The Jepson Desert Manual

The Jepson Desert Manual

Vascular Plants of Southeastern California

Bruce G. Baldwin
Steve Boyd
Barbara J. Ertter
Robert W. Patterson
Thomas J. Rosatti
Dieter H. Wilken

Editors

Margriet Wetherwax

Managing Editor

UNIVERSITY OF CALIFORNIA PRESS

University of California Press
Oakland, California

Library of Congress Cataloging-in-Publication Data

The Jepson desert manual: vascular plants of southeastern California /
Bruce G. Baldwin . . . [et al.], editors
    p.   cm.
  Includes bibliographical references and index.
  ISBN 978-0-520-22775-0 (paper: alk. paper)
  1. Desert plants—California, Southern—Identification. 2. Desert plants—
California, Southern—Pictorial works. I. Baldwin, Bruce G., 1957–

QK.149J58 2002
581.78548097949—dc21   2001037795

26 25 24 23 22
11 10 9 8 7 6 5

The paper used in this publication meets the minimum requirements of
ANSI/NISO Z39.48-1992 (R 1997) (*Permanence of Paper*).

The Jepson Desert Manual is dedicated to

JAMES C. HICKMAN, ROBERT ORNDUFF, AND LINCOLN CONSTANCE

As Editor of *The Jepson Manual: Higher Plants of California*, James C. Hickman (1941–1993) was instrumental in fulfilling one of Jepson's primary wishes and in providing amateur and professional botanists with an updated resource for studying and enjoying California's wild plants. *The Jepson Desert Manual* is based directly on the book that Jim worked so hard to help create.

Professor Robert Ornduff (1932–2000), life-long devotee of the California flora and Trustee of the Jepson Herbarium, inspired *The Jepson Desert Manual* with his vision of small, regionally focused works based on *The Jepson Manual*. Bob left a legacy of helping others to understand, enjoy, and conserve California plant life.

Professor Lincoln Constance (1909–2001), Jepson's protégé and one of the original Trustees of the Jepson Herbarium, was an important source of guidance for the Jepson Manual Project and helped chart the course for systematic botany in California and throughout the world. Lincoln was a mentor and role model for generations of botanists and contributed greatly to knowledge of umbels (Apiaceae) and the California flora in general.

CONTENTS

Willis Linn Jepson and giant saguaro
(Carnegiea gigantea), *lower Colorado River, 1912.*

PREFACE

Willis Linn Jepson (1867-1946) dedicated his life to the study of the California flora and endowed the Jepson Herbarium, a unit of the University of California, Berkeley, which continues his efforts to increase botanical knowledge, education, and conservation in California. Among his more than 200 articles and books, Jepson's well-known *Manual of the Flowering Plants of California* (1925) was the first published work designed for field identification of all native and naturalized vascular plants in the state and was the inspiration for *The Jepson Manual: Higher Plants of California* (1993). He founded the California Botanical Society and helped to create the Save-the-Redwoods League and the Sierra Club.

Despite the demands of being a Professor of Botany at U.C. Berkeley, Jepson regularly immersed himself in California's wild plants and wild places, including the southeastern California deserts. His first major botanical forays to the California deserts occurred in 1912, with exploration of the Mojave Desert near Barstow and an expedition by boat down the lower Colorado River. He made frequent visits to the deserts in subsequent years, culminating in 1941 with a collecting trip with Carl Wolf to the Granite Mountains and Kelso Dunes of the eastern Mojave Desert. Among the plants of southeastern California described by Jepson are *Allium fimbriatum* var. *mohavense*, *Astragalus lentiginosus* var. *semotus*, *Calochortus kennedyi* var. *munzii*, *Nemacladus glanduliferus*, *Salvia eremostachya*, and *Streptanthus cordatus* var. *duranii*.

Jepson recorded in his field notebooks his "irresistible fascination" with the California deserts and the importance of the deserts as a source of renewal. Upon entering the vegetation of Joshua tree, creosote bush, and burro-weed east of Tehachapi Summit, he wrote: "... it caused my spirits to rise and at once I felt better in body as well as cheered and sustained in mind." Jepson's feelings while botanizing at Calico Wash, near Barstow, in spring of 1913 perhaps capture best the lasting appeal of desert botany:

> "Sitting here on the ground studying flower parts under the lens is a pleasant occupation. When one's eyes tire there are the desert ranges stretching one beyond another, and a soft breeze blowing from the west."

BRUCE G. BALDWIN
JULY 2001

ACKNOWLEDGEMENTS

The Jepson Desert Manual would not have been possible without the generous financial support of The Giles W. and Elise G. Mead Foundation, the Homeland Foundation, and the Lawrence R. Heckard Endowment Fund of the Jepson Herbarium. Over 80 authors and numerous other individuals made invaluable contributions to the project. Brent D. Mishler and the Trustees of the Jepson Herbarium offered guidance and support; Betsy Ringrose and Staci Markos helped with fundraising and organizational efforts; Richard L. Moe conducted database research and helped to generate the index; Jeffrey Greenhouse researched herbarium and library records (assisted by Sonia Nosratina); Linda A. Vorobik served as principal illustrator and offered design recommendations; Bobbie Angell, Karen Klitz, Lesley B. Randall, Emily E. Reid, and Sarah A. Young prepared illustrations for *The Jepson Manual* that are used here; Susan Gloystein typeset the book; Elizabeth Painter proofread the book; David J. Keil provided input on a variety of nomenclatural and distributional issues; John L. Strother provided assistance with nomenclatural problems; Tony Morosco helped with database research; Richard G. Beidleman provided archival information; Matthew Blowers provided administrative support; Susan J. Bainbridge offered helpful advice; and Lieske Wetherwax helped with typing. The following individuals provided their color photographs: George Becker, John Game, the late Lawrence R. Heckard, Steve Junak, Scott N. Martens, Donald Myrick, Garry Norvell, the late Robert Ornduff, the late G. T. Robbins, the late James T. Vale, the late Charles S. Webber, Dieter Wilken, and Martin F. Wojciechowski. The following U.C. Berkeley students provided important volunteer and work-study assistance: Susie Hanna, Julian Lim, Maiko Minani, Amy Rusev, and Conrad Sweeting. John T. Kartesz and Christopher A. Meacham donated a copy of their *Synthesis of the North American Flora* for database research. Plant distributional data for California desert regions were provided by Roxanne L. Bittman, Grant Fletcher, Frederic G. Hrusa, Matt Lavin, Timothy Messick, James D. Morefield, Arnie Peterson, Andrew C. Sanders, Dean W. Taylor, Arnold Tiehm, and Dana York. The following individuals lent their expertise on various plant groups: Carlos Aedo (Geraniaceae), James Affolter (Apiaceae), J. Chris Pires (*Brodiaea* and *Dichelostemma*), J. Mark Porter (Polemoniaceae), and Pablo Vargas (*Antirrhinum*). Participants in *The Jepson Desert Manual* workshop held in the Granite Mountains in April 2000 helped to test keys and distributional data. On behalf of the Editors, I offer heartfelt thanks to the above foundations and individuals for helping to make this book a reality.

BRUCE G. BALDWIN

AUTHORS CONTRIBUTING TO
THE JEPSON DESERT MANUAL

Susan G. Aiken, Canadian Museum of Nature, Ottawa, Ontario

Geraldine A. Allen, University of Victoria, British Columbia, Canada

Kelly W. Allred, New Mexico State University, Las Cruces

Barrett H. Anderson, Walnut Creek, CA

Dennis E. Anderson, Humboldt State University (retired)

Edward F. Anderson, Desert Botanical Garden, Phoenix, Arizona (deceased)

Loran C. Anderson, Florida State University, Tallahassee

George Argus, Canadian Museum of Nature, Ottawa, Ontario (retired)

Wayne P. Armstrong, Palomar College, San Marcos

Deborah Engle Averett, Chandler, Arizona

Tina J. Ayers, Northern Arizona University, Flagstaff

John D. Bacon, University of Texas, Arlington

Susan J. Bainbridge, University of California, Berkeley

John Baird, Northern Kentucky University, Highland Heights

Marc A. Baker, Chino Valley, Arizona

Bruce G. Baldwin, University of California, Berkeley

Theodore M. Barkley, Botanical Reserch Institute of Texas, Fort Worth

Mary E. Barkworth, Utah State University, Logan

Jim A. Bartel, United States Fish and Wildlife Service, San Diego

David M. Bates, Cornell University, Ithaca, New York

Randall J. Bayer, Australian National Herbarium, Canberra

Tania Beliz, College of San Mateo, California

Robert Berman, Pacific Grove, California

John Bleck, University of California at Santa Barbara

David Brandenburg, Dawes Arboretum, Newark, Ohio

Gregory K. Brown, University of Wyoming, Laramie

Richard K. Brummitt, Royal Botanic Gardens, Kew, England

Roy E. Buck, University of California, Berkeley

Gerald D. Carr, University of Hawaii, Honolulu, Hawaii

Robert L. Carr, Eastern Washington State University, Cheney

Kenton L. Chambers, Oregon State University, Corvallis (retired)

Anita F. Cholewa, University of Minnesota, St. Paul

Tsan-Iang Chuang, Illinois State University, Normal (deceased)

Curtis Clark, California Polytechnic State University, Pomona

W. Dennis Clark, Arizona State University, Tempe

Lynn Clark, Iowa State University, Ames

J. Travis Columbus, Rancho Santa Ana Botanic Garden, Claremont

Steven A. Conley, Redding

Lincoln Constance, University of California, Berkeley (deceased)

Raymond Cranfill, University of California, Berkeley

William J. Crins, Peterborough, Ontario, Canada

Michael Curto, California Polytechnic State University, San Luis Obispo

Thomas F. Daniel, California Academy of Sciences, San Francisco

Jerrold I Davis, Cornell University, Ithaca, New York

W. S. Davis, University of Louisville, Kentucky

Alva G. Day, California Academy of Sciences, San Francisco (retired)

Lauramay T. Dempster, University of California, Berkeley (deceased)

Melinda F. Denton, University of Washington, Seattle (deceased)

James C. Dice, California Department of Fish and Game, Chino Hills

Patrick E. Elvander, University of California, Santa Cruz (deceased)

Barbara J. Ertter, University of California, Berkeley

Frederick Essig, University of South Florida, Tampa

Wayne R. Ferren, University of California, Santa Barbara

Peggy Fiedler, Oakland, CA

Fred R. Ganders, University of British Columbia, Vancouver, Canada

Craig W. Greene, College of the Atlantic, Bar Harbor, Maine

James R. Griffin, University of California, Berkeley (deceased)

James W. Grimes, New York Botanical Garden, Bronx

Erich Haber, Canadian Museum of Nature, Ottawa, Ontario

J. Robert Haller, Santa Barbara Botanical Garden

Richard R. Halse, Oregon State University, Corvallis

Ronald L. Hartman, University of Wyoming, Laramie

M. J. Harvey, Victoria, British Columbia, Canada

Richard L. Hauke, University of Rhode Island, Kingston (retired)

Frank G. Hawksworth, United States Forest Service, Fort Collins, Colorado (deceased)

Lawrence R. Heckard, University of California, Berkeley (deceased)

Douglass M. Henderson, University of Idaho, Moscow (deceased)

James Henrickson, University of Texas, Austin

James C. Hickman, University of California, Berkeley (deceased)

Steven R. Hill, Center for Biodiversity, Champaign, Illinois

Peter C. Hoch, Missouri Botanical Garden, St. Louis

Carol A. Hoffman, University of Georgia, Athens

Noel H. Holmgren, New York Botanical Garden, Bronx

Duane Isely, Iowa State University, Ames (deceased)

James K. Jarvie, Utah State University, Logan

Judith Jernstedt, University of California, Davis

Dale E. Johnson, Portland, Oregon

James D. Jokerst, Jones and Stokes Associates, Sacramento (deceased)

Elaine Joyal, Arizona State University, Tempe

Glenn Keator, Sebastopol

David J. Keil, California Polytechnic State University, San Luis Obispo

Ronald B. Kelley, Eastern Oregon University, Le Grande

Walter A. Kelley, Mesa College, Grand Junction, Colorado

Daryl L. Koutnik, County of Los Angeles

Arthur C. Kruckeberg, University of Washington, Seattle (retired)

John C. La Duke, University of North Dakota, Grand Forks

Meredith A. Lane, Academy of Natural Sciences, Philadelphia, Pennsylvania

Thomas Lemieux, University of Colorado, Boulder

Harlan Lewis, University of California, Los Angeles (retired)

Richard A. Lis, California Department of Fish and Game, Redding

R. John Little, Sacramento

Robert I. Lonard, Pan American University, Edinburg, Texas

Scott N. Martens, University of California, Davis

Joy Mastrogiuseppe, Washington State University, Pullman

Niall F. McCarten, Davis, CA

Elizabeth McClintock, California Academy of Sciences, San Francisco (retired)

Katy K. McKinney, Texas A&M University, College Station

Dale W. McNeal, University of the Pacific, Stockton

Michael R. Mesler, Humboldt State University, Arcata

Timothy Messick, Davis, CA

Paul Meyers, Pierce College, Los Angeles

Jennifer Milburn, Missouri Botanical Garden

Richard L. Moe, University of California, Berkeley

Michael Moore, University of Georgia, Athens

John S. Mooring, Santa Clara University

Reid Moran, San Diego Natural History Museum (retired)

James D. Morefield, Nevada Natural Heritage Program, Carson City

Nancy R. Morin, The Arboretum at Flagstaff, Arizona

Michael Nee, New York Botanical Garden, Bronx

Elizabeth Neese, University of California, Berkeley (retired)

Guy L. Nesom, Botanical Research Institute of Texas, Fort Worth

Bryan D. Ness, Pacific Union College, Angwin

Richard Olmstead, University of Washington, Seattle

Robert Ornduff, University of California, Berkeley (deceased)

Elizabeth L. Painter, University of California, Berkeley

Bruce D. Parfitt, University of Michigan, Flint

Robert W. Patterson, San Francisco State University

Willard W. Payne, Sanibel, Florida (retired)

Paul M. Peterson, Smithsonian Institution, Washington, D.C.

C. Thomas Philbrick, Western Connecticut State University, Danbury

Duncan M. Porter, Virginia Polytechnic and State University, Blacksburg

A. Michael Powell, Sul Ross State University, Alpine, Texas

Robert A. Price, University of Georgia, Athens

Barry A. Prigge, University of California, Los Angeles

James S. Pringle, Royal Botanical Gardens, Hamilton, Ontario, Canada

Charles F. Quibell, Sonoma State University, Rohnert Park (retired)

Peter H. Raven, Missouri Botanical Garden, St. Louis

John R. Reeder, University of Arizona, Tucson (retired)

James L. Reveal, University of Maryland (retired)

Rhonda Riggins, California Polytechnic State University, San Luis Obispo (retired)

Warren Roberts, University of California, Davis

Reed C. Rollins, Harvard University, Cambridge, Massachusetts (deceased)

Thomas J. Rosatti, University of California, Berkeley

John O. Sawyer, Jr., Humboldt State University, Arcata

Robert L. Schlising, California State University, Chico

Clifford L. Schmidt, San Jose State University (retired) and Salem, Oregon

Alfred E. Schuyler, Academy of Natural Sciences, Philadelphia, Pennsylvania

John C. Semple, University of Waterloo, Ontario, Canada

James R. Shevock, United States Department of the Interior, San Francisco

Teresa Sholars, College of the Redwoods, Fort Bragg

Leila M. Shultz, Harvard University, Cambridge, Massachusetts

Beryl B. Simpson, University of Texas, Austin

Mark W. Skinner, USDA NRCS National Plant Data Center, Baton Rouge, Louisiana

Alan R. Smith, University of California, Berkeley

Galen Smith, University of Wisconsin, Whitewater (retired)

James P. Smith, Humboldt State University, Arcata

Robert J. Soreng, Smithsonian Institution, Washington, D. C.

Richard Spellenberg, New Mexico State University, Las Cruces (retired)

G. Ledyard Stebbins, University of California, Davis (deceased)

William J. Stone, University of California, Berkeley

John L. Strother, University of California, Berkeley

Scott D. Sundberg, Oregon State University, Corvallis

Janice C. Swab, Meredith College, Raleigh, North Carolina

Fosiée Tahbaz, University of California, Berkeley

Jennifer Talbot, Washington University, St. Louis, Missouri

Dean Taylor, University of California, Berkeley

Mary Susan Taylor, Missouri Botanical Garden, St. Louis

John Thieret, Northern Kentucky University, Highland Heights

David M. Thompson, Cedar Falls, Iowa

Robert F. Thorne, Rancho Santa Ana Botanical Garden, Claremont (retired)

Steven L. Timbrook, Ganna Walska Lotusland Foundation, Santa Barbara

Gordon C. Tucker, New York State Museum, Albany

John M. Tucker, University of California,Davis (retired)

Charles E. Turner, US Department of Agriculture, Albany, California (deceased)

Staria S. Vanderpool, Arkansas State University, State University

Barbara Veno, California State University, Los Angeles

Nancy J. Vivrette, Ransom Seed Laboratory, Carpinteria

Linda Ann Vorobik, University of California, Berkeley

Warren H. Wagner, University of Michigan, Ann Arbor (deceased)

Warren L. Wagner, Smithsonian Institution, Washington, D.C.

Gary D. Wallace, Rancho Santa Ana Botanic Garden, Claremont

Michael J. Warnock, Sam Houston State University, Huntsville, Texas

Grady L. Webster, University of California, Davis (retired)

Robert Webster, US Department of Agriculture, Beltsville, Maryland

Philip V. Wells, University of Kansas, Lawrence (retired)

Thomas L. Wendt, University of Texas, Austin

Margriet Wetherwax, University of California, Berkeley

R. David Whetstone, Jacksonville State University, Alabama

Sherry Whitmore, San Antonio, Texas

Delbert Wiens, University of Utah, Salt Lake City (retired)

Dieter H. Wilken, Santa Barbara Botanic Garden

Michael P. Williams, University of California Reserve System

Martin F. Wojciechowski, Arizona State University, Tempe

Dennis W. Woodland, Andrews University, Berrien Springs, Michigan

George Yatskievych, Missouri Botanical Garden, St. Louis

PHILOSOPHY OF
THE JEPSON DESERT MANUAL

The purpose of *The Jepson Desert Manual* is to provide amateur and professional botanists with a single work focused exclusively on identification of vascular plants in the California deserts. For desert botany, a field manual should be as light and easy to use as possible — a goal that we sought to achieve by simplifying keys and distilling treatments from *The Jepson Manual*: *Higher Plants of California* (J. Hickman, ed., 1993) to include only plants found within the southern Great Basin, the Mojave Desert, and the Sonoran Desert of California. The flora of the California deserts is sufficiently distinct from the much more diverse and complex flora of the California Floristic Province to warrant separate treatment in a substantially smaller, simpler, and more extensively illustrated manual than would be possible to produce for the California flora as a whole. By confining our focus to the California deserts, we were able to augment illustrations of desert plants from *The Jepson Manual* with ~300 new, partial or full illustrations and 130 photographs without sacrificing portability of the book.

The area covered by *The Jepson Desert Manual* corresponds with some but not all interpretations of "the California deserts". The decision to include the southern Great Basin but not the northern Great Basin (the Modoc Plateau) in *The Jepson Desert Manual* was based on floristic and practical considerations. The southern Great Basin, i.e., the region east of the Sierra Nevada (including the White-Inyo Range), is much more similar floristically to the Mojave and Sonoran deserts than is the Modoc Plateau. Indeed, some botanists regard the Mojave Desert and southern Great Basin to constitute a single floristic unit apart from the northern Great Basin and the Sonoran Desert. In addition, the southern Great Basin and the Mojave and Sonoran deserts constitute a contiguous area of special interest to desert botanists, who may visit all three regions during a single trip to the field.

Much of the effort that went into *The Jepson Desert Manual* was to improve accuracy of distributional information based on vouchered records. We adhered to the geographic subdivisions of *The Jepson Manual*, which have become widely adopted by Californian naturalists over the last eight years. Finer-scale resolution of plant distributions is offered for plants of highly limited occurrence. We retained information on state-wide and out-of-state distributions for plants found both within and outside the California deserts to give users of the manual an appreciation for the ecological and geographic range of desert plants (desert distributions are highlighted in bold-face type). In the interest of meeting our goal of producing a compact, portable field manual, we chose not to present more detailed distributional data for widespread desert plants than is possible using the geographic subdivisions adopted here.

One of the important decisions faced during production of *The Jepson Desert Manual* was whether to include or exclude taxa that occur along the western edge of the California deserts. Although the spine of mountains separating the California deserts from the rest of California is a dramatic and relatively abrupt geographic boundary, plants characteristic of the California Floristic Province often spill out onto the western edge of the deserts, especially in areas where the transition between Mediterranean and desert conditions is relatively gradual. Strict adherence to a policy of inclusion of all taxa recorded from at least one site inside a sharply delimited western boundary for the deserts would have defeated our goal of producing a small, simple manual for desert botanists. We trust that the values we chose to emphasize offset the limitations of *The Jepson Desert Manual* for identifying all vascular plants in transitional areas at the western desert edge (e.g., western Antelope Valley in the Mojave Desert, Morongo Valley in the Sonoran Desert).

Another challenge we faced was to provide updated information on taxonomy and nomenclature of desert plants in a way that allows easy cross-referencing of taxa and plant names between *The Jepson Desert Manual* and *The Jepson Manual*. The compromise we adopted allows users of *The Jepson Desert Manual* to learn about changes in plant names and plant classification without introducing the confusion of unfamiliar names and circumscriptions of taxa in keys and descriptions. With minor exceptions, the taxonomy and nomenclature followed in keys and descriptions in *The Jepson Desert Manual* are the same as in *The Jepson Manual*; taxonomic and name changes for families, genera, species, subspecies, and varieties are discussed or noted at the end of descriptions, generally with reference to relevant literature. Name changes of species, subspecies, and varieties are highlighted in bold and noted with an asterisk. Taxa new to the state, newly described, or recently recognized are noted in the treatments.

The spirit and goals of the original Jepson Manual Project guided production of *The Jepson Desert Manual* and we are greatly indebted to the late James C. Hickman, the late Lawrence R. Heckard, and others who were involved in editing of *The Jepson Manual*, on which *The Jepson Desert Manual* is based.

BRUCE G. BALDWIN

CONVENTIONS USED IN
THE JEPSON DESERT MANUAL

The Jepson Desert Manual follows the principles and practices adopted in *The Jepson Manual* to achieve the goals of comprehensiveness, conciseness, and accessibility of floristic information. In a field manual intended for a wide group of users, precision, conciseness, and clarity are all of utmost importance but can be difficult to reconcile. To balance these goals, assumptions and conventions were adopted for *The Jepson Manual* and *The Jepson Desert Manual*. The most important conventions are outlined below.

Overall Design Conventions

Comprehensiveness. All native, validly named taxa recognized by the authors are included (no new names are published in *The Jepson Desert Manual*). In addition, all aliens that are known to have become naturalized in the California deserts are included. The general policy was not to include (or to note only in passing) waifs or non-reproducing but long-persisting individuals or clones. As in other areas, the desired line was impossible to draw cleanly; strong views of authors about inclusion or exclusion of taxa normally were accommodated. Validly named taxa not treated in *The Jepson Manual* but now known to occur in the California deserts are noted within the corresponding treatments in *The Jepson Desert Manual*.

Uniformity. Achieving consistency among treatments of the nearly 200 contributing authors was an important goal in *The Jepson Desert Manual*, as in *The Jepson Manual*. Some compromises proved necessary but ultimately a high level of uniformity across treatments was realized.

Book Organization and Design. Ease of accessibility to information is the primary basis for the organization and design of *The Jepson Desert Manual*. The four traditionally recognized groups of higher plants are arrayed in the following sequence: ferns and their allies, gymnosperms, dicots, and monocots. Families are arrayed alphabetically within each of the four major groups, as are genera within families, species within genera, and infraspecific taxa within species.

Measurements. All linear measurements are given in metric units. Metric conversion scales (centimeters to inches, meters to feet) are provided on the inside front cover.

Illustrations. Illustrations are provided to aid identification of plants. Except for some aliens, each genus is represented, usually by a habit drawing. Because all taxa could not be illustrated, priority was given to those commonly encountered, those uncommon or threatened enough to warrant monitoring, and those with unusual or difficult diagnostic features. All illustration plates appear on right-hand pages. Illustrations of taxa generally follow corresponding descriptions (page numbers for illustrations are given at the beginning of taxon descriptions) and are in alphabetical order.

Index. All family names, generic names, common names, and names here considered to be synonyms or to have been misapplied are indexed in a single alphabetical listing. Because accepted specific and infraspecific names are all ordered alphabetically within the text, under the appropriate genus, they can be found easily and are not listed in the index.

Keys and Their Conventions

All Keys

Keys in *The Jepson Desert Manual* are artificial means for identifying plants (keys that are "natural", in the sense that they divide groups according to their relationships, can be more difficult to use than artificial keys). Each key is a series of paired, mutually exclusive statements that divides a set of taxa into progressively smaller subsets until all possibilities but one have been eliminated (hence, they are commonly called "dichotomous" keys). Keys are used at all levels, from the separation of groups of families to the separation of varieties of a species.

Keys in *The Jepson Desert Manual* are made up of numbered, paired, alternative statements about plant features; only one statement or "lead" of the pair (or "couplet") should be true for a given plant (except for taxa that can be identified in more than one part of a key — see below). In each couplet, the number of the first statement is followed by a period (.), while that of the second is followed by a prime ('). Couplets are arranged in a series of indented steps, like sequential clues that allow the user to see where any decision leads. Each decision between statements of a couplet eliminates all possibilities included under the rejected statement and leads in turn to the lowest-numbered couplet included under the accepted statement (and thereby to another decision). This process is repeated until the end point is reached, that is, when all inappropriate possibilities have been eliminated and the probable identity of the plant is revealed. Identifications should be confirmed by reading the entire description, considering the statements of habitat and geography, and studying available illustrations. In critical cases, material should be compared to specimens in an herbarium, ideally to specimens identified by an expert in the group being considered.

All statements have a morphological basis, but geographic, habitat, host, and other corroborative information is sometimes provided. All keys are constructed to differentiate the taxa as they appear in California in the simplest way; a corollary is that plants of a taxon represented in the California desert flora but collected outside California may not key well in *The Jepson Desert Manual*.

Complex or variable taxa often may be keyed in more than one place in a key. A superscript integer preceding a taxon name in the key indicates the number of places where the taxon can be reached in that key. This convention is not followed in the key to families (to avoid confusion — see "Family Key" below), even though members of a family sometimes "key out" in more than one place in a key.

Keys were constructed to emphasize both highly distinctive and readily assessed features. Conveniently, sometimes a feature has both attributes. Other times, the two attributes are at odds and compromise was necessary. For example, easily assessed features may be given first within a key lead, with unique but more difficult-to-observe characters (or those less likely to be present on an average specimen) listed later in the key lead, to allow corroboration or correction of an impression. For some taxa, only a single feature (such as fruit type) accurately separates groups, so, by default, it is all that is used. For example, the taxonomy of various groups (e.g., Apiaceae, Boraginaceae, Ericaceae, *Carex*) relies on features of mature fruit.

Sometimes features unique to one part of a couplet (that is, those that cannot be readily compared in the other key lead) nevertheless provide useful corroborating evidence for a decision. When such unique information is included, it is set off from the rest of the statement by a dash (—), indicating that there is no comparable information for the other half of the couplet.

Family Key. The first portion of the *Key to California Desert Plant Families* is a key to groups of families that share taxonomically important or easily seen characters. Users of the family key may wish to become especially familiar with the kinds of information asked for in the Key to Groups. The Key to Groups will be easier to use after learning to distinguish ferns and fern allies, gymnosperms, dicots, or monocots and learning to examine and identify flower parts.

Two other important conventions are followed in the key to families. Members of variable families may "key out" in two or more places, often with the end point being a genus or species rather than the entire family. Attempts were made to anticipate common mistakes and misinterpretations, and to guide the user past such places in the keys, thus minimizing technical errors leading to misidentifications (Euphorbiaceae, sometimes with inflorescences that appear to be flowers, is an example of such a case).

Other Keys. A key to the genera of a family follows each family description (if the family includes more than one genus in the California deserts); a key to species (generally including infraspecific taxa as well) follows each genus description (if the genus includes more than one species in the California deserts). For a few species-rich genera (e.g., *Astragalus*), infraspecific taxa are keyed after the relevant species description rather than in the main key to species.

Descriptions and Their Conventions

All Descriptions

Contents of descriptions are comparable among all members of a taxon and are constructed to focus attention on diagnostic features. Characters included in descriptions are treated in the same sequence and with comparable modifiers in the descriptions of all related taxa at the same rank.

Characters are described at the highest rank at which they apply and are not repeated in lower level descriptions. Consequently, family descriptions must be consulted in order to understand the important characters of included genera. Likewise, generic descriptions must be consulted for characters of included species and infraspecific taxa. For character states that generally but not always occur in a taxon, the abbreviation "gen" is used. Then, at lower ranks, only exceptions to the general statement are given.

Descriptions are composed of several statements, most of them highlighted by an all-capital, bold-face abbreviation of a major plant part (stem, leaf, inflorescence, flower, fruit, seed). Statements describing major plant parts are separated by periods. Consistently different, complex forms of inflorescences or flowers generally are separated into different statements, e.g., staminate catkins, disk flowers.

Nouns begin statements and, generally, phrases within statements. Plant parts are described sequentially from basal to apical on the plant axis. Nouns are followed by a specific sequence of adjectives (e.g., position, number, size, shape, color, hair types). Adjectives always modify the immediately preceding noun.

First Descriptive Statement. The first statement of a description covers any important general aspects of the plant, including growth habit, size and shape, sexuality, general hairiness, etc. Such statements are the only ones that do not begin with bold-face abbreviations.

Other Descriptive Statements. Several conventions are followed regarding punctuation within statements of a description. If the entire organ to which the statement pertains is described, the adjectives immediately following are not separated from the main noun by punctuation (e.g., "**LVS** alternate, compound...."). If the main noun is immediately broken into parts that are described sequentially, it is followed by a colon (e.g., "**LF**: petiole 1–3 cm, winged; blade ± 4 cm, round, entire...."). The secondary, noun-initiated phrases of a statement are separated by semicolons; within these secondary phrases, adjectives (and sometimes tertiary nouns) are separated by commas.

Descriptions of Families

All families are given names based on genera and end in "-aceae". Among family names used here are alternate names for families that may be better known by other names: Apiaceae (not Umbelliferae), Asteraceae (not Compositae), Brassicaceae (not Cruciferae), Fabaceae (not Leguminosae), Hypericaceae (not Guttiferae), Lamiaceae (not Labiatae), Arecaceae (not Palmae), and Poaceae (not Gramineae). A common (English, colloquial) name is given to each family for ease of association. Authors of family names are not given.

Every family with native or naturalized representatives in the California deserts is described. Family descriptions account for variability throughout the world. The general form of the description itself is the same as for genera and species.

Approximate numbers are given, on a worldwide basis, for genera and species in each family. Overall range is also summarized. Where appropriate, notes on cultivated, useful, or toxic forms are included. A reference may be given [in brackets] that supplies an appropriate entry into literature on the family. (The same strategy is used for genera and sometimes for species.)

Descriptions of Genera

If a family comprises only one genus worldwide, the family description serves as the genus description (e.g., Krameriaceae). In such cases, the number of species and their overall range are given under the family description, together with an appropriate reference [in brackets], and sometimes additional notes. The derivation of the genus name is given under the genus description. If a family is made up of more than one genus, the genera that are native or naturalized in the California deserts are treated.

Genera are assigned a common name only if one is believed to be in common usage. As with families, the general form of the description is the same as for species. Authorship is not given for genus names.

Descriptions of Species, Subspecies, and Varieties

Scientific Names. Names of genera, species, subspecies, and varieties appear in *italics*. Genus names are always capitalized; specific, subspecific, and varietal epithets are never capitalized, regardless of their origin. Names of taxa considered native to California are printed in bold-face italic **Times**. Names of alien taxa are in italic Helvetica, a sans-serif type with a lighter appearance than the names of native taxa.

Authors of Names. Specific, subspecific, and varietal epithets are followed immediately by the name of the person or persons (often abbreviated, see below) who validly published the name (except that no authors are given for the typical subspecies or variety within a species; e.g., *Phacelia campanularia* A. Gray ssp. *campanularia*). An indication of a scientific name is not strictly accurate and complete without the name(s) of the validating author(s).

The Jepson Desert Manual follows *The Jepson Manual* in using the 1980 *Draft Index of Author Abbreviations compiled at The Herbarium, Royal Botanic Gardens, Kew*. Author names that are not accounted for in the *Draft Index* generally are given here for clarity as one or two initials plus an unabbreviated surname. Mabberley's useful and readily available *The Plant Book* (Cambridge University Press, 1987) includes a list of author abbreviations that is somewhat abridged from the *Draft Index*.

Common Names. Common names for species are printed in Times small capitals. Common names are given only if a colloquial name is in common usage. Common names are also given to provide a cross reference to a taxon that is rare enough to be "sensitive". To facilitate communication among all users, common names accepted in the California Native Plant Society's *Inventory of Rare and Endangered Plants of California* are used in *The Jepson Desert Manual*.

Chromosome Numbers. Chromosome reports often are presented but have not been verified for this work and should not be taken as definitive. Groups without reported counts may be excellent candidates for chromosomal study.

Habitats and Elevations. Habitats, elevational ranges (together with geographic ranges, below) combine to provide a reasonable prediction of where a plant taxon might be found. Habitats and elevations are given for each taxon at rank of species or below. Plant taxa are highly variable in the specificity of their habitat and elevational requirements. Many plant taxa grow in a wide range of plant communities, as long as certain habitat requirements are met. Others share a very specific habitat with a highly predictable group of other taxa. Habitat specifications may include degree of exposure to sunlight, soil parent material, soil texture and soil moisture conditions. Elevational ranges are given in meters (a conversion scale to feet is found on the inside front cover) and should be regarded as a minimum range.

Geographic Range Descriptions. A hierarchical system is used to describe geographic ranges in the California deserts (see Geographic Subdivisions of California, p. 33).

All range specifications assume the restrictions of cited habitats and elevational ranges. At the most general, "CA" means that a taxon occurs in all three floristic provinces in California; it may be expected throughout the state.

A combined habitat and range statement might read "Sandy washes; < 300 m. D". This statement implies that the taxon might be found in washes throughout the Mojave Desert (DMoj) and the Sonoran Desert (DSon). However, elevational range excludes it from washes in the higher portions of DSon and DMoj, including all of the Desert Mountains (DMtns).

Authors based taxon distributions on documented occurrences (normally herbarium specimens). Two conventions are followed to assure that ranges are described as completely as possible. Alien taxa with expanding ranges are generally specified as "expected elsewhere" or "expected more widely". For native taxa reported from a region from which no documentation had been seen by authors, a different wording is used: regions of questionable occurrence are listed, but followed by a "?" (e.g., SNE, w DMoj?).

Synonyms and Misapplied Names. Complete synonymies that show the entire nomenclatural histories of taxa are not given because of space constraints. Names accepted by Munz (1959 *A California Flora*, as corrected by his 1968 *Supplement*) and by the fifth edition of the California Native Plant Society's *Inventory of Rare and Endangered Plants of California* are all presented, either as names for recognized taxa or as synonyms, misapplied names, or unclearly differentiated forms (see below).

Synonyms and misapplied names appear in brackets. The two are differentiated as follows: "[var. *aggregata* Rydb.; *A. rubra* Michaux; *A. coccinea* Rottb. misapplied]". If no species name is specified for a synonym or misapplied name, then the species just described is assumed. Included, lower-rank taxa that are not recognized are listed first; other synonyms are then listed in alphabetical order. Misapplied names appear last.

Unclearly differentiated forms are normally diagnosed in a note following the synonymy, e.g., "Pls from n SNE with reddish hairs have been called *P. rubrotincta* E. Greene". Such a statement reserves judgment about whether an entity should be considered taxonomically distinct or merely a minor, taxonomically insignificant variant of a recognized taxon.

Highly variable taxa are so noted in a sentence appended to the description. Any known problems concerning the variation pattern are specified. The phrase "more study needed" is used if additional work might result in a more satisfactory taxonomic disposition.

Toxicity. Some California desert plants are seriously toxic. Fuller & McClintock's *Poisonous Plants of California* (University of California Press, 1986) provides an excellent overview of plants that are both major and minor sources of poisoning. Plants that have been toxic to animals or people in California (or are expected to be toxic) are cued with an all-capital "TOXIC...". Usually, some specifics are included, e.g., "TOXIC to livestock but rarely eaten"; "TOXIC: causes severe contact dermatitis"; or "TOXIC: ingestion of a single seed has been fatal to humans".

Weediness. The California deserts support an increasing number of aggressive weeds that were originally native to other parts of the world (i.e., are aliens). Both state and federal governments have enacted legislation designating the worst alien weeds as "noxious weeds" that should be eradicated wherever they are found. Listed alien weeds are specified as "NOXIOUS WEED" in *The Jepson Desert Manual*.

Other, non-listed weeds may be seriously aggressive in some circumstances (such as disturbed roadsides and urban gardens) but do not pose an immediate threat to agriculture or to natural habitats. Such plants are noted as aggressive and pernicious weeds.

Horticultural Value. The final segment of many descriptions of native taxa is a summary of horticultural value and requirements for growth. Such entries are preceded by the symbol ❀. See the chapter "Horticultural Information" in *The Jepson Desert Manual*, p. 29, for more information.

GLOSSARY

Illustrations by Dr. Linda Ann Vorobik, Susan Stanley, & Sarah A. Young

The vocabulary, definitions, and illustrations used here are mostly adopted from *The Jepson Manual* (Hickman 1993). Terminology was chosen for ease of communication to a broad community of users. Please see *The Jepson Manual* for additional discussion of the philosophy and decisions that guided preparation of the glossary.

The page locations of glossary illustrations are given in parentheses before the definition. Some terms are not illustrated in the glossary itself, but supplementary references are given to examples found in illustrations in the main text. These references are to taxa (not pages) and are found at the ends of definitions.

abundant. Very likely to be encountered; nearly always found in appropriate habitats, sometimes forming dense stands. (see common) (see pp. 27-28 for more information)

achene. (p. 21) Dry, indehiscent, 1-seeded fruit from a 1-chambered ovary, often appearing to be a naked seed.

acuminate. (p. 14) Having a long-tapered, sharp tip, the sides of which are concave. (see acute, awl-like)

acute. (p. 14) Having a short-tapered, sharp tip, the sides of which are convex or straight and converge at less than a right angle. (see acuminate, obtuse)

adherent. Superficially appearing fused to another organ (of like or unlike type) but separable from it, as "perianth adherent to fruit".

adventitious. Arising at unusual times or places. Said of plant structures such as roots on aerial stems.

aggressive. Growing or spreading rapidly, outcompeting other plants, difficult to control. Said especially of weeds. (see weed)

alien. Not native; introduced purposely or accidentally into an area. (see native, naturalized, waif, weed)

alpine. Found above timberline. (see subalpine)

alternate. (p. 13) 1. Arranged singly, often spirally, along an axis — e.g., one leaf per node. (see opposite, whorled) 2. Occurring between structures, or in different ranks, as "stamens alternate petals". (see rank)

angiosperm. Plant that bears true flowers, made up of two major groups, dicots and monocots. (see flower, dicot, monocot)

annual. (p. 10) Completing life cycle (germination through death) in one year or growing season, essentially non-woody. (see biennial, herb, perennial)

annulus (annuli). On the sporangium of most ferns, a row of cells with partly thickened walls that functions in the often catapult-like release of spores.

anther. (p. 17) Pollen-forming portion of a stamen.

appressed. (pp. 11, 19) Pressed against. Said especially of hairs that are parallel or nearly parallel to and often in contact with the surface or axis of origin.

aquatic. Growing under, in, or on water, rooted in bottom sediment or floating. Does not include plants of seeps or wet rocks, but does include those with part of the shoot submerged, even though other parts may be above water. (see emergent) (example, *Potamogeton gramineus*)

areole. In Cactaceae, a well defined, axillary area bearing one to many spines and generally other, shorter structures. (example, *Ferocactus cylindraceus*)

aril. Fleshy, corky, or bony appendage arising at or near the point of seed attachment, sometimes completely covering the seed.

armed. General term meaning bearing prickles, spines, or thorns. (see spine)

ascending. (pp. 10, 11) Curving or angling upward from base (generally 30–60° less than vertical or away from axis of attachment). (see decumbent, erect)

asymmetric. (p. 16) Irregular in shape; in no way divisible into identical or mirror-image halves. (see bilateral, biradial, radial)

awl-like. (p. 15) Narrow throughout, but broader at the base and tapered to a sharp tip. (see acuminate)

awn. (pp. 14, 22) 1. Bristle-like appendage or elongation, generally at the tip of a larger structure. 2. Stiff, needle-like pappus element in Asteraceae.

axil. (p. 13) The upper angle between axis and branch or appendage — e.g., between stem and leaf.

axile. (p. 17) Pertaining to an axis, as of a placenta along the central axis in a compound ovary with more than one chamber.

axillary. (pp. 13, 20) Pertaining to or within an axil, especially a leaf axil.

axis (axes). (pp. 17, 22) Line of direction, growth, or extension; structure occupying such a position — e.g., the main stem of a plant or inflorescence, the midrib of a leaf.

banner. (p. 16) Uppermost, often largest petal of many members of Fabaceae.

barbed. (p. 19) Having sharp, normally downward- or backward-pointing projections. Said of an awn, bristle, or other structure.

bark. All tissues outside or covering the wood (hardened xylem tissue) of non-herbaceous plants. Bark patterns are important for identification in many trees and shrubs.

basal. (p. 10) Found at or near the base of a plant or plant part. Especially said of leaves clustered near the ground or of a placenta confined to the base of an ovary.

bell-shaped. (p. 16) Widening more or less abruptly at the base and then generally more gradually above. Generally said of a fused calyx or corolla.

berry. Fleshy, indehiscent fruit in which the seeds are not encased in a stone and are generally more than 1. (see drupe, pome) (example, *Solanum americanum* fruit)

biennial. Completing life cycle (germination through death) in two years or growing seasons (generally flowering only in the second) and non-woody (at least above ground), often with a rosette the first growing season. (see annual, herb, perennial)

bilateral. (p. 16) Divisible into mirror-image halves in only one way. (see asymmetric, biradial, radial)

biradial. (p. 16) Divisible into mirror-image halves in two ways; isobilateral. (see asymmetric, bilateral, radial)

bisexual. (p. 17) Flowers with both fertile stamens and fertile pistils.

blade. (pp. 13, 17, 20, 22) Expanded portion of a leaf, petal, or other structure, generally flat but sometimes rolled or cylindric.

brackish. Somewhat salty. Said generally of a mixture of marine and fresh water.

bract. (pp. 15, 16) Small, leaf- or scale-like structure associated with an inflorescence or cone. Generally subtends a branch, peduncle, pedicel, flower, or cone scale. (see bractlet)

bractlet. (p. 16) 1. Relatively small, generally secondary bract within an inflorescence. 2. Bract-like structure on a pedicel that often does not directly subtend another structure. (see bract)

bristle. 1. Relatively large, generally stiff, more or less straight hair. (example, *Navarettia breweri*) 2. Fine, generally cylindric pappus element in Asteraceae. (example, *Calycoseris parryi*)

bud. (pp. 12, 16, 20) 1. An incompletely developed shoot or leaf. 2. an unopened flower. Sometimes protected by bud scales or sepals, generally axillary or at a stem tip.

bulb. (p. 10) Short underground stem and the fleshy leaves or leaf bases attached to and surrounding it — e.g., an onion. (see stem)

bulblet. (p. 10) 1. Small bulb generally produced at the base of a bulb. 2. Any small, bulb-like structure that propagates a plant, often in a leaf or bract axil.

bur. Fruit or fruiting inflorescence with awns or bristles, often barbed, that attaches to and is dispersed by a passing animal. (example, *Xanthium strumarium*)

callus. (p. 22) In some Poaceae, enlarged base of floret. Sometimes hairy.

calyx (calyces). (p. 17) Collective term for sepals; outermost or lowermost whorl of flower parts, generally green and enclosing remainder of flower in bud. Sometimes indistinguishable from corolla.

canescent. Covered with dense, fine, generally grayish white hairs. (example, *Phoenicaulis cheiranthoides* leaf)

capsule. (p. 18) Dry, generally many-seeded fruit from compound pistil, nearly always dehiscent (irregularly or by pores, slits, or lines of separation). (see circumscissile, loculicidal, septicidal)

carpel. (p. 17) The basic female structure of flowering plants, in concept derived from a leaf; an evolutionary term rarely used for identification in *The Jepson Desert Manual*. (see pistil)

catkin. (p. 16) Spike of unisexual flowers with inconspicuous perianths, sometimes pendent and often with conspicuous bracts.

caudex (caudices). (p. 10) Short, sometimes woody, more or less vertical stem of a perennial, at or beneath ground level. (see stem)

cauline. (p. 10) Borne on a stem; not basal. Said especially of leaves borne along an above-ground stem.

centimeter. One-hundredth of a meter; 10 millimeters (abbreviation: cm). (see ruler on inside front cover)

cespitose. (p. 10) Having a densely clumped, tufted, or cushion-like growth form, with the flowers held above the clump or tuft.

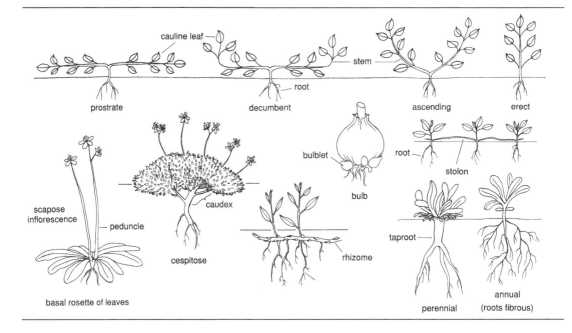

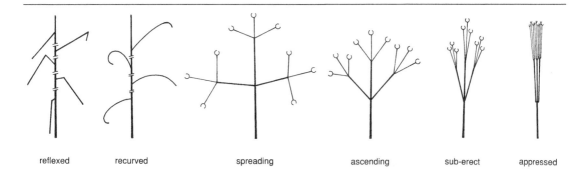

reflexed   recurved   spreading   ascending   sub-erect   appressed

chaff. (p. 21) Dry bracts; in Asteraceae, dry, generally papery or scaly, often persistent bracts on a receptacle.

chamber. (p. 17) Compartment or cavity within an ovary, capsule, or other hollow structure.

ciliate. (p. 19) Having generally straight hairs along the margin or edge.

circumboreal. Found around the world at northern latitudes.

circumscissile. (p. 18) Dehiscent by a transverse line, the top coming off as a lid. Generally said of a capsule. (see loculicidal, septicidal)

claw. (p. 17) Stalk-like base of some free sepals or petals.

cleistogamous. Bud-like, unopening flowers that are generally self-pollinated.

clone. Genetically identical individuals resulting from asexual reproduction (fragmentation of rhizomes or stolons, budding, etc.). Often used for an apparent population, the members of which are or were connected — e.g., aspens, cattails, duckweeds.

collar. (p. 22) The back (generally the outer surface) of a grass leaf at the junction of sheath and blade.

column. (p. 17) Structure at the center of an orchid flower formed by fusion of stamen(s) and style.

common. Likely to be encountered. (see abundant, rare, uncommon) (see p. 27 for more information)

compound. (p. 12) 1. Composed of two or more parts, as a *compound leaf* composed of leaflets (see compound leaf) or a *compound pistil* composed of fused or partly fused carpels. 2. Repeating a structural pattern (a compound umbel is an umbel of umbels). (see simple)

compound leaf. (p. 12) A leaf divided into distinct parts. In a *1-compound leaf*, the blade is divided into primary leaflets connected by an axis but no blade material (if there is connecting blade material, the leaf is lobed or dissected); in a *2-compound leaf*, the primary leaflets are so divided into secondary leaflets (if there is connecting blade material, primary leaflets are lobed); etc. (see palmate, pinnate)

compressed. (p. 19) Flattened side-to-side or front-to-back. (see depressed)

concave. Hollowed or indented, as the interior of a curved surface. (see convex)

cone. Reproductive structure composed of an axis, scales, and sometimes bracts. 1. Non-woody structure producing spores (e.g., clubmosses, horsetails) or pollen (e.g., conifers). 2. Generally woody structure producing seeds (e.g., most conifers, alders). (example, *Abies concolor*)

conic. Having a 3-dimensional shape defined by a wide, more or less round base, the sides evenly tapered to a narrow tip.

continuous. Having parts spaced evenly and without interruption, not clumped. Generally said of inflorescences. (see interrupted)

convex. Rounded outward, as the exterior of a curved surface. (see concave)

cordate. (p. 14) Heart-shaped, as of a leaf. Sometimes said of a leaf base with rounded lobes of which the sides adjacent to the petiole are convex. (see reniform)

corm. Short, thick, unbranched, underground stem often surrounded by dry (not fleshy) leaves or leaf bases. (see bulb, stem) (example, *Muilla maritima*)

corolla. (pp. 17, 21) Collective term for petals; whorl of flower parts immediately inside or above calyx, often large and brightly colored. Sometimes indistinguishable from calyx.

cotyledon. Seed-leaf; a modified leaf present in the seed, often functioning for food storage. Persistent in some annuals and of aid in their identification. (example, *Lupinus microcarpus*)

crenate. (p. 14) Scalloped — e.g., margins with gen acute sinuses between shallow, rounded teeth.

cylindric. (p. 14) Elongate, with parallel sides and, at any point, round in transverse section.

cyme. (p. 16) Branched inflorescence in which the central or uppermost flower opens before the peripheral or lowermost flowers on any axis. (see panicle)

deciduous. Falling off naturally at the end of a growing period. Generally said of leaves that fall seasonally and all together or of plants that are seasonally leafless. (see evergreen)

decimeter. One-tenth of a meter; 10 centimeters (abbreviation: dm). (see ruler on inside front cover)

decumbent. (p. 10) Mostly lying flat on the ground but with tips curving up. (see ascending)

decurrent. (p. 14) Having a wing-like or ridge-like extension beyond the actual or apparent point of attachment. Said especially of a leaf base that seems to continue down its stem.

dehiscent. (p. 18) Splitting open at maturity to release contents. Said especially of fruit or anthers. (see indehiscent)

deltate. (p. 14) More or less equilaterally triangular, with basal corners generally rounded.

dense. Congested or compact. Especially said of disposition of flowers in an inflorescence. (see open)

dentate. (p. 14) Having margins with sharp, relatively coarse teeth pointing outward, not tipward. (see serrate)

depressed. (p. 19) 1. Flattened from above and below. 2. with the center lower than the margins. (see compressed)

dicot. A member of the larger main subgroup of flowering plants; generally having two cotyledons, flower parts in 4's, 5's, or spirals, pinnate or palmate leaf venation, stem veins in rings (but often not all of these) — e.g., poppy, cactus, rose, sunflower.

dioecious. Male and female (or staminate and pistillate) plants separate. Said of a taxon in which individual plants produce either kind of unisexual fertile reproductive structures, but not both. (see monoecious) (example, *Salix laevigata*)

diploid. Having two sets of chromosomes (maternal and paternal), the normal complement in plant cells (except spores, sperm, eggs, some others); 2*n*. (see haploid, *n*, polyploid)

disciform head. In Asteraceae, a head composed of disk flowers and marginal pistillate flowers with minute or missing ligules, superficially similar to discoid head. (see ligulate head, radiate head)

discoid head. (p. 21) In Asteraceae, a head composed entirely of disk flowers. (see disciform head, ligulate head, radiate head)

disk. (p. 17) 1. Fleshy, often nectar-secreting structure near (often surrounding) an ovary base. 2. In Asteraceae, the part of a head made up of disk flowers.

disk flower. (p. 21) In Asteraceae, the generally bisexual (never pistillate), generally radial, ligule-less flower with a 5- (rarely 4-)lobed corolla. Appearing without other flower types (discoid head) or with marginal ray or pistillate flowers (radiate or disciform heads, respectively). (see ligulate flower, ray flower)

dissected. Irregularly, sharply, and deeply cut but not compound. Said especially of leaves. (see compound leaf, lobe) (example, *Cymopteris deserticola*)

distal. Farther away from the origin or point of attachment, more toward the edge or tip. (see proximal)

drupe. Fleshy or pulpy, indehiscent, superficially berry-like fruit with one seed encased in a hardened stone that is derived from inner ovary tissue (sometimes several seeds are encased separately or together). Ovary outside stone sometimes edible but stone and contents generally inedible. (see berry, nut, pome, stone) (example, *Prunus emarginata* fruit)

elliptic. (p. 14) In the shape of an ellipse (flattened circle). (see oblong)

emergent. 1. Plant normally rooted underwater and extending above the water surface. 2. Any part of a plant normally held above the water surface. (see aquatic) (example, *Polygonum amphibium*)

endangered. Survival is in immediate jeopardy. Used only for taxa given such status by law. (see extant, extinct, extirpated, rare, threatened) (see pp. 27, 28 for more information)

endemic. Native to a well defined geographic area and restricted to that area.

entire. (p. 14) Having margins that are continuous and smooth (i.e., without teeth, lobes, etc.).

ephemeral. 1. Lasting a short time. 2. Completing the life cycle (germination through death) or growth cycle in much less than one year, as most desert herbs. Said also of plant parts that are functional for a relatively short time or fall early.

epidermis. Outermost cell layer (or layers) of non-woody plant parts.

epipetalous. (p. 17) Stamens that are partly fused to the petals and therefore appear to arise from them.

erect. (p. 10) Upright; vertically oriented.

evergreen. 1. Never lacking green leaves. 2. Having leaves that remain green and on the plant for more than one season and do not fall all together. (see deciduous)

exceeding. Surpassing another structure tipward. (see exserted)

exserted. (p. 17) Protruded out of surrounding structure(s). (see exceeding, included)

extant. Surviving, in existence; not completely died out or destroyed. (see extinct, extirpated)

extinct. No longer living anywhere; completely died out. (see extant, extirpated)

extirpated. Destroyed or no longer surviving in the area being referred to (may survive outside that area). (see extant, extinct)

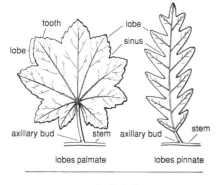

simple leaf

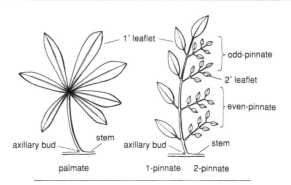

compound leaf

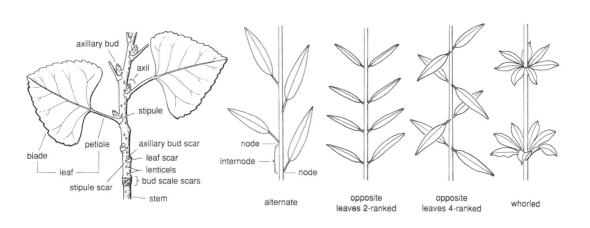

axillary bud
axil
stipule
petiole
axillary bud scar
blade
leaf scar
leaf
lenticels
bud scale scars
stipule scar
stem
node
internode
node
alternate
opposite
leaves 2-ranked
opposite
leaves 4-ranked
whorled

exudate. Fluid or solid material discharged from a plant surface that may take on a characteristic color or texture.

fertile. Reproductively functional. Said of a plant or plant part that produces or is associated with the production of functional spores, pollen, ovules, or seeds. (see sterile)

fibrous. (p. 10) 1. Composed of fine or slender structures. 2. Having a root system composed of many roots similar in length and thickness, as in grasses. (see taproot)

filament. (p. 17) Anther-stalk; the often thread-like portion of a stamen.

fleshy. Thick and juicy; succulent. (example, *Sesuvium verrucosum*)

floret. (p. 22) In Poaceae, a single flower and its immediately subtending bracts (generally lemma and palea).

follicle. (p. 18) Dry, generally many-seeded fruit from a simple pistil, dehiscent on only one side, along a single suture. A flower may have a simple fruit of 1 follicle or an aggregate fruit of several follicles. (see fruit)

forked. (p. 19) 1. Branching into two parts of about equal size. 2. Hair with branches that do not radiate from a common point. (see stellate)

free. Not fused to other parts; distinct, separate. (see adherent, fused)

free-central. (p. 17) Pertaining to a placenta along the central axis in a compound ovary with only one chamber. (see axile, basal, parietal)

fringed. (p. 19) Having ragged or finely cut margins.

fruit. (p. 18) A ripened ovary and sometimes associated structures. A *simple fruit* develops from one ovary — e.g., cherry, apple, the latter derived largely from the hypanthium; *aggregate* and *multiple fruits* develop from ovaries of one and more than one flower, respectively, held together as a unit — e.g., a strawberry is an aggregate fruit of achenes held together by a juicy, red flower receptacle; a fig is a multiple fruit of achenes surrounded by a fleshy inflorescence receptacle. (see achene, berry, capsule, drupe, follicle, legume, nut, nutlet, pome, utricle)

funnel-shaped. (p. 16) Widening from the base more or less gradually through the throat into an ascending, spreading, or recurved limb. Said usually of a fused calyx or corolla.

fused. (p. 21) United, as the petals together into a corolla tube or stamens onto petals; not free. (see adherent, free)

fusiform. (p. 19) Elongate, widest at the middle, tapered to both ends.

glabrous. Without hairs.

gland. A small, often spheric body that exudes a generally sticky substance, on (or embedded in) epidermis or at the tip of a hair. (example, *Psorothamnus arborescens*)

glaucous. Covered with a generally whitish or bluish, waxy or powdery film that is sometimes easily rubbed off.

glume. (p. 22) In Poaceae, each of generally two sheathing bracts that are the lowermost parts of a spikelet. (see lemma, palea)

granular. (p. 19) Covered with minute bumps. (see papillate, tubercle)

gymnosperm. Woody plant with seeds that are not borne in ovaries but in cones or naked on branches — e.g., pine, sequoia, ephedra, yew.

habit. Characteristic mode of growth; general form or shape of a plant — e.g., cespitose, herb, scapose, shrub.

habitat. Natural setting or abode of a plant, generally specified as a plant community or set of environmental features.

hair. Thread-like epidermal outgrowth. (see puberulent, strigose, trichome)

haploid. Having one set of chromosomes (maternal or paternal), the normal complement in spores, sperm, eggs, and some other cells that are derived from those; *n*. (see diploid, *n*, polyploid)

hastate. (p. 14) Arrowhead-shaped, with two basal lobes oriented more or less perpendicularly to the long axis. (see sagittate)

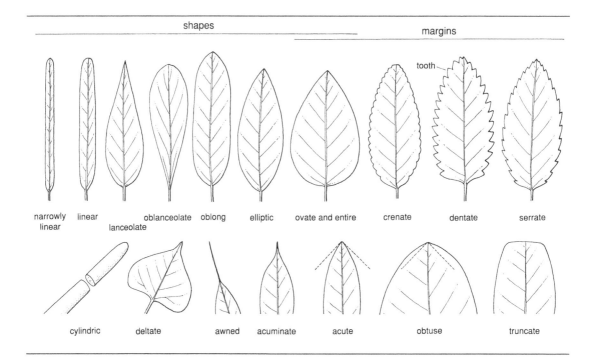

shapes   margins

narrowly   linear   oblanceolate   oblong   elliptic   ovate and entire   crenate   dentate   serrate
linear   lanceolate   tooth

cylindric   deltate   awned   acuminate   acute   obtuse   truncate

head. (p. 16) Dense, often spheric inflorescence of sessile or subsessile flowers.

hemispheric. Shaped like a dome or half sphere.

herb. Plant with little or no wood above ground; above-ground parts are of less than one year or growing season duration. All plants called annual, biennial, or perennial in *The Jepson Desert Manual* are herbs. (see annual, biennial, perennial, subshrub)

herbaceous. Lacking wood; having the characteristics of an herb.

herbage. The non-woody, above ground parts of a plant, especially the leaves and young stems taken together.

heterostylous. Having different kinds of style (and stamen) lengths. Said of a species in which individuals produce only one of two or more flower types, each differing in style (and generally stamen) length.

hypanthium (hypanthia). (p. 17) Structure derived from the fused lower portions of sepals, petals, and stamens and from which these parts seem to arise, the whole generally in the shape of a tube, cup, or plate. An inferior ovary is fused to part or all of a ahypanthium. (see inferior ovary)

included. (p. 17) Not protruding out of surrounding structure(s). (see exserted)

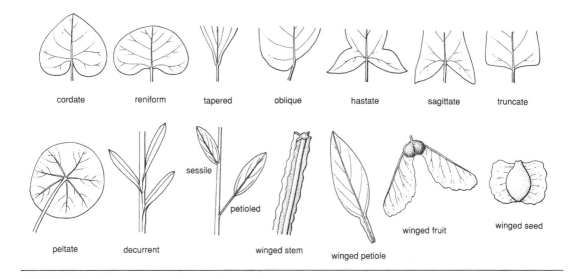

cordate   reniform   tapered   oblique   hastate   sagittate   truncate

peltate   decurrent   sessile   petioled   winged stem   winged petiole   winged fruit   winged seed

indehiscent. (p. 18) Not opening to release contents. Generally said of fruits. (see dehiscent)

indusium (indusia). (p. 20) In many ferns, a veil- or scale-like outgrowth of the leaf surface or margin that covers a sorus (cluster of sporangia).

inferior ovary. (p. 17) An ovary that appears to be beneath its flower or to bear the sepals, petals, and stamens at or above its summit, owing to its fusion to the hypanthium. (see hypanthium, superior ovary)

inflorescence. (pp. 10, 16, 22) An entire cluster of flowers and associated structures — e.g., axes, bracts, bractlets, pedicels. Often difficult to define as to type and boundaries but generally excluding full-sized foliage leaves.

infraspecific. Below the species level. Said especially of variation within a species, whether taxonomically significant (i.e., characterizing subspecies or varieties) or not.

intergrade. To merge gradually from one extreme to another through a more or less continuous series of intermediates.

intermediate. Between extremes in form and sometimes in other ways.

lanceolate. (p. 14) Narrowly elongate, widest in the basal half, often tapered to an acute tip.

lateral. Referring to the sides(s) of a structure — e.g., laterally compressed (flattened side-to-side), lateral branch (from "side" of stem).

leaf. (pp. 12, 13, 20, 22) Stem appendage with a structure such as a bud, branch, or flower in its axil, generally green and often composed of a stalk (petiole) and a flat, expanded, photosynthetic area (blade).

leaflet. (pp. 12, 20) One leaf-like unit of a compound leaf, which may be primary, secondary, etc. (see compound leaf)

legume. (p. 14) 1. In Fabaceae, a dry or somewhat fleshy, one- to many-seeded fruit from a simple pistil, typically dehiscent longitudinally along two sutures and splitting into halves that remain joined at the base, sometimes indehiscent or breaking crosswise into one-seeded segments. 2. A plant with such a fruit.

lemma. (p. 22) In Poaceae, the lower, generally larger of two sheathing bracts that directly subtend a flower; the lowermost part of a floret. (see glume, palea, spikelet)

lenticel. (p. 13) A spongy area (pore), most common on surfaces of twigs or fruits.

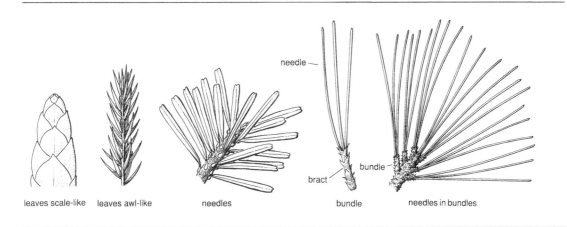

leaves scale-like   leaves awl-like   needles   bundle   needles in bundles

needle —

bundle —

bract

internode. (p. 13) Stem segment between leaves; segment of an axis between two successive attachment points for appendages. (see node)

interrupted. Having parts spaced unevenly. Generally said of inflorescences in which the axis is elongated between flower clusters. (see continuous)

involucel. (p. 16) A secondary involucre (group of bracts) within an inflorescence — e.g., those subtending the secondary umbels in members of Apiaceae.

involucre. (pp. 16, 21) Group of bracts more or less held together as a unit, subtending a flower, fruit (acorn cup), or inflorescence (the combined phyllaries of a daisy).

keel. (p. 16) 1. Ridge or crease more or less centrally located on the long axis of a structure, generally on the under or outer side. 2. The two lowermost, fused petals of many members of Fabaceae.

lenticular. Lens- or discus-shaped, with both sides convex.

ligulate flower. (p. 21) In Asteraceae, a bisexual, bilateral flower with the long, outer portion of the corolla (the ligule) 5-lobed. Appears only with other ligulate flowers in a ligulate head. (see disk flower, ray flower)

ligulate head. (p. 21) In Asteraceae, a head composed entirely of ligulate flowers. (see disciform head, discoid head)

ligule. (pp. 21, 22) 1. In Asteraceae, the strap- or blade-like outer portion of the corolla in ligulate and ray flowers. 2. In Poaceae and other grass-like plants, an appendage at the juncture of leaf sheath and blade, generally with a membranous or fringed margin.

limb. (p. 17) In calyces or corollas with fused parts, the expanded, often lobed portion above the tube or throat.

linear. (p. 14) Elongate, with nearly parallel sides, and narrower than oblong.

lip. (p. 16) 1. Upper or lower of two parts in an unequally divided calyx or corolla. 2. In Orchidaceae, generally the largest, lowest, most highly modified perianth part.

lobe. (pp. 12, 17, 20–22) 1. A major expansion or bulge, such as on the margin of a leaf or petal or on the surface of an ovary. 2. The free tips of otherwise fused structures, such as sepals or petals.

loculicidal. (p. 18) A capsule, longitudinally dehiscent through the ovary wall at or near the center of each chamber. (see circumscissile, septicidal)

longitudinal. Pertaining to length or the lengthwise dimension; parallel to the axis. (see transverse)

margin. (pp. 14, 20) The edge, generally of a leaf or perianth part.

membranous. Thin, pliable, sometimes somewhat translucent, sometimes green. (see scarious) (example, *Leymus cinereus* ligule)

meter. Unit of length in the metric system, equal to 39.4 inches, slightly more than a yard (abbreviation: m). (see ruler on inside front cover)

millimeter. One-thousandth of a meter; one-tenth of a centimeter. The smallest unit of size used in *The Jepson Desert Manual* (abbreviation: mm). (see ruler on inside front cover)

monocot. The smaller main subgroup of flowering plants; generally having one cotyledon, flower parts in 3's, parallel leaf venation, stem veins scattered (but often not all of these) — e.g., lily, orchid, grass, cattail, palm.

monoecious. Male and female (or staminate and pistillate) unisexual structures (flowers) on the same plant. Said of a taxon having only unisexual fertile reproductive structures. (see dioecious) (example, *Alnus rhombifolia*)

montane. Pertaining to mountains; the region between foothills and subalpine.

n. Number of chromosomes in sperm and egg cells. The number in other cells of plants treated in *The Jepson Desert Manual* is generally 2*n*. (see diploid, haploid, polyploid)

native. Occurring naturally in an area, not as either a direct or indirect consequence of human activity; indigenous; not alien.

naturalized. Alien (not native) but reproducing without human fostering. (see native, waif)

nectary. Structure that secretes nectar, often near the base of an ovary or in a perianth spur. Nectar is a nutritive solution consumed by animal visitors that are often pollinators. (example, *Symphoricarpos rotundifolius*)

needle. (p. 15) A narrowly linear, often waxy, generally evergreen leaf, especially of conifers.

node. (pp. 13, 22) Position on an axis (generally a stem) from which one or more structures (especially leaves) arise.

nut. A mostly dry, indehiscent fruit in which a single seed is encased in a hard shell that is derived from inner ovary tissue. Ovary tissue outside shell sometimes

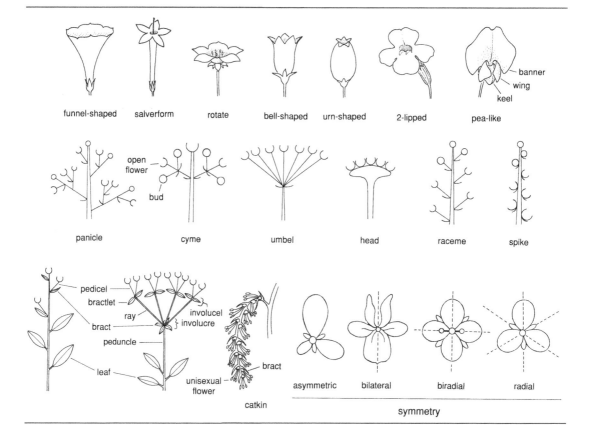

funnel-shaped salverform rotate bell-shaped urn-shaped 2-lipped pea-like banner wing keel

panicle open flower bud cyme umbel head raceme spike

pedicel bractlet ray involucel involucre bract peduncle leaf unisexual flower bract catkin

asymmetric bilateral biradial radial

symmetry

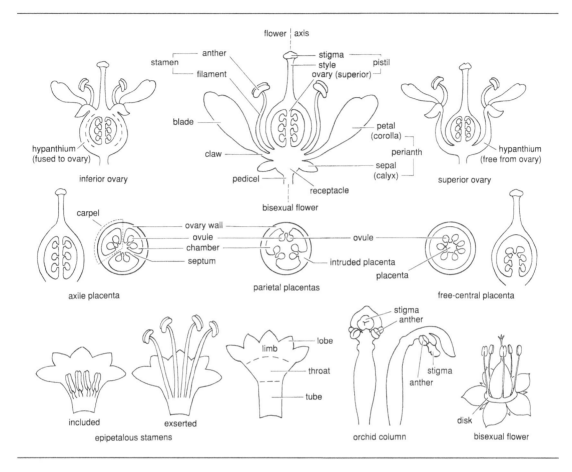

fleshy (generally inedible) but seed within shell often edible. (see drupe) (example, *Quercus palmeri*)

nutlet. Small, dry nut (or nut-like fruit), generally one of several produced by a single flower. (see nut, drupe) (example, Boraginaceae, Lamiaceae)

ob-. (p. 14) A prefix indicating inversion of shape — e.g., lanceolate and oblanceolate leaf blades are widest below and above the middle, respectively.

oblique. (p. 14) Having unequal sides or an asymmetric base.

oblong. (p. 14) Longer than wide, with nearly parallel sides and rounded corners; wider than linear. (see elliptic, linear)

obtuse. (p. 14) Having a short-tapered, blunt tip or base, the sides convex or straight and converging at more than a right angle. (see acute)

open. Uncongested or diffuse. Said especially of the disposition of flowers in an inflorescence. (see dense)

opposite. (p. 13) 1. Located directly across from. 2. Two structures (generally leaves) per node. 3. Superimposed structures that are in the same rank — e.g., "stamens opposite petals". (see alternate)

or. Unless stated otherwise, in *The Jepson Desert Manual*, defined as "one or the other or both"; for example, "leaves toothed or lobed" does not exclude leaves that are both toothed and lobed.

ovary. (pp. 17–19) Ovule-bearing, generally wider, portion of pistil, normally developing into a fruit as ovules become seeds. (see pistil)

ovate. (p. 14) Egg-shaped in two dimensions, widest below the middle, as of a leaf.

ovoid. Egg-shaped in three dimensions, widest below the middle, as of a fruit.

ovule. (p. 17) Structure containing an egg; a seed prior to fertilization.

palea. (p. 22) In Poaceae, the upper and generally smaller of two sheathing bracts subtending a flower, itself generally ensheathed by the lemma. (see glume, lemma, spikelet)

palmate. (p. 12) Radiating from a common point. Generally said of veins, lobes, or leaflets of a leaf.

panicle. (p. 16) Branched inflorescence in which the basal or lateral flowers (or some of them) open before the terminal or central flowers on any axis. (see cyme)

papillate. (p. 19) Bearing small, rounded or conic protuberances (papillae). Said especially of a leaf or fruit surface.

pappus. (p. 21) In Asteraceae, the aggregate of structures such as awns, bristles, or scales arising from the top of the inferior ovary, in the place sepals would be expected.

parasite. A plant that benefits from a physical connection to a host plant of another species and often in time harms the host. *Green parasites* derive water and dissolved substances and often are able to survive without the connection, while *non-green parasites* obtain in addition energy-rich products of photosynthesis and require the connection to survive.

parietal. (p. 17) Pertaining to placentas on the inside surface of the ovary wall in a compound ovary with one or more chambers.

pedicel. (pp. 16, 17) Stalk of an individual flower or fruit. (see peduncle, ray)

peduncle. (pp. 10, 16, 21) Stalk of an entire inflorescence or of a flower or fruit not borne in an inflorescence. (see pedicel, ray)

peltate. (p. 14) With the stalk (of a leaf, scale, or other flat structure) attached toward the middle, not at a margin.

petiole. (pp. 13, 14, 20) Leaf stalk, connecting leaf blade to stem.

phyllary. (p. 21) In Asteraceae, a bract of the involucre that subtends a head.

pinnate. (p. 12) Feather-like, with two rows of structures on opposite sides of an axis. Generally said of veins, lobes, or leaflets arranged in two dimensions along either side of an axis. A leaf is *odd-pinnate* if there is a terminal leaflet, *even-pinnate* if there is not, and either may be *1-pinnate* (blade divided into primary leaflets), *2-pinnate* (primary leaflets divided into secondary leaflets), etc. (see compound leaf)

pistil. (p. 17) Female reproductive structure of a flower, composed of an ovule-containing ovary at the base, one or more pollen-receiving stigmas at the tip, and generally one or more styles between ovary and stigma. A flower may have one or more *simple pistils* (each a single, free carpel with a single ovary chamber, placenta, and stigma) or one *compound pistil* (two or more fused

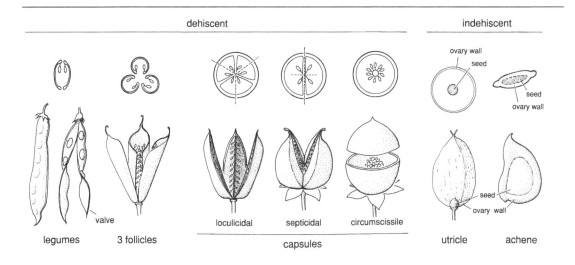

pendent. Drooping, hanging, or suspended from a point of attachment above. (example, *Amelanchier utahensis* fruit)

perennial. (p. 10) Living more than two years or growing seasons; restricted in *The Jepson Desert Manual* to plants that are essentially non-woody aboveground. (see annual, biennial, herb, subshrub)

perianth. (p. 17) Calyx and corolla collectively, whether or not they are distinguishable.

perianth part. An individual member of a perianth, whether or not calyx and corolla are distinguishable. Normally used when they are not distinguishable.

perigynium. (p. 19) Sac-like structure enclosing the ovary and achene in *Carex*, of diverse form and critical to identification.

persistent. Not falling off; remaining attached. (see deciduous, ephemeral)

petal. (p. 17) Individual member of the corolla, whether fused or not; often conspicuously colored. (see sepal)

or partially fused carpels, the exact number often equaling the number of ovary lobes, ovary chambers, placentas, styles, or stigmas).

pistillate. Having fertile pistils but sterile or missing stamens. Said of flowers, inflorescences, or plants. (example, *Salix laevigata* flower)

placenta. (p. 17) Structure or area to which ovules are attached in an ovary, variously shaped and positioned.

planoconvex. Solid shape, with one side nearly flat, the other rounded. (example, *Carex leporinella* perigynium)

pleated. Having accordion-like folds.

plumose. (p. 19) Plume-like; generally with fine appendages arrayed in three dimensions around an axis, or in tufts held together at the base. Said especially of certain stigmas and pappus elements.

pollination. Placement, in any way, of pollen on a stigma (or other floral surface through which fertilization may be achieved).

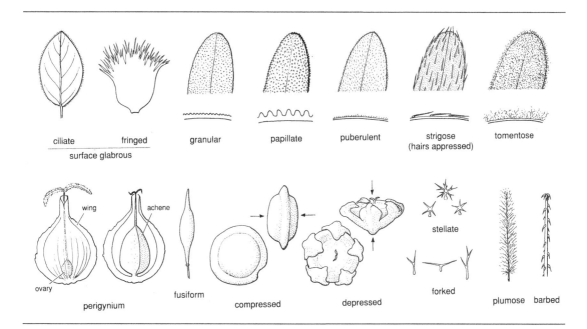

ciliate fringed granular papillate puberulent strigose (hairs appressed) tomentose

surface glabrous

wing achene

ovary

perigynium fusiform compressed depressed stellate forked plumose barbed

polyploid. Having three or more sets of chromosomes; 3*n*, 4*n*, etc. (see diploid, haploid, *n*)

pome. In Rosaceae, a fleshy, indehiscent fruit, such as an apple or pear. Derived from a compound, inferior ovary (the core and inner fleshy material) and its surrounding hypanthium (outer fleshy material and skin). (see berry, drupe) (example, *Amelanchier utahensis* fruit)

prickle. Superficial, sharp-pointed projection, derived from epidermis, bark, etc. (see armed, spine, thorn) (example, *Rosa woodsii* stem)

prostrate. (p. 10) Lying flat on the ground. (see decumbent)

protandrous. Releasing pollen first. Said of a flower (or plant with unisexual flowers) in which pollen release precedes and does not overlap stigma receptivity.

protogynous. Receiving pollen first. Said of a flower (or plant with unisexual flowers) in which stigma receptivity precedes and does not overlap pollen release.

proximal. Closer to the origin or point of attachment (or farther away from the edge or tip). (see distal)

puberulent. (p. 19) Having hairs normally visible only when magnified.

raceme. (p. 16) Unbranched inflorescence of pediceled flowers that open from bottom to top. (see panicle, spike)

radial. (p. 16) Divisible into mirror-image halves in three or more ways. (see asymmetric, bilateral, biradial)

radiate head. (p. 21) In Asteraceae, a head composed of central disk flowers and marginal ray flowers.

rank. (p. 13) 1. A row or column of parts of the same orientation along an axis — e.g., leaves on an erect stem that are arranged in four vertical rows are 4-ranked. (see alternate, opposite) 2. In classification, a level — e.g., family, genus, species. (see taxon)

rare. Extremely unlikely to be encountered, often not present in appropriate habitats and often restricted to a small number of sites. Used only for certain taxa included in *Index of Rare and Endangered Vascular Plants of California* (CNPS), some of which have been accorded such legal status by the State of California. (see endangered, threatened, uncommon)

ray. (p. 16) A primary, radiating axis, as a primary branch in a compound umbel. (see pedicel, peduncle)

ray flower. (p. 21) In Asteraceae, a generally pistillate or sterile, bilateral flower with the long, outer portion of the corolla (ligule) often 3-lobed, appearing on the margin of a head and accompanied by more central disk flowers. (see ligulate flower, disk flower)

receptacle. (pp. 17, 21) 1. In individual flowers, the structure to which flower parts are attached. 2. In heads or head-like inflorescences, especially in Asteraceae, the structure to which flowers or sometimes heads are attached.

recurved. (p. 11) Gradually curved downward or backward.

reduced. Gradually smaller; often narrower, less lobed, etc.

reflexed. (p. 11) Abruptly bent or curved downward or backward.

reniform. (p. 14) Kidney-shaped, as of a leaf; sometimes a leaf base having rounded lobes with the sides adjacent to the petiole clearly concave. (see cordate)

rhizome. (pp. 10, 20, 22) Underground, often elongate, more or less horizontal stem. Distinguished from root by presence of leaves, leaf scars, scales, buds, etc. (see stem)

rib. 1. Ridge, as on a fruit. 2. Raised vein, as on a leaf or perianth part. (example, *Carex pellita* perigynium)

root. (pp. 10, 20, 22) Underground structure of a plant, generally branched, without appendages, generally

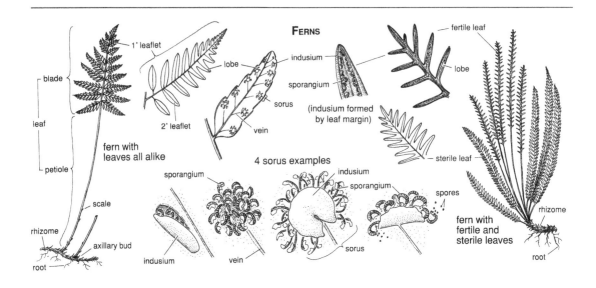

FERNS

growing into the ground from the base of a stem. Its functions include anchorage, absorption of water and nutrients, and food storage. (see bulb, corm, rhizome, stem)

rosette. (p. 10) A radiating cluster of leaves generally at or near ground level.

rotate. (p. 16) Wheel-shaped, spreading, or saucer-shaped. Said of a fused corolla with a short or nonexistent tube and a spreading limb.

sagittate. (p. 14) Arrowhead-shaped, with two basal lobes oriented nearly parallel to the long axis. (see hastate)

salverform. (p. 16) Having a slender tube and an abruptly spreading, flat limb. Said especially of a fused corolla.

scabrous. Rough to the touch, generally owing to short stiff hairs. (example, *Leersia oryzoides*)

scale. 1. Wide, appressed, membranous, epidermal outgrowth (example, *Cheilanthes covillei*). 2. Structure partially or entirely covering an over-wintering bud (bud scale) (example, *Salix gooddingii* bud). 3. In gymnosperms, a woody, seed-bearing structure attached to the cone axis (cone scale) (example, *Abies magnifica*). 4. In Asteraceae, a flat, membranous pappus element (example, *Dugaldia hoopesii*). Leaves or bracts may be scale-like in one or more of the preceding ways.

scapose. (p. 10) Pertaining to a plant or an inflorescence having a relatively long peduncle that arises from ground level, often from a rosette, sometimes bearing bracts but without leaves.

scar. (p. 13) Mark left by the natural separation of two structures, as a leaf scar on a stem.

scarious. Thin, dry, pliable, dark-colored or translucent but not green. Often like dry onion peel. (see membranous) (example, *Carex incurviformis* pistillate flower bract)

scree. Relatively unstable, sloping accumulation of small rock fragments, often at a cliff base. (see talus)

sculpture. Surface ornamentation, often visible only when magnified, as on a seed. (example, *Plagiobothrys nothofulvus* nutlet)

seed. (p. 18) A fertilized ovule, the earliest product of sexual reproduction in plants. In descriptions, the fully mature form (at full fruit maturation) is assumed unless noted.

segment. One of the repeated components of an organ, such as a perianth, fruit, or leaf (example, *Eremalche rotundifolia* fruit). A leaf segment is one of the ultimate or smallest divisions of the blade (not a marginal lobe, tooth, bristle, etc.).

sepal. (p. 17) Individual member of the calyx, whether fused or not, generally green. (see petal)

septicidal. (p. 18) Pertaining to a capsule, dehiscent longitudinally through the ovary wall at or near the center of each septum, such that each resulting valve or segment corresponds to a single chamber. (see circumscissile, loculicidal)

septum (septa). (p. 17) Wall between chambers in a compound ovary.

series. (p. 21) A group of structures of similar size or shape, generally more or less in a whorl — e.g., involucre bracts may be in one or more series.

serpentine. General term for rocks with unusually high concentrations of magnesium and iron or the soils derived from them. Both are characterized by low levels of calcium and other nutrients and high levels of magnesium, iron, and certain toxic metals. Many plant taxa are restricted to or excluded from serpentine.

serrate. (p. 14) Having margins with sharp teeth generally pointing tipward, not outward. (see dentate)

sessile. (p. 14) Without a petiole, peduncle, pedicel, or other kind of stalk.

sheath. (p. 22) Structure that surrounds or partly surrounds another structure, often tubular, as a leaf base in Apiaceae or Poaceae.

COMPOSITE FAMILY

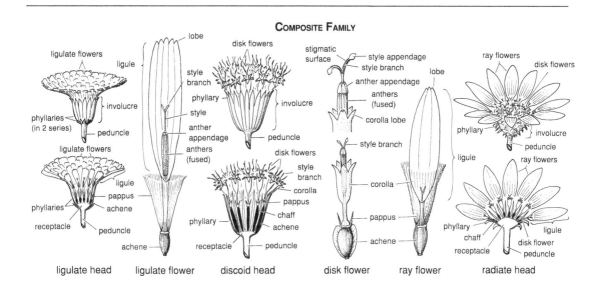

ligulate head   ligulate flower   discoid head   disk flower   ray flower   radiate head

shoot. 1. A stem and its appendages collectively. 2. Sometimes used for all aboveground parts of a plant.

shrub. 1. A woody plant of relatively short maximum height. 2. A woody plant much-branched from the base. (see tree, subshrub)

simple. (p. 12) Composed of a single part; undivided; unbranched. (see compound)

sinus. (p. 12) An indentation, as between adjacent lobes of a margin.

sorus (sori). (p. 20) In many ferns, a distinct cluster of sporangia.

spheric. Globe- or ball-shaped; round in three dimensions.

spike. (p. 16) Unbranched inflorescence of sessile flowers, nearly always opening from bottom to top.

spikelet. (p. 22) 1. In Poaceae, the smallest aggregation of florets plus any (generally 2) subtending glumes. 2. In Cyperaceae, the smallest aggregation of flowers (i.e., generally more than 2) and associated bracts.

spine. Sharp-pointed projection, derived from leaf (often vein tip) or other organ, such as ovary wall. Sometimes used for any sharp projection. (see armed, prickle, thorn) (example, *Cirsium arvense* leaf)

sporangium (sporangia). (p. 20) In non-seed plants (fern allies and ferns), a spore-producing organ (some ferns, such as *Marsilea*, bear sporangia in hard cases).

spore. (p. 20) The minute, dispersing, reproductive unit of non-seed plants (fern allies and ferns); one of very many haploid cells dispersed from a diploid parent plant, normally developing into a small haploid plant that produces eggs, sperm, or both, the fusion of which results in new diploid offspring.

spreading. (p. 11) Oriented more or less perpendicularly to the axis of attachment; often, more or less horizontal.

spur. Hollow, often conic, projection or expansion, generally of a perianth part and containing nectar. (example, *Aquilegia formosa* flower)

stamen. (p. 17) Male reproductive structure of a flower, typically composed of a stalk-like filament and a terminal, pollen-producing anther. Filaments sometimes partly fuse to the corolla, or to other filaments to form a tube. (see anther, filament, pistil)

staminate. Having fertile stamens but sterile or missing pistils. Said of flowers, inflorescences, or plants. (see pistillate) (example, *Salix laevigata*)

staminode. Sterile stamen, often modified in appearance, sometimes petal-like or elaborate in structure. (example, *Penstemon palmeri*)

stellate. (p. 19) Star-like. Generally said of a hair with three or more branches radiating from a common point. (see forked)

stem. (pp. 10, 12, 13) Axis or axes of a plant, bearing appendages such as leaves, axillary buds, and flowers. Sometimes below ground. (see bulb, caudex, corm, rhizome, root, stolon, tuber)

sterile. Not reproductively functional. Said of a plant or plant part that does not produce or is not associated with the production of functional spores, pollen, ovules, or seeds. (see fertile)

stigma. (pp. 17, 21, 22) The part of a pistil on which pollen is normally deposited, generally terminal and elevated above the ovary on a style, generally sticky or hairy, sometimes lobed.

stipule. (p. 13) Appendage at base of petiole, generally paired, variable in form but often leaf- or scale-like, sometimes a spine.

stolon. (pp. 10, 22) Runner; a normally thin, elongate stem lying more or less flat on the ground and forming roots as well as erect stems or shoots (i.e., ultimately new plants) at generally widely spaced nodes. (see shoot, stem)

stomate. A minute pore on a leaf or stem through which gases such as carbon dioxide, oxygen, and water vapor pass by diffusion. Features of stomates help identify some plants.

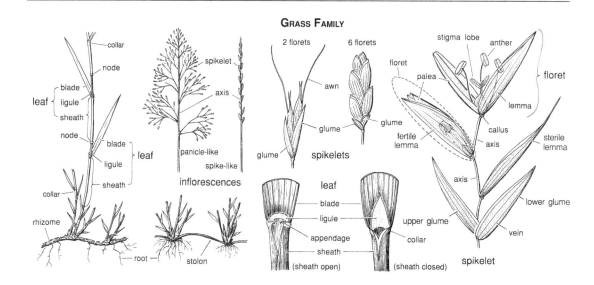

GRASS FAMILY

stone. In a drupe, the very hard inner ovary wall and the generally single seed it surrounds; occurring one or more per flower, free or variously fused. (example, *Prunus emarginata* fruit)

stout. Thick, sturdy, not slender.

striate. With fine, longitudinal channels, lines, or ridges.

strigose. (p. 19) With stiff, straight, sharp, appressed hairs.

style. (pp. 17, 21) Stalk-like portion that connects ovary to stigma in many pistils.

sub-. A prefix meaning almost, just below, or somewhat imperfectly.

subalpine. Just below timberline; between montane and alpine.

submersed. 1. A plant normally rooted and remaining underwater. 2. The part of such a plant normally held underwater.

subshrub. A plant with the lower stems woody, the upper stems and twigs not woody (or less so) and dying back seasonally. (see perennial, shrub)

subtend. Occurring immediately below, as sepals subtending petals or leaves subtending axillary buds.

superior ovary. (p. 17) An ovary that is free from the perianth or hypanthium and appears to sit on top of the receptacle. Lower parts of sepals, petals, and stamens (or hypanthium) arise from near ovary base instead of its top. (see inferior ovary)

suture. Groove or line of dehiscence or fusion.

talus. Relatively stable, sloping accumulation of large rock fragments, often at a cliff base. (see scree)

tapered. (p. 14) Gradually (not abruptly) narrower or smaller at base or tip. (see truncate)

taproot. (p. 10) Main, tapered root that generally grows straight down into soil and has smaller, lateral branches.

taxon (taxa). In classification, a group of organisms (such as plants) at any rank — e.g., species, family. (see rank)

tendril. A slender, coiling structure (generally stem, stipule, or leaf tip) by which a climbing plant becomes attached to its support. (example, *Lathyrus lanszwertii* leaf)

terminal. At the tip of a structure.

ternate. Once or repeatedly lobed or compounded into three parts, as a clover leaf. (example, *Trifolium wormskioldii* leaf)

thorn. Sharp-pointed branch. (see armed, prickle, spine) (example, *Castela emoryi*)

threatened. Survival is in jeopardy but not the extreme jeopardy implied by "endangered". May be used in a general sense, as well as to indicate such status accorded by law (see endangered, rare)

throat. (p. 17) In flowers with fused sepals or petals, the expanded, fused portion above the tube and below the limb.

tomentose. (p. 19) Covered with densely interwoven, generally matted hairs.

tooth (teeth). (pp. 12, 14) A small, pointed projection of a margin. (see dentate, serrate)

transverse. Pertaining to width or the widthwise dimension; perpendicular to the axis. (see longitudinal)

tree. A woody plant of medium to tall maximum height, with generally one relatively massive trunk at the base. (see shrub) (example, *Carnegiea gigantea*)

trichome. Any epidermal outgrowth of a plant. Not used in *The Jepson Desert Manual*. (see hair, scale)

truncate. (p. 14) Abruptly (not gradually) narrower or smaller at base or tip, as if cut straight across or nearly so. (see tapered)

tube. (p. 17) In flowers with fused sepals or petals, the more or less cylindric, fused portion at the base.

tuber. Short, thick, fleshy, underground stem for storage (of water, food, or both) and sometimes propagation — e.g., potato. (see stem)

tubercle. Small, wart-like projection. (example, *Cryptantha muricata* nutlet)

twig. In woody plants, a terminal stem segment, produced during the current or most recent growth period.

twining. Twisting or coiling, normally for the purpose of climbing. Generally said of stems or tendrils that wind around a support. (example, *Antirrhinum filipes*)

ultimate. Last, most distal, or smallest, as all the tips of a branching stem or the smallest divisions (segments) of a compound leaf.

umbel. (p. 16) Inflorescence in which three to many pedicels radiate from a common point. May be compound, in which case larger inflorescence branches (rays) also radiate from a common point. Characteristic of but not confined to Apiaceae.

uncommon. Unlikely to be encountered and sometimes not present in appropriate habitats. Used in a general sense and (in all capital letters) also for some taxa included in *Inventory of Rare and Endangered Vascular Plants of California* (CNPS). (see common, rare)

unisexual. Having flowers in which either stamens or pistils, but not both, are fertile. (see bisexual, pistillate, staminate)

urn-shaped. (p. 16) Pertaining to a fused calyx or corolla that is gradually or abruptly narrowed toward the tip.

utricle. (p. 18) Mostly dry, generally indehiscent fruit from a generally compound pistil in which a single seed is loosely enclosed by a balloon- or bladder-like ovary wall.

valve. (p. 18) One of the parts into which a capsule or legume splits.

vascular. Pertaining to plant veins or to plants with veins. Only vascular plants are treated in *The Jepson Desert Manual*.

vein. (pp. 20, 22) 1. Tissue specialized for transport of substances within a plant (xylem for water and dissolved substances, phloem for energy-rich, organic compounds). 2. A strand of such tissue, often seen as a bundle in transverse section.

vernal. Pertaining to the spring season.

vestigial. Rudimentary. Said of a structure that is undeveloped, poorly developed, or degenerate and therefore non-functional.

vine. A trailing or climbing plant, sometimes attaching to its support by tendrils. (example, *Phaseolus filiformis*)

waif. Alien (not native) and either not reproducing without human intervention or not persisting for more than a few generations and therefore incompletely naturalized. Most known waifs are not treated in *The Jepson Desert Manual*.

weed. A generally alien, generally undesired (sometimes attractive) plant, often adapted to disturbed places, often aggressive, occurring in or near settlements, in fields or gardens, along roadsides, and in relatively undisturbed communities of native plants. Weeds considered legally noxious by the State of California are so noted.

whorl. (p. 13) Group of three or more structures of the same kind (generally leaves or flower parts) at one node.

wing. (pp. 14, 16, 19) 1. Thin, flat extension or appendage of a surface or margin. 2. In many members of Fabaceae, each of two lateral petals.

wiry. Slender, stiff, and tough. Used for stems, etc.

wood. Hard, thickened, vascular tissue (xylem) that develops especially in shrubs, trees, and some vines, generally in concentric rings.

ABBREVIATIONS AND SYMBOLS

The abbreviations adopted in *The Jepson Desert Manual* are the same abbreviations used in *The Jepson Manual: Higher Plants of California* (J. Hickman, ed., 1993). The philosophy behind choosing abbreviations for *The Jepson Manual* was based on a critical need to conserve space (as in *The Jepson Desert Manual*). This need ran counter to that of maximizing ease of understanding. The abbreviations used were selected or designed to be unambiguous, self-explanatory, and easily remembered. They are used throughout *The Jepson Desert Manual,* with the exception of most introductory material, such as the glossary.

Most of the abbreviations are used for describing geographic ranges. About half of the geographic abbreviations were designed during the Jepson Manual Project specifically for geographic units of California. These geographic abbreviations are preceded by asterisks (*) and are discussed more fully in the chapter "Geographic subdivisions of the California deserts" (p. 33). Other geographic abbreviations are in wide use.

Fewer than a dozen plant features are abbreviated. The most commonly encountered abbreviations of plant features are highlighted in **BOLDFACE CAPITALS** at the beginning of sentences within a plant description ("st", "lf", "fl", etc.). These abbreviations serve as visual guides to the structure of the description. A few other abbreviations (e.g., "lfless") are derived from them. Four special abbreviations are used only in certain illustration captions to save space and one is used only to cross-reference terms in the glossary. Thirteen abbreviations are restricted to summaries of horticultural potential; these summaries follow some descriptions.

Abbreviations

Afr = Africa
AK = Alaska
Am = Americas, western hemisphere
ann = annual
AZ = Arizona

b = back (of an organ, used only in some illustration captions)
B.C. = British Columbia
bien = biennial
br = bract (used only in some illustration captions)

c = central
CA = California
*CA-FP = California Floristic Province
C.Am = Central America
Can = Canada
*CaR = Cascade Range
*CaRF = Cascade Range Foothills
*CaRH = High Cascade Range
*CCo = Central Coast
*ChI = Channel Islands
cm = centimeter
Co. = County
cos. = counties
cult = cultivated
CVS = horticultural entries only: cultivars are available in the trade
*CW = Central Western California

*D = Desert Province
DFCLT = horticultural entries only: difficult; needs special care in gardens
diam = diameter
dm = decimeter
*DMoj = Mojave Desert
*DMtns = Desert Mountains
DRN = horticultural entries only: requires excellent drainage
DRY = horticultural entries only: intolerant of frequent summer water
*DSon = Sonoran (Colorado) Desert

e = east(ern)
e-c = east-central
e.g. = for example
esp = especially
et al = and others
etc = and so on
Eur = Europe
exc = except, excluding

f = front (of an organ, used only in some illustration captions)
fl, fls (**FL, FLS**) = flower(s), floral, flowering
fld = flowered
fr (**FR**) = fruit

*GB = Great Basin Province
gen = generally, usually (also some senses of "mostly")
geog = geographic (-al, -ally)
GRCVR = horticultural entries only: good groundcover
*GV = Great Central Valley

ID = Idaho
i.e. = that is to say
incl = including, included (in)
INV = horticultural entries only: invasive; may displace or overrun other plants
IRR = horticultural entries only: requires moderate summer watering (irrigation)

*KR = Klamath Ranges
KS = Kansas

lf (**LF**) = leaf
lfless = leafless
lflet = leaflet
lvs (**LVS**) = leaves

m = meter
Medit = Mediterranean
Mex = Mexico
mm = millimeter
*MP = Modoc Plateau
MT = Montana
Mtn(s) = Mountain(s) (proper name)
mtn(s) = mountain(s) (not a proper name)

n = north(ern)
n-c = north-central
N.Am = North America
*NCo = North Coast
*NCoR = North Coast Ranges
*NCoRH = High North Coast Ranges
*NCoRI = Inner North Coast Ranges
*NCoRO = Outer North Coast Ranges
NE = Nebraska
ne = northeast(ern)
NM = New Mexico
NV = Nevada
*NW = Northwestern California
nw = northwest(ern)

OK = Oklahoma
OR = Oregon
orn = ornamental

per = perennial herb
peri = perigynium (used only in some *Carex* illustrations)
pl(s) = plant(s)
*PR = Peninsular Ranges

s = south(ern)
S.Am = South America
s-c = south-central
*SCo = South Coast
*SCoR = South Coast Ranges
*SCoRI = Inner South Coast Ranges
*SCoRO = Outer South Coast Ranges
*ScV = Sacramento Valley
SD = South Dakota
se = southeast(ern)
sect(s). = section(s) (abbreviated only as taxonomic rank)
SHD = horticultural entries only: does best in full or part shade
*SnBr = San Bernardino Mountains
*SnFrB = San Francisco Bay Area
*SnGb = San Gabriel Mountains
*SnJt = San Jacinto Mountains
*SnJV = San Joaquin Valley
*SN = Sierra Nevada
*SNE = East of Sierra Nevada (e.g., Mono Valley, Owens Valley)
*SNF= Sierra Nevada Foothills
*SNH = High Sierra Nevada
sp. = species (singular)
spp. = species (plural)
ssp. = subspecies (singular)
sspp. = subspecies (plural)
st, sts, (**ST, STS**) = stem, stems
STBL = horticultural entries only: stabilizer; good for restoring degraded areas
subg. = subgenus, subgenera
subsect(s). = subsection(s) (abbreviated only as taxonomic rank)
SUN = horticultural entries only: does best in ± full sun
*SW = Southwestern California
sw = southwest(ern)

*Teh = Tehachapi Mountain Area
temp = temperate
*TR = Transverse Ranges
trop = tropical
TRY = horticultural entries only: ± untested but worth pursuing
TX = Texas

US = United States
UT = Utah

var(s). = taxonomic variety (taxonomic varieties)
vs = versus

w = west(ern)
WA = Washington (the state)
*W&I = White and Inyo Mountains

w-c = west-central
WET = horticultural entries only: roots need to be in
   continually moist or wet soil

*WTR = Western Transverse Ranges
*Wrn = Warner Mountains
WY = Wyoming

Symbols

The following symbols are used often. Most are quantitative, referring to number, height, length, width, etc.; however, "±" may be qualitative as well, referring to color, fusion, symmetry, etc. Mathematically, the symbol "<" means "up to" (or "approaches as a limit"), so it is generally also equivalent to "=". For example, "ST < 5 m" includes those stems that are exactly 5 m. In cases where equivalency is emphasized, the construction "< or =" is used.

<<   much less than (in size); greatly exceeded by
<   fewer than (in number); less than or up to (in size); exceeded by
=   equal to (generally in size or exsertion)
>   more than (in number); greater than or equal to (in size); exceeding
>>   much greater than (in size); greatly exceeding
0   none, absent
×   multiplication sign, meaning "times" or indicating hybridity

°   degree of compoundness, branching, or angle [e.g., $1°$ (primary), $2°$ (secondary); 45–$60°$ angle]
❀   precedes a statement of horticultural potential
±   more or less, approximately
—   in keys, indicates there is no comparable information for the other half of the couplet
( )   rarely can be this number [e.g., tree (2)4–10]

COMMONNESS AND RARITY

Specifying an accurate "likelihood of encounter" for each plant taxon that grows in the California deserts is impossible. In *The Jepson Desert Manual,* the most widespread and commonly encountered taxa are termed "abundant". Those that are obvious but somewhat less likely to be encountered or less widespread are considered "common". Lack of a "likelihood of encounter" designation generally indicates that a taxon is neither especially common nor especially rare. Taxa designated as "rare" or "uncommon" are not very likely to be encountered but are, nevertheless, not rare enough to be considered sensitive or to be included in official lists of rare and endangered plants.

For plant taxa officially listed as warranting conservation attention, *The Jepson Desert Manual* provides rarity designations that reflect those in the California Native Plant Society's (CNPS's) *Inventory of Rare and Endangered Vascular Plants of California,* fifth edition (address for copies: CNPS Publications, 1722 J St., Suite 17, Sacramento, CA 95814). Discrepancies between rarity designations in *The Jepson Desert Manual* and in the CNPS *Inventory* generally reflect minor differences in the taxonomies followed in the two works. For example, some plant taxa recognized by CNPS are regarded as minor variants that do not warrant taxonomic recognition in *The Jepson Desert Manual.* For taxa given special protection by California or Federal law, the legal designations are provided instead of the CNPS designations. The conventions followed here are the same as those followed in *The Jepson Manual: Higher Plants of California* (J. Hickman, ed., 1993).

The less sensitive taxa included in the CNPS *Inventory* on List 3 (more information needed) or List 4 (watch list) are designated "UNCOMMON" (all capital letters; not to be confused with "uncommon") in *The Jepson Desert Manual.* The "UNCOMMON" designation means that the taxon is of special interest because of its potential rarity; the reader is thereby referred to the CNPS *Inventory* for more information.

The Jepson Desert Manual restricts application of the word "RARE" (all capital letters; not to be confused with "rare") to those taxa included in or proposed for inclusion in CNPS *Inventory* List 1B (rare, threatened, or endangered throughout range) or List 2 (rare, threatened, or endangered in California but more common outside the state). In *The Jepson Desert Manual,* taxa on *Inventory* List 2 are noted to be "RARE in CA".

Highlighting is used to note special protection accorded by either California or Federal law. Boldface type indicates legal status, followed by "US" or "CA" to indicate protection under Federal or California law ("**ENDANGERED** US", "**THREATENED** US", "**ENDANGERED** CA", "**THREATENED** CA", or "**RARE** CA").

Taxa for which living representatives have not been found for many years, despite much effort, make up List 1A of the CNPS *Inventory.* List 1A taxa are designated "**PRESUMED EXTINCT**" (or "**PRESUMED EXTIRPATED** in CA", if at least one population survives outside the state). Although often without legal status (very few are listed as "Endangered" by the State of California), taxa that are presumed to be extirpated in California or extinct are highlighted with boldface designations to draw attention to the importance of finding and protecting any remaining populations.

The information presented above on commonness and rarity is summarized in Table 1 (following page).

TABLE 1. RARITY LISTINGS IN *THE JEPSON DESERT MANUAL*

CNPS *Inventory* List	Legal Status (as abbreviated in *Inventory*)	*The Jepson Desert Manual*
1A		**"PRESUMED EXTINCT"** **"PRESUMED EXTIRPATED** in CA (extant in...)"
1B	(none)	"RARE"
	CE/FE	**"ENDANGERED** CA, US"
	CE/—	**"ENDANGERED** CA"
	—/FE	**"ENDANGERED** US"
	—/FT	**"THREATENED** US"
	CT/—	**"THREATENED** CA"
	CR/—	**"RARE** CA"
2		"RARE in CA"
3,4		"UNCOMMON"
—		No designation (most taxa) or "Rare", "Uncommon", "Common", or "Abundant"

Legal Status Abbreviations

CE = Endangered status provided by California law

FE = Endangered status provided by United States law

CT = Threatened status provided by California law

FT = Threatened status provided by United States law

CR = Rare status provided by California law

HORTICULTURAL INFORMATION

Modified from James C. Hickman & Warren L. Roberts, *The Jepson Manual*

The Jepson Desert Manual adopts the horticultural information provided *in The Jepson Manual: Higher Plants of California* (J. Hickman, ed., 1993). The following information on uses of native plants in California gardens was gathered by a statewide Horticultural Advisory Council (see *The Jepson Manual* for additional details). The Council drew on the expertise and experience of Council members and many other native-plant growers to generate the horticultural recommendations. Taxonomic authors are not responsible for the content of horticultural statements.

Most native plants are restricted in where they can be grown with ease. To help distinguish areas where plants can be expected to perform well without undue effort, the system of numbered climate zones used in *The Jepson Manual* was adopted. The system was based on an important element of Sunset Publishing Corporation's *Western Garden Book*. Sunset Publishing Corporation graciously allowed us to use an adaptation of these zones and an outline map showing their general locations.

Interpreting Horticultural entries

1. Horticultural entries, when present, are found at the end of a species treatment, following the statements of habitat and of elevational and geographic ranges (which are themselves horticulturally useful). Each entry is preceded by the special symbol ❀ and is made up of abbreviations and numbers.
2. Each entry includes capital-letter terms or abbreviations that represent appropriate growing conditions (see list below).
3. Sets of terms or abbreviations are generally followed or preceded by a numerical list of Sunset Publishing Corporation's *Western Garden Book* climate zones in which members of the taxon can be expected to grow if their requirements are met.
4. Zones that are especially appropriate are indicated by bold-face numbers
5. If a plant has different requirements in different zones, the entry is divided into two or more contrasting or additive parts.

Terms and Abbreviations Used in Horticultural Entries

CVS   Cultivar(s) available in the horticultural trade.

DFCLT   Difficult; needs special care in all zones; has complex requirements.

DRN   Requires excellent drainage. Compacted or other water-holding soils may need to be modified.

DRY   Intolerant of frequent summer water; should not be planted near lawns or other moisture-loving plants.

GRCVR   Good groundcover.

INV   Invasive; used for plants that, once established, tend to outcompete, displace, or overrun others.

IRR   Requires moderate summer watering (irrigation), generally 1–4 times per month depending upon the absorption rate and water retention capacity of the soil.

SHD   Does best in full or part shade; may tolerate morning and winter sun.

STBL　Native plants especially good for stabilizing or restoring disturbed or degraded (including logged or burned) areas, for erosion and slope control, for wildlife food or cover, etc. May be less suitable for general garden use.

SUN　Does best in full or nearly full sun; tolerates summer afternoon sun.

TRY　Insufficiently tested but worth pursuing, especially within its natural range.

WET　Roots need to be in continually moist or wet soil.

Example

The horticultural statement for *Prosopis glandulosa* var. *torreyana*, mesquite, is , "❀ SUN, DRN:7,**9**, 10,**12,14–16**,17,18,**19–24**&IRR:**8,11,13**; also STBL". This statement is intended to be read as follows:

"Given well drained soil and full sun, mesquite can be grown in zones 7,9,10,12,14–16,17,18,19–24 but is especially successful in zones 9,12,14–16,19–24, and can be grown successfully with moderate summer water in zones 8, 11, and 13. This species is also used for stabilizing disturbed or degraded areas."

Cautions

Although *The Jepson Desert Manual* summarizes garden values for a broad range of native desert plants, the information presented is general and is intended only as a guide. More detailed guidelines should be sought from experienced gardeners, botanical gardens and arboreta, special classes and seminars, extension agents, nursery managers, and suppliers concerned with growing native plants. Local chapters of the California Native Plant Society are especially valuable sources of information and can help beginning native-plant gardeners in many ways, including recommending and interpreting standard and specialized horticultural library references, books, guides, and journal articles about California desert plants.

Availability of native desert plants for gardens is highly variable. Nursery operators, propagators, and seed distributors respond to demand. Asking them to obtain nursery stock of plants you wish to grow will eventually increase their availability. The highly popular and successful sales of native plants by chapters of the California Native Plant Society and by various arboreta and botanical gardens provide additional excellent, diverse sources of garden material propagated from plants already in cultivation. Keep in mind that plants propagated from locally native material will often be better adapted to local conditions than plants that originate from the same species in other parts of its range. Also, the potential for genetic contamination of adjacent natural populations is reduced if only locally obtained natives are used in gardens.

Limitations of space and format reduce exactness in the presentation of horticultural parameters, especially for those plants designated "difficult" ("DFCLT"). Many beautiful and useful plants have exacting requirements regarding the following: 1) seasonality and position of sun and shade; 2) soil factors such as texture and composition, alkalinity or acidity, salinity or other mineral toxicities or deficiencies; 3) the need for fertilizers and "feeding"; 4) biological factors such as mycorrhizal and semi-parasitic associations, mulch, competition, and predation; 5) tolerance of atmospheric conditions such as wind (dry, hot, freezing, salty) and pollution. The importance of these criteria to the successful cultivation of these plants often varies widely in relation to the factors selected for horticultural entries.

We discourage the collection of plants growing in the wild, especially of cacti and other overharvested succulents. On public lands, plants may be collected only with a permit from the governing agency. Plants designated "ENDANGERED", "THREATENED", or "RARE" are afforded special legal protection at Federal or State levels. Reduction of wild populations of these taxa for any reason, including horticultural "take", is against the law. No detailed horticultural information is included in *The Jepson Desert Manual* for any taxon protected by either Federal or State law or for

any taxon specified as rare in the CNPS *Inventory*. Some rare taxa are available in cultivation; they are indicated by the statement "In cult". For taxa in cultivation, more information on growing requirements is available from nurseries, CNPS chapters, and botanical gardens.

Under no conditions may readers assume that the inclusion or horticultural information is license or encouragement to remove native plants or plant parts from the wild in any form, including seeds. Neither should they assume that information herein guarantees horticultural success or failure within the designated areas or conditions.

Descriptions of Climate Zones

The *Western Garden Book* system has 19 horticultural climate zones in California; 5 of these are in the California deserts. Adaptations of the original descriptions of these 5 zones are given below, with cross-references to the geographic subdivisions of California used to describe the natural distributions of plants. (For more information on those geographic subdivisions, see pp. 33–35.)

Boundaries of climate zones are arbitrary, because climates vary gradually. If your garden is well inside the boundaries of a climate zone (see map, p. 32, and for more detail, those in Sunset Publishing Corporation's *Western Garden Book*), you can assume that the characteristics of the zone hold nearly all the time. If your garden is near a border, your climate will sometimes resemble that across the line. Very local conditions like fence-rows, south-facing walls, slope, or even soil type may produce "microclimates" that are different from the prevailing climate of your region (the same is true of natural communities), perhaps offsetting the effective climate to that of a geographically adjacent zone.

Zones 1–3. Snowy Parts of the West

ZONE 1. Extreme winter cold associated with northern latitudes, continental air masses, or high elevations. Growing season averages 100 days; frosts can occur at any time. High and northern parts of the Great Basin, highest desert mountains.

ZONE 2. Less cold than Zone 1, but soil freezes in winter. Growing season averages 150 days. Edges of Owens Valley, most desert mountains.

ZONE 3. Mildest high-elevation or cold-interior climates. Growing season averages 160 days. Lowest high-mountain areas.

Zones 10–13. Deserts

[ZONE 10. High deserts beyond California: southern Nevada, southwest Utah, Arizona; more rain (especially in summer) and less wind than Zone 11.]

ZONE 11. Medium to high deserts of California and southern Nevada. Wide swings in temperature: cold winters and nights; very hot summers and days; late spring frosts are likely; windy; combinations of winter wind and sun may be harmfully desiccating. Owens Valley, all but lowest Mojave Desert, highest parts of Sonoran Desert, western desert edge.

[ZONE 12. Intermediate desert beyond California; confined to Arizona; colder, with more summer rain than Zone 13.]

ZONE 13. Low, essentially subtropical, deserts of California and Arizona. Much warmer winters than Zone 11 (few nights below freezing); summer storms more common than in Zone 11. Most of Sonoran Desert, lowest parts of eastern Mojave Desert, eastern edge of Peninsular Ranges.

Climate Zones of California

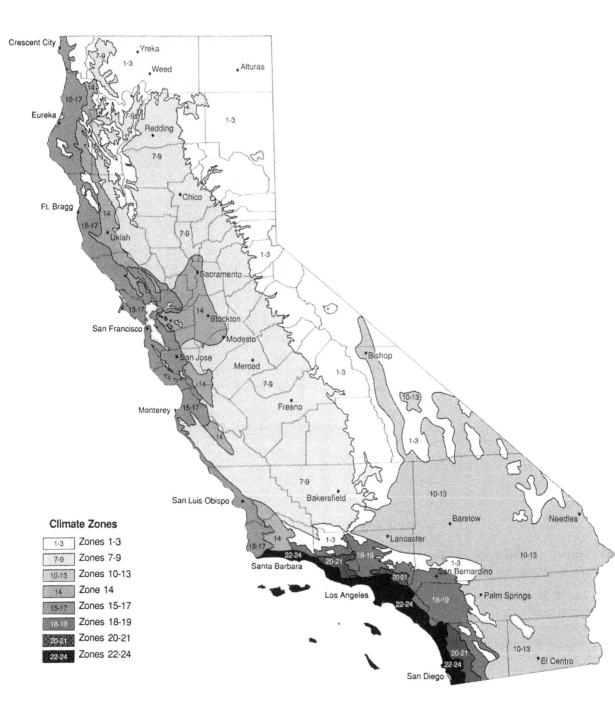

Climate Zones

1-3	Zones 1-3
7-9	Zones 7-9
10-13	Zones 10-13
14	Zone 14
15-17	Zones 15-17
18-19	Zones 18-19
20-21	Zones 20-21
22-24	Zones 22-24

GEOGRAPHIC SUBDIVISIONS OF
THE CALIFORNIA DESERTS

The Jepson Desert Manual follows the geographic system developed for describing plant distributions in *The Jepson Manual: Higher Plants of California* (J. Hickman, ed., 1993) and subsequently adopted in other books on California plants. The geographic system was developed in the interest of providing a biologically meaningful, predictive, and concise means of representing plant distributions within California. The 50 units constituting the system each have a unique abbreviation (often with directional modifiers, e.g., "s DMoj" for southern Mojave Desert) and are arranged in a hierarchy of provinces, regions, subregions, and districts (outlined on the map of geographic subdivisions on the inside back cover). Inclusion of elevational and habitat data in descriptions refine the distributional field within geographic units. For taxa of highly limited occurrence, additional distributional details are often indicated parenthetically. The complete within-state distribution of California desert taxa is provided in *The Jepson Desert Manual,* as in *The Jepson Manual,* but distributional data for the deserts are highlighted in bold.

The three primary geographic units used here for California are the California Floristic Province (CA-FP), the Great Basin Province (GB), and the Desert Province (D). All three units extend outside California.

The focus of *The Jepson Desert Manual* is on the Desert Province (D), comprising the Mojave Desert region (DMoj) and the Sonoran Desert region (DSon), and on the southern Great Basin, i.e., the region east of the Sierra Nevada (SNE). Distributions given as "GB" indicate occurrence in both the southern Great Basin (SNE) and the Modoc Plateau (MP — outside the area covered in *The Jepson Desert Manual*). Distributions given as "D" indicate occurrence in both the Mojave Desert (DMoj) and the Sonoran Desert (DSon). As discussed in more detail elsewhere (Philosophy; Floristic Diversity in the California Deserts), the rationale for delimiting the California deserts to include the southern Great Basin region (SNE) as well as the Desert Province (D) was based on floristic and practical considerations.

The southern Great Basin region (SNE) and the Mojave Desert region (DMoj) each includes two subregions; the Sonoran Desert region (DSon) is not divided into subregions. The White-Inyo Range (W&I) is a subregion within the southern Great Basin (SNE); the Desert Mountains (DMtns) are a subregion within the Mojave Desert (DMoj). A distribution given as "W&I" indicates that a plant found in the White-Inyo Range is known from nowhere else in the southern Great Basin. Similarly, a distribution given as "DMtns" indicates that a plant found in the Desert Mountains is known from nowhere else in the Mojave Desert. A distribution of "SNE" indicates that a plant occurs somewhere in the southern Great Basin (SNE) outside the White-Inyo Range (W&I) and is expected to occur in the White-Inyo Range. Similarly, a distribution of "DMoj" indicates that a plant occurs somewhere in the Mojave Desert (DMoj) outside the Desert Mountains (DMtns) and is expected to occur in the Desert Mountains.

Boundaries among the geographic subdivisions adopted here and in *The Jepson Manual* generally cannot be drawn as sharply as political boundaries, especially from the perspective of natural transitions in vegetation and flora. Even along the dramatically steep eastern slope of the Sierra Nevada, the transition from Sierra Nevada (SN) to southern Great Basin (SNE) or Mojave Desert (DMoj) vegetation and flora is gradual. The following delimitations of provinces, regions, and subregions covered in *The Jepson Desert Manual* reflect the "fuzziness" that often characterizes biologically meaningful boundaries.

Great Basin Province (GB)

The Great Basin Province (GB) lies to the east of the California Floristic Province (CA-FP) in the northern two-thirds of California and abuts the Desert Province (D) at its southern margin. The southern Great Basin, east of the Sierra Nevada (SNE), is the only region of the Great Basin Province (GB) covered in *The Jepson Desert Manual.*

EAST OF THE SIERRA NEVADA (SNE)

This region (entirely south of Lake Tahoe) has a wide elevational range, from Owens Lake at 1100 m to White Mountain Peak at 4330 m. The part of the southern Great Basin (SNE) falling outside of the White-Inyo Range (W&I) supports primarily a mosaic of sagebrush scrub, pinyon/juniper woodland, and cottonwood-dominated riparian vegetation. There are also extensive areas of Jeffrey-pine forest in the Mono Craters area, subalpine-fir/pine forest on Glass Mountain (3400 m), and alpine vegetation at the top of the Sweetwater Mountains (3550 m). The southern Great Basin (SNE) extends along the eastern edge of the Sierra Nevada (SN) to the southern limit of Owens Valley and the White-Inyo Range (W&I), where there is a gradual transition to Mojave Desert (DMoj) creosote-bush/burro-weed-dominated scrub vegetation. To the east of the junction of the White-Inyo Range (W&I) at Westgard Pass lies a low (1500-2000 m) outlier of the southern Great Basin (SNE) that includes the Deep Springs and Fish Lake valleys.

The boundary of the southern Great Basin (SNE) with the California Floristic Province (CA-FP) along the eastern edge of the Sierra Nevada (SN) is generally defined by an indefinite break between either upper montane (red-fir/lodgepole-pine) forest or Jeffrey-pine forest on the California Floristic Province (CA-FP) side and either pinyon/juniper woodland or sagebrush scrub on the southern Great Basin (SNE) side. As noted above, there is also Jeffrey-pine forest in the southern Great Basin (SNE), e.g., Mono Craters area. In some places, U.S. Highway 395 approximates the boundary; south of Bishop, the boundary lies to the west of that highway, farther up the east slope of the Sierra Nevada (SN).

White and Inyo Mountains (W&I). The White-Inyo Range (W&I) is considered a separate subregion because it supports subalpine bristlecone-pine and limber-pine woodlands as well as unique, treeless, alpine vegetation. (White Mountain Peak 4330 m; Inyo and Waucoba peaks both ± 3400 m).

Desert Province (D)

The Desert Province (D) encompasses the Mojave Desert (DMoj) and Sonoran Desert (DSon) of southeastern California. This province lies east of the California Floristic Province (CA-FP) and south of the Great Basin Province (GB). A matrix of creosote-bush/burro-weed-dominated scrub vegetation occurs throughout much of the lowlands, with saltbush scrub characteristic of alkaline basins.

The boundary of the Desert Province (D) with the southern Great Basin (SNE), in the north, is the transition from sagebrush scrub or pinyon/juniper woodland vegetation (GB) to creosote-bush/burro-weed-dominated vegetation (D). Deep Springs and Fish Lake valleys are in the Great Basin Province (GB); Eureka and Saline valleys are in the Desert Province (D). Southward, the mixed vegetation of Owens Valley is included in the southern Great Basin (SNE). South of Owens Valley, the California Floristic Province (CA-FP) abuts the Desert Province (D), and the boundary lies between the chaparral or pinyon/juniper woodland on the California Floristic Province (CA-FP) side and creosote-bush/burro-weed-dominated vegetation in the Desert Province (D); the Little San Bernardino Mountains are considered part of the Desert Province (D). Montane vegetation of the eastern Sierra Nevada (SN), northern Transverse (WTR), and eastern Peninsular (PR) ranges tends to grade into desert vegetation on the lower slopes of these mountains. Some taxa are limited to this interface, which may be specified as "w edge DSon". In Riverside, San Diego, and southwesternmost Imperial counties, the Santa Rosa, Volcan, Laguna, and Jacumba mountains make up the eastern edge of the California Floristic Province (CA-FP).

MOJAVE DESERT (DMoj)

The Mojave Desert (DMoj) occupies the northern two-thirds of the Desert Province (D). The region exhibits greater temperature ranges and more extreme elevational relief than the Sonoran Desert (DSon) to the south. Joshua tree (*Yucca brevifolia*) and Mohave yucca (*Y. schidigera*) are conspicuous, widespread members of Mojave Desert (DMoj) vegetation that are absent from the Sonoran Desert (DSon).

Desert Mountains (DMtns). Although the entire Mojave Desert (DMoj) is a series of mountains and intervening (often wide) valleys, some ranges reach sufficient elevation to support pinyon/juniper woodland vegetation and are therefore recognized as a distinct subregion, the Desert Mountains (DMtns). These high ranges include the Last Chance, Grapevine, Panamint, Coso, Argus, Kingston, Clark, Ivanpah, New York, Providence, Granite, Old Woman, and Little San Bernardino mountains. Four of these ranges (Panamint, Kingston, Clark, and New York mountains) also support white fir or limber pine at their highest elevations. The Desert Mountains (DMtns) have unique elements but also overlap floristically with pinyon/juniper woodland vegetation of the adjacent California Floristic Province (CA-FP). Some of the eastern Desert Mountains (DMtns) support taxa that occur more widely (in the Desert Province (D) or Great Basin Province (GB) outside the state) but are otherwise unknown in California.

SONORAN DESERT (DSon)

The Sonoran Desert (DSon), also known as the Colorado Desert, occupies the southern one-third of the Desert Province (D), south of the Mojave Desert (DMoj). The physiographic line separating the two desert regions is not always clear, but overall the Sonoran Desert (DSon) is lower, warmer, and somewhat distinct floristically. Conspicuous members of the Sonoran Desert (DSon) flora that are absent from the Mojave Desert (DMoj) or confined to its southeastern limits include blue palo verde (*Cercidium floridum*), ocotillo (*Fouquieria splendens*), chuparosa (*Justicia californica*), and ironwood (*Olneya tesota*).

The approximate boundary between the Mojave Desert (DMoj) and the Sonoran Desert (DSon), from west to east, is along the south edge of the Little San Bernardino, Cottonwood, and Eagle mountains (all in DMoj), then north along the eastern edge of the Coxcomb Mountains (DMoj) and around the Old Woman, Turtle, and Chemehuevi mountains (all in DMoj) to the Colorado River. The Chuckwalla and Whipple mountains are in the Sonoran Desert (DSon).

THE CALIFORNIA DESERTS:
SETTING, CLIMATE, VEGETATION, AND HISTORY

Bruce G. Baldwin and Scott N. Martens

A visitor to southeastern California in mid-summer may be struck by the impression of an ancient desert, desolate of life. Yet, the desert conditions that exist today are young and a visit in spring would reveal surprising botanical richness.

The California deserts (including the southern Great Basin) and the desert flora are in part the products of rapid climatic and geologic changes, which have also shaped California and the California flora in general. The development of more extreme aridity in the deserts than in the rest of California stimulated the rise of a diverse flora that is more similar to the floras of adjacent desert regions outside the state than to that of the California Floristic Province. The diversity of the desert flora also reflects the diversity of environments that have arisen in the California deserts.

Setting and Climate

The California deserts are abruptly separated from the rest of California to the west by the steep escarpments of the Sierra Nevada, Transverse Ranges, and Peninsular Ranges. Extending throughout southeastern California and beyond the state to the east and south is a vast expanse of desert terrain dominated by mostly north-south trending mountain ranges separated by deep basins. Extensive faulting has produced some extreme topography. For example, Telescope Peak (3368 m) in the Panamint Range towers above the floor of Death Valley, where Badwater is the lowest point in North America (– 86 m). The upper reaches of the ranges are often composed predominantly of exposed bedrock of diverse types (e.g., granite, volcanics, limestone, dolomite). From the canyons emanate broad fans of alluvial debris that coalesce and surround the ranges. Ephemeral watercourses (washes) dissect the slopes and usually terminate in basins that have no outlet and often contain dry lakes (playas) with extensive salt accumulation. Some perennial streams feed basins that contain permanent saline lakes (e.g., Mono Lake and, prior to water diversion, Owens Lake). At basin edges, dunes formed from windblown sand are sometimes found. At the southeastern edge of the California deserts, the Colorado River serves as the only drainage to the ocean.

Although the winter-wet/summer-dry climate of California prevails throughout much of the California deserts, winter storms from the Pacific are largely restricted from entering the deserts by high ranges along the western boundary. This rain-shadow effect extends to monsoonal summer precipitation, which is partly limited from entering the deserts by mountains to the east and distance from moist subtropical air masses.

Climatic variation within the California deserts is evident along north-south and east-west gradients, as well as elevational gradients. Monsoonal summer rainfall enters the deserts from the southeast and decreases in amount to the west and north. Freezing temperatures and snowfall are more frequent and of longer duration to the north. Superimposed on these regional climatic patterns, precipitation generally increases with increasing elevation. Effectiveness of precipitation for plant growth also increases with increasing elevation because temperatures and evaporation decrease with increasing elevation.

These regional and elevational climatic gradients are reflected by the geographic subdivisions of the California deserts (see inside back cover). The Great Basin, to the north, generally experiences longer periods of freezing temperatures than the Mojave and Sonoran deserts, to the south. The Mojave Desert generally experiences longer periods of freezing temperatures and receives less summer rainfall than the lower-elevation Sonoran Desert, to the south. The Desert Mountains and the White-Inyo Range receive more precipitation and experience lower temperatures than the surrounding Mojave Desert and Great Basin lowlands. These climatic differences among geographic subdivisions of the California deserts significantly influence vegetation and flora.

Vegetation

Vegetation patterns in the California deserts are associated with variation in climate, landforms, elevation, and substrates. Hot, dry lowlands are characterized by low-cover, shrub-dominated (scrub) vegetation of simple vertical structure and low perennial diversity. Density and complexity of woody vegetation increases with increasing elevation and with increasing availability of water in general (e.g., along watercourses). Woodlands of small trees are generally restricted to the wetter environments. Vegetation is absent or nearly so on substrates that are physically or chemically inhospitable to plant growth, such as open dunes and salt-encrusted playas. Seasonal, rather than spatial, differences in availability of water are of primary importance to the annual flora, which constitutes much of the floristic diversity in the Mojave and Sonoran deserts. Most of the annuals germinate during the winter and flower in spring; a small minority of annuals are active after summer rains.

Scrub vegetation dominates the lowlands of the Desert and Great Basin Provinces. Across the vast, hot, dry lowlands of the Mojave and Sonoran deserts, creosote bush (*Larrea tridentata*) and burro-weed (*Ambrosia dumosa*) are conspicuous shrubby dominants. Creosote bush is an evergreen that is characteristic of warm deserts throughout North America. Burro-weed is drought-deciduous, losing leaves in the dry summer season, and leafing out rapidly after rains. Other shrubs (including cacti and yuccas) that are often important members of lowland desert scrub vegetation include Pima rhatany (*Krameria erecta*), box thorn (*Lycium andersonii*), beavertail cactus (*Opuntia basilaris* var. *basilaris*), pencil cactus (*O. ramosissima*), and Mohave yucca (*Y. schidigera*). Vegetation density and height are reduced in areas with poor water infiltration resulting from either a flat, stone-covered surface (desert pavements) or subsurface hardpans (e.g., caliche).

In the Great Basin, the prevailing scrub vegetation of lower elevations is dominated by big sagebrush (*Artemisia tridentata* ssp. *tridentata*). Big sagebrush is an evergreen shrub that can withstand long periods of freezing temperatures. Big sagebrush and other common Great Basin shrubs, including blackbush (*Coleogyne ramosissima*), Mormon tea (*Ephedra nevadensis*), hop-sage (*Grayia spinosa*), and winter fat (*Krascheninnikovia lanata*), are also components of Mojave Desert scrub vegetation at higher elevations.

Joshua tree (*Yucca brevifolia*) is frequently a conspicuous component of lowland and higher elevation vegetation (from 500 to 2000 m) in the Mojave Desert. These bizarre yuccas are distributed throughout the Mojave Desert of California and other western states and have been regarded by some plant geographers and ecologists as an indicator of Mojave Desert vegetation.

Conifer woodlands are restricted to some of the high ranges of the Mojave Desert and Great Basin above ~1300 m. Pinyon pines and junipers are the most widespread conifers and often form extensive woodlands. Four species of pinyons and junipers are found in the California deserts: Colorado pinyon (*Pinus edulis*), singleleaf pinyon (*P. monophylla*), California juniper (*Juniperus californica*), and Utah juniper (*J. osteosperma*). Subalpine woodlands of limber pine (*Pinus flexilis*), western bristlecone pine (*Pinus longaeva*), or both occur in the Panamint Mountains and White-Inyo Range. Some western bristlecone pines, most commonly found on dolomite or limestone, are the oldest trees in the world (up to ~5000 years). The subalpine woodlands are sometimes separated from lower elevation woodlands of pinyon and juniper by a zone of sagebrush-dominated scrub. Where this phenomenon occurs and where subalpine woodland is capped by alpine vegetation, as in the White-Inyo Range, an unusual double tree-line is evident. Relict white fir (*Abies concolor*) is found in small stands in the highest ranges of the Mojave Desert (Clark Mountain, Kingston Range, and New York Mountains). The northwestern-most ranges of the California deserts (e.g., Sweetwater Mountains) have a conifer flora similar to that of the nearby Sierra Nevada.

Various fine-scale physical features exert a strong influence on vegetation and flora. Bedrock exposures often have a vegetation of distinctive composition that is influenced by the interaction of elevation and lithology. At lower elevations, widespread members of vegetation on bedrock include brickellbush (*Brickellia arguta* var. *arguta*), brittlebush (*Encelia farinosa*), hedgehog cactus (*Echinocereus engelmannii*), arrow-leaf (*Pleurocoronis pleuriseta*), and rock nettle (*Eucnide urens*). At higher elevations, calcareous exposures support a unique diversity of perennial vegetation, as in other regions of California, including oreganillo (*Aloysia wrightii*), Panamint butterfly bush (*Buddleja utahensis*), yerba desierto (*Fendlerella utahensis*), and rock spiraea (*Petrophytum caespitosum* ssp. *caespitosum*).

Washes (ephemeral watercourses) in the Mojave and Sonoran deserts have a characteristic vegetation comprising deep-rooted shrubs and trees that tolerate flash floods. Common members of wash vegetation are catclaw (*Acacia greggii*), desert-willow (*Chilopsis linearis*), cheesebush (*Hymenoclea salsola*), and mesquite (*Prosopis glandulosa* var. *torreyana*). Sonoran Desert washes commonly have additional tree and shrub diversity including blue palo verde (*Cercidium floridum* ssp. *floridum*), desert-lavender (*Hyptis emoryi*), ironwood (*Olneya tesota*), and smoke tree (*Psorothamnus spinosus*).

True riparian vegetation, comprising dense stands of trees and shrubs, is limited to the relatively few perennial watercourses (e.g., Colorado River, Mojave River). Common trees of desert riparian areas are Fremont cottonwood (*Populus fremontii* ssp. *fremontii*), Goodding's black willow (*Salix gooddingii*), and red willow (*S. laevigata*). Riparian shrubs include mule fat (*Baccharis salicifolia*), arrow weed (*Pluchea sericea*), and narrow-leaved willow (*Salix exigua*). Invasive tamarisks (*Tamarix chinensis* and *T. ramosissima*) and giant reed (*Arundo donax*) have displaced much of the native riparian vegetation in the California deserts.

On a smaller scale, isolated seeps and springs give rise to oases that support trees and shrubs associated with riparian vegetation. Some oases in the Sonoran Desert harbor the only palm native to California, California fan palm (*Washingtonia filifera*).

In basins where water collects, evaporation results in alkaline or saline conditions unfavorable to plant growth. Barren salt flats or dry lakes (playas) are ringed by distinctive vegetation composed of shrubby members of Chenopodiaceae (in the broad sense), including iodine bush (*Allenrolfea occidentalis*), saltbushes (e.g., *Atriplex confertifolia*, *A. lentiformis*), and greasewood (*Sarcobatus vermiculatus*). Rarely, basins with standing water supplemented by seeps or springs support submerged aquatic vegetation, such as holly-leaved water-nymph (*Najas marina*), ditch-grass (*Ruppia cirrhosa*), and horned-pondweed (*Zannichellia palustris*). Alkaline meadows surrounding desert springs harbor various sedges, rushes, and grasses. Examples of alkaline meadow taxa are saw-grass (*Cladium californicum*), Cooper's rush (*Juncus cooperi*), saltgrass (*Distichlis spicata*), and alkali sacaton (*Sporobolus airoides*).

On the leeward margins of basins, dunes may form from windblown sand. Where the sand is not actively shifting, vegetation consists of a wide diversity of herbaceous plants that are mostly restricted to dune habitats. Examples of common dune herbs are sand verbena (*Abronia villosa*), desert-marigold (*Baileya pauciradiata*), desert lily (*Hesperocallis undulata*), basket evening primrose (*Oenothera deltoides* ssp. *deltoides*), and tiquilia (*Tiquilia plicata*). Perennial grasses, such as panicgrass (*Panicum urvilleanum*), big galleta (*Pleuraphis rigida*), and Eureka Valley dune grass (*Swallenia alexandrae*), may achieve dominance, and deep-rooted trees and shrubs, including desert-willow (*Chilopsis linearis*) and mesquite (*Prosopis glandulosa* var. *torreyana*), may occur.

History

The modern California deserts originated less than 10,000 years bp (before present), since humans first arrived in North America. The region occupied by our modern deserts has undergone dramatic changes over the last 15 million years. Fossil evidence indicates that prior to 15 million years bp the vegetation of much of western North America included relatives of plants that today are characteristic of summer-wet vegetation in eastern North America and Asia. Fossil and paleoclimatic data provide evidence for a drying trend, commencing in mid-Miocene (~15 million years bp) and continuing into the Pliocene (< 5 million years bp), that involved a dramatic reduction in summer rainfall. During this time, plants exhibiting characteristics associated with drought-resistance appeared in the desert region. Some plant families with extreme modifications related to drought-resistance, such as the cactus family (Cactaceae) and ocotillo family (Fouquieriaceae), have a poor fossil record and were once thought to be extremely ancient, possibly dating back to the age of the dinosaurs (> 65 million years bp). However, recent evidence indicates that these groups are more contemporary and probably diversified most extensively since the mid-Miocene.

Dramatic changes in the desert vegetation during the last ~22,000 years are well-recorded in fossil packrat middens. Pinyons, junipers, and other species now associated with desert uplands were evidently typical of desert lowlands from ~22,000 to 12,000 years bp, during the last major glaciation of the Pleistocene. Concurrently, elevated precipitation and cool temperatures sustained large, freshwater lakes where dry-lake basins exist today. Desert scrub vegetation was restricted to the lowest elevations (e.g., the lower Colorado River Valley) and areas to the south in Mexico. Following the end of the Pleistocene, an increase in summer temperature and decrease in precipitation led to rapid conversion of lowland woodlands to modern scrub vegetation, and the retreat of woodland vegetation to higher elevations and latitudes. Creosote bush, burro-weed, and other plants that were confined to low-elevation pockets and southerly areas during the Pleistocene underwent distributional expansion in the Holocene (< 12,000 years bp) to become dominant components of lowland vegetation throughout the Mojave and Sonoran deserts. Some creosote bushes that date back to early Holocene, when lowland woodlands were giving way to modern scrub, have survived to the present by clonal reproduction. Climatic fluctuations during the mid to late Holocene, such as the Altithermal drought at 7000 to 4500 years bp, resulted in evident vegetation shifts in the Great Basin and Mojave Desert. The transition to modern desert vegetation was essentially complete by ~8500 to ~5000 years bp.

Some books pertinent to California desert vegetation

Barbour, M. G. and J. Major (eds.). 1988. Terrestrial vegetation of California. New Expanded Edition. California Native Plant Society, Special Publication Number 9.

Betancourt, J. L., T. R. Van Devender, and P. S. Martin (eds.). 1990. Packrat middens : The last 40,000 years of biotic change. University of Arizona Press, Tucson.

Grayson, D. K. 1993. The desert's past: A natural prehistory of the Great Basin. Smithsonian Institution Press, Washington D. C.

Hall, C. A., Jr. (ed.). 1991. Natural history of the White-Inyo Range, eastern California. University of California Press, Berkeley, CA.

Hall, C. A., Jr. and V. Doyle-Jones (eds.). 1988. Plant biology of eastern California: The Mary DeDecker Symposium. White Mountain Research Station, University of California, Los Angeles.

Jaeger, E. C. 1965. The California deserts. Fourth Edition. Stanford University Press, Stanford, CA.

Latting, J. and P. G. Rowlands (eds.). 1995. The California desert: An introduction to natural resources and man's impact. June Latting Books, Riverside, CA.

Raven, P. H. and D. I. Axelrod. 1995. Origin and relationships of the California flora. California Native Plant Society, Sacramento, CA.

Thorne, R. F., B. A. Prigge, and J. Henrickson. 1981. A flora of the higher ranges and the Kelso Dunes of the eastern Mojave Desert in California. Southern California Botanists, Claremont, CA.

FLORISTIC DIVERSITY IN
THE CALIFORNIA DESERTS

Bruce G. Baldwin and Richard L. Moe

Comparisons of floristic diversity across different geographic regions often yield insights of ecological or evolutionary significance. Results of such broad-scale comparisons must be interpreted with caution, however. A focus on areas that are demarcated at least in part by unnatural (e.g., political) boundaries may hamper interpretation of natural patterns of diversity. Also, the basic units for comparison in floristic analyses, i.e., minimal-rank taxa (species, subspecies, and varieties), may not be comparable in age, biology, or diversity of component evolutionary lineages. Lack of comparability may be even more pronounced for higher-level taxa (e.g., genera and families). Floristic plant geography and historical biogeography in general should benefit as detailed phylogenetic data become available for more of the world's flora, including the flora of the California deserts. With the above provisos in mind, a preliminary, traditional perspective on floristic patterns is offered here as a general orientation to plant diversity in the California deserts. Data are from *The Jepson Desert Manual* and *The Jepson Manual: Higher Plants of California* (J. Hickman, ed., 1993).

Overall patterns of diversity

The California desert flora encompasses 2267 (37%) of the 6200 minimal-rank vascular-plant taxa (hereafter, taxa) native to California. Naturalized alien taxa constitute a substantially smaller percentage of the California desert flora (9%, ~232 taxa) than of the California flora in general (15%, ~1064 taxa). Approximately one-third (785) of the native taxa in the California deserts do not occur elsewhere in the state. The desert regions with the greatest climatic and elevational diversity also contain the most vascular plant diversity; the Mojave Desert (1409 native taxa) and the southern Great Basin (1363 native taxa) each encompass about twice as many taxa as the Sonoran Desert (720 native taxa).

Endemism

The California flora is extraordinary among continental floras for having such a high level of diversity and endemism. The California Floristic Province (i.e., the Mediterranean-climatic region of western California and adjacent southwestern Oregon in the United States and northwestern Baja California in Mexico) is the area with the most distinctive flora, ~35% of Californian taxa that occur within the California Floristic Province are endemic to the state (endemism is even higher within the California Floristic Province as a whole). Like other areas on earth with a Mediterranean climate (winter-wet, summer-dry), the California Floristic Province is not only rich in plant diversity but also is believed to be an area of extensive evolutionary diversification or speciation.

The California desert flora has proportionately fewer endemics than are found in the Californian portion of the California Floristic Province. Only small parts of the Great Basin and the Mojave and Sonoran deserts lie within California whereas most of the California Floristic Province occurs within the state. The eastern, southern, and northern boundaries of the California deserts are political borders that mostly do not correlate with major ecological boundaries (except, to some extent, the lower Colorado River Valley). Much of the California desert flora extends into the adjacent deserts of Arizona and Nevada in the United States, and Baja California in Mexico. A broad geographic view reveals conspicuous endemics of the western North American deserts that are not endemic to California, such as Mohave yucca, *Yucca schidigera* (Liliaceae, in the broad sense), in the Mojave Desert and giant saguaro, *Carnegiea gigantea* (Cactaceae), in the Sonoran Desert. Low levels of endemism in the California deserts do not necessarily reflect lack of diversification or speciation in the region; taxa that originated in the California deserts may have spread outside the state or into other regions of California. For example, some widespread desert annuals that appear to have descended from ancestors in the California Floristic Province probably originated in the adjacent California deserts prior to dispersal elsewhere, e.g., the desert pincushions, *Chaenactis fremontii* and *C. stevioides* (Asteraceae).

Within the California deserts, the region of greatest floristic endemism is the Mojave Desert, with more than 40 taxa found nowhere else. The majority of Californian Mojave Desert endemics are confined to the Desert Mountains, including (1) the Argus, Coso, Grapevine, Last Chance, and Panamint mountains of the Death Valley region, (2) the Clark, Granite, Ivanpah, Kingston, New York, Old Woman, and Providence mountains of the eastern Mojave Desert, and (3) the Little San Bernardino Mountains of the southwestern Mojave Desert. A concentration of endemics in the Desert Mountains is associated with higher precipitation than in other areas of the Mojave Desert. The high level of endemism in the Desert Mountains also may reflect geographic isolation of the ranges from one another and other areas of similar climate. At least some of the endemics are probably paleo-endemics that were once more widely distributed in the deserts during glacial episodes of the Pleistocene; others may be neo-endemics that have originated recently in isolation, since the end of the Pleistocene (< 12,000 years before present).

Endemics recorded for the Desert Mountains in the Californian Mojave Desert

Apiaceae
    Cymopterus panamintensis var. *panamintensis*   Panamint cymopterus
Asteraceae
    Enceliopsis covillei   Panamint daisy
Crassulaceae
    Dudleya saxosa ssp. *saxosa*   Panamint dudleya
Euphorbiaceae
    Tetracoccus ilicifolius   Holly-leaved tetracoccus
Fabaceae
    Astragalus panamintensis   Panamint milkvetch
    Lotus argyraeus var. *notitius*   Providence Mountains lotus
Malvaceae
    Sphaeralcea rusbyi var. *eremicola*   Panamint mallow
Polygonaceae
    Eriogonum ericifolium var. *thornei*   Thorne's buckwheat
    E. gilmanii   Gilman's buckwheat
    E. hoffmannii var. *hoffmannii*   Hoffmann's buckwheat
    E. hoffmannii var. *robustius*   Robust Hoffmann's buckwheat
    E. intrafractum   Jointed buckwheat
Rosaceae
    Ivesia patellifera   Kingston Mountains ivesia
Rubiaceae
    Galium argense   Argus bedstraw
    G. hilendiae ssp. *carneum*   Panamint Mountains bedstraw
    G. hypotrichium ssp. *tomentellum*   Telescope Peak bedstraw
Scrophulariaceae (in the broad sense)
    Mimulus rupicola   Death Valley monkeyflower
    Penstemon clevelandii var. *mohavensis*   Mojave beardtongue
    P. stephensii   Stephens' beardtongue
Liliaceae (in the broad sense)
    Calochortus panamintensis   Panamint mariposa lily

The Death Valley region, including the northern Desert Mountains, has been noted for having a flora with an unusually high number of endemics. Although the Death Valley region extends well into Nevada, some of the plants endemic to the area are restricted to California, including endemics of the northern Desert Mountains (some of the taxa listed above) and those listed below that occur outside the Desert Mountains. Two of the Californian Death Valley region endemics have been treated for decades in monotypic genera: *Gilmania* (Polygonaceae), a rarely seen, annual buckwheat from Death Valley, and *Swallenia* (Poaceae), a perennial grass from the Eureka Dunes. Rock lady, *Maurandya petrophila* (Scrophulariaceae, in the broad sense), a tufted perennial from limestones in the Grapevine Mountains, has recently been treated as the sole representative of the genus *Holmgrenanthe*.

Endemics recorded for the Californian Death Valley region
(not endemic to the northern Desert Mountains)

Brassicaceae
    Sibara rosulata   East Mojave rock cress
Fabaceae
    Astragalus lentiginosus var. *micans*   Shining milkvetch
Hydrophyllaceae
    Nama demissum var. *covillei*   Coville's purple mat
Loasaceae
    Petalonyx thurberi ssp. *gilmanii*   Death Valley sandpaper plant
Onagraceae
    Camissonia cardiophylla ssp. *robusta*   Big heartleaf sun cup
    Oenothera californica ssp. *eurekensis*   Eureka Dunes evening primrose
Polygonaceae
    Gilmania luteola   Golden carpet
Scrophulariaceae (in the broad sense)
    Maurandya [*Holmgrenanthe*] *petrophila*   Rock lady
Poaceae
    Swallenia alexandrae   Eureka Valley dune grass

Other endemics of the Death Valley region that have an extremely limited or uncertain occurrence in Nevada include Death Valley ringstem, *Anulocaulis annulatus* (Nyctaginaceae); black milkvetch, *Astragalus funereus* (Fabaceae); Death Valley sage, *Salvia funerea* (Lamiaceae); and limestone beardtongue, *Penstemon calcareus* (Scrophulariaceae, in the broad sense).

The western and central parts of the Mojave Desert constitute an endemic area that receives minimal summer precipitation and therefore can be regarded as having an extremely dry Mediterranean climate. In contrast, some parts of the eastern Mojave Desert and the Sonoran Desert receive a substantial proportion of annual precipitation in the summer. At least one of the taxa endemic to the western Mojave Desert, Red Rock tarplant, *Hemizonia* [*Deinandra*] *arida* (Asteraceae), is an example of colonization of the western Mojave Desert by a lineage from the California Floristic Province, where a true Mediterranean climate prevails. Taxa endemic to the western and/or central Mojave Desert are indicated below.

Endemics recorded for the western and/or central Mojave Desert in California

Apiaceae
    Cymopterus deserticola   Desert cymopterus
Asteraceae
    Eriophyllum mohavense   Barstow woolly sunflower
    Hemizonia [*Deinandra*] *arida*   Red Rock tarplant

Boraginaceae
  Cryptantha clokeyi   Clokey's cryptantha
Brassicaceae
  Lepidium nitidum var. *howellii*   Howell's peppergrass
Fabaceae
  Astragalus jaegerianus   Lane Mountain milkvetch
Loasaceae
  Mentzelia eremophila   Desert-loving blazing star
Polemoniaceae
  Gilia latiflora ssp. *elongata*   Broad-flowered gilia
  G. ochroleuca ssp. *ochroleuca*   Desert gilia
Polygonaceae
  Chorizanthe spinosa   Mojave spineflower
Scrophulariaceae (in the broad sense)
  Castilleja exserta ssp. *venusta*   Purple owl's clover
  Mimulus mohavensis   Mojave monkeyflower

The Sonoran Desert of California contains a substantial diversity of conspicuous shrubs and trees not found elsewhere in the state. Distinctive Sonoran Desert taxa include chuparosa, *Justicia californica* (Acanthaceae); ironwood, *Olneya tesota* (Fabaceae); ocotillo, *Fouquieria splendens* (Fouquieriaceae); crown-of-thorns, *Koeberlinia spinosa* ssp. *tenuispina* (Koeberliniaceae); desert-lavender, *Hyptis emoryi* (Lamiaceae); and California fan palm, *Washingtonia filifera* (Arecaceae). A low level of endemism in the Californian part of the Sonoran Desert strongly reflects the similarity of desert habitats to the east in Arizona and to the south in Baja California, Mexico. Relatively few Sonoran Desert endemic taxa are restricted to California.

Endemics recorded for the Californian Sonoran Desert

Asteraceae
  Xylorhiza cognata   Mecca-aster
Cactaceae
  Opuntia ×*munzii*   Munz's cholla
Fabaceae
  Astragalus lentiginosus var. *coachellae*   Coachella Valley milkvetch
  Lupinus excubitus var. *medius*   Mountain Springs bush lupine
Hydrophyllaceae
  Phacelia campanularia ssp. *campanularia*   Desert bells
Lamiaceae
  Salvia greatae   Orocopia sage
Rubiaceae
  Galium angustifolium ssp. *borregoense*   Borrego bedstraw

Taxa endemic to California that are confined to the Sonoran Desert and Mojave Desert include California ditaxis, *Ditaxis californica* (Euphorbiaceae); Providence Mountain milkvetch, *Astragalus nutans* (Fabaceae); tapering desert bells, *Phacelia campanularia* ssp. *vasiformis* (Hydrophyllaceae); and Little San Bernardino Mountains gilia, *Gilia maculata* (Polemoniaceae). The distribution of another California desert endemic, desert peppergrass, *Lepidium fremontii* var. *stipitatum* (Brassicaceae), spans the Mojave and Sonoran deserts and the White-Inyo Range in the southern Great Basin.

Endemics to the southern Great Basin of California are mostly concentrated in the White-Inyo Range, which is exceeded in elevation within California only by the Sierra Nevada. Unlike other ranges of the California deserts, alpine habitat in the White-Inyo Range is extensive and approximately half of the endemics to the range are alpine taxa. The White-Inyo Range extends from California into Nevada without any ecological barriers between the states. Consequently, most plants endemic to the White-Inyo Range occur in both California and Nevada.

Endemics recorded for the Californian part of White-Inyo Range

Asteraceae
    Perityle inyoensis   Inyo laphamia
Boraginaceae
    Cryptantha roosiorum   Bristlecone cryptantha
    Hackelia brevicula   Poison Canyon stick-seed
Brassicaceae
    Caulostramina jaegeri   Jaeger's caulostramina
Fabaceae
    Astragalus cimae var. *sufflatus*   Bladdery Cima milkvetch
Hydrophyllaceae
    Phacelia amabilis   Saline Valley phacelia
Rosaceae
    Horkelia hispidula   White Mountains horkelia
    Potentilla morefieldii   Morefield's cinquefoil
Scrophulariaceae (in the broad sense)
    Penstemon monoensis   Mono beardtongue

About seven taxa of California endemics found in the White-Inyo Range are otherwise known in the state only from the Desert Mountains of the Mojave Desert, generally from the mountains of the Death Valley region. The highly distinctive, shrubby buckwheat *Dedeckera* (Polygonaceae) is a particularly prominent example of endemism in the White-Inyo Range and Desert Mountains of California.

Endemics recorded for the White-Inyo Range and Desert Mountains of California

Asteraceae
    Ericameria gilmanii   Gilman's ericameria
    Perityle megalocephala var. *oligophylla*   Limestone laphamia
    Perityle villosa   Hanaupah laphamia
Polygonaceae
    Dedeckera eurekensis   July gold
    Eriogonum eremicola   Wildrose Canyon buckwheat
    E. microthecum var. *panamintense*   Panamint Mountains buckwheat
Scrophulariaceae (in the broad sense)
    Penstemon scapoides   Pinyon beardtongue

Endemics to the southern Great Basin of California that occur outside the White-Inyo Range are mostly associated with saline soils of the Owens Valley and vicinity or with volcanic pumice and gravel of the Mono Craters area.

Endemics recorded for the southern Great Basin (east of the Sierra Nevada) in California

Asteraceae
    Erigeron calvus   Bald daisy

Brassicaceae
    Draba incrassata   Sweetwater Mountains draba
Fabaceae
    Astragalus lentiginosus var. *piscinensis*   Fish Slough milkvetch
    Astragalus monoensis   Mono milkvetch
    Lupinus duranii   Mono Lake lupine
Malvaceae
    Sidalcea covillei   Owens Valley checkerbloom
Onagraceae
    Camissonia claviformis ssp. *lancifolia*   Clavate-fruited evening primrose
Liliaceae
    Calochortus excavatus   Inyo County star-tulip

Some California endemics that occur outside the White-Inyo Range in the southern Great Basin are otherwise found in California only in the adjacent Mojave Desert. Southern Great Basin and Mojave Desert endemics known only from California include Parish's popcorn flower, *Plagiobothrys parishii* (Boraginaceae); Death Valley spurge, *Chamaesyce vallis-mortae* (Euphorbiaceae), Panamint Mountains lupine, *Lupinus magnificus* (Fabaceae); Mojave woollystar, *Eriastrum densifolium* ssp. *mohavense* (Polemoniaceae); and western Mojave buckwheat, *Eriogonum mohavense* (Polygonaceae).

Rarity

More than one-tenth of all native taxa in the California deserts, including most endemics, are listed as warranting conservation attention by the California Native Plant Society (CNPS), the State of California, or the United States government. Approximately one-third of the native taxa restricted within California to the deserts (but not necessarily endemic to California) similarly are listed by CNPS or the state or federal government. The proportion of rare or uncommon taxa in the California deserts is nonetheless substantially lower than in California as a whole, wherein more than one-fifth of all native taxa are listed. The high diversity of taxa restricted to threatened or dwindling lowland habitats in the California Floristic Province accounts in part for the high level of plant rarity and endangerment on a state-wide level.

Floristic comparisons within California

Floristic comparisons among the California deserts illustrate the absence of a sharp boundary between the Great Basin and the Mojave and Sonoran deserts. Pairwise comparisons of the floras of the southern Great Basin, Mojave Desert, and Sonoran Desert (using Sorensen's index to correct for differences in regional diversity) show that floristic similarity between the Mojave Desert and the Sonoran Desert is approximately the same (~50%) as floristic similarity between the Mojave Desert and the southern Great Basin. Floristic similarity between the Mojave Desert and the southern Great Basin is about as high as floristic similarity between the southern Great Basin and northern Great Basin, i.e., the Modoc Plateau, in California. Floristic similarity between the southern Great Basin and Sonoran Desert of California is only ~25%. These findings help to understand why various plant geographers have treated the Californian Mojave Desert (at least in part) within floristic units that include either the southern Great Basin or the Sonoran Desert.

The Desert Mountains contribute substantially to floristic similarity between the Mojave Desert and southern Great Basin. If taxa restricted to the Desert Mountains are excluded from the Mojave Desert flora, then the pairwise similarity between the Mojave Desert and the Sonoran Desert floras is much higher than the similarity between the Mojave Desert and southern Great Basin floras, regardless of whether the White-Inyo Range is included or excluded from the southern Great Basin flora.

In comparisons with each of the California deserts, the California Floristic Province shows highest floristic similarity with the southern Great Basin and lowest floristic similarity with the Sonoran Desert.

Similarity between the high montane floras of the California deserts and the California Floristic Province appears to account for much of the disparity in floristic similarity between the California Floristic Province and each of the deserts. Floristic comparisons of each of the desert regions with the California Floristic Province yield similar values if the montane floras of the White-Inyo Range and the Desert Mountains are removed from the southern Great Basin and Mojave Desert floras.

To obtain a better understanding of patterns of diversity, endemism, and rarity within the deserts, we encourage collaboration among floristicians, plant geographers, and systematists focused on different geographic areas and plant lineages throughout western North America and beyond.

KEY TO CALIFORNIA DESERT PLANT FAMILIES

David J. Keil

Key To Groups

1. Specimens at hand without both stamens and pistils available for examination; fls either not produced at all or, if present, fls unisexual and either staminate or pistillate fls unavailable
  2. Pls aquatic, available specimens reproducing mainly by vegetative means; fls, frs, or spores 0 or not readily apparent . . . . **Group 1** (p. 48)
  2' Pls aquatic or terrestrial; fls, frs, seeds, spores, or other specialized reproductive structures present and readily apparent
    3. Pls bearing only spores or pollen; available specimens without ovules, ovaries, seeds, frs, or bulblets
      4. Ferns and their relatives; reproductive spores produced and released directly from sporangia borne on surfaces or in axils of some or all lvs or borne in terminal cones; fls and seeds never formed; herbs only [FERNS & FERN-ALLIES] . . . . **Group 2** (p. 50)
    4' Angiosperms and gymnosperms in pollen-producing condition only; pollen borne in anthers or pollen cones; staminate fls or other pollen-producing structures; pistils and other ovule-producing structures 0; herbs, shrubs, and trees
      5. Pls conspicuously woody; tree and shrubs [some ANGIOSPERMS and GYMNOSPERMS] **Group 3** (p. 50)
      5' Pls herbaceous or woody only at base; ann and per herbs [some ANGIOSPERMS] . . . . . **Group 4** (p. 51)
  3' Pls bearing ovaries, ovules, seeds, frs, or bulblets; ovules borne in ovaries, naked on branches, in cones, or cone-like structures; pollen-producing structures present or 0
    6. Pls herbaceous, sometimes woody at base [some ANGIOSPERMS]
      7. Fls 0, replaced in infls by bulblets that are dispersed in place of seeds or frs . . . . **Group 5** (p. 52)
      7' Fls present, unisexual, bearing only ovaries . . . . **Group 6** (p. 52)
    6' Pls woody [some ANGIOSPERMS and GYMNOSPERMS]
      8. Ovules not enclosed in ovaries, exposed to air at time of pollination; pollen deposited directly on ovules; stigmas never present; seeds borne between scales of cone or naked on branches [GYMNOSPERMS] . . . . **Group 7** (p. 53)
      8' Ovules enclosed in ovaries at time of pollination; pollen deposited on stigmas; seeds borne inside frs derived from ovaries [mostly ANGIOSPERMS] . . . . **Group 8** (p. 53)
1' Specimens at hand with both stamens and pistils available for examination; fls bisexual or, if unisexual, both staminate and pistillate fls present
  9. Fl parts (esp perianth) gen in 3s (rarely 1 or in 2s or 4s); principal lf veins gen parallel (rarely pinnate or palmate); vascular bundles of st (in cross-section) scattered or in diffuse circles; sts not increasing in diam through secondary growth exc in some woody Liliaceae; 1st-year taproot gen 0; cotyledon gen 1 [mostly MONOCOTS]
    10. Perianth conspicuous, at least inner whorl corolla-like in color and texture; infl various . . . . **Group 9** (p. 55)
    10' Perianth inconspicuous, individual perianth parts gen scale-like or 0, never petal-like; infl unit most commonly a spike, spikelet, or head . . . . **Group 10** (p. 55)
  9' Fl parts (esp perianth) gen in 4s or 5s (rarely in 3s) or in a spiral with number of parts indefinite; principal lf veins gen pinnate or palmate (rarely parallel); vascular bundles of young st (in cross-section) gen in a well defined ring; sts gen increasing in diam through secondary growth, often becoming stiffly woody in age; 1st-year taproot gen present; cotyledons gen 2 [mostly DICOTS]
    11. Pistils 2 or more per fl . . . . **Group 11** (p. 56)
    11' Pistils 1 per fl
      12. Perianth 0 or in a single whorl (appearing to be sepals or petals but not both)
        13. Pls conspicuously woody; trees or shrubs
          14. Infl, at least the staminate, of catkins or catkin-like spikes . . . . **Group 12** (p. 56)
          14' Infl various but not catkin-like . . . . **Group 13** (p. 57)
        13' Pls herbaceous (ann or per) or woody only at base
          15. Ovary inferior . . . . **Group 14** (p. 58)
          15' Ovary superior . . . . **Group 15** (p. 58)
      12' Perianth in 2 or more whorls (gen both sepals and petals) or perianth parts spiraled 2 or more times around floral axis

16. Petals fused at least at base, forming a ring or a tube, falling as a unit
16' Petals gen free at least at base, falling singly (in a few families joined and falling in 2s or 3s, but
    not forming a ring or a tube)
18' Ovary superior
19' Stamens twice as many as petals or fewer
20' Lvs simple, sometimes much reduced

Group 1: Aquatic Plants in Vegetative Condition

1. Pls of saltwater habitats, e.g., in brackish to saline lakes, ponds, pools, etc.
  2. Pls not differentiated into lvs or lf-like structures and sts or st-like structures [algal groups, not further
     treated in this book]
  2' Pls differentiated into lvs or lf-like structures and sts or st-like structures; nodes and internodes readily
     apparent
    3. Pls with a central axis and whorled, cylindric branches [CHAROPHYTA, algal group not further
       treated here]
    3' Pls with a short to elongate st bearing alternate or opposite lvs
      4. Pls submersed; lvs and sts weak
        5. Lf tips shallowly 2–3-toothed . . . . **CYMODOCEACEAE**
        5' Lf tips entire
          6. Lvs gen alternate; lf bases sheathing; free stipules 0 . . . . **POTAMOGETONACEAE** (*Ruppia*)
          6' Lvs gen opposite; lf bases not sheathing; free membranous stipules present . . . **ZANNICHELLIACEAE**
      4' Pls emergent; lvs and sts able to support themselves out of water
        7. Lvs all basal
          8. Lf blades wiry, tough, or stiffly rigid; internal cross partitions gen present and evident to touch;
             lvs ± 3-ranked . . . . **JUNCACEAE**
          8' Lf blades soft, not wiry and tough or stiffly rigid; internal cross partitions 0; lvs ± 2-ranked
            9. Lf blades cylindric or ± flat, thickish . . . . **JUNCAGINACEAE** (*Triglochin*)
            9' Lf blades flat, thin . . . . **POACEAE**
        7' Lvs basal and cauline
          10. Sts ± triangular . . . . **CYPERACEAE**
          10' Sts round
            11. Lvs cylindric . . . . **JUNCACEAE**
            11' Lvs flat . . . . **POACEAE**
1' Pls of freshwater habitats
  12. Pl body consisting of a jointed central axis and whorled branches; lvs 0 [algal groups not further
      treated here]
  12' Pl body various, but not consisting of a jointed central axis and whorled branches; lvs various
    13. Individual pls gen 1–20 mm in length or diam, free-floating or stranded along shore, sometimes not
        differentiated into lvs and sts
      14. Pl body forked or with forked grooves on upper surface [RICCIACEAE, liverworts not further
          treated here]
      14' Pl body neither forked nor with forked grooves
        15. Pl body differentiated into a short, often branched st covered with tiny, overlapping, scale-like,
            green or red-purple-tinged lvs . . . . **AZOLLACEAE**
        15' Pl body spheric to disk-shaped or oblong, not differentiated into sts and lvs . . . . **LEMNACEAE**
    13' Individual pls often >> 20 mm, anchored or sometimes free-floating, most often clearly differentiated
        into sts and lvs
      16. Pls not differentiated into lvs or lf-like structures and sts or st-like structures [various algal groups,
          not further treated here]
      16' Pls differentiated into lvs or lf-like structures and sts or st-like structures; nodes and internodes
          readily apparent (or in *Utricularia* producing dissected, lf-like underwater sts)
        17. Lvs all scale-like or narrowly linear, gen very thin, entire or minutely toothed, densely overlapping,
            all alternate; roots never present [MUSCI, true mosses, not further treated here]
        17' Lvs linear to ovate, very thin to thick and ± fleshy, entire to variously toothed, lobed, or compound,
            gen not densely overlapping exc sometimes in new growth, alternate, opposite, whorled, or
            sometimes all basal; roots gen present though sometimes few
          18. Pls raft-like, free-floating on water surface, breaking apart into individuals or small clumps;
              rhizomes short or 0 — lf blades palmately veined, light green, velvety-hairy . . . . **ARACEAE** (*Pistia*)

18' Pls gen anchored in bottom sediments or, if free-floating, most of pl body below water surface,
    mostly not readily breaking apart; rhizomes present or 0, sometimes elongated
  19. Sts gen erect; lvs all basal or alternate on an erect, emergent st or borne on a rhizome anchored
    in bottom sediments
    20. Lvs petioled with expanded blades or lower ones narrowly strap-shaped . . . . **ALISMATACEAE**
    20' Lvs sessile, linear or reduced to bladeless sheaths
      21. Lvs vertically folded and attached "edge-on" to st, obviously 2-ranked . . . . **JUNCACEAE** (*Juncus*)
      21' Lvs cylindric, ± angled, with ordinary flat blades, or reduced to bladeless sheaths, not
        obviously 2-ranked
        22. Lvs with flat blades or reduced to bladeless sheaths
          23. Lvs bladeless or nearly so, composed mainly of a tubular sheath
            24. Sts ± triangular . . . . **CYPERACEAE**
            24' Sts round, flattened, or several-angled
              25. Lf sheaths closed, forming a continuous cylinder around st . . . . **CYPERACEAE**
              25' Lf sheaths open on one side, with overlapping margins . . . . **JUNCACEAE**
          23' Lvs with ordinary flat blades and gen also ± tubular sheaths
            26. Lf blade with a well developed midvein or keel; st gen triangular . . . . **CYPERACEAE**
            26' Lf blade rounded on back or flat, lacking a prominent midvein or keel; st cylindric
              27. Lf width < or = 5 mm; remnants of old infl a raceme; inner surface of sheaths not
                mucilaginous . . . . **JUNCAGINACEAE** (*Triglochin*)
              27' Lf width gen > 5 mm; remnants of old infl a dense spike; inner surface of lf sheaths
                mucilaginous . . . . **TYPHACEAE**
        22' Lvs cylindric or angled
          28. Rhizomes 0; lvs densely tufted
            29. Pls per; lvs often with internal cross-partitions . . . . **JUNCACEAE** (*Juncus*)
            29' Pls ann; lvs without internal cross-partitions or partitions obscure . . . **JUNCAGINACEAE** (*Lilaea*)
          28' Rhizomes present; lvs tufted or borne along rhizome
            30. Lvs sharply angled . . . . **CYPERACEAE**
            30' Lvs cylindric
              31. Lvs with internal cross-partitions . . . . **JUNCACEAE** (*Juncus*)
              31' Lvs without internal cross-partitions . . . . **JUNCAGINACEAE** (*Triglochin*)
  19' Sts gen weak, often taking form of underwater rhizomes or stolons, unable to support pl
    body outside of water; lvs gen all cauline on submersed sts
    32. Lvs (or lf-like branches) compound or very deeply divided
      33. Lvs palmately compound — lflets 4, floating or emergent . . . . **MARSILEACEAE**
      33' Lvs pinnately compound or dissected, repeatedly forked, or irregularly divided
        34. Lflets lanceolate or wider; lvs mostly aerial
          35. Lflets coarsely serrate or dentate; petioles sheathing . . . . **APIACEAE**
          35' Lflets entire or teeth rounded; petioles not sheathing . . . . **BRASSICACEAE** (*Rorippa*)
        34' Lflets or lobes linear, at least on submersed lvs; lvs all or many submersed
          36. Underwater parts bearing bladder-like traps that capture aquatic invertebrates and
            microorganisms; pls carnivorous . . . . **LENTIBULARIACEAE** (*Utricularia*)
          36' Underwater parts lacking bladder-like traps; pls not carnivorous
            37. Lvs alternate . . . . **RANUNCULACEAE** (*Ranunculus*)
            37' Lvs opposite or whorled
              38. Lvs repeatedly forked . . . . **CERATOPHYLLACEAE**
              38' Lvs pinnately divided . . . . **HALORAGACEAE** (*Myriophyllum*)
    32' Lvs simple, margins entire, toothed, or shallowly lobed
      39. Lvs long, narrow, strap-shaped or ribbon-like, often >> 20 cm, all basal . . . . . . **ALISMATACEAE**
      39' Lvs linear to ovate, gen < 20 cm, gen all cauline
        40. Lvs with stipules or sheathing bases; margins entire
          41. Principal veins of lf pinnate . . . . **POLYGONACEAE** (*Polygonum*)
          41' Principal veins of lf parallel or pinnate-parallel or only 1 lf vein apparent
            42. Lvs gen alternate; floating lvs sometimes present . . **POTAMOGETONACEAE** (*Potamogeton*)
            42' Lvs mostly or all opposite; lvs all submersed . . . . **ZANNICHELLIACEAE**
        40' Lvs without stipules or sheathing bases; margins entire, toothed, or shallowly lobed
          43. Lvs alternate . . . . **ONAGRACEAE** (*Ludwigia peploides*)
          43' Lvs opposite or whorled
           44. Lvs mostly aerial
            45. Lf veins palmate . . . . **SCROPHULARIACEAE** (*Bacopa, Mimulus*)
            45' Lf veins pinnate
              46. Lvs entire . . . . **ONAGRACEAE** (*Ludwigia repens*)
              46' Lvs toothed . . . . **SCROPHULARIACEAE** (*Mimulus, Veronica*)
          44' Lvs mostly or all submersed
           47. Lvs whorled . . . . **HYDROCHARITACEAE**

47' Lvs opposite
    48. Lvs entire, linear to ovate, not enlarged at base, sometimes with floating upper lvs different from submersed lvs . . . . **CALLITRICHACEAE**
    48' Lvs minutely toothed to coarsely lobed, gen enlarged at base, all similar in form and submersed . . . . **HYDROCHARITACEAE** (*Najas*)

Group 2: Ferns and Fern Allies; Seeds Absent, Plants Reproducing by Spores

1. Pl aquatic or in dried pools
  2. Pl floating or stranded on mud; lvs sessile, scale-like, 2-lobed; pl body gen < 1 cm . . . . **AZOLLACEAE**
  2' Pl rooted on bottom of pond; lvs petioled, palmately compound, lflets 4; pl body >> 1 cm . . **MARSILEACEAE**
1' Pl terrestrial or growing on other pls
  3. Lvs sessile, scale-like, 1-veined
    4. Lvs whorled; internodes of st long, green, forming main photosynthetic surface of pl; sporangia on undersides of peltate scales that are clustered into a terminal cone . . . . **EQUISETACEAE**
    4' Lvs crowded on sts; internodes of st very short; sporangia in axils of 4-ranked upper lvs
. . . . **SELAGINELLACEAE**
  3' Lvs petioled, simple or 1–4-pinnate
    5. Sporangia ± 1 mm thick, sessile in panicle-like clusters on stalks fused to petiole, annulus 0; new lvs never coiled . . . . **OPHIOGLOSSACEAE**
    5' Sporangia << 1 mm thick, slender-stalked, gen in sori on underside of lf, often near margin, annulus present; new lvs coiled, unrolling as they develop
      6. Sori borne along margin of lf; indusia 0 or sori covered by reflexed lf margin . . . . **PTERIDACEAE**
      6' Sori borne away from margin on underside of lvs or lflets; indusia present or 0
        7. Sporangia scattered along veins, not clustered into distinct sori; indusia 0 . . . . **PTERIDACEAE**
        7' Sporangia in distinct sori; indusia sometimes present
          8. Lvs 1–2-pinnately compound; indusia present . . . . **DRYOPTERIDACEAE**
          8' Lvs simple, pinnately lobed; indusia 0 . . . . **POLYPODIACEAE**

Group 3: Woody Gymnosperms and Angiosperms in Pollen-Producing Condition Only; Ovulate Cones or Pistillate or Bisexual Flowers Absent or Not Evident

1. Pls parasitic on sts of woody host pls . . . . **VISCACEAE**
1' Pls free-living
  2. Trees; trunk unbranched, covered with persistent woody lf bases; lvs 1–several m, pinnately compound (*Phoenix*) . . . . **ARECACEAE**
  2' Trees or shrubs; trunk(s) gen branched, gen not covered with persistent lf bases; lvs gen smaller, simple or compound
    3. Lvs stiff and sword-like, 0.5–1.5 m; infl a large panicle . . . . **LILIACEAE** (*Nolina*)
    3' Lvs not sword-like, mostly smaller; infls various
      4. Lvs opposite or whorled
        5. Lvs compound
          6. Lvs palmate
            7. Lflets 3; infl few-fld ± flat-topped . . . . **ACERACEAE** (*Acer glabrum*)
            7' Lflts gen 5 or more; infl many-fld, cylindric to conic . . . . **HIPPOCASTANACEAE**
          6' Lvs pinnate
            8. Trees or large shrubs; sepals gen 4, very small . . . . **OLEACEAE** (*Fraxinus*)
            8' Woody vines; sepals gen 4, conspicuous, petal-like . . . . **RANUNCULACEAE** (*Clematis*)
        5' Lvs simple
          9. Lvs and sometimes sts thick and fleshy — lvs sometimes reduced to fleshy scale . . . **CHENOPODIACEAE**
          9' Lvs and sts not fleshy
            10. Margins of lvs toothed or lobed
              11. Lvs palmately veined and lobed . . . . **ACERACEAE**
              11' Lvs pinnately veined and toothed . . . . **EUPHORBIACEAE** (*Tetracoccus*)
            10' Margins of lvs entire
              12. Weak-stemmed subshrub . . . . **RUBIACEAE** (*Galium*)
              12' Stouter shrub or tree
                13. Fls in catkins or pollen produced in ± catkin-like cones
                  14. Catkins elongate, pendent; lf blades well developed . . . . **GARRYACEAE**
                  14' Catkin-like pollen cones short, not pendent; lvs scale-like
                    15. Internodes very short; st concealed by persistent, overlapping, green, scale-like lvs; pollen sacs sessile beneath scales of small cone . . . . **CUPRESSACEAE**
                    15' Internodes elongated; st green, bearing dry, scale-like, often early-deciduous lvs; pollen sacs borne on filaments . . . . **EPHEDRACEAE**
                13' Fls in head-like or raceme-like clusters

16. Lvs thick, leathery, evergreen; fls in peduncled, head-like clusters; sepals well developed
. . . . **SIMMONDSIACEAE**

16' Lvs thin, deciduous; fls in sessile, axillary clusters or short-peduncled racemes; sepals
very small

  17. Pl not spiny; bark of young sts reddish; fls in short, axillary racemes
. . . . **EUPHORBIACEAE** (*Tetracoccus*)

  17' Pl spiny; bark smooth, gray; fls sessile, in axillary clusters . . . . **OLEACEAE** (*Forestiera*)

4' Lvs alternate

  18. Fls not in catkins or spikes; pollen never produced in cones

    19. Lvs palmately lobed . . . . **PLATANACEAE**

    19' Lvs not palmately lobed

      20. Sts and lvs densely covered with stellate hairs . . . . **EUPHORBIACEAE**

      20' Sts and lvs glabrous or hairy but not stellate

        21. Lvs compound

          22. Lvs once compound, lflets 3 . . . . **ANACARDIACEAE**

          22' Lvs 2-pinnate; stamens many . . . . **FABACEAE** (*Acacia*)

        21' Lvs simple

          23. Fls in involucred heads, these gen secondarily clustered . . . . **ASTERACEAE**

          23' Fls solitary or variously clustered, not in involucred heads

            24. Lvs gen grayish, either soft-hairy or covered with powdery scales . . . . . **CHENOPODIACEAE**

            24' Lvs green

              25. Lvs narrow, bases symmetric; shrubs . . . . **EUPHORBIACEAE** (*Tetracoccus*)

              25' Lvs wide, bases oblique; trees . . . . **ULMACEAE** (*Celtis*)

  18' Fls in catkins or catkin-like spikes or pollen produced in ± catkin-like cones

    26. Shrubs of ± saline desert habitats; lvs thick, often ± fleshy; spikes erect or very short and
axillary . . . . **CHENOPODIACEAE**

    26' Shrubs or trees, gen of moister habitats; lvs mostly thinner; spikes, catkins, or pollen cones erect
to pendent

      27. Lvs needle-like; pls evergreen . . . . **PINACEAE**

      27' Lvs with well developed blades; pls evergreen or deciduous

        28. Bractlets of infl 0 or very inconspicuous at time of flowering

          29. Evergreen shrubs or trees; calyx present, of 4–6 sepals . . . . **FAGACEAE**

          29' Deciduous shrubs or trees; calyx 0 . . . . **SALICACEAE**

        28' Bractlets of infl well developed and readily visible at time of flowering

          30. Catkin-bractlets fringe-margined; trees; petioles long . . . . **SALICACEAE** (*Populus*)

          30' Catkin-bractlets entire or appearing 3-lobed; shrubs or trees; petioles various

            31. Fls inserted individually along catkin axis, each subtended by 1 bractlet (this sometimes
early deciduous) . . . . **SALICACEAE** (*Salix*)

            31' Fls inserted in groups subtended and often somewhat concealed by 1–3 bractlets

              32. Staminate infl pendent . . . . **BETULACEAE**

              32' Staminate infl stiff, erect or spreading . . . . **EUPHORBIACEAE** (*Acalypha*)

Group 4: Herbaceous Monocots and Dicots in Staminate Condition, Producing Only Pollen; Pistillate or Bisexual Flowers Absent or Not Evident

1. Pls fully aquatic, submersed, floating in water, or stranded on mud

  2. Pls gen free-floating, disk-like, whole pl <15 mm diam, often < 5 mm diam; lvs and sts not
differentiated; roots 0 or 1–few, unbranched — stamens emerging from tiny flap of tissue . . **LEMNACEAE**

  2' Pls rooted in bottom sediments or free-floating, gen >> 15 mm; lvs and sts clearly differentiated; roots
often branched

    3. Lvs alternate . . . . **CYMODOCEACEAE**

    3' Lvs opposite or whorled

      4. Lf blades entire, toothed, or shallowly lobed . . . . **HYDROCHARITACEAE**

      4' Lf blades (at least those of submersed lvs) divided into linear lobes

        5. Blades of lvs repeatedly forked . . . . **CERATOPHYLLACEAE**

        5' Blades of lvs pinnately divided . . . . **HALORAGACEAE** (*Myriophyllum*)

1' Pls terrestrial or, if growing in wet places, rooted in place and extending well above water surface

  6. Pls parasitic on sts of woody host pls

    7. Fls of parasite borne directly on sts of host; remainder of parasite internal within tissues of host; sts
and lvs not differentiated . . . . **RAFFLESIACEAE**

    7' Fls borne on leafy branches of parasite; shoots of parasite external; lvs differentiated though sometimes
reduced to scales . . . . **VISCACEAE**

  6' Pls free-living

    8. Lvs opposite or whorled, not all basal

      9. Sts or lvs thick and fleshy; infls terminal spikes . . . . **CHENOPODIACEAE**

      9' Sts and lvs of normal texture, not thick and fleshy; infls various . . . . **RUBIACEAE**

8' Lvs alternate or all basal
   10. Lvs stiff and sword-like, 5–15 dm; infl a large panicle; perianth parts 6 . . . . **LILIACEAE** (*Nolina*)
   10' Lvs not sword-like, often smaller, infls various; perianth parts mostly other than 6
      11. Blades of lvs linear or narrowly lanceolate, simple and entire, veins parallel; lf bases sheathing sts;
         fls in spikelets, these gen in secondary clusters
         12. St triangular, nodes not swollen; lf blades often channeled . . . . **CYPERACEAE**
         12' St round; nodes gen swollen and knot-like; lf blades gen flat . . . . **POACEAE**
      11' Blades of lvs variously shaped, sometimes toothed, lobed, or compound, veins mostly pinnate or
         palmate; lf bases often not sheathing sts; fls not in spikelets
         13. Lvs all basal; fls with 3 petals and many stamens . . . . **ALISMATACEAE**
         13' At least some lvs cauline; fls not with both 3 petals and many stamens
            14. Pls vines; tendrils present . . . . **CUCURBITACEAE**
            14' Pls prostrate to erect herbs; tendrils 0
               15. Fls in heads or umbels
                  16. Infl a simple or compound umbel . . . . **APIACEAE**
                  16' Infl of 1 or more heads
                     17. Heads without involucres . . . . **CHENOPODIACEAE**
                     17' Each head subtended by an involucre
                        18. Stamens free; petals free or 0 . . . . **APIACEAE**
                        18' Stamens fused by anthers; petals fused or 0 . . . . **ASTERACEAE**
               15' Fls in axillary clusters, racemes, or panicles
                  19. Lvs 1–4-ternately compound . . . . **RANUNCULACEAE** (*Thalictrum*)
                  19' Lvs entire or toothed
                     20. Lvs covered with stellate hairs . . . . **EUPHORBIACEAE** (*Croton*)
                     20' Lvs without stellate hairs
                        21. Lvs bearing bead-like, sessile hairs or powdery scales . . . . . . **CHENOPODIACEAE** (*Atriplex*)
                        21' Lvs glabrous
                          22. Stipules 0; sepals 3–5, in a single series or strongly overlapping
                        . . . . **AMARANTHACEAE** (*Amaranthus*)
                        22' Stipules present, forming a sheath around st; sepals 6, in two series
                        . . . . **POLYGONACEAE** (*Rumex*)

Group 5: Monocots and Dicots; Flowers Replaced by Bulblets that are Dispersed in Place of Seeds and Fruits

1. Lvs linear, sheathing at base . . . . **POACEAE** (*Poa bulbosa*)
1' Lvs wide, palmately lobed, not sheathing at base . . . . **RANUNCULACEAE** (*Aconitum columbianum*)

Group 6: Herbaceous Monocots and Dicots in Pistillate Condition, Producing Ovules within Ovaries; Staminate or Bisexual Flowers Absent or Not Evident

1. Pls fully aquatic, submersed, floating in water, or stranded on mud
  2. Pls gen free-floating, disk-like, whole pl <15 mm diam, often < 5 mm diam; lvs and sts not
    differentiated; roots 0 or 1–few, unbranched — pistil emerging from tiny flap of tissue . . . . . **LEMNACEAE**
  2' Pls rooted in bottom sediments or free-floating, gen >>15 mm; lvs and sts clearly differentiated; roots
    often branched
    3. Lvs alternate
      4. Lf tips shallowly 2-toothed; frs paired, axillary; pls of inland saltwater habitats . . **CYMODOCEACEAE**
      4' Lf tips entire; frs borne in terminal, umbel-like clusters; pls of freshwater or inland saltwater
        habitats . . . . **POTAMOGETONACEAE** (*Ruppia*)
    3' Lvs opposite or whorled or all basal
      5. Lvs all basal; fls solitary in lf-axils; style very elongated, often > 5 cm . . . . **JUNCAGINACEAE** (*Lilaea*)
      5' Lvs cauline; fls axillary or variously clustered; styles gen << 5 cm (exc in some Hydrocharitaceae)
        6. Lf blades (at least those of submersed lvs) divided into linear lobes
          7. Blades of lvs repeatedly forked . . . . **CERATOPHYLLACEAE**
          7' Blades of lvs pinnately divided . . . . **HALORAGACEAE** (*Myriophyllum*)
        6' Lf blades entire, toothed, or shallowly lobed
          8. Petals and sepals both 3, evident; perianth and stigmas borne at water surface at end of long, tubular
            hypanthium; ovary inferior, sessile in lf axil . . . . **HYDROCHARITACEAE**
          8' Petals and sepals both 0; stigmas submersed, borne at end of short styles; ovary or ovaries superior,
           sessile or short-stalked in lf axil
            9. Pistil 1, compound, bearing 2–4 slender stigmas; lvs subentire to finely or coarsely toothed
            . . . . **HYDROCHARITACEAE** (*Najas*)
            9' Pistils 2–10, simple, each with a cup-like stigma; lvs entire . . . . **ZANNICHELLIACEAE**

1' Pls terrestrial or, if growing in wet places, rooted in place and extending well above water surface
  10. Pls parasitic on sts of woody host pls
    11. Fls of parasite borne directly on sts of host; remainder of parasite internal within tissues of host; sts
        and lvs not differentiated . . . . **RAFFLESIACEAE**
    11' Fls borne on leafy branches of parasite; shoots of parasite external; lvs differentiated, though
        sometimes reduced to scales . . . . **VISCACEAE**
  10' Pls free-living
    12. Lvs opposite or whorled, not all basal
      13. Sts or lvs thick and fleshy; infls terminal spikes; fr unlobed, 1-seeded . . . . **CHENOPODIACEAE**
      13' Sts and lvs of normal texture, not thick and fleshy; infls various; fr 2-lobed, 2-seeded . . . . **RUBIACEAE**
    12' Lvs alternate or all basal
      14. Blades of lvs linear or narrowly lanceolate, simple and entire; veins parallel; lf bases sheathing st
        15. Lvs stiff and sword-like, 0.5–1.5 m; infl a large panicle; perianth parts 6, well developed
            . . . . **LILIACEAE** (*Nolina*)
        15' Lvs not sword-like, often smaller; infls various; perianth parts 0 or reduced to bristles or minute scales
          16. Lvs all basal; fls solitary in lf axils; styles very elongated, often > 5 cm . . **JUNCAGINACEAE** (*Lilaea*)
          16' Lvs basal and cauline or all cauline; fls in spikes or spikelets, these gen in secondary clusters;
              styles gen << 5 cm
            17. Sts triangular; nodes not swollen; lf blades often channeled . . . . **CYPERACEAE**
            17' Sts round; nodes gen swollen and knot-like; lf blades gen flat . . . . **POACEAE**
      14' Blades of lvs variously shaped, sometimes toothed, lobed, or compound; veins mostly pinnate or
          palmate; lf bases often not sheathing sts — fls not in spikelets
        18. Lvs all basal; fls with 3 petals and many free pistils . . . . **ALISMATACEAE**
        18' At least some lvs cauline; fls not with both 3 petals and many free pistils
          19. Pls vines; tendrils present . . . . **CUCURBITACEAE**
          19' Pls prostrate to erect herbs; tendrils 0
            20. Fls in heads or umbels
              21. Infl a simple or compound umbel . . . . **APIACEAE**
              21' Infl of 1 or more heads
                22. Heads without involucres; ovaries superior . . . . **CHENOPODIACEAE**
                22' Each head subtended by an involucre; ovaries inferior
                  23. Heads borne in umbels . . . . **APIACEAE**
                  23' Heads borne in cymes, panicles, or racemes . . . . **ASTERACEAE**
            20' Fls in axillary clusters or in racemes or panicles
              24. Lvs 1–4-ternately compound . . . . **RANUNCULACEAE** (*Thalictrum*)
              24' Lvs entire or toothed
                25. Lvs densely stellate; ovary chambers gen 3; fr a capsule . . . . **EUPHORBIACEAE** (*Croton*)
                25' Lvs not stellate; ovary chamber 1; fr an achene or utricle
                  26. Lvs bearing bead-like, sessile hairs or powdery scales . . . . **CHENOPODIACEAE** (*Atriplex*)
                  26' Lvs glabrous
                    27. Stipules 0; ovary ovoid, not angled; sepals 3–5, in a single series or strongly overlapping
                        . . . . **AMARANTHACEAE** (*Amaranthus*)
                    27' Stipules present, forming a sheath around st; ovary triangular; sepals 6, in 2 series
                        . . . . **POLYGONACEAE** (*Rumex*)

Group 7: Gymnosperms; Seeds Produced But Not Enclosed in Pistils; Flowers Not Produced

1. Lvs needle-like, borne singly or in bunches of 2–5 . . . . **PINACEAE**
1' Lvs scale-like, opposite, alternate, or whorled
  2. Sts jointed at nodes, internodes long, green; lvs dry, scale-like, often early deciduous, not overlapping;
     seed(s) surrounded by perianth-like scales; pollen cones with stamen-like structures, pollen sacs
     borne on stout filaments . . . . **EPHEDRACEAE**
  2' Sts not jointed at nodes, internodes gen very short; lvs green, scale-like or awl-like, persistent,
     crowded, sometimes overlapping; seeds enclosed in berry-like fleshy cones or enclosed by woody
     cone scales; pollen cones with sessile pollen sacs . . . . **CUPRESSACEAE**

Group 8: Woody Plants, Mostly Angiosperms in Ovule-Producing Condition Only; Staminate or Bisexual Flowers Absent or Not Evident

1. Pls parasitic on sts of woody host pls . . . . **VISCACEAE**
1' Pls free-living
  2. Trees; trunk unbranched, covered with persistent woody lf bases; lvs to several m . . . . **ARECACEAE**
  2' Trees or shrubs; trunk(s) gen branched, gen not covered with persistent lf bases; lvs gen much smaller,
     simple or compound

3. Lvs stiff and sword-like, 5–45 dm; infl a large panicle .... **LILIACEAE** (*Nolina*)
3' Lvs not sword-like, mostly smaller; infls various
  4. Lvs opposite or whorled
    5. Lvs compound
      6. Lvs palmate; lflets 3 .... **ACERACEAE** (*Acer glabrum*)
      6' Lvs pinnate; lflets often > 3
        7. Tree or large shrubs; sepals gen 5 or number difficult to determine, very small; pistil 1; styles never plumose; fr winged .... **OLEACEAE** (*Fraxinus*)
        7' Woody vines; sepals gen 4, conspicuous, petal-like; pistils many; styles elongated and plumose in fr .... **RANUNCULACEAE** (*Clematis*)
    5' Lvs simple
      8. Lvs and sometimes sts thick and fleshy; lvs sometimes reduced to fleshy scales .. **CHENOPODIACEAE**
      8' Lvs and sts not fleshy
        9. Margins of lvs toothed .... **EUPHORBIACEAE** (*Tetracoccus*)
        9' Margins of lvs entire
          10. Weak-stemmed subshrubs .... **RUBIACEAE** (*Galium*)
          10' Stouter shrub or trees
            11. Fls in elongate, pendent catkins .... **GARRYACEAE**
            11' Fls not in catkins
              12. Sts green, jointed; lvs thin, dry, scarious, scale-like, often early-deciduous ... **EPHEDRACEAE**
              12' Sts gray or brown or only very young sts greenish, not jointed; lvs thin or thick, green, not dry and scarious, not scale-like, persistent or deciduous
                13. Lvs thick, leathery, evergreen; fls solitary; sepals well developed ...... **SIMMONDSIACEAE**
                13' Lvs thin, deciduous; fls in sessile, axillary clusters or short-peduncled racemes; sepals very small
                  14. Fls subsessile, solitary or few in axils; styles 3-4; ovary 3-4-lobed; fr a capsule .... **EUPHORBIACEAE** (*Tetracoccus*)
                  14' Fls pedicelled, in axillary clusters; style 1; ovary unlobed; fr a drupe . **OLEACEAE** (*Forestiera*)
  4' Lvs alternate
    15. Fls in catkins or catkin-like spikes
      16. Shrubs of ± saline desert habitats; lvs ± thick and fleshy or very narrow and densely hairy; spikes erect or very short and axillary .... **CHENOPODIACEAE**
      16' Shrubs or trees, gen of moister habitats; lvs thin and wide; spikes or catkins erect or pendent
        17. Stipules 0; fls individually enclosed by a pair of appressed (and sometimes fused) bractlets .... **CHENOPODIACEAE**
        17' Stipules present, evident at least on new growth, sometimes deciduous but leaving evident scars; fls not individually enclosed by a pair of appressed bractlets
          18. Styles repeatedly branched into many fine, thread-like divisions; ovary 3-lobed, maturing as a capsule; brittle-stemmed shrub .... **EUPHORBIACEAE** (*Acalypha*)
          18' Styles or style branches 2–4, gen not further divided; ovary gen unlobed, frs various; sts mostly not brittle
            19. Catkins very dense at time of flowering; only stigmas exserted beyond tips of tightly appressed bractlets; at maturity, catkin ± cone-like with ovaries ripening as winged achenes .... **BETULACEAE**
            19' Catkins looser at time of flowering; bractlets, if present, not tightly appressed, ovaries gen visible on close inspection; at maturity, catkin becoming a ± open cluster of capsules with hair-tufted seeds .... **SALICACEAE**
    15' Fls not in catkins or spikes — ovules never produced in cone-like catkins
    20. Lvs palmately lobed .... **PLATANACEAE**
    20' Lvs not palmately lobed
      21. Sts and lvs stellate-hairy
        22. Ovary clearly superior, 2-4-lobed; fr a capsule; involucre 0; subshrub with entire lvs or desert shrub with crenate lvs .... **EUPHORBIACEAE**
        22' Ovary inferior but sepals very small and often concealed by bractlets of involucre; fr an ovoid nut, individually subtended by a cup-like involucre; shrub or tree with lvs entire or toothed, but not crenate .... **FAGACEAE**
      21' Sts and lvs glabrous or hairy but not stellate
        23. Corolla present; perianth parts in 2 series
          24. Fls in involucred heads, these gen secondarily clustered; ovary 1 per fl, inferior; calyx modified as a white to brownish pappus .... **ASTERACEAE**
          24' Fls in racemes; ovaries 1-5, superior; calyx green, not modified .... **ROSACEAE**
        23' Corolla 0; perianth parts 0 or in 1 series
          25. Fls subtended or surrounded at base by an involucre of tiny scale-like bractlets; ovary inferior to minute calyx; fr an ovoid nut, subtended by a cup-like involucre ...... **FAGACEAE** (*Quercus*)

25' Fls without an involucre; ovary superior or seemingly inferior between a pair of tightly
    appressed bractlets; fr an achene, utricle, capsule, or drupe
  26. Each fl individually enclosed by a pair of tightly appressed bractlets; fr an achene or
    utricle . . . . **CHENOPODIACEAE**
  26' Fls not enclosed by paired bractlets; fr a capsule or drupe
    27. Lvs narrow; bases symmetric; shrubs . . . . **EUPHORBIACEAE** (*Tetracoccus*)
    27' Lvs wide; bases oblique; trees . . . . **ULMACEAE** (*Celtis*)

Group 9: Monocots and some Dicots; Flower Parts in Threes;
At Least Inner Perianth Parts Petal-like

1. Pls aquatic, floating on water surface, stranded on mud, or rooting in bottom sediments; sts often weak
  and unable to support pls out of water, not extending much above water surface
  2. Pistils many; lvs all basal, strap-like or with sagittate blade . . . . **ALISMATACEAE**
  2' Pistil 1; lvs opposite or whorled, linear or oblong . . . . **HYDROCHARITACEAE**
1' Pls terrestrial or, if growing in water, extending well above water surface; sts able to support pls out
  of water
  3. Trees, shrubs, or fleshy rosette pls; lvs large, stiff
    4. Lvs pinnately compound or palmately lobed, never fleshy; trees; trunk unbranched . . . . **ARECACEAE**
    4' Lvs simple, entire to coarsely toothed or with marginal fibers, often fleshy; trees, shrubs, or fleshy
    rosette pls; trunk, if present, branched or unbranched . . . . **LILIACEAE**
  3' Herbs; lvs gen thinner
    5. Ovary or ovaries superior
      6. Pistils many; sepals green; petals white . . . . **ALISMATACEAE**
      6' Pistil 1, entire to ± 3-lobed; sepals and petals gen similar, both gen petal-like, colors various
      . . . . **LILIACEAE**
    5' Ovary inferior or partly so
      7. Fertile stamen 1, anther, filament, and sterile stamens wholly fused above ovary to stigma and style,
      forming a column; ovary with a half twist; fls bilateral, lower petal very different from other 2 in
      size, shape, or color . . . . **ORCHIDACEAE**
      7' Fertile stamens 3 or more, anthers free, filaments free or fused to hypanthium; ovary not twisted;
      fls radial or bilateral
        8. Stamens 3; lvs 2-ranked, gen folded lengthwise and apparently attached "on edge", only bottom
        side of lf visible . . . . **IRIDACEAE**
        8' Stamens gen 6; lvs not 2-ranked and folded, top and bottom sides both visible or lvs ± cylindric
        . . . . **LILIACEAE**

Group 10: Monocots; Perianth Inconspicuous or Absent

1. Lf blade wide; veins pinnate or palmate
  2. Floating herbs with simple, entire, sessile lvs; infl a short, fleshy spike subtended by a small, whitish
  bract . . . . **ARACEAE**
  2' Trees with large, pinnately compound or palmately lobed, long-petioled lvs; infl a panicle subtended
  by a ± woody bract . . . . **ARECACEAE**
1' Lf blade linear to narrowly lanceolate (wider if floating) or 0; veins parallel
  3. Pls aquatic, submersed or floating
    4. Pl < 1 cm, floating or stranded on shore, not differentiated into sts and lvs; roots unbranched or 0
    . . . . **LEMNACEAE**
    4' Pl >> 1 cm, gen rooted, differentiated into sts and lvs; roots gen present, often branched
      5. Lvs opposite or whorled
        6. Lvs coarsely to finely toothed, stipules 0 (lf base may have ear-like lobes); pistil 1, compound,
        with 3–4 stigmas; fr without a stiff beak . . . . **HYDROCHARITACEAE** (*Najas*)
        6' Lvs entire, stipules free, membranous, entire; pistils > 1, simple, each with 1 stigma; fr with a
        stiff beak . . . . **ZANNICHELLIACEAE**
      5' Lvs alternate
        7. Lvs gen with 2–3 teeth near tip; fls solitary in axils, unisexual, pls dioecious; perianth 2
        minute protuberances or 0; stamens 2, fused together; pistils 2 . . . . **CYMODOCEACEAE**
        7' Lvs entire; fls in cylindric spikes, bisexual; perianth segments 4; stamens 2–4, separate;
        pistils 4 or more . . . . **POTAMOGETONACEAE**
  3' Pls terrestrial or strongly emergent
    8. Fls all unisexual, infl a dense spike 1–4 cm diam with staminate fls in distal half . . . . **TYPHACEAE**
    8' Fls bisexual or, if some or all unisexual, infls gen < 1 cm diam; infls various
      9. Fr a capsule or breaking into segments; ovules 2–many; perianth segments 4–6 in 2 whorls, well
      developed, all scale-like
        10. Infl a panicle or group of head-like clusters or fl solitary; stigmas slender, spreading; fr a capsule
        . . . . **JUNCACEAE**

      10' Infl a bractless raceme; fr breaking into segments . . . . **JUNCAGINACEAE** (*Triglochin*)
    9' Fr an achene or grain; ovule 1; perianth inconspicuous or 0
      11. Lvs cylindric, all basal; pistillate fls axillary, with long, slender styles; bisexual fls in bractless
         spikes . . . . **JUNCAGINACEAE** (*Lilaea*)
      11' Lvs flat, cylindric, or scale-like, basal or cauline; infl unit a spikelet of minute fls subtended by
         scale-like bractlets
        12. Sts lfless, lvs reduced to bladeless basal scales; spikelet 1, terminal on st
        . . . . **CYPERACEAE** (*Eleocharis*)
        12' Sts ± leafy; spikelets gen >1
         13. Sts gen 3-angled, solid; nodes not swollen; lvs gen 3-ranked; each fl subtended by 1 bract
           (in pistillate fl); perianth of bristles or 0; fr an achene . . . . **CYPERACEAE**
         13' Sts cylindric, internodes often hollow; nodes swollen, knot-like; lvs 2-ranked or not in
           obvious ranks; each spikelet gen with 2 bractlets (glumes) at base; each fl gen enclosed by
           2 additional bractlets (lemma and palea); perianth of 2 tiny scales or 0; fr a grain . . . . . . **POACEAE**

Group 11: Dicots; Pistils Two or More, Simple

1. Perianth 0 or of only one whorl, gen called sepals even when petal-like
  2. Hypanthium present . . . . **ROSACEAE**
  2' Hypanthium 0
    3. Trees; lvs palmately lobed; axillary bud covered by petiole base; perianth 0 or reduced to tiny scales;
      infl a cluster of spheric heads; monoecious; fr a dense head of hairy achenes . . . . **PLATANACEAE**
    3' Herbs or woody vines; lvs various; axillary bud gen exposed; perianth present, gen well developed;
      infls various but not spheric heads; fr a cluster of achenes or follicles . . . . **RANUNCULACEAE**
1' Perianth of 2 or more whorls or spirals, outer gen called sepals, inner called petals
  4. Stamens >2 times as many as petal-like perianth segments or >15
    5. Pls obviously woody; shrubs or trees
      6. Lvs glabrous, entire; seeds with arils . . . . **CROSSOSOMATACEAE** (*Crossosoma*)
      6' Lvs hairy, toothed, lobed, or compound; seeds without arils . . . . **ROSACEAE**
    5' Pls herbaceous or slightly woody at base
      7. Hypanthium present . . . . **ROSACEAE**
      7' Hypanthium 0
        8. Sepals 2–3, falling as fl opens, not petal-like; pistils weakly fused in fl, separating in age
        . . . . **PAPAVERACEAE** (*Platystemon*)
        8' Sepals 4–many, present in open fls, sometimes petal-like; pistils gen wholly free . . **RANUNCULACEAE**
  4' Stamens 2 times as many as petals or fewer
    9. Petals fused together, at least at base
      10. Pistils 5; lvs fleshy; sap clear . . . . **CRASSULACEAE**
      10' Pistils 2; lvs gen not fleshy; sap gen milky
        11. Styles 1 or 2; stigmas fused; anthers free from but sometimes lying against style(s) or stigmas
        . . . . **APOCYNACEAE**
        11' Styles 2; stigmas fused together and to anthers, entire unit (style-stigma-anther head)
        drum-shaped . . . . **ASCLEPIADACEAE**
  9' Petals free but sometimes attached to a disk-like to tubular hypanthium
    12. Lvs and often sts thick and fleshy . . . . **CRASSULACEAE**
    12' Lvs and sts not very thick and fleshy
      13. Hypanthium well developed; herbs . . . . **ROSACEAE**
      13' Hypanthium 0 or inconspicuous; shrubs
        14. Sts slender, ± lfy, brown or gray; fls solitary in lf axils; fr a cluster of follicles
        . . . . **CROSSOSOMATACEAE** (*Glossopetalon*)
        14' Sts stout, leafless, green; fls in dense panicles; fr a cluster of ± dry drupes . . . . . . **SIMAROUBACEAE**

Group 12: Woody Dicots; Perianth in One Whorl or Absent;
Staminate Flowers in Catkins or Catkin-like Spikes

1. Pls parasitic, attached to branches of trees or shrubs, not rooted in soil . . . . **VISCACEAE**
1' Pls not parasitic, gen rooted in soil
  2. Pls fleshy, ± woody at base; lvs fleshy, scale-like or linear to oblanceolate; catkins erect
  . . . . **CHENOPODIACEAE**
  2' Pls non-fleshy shrubs or trees; lvs with well developed flat blades; catkins often drooping
    3. Lvs opposite . . . . **GARRYACEAE**
    3' Lvs alternate
      4. Fr a capsule; seeds several–many, each bearing a tuft of hairs; fls subtended by minute glands or
        fringed bractlets . . . . **SALICACEAE**
      4' Fr dry or fleshy, indehiscent; seed 1, without a tuft of hairs; bractlets, if any, not fringed

5. Frs many, small, flattened, and often winged, clustered in a cone-like infl; pls deciduous; pistillate fls without sepals . . . . **BETULACEAE**
5' Frs 1–3, each a well developed nut, subtended or surrounded by an involucre; pls deciduous or evergreen; pistillate fls with minute sepals — ovary inferior . . . . **FAGACEAE**

Group 13: Woody Dicots; Perianth in One Whorl or Absent; Inflorescences Not Catkins

1. Infl a dense spike, spheric head, or fls enclosed within a fleshy, hollow receptacle
  2. Lvs bipinnately compound or modified into linear or oblong, entire, flattened axes without lflets — fr a legume . . . . **FABACEAE** (Mimosoideae)
  2' Lvs simple, ovate or wider in overall outline, palmately lobed . . . . **PLATANACEAE**
1' Infl not a dense spike, spheric head, nor enclosed in a fleshy receptacle
  3. Pls parasitic, attached to branches of trees or shrubs, not terrestrial . . . . **VISCACEAE**
  3' Pls not obviously parasitic, gen rooted in soil (sometimes root-parasitic)
    4. Ovary inferior, partly so, or appearing so
      5. Sts and lvs covered with silvery scales . . . . **ELAEAGNACEAE**
      5' Sts and lvs without silvery scales
        6. Infl unit an involucred head (superficially resembling a single fl) . . . . **ASTERACEAE**
        6' Infl open or fls solitary or paired
          7. Lvs opposite . . . . **CAPRIFOLIACEAE** (*Lonicera*)
          7' Lvs alternate . . . . **SANTALACEAE**
    4' Ovary superior
      8. Lvs opposite
        9. Fr winged, indehiscent; lvs gen lobed or compound
          10. Stigma 1, entire or slightly 2-lobed; fr wings 2 or sometimes 3; lvs palmately lobed . . **ACERACEAE**
          10' Stigmas 2, elongate; fr wing 1; lvs gen pinnately compound or appearing simple (lflet 1, sometimes 3–5 on twigs) . . . . **OLEACEAE** (*Fraxinus*)
        9' Fr not winged; lvs simple, entire or toothed
          11. Fls bisexual; stamens > 25 . . . . **ROSACEAE** (*Coleogyne*)
          11' Fls unisexual; stamens gen 4–12
            12. Staminate and pistillate fls in sessile umbels . . . . **OLEACEAE** (*Forestiera*)
            12' Staminate fls in stalked clusters; pistillate fls solitary
              13. Staminate infl head-like, fls subsessile; pistillate fls stalked; fr unlobed . . . . **SIMMONDSIACEAE**
              13' Staminate infl open, raceme or panicle; staminate and pistillate fls clearly stalked; fr 4–5-lobed . . . . **EUPHORBIACEAE** (*Tetracoccus*)
      8' Lvs alternate or whorled
        14. Infl fl-like with a cup-like involucre bearing colored, often petal-like nectaries; each stamen (actually a staminate fl) with a jointed stalk (composed of a filament and a pedicel); pistil (actually a pistillate fl) 3-lobed, exserted from involucre; sap milky . . . . **EUPHORBIACEAE** (*Euphorbia*)
        14' Infls various, but not fl-like; stamens gen > 1 per fl, filaments not joined; pistils various; sap mostly clear
          15. Fls strongly bilateral; fr an indehiscent, spiny pod . . . . **KRAMERIACEAE**
          15' Fls radial or nearly so; fr not spiny
             16. Pls covered with grayish or silvery scales
               17. Fls tubular, some or all bisexual; trees or large shrubs; lvs linear-lanceolate . . . . **ELAEAGNACEAE** (*Elaeagnus*)
               17' Fls not tubular, all unisexual; dioecious shrub or subshrub; lvs various
                 18. Ovary of pistillate fl enclosed by 2 appressed bractlets; lf surface powder-like, hairs not stellate . . . . **CHENOPODIACEAE** (*Atriplex*)
                 18' Ovary of pistillate fl subtended by 5 sepals; hairs scale-like, stellate **EUPHORBIACEAE** (*Croton*)
            16' Pls without scales
             19. Fl parts in 3s . . . . **POLYGONACEAE** (*Eriogonum*)
             19' Fl parts gen in 4s or 5s
               20. Stamens as many as and alternating with sepals . . . . **RHAMNACEAE**
               20' Stamens opposite sepals or more numerous
                 21. Style and stigma 1; style feathery with long, stiff hairs; fr an achene surrounded by a tubular hypanthium . . . . **ROSACEAE** (*Cercocarpus*)
                 21' Styles or stigmas 2–4
                   22. Fls bisexual
                     23. Shrub; lvs mostly narrow, base symmetric; fr an achene or utricle, not flattened or winged . . . . **CHENOPODIACEAE**
                     23' Tree; lvs wide, flat, base gen oblique; fr a drupe or round to ovate, winged achene . . . . **ULMACEAE**
                 22' Fls unisexual

24. Large shrubs or small trees; lvs 3–10 cm, green; lf bases oblique; fr a drupe
. . . . **ULMACEAE** (*Celtis*)
24' Shrubs 2.5 m or less; lvs gen < 3 cm, sometimes gray; lf bases gen symmetric; fr dry
   25. Calyx of pistillate fls 0, ovary enclosed by a pair of tightly appressed bractlets; ovary
      unlobed; fr indehiscent, seed 1 . . . . **CHENOPODIACEAE**
   25' Calyx of pistillate fls well developed; ovary 2–3-lobed; fr dehiscent, seeds 2 or more
   . . . . **EUPHORBIACEAE**

Group 14: Herbaceous Dicots; Perianth in One Whorl or Absent; Ovary Inferior

1. Pls internal st-parasites on desert shrubs . . . . **RAFFLESIACEAE**
1' Pls free-living or root-parasites
  2. Perianth parts free or nearly so
   3. Infl a simple or compound umbel . . . . **APIACEAE**
   3' Infl various
    4. Pls aquatic, floating or stranded on mud . . . . **HALORAGACEAE**
    4' Pls terrestrial or sometimes growing in damp soil
     5. Lvs coarsely toothed or lobed; infl of axillary clusters . . . . **DATISCACEAE**
     5' Lvs entire; infl a dense spike with large, petal-like bracts . . . . **SAURURACEAE**
  2' Perianth parts fused
   6. Infls of 1–many involucred heads
    7. Ovary truly inferior; anthers fused around style or, if free, fls unisexual; stigma lobes 2
    . . . . **ASTERACEAE**
    7' Ovary actually superior, but appearing inferior (base tightly enwrapped by perianth base); anthers free;
     stigma unlobed . . . . **NYCTAGINACEAE**
   6' Infl various but not of involucred heads
    8. Lvs alternate or all basal
     9. Tendrils present; fls unisexual . . . . **CUCURBITACEAE**
     9' Tendrils 0; fls bisexual . . . . **SANTALACEAE**
    8' Lvs opposite or whorled
     10. Perianth lobes 4 or fewer . . . . **RUBIACEAE**
     10' Perianth lobes 5 or more
      11. Fls gen radial or nearly so; ovary actually superior (tightly enwrapped by hardened or winged
       calyx base); lvs of a pair often unequal . . . . **NYCTAGINACEAE**
      11' Fls bilateral; ovary truly inferior; lvs of a pair equal . . . . **VALERIANACEAE**

Group 15: Herbaceous Dicots and Some Monocots;
Perianth in One Whorl or Absent; Ovary Superior

1. Lvs alternate or all basal
  2. Pls aquatic, floating on water surface — lvs in rosettes, sessile, velvety, palmately veined; fls enclosed
   by small, sheathing bracts . . . . **ARACEAE** (*Pistia*)
  2' Pls terrestrial or rooted in shallow water
   3. Lvs with stipules, gen well developed
    4. Style 1, unbranched, or stigma sessile
     5. Ovule 1; fr an achene . . . . **ROSACEAE**
     5' Ovules several–many; fr a capsule . . . . **VIOLACEAE**
    4' Styles or style branches 2 or more
     6. Stipules distinct, not fused around st, not shredded as branches develop . . . . **EUPHORBIACEAE**
     6' Stipules fused into a sheath around st, sometimes shredded as branches develop . . . **POLYGONACEAE**
   3' Lvs without stipules, sometimes reduced to scale-like bracts
    7. Perianth parts 6, ± petal-like . . . . **POLYGONACEAE**
    7' Perianth parts 5 or fewer, sometimes 0, gen not petal-like
     8. Infl fl-like with a cup-like involucre bearing colored, often petal-like nectaries; each stamen
      (actually a staminate fl) with a jointed stalk (composed of a filament and a pedicel); pistil (actually
      a pistillate fl) 3-lobed, exserted from involucre; sap milky . . . . **EUPHORBIACEAE**
     8' Infl various, but not flower-like; stamens gen > 1 per fl, filament not jointed; pistils various;
      sap not milky
      9. Ovules 2–many; infl a raceme
       10. Sepals 5 . . . . **POLYGALACEAE**
       10' Sepals 4
        11' Stamens 2, 4, or 6 . . . . **BRASSICACEAE**
        11' Stamens many . . . . **PAPAVERACEAE**
      9' Ovule 1; infls various
       12. Herbage ± densely covered with branched hairs

13. Fls bisexual; hairs soft, irregularly branched, branches all short; fr indehiscent
. . . . **AMARANTHACEAE** (*Tidestromia*)
13' Fls unisexual; hairs harshly stellate, central branches often long, spreading, bristle-like;
fr dehiscent . . . . **EUPHORBIACEAE** (*Eremocarpus*)
12' Herbage glabrous or ± hairy, hairs not branched, not stellate
14. Stigma 1, sessile . . . . **URTICACEAE** (*Parietaria*)
14' Stigmas > 1, sometimes borne on slender styles
15. Bracts subtending fls dry, scarious; pls neither fleshy nor with powdery or beaded surface;
habitats gen not saline . . . . **AMARANTHACEAE**
15' Bracts subtending fls fleshy, herbaceous, or 0; pls fleshy or surface powdery or with bead-like
hairs; habitats often saline or alkaline . . . . **CHENOPODIACEAE**
1' Lvs opposite or whorled
16. Pls aquatic, weak-stemmed; submersed, floating, or stranded on mud
17. Lvs opposite; ovary 2- or 4-lobed, 4-seeded . . . . **CALLITRICHACEAE**
17' Lvs whorled; ovary unlobed; seed 1 . . . . **CERATOPHYLLACEAE**
16' Pls terrestrial, sometimes in damp soil
18. Ovary chambers 2 or more (doubtful cases should be keyed both ways)
19. Fls unisexual . . . . **EUPHORBIACEAE**
19' Fls bisexual
20. Lvs opposite; fr circumscissile . . . . **AIZOACEAE**
20' Lvs appearing whorled; fr splitting lengthwise through chambers . . . . **MOLLUGINACEAE**
18' Ovary chamber 1
21. Ovules 3–many; fr a capsule
22. Placentas parietal . . . . **PAPAVERACEAE**
22' Placentas free-central or basal
23. Capsule circumscissile . . . . **AIZOACEAE**
23' Capsule splitting lengthwise, at least at tip
24. Sepals free . . . . **CARYOPHYLLACEAE**
24' Sepals fused . . . . **PRIMULACEAE** (*Glaux*)
21' Ovule 1; fr an achene, 1-seeded capsule, or utricle
25. Style 1, undivided or stigma 1, sessile
26. Sepals petal-like, fused into a tube; lvs entire or slightly lobed, stinging hairs 0 . . . **NYCTAGINACEAE**
26' Sepals green, small, inconspicuous; lvs toothed or entire (sometimes with stinging hairs)
27. Herbage glabrous or ± hairy, hairs not branched . . . . **URTICACEAE**
27' Herbage densely covered with branched hairs
28. Fls bisexual; hairs soft, irregularly branched, branches all short; fr indehiscent
. . . . **AMARANTHACEAE** (*Tidestromia*)
28' Fls unisexual; hairs harshly stellate, central branches often long, spreading, bristle-like;
fr dehiscent . . . . **EUPHORBIACEAE** (*Eremocarpus*)
25' Styles or style branches 2 or more
29. Fr a triangular achene; sepals 6 in two similar to very dissimilar whorls or in a single series
. . . . **POLYGONACEAE**
29' Fr not triangular; sepals 5 or fewer
30. Lvs with stipules . . . . **CARYOPHYLLACEAE**
30' Lvs without stipules
31. Lvs reduced to fleshy scales; infl a fleshy spike . . . . **CHENOPODIACEAE**
31' Lvs linear to ovate; infls various
32. Bracts subtending fls dry, scarious; pls neither fleshy nor with powdery or beaded surface;
habitats gen not saline . . . . **AMARANTHACEAE**
32' Bracts subtending fls lf-like or fleshy; pls often fleshy or surface powdery or with bead-like
hairs; habitats often ± saline . . . . **CHENOPODIACEAE**

Group 16: Dicots; Petals Fused; Ovary Superior

1. Stamens more numerous than corolla lobes
2. Lvs 2-pinnate—fr a legume; stamens many, very conspicuous . . . . **FABACEAE** (Mimosoideae)
2' Lvs simple, sometimes reduced to linear or scale-like bracts
3. Fl parts in 3s; fr a triangular achene . . . . **POLYGONACEAE**
3' Fl parts in 5s; fr a capsule or berry
4. Filaments fused, forming a tube aroundstyle
5. Fls radial; lf veins gen palmate; stamens many; filament tube a continuous cylinder . . . . **MALVACEAE**
5' Fls bilateral; lf veins pinnate or nearly parallel; stamens 8; filament tube open on one side
. . . . **POLYGALACEAE**
4' Filaments free

6. Sts bearing a stout, petiolar spine at each node; corolla tubular, bright red; filaments fused to base of corolla . . . . **FOUQUIERIACEAE**
6' Sts spineless; corolla urn-shaped to widely funnel-shaped, variously colored; filaments free from corolla or weakly adherent at base
  7. Shrubs; fls often urn-shaped; anthers releasing pollen through pores; ovules several–many; fr a berry or drupe . . . . **ERICACEAE**
  7' Herbs or weak-stemmed subshrubs; fls not urn-shaped; anthers dehiscent by lateral slits; ovule 1; fr an achene tightly enwrapped by hardened or winged base of corolla-like calyx
. . . . **NYCTAGINACEAE**
1' Stamens equal in number to corolla lobes or fewer
 8. Fertile stamens fewer than corolla lobes; fls mostly bilateral — 1 or more sterile stamens sometimes present
  9. Perianth parts in 6s; fls radial; stigmas and style branches 3 . . . . **POLYGONACEAE**
  9' Perianth parts in 4s or 5s; fls radial or bilateral; stigmas 1 or 2, style unbranched or branches 2
   10. Pls entirely non-green, ± fleshy . . . . **OROBANCHACEAE**
   10' Pls green and photosynthetic
    11. Pl a carnivorous aquatic with finely divided, submersed lf-like branches bearing small, hollow traps; placentas free-central . . . . **LENTIBULARIACEAE**
    11' Pls not carnivorous; terrestrial or aquatic but lacking finely divided, submersed, lf-like branches; placentas parietal or axile
     12. Ovary deeply 4-lobed; style gen arising from base of ovary lobes — fr 4 nutlets . . . . . **LAMIACEAE**
     12' Ovary entire or ± shallowly lobed; style arising from tip of ovary
      13. Ovules many per ovary chamber
       14. Trees or shrubs
        15. Capsule 15–30 cm, linear; trees or large shrubs; lvs long, linear . . . . **BIGNONIACEAE**
        15' Capsule << 15 cm, variously shaped; gen smaller shrubs; lvs various . . **SCROPHULARIACEAE**
       14' Herbs
        16. Capsule > 5 cm, densely glandular, tipped with a hooked beak several cm long; placentas parietal . . . . **MARTYNIACEAE**
        16' Capsule < 2 cm, mostly not glandular, beak 0 or < 1 cm; placentas axile **SCROPHULARIACEAE**
      13' Ovules 1–4 per ovary chamber
       17. Apparent corolla actually a petal-like calyx, narrowed above ovary; calyx base persistent around ovary, becoming hardened or winged at maturity; stigma, ovary chamber, and ovule 1; fr an achene . . . . **NYCTAGINACEAE**
       17' True corolla present, not narrowed above ovary; calyx base not persistent around ovary, neither hardened nor winged; stigmas, ovary chambers, and ovules > 1; frs various
        18. Ann . . . . **SCROPHULARIACEAE** (*Collinsia*)
        18' Per herbs or shrubs
         19. Lvs toothed, lobed, or compound; fr 2–4 nutlets . . . . **VERBENACEAE**
         19' Lvs entire; fr a capsule
          20. Corolla strongly bilateral, red or white; capsule wall thick, stiff . . . . **ACANTHACEAE**
          20' Corolla radial or nearly so, yellow or white; capsule ± inflated, its wall thin, membranous
. . . . **OLEACEAE** (*Menodora*)
8' Fertile stamens equal in number to corolla lobes; fls mostly radial
 21. Pls entirely non-green, parasitic
  22. Pl a thread-like, lfless, twining vine . . . . **CUSCUTACEAE**
  22' Pl an erect or mound-shaped, fleshy root-parasite with scale-like lvs . . . . **LENNOACEAE**
 21' Pls green and photosynthetic
  23. Ovaries 2, free, sometimes fused by a single style or by a complex stigmatic structure; fr of 1–2 follicles; sap milky
   24. Anthers free from stigma or slightly adherent, sometimes forming a cone or a dome over stigma complex; fls solitary or in cymes . . . . **APOCYNACEAE**
   24' Anthers fused to stigma and top of style, forming a drum-shaped cylindric structure with 5 vertical slits on sides; fls mostly in umbels . . . . **ASCLEPIADACEAE**
  23' Ovary 1, sometimes deeply lobed; frs various but not follicles; sap mostly clear
   25. Ovary deeply 2–4-lobed, esp in fr; fr breaking apart into 1-seeded nutlets or segments (sometimes only one nutlet maturing)
    26. Lvs opposite throughout; sts 4-angled; herbage strongly aromatic; infl never coiled . . . **LAMIACEAE**
    26' Lvs alternate above or throughout; sts ± round; herbage not aromatic; infl gen coiled when young, unrolling in fl or fr . . . . **BORAGINACEAE**
   25' Ovary entire or shallowly lobed; fr gen a capsule, berry, drupe, or achene
    27. Sepals 2; sts and lvs ± fleshy . . . . **PORTULACACEAE**
    27' Sepals 4 or more; sts and lvs gen not fleshy
     28. Perianth parts in 6s — fr a triangular achene . . . . **POLYGONACEAE**
     28' Perianth parts in 4s or 5s

29. Apparent corolla actually a petal-like calyx, narrowed above the ovary; calyx base persistent around ovary, becoming hardened or winged at maturity; stigma, ovary chamber, and ovule 1; fr an achene . . . . **NYCTAGINACEAE**
29' True corolla present, calyx gen not narrowed above ovary; calyx base not persistent around ovary, not hardened or winged; stigmas 1–3, ovules mostly >1; frs various
  30. Stamens opposite corolla lobes; placentas free-central . . . . **PRIMULACEAE**
  30' Stamens alternate with corolla lobes; placentas parietal or axile (rarely basal)
    31. Infl or branches of infl coiled when young, unrolling and ± open when mature, gen ± 1-sided
      32. Stigmas 2, gen not expanded, styles 2 or 1 and ± divided, at least near tip
        . . . . **HYDROPHYLLACEAE**
      32' Stigma 1, ± expanded and disk- or head-like, sessile or borne on undivided style
        33. Lvs entire, fleshy, and glabrous or ± bristly; fr breaking into 2 or 4 nutlets
          . . . . **BORAGINACEAE** (*Heliotropium*)
        33' Lvs gen toothed or lobed, gen not fleshy, variously hairy; fr a capsule . **HYDROPHYLLACEAE**
    31' Infl not coiled and unrolling, not 1-sided
      34. Calyx lobes gen ± fused by a much thinner, transparent or translucent membrane; stigmas and ovary chambers mostly 3 (less commonly 2 or 1) . . . . **POLEMONIACEAE**
      34' Calyx lobes without a transparent or translucent marginal membrane; stigmas mostly 2 (less commonly 1); ovary chambers mostly 1–2
        35. Lvs gen alternate or of unusual arrangement
          36. Stigmas 3
            37. Twining vines; lvs wide, palmately veined and often palmately lobed
              . . . . **CONVOLVULACEAE** (*Ipomoea*)
            37' Erect or prostrate herbs; lvs mostly pinnately compound or lobed, if palmately veined then lobes narrow . . . . **POLEMONIACEAE**
          36' Stigmas 1 or 2
            38. Style 1 unbranched; stigma 1 . . . . **SOLANACEAE**
            38' Styles 2 or style 1 and 2-branched; stigmas 2
              39. Fls 1(–5) per peduncle, gen subtended by a pair of opposite or subopposite bracts appressed against calyx or borne below fl(s) on peduncle; lvs often cordate, sagittate, or hastate; sts often trailing or twining; corolla 2–6 cm . . . . **CONVOLVULACEAE**
              39' Fls 1–many, not subtended by paired bracts; lvs entire to toothed, lobed, or compound, mostly not cordate, sagittate, or hastate; sts often erect; corolla gen < 2.5 cm
                40. Fls solitary in upper lf axils; per . . . . **CONVOLVULACEAE** (*Cressa*)
                40' Fls in few–many-fld clusters; ann or per . . . . **HYDROPHYLLACEAE**
        35' Lvs opposite, whorled, or all basal
          41. Corolla lobes 4
            42. Shrubs . . . . **BUDDLEJACEAE**
            42' Herbs
              43. Corolla of a normal, petal-like texture; infl often a cyme or fls solitary; lvs often cauline . . . . **GENTIANACEAE**
              43' Corolla dry, scarious; infl a dense spike (rarely fl only 1); lvs in most spp. all basal
                . . . . **PLANTAGINACEAE**
          41' Corolla lobes 5
            44. Fls solitary, scapose . . . . **HYDROPHYLLACEAE** (*Hesperochiron*)
            44' Fls 1–many, borne on a ± leafy st
              45. Style branches or stigmas 3; fr a capsule . . . . **POLEMONIACEAE**
              45' Style branches or stigmas 2 or style undivided and stigma 1; fr a capsule or a cluster of 4 nutlets
                46. Lvs toothed, lobed, or compound . . . . **HYDROPHYLLACEAE**
                46' Lvs entire
                  47. Pls glandular-hairy . . . . **SOLANACEAE** (*Petunia*)
                  47' Pls glabrous or hairy, but not glandular
                    48. Sts spreading or mat-forming; sts and lvs densely strigose; fls sessile or subsessile in forks of st; lvs of a pair often unequal . . . . **BORAGINACEAE** (*Tiquilia*)
                    48' Sts mostly erect or ascending; sts and lvs glabrous or short-hairy; fls axillary or in terminal infls; lvs of a pair equal . . . . **GENTIANACEAE**

Group 17: Dicots; Petals Fused; Ovary More or Less Inferior

1. Stamens many
  2. Spiny per with fleshy sts; petals many, overlapping in several series . . . . **CACTACEAE**
  2' Bristly, slender-stemmed ann; petals 5, in 1 series . . . . **LOASACEAE**
1' Stamens 8 or fewer
  3. Tendrils gen present; fls unisexual—monoecious vines . . . . **CUCURBITACEAE**

3' Tendrils 0; fls bisexual or unisexual
    4. Ovary actually superior, surrounded by but not fused to hardened or winged base of corolla-like
       calyx; style passing through constriction separating calyx base from petal-like portion and joining
       to top of ovary . . . . **NYCTAGINACEAE**
    4' Ovary truly inferior, fused to bases of surrounding perianth parts; style not passing through
       calyx constriction
      5. Fls borne in involucred heads of 1–many fls, these often secondarily clustered; sepals never green,
         the calyx modified as a white to brownish pappus of 1–many dry scales, bristles, or awns, or
         sometimes 0 . . . . **ASTERACEAE**
      5' Fls borne in cymes, racemes, or panicles; sepals gen ± green (forming a pappus in Valerianaceae)
        6. Lvs alternate; filaments ± free from corolla . . . . **CAMPANULACEAE**
        6' Lvs opposite or whorled; filaments strongly fused to corolla tube or throat
          7. Stamens fewer than corolla lobes . . . . **VALERIANACEAE**
          7' Stamens as many as corolla lobes
           8. Shrubs and vines; lvs opposite, simple or compound; fr a berry; corolla lobes gen 5; fls radial or
            bilateral . . . . **CAPRIFOLIACEAE**
           8' Herbs and brittle-stemmed subshrubs; lvs opposite or whorled, simple; fr 2-lobed, gen breaking
            into 2 one-seeded segments at maturity; corolla lobes 3–4; fls radial . . . . **RUBIACEAE**

Group 18: Dicots; Petals Free; Ovary Inferior

1. Pls aquatic . . . . **HALORAGACEAE**
1' Pls terrestrial
  2. Stamens opposite and equal in number to petals . . . . **RHAMNACEAE**
  2' Stamens alternate with petals or different in number
    3. Styles > 1
      4. Ovules and seeds gen 1–2 per chamber
        5. Herbs; stamens 5; fls in umbels or heads . . . . **APIACEAE**
        5' Shrubs or trees; stamens 10–many; fls gen not in umbels or heads
          6. Lvs opposite; stamens 10; hypanthium very short or 0; fr a capsule
           . . . . **PHILADELPHACEAE** (*Fendlerella*)
          6' Lvs alternate; stamens many; hypanthium well developed; fr a pome . . . . **ROSACEAE**
      4' Ovules and seeds gen several–many per chamber
        7. Sepals 2; petals 4–6; lvs ± fleshy . . . . **PORTULACACEAE**
        7' Sepals 5; petals 5, lvs not fleshy
          8. Herbs . . . . **SAXIFRAGACEAE** (*Heuchera*)
          8' Shrubs
           9. Stamens 5; lvs alternate . . . . **GROSSULARIACEAE**
           9' Stamens 10–many; lvs opposite . . . . **PHILADELPHACEAE**
    3' Style 1
      10. Sts fleshy, spiny; sepals and petals indefinite in number and not sharply differentiated, in spirals
        . . . . **CACTACEAE**
      10' Sts not fleshy, not spiny; sepals and petals in definite whorls
        11. Pls tendril-bearing vines; monoecious . . . . **CUCURBITACEAE**
        11' Pls not tendril-bearing; fls gen bisexual
          12. Lvs armed with rough, barbed, or stinging hairs . . . . **LOASACEAE**
          12' Lvs glabrous or ± soft-hairy to glandular
           13. Herbs . . . . **ONAGRACEAE**
           13' Shrubs or trees
            14. Lvs opposite, entire; petals 4; hypanthium 0 . . . . **CORNACEAE**
            14' Lvs alternate, palmately lobed; petals 5; hypanthium well developed. . . . . **GROSSULARIACEAE**

Group 19: Dicots; Petals Free; Stamens Many; Ovary Superior

1. Ovary chambers 2–many
  2. Fls unisexual—ovary gen 3-lobed. . . . . **EUPHORBIACEAE**
  2' Fls bisexual
    3. Filaments fused into tube around style; anther chamber 1
      4. Fls radial; petals 5; stamens many; filament tube cylindric . . . . **MALVACEAE**
      4' Fls bilateral; petals 3; stamens 8; filament tube open on one side . . . . **POLYGALACEAE**
    3' Filaments free or fused at base into groups; anther chambers 2
      5. Hypanthium 0; lvs opposite; stamens in 3(–5) bunches . . . . **HYPERICACEAE**
      5' Hypanthium present; lvs alternate; stamens not in bunches. **ROSACEAE**
1' Ovary chamber 1
  6. Ovules 1–2 per ovary; fr a drupe, follicle, or achene
    7. Fl parts in 3s; fr triangular . . . . **POLYGONACEAE**

7' Fl parts in 5s; fr ± round . . . . **ROSACEAE**
6' Ovules 2–many per ovary; fr a capsule, follicle, legume, or berry
  8. Lvs palmately compound, densely covered with stalked glands — fl showy, parts in 4s . . **CAPPARACEAE**
  8' Lvs various, glabrous to densely hairy but not glandular
    9. Infl a spike, or head
      10. Shrubs or trees; fr a legume . . . . **FABACEAE** (Mimosoideae)
      10' Herbs; fr a follicle or berry . . . . **RANUNCULACEAE**
    9' Infl a panicle or a cyme or fl solitary
      11. Fr a follicle; hypanthium present . . . . **ROSACEAE**
      11' Fr a capsule; hypanthium gen 0
        12. Sepals gen 2–3, sometimes fused into a conical cap, falling as fl opens; sap (at least in roots)
          gen milky or colored . . . . **PAPAVERACEAE**
        12' Sepals 2–9, persistent, never fused to tip; sap clear — pl gen ± fleshy . . . . **PORTULACACEAE**

Group 20: Dicots; Petals Generally Free; Stamens Twice as Many as Petals or Fewer; Leaves Compound or Nearly So; Ovary Superior

1. Lvs opposite
  2. Trees or well developed shrubs
    3. Lvs simple, palmately lobed; fls ± radial; ovary 2-winged . . . . **ACERACEAE**
    3' Lvs compound, lflets 5 or more; fls strongly bilateral; ovary not winged . . . . **HIPPOCASTANACEAE**
  2' Herbs or subshrubs
    4. Lflets toothed or dissected; infl gen an umbel; fruit long-beaked, segments coiled when dry
    . . . . **GERANIACEAE**
    4' Lflets entire; fls solitary or in pairs; ovary not beaked . . . . **ZYGOPHYLLACEAE**
1' Lvs alternate
  5. Petals 4 or 6
    6. Petals 6 . . . . **BERBERIDACEAE**
    6' Petals 4
      7. Corolla strongly bilateral
        8. Stamens 10, all filaments, or 9 of them fused, forming tube around ovary—petals actually 5, but
          lower 2 ± fused; petals not spurred . . . . **FABACEAE** (Papilionoideae)
        8' Stamens 6, free; 1 petal prolonged into a spur . . . . **PAPAVERACEAE** (*Corydalis*)
      7' Corolla radial or ± weakly bilateral
        9. Lvs pinnately dissected or compound . . . . **BRASSICACEAE**
        9' Lvs palmately compound . . . . **CAPPARACEAE**
  5' Petals 5
    10. Lvs 2 or more times compound or divided . . . . **FABACEAE**
    10' Lvs once compound
      11. Fls strongly bilateral; stamens gen 10, all filaments or 9 of them fused, forming tube around ovary
      . . . . **FABACEAE** (Papilionoideae)
      11' Fls radial or weakly bilateral; stamens 5–12, gen free
        12. Lflets 3
          13. Shrubs . . . . **ANACARDIACEAE**
          13' Herbs . . . . **OXALIDACEAE**
        12' Lflets 4 or more
          14. Herbs or subshrubs . . . . **FABACEAE**
          14' Shrubs or trees
            15. Petals ± white, ± 4 mm; fls solitary or few; lflets gland-dotted; herbage aromatic . . **BURSERACEAE**
           15' Petals yellow, 8–12 mm; fls in racemes or panicles, often many; lflets not gland-dotted; herbage
              not aromatic . . . . **FABACEAE** (*Senna*)

Group 21: Dicot Shrubs and Trees; Petals Free; Stamens Twice as Many as Petals or Fewer; Leaves Simple; Ovary Superior

1. Lvs opposite or whorled
  2. Style 1, unbranched, stigma 1, entire or nearly so—lvs deeply 2-lobed; fls bright yellow; fr densely
    white-hairy . . . . **ZYGOPHYLLACEAE** (*Larrea*)
  2' Styles or style branches 2–5, each terminated by a stigma
    3. Stamens more numerous than petals; petals flat; lvs palmately veined . . . . **ACERACEAE**
    3' Stamens equal in number to and opposite petals; petals gen ± cupped; lvs pinnately veined
    . . . . **RHAMNACEAE**
1' Lvs alternate, sometimes reduced to minute scales or so quickly deciduous that pls gen lfless
  4. Lvs palmately veined, blade well developed — vine with tendrils . . . . **VITACEAE**
  4' Lvs pinnately veined, 1-veined, or reduced to bladeless scales

    5. Fls strongly bilateral
      6. Stamens 4, all inserted on 1 side of ovary, free or nearly so; sepals rose-purple, petal-like; petals 5, lower 2 reduced to fleshy scales; fr bearing slender, barb-tipped spines . . . . **KRAMERIACEAE**
      6' Stamens 8 or 10, all filaments or 9 of them fused into tube around ovary; sepals variously colored, sometimes petal-like; petals 3 or 5, not reduced to fleshy scales; fr a capsule or legume
        7. Sepals all fused, at least at base, gen not petal-like; petals 5, odd petal (banner) uppermost; lower 2 lateral petals gen free at base but fused toward tip, forming a keel that enclosed stamens and ovary; fr a legume . . . . **FABACEAE** (Papilionoideae)
        7' Sepals free, petal-like, 2 spreading and very different from other 3; petals 3, odd petal lowermost, banner petal never present; fr a capsule . . . . **POLYGALACEAE**
    5' Fls radial or nearly so
      8. Stamens equal in number to petals
        9. Lvs all reduced to scales; twigs very slender, jointed, green . . . . **TAMARICACEAE**
        9' Lvs with expanded blades, linear to ovate, sometimes early deciduous; twigs not jointed, green or brown
          10. Stamens opposite petals; petals gen with cupped blade . . . . **RHAMNACEAE**
          10' Stamens alternate petals; petals mostly flat
            11. Stigmas 3–5, style 1, unbranched, or style branches 3
              12. Lvs > 25 mm, petioled, margin not thickened; style branches or styles 3; fr a drupe
. . . . **ANACARDIACEAE**
              12' Lvs < 15 mm, sessile or subsessile, margin thickened; style 1, stigmas 4–5; fr a 1-seeded capsule . . . . **CELASTRACEAE** (*Mortonia*)
            11' Stigma 1, sessile or borne on an unbranched style
              13. Lvs entire; fr a follicle . . . . **CROSSOSOMATACEAE** (*Glossopetalon*)
              13' Lvs toothed; fr dry, roughened, breaking into 5 one-seeded pieces . . . **STERCULIACEAE** (*Ayenia*)
      8' Stamens more numerous than petals
        14. Fls dark blue-purple; sts and lvs gland-dotted, very strongly scented . . . . . . **RUTACEAE** (*Thamnosma*)
        14' Fls white to yellow or green; sts and lvs not scented
          15. Fls unisexual
            16. Ovary unlobed, chamber 1, containing 1 or 2 seeds; hairs 0 or attached at one end; stiffly branched shrub . . . . **CROSSOSOMATACEAE** (*Glossopetalon*)
            16' Ovary 3-lobed, chambers 3, each 1-seeded; hairs 2-branched, attached in middle; weak-stemmed subshrub . . . . **EUPHORBIACEAE** (*Ditaxis*)
           15' Fls bisexual
            17. Anthers opening by small round pores; petals 5 . . . . **ERICACEAE** (*Ledum*)
            17' Anthers opening along sides by elongated slits; petals 4–6
              18. Fls in elongated, terminal, bractless racemes; sepals and petals 4; stamens 6; sts not ending in spines . . . . **BRASSICACEAE**
              18' Fls solitary or in small, axillary clusters, if in racemes, fls subtended by bracts; sepals and petals 4–6; stamens 8–12; sts often terminating in stout spines
                19. St grooved; ovary sessile on receptacle, chamber 1; lvs linear or narrowly elliptic, ± persistent; fr a follicle . . . . **CROSSOSOMATACEAE** (*Glossopetalon*)
                19' St smooth, gen lfless most of year; ovary raised ± 1 mm above receptacle on short stalk, chambers 2; lvs scale-like, soon deciduous; fr a berry . . . . **KOEBERLINIACEAE**

Group 22: Mostly Dicot Herbs; Petals Generally Free; Stamens Twice as Many as Petals or Fewer; Leaves Simple; Ovary Superior

1. Fls bilateral
  2. Petals 2— developing fr open at tip . . . . **RESEDACEAE**
  2' Petals 4 or 5
    3. Petals 4, not strongly overlapping; stamens 6, 4 with long filaments, 2 with short filaments
. . . . **BRASSICACEAE**
    3' Petals 5, strongly overlapping; stamens 5–10, ± equal in length
      4. Odd petal uppermost in fl; leaves gen compound; ; fr a legume or indehiscent pod
. . . . **FABACEAE** (Papilionoideae)
      4' Odd petal lowermost in fl; lvs simple; fr a capsule
        5. Lvs sessile, without stipules; filaments fused, forming a U-shaped tube around ovary; fr flat
. . . . **POLYGALACEAE**
        5' Lvs petioled, with stipules; filaments free; fr ovoid . . . . **VIOLACEAE**
1' Fls radial
  6. Petals 4 or fewer
    7. Lvs alternate, sometimes all basal
      8. Petals 4 . . . . **BRASSICACEAE**
      8' Petals 2–3 . . . . **POLYGONACEAE**

7' Lvs opposite or whorled
  9. Sepals free or nearly so . . . . **ELATINACEAE**
  9' Sepals fused, at least toward base
    10. Petals borne on receptacle; style 2–3 branched — calyx cylindric . . . . **FRANKENIACEAE**
    10' Petals borne on inner face of tubular hypanthium; style unbranched . . . . **LYTHRACEAE**
6' Petals 5 or more
  11. Fls unisexual . . . . **EUPHORBIACEAE**
  11' Fls bisexual
    12. Sepals 2 or 3; lvs ± fleshy . . . . **PORTULACACEAE**
    12' Sepals 5 or more; lvs gen not fleshy
      13. Stamens more numerous than petals
        14. Lvs alternate, sometimes all basal
          15. Lvs toothed or lobed; style branches and stigmas 5; ovules 1 per ovary chamber; fr breaking
            into 5 segments, each tipped with coiled beak segment . . . . **GERANIACEAE**
          15' Lvs entire; styles or style branches 1–3, stigmas 1–3; ovules several–many per ovary chamber;
            fr a capsule without an elongate beak
        14' Lvs opposite or whorled
          16. Stigma and style 1 — fls with a cylindric hypanthium; lvs entire . . . . **LYTHRACEAE**
          16' Stigmas 2–6
      13' Stamens equal in number to petals or fewer
        17. Style 1, unbranched
          18. Stamens alternate with petals . . . . **GENTIANACEAE** (*Swertia*)
          18' Stamens opposite petals . . . . **PRIMULACEAE**
        17' Styles or style branches 2 or more
          19. Lvs opposite
            20. Sepals fused . . . . **FRANKENIACEAE**
            20' Sepals free
              21. Lvs entire . . . . **CARYOPHYLLACEAE**
              21' Lvs toothed, lobed, or compound . . . . **GERANIACEAE**
          19' Lvs alternate, sometimes all basal
            22. Lvs linear, entire . . . . **LINACEAE**
            22' Lvs wider, often toothed or lobed
              23. Styles or style branches 5; fr breaking into 5 one-seeded segments, each tipped with coiled
                beak segment . . . . **GERANIACEAE**
              23' Styles, style branches, or stigmas 2–4; fr a capsule with many small seeds
                . . . . **SAXIFRAGACEAE**

AZOLLACEAE MOSQUITO FERN FAMILY

Alan R. Smith

Pl free-floating or stranded on mud, gen 1–5 cm, often fan-shaped; roots pendent from st forks, unbranched. **ST** forked repeatedly or pinnate, thread-like, easily fragmented at joints. **LVS** alternate, in 2 rows, sessile, often over-lapped, 0.5–1.5 mm, seemingly paired but actually of 2 roundish to ovate lobes; upper lobe floating or emergent, thick, greenish or reddish, margin whitish; lower lobe submersed, gen slightly larger, thinner, whitish. **SPORAN-GIA** in seemingly axillary cases of 2 kinds, cases gen in pairs of 1 kind. **MALE SPORANGIUM CASE** 1.2–2 mm diam, spheric; tip dark-pointed; wall transparent; sporangia gen 20–100+, long-stalked; spores 32 or 64, spheric, in gen 3–6 barbed masses. **FEMALE SPORANGIUM CASE** 0.2–0.4 mm diam, hemispheric or spheric; tip obtuse, covered by dark, conic, spongy structures that aid in flotation; wall ± opaque; sporangium 1, sessile; spore 1, spheric. 1 genus, ± 7 spp.: ± worldwide. *Salvinia molesta* D.S. Mitch. has been tentatively identified from Walter's Camp, Imperial Co.; more study needed.

AZOLLA

(Greek: dry kill, from pl death in dried habitats) [Perkins et al. 1985 Scanning Electron Microscopy 1985(IV):1719–1734] Used as green manure in rice paddies because of nitrogen-fixing algae in upper lf lobe; spp. identification gen requires female sporangium cases (gen 0 on herbarium specimens), often lf sectioning, compound microscope.

A. filiculoides Lam. (p. 71) Pl green to reddish, fertile only when ascending. **STS** < 5+ cm: immature prostrate, internodes < 5 mm; mature ascending, internodes < 1 mm. **SPORANGIUM** **CASES** gen male. Common. Ponds, slow streams, wet ditches; 0–1600 m. CA-FP, **GB**; to WA, AZ, S.Am, Eurasia, Afr; also e US. ❀still water [not saline] or mud:**4–9**,10,**14–24**.

DRYOPTERIDACEAE WOOD FERN FAMILY

Alan R. Smith & Thomas Lemieux

Per, in soil or rock crevices; rhizome gen short-creeping, suberect, or erect, scales large, gen tan to brown, gen 1-colored. **LVS** gen tufted, 5–200+ cm, gen ± alike; petiole gen firm, base gen darker, with 2–many vascular strands; blade 1–4-pinnate, often with scales, hair-like scales, hairs (exc clear, needle-like hairs gen 0), or short-stalked glands on axes, sometimes between veins, veins free to netted; 1° and 2° axes gen grooved on upper side. **SPO-RANGIA**: sori round, less often oblong or J-shaped, along or at tips of veins; indusia peltate, round-reniform, oblong to linear, J-shaped, hood-like, or cup-like, rarely 0; spores elliptic, winged, ridged, or spiny, scar linear. ± 60 genera, > 1000 spp.: worldwide, esp trop, wooded areas. *Woodsia* sometimes in Woodsiaceae; *Athyrium, Cystopteris* sometimes in Athyriaceae.

1. Indusium cup-like, of many segmented hair- or scale-like fragments or lobes encircling sorus from
   below, often obscure in age . . . . **WOODSIA**
1' Indusium peltate, round-reniform, oblong, J-shaped, or hood-like, ± covering sorus from above or 1 side,
   gen conspicuous in age
   2. Indusium oblong or J-shaped — petiole base in ×-section with 2 crescent-shaped vascular strands
   . . . . **ATHYRIUM**
   2' Indusium peltate, round-reniform, or hood-like
      3. Blade 1-pinnate, lower lflets only slightly lobed, margin with bristle-like tips; indusium peltate,
         centrally attached, without a sinus . . . . **POLYSTICHUM**
      3' Blade 2–4-pinnate, lflet margin lacking bristle-like tips; indusium hood-like or round-reniform,
         ± centrally attached and with a sinus
         4. Indusium hood-like; petiole < 1.5 mm wide, base with 2 vascular strands; blade gen 10–20 cm,
            5–8 cm wide . . . . **CYSTOPTERIS**
         4' Indusium round-reniform; petiole >1.5 mm wide, base with many vascular strands; blade gen
            20–75 cm, 10–20 cm wide . . . . **DRYOPTERIS**

ATHYRIUM LADY FERN

Rhizome short-creeping to suberect, stout. **LF**: petiole stout, fleshy, easily crushed, straw-colored exc base gen black-ened, base scaly, in ×-section with 2 crescent-shaped vascular strands; blade gen 2-pinnate or more, ± glabrous, veins free. **SPORANGIA**: sori ± round, ± oblong, or J-shaped; indusium oblong or J-shaped, laterally attached, or 0. ± 100 spp.: gen n temp, esp e Asia. (Greek: doorless, from enclosed sori)

A. filix-femina (L.) Roth var. *cyclosorum* Rupr. (p. 71) **LF**: blade elliptic to lanceolate, 1–2-pinnate, segments deeply pin-nately lobed to ± toothed, margin not reflexed over sorus. 2*n*=80. Woods, along streams, seepage areas; 0–3200 m. CA-FP (exc GV, PR), **GB**; to AK, w Can, ID, Colorado, n Mex. [var. *californicum* F.K. Butters; var. *sitchense* Rupr.] Highly variable, but named vars. in w N.Am seem indistinct; other vars. worldwide. ❀IRR or WET:**4–6**,**17**&SHD:1–3,7,14,**15,16**,18–23,**24**.

CYSTOPTERIS FRAGILE FERN

Rhizome gen short-creeping. **LF**: petiole ± fleshy, often with few scales, base in ×-section with 2 vascular strands; blade 2–4-pinnate, veins free. **SPORANGIA**: sori round; indusium hood-like, arched over sorus, attached on side away from margin. ± 10 spp.: gen temp. (Greek: bladder fern, from indusium)

C. fragilis (L.) Bernh. (p. 71) Rhizome 2–4 mm diam; scales at tip, lanceolate, brownish, shining, glabrous, entire. **LF** 8–30(37) cm; petiole gen < blade, < 1.5 mm wide, base straw-colored to reddish brown; blade gen 10–24 cm, 5–9 cm wide, ovate-lanceolate, lowest 2–4 1° lflets ± < others. $2n=168$. Shady, moist rock crevices, meadows, banks, streamsides; 50–3800 m. KR, NCoRO, NCoRH, CaR, n&c SNF, SNH, SnFrB, SCoR, TR, PR, **GB, DMtns**; worldwide. ❀WET or IRR:**4–6,17**&SHD:1–3,7, 14,**15,16**,18–23,**24**.

DRYOPTERIS WOOD FERN

Rhizome short-creeping or ascending to suberect, stout. **LF**: petiole > 1.5 mm wide, firm, more densely scaly than midrib, base in ×-section with many round vascular strands in an arc; blade 1–3-pinnate or more, veins free, simple or forked; segments deeply pinnately lobed or not. **SPORANGIA**: sori round; indusium round-reniform, ± centrally attached at a sinus, gen persistent. ± 100 spp.: ± worldwide, esp e Asia. (Greek: oak, fern) Hybrids unknown in CA, frequent in e N.Am. [Montgomery & Paulton 1981 Fiddlehead Forum 8:25–31]

D. filix-mas (L.) Schott (p.71) MALE FERN **LF** ± 40–70(100+) cm, 15–25(30+) cm wide; petiole, midrib nonglandular; blade elliptic, 2-pinnate, segments deeply pinnately lobed or not, teeth ± without bristle-like tips, veins gen ending short of margin; long-est 1° lflets near middle, rarely near base; scales of 1° lflet mid-ribs ± linear or hair-like. $2n=164$. RARE in CA. Granitic cliffs; 2400–3100 m. SnBr, **W&I**; to B.C., ne N.Am, Eur, Afr. In cult.

POLYSTICHUM SWORD FERN

Rhizome gen suberect to erect, often stout. **LF**: petiole stout, firm, gen densely scaly, in ×-section with many round vascular strands in an arc; blade gen 1–3-pinnate, thin to leathery, scaly, veins gen free, rarely casually joined; 1° lflet bases often wider on distal side; teeth gen incl bristle-like tips that are < 2 mm. **SPORANGIA**: sori round; indusium gen peltate, sinus 0. 175+ spp.: ± worldwide (Greek: many rows, from rows of sori on type sp.)

P. scopulinum (D.C. Eaton) Maxon (p. 71) **LF** 10–50 cm; petiole 1/4–1/2 blade length, base scales 1.5–2(3) mm wide, lanceolate to elliptic; blade narrow-lanceolate, 1-pinnate, 1° lflets shallowly lobed, oblong-lanceolate, longest 1.5–3 cm. **SPORANGIA**: indusium entire. $2n=164$. Serpentine to acidic soils, gen full sun, rock crevices, boulder bases; 400–3200 m. KR, NCoRO, NCoRH, CaRH, n&c SNH, SnBr, SnJt, **DMtns (Surprise Can-yon, Panamint Mtns)**; to B.C., Rocky Mtns, AZ. Probably fertile hybrid between *P. imbricans* and *P. lemmonii*. DMtns distribution based on an RSA herbarium specimen, Rompert 229, 1977; 1977; this specimen differs from others in the range in having lflets only shallowly, if at all, lobed ❀DRN:**4–6** &IRR, SHD:1–3,7,14–18

WOODSIA CLIFF FERN

Rhizome gen ascending to suberect, short, with many old petiole bases. **LF** often glandular or hairy; petiole base with 2 vascular strands; blade 1–2-pinnate, segments ± toothed to pinnately lobed, veins free. **SPORANGIA**: sori round, gen not at margins; indusium cup-like, often of many segmented hair- or scale-like fragments or lobes encircling sorus from below, often of crusty, whitish beads, often obscure in age. ± 30 spp.: gen n temp. (J. Woods, Britain, b 1776) [Brown 1964 Beih Nova Hedwigia 16:1–154] ❀TRY.

1. Hairs on lower surface lf axes ± flat, segmented, ± 0.5–1 mm and nonsegmented, glandular, ± 0.1 mm
. . . . *W. scopulina*
1' Hairs on lower surface lf axes 0 or nonsegmented, glandular, ± 0.1 mm
  2. Indusium of segmented hairs . . . . *W. oregana*
  2' Indusium of scale-like fragments or lobes . . . . *W. plummerae*

W. oregana D. Eaton (p. 71) **LF** 5–25 cm, 1–3.5 cm wide; blade tip ± acute, nonforked; blade lower surface hairs 0 or ± 0.1 mm, nonsegmented, glandular; 1° lflets 0.5–2.5 cm, 0.3–1.3 cm wide, pinnately lobed to 1-pinnate, margin fine-toothed. **SPORANGIA**: indusium of segmented hairs. $2n=76,152$. Crevices, rock bases; 900–2800 m. KR, CaRH, n&s SNH, SnBr, PR, MP?, **W&I, DMtns**; to B.C., e Can, n US, OK, AZ. Sierra Co. citation by Brown a mislabeled specimen.

W. plummerae Lemmon (p. 71) PLUMMER'S WOODSIA **LF** < 25 cm, < 4 cm wide; blade tip often blunt or forked; blade lower surface hairs ± 0.1 mm, nonsegmented, glandular; 1° lflets < 3 cm, < 1.5 cm wide, pinnately lobed to 1-pinnate, margin toothed to shallow-lobed. **SPORANGIA**: indusium of scale-like fragments or lobes ending in hairs or not. $2n=152$. RARE in CA. Crevices, rock bases; 1600–2000 m. **DMtns**; to TX, n Mex. San Diego Co. citation by Brown a mislabeled specimen.

W. scopulina D. Eaton (p. 71, pl. 1) **LF** < 36 cm, 5–8 cm wide; blade tip ± acute, nonforked; blade lower surface hairs ± 0.5–1 mm, ± flat, segmented and ± 0.1 mm, nonsegmented, glandular; 1° lflets < 4 cm, 2.5 cm wide, pinnately lobed to 1-pinnate, margin toothed to shallow-lobed. **SPORANGIA**: indusium of narrow scale-like lobes. $2n=76,152$. Crevices, rock bases; 1400–3500 m. KR (Trinity Co.), CaRH, SNH, SnBr, **W&I**; to AK, e Can, w US.

EQUISETACEAE HORSETAIL FAMILY

Richard L. Hauke

Per from rhizome (above ground st ann or per). **ST** gen erect, ridged lengthwise, hollow exc at nodes, sometimes of 2 kinds (sterile, fertile); branches 0 or whorled and alternate lvs, sometimes solid. **LVS** scale-like, whorled, fused into nodal sheath with as many teeth as lvs, gen not green. **SPORANGIA** several on inner surface of peltate scales that are clustered into a terminal cone; spores of 1 kind per sp., spheric, green, unmarked, with 4 strap-like appendages. 1 genus, 15 spp.: world-wide exc Australia, New Zealand. [Hauke 1978 Nova Hedwigia 30:385–455]

EQUISETUM HORSETAIL, SCOURING RUSH

(Latin: horse, bristle, from roots of *E. fluviatile* L.)

1. Green sts regularly branched, lacking cones . . . . *E. arvense*
1' Green sts unbranched, bearing cones . . . . *E. laevigatum*

E. arvense L. (p. 71) COMMON HORSETAIL **STS** ann, of 2 kinds. **STERILE ST** 10–60 cm, green; basal internode of branch > subtending sheath; sheath 3–8.5 mm, ± as long as wide, teeth 6–14, 1.5 –3.5 mm, dark, often joined but not fused; branch with 3–4 rounded ridges, solid. **FERTILE ST** 11–32 cm, unbranched, fleshy, brown, ephemeral; sheath 5–11 mm, > that of sterile st, teeth 6–10, 3–7.5 mm. Moist, disturbed areas; < 3000 m. CA-FP, MP, **W&I**; N.Am, Eur, Asia. ❀ IRR:**1,4–7,15–17**,24&SHD:**2,3,8,9,14**,18–23; INV.

E. laevigatum A. Braun (p. 71) SMOOTH SCOURING RUSH **ST** ann (or per in s), of 1 kind, 30–180 cm, green, unbranched; sheath 6–15 mm, longer than wide, gen with 1 dark band at tip, teeth 10–26, gen deciduous. **SPORANGIA**: cone tip rounded. Moist, sandy or gravelly areas; < 3000 m. CA-FP, MP, **W&I**, **DMtns**; to B.C., e US. [*E. funstoni* A.A. Eaton; *E. kansanum* J. Schaffner] ❀IRR:**1–7,14–17,22–24**&SHD:**8,9,18–21**;INV.

MARSILEACEAE MARSILEA FAMILY

Alan R. Smith & Thomas Lemieux

Pl gen aquatic, gen rooted in, often stranded on mud; rhizome creeping, slender, branched. **LVS** floating, emersed, or out of water, ± alike; blade 1-palmate or 0, << petiole; veins not or repeatedly forked, free or netted. **SPORANGIA** in stalked, spheric or ± flat-ovoid, hard cases of 1 kind, near petiole base. **SPORES** large (female) and small (male), in separate sporangia. 3 genera, ± 70 spp.: esp temp.

MARSILEA

LF like that of clover or wood sorrel; blade 1-palmate, lflets 4, wedge-shaped, hairy. **SPORANGIUM CASE** fused to stalk 0.8–1.7 mm, ± flat-ovoid, hairs long, dense, deciduous or not, teeth 1–2, near base. > 60 spp.: esp temp. (L.F. Marsigli, Italian botanist, 1656–1730) [Johnson 1986 Syst Bot Monogr 11: 1–87]

M. vestita Hook. & Grev. ssp. *vestita* (p. 71) **LVS**: petioles of floating lvs weak, 6–35 cm, others 3–8 cm; lflet sides (or 1) often concave, distal margin convex, ± entire. **SPORANGIUM CASE** 3–8 mm, 3–7 mm wide; stalk not bent. Creek beds, flood basins; < 2200 m. KR, NCoRI, CaR, s SNF, SNH, GV, CCo, SnFrB, SCoRO, SCo, WTR, SnBr, PR, MP, **DSon**; to w&c Can, Mex, also Peru. ❀IRR:**4–7,14–17,24**,WET:1–3,8–13,**18–23**; GRCVR.

OPHIOGLOSSACEAE ADDER'S-TONGUE FAMILY

Warren H. Wagner, Jr.

Per, small, fleshy, gen glabrous; caudex gen underground, unbranched; roots glabrous, with bulblets or plantlets or not. **LF** gen 1 per caudex per year, divided into 2 facing parts with a common stalk; sterile part separated from fertile at to well above ground, blade simple to compound, veins free and forked (or netted, with incl veinlets); fertile part bladeless, bearing sporangia, simple to compound. **SPORANGIA** dehiscent into 2 valves, ± 1 mm wide, thick-walled. 3 genera, 70–85 spp.: ± worldwide, gen rare or overlooked. Fern-like pls with many traits of seed pls. Specimens must be carefully spread and pressed for identification; haploid generation underground, fleshy, non-green, associated with fungi.

BOTRYCHIUM GRAPE-FERN, MOONWORT

Roots smooth, pale or cork-ridged, dark gray, without bulblets or plantlets. **LF** gen deciduous; bud glabrous or hairy; sterile part gen ± 1–3-pinnate (rarely simple or entire), linear to deltate, segments linear to oblong and midribbed or spoon- to wedge- or fan-shaped and not midribbed, veins free, forked, margins entire to dentate or irregularly cut; fertile part 1–3-pinnate, < to > sterile. **SPORANGIA** not sunken in axis; stalk 0 or short. 40–50 spp.: gen temp to arctic or alpine. (Greek: bunch of grapes, from clusters of sporangia) [Wagner & Wagner 1983 Amer Fern J 73:53–62] Difficult, needing careful study; most spp. very uncommon, sporadic; good sampling of populations highly desirable in specimens. ❀TRY;DFCLT.

1. Blades gen narrowly oblong; basal lflets ± equal in size and cutting to next pair; fertile lf part arising near or gen above middle of common lf stalk . . . . **B. lunaria**
1' Blades ovate to often ternate; basal lflets gen much larger and more complex than next pair; fertile lf part arising near to below middle of common lf stalk . . . . **B. simplex**

B. lunaria (L.) Sw. MOONWORT **LF:** sterile part separated from fertile near or gen above middle of lf, stalk < 1.5 mm, blade 1-pinnate, gen 6–10 cm, 2–4 cm wide, oblong, thick, dark green, segments gen touching to overlapped, 4–9 pairs, widely fan-shaped, not midribbed, lower with margins at base meeting at 120–160°, outer margins gen entire; fertile part mostly 1-pinnate, 0.8–2 × sterile. 2*n*=90. RARE in CA. Fields, meadows; 3000–3400 m. c SNH (e slope, Tuolumne Co.), **n SNE (Mono Co.);** n hemisphere, s S.Am. Australia, New Zealand.

B. simplex Hitchc. (p. 71) YOSEMITE MOONWORT **LF:** stalk < 1 × blade; sterile part separated from fertile well below middle of lf, gen at top of lf sheath (well above ground in ± young pls), blade simple, deeply lobed, to 2-pinnate, < 12 cm, oblong to ovate, firm, dull green, segments touching to well separated, fan- to wedge-shaped, ± oblique, not midribbed, outer margins entire to slightly crenate; fertile part 1-pinnate, 3–8 × sterile. Uncommon. Open marshes, damp meadows; 2200–3300 m. NCoRH, CaR, SN, SnBr, **W&I;** to e N.Am, Eur, Japan. W N.Am form probably warrants ssp. or sp. status. W N.Am pls with sterile lf part ternate-pinnate have been called var. *compositum* (Lasch) Milde.

POLYPODIACEAE POLYPODY FAMILY

Sherry Whitmore & Alan R. Smith

Per, on pls, rocks, in rock crevices, less often in soil, humus, or on dunes; rhizome short- to long-creeping, branched, glaucous to not, scaly. **LVS** ± alike or of 2 kinds, fertile and sterile; petiole thin to thick, straw-colored or green to brown or black, gen jointed to persistent knob on rhizome; blade gen simple to 1-pinnate, membranous to fleshy or leathery; veins free to fused, gen forked. **SPORANGIA:** sori round to elongate, rarely linear, gen 1 per areole, in 1–several rows on each side of segment midrib; indusium 0; spores gen ± elliptic, ± smooth to coarse-tubercled or -ridged, scar linear. ± 46 genera, ± 650 spp.: worldwide, esp Old World trop. Numbers of genera, spp. depend on treatment; many spp. cult.

POLYPODIUM POLYPODY

Sherry Whitmore

Rhizome long-creeping; scales lanceolate, ± brownish, 1-colored or often with darker central area or midstripe. **LVS** 0.2–10(20) dm, ± alike or fertile > sterile; petiole glabrous to scaly; blade 1-pinnate to gen deeply pinnately lobed, rarely simple and unlobed, glabrous to hairy, glandular or not, scales on lower surface midrib near base gen lanceolate or linear-lanceolate, gen ± brown; veins free to fused. **SPORANGIA:** sori in 1 row on each side of segment midrib, gen raised, sometimes incl branched or unbranched, glandular hairs, sporangium-like structures, or shriveled sporangia; spores yellow. ± 160 spp.: gen New World, trop, some temp, few boreal. (Latin: many feet, from rhizome) [Whitmore & Smith 1991 Madroño 38:233–248] 50% or more malformed spores indicates hybrid involving 2 or more spp. in CA.

P. hesperium Maxon (p. 71) WESTERN POLYPODY Rhizome 3–6 mm diam, whitish glaucous or not, taste acrid to sweet; scales 1-colored or gen with ± darker central area. **LVS** alive until new lvs formed; blade 2–25 cm, oblong to oblong-ovate, ± membranous to ± thick, ± firm, upper surfaces of midribs glabrous, segments entire to serrate, tips gen obtuse to acute, veins free. **SPO-**RANGIA: sori 1–2.5 mm, ovate, each with 0–5(10) dark brown or reddish black, shriveled, glandular sporangia. 2*n*=148. Rock crevices, talus slopes, under rock ledges; 1400–2980 m. KR, n&c SNH, SnBr, SnJt, **W&I, e DMtns (New York Mtns);** to B.C., Rocky Mtns, n Mex. [*P. vulgare* L. var. *columbianum* Gilbert] ✸TRY;DFCLT.

PTERIDACEAE BRAKE FAMILY

Alan R. Smith & Thomas Lemieux

Per, in soil or on or among rocks; rhizome creeping to erect, scaly. **LVS** gen all ± alike (or of 2 kinds, fertile and sterile), gen < 50 cm, often < 25 cm; petiole gen thin, wiry, often dark, in transverse section with vascular strands gen 1–3, less often many in a circle; blade gen pinnate or ± palmate-pinnate (see *Adiantum*), often 2 or more compound, lower surface often with glands, ± powdery exudate, hairs, or scales; segments round, oblong, fan-shaped, or otherwise, veins gen free. **SPORANGIA** in sori or not, marginal, submarginal, or along veins, sometimes covered by recurved, often modified segment margins (false indusia); true indusia 0; spores spheric, sides sometimes flat, scar with 3 radiating branches. ± 40 genera, 500 spp.: worldwide, esp dry areas. Definition of *Cheilanthes* and related genera problematic; traditional limits often untenable.

1. Lf segment margin gen not recurved, ± unmodified, not covering sporangia
  2. Lf 1-pinnate; upper surface with stellate scales, margins sometimes shallow (not deeply) pinnately lobed or dissected . . . . **ASTROLEPIS**
  2' Lf either 1-pinnate with lflets pinnately dissected or lf gen more divided; upper surface of lf glabrous or glandular, scales 0

3. Sporangia along veins for outer 1/3–2/3; lf segments narrowed at base . . . . ²**ARGYROCHOSMA**
3' Sporangia along veins ± throughout (best seen on immaturre, fertile lf); lf segments not narrowed at base . . . . **PENTAGRAMMA**
1' Lf segment margin gen recurved at least partly, often modified, gen covering sporangia at least partly
  4. Sporangia borne on and covered by highly modified, recurved part of segment margin; segments fan-shaped, thin textured . . . . **ADIANTUM**
  4' Sporangia borne on unmodified segment surface, gen covered at least partly by recurved part of margin; segments lanceolate, round, or otherwise, gen thick-textured
    5. Lf lower surface with scales, hairs, or glands . . . . **CHEILANTHES**
    5' Lf lower surface glabrous or covered with colored exudate
      6. Lf lower surface densely covered with white or yellow exudate, upper sparsely dotted with same . . . . **NOTHOLAENA**
      6' Lf without exudate (exc lower surface in *Argyrochosma limitanea*)
        7. Sporangia along veins for outer 1/3–2/3; rhizomes scales without darker mid-stripe, lanceolate; false indusium 0 . . . . ²**ARGYROCHOSMA**
        7' Sporangia along veins only at tips; rhizome scales gen with darker mid-stripe, if this lacking then scales hair-like; false indusium present . . . . **PELLAEA**

ADIANTUM

Pl in soil or rock crevices; rhizome short-creeping, scales variously colored. **LF** < ± 1 m; petiole cylindric, gen dark reddish brown to blackish, shiny, ± scaly at base; blade 2–3-pinnate or ± palmate-pinnate (1st division ± palmate, subsequent ones pinnate), segments stalked, fan-shaped or oblong, gen lobed, toothed, or both; axes, blades lacking colored exudate. **SPORANGIA** borne along veins on and covered by highly modified, recurved part of segment margin, appearing to run together at maturity; false indusia ± semi-circular to linear; spores gen smooth, tan. ± 200 spp.: trop, temp. (Greek: unwettable) Widely cult.

A. capillus-veneris L. (p. 71) SOUTHERN MAIDEN-HAIR **LF** gen 20–40(7–50+) cm; petiole dark brown to blackish; blade 2–3-pinnate; segments cut or lobed often > 1/4 way to base, often with < 4 ± irregular lobes, margins at base converging at 45–90°, stalk color often extending gradually into base, midvein often extending part way along 1 margin. **SPORANGIA**: sori (and false indusia) 3–11, rarely 2 per segment, gen < 5 mm. 2*n*=60.

Uncommon (or locally common). Shaded, rocky or moist banks, exposed sites or not; < 2000 m. NCoR, CaRF, n SNF, s SNH, CCo, ScoRO, SW (exc SCo), **SNE, D**; gen s US, worldwide, esp temp. Widely cult (incl many cultivars); recency of collections, erratic distribution suggest sp. alien in CA. ❀IRR or WET, DRN:4–6&SHD:2,3,**7**,8–12,**13–24**.

ARGYROCHOSMA

Pl in soil or rock crevices; rhizome short-creeping, scales linear-lanceolate, tan to ± reddish throughout. **LF** < 40 cm; hairs 0; scales 0; petiole cylindric, dark, glabrous or ± scaly at base; blade 2–5-pinnate, segments stalked, gen < 5 mm, round to oblong, blue- to gray-green, gen thick, veins obscure; axes, blades covered with whitish exudate on lower side or not. **SPORANGIA** along veins for outer 1/3–2/3 of segments; segment margin unmodified, often only slightly recurved; spores tan, coarsely ridged. ± 20 spp.: Am. (Greek: silver ornament) Considered closer to *Pellaea* than to *Notholaena*. [Windham 1987 Amer Fern J 77:37–41]. ❀DFCLT.

1. Lf blade without whitish exudate on lower surface, 2–3-pinnate; basal 1° lflets spreading to ± ascending . . . . *A. jonesii*
1' Lf blade covered with whitish exudate on lower surface, 3–5-pinnate; basal 1° lflets ± strongly ascending . . . . *A. limitanea* var. *limitanea*

A. jonesii (Maxon) M.D. Windham **LF** 5–15 cm; petiole dark brown; blade 2–3-pinnate, ± ovate, exudate 0; basal 1° lflets spreading to ± ascending, stalks < 5 mm; 2° lflet stalk 0–1.5 mm; segments gen 2–5 mm. **SPORANGIA** 64-spored. 2*n*=54,108. Gen calcareous rock crevices, cliff bases; 400–1800 m. s SNH, SCoRO, SnGb, **W&I, DMtns**; to sw UT, AZ. [*Cheilanthes j.* (Maxon) Munz, *Notholaena j.* Maxon]

A. limitanea (Maxon) M.D. Windham var. *limitanea* (p. 75) CLOAK FERN **LF** 10–25 cm; petiole dark brown to black; blade 3–5-pinnate, ovate to triangular, covered with whitish exudate on lower surface; basal 1° lflets ± strongly ascending, stalks 5–10 mm; 2° lflet stalk 3–6 mm; segments gen 1.5–3 mm. **SPORANGIA** 32-spored. RARE in CA. In crevices, esp bases of calcareous rocks; 1800 m. e DMtns (**New York Mtns, San Bernardino Co.**) ; to UT, NM, nw Mex. [*Cheilanthes l.* (Maxon) Mickel var. *l.*; *Notholaena l.* Maxon] Produces spores asexually; common name, even more so than others, is applied to ferns in other groups.

ASTROLEPIS

Pl in soil or rock crevices; rhizome ± short-creeping-decumbent, scales gen linear to linear-lanceolate, toothed, pale to reddish brown, older with darker, irregular central area or not. **LF**: axes gen orange to reddish brown, scaly; blade 1-pinnate, linear, lflets sometimes shallowly but not deeply pinnately lobed or dissected, upper surface with stellate scales. **SPORANGIA** along veins, obscured by dense scales; segment margin unmodified, not recurved. ± 6 spp.: sw US through S. Am. (Greek: star scale) [Benham et al. 1988 Amer J Bot 75(6:2):138]

Azolla filiculoides
Azollaceae

Athyrium filix-femina var. cyclosorum
Dryopteridaceae

Cystopteris fragilis

Dryopteris filix-mas

Polystichum scopulinum

Woodsia oregana

W. plummerae

W. scopulina

Equisetum arvense
Equisetaceae

E. laevigatum

Marsilea vestita
Marsileaceae

Botrychium simplex
Ophioglossaceae

Polypodium hesperium
Polypodiaceae

Adiantum capillus-veneris
Pteridaceae

A. cochisensis (Goodd.) D.M. Benham & M.D. Windham (p. 75) SCALY CLOAK FERN Rhizome short; scales ± 10 mm, 0.1–0.5 mm wide, linear; teeth ± sparse, more pronounced in upper half or not. **LF**: petiole 3–6(10) cm, 1–1.5 mm wide, scales appressed, 0.5 mm, whitish; blade 1-pinnate, 8–15(20) cm, tapered to tip; 1° lflets < 0.5 cm, with jointed stalk and 0–3 pairs of lobes, gen obtuse at tip, upper surface with persistent, stellate scales, lower surface with lanceolate, whitish to tan, abundantly toothed and finely dissected scales covering small (< 0.1 mm) glandular hairs. **SPORANGIA** in a submarginal band when mature, ± visible, erupting through scales. RARE in CA. Limestone slopes, crevices; 900–1800 m. **DMtns**; to TX, Mex, incl Baja CA. [*Cheilanthes c.* (Goodd.) Mickel; *C. sinuata* (Sw.) Domin var. *c.* (Goodd.) Munz]

CHEILANTHES

Pl in soil or rock crevices; rhizome short- to long-creeping-decumbent, gen many-branched, scales gen linear-lanceolate, pale to dark, with darker mid-stripe or not. **LF** < 75 cm; petiole cylindric, reddish brown to blackish; blade gen 2–3-pinnate, gen oblong to narrowly triangular; segments gen small, ± flat or lower side concave (from recurved margins). **SPORANGIA** along margin, in discrete patches to continuous, partly to completely covered by recurved margin (gen not recurved in *C. cooperae*). 150+ spp.: gen Am, gen dry areas. (Greek: lip fl, from location of sporangia) ✺DFCLT.

1. Lf blade without scales, with or without hairs, with or without glands
  2. Rhizome scales without darker midstripe; lf surface with ± clear, sticky exudate, gen with sessile or short-stalked glands . . . . ***C. viscida***
  2' Rhizome scales with darker midstrip; lf surface without sticky exudate
    3. Hairs on lf segment upper surface sparse, gen not intertwined; darker midstripe of rhizome scales not thread-like, > 0.1 mm wide at base; 1° lflet stalk ± 0–1 mm; 2° lflet stalk greenish on upper surface, brownish on lower . . . . ***C. feei***
    3' Hairs on lf segment upper surface very dense, intertwined; darker midstripe of rhizome scales thread-like, < 0.1 mm wide at base; 1° lflet stalk 1–2(5) mm; 2° lflet stalk gen brownish on upper and lower surfaces . . . . ***C. parryi***
1' Lf blade with scales, with or without hairs, without glands
  4. Rhizome long-creeping, lvs well separated . . . . ***C. wootonii***
  4' Rhizome short-creeping, lvs crowded
    5. Lf blade linear oblong; lf segments ± round or gen oblong, scales on lower surface deeply dissected, long-ciliate . . . . ***C. gracillima***
    5' Lf blade oblong to narrowly triangular; lf segments ± round, scales on lower surface entire to ciliate, gen covering more dissected scales
      6. Lf segments glabrous on upper side, scales in immediate contact with lower surface dissected but with an obvious and relatively large body, scales of main axes gen > 1 mm wide . . . . ***C. covillei***
      6' Lf segments, or at least some, with deeply dissected scales on upper surface, scales in immediate contact with lower surface deeply dissected, with a body only 1–2 cells wide, scales of main axes gen < 1mm wide . . . . ***C. intertexta***

C. covillei Maxon (p. 75) Rhizome short-creeping; scales gen reddish brown, with darker mid-stripe. **LF** 8–22(30+) cm, 2–4(6) cm wide, dark green; petiole < 2 mm wide, scales linear-lanceolate, whitish to reddish brown; blade 3–4-pinnate; segments small, ± round, upper side glabrous, lower concave, obscured by scales that exceed margin, scales originating from axes, > 2 mm, 1 mm wide, ± entire, covering gen more dissected scales. **SPORANGIA** obscured by recurved segment margin and scales. 2*n*=60. Crevices, bases of rocks, sun or shade; 600–2400 m. NCoR, SN, SnFrB, SCoR, TR, PR, **SNE**, **DMtns**; to UT, AZ, Baja CA. Hybridizes with *C. parryi* (*C. ×parishii* Davenp.) in s CA; also with *C. intertexta.*

C. feei T. Moore (p. 75) Rhizome short-creeping; scales light to reddish brown, gen with darker mid-stripe. **LF** 6–15(18) cm, 1.5–3 cm wide, pale green, scales 0; petiole ± 1 mm wide, hairs < 2 mm, pale or tan with ± orange constrictions; blade gen 3-pinnate; segments small, ± round, lower surface concave, hairs tangled, long, whitish to brownish, ± sparse on upper surface, very dense on lower surface. **SPORANGIA** partly obscured by hairs, less so by segment margin. Gen limestone crevices, slopes, cliffs; 1200–3000 m. **W&I**, **DMtns**; to B.C., MT, c US, Mex.

C. gracillima D. Eaton (p. 75) Rhizome short-creeping; scales light brown, with darker (sometimes reddish brown) mid-stripe or not. **LF** 6–18(30) cm, 1–2(3) cm wide, dark green; petiole ± 1 mm wide, scales linear-lanceolate, ciliate at base; blade linear-oblong, gen 3–5 × longer than wide, axes with long, narrow scales; segments small, ± round to gen oblong, lower surface concave, with dense, pale, deeply dissected, long-ciliate scales, upper surface with similar scales or glabrous. **SPORANGIA** on young lvs often entirely obscured by scales and recurved segment margin. Gen granite cliffs, crevices; 400–3200 m. NW, CaR, SN, CW, **SNE**; to B.C., MT. Hybridizes with *C. intertexta.*

C. intertexta (Maxon) Maxon Rhizome short-creeping; scales pale with darker (reddish brown to blackish) mid-stripe extending ± to margin or not. **LF** 6–14(20) cm, 1.5–3 cm wide, dark green; petiole < 1 mm wide, scales lanceolate, ciliate at base, pale; blade 3-pinnate; segments small, ± round, lower surface concave, with pale to reddish brown scales, scales < 1 mm wide, ciliate at base, covering more deeply dissected scales, barely exceeding segment margin, upper surfaces glabrous or at least some with deeply dissected scales. **SPORANGIA** gen obscured by scales and recurved segment margin. Crevices, bases of rocks; 300–2800 m. NCoRO, NCoRH, SNH, SnFrB, SCoR, **W&I**; NV. Hybridizes with *C. covillei*, *C. gracillima.*

C. parryi (D. Eaton) Domin (p.75, pl. 2) Rhizome short-creeping, > 6 cm; scales medium brown, most with darker (thread-like) mid-stripe. **LF** 6–15(25) cm, 1–2(3) cm wide; scales 0; petiole < 1 mm wide, hairs short to long, bent, appressed to ± spreading, glandular and not, pale; segments small, ± round, ± flat, hairs long (4+ mm), tangled, gen non-glandular, very dense on both surfaces, making upper surface silver-whitish, lower tan to brown or golden. **SPORANGIA** ± visible through hairs at segment margin; spores blackish. 2*n*=60. Limestone, granite crevices, rocks; 100–1500 m. PR, **SNE**, **D**; to UT, AZ, Baja CA.

C. viscida Davenp. (p. 75) VISCID LACE FERN Rhizome short-creeping; scales reddish brown, without darker mid-stripe. **LF** 10–15(25) cm, 2(3) cm wide, pale to dull green; petiole < 1 mm wide, scales sparse, at base, gland stalks 0–0.3 mm; segments small, ± oblong, ± flat, upper and lower sides covered with ± clear, sticky exudate from glands. **SPORANGIA** visible at recurved segment margins; spores brownish. UNCOMMON. Limestone, granite crevices, rocks; 100–1600 m. e edge SnBr, SnJt, e PR, **DMtns**, **w edge DSon**; Baja CA.

C. wootonii Maxon (p. 75) WOOTON'S LACE FERN Rhizome long-creeping; scales light brown, without darker mid-stripe. **LF** 10–20 cm, 2–3 cm wide; petiole 1–2 mm wide; blade 3–4-pinnate; segments small, ± round, lower surface concave, densely covered with ciliate, linear-lanceolate scales, upper side glabrous. **SPORANGIA** gen obscured by dense, overlapping scales. RARE in CA. Rocky outcrops; 1600–1800 m. **SNE**, **DMtns**; to TX, Baja CA.

NOTHOLAENA

Pl in soil or often in granite rock crevices; rhizome short-creeping to suberect, scales linear-lanceolate. **LF**: petiole gen cylindric, dark brown to black, glabrous to ± scaly; blade 2–4-pinnate, axes and segments with white to yellow ± powdery exudate on lower, often upper sides, segments gen sessile, sometimes slightly narrower at base. **SPORANGIA** in ± continuous, marginal bands at maturity; segment margin recurved and partly covering sporangia, unmodified; spores finely ridged or granular, often blackish. ± 25 spp.: gen Mex, sw US, few in Caribbean, S.Am. (Greek: false cloak, from lf blade margin not reflexed as it is in *Cheilanthes*) [Tryon 1956 Contr Gray Herb 179:1–106]

N. californica D. Eaton (p. 75) CALIFORNIA CLOAK FERN Rhizome scales rigid, with darker (blackish) midrib extending nearly to margins, finely ciliate. **LF** 3-pinnate, ± 3–13 cm; blade axes brown to black, glabrous or with white to yellow exudate; lowermost 1° lflets each strongly asymmetric (more developed on basal side); segment lower surface covered with white to yellow exudate, hairs 0, scales 0, upper surface sparsely dotted with white to yellow exudate. **SPORANGIA** 32-spored. UNCOMMON. Dry, rocky slopes, in rock crevices, under rock ledges; 200–1300 m. s ChI, SnGb, SnBr, PR, **DMtns**, **DSon**; AZ, nw Mex. [ssp. *nigrescens* Ewan; *Aleuritopteris cretacea* (Liebm.) Fourn. misapplied] At least 2 forms in CA are chemically distinct: one with pale to bright yellow exudate on lf upper surface; the other with white exudate on lf upper surface. ✿DFCLT.

PELLAEA CLIFF-BRAKE

Pl in soil or rock crevices; rhizome short- to long-creeping, scales overlapping, narrowly linear, light- to reddish or medium-brown, often with darker mid-stripe. **LVS** erect, persistent, < 1 m; petioles ± cylindric, gen dark or reddish brown to blackish, ± shiny, glabrous; blade 1–4-pinnate; segments gen stalked, gen free, linear to rounded, lobed or not, often folded lengthwise when dried; veins gen free. **SPORANGIA** in ± continuous, submarginal bands, among a whitish to yellowish exudate or not; segment margin gen recurved, gen modified; spores tan to light yellow. ± 35 spp.: trop, temp, few in Eur, 0 in Asia. (Greek: dusky, from bluish gray lvs) [Tryon 1957 Ann Missouri Bot Gard 44(2):125–193] Not commonly cult.

1. Lf 1-pinnate, 1° lflets unlobed or deeply 2(3)-lobed; fracture lines at base of petiole many; rhizome scales
    without darker mid-stripe . . . . *P. breweri*
1' Lf 2–3-pinnate, 1° lflets compound; fracture lines at base of petiole 0
    2. Lf 2–3-pinnate, at least some basal segments 3-parted . . . . *P. mucronata* var. *mucronata*
    2' Lf 2-pinnate, basal segments not 3-parted
        3. Fertile segment not appearing folded in half, recurved margins not meeting, lower surface therefore
            visible; DMtns (Providence, New York Mtns) . . . . *P. truncata*
        3' Fertile segment gen appearing folded in half, recurved margins ± meeting, lower surface therefore
            not visible; SNE, DMtns (exc Providence, New York Mtns) . . . . *P. mucronata* var. *californica*

P. breweri D. Eaton (p. 75) Rhizome short-creeping, branched, > 10 cm, 5(7) mm wide; scales very narrowly linear (hair-like), reddish brown, without darker mid-stripe. **LVS** clustered, 8–20(25) cm, 2–3(4) cm wide, pale greenish; petiole < 2 mm wide, fracture lines at base many; blade 1-pinnate, oblong, main axis green at tip; 1° lflets < 2 cm, < 1.5 cm wide, lanceolate-ovate, deeply 2(3)-lobed. **SPORANGIA** 64-spored; spores dark to light brown. 2*n*=58. Gen n-facing granite rock crevices, slopes; 1500–3700 m. KR, SNH, **GB**, **DMtns**; to WA,ID, Colorado. ✿DFCLT.

P. mucronata (D. Eaton) D. Eaton BIRD'S-FOOT FERN Rhizome short-creeping, branched, > 8 cm, 0.5–1 cm wide; scales brownish with darker mid-stripe. **LVS** ± clustered, greenish to purplish; petiole < 2(3) mm wide; blade 2–3(4)-pinnate, narrowly triangular to oblong; segments 2–6(8) mm, 0.5–2(4) mm wide, linear to oblong, with a small point at tip. **SPORANGIA** 64-spored. 2*n*=58. Rocky or dry areas; 20–3000 m. NCoR, CaR, SN, GV (Sutter Buttes), CW, SW, **SNE**, **DMtns**; Baja CA. Hybrids with *P. truncata* uncommon.

   var. *californica* (Lemmon) Munz & I.M. Johnston (p. 75) **LF** 15–25(33) cm, 2–4(8) cm wide; blade 2-pinnate; 1° lflets often overlapping, ascending; fertile segments gen appearing folded in half, recurved margins ± meeting, lower surface therefore not visible. Habitat of sp.; 1800–3000 m. c SNH, TR, PR, **SNE**, **DMtns**. [*P. compacta* (Davenp.) Maxon] ✿DFCLT.

   var. *mucronata* **LF** 20–40(60) cm, 5–15 cm wide; blade 2–3(4)-pinnate; 1° lflets not overlapping, gen ± spreading to widely ascending; fertile segments not appearing folded in half, recurved margins not meeting, lower surface therefore visible. Habitat of sp.; 20–2400 m. NCoR, CaR, SN, GV (Sutter Buttes), CW, SW, **DMtns**;Baja CA. ✿DRN,DRY;2,**7**,9,**14–18,23, 24**;DFCLT.

P. truncata Goodd. (p. 75) CLIFF-BRAKE Rhizome short-creeping, branched, to 8+ cm, 2–3 mm wide; scales brownish with darker mid-stripe. **LVS** clustered, 15–30(36) cm, 4–11 cm wide, olive-green; petiole 1(2) mm wide, ± flat or upper surface grooved; blade 2-pinnate, narrowly triangular to oblong-triangular; 1° lflets gen not overlapping, ± spreading; segments 5–8 mm, 1–4 mm wide, linear to oblong, with a small point at tip, margins wavy-crenate, often (esp sterile) whitish, fertile not appearing folded in half, recurved margins not meeting, lower surface therefore visible. **SPORANGIA** 64-spored. 2*n*=58. RARE in CA. Gen in crevices of or at bases of granite (in CA) or igneous rock; 1200–1900 m. **e DMtns (Providence, New York mtns)**; to Colorado, TX, Baja CA. [*P. longimucronata* Hook. misapplied]

PENTAGRAMMA GOLDBACK or SILVERBACK FERN

Pl in soil or rock crevices; rhizome short-creeping-decumbent, gen 3–5(8) mm wide, scales linear-lanceolate, with darker mid-stripe. **LF**: petiole 5–20(32) cm, 0.5–2(3) mm wide; blade gen 2–3-pinnate, 2–8(15) cm, triangular or gen

5-sided, with white or yellow exudate on lower surface, with or without exudate on upper surface, main axis shallowly to deeply grooved on upper surface; lowermost 1° lflets each strongly asymmetric (more developed on basal surface); veins free. **SPORANGIA** along veins ± throughout; segment margins unmodified, recurved or not. 2 spp.: w N.Am. A puzzling complex of intergrading chemical, chromosomal, and morphological variants (see Yatskievych et al. 1990 Amer Fern J 80:9–17); for these we prefer the rank of var. but have used ssp. here because combinations at that rank already exist. ❀DFCLT

P. triangularis (Kaulf.) G. Yatskievych, M.D. Windham & E. Wollenweber Rhizome tip, scales without exudate. **LF**: petiole brown to reddish brown, with or without exudate; blade 3–10(18) cm, gen pale to dark green, upper surface gen without exudate. 2*n*=60,90,120,150. Common. Gen shaded slopes or rocky areas; < 2300 m. CA-FP, **SNE**, **DMtns**; to B.C., NV, Baja CA. [*Pityrogramma t.* (Kaulf.) Maxon] 3 sspp. in CA.

1. Lf blade upper surface gen glabrous . . . . . ssp. *triangularis*
1' Lf blade upper surface with sparse, minute, yellowish glands . . . . ssp. *maxonii*

ssp. ***maxonii*** (Weath.) G. Yatskievych, M.D. Windham & E. Wollenweber **LF**: blade upper surface gen with sparse, minute (0.1 mm), yellowish glands, lower surface often with many yellowish or reddish glands, margins not recurved; lower 1° lflets 2–5(7) cm; upper 1° lflets and 2° lflets on basal surface of lowermost 1° lflets deeply pinnately lobed to nearly 1-pinnate. Gen ± shaded, near rocks, boulders; 300–1400 m. SnBr, SnJt, **DMtns**; AZ, Baja CA. [*Pityrogramma t.* (Kaulf.) Maxon var. *m.* Weath.]

ssp. ***triangularis*** (p. 75) **LF**: blade upper surface gen glabrous, not sticky, margins not recurved; lower 1° lflets 2–6(11) cm; upper 1° lflets and 2° lflets on basal surface of lowermost 1° lflets deeply pinnately lobed to nearly 1-pinnate. Common. Gen shaded, sometimes rocky or wooded areas; < 2300 m. CA-FP, **SNE**, **DMtns**; to B.C., NV, Baja CA. Ssp. *semipallida* (J. Howell) G. Yatskievych, M.D. Windham & E. Wollenweber is chemically distinct, morphologically and geog less distinct, sometimes recognized taxonomically.

SELAGINELLACEAE SPIKE-MOSS FAMILY

Dieter H. Wilken

Pls on soil, rocks, or other pls, cespitose, mat-like (± flat), or cushion-like (rounded). **STS** pendent to erect, short to widely spreading, rooting at base or in branch fork; branches intricately intertwined or not. **LVS** many, simple, overlapping, appressed, small, ± scale-like, 1-veined, smooth to grooved on back, sessile to decurrent, gen 4-ranked; fertile lvs at same node of prostrate sts equal or not, if unequal, lvs below st gen appressed, lvs above st ascending to spreading. **CONES** gen terminal; lvs like those on sterile sts or not, gen strongly overlapping, triangular in ×-section. **SPORANGIA** 1 per lf axil, spheric to reniform; lower gen with (1–3)4 large, 3-ridged, yellow to orange spores; upper gen with many, small, gen pale-colored spores.

SELAGINELLA SPIKE-MOSS

Only genus. ± 700 spp.: worldwide, gen trop & warm temp. (Latin: small *Selago*, ancient name for some *Lycopodium*). Some cult as groundcover & curiosity (*S. kraussiana, S. lepidophylla*, resurrection plant). Hand lens required to observe lf shape, margin, bristle at tip, cones. ❀TRY;DFCLT.

1. Main sts above ground erect, rooting at or near st base; rhizomes present . . . . ***S. bigelovii***
1' Main sts above ground prostrate to spreading, rooting at branch forks; rhizome 0 or inconspicuous
  2. Lvs ± lanceolate, unequal; lvs below wider and > lvs above at same nodes as viewed at 20×; bristle of young lf tip twisted, deciduous . . . . ***S. eremophila***
  2' Lvs linear, narrowly lanceolate, or oblong; lvs below = lvs above at same nodes as viewed at 20×; lf tip acute or bristle rigid
    3. Sts strongly intertwined, fragile when dry; cone lf margins ciliate to minutely dentate . . . . . ***S. leucobryoides***
    3' Sts not strongly intertwined, not fragile when dry; cone lf margins gen entire . . . . ***S. watsonii***

S. bigelovii L. Underw. (p.75) Pl cespitose. **STS** ascending to erect; main sts clumped, 5–15(20) cm, ± 1 mm wide, rooting at base; lateral branches 1–4 cm, ascending to erect. **LVS** sessile; upper and lower equal, 1–3.5 mm, linear to narrowly lanceolate, bristle-tipped, bristle < 1 mm, ± rigid, marginal hairs spreading to ascending. **CONE** gen < 1 cm; lvs 1.5–2.5 mm, lanceolate to ovate; large spores yellow. Open sites, rocks, crevices, shrubland, woodland; < 2000 m. s NW, s SN, CW, SW, **sw edge DMoj**; Baja CA. ❀ DRN,DRY: **15–17**&SHD:**7**,14,**18**,19;DFCLT.

S. eremophila Maxon (p. 75) DESERT SPIKE-MOSS Pl mat-like, dense. **STS** prostrate to spreading; main sts 5–14 cm; lateral branches < 1 cm, ascending. **LVS** decurrent; upper and lower lvs of main sts unequal (lower 1.5–3 mm, upper ± 1.5 mm), ± lanceolate, marginal hairs spreading; young lvs bristle-tipped, bristle < 1 mm, soft, twisted, deciduous; mature tip gen acute. **CONE** 3–10 mm; lvs ovate to deltate; large spores yellow. UNCOMMON. Shaded sites, gravelly soils, crevices, among rocks; < 900 m. e PR, **DSon**; to TX, Baja CA.

S. leucobryoides Maxon MOJAVE SPIKE-MOSS Pl cushion-like. **STS** prostrate, strongly intertwined, fragile when dry; main sts 1–2(4) cm; lateral branches ascending to erect, < 1 cm. **LVS** decurrent; upper and lower equal, 1.5–3(4) mm, linear to lanceolate, bristle-tipped, bristle < 0.5 mm, rigid, margin sparsely ciliate to minutely dentate. **CONE** 5–12 mm; lvs lanceolate to ovate; large spores yellow. UNCOMMON. Among rocks, crevices, gen limestone, shrubland, woodland; 600–2300 m. **DMtns (Panamint, Kingston, Providence mtns)**; sw NV.

S. watsonii L. Underw. (p. 75) Pl cushion-like, ± open to dense. **STS** prostrate to decumbent; main sts 2–12 cm; lateral branches ascending to erect, 1–3(5) cm. **LVS** decurrent; upper and lower equal, 1.5–3.5 mm, linear-lanceolate to oblong, bristle-tipped or acute, bristle << 0.5 mm, rigid, margin entire to sparsely ciliate, hairs ± spreading. **CONE** 10–30 mm; lvs lanceolate to ovate; large spores yellow. Open rocky sites, coniferous forest, alpine; 2300–4100 m. KR (Trinity Alps), SNH, TR, PR, **n SNE**, **W&I**; to OR, MT, UT.

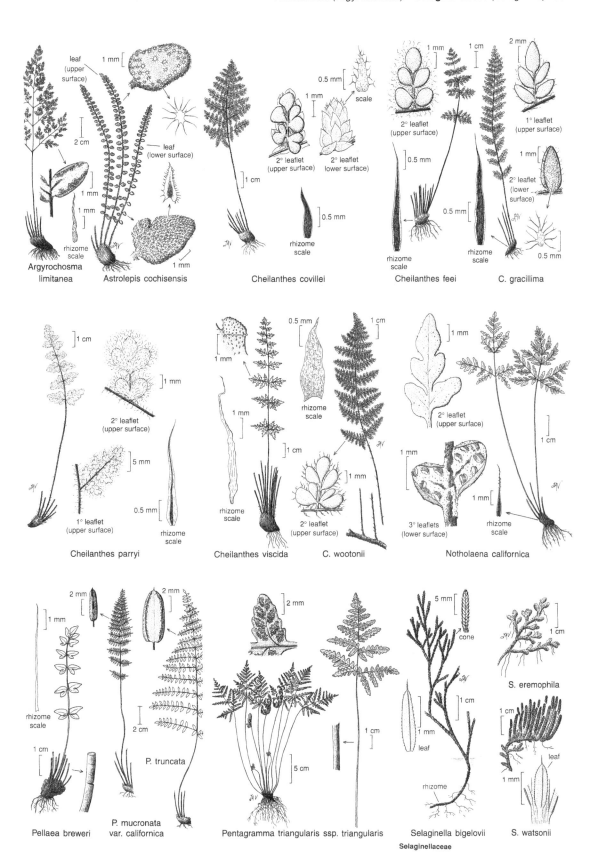

Argyrochosma limitanea

Astrolepis cochisensis

Cheilanthes covillei

Cheilanthes feei

C. gracillima

Cheilanthes parryi

Cheilanthes viscida

C. wootonii

Notholaena californica

Pellaea breweri

P. mucronata var. californica

P. truncata

Pentagramma triangularis ssp. triangularis

Selaginella bigelovii

S. eremophila

S. watsonii

Selaginellaceae

CUPRESSACEAE CYPRESS FAMILY

Jim A. Bartel

Shrub, tree, evergreen, monoecious or dioecious. **LVS** cauline, opposite and 4-ranked or whorled in 3's and 6-ranked, gen scale-like, decurrent, completely covering young sts. **POLLEN CONE** small, axillary or terminal. **SEED CONE** ± fleshy to woody, gen hard at maturity; scales opposite or whorled. **SEEDS** 1–many per scale, gen angled or winged, gen wind-dispersed. 17 genera, ± 120 spp.: worldwide; all N.Am genera cult. [Elias 1980 Complete Trees N.Am] Juvenile lvs needle- or awl-like, sometimes present in ± mature pls, esp in response to grazing or infection, esp in *Cupressus, Juniperus*. Recently treated to include Taxodiaceae (in CA, *Sequoia, Sequoiadendron*) [Gadek et al. 2000 Amer J Bot 87: 1044-1057]

JUNIPERUS JUNIPER

Shrub, tree, gen dioecious. **ST**: bark thin, peeling in strips; young shoots 4-angled to cylindric. **LVS** opposite and 4-ranked or whorled in 3's and 6-ranked, scale-like to less often awl- or needle-like. **POLLEN CONE**: pollen sacs 2–6 per scale. **SEED CONE** 5–18 mm, ± fleshy, berry-like, glaucous or not, dry or resinous, sweet, formed by fusion of scales, ± spheric, surrounded at base by minute scale-like bracts, gen maturing 2nd year; scales 3–8, opposite or whorled in 3's. **SEEDS** 1–3(12) per cone, ± flat, unwinged, often not angled, gen animal-dispersed over 2 years; cotyledons 2–6. ± 60 spp.: n hemisphere. (Latin: juniper) [Vasek 1966 Brittonia 18:350–372]

1. Seed cone maturing blue-black; tree 5–15 m . . . . *J. occidentalis* var. *australis*
1' Seed cone maturing red-brown; shrub, tree 1–8(10) m
  2. Lf gland obvious; trunks several at base; dioecious . . . . *J. californica*
  2' Lf gland obscure; trunk gen 1 at base; monoecious . . . . *J. osteosperma*

J. californica Carrière (p. 81) CALIFORNIA JUNIPER Shrub or tree 1–4(10) m, dioecious. **ST**: trunks several at base; bark gray, thin, outer layers persistent. **LVS** gen whorled in 3's, 6-ranked, closely appressed, scale-like; gland obvious. **POLLEN CONE** 2–3 mm, oblong. **SEED CONE** 7–12 mm, spheric to ovoid, bluish maturing red-brown, dry. **SEEDS** 1–3 per cone, 5–7 mm, pointed, angled, brown. Dry slopes, flats, pinyon/juniper woodlands; 50–1500 m. NCoRI, SNF, SnFrB, SCoRI, TR, PR, **DMtns**; s NV, nw AZ, Baja CA (Cedros, Guadalupe islands). ❀DRN: 3,6,**7**,10,**14–16**,17,**18–23**,24&IRR:8,9,11,12.

J. occidentalis Hook. Tree 5–15 m. **ST**: bark brown to red-brown. **LVS** opposite and 4-ranked or whorled in 3's and 6-ranked, closely appressed, scale-like. **POLLEN CONE** 2–3 mm, oblong. **SEED CONE** 5–12 mm, blue-green maturing blue-black, resinous. **SEEDS** 2–3 per cone, 6 mm, ovoid, acute, grooved or pitted. 2n= 22. Dry slopes, flats, forests, woodlands; 100–3100 m. NCoRH, CaRH, SNH, SnGb, SnBr, **SNE**, **DMtns**; to WA, ID, w NV. 2 vars in CA.

var. ***australis*** (Vasek) A. Holmgren & N. Holmgren (p. 81) SIERRA JUNIPER Gen dioecious. **ST**: bark red-brown. **LVS** gen whorled in 3's, 6-ranked. **SEED CONE** 5–9 mm. **SEED**: cotyledons 2–4. Exposed, dry, rocky slopes, flats, forests, pinyon/juniper woodlands; 100–3100 m. NCoRH, SNH, SnGb, SnBr, **n SNE, W&I, DMtns**; w NV. ❀DRN:4,5,**6**,15–17,24&IRR:**1–3**,**7**,8–10,**14**,18–23.

J. osteosperma (Torrey) Little (p. 81, pl. 3) UTAH JUNIPER Tree < 8 m, monoecious. **ST**: trunk gen 1; bark thin, gray-brown aging ash-white. **LVS** gen opposite, 4-ranked, closely appressed, scale-like; gland obscure. **POLLEN CONE** 2–3 mm, cylindric. **SEED CONE** 5–13 mm, spheric, brown maturing red-brown, dry. **SEEDS** 1(2) per cone, 3–4 mm, ovoid, strongly angled. Pinyon/juniper woodlands; 1300–2600 m. SnGb, SnBr, **GB, DMtns**; to MT, NM. ❀DRN:1,**2**,4–6,15,**16**,22,24& IRR:**3**,**7**,8–12,**14**,18–21,23.

EPHEDRACEAE EPHEDRA FAMILY

James R. Griffin

Shrub or tree-like, densely branched, gen dioecious. **ST** gen erect, < 2 m, jointed; node conspicuous, internode > lf; bark with irregular, longitudinal cracks, gen gray; twigs whorled, grooved, greenish and photosynthetic when young, glaucous, glabrous to scabrous, sometimes thorn-like. **LVS** 2–3 per node, not green; bases ± fused into sheath, thickening with age; tips often ± deciduous. **POLLEN CONES** 1–5 per node, gen short-stalked, gen spheric; bracts gen flexible, lower sterile, upper subtending 2–8 stamen-like structures. **SEED CONES** gen 1–3 per node, sessile or stalk < 5 mm; bracts gen flexible, lower sterile, upper enclosing 1–3 seeds. **SEED** spheric to cylindric, smooth or furrowed, gen angled at top, gen brown. 1 genus, 42 spp.: N.Am, S.Am, Medit, Asia. [Cutler 1939 Ann Mo Bot Gar 26:373–424]

EPHEDRA EPHEDRA, MORMON TEA

(Greek: *Equisetum*, for resemblance to those pls)

1. Lvs gen 3 per node
  2. Twig tip thorn-like; lf sheath fibrous, persistent . . . . *E. trifurca*
  2' Twig tip narrowed but not thorn-like; lf sheath not fibrous, not persistent
    3. Seed not or faintly angled; twig yellow-green when young; lf base recurved, thickened when old
    . . . . *E. californica*
    3' Seed clearly angled at top; twig gray-green when young; lf base not recurved, not thickened when old
    . . . . *E. funerea*

1' Lvs gen 2 per node
  4. Lvs deciduous exc bases gen persistent; seeds gen 2
    5. Twig pale green, glaucous when young; lf base gray; seed cone stalk 1–5 mm . . . . *E. nevadensis*
    5' Twig bright green to yellow-green when young; lf base brown; seed cone sessile or stalk < 5 mm . . *E. viridis*
  4' Lvs persistent; seed gen 1
    6. Seed gen smooth . . . . *E. aspera*
    6' Seed furrowed . . . . *E. fasciculata*
      7. Seed < 8 mm . . . . var. *clokeyi*
      7' Seed > 8 mm . . . . var. *fasciculata*

E. aspera S. Watson   **ST** < 1.5 m; twig greenish when young, aging yellow, tip not thorn-like. **LVS** 2 per node, < 6 mm, gen persistent, sometimes deciduous but leaving thickened bases. **POLLEN CONES** gen 2 per node, < 7 mm. **SEED CONES** gen 2 per node, < 10 mm, ovoid, sessile. **SEED** 1, 5–8 mm, ovoid, gen smooth. Creosote-bush scrub, Joshua-tree woodland; < 1500 m. **D**; to TX, Mex. [*E. nevadensis* var. *a.* L. Benson] ❀TRY.

E. californica S. Watson (p. 81)   DESERT TEA   **ST** < 1.5 m; twig yellow-green when young, aging gray-brown, tip not thorn-like. **LVS** 3 per node, < 6 mm, with whitish margins wearing away to leave brown, thickened bases with recurved tips. **POLLEN CONES** 1–3 per node, < 9 mm. **SEED CONES** < 12 mm, ovoid, short-stalked. **SEED** gen 1, < 10 mm, ± spheric, smooth, not or faintly 4-angled. Scattered in arid grassland, chaparral, creosote-bush scrub; < 900 m. s SNF, Teh, w SnJV, SCoR, SW, **D**; to w AZ, Baja CA. ❀DRN,DRY,SUN:**7–12,14**,15–17,**18–21**,22–24.

E. fasciculata Nelson   **ST** gen low or prostrate, < 1 m; branch ± flexible; twig pale green when young, aging yellow, tip not thorn-like. **LVS** 2 per node, < 3 mm, persistent; bases slightly thickened, whitish. **POLLEN CONES** gen 2–3 per node, < 8 mm. **SEED CONES** 6–13 mm, sessile or short-stalked. **SEED** gen 1, 6–13 mm, obovoid-ellipsoid, furrowed. Creosote-bush scrub; < 1500 m. **D**; to AZ, s UT.

  var. ***clokeyi*** (Cutler) Clokey   **SEED CONE** 6–10 mm, obovoid. **SEED** 5–8 mm. **D**; to s UT. [*E. clokeyi* Cutler] ❀TRY.

  var. ***fasciculata*** (p. 81)   **SEED CONE** 6–13 mm, elliptic. **SEED** 8–13 mm. **D**; to AZ. ❀TRY.

E. funerea Cov. & C. Morton   DEATH VALLEY EPHEDRA   **ST** < 1.5 m; twig pale gray-green when young, aging gray, tip not thorn-like. **LVS** gen 3 per node, < 5 mm; bases not thickened, persis-tent, not recurved. **POLLEN CONES** 1–3 per node, 5–8 mm. **SEED CONES** < 15 mm, elliptic, sessile or short-stalked. **SEED** gen 1, sometimes 2–3, 6–9 mm, ovoid, 4-angled, smooth to sca-brous. Creosote-bush scrub; < 1700 m. **DMoj**; to w NV. ❀DRN,DRY, SUN:7,9,**10–12**,14,16,18–23.

E. nevadensis S. Watson (p. 81)   **ST** < 1.3 m; twig pale green, glaucous when young, aging yellow to gray, tip not thorn-like. **LVS** 2, sometimes 3 per node, 2–8 mm; bases thickened, persis-tent, gray. **POLLEN CONES** 1–3 per node, 4–8 mm, elliptic. **SEED CONES** 5–11 mm, spheric; stalk 1–5 mm. **SEEDS** gen 2, 6–9 mm, hemispheric when 2, spheric when 1, smooth. Creosote-bush scrub, Joshua-tree woodland; < 1100 m. s SN, SNE, **D**; to OR, UT. ❀DRN,DRY,SUN:**3,7,9–12,14,18–23**.

E. trifurca Torrey (p. 81)   **ST** < 2 m; twig pale green when young, aging yellow to gray-green, rigid, tip thorn-like. **LVS** gen 3 per node, 5–15 mm; tips ± spine-like; sheath fibrous, persis-tent, aging gray. **POLLEN CONES** 1–4 per node, 6–10 mm. **SEED CONES** 10–14 mm, obovoid, sessile or short-stalked. **SEED** gen 1, sometimes 2–3, 9–15 mm, 4-angled, smooth, light brown. Creosote-bush scrub; < 2100 m. **DSon**; to TX, Mex. Reportedly uniform, but pls in DMoj may be intermediate to *E. funerea*. ❀DRN, DRY,SUN:**2**,3,**7–12**,13,**14**,15,16,**18–23**.

E. viridis Cov. (p. 81, pl. 4)   GREEN EPHEDRA   **ST** < 1.5 m; twig bright green to yellow-green when young, aging yellow, tip not thorn-like. **LVS** gen 2 per node, < 6 mm; bases thickened, persis-tent, brown. **POLLEN CONES** 2–5 per node, 5–7 mm. **SEED CONES** 2–6 per node, 6–10 mm, obovoid, sessile or stalk < 5 mm. **SEEDS** 2, 5–8 mm, 3-angled, smooth. 2*n*=14. Sagebrush scrub, creosote-bush scrub, pinyon/juniper woodland; 900–2300 m. s SNF, Teh, SnJV, SCoR, WTR, **GB, DMtns**; to Colorado. [*E. nevadensis* var. *v.* (Cov.) M.E. Jones] ❀DRN,DRY, SUN:1,**2,3,7–12**,13,**14–24**.

PINACEAE   PINE FAMILY

James R. Griffin

Tree or shrub, monoecious, evergreen. **ST**: young crown conic; twig not grooved, resinous, gen persistent. **LVS** simple, gen alternate, sometimes in bundles or appearing ± 2-ranked, linear or awl-like; bases decurrent, sometimes woody, persistent several years. **POLLEN CONE** gen < 6 cm, not woody, deciduous. **SEED CONE** gen woody; bracts, scales gen persistent; scale not peltate, fused to or free from subtending bract. **SEEDS** 2, on upper side of scale base. 10 genera, 193 spp.: mostly n hemisphere; many of great commercial value, supplying > half of world's timber. [Price 1989 J Arnold Arbor 70:247–305] Keys adapted by J. Robert Haller.

1. Lvs gen in bundles of 2–5 (1 in *P. monophylla*), gen 2.5–35 cm; seed cone bract fused to scale
  inconspicuous . . . . **PINUS**
1' Lvs not in bundles, gen 1–6 cm; seed cone bract free from scale, conspicuous
  2. Lf scar ± flush with twig surface, smooth, round; seed cone erect, > 7 cm, scales and bracts deciduous
    at maturity, axis persistent on st . . . . **ABIES**
  2' Lf scar raised above twig surface on persistent woody base, not smooth, not round; seed cone pendent,
    < 8 cm, whole cone ± persistent on st . . . . **TSUGA**

ABIES   FIR

ST: young bark smooth, with resin blisters, mature bark thick, deeply furrowed; young branches appearing whorled; twig glabrous or hairy; lf scars smooth, round, flush with surface; bud gen ± spheric, gen < 1 cm, ± resinous. **LVS** 2–9 cm, sessile, twisted at base to become 2-rankcd, often curved upward on upper twigs, gen ± flat; upper surface with 2 longitudinal, whitish bands, midrib sometimes depressed; lower surface with or without whitish bands, midrib sometimes ridge-like. **SEED CONE** erect, < 23 cm, maturing 1st season; stalk gen 0; bracts, scales deciduous; bract incl or exserted, free from scale; axis persistent on st. **SEED** with obvious resin deposits on surface; wing < 2.5 cm. $2n=24$ for all reports. 39 spp.: n hemisphere. (Latin: silver fir) [Vasek 1985 Madroño 32:65–77]

1. Lf ± flat, light green; mature bark dark gray to ± black; seed cone < 13 cm . . . . *A. concolor*
1' Lf ± 3–4-angled, light blue-green; mature bark dark red; seed cone > 12 cm . . . . . . *A. magnifica* var. *magnifica*

A. concolor (Gordon & Glend.) Lindley (p. 81) WHITE FIR **ST**: trunk < 61 m, < 2.7 m wide; mature crown rounded; young bark white-gray, old bark gray-brown to ± black, thick, deeply furrowed, with alternate dark and light layers; twig glabrous; bud resinous. **LVS** ± 2-ranked on lower branches, twisted upward on higher branches, 3–9 cm, ± flat; upper surface without white bands; tip gen blunt or acute. **SEED CONE** 7–13 cm; stalk < 5 mm; bract incl. Mixed-conifer to lower red-fir forest; 900–3100 m. KR, NCoR, CaRH, SNH, Teh, TR, PR, MP, **n SNE, DMtns**; to c OR, ID, Colorado, n Baja CA. Most CA pls and some Rocky Mtn pls may be called var. *lowiana* (Gordon) Andr. Murray. Relationship of s CA pls (esp from DMtns) to gen Rocky Mtn var. *concolor* is under study. ❀DRN:1,**4**,5,**6**&IRR:**2,3,7**,15–17,24&SHD:**10,14**,18–23;CVS.

A. magnifica Andr. Murray CALIFORNIA RED FIR **ST**: trunk < 57 m, < 2.5 m wide; mature crown ± cylindric, top rounded; young bark gray, with resin blisters, mature bark deeply furrowed, with dark reddish ridges; twig hairy 1st season; bud ± resinous. **LVS** < 3.5 cm, ± 3–4-angled; upper surface with 2 whitish bands; tip notched or blunt. **SEED CONE** 12–23 cm; bract incl or ± exserted. mixed-conifer to subalpine forests; 1200–2800 m. KR, NCoRH, CaRH, SNH, **SNE**; s OR (Cascade Mtns), w-c NV. 2 vars in CA.

var. ***magnifica*** (p. 81) Cotyledons 6–13. **SEED CONE**: bract incl. Mixed-conifer to subalpine forests; 1200–2600 m. KR, NCoRH, CaRH, n&c SNH, **SNE**; w-c NV. ❀DRN:**1**,4,5, 15&IRR:2,**3,6**,16 &SHD:7,14.

PINUS PINE

ST: young crown conic, mature crown often rounded or flat; branches ± whorled in young pls; young bark smooth, mature bark furrowed; bud ± conic, gen resinous. **LVS** gen 2.5–35 cm, gen sessile, in bundles of 1–5; bundles solitary in axils of alternate, awl-like bracts, each bundle enclosed at base in a sometimes deciduous sheath of bracts, gen persistent several seasons. **SEED CONES** often whorled, gen maturing and opening 2nd season, sometimes persistent on st; stalk 0 or < 16 cm; bract incl, fused to scale, minute; scale tip reflexed and elongated 3–7 cm, or often with a rounded or angled, often prickled knob < 3 cm. **SEED**: coat hard, sometimes woody. $2n=24$ for all reports. 94 spp.: n hemisphere. (Latin: pine) [Millar & Critchfield 1988 Madroño 35:39–53]

1. Lf-bundle sheath deciduous or mostly so; seed cone scale tip gen without prickle
  2. Lvs 1–2 per bundle
    3. Lvs gen 2 per bundle . . . . *P. edulis*
    3' Lvs gen 1 per bundle . . . . *P. monophylla*
  2' Lvs 5 per bundle
    4. Seed cone gen < 8 cm, gen torn apart by animals, gen not on ground intact; seed wings persistent on
      scale . . . . *P. albicaulis*
    4' Seed cone gen > 8 cm, gen not torn apart by animals, gen on ground intact; seed wings persistent on
      scale or not
      5. Seed cone scale thickened to a prominent knob at tip, with 2–6 mm prickle . . . . *P. longaeva*
      5' Seed cone scale not thickened to a prominent knob at tip, without prickle
        6. Seed cone stalk < 2 cm; scale moderately thickened near tip; seed wings narrow, gen persistent
          on scale . . . . *P. flexilis*
        6' Seed cone stalk > 2 cm; scale not thickened near tip; seed wings broad, deciduous from scale
          7. Seed cone gen > 20 cm; mature bark thick, dark purplish red . . . . *P. lambertiana*
          7' Seed cone gen < 20 cm; mature bark thin, dark gray to bright reddish brown . . . . *P. monticola*
1' Lf-bundle sheath persistent; seed cone scale tip with prickle, at least when immature (sometimes 0 at maturity)
  8. Lvs 2 per bundle, < 9 cm; seed cone < 6 cm . . . . *P. contorta* ssp. *murrayana*
  8' Lvs 3 per bundle, > 9 cm; seed cone > 6 cm
    9. Basal seed cone scale tips recurved, elongated 3–7 cm; seed cone scale prickles gen worn off by maturity;
      seed > wing; lf gray-green, gen drooping . . . . *P. sabiniana*
    9' Basal seed cone scale tips rarely recurved, then elongated < 3 cm; seed cone scale prickles evident
      at maturity; seed < wing; lf grayish blue-green to dark green, gen straight
      10. Seed cone gen 13–25 cm, lower surface of cone scale gen not darker than upper; lf grayish blue-
        green, glaucous; bark odor banana-, pineapple-, or vanilla-like . . . . *P. jeffreyi*
      10' Seed cone gen 7–16 cm, lower surface of cone scale gen darker than upper; lf deep green; bark
        odor not banana-, pineapple- or vanilla-like . . . . *P. ponderosa*

P. albicaulis Engelm. (p. 81) WHITEBARK PINE **ST** in most exposed places shrubby to prostrate; trunks sometimes multiple, < 26 m, < 1.5 m wide, much wider at base; bark when mature graywhite, smooth, thin; mature crown often deformed by wind. **LVS** 5 per bundle, 3–7 cm, ± curved, dark green, stiff; sheath deciduous. **SEED CONE** sessile, erect, 3.5–9 cm, ovate, purple-brown, gen torn apart (and seeds dispersed) by animals; scale tip knobs angled, prickled. **SEED**: wing persistent on scale. Upper red-fir

forest to timberline, esp subalpine forest; < 3700 m. KR, CaRH, SNH, Wrn, **SNE**; to B.C., WY. ❀DRN,SUN:**1**,6,15–17&IRR:**2,3**,7;DFCLT.

P. contorta Loudon LODGEPOLE PINE **ST**: mature bark scaly, thin; trunk 2–34 m (extremely variable at maturity). **LVS** 2 per bundle, 2.5–8.6 cm; sheath persistent. **SEED CONE** pendent, < 6 cm, brown; stalk ± 0; scale tip knobs < 5 mm, angled, prickled. Coastal to subalpine forest; < 3500 m. NCo, KR, CaRH, SNH,

SnGb, SnBr, SnJt, **GB**; to B.C., SD, n Baja CA. 3 sspp in CA.

ssp. ***murrayana*** (Grev. & Balf.) Critchf. (p. 81) LODGEPOLE PINE **ST**: trunk < 34 m. **SEED CONE** ± symmetric, deciduous within several years. Lodgepole forest, wet meadows, cold places in mixed-conifer forest; < 3500 m. KR, CaRH, SNH, SnGb, SnBr, SnJt, **GB**; OR, n Baja CA. [*P. murrayana* Grev. & Balf.] ❁ SUN: 4–6,15,16&IRR:**1,2,7,17**&SHD:**3**,14.

P. edulis Engelm. (p. 81) COLORADO PINYON PINE **ST**: trunk < 15 m, < 1.1 m wide; bark shallowly furrowed, red-brown; mature crown ± rounded. **LVS** gen 2 per bundle, 2–6 cm, dark green; sheath deciduous. **SEED CONE** erect, 3–7 cm, ovoid, yellow-brown; stalk ± 0; scale tip knobs < 1 cm, angled, truncate. **SEED**: wing persistent on scale. RARE in CA. Pinyon/juniper woodland; 1300–2700 m. **DMtns (New York Mtns)** to s WY, w TX, n Mex. Extent of hybridization with *P. monophylla* under study.

P. flexilis James (p. 81) LIMBER PINE **ST**: trunk < 20 m, 2.9 m wide; mature bark dark brown, deeply furrowed, forming rectangular plates; mature crown branches sometimes reaching ground. **LVS** 5 per bundle, 2–9 cm, in dense tufts of bundles at branch ends, stiff, gen curved; sheath deciduous. **SEED CONE** pendent, 7–20 cm, oblong, yellow-brown, opening late 2nd season; stalk < 2 cm; scales ± thickened near tips, angled. **SEED**: wing gen persistent on scale. Lodgepole, subalpine, bristlecone forests; < 3700 m. SNH, TR, PR, **SNE**, **n DMtns**; to w Can, SD, NM. ❁DRN:**1**,4,6,15–17&IRR:**2,3**,7,10,14, **18**,19.

P. jeffreyi Grev. & Balf. (p. 81) JEFFREY PINE **ST**: trunk < 53 m, 2.3 m wide; mature bark gen red-brown, furrows closely spaced, deep, forming ridges, odor banana-, pineapple-, or vanilla-like; mature crown rounded; bud not resinous, scales light brown, white-hairy. **LVS** 3 per bundle, 12–22 cm, grayish blue-green, glaucous. **SEED CONE** spreading or recurved, gen 13–25 cm, ± oblong, brown; stalk < 3 cm, persistent with basal scales; scale surfaces gen of similar color, tips reflexed, elongated < 3 cm, with prickled knobs. Upper mixed-conifer, red-fir forests, elsewhere on serpentine; 450–3100 m. KR, NCoR, CaR, SN, SCoRI, TR, PR, **GB**; sw OR, w NV, n Baja CA. ❁DRN,SUN: 1,**4–6,15–17**&IRR:**2,3,7**,8, 9,**14,18,19**.

P. lambertiana Douglas (p. 81) SUGAR PINE **ST**: trunk < 70 m, < 3.3 m wide; mature bark thick, dark purplish red, irregularly furrowed into plate-like ridges with deciduous platelets; mature crown flattened, with large, ± horizontal branches. **LVS** 5 per bundle, 5–11 cm, gen stiff, sometimes twisted; sheath deciduous. **SEED CONE** pendent, 15–63 cm, cylindric, yellow-brown; stalk 5–16 cm; scales not thickened near tips. **SEED** < wing. Mixed-conifer, mixed-evergreen forests; < 3200 m. NW, CaR, SN, SW, **w GB**; OR, n Baja CA. ❁DRN:1,**4–6,15**,16,17 &IRR:2,3,7.

P. longaeva D. Bailey (p. 81, pl. 5) WESTERN BRISTLECONE PINE **ST**: trunks often multiple, twisted, strongly tapered upward, < 16 m, < 2 m wide; bark reddish brown; mature crown bushy, irregular. **LVS** 5 per bundle, 1–4 cm, often with white resin spots; sheath deciduous. **POLLEN CONE** when mature deep red.

SEED CONE ± spreading or pendent, < 5–14 cm, ovoid-oblong, dark red-brown; stalk 0–2 cm; scale tips thickened to prominent knob, prickle < 6 mm, slender, reflexed. **SEED** < wing. UNCOMMON. Bristlecone-pine forest; 2200–3700 m. **W&I, n DMtns**; to e UT. [*P. aristata* Engelm., in part]. ❁DRN:1,4–6,15–17& SHD,IRR:2,3,7,18,19.

P. monophylla Torrey & Frémont (p. 81, pl. 6) SINGLELEAF PINYON PINE **ST**: trunk < 15 m, < 40 cm wide; mature crown much-branched, rounded. **LF** gen 1 per bundle, 2–7 cm, often curved, gray or blue-green; sheath deciduous. **SEED CONE** erect, 3–12 cm, spheric-ovoid, light or reddish brown; stalk < 1 cm; scale tip knobs < 1 cm, angled, truncate. **SEED** wing persistent on scale. Pinyon/juniper woodland; < 2800 m. c&s SNH, Teh, se SCoRO, TR, PR, **SNE, DMtns**; to se ID, n Baja CA. [*P. californiarum* D. Bailey] Extent of hybridization with *P. edulis* under study. ❁DRN:1,**2,3**,4–6,7,10,**14–16**,17,24,**18–23**&IRR: 8–12.

P. monticola Douglas (p. 81) WESTERN WHITE PINE **ST**: trunk < 73 m, < 2 m wide; mature bark dark gray to bright reddish brown, in ± square blocks, relatively thin; mature crown narrowly conic. **LVS** 5 per bundle, 3–10 cm, gen persistent < 4 years, gen straight, flexible, blue-green, glaucous; sheath deciduous. **SEED CONE** pendent, 9–25 cm, cylindric, yellow-brown; stalk 2–5 cm; scales thinnest at tips. **SEED** < wing. Upper mixed-conifer to subalpine forests; 150–3400 m. KR, CaRH, SNH, **GB**; to B.C., MT. ❁DRN:1,**4–6**&IRR:**2,3**,7,15,16; DFCLT.

P. ponderosa Laws. (p. 81) PACIFIC PONDEROSA PINE **ST**: trunk < 68 m, 2.2 m wide; branches sometimes lacking in lower half when mature; mature bark gen yellow-brown, furrows shallow, well spaced, forming plates, odor not banana-, pineapple-, or vanilla-like; mature crown short, conic or flat-topped; bud resinous, scales red-brown, dark-hairy. **LVS** gen 3 per bundle, 12–26 cm, < 2 mm thick, not glaucous, deep green; sheath persistent. **SEED CONE** ± spreading or recurved, gen 9–18 cm, ovoid, dark brown; stalk < 2 cm, persistent with basal scales; scales gen blackish on lower surfaces; knob prickles < 3 mm. **SEED** < wing. mixed-conifer, lower mixed-evergreen forests, gen absent from serpentine; 150–2300 m. CA-FP (exc NCo, GV, SCo), **SNE**; to B.C., MT, Nebraska, n Mex. ❁ DRN,SUN:1,**4–6,15,16**,17& IRR:**2,3,7**,8,9,**14**, **18**,19.

P. sabiniana D. Don (p. 81) GRAY or FOOTHILL PINE **ST**: trunk < 38 m, < 2 m wide, often leaning; several major branches after 20–30 years; bark dark gray with irregular furrows, forming yellow plates when very old. **LVS** 3 per bundle, 9–38 cm, gray-green, fragrant; sheath persistent. **SEED CONE** pendent, 10–28 cm, ovate-oblong, brownish, opening slowly 2nd season, then persistent several years; stalk < 7 cm, persistent (with basal cone scales) several years; scale tip reflexed, elongated 3–7 cm, angled. **SEED** > wing. Foothill woodland, n oak woodland, chaparral, infertile soils in mixed-conifer and hardwood forests; 150–1500 m. CA-FP (exc n NW, n CaR, SnJV), **w GB, w D**. Common name 'digger pine' is pejorative in origin, so best avoided. ❁ DRN,SUN: 3,**4–7,14–24** &IRR:8–11.

TSUGA HEMLOCK

ST: crown conic, top slender, gen nodding; branches ± drooping; bark reddish brown; twig hairy or puberulent, lf bases ascending, wedge- or scale-like, persistent; bud ovoid. **LVS** 5–25 mm, tapered to a short petiole above persistent base, spreading, ± 2-ranked or not, ± flat, sometimes grooved on upper surface, ridged on lower. **SEED CONE** pendent, 1.2–7.5 cm, ovoid to oblong, opening 1st season; stalk ± 0; bract incl, free from scale. **SEED** plus wing < 2 cm. $2n=24$ for all reports. 10 spp.: n hemisphere. (Japanese: hemlock)

T. mertensiana (Bong.) Carrière (p. 81) MOUNTAIN HEMLOCK **ST**: trunk < 35 m, 2.2 m wide, prostrate at timberline; mature bark red- or purple-brown, with narrow grooves between narrow ridges; twig puberulent. **LVS** more crowded above twig, 10–25 mm, glaucous and with white bands on both surfaces, persistent

< 7 years, tip rounded. **SEED CONE** 3–7.5 cm. Subalpine, some scattered in cold areas of red-fir, mixed-conifer forest; 1200–3500 m. KR, NCoRH, CaRH, SNH, MP, **n SNE**; to AK, MT. ❁DRN,IRR:1,**2**,4,5&SHD:**6**,7,14–17.

ACANTHACEAE ACANTHUS FAMILY

Lawrence R. Heckard

Ann, per, shrubs. **LVS** simple, gen opposite; stipules 0. **INFL**: cyme, spike, or raceme, bracted. **FL** bisexual; calyx deeply (3)4–5-lobed; corolla 4–5-lobed, nearly radial to 2-lipped; stamens (2)4, epipetalous, anther sacs sometimes dissimilar in size or placement; ovary superior, chambers 2, ovules 4–many, placentas axile, style long, stigmas 1–2. **FR**: capsule, loculicidal, gen dehiscing explosively. **SEEDS** gen 2 per chamber, ejected by specialized hook-like stalks. 250 genera, 3000 spp.: esp trop; some orn: *Justicia* (Beloperone, shrimp-plant), *Acanthus*, *Thunbergia*. [Daniel 1984 Desert Plants 5:l62–l79]

1. Corolla resembling pea fl, < 2 cm, tube < lips, white with yellow, maroon-streaked eye on upper lip; anther sacs ± equally inserted on filament, not spurred; fr glabrous . . . . **CARLOWRIGHTIA**
1' Corolla long-tubular, > 2 cm, tube > lips, reddish (rarely yellow) throughout; anther sacs unequally inserted on filament, lower sac spurred; fr canescent . . . . **JUSTICIA**

CARLOWRIGHTIA

Subshrub, shrub. **INFL** gen spike- or panicle-like. **FL**: calyx 5-lobed; corolla appearing 4-lobed, tube < lobes, slender, barely wider at throat, uppermost lobe gen notched (actually 2 lobes ± entirely fused), 2 lateral lobes ascending or reflexed, lowest lobe ± keeled, containing anthers and style; stamens 2, anther sacs ± equally inserted on filament, opening toward uppermost lobe, not spurred. **FR** compressed-ovoid, on flattened stalk. **SEEDS** 4, disk-like, notched at base. 23 spp.: esp sw US, Mex. (Charles Wright 1811–85, Am botanical collector) [Daniel 1988 Brittonia 40: 245–255]

C. arizonica A. Gray (p. 85) ARIZONA CARLOWRIGHTIA **ST** multi-branched, < 1 m, puberulent; hairs erect or recurved. **LVS** variable in size and shape; blade gen < 5 cm, lanceolate to elliptic, sometimes cordate, puberulent; petiole 0–8 mm. **INFL**: spike; fls 1–several per bract. **FL**: calyx 1.5–5 mm, lobes narrow-triangular; corolla 10–18 mm, white with yellow, maroon-streaked eye on upper lip. **FR** 7–11 mm, glabrous. **SEED** 3–4 mm, tubercled; margin finely dentate. 2*n*=18. RARE in CA. Rocky slopes; ± 300 m. **w DSon (Palm Canyon, Borrego Springs)**; to TX, C.Am.

JUSTICIA

Ann, per, shrub. **INFL**: gen spike or raceme. **FL**: corolla 2-lipped, tube > lips, wider upward, upper lip notched or 2-lobed, lower lip 3-lobed; stamens 2, gen appressed to upper lip, anther sacs unequally placed on filament, opening toward lower lip, gen at least lower sac spurred. **FR** club-shaped; stalk flattened. **SEEDS** 4, outlined on fr surface. 300 spp.: trop, subtrop. (James Justice, 18th century Scottish horticulturist)

J. californica (Benth.) D. Gibson (p. 85, pl. 7) BELOPERONE, CHUPAROSA Shrub. **ST** < 2 m, gen lfless in fl, canescent. **LF**: petiole < 20 mm; blade 1–6 cm, ovate, triangular, or ± round, puberulent. **INFL**: raceme; bracts <1 cm, lanceolate-elliptic, falling early; bractlets 2–5 mm, narrow-tapered. **FL**: calyx 5–8 mm, lobes 5, lanceolate; corolla 2–4 cm, puberulent, dull scarlet (yellow), lips 1–2 cm, < tube, lobes of lower lip 1–4 mm. **FR** 1.5–2 cm, canescent. **SEED** 2.5–3.5 mm, ± round, mottled. 2*n*=28. Dry, sandy or rocky soils, esp washes; < 800 m. **w PR** (San Diego River e of Wildcat Canyon), **DSon**; AZ, nw Mex. [*Beloperone c.* Benth.] ❀SUN,DRN:18–22, **23**,24&IRR12, **13**,14;DFCLT < 25° causes damage; CVS. Mar–Jun

ACERACEAE MAPLE FAMILY

James R. Shevock

Shrubs, trees, sometimes monoecious, dioecious, or with staminate and bisexual fls. **LVS** opposite, gen simple, gen palmately lobed (rarely pinnate), gen deciduous; stipules 0. **INFL**: panicle, raceme, or umbel-like. **FL** small; perianth gen ± yellowish green; sepals (4)5, free; petals gen 5 (sometimes 0, 4, or 6), free, gen sepal-like; stamens gen 8 (sometimes 5, 10, or 12), gen attached to edge of nectary disk; ovary superior, chambers 2, each 2-ovuled. **FR**: gen pair of achenes, conspicuously winged. **SEED** gen 1 per achene. 2 genera, ± 120 spp.: n temp, trop mtns (*Dipteronia*: 2 spp.: China). Some *Acer* important as timber or orn, often has bright autumn colors. [Ogata 1967 Bull Tokyo U For 63:89–206] Recently treated, along with Hippocastanaceae (in CA, *Aesculus*), in expanded Sapindaceae [Gadek et al 1996 Amer J Bot 83: 802-811]

ACER MAPLE

LF simple or pinnately compound. **INFL**: fl clusters drooping, gen appearing before or with emerging lvs. **FR** paired, each with elongate wing ribbed on proximal side. ± 118 spp.: n hemisphere. (Latin name for maple)

A. glabrum Torrey MOUNTAIN MAPLE Shrub or small tree, 2–6 m, ± erect, dioecious (or staminate pl with some bisexual fls). **LF** 2–4 cm wide, palmately lobed. **INFL** dense, gen < 10-fld. **FL**: petals present. **FR**: body glabrous; wings overlapping or spreading < 45°. Moist to fairly dry, montane, rocky slopes, canyons; 1500–2800 m. KR, CaRH, SN, SnJt, MP, **SNE, DMtns**; to AK, c US, NM. Complex of 6 vars. (var. *glabrum* in US Rocky Mtns, c US). 3 vars in CA.

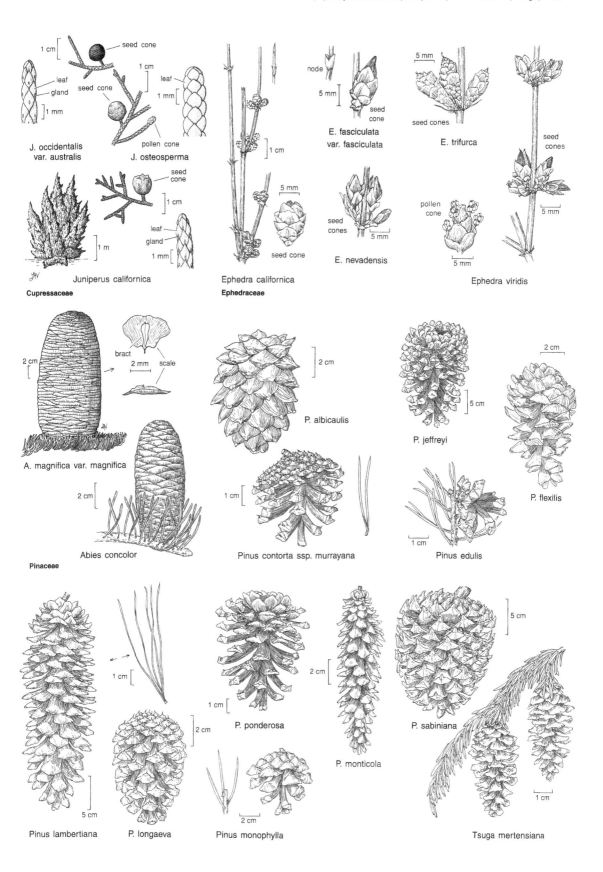

J. occidentalis
var. australis

J. osteosperma

Juniperus californica

Cupressaceae

Ephedra californica

Ephedraceae

E. fasciculata
var. fasciculata

E. nevadensis

E. trifurca

Ephedra viridis

A. magnifica var. magnifica

Abies concolor

Pinaceae

P. albicaulis

Pinus contorta ssp. murrayana

P. jeffreyi

Pinus edulis

P. flexilis

Pinus lambertiana

P. longaeva

P. ponderosa

Pinus monophylla

P. monticola

P. sabiniana

Tsuga mertensiana

var. *diffusum* (E. Greene) F.J. Smiley (p. 85) **ST**: twigs gen whitish. **LF** gen 1.2–2.8 cm wide; teeth few, blunt. **INFL**: peduncle + pedicel 1–2 cm. **FR**: wings spreading gen 45°. Habitats of sp. SnJt, **SNE, DMtns**; to UT. ⚘DRN,IRR:1–7. Apr–May

AIZOACEAE FIG-MARIGOLD FAMILY

John Bleck, Wayne R. Ferren Jr., Nancy J. Vivrette

Ann, per, shrub, gen fleshy. **ST** underground or prostrate to erect. **LVS** gen simple, gen cauline, gen opposite; stipule gen 0; blade gen glabrous, often glaucous. **INFL**: cyme or fl solitary. **FL** gen bisexual, radial; hypanthium present; sepals 3–8; petals gen many in several whorls, free or fused at base, linear, sometimes 0; stamens 1–many, free or fused in groups, outer often petal-like; nectary a ring or separate glands; pistil 1, ovary superior to inferior, chambers 1–20, placentas gen parietal, styles 0–20, stigmas 1–20. **FR**: gen capsule, opening by flaps or circumscissile, or berry or nut. **SEEDS** 1–many per chamber, often with aril. 130 genera, 2500 spp.: gen subtrop, esp s Afr; many cult, some waifs in CA (e.g., *Disphyma australe* (Aiton) J. Black: ovary glands convex and minutely crenate, stigmas densely plumose, fr chambers 5, seeds ovate, ± smooth; *Lampranthus* spp.: ovary glands fused, fr chambers 5, seeds pear-shaped, ± black, rough; both genera members of Ruschieae; *Mesembryanthemum crystallinum*, petals many, has been reported as a roadside occurrence in DSon). [Ferren et al. 1981 Madroño 28:80–85] *Mollugo* is in Molluginaceae. Key to genera adapted by Margriet Wetherwax.

1. Styles 2–5; ovary half-superior; calyx lobes 4–10 mm; lvs of a pair equal . . . . **SESUVIUM**
1' Styles 1–2; ovary fully superior; calyx lobes ± 2.5 mm; lvs of a pair unequal . . . . **TRIANTHEMA**

SESUVIUM SEA-PURSLANE

Wayne R. Ferren Jr.

Ann, per, shrub, glabrous, gen papillate. **ST** prostrate to erect, forming mats < 2 m diam; nodes sometimes rooting. **LVS** gen < 6 cm; stipule 0; petiole base gen wide with scarious margins; blade linear to ovate, entire. **INFL**: cyme, cluster or fl solitary; bracts 0 or 2. **FL**: hypanthium obconic; calyx lobes 5, gen hooded near tip, reddish within; petals 0; stamens 1–many, often fused at base; ovary half-superior, chambers 2–5, placentas axile, styles 2–5, papillate. **FR**: capsule, circumscissile, ovoid to conic, thin-walled. **SEEDS** many, ± reniform, gen smooth, shiny, black or brown; aril present. 8 spp.: gen trop, subtrop coasts, deserts.

S. verrucosum Raf. (p. 85) WESTERN SEA-PURSLANE Per, branched from base, minutely papillate. **LF** 0.5–4 cm, linear to widely spoon-shaped; base clasping st. **INFL**: fl solitary, axillary, sessile or peduncle short. **FL**: calyx lobes 4–10 mm, margins scarious, hooded or beaked, outer surface papillate; stamens many, filaments fused to midlength, reddish. **FR** 4–5 mm. **SEED** 0.8–1 mm, smooth. Uncommon. Moist or seasonally dry flats, margins of gen saline wetlands; < 1400 m. GV, SCoRO, SCo, WTR, PR, SNE, D; to OR, KS, S.Am. Apr–Nov

TRIANTHEMA HORSE-PURSLANE

Wayne R. Ferren Jr.

Ann, per, shrub, branched from base, glabrous, hairy, or papillate, ± fleshy. **ST** gen prostrate. **LVS** ± opposite, unequal; stipules papery; blade linear to round, base tapered, margin entire. **INFL**: fl gen solitary; bracts 2. **FL**: calyx lobes 5, tip pointed; petals 0; stamens 5–10; ovary superior, chambers 1–2, ovules 1–few, placentas basal, style 1–2. **FR**: capsule, papery or leathery, circumscissile; lid winged. **SEED** reniform, rough; aril present. 20 spp.: trop, subtrop, esp Australia. (Greek: 3-fld)

T. portulacastrum L. (p. 85) Ann. **ST** < 10 dm; young branches with lines of hairs below petioles. **LF**: stipules reduced to 2 teeth on petiole; petiole gen = blade; blade < 4 cm, smaller on twigs, elliptic to ± round, base tapered, tip often notched. **INFL**: fl sessile, ± covered by stipule. **FL**: calyx 3–5 mm, lobes 2.5 mm, lanceolate, purplish within; ovary chamber 1, stigmas 2. **FR** 4–5 mm, cylindric, ± curved; wings of lid 2, prominent, erect. **SEED** 1.5–2 mm, ridged, reddish brown to black. Uncommon. Moist or seasonally dry wetlands, waste places; < 1000 m. SnJV, D; to TX, e N.Am., S.Am., trop Old World. Jun–Nov

AMARANTHACEAE AMARANTH FAMILY

James Henrickson

Ann to tree. **ST** prostrate to erect. **LVS** alternate or opposite, simple, gen entire, bract-like upward; stipules 0. **INFL**: cymes, often arrayed ± in spikes or panicles; bracts 1–3 per fl, gen membranous-scarious, tip often short-pointed or spine-like. **FL** gen small, uni- or bisexual, radial; sepals 0–5, fused at base or free, often ± scarious; petals 0; stamens 0–5, opposite sepals (staminodes sometimes alternating), filaments sometimes fused at base; ovary superior, 1-chambered, ovules 1–several, erect or pendent on ± basal stalks, styles 0–3, stigma head-like or 2–3-lobed. **FR**: utricle to circumscissile capsule. **SEED** lenticular to spheric, hard. ± 65 genera, ± 900 spp.: trop, subtrop. [Robertson 1981 J Arnold Arbor 62:267–313] Recently treated to include Chenopodiaceae [Downie et al. 1997 Amer J Bot 84: 253–273]. Key to genera adapted by Margriet Wetherwax.

1. Cauline lvs alternate; fl unisexual; bract midrib thick, margin scarious, tip short-pointed or spine-like
. . . . **AMARANTHUS**

1' Cauline lvs opposite; fl bisexual; bract scarious throughout, tip gen not short-pointed or spine-like
· **TIDESTROMIA**

AMARANTHUS PIGWEED, AMARANTH

Ann, monoecious or dioecious. **LVS** alternate; blade linear to ovate. **INFL**: cymes in dense, spike-like clusters; bract 1, tip gen short-pointed or -spined, gen scarious; bractlets 0–2. **STAMINATE FL**: sepals 3–5, ± equal, ± as bracts; stamens (1)3–5, filaments free; staminodia 0. **PISTILLATE FL**: sepals (3–)5, ± equal, scarious exc midvein, fused at base, falling with fr; ovary compressed-ovoid, styles (2)3, stigmas slender, papillate, ovule 1, erect. **FR** circumscissile or indehiscent, smooth or inflated-wrinkled; walls membranous to spongy-hardened. **SEED** 1, lenticular, smooth, reddish to black. ± 60 spp.: worldwide; some potherbs, some cult for seed. (Greek: unfading, from persistent bracts and sepals) Some spp. (esp *A. cruentus, A. powellii, A. retroflexus*) hybridize complexly. [Tucker & Sauer Madroño 1958 14:252–261] Key to species adapted by Margriet Wetherwax.

1. Dioecious
  2. Pl in hand pistillate
    3. Outermost sepal > others, tip acute, spine-like; bracts >> inner sepals · ²*A. palmeri*
    3' Outermost sepal slightly > others, obtuse to notched, short-pointed; bracts slightly > inner sepals ²*A. watsonii*
  2' Pl in hand staminate
    4. Outer sepal spine-tipped, > others; bracts = or > sepals · ²*A. palmeri*
    4' Outer sepal short-pointed, slightly > others; bracts < or = sepals · · · · · · · · · · · · · · · · · · · ²*A. watsonii*
1' Monoecious (staminate fls sometimes only near infl tip)
  5. Most nodes with 2 rigid spines · *A. spinosus*
  5' Nodes without spines
    6. Infl clusters axillary throughout pl, not clearly terminal
      7. Sts ascending to erect, pl spheric; pistillate sepals 3, 0.7–2.2 mm · *A. albus*
      7' Sts prostrate; pistillate sepals (1)3–5, 1–3.5 mm
        8. Seed 1.2–1.7 mm diam; pistillate sepals 4–5, outer green along midrib, 1.5–3.5 mm · · · · · · · · *A. blitoides*
        8' Seed 0.8–1.1 mm diam; pistillate sepals 1–3, white and scabrous throughout, 1–1.2 mm · · · *A. californicus*
    6' Infl clearly terminal (sometimes also axillary)
      9. Axillary clusters 0 in lower 1/2 of pl · *A. powellii*
      9' Axillary clusters scattered gen to pl base
        10. Pistillate bracts < sepals; pistillate sepals widely fan-shaped, < or = 2.5 mm wide, margin with
          small, finger-like projections · *A. fimbriatus*
        10' Pistillate bracts gen > sepals; pistillate sepals ± spoon-shaped, 0.7–1.5 mm wide, irregularly toothed
        · *A. torreyi*

A. albus L. (p. 85) TUMBLEWEED Pl ascending to erect, 1–7 dm, bushy, monoecious. **LVS**: cauline petiole 5–50 mm, blade 8–30(70) mm, elliptic to obovate, gen early deciduous; axillary lvs gen smaller, persistent. **INFL**: clusters axillary, throughout pl; bracts 2–3.5 mm, green, spined, margin scarious. **STAMINATE FL**: sepals 3, ± 1 mm; stamens 3. **PISTILLATE FL**: sepals 3, 0.7–2.2 mm, subequal, oblong-lanceolate, tip acute to obtuse, margin scarious. **FR** 1.5–2 mm, circumscissile; surface rough throughout. **SEED** ± 1 mm wide, red-brown. 2*n*=32. Common. Weed of waste places, roadsides, fields; < 2200 m. **CA**; widespread N.Am, to Eurasia; native to trop Am. Jun–Oct

A. blitoides S. Watson Pl prostrate, 3–7 dm, monoecious. **LVS**: cauline petiole 4–30 mm, blade 5–40 mm, elliptic to widely (ob)ovate, early-deciduous; axillary lvs smaller, persistent. **INFL**: clusters axillary, throughout pl; bracts 1–3.5 mm, < or = sepals, midrib thick, spine-tipped, green, margins scarious esp near base. **STAMINATE FL**: sepals 4–5, 1.5–2.5 mm; stamens (3)4–5. **PISTILLATE FL**: sepals 4–5, 1–3.5 mm, oblong to ovate, outer 2 > others, more reflexed, tips more tapered and spiny. **FR** ± 2.5 mm, circumscissile; lid often rough below. **SEED** 1.3–1.7 mm wide, black; margin acute. Waste places; < 2200 m. CA-FP, **W&I, D**; to WA, e US, c&s Eur. [*A. graecizans* L. misapplied] Jul–Nov

A. californicus (Moq.) S. Watson (p. 85) Pl prostrate, monoecious. **ST** 1–5 dm. **LVS**: cauline petiole 1–10(20) mm, blade 3–12 (30) mm, spoon-shaped or obovate, gen early deciduous; axillary lvs much smaller, clustered, persistent. **INFL**: clusters axillary, scattered; bracts 1–2 mm, scarious at margins or ± throughout, often inconspicuous, tip acute, short-spined. **STAMINATE FL**: sepals (2)3, 1–1.5 mm; stamens 3. **PISTILLATE FL**: sepals (1–)3, 1–1.2 mm, subequal, narrowly lanceolate. **FR** 1 mm,

circumscissile; lid smooth or wrinkled at base. **SEED** 0.8–1.1 mm wide, red-brown. Seasonally moist flats, lake margins; < 2800 m. **CA**; to s Can, w TX. ✿TRY:STBL only. Jul–Oct

A. fimbriatus (Torrey) Benth. (p. 85, pl. 8) Pl erect, 4–10 dm, often reddish, monoecious. **LF**: petiole 5–20 mm; blades 20–100 mm, linear to narrowly lanceolate. **INFL**: lower clusters axillary, scattered; terminal clusters spike-like, 10–30 cm, 1–2 cm wide, crowded, leafy; bracts 1–1.8 mm, scarious (exc midrib). **STAMINATE FL**: sepals 5, 1–1.5 mm; stamens 3. **PISTILLATE FL**: sepals 5, 1.5–3.3 mm, < or = 2.5 mm wide, subequal, fan-shaped, tip long-toothed, fringed, reflexed. **FR** ± 1.5 mm, circumscissile; lid strongly inflated-wrinkled near base. **SEED** 0.9–1 mm wide. *n*=17. Common. Sandy, gravelly slopes, washes, after summer rains; 600–1700 m. **D**; to s Can, MT, w TX, Baja CA. Aug–Nov

A. palmeri S. Watson (p. 85) Pl erect, 2–20 dm, dioecious. **LF**: petiole 10–90 mm; blade 15–170 mm, lanceolate to ovate. **INFL** spike-like, terminal, solitary or with axillary ones at base, 10–50 cm, 7–15(25) mm wide; bracts ± lanceolate, scarious below, spine-tipped, staminate bracts 2.5–5 mm, pistillate bracts 2.5–6 mm, recurved in fr. **STAMINATE FL**: sepals 5, 2.5–6 mm, outer ± > inner; stamens (3)5. **PISTILLATE FL**: sepals 5, 1.7–3.8 mm, narrowly spoon-shaped, reflexed in fr, tips of inner gen rounded to slightly notched, tips of outer acute to spined. **FR** ± 1.5 mm, circumscissile; lid smooth or rough. **SEED** 0.8–1.4 mm wide, red-brown. Abundant. Roadsides ditches, fields, arroyos; < 1200 m. SnJV, CW, **D**; to c US, c Mex. Aug–Nov

A. powellii S. Watson (p. 85) Pl erect, 3–20 dm, monoecious. **LF**: petiole 10–50 mm; blade 15–90 mm, lower diamond-shaped to ovate, upper ± lanceolate. **INFL**: panicle of spike-like clusters

1–1.8 cm wide; axillary clusters 1–6 cm; terminal clusters < or = 25 cm; bracts 3–6(8) mm, spine-tipped, midrib green, thick, margin scarious in lower 2/3. **STAMINATE FL**: sepals 4–5, 2–3 mm; stamens 3–5. **PISTILLATE FL**: sepals 5, 1.7–3.6 mm, unequal, ± oblong, tip short-pointed. **FR** 2.5–3.6 mm, circumscissile, smooth throughout. **SEED** 0.8–1.4 mm wide, reddish black. Uncommon. Waste places; < 800 m. SNF, GV, CW, SW, **W&I**; to WA, n-c US, TX, Mex, w S.Am. Jul–Oct

A. spinosus L. Pl erect, 2–10 dm, ± spheric, monoecious; spines at lower nodes gen 2, 8–20 mm, slender. **LF**: petiole 12–70 mm; blade 14–80 mm, ± ovate. **INFL** dense; axillary clusters below middle ± spheric, each with 2 bracts elongated, stiff, spiny; terminal clusters spike-like, 3–17 cm, 4–13 mm wide, cylindric; upper bracts 1.5–6 mm, spine-tipped, midrib thick, green, margin scarious below. **STAMINATE FL**: sepals 5, 1.5–2 mm; anthers 5. **PISTILLATE FL**: sepals 5, 1.5–2 mm, oblanceolate to spoon-shaped, tip obtuse to spined. **FR** 1.5–2 mm, irregularly dehiscent; lid smooth throughout or base rough. **SEED** ± 1 mm wide, reddish black. $2n=34$. Very uncommon as waif. Roadsides, waste places; < 700 m. SCo, **w DMoj**; to e US, Mex; native to trop Am.

A. torreyi (A. Gray) Benth. (p. 85) Pl erect, 1–7 dm, monoecious. **LF**: petiole 4–16 mm; blade 12–50 mm, oblanceolate to ovate. **INFL**: axillary clusters scattered to st base, crowded above into terminal, leafy spike < or = 10 cm, 7–15 mm wide; bracts 1.3–3.5 mm, midrib green, margins scarious below, sparsely glandular-hairy. **STAMINATE FL**: sepals 5, 2.5–4 mm; stamens 3. **PISTILLATE FL**: sepals 5, 1.4–2.5 mm, narrowly spoon-shaped, tips rounded to slightly notched, jagged. **FR** 1.4–2 mm, circumscissile; lid coarsely wrinkled-inflated near base. **SEED** ± 1 mm wide, black. Uncommon. Sandy flats, arroyos, after late summer rains; 1200–1700 m. **DMoj**; w AZ.

A. watsonii Standley (p. 85) Pls erect, 1–10 dm, dioecious. **ST** glandular-hairy. **LF**: petiole 3–90 mm; blade 10–80 mm, ± oblong to (ob)ovate. **INFL**: axillary clusters few, scattered below; terminal cluster 2–15 cm, 8–20 mm wide, spike-like; bracts 2.5–4 mm, spine-tipped, midrib ± thick, green, with stalked glands, margins scarious below. **STAMINATE FL**: sepals 5, 1.5–2 mm; stamens 3–5. **PISTILLATE FL**: sepals 5, 1.7–2.2 mm, ± fan-shaped, margin jagged, tip obtuse to slightly notched. **FR** 1.5–2 mm, circumscissile, smooth. **SEED** ± 1 mm wide, reddish black. Very uncommon. Depressions, waste places, after winter rains; < 500 m. **s DSon (Imperial Co.)**; nw Mex. Aug–Sep

TIDESTROMIA

Ann to subshrub, ± mounded; hairs short, branched. **LVS** entire, often asymmetric, petioled; lowest alternate to opposite; upper opposite or whorled in 3'0s, gen reduced, larger falling early. **INFL**: clusters of 1–5 fls, axillary, sessile, subtended and ± enclosed by involucres of 2–3 bract-like lvs that become hardened; bracts scarious. **FL** bisexual; sepals 5, 1–1.5 mm, 1-veined, glabrous inside, sometimes canescent outside, reflexed, inner scarious-margined; stamens 5, filaments fused below into a cup, alternate 5 short staminodes; ovary spheric, style 1, short, stigma head-like or 2-lobed. **FR** indehiscent; wall membranous. **SEED** obovoid, brown. 7 spp.: w N.Am deserts. (Ivar T. Tidestrom, Swedish-born botanist of sw US, 1864–1956) Key to species adapted by Margriet Wetherwax.

1. Ann; young lvs whiter (or grayer) than older lvs; sepals 1.8–2.6 mm; filaments 0.6–1.6 mm . . . . . *T. lanuginosa*
1' Per; young and old lvs ± equally white or gray; sepals 1.3–1.7 mm; filaments < or = 0.5 mm . . . *T. oblongifolia*

T. lanuginosa (Nutt.) Standley Ann 1–9 dm, < or = 20 dm wide. **LVS**: upper << lower; young terminal lvs white-gray, tomentose-canescent, hairs short, much-branched, wearing off, older lvs green above, more whitish below; blades 6–22 mm, (ob)ovate, base wedge-shaped. **FL**: sepals 1.8–2.6 mm; filament tube 0.3–0.4 mm, free filaments 0.6–1.6 mm. Slopes; ± 1200 m. **e DMtns (Granite Mtns)**; to Colorado, TX, Mex. Jul–Oct

T. oblongifolia (S. Watson) Standley (p. 91) Per 1–9 dm, < or = 15 dm wide. **LVS**: upper, persistent lvs 2–4(8) mm; blade 10–30(65) mm, narrowly ovate to round, base ± obliquely wedge-shaped, persistently grayish white-canescent. **FL**: sepals 1.3–1.7 mm; filament tube 0.2 mm, free filaments < or = 0.5 mm. Common. Washes, rocky hillsides; < 1200 m. **D**; NV, AZ, n Baja CA. [ssp. *cryptantha* (S. Watson) Wiggins] ✿TRY:STBL only. Apr–Dec

ANACARDIACEAE SUMAC OR CASHEW FAMILY

Dieter H. Wilken

Shrub, tree, gen dioecious or fls bisexual and unisexual, ± resinous, sometimes milky, gen aromatic. **LVS** simple or compound, alternate, deciduous or evergreen; stipules 0. **INFL**: raceme or panicle; fls gen many. **FL** gen unisexual, radial; sepals 5, base gen ± fused; petals 5, gen > sepals, free; stamens 5 or 10, reduced and sterile in pistillate fls; ovary superior, vestigial or 0 in staminate fls, subtended by ± lobed, disk-like nectary, chamber gen 1, ovule gen 1, styles 1–3. **FR** drupe-like, glabrous, sticky, or short-hairy; pulp ± resinous, sometimes aromatic. 70+ genera, ± 850 spp.: trop, warm temp; some orn (*Rhus, Schinus*), some cult for fr (*Anacardium*, cashew; *Mangifera*, mango). [Brizicky 1962 J Arnold Arbor 43:359–375] TOXIC: many genera produce contact dermatitis.

1. Infl terminal, branches stiff; fls sessile; fr red; DMtns . . . . *Rhus trilobata*
1' Infl axillary, branches slender, loose; fls pedicelled; fr creamy white;
   sw edge DMoj . . . . *Toxicodendron diversilobum*

RHUS

Shrub, tree, dioecious or bisexual and pistillate. **LVS** simple or compound, deciduous or evergreen, entire, toothed, or lobed. **INFL**: panicle, terminal on short twigs, open to dense; fls gen sessile. **FL**: stamens 5; styles 3, free or ± fused. **FR** spheric or ± compressed, glabrous or glandular-hairy, gen reddish; pulp thin or thick, ± resinous. ± 150 spp.: warm temp. (Greek: ancient name for sumac) [Brizicky 1963 J Arnold Arbor 44:60–80]

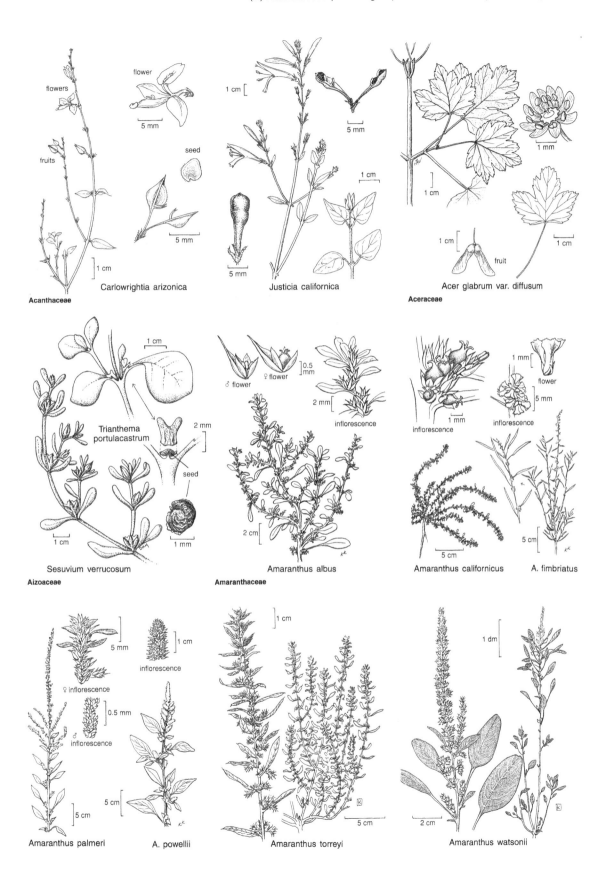

Carlowrightia arizonica

Acanthaceae

Justicia californica

Acer glabrum var. diffusum

Aceraceae

Trianthema portulacastrum

Sesuvium verrucosum

Aizoaceae

Amaranthus albus

Amaranthaceae

Amaranthus californicus   A. fimbriatus

Amaranthus palmeri   A. powellii

Amaranthus torreyi

Amaranthus watsonii

R. trilobata Torrey & A. Gray (p. 91, pl. 9) SKUNKBRUSH Shrub, 0.5–2.5 m. **LF** deeply lobed to compound, deciduous, thin, flat; petiole 5–15 mm; lobes or lflets gen 3, margins crenate to slightly lobed, lower surfaces tomentose to ± glabrous; terminal lobe or lflet 10–35 mm, ± diamond-shaped; lateral 5–18 mm, gen ovate. **INFL**: branches short, stiff; fl before lvs appear. **FL**: sepals yellow-green to reddish; petals gen yellow. **FR** 5–8 mm diam, sparsely hairy, sticky, gen bright red-orange. Slopes, washes, shrubland; < 2200 m. CA-FP, **DMtns**; to s Can, c US, n Mex. [var. *anisophylla* (E. Greene) Jepson; var. *malacophylla* (E. Greene) Jepson; var. *quinata* Jepson] Geog variation in w N.Am needs further study. ❀4,5,**6,7**, **14**,15–17,22–24;IRR:1–3,8–10,**18–21**;also STBL. Mar–Apr

TOXICODENDRON POISON OAK, POISON IVY

Shrub, tree, vine, gen dioecious. **LVS** gen ternate or pinnately compound, ± resinous; lflets 3–9, lateral gen opposite, thin to ± leathery, entire, toothed,or lobed. **INFL**: raceme or panicle, axillary, ± open; fls pedicelled. **FL**: stamens 5, sterile or reduced in pistillate fls; styles ± fused, stigmas 3. **FR** gen spheric, becoming papery or leathery, cream to brown; pulp resinous. 6 spp.: Am, e Asia. (Latin: poisonous tree) [Gillis 1971 Rhodora 73:161–237,370–443] TOXIC: resin on lvs, sts, frs causes severe contact dermatitis; one of the most hazardous pls in CA.

T. diversilobum (Torrey & A. Gray) E. Greene (p.91) WESTERN POISON OAK Shrub (sometimes tree-like) 0.5–4 m or vine < 25 m. **ST**: twigs glabrous to sparsely hairy, gray- to red-brown. **LF**: petiole 1–10 cm; lflets gen 3(5), ± round to oblong, thin to ± leathery, becoming bright red in autumn, base truncate to rounded, tip obtuse to rounded, margin entire, wavy, or slightly lobed, upper surface glabrous, ± shiny, lower sparsely short-hairy; terminal lflet 1–13 cm, 1–8 cm wide; lateral lflets 1–7 cm, 1–6 cm wide. **INFL** drooping, spreading, or erect; pedicels 2–8 mm; bractlets < 1 mm. **FL**: sepals, petals gen ovate, yellow-green. **FR** 1.5–6 mm diam, spheric to slightly compressed, becoming leathery, glabrous to finely bristly, creamy white; pulp white, black-striate. 2*n*=30. Canyons, slopes, chaparral, oak woodland; < 1650 m. CA-FP, **sw edge DMoj**; to B.C., Baja CA. [*Rhus d.* Torrey & A. Gray] Apr–May

APIACEAE [UMBELLIFERAE] CARROT FAMILY

Lincoln Constance

Ann, bien, per (rarely shrub, tree), often from taproot. **ST** often ± scapose, gen ribbed, hollow. **LVS** basal and gen some cauline, gen alternate; stipules gen 0; petiole base gen sheathing st; blade gen much dissected, sometimes compound. **INFL**: umbel or head, simple or compound, gen peduncled; bracts present (in involucres) or not; bractlets gen present (in involucels). **FLS** many, small, gen bisexual (or some staminate), gen radial (or outer bilateral); calyx 0 or lobes 5, small, atop ovary; petals 5, free, gen ovate or spoon-shaped, gen incurved at tips, gen ± ephemeral; stamens 5; pistil 1, ovary inferior, 2-chambered, gen with a ± conic, persistent projection or platform on top subtending 2 free styles. **FR**: 2 dry, 1-seeded halves that separate from each other but gen remain attached for some time to a central axis; ribs on each half 5, 2 marginal and 3 on back; oil tubes 1–several per interval between ribs. 300 genera, 3,000 spp.: ± worldwide, esp temp; many cult for food or spice (e.g., *Carum*, caraway; *Daucus*; *Petroselinum*); some highly toxic (e.g., *Conium*). Underground structures here called roots, but true nature remains problematic. Mature fr gen critical in identification; shapes gen given in outline, followed by shape in ×-section of 2 fr halves together.

1. Fls in simple umbels; fr central axis not an obvious structure . . . . **HYDROCOTYLE**
1' Fls in simple or compound umbels (these sometimes head-like); fr central axis gen an obvious, separate structure
  2. Ann or bien from slender taproot, reproducing solely from seed
    3. Corolla yellow . . . . **BUPLEURUM**
    3' Corolla white
      4. Bracts lf-like, gen dissected . . . . **DAUCUS**
      4' Bracts 0 or inconspicuous, not lf-like
        5. Involucel 0; fr glabrous . . . . **CICLOSPERMUM**
        5' Involucel present; fr conspicuously bristly or sharply scabrous, at least on ribes
          6. Infl exc frs puberulent or minutely hairy; fr oblong-ovate . . . . **AMMOSELINUM**
          6' Infl exc frs glabrous; fr widely ovate . . . . **SPERMOLEPIS**
  2' Per or bien, from various, persistent, often swollen, tuberous underground structures, as well as reproducing from seed
    7. Pl glabrous, gen in wet or moist soil of marshes, streambanks, etc., or where water stands for long periods; corolla white
      8. Sts from well developed single or clustered tubers, gen not rooting at lower nodes
        9. Fr very compressed front-to-back, marginal ribs widely thin-winged . . . . **OXYPOLIS**
        9' Fr slightly compressed side-to-side, ribs subequal, thread-like, not clearly winged . . . . . . **PERIDERIDIA**
      8' Sts from stout taproots or rooting at lower nodes or along rhizomes or stolons
        10. Bracts, bractlets conspicuous; fr halves gen adhering to central fr axis
          11. Fr ribs thread-like, inconspicuous in corky fr wall . . . . **BERULA**
          11' Fr ribs prominent, corky . . . . **SIUM**
        10' Bracts, bractlets inconspicuous to 0; fr halves not adhering to fr central axis
          12. Bien, per; lvs 1-pinnate, lflets lanceolate to ± round; fr axis notched at tip; fr not conspicuously corky . . . . **APIUM**

12' Per; lvs 1–3-pinnate or 1–3-ternate-pinnate, lflets mostly lanceolate; fr axis divided to base; fr conspicuously corky . . . . **CICUTA**
7' Pl glabrous or variously hairy, not restricted to wet places; corolla white, yellow, or purple
  13. Herbage anise- or licorice-scented; lf segments thread-like . . . . **FOENICULUM**
  13' Herbage scented or not, but not anise-scented; lf segments wider than threadlike
    14. Fr ribs subequal, none expanded into definite wings
      15. Pl dwarfed, cushion-forming, 2–5 cm; lf segments not sharply serrate; alpine . . . . **PODISTERA**
      15' Pl gen not dwarfed, not cushion-forming, 10–120(300) cm, if < 20 cm, lf segments sharply serrate (st 0 in *Tauschia*); gen lower than alpine
        16. Fr linear to oblong, 12–22 mm; fr oil tubes obscure; roots licorice-scented . . . . **OSMORHIZA**
        16' Fr oblong to narrowly elliptic, 5–8 mm; fr oil tubes evident, 4–5 per rib-interval; roots not licorice-scented . . . . **TAUSCHIA**
    14' Fr ribs unequal, some expanded into definite wings
      17. Ribs at margins of fr halves winged, others not or more narrowly so; fr very compressed front-to-back
        18. Pl very robust, 1–3 m; outer petals of marginal fls > others; oil tubes extending part way to base of fr . . . . **HERACLEUM**
        18' Pl not very robust, rarely 1 m; outer petals of marginal fls = others; oil tubes extending to base of fr . . . . **LOMATIUM**
      17' Ribs at margins of fr halves and some others ± winged; fr cylindric to compressed front-to-back, rarely very much so
        19. Pl low, gen spreading, st 0 or short; rays well developed or not; projection atop ovary 0 . . . . **CYMOPTERUS**
        19' Pl erect, st gen present; rays well developed; projection atop ovary conic
          20. Pedicels distinct to slightly webbed at base, but not reduced to a disk; 2° umbels open, not head-like . . . . **ANGELICA**
          20' Pedicels reduced to a disk; 2° umbels head-like . . . . **SPHENOSCIADIUM**

AMMOSELINUM

Ann, taprooted. **ST** erect or gen loosely branched, glabrous or roughened. **LF**: petiole entirely sheathing; blade oblong to obovate, ternately or ternate-pinnately dissected, segments linear to spoon-shaped. **INFL**: umbels compound, peduncled or some sessile, puberulent; bracts gen 0; bractlets several, narrow; rays, pedicels few, spreading or spreading-ascending, unequal. **FL**: calyx lobes 0; petals ovate, white, tips obtuse, not narrowed, not incurved. **FR** oblong-ovate, compressed side-to-side; ribs subequal, prominent, conspicuously bristly to glabrous; oil tubes per rib-interval 1–3; fr axis notched at tip. **SEED**: face flat to concave. 4 spp.: 3 N.Am, 1 S.Am. (Greek: sand-parsley)

A. giganteum J. Coulter & Rose (p. 91) DESERT SAND-PARSLEY Pl 1–2 dm. **LF**: petiole 3–8 mm; blade 1.5–2.5 cm, obovate, segments 4–13 mm, linear, glabrous or roughened. **INFL**: peduncles 0–4 cm; rays 4–8, 0–2 cm; pedicels 1–10, 0–8 mm. **FR** 3–5 mm, oblong-ovate; ribs corky, sharply scabrous. $2n=38$. RARE in CA. Heavy soil under shrubs; ± 400 m. **DSon (Hayfield Lake, Riverside Co., 1922)**; AZ, n Mex. Possibly alien in CA. Mar–Apr

ANGELICA

Per, taprooted. **ST** erect, leafy, hollow. **LVS**: petioles gen inflated; cauline often bladeless; blades compound, rarely dissected, lflets gen wide, distinct, when lf dissected, segments narrow, connected. **INFL**: umbels compound, peduncled; bracts 0; bractlets 0 or many and conspicuous; rays, pedicels many, spreading-ascending or ascending. **FL**: calyx lobes 0 or minute; petals wide, white, pink, red, or purple. **FR** oblong to round, gen very compressed front-to-back (rarely slightly so or cylindric), glabrous to hairy; ribs gen unequal, winged but marginal gen wider than others; oil tubes per rib-interval 1–several, adhering to fr wall or rarely to seed; fr axis divided to base. **SEED**: face flat. 50–60 spp.: temp N.Am, Asia. (Latin: angelic, for cordial and medicinal properties) [DiTomaso 1984 Madroño 31:69–79]

1. Lf triangular-ovate, 2–3-ternate-pinnately dissected, segments linear, entire; rays 20–40, subequal . . . . *A. lineariloba*
1' Lf oblong, 1–3-pinnate or -ternate-pinnate, lflets lanceolate to widely ovate, serrate or dentate, rarely entire; rays 7–14, unequal . . . . *A. kingii*

A. kingii (S. Watson) J. Coulter & Rose (p. 91) Pl 3–20 dm, glabrous to roughened. **LF** 1.5–4 dm, oblong, 1-ternate-pinnate (rarely ± 2-pinnate); lflets 3–12 cm, ± lanceolate, acute to acuminate, entire to sparsely serrate. **INFL** roughened; bracts, bractlets 0; rays 7–14, 0.5–10 cm, unequal, ascending; rays, pedicels webbed at base. **FL**: petals hairy; ovary minutely bristly. **FR** 4–5 mm, oblong. $2n=44$. Subalpine streambanks; 1900–3000 m. **W&I (White Mtns)**; to UT. ❀WET:1. Jun–Aug

A. lineariloba A. Gray (p. 91) Pl 5–15 dm, nearly glabrous to scabrous. **LF** 1–3.5 dm, triangular-ovate, 2–3-ternate-pinnately dissected; segments 2–10 cm, linear to linear-oblong, acute, entire. **INFL** scabrous; bracts, bractlets 0; rays 20–40, subequal; rays, pedicels not webbed at base. **FL**: petals, ovary roughened to becoming glabrous. **FR** 10–13 mm, oblong to wedge-shaped. Rocky open slopes; 2300–3000 m. c&s SNH, SNE; NV. [var. *culbertsonii* Jepson] ❀TRY. Jun–Aug

APIUM

Ann, bien, per, taprooted or fibrous-rooted from horizontal rhizome. **ST** prostrate to erect, hollow, rooting from lower nodes or not, glabrous. **LF**: blade oblong to obovate, 1-pinnate or ternate-pinnately dissected, lflets paired, lanceolate or ± round (segments linear). **INFL**: umbels compound, peduncled or not; bracts, bractlets conspicuous to 0; rays, pedicels few, spreading-ascending. **FL**: calyx lobes 0 or minute; petals wide, white or greenish white; projection atop ovary sometimes flat. **FR** ovate-oblong to round, compressed side-to-side; ribs subequal, thread-like to obtuse and corky; oil tubes per rib-interval 1; fr axis entire or notched at tip. **SEED**: face flat. $2n=22$. ± 20 spp.: gen s hemisphere (S.Am, s Afr, Australia, New Zealand), also Eurasia, 2 spp. weedy in CA. (Classical name for celery)

A. graveolens L. CELERY   Pl 5–15 dm. **ST** not rooting at nodes. **LF**: petiole 0.3–2.5 cm; blade 7–18 cm, oblong to obovate, lflets 2–4.5 cm, ovate to ± round, gen lobed. **INFL**: rays 7–16, 0.7–2.5 cm, subequal; calyx lobes minute. **FR** 1.5–2 mm diam, elliptic to nearly round; fr axis notched at tip. Wet places; gen < 1000 m. **CA**; temp zones worldwide; native to Eurasia. Cult and naturalized widely. May–Jul

BERULA   CUTLEAF WATER-PARSNIP

1 sp. (Latin: water-cress)

B. erecta (Hudson) Cov. (p. 91)   Per, < 15 dm, stoloned, glabrous. **ST** ascending or erect, branched, hollow, rooting at lower nodes. **LF**: petiole 4–12 cm, narrowly sheathing; blade 1.5 dm, oblong, 1-pinnate, much-dissected if submersed, lflets 7–12 pairs, 1–8 cm, oblong to ovate, sessile, serrate or lobed. **INFL**: umbels compound, peduncled, terminal or opposite lvs; bracts, bractlets lf-like, lanceolate; rays, pedicels ± unequal, spreading. **FL**: calyx lobes minute, persistent; petals wide, white. **FR** 1.5–2 mm, nearly round, compressed side-to-side, glabrous; ribs thread-like, inconspicuous in corky fr wall; oil tubes apparent but deeply embedded in fr wall; fr axis divided to base, its branches adhering to and falling with fr-halves. $2n=18$. Marshy areas, streams; < 1800 m. **CA**; Eurasia, Afr. Possibly TOXIC: implicated in some livestock poisonings. Jul–Oct

BUPLEURUM

Ann, bien, per, shrub, taprooted. **ST** decumbent to erect, branched, gen glaucous. **LVS**: basal petioled; cauline gen sessile, clasping or fused around st at base; blades linear to ovate or obovate, simple, gen parallel-veined, margins entire or minutely serrate. **INFL**: umbels compound; bracts gen present; bractlets gen wide, conspicuous; rays, pedicels few–many, spreading. **FL**: calyx lobes 0; petals narrow to wide, yellow or dark purple, tips narrowed. **FR** oblong to round, slightly compressed side-to-side; ribs subequal, thread-like to narrowly winged; oil tubes per rib-interval gen several; fr axis divided to base. **SEED**: face ± flat. ± 100 spp.: Eurasia, s Afr, n N.Am, 1 reportedly naturalized in CA. (Greek: ox rib)

B. lancifolium Hornem.   Pl 0.5–5 dm, glaucous. **LF**: basal 3–10 cm, linear to oblong-lanceolate; cauline lanceolate to widely ovate, base fused around st, margin scarious. **INFL**: peduncles 1–8 cm; bracts 0; bractlets gen 5, > fls and frs, lf-like, ovate to ± round, fused at base; rays 2–3. **FL**: corolla yellow. **FR** 2–3 mm, ovate to round, roughened. $2n=16$. May be naturalized in gardens; gen < 1000 m. **CA**; e US; native to Medit. Reported as *B. subovatum* Link. May–Jun

CICLOSPERMUM

Ann, taprooted, glabrous. **ST** decumbent to erect, gen loosely branched. **LVS**: sessile or petioled; blade oblong to triangular-ovate, pinnately or ternate-pinnately dissected or compound, segments or lflets gen thread-like to linear. **INFL**: umbels compound or some simple; bracts, bractlets gen 0; rays few; rays, pedicels spreading. **FL**: calyx lobes 0; petals oblong, white, tips acutish but not incurved. **FR** narrowly ovate to elliptic, compressed side-to-side, glabrous or hairy; ribs subequal, thread-like to prominent and corky; oil tubes per rib-interval 1; fr axis shallowly notched. **SEED**: face flat. 3 spp.: S.Am, 1 ± worldwide aggressive weed. (Greek: circular seed)

C. leptophyllum (Pers.) Britton & E. Wilson (p. 91)   Pl 0.5–6 dm. **LF** 3.5–10 cm; petiole 2.5–12 cm, gen 0 on cauline lvs, sheath margin scarious; segments 3–15 mm, thread-like to linear, entire. **INFL** < 2 cm, gen some sessile; rays 1–3; pedicels 6–20, 2–16 mm, spreading. **FR** 1.2–3 mm wide, elliptic to ovate. $2n=14$. Lawns, roadsides; < 150 m. **CA**; ± worldwide, warm temp. Sporadic in CA. [*Apium l.* (Pers.) Benth.] Apr–Aug

CICUTA   WATER HEMLOCK

Per, glabrous; rhizome divided internally into chambers, with sap that oxidizes to reddish brown, bearing fibrous or tuberous roots. **ST** erect, hollow. **LF**: blade oblong to triangular-ovate, 1–3-pinnate or ternate-pinnate, lflets linear to ovate-lanceolate, serrate or irregularly cut. **INFL**: umbels compound; bracts gen 0; bractlets gen inconspicuous; rays, pedicels many, spreading. **FL**: calyx lobes minute; petals wide, white, tips narrowed. **FR** ovate to round, slightly compressed side-to-side; ribs low, corky, sometimes unequally spaced; oil tubes per rib-interval 1; fr axis divided to base. **SEED**: face flat or concave. ± 4 spp.: Eurasia, N.Am. (Ancient Latin name) [Mulligan & Munro 1981 Canad J Plant Sci 61: 93–105] More evidence from ripe fr and chromosomes needed to substantiate proposed cryptic spp. TOXIC: both spp. below contain cicutoxin, a virulent poison; many livestock and human deaths recorded. Most lethally toxic native plants.

1. Fr gen round, ribs much wider than intervals between; areas surrounded by veins on lf lower surface coarse, gen some elongate .... *C. douglasii*
1' Fr gen ovate, ribs gen as wide as to much narrower than intervals between; areas surrounded by veins on lf lower surface fine, gen rounded or square .... *C. maculata* var. *angustifolia*

C. douglasii (DC.) J. Coulter & Rose (p. 91) Pl 15–30 dm. **LF** 1.5–4.5 dm, narrowly ovate to triangular-ovate, 1–2(3)-pinnate; lflets 1–10(15) cm, linear to widely lanceolate, acute or acuminate, subentire to coarsely serrate, areas surrounded by veins on lower surface coarse, gen some elongate. **INFL**: umbels compound, terminal and lateral; peduncles 2–18 cm; rays 15–30(35), 2–8 cm; pedicels 20–30, 2–10 mm. **FR** 2–4 mm, gen round; ribs much wider than intervals between. 2*n*=44. Wet places, often in water; < 2500 m. NCo, CaRH, SNH, CCo, SCo, **GB**; to B.C., MT. Jun–Sep

C. maculata L. Pl 10–15 dm. **LF** 1–4 dm, ovate to triangular-ovate, 1–2-pinnate; lflets 2–10 cm, lanceolate, acute or acuminate, coarsely to sparsely serrate, areas surrounded by veins on lower surface fine, gen rounded or square. **INFL**: umbels compound, terminal and lateral; peduncles 2.5–12 cm; rays 15–30, 2–4.5 cm; pedicels 15–30, 2–10 mm. **FR** 3–4 mm, gen ovate; ribs gen as wide as to much narrower than intervals between. 2*n*=22. Wet places; < 2000 m. CA-FP, **GB**; to AK, c Mex. 2 vars in CA.

var. *angustifolia* Hook. **LF** gen 1–2-pinnate. **FL**: styles < 1 mm. Wet meadows, etc.; 1500–2100 m. SnBr, **GB**. [*C. occidentalis* E. Greene, *C. valida* E. Greene] Jun–Sep

CYMOPTERUS

Per, taprooted, gen glabrous. **ST** gen 0 or short. **LVS** mostly basal, membranous to subleathery or fleshy; blade oblong to widely ovate or round, palmately or pinnately lobed to 1–2-pinnately or -ternate-pinnately dissected or compound, segments or lflets linear to obovate, entire to variously lobed, gen spine-tipped. **INFL**: umbels compound, gen terminal, scapose, open to spheric, dense, peduncled; bracts, bractlets conspicuous and scarious (or rarely 0); rays few–many (rays and pedicels sometimes ± 0). **FL**: calyx lobes prominent to 0; petals oblong to obovate, white, yellow, or purple, tips narrowed; projection atop ovary 0. **FR** oblong to ovate, subcylindric to compressed front-to-back; ribs subequal or unequal, marginal and some or all others thin-or corky-winged, or rarely some or all wingless; oil tubes per rib-interval 1–several; fr axis 0 or divided to base. **SEED**: face flat to longitudinally concave or grooved. ± 50 spp.: w N.Am. (Greek: wave wing) [Mathias 1930 Ann Missouri Bot Gard 17:213–476] Generic boundaries fluctuating. Some spp. outside CA are TOXIC to livestock.

1. Rays ± 0 (umbels dense, spheric); bractlets 0 or poorly developed
2. Lf minutely hairy or roughened; bracts conspicuous, fused below; fr glabrous
.... *C. cinerarius*
2' Lf glabrous; bracts 0; fr hairy
3. Lf round, ternate .... *C. ripleyi*
3' Lf oblong-ovate, ternate–2-pinnately or 2-pinnately dissected
4. Pl sessile on taproot; fr wings unequal .... *C. deserticola*
4' Pl with lfless stalk above taproot; fr wings subequal .... *C. globosus*
1' Rays ± developed (umbels open to ± dense, not spheric); bractlets evident
5. Pl not woody; corolla purplish; lf simple to 2-pinnately dissected or compound, lobes, segments, or lflets wider than linear
6. Bracts 0; lf round-reniform, ternate .... *C. gilmanii*
6' Bracts conspicuous, scarious, veiny; lf oblong-ovate, 1–2-pinnate or -ternate
7. Bracts and bractlets greenish or purplish, with many green or purple veins; pedicals < 1 mm
.... *C. multinervatus*
7' Bracts and bractlets white, with 1–5 green or white veins; pedicels 3–8 mm .... *C. purpurascens*
5' Pl woody at base; corolla ± yellow or white; lf finely dissected, segments gen ± linear
8. Lf finely hairy; corolla white; umbels ± dense .... *C. aboriginum*
8' Lf glabrous; corolla ± yellow; umbels open
9. Rays gen unequal; calyx lobes persistent; bractlets gen < mature pedicels .... *C. terebinthinus*
10. Lf blade ± as long as wide .... var. *californicus*
10' Lf blade much longer than wide .... var. *petraeus*
9' Rays subequal; calyx lobes 0; bractlets > mature pedicels .... *C. panamintensis*
11. Lf segments not crowded, 3–20 mm, flexible .... var. *acutifolius*
11' Lf segments ± crowded, 1–5 mm, rigid .... var. *panamintensis*

C. aboriginum M.E. Jones Pl 1–3.5 cm, finely hairy or scabrous, gray-green. **ST** 0. **LF**: petiole 2–13 cm; blade 1–4.5 cm, oblong, ternate–2-pinnately or 3-ternately dissected, segments 2–8 mm, linear. **INFL**: peduncles 8–30 cm, = or > lvs; bracts gen 0; bractlets linear, subscarious, acute; rays 3–10, 4–20 mm; pedicels 3–7 mm. **FL**: corolla white. **FR** 6–11 mm, oblong to ovate; ribs subequal, wings 2 × body in width; oil tubes per rib-interval 2–8. 2*n*=22. Rocky slopes; 1500–2500 m. **DMtns**, **W&I**; NV. Apr–Jun

C. cinerarius A. Gray Pl 7–8 cm, minutely hairy or rough-

ened, glaucous. **ST** 0. **LF**: petiole 3–5 cm; blade 1–2.5 cm, oblong-ovate, 2-pinnately dissected, segments 1–3 mm. **INFL**: peduncles > lvs; bracts conspicuous, scarious-margined, fused below; bractlets obscure; rays, pedicels ± 0. **FL**: corolla white. **FR** 6 mm, narrowly wedge-shaped; ribs subequal, wings < body in width; oil tubes per rib-interval 5–8. Rocky mtn slopes; 2500–3500 m. SNH, **SNE**; NV. Jun–Jul

C. deserticola Brandegee (p. 91) DESERT CYMOPTERUS Pl ± 15 cm, glabrous. **ST** 0. **LF**: petiole 4–10 cm; blade 2–6.5 cm, oblong-ovate, ternate–2-pinnately dissected, segments 1–4 cm, lanceolate, acuminate. **INFL**: peduncles = or > lvs; bracts, gen bractlets 0; rays, pedicels ± 0. **FL**: corolla purple. **FR** 5–7 mm, oblong-ovate to wedge-shaped; ribs unequal, marginal wings < body in width, others inconspicuous; oil tubes per rib-interval 3–5. 2*n*=22. RARE. Sandy flats, slopes; ± 1500 m. **w DMoj**. Apr

C. gilmanii C. Morton GILMAN'S CYMOPTERUS **ST** 12–23 cm, glabrous, glaucous; base fibrous. **LF**: petiole 8–18 cm; blade 2.5–4.5 cm, round-reniform, ternate, lflets obovate, deeply lobed, ultimate lobes toothed, spine-tipped. **INFL**: peduncles > lvs; bracts 0; bractlets inconspicuous, lf-like, linear to lanceolate; rays ± 8, 1–2 cm; pedicels 2–5 mm. **FL**: corolla purplish. **FR** 7–8 mm, widely ovate; ribs unequal, marginal and 1 or 2 other wings > body in width. RARE in CA. Limestone, gypsum slopes; 1000–2000 m. **ne DMtns**; NV. Apr–May

C. globosus (S. Watson) S. Watson Pl 3–20 cm, with lfless stalk above taproot, glabrous, glaucous. **ST** (above lfless stalk) 0. **LF**: petiole 1–10 cm; blade 1–7 cm, oblong-ovate, ternate–2-pinnately or 2-pinnately dissected, segments 0.5–6 mm, ± indistinct. **INFL**: peduncles = or > lvs; bracts 0; bractlets linear, small; rays, pedicels ± 0. **FL**: corolla white or purple. **FR** 6–11 mm, narrowly wedge-shaped; ribs subequal, wings < body in width; oil tubes per rib-interval gen 1. 2*n*=22. Uncommon. Sandy, open flats; 1200–2100 m. **SNE**; NV, UT. Mar–May

C. multinervatus (J. Coulter & Rose) Tidestrom Pl 4–20 cm, with lfless stalk above taproot or not, glabrous. **ST** (above lfless stalk) 0. **LF**: petiole 2–7 cm; blade 1–8.5 cm, oblong-ovate, 2-pinnately or ternate–pinnately dissected, glaucous, fleshy, segments 0.5–6 mm, ± indistinct. **INFL**: peduncles = or > lvs, 2–14 cm; bracts greenish or purplish, scarious, many-veined, forming a shallow sheath or cup; bractlets like bracts; fertile rays 1–5, 0.5–2.5 cm; pedicels < 1 mm. **FL**: corolla purplish. **FR** 8–17 mm wide, oblong-ovate to ovate; ribs subequal, wings 2–3 × body in width; oil tubes per rib-interval 3–9. Uncommon. Sandy and rocky (limestone?) slopes; 630–1500 m. **DMoj**; to UT, NM. Mar–Apr

C. panamintensis J. Coulter & Rose Pl 0.5–4 dm, glabrous; base woody. **ST** 0. **LF**: petiole 1–10 cm; blade 1–14 cm, oblong-ovate to obovate, ternate–2–3-pinnately dissected, segments 1–5 mm, ± linear, acute, gen ± distinct. **INFL**: peduncles > lvs, 3–25 cm; bracts 0; bractlets linear, acuminate, fused at base; fertile rays 5–15, 1–6.5 cm, subequal; pedicels 4–13 mm. **FL**: corolla greenish yellow. **FR** 6–10 mm, oblong-ovate; ribs subequal, wings = or > body in width; oil tubes per rib-interval 1–5. Uncommon. Rocky slopes, canyon walls; 700–2500 m. **DMoj**.

var. ***acutifolius*** (J. Coulter & Rose) Munz **LF**: segments not crowded, 3–20 mm, flexible. 2*n*=22. Rocky canyon walls;

700–1000 m. **DMoj**. Mar–Apr

var. ***panamintensis*** **LF**: segments ± crowded, 1–5 mm, rigid. Rocky slopes; 800–2500 m. **DMtns**. Mar–May

C. purpurascens (A. Gray) M.E. Jones Pl 3–15 cm, with lfless stalk above taproot or not, with persistent fibers, glabrous. **ST** (above lfless stalk) 0–15 cm. **LF**: petiole 1–4 cm; blade 1.2–5 cm, oblong-ovate, 1–2-pinnately (rarely ternate–pinnately) dissected, glaucous, fleshy, segments 1–8 mm, ± indistinct. **INFL**: peduncles 1.5–7 cm, = or > lvs; bracts white, with 1–5 green or white veins, fused below; bractlets like bracts; fertile rays 3–5, 4–10 mm; pedicels 3–8 mm. **FL**: corolla purplish. **FR** 8–18 mm wide, widely ovate; ribs subequal, wings 2–3 × body in width; oil tubes per rib-interval 3–4. Shrubby slopes; 1300–2200 m. **e DMoj, W&I**; to ID, NV, AZ. ❀TRY. Mar–May

C. ripleyi Barneby RIPLEY'S CYMOPTERUS Pl 10–15 cm, glabrous. **ST** 0. **LF**: petiole 3–10 cm; blade 2–5 cm, round, ternate, lflets wedge-shaped, deeply 3-lobed, lobes again lobed. **INFL**: peduncles > lvs; bracts 0; bractlets small, chaffy; rays, pedicels ± 0. **FL**: corolla purple (rarely white). **FR** 6–7 mm, wedge-shaped to obovate, hairy; ribs unequal, marginal winged, others not; oil tubes minute. 2*n*=22. RARE in CA. Sandy soil; 1000–1600 m. **s SNH, s SNE, n DMtns**; NV. [var. *saniculoides* Barneby]

C. terebinthinus (Hook.) M.E. Jones Pl 1.5–4.5 dm, gray-green, glabrous; base woody. **ST** 0 to very short. **LF**: petiole 2–16 cm; blade 1.5–18 cm, ± ovate, pinnately or ternate–pinnately dissected, segments 1–4 mm, linear, ± rigid, acute. **INFL**: peduncles 1–3.5 cm, gen < lvs; bracts 0; bractlets 2–6 mm, gen linear, acute; rays 3–24, 0.5–8 cm, gen unequal; pedicels 1–8 mm. **FL**: corolla yellow. **FR** 5–10 mm wide, ± ovate; ribs gen subequal, wings often irregularly curled, = or > body in width; oil tubes per rib-interval 3–12. Rocky or sandy slopes; 150–3500 m. KR, CaRH, SNH, **SNE**; WA, OR, to Rocky Mtns. [*Pteryxia t.* (Hook.) J. Coulter & Rose]

var. ***californicus*** (J. Coulter & Rose) Jepson (p. 97) Pl herbage gray-green to bright green. **ST** 0 to very short. **LF**: blades ± as long as wide, appearing ± full from relatively large number and size of segments. 2*n*=22. Sand, rocks; 150–3500 m (lowest on serpentine). KR, CaRH, SNH, **n SNE**; n NV. [*Pteryxia t.* var. *c.* (J. Coulter & Rose) Mathias] May–Jun

var. ***petraeus*** (M.E. Jones) Goodrich Pl herbage gray-green. **ST** very short. **LF**: blades much longer than wide, appearing skeleton-like from relatively small number and size of segments. 2*n*=22. Rocky alpine slopes; 1800–3400 m. **c SNE, W&I**; NV, UT. [*Pteryxia p.* (M.E. Jones) J. Coulter & Rose] May–Jun

DAUCUS

Ann, bien, taprooted, hairy. **ST** decumbent or erect, gen ± branched. **LF**: blade oblong, pinnately dissected, segments linear to lanceolate. **INFL**: umbels compound; bracts, bractlets gen present; bracts conspicuous, gen pinnately lobed; bractlets entire to toothed; rays gen many, spreading, in fr incurving to form a nest-like umbel. **FLS**: outer sometimes ± bilateral; calyx lobes 0 or evident; petals wide, white, tips narrowed, unequally 2-lobed. **FR** oblong to ovate, compressed front-to-back; ribs 10, 1° thread-like and bristly, 2° winged and prickly; oil tubes 1 beneath each 2° ribs; fr axis entire or notched at tip. ± 20 spp.: Am, Eurasia, n Afr, Australia. (Greek: carrot) [Sáenz Laín 1980 Ann Jard Bot Madrid 37:481–533]

D. pusillus Michaux (p. 97) Pl 0.3–9 dm, gen simple or few-branched. **LF**: petiole 4–15 cm; blade 3–10.5 cm, segments 1–5 mm, linear, acute, entire, ± bristly. **INFL**: peduncles 1–4.5 cm, bristles reflexed to spreading; rays 0.4–4 cm; pedicels 2–9 mm. **FR** 3–5 mm, oblong. *n*=22. Rocky or sandy places; 0–1500 m. CA-FP (esp coastal), **DMtns**; to B.C., se US, S.Am. Apr–Jun

FOENICULUM

1 sp. (Latin: fennel)

F. vulgare Miller (p. 97) FENNEL Per, taprooted, 0.9–2 m, glabrous, glaucous, anise- or licorice-scented. **ST** erect, branched, solid. **LF**: petiole 7–14 cm, conspicuously sheathing; blade 3–4 dm wide, triangular-ovate, pinnately finely dissected, segments

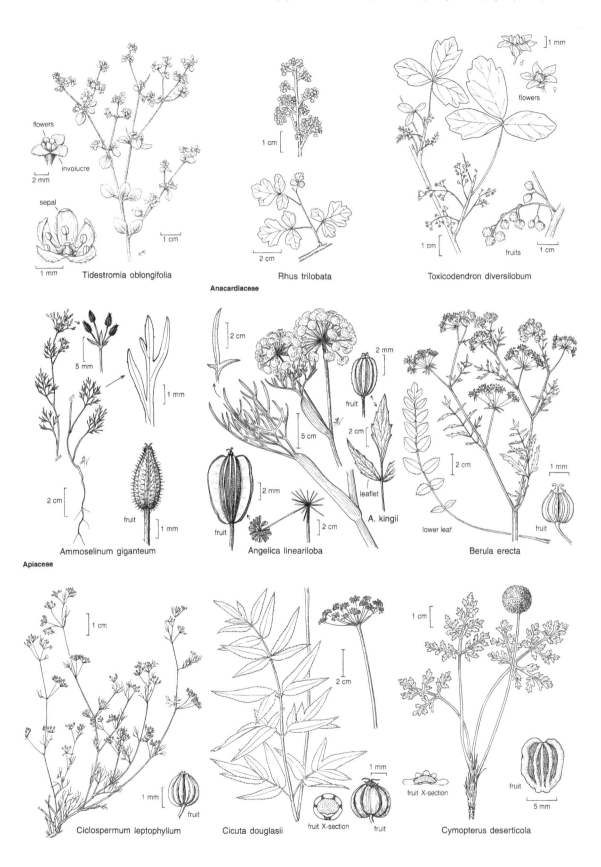

Tidestromia oblongifolia

Rhus trilobata

Anacardiaceae

Toxicodendron diversilobum

Ammoselinum giganteum

Apiaceae

Angelica lineariloba

A. kingii

Berula erecta

Ciclospermum leptophyllum

Cicuta douglasii

Cymopterus deserticola

4–40 mm, thread-like. **INFL**: umbels compound, peduncled; bracts, bractlets 0; rays 15–40, unequal, 1–4 cm, spreading-ascending to ascending; pedicels 18–25, 1–10 mm, subequal. **FL**: calyx lobes 0; petals wide, yellow, tips narrowed. **FR** 3.5–4 mm, oblong-ovate, compressed side-to-side, glabrous; ribs subequal, prominent, acute; oil tubes per rib-interval 1; fr axis divided to base. **SEED**: face gen flat. 2*n*=22. Roadsides, waste places; 0–350 m. CA-FP, **W&I** ; native to s Eur; widely escaped from cult in w hemisphere. Locally abundant and invasive. May–Sep

HERACLEUM

Per from taproot or clustered roots. **ST** stout, erect, gen branched, hollow. **LVS**: blades oblong to round, simple to ternately, pinnately, or palmately compound (rarely simple), lflets large, lobed or serrate; uppermost cauline often reduced to enlarged sheaths. **INFL**: umbels compound, large, often sterile at margins; bracts 0–few, often deciduous; bractlets gen present, persistent; rays, pedicels many, spreading-ascending. **FLS**: marginal bilateral, with outer petals > others, 2-lobed; calyx lobes gen 0; petals wide, white, yellowish, or rosy. **FR** oblong-ovate to round or obcordate, very compressed front-to-back; ribs unequal, marginal thin-winged, veined near outer margin, others thread-like; oil tubes per rib-interval 1–2, unequal in length; fr axis divided to base. **SEED**: face flat. ± 80 spp.: Eurasia, e Afr, 1 in N.Am. (Hercules, presumably from large stature of some spp.)

H. lanatum Michaux (p. 97) COW PARSNIP Pl 1–3 m, stout, strong-scented, tomentose. **LF** round to reniform; petioles 1–4 dm, widely sheathing, upper sheaths enlarged, bladeless; blades 2–5 dm wide, ternate, lflets 1–4 dm wide, ovate to round, cordate, coarsely serrate and lobed. **INFL** tomentose or long-hairy; peduncle 5–20 cm; rays 15–30, 5–10 cm, unequal; pedicels 8–20 mm. **FL**: petals obovate, white. **FR** 8–12 mm, obovate to obcordate, ± hairy. 2*n*= 22. Moist places, wooded or open; < 2600 m. CA-FP, **GB**; to AK, e US, AZ. Relationship to some Eurasian taxa unclear; only native sp. in family occurring on both coasts of N.Am. ❀**4,5**,IRR:**6,15–17,24**&SHD:**7**,14,18–23. Apr–Jul

HYDROCOTYLE

Per, creeping or sprawling, glabrous or hairy; rhizomes or st rooting at nodes. **LVS** simple; petiole scarious-stipuled, not sheathing; blade ± round, peltate or not, entire to deeply lobed. **INFL**: umbels simple, open or dense; bracts 0 or inconspicuous; pedicels 0–many, spreading. **FL**: calyx lobes 0 or minute; petals obtuse or acute, greenish or yellowish white to purplish, tip not incurved. **FR** elliptic to round, very compressed side-to-side; ribs subequal, thread-like, distinct or not; oil tubes 0 (but individual oil cells in fr wall); fr central axis not an obvious structure. **SEED**: face flat to convex. ± 100 spp.: worldwide, esp s hemisphere. (Greek: water cup, apparently from lf shape)

H. verticillata Thunb. (p. 97) Pl low, glabrous. **ST** creeping. **LF**: petiole 0.5–25 cm, slender; blade 1–4 cm wide, round-peltate, crenate or very shallowly, crenately, subequally 8–13-lobed. **INFL**: umbels small, 1–15-fld, in a simple or forked spike; peduncles 1.5–20 cm; pedicels 0 or short. **FR** 1–3 mm, elliptic; ribs acute, evident. Swampy ground, lake margins; < 100 m. CA-FP, **W&I**; to e N.Am, S.Am, Hawaii, s Afr. [var. *triradiata* (A. Rich.) Fern.] ❀WET:**5**, 7,14,**15–24**;INV. Apr–Sep

LOMATIUM

Per from taproot or gen deep-seated tuber, glabrous to tomentose. **ST** 0 or erect, simple or branched; base fibrous (from old lf sheaths) or not. **LF**: blade oblong to triangular-ovate or obovate, ternately, pinnately, or ternate-pinnately dissected or compound, segments or lflets thread-like to wide. **INFL**: umbels compound, peduncled; bracts gen 0; bractlets gen present, 0 to conspicuous; rays, pedicels spreading to erect, often webbed at base. **FL**: calyx lobes gen 0; petals wide, yellow, white, or purple, tips narrowed; projection atop ovary 0. **FR** linear to obovate, very compressed front-to-back; marginal ribs widely to narrowly thin or thick-winged, others thread-like; oil tubes per rib-interval 1–several; fr axis divided to base. **SEED**: face flat to concave. ± 75 spp.: c&s N.Am. (Greek: bordered, from prominent marginal fr wing) [Schlessman 1984 Syst Bot Monogr 4:1–55] Fr wing width expressed as width of 1 wing, not both together.

1. Fls white to cream or red to purple; anthers gen purple
  2. Pl glabrous or nearly so . . . . ***L. nevadense*** var. ***parishii***
  2' Pl densely grayish, finely hairy
    3. Fls white to cream; sts, if present, neither thickened nor fibrous . . . . ***L. nevadense*** var. ***nevadense***
    3' Fls red to purple; sts, if present, gen thickened, fibrous . . . . ²***L. mohavense***
1' Fls yellow, sometimes drying white; anthers gen yellow
  4. Pl glabrous, gen glaucous
    5. Lflets ovate, sharply lobed; pedicels 5–10 mm, webbed . . . . ***L. rigidum***
    5' Lflets linear, entire; pedicels 10–17 mm, not webbed . . . . ***L. parryi***
  4' Pl variously pubescent, gen not glaucous
    6. Sts, st-lvs prominent
      7. Lf divisions short and crowded; lf sheaths and bracts conspicuously scarious-dilated . . . . ***L. utriculatum***
      7' Lf divisions elongate and distinct; lf sheaths and bracts narrow and inconspicuous
        . . . . ***L. triternatum*** var. ***macrocarpum***
    6' St 0–very short, with 0–few lvs
      8. Pl base fibrous; fr wings = to > body . . . . ²***L. mohavense***
      8' Pl base not fibrous; fr wings < body

9. Herbage ± fleshy; rays slightly webbed . . . . *L. plummerae*
9' Herbage not fleshy; rays not webbed
   10. Fr 12–16 mm, glabrous, wings << body in width . . . . *L. dissectum* var. *multifidum*
   10' Fr 4–12 mm, gen hairy, wings gen 1/2 body in width . . . . *L. foeniculaceum*
      11. Petals glabrous; fertile rays gen 1 . . . . ssp. *inyoense*
      11' Petals minutely ciliate; fertile rays 2–14 . . . . ssp. *fimbriatum*

L. dissectum (Torrey & A. Gray) Mathias & Constance Pl 3–14 dm, glabrous to puberulent or minutely scabrous, ± glaucous; taproot stout, thickened. **ST** rarely 0; base with 1 or more scarious sheaths. **LF**: petiole 3–30 cm; blade 15–35 cm wide, triangular-ovate to round, ternate-pinnately dissected, segments 2–22 mm, linear-oblong; cauline lvs gen few, like basal. **INFL** glabrous; peduncle 1.5–6 dm; bractlets several, > or < fls, linear; rays 10–30, 3–10 cm, spreading; pedicels 1–20 mm. **FL**: corolla maroon-red or yellow. **FR** 12–16 mm, oblong-ovate to elliptic, glabrous, pumpkin-seed-like; wings thick, << body in width; oil tubes obscure. Wooded or shrubby slopes; 150–3000 m. CA-FP, **GB**; to w Can, Baja CA. 2 vars in CA.

var. *multifidum* (Torrey & A. Gray) Mathias & Constance **LF**: segments 2–22 mm, 0.5–2 mm wide. **FL**: corolla yellow. **FR**: pedicels gen 5–15 mm. 2*n*=22. Wooded or shrubby slopes, often coniferous forest; 600–3000 m. KR, CaRH, SNH, Teh, SCo, SnGb, SnBr, **GB**; to w Can, Baja CA. ❀TRY. May–Jul

L. foeniculaceum (Nutt.) J. Coulter & Rose Pl 0.3–3 dm, taprooted, densely puberulent or soft-hairy to tomentose. **ST** 0. **LF**: petiole 1–15 cm, < blade, wholly sheathing; blade gen 2.5–18 cm wide, oblong to obovate, pinnately or ternate-pinnately dissected, segments 1–7 mm, linear to obovate, pointed. **INFL** hairy to nearly glabrous; peduncle 0.3–3 dm; bractlets fused or not, < to = fls, linear to linear-lanceolate, entire or lobed, acute; rays 1–30, gen 1–8 cm, spreading-ascending or spreading; pedicels 1–15 mm. **FL**: corolla yellow, rarely purplish. **FR** 4–12 mm, oblong-ovate, gen hairy; wings gen 1/2 body in width; oil tubes per rib-interval 1–7. Sagebrush, pine woodland, open summits, subalpine scrub; 1600–3300 m. **SNE, DMtns (Last Chance Mtns)**; to B.C., UT, TX.

ssp. *fimbriatum* Theob. Pl 7–30 cm. **LF**: segments 1–5 mm, linear. **INFL**: peduncle 10–30 cm; rays 2–14, 0.5–6 cm. **FL**: petals yellow or purplish, minutely ciliate. 2*n*=22. Sagebrush, pine woodland; 1600–3300 m. **SNE, DMtns (Last Chance Mtns)**; to UT. Apr–Jun

ssp. *inyoense* (Mathias & Constance) Theob. INYO LOMATIUM Pl 3–12 cm. **LF**: segments 1–2 mm, linear to obovate. **INFL**: peduncle 3–12 cm; rays gen 1, < 5 cm. **FL**: petals pale yellow, glabrous. UNCOMMON. Open summits, subalpine scrub; < 3000 m. **W&I (Inyo Mtns)**; possibly to ID, NV. [*L. inyoense* Mathias & Constance] May be a form induced by high-elevation conditions. Jun–Jul

L. mohavense (J. Coulter & Rose) J. Coulter & Rose (p. 97, pl. 10) Pl 1–4 dm, densely grayish short-hairy; taproot long, often thickened. **ST** ± 0; base fibrous. **LF**: petiole 2–12 cm, > blade; blade 2–10 cm, oblong to ovate, 3–4-pinnately dissected, segments 2–5 mm, linear-oblong to obovate, ± crowded. **INFL** finely hairy; peduncle 8–22 cm; bractlets 8–12, 2–4 mm, linear to linear-lanceolate, acute, free or fused basally, scarious-margined, often obscured by hairs; rays 8–18, 1–5 cm, spreading-ascending, unequal; pedicels 1–10 mm. **FL**: petals yellow or purple, glabrous. **FR** 4.5–11 mm, ovate to round, ± glabrous to densely short-hairy; wings = to > body in width; oil tubes per rib-interval 1–4. 2*n*=22. Desert flats, slopes, summits, scrub, or woodland; 1000–2000 m. SCoR, WTR, **D**; Baja CA. Apr–May

L. nevadense (S. Watson) J. Coulter & Rose Pl 1–4.5 dm; taproot slender, sometimes swollen below; herbage grayish, gen finely hairy. **ST** 0 or short. **LF**: petiole 4–6 cm; blade 3.5–10 cm, oblong to obovate, gen 2–3-pinnately dissected, segments gen 2–3 mm, linear or oblong, pointed, often crowded; cauline lvs 0 or like basal. **INFL** finely hairy; peduncle 0.5–3 dm; bractlets 1–

10, linear and free to obovate and ± fused (involucel radial to 1-sided), conspicuously scarious or scarious-margined, gen glabrous or nearly so; rays 8–22, 1–2.5 cm, unequal, spreading; pedicels 3–10 mm. **FL**: corolla white to cream. **FR** 6–11 mm, oblong to round or obovate, densely hairy to glabrous; wings < to > body in width; oil tubes per rib-interval 1–9. 2*n*=22. Sagebrush scrub, woodland, desert scrub; 1000–3000 m. CaRH, SNH, SnGb, SnBr, **GB, DMtns**; to OR, UT, NM, n Mex.

var. *nevadense* **LF** 2–3-pinnately dissected; segments gen 2–3 mm. **FL**: ovary (often fr) densely finely hairy. Sagebrush scrub, woodland; 1500–3000 m. CaRH, n&c SNH, **GB, DMtns**; to OR, UT, AZ. Apr–Jul

var. *parishii* (J. Coulter & Rose) Jepson **LF** often 1–2-pinnately dissected; segments < 3 mm. **FL**: ovary (and fr) glabrous or nearly so or ± roughened. Sagebrush, desert scrub, pine woodland; 1000–3000 m. c&s SNH, SnGb, SnBr, **GB, DMtns**; to NV, NM. [var. *holopterum* Jepson; var. *pseudorientale* Munz] Apr–Jul

L. parryi (S. Watson) J.F. Macbr. Pl 2–4 dm, taprooted, glabrous, glaucous. **ST** ± 0; base fibrous. **LF**: petiole 6–10.5 cm, widely sheathing basally; blade 5–20 cm, narrowly oblong, 2–3-pinnate, lflets 2–9 mm, linear, entire, sharp-pointed. **INFL**: peduncle 1–2.5 dm, slightly swollen at top; bractlets 3–8, 3–6 mm, linear, acute, scarious, entire or lobed; rays 8–15, ascending to suberect, 2–4.5 cm, subequal; pedicels 10–17 mm. **FL**: corolla yellow. **FR** 9–15 mm, oblong to ± diamond-shaped; wings = or > body in width; oil tubes per rib-interval 2–3. Rocky slopes, gen in pinyon woodland; 1500–2500 m. **DMtns**; to UT, AZ. May–Jun

L. plummerae J. Coulter & Rose Pl 1.2–3.5 dm; taproot slender; herbage grayish, dull, ± fleshy; hairs dense, fine, soft to ± 0. **ST** short, lvs crowded at base. **LF**: petiole 3–6 cm, gen wholly scarious-sheathing; blade 5–10 cm, oblong to ovate, ternate-pinnately dissected, segments 3–7 mm, linear to oblong, obtuse or acutish; cauline lvs like basal. **INFL** glabrous or finely soft-hairy; peduncle 0.7–3 dm, spreading-ascending; bractlets 5–10, linear-lanceolate to obovate, gen ± fused into 1-sided, scarious, veiny, irregularly cut cup that is = or > fls; rays 10–25, 0.5–7.5 cm, unequal, spreading-ascending, slightly webbed; pedicels 3–8 mm. **FL**: corolla light yellow. **FR** 9–13 mm, oblong to oblong-ovate, glabrous; wings < body in width; oil tubes per rib-interval 1–several. Rocky places, sagebrush, pine woodland; 1500–2300 m. CaRH, n SNH, **GB**; w NV. [var. *austiniae* (J. Coulter & Rose) Mathias; var. *sonnei* (J. Coulter & Rose) Jepson] Extremely variable in hairiness. May–Jun

L. rigidum (M.E. Jones) Jepson STIFF LOMATIUM Pl 1.5–6 dm; taproot massive; herbage glabrous, green. **ST** ± 0; lvs clustered at base. **LF**: petiole 5–15 cm; blade 7–15 cm, oblong to ovate, ternate-pinnate or 2-pinnate, lflets 1–2 cm, ovate, pinnately sharply lobed; cauline lvs 0 or like basal. **INFL**: peduncle 1–5 dm; bractlets 5–8, 3–8 mm, lanceolate, acuminate, ± fused basally, reflexed; rays 10–16, 2.5–5 cm, spreading-ascending, webbed; pedicels 5–10 mm, webbed. **FL**: corolla yellow; calyx lobes < 1 mm, evident. **FR** 6–12 mm, oblong-ovate, glabrous; wings < body in width; oil tubes per rib-interval 3. UNCOMMON. Rocky slopes near streams, sagebrush scrub, pinyon/juniper woodland; 1200–2200 m. **SNE (Big Pine, Bishop creeks, Inyo Co.)**. Vegetatively like *Tauschia parishii*. Apr–May

L. triternatum (Pursh) J. Coulter & Rose Pl 1.5–10 dm, gen finely soft-hairy or puberulent; taproot slender to massive. **ST** prominent. **LF**: petiole 7–20 cm, sheathing ± to middle; blade

7–20 cm, oblong-ovate to triangular-ovate or obovate, 1–2-ternate-pinnate, lflets 1.5–20 cm, gen linear to widely lanceolate, gen entire; cauline lvs 0 or gen wholly sheathing. **INFL:** peduncle gen 1–4.5 dm, spreading to erect; bractlets (0)3–8, 1–5 mm, thread-like to linear-lanceolate, ± scarious; rays 5–20, 2–10 cm, spreading or spreading-ascending, unequal, webbed; pedicels 1–10 mm, webbed. **FL:** corolla yellow; ovary glabrous to densely puberulent. **FR** 6–22 mm, oblong, puberulent or glabrous; wings gen < body in width; oil tubes per rib-interval 1. Uncommon. Sagebrush/juniper, pine woodland or forest, open slopes, meadows, open serpentine ridges, scrub; 200–2000 m. KR, NCoR, **GB**; to w Can, WY. Variable in GB, adjacent areas. Var. *anomalum* (M.E. Jones) Mathias evidently not in CA. 2 vars in CA.

var. ***macrocarpum*** (J. Coulter & Rose) Mathias Pl often stout. **FL:** ovary densely puberulent. 2*n*=44. Sagebrush/juniper, pine woodland, open slopes, meadows; 200–1500 m. KR, NCoR, GB; to WA, ID, NV. [*L. alatum* J. Coulter & Rose, *L. giganteum* J.Coulter & Rose] Apr–Jul

L. utriculatum (Torrey & A. Gray) J. Coulter & Rose Pl 1–5 dm, glabrous to densely puberulent; taproot gen slender. **ST** ± leafy. **LF:** petiole 1.5–10 cm, widely sheathing; blade 5–16 cm, oblong to ovate, pinnately or ternate-pinnately dissected, segments 2–25 mm, gen linear; cauline lvs conspicuously wholly sheathing. **INFL:** peduncle 0.5–3 dm; bractlets 3–12, 3–6 mm, gen obovate or oblanceolate, ± scarious, veiny, gen entire; rays 5–20, 2–12 cm, spreading to ascending, unequal; pedicels 2–9 mm, webbed. **FL:** calyx lobes sometimes evident; corolla yellow; ovary puberulent or glabrous. **FR** 5–15 mm, oblong to obovate, glabrous or nearly so; wings thin, = or > body in width; oil tubes per rib-interval 1–4. 2*n*= 22. Open grassy slopes, meadows, woodland; 50–1550 m. CA-FP, **DMtns**; to B.C. [*L. vaseyi* J. Coulter & Rose] Variable. ✿ TRY. Feb–May

OSMORHIZA

Per, nearly glabrous to hairy; roots thick, clustered, licorice-scented. **ST** branched, leafy. **LF:** blade oblong to triangular-ovate, 2-pinnate or ternate-pinnate or 2–3-ternate, lflets lanceolate to round. **INFL:** umbels compound; bracts 0; bractlets 0–several and conspicuous; rays, pedicels few, spreading-ascending to spreading. **FL:** calyx lobes 0; petals obovate, white, purple, or greenish yellow (white), tips narrowed; disk sometimes present. **FR** linear to oblong, cylindric to club-shaped, slightly compressed side-to-side, bristly to glabrous; base obtuse or long-tapered into tail, tip tapered into beak or obtuse; ribs thread-like; oil tubes per rib-interval obscure; fr axis divided in upper 1/2. **SEED:** face concave or grooved. ± 10 spp.: Am, e&s Asia. (Greek: sweet root) [Lowry & Jones 1985 Ann Missouri Bot Gard 71:1128–1171]

O. occidentalis (Nutt.) Torrey (p. 97) Pl 4–12 dm, glabrous to sparsely fine-hairy. **LF:** petiole 5–25 cm; blade 1–2 dm, oblong to ovate, 2-pinnate, lflets 2–10 cm, oblong-lanceolate to ovate, serrate and gen irregularly cut or lobed. **INFL:** peduncle 6–20 cm; bractlets gen 0; rays 5–12, gen 3–8 cm, ascending to spreading-ascending; pedicels 3–8 mm. **FL:** corolla yellow; styles 0.8–1.4 mm; disk conspicuous. **FR** 12–22 mm, linear-fusiform, not long-tapered at base; tail 0; tip narrowed below; ribs (and intervals) glabrous. 2*n*=22. Coniferous forest, oak woodland; 350–3100 m. KR, NCoR, CaRH, n&c SNH, MP, **n SNE**; to w Can, Colorado. May–Jul

OXYPOLIS

Per, glabrous; tubers clustered. **ST** erect, gen branched. **LF:** blade linear or oblong to triangular-ovate, ternate, pinnate, or simple; main axis segmented, hollow, sometimes bladeless. **INFL:** umbels compound; involucre, involucel extremely variable; bracts, bractlets 0 or 1–several, ± inconspicuous; rays, pedicels few–many, spreading-ascending. **FL:** calyx lobes conspicuous or minute; petals wide, white or purple, tips narrowed. **FR** oblong to obovate, very compressed front-to-back; ribs unequal, marginal widely thin-winged (wings veined on back at inner margin), others thread-like; oil tubes per rib-interval 1; fr axis divided to base. ± 6 spp.: w&e US, Caribbean. (Greek: sharp white)

O. occidentalis J. Coulter & Rose (p. 97) Pl 6–15 dm. **LF:** petiole 1–5 dm; blade 12–30 cm, oblong, 1-pinnate, lflets 5–13, 3.5–9.5 cm, lanceolate to widely ovate, crenate or serrate to irregularly cut; cauline lvs have enlarged petiole, smaller lflets than basal. **INFL:** peduncle 6–30 cm; bracts 0–1(8), 5–20 mm, linear, scarious; bractlets like bracts, < 10 mm; rays 12–24, 2–8.5 cm; pedicels 3–15 mm. **FL:** calyx lobes conspicuous. **FR** 5–6 mm, oblong or ovate. 2*n*=36. Bogs, wet meadows, streamsides, often in coniferous forest; 1200–2600 m. CaRF, SN, c SnBr, **W&I**. Jul–Aug

PERIDERIDIA   YAMPAH

Per, glabrous, often glaucous; roots tuberous, single or clustered, or fibrous, clustered. **ST** erect, branched. **LF:** blade lanceolate to triangular-ovate, gen 1–2-ternate-pinnate or 1–2-pinnately or ternate-pinnately dissected, lflets or segments gen linear to linear-lanceolate. **INFL:** umbels compound; bracts 0–many, conspicuous and reflexed or not; bractlets several–many, narrow, ± scarious; rays, pedicels few–many, gen spreading-ascending; 2° umbels gen convex on top. **FL:** calyx lobes evident; petals gen obovate, white, tips narrowed. **FR** linear-oblong to round, slightly compressed side-to-side or not at all, glabrous; ribs subequal, thread-like to prominent, not winged; oil tubes per rib-interval 1–several; fr axis divided to base. **SEED:** face flat to grooved. ± 12 spp.: gen w Am. (Greek: around the neck, from involucre) [Chuang & Constance 1969 Univ Calif Publ Bot 55] Roots, basal lvs needed for identification. Key to species adapted by Margriet Wetherwax.

1. Basal lvs 1–2-ternate-pinnately dissected; bracts 8–12 — oil tubes per rib interval 2–3
. . . . ***P. bolanderi*** ssp. ***bolanderi***
1' Basal lvs 1–2-ternate or 1–2-pinnate with 1–3 pairs of 1° lflets; bracts 0–2
2. Rays in fr unequal, spreading-ascending; bractlets 7–10, 1–2 mm; oil tubes per rib interval 1 . . . . ***P. lemmonii***
2' Rays in fr subequal, ascending; bractlets 6–8, 3–4 mm; oil tubes per rib interval 3–4 . . ***P. parishii*** ssp. ***latifolia***

P. bolanderi (A. Gray) Nelson & J.F. Macbr. Pl 1.5–9 dm; roots tuberous, single or 2–3-clustered, 1–7 cm. **LF**: basal petiole 2–15 cm; basal blade 10–20 cm, ± ovate, gen 1–2-ternate-pinnately dissected, segments 0.5–6 cm, thread-like to oblong, gen lobed, toothed; cauline lvs ternate-pinnately dissected or 1-ternate. **INFL**: peduncle 2–20 cm; bracts 8–12, 3–12 mm, ± lanceolate, gen acuminate; bractlets 4–10, 3–9 mm, like bracts; rays 9–23, 1–2 cm, subequal, ascending or spreading-ascending; pedicels 2–5 mm; 2° umbels 18–30-fld. **FL**: petals 1-veined; styles 2 mm. **FR** 4–6 mm, oblong; ribs thread-like; oil tubes per rib-interval 2–3. Meadows, scrub, pine forest, blue-oak woodland, summer-dry clay soil; 600–2000 m. NW, SN, MP, **n SNE**; to WY, UT. 2 sspp. in CA.

  ssp. *bolanderi* (p. 97) Pl 1.5–8 dm, green. **LF**: segments 1–4 mm wide, oblong to thread-like, terminal 3–8 cm, lateral 0.5–3 cm. **INFL**: bracts, bractlets wider than linear-lanceolate, deciduous, wholly scarious, margins uneven; 2° umbels 25–30-fld. $2n=38$. Meadows, scrub, pine forest; 1000–2000 m. KR, CaRH, SNH, MP, **n SNE**; to OR, ID, WY, NV. ❀ DFCLT. Jun–Aug

P. lemmonii (J. Coulter & Rose) Chuang & Constance (p. 97) Pl 2.5–9 dm; roots tuberous, gen single, 1.5 cm. **LF**: basal petiole 4–15 cm; basal blade 10–30 cm, ± ovate, 1-ternate or 1–2-pinnate with 1–2 pairs of 1° lflets, 1° lflets 3–10 cm, linear-lanceolate, gen entire; cauline lvs 1-ternate. **INFL**: peduncle 3–20 cm; bracts gen 0; bractlets 7–10, 1–2 mm, linear-lanceolate; rays 10–14, 1–3.5 cm, unequal, spreading-ascending; pedicels 3–5 mm; 2° umbels 20–35-fld. **FL**: petals 1-veined; styles 1–1.5 mm. **FR** 3–4.5 mm, oblong, to ± round; ribs thread-like; oil tubes per rib-interval 1. $2n=36$. Open meadows, coniferous forest edges; 1000–2500 m. SN, **n SNE**; se OR, w NV. Jul–Aug

P. parishii (J. Coulter & Rose) Nelson & J.F. Macbr. Pl 1.5–9 dm, green; roots tuberous, single, 1–2.5 cm, fusiform. **LF**: basal petiole 3–10 cm; basal blade 10–20 cm, ± ovate, gen 1-ternate or 1-pinnate with 1–3 pairs of lflets, lflets 3–15 cm, ± lanceolate, entire; cauline lvs 1-ternate. **INFL**: peduncle 3–20 cm; bracts 0–2, bristle-like; bractlets 3–8, 2–4 mm, linear-lanceolate, scarious-margined; rays 5–20, 1–4.5 cm, subequal or unequal, ascending or spreading-ascending; pedicels 3–4 mm; 2° umbels 13–27-fld. **FL**: petals 1-veined; styles 1–1.5 mm. **FR** 2.5–5 mm wide, oblong to ± round; ribs thread-like; oil tubes per rib-interval 3–4. Moist meadows, open coniferous forests; 2000–3000 m. KR, CaR, SNH, TR, PR, **SNE**; to NV, NM. 2 sspp. in CA.

  ssp. *latifolia* (A. Gray) Chuang & Constance **INFL**: umbels flat or convex on top; rays gen 12–14, subequal; bractlets 6–8, ± = pedicels. **FL**: styles 1.5 mm. **FR** 2.5–3.5 mm, ovate to ± round. $2n=38$. Wet meadows, open coniferous forests; 2000–3000 m. KR, CaR, SNH, TR, PR, **SNE**; NV.

PODISTERA

Per, taprooted, low, fibrous at base, ± puberulent. **ST** 0. **LVS** basal; blades oblong to widely ovate, 1–2-pinnate, lflets linear to round. **INFL**: umbels compound, dense, head-like; bracts 0 or linear; involucel 1-sided; bractlets several, narrow or wide, gen partly fused; rays few, cylindric or flattened to winged; pedicels like rays. **FL**: calyx lobes conspicuous; petals wide, yellow, purplish, or white, tips narrowed. **FR** oblong-ovate to ovate, slightly compressed side-to-side, glabrous; ribs subequal, thread-like to prominently corky, obtuse; oil tubes per rib-interval 2–several; fr axis not seen. **SEED** compressed front-to-back; face flat to slightly concave. 4 spp.: mtns of w N.Am. (Greek: solid foot, from compact habit)

P. nevadensis (A. Gray) S. Watson (p. 101) SIERRA PODISTERA Pl forming compact cushions 2–5 cm, 2–5 dm diam. **LF**: petiole 3–15 mm, conspicuously white scarious-sheathing; blade 3–10 mm, oblong to ovate, 1-pinnate, lflets 1–6 mm, linear to lanceolate, entire, pointed. **INFL**: peduncle 5–30 mm; bracts 0; bractlets 2–4 mm, ± = fls and frs, ovate, strongly fused into a cup; rays winged, very short; pedicels 0–few, < fr. **FL**: corolla yellow. **FR** 4–4.5 mm, oblong-ovate; ribs thread-like; oil tubes per rib-interval 3–4. $2n=22$. UNCOMMON. Unglaciated granitic gravel, scree, crevices above timberline; 3000–4000 m. n&c SNH, SnBr, **W&I**. Jul–Sep

SIUM

Per, glabrous; roots clustered, fibrous or ± tuberous. **ST** erect or ascending, branched. **LF**: blade oblong to ovate, 1-pinnate, sometimes 2-pinnate when submerged, lflets distinct, serrate, irregularly cut, or pinnately lobed. **INFL**: umbels compound, gen opposite a lf; bracts, bractlets lf-like, often reflexed, conspicuous; rays, pedicels many, spreading-ascending. **FL** sometimes ± bilateral; calyx lobes 0 or minute; petals wide, white, narrowed at tips, outer slightly > others. **FR** ovate to round, slightly compressed side-to-side; ribs prominent, subequal, corky; oil tubes per rib-interval 1–3; fr axis entire or divided to base, adhering to fr halves or not. **SEED**: face flat. $2n=12$. ± 10 spp.: N.Am., Eurasia, Afr. (Greek: for some aquatic member of family)

S. suave Walter (p. 101) Pl 6–12 dm, stout. **LF**: petiole 1–8 dm, segmented; blade 6–25 cm, 7–18 cm wide, lflets 1–4 cm, linear or lanceolate, serrate or irregularly cut. **INFL**: peduncle 4–10 cm; bracts 6–10, 3–15 mm, linear or lanceolate, acute, entire or irregularly cut, reflexed; bractlets 4–8, 1–3 mm, linear-lanceolate; rays 10–20, 1.5–3 cm, slender, subequal; pedicels 3–5 mm. **FR** 2–3 mm wide. $2n=12$. Wet soil of swamps, marshes, streambanks; < 2000 m. SnFrB (Suisun Marshes), **GB**; to B.C., e N.Am, e Asia. ❀TRY. Jul–Aug

SPERMOLEPIS

Ann, taprooted, glabrous. **ST** gen spreading, branched. **LF**: blade oblong to ovate, ± ternate-pinnately dissected, segments thread-like to linear. **INFL**: umbels compound, terminal and lateral, peduncled or not; bracts 0; bractlets few, narrow; rays, pedicels few, suberect, gen ± spreading. **FL**: calyx lobes 0; petals oblong to ovate, white, tips not narrowed, not incurved. **FR** ovate, slightly compressed side-to-side, smooth, tubercled, or short-bristly; ribs low, thread-like; oil tubes per rib-interval 1–3; fr axis divided at tip. **SEED**: face grooved. 5 spp.: s US, Hawaii, s S.Am. (Greek: seed scale, from tubercled or bristly fr)

S. echinata (DC.) A.A. Heller (p. 101) Pl low, spreading, 5–40 cm. **LF** ovate; petiole 3–20 mm; blade 7–25 mm wide, segments 2–18 mm, thread-like. **INFL**: peduncles 1–5 cm; bractlets few, thread-like to linear, entire or toothed; rays 5–14, 1–15 mm,

gen ± ascending, very unequal; pedicels gen < 7 mm, those of central fl of each 2° umbel gen 0. **FR** 1.5–2 mm wide, widely ovate; ribs prominent, short-bristly. 2*n*=16,20. Very uncommon. Rocky slopes, sandy flats; 60–1500 m. **DSon (Borrego Valley)**; to se US, n Mex.

SPHENOSCIADIUM

1 sp. (Greek: wedge umbrella, from umbel)

S. capitellatum A. Gray (p. 101) SWAMP WHITE HEADS, RANGER'S BUTTONS Per, ± scabrous; root tuberous. **ST** erect, 5–18 dm, gen branched, leafy. **LF**: petiole 1–4 dm; blade 1–4 dm, oblong to ovate, 1–2-pinnate or ternate then pinnate, lflets 1–12 cm, gen ± lanceolate, acute, sparsely toothed to irregularly cut or pinnately lobed; cauline lf sheaths conspicuously enlarged. **INFL**: umbels compound, tomentose; peduncle 7–40 cm; bracts 0; bractlets many, linear, bristle-like; rays 4–18, 1.5–10 cm, ascending to reflexed; pedicels reduced to a disk; 2° umbels head-like, spheric.

FL: calyx lobes 0; petals obovate, white or purplish, tips narrowed; styles slender. **FR** 5–8 mm, wedge-shaped-obovate, very compressed front-to-back, tomentose; ribs unequally winged, marginal wider than others; oil tubes per rib-interval 1; fr axis divided to base. **SEED**: face ± flat. 2*n*=22. Wet meadows, streamsides, lakeshores; 900–3000 m. NCoRI, CaR, SNH, SW, **GB**; to OR, ID, NV, Baja CA. TOXIC to livestock, but rarely eaten. ✿WET:1,2,6,7,14–16; DFCLT. Jul–Aug

TAUSCHIA

Per, taprooted or roots tuberous, glabrous to hairy. **ST** 0 or low. **LF**: blade oblong to obovate, 1–2-pinnate or -ternate, lflets wide, margins entire to pinnately lobed. **INFL**: umbels compound, terminal; peduncle gen > lf; bracts gen 0; involucel 1-sided; bractlets inconspicuous to lf-like; rays, pedicels few–many, ascending to reflexed, gen few fertile. **FL**: calyx lobes evident to 0; petals wide, yellow, tips narrowed; styles short to slender; projection atop ovary inconspicuous. **FR** oblong to round, ± compressed side-to-side, glabrous; ribs prominent to thread-like, subequal, unwinged; oil tubes per rib-interval several; fr axis gen divided ± to base. **SEED**: face gen grooved or concave. (I.F. Tausch, Czech botanist, 1793–1848) ± 35 spp.: w N.Am., C.Am, n S.Am.

T. parishii (J. Coulter & Rose) J.F. Macbr. (p. 101) Pl 1–4 dm, glabrous, ± glaucous. **ST** 0. **LF**: petiole 5–15 cm; blade 8–15 cm, oblong to ovate, 2-pinnate, lflets 15–40 mm, oblong to ovate, sharply serrate to pinnately lobed. **INFL**: peduncle 10–30 cm; bractlets few, 5–12 mm, linear, entire; rays 12–18, 3–6 cm, subequal, spreading and reflexed; pedicels 2–7 mm. **FL**: calyx

lobes evident; styles slender. **FR** 5–8 mm, oblong to narrowly elliptic; ribs narrow, prominent, acute; oil tubes per rib-interval 4–5; fr axis divided in upper 2/3. 2*n*=22. Rocky or sandy soil, pine woodland; 1200–2400 m. Teh, SnGb, SnBr, SnJt, **SNE**. May–Jul

APOCYNACEAE DOGBANE FAMILY

Lauramay T. Dempster

Per (sometimes ann, shrub, vine, tree); sap milky. **LVS** simple, entire, opposite, alternate, or subwhorled; stipules 0 or small. **INFL**: cyme; fls 1–many, axillary or terminal. **FL** bisexual, radial; perianth parts overlapping, at least in bud; sepals 5, fused at base, persistent; petals 5, fused in ± basal half; stamens 5, attached to corolla tube or throat, alternate lobes; ovaries 2, ± superior, gen free, styles and stigmas fused. **FR**: gen 2 follicles. **SEEDS** many, often with tuft of silky hairs. ± 150–200 genera, 1000–2000 spp.: esp trop; many orn (*Oleander*; *Plumeria*, frangipani); some alkaloids highly toxic, some used in medicine. Recently treated to include Asclepiadaceae [Civeyrel et al 1998 Mol Phylog Evol 9: 517-527] Key to genera by Margriet Wetherwax.

1. Stamens attached near base of corolla tube, apparently below level of stigma; seed with tuft of long hairs
   2. Lvs, sts, roots not fleshy; fl < 8 mm; infl many-fld . . . . **APOCYNUM**
   2' Lvs, sts, roots fleshy; fl > 15 mm; infl few-fld . . . . *Cycladenia humilis* var. *venusta*
1' Stamens attached near top of corolla tube, near level of stigma; seeds glabrous
   3. Lvs alternate to subwhorled; fls several–many . . . . *Amsonia tomentosa*
   3' Lvs opposite; fls gen solitary in axils . . . . *Catharanthus roseus*

AMSONIA

Pl erect. **ST** ± woody. **LVS** alternate to subwhorled. **INFL**: fls several–many. **FL**: corolla salverform; anthers free from each other and stigma; nectary 0 or low ring around ovaries; style ± thread-like, stigma skirted at base. **SEED** glabrous. 5–25 spp.: N.Am, Japan. (Probably for Charles Amson, Virginian physician, 18th century) [McLaughlin 1982 Ann Missouri Bot Gard 69:336–350]

A. tomentosa Torrey & Frémont (p. 101) Pl glabrous or grayish tomentose. **STS** several–many from woody crown, 16–36 cm. **LF** 2–4 cm; petiole short or 0; blade ovate-lanceolate. **FL**: calyx lobes erect, thread-like above base; corolla whitish, blue, or greenish, tube ± 15 mm, inflated above middle, narrower just below

spreading lobes; style spheric just below stigma. **FR** 3–8 cm, narrowed between seeds, often breaking into 1-seeded segments. Desert plains, canyons; 300–1800 m. SnBr (n slope), **s DMtns**, **DSon**; to UT. [*A. brevifolia* A. Gray] Hairiness is probably a simple, genetically recessive trait. ✿TRY. Mar–May

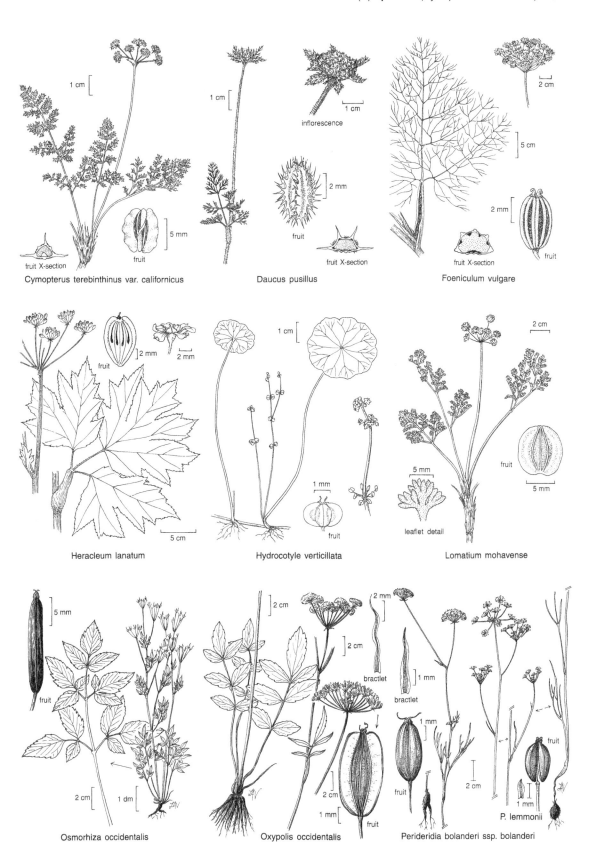

Cymopterus terebinthinus var. californicus

Daucus pusillus

Foeniculum vulgare

Heracleum lanatum

Hydrocotyle verticillata

Lomatium mohavense

Osmorhiza occidentalis

Oxypolis occidentalis

Perideridia bolanderi ssp. bolanderi

P. lemmonii

APOCYNUM DOGBANE, INDIAN HEMP

ST ascending to erect. **LVS** opposite. **INFL**: fls several–many. **FL** small; corolla cylindric to bell-shaped, 5-lobed, with 5 triangular appendages alternate stamens; stamens attached at base of tube, filaments short, wide, anthers forming adherent cone around stigma, each partly sterile, sharply sagittate; nectaries 5, free, around and < ovaries; style ± 0, stigma massive, ovoid, obscurely 2-lobed. **FR** slender, cylindric, pointed. **SEED** with tuft of long hairs. ± 7 spp.: N.Am. (Greek: away from, dog, from ancient use as dog poison). The 2 CA spp. hybridize extensively; many hybrid forms have been named.

1. Pl 1.6–3 dm; lf drooping to spreading, ovate to round, dark green above, pale below; corolla reddish
   purple or pink to white, sometimes with pink stripes, lobes spreading to recurved; calyx << corolla tube
   . . . . *A. androsaemifolium*
1. Pl 3–12 dm; lf ascending, lanceolate to narrowly ovate, yellowish green above, below; corolla
   greenish or white, lobes ± erect; calyx ± = corolla tube. . . . . *A. cannabinum*

A. androsaemifolium L. (p. 101) BITTER DOGBANE **ST** diffusely branched. **LF**: petiole << blade; blade 4–6 cm, base gen round or cordate, tip round or obtuse to ± acute. **FL**: corolla 4–8 mm, ± bell-shaped. **FR** 7–11 cm, pendent to erect. Open slopes, rocky places, with conifers, chaparral; 200–2500 m. KR, NCoRH, NCoRI, CaRH, SNH, SnFrB, SCoRI, SnBr, SnJt, MP, **W&I**; to e N.Am, Can. [var. *glabrum* Macoun; *A. medium* E. Greene var. *floribundum* (E. Greene) Woodson; *A. pumilum* (A. Gray) E. Greene] ❀ DRN: **1**,4–6,15,17&SHD:**2,3**,16&IRR:**7**,14,18,20,21; STBL,INV. Jun–Aug

A. cannabinum L. (p. 101) INDIAN HEMP **ST** stout, ± stiffly erect, branched near top. **LF**: petiole << blade; blade 5–8 cm, base tapered to cordate, sometimes clasping st, tip obtuse to acute. **FL**: corolla 2.5–5 mm, cylindric to urn-shaped. **FR** 6–9 cm, ± pendent. Moist places near streams, springs, etc., or weed in orchards; < 2000 m. KR, NCoR, CaRH, SNH, Teh, SnJV, SnFrB, SCoR, TR, PR, **GB**; to B.C., e N.Am. [var. *glaberrimum* A. DC.; *A. sibiricum* Jacq. var. *salignum* (E. Greene) Fern.] ❀WET: **1–3**,4,5,**6–9**,10, **14–21**;STBL,INV; important traditional source of fiber. Jun–Aug

CATHARANTHUS

Pl ± puberulent. **LVS** ± opposite. **INFL**: fls gen solitary in axils. **FL**: calyx lobes long, slender; corolla pink, tube ± cylindric, lobes asymmetric; stamens attached near top of corolla tube, filaments ± straight, anthers held above, free from stigma, each completely fertile; nectaries 2, alternate and gen > ovaries; style thread-like, stigma skirted at base. **SEED** glabrous. 3–7 spp.: Madagascar, India. (Greek: pure fl) [Taylor & Farnsworth 1975 The *Catharanthus* alkaloids]

C. roseus (L.) G. Don MADAGASCAR PERIWINKLE **ST** erect, 30–60 cm. **LF**: petiole gen < 1 cm; blade ± elliptic. **FL**: corolla 3–5 cm wide, pink. **FR** ± straight. Desert spring; 160 m. **DSon (Borrego Springs, San Diego Co.)**; native to Madagascar. [*Vinca*

rosea L.] Widely cult, naturalized in warm temp to trop regions. Alkaloids used to treat childhood leukemia, Hodgkin's disease, other cancers.

CYCLADENIA

Pl fleshy (incl large root), glabrous to tomentose. **ST** ± erect. **LVS** opposite. **INFL**: fls 2–6 near tips of axillary peduncles. **FL**: calyx lobes slender; corolla funnel-shaped with 5 rounded appendages behind anthers; stamens appearing attached at base of corolla tube but filaments epipetalous up to stigma level, anthers forming cone around but free from stigma, each partly sterile, sharply sagittate; nectary disk 5-lobed, around and < ovaries; style thread-like, stigma skirted at base. **SEED** with tuft of long hairs. 1 sp.: CA. (Greek: ring gland, from nectary)

C. humilis Benth. Pl 6–12 cm, gen glabrous, glaucous. **LF**: pairs 2–5; lf < 9 cm; petiole < to > blade; blade ovate or rounded. **FL**: calyx lobes narrowly triangular; corolla 15–20 mm, rose-purple, lobes obovate or round, margins wavy; filaments hairy. **FR** 3–5 cm. Sandy flats to talus slopes in open pine forest, chaparral; 1200–2800 m. KR, NCoRH, CaRH, n SNH, SCoRO, SnGb, SNE. ❀TRY. 2 vars in CA.

var. *venusta* (Eastw.) Munz (p. 101) Pl glabrous, glaucous, but infl hairy. Talus, loose gravel, dry ground in light shade of pines, chaparral; 1550–2500 m. SCoRO (Santa Lucia Range), SnGb, **SNE**. Pls of Inyo Co. have narrower, darker red corolla lobes. May–Jul

ASCLEPIADACEAE MILKWEED FAMILY

Carol A. Hoffman

Ann, per, shrub, vine; sap milky. **LVS** simple, gen opposite or whorled; stipules 0 or small. **INFL**: cyme, terminal or axillary, umbel- or raceme-like, or fl solitary. **FL** bisexual, radial; sepals 5, gen reflexed; petals 5, gen reflexed or spreading; stamens 5, fused to form filament column and anther head, gen with 5 elaborate appendages on outside of filament column, pollen removed in pairs of massive sacs; ovaries 2, superior, free, style tips gen fused into massive pistil head surrounded by anther head. **FR**: follicle (1 ovary gen aborts). **SEEDS** many, ± flat, with tuft of silky hairs. 50–250 genera, 2000–3000 spp.: esp trop, subtrop S.Am, s Afr; orn (*Asclepias*, *Hoya*, *Stapelia*). Cardiac glycosides produced by some; used as arrow poisons, in medicine to control heart contraction, and by some insects for defense. Recently treated to be included within Apocynacaeae [Civeyrel et al 1998 Mol Phylog Evol 9: 517–527]] Keys adapted by Margriet Wetherwax.

1. St not twining; widespread . . . . **ASCLEPIAS**
1' St twining in CA; gen D
  2. Filament column without appendages . . . . **CYANCHUM**
  2' Filament column with appendages
    3. Filament-column appendages fused margin-to-margin into a lobed, cup- or plate-like structure
      around anther head . . . . **MATELEA**
    3' Filament-column appendages free . . . . **SARCOSTEMMA**

ASCLEPIAS MILKWEED

Per, ann, shrub. **ST** prostrate to erect. **LVS** alternate, opposite, or whorled; blade narrowly linear to ovate or cordate. **INFL**: umbel-like, gen terminal and in (esp upper) axils. **FL**: ring of tissue at base of corolla 0; filament-column appendages (hoods) free, sometimes elevated above corolla base, each often with an elongate projection (horn) attached to inside, solid, margins curved in and meeting or nearly meeting on side adjacent to column but not fused; top of pistil head flat or conic. **FR** gen erect (pedicel gen pendent), narrowly ovoid, smooth or with tubercles. 100 spp.: Am. (Asklepios, ancient Greek physician) [Woodson 1954 Ann Missouri Bot Gard 41:1–211]

1. Lvs ephemeral (st gen lfless), narrowly linear
  2. Hoods < anther head . . . . *A. albicans*
  2' Hoods >> anther head . . . . *A. subulata*
1' Lvs persistent, narrowly linear to broadly ovate
  3. Lf 1 per node (alternate or clustered)
    4. Lvs lanceolate; hood not elvated above corolla base, subreflexed to spreading at base, ± = anther head
      . . . . *A. asperula*
    4' Lvs linear; hoods elevated above corolla base, ± erect at base, > anther head . . . . *A. linaria*
  3' Lvs > 1 per node (opposite or whorled)
    5. Horn 0 or incl in hood
      6. Hoods ± = anther head; pl ± glabrous; SNE (Mono Co.) . . . . *A. cryptoceras*
      6' Hoods < anther head; pl densely hairy; D . . . . *A. californica*
    5' Horn exserted from hood (length sometimes < hood)
      7. Horn > hood; pl gen glabrous; lvs in whorl of 3–5, often with axillary clusters of small lvs . . *A. fascicularis*
      7' Horn < to ± = hood; pl glabrous to very hairy; lvs opposite (or if whorled, then petiole 0 to short)
        8. Hoods > anther head by at least 1/2 of hood length
          9. Corolla greenish white; hoods gen not elevated above corolla base, ± = exserted horns, ± erect, tips
            rounded . . . . *A. nyctaginifolia*
          9' Corolla rose-purple; hoods slightly elevated above corolla base, >> exserted horns, ± ascending,
            tips acute . . . . *A. speciosa*
        8' Hoods not > anther head by at least 1/2 of hood length
          10. Axillary infls ± sessile . . . . *A. vestita*
          10' Axillary infls peduncled
            11. Lvs opposite or whorled, sessile to short-petioled, blade bases rarely cordate; hoods slightly
              < anther head . . . . *A. eriocarpa*
            11' Lvs opposite, sessile, blade bases often (shallowly) cordate; hoods slightly > anther head . . . . *A. erosa*

A. albicans S. Watson WHITE-STEMMED or WAX MILKWEED Shrub, ± hairy, waxy. **ST** ± erect. **LVS** whorled in 3's, ephemeral; blade narrowly linear. **INFL** gen terminal. **FL**: corolla reflexed, greenish white, sometimes tinged brown or pink; hoods elevated above corolla base, < anther head, yellowish; horns > hoods. Dry washes, gravelly slopes; 200–1100 m. **D**; AZ, Baja CA. Mar–May

A. asperula (Decne.) Woodson ssp. *asperula* (p. 101) ANTELOPE HORNS Per, ± hairy. **ST** decumbent to ascending. **LVS** alternate to clustered in 3's, persistent; petiole short; blade narrowly lanceolate. **INFL** terminal. **FL**: corolla spreading to ascending, greenish white; hoods not elevated above corolla base, ± = anther head, purplish; horns incl in hoods. Dry, open, rocky places; 1500–2000 m. **DMoj**; sw U.S., Mex. [*A. capricornu* ssp. *occidentalis* Woodson] ❀TRY. May–Jul, Sep

A. californica E. Greene (p. 113) CALIFORNIA or ROUND-HOODED MILKWEED Per, densely hairy. **ST** decumbent to ± ascending. **LVS** opposite, persistent; petiole 0–short; blade ovate. **FL**: corolla reflexed, purplish; hoods elevated above corolla base or not, < anther head, dark purple; horns 0–minute. Flats, grassy or shrubby hillsides; 200–2100 m. c&s SNF, CW, SW, **D**; n Baja CA. Pls from c&s SNF, CW with hoods elevated above corolla base have been called ssp. *greenei* Woodson. ❀TRY. Apr–Jul

A. cryptoceras S. Watson HUMBOLDT MOUNTAINS MILKWEED Per, ± glabrous. **ST** prostrate to decumbent. **LVS** opposite, persistent, sometimes sessile; blade ovate to nearly round, sometimes cordate, base rarely clasping st. **INFL** gen terminal. **FL**: corolla reflexed, greenish yellow; hoods not elevated above corolla base, ± = anther head, pinkish tan; horns 0 or incl in hoods. **FR**: pedicel erect. Sandy or gravelly slopes, canyon bottoms, arid plains; 1400–1700 m. **SNE (Mono Co.)**; to WA, WY, Colorado, AZ. If the dubious sspp. are recognized, CA pls (with hoods slightly < anther head) are ssp. *davisii* (Woodson) Woodson. Ssp. *cryptoceras* (hoods gen slightly > anther head) has been reported from SNE (Mono Co.), is listed as UNCOMMON. Even if recognized, its occurrence in CA must be doubted. ❀TRY. May–Jun

A. eriocarpa Benth. KOTOLO or INDIAN MILKWEED Per, very hairy (sometimes becoming less so). **ST** erect. **LVS** opposite or in whorls of 3–4, persistent; petiole 0 to short; blade lanceolate, elliptic, or ovate, base tapered or obtuse, rarely cordate. **FL**: corolla reflexed to ascending, cream, sometimes tinged pink; hoods slightly elevated above corolla base, < to ± = anther head, cream, sometimes tinged purple; horns exserted, ± = hoods. Dry, barren areas; 200–1900 m. **CA**; NV, n Baja CA. ❀DRN,DRY,&SUN: **7–9**,10, **14,15**,16,17,**18**,19–24. Jun–Aug

A. erosa Torrey (p. 113, pl. 11) DESERT MILKWEED Per, glabrous to very hairy. **ST** ascending to erect. **LVS** opposite, persistent, sessile; blade lanceolate, elliptic, or ovate, base tapered to cordate, sometimes clasping st. **FL**: corolla reflexed to spreading, pale cream or greenish white; hoods elevated above corolla base, > anther head, cream, yellow, or reddish; horns exserted, < to ± = hoods. Dry slopes, washes; 150–1900 m. SW, **W&I, D**; AZ, Baja CA. May–Jul

A. fascicularis Decne. NARROW-LEAF MILKWEED Per, gen glabrous. **ST** ascending to erect. **LVS** in whorl of 3–5, often with axillary clusters of small lvs, persistent; petiole short; blade narrowly lanceolate, base tapered. **FL**: corolla reflexed, greenish white, sometimes tinged purple; hoods elevated above corolla base, gen < anther head, greenish white; horns gen > hoods (and anther head). **FR**: pedicel erect. Dry ground, valleys, foothills; 50–2200 m. **CA** (exc NCo, CCo, SCo); to WA, UT, Baja **CA**. [not *A. mexicana* Cav.] ❀DRN,DRY,&SUN:**3,7–9**,10,**14,15**, 16,17,**18**,19–24. Jun–Sep

A. linaria Cav. Per or shrub, ± hairy. **ST** erect. **LVS** alternate, persistent, sessile; blade linear, resembling pine needle. **FL**: corolla reflexed, inner surface greenish white, outer surface pinkish or purplish; hoods gen elevated above corolla base, > anther head, greenish white; horns exserted, < hoods. Uncommon. Open woodland, limestone ridges, rocky hills, canyons, arroyos, dry abandoned pastures; 1000–1400 m. **D**; AZ, Mex.

A. nyctaginifolia A. Gray MOJAVE MILKWEED Per, hairy. **ST** decumbent to ascending. **LVS** opposite, persistent, petioled; blade lanceolate, elliptic, or ovate. **FL**: corolla reflexed, greenish white, outer surface sometimes purple; hoods not to slightly elevated above corolla base, >> anther head, greenish white; horns exserted, ± = hoods. Arroyos, dry slopes; 1000–1700 m. **DMoj**; to NM. May–Jun

A. speciosa Torrey GREEK or SHOWY MILKWEED Per, hairy. **ST** ascending to erect. **LVS** opposite, persistent; petiole short; blade elliptic to ovate, base rarely cordate and clasping st. **FL**: corolla reflexed, rose-purple; hoods slightly elevated above corolla base, >> anther head, pink, aging yellow; horns exserted, ± = anther head. Many habitats incl fields, roadsides; 0–1900 m. **CA**; to B.C., c Can, TX. [*A. giffordii* Eastw. (see Gilmartin 1980 Bull Torrey Bot Club 104:496–505)] ❀SUN&DRN: 1,**2,3**,4,5,**6,10,14,15**,16,17&IRR:**7–9,14,15,18**,19–24;STBL,INV. May–Jul

A. subulata Decne. RUSH MILKWEED, AJAMETE Per, gen glabrous exc infl. **ST** erect. **LVS** opposite, ephemeral, sessile; blade narrowly linear. **FL**: corolla reflexed, yellowish white; hoods slightly elevated above corolla base, >> anther head, yellowish white; horns exserted, = or slightly < hoods, >> anther head. **FR** pendent. Arroyos, washes; < 700 m. **D**; NV, AZ, n Mex. Apr–Dec

A. vestita Hook. & Arn. WOOLLY MILKWEED Per, gen densely hairy, sometimes becoming ± glabrous. **ST** ascending. **LVS** opposite, persistent; petiole gen short; blade lanceolate, elliptic, or ovate. **FL**: corolla reflexed, yellowish white or purplish; hoods slightly elevated above corolla base, ± = anther head, yellowish white, sometimes with vertical, brown stripe; horns exserted, ± = hoods. Dry plains, shrubby flats, hillsides, desert canyons; 50–1350 m. GV, CW, TR, **DMoj**. Pls from CW, TR, DMoj with purple corollas have been called var. *parishii* Jepson. May–Jun

CYNANCHUM VINE MILKWEED

Per, shrub. **ST** twining (elsewhere sometimes prostrate to erect). **LVS** opposite; blade linear to ovate. **INFL** axillary, umbel- or raceme-like. **FL**: corolla ± erect, ring of tissue at base 0; filament-column appendages 0 (elsewhere gen free); pistil head flat, conic, or with 2 lobes on top. **FR** gen erect, fusiform to narrowly ovoid. ± 200 spp.: temp, esp trop. (Greek: dog strangle, from ancient supposition of or use as dog poison) [Sundell 1981 Evol Monogr 5:1–63]

C. utahense (Engelm.) Woodson (p. 113) UTAH CYNANCHUM Per. **ST** slender, much-branched, < 1 m. **LF**: blade 1.5–4 cm, linear, becoming reflexed. **INFL** umbel-like. **FL**: corolla 1.5–3 mm, bell-shaped, lobes incurved, hood-like, yellow, becoming orange. **FR** 4–6 cm, with fine longitudinal grooves. UNCOMMON. Dry, sandy or gravelly areas; < 1000 m. **DMoj**; to UT, AZ. Apr–Jun

MATELEA

Per, shrub. **ST** twining (elsewhere sometimes prostrate to erect). **LVS** opposite; blade often ± cordate. **INFL**: fls 1–2, axillary (elsewhere various). **FL**: corolla ± spreading, ring of tissue at base 0; filament-column appendages gen fused margin-to-margin into 5-lobed, cup- or plate-like structure around anther head, fused to base of filament column, each lobe with a vertical, flap-like ridge; pistil head flat. **FR** erect or pendent, fusiform to ± ovoid, smooth, gen tubercled. ± 200 spp.: trop, warm temp Am. [Stevens 1976 Diss Abstr B 37(2):587]

M. parvifolia (Torrey) Woodson (p. 113) SPEARLEAF, TALAYOTE Per. **ST** slender, much-branched, < 0.5 m. **LF**: blade 0.5–2 cm, cordate-sagittate. **FL**: corolla greenish or purple, each sinus with acute, turned-out tooth. **FR** ± 7 cm, with fine longitudinal grooves. RARE in CA. Dry, rocky areas; 700–1000 m. **D**; to NV, TX, Baja CA. Populations widely scattered. Mar–May

SARCOSTEMMA CLIMBING MILKWEED

Per, shrub. **ST** twining (elsewhere sometimes prostrate to ± erect). **LVS** opposite; blade often linear to narrowly lanceolate or hastate. **INFL** axillary, often umbel-like. **FL**: corolla gen ± spreading to ± erect, with ring of tissue at base; filament-column appendages free, ± spheric, attached to base of filament column, projections hollow; pistil head ± conic, 2-lobed, or both. **FR** gen erect, narrowly fusiform to narrowly ovoid, with fine longitudinal grooves. ± 34 spp.: N.Am, Afr to Australia. (Greek: fleshy crown, from sac-like filament-column appendages) [Holm 1950 Ann Missouri Bot Gard 37:477–560] Our spp. now treated in *Funastrum* [Liede 1996 Syst Bot 21: 31–44]

1. Pl sparsely hairy; corolla ± purple; filament-column appendages free from ring of tissue at base of corolla
   . . . . ***S. cyanchoides*** ssp. ***hartwegii***
1' Pl densely hairy; corolla ± greenish white; filament-column appendages fused to ring of tissue at base
   of corolla . . . . ***S. hirtellum***

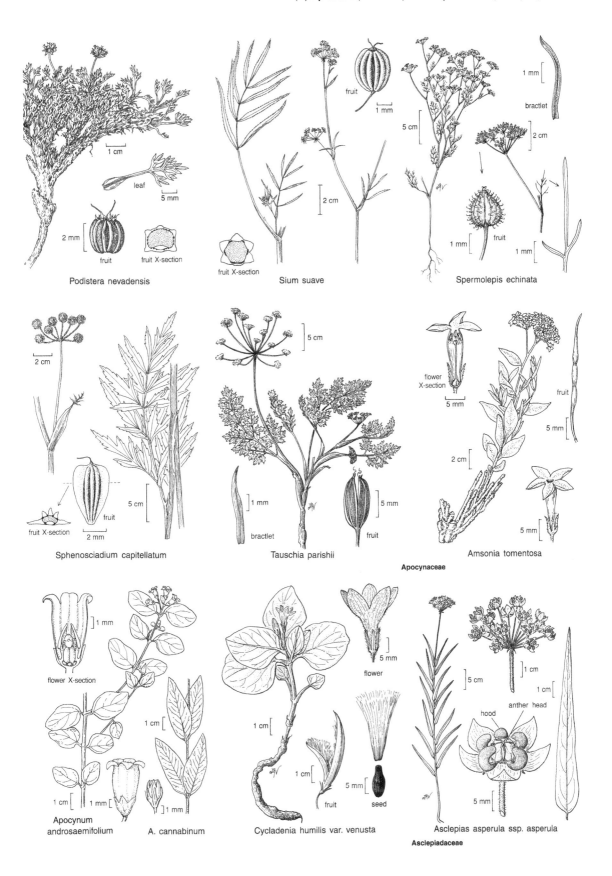

Podistera nevadensis

Sium suave

Spermolepis echinata

Sphenosciadium capitellatum

Tauschia parishii

Amsonia tomentosa

Apocynaceae

Apocynum androsaemifolium

A. cannabinum

Cycladenia humilis var. venusta

Asclepias asperula ssp. asperula

Asclepiadaceae

S. cynanchoides Decne. ssp. *hartwegii* (Vail) R. Holm (p. 113, pl. 12) CLIMBING MILKWEED Pl green; hairs gen sparse, ± appressed. **LF**: blade base gen hastate or truncate. **FL**: corolla pink to purple, or lobes white with purple streak; filament-column appendages free from ring of tissue at base of corolla. **FR** gen 1. Dry, sandy, rocky arroyos or plains, in ditches near cult; 30–1600 m. **D**; to UT, AZ, Mex. ❀TRY. Apr–Jul

S. hirtellum (A. Gray) R. Holm TRAILING TOWNULA Pl gray-green; hairs gen dense, short, erect. **LF**: blade base variously tapered. **FL**: corolla white to greenish white; filament-column appendages fused to ring of tissue at base of corolla. **FR** gen 2, often spreading. Hard desert pavement, washes; 150–1200 m. **D**; NV, AZ. ❀TRY. Mar–May

ASTERACEAE [Compositae] SUNFLOWER FAMILY

David J. Keil, Family Editor and author, except as specified

Ann to tree. **LVS** basal or cauline, alternate to whorled, simple to compound. **INFL**: 1° infl a head, each resembling a fl, 1–many, gen arrayed in cymes, gen subtended by ± calyx-like involucre; fls 1–many per head. **FLS** bisexual, unisexual, or sterile, ± small, of several types; calyx 0 or modified into pappus of bristles, scales, or awns, which is gen persistent in fr; corolla radial or bilateral (rarely 0), lobes gen (0)4–5; stamens 4–5, anthers gen fused into cylinder around style, often appendaged at tips, bases, or both, filaments gen free, gen attached to corolla near throat; pistil 1, ovary inferior, 1-chambered, 1-seeded, style 1, branches 2, gen hair-tufted at tip, stigmas 2, gen on inside of style branches. **FR**: achene, cylindric to ovoid, gen deciduous with pappus attached. ± 1300 genera, 21,000 spp. (largest family of dicots): worldwide. Largest family in CA. Also see tribal key to CA genera: Strother 1997 Madroño 44(1):1-28.

Key to Groups

1. Ligules (strap-shaped, petal-like corollas) 0
  2. Heads disciform [fls of 2 kinds (some pistillate or sterile, others bisexual or staminate) in same or different heads] . . . . **GROUP 1**
  2' Heads discoid [fls bisexual, gen all fertile (sometimes outermost enlarged, ± bilateral)]
    3. Receptacle bearing chaff (scale-like bractlets) or long, stiff hairs among fls
      4. Receptacle chaffy . . . . **GROUP 2**
      4' Receptacle bristly . . . . **GROUP 3**
    3' Receptacle ± naked (sometimes bearing minute scales or short hairs among fls)
      5. Pappus 0 or only a low crown . . . . **GROUP 4**
      5' Pappus well developed
        6. Pappus of bristles (sometimes with an additional series of shorter bristles or scales) . . . . **GROUP 5**
        6' Pappus of flat, ± membranous scales or stiff, ± needle-like awns . . . . **GROUP 6**
1' Ligules present
  7. Heads either ligulate or of 2-lipped disk fls; flowers of head all of same kind; sap milky or clear
    8. Corollas readily withering, 1-lipped, ligules 5-lobed; sap gen milky . . . . **GROUP 7**
    8' Corollas gen not readily withering, 2-lipped, outer lip 3-lobed, spreading, inner lip 2-lobed, shorter, recurved; sap clear . . . . **TRIXIS**
  7' Heads radiate; fls of head of 2 kinds, outer ray fls pistillate or sterile; central disk fls gen bisexual; sap gen clear
    9. Receptacle bearing chaff scales among disk fls (sometimes only 1 ring of chaff scales between ray and disk fls)
      10. Phyllaries in 1 series, each subtending a ray fl. . . . . **GROUP 8**
      10' Phyllaries in 2+ series, not all subtending ray fls . . . . **GROUP 9**
    9' Receptacle naked or bearing minute scales or hairs among fls
      11. Pappus 0 or only a low crown . . . . **GROUP 10**
      11' Pappus well developed on ray or disk frs (or both)
        12. Pappus of flat, ± membranous scales or stiff, ± needle-like awns . . . . **GROUP 11**
        12' Pappus of bristles (sometimes with an additional series of shorter bristles or scales)
          13. Ligules white to purple . . . . **GROUP 12**
          13' Ligules yellow to orange or red . . . . **GROUP 13**

Group 1: Heads disciform; fls of 2 kinds, some pistillate or neuter, others bisexual or staminate; pistillate fls and disk fls in same or different heads

1. Pistillate fls and staminate fls in different heads
  2. Pls dioecious, staminate and pistillate heads on different pls
    3. Herbs, sometimes prostrate, not sticky; pls gen ± woolly . . . . **ANTENNARIA**
    3' Shrubs or erect herbs with ± sticky lvs; pls not woolly . . . . **BACCHARIS**
  2' Pls monoecious, staminate and pistillate heads on same pl

4. Subshrubs or shrubs
  5. Lvs linear to ovate, toothed or variously lobed; bracts of pistillate heads spiny . . . . [2]**AMBROSIA**
  5' Lvs thread-like and entire or divided into thread-like lobes; bracts of pistillate heads flat, scarious
  . . . . **HYMENOCLEA**
4' Herbs
  6. Staminate heads in long, terminal spikes or racemes; burs gen 3–10 mm . . . . [2]**AMBROSIA**
  6' Staminate heads congested; burs gen > 20 mm . . . . **XANTHIUM**
1' Pistillate or sterile fls in same heads as staminate or bisexual fls
  7. Outer fls of head without corollas; heads not embedded in wool
    8. Ann; fr winged . . . . **DICORIA**
    8' Per to shrubs; fr not winged . . . . [2]**IVA**
  7' Outer fls of head with corollas, these sometimes very narrowly tubular; heads sometimes embedded in
    wool
    9. Outer fls of head sterile, corollas sometimes expanded, ± bilateral; heads not very small, not embedded
      in wool
      10. Lvs spiny . . . . **CNICUS**
      10' Lvs not spiny
        11. Pappus of awns with reflexed barbs; phyllaries in 2 series, unarmed . . . . **BIDENS**
        11' Pappus of smooth or minutely rough bristles; phyllaries in 3+ series, often fringed or spine-
          tipped . . . . **CENTAUREA**
    9' Outer fls of head pistillate, corollas often narrowly tubular; heads sometimes very small, embedded in
      wool
      12. Phyllaries (or outermost bracts of head) papery, membranous or scarious; sometimes green below
        middle or in narrow, central band; heads often embedded in wool
        13. Phyllaries gen many, overlapping, inner > fls; receptacle naked; pistillate fls not subtended by bracts
          14. Pappus 0 or a minute crown . . . . [3]**ARTEMISIA**
          14' Pappus of bristles
            15. Lvs ± tomentose, esp below; all phyllaries thin, membranous, margin ± widely scarious
            . . . . **GNAPHALIUM**
            15' Lvs glandular or silky with short, appressed hairs; upper half of outer phyllaries thick,
              leathery, margin narrowly or not scarious . . . . [2]**PLUCHEA**
        13' Phyllaries 0 or 2–6, not overlapping, < fls; outer receptacle chaffy; some or all pistillate fls
          individually subtended by bracts
          16. Disk fls bisexual, pappus gen of > 12 bristles, ± exserted; inner pistillate fls with pappus . . . . . **FILAGO**
          16' Disk fls staminate, pappus 0 or of 1–12 bristles, incl; pistillate fls pappus 0
            17. Chaff scales subtending disk fls (disk chaff) enlarged, rigid, ± spreading, very different from
              scales subtending pistillate fls (pistillate chaff); tips spine-like, hooked . . . . . . **ANCISTROCARPHUS**
            17' Disk chaff 0 or scales ± gradually reduced, scarious, erect; tips not spine-like or hooked
            . . . . **STYLOCLINE**
      12' Phyllaries green gen throughout, sometimes scarious-margined; heads gen not embedded in wool
        18. Pappus well developed
        19. Pappus of flattened scales or barbed awns; lvs opposite . . . . [2]*Lasthenia microglossa*
        19' Pappus of bristles; lvs gen all alternate
          20. Pappus bristles 1–2; lvs coarsely toothed to palmately lobed . . . . [2]*Perityle emoryi*
          20' Pappus bristles gen many; lvs entire to pinnately lobed
            21. Heads ± flat, button-like . . . . **ERIGERON**
            21' Heads gen cylindric or bell-shaped
              22. Ann; disk corollas gen yellow . . . . **CONYZA**
              22' Per or shrub; disk corollas gen pinkish or purple . . . . [2]**PLUCHEA**
        18' Pappus 0 or reduced to a minute crown
        23. Phyllary margins widely scarious or transparent
          24. Heads in spikes, racemes or panicles . . . . [3]**ARTEMISIA**
          24' Heads in wide, flat-topped clusters . . . . **SPHAEROMERIA**
        23' Phyllary margins not scarious or transparent
          25. Ovary of disk fls much reduced; style tip truncate or tack-shaped
            26. Anthers fused . . . . [3]**ARTEMISIA**
            26' Anthers ± free . . . . [2]**IVA**
          25' Ovary of disk fl well developed; style tip ± branched
            27. Lvs wide, coarsely toothed to palmately lobed . . . . [2]*Perityle emoryi*
            27' Lvs linear, entire to ± toothed
              28. Plants not glandular, not strongly scented; lvs opposite . . . . [2]*Lasthenia microglossa*
              28' Plants glandular, ± strongly scented; lvs alternate or clustered at nodes
              . . . . *Madia minima* (segregate genus *Hemizonella*)

Group 2: Heads discoid; receptacle chaffy

1. Shrubs
  2. Heads cylindric to narrowly bell-shaped; lvs linear, often early deciduous; fr 3-angled, not ciliate; pappus of bristles . . . . **BEBBIA**
  2' Heads hemispheric; lvs elliptic or narrowly ovate; fr flattened, ciliate; pappus of 2 scales or 0
  . . . . *Encelia frutescens*
1' Herbs
  3. Corollas white to pinkish . . . . *Chaenactis carphoclinia*
  3' Corollas yellow
    4. Lf margins and phyllaries unarmed; pappus of 1–5 barbed awns . . . . *Bidens frondosa*
    4' Lf margins and phyllaries spiny; pappus of many narrow scales. . . . . **CARTHAMUS**

Group 3: Heads discoid; receptacle bristly

1. Lvs not spiny
  2. Phyllary tips not expanded or fringed, never spine-tipped. . . . . **ACROPTILON**
  2' Phyllary tips (at least inner) prominently expanded, ± fringed with short spines or irregular teeth, sometimes spine-tipped . . . . **CENTAUREA**
1' Lvs spiny
  3. Corollas yellow to orange-red
    4. Pappus of many unequal, narrow scales in several series, gen 0 on outer frs; all fls fertile; fr 4-angled
    . . . . **CARTHAMUS**
    4' Pappus of 20 stiff bristles or bristle-like awns in 2 series, on all frs; outer fls sterile, corollas 3-lobed, ovary vestigial; fr cylindric, 20-ribbed. . . . . **CNICUS**
  3' Corollas white to blue, red, or purple
    5. Pappus of roughened or barbed bristles . . . . **CARDUUS**
    5' Pappus of long-plumose bristles. . . . . **CIRSIUM**

Group 4: Heads discoid; receptacle naked; pappus 0 or reduced to a low crown

1. Fls gen 1–8 per head; heads often grouped in tight, often headlike 2° clusters
  2. Shrubs; 1° lvs ± persistent as spines. . . . . **HECASTOCLEIS**
  2' Herbs; lvs never forming spines
    3 Pl 1–2.5 cm, tufted, ± woolly . . . . *Eriophyllum mohavense*
    3' Pl 15–80+ cm, erect, glandular, ± bristly . . . . *Madia glomerata*
1' Fls 10–many per head; heads not grouped in 2° heads
  4. Fr compressed
    5. Phyllaries in 1 series, weakly fused; lvs long-acuminate . . . . **PERICOME**
    5' Phyllaries in 2–3 series, ± equal, free; lvs not long-acuminate . . . . **PERITYLE**
  4' Fr not compressed
    6. Shrub; heads many in panicles . . . . **ARTEMISIA**
    6' Ann; heads 1–few
      7. Pl densely woolly; lvs 3-lobed at tip . . . . *Eriophyllum pringlei*
      7' Pl glabrous or inconspicuously short hairy; lvs pinnately dissected . . . . *Chamomilla occidentalis*

Group 5: Heads discoid; receptacle naked; pappus of bristles

1. Lvs and phyllaries dotted or streaked with embedded translucent oil glands, otherwise glabrous; odor strong, unpleasant . . . . **POROPHYLLUM**
1' Lvs and phyllaries without embedded oil glands (sometimes glandular-hairy or dotted with sessile resin glands); odor 0 or gen not very unpleasant
  2. Shrubs or subshrubs
    3. Corollas white or cream to dull purple or greenish
      4. Ovaries very reduced, sterile; style incl . . . . **BACCHARIS**
      4' Ovaries well developed, forming frs; style often exserted
        5. Fr 10-ribbed . . . . [2]**BRICKELLIA**
        5' Fr 5-ribbed
          6. Lvs narrowly linear, blade not evident, entire . . . . *Chrysothamnus albidus*
          6' Lvs petioled, blade obvious, gen toothed
            7. Blade gen = or > petiole; pappus of bristles only. . . . . [2]**AGERATINA**
            7' Blade gen << petiole; pappus of bristles and short scales . . . . **PLEUROCORONIS**
    3' Corollas yellow to orange
      8. Pappus bristles 1–2
        9. Lf blades 3–12 cm, petioles 15–50 mm; phyllaries in 1 series, weakly fused . . . . [2]**PERICOME**
        9' Lf blades < 2.5 cm; petioles 1–6 mm; phyllaries in 2 equal series, free . . . . **PERITYLE**

8' Pappus bristles gen many
  10. Heads ± spheric; phyllaries ovate or widely elliptic . . . . *Acamptopappus sphaerocephalus*
  10' Heads cylindric to bell-shaped; phyllaries linear to narrowly ovate
    11. Phyllaries ± equal, in 1–2 series
      12. Phyllaries 9–18; pl glabrous, resinous-glandular; fls 12–21 . . . . **PEUCEPHYLLUM**
      12' Phyllaries 4–7; pl ± woolly; fls 4–8 . . . . **TETRADYMIA**
    11' Phyllaries unequal, graduated in several series
      13. Lvs (at least upper) scale-like
        14. Heads solitary; fr appressed-hairy . . . . *Machaeranthera carnosa*
        14' Heads in panicle-like clusters; fr ± glabrous . . . . *Lepidospartum squamatum*
      13' Lvs well developed, linear to oblong or obovate
        15. Lvs toothed . . . . ²**ISOCOMA**
        15' Lvs entire
          16. Phyllaries in 5 ± distinct vertical ranks . . . . ²**CHRYSOTHAMNUS**
          16' Phyllaries not in distinct vertical ranks
            17. Sts ± tomentose
              18. Phyllaries obtuse to acute . . . . *Lepidospartum latisquamum*
              18' Phyllaries acuminate
                19. Lvs glabrous to ± tomentose . . . . *Chrysothamnus parryi*
                19' Lvs stalked-glandular . . . . *Ericameria discoidea*
            17' Sts glabrous or ± hairy
              20. Corolla throat abruptly expanded above tube . . . . ²**ISOCOMA**
              20' Corolla throat gradually expanded above tube
                21. Fls 2–8 . . . . ²**CHRYSOTHAMNUS**
                21' Fls 9–70
                  22. Dwarf alpine shrubs; gen > 3200 m . . . . *Chrysothamnus viscidiflorus*
                  22' Low to erect non-alpine shrubs; < 2900 m . . . . **ERICAMERIA**
2' Herbs
  23. Corollas white to purple
    24. Pappus of 3 bristles and 3 minute scales . . . . **MALPERIA**
    24' Pappus of gen >> 6 bristles
      25. Phyllaries ± equal, in 1–2 series . . . . ²**AGERATINA**
      25' Phyllaries unequal, graduated in several series . . . . ²**BRICKELLIA**
  23' Corollas yellow to orange
    26. Phyllaries unequal, in 2–several series, often strongly graduated
      27. Pl low, mounded, densely gray-woolly, strongly scented; lvs often wider than long; phyllaries
        in 2 series, outer phyllaries wide, tips spreading . . . . ²**PSATHYROTES**
      27' Pl spreading to erect, not densely woolly, scented or not; lvs gen longer than wide; phyllaries in
        3–several series
        28. Pappus bristles flattened, readily deciduous; involucre conspicuously gummy-resinous . . **GRINDELIA**
        28' Pappus bristles ± cylindric, ± persistent; involucre gen not gummy-resinous
          29. Ann; outer fls of head often bilateral, ± ray-like . . . . **LESSINGIA**
          29' Per; all fls of head radial
            30. Pl glabrous; fls 4–7 per head . . . . *Chrysothamnus gramineus*
            30' Pl ± hairy, sometimes glandular; fls 14–60 per head
              31. Phyllary tips spreading . . . . *Machaeranthera canescens*
              31' Phyllary tips gen appressed or outermost spreading . . . . ²**ERIGERON**
    26' Phyllaries ± equal, in 1–2 series
      32. Lvs all basal . . . . **RAILLARDELLA**
      32' Some or all lvs cauline
        33. Cauline lvs opposite (uppermost sometimes alternate)
          34. Pappus bristles many; phyllaries in 2 series, free . . . . *Arnica parryi*
          34' Pappus bristles 1–2; phyllaries in 1 series, fused at base . . . . ²**PERICOME**
        33' Cauline lvs alternate throughout
          35. Per, often from caudex or rhizome
            36. Pappus double, outer pappus of minute bristles or scales; phyllaries not black-tipped . . ²**ERIGERON**
            36' Pappus single; phyllaries often black-tipped . . . . ²**SENECIO**
          35' Ann or short-lived per from slender taproot
            37. Main phyllaries in 1 series, outer, if present, << inner . . . . ²**SENECIO**
            37' Main phyllaries in 2 series, ± equal
              38. Pappus bristles free; heads short-peduncled; lvs entire or blunt-toothed . . . . ²**PSATHYROTES**
              38' Pappus bristles fused at base in 5 groups; heads long-peduncled; lvs sharply toothed or
                lobed . . . . **TRICHOPTILIUM**

Group 6: Heads discoid; receptacle naked; pappus of scales or awns

1. Heads ± sessile, sometimes in dense 2° clusters
  2. Ann, 1–8 cm, tufted; lvs not spiny; heads 3–25-fld . . . . **ERIOPHYLLUM**
  2' Shrub, 4–7 dm; lvs spine-tipped, weakly spine-margined; heads 1-fld . . . . **HECASTOCLEIS**
1' Heads evidently peduncled
  3. Phyllaries strongly graduated, overlapping in 2–many series, outer << inner
    4. Lf blade diamond-shaped, slender-petioled, gen few-toothed . . . . **PLEUROCORONIS**
    4' Lf blade linear or oblong, sessile, gen entire
      5. Shrub; involucres hemispheric to spheric; pappus of many narrow, persistent, bristle-like scales
      . . . . *Acamptopappus sphaerocephalus*
      5' Ann; involucres cylindric to narrowly bell-shaped; pappus of 3 bristles and 3 minute scales . . **MALPERIA**
  3' Phyllaries subequal, in 1–3 series, not or weakly overlapping, outer ± = inner
    6. Phyllary margins thin, ± scarious, often brownish to purple
      7. Per; pappus scales 12–22 . . . . **HYMENOPAPPUS**
      7' Ann; pappus scales gen 8 . . . . **SCHKUHRIA**
    6' Phyllary margins gen ± herbaceous, not or inconspicuously scarious
      8. Corollas yellow
        9. Outer corollas bilateral, much enlarged, ray-like; pappus scales entire or toothed . *Chaenactis glabriuscula*
        9' Outer corollas radial, not enlarged; pappus scales dissected into bristles . . . . **TRICHOPTILIUM**
      8' Corollas white to purple
        10. Fr 3–9 mm; pappus scale without midrib; lvs often toothed or lobed . . . . **CHAENACTIS**
        10' Fr 10–15 mm; pappus scale midrib well-developed; lvs entire . . . . **PALAFOXIA**

Group 7: Heads ligulate; ligules all 5-lobed; sap gen milky

1. Pappus 0—lvs all basal or nearly so; corollas white . . . . **ATRICHOSERIS**
1' Pappus present
  2. Pappus of scales
    3. Pappus scales < 0.5 mm, blunt . . . . **CICHORIUM**
    3' Pappus scales 5–15 mm, awn-tipped . . . . **UROPAPPUS**
  2' Pappus of bristles and/or awns
    4. Bristles or awns of pappus plumose
      5. Pappus bristles very unequal in length, much longer on 1 side of fr . . . . **ANISOCOMA**
      5' Pappus bristles ± equal in length
        6. Lvs linear to lanceolate, grass-like, entire; involucre 2.5–7 cm . . . . **TRAGOPOGON**
        6' Lvs lanceolate to elliptic or scale-like, larger lvs ± toothed or lobed; involucre 0.5–3 cm
          7. Frs beaked; corollas white or cream, sometimes red-veined . . . . **RAFINESQUIA**
          7' Frs not beaked; corollas pink to pale lavender . . . . **STEPHANOMERIA**
    4' Bristles or awns of pappus smooth or barbed
      8. Fls 3–5 per head
        9. Pappus double, of 5 awns and gen many bristles, all fused at base . . . . **CHAETADELPHA**
        9' Pappus single, of many bristles
          10. Ann; sts not thorny . . . . **PRENANTHELLA**
          10' Subshrub; sts thorny . . . . *Stephanomeria spinosa*
      8' Fls many per head
        11. Frs weakly to strongly flattened
          12. Frs beaked, beak sometimes short and thick; corollas pale yellow or blue . . . . **LACTUCA**
          12' Frs not beaked; corollas yellow . . . . **SONCHUS**
        11' Frs not flattened
          13. Fr widely cylindric or narrowed at base, not beaked
            14. Pappus bristles persistent on mature fr; lower lvs and often entire pl with long glandless hairs
            . . . . **HIERACIUM**
            14' Some or all pappus bristles falling together from mature fr; pls glabrous, hairy or glandular,
              but long, glandless hairs 0 (exc *M. californica*) . . . . [2]**MALACOTHRIX**
          13' Fr fusiform, often beaked
            15. Pappus bristles, at least inner, ± fused at base and readily deciduous from mature fr
            16. Fr not beaked . . . . [2]**MALACOTHRIX**
            16' Fr short-beaked
              17. Fr tapered to beak; lf without hardened margins . . . . **CALYCOSERIS**
              17' Fr abruptly beaked; lf with white, hardened margins . . . . **GLYPTOPLEURA**
           15' Pappus bristles persistent on mature fr
            18. Lvs basal and cauline . . . . **CREPIS**
            18' Lvs all basal
              19. Phyllaries all equal or in several graded series; fr beak < body . . . . *Agoseris glauca*
              19' Phyllaries in 2 distinct series; fr beak slender, > body . . . . **TARAXACUM**

Group 8: Heads radiate; phyllaries in 1 series, each subtending a ray fl; receptacle chaffy, scales sometimes restricted to 1 ring separating ray and disk fls

1. Phyllary margins ± flat, not clasping or enclosing ray ovaries
  2. Lvs woolly; ray corollas ± thin, deciduous; receptacle minutely chaffy in center . . . . . ***Eriophyllum ambiguum***
  2' Lvs scabrous; ray corollas leathery, persistent on frs; receptacle chaffy throughout, scales awn-tipped
  . . . . **SANVITALIA**
1' Phyllary margins folded, ± clasping or enclosing ray ovaries
  3. Phyllary margins not adjacent or overlapping, ± clasping ovary margins but not completely enclosing
    ovary . . . . **HEMIZONIA** (segregate genus *Deinandra*)
  3' Phyllary margins adjacent or overlapping, ± completely enclosing ovary
    4. Disk pappus of basally plumose awns or scabrous bristles . . . . **LAYIA**
    4' Disk pappus 0 . . . . **MADIA** (including segregate genus *Hemizonella*)

Group 9: Heads radiate; phyllaries in 2+ series; receptacle chaffy throughout

1. Chaff scales flat, not folded around disk ovaries
  2. Ray corollas yellow
    3. Pappus of barbed awns . . . . **BIDENS**
    3' Pappus 0 or of flat scales . . . . **COREOPSIS**
  2' Ray corollas white to pink
    4. Disk corollas yellow . . . . **GALINSOGA**
    4' Disk corollas white to pink
      5. Lvs finely dissected, aromatic; heads many in flat-topped clusters; ligules ovate to round . . . . ²**ACHILLEA**
      5' Lvs entire or toothed, not aromatic; heads solitary or in small cymes; ligules narrowly linear . . . . **ECLIPTA**
1' Chaff scales folded around disk ovaries
  6. Ray and disk corollas white . . . . ²**ACHILLEA**
  6' Ray and disk corollas yellow to orange
    7. Ray fls fertile (style present, ovary well developed)
      8. Ann; disk fr compressed, margin winged; pappus of 2 awns . . . . **VERBESINA**
      8' Per; disk fr gen 4-angled; pappus 0 or of scales
        9. Lvs mostly basal, triangular, base cordate and ± hastate, cauline few, linear to oblanceolate; pappus 0
        . . . . **BALSAMORHIZA**
        9' Lvs basal and cauline; oblanceolate to obovate, base acuminate to obtuse; pappus of scales . . **WYETHIA**
    7' Ray fls sterile (style gen 0, ovary vestigial)
      10. Fr not or weakly compressed, margin not thin
        11. Pappus 0 . . . . **HELIOMERIS**
        11' Pappus present
          12. Herbs; pappus scales gen readily deciduous from fr . . . . **HELIANTHUS**
          12' Shrubs; pappus scales often persistent on fr . . . . **VIGUIERA**
      10' Fr strongly compressed, margin ± thin
        13. Lvs all basal or sub-basal . . . . **ENCELIOPSIS**
        13' Lvs mostly or all cauline
          14. Shrubs or subshrubs, woody well above base . . . . **ENCELIA**
          14' Ann . . . . **GERAEA**

Group 10: Heads radiate; receptacle naked; pappus 0 or low crown

1. Lvs opposite throughout
  2. Lf margins without embedded oil glands; not bristly-ciliate near base; herbage not scented; spring fl
  . . . . **LASTHENIA**
  2' Lf margins with embedded oil glands; bristly-ciliate near base; herbage strongly scented; summer–
    autumn fl . . . . **PECTIS**
1' Lvs alternate throughout or lower opposite
  3. Shrubs; ligules 1 mm or less . . . . ***Artemisia bigelovii***
  3' Herbs or subshrubs; ligules > 3 mm
    4. Ray corollas white to purple; pl puberulent or short-stiff-hairy . . . . **MONOPTILON**
    4' Ray corollas yellow when fresh; pl glandular or tomentose
      5. Pl glandular-hairy . . . . **BAHIA**
      5' Pl tomentose, sometimes also sessile glandular
        6. Old ray corollas papery, persistent on frs when dry, reflexed . . . . **BAILEYA**
        6' Old ray corollas withering, deciduous from maturing frs
          7. Ray corollas without very small lobe opposite ligule . . . . **ERIOPHYLLUM**
          7' Ray corollas with very small lobe opposite ligule . . . . **MONOLOPIA**

Group 11: Heads radiate; receptacle naked; pappus of scales or awns

1. Lvs and phyllaries dotted or streaked with embedded translucent oil glands — herbage strongly scented
  2. Ray corollas white to pink-purple . . . . **NICOLLETIA**
  2' Ray corollas yellow to orange or red
    3. Phyllaries in 1 series, falling with ray frs; disk pappus gen of bristles . . . . **PECTIS**
    3' Phyllaries in 2 or 3 series, persistent; pappus scales ± divided into bristles or awns
      4. Lvs toothed or divided into ± flat segments; involucre 10–18 mm . . . . **ADENOPHYLLUM**
      4' Lvs divided into stiff, needle-like segments; involucre 5–6 mm . . . . **THYMOPHYLLA**
1' Lvs and phyllaries without embedded translucent oil glands, sometimes glandular-hairy or dotted with sessile resin glands
  5. Ray corollas white to purple
    6. Per — pappus scales long, narrow, bristle-like . . . . **TOWNSENDIA**
    6' Ann
      7. Lvs wide, palmately lobed or toothed . . . . *Perityle emoryi*
      7' Lvs narrow, entire
        8. Pls puberulent or short-stiff-hairy . . . . **MONOPTILON**
        8' Pls tomentose
          9. Ligules 1.5–2.5 mm; pappus scales 2 . . . . [2]**EATONELLA**
          9' Ligules 3–7 mm; pappus scales ± 10 . . . . [2]**ERIOPHYLLUM**
  5' Ray corollas yellow to red
    10. Phyllaries in several series, strongly graduated
      11. Phyllary tips spreading to recurved; involucre gummy-resinous — pappus of 2–6 slender deciduous awns . . . . **GRINDELIA**
      11' Phyllary tips ± appressed; involucre not or barely gummy-resinous
        12. Heads ± spheric; disk fls 30–80 . . . . *Acamptopappus shockleyi*
        12' Heads narrow; disk fls 1–13
          13. Lvs elliptic to obovate; disk pappus scales many, twisted, bristle-like . . . . **AMPHIPAPPUS**
          13' Lvs linear; disk pappus scales wide, flat, straight . . . . **GUTIERREZIA**
    10' Phyllaries in 1–3 series, ± equal, not strongly graduated
      14. Lvs gen opposite throughout . . . . **LASTHENIA**
      14' Lvs gen alternate (sometimes opposite below) or all basal
        15. Old ray corollas papery, persistent on frs when dry . . . . **PSILOSTROPHE**
        15' Old ray corollas withering, deciduous from maturing frs
          16. Phyllaries in 1 series
            17. Disk fr ± compressed; ann . . . . [2]**EATONELLA**
            17' Disk fr (exc sometimes outermost) ± cylindric or club-shaped, sometimes 3–4-angled; ann to shrub . . . . [2]**ERIOPHYLLUM**
          16' Phyllaries in 2–3 ± equal series
            18. Pls glandular-hairy
              19. Pappus scales gen 4 in 2 unequal pairs; lvs entire to lobed . . . . **HULSEA**
              19' Pappus scales 1–13; lvs 1–several times ternately dissected . . . . **BAHIA**
            18' Pls not glandular-hairy —pappus scales 5–7, ± equal
              20. Lvs gen oblong to elliptic or oblanceolate, entire . . . . **DUGALDIA**
              20' Lvs linear and entire, or if wider, divided into linear lobes . . . . **HYMENOXYS**

Group 12: Heads radiate; ligules white to purple; receptacle naked; pappus of bristles

1. Lvs palmately lobed or coarsely dentate; pappus bristle 1 . . . . *Perityle emoryi*
1' Lvs entire to pinnately lobed; pappus bristles gen > 1
  2. Ligules inconspicuous, not or scarcely exceeding disk
    3. Phyllaries without resin-filled veins
      4. Phyllaries all ± =. . . . . [4]**ERIGERON**
      4' Phyllaries strongly graduated in several series. . . . . [2]*Machaerantha canescens* **var.** *shastensis*
    3' Phyllaries with 1–3 resin-filled veins, these orange or brown when dry
      5. Resin-filled phyllary veins gen 3
        6. Subshrubs; branches often thorny . . . . [3]**CHLORACANTHA**
        6' Herbs; branches not thorny. . . . . **TRIMORPHA**
      5' Resin-filled phyllary vein 1
        7. Disk narrow, disk fls few; pappus single . . . . **CONYZA**
        7' Disk wide, flat, disk fls many; pappus double . . . . [4]**ERIGERON**
  2' Ligules conspicuous, gen much exceeding disk
    8. Pappus bristles alternating with well developed scales — pls pungently ill-scented. . . . . **NICOLLETIA**

8' Pappus of bristles only or of bristles and very short scales
  9. Pappus either of 5 bristles alternating with 5 narrow scales or of 1 apically plumose bristle and a
  low crown . . . . **MONOPTILON**
9' Pappus bristles gen > 20
  10. Bristles flattened, barbed to subplumose . . . . **TOWNSENDIA**
10' Bristles cylindric, smooth to barbed
  11. Main phyllaries ± equal in 1–3 series
    12. Pappus single; rays often > 2 mm wide . . . . ³**ASTER**
    12' Pappus gen double, outer of short bristles or scales, inner of soft bristles; rays gen < 2 mm wide
    . . . . ⁴**ERIGERON**
  11' Main phyllaries unequal, overlapping in 2–several series
    13. Ray fls sterile (style 0; ovary ± vestigial) . . . . ²*Machaeranthera canescens* var. *shastensis*
    13' Ray fls fertile (style present, ovary well developed, fruiting)
      14. Shrubs or subshrubs
        15. Phyllary veins resin-filled, orange or brown when dry; pls often thorny . . . . . . ³**CHLORACANTHA**
        15' Phyllary veins not resin-filled, not orange or brown when dry; pls unarmed
          16. Ray fls 15–40 mm; lvs often sharply toothed . . . . ²**XYLORHIZA**
          16' Ray fls 5–10 mm; lvs entire
            17. Slender subshrub, ± strigose, ± glandular . . . . ²**CHAETOPAPPA**
            17' Stiff shrub, glabrous, sticky-resinous . . . . *Ericameria gilmanii*
      14' Herbs
        18. Pappus double, outer of short bristles or scales, inner of longer bristles
          19. Resin-filled veins of phyllaries gen 3, these orange or brown when dry; sts often modified
          as thorns . . . . ³**CHLORACANTHA**
          19' Resin-filled veins of phyllaries 0–1; sts not forming thorns
            20. Rays often wider than narrowly linear, 8–16 . . . . *Aster scopulorum*
            20' Rays narrowly linear, often very many . . . . ⁴**ERIGERON**
        18' Pappus single, of ± equal bristles
          21. Phyllaries herbaceous throughout (or margin scarious to ± whitish)
            22. Lvs gen 2–25 cm; pls often > 20 cm; style appendages lanceolate to awl-shaped, acute . . ³**ASTER**
            22' Lvs gen 4–12 mm; pls gen < 15 cm; style appendages ovate or oblong, obtuse . . ²**CHAETOPAPPA**
          21' Phyllary tips herbaceous, bases white to straw-colored
            23. Per, gen from rhizome; lvs entire to toothed but teeth not bristle-tipped; phyllary tips gen
            appressed . . . . ³**ASTER**
            23' Ann to per from taproot; lvs gen toothed, teeth sometimes bristle-tipped; phyllary tips
            often spreading
              24. Heads gen < 3 cm diam, often in cymes . . . . **MACHAERANTHERA**
              24' Heads 3.5–7+ cm diam, solitary . . . . ²**XYLORHIZA**

Group 13: Heads radiate; ligules yellow to orange or red; receptacle naked; pappus of bristles

1. Lvs opposite, at least below
  2. Lvs and phyllaries dotted or streaked with embedded oil glands; pls very strongly scented
    3. Subshrub; lvs sharply toothed to pinnately dissected; phyllaries in 3 series ²**ADENOPHYLLUM**
    3' Ann; lvs linear, entire, bristly-ciliate at base; phyllaries 8, in 1 series . . . . **PECTIS**
  2' Lvs and phyllaries without embedded oil glands; pls not or faintly scented
    4. Per; involucre 7–20 mm; pappus bristles >> 2 mm, free . . . . **ARNICA**
    4' Ann; involucre 5–7 mm; pappus bristles 1–2 mm, fused at base . . . . ²**SYNTRICHOPAPPUS**
1' Lvs alternate throughout
  5. Ann or bien
    6. Ray pappus 0 . . . . ²**HETEROTHECA**
    6' Ray (and disk) frs with pappus
      7. Main phyllaries unequal, in 4–6 series
        8. Pappus bristles flattened, readily deciduous; involucre gummy-resinous . . . . ³**GRINDELIA**
        8' Pappus bristles ± cylindric, gen persistent; involucre not gummy-resinous . . . . . *Machaeranthera gracilis*
      7' Main phyllaries ± equal, in 1–3 series
        9. Rays many; pappus double, outer of short scales, inner of bristles . . . . ²**PULICARIA**
        9' Rays 5–13; pappus single
          10. Phyllaries 8+, not enfolding ray fr . . . . ³**SENECIO**
          10' Phyllaries gen 5, each partially enfolding ray fr . . . . ²**SYNTRICHOPAPPUS**
5' Per to shrubs

11. Shrubs or subshrubs
 12. Outer phyllaries with an embedded, translucent, swollen oil gland near tip; plants pungently ill-scented
 . . . . ²**ADENOPHYLLUM**
 12' Outer phyllaries without oil glands; pls scented or not
  13. Main phyllaries in 1 series, equal (often a few, gen much shorter, outer phyllaries present) . . ³**SENECIO**
  13' Main phyllaries in 2–7 series, often graduated, often unequal
   14. Pappus bristles flattened; phyllaries ± ovate
    15. Phyllary tips recurved to coiled; involucre gummy-resinous; ray fls 13–40 . . . . ³**GRINDELIA**
    15' Phyllary tips ± appressed; involucre ± resinous but not visibly gummy; ray fls 1–14
     16. Rays 5–14, >> involucre; disk fls 30–80, forming frs . . . . *Acamptopappus shockleyi*
     16' Rays 1–2, barely > involucre; disk fls 3–7, staminate . . . . **AMPHIPAPPUS**
   14' Pappus bristles ± cylindric; phyllaries gen linear to oblong or narrowly lanceolate
    17. Woody sts ± prostrate, pl cushion-forming . . . . ²**STENOTUS**
    17' Woody sts gen erect
     18. Lvs entire . . . . **ERICAMERIA**
     18' Lvs gen sharply toothed . . . . **HAZARDIA**
11' Herbs
 19. Main phyllaries in 1 series, equal (often with a few, gen much shorter outer phyllaries) . . . . . . ³**SENECIO**
 19' Phyllaries in 2 or more series, subequal to strongly graduated
  20. Disk pappus double, outer < 1mm, of bristles or scales, inner of bristles 2–9 mm
   21. Ligules 4–10 mm; disk fr 2–5 mm, ± flat; pappus 6–9 mm . . . . ²**HETEROTHECA**
   21' Ligules 1.5–2 mm; disk fr ± 1 mm, 5-ribbed; pappus 2–3 mm . . . . ²**PULICARIA**
  20' Disk pappus single
   22. Pappus bristles flat, readily deciduous; involucre strongly gummy-resinous, esp in bud . . ³**GRINDELIA**
   22' Pappus bristles ± cylindric, gen persistent; involucre not or moderately resinous
    23. Lvs toothed or lobed
     24. Involucre 2.5–6 mm; heads gen small, in racemes or panicles, sometimes clustered on 1 side of
       branches . . . . ²**SOLIDAGO**
     24' Involucre 5–22 mm; heads gen not small, not in 1-sided clusters
      25. Pl from stout taproot; basal rosette well developed . . . . **PYRROCOMA**
      25' Pl from slender taproot or caudex; basal rosette 0 or poorly developed
       26. Fr 2–3 mm, obconic; lf teeth or lobes bristle- or minutely spine-tipped; upper lvs scale-like
        . . . . *Machaeranthera pinnatifida*
       26' Fr 5 mm, ± cylindric or sometimes compressed; lf teeth gen not bristle-tipped or spine-tipped;
        upper lvs gen little reduced . . . . **TONESTUS**
    23' Lvs entire
     27. Phyllaries in vertical ranks; disk fls staminate . . . . **PETRADORIA**
     27' Phyllaries not in vertical ranks; disk fls forming frs
      28. Pl low, mat-forming; head solitary; involucres 6–12 mm diameter . . . . ²**STENOTUS**
      28' Pl erect; heads many; involucre 2–4 mm diameter
       29. Lvs resin-dotted . . . . **EUTHAMIA**
       29' Lvs not resin-dotted . . . . ²**SOLIDAGO**

ACAMPTOPAPPUS GOLDENHEAD

Meredith A. Lane

Subshrubs, appearing glabrous. **STS** decumbent, widely spreading or erect, striate; old growth gray, bark sometimes shreddy with age; new growth white below, tips green. **LVS** simple, sometimes in axillary clusters below, spreading-ascending to appressed-erect, linear to narrowly oblanceolate, gen minutely spine-tipped, pale green to light gray-green. **INFL**: heads radiate or discoid, 1–many in rounded to ± flat-topped clusters, very small in bud, expanding rapidly in fr; involucres widely bell-shaped to nearly spheric; phyllaries 20–24 in 2–3 series, ovate to ovate-elliptic, bases cream-yellow, tips green, margins scarious; receptacle deeply pitted, with projections between fls but not chaffy. **RAY FLS** present or 0; ligules yellow. **DISK FLS** many; corollas funnel-shaped, yellow, sinuses deep, lobes spreading to reflexed, style-branch appendages lanceolate. **FR** densely long-hairy, hairs white, bronze, or brownish; pappus of 20–25 wide, stiff, widely spreading bristles, slightly > fr. 2 spp.: sw US. (Greek: unbending pappus) [Lane 1988 Madroño 35:247–265]

1. Ray fls present; involucres bell-shaped to hemispheric . . . . *A. shockleyi*
1' Ray fls 0; involucres hemispheric to spheric . . . . *A. sphaerocephalus*
 2. Sts and lvs rough-puberulent . . . . var. *hirtellus*
 2' Sts and lvs glabrous (or lf margins sparsely rough-puberulent) . . . . var. *sphaerocephalus*

A. shockleyi A. Gray (p. 113) **STS** decumbent to ascending, gen < 4 dm, gen minutely hairy to scabrous. **LVS** < 2.5 cm, < 4 mm wide, gen oblanceolate, densely, minutely hairy to scabrous. **INFL**: heads few, terminal; involucre bell-shaped to hemispheric. **RAY FLS** 5–14; corollas < 2 cm, ligules < 6 mm wide. **DISK FLS** 30–80; corollas 2–5 mm. **FR** < 5 mm. 2*n*=18. Slopes and ridges; 500–2000 m. SNE, **DMoj**; s NV. ❀TRY. Apr–Jun

A. sphaerocephalus (A. Gray) A. Gray (p. 113) **STS** much-branched, ascending to erect, gen < 1 m. **LVS** < 1.5 cm, < 3 mm wide, gen linear to oblanceolate. **INFL**: heads very many, borne singly or clustered; involucre hemispheric to spheric. **RAY FLS**

0. **DISK FLS** 13–27; corollas 2–3.5 mm. **FR** < 3 mm. 2*n*=18. Gravelly or rocky soils on flats or slopes in deserts to juniper woodlands; 60–2200 m. Teh, SnGb, SnBr, PR, **SNE, D**; to UT, s NV, AZ.

var. *hirtellus* S.F. Blake Herbage ± densely rough-puberulent. **STS** gen < 6 dm. Habitats of sp.; < 1600 m. Teh, SnGb, SnBr, **SNE, DMoj**; AZ, s NV. May–Jun

var. *sphaerocephalus* Herbage glabrous (or lf margins sparsely rough-puberulent). **STS** gen < 1 m. Habitats of sp.; Teh, SnGb, SnBr, PR, **D**; to UT, s NV, AZ. < 2200 m. Apr–Jun

ACHILLEA YARROW, MILFOIL

Per, strongly scented. **LVS** alternate, simple to 3-pinnately dissected, ± hairy, ± reduced upward. **INFL**: heads gen radiate, many, in flat-topped clusters; involucre bell-shaped or ovoid; phyllaries graded in 3–4 unequal series, ovate, obtuse; margins membranous; receptacle flat to rounded; chaff scales narrow, transparent. **RAY FLS** few; ligules short, round, white, pink, or yellow. **DISK FLS** ± many; corollas short, white to purple or yellow. **FR** oblong to obovate, compressed, thick-margined, glabrous; pappus 0. ± 85 spp.: N.Am, Eurasia, n Afr. (Greek: Achilles of ancient mythology) [Tyrl 1975 Brittonia 27:187–196]

A. millefolium L. (p. 113) Pl 10–200 cm. **LVS** very finely 3-pinnately divided; cauline lvs ± clasping. **INFL**: phyllaries 4–9 mm. **RAY FLS** gen 3–8; ligules 2.5–4 mm, ovate to round, white to pink. **DISK FLS** 15–40; corollas 2–3 mm, white to pink. **FR** ± 2 mm. 2*n*=36,45,54,63,72. Many habitats; < 3500 m. CA-FP, **GB**; circumboreal. [*A. borealis* Bong. sspp. *arenicola* (A.A. Heller) Keck, *californica* (Pollard) Keck; *A. lanulosa* Nutt. sspp. *alpicola* (Rydb.) Keck, *lanulosa*; *A. m.* vars. *alpicola* (Rydb.) Garrett, *arenicola* (A.A. Heller) Nobs, *borealis* (Bong.) Farwell, *californica* (Pollard) Jepson, *gigantea* (Pollard) Nobs, *lanulosa* (Nutt.) Piper, *littoralis* Nobs, *pacifica* (Rydb.) G.N. Jones, *puberula* (Rydb.) Nobs] Highly variable polyploid complex; lf size and hairiness esp variable. ❀SUN,DRN:**4–6,17**&IRR:**1–3,7–9,14–16,18–24**& part SHD:**10**,11–13;GRNCVR;CVS; rather INV; also STBL. Summer

ACROPTILON RUSSIAN KNAPWEED

David J. Keil & Charles E. Turner

1 sp.: Eurasia. (Greek: feather-tipped, from pappus bristles)

A. repens (L.) DC. (p. 113) RUSSIAN KNAPWEED Per 3–10 dm, from dark rhizome. **STS** erect; branches ascending, ± cobwebby-tomentose. **LVS** below middle 4–10 cm, oblong, 1–2-pinnately lobed, above middle 1–3 cm, linear to narrowly lanceolate, entire to toothed; blades glabrous to ± tomentose. **INFL**: heads discoid in ± flat-topped or panicle-like clusters, leafy; involucre 10–14 mm, ± ovoid; phyllaries in several unequal series, entire, ± soft-hairy, inner narrower, tips widely scarious; receptacle bristly; fls ± 30. **FL**: corolla ± 15 mm, white to blue, tube very slender, abruptly wider, lobes linear; anther bases short-tailed, tips oblong; style with minutely hairy, swollen node above, papillate above node, branches very short, tips triangular. **FR** 3–4 mm, obovoid, slightly compressed, glabrous; pappus bristles many, 6–8 mm, ± deciduous, barbed below, short-plumose above. 2*n*=26. Fields, roadsides, cult ground; < 1900 m. **CA** (exc wettest NW, driest GB, D); N.Am; native to c Asia. [*Centaurea r.* L.] NOXIOUS WEED. May–Sep

ADENOPHYLLUM

Ann, per, subshrubs. **STS** erect. **LVS** simple or pinnate, opposite or alternate, dotted with embedded oil glands. **INFL**: heads radiate or discoid, solitary or in few-headed, often leafy-bracted, often open cymes; peduncles slender, with lf-like or scale-like bracts; involucre bell-shaped to cylindric; phyllaries in 3 series, outer << others, free, 2 inner ± equal, fused or free, gland-dotted; receptacle flat, bearing short, fringed scales. **RAY FLS** 0–few; corollas yellow to red. **DISK FLS** few–many; corollas yellow to orange; style tips tapered. **FR** obconic, ribbed; pappus of scales dissected into slender bristles. 10 spp.: sw US, n Mex. (Greek: gland-leaf) [Strother 1986 Sida 11:371–378]

1. Lvs simple, ovate to oblanceolate, coarsely toothed or shallowly lobed; involucres 15–18 mm; pappus
    scales 15–20 . . . . *A. cooperi*
1' Lvs deeply pinnately divided into 3–5 linear lobes; involucres 10–15 mm; pappus scales 8–12
    . . . . *A. porophylloides*

A. cooperi (A. Gray) Strother (p. 113, pl. 13) Subshrub, glabrous or short-rough-hairy; odor unpleasant. **STS** many, 3.5–6 dm, gen much-branched. **LVS** gen alternate, sessile, 8–25 cm, ovate to oblanceolate, coarsely toothed or shallowly lobed; blade with 1–2 glands near base, 1 near tip. **INFL**: heads radiate (rarely discoid); peduncles 6–15 cm, swollen beneath heads, bracts lf-like; involucre cylindric; outer phyllaries 12–22, 5–8 mm, linear, long-acuminate, each with 1 central gland; inner phyllaries ± 20, 15–18 mm, linear, fused below, gland-dotted. **RAY FLS** (0)7–13; ligule 8–9 mm, yellow to red-orange. **DISK FLS** many; corollas yellow 8–10 mm. **FR** 5–7 mm, hairy; pappus scales 15–20, 8–10 mm, each dissected into 5–9 bristles. 2*n*=26. Dry, sandy slopes and washes; 600–1550 m. **DMoj**; s NV, nw AZ. [*Dyssodia c.* A. Gray] May–Jun, Sep–Nov

A. porophylloides (A. Gray) Strother (p. 113) Subshrub, glabrous; odor unpleasant. **STS** many, 2–6 dm, gen much-branched. **LVS** opposite below, alternate above, 1.5–4 cm, deeply divided;

lobes linear, entire to sharply serrate, each with 1 gland at base, 1 near tip. **INFL**: heads radiate (discoid); peduncles 2–8 cm, naked or with 1–5 narrow bracts; involucre cylindric; outer phyllaries 12–16, 3–8 mm, erect or recurved, each with a central gland; inner phyllaries 12–20, 10–15 mm, lanceolate, fused below, gland-dotted. **RAY FLS** (0)8–12; ligules 2–4 mm, yellow to red-orange. **DISK FLS** many; corollas 7–8 mm, yellow-orange. **FR** 5 mm, sparsely hairy; pappus scales 8–12, 7–8 mm, each dissected into 7–11 bristles. 2*n*=26. Dry, rocky hillsides, washes; 200–1460 m. **s DMoj, DSon**; to AZ, n Mex. [*Dyssodia p.* A. Gray] Mar–Jun

AGERATINA

A. Michael Powell

Per, shrubs. **LVS** gen opposite; blade elliptic to triangular, margin entire to lobed. **INFL**: heads discoid, solitary or in ± flat-topped cymes; phyllaries subequal, in 1–2(3) series; receptacle flat to conic, naked. **FLS** 10–60; corollas ± white or blue to pink-tinged, cylindric (or throat wider). **FR** 5-angled, gen 5-ribbed; pappus of 5–40 slender scabrous bristles, often easily detached. (Latin: resembling *Ageratum*) [King & Robinson 1987 Monogr Syst Bot Missouri Bot Garden 22:428–436]

1. Lvs opposite; heads 6–8 mm . . . . *A. herbacea*
1' Lvs gen alternate; heads 8–10 mm . . . . *A. occidentalis*

A. herbacea (A. Gray) R. King & H. Robinson (p. 119) Per; caudex woody. **ST** 4.5–7 dm, erect or spreading, green, puberulent. **LVS** gen opposite; blade gen triangular to ± cordate, yellowish to light or grayish green, glabrous to puberulent. **INFL**: heads 6–8 mm, in dense clusters; phyllaries puberulent. **FL**: corolla white. **FR** 2–3 mm. 2*n*=34. Common. Rocky pinyon/juniper woodland; 1600–2200 m. **e DMtns (Clark, New York, Providence mtns)**; to Colorado, NM, w TX, n Mex. [*Eupatorium h.* (A. Gray) E. Greene] ❀TRY. May–Jun, Oct–Nov

A. occidentalis (Hook.) R. King & H. Robinson (p. 119) Per; caudex woody, ± rhizomatous. **ST** 1.5–7 dm, erect or ascending, ± green or purple, puberulent. **LVS** gen alternate; blade ± triangular, ± serrate, glandular. **INFL**: heads 8–10 mm, clustered; phyllaries puberulent. **FL**: corollas ± white to blue. **FR** 3–3.5 mm. 2*n*=34. Common. Rocks; 2100–3700 m. NW, CaRH, SN, Wrn, **n SNE, W&I**; to WA, ID, UT. [*Eupatorium o.* Hook.] Pls from SN are smaller than those from NCoR. ❀ DRN,SHD:**2,7**,8–11,14–21;DFCLT. Jul–Sep

AGOSERIS

Kenton L. Chambers

Ann or per, gen scapose; sap milky. **LVS** ± basal, narrow, entire to pinnately lobed. **INFL**: head 1, ligulate, erect, 8–60 mm; phyllaries subequal or overlapping in 2–4 series, inner often elongated in fr; receptacle flat, naked. **FLS** gen many; ligules = to >> involucre, yellow or red-orange, readily withering, outer ± reddish below. **FR** linear to fusiform, ± 10-ribbed; beak often = or >> body; pappus of many fine, simple, white bristles. ± 10 spp.: Am. (Greek: goat chicory) [Q. Jones 1954 PhD Harvard U.] Like *Taraxacum* but closely related to *Nothocalais*. Hybrids and polyploidy complicate variation in some spp.

1. Corollas orange or brick-red, drying purplish; SNE . . . . *A. aurantiaca*
1' Corolla yellow (often reddish on back), drying pinkish; GB . . . . *A. glauca*
 2. Pl 20–50 cm; infl glabrous; lvs entire or few-toothed, ± glabrous; W&I . . . . var. *glauca*
 2' Pl 3–25 cm; upper infl axis (and sometimes phyllaries) hairy; GB
  3. Infl hairs nonglandular, opaque; outer phyllaries lanceolate, often abruptly long-tapered; lvs long-
   tapered, irregular lobes often angled toward base . . . . var. *laciniata*
  3' Some infl hairs glandular, translucent; outer phyllaries ± ovate, evenly tapered; lvs acute, ± entire
   . . . . var. *monticola*

A. aurantiaca (Hook.) E. Greene (p. 119) Per 10–50 cm. **LVS** gen linear to (widely) oblanceolate, (acute to) long-tapered, entire to irregularly lobed in lower 2/3, (sub)glabrous. **INFL**: base of head tomentose; involucre12–30 mm; outer phyllaries gen narrowly oblong-lanceolate, long-tapered, < inner phyllaries in fr, glabrous to soft-hairy, nonglandular. **FLS** many,= to > involucre; ligules orange or brick-red, drying purplish or dark pink. **FR**: body 4–9 mm, fusiform; ribs often minutely hairy; beak sometimes < body. 2*n*=18,36. Meadows, shrubland, streamsides; 1500–3500 m. KR, NCoRH, CaRH, SNH, **GB (exc W&I)**; to AK, w Can, SD, NM. Reported yellow-fld pls from outside CA much like *A.glauca*. ❀ TRY. Jul–Aug

A. glauca (Pursh) Raf. Per 3–50 cm, decumbent to erect, gen ± tomentose. **LVS** linear to (ob)lanceolate, acute to long-tapered, entire to lobed. **INFL**: involucre 10–25 mm; phyllaries overlapping in several series, outer ovate and acute to lanceolate and long-tapered, glabrous to ciliate or soft-hairy. **FLS** many, > involucre; ligules yellow. **FR**: body 4–10 mm, linear-fusiform; ribs straight, smooth; beak < body. Sagebrush scrub, coniferous forest, meadows, alpine slopes; 1400–3800 m. KR, CaRH, SNH, GB; to w Can, n-c US, NM.

var. *glauca* (p. 119) Pl 20–50 cm, glabrous. **LVS** linear to oblanceolate, long-tapered, entire or few-toothed, (sub)glabrous, glaucous. **INFL**: phyllaries lanceolate, long-tapered, often marked reddish, glabrous. 2*n*=36. Meadows, open coniferous woods; 1400–2500 m. **W&I**; range of sp. outside CA. Jul–Aug

var. *laciniata* (D. Eaton) F.J. Smiley (p. 119) Pl 3–25 cm, ± soft-white-opaque-hairy (esp below head), nonglandular. **LVS** ± narrowly lanceolate, long-tapered, gen pinnately (and downwardly) lobed. **INFL**: outer phyllaries lanceolate, ± long-tapered, gen glabrous, red-marked. 2*n*=18. Dry, open sagebrush scrub, coniferous woods; 1400–3800 m. CaRH, SNH, **GB**; to ID, MT, WY, NM. Intergrades with var. *monticola*. May–Aug

var. *monticola* (E. Greene) Q. Jones (p. 119) **STS** 3–40 cm, ± soft-yellow-translucent-hairy (esp below head), ± glandular. **LVS** gen oblanceolate, acute, glabrous, entire to short-lobed. **INFL**: outer phyllaries ± ovate, ± acute, gen hairy and glandular, often red-marked. 2*n*=18,36. Moist meadows, streambanks, coniferous woods, alpine slopes; 1500–3800 m. KR, CaRH, SNH, **GB**; to WA, NV. [var. *dasycephala* (Torrey & A. Gray) Jepson misapplied] Jul–Aug

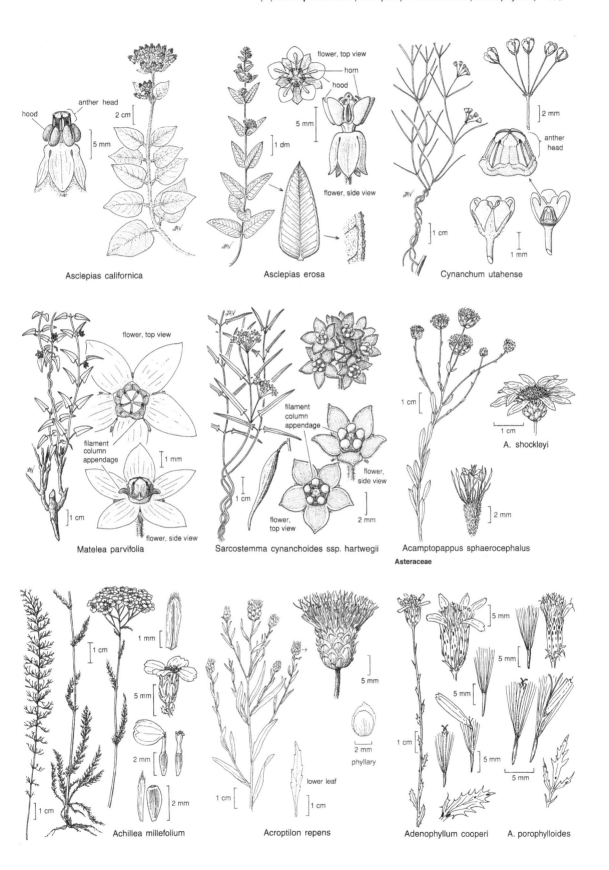

Asclepias californica

Asclepias erosa

Cynanchum utahense

Matelea parvifolia

Sarcostemma cynanchoides ssp. hartwegii

Asteraceae

Acamptopappus sphaerocephalus

A. shockleyi

Achillea millefolium

Acroptilon repens

Adenophyllum cooperi   A. porophylloides

AMBROSIA RAGWEED, BUR-SAGE

Willard W. Payne

Ann to shrub, monoecious. **LVS** often opposite below, gen alternate above, gen petioled, hairy, glandular. **INFL**: staminate heads gen many in ± spikes or racemes, phyllaries fused into shallow cup; pistillate heads gen clustered below staminate, gen spiny, bur-like; involucre ± 0; receptacle chaffy; chaff scales spirally arrayed, fused below, tips gen becoming spiny; each pistillate fl in separate chamber. **STAMINATE FLS** ± many; corolla yellow or translucent; anthers free; style unbranched. **PISTILLATE FLS** 1–5; corolla 0; style branches long. **FR** enclosed in bur; pappus 0. (Greek: early name for aromatic plants; mythic food of the gods) [Payne 1976 Pl Syst Evol 125:169–178] Closely related to (indistinct from) *Hymenoclea* [Miao et al. 1995. Amer J Bot 82: 924-932; Baldwin et al. 1996. Madroño 43: 15-27] Wind-blown pollen often highly allergenic. Key to species adapted by Margriet Wetherwax.

1. Ann from slender tap-root
  2. Spines of bur many, scattered, sharp; distal staminate phyllary tips ± black-lined . . . . ***A. acanthicarpa***
  2' Spines of bur few, below beak, blunt; staminate phyllary tips uniformly green . . . . ***A. artemisiifolia***
1' Subshrub or shrub; sts persisting 2+ seasons
  3. Spines of burs hooked, round in ×-section . . . . ***A. ilicifolia***
  3' Spines of burs straight, flat in ×-section
    4. Cauline lvs ± compound; bur puberulent . . . . ***A. dumosa***
    4' Cauline lvs few-toothed to shallowly lobed; bur long-soft-hairy . . . . ***A. eriocentra***

A. acanthicarpa Hook. (p. 119) ANNUAL BUR-SAGE Ann 4–15 dm, from slender taproot. **ST** gray-green, ± stiffly strigose-bristly. **LVS**: petioles winged; blade pinnately divided, < 8 cm, 7 cm wide. **INFL**: staminate heads 2–5 mm diam, involucre lobes 3–9, tips of longest 3 ± black-lined along midveins; pistillate heads 1-fld. **FR**: body of bur 5–7 mm, ovoid, gen ± golden, glabrous or puberulent; spines 0–30, scattered, flat, straight, sharp. 2*n*=36. Sandy plains, disturbed sites, many communities; < 2200 m. NCoR, SN, SW, **GB, D**; to WA, MT, WY, TX, nw Mex. [*Franseria a.* (Hook.) Cov.] Aug–Nov

A. artemisiifolia L.(p. 119) COMMON RAGWEED Ann < 7 dm, much-branched. **ST** green, red- or black-marked, weakly hairy. **LVS** opposite below; cauline lvs 3–12 cm, widely ovate, gen 2–3-pinnately parted, ± hairy. **INFL**: staminate heads 2–5 mm diam, involucre asymmetric, lobes 5–10, green; pistillate heads 1-fld. **FR**: bur 2–4 mm, widely obconic, green to brown, ± puberulent; spines 4–12, blunt, vestigial, ± in 1 whorl below beak. 2*n*=36. Uncommon. Disturbed sites; < 650 m. NW, e ScV, SCo, **SNE**; native to e US. Aug–Oct

A. dumosa (A. Gray) Payne (p. 119) BURRO-WEED Shrub 2–9 dm, much-branched. **STS** softly gray-white canescent. **LVS** gen ± clustered on short branches, (sub)sessile; blade 0.5–4 cm, ± ovate, 1–3-pinnate, canescent. **INFL**: staminate heads 3–5 mm diam, involucre lobes 5–8; pistillate heads 2-fld. **FR**: bur 5–9 mm, spheric, golden to purple or brown, puberulent; spines 12–

35, scattered, flat, straight, sharp. 2*n*=36,72,108,126,144. Creosote-bush scrub; < 1600 m. SNE, **D**; to sw UT, AZ, nw Mex. [*Franseria d.* A. Gray] Hybridizes with *Hymenoclea salsola*) ❀TRY. Feb–Jun, Sep–Nov

A. eriocentra (A. Gray) Payne (p. 119) WOOLLY BUR-SAGE Shrub 3–18 dm, ± spheric. **STS** gray-brown, ± woolly, becoming glabrous. **LVS**: petioles winged; blades 1–9 cm, ± lanceolate, coarsely toothed or pinnately lobed, ± rolled under, ± green above, **INFL**: staminate heads few, 5–7 mm diam, involucre lobes 5–8; pistillate heads gen 1-fld. **FR**: bur 8–11 mm, greenish brown, densely long-soft-hairy; spines 12–20, gen near middle, straight, flat, stout, sharp, tips ± hair-tufted. 2*n*=36. Dry washes and slopes; 800–1700 m. **e DMoj, DMtns**; to sw UT, nw AZ. Mar–May

A. ilicifolia (A. Gray) Payne (p. 119) Shrub < ± 1 m, matted. **STS** greenish when young, densely glandular and stiffly short-hairy; older bark gray-white. **LVS** 2–10 cm, ovate to round, leathery, brittle, dark gray-green, sticky, veiny below; teeth long, spine-tipped; sessile, ± clasping. **INFL**: staminate heads 10–15 mm diam, involucre lobes 10–15, lanceolate, spine-tipped; pistillate heads gen 2-fld. **FR**: bur < 20 mm, spheric, ± brown, sticky; spines 20–70, scattered, < 6 mm, ± cylindric, strongly hooked. 2*n*=36. Sandy washes, rocky canyons, creosote-bush scrub; < 500 m. SnGb, SnBr, **s D**; w AZ; Baja CA. [*Franseria i.* A. Gray] Feb–Apr

AMPHIPAPPUS CHAFF-BUSH

Meredith A. Lane

1 sp. (Greek: double pappus) [Porter 1943 Amer J Bot 30:481–483]

A. fremontii Torrey & A. Gray (p. 119) Shrub, glabrous. **STS** much-branched, widely spreading to ascending, gen < 6 dm, striate, smooth, white above, grayish below; lfless sts spiny. **LVS** short-petioled, gen < 2 cm, obovate or elliptic, entire, sometimes ± thick, light yellow- or gray-green. **INFL**: heads radiate, in crowded, flat-topped clusters gen 3–5 cm wide; involucres cylindric, ± 5 mm, < 3 mm wide; phyllaries 7–12, ovate, whitish or pale green. **RAY FLS** 1–2, barely > involucre, tips 2- or 3-toothed, yellow. **DISK FLS** 3–7, staminate; corollas narrowly funnel-shaped, sinuses deep, lobes reflexed, style-branch appendages lanceolate to rounded. **FR** of ray fls < or = 3 mm, hairy; pappus of 15–20 stout bristles fused at base, gen 1 mm; ovary of disk fls

< or = 1 mm, glabrous; pappus of 25 flattened, twisted bristles gen < or = 3 mm. 2*n*=18. Rocky or gravelly flats, slopes, canyons; < 1600 m. **c&e DMoj**; to UT, NV, AZ.

1. Lvs glabrous . . . . var. ***fremontii***
1' Lvs scabrous . . . . var. ***spinosus***

var. ***fremontii*** **LVS** glabrous, sometimes gummy. Habitats of sp. **ne DMoj**; NV. Apr–May

var. ***spinosus*** (Nelson) C.L. Porter **LVS** minutely scabrous on surfaces and margins. Habitats of sp. **e-c DMoj**; to UT, NV, AZ. [ssp. *s.* (Nelson) Keck] ❀TRY.

ANCISTROCARPHUS WOOLLY FISHHOOKS

James D. Morefield

1 sp. (perhaps another undescribed). (Greek: fishhook chaff) Often placed in *Stylocline*, but more closely related to *Hesperevax* or Eur *Cymbolaena*.

A. filagineus A. Gray (p. 119) Ann, ± gray, cobwebby to tomentose. **STS** 1–several from base, spreading to erect, < 15 cm, ± forked. **LVS** simple, alternate, ± sessile, < 30 mm, linear-oblanceolate to ± elliptic, entire. **INFL**: heads disciform, ± sessile in leafy-bracted groups of 2–5, < 9 mm; phyllaries 0 or 3–6, vestigial, ± equal, membranous; receptacle 1–2 × longer than wide, expanded at tip, chaffy; outer chaff scales phyllary-like, each enclosing a pistillate fl, falling with a fr, woolly, strongly 3-veined, body hard, tip scarious-winged; innermost chaff scales each subtending a disk fl, 2.7–4.1 mm (> outer), open, concave, persis-tent, thinly woolly, in fr rigid, enlarged, spreading, tip hooked strongly inward, forming a spine. **PISTILLATE FLS** 5–10 in 1–2 series; corollas tubular. **DISK FLS** staminate, 3–5, 1–1.5 mm, 4–6-lobed. **FR** 1.4–2 mm, 0.6–0.9 mm wide, obovoid, com-pressed front-to-back, smooth, dull, black-banded near base; pappus 0. Bare or grassy, often serpentine slopes, road beds, vernally moist places; < 1900 m. CA-FP (exc coast), MP, **sw DMoj**; to OR, ID, NV, nw Baja CA. [*Stylocline f.* (A. Gray) A. Gray incl var. *depressa* Jepson] Apr–May

ANISOCOMA SCALE BUD

G. Ledyard Stebbins

1 sp.: sw US, n Mex. (Greek: unequal pappus)

A. acaulis Torrey & A. Gray (p. 119) Ann; sap milky. **LVS** all basal, 3–5 cm, pinnately lobed; lobes toothed, ± hairy. **INFL**: head ligulate, solitary, 2–3 cm; peduncle 5–20 cm, glabrous; phyl-laries papery-transparent on margins, often with reddish tips and dots, in several series, outer short, oblong, blunt, inner linear, pointed; receptacle with long, narrow chaff scales. **FLS** many; ligules pale yellow, readily withering. **FR** 10–15-veined; pappus of plumose bristles in 1 series, white, those on inner side of achene < on outer. 2*n*=18. Sandy washes, dry slopes; 600–2400 m. s SN, Teh, SnJV, SCoRO, TR, PR, **SNE, D**; NV, AZ, Baja CA. ❀TRY. Apr–Jun

ANTENNARIA PUSSY-TOES, SPRING EVERLASTING

G. Ledyard Stebbins & Randall J. Bayer

Per, often matted, dioecious; staminate pls gen present. **LVS** alternate, entire, gen ± tomentose. **INFL**: heads discoid or disciform, 1 or in cyme-like clusters; phyllaries many in several series, papery or membranous (staminate wider, more conspicuous); receptacle naked. **STAMINATE FLS** 2–5 mm; corollas white, yellow, or red; pappus bristle tips gen enlarged. **PISTILLATE FLS** 2–10 mm; corollas barely lobed, white, yellow or red. **FR** 0.5–3.5 mm, ± elliptic; pappus bristles many, soft, weakly barbed. ± 40 spp.: Am, n Eurasia. (Latin: antenna) [Bayer 1990 Can J Bot 68:1389–1397 & Madroño 37:171–183] Many races reproduce by asexual seeds, their populations entirely pistillate pls. Key to species adapted by Randall J. Bayer.

1. Head 1 per fl st . . . . *A. dimorpha*
1' Heads 2–3 per fl st
  2. Phyllaries dark brown-black; sts 3-13 cm
    3. Lowest cauline lf > 11 mm; herbage gen not glandular; pistillate corolla > 3.0 mm in fr; staminate
      corolla gen > 2.8 mm in fl . . . . *A. media*
    3' Lowest cauline lf < or = 11 mm; herbage gen glandular; pistillate corolla < or = 3.0 mm in fr; staminate
      corolla gen < or = 2.8 mm in fl. . . . . *A. pulchella*
  2' Phyllaries white, rose, straw-colored, or pale brown; sts 6-40 cm
    4. Stolons ascending, slightly woody; phyllaries pale yellow to pale brown; staminate pls present —
      W&I . . . . *A. umbrinella*
    4' Stolons spreading horizontally or ascending, not woody; phyllaries various colors; staminate pls ± 0 . . *A. rosea*
      5. Longest lf of fl rosettes < 20 mm; phyllaries sometimes brown . . . . ssp. *confinis*
      5' Longest lf of fl rosettes > 20 mm; phyllaries gen not brown . . . . ssp. *rosea*

A. dimorpha (Nutt.) Torrey & A. Gray (p. 119) **STS** < 4 cm, cespitose from much-branched caudex; stolons 0. **LVS** basal, 8–11 mm, linear to narrowly spoon-shaped, 1-veined, ± gray-tomentose. **INFL**: head 1; involucres 6–8 mm (staminate) or 10–11 mm (pistillate), base hairy; phyllaries narrow, acute-acuminate, dingy brown. **STAMINATE FLS**: corollas 3–5 mm; pappus bristle tips slender. **PISTILLATE FLS**: corollas 8–10 mm. **FR** 2–3.5 mm, hairy; pappus 10–12 mm. 2*n*=28,56. Dry places; 800–2400 m. KR, NCoRH, SNH, TR, **GB**; to sw Can, MT, WY, NE. May–Jul

A. media E. Greene (p. 119) **STS** 5–13 cm; stolons many, matted. **LVS**: basal 6–19 mm, ± linear to spoon-shaped, 1-veined, densely woolly; cauline 5–20 mm, linear. **INFL**: heads 2–7; in-volucres 4–8 mm, base woolly; phyllaries narrow and acute (pistillate) or wide and blunt (staminate), upper part dark brown-ish black. **STAMINATE FLS**: corollas 2.8–4.5 mm. **PISTIL-LATE FLS**: corollas 3–4.5 mm. **FR** 0.6–1.6 mm, papillate or smooth; pappus 4–5.5 mm. 2*n*=56,98,112. High meadows, snow basins, ridges; 1800–3900 m. KR, NCoRH, CaRH, SNH, SnBr, **W&I**; to sw Can, MT, Colorado, NM. [*A. alpina* (L.) Gaertner var. *m.* (E. Greene) Jepson] ❀DRN,IRR,SUN:1–3,7,14–17;DFCLT. Jul–Aug

A. pulchella E.Greene BEAUTIFUL PUSSY-TOES **STS** 3–12 cm; stolons many, matted. **LVS**: basal 6–12 mm, ± linear to spoon-shaped,1-veined, densely woolly, often purple-glandular; cauline 3–11 mm, linear. **INFL**: heads 4–6; involucres 3.5–5 mm, base woolly; phyllaries narrow and acute (pistillate) or wide and blunt (staminate),upper part gen dark brownish black (to whitish at very tip). **STAMINATE FLS**: corollas 1.9–2.8. **PISTILLATE FLS**: corollas 2–3 mm. **FR** 0.7–1.3 mm, papillate or smooth; pappus 2.5–3.5. 2*n*=28.UNCOMMON. High meadows,snow basins, ridges; 2800–3700 m. SNH, **n SNE (Sweetwater Mtns)**. [*A. alpina* (L.) Gaertnervar. *scabra* Jepson] Jul–Aug

A. rosea E. Greene Pls gen all pistillate. **STS** 9–40 cm; stolons short, horizontal or ascending. **LVS**: basal 8–40 mm, spoon- to wedge-shaped, 1-veined, ± gray-tomentose; cauline 6–36 mm. **INFL**: heads 3–16; involucre 3.5–7.5 mm, base hairy; phyllaries wide, acute, tips white, rose, yellowish, or brownish. **PISTILLATE FLS**: corollas 2.5–5.5 mm. **FR** 0.7–1.8 mm, papillate or smooth; pappus 3.5–6.5 mm. 2*n*=42,56,70. Woods, meadow edges, rock barrens, dry ridges; 1200–3700 m. KR, CaRH, SNH, SnGb, SnBr, SnJt, **GB**; to AK, e Can, NM. [*A. microphylla* Rydb. misapplied, in part] Highly variable; sspp. intergrade extensively.

ssp. *confinis* (E. Greene) R. Bayer(p. 119) **STS** 9–25 cm; stolons decumbent, 15–45 mm. **LVS**: basal 8–20 mm; cauline 6–20 mm. **INFL**: heads 4–11; involucre 4–6.5 mm; phyllaries often brown. **FL**: corollas 2.5–4 mm. **FR**: pappus 3.5–5 mm. Habitat and range of sp. Jun–Aug

ssp. *rosea* **STS** 10–40 cm; stolons decumbent, 20–70 mm. **LVS**: basal 20–40 mm; cauline 8–36 mm. **INFL**: heads 6–20; involucre 5–8 mm; phyllaries not brown. **FL**: corollas 3–4.5 mm. **FR**: pappus 4–6 mm. Habitats and range of sp. ❀DRN,IRR,SUN: 1–3,6,7,15–17;DFCLT. Jun–Aug

A. umbrinella Rydb. **STS** 7–16 cm; base ± woody; stolons or sterile shoots ascending, slightly woody. **LVS**: basal 10–17 mm, spoon- to wedge-shaped, 1-veined, ± gray-tomentose; cauline 8–18 mm. **INFL**: heads 3–8; involucres 3–6.5 mm, base woolly; phyllaries narrow and acute (pistillate) or wide and blunt (staminate), upper part whitish to pale brownish. **STAMINATE FLS**: corollas 2.5–3.5 mm. **PISTILLATE FLS**: corollas 2.5–3.5 mm. **FR** 0.5–1.2 mm, glabrous; pappus 3–5 mm. 2*n*=28,56. Uncommon. Dry sagebrush scrub, open yellow-pine forest; 1800–2000 m. n SNH (Plumas Co.), **W&I**; to sw Can, MT, Colorado, NV.

ARNICA

Theodore M. Barkley

Per gen from long, naked rhizome. **LVS**: basal 0 or gen withered by fl; cauline opposite. **INFL** ± flat-topped; heads radiate or discoid, 1–many; involucre hemispheric to obconic; phyllaries gen in 2 ± equal series; receptacle ± flat, naked. **RAY FLS** (0)6–21; ligules (orange-)yellow. **DISK FLS** many; corolla gen soft-hairy, colored like ligules; anther bases entire or slightly sagittate, tips triangular; style branches flat, tips truncate, very short, hair-tufted. **FR** ± cylindric, 5–10-veined; pappus of many barbed to subplumose bristles, white to red-brown. ± 27 spp.: N.Am, Eurasia. (Latin or Greek: ancient name) [Downie & Denford 1988 Rhodora 90:245–275] Diploid spp. sexual; polyploid spp. gen form seeds asexually.

1. Cauline lvs gen 5–12 pairs, ± equal or gradually reduced upward; heads 4–12
  2. Phyllaries ± obtuse; inner face of tip hair-tufted; tube of disk corolla 3–4.5 mm . . *A. chamissonis* var. *foliosus*
  2' Phyllaries acute; tip no more hairy than body; tube of disk corolla 2–3 mm
    3. Sts 1–few from rhizome; lvs toothed . . . . *A. amplexicaulis*
    3' Sts clustered from caudex-like rhizome; lvs ± entire . . . . *A. longifolia*
1' Cauline lvs gen 2–4(5) pairs, often strongly reduced upwards; heads gen < 5
  4. Pappus white, bristles short-barbed . . . . *A. sororia*
  4' Pappus tawny or brownish yellow, bristles subplumose
    5. Ray fls 12–18, ligules 15–25 mm; lower herbage puberulent to long-hairy but not at all woolly . . . . *A. mollis*
    5' Ray fls (0)6–10, ligules 7–15 mm; lower herbage woolly . . . . *A. parryi*

A. amplexicaulis Nutt. (p. 119) Pl 3–8 dm from short rhizome or caudex, gen hairy and glandular, esp upward. **STS** 1–few, sometimes branched above middle. **LVS**: basal petioled; cauline 5–10 pairs, ± sessile, 4–12 cm, ± elliptic to ovate, ± toothed. **INFL**: heads radiate, (3)4–9+; involucre 9–15 mm, bell-shaped or widely obconic; phyllaries acute, tip hairier. **RAY FLS** 8–14; ligules 1–2 cm. **FR** 4–6 mm, sparsely hairy, sometimes glandular; pappus subplumose, brownish. 2*n*=38,57,76. Moist, open woodlands, streambanks; 2200–3500 m. KR, CaRH, SNH, **SNE**; to s AK, MT. Gen asexual. ❀TRY. Jul–Aug

A. chamissonis Less. ssp.*foliosa* (Nutt.) Maguire (p. 119) Pl 3–8 dm, ± hairy, gen glandular upward. **ST** 1, branched above. **LVS**: cauline 5–10 pairs, sessile or short-petioled, 5–20(30) cm, blade (ob)lanceolate, (sub)entire, upper ± reduced. **INFL**: heads radiate, 5–10; involucre 8–15 mm, bell-shaped; phyllaries ± obtuse, inner face gen hair-tufted near tip. **RAY FLS** gen 13; ligules 1–2 cm. **FR** 4–5 mm, ± glabrous to short-hairy, glandular; pappus short-barbed or weakly subplumose, dirty white to straw-colored. 2*n*= 38,57,106–108. Damp meadows, rocky places, coniferous forest; 1800–3500 m. KR, CaR, SN, SnBr, Wrn, **n SNE, W&I**; to B.C., MT, NM. [vars. *bernardina* (E. Greene) Maguire, *incana* (A. Gray) Hultén, *jepsoniana* Maguire] Gen asexual. ❀IRR,SHD:1–3,**4–7**,10,14, **15–18**,19–24. Jul–Aug

A. longifolia D. Eaton Pl 3–6 dm from caudex-like rhizome, scabrous, short-hairy upward, often ± sticky. **STS** clustered, often forming large patches. **LVS**: basal few, < cauline; cauline 5–7 pairs, lowest pairs often fused around st, blade 5–12 cm, ± lanceolate, ± entire, upper ± reduced. **INFL**: heads radiate, 3–20; involucre 7–10 mm, bell-shaped or widely obconic; phyllaries acute, glandular, ± long-hairy, (esp tip). **RAY FLS** 8–13; ligules 1–2 cm. **FR** 4–6 mm, subglabrous to glandular and hairy; pappus short-barbed or subplumose, red- to yellow-brown. 2*n*=38,±50. Gen wet meadows or open coniferous forest; 1800–3500 m. KR, s NCoRH, CaRH, SNH, MP, **n SNE**, **W&I**; to w Can, MT, Colorado. [ssp. *myriadenia* (Piper) Maguire] Gen asexual in CA. ❀ IRR,DRN,SHD:**1**,2–5,**6**,7,14,**15–17**, 18–24. Jul–Aug

A. mollis Hook. Pl 2–6 dm from short rhizome or loose caudex, ± hairy, glandular. **STS** 1–several, little-branched. **LVS**: cauline 3–5 pairs, (sub)sessile, lowest gen largest, blade 4–20 cm, (ob)lanceolate to (ob)ovate, entire to unevenly dentate. **INFL**: heads radiate, 1–3(7); involucre 10–16 mm, hemispheric to spreading-bell-shaped; phyllaries acute, sparsely soft-hairy. **RAY FLS** 12–18; ligules 1–3 cm. **FR** 6–7 mm, forked-hairy, gen glandular; pappus subplumose, (yellow-)brown. 2*n*=38,57,76,152. Meadows, streambanks in subalpine zone; 2500–3500 m. KR, CaRH, SNH, **n SNE, W&I**; to WA, w Can, MT, Colorado. Sexual or not. ❀ TRY. Jul–Sep

A. parryi A. Gray Pl 1–6 dm from short rhizome; hairs various. **ST** 1, gen unbranched. **LVS**: basal ± sessile, ± ovate, < cauline; cauline 3–4 pairs, lower pairs ± crowded, petioled, blade 5–20 cm, lanceolate to ± ovate, (sub)entire, upper pairs sessile, reduced. **INFL**: heads gen radiate, (1)3–9, gen nodding in bud; involucre 15–20 mm, ± obconic; phyllaries acute, sparsely softhairy. **RAY FLS** (0)6–10; ligules < 1.5 cm. **FR** 4–6 mm, glabrous to unevenly hairy and glandular; pappus subplumose, brownish yellow. 2*n*=38,57,76,±97. Slopes in coniferous forest; 600–2400 m. s NCoRH, n&c SNH, **W&I**; to nw Can, MT, Colorado. CA pls gen radiate, have been called ssp. *sonnei* (E. Greene) Maguire. Jul–Aug

A. sororia E. Greene (p. 119) TWIN ARNICA Pl 1–5 dm from short, sparsely scaly rhizome, stalked-glandular, unevenly hairy upward; axils often sparsely white-hairy. **STS** 1–several, branched or not. **LVS**: cauline 3–6 pairs, lower petioled, blades 4–14 cm, ± oblanceolate, (sub)entire; upper lvs sessile, reduced. **INFL**: heads radiate, 1–5; involucre 10–15 mm, widely hemispheric; phyllaries acute, stalked-glandular, hairy, ciliate, inner face glabrous. **DISK FLS**: corollas stalked-glandular. **RAY FLS** 8–17; ligules 1–3 cm. **FR** 3–5.5 mm, hairy, sometimes sparsely glandular; pappus short-barbed, white. 2*n*=38. RARE. Open sagebrush scrub; 1400–1800 m. **GB**; to w Can, WY. May–Aug

ARTEMISIA SAGEBRUSH

Leila M. Shultz

Ann to shrubs, gen aromatic. **LVS** entire to ± lobed, glabrous to densely hairy; hairs glandular (resin-filled) or T-shaped, hollow. **INFL**: gen panicle; heads gen discoid or disciform, in racemes or panicles; involucre ovoid to hemispheric, gen concealing fls; phyllaries in several series, margins scarious; receptacle conic, gen naked. **PISTILLATE FLS** 0–many; corollas gen < 2 mm. **DISK FLS** 4–many, gen forming frs, sometimes staminate; corollas < 2 mm, pale yellow; anther tips acute to awl-shaped; style branches flat, fringed or blunt (sometimes simple, tack-shaped in staminate fls). **FR** < 2 mm, obovoid or fusiform, ribbed or smooth, glabrous, hairy, or resinous; pappus gen 0 or minute crown. ± 300 spp.: esp n hemisphere. (Greek: Artemis, goddess of the hunt, and noted herbalist, Queen of Anatolia) [Keck 1946 Proc Calif Acad Sci (4)25:421–468; Shultz 1983 PhD thesis Claremont Graduate School]

1. Herbs, sometimes woody at base but not shrubby
  2. Stem lvs gen entire, linear, gen glabrous (exc in D), gen anise- or tarragon-scented . . . . ***A. dracunculus***
  2' Stem lvs gen lobed or divided, not linear, variously hairy, not anise- or tarragon-scented
    3. Lvs pinnately divided, phyllaries strongly scarious; rhizome 0
      4. Pl annual or biennial; st glabrous, simple; basal lvs 0; infl ± dense. . . . . ***A. biennis***
      4' Pl perennial; st hairy, branched from base, from basal lf cluster; infl ± open . . . . . ***A. norvegica*** ssp. *saxatilis*
    3' Lvs entire or lobed, not divided to midrib; phyllary margins obscured by dense tomentum; rhizome present
      5. Lvs gen 1–8 cm wide, ± glabrous above, glabrous or hairy below; sts 3–25 dm
        6. Lvs entire to coarsely 3–5-lobed near tip, lateral lobes obtuse to acute, not glandular above, densely
          tomentose below; ± scented (not lemony) . . . . ***A. douglasiana***
        6' Lowest lvs1–2-pinnately divided, lateral lobes acute; lvs gland-dotted above, tomentose (or
          becoming glabrous) below; strongly lemon-scented . . . . ***A. michauxiana***
      5' Lvs gen < 1 cm wide, densely gray-hairy (or becoming ± glabrous above); sts gen < 8 dm . . . ***A. ludoviciana***
        7. Infl open; lvs gen 1–2 cm . . . . ssp. *albula*
        7' Infl narrow; lvs gen 2–11 cm
          8. Lvs deeply lobed . . . . ssp. *incompta*
          8' Lvs entire to few-toothed . . . . ssp. *ludoviciana*
1' Subshrubs or shrubs
  9. Pl spiny, compact (hairy); infl short, heads ± hidden by lvs; lvs palmately 3–5-divided; fl spring/early
    summer . . . . ***A. spinescens***
  9' Pl unarmed, compact to tall; infl elongate, lvs narrow and entire or wedge-shaped, lobes 0–5; fl
    late summer/autumn
    10. Heads usually nodding; lvs not clustered, entire or teeth 3, sharply pointed — W&I, DMtns . . . ***A. bigelovii***
    10' Heads mostly erect; lvs of vegetative shoots in axillary clusters, entire or teeth 3, rounded to acute
      11. Lvs gen linear, entire, winter-deciduous; heads gen > 5 mm wide . . . . ***A. cana*** ssp. *bolanderi*
      11' Lvs gen wedge-shaped, 3-toothed, persistent in winter; heads gen < 5 mm wide
        12. Pl gen < 3 dm; infl narrow, gen < 3 cm wide
          13. Lvs of fl-sts shallowly lobed; heads gen 3–4 mm diam, sessile, 2–4 per cluster; phyllaries gray-hairy
           . . . . ***A. arbuscula*** ssp. *arbuscula*
          13' Lvs of fl-sts entire; heads < 3 mm diam, solitary on slender peduncle; phyllaries shiny-resinous,
           straw-colored . . . . ***A. nova***
        12' Pl gen > 3 dm; infl > 3 cm wide
          14. Pl sticky-resinous; heads > 3 mm diam; meadows, n SNE, W&I . . . . ***A. rothrockii***
          14' Pl not sticky; heads gen < 3 mm diam (4.5–6 mm in *A. spiciformis*); widespread in mtns, dry valleys
           15. Lvs entire to irregularly 3–6-lobed, partly deciduous; pl sprouting from roots; high mtn meadows
             . . . . ***A. spiciformis***
           15' Lvs gen regularly 3-lobed, wedge-shaped, evergreen; pl not root-sprouting; widespread . . . . ***A. tridentata***
             16. Infl narrow, branches gen erect . . . . ssp. *vaseyana*
             16' Infl wide, branches erect to spreading or drooping
               17. Infl branches drooping; fr hairy . . . . ssp. *parishii*
               17' Infl branches erect to spreading; fr glandular . . . . ssp. *tridentata*

A. arbuscula Nutt. LOW SAGEBRUSH Shrub < 3 dm, mounded, gray, evergreen. **STS** much-branched. **LVS** 3–9 mm, wedge-shaped, 3-lobed, hairy, gray-green. **INFL** spike-like; heads discoid, 2–4 mm diam, gen 2–4 per cluster, ± sessile; phyllaries ovate, densely hairy, margins transparent. **PISTILLATE FLS** 0. **DISK FLS** 4–6. **FR** 0.7–0.8 mm, light brown, finely resinous. $2n=18,36$. Clay soils, valleys, slopes; 1500–3800 m. NCoRH, SNH, MP, **W&I**; to WA, MT, WY, UT. 2 sspp. in CA.

ssp. **arbuscula** **LVS** of fl-sts shallowly 3-lobed. **INFL**: heads 3–4 mm diam. $2n=18,36$. Habitats and range of sp. ❀SUN:1,**2**, **17**&IRR,DRN:**3**,**7**,8–10,**14–16**,18–**24**. Jul–Aug

A. biennis Willd. (p. 127) Ann or bien gen 3–20 dm, glabrous, green, unscented. **ST** 1, erect, finely striate, often reddish. **LVS** 4–13 cm, widely lanceolate, 2-pinnately divided ± to midrib; lobes sharply toothed. **INFL** ± dense, leafy; heads 2–4 mm diam, erect, 1–several in tight groups; phyllaries widely elliptic to obovate, green, margins widely scarious. **PISTILLATE FLS** 0. **DISK FLS** 15–40. **FR** 0.2–0.9 mm, 4–5-veined, glabrous. $2n=18$. Disturbed moist sites; < 2200 m. NCoRO, SnJV, CW, SCo, WTR, SnBr, **GB**, **DMtns**; N.Am; native to Eur. Aug–Oct

A. bigelovii A. Gray BIGELOW SAGEBRUSH Shrub 4–6 dm, rounded, branched from base. **STS** many, slender, curved, silvery-canescent. **LVS** 0.5–3 cm, entire to sharply 3-toothed, densely hairy. **INFL** 6–23 cm, 1–4 cm wide; heads 2–2.5 mm diam, nodding; phyllaries 12–15, ovate, obtuse, densely hairy, margins narrowly scarious. **PISTILLATE FLS** 0–2; ligules < 1 mm. **DISK FLS** 2–3; corollas < 1.5 mm. **FR** elliptic in outline, 5-ribbed, glabrous. $2n=18,36$. Sandy, often limestone soils; 1300–1900 m. **W&I**, **DMtns**; to Colorado, TX. [*A. petrophila* Wooton & Standley] ❀TRY. Aug–Sep

A. cana Pursh ssp. **bolanderi** (A. Gray) G. Ward (p. 127) SILVER SAGEBRUSH Shrub < 9 dm, from woody trunk. **STS** white-felty. **LVS** (1.5)3–4 cm, linear to narrowly lanceolate, gen entire, winter-deciduous. **INFL** 12–20 cm, 1–2 cm wide, leafy; heads gen sessile, 2–3 per cluster, gen > 5 mm diam, erect; phyllaries elliptic to ovate, outer acute, inner obtuse, membranous, margins scarious. **PISTILLATE FLS** 0. **DISK FLS** 8–16. **FR** < 1.2 mm, resinous. $2n=18,36$. Gravelly soils, meadows, streambanks; 1600–3300 m. CaRH, s SNF, SNH, **GB**; to OR, NV. [*A. tridentata* ssp. *b.* (A. Gray) H.M. Hall & Clements] ❀IRR,SUN,DRN:**1**–**3**,**7**,14, **15–17**,18–21. Sep–Oct

A. douglasiana Besser(p. 127) MUGWORT Per 5–25 dm, from rhizome. **STS** many, erect, brown to gray-green. **LVS** evenly spaced, 1–11(15) cm, narrowly elliptic to widely oblanceolate, entire or coarsely 3–5-lobed near tip, sparsely tomentose above, densely white-tomentose below. **INFL** 10–30 cm, 3–9 cm wide, leafy; branches widely spreading; heads 2–4 mm diam, bell-shaped, gen nodding; phyllaries ± widely (ob)ovate, gray-tomentose, margins wide, transparent. **PISTILLATE FLS** 6–9. **DISK FLS** 9–25, staminate. **FR** < 1 mm, glabrous. $2n=18,36,54$. Common. Open to shady places, often in drainages; < 2200 m. CA-FP, **n SNE**; to WA, ID, Baja CA. ❀4–6,17&IRR:1–3,**7**–9,10,**14–16**,18–24; may be INV; STBL. Jun–Oct

A. dracunculus L. TARRAGON Per 5–15 dm, from rhizome and woody caudex, gen glabrous, odorless or tarragon-scented. **STS** many, stiff, erect, brown. **LVS** basal and cauline, 1–7 cm, linear, entire or with few linear lobes, bright green, glabrous (exc D). **INFL** 15–45 cm, leafy; heads 2–3.5 mm diam, gen nodding; phyllaries widely ovate, glabrous, light brown, membranous, margins widely transparent. **PISTILLATE FLS** 14–25. **DISK FLS** 8–20, staminate. **FR** 0.5–0.8 mm, glabrous. $2n=18$. Common. Meadows, disturbed sites; < 3400 m. **CA**; to AK, n Can, n-c US, n Mex, Eurasia. ❀DRN,SUN:4,**5**,**6**;IRR:1,**2**,3,**7**,8,9,**14–24**; [CVS non CA]. Aug–Oct

A. ludoviciana Nutt. SILVER WORMWOOD Per 3–10 dm, from rhizome. **STS** many, simple, gray- to white-tomentose. **LVS** 1–11 cm, linear to narrowly elliptic, entire to deeply lobed, densely tomentose. **INFL** 5–30+ cm, narrow, open to dense; heads < 7 mm diam, gen nodding; phyllaries lanceolate to (ob)ovate, densely tomentose, margins narrowly transparent. **PISTILLATE FLS** 5–12. **DISK FLS** 6–30. **FR** < 0.5 mm, glabrous. $2n=36,54$. Common. Gen dry, sandy to rocky soils; < 3500 m. NCoRO, SN, SnJV, SW (exc ChI), **GB**, **D**; to WA, e Can, TX, n Mex. 4 sspp. in CA.

ssp. **albula** (Wooton) Keck **LVS** gen 1–2 cm, linear to obovate, entire to ± serrate or lobed, ± persistently white-tomentose on both surfaces, often ± curled under. **INFL** open; involucre ± 3 mm. **DISK FLS** 8–13. Dry, sandy soils, shrubland, forest; < 1400 m. PR, **D**; to Colorado, w TX, nw Mex. ❀SUN:**4**–**6**&IRR:2,3,**7–10**,11,**14–24**;CVS;INV;STBL. May–Oct

ssp. **incompta** (Nutt.) Keck **LVS**: lower gen 2–8 cm, 1–2-divided into lance-linear lobes; upper entire to divided, white-tomentose below, becoming ± glabrous above, ± flat. **INFL** narrow; involucre 2.5–3.5 mm. **DISK FLS** 15–30. $2n=36$. Shrubland, woodland, coniferous forest; < 3500 m. SN, TR, PR, **GB**, **DMtns**; to OR, MT, Colorado. ❀STBL. Jul–Sep

ssp. **ludoviciana** **LVS**: lower gen 3–11 cm, entire to few-toothed ± near tip, white-tomentose on both surfaces or becoming ± glabrous above, flat. **INFL** narrow; involucre 3–4 mm. **DISK FLS** 6–20. $2n=36$. Rocky soils, shrubland, coniferous forest; < 2600 m. NCoRO, SN, SnJV, SW (exc ChI), **GB**, **DMoj**; range of sp. outside CA. ❀STBL. Jul–Sep

A. michauxiana Besser LEMON SAGEWORT Per 3–10 dm, from rhizome, lemon-scented. **STS** many, unbranched, green. **LVS** 1.5–11 cm, linear to narrowly elliptic, 1–2-pinnately divided, green, glabrous and dotted with yellow glands above, white-tomentose or becoming ± glabrous below. **INFL** 8–15 cm, narrow; heads < 5.5 mm diam, nodding; phyllaries elliptic to widely (ob)ovate, often purplish and dotted with yellow glands, gen ± glabrous, margins widely scarious. **PISTILLATE FLS** 9–12. **DISK FLS** 15–35. **FR** 0.5–1 mm, glabrous. $2n=18,36$. Subalpine to alpine scree, talus, drainages; 3000–3500 m. **W&I**; to B.C., MT, Colorado. ❀TRY; DFCLT. May–Aug

A. norvegica Fries ssp. **saxatilis** (Besser) H.M. Hall & Clements BOREAL SAGEWORT Per 4–6 dm, from branched caudex, mildly fragrant. **STS** erect, loosely tomentose. **LVS**: basal 6–15 cm; cauline < 10 cm; 1–2-pinnately divided; lobes 1–2 mm wide. **INFL** open; peduncles 0.3–10 cm, slender; heads ± 9–11 mm diam, nodding; phyllaries widely (ob)ovate, sparsely hairy, margins dark, scarious. **PISTILLATE FLS** 6–12. **DISK FLS** 30–80. **FR** glabrous. Rocky slopes; 2300–3800 m. KR, SNH, SnJt, **W&I**; to AK, n Can, MT, Colorado. ❀TRY. Jul–Sep

A. nova Nelson BLACK SAGEBRUSH Shrub 1–3 dm, loosely branched from short trunk. **STS** canescent, becoming ± glabrous. **LVS** 0.5–2 cm, wedge-shaped, gen 3-toothed at tip (gen entire on fl-sts), evergreen. **INFL** slender; branches erect; heads 2–3 mm diam; phyllaries elliptic to ovate, inner gen glabrous, shiny-resinous, straw-colored, margins transparent. **PISTILLATE FLS** 0. **DISK FLS** 3–6. **FR** 0.8–1 mm, ribbed, glabrous or resinous. $2n=18,36$. Shallow, rocky soils in desert valleys, dry slopes; < 2300 m. CaRH, SnJt, **GB**, **DMtns**; to OR, MT, NM. [*A. arbuscula* ssp. *n.* (Nelson) G. Ward] Often confused with *A. arbuscula*. ❀DRN:1, **2**,**3**&IRR:**7**,**15–21**. Sep–Nov

A. rothrockii A. Gray ROTHROCK SAGEBRUSH Shrub 2–5 dm from narrow trunk, sticky-resinous, dark green throughout (sts sometimes ± white-hairy), pungently aromatic. **LVS** 1–1.5(2) cm, (0)3-lobed (gen entire on fl-sts), evergreen, canescent, sometimes becoming ± glabrous. **INFL** narrow; heads 3–5 mm diam, erect, sessile to short-peduncled; phyllaries ± ovate, sparsely hairy, straw-colored, shiny, margins wide, scarious. **PISTILLATE FLS** 0. **DISK FLS** gen 10–16. **FR** 0.8–2 mm, smooth, resinous; pappus present on outer frs. $2n=18,36,54$. Clay soils, meadows; 2000–3100 m. c&s SNH, SnGb, SnBr., **n SNE**, **W&I**. ❀IRR,SUN,DRN:**1**,**2**,3,**7**,14,**15**, **16**,17–21. Aug–Sep

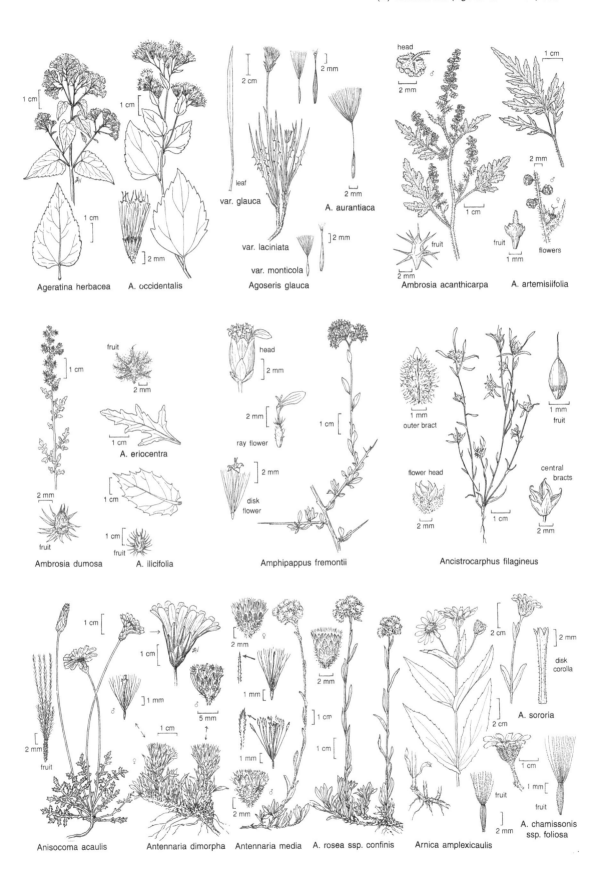

var. glauca

leaf

A. aurantiaca

var. laciniata

var. monticola

Agoseris glauca

head

fruit

fruit

flowers

Ageratina herbacea   A. occidentalis   Ambrosia acanthicarpa   A. artemisiifolia

fruit

A. eriocentra

head

ray flower

disk flower

outer bract

fruit

flower head

central bracts

fruit

Ambrosia dumosa   A. ilicifolia   Amphipappus fremontii   Ancistrocarphus filagineus

fruit

disk corolla

A. sororia

fruit

fruit

fruit

Anisocoma acaulis   Antennaria dimorpha   Antennaria media   A. rosea ssp. confinis   Arnica amplexicaulis   A. chamissonis ssp. foliosa

A. spiciformis Osterh. SNOWFIELD SAGEBRUSH Shrub 3–8 dm, gray-tomentose, widely branched, root-sprouting. **LVS** 2.5–5.5 cm, entire to irregularly 3–6-toothed (gen entire on fl-sts), turning yellow, ± winter-deciduous. **INFL** 5–20 cm, narrow, leafy; heads 4.5–6 mm diam, erect; phyllaries oblanceolate to obovate, ± hairy, margins brownish scarious. **PISTILLATE FLS** 0. **DISK FLS** 8–23. **FR** ± 1 mm, glabrous. 2*n*=18,36,54. Common. Moist open slopes to rocky meadows; 2100–3700 m. c&s SNH, **W&I**; to WY, Colorado. Often confused with *A. rothrockii*; perhaps hybrid of *A. cana* & *A. tridentata*. ✿TRY.

A. spinescens D. Eaton (p. 127) BUDSAGE Shrub < 3 dm, stout, mound-like, pungently aromatic. **STS** canescent; old branches forming thorns. **LVS** < 2 cm, round, palmately 3–5-divided into narrow segments, densely soft-hairy. **INFL** spike-like, ± surrounded by lvs; heads < 5 mm diam; phyllaries few, widely obovate, densely soft-hairy, margins obscurely scarious. **PISTILLATE FLS** 2–8. **DISK FLS** 2–15, densely soft-hairy, staminate; corolla widely bell-shaped; style unbranched. **FR** < 0.5 mm, lightly veined, long-hairy. 2*n*=18,36. Clay or gravelly, often saline soils, shrubland; 900–1600 m. SnBr, **SNE**, **DMoj**; to OR, MT, NM. Apr–May

A. tridentata Nutt. (p. 127) Shrub < 30 dm, from thick trunk, gray-hairy. **STS** gen glabrous. **LVS** 1–3(6) cm, gen wedge-shaped, gen 3(0–5)-toothed at tip, often in axillary clusters, persistent, gray-green, densely hairy. **INFL**: heads 2–2.5 mm diam, gen erect; phyllaries oblanceolate to widely obovate, densely tomentose, margins ± transparent. **PISTILLATE FLS** 0. **DISK FLS** 4–6. **FR** 1–2 mm, glandular or hairy. 2*n*=18,36. Common. Dry soils, valleys, slopes; 300–3000+ m. CaRH, SNH, SCoRI, s SnJV, SCo, TR, **GB**, **DMoj**; to WA, n-c US, NM. 4 sspp. in CA.

ssp. **parishii** (A. Gray) H.M. Hall & Clements Pl 10–30 dm. **LVS** gen 3–6 cm, linear to narrowly wedge-shaped. **INFL** < 30 cm, wide, often surrounded by lvs; branches gen drooping. **FR** hairy. Uncommon. Dry, sandy soils; 300–2000 m. SCoRI, SCo, WTR, **W&I**, **DMoj**. Oct–Nov

ssp. **tridentata** BIG SAGEBRUSH Pl < 20 dm. **LVS** gen 1.2–4 cm, narrowly wedge-shaped. **INFL** < 30 cm, 5–15 cm wide, often surrounded by lvs; branches gen spreading. **FR** glandular. 2*n*=18,36. Valleys, benches, sandy to coarse gravelly soils; 800–1900 m. CaRH, s SnJV, SCoRI, WTR, **GB**; to WA, MT, NM. ✿DRN, SUN:1,**2,3**,6,**7–10**,11,**14–16**,17,**18–21**,22–24;may be INV. Aug–Oct

ssp. **vaseyana** (Rydb.) Beetle MOUNTAIN SAGEBRUSH Pl gen < 10 dm. **LVS** gen 2–3 cm, wedge-shaped. **INFL** < 30 cm, narrow, exposed; branches erect, exceeding vegetative branches. **FR** glandular. 2*n*=18,36. Common. Dry slopes; 1800–3000+ m. CaRH, SNH, TR, **n SNE**, **W&I**; to ID, ND, Colorado, NM. ✿TRY.

ASTER ASTER

Geraldine A. Allen

Ann or per from caudex or rhizome. **ST** gen erect, 1–20 dm. **LVS** basal, cauline, or both, alternate, gen entire; basal gen petioled. **INFL**: heads gen radiate, solitary or in a cyme or panicle; involucre obconic to hemispheric; phyllaries in 2–6 series, outer gen < inner, free, at least inner with pale, papery margins; receptacle ± flat, naked. **RAY FLS** 0–many; corolla violet to pink or white. **DISK FLS** many; corolla and anthers gen yellow, tube gen < throat; anther tips ± triangular; style branches flat on inner face, base ± warty, tip acute, hairy. **FR** gen rounded, ± ribbed, ± brown; pappus of bristles, white to brownish. ± 250 spp.: N.Am, Eurasia, Afr. (Greek: star) [Allen 1984 Syst Bot 9:175–19] *See also revised taxonomy of Nesom, G. 1994 Phytologia 77: 141–297.

1. Ann, taprooted; rays < 8 mm, gen > disk fls by 2 mm or less, pink to violet
  2. Phyllaries obtuse to rounded; lvs elliptic to obovate, gen obtuse . . . . **A. frondosus**
  2' Phyllaries acute to acuminate; lvs linear to oblanceolate, acute . . . . **A. subulatus** ssp. **ligulatus**
1' Per, gen from rhizome or caudex; rays > 8 mm, conspicuous, white to pink or violet
  3. Heads solitary; lvs mostly basal, cauline lvs 0 or < 1/3 as long as basal lvs . . . . . . **A. alpigenus** var. **andersonii**
  3' Heads gen > 1 per st; cauline lvs well developed, > 1/3 as long as basal lvs, or basal lvs 0
    4. Outer pappus bristles < 2 mm, << inner; pl 12 cm or less; lvs all cauline, < 1.5 cm . . . . **A. scopulorum**
    4' Outer pappus bristles gen ± = inner; pl > 12 cm, or basal lvs present and >> 1.5 cm
      5. Involucre and peduncles ± glandular
        6. Lvs ± similar in size; pl with both glandular and nonglandular hairs . . . . **A. campestris**
        6' Lvs smaller upward; pl glabrous except for glandular involucre and upper st . . . . **A. pauciflorus**
      5' Involucre glabrous to ± hairy, not glandular
        7. Basal or lower cauline lvs largest; basal lvs gen present at fl; st 2–8 dm
          8. Outer phyllaries obtuse to rounded, < inner ones, ± pale-margined at base, with green area at tip gen < 2.5 × longer than wide; pl gen densely and uniformly strigose, at least beneath heads
           . . . . **A. ascendens**
          8' Outer phyllaries acute to obtuse, < or = inner ones, ± pale-margined at base, with green area at tip > 2.5 × longer than wide; pl glabrous to ± hairy . . . . **A. occidentalis** var. **occidentalis**
        7' Lvs equal or middle cauline lvs largest; basal lvs gen 0 at fl; st 4–16 dm
          9. Rays ± white to pink; outer phyllaries acute to obtuse, ± greenish throughout, gen < 3 × longer than wide . . . . **A. eatonii**
          9' Rays ± white to violet; outer phyllaries acute to acuminate, pale-margined > 1/2 length, gen > 3 × longer than wide . . . . **A. lanceolatus** ssp. **hesperius**

A. alpigenus (Torrey & A. Gray) A. Gray var. **andersonii** (A. Gray) M. Peck (p. 127) Per; caudex taprooted. **STS** decumbent to ± erect, 1–4 dm, ± hairy above. **LVS** mainly basal, sessile, 3–25 cm, linear to narrowly elliptic, acute, ± glabrous; cauline 0 or < 1/3 as long. **INFL**: heads solitary; phyllaries subequal or outer < inner, lanceolate to oblong, acute, green to ± purple-tinged, pale-margined at base. **RAY FLS** many; corollas gen 10–16 mm, violet to light purple. **FR** ± hairy throughout. 2*n*=18,36. Meadows; 1500–3700 m. KR, NCoR, CaR, SN, SnJt, Wrn, **n SNE**, **W&I**; OR, NV. Var. *alpigenus* farther n *Oreostemma* *alpigenus* (Torrey & A. Gray) E. Greene var. *andersonii* (A. Gray) G.L. Nesom ❀4–6;IRR:1,15–17&SHD;2,3,7,14,18;DFCLT. Jul–Aug

A. ascendens Lindley (p. 127) Per; rhizomes elongate. **STS** 2–6 dm, strigose at least above. **LVS** basal and cauline, smaller upward, 4–15 cm, oblong to oblanceolate, acute, glabrous to strigose. **INFL**: heads in an open cyme; phyllaries oblong to narrowly obovate, obtuse (outer) to ± acute (inner), green at tip, pale-margined at base. **RAY FLS** many; corollas 8–15 mm, violet. **FR** hairy. 2*n*=26,52. Meadows, disturbed places; 500–2000 m. SN (e slope), Teh, SnGb, SnBr, **GB**; to w Can, Colorado, n AZ. **Symphyotrichum ascendens* (Lindl.) G.L. Nesom ❀TRY. Jul–Sep

A. campestris Nutt. Per; rhizomes elongate. **STS** 1–5 dm, glandular and gen strigose. **LVS** cauline, 2–5 cm, linear to narrowly oblong, acute to obtuse, glabrous to ± short-hairy. **INFL**: heads in a narrow cyme; phyllaries linear to lanceolate, acute, green, ± pale-margined at base, ± glandular. **RAY FLS** many; corollas 6–10 mm, violet. **FR** hairy. 2*n*=10. Dry meadows; 1800–2700 m. **SNE**; to e WA, UT. **Symphyotrichum campestre* (Nutt.) G.L. Nesom ❀ TRY. Jul–Aug

A. eatonii (A. Gray) Howell (p. 127) Per; rhizomes short. **STS** 4–10 dm, ± hairy above. **LVS** basal and cauline (basal 0 at fl), sessile, 5–15 cm, narrowly lanceolate, acute, gen entire, glabrous to ± hairy. **INFL**: heads in a narrow, many-headed cyme; phyllaries ± subequal, oblong to ovate, inner pale-margined at base, acute, outer ± green throughout, obtuse to acute. **RAY FLS** many; corollas 6–12 mm, white to pink or pink-purple. **FR** hairy. 2*n*=16,32,48,64. Wet places; 500–2000 m. CaR, SN, **GB**; to B.C., Colorado. **Symphyotrichum eatonii* (A. Gray) G.L. Nesom ❀TRY. Jul–Aug

A. frondosus (Nutt.) Torrey & A. Gray (p. 127) Ann. **STS** decumbent to erect, 2–6 dm, ± glabrous. **LVS** basal and cauline, sessile, gen 2–5 cm, elliptic to obovate, ± obtuse, glabrous to finely ciliate. **INFL**: heads in a narrow cyme; phyllaries oblanceolate to obovate, obtuse to rounded, green, inner ± pale-margined at base. **RAY FLS** many; corollas < 8 mm, pink-purple, slightly > disk corollas. **FR** ± hairy. 2*n*=14. Marshes, lake edges, often alkaline; 700–2200 m. CaR, SN, TR, PR, **GB**; to B.C., WY, AZ. **Symphyotrichum frondosum* (Nutt.) G.L. Nesom Aug

A. lanceolatus Willd. ssp. *hesperius* (A. Gray) Semple & J. Chmielewski (p. 127) Per; rhizomes long. **STS** 6–16 dm; hairs ± in lines throughout. **LVS** basal and cauline (basal 0 at fl), 8–16 cm, lanceolate, acute, entire to ± serrate, ± hairy. **INFL**: heads in an open, leafy-bracted cyme; phyllaries linear to oblong, acute to acuminate, pale-margined at base to > 1/2 length. **RAY FLS** many; corollas 6–12 mm, ± white to violet. **FR** hairy. 2*n*=64.

Wet places; < 1500 m. SCo, PR, **SNE**; to Alberta, n-c US. [*A. h.* A. Gray] **Symphyotrichum lanceolatum* (Willd.) G.L. Nesom ssp. *hesperium* (A. Gray) G.L. Nesom ❀TRY. Jul–Aug

A. occidentalis (Nutt.) Torrey & A. Gray Per; rhizomes long. **STS** 2–8 dm, glabrous to ± hairy. **LVS** basal and cauline, 5–15 cm, ± smaller upward, ± linear to narrowly elliptic, acute, ± entire, ± glabrous. **INFL**: heads in an open cyme; phyllaries linear to oblong, acute or outer ± obtuse, green at tip, ± pale-margined at base. **RAY FLS** many; corollas 8–12 mm, violet. **FR** hairy. 2*n*=16,32, 48,64. Meadows; 1200–2800 m. NW, CaR, SN, TR, SnJt, **GB**; to B.C., Colorado, Baja CA. **Symphyotrichum spathulatum* (Lindl.) G.L. Nesom

var. *occidentalis* (p. 127) **STS** 2–8 dm. **LVS** lanceolate to narrowly elliptic. 2*n*=16,32,48,64. Habitats and range of sp. [vars. *delectabilis* (H.M. Hall) Ferris, *parishii* (A. Gray) Ferris] Larger pls have been called var. *intermedius* A. Gray. **Symphyotrichum spathulatum* (Lindl.) G.L. Nesom var. *spathulatum*, in part. ❀4–6&IRR:1–3,**7**,14,**15–17**, 18. Jun–Aug

A. pauciflorus Nutt. Per; rhizomes long. **STS** 3–12 dm, glabrous at base, glandular above. **LVS** basal and cauline, smaller upward (bract-like in infl), sessile, 6–10 cm, ± linear, acute to acuminate, ± glabrous. **INFL**: heads in an open cyme with heads at ends of branches; phyllaries ± oblong, acute, ± green, ± glandular. **RAY FLS** many; corollas 5–8 mm, ± white to pale purple. **FR** hairy. 2*n*= 18. Uncommon. Damp alkaline places; 200–700 m. n **DMoj** (**Inyo Co**); to c Can, TX, Mex. **Almutaster pauciflorus* (Nutt.) Löve & Löve. Jun–Oct

A. scopulorum A. Gray (p. 127) Per; caudex fibrous-rooted. **STS** ascending to erect, 4–12 cm, ± hairy. **LVS** cauline, crowded on lower 2/3 of st, 0.3–1.2 cm, narrowly oblanceolate to elliptic, firm, ± abruptly pointed, densely short-hairy. **INFL**: heads solitary; phyllaries lanceolate to oblong, acute to acuminate, green to ± purple, inner pale-margined below. **RAY FLS** gen 8–16; corollas 7–12 mm, violet to purple. **FR** hairy; pappus outer bristles ± 1 mm, << inner. 2*n*=18. Dry, rocky places; 1500–3000 m. Wrn, n **SNE**, **W&I**; to e OR, WY, NV. **Ionactis alpina* (Nutt.) E. Greene ❀TRY;DFCLT. May–Jul

A. subulatus Michaux var. *ligulatus* Shinn. (p. 127) Ann. **STS** erect, 2–8 dm, glabrous. **LVS** basal and cauline, sessile, gen 3–6 cm, linear to oblanceolate, acute, glabrous. **INFL**: heads in an open cyme; phyllaries linear to lanceolate, acute to long-tapered, green, ± pale-margined. **RAY FLS** many; corollas gen < 7 mm, pink to violet, barely > disk corollas. **FR** ± hairy. 2*n*=10. Wet places, often alkaline; < 200 m. GV, CW, SW, **DSon**; to c&se US, Mex. [*A. exilis* Elliott] **Symphyotrichum divaricatum* (Nutt.) G.L. Nesom Jul–Oct

ATRICHOSERIS TOBACCO-WEED, GRAVEL-GHOST

G. Ledyard Stebbins

1 sp.: sw US. (Greek: chicory-like pl without pappus)

A. platyphylla A. Gray (p. 127, pl. 14) Ann, 2–18 dm, glabrous; sap milky. **ST** erect. **LVS**: basal and lowest cauline gen flat against soil, sessile or tapering to a short, winged petiole, 3–10 cm, obovate, obtuse, finely dentate, gray-green, often purple-tinged, esp below; cauline few, reduced to inconspicuous, triangular scales. **INFL**: heads ligulate, 1.5–2.5 cm diam, few–many

in an open scapose cyme; involucre 6–10 mm; phyllaries in 2–4 series, outer short, triangular to lanceolate, inner 10–15, ± equal; receptacle naked. **FLS** fragrant; ligules white, readily withering. **FR** 4–4.5 mm with 5 thick, corky ribs, whitish; pappus 0. 2*n*=18. Desert valleys and washes; < 1400 m. **D**; to UT, AZ. ❀TRY. Mostly Mar–May

BACCHARIS

Scott Sundberg

Per to shrub, dioecious, sometimes aromatic, often ± sticky-resinous. **STS** erect, channeled. **LVS** cauline, alternate, simple, reduced to bracts above. **INFL**: heads discoid and disciform, borne in terminal or lateral racemes, panicles or cymes; phyllaries overlapping in several series; receptacle naked or chaffy. **DISK FLS** gen many, functionally staminate; corollas white to pink-tinged; ovary much reduced; pappus of bristles < involucre. **PISTILLATE FLS** gen many; corollas thread-like, ± whitish. **FR** ± cylindric, 4–10-ribbed; pappus of many bristles > involucre. 250–400 spp.: Am. (Latin: Bacchus, god of wine) [Boldt 1989 *Baccharis* TX Agric Exp Sta, College Station] Key to species adapted by Dieter Wilken.

1. Sts finely glandular-puberulent; fr glandular-puberulent . . . . *B. brachyphylla*
1' Sts glabrous, often sticky; fr glabrous
  2. Fr ribs 5; lvs lanceolate; pistillate fls > 50 per head . . . . *B. salicifolia*
  2' Fr ribs gen 10; lvs oblanceolate, linear-oblanceolate, or obovate; pistillate fls 15–40 per head
    3. Lvs gen coarsely toothed, 35–70 mm, gen persistent at fl, principal veins 3; pls without many
      erect branches . . . . *B. emoryi*
    3' Lvs gen entire, < 35 mm, gen 0 at fl, principal lf veins 1; pls with many erect branches
      4. Lvs linear-oblanceolate; pistillate involucres > 5mm; staminate involucres > 4 mm, corollas >
        4 mm; fr > 2 mm . . . . *B. sarothroides*
      4' Lvs oblanceolate to obovate; pistillate involucres < 4.5 mm; staminate involucres < 4 mm,
        corollas < 4 mm; fr < 1.5 mm . . . . *B. sergiloides*

B. brachyphylla A. Gray Shrub < 1 m, glandular. **STS** much-branched, wand-like. **LVS** sessile, < 17 mm, linear-lanceolate, entire; principal vein 1. **INFL**: heads in open panicles; involucre funnel- or bell-shaped, of staminate heads 3.8–5.2 mm, of pistillate heads 4.5–6 mm; phyllaries in 4–5 series, lanceolate, glandular-puberulent, acute to long-tapered; receptacle convex, honeycombed, puberulent, chaff 0. **STAMINATE FLS** (8)12–18(29); corollas 3.3–4.2 mm; pappus 3–4.4 mm. **PISTILLATE FLS** 8–18; corollas 2–2.8 mm. **FR** 2–3 mm, glandular-puberulent; ribs 5; pappus 4.5–6.5 mm. 2*n*=18. Canyon bottoms, dry washes; 300–1200 m. SnBR, SnJt, **D**; to TX, n Mex. ❀TRY;STBL. Jul–Nov

B. emoryi A. Gray (p. 127) Shrub < 4 m, glabrous, gen sticky. **STS** erect; branches ascending, sometimes lfless at fl. **LVS** sessile or with winged petiole, 35–70 mm, oblanceolate, 0–8-toothed or -lobed; base wedge-shaped; principal veins 3. **INFL**: heads in clusters of 3–5 in a panicle; involucre cylindric to bell-shaped, of staminate heads 4–6 mm, of pistillate heads 6–8 mm; phyllaries in 5–7 series, linear-lanceolate, glabrous, gen sticky, tip acute; receptacle flat, honeycombed, gen hairy (sometimes long-bristly), chaff 0. **STAMINATE FLS** 17–38; corollas (3.2)4–5 mm; pappus 3–5 mm. **PISTILLATE FLS** 20–40; corollas 3.5–4 mm. **FR** 1.2–2 mm, glabrous; ribs 10; pappus 8–12 mm. Sandy edges of rivers and washes, salt marshes; 0–600 m. SCoRI, SCo, WTR, PR, **D**; to UT, TX, n Mex. ❀TRY;STBL. Aug–Dec

B. salicifolia (Ruiz & Pav.) Pers. (p. 127) MULE FAT, SEEP-WILLOW, WATER-WALLY Shrub < 4 m, glabrous to minutely puberulent, often ± sticky. **STS**: main sts gen 1–few; branches few–many, short, spreading or ascending. **LVS**: petioles winged; blades < 150 mm, lanceolate, entire to toothed, principal veins 1–3. **INFL**: heads in a pyramid-shaped to rounded panicle; involucre hemispheric, of staminate heads (3)4–6 mm, of pistillate heads 3–6 mm; phyllaries in 4–5 series, awl-shaped to lanceolate, irregularly toothed, gen tinged red, glabrous (exc ciliate margins), tip obtuse to long-tapered; receptacle flat to convex, smooth, glabrous to tomentose, chaff 0. **STAMINATE FLS** (10)17–48; co-

rollas (3)4–6 mm, pappus (3)3.6–5 mm. **PISTILLATE FLS** 50–150; corollas 2.2–3.5 mm. **FR** 0.8–1.3 mm, glabrous; ribs 5; pappus 4.2–6 mm. 2*n*=18. Canyon bottoms, moist streamsides, irrigation ditches, often forming thickets; < 1250 m. NW, CaRF, SNF, GV, Teh, CW, SW, **D**; to TX, Mex, S.Am. Summer forms (infl terminal, lvs mostly toothed) formerly separated from winter forms (infls lateral, lvs entire). [*B. glutinosa* Pers.; *B. viminea* DC.] ❀STBL. Feb–Dec

B. sarothroides A. Gray (p. 127) BROOM BACCHARIS Shrub < 4 m, glabrous, sticky. **STS**: branches dense, erect, often lfless at fl. **LVS** sessile, < 35 mm, linear-oblanceolate, thick, entire, ± rolled under; principal vein 1. **INFL**: heads in a panicle; involucre cylindric to hemispheric, of staminate heads 4.2–5.2 mm, of pistillate heads 5.3–7 mm; phyllaries in 5–6 series, lanceolate to ovate, hard, glabrous, gen sticky, tip rounded to acute; receptacle flat to concave, honeycombed, chaff 0. **STAMINATE FLS** 18–35; corollas 4.2–5 mm; pappus 3–4.5 mm. **PISTILLATE FLS** 19–31; corollas 2.5–3.5 mm. **FR** 2–2.6 mm, glabrous; ribs 10; pappus 7–10.8 mm. 2*n*=18. Gravelly and sandy washes, roadsides; < 850 m. PR, **D**; to TX, n Mex. *B. emoryi* without st lvs may key here. ❀SUN, DRN:15–18,**19–24**&IRR:**7–14**;male CV. Jun–Oct

B. sergiloides A. Gray DESERT BACCHARIS Shrub < 2 m, glabrous, gen sticky. **STS**: branches gen many, erect, often lfless at fl. **LVS** barely petioled, < 35 mm, oblanceolate to obovate, 0–4-toothed; principal vein 1. **INFL**: heads in a panicle; involucre funnel- to bell-shaped, of staminate heads 2.2–4 mm, of pistillate heads 3–4.5 mm; phyllaries in 4–6 series, lanceolate, hard, glabrous, gen sticky, tip rounded to acute; receptacle conic to convex, gen smooth, glabrous to puberulent, chaff sometimes 0 in staminate pls. **STAMINATE FLS** (11)24–33; corollas 2.2–3.3 mm; pappus 2–3 mm. **PISTILLATE FLS** 15–30; corollas 1.6–2.7 mm. **FR** 1–1.5 mm, glabrous; ribs 10; pappus 1.7–3.1 mm. 2*n*=18. Gravelly or sandy stream beds; 600–1575 m. Teh, PR, **D**; NV, AZ, n Mex. ❀TRY. May–Oct

BAHIA

Ann, bien, per, subshrubs. **LVS** simple or ternately dissected, opposite or alternate, sessile or petioled, dotted with resin glands. **INFL**: heads radiate, in few–many-headed cymes; involucre hemispheric or bell-shaped; phyllaries in 1–3 series, free, ± equal, reflexed in age; receptacle flat or slightly rounded, naked. **RAY FLS** 5–20; ligules yellow or white. **DISK FLS** gen many; corollas yellow, radial or outer bilateral; style tips obtuse. **FR** 4-sided, narrowly obpyramidal; pappus scales (0)8–15, lanceolate to obovate. 12 spp.: sw US, Mex, S.Am. (J.F. Bahi, botany professor, Barcelona) [Ellison 1964 Rhodora 66:67–86,177–215,281–311]

B. dissecta (A. Gray) Britton (p. 127) Ann, bien. **ST** gen 1, erect, 2–12 dm, openly branched above, glandular. **LVS** alternate; petiole 1–6 cm or upper lvs sessile; blade 1–3.5 cm, 1–several × ternately divided into linear to oblong segments 1–5

mm wide, strigose or hairs short, spreading. **INFL** many-headed, panicle-like; peduncles 0.7–7 cm; involucre 7–10 mm diam, hemispheric to widely bell-shaped; phyllaries 12–24 in 2–3 series, 4.5–6 mm, oblanceolate or lanceolate, acuminate, soft-hairy, glandular. **RAY FLS** 10–20; ligules 5–7.5 mm, yellow. **DISK FLS**: corollas radial or bilateral, 2.5–4.5 mm; stamens often abortive. **FR** 3–4.5 mm, dark brown or black, glabrous to short-rough-hairy; pappus 0 (rarely scales 1–13, 1.5–3 mm, lanceolate).

$2n$=36. Dry, open, forest slopes; 1800–2650 m. SnBr, e PR (Santa Rosa Mtns), **e DMtns**; to WY, Colorado, TX, n Mex. Triploid, reproducing asexually. Aug–Sep

BAILEYA DESERT-MARIGOLD

Ann, per, ± tomentose throughout. **STS** 1–many from base. **LVS** basal and alternate, simple, entire to deeply lobed, ± reduced upward, petioled or upper sessile, tomentose and glandular. **INFL**: heads radiate, solitary or in few-headed cymes; peduncles short to very long; involucre cylindric to bell-shaped or hemispheric; phyllaries in 1–2 ± equal series, linear-lanceolate; receptacle flat to slightly rounded, naked or with scattered, narrow chaff. **RAY FLS** 4–many; ligules ovate, 3-lobed, sessile on ovary, yellow, drying cream, papery, reflexed and persistent on fr when dry. **DISK FLS** 8–many; corollas yellow, gland-dotted, lobes triangular, long-hairy; anther tips triangular; style tips short-triangular. **FR** linear to club-shaped, cylindric or ± angled, short-rough hairy and gland-dotted to glabrous; pappus 0. 3 spp.: sw US, Mex. (J.W. Bailey, Am microscopist, born 1811) [Brown 1973 PhD dissertation, AZ State Univ; Turner 1993 Sida 15: 491–508]

1. Ray fls 4–8, pale yellow; involucre gen 4–6 mm diam; heads 2–3 . . . . *B. pauciradiata*
1'. Ray fls 20–many, bright yellow; involucre 7+ mm diam; heads solitary
  2. Ligules 10–20 mm, widely linear or oblong, prominently 3-lobed; fr cylindric or barely angled;
    ribs ± equal; peduncles gen 10–20+ cm, ± scape-like . . . . *B. multiradiata* var. *multiradiata*
  2'. Ligules 6–10 mm, widely elliptic to obovate, shallowly 3-lobed; fr ribbed and angled; ribs of angles
    most prominent; peduncles gen 2–10 cm, gen leafy-bracted . . . . *B. pleniradiata*

B. multiradiata A. Gray var. ***multiradiata*** (p. 131) Ann, per, canescent-tomentose. **STS** 2–5 dm, gen branched only at base. **LVS** mostly basal and on lower sts, these 2–10 cm; petioles winged; blades 1–3-pinnately divided, lobes linear to ovate; mid-st lvs 0 or reduced to linear, entire bracts. **INFL**: heads solitary, showy; peduncles 1–3 dm, ± naked; involucre 10–25 mm diam, hemispheric; phyllaries 5–8 mm. **RAY FLS** 50–60 in > 1 series; ligule 10–20 mm, widely linear or oblong, bright yellow; lobes prominent, lanceolate to ovate. **DISK FLS** many; corollas 3–4 mm. **FR** 2.5–4 mm, cylindric, not or only slightly angled, ± equally ribbed. $2n$=32. Desert roadsides, flats, washes, hillsides; 600–1600 m. **DMoj**; to NV, AZ, NM, n Mex. ❀SUN,DRN,DRY: 7–9,**10–13**,14,**18–21**,22,23. Apr–Jul, Oct

B. pauciradiata A. Gray (p. 131) Ann (rarely per), ± loosely tomentose. **STS** 1–5 dm, gen branched above base. **LVS** mostly cauline; basal soon withering; cauline lvs 4–14 cm, not markedly reduced upward, linear-oblong to lanceolate or oblanceolate, entire or divided into linear lobes. **INFL**: heads 2–3 in cymes; peduncles 2–5 cm; involucre 4–6 mm diam, cylindric to bell-shaped; phyllaries 8–10, 4–6 mm. **RAY FLS** 4–8; ligules 4–10 mm, obovate, shallowly 3-lobed, pale yellow. **DISK FLS** 8–20; corolla 2.5–3 mm. **FR** 4–5 mm, narrowly club-shaped, evenly ribbed. $2n$=32. Sandy desert soils, esp dunes; < 1100 m. **D**; w AZ, n Mex. Feb–Jun, Oct

B. pleniradiata A. Gray (p. 131, pl. 15) Ann, canescent-tomentose. **STS** 1–5 dm, branched mostly in basal half. **LVS** mostly cauline; basal gen withering by time of fl; basal and lower cauline similar, 2–8 cm, petioled; blade 1–3 × divided into oblong to ovate lobes; upper cauline sessile, simple, linear to narrowly oblanceolate, gen entire. **INFL**: heads solitary; peduncles 2–10 cm; involucre 7–10 mm diam, hemispheric; phyllaries 20–30, 4–6 mm. **RAY FLS** 20–60, in 2 or more series; ligules 6–10 mm, widely elliptic to obovate, ± entire to shallowly 3-lobed, bright yellow. **DISK FLS** many; corollas 3 mm. **FR** 3–4 mm, cylindric, distinctly angled; ribs of angles most prominent. $2n$=32. Desert roadsides, sandy soils; < 1500 m. **SNE, D**; to UT, AZ, n Mex. ❀TRY. Mar–Jun, Oct–Nov

BALSAMORHIZA BALSAM-ROOT

Per from fleshy taproot; caudices 1–many. **STS** erect. **LVS** ± basal, few cauline, long-petioled; blade entire to 1–3-pinnately lobed. **INFL**: heads 1–few, radiate; peduncles long, bracts 0–few; involucre hemispheric to bell-shaped; phyllaries in 2–4 series; receptacle flat; chaff folded around frs. **RAY FLS** showy; ligules yellow. **DISK FLS** many; corollas yellow, tube short, throat cylindric to narrowly bell-shaped; style branches tapered. **FR** oblong, 3–4-angled; pappus 0. ± 12 spp.: w N.Am. (Greek: balsam root, from sticky sap of taproot) [Weber 1982 Phytologia 50:357–359] Hybrids common. ❀TRY.

B. sagittata (Pursh) Nutt. (p. 131) **STS** 2–6 dm, ± short-tomentose, minutely glandular. **LVS**: basal 20–50 cm, blade widely triangular, entire, acute or obtuse, base cordate and ± hastate, upper surface soft-hairy, lower surface short-tomentose to finely strigose, ± canescent; cauline gen several, linear to oblan-ceolate. **INFL**: heads 1–few; outer phyllaries 10–25 mm, 4–9 mm wide, oblong-lanceolate to ovate, obtuse to acute, ± tomentose. **RAY FLS**: ligules 2.5–4 cm. **DISK FLS**: corollas 6–8 mm. **FR** 7–9 mm. Open forest, scrub; 1400–2600 m. SNH, **GB**; to B.C., Rocky Mtns. May–Jul

BEBBIA SWEETBUSH

Subshrubs, shrubs, ± strongly scented. **STS** many, slender, from thick, woody root-crown, short-lived, very brittle, often lfless. **LVS** simple, opposite (or upper alternate), sessile or petioled; blades linear to triangular, entire to dentate or irregularly lobed. **INFL**: heads discoid, solitary or in open, rounded cymes; peduncles slender; involucre cylindric to narrowly bell-shaped; phyllaries graduated in several series; receptacle rounded, chaffy, scales folded around frs. **FLS** many; corollas yellow; anther tips ovate, acute; style tips tapered, acute. **FR** club-shaped, compressed, 3-angled, brown to black; hairs ascending, white; pappus of 15–30 subplumose bristles. 2 spp.: sw US, nw Mex. (M.S. Bebb, Am botanist, 1833–1895) [Whalen 1977 Madroño 24:112–123]

B. juncea (Benth.) E. Greene var. ***aspera*** E. Greene (p. 131) **STS** 5–15 dm, much-branched, forming a rounded bush < or = 3 m diam, glabrous or short-bristly. **LVS** 1–3(9) cm, linear and entire or with few, sharp, pinnate lobes, drought-deciduous. **INFL**: heads few; peduncles 1.5–6 cm; involucre 4–15 mm diam; phyl-laries 1–7 mm, lanceolate to linear, acute. **FLS**: corollas 6.5–10 mm. **FR** 2–3.5 mm; pappus 6–10 mm. 2*n*=18. Common. Dry, rocky slopes, desert plains, washes; < 1500 m. SW, SNE, **D**; to s NV, TX, nw Mex. ❀DRN:7–11,**12,13**,14,18–24;bright green sts. Apr–Jul

BIDENS STICKTIGHT, SPANISH-NEEDLES,

Ann, per, shrubs. **STS** prostrate to erect. **LVS** simple or pinnate, gen opposite, sessile or petioled. **INFL**: heads radiate or discoid, gen few in CA; involucre cylindric to bell-shaped; phyllaries in 2 dissimilar series, outer gen ± lf-like in texture, inner thinner, with transparent or scarious margins; receptacle chaffy; chaff scales narrow, flat. **RAY FLS** 0 or few; ligules yellow or white. **DISK FLS** gen many; corollas yellow, radial (or outermost white, bilateral). **FR** nar-rowly club-shaped, thick or compressed front-to-back; pappus 0 or awns 1–several, gen barbed. ± 230 spp.: world-wide. (Latin: 2 teeth) [Sherff & Alexander 1955 N.Am Flora 2(2):70–129]

1. Lvs pinnately compound, petioled; ligules 0 or 2–3.5 mm . . . . *B. frondosa*
1' Lvs simple, sessile; ligules 15–30 mm . . . . *B. laevis*

B. frondosa L. (p. 131) STICKTIGHT Ann, ± glabrous. **STS** 5–12 dm, square. **LVS** compound, petioled; lflets 2–8 cm, lanceolate, gen acuminate, serrate. **INFL**: heads radiate or discoid, erect; peduncles 2–10 cm; involucre ± 1 cm diam, hemispheric; outer phyllaries 5–8, 1–5 cm, ± linear, ciliate; inner phyllaries 5–7 mm, ovate; chaff scales 5–7 mm, brownish. **RAY FLS** 0–few; corol-las 2–3.5 mm. **DISK FLS**: corollas ± 2 mm, orange. **FR** 6–10 mm, blackish, narrowly wedge-shaped, flat, ± glabrous to stiffly hairy; pappus awns gen 2, 3–4.5 mm. 2*n*=48. Uncommon. Damp soil, esp disturbed sites; gen < 1600 m. NCoR, GV, SW, **GB**; to e N.Am. Aug–Oct

B. laevis (L.) Britton, Sterns & Pogg. (p. 131) BUR-MARIGOLD Ann, per, gen ± glabrous. **STS** 2–25 dm, ± decumbent to erect, ± cylindric. **LVS** simple, sessile; bases sometimes fused around st; blades 5–15 cm, ± lanceolate, acute to acuminate, serrate. **INFL**: heads radiate, erect in fl, often nodding in fr; peduncle 2–10 cm; involucre 1–2 cm diam, bell-shaped; outer phyllaries 6–8, 1–2 mm, linear-lanceolate, sparsely ciliate; inner phyllaries 8–16 mm, obovate; chaff scales reddish tipped. **RAY FLS** 7–8; ligules 1.5–3 cm, yellow. **DISK FLS**: corollas 4–6 mm, yellow. **FR** nar-rowly wedge-shaped, flat or 3–4-angled; angles thin, barbed; pappus awns 2–4, 3–5 mm. 2*n*=22. Freshwater wetlands; gen < 300 m. GV, SCoR, SW, **SNE, DMoj (Mojave River)**; to e N.Am. ❀WET,SUN:7,**8,9**,10,11,**12,14–16,18–24**. Aug–Nov

BRICKELLIA BRICKELLBUSH

A. Michael Powell

Per from ± woody caudex or shrub (rarely ann). **LVS** alternate or opposite, simple, veiny, gen resinous-dotted. **INFL**: heads discoid, gen clustered; involucre cylindric to bell-shaped; phyllaries overlapping, strongly nerved; receptacle gen flat, chaff 0. **FLS**: corollas cylindric, ± white or tinged red; anther bases rounded or slightly cordate, tips ovate; style branches long, club-shaped, tips rounded. **FR** 10-ribbed, gen cylindric, hairy; pappus gen of many, gen scabrous bristles. (John Brickell, early botanist in Georgia) [King & Robinson 1987 Monogr Syst Bot Missouri Bot Gard 22:220–224]

1. Per or subshrubs from woody caudex . . . . *B. oblongifolia* var. *linifolia*
1' Shrubs — 2–20 dm
  2. Pl gen 10 dm or more
    3. Heads 8–18-fld
      4. Heads 12–14 mm; phyllaries ± glabrous . . . . *B. californica*
      4' Heads 8–10 mm; phyllaries puberulent . . . . *B. desertorum*
    3' Heads 3–7-fld
      5. Lvs linear; phyllaries 10–12 . . . . *B. longifolia*
      5' Lvs lanceolate to ovate; phyllaries ± 20 . . . .
        6. Lvs gen serrate . . . . *B. knappiana*
        6' Lvs ± entire . . . . *B. multiflora*
  2' Pl gen 2–6 dm
    7. Phyllary tips spreading or recurved
      8. Lvs white-tomentose . . . . *B. nevinii*
      8' Lvs ± green
        9. Sts glandular-hairy . . . . *B. microphylla*
        9' Sts tomentose-puberulent . . . . *B. watsonii*
    7' Phyllary tips erect
      10. Pls white-tomentose; heads ± 24 mm . . . . *B. incana*
      10' Pls green; heads gen 12–15 mm
        11. Outer phyllaries ovate . . . . *B. atractyloides*
        11' Outer phyllaries ± linear to widely lanceolate
          12. Lvs oblong to spoon-shaped, not leathery, entire; outer phyllaries linear-oblong, acute, ± 3 mm,
            puberulent to minutely tomentose . . . . *B. frutescens*

12' Lvs ovate, leathery, gen sharply toothed; outer phyllaries widely lanceolate, long-tapered, ± 10 mm, gen minutely scabrous . . . . *B. arguta*
   13. Outer phyllaries ± entire . . . . var. *arguta*
   13' Outer phyllaries clearly dentate . . . . var. *odontolepis*

B. arguta Robinson (p. 131) Shrub 2–4 dm. **STS** much-branched, zigzag, glandular-puberulent. **LVS** alternate, subsessile, 1–2 cm, ovate, leathery, prominently toothed or entire, scabrous, gen glandular-puberulent. **INFL:** heads 13–15 mm, solitary; involucre bell-shaped; phyllaries 26–33, overlapping to subequal; outer lance-ovate, ± entire to dentate, inner linear, 4–13-nerved, greenish, gen minutely scabrous, tips erect. **FLS** 40–55. **FR** 4 mm. Rocky places; < 1500 m. **D;** n Baja CA. Perhaps same as *B. atractyloides.*

var. ***arguta*** **INFL:** outer phyllaries ± entire. $2n=18$. Common. Habitats and range of sp. ❀DRN,SUN:7–16,18–21;STBL. Mar–Jun

var. ***odontolepis*** Robinson **INFL:** outer phyllaries clearly dentate. Uncommon. Habitats of sp. **DSon.**

B. atractyloides A. Gray Shrub 1.7–3.5 dm. **STS** much-branched, gen glandular-puberulent. **LVS** alternate, gen short-petioled, 1.5–3 cm, ± lanceolate to ovate, leathery, clearly toothed, scabrous, gen glandular-puberulent. **INFL:** heads 12–14 mm, solitary; involucre bell-shaped; phyllaries 24–32, overlapping to subequal, outer ± ovate, innermost linear, 4–15-veined, greenish to tan, minutely scabrous, tips gen erect. **FLS** 50–65. **FR** 4–5 mm. $2n=18$. Uncommon. Rocks, dry slopes, washes; 600–1400 m. **DMoj;** NV, AZ, UT. Intergrading with *B. arguta* in Inyo Co. Apr–May

B. californica (Torrey & A. Gray) A. Gray (p. 131) Shrub 5–20 dm. **STS** many, branched, puberulent to weakly tomentose. **LVS** alternate, short-petioled, 1–6 cm, triangular-ovate, ± serrate, gland-dotted, puberulent, reduced upward. **INFL:** heads 12–14 mm, in a leafy panicle with small clusters on short lateral branchlets; involucre cylindric; phyllaries 21–35, overlapping, 3–5-veined, green or ± purple, ± glabrous, outer ± lanceolate to ovate, inner linear to oblong, tips rounded to ± acute, erect. **FLS** 8–18. **FR** 3 mm. $2n=18$. Common. Diverse dry habitats; < 2700 m. NW, CaRF, SN, SnJV, CW, SW, PR, **W&I, D;** to ID, Colorado, TX, n Mex. Variable. ❀DRN, SUN:1–3,**7**,8–13,**14–16,18–24**;STBL. May, Aug–Oct

B. desertorum Cov. Shrub 8–15 dm. **STS** intricately branched, puberulent to short-tomentose. **LVS** opposite or alternate, short-petioled, 0.5–1.2 cm, ovate, ± serrate, minutely tomentose. **INFL:** heads 8–10 mm, in small clusters on short lateral branchlets; involucre cylindric; phyllaries ± 21, overlapping, linear to oblong, 3–5-veined, ± green, purple, or yellow, ± minutely tomentose, tips erect. **FLS** 8–12. **FR** 2–3 mm. $2n=18$. Common. Rocky places, desert scrub; 200–1400 m. **D;** NV, AZ. ❀TRY;STBL. Mar–May, Sep–Dec

B. frutescens A. Gray (p. 131) Shrub 3–6 dm, aromatic. **STS** slender, rigid, spreading, ± short-tomentose. **LVS** alternate, short-petioled, 0.5–2 cm, oblong to spoon-shaped, puberulent, entire. **INFL:** heads 13–15 mm, gen 1(3), terminal; involucre cylindric; phyllaries ± 21, overlapping, linear-oblong, 4-veined, green or purplish-tinged, ± minutely tomentose, tips erect. **FLS** 20–30. **FR** 3–4.8 mm. Common. Rocky slopes, desert scrub; 600–1200 m. **DSon;** NV, Baja CA. ❀TRY;STBL. Apr–May

B. incana A. Gray (p. 131) Shrub 4–10 dm, ± spheric. **STS** many, simple or branched, white-tomentose. **LVS** alternate, sessile or short-petioled, 1–3 cm, ovate, entire to minutely serrate, tomentose. **INFL:** heads ± 24 mm, solitary; involucre bell-shaped, tomentose; phyllaries ± 40, overlapping, ± gray (greenish to purple beneath hairs), outer oblong-ovate, inner linear-lanceolate, 5–9-veined, hidden by dense hairs, tips erect. **FLS** ± 60. **FR** ± 1 cm. $2n=$ 18. Uncommon. Sandy washes, flats; < 1700 m. **D;** NV. ❀DRN, DRY:2,3,7–12,**14**,18–24;DFCLT. Apr–Jun, Oct

B. knappiana Drew (p. 131) KNAPP'S BRICKELLIA Shrub 10–20 dm. **STS** slender, willow-like, ± sticky, puberulent. **LVS** alternate; petioles 4–5 mm; blades 2.5–3.5 cm, lanceolate or narrowly ovate, gen serrate, gland-dotted. **INFL:** heads ± 7 mm, in a panicle-like cluster with most heads ± lateral; involucre cylindric; phyllaries ± 20, overlapping, linear-elliptic, 4-veined, greenish, puberulent, tips erect. **FLS** 5–7. **FR** 2.5 mm. RARE. Desert scrub; 800–1700 m. **n&e DMtns (Panamint, n Kingston mtns).** Origin possibly *B. multiflora* × *B. californica* or *B. desertorum*; needs further study.

B. longifolia S. Watson (p. 131) Shrub 10–15 dm. **STS** much-branched, ± glabrous. **LVS** alternate, ± sessile, 3–10 cm, linear, entire, appearing varnished, subglabrous to puberulent. **INFL:** heads ± 5 mm, in small raceme-like clusters; involucre cylindric; phyllaries 10–12, overlapping, 4-veined, outer ovate, inner ± lanceolate, greenish, glabrous or minutely puberulent, tips erect. **FLS** 3–5. **FR** 1.8 mm. Uncommon. Washes; 1000–1700 m. **n&w DMoj;** to UT, AZ. Indistinct from *B. multiflora.* Apr–May

B. microphylla (Nutt.) A. Gray Shrub 3–6 dm. **STS** erect, branched from base, glandular-hairy. **LVS** green, short-petioled; blade 0.7–2 cm, ovate to round, entire to serrate-lobed, glandular. **INFL:** heads 10–12 mm, 1–3 per panicle on short branches; involucre cylindric to narrowly bell-shaped; phyllaries ± 45, overlapping, outer obovate, inner linear-oblong, 3–5-veined, greenish, glandular, tips darker, recurved or spreading. **FLS** ± 22. **FR** 4–4.5 mm. $2n=18$. Uncommon. Among rocks; 1000–2700 m. s SNF, SnGb, Wrn, **SNE, DMoj;** to ID, NV. Sep–Oct

B. multiflora Kellogg Shrub 10–20 dm. **STS** erect, branched. **LVS** alternate, short-petioled; blade 3–8 cm, lanceolate, ± entire, gummy, gland-dotted, sparsely bristly. **INFL:** heads ± 7 mm, in small raceme-like clusters; involucre cylindric; phyllaries ± 20, overlapping, lanceolate to ovate, 4-veined, greenish, glabrous or minutely puberulent, tips erect. **FLS** 3–5. **FR** 1.7–2.2 mm. Uncommon. Rocks, washes; 600–2400 m. **W&I, DMoj;** NV. May–Jul, Sep–Oct

B. nevinii A. Gray (p. 131) NEVIN'S BRICKELLBUSH Shrub 3–5 dm. **STS** erect, branched, dense, white-tomentose. **LVS** alternate, ± sessile, 0.6–1.8 cm, ovate to ± cordate, dentate-serrate, tomentose. **INFL:** heads ± 1.5 cm, few in panicle-like clusters on short branches; involucre subcylindric; phyllaries ± 30, overlapping, linear-oblong, 3–6-veined, whitish woolly, tips spreading or recurved. **FLS** ± 23. **FR** ± 4 mm. $2n=18$. UNCOMMON. Desert scrub; 300–1900 m. SCoRI, WTR, SnGb, **DMoj;** NV. Related to *B. microphylla, B. watsonii.* ❀DRN,DRY:7–12,14,18–24;DFCLT. Sep–Nov

B. oblongifolia Nutt. var. ***linifolia*** (D. Eaton) Robinson (p. 131) Per or subshrub 2–4 dm, rounded. **STS** much-branched, gray-hairy, gen glandular. **LVS** alternate, ± sessile, 1–3 cm, linear to ± ovate, 0–2-toothed, ± densely puberulent, gen glandular. **INFL:** heads 15–17 mm, gen solitary; involucre cylindric to bell-shaped; phyllaries 25–35, overlapping, outer shorter, lance-oblong, inner linear, gen 4-veined, green, glandular-puberulent, tips erect. **FLS** 35–45. **FR** 4–6 mm. $2n=18$. Common. Desert woodland; < 2800 m. SnJt, **SNE, D;** to B.C., MT, UT, Colorado. Variable; 2 other vars.. ❀TRY;STBL. May–Jul

B. watsonii Robinson Shrub or subshrub 2–3 dm, aromatic. **STS** much-branched, tomentose-puberulent. **LVS** alternate, short-petioled; blades 5–15 mm, ovate, light green, > tomentose, ± dentate. **INFL:** heads 9–11 mm, in panicle-like clusters with 1–3 heads at branchlet ends; involucre cylindric; phyllaries ± 25, ± spreading, ± ovate, 3–5-veined, glandular, tips green. **FLS** 15–18. **FR** 3.5–4.5 mm. $2n=18$. Uncommon. Among rocks, on rock walls, gen in woodland; 1500–2400 m. **DMoj;** to UT. May, Aug–Oct

CALYCOSERIS

G. Ledyard Stebbins

Ann; sap milky. **ST** 1–3 dm, branched, glabrous below, with tack-shaped glands above. **LVS**: basal and lower cauline petioled, 1–2-pinnately divided into long, linear lobes, glabrous; cauline alternate, gradually or abruptly reduced, upper sessile. **INFL**: heads ligulate, showy, solitary or in open, few-headed cymes; phyllaries in 2 series, scarious-margined, outer short, wide, inner many, linear; receptacle minutely bristly, otherwise naked. **FLS** many; ligules yellow or white, readily withering. **FR** 5–6-ribbed, tapered to short beak; pappus white, of many slender bristles that fall together. 2 spp.: sw US, nw Mex. (Greek: cup-like chicory)

1. Ligules yellow; fr smooth or nearly so . . . . *C. parryi*
1' Ligules white; fr roughened on ribs . . . . *C. wrightii*

C. parryi A. Gray (p. 137) **ST**: glands dark colored. **LF** 3–12 cm. **INFL**: heads 2–4 cm diam; involucre 10–15 mm. **FL**: ligules yellow. **FR** 7–9 mm, smooth or nearly so; pappus 6–8 mm. $2n=14$. Sandy to gravelly soils, washes, slopes; < 2000 m. Teh, PR, SNE, **D**; to UT, AZ, Baja CA. Apr–May

C. wrightii A. Gray **ST**: glands pale. **LF** 4–10 cm. **INFL**: heads 3–4 cm diam; involucre 10–15 mm. **FL**: ligules white, often purple-tinged, esp below. **FR** 5–8 mm, roughened on ribs; pappus 5–6.5 mm. $2n=14$. Desert washes, gravelly slopes, desert plains; 150–1150 m. **W&I**, **D**; to UT, TX. Mar–May

CARDUUS PLUMELESS THISTLE

David J. Keil & Charles E. Turner

Ann to per. **STS** erect. **LVS** alternate, reduced upward, decurrent as spiny wings, spiny-dentate and pinnately lobed, glabrous to tomentose; basal tapered to winged petiole; cauline sessile. **INFL**: heads discoid, 1–20 at branch tips; involucre cylindric to spheric; phyllaries overlapping in several series, spine-tipped; receptacle flat, whitish bristly. **FLS**: corollas white to pink or purple, tube long, slender, throat abrupt, short, lobes linear; anther bases short-sagittate, tips oblong; style with slightly swollen node, cylindrical above node, branches very short. **FR** ovoid, slightly compressed, glabrous; base slightly angled; pappus of many flat, minutely barbed, persistent bristles. ± 90 spp.: Eurasia, e Afr. (Latin: ancient name) [Howell 1959 Leaflets W Bot 9:17–29]

C. nutans L. (p. 137) MUSK THISTLE Bien. **STS** 4–15 dm, glabrous to woolly, narrowly spiny-winged. **LVS**: basal 10–40 cm, 1–2-pinnately lobed; cauline glabrous or sparsely hairy. **INFL**: heads gen solitary, often nodding; involucre 2–7 cm diam, spheric; phyllaries spreading, lanceolate to ovate, spine-tipped. **FLS**: corollas 20–25 mm, purple; tube 12–14 mm, throat 4–5 mm, lobes 4–6 mm. **FR** 4–5 mm, golden to brown; pappus 13–25 mm. $2n=16$. Roadsides, pastures, waste areas; 100–1200 m. KR, CaR, n SNH, SCo, MP, **DMoj**; N.Am.; native to Eur. [var. *leiophyllus* (Petrovic) Arènes] NOXIOUS WEED. Jun–Jul

CARTHAMUS DISTAFF THISTLE

David J. Keil & Charles E. Turner

Ann in CA. **ST** gen erect, leafy, branched above or throughout. **LVS** alternate, gen pinnately lobed, ± spiny; basal often 0 by fl; cauline gen clasping, gen spreading to recurved, lanceolate to ovate, rigid. **INFL**: heads discoid, solitary; involucre ± urn-shaped; outer phyllaries ± lf-like, inner with spiny appendages in CA; receptacle convex to conic, chaffy. **FLS**: corolla tube very slender, throat abruptly expanded, lobes linear; filaments gen densely hairy, anther bases short-tailed, tips oblong; style tip with minutely hairy node and terminal segment minutely papillate, barely notched. **FR** oblong to obpyramidal, ± 4-angled, glabrous, attached at side; outer frs gen ± roughened, pappus gen 0; inner frs smooth, pappus 0 or of many, narrow, gen unequal scales. 14 spp.: Medit. (Arabic: frs from fl color)

C. baeticus (Boiss. & Reuter) Nyman (p. 137) SMOOTH DISTAFF THISTLE Pl 4–10 dm, ± sparsely hairy. **ST** white. **LVS** very spiny. **INFL**: involucre 20–25 mm, gen becoming glabrous; outer phyllaries 35–55 mm, recurved, gen 2 × inner, tip appendages prominently spiny. **FLS**: corollas 25–35 mm, yellow. **FR** 4–6 mm, brown; pappus 8–10 mm. $2n=64$. Disturbed ground; < 500 m. c SNF, c GV, SnFrB, s SCoR, SW, **w DMoj**; native to Medit. NOXIOUS WEED. Jul–Aug

CENTAUREA KNAPWEED, STAR-THISTLE

David J. Keil & Charles E. Turner

Ann to per, ± branched. **LVS** alternate; lower gen deeply 1–2-lobed, segments gen narrow; upper reduced. **INFL**: heads discoid (sterile outer fls sometimes ± ray-like); involucre cylindric to hemispheric; phyllaries many, graded, gen ± ovate, scarious-margined, tip appendages fringed to spiny; receptacle flat, long-bristly. **FLS**: inner fruiting; anther bases tailed, tips oblong; style top minutely hairy, tips minutely branched. **FR** ± barrel-shaped, ± compressed, attached ± at side; pappus gen of stiff, unequal bristles or narrow scales. ± 500 spp.: esp Eurasia, n Afr (± 2 N.Am); some cult (waifs may incl *C. cineraria* L., *C. eriophora* L., *C. jacea* L., *C. moschata* L., *C. muricata* L., *C. salmantica* L.). (Greek: ancient name) Many NOXIOUS or invasive weeds.

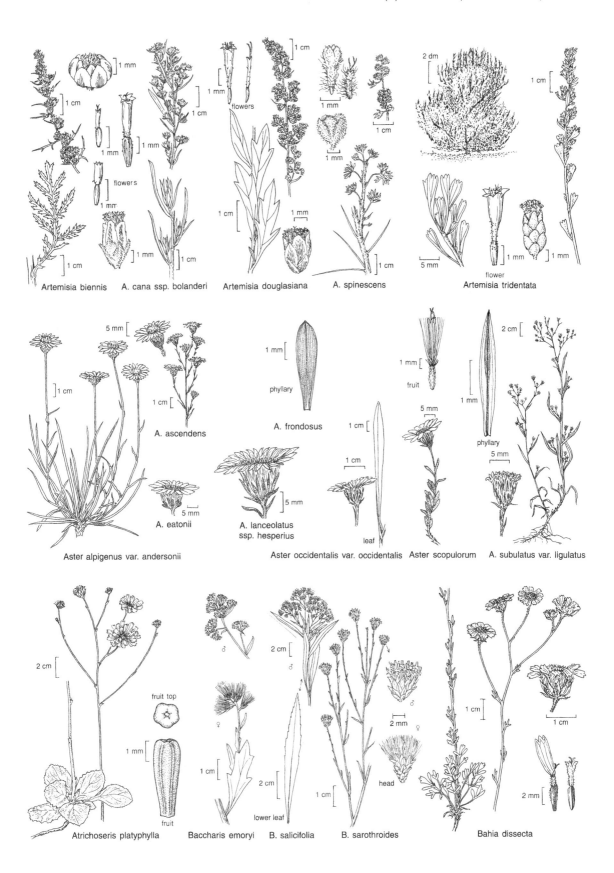

Artemisia biennis A. cana ssp. bolanderi Artemisia douglasiana A. spinescens Artemisia tridentata

flowers

flower

Aster alpigenus var. andersonii A. ascendens A. frondosus A. eatonii A. lanceolatus ssp. hesperius Aster occidentalis var. occidentalis Aster scopulorum A. subulatus var. ligulatus

phyllary

fruit

phyllary

leaf

Atrichoseris platyphylla Baccharis emoryi B. salicifolia B. sarothroides Bahia dissecta

fruit top

fruit

lower leaf

head

1. Corollas yellow
  2. Central spine of main phyllaries 5–10 mm, ± purplish . . . . *C. melitensis*
  2' Central spine of main phyllaries 10–25 mm, straw-colored or brownish . . . . *C. solstitialis*
1' Corollas white to pink or purple
  3. Phyllary appendages fringed with straw-colored spines; central spine of main phyllaries 1–3 mm;
    pappus 0 or of scales < 1 mm . . . . *C. diffusa*
  3' Phyllary appendages fringed with slender, dark teeth; main phyllaries not spine-tipped; pappus of
    bristles 1–2 mm . . . . *C. maculosa*

C. diffusa Lam. DIFFUSE KNAPWEED Ann to per 2–8 dm, much-branched, puberulent, ± gray-tomentose. **LVS**: lower 10–20 cm, ± deeply 2-lobed. **INFL** panicle-like; involucre 10–13 mm, cylindric to narrowly ovoid; main phyllaries pale green, prominently parallel-veined, appendages fringed with slender, straw-colored spines, central spine 1–3 mm. **FLS** few; corollas 12–13 mm, equal, white to pink or pale purple, sterile corollas slender. **FR** ± 2.5 mm, dark brown; pappus 0 or scales < 1 mm, white. 2*n*=18. Fields, roadsides; < 2300 m. NW, CaR, n&c SN, SCoR, MP, **W&I**; native to se Eur. NOXIOUS WEED.

C. maculosa Lam. (p. 137) SPOTTED KNAPWEED Bien 3–10 dm, ± gray-tomentose. **LVS** resin-dotted; lower 10–15 cm, ± deeply 1–2-lobed. **INFL** open; heads gen many; involucre 10–13 mm, ovoid; phyllaries pale green or pink-tinged, prominently parallel-veined, appendages fringed with slender, dark teeth. **FLS** 30–40; corollas 12–25 mm, white to pink or purple, sterile corollas ± slender. **FR** 3–3.5 mm, ± pale brown, finely hairy; pappus bristles 1–2 mm, white. 2*n*=18,36. Disturbed areas; < 2000 m. NW, CaR, SN, n ScV, n CW, s PR, MP, **n SNE**; native to Eur. NOXIOUS WEED. Summer

C. melitensis L. TOCALOTE Ann 1–10 dm, ± gray-hairy. **LVS** resin-dotted; ± scabrous; lower 2–15 cm, entire to lobed, gen 0 at fl; cauline long-decurrent. **INFL**: heads 1–few; involucre 10–15 mm, ovoid, ± cobwebby or becoming glabrous; main phyllaries ± straw-colored, appendage purplish, base spine-fringed, central spine 5–10 mm, slender. **FLS** many; corollas 10–12 mm, ± equal, yellow, sterile corollas slender. **FR** ± 2.5 mm, ± light brown, finely hairy; pappus bristles 2.5–3 mm, white. 2*n*=18,24,36. Disturbed fields, open woods; < 2200 m. CA-FP, **D (uncommon)**; native to s Eur. Invasive. May–Jun

C. solstitialis L. (p. 137) YELLOW STAR-THISTLE Ann 1–10 dm, ± rounded, gray-tomentose. **LVS** ± scabrous-bristly; lower 5–15 cm, 1–2-lobed, gen 0 at fl; cauline long-decurrent. **INFL** open; heads 1–many; involucre 13–17 mm, ovoid; main phyllaries pale green to straw-colored or brown, appendages palmately spiny, central spine 10–25 mm, stout. **FLS** many; corollas 13–20 mm, ± equal, yellow, sterile corollas slender. **FR** 2–3 mm, glabrous; outer fr dark brown (pappus 0); inner fr ± mottled light brown (pappus bristles 2–4 mm, fine, white). 2*n*=16. Pastures, roadsides, disturbed grassland or woodland; < 1300 m. CA-FP, **DMoj**; native to s Eur. NOXIOUS WEED. Cumulatively TOXIC to horses. May–Oct

CHAENACTIS PINCUSHION

James D. Morefield

Ann to subshrubs, gen ± hairy. **LVS** alternate or basal, gen petioled, reduced upward, entire and linear or gen elliptic to ovate or obovate and 1–4-pinnately lobed; 1+ lobes longest near middle or base of blade. **INFL**: heads discoid (but outer fls often enlarged, ± ray-like), 1–many per st, gen in terminal cymes; peduncle gen hairy like phyllary bases; involucre gen < 15 mm diam, cylindric to obconic or hemispheric; phyllaries in 1–2 ± equal series, gen linear to lanceolate, tips gen ± flat, gen ± green; receptacle flat to rounded, gen naked. **FLS** 10–many; corollas radial (outer, if enlarged, ± bilateral), gen white to pinkish or yellow, gen opening in daytime; anthers gen exserted. **FR** club-shaped, gen not compressed, stiffly hairy; pappus 0 or of 4–20 fringed scales in 1–few series. 18 spp.: w N.Am. (Greek: gaping ray, from enlarged outer corollas of some) [Mooring 1980 Amer J Bot 67:1304–1309] Spp. of sect. *Chaenactis* hybridize.

1. Per, rarely bien or fl 1st year; pappus scales 8–20 per fr in indistinct series, ± equal (sect. *Macrocarphus*)
  2. Phyllaries and peduncles ± tomentose, glands 0; lvs gen 1-pinnately lobed; pls scapose, lfless above base;
    head 1 per st . . . . *C. alpigena*
  2' Phyllaries and peduncles ± cobwebby, ± glandular-hairy; lvs gen 2-pinnately lobed; pls and heads
    various . . . . *C. douglasii*
    3. Pls scapose, lfless above base; heads 1(–2) per st; alpine . . . . var. *alpina*
    3' Pls not scapose, ± leafy above base; heads (1)2–many per st; gen below alpine . . . . var. *douglasii*
1' Ann; pappus scales 4–8 in 1 or 2 distinct series, outer series (when present) gen << inner
  4. Tips of longest phyllaries acuminate, needle-like, cylindric, gen reddish at fl time; lower st and lvs
    whitish scaly- to granular-puberulent, not cobwebby (sect. *Acarphaea*) . . . . *C. carphoclinia*
    5. Lvs basal and cauline, < 7 cm; bases of petioles not strongly enlarged . . . . var. *carphoclinia*
    5' Lvs gen basal, longest 7–10 cm; bases of petioles strongly enlarged . . . . var. *peirsonii*
  4' Tips of longest phyllaries obtuse to acute, ± flattened, gen greenish at fl time; lower st and lvs ±
    cobwebby or glabrous (sect. *Chaenactis*)
    6. Corollas ± bright to deep yellow . . . . *C. glabriuscula*
      7. Pls not scapose, gray-cobwebby; heads 2–many per st; longest peduncles < 5 cm; longest pappus
        2–4 mm; basal rosette gen withering . . . . var. *glabriuscula*
      7' Pls ± scapose, ± white-woolly; heads 1(–3) per st; longest peduncles > 5 cm; longest pappus
        4–6 mm; basal rosette ± persistent . . . . var. *lanosa*
    6' Corollas white to pinkish, rarely pale yellow
      8. Peduncles not glandular-hairy near heads, cobwebby to glabrous; corollas all radial, ± equal;
        head length > width, longest phyllaries (10)12–18 mm, tips gen recurved, soft; pappus scales ± 8,
        outer 4 << inner 4

9. Sts and lvs thinly tomentose; lvs not fleshy, lobes flattened; phyllaries ± tomentose; corollas ± 2 × fr, gen > 9 mm, anthers ± included . . . . *C. macrantha*
9' Sts and lvs gen glabrous; lvs ± fleshy, lobes ± cylindric; phyllary bases glabrous, tips densely puberulent; corollas ± = fr, gen 7–9 mm, anthers exserted . . . . *C. xantiana*
8' Peduncles ± glandular-hairy near heads, sometimes also cobwebby; outer corollas strongly bilateral, >> inner; head length ± = width, longest phyllaries < 10(–12) mm, tips gen erect, rigid; pappus scales (at least of central fls) ± 4, ± equal
   10. Outer phyllaries sharp, becoming glabrous; st base glabrous at fl time; pappus of inner fls > buds, tips visible; lvs entire or 1-pinnately lobed, lobes 1–2(5) distant pairs . . . . *C. fremontii*
   10' Outer phyllaries gen blunt, persistently glandular hairy or cobwebby; st base at least thinly cobwebby at fl time; pappus < buds, tips hidden; lvs 1–2-pinnately lobed, 1° lobes gen 4–8 ± crowded pairs . . . . *C. stevioides*

C. alpigena Sharsm. (p. 137) Per, scapose, cespitose to ± matted. **STS** gen several, erect to prostrate, < 10 cm, tomentose near base, hairs gen thinning with age. **LVS** < 4 cm, densely tomentose, rarely becoming glabrous, not fleshy; basal rosette persistent; largest blades linear to obovate, curled and 1–2-pinnately lobed (s pls) to flat and ± palmately lobed or entire (n pls), 1° lobes 2–7 pairs, ± crowded, longest near middle. **INFL:** heads 1 per st; peduncles < 10 cm; involucre obconic to cylindric, ± tomentose, glandless; longest phyllaries 9–14 mm, tips erect, ± rigid, blunt. **FLS:** corollas radial, 5.5–8 mm, ± equal, white to pinkish. **FR** 5–8 mm; pappus scales 8–20 in indistinct series, ± equal, longest 5–8.5 mm. Loose sand or gravel; 2300–4000 m. c&s SNH, **W&I (n White Mtns.);** w NV. Summer

C. carphoclinia A. Gray PEBBLE PINCUSHION Ann, branched above middle. **ST** gen 1, ± erect, < 7 dm, whitish granular-puberulent below and gen on lvs; hairs thinning with age. **LVS** < 11 cm, gen not fleshy; basal rosette withering; largest blades (2)3–4-pinnately lobed, 1° lobes 2–7(10) pairs, ± crowded, longest near base, tips ± cylindric. **INFL:** heads gen several per st; peduncles < 7 cm; involucre cylindric to hemispheric, glandular-or wavy-hairy; longest phyllaries 7–10 mm, tips (at least of inner) erect, rigid, acuminate, needle-like, cylindric, gen reddish; receptacle gen ± chaffy, chaff resembling phyllaries. **FLS:** corollas radial, 4–6 mm, white to pinkish, outer spreading but barely enlarged. **FR** 3–4 mm; pappus scales in 1 series, scales of outer fr (0–)4, unequal, < inner, scales of inner fr 4, equal, longest 3–5 mm. Gen open rocks or gravel; < 1900 m. e PR, **SNE, D;** to UT, AZ, nw Mex. Often mistaken for *C. stevioides.* Spring.

   var. *carphoclinia* (p. 137) **ST** < 4 dm. **LVS** basal and cauline, all < 7 cm; petiole narrow, soft, base not strongly enlarged. 2n=16. Habitats and range of sp. [var. *attenuata* (A. Gray) M.E. Jones] Mar–Jun

   var. *peirsonii* (Jepson) Munz PEIRSON PINCUSHION **ST** gen 4–6 dm. **LVS** gen basal, longest 7–10 cm; petiole stout, hard, base strongly enlarged. RARE. Habitats of sp; < 500 m. e PR (e Santa Rosa Mtns), **adjacent w DSon?** Mar–Apr

C. douglasii (Hook.) Hook. & Arn. Per, bien(?), sometimes fl lst year. **STS** 1–several, erect to spreading, < 50 cm, gen thinly grayish cobwebby; hairs thinning with age. **LVS** < 15 cm, gen cobwebby to ± tomentose, not fleshy; basal rosette ± persistent; largest blades gen 2-pinnately lobed, 1° lobes 3–7 pairs, ± crowded, longest near middle, tips curled. **INFL:** heads 1–many per st; peduncles < 10 cm; involucre obconic to hemispheric, gen glandular-hairy; longest phyllaries 9–14 mm, tips erect, ± rigid, blunt. **FLS:** corollas radial, 5–8 mm, white to pinkish, outer somewhat enlarged. **FR** 5–8 mm; pappus scales 8–20 in indistinct series, ± equal, longest gen 3–6 mm. Dry, open, often disturbed areas, alpine crevices; 1000–3500 m. NW, CaR, SN, **GB, n DMtns;** to B.C., Colorado, AZ.

   var. *alpina* A. Gray ALPINE DUSTY MAIDENS Per, scapose, cespitose to ± matted. **STS** < 10 cm. **LVS** < 7 cm, sometimes becoming glabrous. **INFL:** heads 1(–2) per st; longest phyllaries 9–12 mm. 2n=12. RARE in CA. Open, subalpine to alpine gravel and crevices; 3000–3400 m. n SNH (Alpine, El Dorado cos., esp Freel Peak area), **n DMtns (Panamint Mtns);** to ID, MT, Colorado. [*C. a.* (A. Gray) M.E. Jones] Intergrades

downslope with var. *douglasii;* more uniform outside CA (CA pls are ± atypical). Summer

   var. *douglasii* (p. 137) DUSTY MAIDENS Per, bien(?), sometimes fl lst year, gen branched below middle. **STS** < 50 cm. **LVS** < 15 cm. **INFL:** heads (1)3–many per st in a ± leafy cyme; longest phyllaries 11–14 mm. 2n=12,24,36 (extra chromosomes also reported). Dry, open, often disturbed areas; 1000–3500 m. NW, CaR, SN, **GB, n DMtns;** to B.C., MT, Colorado, n AZ. [vars. *achilleifolia* (Hook. & Arn.) A. Gray, *montana* M.E. Jones, *rubricaulis* (Rydb.) Ferris; *C. ramosa* Stockw.] More distinctive variants occur outside CA. ❀DRN:1,15,16&IRR:2,3,7,18; DFCLT. Summer

C. fremontii A. Gray (p. 137) DESERT PINCUSHION Ann, branched below middle, sometimes ± scapose. **STS** 1–many, ± erect, < 40 cm, becoming glabrous before fl time exc near heads. **LVS** < 9 cm, gen glabrous, ± fleshy; basal rosette withering; largest blades entire or 1-pinnately lobed, lobes 1–2(5) pairs, well separated, longest near middle, tips cylindric. **INFL:** heads 1–few per st; peduncles < 11 cm, gen glandular-hairy near heads; involucre obconic to hemispheric, ± truncate at base, becoming glabrous; longest phyllaries 8–10(12) mm, tips erect, rigid, sharp. **FLS:** corollas white to pinkish, outer bilateral, greatly enlarged, inner radial, 5–7 mm. **FR** (3)6–8 mm; pappus scales in 1 series, smaller and unequal on outer fr, on inner fr 4(5), equal, longest gen 6–8.5 mm, > buds, tips visible. 2n=10. Open sand or gravel; < 1600 m. Teh, s SnJV, SCoRI, more common in **s SNE, D;** to sw UT, w AZ, n Baja CA. Scattered. ❀TRY. Spring

C. glabriuscula DC. YELLOW PINCUSHION Ann. **STS** gen 1–few, erect to spreading, < 50 cm; hairs thinning with age. **LVS** < 11 cm, ± cobwebby, fleshy or not; largest entire or 1–2-pinnately lobed, 1° lobes 1–7 pairs, longest near blade middle, tips flat to curled or cylindric. **INFL:** heads 1–several per st; peduncle < 20 cm; involucre widely cylindric to obconic or hemispheric, gen ± tomentose or glandular-hairy; longest phyllaries 4.5–9 mm, tips erect, ± rigid, gen blunt. **FLS:** corollas ± bright to deep yellow, outer bilateral, greatly enlarged, inner radial, 4–8 mm. **FR** 3–9 mm; pappus scales (1)4–8 in 1–2 series, scales of outer fr gen < inner, unequal, scales of inner fr gen equal, longest 1–8 mm. 2n=12. Gen dry, open places; < 1600 m. CA-FP, **w edge D;** n Baja CA. Highly variable; some forms like *C. stevioides* exc fl color. 5 vars. in CA.

   var. *glabriuscula* Pls branched above or below middle, thinly grayish cobwebby. **STS** < 50 cm, ± erect. **LVS** < 9 cm, gen not fleshy; basal rosette withering; largest blades 1(2)-pinnately lobed, 1° lobes 2–7 pairs, well separated to ± crowded, tips flat to ± cylindric. **INFL:** heads gen several per st; peduncles < 5 cm; involucre obconic to hemispheric; longest phyllaries 5–7 mm, < 2 mm wide, gen ± grayish tomentose at least on midrib. **FLS:** inner corollas 4–6 mm. **FR** 3–5.5 mm; pappus scales 4 in 1 series, longest 2–4 mm, sometimes 0 on outer fr. 2n=12. Open slopes, sandy places; < 1600 m. s SNF, Teh, CW, SW, **w edge D;** n Baja CA. [vars. *curta* (A. Gray) Jepson, *tenuifolia* (Nutt.) H.M. Hall] ❀TRY. Apr–Jun

   var. *lanosa* (DC.) H.M. Hall Pls ± scapose, ± densely white-woolly. **STS** < 10 cm, decumbent to erect. **LVS** < 11 cm, gen not fleshy; basal rosette ± persistent; largest blades entire or 1-pinnately lobed, lobes 1–2(5) pairs, well separated, tips flat to cylindric. **INFL**: heads 1 per st; peduncles < 20 cm; involucre obconic to hemispheric; longest phyllaries 6–8 mm, < 2 mm wide, whitish tomentose. **FLS**: inner corollas 5–6.5 mm. **FR** 4–6 mm; pappus scales 4 in 1 series, longest 4–6 mm. 2n=12. Loose sand; < 700 m. SCoRI, SCo, TR, **nw edge DSon**. [var. *denudata* (Nutt.) Munz] ❀DRN,DRY,SUN:**7**,8,9,11,**14–17,19–24**. Apr–Jul

C. macrantha D. Eaton (p. 137) MOJAVE PINCUSHION Ann, branched below middle. **STS** 1–few, erect, < 35 cm, thinly tomentose; hairs thinning with age. **LVS** < 6 cm, not fleshy; basal rosette withering; largest blades 1-pinnately lobed, lobes gen 2–5 pairs, ± well separated, longest near middle to ± near base, tips flat. **INFL**: heads 1–few per st; peduncles < 9 cm; involucre widely cylindric, ± tomentose, glands 0; longest phyllaries 12–18 mm, tips gen recurved, soft, blunt. **FLS**: corollas all radial, 9–11 mm, ± equal, white to pinkish, gen opening at night; anthers ± incl. **FR** 5–6 mm; pappus scales 8 in 2 series, longest 5–7 mm, outer << inner. 2n=12. Open (often calcareous) sand or gravel; 1000–2000 m. **GB, n DMoj**; to se OR, sw ID, w UT, n AZ. Spring

C. stevioides Hook. & Arn.(p. 137) DESERT PINCUSHION Ann, branched above or below, sometimes ± scapose. **STS** 1–many, erect, < 45 cm, cobwebby (at least thinly so near base); hairs thinning with age. **LVS** < 11 cm, ± cobwebby, gen not fleshy; basal rosette gen withering; largest blades 1–2-pinnately lobed, lobes 4–8 pairs, crowded to well separated, longest near middle, tips gen curled. **INFL**: heads gen several per st; peduncles < 11

cm; involucre obconic to hemispheric, glandular-hairy or ± cobwebby; longest phyllaries 5.5–8(10) mm, tips erect, ± rigid, gen blunt. **FLS**: corollas white to pink, rarely pale yellow, outer bilateral, greatly enlarged, inner radial, 4.5–6.5 mm. **FR** 4–6.5 mm; pappus scales 4–5 in 1(–2) series, scales of outer fr < inner, unequal, scales of inner fr equal to unequal, longest 1.5–6 mm. 2n=10. Common. Open flats, slopes; < 2100 m. SCoRI, **SNE, D**; to se OR, sw ID, w Colorado, sw NM, nw Mex. [var. *brachypappa* (A. Gray) H.M. Hall; *C. mexicana* Stockw.] Highly variable, but not clearly divisible. See also *C. carphoclinia*. ❀TRY. Spring

C. xantiana A. Gray (p. 137) Ann, branched above middle. **STS** 1–several, erect, < 50 cm; base gen becoming glabrous before fl time. **LVS** < 7 cm, gen glabrous, ± fleshy; basal rosette withering; largest blades entire or 1-pinnately lobed, lobes 1–2(5) pairs, well separated, longest near middle, tips cylindric. **INFL**: heads few–many per st; peduncles < 6 cm; involucre widely obconic, base (at least of outer phyllaries) glabrous, tips densely puberulent, glandless; longest phyllaries 10–18 mm, gen recurved, tips soft, blunt. **FLS**: corollas white to pinkish, all radial, 7–9 mm, outer somewhat enlarged. **FR** 5–9 mm; pappus scales 8 in 2 series, outer << inner, longest 5–9 mm. 2n=14. Sand; 300–2500 m. s SN, SCoR, TR, **GB, w DMoj**; se OR, w NV, nw AZ. ❀DRN,SUN,DRY:1–3,**7**,8–12,**14–16,18–24**. Spring–early summer

CHAETADELPHA

G. Ledyard Stebbins

1 sp.: w US. (Greek: fused bristles) [Tomb 1972 Madroño 21:459–462]

C. wheeleri A. Gray (p. 137) Per from branched caudex; sap milky; herbage glabrous. **STS** 1–4 dm, much-branched. **LVS** all cauline, alternate, 1–5 cm, linear to narrowly lanceolate (or scale-like), acute, entire. **INFL**: heads ligulate, solitary on branchlets 2–5(7) cm; involucre 11–15 mm, narrowly cylindric; phyllaries in 2 series, outer 7–10 short, wide, inner 5 long, linear-oblong; receptacle naked. **FLS** 5 per head; ligules 6–12 mm, pale lavender-white, readily withering. **FR** 8–12 mm, columnar, 4–5-ridged, glabrous; pappus double, of 5 stiff awns and gen many bristles, all fused at base, yellowish or tan. 2n=18. Sand dunes, desert scrub; 900–1900 m. **W&I, n DMoj (Eureka Valley)**; se OR, w NV. ❀TRY. May–Sep

CHAETOPAPPA

Geraldine A. Allen

Ann to subshrub. **LVS** alternate, linear to oblanceolate, entire. **INFL**: heads radiate, solitary; phyllaries in 2–6 series, outer < inner, margins translucent, papery; receptacle ± flat, chaff 0. **RAY FLS** 5–24; ligules coiled at maturity, white to blue or pink-purple. **DISK FLS** 5–many; corollas yellow; anther tips triangular; style tips triangular. **FR** gen linear-fusiform, round or ± compressed; pappus 0 or of scales, bristles, or both. ± 10 spp.: sw US, n Mex. (Greek: bristle-pappus) [Nesom 1988 Phytologia 64:448–456]

C. ericoides (Torrey) G. Nesom (p. 137) Per to subshrub; caudex ± woody. **STS** ascending to erect, 5–15 cm, gen branched, strigose, ± glandular. **LVS** basal and cauline (basal 0 at fl), 4–12 mm, linear to obovate, obtuse to abruptly pointed, ± glandular, bristly-hairy below. **INFL**: phyllaries lanceolate to oblong, acute to acuminate, green, ± purple-tipped. **RAY FLS** gen 12–21; corolla 5–10 mm, white to ± pink. **FR** ± rounded, brown, hairy; pappus of minutely barbed bristles, ± white. 2n=16,32. Dry slopes; 1300–2900 m. SN (e slope), **W&I, DMtns**; to NE, TX, n Mex. [*Leucelene e.* (Torrey) E. Greene] ❀TRY. Apr–May

CHAMOMILLA

Elizabeth McClintock

Ann, sometimes aromatic. **STS** branched, erect or decumbent. **LVS** alternate, irregularly 2–3-pinnately lobed; segments linear; petiole short or 0. **INFL**: heads discoid, solitary or 2–3; receptacle conic, naked; phyllaries in 2–3 unequal series, margins scarious. **FLS** many, yellow, tubular, 4-lobed, narrowed above; anthers very small, tips ovate, bases rounded or ± cordate; style short, branches truncate with shrub-like tips. **FR** cylindric, sometimes gelatinous when wet, ribbed; pappus a narrow crown or 0. ± 5 spp.: Eur, N.Am. (Derivation of name not known) [Moe 1977 Dissertation, Univ CA, Berkeley] *C. suaveolens* has been reported from Independence, CA. *For revised taxonomy of *Chamomilla*, see Bremer & Humphries 1993 Bull Nat Hist Mus Lond (Bot) 23: 71-177

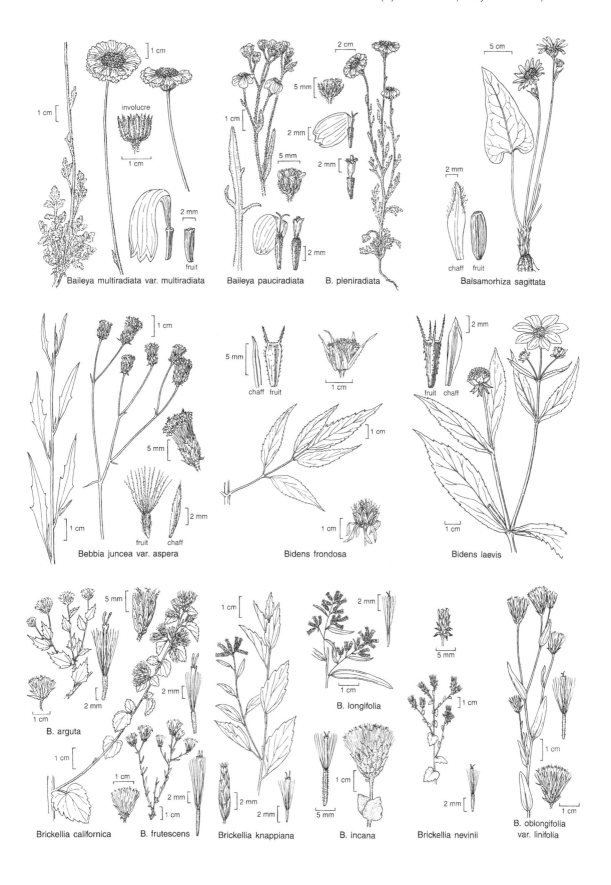

Baileya multiradiata var. multiradiata

involucre

fruit

Baileya pauciradiata

B. pleniradiata

Balsamorhiza sagittata

chaff   fruit

Bebbia juncea var. aspera

fruit   chaff

chaff   fruit

Bidens frondosa

fruit   chaff

Bidens laevis

B. arguta

Brickellia californica

B. frutescens

Brickellia knappiana

B. longifolia

B. incana

Brickellia nevinii

B. oblongifolia
var. linifolia

C. occidentalis (E. Greene) Rydb. (p. 137) Pls 15–45(70) cm; herbage not strongly scented. **STS** often branched only above. **LVS** sessile, < 7 cm, glabrous. **INFL**:heads gen < 1.5 cm diam, ± conic to spheric, remaining intact at maturity. **FLS**: corollas 1–2 mm. **FR** angled, gelatinous when wet; tip and pappus crown 2-lobed, with wide brown gland extending from tip of each lobe to ± middle of achene. Common. Undisturbed alkali flats, vernal pools, edges of salt marshes; < 2400 m. NCoRO, CaRH, SNH, SnJV, SnFrB, SCoRO, SCo, **DMoj**. ***Matricaria occidentalis** E. Greene] ✿SUN:5,**7–9,14–24**;used as substitute for chamomile. May–Aug

CHLORACANTHA

Guy L. Nesom

1 sp.: sw US, Mex, C.Am. (Latin: green thorns) [Nesom et al. 1991 Phytologia 70:371–381]

C. spinosa (Benth.) G. Nesom var. *spinosa* (p. 137) Subshrub from stout rhizome, ± glabrous. **STS** erect, 5–15(25) dm; branches sometimes thorn-like. **LVS** alternate, 10–50 cm, oblanceolate, gen entire, 1-veined, early deciduous. **INFL**: heads radiate, few–many in loose clusters; phyllaries in (3)4–5 series, inner 4.5–7.5 mm, ± lanceolate, veins (1)3(5), parallel, orange-resinous, tips gen rounded, margins transparent; receptacle naked. **RAY FLS** 10–33; corollas 4–8(11) mm, white, ligules 1–2 mm wide, coiled at maturity. **DISK FLS** many; corollas 3–6 mm, yellow; style appendages ± deltate. **FR** 1.5–3.5 mm, slightly compressed, glabrous; veins 5(6), whitish to golden-brown; pappus of many barbed bristles in ± 2 series, outer gen << inner. 2*n*=18. Seeps, moist streamsides, ditches, sometimes saline or drier areas; < 1250 m. PR, **DSon**; to UT, s-c US, Mex. [*Aster s.* Benth.] Jun–Dec

CHRYSOTHAMNUS RABBITBRUSH

Loran C. Anderson

Per to shrub. **STS** erect, often densely clustered. **LVS** alternate, sessile, entire. **INFL**: heads discoid, arrayed in ± dense cymes, peduncled or subsessile; involucre gen cylindric; phyllaries gen in 3–5 series (± 5 vertical ranks), free, overlapping, keeled; receptacle naked. **FLS** 2–20 (often 5) per head; corollas gen yellow, lobes 0.5–3 mm, gen spreading; style branches long, slender, gen exserted. **FR** narrowly cylindric, ± 5-ridged, gen light brown; pappus of many white to brownish bristles. 16 spp.: sw Can to n Mex. (Greek: golden shrub) Relationships to *Ericameria* warrant further study [Anderson 1995 Great Basin Naturalist 55:84–88; Neson & Baird 1995 Phytologia 78:61–68]

1. Sts glabrous, puberulent, or gland-dotted
  2. Herbage gland-dotted; corollas yellow
    3. Involucres 6–8 mm, obconic; phyllaries weakly vertically ranked, tips not swollen . . . . *C. paniculatus*
    3' Involucres 7–9.5 mm, cylindric; phyllaries strongly vertically ranked, tips swollen, green . . . . . *C. teretifolius*
  2' Sts (and gen lvs) not gland-dotted; corollas gen yellow (white)
    4. Corollas white; lvs gland-dotted . . . . *C. albidus*
    4' Corollas yellow; lvs not gland-dotted
      5. Fr glabrous; lvs flat, rigid, persistent
        6. Involucres gen 10–13 mm; phyllaries strongly keeled . . . . *C. depressus*
        6' Involucres gen 12–15 mm; phyllaries not keeled . . . . *C. gramineus*
      5' Fr (gen densely) hairy; lvs gen twisted
        7. Phyllaries acuminate to bristle-tipped; lvs < 1 mm wide . . . . *C. greenei*
        7' Phyllaries obtuse to acute; lvs 1–10 mm wide . . . . *C. viscidiflorus*
          8. Upper sts hairy; lvs glabrous . . . . ssp. *puberulus*
          8' Sts glabrous; lvs glabrous (exc gen ciliate)
            9. Lvs ± 1 mm wide; fls 3–4(5) per head; involucre ± obconic . . . . ssp. *axillaris*
            9' Lvs 1–10 mm wide; fls gen 4+ per head; involucre cylindric . . . . ssp. *viscidiflorus*
1' Sts tomentose to felted
  10. Phyllaries long-acuminate, weakly keeled; infl gen raceme-like; heads 5–10-fld . . . . *C. parryi*
    11. Lvs oblanceolate, glands short-stalked . . . . ssp. *asper*
    11' Lvs linear to oblanceolate, stalked glands 0
      12. Heads 1–3 per infl; corolla 8–9 mm . . . . ssp. *monocephalus*
      12' Heads many per infl; corolla > 9 mm . . . . ssp. *nevadensis*
  10' Phyllaries obtuse to acute, strongly keeled; infl gen panicle-like; heads 5-fld . . . . *C. nauseosus*
    13. Involucre subglabrous to tomentose; sts gen whitish; lvs gen dark green to grayish white
      14. Involucre 10–14 mm; lvs gen green . . . . [2]ssp. *bernardinus*
      14' Involucre 7–10(11) mm; lvs gen gray or white
        15. Corolla lobes 1–2 mm; style appendage > stigma . . . . ssp. *albicaulis*
        15' Corolla lobes 0.5–1 mm; style appendage < stigma . . . . ssp. *hololeucus*
    13' Involucre glabrous (outer phyllaries sometimes ciliate or scaly-margined); sts greenish or brownish;
      lvs greenish yellow or 0
      16. Fr glabrous . . . . ssp. *leiospermus*
      16' Fr hairy
        17. Lvs 1–3 mm wide, 1–5-veined; involucre > 10 mm . . . . [2]ssp. *bernardinus*
        17' Lvs 1 mm wide or less, 1-nerved; involucre < 10 mm

18. Phyllaries abruptly pointed, recurved . . . . ssp. *ceruminosus*
18' Phyllaries acute, erect
    19. Involucres 6–10 mm; sts gen leafy . . . . ssp. *consimilis*
    19' Involucres 8.5–12 mm; sts often lfless . . . . ssp. *mohavensis*

C. albidus (A. Gray) E. Greene (p. 139) Shrub 3–12 dm. **STS** brittle, glabrous. **LVS** 2–3.5 cm, linear, ± cylindric, very resinous. **INFL:** heads in cymes; involucres 7.4–9.5 mm, ± cylindric; phyllaries ± lanceolate, in poorly defined ranks, barely keeled, straw-colored, resinous, tips abruptly pointed, green. **FLS** 5–6; corollas 6–8 mm, white, lobes gen recurved; style appendage > stigma. **FR** 4–5 mm, hairy; pappus slightly < corolla. 2*n*=18. Uncommon. Saline or alkaline soils; 300–1300 m. **SNE, n DMoj**; to w UT. Aug–Nov

C. depressus Nutt. (p. 139) Subshrub 1–2 dm; caudex much-branched. **STS** unbranched below infl, brittle, rough, ± gray-hairy. **LVS** 2–3 cm, oblanceolate, rough-hairy. **INFL:** heads in small, dense cymes; involucre 10–15 mm, ± cylindric; phyllaries in well defined vertical ranks, strongly keeled, acute to acuminate, straw-colored or purplish. **FLS** 5–6; corollas 8–11 mm. **FR** 5–6.5 mm, gen glabrous (sparsely glandular); pappus ± = corolla. 2*n*=18. Rocky crevices; 1000–2100 m. **DMtns**; to Colorado, NM. Aug–Oct

C. gramineus H.M. Hall (p. 139) Per 2–5 dm. **STS** ± glabrous. **LVS** < 9 cm, ± lanceolate, firm, glabrous exc finely rough-hairy margins. **INFL:** heads few; involucre 11–17.5 mm, cylindric; phyllaries weakly vertically ranked, oblong to ovate, firm, not keeled, tips irregularly toothed or notched, abruptly pointed. **FLS** 4–7; corollas 9.5–11.5 mm; style appendage > stigma. **FR** 7–9 mm, glabrous; pappus ± = corolla. 2*n*=18. Uncommon. Pinyon/juniper woodland, bristlecone-pine forest; 2200–2900 m. **s W&I, n DMtns (Panamint Mtns)**; s NV. [*Petradoria discoidea* L. Anderson] Jul–Aug

C. greenei (A. Gray) E. Greene (p. 139) Shrub 1–2 dm. **STS** gen glabrous. **LVS** 1–2 cm, thread-like, straight or twisted. **INFL:** heads in dense cymes; involucre 5–8 mm, cylindric; phyllaries widely lanceolate, clearly vertically ranked, weakly keeled, tips (at least outer) gen ascending, long-tapered, sticky. **FLS** 4–5; corollas 4–5 mm; style appendage < stigma. **FR** 3–3.5 mm, hairy; pappus < corolla. 2*n*=18. Uncommon. Sandy washes; 1340–1830 m. **SNE (Fish Lake Valley), n DMtns (Cottonwood Mtns)**; to Colorado, NM.

C. nauseosus (Pallas) Britton RABBITBRUSH Shrub < 28 dm, ± tomentose; odor strong. **STS** whitish to green, ± flexible, very leafy or naked at fl. **LVS** 1–7 cm, thread-like to narrowly (ob)lanceolate. **INFL** dense, flat-topped or rounded, panicle-like; involucre 6–14.5 mm, cylindric; phyllaries ± lanceolate to ovate, in vertical ranks, ± strongly keeled, firm, obtuse to acute. **FLS** gen 5; corollas 6–12 mm; style appendage gen > stigma. **FR** 3–8 mm, gen hairy; pappus gen = corolla. 2*n*=18. Diverse habitats; 50–3300 m. NW, CaR, SN, SCoR, TR, PR, **GB**, **DMoj**; to B.C., MT, Colorado, Baja CA. Highly variable; 22 sspp. in w US. 8 sspp. in CA.

  ssp. ***albicaulis*** (Nutt.) H.M. Hall & Clements (p. 139) Pl 2–10 dm, gray-white to dark green. **LVS** 3–6 cm, ± narrowly lanceolate. **INFL:** involucres 7–10(11) mm, ± tomentose. **FLS:** corollas gen 9–11 mm. Common. Diverse dry habitats; 50–3300 m. NW, CaR, SN, **W&I**; to B.C., MT, Colorado. Pls with large, wide lvs have been called var. *macrophyllus* Howell or *C. californicus* E. Greene. Intergrades with sspp. *consimilis, hololeucus* ❀ DRN,SUN, DRY:1–3,**7**,8,9,**10**,11–16,**18–21**,22–24.

  ssp. ***bernardinus*** (H.M. Hall) H.M. Hall & Clements Pl 5–15 dm, white-tomentose to green. **LVS** 2.5–5 cm, ± narrowly lanceolate. **INFL:** involucres 10–14 mm, glabrous to hairy. **FLS:** corollas 10–12 mm. Open yellow-pine forest; 1200–2900 m. s SN, TR, n PR, **W&I**. Much like robust ssp. *albicaulis*. Sometimes also confused with sspp. *consimilis, mohavensis*. ❀ TRY.

  ssp. ***ceruminosus*** (Durand & Hilg.) H.M. Hall & Clements (p. 139) Pl 6–15 dm, gray- or yellow-green. **STS** ± lfless. **LVS** (if any) 1–3 cm, thread-like, curved. **INFL:** involucres 7–8.2 mm, glabrous, sticky; phyllary tips abruptly narrowed, recurved. **FLS:** corollas 6–7 mm. Gravelly arroyos; 700–1700 m. **DMoj**. Locally common.

  ssp. ***consimilis*** (E. Greene) H.M. Hall & Clements Pl 5–25 dm. **STS** leafy, yellowish green. **LVS** 2–6 mm, ± thread-like. **INFL** ± narrow; involucres 6–10 mm, glabrous; phyllary midribs brownish. **FLS:** corollas 6–7 mm. Common. Gen alkaline soils; 1000–2900 m. SN (e slope), Teh, s SCoRI, SnBr, PR, **GB**; to OR, MT, WY, NM, Baja CA. Sometimes intergrades with sspp. *albicaulis* (gen higher elevation) and *mohavensis*. Robust pls from s SNE have been called ssp. *viridulus* (H.M. Hall) H.M. Hall & Clements. ❀ TRY.

  ssp. ***hololeucus*** (A. Gray) H.M. Hall & Clements Pl 3–25 dm, white to gray- or yellow-green. **LVS** 3–7 cm (shorter in infl), ± narrowly lanceolate, sometimes reflexed. **INFL:** involucres gen 8–9 mm, straw-colored, less hairy inward; phyllary midribs brownish. **FLS:** corollas 8–9.5 mm, lobes erect or bent inward. Common. Well drained granitic or limestone soils in scrub or woodland; 150–2500 m. SNH, Teh, SCoRO, WTR, **SNE**, **DMtns**; to s OR, UT, n AZ. Yellow-green pls have been called ssp. *gnaphalodes* (E. Greene) H.M. Hall & Clements. ❀ TRY.

  ssp. ***leiospermus*** (A. Gray) H.M. Hall & Clements (p. 139) Pl gen 3–6 dm. **STS** gen lfless, yellowish. **LVS** (if any) 1–3 cm, thread-like. **INFL:** involucres 8–9.5 mm, gen straw-colored (purplish), glabrous. **FLS:** corollas 6–7.5 mm, lobes erect or bent inward. **FR** glabrous. Common. Dry sand, gravel, rocky crevices; 700–2400 m. **SNE, DMtns**; to UT, n AZ. ❀ TRY.

  ssp. ***mohavensis*** (E. Greene) H.M. Hall & Clements Pl 5–28 dm. **STS** few-branched, often ± lfless. **LVS** (if any) 1.5–3 cm, thread-like. **INFL:** involucres 8.5–12 mm, straw-colored, glabrous. **FLS:** corollas 7–10.5 mm. Common. Dry scrub; 400–2400 m. SCoRO, TR, **DMoj**; s NV. Sometimes intergrades with sspp. *consimilis* and *hololeucus*. ❀ DRN,SUN,DRY:1–3,7–10,**11**,12–16,18–21.

C. paniculatus (A. Gray) H.M. Hall (p. 139) BLACK-STEM Shrub 5–10(18) dm. **STS** erect, clustered, gland-dotted, gen ± black-banded (from fungal or insect attack). **LVS** 1.5–3.5 cm, thread-like, light green, resinous. **INFL:** heads in elongate cymes; involucre 6–8 mm, obconic; phyllaries oblong to ovate, weakly keeled, weakly 5-ranked, obtuse, resinous. **FLS:** corollas 5.5–7 mm; style appendage ± = stigma. **FR** 3.5–4 mm, hairy; pappus ± = corolla. 2*n*= 18. Common. Gravelly washes; 400–1600 m. **W&I, D**; to sw UT, nw AZ. ❀ TRY. Jun–Dec

C. parryi (A. Gray) E. Greene Subshrub or shrub, 1–9 dm. **STS** prostrate to erect, white to green, tomentose. **LVS** 1–7 cm, gen < 4 mm wide, thread-like to ± oblanceolate, ± glabrous to tomentose. **INFL:** heads gen ± many, in long or rounded cymes; involucre 10–18 mm, widely cylindric; phyllaries ± lanceolate, weakly 5-ranked, keeled, ± membranous, tips green. **FLS** 5–18; corollas 8–12.5 mm; style appendage > stigma. **FR** 4–7 mm, hairy; pappus gen < corolla. 2*n*=18. Dry places, gen open forest; 700–3700 m. KR, NCoR, CaR, SNH, SnBr, **GB**, **DMtns**; to s OR, Colorado, AZ. 12 sspp. in w US. 6 sspp. in CA.

  ssp. ***asper*** (E. Greene) H.M. Hall & Clements **LVS** 1.5–4 cm, straight or curved, gray or green, short-glandular. **INFL** spike-like or branched, ± dense; phyllaries straight or spreading, whitish, straw-colored, or purplish. **FLS** 5–10. Dry forest to alpine barrens, often in pumice or gravel; 1900–3300 m. c&s SNH, SnBr, **SNE**, **DMtns**; NV.

  ssp. ***monocephalus*** (Nelson & Kenn.) H.M. Hall & Clements **LVS** 1–2.5 cm, green; glands 0. **INFL:** heads gen 1(–3), gen overtopped by lvs; phyllaries erect or spreading, straw-colored. **FLS** 5–8. Common. Open subalpine forest, talus, alpine barrens; 2800–3700 m. c SNH, **n SNE (Sweetwater Mtns)**; w NV. Probably derived from ssp. *nevadensis*.

ssp. *nevadensis* (A. Gray) H.M. Hall & Clements (p. 139) **STS** stout. **LVS** 2–4 cm, green; glands 0. **INFL** raceme-like; phyllaries long-tapered, spreading or reflexed. **FLS** 5–6. Scrub, open yellow-pine forest, rarely on serpentine; 1100–2700 m. n&c SNH, **GB**; to s OR, sw UT, n AZ. Intergrades with ssp. *monocephalus* s. ❀TRY.

C. teretifolius (Durand & Hilg.) H.M. Hall (p. 139) Shrub 2–12 dm. **STS** much-branched, brittle, green, gland-dotted. **LVS** 1–3.5 cm, thread-like, subcylindric, dark green, resinous. **INFL**: heads scattered or in long, dense cymes; involucre 7–9.5 mm, narrowly cylindric; phyllaries strongly ranked, straw-colored, ± keeled, tips enlarged, green. **FLS** 5–7; corollas 6.5–8 mm; style appendage < stigma. **FR** 4–5 mm, hairy; pappus slightly < corolla. 2*n*=18. Rocky flats, slopes; 600–2400 m. Teh, TR, SnJt, **SNE, DMoj**; s NV, nw AZ. Sep–Nov

C. viscidiflorus (Hook.) Nutt. YELLOW RABBITBRUSH Shrub 1–15 dm; **STS** gen erect, brittle, white (greener upward). **LVS** 1–7.5 cm, 1–10 mm wide, thread-like to oblong, flat or twisted, (gray-) green, ± sticky. **INFL**: heads in dense, flat-topped or rounded cymes; involucre 5–10 mm, gen cylindric; phyllaries gen ± lanceolate, in ± 5 vertical ranks, keeled, yellow-green, ±

sticky, tips obtuse to acute. **FLS** 3–13; corolla 3.5–7.5 mm; style exserted, appendage gen > stigma. **FR** 3–5 mm, hairy; pappus ± = corolla. 2*n*=18, 36,54. Sagebrush scrub, pinyon/juniper woodland; 900–4000 m. KR, CaR, SN, TR, **GB, DMtns**; to s Can, MT, NM. Highly variable; 5 sspp. in w US. 4 sspp. in CA.

ssp. *axillaris* (Keck) L. Anderson **STS** glabrous. **LVS** 1–3 cm, thread-like, glabrous (exc ± ciliate). **INFL**: involucre ± obconic. **FLS** 3–5; corolla 3.5–4.5 mm. 2*n*=18. Uncommon. Gravelly washes; 1300–2000 m. **s W&I, ne DMtns**; to UT, n AZ. [*C. a.* Keck] Jul–Oct

ssp. *puberulus* (D.C. Eaton) H.M. Hall & Clements (p. 139) Pl gray-green, densely puberulent. **LVS** 1–3 cm, gen 1–2 mm wide, thread-like to ± oblanceolate. **FLS** 4–7; corolla 4.5–5.5 mm. 2*n*=18,36 (tetraploids larger, lower elevations). Sagebrush scrub, pinyon/juniper woodland, subalpine slopes; 1500–3000 m. SN (e slope), SnBr, **GB, DMtns**; to s ID, UT, n AZ.

ssp. *viscidiflorus* (p. 139) **STS** glabrous. **LVS** 0.5–7.5 cm, linear to ± lanceolate, glabrous (exc ± ciliate). **FLS** 4–13; corolla 4–7.5 mm. 2*n*=18,36,54. Common. Sagebrush scrub, pinyon/juniper woodland, alpine talus; 900–4000 m. KR, CaR, SN, n WTR, SnBr, **GB, s DMtns**; to WA, MT, NM. Intergrading forms have been called sspp. *latifolius* (D.C. Eaton) H.M. Hall & Clements (wide lvs); *pumilus* (Nutt.) H.M. Hall & Clements (small pls); *stenophyllus* (A. Gray) H.M. Hall & Clements (narrow lvs). ❀TRY. Jul–Sep

CICHORIUM CHICORY

G. Ledyard Stebbins

Ann, bien, per; sap milky. **STS** 3–10 dm, branched. **LVS** basal and cauline, reduced upward; lower 1–2 dm, toothed or pinnately lobed, wing-petioled; middle sessile, sometimes clasping; upper greatly reduced. **INFL**: heads ligulate, showy, terminal and axillary, lateral sessile; involucre ± cylindric; phyllaries in 2 series, hardened in basal half, outer short, spreading, inner elongate, erect; receptacle naked. **FLS** many; ligules blue to purple. **FR** oblong, glabrous, 5-angled; pappus of minute blunt scales. 8 spp.: Eur, Medit, Afr. (Old Arabic name)

C. intybus L. (p. 139) Per < 1 m, from deep, woody taproot; herbage glabrous to short-bristly, esp near base. **STS** erect; branches stout, ascending. **LVS**: lower oblong to elliptic in outline, subentire to coarsely pinnately lobed. **INFL**: peduncles of terminal heads not thickened. **FLS**: ligules blue (pink, white). **FR** 1.5–2.5 mm; pappus scales < 0.5 mm. 2*n*=18. Common. Roadsides, waste places; < 1500 m. CA-FP, **n DMtns (Panamint Mtns)**; widespread in N.Am; native to Eur. Roasted roots used as coffee flavoring or substitute. Jun–Oct

CIRSIUM THISTLE

David J. Keil & Charles E. Turner

Ann to per (sometimes short-lived, dying after fl once). **STS** gen erect. **LVS**: lower gen tapered or petioled, often wavy-margined, gen pinnately lobed, ± dentate, lobes and teeth spine-tipped, margin gen spiny-ciliate, glabrous to tomentose; upper gen sessile, ± reduced. **INFL**: heads discoid, 1–many; involucre cylindric to spheric; phyllaries many, graduated in several series, outer spine-tipped; receptacle flat, long-bristly. **FLS** gen many; corollas ± bilateral, white to red or purple, tube long, slender, lobes linear; anther bases sharply sagittate, tips oblong; style tip with slightly swollen node, appendage (above node) long, cylindric, branches very short. **FR** ovoid, glabrous; scar slightly angled; pappus bristles many, plumose, ± persistent or falling in ring. ± 200 spp.: N.Am, Eurasia. (Greek: thistle) Taxa difficult, incompletely differentiated, hybridize.

1. Fls unisexual, pls dioecious; heads gen 1–1.5 cm . . . . ***C. arvense***
1' Fls bisexual; heads gen > 1.5 cm
  2. Upper lf surface harshly bristly; sts spiny-winged . . . . ***C. vulgare***
  2' Upper lf surface glabrous to tomentose; sts wingless or ± short-winged
    3. Corollas rose-purple, bright pink to red
      4. Per; sts often > 1 from base; heads on short leafy peduncles
        5. Corollas bright pink to red; corolla tube < throat; W&I . . . . ***C. arizonicum***
        5' Corollas rose-purple; corolla tube > 2 × throat; DMtns . . . . ***C. nidulum***
      4' Bien; st gen 1; heads gen long-peduncled, often in ± open cymes
        6. Corolla throat abruptly expanded from tube; e DMoj . . . . ²***C. neomexicanum***
        6' Corolla throat gradually expanded from tube; W&I, w DMoj . . . . ***C. occidentale*** var. ***venustum***

3' Corollas gen white to ± pale purple
  7. Lvs all basal or crowded on very short, densely leafy st; heads ± sessile, closely subtended by rosette
    lvs . . . . ***C. scariosum***
  7' Lvs basal and cauline; heads sessile or peduncled, clearly raised above rosette lvs
    8. Corollas 29–37 mm; per from creeping roots . . . . ***C. ochrocentrum***
    8' Corollas 16–27 mm; bien from taproot
      9. Middle and outer phyllaries tightly appressed, only spine tip spreading . . . . ***C. mohavense***
      9' Middle and outer phyllaries ascending to spreading or reflexed
        10 Lvs thinly tomentose on both surfaces, becoming ± glabrous, esp above; n SNE . . . . . . ***C. canovirens***
        10' Lvs densely and gen persistently gray tomentose on both surfaces, esp below; e DMoj
        . . . . ²***C. neomexicanum***

C. arizonicum (A. Gray) Petrak ARIZONA THISTLE Per 2–10 dm; taproot ± woody. **STS** often > 1 from base, thinly tomentose or ± glabrous. **LVS** ± tomentose (esp below), becoming ± glabrous; lower 1–2 dm, tapered to spiny-winged petioles, oblong-obovate, ± lobed, lobes gen further lobed or toothed, main spines 5–15 mm; middle and upper not strongly reduced, clasping or short-decurrent. **INFL:** heads gen few, short-peduncled, ± closely subtended by uppermost lvs; involucres 3–4 cm, 1.5–2 cm diam when fresh, cylindric or narrowly ovoid, sparsely tomentose; phyllaries linear or linear-lanceolate, entire, gen erect or ascending, outer and middle tipped with spines 10–20+ mm, inner with tips flat or short-spined, straight, often red or purple, puberulent. **FLS** exserted; corollas 30–34 mm, red, tube 7–9 mm, throat 11–14 mm, lobes 12–13 mm. **FR** 4–6 mm, ± compressed, shiny brown; pappus 24–28 mm. 2*n*=32. Open forests, sagebrush scrub; 2400–3700 m. s SNH, **W&I**; to UT, NM. [*C. eatonii* (A. Gray) Robinson and *C. nidulum* (M.E. Jones) Petrak misapplied] Variable, needs further study; close to *C. nidulum*. ❀TRY. Jul–Aug

C. arvense (L.) Scop. (p. 139)) CANADA THISTLE Per 5–10 dm, dioecious; rootstock creeping; herbage green. **STS** colonial, very leafy. **LVS** 5–20 cm, mostly cauline, gradually reduced upward, sessile, tapered at base, sometimes decurrent as spiny wings, subentire to coarsely dentate or 1–2 × lobed; main spines 3–5 mm. **INFL:** heads several–many, cymes tight to ± open, rounded or flat-topped; peduncles 0–4 cm; involucre hemispheric to ovoid, 1–2 cm, 1–2 cm diam, gen ± purplish, ± tomentose when young; outer phyllaries ovate, tipped by spines ± 1 mm, inner lanceolate, tips flat, membranous. **FLS:** corollas gen purplish, sometimes white or pink. **STAMINATE FL:** corollas 12–13 mm, > pappus, tube 8 mm, throat 1–1.5 mm, lobes 3–4 mm. **PISTILLATE FL:** corollas 14–20 mm, gen < pappus, tube 10–16 mm, throat ± 1 mm, lobes 2–3 mm. **FR** 2–3 mm; pappus 13–23 mm. 2*n*=34,36. Disturbed places; < 1800 m. CA-FP (exc SnJV, s SN), **SNE, w DMoj;** N.Am.; native to Eur. NOXIOUS WEED. Jun–Sep

C. canovirens Rydb. GRAY-GREEN THISTLE Bien or short-lived per 3–10 dm, from taproot. **ST** gen 1, gen simple below, ± cobwebby and soft-hairy (hairs jointed, multicellular). **LVS** ± tomentose (esp lower surface), sometimes becoming ± glabrous; lower ± petioled, 1–2(3) dm, oblong or narrowly oblanceolate, gen lobed, main spines 5–10 mm; middle and upper smaller, decurrent as spiny wings, toothed or shallowly lobed. **INFL:** heads 1–many in flat-topped to raceme-like clusters, sessile or peduncles gen 1–10 cm; involucres 1.5–2.5 cm, 1.5–3 cm diam, ovoid to hemispheric, ± loosely tomentose; phyllaries linear-lanceolate to ovate, ascending to spreading, midribs of outer and middle often with narrow glandular area, spines 5–10 mm, inner with tips flat, straight. **FLS:** corollas 20–27 mm, dull white to pale purple, tube 7–10 mm, throat 7–10 mm, lobes 4–7 mm. **FR** 4–6 mm, dark brown, ± compressed; pappus 15–25 mm. 2*n*=34. Shrubby areas, open forests, roadsides; 1600–3400 m. e SNH, MP, **n SNE.** [*C. utahense* Petrak misapplied] Needs further study..

C. mohavense (E. Greene) Petrak (pl. 16) MOJAVE THISTLE Bien (or short-lived per) 5–25 dm. **ST** gen 1, gen simple below, openly branched above, ± white-tomentose. **LVS** ± densely tomentose, lighter below; lower 1.5–6 dm, spiny-petioled, elliptic to oblanceolate, toothed to deeply lobed, lobes gen rigidly spreading, simple or with 1–2 pairs of coarse teeth or 2+ lobes,

main spines 3–20 mm; middle and upper smaller, narrower, decurrent as spiny wings, upper much reduced, often long-acuminate, gen very spiny, spines 4–25 mm. **INFL:** heads in loose to crowded cyme- or panicle-like clusters (sometimes on short axillary branches); peduncles leafy, 0–10 cm; involucres 1.5–2.5 cm, 1.5–2 cm diam, ± ovoid, ± loosely tomentose, becoming glabrous; phyllaries strongly graduated (outer ovate, inner oblong), entire, tightly appressed, midribs of middle often with glandular area, spines 3–7 mm, ascending or spreading, inner with erect, ± twisted flat tips. **FLS:** corollas 16–25 mm, white to lavender or pink, tube 7–11 mm, throat 4–7 mm, lobes 4–8 mm. **FR** 3.5–6 mm, straw-colored to brown, gen not compressed; pappus ± 15 mm. 2*n*=30,32. Damp soil around springs, canyons, streams, ditches; 400–2800 m. **SNE, DMoj;** NV. Jul–Oct

C. neomexicanum A. Gray DESERT THISTLE Bien (or short-lived per) 4–29 dm. **ST** gen 1, gen simple below; branches few above, ascending, ± white-cobwebby-tomentose, puberulent. **LVS** ± persistently gray-tomentose (both surfaces), lighter below; lower 6–35 cm, petioled or tapered to spiny-winged base, oblong-elliptic to oblanceolate, ± lobed, lobes gen rigidly spreading, simple or with 2–4 coarse teeth or 2+ lobes, main spines 5–15 mm; middle and upper gen smaller, narrower, decurrent as spiny wings, uppermost well separated, much reduced, ± bract-like, sometimes barely a cluster of long spines. **INFL:** heads 1–few in open cyme-like clusters (sometimes on short axillary branches); peduncles 2.5–30 cm, leafy; involucres 2–2.5 cm, 2.5–5 cm diam, hemispheric or bell-shaped, ± loosely tomentose, sometimes glabrous; phyllaries linear-lanceolate, entire, sometimes midveins with glandular area, outer and middle spreading to reflexed, spines 4–15 mm, inner with tips ± erect, flat. **FLS:** corollas 18–27 mm, white to pale lavender or pink, tube 8–14 mm, throat 4–7 mm, lobes 5–9 mm. **FR** 5–6 mm, dark brown, ± flattened; pappus 15–20 mm. 2*n*=30. Canyons, slopes, roadsides; 800–2100 m. **e DMoj;** to Colorado, NM. Closely related to *C. occidentale.* Apr–May

C. nidulum (M.E. Jones) Petrak Per 2–10 dm; taproot woody. **STS** often > 1 from base, thinly tomentose, sometimes ± glabrous. **LVS** ± tomentose (esp below), becoming ± glabrous; lower 1–3 dm, tapered to spiny-winged petioles, oblong-elliptic, long-acute, divided ± to midvein, lobes with 2–4 narrow 2+ lobes or coarse teeth, main spines 10–30 mm; middle and upper sessile, not strongly reduced, clasping or short-decurrent, exceedingly spiny. **INFL:** heads gen few, short-peduncled, ± closely subtended by uppermost lvs; involucres ± 3 cm, 1.5–2 cm diam when fresh, cylindric or narrowly ovoid, sparsely tomentose or becoming glabrous; phyllary bodies lanceolate to ± ovate, entire, tip spreading to erect, spines 20–25 mm, inner with tips straight, flat or short-spiny, often red or purple, puberulent. **FLS** exserted; corollas 28–30 mm, ± rose-purple, tube 11–14 mm, throat 4–6 mm, lobes 12–13 mm. **FR** 5–6 mm, ± flattened, shiny brown; pappus 18–20 mm. Uncommon. Pinyon/juniper woodland; 1500–2300 m. **e DMtns (Clark, New York Mtns).** Most spiny thistle in CA. Related to *C. arizonicum.* Jul–Oct

C. occidentale (Nutt.) Jepson Bien 1–30 dm, erect or low, mound-like. **ST** gen 1, branched above (near base in dwarf pls), ± tomentose. **LVS** ± densely gray- or whitish tomentose, esp below; lower 1–4 dm, petioles spiny-winged, blade oblanceolate, lobed 1/2+ to midvein, lobes widely triangular, dentate or further lobed, main spines 1–10 mm; upper gradually reduced, sessile, ± clasping or short-decurrent, linear or oblong, often entire, often spinier than lower, uppermost bract-like. **INFL**: heads 1–several in loose to tight cluster (barely raised above rosette in dwarf pls); peduncles 1–30 cm; involucres 1.5–5 cm, 1.5–8 cm diam, ovoid to spheric; phyllaries ± equal to strongly graduated, linear or linear-lanceolate, straight, ascending and appressed to widely radiating, often connected side-to-side by conspicuous cobwebby hairs, spines 3–10+ mm, inner with tips flat, straight. **FLS**: corollas 18–40 mm, white to purple or red, tube 8–18 mm, throat 5–7 mm, lobes 5–10 mm. **FR** 5–6 mm, shiny, ± brown; pappus 15–30 mm. 2*n*=30. Many habitats; < 3600 m. CA-FP, MP, **W&I**, w **DMoj**; to s OR, sw ID, w NV. Variable; ± distinctive, intergrading races often treated as sp. 5 vars.. in CA.

var. *venustum* (E. Greene) Jepson (p. 139) VENUS THISTLE
Pls gen 5–30 dm, erect. **INFL**: heads gen long-peduncled, sometimes in tight clusters at ends of peduncles, well elevated above lower lvs; involucre 2–6 cm diam, subglabrous to densely cobwebby; middle phyllary tips 5–20+ mm, gen 2–3 mm wide, ascending to rigidly spreading or reflexed. **FLS**: corollas 23–35 mm, gen ± red (white, pink, purple). 2*n*=30. Disturbed places, grassland, woodland; < 3600 m. NCoR, s SN, SnFrB, SCoR, WTR, **W&I**, w **DMoj**. [*C. coulteri* Harvey & A. Gray misapplied; *C. proteanum* J. Howell] ❀DRN,SUN:**7,14–16**,17&IRR:1,2,**3,18–21**,22–24. May–Jul

C. ochrocentrum A. Gray (p. 139) YELLOWSPINE THISTLE Per 2.5–10 dm; roots creeping. **STS** gen simple below, few-branched above, white-tomentose. **LVS** thinly gray-tomentose above, white-tomentose below; lower 10–25 cm, tapered to spiny petioles, elliptic to oblanceolate, deeply lobed, lobes gen rigidly spreading, simple or with 2–4 ± narrow 2+ lobes or coarse teeth, main spines 3–10 mm; middle and upper gradually reduced, decurrent as spiny-margined wings, gen very spiny, spines 5–15 mm. **INFL**: heads 1–few in cymes; peduncles 0–10 cm, leafy; involucres 2.5–3.5 cm, 2–3.5 cm diam, ± ovoid to bell-shaped, ± loosely tomentose, becoming glabrous; phyllaries strongly graduated (outer ovate, inner oblong), minutely roughened or toothed, tightly appressed, midribs of middle phyllaries often with glandular area, spines 5–12 mm, stout, spreading to reflexed, inner with tips erect or recurved, ± twisted, flat, sometimes ± expanded and fringed. **FLS**: corollas 29–37 mm, white to pale lavender or pink, tube 14–17 mm, throat 7–10 mm, lobes 8–11 mm. **FR** 7–8 mm, light brown, ± thick; pappus 25–30 mm. 2*n*=15,16,17.

Disturbed places, fields; < 1700 m. n NCo, KR, e SN, n ChI, TR, PR, MP, **W&I**; native to c US. NOXIOUS WEED. Apr–Jul

C. scariosum Nutt. (p. 139) ELK THISTLE Bien (or short-lived per) 0.5–10+ dm. **ST** often 0 or very short, sometimes erect, often ± fleshy, ridged, glabrous to loosely tomentose, sometimes coarse-hairy. **LVS** often basal, glabrous to loosely tomentose above, glabrous to densely tomentose below; lower 1–4 dm, tapered or spiny-petioled, oblong to oblanceolate, subentire to deeply lobed, sometimes with 2+ lobes or teeth, main spines gen 2–7 mm; cauline 0 or well distributed, ± petioled. **INFL**: heads gen ± sessile in basal lf rosette or in clusters at st tips, gen closely subtended by lvs; involucres 2–4.5 cm, 2–5 cm diam, ovoid to bell-shaped, glabrous; phyllaries linear-lanceolate to ovate, entire to minutely toothed, outer tipped by ascending spines 2–5 mm, inner with tips entire or toothed, flat or crinkled. **FLS**: corollas 23–32 mm, ± white to purple, tube 10–16 mm, throat 6–10 mm, lobes 5–6 mm. **FR** 3–5 mm, oblong-obovate, ± compressed; pappus 15–30 mm. 2*n*=34. Moist places, meadows; (400)700–3400 m. KR, CaR, SN, SCoR (very uncommon), TR, PR, **GB**; to B.C., MT, Colorado, NM, Baja CA. [*C. drummondii* Torrey & A. Gray and *C. foliosum* (Hook.) DC., misapplied; *C. tioganum* Congdon] Extremely variable, needing study; dwarf and tall pls sometimes occur together. Pls from SNE & SnBr, with sts 0, heads small, corollas purple, have been called *C. congdonii* R. Moore & Frankton. Jun–Aug

C. vulgare (Savi) Ten. (p. 139) BULL THISTLE Bien 3–20 dm. **ST** gen 1, ± openly branched above middle, loosely tomentose, often glandular-hairy. **LVS** harshly bristly above, sometimes ± tomentose when young, ± tomentose below; main veins prominently raised on lower surface, ± glandular; lower 10–40 cm, sessile or wing-petioled, shallowly to deeply 1–2 × lobed; cauline gradually reduced, long-decurrent as spiny wings, gen spinier than lower, main lobes gen rigidly spreading, spine-margined, otherwise entire, tip prolonged, main spines < 15 mm. **INFL**: heads 1–several, ± clustered, closely subtended by bract-like uppermost lvs; peduncles 1–6 cm; involucres 3–4 cm, 2–4 cm diam, hemispheric or bell-shaped; phyllaries graduated in 5–10 series, tips linear to linear-lanceolate, spreading to reflexed, spines 1–5 mm. **FLS**: corollas 25–35 mm, purple, tube 18–25 mm, throat 5–6 mm, lobes 5–6 mm. **FR** 3.5–4.5 mm, light-brown or tan; pappus 20–30 mm. 2*n*=68. Common. Disturbed areas; < 2300 m. CA-FP, **GB**; N.Am; native to Eur. Jun–Sep

CNICUS BLESSED THISTLE

David J. Keil & Charles E. Turner

1 sp.: Eur. (Latin: from Greek name for safflower)

C. benedictus L. Ann < 6 dm. **STS** gen branched throughout, gen reddish, ± loosely tomentose. **LVS** ± cauline, alternate, sessile, short decurrent or tapered to winged petiole, 6–25 cm, (ob)lanceolate, spine-toothed or pinnately lobed, strongly veined, sparsely hairy, gland-dotted. **INFL**: heads disciform, sessile among lf-like bracts; involucre 2–4 cm, ± spheric; outer phyllaries ovate, base appressed, spine tips spreading, inner phyllaries lanceolate, spine tips pinnately divided; receptacle bristly. **STERILE FLS** few, marginal; corollas very slender, lobes 3, linear.

DISK FLS many; corollas ± 2 cm, yellow, lobes linear; anther bases sagittate, tips long-appendaged; style tip minutely hair-ringed just below rounded lobes. **FR** ± 8 mm, cylindric, ribbed, attached laterally, glabrous; top with 10-toothed rim; pappus 2 series of awns, outer ± 10 mm, smooth or rough, inner 2–5 mm, rough from short, spreading hairs. 2*n*=22. Roadsides, fields, waste places; < 800 m. NCoR, GV, CW, SCo, **w DMoj**; native to Eur. Scattered. Apr–Jul

CONYZA

Ann or per. **STS** gen erect, leafy. **LVS** alternate, linear to ± (ob)lanceolate, entire to pinnately dissected, obtuse to acute. **INFL**: heads gen disciform (or minutely radiate), gen many in raceme- or panicle-like clusters; phyllaries in 2–3 equal or unequal series, free, linear to lanceolate, often narrowly scarious-margined, reflexed in age; receptacle naked. **PISTILLATE FLS** gen many; corollas white, pink, or cream, narrowly cylindric. **DISK FLS** ± few; corollas gen yellow, lobes short-triangular; style tips lanceolate, incl or short-exserted. **FR** elliptic, compressed, puberulent; pappus of bristles. ± 50 spp.: esp trop. (Greek: flea; name of Pliny & Dioscorides for a fleabane) Descended from within *Erigeron* [Noyes 2000 Plant Syst. Evol. 220:93–114]

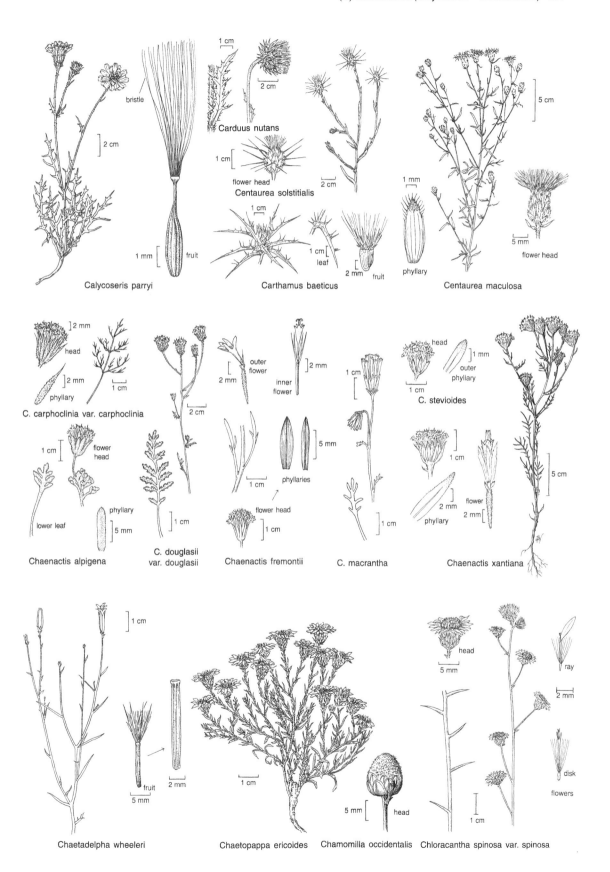

Calycoseris parryi

bristle

fruit

Carduus nutans

Centaurea solstitialis

flower head

Carthamus baeticus

leaf

fruit

Centaurea maculosa

phyllary

flower head

C. carphoclinia var. carphoclinia

head

phyllary

Chaenactis alpigena

flower head

lower leaf

phyllary

C. douglasii var. douglasii

outer flower

inner flower

Chaenactis fremontii

phyllaries

flower head

C. macrantha

C. stevioides

head

outer phyllary

phyllary

flower

Chaenactis xantiana

Chaetadelpha wheeleri

fruit

Chaetopappa ericoides

Chamomilla occidentalis

head

Chloracantha spinosa var. spinosa

head

ray

disk

flowers

1. Midvein of phyllary resin-filled, orange when dry; herbage strigose or stiff-spreading hairy; ligules present, < 1 mm . . . . *C. canadensis*
1' Midvein of phyllary not resin-filled, green to purple; herbage densely soft-hairy and glandular; ligules 0 . . . . *C. coulteri*

C. canadensis (L.) Cronq. (p. 147) HORSEWEED Ann 1–20(30) dm. **ST** gen simple below, much-branched above, strigose or stiff-spreading-hairy. **LVS** 1–10 cm, entire to shallowly few-lobed, glabrous to strigose, often ciliate. **INFL**: heads obscurely radiate; lateral clusters not overtopping central; fresh involucre gen 2.5–4 mm diam; phyllaries 1–3.5 mm, glabrous to strigose, midvein brown, resin-filled. **PISTILLATE FLS** 20–40; corollas 2.5–3 mm, ligule < 1 mm. **DISK FLS** 7–13; corollas 2.5–3 mm. **FR** ± 1.5 mm; pappus 2.5–3 mm, gen whitish in age. $2n$=18,36,54. Common. Waste ground; gen < 2000 m. **CA**; ± worldwide. Jun–Sep

C. coulteri A. Gray (p. 147) Ann 2–10(20+) dm, densely soft-hairy and glandular. **ST** gen simple below, much-branched above. **LVS** 2–8 cm, sessile, ± clasping, gen coarsely dentate or shallowly spreading-lobed. **INFL**: lateral clusters gen not overtopping central; peduncles 3–10 mm; fresh involucres 4–5 mm diam; phyllaries 2–4 mm, densely glandular, midvein ± green, not resin-filled. **PISTILLATE FLS** many; corollas 2 mm, very slender, cream; style long-exserted, branches unequal. **DISK FLS** 5–15; corollas 3–3.5 mm. **FR** 1 mm; pappus 3–4 mm, gen white in age. $2n$=18. Disturbed places; gen < 1000 m. s SNF, SnJV, CW, SW, **W&I, D**; to Colorado, TX, Mex. May–Oct

COREOPSIS TICKSEED

Ann, per, shrubs. **STS** slender to stout and fleshy. **LVS** simple to several times pinnately dissected, basal or cauline, opposite or less commonly alternate, sessile or petioled. **INFL**: heads radiate, solitary or in few–many-headed cymes; peduncles short to long; involucre hemispheric or bell-shaped; phyllaries in 2 series, outer ± spreading, thick, green, inner thin, membranous; receptacle flat to rounded, chaffy; scales flat, scarious. **RAY FLS** fertile or sterile; ligules gen yellow, showy. **DISK FLS** many; corollas 4–5-lobed, yellow; style tips truncate to long-tapered. **FR**: ray and disk achenes alike or different, gen compressed front-to-back, often winged; pappus 0 or of 2 awns or scales. ± 114 spp.: Am, Afr. (Greek: bedbug-like, from fr) [Smith 1984 Sida 10:276–289]

1. Disk and ray frs similar, never ciliate; disk pappus gen 0 . . . . *C. californica* var. *californica*
1' Disk and ray frs different, disk frs ciliate, pappus scales 2; ray frs glabrous, pappus 0
  2. Outer phyllaries linear-oblong; pappus scales 1.7–2.8 mm . . . . *C. bigelovii*
  2' Outer phyllaries triangular-ovate; pappus scales 2.5–5 mm . . . . *C. calliopsidea*

C. bigelovii (A. Gray) H.M. Hall (p. 147)) Ann gen 1–3 dm, glabrous. **STS** 1–many, erect. **LVS** basal (or few cauline, alternate); petiole 1–5 mm; blade 2–8 cm, 1–2-pinnately divided into linear segments 1–2 mm wide, grooved above. **INFL**: heads solitary, ± scapose; involucre cylindric, base truncate; outer phyllaries 4–7, 5–12 mm, linear; inner phyllaries 6–8, 6–10 mm, ovate, acute, margin scarious; chaff scales gen 5–8 mm, lanceolate to oblanceolate, fused to base of disk achenes. **RAY FLS** 5–10, fertile; ligules 5–25 mm, obovate, spreading, yellow. **DISK FLS** 20–50; corollas ± 4 mm, yellow. **FR**: ray achenes 3–5 mm, oblong to obovate, brown or splotched with tan, rough, glabrous, wing narrow, pappus 0; disk achenes 4–6 mm, oblong to oblanceolate, dark brown or splotched with tan, shiny, outer face glabrous, inner face with central row of hairs, margins ciliate, hairs 1–1.5 mm; pappus scales 1.7–2.8 mm, lanceolate. $2n$=24. Open woodlands, grasslands, deserts; 150–1500 m. SCoRI, TR, Teh, s SNF, **DMoj, n DSon**. ❀DRN,SUN:**8, 9**,10,11,**12,13,18–24**&DRY:7,14–16. Mar–May

C. californica (Nutt.) H. Sharsm. var. *californica* (p. 147) Ann gen 5–30 cm, glabrous. **STS** 1–many, erect. **LVS** all basal or few cauline, gen erect, 2–10 cm, 0.5 mm wide, linear, thread-like, entire or lobes 1–2, short, linear, ± cylindric, tip obtuse, red. **INFL**: heads solitary, ± scapose; involucre widely cylindric, base rounded; outer phyllaries 2–7, 4–7 mm, narrowly lanceolate, hairs at base yellow or red, glandular, tip red; inner phyllaries 5–8, 6–

10 mm, widely lanceolate, acute, margin narrowly scarious; chaff scales 4–5.5 mm, linear to oblanceolate, free from disk achenes. **RAY FLS** 5–12, fertile; ligules 5–15 mm, obovate, yellow. **DISK FLS** 10–30; corollas 3.5–5 mm, yellow. **FRS** alike, 2.5–4.3 mm, obovate, rusty-tan to light brown, often red- or black-spotted near margin, puberulent (hairs club-shaped), wing irregularly thickened; pappus 0 (or scales 2). $2n$=24. Desert plains, washes; 30–600 m. s SnJV, s SCoRI, TR, **D**; Baja CA. ❀TRY. Mar–May

C. calliopsidea (DC.) A. Gray (p. 147) Ann gen 1–4(6+) dm, glabrous. **STS** 1–many, erect, simple or few-branched. **LVS** basal and alternate; petiole 1–5 mm; blade 1–5 cm, 1–2-pinnately divided into linear segments 0.5–2 mm wide, grooved above; upper lvs sometimes simple. **INFL**: heads solitary; involucre bell-shaped; outer phyllaries 4–6, 3–8 mm, triangular-ovate, fused at base; inner phyllaries gen 8, 8–10 mm, ovate, acute, margin narrowly scarious; chaff scales 6–7 mm, lanceolate to oblanceolate, fused to base of disk achene. **RAY FLS** gen 8, fertile; ligules 10–35 mm, obovate, yellow. **DISK FLS** 15–50; corollas ± 5 mm, yellow. **FR**: ray achenes 5–6 mm, ovate, tan or brown, glabrous, wing smooth, flat, pappus 0; disk achenes 6–7 mm, linear to oblanceolate, outer face dark brown, shiny, glabrous, inner face covered with white hairs, margin ciliate, hairs 2–3 mm; pappus scales 2.5–5 mm, lanceolate. $2n$=24. Deserts, dry grassy areas; 200–1100 m. SCoRI, TR, **w DMoj**. ❀TRY. Mar–May

CREPIS HAWKSBEARD

G. Ledyard Stebbins

Ann, bien, per from taproot; sap milky. **STS** erect, < 8 dm. **LVS** basal or cauline, entire to pinnately lobed. **INFL**: heads ligulate, clustered in cymes; phyllaries in 2 distinct series; receptacle naked. **FLS** 5–60; ligules yellow, readily withering. **FR** tapered at both ends, sometimes beaked; pappus of many soft, hair-like bristles. ± 200 spp.: esp n hemisphere. (Greek: sandal, for unknown reason) Sexual forms of native spp. are distinct but (exc *C. nana, C. runcinata*) connected by many asexually reproducing forms of hybrid origin that obscure boundaries. Asexual forms are all placed in same sp. as sexual forms. ❀TRY. Key to species adapted by David J. Keil.

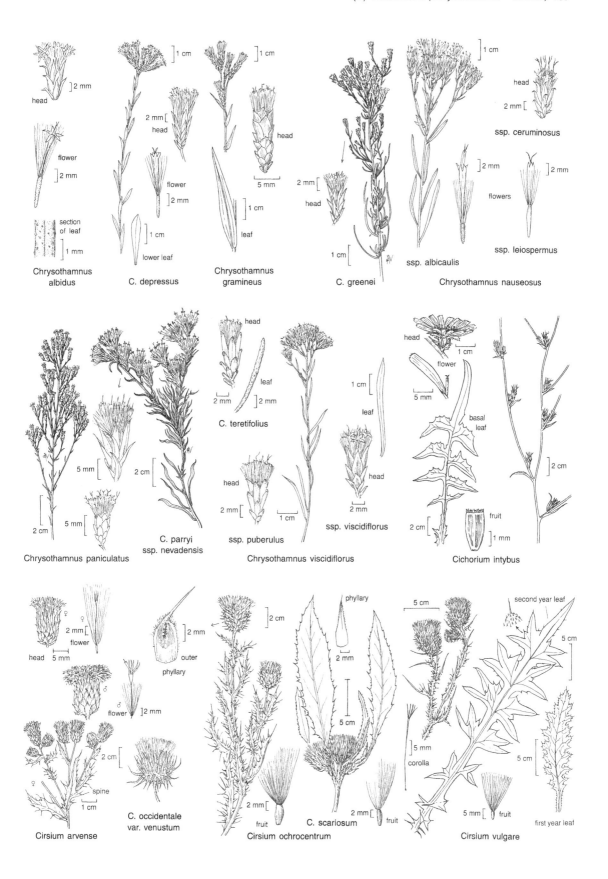

Chrysothamnus albidus

C. depressus

Chrysothamnus gramineus

C. greenei

ssp. albicaulis

Chrysothamnus nauseosus

ssp. ceruminosus

ssp. leiospermus

Chrysothamnus paniculatus

C. parryi ssp. nevadensis

C. teretifolius

ssp. puberulus

ssp. viscidiflorus

Chrysothamnus viscidiflorus

Cichorium intybus

Cirsium arvense

C. occidentale var. venustum

Cirsium ochrocentrum

C. scariosum

Cirsium vulgare

1. Sts glabrous or bearing long, straight bristles
  2. Sts 2–7 cm, much-branched; heads 2–4, borne among lvs . . . . ***C. nana***
  2' Sts 2.5–8 dm, unbranched; heads 3–20 in open cyme . . . . ***C. runcinata*** ssp. *hallii*
1' Sts tomentose
  3. Inner phyllaries glabrous or lightly tomentose; involucres 9–10 mm . . . . ***C. acuminata***
  3' Inner phyllaries ± densely tomentose; involucres 10–21 mm
    4. Fls 7–12 per head; involucres 10–16 mm; outer phyllaries lanceolate-deltate . . . . ***C. intermedia***
    4' Fls 9–40 per head; involucres 11–19 mm; outer phyllaries linear to deltate . . . . ***C. occidentalis***

C. acuminata Nutt. (p. 147) Per from deep taproot. **STS** 2–7 dm, branched, tomentose. **LVS**: basal 12–40 cm, tomentose, gray-green, lobes narrowly triangular, acute, sometimes with short secondary lobes; cauline similar, smaller. **INFL**: heads many in a compound cyme; involucre 9–10 mm, cylindric to obconic; outer phyllaries narrowly triangular; inner phyllaries 5–8, lanceolate, smooth and shining or lightly tomentose. **FLS** gen 5–6(10). **FR** 6.5–8 mm, tapered to both ends, ± 12-ribbed, pale yellow, ± brown, or dusky white. $2n$=22,33,44,55,88. Open, rocky places; 1200–3300 m. KR, NCoRH, CaRH, SNH, Teh, TR, **GB**; to WA, MT, WY, Colorado. Jun–Aug

C. intermedia A. Gray Per from stout taproot. **STS** 3–7 dm, stout, branched near middle or above, tomentose. **LVS**: basal 15–40 cm, tomentose, lobes narrowly triangular, often bearing small 2° lobes; cauline similar, smaller. **INFL**: heads many; involucre 10–16 mm, cylindric to bell-shaped; outer phyllaries small, lanceolate-deltate; inner phyllaries 7–10, densely and evenly tomentose, sometimes glandular. **FLS** 7–12. **FR** 5.5–9 mm, narrowed at both ends, 10–12-ribbed, yellow, buff, or brown; pappus dusky to white. $2n$=33,44,55,88. Dry slopes, open forest; 1400–3300 m. KR, NCoRH, CaRH, NCoRH, n&c SN, **GB**; to WA, w Can, WY, Colorado. Complex series of asexually reproducing forms, probably of hybrid origin, combining characters of *C. acuminata, C. modocensis, C. occidentalis,* & *C. pleurocarpa.* Jun–Aug

C. nana A. Richards (p. 147) Per from taproot or creeping rhizome, dwarf, purplish green. **STS** many, 2–7 cm, much-branched, in dense clumps. **LVS** mostly basal, glabrous, oblanceolate to obovate or elliptic, entire or shallowly lobed. **INFL**: heads in clusters of 2–4 among lvs; involucre 10–13 mm; outer phyllaries lanceolate or ovate, acute; inner phyllaries 10, oblong, obtuse, short-hairy at tip. **FLS** 9–12. **FR** 4–6 mm, golden brown; tip acuminate or obtuse and beak short; pappus < or = fr. $2n$=14. Stony or gravelly scree; 2600–3700 m. c&s SNH, SnGb, **SNE**

(Sweetwater Mtns), n DMtns (**Panamint Mtns**); to AK, MT, WY, Colorado, ne N.Am, Asia. [ssp. *ramosa* Babc.] Jul–Aug

C. occidentalis Nutt. (p. 147) Per from deep taproot; herbage densely gray-tomentose, without spreading, glandless or glandular hairs. **STS** 1.5–4 dm, branched from base or middle. **LVS** 10–30 cm, toothed to deeply lobed. **INFL**: heads 10–30, clustered in cymes; involucre 11–19 mm; outer phyllaries linear to deltate; inner phyllaries lanceolate, acute, tomentose and sometimes with short, gland-tipped hairs. **FLS** 9–40. **FR** 6–10 mm, tapered to both ends, beakless, strongly 10–18-ribbed, light to dark brown; pappus dusky to white. $2n$=22,33,44,55,66,77,88. Dry terraces, mtn slopes; 800–2700 m. KR, NCoRH, CaRH, SNH, Teh, se SnFrB (Mount Hamilton), WTR, SnBr, **GB, DMtns**; to w Can, MT, WY, NM. Pls called sspp. *conjuncta* (Jepson) Babc. & Stebb., *costata* (A. Gray) Babc. & Stebb., *pumila* (Rydb.) Babc. & Stebb. intergrade extensively. Highly variable; incl genes from most or all dry-land montane spp. Jun–Aug

C. runcinata Torrey & A. Gray Per from taproot. **STS** 2.5–8 dm, ± lfless, glabrous; branches 0. **LVS**: basal 7–27 cm, oblanceolate or elliptic, pinnately lobed, minutely dentate, or entire, glabrous; cauline 0–few, much reduced. **INFL**: heads (in CA sspp.) 3–20 in open cyme; involucre 8–21 mm; outer phyllaries linear to lanceolate; inner phyllaries 10–16, narrowly to widely lanceolate, glandular-hairy. **FLS** 20–50. **FR** 3.5–7.5 mm, 10–13-ribbed, light to dark-brown; tip acuminate or beak short; pappus white. $2n$=22. Moist depressions, streambanks; 1250–1500 m. s MP, **SNE**; to WA, s-c Can, n-c US. 2 sspp. in CA.

ssp. ***hallii*** Babc. & Stebb. (p. 147) HALL'S MEADOW HAWKSBEARD **LVS** glaucous, winged at base, closely dentate. **FR** tapered or beak short. $2n$=22. RARE in CA. Moist, alkaline valley bottoms; 1250–1450 m. **SNE (Benton, s Mono Co. to Bishop, n Inyo Co.)**; NV. Declining from grazing, habitat drainage. Jun–Jul

DICORIA

Willard W. Payne

Ann, taprooted, much-branched, gen ± white-hairy. **LVS** petioled, gen opposite below, alternate above. **INFL**: heads disciform, many; involucre cup- to bell-shaped; phyllaries few in 1 series, small, free; receptacle chaffy. **PISTILLATE FLS** 0–2; corolla 0; style branches long. **STAMINATE FLS** 5–20; corolla greenish to dull purple; anthers free, exserted, filaments fused, attached at corolla tube base. **FR** compressed, winged, falling with chaff scale; pappus 0. 3–4 spp.: sw US, nw Mex. (Greek: 2 bugs, from 2-fr heads)

D. canescens A. Gray (p. 147) Pl 3–9 dm, gen white-hairy. **LVS** gen ± densely canescent; lower 3–5 cm, lanceolate to triangular-ovate, toothed, gen 3-veined; upper reduced, ± round, ± entire. **INFL**: phyllaries gen 3–5, 2–5 mm, reflexed in age. **PISTILLATE FLS** gen 2. **STAMINATE FLS**: corolla ± 3 mm.

FR 3–6 mm, ± keeled, glabrous, wings toothed. $2n$=36. Sandy soil; < 1300 m. **D**; to s UT, w AZ, nw Mex. Highly variable; features vary ± independently. [sspp. *clarkiae* (Kenn.) Keck, *hispidula* (Rydb.) Keck] ❀TRY;STBL. Sep–Jan

DUGALDIA

Per, glabrous to densely hairy. **STS** erect. **LVS** simple, basal and cauline, alternate; lower wing-petioled; upper sessile; blade 3-veined. **INFL**: heads large, radiate, 2–15 in ± flat-topped cymes; peduncles short to long; involucre disc-like or widely bell-shaped; phyllaries in 2 series; receptacle ovoid to spheric, naked. **RAY FLS** 14–35; ligules yellow to orange, elongated, 3–4-lobed. **DISK FLS** many; corollas yellow; tips of style branches truncate, shrub-like. **FR** oblong, 4-angled; hairs straight; pappus of 5–7 lanceolate scales. 3 spp.: w US to C.Am. (Dugald Stewart, Scotland, 1753–1828) [Bierner 1974 Brittonia 26:385–392] *For revised taxonomy see Bierner 1994 Sida 16:1–8.

D. hoopesii (A. Gray) Rydb. (p. 147) **STS** 3–9 dm, glabrous to ± hairy. **LVS** entire, obtuse to acute; lower cauline and basal oblong to oblanceolate; middle oblong to elliptic; upper linear to lanceolate. **INFL**: peduncles 3–16 cm, white-tomentose below heads; disk 12–17 mm, 19–26 mm diam; phyllaries in 2 series, outer lanceolate to ovate, ± fused, ± hairy, inner smaller, elliptic, acuminate, free, glabrous. **RAY FLS** 14–26; ligules 1.5–3.5 cm, yellow to orange. **DISK FLS**: corollas 4–5.5 mm. **FR** 3.5–4.5 mm; pappus 3–4 mm. 2*n*=30. Meadows; 1500–3000 m. KR, SN, Wrn, **n SNE**; to OR, WY, NM. [*Helenium h.* A. Gray] ******Hymenoxys hoopesii*** (A. Gray) Bierner ✿ IRR or WET,SUN:**1,2**,3,6,7,15,16,18. Jul–Sep

EATONELLA

1 sp.: w US. (Daniel C. Eaton, American botanist, 1834–1895)

E. nivea (D.C. Eaton) A. Gray (p. 147) Ann 1–4 cm, densely tomentose. **STS** congested. **LVS** basal and alternate, very crowded, gen 1 cm or less, linear to obovate, entire. **INFL**: heads radiate, solitary, terminal, subsessile, in fl ± concealed by lvs; peduncle in fr elongating to ± 1 cm; involucre 4–5 mm diam, bell-shaped; phyllaries 8–12 in 1 series, 4–5 mm, linear-oblong, free, in fr reflexed; receptacle flat, naked. **RAY FLS** 1 per phyllary, inconspicuous; ligules 1.5–2.5 mm, light yellow, often drying ± purple. **DISK FLS** many; corollas 2 mm, 4–5-lobed, yellow. **FR** ± 3 mm, oblanceolate, flattened; surfaces glabrous, black and shiny; margins ciliate with long, white hairs; pappus of 2 fringed scales 1–2 mm. 2*n*=19. Sandy soil; 1350–2900 m. **GB**; se OR, w NV. ✿TRY. May–Jun

ECLIPTA

Ann. **STS** prostrate to erect, often rooting below, simple to much-branched. **LVS** simple, opposite, sessile. **INFL**: heads radiate, small, solitary or in small cymes; peduncles short or 0; involucre hemispheric; phyllaries in 1–2 ± equal series, free, ovate; receptacle rounded, chaffy, scales narrowly linear, bristle-like. **RAY FLS** many; corollas white; ligules short, narrowly linear. **DISK FLS** many; corollas white; style branches very short, incl; anthers brown, incl. **FR** 4-angled, ± flat, brown, glabrous; pappus a crown of minute bristles or 0. 4 spp.: esp trop. (Greek: deficient, from absence of pappus)

E. prostrata (L.) L. (p. 147) Pl ± strigose throughout. **STS** 1–10 dm. **LVS**: blades linear, lanceolate, or narrowly elliptic, entire or short-toothed, tip acute. **INFL**: peduncles 0–15 mm; involucre 4–10 mm diam; phyllaries 4–5 mm, acute. **RAY FLS**: corolla 1.5–3 mm. **DISK FLS**: corolla 1.5–2 mm. **FR** 1.7–2.2 mm, obovate, smooth or ± warty; pappus < or = 0.2 mm. 2*n*=12. Damp places; < 300 m. GV, SCoR, SW, **DSon**. [*E. alba* (L.) Hassk.] A weed on all continents. Source of dark dye, medicine against roundworm parasites. All year

ENCELIA

Curtis Clark

Shrubs. **STS** gen many from base. **LVS** alternate, gen drought-deciduous, simple, petioled, entire or rarely toothed. **INFL**: heads radiate or discoid, solitary or in cyme-like panicles; peduncles gen long; involucre hemispheric; phyllaries in 2–3 series, free; receptacle chaffy, scales folded around frs and falling with them. **RAY FLS** sterile; style 0; ligules yellow. **DISK FLS** many; corollas yellow or brown-purple, tube slender, throat abruptly expanded, lobes triangular; anther tips ovate, ± acute; style tips triangular. **FR** strongly compressed, obovate or wedge-shaped; edges long-ciliate; faces glabrous or short-hairy; pappus of 2 narrow scales or 0. 13 spp.: w N.Am, w S.Am. (Christopher Encel, 16th century) Commonly hybridizing, esp in disturbed areas; *E. farinosa* × *E. frutescens* is common; *E. farinosa* × *E. actoni*, *E. actoni* × *E. frutescens*, *E. frutescens* × *E. virginensis*, *E. farinosa* × *Geraea canescens* have been reported. Key to species adapted by Margriet Wetherwax.

1. Heads in panicles, always radiate; lvs ± hairy, with curled hairs
  2. Heads in tight panicles; ray corollas well developed; lvs densely hairy, without strigose hairs . . . . . ***E. farinosa***
  2' Heads in loose panicles; ray corollas often short, few, deeply lobed; lvs moderately hairy, often with
    some strigose hairs . . . . ***E. farinosa*** × ***E. frutescens***
1' Heads solitary, rayed or rayless; lvs strigose, canescent, or both
  3. Rays 0; lvs strigose but not canescent . . . . ***E. frutescens***
  3' Rays present; lvs canescent
    4. Lvs silvery canescent, with no strigose hairs; rays 15–25, with ligules > 10 mm and shallowly toothed;
      W&I, w D, DMtns . . . . ***E. actoni***
    4' Lvs with some strigose hairs intermixed with a softer canescence; rays gen < 21, with ligules < 15 mm
      and deeply toothed; e DMoj, DMtns . . . . ***E. virginensis***

E. actoni Elmer (p. 147) Shrub 5–15 dm, with many slender branches from base. **STS** branched below; young sts hairy; older sts with fissured bark. **LVS** scattered along sts; petioles 6–12 mm; blades 2.5–4 cm, ovate to deltate, acute, silvery green, canescent. **INFL**: heads radiate, solitary; peduncles canescent; involucre 8–14 mm; phyllaries ovate. **RAY FLS** 15–25; ligules 10–25 mm. **DISK FLS**: corollas 5–6 mm, yellow. **FR** 5–7 mm; pappus gen 0. 2*n*=36. Open areas, rocky slopes, roadsides; 800–1500 m. sw SnJV and adjacent WTR (Cuyama Valley), **w D** and adjacent CA-FP, **W&I, DMtns**; sw NV, n Baja CA. [*E. virginensis* Nelson ssp. *a.* (Elmer) Keck] ✿DRY,SUN,DRN:7,**8,9**, 10,**11**,12,13,**14**,15,16,18,**19–21**,22–24. Feb–Jul

E. farinosa Torrey & A. Gray (p. 147, pl. 17) BRITTLEBUSH. INCIENSO Shrub 3–15 dm, from 1 or several trunks; sap fragrant. **STS** much-branched above; young sts tomentose; older sts with smooth bark. **LVS** clustered near st tips; petioles 10–20 mm; blades 2–7 cm, ovate to lanceolate, obtuse or acute, silver- or gray-tomentose. **INFL**: heads radiate, 3–9 in panicles; peduncles ± yellow, glabrous exc just below heads; involucre 4–10 mm; phyllaries lanceolate. **RAY FLS** 11–21; ligules 8–12 mm. **DISK FLS**: corollas 5–6 mm, yellow or brown-purple. **FR** 3–6 mm; pappus 0. $2n=36$. Coastal scrub, stony desert hillsides; < 1000 m. e SCo and adjacent PR, **D**; to sw UT, AZ, nw Mex. Dried resin used as incense. [var. *phenicodonta* (S.F. Blake) I.M. Johnston] ❀SUN,DRN:**8–14**,15,16,18, **19–24**. Mar–May

E. farinosa× **E. frutescens** Shrub 5–12 dm with few–many branches from 1–several short trunks. **STS** branched along their length; young sts hairy; older with rough bark. **LVS** scattered along sts; petioles 5–15 mm; blades 1–5 cm, elliptic, lanceolate, or narrowly ovate, obtuse, gray-green, lightly tomentose with some strigose hairs. **INFL**: heads radiate or sometimes discoid, 2–5 in loose panicles; peduncles glabrous or strigose; involucre 5–12 mm; phyllaries lanceolate. **RAY FLS** 1–11; ligules 5–20 mm, deeply 3-lobed. **DISK FLS**: corollas 5–6 mm, yellow. **FR** 5–8 mm; pappus 2 slender scales or 0. $2n=36$. Roadsides, waste places, desert washes, flats; < 800 m. **D**; w AZ, n Baja CA. Hybrids and backcrosses commonly found with parent spp. Many herbarium specimens of "*E. virginensis*" are these.

E. frutescens (A. Gray) A. Gray (p. 147)   Shrub 5–15 dm, with many slender branches from 1–several short trunks. **STS** branched below; young sts glabrous; older sts with fissured bark. **LVS** scattered along sts; petioles 2–7 mm; blades 1–2.5 cm, elliptic or narrowly ovate, obtuse, green, strigose. **INFL**: heads discoid, solitary; peduncles strigose; involucre 6–12 mm; phyllaries lanceolate. **RAY FLS** 0. **DISK FLS**: corollas 5–6 mm, yellow. **FR** 6–9 mm; pappus 2 slender scales or 0. $2n=36$. Desert washes, flats, slopes, roadsides; < 800 m. **D**; s NV, w AZ, Baja CA. ❀SUN,DRN:**8,9**,10–13,**14**,15,16,18,**19–24**. Feb–May

E. virginensis Nelson   Shrub 5–15 dm, with many slender branches from base. **STS** branched below; young sts hairy; older sts with fissured bark. **LVS** scattered along sts; petioles 2–7 mm; blades 1.2–2.5 cm, narrowly ovate to deltate, acute or obtuse, gray-green, lightly canescent, with some strigose hairs. **INFL**: heads radiate, solitary; peduncles canescent; involucre 9–13 mm; phyllaries narrow ovate. **RAY FLS** 11–21; ligules 8–15 mm. **DISK FLS**: corollas 5–6 mm, yellow. **FR** 5–8 mm; pappus gen 0. $2n=36$. Desert flats, rocky slopes, roadsides; 500–1500 m. **e DMoj, DMtns**; to sw UT, nw AZ. ❀SUN,DRN:2,**7–9**,10,**11**,12,13,**14**,15,16,**18–24**. Apr–May, Dec

ENCELIOPSIS

Curtis Clark

Per from stout caudex, subscapose. **STS** densely leafy at base, lfless above. **LVS** basal and closely alternate, simple, petioled or sessile, entire, 3-veined. **INFL**: heads radiate or discoid, solitary; peduncles long; involucre hemispheric; phyllaries in 2–3 series, free; receptacle chaffy, scales folded around frs and falling with them. **RAY FLS** sterile; style 0; ligules yellow. **DISK FLS** many; corollas yellow, tube slender, throat abruptly expanded, lobes triangular; anther tips ovate, ± acute; style tips triangular. **FR** strongly compressed, wedge-shaped; edges ± white, corky, glabrous or long-ciliate; faces black, glabrous or ± hairy; pappus of 2 narrow awns and a crown of shorter scales. 3 spp.: w N.Am. (Greek: like *Encelia*)

1. Petioles winged, wings merging with blades, blades diamond-shaped or widely elliptic; herbage silvery
   . . . . *E. covillei*
1' Petioles not or barely winged, blades ovate; herbage dull gray . . . . *E. nudicaulis*

E. covillei (Nelson) S.F. Blake (p. 149, pl. 18) PANAMINT DAISY   Pl 1.5–8(10+) dm; herbage silvery canescent, hairs fine, ± appressed. **STS** woody at base. **LVS**: petioles winged, wings merging with blades; blades 4–10 cm, 2–8 cm wide, diamond-shaped or widely elliptic, 3-veined. **INFL**: heads radiate, 9–13 cm diam; peduncles 3–10 dm, gray-puberulent; involucre 1.8–3 cm; phyllaries in 3 series, lanceolate to ovate, acuminate, densely gray-puberulent. **RAY FLS** 20–35; ligules 3–5 cm. **FR** ± 10 mm, 6.5 mm wide, glabrous or puberulent; pappus awns ± 1 mm, smooth. $2n=36$. RARE. Stony hillsides, canyons; 400–1250 m. **n DMtns (w side Panamint Mtns)**. [*E. argophylla* (D.C. Eaton) Nelson var. *grandiflora* (M.E. Jones) Jepson] Apr–Jun

E. nudicaulis (A. Gray) Nelson (p. 149) NAKED-STEMMED DAISY   Pl 1–4 dm; herbage dull gray, hairs short, ± spreading. **STS** woody at base. **LVS**: petioles not or only slightly winged; blades 2–6 cm, 2–6 cm wide, ovate, 3-veined. **INFL**: heads radiate, 4–9 cm diam; peduncles 1.5–4.5 dm, gray-puberulent; involucre 1–2 cm; phyllaries in 3 series, narrowly lanceolate from ovate base, acute, densely gray-puberulent. **RAY FLS** ± 21; ligules 2–4 cm. **FR** ± 9 mm, 3.5 mm wide, silky-hairy; pappus awns 1–1.5 mm, smooth. $2n=36$. UNCOMMON. Stony hillsides and canyons; 950–2000 m. **W&I, DMtns**; to ID, UT, n AZ. ❀TRY;DFCLT. May

ERICAMERIA GOLDENBUSH

Gregory K. Brown & David J. Keil

Shrubs < 50 dm, resinous, gen gland-dotted. **LVS** < 10 cm, thread-like to wedge-shaped, entire. **INFL** various; heads radiate or discoid; involucre 3–14 mm, obconic to hemispheric; phyllaries in 2–6 series, ± lanceolate to ovate, gen resinous, tips erect to recurved, obtuse to acuminate or tailed, midrib often thickened with a resin gland. **RAY FLS** 0–30; corollas 2–12 mm, gen yellow. **DISK FLS** 4–70+; corollas 3–11 mm, yellow. **FR** 2–8 mm, ribbed; pappus white to brown. ± 27 spp.: w N.Am. [Nesom 1990 Phytologia 68:144–155] Gen fls summer/autumn. Some spp. hybridize with *Chrysothamnus nauseosus*; relationship to spp. of *Chrysothamnus* warrants further study. [Anderson 1995 Great Basin Naturalist 55:84–88; Nesom & Baird 1995 Phytologia 78: 61–68] Key to species adapted by Margriet Wetherwax.

1. Lvs narrowly oblong to obovate, widest gen > 2.5 mm wide
   2. Involucre gen 5–7 mm wide . . . . *E. cuneata*
      3. Lvs wedge-shaped, sessile, largest 3–18 mm . . . . var. *cuneata*
      3' Lvs oblanceolate, blunt, gen distinctly petioled, largest 9–25 mm . . . . var. *spathulata*
   2' Involucre gen 5–15 mm wide

4. Heads discoid . . . . *E. discoidea*
4' Heads radiate
  5. Lvs 6–12 mm; involucre 7–9 mm; outer phyllaries parchment-like, only tips green; disk fls 15–18
    . . . . *E. gilmanii*
  5' Lvs 15–40 mm; involucre 8–15 mm; outer phyllaries green, lf-like; disk fls 13–40 . . . . *E. suffruticosa*
1' Lvs linear to thread-like and cylindric, widest < 2.5 mm wide
  6. Involucre 6–18 mm wide, obconic to hemispheric
    7. Involucre 8–14 mm; phyllary margins cut-ciliate . . . . *E. linearifolia*
    7' Involucre 5–10 mm; phyllary margins woolly-ciliate . . . . *E. pinifolia*
  6' Involucre 3–6 mm wide, ± narrowly obconic
    8. Lvs not gland-dotted, glabrous, ± sticky . . . . *E. nana*
    8' Lvs gland-dotted, resinous, glabrous or puberulent
      9. Young herbage glabrous; lvs 10–60 mm; pl < 3 m . . . . *E. laricifolia*
      9' Young herbage puberulent; lvs 3–40 mm; pl < 1 m
        10. Involucre 4–5 mm; disk corollas 3–5 mm; ray fls (0)1–2 . . . . *E. cooperi*
        10' Involucre 6–7 mm; disk corollas 5–8 mm; ray fls 1–6 . . . . *E. palmeri* var. *pachylepis*

E. cooperi (A. Gray) H.M. Hall var. ***cooperi*** (p. 149) Pl 3–6 dm, puberulent, gland-dotted. **LF** 3–15 mm, ± linear, acute. **INFL**: heads gen radiate, in open cymes; involucre 4–5 mm, 3–4 mm diam, narrowly bell-shaped; phyllaries 9–15 in 3–4 series, 2.5–4 mm, oblong to ovate, obtuse or outer acute. **RAY FLS** 0–2; corollas 4–9 mm. **DISK FLS** 4–12; corollas 3–5 mm. **FR** 3–3.5 mm, obconic to subcylindric, softly silky-hairy, veins 10–12, thin; pappus 3–4.5 mm, white. 2*n*=18. Rocky slopes, valleys, in creosote-bush scrub, Joshua-tree woodland; 800–2000 m. **SNE, DMoj**; s NV. [*Haplopappus c.* (A. Gray) H.M. Hall] Hybridizes with *E. linearifolia*. ✿TRY;DFCLT. Mar–Jun

E. cuneata (A. Gray) McClatchie Pl 1–10 dm, glabrous, ± gland-dotted. **LF** 2–25 mm, ± oblanceolate or obovate, obtuse. **INFL**: heads radiate or discoid in small compact cymes; involucre 6–12 mm, 4–14 mm diam, obconic; phyllaries 20–30 in 4–6 series, lanceolate to obovate, glabrous, sometimes resinous. **RAY FLS** 0–3; corollas < 5 mm. **DISK FLS** 7–70; corollas ± 5.5 mm. **FR** 2.5–3 mm, 5-ribbed, silky-hairy; pappus < corolla, sparse, brown. Outcrops, slopes, cliffs; 100–2800 m. SN, SCoR, TR, PR, **SNE, D**; s NV, AZ, nw Mex. [*Haplopappus c.* A. Gray] 3 vars. in CA.

  var. ***cuneata*** **LVS**: largest 3–14(18) mm, 2–9(12) mm wide, wedge-shaped, sessile. **INFL**: heads radiate or discoid, 8–11 mm, 5–7 mm diam. **DISK FLS** 12–33. 2*n*=18. Granite outcrops; 1000–2800 m. SN, WTR, SnGb, PR, **SNE**. ✿DRN,DRY,SUN: 1–3,**7**, 14,**15,16,18**,19–24;GRNCVR. Sep–Nov

  var. ***spathulata*** (A. Gray) H.M. Hall (p. 149) **LVS**: largest (9)12–25 mm, 4–16 mm wide, ± obovate; base petiole-like; tip gen widely obtuse or notched. **INFL**: heads gen discoid, 8–11 mm, 5–7 mm diam. **DISK FLS** 7–15. 2*n*=18. Rock outcrops; 100–1900 m. Teh, SCoRI, **D**; s NV, AZ, nw Mex. Intergrades with var. *cuneata* in Kern Co. [*Haplopappus c.* var. *s.* (A. Gray) S.F. Blake] ✿TRY; DFCLT.

E. discoidea (Nutt.) G. Nesom Pl 1–4 dm. **ST** densely white-tomentose. **LF** 10–30 mm, oblong to oblanceolate, sessile, obtuse to acute, stalked-glandular. **INFL**: heads discoid, 1–few in terminal clusters; involucre 9–13 mm, 8–12 mm diam, obconic to bell-shaped; phyllaries in 2–3 series, lanceolate, acuminate, scarious, grading into upper lvs. **DISK FLS** 10–26; corollas 9–11 mm. **FR** 5–6 mm, narrowly obconic, hairy; pappus brownish. 2*n*=18. Rocky slopes; 2700–3700 m. SNH, Wrn, **n SNE (Sweetwater Mtns)**; to OR, ID, WY, Colorado, UT. [*Haplopappus macronema* A. Gray] Jul–Sep

E. gilmanii (S.F. Blake) G. Nesom (p. 149) GILMAN'S ERICAMERIA Pl 2–4 dm, aromatic, glabrous. **LF** 6–12 mm, oblanceolate, obtuse, often folded. **INFL**: heads radiate, 1–few in cymes; involucre 7–9 mm, ± 5 mm diam, narrowly bell-shaped; phyllaries ± 25 in 4–6 series, outer widely lanceolate, tip often thickened, green, recurved, inner phyllaries oblong, parchment-like, resinous. **RAY FLS** 4–6; corollas 8–10 mm, white or pale

yellow. **DISK FLS** 15–18; corollas 7–7.5 mm. **FR** 3–4 mm, ± cylindric, 5-ribbed, silky-hairy; pappus < 6.5 mm, whitish. 2*n*=18. RARE. Open coniferous forests, gen on limestone; 2100–3400 m. **W&I, n DMtns (Panamint Mtns)**. [*Haplopappus g.* S.F. Blake] Aug–Sep

E. laricifolia (A. Gray) Shinn. TURPENTINE-BRUSH Pl 3–10 dm, glabrous, aromatic. **LF** 10–30 mm, gen subcylindric, ± acute. **INFL**: heads radiate, in cymes; involucre 3–5 mm, 3–5 mm diam, obconic; phyllaries 12–20 in 3–4 series, ± linear, acute, glabrous, midrib a brownish to yellowish gland. **RAY FLS** 3–11; corollas 8–11 mm. **DISK FLS** 10–18; corollas 5.5–6.5 mm. **FR** 3.5–4 mm, narrowly obconic, obscurely 4-ribbed, densely white-soft-hairy; pappus 5–8 mm, tan. 2*n*=18. Rocky canyons, pinyon/juniper woodland, creosote-bush scrub; 1000–2000 m. **DMtns**; to TX, n Mex. [*Haplopappus l.* A. Gray] ✿TRY;DFCLT. Sep–Oct

E. linearifolia (DC.) Urb. & J. Wussow (p. 149) INTERIOR GOLDENBUSH Pl 4–15 dm, glabrous to ± puberulent. **LF** 10–55 mm, linear, acute; base narrowed. **INFL**: head radiate, 1, on ± lfless peduncles; involucre 8–14 mm, 10–18 mm diam, hemispheric; phyllaries in 2–3 series, linear to lanceolate, acuminate, stalked-glandular, center green, margin cut-ciliate, scabrous. **RAY FLS** 13–18; corollas 9–20 mm. **DISK FLS** many; corollas 6–10 mm. **FR** 4–5 mm, compressed, 6–8-veined, densely silky-hairy; pappus 5.5–7 mm, white. 2*n*=18. Dry slopes, valleys; < 2000 m. s SNF, ScV (Sutter Buttes), s SnJV, e CW, WTR, **DMoj, w DSon**; NV, sw UT, w AZ. [*Haplopappus l.* DC.] Hybridizes with *E. cooperi*. ✿DRN,DRY,SUN:2,3,**7**,9–12,14–16,18–24;DFCLT. Mar–May

E. nana Nutt. Pl 1–5 dm, glabrous. **LF** 10–15 mm, linear to narrowly oblanceolate, gen curved, acute, not gland-dotted, ± sticky. **INFL**: heads radiate, in dense leafy cymes; involucre 5.5–7.5 mm, 3–4 mm diam, obconic; phyllaries 20–30 in 4–5 series, lanceolate, acute to acuminate, glabrous. **RAY FLS** 1–7; corollas 2–3 mm. **DISK FLS** 4–8; corollas 4.5–6.5 mm. **FR** 4.5–5.5 mm, cylindric, faintly 5-angled, glabrous to densely hairy; pappus = disk corollas, light brown. 2*n*=18. Rocky soils, cliffs; 2100–2800 m. **c SNE, DMtns**; to WA, ID, UT. [*Haplopappus n.* (Nutt.) D. Eaton] Jul–Nov

E. palmeri (A. Gray) H.M. Hall (p. 149) Pl 5–40 dm, glabrous to puberulent, gland-dotted when young. **LF** < 40 mm, linear, acute. **INFL**: heads radiate, many; involucre 5–8.5 mm, 3.5–4.5 mm diam, obconic to ± cylindric; phyllaries 16–24 in 4–5 series, < disk fls, oblong, tips greenish, oblong or acute, margins narrow, white, ciliate. **RAY FLS** 1–8; corollas 4–6 mm. **DISK FLS** 5–20; corollas 5–8 mm. **FR** 3–4 mm, subcylindric, 4–7-angled, hairy; pappus of disk fls > corollas, brown. Plains, foothills; < 800 m. SCo, **w DSon**; n Baja CA. [*Haplopappus p.* A. Gray] 2 vars. in CA.

  var. ***pachylepis*** (H.M. Hall) G. Nesom Pl 5–15 dm, very

leafy, puberulent. **LF** 5–16 mm, thread-like. **INFL**: involucre 6–7 mm. **RAY FLS** 1–6. Coastal scrub, disturbed chaparral; < 800 m. n SCo, ChI, **w DSon**. [*Haplopappus palmeri* A. Gray var. *p.* H.M. Hall] ✿TRY. Aug–Dec

E. pinifolia (A. Gray) H.M. Hall PINE-BUSH Pl 6–25 dm, ± glabrous. **LF** 10–40 mm, thread-like, subacute; short lvs clustered in axils. **INFL**: heads radiate; head in spring 1, large; heads in autumn many, smaller; involucre 5–10 mm, 5–15 mm diam, obconic to hemispheric; phyllaries 20–26 in ± 4 series, lanceolate-acuminate to oblong, woolly-ciliate, tips green, outer often tailed, inner acute, margin narrow. **RAY FLS** 15–30 in spring, 5–10 in autumn; corollas 3.5–5 mm. **DISK FLS** 12–many, corollas 6–7.5 mm. **FR** 4.5–5 mm, subcylindric, striate, lightly hairy; pappus > disk corollas, reddish or tan. 2n=18. Chaparral, oak wood-

land, scrub away from coast; < 1700 m. WTR, PR, **w DSon**. [*Haplopappus p.* (A. Gray) H.M. Hall] ✿DRN,DRY,SUN:7,14–17,19–21. Apr–Jul, Sep–Jan

E. suffruticosa (Nutt.) G. Nesom Pl 1.5–4 dm, glandular-puberulent. **LF** 15–40 mm, ± oblanceolate, acute. **INFL**: heads radiate, 1–few in small cymes; involucre 8.5–15 mm, 10–15 mm diam, bell-shaped; phyllaries lanceolate, outer green and leafy throughout, inner green above, acuminate, glandular. **RAY FLS** 1–6; corollas 7–12 mm. **DISK FLS** 13–40; corollas 8.5–10.5 mm. **FR** 5.5–6.5 mm, narrowly obconic, angled, hairy; pappus < disk corollas, white to pale yellow. 2n=18. Coniferous forest, alpine places; 2400–3700 m. SNH, **n SNE (Sweetwater Mtns)**, **W&I**; to OR, MT, WY, AZ. [*Haplopappus s.* (Nutt.) A. Gray] Jul–Sep

ERIGERON FLEABANE DAISY

Guy L. Nesom

Ann to per (subshrub). **STS** gen erect. **LVS** alternate, gen entire. **INFL**: heads gen radiate, 1–many in loose, panicle-like or flat-topped clusters; involucre hemispheric; phyllaries narrowly lanceolate, in 2–several equal to strongly graded series; receptacle flat to steeply conic, naked, smooth to shallowly pitted. **RAY FLS** (0) gen 10–many; ligules gen white, pink, or blue (yellow). **DISK FLS** many; corollas gen narrowly funnel-shaped, yellow; style tips 0.1–0.8 mm, ± triangular. **FR** 0.5–3 mm, gen ± oblong, compressed to ± cylindric, gen 2-ribbed, gen sparsely hairy; pappus (0) gen of 6–50 longer, inner bristles and shorter outer bristles, narrow scales, or short crown. ± 375 spp.: wordwide. (Greek: early old age) [Nesom 1992 Phytologia 72:157–208]

1. Heads discoid or disciform; pistillate fls 0 or << involucre
  2. Pistillate fls 0 . . . . ***E. bloomeri*** var. ***bloomeri***
  2' Pistillate fls present, corollas inconspicuous
    3. Heads gen 1–many; sts gen branched near mid-st or below
      4. Per — pappus bristles 12–17 . . . . ***E. aphanactis*** var. ***aphanactis***
      4' Ann or bien
        5. Pappus bristles 15–20 . . . . ***E. calvus***
        5' Pappus bristles 6–9(12) . . . . ³***E. divergens***
    3' Heads solitary; sts unbranched
      6. Per; lvs (1)2–3-ternately dissected; pappus bristles 12–25 . . . . ²***E. compositus***
      6' Ann or bien; lvs entire to lobed; pappus bristles 6–9(12) . . . . ³***E. divergens***
1' Heads radiate; pistillate fls with obvious ligules
  7. Lvs mostly 1–3-ternately dissected
    8. Lvs (1)2–3-ternately dissected; ligules <1 mm wide; caudex branches thick, ascending . . . . . . ²***E. compositus***
    8' Lvs deeply 3-lobed at tip; ligules 1–2 mm wide; caudex branches thin, rhizome-like . . . . ***E. vagus***
  7' Lvs entire to shallowly lobed
    9. Fr ribs 4–8; herbage gen ± silvery or gray, hairs dense, appressed
      10. Fr ribs 6–8; basal lvs tufted, persistent . . . . ***E. argentatus***
      10' Fr ribs 4(6); basal lvs gen 0 at fl . . . . ***E. utahensis***
    9' Fr ribs gen 2 (if more, herbage not silvery)
      11. Phyllaries strongly graded; lvs entire, not clasping, basal 0, cauline gen evenly sized and spaced; infls arising near st tips . . . . ***E. breweri***
        12. Sts (30)40–75 cm, not wiry or brittle; phyllary hairs short, white, ± appressed, glandless . . . . var. ***covillei***
        12' Sts 20–30 cm, wiry, brittle; phyllary hairs long, stiffly spreading, translucent (also some glandular)
        . . . . var. ***porphyreticus***
      11' Phyllaries gen ± equal; lvs entire to lobed, sometimes clasping, basal sometimes present, cauline often strongly reduced upwards; infls arising near mid-st
        13. Pls ann or bien
          14. Cauline lvs clasping; pls fibrous-rooted . . . . ***E. philadelphicus***
          14' Cauline lvs not clasping; pls taprooted
            15. Sts and phyllaries densely, evenly puberulent (hairs < 0.5 mm) . . . . ³***E. divergens***
            15' Sts and phyllaries glandular-puberulent, also sparsely spreading-hairy (hairs 0.5–2 mm) . . . ***E. lobatus***
        13' Pls distinctly per
          16. Cauline leaves ± clasping; pls fibrous-rooted, gen from obvious lateral rhizomes
            17. Phyllaries blackish purple, sts with spreading hairs . . . . ²***E. algidus***
            17' Phyllaries greenish, sts ± appressed-hairy . . . . ***E. peregrinus***
              18. Herbage strigose . . . . var. ***callianthemus***
              18' Herbage spreading-hairy . . . . var. ***hirsutus***

16' Cauline lvs not at all clasping; pls taprooted
  19. Sts glabrous or appressed-hairy
    20. Basal lvs linear; fr ribs densely ciliate, faces glabrous . . . . *E. compactus*
    20' Basal lvs narrowly oblanceolate to spoon-shaped; fr ribs and faces sparsely and evenly hairy
      21. Pls cespitose; caudex stout, branched; basal lvs spoon-shaped, petioled; cauline lvs abruptly
        reduced . . . . *E. tener*
      21' Pls not cespitose; caudex gen slender; basal lvs narrowly oblanceolate, petiole indistinct;
        cauline lvs gradually reduced . . . . *E. eatonii* var. *sonnei*
  19' Sts spreading-hairy
    22. Lf blades clearly petioled
      23. Basal leaves 20–70 mm; cauline lvs much reduced; sts 2–30 cm; phyllaries blackish purple;
        ray fls (30)50–125, ligules 7–13 mm . . . . ²*E. algidus*
      23' Basal lvs 5–25 mm; cauline lvs 0; sts 1–4 cm; phyllaries green; ray fls 15–40, ligules 4–6 mm
        . . . . *E. uncialis* var. *uncialis*
    22' Lvs linear or gradually tapered to base, petiole indistinct
      24. Margins of lower lvs gen hairy but not stiffly spreading-ciliate; > 2200 m
        25. Basal lvs 20–80 mm; sts ± nonglandular; phyllaries green; ray fls 25–55, ligules 6–11 mm
          . . . . *E. clokeyi*
        25' Basal lvs 6–35 mm; sts densely glandular; phyllaries blackish purple; ray fls 20–37,
          ligules 4–7(10) mm . . . . *E. pygmaeus*
      24' Margins of lower lvs stiffly long-spreading-ciliate; 1200–1800 m
        26. Disk corollas (sub)glabrous; pappus bristles 12–16; outer pappus of bristles or narrow scales;
          ray fls (40)55–115, ligules 8–12 mm . . . . *E. pumilus* var. *intermedius*
        26' Disk corollas sharply scabrous; pappus bristles 7–15; outer pappus of wide scales; rays 40–60,
          ligules 7–9 mm . . . . *E. concinnus*
          27. Sts leafy; heads many . . . . var. *concinnus*
          27' Sts ± lfless; head 1 . . . . var. *condensatus*

E. algidus Jepson (p. 149) Per 2–30 cm, from (sub)simple caudex and fibrous roots, unbranched, sparsely and loosely spreading-hairy and glandular. **LVS**: basal 2–7 cm, oblanceolate to spoon-shaped; cauline much reduced. **INFL**: head 1, 8–16 mm diam; phyllaries ± equal, blackish purple, tips spreading to reflexed. **RAY FLS** 30–125; corollas 7–13 mm, ligules blue or pink to white, coiled. **FR**: pappus bristles 12–20. Alpine meadows, talus; 2600–3700 m. SNH, SNE; w NV. [*E. petiolaris* E. Greene] ❀DRN, SUN:**1**&IRR:**2**,3,15–18;DFCLT. Jul–Aug

E. aphanactis (A. Gray) E. Greene (p. 149) Per 8–25 cm, from taproot and short-branched caudex, often densely cespitose, stiffly spreading-hairy, sessile-glandular. **LVS**: basal 4–7 cm, sometimes long-petioled, linear-oblanceolate, gradually reduced upwards. **INFL** flat-topped; heads disciform, 1–many, 8–12 mm diam; phyllaries ± equal. **PISTILLATE FLS** many; ligules 0 or < involucre. **DISK FLS**: corollas gen abruptly inflated to throat. **FR**: pappus bristles 12–17. Sagebrush or juniper scrub; 1300–2600 m. SnBr, SNE, DMtns; to OR, Colorado, NM. 2 vars. in CA.

  var. **aphanactis** STS gen branched in lower 1/2. **INFL**: heads gen 2–many. 2*n*=18. Habitats and ± range of sp. Apr–Sep

E. argentatus A. Gray (p. 149) Per 10–40 cm, from woody taproot and short-branched caudex, often cespitose, densely silvery-hairy. **LVS**: basal many, erect, gen 2–5 cm, narrowly oblanceolate; cauline scattered on lower 2/3 of st. **INFL**: head 1, 12–22 mm diam; phyllaries ± equal. **RAY FLS** 25–48; corollas 10–16 mm, ligules blue, coiled. **FR** 6–8-ribbed, densely hairy; pappus bristles 25–40. 2*n*=18. Rocky slopes, pinyon/juniper woodland; 2000–2300 m. **W&I, ne DMtns (Last Chance Mtns)**; to UT, w AZ. ❀TRY;DFCLT. May–Jul

E. bloomeri A. Gray Per 4–15 cm, from taproot and short-branched caudex, cespitose, ± glabrous to densely white-strigose. **LVS**: basal 2–7 cm, ± linear; cauline lvs on lower st. **INFL**: head discoid, 1, 7–20 mm diam; phyllaries ± equal, hairy, sparsely sessile-glandular. **FR**: pappus bristles 25–40. Rocky slopes, lava beds, meadows; 600–2300 m. KR, CaR, n SNH, **GB**; to WA, ID, NV. 2 vars. in CA.

  var. **bloomeri** Herbage strigose. 2*n*=18. Habitats of sp.; 800–2000 m. Range of sp. (exc KR). [var. *pubens* Keck] ❀DRN, SUN:1&IRR:2,7,14–17;DFCLT. May–Jul

E. breweri A. Gray Per 7–75 cm, from woody roots and slender-branched caudex, gen not wiry or brittle, gen densely short-spreading-hairy; glands gen 0. **LVS** cauline, 5–40 mm, linear to oblanceolate, evenly sized and spaced. **INFL**: heads gen radiate, 8–15 mm diam; phyllaries strongly graded in 3–5 series, glandular or not, tip gen ± like body. **RAY FLS** 12–45, 4–7 mm, ligules white to pink or drying blue, weakly coiled. **FR**: pappus bristles 22–46. Many habitats; 300–3100 m. KR, SNH, TR, SnJt, **SNE, DMoj**; OR, w NV. 6 vars. in CA.

  var. **covillei** (E. Greene) G. Nesom **ST** 30–75 cm; base thicker. **LF** 15–30 mm; hairs 0.2–0.4 mm. **INFL**: phyllary hairs nonglandular, short, white, thick-based, ascending, much sparser inward. Open, rocky sagebrush scrub, chaparral, juniper woodland; ± 1000–1900 m. s SNH (e slope), SnGb, SnBr, SnJt, **DMoj**. [*E. foliosus* var. *c*. (E. Greene) Compton] May–Sep

  var. **porphyreticus** (M.E. Jones) Cronq. **ST** 20–30 cm, wiry, brittle. **LF** 5–30 mm; hairs 0.1–0.2 mm. **INFL**: phyllaries thick-margined, hairs glandular and not, long, thick-based, stiffly spreading, translucent. 2*n*=18. Open, rocky sagebrush scrub to yellow-pine forest; 1200–2600 m. SnBr, **SNE, DMoj**; sw NV. May–Aug(Sep)

E. calvus Cov. BALD DAISY Bien or short-lived per 10–14 cm, from taproot, ± long-spreading-hairy; base much-branched. **LVS**: basal 3–5 cm, spoon-shaped; cauline abruptly reduced. **INFL**: heads disciform or obscurely radiate, 1–few, 13–14 mm diam; phyllaries ± equal, sessile-glandular, hairs thick-based. **RAY FLS** 0–many, < phyllaries. **FR**: pappus bristles 15–20. RARE. Sagebrush and desert scrub; ± 1200 m. **s SNE (w base Inyo Mtns)**. Closely related to *E. divergens*; also confused with *E. aphanactis*. May

E. clokeyi Cronq. (p. 149) Per 5–20 cm, from stout taproot and (sub)simple caudex, ascending to erect, unbranched; hairs short, stiffly spreading to reflexed, glandless. **LVS**: basal 2–8 cm, ± oblanceolate; cauline not clasping, gen strongly reduced by mid-st. **INFL**: head 1, 8–12 mm diam; phyllaries ± equal, sessile-glandular, hairs short, stiffly spreading to reflexed. **RAY FLS** 25–55; corollas 6–11 mm, ligules white to blue, reflexed. **FR**: pappus bristles 13–22. 2*n*=18. Sagebrush scrub to alpine talus; 2200–3400 m. s SNH, **SNE, DMtns**; to UT. ❀TRY;DFCLT. Jun–Sep

E. compactus S.F. Blake (p. 149) Per 2–8 cm, from taproot and short-branched caudex, cespitose, unbranched, short-white-appressed-hairy. **LVS** basal, 5–25 mm, linear. **INFL**: head 1, 10–14 mm diam; phyllaries ± equal. **RAY FLS** 15–32; corollas 7–11 mm, ligules white, sometimes striped lilac, coiled. **FR**: ribs densely ciliate; faces glabrous; pappus bristles 30–40. Rocky slopes, pinyon/juniper woodland; 1800–2300 m. **W&I**; to UT. ❀TRY;DFCLT. May–Jun

E. compositus Pursh (p. 149) Per 3–15 cm, from stout taproot and short-branched caudex, cespitose, unbranched, densely sessile-glandular, sometimes also sparsely soft-bristly. **LVS**: gen basal, 1–5 cm, oblanceolate to spoon-shaped, (1)2–3-ternately divided; cauline lvs (if any) much reduced, gen entire. **INFL**: head ± radiate or disciform, 1; disk 10–18 mm diam; phyllaries ± equal. **RAY FLS** (0) ± 30–60; corollas 1–5 mm, ligules gen < styles in CA, blue to pinkish or white, weakly coiled. **FR**: pappus bristles 12–15. 2*n*=18, 27,36,45,54,63. Rocky slopes, crevices, talus; 2000–4300 m. SNH, Wrn, **SNE**; to AK, e Can, Colorado, AZ. [vars. *discoideus* A. Gray, *glabratus* Macoun] ❀DRN,IRR:1,2,**3**,6,**7,14–16,18**. May–Sep

E. concinnus (Hook. & Arn.) Torrey & A. Gray Per 6–16 cm, from woody taproot and thick, short-branched caudex, cespitose, 1–4-branched below mid-st; hairs spreading or reflexed, soft-bristly and minutely stalked-glandular. **LVS**: basal erect, crowded, 2–6 cm, ± linear to oblanceolate; cauline 0 or gradually reduced upward. **INFL** flat-topped; heads 1–many, 7–11 mm diam; phyllaries ± equal. **RAY FLS** 40–60; corollas 7–9 mm, ligules blue to white or pink, reflexed or weakly coiled. **DISK FLS**: corollas ± widely funnel-shaped. **FR**: pappus bristles 7–15, outer series of prominent, wide scales 0.2–0.5 mm. Sandy to rocky slopes, crevices; 1200–1800 m. **DMtns**; to ID, WY, Colorado, NM.

var. *concinnus* (p. 149) **ST** leafy. **INFL**: heads many. 2*n*=18. Habitats and range of sp. [*E. pumilus* ssp. *concinnoides* Cronq.] Apr–June(Aug)

var. *condensatus* D. Eaton **ST** ± lfless. **INFL**: head 1. Habitats and range of sp. Scattered variant. [Nesom 1983 Sida 10:159–166] May–Jul

E. divergens Torrey & A. Gray Ann 10–45 cm, from slender taproot, branched near mid-st, ± densely reflexed- to spreading-hairy, minutely sessile-glandular near and on heads. **LVS**: basal and lower cauline 2–6 cm, ± obovate, entire to lobed, gradually reduced upward, not clasping. **INFL** often flat-topped; heads gen radiate, gen many, 7–11 mm diam; phyllaries ± equal. **RAY FLS** gen 75–150; corollas (0)5–10 mm, white to purple. **DISK FLS**: corollas abruptly wider at throat. **FR**: pappus bristles 6–9(12), outer series short bristles or narrow scales. 2*n*=18,27,36. Desert scrub to yellow-pine forest; 500–2600 m. s SN, SnGb, SnBr, SnJt, **GB, D**; to B.C., TX, nw Mex. Variable. ❀DRN,SUN:1,**2,3,7,10**,12,**14**,15, 16,**22–24**&IRR: **8,9**,11,**18–21**. Apr–Aug(Sep)

E. eatonii A. Gray Per 4–33 cm, from taproot and (sub)simple caudex, prostrate to erect, 0–few-branched below mid-st, ± ascending-hairy. **LVS** linear to narrowly oblanceolate, 3-veined; basal gen present at fl; middle cauline 1–3(5) cm, strigose. **INFL**: heads 1(4), long-peduncled, 8–23 mm diam; phyllaries ± equal. **RAY FLS** 15–39; corollas 7–15 mm, ligules white, gen bluish or pinkish beneath, weakly coiled. **FR**: pappus bristles 16–30. Open grassland, rocky flats, gen in sagebrush or pinyon/juniper scrub; 1000–2900 m. CaR, n&c SNH, **GB**; to OR, WY, Colorado, AZ. CA vars. intergrade; other vars. in Rocky Mtns. 3 vars. in CA.

var. *sonnei* (E. Greene) G. Nesom **STS** 4–21 cm. **INFL**: head 8–16 mm diam; longest phyllaries 5–8 mm. **DISK FLS**: corollas 3.5–5 mm. **FR**: pappus bristles 18–30, 3.5–5 mm. 2*n*=18. Rocky grassland or sagebrush scrub; 1800–2800 m. n&c SNH, **SNE**; c–w NV. [*E. s.* E. Greene] May–Sep

E. lobatus Nelson Ann gen 14–50 cm, from slender taproot, branched at base, glandular-puberulent. **LVS**: basal 5–10 cm, ± obovate, pinnately 4–8-lobed; cauline gradually reduced upwards.

INFL: heads radiate, 6–10 mm diam; phyllaries glandular-puberulent. **RAY FLS** 85–110; corollas 6–8 mm, drying blue. **FR**: pappus bristles 11–12. Sandy soil, creosote-bush scrub; ± 550 m. **DMoj** (e San Bernardino Co.); NV, AZ, n Mex Mar–Apr

E. peregrinus (Pursh) E. Greene Per 8–45 cm, from short rhizome, branched on upper sts, glabrous or strigose above. **LVS**: basal 5–20 cm, oblanceolate to spoon-shaped, glabrous or sparsely spreading-hairy; cauline ± reduced upward, gen lanceolate to ovate, subclasping. **INFL** flat-topped; heads 1–4, 10–21 mm diam; phyllaries ± equal, with long-acuminate, loosely spreading tips, densely stalked-glandular. **RAY FLS** 30–105; corollas 8–15 mm, ligules white to purple, coiled. **FR** 4–7-ribbed; pappus bristles 20–30. Clearings, talus, alpine meadows; 1300–3400 m. KR, CaRH, SNH, Wrn, **SNE**; to AK, MT, Colorado, NM, e Asia. ❀TRY;DFCLT.

var. *callianthemus* (E. Greene) Cronq. (p. 149) Herbage strigose. 2*n*=18. Habitats and range of sp. (exc n Can to e Asia). [var. *angustifolius* (A. Gray) Cronq.] Jul–Sep

var. *hirsutus* Cronq. Herbage spreading-hairy. Habitats of sp.; 2200–3200 m. c&s SNH, **SNE**. Jul–Aug

E. philadelphicus L. (p. 149) Bien or short-lived per 25–80 cm, from fibrous roots and often short rhizome, branched ± near st tips, wider below heads, sparsely loose-spreading-hairy. **LVS**: basal 8–15 cm, oblong-obovate to spoon-shaped, gen coarsely toothed; cauline lanceolate to ovate, clasping, little reduced. **INFL** flat-topped; heads 1–many, 6–15 mm diam; phyllaries ± equal, bases often ± fused. **RAY FLS** ± 150–400; corollas 6–9 mm, ligules white or pinkish, coiled. **FR**: pappus bristles 20–30. 2*n*=18. Streamsides, other moist habitats; < 1200 m. CA (exc GV, DMtns); to e US. Scattered. ❀WET or IRR:1,2,**14,18–24**&SUN:**3–7,8**–10,**15–17**;INV. May–Jun

E. pumilus Nutt. var. *intermedius* Cronq. (p. 149) Per 8–35 cm, from woody taproot and thick-branched caudex, cespitose, 0–4-branched below mid-st, stiffly spreading- or reflexed-hairy, minutely stalked-glandular. **LVS**: basal erect, crowded, 2–8 cm, linear to oblanceolate; cauline gradually reduced. **INFL** flat-topped; heads 9–14 mm diam; phyllaries ± equal. **RAY FLS** 40–115; corollas 8–12 mm, ligules blue to white or pink, reflexed, sometimes weakly coiled. **DISK FLS**: corollas abruptly wider at throat. **FR**: pappus bristles 12–16, outer series inconspicuous scales. 2*n*=18, 36. Open slopes, meadows; 1200–1800 m. CaR, n SNH, MP, **W&I**; to B.C., MT, WY, UT. Var. *pumilus* on e side Rocky Mtns. ❀TRY. May–Aug

E. pygmaeus (A. Gray) E. Greene (p. 149) Per 1–6 cm, from taproot and short-branched caudex, gen cespitose, unbranched, ± spreading-hairy, densely sessile-glandular. **LVS** gen basal, < 4 cm, linear to narrowly oblanceolate; cauline (if any) below mid-st. **INFL**: head 1, 6–15 mm diam; phyllaries ± equal, purple-black. **RAY FLS** 20–37; corollas 4–10 mm, ligules blue or purple, rarely white, not coiled or reflexed. **FR**: pappus bristles 15–25. 2*n*=18. Rocky sites, subalpine forest to alpine talus; 2900–4100 m. c SNH, **SNE**; w-c NV. ❀TRY;DFCLT. Jul–Aug

E. tener A. Gray Per 2–15 cm, from taproot and branched caudex, cespitose, gen simple, short-white-appressed-hairy, glandless. **LVS**: basal 1–8 cm, long-petioled, oblanceolate to elliptic; cauline much reduced. **INFL**: head 1(2), 5–12 mm diam; phyllaries ± equal, spreading-hairy, sessile-glandular. **RAY FLS** 18–40; corollas 4–8 mm, ligules bluish, not coiled or reflexed. **FR**: pappus bristles 15–30. Crevices or ledges from sagebrush scrub to yellow-pine forest; 2300–3400 m. KR, SNH, MP, **W&I**; to se OR, WY, UT. Jun–Sep

E. uncialis S.F. Blake var. *uncialis* (p. 149) LIMESTONE DAISY Per 1–4 cm, from taproot and slender-branched caudex, simple, spreading-hairy. **LVS** basal, 5–25 mm, long-petioled, widely elliptic to obovate. **INFL**: head 1, 6–11 mm diam; phyllaries ± equal, densely sessile-glandular. **RAY FLS** 15–40; corollas 4–6 mm, ligules white to pinkish, not coiled or reflexed. **FR**: pappus bristles 13–22. UNCOMMON. Limestone crevices

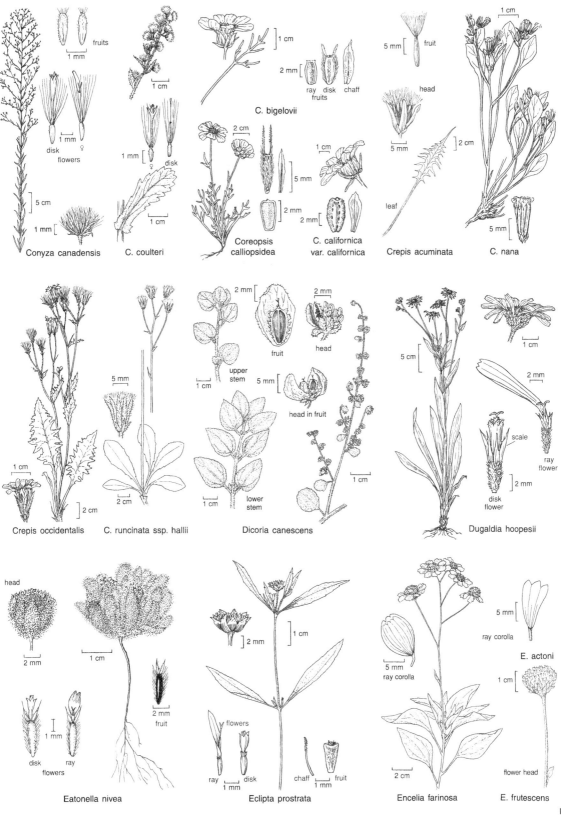

fruits
1 mm

disk
flowers
1 mm ♀

5 cm

1 mm

Conyza canadensis

C. coulteri
1 cm

1 mm ♀ disk
1 cm

C. bigelovii
1 cm
ray disk chaff
2 mm fruits

Coreopsis
calliopsidea
2 cm
5 mm
2 mm

C. californica
var. californica
1 cm
2 mm

fruit
5 mm
head
5 mm
leaf
2 cm

Crepis acuminata

C. nana
1 cm
5 mm

Crepis occidentalis
1 cm
2 cm

C. runcinata ssp. hallii
5 mm
2 cm

Dicoria canescens
upper stem
1 cm
fruit
2 mm
head
2 mm
head in fruit
5 mm
lower stem
1 cm
1 cm

Dugaldia hoopesii
5 cm
1 cm
scale
2 mm
ray flower
disk flower
2 mm

Eatonella nivea
head
2 mm
1 cm
2 mm
fruit
1 mm
disk ray
flowers

Eclipta prostrata
2 mm
1 cm
flowers
ray disk
1 mm
chaff fruit
1 mm

Encelia farinosa
5 mm
ray corolla
2 cm

E. actoni
5 mm
ray corolla

E. frutescens
1 cm
flower head

from sagebrush scrub to subalpine forest; 2100–2900 m. **W&I, DMoj**; c NV. Var. *conjugans* S.F. Blake in s NV. Jun–Jul

E. utahensis A. Gray (p. 149) Per 10–50 cm, from thick, peeling taproot and branched caudex, 0–few-branched near mid-st, densely silvery-hairy. **LVS**: basal 6–8 cm, narrowly oblanceolate, gen 0 by fl; cauline ± reduced. **INFL**: heads 1–4(10), 10–15 mm diam; phyllaries ± equal, densely white-strigose, minutely glandular. **RAY FLS** 16–28; corollas 12–15 mm, ligules bluish, coiled. **FR** 4(6)-ribbed; pappus bristles 20–35. 2*n*=18. Limestone slopes; ± 1500 m. **e DMtns (Providence Mtns)**; to Colorado, n AZ.

May–Jul(Sep)

E. vagus Payson (p. 149) Per 2–5 cm, from (taproot and) branched, spreading caudex, simple, spreading-hairy, sometimes glandular. **LVS** gen 1–3 cm, oblanceolate to spoon-shaped, with 3 long-obovate lobes at tip; cauline gen 0 or much reduced. **INFL**: head 1, 8–16 mm diam; phyllaries ± equal, purple at least distally. **RAY FLS** 25–35; corollas 4–7 mm, ligules white to pink, sometimes with midstripe, coiled. **FR**: pappus bristles 16–20. 2*n*=18. Alpine talus; 3300–4400 m. c SNH, **W&I**; to se OR, Colorado. Jun–Aug

ERIOPHYLLUM WOOLLY SUNFLOWER

John S. Mooring & Dale E. Johnson

Ann to shrubs, ± woolly. **LVS** gen alternate, entire to nearly compound. **INFL**: heads 1–many, gen radiate; cluster often ± flat-topped; involucre obconic to hemispheric; phyllaries in 1 series, free or ± fused; receptacle flat to columnar, gen naked. **RAY FLS** gen ± 1 per phyllary; ligules entire to lobed, gen yellow (white). **DISK FLS** (3)10–300; corolla yellow; anther tips ovate, deltate or awl-shaped. **FR** 4-angled or flattened in outer fls, gen club-shaped in inner fls; pappus 0–15 ± jagged or fringed scales. 14 spp.: w N.Am. (Greek: woolly lf) [Mooring 1991 Madroño 38:213–226] Ann spp. by Dale E. Johnson.

1. Shrubs or subshrubs, 10–100 cm
  2. Shrubs 20–100 cm; heads 10–30+; peduncles gen < 1 cm; ligules (0) 2–5 mm
    . . . . ***E. confertiflorum* var. *confertiflorum***
  2' Subshrubs 10–20(40) cm; head 1, peduncles 3–10 cm; ligules 6–10 mm . . . . . . ***E. lanatum* var. *integrifolium***
1' Ann, 1–30 cm
  3. Heads discoid
    4. Phyllaries 3–4; lf entire or 2–3-lobed (lobes pointed), flat or weakly rolled under; rare . . . . ***E. mohavense***
    4' Phyllaries 6–8; lf gen 3-lobed (lobes rounded), strongly rolled under; common . . . . ***E. pringlei***
  3' Heads radiate
    5. Ligules white; pappus 0.5–2.5 mm, of alternating long and short scales . . . . ***E. lanosum***
    5' Ligules yellow or cream; pappus 0 or of ± equal scales
      6. Anther tips deltate; disk corolla lobes non-glandular; pappus 0 or < 0.2 mm, translucent
        . . . . ***E. ambiguum* var. *paleaceum***
      6' Anther tips awl-like; disk corolla lobes glandular; pappus (0) 0.4–0.8 mm, opaque . . . . ***E. wallacei***

E. ambiguum (A. Gray) A. Gray Ann 5–30 cm, decumbent to ascending. **LF** < 4 cm, oblong-oblanceolate, entire to shallowly lobed. **INFL**: head 1; peduncle 1–8 cm; involucre 3–6 mm, obconic to hemispheric; phyllaries 6–10, acuminate; receptacle conic. **RAY FLS** 6–10; ligules 2–10 mm. **DISK FLS** many; corollas 1.3–3 mm, tube and throat hairy; anther tips deltate, smooth. **FR** 2.2–3 mm, ± strigose; pappus 0 or < 0.5 mm. 2*n*=14. Woodland, desert; < 2800 m. s SNF, Teh, **SNE, D**; s NV. 2 vars. in CA.

var. *paleaceum* (Brandegee) Ferris (p. 155) **LF** entire or 3-lobed near tip. **INFL**: involucre 5–7 mm; phyllaries free; receptacle tip sometimes scaly. **DISK FLS**: corolla lobes with 1-celled hairs. **FR**: pappus 0 or scales entire, < 0.2 mm. Desert scrub or woodland; 100–2800 m. s SNF, Teh, **SNE, D**; s NV. ✿TRY. Apr–Jun

E. confertiflorum (DC.) A. Gray GOLDEN-YARROW Subshrub or shrub gen 2–7 dm. **LF** 1–5 cm, ± obovate, deeply 3–5-lobed to nearly 2-pinnately compound, rolled under, becoming ± glabrous above. **INFL**: heads 3–30+; peduncles 0–25 cm; involucres 3–7 mm, bell-shaped, tomentose; phyllaries 4–7, obtuse, keeled, strongly overlapping, ± free; receptacle ± convex. **RAY FLS** (0)4–6; ligules 2–5 mm. **DISK FLS** 10–75; corollas 2–4 mm, puberulent to glandular. **FR** 2–4 mm; pappus gen < 1 mm. Many dry habitats; < 3000 m. NCoR, SN, CW, SW, **w edge D**; Baja CA. Highly variable, intergrading complex. Hybridizes with *E. lanatum*. 2 vars. in CA.

var. *confertiflorum* (p. 155) Pl persistently tomentose. **INFL**: heads gen 10–30+; clusters dense; peduncles gen < 10 mm; involucres 3–5 mm. **RAY FLS** sometimes 0. **DISK FLS** gen 10–35; corollas 2–3 mm. **FR** 2–3 mm; pappus scales 5–14,

± equal. 2*n*=16,32,48,64. Habitats and range of sp. [var. *discoideum* E. Greene; var. *laxiflorum* A. Gray] ✿ DRN,SUN:5,**15–17**&IRR:7, 8,9,**14,18–24**;also STBL. Apr–Aug

E. lanatum (Pursh) James Forbes Gen ± subshrub 1–10 dm. **LF** 1–8 cm, linear to ovate, entire to ± 2-pinnately compound, gen becoming glabrous above. **INFL**: heads 1–5+; peduncles 3–30 cm; involucres 5–12 mm, bell-shaped to hemispheric; phyllaries 5–15; receptacle ± flat to ± conic. **RAY FLS** (0) gen 8–13; ligules 6–20 mm, oblong to elliptic. **DISK FLS** 20–300; corollas 2.5–5 mm, tube gen glandular. **FR** gen variable, 2–5 mm, glandular or hairy; pappus scales 0 or 6–12, 0–2 mm, translucent. Many (gen dry) habitats; < 4000 m. **CA** (exc SnJV, D); to B.C., MT, WY, NV. Polyploid pillar complex of intergrading races; key is to modal populations; some vars. hybridize with *E. confertiflorum*. 8 vars. in CA.

var. *integrifolium* (Hook.) F.J. Smiley OREGON SUNSHINE **LF** 1–4 cm, wedge-shaped to obovate, gen entire (to 5-lobed at tip); margins flat. **INFL**: head 1; peduncle 3–10 cm; involucres 6–8 mm; phyllaries gen 8 (5–10), strongly keeled. **RAY FLS** gen 8(5–10); ligules 6–10 mm. **DISK FLS**: corollas ± 4 mm. **FR** 3–4 mm; pappus < 2 mm. 2*n*=16,32,48,64. Cold, dry sites; 1400–3500 m. KR?, CaRH, n&c SNH, **GB**; to WA, WY, NV. [var. *monoense* (Rydb.) Jepson] Variable and complex. ✿TRY;DFCLT. Jul–Aug

E. lanosum (A. Gray) A. Gray (p. 155) Ann 1–15 cm, decumbent-ascending, often reddish, sparsely woolly. **LF** 5–20 mm, linear-oblanceolate, entire or lobed at tip. **INFL**: head 1; peduncles 1–5 cm; involucre 5–7 mm, ± cylindric; phyllaries 8–10, acuminate, free; receptacle conic. **RAY FLS** 8–10; ligules

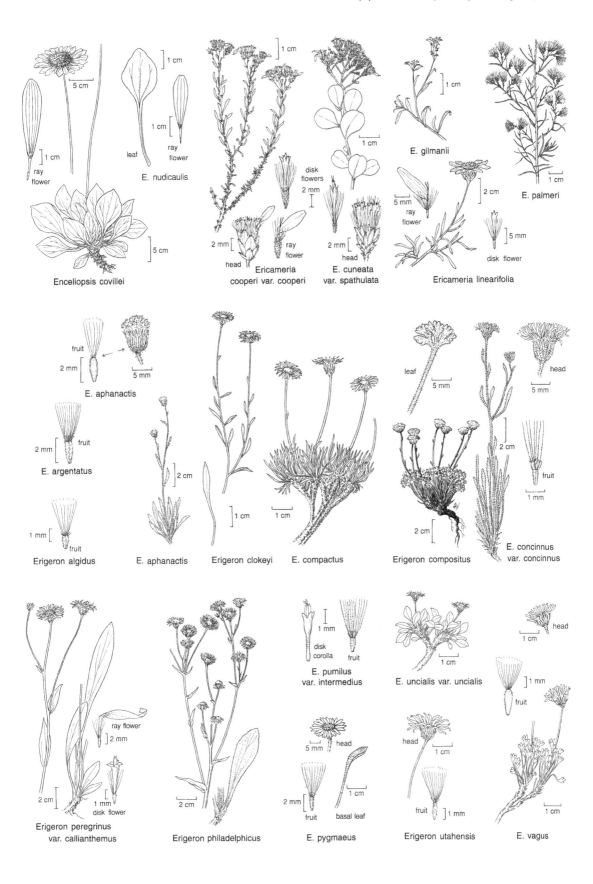

Enceliopsis covillei

ray flower

leaf

ray flower

E. nudicaulis

Ericameria cooperi var. cooperi

disk flowers

ray flower

head

E. cuneata var. spathulata

head

E. gilmanii

E. palmeri

ray flower

disk flower

Ericameria linearifolia

fruit

E. aphanactis

E. argentatus

fruit

Erigeron algidus

fruit

E. aphanactis

Erigeron clokeyi

E. compactus

leaf

head

Erigeron compositus

fruit

E. concinnus var. concinnus

Erigeron peregrinus var. callianthemus

ray flower

disk flower

Erigeron philadelphicus

disk corolla

fruit

E. pumilus var. intermedius

head

fruit

basal leaf

E. pygmaeus

E. uncialis var. uncialis

head

head

fruit

Erigeron utahensis

head

fruit

E. vagus

3–7 mm, oblong, white, sometimes red-veined. **DISK FLS** many; corollas 2–3 mm, glabrous; anther tips awl-like, glabrous. **FR** 2.5–4.5 mm, linear or narrowly club-like, glabrous to minutely strigose; pappus scales 0.5–2.5 mm, gen very unequal, longest awned. $2n=8$. Desert scrub; < 1400 m. **D**; to sw UT, AZ, nw Mex. ❀TRY. Feb–May

E. mohavense (I.M. Johnston) Jepson (p. 155) BARSTOW WOOLLY SUNFLOWER Ann 1–2.5 cm, tufted, spreading, loosely white-woolly. **LF** 3–10 mm, spoon- to wedge-shaped, entire or 2–3-lobed (lobes pointed). **INFL**: head 1, discoid, ± sessile; involucre 3–4 mm, ± cylindric; phyllaries 3–4, acute, free; receptacle ± pointed-columnar, flanges 3, protruding between frs, spine-tipped. **RAY FLS** 0. **DISK FLS** ± 3; corollas ± 2 mm, throat minutely puberulent; anther tips narrowly deltate. **FR** 2–2.5 mm, narrowly obconic, strigose; pappus ± 1.5 mm. RARE. Creosote-bush scrub; 500–800 m. **c DMoj** (**wc San Bernardino Co.**). Apr–May

E. pringlei A. Gray (p. 155) Ann 1–8 cm, ± tufted-spreading, white-woolly. **LF** 3–10 mm, wedge-shaped, gen 3-lobed, very woolly; margins rolled under. **INFL**: heads discoid, ± sessile in leafy clusters at branch tips; involucre 3–6 mm, hemispheric; phyllaries 6–8, acuminate, free; receptacle convex. **RAY FLS** 0. **DISK FLS** 10–25; corollas ± 2 mm, minutely glandular; anther tips deltate. **FR** 1.5–2 mm, strigose; pappus ± 1 mm. $2n=14+$. Chaparral, sagebrush or desert scrub or woodlands; 300–2200 m. s SNF, Teh, s SCoRO, SCoRI, TR, **SNE**, **D**; to s NV, AZ. ❀TRY. Apr–Jun

E. wallacei (A. Gray) A. Gray (p. 155) Ann 1–15 cm, often tufted, woolly. **LF** 7–20 mm, spoon-shaped to obovate, entire or 3-lobed. **INFL**: head 1; peduncles 1–3 cm; involucre 5–7 mm, bell-shaped; phyllaries 5–10, acute, free; receptacle hemispheric. **RAY FLS** 5–10; ligules 3–4 mm, sometimes cream-white. **DISK FLS** many; corollas 2–3 mm, throat minutely puberulent; anther tips awl-like, glabrous. **FR** ± 2 mm, narrowly club-shaped, glabrous or minutely strigose; pappus (0)0.4–0.8 mm. $2n=10+$. Chaparral, sagebrush or desert scrub or woodlands; < 2400 m. e SnFrB, SnGb, SnBr, PR, **SNE**, **D**; to sw UT, nw AZ, n Baja CA. [var. *rubellum* (A. Gray) A. Gray] ❀TRY. Mar–May

EUTHAMIA GRASS-LEAVED GOLDENROD

John C. Semple

Per from rhizome, ascending to erect, branched above. **LVS** alternate, sessile, linear-lanceolate, 3–5-veined, entire, resin-dotted; margins finely scabrous. **INFL** dense, sometimes flat-topped; heads radiate, subsessile; involucre ± ovoid; phyllaries in graded series, midrib gen ± swollen, translucent; receptable convex, naked, pitted. **RAY FLS**: ligules yellow. **DISK FLS**: corollas yellow, ± glabrous; style branches finely papillate, appendages narrowly triangular. **FR** fusiform; pappus bristles 25–45, long, in 1 whorl. ± 8 spp.: N.Am. (Greek: well crowded, from dense infl) [Sieren 1981 Rhodora 83: 551–579]

E. occidentalis Nutt. (p. 155) WESTERN GOLDENROD **STS** < 2 m, smooth, sometimes ± white. **LVS** < 10 cm, < 6 mm wide; lower deciduous; middle largest. **INFL** large, panicle-like, ± resinous; branches ascending; involucre 3–5 mm; phyllaries in 3–4 series. **RAY FLS** 15–25; ligules 1.5–2.5 mm. **DISK FLS** 6–15; corollas 3–4 mm. **FR** 1 mm, strigose. $2n=18$. Ditches, marshes, streambanks, meadows; < 2300 m. **CA** (exc D); to w Can, n-c US, NM, n Baja CA. [*Solidago o.* (Nutt.) Torrey & A. Gray] ❀IRR:1,**2,3**,4, 5,**6,7**,8–10,**14–24**;INV. Jul–Nov

FILAGO COTTONROSE, FILZKRAUT

James D. Morefield

Ann, grayish, cobwebby to tomentose. **STS** gen ± evenly leafy below, ± lfless between upper forks. **LVS** simple, alternate or seeming whorled, ± sessile, entire. **INFL**: heads disciform, ± sessile, gen in groups of 2–10(20), ± ovoid to conic until mature; bracts lf-like; phyllaries ± 0; receptacle gen < 2 × longer than wide, gen expanded at tip, chaffy; chaff scales gen 10–20, ± phyllary-like, each subtending a pistillate fl, gen evenly curved inward; outer scales each ± folded around a fl, gen falling with a fr, ± woolly, back gen rounded, tip gen narrowly obtuse to acute, ± scarious-winged; innermost chaff scales gen > outer, open, boat-shaped, persistent, ± glabrous, ± rigid throughout, gen spreading at maturity. **PISTILLATE FLS** in (3)4–8 series, all or outer subtended by chaff scales; corollas tubular. **DISK FLS** bisexual, not subtended by chaff scales; corolla lobes 4–5. **FR** ± obovoid, gen ± compressed side-to-side; outer fr enfolded by chaff scale, gen erect, straight, smooth, shiny, pappus 0; inner fr not enfolded by chaff scale, slightly < outer, rougher or papillate, dull, pappus gen of 16–30 bristles, ± deciduous, gen cohering in a ring. 24 spp.: Eur, n Afr, sw Asia, sw N.Am. (Latin: with threads, from woolly hairs) [Wagenitz 1976 Sida 6:221–223] Characters may be unreliable in dwarf pls. Subg. *Oglifa* sometimes treated as genus *Logfia* in Eur. 3 Eur. aliens occur near CA: *F. arvensis, minima, vulgaris*.

1. Sts gen lfless between lower forks, ± glabrous, purplish to black at fl time; heads gen in groups of
    4–10, restricted to forks and tips of branches, subtending lvs ± linear, gen 2 × heads or longer; fls
    inside innermost chaff scales 4–12, 0–2 pistillate . . . . *F. arizonica*
1' Sts ± evenly leafy between lower forks or branches, cobwebby, grayish to greenish at fl time; heads gen
    in groups of 1–4, not restricted to forks and tips of branches, subtending lvs elliptic to obovate, gen <
    1.5 × heads; fls inside innermost chaff scales 12–40, 9–33 pistillate
    2. Sts gen erect, gen not forked, central axis gen dominant; upper lvs gen oblanceolate, acute; body of
        outer chaff scales firm, thickened; lobes of disk fl gen 4, gen reddish purple; pappus bristles of
        disk fr 17–23, detaching in a ring . . . . *F. californica*
    2' Sts gen spreading, ± evenly forked (no dominant central axis); upper lvs gen elliptic to obovate, obtuse;
        body of outer chaff scales soft, membranous; lobes of disk fl gen 5, gen yellowish to brownish; pappus
        bristles of disk fr 11–15, detaching singly or in 2's . . . . *F. depressa*

F. arizonica A. Gray (p. 155) **STS** gen several from base, spreading, forked, < 20 cm, gen lfless between lower forks, purplish to black, ± glabrous; central st gen 0 or not dominant. **LVS** < 25 mm, linear to narrowly oblanceolate, acute, flexible, grayish to green, cobwebby; uppermost gen 2 × heads or longer. **INFL**: heads in loose, ± hemispheric groups of 4–10, restricted to forks and tips of branches, longest ± 4 mm, ± 3 mm wide, grayish, largest groups 8–10 mm diam; chaff scales in 5 vertical ranks, longest 2.2–2.7 mm, body gen ± hard. **PISTILLATE FLS** in 3–4 series; each fl subtended by a chaff scale. **DISK FLS** 4–10; corollas 1.2–1.7 mm, lobes 5, gen brownish to yellowish. **FR**: outer fr ascending, 0.9–1 mm, ± bent, compressed front-to-back; inner fr densely and minutely papillate, pappus bristles 17–23, 1.3–2 mm, falling in a ring. *n*= 14. Locally or seasonally moist, gen clay soils; 0–800 m. SCo, s ChI, PR, **w-most DSon**; s-c AZ, nw Mex. Mar–Apr

F. californica Nutt. (p. 155) **STS** 1–several from base, ± erect, gen not forked, < 55 cm, grayish to green, cobwebby; central axis gen dominant. **LVS** < 20 mm, gen oblanceolate, acute, flexible, grayish to green, cobwebby; uppermost gen < 1.5 × heads. **INFL**: heads solitary or in loose, ± hemispheric groups of 2–4, not restricted to forks and tips of branches, longest 3.5–4.5 mm, 2.5–3 mm wide, grayish, largest groups 5–9 mm diam; chaff scales in spiral ranks, longest 2.7–3.3 mm, body gen ± hard. **PISTILLATE FLS** in 4–8 series; inner 9–30 fls not subtended by chaff

scales. **DISK FLS** 4–7; corollas 1.9–2.8 mm, lobes 4, gen bright reddish purple. **FR**: outer fr erect, 0.9–1 mm, straight, compressed side-to-side; inner fr gen sparsely papillate, pappus bristles 17–23, 1.9–3 mm, falling in a ring. *n*=14. Mostly rocky places, crevices, or drainages; 0–1800 m. CA-FP (most common s), SNE, **D (esp DMtns)**; to sw UT, w TX, nw Mex. Dwarf pls like *F. depressa* exc adherent pappus bristles, 4-lobed disk corollas. Mar–May

F. depressa A. Gray (p. 155) **STS** gen several from base, ± spreading, forked, < 11 cm, grayish to whitish, tomentose; central axis not dominant. **LVS** < 11 mm, elliptic to obovate, obtuse, rarely acute, flexible, grayish to whitish; uppermost ± equaling heads. **INFL**: heads in loose, ± hemispheric groups of 2–5, not restricted to forks and tips of branches; longest 3–4 mm, ± 2–2.5 mm wide, grayish to whitish, largest groups 5–9 mm diam; chaff scales in spiral ranks, longest 2.1–3.1 mm, body (exc central vein) of outer scales soft, membranous. **PISTILLATE FLS** in 4–7 series; inner 4–20 fls not subtended by chaff scales. **DISK FLS** 3–5; corollas 1.3–2 mm, lobes 5, gen brownish to yellowish. **FR**: outer fr erect, 0.7–0.9 mm, straight, compressed side-to-side; inner fr smooth (rarely sparsely papillate), dull; pappus bristles 11–15, 1.3–2.4 mm, falling singly or in 2's. Sandy washes or open alluvium; 0–1500 m. s SNE, D (rare s SnJV margin, SCo); s NV, se AZ, nw Mex. See *F. californica.* Feb–May

GALINSOGA

Ann. **STS** gen erect. **LVS** simple, opposite, petioled; blade 3-veined. **INFL**: heads radiate or discoid, gen small, in leafy-bracted cymes; peduncles slender; involucre bell-shaped; phyllaries in 2 series, free; receptacle conic, chaffy, scales of 2 kinds, outer fused in groups of 2–3 together with a phyllary around a ray fl, inner narrower, subtending disk fls. **RAY FLS** (0)5–8, ligule short, white. **DISK FLS** 8–50; corolla yellow; style tips acute. **FR** obconic, round to ± angled; pappus of fringed scales or 0. ± 15 spp.: Am trop; some widespread weeds. (D. Mariano Martinez de Galinsoga, Spanish physician, 18th century) [Canne 1977 Rhodora 79:319–389]

G. parviflora Cav. var. *parviflora* (p. 155) Ann, glabrous or sparsely soft-hairy, sometimes also glandular. **ST** 10–60 cm, simple or much-branched. **LF**: petiole < or = 2.5 cm; blade 1–11 cm, ± ovate, acute, finely dentate to coarsely serrate. **INFL**: heads 2–6 mm diam; cymes round- to ± flat-topped; peduncles 1–40 mm; phyllaries and chaff persistent; outer phyllaries 2–4, 1.2–2.2 mm, 0.6–1.5 mm wide, margins scarious; inner phyllaries

gen 5, 2.5–3.5 mm, 1.3–2.6 mm wide; inner chaff scales deeply 3-lobed. **RAY FLS** gen 5; ligules 1–1.5 mm. **DISK FLS** 8–50; corollas 1.3–1.8 mm. **FR** 1.2–2.5 mm, glabrous or strigose; pappus of ray achenes of 5–8 unequal scales < or = 1 mm; pappus of disk achenes of 15–20 obtuse to acute scales < or = 2 mm. 2*n*=16. Gardens, fields; gen < 1000 m. SnFrB, SW, **SNE**; worldwide; native to S.Am.

GERAEA

Curtis Clark

Ann, per. **STS** erect; branches ascending. **LVS** basal and alternate, simple, sessile or petioled, entire or toothed, 3-veined from base. **INFL**: heads radiate or discoid, solitary or in few-headed panicles; peduncles ± elongated; involucre hemispheric; phyllaries in 2–3 series, free; receptacle chaffy, scales folded around frs and falling with them. **RAY FLS** sterile; style 0; ligules yellow. **DISK FLS** many; corollas yellow, tube slender, throat gradually expanded, lobes triangular; anther tips ovate, ± acute; style tips triangular. **FR** strongly compressed, narrowly wedge-shaped; edges ± white, long-ciliate; faces black, ± hairy; pappus of 2 narrow awns. 2 spp.: sw US, nw Mex. (Greek: old, from white-haired involucre)

G. canescens A. Gray (p. 157) DESERT-SUNFLOWER Ann, taprooted; herbage bristly or soft-hairy. **STS** 1–8 dm, simple to openly much-branched. **LVS** 1–10 cm, sessile above, wing-petioled below; blade lanceolate or ovate to elliptic or oblanceolate, green or ± canescent, tip acute, base tapering to wing, margin entire or dentate. **INFL**: heads radiate, solitary or few–

many in panicles; involucre 7–12 mm; phyllaries narrowly lanceolate, acute, green, ciliate. **RAY FLS** 10–21; ligules 1–2 cm. **DISK FLS**: corollas 4–5 mm. **FR** 6–7 mm; pappus awns 3–4 mm. 2*n*=36. Sandy desert soils; < 1300 m. **D**; to sw UT, w AZ, n Mex. Sometimes hybridizes with *Encelia farinosa.* ✤TRY. Feb–May, Oct–Nov

GLYPTOPLEURA

G. Ledyard Stebbins

1 sp.: w N.Am. (Greek: carved side, from sculptured fr)

G. marginata D. Eaton (p. 157, pl. 19) Ann, taprooted, forming small tufts; sap milky. **STS** many, semi-prostrate, 2–5 cm.

LVS 2–5 cm, lobed; lobes rounded, toothed, margins whitish, hard. **INFL**: heads ligulate, gen 1 (2–3 in cymes); involucre

10–12 mm, cylindric; phyllaries in 2 series, outer few, linear, inner 7–12, equal, narrow; receptacle naked. **FLS** 7–16; ligules cream to pale yellow, readily withering. **FR** 4 mm, oblong, often curved, obtusely 5-angled; ribs alternating with 5 rows of pits, abruptly short-beaked; pappus ± 8 mm, of many white bristles, outer falling separately, inner persistent. 2*n*=18. Local on sandy flats; 600–2100 m. s SNH, **SNE, DMoj**; to OR, UT, AZ. [*G. setulosa* A. Gray] ❀TRY. Apr–Jun

GNAPHALIUM CUDWEED, EVERLASTING

G. Ledyard Stebbins

Ann or per, gen ± woolly or tomentose. **LVS** alternate, sessile, entire. **INFL**: heads disciform, many, small, ± sessile in clusters; involucres ± cylindric to spheric, often bell-shaped when pressed; phyllaries graded in several series, transparent to opaque at tips or scarious ± throughout; receptacle flat, naked. **PISTILLATE FLS** many, in several series; corollas very slender, minutely lobed, cream to pale yellow or tip reddish. **DISK FLS** few; corollas ± cylindric to funnel-shaped, whitish to purplish; anther bases short-tailed; style branches wider at tip, truncate. **FR** < 1 mm, oblong; pappus of many fine bristles, bases sometimes fused. ± 120 spp.: worldwide. (Greek: lock of wool). Desert spp. except *G. palustre* now treated in *Pseudognaphalium*. For revised taxonomy see Anderberg 1991 Opera Bot 104: 1-195 Key to species adapted by David J. Keil.

1. Per (bien); basal tufts of lvs or sterile shoots present at fl . . . . ***G. canescens***
  2. Heads 5–6 mm; lvs of sterile shoots linear to oblanceolate . . . . ssp. *canescens*
  2' Heads 4–5 mm; lvs of sterile shoots narrowly spoon-shaped . . . . ssp. ***thermale***
1' Ann or bien; basal tufts of lvs or sterile shoots 0 at fl
  3. St 1–30 cm; phyllaries brown; tips short, whitish; fls 40–60 per head . . . . ***G. palustre***
  3' St 30–70 cm in well developed pls; phyllary upper 1/2 white or pale yellow, transparent or opaque; fls 50–120 per head
    4. Heads 3–4.5 mm; pistillate corollas 1.5–2 mm; pappus bristles falling in clusters . . . . ***G. luteo-album***
    4' Heads 3.8–5.5 mm; pistillate corollas 1.8–2.5 mm; pappus bristles falling singly . . . . ***G. stramineum***

G. canescens DC. Bien or short-lived per 20–110 cm, much-branched above, scented or not, gray-tomentose throughout; basal lf-tufts gen present. **LVS** 15–80 mm, linear to (ob)lanceolate, sometimes decurrent. **INFL** panicle-like; heads many; involucre 4.5–6 mm, ovoid; phyllaries white or pale straw-colored, outer ± tomentose (esp base). **FLS** 25–45; pistillate corollas 2.5–4 mm. **FR**: pappus bristles free. Many habitats; < 2500 m. **CA (exc DSon)**; to B.C., WY, n Mex. 4 sspp. in CA.

  ssp. ***canescens*** DC. Pl 25–70 cm, barely scented or not. **LVS** 20–55 mm, narrowly oblanceolate; cauline spreading, gen not decurrent. **INFL** open. **FLS** 30–45; pistillate corollas ± 3.5 mm. 2*n*= 14. Canyons, rocky slopes; < 2000 m. **DMtns**; to NM, n Mex.

  ssp. ***thermale*** (E. Nelson) Stebb. & Keil (p. 157) Pl 20–70 cm, ± scented. **LVS** 15–80 mm, linear to ± spoon-shaped (basal tufts); cauline ascending, upper decurrent. **INFL**: branches short, ascending. **FLS** 20–35; pistillate corollas 3–3.5 mm. Dry woods, roadsides; 1000–2500 m. KR, CaRH, SNH, SnBr, SnJt, MP, **W&I**; to B.C., WY, Colorado.

G. luteo-album L. (p. 157) Ann 15–60 cm, gen white-(yellow-) woolly, unscented. **STS** 1–several from base, branched below or above, densely leafy below. **LVS** 10–60 mm, linear to spoon-shaped, short-decurrent. **INFL**: heads many in spheric clusters; involucre 3–4.5 mm, widely cylindric to ovoid; phyllaries smooth, transparent, whitish to yellowish above, ± glabrous. **FLS** 40–100; pistillate corollas 1.5–2 mm. **FR**: pappus bristles falling together or in clusters. 2*n*=14. Fields, waste places; < 2100 m. NCoRO, c&s SN, GV, w CW, SW, **SNE, D**; widespread weed; native to Eurasia. All year

G. palustre Nutt. (p. 157) Ann 1–30 cm, unscented, ± tomentose throughout. **STS** gen branched at base, leafy. **LVS** 4–30 mm, oblong to spoon-shaped, clasping or short-decurrent. **INFL**: clusters small, terminal or axillary; involucre 3–3.5 mm, ± cylindric to ± ovoid; phyllaries smooth and brown with whitish tips above, tomentose below. **FLS** 40–60; pistillate corollas 1.5–2 mm. **FR**: pappus bristles free. 2*n*=14. Common. Moist places; < 2700 m. **CA (exc MP)**; to w Can, MT, NM. May–Oct

G. stramineum Kunth Ann or bien 8–70 cm, unscented, ± tomentose throughout. **STS** 1–several from caudex. **LVS** 10–70 mm, lanceolate to oblong or narrowly spoon-shaped, decurrent. **INFL**: heads in dense terminal clusters; involucre 3.8–5.5 mm, ovoid; phyllaries transparent to opaque, white to straw-colored in upper part, base glabrous or loosely tomentose. **FLS** 65–110; pistillate corollas 1.8–2.5 mm. **FR**: pappus bristles free. 2*n*=28. Moist, disturbed places; < 2000 m. **CA**; to B.C., TX, Mex. [*G. chilense* Sprengel] Jun–Oct

GRINDELIA GUMPLANT

Meredith A. Lane

Bien to subshrub from taproot or woody caudex, glabrous to tomentose or glandular-sticky. **LVS** entire to pinnately lobed, gen clasping, gland-dotted. **INFL**: heads gen radiate, 1–many; involucres obconic to hemispheric, gen gummy; phyllaries in 4–10 series, bases gen tough, tips green; receptacle flat to convex, naked, ± pitted. **RAY FLS** 0–many; ligules yellow. **DISK FLS**: corollas yellow; style appendages linear to lanceolate, gen = or > stigmatic portion. **FR** cylindric or swollen-obconic, shiny white to ± brown, glabrous, smooth to ridged; pappus of 1–6 awns ± < disk corollas, gen < 0.2 mm wide, gen U-shaped in ×-section, gen entire, deciduous. ± 80 spp.: c&w N.Am, S.Am. (D.H. Grindel, 1776–1836, Latvian botanist) Hybrids common.

1. Pl white to yellowish; outer phyllaries long acuminate, gen coiled 360° — lf gray-green
  . . . . ***G. squarrosa*** var. ***serrulata***
1' Pl tan to reddish or white to coppery; outer phyllaries acute to acuminate, erect to reflexed (not tightly coiled)

2. Heads in panicles; involucres obconic; phyllaries ± erect; lf dark green; e DMoj (Ash Meadows)
. . . . *G. fraxino-pratensis*
2' Heads in cymes; involucres bell-shaped or hemispheric; phyllaries erect to ± recurved; lf light yellowish
green; w DSon (San Diego Co.) . . . . *G. hirsutula* var. *hallii*

G. fraxino-pratensis Rev. & Beatley (p. 157) ASH MEADOWS GUMPLANT Per 5–12 dm, erect, branched above, tan to reddish, glabrous, resinous. **LVS**: cauline 1–7 cm, gen oblanceolate to oblong, entire to serrate toward tip, acute, gen dark green, densely gland-dotted, bases narrowly clasping. **INFL**: heads 1–4; involucres 5–10 mm diam, ± obconic; phyllaries 4–5 series, ± erect. **RAY FLS** ± 13; corollas 4–4.5(7) mm. **DISK FLS** ± 15; throat ± narrow. **FR** 2.5–4 mm, golden-brown, oblong; top gen truncate; pappus awns 2, ± 3/4 disk corolla. $2n=24$. **THREATENED** US. Wet clay of meadows, woodland borders; ± 700 m. **e DMoj (Ash Meadows, Inyo Co.)**; NV. Threatened by water diversion.

G. hirsutula Hook. & Arn. Per 2–15 dm, erect, few-branched above, green to red-purple or -brown, glabrous to tomentose. **LF** 1–10 cm, oblong to lanceolate, entire to lobed (less so upward) yellow-, red-, or gray-green. **INFL**: heads often subtended by phyllary-like bracts; involucres 7–32 mm diam, hemispheric to bell-shaped; phyllaries 4–5 series, gen lanceolate-acute, outer erect to reflexed. **RAY FLS** 10–60+; ligules 8–20 mm. **DISK FLS** many; throat narrow. **FR** 2.5–5.5 mm, golden- to red-brown, smooth to ridged; pappus awns flat. $2n=12,24$. Sandy, clay, or serpentine slopes or roadsides; < 1700 m. NCoR, n&c SNF, ScV, CW, WTR, PR, **DSon**; to B.C.? 4 vars. in CA.

var. ***hallii*** (Steyerm.) M.A. Lane (p. 157) SAN DIEGO GUMPLANT Pl 2–6 dm, gen white to coppery. **LF** light yellowish green. **INFL**: heads gen not subtended by bracts; involucre gen 8–12 mm diam; phyllaries erect to ± recurved. **RAY FLS** gen 12–20; ligules 8–9 mm. **FR** 4.5–5 mm. $2n=6$. RARE. Meadows, dry slopes, open pine/oak woodlands; 800–1700 m. PR, **w DSon (San Diego Co.)**. [*G. hallii* Steyerm.] In cult. Jul–Oct

G. squarrosa (Pursh) Dunal var. ***serrulata*** (Rydb.) Steyerm. (p. 157) Bien 1–6 dm, decumbent to erect, branched, white to yellowish, glabrous. **LF** 1.5–7 cm, oblong to ovate, dentate (teeth with rounded, swollen tips), gray-green, glabrous. **INFL**: heads sometimes subtended by bracts; involucres 12–20 mm diam, bell-shaped; phyllaries 5–6 series, outer coiled 360+. **RAY FLS** 0 or 24–36; corollas 8–10 mm. **DISK FLS** gen > 120; corolla throat abruptly wider. **FR** 2.3–3 mm, light brown to yellowish, smooth to 2–3-ribbed; top truncate; pappus awns 2–3(6). $2n=12,24$. Disturbed roadsides, streamsides; < 1000 m. SN (e slope), c ScV, SCo, **SNE, DMoj**; native from WY to NM. TOXIC: concentrates selenium. Jul–Sep

GUTIERREZIA SNAKEWEED, MATCHWEED

Meredith A. Lane

Ann to subshrubs, appearing glabrous. **STS** single and branching above or branching from base, ascending, < 1.5 m, ± striate, gen fibrous, gummy, minutely scabrous, yellow to tan or gray. **LVS** alternate, sometimes in axillary clusters, entire, gland-dotted, sometimes gummy, glabrous or minutely scabrous, dark gray-green. **INFL**: heads radiate, solitary or in short-peduncled clusters; involucres narrowly to widely obconic; phyllaries in 3–4 series, whitish yellow, tips green; receptacle naked, minutely hairy. **RAY FLS** 1–13; corollas yellow. **DISK FLS** 1–13 (in CA spp.); corollas yellow, club- or narrowly funnel-shaped, lobes short, recurved; style appendages lanceolate. **FR** narrowly obconic, light tan, hairy; hairs appressed, white; pappus of 1–2 series of finely toothed, white or yellowish scales gen 1/2 fr length (in CA spp.) or much reduced. 25 spp.: 10 w N.Am, 15 S.Am. (Gutierrez, surname of a noble Spanish family) [Lane 1985 Syst Bot 10:7–28] TOXIC to livestock, fresh or dried in hay.

1. Involucres cylindric; phyllaries 4–6 . . . . *G. microcephala*
1' Involucres obconic; phyllaries 8–21 . . . . *G. sarothrae*

G. microcephala (DC.) A. Gray (p. 157) STICKY SNAKEWEED Subshrub 2–6 dm, much-branched, often nearly spheric. **STS** brown below, yellow or green above. **LVS** linear to thread-like. **INFL**: heads 1–3-fld, in groups of 5–6, sessile; involucres gen < 3.2 mm, < 1.2 mm diam, cylindric; phyllaries 4–6 in 2 series. **RAY FLS** 1–2; corollas 2.1–3.5 mm. **DISK FLS** 1–2, functionally staminate; corollas 2.2–3.3 mm. **FR** 1–1.5 mm. $2n=8,16,24,32$. Grasslands, sand dunes; 1800–2500 m. SCo, SnBr, PR, **SNE, D**; to Colorado, c Mex. ❀DRN,DRY,SUN:**8–11,14–24**. Jul–Oct

G. sarothrae (Pursh) Britton & Rusby (p. 157) BROOM SNAKEWEED, MATCHWEED Subshrub 1–6 dm. **STS** sprawling or upright, brown below, green or tan above. **LVS** lance-linear if single, thread-like if clustered. **INFL**: heads 6–14-fld, in clusters of 5 or fewer, on peduncles < 1.5 mm or sometimes sessile; involucre gen < or = 4.5 mm, < or = 2.5 mm diam, narrowly obconic; phyllaries 8–21 in 2–3 series. **RAY FLS** 2–8; corollas 3–5.4 mm. **DISK FLS** 2–9, fertile; corollas 2.3–3.5 mm. **FR** 0.9–1.6 mm. $2n=8,16,32$. Grasslands, deserts, montane areas; 50–2900 m. SCo, WTR, PR, **D**; to WA, s-c Can, c US, n Mex. ❀DRN, DRY,SUN:**7–11,14–16**,17,**18–24**. May–Oct

HAZARDIA

Gregory K. Brown & W. Dennis Clark

Per or shrub, gen resinous. **STS** < 2.5 m, leafy. **LVS** gen sessile or short-petioled. **INFL**: heads radiate or discoid; involucres obconic to bell-shaped; phyllaries in several overlapping series, linear to oblanceolate, gen recurved. **RAY FLS** 0–25; corollas < 9 mm, yellow. **DISK FLS** 4–60; corollas 4–10 mm, gradually flared from middle. **FR** 1–10 mm, 4–5-angled; pappus of 20–60 bristles, 2.5–12 mm, white to reddish brown. 13 spp.: w N.Am. (Barclay Hazard, 19th Century CA botanist) [Clark 1979 Madroño 26:105–127]

H. brickellioides (S.F. Blake) W. Clark (p. 157) Shrub 2–8 dm, yellow-glandular, hairy to scabrous. **LVS** 10–35 mm, elliptic to obovate, leathery, teeth (0)2–8, teeth and tip spiny. **INFL**: heads radiate; involucre 4–5 mm wide, cylindric to obconic; phyllaries 15–25, 3–7 mm, lanceolate, acute, recurved (or inner erect), bristly-glandular. **RAY FLS** 5–8; tube 4–5 mm; ligules 2–4 mm.

DISK FLS 8–12; corollas 6–8 mm. **FR** 2–3 mm, hairy; veins 5, white; pappus 5–7 mm, white to brownish. 2*n*=12. Limestone outcrops, cliffs; 700–2100 m. **DMoj**. [*Haplopappus b.* S.F. Blake] Locally common. Apr–Sep

HECASTOCLEIS

1 sp. (Greek: each enclosed, from 1-fld heads)

H. shockleyi A. Gray (p. 157) Shrub. **STS** 4–7 dm, stiff, much-branched, glandular-puberulent or becoming glabrous exc for tufts of soft hair in axils of persistent lf bases. **LVS** simple, alternate, some also clustered in axils of older sts; primary lvs 1–3 cm, linear to linear-lanceolate, sessile, spine-tipped, sparsely spiny-dentate, ± persistent as spines when dry; clustered axillary lvs narrower, obtuse to acute, gen not toothed. **INFL**: heads 1-fld, sessile, in dense, head-like clusters surrounded by involucre of persistent, ovate, spiny-toothed, net-veined bracts 1–2 cm; true involucre narrowly cylindric; phyllaries ± 15 in several unequal series, 4–10 mm, linear, acuminate, loosely soft-hairy; receptacle naked. **FL** 1 per head; corolla pink to reddish purple in bud, greenish white in fl, lobes linear, equal; filaments inserted near base of corolla, anthers purple, exserted, bases with stiff, bristle-like tails, tips short-triangular; style tips very shallowly lobed. **FR** cylindric, glabrous; pappus a crown of fringed scales < or = 1 mm. 2*n*=16. Dry, rocky slopes; 1200–2200 m. **SNE, DMtns**; w NV. May–Jun

HELIANTHUS SUNFLOWER

Ann or per. **STS** gen erect. **LVS** opposite or alternate, gen reduced upward, often 3-veined from near base, gen rough-hairy. **INFL**: heads radiate, solitary or in cymes; involucre bell-shaped to hemispheric; phyllaries in 1–3 gen ± equal series, free; receptacle flat to rounded; chaff scales 0–3-lobed. **RAY FLS** 10–many, sterile; ligules yellow. **DISK FLS** many; corollas yellow to red or purple, tube short, throat base often swollen, lobes triangular; style appendages triangular. **FR** oblanceolate to obovate, ± compressed; sides rounded; pappus gen of 2 deciduous, lanceolate to ovate scales (sometimes also 1–several shorter scales). 67 spp.: Am. (Greek: sun fl) [Heiser 1969 Mem Torrey Bot Club 22(3):1–218] Key to species adapted by Bruce G. Baldwin.

1. Lvs ± sessile; per
  2. Disk corolla lobes red to purple; pls < 15 dm, from horizontal creeping roots or taprooted . . . . ***H. ciliaris***
  2' Disk corolla lobes yellow; pls 5–40 dm, from clustered tuber-like roots . . . . ***H. nuttallii*** ssp. ***nuttallii***
1' Lvs ± long-petioled; ann or per
  3. Phyllaries gen > 4 mm wide, abruptly acuminate; lf bases cordate . . . . ***H. annuus***
  3' Phyllaries gen < 4 mm wide, acute or gradually acuminate; lf base gen truncate to wedge-shaped . . . ***H. niveus***
    4. Central chaff scales stiff-white-hairy; lvs densely stiff-gray-hairy . . . . ssp. *canescens*
    4' Central chaff scales glabrous or fine-appressed-hairy; lvs densely soft-white-hairy . . . . ssp. *tephrodes*

H. annuus L. Ann < 3 m. **STS** ± rough-hairy. **LVS** long-petioled; blade 10–40 cm, widely lanceolate to widely ovate, base gen ± cordate, tip obtuse to acute, margin serrate. **INFL**: heads gen few–many; peduncles 2–20 cm; involucre 1.5–3(20) cm diam; phyllaries 15–25 mm, lanceolate to widely ovate, abruptly acuminate, glabrous to rough-hairy, gen ciliate; chaff scales deeply 3-lobed, middle lobe long-acuminate. **RAY FLS** 15–many; ligule gen > 2.5 cm. **DISK FLS**: corollas 5–8 mm, lobes red to purple or yellow. **FR** 3–15 mm; pappus scales 2–3.5 mm. 2*n*=34. Disturbed areas, shrubland, many other habitats; < 1900 m. **CA**; to e N.Am. Highly variable; hybridizes with several other spp. [ssp. *jaegeri* (Heiser) Heiser; ssp. *lenticularis* (Douglas) Cockerell; var. *macrocarpus* (DC.) Cockerell] Head 1, very large in cult forms. ❀SUN: 2–5,**6**,**17**&IRR:**7–16**,**18–24**. Jul–Oct

H. ciliaris DC. BLUEWEED Per 4–7 dm, from rhizome-like, horizontal roots. **STS** ± glabrous, glaucous. **LVS** gen opposite, sessile, 3–7.5 cm, oblong or lanceolate, acute, entire to shallowly lobed, wavy, glabrous, sparsely short-rough-hairy, glaucous, sometimes ciliate. **INFL**: heads few; peduncles 3–13 cm; involucre 12–25 mm diam; phyllaries gen unequal, 3–8 mm, < disk, oblong to ovate, tip obtuse or abruptly pointed to acute, ciliate, (sub)glabrous; chaff scales 8–10 mm, entire or 3-toothed, tip hairy. **RAY FLS** (0)10–18; ligules ± 1 cm. **DISK FLS**: corollas 4–5 mm, lobes red. **FR** ± 3 mm; pappus scales ± 2 mm. Irrigated fields, roadsides; gen < 300 m. 2*n*=68,102. ScV, SnFrB, SCo, **w D**; native to sc US, n Mex. **NOXIOUS WEED**. Jun–Nov

H. niveus (Benth.) Brandegee Ann or per < 15 dm, from taproot. **LVS** gen alternate, ± long-petioled; blade lanceolate to ovate, wedge-shaped to ± cordate at base, obtuse to acute, densely canescent to long-silky. **INFL**: heads 1–few; involucre 8–28 mm diam; phyllaries 8–12 mm, ± lanceolate, acute, canescent to soft-hairy, ± = disk; chaff scales 10–11 mm, entire to deeply 3-lobed.

RAY FLS 13–21; ligules 12–25 mm. **DISK FLS**: corollas 5–7 mm, lobes red to dark purple in CA. **FR** 3–8 mm; pappus scales 2–3 mm (gen also several shorter scales). 2*n*=34. Open, sandy places; gen < 300 m. SCo, **DSon**; to TX, n Mex.

ssp. ***canescens*** (A. Gray) Heiser Ann or per. **STS** strigose and stiff-spreading-hairy. **LVS**: blades 2–12 cm, lanceolate to ovate, obtuse, densely stiff-hairy. **INFL**: phyllaries gen slightly > disk; central chaff scales gen stiff-white-hairy. **FR** 3–4 mm, short-hairy. Habitats and range of sp. [*H. petiolaris* Nutt. var. *c.* A. Gray] ❀DRN,SUN:**15**,**16**,17,&IRR:**8–14**,**19–24**. Mar–Jun

ssp. ***tephrodes*** (A. Gray) Heiser (p. 157) ALGODONES DUNES SUNFLOWER Gen per, ± shrubby. **STS** soft-white-appressed-hairy. **LVS**: blades gen 3–7 cm, triangular-ovate, densely soft-hairy. **INFL**: phyllaries not > disk; central chaff scales glabrous or fine-appressed-hairy. **FR** 4–8 mm, densely long-hairy. 2*n*=34. **ENDANGERED** CA. Sand dunes; gen < 100 m. **s DSon (Imperial Co.)**; sw AZ, n Mex. [*H. tephrodes* A. Gray] Mar–May, Oct–Jan

H. nuttallii Torrey & A. Gray Per 5–40 dm, from clustered, tuber-like roots; rhizome short. **STS** glabrous or hairy. **LVS** alternate or opposite, subsessile; blade 10–20 cm, narrowly lanceolate to ovate, acute to acuminate, entire or serrate, glabrous to hairy. **INFL**: heads few–many; peduncles 1–18 cm; involucre 1–2 cm diam; phyllaries ± erect, 8–16 mm, gen < 3 mm wide, ± linear, = or slightly > disk, glabrous or tomentose; chaff scales 8–12 mm, entire or 3-toothed, acute, short-rough-hairy. **RAY FLS** 12–20; ligules 15–25 mm. **DISK FLS**: corollas 5–6 mm, lobes yellow. **FR** 3–4 mm; pappus scales 3–4 mm (sometimes also with shorter scales). Damp meadows, marshes; < 500 (c-w SW) or 1200–2500 m. c-w CW, SnGb, SnBr, **GB**; to B.C., e Can, NM. 2 sspp. in CA.

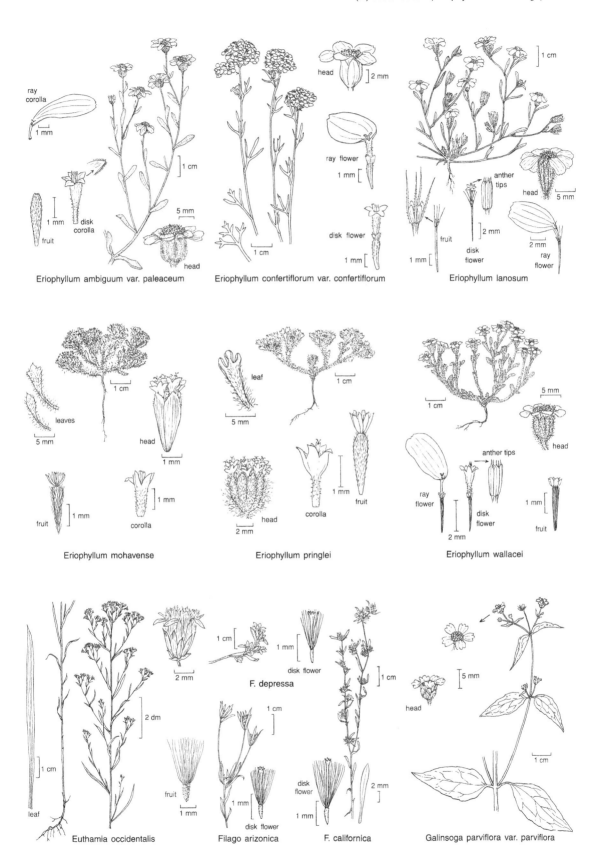

Eriophyllum ambiguum var. paleaceum

Eriophyllum confertiflorum var. confertiflorum

Eriophyllum lanosum

Eriophyllum mohavense

Eriophyllum pringlei

Eriophyllum wallacei

Euthamia occidentalis

Filago arizonica

F. californica

Galinsoga parviflora var. parviflora

ssp. *nuttallii* (p. 157) **STS** glabrous to scabrous. **LVS**: blade scabrous above, stiffly hairy below. **INFL**: peduncles glabrous; phyllaries glabrous or strigose. 2*n*=34. Damp meadows; 1200– 2500 m. SnGb, SnBr, **GB**; to B.C., e Can, NM. ⚘IRR,SUN:1–5, **6**,7–10,**14–24**. Jul–Sep

HELIOMERIS GOLDEN-EYE

Ann, per. **STS** 1–many from base, slender, spreading or erect. **LVS** simple, opposite or upper alternate, linear to narrowly elliptic, entire. **INFL**: heads radiate, solitary or in few-headed cymes; peduncles slender, bracts 0–few, linear; involucre hemispheric, appearing disk-shaped when pressed; phyllaries in 2–3 ± equal series, linear; receptacle conic, chaffy, scales linear-lanceolate. **RAY FLS** sterile; corollas yellow-orange; ligules showy, oblong, entire or nearly so. **DISK FLS** many; corollas yellow-orange; style tips triangular. **FR** oblanceolate, ± 4-angled, flat, glabrous; pappus 0. 6 spp.: N.Am. (Greek: sun part, from showy heads) [Yates & Heiser 1979 Proc Indiana Acad Sci 88:364–372]

H. multiflora Nutt. var. *nevadensis* (Nelson) Yates (p. 161) Per from branched, woody rootstock. **STS** 3–9 dm, slender, erect or spreading, glabrous or finely strigose or puberulent. **LVS** 2–6 cm, linear to oblong or narrowly elliptic, ± stiffly hairy near base; tip obtuse or acute; surfaces strigose. **INFL**: peduncles 5–15 cm; involucre 7–13 mm diam; phyllaries 4–7 mm, linear, strigose. **RAY FLS** 8–15; ligules 15–20 mm, oblong to ovate. **DISK FLS**: corollas 3 mm; anthers yellow. **FR** 2 mm. 2*n*=16. Dry, rocky slopes, upland valleys; 1200–2400 m. **SNE, DMtns**; to UT, AZ. [*Viguiera m.* (Nutt.) S.F. Blake var. *n.* (Nelson) S.F. Blake] Other vars. widespread in mtns of w N.Am. ⚘TRY. May–Sep

HEMIZONIA TARPLANT, TARWEED

Ann to shrub, gen glandular, aromatic. **STS** gen branched above middle or throughout. **LVS** gen cauline (some also basal and cauline, gen alternate, gen linear to (ob)lanceolate, entire to pinnately lobed, gen not spine-tipped; lower gen toothed to lobed; upper gen entire. **INFL**: heads radiate, gen 1–many in open cymes; involucre gen hemispheric; phyllaries gen linear to lanceolate, half-enclosing ray frs; chaff scales gen in 1 ring between ray and disk fls (scattered). **RAY FLS** 3–many; ligules gen 3-lobed, white to yellow. **DISK FLS** 3–many, staminate or fruiting; corollas white to yellow, becoming red; anther tips ovate; style branches long, tips bristly. **FR**: ray achenes ± 3-angled; pappus 0; disk achenes cylindric or obconic, pappus 0 or scales gen linear to lanceolate. ± 25 spp.: CA, OR, w AZ, n Baja CA. (Greek: half girdle, from sheathing phyllaries) [Tanowitz 1982 Syst Bot 7:314–339; Venkatesh 1958 Amer J Bot 45:77–84] *See revised taxonomy of Baldwin 1999 Novon 9: 462–471. Desert taxa below now part of segregate genus *Deinandra*. *Centromadia fitchii* (A. Gray) E. Greene (*Hemizonia fitchii* A. Gray in The Jepson Manual) has been reported as a weed along upper Mojave River. Key to species adapted by Bruce G. Baldwin.

1. Ray fls 5–10 per head; disk fls 18–25 per head; disk pappus gen 0 . . . . *H. arida*
1' Ray fls 5 per head; disk fls 6 per head; disk pappus scales 5–12
  2. Basal lvs clearly toothed to lobed; pappus scales ± linear, entire or fringed, ± uniform; dry sites;
    fl spring to summer . . . . *H. kelloggii*
  2' Basal lvs subentire or entire; pappus scales lanceolate to rectangular, deeply cut, often highly irregular;
    moist sites; fl summer to fall . . . . *H. mohavensis*

H. arida Keck (p. 161) RED ROCK TARPLANT Ann 2–10 dm. **STS** erect; branches ascending, ± bristly, ± sparsely glandular. **LVS**: lower 4–8 cm, linear-oblanceolate, dentate, glabrous; upper widely lanceolate, entire, ± bristly. **INFL** ± flat-topped; heads short-peduncled; involucre 4.5–6.5 mm; phyllaries soft-bristly, densely glandular. **RAY FLS** 5–10; ligules 5–9 mm, pale yellow. **DISK FLS** 18–25, gen staminate; corollas and anthers yellow. **FR** ± 2.5 mm; beaked; disk pappus ± 0. 2*n*=24. **RARE** CA. Clay soil of washes; 300–950 m. **w DMoj (Red Rock Canyon, e Kern Co.)**. Threatened by vehicles. **Deinandra arida* (Keck) B.G. Baldwin May–Nov

H. kelloggii E. Greene (p. 161) Ann 1–10 dm. **STS** soft-hairy to bristly below, bristly and glandular-puberulent above. **LVS**: bristly; lower 3–9 cm, oblong-oblanceolate, dentate to lobed; upper linear. **INFL** open; heads long-peduncled; involucre 4.5– 5 mm; phyllaries soft-hairy to densely glandular. **RAY FLS** 5; ligule 4–8 mm, pale yellow. **DISK FLS** 6, gen staminate; corollas and anthers yellow. **FR** 2.5–3 mm, beaked; disk pappus scales 6–12. 2*n*=18. Common. Open areas; < 700 m. s SNF, SnJV, e SnFrB, SW, alien in s NCoRI, **w edge DMoj**; n Baja CA. **Deinandra kelloggii* (E. Greene) E. Greene Apr–Jul

H. mohavensis Keck MOJAVE TARPLANT Ann 1.5–3 dm. **STS** erect, soft-hairy and sticky-glandular. **LVS**: lower 3–5 cm, oblanceolate, (sub)entire, bristly; upper oblong-oblanceolate, glandular-puberulent. **INFL**: heads sessile, clustered; involucre 4.5–6 mm; phyllaries soft-hairy to densely glandular. **RAY FLS** 5; ligule ± 5 mm, yellow. **DISK FLS** 6, staminate; corollas and anthers yellow. **FR** 2.5–3 mm, beaked; disk pappus scales 5–9, ± fused. 2*n*=22. **ENDANGERED** CA. 900–1300 m. SnJt, s PR, **sw edge DMoj**. **Deinandra mohavensis* (Keck) B.G. Baldwin Jul–Sep

HETEROTHECA GOLDENASTER, TELEGRAPH WEED

John C. Semple

Ann to per, taprooted, branched above, strigose-bristly; hairs minutely knobby. **LVS** gen ± cauline, alternate; lower oblanceolate to ovate, base or petiole ± spreading-hairy; upper reduced, glandular. **INFL** ± flat-topped; heads gen radiate; involucre ± bell-shaped; phyllaries in 3–5 graded series; receptacle naked, pitted. **RAY FLS** (0)10–30(40); ligules yellow. **DISK FLS** many; corollas yellow; style branches finely papillate, appendage narrowly triangular. **FR** obconic; ray fr ± 3-angled, pappus 0 or of bristles; disk fr compressed, outer pappus of narrow scales < 1 mm, inner of

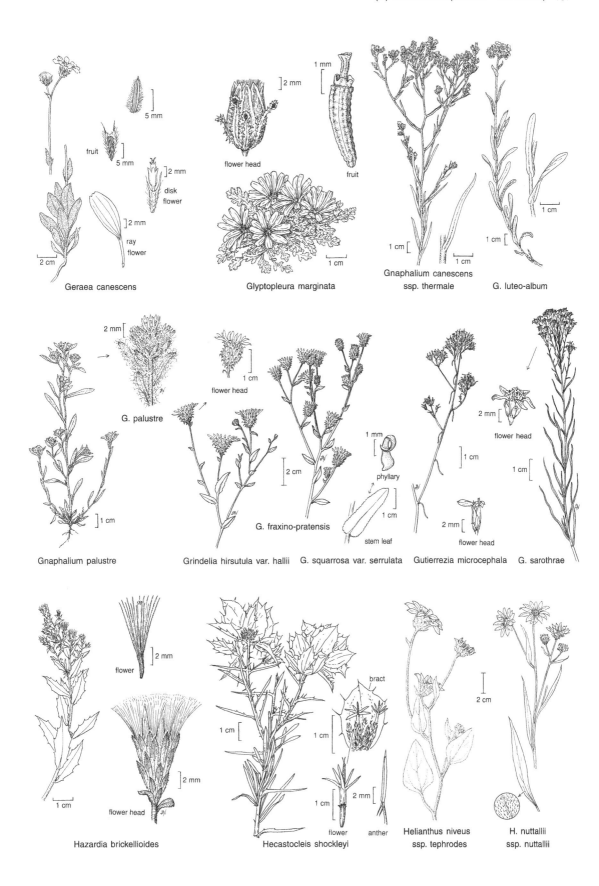

Geraea canescens

flower head

fruit

disk flower

ray flower

fruit

Glyptopleura marginata

Gnaphalium canescens ssp. thermale

G. luteo-album

G. palustre

Gnaphalium palustre

flower head

Grindelia hirsutula var. hallii

phyllary

stem leaf

G. fraxino-pratensis

G. squarrosa var. serrulata

flower head

flower head

Gutierrezia microcephala

G. sarothrae

flower

flower head

Hazardia brickellioides

bract

flower   anther

Hecastocleis shockleyi

Helianthus niveus ssp. tephrodes

H. nuttallii ssp. nuttallii

30–45 bristles 3–7 mm. ± 30 spp.: esp w N.Am. (Greek: different cases, from ray and disk frs) [Semple 1990 Brittonia 42:221–228] *H. sessiliflora* var. *fastigiata* possibly introduced in desert areas.

1. Per; ray frs hairy; ray pappus present (sect. *Phyllotheca)* — Little San Bernardino Mtns . . *H. villosa* var. *scabra*
1' Ann to short-lived per; ray frs ± glabrous; ray pappus 0 (sect. *Heterotheca)* . . . .
  2. Upper lvs not clasping;widespread, possibly introduced in D . . . . *H. grandiflora*
  2' Upper leaves clasping; e DSon, uncommon **. . . . .** *H. psammophila*

H. grandiflora Nutt.(p. 161) TELEGRAPH WEED Ann to short-lived per < 20 dm, erect, ± densely bristly, glandular. **LVS**: lower lvs petioled, basal lobes ear-like, clasping; middle lvs densely appressed-hairy, lanceolate; upper lvs sessile, ascending, less hairy and more glandular upward. **INFL** ± panicle-like, densely glandular; involucre 6–9 mm; phyllaries in 4–6 series. **RAY FLS** 25–40; ligules 5–8 mm. **DISK FLS** 30–75; corollas 4–6 mm. **FR** 2–5 mm: rays frs ± glabrous, pappus 0; disk frs strigose, outer pappus 0.2–0.7 mm, inner 3–5 mm. 2*n*=18. Disturbed areas, dry streams, sand dunes; gen < 300 mm. s NW, SNF, s Teh, GV, CW, **D (uncommon)**; nw Mex; introduced into AZ, UT. ❀ SUN:7,8,**9**,10,**14–17**,18,**19–24**;INV. All year

H. psammophila B. Wagenkn. (p. 161) CAMPHOR WEED Ann 3–15 dm, ascending to erect, ± bristly-scabrous, glandular (densely so above); large pls much-branched. **LVS**: lower wing-petioled, basal ear-like lobes clasping; middle ± sessile, ovate, clasping; upper spreading. **INFL**: involucre 6–12 mm; phyllaries in 4–6 series. **RAY FLS** 20–30; ligules 4–10 mm. **DISK FLS** 40–75; corollas 4–6 mm. **FR** 2–4 mm; ray frs (sub)glabrous, pappus 0; disk frs densely strigose, outer pappus 0.2–0.6 mm,

inner 6–9 mm. 2*n*=18. Uncommon. Disturbed, sandy soils; < 300 m. SCo, **e DSon**; to TX, nw Mex. [*H. subaxillaris* (Lam.) Britton & Rusby misapplied] Aug–Oct

H. villosa (Pursh) Shinn. Per 1–13 dm, decumbent to erect, ± bristly-strigose, sometimes densely glandular above. **LVS** ± (ob)lanceolate, gen flat; hairs gen < 1 mm; lower tapered to base; upper tapered to subclasping. **INFL**: heads 1–many. **RAY FLS** 7–30; ligules 4–12 mm. **DISK FLS** 30–80; corollas 4–8 mm, lobes (sub)glabrous (any hairs < 0.2 mm), glands few. **FR** 2–4 mm. Crevices, open sand and gravel, lava flows; 400–3100 m. KR, CaR, n&c SNH, s SNF, MP, **sw DMtns**; w N.Am. Highly variable; local forms often ± distinct. 4 vars. in CA.

var. ***scabra*** (Eastw.) Semple Pl < 5 dm. **LVS**: mid-st lvs narrowly triangular, flat, ± sparsely scabrous to bristly, ± densely glandular (esp later growth). **RAY FLS** 9–14; ligules 5–6 mm. **DISK FLS** 30–50; corollas 5–7 mm. **FR** 2–3 mm. 2*n*=18,36. Uncommon. Rock crevices; 1200–1300 m. **sw DMtns (Little San Bernardino Mtns)**; to UT, n AZ. [*Chrysopsis v.* (Pursh) Nutt. var. *hispida* (Hook.) D. Eaton misapplied, in part] Possibly alien in CA. Apr–May, Oct–Nov

HIERACIUM HAWKWEED

G. Ledyard Stebbins

Per; sap milky; herbage gen long-hairy. **STS** erect, 1–10 dm. **LVS** basal or cauline, alternate. **INFL**: heads ligulate, few–many in cymes or panicles; involucre cylindric; phyllaries in 2–4 series of different lengths; receptacle naked. **FLS** few–many; ligules yellow, white, or orange, readily withering. **FR** cylindric, slender; pappus of many slender bristles, brittle, dull white, tawny, or brownish. ± 250 spp.: ± worldwide. (Greek: hawk) Many reproduce only by asexual seeds.

1. Sts gen 1–3.5 dm; lvs densely hairy; fls 6–15 per head . . . . *H. horridum*
1' Sts gen 3–7 dm; lvs sparsely to moderately hairy; fls 15–50 per head . . . . *H. scouleri*

H. horridum Fries (p. 161) Stolons 0; herbage densely long-hairy; hairs whitish or brownish. **STS** 1–several from taproot, 1–3.5 dm, much-branched. **LVS** mostly cauline, 3–10 cm, oblong, entire. **INFL**: heads many, in open cymes or panicles; involucre 6–9 mm, narrow; phyllaries glandular and long-hairy. **FLS** 6–15; ligules yellow. **FR** 3 mm; pappus brownish. 2*n*=18. Rocky places, crevices; 1350–3300 m. KR, CaRH, SNH, SnJt, **GB**; OR. ❀TRY. Jul–Aug

H. scouleri Hook. Stolons 0. **STS** 3–7 dm, simple, ± glabrous

to bristly. **LVS**: gen basal and lower cauline, 10–20 cm, lanceolate to oblanceolate, entire, sparsely to moderately long-hairy. **INFL** glabrous to hairy or glandular; heads many, in cymes; involucre 8–10 mm, glandular-hairy. **FLS** 15–50; ligules yellow. **FR** 3 mm; pappus dull brown. 2*n*=18. Open woods, shrubby places; 450–2250 m. KR, NCoRH, CaRH, n SNH, **GB**; to B.C., MT. Pls with naked sts and many dark, glandular hairs have been called var. *nudicaule* (A. Gray) Cronq. [*H. cynoglossoides* Arv.-Touv. misapplied to CA pls] ❀DRN:4,5,**6,15–17**&SHD:1,2,**7**,8–10,**14**,18–24. May–Jul

HULSEA

Dieter H. Wilken

Ann to per. **STS** 1–5, 1–15 dm, ± hairy, glandular. **LVS** alternate; petioles gen ciliate; blades gen ± oblanceolate, entire to lobed, ± reduced upward. **INFL** raceme-like; heads radiate, 1–many; bracts ± narrowly lanceolate; involucre hemispheric to obconic; phyllaries many, in 2–3 series, linear to obovate, green, ± glandular; receptacle naked. **RAY FLS**: ligules yellow to red. **DISK FLS** many; corollas 5–9 mm, yellow to orange, gen glabrous. **FR** 4–10 mm, linear to club-like, black, ± hairy; pappus scales gen 2 pairs, 4–10 mm, gen deeply cut, gen translucent. ± 8 spp.: w US. (G.W. Hulse, US Army surgeon, botanist, 1807–1883) [Wilken 1977 Madroño 24:48–55] Self-sterile.

1. Pl leafy; ray fls linear, ligules < 2 mm wide, red . . . . *H. heterochroma*
1' Pl ± scapose; ray fls oblong to elliptic, ligules > 3 mm wide, yellow to red-tinged
  2. Lower lvs glandular-hairy, green, coarsely toothed — subalpine to alpine talus . . . . *H. algida*
  2' Lower leaves woolly to long-soft-wavy-hairy, gen grayish, entire to lobed . . . . *H. vestita*

3. Ray fls 18–32, corollas 12–18 mm, ligules yellow — SNE, n DMtns . . . . ssp. *inyoensis*
3' Ray fls 9–16, corollas 5–9 mm, ligules yellow, orange, or reddish
    4. Basal lf blades sparsely woolly, somewhat greenish; fls orange to red-tinged; sw DMtns . . . . ssp. *parryi*
    4' Basal lf blades densely woolly, grayish; fls gen yellow, sometimes red-tinged below; n SNE . . . . ssp. *vestita*

H. algida A. Gray (p. 161) Per < 4 dm, ± long-soft-hairy, ± glandular. **LVS**: basal < 10 cm, ± coarsely toothed. **INFL**: heads 10–20 mm, 10–25 mm wide, phyllaries 8–15 mm, 1–3 mm wide, narrowly oblong, long-tapered. **RAY FLS** 25–60; corollas 10–15 mm, 2–4 mm wide, yellow, puberulent. **FR** 6–10 mm, sparsely hairy; pappus scales < 1.5 mm, ± equal. 2*n*=38. Subalpine to alpine talus; 3000–4000 m. SNH, **n SNE, W&I**; to OR, MT, NV. Jul–Aug

H. heterochroma A. Gray (p. 161)) Ann or per < 15 dm, ± densely glandular. **LVS**: basal 10–20 cm, coarsely toothed. **INFL**: heads gen 20 mm, 15 mm wide; phyllaries 10–14 mm, linear to lanceolate, long-acuminate. **RAY FLS** 30–60; corollas 6–10 mm, linear, red, hairy. **FR** 6–8 mm, very hairy; pappus scales 1–3 mm, unequal. 2*n*=38. Open sites, recent burns; 300–2500 m. SN, SCoRO, TR, n PR, **n DMtns (Panamint Mtns)**; to UT. Hybridizes with *H. vestita* ssp. *parryi*. ❀ DRN,SUN:1–7,14–24;DFCLT. Jun–Aug

H. vestita A. Gray Per gen 1–10 dm. **STS** gen leafy in lower 1/3–1/2. **LVS**: basal < 8 cm, 1–3 cm wide, spoon-shaped, entire to lobed, woolly above; cauline few. **INFL**: heads < 15 mm, < 12 mm wide; bracts lanceolate to ovate, glandular, ± long-soft-hairy; phyllaries 8–11 mm, oblong to obovate, acuminate, hairy. **RAY**

FLS 9–32; corollas 5–18 mm, 2–5 mm wide, yellow to red, puberulent. **FR** 5–7 mm, moderately hairy; pappus 1–2 mm, gen ± equal. Open gravel, talus slopes; 1300–3900 m. c&s SNH, TR, PR, **SNE, DMtns**; NV. Sspp. gen geog separated. 6 sspp. in CA.

ssp. ***inyoensis*** (Keck) Wilken INYO HULSEA Pl < 7 dm. **LVS**: petioles gen > blade, green; basal lvs lobed. **INFL**: bracts barely woolly; phyllary tips ± green. **FLS** yellow; ray corollas 12–18 mm. 2*n*=38. RARE in CA. Rocky pinyon/juniper woodland; 1700–3000 m. **SNE, n DMtns**; w NV. [*H. i.* (Keck) Munz] May–Jun

ssp. ***parryi*** (A. Gray) Wilken (p. 161) Pl gen 2–6 dm. **LVS**: petiole gen > blade, ± green; basal lvs deeply lobed, glandular and non-glandular below. **INFL**: bracts sparsely woolly; phyllary tips red-tinged. **FLS** orange to red; ray corollas 5–7 mm, orange or reddish. 2*n*=38. Sagebrush scrub to fir forest ; 2000–2500 m. SnBr, **sw DMtns (Little San Bernardino Mtns)**. [*H. p.* A. Gray] May–Aug

ssp. ***vestita*** (p. 161) Pl 1–4 dm. **STS** sometimes ± lfless. **LVS**: petiole = blade, green; basal lvs entire, woolly. **INFL**: bracts densely woolly; phyllary tips green or red-tinged. **FLS** gen yellow; ray corollas 5–9 mm. 2*n*=38. Montane sagebrush scrub to fir forest; 2400–3000 m. c&s SNH, **n SNE**. May–Jul

HYMENOCLEA WINGED RAGWEED

Willard W. Payne

Subshrub to small tree, monoecious. **STS** slender. **LVS** alternate, entire or narrowly lobed, white-hairy above; margins curled upward. **INFL**: staminate heads many, in spike- or panicle-like clusters; phyllaries 4–8, fused into shallow cup, greenish; receptacle chaffy; pistillate heads ± spheric, beaked, in fr winged, 1-fld, ± bur-like, phyllaries 0–few, chaff scales ± fused, free tips winged, papery. **STAMINATE FLS** 4–18; corolla translucent; filaments and anthers free; style unbranched. **PISTILLATE FL**: corolla 0; style branches long; pappus 0. **FR** enclosed in bur, dispersed by wind or water. 2–4 spp.: sw US, n Mex. (Greek: membrane-enclosed, from bur) [Peterson & Payne 1974 Brittonia 26:397] Like *Ambrosia, Xanthium*.

1. Wings of bur < 1.5 mm wide, in 1 whorl; fls autumn . . . . ***H. monogyra***
1' Wings of bur 1.5–8 mm wide, often scattered; fls spring. . . . . ***H. salsola***
  2. Wings ± 7, in ± 1 central whorl . . . . var. ***pentalepis***
  2' Wings ± 10–13, scattered
    3. Wings appressed, enclosing body of bur . . . . var. ***fasciculata***
    3' Wings spreading perpendicular to body of bur . . . . var. ***salsola***

H. monogyra A. Gray (p. 161) Shrub or small tree 1–4 m, much-branched above. **STS** straw-colored to dark brown; old bark gray-brown. **INFL**: fr heads 2–4 mm, fusiform; wings 7–12, 1.5–2 mm, < 1.5 mm wide, oblanceolate, in 1 central whorl. Washes, dry riverbeds; < 500 m. SW, **SNE, DSon**; to w NV, TX, n Mex. Aug–Nov

H. salsola A. Gray CHEESEBUSH Subshrub < 2 m, branched throughout. **STS** pale straw-colored. **INFL**: fr heads 2.8–6.3 mm, widely fusiform; wings 5–19, 2–7 mm, 1.5–8 mm wide, often reniform, in 1 whorl or scattered. Dry flats, washes, fans; < 1800 m. s SnJV, s SCoRI, SW, **SNE, D**; to sw UT, AZ, nw Mex. Fls spring. Vars. intergrade. Hybridizes with *Ambrosia dumosa*.

var. ***fasciculata*** (Nelson) Peterson & Payne **INFL**: bracts

subtending bur gen several, widely ovate, gen green, hairy. **FR**: bur wings (10–)13(–19), scattered, appressed, base pitted. Habitats of sp. s SnJV, SCoRI, SCo, WTR, **DMoj**; to sw UT, w AZ. [var. *patula* (Nelson) Peterson & Payne] Mar–Jun

var. ***pentalepis*** (Rydb.) L. Benson **INFL**: bract subtending bur gen 1, obovate, transparent. **FR**: bur wings (5–)7(–13), whorled at center, spreading. Habitats of sp. SW, **D**; s NV, sw AZ, nw Mex.

var. ***salsola*** (p. 161, pl. 20) **INFL**: bracts subtending bur gen several, ovate, green or transparent, hairy. **FR**: bur wings (5–)10 (–16), scattered, spreading. Habitats of sp. s SnJV, s SCoRI, SW, **SNE, D**; to sw UT, w AZ, Baja CA. Mar–Jun

HYMENOPAPPUS

Bien, per from taproot bearing 1–several caudices. **STS** erect. **LVS** simple to 2-pinnately dissected; basal and alternate, reduced upward, gland-dotted. **INFL**: heads radiate or discoid, in few–many-headed panicles; involucre obconic to widely bell-shaped; phyllaries in 2–3 series, ± equal, scarious-margined; receptacle flat or rounded, gen naked. **RAY FLS** 0 (in CA) or 8; ligules white. **DISK FLS** 10–many; corollas yellow, white, or reddish purple, tube slender, throat abruptly enlarged, lobes triangular, reflexed; style branches obtuse. **FR** obpyramidal, 4-angled; pappus 0 or of many thin, transparent, linear-oblong to ovate, obtuse scales. 10 spp.: N.Am. (Greek: membranous pappus) [Turner 1956 Rhodora 58:163–186, 208–242, 250–269, 295–308]

H. filifolius Hook. Per 0.5–10 dm, ± glabrous to densely tomentose. **LVS**: basal 3–20 cm, 2-pinnately dissected into 2–50 mm linear or thread-like segments, minutely gland-dotted; cauline 0 or few, gen much reduced upwards. **INFL**: heads discoid, 1–many; peduncles 0.5–16 cm; principal phyllaries 3–14 mm, 2–5 mm wide, margin white or yellowish scarious for 1–4 mm from acute to obtuse tip; fls 10–70 per head. **FLS**: corollas 2–7 mm, yellow or white. **FR** 3–7 mm, densely short-hairy; pappus scales 12–22, gen linear-oblong. Many ± dry habitats, sometimes on limestone; ± 1000–3000 m. TR, PR, **W&I**, **DMtns**; to WA, SD, AZ, n Mex. 4 vars. in CA.

1. Corollas 3–4 mm; anthers 2–3 mm; fr 4.5–5.5 mm
   . . . . var. ***nanus***
1' Corollas 4–7 mm; anthers 3–4 mm; fr 5–7 mm
   2. Corollas white; peduncles 8–16 cm . . . . . var. ***eriopodus***
   2' Corollas yellow; peduncles 2–10 cm . . var. ***megacephalus***

var. ***eriopodus*** (Nelson) B. Turner (p. 161) Pl 4–8 dm, ± tomentose below, becoming glabrous above. **LVS**: basal 10–20 cm, divisions 10–20 mm, 0.4–1 mm wide, thread-like; cauline 2–7, gen glossy green. **INFL**: heads 3–8; peduncles 8–16 cm; main phyllaries 7–10 mm, 2–4 mm wide. **FLS**: corollas 4–5 mm, ± white; anthers 3–4 mm. **FR** 5.5–6 mm, evenly hairy; hairs 0.5–1.5 mm. Limestone soil, with pines, junipers; 1600–1700 m. **e DMtns (Clark, New York mtns)**; s NV, sw UT. May–Jun, Oct

var. ***megacephalus*** B. Turner (p. 161) Pl 3–7 dm, tomentose through-out. **LVS**: basal 8–20(30) cm, divisions 8–30 mm, 1–2 mm wide, flat, linear; cauline 2–6. **INFL**: heads 3–14; peduncles 2–10 cm; main phyllaries 8–14 mm. **FLS**: corollas 4–7 mm, yellow; anthers 3–4 mm. **FR** 5–7 mm, evenly hairy; hairs 1–2 mm. Sandy or gravelly desert soils in valleys, washes; ± 1000–1500 m. **e DMtns (Providence Mtns)**; to w Colorado, AZ.

var. ***nanus*** (Rydb.) B. Turner (p. 161) Pl 0.5–5 dm, ± evenly, sparsely tomentose throughout. **LVS**: basal 2–12 cm, divisions 5–15 mm, 0.5–1 mm, thread-like; cauline 0–3. **INFL**: heads 1–6; peduncles 3–15 cm; main phyllaries 6–9 mm. **FLS**: corollas 3–4 mm, pale yellow; anthers 2–3 mm. **FR** 4.5–5.5 mm, evenly hairy; hairs gen 0.2–1 mm. $2n$=34. Limestone soil, often with juniper; 1500–3000 m. **W&I (Inyo Mtns)**; also c&e NV to w UT, nw AZ. Jul–Aug

HYMENOXYS

Ann, bien, per. **STS** erect, simple to much-branched. **LVS** simple, basal or alternate, sessile or petioled, entire or divided 1 or more times into linear lobes, dotted with sunken resin-glands. **INFL**: heads gen radiate, solitary or in few–many-headed cymes; involucre hemispheric; phyllaries in 2–3 similar or dissimilar series; receptacle flat to rounded, naked. **RAY FLS** few–many; corolla yellow, fan-shaped, 3–5-lobed. **DISK FLS** many; corolla yellow. **FR** obpyramidal, gen 5-angled, hairy; pappus of ± 5 membranous, often awn-tipped scales. 28 spp.: w N.Am, S.Am. (Greek: sharp membrane, from pappus) For revised taxonomy of *H. acaulis* see M.W. Bierner and R.K. Jansen 1998 Lundelia 1:17–26.

1. Lvs entire, gen all basal; head solitary, scapose (segregate genus *Tetraneuris*) . . . . ***H. acaulis*** var. ***arizonica***
1' Lvs divided into linear lobes, basal and cauline; heads gen several–many in a cyme
   2. Involucre diam gen 15 mm or more . . . . ***H. lemmonii***
   2' Involucre diam gen 10 mm or less
      3. Bien or short-lived per; basal lvs gen present at time of fl; involucre diam gen 9–10 mm . . . . ***H. cooperi***
      3' Ann; basal lvs 0–few at time of fl; involucre diam 5–7 mm . . . . ***H. odorata***

H. acaulis (Pursh) K. Parker var. ***arizonica*** (E. Greene) K. Parker (p. 161) Per 8–20 cm from branched caudex. **STS** unbranched. **LVS** gen all basal, 2–6 cm, linear to narrowly oblanceolate, entire, soft-hairy, esp at base, becoming ± glabrous in age. **INFL**: heads solitary, scapose; involucre tomentose; outer phyllaries = inner, 5–6 mm, oblanceolate to obovate, free. **RAY FLS** 8–13; ligule 10–15 mm, bright yellow, fading to cream and ± persistent on fr. **DISK FLS**: corolla 3.5–4 mm. **FR** 3–4 mm; pappus scales 5–7, 2–3 mm, obovate, entire or awn-tipped. $2n$=30,60. Dry, rocky slopes; 1400–2800 m. SnJt, **DMtns**; to WY, Colorado, AZ. Apr–Jun

H. cooperi (A. Gray) Cockerell Bien or short-lived per 15–90 cm, lightly canescent with soft, white, jointed hairs, ± glabrous in age. **ST** gen branched above middle. **LVS** basal and cauline, 4–9 cm, divided into 1–2 mm wide linear lobes. **INFL**: heads in ± flat-topped cyme; peduncles > 10 cm; involucre gen 9–10 mm diam; outer phyllaries fused at base, 5–7 mm, < or = inner series. **RAY FLS** 8–14; ligules 10–13 mm. **DISK FLS**: corolla 3.5–4.5 mm. **FR** ± 3 mm; pappus scales, 1–2 mm, ovate, entire or short-awn-tipped. $2n$=30. Dry, rocky slopes; 1200–2000 m. **DMtns**; NV, AZ. May–Jun, Sep–Oct

H. lemmonii (E. Greene) Cockerell (p. 161) Bien or per 4–50 cm, persistently tomentose or glabrous in age. **STS** often > 1, gen branched above middle. **LVS** basal and cauline, 2–9 cm, divided into 1–5 mm wide linear lobes. **INFL**: heads in ± flat-topped cyme; peduncles gen < 8 cm; involucre gen 15 mm diam or more; outer phyllaries fused at base, 5–7 mm, < or = inner series. **RAY FLS** 8–14; ligules 10–15 mm. **DISK FLS**: corolla 3.5–4.5 mm. **FR** ± 3 mm; pappus scales, 1–2 mm, lanceolate to ovate, entire. $2n$=30. Meadows or dry, rocky slopes; 700–3500 m. e KR, **GB**; se OR, s ID, NV, UT. Dwarf, subalpine form has been called *H. cooperi* (A. Gray) Cockerell var. *canescens* (D. Eaton) K. Parker. Jun–Aug

H. odorata DC. Ann 15–60 cm, soft-hairy, ± glabrous in age. **STS** branched above. **LVS**: basal gen 0 at time of fl; cauline 1–5 cm, divided into linear lobes 1 mm or less wide. **INFL**: heads in flat-topped cyme; peduncles 2–15 cm; involucre gen 5–7 mm diam; outer phyllaries 3–5 mm, < inner phyllaries, thickened and fused at base. **RAY FLS** ± 8; ligules gen 1 cm or less, yellow, fading to cream and ± persistent on fr. **DISK FLS**: corolla 3–4 mm. **FR** ± 2 mm; pappus scales 2 mm, lanceolate. $2n$=11,12,14,15. Sandy flats near Colorado River; < 150 m. **e DSon**; to UT, TX, n Mex. TOXIC range pl outside CA. Feb–May

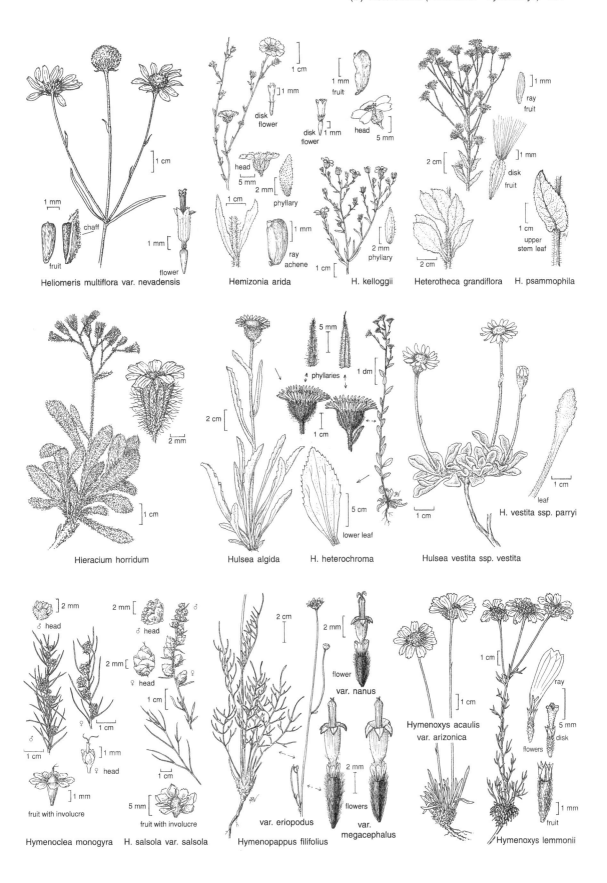

Heliomeris multiflora var. nevadensis

chaff

fruit

flower

1 mm

1 cm

Hemizonia arida

disk flower

1 mm

disk flower

head

phyllary

ray achene

1 cm

5 mm

2 mm

1 mm

fruit

1 mm

head

5 mm

H. kelloggii

2 mm

phyllary

1 cm

2 mm

Heterotheca grandiflora

ray fruit

1 mm

disk fruit

1 mm

2 cm

H. psammophila

upper stem leaf

1 cm

2 cm

Hieracium horridum

2 mm

1 cm

Hulsea algida

phyllaries

5 mm

1 dm

2 cm

1 cm

lower leaf

5 mm

H. heterochroma

Hulsea vestita ssp. vestita

H. vestita ssp. parryi

leaf

1 cm

1 cm

Hymenoclea monogyra

♂ head

2 mm

♀

♀ head

1 mm

fruit with involucre

1 mm

1 cm

H. salsola var. salsola

♂ head

2 mm

♀ head

2 mm

1 cm

fruit with involucre

5 mm

1 cm

Hymenopappus filifolius

var. eriopodus

2 cm

flower

var. nanus

2 mm

var. megacephalus

flowers

2 mm

Hymenoxys acaulis var. arizonica

1 cm

2 mm

Hymenoxys lemmonii

ray

disk

flowers

fruit

1 cm

5 mm

1 mm

ISOCOMA GOLDENBUSH

Meredith A. Lane

Subshrubs, glabrous to scabrous or hairy. **STS** prostrate or decumbent to ascending or erect, < 3 m, ± striate below, minutely scabrous, yellow-white or gray to red-brown. **LVS** alternate, sometimes clustered in axils, entire or toothed, gland-dotted, sometimes gummy, glabrous, minutely scabrous, or tomentose, light to dark gray-green. **INFL**: heads discoid, in loose to tight clusters, these borne in flat-topped or ± spheric cymes; involucres obconic; phyllaries yellow-white below, texture cartilage-like, tips green; receptacles flat, naked. **FLS** yellow; tubes narrowly cylindric, abruptly expanded into larger cylindric throat; sinuses shallow, lobes erect; style branch appendages triangular. **FR** narrowly obconic, light tan, silky-hairy; hairs white, yellow, tan, or light red-tan; pappus of 1–2 series of white, yellowish, or red-tan bristles ± 2 × fr. ± 10 spp.: sw N.Am, Mex. (Greek: equal hair-tuft, from fls) [Nesom 1991 Phytologia 70: 69–114]

I. acradenia (E. Greene) E. Greene Pl < 1.3 m, rounded or open. **STS** erect or ascending, branched from ground or above, glabrous or minutely scabrous, yellow-white, varnished, shiny, becoming yellow-tan or gray with age. **LVS** 1.5–6 cm, 1.5–15 mm wide, linear, obovate or spoon-shaped, entire or toothed, gland-dotted, glabrous or minutely scabrous, gen light gray-green. **INFL**: heads in loose to tight clusters of 4–5; involucre 4–5 mm, 4–5 mm diam; phyllaries 22–36 in 3–6 series, oblong, tips blunt, rounded, or acute, green or tan to 1/4 total length of phyllary, swollen by glandular exudate below surface, appearing wart-like. **FLS** 6–17. **FR** 2–3.5 mm; pappus 3–5.5 mm, white-yellow, bristles unequal. $2n=12$. Sandy or clay soils in alkaline or gypsum flats or slopes; < 1300 m. Teh, SnJV, SCoR, SnBr, **D**; NV, AZ, Baja CA. [*Haplopappus a.* (E. Greene) S.F. Blake] Vars. intergrade somewhat. Key to vars.adapted by Margriet Wetherwax.

1. Lvs conspicuously reduced upward, grading into
   bracts below heads, these grading into phyllaries;
   fls 10–17 . . . . var. *bracteosa*
1'. Lvs uniform throughout, or only slightly reduced
   upward, not grading into bracts below heads; fls 6–14
   2. Lvs narrowly oblanceolate, entire; involucres 4–5 mm
   . . . . var. *acradenia*

2'. Lvs narrowly obovate, with long-tapering base,
   to obovate, with 4–6 teeth per side; involucres
   5–7 mm . . . . var. *eremophila*

var. *acradenia* (p. 169) **STS** < 0.8 m. **LVS** < 5 cm, not much reduced above, entire. **INFL**: involucre not closely subtended by bracts; phyllaries 22–28 in 3–4 series, tips blunt, sometimes abruptly soft-pointed. **FLS** 6–12. Alkaline soils; < 1000 m. SnJV, SCoR, SnBr, **DMoj**. Aug–Nov

var. *bracteosa* (E. Greene) G. Nesom (p. 169) **STS** < 0.9 m. **LVS** < 5 cm, much reduced above, grading into bracts, sometimes toothed. **INFL**: involucre closely subtended by bracts grading into phyllaries; phyllaries 25–36 in 4–6 series, tips blunt to acute (inner ones), abruptly soft-pointed. **FLS** 10–17. Sandy, alkaline soils; < 900 m. SnJV, SCoRI, **nw DMoj**. [*Haplopappus a.* ssp. *b.* (E. Greene) H.M. Hall]

var. *eremophila* (E. Greene) G. Nesom (p. 169) **STS** < 1.3 m. **LVS** < 6 cm, not much reduced above; margins with 4–6 shallow to deep, abruptly soft-pointed teeth per side. **INFL**: involucre not closely subtended by bracts; phyllaries < 28 in 3–5 series, tips widely rounded. **FLS** 7–14. Alkali or gypsum silt on flats or slopes; < 1300 m. Teh, SCoRI, **D**; AZ, NV, Baja CA. [*Haplopappus a.* ssp. *e.* (E. Greene) H.M. Hall]

IVA WORMWOOD

Willard W. Payne

Ann to shrub. **LVS** gen opposite below, ± alternate above. **INFL**: heads many, disciform; involucre cup-shaped, shallow; phyllaries few, free or fused; receptacle chaffy. **PISTILLATE FLS** 0–6; corolla 0 or cylindric. **STAMINATE FLS** 5–20; corolla translucent; anthers free, yellow; style tip truncate or tack-shaped. **FR** obovate, ± compressed; pappus 0. (Latin: from mint *Ajuga iva*, with similar aroma) [Jackson 1960 Univ KS Sci Bull 41:793–876] Generic limits need revision, see Miao et al 1995 Amer J Bot 82: 924-932. Key to species adapted by Margriet Wetherwax.

1. Ann; lvs 2-pinnately dissected — < 3 dm . . . . *I. nevadensis*
1'. Per or shrub; lvs entire or pinnately linear-lobed
   2. Shrub 10–20 dm; phyllaries free . . . . *I. acerosa*
   2'. Per 1–6 dm; phyllaries ± fused . . . . *I. axillaris* ssp. *robustior*

I. acerosa (Nutt.) R. Jackson (p. 169) Shrub 1–2 m, sometimes appearing rush-like. **STS** ± green. **LVS** 5–15 cm, ± thick, ± linear, entire to pinnately linear-lobed, finely canescent. **INFL** panicle-like; head erect; involucre 3–4 mm; phyllaries free; chaff scales ± cut, tips ± wider. **PISTILLATE FLS**: corolla 0. **FR** plump, densely long-soft-hairy. $2n=36$. Saline soils; < 700 m. **n DMoj**; to Colorado, NM. [*Oxytenia a.* Nutt.; Bolick 1983 Adv Clad 2:125–141] Aug–Dec

I. axillaris Pursh ssp. *robustior* (Hook.) Bassett (p. 169) POVERTY WEED Per 1–6 dm, from rhizome. **STS** green to brown. **LVS** 1–4 cm, linear to obovate, entire, subsessile, ± red-gland-dotted. **INFL** raceme-like; heads short-peduncled, nodding;

involucre 5–6 mm; phyllaries ± fused; chaff scales linear. **PISTILLATE FLS**: corolla vestigial. **FR** plump, glabrous. $2n=36,54$. Common. Many (saline) habitats; < 2500 m. **CA**; to B.C., MT, c US, TX. May–Sep

I. nevadensis M.E. Jones (p. 169) Ann < 3 dm, from stout taproot, ill-scented. **STS** yellow. **LVS** 8–18 mm, 2-pinnately dissected (segments ± obtuse), ± glandular and white-strigose. **INFL**: small cymes, leafy-bracted; heads erect; involucre 2–3 mm; phyllaries gen 3, free outer chaff scales wide, inner linear. **PISTILLATE FLS**: corolla short. **FR** glabrous; inner face flat or concave. Alkaline, sandy plains; 1000–2000 m. **nw W&I**; w NV. [Bolick 1983 Adv in Clad 2:125–141] Jun–Oct

LACTUCA LETTUCE

G. Ledyard Stebbins

Ann to per; sap milky. **STS** erect, 0.5–1.5+ m. **LVS** basal and cauline, alternate, entire to pinnately lobed. **INFL**: heads ligulate, many, in panicles; involucre cylindric; phyllaries in 2–several series; receptacle naked. **FLS** few–many; ligules yellow or cream to blue, readily withering. **FR** flattened, short- or long-beaked; pappus of many bristles, falling separately. ± 100 spp.: ± worldwide temp. (Latin: milky) Key to species adapted by David J. Keil.

1. Corollas yellow; ann from taproot; lf midrib and st prickly . . . . *L. serriola*
1' Corollas blue; per from deep rhizome; lf midrib and st glabrous . . . . *L. tatarica*

L. serriola L. (p. 169) PRICKLY LETTUCE Ann from taproot, 0.5–1.5 m. **STS** erect, prickly-bristly. **LVS** few–many, oblanceolate to obovate, dentate to coarsely lobed, clasping, prickly-bristly on midvein. **INFL** open; branches often widely spreading; heads in fl 4–6 mm wide; involucre in fr 10–12 mm. **FLS** 14–20; corolla pale yellow. **FR** 6–7 mm (incl beak), light to dark brown, rough-hairy, several-veined on each face, unwinged, beak = or > body; pappus 4–5 mm, white. 2*n*=18. Abundant. Weed of disturbed places; < 2000 m. **CA**; native to Eur. [var. *integra* Gren. & Godron] May–Sep

L. tatarica (L.) C. Meyer ssp. *pulchella* (Pursh) Stebb.(p. 169) Per from extensive rhizomes, 3–10 dm. **STS** erect, glabrous. **LVS** many, 5–15 cm, linear to elliptic, entire to lobed. **INFL**: heads in fl 2–3 cm wide; involucre in fr 1.4–1.8 mm. **FLS** 10–15; corolla bright blue, conspicuous. **FR** 5–6 mm (incl beak), pale brown, glabrous, several-veined on each face, unwinged; beak very short; pappus 8–11 mm, white. 2*n*=18. Dry to moist alluvial valleys; 1150–2000 m. CaRH, **GB**; to AK, e Can, c US, NM. ❀TRY. Jun–Sep

LASTHENIA GOLDFIELDS

Robert Ornduff

Ann, per, glabrous or hairy. **ST** gen branched, gen erect, < 60 cm. **LVS** opposite, < 20 cm, entire to pinnately cut. **INFL**: heads radiate, solitary or in cymes; phyllaries in 1 or 2 series, free or partly fused; receptacle narrowly conic to hemispheric, naked, smooth, pitted, or rough. **RAY FLS** 4–21; ligules gen yellow. **DISK FLS** gen many; corollas gen 5-lobed, gen yellow; anther tips acuminate to triangular; style tips triangular or round, gen hair-tufted. **FR** < 5 mm, cylindric to obovoid, black or gray; pappus of awns, scales, or 0. 17 spp.: w N.Am, Chile. (Greek: female pupil of Plato) [Ornduff 1966 Univ Calif Publ Bot 40:1–92] Gen self-incompatible (cross-pollinated).

1. Phyllaries fused > 2/3 their length; pappus 0; fr hairs wart-like . . . . *L. glabrata* ssp. *coulteri*
1' Phyllaries free; pappus present or 0; fr hairs slender or 0
  2. Ray ligules < 2 mm; phyllaries gen 4 . . . . *L. microglossa*
  2' Ray ligules > 2 mm; phyllaries gen > 6
    3. All lvs ± entire; corollas dark red in 5% aqueous KOH; pl not glandular . . . . *L. californica*
    3' Middle cauline lvs gen pinnately divided; corollas yellow in 5% aqueous KOH; pl gen glandular
    . . . . *L. coronaria*

L. californica Lindley (p. 169) Ann < 40 cm. **ST** simple or freely branched, ± hairy. **LVS** 0.8–7 cm, linear to oblanceolate, entire, hairy, ± fleshy in coastal forms. **INFL**: involucre 5–10 mm, bell-shaped or hemispheric; phyllaries 4–13, free, hairy; receptacle conic, rough, glabrous. **RAY FLS** 6–13; ligules 5–10 mm. **DISK FLS** gen many; anther tips triangular; style tips triangular. **FR** < 3 mm, linear to ± club-shaped, glabrous or hairy; pappus of 1–7 narrow awns, wider awned scales, or 0. 2*n*=16,32. Abundant. Many habitats; < 1500 m. CA-FP, **w DMoj**; sw OR, AZ, Mex. Highly variable; needs further study. [*L. chrysostoma* (Fischer & C. Meyer) E. Greene] ❀SUN:**7–10**,11,12,**14–24**. Feb–Jun

L. coronaria (Nutt.) Ornd. (p. 169) Ann < 40 cm; herbage sweetly scented. **ST** simple or much-branched; hairs short, glandular, or long, non-glandular, or mixed. **LVS** 1.5–6 cm, linear, entire or pinnately lobed or cut, gen glandular-hairy. **INFL**: involucre 4–7 mm, obconic to hemispheric; phyllaries 6–14, free, gen glandular-hairy; receptacle conic, hairy. **RAY FLS** 6–15; ligules 3–10 mm. **DISK FLS** many; anther tips elliptic; style tips triangular or dome-shaped. **FR** < 2.5 mm, linear to narrowly club-shaped, hairy; pappus gen a mixture of lanceolate or oblong, truncate scales or 0, gen different in ray and disk frs. 2*n*=8,10. UNCOMMON. Sunny, open places; < 700 m. SCo, PR, **w D**; nw Baja CA, Guadalupe Island, Mex. ❀TRY. Mar–May

L. glabrata Lindley Ann < 60 cm. **ST** erect, simple or branched, glabrous or slightly hairy. **LVS** 4–15 cm, linear or awl-shaped, entire, glabrous. **INFL**: involucre 5–10 mm, hemispheric; phyllaries 10–14, fused, glabrous; receptacle conic, papillate, glabrous or sparsely hairy. **RAY FLS** 7–15; ligules 4–14 mm. **DISK FLS** many; anther tips ovate or triangular; style tips triangular, hair-tufted. **FR** < 3.5 mm, club-shaped or ovoid, glabrous or papillate; pappus 0. 2*n*=14. Saline places, vernal pools; < 1000 m. NCoRI, Teh, ScV, n SnJV, SnFrB, SCoRO, SCo, n Chl (Santa Rosa Island), PR, **w DMoj**. 2 sspp. in CA.

  ssp. *coulteri* (A. Gray) Ornd. (p. 169) COULTER GOLDFIELDS **FR** warty-hairy. 2*n*=14. RARE. Habitats of sp. Teh (1 station), s SCoRO, SCo, n Chl (Santa Rosa Island), PR, **w DMoj**. [var. *c.* A. Gray] Apr–May

L. microglossa (A.DC.) E. Greene (p. 169) Ann < 25 cm. **ST** sprawling or erect, simple or much-branched, hairy. **LVS** 1.5–8 cm, linear or awl-shaped, ± entire, hairy. **INFL**: involucre 6–8.5 mm, cylindric to narrowly obconic; phyllaries ± 4, hairy; receptacle narrowly conic, glabrous. **RAY FLS** ± 4; ligules < 1 mm or sometimes 0. **DISK FLS** few; corolla gen 4-lobed; anther tips narrowly tapered; style tips lanceolate, glabrous. **FR** < 5 mm, ± linear, hairy; pappus of < 4 scales, each lanceolate, yellowish or white, awn-tipped, or 0. 2*n*=24. Shaded slopes; < 1000 m. NCoRI, ScV (1 collection), SnFrB, SCoR, TR, PR, **DMoj**. Mostly self-pollinated. Mar–May

LAYIA

Bruce G. Baldwin & Susan J. Bainbridge

Ann, gen ascending to erect, often black-glandular, gen ± purplish or brownish. **LVS** alternate, gen linear to (ob)lanceolate, sessile, gen pinnately lobed, reduced upward. **INFL**: heads gen radiate; involucre obconic to urn-shaped; phyllaries gen folded completely around ray fr, falling with fr, gen ± hairy; receptacle flat to slightly convex, chaff scales free. **RAY FLS** (0)3–27; ligules white (often aging pinkish) to yellow, tubes hairy. **DISK FLS** 5–many; corollas yellow, puberulent; anther tips acute, long-tapered; style branches long, bristly. **FR** gen 2–5 mm, gen club-shaped, black; ray fr compressed back to front, gen ± glabrous, pappus 0; disk fr ± straight, gen ± hairy, pappus various. 14 spp.: w N.Am. (George T. Lay, early 19th century English pl collector) [Kyhos et al. 1990 Ann Missouri Bot Gard 77:84–95]

1. Pappus of 10–15 flattened, linear, plumose awns; ligules white or yellow; D, GB . . . . *L. glandulosa*
1' Pappus of (0)S14–32 densely woolly or scabrous bristles; ligules yellow or white-tipped; w edge DMoj
. . . . *L. platyglossa*

L. glandulosa (Hook.) Hook. & Arn. (p. 169) WHITE LAYIA Pl 4–60 cm, glandular, sometimes spicy-scented. **LVS** < 10 cm, linear to obovate, thin, gen ± irregularly lobed below. **INFL**: peduncles 0–7 cm; phyllaries 4–11 mm, basal margins interlocked by cobwebby hairs; chaff scales between ray and disk fls. **RAY FLS** 3–14; ligules 3–22 mm, white to yellow. **DISK FLS** 17–105; corollas 3.5–6.5 mm; anthers yellow. **FR** gen ± hairy; disk pappus of 10–15 awns, 2–5 mm, linear, flat, white, plumose below, often woolly on inner surface, scabrous above. 2*n*=16. Open, sandy soils, < 2700 m. CaRH, SNH, Teh, SnJV, CW, SW (exc ChI), **GB**, **D**; to WA, ID, UT, NM. [ssp. *lutea* Keck] ✿DRN, SUN:1–3,**7–10**,11,12,**14–24**. Mar–Jun

L. platyglossa (Fischer & C. Meyer) A. Gray (p. 169) TIDY-TIPS Pl 3–70 cm, decumbent to erect, glandular, not scented. **LVS** 4–100 mm, linear to (ob)lanceolate, sometimes fleshy; lower lvs lobed ± 1/2 to midvein. **INFL**: peduncles < 13 cm; phyllaries 4–18 mm, basal margins interlocked by cottony hairs; chaff scales between ray and disk fls. **RAY FLS** 5–18; ligules 3–20 mm, yellow, gen white-tipped. **DISK FLS** 6–124; corollas 3.5–6 mm; anthers gen purple. **FR** 2.5–7 mm; ray fr dull; disk pappus of (0)14–32 bristles, 1–6 mm, whitish, gen scabrous throughout, not woolly on inner surface (in SW, rarely short-plumose below and woolly inside). 2*n*=14. Common. Many habitats; < 2000 m. NW, GV, CW, SW, **w edge DMoj**. [ssp. *campestris* Keck; *L. ziegleri* Munz, Ziegler's tidy-tips] ✿ SUN:1–6,**7–9**,14–23,24. Mar–Jun

LEPIDOSPARTUM SCALE-BROOM

Theodore M. Barkley

Shrubs or small trees gen < 3 m, broom-like. **LVS** alternate, entire, thread- to needle- or scale-like. **INFL**: heads discoid, in panicle-like clusters at branch tips; involucre cylindric to obconic; phyllaries overlapping; receptacle naked. **RAY FLS** 0. **DISK FLS** 4–17; corollas yellow, tube long, throat abruptly wider, lobes long; anther bases sagittate to tailed, tips ± lanceolate; style branches long, tips conic or hair-tufted. **FR** ± fusiform; pappus many bristles in 3–4 series. 3 spp.: w N.Am. (Greek: scale-broom) [Strother 1978 N.Am Fl II 10:171–173]

1. Lvs of fl-sts 20–30 mm, thread- or needle-like; fls gen 5; fr densely hairy . . . . *L. latisquamum*
1' Lvs of fl-sts 2–3 mm, scale-like; fls 9–17; fr ± glabrous . . . . *L. squamatum*

L. latisquamum S. Watson (p. 169) Pl narrow. **STS** striate; ribs glabrous, grooves felted-tomentose. **LVS** of fl-sts gen 20–30 mm, thread- or needle-like, short-hairy or becoming glabrous. **INFL**: heads 3–5; main inner phyllaries 3–5, 6–8 mm, outer grading into subtending bracts. **FLS** gen 5; corollas pale yellow. **FR** 5–6.5 mm, 5-veined, ± long-white-hairy between veins; pappus bristles < 11 mm, white to brownish. 2*n*=60. Sandy or gravelly pine/juniper woodlands, open scrubland; 1400–1500 m. SnGb (n slope), **W&I**, **DMtns**; to UT. ✿DRN,SUN:7,8–11,14,**19–21**,22–24. Jun–Oct

L. squamatum (A. Gray) A. Gray (p. 169) Pl spreading, round-topped, woolly, soon becoming glabrous. **LVS** of fl-sts 2–3 mm, scale-like, appressed; axils often woolly-tufted. **INFL**: heads 1–5; main inner phyllaries 7–23, 4–7 mm, outer grading into subtending scale-like bracts. **FLS** gen 9–17; corollas yellow. **FR** 3.5–5 mm, 10–15-veined, ± glabrous; pappus bristles 5–8 mm, whitish brown. 2*n*=±90. Sandy or gravelly washes, stream terraces; < 1800 m. SNF, SCoRI, SW, **D**; Baja Ca. [var. *palmeri* (A. Gray) L. Wheeler] Apparently TOXIC but unpalatable. ✿DRN,SUN:**7–9**,10,11,**14–16**,**18–21**,22–24. Aug–Oct

LESSINGIA

Meredith A. Lane

Ann, per, subshrubs, decumbent to erect, taprooted. **STS** simple or branched from base or above. **LVS** simple, entire to pinnately lobed; basal petioled, ovate to spoon-shaped; cauline reduced upward, sessile, (ob)lanceolate to (ob)ovate. **INFL**: heads radiate or discoid, terminal, solitary or clustered; involucres hemispheric, bell-shaped, obconic, or cylindric; phyllaries in 4–9 series, thin to tough but flexible, tips green or tinged purplish; receptacle concave, naked, shallowly pitted. **RAY FLS** present or 0, fertile or sterile; corollas yellow or white to purple. **DISK FLS** gen fertile (exc *L. occidentalis*); corollas yellow, violet, purple, pink, or white, funnel-shaped to cylindric, marginal ones in rayless heads often enlarged and bilateral, deeply lobed on inner side, lobes spreading from head center; style appendages flat, triangular, awl-shaped, or cusped. **FR** obconic, mottled purple-brown; hairs dense, appressed, silky; pappus 0 in ray fls, in disk fls of many bristles, these free or fused at base, or fused throughout into awns, white, tan, or red-brown. 14 spp.: CA; NV, AZ, n Baja CA. (C.F. Lessing, 1809–1862, German specialist in Asteraceae) Incl CA sp. previously treated in *Benitoa* and *Corethrogyne*.

L. lemmonii A. Gray Ann; herbage grayish tomentose, sometimes becoming glabrous with age. **STS** decumbent or erect, 0.3–4 dm. **LVS**: basal deciduous, < 6 cm, oblanceolate or long-tapered obovate, entire to pinnately lobed; cauline < 2 cm, awl-shaped or linear to obovate, entire or tip with few teeth. **INFL**: heads discoid, solitary, terminal; involucres 4–6 mm, narrowly obconic to bell-shaped; phyllaries oblong, obtuse or acute, puberulent or tomentose, margins with large sessile glands. **RAY FLS** 0. **DISK FLS** 10–25; corollas funnel-shaped to tubular, marginal corollas funnel-shaped in small heads, yellow, with or without white band in throat; style branches 1–2 mm, appendages 0.7–1.3 mm, with long, abrupt point. **FR** 1.5–3.5 mm; pappus bristles many, free to base, white or tannish white. 2*n*=10. Sandy soils; 200–1850 m. TR, **W&I**, DMoj; w NV, nw AZ. Vars. intergrade. Key to vars. adapted by Margriet Wetherwax.

1. Phyllaries tomentose . . . . var. *peirsonii*
1' Phyllaries puberulent
  2. Cauline lvs fewer above, not clasping base of involucre . . . . var. *lemmonii*

2' Cauline lvs many, clasping base of involucre . . . . var. *ramulosissima*

var. ***lemmonii*** (p. 169) Herbage tomentose, partly glabrous with age. **LVS**: cauline fewer above, not clasping base of involucre. **INFL**: phyllaries puberulent. Dry, sandy soil; < 1900 m. TR, **W&I**, DMoj; nw AZ. [*L. germanorum* Cham. var. *l.* (A. Gray) J. Howell] May–Oct

var. ***peirsonii*** (J. Howell) Ferris (p. 169) Herbage densely and persistently tomentose. **LVS**: cauline few, clasping base of involucre. **INFL**: phyllaries densely and persistently tomentose. Open slopes in sandy soil; 300–1850 m. e WTR, **w DMoj**; w NV. [*L. germanorum* Cham. var. *p.* J. Howell] July

var. ***ramulosissima*** (Nelson) Ferris (p. 169) Herbage tomentose, becoming less so with age. **LVS**: cauline many, clasping base of involucre. **INFL**: phyllaries puberulent. Very dry, sandy soil; 800–1200 m. TR, **W&I**, DMoj; w NV. Sometimes forms tumbleweeds.

MACHAERANTHERA

David J. Keil & Gregory K. Brown

Ann to subshrubs. **STS** from taproot or ± branched caudex. **LVS** simple, alternate, entire to pinnately dissected; teeth or lobes often ± bristle-tipped. **INFL**: heads radiate or discoid, solitary or cymosely clustered; involucre bell-shaped, hemispheric, or obconic; phyllaries in 2–several series of unequal length, basal portion straw-colored to purplish, tips green; receptacle convex, naked or with short, triangular scales (not chaff). **RAY FLS** 8–many; corollas yellow, white, pink, blue, or purple. **DISK FLS** 10–many; corollas yellow; style tips triangular to linear, acute. **FR** linear to club-shaped or obovoid, smooth or several–many-ribbed, glabrous to densely hairy; pappus of many unequal bristles (ray pappus sometimes 0). ± 35 spp.: temp w N.Am. (Greek: sword-like anthers) [Hartman 1990 Phytologia 68:439–465] Key to species adapted by David J. Keil.

1. Heads discoid
  2. Ann to short-lived per, < 5 dm; fl-sts ± hairy and glandular . . . . [2]*M. canescens* var. *shastensis*
  2' Subshrub, 5–9 dm; fl-sts glabrous, glaucous . . . . *M. carnosa*
1' Heads radiate
  3. Ray corollas yellow
    4. Ann; ray fls 16–18 . . . . *M. gracilis*
    4' Per; ray fls 30–45 . . . . *M. pinnatifida* var. *gooddingii*
  3' Ray corollas white to purple
    5. Ray pappus 0 or obscure . . . . *M. arida*
    5' Ray pappus conspicuous
      6. Lvs 1–2-pinnately dissected . . . . *M. tanacetifolia*
      6' Lvs entire or toothed
        7. Phyllary tips narrowly awl-shaped, gen hairy throughout . . . . *M. asteroides* var. *asteroides*
        7' Phyllary tips acute to awl-shaped, gen ± glandular . . . . *M. canescens*
          8. Sts glandular, but not canescent-puberulent . . . . var. *leucanthemifolia*
          8' Sts either canescent-puberulent throughout, without glandular hairs, or sts canescent and glandular
            9. Ray fls with styles, fertile, well developed; involucre gen 8–12 mm; phyllaries gen in 5–10 series
            . . . . var. *canescens*
            9' Ray fls without styles, sterile, often ± reduced; involucre 6–9 mm; phyllaries gen in 3–5 series
            . . . . [2]var. *shastensis*

M. arida B. Turner & D. Horne (p. 171) SILVER LAKE DAISY Ann. **STS** branched from base, 5–30 cm, glandular with nonglandular hairs interspersed. **LVS** gen sessile, 1–30 mm, 1–10 mm wide, oblong, toothed to pinnately lobed; lobes and teeth bristle-tipped; upper lvs reduced, appressed. **INFL**: heads radiate, solitary or in cymes; involucre 3–6 mm, 5–10 mm wide, hemispheric; phyllaries in 2–3 series, oblong to oblanceolate, glandular. **RAY FLS** 25–35, < or = 7 mm, white to lavender. **DISK FLS** 28–45, 2.5–3 mm. **FR** 1.4–1.9 mm, hairy; ray pappus 0 or obscure; disk pappus of many bristles, 2–3 mm, white. 2*n*=10. Uncommon. Riverbanks, sandy, alkaline flats, roadsides;

30–300 m. **D**; s NV, s AZ, n Mex. [*Psilactis coulteri* A. Gray misapplied] Mar–Jun

M. asteroides (Torrey) E. Greene Bien, per < 10 dm, gen canescent-puberulent and nonglandular. **STS** 1–several from base, gen branched above and ± bushy. **LVS** gen 3–10 cm, gen 5–25 mm wide, lanceolate to oblanceolate, irregularly dentate to minutely serrate or subentire; lower tapered; upper clasping. **INFL**: heads radiate; phyllaries gen in 5–12 series, tips short-triangular to elongate, acuminate, spreading to bent backward, puberulent. **RAY FLS** many; corollas blue-purple; ligules 1–2 cm. **DISK**

FLS many; corollas 5.5–8 mm. **FR** 2.5–3.5 mm, narrowly obovate, weakly curved and ± flattened with 5–7 ribs on each face, glabrous or ± silky; pappus 6–8 mm. Chaparral, woodland, shrubland; ± 100 m; 800–2400 m. PR, **DSon**; to s NV, sw NM, n Mex. Perhaps not a different sp. from *M. canescens*. 2 vars. in CA.

var. *asteroides* (p. 171) **LVS**: middle cauline lvs 6–15(25) mm wide, minutely serrate. **INFL**: involucre hemispheric; phyllary tips 3–6 mm, narrowly acuminate. 2*n*=8. Shrubland; ± 100 m. **e DSon (near Colorado River)**; to sw NM, n Mex.

M. canescens (Pursh) A. Gray   HOARY-ASTER   Ann to per < 12 dm, gen canescent-puberulent and often glandular. **STS** 1– several from base, gen branched above and ± bushy. **LVS** gen 3– 10 cm, gen 2–6 mm wide, linear to obovate, subentire to dentate or minutely serrate; lower tapered; upper sessile, clasping. **INFL**: heads radiate; phyllaries gen in 3–10 series, tips short-triangular to elongate, acuminate, spreading to bent backward, gen ± glandular or glabrous. **RAY FLS** many (0 in var. *shastensis*); corollas blue-purple; ligules 1–2 cm. **DISK FLS** many; corollas 5.5–8 mm. **FR** 2.5–3.5 mm, narrowly obovate, weakly curved and ± flattened with 5–7 ribs on each face, glabrous or ± silky; pappus 6–8 mm. Common. Many habitats; 300–3400 m. KR, CaR, SN, TR, PR, **GB, DMtns**; w N.Am. Variable; perhaps not a different sp. from *M. asteroides*. 5 vars. in CA.

var. *canescens* (p. 171) Ann to short-lived per. **STS** 1–5 dm, spreading to erect; branches ascending to loosely spreading. **INFL**: heads radiate, 6–12(14) mm, 10–15 mm wide (when pressed); involucres 8–12 mm. **RAY FLS** present, fertile (rarely reduced or 0); style well developed. Open montane habitats; 2000–3000 m. CaR, SN, TR, **n SNE, W&I**; to WA, c Can, Colorado, AZ. [*M. shastensis* A. Gray vars. *glossophylla* (Piper) Cronq. & Keck, *montana* (E. Greene) Cronq. & Keck; *M. tephrodes* (A. Gray) E. Greene] ❀ DRN,SUN:1–3,**15, 16**. Jul–Sep

var. *leucanthemifolia* (E. Greene) Welsh Ann to short-lived per. **STS** 1–5 dm, spreading or erect; branches ascending to loosely spreading. **INFL**: heads radiate; 6–12(14) mm, 10–15 mm wide (when pressed); involucres 5–9 mm. **RAY FLS** present, fertile; style well developed. Desert scrub; 1000–2000 m. e PR, **W&I, DMtns**; to e OR, UT. [*M. l.* (E. Greene) E. Greene] May–Jun

var. *shastensis* (A. Gray) B. Turner Ann to short-lived per. **STS** 1–5 dm, spreading or erect; branches loosely spreading to ascending. **INFL**: heads discoid or radiate; 6–12(14) mm, 10–15 mm wide (when pressed); involucres 6–9 mm. **RAY FLS** 0 or reduced, often sterile; style 0. Various montane habitats; 1500–3400 m. KR, CaR, n SN, MP, **W&I**; s OR, w NV. [*M. s.* A. Gray, incl var. *eradiata* (A. Gray) Cronq. & Keck]

❀ DRN,SUN:1–3,**6,7,15,16**. Jul–Sep

M. carnosa (A. Gray) G. Nesom (p. 171)   SHRUBBY ALKALI AS-TER Subshrub 5–9 dm, much-branched, ± glabrous, glaucous. **LVS**: lower 1–2 cm, linear, entire, fleshy; upper reduced to appressed, awl-shaped scales 1–4 mm. **INFL**: heads discoid, solitary at branch tips; involucre 5.5–7 mm; phyllaries gen in 4–5 series, with tips acute or acuminate, appressed or outer spreading. **RAY FLS** 0. **DISK FLS** 12–18; corollas ± 6 mm. **FR** 2–3 mm, subcylindric, many-ribbed, appressed-hairy; pappus ± 6 mm. Alkaline soils; 100–1600 m. SnJV, **SNE, DMoj**; s NV, w&s AZ. [*Aster intricatus* (A. Gray) S.F. Blake] Jun–Oct

M. gracilis (Nutt.) Shinn. (p. 171) Ann. **STS** erect, 3–25 cm, leafy, branched at or above base, bristly throughout. **LVS** 1–3 cm, 3–7 mm wide; lower oblanceolate, elliptic, or oblong in outline, 1–2-pinnately lobed; upper linear, reduced, lobes and teeth bristle-tipped. **INFL**: heads radiate, solitary or in cymes; involucre 6–7 mm, 7–12 mm wide, hemispheric; phyllaries in 4–6 series, linear-lanceolate, bristle-tipped, hairy. **RAY FLS** 16–18; ligules 7–12 mm, yellow. **DISK FLS** 44–65; corollas 4.5–5.5 mm. **FR** 2.2–2.8 mm, canescent; pappus < or = 5 mm, bristles unequal, slightly wider at base, white to reddish brown. 2*n*=4,8. Sandy or rocky places; < 1500 m. **D**; to Colorado, TX, n Mex. [*Haplopappus g.* (Nutt.) A. Gray; *H. ravenii* R. Jackson] Apr–Jun

M. pinnatifida (Hook.) Shinn. var. *gooddingii* (Nelson) B. Turner & R. Hartman Per. **STS** erect or ascending, 2–6 dm, gen leafy, at least in lower half, ± glandular-puberulent to canescent. **LVS** 2–5 cm, pinnately lobed; lobes linear, bristle-tipped; upper reduced, entire. **INFL**: heads radiate, solitary at tips of long branches; involucre 6–9 mm, 10–18 mm wide, hemispheric; phyllaries linear-lanceolate, tip with short bristle, greenish, prominently glandular-puberulent and scabrous. **RAY FLS** 30–45; ligules 6–16 mm, yellow. **DISK FLS** many; corollas 4–6 mm. **FR** 2–3 mm, appressed-hairy; pappus of many unequal bristles 3–5 mm, tan. Rocky places; < 625 m. **D**; NV, AZ, nw MEX. [*Haplopappus gooddingii* (Nelson) Munz & I.M. Johnston] ❀TRY. Feb–May

M. tanacetifolia (Kunth) Nees Ann, bien 1–7 dm, puberulent to densely glandular. **STS** 1–several from base, gen branched above and ± bushy. **LVS** gen 3–12 cm, 1–2-pinnately dissected. **INFL**: heads radiate; phyllaries gen in 3–5 series, tips elongate, acuminate, spreading to bent backward. **RAY FLS** many; corollas blue-purple; ligules 1–2 cm. **DISK FLS** many; corollas 5–7 mm. **FR** 3–4 mm, narrowly obovate, ± flattened; ribs 4–6 on each face, silky; pappus 4–6 mm. 2*n*=8. Uncommon. Desert scrub, pinyon/juniper woodland; ± 1700 m. **e DMtns (New York Mtns.)**; to MT, SD, TX, n-c Mex.

MADIA TARWEED

Ann or per, gen densely glandular, aromatic. **STS** 1–several, gen simple below, ± branched above. **LVS** gen opposite below, alternate above, gen linear to lanceolate, entire to slightly toothed. **INFL**: heads gen radiate, gen peduncled, few–many; phyllaries gen 1–20, free, enclosing (and falling with) ray achenes; receptacle ± flat, gen glabrous; chaff scales gen ± fused, in ring between ray and disk fls. **RAY FLS** gen 1–20, sometimes minute; ligules 2–3-lobed, gen yellow. **DISK FLS** 1–many, sometimes staminate; corollas yellow or maroon; anther tips triangular-ovate; style tips linear to oblong, acute, bristly. **FR** club-shaped or obovoid; ray achenes compressed, thickened, or 3-angled (1 angle toward center of head), ridged, sometimes beaked; pappus 0 or of short scales; disk achenes ± symmetric; pappus 0 or of 4–10 scales or bristles. 21 spp.: w N.Am, sw S.Am. (Chilean name) [Nelson & Nelson 1980 Brittonia 32:323–325] *See revised taxonomy of Baldwin 1999 Novon 9:462–471.

1. Disk fls gen 25–50+, staminate, not forming frs; ray fls 5–21; ligules 6–20 mm, often maroon-spotted at base . . . . *M. elegans*
  2. Basal rosette well developed; sts stout, very leafy below; phyllary tips = or > body; disk corollas yellow; gen < 1000 m . . . . ssp. *densifolia*
  2' Basal rosette weakly developed or 0; sts slender to ± stout, lvs crowded or well separated; phyllary tips gen = body; disk corollas yellow or maroon; 900–2500 m . . . . ssp. *elegans*
1' Disk fls 1–12, fertile, forming frs; ray fls 0–9; ligules 0.5–8 mm, not maroon-spotted at base

3. Disk corollas glabrous; anthers yellow; phyllary tips not flat; sts threadlike (segregate genus *Hemizonella*)
.... *M. minima*
3' Disk corollas puberulent; anthers black; phyllary tips ± flat; sts gen not threadlike
    4. Involucre narrowly ovoid or ellipsoid; ray fls 0–3; ray frs widest at middle, ends truncate ..... *M. glomerata*
    4' Involucre depressed-spheric to urn-shaped; ray fls 3–9; ray frs widest toward top, tapered to base ... *M. gracilis*

M. elegans Lindley COMMON MADIA Ann 1–25 dm, strongly scented. **STS** simple to branched throughout, often very leafy below, soft-hairy below, sparsely to densely stalked-glandular esp above; glands yellow to black. **LVS** 2–20 cm, linear to widely lanceolate, entire or ± serrate, soft-hairy to bristly, ± glandular. **INFL**: heads in open, rounded to ± flat-topped cymes (sometimes panicle-like); involucre 4.5–12 mm, bell-shaped to hemispheric; phyllaries ± bristly, glandular or not, tips flat; receptacle hairy; chaff scales strongly fused. **RAY FLS** 5–21; corolla tubes 0.5–1 mm, ligules (2.5)4–20 mm, fan-shaped, deeply lobed, yellow, base gen maroon-spotted. **DISK FLS** 25–50+, staminate; corollas 2.5–5 mm, yellow or maroon; anthers yellow or black. **FR**: ray achenes 2.5–5 mm, compressed side-to-side or ± 3-angled, glabrous, black or dark brown, sometimes mottled, beak 0, pappus 0; disk ovaries slender, much reduced, pappus 0. Grassland, open forest; < 3350 m. CA-FP (exc ChI), **GB**; to n OR, Baja CA. Highly variable; intermediates blur distinctions among extremes. 4 sspp. in CA.

  ssp. **densifolia** (E. Greene) Keck (p. 171) Pl 2.5–25 dm. **LVS**: basal rosetted; cauline densely overlapping, densely soft-hairy or bristly, upper (sometimes lower) strongly glandular, hairy. **INFL**: involucre 6–12 mm; phyllary tips often > body. **RAY FLS** (5)12–20; ligules gen 10–20 mm. **DISK FLS** 15–many; corollas yellow; anthers purple-black. **FR**: ray achenes compressed. 2*n*=16. Grassy slopes, valleys; gen < 1000 m. CA-FP (exc ChI), **GB**; to OR. ❀SUN:2,3,5,6,**7–9,14–24**. Aug–Nov

  ssp. **elegans** Pl 2–9 dm. **LVS**: basal few, rosette small; lower cauline ± crowded, soft-hairy, upper ± strongly glandular. **INFL**: involucre 4–10 mm; phyllary tips < or = body. **RAY FLS** 8–16; ligules 6–15 mm. **DISK FLS** few–many; corollas yellow or maroon; anthers purple-black. **FR**: ray achenes compressed. 2*n*=16. Meadows, dry slopes, open forests; 900–2500 m. CA-FP (exc GV, ChI), **SNE (uncommon)**; OR. ❀TRY. Jun–Aug

M. glomerata Hook. (p. 171) MOUNTAIN TARWEED Ann 1.5–8 (12) dm, strongly ill-scented. **STS** simple below or with stiff ascending branches, very leafy, soft-hairy to ± bristly, esp below, stalked-glandular above; glands yellow. **LVS** 2–10 cm, mostly cauline, often with axillary clusters, gen ascending, loosely strigose, often bristly-ciliate; upper glandular. **INFL**: heads discoid or inconspicuously radiate, ± sessile to short-peduncled, in dense cymes or panicle-like clusters; involucre (incl chaff) 5.5–9 mm, narrowly ovoid or ellipsoid, sometimes curved; phyllaries 0–3, soft-hairy or ± bristly, ± glandular, tips flat; chaff scales 1–few, free, ± like phyllaries. **RAY FLS** 0–3; corolla tubes 1–3.5 mm, ligules 1.5–3 mm, greenish yellow or purple-tinged. **DISK FLS** 1–5(12), fertile; corollas 3–4.5 mm; anthers black. **FRS** alike, 4–6 mm, oblanceolate, compressed side-to-side, glabrous, black; beak 0; pappus 0. 2*n*=28. Forest openings; 1050–2700 m. s KR (Trinity Alps), CaR, c&n SNH, SnBr, **GB**; to AK, SD, Colorado, NM. Jul–Sep

M. gracilis (Smith) Keck SLENDER TARWEED Ann 1–10 dm, ± fragrant. **STS**: branches ascending, gen not overtopping main st, ± bristly to long-soft-hairy below, gen glandular above. **LVS** 3–10 cm, soft-hairy, sessile to long-peduncled, in raceme- or panicle-like clusters; involucre 6–10 mm, depressed-spheric to urn-shaped; phyllaries ± densely stalked-glandular, tips flat; chaff scales fused > 1/2 length, easily separated. **RAY FLS** 3–9; corolla tubes 2–3 mm, ligules 1.5–8 mm, lemon-yellow. **DISK FLS** 2–12, fertile; corollas 3–5 mm; anthers black. **FRS** alike, 2.8–5 mm, obovate, flat, ± bowed outward, glabrous, often mottled; beak ± 0; pappus 0. 2*n*=32,48. Grassy areas, woodlands, open forests, many habitats; < 2400 m. CA-FP, **GB**; to B.C., MT, UT, n Baja CA; also s S.Am. [sspp. *collina* Keck, *pilosa* Keck] Highly variable. Apr–Aug

M. minima (A. Gray) Keck (p. 171) Ann 2–15 cm. **STS** very slender, openly branched above or throughout, glandular (esp below), glandular puberulent above. **LVS** 1–2.5 cm, often clustered at nodes, sometimes ± toothed, finely appressed- to spreading-bristly. **INFL**: heads solitary or few, in dense cymes; peduncles 1–12 mm, ± thread-like, gen bracted below; involucre 2–4 mm, widely top-shaped or obovoid; phyllaries loosely appressed, tips not flat, backs minutely stalked-glandular (glands golden to black); chaff scales strongly fused. **RAY FLS** 3–5, inconspicuous; corolla tubes 0.5–1 mm, ligules 0.5–1 mm, pale yellow. **DISK FLS** 1–2, fertile; corolla 1–2.3 mm; anthers yellow. **FR**: ray achenes 1.8–2.8 mm, compressed front-to-back, strongly bowed out, not angled, hairy, black, beaked, pappus 0; disk achenes ± cylindric or club-shaped, pappus 0. 2*n*=32. Open forest, shrubland; 550–2600 m. KR, NCoR, CaR, SN, se SnFrB, n WTR, SnBr, PR, MP, **w DMoj**; to B.C. **Hemizonella minima* A. Gray May–Jul

MALACOTHRIX

W.S. Davis

Ann or per < 60 dm, gen ± branched, gen erect; sap milky. **LVS** gen basal and cauline, alternate, sessile, gen reduced upward. **INFL**: heads ligulate, 2–10 mm diam; involucre 5–20 mm, gen bell-shaped; phyllaries in 3–6 series; receptacle naked or with fragile bristles < 5 mm. **FLS**: corollas readily withering; ligules yellow or white, gen ± purple-striped below. **FR** gen fusiform, straw-colored to purple-brown, truncate; veins 15 (5 gen prominent, 10 sometimes obscure); outer pappus ± 0 or of 0–6 smooth, persistent bristles, inner 12–32 bristles fused at base, readily deciduous, minutely barbed below. 21 spp.: w N.Am, s S.Am. (Greek: soft hair) [Williams 1957 Am Midland Naturalist 58:494–517] Key to species adapted by Margriet Wetherwax.

1. Cauline lvs with ear-like basal lobes; involucre gen spheric; outer phyllaries round .... *M. coulteri*
1' Cauline lvs gen without ear-like basal lobes; involucre bell-shaped; outer phyllaries lanceolate to ovate
  2. Outer ligules gen exserted 1–2 mm .... *M. stebbinsii*
  2' Outer ligules exserted > 5 mm
    3. St scapose; lvs gen long-hairy at base .... *M. californica*
    3' Sts leafy, ± branched; basal lvs gen not long-hairy at base
      4. Basal lvs gen not fleshy, narrowly toothed or lobed .... *M. glabrata*
      4' Basal lvs ± fleshy, with 3–8 pairs of ± equal, short, wide, gen toothed lobes

5. Cauline lvs widest at base; outer pappus a scalloped crown, bristles deciduous . . . . *M. sonchoides*
5' Cauline lvs gen narrowed to strap-like base; outer pappus of irregular teeth, persistent bristles 0–6
. . . . *M. torreyi*

M. californica DC. (p. 171) Ann 5–45 cm, scapose. **ST** gen glabrous. **LVS** basal, gen linear to oblanceolate; bases conspicuously long-hairy; teeth or narrow lobes well spaced. **INFL**: heads 5.2–6 mm diam; involucre gen 10–15 mm; outer phyllaries ± 1/2 inner, lanceolate, tangled-hairy. **FLS**: corollas 17–22 mm, gen yellow (white), outer corollas exserted 11–13 mm. **FR** 2–3.4 mm, ± light brown; outer pappus teeth irregular, bristles 2. *2n*=14. Open, sandy soil in grassland, oak woodland, chaparral, desert margins; < 1700 m. SnJV, CW, SW, **DMoj**; Mex. ❀TRY. Mar–May

M. coulteri A. Gray (p. 171) SNAKE'S-HEAD Ann 5–50 cm, ± branched, ascending or erect, ± glabrous, glaucous. **LVS**: basal oblanceolate to obovate, entire, toothed or short-lobed; cauline ear-like at base. **INFL**: heads 6–10 mm diam; involucre gen spheric, 10–20 mm; outer phyllaries widely overlapping, ± = inner, widely ovate to round. **FLS**: corollas 8–12 mm, gen pale yellow (white), outer corollas exserted 2–5 mm. **FR** 1.6–3.2 mm, ± light brown; 5 veins ± winged; outer pappus teeth uneven, bristles 2–6. *2n*=14. Sandy, open areas, in coastal sage scrub, grassland, deserts; < 1500 m. SnJV, CW, (formerly n ChI), WTR, PR, **SNE**, **DMoj**. ❀SUN,DRN:**7–11,14–16**,17,18,**19–24**. Mar–May

M. glabrata A. Gray (p. 171, pl. 21) DESERT DANDELION Ann 6–40 cm, ± glabrous. **LVS**: basal oblanceolate to obovate, with well spaced teeth or long narrow lobes, base sometimes ± hairy; cauline gen long-lobed. **INFL**: heads gen 4–7 mm diam; involucre 9–17 mm; outer phyllaries gen 1/2 inner, lanceolate, sometimes short-white-hairy. **FLS**: corollas 15–23 mm, gen pale yellow (white), outer corollas gen exserted 9–13 mm. **FR** 2–3.3 mm; outer pappus of irregular teeth, bristles 1–5. *2n*=14. Coarse soils in open areas or among shrubs; < 2000 m. SnJV, **SNE**, **D**; to OR, ID, UT, Mex. ❀SUN,DRN:1,2,**3,7–10**,11,**14–16,18–21**,22,23. Mar–Jun

M. sonchoides (Nutt.) Torrey & A. Gray (p. 171) Ann 6–40 cm, ± glabrous. **LVS**: basal obovate, ± fleshy, ± equally and widely 6–16-lobed; cauline widely lobed at base. **INFL**: heads 4–6 mm diam; involucre 7–13 mm; outer phyllaries ± 1/2 inner, ovate, margins sometimes with tack-shaped hairs. **FLS**: corollas 10–14 mm, lemon yellow, outer corollas exserted 6–10 mm. **FR** 1.8–3 mm, dark or red-brown; outer pappus a scarious, rounded-toothed crown, bristles 0. *2n*=14. Gen deep, sandy soils; 400–1400 m. **SNE**, **DMoj**; to WY, NM. ❀TRY. Apr–Jun

M. stebbinsii W. Davis & Raven (p. 171) Ann 5–60 cm, gen glabrous. **LVS**: basal gen obovate, ± narrowly lobed; upper cauline ± 2–4-lobed near base. **INFL**: heads ± 3–4 mm diam; involucre 7–10 mm; outer phyllaries ± 1/2 inner, lanceolate to ovate. **FLS**: corollas 6–8 mm, white or pale yellow, outer corollas exserted 1–2 mm. **FR** 1.7–2.3 mm, ± light brown; veins ± equal, very tip veinless; outer pappus of narrow teeth, bristles 1(2). *2n*=28. Uncommon. Gravelly soils beneath shrubs; 300–1300 m. **SNE**, **D**; OR, NV, AZ. *M. similis* W. Davis & Raven (fr veined to very tip), mostly of Mex, collected twice in CA (n SCo, n ChI).

M. torreyi A. Gray (p. 171) Ann 4–40 cm, often with some tack-shaped hairs. **ST** ± glaucous. **LVS**: basal obovate, ± fleshy, ± equally and widely 6–16-lobed; cauline sometimes entire, gen narrowed to strap-like base. **INFL**: heads 4–5 mm diam; involucre 8–14 mm; outer phyllaries ± 1/2 inner, gen lanceolate, with tack-shaped hairs. **FLS**: corollas 14–20 mm, medium yellow, outer corollas exserted 7–10 mm. **FR** ± 2.5–4 mm, brown; outer pappus of irregular teeth, bristles 0–6. *2n*=14. Coarse soils in desert scrub; 1500–1800 m. **GB**; to OR, ID, WY, Colorado, AZ. ❀TRY.

MALPERIA

A. Michael Powell

1 sp. (Anagram formed from name of collector of type material, E. Palmer, 1831–1911) [King & Robinson 1987 Monogr Syst Bot Missouri Bot Garden 22:235–237]

M. tenuis S. Watson (p. 171) BROWN TURBANS Ann < 40 cm, glabrous to puberulent. **LVS** opposite below, alternate above, gen sessile, < 5 cm, linear, gen entire. **INFL** open; heads discoid, many, 8–14 mm, 5–6 mm wide; involucre cylindric to narrowly bell-shaped; phyllaries 20–25, unequal, in several series, lanceolate; receptacle flat, naked. **FLS** ± 30; corollas 5–6 mm, cylindric, ± white, pink-tinged; anther bases rounded, tips oblong; style branches long, club-shaped, rounded. **FR** 3 mm, slender, 5-ribbed; pappus of 3 bristles 5 mm and 3 scales 0.5 mm. *2n*=20. RARE in CA. Sandy creosote-bush scrub; < 500 m. **DSon**; n Mex. Apr, Dec

MONOLOPIA

Dale E. Johnson

Ann 1–8 dm, sometimes decumbent, ± woolly. **LVS** opposite below, alternate above, linear to (ob)lanceolate, entire to wavy-dentate. **INFL**: head solitary, radiate; involucre hemispheric; phyllaries in 1 series, free or fused into lobed cup, tips black-hairy; receptacle conic, naked. **RAY FLS** 1 per phyllary; corolla gen yellow, with small lobe opposite ligule. **DISK FLS** yellow; lobe hairs large, nonglandular. **FR**: ray frs 3-angled (disk frs 4-angled), ± obconic; pappus 0. 4 spp.: CA. (Greek: single husk, from phyllaries) [Johnson 1991 Novon 1:119–124]

M. lanceolata Nutt. (p. 171) STS simple or branched above and below. **INFL**: peduncles 1–13 cm; involucre 6–10 mm; phyllaries ± 8, elliptic-oblanceolate, free or fused to ± 1/2. **RAY FLS**: ligules 10–20 mm, 3-lobed. **FR** 2–4 mm, compressed front-to-back, strigose. *2n*=20. Grassland, bare clay, open chaparral, woodland; 50–1600 m. s SNF, Teh, c SCo, WTR, SnGb, nw PR, **w edge DMoj**. ❀ SUN,DRN, IRR :**7,14–24**. Mar–May

MONOPTILON DESERT STAR

Ann, gen prostrate, often minute, gen ± short-bristly. **LVS** alternate, often tufted below heads, linear or oblanceolate, entire. **INFL**: heads radiate, ± subsessile; involucre bell-shaped or hemispheric; phyllaries many in 1 series, equal,

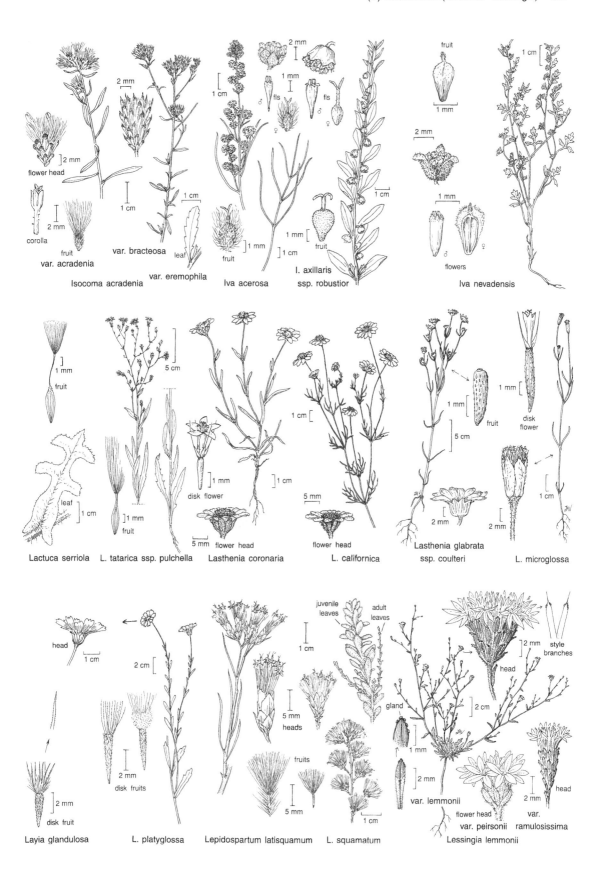

2 mm

flower head

2 mm

corolla

fruit

var. acradenia

Isocoma acradenia

var. bracteosa

1 cm

var. eremophila

leaf

1 cm

Iva acerosa

2 mm

1 mm

♂ fls

♀ fls

♂ ♀

1 mm

fruit

1 mm

fruit

I. axillaris

ssp. robustior

fruit

1 mm

2 mm

1 mm

♂ ♀

flowers

1 cm

Iva nevadensis

1 mm

fruit

leaf

1 cm

Lactuca serriola

5 cm

1 mm

fruit

L. tatarica ssp. pulchella

disk flower

1 mm

1 mm

5 mm

5 mm flower head

Lasthenia coronaria

1 cm

1 cm

5 mm

flower head

L. californica

1 mm

fruit

5 cm

2 mm

Lasthenia glabrata

ssp. coulteri

disk flower

1 mm

2 mm

1 cm

L. microglossa

head

1 cm

2 cm

disk fruits

2 mm

disk fruit

2 mm

Layia glandulosa

L. platyglossa

juvenile leaves

adult leaves

1 cm

5 mm

heads

fruits

5 mm

1 cm

Lepidospartum latisquamum

L. squamatum

gland

1 mm

2 mm

var. lemmonii

flower head

var. peirsonii

head

2 cm

style branches

2 mm

head

2 mm

var. ramulosissima

Lessingia lemmonii

0± folded, gen acuminate, purplish; receptacle convex, naked. **RAY FLS** many; ligules white to purple, often dark-veined. **DISK FLS** many; corollas yellow; style tips short-triangular. **FR** compressed, oblong to obovate, finely appressed-hairy, light brown; pappus of scales and gen 1 or more slender bristles. 2 spp.: sw N.Am. (Greek: 1 feather, from pappus)

1. Pappus a minute crown of fused scales 0.1–0.2 mm and 1 plume-tipped bristle . . . . *M. bellidiforme*
1' Pappus of 0–12 non-plumose bristles and several narrow scales ± dissected into bristles . . . . *M. bellioides*

M. bellidiforme A. Gray (p. 171) **STS** < 6 cm. **LVS** 4–10 mm, gen oblanceolate, obtuse. **INFL**: phyllaries 4–4.5 mm. **RAY FLS**: corolla 5–7 mm; ligule 3–5 mm. **DISK FLS**: corollas 3–4.5 mm. **FR** 1.5–2 mm; pappus a minute crown of fused scales 0.1–0.2 mm and 1 plume-tipped bristle 3–4 mm. 2n=16. Sandy deserts, washes; 660–1000 m. **DMoj**; to sw UT, w AZ. Mar–Jun

M. bellioides (A. Gray) H.M. Hall (p. 171, pl. 22) **STS** < 25 cm. **LVS** 5–10 mm, gen linear, obtuse to subacute. **INFL**: phyllaries 4–6 mm. **RAY FLS**: corollas 6–11.5 mm; ligule 5–8.5 mm. **FR** 1.5–2 mm; pappus of 0–12 bristles 1–2 mm and several slender scales 0.5–1 mm, ± dissected into bristles. 2n=16. Sandy deserts, washes; 200–1200 m. **D**; to sw UT, w AZ, nw Mex. Feb–May, Sep

NICOLLETIA

Ann, per; herbage ill-scented, glabrous, often glaucous. **LVS** gen alternate, simple, pinnately lobed; each lobe with an embedded oil gland, bristle-tipped. **INFL**: heads radiate, solitary or in cymes; peduncles short, stout; involucre hemispheric to bell-shaped; phyllaries gen in 2 series, free, outer 0–6, short, lanceolate to triangular, inner 8–12, oblong, membrane-margined, gland-dotted; receptacle convex, naked. **RAY FLS** gen 8; ligules showy, white to pink.-purple. **DISK FLS** many; corollas yellow, sometimes purple-tipped; style-tips long, thread-like. **FR** narrowly club-shaped, short-hairy; pappus of 5 bundles of bristles alternating with 5 lanceolate scales. 3 spp.: sw US, n Mex. (J.N. Nicollet, French astronomer, explorer, 1786–1843) [Strother 1978 Sida 369–374]

N. occidentalis A. Gray (p. 175, pl. 23) Per from deep taproot, glaucous. **STS** erect, 12–30 cm. **LVS** ± fleshy, 2–7 cm, divided into 5–11 pairs of short lobes. **INFL**: heads solitary; peduncles gen 2–10 mm; involucre ± cylindric; outer phyllaries 4–8 mm; inner phyllaries 14–18 mm, linear to ovate, acute to acuminate, gland-dotted. **RAY FLS** 8–12; ligules 4.5–8.5 mm. **DISK FLS**: corollas 8–9.5 mm. **FR** 7–9 mm; pappus bristles 3–7 mm, scales 6–8 mm. 2n=20. Sandy desert soils, often dunes, washes; 600–1400 m. s SNH, **DMoj**; also n Baja CA. Apr–Jun

PALAFOXIA

Ann, per, subshrubs. **STS** ascending or erect. **LVS** simple, opposite below, alternate above, gen petioled; blades entire. **INFL**: heads radiate or discoid, in cymes; involucre ± cylindric; phyllaries in 2–3 ± equal series; receptacle flat, naked. **RAY FLS** 0–few; corolla ± purple. **DISK FLS** few–many; corollas radial or ± bilateral, white to ± purple, lobes narrowly triangular; anther tips triangular; style branches linear. **FR** 4-angled, narrowly obpyramidal; pappus of 4–10 scales. 12 spp.: sw US, Mex. (J. Palafox, Spanish general, 1776–1847) [Turner & Morris 1976 Rhodora 78:567–628]

P. arida B. Turner & M. Morris (p. 175) Ann. **STS** gen erect, much-branched. **LVS** 2–12 cm; petioles 5–15 mm; blades linear to linear-lanceolate. **INFL**: heads discoid; cymes ± flat-topped; involucre cylindric or ± obconic; phyllaries linear, scabrous to densely glandular. **FLS** 9–40; corollas white to pink; anthers pink to purple. **FR** 10–15 mm, ± strigose; pappus of outer achenes 0 or of 3–8 scales of varying length; pappus of inner achenes of 4 scales 8–12 mm and 4 shorter scales. Sandy places; < 1000 m. **D**; NV, AZ, n Mex. [*P. linearis* (Cav.) Lagasca misapplied]

1. Pl gen 1–7 dm; main st < or = 5 mm diam; heads gen 20–25 mm . . . . var. *arida*
1' Pl gen 9–20 dm; main st 5–10 mm diam; heads 28–35 mm . . . . var. *gigantea*

var. *arida* **STS** gen 1–7 dm; axis and branches < or = 5 mm diam, ± rough-hairy, glandular above. **LVS** 2–10 cm, ± rough-canescent. **INFL**: heads gen 5–40, gen 20–25 mm (incl fls); involucr, 5–10 mm diam, cylindric; phyllaries 10–20 mm. **FLS** 9–20; corollas 9–11 mm, outer ± bilateral. **FR** 10–15 mm. 2n=24. Common. Habitats and range of sp. [*P. linearis* var. *l.*] ⊛STBL. Jan–Sep

var. *gigantea* (M.E. Jones) B. Turner & M. Morris GIANT SPANISH-NEEDLE **STS** gen 9–20 dm, ± glabrous; axis and branches gen 5–10 mm diam, ± glabrous. **LVS** 6–12 cm, gen ± glabrous. **INFL**: heads gen 10–20, 28–35 mm (incl fls); involucre 10–20 mm diam, obconic; phyllaries 16–25 mm. **FLS** 18–40; corollas 10–13 mm, radial. **FR** 12–16 mm. 2n=24. RARE. Desert sand dunes; < 100 m. **DSon (se Imperial Co.)**; sw AZ. [*P. linearis* var. *g.* M.E. Jones] Mar–May

PECTIS

Ann, per, often scented. **STS** prostrate to erect. **LVS** simple, opposite, sessile, narrow, bristly-ciliate, dotted with embedded oil glands. **INFL**: heads radiate, peduncled, solitary or in leafy cymes; involucre cylindric to bell-shaped; phyllaries in 1 series, free, gland-dotted; receptacle naked. **RAY FLS** as many as and subtended by phyllaries; corollas yellow. **DISK FLS** few–many; corollas yellow, 4–5-lobed, gen 2-lipped. **FR** cylindric, gen puberulent; pappus of bristles, scales, or awns. ± 85 spp.: w US, Caribbean, S.Am. (Greek: comb, from ciliate lvs) [Keil 1977 Rhodora 79:32–78]

P. papposa Harvey & A. Gray var. *papposa* (p. 175) CHINCH-WEED Ann 1–20 cm, mound-shaped; herbage spicy-scented. **STS** 1–several from base, simple or much-branched. **LVS** narrowly linear, gland-dotted on margins. **INFL**: heads in dense cymes, 6–10 mm diam; peduncles 3–10 mm; phyllaries 8, 3–5 mm, linear, each with subterminal gland and several smaller

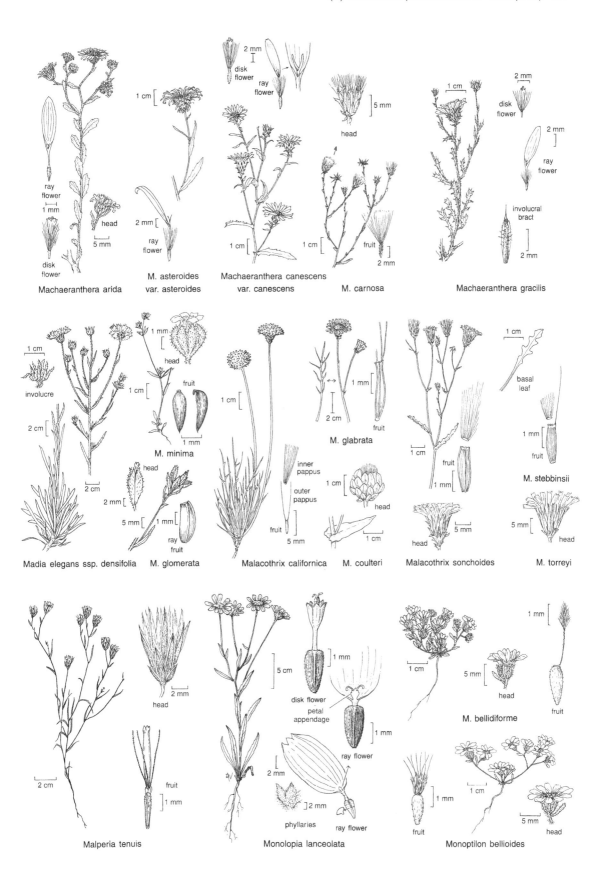

Machaeranthera arida

M. asteroides
var. asteroides

Machaeranthera canescens
var. canescens

M. carnosa

Machaeranthera gracilis

Madia elegans ssp. densifolia

M. glomerata

M. minima

M. glabrata

M. stebbinsii

Malacothrix californica

M. coulteri

Malacothrix sonchoides

M. torreyi

Malperia tenuis

Monolopia lanceolata

M. bellidiforme

Monoptilon bellioides

submarginal glands. **RAY FLS** 8; corollas 3–6 mm. **DISK FLS** 6–14; corollas 2–3.5 mm. **FR** 2–4.5 mm; pappus of rays a low crown; pappus of disk ± 20 subplumose bristles. Arid plains, rocky slopes; < 1500 m. **SNE, D**; to sw UT, sw NM, nw Mex. ❀STBL. After summer rains, Jun, Sep–Nov

PERICOME

Per, subshrub. **STS** many from base. **LVS** simple, gen opposite (or uppermost alternate), petioled, deltate-ovate, puberulent, gland-dotted; tip long-acuminate. **INFL**: heads discoid, small, few–many in ± flat-topped cymes, these often arrayed in leafy-bracted, compound clusters; peduncles slender; involucre cylindric to bell-shaped; phyllaries in 1 series, linear, ± fused; receptacle rounded, naked. **FLS** many; corollas 4-lobed, creamy yellow; anther tips triangular; style tips linear, tapered. **FR** oblanceolate, flat; surfaces black, puberulent, margins ± thickened, densely ciliate; pappus a low crown of fringed scales, sometimes with 1–2 bristles. 2 spp.: sw US, n Mex. (Greek: hairs around, from ciliate fr margin) [Powell 1973 Southw Naturalist 18:335–339]

P. caudata A. Gray (p. 175) **STS** < or = 2 m, much-branched, ± puberulent, resin-dotted. **LVS** many; petioles 1.5–5 cm; blades 3–12 cm, base rounded to cordate or hastate, margin entire or basal half-toothed or shallowly few-lobed. **INFL**: peduncles 5–30 mm; involucre 4.5–6 mm diam; phyllaries 4.5–7 mm, fused in lower half, margins transparent, tips soft-hairy. **FLS**: corollas 3–5 mm. **FR** 3.5–5 mm; pappus scales ± 1 mm, bristles 0–2, 1–4.5 mm, gen unequal. 2n=36. Dry, rocky slopes; 1200–2400 m. s SN, **SNE**; to Colorado, TX, n Mex. Jul–Oct

PERITYLE

Ann, per, subshrubs, shrubs. **LVS** opposite or alternate, simple to deeply divided or compound, sessile or petioled. **INFL**: heads radiate, discoid or disciform, solitary or in cymes; peduncles short or long; involucre cylindric, hemispheric, or bell-shaped; phyllaries in 2–3 ± equal series, linear to ovate; receptacle flat to conic, naked. **RAY FLS**: ligules yellow or white. **DISK FLS** many; corollas yellow or white, 4-lobed; anther tips triangular; style tips tapered. **FR** linear to oblanceolate, very flat, sometimes weakly 3–4-angled; surface dark brown or black, glabrous or puberulent; margins gen ± thick, puberulent to strongly ciliate; pappus 0 or a crown of fringed scales and 0–2 slender bristles. ± 75 spp.: sw N.Am. (Greek: around the margin, from thick fr margin) [Powell 1974 Rhodora 76:229–306] ❀STBL.

1. Ann; heads gen radiate . . . . *P. emoryi*
1' Subshrubs; heads discoid
  2. Herbage gen long-hairy
    3. Lf margins serrate to lobed; lvs opposite or alternate . . . . *P. inyoensis*
    3' Lf margins entire or lobes 1–3 per margin; lvs alternate . . . . *P. villosa*
  2' Herbage short-rough-hairy . . . . *P. megalocephala*
    4. Lvs many, narrowly ovate to ± round, 4–9 mm wide . . . . var. *megalocephala*
    4' Lvs few, linear to lanceolate, 1–4 mm wide . . . . var. *oligophylla*

P. emoryi Torrey (p. 175) Ann 2–60 cm, puberulent to rough-hairy and glandular. **STS** simple to much-branched. **LVS** gen alternate, petioled; blades 2–10 cm, ovate, round, or triangular, coarsely toothed to palmately lobed, teeth and lobes gen again toothed or lobed. **INFL**: heads radiate (rarely disciform), 1–many; peduncles 0.1–7 cm; involucre hemispheric to bell-shaped; phyllaries many, 5–6 mm, lanceolate or oblanceolate to ovate. **RAY FLS** gen 8–12; ligules 1.5–4 mm, white, rarely vestigial. **DISK FLS**: corollas 2–2.5 mm, yellow. **FR** gen 2–3 mm; margins thin, ciliate; surfaces of ray achenes gen ± puberulent; surfaces of disk achenes gen glabrous; pappus scales well developed or vestigial, bristle 0 or 1, 1–2.5 mm. 2n=64–72, 100–116. Common. Desert plains, slopes, and washes; < 1000 m. **D** (uncommon SCo, ChI); NV, AZ, n Mex; also in w S.Am. Feb–Jun

P. inyoensis (Ferris) A. Powell (p. 175) INYO LAPHAMIA Subshrub 12–25 cm. **STS** many; hairs soft, spreading. **LVS** opposite or alternate; petioles 0.5–2 mm; blade 1–2 cm, ovate to triangular or round, tip ± acute, margin serrate to lobed, surface soft-hairy and glandular. **INFL**: heads discoid, 1–3; peduncles 1–4 cm; involucre bell-shaped; phyllaries 14–21, 5.6–6.5 mm, linear-lanceolate. **RAY FLS** 0. **DISK FLS**: corollas 4–5 mm. **FR** 3–3.5 mm, 1 or both surfaces rounded or angled, puberulent; margins hairy; pappus 0 or a crown of vestigial scales. 2n=±36. RARE. Dry, rocky slopes; 1800–2600 m. **W&I (s Inyo Mtns)**. [*Laphamia i.* Ferris]

P. megalocephala (S. Watson) J.F. Macbr. Subshrub 15–55 cm, short-rough-hairy. **STS** many, much-branched. **LVS** gen alternate; petiole 1–6 mm; blade 7–15 mm, 1–10 mm wide, linear to widely ovate, tip acute to obtuse, margin gen entire, rarely serrate or lobed. **INFL**: heads discoid, 1–few, loosely clustered; peduncles gen 1–5 cm, bracts lf-like; involucre 3.5–7.5 mm diam, bell-shaped; phyllaries 14–19, 5–6 mm, lanceolate or oblong. **RAY FLS** 0. **DISK FLS**: corollas 3.5–4.2 mm. **FR** 2.5–3 mm; 1 or both surfaces rounded or ± angled, puberulent; margin hairy; pappus 0 or a crown of vestigial scales, rarely with 1 bristle. Rocky places; 1300–2800 m. **W&I, DMtns**; NV.

var. *megalocephala* Pl gen 30–60 cm. **LVS** many, 7–15 mm, 4–9 mm wide, lance-ovate to round. **INFL**: heads gen 5–6 mm diam. **FR**: pappus bristle 0. 2n=34. Rocky slopes; 1500–2800 m. **W&I, DMtns**; NV. [*Laphamia megalocephala* S. Watson var. *m*] Jun–Nov

var. *oligophylla* A. Powell Pl gen 15–35 cm. **LVS** few, 7–17 mm, gen 1–4 mm wide, linear to lanceolate. **INFL**: heads gen 3.5–7.5 mm diam. **FR**: pappus bristle 0 or rarely 1. 2n=68. Rocky slopes; 1300–2600 m. **W&I, DMtns**. [*Laphamia m.* S. Watson var. *intricata* (Brandegee) Keck misapplied] Jul–Aug

P. villosa (S.F. Blake) Shinn. HANAUPAH LAPHAMIA Subshrub 13–20 cm, soft-hairy. **LVS** alternate; petiole 3–6 mm; blade 12–22 mm, ovate to widely wedge-shaped, tip acute, margins entire or with 1–3 short, pointed lobes, surface soft-hairy. **INFL**: heads discoid, 1–3; peduncle gen 1–2 cm; involucre 5–7 mm diam, bell-shaped; phyllaries 13–23, 6 mm, linear-lanceolate to oblong. **RAY FLS** 0. **DISK FLS**: corollas 4–5 mm, yellow. **FR** 3–3.5 mm, puberulent; surfaces rounded; pappus 0 or a vestigial crown, sometimes with 1–2 bristles 1–2 mm. 2n(=3x)=51. RARE. Dry, rocky slopes; 1700–2600 m. **W&I (Inyo Mtns), n DMtns (Panamint, Grapevine mtns)**. [*Laphamia v.* S.F. Blake] Jul–Sep

PETRADORIA   ROCK GOLDENROD

Meredith A. Lane

1 sp.: w N.Am. (Greek: rock goldenrod)

P. pumila (Nutt.) E. Greene ssp. ***pumila*** (p. 175)   Per < 3 dm, light green, glabrous, taprooted. **STS** several, erect, striate, gummy. **LVS** alternate, 2–12 cm, linear to (ob)lanceolate, leathery, 3- veined, resin-dotted, entire; margins scabrous. **INFL**: heads radiate, many, in dense, flat-topped clusters; involucres < 10 mm, < 3 mm diam, cylindric; phyllaries 10–21 in 3–6 series, oblong to ovate, light yellow, tips green; receptacle naked. **RAY FLS** 2– 3; ligules yellow. **DISK FLS** 2–4, staminate; corollas < 6.2 mm, yellow. **FR** 4–5 mm, 6–9-veined, compressed, glabrous; pappus bristles thread-like, ± twisted, brownish. 2*n*=18. Rocky soils, pine forest to juniper scrub, often on limestone; 2300–3400 m. **DMtns**; to ID, WY, NM. [Anderson 1963 Trans Kansas Acad Sci 66:632–684] ✺TRY. Jul–Oct

PEUCEPHYLLUM

1 sp.: sw N.Am. (Greek: fir-leaf) [Strother 1978 N.Am Fl II 10:160–173]

P. schottii A. Gray (p. 175, pl. 24)   PYGMY-CEDAR   Shrub or small tree < 3 m, rounded. **ST** densely leafy, green. **LVS** alternate, gen 1–2 cm, narrowly linear, thick, gen entire, glabrous, gland-dotted, resin-varnished. **INFL**: heads solitary, discoid; peduncle 8–25 mm, leafy-bracted; involucre ± obconic; phyllaries 9–18 in 1 series, 8–12 mm, linear to lanceolate, thick, acuminate, gland-dotted near tip, margins often scarious; receptacle flat, naked. **DISK FLS** 12–21; corollas 6.5–8.5 mm, pale yellow, tube << throat; anther bases weakly tailed, tips lanceolate to ovate; style branches minutely papillate, rounded-truncate. **FR** 3–4 mm, narrowly obconic, weakly angled, blackish, bristly; pappus of many fine bristles, 2–5 mm (sometimes also 15–20 slender scales 4–6 mm), straw-colored to red-brown. Rocky slopes, often among boulders; < 1400 m. **D**; AZ, nw Mex. ✺TRY; DFCLT. Dec–May

PLEUROCORONIS   ARROW-LEAF

A. Michael Powell

Per or subshrub. **LVS** opposite below, alternate above. **INFL** open; heads discoid, few; phyllaries in several series, outer short, ovate, inner lanceolate; receptacle ± flat, naked. **FLS** 25–30; anther bases ± rounded, tips ovate or oblong; style branches, ± 1.5 mm, club-shaped. **FR** ellipsoid to obconic, 4–5-ribbed; sides densely hairy; pappus of bristles and scales. 3 spp.: sw US, nw Mex. (Greek & Latin: side crown, from pappus) [King & Robinson 1987 Monogr Syst Bot Missouri Bot Garden 22:237–239]

P. pluriseta (A. Gray) R. King & H. Robinson (p. 175)   Subshrub < 60 cm. **LVS** thin; blades 3–10 mm (<< glandular petioles), lanceolate to ± diamond-shaped, gen few-toothed, glabrous or ± glandular. **INFL**: heads 6–11 mm; phyllaries ± 25, 2.5–6 mm, glandular, tips darker, gen recurved. **FLS**: corollas 4–5 mm. **FR** 3–4 mm; pappus bristles 10–16, 2.5–5 mm, scales 10–12, 1–2 mm. 2*n*=18. Common. Rocky creosote-bush scrub; < 1300 m. **D**; to UT, AZ, nw Mex. [*Hofmeisteria p.* A. Gray] Mar–May, Oct–Jan

PLUCHEA

G. Ledyard Stebbins

Ann to shrub, stiff. **LVS** alternate. **INFL**: heads disciform, many; involucre ± hemispheric; phyllaries in 3–5 unequal series, outer leathery above, inner narrower, ± membranous; receptacle naked. **PISTILLATE FLS** many; corolla very slender, 4–5-lobed. **DISK FLS** few, fruiting or staminate; corolla 5-lobed, pink or purple in CA; anther bases short-tailed, tips ± ovate; style branches 0 to short, obtuse. **FR** ± cylindric, grooved; pappus 1 series of slender bristles. ± 40 spp.: trop, warm temp. (N.A. Pluche, 18th century French naturalist)

1. Ann or per, glandular; lvs 4–12 cm, ovate, toothed, not crowded . . . . ***P. odorata***
1' Shrub, nonglandular, silky-hairy; lvs 1–4 cm, linear to lanceolate, entire, crowded . . . . ***P. sericea***

P. odorata (L.) Cass. (p. 175)   SALT MARSH FLEABANE   Ann or per 5–12 dm, coarse, glandular, ill-scented. **LVS** 4–12 cm, ovate, not crowded, toothed. **INFL**: involucre 4.5–5.5 mm. **PISTILLATE FLS**: corollas 3.5–4 mm, purple. **DISK FLS**: corollas 4–5 mm, purple. **FR** ± 1 mm, minutely rough-hairy; pappus bristles slender to tip. 2*n*=20. Moist, often saline valley bottoms; 0–300 m. s SNF, GV, SnFrB, SCo, ChI, **D**; to e US, Caribbean, n S.Am. [*P. purpurascens* (Sw.) DC. and *P. camphorata* (L.) DC. misapplied] ✺TRY. Jul–Nov

P. sericea (Nutt.) Cov. (p. 175)   ARROW WEED   Shrub 1–5 m, finely silky, not scented, nonglandular. **LVS** 1–4 cm, linear to lanceolate, crowded, entire. **INFL**: involucre 3.5–4.5 mm. **PISTILLATE FLS**: corollas 5–6.5 mm, pink to deep rose. **DISK FLS**: corollas 5–6 mm, pink to deep rose. **FR** 0.5–1 mm, smooth. 2*n*=20. Forming thickets in stream bottoms, washes, canyons, around springs, sometimes in saline areas; < 600 m. SnJV, SCoRI, SCo, s ChI, TR, PR, **D**; to TX, nw Mex. ✺STBL;INV. Mar–Jul

POROPHYLLUM

Ann, per, subshrubs; odor strong. **STS** erect. **LVS** simple, alternate or opposite; blade dotted with embedded oil glands. **INFL**: heads discoid, peduncled, solitary or in ± leafy cymes; involucre ± cylindric; phyllaries in 1 series, equal, free or fused at base, gland-dotted; receptacle naked. **FLS**: corollas white to greenish yellow or purple; style branches long, slender. **FR** cylindric; pappus of many bristles. ± 30 spp.: N.Am., S.Am. (Greek: pore-leaf, from gland-dotted lvs) [Johnson 1969 Univ Kansas Sci Bull 48:225–267]

P. gracile Benth. (p. 179) ODORA Subshrub; herbage ill-scented. **STS** 1–many, 3–7 dm, glaucous, glabrous; branches slender, ascending. **LVS** 1–5 cm, narrowly linear, entire, glaucous. **INFL**: heads 1–few; involucre 4–8 mm diam; phyllaries 5, free, 10–16 mm, oblong, glaucous, dotted or streaked with glands. **FLS** 20–30; corolla 7–9 mm, purplish or whitish. **FR** 8–10 mm; pappus 6–7 mm, dull white to brownish. 2*n*=48. Rocky slopes; < 1500 m. **D**; to NV, TX, n Mex. ❀STBL. Oct–Jun

PRENANTHELLA

G. Ledyard Stebbins

1 sp.: w N.Am. (Latin: diminutive of *Prenanthes*)

P. exigua (A. Gray) Rydb. (p. 179) Ann, 1–4 dm; sap milky. **STS** slender, gen openly much-branched, sparsely glandular-puberulent. **LVS**: basal, 2–4 cm, oblanceolate, entire to pinnately lobed, ciliate, otherwise glabrous, often early-deciduous; cauline few, reduced, upper scale-like. **INFL**: heads ligulate, many, in open panicles; peduncles very slender; involucre cylindric, 4–5 mm; phyllaries in 2 series, outer 2–3, < 1 mm, inner 3–4, 4–5 mm, lanceolate; receptacle naked. **FLS** 3–4; ligules light pink or white, readily withering. **FR** 3–4 mm, cylindric, 5-ribbed, white; pappus of fine bristles, 2–4 mm, white. 2*n*=14. Desert canyons and valleys, juniper woodlands; < 1850 m. **SNE**, **D**; to UT, Colorado, w TX. [*Lygodesmia e*. A. Gray] [Tomb 1972 Brittonia 24:223–228] Mar–May

PSATHYROTES

Theodore M. Barkley

Ann to low, dense subshrub (< 40 cm diam in CA), much-branched, ± hairy and scaly; odor turpentine-like. **LVS** alternate, hairy; blade ± ovate to reniform. **INFL**: heads discoid, peduncled in axils; involucre ± obconic; phyllaries in 2 series; receptacle flat, naked. **DISK FLS** 9–50; corollas cylindric, in CA light yellow and often fading reddish, glandular, ± soft-hairy; anther bases ± sagittate, tips acute to blunt; style branches ± shaggy-papillate, ± truncate or with tapered appendage. **FR** cylindric to obconic, weakly 10-ribbed, densely hairy in CA; pappus of many bristles in 1–4 series. 5 spp.: sw N.Am. (Greek: brittleness, from sts) [Strother 1978 N.Am Fl II 10:142–146]

1. Lvs sparsely tomentose; outer phyllaries erect, like inner; fls gen 13–16; pappus bristles < 50, in 1 series
. . . . *P. annua*
1' Lvs velvety; outer phyllaries recurved, wider than inner; fls gen 21–26; pappus bristles > 120, in 3–4 series
. . . . *P. ramosissima*

P. annua (Nutt.) A. Gray (p. 179) **STS** often purplish. **LVS** short-petioled; blade 8–16 mm, gen ± toothed upward, (gray-)green. **INFL**: outer phyllaries 4–5 mm, erect; inner phyllaries gen 8 or 13. **FLS** 13–16; corollas 4–4.5 mm. **FR** ± 2.5 mm; pappus bristles 35–50 in 1 series, coarse, red-brown. 2*n*=34. Dry, sandy shadscale scrub, alkali flats; gen 800–2000 m. **SNE**, **DMoj**; to ID, UT, nw AZ. Jun–Oct

P. ramosissima (Torrey) A. Gray (p. 179) TURTLEBACK **STS** woolly, becoming glabrous and shiny. **LVS** ± long-petioled; blade 8–20 mm, prominently few-toothed, brown- to gray-green, velvety and woolly-scaly. **INFL**: outer phyllaries 5–6 mm, recurved, wider than inner; inner phyllaries 12–15, deciduous. **FLS** 16–32; corollas 4.5–5 mm. **FR** 2–3+ mm; pappus bristles 120–140 in 3–4 series, fine, brownish. 2*n*=34. Sandy creosote-bush scrub; gen < 1000 m. **s SNE**, **D**; s NV, AZ, nw Mex. ❀DRN,SUN: 10,11,**12,13**& DRY: 1–3,8, 9,19–21;DFCLT. Mar–Jun (sometimes in winter)

PSILOSTROPHE PAPER-DAISY

Per, subshrubs. **STS** 1–many, erect, ± hairy. **LVS** simple, alternate, entire or pinnately lobed, ± soft-hairy; upper often ± glandular. **INFL**: heads radiate, solitary or in cymes; peduncles short to long; involucre cylindric to bell-shaped; phyllaries 5–13 in 1–2 subequal series; receptacle flat, naked. **RAY FLS** 2–6; corollas yellow, fading to cream, persistent on fr; ligules ovate, 3-lobed, reflexed when dry. **DISK FLS** few–many; corollas yellow, densely hairy; anther tips triangular; style tips truncate. **FR** cylindric (or ray achenes slightly flat), ribbed, glabrous, glandular or soft-hairy; pappus of 4–6 unequal transparent scales. 6 spp.: sw US, n Mex. (Greek: naked turn) [Brown 1973 PhD dissertation AZ State Univ]

P. cooperi (A. Gray) E. Greene (p. 179) Per, subshrub 2–6 dm. **STS** densely white-tomentose, openly branched. **LVS** 1–8 cm, linear, entire, tomentose or becoming glabrous. **INFL**: heads 1–few; peduncles 3–8 cm; involucre 3–5 mm diam, cylindric; phyllaries in 2 series, outer 5–8, 6–8 mm, lanceolate, ± soft hairy, inner 4–5, shorter, membranous. **RAY FLS** 3–6; ligules 8–18 mm. **DISK FLS** 10–25; corollas 4–5 mm. **FR** 2–3 mm, glabrous or sparsely glandular; pappus scales 1–2 mm. 2*n*=32. Dry plains, hillsides, washes; 150–1500 m. **e DMoj**, **DSon**; to UT, AZ, n Mex. ❀DRN:10,11, **12,13**,19–21&DRY,SUN:7–9,14,18. Apr–Jun, Oct–Dec

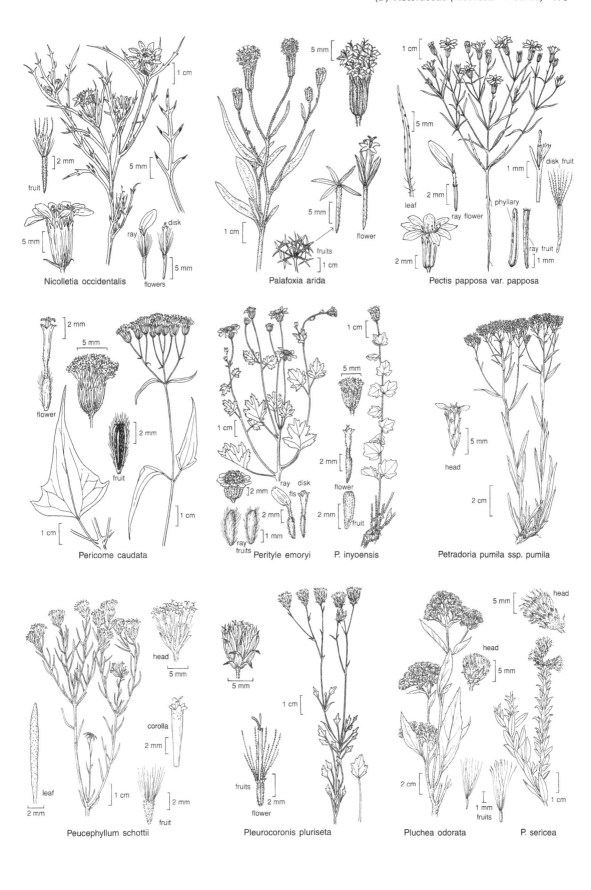

Nicolletia occidentalis

Palafoxia arida

Pectis papposa var. papposa

Pericome caudata

Perityle emoryi

P. inyoensis

Petradoria pumila ssp. pumila

Peucephyllum schottii

Pleurocoronis pluriseta

Pluchea odorata

P. sericea

PULICARIA

G. Ledyard Stebbins

Ann to per. **LVS** basal or some cauline, alternate. **INFL**: heads radiate, 1–many; involucre ± hemispheric; phyllaries in 2–several series, narrowly membranous-margined; receptacle naked. **RAY FLS** many; ligules linear, yellow. **DISK FLS** many; corolla yellow; anther base appendages bristle-like; style branches ± club-like, spreading. **FR** subcylindric, 5-ribbed, short-hairy; outer pappus a short crown of ± fused scales, inner of barbed bristles. (Latin: flea-like) [Raven 1963 Aliso 5:251–253]

P. paludosa Link Pl 6–12 dm, from short rhizome, stiff, ± soft-hairy. **LVS** 1–3(8) cm, linear to oblong, entire, often rolled under, ± clasping. **INFL**: heads many; peduncles 1–3 cm; involucre 4–5 mm, 6–10 mm diam; phyllaries 2.7–4.8 mm, subequal in 2–3 series, narrowly linear. **RAY FLS**: corollas 3.5–4 mm, tube 1.5–2 mm, ligule 1.5–2 mm. **DISK FLS**: corollas 2.2–3 mm. **FR** ± 1 mm; outer pappus < 0.4 mm; inner pappus bristles 8–16, 2–3 mm. 2*n*= 18. Uncommon. Damp sand; < 700 m. w SnJV, SW, **e** DSon; native to Eur. [*P. hispanica* (Boiss.) Boiss.]

PYRROCOMA

Gregory K. Brown

Per from woody taproot. **STS** 1–many, decumbent to erect, gen red-tinged. **LVS** alternate, simple, glabrous to tomentose or glandular; basal petioled; cauline gen clasping, reduced. **INFL**: heads gen radiate, 1–many; involucre hemispheric to bell-shaped; phyllaries in 2–6 ± graduated series, herbaceous. **RAY FLS** 10–80; corollas 2–35 mm, yellow. **DISK FLS** 20–100+; corollas 5–15 mm, cylindric to funnel-shaped, yellow. **FR** 3–4-angled, gen hairy; pappus bristles 15–60, gen rigid, unequal. ± 10 spp.: w N.Am. (Greek: reddish pappus) [Mayes 1976 PhD Univ TX] Formerly incl in *Haplopappus*. Key to species adapted by Margriet Wetherwax.

1. Heads gen 1(–4)
  2. Phyllaries oblong to oblanceolate, in 3–4 graduated series, glabrous; fr glabrous . . . . **P. apargioides**
  2' Phyllaries ± linear, in ± 2 series, barely or not overlapping, long-soft-hairy; fr silky . . **P. uniflora** var. **uniflora**
1' Heads gen 3+
  3. Heads 4–25, in flat-topped clusters, panicle-like . . . . **P. lanceolata** var. **lanceolata**
  3' Heads 3–15+, in narrow clusters, raceme- or spike-like . . . . **P. racemosa**
    4. Infl a raceme-like cluster, not crowded; involucre 8–12 mm diam . . . . var. **paniculata**
    4' Infl spike-like, gen crowded; involucre 5–7 mm diam. . . . . var. **sessiliflora**

P. apargioides (A. Gray) E. Greene (p. 179) **STS** 5–18 cm, 0-few-lvd, glabrous to sparsely tomentose. **LVS**: basal ± 4–10 cm, ±(ob)lanceolate, leathery, gen ± coarsely dentate or cut, glabrous; cauline reduced, gen entire. **INFL**: head gen 1; involucre 13–20 mm wide, hemispheric; phyllaries in 3–4 graded series, 6–10 mm, ± oblong to oblanceolate, green toward tip, glabrous. **RAY FLS** 11–40; ligules 7–16 mm. **DISK FLS** 45–90; corollas 5–7 mm. **FR** 2.5–6 mm, 3-angled, glabrous; pappus 5–7.5 mm, tan. 2*n*=12. Rocky slopes, meadows, forest openings; 2200–3700 m. SN, **SNE**; w NV. [*Haplopappus a.* A. Gray] ✿TRY. Jul–Sep

P. lanceolata (Hook.) E. Greene Pl gen ± (sub)glabrous (densely glandular upward). **STS** 15–50 cm. **LVS** ± (ob)lanceolate, sometimes glandular; basal petioled, 6–30 cm, dentate; cauline sessile, reduced. **INFL**: heads 4–25, in flat-topped clusters; involucre 9–22 mm wide, hemispheric; phyllaries overlapping in 3–4 series, 6–11 mm, unequal, ± lanceolate. **RAY FLS** 18–40; ligules 6–12 mm. **DISK FLS** many; corollas 5–7 mm. **FR** ± 4 mm, 4-angled, silky; pappus ± 5–7 mm, tan. Alkaline meadows, marsh edges, open places; 1300–2800 m. MP, **n SNE**; to OR, ID, s-c Can, c US. [*Haplopappus l.* (Hook.) Torrey & A. Gray] Variable; needs study. 2 vars. in CA.

  var. *lanceolata* Pl (sub)glabrous. 2*n*=12,24,36. Habitats and range of sp. [*H. l.* (Hook.) Torrey & A. Gray incl ssp. *tenuicaulis* (D. Eaton) H.M. Hall] ✿TRY.

P. racemosa (Nutt.) Torrey & A. Gray **STS** 15–90 cm, gen glabrous. **LVS**: basal 5–36 cm, (ob)lanceolate to widely elliptic, entire to serrate, petioles tomentose; cauline clasping, reduced, gen serrate, glabrous. **INFL**: heads 3–15+, in ± narrow clusters; involucre 5–18 mm diam, hemispheric or bell-shaped; phyllaries overlapping in 4–5 series, 6–13 mm, (ob)lanceolate to oblong, (sub)glabrous. **RAY FLS** 7–28; ligules 5–10 mm. **DISK**

FLS 20–65; corollas 5–8 mm. **FR** 2.5–5.5 mm, 4-angled, glabrous to densely tomentose; pappus 6–9 mm, tan to brownish. Many habitats; < 2500 m. KR, NCoR, c&s SNH, Teh, s ScV, SnFrB, SCoRI, SCo, **GB**, **DMoj**; to OR, UT. [*Haplopappus r.* (Nutt.) Torrey] 5 vars. in CA.

  var. *paniculata* (Nutt.) J. Kartesz & K. Gandhi **STS** glaucous. **INFL** narrow; involucre 5–8.5 mm, 8–12 mm diam. 2*n*=12, 24. Alkaline flats, saline meadows; 150–2500 m. c&s SNH, Teh, **GB**, **DMoj**; to OR, ID, UT. Highly variable. [*H. r.* ssp. *glomeratus* (Nutt.) H.M. Hall] Jul–Oct

  var. *sessiliflora* (E. Greene) G. Brown & Keil (p. 179) **STS** glaucous. **LVS**: basal linear-oblanceolate. **INFL** gen crowded, spike-like; involucre 5–8.5 mm, 5–7 mm diam; phyllary base pale, leathery, margin translucent, tip green, recurved. 2*n*=12. Dry alkaline flats or saline meadows; 300–2200 m. **se SNE**, **DMoj**; w NV. [*H. r.* ssp. *s.* (E. Greene) H.M. Hall] Jul–Oct

P. uniflora (Hook.) E. Greene **STS** 7–38 cm. **LVS** ± tomentose or woolly; basal 3–12 cm, (ob)lanceolate, sharply dentate to cut; cauline few, clasping, reduced. **INFL**: heads 1(–4), in raceme-like cluster; involucre 6–13 mm, 11–20 mm diam, hemispheric; phyllaries barely or not overlapping in ± 2 series, 6–12 mm, ± linear, herbaceous, gen tomentose to woolly. **RAY FLS** 25–45; ligules 7–11 mm. **DISK FLS** 35–60; corollas 5–8 mm. **FR** ± 3–4 mm, 3–4-angled, silky; pappus ± 5–8 mm, tan. Alkaline soils of mtn meadows, open forest, near hot springs; 1400–2900 m. SnBr, **GB**; to OR, ID, MT, WY, Colorado. [*Haplopappus u.* (Hook.) Torrey & A. Gray] 2 vars. in CA.

  var. *uniflora* (p. 179) Herbage tomentose or becoming glabrous. **INFL**: involucre 6–9 mm; phyllaries equal. 2*n*=12. Habitats and range (exc SnBr) of sp. ✿TRY.

RAFINESQUIA

G. Ledyard Stebbins

Ann, glabrous; sap milky. **STS** erect, gen branched above. **LVS** basal and cauline, alternate, oblong to widely elliptic or oblanceolate, dentate or pinnately lobed; lower ± petioled or sessile, clasping; upper bract-like. **INFL**: heads ligulate, solitary or in ± flat-topped or panicle-like clusters; peduncles with scale-like or reduced lf-like bracts; involucre cylindric or obconic; phyllaries in 3–4 series, outer 2–3 series unequal, lanceolate to ovate, acute to acuminate, tips spreading, innermost series ± equal, linear-acuminate, >> outer, erect, gen ± membrane-margined; receptacle flat or convex, naked. **FLS** many; ligules white or cream, often rose-tinged, esp beneath, readily withering. **FR** narrowly elliptic; body smooth or tubercled, weakly ribbed, tapered to a beak; pappus a ring of plumose bristles. 2 spp.: sw US, n Mex. (C.S. Rafinesque, eccentric US naturalist, 1783–1840)

1. Ligules 5–8 mm, 3–5 mm > phyllaries; fr beak very slender, ± = and clearly different from body; side hairs of pappus bristles stiff, straight . . . . ***R. californica***
1' Ligules 15–20 mm, 10+ mm > phyllaries; fr beak stout, < and unclearly different from body; side hairs of pappus bristles soft, ± matted . . . . ***R. neomexicana***

R. californica Nutt. CALIFORNIA CHICORY **ST** gen 1 from base, erect, branched chiefly in upper half, 2–15+ dm, in larger pls 10+ mm diam near base. **LVS** 3–15 cm, oblong, dentate or coarsely and ± widely lobed. **INFL**: heads gen several–many; peduncles 1–8 cm; involucres 14–20 mm. **FLS**: ligules 5–8 mm, slightly > phyllaries. **FR** 9–11 mm (incl beak); body 4–5 mm, ± glabrous to short-rough-hairy; beak slender, ± = body; pappus bristles 6–10 mm, very fine, plumose to tip with straight hairs, dull white to brownish; bristles bearing straight hairs. 2*n*=16. Shrubby slopes, open woods, deserts, often common after fires; < 1500 m. NCoR, SNF, SnFrB, SCoR, SW, **W&I, D**; to sw UT, AZ, Baja CA. ❀TRY. Apr–Jul

R. neomexicana A. Gray (p. 179) DESERT CHICORY **STS** 1– several from base, 1–4 dm, in larger pls gen < 5 mm diam near base. **LVS** 3–15 cm, dentate or ± narrowly lobed. **INFL**: heads gen 1–5 per st; peduncles 2–8(15) cm; involucres 17–30 mm. **FLS**: ligules 15–20 mm, >> phyllaries. **FR** 12–14 mm (incl beak); body 8–10 mm, puberulent or appressed-scaly; beak stout, gen < body; pappus bristles 12–17 mm, stiff, wider at base, plumose with tangled hairs, often merely barbed at tip, white. 2*n*=16. Gravelly and sandy desert soils, often partially supported in branches of shrubs; gen < 1400 m. **D**; to sw UT, w TX, n Mex. ❀TRY. Feb–May

RAILLARDELLA

Bruce G. Baldwin

Per, ± scapose from gen branched rhizome; rosettes often clumped. **LVS** simple, sessile, ± basal. **INFL** glandular, sometimes hairy; heads radiate or discoid, gen solitary; phyllaries gen as many as ray fls, ± folded around ray ovaries; chaff scales in 1 involucre-like series, ± equal, ciliate-hairy. **RAY FLS** 0–13; corollas yellow to red; ligules often deeply lobed. **DISK FLS** 7–84; corollas yellow to red; style branches long, tips bristly. **FR** ± cylindric, ± straight, ascending-hairy, black; pappus of flat, ciliate-plumose bristles. 3 spp.: montane CA, w NV, OR. (Latin: small *Raillardia* or *Railliardia*) [Baldwin et al. 1991 Proc Natl Acad Sci USA 88:1840–1843]

R. argentea (A. Gray) A. Gray (p. 179) SILKY RAILLARDELLA Pl 1–14.5 cm. **LVS** basal, entire, 0.7–8 cm, gen oblanceolate, silky-hairy, minutely glandular or nonglandular. **INFL**: heads discoid, solitary, cylindric to bell-shaped; peduncles 0.1–12.5 cm, sub-tending bracts 0–1; phyllaries 0; chaff scales 5–15, 6–16 mm, ±

fused. **RAY FLS** 0. **DISK FLS** 7–26; corollas 6–11 mm, yellow. **FR** 5–9.5 mm, ± linear; pappus bristles 16–30, 6–11 mm. 2*n*=34,36. Dry, open, gen gravelly sites; 2200–3900 m. CaRH, SNH, SnBr, Wrn, **SNE**; w NV, OR. ❀DRN,IRR,SUN: 1–3,6,7,15,16,18;DFCLT. Jul–Aug

SANVITALIA

Ann, per. **STS** simple to much-branched, prostrate to erect. **LVS** simple, opposite. **INFL**: heads radiate, solitary or in few-headed cymes; peduncle slender; involucre disk-like to hemispheric; phyllaries in 1–2 series; receptacle conic, chaffy; chaff scales lanceolate, ± awn-tipped. **RAY FLS** 5–13; tube 0; ligules 2–3-lobed, cream to orange, persistent on fr. **DISK FLS** many; corollas cream to yellow or brown; lobes very small; style tips triangular. **FR** glabrous; ray achenes thick, pappus of short, stout awns; disk achenes short, pappus 0 or of 2 awns. 5 spp.: sw US, Mex, n C.Am, S.Am. (either Sanvital, a Spanish botanist, or the Italian Sanvital family) [Strother 1979 Madroño 26:173–179]

S. abertii A. Gray (p. 179) ABERT'S SANVITALIA Ann 2–29 cm. **STS** spreading or erect, simple to much-branched, strigose. **LVS** sessile or short-petioled, 2–5 cm, linear to lanceolate or narrowly elliptic, acute, scabrous. **INFL**: heads gen in cymes; peduncle 0–30 mm; phyllaries 5–11, prominently veined, acute, ± glabrous; awn-tips of chaff scales > disk fls. **RAY FLS** 1 per phyllary;

corollas yellow, drying cream; ligules thick, 2–3 mm, ± leathery, gen 2-lobed. **DISK FLS**: corollas 1–2 mm, cylindric, yellow, drying cream. **FR**: ray achenes 3–4 mm, straw-colored, pappus awns 3, < or = 1 mm, stout; disk achenes 2.5–3.5 mm, brown, ± 4-angled, warty, pappus 0. 2*n*=22. RARE. Dry slopes; 1800 m. **DMtns (Clark, New York mtns)**; to TX, n Mex. Aug–Sep

SCHKUHRIA

Ann, per. **STS** often much-branched. **LVS** simple or pinnately divided, gen gland-dotted; lower opposite; upper alternate. **INFL**: heads radiate or discoid, in open cymes; peduncles short to long; involucre obconic to bell-shaped;

phyllaries 4–18, oblanceolate to obovate, obtuse, margins scarious; receptacle rounded, naked. **RAY FLS** 0–3; corollas yellow or white; ligules short, inconspicuous. **DISK FLS** few–many; corollas yellow; style tips acute, triangular. **FR** narrowly obpyramidal, gen 4-angled; angles hairy; pappus gen of 8 scales. 6 spp.: s N.Am, n S.Am (C. Schkuhr, 1741–1811, German botanist) [Heiser 1945 Ann Missouri Bot Gard 32:265–278]

S. multiflora Hook. & Arn. var. **multiflora** (p. 179) Ann. **STS** decumbent or erect, 5–25 cm, glandular, strigose, or becoming glabrous. **LVS** 2–4 cm, dissected into thread-like lobes 0.5–1 mm wide. **INFL**: heads discoid; cymes few-headed; peduncles 5–30 mm, glandular; involucre obconic; phyllaries 7–9, 5–6 mm, oblanceolate, green-centered, margins often red or yellow. **RAY FLS** 0. **DISK FLS** 15–30; corollas 1.5–2 mm. **FR** 3–4 mm; pappus scales 1–2 mm, obtuse to acute. $2n$=22. Dry, sandy soils; 1500–1700 m. **e DMoj**; to TX, n Mex; also in S.Am. [*Bahia neomexicana* (A. Gray) A. Gray] Sep

SENECIO GROUNDSEL, RAGWORT, BUTTERWEED

Theodore M. Barkley

Ann to tree-like, gen ± loosely hairy, becoming ± subglabrous; roots gen fibrous, branched. **STS** gen 1–few per rosette. **LVS** alternate; lower gen petioled; middle gen weakly clasping; uppermost gen bract-like. **INFL** often ± flat-topped; heads radiate or discoid, 1–many; involucre cylindric to hemispheric; main phyllaries in 1 equal series (often subtended by few, much-reduced outer phyllaries). **RAY FLS** 0–13; ligules gen yellow to orange. **DISK FLS** gen < 40; corolla gen ± yellow; style tips truncate to obtuse, gen hair-tufted. **FR** cylindric; ribs shallow, often stiff-hairy; pappus of thin, minutely barbed, deciduous bristles ± = fr body. ± 1500 spp.: worldwide; some cult, some of unusual form. (Latin: old man, from white pappus) *For revised taxonomy of *Senecio*, see Barkley 1999 Sida 18(3): 661–672.

1. Weak shrub or subshrub; branching aspect shrub-like; lvs or segments linear or nearly so
  2. Main phyllaries ± (13)21, outermost phyllaries prominent, some 1/2 × main phyllaries
    . . . . ***S. flaccidus* var. *monoensis***
  2' Main phyllaries ± 8(13), outer phyllaries 0 or inconspicuous . . . . ***S. spartioides***
1' Ann or per; lvs gen wider than linear
  3. Ann from slender taproot
    4. Ligules conspicuous, ± 10 mm . . . . ***S. californicus***
    4' Ligules 0 or minute
      5. Main lvs cordate-clasping, bases 1–2 cm wide . . . . ***S. mohavensis***
      5' Main lvs petiolate, at most weakly clasping — introduced weed . . . . ***S. vulgaris***
  3' Per from taproot, caudex, or rhizome
    6. Lvs ± equal, evenly spaced, or crowded upward at fl
      7. Pls 5–20 dm, erect; montane
        8. Lvs linear-lanceolate, tapered . . . . ***S. serra* var. *serra***
        8' Lvs narrowly triangular, truncate or hastate . . . . ***S. triangularis***
      7' Pls gen < 2 dm**,** spreading; alpine and subalpine
        9. Pls tap-rooted; lvs oblanceolate to ovate . . . . ***S. fremontii* var. *occidentalis***
        9' Pls rhizomed; lvs narrowly lanceolate to linear . . . . ***S. pattersonensis***
    6' Lowermost lvs largest; cauline lvs progressively reduced upward
      10. St single from button-like caudex; roots many, fleshy-fibrous, unbranched
        11. Pl glabrous or minutely hairy among heads; herbage glaucous . . . . ***S. hydrophilus***
        11' Pls hairy and not glaucous; sometimes becoming subglabrous . . . . ***S. integerrimus***
          12. Main phyllaries 5–10 mm, strongly black-tipped; becoming subglabrous; open places in sagebrush
            scrub, seasonally wet meadows . . . . var. ***exaltatus***
          12' Main phyllaries (8)10–12(+) mm, gen greenish or minutely black-tipped; ± hairy; open places
            in coniferous forests . . . . var. ***major***
      10' Sts single or clustered from taproot, rhizome, or non-button-like caudex; roots branched or unbranched
        13. Sts from thick, short, creeping rhizome; roots fleshy, unbranched; lf margins with numerous
          small, hard, translucent teeth . . . . ***S. scorzonella***
        13' Sts from taproot, thin rhizome, or caudex; lateral roots thin, branched; lf margins various but
          without small, translucent teeth (segregate genus *Packera*)
          14. Basal and lower cauline lvs pinnately divided or lobed; pl glabrous at fl . . . . ***S. multilobatus***
          14' Basal and lower cauline lvs entire to dentate; pl variously hairy
            15. Heads gen > 6; herbage permanently hairy; 1300–3600 m, GB . . . . ***S. canus***
            15' Heads 1–6; herbage early hairy, becoming ± glabrous at fl; 3000–4000 m, n W&I . . . ***S. werneriifolius***

S. californicus DC. Ann 1–4 dm, from taproot, slender, ± glabrous. **ST**: branches arched upward. **LVS** thin (to fleshy near ocean); lower weakly petioled, 2–7 cm, ± linear to ovate, subentire to deeply dentate. **INFL**: heads radiate, (1)3–10+; main phyllaries ± 21, 5–7 mm, tips black. **RAY FLS** 13; ligules < 10 mm. **DISK FLS** < 40. **FR** soft-hairy. $2n$=40. Coastal strand to shrubland; < 1200 m. s SN, Teh, CW, SW, **w DSon**; Baja CA. ❀SUN,DRN:**7**, 12,13,**14–24**. Mar–May

S. canus Hook. Per 1–4 dm, from caudex or rhizome, ± felty-gray-tomentose. **LVS**: basal 2.5–5 cm (< petiole), ± lanceolate to ovate, entire to weakly dentate. **INFL**: heads radiate, 6–12; main phyllaries ± 13, sometimes 21, 5–6+ mm, tips green. **RAY FLS** 8(13); ligules (5)8–10 mm. **DISK FLS** < 40. **FR** glabrous. $2n$=46, 92, + polyploids. High rocky plains, sagebrush scrub; 1300–3600 m. e KR, CaR, SN, **GB**; to Can, c US. Locally abundant.

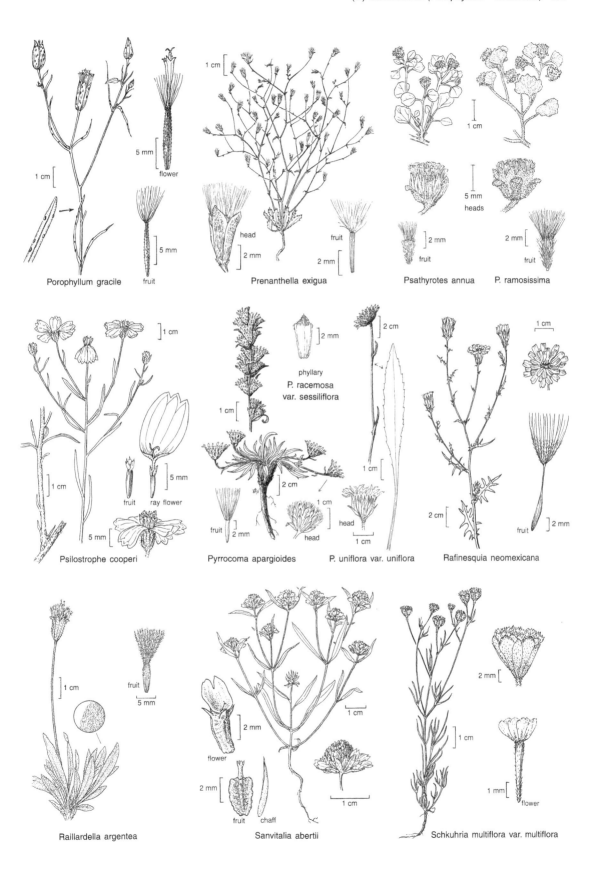

Porophyllum gracile

Prenanthella exigua

Psathyrotes annua   P. ramosissima

Psilostrophe cooperi

Pyrrocoma apargioides   P. uniflora var. uniflora   Rafinesquia neomexicana

P. racemosa var. sessiliflora

Raillardella argentea

Sanvitalia abertii

Schkuhria multiflora var. multiflora

*****Packera cana** (Hook.) W. A. Weber & Á. Löve ❀SUN, DRN:**17** &IRR:1–3,**7**,10,11,**14–16**,18,24. May–Aug

S. flaccidus Less. Subshrub (may fl 1st year) 3–15+ dm, ± taprooted, arching-branched above, glabrous to woolly. **LVS** thread-like, sometimes deeply divided into narrow segments, sometimes with axillary lf clusters. **INFL:** heads radiate, 3–10(20) per cluster; main phyllaries ± 13, sometimes 21, 5–10+ mm; outer phyllaries ± prominent. **RAY FLS** 8(13); ligules 10–20 mm, ± yellow. **DISK FLS** < 40. **FR** soft-hairy. 2*n*=40. Dry, rocky or sandy sites; < 2000 m. CA-FP (exc NCo, KR), **SNE, D**; to Colorado, TX, w&c Mex. 2 vars. in CA.

var. **monoensis** (E. Greene) B. Turner & T. Barkley Pl often not appearing shrubby, ± glabrous. **INFL:** main phyllaries ± 21, sometimes 13. 2*n*=40. Exposed basins, foothills; 500–2000 m. **SNE, D**; to TX, nw Mex. [*S. douglasii* var. *monoensis* (E. Greene) Jepson] ❀TRY. Mar–May (sometimes in fall)

S. fremontii Torrey & A. Gray Per 1–2 dm, ± prostrate to arched upward from caudex, glabrous, often purplish below. **LVS** thickish, tapered to petiole or sessile and weakly clasping; larger 2–4.5 cm, oblanceolate to ovate, dentate. **INFL:** heads radiate, 1–5; main phyllaries ± 13, sometimes 8, 5–10 mm, tips green. **RAY FLS** 8; ligules 8–12 mm. **DISK FLS** < 40. **FR** glabrous or angles hairy. 2*n*=40,40+,80. Talus, other rocky places; 2600–3600 m. CaRH, SN, SnBr, Wrn, **SNE**; to sw Can, WY, Colorado. 2 vars. in CA.

var. **occidentalis** A. Gray (p. 185) **LVS** few and much reduced above; lower gen < 3.5 cm. **INFL:** main phyllaries 5–8 mm. Habitats of sp. SN, SnBr, **SNE**; w NV. ❀TRY. Jul–Aug

S. hydrophilus Nutt. Bien or per 5–20 dm, from button-like caudex, glabrous, glaucous, light green; roots fleshy, unbranched. **ST** hollow. **LVS** thick; lower blades 5–20+ cm, > petiole, elliptic to oblanceolate, entire or shallowly toothed. **INFL:** heads radiate or discoid, 20–80+; main phyllaries 8(13), 5–8 mm, tips often black. **RAY FLS** (0)5; ligules 3–8 mm. **DISK FLS** < 40. **FR** glabrous. 2*n*=40. Swamps, muddy sites, tolerant of standing saltwater; < 2300 m. s NCoR, CaR, SN, deltaic GV, n CW, **GB**; to B.C., Colorado. Reduced from wetland development. ❀TRY;STBL. May–Aug

S. integerrimus Nutt. Bien or per 2–7 dm, from button-like caudex; roots fleshy, unbranched. **LVS:** lower petioled or tapered to base, 6–25 cm; blade gen lanceolate, subentire. **INFL:** heads radiate, 6–20(30), central head often largest, peduncle shorter; main phyllaries ± 21, sometimes 13. **RAY FLS** (8)13; ligules 6–15 mm, yellow or off-white. **DISK FLS** 20–50. **FR** glabrous. 2*n*=40,80. Grassland, open forest; 150–3600 m. KR, NCoRI, CaR, SN (exc Teh), **GB**; to sw Can, n-c US, Colorado. 3 vars. in CA.

var. **exaltatus** (Nutt.) Cronq. (p. 185) **LVS:** petiole often indistinct. **INFL:** main phyllaries gen < 10 mm, ± strongly black-tipped. **RAY FLS** rarely 0; ligules yellow. Habitats of sp.; 1400–3200 m. CaR, SN (exc Teh), **GB**; to sw Can, WY, Colorado. ❀TRY. May–Jul

var. **major** (A. Gray) Cronq. (p. 185) Pl persistently softhairy. **LVS:** petiole often indistinct. **INFL:** main phyllaries 8–12 mm, greenish or minutely black-tipped. **RAY FLS** rarely 0; ligules yellow. Forest; 150–3600 m. KR, NCoRI, CaRH, SN, **W&I**; s OR. ❀ TRY. May–Aug

S. mohavensis A. Gray (p. 185) Ann 1–4 dm, from short, often twisted taproot, much-branched upward, glabrous, often purplish. **LVS:** lower blades 2–6 cm (petiole poorly defined), (ob)ovate, unevenly dentate to lobed; upper blades clearly clasping. **INFL:** heads radiate or disciform, 3–10; main phyllaries ± 8, sometimes 13, 6–7 mm, tips green. **RAY FLS** few, inconspicuous; ligules short, ± = phyllaries. **DISK FLS** < 40. **FR** soft-hairy. 2*n*=40. Sandy washes, flats; < 1000 m. **D**; s NV, w AZ, nw Mex. [Liston et al 1989 Amer J Bot 76:383–388]. Mar–May

S. multilobatus A. Gray Ann to per 2–5 dm, from taprooted caudex. **LVS:** lower blade 3–9+ cm, ± (ob)ovate, deeply pin-

nately dissected. **INFL:** heads gen radiate, 8–25; main phyllaries ± 13, sometimes 21, 4–8+ mm, tips green or yellow. **RAY FLS** (0)8(13); ligules < 10 mm. **DISK FLS** < 40. **FR** gen glabrous (or angles hairy). 2*n*=46,92. Rocky or sandy soils, sagebrush scrub or open woodland; 1400–3200 m. SNH (e slope), **SNE, DMtns**; to ID, WY, NM. [*S. stygius* E. Greene] *****Packera multilobata** (A. Gray) W. A. Weber & Á. Löve ❀TRY. May–Jul

S. pattersonensis Hoover (p. 185) MONO RAGWORT Per gen < 1 dm, from rhizome, arched upward, glabrous, sometimes reddish. **LVS** thick to ± fleshy, evenly spaced, 2–4 cm, ± narrowly lanceolate, sometimes with 1–2 lateral lobes, margins entire to wavy, often rolled under, sometimes decurrent. **INFL:** heads radiate, 1(–4); main phyllaries ± 13, 5–8 mm, tips green. **RAY FLS** 8; ligules 5–10 mm. **DISK FLS** < 40. **FR** glabrous. UNCOMMON. Talus slopes; 2900–3700 m. c SNH, **SNE**. Jul–Aug

S. scorzonella E. Greene (p. 185) Per 2–5 dm, from stout rhizome or caudex, gen evenly short-woolly; roots fleshy, unbranched. **LVS:** basal blades 10–24 cm, (ob)lanceolate, teeth fine, hard, dark, base tapered. **INFL:** heads gen radiate, 10–24; main phyllaries ± 13, 3–5 mm, tips often black. **RAY FLS** (0)5+; ligules 5–8+ mm. **DISK FLS** < 40. **FR** glabrous. Open forest, meadow edges; 1600–3500 m. CaRH, SNH, n **SNE**, n **W&I**. [*S. covillei* E. Greene] ❀TRY. Jul–Aug

S. serra Hook. var. **serra** Per 5–20 dm, from branched woody caudex, sometimes red-tinged below. **LVS** evenly spaced, ± equal; blades 5–20 cm, > petiole ± lanceolate, subentire to dentate; lowermost soon deciduous. **INFL:** heads radiate, 30–60+; main phyllaries ± 13, sometimes 8, 4–6 mm, tips often black. **RAY FLS** 8; ligules 5–7 mm. **DISK FLS** < 40. **FR** (sub)glabrous. 2*n*=40. Damp, open coniferous forest or sagebrush scrub; 1300–3200 m. s SNF, **SNE**; to WA, MT, WY. ❀ TRY. Jul–Aug

S. spartioides Torrey & A. Gray Subshrub 2–8 dm, arched upward from taprooted crown. **LVS** 5–10 cm, linear, entire, ± toothed, or 2–4-lobed. **INFL:** heads radiate, 10–20+; main phyllaries ± 8, sometimes 13, 6–9 mm, tips green. **RAY FLS** 5(8); ligules 10+ mm, ± yellow. **DISK FLS** < 40. **FR** soft-hairy. 2*n*=40. Dry, open, rocky places; 1800–3200 m. SNH (e slope), SnBr, **SNE, n DMtns**; to SD, NM, nw Mex. May hybridize with *S. fremontii* var. *occidentalis* in n SNE (Sweetwater Mtns). Jul–Aug

S. triangularis Hook. (p. 185) Per 5–12+ dm, from branched, woody caudex. **LVS** evenly spaced, ± equal; blades 4–14 cm, < petiole ± triangular, subentire to dentate, base truncate to hastate. **INFL:** heads radiate, 10–30+; main phyllaries ± 13, sometimes 8 or 21, 6–10 mm, tips barely black. **RAY FLS** 8; ligules < 15 mm. **DISK FLS** < 40. **FR** glabrous. 2*n*=40,80. Wet meadows, streambanks in open, coniferous forest; 1000–3500 m. KR, CaR, SN, SnGb, SnBr, PR, **GB**; to AK, Colorado. ❀ TRY. Jul–Sep

S. vulgaris L. Ann 1–6 dm, arched upward, ± glabrous. **LVS:** blades 2–10 cm (petiole weakly defined), oblanceolate to obovate, deeply and unevenly dentate to lobed. **INFL:** heads discoid, 8–10; main phyllaries ± 21, 4–6 mm, tips often black; outer phyllaries 2–6, short, strongly black tipped. **RAY FLS** 0. **DISK FLS** < 40. **FR** subglabrous. 2*n*=40,80. Abundant. Gardens, farmlands, other disturbed sites; < 1500 m. CA-FP, **GB**; native to Eurasia. Most of year

S. werneriifolius A. Gray Per gen 1–2 dm, ± tufted from branched, rhizome-like caudex, ± scapose. **LVS** thick, ± basal; blade < 4 cm, < (sometimes indistinct) petiole, linear-oblong, rolled under. **INFL:** heads gen radiate, 1–6; main phyllaries ± 13(21), 4–10 mm, tips green. **RAY FLS** (0)8(13); ligules 5–10 mm. **DISK FLS** < 40. **FR** glabrous. 2*n*=44,46. Talus, open sites, among trees near timberline, in loose soil; 3000–4000 m. n&c SNH, **n SNE, n W&I**; to ID, MT, Colorado, n AZ. CA pls have narrower, more rolled lvs than pls from Rocky Mtns but intergrade completely. [*S. muirii* Greenman] *****Packera werneriifolia** (A. Gray) W. A. Weber & Á. Löve ❀TRY;DFCLT. Jul–Aug

SOLIDAGO GOLDENROD

John C. Semple

Per from woody caudex or rhizome, branched above. **LVS** alternate, resinous, often sessile. **INFL**: heads radiate, few–many, in ± flat-topped to panicle-like, often ± 1-sided clusters; involucre cylindric to bell-shaped (wider when dry); phyllaries in 3–5 graduated, overlapping series, midrib gen ± swollen, translucent. **RAY FLS** few–many; ligules yellow. **DISK FLS** few–many; corollas yellow, gen glabrous; style branches finely papillate, appendage triangular. **FR** obconic, compressed; pappus of 25–45 long-barbed bristles in 1 series. ± 150 spp.: esp N.Am (S.Am, Eurasia). (Greek: make-well, from purported medicinal value) [Semple et al. 1990 Can J Bot 68:2070–2082]

1. Middle cauline lvs largest
  2. Infl ± widely pyramid-shaped; involucre 3–5 mm; lvs weakly serrate; escaped from cultivation . . *S. altissima*
  2' Infl ± club-shaped; involucre 2.5–3.5 mm; lvs gen serrate; SNE . . . . *S. canadensis* ssp. *elongata*
1' Lower cauline (or basal) lvs largest
  3. Infl flat- to round-topped; phyllaries often long-acuminate, not strongly graded; lvs gen long-ciliate;
    alpine . . . . *S. multiradiata*
  3' Infl panicle- or raceme-like or pyramid-shaped; phyllaries rounded to barely acuminate, ± strongly
    graded; lvs not long-ciliate; up to subalpine
    4. Herbage ± short-hairy; upper nodes without axillary lf-clusters
      5. Herbage ± densely hairy; lvs gen not 3-veined; phyllaries strigose; disk corolla throat obscure . . *S. californica*
      5' Herbage ± sparsely hairy; lvs ± 3-veined; phyllaries subglabrous; disk corolla throat obvious . . . *S. sparsiflora*
    4' Herbage ± glabrous; upper nodes often with axillary lf-clusters
      6. Phyllaries very narrowly triangular, inrolled near tip, sharply acute; rays gen 5–10; n DMtns . . . *S. confinis*
      6' Phyllaries lanceolate to ± ovate, not inrolled, obtuse to acuminate; rays gen 8–12; GB . . . . *S. spectabilis*

S. altissima L. var. *altissima* LATE GOLDENROD **STS** from rhizome, 2–15 dm, hairy to base. **LVS** 4–15 cm; mid-st lvs largest, (ob)lanceolate, strongly 3-veined, ± densely scabrous to short-hairy. **INFL** gen large, dense, pyramid-shaped; heads many; involucre 3–5 mm; phyllaries lanceolate, outer 1/4–1/3 length inner, margins ± strigose. **RAY FLS** 10–15; ligules 1–2 mm. **DISK FLS** 3–7; corollas 2.5–4.5 mm. **FR** 1–1.5 mm, ± strigose. 2n=36,54. Uncommon. Disturbed sites; < 2800 m. NW, CaR, SN, CW, **GB**; to B.C.; native to e US. Oct–Nov

S. californica Nutt. (p. 185) CALIFORNIA GOLDENROD Herbage gen ± densely short-soft-hairy. **STS** 2–15 dm, short-rhizomed. **LVS**: lower < 14 cm, oblanceolate to obovate, serrate, sometimes 3-veined, base tapered; upper much reduced. **INFL** long, 1-sided, pyramid- or wand-shaped, many-headed (or short, raceme-like, few-headed); involucre 3–5 mm; phyllaries gen narrow, acute, strigose, outer 1/4–1/3 length inner. **RAY FLS** 6–11; ligules 3–5 mm. **DISK FLS** 6–17; corollas 3–5 mm. **FR** 0.7–1.5 mm, ± densely strigose. 2n=18,36. Woodland margins, grassland, disturbed soils; < 2300 m. CA-FP, MP, **W&I**; OR, Baja CA. ❀ SUN:**4–6**&IRR:1–3,**15–23**&SHD:**7**,8,9,**14**. Jul–Oct

S. canadensis L. ssp. *elongata* (Nutt.) Keck (p. 185) CANADA GOLDENROD Herbage ± sparsely strigose (more so above). **STS** 25–150 cm, rhizomed. **LVS**: middle cauline largest, 5–15 cm, ± lanceolate, gen 3-veined, toothed. **INFL** panicle-like; heads many; involucre 2.5–3.5 mm; phyllaries lanceolate, outer 1/4–1/3 length inner. **RAY FLS** 8–15; ligules 1.5–2 mm. **DISK FLS** 5–12; corollas 2.5–3.5 mm. **FR** 1–1.5 mm, ± strigose. 2n=18,36. Meadows, thickets; < 2800 m. NW, CaR, SN, CW, **GB**; to B.C. Pls in CW have thicker, veinier lvs. *S. c.* incls many vague races across N.Am; one of the most difficult taxonomic problems in N.Am. ❀SUN:**4–6**&IRR:1–3,**7**,8–10,**14–24**;INV. May–Sep

S. confinis A. Gray (p. 185) SOUTHERN GOLDENROD Pl ± glabrous. **STS** < 21 dm, often stout, from short, branched caudex. **LVS**: lowest largest, 5–25 cm, gen < 10 × longer than wide, entire, often ± fleshy, base nearly sheathing; uppermost sometimes scale-like or with axillary lf-clusters. **INFL** panicle-like; heads gen many; involucre 2.5–4 mm; phyllaries very narrowly triangular, inrolled near tip, sharply acute, outer 1/3–2/3 length inner;

midrib gen enlarged, translucent. **RAY FLS** 3–13; ligules 1–2.5 mm. **DISK FLS** 10–20; corollas ± 3–4 mm. **FR** 1–1.5 mm, ± strigose. 2n=18. Wet streambanks, springs, marshes; gen < 2500 m. Teh, CCo (formerly SnFrB), SCoR, SW, **W&I, n DMtns**. Involucres of n DMtns pls like those of *S. spectabilis*. Jul–Oct

S. multiradiata Aiton (p. 185) NORTHERN GOLDENROD Pl ± (sub)glabrous, hairier above. **STS** < 5 dm, from woody caudex. **LVS**: lower cauline < 12 cm, ± linear to spoon-shaped; upper reduced, ± linear to ovate, subclasping, gen long-ciliate. **INFL** flat- to round-topped; heads few–many; involucre 4–7 mm; phyllaries ± lanceolate, acute to long-acuminate, outer gen 2/3 to = length inner. **RAY FLS** 12–18; ligules 2–4 mm. **DISK FLS** 12–35; corollas 3–5 mm. **FR** 2–3 mm, ± sparsely strigose. 2n=18,36. Alpine slopes, meadows; 2600–3700 m. KR, CaRH, SN, **n SNE, W&I**; to AK, arctic Can, NM. ❀IRR,DRN,SUN:**1**,2–6,**7**,14,**15–18**,19–21. Jun–Sep

S. sparsiflora A. Gray Pl ± sparsely short-hairy. **STS** 2–15 dm, short-rhizomed. **LVS**: lower largest, < 14 cm, oblanceolate to obovate, tapered at base, serrate, 3-veined, base tapered; upper much reduced, entire. **INFL** pyramid-shaped, ± sparsely branched, 1-sided at tip; heads gen many; involucre 3–6 mm; phyllaries lanceolate, rounded or obtuse, ± resinous. **RAY FLS** 6–9; ligules 4–6 mm. **DISK FLS** 5–12; corollas 3.5–5 mm, throat ± obvious. **FR** ± 1–2.5 mm, ± strigose. 2n=18,36,54. Margins of dry woodlands, grasslands, disturbed soils; 500–2200 m. **GB**; to NM, Mexico.

S. spectabilis (D. Eaton) A. Gray (p. 185) SHOWY GOLDENROD Pl ± glabrous. **STS** < 18 dm, from short caudex. **LVS**: basal largest, < 25 cm, oblanceolate, tapered, entire or serrate toward tip, ± fleshy; cauline reduced, entire, axils often with lf-clusters. **INFL** ± panicle-like, ascending to arching, tip sometimes 1-sided; heads gen many; involucre 3–4 mm; phyllaries lanceolate to narrowly ovate, obtuse to acuminate. **RAY FLS** 6–21; ligules 1.5–3.5 mm. **DISK FLS** 8–22; corollas 2.5–4.5 mm. **FR** 1–2 mm, sparsely strigose. 2n=18. Bogs, alkaline meadow; 300–2300 m. CaRH, n SNH (e slope), **GB**; to OR, NV. [*S. missouriensis* Nutt., Missouri goldenrod, misapplied] ❀DFCLT. Jul–Sep

SONCHUS SOW THISTLE

G. Ledyard Stebbins

Ann to shrubs; sap milky. **STS** erect, smooth, leafy. **LVS** basal and cauline, alternate, ± entire to toothed and coarsely pinnate-lobed; cauline gen sessile, clasping. **INFL**: heads ligulate, in cymes; involucre swollen at base; phyllaries gen in 3 series, outer many, short-triangular, those of inner 2 series linear, tapered; receptacle ± flat, naked. **FLS** many; ligules yellow, readily withering. **FR** gen ± flat, beakless; pappus of many fine, white bristles. 54 spp.: Eurasia, Afr. (Ancient Greek name) [Boulos 1972–74 Bot Not 125:287–319, 126:155–196; 127:7–37,402–451] Key to species adapted by David J. Keil.

1. Basal clasping lobes of lvs rounded, curled or coiled; fr 3-ribbed on each side, otherwise smooth; ligule < corolla tube . . . . ***S. asper***
1' Basal clasping lobes of lvs acute, not curled or coiled; fr 2–4-ribbed and transversely wrinkled; ligule ± = corolla tube . . . . ***S. oleraceus***

S. asper (L.) Hill ssp. ***asper*** (p. 185) PRICKLY SOW THISTLE Ann 1–12 dm. **STS** mostly unbranched below infl. **LVS**: basal tapered to gen sessile bases; cauline sessile, clasping, upper often widest near base, basal clasping lobes rounded, strongly curved to coiled; blades dentate, sometimes ± lobed, teeth and lobes tipped with soft spines. **INFL**: peduncles 0.5–5 cm, gen ± bristly-glandular; involucre 10–13 mm. **FLS**: ligule ± 1/3 < tube. **FR** 2–3 mm, very flat, 3- ribbed on each face, otherwise smooth; pappus bristles ± 3 × fr. 2*n*= 18. Common. Weed in slightly moist waste places, gardens, along streams; < 1900 m. **CA**; to e US; native to Eur. Much like *S. oleraceus*. Most of year

S. oleraceus L. (p. 185) COMMON SOW THISTLE Ann 1–14 dm. **STS** often branched from below middle. **LVS**: basal gen < cauline, gen tapered or abruptly wing-petioled; cauline 5–35 cm, upper sessile, clasping, often widest at base, basal clasping lobes acute, not curled or coiled; blades nearly all lobed exc in dwarfed pls, lobes variable in width, terminal lobe often widely arrowhead-shaped. **INFL**: peduncles 0.5–7 cm, glabrous to bristly-glandular, sometimes cottony-tomentose just below heads; involucre 10–13 mm. **FLS**: ligule ± = tube. **FR** 2.5–3.8 mm, flat, 2–4-ribbed, cross-wrinkled on each face; pappus bristles ± 2 × fr. 2*n*=32. Abundant. Weed in waste places, gardens, etc; < 1500 m. **CA**; to e US; native to Eur. Much like *S. asper*. Most of year

SPHAEROMERIA

Elizabeth McClintock

Per, subshrub, often from long, thick caudex, silky; hairs basally forked, attached off-center; herbage dotted with resin glands. **LVS** alternate, often crowded at base; simple and entire or 3-lobed to 1–3-pinnately dissected. **INFL**: heads disciform, few–many; involucre hemispheric; phyllaries in 2–3 overlapping series, scarious-margined and -tipped; receptacle hemispheric, naked or hairy. **PISTILLATE FLS** few, marginal; corolla cylindric to lance-ovoid. **DISK FLS** more numerous; corolla tubular to bell-shaped; anther tips ovate, bases rounded or ± cordate; style branches truncate, tips shrub-like. **FR** cylindric, gen 5–10-ribbed; pappus gen 0 or a narrow crown. 9 spp.: w. N.Am. (Greek: spherical division) [Holmgren et al 1976 Brittonia 28:255–272] Segregated from *Tanacetum*.

S. cana (D.C. Eaton) A.A. Heller (p. 185) Subshrub. **STS** 15–30 cm, leafy, gen branched from base. **LVS** 5–12 mm, basal and cauline, sessile; basal and lower cauline gen 3–4-lobed, upper entire. **INFL**: heads gen 3–8, in dense clusters (rarely solitary); receptacle glabrous. **FR** < 2 mm, 10-ribbed; pappus 0. 2*n*=18.

Uncommon. Rocky places, ledges, trails, talus near or above timberline; 3000–4000 m. c&s SNH, **SNE, n DMtns**. [*Tanacetum c.* D.C. Eaton] Sage-like smell sometimes reported. ✿DRN,DRY,SUN:1–3,15,16;DFCLT. Jul–Aug

STENOTUS

Gregory K. Brown & David J. Keil

Per < 5 m diam, ± mat-forming. **LVS** gen crowded at branch tips, persistent for 2–3 years, < 10 cm, linear to oblanceolate, gen rigid, entire, scabrous; petiole indefinite. **INFL**: head radiate, solitary; peduncles < 15 cm, naked; involucre 5–10 mm, hemispheric; phyllaries in 2–3 series, linear to ± ovate, acute to acuminate. **RAY FLS** 6–15; ligules 7–12 mm, yellow. **DISK FLS** 25–50; corollas 6–7.5 mm, funnel-shaped, yellow. **FR** 3.5–7 mm, gen densely silky; pappus of soft bristles. ± 5 spp.: w N.Am. (Greek: narrow ear)

S. acaulis Nutt. (p. 185) **LVS** 2–10 cm, oblanceolate, gen 3-veined, glabrous to short-bristly, sometimes sticky. **INFL**: involucre 6–10 mm. **RAY FLS** 6–15, 8–12 mm. **DISK FLS**: corollas 6–7.5 mm, tube glabrous; **FR** glabrous or densely silky;

pappus white or tan. Dry, rocky, open shrubland; 1900–3200 m. CaR, SN, **GB, DMtns**; to WA, MT, WY, Colorado, UT. [*Haplopappus a.* (Nutt.) A. Gray] ✿TRY;DFCLT. May–Aug

STEPHANOMERIA

G. Ledyard Stebbins

Ann to shrubs, glabrous or hairy; sap milky. **STS** branched, 0.5–3 m. **LVS** cauline or some basal, alternate. **INFL**: heads ligulate; involucre cylindric; phyllaries in 2–several series; receptacle naked. **FLS** few–many; ligules lavender, pink, or whitish, readily withering. **FR** linear, club-shaped, or oblong; pappus gen of 9–30 stiff, plumose bristles.

24 spp.: w N.Am. (Greek: wreath division) [Gottlieb 1972 Madroño 21:463–481] Ann, tap-rooted, rosette-forming spp. highly variable and complexly interrelated. Key to species adapted by David J. Keil.

1. Ann
  2. Heads solitary or clustered on short. stiffly spreading branches; peduncles < 10 mm . . . *S. exigua* ssp. *coronaria*
  2' Heads in an open panicle; peduncles gen 10–40 mm . . . . *S. exigua* ssp. *exigua*
1' Per or subshrub
  3. Cauline lvs well developed at fl time, linear-lanceolate or wider; heads 12–15 mm; fls 10–14 . . . . *S. parryi*
  3' Cauline lvs small, linear or scale-like, often 0 at fl time; heads 8–10 mm; fls 3–5
    4. Sts forming sharp thorns; pappus of many non-plumose bristles . . . . *S. spinosa*
    4' Sts not forming thorns; pappus of 20–30 plumose bristles
      5. Branches slender, flexible, in fl gen bearing thin, linear, flexible lvs; dry montane forests, open subalpine
        slopes . . . . *S. tenuifolia*
      5' Branches stout, rigid, gen lfless exc for a few short scales; low desert-like habitats . . . . *S. pauciflora*
      6. Herbage canescent . . . . var. *parishii*
      6' Herbage glabrous, glaucous . . . . var. *pauciflora*

S. exigua Nutt. Ann 2–6 dm, glabrous to minutely glandular. **STS** much-branched. **LVS:** basal rosette gen withered at fl; cauline small, linear, often 0 at fl. **INFL** various; involucre 6–7 mm; outer phyllaries << inner, reflexed or erect and appressed; inner phyllaries 4–6. **FLS** 5–6; ligules 6–8 mm, white to pink. **FR** 2–6.8 mm, 5-angled; pappus bristles variously plumose, white. Common. Desert scrub, dry disturbed ground; < 2000 m. **CA;** to WA, ID, Colorado, Tex, Baja CA. Highly variable. 5 sspp. in CA.

  ssp. **coronaria** (E. Greene) Gottlieb **INFL** glabrous or sparsely hairy; peduncles 3–5 mm; outer phyllaries appressed. **FR** 2.3–3.1 mm; pappus bristles plumose on upper 60–85%. 2*n*=16. Dry steppes, shrubby slopes; 200–800 m. KR, SNH, Teh, SnJV, CCo, ChI, TR, MP, **SNE;** to OR, sw ID, w NV. [var. *c.* (E. Greene) Jepson]

  ssp. **exigua** (p. 185) **INFL** glabrous or sparsely glandular; peduncles 10–40 mm; outer phyllaries erect, appressed. **FR** 2.5–3.5 mm; pappus bristles plumose on upper 45–55%. 2*n*=16. Deserts, dry slopes; 100–2000 m. **GB, D;** to Colorado, TX, Baja CA [var. *pentachaeta* (D.C. Eaton) H.M. Hall] May–Sep

S. parryi A. Gray Per from strong taproot, 2–4 dm, glabrous. **STS** 1–few. **LVS** 2–8 cm, lobed. **INFL:** heads in open cyme; peduncles 4–8 mm; involucre 12–16 mm; inner phyllaries 7–10; outer phyllaries erect to reflexed, << inner. **FLS** 10–14; ligules 11–15 mm, whitish. **FR** 3–4 mm; pappus bristles plumose above, minutely barbed at base, brownish or pale yellowish. 2*n*=32. Deserts; 680–2000 m. **SNE, DMoj.** May–Jun

S. pauciflora (Torrey) Nelson WIRE-LETTUCE Per or subshrub, 3–6 dm, glabrous or densely short-tomentose. **STS:** branches many, ± stout, rigid. **LVS:** cauline often reduced to scales. **INFL:** heads at tips of side branches; peduncles 1–13 mm (or 0 above); involucre 8–10 mm; outer phyllaries << inner, appressed or spreading; inner phyllaries 5. **FLS** 3–5; ligules lavender-pink. **FR** 3–4 mm; pappus plumose but not to base, pale brownish. 2*n*=16. Dry flats, deserts; < 2400 m. s SNH, Teh, SnJV, SCoRI, TR, **SNE, D;** to UT, TX, nw Mex; also in KS.

  var. **parishii** (Jepson) Munz Herbage densely short-tomentose. Gravelly soil; < 1000 m. **DMoj;** w NV. ❀TRY.

  var. **pauciflora** Herbage glabrous, glaucous. Habitat and range of sp. [*S. myrioclada* D.C. Eaton] ❀TRY. May–Aug

S. spinosa (Nutt.) Tomb Subshrub 1–4 dm, from thick caudex, woolly at base and in lower axils, otherwise glabrous. **STS** few–several; branches intricate, becoming sharp thorns. **LVS** linear to scale-like. **INFL:** heads terminal on branchlets; peduncles < 7 mm; involucre 8–9 mm; outer phyllaries unequal, ± triangular, ascending; inner phyllaries 4–6. **FLS** 3–5; ligules pink. **FR** 4 mm; pappus of many non-plumose bristles, white to tawny. 2*n*=16. Deserts, scrub, dry mtn slopes; 1200–3300 m. SNH, SnGb, SnBr, MP, **n SNE, W&I, DMtns;** NV, s OR. [*Lygodesmia s.* Nutt.] Jul–Sep

S. tenuifolia (Torrey) H.M. Hall WIRE LETTUCE Per 1–6 dm, from woody caudex, glabrous. **STS** several; branches many, slender, ascending. **LVS** 1.5–5 cm, linear or thread-like, gen present at fl. **INFL:** heads solitary, nearly sessile at branch tips; involucre 8–10 mm; outer phyllaries << inner, erect or ascending; inner phyllaries gen 5. **FLS** 5. **FR** 2–3 mm; pappus bristles plumose throughout, white to tawny. 2*n*=16. Scrub, dry mtn slopes; 1100–3360 m. CaRH, SNH, MP, **n DMtns (Argus Mtns);** to WA, ID, UT, NM, nw Mex. Jul–Aug

STYLOCLINE NEST STRAW

James D. Morefield

Ann, gen grayish, cobwebby to tomentose. **STS** 1–several from base, ± spreading to ascending, ± forked, gen ± evenly leafy below, ± lfless between upper forks. **LVS** simple, alternate, ± sessile, gen elliptic to oblanceolate, entire. **INFL:** heads disciform, ± sessile in groups of 2–10; bracts like lvs; phyllaries 0 or vestigial (or 1–4 reduced, unequal scales); receptacle (2.8)4–8 × > wide, ± cylindric to club-shaped; chaff scales subtending pistillate fls phyllary-like, each gen enclosing a fl (or outermost open), falling with a fr, gen woolly, ± scarious-winged, wing gen elliptic to obovate, base gen acute; chaff scales subtending disk fls reduced, open or folded, glabrous to cobwebby, scarious. **PISTILLATE FLS** in 3–several series; corollas tubular. **DISK FLS** 2–6, staminate; ovary vestigial. **FR** obovoid, smooth, shiny; style ± at tip; pappus 0; disk pappus gen of 1–12 bristles. 7 spp.: sw N.Am. (Greek: column bed, from long receptacle) [Morefield 1992 Madroño 39: 114–130] Close to *Filago* subg. *Oglifa*, esp *F. depressa*.

1. Longest chaff scales scarious-winged throughout, wing = entire scale, widest below middle of scale; phyllaries 2–4, ± persistent, like chaff wings, completely scarious . . . . *S. gnaphaloides*
1' Longest chaff scales scarious-winged only toward tip, wing << entire scale, widest well above middle of scale; phyllaries 0 or 1–3, deciduous, unlike chaff wings, either vestigial or only edges scarious

2. Receptacle club-shaped, length 2.8–3.5 × width; outer fr 0.6–0.8 mm; lower lvs gen obtuse; heads ± spheric, 3–4 mm wide; longest chaff scale < 3.4 mm . . . . *S. sonorensis*
2' Receptacle ± cylindric in outline, length 4–8 × width; outer fr 1–1.6 mm; lower lvs gen acute; heads and chaff various
   3. Heads ovoid, largest 2.5–4 mm wide; longest chaff scales 2.8–3.3 mm, body firm, thickened; outermost chaff scales open, nearly glabrous; fr compressed front-to-back . . . . *S. psilocarphoides*
   3' Heads spheric, largest 5–9 mm wide; longest chaff scales 3.4–4.5 mm, body various; outermost chaff scales each enclosing a fl, woolly; fr various
      4. Largest uppermost lvs gen 4–11 mm, all ± elliptic to oblanceolate or obovate; fr compressed front-to-back; body of longest chaff scale firm, thickened, splitting lengthwise if forced . . . . . . *S. intertexta*
      4' Largest uppermost lvs gen 11–18 mm, at least some awl-like to lanceolate; fr compressed side-to-side; body of longest chaff scale soft, membranous, easily tearing when wool is pulled . . . . *S. micropoides*

S. gnaphaloides Nutt. (p. 189) EVERLASTING NEST STRAW **STS** < 24 cm. **LVS** < 14 mm, gen obtuse; uppermost < 13 mm, ± elliptic to obovate. **INFL:** heads 3–6 mm, spheric; phyllaries 2–4, 1–3.5 mm, ± ovate, scarious throughout, ± persistent; receptacle cylindric, length 5–8 × width; longest chaff scale 1.8–4.5 mm, widely ovate, body (exc midvein) membranous, winged throughout, wing widely ovate, widest below middle, rounded to cordate at base; outermost chaff scales ± closed, cobwebby. **DISK FLS** 1–2 mm. **FR** 0.8–1 mm, compressed side-to-side; disk ovary 0–0.2 mm, pappus bristles gen 1–5, 1.3–1.9 mm. *n*=14. Open, gen sandy soil; < 1200 (1700 m). s NCoRI, SNF, Teh, SnJV, CW, SW, **sw D**; to AZ, nw Mex. [Misspelled *S. gnaphalioides*] Mar–May

S. intertexta Morefield (p. 189) **STS** < 11 cm. **LVS** < 15 mm, gen acute; largest uppermost gen 4–11 mm, ± elliptic to obovate. **INFL:** largest head 5–6 mm, spheric; phyllaries 0 or vestigial; receptacle ± cylindric, length 4–7 × width; longest chaff scale 3.4–4.2 mm, winged toward tip, body hard; outermost chaff scales ± closed, woolly. **DISK FLS** 1.1–2.3 mm. **FR** 1–1.4 mm, compressed front-to-back; disk ovary 0–0.3 mm, pappus bristles 0 or 1–4(8), 1.1–2 mm. Stable, rocky or sandy, often calcareous soils; 50–1400 m. **ne DMoj, nw DSon**; to sw UT, w-most AZ. Perhaps derived from *S. micropoides × S. psilocarphoides*. Feb–May

S. micropoides A. Gray (p. 189) DESERT NEST STRAW **STS** < 21 cm. **LVS** < 20 mm, acute; largest uppermost gen 11–18 mm, awl-like to lanceolate. **INFL:** largest head 5–9 mm, spheric; phyllaries 0 or vestigial; receptacle ± cylindric, length 5–8 × width; longest chaff scale 3.4–4.5 mm, winged toward tip, body (exc midvein) membranous; outermost chaff scales ± closed, woolly. **DISK FLS** 1.4–2 mm. **FR** 1–1.4 mm, compressed side-to-side;

disk ovary 0–0.3 mm, pappus bristles gen 2–5(10), 1.1–2 mm. *n*=14. Stable, rocky or sandy, often calcareous soils; 50–1600 m. **s SNE, ne DMoj, nw DSon**; to w TX, n-most Mex. Feb–May

S. psilocarphoides M. Peck (p. 189) PECK NEST STRAW **STS** < 18 cm, ± lfless between lower forks. **LVS** gen acute, < 18 mm; uppermost < 10 mm, ± elliptic to obovate. **INFL:** heads 3.5–5 mm, 2.5–4 mm wide, ± ovoid; phyllaries 0 or vestigial (or 1–3, 1.5–2.5 mm, obovate, deciduous), body hard, not scarious; receptacle ± cylindric, length 5–8 × width; longest chaff scale 2.8–3.3 mm, winged toward tip; outermost chaff scales open, nearly glabrous. **DISK FLS** 1.2–2 mm. **FR** 1.1–1.6 mm, compressed front-to-back; disk ovary 0.1–0.4 mm, pappus bristles 0 or 1–3, 1.1–1.5 mm. Stable, sandy or rocky soils; (150)600–1800 m. **SNE, DMoj, w edge DSon**; to se OR, sw ID, sw UT. CA pls long mistaken for other *Stylocline* or *Filago* spp. ❀TRY. Feb–May

S. sonorensis Wiggins (p. 189) MESQUITE NEST STRAW **STS** < 15 cm. **LVS** < 13 mm, gen obtuse; uppermost < 10 mm, elliptic to ± lanceolate, acute. **INFL:** heads 3.5–4.5 mm, 3–4 mm wide, ± spheric; phyllaries 0 or vestigial; receptacle club-shaped, length 2.8–3.5 × width; longest chaff scale 1.9–3.1 mm, winged toward tip, body (exc midvein) gen membranous; outermost chaff scales ± closed, woolly. **DISK FLS** 1.2–1.9 mm. **FR** 0.6–0.8 mm, ± compressed side-to-side; disk ovary 0.3–0.6 mm, pappus bristles 3–8, 0.9–1.3 mm. RARE in CA. Open, sandy drainages; ± 400 m. **DSon** (Hayfields, Riverside Co., April 1930, perhaps extirpated); se AZ, ne Sonora. Like *Filago depressa*, which may have been involved in its origin. Mar–Apr

SYNTRICHOPAPPUS

Dale E. Johnson

Ann 2–10 cm, gen loosely woolly. **LVS** alternate (or lowest opposite), simple. **INFL:** heads solitary, radiate; involucre subcylindric; phyllaries in 1 series, spreading in fr, oblanceolate, acute, alternate phyllaries scarious-margined; receptacle convex, naked. **RAY FLS** 1 per phyllary; ligules 3-lobed, yellow (or white and red-veined). **DISK FLS** narrowly funnel-shaped, yellow, glabrous; stamen tips narrowly triangular. **FR** narrowly obconic; pappus 0, or of many bristles fused at base. 2 spp.: sw US. (Greek: fused bristly pappus) [Johnson 1991 Novon 1:119–124] ❀TRY.

S. fremontii A. Gray (p. 189) Pl ± decumbent. **LVS** 5–20 mm, wedge-shaped below, spoon-shaped above, 3-lobed (margins rolled under) or entire. **INFL:** peduncles 3–25 mm; involucre 5–7 mm; phyllaries gen 5. **RAY FLS:** ligules 3–5 mm, yel-

low. **FR** 2–3.5 mm, strigose; pappus of 30–40 bristles, ± 2 mm. 2*n*=12. Open, sandy to gravelly areas; 600–2500 m. **DMoj**; to sw UT, nw AZ. Apr–Jun

TARAXACUM

G. Ledyard Stebbins

Per from taproot; sap milky. **STS** naked, hollow. **LVS** all basal, toothed or lobed; lobes acute. **INFL:** heads ligulate, solitary, scapose; phyllaries many, outer ovate to lanceolate, gen reflexed, inner erect, linear; receptacle convex, naked. **FLS** many; ligules yellow, readily withering. **FR** fusiform; ribs rough; beak slender, >> body; pappus of many, white, slender bristles, not plumose. Many named spp., mostly reproducing clonally by asexual seeds: Eurasia, CA. (Greek: ancient name) [Taylor 1987 Bull Torrey Bot Club 114:109–120] *T. lyratum* has been recorded from e SNE.

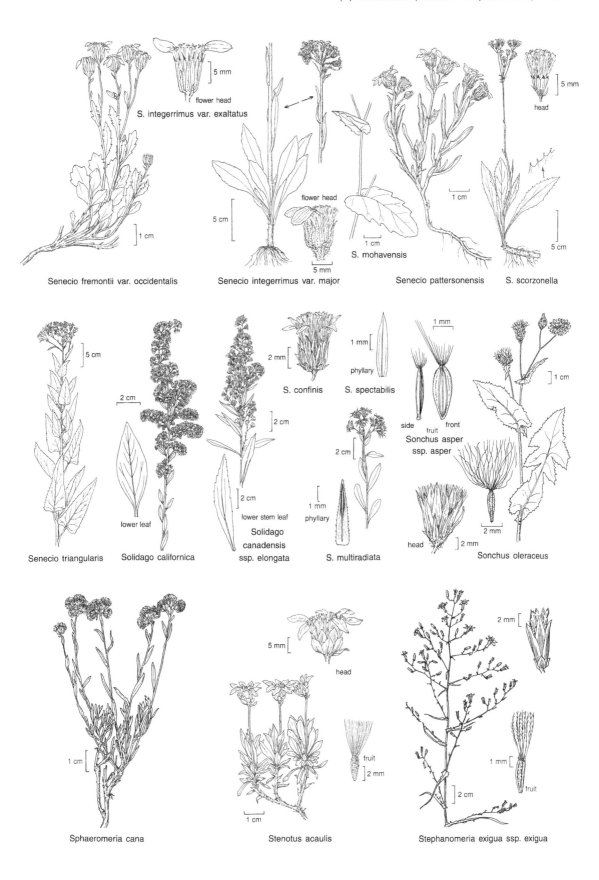

S. integerrimus var. exaltatus

flower head

5 mm

Senecio fremontii var. occidentalis

1 cm

5 cm

flower head

Senecio integerrimus var. major

5 mm

S. mohavensis

1 cm

Senecio pattersonensis

1 cm

head

5 mm

S. scorzonella

5 cm

Senecio triangularis

5 cm

2 cm

lower leaf

Solidago californica

2 cm

2 cm

lower stem leaf

Solidago canadensis ssp. elongata

2 cm

S. confinis

2 mm

S. spectabilis

phyllary

1 mm

S. multiradiata

2 cm

phyllary

1 mm

side   front

fruit

Sonchus asper ssp. asper

1 mm

1 cm

head

2 mm

2 mm

Sonchus oleraceus

Sphaeromeria cana

1 cm

Stenotus acaulis

head

5 mm

fruit

2 mm

1 cm

Stephanomeria exigua ssp. exigua

2 mm

2 cm

fruit

1 mm

T. officinale Wigg. (p. 189) **LVS** variously toothed or lobed. **INFL**: heads 2–3.3 cm diam; outer phyllaries reflexed, 5–8 mm; tips of inner phyllaries acute to minutely truncate. **FR** grayish or olive-brown; body tubercled near tip. 2*n*=16,24,48. Abundant. Weed, esp of lawns, meadows; 0–3300 m. CA-FP, **GB**; native to Eur. [*T. laevigatum* (Willd.) DC.] Most of year

TETRADYMIA COTTON-THORN, HORSEBRUSH

Shrubs. **STS** ± tomentose. **LVS** alternate and gen clustered in axils, linear to (ob)lanceolate, sometimes persisting as stiff spines, glabrous to tomentose. **INFL**: heads discoid, axillary or in ± rounded, terminal clusters; involucre cylindric to hemispheric; phyllaries in 1–2 ± equal series, often keeled; receptacle naked. **DISK FLS** gen 4–8; corollas cream to yellow, lobes long, spreading; anther bases ± sagittate, tips obtuse or acute; style branches papillate to short-bristly, tips truncate to conic. **FR** obconic or fusiform, often angled; pappus 0 or of gen many bristles or slender scales. 10 spp.: w N.Am. (Greek: 4 together, from 4-fld heads of some) [Strother 1974 Brittonia 26:177–202] Esp fl buds TOXIC to sheep (toxicity poorly understood).

1. Sts unarmed (lvs not forming spines)
   2. Main lvs densely tomentose or silvery; clustered lvs lanceolate to ± oblanceolate . . . . **T. canescens**
   2' Main lvs glabrous or sparsely tomentose; clustered lvs narrowly linear
      3. Main lvs narrowly awl-shaped, stiffly ascending or appressed, gen 5–10 mm; clustered lvs gen 3–10 mm, glabrous; fr gen 3–4 mm, short-stiff-hairy; pappus of ± 100 bristles . . . . **T. glabrata**
      3' Main lvs ± threadlike, soft, gen 10–30 mm; clustered lvs gen 10–20 mm, loosely tomentose; fr 5–6 mm, long-soft-hairy; pappus of ± 20 stiff bristles or slender scales (± hidden by fr hairs) . . **T. tetrameres**
1' Sts armed with spines derived from main lvs
   4. Fr glabrous, 2.5–3.5 mm . . . . **T. argyraea**
   4' Fr ± long-soft-hairy, 5–8 mm
      5. Clustered lvs silvery-hairy; fr ± short-hairy; pappus of many fine bristles 9–12 mm . . . . **T. stenolepis**
      5' Clustered lvs ± glabrous; fr densely long-hairy; pappus of ± 25 slender scales 6–9 mm
         6. Spines sharply recurved, < 2 cm, persistently tomentose; involucres gen 8–12 mm, bell-shaped; fr gen 6–8 mm . . . . **T. spinosa**
         6' Spines straight, gen 2–4 cm, becoming glabrous; involucres gen 7–9 mm, obconic; fr gen 4–5 mm . . . . **T. axillaris**
            7. Phyllaries and peduncles glabrous; fr hairs gen 6–8 mm . . . . var. **axillaris**
            7' Phyllaries and peduncles tomentose; fr hairs gen 9–11 mm . . . . var. **longispina**

T. argyraea Munz & Roos STRIPED HORSEBRUSH Pl < 20 dm, spiny. **STS** becoming ± glabrous in stripes below spines. **LVS**: main lvs 1–2(3) cm, canescent or becoming glabrous, linear, forming straight or ± upturned spines; clustered lvs 3–8(20) mm, thread- to club-like, ± glabrous. **INFL**: heads gen 2–5; peduncles gen 1–4 mm, tomentose, bracts 0; involucre ± 7 mm, obconic; phyllaries 5, narrowly elliptic, tomentose. **FLS** 5; corollas ± 9 mm, pale yellow. **FR** 2.5–3.5 mm, glabrous; pappus of many fine bristles, ± 8 mm. 2*n*=60. UNCOMMON. Pinyon/juniper woodland; 1400–2100 m. **DMtns**. Aug–Sep

T. axillaris Nelson Pl < 15 dm, spiny. **STS** evenly tomentose. **LVS**: main lvs 1–5 cm, forming straight spines, tomentose, becoming glabrous; clustered lvs 2–12(20) mm, thread- to club-like, ± glabrous. **INFL**: heads 1–3 in axils of previous year's growth; peduncles gen 4–15 mm, gen bracted; involucre 7–9 mm, obconic; phyllaries 5, ovate. **FLS** 5–7; corollas 7.5–9 mm, pale yellow. **FR** gen 4–5 mm, densely long-white-hairy; pappus of ± 25 slender scales, 6–7.5 mm. Sagebrush or saltbush scrub; 1200–2300 m. **s SNE**, **DMoj**; to sw UT.

var. *axillaris* **INFL**: peduncles and phyllaries glabrous. **FR**: hairs 6–8(10) mm. 2*n*=60. Habitats of sp. **ne DMoj**; s NV. Apr–May

var. *longispina* (M.E. Jones) Strother (p. 189) **INFL**: peduncles and phyllaries tomentose. **FR**: hairs 9–11(14) mm. 2*n*= 60,62. Habitats of sp. **s SNE**, **w DMoj**; to sw UT.

T. canescens DC. (p. 189) Pl 1–8 dm, unarmed. **STS** unevenly tomentose, becoming ± glabrous in stripes below nodes. **LVS**: main lvs < 4 cm, ± (ob)lanceolate, sparsely tomentose to silvery; clustered lvs like (gen <) main lvs. **INFL**: heads gen 3–6 in flat-topped clusters; peduncles 5–15(25) mm, bracts 0; involucre 6–8(12) mm, cylindric to obconic; phyllaries 4, oblong to ovate. **FLS** 4; corollas 7–15 mm, creamy to bright yellow. **FR** 2.5–5 mm, glabrous or short-stiff-hairy; pappus of many fine bristles, 6–11 mm. 2*n*=60, 90,120. Sagebrush scrub, pinyon/juniper woodland, forest; (400) 1600–3300 m. TR, s PR, **GB**, **DMoj**; to B.C., MT, WY, NM. ❀TRY. Jul–Aug

T. glabrata Torrey & A. Gray (p. 189) Pl < 12 dm, unarmed. **STS** unevenly tomentose, becoming ± glabrous in stripes below nodes. **LVS**: main lvs 5–10(20) mm, narrowly awl-shaped, ascending to appressed, sparsely woolly or becoming glabrous; clustered lvs 3–10(15) mm, thread- to ± club-like, glabrous. **INFL**: heads gen 3–7; peduncles gen 5–10 mm, bracts 0; involucre 7–10 mm, obconic; phyllaries 4, lanceolate to obovate. **FLS** 4; corollas 9–10 mm, cream to golden yellow. **FR** 3–5 mm, short-stiff-hairy; ribs glandular; pappus of many fine bristles, 6–8 mm. 2*n*=60,62,120,180. Sagebrush scrub, pinyon/juniper or Joshua-tree woodland; 800–2400 m. **GB**, **DMoj**; to OR, ID, UT. ❀TRY. May–Jul

T. spinosa Hook. & Arn. Pl < 10 dm, spiny. **STS** tomentose. **LVS**: main lvs 5–25 mm, tomentose, forming rigid, recurved spines; clustered lvs 3–15(25) mm, thread-like to ± oblanceolate, ± glabrous. **INFL**: heads gen 1–2 in axils of previous year's growth; peduncles gen 5–30 mm, tomentose, bracted; involucre 8–12 mm, ± bell-shaped; phyllaries 4–6, oblong to ovate. **FLS** 5–8; corollas 6–10 mm, ± yellow. **FR** 6–8 mm, densely long-white-hairy; pappus of ± 25 slender scales, 6–9 mm. 2*n*=60. Gen saltbush scrub; 800–2400 m. s MP, **n SNE**; to OR, ID, MT, WY, n NM.

T. stenolepis E. Greene (p. 189) Pl < 12 dm, spiny. **STS** unevenly tomentose, becoming ± glabrous in stripes below spines. **LVS**: main lvs 2–3 cm, tomentose or becoming glabrous, forming ± straight spines; clustered lvs 10–30 mm, ± oblanceolate, tomentose or silvery-hairy. **INFL**: heads gen 4–7; peduncles gen 5–12 mm, tomentose, bracts 0; involucre 8–10 mm, narrowly obconic; phyllaries (4)5, narrowly ovate. **FLS** (4)5; corollas 10–12 mm, pale yellow. **FR** 5–8 mm, ± short-soft-hairy; pappus of many fine bristles, 9–12 mm. 2*n*=60. Joshua-tree woodland, creosote-bush scrub; 600–1500 m. **SNE**, **DMoj**; s NV May–Aug

T. tetrameres (S.F. Blake) Strother Pl < 20 dm, unarmed. **STS** tomentose. **LVS** sparsely tomentose; main lvs 1–4 cm, linear, thread-like, soft; clustered lvs 10–20 mm, thread-like to linear-oblanceolate. **INFL**: heads gen 4–6 on short side-branches; peduncles gen 1–3 mm, tomentose, bracts 0–2; involucre 8–9 mm, obconic; phyllaries 4(5), widely elliptic. **FLS** 4(5); corollas ± 8 mm, pale yellow. **FR** 5–6 mm, densely long-soft-hairy; pappus of ± 20 stiff bristles or slender scales, 3–5 mm, ± hidden by fr hairs. 2*n*=60. Dunes, deep sand, sagebrush scrub; 1200–2100 m. **n SNE**; w NV.

THYMOPHYLLA

Ann, per, subshrubs. **LVS** simple or pinnately divided, opposite or alternate, dotted with embedded oil glands. **INFL**: heads radiate or discoid, sessile or peduncled, solitary or in few-headed cymes; involucre hemispheric to bell-shaped; phyllaries in 2–3 series (outer, if present, few, free, inner fused, gland-dotted); receptacle flat to rounded, naked. **RAY FLS** 0–few; corollas yellow or white. **DISK FLS** many; corollas yellow; style tips truncate or conic. **FR** obpyramidal, obconic, or cylindric; pappus of scales, each awn-tipped or dissected into bristles. ± 17 spp.: s N.Am, Caribbean. (Greek: thyme-leaved) [Strother 1986 Sida 11:371–378]

T. pentachaeta (DC.) Small var. ***belenidium*** (DC.) Strother (p. 189) Subshrub 1–3 dm. **STS** gen many, slender, very leafy. **LVS** opposite, 1–2 cm, pinnately divided into 3–5 stiff, linear lobes, dotted with tiny glands, puberulent. **INFL**: heads radiate, solitary; peduncles 2.5–4.5 cm; involucre 3.5–6 mm diam; outer phyllaries 3–5, short-triangular; inner phyllaries 13 in 2 series, 5–6 mm, gland-dotted, outer linear, ciliate, inner wider. **RAY FLS** gen 13; ligules 2–3.5 mm, yellow. **DISK FLS**: corollas 2.5–3 mm. **FR** 2–3 mm, puberulent; pappus of 10 scales, each 2.5–3 mm, dissected into 3 awns (or 5 of the 10 truncate, ± 1 mm). 2*n*=26. Dry roadsides, gravelly slopes; 900–1800 m. SnJt, **e DMoj**; to TX, n Mex; also in s S.Am. [*Dyssodia p.* (DC.) Robinson var. *b.* (DC.) Strother; *D. thurberi* (A. Gray) Robinson] ✿TRY. Apr–Jun

TONESTUS

Gregory K. Brown

Per from taproot or branched caudex, glandular. **LVS**: basal persistent; cauline alternate, reduced above, oblanceolate, 1–3-veined. **INFL** loose; heads, radiate, 1–5; involucre bell-shaped; phyllaries in 3–4 ± equal series, lanceolate, 1-veined, outer graduated into upper lvs; receptacle convex, naked. **RAY FLS** 10–25; ligules < 12 mm, yellow. **DISK FLS** many; corolla flaring slightly, yellow. **FR** narrowly oblong, ± cylindric, fusiform, or compressed, hairy; pappus of many white bristles in 1 series. ± 4 spp.: w N.Am. (Anagram of *Stenotus*) [Nesom & Morgan 1990 Phytologia 68:174–180]

T. peirsonii (Keck) G. Nesom & R. Morgan (p. 189) **STS** < 20 cm. **LVS** < 8 cm, toothed. **INFL**: involucre 15–20 mm; phyllaries green. **DISK FLS**: corollas 8–10 mm. **FR** 5 mm; pappus < disk corolla. 2*n*=90. Rocky alpine and subalpine slopes; 2900–3700 m. c SNH (Inyo, Fresno cos.), **W&I**. [*Haplopappus p.* (Keck) J. Howell] ✿ TRY:DFCLT. Jul–Aug

TOWNSENDIA

Geraldine A. Allen

Ann to per. **STS** 0 to erect, < 30 cm. **LVS** alternate, entire, petioled. **INFL**: head, radiate gen solitary; involucre conic to hemispheric; phyllaries in 2–6 series, outer gen < inner, free, margins scarious to ciliate; receptacle flat, naked. **RAY FLS** many; ligules white, pink, blue, or yellow. **DISK FLS** many; corollas ± yellow; style branches flat, tip hairy. **FR** ± compressed, brown, glabrous to hairy; pappus of minutely barbed, gen flat bristles (ray frs sometimes also with small outer series). 21 spp.: w N.Am. (D. Townsend, amateur US botanist, 1787–1858) [Reveal 1970 GB Naturalist 30:23–52] Some spp. reproduce by asexual seed.

1. Lvs long-soft-woolly; pappus readily deciduous . . . . ***T. condensata***
1' Lvs ± finely strigose; pappus ± firmly attached to fr
  2. Fr glabrous; lvs gen linear to oblanceolate, acute . . . . ***T. leptotes***
  2' Fr ± spreading-hairy; lvs oblanceolate to obovate, ± obtuse
    3. Sts (2)5–30 cm, leafy to middle or above . . . . ***T. parryi***
    3' Sts 2–8 cm, leafy in lower half . . . . ***T. scapigera***

T. condensata Eaton (p. 189) Bien or per < 5 cm; caudex taprooted. **LVS** basal, gen 1–1.5 cm, ± narrowly obovate, rounded, entire, long-soft-woolly. **INFL**: phyllaries ± subequal, lanceolate, ± acuminate, scarious-margined, ± long-hairy. **RAY FLS**: ligules white to pink or violet. **FR** short-hairy; pappus readily deciduous. 2*n*=18. Gravel slopes; 3200–3500 m. **n SNE (Mono Co)**; to sw Can, WY, Colorado. May–Jun

T. leptotes (A. Gray) G.E. Osterh. (p. 189) Per < 3 cm; caudex taprooted. **LVS** basal, gen 1–2 cm, ± linear to oblanceolate, gen acute, entire, gen strigose. **INFL**: phyllaries lanceolate, acuminate, margins scarious-ciliate, glabrous to ± hairy. **RAY FLS**: ligules white to pink or blue. **FR** glabrous; pappus ± persistent. 2*n*=18. Rocky slopes; 3500–3700 m. **W&I**; to MT, NM. Jul–Aug

T. parryi Eaton PARRY'S TOWNSENDIA Per (2)5–30 cm; caudex taprooted. **LVS** basal and cauline, gen 2–6 cm, oblanceolate to obovate, obtuse to rounded, entire, gen strigose. **INFL**: phyllaries lanceolate, acuminate, margins scarious-ciliate, glabrous to ± hairy. **RAY FLS**: ligules violet to blue. **FR** ± spreading-hairy; pappus ± persistent. 2*n*=18,36. RARE in CA. Open places; 3000–3500 m. **n SNE (Sweetwater Mtns, Mono Co.)**; to sw Can, Colorado. May–Jul

T. scapigera Eaton (p. 189, pl. 25) Per 2–8 cm; caudex taprooted. **LVS** basal and cauline, 1–5 cm (< petiole), obovate, ± obtuse to rounded, entire, strigose. **INFL**: phyllaries narrowly ovate to oblong, ± acute, scarious-margined, ± hairy. **RAY FLS**: ligules white to pink or violet. **FR** ± spreading-hairy; pappus ± persistent. Rocky slopes; 1400–3500 m. **SNE**; NV. ✿DRN, SUN:1–3,7,18;DFCLT. May–Aug

TRAGOPOGON   GOAT'S BEARD

G. Ledyard Stebbins

Bien or per from strong taproot; herbage ± glabrous; sap milky. **STS**: branches few, strongly ascending. **LVS** basal and cauline, alternate, entire, grass-like, parallel-veined. **INFL**: heads ligulate, solitary at branch tips; peduncles long, naked; involucre cylindric or narrowly conic; phyllaries in 1 series; receptacle naked. **FLS** yellow to bronze or purple; ligules readily withering. **FR** 2.5–3 cm; beak stout, > body; pappus of stout plumose bristles, 2+ bristles tangled, tips of a few bristles > the rest, unbranched; frs spreading, forming a spheric head 4–5 cm diam. ± 45 spp.: Eurasia. (Greek: goat's beard)

T. dubius Scop. Ann, bien 3–10 dm. **LVS** 2–5 dm. **INFL**: peduncle much wider toward tip; phyllaries 8–13, 2.5–4 cm in fl heads, >> fls, 4–7 cm in fr heads. **FLS**: ligules pale lemon-yellow. **FR** 25–35 mm; pappus ± white. $2n=12$. Uncommon. Weed in waste places; 0–2700 m. n SN, SnJV, SnFrB, SCoRO, SnBr, **GB**; native to Eur. May–Jul

TRICHOPTILIUM

1 sp. (Greek: feathery bristle, from pappus)

T. incisum A. Gray (p. 193, pl. 26) Ann, short-lived per 5–25 cm, gen ± tomentose. **STS** 1–several from base. **LVS** simple, basal and cauline, alternate or subopposite, mostly clustered in lower half of pl, 1–3 cm, sessile or tapered to a short, winged petiole; blade acute, sharply dentate or shallowly lobed, densely tomentose, resin-dotted. **INFL**: heads discoid, solitary; peduncles 3–11 cm, slender, bractless, glandular-puberulent; involucre 6–12 mm diam, hemispheric or bell-shaped, often appearing ± disk-like when pressed; phyllaries in 2 ± equal series, free, 5–7 mm, linear-elliptic, acute; receptacle rounded, naked. **FLS** many; corollas 4.5–7 mm, yellow, outermost sometimes enlarged and ± bilateral; style-tips truncate. **FR** 2–3 mm, obpyramidal, glabrous to densely strigose; pappus of 5 scales, each 5–7 mm, dissected into many bristles. $2n=26$. Dry slopes, plains; < 1000 m. s DMoj, DSon; s NV, sw AZ, n Baja CA. ✿TRY. Feb–May, Oct–Nov

TRIMORPHA

Guy L. Nesom

Ann, bien, per, fibrous-rooted, gen from short rhizome. **LVS** oblanceolate to obovate or spoon-shaped, entire. **INFL**: heads disciform or inconspicuously radiate, solitary or in few-headed clusters; buds erect; involucre hemispheric; phyllaries ± graduated, narrowly lanceolate outer and middle with 3 orange-resinous veins, at least near base; receptacle naked. **PISTILLATE FLS** in ± 2 series; outer gen with very narrow, sometimes coiled ligules; inner gen without ligules (exc *T. lonchophylla*). **DISK FLS**: corollas yellow, narrowly tubular, not hard or inflated; veins orange-resinous; style tips deltate. **FR** narrowly oblong, 2–4-ribbed, ± flat; pappus of barbed bristles in 1–2 series, outer 0 or bristles few, short, inner at maturity > involucre. ± 45 spp.: N.Am, Eurasia. (Latin & Greek: 3 forms, from fl types) [Nesom 1989 Phytologia 67:61–66] Formerly treated as *Erigeron* sect. *Trimorpha.*

T. lonchophylla (Hook.) G. Nesom (p. 193) Ann, bien. **STS** 4–20 cm, sparsely to densely bristly or soft-hairy; glands 0. **LVS**: basal 2–8 cm, gradually reduced and becoming linear upward. **INFL** gen raceme-like or of 1 head; phyllaries ± rough-hairy, without glands, gen purple-tipped. **PISTILLATE FLS** in 2–3 series, all with very narrow ligules 2–3 mm, barely extending beyond involucre. **DISK FLS**: corollas 3–4.5 mm. **FR** 1.3–1.5 mm. $2n=18$. Meadows, creek and ditch banks; 1800–3550 m. SN, SnBr, **W&I**; to AK, e Can, NM. [*Erigeron l.* Hook.] Jul–Aug

TRIXIS

Shrubs. **LVS** simple, alternate, sessile or petioled; blade long, entire or toothed, gen ± glandular. **INFL**: heads discoid (but often appearing radiate), sessile or short-peduncled, in flat-topped or panicle-like cymes; involucre cylindric; phyllaries gen in 1 series, linear, keeled; receptacle flat, short-hairy. **FLS** few–many; corolla yellow, 2-lipped, outer lip spreading, ligule-like, 3-lobed, sometimes recurved, inner lip 2-lobed, recurved or coiled; anthers exserted, base with tail-like appendages, tip oblong, acute; style tips truncate. **FR** ± cylindric, 5-ribbed, black or brown, short-hairy and glandular; pappus of many bristles. ± 65 spp. N.Am, S.Am. (Greek: 3-fold, from outer corolla lip) [Anderson 1972 Mem NY Bot Gard 22(3):1–68]

T. californica Kellogg var. ***californica*** (p. 193, pl. 27) Pl glandular and short-hairy. **STS** 2–20 dm, erect, stiff, much-branched. **LVS** many; petiole 1.5–5 mm, winged; blade 2–11 cm, linear-lanceolate to lanceolate, acute at both ends, entire or serrate, margin often rolled under. **INFL**: peduncles gen < or = 5 mm, bearing 5–7 bracts; phyllaries 8–10, 8–14 mm, green, tips acute. **FLS** 11–25; corolla glandular, tube 6–9 mm, outer lip 5–8 mm, spreading, inner lip 4–5.5 mm, gen coiled. **FR** 6–10.5 mm; pappus 7.5–12 mm, dull white. $2n=54$. Dry, rocky slopes, washes, desert flats; gen < 1000 m. **D**; to TX, n Mex. ✿DRN, DRY,SUN: 10,11,**12,13,19–21**,22–24.

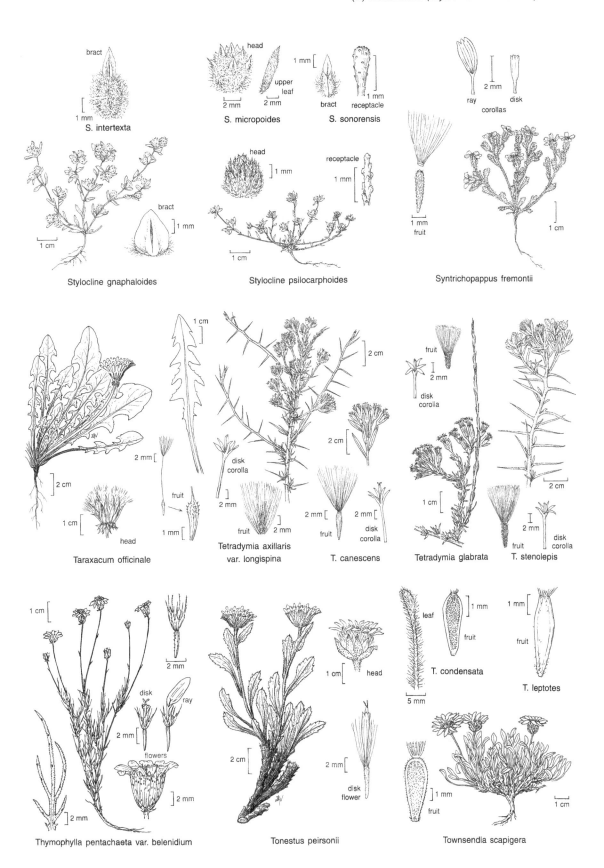

S. intertexta

S. micropoides

S. sonorensis

ray   disk
corollas

Stylocline gnaphaloides

Stylocline psilocarphoides

Syntrichopappus fremontii

Taraxacum officinale

Tetradymia axillaris
var. longispina

T. canescens

Tetradymia glabrata

T. stenolepis

Thymophylla pentachaeta var. belenidium

Tonestus peirsonii

T. condensata

T. leptotes

Townsendia scapigera

UROPAPPUS SILVER PUFFS

Kenton L. Chambers

1 sp. (Latin: tailed pappus, from awn-tipped scales)

U. lindleyi (DC.) Nutt. (p. 193) Ann 5–70 cm, ± scapose; sap milky. **LF** 5–30 cm, ± basal, ± linear, long-tapered, entire to narrowly lobed, ± soft-hairy (esp petiole base). **INFL**: head, ligulate solitary, erect; involucre 10–40 mm, glabrous; phyllaries narrowly lanceolate, outer progressively shorter; receptacle naked. **FLS** 5–many; corollas yellow (often reddish below), readily withering, < or = involucre. **FR** 7–17 mm, slender, tapered to tip in CA, obscurely 10-ribbed, gen blackish; outermost frs scabrous; pappus scales 5, 5–15 mm, deciduous, smooth, silvery, bristle-tip 4–6 mm, slender, smooth, from notched scale tip. $2n=18$. Common. Open grassland, woods, chaparral, deserts, gen in loose soils; < 1800 m. **CA (exc NCo)**; to WA, ID, UT, w TX, n Mex. [*Microseris l.* (DC.) A. Gray; *M. linearifolia* (DC.) Schultz-Bip.] ❀TRY. Apr–May

VERBESINA CROWNBEARD

Ann, per, shrubs. **LVS** simple, opposite or alternate. **INFL**: heads radiate or discoid, solitary or in few–many-headed cymes; phyllaries in 2–6 equal to very unequal series, free; receptacle flat to conic, chaffy; chaff scales entire, folded around ovaries. **RAY FLS** fertile or sterile; corollas yellow to white. **DISK FLS** many; corollas yellow to orange; style branches acute to acuminate. **FR**: ray achenes triangular or 0; disk achenes flattened, wing-margined, obovate; pappus 0 or of 2 awns. (Derived from *Verbena*) [Coleman 1966 Madroño 18:129–137; Coleman 1966 Amer Midl Naturalist 76:475–481]

V. encelioides (Cav.) A. Gray ssp. **exauriculata** (Robinson & Greenman) J. Coleman (p. 193) GOLDEN CROWNBEARD Ann 15–130 cm; odor unpleasant. **STS** densely short-hairy. **LVS** petioled; blade lanceolate to triangular-ovate, 3-veined from base, coarsely dentate, dull green, ± strigose-canescent. **INFL**: heads 1–many; peduncles subtended by lf-like bracts; phyllaries in 2 or 3 ± equal series, 8–10 mm, linear-lanceolate, acute, ± strigose; chaff scales 6–8 mm, abruptly acuminate. **RAY FLS** fertile; ligule orange-yellow, ± 3-lobed; style present. **DISK FLS**: corolla 5–6 mm; anthers yellow to light brown. **FR**: ray achenes 3–4 mm, obovoid, triangular, wing 0, pappus 0; disk achene 4–6.5 mm, obovate, flattened, brown or black, soft-hairy, wing wide, ± white, corky, pappus awns 2, 1–2.5 mm. $2n=34$. Disturbed areas, roadsides, fields; gen < 200 m. SnJV, SCoR (Salinas Valley), SCo, **DSon**; w N.Am; native AZ to Great Plains, Mex. [*V. en.* var. *ex.* Robinson & Greenman] TOXIC to livestock but unpalatable. May–Dec

VIGUIERA

Ann, per, shrubs. **STS** gen several from base. **LVS** simple, alternate or opposite, sessile or petioled. **INFL**: heads radiate or discoid, solitary or in few-headed, terminal cymes; peduncles long or short; involucre hemispheric or bell-shaped; phyllaries in several series, equal to very unequal; receptacle rounded to conic, chaffy; chaff scales entire or 3-lobed, folded around frs. **RAY FLS** 0–many, sterile; corolla yellow; ligules entire to 3-lobed. **DISK FLS** many; corollas yellow or orange; anther tips triangular; style tips triangular. **FR** ± flattened, obovate, glabrous or ± hairy; pappus of scales, gen 1 or more lanceolate. ± 150 spp.: New World. (L.G.A. Viguier, 1790–1867, French physician, botanist) [Shilling 1990 Madroño 37:149–170]

1. Lvs short-hairy on both surfaces, blade 1–3.5 cm, triangular ovate, often toothed, veins not strongly raised beneath . . . . *V. parishii*
1' Lvs densely canescent-tomentose above, dull green and loosely tomentose beneath, blade 2–9 cm, ovate, entire, veins prominently raised beneath . . . . *V. reticulata*

V. parishii E. Greene (p. 193) Shrub < 2 m diam, short-rough-hairy throughout. **ST** 6–13 dm, much-branched. **LVS** opposite below, alternate above; petiole 2–8 mm; blade gen 1–3.5 cm, triangular-ovate, 3-veined from obtuse to truncate or subcordate base, tip obtuse to acute, entire or teeth few, short, surfaces green to lightly canescent. **INFL**: heads solitary or in open, few-headed cymes; peduncles 3–15 cm, slender, bracts 0 or few and lf-like; heads radiate; involucre 10–13 mm diam, hemispheric or appearing disk-like when pressed; phyllaries in 2–3 equal or unequal series, 3–9 mm, lance-oblong, tips abruptly narrowed, surface green to canescent; chaff scales 5–6 mm. **RAY FLS** 8–15; ligules 1–1.5 cm. **DISK FLS**: corollas 3.5–5 mm. **FR** 2.7–3.8 mm; pappus of 2 fringed scales (each 0.5–1 mm) and 2–3 lanceolate scales (each 2–3 mm). $2n=36$. Common. Washes, dry, rocky slopes; < 1500 m. **D**; AZ, NV, nw Mex. [*V. deltoidea* A. Gray var. *p.* (E. Greene) Vasey & Rose] ❀DRN:10,11,**12–14**,18,**19–**24&DRY,SUN:15–17; continued fls & lvs with IRR;GRNCVR; also STBL Feb–Jun, Sep–Oct

V. reticulata S. Watson (p. 193) Shrub < 1.5 m diam. **STS** many, 5–15 dm, soft-hairy; bark peeling in age. **LVS** opposite below, alternate above; petioles 3.5–30 mm; blades 2–9 cm, ovate, 3-veined from truncate to cordate base, tip acute, margin entire, upper surface densely canescent-tomentose, undersurface with veins prominently raised, gray-green, loosely tomentose. **INFL**: cyme few-headed, on long, ± naked branch; bracts reduced to scales 3–10 mm; peduncles 5–50 mm; heads radiate; involucre hemispheric or appearing disk-like when pressed; phyllaries in 2–3 unequal series, 2–5 mm, oblong to ovate, obtuse, short-white-hairy; chaff scales 4–5.5 mm. **RAY FLS** 10–15; ligules 7–15 mm. **DISK FLS**: corollas 3–4 mm. **FR** 2.5–4 mm, obovate; pappus of 2 scales (each 0.5–1 mm) and 1–2 lanceolate scales (each 1–2.8 mm). Arid slopes; < 1500 m. **DMoj**; w NV. ❀TRY. Feb–Jun

WYETHIA MULES EARS

Per from stout, taprooted caudex, gen ± unbranched. **LVS** alternate, gen reduced upward; petiole gen << blade, often winged. **INFL**: heads gen radiate, 1–few, ± large; involucre hemispheric or bell-shaped; phyllaries ± many in 2–3 series, outer often ± lf-like; receptacle ± convex; chaff scales ± linear, entire, gen ± hairy. **RAY FLS** (0)5–20; ligules yellow. **DISK FLS** many; corollas yellow, lobes short; anthers brown or yellow, tips triangular; style tips linear, tapered. **FR**: pappus 0 or crown of scales; ray achenes 3-angled; disk achenes 4-angled, sometimes ± compressed. (Nathaniel J. Wyeth, US explorer, 1802–1856) [Weber 1946 Amer Midl Naturalist 35:400–452]

W. mollis A. Gray (p. 193) Pls 3–5 dm, ± tomentose, often becoming glabrous. **LVS**: basal > cauline; blade 20–40 cm, oblanceolate to widely obovate, base acuminate to obtuse. **INFL**: peduncle 1–10 cm, leafy-bracted; involucre ± 2 cm diam, ± bell-shaped; phyllaries gen few, ± equal, 15–35 mm, oblong-lanceolate, acute to obtuse; chaff scales 15–16 mm. **RAY FLS** 5–11; ligules 30–45 mm. **DISK FLS**: corollas 10 mm. **FR** 9–11 mm; upper half puberulent to tomentose; pappus of short scales, < 1 mm, sometimes also 1–few lanceolate scales < 8 mm. $2n=38$. Open forest, dry, rocky slopes; 1200–3400 m. KR, CaR, n&c SN, **n SNE**; se OR, w NV. ❀ TRY: DFCLT. May–Aug

XANTHIUM COCKLEBUR

Willard W. Payne

Ann. **LVS** alternate, petioled, gen ± lobed, ± hairy. **INFL**: staminate heads in clusters, involucre 0, receptacle chaffy; pistillate heads 2-fld, 2-beaked, spiny, bur-like; phyllaries 0 or minute; chaff scales many, spirally arrayed, fused below, free tips spiny, hooked. **STAMINATE FLS** many; corolla translucent; filaments fused, attached to corolla tube base, anthers free; ovary slender. **PISTILLATE FLS** gen 2 per head; corolla 0; style branches long. **FR** 2, enclosed in bur, germinating in successive years; pappus 0. 2 spp.: worldwide. (Greek: yellow, from fr-extract dye)

X. strumarium L. (p. 193) COCKLEBUR **STS** < 15 dm, thick, ± fleshy, gen ± red- or black-spotted, unarmed. **LVS** long-petioled; blades < 15 cm, widely triangular, gen ± 3-lobed, coarsely toothed, green below and above, ± glandular, scabrous; base gen 3-veined. **INFL**: pistillate heads clustered below staminate heads. **FR**: bur cylindric to barrel-shaped; spines gen stout, ± glandular. $2n=36$. Common. Disturbed areas; gen < 500 m. **CA**; worldwide. [vars. *canadense* (Miller) Torrey & A. Gray and *glabratum* (DC.) Cronq.] Highly variable; populations show founder effects. Jul–Oct

XYLORHIZA DESERT-ASTER

Per, subshrubs, shrubs. **STS** gen white, glabrous or hairy. **LVS** simple, alternate, entire or toothed; midrib white. **INFL**: heads radiate, solitary, peduncled; involucre bell-shaped or hemispheric; phyllaries graduated in several series; receptacle convex, naked. **RAY FLS** gen many; corollas white to blue or purple. **DISK FLS** many; corolla yellow; style tips linear, acute. **FR** linear to club-shaped, weakly compressed, covered with long, appressed hairs; pappus of many, unequal bristles. 8 spp: w N Am (Greek: woody root) [Watson 1977 Brittonia 29:199–216]

1. Herbs or subshrubs; sts gen branched only in basal half; peduncles 8–22 cm . . . . *X. tortifolia* var. *tortifolia*
1' Shrubs; sts gen branched throughout; peduncles 0–11 cm
  2. Younger sts and peduncles glandular; phyllaries loosely appressed, outermost glandular, innermost < immediately preceding series . . . . *X. cognata*
  2' Younger sts and peduncles glabrous; phyllaries tightly appressed, essentially glabrous; innermost phyllaries = or > immediately preceding series . . . . *X. orcutii*

X. cognata (H.M. Hall) T.J. Watson (p. 193) MECCA-ASTER Shrub < 1.5 m. **STS** gen short-glandular, glabrous in age. **LVS** 1–5 cm, oblanceolate to ovate, glabrous to glandular, obtuse or acute, ± spiny-dentate or entire, not reduced upward. **INFL**: peduncles 0–11 cm; phyllaries 8–19 mm, 0.8–2.2 mm wide, subglabrous to densely glandular, innermost < immediately preceding series. **RAY FLS** 20–30; tube 5–8 mm; ligule 1.8–2.5 cm, light blue. **DISK FLS** 40–80; corolla 7–9 mm. **FR** 3–4.5 mm; pappus bristles < or = 9.5 mm. $2n=12$. RARE. Arid canyons; 20–240 m. **n DSon (Riverside Co.)**. [*Machaeranthera c.* (H.M. Hall) Cronquist & Keck] Jan–Jun

X. orcutii (Vasey & Rose) E. Greene (p. 193) ORCUTT'S WOODY-ASTER Shrub < 1.5 m. **STS** glabrous (rarely puberulent below heads). **LVS** 2–6 cm, oblanceolate to oblong or lanceolate, glabrous or sparsely hairy on margins, obtuse or acute, ± spiny-dentate or entire, not much reduced upward. **INFL**: heads sessile (or peduncle < or = 11 cm); phyllaries 5–12 mm, 1.5–3.5 mm wide, glabrous, innermost = or > immediately preceding series. **RAY FLS** 25–40; tube 4–6 mm; ligule 1.2–3.2 cm, light blue.

DISK FLS 55–140; corolla 8–10.5 mm. **FR** 3–4 mm; pappus bristles < or = 12 mm. $2n=12$. RARE. Arid canyons; 20–300 m. **s DSon (Imperial, San Diego cos.)**. [*Machaeranthera o.* (Vasey & Rose) Cronq. & Keck] Mar–Apr

X. tortifolia (Torrey & A. Gray) E. Greene var. *tortifolia* (p. 193, pl. 28) MOJAVE-ASTER Per, subshrubs 2–6 dm from much-branched caudex. **STS** with long, non-glandular hairs and shorter, stalked glands. **LVS** 2.5–10 cm, linear to lanceolate, oblanceolate or elliptic, gen soft-hairy and glandular, acute to spine-tipped, ± spiny-dentate, reduced upward. **INFL**: peduncles 8–22 cm; phyllaries 5–25 mm, 0.7–2.5 mm wide, soft-hairy and glandular, innermost > or < immediately preceding series. **RAY FLS** 25–60; tube 4–6 mm; ligule 1–3.3 cm, light blue or white. **DISK FLS** 70–110; corolla 5.5–8.5 mm. **FR** 3–6 mm; pappus bristles < or = 9 mm. $2n=12, 24$. Desert slopes, canyons; 240–2000 m. **s SNE, DMoj, n DSon**; to sw UT, w AZ. [*Machaeranthera t.* (Torrey & A. Gray) Cronq. & Keck] ❀DRN, SUN:2,7–10,11–13,18–24. Mar–May, Oct

BERBERIDACEAE BARBERRY FAMILY

Michael P. Williams

Per, shrub, gen from rhizomes; caudex sometimes present, glabrous, glaucous, or hairy. **STS** spreading to erect, branched or not. **LVS** simple, 1–3-ternate, or pinnately compound, basal and cauline, gen alternate, deciduous or evergreen, petioled. **INFL**: gen raceme, spike, or panicle, scapose, terminal, or axillary. **FL**: sepals 6–18 or 0, gen in whorls of 3; petals gen 6, in 2 whorls of 3, or 0; stamens 6–12, free or fused at base, 2-whorled or not, anthers dehiscent by flap-like valves or longitudinal slits; ovary superior, chamber 1, ovules gen 1–10, style 1 or 0, stigma flat or spheric. **FR**: berry, capsule, or achene. 16 genera, ± 670 spp.: temp, trop worldwide; some cult (*Berberis, Epimedium, Nandina* (Heavenly bamboo), *Vancouveria*). [Ernst 1964 J Arnold Arbor 45:1–35]

BERBERIS OREGON-GRAPE, BARBERRY

Shrub, gen from rhizomes. **STS** spreading to erect, branching, spiny or not, sometimes vine-like; inner bark, wood gen bright yellow; bud bracts deciduous or persistent. **LVS** simple or pinnately compound, cauline, alternate, deciduous or evergreen; lflets gen 3–11, ± round to lanceolate, gen spine-toothed. **INFL**: raceme, axillary or terminal. **FL**: sepals 9 in 3 whorls of 3; petals 6 in 2 whorls of 3, base gen glandular; stamens 6, anther valves pointed down to ± spreading; ovules 2–9, stigma ± spheric. **FR**: berry, spheric to elliptic, gen purple-black. ± 600 spp.: temp worldwide. (Latin: ancient Arabic name for barberry) [see Moran 1982 Phytologia 52:221–226 for relationship between *Berberis* and *Mahonia*] Roots often TOXIC; spines may inject fungal spores into skin. Key to species adapted by Margriet Wetherwax.

1. Infl dense, fls > 10; largest lflets gen > 4 cm, > 2 cm wide; petiole gen 1–6 cm; terminal lflets gen ovate to widely elliptic . . . . ***B. aquifolium*** var. ***repens***
1' Infl open, fls gen < 10; largest lflets < 4 cm, < 2 cm wide; petiole < 1 cm; terminal lflets gen lanceolate, oblong, or narrowly elliptic
   2. Lflets 3–7(9); terminal lflet gen ovate-lanceolate, length gen < 3 × width; fl 8–12; fr yellowish red to reddish purple . . . . ***B. fremontii***
   2' Lflets 3–5; terminal lflet gen narrowly lanceolate, length gen > 3 × width; fl 3–5; fr reddish brown to dark purple . . . . ***B. haematocarpa***

B. aquifolium Pursh **STS** spreading to erect, 0.1–2 m; bud scales gen deciduous. **LVS** cauline, not crowded, 8–24 cm; petiole 1–6 cm; lflets 5–9, 2–7.5 cm, 1.5–4.5 cm wide, ± round to elliptic, ± flat to strongly wavy, base slightly lobed to wedge-shaped, tip acute to obtuse (exc tooth), margin serrate, spine-tipped teeth 6–24(40), 2–5 mm. **INFL** 3–6 cm, dense; axis internodes 2–4 mm in fl, fr. **FR** 4–7 mm diam, ovoid to obovoid, glaucous, dark blue to purple. **SEEDS** 4–5 mm. Slopes, canyons, coniferous forest, oak woodland, chaparral; < 2200 m. CA-FP, **GB, DMoj**; to Can, w Great Plains, n Mex. Vars. intergrade; variation needs study. 3 vars. in CA.

var. ***repens*** (Lindley) H. Scoggan (p. 201) **STS** spreading to erect, gen < 0.8 m. **LVS** 8–18 cm; petiole 1–6 cm; lflets gen 5–7, 3–7 cm, 2.5–4.5 cm wide, round, widely elliptic, or ovate, ± flat, base oblique to slightly rounded; marginal teeth 11–15, 0.5–1 mm. **FR** obovoid to elliptic, blue to dark blue. 2*n*=28. Habitats of sp.; 300–2200 m. KR, NCoR, CaR, SN, PR, **GB, DMoj**; to B.C., Great Plains. [*B. amplectens* (Eastw.) Wheeler; *B. pumila* E. Greene; *B. repens* Lindley; *B. sonnei* (Abrams) McMinn; *Mahonia sonnei* Abrams, Truckee barberry **ENDANGERED** CA, US] In cult. Apr–Jun

B. fremontii Torrey (p. 201) **STS** erect, 0.1–4(5) m; bud bracts < 5 mm, gen deciduous. **LVS** 3–6 cm, crowded on short lateral sts; petiole < 1 cm; lflets 3–7(9); terminal lflet 1.5–2.5 cm, 1–1.5 cm wide, ovate, oblong, or lanceolate, wavy, gen folded along midrib, base truncate to wedge-shaped, tip gen acute, margin ± lobed, spine-tipped teeth 3–8, 2–3 mm. **INFL** 4–5.5 cm, open; axis internodes 2–10 mm, 5–10 mm in fr; fls 8–12. **FR** 6–15 mm diam, ± spheric, glaucous, yellowish or purplish red to dark purple. **SEEDS** 3–4 mm. Rocky slopes, pinyon/juniper woodland, chaparral; 900–1850 m. PR, **e&s DMoj**; to Colorado, NM, Mex. [*B. higginsiae* Munz; *Mahonia higginsiae* (Munz) Ahrendt, Higgins' barberry] Intergrades with *B. haematocarpa*, esp in e DMoj. ✿4–6,**7,14–16**,17,**23,24**&IRR:2,3,**8–10**,11–13,**18–22**. Apr–Jun

B. haematocarpa Wooton **STS** erect, 0.5–4 m; bud bracts < 5 mm, gen deciduous. **LVS** 3–6 cm, crowded on short lateral sts; petiole < 1 cm; lflets 3–5; terminal lflet 3–3.5 cm, 0.8–1.2 cm wide, narrowly lanceolate, wavy, gen folded along midrib, base truncate to wedge-shaped, tip gen acuminate, margin ± lobed, spine-tipped teeth 3–8, 2–3 mm. **INFL** 2–3.5 cm, open; axis internodes 2–10 mm, 5–10 mm in fr; fls 3–5. **FR** 8–10 mm diam, ± spheric, reddish brown to dark red. **SEEDS** 3–4 mm. Rocky slopes, pinyon/juniper woodland, chaparral; 1000–1700 m. **e&s DMoj**; to TX, Mex. Perhaps best treated as part of *B. fremontii*. ✿4–6,**7,14–16,23,24**& IRR:2,3,**8–10**,11–13,**18–22**. May–Jun

BETULACEAE BIRCH FAMILY

John O. Sawyer, Jr.

Tree, shrub, monoecious. **ST**: trunk < 35 m; bark ± smooth; lenticels present. **LVS** simple, alternate, petioled, deciduous; stipules deciduous; blade ovate to elliptic, gen serrate, ± doubly so. **INFL**: catkin, gen appearing before lvs, often clustered; bracts each subtending 2–3 fls and 3–6 bractlets. **STAMINATE INFL** pendent, ± elongate. **PISTILLATE INFL** pendent or erect, developing variously in fr (see key to genera). **STAMINATE FL**: sepals 0–4, minute; petals 0; stamens 1–10; pistil vestigial or 0. **PISTILLATE FL**: sepals 0–4; petals 0; stamens 0; pistil 1, ovary inferior, chambers 2, each 1-ovuled, stigmas 2. **FR**: nut or nutlet, sometimes winged, subtended or enclosed by 1–2 bracts. 6 genera, 105 spp.: gen n hemisphere; some cult.

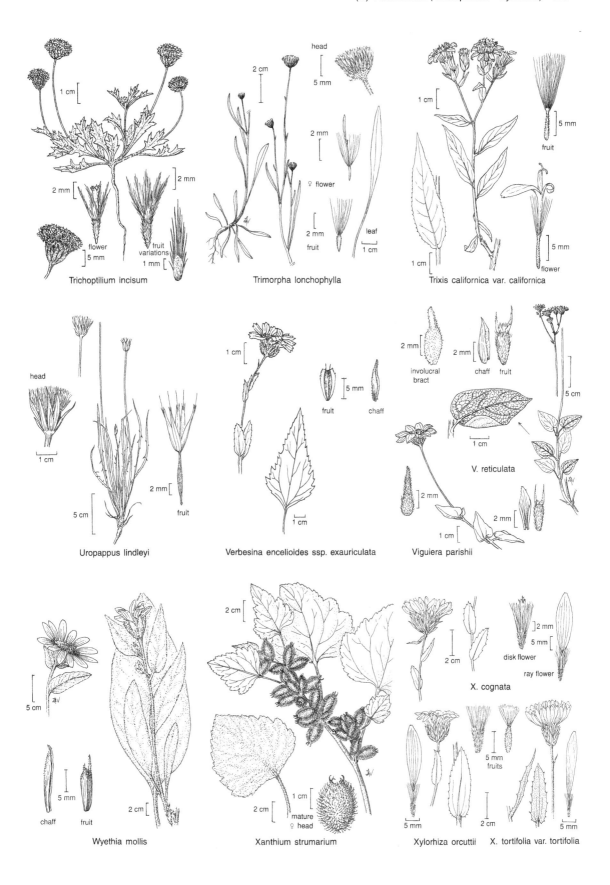

Trichoptilium incisum

Trimorpha lonchophylla

Trixis californica var. californica

Uropappus lindleyi

Verbesina encelioides ssp. exauriculata

Viguiera parishii

V. reticulata

Wyethia mollis

Xanthium strumarium

Xylorhiza orcuttii X. tortifolia var. tortifolia

X. cognata

1. Pistillate catkin cone-like, with woody bracts remaining attached after fr release . . . . **ALNUS**
1' Pistillate catkin not cone-like, with papery, lobed bracts released with, but not attached to fr. . . . . **BETULA**

ALNUS ALDER

Tree, shrub. **ST**: trunk < 35 m; bark smooth, gray to brown; twigs glabrous to finely hairy, reddish gray; lenticels small; winter buds stalked, 2-scaled. **LF** glabrous to finely hairy; blade 3–15 cm, elliptic to ovate, base ± truncate to tapered, sometimes subcordate. **STAMINATE INFL** 5–20 cm; bracts each subtending 3 fls and 4 bractlets. **PISTILLATE INFL** 5–20 mm; bracts each subtending 2 fls and 4 fused bractlets. **STAMINATE FL**: sepals 4; stamens 1–4. **PISTILLATE FL**: sepals 0. **FRS** many, in cone-like catkin, bracts 3 mm, woody, winged. 30 spp.: n hemisphere, S.Am. (Latin: alder) [Furlow 1979 Rhodora 81:1–121, 151–248] Root nodules contain nitrogen-fixing bacteria; wood used for interior finishing, to smoke fish, meats.

A. rhombifolia Nutt. (p. 201) WHITE ALDER Tree. **ST**: trunk < 35 m. **LF**: blade thick, base tapered to round, tip round to acute, margin gen ± flat, upper surface green, midrib and major veins not indented, lower surface yellow-green. Along permanent streams; 100–2400 m. CA-FP, MP (uncommon), **w DSon**; to WA, ID. ❀ SUN:**1–3**, 4–6,**7**,**8**,**9**,10,**14–18**,19,**22–24**;STBL. Jan–Apr

BETULA BIRCH

Tree, shrub. **ST**: trunk < 30 m; bark smooth or scaly, aromatic, often peeling in thin layers; twigs puberulent, glandular, or both; lenticels prominent; winter buds sessile, 3-scaled. **LF** glandular-hairy; blade 2–5 cm, widely elliptic, base ± truncate to tapered. **STAMINATE INFL** 2–7 cm; bracts each subtending 3 fls and 3 bractlets. **PISTILLATE INFL** 2–3 cm; bracts each subtending 3 fls and 3 bractlets. **STAMINATE FL**: sepals 4; stamens 2. **PISTILLATE FL**: sepals 0. **FRS** many, in a non-cone-like catkin, winged; bracts lobed, papery, released with but not attached to fr. 50 spp.: circumboreal. (Latin: birch) Important wildlife food; wood used for interior finishing; many spp. cult.

B. occidentalis Hook. (p. 201) WATER BIRCH Tree, shrub. **ST**: trunks < 10 m; bark black, red-brown, not peeling; twigs with large resin glands, hairy. **LF**: petiole < 15 mm, hairy; blade 2–5 cm, widely ovate, thin, glands esp on upper surface, base ± truncate to tapered, tip acute, margin doubly serrate exc at base. **PISTILLATE INFL** 3–5 cm; bract fringed with hairs. Streamsides, springs; 600–2500 m. KR, CaRH, SNH, **GB**, **DMtns**; scattered in w N.Am. ❀WET:**1–3**,4–6,**7**,**9**,**10**,14–17,&SHD:**18**,**19**,20–24;STBL. Apr–May

BIGNONIACEAE BIGNONIA FAMILY

Lawrence R. Heckard

Per to tree (many woody vines). **LVS** gen 1–3-pinnately or -palmately compound, rarely simple, gen opposite or whorled. **FL** bisexual, showy; calyx gen 5-lobed, sometimes 2-lipped; corolla funnel- or bell-shaped, 5-lobed, gen 2-lipped; stamens epipetalous, gen 4, paired, a 5th vestigial or 0; ovary superior, chambers gen 2, placentas 4, axile (or chamber 1, placentas 2–4, parietal), style long, stigma 2-lobed. **FR**: gen capsule, long, cylindric, 2-valved. **SEEDS** many, flat, gen winged. 110 genera, 800 spp.: gen trop, esp S.Am.; many orn (*Campsis*, trumpet creeper; *Catalpa*; *Jacaranda*).

CHILOPSIS DESERT-WILLOW

1 sp. (Greek: resembling lips, from fl shape) [Henrickson l985 Aliso 11:179–197]

C. linearis (Cav.) Sweet ssp. *arcuata* (Fosb.) Henrickson (p. 201, pl. 29) Shrub or tree 1.5–7 m, willow-like. **LVS** deciduous, gen alternate (often some opposite to whorled on same pl); blade 10–26 cm, ± linear, curved. **INFL**: panicle or raceme, terminal. **FL**: calyx 8–14 mm, inflated, 2-lipped, gen soft-hairy, purplish; corolla 2–5 cm, sweetly fragrant, gen light pink to lavender with yellow ridges and purple lines on throat and lower lobes, lobes spreading, margins jagged, wavy; stamens incl; stigma lobes closing when touched. **FR** < 35 cm, linear, round in ×–section. **SEED** 6–12 mm, oblong; both ends long-hairy. 2*n*=40. Common. Sandy washes; < 1500 m. **D**, adjacent TR, PR; to UT, NM, n Mex. ❀SUN,DRN:7,**14**,15–17,24 &IRR:**8**,**9**,10,11,**12**,**13**,**18**–23. May–Sep

BORAGINACEAE BORAGE FAMILY

Ann, per, shrubs, gen bristly or sharply hairy. **ST** prostrate to erect. **LVS** cauline, often with basal rosette, gen simple, alternate; lower sometimes opposite, entire. **INFL**: cyme, gen elongate, panicle-, raceme-, or spike-like, coiled in fl, gen uncoiled in fr or fls 1–2 per axil. **FLS** gen bisexual, gen radial; sepals 5, free or fused in lower half; corolla 5-lobed, gen salverform, top of tube gen appendaged, appendages 5, alternating with stamens, sometimes arching over tube; stamens 5, epipetalous; ovary superior, gen 4-lobed, style gen entire. **FR**: nutlets 1–4, smooth to variously roughened, sometimes prickly or bristled. ± 100 genera, ± 2000 spp.: trop, temp, esp w N.Am, Medit; some cult (*Borago, Echium, Myosotis, Symphytum*). Almost all genera may be TOXIC from alkaloids or accumulated nitrates. Recently treated to include Hydrophyllaceae [Olmstead et al 2000 Mol Phylog Evol 16: 96-112] Family description by Timothy Messick, key to genera by Ronald B. Kelley.

1. Style deeply divided; stigmas 2 . . . . **TIQUILIA**
1' Style simple or 0; stigma 1
  2. Style attached atop ovary, deciduous in fr . . . . **HELIOTROPIUM**
  2' Style attached to receptacle, gen persistent
    3. Mature nutlets spreading widely
      4. Sepals in fr very unequal, upper 2 >> others, partly fused, arched over 1 nutlet, ± bur-like,
        with 5–10 stout spines each with hooked bristles, lower 3 sepals distinct; nutlets 2 . . . . **HARPAGONELLA**
      4' Sepals in fr ± equal or, if ± unequal, upper 2 > others, without spines but with hooked or
        straight bristles, distinct, not arched over 1 nutlet; nutlets gen 4 . . . . **PECTOCARYA**
    3' Mature nutlets ± erect
      5. Receptacle ± flat; nutlet scar gen basal
        6. Corolla yellow . . . . **LITHOSPERMUM**
        6' Corolla blue . . . . **MERTENSIA**
      5' Receptacle ± conic or elongate; nutlet scar gen lateral
        7. Nutlet margins prickled, prickles barbed
          8. Gen per; pedicel recurved in fr, receptacle wide, pyramidal, ± 1/2 nutlet length . . . . **HACKELIA**
          8' Ann; pedicel erect in fr, receptacle narrow, tapered, ± = nutlet . . . . *Lappula redowskii*
        7' Nutlet margins prickled or not, prickles if present not barbed
          9. Corolla limb and tube bright yellow or orange
            10. Ann . . . . **AMSINCKIA**
            10' Per . . . . *Cryptantha confertiflora*
        9' Corolla limb and tube white to creamy yellow
          11. Nutlet wall grooved above scar, groove flared open to 2-forked at base; scar gen recessed or
            depressed . . . . **CRYPTANTHA**
          11' Nutlet wall keeled above scar; scar gen elevated . . . . **PLAGIOBOTHRYS**

AMSINCKIA FIDDLENECK

Fred R. Ganders

Ann; hairs gen bristly, often with bulbous bases. **ST** gen erect, 2–12 dm, gen green. **LVS** basal and cauline, alternate, sessile or lower short-petioled, gen linear to narrowly lanceolate or oblong, gen ± entire. **INFL** spike-like, gen ± terminal; tip coiled. **FL** gen radial; calyx lobes 5, sometimes appearing to be 2–4 from fusion; corolla orange or yellow, limb gen with 5 red-orange marks. **FR**: nutlets erect, ± triangular, gen with oval lateral scar, gen with round or sharp tubercles. 10 spp.: w N.Am, sw S.Am, widely alien elsewhere. (W. Amsinck, patron of Hamburg Botanic Garden, early 19th century) [Ray & Chisaki 1957 Amer J Bot 44:529–554] Self-compatible; often heterostylous; large-fld taxa gen cross-pollinated, small-fld self-pollinated. Seeds and herbage TOXIC to livestock (esp cattle) from alkaloids and high nitrate concentrations. Sharp plant hairs irritate human skin. Key to species adapted by Ronald B. Kelley.

1. Calyx lobes 5; corolla tube 10-veined near base; nutlet surface gen sharp-tubercled, often ridged . . *A. menziesii*
  2. Corolla 7–11 mm, 4–10 mm wide at top, ± orange . . . . var. *intermedia*
  2' Corolla 4–7 mm, 2–3 mm wide at top, pale yellow . . . . var. *menziesii*
1' Calyx lobes 2–4; corolla tube 20-veined near base; nutlet surface smooth, cobblestone-like, or round-tubercled
  3. St (exc infl) gen ± glabrous, glaucous, ± pink below; lf glaucous; nutlet surface smooth, with a
    longitudinal groove, scar not obvious . . . . *A. vernicosa* var. *vernicosa*
  3' St hairy, not glaucous, green; lf not glaucous; nutlet surface cobblestone-like or round-tubercled,
    ungrooved, with lateral scar . . . . *A. tessellata*
    4. Corolla 12–16 mm, 6–10 mm wide at top; anthers not appressed to, often below stigma . . . . var. *gloriosa*
    4' Corolla 8–12 mm, 2–6 mm wide at top; anthers appressed to stigma . . . . var. *tessellata*

A. menziesii (Lehm.) Nelson & J.F. Macbr. RANCHER'S FIREWEED **FL**: calyx lobes 5; corolla 4–11 mm, 2–10 mm wide at top, tube 10-veined near base, limb with red-orange marks or not; anthers appressed to stigma in corolla throat. **FR** 2–3.5 mm; surface tubercled, sometimes ridged. Abundant. Open, gen disturbed places; < 1700 m. CA; to B.C., AZ, ID, UT; also in S.Am; alien in e US, e hemisphere. 100+ named, mostly indistinct variants; self-pollinated; different variants may grow together and remain distinct but intergrade over their ranges.

  var. ***intermedia*** (Fischer & C. Meyer) Ganders **FL**: corolla 7–11 mm, 4–10 mm wide at top, ± orange, limb gen with 5 red-orange marks. 2*n*=30,34,38. Habitat and elevation of sp. **CA**; to B.C., AZ, ID, Baja CA. [*A. i.* Fischer & C. Meyer, incl var. *echinata* (A. Gray) Wiggins] ❀STBL;SUN:2,3,**7–10,14**,15–

17,**18–23**,24. Mar–Jun

  var. ***menziesii*** (p. 201) **FL**: corolla 4–7 mm, 2–3 mm wide at top, pale yellow, limb gen without red-orange marks. 2*n*=16,26,34. Habitat and elevation of sp. **CA**; to B.C., AZ, ID; naturalized in c&e US; also in S.Am. [*A. helleri* Brand; *A. micrantha* Suksd.; *A. retrorsa* Suksd.] Apr–Jun

A. tessellata A. Gray DEVIL'S LETTUCE **FL**: calyx lobes 2–4; corolla 8–16 mm, 2–10 mm wide at top, yellow or orange, tube 20-veined near base. **FR** 2.5–4 mm; surface cobblestone-like or round-tubercled, ridged or not. Common. Often disturbed places; 50–2200 m. NCoRI (Colusa Co.), GV, SnFrB, SCoR, WTR, **GB**, **D**; to WA, ID, AZ, S.Am.

var. *gloriosa* (Suksd.) Hoover (p. 201) **FL**: corolla 12–16 mm, 6–10 mm wide at top, orange; anthers not appressed to, gen below stigma. 2*n*=24. Sandy or shaly soils; 50–1700 m. NCoRI, ScV (Colusa Co.), SnFrB, SCoR, WTR, **w DMoj**. [*A. g.* Suksd.] ❀ SUN: TRY.

var. *tessellata* (p. 201) **FL**: corolla 8–12 mm, 2–6 mm wide at top, yellow; anthers appressed to stigma. 2*n*=24. Rocky or sandy soils; 50–2200 m. SnJV, SnFrB, SCoR, **GB, D**; to WA, ID, AZ, S.Am. [*A. t.* var. *elegans* (Suksd.) Hoover]

A. vernicosa Hook. & Arn.   **ST** (exc infl) gen ± glabrous, glaucous, ± pink below. **LF** glaucous. **FL**: calyx lobes 2–4; corolla 8–22 mm, 2–14 mm wide at top, yellow or orange, tube 20-veined near base. **FR** 3.5–6 mm; surface smooth, shiny, with a longitudinal groove, scar not obvious. Uncommon. Loose, shaly slopes; 50–1400 m. s SNF, w SnJV, SnFrB, SCoR, **DMoj**. 2 vars. in CA.

var. *vernicosa* (p. 201) **FL**: corolla 8–12 mm, 2–8 mm wide at top, yellow. **FR**: groove unforked. 2*n*=14. Habitat of sp.; 50–1400 m. s SNF, w SnJV, SnFrB, SCoR, **DMoj**. Mar–May

CRYPTANTHA

Walter A. Kelley and Dieter H. Wilken

Ann, bien, per. **ST** simple or branched; branches gen ascending to erect, hairy. **LVS** strigose, rough-hairy, or bristly, largest bristles (esp lower surface) bulbous-based; basal whorled; cauline gen opposite below, alternate above. **INFL** gen terminal, gen elongated in fr, open (fls in fr not overlapping or touching side to side) or dense (fls in fr overlapping or touching side to side). **FL**: sepals ± free; corolla gen white, tube gen 1–13 mm, appendages 5, white to yellow, limb 1–5 mm wide in ann, 6–12 mm wide in per; anthers incl; ovary gen 4-lobed. **FR**: nutlets 1–4, back gen grayish brown, smooth and shiny or granular, tubercled, or rough at 10×, margin rounded, sometimes sharp-edged, groove on inside surface narrow, open to closed, sometimes raised, edges inrolled to sharp-angled, gen forked or flared open at base. ± 160 spp.: w Am. (Greek: hidden fls, from cleistogamous fls of some spp.) [Higgins 1971 Brigham Young Sci Bull Biol Ser 13(4):1–63] Ann spp. gen self-pollinating; per spp. homostylous or heterostylous. Many ann spp. difficult to separate; observation of nutlets and hairs requires magnification at 20×.

Keys to Groups

1. Bien or per, gen with persistent basal rosette or tuft of lvs; caudex present in per, gen woody
     (exc C. *virginensis*) . . . . **Group 1**
1' Ann, basal lvs gen not persistent, gen not green in fr; caudex 0
  2. Most fls with 1 nutlet (sometimes 2) . . . . **Group 2**
  2' Most fls with 4 nutlets (sometimes 3) . . . . **Group 3**

Group 1: Bien or Per

1. Sepals 1–2 mm, gen 2–3 mm in fr; infl gen open in fr, fls pedicelled to subsessile; basal lvs not persistent, not green in fr; woody caudex 0
  2. Infl dense in fr; fls subsessile, ascending to erect in fr; nutlets gen equal . . . . ⁴*C. holoptera*
  2' Infl open in fr; fls pedicelled, spreading in fr; nutlets gen of 2 kinds, 1 > other 2–3 . . . . ³*C. racemosa*
1' Sepals 2–10 mm, 3.5–15 mm in fr; infl ± dense in fr, fls subsessile; basal lvs persistent, gen green in fr; caudex present, gen woody (not so in C. *virginensis*)
  3. Corolla tube > sepals
    4. Corolla and appendages yellow; nutlet back smooth, groove closed, groove edges ± overlapping
         . . . . *C. confertiflora*
    4' Corolla white, appendages light yellow; nutlet back rough, groove open, narrowed near middle,
         groove edges elevated . . . . *C. flavoculata*
  3' Corolla tube = or < sepals
    5. Sts prostrate to ascending; nutlet back clearly curved, surface smooth; groove closed, edges gen
         overlapping . . . . *C. cinerea* var. *abortiva*
    5' Sts ± erect (gen cespitose in C. *humilis*); nutlet back straight, surface smooth or rough; groove
         closed or open, edges not overlapping
      6. Sts 1–2 cm, < or = basal lvs; basal lvs gen < 1 cm . . . . *C. roosiorum*
      6' Sts > 2 cm, > basal lvs; basal lvs gen >> 1 cm
        7. Nutlet lanceolate to narrowly ovate, inner surface (between groove and margin) gen smooth, shiny
          8. Corolla tube 2–3 mm; nutlet 2–3 mm; W&I, n DMtns . . . . *C. nubigena*
          8' Corolla tube 3–4 mm; nutlet 3–4 mm; n SNE (Sweetwater Mtns) . . . . *C. sobolifera*
        7' Nutlet ovate, inner surface gen rough or wrinkled
          9. Bien or short-lived per; caudex not woody . . . . ³*C. virginensis*
          9' Long-lived per; caudex woody
            10. Edge of nutlet groove not elevated; GB . . . . *C. humilis*
            10' Edge of nutlet groove elevated; DMtns . . . . *C. tumulosa*

Group 2: Ann; nutlet gen 1(2)

1. Basal lvs tufted, some green and persistent in fr; st hairs soft-downy under densely spreading bristles
     . . . . ³*C. virginensis*

1' Basal lvs not tufted, gen withered in fr; st hairs strigose, rough-hairy, or bristly, not soft-downy
  2. Nutlets 2 and not identical (1 nutlet smooth on back, other nutlet fine-granular on back) . . . . ²*C. maritima*
  2' Nutlet 1 or 2 and identical (both with back smooth or roughened)
    3. Nutlet back granular, tubercled, or papillate, shiny or dull, not smooth
      4. Sepals 2–6 mm, 3.5–10 mm in fr
        5. Sepals 4–6 mm, 5–10 mm in fr; corolla limb gen < 3 mm wide; appendages white to light yellow
          . . . . ²*C. barbigera*
        5' Sepals 2–3 mm, 3.5–5 mm in fr; corolla limb 3–6 mm wide; appendages gen bright yellow . . ²*C. intermedia*
      4' Sepals 1.5–2.5 mm, 2–4 mm in fr
        6. Fl spreading to recurved in fr; sepal tips recurved in fr; nutlet curved . . . . *C. recurvata*
        6' Fl ascending to ± erect in fr; sepal tips ± straight, erect in fr; nutlet ± straight
          7. Sepals ovate to elliptic in fr; nutlet margin sharp-angled to narrowly winged near tip; throat
            appendages gen bright yellow; style gen = nutlet . . . . *C. utahensis*
          7' Sepals lanceolate to linear-oblong in fr; nutlet margin rounded; throat appendages white to light
            yellow; style < nutlet . . . . *C. decipiens*
    3' Nutlet back smooth throughout, shiny
      8. Sepals gen soft-hairy, not bristly, hairs appressed to stiffly ascending — GB, D . . . . *C. gracilis*
      8' Sepals strigose and bristly, bristles reflexed, spreading, or ascending
        9. Sepals 6–10 mm in fr; longest sepal bristles 3–4 mm in fr — w DSon (Borrego Valley) . . . . *C. ganderi*
        9' Sepals 1–6 mm in fr; longest sepal bristles < 3 mm in fr
          10. Nutlet groove clearly off-center (nearer one margin than center), closed (edges overlapping),
            abruptly flared open at base; infls axillary and terminal, ± spheric in fl . . . . *C. glomeriflora*
          10' Nutlet groove ± central, open to closed, gen forked at base; infl terminal, gen elongate in fl . . ²*C. maritima*

Group 3: Ann; nutlets gen 4(3)

1. Pls gen wider than tall, rounded to cushion-like branches many, dense, spreading to ascending
  2. Calyx circumscissile below middle in fr, limb and upper tube falling as 1 unit, lower part of tube scarious,
    persistent, cup-like in fr . . . . *C. circumscissa*
  2' Calyx intact in fr, not circumscissile . . . . ²*C. micrantha*
1' Pls gen taller than wide, not rounded or cushion-like
  3. Basal lvs tufted, some green, persistent in fr; st hairs soft-downy under densely spreading bristles
    . . . . ³*C. virginensis*
  3' Basal lvs not tufted, gen withered in fr; st hairs strigose, rough-hairy, or bristly, not soft-downy
    4. Nutlet margin sharp-angled in part or throughout (acute in ×-section near tip) or winged
      5. Nutlet back smooth or minutely rippled to weakly tubercled, shiny
        6. Nutlet back smooth
          7. Corolla limb 3–5 mm wide . . . . *C. mohavensis*
          7' Corolla limb 1–2 mm wide . . . . *C. watsonii*
        6' Nutlet back minutely rippled to weakly tubercled
          8. Nutlets gen of 2 kinds, 1 gen > other 2–3 . . . . ³*C. racemosa*
          8' Nutlets gen equal
            9. Nutlet inner surface smooth, flat; groove edges not elevated . . . . *C. costata*
            9' Nutlet inner surface weakly tubercled, not flat; groove edges gen elevated . . . . ⁴*C. holoptera*
      5' Nutlet back clearly tubercled to granular, dull or shiny
        10. Nutlet margin sharp-angled, not clearly winged (exc sometimes at tip in *C. racemosa*)
          11. Nutlets gen equal; infl dense in fr; fls subsessile in fr . . . . ⁴*C. holoptera*
          11' Nutlets gen of 2 kinds, 1 gen > other 2–3; infl open in fr; fls pedicelled in fr . . . . ³*C. racemosa*
        10' Nutlet margin clearly winged (1 nutlet gen not winged in *C. pterocarya*)
          12. Nutlets gen of 2 kinds, 3–4 winged, 1 sometimes not winged; nutlet wing gen lobed at tip,
            0.5–1 mm wide; sepals ovate to lanceolate and 4–6 mm in fr, densely strigose, bristles few or 0
            . . . . *C. pterocarya*
          12' Nutlets all winged; nutlet wing not lobed, < 0.5 mm wide; sepals oblong to linear and 2.5–4 mm
            in fr, bristly to rough-hairy, hairs spreading to ascending
            13. Corolla limb 1–2 mm wide; nutlet margin evenly and narrowly winged, back sparsely
              tubercled, groove opened widely below middle; e DMoj, DSon . . . . ⁴*C. holoptera*
            13' Corolla limb 4–6 mm wide; nutlet margin narrowly winged near base, widely winged above
              middle, back fine-tubercled to white-granular; groove narrowly open at base; w DMoj . . . *C. oxygona*
    4' Nutlet margin rounded throughout (obtuse in ×-section near tip); not sharp angled or winged
      14. Nutlets clearly of 2 kinds, 1 gen > other 2–3
        15. Distal st part ascending to erect, hairs appressed and spreading; infl gen dense, fls spreading to
          ascending in fr . . . . *C. angustifolia*
        15. Distal st part spreading, prostrate, or sprawling over rocks, other pls, hairs appressed to ascending;
          infl open, fls ± appressed in fr . . . . *C. dumetorum*
      14' Nutlets gen equal
        16. Infls axillary and terminal, fls ± evenly distributed along st

17. Style > nutlet; sepals 1–3 mm in fr; infl clearly bracted, bracts gen = sepals . . . . ²*C. micrantha*
17' Style = or < nutlet; sepals 5–6 mm in fr; infl bracts 0 . . . . ²*C. echinella*
16' Infl mostly terminal; lower fls sometimes axillary, subtended by reduced lvs
  18. Nutlet back smooth, shiny . . . . *C. torreyana*
  18' Nutlet back papillate, granular, or tubercled, shiny or dull
    19. Corolla limb 3–6 mm wide, appendages bright yellow
      20. Style = or < nutlet . . . . ²*C. intermedia*
      20' Style > nutlet . . . . ²*C. muricata*
    19' Corolla limb gen < 3 mm wide, appendages white to light yellow
      21. Most st hairs ascending or appressed, spreading hairs 0 or few
        22. Nutlets lanceolate, tip tapered; sepals gen 6–10 mm in fr . . . . *C. nevadensis*
        22' Nutlets ovate to deltate; tip acuminate; sepals 3–7 mm in fr
          23. Infl dense in fr, fls gen clustered near axis tip . . . . ²*C. simulans*
          23' Infl open in fr, fls well spaced on axis
            24. Nutlet back densely high-tubercled, tubercle tips translucent; style ± > nutlet; c DMoj
              (near Barstow) . . . . ²*C. clokeyi*
            24' Nutlet back sparsely low-tubercled and densely fine-granular; style = or < nutlet; mtns >
              650 m . . . . ²*C. simulans*
      21' Most st hairs spreading, sometimes also strigose
        25. Nutlet back densely-fine-papillate . . . . ²*C. echinella*
        25' Nutlet back tubercled (sometimes weakly so in *C. ambigua*)
          26. Sepals 1–2 mm, 2–4 mm in fr . . . . ²*C. muricata*
          26' Sepals 2–6 mm, 4–10 mm in fr
            27. Style > nutlet; c DMoj (near Barstow) — nutlet tubercle tips translucent . . . . ²*C. clokeyi*
            27' Style = or < nutlet; GB, D
              28. Sepals 2–3 mm, 4–7 mm in fr; GB . . . . *C. ambigua*
              28' Sepals 4–6 mm, 5–10 mm in fr; SNE, D . . . . ²*C. barbigera*

C. ambigua (A. Gray) E. Greene Ann 10–35 cm. **ST** branched throughout, strigose and spreading-hairy. **LF** 1–4 cm, linear to oblong, strigose to bristly; bristles ascending to spreading. **INFL** dense to open below in fr. **FL**: sepals 2–3 mm, 4–7 mm and linear in fr, densely strigose and spreading-bristly; corolla 3–4 mm wide. **FR**: nutlets 4, 1.5–2.5 mm, ± ovate, back tubercled, groove ± closed, flared open or forked at base. Open, sandy or gravelly areas, sagebrush scrub, coniferous forest; 1300–2400 m. CaR, n&c SN, **GB**; to B.C., MT, Colorado. Jun–Jul

C. angustifolia (Torrey) E. Greene Ann 5–60 cm. **ST** simple to branched throughout, strigose and rough-hairy to bristly; hairs spreading. **LF** 1–4 cm, linear to oblong; bristles spreading. **INFL** dense to open below in fr. **FL**: sepals 1–2 mm, 2.5–4 mm and ± linear in fr, densely spreading-bristly; corolla 1–3(4) mm wide. **FR**: nutlets 4 (1 slightly > other 3), ± 1 mm, lanceolate to narrowly ovate, back brown, white-granular, dull, groove ± open throughout to flared open at base. Sandy to rocky soils, creosote scrub, desert woodland; < 1400 m. **SNE, D**; to TX, n Mex. [*C. inaequata* I.M. Johnston] Mar–May

C. barbigera (A. Gray) E. Greene Ann 10–50 cm. **ST** simple to branched gen below middle, strigose and densely spreading-bristly. **LF** 1–5(8) cm, linear-oblong to narrowly lanceolate; bristles spreading. **INFL** gen dense to open in fr. **FL**: sepals 4–6 mm, 5–10 mm and narrowly oblanceolate to oblong in fr, gen spreading-bristly esp at base; corolla tube 3–5 mm, limb 1–2 mm wide. **FR**: nutlets 1–4, 1.5–2 mm, lanceolate to ovate, back white-tubercled, groove ± open, esp at forked base, to slightly flared open at base. Open, sandy to rocky soils; 300–2250 m. s SN, Teh, **SNE, D**; to UT, NM, Baja CA. Feb–May

C. cinerea (E. Greene) Cronq. var. ***abortiva*** (E. Greene) Cronq. (p. 201) BOWNUT CRYPTANTHA Per 3–20 cm; caudex woody. **ST** branched from base, prostrate to ascending, ± strigose. **LF** 1.5–9 cm, linear to linear-oblanceolate; upper surface ± strigose; lower surface also bulbous-based bristly. **INFL** gen dense, cylindric; branches 3–many, elongate and straight in fr. **FL**: sepals 2–4 mm, 4–7 mm in fr, hairs gen soft, bristles few; corolla tube 2–4 mm, limb 5–9 mm wide, throat gen yellow. **FR**: nutlets 1–4, 1.5–2.5 mm, ± ovate, slightly curved inward, back smooth, groove closed, edges overlapping. Common. Sandy soils; 1800–3300 m. **SNE, DMtns**; s NV. [*C. jamesii* (Torrey) Payson var. *a.*

(E. Greene) Payson] Other vars. in Great Basin, Rocky Mtns. ❀ TRY,DFCLT. May–Aug

C. circumscissa (Hook. & Arn.) I.M. Johnston (p. 201, pl. 30) Ann < 10 cm, cushion-like; taproot gen red, purple when dry. **ST** much-branched throughout, strigose and bristly or rough-hairy; hairs gen ascending. **LF** 0.3–1.5 cm, linear to narrowly oblanceolate, bristly to rough-hairy; hairs ± ascending. **INFL** axillary or in branch forks; fls 1–5 per cluster, dense in fr. **FL**: sepals 1.5–2 mm, 2.5–4 mm and circumscissile below middle in fr, hairs ± like lvs; corolla limb 1–6 mm wide, appendages gen yellow. **FR**: nutlets 3–4, ± 1.5–2 mm, ovate, back gen smooth, shiny, gen mottled gray and brown, groove ± open below middle, forked at base. 2*n*=24,36. Sandy soils; 300–3700 m. SN, s SnJV, SCoRI (se Monterey Co.), e SCo, TR, e PR, **GB, D**; to WA, Colorado; also in s S.Am. [vars. *hispida* (J.F. Macbr.) I.M. Johnston, *rosulata* J. Howell] Pls from n SnGb, w SnBr, sw DMoj with corolla limb 3.5–6 mm wide, throat yellow, 2*n*=12, have been called *C. similis* K. Mathew & Raven. Corolla limb width intergrades in some places. Jul–Aug

C. clokeyi I.M. Johnston(p. 201) CLOKEY'S CRYPTANTHA Ann 8–15 cm. **ST** gen branched throughout, mostly strigose. **LF** 0.5–3 cm, linear-lanceolate to oblong, stiffly strigose to spreading-rough-hairy; hair bases gen bulbous. **INFL** gen dense above middle; lower fls not touching in fr. **FL**: sepals 2–3 mm, 5–7(10) mm and ± linear in fr, densely strigose and spreading-bristly; corolla limb 1–2 mm wide. **FR**: nutlets 4, 2–2.5 mm, ± ovate, back, inside surface tubercled, tubercle tips translucent, groove open, flared at base, edge slightly elevated. RARE. Sandy or gravelly soils; 800–900 m. **c DMoj (near Barstow)**. [*C. muricata* (Hook. & Arn.) Nelson & J.F. MacBr. var. *c.* (I.M. Johnston) Jepson] Apr–Jun

C. confertiflora (E. Greene) Payson YELLOW CRYPTANTHA Per 13–44 cm; caudex woody. **ST** simple, erect, tomentose below, strigose and sparsely bristly above. **LF** 3–12 cm, ± oblanceolate; upper surface strigose; lower surface also with bulbous-based bristles. **INFL** gen dense, ± head-like at tip; axillary clusters below. **FL**: sepals 6–10 mm, 9–14 in fr, strigose and spreading-bristly; corolla yellow, tube 9–13 mm, limb 8–12 mm wide. **FR**: nutlets gen 4, 3–4 mm, ovate to deltate, back smooth, groove closed, edges ± overlapping. Common. Dry, rocky soils; 1200–

2700 m. **SNE**, **DMtns**; to sw UT, ne AZ. Heterostylous. ❀TRY; DFCLT. May–Jul

C. costata Brandegee RIBBED CRYPTANTHA Ann 10–20 cm. **ST** branched throughout, densely strigose and spreading-bristly. **LF** 1–3(4) cm, ± linear, bristly; longest bristles spreading to ascending. **INFL** dense in fr. **FL:** sepals 3–4 mm, 4–6 mm and ± oblong in fr, hairs like lvs; corolla limb 1–2 mm wide. **FR:** nutlets gen 4, 1.5–2 mm, lanceolate, back smooth to minutely rippled, inner surface smooth, flat, margin sharp-angled, groove gen open, flared at base, edge flat. UNCOMMON. Sandy soils, creosote scrub; < 500 m. **e DMoj**, **DSon**; AZ, Baja CA. Feb–May

C. decipiens (M.E. Jones) A.A. Heller (p. 201) Ann 10–40 cm. **ST** simple to branched throughout, strigose to rough-hairy or bristly; hairs appressed to spreading. **LF** 0.5–5 cm, linear to narrowly lanceolate; bristles appressed to spreading. **INFL** dense to open in fr; lower fls not touching, appressed to ascending in fr. **FL:** sepals 1.5–2.5 mm, 2–4 mm and linear-oblong in fr, densely bristly to rough-hairy, hairs ascending to spreading; corolla limb 1–5 mm wide. **FR:** nutlets 1–2, 1.5–2 mm, lanceolate, gen white-granular throughout, sometimes smooth, shiny, white-granular near tip, groove closed to ± open, sometimes flared open at base, raised, edges sharp-angled. Open, sandy areas, grassland, shrubland; < 1500 m. Teh, SnFrB, s SnJV, SCoR, WTR, **W&I**, **DMoj, n DSon**; to sw UT, AZ. Pls from w of D, with corolla limb 2–3 mm wide and st gen hairs gen appressed, have been called *C. corollata* (I.M. Johnston) I.M. Johnston. Mar–May

C. dumetorum (A. Gray) E. Greene (p. 201) Ann 10–60 cm. **ST** branched throughout; branches prostrate to ascending, gen sprawling over rocks, other pls, ± fragile, strigose to appressed-stiff-hairy. **LF** 0.5–3 cm, linear to lanceolate; bristles ascending to ± appressed. **INFL** axillary and terminal, open in fr; fls ± appressed. **FL:** calyx assymetrical, sepals ± 1 mm, 2–3 mm, folded, oblong in fr, gen bristly, some bristles > sepal; corolla limb 0.5–1 mm wide. **FR:** nutlets 4 (1 gen > other 3), 2–2.5 mm, lanceolate to narrowly ovate, back gen white-granular, groove ± open throughout, not gen flared or forked at base. Sandy or gravelly soils, gen under other pls; 200–1500 m. SNH (e slope), TR (n slope), **SNE**, **DMoj, n DSon**; to sw UT. Apr–May

C. echinella E. Greene Ann 5–35 cm. **ST** simple or branched throughout, strigose and spreading-bristly. **LF** 0.5–5 cm, oblong to oblanceolate, strigose and bristly; bristles ascending. **INFL** dense to open. **FL:** sepals 2–3 mm, 5–6 mm and ± linear in fr, strigose and spreading-bristly; corolla limb 1–1.5 mm wide. **FR:** nutlets gen 4, 1.5–2 mm, ovate, back densely fine-papillate, groove closed to ± open, gen forked at base. Open, gravelly or rocky soils, woodland, coniferous forest; 700–2800 m. SN, TR, **GB, n DMtns**; to OR, ID. Jun–Aug

C. flavoculata (Nelson) Payson (p. 201) Per 5–35 cm; caudex woody. **ST** simple, erect; bristles ± soft, spreading. **LF** 3–11 cm, linear-oblanceolate to spoon-shaped; upper surface densely strigose and with bulbous-based bristles or only silky-strigose; lower surface strigose and with bulbous-based bristles or only silky-strigose. **INFL** dense, cylindric or ± head-like, elongate in fr. **FL:** sepals 5–7 mm, 8–10 mm in fr, spreading-bristly; corolla tube 7–12 mm, limb 7–12 mm wide, appendages yellow. **FR:** nutlets 4, 2.5–4 mm, lanceolate to ± ovate, back rough, groove open, narrowed near middle, edges elevated. Common. Loose soils; 1300–3200 m. **SNE**, **DMtns**; to ID, Colorado. Heterostylous. ❀TRY;DFCLT. May–Jul

C. ganderi I.M. Johnston (p. 201) GANDER'S CRYPTANTHA Ann 10–40 cm. **ST** simple to branched throughout, strigose and rough-hairy to bristly; hairs spreading. **LF** 1–3(4) cm, linear to narrowly lanceolate; bristles spreading, some bulbous-based. **INFL** open in fr. **FL:** sepals 3–4 mm, 6–10 mm and linear-oblong in fr, densely strigose to rough-hairy, midvein densely spreading-long-bristly; corolla limb 1–2 mm wide. **FR:** nutlets 1–2, 2.5–3 mm, lanceolate, back smooth, shiny, with a faint longitudinal ridge, gen mottled gray-brown, groove ± open, clearly forked at base.

RARE. Open, sandy soils, creosote scrub; < 400 m. **w DSon (Borrego Valley)**; nw Mex. Feb–May

C. glomeriflora E. Greene Ann < 15 cm. **ST** gen branched throughout, gen strigose. **LF** 0.5–2 cm, linear to oblong-lanceolate, strigose to appressed-bristly; some bristles bulbous-based. **INFL** axillary and terminal, dense, spheric in fl, open, elongate in fr. **FL:** sepals 1.5–2 mm, 2–2.5 mm and linear-lanceolate in fr, strigose at base, ascending-bristly near tip; corolla limb gen < 1 mm wide. **FR:** nutlet gen 1, 1.5–2 mm, ± ovate, back smooth, shiny, groove closed, abruptly flared open at base, edges strongly overlapping esp near base. Open slopes, meadows; 1800–3400 m. SNH (e slope), **W&I**. Jun–Sep

C. gracilis Osterh. Ann 10–35 cm. **ST** simple or branched throughout; branches decumbent to erect, strigose and rough-hairy to short-bristly, hairs spreading. **LF** 1–3.5 cm, linear to narrowly oblanceolate, densely bristly. **INFL** gen dense in fr; lowest fls sometimes not touching. **FL:** sepals < 2 mm, 2–3 mm and lanceolate to narrowly ovate in fr, gen soft-hairy, hairs appressed to stiffly ascending; corolla limb 1–2 mm wide. **FR:** nutlet gen 1, ± 2 mm, lanceolate to narrowly ovate, back smooth, shiny, groove ± closed, ± flared open at base, edges ± overlapping. Sandy to rocky soils, slopes; 900–2150 m. **GB, e DMoj**; to OR, ID, Colorado. Apr–Jun

C. holoptera (A. Gray) J.F. Macbr. (p. 201) WINGED CRYPTANTHA Ann (per), 10–50 cm. **ST** gen branched throughout, strigose and bristly; bristles spreading to ascending, some bulbous-based. **LF** 1–3.5(5) cm, linear, narrowly lanceolate, or oblong, short-bristly; bristles spreading. **INFL** gen dense in fr; lowest fls not touching. **FL:** sepals < or = 2 mm, 2.5–3(4) mm and linear-oblong in fr, densely bristly to rough-hairy, hairs spreading; corolla limb 1–2 mm wide. **FR:** nutlets 4, ± 1–2 mm, subequal, ovate to triangular, back and gen inner face sparsely tubercled, margin gen narrowly winged, groove opened widely below middle, edge gen elevated. UNCOMMON. Sandy to rocky soils; 100–1200 m. **e DMoj, DSon**; w NV, AZ. Mar–Apr

C. humilis (E. Greene) Payson (p. 201) Per 5–30 cm, gen cespitose; caudex woody. **ST** simple, erect, spreading bristly. **LF** 1.5–5 cm, oblanceolate to spoon-shaped, strigose to ± tomentose and bristly; bristles appressed, bulbous-based. **INFL** dense, cylindric. **FL:** sepals 2.5–5 mm, 5–10 mm in fr; corolla limb 4–10 mm wide, appendages white to yellow. **FR:** nutlets 1–4, 2.5–4.5 mm, ovate, back rough, groove open, narrow to triangular, edges not elevated. Common. Gravelly soils; 1700–3600 m. **GB**; to ID, Colorado. [*C. alpicola* Cronq.] Variable; further study needed. ❀TRY;DFCLT. Jun–Aug

C. intermedia (A. Gray) E. Greene (p. 201) Ann 10–60 cm. **ST** branched throughout; branches gen ascending, strigose and rough-hairy to bristly, hairs spreading. **LF** 1.5–5 cm, linear to lanceolate, spreading-bristly. **INFL** open in fr. **FL:** sepals 2–3 mm, 3.5–5 mm and linear-oblong in fr, rough-hairy and bristly, bristles ascending to spreading; corolla limb 3–6 mm wide, throat appendages bright yellow. **FR:** nutlets 1–4, 1.5–2 mm, widely lanceolate to ovate, back gen tubercled to sparsely rough-granular, shiny, groove ± open, esp at forked base. Sandy to rocky soils, oak woodland, coniferous forest; 300–2800 m. 2*n*=24. **CA** (exc coast, GV, D); to B.C., ID, Baja CA. [*C. hendersonii* (Nelson) Piper] Mar–Jul

C. maritima (E. Greene) E. Greene (p. 205) Ann 10–30(40) cm. **ST** simple to branched throughout; branches ascending to erect, gen strigose and spreading-rough-hairy. **LF** 1–4 cm, linear to narrowly oblanceolate, appressed bristly to rough-hairy, some or all hairs bulbous-based. **INFL** gen dense in fr; lowest fls sometimes not touching, bracted. **FL:** sepals 1.5–2 mm, 2–3 mm and linear in fr, densely rough- to long-soft-hairy, hairs ascending, sometimes spreading-bristly. **FR:** nutlets 1–2, 1.5–2 mm, lanceolate, smooth, shiny (if 2, 1 with back smooth, shiny, other fine-granular), groove ± closed to ± near base, edges overlapping. Sandy to gravelly soils; < 1500 m. SW, **D**; AZ, Baja CA. [var. *pilosa* I.M. Johnston] Mar–May

C. micrantha (Torrey) I.M. Johnston (p. 205)   Ann < 15 cm, rounded to cushion-like or taller than wide; taproot gen red, gen purple when dry. **ST** branched throughout; branches spreading to erect, strigose. **LF** < 1 cm, linear to narrowly oblanceolate, short-bristly; hairs ± ascending. **INFLS** axillary and terminal, gen dense in fr, bracted. **FL**: sepals ± 1–1.5 mm, ± 2 mm and linear-oblong in fr, hairs like lvs; corolla limb 0.5–3.5 mm wide. **FR**: nutlets gen 4, ± 1 mm, ± lanceolate, back white-granular to tubercled, sometimes smooth, shiny, groove gen flared open only at base. 2*n*=24. Sandy soils; 200–2300 m. c&s SN (e slope), SW (exc ChI), **SNE, D**; to UT, TX, n Mex. [var. *lepida* (A. Gray) I.M. Johnston] Mar–Jun

C. mohavensis (E. Greene) E. Greene   Ann 10–40 cm. **ST** simple to branched throughout, rough-hairy; hairs ascending to spreading. **LF** 0.5–4 cm, linear to oblong, densely bulbous based bristly. **INFL** gen dense in fr. **FL**: sepals ± 2 mm, 3–5 mm and lanceolate in fr, densely rough-hairy and sparsely spreading-bristly; corolla limb 3–5 mm wide. **FR**: nutlets 4, ± 2 mm, widely lanceolate to ovate, back smooth, shiny, margin sharp-angled (esp above middle), groove ± closed, 2-forked at base, scar open, ± triangular. Open areas; 600–2800 m. c&s SNH (e slope), Teh, **w DMoj**. May–Jul

C. muricata (Hook. & Arn.) Nelson & J.F. Macbr. (p. 205)   Ann 10–100 cm. **ST** branched throughout or above middle, thinly strigose and spreading-soft- to rough-hairy. **LF** 0.5–4 cm, ± linear, rough-hairy to bristly; some bristles bulbous-based. **INFL** gen open in fr; upper fls sometimes touching or overlapping. **FL**: sepals 1–2 mm, 2–4 mm and ± lanceolate in fr, strigose and spreading-bristly; corolla limb 2–6 mm wide. **FR**: nutlets 4, 1–2 mm, ovate to deltate, back tubercled, groove ± closed, forked to flared open at base. Sandy or gravelly, open areas; 150–2700 m. CA-FP (exc NCo, KR), **W&I (Inyo Mtns)**; Baja CA. [vars. *denticulata* (E. Greene) I.M. Johnston, *jonesii* (A. Gray) I.M. Johnston] Apr–Jun

C. nevadensis Nelson & Kenn.   Ann 10–60 cm. **ST** simple to branched throughout, strigose; some hairs ascending. **LF** 1–4(5) cm, linear to oblong, gen bristly; bristles ± ascending. **INFL** gen dense in fr; lowest fls sometimes not touching. **FL**: sepals 3–3.5 mm, (4)6–10 mm and linear in fr, densely rough-hairy and bristly, hairs ascending, bristles spreading; corolla limb 1–2 mm wide. **FR**: nutlets 4, 2–2.5 mm, lanceolate, back densely tubercled, groove ± closed, forked or flared open at base. Sandy, gravelly soils; < 2100 m. s SNF, SnFrB, Teh, SnJV, SCoR, **W&I, D**; to s UT, Baja CA. [var. *rigida* I.M. Johnston] Pls from ne DMoj, with sepals ± 4–6 mm in fr and nutlet back ± fine-tubercled have been called *C. scoparia* A. Nelson, desert cryptantha. Mar–May

C. nubigena (E. Greene) Payson   Per 2–30 cm; caudex gen woody. **ST** simple, erect, spreading-bristly. **LF** 2–5 cm, linear-oblanceolate to spoon-shaped; upper surface spreading-bristly; lower surface bristles bulbous-based. **INFL** dense, ± head-like to cylindric, not elongated in fr. **FL**: sepals 2–3 mm, 4–7 mm in fr, strigose and spreading-bristly; corolla tube 2–3 mm, 4–7 mm in fr, limb 3.5–4 mm wide, appendages gen yellow. **FR**: nutlets gen 4, 2–3 mm, lanceolate to narrowly ovate, back rough, inner surface gen smooth, groove ± closed. Common. Volcanic gravel, talus; 2400–3800 m.   c&s SNH (e slope), **W&I, n DMtns.** ✿TRY;DFCLT. Jul–Aug

C. oxygona (A. Gray) E. Greene   Ann 10–40 cm. **ST** branched throughout, strigose and sparsely rough-hairy; hairs spreading. **LF** 0.5–3(4) cm, narrowly oblanceolate to oblong, bristly; bristles ascending. **INFL** gen dense in fr; lowest fls not touching. **FL**: calyx ± 2 mm, 3–4 mm and narrowly lanceolate in fr, rough-hairy to few-bristled, hairs gen ascending; corolla scented, limb 4–6 mm wide. **FR**: nutlets 4, ± 2 mm, ± ovate, back finely tubercled to white-granular, sometimes mottled gray-brown, margin winged, groove narrowly open at base. Open sites, slopes; < 1950 m. c&s SNF, SNH (e slope), Teh, SnJV, SCoRI, SW (exc ChI), **SNE, w DMoj**. Mar–May

C. pterocarya (Torrey) E. Greene (p. 205)   Ann 10–40(50) cm. **ST** gen branched throughout; branches few, strigose and rough-hairy, hairs gen ascending. **LF** 0.5–5 cm, linear to oblong, bristly; bristles ascending. **INFL** gen open in fr; lowest fls not touching. **FL**: sepals 2–2.5 mm, 4–6 mm and widely lanceolate to ovate in fr, densely strigose to rough-hairy, hairs ascending, bristles few or 0. **FR**: nutlets 3–4, ± 2 mm, winged or 3 clearly winged, 1 not, back tubercled to white-granular, groove flared open to 2-forked at base; winged nutlet widely ovate to ± round, wing gen minutely white-lobed; wingless nutlet narrowly ovate. Sandy to gravelly soils; < 2500 m. s SN, s SnJV, e PR, **GB, D**; to WA, ID, Colorado, TX. Pls with all 4 nutlets winged are called var. *cycloptera* (E. Greene) J.F. Macbr. [var. *purpusii* Jepson] Mar–Jun

C. racemosa (S. Watson) E. Greene (p. 205)   Per 20–100 cm, sometimes rounded, sometimes shrubby, gen fl first year. **ST** branched intricately throughout; branches spreading to ascending; distal branches ± wiry, canescent and sparsely spreading-bristly. **LF** 0.5–3.5 cm, narrowly oblanceolate to linear, bristly. **INFL** open in fr; fls in fr pedicelled. **FL**: sepals 1–2 mm, 2–3 mm and narrowly lanceolate in fr, bristly in lower 1/2, bristles spreading to reflexed; corolla limb 0.5–1 mm wide. **FR**: nutlets gen 4 (1 gen slightly > other 2–3), ± 1–1.5 mm, lanceolate-ovate, back sparsely tubercled to white-granular, margin sharp-angled, minutely winged at tip, groove opening widely, scar triangular. Rocky slopes, washes, canyons; < 1500 m. SnBr (n slope), PR, **s SNE, D**; s NV, AZ, Baja CA. Mar–May

C. recurvata Cov. (p. 205)   Ann 5–35 cm. **ST** branched from base or throughout, strigose. **LF** 1–2(3) cm, linear to narrowly oblanceolate, with many bulbous-based bristles. **INFL** open in fr; fls spreading to recurved. **FL**: sepals 1.5–2 mm, 2–3.5 mm and curved in fr, densely spreading-bristly, hairs collectively light yellow-brown; corolla limb 1–2 mm wide. **FR**: nutlet 1–2 (if 2, 1 > other), 1.5–2 mm, lanceolate, curved, back fine-granular, brown, groove ± closed, flared open at base. Sandy to rocky soils; 700–2500 m. **SNE, DMoj**; to OR, ID, Colorado. Apr–Jun

C. roosiorum Munz (p. 205)   BRISTLECONE CRYPTANTHA   Per 1–2 cm, cespitose; caudex woody. **ST** simple, erect, barely > basal lvs, spreading-soft-bristly. **LF** 0.5–1.2 cm, oblanceolate to spoon-shaped, densely strigose to tomentose; bristles ± appressed, bulbous-based. **INFL** dense, ± head-like. **FL**: sepals 2.5–3 mm, 3.5–4.5 mm in fr, densely strigose and spreading-bristly; corolla tube ± 2.5–3 mm, limb 4.5–5.5 mm wide, appendages yellow. **FR**: nutlets gen 4, 2.5–3 mm, lanceolate-ovate, back rough, groove ± open, ± triangular. RARE CA. Rocky soils, high ridges; > 3000 m. **W&I (n Inyo Mtns)**. Jun–Jul

C. simulans E. Greene   Ann 5–40 cm. **ST** branched throughout, strigose and rough-hairy; hairs gen ascending. **LF** 0.5–3.5 cm, linear to narrowly oblanceolate, rough-hairy to bristly; hairs ascending to ± appressed. **INFL** gen open; fls ascending to erect. **FL**: sepals 2.5–3 mm, 3–6 mm and linear-lanceolate in fr; corolla limb 1–2 mm wide. **FR**: nutlets gen 4, ± 2 mm, ovate, tip acuminate, back densely granular, sparsely tubercled, groove ± closed, forked at base. Open sites, coniferous forest; 650–2300 m. KR, NCoRH, CaR, SN, TR, PR, **GB**; to WA, ID. May–Jul

C. sobolifera Payson (p. 205)   Per 5–20 cm; caudex woody. **ST** sparsely to densely bristly. **LF** 1–4 cm, oblanceolate to spoon-shaped, strigose to tomentose and bristly; lower surface bristles bulbous-based. **INFL** dense, head-like to cylindric. **FL**: sepals 3–4 mm, 6–9 mm in fr, densely bristly; corolla tube 3–4 mm, limb 4–9 mm wide, throat yellow. **FR**: nutlets 2–4, 3–4 mm, lanceolate, back rough to smooth, inner surface gen smooth, groove closed to ± open. Common. Volcanic soils, pumice; 1300–3100 m. KR, CaRH, MP, **n SNE (Sweetwater Mtns)**; to e OR, MT, nw NV. [*C. schoolcraftii* Tiehm, *C. subretusa* I.M. Johnston, Mount Eddy cryptantha] Variable; pls in n SNE like *C. nubigena*; needs study. Jul–Aug

C. torreyana (A. Gray) E. Greene (p. 205)   Ann 10–40 cm. **ST** simple or branched throughout, weakly strigose to rough-hairy

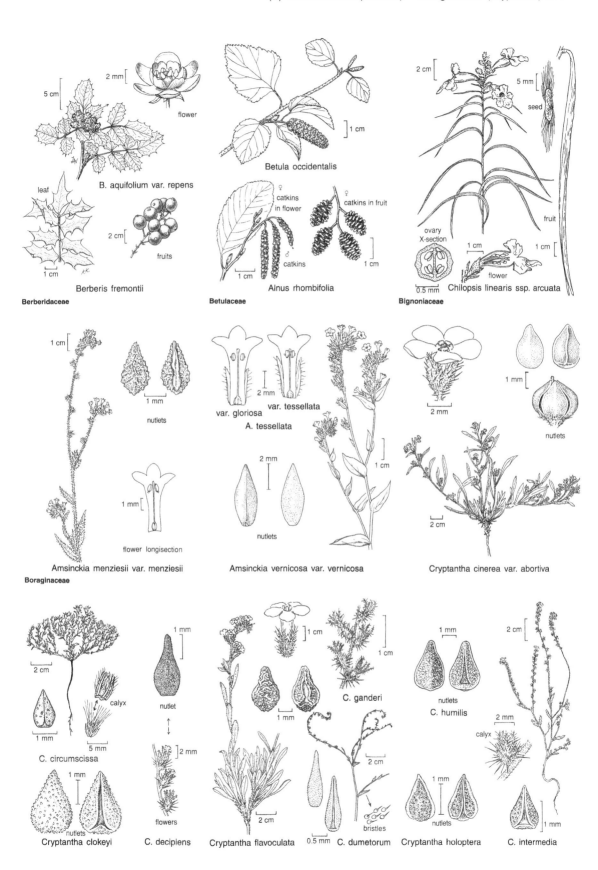

B. aquifolium var. repens

flower

leaf

fruits

Berberis fremontii

Berberidaceae

Betula occidentalis

♀ catkins in flower

♂ catkins

♀ catkins in fruit

Alnus rhombifolia

Betulaceae

seed

fruit

ovary X-section

flower

Chilopsis linearis ssp. arcuata

Bignoniaceae

nutlets

Amsinckia menziesii var. menziesii

flower longisection

Boraginaceae

var. gloriosa

var. tessellata

A. tessellata

nutlets

Amsinckia vernicosa var. vernicosa

nutlets

Cryptantha cinerea var. abortiva

C. circumscissa

calyx

nutlets

Cryptantha clokeyi

nutlet

flowers

C. decipiens

Cryptantha flavoculata

C. ganderi

bristles

C. dumetorum

nutlets

C. humilis

nutlets

Cryptantha holoptera

calyx

nutlets

C. intermedia

or bristly; hairs ascending to spreading. **LF** 0.3–3(4) cm, linear to oblanceolate, strigose to rough-hairy, sometimes bristly; hairs ascending. **INFL** open in fr. **FL**: sepals 2–2.5 mm, 3.5–6 mm and linear-lanceolate in fr, strigose and spreading-bristly; corolla limb 1–2 mm wide. **FR**: nutlets 4, 1.5–2 mm, ± ovate, back flat, smooth, shiny, gen mottled, groove closed, slightly raised, forked and closed at base. Open areas, slopes, gen coniferous forest; 350–2000 m. KR, NCoR, CaRF, SN, SnFrB, **GB**; to B.C., MT, Colorado. [var. *pumila* (A.A. Heller) I.M. Johnston] May–Aug

C. tumulosa (Payson) Payson NEW YORK MOUNTAINS CRYPTANTHA Per 7–25 cm; caudex woody. **ST** simple, erect, densely stiff- to soft-hairy. **LF** 3–6 cm, oblanceolate to ± spoon-shaped, tomentose to bulbous-based bristly. **INFL** cylindric, not gen elongated in fr. **FL**: sepals 3.5–5 mm in fl, 7–10 mm in fr, linear-lanceolate, densely bristly; corolla tube 3.5–4.5 mm, limb 6–8 mm wide, appendages yellow. **FR**: nutlets 1–3, 3–4.5 mm, ovate, both surfaces rough, back ridged down middle, groove open-triangular, edges elevated. UNCOMMON. Gravel or clay, granitic or limestone soils; 1400–2100 m. **n&e DMtns**; sw NV (Spring Mtns). Like per form of *C. virginensis*. Apr–Jun

C. utahensis (A. Gray) E. Greene Ann 10–30 cm. **ST** branched throughout; branches gen ascending, strigose. **LF** 0.3–3(5) cm, linear to oblong, ± appressed-bristly; some bristles on lower surface bulbous-based. **INFL** gen dense in fr; lower fls not touching. **FL**: sepals 2–2.5 mm, 2.5–3 mm and elliptic to ovate in fr, densely strigose to rough-hairy near margin, hairs ascending; corolla scented, tube 2–2.5 mm, limb 2–4 mm wide, throat appendages gen bright yellow. **FR**: nutlets 1(2), ± 2 mm, lanceolate, ± 3-sided, margin sharp-angled to narrowly winged distally, back finely tubercled to white-granular, groove flared open at base. Sandy to gravelly soils, creosote scrub, pinyon/juniper woodland; < 2000 m. s SNE, **D**; to sw UT, AZ. Mar–May

C. virginensis (M.E. Jones) Payson (p. 205) Ann to short-lived Per 10–40 cm; caudex not woody. **ST** simple, erect, downy-hairy under densely spreading, ± stiff hairs. **LF** 2–12 cm, oblanceolate to spoon-shaped, strigose to tomentose and bulbous-based bristly. **INFL** dense, gen cylindric, gen elongated in fr. **FL**: sepals 3–5 mm, 5–11 mm in fr, densely bristly; corolla tube 3–5 mm, limb 5–10 mm wide, appendages yellow. **FR**: nutlets 1–4, 2.5–4.5 mm, ovate, back rough, gen ridged down middle, groove open, edges elevated. Common. Loose soils; 1900–3100 m. **W&I, DMtns**; to sw UT, nw AZ. Ann or bien pls from high W&I, with 1 st and infl not elongated in fr, have been called *C. hoffmannii* I.M. Johnston. ✿TRY;DFCLT. Apr–Jun

C. watsonii (A. Gray) E. Greene Ann 10–40 cm. **ST** branched throughout, thinly strigose, spreading-rough-hairy. **LF** 0.5–3.5 cm, ± linear, densely strigose to appressed-rough-hairy above; lower surface bristly. **INFL** dense to open in fr. **FL**: sepals ± 2 mm, 2.5–3.5 mm and lanceolate in fr, densely strigose, spreading bristly; corolla limb 1–2 mm wide. **FR**: nutlets 4, 1.5–2 mm, lanceolate to narrowly ovate, back smooth, shiny, margin sharp-angled, groove ± closed, short-forked at base. Rocky areas, pinyon/juniper woodland, coniferous forest; 1500–3200 m. e SNH, **GB**; to WA, ID, Colorado. Jun–Aug

HACKELIA STICKSEED

Robert L. Carr

Per (rarely bien); hairs appressed to spreading; caudex gen branched in age, often ± woody, taprooted. **ST** ascending or erect. **LVS**: lowest gen with petioles ± = blades, ± winged; other lvs gen sessile, becoming bract-like toward infl. **INFL**: cymes, gen ± terminal, gen > 3, gen arrayed in panicles, coiled at tips. **FL** radial; corolla rotate-salverform, with appendage near base of each lobe. **FR**: nutlets erect, > style, gen with lateral-medial scar, gen with barb-tipped prickles on margin and exposed face. 40 spp.: gen w N.Am, se Asia. (J. Hackel, Czech botanist, born 1783) [Gentry & Carr 1976 Mem New York Bot Gard 26:121–227] Difficult genus needing much work, esp in n CA, se Asia; sometimes merged with *Lappula*. ✿TRY;DFCLT. Key to species adapted by Ronald B. Kelley.

1. Corolla pale blue; hairs at mid-st gen ± appressed downward . . . . ***H. brevicula***
1' Corolla blue; hairs at mid-st gen ± spreading
  2. Nutlet facial prickles gen 0, rarely 3; infl elongate, narrow; robust bien to short-lived per . . . . . ***H. floribunda***
  2' Nutlet facial prickles gen 4–10; infl gen open, wide; per . . . . ***H. micrantha***

H. brevicula (Jepson) J. Gentry (p. 205) POISON CANYON STICKSEED **ST** 2–6 dm; hairs ± strongly appressed downward, ± stiff, coarse. **LVS**: caudex 6–18 cm, 5–18 mm wide, narrow-elliptic; cauline ± similar but gen smaller, esp above. **FL**: corolla pale blue, limb 5–8 mm wide. **FR**: nutlets 2–3 mm, facial prickles 0–7, < marginal. RARE. Open hillsides, dry stream beds, open aspen stands; 2600–3200 m. **W&I (Mono Co.)**. [*H. patens* (Nutt.) I.M. Johnston misapplied] Jul–Aug

H. floribunda (Lehm.) I.M. Johnston (p. 205) Bien to short-lived per. **ST** gen 4–12 dm; hairs gen spreading, ± coarse. **LVS**: caudex gen < cauline, ephemeral; lower cauline gen 5–24 cm, 5–35 mm wide, oblanceolate to narrow-elliptic. **INFL** elongate, narrow. **FL**: corolla blue, limb gen 4–8 mm wide, sometimes smaller. **FR**: nutlets 2–4 mm, marginal prickles sometimes fused basally, ± forming a wing, facial prickles gen 0–3, < marginal. *n*=±12. Meadows, stream banks, other vernally wet areas, less often open slopes, forests; 700–3100 m. SN, Wrn, **SNE**; w N.Am. Jun–Aug

H. micrantha (Eastw.) J. Gentry (p. 205) **ST** 3–11 dm; hairs gen ± 0 to ± sparse, gen ± spreading, often > 1 mm, ± strigose in infl. **LVS**: caudex 6–33 cm, 7–37 mm wide, narrowly elliptic to oblanceolate; lower cauline gen 5–23 cm, 6–24 mm wide. **FL**: corolla blue, limb gen 5–11 mm wide, sometimes smaller. **FR**: nutlets 3–5 mm, facial prickles gen 4–10, < marginal. *n*=12. Meadows, along streams, open slopes, forests; 700–3400 m. KR, NCoR, CaR, SN, MP, **n SNE (Sweetwater Mtns)**; w N.Am. [*H. jessicae* (MacGregor) Brand] Jul–Aug

HARPAGONELLA

Timothy C. Messick

1 sp. (Latin: small grappling hook, from calyx spines)

H. palmeri A. Gray (p. 205) PALMER'S GRAPPLING HOOK **ST** ascending to erect, 3–30 cm. **INFL**: pedicels in fr 0.5–1 mm, twisted. **FL**: sepals in fr > nutlets, upper 2 > others, partly fused, arched over 1 nutlet, ± bur-like, with 5–10 stout spines, each with hooked bristles, lower 3 sepals distinct. **FR**: nutlets 2, 1–4 mm, dissimilar, ± oblanceolate, margins entire. *n*=12. RARE in

CA. Dry sites in chaparral, coastal scrub, grassland; < 450 m. SCo, PR, **sw DSon**; AZ, nw Mex. Mar–Apr

HELIOTROPIUM HELIOTROPE

Dieter H. Wilken

Ann, per, shrub, glabrous to bristly or strigose. **STS** prostrate to erect, branched. **LVS** gen cauline, petioled to sessile, gen entire. **INFL**: fl 1 and axillary or terminal spikes with many fls, coiled in fl. **FL**: corolla rotate to bell-shaped, white to purple; stamens inserted on upper tube, incl, anthers ± sessile; style attached atop ovary, stigma linear to disk-like. **FR**: nutlets 2 or 4, erect, gen ovoid to spheric, smooth, roughened or hairy, scar gen lateral. ± 250 spp.: temp, trop. Orn, cult for medicinal drugs. (Greek: sun turning, because some spp. fl at summer solstice)

1. Ann; fls axillary, solitary . . . . *H. convolvulaceum*
1' Per; fls many in terminal coiled spikes. . . . . *H. curassavicum*

H. convolvulaceum (Nutt.) A. Gray var. ***californicum*** (E. Greene) I.M. Johnston (p. 205) Ann, taprooted. **ST** ascending to erect, 7–18 cm, canescent. **LF** 1–4 cm, elliptic to ovate, gen petioled, acute, densely strigose. **INFL**: fls 1, axillary. **FL**: calyx lobes lanceolate, long-tapered, densely bristly; corolla 7–10 mm, ± rotate, limb 8–12 mm wide, papery, white. **FR**: nutlets 4, long-soft-hairy. 2*n*=42. Sandy soils; < 700 m. **D**; to Great Plains, n Mex. ❀TRY. Apr–May

H. curassavicum L. (p. 205) Per, sometimes from rhizome-like root. **ST** prostrate to weakly ascending, 1–6 dm, fleshy, glabrous. **LF** 1–6 cm, gen oblanceolate, short-petioled to subsessile, acute to obtuse, fleshy, glabrous. **INFL**: spikes 2–4, terminal, coiled in fl. **FL**: calyx lobes oblong to narrowly ovate, glabrous; corolla 3–5 mm, bell-shaped, limb 3–4 mm wide, white to bluish. **FR**: nutlets 4, smooth. Moist to dry, saline soils; < 2100 m. **CA**; NV, AZ, subtrop, trop Am. [*H. c.* var. *oculatum* (A.A. Heller) I.M. Johnston] ❀6,**7**, **14–17,24**IRR:2,3,**8,9**,10–13,**18–23**;STBL;INV. Mar–Oct

LAPPULA STICKSEED

Ronald B. Kelley

Ann, hairy, taprooted. **ST** ± erect; branches 0–many. **LVS** basal and cauline, sessile, entire. **INFL**: raceme, long, ± terminal; bracts lf-like. **FL**: calyx ± deeply 5-lobed, enlarging in fr; corolla 5-lobed, funnel-shaped, appendages present; style entire. **FR**: nutlets gen 4, ovate, covered with ± long, barbed prickles. 12–14 spp.: n hemisphere. (Latin: little bur)

L. redowskii (Hornem.) E. Greene **ST** 0.5–3.5 dm. **LF** 1–4 cm, linear to lanceolate. **INFL**: pedicel 1–2 mm. **FL**: calyx 3–3.5 mm in fr, lobes lanceolate, ± erect in fr; corolla 1.5–2.5 mm wide, blue to white. **FR**: nutlets 2–3 mm. Dry, open, rocky, often disturbed sites; 600–3300 m. SNH, SnBr, SnJt, **GB, DMoj**; w N.Am, Eurasia. Var. *c.* may occupy warmer, drier habitats than var. *r.*

1. Nutlet marginal prickles much wider at base, fused to form a crown . . . . var. ***cupulata***
1' Nutlet marginal prickles slightly wider at base, slightly fused, not forming a crown . . . . var. ***redowskii***

var. ***cupulata*** (A. Gray) M.E. Jones (p. 205) **FR**: nutlet margin prickles much wider at base, fused to form a crown. Habitat of sp.; 600–2100 m. **GB, DMoj**; w N.Am. [var. *desertorum* (E. Greene) I.M. Johnston]

var. ***redowskii*** (p. 205) **FR**: nutlet marginal prickles slightly wider at base, slightly fused, not forming a crown. 2*n*=48. Habitats of sp.; 1300–3300 m. SNH, SnBr, SnJt, **GB, DMtns**; w N.Am, Eurasia. Apr–Jul

LITHOSPERMUM STONESEED

Ronald B. Kelley

Per, ann, hairy, taprooted. **ST** erect. **LVS** gen cauline, ± sessile, entire. **INFL**: bracted cymes, open panicles, or fls solitary in upper lf axils. **FL**: calyx deeply 5-lobed, enlarging in fr; corolla 5-lobed, funnel-shaped or salverform, gen ± yellow, appendages present or not; style entire. **FR**: nutlets 1–4, 3–6 mm, smooth to pitted or wrinkled. 75 spp.: worldwide, gen temp or mtn. (Greek: stone seed) [Baker 1961 Rhodora 63:229–235]

L. incisum Lehm. (p. 205) Per, strigose; caudex woody. **STS** few–several, 1–3 dm, clustered, ± unbranched. **LVS** many; blade 1.5–6 cm, linear to linear-oblong. **INFL**: cymes many, in upper axils; pedicels 2–5 mm, ± recurved in fr. **FL**: corolla 15–35 mm, 2–3.5 × calyx, 10–20 mm wide, salverform, yellow, appendaged.

FR: nutlets ± pitted, shiny, gray. 2*n*=24,36. Sandy, rocky slopes, pinyon/juniper woodland; 1650–1700 m. **DMtns (Keystone Canyon, New York Mtns, San Bernardino Co.)**; to B.C., MT, Great Plains, s NV. Not heterostylous; cleistogamous fls present. ❀DRN:1–3, 16,17;DFCLT.

MERTENSIA BLUEBELLS, LUNGWORT

Elaine Joyal

Per from branched caudex, glabrous to coarsely hairy. **ST** ± erect. **LVS** cauline and gen basal, alternate, gen petioled (upper gen sessile). **INFL**: cyme, gen panicle- or raceme-like; bracts 0. **FL**: calyx gen deeply lobed; corolla blue, gen abruptly expanded at throat, limb often ± cylindric or flared; filaments often ± flat, gen attached ± below obvious corolla appendages, anthers incl. **FR**: nutlets gen wrinkled, each attached near or below middle to convex receptacle. ± 50 spp.: N.Am, temp Eurasia. (F.C. Mertens, Germany, 1764–1831) [Milek 1988 PhD U Northern Colorado; Strachan 1988 PhD U Montana] Hybrids common; identification sometimes difficult, esp with MP. ❀TRY;DFCLT Key to species adapted by Ronald B. Kelley.

1. Pl 4–15 dm; lateral veins of cauline lvs conspicuous; wet places in mtns, fls late spring, summer .... *M. ciliata*
1' Pl gen < 2 dm; lateral veins of cauline lvs obscure; gen spring-moist, drying places with sagebrush
.... *M. oblongifolia* var. *nevadensis*

M. ciliata (Torrey) G. Don (p. 205) STREAMSIDE BLUEBELLS Pl 4–15 dm, glabrous, sometimes glaucous. **STS** clustered on thick, branched caudex, leafy. **LVS**: basal gen > cauline; cauline with conspicuous lateral veins, lower petioled; blades lanceolate to ovate, acute. **INFL** panicle-like, open. **FL**: corolla 10–17 mm, tube ± = or > limb, often with ring of hairs below middle inside; filaments wide, gen > anthers; style exserted 2–5 mm. 2*n*=24,48. Streamsides, wet meadows, damp thickets, wet cliffs; 1700–3600 m. n SNH, MP, **W&I**; to s OR, NV, UT. Pls with style ± incl have been called var. *stomatechoides* (Kellogg) Jepson. ❀DRN, IRR:1,2,17&WET:14–16;DFCLT. May–Aug

M. oblongifolia (Nutt.) G. Don SAGEBRUSH BLUEBELLS Pl gen < 4 dm, glabrous to strigose. **STS** many, firmly attached to stout, deep, sometimes fleshy caudex. **LVS**: basal gen well developed; cauline gen 2.5–7 × longer than wide, lateral veins obscure, lower lvs gen petioled. **INFL** ± panicle-like, gen dense. **FL**: corolla 10–20 mm, tube gen 1.3–2 × longer than limb, sometimes with ring of hairs inside; filaments wide, ± = anthers; style ± incl. Open slopes, drier meadows, gen spring-moist places, esp with sagebrush; 1000–3000 m. CaRH, n&c SNH, MP, n SNE **(Sweetwater Mtns)**; to WA, WY, Colorado. Polyploid complex (2*n*=24,48). 3 vars. in CA.

var. *nevadensis* (Nelson) L.O. Williams (p. 205) **ST** gen < 20 cm. **LF** glabrous, sometimes bumpy. **FL**: corolla tube glabrous inside. 2*n*=24. Habitats and elev. gen of sp. CaRH, n&c SNH, Wrn, **n SNE (Sweetwater Mtns)**; to e OR, ID, Colorado. Most common var. in CA. Apr–Jun

PECTOCARYA

Timothy C. Messick and Barbara Veno

Ann. **ST** 2–40 cm, strigose, breaking apart at nodes or not. **LVS** gen alternate, gen 0.5–4 cm, ± linear, strigose to sharp-bristled. **INFL**: pedicel in fr gen free from nutlets, gen recurved. **FL**: sepals gen < fr, upper 2 in fr gen > others; corolla 0.8–3 mm, white; style attached to receptacle, unbranched, gen persistent, stigma 1, head-like. **FR**: nutlets gen 4, spreading, 1–4.5 mm, gen paired, gen compressed, marginal prickles hooked at tip, not barbed; all 4 often dissimilar in shape, ornamentation, margin width. 15 spp.: CA to B.C., WY, TX, n Mex; also S.Am. (Greek: comb nut, from dentate nutlet margins in some spp.) [Veno 1979 PhD dissertation UCLA] ❀STBL. Key to species adapted by Ronald B. Kelley.

1. Calyx radial, sepals ± equal, > nutlets; nutlet margins ± entire, not dentate; lower cauline lvs opposite, fused at base (sect. *Gruvelia*) .... *P. setosa*
1' Calyx bilateral, upper 2 sepals > others, all < nutlets; nutlet margins entire to gen dentate; lower cauline lvs mostly alternate, free (sect. *Pectocarya*)
 2. Basal fls cleistogamous; basal nutlet margins less ornamented than cauline
  3. Pedicel in fr partly fused to 1 nutlet; cauline nutlets gen curved; lower 3 sepals unequal...... *P. heterocarpa*
  3' Pedicel in fr free from nutlets; cauline nutlets gen straight; lower 3 sepals ± equal .... *P. peninsularis*
 2' Basal fls not cleistogamous; basal and cauline nutlet margins similarly ornamented
  4. Nutlet margins ± entire, bristled but not or barely dentate .... *P. penicillata*
  4' Nutlet margins dentate
   5. Nutlet strongly recurved to coiled, linear; nutlet margin teeth distinct ± to base............. *P. recurvata*
   5' Nutlet straight to moderately recurved, ± oblanceolate; nutlet margin teeth distinct ± to or fused at base
    6. Sts prostrate to decumbent; nutlet margin teeth gen narrower at base than length, distinct ± to base; nutlets straight or slightly recurved at tip .... *P. linearis* ssp. *ferocula*
    6' Sts ascending to erect; nutlet margin teeth ± as wide at base as length, fused at base; nutlets slightly to moderately recurved ± throughout .... *P. platycarpa*

P. heterocarpa (I.M.Johnston) I.M. Johnston (p. 209) **ST** prostrate to ascending, 2–25 cm. **INFL**: pedicel in fr 1.3–2.8 mm, partly fused to 1 nutlet. **FLS**: basal cleistogamous; in fr lower 3 sepals unequal, < others. **FR**: nutlet 1.2–3 mm, oblong to oblanceolate; basal nutlets 2–4 per fl, not paired, reflexed, 1 unmargined, others narrowly entire- to ± dentate-margined; cauline nutlets paired, curved, margins ± entire (and narrowly to widely membranous) to dentate. *n*=12. Washes, roadsides, openings in creosote-bush scrub, Joshua-tree woodland; < 1400 m. SW, **W&I, D**; to UT, TX, nw Mex. Feb–May

P. linearis (Ruiz & Pav.) DC. ssp. *ferocula* (I.M. Johnston) Thorne (p. 209) **ST** prostrate to decumbent, 6–26 cm. **INFL**: pedicel in fr 1.5–3 mm. **FR**: nutlet body 2–3.8 mm, straight or slightly recurved near tip, linear-oblanceolate; margin teeth distinct ± to base, width at base gen < length. *n*=24. Roadsides, grassy slopes, clearings; 5–2000 m. GV, CW, SW, **DMoj**; Baja CA; also s S.Am. Mar–May

P. penicillata (Hook. & Arn.) A. DC. (p. 209) **ST** prostrate to decumbent, 2–25 cm. **INFL**: pedicel in fr ascending, 1.3–2.5 mm. **FR**: nutlets straight, 1.1–3.3 mm, oblanceolate; margins erect to incurved, ± entire, all in a fl ± equal in width, bristled only above ± middle. *n*=12. Disturbed sites, roadsides in many communities; 90–2100 m. CA-FP, **W&I, D**; to B.C., WY, AZ, n Baja CA. Pls with margins of 1–2 nutlets narrower than others in same fl, bristles from nutlet base to tip, *n*=24, may be distinct sp. Mar–May

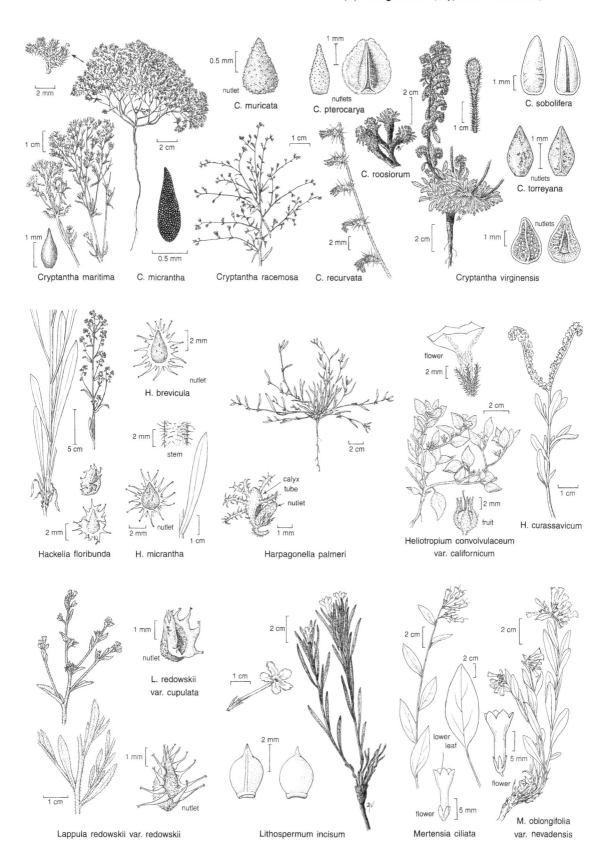

2 mm

0.5 mm

nutlet

C. muricata

1 mm

nutlets

C. pterocarya

2 cm

1 cm

1 mm

C. sobolifera

1 mm

nutlets

C. torreyana

C. roosiorum

1 cm

1 cm

2 cm

1 mm

nutlets

1 cm

2 mm

1 mm

Cryptantha maritima   C. micrantha   Cryptantha racemosa   C. recurvata   Cryptantha virginensis

0.5 mm

2 cm

2 mm

H. brevicula

2 mm

stem

2 mm

nutlet

5 cm

2 mm

2 mm

nutlet

1 cm

Hackelia floribunda   H. micrantha

2 cm

2 cm

calyx
tube

nutlet

1 mm

Harpagonella palmeri

flower

2 mm

2 cm

2 mm

fruit

1 cm

Heliotropium convolvulaceum
var. californicum

H. curassavicum

1 mm

nutlet

L. redowskii
var. cupulata

1 mm

nutlet

1 cm

Lappula redowskii var. redowskii

2 cm

1 cm

2 mm

Lithospermum incisum

2 cm

2 cm

lower
leaf

flower   5 mm

Mertensia ciliata

2 cm

5 mm

flower

M. oblongifolia
var. nevadensis

P. peninsularis I.M. Johnston (p. 209) **ST** prostrate to ascending, 2–24 cm. **INFL**: pedicel in fr 1–1.5 mm. **FLS**: basal cleistogamous; in fr lower 3 sepals ± equal. **FR**: nutlets 1.1–2 mm, elliptic to ovate; basal nutlets 2–4 per fl, not paired, reflexed, 1–3 unmargined, others narrowly ± dentate-margined; cauline nutlets paired, gen straight, margins narrowly to widely membranous, ± dentate. *n*=12. Washes, roadsides, clearings; 30–300 m. **DSon**; Baja CA.

P. platycarpa (Munz & I.M. Johnston) Munz & I.M. Johnston (p. 209) **ST** ascending to erect, 4–25 cm. **INFL**: pedicels in fr 2.5–4 mm. **FR**: nutlet slightly to moderately recurved ± throughout, 2.5–4.5 mm, oblanceolate to narrowly obovate; margin teeth fused at base, width at base ± = length. *n*=24. Washes, roadsides in creosote-bush scrub, Joshua-tree woodland; 150–1500 m. **SW, W&I, D**; s NV, AZ, nw Mex. Mar–May

P. recurvata I.M. Johnston (p. 209) **ST** ascending to erect, 3.5–21 cm. **FR**: pedicels in fr 2–3 mm. **FR**: nutlets strongly recurved to coiled, 2.5–4 mm, linear; margin teeth distinct ± to base, linear (or width at base < length). *n*=12. Shelter of rocks, bases of shrubs, sometimes roadsides, creosote-bush scrub, Joshua-tree woodland; 10–1600 m. **W&I, D**; to s NV, sw NM, nw Mex. Mar–May

P. setosa A. Gray (p. 209) **ST** ascending to erect, 2–23 cm. **LVS**: lower opposite, fused at base; upper alternate. **INFL**: pedicels in fr ascending or reflexed, ± 0.5 mm. **FL**: sepals in fr ± equal, > nutlets, with appressed, short, and several spreading, long, stiff, hairs (incl bristles), tips with straight bristles. **FR**: nutlets 1.5–4 mm, ± obovate to ± round; margins ± entire, membranous-winged, wide on 3 nutlets, narrow on 1. *n*=12. Clearings in sagebrush scrub, creosote-bush scrub, pinyon/juniper woodland, grassland; 150–2300 m. s **SN, CW, SW, SNE, D**; to WA, ID, UT, AZ, n Baja CA. Apr–May

PLAGIOBOTHRYS POPCORNFLOWER

Timothy C. Messick

Ann, per, gen strigose. **ST** prostrate to erect, branched at base or above, < 5 dm. **LVS** simple, 0.5–10 cm, gen smaller upward; all cauline (lower opposite, linear to oblong, upper gen alternate) or both basal (often in rosettes) and cauline (alternate, linear to oblanceolate). **INFL**: raceme or spike, coiled in bud, gen elongate in fr; bracts 0–many; pedicels gen 0–1 mm. **FL** bisexual; sepals fused below middle, 2–10 mm in fr; corolla 1–12 mm wide, all white or yellow inside tube. **FR**: nutlets gen 4, 1–3.5 mm; back gen with midrib, lateral ribs, cross-ribs, interspaces, gen tubercled, sometimes prickled or bristled; scar gen lateral (on side) near middle or base, sometimes basal (on bottom) or oblique (between side and bottom), sometimes on a stalk or short peg, gen ovate to triangular. ± 65 spp.: temp w N.Am, w S.Am. (Greek: sideways pit, from position of nutlet attachment scar) [Higgins 1974 Great Basin Natur 34(2):161–166; Johnson 1932 Contr Arnold Aboretum 3:1–102] Fully mature nutlets critical for identification; intergradation common in some spp. groups; sect. *Allocarya* often treated as a separate genus; many spp. need further study. Key to species adapted by Ronald B. Kelley.

1. Nutlet scar at tip of oblique stalk (sect. *Echidiocarya*) .... ***P. collinus***
  2. Corolla 4–7 mm wide; st hairs gen ± fine, appressed................................ var. ***californicus***
  2' Corolla 1–3 mm wide; st hairs gen coarse, spreading .... var. ***fulvescens***
1' Nutlet scar sessile, lateral to oblique, or on short, basal to rarely ± oblique peg (gen < stalks in 1.), not on oblique stalk
  3. Lower cauline lvs alternate, sometimes ± paired, upper alternate; basal rosette present or 0; corolla tube white; nutlet scar lateral near middle; habitat gen dry, upland
    4. Scar circular, infl short, coiled in fr (sect. *Sonnea*) (see also 6.) .... ***P. hispidus***
    4' Scar linear; infl gen elongate in fr
      5. Nutlet scar long, narrow, along crest of keel; basal rosette 0 (sect. *Amsinckiopsis*)
        6. Corolla 1–3 mm wide; nutlet covered with cobblestone-like bulges, tubercles 0 .... ***P. jonesii***
        6' Corolla 4–7 mm wide; nutlet irregularly ribbed, tubercled .... ***P. kingii***
          7. Infl ± short, coiled in fr; st 0.5–1.5 dm .... var. ***harknessii***
          7' Infl ± elongate in fr; st 1–4 dm.. var. ***kingii***
      5' Nutlet scar short, ± round, below keel; basal rosette gen conspicuous (sect. *Plagiobothrys*)
        8. Interspaces of ± ovoid-shaped nutlet back gen wide, flat, between narrow, sometimes ± toothed cross-ribs
          9. Bracts few, near base of infl; corolla 3–9 mm wide; nutlet slightly arched in profile ...... ***P. nothofulvus***
          9' Bracts many, throughout infl; corolla 2–3 mm wide; nutlet gen strongly arched in profile
            10. Calyx circumscissile in fr; nutlets gen 2, strongly attached .... ***P. arizonicus***
            10' Calyx not cicumscissile in fr; nutlets gen 4, weakly attached .... ***P. canescens***
        8' Interspaces of ± cross-shaped nutlet back gen narrow, groove-like, sometimes obscure, between wide, sometimes ± tubercled cross-ribs .... ***P. tenellus***
3' Lower cauline lvs opposite, upper gen alternate; basal rosette 0; corolla tube often yellow inside; nutlet scar lateral near base or basal, sometimes oblique; habitat gen seasonally wet (sect. *Allocarya*)
  11. Nutlet scar basal or rarely ± oblique, often on short peg; sepal midrib ± fleshy
    12. St decumbent; calyx gen strongly bent... ***P. leptocladus***
    12' St ascending to erect; calyx straight or slightly bent .... ***P. stipitatus*** var. ***micranthus***
  11' Nutlet scar lateral near base or sometimes oblique, sessile; sepal midrib gen slender
    13. St hairs spreading
      14. Nutlets < 2 mm, cross-ribs prominent, scar lateral near base; bracts few, near base of infl .... ***P. parishii***
      14' Nutlets gen > 2 mm, cross-ribs low, scar gen oblique; bracts many, throughout infl .... ***P. salsus***
    13' St hairs appressed
      15. Nutlet scar gen oblique, triangular .... ***P. cognatus***

15' Nutlet scar gen lateral, elongate

16. Nutlet not bristled, not scabrous, scar elongate with thin, incurved margins . . . . *P. cusickii*

16' Nutlet minutely bristled or scabrous, scar narrowly elliptic . . . . *P. hispidulus*

P. arizonicus (A. Gray) A. Gray (p. 209) Ann; sap purple. **ST** ascending to erect, 1–4 dm; hairs rough, sharp, some spreading. **LVS:** basal in rosette, 1.5–5 cm; cauline alternate. **INFL** bracted throughout. **FL:** calyx ± 3 mm, gen circumscissile in fr; corolla 2–2.5 mm wide. **FR:** nutlets gen 2, ± 2 mm, ovoid, strongly arched in profile, strongly attached; midrib, lateral ribs, cross-ribs narrow; interspaces wide; scar lateral near middle, round. Common. Dry, coarse soils in scrub or woodlands; < 2100 m. Teh, e SnFrB, SCoRI, TR, **W&I, D**; to NM, n Mex. Intergrades with *P. canescens, P. nothofulvus.* 🌣TRY. Mar–May

P. canescens Benth. (p. 209) Ann; sap purple. **ST** prostrate to erect, 1–6 dm, purple-dyed; hairs long, rough or bristled. **LVS:** basal in rosette, 1.5–5 cm; cauline alternate. **INFL** bracted throughout. **FL:** calyx 4–6 mm, gen not circumscissile in fr; corolla 2–3 mm wide. **FR:** nutlets ± 2 mm, round-ovoid, gen strongly arched in profile, gen weakly attached; midrib, lateral ribs, cross-ribs narrow; interspaces wide; scar lateral near middle, round. Common. Grasslands, woodlands, coastal scrub; < 1400 m. CaRF, SNF, GV, SW, **w DMoj**. Intergrades with *P. arizonicus, P. nothofulvus.* 🌣TRY. Mar–May

P. cognatus (E. Greene) I.M. Johnston (p. 209) Ann, strigose. **ST** decumbent to erect, 0.5–3 dm. **LVS** cauline; lower 2–7 cm. **INFL** bracted below middle. **FL:** calyx 2–4 mm; corolla 1–1.5 mm wide. **FR:** nutlet 1–2 mm, lanceolate-ovate, scabrous-bristled to ± papillate; midrib, lateral ribs short, near tip; cross-ribs irregular, rounded; scar oblique, ± triangular. Moist places in meadows, forests; < 2100 m. c SN, NCoRI, **SNE**; to WA, Rocky Mtns, AZ. [*P. scouleri* (Hook. & Arn.) I.M. Johnston var. *penicillatus* (E. Greene) Cronq., in part] With *P. cusickii, P. hispidulus, P. reticulatus, P. scouleri* (OR, WA to Rocky Mtns, Can), and possibly others, forming a widespread, highly variable complex of poorly defined taxa in need of further study. Jun–Aug

P. collinus (Phil.) I.M. Johnston (p. 209) Ann; hairs appressed to spreading, fine or coarse. **ST** prostrate to ascending, gen 1–4 dm. **LVS** cauline; lower gen opposite, 1–4 cm; upper alternate. **INFL** gen elongate, bracted throughout. **FL:** calyx ± 3 mm; corolla 1–7 mm wide. **FR:** nutlet 1–2 mm, ovoid; midrib, lateral ribs, cross-ribs sharp; scar at tip of oblique stalk. Dry places, many habitats; < 2300 m. s SN, s SnJV, CW, SW, **nw edge DSon (San Gorgonio Pass)**; Mex, Chile. 4 vars. in CA.

var. ***californicus*** (A. Gray) Higgins Pl not cespitose. **ST:** hairs gen ± fine, appressed. **LF** 1–3 cm, 2–5 mm wide, oblancelate. **INFL** > lvs. **FL:** corolla 4–7 mm wide. Grassland, coastal scrub; < 2300 m. s SN, s SnJV, CW, SW, **nw edge DSon (San Gorgonio Pass)**; Mex. [*P. californicus* (A. Gray) E. Greene] Mar–May

var. ***fulvescens*** (I.M. Johnston) Higgins Pl not cespitose. **ST:** hairs gen coarse, spreading. **LF** 1–3 cm, 3–5 mm wide, oblanceolate. **INFL** > lvs. **FL:** corolla ± 2 mm wide. Dry places, chaparral, coniferous forest; gen 600–2000 m. SCoRO, SW, **w DSon**; AZ, Mex, Chile. [*P. californicus* var. *f.* I.M. Johnston]

P. cusickii (E. Greene) I.M. Johnston (p. 209) Ann, strigose. **ST** prostrate to ascending, 0.5–2 dm. **LVS** cauline; lower 3–10 cm. **INFL** bracted below middle. **FL:** calyx 1.5–4 mm; corolla 1–1.5 mm wide. **FR:** nutlet 1–2 mm, lanceolate-ovate; midrib, lateral ribs short, near tip; cross-ribs irregular, rounded; scar lateral near base, elongate, with thin, incurved margins. Montane flats, meadows; 1200–2100 m. **GB**; to WA, NV. [*P. scouleri* (Hook. & Arn.) I.M. Johnston var. *penicillatus* (E. Greene) Cronq., in part]. May–Aug

P. hispidulus (E. Greene) I.M. Johnston (p. 209) Ann, strigose. **ST** prostrate or ± ascending, 0.5–4 dm. **LVS** cauline; lower 1–5 cm. **INFL** bracted below middle. **FL:** calyx ± 3 mm; corolla

1–2 mm wide. **FR:** nutlet 1.5–2 mm, ± ovoid, minutely bristled or scabrous; midrib, cross-ribs narrow, low; interspaces wide; scar lateral near base, ± narrowly elliptic. Moist or drying sites; 1200–3400 m. KR, SN, TR, PR, **W&I**; to WA, WY, NV. [*P. scouleri* (Hook. & Arn.) I.M. Johnston var. *penicillatus* (E. Greene) Cronq., in part] Jun–Aug

P. hispidus A. Gray (p. 209) Ann, sharply hairy and sparsely short-tomentose. **ST** erect, 0.5–2 dm. **LVS** mostly cauline, alternate, 1.5–4 cm, ± blistered. **INFL** short, coiled in fr, bracted below middle. **FL:** calyx 2 mm; corolla 1 mm wide. **FR:** nutlet gen 1, 1–1.5 mm, ovoid; midrib, lateral ribs low, wide; cross-ribs 0; scar lateral gen above middle, ± round. Dry places, gen in sandy soil; 1200–2800 m. CaR, **GB**; OR, w NV. Jun–Aug

P. jonesii A. Gray (p. 209) Ann; hairs sharp, bristled, spreading. **ST** ascending to erect, < 4 dm. **LVS** cauline, alternate; lower 2–10 cm. **INFL** bracted near base. **FL:** calyx 4–8 mm; corolla 1–3 mm wide. **FR:** nutlets 3–4, 2–3 mm, triangular-ovate, covered with cobblestone-like bulges; tubercles 0; midrib, lateral ribs weak; cross-ribs 0; scar lateral near middle, along crest of keel, long, narrow. Sandy, gravelly, or rocky slopes, creosote-bush scrub to pinyon/juniper woodland; < 1800 m. SNE, **D**; to UT, w AZ, Mex. Apr–May

P. kingii (S. Watson) A. Gray (p. 209) GREAT BASIN POPCORNFLOWER Ann; hairs coarse, stiff, spreading, unequal. **ST** ascending to erect, 0.5–4 dm. **LVS** cauline, alternate; lower crowded, < 6 cm; upper scattered. **INFL** bracted near base. **FL:** calyx 3–4 mm; corolla 4–7 mm wide. **FR:** nutlet 2–3 mm, ovoid, arched in profile; midrib, lateral ribs, cross-ribs irregular, coarse; scar lateral near middle, along crest of keel, long, narrow. Dry, open slopes, sagebrush scrub, saltbush scrub, juniper woodland; 1200–2300 m. **GB, DMoj**. Vars. intergrade.

var. ***harknessii*** (E. Greene) Jepson STS all ascending, 0.5–1.5 dm. **LF** < 1 cm wide. **INFL** ± short, coiled in fr. Habitat and elevation as in sp. **GB**; se OR, NV.

var. ***kingii*** STS ascending at pl sides to erect at pl center, 1–4 dm. **LF** < 2 cm wide. **INFL** ± elongate in fr. Habitat and elevation as in sp. **SNE, DMoj**; to se OR, w UT.

P. leptocladus (E. Greene) I.M. Johnston (p. 209) ALKALI PLAGIOBOTHRYS Ann, sparsely strigose to ± glabrous. **ST** decumbent, 1–3 dm. **LVS** cauline; lower 3–10 cm. **INFL** bracted below middle. **FL:** calyx 4–8 mm, gen strongly bent, turning corolla skyward, sepal midribs ± fleshy; corolla 1–2 mm wide. **FR:** nutlet 1.5–2.5 mm, ± lanceolate; midrib above middle; cross-ribs 0 or few, weak, above middle; back often minutely bristled; scar basal, often on short peg. Gen alkaline clay soils in vernal pools, wet places; < 2500 m. SW, **w DMoj**; to AK, c Can, Mex. Mar–May

P. nothofulvus (A. Gray) A. Gray (p. 209) POPCORNFLOWER Ann; hairs spreading, rough, sharp; sap purple. **ST** ± erect, 2–7 dm. **LVS:** basal in rosette, 3–10 cm; cauline few, alternate. **INFL** bracted near base. **FL:** calyx 2–3 mm, often circumscissile in fr; corolla 3–9 mm wide. **FR:** nutlets gen 1–3, ± 2 mm, round-ovoid, abruptly narrowed below acute tip, slightly arched in profile, sometimes strongly attached; midrib, lateral ribs, cross-ribs narrow; interspaces wide; scar lateral near middle, round. $2n=24$. Abundant. Grasslands, woodlands; gen < 800 m. CA-FP, **rarely edge of D**; to WA, Mex. Intergrades with *P. arizonicus, P. canescens.* 🌣TRY. Mar–May

P. parishii I.M. Johnston (p. 209) PARISH'S POPCORNFLOWER Ann; hairs short, spreading. **ST** prostrate, 0.5–3 dm. **LVS** cauline; lower 1–5 cm, lower surface blistered. **INFL** bracted near base. **FL:** calyx 2–3 mm; corolla 3–5 mm wide. **FR:** axial nutlet 1–1.8 mm, > others, ovoid or lance-ovoid; midrib, lateral ribs, cross-ribs narrow, irregular; scar lateral near base, gen linear to narrowly triangular. Uncommon. Wet, alkaline soil around desert springs; 750–1400 m. **SNE, DMoj**. Apr–Jun

P. salsus (Brandegee) I.M. Johnston Ann; hairs stiff, spreading. **ST** decumbent to erect, 0.6–1.6 dm. **LVS** cauline; lower 3–6 cm. **INFL** bracted throughout. **FL**: calyx 2–5 mm; corolla 2–4 mm wide. **FR**: nutlet 1.5–2.5 mm, lanceolate; cross-ribs few, low, above middle, arched or irregular; scar gen oblique, narrowly ovate. Moist, alkaline mud flats; ± 700 m. **ne DMoj (e Inyo Co)**; se OR, NV.

P. stipitatus (E. Greene) I.M. Johnston Ann, short-strigose. **ST** ascending to erect, 1–5 dm, often fleshy, hollow. **LVS** cauline; lower 2–11 cm. **INFL** bracted below middle; pedicels gen solid. **FL**: calyx 5–8 mm, sepal midribs ± fleshy; corolla 2–12 mm wide. **FR**: nutlet 1.5–2.5 mm, lanceolate to lance-ovate; midrib, lateral ribs low, rounded, below middle obscure; cross-ribs above

middle, rounded, oblique or arched, sometimes bristled; scar basal or rarely ± oblique, gen on short peg. Vernal pools, wet sites; < 1500 m. CA-FP, **GB**; OR. 2 vars. in CA.

var. ***micranthus*** (Piper) I.M. Johnston (p. 209) **FL**: corolla 2–3 mm wide. Common. Vernal pools, wet sites in grasslands to conifer forests; < 1500 m. CA-FP, **GB**; se OR. Apr–Jul

P. tenellus (Nutt.) A. Gray (p. 209) Ann; hairs spreading, shaggy. **STS** several, erect, 0.5–3 dm. **LVS**: basal in rosette, 1–5 cm; cauline few, alternate. **INFL** ± strongly coiled toward tip, bracted near base; fls gen > 10 per branch. **FL**: calyx 3–5 mm; corolla 1–3 mm wide. **FR**: nutlet 1–2 mm, thickly cross-shaped; cross-ribs wide, gen tubercled; interspaces narrow, groove-like, often obscure; scar lateral near middle. Common. Dry slopes in grassland, scrub, woodland, or forest; < 1700 m. CA-FP, **GB (uncommon), D**; to B.C., ID, UT, AZ, Mex. Mar–May

TIQUILIA

Ronald B. Kelley

Ann, per, subshrub, variously hairy, ± taprooted; rhizome gen 0. **ST** spreading to prostrate. **LVS** cauline, alternate, gen clustered, evergreen, petioled; margin rolled under, entire or ± crenate. **INFL** ± axillary; fls solitary or clustered, sessile. **FL**: calyx ± deeply 5-lobed, not enlarging in fr; corolla 5-lobed, gen ± funnel-shaped, tube yellow when young, appendages 0; style branches 2. **FR**: nutlets 1–4, sometimes ± tubercled. 27 spp.: w hemisphere deserts. (native S.Am. name for flower) [Richardson 1977 Rhodora 79:467–572] Separated from *Coldenia* of e hemisphere.

1. Branches alternate; per; lf veins obscure, blade ovate to narrowly elliptic; fls gen ± solitary; style branched < 1/3 from tip; fr 4-grooved (sect. *Stegnocarpus*) . . . . **T. canescens**
  2. Corolla 4–7.5 mm, at top 2.5–4.5 mm wide . . . . var. **canescens**
  2' Corolla 8–12 mm, at top 5–8 mm wide . . . . var. **pulchella**
1' Branches opposite; ann or per; lf veins obvious, ± sunken, blade ovate, round, or obovate; fls clustered; style branched 1/3–4/5 from tip; fr 4-lobed (sect. *Tiquiliopsis*)
  3. Rhizome present; st ± glandular; lf veins deeply sunken, 4–7 pairs; hairs within calyx long . . . . **T. plicata**
  3' Rhizome 0; st ± non-glandular; lf veins shallowly sunken, 2–3 pairs; hairs within calyx short or 0
    4. Ann; style < calyx; lf margin entire, lateral veins ± 30° from midvein; corolla pink to white; seed oblong-ovoid . . . . **T. nuttallii**
    4' Per; style > calyx; lf margin ± crenate, lateral veins ± 45° from midvein; corolla blue, purple, or lavender; seed spheric . . . . **T. palmeri**

T. canescens (DC.) A. Richardson Per, subshrub. **ST**: branches alternate; hairs ± spreading. **LVS** sometimes clustered, white-tomentose; blade 5–13 mm, ovate to narrowly elliptic, veins obscure, margin entire, spiny-ciliate. **INFL**: fls ± solitary; bracts 0. **FL**: calyx 3–5 mm, free 2/3–3/4 length; style branched < 1/3 from tip, shortly exserted from calyx. **FR** spheric, 4-grooved, not lobed. **SEED** 2–2.5 mm, ovoid, minutely tubercled, hairy or not. Slopes, ridges of broken granite, limestone, gneiss; 500–1500 m. **DMtns, DSon**; sw N.Am., n Mex. [*Coldenia c.* DC.]

var. ***canescens*** (p. 209) **FL**: corolla 4–7.5 mm, 2.5–4.5 mm wide at top, lavender, pink, or white. *n*=9. Habitats and range of sp. ❀DFCLT;TRY.

var. ***pulchella*** (I.M. Johnston) A. Richardson **FL**: corolla 8–12 mm, 5–8 mm wide at top, blue or lavender. Habitats of sp. **DSon (Imperial, Riverside cos.)**; sw AZ. ❀TRY.

T. nuttallii (Hook.) A. Richardson (p. 209) Ann. **ST**: branches opposite; hairs ± appressed. **LVS** clustered; hairs ± spreading; blade 3.5–9 mm, ovate to round, margin entire, veins 2–3 pairs, shallowly sunken, ± 30° from midvein. **INFL**: fls clustered, bracted. **FL**: calyx 3–5 mm, free 2/3–3/4 length, hairs within short; corolla 3–4 mm, 2–2.5 mm wide, pink to white; style < calyx, branched 1/3–1/2 from tip. **FR** deeply 4-lobed. **SEED**

oblong-ovoid, smooth, shiny. *n*=8. Sandy plains, washes, slopes, saline flats; < 2400 m. Teh, e MP, **SNE, DMoj**; to OR, UT; also in Argentina. [*Coldenia n.* Hook.] May–Aug

T. palmeri (A. Gray) A. Richardson (p. 209) Per, ± woody; bark white. **ST**: branches opposite; hairs ± shaggy. **LVS** clustered, grayish strigose; blade 3.5–11 mm, ovate to round, margin ± crenate, veins 2–3 pairs, shallowly sunken, ± 45° from midvein. **INFL** bracted; fls clustered. **FL**: calyx 2–3.5 mm, free ± 1/2 length, hairs within short or 0; corolla 5–9 mm, 4–5 mm wide, blue, purple, or lavender; style > calyx, branched 1/2 from tip. **FR** deeply 4-lobed. **SEED** spheric, smooth, shiny. *n*=8,9. Sandy gravel soils; < 900 m. **D (esp w edge DSon and near Colorado River)**; sw NV, w AZ, n Mex. [*Coldenia p.* A. Gray] Apr–Jun

T. plicata (Torrey) A. Richardson (p. 209, pl. 31) Per, ± woody; rhizome present. **ST**: branches opposite, ± glandular. **LVS** clustered, white-canescent; blade 3–12 mm, obovate to widely ovate, margin entire, veins 4–7 pairs, deeply sunken. **INFL**: fls clustered; bracts 0. **FL**: calyx 2–3 mm, free ± total length, hairs within long; corolla 4–6 mm, 2–3 mm wide, blue to lavender; style > calyx, branched 1/2–4/5 from tip. **FR** deeply 4-lobed. **SEED** ovoid, smooth, shiny. *n*=8. Dune sand, sandy gravel flats; < 900 m. **D**; w AZ, s NV, n Mex. [*Coldenia p.* (Torrey) Cov.] Apr–Jun

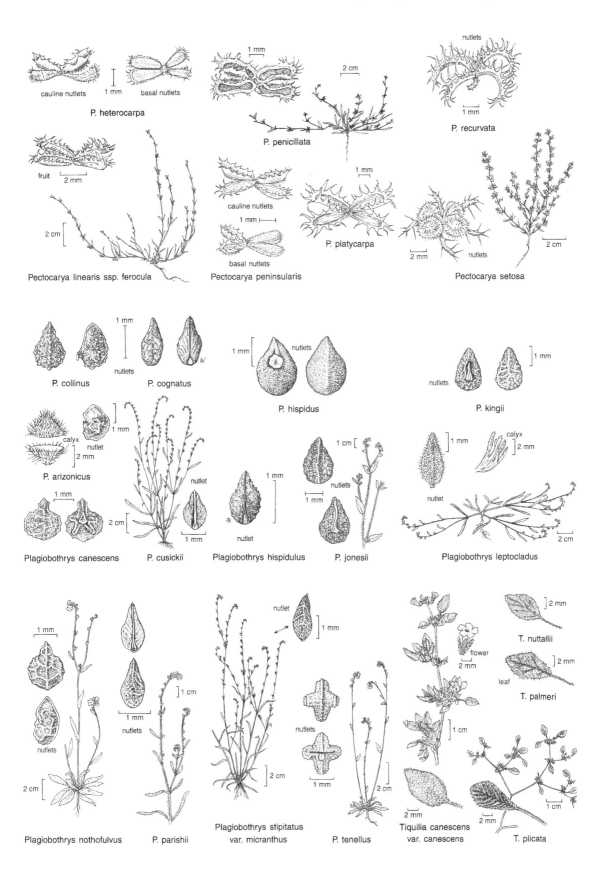

cauline nutlets   basal nutlets   1 mm

P. heterocarpa

fruit   2 mm

2 cm

Pectocarya linearis ssp. ferocula

1 mm

2 cm

P. penicillata

cauline nutlets

1 mm

basal nutlets

Pectocarya peninsularis

1 mm

P. platycarpa

nutlets

1 mm

P. recurvata

2 mm   nutlets

2 cm

Pectocarya setosa

P. collinus   P. cognatus

1 mm

nutlets

1 mm   nutlets

P. hispidus

nutlets

1 mm

P. kingii

calyx

nutlet

1 mm

2 mm

P. arizonicus

1 mm

nutlet

2 cm

1 mm

Plagiobothrys canescens   P. cusickii

1 mm

nutlet

Plagiobothrys hispidulus

1 cm

nutlets

1 mm

nutlet

P. jonesii

1 mm

nutlet

calyx

2 mm

nutlet

2 cm

Plagiobothrys leptocladus

1 mm

nutlets

2 cm

Plagiobothrys nothofulvus   P. parishii

1 cm

nutlet

1 mm

nutlets

1 mm

2 cm

Plagiobothrys stipitatus
var. micranthus

nutlets

1 mm

2 cm

P. tenellus

flower

2 mm

1 cm

2 mm

Tiquilia canescens
var. canescens

2 mm

T. nuttallii

2 mm

leaf

T. palmeri

1 cm

2 mm

T. plicata

BRASSICACEAE [CRUCIFERAE] MUSTARD FAMILY

Ann to subshrub. **LVS** gen basal and cauline, alternate, gen simple; stipules 0. **INFL**: gen raceme. **FL** bisexual; sepals 4, free; petals (0)4, free, gen white or yellow, often clawed; stamens gen (2,4)6, gen 4 long, 2 short; ovary 1, superior, chambers gen 2, septum membranous, connecting 2 parietal placentas, style 1, stigma simple or 2-lobed. **FR**: gen capsule ("silique") with 2 deciduous valves, sometimes breaking transversely or indehiscent. **SEEDS** 1– many per chamber in 1 or 2 rows. 300+ genera, 3000+ spp.: worldwide, esp cool regions; some cult for food (esp *Brassica, Raphanus*) and orn. Recently treated to include Capparaceae [Rodman et al 1993 Ann Missouri Bot Gard 80: 686-699; Rollins 1993 Cruciferae of Continental North America. Stanford Univ Press] Family description by Robert A. Price. Key to genera by Roy Buck.

Reed C. Rollins, except as specified

1. Fr gen at least 4 × longer than wide, gen linear
  2. Lf hairs branched (sometimes also simple)
    3. Lf deeply lobed or compound; fr < 3 cm
      4. Fr ± cylindric, valves midribbed ± to tip; pedicel gen thread-like; petals yellow or whitish
      . . . . [2]**DESCURAINIA**
      4' Fr gen flat, valves midribbed only at base; pedicel linear, width nearly = fr width; petals white to
      purplish . . . . [2]**SIBARA**
    3' Lf entire to gen shallowly lobed; fr often > 3 cm
      5. Fr flat
        6. Seeds 1 row per chamber
          7. Fr not long-tapered; taproot often ± slender; basal lvs entire or dentate, gen < 10 cm; seed often
          winged . . . . [3]**ARABIS**
          7' Fr long-tapered to tip; taproot very thick; basal lvs entire, 3–15 cm; seed unwinged . . **PHOENICAULIS**
        6' Seeds 2 rows per chamber
          8. Fr 5–7 mm wide, tapered, sharp-pointed; seed coat silvery . . . . [2]**ANELSONIA**
          8' Fr < 5 mm wide (or tip not tapered and sharp-pointed); seed coat not silvery
            9. Fr gen > 3 cm, gen > 10 × longer than wide; petals not yellow . . . . [3]**ARABIS**
            9' Fr gen < 2 cm, < 10 × longer than wide; petals often yellow . . . . [3]**DRABA**
      5' Fr ± 4-sided or round in ×–section
        10. Petals gen yellow to orange; st solitary or few-branched . . . . **ERYSIMUM**
        10' Petals white to rose-violet; st many-branched
          11. Petals 3–5 mm, white; fr gen < 3 cm . . . . **HALIMOLOBOS**
          11' Petals 6–9 mm, rose-violet; fr 4–6 cm . . . . **MALCOLMIA**
1'. Lf hairs 0 or simple (rarely branched in *Caulanthus*)
  12. Fr stalk above pedicel 1–3 cm; stamens ± equal . . . . **STANLEYA**
  12' Fr stalk above pedicel 0–1 cm; stamens gen 4 long, 2 short or in 3 unequal pairs
    13. Calyx ± urn- or flask-shaped, nearly closed at full fl, often not green; petal claw often ± wider than
    blade, blade gen strap-shaped, gen channeled
      14. Fr gen ± cylindric or slightly flattened parallel to septum; seed gen unwinged . . . . **CAULANTHUS**
      14' Fr gen strongly flattened parallel to septum; seed gen winged
        15. Fr pendent, narrowed to indehiscent beak-like apex; sepals gen < 4 mm . . . . **STREPTANTHELLA**
        15' Fr spreading to erect, dehiscent to tip, beak-like apex 0; sepals gen > 7 mm . . . . **STREPTANTHUS**
    13' Calyx not urn- or flask-shaped, ± open at full fl, gen greenish; petal claw 0 or narrower than blade,
    blade gen spoon-shaped or oblanceolate to obovate (sometimes linear), not channeled
      16. Fls gen subtended by bracts; fr flattened perpendicular to septum . . . . **TROPIDOCARPUM**
      16' Bracts gen 0 (may subtend a few lower fls on main branches); fr ± cylindric or ± flattened parallel
      to septum
        17. Fr not opening by valves
          18. Pl glandular; fr often upcurved, not grooved . . . . **CHORISPORA**
          18' Pl not glandular; fr straight, grooved below . . . . **RAPHANUS**
        17' Fr opening by valves
          19. Fr beak prominent, gen > 5 mm
            20. Seeds 2 rows per chamber; petals gen prominently dark-veined . . . . **ERUCA**
            20' Seeds 1 row per chamber; petals not dark-veined . . . .
              21. Fr erect to spreading, valves with 1 midvein . . . . **BRASSICA**
              21' Fr appressed to st, valve with 3–7 veins, esp when young . . . . **HIRSCHFELDIA**
          19' Fr beak gen 0 or obscure
            22. Fr gen ± flat
              23. Fr clearly stalked above receptacle . . . . [2]**THELYPODIUM**
            23' Fr ± not stalked above receptacle
              24. Fr valves with prominent midrib . . . . [3]**ARABIS**

24' Fr valves without prominent midrib
    25. Fr valves gen opening elastically, sometimes coiling from base . . . . **CARDAMINE**
    25' Fr valves not opening elastically . . . . ²**SIBARA**
  22' Fr ± 4-sided or round in ×-section . . . .
    26. Seeds 2 rows per chamber . . . . *Rorippa nasturtium-aquaticum*
    26' Seeds 1 row per chamber
      27. Petals white to purple (rarely pale yellow in *Guillenia*)
        28. Pl woody at base; fr ± not stalked above receptacle . . . . **CAULOSTRAMINA**
        28' Pl gen not woody at base; fr ± stalked above receptacle, stalk gen > 0.5 mm
          29. Ann; fr stalk < 1 mm, ± as wide as body; fr gen reflexed, not densely clustered . . **GUILLENIA**
          29' Gen bien to per (rarely ann); fr stalk < to > 1 mm, narrower than body; fr not reflexed,
            often densely clustered . . . . ²**THELYPODIUM**
      27' Petals ± yellow
        30. Cauline lvs not clasping st . . . . **SISYMBRIUM**
        30' Cauline lvs clasping st
          31. Lf pinnately lobed to ± compound; pl not glaucous; fr 1–5 cm . . . . **BARBAREA**
          31' Lf ± entire; pl glaucous; fr 8–13 cm . . . . **CONRINGIA**
1' Fr < 4 × longer than wide, linear or not
32. Fr round to somewhat flattened in ×-section, sometimes inflated
  33. Lf gen deeply lobed to compound . . . . ²**DESCURAINIA**
  33' Lf entire to shallowly lobed
    34. Hairs 0 or simple
      35. Lf entire, < 5 mm wide; fr dehiscent . . . . ²**CUSICKIELLA**
      35' Lf gen ± toothed (sometimes obscurely so) or lobed (rarely entire), gen > 5 mm wide; fr dehiscent
        or not
        36. Fr ± indehiscent, gen < 2 × longer than wide; seeds 2–4 . . . . **CARDARIA**
        36' Fr dehiscent, gen > 2 × longer than wide; seeds several–many . . . . **RORIPPA**
    34' Some hairs gen forked, stellate, or multibranched, esp near st base
      37. Herbage hairs mostly simple or multibranched; seed gen 1 per chamber . . . . ²**CUSICKIELLA**
      37' Herbage hairs mostly stellate; seeds gen 2+ per chamber
        38. Fr not notched . . . . ³**LESQUERELLA**
        38' Fr tip (and often base) ± deeply notched . . . . **PHYSARIA**
32' Fr ± strongly flattened, not inflated
  39. Fr 1-seeded, 1-chambered, indehiscent . . . . **THYSANOCARPUS**
  39' Fr gen 2+-seeded, 2-chambered, gen dehiscent
    40. Fr flattened perpendicular to septum
      41. Infl axillary; fr of 2 nutlet-like halves, wrinkled or ridged, spiny . . . . **CORONOPUS**
      41' Infl often terminal; fr not of nutlet-like halves, not wrinkled, ridged, or spiny
        42. Seed 1 per chamber
          43. Fr deeply lobed at tip and base between ± round halves; herbage hairs multibranched, stellate
           . . . . **DITHYREA**
          43' Fr sometimes notched at tip but not deeply lobed; herbage hairs 0 or simple . . . . **LEPIDIUM**
        42' Seeds 2+ per chamber
          44. Upper cauline lvs ± clasping st
            45. Fr obtriangular-obcordate, corners ± acute; basal lvs gen lobed to deeply dissected . . . . **CAPSELLA**
            45' Fr widely oblong to round, corners rounded; basal lvs entire to dentate . . . . **THLASPI**
          44' Upper cauline lvs 0 or not clasping st
            46. Lf deeply lobed to compound, segments linear, often sharp-pointed . . . . **POLYCTENIUM**
            46' Lf entire to deeply lobed, lobes gen wider than linear, not sharp-pointed
              47. Lf hairs 0 or sparse, branched; petals ± 1 mm, white . . . . **HUTCHINSIA**
              47' Lf hairs dense, gen stellate or multibranched; petals > 5 mm, gen not white
                48. Lvs entire, hairs gen stellate (sometimes also simple); petals < 15 mm, yellowish; fr. < 6 mm
                  wide . . . . ³**LESQUERELLA**
                48' Lvs gen pinnately lobed, hairs ± multibranched; petals 15–25 mm, purplish to brownish;
                  fr 10–20 mm wide . . . . **LYROCARPA**
    40' Fr flattened parallel to septum
      49. Fr > 2 × longer than wide
        50. Fr 5–7 mm wide, tip ± tapered, ± sharp-pointed; seed 2.5–3 mm, coat silvery . . . . ²**ANELSONIA**
        50' Fr gen < 5 mm wide, tip not tapered or sharp-pointed; seed gen < 2 mm, coat not silvery . . . . . ³**DRABA**
      49' Fr < 2 × longer than wide (often ± round)
        51. Lvs basal and cauline . . . . ³**LESQUERELLA**
        51' Lvs basal (rarely a few cauline)
          52. Fls in racemes; seed wing 0 . . . . ³**DRABA**
          52' Fls solitary; seed wing ± 1 mm wide . . . . **IDAHOA**

ANELSONIA

1 sp. (A. Nelson, Rocky Mountain botanist, 1859–1952)

A. eurycarpa (A. Gray) J.F. Macbr. & Payson (p. 219) Per; roots deep; caudex loosely branched; hairs branched, dense on lvs, infl axes, sepals. **LVS** basal, dense, overlapped, entire, canescent; base tapered. **INFL** scapose, umbel-like. **FL**: sepals early deciduous, bases not sac-like; petals small, white; stigma 2-lobed. **FR** 2–3 cm, 5–7 mm wide, lanceolate-elliptic, flat parallel to septum, glabrous, gen purplish; base ± obtuse or rounded; tip ± tapered, sharp-pointed; valves leathery; pedicel erect to ascending; style 1–2 mm. **SEEDS** several–many, 2 rows per chamber, 2.5–3 mm, ± flat, oblong; coat silvery; wing 0; hairs dense, minute, club-shaped; embryonic root at edges of both cotyledons. Broken rock, talus, slopes, ridges; > 3000 m. SN, **SNE**; to ID, NV. [*Phoenicaulis e.* (A. Gray) Abrams] Jul–Aug

ARABIS   ROCK CRESS

Bien, per; base woody or not; hairs 0 to dense, simple, forked, stellate, or multibranched; caudex branched or not. **ST** branched or not, cylindric, leafy. **LVS**: basal petioled, entire or dentate; cauline gen sessile, entire or dentate, base often lobed, often clasping st. **INFL**: bracts 0. **FL** erect to reflexed; sepals erect; petals spoon-shaped to oblong and narrowed at base or narrowly obovate, white to deep purple, rarely straw-colored. **FR** erect to reflexed, linear, straight to curved, flat parallel to septum, rarely ± cylindric. **SEEDS** ± many, gen 1 row per chamber, flat or plump, winged or not; embryonic root at edges of both cotyledons. ± 120 spp.: temp N.Am, Eurasia, Afr. (Latin: of Arabia) *A. pendulina* has been collected once from e SNE. Key to species adapted by Roy Buck.

1. Seed 2.5–5 mm wide incl 1–3 mm wing; fr gen > 3 mm wide
   2. Pedicel in fr reflexed or recurved
      3. Fr tip obtuse; seeds 2 rows per chamber; lf, st hairs dense throughout . . . . *A. glaucovalvula*
      3' Fr tip acute; seeds 1 row per chamber; lf, st hairs 0 to dense below, 0 to sparse above
          . . . . **A. suffrutescens** var. **suffrutescens**
   2' Pedicel in fr erect to ± spreading
      4. Lf, lower st whitish hairy; pedicel hairy . . . . ²**A. dispar**
      4' Lf, lower st glabrous or hairy but not whitish; pedicel glabrous
         5. Basal lvs < 3 mm wide; hairs multibranched, dense on basal lvs, 0 to sparse on upper cauline . . . ²**A. pinzlae**
         5' Basal lvs > 3 mm wide; hairs gen 3–5-rayed, 0 to ± dense on basal lvs, 0 on upper cauline . . . **A. platysperma**
            6. Basal lvs glabrous; pl 0.5–2(3) dm; 3000–3600 m . . . . var. **howellii**
            6' Basal lvs hairy; pl 1–4 dm; 1300–3350 m . . . . var. **platysperma**
1' Seed gen < 2 mm wide incl 0–1 mm wing; fr gen < 3 mm wide
   7. Basal lvs obovate to widely oblanceolate (rarely oblong), rosetted, gen thin, tips gen obtuse to rounded; fr erect to ± spreading . . . . **A. hirsuta**
      8. Petals 5–9 mm, white to pinkish; cauline lvs gen spaced, lower ± glabrous; fr ± ascending to ± spreading; outer sepals strongly sac-like at base . . . . var. **glabrata**
      8' Petals < 5 mm, white to cream-white; cauline lvs gen overlapped, lower ± hairy; fr erect; outer sepals weakly sac-like at base . . . . var. **pycnocarpa**
   7' Basal lvs linear to oblanceolate, if wider then hairy or fr reflexed or both, sometimes clustered but not rosetted, ± thick, tips gen acute; outer sepals gen not sac-like at base
      9. Sts, pedicels, lvs gray to grayish white with hairs dense, minute, multibranched
         10. Pedicel erect to spreading
            11. Seeds 2 rows per chamber, ± unwinged, plump, ± 1 mm wide; cauline lvs overlapped . . . . . . *A. shockleyi*
            11' Seeds 1 row per chamber, winged, flat, 1–2.5 mm wide; cauline lvs gen spaced
               12. Seed wing > 0.5 mm wide; fr 2.5–3.5 mm wide, spreading-ascending . . . . ²**A. dispar**
               12' Seed wing < 0.5 mm wide; fr ± 2 mm wide, gen spreading . . . . ³**A. inyoensis**
         10' Pedicel recurved to pendent
            13. Seeds 1 row per chamber; cauline lvs oblong to lanceolate, ± crowded, often ± toothed or ± lobed
               . . . . **A. puberula**
            13' Seeds 2 rows per chamber; cauline lvs gen linear, not crowded, entire . . . . ²**A. pulchra**
               14. Fr reflexed-appressed, hairs dense, pedicel reflexed . . . . var. **pulchra**
               14' Fr pendent to spreading, hairs dense or ± 0, pedicel spreading-recurved
                  15. Fr, upper st, pedicel ± glabrous . . . . var. **gracilis**
                  15' Fr, upper st, pedicel densely hairy . . . . var. **munciensis**
      9' Sts, pedicels, at least cauline lvs greenish with hairs 0 to dense, fine to coarse but gen larger than minute, simple to multibranched
         16. Pedicel in fr spreading to reflexed-appressed
            17. Seeds 2 rows per chamber; cauline lvs linear; petals 8–20 mm, gen showy, limbs spreading; fr hairs dense . . . . ²**A. pulchra** (see 14. for vars)
            17' Seeds 1 row per chamber; cauline lvs gen lanceolate or ovate to oblong, rarely linear or narrowly lanceolate; petals < 12 mm, limbs ± erect; fr hairs 0 or very sparse
               18. Basal lvs gen linear; petals white, ± 4 mm; pedicel widely arched with ± pendent fr . . . . *A. cobrensis*

18' Basal lvs widely spoon-shaped to oblanceolate (linear to narrowly oblanceolate in
   A. microphylla var. *microphylla*); petals gen purplish to pinkish, 4 mm; pedicel gen not widely
   arched with pendent fr (exc in *A. holboellii* var. *pinetorum*)
   19. Pedicel 2–4(6) mm; cauline lvs oblong to ± ovate, hairy or not . . . . ²*A. lemmonii* (see 35. for vars)
   19' Pedicel (4)6–20 mm; cauline lvs linear-lanceolate to oblong, gen hairy
      20. Pedicel in fr recurved to reflexed, ± straight; fr gen ± straight, reflexed to pendent
         (exc var. *pinetorum*) . . . . *A. holboellii*
         21. Cauline lf base tapered, not clasping st; pl 1–2 dm; basal lvs < 3 mm wide . . . . . . var. *pendulocarpa*
         21' Cauline lf base lobed, often ± clasping st; pl gen 2–9 dm; basal lvs > 3 mm wide
            22. Pedicel in fr arched to gently recurved; fr ± curved . . . . var. *pinetorum*
            22' Pedicel in fr reflexed, gen straight; fr gen straight . . . . var. *retrofracta*
      20' Pedicel in fr spreading, straight or spreading-recurved; fr straight, spreading to widely recurved
         23. Sts 1–2(3) dm, many, slender; cauline lvs gen few, spaced . . . . ³*A. microphylla* var. *microphylla*
         23' Sts 2–9 dm, 1–several, ± stout; cauline lvs gen many, overlapped below (exc sometimes *A. perennans*)
            24. Basal lvs entire, grayish, hairs fine, dense; st hairs dense at least below, appressed . . . ³*A. inyoensis*
            24' Basal lvs, or at least outer, dentate, rarely entire, greenish, hairs coarse, sparse to
               dense; st hairs spreading, rarely appressed
               25. Outer basal lvs oblanceolate, tip ± obtuse; pedicel slender, 10–20 mm, glabrous in fr;
                  petals 6–9 mm, 1.5–2.5 mm wide . . . . *A. perennans*
               25' Outer basal lvs linear-oblanceolate, tip acute; pedicel stout, 5–15 mm, glabrous to hairy
                  in fr; petals 9–12 mm, 2–4 mm wide . . . . ²*A. sparsiflora* (see 32' for vars)
16' Pedicel in fr erect to spreading
   26. Basal, lower cauline lvs greenish, hairs gen 0 or sparse, simple to multibranched; sts gen 1–several
      27. Fr ± 1mm wide, straight to downcurved; seeds round, ± 1 mm wide; st hairy below
         . . . . ³*A. microphylla* var. *microphylla*
      27' Fr 1.5–3.5 mm wide, gen straight; seeds round to oblong, 1.2 mm wide; st hairy below or not
         28. Seeds 2 rows per chamber, gen oblong, wing gen on 1 side and 1 end, narrow or 0 elsewhere;
            fr erect, tip obtuse, rarely subacute; basal lf narrowly boat-shaped, centrally attached, rarely 0;
            petals white, rarely pinkish . . . . *A. drummondii*
         28' Seeds 1 row per chamber, oblong to round, wing on 1 end or gen all around; fr spreading to erect,
            tip gen acute; basal lf hairs 0 or multibranched, not boat-shaped; petals pink to rose or purplish,
            rarely white
            29. Sts 3–9 dm, 1–several from simple or branched caudex ; fr ascending to spreading, rarely ±
               reflexed . . . . *A. ×divaricarpa*
            29' Sts < 3 dm, several–many from gen branched caudex; fr erect to ± spreading . . . *A. lyallii* var. *nubigena*
   26' Basal, lower cauline lvs grayish or greenish, hairs dense, stellate or mulitbranched; sts gen many
      30. Petals 7–14 mm; style gen < 1 mm, sometimes 0
         31. Basal lvs linear-oblanceolate, minutely hairy, grayish, tip obtuse; fr ± straight; caudex branches
            not elongate . . . . ³*A. inyoensis*
         31' Basal lvs linear to linear-oblanceolate, ± coarsely hairy, greenish, tip acute; fr gen curved; caudex
            branches elongate . . . . ²*A. sparsiflora*
            32. Basal lvs entire; pedicel ascending in fr, hairs gen ± 0; st gen branched above . . . . var. *sparsiflora*
            32' Basal lvs gen ± dentate; pedicel spreading in fr, hairs gen spreading; st rarely branched above
               . . . . var. *subvillosa*
      30' Petals < 7 mm; style gen ± 0, rarely < 1 mm
         33. Basal lvs, lower sts greenish, hairs not matted; cauline lvs linear to narrowly lanceolate, ±
            spaced; pedicel ± glabrous . . . . ³*A. microphylla* var. *microphylla*
         33' Basal lvs, lower sts grayish, hairs ± matted; cauline lvs oblong to narrowly ovate, crowded at
            base (exc in *A. fernaldiana* var. *stylosa*); pedicel gen hairy
            34. Fr tip obtuse; style ± 0; cauline lvs with basal lobes . . . . ²*A. lemmonii*
               35. Fr ascending to ± spreading; basal lvs narrowly oblanceolate to lanceolate; infl in fr not
                  1-sided . . . . var. *depauperata*
               35' Fr spreading to ± reflexed; basal lvs obovate to widely oblanceolate; infl gen 1-sided . . . var. *lemmonii*
            34' Fr tip acuminate; style < 1 mm; cauline lvs with 0 or inconspicuous basal lobes
               36. Frs ± curved to ± straight, many per st, spreading to ± ascending . . . . *A. bodiensis*
               36' Fr ± straight, few per st, ± erect
                  37. Pl 10–30 cm; basal lvs 10–20 mm, spoon-shaped to linear-oblanceolate; pedicels 5–10 mm;
                     fr 4–6 cm . . . . *A. fernaldiana* var. *stylosa*
                  37' Pl 3–8 cm; basal lvs 610 mm, linear-lanceolate or narrower; pedicels 3–5 mm; fr 2–4 cm
                     . . . . ²*A. pinzlae*

A. bodiensis Rollins BODIE HILLS ROCK CRESS Per; caudex
branched; hairs minute, multibranched, some large, simple. **STS**
several–many, erect, 1.5–3.5 dm; hairs dense below, 0 above.
LVS acute at tip; basal petioled, 1–3 cm, linear to linear-
oblanceolate, grayish with dense hairs, entire; cauline sessile,
1–2.5 cm, oblong, base lobes 0 or inconspicuous. **FR** spreading
to ± ascending, 3–5 cm, ± curved to ± straight; pedicel 3–6 cm,
ascending to ± spreading. **SEED** widely oblong to ± round; wing
narrow. RARE. Rock crevices, open slopes; 2500–3100 m. **SNE**
(**n Mono Co.**); w NV.

A. cobrensis M.E. Jones (p. 219) MASONIC ROCK CRESS Per; caudex branched; hairs multibranched, minute. **STS** several–many, simple or gen branched above, 2–5 dm, slender. **LVS:** basal many, 2–5 cm, gen linear, entire, gray, hairs dense, fine, tip acute; cauline few, sessile, base ± lobed, ± clasping st. **FL:** petals ± 4 mm, white. **FR** ± pendent, 2–4 cm, ± straight; tip obtuse; hairs 0 or very sparse; pedicel hairs sparse; style very short or 0. **SEED** ± oblong to ± round; wing ± wide. $2n=14$. RARE in CA. Sandy soils, sagebrush scrub; 1375–2800 m. **n SNE (Mono Co.), n DMtns (Panamint Mtns);** to WY. Jun–Jul

A. dispar M.E. Jones Per; caudex branched; hairs minute, multibranched. **STS** several, simple or branched above base, 1–2.5 dm; hairs below, dense. **LVS** whitish hairy; basal many, erect, 1.5–2.5 cm, slender-petioled, linear-oblanceolate to oblanceolate, entire; cauline 1–2 cm, sessile, widely linear. **FL:** petals > sepals, obovate, purplish. **FR** ascending, 5–7 cm, 2.5–3.5 mm wide, glabrous; tip acute; pedicel ± erect to ascending, 1–2 cm, hairy. **SEED** ± 2.5 mm, round; wing wide. Loose gravelly slopes, compact talus; 1200–2400 m. s SNH, n SnBr, **SNE, DMtns;** sw NV. Apr–May

A. ×divaricarpa Nelson (p. 219) Bien, rarely per; caudex branched or not; hairs branched, appressed, gen 3–5-rayed. **STS** 1–3, simple or branched above, 3–9 dm; hairs 0 or below. **LVS:** basal 2–6 cm, widely oblanceolate to narrowly spoon-shaped, dentate to subentire, hairs sparse, tip ± acute; cauline sessile, narrowly oblong to lanceolate, often sagittate, clasping st, entire or lower ± dentate, hairs 0 or sparse. **FL:** petals spoon-shaped, pink to purplish. **FR** ascending to spreading, rarely ± reflexed, 2–8 cm, straight; pedicel ascending to ± reflexed, 6–12 mm; style very short or 0. **SEED** ± round; wing narrow. $2n=14,13+2,21,20+2,28$. Gravelly, calcareous soils; 2100–3300 m. CaR, SN, **GB;** to Rocky Mtns, e Can. [var. *interposita* (E. Greene) Rollins] (see Rollins 1983 Amer J Bot 70:625–634 for discussion of hybridity) Jul–Aug

A. drummondii A. Gray (p. 219) Bien, per; caudex gen simple; hairs sessile, narrowly boat-shaped, centrally attached. **STS** 1–few, simple or branched above, 3–9 dm; hairs 0 or below, sparse. **LVS:** basal petioled, 2–8 cm, narrowly oblanceolate, entire to sparsely dentate, hairs rarely 0, tip gen acute; cauline sessile, clasping st, 2–7 cm, oblong to oblong-lanceolate, glabrous, tip acute. **FL:** petals white, rarely pinkish. **FR** erect, 4–10 cm, gen crowded, straight, glabrous; tip gen obtuse, rarely subacute; pedicel erect, 1–2 cm, glabrous; style short or 0. **SEEDS:** 2 rows per chamber, gen oblong; wing gen on 1 end and 1 side, narrow or 0 elsewhere. $2n=14,20,28$. Calcareous gravels, talus, open disturbed areas; 1800–3200 m. SN, **GB;** to AK, e N.Am. [*A. confinis* S. Watson] Jun–Jul

A. fernaldiana Rollins var. **stylosa** (S. Watson) Rollins STYLOSE ROCK CRESS Per; caudex branched; hairs gen minute, many-branched. **STS** several–many, gen simple, 1–3 dm. **LVS:** basal many, often in sterile clusters, 1–2 cm, spoon-shaped to linear-oblanceolate, entire, grayish, hairs dense, petioles short, ± ciliate; cauline sessile, lobes at base 0 or inconspicuous, hairs dense on lower, gen 0 on upper. **FL:** petals 5–7 mm, pink to lavender. **FR** ± erect, 4–6 cm, ± straight, glabrous, sessile above receptacle; pedicel spreading-ascending, 5–10 mm, hairy; style < 1 mm. **SEED** oblong; wing narrow or 0. RARE. Ridges, rocky, limestone soils, sagebrush scrub; 2300–2800 m. **GB;** NV. var. *f.* in NV, UT, Colorado. Jun–Jul

A. glaucovalvula M.E. Jones Per; caudex branched; hairs minute, multibranched. **STS** 1–several, simple or branched above, 1.5–4 dm, densely canescent. **LVS:** basal 2–5 cm, linear to ± wider, entire, tip obtuse to ± acute, hairs dense; cauline 1–4 cm, sessile, lanceolate to linear-lanceolate, tapered below, gray. **FL:** petals white to pinkish. **FR** reflexed, 2–4.5 cm, 5–8 mm wide, oblong, glabrous, glaucous; base, tip obtuse; pedicel stout, recurved; hairs dense; style < 1 mm. **SEEDS:** 2 rows per chamber, round; wing very wide. $2n=14$. Rocky, limestone soils, open sites, summits, hill slopes, desert shrubland; 1000–1300 m. **e SNE, DMoj;** NV. ❀TRY. Mar–May

A. hirsuta (L.) Scop. Bien, weak per; caudex branched or not; hairs simple or forked, coarse, spreading. **STS** 1–several, simple or branched above, erect, 2–7 dm. **LVS:** basal short-petioled, 2–8 cm, oblong to obovate, entire to dentate, tip obtuse; cauline sessile, 1–5(7) cm, lanceolate to obovate, sagittate, clasping st. **FR** 3–6 cm, flat, glabrous; pedicel 0.5–1.5 cm; hairs 0, rarely sparse; style 0.5–1 mm. **SEED** ± round to ± rectangular; wing prominent or narrow on 1 end to 0. Gravelly soils, swales, disturbed sites; 1000–2600 m. KR, NCoRO, CaR, SN, SnBr, **W&I;** to AK, e Can, Colorado. Var. *eschscholtziana* (Andrz.) Rollins OR to AK, e Asia; var. *hirsuta* native of Eur, Asia, not in N.Am.

var. **glabrata** Torrey & A. Gray (p. 219) **ST:** hairs sparse below, 0 above. **LVS:** basal with hairs sparse to 0; cauline gen well spaced, lower ± glabrous. **FL:** outer sepals strongly sac-like at base; petals 5–9 mm, white to pinkish. **FR** ± ascending to ± spreading. Gravelly soils, swales, open woods; 1200–2600 m. KR, NCoRO, CaR, SN, SnBr, **W&I;** to B.C., Colorado. May–Jul

var. **pycnocarpa** (M. Hopk.) Rollins **ST:** hairs ± throughout. **LVS:** basal hairy; cauline gen overlapped, lower ± hairy. **FL:** outer sepals weakly sac-like at base; petals < 5(–8) mm, white to cream-white. **FR** erect. $2n=16,32$. Meadows, gravel bars, shady slopes, moist, disturbed sites; 1000–1900 m. CaR, SN, SnBr; **W&I;** to AK, e Can. [*A. p.* M. Hopk.]

A. holboellii Hornem. Bien, per; caudex branched or not; hairs gen multibranched. **STS** 1–several, erect. **LVS:** basal 1–5 cm, linear-oblanceolate to widely spoon-shaped, tip acute, hairs dense; cauline sessile, 2–4 cm, oblong to lanceolate, entire, lower surface densely hairy, upper surface ± hairy to not. **FL:** petals purplish pink to whitish. **FR** 3–7 cm, gen ± straight, glabrous; pedicel arched or recurved to ± reflexed, (4)6–16 mm, straight to ± curved, slender, hairy or not. **SEED** round; wing narrow all around. Rocky slopes, open sites; 500–3500 m. NW, CaR, SN, SnFrB, **GB;** to AK, e Can, Colorado. 5 vars. in N.Am, difficult to distinguish, esp in fl.

var. **pendulocarpa** (Nelson) Rollins **ST** gen simple, 1–2 dm, slender; hairs present below, ± 0 above. **LVS:** basal < 3 mm wide, entire; cauline base tapered. **FR** pendent. $2n=14$. Rocky slopes; 1800–3000 m. c SNH, **GB;** to B.C., Colorado. May–Jul

var. **pinetorum** (Tidestrom) Rollins **ST** gen branched above, 3–9 dm, hairy below. **LVS** minutely hairy; basal oblanceolate. **FR** 4–7 cm, ± curved, glabrous; pedicel arched to gently recurved. $2n= 21$. Rocky slopes; 2000–3500 m. CaR, SN, **SNE (Sweetwater Mtns);** to B.C., NE. May–Jul

var. **retrofracta** (Graham) Rydb. (p. 219) **ST** branched above, 2–9 dm; hairs throughout or ± 0 above. **LVS** minutely hairy; basal > 3 mm wide, entire to ± dentate; cauline base lobed, often ± clasping st. **FR** gen straight; pedicel reflexed, gen straight. $2n=14,21+1$. Common. Rocky slopes, open areas; 500–2400 m. NW, CaR, SN, SnFrB, **GB;** to AK, e Can. May–Jul

A. inyoensis Rollins Per; caudex branched; hairs minute, multibranched. **STS** several, erect, 2–5 dm, rigid; hairs dense, esp below. **LVS:** basal many, 2–3 cm, linear-oblanceolate to spoon-shaped, entire, grayish, hairs dense, tip acute or obtuse; cauline sessile, 1–2.5 cm, gray, base lobed, clasping st, or tapered. **FL:** petals spoon- to tongue-shaped, pink to purplish. **FR** gen spreading, ± 2 mm wide, ± straight, glabrous; pedicel gen spreading, 6–12 mm. **SEED** round; wing < 0.5 mm wide. $2n=21,23$. Rocky ridges, slopes; 1500–3500 m. SNH, **SNE;** NV. ❀TRY. May–Jul

A. lemmonii S. Watson Per; caudex branched; hairs multibranched, minute. **STS** several–many, simple, gen prostrate or decumbent, 6–20 cm, slender, hairy or gen glabrous above. **LVS:** basal 1–2 cm, entire or few-toothed, hairs dense, tip gen acute; cauline sessile, 4–10 mm, oblong-lanceolate to ± ovate, hairy or not, base lobed, gen clasping st. **FL:** petals pink to purplish. **FR** 2–4 cm, glabrous; pedicel 2–4(6) mm, glabrous or not; style ± 0.

SEED round; wing narrow. Rocky to gravelly soils; 2400–4300 m. CaR, SNH, **SNE**; to Yukon, Colorado. 4 vars. total.

var. *depauperata* (Nelson & Kenn.) Rollins **LVS**: base narrowly oblanceolate to lanceolate. **INFL** in fr not 1-sided. **FR** ascending to ± spreading. Talus, rocky slopes, gravelly soils; 2400–4300 m. SNH, **SNE**; NV. [*A. d.* Nelson & Kenn.] Jul–Aug

var. *lemmonii* **LVS**: base obovate to widely oblanceolate. **INFL** in fr gen 1-sided. **FR** spreading to reflexed. 2*n*=14. Talus, rock fields, ridges, outwash gravels; 2400–3700 m. CaR, SNH, **W&I**; to Yukon, Colorado.

A. lyallii S. Watson Per; caudex branched; hairs multibranched, ± minute. **STS** few–many, < 15 cm, glabrous. **LVS**: basal narrowly petioled, 1–3 cm, entire; cauline few, well spaced, sessile, 1–2 cm, lanceolate to oblong, base gen tapered, tip acute. **FL**: petals spoon-shaped, rose to purplish. **FR** erect to ± spreading, 3–5 cm, straight, glabrous; tip tapered; style short or 0. **SEED** round, winged. Rock crevices, slopes, ridges; 2400–3800 m. CaRH, SNH, Wrn, **W&I**; to B.C., WY. 2 vars. in CA.

var. *nubigena* (J.F. Macbr. & Payson) Rollins **ST** ± prostrate to ascending, slender. **LVS**: basal 1–2.4 mm wide, linear to linear-oblanceolate, tip acute to acuminate. Rock slides, open ridges; 2900–3800 m. SNH, **W&I**; to ID, WY. Jul–Aug

A. microphylla Nutt. var. *microphylla* SMALL-LEAVED ROCK CRESS Per; caudex branched, underground; hairs gen multibranched, some simple. **STS** several–many, branched or gen not, 1–2(3) dm in CA, slender; hairs spreading below, 0 above. **LVS**: basal 5–29 mm, linear to narrowly oblanceolate, entire, greenish, tip acute, hairs minute; cauline few, sessile, 1–2 cm, narrowly lanceolate, base lobed, ± clasping st. **FL**: petals spoon-shaped, pale rose to purplish. **FR** ± ascending to spreading, 2–6 cm, ± 1 mm wide, straight to downcurved, glabrous; pedicel spreading to ascending, 5–15 mm, slender, ± glabrous; style < 1 mm or ± 0. **SEED** ± 1 mm, round; wing narrow. 2*n*=14. UNCOMMON. Rock crevices, basaltic or granitic outcrops; 1700–2700 m. n SNH, **GB**; to WA, MT, WY.

A. perennans S. Watson (p. 219) Per; caudex branched or not, gen above ground; hairs coarse, multibranched. **STS** several, simple or branched above, 2–6 dm, hairy below, glabrous above. **LVS**: basal many, petioled, 2–6 cm, oblanceolate to widely oblanceolate, dentate, rarely entire, hairs dense; cauline sessile, 1–3 cm, lanceolate, base lobed, ± clasping st. **FL**: petals 6–9 mm, spoon-shaped, purple to pinkish. **FR** spreading or recurved, 4–6 cm; pedicel spreading-recurved, 1–2 cm, slender, gen glabrous. **SEED** round; wing all around. 2*n*=14. Calcareous rocks, canyon walls, gravelly slopes, pinyon/juniper woodland; 300–2200 m. SnGb, SnBr, e PR, **SNE**, **DMtns**; to TX. ✿TRY. Apr–Jul

A. pinzlae Rollins PINZL'S ROCK CRESS Per; caudex branched or not; hairs multibranched, minute. **ST** simple, 3–8 cm, slender; hairs below, sparse to 0 above. **LVS**: basal tufted, erect, 6–10 mm, linear-lanceolate, densely gray-hairy; cauline 3–5, sessile, 4–6 mm, narrowly oblong, hairy. **FL**: petals erect, ± 5 mm, spoon-shaped, purple at and whitish below tip. **FR** erect, 2–4 cm, 2–3 mm wide, linear, glabrous; margins uneven; tip acuminate; pedicel erect to spreading-ascending. **SEED** ± round; wing wide all around. RARE. Rocky slopes 3000–3350 m. **W&I (White Mtns)**; NV.

A. platysperma A. Gray Per; caudex branched; hairs gen 3–5-rayed. **STS** several–many, simple or often branched above, ± decumbent to erect, hairy to glabrous. **LVS**: basal many, 2–6 cm, oblanceolate, entire, tip acute to obtuse; cauline few, well spaced, sessile, 1–1.5 cm, oblong to linear-lanceolate, upper glabrous. **FL**: petals spoon-shaped, pink, rarely whitish. **FR** erect to ascending, 3–7 cm, 3–5 mm wide, straight, flat; tip acuminate; pedicel ascending to ± spreading, 5–15 mm, glabrous; style < 1 mm to 0. **SEED** round; wing wide. Rocky slopes, ridges; 1300–3600 m. NCoRO, CaR, SNH, **GB**; OR, NV.

var. *howellii* (S. Watson) Jepson **ST** simple, 0.5–2(3) dm. **LVS**: basal glabrous; cauline bases clasping st or not. Rocky slopes, ridges; 3000–3600 m. NCoRO, CaR, SNH, **GB**; OR, NV. Jul–Aug

var. *platysperma* (p. 219) **ST** often branched above, 1–4 dm. **LVS**: basal hairy; cauline base tapered, not clasping st. Slopes, granitic outcrops, limestone ridges; 1300–3350 m. s SNH, **SNE**; OR, NV. ✿TRY.

A. puberula Torrey & A. Gray Bien, per; caudex simple; hairs dense, multibranched, fine. **STS** 1 or few, simple or branched above, 1.5–5 dm, often stout, gray. **LVS**: basal 1–3 cm, oblanceolate to linear-oblanceolate, gray, entire or few-toothed, tip acute; cauline many, ± crowded, sessile, 1–3 cm, oblong to lanceolate, often ± toothed or ± lobed, base ± lobed, ± clasping st. **FL**: petals 7–10 mm, spoon-shaped to narrower, rose to purplish or ± white. **FR** pendent to reflexed, 3–6 cm; tip gen obtuse; hairs ± dense; pedicel recurved or reflexed, 4–8 mm, hairs dense; style 0. **SEED** round, plump; wing narrow. Rocky sites, sagebrush scrub, juniper woodland; 1200–3200 m. **GB**; to WA, ID, UT. Jun–Jul

A. pulchra M.E. Jones Per; base subshrubby; caudex branched or not, elevated; hairs multibranched, often appressed, minute. **STS** 1–several, branched or not, 2–6 dm; hairs dense or 0 above. **LVS**: hairs dense; basal 4–8 cm, linear, entire; cauline sessile, 2–6 cm, linear, entire, base lobed or not, clasping st or not. **FL**: petals 8–20 mm, widely spoon-shaped, purple to reddish purple. **FR** 4–7 cm; pedicel 8–20 mm, hairs dense; style very short or 0. **SEEDS**: 2 rows per chamber, ± round; wing prominent. Canyons, slopes, washes; 600–2200 m. SN, **GB**, **D**; to UT, AZ, Mex. 5 vars. total.

var. *gracilis* M.E. Jones **ST**: upper ± glabrous. **FR** pendent to ± spreading, ± glabrous; pedicel spreading-recurved, ± glabrous. 2*n*=14. Limestone soils; 850–1850 m. SN, **W&I**, **D**; AZ, n Mex. [vars. *glabrescens* Wiggins and *viridis* Jepson]. ✿TRY. Apr–May

var. *munciensis* M.E. Jones DARWIN ROCK CRESS **ST**: upper densely hairy. **FR** pendent to spreading; hairs dense; pedicel spreading-recurved, hairs dense. 2*n*=21. RARE in CA. Canyon slopes, ledges, rock outcrops, desert shrublands; 1100–2000 m. **GB**, **DMoj**; to UT. Apr

var. *pulchra* (p. 219) **ST**: upper densely hairy. **FR** reflexed-appressed; hairs dense; pedicel reflexed; hairs dense. 2*n*=14. Canyons, slopes, rocky washes; 600–2200 m. **GB**, **DMoj**; NV, Mex. ✿TRY.

A. shockleyi Munz SHOCKLEY'S ROCK CRESS Per; caudex simple; hairs multibranched, fine, dense. **STS** 1–few, simple or branched above, 1.5–4 dm, stout, grayish white. **LVS** grayish white; basal crowded, 1–2 cm, spoon-shaped to obovate, tip acute; cauline many, overlapped, 1–2 cm, widely lanceolate, base ± lobed or not, clasping st or not. **FL**: petals 8–11 mm, ± linear, pink. **FRS** crowded, ascending, 5–8 cm, straight to outcurved; hairs sparse to 0; pedicel ascending, 8–12 mm, straight, hairs dense; style ± 0. **SEEDS**: 2 rows per chamber, oblong, plump; wing ± 0. RARE in CA. Limestone and quartzite ridges, gravel; 1000–2000 m. SnBr, **DMoj**; UT. May–Jun

A. sparsiflora Torrey & A. Gray Per; caudex branched or not; hairs multibranched to simple, spreading or appressed, coarse. **STS** 1–several, simple or branched above, 3–9 dm, gen stout; hairs below, sometimes above. **LVS**: basal many, 3–10 cm, linear-oblanceolate to wider, entire or ± dentate, hairy, tip acute; cauline many, sessile, 2–8 cm, linear-lanceolate or wider, base lobed, gen sagittate, clasping st. **FL**: petals 9–12 mm, 2–4 mm wide, spoon-shaped, pink to purple. **FR** ascending to recurved, 6–12 cm, straight or curved, glabrous; pedicel ascending to spreading-recurved, 5–15 mm, often stout, glabrous to hairy. **SEED** round; wing narrow. Rocky slopes, valleys; < 2800 m. CaR, SNF, SW, **GB**; to ID, UT. 4 vars. in CA.

var. ***sparsiflora*** (p. 219) **ST** gen branched above; hairs spreading, simple below, ± 0 to very sparse above. **LVS**: basal linear-oblanceolate, entire. **FR**: pedicel ascending; hairs ± 0 (to sparse, spreading). Steep, gravelly slopes, grassy sagebrush scrub, basaltic talus; 1300–2800 m. **GB**; to ID, UT. May–Jul

var. ***subvillosa*** (S. Watson) Rollins **ST** rarely branched above; hairs spreading, simple or forked, ± 0 above. **LVS**: basal oblanceolate to wider, gen ± dentate. **FR**: pedicel spreading; hairs gen spreading. 2n=21,21+1,22. Canyon slopes, talus, among granitic boulders, sagebrush; 500–1100 m. CaR, n SNF, **n SNE**; to B.C., MT, WY. May–Jun

A. suffrutescens S. Watson Per; base woody; caudex widely branched. **STS** several–many, simple, rarely branched above;

hairs ± 0 or below, multibranched or stellate. **LVS**: hairs 0 to dense; basal 1–4 cm, linear to ± spoon-shaped; cauline few, sessile, 1–3 cm, lanceolate, base lobed, clasping st. **FL**: petals 5–7 mm, spoon-shaped, rose to purplish. **FR** 3–6 mm wide, glabrous; tip acute; pedicel 4–10 mm, slender, glabrous; style < 1 mm or 0. **SEED** round; wing all around, wide. Gravelly to rocky slopes, basalt, pumice, serpentine; ± 1500 or 1700–2800 m. e KR, n NCoRH, CaR, SN, **n SNE (Sweetwater Mtns)**; to OR, ID. 2 vars. in CA.

var. ***suffrutescens*** **ST** 2–5 dm; hairs sparse to ± 0. **LVS**: basal hairs sparse to ± 0. **FR** pendent to reflexed, 4–7 cm. Dry, rocky slopes, basaltic, serpentine outcrops; 1700–2800 m. e KR, CaR, SN, **n SNE (Sweetwater Mtns)**; to OR, ID. [var. *perystylosa* Rollins] Jun–Jul

BARBAREA

Bien, per, erect; hairs simple, sparse, stiff, or 0. **LVS**: basal, lower cauline pinnately lobed to ± compound, terminal lobe > lateral; cauline smaller upward, upper sessile, base ± lobed, clasping st. **INFL** terminal; bracts 0. **FL**: petals yellow. **FR** linear, ± cylindric to ± 4-sided; valves strongly 1-veined base to tip; style 0.5–3 mm, beak-like. **SEEDS** many, 1 row per chamber; wing 0; embryonic root at edges of both cotyledons. ± 20 spp.: e Eur, sw Asia. (Saint Barbara) Key to species adapted by Roy Buck.

1. Style in fr stout, < 1mm; upper cauline lvs pinnately lobed . . . . ***B. orthoceras***
1' Style in fr slender, 1.5–3 mm; upper cauline lvs entire to coarsely dentate . . . . ***B. vulgaris***

B. orthoceras Ledeb. (p. 219) Bien, per; caudex simple. **ST** 1–6 dm, branched, stiff, angled. **LVS** pinnately lobed; basal < 12 cm, with 2–3 pairs of lateral lobes, terminal lobe ovate, entire or irregularly toothed; cauline less lobed upward, often clasping st. **FL**: petals bright yellow. **FR** erect to spreading-ascending, (1.5)2.5–5 cm, straight; pedicel 2–3 mm, < 1 mm thick; style 0.5–1 mm. **SEED** ± 1.5 mm. 2n=16. Damp meadows, wet rocks, streambanks, moist woods; 700–3350 m. CA-FP (exc GV), **n SNE, W&I**; to AK, e N.Am, Mex, also in temp Asia. [var. *dolichocarpa* Fern.] May–Sep

B. vulgaris R.Br. (p. 219) COMMON WINTER CRESS, YELLOW ROCKET Bien. **ST** 2–8 dm, branched above, coarse. **LVS**: basal with 1–4 pairs of lateral lobes; upper cauline entire to coarsely dentate, glabrous. **FL**: petals 6–8 mm, bright yellow. **FR** erect to ascending, 1–3 cm, straight; pedicel 3–5 mm, < 1 mm thick; style gen 2–3 mm. **SEED** 1–1.5 mm, oblong. 2n=16. Uncommon. Disturbed sites; < 1000 m. **GB**, expected elsewhere; native to Eurasia. Apr–May

BRASSICA MUSTARD, TURNIP

Ann, per; hairs simple. **ST** erect, branched, glabrous above. **LVS**: basal and lower cauline petioled, dentate to pinnately lobed, lateral lobes < terminal. **INFL** terminal; bracts ± 0. **FL**: sepals erect; petals gen yellow. **FR** linear; valves 1-veined; beak conic or cylindric, with seeds 0 or rarely 1–2. **SEEDS** many, 1 row per chamber, spheric, finely to coarsely netted. ± 35 spp.: Medit, Eurasia, some naturalized ± worldwide. (Latin: cabbage) Naturalizing CVS soon lose desirable food properties. Key to species adapted by Roy Buck.

1. Upper cauline lvs sessile, base lobed, partly or fully clasping st . . . . ***B. rapa***
1' Upper cauline lvs short-petioled or sessile, base tapered . . . . ***B. tournefortii***

B. rapa L. (p. 219) TURNIP, FIELD MUSTARD Ann, erect; hairs 0 or very sparse, not stiff. **ST** simple to freely branched, 2–10 dm. **LVS**: lower cauline ± pinnately lobed, lateral lobes 2–4, terminal lobe obovate, wavy-dentate; middle, upper lvs sessile, base lobed, ± clasping st. **FL**: petals 6–11 mm, yellow. **FR** ascending to ± spreading, 3–7 cm; pedicel ± ascending, 7–25 mm; beak (8)10–15 mm, narrowed to a slender style. **SEED** ± 1.5 mm wide, very finely netted. 2n=20. Grainfields, orchards, disturbed areas; < 1500 m. CA-FP, **SNE**; widespread US, native to Eur. [*B.*

campestris L.] Mostly Jan–May

B. tournefortii Gouan (p. 219) Ann, branched ± from base, widely above. **ST** < 7 dm; hairs on lower st ± dense, stiff, white. **LVS**: basal rosetted, persistent, petioled, pinnately lobed, serrate-dentate; cauline few, base tapered, uppermost bract-like. **FR** 3–7 cm, cylindric, narrowed between seeds; pedicels spreading, lower 8–15 mm; beak 1–1.5 cm, stout. **SEED** ± 1 mm wide, finely netted. 2n=20. Roadsides, washes, open areas; < 800 m. SW, **D**; to s NV, TX; native to Medit. Locally abundant. Jan–Jun

CAPSELLA SHEPHERD'S PURSE

Ann, bien; hairs 0, simple, or branched. **LVS**: basal clustered or rosetted, entire to dissected or lobed; cauline sessile, sagittate, ± clasping st. **FL**: sepals green or reddish, bases not sac-like; petals ± 2 mm, obovate to spoon-shaped, white or pinkish. **FR** obtriangular-obcordate, flattened perpendicular to septum; valves keeled. **SEEDS** many; wing 0; embryonic root at back of 1 cotyledon. ± 4 spp.: Eur. (Latin: little box)

C. bursa-pastoris (L.) Medikus (p. 219) Ann; hairs simple and stellate. **ST** erect, 1–5 dm, branched or not. **LVS**: basal rosetted, petioled, 3–6 cm, oblanceolate, subentire to pinnately lobed or dis-sected. **INFL** many-fld. **FL**: petals white, distinctly clawed. **FR** 4–8 mm; pedicel spreading to ascending. $2n$=32. Disturbed sites, gardens; < 2300 m. **CA**; N.Am; native to Eur. All year

CARDAMINE BITTER-CRESS, TOOTHWORT

Ann, bien, per, from taproots, fibrous roots, or tuber-like rhizomes; hairs 0 or simple. **LVS** entire or palmately or pinnately lobed to compound; rhizome lvs often present, separate from others. **INFL** bracted or not. **FL**: sepals equal at base; petals white to pink or rose. **FR** linear, gen flat; valves gen opening elastically, sometimes by coiling from base; septum margins intruding on valves. **SEEDS** many, 1 row per chamber, wingless (± margined in *C. oligosperma*); embryonic root at edges of both cotyledons. ± 170 spp.: most temp parts of world. (Greek: for a cress with medicinal uses) Key to species adapted by Roy Buck.

1. Per, rhizomed; lvs simple or lflets gen 3 . . . . *C. breweri* var. *breweri*
1' Ann or bien, taprooted; lvs pinnate, lateral lflets > 2 pairs . . . . *C. oligosperma*

C. breweri S. Watson Per. **ST** 15–50(70) cm; hairs gen 0 or below. **LF** simple or lflets 3 or 5; simple lvs and terminal lflets > lateral lflets, ovate to cordate, wavy-margined to shallowly lobed. **FL**: petals ± 5 mm, white. **FR** erect, 15–30 mm; pedicel ascending, 5–20 mm; style 1.2 mm. **SEED** 1.5–2 mm. Wet sites, coniferous forest; < 3200 m. n NW, CaRH, SNH, WTR, SnBr, **SNE**; to B.C., MT, Colorado. 2 vars. in CA.

var. *breweri* (p. 219) Rhizome ± elongate, spreading, 1–4 mm thick. **STS** 1 or few, decumbent or erect, rarely prostrate and rooting at nodes. **LVS**: lower cauline lflets gen 3, wedge-shaped to truncate at base. $2n$=42–48. Margins of streams, seeps, lakes, swamps; 1200–3200 m. CaRH, SNH, WTR, SnBr, **SNE**; to B.C.,

MT, Colorado. ❀IRR,DRN:1,4,5,6,15–17&SHD:**2,3**,7,14,**18**.

C. oligosperma Torrey & A. Gray (p. 219) Ann, bien, taprooted. **STS** erect to ascending, 1–several, gen > 20 cm, branched; hairs short or 0. **LVS** many; basal ± rosetted, 1-pinnate, lateral lflets 5–9, 3–20 mm, short-stalked, widely obovate to ± round, sparsely, shallowly dentate, < terminal. **INFL** 3–10 cm, elongate. **FL**: petals 2–4 mm, white. **FR** ascending to erect, 15–25 mm, 1–2 mm wide; hairs 0 or sparse; pedicel ascending to erect, 5–15 mm; style < 0.5 mm. **SEED** oblong-ovate, narrowly margined. $2n$=16. Wet meadows, shady banks, creek bottoms; < 1100 m. CA-FP, **W&I**; B.C., MT, Colorado. Mar–Jul

CARDARIA WHITE-TOP, HOARY CRESS

Per, gen strongly rhizomed; hairs 0 to dense, simple, fine. **STS** 1–several, gen erect. **LVS** simple, toothed; base often lobed, clasping st. **INFL** ± flat-topped, wider than long, gen dense. **FL**: petals white. **FR** cordate or ovate to ± round to obovate, indehiscent or tardily dehiscent, partly inflated or flattened perpendicular to septum; style 1–2 mm, slender. **SEEDS** 1–2 per chamber, ovate, ± flat; wing 0; embryonic root at back of 1 cotyledon. 5 spp.: Eurasia. (Greek: heart-shaped, from fr of *C. draba*) [Mulligan & Findley 1974 Canad J Pl Sci 54:149–160]

1. Sepals, fr hairy . . . . *C. pubescens*
1' Sepals, fr glabrous
  2. Fr widely ovate or widely obovate to ± round, not narrowed at septum . . . . *C. chalepensis*
  2' Fr widely cordate or widely ovate, gen narrowed at septum . . . . *C. draba*

C. chalepensis (L.) Hand.-Mazz. (p. 223) LENS-PODDED HOARY CRESS **ST** 2–4 dm; hairs below, sparse to ± 0 above. **LVS** widely oblanceolate to obovate; basal short-petioled, ± toothed to entire; middle and upper cauline sessile, base lobed, clasping st. **FL**: sepals glabrous; petals 3–4 mm. **FR** 2.5–6(8) mm, ± round to widely ovate or widely obovate, inflated, not narrowed at septum, glabrous. $2n$=80. Disturbed, gen saline soils, fields; < 1500 m. CA-FP, **GB**; to Can, c US; native to e Asia. NOXIOUS WEED.

C. draba (L.) Desv. (p. 223) HEART-PODDED HOARY CRESS **ST** 2–5 dm; hairs below, sparse or ± 0 above. **LVS** widely oblanceolate to obovate; basal lvs short-petioled, ± toothed to entire; middle and upper cauline lvs sessile, base lobed, clasping st. **FL**:

sepals glabrous, margin white; petals 3–4 mm. **FR** 3–5 mm wide, widely cordate or ovate, gen narrowed at septum, glabrous. $2n$=62,64. Disturbed, gen saline soils, fields, roadsides; < 1200 m. CA-FP, **GB**; to Can, c US; native to Eurasia. NOXIOUS WEED. Pls called var. *repens* (Shrenk) O. Schulz are apparent hybrids with *C. chalepensis*. Mar–Jun

C. pubescens (C. Meyer) Jarmol. (p. 223)) WHITE-TOP **ST** 1–4 dm, hairy. **LVS** ± dentate; basal short-petioled; cauline sessile, oblong-linear to -lanceolate, base lobed, clasping st. **FL**: sepals hairy, margin white; petals 2–3.5 mm. **FR** 3–4.5 mm, ovate, round or obovate, inflated, not narrowed at septum, hairy. $2n$=16. Saline soils, fields, ditchbanks; < 2000 m. **CA**; to e N.Am; native to Asia. NOXIOUS WEED. [var. *elongata* Rollins] Apr–Jul

CAULANTHUS JEWELFLOWER

Roy E. Buck

Ann to per, gen tapered-hairy on lvs and lower st. **ST** gen ascending to erect, ± glaucous. **LVS** ± entire to deeply cut; basal gen rosetted, withering, gen oblanceolate to obovate; cauline gen linear to obovate, clasping, reduced. **INFL** becoming more open; bracts gen 0. **FL** biradial to ± bilateral; calyx ± urn-shaped, sepals often ± pouched below, gen not green, gen not darker in bud, gen erect after fl; petal (and sepal) margins often scarious, wavy or not; filaments gen in 3 pairs, gen free (or longest 1–2 pairs ± fused below); style < 4 mm, stigma gen 2-lobed. **FR** ascending to reflexed, gen cylindric. **SEED** gen ± oblong, gen compressed, gen ± brown. ± 14 spp.: ± sw N.Am. (Greek: st fl, from use of some as cauliflower-like vegetable) Key to species adapted by Roy Buck.

1. Middle and upper cauline lvs clasping, sessile
    2. Sepals gen obviously darker in bud (exc rarely *C. coulteri*); longest filament pair gen ± fused at least at base
        3. St not inflated; pl obviously hairy below; stigma whitish . . . . ***C. coulteri*** var. ***coulteri***
        3' St ± inflated; pl glabrous or sparsely hairy below; stigma purplish . . . . ***C. inflatus***
    2' Sepals not or barely darker in bud; longest filament pair free
        4. Pl glabrous or st inconspicuously appressed-hairy; upper lvs ± hastate at base . . . . ***C. cooperi***
        4' Pl ± conspicuously spreading-hairy below; upper lvs clasping but not hastate . . . . ***C. simulans***
1' Middle and upper cauline lvs not clasping, petioled or not
    5. Sepal hairs ± dense; ann . . . . ***C. hallii***
    5' Sepals gen glabrous or hairs sparse (if dense, per from woody caudex)
        6. Ann to weak per, caudex gen 0 or weak; lvs and lower st conspicuously hairy . . . . ***C. pilosus***
        6' Per from woody caudex; lvs and lower st ± glabrous
            7. Basal and lower cauline lvs gen < 3 × longer than wide, narrowed abruptly to petiole; basal ± not
               rosetted; seed yellowish or light brown . . . . ***C. glaucus***
            7' Basal and lower cauline lvs gen > 3 × longer than wide, tapered gradually to petiole (unless base
               deeply lobed), basal rosetted; seed dark brown
                8' St gen not inflated; sepals glabrous or tips sparsely hairy; stigma lobes < 0.5 mm . . . . . ***C. major*** var. ***major***
                8' St ± inflated; sepals glabrous or densely hairy; densely hairy; stigma lobes > 0.5 mm . . . . . . ***C. crassicaulis***
                9. Sepals ± densely hairy . . . . var. ***crassicaulis***
                9' Sepals glabrous . . . . var. ***glaber***

C. cooperi (S. Watson) Payson Ann. **ST** often irregularly twisted or weakly twining, gen branched, glabrous or inconspicuously branched-appressed-hairy. **LVS** < 6 cm, glabrous; basal entire or shallowly cut, tapered to short, winged petiole; cauline ± lanceolate-oblong, entire to dentate, sessile, clasping, hastate, tips acute. **FL:** sepals 5–7 mm, not pouched below; petals 7–10 mm, yellowish or purplish, margins scarious, not wavy; style 1–2 mm. **FR** spreading to reflexed, 2–4.5 cm, often curved. Common. Open, sandy or gravelly soil, gen among shrubs; 300–2500 m. n TR, e PR, **s SNE, D**; to sw UT, w AZ, Baja CA. Pls with st and fr greenish or straw-colored often mixed with pls with st and fr dark purple. Mar–Apr

C. coulteri S. Watson Ann, ± glabrous or ± bristly below. **ST** gen branched above. **LVS** < 13 cm; hairs branched or not; basal ± entire to deeply cut, tapered to short, winged petiole; cauline oblong to ovate, entire to cut, sessile, clasping. **FL:** sepals erect or spreading, (5)7–18 mm, ± pouched below, keeled, glabrous or ± bristly, gen ± darker in bud; petals 8–31 mm, whitish, cream and purple-veined, purplish, or brownish, margins wavy, gen scarious; longest 1–2 pairs of filaments ± fused; style < 1 mm. **FR** erect to reflexed, 4–13 cm. Dry, exposed slopes; 80–2000 m. s SNF, Teh, sw SnJV, se SnFrB, e SCoRO, SCoRI, nw WTR, **sw edge DMoj (Kern Co.).** 2 vars. in CA.

var. ***coulteri*** (p. 223) Gen densely hairy; gen some lf and sepal hairs branched. **LVS:** basal sometimes deeply cut; lower cauline gen coarsely dentate to deeply cut. **FL:** gen only longest filament pair ± fused; stigma lobes 0.5–2 mm. **FR** gen reflexed. Habitats of sp. ± SNF, Teh, se SCoRO, nw WTR, **sw edge DMoj (Kern Co.).** Variable; some distinct local races. ❀SUN,DRN: **7,8,9,11,14–16,17,18,**19–24. Mar–May

C. crassicaulis (Torrey) S. Watson THICK-STEM WILD CABBAGE Per from woody caudex; herbage ± glabrous. **ST** ± inflated, branched or not. **LVS** < 10 cm, petioled; rosette persistent; basal and lower cauline oblanceolate, entire to deeply cut; upper much-reduced, ± linear, entire. **FL:** sepals 8–15 mm, ± pouched below; petals 10–20 mm, purplish or brownish, margins scarious, not wavy; style < 0.5 mm, stigma lobes > 0.5 mm. **FR** erect to ascending, 6–13 cm. Dry sagebrush scrub, pinyon/juniper woodland; 900–2900 m. **W&I, n&c DMtns;** to ID, WY, Colorado, AZ.

var. ***crassicaulis*** (p. 223) **FL:** sepals greenish, gen purplish near tips, ± densely hairy. Uncommon. Habitats and range of sp.

var. ***glaber*** M.E. Jones **FL:** sepals cream in CA, glabrous. Uncommon. Habitats of sp.; 900–2200 m. **e DMtns (Providence Mtns);** to UT. CA pls with cream sepals may be distinct from more e pls with greenish or purplish sepals. [*C. g.* (M.E. Jones) Rydb.]

C. glaucus S. Watson Per from woody caudex, glabrous. **ST** gen branched above. **LVS:** basal ± not rosetted, blades 2–10 cm (± > petioles), widely oblong to widely obovate, entire or base few-lobed, abruptly narrowed to petiole; lower cauline sometimes ± reduced; upper linear to lanceolate, petioled. **FL:** sepals 6–10 mm, not pouched below; petals 9–18 mm, yellowish green or purplish, margins scarious, not wavy; style < 1.3 mm. **FR** spreading, 4–15 cm, gen curved. **SEED** yellowish or light brown. Uncommon. Open, rocky slopes, often in crevices; 1400–2500 m. **W&I, n DMtns (Grapevine, Last Chance Mtns);** w NV. May–Jun

C. hallii Payson Ann. **ST** gen not inflated, hollow, glabrous, sometimes branched above. **LVS** ± short-petioled; basal blades 3–18 cm, oblanceolate or oblong, ± cut, ± glabrous or bristly; cauline blades oblong, ± entire to cut, ± glabrous. **FL:** sepals 6–8 mm, ± pouched below, ± densely white-hairy; petals 8–9 mm, whitish or pale yellowish, purple-veined or not, margins not or barely wavy; style < 2 mm. **FR** ascending to spreading, 6–11 cm. Uncommon. Dry, open areas; 150–1800 m. e PR, **s DMtns (Little San Bernardino Mtns),** w edge DSon; n Baja CA. Apr–May

C. inflatus S. Watson (p. 223, pl. 32) DESERT CANDLE Ann, glabrous or sparsely hairy below. **ST** ± inflated, sometimes branched above. **LVS:** basal 2–7 cm, ± entire to finely dentate esp toward tips, tapered to short, winged petiole; cauline widely oblong to ovate, entire or ± dentate, sessile, clasping. **FL:** sepals erect or spreading, 8–10 mm, ± pouched below, darker in bud; petals 8–14 mm, purplish esp on veins, margins scarious, wavy; longest 1–2 filament pairs ± fused; style < 0.5 mm. **FR** ascending, 5–11 cm. Open, sandy plains to rocky slopes; 150–1500 m. s SNF (Kern Co.), c&s SnJV (w edge), SCoRI, n WTR (± uncommon in preceding), **sw DMoj (common).** Only sp. with ± purplish stigma. ❀TRY;DFCLT. Mar–May

C. major (M.E. Jones) Payson Per from woody caudex, ± glabrous. **ST** erect, gen not inflated, hollow, sometimes branched. **LVS:** rosette persistent; basal and lower cauline blades 1–9 cm, oblanceolate or elliptic, entire to deeply cut, tapered to long petiole; upper lvs much reduced, ± linear, entire. **FL:** sepals sometimes sparsely hairy at tips; petals 9–16 mm, darker on veins, margins scarious, not wavy; style < 0.5 mm, stigma lobes < 0.5 mm. **FR** erect to ascending, 4–13 cm. Dryish, often rocky slopes, sometimes in shade; 1500–2500 m. n SNH (Alpine Co.), SnGb, SnBr, s MP, **e DMtns (Providence, New York mtns);** se OR, n&w NV; also e UT. Possibly not distinct from *C. crassicaulis.* 2 vars. in CA.

var. ***major*** (p. 223) **ST** gen ± stout. **INFL:** pedicel gen glabrous (rarely bristly). **FL:** sepals 8–10 mm, pouched below,

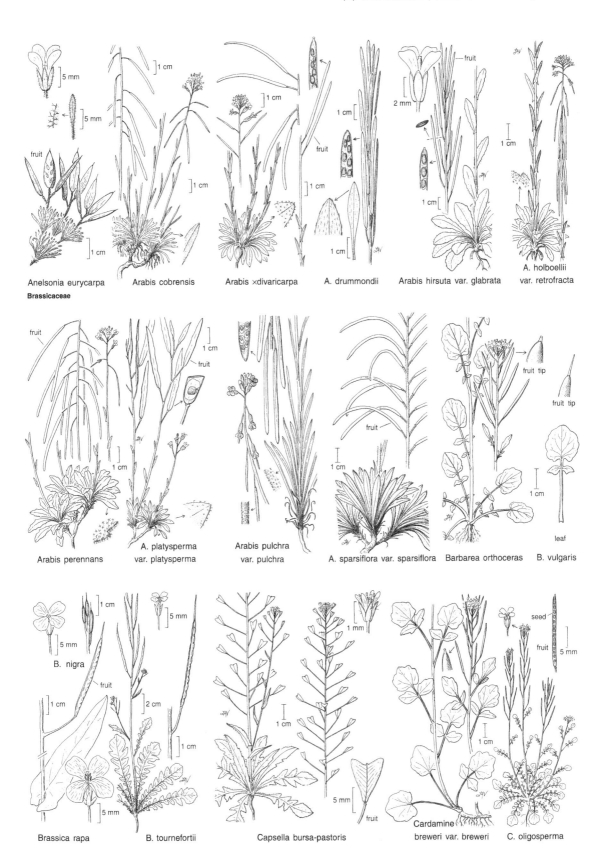

Anelsonia eurycarpa
Brassicaceae

Arabis cobrensis

Arabis ×divaricarpa

A. drummondii

Arabis hirsuta var. glabrata

A. holboellii var. retrofracta

Arabis perennans

A. platysperma var. platysperma

Arabis pulchra var. pulchra

A. sparsiflora var. sparsiflora

Barbarea orthoceras

B. vulgaris

B. nigra

Brassica rapa

B. tournefortii

Capsella bursa-pastoris

Cardamine breweri var. breweri

C. oligosperma

greenish white or cream; petals brownish. Habitats and elevations of sp. SnGb, SnBr, **e DMtns (Providence, New York mtns)**; also e UT. May–Jul

C. pilosus S. Watson (p. 223) CHOCOLATE DROPS, HAIRY WILD CABBAGE Ann? to weak per; hairs slender, barely tapered. **ST:** branches 0–many. **LVS:** lower blades 1–25 cm (± > petioles), oblanceolate to oblong, ± cut; upper blades linear to oblanceolate, entire or few-lobed, petioled. **INFL:** lowest 1–4 fls often bracted. **FL:** sepals 3–10 mm, 2 or 4 pouched at base, greenish to purple, darker in bud or not; petals 5–13 mm, whitish or purplish, margins wavy, scarious or not; style gen < 0.5 mm. **FR** ascending to spreading, 2–18 cm, often curved. **SEED** 1–3 mm. Uncommon.

Open, dry areas; 600–2800 m. c&s SNH (e slope), s MP (Honey Lake), **SNE, n DMtns**; to se OR, s ID, w UT. Apr–Jul

C. simulans Payson PAYSON'S JEWELFLOWER Ann, ± conspicuously spreading-bristly below. **ST** gen branched above. **LVS** < 10 cm; basal coarsely dentate to shallowly cut, petiole 0 or short, winged; cauline oblong to ovate, entire to coarsely dentate, sessile, clasping. **FL:** sepals 3–7 mm, barely pouched at base; petals 5–10 mm, cream-yellow, margins not scarious, not wavy; style 0 or < 4 mm. **FR** ± reflexed, 2–8 cm, often curved. RARE. Open, dry areas; 400–2200 m. e SCo (w Riverside Co.), e PR, **w edge DSon**. Apr–Jun

CAULOSTRAMINA

1 sp. (Latin: straw-colored stem)

C. jaegeri (Rollins) Rollins (p. 223) JAEGER'S CAULOSTRAMINA Per, deeply rooted; base woody. **STS** few–many, 1–3 dm, branched at base, wavy, glabrous, gray-green. **LVS** all similar, glabrous, grayish; basal deciduous; petioles 1–2 cm, slender; blades 1.5–4 cm, elliptic to ovate, wavy to few-lobed, base tapered to truncate, reniform, or cordate, veins conspicuous. **INFL** open; bracts 0 or rarely below. **FL:** petals 8–14 mm, spoon-shaped, white to pale lavender, purple-veined. **FR** spreading to ± erect, 3–5 cm, linear, ± cylindric, glabrous; pedicel spreading to ascending, 8–10 mm; style < ± 1 mm. **SEEDS:** 1 row per chamber, < ± 1 mm; wing 0. RARE. Rock crevices, cliffs; 1800–2400 m. **s W&I (Inyo Mtns)**. [*Thelypodium j.* Rollins] May–Jun

CHORISPORA

Ann to per; hairs simple, gen glandular and not. **LVS** entire to pinnately lobed; basal not rosetted. **INFL:** bracts 0. **FL:** inner sepals sac-like at base; petals purplish blue to white or yellow; anthers ± exserted. **FR** cylindric, sessile above receptacle, tapered to sharp beak, indehiscent but breaking into 1-seeded segments. **SEEDS:** 1 row per chamber, ± flat, embedded in cavities of septum; embryonic root at edges of both cotyledons. ± 12 spp.: Eurasia. (Latin: from breaking between seeds)

C. tenella (Pallas) DC. (p. 223) Ann. **ST** 1–5 dm, branched from near base. **LVS** elliptic-oblong to lanceolate or oblanceolate; basal and lower cauline petioled, blades 3–8 cm, wavy-dentate to pinnately lobed; upper cauline sessile, ± entire to dentate. **FL:** sepals erect, 6–8 mm, free but forming a tube; petals narrowly clawed, magenta. **FR** spreading to ascending, 35–45 mm, lanceolate, often upcurved; beak 7–20 mm; pedicel ascending, 2–4 mm, stout; style 0, stigma minute, entire. 2*n*=14. Grainfields, roadsides, waste places; < 1300 m. CaR, GV, SCo, **GB**; widespread to c US; native to Eur.

CONRINGIA HARE'S EAR

Ann, per, glabrous, gen glaucous. **ST** erect. **LVS:** basal gen entire, ± fleshy; cauline lanceolate or oblong to ± round, base cordate, clasping st. **INFL** ± flat-topped; bracts 0. **FL:** outer 2 sepals linear to narrowly oblong, inner 2 wider; petals yellow or white. **FR** linear; stigma head-like, entire or 2-lobed. **SEEDS:** 1 row per chamber, oblong, plump; wing 0; embryonic root at back of 1 cotyledon or twisted. 6 spp.: Eurasia. (H. Conring, professor at Helmstedt, Germany, 1606–1681)

C. orientalis (L.) Dumort. (p. 223) Ann. **ST** 3–7 dm, gen simple. **LVS:** basal 5–9 cm, oblanceolate to obovate, narrowed to base, ± entire; cauline sessile, oblong-lanceolate. **FL:** sepals erect, 6–8 mm, acute; petals 7–12 mm, lemon to creamy yellow, claws slender. **FR** 8–13 cm, ± beaded, 4-angled to ± cylindric; pedicel ascending, 10–15 mm; style ± 1 mm, thick. 2*n*=14. Fields, roadsides, open, disturbed places; < 900 m. SCo, **SNE, DSon**, expected elsewhere; c US; native to Eurasia. Apr–Jun

CORONOPUS WART or SWINE CRESS

Ann, bien; odor often foul; hairs 0 or simple. **STS** several–many. **LVS** entire to very deeply pinnately lobed; basal and lower cauline petioled; upper subsessile. **INFL** from lf axils, dense. **FL** minute: sepals spreading; petals small or 0, white or purplish; stamens 2 or 4. **FR** of 2 nutlet-like halves, inflated exc at septum, flat perpendicular to septum, sessile above receptacle, indehiscent; valves hardened, each ± surrounding 1 seed. **SEED** oblong to ± reniform; embryonic root at back of 1 cotyledon. ± 10 spp.: gen Medit, few n S.Am. (Greek: crow foot, from lf shape)

1. Style 0 or incl in fr tip notch; fr base cordate, tip notched; valve surface net-wrinkled . . . . **C. didymus**
1' Style exceeding fr tip; fr base cordate to ± truncate, tip acute; valve surface with coarse ridges, spine-like projections . . . . **C. squamatus**

C. didymus (L.) Smith (p. 223) Ann. **ST** prostrate to decumbent, much-branched, 20–50 cm. **LVS** many, ovate-oblong, very deeply pinnately lobed; lobes narrow, entire to dentate or dissected. **FL** ± 0.5 mm; petals linear, white. **FR:** pedicel 1.5–2.5 mm. 2*n*=32. Common. Disturbed areas, gardens, fields; < 2000 m. **CA**; N.Am; native to Eurasia. Mar–Jul

C. squamatus (Forsskal) Asch. (p. 223) SWINE CRESS Ann, bien. **ST** prostrate to decumbent, gen much-branched, 5–30 cm. **LVS** many, oblanceolate to obovate, very deeply pinnately lobed; lobes narrow, gen coarsely toothed or ± dissected. **FL** 1– 1.5 mm; petals obovate, white. **FR**: pedicel 1–2 mm. 2*n*=32. Disturbed areas, fields; < 300 m. SnFrB, ScV, **DSon**; native to Eur. NOXIOUS WEED.

CUSICKIELLA

Per, low-cespitose, taprooted. **LVS** gen basal, clustered at tips of caudex branches, sessile, entire. **FL**: sepals hairy or not, bases not sac-like; petals erect, white or yellowish, distinct claws 0; ovules 2 per chamber. **FR** sessile above receptacle, ovate to ± oblong or fusiform, not or ± flat toward tip; valves rounded or keeled; pedicel ascending. **SEED** gen 1, 2–3 mm, brown, plump, ± round or ovate-oblong; wing 0; embryonic root at back of 1 cotyledon. 2 spp.: w US. (W.C. Cusick, Oregon pl collector, 1842–1922) [Rollins 1988 J Jap Bot 63:65–69]

1. Bracts subtending pedicels 0; fr valves rounded; lf 5–12 mm; petals white . . . . *C. douglasii*
1' Bracts subtending lower pedicels, lf-like; fr valves keeled; lf 2–4 mm; petals yellowish . . . . . . *C. quadricostata*

C. douglasii (A. Gray) Rollins (p. 223) Caudex branches underground. **ST** 3–5 cm, hairy or not. **LF** oblanceolate; hairs gen on margins, some not, gen simple. **FR** 3–7 mm, ± flat; hairs gen simple or 0; pedicel hairy or not; style 0.5–1.5 mm. Rocky ridges, slopes; 1500–2450 m. KR, NCoR, SN, SnBr, **GB**; to WA, ID, UT. [*Draba douglasii* A. Gray var. *crockeri* (Lemmon) C. Hitchc.] May–Jun

C. quadricostata (Rollins) Rollins (p. 223) Caudex branches gen underground. **ST** 2–5 cm, hairy. **LF** linear to oblong; hairs on margins, surfaces, simple or multibranched. **FR** 3–4 mm, 4-sided; hairs simple or forked; pedicel hairy; style 0.5–1 mm. Rocky flats, sagebrush scrub, slopes; 2400–2800 m. SNE (**Mono Co.**); w NV. [*Draba q.* Rollins] Jul

DESCURAINIA TANSY MUSTARD

Ann, bien, rarely per; hairs gen minute, gen multibranched, fewer simple, some glandular. **ST** branched. **LVS** 1–many-pinnately lobed to compound; basal often rosetted but withering. **INFL** elongating; bracts gen 0. **FL**: petals < 3 mm, yellow or whitish, blades obovate, obtuse. **FR** linear to ± obovate, straight or uneven-margined, gen ± cylindric; style < 0.8 mm or 0, stigma gen head-like, entire. **SEEDS**: 1–2 rows per chamber, gen < 1 mm, elliptic, plump, gelatinous when wet; wing 0; embryonic root at back of 1 cotyledon. ± 40 spp.: worldwide temp. (F. Descourain, French botanist, 1658–1740) [Detling 1939 Amer Midl Naturalist 22: 481–520] Relationships, characters of spp. difficult. May be TOXIC to livestock.

1. Fr fusiform, base as acute as tip; style gen 0.5–0.8 mm . . . . *D. californica*
1' Fr linear, oblong, or ± widely elliptic to ± club-shaped or ± obovate, base as acute as to more acute than tip; style < 0.8 mm or 0
  2. Fr oblong or ± widely elliptic to ± club-shaped to ± obovate, tip obtuse or rounded; style < 0.5 mm or 0; seeds gen in 2 rows per chamber
    3. Fr 2.5–4 mm, ± widely elliptic to ± obovate; basal lvs not rosetted, not shed early . . . *D. paradisa* ssp. *paradisa*
    3' Fr 4–20 mm, oblong to ± club-shaped; basal lvs rosetted, shed early . . . . *D. pinnata*
      4. Pedicel in fr ± 45° from infl axis; herbage hairs ± not dense . . . . ssp. *intermedia*
      4' Pedicel in fr gen ± 60–110° from infl axis; herbage hairs dense
        5. Lf lobes or lflets obtuse at tip, oblanceolate to obovate . . . . ssp. *glabra*
        5' Lf lobes of lflets gen acute at tip, linear to obovate
          6. Petals 1–2 mm, pale yellow to whitish; branches 0, below, or above; basal and cauline lvs canescent
          . . . . ssp. *halictorum*
          6' Petals 2–3.5 mm, bright yellow; branches 0 or above; lvs ± greenish . . . . ssp. *menziesii*
  2' Fr linear to narrowly oblong, tip acute to acuminate; style gen 0.5–0.8 mm; seeds gen in 1 row per chamber
    7. Pedicel in fr erect to ascending . . . . *D. incana*
    7' Pedicel in fr ascending to spreading
      8. Lf very deeply 2–3-pinnately lobed to compound; fr 10–35 mm, septum with (2)3 veins . . . . *D. sophia*
      8' Lf 1–2-pinnately lobed to ± 1-compound, 1° lobes or 1° lflets dentate to deeply lobed; fr 5–16 mm, septum with 0 or 1 indistinct vein
        9. Pl whitish; fr gen straight . . . . *D. obtusa* ssp. *adenophora*
        9' Pl greenish; fr gen ± curved . . . . *D. incisa*
          10. Pedicels ± ascending, (10)12–20 mm, lower > upper; fr ascending to gen erect . . . . ssp. *filipes*
          10' Pedicels ± spreading, 5–10 mm, lower ± = upper; fr ascending to ± spreading . . . . ssp. *incisa*

D. californica (A. Gray) O. Schulz (p. 223) Winter ann to bien. **ST** 1, 3–8 dm; glands 0; branches above, slender. **LVS** oblanceolate to obovate, greenish, 1-pinnately lobed; lobes 2–4 pairs, 1.5–6 cm, lanceolate to oblanceolate, tip acute or obtuse, margins serrate to dissected; lower cauline lvs ± hairy, upper ± glabrous. **INFL**: bracts subtending branches, lf-like. **FL**: petals spreading, < 2 mm, barely > sepals, bright yellow. **FR** erect to spreading, 2–5 mm, fusiform, ± straight or curved; pedicel ascending to spreading, 3–7 mm, slender; style gen 0.5–0.8 mm.

SEEDS 1–3, 1 row per chamber. 2*n*= 14. Open sites, sagebrush scrub, shrubland, aspen groves, open woodlands; 2100–3400 m. SNH, **GB**, DMtns; to WY, NM. May–Aug

D. incana (Fischer & C. Meyer) Dorn (p. 223) Bien, canescent to greenish. **ST** 3–12 dm, slender, branched, with glandular hairs or not. **LVS** 1.5–10 cm, widely lanceolate or oblanceolate to ovate; basal and lower cauline lvs 1–2-pinnately lobed or -compound, upper simple to 1-pinnate, entire. **FL**: petals ± 2 mm, yellow. **FR** 5–25 mm, 0.5–1.5 mm wide, linear, gen straight; both ends acute; pedicel erect to ascending, 2–4(7) mm; style < 1 mm. **SEEDS** 4–8, 1 row per chamber. $2n=14$. Open sites, meadows, sagebrush scrub, open aspen groves; 1500–3400 m. KR, SN, SnBr, **GB**; to AK, e N.Am. [*D. richardsonii* (Sweet) O. Schulz; *D. r.* ssp. *viscosa* (Rydb.) Detl.] May–Aug

D. incisa (A. Gray) Britton   Ann; hairs gen nonglandular. **ST** 1, gen branched above, rarely below. **LVS**: basal ± 1–2-pinnately lobed or -compound, 5–10 cm, obovate, soon withering; cauline reduced, gen simpler upward. **INFL** elongate; hairs on axes minute, glandular and not. **FL**: petals ± 2 mm, spoon-shaped, yellow. **FR** < 1 mm wide, linear to narrowly oblong, gen ± curved, cylindric, glabrous; both ends acute to acuminate; pedicel slender. **SEEDS**: 1 row per chamber. Open sites, shrubland; 1000–3100 m. CaR, TR, PR, **GB**; to B.C., NM. 4 sspp. total.

ssp. *filipes* (A. Gray) Rollins   Pl hairs sparse to moderate, minute, multibranched. **FR** ascending to gen erect, 10–16 mm; pedicels ± ascending, (10)12–20 mm, lower > upper. Shaly soils at cliff bases, dry washes, often sagebrush scrub, pinyon/juniper woodland; 1200–2500 m. CaR, **GB**; to WA, WY, NM. [*D. pinnata* ssp. *f.* (A. Gray) Detl.]

ssp. *incisa*   Pl hairs ± dense to ± 0, multibranched, minute. **FR** ± spreading to ascending, 6–12 mm; pedicels ± spreading, 5–10 mm, lower ± = upper. Hillsides, dry water courses, granitic sands, often with sagebrush, juniper woodland; 1000–3100 m. CaR, TR, PR, **GB**; to B.C., NM. [*D. richardsonii* (Sweet) O. Schulz ssp. *i.* (A. Gray) Detl.] Jun–Aug

D. obtusa (E. Greene) O. Schulz ssp. *adenophora* (Wooton & Standley) Detl.   Bien, erect, coarse, whitish. **ST** gen 1, simple or branched above. **LF** 1–6 cm, 1-pinnately lobed or -compound, canescent; hairs minute, multibranched, rarely glandular; lobes or lflets 2–5 pairs, linear to oblanceolate, tip gen obtuse. **FL**: sepals 2–2.5 mm, ± < petals, sparsely glandular; petals yellowish. **FR** 5–15 mm, linear, straight, gen glabrous; both ends tapered abruptly; pedicel ascending to spreading, 12–25 mm, sparsely glandular. **SEEDS** 24–32 per chamber, 1–2 rows per chamber, closely packed, 0.7–1 mm. Gravelly flats, open woods, lake margins; 900–2200 m. e PR, **DMoj**; to NM, Baja CA. May–Jun

D. paradisa (Nelson & Kenn.) O. Schulz ssp. *paradisa*   Ann, greenish to gray-green; hairs gen dense, minute, multibranched. **ST** 1, 1.5–2.5 (3.5) dm, branched from base. **LF** 2–3 cm, canescent to greenish, deeply 1-pinnately lobed to ± compound; lobes or lflets linear to oblong. **INFL**: axis hairs glandular and not. **FL**: petals ± 1 mm, pale yellow. **FR** 2.5–4 mm, ± widely elliptic to ± obovate, glabrous to sparsely hairy; tip rounded; pedicel ± ascending, 3–5 mm. **SEEDS** 2–4 per chamber, 2 rows per chamber. Sandy washes, dunes, sagebrush scrub; 900–2300 m. **GB**; to OR, NV. [*D. pinnata* ssp. *paradisa* (Nelson & Kenn.) Detl.] Ssp. *nevadensis* Rollins occurs in Nevada.

D. pinnata (Walter) Britton   Ann; hairs sparse to dense, multibranched. **ST** 1–7 dm; branches 0, below, or above. **LVS** lanceolate or oblanceolate to ovate; lower cauline lvs 2-pinnately lobed; upper lvs 1–10 cm, ± 1–2-pinnately lobed or 1-compound; lobes or lflets linear to widely obovate, hairs glandular or not. **FL**: petals 1–3.5 mm, bright yellow to cream. **FR** 4–20 mm, oblong to ± club-shaped; pedicel ascending to spreading, 4–20 mm. **SEEDS** 5–20 per chamber, 2 rows per chamber, 0.5–1 mm. Washes, slopes, often saline soils; < 2500 m. CaR, s SNH, GV, CCo, SnFrB, SCoR, SW, **GB**, **DMoj**; widespread N.Am. 8 difficult sspp.

ssp. *glabra* (Wooton & Standley) Detl.   **ST** 1–4 dm, branched from base, ± canescent below, ± glabrous above. **LF** ± 1–2-pinnately lobed; lobes oblanceolate to obovate, obtuse at tip, greenish; hairs dense. **FL**: sepals 0.5–1.5 mm, ± < petals, yellow or rose; petals yellow. **FR** 5–8 mm, oblong to ± club-shaped; pedicel ± 60–110° from infl axis, 4–12 mm. **SEEDS** 8–12 per chamber, ± 0.5 mm. $2n=28$. Sandy or saline soils, dry washes, desert shrubland; < 1500 m. s SnJV, SCoR, **W&I**, **DMoj**; to NM, n Mex.

ssp. *halictorum* (Cockerell) Detl. (p. 223)   **ST** 1.5–5 dm, often glandular above; branches 0, below, or above. **LVS** lanceolate to oblanceolate; basal and lower cauline deeply 2-pinnately lobed, rarely 1-compound, canescent, lobes or lflets gen acute at tip, linear to narrowly oblong, upper simpler, hairs gen sparser. **FL**: sepals 1–2 mm, ± < petals, yellowish or rose; petals < 2 mm, pale yellow to whitish. **FR** ± ascending, 5–20 mm, ± club-shaped; pedicel ± 60–110° from infl axis, 8–12 mm. **SEEDS** 8–20 per chamber, 0.7–1 mm. $2n=14,28,42$. Dry streambeds, stable dunes, juniper woodland; < 1800 m. CaR, SW, **GB**, **DMoj**; to TX, Baja CA.

ssp. *intermedia* (Rydb.) Detl.   **ST** 2–6 dm, gen branched above, gen glabrous above. **LVS**: basal, lower cauline 1–2-pinnately lobed, linear to obovate, hairy; upper lvs 1-pinnate to ± entire, hairs gen 0, rarely glandular. **FL**: sepals 1.5–2.5 mm, yellow; petals erect, 2–3 mm, exceeding sepals by ± 0.5 mm, yellow. **FR** 8–12 mm, ± club-shaped, rarely oblong; pedicel 6–12 mm, ± 45° from infl axis. **SEEDS** 5–13 per chamber, crowded or not, appearing as 1 row per chamber, ± 1 mm. $2n=28$. Dry streambeds, meadows, slopes, sagebrush scrub; 900–2200 m. **GB**; to B.C., s Colorado.

ssp. *menziesii* (DC.) Detl.   **ST** 1–6 dm; branches 0 or above; hairs glandular and not. **LVS** ± greenish; hairs dense; basal and lower cauline 2(3)-pinnately lobed, lobes gen obovate; upper lvs 1(2)-pinnately lobed, lobes acute at tip, linear to oblanceolate. **FL**: sepals 1.5–2.5 mm, ± < petals, bright yellow; petals erect, 2–3.5 mm, bright yellow. **FR** 5–12(15) mm, ± club-shaped; pedicel 5–10 mm, ± 60–110° from infl axis. **SEEDS** 6–8, 0.7–1 mm. $2n=28$. Rocky flats, disturbed areas, washes, chaparral; < 2500 m. s SNH, GV, CCo, SnFrb, SW, **DMtns**. Mar–Jun

D. sophia (L.) Webb   Ann, bien. **ST** 2.5–7.5 dm, short-branched above; hairs sparse to dense, minute, branched, nonglandular, sometimes also larger, simple. **LF** 1–9 cm, oblanceolate to widely ovate, very deeply 2–3-pinnately lobed to ± 2–3-compound; ultimate lobes of lflets gen linear (rarely obovate). **FL**: sepals 2–2.5 mm, yellow; petals erect, ± = sepals, yellow. **FR** 10–35 mm, ± 1 mm wide, linear, straight to ± curved, ± cylindric; pedicel ± ascending, 8–15 mm; septum (2)3-veined. **SEEDS** 10–20, 1 row per chamber, 0.7–1.5 mm, oblong-elliptic. $2n=20,28,38$. Common. Disturbed areas, fields, roadsides, canyon bottoms, desert; < 2600 m. **CA**; native to Eurasia. May–Aug

DITHYREA   SPECTACLE-POD

Ann, per; herbage hairs dense, stellate, multibranched. **LVS** oblanceolate to widely obovate, entire to and shallowly dentate to lobed; basal and lower cauline petioled; middle and upper cauline petioles 0 or short. **INFL**: bracts 0. **FL** fragrant; sepals overlapped; petals erect, narrowly tongue-shaped, white to lavender. **FR** indehiscent, flat perpendicular to septum; 1-seeded halves ± round, each bordered by a raised rim; hairs simple, forked, club-like or not, esp dense on rim; pedicel < 3 mm, stout. **SEED** widely oblong, flat; wing 0; embryonic root at edges of both cotyledons. 2 spp.: CA, Mex. (Greek: with 2 shields, from fr) [Rollins 1979 Publ Bussey Inst Harvard 3–32]

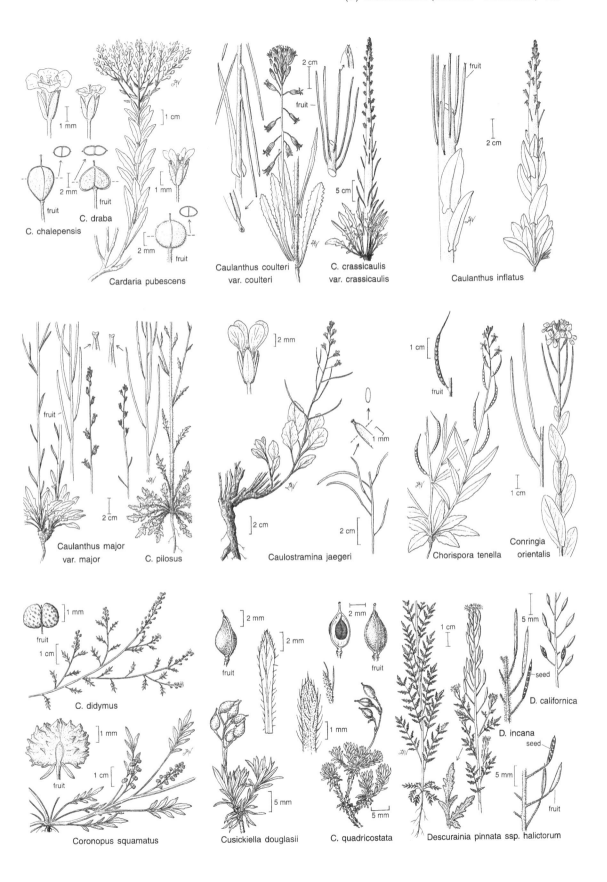

C. chalepensis

C. draba

Cardaria pubescens

Caulanthus coulteri
var. coulteri

C. crassicaulis
var. crassicaulis

Caulanthus inflatus

Caulanthus major
var. major

C. pilosus

Caulostramina jaegeri

Chorispora tenella

Conringia
orientalis

C. didymus

Coronopus squamatus

Cusickiella douglasii

C. quadricostata

Descurainia pinnata ssp. halictorum

D. californica

D. incana

D. californica Harvey (p. 227)   Ann canescent; bases of cauline lvs, pedicels with paired glands. **STS** arising below lf clusters, decumbent to erect, 1–7 dm; branches 0 below, sparse or 0 above. **LVS** not fleshy, canescent; basal 3–15 cm, oblanceolate to widely obovate, dentate to shallowly lobed, tip obtuse; cauline 1–4 cm, widely oblong to ovate. **FL**: petals 12–15 mm, prominently 3-veined, whitish to light lavender. **FR**: hairs simple, club-like or not; pedicel spreading. **SEED** 3–4 mm. 2*n*=20. Abundant. Sandy places, washes, shrubland; < 1400 m. **D**; w AZ, nw Mex. ❀TRY. Mar–May

DRABA

Robert A. Price

Ann to per, often cushion- or mat-forming; hairs often branched. **LVS** basal and sometimes cauline, entire or shallowly toothed. **FL**: sepal bases equal; petals < 10 mm, yellow or white, claw and limb gen distinct. **FR** < 30 mm, gen lanceolate to ovate, gen flat parallel to septum, less often partially inflated, sometimes twisted or wavy. **SEEDS:** 2 rows per chamber; wing gen 0. 350+ spp.: n hemisphere, mtns of S.Am.   (Greek: acrid) [Rollins & Price 1988 Aliso 12:17–27] Key to species adapted by Roy Buck.

1. Ann (may germinate in winter); styles < 0.2 mm; < 2500 m
  2. Lf gen dentate; frs lanceolate to oblong, often loosely clustered; infl axes gen hairy . . . . *D. cuneifolia*
  2' Lf gen entire (to slightly dentate); frs ± linear, densely clustered; infl axes glabrous . . . . *D. reptans*
1' Bien to per; styles gen conspicuous, often > 0.2 mm (exc. *D. albertina*); gen > 2000 m
  3. Petals yellow; lvs basal; sts ± cushion- or mat-forming; seeds gen > 1 mm
    4. Lf obovate, thick, ± fleshy, gen > 3 mm wide . . . . *D. incrassata*
    4' Lf linear to oblanceolate, ± thin, 0.5–3 mm wide
      5. Lf margin prominently stiff-hairy. . . . . *D. densifolia*
      5' Lf margin glabrous, or hairs not prominent
        6. St hairs 0 or stellate; lf hairs not bush-like, main axis obvious . . . . *D. oligosperma* var. *oligosperma*
        6' St hairs bush-like; lf hairs ± bush-like or stellate, main axis not obvious . . . . *D. subumbellata*
  3' Petals gen white to cream (exc. *D. albertina*); lvs basal and gen also cauline; sts not mat-forming (sometimes ± tufted); seeds gen < 1 mm
    7. Stellate st, lf hairs with 4 or fewer branches
      8. Lf lower surface with forked and cross-shaped hairs; petals yellow; st 1–40 cm; frs 4–20 mm, not in umbel-like clusters; seed 0.7–1 mm . . . . *D. albertina*
      8' Lf lower surface with simple and forked hairs; petals white; st < 10 cm; frs 3–5 mm, in umbel-like clusters; seed 0.6–0.8 mm . . . . *D. monoensis*
    7' Some stellate st, lf hairs with > 4 branches
      9. Lf lower surface with dense, many-branched, stellate hairs, gen whitish; lower pedicels gen < 3 mm, appressed; fr gen twisted, with gen > 3-branched, stellate hairs . . . . *D. breweri*
      9' Lf lower surface with 4–8-branched, stellate hairs, gen gray-green; lower pedicels > 3 mm, appressed; fr not twisted, glabrous or hairs simple and forked . . . . *D. californica*

D. albertina E. Greene   Bien to short-lived per. **STS** 1–several from base, 1–40 cm, often branched from lf axils; hairs near base, coarse, simple and forked. **LVS**: basal 3–40 mm, oblanceolate to obovate, entire to minutely dentate; lower surface hairs forked and cross-shaped, upper surface hairs 0 or simple and forked; cauline 0–7. **INFL** < 35-fld, gen glabrous. **FL**: petals 2–3 mm, yellow. **FR** 4–20 mm, ± linear to narrowly ovate; hairs 0, rarely short; style < 0.2 mm. **SEEDS** < 40, 0.7–1 mm; wings 0. 2*n*=24. Moist meadows, streambanks; > 1500 m. CaRH, SNH, SnGb, MP, **n SNE, W&I**; w US, w Can. [*D. stenoloba* Ledeb. vars. *nana* (O. Schulz) C. Hitchc. and *ramosa* C. Hitchc. (branched draba)] Small, lfless, high-elevation pls often confused with *D. crassifolia* Graham (esp var. *nevadensis* C. Hitchc., dolomite draba) of n&w N.Am outside CA, sparsely hairy, lf hairs not cross-shaped. Jun–Aug

D. breweri S. Watson (p. 227)   Per. **STS** gen several from base, < 15 cm; hairs dense, stellate. **LVS**: basal 5–15 mm, oblanceolate to obovate, gen whitened, hairs dense, stellate, main axis obscure (branches again branched); cauline 1–10, gen entire. **INFL** < 30-fld, elongate; lower pedicels gen < 3 mm, appressed, not bracted. **FL**: petals 2–4 mm, white. **FR** 4–10 mm, ± lanceolate, gen twisted; hairs > 3-branched, stellate; style < 0.5 mm. **SEEDS** < 35, 0.6–1 mm; wings 0. Open, rocky areas, gen above timberline; > 2500 m. CaRH, SNH, **n SNE, W&I**; NV. Jul–Aug

D. californica (Jepson) Rollins & R.A. Price (p. 227)   CALIFORNIA DRABA   Per. **STS** 1–several, < 10(15) cm; hairs dense, gen forked and stellate. **LVS**: basal 10–50 mm, ± lanceolate, gen entire, lower surface gen grayish green, hairs 4–7-branched, stellate; cauline 0–3, overlapped. **INFL** < 35-fld; lower pedicels ascending, > 3 mm. **FL**: petals 2–4 mm, white. **FR** ± throughout st, 5–12 mm, narrowly elliptic to lanceolate, not twisted, not wavy; hairs simple and forked or 0; style < 0.5 mm. **SEEDS** < 25, 0.8–1.2 mm, unwinged (tip gen with small extension). UNCOMMON. Open, rocky areas; > 3000 m. **n W&I (White Mtns, Mono Co.).** [*D. cuneifolia* Torrey & A. Gray var. *californica* Jepson] Locally common.

D. cuneifolia Torrey & A. Gray (p. 227)   Ann. **STS** 1–few from base, < 40 cm; hairs short, simple and branched, often stellate, rarely 0. **LVS** gen dentate; basal 5–70 mm, oblanceolate to obovate, surface hairs stellate; cauline 1–4. **INFL** < 75-fld; axes gen hairy; lowest pedicel < 2 × fr. **FL**: petals < 5 mm, often smaller or 0 in lateral infls, divided < 1/8 to base, white. **FRS** often loosely clustered, 3–12 mm, lanceolate to oblong; hairs 0 or simple, forked, and cross-shaped; style < 0.2 mm. **SEEDS** < 100, 0.5–0.7 mm; wings 0. 2*n*=16,32. Open or disturbed places; < 2000 m. s SN, SnJV, SW, **D**; w US, n Mex. Vars. *cuneifolia, integrifolia* S. Watson, *sonorae* (E. Greene) Parish recognized by Hartman et al [1976 Brittonia 27: 317–327] but ± intergrade in CA. May

D. densifolia Nutt. (p. 227) Per. **STS** cushion-forming, < 15 cm, glabrous to hairy. **LVS** basal, 2–15 mm, linear to ± oblanceolate, entire; midrib prominent below; surface hairs 0 or sparse, short, branched; margin hairs prominent, stiff, simple. **INFL** < 20-fld. **FL**: petals 2–6 mm, yellow. **FR** 2–7 mm, ± ovate; hairs 0 or short, simple to stellate; style 0.5–1 mm. **SEEDS** < 15, 1.2–1.8 mm; wings 0. 2*n*=36. Alpine barrens, rocky slopes; > 2000 m. SNH, **n SNE, W&I**; nw US, w Can. Reproduces by asexual seed. Jul–Aug

D. incrassata (Rollins) Rollins & R.A. Price SWEETWATER MOUNTAINS DRABA Per. **STS** ± mat-forming, < 15 cm; hairs 0 or sparse, simple. **LVS** basal, < 15 mm, obovate, thick, entire; surface hairs 0 or sparse, simple and forked; margin hairs gen prominent. **INFL** < 30-fld. **FL**: petals ± 3–5 mm, yellow. **FR** 3–12 mm, ± ovate, gen not wavy, not twisted; hairs 0 or sparse, marginal; style 0.2–0.8 mm. **SEEDS** < 20, 1.2–1.8 mm; wings 0. UNCOMMON. Alpine barrens, rocky slopes; > 2500 m. **n SNE (Sweetwater Mtns, Mono Co.)**. [*D. lemmonii* S. Watson var. *i.* Rollins]

D. monoensis Rollins & R.A. Price (p. 227) WHITE MOUNTAINS DRABA Per. **STS** 1–few from base, < 10 cm; hairs gen dense, simple and forked. **LVS**: basal 5–20 mm, oblanceolate, entire or teeth few; surface hairs simple and forked; cauline 0–2. **INFL** 10–20-fld, umbel-like in fr. **FL**: petals 2–3 mm, white. **FR** 3–5 mm, ovate; hairs 0 or sparse, simple; style < 0.3 mm. **SEEDS** 10–20, 0.6–0.8 mm; wings 0. RARE. Moist gravel, rock crevices; > 3000 m. **n W&I (White Mtns, Mono Co.)**, probably also in SNH. [*D. fladnizensis* Wulfen misapplied] Aug

D. oligosperma Hook. var. *oligosperma* (p. 227, pl. 33) Per. **STS** cushion-forming, < 12 cm; hairs 0 or stellate. **LVS** basal, 3–10 mm, linear-oblanceolate, entire; midrib prominent on lower surface; surface hairs sparse to dense, gen appressed; main axis obvious, branches shorter; margin hairs 0 or not very prominent. **INFL** < 20-fld. **FL**: petals 2.5–4.5 mm, yellow. **FR** 2.5–7 mm, ± ovate; hairs 0 or simple, hooked, or branched like lf hairs; style < 1 mm. **SEEDS** < 12, 1.3–1.8 mm. 2*n*=64. Common. Alpine barrens, dry slopes; > 2000 m. SNH, **n SNE, W&I**; w US, w Can. Dwarfed pls of high elevations have been called var. *subsessilis* (S. Watson) O. Schulz; gen reproduces by asexual seed. Jul–Aug

D. reptans (Lam.) Fern. Ann. **STS** 1–several from base, < 15 cm; hairs near base simple, forked, and stellate, near infl 0. **LVS**: basal 5–20 mm, oblanceolate to obovate, gen entire; lower surface hairs forked and stellate; upper surface hairs simple and forked; cauline 1–5. **INFL** < 30-fld; axes glabrous; lowest pedicel < 2 × fr. **FL**: petals 0.5–5 mm, white, often smaller or 0 in lateral infls. **FRS** densely clustered in upper 1/3 of st, 5–20 mm, ± linear; hairs 0 or short, stiff, simple; style < 0.2 mm. **SEEDS** < 75, 0.5–0.7 mm; wings 0. 2*n*=16,30,32. Very uncommon. Open or disturbed areas; < 2500 m. n SNH (Emigrant Gap, n Placer Co.), **n DMtns (w Panamint Mtns)**; US, s-c Can. [*D. micrantha* Nutt.] Probably a waif in CA.

D. subumbellata Rollins & R.A. Price (p. 227) WHITE MOUNTAINS CUSHION DRABA Per. **STS** cushion- or mat-forming, < 5 cm; hairs dense, bush-like. **LVS** basal, 2–6 mm, oblong to obovate, entire; surface hairs dense, bush-like; margin hairs 0 or not very prominent. **INFL** 2–10-fld. **FL**: petals 3–5 mm, yellow. **FR** 2–7 mm, ovate, inflated at base, not twisted; hairs branched; style 0.2–0.7 mm. **SEEDS** < 10, ± 1.5 mm; wings 0. RARE. Scree, talus, among rocks; > 3000 m. s SNH (e slope, nw Inyo Co.), **n W&I (White Mtns, Mono Co.)**.

ERUCA SALAD-ROCKET

Ann, per. **LVS**: basal petioled, deeply pinnately lobed to ± compound, rarely simple, entire; cauline short-petioled to sessile, shallowly lobed to entire. **INFL** gen dense, many-fld; bracts 0. **FL**: sepals erect, linear to oblong; petals widely obovate to oblanceolate, veins prominent, dark. **FR** linear to oblong or elliptic, beaked; pedicel ascending to erect, subappressed. **SEEDS:** 2 rows per chamber, orange or brown; cotyledons doubly folded. 3 spp.: Eurasia. (Latin: perhaps burn, from pl taste)

E. vesicaria (L.) Cav. ssp. *sativa* (Miller) Thell. (p. 227) **ST** gen branched from base, 2–10 dm; hairs toward base, simple, often reflexed. **LF** 5–15 cm, gen widely oblanceolate; hairs on upper surface 0, on lower along midrib, few, simple. **FL**: petals erect to ascending, 1.5–2 cm, ± 2 × sepals, white to yellowish. **FR** 1–2.5 cm, ± cylindric, 4-ribbed; pedicel 2–5 mm; beak ± flat, 1/2 to = valve. Disturbed areas, fields, roadsides; < 400 m. CaR, GV, SCoR, **DSon**; N.Am.; native to Eur. [*E. s.* Miller] May–Jul

ERYSIMUM WALLFLOWER

Robert A. Price

Ann to subshrub; hairs appressed, forked to many-branched. **LVS** in basal rosettes and cauline, simple, entire to lobed. **FL**: petals clawed, gen cream to orange; stigma 2-lobed. **FR** narrow, round, 4-sided, or ± flattened (gen parallel to septum); style 0.2–5 mm. **SEEDS** 1–2 rows per chamber, 1–4 mm, often ± winged. 160+ spp.: temp n hemisphere, esp Eurasia. (Greek: to help, from medicinal uses) [Rossbach 1958 Madroño 14:261–267] Incl *Cheiranthus*; native CA taxa all related to *E. capitatum*.

E. capitatum (Hook.) E. Greene WESTERN WALLFLOWER Bien or short-lived per. **STS** 1–few, 0.5–100+ cm. **LVS** ± linear to spoon-shaped, entire to toothed; tip acute; lower ± 2–25 cm; hairs 2–several-branched. **FL**: petals 12–30 mm, gen orange to yellow (cream to reddish). **FR** gen ascending, 3–15 cm, 1–4 mm wide, slightly fleshy when immature, ± 4-sided or slightly flattened when mature; style 0.2–5 mm. **SEED** 1–4 mm, 0.7–2 mm wide; wing gen 0 or at tip, rarely along sides. *n*=18. Common. Many habitats, gen inland; 0–4000 m. **CA (exc GV)**; to e-c US. Highly variable, with many intergrading local variants. 4 sspp. in CA. Key to vars. adapted by Margriet Wetherwax.

1. Fr flattened between seeds; basal lvs gen
   (ob)lanceolate . . . . var. *capitatum*
1' Fr gen 4-sided; basal lvs spoon-shaped . . . . . . var. *perenne*

var. *capitatum* (p. 227) Caudex gen short. **LVS** (ob)lanceolate, entire or dentate; tip sometimes obtuse; hairs 2–several-branched; lower lvs 2–10(25) cm. **FL**: petals orange to yellow. **FR** ascending, gen 4-sided, straight or slightly curved. Common. Many habitats; 0–4000 m. **CA (exc GV, SNH)**; w N.Am. [*E. asperum* (Nutt.) DC. var. *stellatum* J. Howell; *E. moniliforme* Eastw.] In CA gen bien, orange-fld, hairs 2–4-branched. Pls from DMoj with 2–3-branched hairs, ± entire lvs, ± flattened frs have been called var. *bealianum* (Jepson) Rossbach; yellow-fld GB pls with gen 2-branched hairs have been called *E. argillosum* (E. Greene) Rydb. ✿SUN,DRN:**4–6,15–17**&IRR: 1–3,**7,14,22–24**&SHD:8–10,12,**18–21**. Mar–Jul

    var. *perenne* (Cov.) R.J. Davis (p. 227)   Caudex short. **LVS**

gen oblanceolate, subentire to dentate; tip obtuse; hairs gen 2–3-branched; lower lvs 2–10 cm, ± spoon-shaped. **FL**: petals yellow. **FR** ascending, partly flattened, narrowed between seeds. Montane slopes to alpine barrens; 2000–4000 m. KR, CaRH, SNH, **n SNE (Sweetwater Mtns)**; OR, NV. [*E. p.* (Cov.) Abrams] ✿ IRR,DRN,SUN:1–3,**4–7**,8,9, **14–24**. Jun–Aug

GUILLENIA

Roy E. Buck

Ann, glabrous to ± hairy below. **ST** often hollow, ± glaucous. **LVS**: basal rosetted, often withering in fl, gen ± oblanceolate, entire to deeply cut, petioles < blades; upper lvs reduced. **INFL** longer in fr; bracts gen 0. **FL**: sepals pouched at base or not, greenish or not; petals ± linear to obovate; anthers coiled or ± curved when open; style gen ± tapered, stigma small, entire or shallowly 2-lobed. **FR** ascending to reflexed, ± cylindric; stalk-like base < 1 mm, = body width. **SEED** ± oblong, brownish or yellowish; wing 0. 3 spp.: w N.Am. (Father C. Guillen, Jesuit missionary, Mexico, born 1677)

G. lasiophylla (Hook. & Arn.) E. Greene (p. 227) CALIFORNIA MUSTARD **LVS**: lower blades < 22 cm, lanceolate to ± oblong, entire to cut; lower cauline gen > basal; uppermost lvs ± subsessile. **INFL**: pedicel in fr 0.5–4 mm. **FL**: parts ± erect; sepals 1.5–4 mm, not pouched at base, greenish (pinkish), narrowly scarious-margined; petals 3–6 mm, ± oblanceolate, gen white or pale yellow (pinkish), blade not channeled, not wavy-margined, narrowed to claw; style 0.1–2.4 mm. **FR** gen reflexed, 1–7 cm, straight or outcurved. **SEED** ± 1 mm, yellowish or brownish. *n*=14. Common. Dry, open, sometimes disturbed areas; < 2500 m. **CA (exc MP)**; to B.C., UT, nw Mex. [*Thelypodium l.* (Hook. & Arn.) E. Greene incl vars. *inalienum* Robinson, *rigidum* (E. Greene) Robinson, *utahense* (Rydb.) Jepson] Highly variable; needs study. Mar–Jun

HALIMOLOBOS

Bien, per, hairy or not. **ST** erect, herbaceous. **LVS** simple, entire to deeply lobed; basal often shed early. **FL**: sepals erect, bases not sac-like. **FR** 0.5–3 cm, linear, cylindric, hairy or not. **SEEDS** many, (1)2 rows per chamber, 0.4–1.5 mm, elliptic; wing 0; embryonic root at back of 1 cotyledon. 19 spp.: gen Mex. (Greek: sea pod, from resemblance to *Alyssum halimifolium*) [Rollins 1943 Contr Dudley Herb 3:241–265]

1. Cauline lf base tapered, not eared; fr densely hairy; basal lvs gen 0 . . . . *H. jaegeri*
1' Cauline lf base eared, not tapered; fr glabrous; basal lvs gen present . . . . *H. virgata*

H. jaegeri (Munz) Rollins (p. 233) Per; hairs multibranched; base woody. **STS** several–many, many-branched, 2–6 dm. **LVS**: basal gen 0; cauline 1–6 cm, oblanceolate to ovate, coarsely toothed to deeply lobed, lower short-petioled or sessile, upper sessile. **INFL** dense. **FL**: petals 4–5.5 mm, linear-oblanceolate to spoon-shaped, white; stamens not exserted. **FR** spreading, 1.5–2.5 cm, < 1 mm wide, gen slightly upcurved; pedicels spreading, 2–7 mm, densely hairy; style 0.5–2.0 mm. **SEED** ± 1 mm, embedded in septum. Limestone cliffs, steep rock outcrops, sagebrush/juniper areas; 1500–2500 m. **W&I, DMoj**; NV. [*H. diffusa* (A. Gray) O. Schulz var. *j.* (Munz) Rollins] May–Sep

H. virgata (Nutt.) Schulz VIRGATE HALIMOLOBOS Bien, per; hairs simple or multibranched; base not woody. **ST** gen 1, simple below, branched above, 1–4 dm. **LVS**: basal gen present, 2–6 cm, oblanceolate, entire or toothed; cauline 1.5–4 cm, lanceolate to oblong, entire to toothed, sessile. **INFL** open to dense. **FL**: petals 3–4 mm, spoon-shaped, white; stamens not exserted. **FR** erect, 2–4 cm, ± 1 mm wide, straight; pedicels ascending, 5–12 mm, ± hairy; style ± 0.5 mm. **SEED** ± 1 mm, not embedded in septum. RARE in CA. Meadows, near aspen groves, pinyon/juniper woodland; 2000–3000 m. **SNE**; to w Can, Colorado.

HIRSCHFELDIA

Ann, bien, per; hairs simple, dense. **ST** erect. **LVS** basal, cauline, lobed. **INFL** narrow, greatly elongating in fr; bracts 0. **FL**: sepals spreading, bases not or barely sac-like; petals obovate, clawed, pale yellow to white; stigma expanded. **FR** cylindric; beak abruptly swollen at base. 2 spp.: s Eur. (C. Hirschfeldt, horticulturist, 1742–1792)

H. incana (L.) Lagr.-Fossat (p. 233) Bien, per, canescent. **ST** branched from base and above, 2–10 dm. **LVS**: basal rosetted, flat on ground, pinnately lobed, terminal lobe > lateral, ± crenate-dentate; cauline ± sessile, ± simple, not clasping st. **FR** erect, appressed, 1–1.5 cm, glabrous; valves 3–7-veined; pedicel erect, 3–4 mm, stout, club-like, beak 3–6 mm, flat. **SEEDS** 1 row per chamber, spheric, reddish brown. 2*n*=14. Roadsides, creek bottoms, disturbed areas; < 1600 m. NCo, SNF, GV, CW, SCo, **DMoj**; native to Medit. [*Brassica geniculata* (Desf.) Ball; *H. adpressa* Moench]

HUTCHINSIA

Ann; hairs 0 or sparse, minute, branched. **INFL** open; bracts 0. **FL**: petals white. **FR** ± elliptic, ± flat perpendicular to septum; wing 0; notch 0. **SEEDS** several per chamber; wing 0; embryonic root at back of 1 cotyledon. 3 spp.: Eur, 1 apparently native in N.Am as well. (E. Hutchins, Irish botanist, 1785–1815)

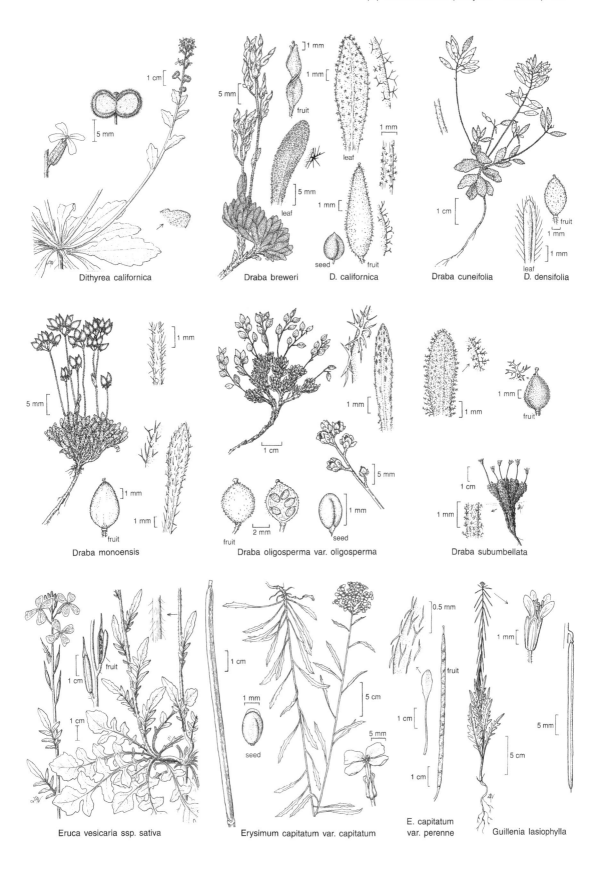

Dithyrea californica

Draba breweri

D. californica

Draba cuneifolia

D. densifolia

Draba monoensis

Draba oligosperma var. oligosperma

Draba subumbellata

Eruca vesicaria ssp. sativa

Erysimum capitatum var. capitatum

E. capitatum var. perenne

Guillenia lasiophylla

H. procumbens (L.) Desv. (p. 233) **ST** ± decumbent to ± erect, (3)5–10(15) cm, many-branched. **LVS**: basal and lower cauline petioled, 5–20 mm, obovate, entire to lobed; upper lvs fewer, sessile, reduced. **FL**: sepals ± 1 mm; petals ± 1 mm, obovate or narrower.

FR 2.5–3.5 mm, elliptic to ± obovate; valve veins prominent, netted; style < 0.2 mm. $2n=12$. Alkali flats, saline seeps, shaded sites, sagebrush scrub, juniper woodland; < 2600 m. **CA** (exc KR, SNH); to B.C., Labrador, Baja CA, also Eur. Mar–Jul

IDAHOA FLAT-POD

1 sp. (State of Idaho)

I. scapigera (Hook.) Nelson & J.F. Macbr. (p. 233) Ann, scapose, glabrous. **LVS** rosetted, several–many, petioled; blades 1–3 cm, ovate, entire to deeply lobed, lobes few, oblanceolate, terminal > others. **INFL**: peduncles many, (2)3–13 cm, 1-fld, slender, bractless. **FL**: sepals 1.4–2 mm, gen reddish or purplish; petals = or ± > sepals, white. **FR** 6–12 mm, round to widely ovate or elliptic, flattened parallel to septum; beak < 1 mm; style ± 0. **SEED** 3.5–5 mm wide incl ± 1 mm wide wing, ± round, flat; embryonic root at edges of both cotyledons. $2n=16$. Moist ledges, slopes, meadows, foothills; 600–1900 m. CaR, SnFrB (Mount Hamilton), NCoRI (Mount Saint Helena), n SN, s SNF, SCoRI, **GB**; to B.C., MT. ❀TRY. Feb–Apr

LEPIDIUM PEPPERGRASS, PEPPERWORT

Ann to shrub; hairs 0 or simple. **LVS**: basal not rosetted, gen petioled, entire to pinnately lobed; cauline short-petioled to sessile, sometimes clasping or surrounding st. **FL** small; sepals erect or spreading, oblong to ovate, shed early or persistent; petals linear to obovate, gen white, rarely yellowish, sometimes bristle-like or 0; stamens 6, 4, or 2. **FR** dehiscent, oblong to elliptic or obcordate, flat perpendicular to septum; pedicel cylindric or flat, winged or not. **SEEDS** 1 per chamber, gelatinous when wetted; wing narrow or 0; embryonic root at back of 1 cotyledon, rarely at edges of both. ± 175 spp.: ± worldwide. (Greek: little scale, from fr) [Hitchcock 1936 Madroño 3:265–300] Key to species adapted by Roy Buck.

1. Upper cauline lvs surrounding st . . . . *L. perfoliatum*
1' Upper cauline lvs not surrounding st
  2. Stigma gen exceeding fr; fr tip notched or not
    3. St prostrate to ascending; fr valves with prominent, ascending-spreading tips; style < to > 1/2 fr; petals yellow . . . . *L. flavum*
     4. Fr 3–4.5 mm, 3–3.5 mm wide; style < 1/2 fr . . . . var. *felipense*
     4' Fr 2–3 mm, 1.5–2.2 mm wide; style > 1/2 fr . . . . var. *flavum*
    3' St gen erect, outer decumbent or not; fr valves without prominent tips; style < 1/2 fr; petals white, rarely pale cream
     5. Pls rhizomed, individuals often in contact, in colonies; basal lvs < 3 dm, toothed . . . . *L. latifolium*
     5' Pls not rhizomed, individuals not in contact, not in colonies; basal lvs gen < 1 dm, gen lobed or compound
      6. Pl glabrous, grayish, ± woody; fr obovate to round . . . . *L. fremontii*
       7. Fr widely ovate to ± round or obovate, stalk above receptacle ± 0 . . . . var. *fremontii*
       7' Fr widely obovate to ± round, stalk above receptacle ± 1 mm . . . . var. *stipitatum*
      6' Pl with at least some hairs, not grayish, base rarely woody; fr elliptic to round
       8. St hairs long, flat and short, club-shaped; upper cauline lvs gen pinnately lobed; winter ann, bien . . . . *L. thurberi*
       8' St hairs all oblong, club-shaped or scale-like, short, rarely ± 0; upper cauline lvs gen entire; bien to per . . . . *L. montanum*
        9. Hairs of sts, pedicels ± dense, oblong; per, rarely bien . . . . var. *cinereum*
        9' Hairs of sts, pedicels sparse to ± dense, club-shaped or scale-like, or ± 0; per, ± woody at base or not . . . . var. *montanum*
  2' Stigma exceeded by fr; fr tip notched
    10. Pedicel in fr very flat, > (2)3 × wider than thick, gen = or < fr
     11. Lf oblanceolate to oblong, toothed or with obovate to oblong lobes; fr margin not upturned . . . . *L. lasiocarpum* var. *lasiocarpum*
     11' Lf linear or with linear lobes; fr margin upturned or not
      12. Fr hairs ± throughout, rarely 0; pedicel suberect to ascending, rarely ± reflexed . . . . *L. dictyotum* var. *dictyotum*
      12' Fr hairs 0 or marginal; pedicel ± recurved to ± ascending . . . . *L. nitidum*
       13. Fr gen fringed with hairs; st hairs dense . . . . var. *howellii*
       13' Fr glabrous; st hairs 0 to ± dense . . . . var. *nitidum*
    10' Pedicel in fr cylindric to flat, gen < or ± 2 × wider than thick, gen = or > fr (< fr in *L. strictum*)
     14. Lvs lobed or divided below and gen above; st gen prostrate, branched, 0.5–2 dm . . . . *L. strictum*
     14' Lvs entire to lobed below, entire to toothed, rarely ± lobed above; st erect, branched or not, gen > 2 dm
      15. Petals = to 2(3) × sepals; fr hairs 0 . . . . *L. virginicum*
       16. Pedicel in fr very flat, hairs gen dense; embryonic root at back of 1 cotyledon . . . . var. *pubescens*
       16' Pedicel ± cylindric, hairs gen sparse; embryonic root at edges of both cotyledons . . . . . . var. *virginicum*
      15' Petals 0 (rarely vestigial, < sepals); fr hairs 0 or minute . . . . *L. densiflorum*

17. Fr ± 2.5 mm, glabrous; pedicels very slightly flat, > 9 per cm . . . . var. *densiflorum*
17' Fr ± = or > 3 mm, hairy at least on margin or glabrous; pedicels flat at least on one side, < 9 per cm
  18. Pedicel flat on both sides, ± 2 × wider than thick; fr glabrous . . . . var. *ramosum*
  18' Pedicel flat gen on lower side, < ± 2 × wider than thick; fr hairy at least on margin or glabrous
    19. Fr glabrous . . . . var. *macrocarpum*
    19' Fr hairy at least on margin
      20. Fr hairy on margin . . . . var. *elongatum*
      20' Fr hairy ± throughout . . . . var. *pubicarpum*

L. densiflorum Schrader Ann, bien. **ST** gen 1, erect, branched above, 1–3(4) dm; hairs minute, flat, obtuse. **LVS:** basal rosetted, 3–10 cm, toothed to pinnately lobed; cauline entire to lobed, upper sessile, not clasping st. **FL:** sepals ± 1 mm; petals 0, rarely vestigial, white; stamens 2(4). **FR** ± 2.5–3.5 mm, oblong-obovate; hairs 0 or minute; valve tips rounded; pedicel ± ascending, slender, gen < ± 2 × wider than thick; style ± 0. **SEED** gen margined or winged. Sandy soils, plains, slopes, gen disturbed sites; < 1800 m. KR, CaR, n SNH, GV, SCo, **GB**, **DMoj**; to AK, Can, e US. Vars. difficult.

var. **densiflorum** (p. 233) Ann. **FR** ± 2.5 mm, 2 mm wide, widest at middle or just above, glabrous; pedicel very slightly flat. 2*n*=32. Grassy slopes, flood plains, disturbed areas; < 1200 m. CaR, GV, **DMoj**, expected elsewhere; to AK, se Can, e US.

var. **elongatum** (Rydb.) Thell. Bien. **FR** 3–3.5 mm, 2.5–3 mm wide, hairy on margins; pedicel ± flat. 2*n*=32. Sandy banks, alkaline soils, open areas; 700–1800 m. SCo, **GB**; to AK, ID.

var. **macrocarpum** G. Mulligan. Bien. **FR** ± 3 mm, 2.5–3 mm wide, glabrous; pedicel flat. 2*n*=32. Roadsides, rocky knolls, sagebrush scrub; < 2200 m. KR, n SNH, ScV, **SNE**, expected elsewhere; to B.C., n-c US.

var. **pubicarpum** (Nelson) Thell. Ann, winter ann. **FR** 3–3.5 mm, 2.5–3 mm wide, hairy ± throughout; pedicel ± flat. 2*n*=32. Roadbanks, meadows; < 1200 m. KR, CaR, **GB**, **DMtns**, expected elsewhere; to s-c Can, UT.

var. **ramosum** (Nelson) Thell. Ann. **FR** ± 3.5 mm, glabrous; pedicel flat on upper, lower sides, ± 2 × wider than thick. Sandy soils, slopes, flats, sagebrush scrub; < 800 m. **SNE**, **DMoj**; to WY, AZ.

L. dictyotum A. Gray Ann, low; hairs dense, spreading. **ST:** branched from base; outer branches ± decumbent, 2–20 cm. **LVS:** basal pinnately lobed, lobes linear; cauline 1–2.5 mm wide, linear, gen entire. **FL:** sepals 0.7–1 mm; petals ± > sepals, white, gen 0; stamens 4 or 6. **FR** 3–4.5 mm, ± 2.3 mm wide, notch < 1/5 × seed pouch; hairs ± throughout, rarely 0; pedicel suberect to ascending, rarely ± reflexed, 1.5–3.5 mm, very flat; style 0. **SEED** ± 2 mm. Saline soils, dry streambeds, fields; < 1000 m. **CA** (exc NW, SN); to WA, UT, Baja CA. 2 vars. in CA.

var. **dictyotum** (p. 233) **FR** ± 3 mm; valve tips erect, < 1 mm, rounded to acute, unwinged, notch ± closed or narrowly U- or V-shaped. Saline areas, playas, alkaline soils; < 1000 m. CaR, GV, SnFrB, SCoR, SW, MP, **W&I**, **DMoj**; to WA, UT, Baja CA. Mar–May

L. flavum Torrey Ann, glabrous. **ST** prostrate to ascending, 1–4 dm, branched from base. **LVS:** basal rosetted, 2–5 cm, oblong-lanceolate to spoon-shaped, margin irregularly to pinnately lobed, lobes ± rounded; cauline often toothed or entire toward tip. **FL:** petals 2–3 mm, yellow; stamens 6. **FR** ovate or ± round, ± winged; tip notched; pedicel cylindric or ± flat; style 1–1.5 mm, stigma exceeding fr. Alkaline or sandy soils; < 1400 m. **W&I**, **D**; n NV, Baja CA.

var. *felipense* C. Hitchc. BORREGO VALLEY PEPPERGRASS **FR** 3–4.5 mm, 3–3.5 mm wide; hairs ± 0; style < 1/2 fr. RARE. Sandy soils, creosote-bush scrub; ± 100 m. **DSon (Borrego Valley)**.

var. *flavum* (p. 233) **FR** 2–3 mm, 1.5–2.2 mm wide; hairs sac- or scale-like or 0; style > 1/2 fr. 2*n*=32. Common. Alkaline soils, flats; < 1400 m. **W&I**, **D**; NV, Baja CA. ❀TRY;DFCLT. Mar–May

L. fremontii S. Watson Per, gen shrubby, 4–10 dm, glabrous, gray. **ST** many-branched. **LVS** 3–10 cm, linear, gen pinnately lobed; lobes 3–9, 1–3 mm wide; upper 2–3 mm wide, entire or not. **INFL** many-branched, ± leafy. **FL:** petals ± 3 mm, obovate to spoon-shaped, white, claw slender. **FR** 5–8 mm; valves thin; veins faint. Sandy washes, barren knolls, gravelly soils, rocky slopes, ridges; < 1600 m. s **SNE**, **D**; to s UT, w AZ.

var. *fremontii* (p. 233) **FR** widely ovate to ± round or obovate; stalk above receptacle ± 0; style 0.5–0.8 mm. 2*n*=64. Sandy washes, barren knolls, gravelly soils; < 1600 m. Range of sp. ❀TRY. Mar–May

var. *stipitatum* Rollins **FR** widely obovate to ± round; stalk above receptacle ± 1 mm; style ± 1 mm. Uncommon. Rocky slopes, ridges; 1200–1500 m. **W&I, ne DMtns (Last Chance Mtns), DSon**.

L. lasiocarpum Torrey & A. Gray var. *lasiocarpum* (p. 233) Ann; hairs spreading, rigid. **ST** prostrate to erect, 1–2(3) dm, gen branched from base. **LVS** 2–15 cm, oblanceolate to oblong, toothed or with obovate to oblong lobes; upper sometimes subentire. **FL:** sepals 1–1.5 mm, widely oblong, purplish, margin thin, white, lower side gen hairy; petals minute or 0; stamens 2, 4, or 6. **FR** 2.5–4 mm, (ob)ovate to oblong-elliptic, gen hairy on surfaces or margins; notch < 1/5 × seed pouch; valve tips gen winged; pedicel gen < fr, > 2 × wider than thick, hairs on gen both sides, or 0 on lower; style 0. Dry flats, washes, roadsides, sagebrush; < 600 m. CCo, SW, **W&I**, **D**; to Colorado, AZ, Baja CA. [var. *georginum* (Rydb.) C. Hitchc. misapplied] 3 other vars. outside CA. Apr–May

L. latifolium L. Per 4–10(20) dm, ± glabrous, grayish, rhizomed. **LVS:** basal < 3 dm, 6–8 cm wide, toothed, long-petioled; cauline reduced but many 1–4 cm wide, lower petioled, upper sessile. **INFL:** panicle; hairs sparse or 0. **FL:** sepals < 1 mm, margins wide, white; petals white; stamens 6. **FR** ± 2 mm, ± round; notch 0; hairs sparse; pedicel >> fr, slender, cylindric, hairs sparse or 0; style 0. 2*n*=24. Beaches, tidal shores, saline soils, roadsides; < 1900 m. **CA** (exc KR, D); widespread US; native to Eurasia. Jun–Aug

L. montanum Nutt. Bien to shrub-like, rounded; root crown simple or branched. **STS** gen ± erect, 1–several, (1)2–4 dm; branches 0 or many. **LVS:** basal 3–15 cm, pinnately lobed, lobes gen dentate or dissected; cauline reduced, upper gen entire, petioled to not. **INFL** 2–4 cm, many-fld. **FL:** sepals 1.1 mm, hairs 0 to sparse; petals ± 2 mm, white to pale cream; stamens (2)6. **FR** 2.5–4 mm, ± ovate, glabrous; pedicel ± cylindric, slender; style > notch, 0.3–1 mm. Sandy, gravelly, often saline soils; 800–2100 m. CaR, **GB**, **DMoj**; to OR, MT, AZ, n Mex. 11 often indistinct vars. 3 vars. in CA.

var. *cinereum* (C. Hitchc.) Rollins Per, rarely bien; hairs of sts, pedicels ± dense, minute, short, oblong. **LVS:** basal pinnately lobed, lobes often dentate; upper gen linear, entire. **FR** hairy or not. Sandy areas, saline soils, ravines; 800–1900 m. **DMoj**; NV. [ssp. *c*. C. Hitchc.] Apr–May

var. *montanum* Per (± bien outside CA), ± woody at base or not; hairs of sts, pedicels sparse to dense, short, club-shaped, scale-like, or ± 0. **LVS:** basal pinnately divided, segments lobed or toothed. **FR** 2.5–3 mm, elliptic to ovate, glabrous. Clay or gravelly soils, washes, greasewood, saltbush scrub; 1400–1850 m. **GB**; to AZ, MT.

L. nitidum Torrey & A. Gray Ann. **ST** erect to spreading, 1–4 dm, slender, puberulent near infl, ± glabrous below; branches 0–many. **LVS**: basal 3–10 cm, deeply pinnately divided, segments 6–14, linear, entire to coarsely toothed or lobed; cauline less divided to simple, entire. **FL**: sepals ± 1 mm, ovate, hairy or not, not persistent in fr; petals < 1.5 mm, spoon-shaped, white; stamens gen 6. **FR** 2.5–4(6) mm, ovate-elliptic to ± round, smooth, shiny; notch narrow, 0.2–0.5 mm deep; hairs 0, rarely few, marginal, minute; pedicel ± recurved to ascending, very flat, ± densely puberulent; style 0. Alkaline soils, flats, slopes; < 1500 m. **CA** (exc e D); to c US. 3 vars. in CA.

var. *howellii* C. Hitchc. **ST**: hairs dense. **INFLS** many, dense. **FR** gen fringed with hairs; valves 3.5–5 mm, tips erect or ± ascending, not beak-like. Slopes, flats; 750–1500 m. **w DMoj**.

var. *nitidum* (p. 233) **ST**: hairs 0 to ± dense. **FR** gen glabrous; valves 3.5–6 mm, tips erect or ± ascending, not beak-like. Meadows, alkaline flats, vernal pools; < 1500 m. Range of sp. Feb–May

L. perfoliatum L. (p. 233) Ann, bien. **ST** (1)2–6 dm, ± puberulent below, glabrous above; branches 0–many. **LVS**: basal 2–3-pinnately lobed or divided, segments linear; upper cauline ovate to round, entire or minutely dentate, surrounding st. **FL**: petals ± 1.5 mm, narrowly spoon-shaped, yellowish; stamens 6. **FR** ± 4 mm, diamond-shaped-ovate; hairs 0, rarely few; pedicel gen > fr, slen- der, cylindric, glabrous; style ± 0.2 mm, ± = notch. $2n=16$. Roadsides, fields; 600–1500 m. CaR, GV, SCoRI, **GB**, **D**; widespread N.Am; native to Eurasia. Mar–Jun

L. strictum (S. Watson) Rattan (p. 233) Ann, prostrate; hairs spreading, simple, pointed. **ST** branched from base, 5–20 cm. **LVS**: basal, lower cauline 1–2-pinnately lobed or divided, 3–7 cm, 1–2 cm wide, segments gen oblanceolate or oblong; upper

cauline gen lobed to ± entire. **INFL** many, crowded. **FL**: sepals 1–1.5 mm, persistent in fr; petals 0 or vestigial; stamens 2. **FR** 2.5–3.5 mm, 2–3 mm wide, ovate to oblong-ovate, glabrous or sparsely ciliate; veins prominent, netted; valve tips ascending, gen winged, gen rounded; notch open, ± 0.4 mm; pedicel < fr, ± 2 × wider than thick; style 0. $2n=\pm32$. Uncommon. Disturbed sites, gen urban; < 300 m. **CA-FP**, **GB**; to OR, Colorado. Mar–May

L. thurberi Wooton Ann, bien. **ST** erect, 1–6 dm, branched; hairs long, flat, and short, club-shaped. **LVS**: basal 3–6 cm, petioled, pinnately divided, segments lobed or toothed; upper cauline gen pinnately lobed. **FL**: sepals 1–1.5 mm, hairs soft; petals 2–3 mm, white; stamens 6. **FR** 2–3 mm, 2–2.5 mm wide, elliptic to round, glabrous; pedicel gen < 3 × fr; style 0.4–0.7 mm, > notch. Saline flats, clay soils, grassland; < 1000 m. **DMoj**; to NM, Mex.

L. virginicum L. Ann; hairs 0 to dense. **ST** erect; branches above, gen 0 below. **LVS**: basal 5–15 cm, obovate, ± pinnately lobed to dissected; upper cauline reduced, entire, rarely dentate. **FL**: sepals ± 1 mm, hairs 0 or few, on lower surface; petals 1–2(3) mm, obovate, white; stamens 2(4). **FR** 2.5–4 mm, ± round, glabrous; notch shallow, gen > style; pedicel = or > fr, slender, cylindric to flat, puberulent or hairs large, gen 0 on lower surface. **SEED**: embryonic root gen at back of 1 cotyledon. Dry, gen open areas; < 2400 m. **CA** (exc KR, SNH); widespread N.Am. 9 vars. total, difficult. 4 vars. in CA.

var. *pubescens* (E. Greene) Thell. **ST** 2–7 dm; hairs rigid. **LVS**: cauline entire to shallowly dissected. **INFL**: pedicel ± flat, gen ± winged, hairy. **FR** 3–4 mm. Common. Disturbed areas, abandoned fields, meadows, roadsides; < 2400 m. Range of sp. Mar–Aug

var. *virginicum* **ST** 2–6 dm; hairs sparse to ± dense, minute, rounded. **LVS**: cauline dissected to ± entire. **INFL**: pedicel ± cylindric, hairs sparse. **SEEDS**: embryonic root at edges of both cotyledons. Old fields, roadsides; < 2400 m. Range of sp.; probably alien in CA.

LESQUERELLA BLADDERPOD

Bien, per; hairs ± dense, stellate, rarely simple, often silvery. **ST** gen arising laterally from basal lf cluster. **LVS** simple; basal petioled, linear to round, entire to pinnately lobed; cauline sessile or lower short-petioled, base tapered. **FL**: sepals erect or spreading, oblong to elliptic; petals widely obovate, entire, gen yellow; stamens 6. **FR** elliptic to ± round, plump or ± flattened parallel or perpendicular to septum; hairs stellate or 0; pedicel slender, straight or not. **SEEDS** 2–10(14) per chamber, ± round, ± plump or flat; margin gen 0; embryonic root at edges of both cotyledons. ± 95 spp.: gen N.Am, ± 12 S.Am. (L. Lesquereux, Am botanist, 1805–1889) [Rollins & Shaw 1973 Harvard Univ Press: 1–228] Key to species adapted by Roy Buck

1. Ann; basal rosette 0; caudex 0 . . . . *L. tenella*
1' Per; basal rosette present; caudex present . . . . *L. kingii*
　　2. Fr obovate to obcordate, often ± flat perpendicular to septum, sessile or short-stalked above receptacle, top truncate or shallowly notched; valves gen hairy inside; seeds 2–4 per chamber; petals 5.5–9.5 mm
　　. . . . . ssp. *kingii*
　　2' Fr ± round or elliptic to obovate, plump or ± flat parallel to septum, often short-stalked above receptacle, top obtuse; valves gen glabrous inside; seeds 2–8 per chamber; petals 9–13 mm . . . . . . ssp. *latifolia*

L. kingii (S. Watson) S. Watson Per, caudexed; hairs dense, 5–7-rayed. **STS** prostrate, decumbent or erect, few–many, 0.5–1.5(4) dm. **LVS**: basal blades 2–6 cm, widely elliptic or diamond-shaped to round, entire to ± lobed; cauline 0.5–2 cm, elliptic to obovate, lower short-petioled. **FL**: petals 5.5–13 mm. **FR** 3.5–9 mm; valve hairs dense outside, 0 to dense inside; pedicel 5–10(15) mm, gen S-shaped, rarely straight or 1-curved; style (2)4–9 mm. **SEEDS** 2–8 per chamber, flat, not margined. Dry soils, rocky sites; 1500–2750 m. SnBr, **SNE**, **DMtns**; to UT, Mex. 4 sspp. total, 3 sspp. in CA.

ssp. *kingii* (pl. 34) Pl prostrate to erect. **LVS**: basal entire to ± lobed, petioles 2–3 × blades. **FR** obovate to obcordate, ± wider than long, often ± flattened perpendicular to septum, often short-

stalked above receptacle; top truncate or shallowly notched; valve hairs appressed outside, gen present inside; septum ± not entire. **SEEDS** 2–4 per chamber. Dry, rocky soils, pinyon/juniper woodland; 1500–2750 m. **SNE**, **DMtns**; NV. [var. *cordiformis* (Rollins) Maguire & Holmgren misapplied] ❀TRY. Jul

ssp. *latifolia* (Nelson) Rollins & E. Shaw Pl prostrate to erect. **LVS**: basal widely elliptic or ± round, petioles = to 2 × blades. **FR** elliptic to round, ± flattened parallel to septum, often short-stalked above receptacle; top obtuse; valve hairs dense outside, 0 inside; septum entire; style 2–6 mm, stout. **SEEDS** 4–8 per chamber. $2n=10$. Gravelly soil, limestone outcrops, ridges; 1500–2300 m. **DMtns**; to UT, Mex. ❀TRY.

L. tenella Nelson (p. 233) Ann; hairs ± dense, 4–7-rayed, some often simple. **STS** several, decumbent to erect, 1.5–6 dm, often stout, much-branched. **LVS**: basal not rosetted, blades (1.5)3–6.5 cm, elliptic, entire to dentate; cauline linear to obovate, lower short-petioled. **FL**: petals 6.5–9(11) mm, yellow to orange. **FR** obovate to round, plump, ± sessile above receptacle; valve hairs sparse, stellate outside, dense, simple or stellate inside; pedicel 5–15 mm, spreading to recurved-S-shaped; style 2–4.5 mm. **SEEDS** 2–6 per chamber, flat, margined. 2*n*=10. Sandy soils, washes, slopes; < 1200 m. **D**; to UT, Mex. [*L. palmeri* S. Watson misapplied] Mar–May

LYROCARPA

Ann, per, woody at base or not; hairs dense ± throughout, multibranched. **ST** branched. **LVS** petioled, deeply pinnately lobed to wavy-margined. **INFL** open, elongate. **FL** fragrant; calyx cylindric, sepals erect, converging above, << petals, outer pair ± sac-like at base. **FR** obcordate to lyre-shaped, flat perpendicular to septum. **SEEDS** 3–10 per chamber; embryonic root at edges of both cotyledons. 3 spp.: sw US, Mex. (Greek: lyre fr, from fr shape)

L. coulteri Hook. & Harvey var. *palmeri* (S. Watson) Rollins (p. 233) COULTER'S LYREPOD Per; caudex ± woody, irregularly branched. **STS** several–many, 3–8 dm, straw-colored below, grayish above. **LF** 1–15 cm, gen < 1 cm wide, linear to ± ovate; lobes linear to oblong. **FL**: sepals 8–11 mm, ± 1.5 mm wide; petals 1.5–2.5 cm, linear to oblanceolate, brownish to dull purple, blades 1–3 mm wide, gen twisted, tip acute to acuminate. **FR** 1–2 cm, ± as wide as long; pedicel spreading or ascending, 3–7 mm; style ± 0, stigma with 2 large spreading lobes. **SEEDS** 3–5 per chamber, 2–3 mm, round; wing 0. UNCOMMON. Dry slopes, gravelly flats, washes; < 600 m. **DSon**; Baja CA. Dec–Apr

MALCOLMIA

Ann, per; hairs simple or gen branched. **ST** prostrate to erect, gen branched. **LF** simple to 2-pinnate. **INFL** bracted or not. **FL**: sepals erect, inner pair gen sac-like at base; petals prominently clawed, reddish or purple, rarely white. **FR** linear, sessile above receptacle, dehiscent; style tapered or 0; stigma lobes decurrent, gen fused. **SEEDS** many, 1 row per chamber; margins 0. ± 35 spp.: s Eurasia, n Afr. (W. Malcolm, London nurseryman, 1769–1820)

M. africana R.Br. (p. 233) Ann, often ± prostrate, stiff; branches many. **ST** 1.5–5 dm; hairs dense, small. **LVS** simple; lower 3–6 cm, oblanceolate, sparsely dentate, petioled; upper reduced, graduated to bracts. **FL**: petals 6–9 mm, rose-violet to pink. **FR** ascending, 4–6 cm, 1–1.5 mm wide, cylindric to ± 4-sided, ± narrowed between seeds; pedicel 1–2 mm; style 0, stigma lobes pointed. 2*n*=28. Disturbed areas, desert shrubland; 1250–2000 m. **SNE**; to Rocky Mtns; native to Medit.

PHOENICAULIS

1 sp. (Greek: visible stem)

P. cheiranthoides Torrey & A. Gray (p. 233) Per, ± cespitose; taproot thick, deep; caudex branches covered with old lf bases. **ST** ± scapose, 0.5–2 dm. **LVS** simple, entire, gray or whitish; hairs dense, multibranched; basal densely clustered, 3–15 cm, petioled. **INFL** many-fld. **FL**: sepals erect, 3–6 mm, often pink or purplish, esp on margin; petals 6–15 mm, obovate-oblanceolate, blade narrow, clawed, pinkish to reddish purple; anthers 1.5–2 mm. **FR** ± spreading, 1.5–8 cm, 2–5 mm wide, ± lanceolate, flat parallel to septum, glabrous; pedicel ± spreading, 1–3 cm; valve midrib prominent; style ± 1 mm, stigma ± unlobed. **SEEDS** 3–6, 2 rows per chamber, 3–4 mm, smooth; wing 0; embryonic root at edges of both cotyledons, rarely at edge of one. Common. Basalt outcrops, clay soils, slopes; 1500–3200 m. NW, CaR, **GB**; to WA, ID. [ssp. *glabra* (Jepson) Abrams] ❀TRY;DFCLT. May–Jul

PHYSARIA DOUBLE BLADDERPOD

Per, silvery; taproot enlarged; hairs stellate, gen ± sessile, appressed. **STS** several–many, gen from below rosette; outer prostrate to decumbent. **LVS**: basal obovate to ± round, < petioles; cauline entire, upper gen sessile. **FL**: petals yellow. **FR** widely obovate to ± round, inflated; tip notched to ± middle, gen also notched from base; valves shed with seeds enclosed; style persistent. **SEEDS** 2–6 per chamber, ± flat; margins 0; embryonic root at edges of both cotyledons. 22 spp.: w N.Am, gen Rocky Mtns, Great Basin. (Greek: bellows, from inflated fr)

P. chambersii Rollins (p. 237) Pl cespitose. **ST** unbranched, 5–15 cm. **LVS**: basal 3–6 cm, 1–2 cm wide, entire to dentate, tip obtuse; cauline 1–2 cm, spoon-shaped, tip gen acute. **INFL** 2–10 cm, dense, ± umbel-like. **FR** 1–1.5 cm, hairy; notch in base shallow or ± 0; septum 4–6 mm; pedicel ascending to spreading, 8–15 mm; style 6–8 mm, exserted from notch. **SEEDS** gen 4 per chamber. 2*n*=8, 10,16,24. Limestone soils; 1500–2500 m. **n DMtns (Clark, Grapevine mtns)**; to OR, UT, AZ. ❀TRY;DFCLT. May

POLYCTENIUM

1 sp. (Greek: many combs, from lvs)

P. fremontii (S. Watson) E. Greene Per, cespitose, often glaucous. **STS** erect, few–several, gen simple, 5–15 cm; caudex without persistent lf bases, branched. **LF** 1–2 cm, ± sessile, deeply pinnately lobed to compound, rigid; lobes or lflets ± 1–5 mm, linear, often sharp-pointed with terminal hair; hairs stiff, gen branched. **INFL** ± dense, ± umbel-like; bracts 0. **FL**: sepals erect, 2–3 mm, oblong; petals 5–6 mm, white to pale purple. **FR** ascending to erect, often flat perpendicular to septum, glabrous;

pedicel ascending, 4–6 mm, straight. **SEEDS** 12–28, 1 row per chamber, ± 1 mm, oblong; wing 0; embryonic root at back of 1 cotyledon. Saline soils, playas, lake margins, wet meadows; 1000–2000 m. **GB**; to ID, NV. 2 vars. in CA.

var. *confertum* Rollins (p. 237) **FR** flat perpendicular to septum. Uncommon. Playas, lake margins; ± 2000 m. **SNE**; NV. ❀TRY.

RAPHANUS WILD RADISH

Ann, bien, erect, gen scabrous, taprooted; hairs simple, rigid. **ST** gen branched, esp above. **LVS**: basal and lower cauline petioled, gen pinnately lobed to compound, terminal lobe or lflet widely ovate to round, >> lateral, toothed; upper cauline short-petioled to sessile, ± dentate or few-lobed. **INFL** many-fld; bracts 0. **FL**: sepals erect, inner pair sac-like at base; petals long-clawed. **FR** indehiscent, longitudinally grooved, esp below, transversely jointed; lower part seedless, very short, or 0; upper part seeded, linear to ovate or dagger-shaped, beaked. **SEEDS**: 1 row per chamber, ± spheric; wing 0; cotyledons doubly folded. 3 spp.: Medit. (Greek: appearing rapidly, from seed germination)

R. raphanistrum L. (p. 237) JOINTED CHARLOCK Pl hairs sparse. **ST** gen 1, 3–8 dm. **LVS**: basal 6–20 cm. **FL**: petals 15–20 mm. **FR** 4–8 cm (incl beak), 3–6 mm wide; pedicel ascend-ing, 10–25 mm. 2*n*=18. Disturbed areas, fields, roadsides; < 800 m. C-FP, **GB**; N.Am; native to Medit Eur. Apr–Jun

RORIPPA YELLOW or WATER CRESS

Ann, bien, per; hairs simple, rarely sac-like. **STS** 1–many, prostrate to erect, branched or not, often from center of basal rosette, rooting at nodes or not. **LVS**: basal, lower cauline sessile or short-petioled, entire to pinnately compound; cauline reduced upward. **INFL** terminal and lateral. **FL** gen small; petals 0 or obovate to narrowly spoon-shaped, pale to bright yellow or white. **FR** linear to round, plump; valves 2 or 4(6); pedicel ascending to recurved, slender, gen with 2 minute glands at base; style 0 or prominent, persistent. **SEEDS** 10–200, (1)2 rows per chamber, dense, gen plump; wing 0; embryonic root at edges of both cotyledons. ± 75 spp.: worldwide. (Old Saxon: for these, perhaps other crucifers) [Stuckey 1972 Sida 4:277–340] *R. curvisiliqua* may be found along the Colorado River, near Yuma.

1. Petals white; cauline lvs 1-pinnate; pl in aquatic to very wet places, gen rooting at node
   . . . . **R. nasturtium-aquaticum**
1' Petals yellow; cauline lvs simple, entire to deeply pinnately lobed, rarely 1-pinnate; pl often in wet
   but gen not aquatic places, rarely, if ever, rooting at nodes
   2. Per, from creeping underground roots . . . . **R. sinuata**
   2' Ann, bien, rarely ± short-lived per, gen from taproot
      3. Pl 1–5 dm, prostrate to decumbent, rarely ascending, often with several–many sts from base
         . . . . **R. curvipes** var. **curvipes**
      3' Pl 4–10(14) dm, ± erect, with 1 dominant st from base . . . . **R. palustris** var. **occidentalis**

R. curvipes E. Greene (p. 237) Ann, rarely ± short-lived per, gen ± glabrous. **ST** 1–5 dm, branched, often from near base; hairs often below, ± sparse. **LVS**: basal ± entire to deeply pinnately lobed, rarely partly ± 1-pinnate, gen petioled; cauline 0 or 4–8 cm, 5–15 mm wide, oblong, obovate, or oblanceolate, clasping st, entire to ± pinnately lobed. **FL**: petals ascending, oblong to ± tongue-shaped, yellow. **FR** straight to upcurved, gen narrowed near middle, smooth, glabrous; pedicel 2–5 mm; style 0.5–1 mm. **SEED** ± reniform, finely papillate. Wet sites; < 2300 m. CA-FP, **SNE**; N.Am. 2 vars. in CA.

var. *curvipes* Pl prostrate to decumbent, rarely ascending. **FL**: petals 0.5–1 mm, gen < sepals. **FR** 1.5–5 mm, gen 2 × longer than wide, lanceolate to ovate; tip acute to ± obtuse; pedicel gently recurved or spreading. Mud flats, stream beds, hillside seeps; 800–2300 m. SN, SNE; to c US.

R. nasturtium-aquaticum (L.) Hayek (p. 237) WATER CRESS Per, ± glabrous. **ST** submersed, ± floating, or prostrate on mud, 1–6 dm, rooting at nodes. **LVS** many, 1-pinnate; lflets gen 3–7, widely oblong to ovate, ± entire or wavy-margined. **FL**: petals 3–4 mm, white. **FR** spreading to ± erect, 10–15 mm, narrowly oblong, straight to upcurved; pedicels ± spreading, ± straight, not bracted, junctions not flat, lower 8–15 mm; style < 0.5 mm or 0. **SEEDS** 2 rows per chamber, ± 1 mm wide, ± round. 2*n*=32. Streams, springs, marshes, lake margins; < 2700 m. CA-FP, **n SNE, W&I, DMtns**; temp worldwide. [*Nasturtium officinale* R.Br.] ❀ WET-fresh water: 1, **2–9**, 10–13, **14–24**; widely cult for edible greens. Mar–Nov

R. palustris (L.) Besser (p. 237) Ann, bien, short-lived per, ± erect, with 1 dominant st from base. **ST** > 3 mm diam; branches 0 or ± from base. **LVS**: basal and lower cauline short-petioled to sessile, ± clasping st or not, 5–20(30) cm, oblong to oblanceolate, irregularly dentate to deeply pinnately lobed, yellow. **FL**: petals 1–2.5(3.5) mm, widely spoon-shaped, yellow. **FR** gen > 2 mm wide, straight to ± upcurved; pedicel spreading to ascending, 2–14 mm; style 0.2–1 mm, abruptly attached to fr tip. **SEED** 0.5–1 mm, widely reniform. Gen wet areas; < 2000 m. **CA**; temp N.Am, Eurasia. 10 vars. total. 2 vars. in CA.

var. *occidentalis* (S. Watson) Rollins **ST** 4–10(14) dm; hairs 0 or below, sparse. **LF**: hairs 0. **FR** 7–15 mm, oblong. Stream beds, sand bars, wet depressions; < 2000 m. **CA**; to NM, Mex, AK. [*R. islandica* (Oeder) Borbás vars. *fernaldiana* F.K. Butters & Abbe, *o.* (S. Watson) F.K. Butters & Abbe] Apr–Jul

R. sinuata (Torrey & A. Gray) A. Hitchc. (p. 237) Per. **ST** prostrate to ascending, 1–4(5) dm; hairs sparse to dense, sac-like. **LVS**: basal and lower cauline short-petioled, oblanceolate, deeply dentate to deeply pinnately lobed, rarely 1-pinnate; middle, upper sessile, ± clasping st, 3–6(8) cm, with sac-like hairs on midribs. **FL**: sepals not persistent in fr; petals 3.5–5.5 mm, oblanceolate to spoon-shaped, yellow. **FR** 5–13 mm, narrowly oblong, upcurved; pedicel ascending to recurved, 5–11(15) mm, S-shaped or not; style 1–2 mm. **SEED** 0.7–1 mm, ± widely reniform, angled. Lake shores, playas, wet depressions; 900–1900 m. **GB**; to Can, c US, TX. May–Sep

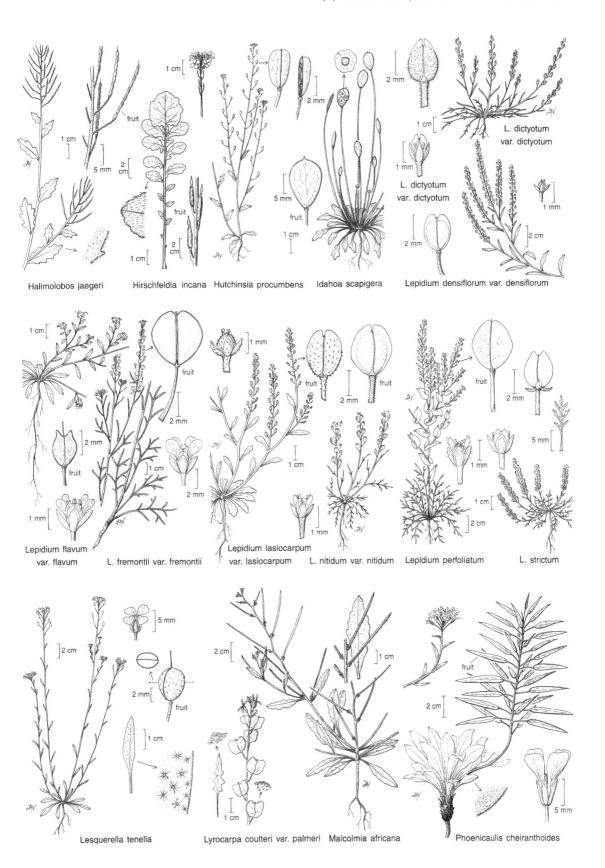

Halimolobos jaegeri   Hirschfeldia incana   Hutchinsia procumbens   Idahoa scapigera   Lepidium densiflorum var. densiflorum

L. dictyotum
var. dictyotum

L. dictyotum
var. dictyotum

Lepidium flavum
var. flavum   L. fremontii var. fremontii   Lepidium lasiocarpum
var. lasiocarpum   L. nitidum var. nitidum   Lepidium perfoliatum   L. strictum

Lesquerella tenella   Lyrocarpa coulteri var. palmeri   Malcolmia africana   Phoenicaulis cheiranthoides

SIBARA

Ann, bien. **STS** 1–several from base, gen erect; hairs 0 or simple or branched. **LVS**: basal and lower cauline pinnately lobed to 1-pinnate; upper often simpler, grayish green. **INFL** elongate, open. **FL** small; sepals ± oblong to ovate, bases gen not sac-like; petals spoon-shaped to ± oblong, white to purplish. **FR** linear, flat parallel to septum or ± cylindric; valve veined or not. **SEED** oblong or ± round, winged or not; embryonic root at back of 1 cotyledon or at edges of 1 or both. 10 spp.: N.Am. (Anagram of Arabis) Key to species adapted by Roy Buck.

1. Fr spreading to ± pendent, 1–1.5 cm, hairs sparse . . . . ***S. deserti***
1' Fr ascending, 1.5–3 cm, hairs 0 . . . . ***S. rosulata***

S. deserti (M.E. Jones) Rollins (p. 237) Ann; hairs minute, branched. **ST** 1, 1–3 dm; branches 0 or above. **LVS**: basal shed early; cauline 2–4 cm, hairs sparse; lower cauline deeply pinnately divided, segments 4–8 mm. **FL**: sepals 1.5–2 mm, oblong; petals 2–3 mm, spoon-shaped, white. **FR** spreading to ± pendent, 10–15 mm, < 1.5 mm wide, flat, ± curved; hairs sparse; pedicel spreading to reflexed, 3–4 mm, hairy or not; style 1–1.5 mm. **SEED** < 1 mm wide, oblong; wing 0; embryonic root at edges of both cotyledons. $2n=26$. Washes, steep hillsides, dry flats; 350–1300 m. **n DMoj**; NV. Mar–Apr

S. rosulata Rollins (p. 237) Ann. **STS** 1–few, 1–3 dm; branches above. **LVS**: basal rosetted, 3–5 cm, persistent, deeply pinnately lobed, lobes 4–8 mm, 1–2 mm wide, hairs 0 or sparse, simple or branched; lower cauline ± lobed, upper entire. **FL**: sepals 1.5–2 mm, oblong; petals 2.5–3 mm, narrowly spoon-shaped, white. **FR** ascending, 15–30 mm, < 1.5 mm wide, flat, glabrous; pedicel ascending, 2–3 mm, upcurved, glabrous; style 2–3 mm, expanded toward tip. **SEED** < 1 mm wide, oblong; wing 0; embryonic root at edges of both cotyledons. Sandy or gravelly washes, scree, calcareous rubble; 250–950 m. **e DMoj**. Mar–Apr

SISYMBRIUM

Ann, bien; hairs simple or 0. **ST** gen erect, branched. **LVS** petioled or sessile, variously lobed or dissected, green or glaucous. **INFL** many-fld; bracts 0. **FL**: sepals erect to ± spreading; petals yellow, clawed. **FR** ascending to erect, linear to ± awl-shaped, gen cylindric, straight or ± curved, hairy or not; valves prominently veined; style conic or 0, stigma 2-lobed. **SEEDS** many, gen 1 row per chamber; margin 0; not or ± gelatinous when wet; embryonic root at back of 1 cotyledon, sometimes obliquely so. ± 90 spp.: most continents. (Greek: for various mustards) Key to species adapted by Roy Buck.

1. Upper cauline lvs with lobes or lflets thread-like to linear, terminal ± = lateral; pedicel in width ± = or >
   fr; outer 2 sepals with erect horns at tip . . . . ***S. altissimum***
1' Upper cauline lvs ± entire or with lobes or lflets lanceolate to triangular, terminal > lateral; pedicel in
   width ± = to << fr; sepal horns 0
   2. Pedicel width ± = fr width . . . . ***S. orientale***
   2' Pedicel width < fr width
      3. St branched from near base; frs gen overtopping fls, gen 3–4 cm; hairs 0 or above, few, ± short . . . . . . ***S. irio***
      3' St branched esp above; frs gen not overtopping fls, 2–3.5 cm; hairs at least on lower sts, long . . . . . ***S. loeselii***

S. altissimum L. (p. 237) TUMBLE or JIM HILL MUSTARD Ann. **ST** 30–150 cm; branches many, esp above. **LVS** petioled, < 15 cm, widely lanceolate; basal, lower cauline ± pinnately lobed to 1-pinnate, lobes or lflets ± lanceolate, dentate; upper with thread-like to linear lobes or lflets, terminal ± = lateral. **FL**: sepals ± 4 mm, outer 2 with erect horns at tip; petals 6–8 mm, pale yellow. **FR** 5–10 cm, ± 1 mm wide, linear, rigid, branch-like; beak 0; pedicel gen spreading, 4–10 mm, width ± = or > fr width. **SEEDS** many, ± 1 mm; embryonic root obliquely at back of 1 cotyledon. $2n=14$. Disturbed areas, fields, roadsides; < 2500 m. **CA**; N.Am; native to Eur. May–Jul

S. irio L. (p. 237) LONDON ROCKET Ann. **ST** 15–50 cm, branched from near base; hairs 0 or above, few, ± short, thin. **LVS**: basal not clustered, petioled, pinnately lobed, terminal lobe > lateral, often hastate; upper cauline pinnately lobed to ± entire. **FL**: petals 2.5–4 mm, barely > sepals, narrowly oblong, pale yellow, claws long. **FR** gen overtopping fls, gen 3–4 cm, ± 1 mm wide; pedicel ascending, 5–11 mm, width < fr width; style ± 0.5 mm. **SEED** < 1 mm, oblong, ± papillate; embryonic root obliquely at back of 1 cotyledon. $2n=14$. Disturbed areas, orchards, roadsides; < 800 m. GV, SW, **W&I, DMtns**; to TX, Baja CA; native to Eur. Jan–Apr

S. loeselii L. (p. 237) Ann. **ST** 40–120 cm; branches esp above; hairs at least below, sparse to dense, spreading or reflexed, long. **LF** petioled, < 1.5 dm, widely deltate-lanceolate, pinnately to irregularly lobed; lobes ± linear to lanceolate, few-toothed. **FL**: sepals 2.5–3.5 mm, lanceolate; petals 6–8 mm, widely obovate, yellow, claws long, narrow. **FR** gen not overtopping fls, 2–3.5 cm, linear, straight or ± incurved; pedicel spreading to ascending, 1–2 cm, width < fr width; style ± 0.5 mm. **SEED** ± 0.7 mm; embryonic root at back of 1 cotyledon, sometimes obliquely so. $2n=14$. Disturbed areas, fields, roadsides; 1200–2000 m. **GB**; to Can, n US; native to Eur.

S. orientale L. (p. 237) Ann; hairs ± soft, of different sizes. **ST** 3 dm, branched. **LVS**: basal clustered, deeply pinnately lobed or compound; cauline lanceolate, with 2 basal, lanceolate, spreading lobes, margins gen entire or few-toothed. **FL**: petals 8–10 mm, pale yellow. **FR** 3–10 cm, linear, straight; beak 0; hairs 0 or sparse; pedicel ascending, 3–6 mm, width ± = fr width; style 1–3 mm, club-shaped; embryonic root at back of 1 cotyledon, sometimes obliquely so. $2n=14$. Disturbed areas, fields; < 1000 m. **CA** (exc NW, CaR, SN); to TX, Baja CA; native to Eur. May

STANLEYA PRINCE'S PLUME

Ann, per, shrub, often glaucous; hairs 0 or simple. **ST** 2–15 dm, branched or not. **LVS**: basal clustered or not; cauline petioled or not, entire to deeply lobed. **INFL** dense, gen > 1 dm; buds club-shaped. **FL**: sepals spreading to reflexed, linear-oblong; petals gen conspicuous, yellow to white; filaments ± equal, >> petals. **FR** linear, flat parallel to septum or ± cylindric; stalk above receptacle 1–3 cm; style ± 0 or short. **SEED** oblong; margin 0; embryonic root at edges of 1 or both cotyledons. 6 spp.: w US. (E. Stanley, English ornithologist, 19th century) Concentrates selenium to TOXIC levels, but rarely eaten. Key to species adapted by Roy Buck.

1. Lower cauline lvs entire to sharp-toothed; petal hairs 0, blades yellow to whitish . . . . *S. elata*
1' Lower cauline lvs deeply pinnately lobed; petal hairs on inner side of claws, dense, blades yellow . . *S. pinnata*
  2. Shrub; trunk short, 4–8 cm wide; lvs yellow-green . . . . var. *inyoensis*
  2' Per to subshrub; trunk 0, sts several–many from base; lvs gray-green, gen glaucous . . . . var. *pinnata*

S. elata M.E. Jones (pl. 35) Per, erect, coarse. **STS** 1–several, 6–15 dm, glabrous; branches 0 or above. **LF** petioled, thick; hairs 0 or sparse, minute; blade 8–15 cm, 1.5–2.5 cm wide, widely oblong to lance-ovate, entire to sharply toothed, tip obtuse to acute. **INFL** 6–20 cm, dense. **FL**: sepals reflexed, 8–12 mm, linear-oblong, yellowish, glabrous; petals 8–10 mm, ± 1 mm wide, yellow to whitish, glabrous, claw bases wide, blades reduced, ± 1 mm wide; filament bases enlarged, papillate. **FR** spreading to ± recurved, 5–10 cm, ± cylindric; stalk above receptacle ± 20 mm; pedicel spreading, 5–10 mm, glabrous; style ± 1 mm. **SEED** ± 2 mm, ± 1 mm wide, brown; embryonic root obliquely at back of 1 cotyledon. $2n=28$. Among boulders, canyons, shrubland; 1300–2000 m. **SNE**, **DMtns**; NV. May–Jul

S. pinnata (Pursh) Britton (p. 237) **STS** several–many, 4–15 dm, glaucous; hairs 0 or sparse; base branched, woody. **LVS** petioled; basal and lower cauline 5–15 cm, 2–5 cm wide, widely lanceolate, deeply pinnately lobed, hairs 0 or short; upper cauline linear-lanceolate to ovate, entire or few-lobed. **INFL** 1–3 dm, dense; buds yellowish. **FL**: sepals spreading or reflexed, 10–15 mm; petals 10–18 mm, yellow, claws with dense, long, wavy hairs on inner side; stamens >> petals, ± equal. **FR** spreading to ± downcurved, 3–8 cm, ± cylindric; stalk above receptacle 10–25 mm; pedicel spreading, 6–12 mm, hairs 0 or few. **SEED** ± 2 mm, oblong, plump; embryonic root at back of 1 cotyledon. Gen open sites, slopes, canyons; < 1850 m. SCo (Conejo Valley), WTR, **GB**, **DMoj**; to c US.

var. *inyoensis* (Munz & Roos) Rev. Shrub. **ST**: trunk short, 4–8 cm wide. **LF** yellow-green. Sandy areas, creosote-bush scrub; 850–1000 m. **W&I**, **n DMoj**; NV. ❀TRY;DFCLT.

var. *pinnata* Per to subshrub. **STS** several–many from base; trunk 0. **LF** gray-green, gen glaucous. $2n=24$. Open areas, chaparral, desert shrubland, woodland, seashore dunes; < 1850 m. Range of sp. ❀SUN,DRN,DRY:7–12,14–16,**18**,19–24;DFCLT.

STREPTANTHELLA

1 sp. (Latin: small *Streptanthus*)

S. longirostris (S. Watson) Rydb.(p. 241) Ann, gen glaucous; hairs 0 or below, sparse, minute, simple. **ST** gen 1; branches many, rarely 0. **LVS**: basal, lower cauline (2)3–6 cm, narrowly oblanceolate, wavy-margined, entire to dentate, rarely lobed; upper cauline ± linear, clasping st or not, gen entire. **FL**: sepals 2–3 mm, upper surface greenish to purple, lower surface white; petals 3–4 mm, narrowly spoon-shaped, white to yellowish with purple veins, blades wavy; anthers < 1 mm, gen ± exserted. **FR** pendent, 35–45 mm, ± 1.5 mm wide, linear, flat parallel to septum; pedicel downcurved to reflexed, 1.5–3 mm; valves ± 3.5 mm, narrowed to a beak-like, indehiscent tip; style ± 0. **SEED** flat; wing narrow; embryonic root at back of 1 cotyledon. $2n=28$. Common. Sandy soils, desert shrubland, woodland; < 2000 m. s SnJV, SCoRI, SW, **W&I**, **D**; to WA, Colorado, NM, Baja CA. [var. *derelicta* Howell] Mar–May

STREPTANTHUS JEWELFLOWER

Roy E. Buck, Dean W. Taylor, and Arthur R. Kruckeberg

Ann to per, glabrous to bristly, gen ± glaucous. **LVS** ± entire to pinnately compound; basal gen rosetted, gen ± petioled; cauline linear to (ob)ovate, often clasping. **INFL** gen ± open; bracts gen 0. **FL** biradial or bilateral; calyx gen ± urn-shaped, sepals erect, gen not green, bases ± pouch-like, gen keeled; petals gen exserted, blade gen narrower than claw, ± channeled, margins ± wavy, gen ± scarious; stamens gen in 3 free pairs; style 0 or short, stigma gen ± entire, blunt. **FR** long, gen strongly compressed parallel to septum. **SEEDS** gen compressed, gen ± winged. ± 40 spp.: sw US, n Mex. (Greek: twisted fl, from wavy-margined petals) [Dolan & LaPré 1989 Madroño 36:33–40; Kruckeberg & Morrison 1983 Madroño 30:230–244] *Caulanthus* sometimes incl here. Calluses on lf margins of some mimic pierid butterfly eggs, reducing larval herbivory. Variable, complex; needs study. Key to species adapted by Roy Buck.

1. Stigma lobes gen 0; pl rhizomed . . . . *S. oliganthus*
1' Stigma ± 2-lobed; rhizomes 0 . . . . *S. cordatus*
  2. Fr gen 4–6 mm wide; n SNE, DMtns . . . . var. *cordatus*
  2' Fr 2.5–4 mm; s SNE, W&I . . . . var. *duranii*

S. cordatus Nutt. Per 2–10 dm, gen simple, glabrous. **LVS**: basal widely obovate, toothed above middle, teeth often bristly, petioles = lvs, often ciliate; cauline few-toothed to entire, gen acute. **FL**: calyx biradial, sepals 8–13 mm, yellowish green in bud becoming purple in fl, tips gen bristly; petals exserted, 10–14 mm, linear, purple; stamens free, equal; stigma 2-lobed. **FR** ascending to ± erect, 5–10 cm, 2.5–6 mm wide, straight. $n=12$. Rocky or sandy sagebrush scrub, pinyon/juniper woodland, ponderosa-pine forest; 1200–3100 m. e CaRH, s SNH, **GB**, **DMtns**; to se OR, WY, n NM. 3 vars. in CA.

var. *cordatus* (p. 241) Pl 2–6 dm. **LVS**: upper cauline widely ovate, acute to obtuse, clasping. **FR** 5–8 cm, 4–6 mm wide. Common. Rocky or sandy sagebrush scrub or pinyon/juniper woodland; 1400–2800 m. e CaRH, MP, **n SNE, e DMtns**; to se OR, WY, n NM. ❀TRY;DFCLT.

var. *duranii* Jepson Pl 2–7 dm. **LVS**: upper cauline widely ovate, acute to obtuse, clasping. **FR** 5–9 cm, 2.5–4 mm wide. Uncommon. Talus, calcareous outcrops; 1800–3100 m. **s SNE, W&I**. May–Jul

S. oliganthus Rollins MASONIC MOUNTAIN JEWELFLOWER Per 2–5 dm, rhizomed, gen simple, glabrous, glaucous. **LVS** 2–8 cm; basal (ob)lanceolate, entire, finely ciliate, petiole > blade; cauline lanceolate to oblong. **FL**: calyx biradial, sepals 7–10 mm, yellow in bud, purple in fl; petals 10–13 mm, tips purple, margins curled; upper stamens free; stigma flat. **FR** spreading to ascending, 5–8 cm, 2.5–3.5 mm wide, straight or slightly curved, wide, flat. 2*n*= 28. RARE. Rocky sites or talus; 2100–2800 m. **n&c SNE (Sweetwater, Masonic mtns), n W&I (n White Mtns)**; w-c NV. [*S. cordatus* var. *exiguus* Jepson] Jun–Jul

THELYPODIUM

Winter ann to bien, per; hairs 0 or simple. **ST** gen erect, branched or not. **LVS**: basal gen rosetted, petioled, gen shed early, entire to pinnately lobed; cauline petioled or sessile, often clasping. **FL**: sepals erect to reflexed, greenish, white, lavender, or purplish, bases sac-like or not; petals linear to oblanceolate, white, lavender, or purple; stamens equal or 4 long, 2 short, paired filaments rarely ± fused. **FR** erect or spreading, narrowly linear, ± narrowed between seeds, cylindric or ± flat parallel to septum, gen stalked above receptacle; pedicel ± flat at base or not, gen expanded at tip; stigma in width < style tip, gen entire. **SEEDS** 1 row per chamber, ± flat; wing gen 0; embryonic root at edge or toward back of 1 cotyledon. 20 spp.: w N.Am. (Greek: female foot, from fr stalk above receptacle) [Al-Shehbaz 1973 Contr Gray Herb 204:1–148] Key to species adapted by Roy Buck.

1. Cauline lvs petioled, lower and middle pinnately lobed, rarely just dentate; fr ± flat parallel to septum to cylindric, rarely ± 4-sided; style ± club-shaped or obconic
   2. Fr pedicel ± straight, spreading, rarely ascending; st solid, not inflated; fr spreading; petal blades linear . . . . ***T. laciniatum***
   2' Fr pedicel upcurved, gen erect at tip; st gen ± hollow, inflated or not; fr ± erect, often appressed; petal blades oblanceolate to spoon-shaped . . . . ***T. milleflorum***
1' Cauline lvs sessile, lower and middle gen entire, rarely dentate; fr cylindric; style cylindric or tapered to tip
   3. Cauline lvs narrowed to base, ± without basal lobes, ascending . . . . ***T. integrifolium***
      4. Petals white; fr pedicel whitish, 6–13 mm, stalk above receptacle 1–3 mm . . . . ssp. *affine*
      4' Petals lavender to purple, rarely whitish; fr pedicel not whitish, 2–5(6) mm, stalk above receptacle 0.5–1 mm . . . . ssp. *complanatum*
   3' Cauline lvs expanded to base, with basal lobes, erect-appressed to ascending
      5. Ann or bien, rhizomes 0 . . . . ***T. crispum***
      5' Per, rhizomed . . . . ***T. brachycarpum***

T. brachycarpum Torrey SHORT-PODDED THELYPODIUM Per, rhizomed. **ST** 3–8(12) dm, branched or not; hairs ± 0 or near base, sparse. **LVS**: basal 4–14(20) cm, ± entire to pinnately lobed, thickish, glaucous, hairy or not; cauline sessile, 2–6 cm, sagittate, clasping st, entire, glaucous. **INFL** spike-like, dense. **FL**: petals white, crinkled. **FR** 1–3 cm, cylindric, narrowed between seeds; stalk above receptacle 2–5(5) mm, slender; pedicel spreading, rarely ± ascending, 1–2 mm, stout, base flat; style 0.5–1 mm, slender. **SEED** plump; embryonic root at back or toward edge of 1 cotyledon. UNCOMMON. Alkaline soils, adobe flats, pond margins; 800–2320 m. KR, NCoR, CaR, **GB**; s OR. ❀TRY. Often confused with *T. crispum*, more study needed.

T. crispum Payson (p. 241) Winter ann, bien. **ST** 1–7 dm, branched; hairs 0 or toward base, sparse. **LVS**: basal 2–15 cm, ± oblanceolate to spoon-shaped, ± pinnately lobed, rarely entire, thickish, glaucous, hairy or not; cauline sessile, 1–5 cm, sagittate, clasping st, entire, glaucous. **INFL** spike-like, dense. **FL**: petals gen white, rarely light lavender with purplish veins, often crinkled. **FR** 1–2.5 cm, cylindric, incurved or straight, narrowed between seeds; stalk above receptacle 0.5–1.5(3.5) mm; pedicel erect to ascending-erect, partly or fully appressed, 2–6 mm, slender, base not flat; style 0.5–1(2.5) mm, slender. **SEED** ± flat; embryonic root at edge or back of 1 cotyledon. 2*n*=26. Alkaline or sandy soils, lake margins, shrubland; 1300–3000 m. SNH, **GB**; NV. Jun–Jul

T. integrifolium Nutt. Bien. **ST** 4.5–17 dm, straight, glabrous, glaucous; branches 0 or above. **LVS**: basal 5–31 cm, oblong to obovate, ± entire, thickish, ± glaucous; cauline sessile, 2–8 cm, narrowed to base, ± without basal lobes, entire to ± dentate, thickish, glabrous. **FL**: petals 6–9 mm, not crinkled. **FR** spreading to ascending, cylindric to ± flat, ± narrowed between seeds; pedicel straight, rarely ± curved; style 0.5–1.5 mm, slender. **SEED**: embryonic roots ± near edge on back of 1 cotyledon. Gen sandy,

silty, or alkaline soils; 700–2500 m. **GB**, **DMoj**; to WA, Rocky Mtns.

　ssp. *affine* (E. Greene) Al-Shehbaz **FL**: petals white. **FR** 2–4 cm, upcurved; stalk above receptacle 1–3 mm; pedicel gen spreading, 6–13 mm, straight to ± curved, whitish, base ± flat. Among shrubs, low dunes, meadows; 700–1100 m. **SNE, DMoj**; to UT.

　ssp. *complanatum* Al-Shehbaz (p. 241) **FL**: petals lavender to purple, rarely whitish. **FR** 1.5–3 cm, straight or curved; stalk above receptacle 0.5–1 mm; pedicel ± spreading, 2–5(6) mm, ± stout, not whitish, base flat. Alkaline or silty soils, woodland; 1100–2500 m. **GB**; to OR, UT.

T. laciniatum (Hook.) Endl. (p. 241) Bien. **ST** 2.5–10(14) dm, branched, glabrous, solid, not inflated. **LVS**: basal and lower cauline petioled, lanceolate to deltate-lanceolate, gen pinnately lobed, thickish; middle, upper cauline entire to dentate, base tapered, tip often acuminate. **INFL** dense. **FL**: petals linear, white to purplish, ± crinkled. **FR** spreading, 3.5–10 cm, flat parallel to septum to cylindric, rarely ± 4-sided, ± narrowed between seeds; stalk above receptacle 1–5 mm; pedicel spreading, rarely ascending, 3–7(15) mm, ± straight, stout; base flat; style 0.7–2.5 mm, stout. **SEED**: embryonic root at edge of 1 cotyledon. Rocky hillsides, basaltic cliffs; 600–1900 m. CaR, **GB**; to B.C., ID. May–Jun

T. milleflorum Nelson Bien. **ST** 1, 4.5–13 dm, gen branched above, gen ± hollow to infl, inflated or not, glabrous, glaucous. **LVS**: basal and lower cauline petioled, 6–23 cm, ± narrowly oblong to lanceolate or ovate, pinnately lobed, rarely dentate or dissected, tip acute. **INFL** dense. **FL**: petals white, blades oblanceolate to spoon-shaped. **FR** erect to erect-ascending, often appressed, 3.5–8.5 cm, linear, ± flat parallel to septum, rarely cylindric, ± narrowed between seeds; stalk above receptacle 1–4

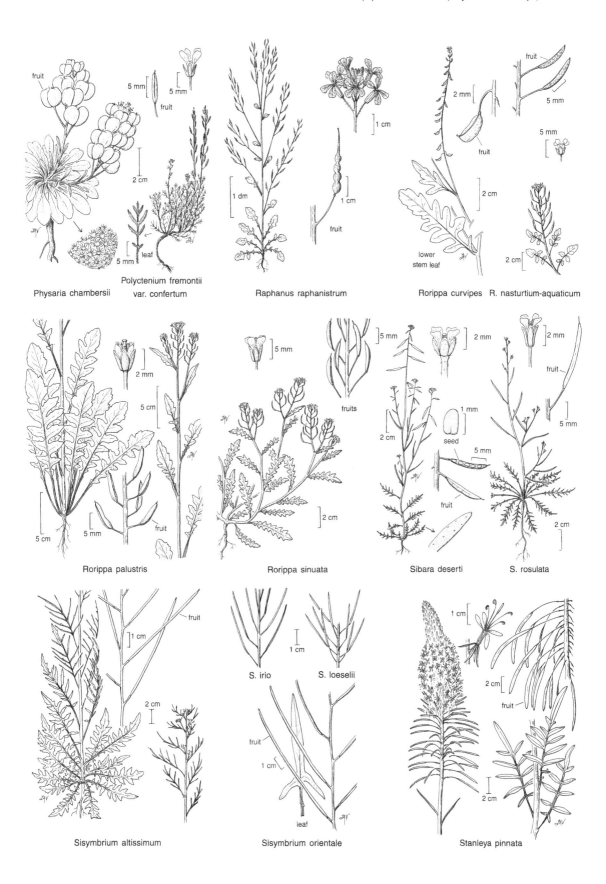

fruit

5 mm
fruit

5 mm

2 cm

5 mm leaf

Physaria chambersii

Polyctenium fremontii
var. confertum

1 dm

1 cm

fruit

Raphanus raphanistrum

2 mm

fruit
5 mm

fruit

5 mm

2 cm

lower
stem leaf

2 cm

Rorippa curvipes R. nasturtium-aquaticum

2 mm

5 cm

5 mm fruit

5 cm

Rorippa palustris

5 mm

fruits

2 cm

Rorippa sinuata

5 mm

2 mm

2 cm

1 mm

seed

5 mm

fruit

Sibara deserti

2 mm

fruit

5 mm

2 cm

S. rosulata

fruit

1 cm

2 cm

Sisymbrium altissimum

1 cm

S. irio

1 cm

S. loeselii

fruit

1 cm

leaf

Sisymbrium orientale

1 cm

2 cm

fruit

2 cm

fruit

Stanleya pinnata

mm, stoutish; pedicel upcurved, gen erect at tip, 2.5–5 mm, gen stout, base ± flat, tip erect; style 0.5–0.8 mm, stout. **SEED**: embryonic root at edge of 1, rarely both cotyledons. $2n=26$. Sandy soils, shrubland; 1300–2500 m. **GB**; to WA, UT. [*T. laciniatum* (Hook.) Endl. var. *m.* (Nelson) Payson]

THLASPI PENNY-CRESS

Ann, bien, per; hairs simple, gen 0. **ST** branched or not. **LVS** simple, entire to dentate; basal ± petioled; cauline sessile, clasping st. **FL**: sepals green or purple-tinged, bases not sac-like; petals 1–2 × sepals, white to purplish. **FR** obcordate or obovate to round, flat perpendicular to septum, tip rounded or notched; valves keeled, often winged. **SEEDS** 2–8 per chamber, ± striate; wing 0; embryonic root at edges of both cotyledons. ± 75 spp.: temp, gen n hemisphere. (Greek: to crush shield, from flat fr or perhaps use of crushed seeds as mustard) [Holmgren 1971 Mem NY Bot Gard 21:1–106]

T. arvense L. (p. 241) FAN-WEED Ann; caudex 0. **ST** 1–5 dm; branches 0–many. **LVS**: basal few, 2–6 cm, short-petioled, oblanceolate, shed early; upper cauline sessile, wavy-margined to dentate, base lobed, clasping st. **FL**: sepals 1.5–2 mm; petals 3–4 mm, white. **FR** 1–1.5 cm wide, widely oblong to round, winged; tip notch 1.5–2.5 mm; pedicel spreading to upcurved, 7–15 mm, slender. **SEED** ± 2 mm, striate; wing 0. $2n=14$. Disturbed areas, fields, roadsides; < 500 m. **CA**; N.Am; native to Eur. May–Aug

THYSANOCARPUS LACEPOD, FRINGEPOD

Ann; hairs 0 or simple. **ST** erect to ascending, slender; branches 0–many. **LVS** simple, sessile, entire, dentate, or pinnately lobed. **INFL** elongate; bracts 0. **FL**: sepals ascending, bases not sac-like; petals white or purple; stamens 6, ± equal, ± exserted. **FR** gen elliptic to round, very flat, 1-chambered (septum 0), indehiscent; wing entire to lobed, often perforated; rays 0 to very distinct; pedicel very slender. **SEED** 1, smooth; embryonic root at edges of both cotyledons. 5 spp.: N.Am. (Greek: fringe fr) Key to species adapted by Roy Buck.

1. Cauline lvs lobed at base, clasping st, often sagittate, (ob)lanceolate, lower dentate to shallowly lobed;
    style incl in or exserted from fr sinus; fr elliptic to round . . . . *T. curvipes*
1' Cauline lvs gen tapered at base, gen not clasping st, linear, lower ± entire to ± deeply lobed; style gen
    exserted from fr sinus; fr elliptic to obovate . . . . *T. laciniatus*

T. curvipes Hook.(p. 241) **ST** 1.5–8 dm, branched or not; hairs gen below. **LVS** (ob)lanceolate; basal, lower cauline ± petioled or not, 1.5–5(7) cm, dentate to shallowly lobed; middle and upper cauline sessile, entire to dentate, base lobed, clasping st. **FL**: sepals ± 1 mm, often purplish, margin white; petals ± = sepals, narrow, white or purple-tinged. **FR** 5–8 mm, elliptic to round, hairy or not; wing entire, wavy-margined, or crenate, often perforated; rays 0 or ± indistinct; pedicel recurved, 4–7 mm; style 0.5–1.5(2) mm, incl in or exserted from sinus. $2n=28$. Common. Slopes, washes, moist meadows; < 1800 m. CA-FP, **D**; to B.C., Rocky Mtns, Baja CA. Vars. *elegans* (Fischer & C. Meyer) Robinson, *eradiatus* Jepson, and *longistylus* Jepson indistinct. ❀ SUN,DRN:6,**7–9,14–24**&IRR:1–3, 10–12. Mar–May

T. laciniatus Torrey & A. Gray (p. 241) Pl gen glaucous; hairs ± 0. **ST** 1–6 dm, slender, branched or not. **LF** 1–4 cm, linear, gen tapered at base, gen not clasping st, ± entire to deeply linear lobed. **FR** 3–6 mm, elliptic, rarely obovate or round; hairs 0 or ± club-shaped; wing entire, wavy-margined, or crenate, perforated or not; rays 0 or ± indistinct; pedicel recurved, 3–6 mm; style 0.4–0.6 mm, gen exserted from sinus. Common. Slopes, rocky ridges, shaded sites; < 2400 m. s SNF, Teh, SnFrB, SCoR, SW, **SNE**, **D**; Baja CA. Vars. *crenatus* (Nutt.) Brewer, *hitchcockii* Munz, *ramosus* (E. Greene) Munz, *rigidus* Munz indistinct. ❀ SUN,DRN:**15–17**&SHD:1–3,7,8–12,**14,18–24**. Mar–May

TROPIDOCARPUM

Ann; hairs gen simple, spreading or reflexed. **ST** prostrate to ± erect, branched. **LVS** deeply pinnately lobed; segments linear to oblong, entire to dentate; basal ± present, not rosetted. **INFL** open; bract lf-like. **FL** < 6 mm; sepals gen spreading, ± 3 mm, ovate-oblong, bases not sac-like; petals obovate to spoon-shaped, yellow or yellowish, purplish-tinged or not; stamens 4 long, 2 short; style slender, stigma obscurely lobed. **FR** linear to oblong, flat perpendicular to narrow septum (or septum ± 0); valves 2 or 4; hairs reflexed. **SEEDS** 1 or 4 rows per chamber, oblong, compressed, brown; wing 0; embryonic root at back of 1 cotyledon. 2 spp.: CA, Baja CA. (Greek: keeled fr)

T. gracile Hook. (p. 241) **STS** prostrate to decumbent, several, 1–5 dm, branched below, above. **LF** 1–5 cm. **FL**: sepals greenish yellow or purplish; petals ± 4 mm, obovate, yellow, purplish tinged or not. **FR**: pedicels ascending to ± erect, gen straight, lower 1–1.5 cm; valves keeled; hairs ± 0–few; style 0.5–3 mm. $2n=16$. Common. Grassy banks, open fields, roadsides, pastures; < 1150 m. NCoRI, CaRF, SNF, Teh, GV, SnFrB, SCoRI, SW, **w DMoj**; Baja CA. [var. *dubium* (Davidson) Jepson] Mar–May

BUDDLEJACEAE BUDDLEJA FAMILY

Elizabeth McClintock

Shrub, tree, rarely herb; hairs gen stellate, branched, scale-like or glandular. **LVS** simple, gen opposite, entire, toothed or lobed; stipules at least partially on st, often ridge-like. **INFL**: cyme but appearing to be a panicle, head, raceme, or spike, terminal or axillary, gen dense. **FL** bisexual, sometimes functionally unisexual, ± radial; calyx lobes 4–5; corolla lobes 4–5; stamens 4–5, attached to corolla tube; ovary superior or half inferior, chambers 2 or 4, style 1, stigma elongate or ± spheric. **FR**: capsule, rarely berry. **SEEDS** small, often winged. ± 10 genera, 150 spp.: trop, subtrop; some cult for orn. [Rogers 1986 J Arnold Arb 67:143-185] Recently treated as included in a reduced Scrophulariaceae (with *Scrophularia, Verbascum*) [Olmstead et al 2001 Amer J Bot 88:348-361].

BUDDLEJA   BUTTERFLY BUSH

Shrub, tree, deciduous or evergreen. **LVS** rarely alternate or ± whorled, lanceolate, oblong, or linear, short-petioled or sessile. **FL** gen fragrant; calyx ± bell-shaped, lobes gen 4, ± = or < tube; corolla bell-shaped, funnel-shaped, or salverform, lobes gen 4, < to << tube, abruptly spreading; stamens gen 4, anthers ± sessile. **FR** 2-parted; calyx persistent. **SEEDS** many, often winged. ± 100 spp.: Am, Afr, Asia. (Rev. Adam Buddle, England, 1660–1715) [Norman 1967 Gentes Herb 10:47–114] Often spelled *Buddleia*, perhaps incorrectly.

B. utahensis Cov. (p. 241) PANAMINT BUTTERFLY BUSH   Shrub < 5 dm, densely branched, deciduous, dioecious; hairs dense, stellate or branched ± throughout. **LF** 1.5–3 cm, linear-oblong, thickish; margin entire to wavy, rolled under. **INFL** paired at upper nodes into single, head-like, spheric, dense clusters ± 10–15 mm wide. **FL** unisexual; calyx 3–4 mm; corolla 4–5 mm, salverform, creamy yellow, becoming purplish or brown-purple, lobes ± 1 mm; stamens incl. **FR** spheric to oblong. **SEEDS** not winged. Uncommon. Slopes, often on dolomite, volcanic rocks, limestone; 900–1700 m. **DMoj**; to UT. May–Oct

BURSERACEAE   TORCHWOOD FAMILY
Niall F. McCarten

Trees, shrubs, dioecious or monoecious. **ST** gen erect, < 15 m. **LVS** simple or compound; cauline, gen alternate, deciduous, petioled; lf axis often winged, glabrous to densely hairy. **INFL**: panicle or fl solitary. **FL** radial; disk ring- or cup-shaped; sepals 3–5; petals 0–5; stamens gen 1–2 × number of petals; ovary superior, chambers 2–5, style 0 or 1. **FR**: drupe or capsule; stones 1–5, each 1-seeded. 17 genera; 500 spp.: worldwide esp trop; some cult (*Boswellia,* frankincense; *Commiphora,* myrrh; *Bursera*).

BURSERA   ELEPHANT TREE, TOROTE

Trees, aromatic. **ST** < 10 m; bark smooth, shedding. **INFL**: panicle. **FL**: sepals 5; petals 5, arising from disk; stamens 10; ovary chambers 3. **FR**: valves 2–3. 60 spp.: trop Am. (J. Burser, born 1500's)

B. microphylla A. Gray (p. 245) ELEPHANT TREE   **ST** < 4 m; branches spreading, gen red; mature bark white. **FL**: sepals ± 5 mm; petals ± 4 mm, white to cream. **LF** pinnately compound, 2–8 cm, glabrous; lflets 7–33, 5–10 mm. **FR**: valves 3; stone yellow. RARE in CA. Rocky slopes; < 700 m. **w edge DSon (San Diego Co.)**; to AZ, Mex. Populations highly localized. Early Summer

CACTACEAE   CACTUS FAMILY
Edward F. Anderson (except *Opuntia*)

Per, shrub, tree, gen fleshy. **ST** cylindric, spheric, or flat; surface smooth, tubercled, or ribbed (fluted); nodal areoles bear fls, gen bear spines from center ("central spines") and margin ("radial spines") (*Opuntia* areoles bear small, barbed, deciduous bristles sometimes called glochids, gen also bear spines). **LF** gen 0. **FL** gen solitary, bisexual, sessile, ± radial; perianth parts gen many, grading from scale-like to petal-like; stamens many; ovary appearing inferior, ± submerged in st, so gen with areoles on surface, style 1, stigma lobes gen many. **FR** gen fleshy, gen indehiscent, spiny, scaly, or smooth. **SEEDS** many. 93 genera, ± 2000 spp.: esp Am deserts; many cult. (Greek: thorny pl) [Benson 1982 Cacti of US & Can; Hunt & Taylor eds 1990 Bradleya 8:85–107]

1. St clearly jointed; small barbed bristles present in areoles; seed covered by white, bone-like aril . . . . **OPUNTIA**
1' St not clearly jointed; barbed bristles 0; seed black or brown
  2. St ribs 0 or inconspicuous, tubercles prominent
    3. Tubercle longitudinally grooved on top (indented in ×-section); central spine not hooked . . . . . **ESCOBARIA**
    3' Tubercle round in ×-section (not grooved); some central spines of areole hooked . . . . **MAMMILLARIA**
  2' St ribs prominent, tubercles 0 to prominent
    4. Pl > 3 m; st > 30 cm diam, gen branching above 1.5 m; fl creamy white . . . . **CARNEGIEA**
    4' Pl < 3; st < 30 cm diam, branching near ground or unbranched; fl yellow to red or magenta
      5. Ovary and young fr spiny, glabrous; st soft-fleshy; branches gen few–many . . . . **ECHINOCEREUS**
      5' Ovary and young fr either spineless or woolly; st firm-fleshy; branches gen 0 (if present, then larger spines with ring-like ridges)
        6. Fr and st tip densely woolly; bracts sharp-tapered . . . . **ECHINOCACTUS**
        6' Fr and st tip not woolly; bracts wide, obtuse to acute
          7. St > 15 cm diam; seed pitted . . . . **FEROCACTUS**
          7' St < 15 cm diam; seed smooth or weakly tuberculed . . . . **SCLEROCACTUS**

CARNEGIEA SAGUARO

Tree. **ST** erect, 3–16 m, 30–75 cm diam, cylindric, gen few-branched; ribs 12–30, prominent; tubercles indistinct; spines 15–30, ± dense, spreading, gray. **FL** 5–6 cm diam, nocturnal; perianth creamy white. **FR** scaly, 25–45 mm diam, obovoid, dehiscing vertically, red, edible. **SEED** 2 mm, obovoid, black. 1 spp.: CA, AZ, Mex. (Andrew Carnegie, Am industrialist, philanthropist, 1835–1919)

C. gigantea (Engelm.) Britton & Rose (p. 245) **ST** columnar, massive, fleshy. **FL** 8.5–12.5 cm; outer parts green, margins lighter; inner parts petal-like, white; ovary green, scaly, style 10– 15 mm. 2n=22. RARE in CA. Rocky hills, plains; < 1500 m. **e DSon**; AZ, Mex. [*Cereus g.* Engelm.] In cult. May–Jun

ECHINOCACTUS CLUSTERED BARREL CACTUS

STS 1–30, often in ± 1 m diam clumps, each 3–6 dm, 10–20 cm diam, spheric to columnar; ribs 13–21, prominent; tubercles indistinct; spines dense. **FL** 4–5 cm diam; perianth yellow tinged with pink. **FR** dry, densely woolly. **SEED** 2–2.5 mm, 2–3.5 mm diam, black. 6 spp.: sw US, Mex. (Greek: hedgehog cactus)

E. polycephalus Engelm. & J. Bigelow var. **polycephalus** (p. 245) **ST** < 1 m; central spines 4, 6–7.5 cm, red or yellow, canescent at first, spreading, lowermost curved slightly; radial spines 6–8, 3–4.5 cm, spreading, slightly curved, red or yellow. **FL**: ovary densely woolly, with sharp-tapered bracts, style ± 20 mm. 2n=22. Rocky hills, silty valleys; < 1000 m. **e** SnBr, **DMoj, n DSon**; to AZ, Mex. ❀ DRN,DRY:2,3,**10–12**,13,19&SUN:20,21. Mar–May

ECHINOCEREUS HEDGEHOG CACTUS

STS 1–many, often densely clumped, each < 1 m, 2–10 cm diam, cylindric; ribs 5–13, prominent; tubercles ± indistinct; spines straight or curved. **FLS** on old growth, often near upper margin of spine-bearing areoles; ovary spiny. **FR** spheric to ovoid, glabrous; spines deciduous. **SEED** ovoid to ± spheric, tubercled, gen black. 47 spp.: sw US, Mex. (Greek: hedgehog candle) [Taylor 1985 Genus *Echinocereus* Royal Botanical Gardens, Kew]

1. Perianth purplish to lavender, closing at night, inner parts gen acuminate or bristle-tipped; anthers yellow; spines glabrous, at least 1 central spine gen angled to flat . . . . *E. engelmannii*
1' Perianth orange to red, remaining open at night, inner parts round to notched; anthers pink to lavender; spines < 1 year old puberulent near tip, all ± angled . . . . *E. triglochidiatus*

E. engelmannii (Engelm.) Lemaire (p. 245, pl. 36) Pl branched, forming clumps or mounds < 1 m diam. **STS** < 60, ± erect, 5–60 cm, 4–9 cm diam, cylindric, green; ribs 10–13; tubercles indistinct; areole wool present in first year only; spines variable in color and shape, always present, glabrous; central spines 2–7, < 8 cm, spreading, straight to twisted; radial spines 6–14, 2–20 cm. **FL** 5–7.5 cm diam, short-funnel-shaped; perianth purplish to magenta or lavender, inner parts gen acuminate or bristle-tipped (sometimes round); anthers yellow. **FR** fleshy, 20–30 mm, spheric, red, edible; spine clusters deciduous. 2n=44. Many dry habitats; < 2400 m. SnBr, PR, **W&I, D**; to UT, AZ, Mex. [*E. munzii* (Parish) L. Benson] Highly variable; sometimes divided into vars, but not satisfactorily. More study needed. ❀ DRN,DRY:2,3,7,10–24. May–Jun

E. triglochidiatus Engelm. (p. 245) Pl gen forming dense mounds. **STS** 1–500, 5–40 cm, 5–15 cm diam, ± spheric to cylindric, light- to bluish green; ribs 5–12; tubercles ± obvious; areole wool persistent; spines highly variable, ± angled, gray, those < 1 year old puberulent near tip; central spines 1–6, difficult to distinguish from radial spines. **FL** < 9 cm, funnel-shaped; perianth orange to red, inner parts round or evenly notched. **FR** 20–25 mm, 10–15 mm diam, pink to red; spines deciduous. 2n=22. Many habitats; 150–3000 m. **W&I, D**; to Rocky Mtns, TX, Mex. [*E. mojavensis* (Engelm. & J. Bigelow) Ruempler] Highly variable; sometimes divided into vars, but not satisfactorily. More study needed. ❀ DRN,DRY:2,3,10,11,14,18–23. Apr–Jun

ESCOBARIA BEEHIVE CACTUS

STS 1–200, gen in ± 50 cm clumps, each 2–15 cm, 2–15 cm diam, ± depressed to cylindric; ribs inconspicuous; tubercles grooved on upper surface from areole to base; central spines straight, ± following tubercle axis; radial spines wide-spreading ± in 1 plane from tubercle tip. **FL** 1–3(6) cm diam; outer perianth parts ciliate. **FR** becoming dry, spheric to club-shaped, red or green; perianth persistent. **SEED** reniform, black or brown, pitted. 16 spp.: w US, Mex. (R. & N. Escobar, Mexico) [Taylor 1986 Cact Succ J Gr Brit 4:36–44]

E. vivipara (Nutt.) F. Buxb. **ST** 2–15 cm, 2–15 cm diam; tubercles 6–9 mm; spines dense, central spines 3–12, white, tip darker; radial spines 12–40, 9–25 mm, straight, white. **FL** 2.5–5 cm diam, straw-yellow, yellow-green, pink, magenta, to purplish. **FR** 12–25 mm, elliptic in outline, green, sometimes with a few scales. Sandy to rocky soils; 75–2700 m. **D**; to UT, AZ, Mex. [*Coryphantha v.* (Nutt.) Britton & Rose; *Mammillaria v.* (Nutt.) Haw.]

1. Central spines < 8; perianth straw-yellow, yellow-green, or pink . . . . var. *deserti*
1' Central spines > 8; perianth magenta, purplish, or pink

2. St 6–8 cm diam, cylindric; fl 3 cm diam, magenta to pink; < 1500 m, s DMoj, DSon . . . . . . var. *alversonii*
2' St 7–15 cm diam, ovoid-spheric; fl 3–5 cm diam, magenta to purplish; > 1500 m, e DMoj . . . . var. *rosea*

var. **alversonii** (J. Coulter) D. Hunt (p. 245) FOXTAIL CACTUS **ST** gen 1, 10–15 cm, 6–8 cm diam, cylindric; central spines 8–10; radial spines 12–18. **FL** ± 3 cm diam, magenta to pink. RARE. Sandy or rocky areas, creosote-bush scrub; 75–600 m. **s DMoj, DSon**. [*Coryphantha v.* var. *a.* (J. Coulter) L. Benson; *Mammillaria a.* (J. Coulter) H. Zeissold] Threatened by collecting. In cult. May–Jun

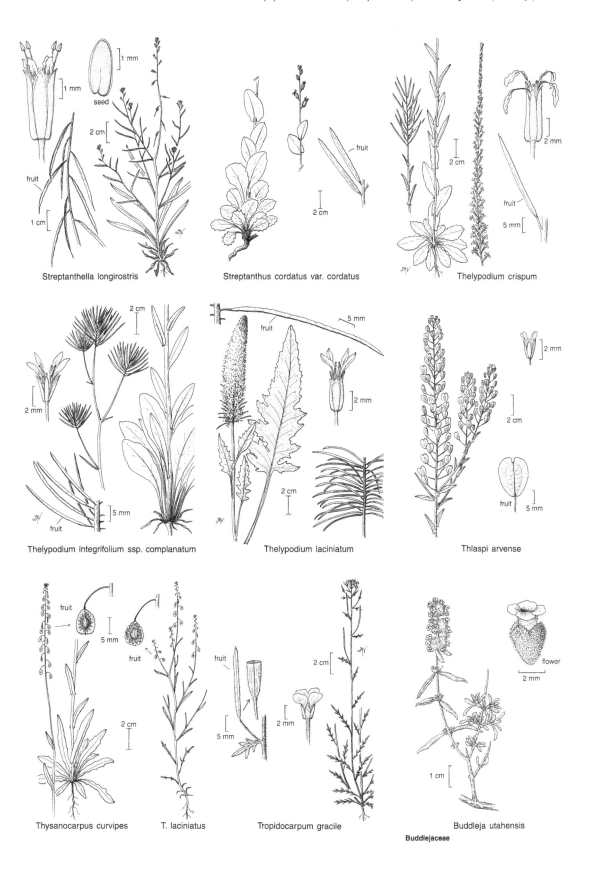

Streptanthella longirostris

Streptanthus cordatus var. cordatus

Thelypodium crispum

Thelypodium integrifolium ssp. complanatum

Thelypodium laciniatum

Thlaspi arvense

Thysanocarpus curvipes

T. laciniatus

Tropidocarpum gracile

Buddleja utahensis

Buddlejaceae

var. *deserti* (Engelm.) D. Hunt **STS** 1–few, 7–15 cm, 7–9 cm diam, cylindric to ovoid; central spines 4–6; radial spines 12–20. **FL** 2–3 cm diam, straw-yellow, yellow-green, or pink. Limestone soils; 1000–2400 m. **DMtns (e San Bernardino Co.)**; to sw UT, nw AZ. [*Coryphantha d.* (Engelm.) Britton & Rose; *Mammillaria d.* Engelm.] ✿ DRN,DRY;DFCLT. Apr–May

var. *rosea* (Clokey) D. Hunt (p. 245) VIVIPAROUS FOXTAIL CAC-TUS **STS** 1–several, 7–18 cm, 7–15 cm diam, ovoid-spheric; central spines 10–12; radial spines 12–18. **FL** 3–5 cm diam, magenta to purplish. 2*n*=22. RARE. Limestone slopes, hills; 1500–2700 m. **DMtns (ne San Bernardino Co.)**; s NV, nw AZ. [*Coryphantha v.* var. *r.* (Clokey) L. Benson] Threatened by collecting. In cult.

FEROCACTUS BARREL CACTUS, VISNAGA

ST gen 1, < 3 m, 20–35 cm diam, depressed-spheric to columnar; ribs 15–27; tubercles inconspicuous; spines dense. **FL** 3–6 cm diam; tube poorly developed; ring of hairs separating perianth and stamens; ovary densely scaly. **FR** 1–2 cm diam, scaly, gen yellow, opening by a pore near base. **SEED** 1.5–3 mm, black, pitted; scar basal. 23 spp.: sw US, Mex. (Latin: fierce cactus) [Taylor 1984 Bradleya 2:19–38]

F. cylindraceus (Engelm.) Orc. CALIFORNIA BARREL CACTUS **ST** 10–30 dm, gen taller than wide, spheric or columnar; ribs 18–27; spines 10–18, erect and spreading, longest recurved, gen with some red, becoming gray. **FL**: perianth parts yellow with red base; ovary 9–12 mm, scales fringed, style 12–20 mm. **FR** yellow. **SEED** 1–3 mm. 2*n*=22. UNCOMMON. Gravelly, rocky, or sandy areas; 60–1500 m. **D (esp e DMoj, w DSon)**; to UT, Mex. [*Echinocactus c.* Engelm.; *F. acanthodes* (Lemaire) Britton & Rose misapplied (rejected name)] Threatened by collecting; monitoring needed.

1. Central spines 7.5–17 cm; seed 2–3 mm; < 600 m
. . . . var. *cylindraceus*

1' Central spines 5–7 cm; seed 1–2 mm; > 700 m
. . . . var. *lecontei*

var. *cylindraceus* (p. 245, pl. 37) **ST**: central spines 7.5–17 cm. **SEED** 2–3 mm. Gravelly or rocky places; < 600 m. **D**; sw AZ, Baja CA. ✿ DRN,DRY:**12**,13,18–21&SUN:7,14,16,17,22–24. Apr–May

var. *lecontei* (Engelm.) H. Brav.-Holl. **ST**: central spines 5–7 cm. **SEED** 1–2 mm. Gravelly, rocky, or sandy places; > 700 m. **D**; to sw UT, AZ, n Mex. ✿ DRN,DRY:**10–12**,13,18–21&SUN:7,14, 16,17,22–24.

MAMMILLARIA NIPPLE CACTUS, FISH-HOOK CACTUS

STS 1–many, 2–30 cm, 2–20 cm diam, spheric to cylindric; ribs inconspicuous; tubercles round in ×-section; central spines gen hooked. **FL** 1–5 cm diam; ovary glabrous. **FR** ovoid to cylindric, red, glabrous. **SEED** black, pitted. 150 spp.: N.Am. (Latin: nipple) [Hunt 1984–87 Bradleya 2:65–96; 3:53–66; 5:17–48]

1. Radial spines > 30; seed with corky aril . . . . ***M. tertrancistra***
1' Radial spines < 30; seed without corky aril
 2. Radial spines white; perianth yellow to white; bristles present in axils of tubercles . . . . ***M. dioica***
 2' Radial spines light brown to red; perianth lavender to reddish purple; axillary bristles 0 . . . . ***M. grahamii***

M. dioica M.K. Brandegee (p. 245, pl. 38) Pl with either all bisexual or all pistillate fls. **ST** gen 1(–many), 5–30 cm, 3–7 cm diam; bristles present in axils of tubercles; central spines 1–4, 8–15 mm; radial spines 11–22 in 1 rank, 4–10 mm, white. **FL** 10–22 mm, 20–40 mm diam; perianth yellow to white. **FR** 10–25 mm, club-shaped to ovoid. **SEED**: aril 0. 2*n*=44,66. Hillsides, washes, coastal scrub to creosote-bush scrub; 10–1500 m. SCo, **w edge DSon**; Baja CA. [var. *incerta* (Parish) Munz] ✿ DRN,DRY:17,19–24. Feb–Apr

M. grahamii Engelm. var. *grahamii* Boed. (p. 245) **STS** gen several (1–many), 7–15 cm, 4–7 cm diam; axillary bristles 0; central spines 1–2, 12–15 mm, longer hooked; radial spines 18–28 in 1 rank, 6–12 mm, light brown to red. **FL** 15–25 mm, 20–30 mm diam; perianth lavender to reddish purple. **FR** 12–25 mm,

cylindric, green, becoming red. **SEED**: aril 0. 2*n*=22. Uncommon. Sandy or rocky canyons, washes, plains, creosote-bush scrub; 300–900 m. **ne DSon (se San Bernardino Co.)**; AZ, n Mex. [*M. milleri* (Britton & Rose) Boed.; *M. microcarpa* Engelm. misapplied (invalid name)] ✿ DRN,DRY;DFCLT. Apr

M. tetrancistra Engelm. (p. 245) **ST** gen 1, 7–25 cm, 3.5–7.5 cm diam, cylindric; axillary bristles 0; central spines 3–4, 18–25 mm, gen hooked, tips dark; radial spines 30–60 in 2–3 ranks, 10–25 mm, white or tips dark. **FL** 30–40 mm, 25–40 mm diam; perianth deep pink to lavender. **FR** 15–32 mm, cylindric. **SEED**: aril corky. 2*n*=22. Uncommon. Sandy hills, valleys, plains, creosote-bush scrub; 130–1400 m. **D**; to UT, AZ, n Mex. ✿ DRN,DRY; DFCLT. Apr

OPUNTIA PRICKLY-PEAR, CHOLLA

Bruce D. Parfitt & Marc A. Baker

Shrubs, trees; roots fibrous. **ST** gen erect, < 12 m; segments flat to cylindric, gen firmly attached; tubercles gen elongate along st; ribs sometimes present; spines 0–many, sometimes flat, tip smooth or barbed, epidermis persistent or separating as a papery sheath; small, barbed deciduous bristles gen many. **LF** small, conic, fleshy, deciduous, obvious on young sts and ovaries. **FR** juicy, fleshy or dry; wall thick, bearing areoles. **SEED** dark brown, encased in a bony, whitish aril. 200 spp.: Am; *O. ficus-indica* cult for food, others for orn. (Possibly from Papago Indian name ("opun") for this food pl; or named for a spiny pl of Opus, Greece) Spines smaller, fewer in shade forms; when yellow, blacken with age. Hybridization common within subgenera.

1. St segments cylindric to club-like, tubercled; spine epidermis sheath-like, gen deciduous
  2. Major spines distinctly flat, epidermal sheath separating only from tip (subg. *Corynopuntia*)
    3. Perianth yellow; largest spine with cross-rows of rough papillae; ovary bristles stiff, with reflexed barbs
    . . . . *O. parishii*
    3' Perianth pink-purple; largest spine smooth; ovary bristles hair-like, with ascending barbs . . . . . . *O. pulchella*
  2' Spines not flat, epidermal sheath separating from entire spine (subg. *Cylindropuntia*)
    4. Fr spineless (sometimes with a few deciduous bristles)
      5. Tubercle length ± = width . . . . *O. bigelovii*
      5' Tubercle length ± 2 × width
        6. Spines 7–10 per tubercle; fr < 2.5 cm; seeds irregularly shaped, < 3 mm . . . . *O.×fosbergii*
        6' Spines 9–16 per tubercle; fr 2.5–3.5 cm; seeds ± spheric, ± 3 mm . . . . *O. ×munzii*
    4' Fr densely spiny
      7. St < 1 cm diam, tubercles < 2 mm high . . . . *O. ramosissima*
      7' St > 1.5 cm diam, tubercles > 3 mm high
        8. Terminal st segment gen < 2 dm, tubercle length < 2 × width; filaments green . . . . *O. echinocarpa*
        8' Terminal st segment gen > 2 dm, tubercle length > 3 × width; filaments red- to green-purple
          9. St gen branched only above; trunk 1 . . . . *O. acanthocarpa* var. *coloradensis*
          9' St gen branched near base; trunks several–many . . . . *O. wolfii*
1' St segments flat, not tubercled; spine epidermis not separable (subg. *Opuntia*)
  10. St minutely papillate-hairy, spines 0 (barbed bristles present); perianth pink to magenta; filaments
    red; fresh stigma white . . . . *O. basilaris* var. *basilaris*
  10' St glabrous, bearing spines; perianth gen yellow; filaments white or red-magenta; fresh stigma green
    11. Fr becoming dry, tan, gen spiny; perianth yellow to pink-magenta; filaments white or red-magenta;
      style slender, white, base slightly thicker . . . . *O. erinacea*
      12. Ovary spiny; spines gen in all st areoles, smaller spines at lower edge of st areoles 3 or more, strongly
        reflexed . . . . var. *erinacea*
      12' Ovary spineless; spines in < 50% of st areoles, smaller spines at lower edge of st areoles 0–4, ±
        reflexed . . . . var. *utahensis*
    11' Fr juicy, gen with some deep purple, spines 0; perianth yellow (base sometimes red); filaments
      white; style thick, white, base much thicker
      13. Trunk 1, erect; areoles gen > 38 per ovary . . . . *O. chlorotica*
      13' Basal branches several, ± decumbent to ascending; areoles < 34 per ovary
        14. St segment > 15 cm wide; perianth base yellow; fr red-purple inside . . . . *O. engelmannii*
        14' St segment < 15 cm wide; perianth base red; fr green inside . . . . *O. phaeacantha*

O. acanthocarpa Engelm. & J. Bigelow var. *coloradensis* L. Benson(p. 245) BUCKHORN CHOLLA **ST** ± tree-like, < 4 m; segments cylindric, terminal < 40 cm, 2–2.5 cm diam; tubercle 20–40 mm, 5–7 mm high; spines 12–21, < 5 cm, pale yellow- to red-brown, sheath pale yellow-brown. **FL:** inner perianth < 3.5 cm, yellow with purple to brown-red tint; filaments purple-red. **FR** dry, tubercled; base acute; lower tubercles >> upper. **SEED** < 6 mm. 2*n*=22. Creosote-bush scrub, Joshua-tree woodland; < 1300 m. **D**; NV, AZ. *O. deserta* Griffiths is probably *O. a.* × *O. echinocarpa*. ✿ DRN, DRY,SUN:**10**,11,**12**,14,18,**19–24**. May–Jun

O. basilaris Engelm. & J. Bigelow BEAVERTAIL CACTUS **ST** clumped, ascending to erect, 7–40 cm; segments flat, gen erect, 5–21 cm, gen puberulent; spines gen 0(–8); bristles many. **FL:** inner perianth ± 4 cm long, pink- magenta; filaments deep magenta-red; style white or pink, stigma white. **FR** 2–4 cm, becoming dry, green and purple becoming tan, gen puberulent; areoles 24–76. **SEED** 6.5–9 mm, ± spheric. Desert, chaparral, pinyon/juniper woodland; 120–2200 m. s SN, Teh, se SnJV, SnGb, SnBr (and adjacent SCo), e PR, **s SNE, D**; to UT, AZ, Mex. 3 vars.in CA.

    var. *basilaris* (p. 249, pl. 39) BEAVERTAIL CACTUS **ST:** segment 8–21 cm, 5–13 cm wide, flat, ± obovate; spines 0. 2*n*=22. Desert to pinyon-juniper woodland; 150–2200 m (higher n). Range of sp. exc SnJV. [var. *ramosa* Parish, *O. whitneyana* E. Baxter] ✿ DRN,DRY,SUN:2,**3,7–14**,16,17,**18–24**. Mar–Jun

O. bigelovii Engelm. (p. 249) TEDDY-BEAR CHOLLA **ST** 1–2 m; trunk 1; main branches few, short, spreading, becoming black; segments cylindric, terminal gen < 10 cm, 4–6 cm diam, easily detached; tubercle 4–11 mm, 3 mm high; spines 4–10, < 2.5 cm, very pale yellow-brown, sheath translucent to pale brown. **FL:** inner perianth 1.5–4 cm, yellow; filaments green. **FR** 1–2 cm,

leathery, tubercled, yellow; spines 0 or bristles few. **SEED** 3–4 mm, gen sterile. 2*n*=33 (rarely 22). Rocky fans, benches, with creosote bush; < 1000 m. **DMoj (Kelso Dunes), DSon**; s NV, AZ, n Mex. Reproduces primarily by rooting of detached segments. ✿ DRN,DRY,SUN:10,11,**12,13,19–21**,22–24. Apr

O. chlorotica Engelm. & J. Bigelow PANCAKE PRICKLY-PEAR **ST** ± tree-like, 1.5–2 m; segments flat, 13–20 cm, gen round; spines 3–8 in at least upper areoles, longest 2.5–5 cm, gen straight, flat, ± reflexed, translucent yellow. **FL:** inner perianth 2–2.5 cm long, yellow; filaments white; style white, stigma yellow-green. **FR** ± 4 cm, juicy; exterior purple-red; middle layer white, seed-pulp pink; areoles 40–70. **SEED** 3 mm. 2*n*=22. Uncommon. Pinyon/juniper woodland, desert-scrub edge, chaparral; 600–1300 m. e PR, **DMtns**; to NV, NM, n Mex. Distinctive stabilized hybrids with *O. phaeacantha* in DMtns (New York Mtns, e San Bernardino Co.), s NV, w AZ have been called *O. curvospina* Griffiths (2*n*=44). ✿ DRN,DRY,SUN:3,**7–12**,13,**14,16,18–21**,22–24. May–Jun

O. echinocarpa Engelm. & J. Bigelow (p. 249) SILVER or GOLDEN CHOLLA **ST** ± tree-like, < 3 m; segments cylindric, terminal gen < 10 cm, 2–3 cm diam; tubercle 6–15 mm, 3–8 mm high; spines 9–20, < 4 cm, pale gray to translucent yellow, sheath gen same color. **FL:** inner perianth < 2.5 cm, green-yellow, rarely red-brown; filaments pale green. **FR** dry, odor of rancid butter, tubercled; base obtuse; spines dense above. **SEED** 4–6 mm. 2*n*=22. Dry habitats; 300–1400 m. **W&I, D**; to NV, UT, AZ. Hybridizes with most co-occurring chollas. ✿ DRN,DRY,SUN:**10**,11,**12**,14,16,18,**19–21**,22–24. Apr–May

O. engelmannii Engelm. var. *engelmannii* ENGELMANN PRICKLY PEAR Shrub, mound-shaped. **ST** gen < 1 m; lower branches gen decumbent, upper spreading to ascending; segments flat, 15–25 cm, gen obovate; spines 3–12 in all areoles, longest 4–5 cm, straight, spreading from areole, ± appressed to st, ± flat, yellow, coated chalky white, base often red-brown. **FL**: inner perianth 4–5 cm long, yellow; filaments white; style white, stigma yellow-green to green. **FR** 4–6.5 cm, juicy, red-purple throughout; areoles 20–32. **SEED** 4–6 mm. 2*n*=66. Uncommon in CA. Desert scrub, dry oak woodland, etc; 900–1500 m. SnJt, **DMtns**; to NV, TX, Mex. [*O. phaeacantha* var. *discata* (Griffiths) L. Benson & Walkington] Hybridizes with *O. phaeacantha*. ✿ DRN,DRY,SUN:2,3,7–9,**10**,11,14,**18**,19–21.

O. erinacea Engelm. & J. Bigelow **ST** clumped; branches gen ascending to erect, gen < 0.5 m tall; segments, 5.5–18 cm, flat, elliptic to obovate; spines 1–24, longest 1.7–13 cm, flat to round, straight to wavy, gen whitish, base yellow-brown, surrounded by shorter, gen reflexed, whiter spines. **FL**: inner perianth 2–2.5 cm long, yellow to pink-magenta; filaments gen white (magenta); style white, stigma green. **FR** 2.5–4 cm, gen spiny, green, tinted red, becoming dry, tan; areoles 14–68. **SEED** 5–6.5 mm. Creosotebush scrub to pine scrub; 900–3300 m. se SNH, SnBr, SnJt, **SNE, DMoj (esp DMtns)**; to WA, Rocky Mtns, NM.

var. ***erinacea*** MOJAVE PRICKLY-PEAR **ST**: spines 4–24 per areole (0–3 in SnJt), gen in all areoles, longest 1.7–13 cm, spines on lower edge of areole 3 or more, straight to slender and wavy, strongly reflexed. **FR** spiny exc sometimes in SnJt; areoles 24–68. 2*n*= 44. Desert scrub, Joshua-tree woodland, pinyon woodland; 900–2200 m. Range of sp. in CA; to UT, AZ. [*O. ursina* A. Weber; not *O. rufispina* Engelm. & J. Bigelow] ✿ DRN,DRY, SUN:2,**3**,7–9,**10**,11,14, **18**,19–21. May–Jun

var. ***utahensis*** (Engelm.) L. Benson **ST**: spines 1–7 per areole, gen in < 50% of areoles, longest 1.7–5.5 cm, spines on lower edge of areole 0–4, straight, ± reflexed. **FR**: spines 0(–4); areoles 14–26. 2*n*=88. Sagebrush scrub to Jeffrey-pine woodland; 2000–2800 m. **GB (Mono, nw Inyo cos.)**; to ID, UT, NM. [*O. rhodantha* Schumann, *O. xanthostemma* Schumann] ✿ DRN,DRY,SUN:1,**2,3,7**, 10,18. Jun–Jul

O.* ×*fosbergii C. Wolf (p. 249) PINK TEDDY-BEAR CHOLLA **ST** 1.5–2.5 m; trunk 1; branches gen several, long; segments cylindric, terminal gen < 10 cm, 4–6 cm diam, easily detached; tubercle 10–20 mm, 3–5 mm high; spines 7–10, < 2.5 cm, pale red-brown, sheath pale yellow-brown. **FL**: inner perianth < 2.5 cm, pale red-brown; filaments green. **FR** < 2.5 cm, dry to leathery, tubercled; spines 0 or bristles few. **SEED** sterile, < 3 mm. 2*n*=33. Uncommon. Valley floors, alluvial fans; 300–450 m. **DSon (e San Diego Co.).** 2 sets of genes probably come from *O. bigelovii*. If other parent sp. is *O. echinocarpa*, *O.* ×*munzii* should be considered a diploid form. [*O. bigelovii* Engelm. var. *hoffmannii* Fosb.] ✿ DRN,DRY,SUN:8–11,**12,13**,14,16,**19–21**,22–24. Apr

O.* ×*munzii C. Wolf (p. 249) MUNZ'S CHOLLA **ST** < 2.4 m; trunk 1; branches several; segments cylindric, terminal easily detached, gen < 10 cm, 3–5 cm diam; tubercle 10–16 mm, 3–4 mm high; spines 9–16, < 4 cm, pale red-brown, sheath yellow-brown. **FL**: inner perianth 1.5–2 cm, yellow-green to red-brown; filaments pale green. **FR** 2.5–3.5 cm, dry, low tubercles, often with deciduous spines. **SEED** 3 mm. 2*n*=22. RARE. Gravelly or sandy soils of washes, canyon walls; 150–600 m. **DSon (Chocolate, Chuckwalla mtns, Imperial, Riverside cos.)** Probable hybrid of *O. bigelovii* and *O. echinocarpa* (see *O.* ×*fosbergii*).

Most localities ± inaccessible. In cult. May

O. parishii Orc. CLUB or MAT CHOLLA **ST** decumbent; branches ascending, in clumps, < 2 m diam, 10–20 cm; segments ± obovoid, terminal 5–7.5 cm, 2–3 cm diam; tubercles 12–25 mm, 3–8 mm high; spines < 21, < 5 cm, largest distinctly flat, with cross-rows of rough papillae exc at tip, gray to brown, margin white, thick, sheath separating only from tip. **FL**: inner perianth 1.5–2.5 cm, yellow; filaments green. **FR** 4.5–8 cm, fleshy, smooth, yellow; spines 0 or easily detached; bristles many. **SEED** 3–4.5 mm. 2*n*=22. Uncommon. Sandy flats; 900–1200 m. **e&s DMoj**; s NV, w AZ. [*O. stanlyi* Engelm. var. *parishii* (Orc.) L. Benson] ✿ DRN,DRY,SUN:**10**,11,**12**,18,19. May–Jun

O. phaeacantha Engelm. (p. 249) Shrub. **ST** decumbent to ± spreading, 0.3–1 m; segments flat, 11–30 cm, gen obovate; spines 1–4(6) per areole on upper 30–70% of segment, largest 3–8 cm, gen flat, spreading, upper 1–2 red-brown near base, white or straw-colored exc at base, smaller 1–3 ± reflexed, gen white or gray. **FL**: inner perianth 3.5–4 cm long, yellow, base red; filaments white; style white, stigma yellow-green to green. **FR** 2.5–6.5 cm, juicy, red-purple; interior gen green; areoles 15–32. **SEED** 3–6 mm. 2*n*=66. Many habitats; 45–2220 m. SCoRO, SnBr, e PR, **DMtns, DSon**; to SD, KS, OK, TX, Mex. [var. *major* Engelm.; *O. littoralis* var. *piercei* (Fosb.) L. Benson & Walkington; probably *O. mojavensis* Engelm. & J. Bigelow] ✿ DRN, DRY:2,7–9,**10**,11,12,**14**,16,**18–24**.

O. pulchella Engelm. (p. 249) SAND CHOLLA **ST** clumped, 10–20 cm, arising from a bristle-covered tuber; segments narrowly club-shaped to cylindric, terminal < 10 cm, 0.5–2.5 cm diam; tubercle 6–9 mm, < 1.5 mm high; spines 8–15, densest near st tip, < 6 cm, bulbous at base, largest flat, sharply angled, without cross-rows of rough papillae; sheath separating only near tip; barbed bristles of tuber gen 1–1.5 cm. **FL**: inner perianth 1.5–2.5 cm, pink-magenta; filaments green to yellow. **FR** 2–3 cm, fleshy, smooth, red, with soft, upwardly barbed bristles. **SEED** 3–6 mm. RARE in CA. Dry-lake borders, sandy flats; 1500–1700 m. **SNE**; to UT. Highly variable; juvenile forms sometimes fl. In cult.

O. ramosissima Engelm. (p. 249, pl. 40) PENCIL CACTUS, DIAMOND CHOLLA **ST** decumbent to ± tree-like, < 1.5 m; segments cylindric, terminal < 10 cm, 4–6 mm diam; tubercle 4.5–7.5 mm, 0–1 mm high; spines 0–3 (gen 1, spreading, long, dark, straight), < 4 cm, pink-gray to dark brown, sheath ± white to pale yellow, translucent, tip darker. **FL**: inner perianth < 6 mm, orange-pink to red-brown; filaments pale green. **FR** < 2 cm, dry; spines dense (rarely 0). **SEED** 2–4 mm. 2*n*=22,44. Desert flats; < 1100 m. **D**; NV, AZ. Variable. *O. wigginsii* L. Benson (sw AZ, reported from DSon) is possibly *O. ramosissima* × *O. echinocarpa*. ✿ DRN,DRY,SUN:8–11,**12,13**,14,19–21. Apr–May

O. wolfii (L. Benson) M. Baker WOLF'S CHOLLA **ST** gen ± erect, < 2 m, gen branched from base; segments cylindric, terminal < 40 cm, 2.5–4 cm diam; tubercle 15–25 mm, 9 mm high; spines 12–22, < 3 cm, pale to dark brown, sheath translucent to pale brown. **FL**: inner perianth 2–3.5 cm, pale purple-brown; filaments red-purple. **FR** 2.5–3 cm, dry, strongly tubercled; spines dense. **SEED** gen sterile. 2*n*=66. Locally common. Dry places above valley floors; 300–1200 m. **w edge DSon**; Baja CA. ✿ DRN,DRY,SUN:**10**,11,**12**, 14,18,**19–21**,22–24.

SCLEROCACTUS PINEAPPLE CACTUS, DEVIL-CLAW

ST gen 1, 5–15 cm, 2–12 cm diam, ovoid to cylindric; ribs 8–21, prominent; tubercles distinct; central spines 1–11, straight or hooked; radial spines gen 3–30, 6–30 mm, straight. **FL** 25–75 mm diam; perianth greenish yellow to magenta; ovary scaly. **FR** becoming dry, 6–25 mm, scaly. **SEED** reniform, tubercled, black. 19 spp.: sw US, Mex. (Greek: hard cactus)

1. Central spines 4–8, straight to curved but not hooked . . . . ***S. johnsonii***
1' Central spines 9–11, some hooked at tip . . . . ***S. polyancistrus***

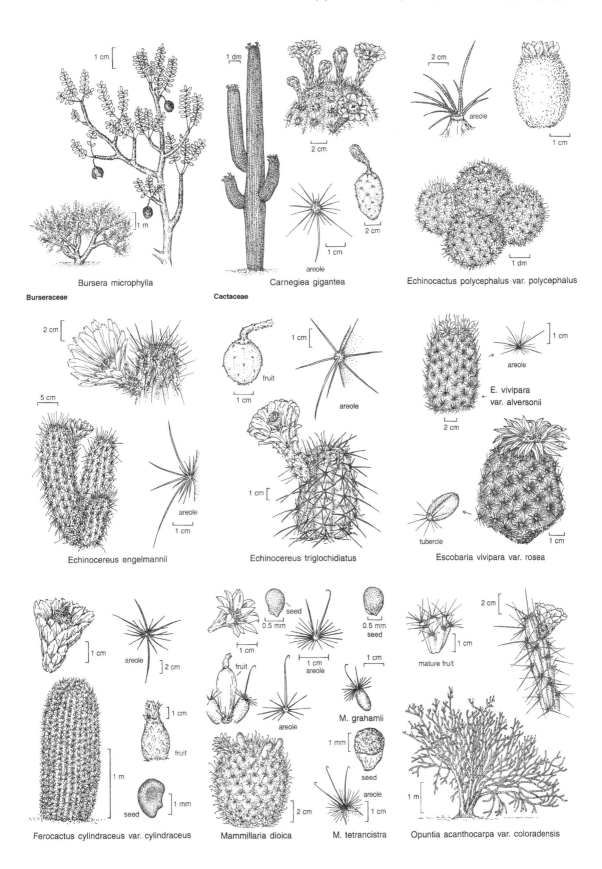

Bursera microphylla

Burseraceae

Carnegiea gigantea

Cactaceae

Echinocactus polycephalus var. polycephalus

Echinocereus engelmannii

Echinocereus triglochidiatus

E. vivipara var. alversonii

Escobaria vivipara var. rosea

Ferocactus cylindraceus var. cylindraceus

Mammillaria dioica

M. grahamii

M. tetrancistra

Opuntia acanthocarpa var. coloradensis

S. johnsonii (Engelm.) N.P. Taylor (p. 249) **STS** gen 1, 10–25 cm, 5–10 cm diam, ovoid to cylindric; ribs 17–21; spines yellow or pink to reddish, central 4–8, radial 9–10. **FL** 5–8 cm diam, greenish yellow, pink, or magenta. **FR** 7–15 mm, 3–5 mm diam; scales widely cordate, ciliate. 2*n*=22. Granitic areas, creosote-bush scrub; 500–1200 m. **n DMoj (Inyo Co.)**; to sw UT, nw AZ. [*Echinocactus j*. Engelm.; *Neolloydia j*. (Engelm.) L. Benson] ✿ DRN,DRY, SUN;DFCLT. Apr–May

S. polyancistrus (Engelm. & J. Bigelow) Britton & Rose (p. 249) MOJAVE FISH-HOOK CACTUS **ST** 1, 10–25 cm, 5–8 cm diam, cylindric; ribs 13–17; central spines 9–11, all but 1–2 hooked, red-brown or white; radial spines 10–15, white. **FL** 4–6 cm diam, rose-purple to magenta. **FR** 20–30 mm, 15–20 mm diam; scales narrow, ciliate near tip. UNCOMMON. Limestone areas, hills and canyons, creosote-bush scrub, Joshua-tree woodland; 750–2100 m. **W&I**, **DMoj**; NV. [*Echinocactus p*. Engelm. & J. Bigelow] ✿ DRN,DRY,SUN; DFCLT. Apr–Jun

CALLITRICHACEAE WATER-STARWORT FAMILY

C. Thomas Philbrick

Ann, in water or on wet ground, monoecious. **ST** slender, gen ascending under water, floating on surface, or prostrate on ground, gen much-branched. **LVS** simple, gen opposite, 4-ranked, linear-lanceolate to spoon-shaped, entire; stipules 0. **INFL**: fls 1–3 per lf axil, the group subtended by gen 2 whitish, inflated bracts. **FLS** minute, unisexual; perianth 0. **STAMINATE FL**: stamen gen 1, filament elongate. **PISTILLATE FL**: ovary superior, ± obcordate, chambers 4, styles 2, thread-like. **FR** 0.6–1.6 mm (width gen ± = length), ± dry, ± grooved longitudinally, splitting into 4 achene-like units. Recently treated in expanded Plantaginaceae (= Veronicaceae sensu Olmstead et al) that includes genera removed from Scrophulariaceae [Angiosperm Phylogeny Group 1998 Ann Missouri Bot Gard 85:531–553; Olmstead et al 2001 Amer J Bot 88:348–361] ✿shallow fresh water: TRY.

CALLITRICHE WATER-STARWORT

The only genus, ± 40 spp.: trop, temp. (Greek: beautiful hair, from slender sts) [Fassett 1951 Rhodora 53:137–155,161–182, 185–194,209–222; Philbrick & Jansen 1991 Syst Bot 16:478–491] Taxonomically difficult; mature fr and 10× magnification needed for identification.

C. verna L. (p. 249) **LVS**: submersed lvs linear-lanceolate; floating, emergent, or terrestrial lvs sometimes present and spoon-shaped. **INFL**: bracts subtending fls 2, whitish, inflated, persistent in fr. **FR** 0.9–1.4 mm, 0.8–1.3 mm wide (wider above middle); wing from base to tip, gen wider above middle, transition to fr wall abrupt; pedicel 0–0.5 mm. Becoming stranded at water edge or submersed < ± 15 dm; < 4000 m. CA-FP, **GB**; to e N.Am, Eur. May–Aug

CAMPANULACEAE BELLFLOWER FAMILY

Nancy Morin, except as specified

Ann to tree. **LVS** gen cauline, gen simple, gen alternate, petioled or not; stipules 0. **INFL**: panicle, raceme, spike, or fls solitary in axils, gen open; bracts lf-like or not. **FL**: bisexual, radial or bilateral, sometimes inverted (pedicel twisted 180°; hypanthium gen present, ± fused to ovary; sepals gen 5; corolla radial to 2-lipped, gen fused (tube sometimes split down back), lobes gen 5; stamens 5, free or ± fused (anthers and filaments fused into tube or filaments fused above middle); ovary inferior, sometimes half inferior, chambers 1–3, placentas axile or parietal, ovules many, style gen 1, 2–5-branched. **FR**: gen capsule, dehiscing on sides or at tip by pores or short valves. **SEEDS** many. ± 70 genera, ± 2000 spp.: worldwide. Some cult for orn (*Campanula, Jasione, Lobelia*). Subfamilies sometimes treated as different families.

1. Anthers and filaments fused; pls of wet areas
  2. Fr spheric; per; corolla red . . . . **LOBELIA**
  2' Fr obovate or cylindric; ann; corolla blue . . . . **PORTERELLA**
1' Anthers free, filaments free at base, fused above middle; pls of dry areas
  3. Sepals linear to triangular; fr fusiform, obconic, or hemispheric, dehiscing by valves; fls inverted
  . . . . **NEMACLADUS**
  3' Sepals oblanceolate; fr hemispheric, top dome-like, circumscissile; fls not inverted . . . . **PARISHELLA**

LOBELIA

Tina Ayers

Per, glabrous or hairy. **LVS** mostly basal or all cauline, 0.5–1.5 cm wide, linear-lanceolate to elliptic, sessile; margin with small, gland-tipped teeth. **INFL**: raceme. **FL** inverted in full bloom by twisted pedicel; corolla red or blue, rarely white, tube entire or with an upper sinus, limb strongly 2-lipped, 2 lobes of upper lip < 3 of lower; stamens fused, gen 2 smaller anthers each with terminal tuft of bristles, 1 sometimes triangular or horn-like, others linear, shorter; ovary ± spheric, chambers 2, placentas 2, axile. **FR** dehiscent by 2 valves at tip. ± 350 spp.: ± worldwide. (Matthias de l'Obel, Flemish botanist, 1538–1616) Fl part positions (upper, next to st; lower, away from st) given at full bloom.

L. cardinalis L. var. *pseudosplendens* McVaugh (p. 249) CARDINAL FLOWER **ST** erect, 4–20 dm, < 1.5 cm diam, purple-red. **FL**: corolla red, rarely white, glabrous, tube 15–20 mm, from upper sinus to base; anther tube 3.5–4.5 mm, triangular bristle at tips of 2 shorter anthers 0. *n*=7. Stream bottoms; 450–1600 m. SnGb, SnBr, PR, **DMtns (Panamint Mtns)**; to w TX, Mex. Incl by McVaugh in ssp. *graminea* (Lam.) McVaugh, with 3 other vars. incl var. *multiflora* (Paxton) McVaugh (pls with dense, short hairs throughout; lvs lanceolate to ovate, probably not in CA). Seriously TOXIC, esp when used as a home remedy. ✸ TRY. Aug–Oct

NEMACLADUS

Nancy R. Morin & Jennifer Milburn

Ann; roots fibrous. **STS** erect or spreading, simple or branched at base or below middle. **LVS** basal; petiole short or 0. **INFL** ± raceme-like; bract 1 per fl, small; pedicel gen thread-like. **FL** inverted; sepals linear to triangular; corolla nearly radial and 5-lobed or 2-lipped (upper lip 3-lobed, lower lip 2-lobed); filaments free at base, fused into tube above, sometimes appendaged at tube base, anthers free, all alike; ovary gen half-inferior in fr, sometimes 0, hemispheric to obconic, sometimes glandular, stigma 2-lobed, papillate. **FR** gen > hypanthium, hemispheric to fusiform; tip pointed or rounded, dehiscing at tip by 2 valves; chambers 2. **SEED** elliptic to oblong. 13 spp.: sw US, nw Mex. (Greek: thread-like branch) [McVaugh 1942 N Amer Flora 32A: 1–134]

1. Ovary superior; fr 2–3 × sepals . . . . *N. longiflorus* var. *breviflorus*
1' Ovary partly to completely inferior; fr < or slightly > sepals
  2. St below branches silver-gray, shiny, sometimes dark; corolla yellow with brown marks . . . . *N. rubescens*
  2' St below branches reddish, brown, or dark purple; corolla white or purplish
    3. Sts spreading to decumbent . . . . *N. rigidus*
    3' Sts gen erect to stiffly ascending (spreading in *N. gracilis, N. sigmoideus*)
      4. Infl axis straight or weakly zigzag
        5. Bracts 2–4 mm, enfolding pedicel base; corolla white with lavender lines; fr hemispheric . . . . . *N. gracilis*
        5' Bracts 2–9 mm, flat, not enfolding pedicel base; corolla white; fr bell-shaped . . . . *N. ramosissimus*
      4' Infl axis strongly zigzag
        6. Corolla divided 1/2 or less to base . . . . *N. pinnatifidus*
        6' Corolla divided nearly to base
          7. Pedicels clearly S-curved; bracts ovate; fls spreading . . . . *S. sigmoideus*
          7' Pedicels straight, slightly curved, or reflexed; bracts linear, elliptic, or lanceolate (sometimes ovate in *N. glanduliferus);* fls spreading or reflexed
            8. Fr obconic (base acute, tip rounded); sepals erect . . . . *N. capillaris*
            8' Fr ± hemispheric (base and tip rounded); sepals spreading or reflexed . . . . *N. glanduliferus*
              9. Pedicels spreading, tip curved; sepals 1.5–2.5 mm . . . . var. *glanduliferus*
              9' Pedicels stiffly ascending, tip straight; sepals ± 1–1.5 mm . . . . var. *orientalis*

N. capillaris E. Greene (p. 249) **STS** stiffly ascending, 7–18 cm; base brownish or purplish. **LF** 3–15 mm, ovate, narrowed abruptly to short petiole, entire, glabrous or hairy. **INFL**: axis zigzag, esp in fr; bracts 1–3 mm, lanceolate to elliptic; pedicels 8–12 mm, 0.1–0.15 mm diam, spreading, straight or slightly curved, tip not curved. **FL**: hypanthium ± 1 mm; sepals ± 0.5 mm, elliptic to lanceolate, erect; corolla 0.7–1.3 mm, divided ± to base, white, glabrous, anthers 0.1–0.2 mm. **FR** 1.5–2.5 mm, obconic (base acute, tip rounded). **SEED** 0.5–0.7 mm, widely elliptic, narrowly ridged. *n*=9. Dry slopes, burned areas; 400–2100 m. NCoR, CaRH, n SNF, SNH, MP, **DMoj**; OR. May–Jul

N. glanduliferus Jepson **STS** stiffly ascending, 5–25 cm; base brownish or purplish. **LF** 3–16 mm, oblanceolate to elliptic, narrowed gradually to petiole, toothed or pinnately lobed, hairy. **INFL**: axis strongly zigzag; bracts 1–3 mm, spreading, linear to ovate; pedicels 6–16 mm, 0.1–0.2 mm diam, spreading, straight or slightly curved, tip gen curved. **FL**: hypanthium ± 1 mm; sepals 1.5–2.5 mm, linear-elliptic to ± deltate, spreading; corolla 2–2.5 mm, divided ± to base, white, upper lobes erect, lower lobes ciliate; filament tube 1–2.3 mm, tip slightly curved, glabrous, anthers 0.2–0.4 mm. **FR** ± 2–4 mm, ± hemispheric (base and tip rounded). **SEED** ± 0.5 mm, cylindric; surface with impressed, vertical lines crossed by fine transverse lines. Dry slopes, sandy soils, washes; < 2400 m. **SNE, D**; to UT, NM, Baja CA.

var. *glanduliferus* (p. 249) **FL**: pedicels spreading, tip curved; sepals 1.5–2.5 mm; filament tube 1.6–2.3 mm. Sandy or gravelly soils, canyons; 150–1900 m. s **SNE, D**; NV, AZ, Baja CA. Mar–May

var. *orientalis* McVaugh **FL**: pedicels stiffly ascending, tip straight; sepals ± 1–1.5 mm; filament tube 1–2 mm. Sandy soils, rocky slopes, washes; < 2400 m. **SNE, D**; to UT, NM, Baja CA. Mar–May

N. gracilis Eastw. (p. 249) SLENDER NEMACLADUS **STS** spreading to ascending, 2.5–10 cm; base dull reddish brown. **LF** 2.5–8 mm, oblanceolate to oblong, narrowed to wide petiole, irregularly dentate to ± pinnately lobed, hairy. **INFL**: axis straight or weakly zigzag; bracts 2–4 mm, linear-oblong, tip recurved; pedicels 5–12 mm, 0.1 mm diam, reflexed, slightly S-curved, tip erect. **FL**: hypanthium 0.5 mm; sepals ± 0.5 mm, linear, spreading; corolla 1.5–2 mm, ± 1/2 divided, white with lavender veins, sparsely hairy, lobes erect; filament tube ± 1 mm, tip curved, fine-hairy; anthers 0.5 mm. **FR** ± 1.5 mm, ± hemispheric (base acute, tip rounded). **SEED** ± 0.5 mm, widely elliptic; surface with vertical zigzag ridges alternating with clearly pitted rows. UNCOMMON. Rocky slopes, sandy washes; < 1900 m. Teh, SnJV (sw Merced Co.), s SCoRO, WTR, **w DMoj (Los Angeles Co.).** Mar–Apr

N. longiflorus A. Gray **STS** erect, 5–21 cm; base brownish or purplish. **LF** 3–12 mm, oblanceolate to ovate or oblong, narrowed to a winged petiole, entire to finely crenate, hairy. **INFL**: axis ± zigzag; bracts 2–4 mm, elliptic to ovate; pedicels 6–23 mm, 0.1 mm diam, ascending, S-curved, tip abruptly curved. **FL**: hypanthium ± 0; sepals ± 1 mm, elliptic, erect; corolla 3–8 mm, divided < 1/2 to base, white, tube cylindric, lobes erect, upper lip puberulent, lower lip glabrous; filament tube 2–7.5 mm, tip abruptly curved, fine-hairy, anthers 0.2–0.6 mm diam; ovary superior in fr. **FR** ± 2.5–5 mm, fusiform (base and tip acute). **SEED** 0.2–0.5 mm, widely elliptic to ± round; surface with obscure, lengthwise, wavy ridges. Sandy or gravelly slopes and washes; 300–2400 m. s SNH, SCoR, SCo, SnGb, SnBr, PR, **sw DMtns (Little San Bernardino Mtns)**; Baja CA. 2 vars. in CA.

var. *breviflorus* McVaugh **FL**: corolla 3–3.5 mm, tube 2–2.5 mm; filament tube ± 2–3 mm. **SEED** < 0.5 mm, ± round. Habitat of sp.; 800–1200 m. s SNH, SnGb, SnBr, PR, **sw DMtns (Little San Bernardino Mtns)**; Baja CA.

N. pinnatifidus E. Greene (p. 255) **STS** stiffly erect, 6–20 cm, brownish or purplish. **LF** 5–20 mm, oblanceolate, narrowed to a long petiole, deeply pinnately lobed; lobes toothed or entire, glabrous. **INFL**: axis zigzag; bracts 2–5 mm, linear to elliptic; pedicels 5–15 mm, < 0.1 mm diam, spreading, S-curved, tip abruptly curved. **FL**: hypanthium 0.5–0.7 mm; sepals ± 1 mm, narrowly triangular, erect; corolla 1.5–2 mm, divided ± 1/3 to base, white or rose-purple, glabrous, lobes erect; filament tube ± 0.7 mm, tip slightly curved, glabrous, anthers 0.1–0.2 mm. **FR** 3–4 mm, fusiform (base and tip acute). **SEED** 0.5 mm, elliptic, with vertical, clearly pitted rows. Dry washes, burned areas, chaparral; 300–1300 m. SCo, SnGb, PR, **DSon**; Baja CA. May–Jun

N. ramosissimus Nutt. **STS** erect, 5–32 cm; base brownish or purplish. **LF** 3–18 mm, oblanceolate, narrowed to slender or wide petiole, irregularly toothed or ± pinnately lobed, margins and base hairy. **INFL**: axis straight; bracts 2–9 mm, linear, flat; pedicels 6–22 mm, 0.7 mm diam, ± spreading, slightly S-curved, tip ± erect. **FL**: hypanthium 0.5 mm; sepals lobes ± 0.5 mm, deltate, erect; corolla 1.5–2 mm, divided > 1/2 length, white, glabrous, lobes erect; filament tube 1–2 mm, tip curved, anthers 0.2–0.3 mm. **FR** ± 1.5–2.5 mm, bell-shaped (base narrowed, tip rounded). **SEED** ± 0.5 mm, subspheric; surface with clearly pitted rows. Dry, sandy or gravelly soils; 150–1600 m. SnFrB (Mount Hamilton), SCoRO, SCo, WTR, SnGb, PR, **DMoj**; to UT, NM, Baja CA. Apr–May

N. rigidus Curran (p. 255) **STS** spreading to decumbent, 4–9 cm; base shiny, purple. **LF** 5–10 mm, elliptic to oblanceolate, narrowed to wide petiole, entire to scalloped, hairy. **INFL**: axis strongly zigzag; bracts 2–3 mm, widely elliptic; pedicels 5–12 mm, 0.2 mm diam, spreading, straight, becoming curved in age.

FL: hypanthium 1 mm; sepals ± 1.5 mm, triangular, erect; corolla 1–1.5 mm, deeply divided, white or purplish, veins and margins red, lobes ovate, erect, glabrous; filament tube 1.2–1.6 mm, tip slightly curved, glabrous, anthers 0.2–0.3 mm. **FR** 3–4 mm, ± hemispheric (base oblique, tip pointed). **SEED** 0.6–0.7 mm, elliptic; surface with wide ridges alternating with pitted rows. Bare soil or sand; 200–2500 m. **GB**; e OR, ID, NV. May–Jun

N. rubescens E. Greene **STS** erect, 5–20 cm; base shiny, silver-gray. **LF** 5–20 mm, elliptic to oblanceolate, narrowed abruptly to winged petiole, entire, toothed, or ± pinnately lobed, glabrous or coarsely hairy. **INFL**: axis weakly zigzag; bracts 1–2.5 mm, widely lanceolate; pedicels 8–15 mm, 0.1–0.2 mm diam, horizontal to ascending, slightly S-curved, tip slightly curved or not. **FL**: hypanthium 0.2–0.3 mm; sepals ± 1 mm, elliptic to deltate, erect; corolla 1.5–2 mm, divided > 1/2 length, yellow with purple or brown marks, upper lobes reflexed, slightly ciliate, lower lobes erect, purple-tipped, densely ciliate; filament tube 2–3 mm, straight or tip slightly curved, anthers 0.6 mm; ovary 1/4–1/2 inferior. **FR** 2–2.5 mm, ± bell-shaped (base narrowed, tip rounded). **SEED** 0.4 mm, widely elliptic; surface with wavy ridges alternating with weakly pitted rows. Dry, sandy or gravelly soils; < 1600 m. PR (e slope), SNE, **D**; NV, AZ, Baja CA. Pls from PR (e slope) & D, with narrow, deeply toothed lvs, filament appendage stalks >> processes have been called var. *tenuis* McVaugh. Apr–May

N. sigmoideus G. Robb. (p. 255, pl. 41) **STS** widely spreading, 4–12 cm; base purplish brown. **LF** 1.5–10 mm, ovate to elliptic, sessile, entire or irregularly dentate, short-hairy. **INFL**: axis strongly zigzag; bracts 0.8–1.5 mm, ovate; pedicels 10–18 mm, < 0.1 mm diam, spreading, S-curved, tip erect. **FL**: hypanthium 0.5 mm; sepals ± 1.5 mm, lanceolate-deltate, erect, spreading in fr; corolla 2.5–3.5 mm, deeply divided, white, yellow at tips, hairy, upper lobes erect, lower lobes spreading; filament tube 1.5 mm, tip curved, fine-hairy, anthers ± 0.3 mm; ovary nearly superior. **FR** ± 2 mm, widely fusiform (base and tip acute). **SEED** ± 0.5 mm, widely elliptic; surface with zigzag ridges alternating with clearly pitted rows. Sandy or gravelly soils, Joshua-tree woodland; 50–2300 m. s SNH (e slope), Teh, e PR, SNE, **DMoj**, **nw DSon**; NV, AZ, Baja CA. Apr–Jun

PARISHELLA

1 sp. (brothers Samuel B. (1838–1928) & William F. Parish, botanical collectors)

P. californica A. Gray (p. 255) Ann, glabrous to hairy. **STS** branched from base, reclining, 1–10 cm. **LVS**: basal rosette, persistent; petiole < 7 mm; blade 6–15 mm, 2–5 mm wide, oblanceolate, leathery, tip rounded, margin narrowly translucent. **INFL**: clusters head-like, terminal; bracts oblanceolate, spreading to erect; pedicels in fr 2–5 mm. **FL** not inverted; hypanthium ± 3.5 mm; sepals 2–6 mm, oblanceolate; corolla white, tube 2–3 mm, flaring, lobes 5, 1–1.5 mm, subequal, oblong; filaments 2.5–3.5 mm, free at base, fused above middle, anthers equal, ± round, lower 4 appendaged at junction with tube, upper not; appendages rod-like, minute, translucent; ovary ± 2-chambered, stigma head-like, smooth. **FR** 3–3.5 mm, 2–3 mm wide; top dome-like, circumscissile. **SEED** ± 0.7 mm, elliptic; surface with 8–10 vertical, pitted rows, pits 10–12 per row. Sandy or gravelly soils; 650–1500 m. sw SnJV (Caliente Mtns), se SCoRO, **DMoj (San Bernardino Co.)**. Apr–May

PORTERELLA

1 sp. (Thomas C. Porter, botanist, 1822–1901)

P. carnosula (Hook. & Arn.) Torrey (p. 255) Ann 2–30 cm, emergent or terrestrial, glabrous. **ST** erect, branched from base or not. **LVS** cauline, sessile; blade (5)12–15 mm, (1)3–5 mm wide, narrowly ovate (aerial) to narrowly triangular (submersed), entire, sometimes few-toothed. **INFL**: raceme; bract 1 per fl, lf-like, ascending; pedicels in fr 1–3 cm. **FL** inverted, fragrant; hypanthium 1–2.5 mm; sepals = or > fr, narrowly triangular; corolla tube 4–5 mm, linear, blue, throat with 2 folds, yellow, limb 2-lipped, upper lip 2-lobed, lobes 1–2 mm, narrowly triangular, erect, deep blue, lower lip 3-lobed, lobes 3–8 mm, ± round, spreading, blue, base yellow or white; stamens fused, anthers ± 2 mm, 2 short anthers appendaged, horn-like appendage ± 0.5 mm, comb-like appendage 4-pronged; ovary 2-chambered, stigma cup- or plate-like, papillate. **FR** 5–10 mm, hemispheric to cylindric; tip conic, dehiscing by valves. **SEED** 1 mm, smooth, fine striate. *n*=12. Moist, grassy roadsides, lake and pond edges; 1500–3100 m. CaRH, n SNH, **GB**; to WY, AZ. ❀ TRY. Jun–Aug

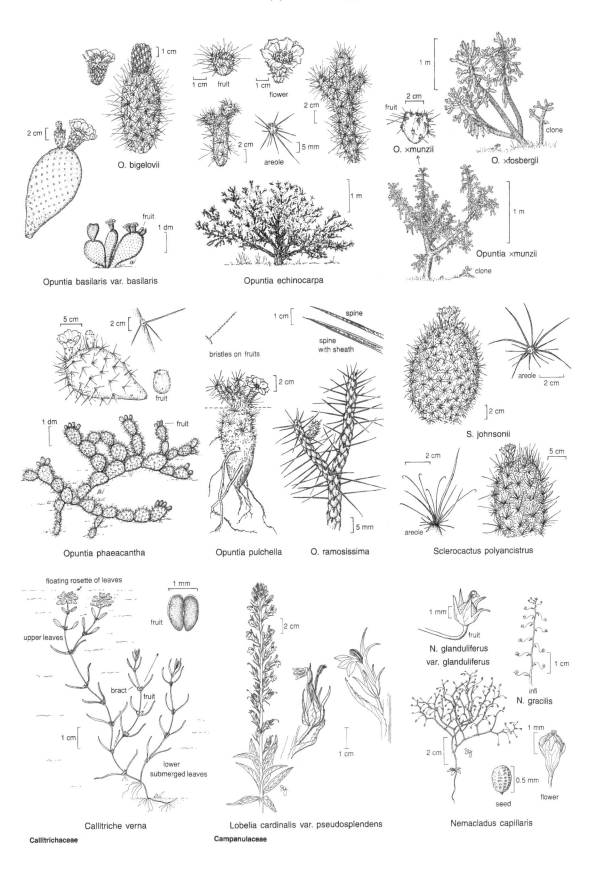

O. bigelovii

Opuntia basilaris var. basilaris

fruit

areole

O. ×munzii

O. ×fosbergii

clone

Opuntia ×munzii

clone

Opuntia echinocarpa

bristles on fruits

spine

spine
with sheath

Opuntia phaeacantha

fruit

Opuntia pulchella

O. ramosissima

S. johnsonii

areole

areole

Sclerocactus polyancistrus

floating rosette of leaves

upper leaves

fruit

bract

fruit

lower
submerged leaves

Callitriche verna

Callitrichaceae

Lobelia cardinalis var. pseudosplendens

Campanulaceae

N. glanduliferus
var. glanduliferus

fruit

infl

N. gracilis

seed

flower

Nemacladus capillaris

CAPPARACEAE CAPER FAMILY

Staria S. Vanderpool

Ann, shrub, tree, ill-smelling. **LVS** gen 1-palmate, gen alternate, gen petioled; stipules gen minute, often bristle-like or hairy; lflets 3–7. **INFL**: raceme, head, or fls solitary, gen longer in fr; bracts gen 3-parted below, simple above, or 0. **FL** gen bisexual, radial to ± bilateral; sepals gen 4, free or fused, gen persistent; petals gen 4, free, ± clawed; stamens gen 6, free, exserted, anthers gen coiling at dehiscence; ovary superior, gen on stalk-like receptacle, chamber gen 1, placentas gen 2, parietal, style 1, persistent, stigma gen minute, ± head-like. **FR**: gen capsule, septicidal; valves gen 2, deciduous, leaving septum (frame-like placentas) behind; pedicel gen ± reflexed to spreading. 45 genera, 800 spp.: widespread trop to arid temp; some cult (*Capparis spinosa*, caper bush). [Ernst 1963 J Arnold Arbor 44:81–93] CA members placed in subfamily Cleomoideae. Alternate family name: Capparidaceae. Recently treated to be included within Brassicaceae [Rodman et al 1993 Ann Missouri Bot Gard 80: 686-699] Keys adapted by Margriet Wetherwax.

1. Shrub; fr an inflated capsule; petals > 4 mm wide . . . . **ISOMERIS**
1' Ann; fr a slightly inflated capsule or pair of nutlets; petals < 4 mm wide
  2. Fls in axillary heads; style in fr stout, spine-like . . . . **OXYSTYLIS**
  2' Fls in terminal racemes or solitary in lf axils; style in fr sometimes elongate but not spine-like
    3. Fr a pair of 1–3-seeded nutlets; septum < 2 mm wide . . . . **WISLIZENIA**
    3' Fr a 2–many-seeded capsule; septum > 2 mm wide
      4. Fr gen 15–45 mm, longer than wide; septum linear to oblong . . . . **CLEOME**
      4' Fr 2–6 mm, often wider than long; septum elliptic to round . . . . **CLEOMELLA**

CLEOME

Ann, gen ± glabrous. **ST** gen branched from upper nodes. **LF**: petiole 5–45 mm; lflets gen 3. **INFL**: raceme, terminal, gen 1–4 cm in fl, gen 5–40 cm in fr; pedicels 4–20 mm. **FL** often ± unisexual (stamens or pistils vestigial), ± bilateral, most parts gen yellow; sepals free or fused; petals sessile to short-clawed. **FR**: capsule, longer than wide; septum linear to oblong; receptacle stalk-like, reflexed to ascending. **SEEDS** 10–40. 150–170 spp.: esp trop, subtrop Am, Afr; some trop weeds. (Early Eur name for a mustard-like pl)

1. Infl dense, many-fld; sepals fused in basal half, persistent; anthers 1.9–2.6 mm . . . . *C. lutea*
1' Infl open, few-fld; sepals free, deciduous; anthers 3–6 mm . . . . *C. sparsifolia*

C. lutea Hook. (p. 255, pl. 42) YELLOW BEE PLANT Pl 2.5–13 dm, ± glabrous. **LF**: lflets gen 5, 1.5–6 cm, linear to elliptic. **FL**: sepals fused in basal half, persistent, 1.6–2.6 mm, lanceolate, minutely dentate, yellow; petals 5–8 mm, oblong to ovate, yellow; stamens 10–20 mm, yellow, anthers 1.9–2.6 mm. **FR** 15–40 mm, 2–5 mm wide, ± round in transverse section, striate; receptacle 5–17 mm. 2*n*=34. Dry, sandy flats, desert scrub, weedy roadsides; 1100–2400 m. **SNE, DMtns**; to n WA, e Colorado. ❀ SUN,DRN,IRR:TRY. May–Aug

C. sparsifolia S. Watson Pl 1–9 dm, densely branched, glabrous, glaucous. **LVS** ± sparse; lflets 3 below, 1 (lvs simple) above, 0.4–1.5 cm, obovate. **INFL** open, few-fld, 2–10 cm, not much expanding in fr. **FL**: sepals free, deciduous, 1.4–3 mm, ovate, acuminate, minutely serrate, brown-green; petals 9–13 mm, strap-shaped, recurved, yellow with brown central streak; stamens 9–15 mm, yellow, anthers 3–6 mm, brown; style 0.1–0.4 mm. **FR** 15–45 mm, 1–3 mm wide, ± round in transverse section, smooth; receptacle 2–5 mm. 2*n*=32. Sand dunes, beaches; 900–2000 m. **SNE, DMoj**; w NV. May–Aug

CLEOMELLA

Ann, gen glabrous. **ST** gen ascending to erect, gen branched from base, often red-tinged. **LVS** gen many; petiole gen 7–20 mm; lflets gen 3. **INFL**: raceme, ± terminal; fls solitary in lf axils, or both; pedicel gen 4–25 mm. **FL** radial to bilateral; parts gen yellow; sepals fused in basal third, gen entire; petals ± sessile, upper 2 often recurved. **FR**: capsule, often wider than long; septum elliptic to round; receptacle stalk-like. **SEEDS** < 10. ± 10 spp.: arid w N.Am. (Diminutive of *Cleome*) [Payson 1922 Univ Wyoming Publ Sci Bot 1:29–46] *C. hillmanii* Nelson, known from near Reno, NV, may be found in adjacent CA.

1. Pl hairy; older sts prostrate; receptacle in fr 6–8 mm, reflexed . . . . *C. obtusifolia*
1' Pl glabrous; older sts ascending to erect; receptacle in fr < 6 mm or 6–10 mm and spreading to ascending
  2. Fls solitary in lf axils; receptacle reflexed in fr . . . . *C. brevipes*
  2' Fls in terminal racemes, sometimes also solitary in lf axils; receptacle spreading to ascending in fr
    3. St branched gen from base; petals 1.8–2.2 mm; receptacle in fr 0.3–0.8 mm . . . . *C. parviflora*
    3' St branched gen from upper nodes; petals 3.5–7 mm; receptacle in fr 6–10 mm . . . . *C. plocasperma*

C. brevipes S. Watson Pl glabrous, glaucous. **ST** 5–45 cm, rough. **LF**: petiole 0.5–3 mm; lflets 5–15 mm, linear to obovate, fleshy. **INFL**: fls solitary in lf axils, incl those near base; pedicels 1.5–3 mm. **FL**: sepals 0.8–1.2 mm, ovate, acuminate; petals 1.5–2 mm, pale yellow; stamens 1.5–2.2 mm, anthers 0.3–0.5 mm; style 0.1–0.3 mm. **FR** 2–3 mm, 2–3.2 mm wide, round; valves slightly conic; receptacle 0.5–3 mm, reflexed. Alkaline marsh, wet, salt-encrusted soil around thermal springs; 400–1400 m. **SNE, DMoj**; w NV. May–Oct

C. obtusifolia Torrey & Frémont (p. 255)   MOJAVE STINKWEED
Pl hairy. **ST** 1–9 dm, rough; younger ascending to erect; older
prostrate, forming circular mat < 9 dm wide. **LF**: lflets 5–15 mm,
obovate. **INFL**: raceme on older sts, 1–10 cm, terminal, dense,
on younger sts fls solitary in lf axils. **FL**: sepals 1–1.5 mm, ovate,
green, hairy, margin with long hairs; petals 4–6(9) mm, dark yellow, lower surface hairy; stamens 8–14 mm, anthers 1.5–2.3
mm; style 1.5–5 mm. **FR** 3–4 mm, hairy, striate; valves conic
to horn-shaped; receptacle 6–8 mm, reflexed. Desert scrub,
sandy, rocky alkaline flats; 300–1200(2000) m. **SNE**, **D**; w NV.
[*C. o.* var. *pubescens* Nelson] Variable; deserves additional study.
Apr–Oct

C. parviflora A. Gray Pl glabrous. **ST** 3–45 cm, smooth. **LVS**
few; lflets 5–35 mm, linear-elliptic, ± fleshy. **INFL**: raceme, 0.5–
30 cm, terminal; fls sometimes also solitary in lf axils. **FL**: sepals 0.5–1 mm, lanceolate; petals 1.8–2.2 mm, pale yellow; sta-

mens 1.9–2.5 mm, anthers 0.4–0.5 mm; style < 0.2 mm, stigma
2-lobed, 0.3 mm, purple. **FR** 3–4 mm; valves slightly conic; receptacle 0.3–0.8 mm, spreading to ascending. Wet, alkaline
meadows about thermal springs in sagebrush desert; 1200–2000
m. **GB**, **DMoj**; w NV. Often occurs with *C. brevipes*, *C.
plocasperma*. May–Aug

C. plocasperma S. Watson (p. 255) Pl glabrous. **ST** branched
gen from upper nodes, 10–55(80) cm, smooth. **LF**: lflets 15–45
mm, linear-elliptic. **INFL**: raceme, 1–20 cm, terminal. **FL**: sepals 0.9–2.2 mm, lanceolate; petals 3.5–7 mm; stamens 8–12 mm,
anthers 1.5–1.9 mm; style 0.8–1.2 mm. **FR** 4–5 mm; valves ±
hemispheric to horn-shaped; receptacle 6–10 mm, spreading to
ascending. Wet, alkaline meadows, greasewood flats, around
thermal springs; 800–1400 m. **GB**, **DMoj**; to OR, ID, UT. Vars.
mojavensis (Payson) Crum, *stricta* Crum of uncertain taxonomic
status. May–Jul

ISOMERIS

1 sp. (Greek: equal part)

I. arborea Nutt. (p. 255) BLADDERPOD Shrub, profusely
branched, gen 5–20 dm, minutely hairy. **LF**: petiole 1–3 cm; lflets
gen 3, 15–45 mm, oblong-elliptic. **INFL**: raceme, 1–30 cm, terminal; pedicels 8–15 mm, thicker in fr. **FL**: sepals fused in basal
half, 4–7 mm, ± entire, green; petals 8–14 mm, 4–5 mm wide,
sessile, yellow; stamens 15–25 mm, yellow, anthers 2–2.5 mm;
style 0.9–1.2 mm or pistil aborting in bud. **FR**: capsule, tardily
dehiscent, 3–4 cm, inflated, oblong to ± spheric, smooth, leath-

ery, light brown; valves 2–3; receptacle stalk-like, 1–2 cm, stout,
reflexed. $2n=$ 34,40. Common. Coastal bluffs, hills, desert
washes, flats; 0–1300 m. s SNF, Teh, SnJV, CCo, SCo, ChI, **D**;
to Baja CA. [*Cleome isomeris* E. Greene] Vars. *angustata* Parish, *globosa* Cov., *insularis* Jepson, have been recognized based
on fr variation that may be loosely correlated with geography. ❀
DRN,SUN&DRY:8,9,**14**, 15–17,**19–24**;IRR:**12,13**;STBL. Most
of year

OXYSTYLIS

1 sp. (Greek: sharp style)

O. lutea Torrey & Frémont (p. 255) Ann, ± glabrous. **ST**
branched from base, 5–15 dm. **LF**: petiole 2.5–7 cm; lflets 3, 2–
6 cm, ± elliptic, thick, firm. **INFL**: head, 0.5–3 cm, axillary, ±
spheric, dense, ± sessile, not elongating in fr; pedicels 1–3 mm,
thicker and reflexed in fr. **FL** radial; sepals free, 1–2 mm, lanceolate, green; petals 2–3 mm, elliptic, straw-yellow, ± sessile;
stamens 3–5 mm, yellow, anthers 1–1.2 mm; ovary 1–1.5 mm, ±

sessile, lobes 2, nearly separate, each 1-ovuled; style 2–3 mm,
wide, fleshy at base. **FR**: nutlets 2, 2.5 mm, ± spheric, smooth,
white to deep purple; receptacle < 2 mm, ± stalk-like; style 4–11
mm, stiff, spine-like. **SEED** 1 per nutlet. $2n=$20,40. Rocky or
sandy alkaline flats; > 600 m. **DMoj**; w NV (Amargosa Desert).
Mar–Oct

WISLIZENIA

1 sp. (A. Wislizenus, pl collector in sw US, born 1810)

W. refracta Engelm. JACKASS CLOVER Ann or per, glabrous
to puberulent. **ST** profusely branched from base, 0.5–24 dm. **LF**:
petiole 3–25 mm; lflets gen 3. **INFL**: raceme, 1–3 cm, dense,
terminal, in fr 4–20 cm; pedicels 5–10 mm. **FL** radial; sepals
free, ± 2 mm, ± entire, green; petals 2.5–6.3 mm, elliptic, yellow,
± sessile but tapered to base; stamens 8–14 mm, yellow; ovary
0.3–0.6 mm, gen exserted, lobes 2, nearly separate, each gen 1-
ovuled, style 2–5.5 mm. **FR**: nutlets 2; valves deciduous; receptacle stalk-like, reflexed; style elongate but not spine-like.
SEEDS gen 1 per nutlet. $2n=$40. Desert washes and flats, fields,
roadsides, esp alkaline soils; 0–800 m. c&s SNF, SnJV, **D**; to
TX, nw Mex. [Keller 1979 Brittonia 31:333–351] Valuable honey
plant. TOXIC but seldom eaten. 3 sspp. in CA.

1. Gen per; lflets 3 below, 1 (lvs simple) above, linear-
    elliptic, 3–8 × longer than wide; sepals ovate; anthers
    1.5–2.3 mm . . . . ssp. *palmeri*

1′ Ann; lflets 3, ovate to obovate, 2–3 × longer than wide;
    sepals lanceolate; anthers 0.9–1.2 mm . . . . . . ssp. *refracta*

ssp. *palmeri* (A. Gray) C.S. Keller (p. 255) Per, short-lived.
ST brown-gray. **LF**: lflets 3 below, 1 (lf simple) above, 17–35
mm, linear-elliptic. **FL**: receptacle 5–14 mm; sepals 1–1.7 mm,
ovate; anthers 1.5–2.3 mm; style 2.5–5.5 mm. Sandy washes,
beach dunes, desert scrub; 0–200 m. **DSon**; nw Mex.

ssp. *refracta* (p. 255) JACKASS CLOVER Ann. **ST** green, tan.
LF: lflets 3, 7–30 mm, ovate. **FL**: receptacle 3–6 mm; sepals
1–1.5 mm, lanceolate; anthers 0.9–1.2 mm; style 2–5 mm. UN-
COMMON. Sandy washes, roadsides, alkaline flats; 600–800
m. **DMoj**, n **DSon**; to w TX. Apr–Nov

CAPRIFOLIACEAE HONEYSUCKLE FAMILY

Lauramay T. Dempster

Subshrub, shrub, vine, or small tree. **LVS** opposite, simple or compound; stipules gen 0. **FL**: calyx tube fused to
ovary, limb gen 5-lobed; corolla radial or bilateral, rotate to cylindric, gen 5-lobed; stamens gen 5, epipetalous,

alternate corolla lobes; ovary inferior, 1–5-chambered, style l. **FR**: berry, drupe, or capsule. ± 12 genera, 450 spp.: esp n temp. Recently treated to include Valerianaceae (and Dipsacaceae); *Sambucus* (and *Viburnum*) recently treated in Adoxaceae [Donoghue et al 1992 Ann Missouri Bot Gard 79: 333-345] Keys adapted by Margriet Wetherwax.

1. Corolla rotate, radial; fls many, in flat-topped or dome-shaped infl; lf compound . . . . **SAMBUCUS**
1' Corolla cylindric, funnel-shaped, or 2-lipped; fls paired or several in elongated, raceme-like infl; lf simple
  2. Corolla ± bilateral, cylindric, 2-lipped; fr black; seeds gen > 2 . . . . **LONICERA**
  2' Corolla ± radial, bell-shaped to salverform; fr white; seeds 1–2 . . . . **SYMPHORICARPOS**

LONICERA HONEYSUCKLE

Shrub, erect or twining. **LVS** simple, entire, short-petioled; 1–2 pairs beneath infl often fused around st. **INFL**: spikes, interrupted, at ends of branches, or fls paired on axillary peduncles and subtended by 0–2 sets of bracts. **FL**: calyx-limb 0 or gen 5-toothed, gen persistent; corolla 5-lobed, ± radial or strongly 2-lipped (4 upper lobes, 1 lower), tube pouched at base; ovary chambers 2–3. **FR**: berry, gen round. ± 200 spp.: temp, subtrop N.Am, Eur, Asia, n Afr. (Adam Lonitzer, German herbalist, l6th century) [Rehder 1903 Rep Missouri Bot Gard 14:27–231]

L. japonica Thunb. JAPANESE HONEYSUCKLE Vine, climbing; herbage glabrous or soft-hairy. **LF** gen 3–8 cm; blade oblong to ovate, base rounded, tip ± acute. **INFL**: fls paired, each pair subtended by 2 lf-like bracts and 4 ± round bractlets that are ± 1/2 ovary length; peduncles short, axillary. **FL**: corolla 25–40 mm, strongly 2-lipped, white turning yellow, often tinged purplish, tube hairy; stamens, style and stigma exserted. **FR** black. 2n=18. Disturbed places; gen < 1000 m. **CA**; abundant in se US; native to Asia. Sporadic escape from cult. Spring & summer

SAMBUCUS ELDERBERRY

Large shrub (or small tree gen lacking main trunk), deciduous. **ST**: pith conspicuous, spongy. **LVS** pinnately (rarely bipinnately) compound, with terminal lflet; lflets serrate. **INFL**: panicle made up of cymes, terminal. **FL** small; calyx 5-toothed; corolla radial, rotate, 5-lobed, white or cream; ovary chambers 3–5, ovules 1 per chamber and suspended from its top, style short, stigmas 3–5. **FR**: drupe. **SEEDS** 3–5. (Greek, the name of a musical instrument made from wood of this genus) 20 spp.: temp, subtrop; some cult as orn. TOXIC in quantity (exc cooked frs).

1. Fr appearing blue (black and densely glaucous); infl flat-topped, central axis gen abruptly shorter and
  weaker than branches . . . . *S. mexicana*
1' Fr bright red, not glaucous; infl ± dome-shaped, central axis dominant . . . . *S. racemosa* var. *microbothrys*

S. mexicana C. Presl (p. 255) BLUE ELDERBERRY Shrub 2–8 m, gen as wide as tall, lacking main trunk. **LF**: lflets 3–9, 3–20 cm, elliptic to ovate, glabrous to hairy, axis often curved or bowed, base often asymmetric, tip acute to acuminate. **INFL** 4–33 cm diam, ± flat-topped; central axis gen abruptly shorter and weaker than branches. **FR** nearly black and densely white glaucous, thus appearing bluish. 2n=36. Common. Streambanks, open places in forest; < 3000 m. CA-FP, **GB**, **DMtns**; to B.C., UT, NM. [*S. caerulea* Raf.] Variable, currently impossible to split into unified subgroups; detailed study warranted. ❀ 4,5,**6,7,14–17,24**,IRR:1–3,**8,9**,10,**18–23**. Mar–Sep

S. racemosa L. RED ELDERBERRY Shrub 2–6 m. **LF**: lflets 5–7, 6–16 cm, lanceolate to oblong-ovate, base gen asymmetric, tip gradually acuminate. **INFL** 6–10 cm diam, ± dome-shaped; central axis remaining dominant throughout development. **FR** gen bright red, not glaucous. Moist places; < 3300 m. CA-FP, n **SNE (Sweetwater Mtns)**; to AK, Colorado, AZ. 2 vars. in CA.

  var. *microbotrys* (Rydb.) Kearney & Peebles (p. 255) **LF**: lflets gen glabrous. Common. Moist places, ± montane; 1800–3300 m. NW, CaR, SN, SnBr, **n SNE (Sweetwater Mtns)**; to Colorado, AZ. [*S. microbotrys* Rydb.] ❀ 4,5,WET:1,2,6,7,15–17;DFCLT Jun–Aug

SYMPHORICARPOS WAXBERRY, SNOWBERRY

Shrub. **ST** decumbent to erect, slender. **LF** simple, deciduous, small, short-petioled; blade gen elliptic to round, some often ± lobed. **INFL**: gen raceme, gen ± terminal, gen few-fld; fl subtended by 2 fused bractlets. **FL** ± radial; hypanthium ± spheric; calyx with 5-toothed, persistent limb; corolla bell-shaped to ± salverform, gen 5-lobed, white or pink, often ± hairy inside; nectaries 1–5, ± basal; stamens gen incl; ovary chambers 4, styles gen incl, stigma head-shaped. **FR** gen berry-like, gen white. **SEEDS** 2 (1 per lateral ovary chamber), ± oblong, planoconvex. ± 10 spp.: N.Am, 1 in China. (Greek: to bear fr together, the berries borne in clusters) [Jones 1940 J Arnold Arbor 21:201–252]

1. Corolla salverform, 8–15 mm, tube slender, glabrous inside, 3–4 × length of spreading lobes; nectary 1;
  lf similar above and below, blade gen 4–12 mm . . . . *S. longiflorus*
1' Corolla bell-shaped, 6–10 mm, tube wide, ± hairy inside, 2–3 × length of erect lobes; nectaries 5; lf more
  obviously veined below, blade gen 8–20 mm . . . . *S. rotundifolius*
  2. Pl trailing; upper 2/3 corolla throat sparsely hairy inside . . . . var. *parishii*
  2' Pl erect, spreading; upper 1/3 of corolla tube glabrous inside, middle 1/3 hairy . . . . var. *rotundifolius*

S. longiflorus A. Gray (p. 255) FRAGRANT SNOWBERRY Pl 9–12 dm, stiff, glabrous or puberulent, often dotted with minute glands. **ST**: branches often spiny; young bark red or brown, old whitish, shredding. **LF**: blade 0.5–2 cm, entire, ± thick, sometimes lanceolate, bluish, veins below not prominent. **INFL**: fls sometimes solitary in axils. **FL** very fragrant; calyx limb ± erect, unevenly and often shallowly lobed, sinuses often round; corolla 8–15 mm, ± salverform, pink or cream, tube slender, often red or purple outside, glabrous inside, lobes 1/5–1/4 corolla length, spreading, ovate; nectary 1, long, slender; style gen hairy above middle. **FR** ± 7 mm, narrowly elliptic, dry. **SEED** ± 5 mm. Among rocks; 1350–1600 m. **GB, DMtns**; to Colorado, TX. ❀ 4,6,IRR:1,15& SHD:2,3,7,16;DFCLT. May–Jun

S. rotundifolius A. Gray Pl 6–12 dm, stiff, puberulent. **ST**: old bark shredding. **LF**: blade 8–20 mm, paler and more prominently veined below. **INFL**: fls 1–2 in axils. **FL**: calyx limb flaring, lobes deep, irregular, margin gen transparent; corolla 6–10 mm, narrowly bell-shaped, pink or white, middle 1/3 of tube lightly hairy, upper 1/3 glabrous, lobes ± erect, ± 1/5–1/3 corolla length; nectary below all 5 lobes; style glabrous. **FR** 8–12 mm, ovoid. **SEEDS** 4–6 mm. Slopes, ridges, open places in forest; 1100–3300 m. CaR, SN, SW, **GB**, **DMtns**; to WA, WY, Colorado, w TX. [*S. vaccinioides* Rydb.]

var. *parishii* (Rydb.) Dempster Pl trailing, 3–6 dm. **ST**: branches often arched, tips rooting; twigs gen ± straight-hairy. **FL**: corolla 6–9 mm, inside of tube sparsely hairy throughout. Slopes, ridges; 1100–3300 m. s SNH, SW, **SNE**, **DMtns**; NM. [*S. parishii* Rydb.] Hard to distinguish from var. *rotundifolius* in Mono Co. ❀ TRY. Jun–Aug

var. *rotundifolius* (p. 255) Pl ± erect, 6–12 dm. **ST**: branches not arched; twigs finely puberulent. **FL**: corolla 7–10 mm, inside of tube hairy in middle 1/3. Rocky or sandy slopes, open places in coniferous forest; 1200–3200 m. CaR, SN, **GB**; to WA, WY, Colorado, w TX. ❀ DRN:4–6,15,16&IRR:1–3,7,14,18; DFCLT. Jun–Aug

CARYOPHYLLACEAE PINK FAMILY

Ronald L. Hartman (except *Silene*)

Ann, bien, per, rarely dioecious, taprooted or rhizome gen slender. **LVS** simple, gen opposite; stipules gen 0; petiole gen 0; blade entire, sheath gen 0. **INFL**: cyme, gen open; fls few–many or fl solitary and axillary; involucre gen 0. **FL** gen bisexual, radial; hypanthium sometimes present; sepals gen 5, ± free or fused into a tube, tube gen herbaceous between lobes or teeth; awns gen 0; petals gen 5 or 0, gen tapered to base (or with claw long, blade expanded), entire to 2–several-lobed, blade gen without scale-like appendages (inner surface), gen without ear-like lobes at base; stamens gen 10, gen fertile, gen free, gen from ovary base; nectaries gen 0; ovary superior, gen 1-chambered, placentas basal or free-central, styles 2–5 or 1 and 2–3-branched. **FR**: capsule or utricle (rarely modified, dehiscent), gen sessile. **SEEDS**: appendage gen 0. 85 genera, 2400 spp.: widespread, esp arctic, alpine, temp, n hemisphere; some cult (*Agrostemma, Arenaria, Cerastium, Dianthus, Gypsophila, Lychnis, Saponaria, Silene, Vaccaria*).

1. Fr a ± modified utricle; stamens arising from hypanthium rim, hypanthium in fr prominent, cylindric to conic (subfamily Paronychioideae)
  2. Ann; base ± glabrous; st prostrate to ascending; sterile stamens 14–19, ± 0.5 mm, thread-like, nectary 0; style 2-branched . . . . **ACHYRONYCHIA**
  2' Per; base densely woolly; st erect; sterile stamens 5; 1–1.5 mm, oblong, petal-like, nectary wide; style 3-branched . . . . **SCOPULOPHILA**
1' Fr a capsule; stamens from ovary base or on disk around ovary; hypanthium obscure
  3. Sepals fused, tube prominent, lobes or teeth < tube; petals long-clawed, gen appendaged near junction with claw (subfamily Silenoideae)
    4. Styles 3–4; fr valves or teeth 3, 6, or 10 . . . . **SILENE**
    4' Styles 2; fr valves gen 4
      5. Calyx 1.5–2.5 mm cup- or bell-shaped, tube white, scarious between teeth . . . . **GYPSOPHILA**
      5' Calyx 7.5–25 mm, ovoid to widely tubular, tube green, herbaceous between teeth
        6. Infl very dense, pedicels 0–3 mm; calyx in fl 20–25 mm, rounded, keels 0; petal appendages 2 . . . . **SAPONARIA**
        6' Infl open, pedicels 5–40+ mm; calyx in fl 7.5–17 mm, 5-angled or 5-keeled; petal appendages 0 . . . . **VACCARIA**
  3' Sepals ± free; petals not long-clawed, not appendaged at junction with claw (subfamily Alsinoideae)
    7. Stipules 0.5–11 mm, ovate to bristle-like, scarious
      8. Fls axillary, 1–2; sepals spine-tipped; petals 0; stipules bristle-like . . . . **LOEFLINGIA**
      8' Fls in cyme, few–many; sepals awnless; petals 5; stipules ovate to lanceolate . . . . **SPERGULARIA**
    7' Stipules 0
      9. Petals 2-lobed, often ± to base
        10. Fr cylindric, often curved, tip 10-toothed . . . . **CERASTIUM**
        10' Fr spheric to ± ovoid, tip not toothed, valves 6, sometimes 8 or 10 . . . . **STELLARIA**
      9' Petals entire or ± notched (0 in *Minuartia stricta*)
        11. Styles 4–5, alternate with sepals; fr valves 4–5 . . . . **SAGINA**
        11' Styles 3, opposite sepals; fr valves 3 or 6
          12. Ovary sutures 6; fr valves 6 . . . . **ARENARIA**
          12' Ovary sutures 3; fr valves 3 . . . . **MINUARTIA**

ACHYRONYCHIA

1 sp.   (Greek: chaff fingernail, from silvery, chaffy sepals)

A. cooperi Torrey & A. Gray (p. 259, pl. 43)   ONYX FLOWER.
FROST-MAT   Ann, prostrate to ascending, glabrous to ± hairy,
taprooted. **STS** many, 3–17 cm. **LF:** stipules 0.1–0.4 mm, ± ovate,
scarious, ± fringed, white; blade 3–20 mm, oblanceolate; vein 1.
INFL: cyme, axillary; fls 20–60+; pedicels 0.5–2.5 mm. **FL** 2.5–
3 mm; hypanthium in fr ± cylindric; calyx abruptly expanded
above; sepals 5, free, ± 1.2–1.5 mm, ovate to reniform, green,
fleshy, margin wide, scarious, ± jagged, white, deciduous; petals
0; fertile stamens 1–2, sterile stamens 14–19, ± 0.5 mm, thread-
like, arising from hypanthium rim; ovary superior, style 2-
branched in upper 1/2, 0.3–0.4 mm. **FR:** utricle, ovoid; teeth 8–
10, minute. **SEED** 1, ± 1 mm, ovoid, ± compressed, tan, red dot
near narrow end. Sandy slopes, flats, washes; 50–700 m. **D**;
AZ, Mex. Closely related to and possibly same as *Scopulophila*.
Jan–May

ARENARIA   SANDWORT

Ann, per, erect to mat-forming, taprooted. **ST** gen round in ×-section. **LF:** blades thread-like to ovate; veins 1–5.
INFL: cyme, terminal or axillary, open to head- or umbel-like; fls 1–many; peduncles and pedicels 0–50+ mm. **FL:**
hypanthium barely present; sepals 5, ± free, 1.5–8 mm, ± lanceolate to widely ovate, glabrous to glandular-hairy;
petals 0 or 5, 1.5–10 mm, entire or notched; stamens inserted on obscure to prominent disk; ovary ± superior, styles 3,
0.5–2 mm. **FR:** capsule, ovoid to urn-shaped; teeth 6, ascending to recurved. **SEEDS** 1–15+, grayish, dark brown,
reddish brown, yellowish tan, blackish purple, or blackish. 150 spp.: n temp, esp mtns, arctic Am, Eurasia. (Latin:
sand, a common habitat) [McNeill 1980 Rhodora 82:495–502]

1. Infl head-like cyme, dense to ± open; pedicels < 7 mm . . . . *A. congesta*
  2. Pedicels ± 0; infl tightly dense . . . . var. *charlestonensis*
  2' Pedicels obvious; infl ± open . . . . var. *subcongesta*
1' Infl ± open cyme or fls 1–2; pedicels some or all > 7 mm
  3. St gen 2–20 cm; lvs 0.5–2 cm; nectaries rounded, < 0.5 mm . . . . *A. kingii* var. *glabrescens*
  3' St gen 20–40 cm; lvs 2–6 cm; nectaries 2-lobed, 0.7–1.5 mm . . . . *A. macradenia*
    4. Sepals 3–4.3 mm, in fr < 5.5 mm; infl open, branches spreading . . . . ssp. *ferrisiae*
    4' Sepals 4.5–7.2 mm, in fr < 8 mm; infl ± compact, branches erect or ascending . . . . var. *macradenia*

A. congesta Nutt.   Per, tufted, ± green. **ST** 8–40 cm, ± dull,
often glandular-hairy. **LF** 10–80 mm, 0.5–2 mm wide, thread- to
needle-like, herbaceous or ± fleshy, sharply acute to spine-tipped;
vein 1. **INFL:** cyme, terminal, head- or umbel-like; fls few–many,
dense to ± open; pedicels < 7 mm. **FL:** sepals 3–6 mm, in fr < 6.5
mm, rounded to acute; petals 5–8 mm; nectaries < 0.2 mm,
rounded. **SEEDS** 4–8, 1.4–3 mm, widely elliptic to ovate, compressed, reddish brown; tubercles low, rounded, often elongate.
Dry, rocky or sandy slopes, ridges, rock crevices; 1200–3300 m.
KR, NCoRH, CaRH, SNH, **GB, DMtns**; to WA, MT, Colorado.
Vars. often intergrade. 6 vars. in CA.

  var. *charlestonensis* Maguire   **LF** 10–20 mm, < 1 mm wide,
± herbaceous, needle-like. **INFL:** cyme, dense; bracts closely
enveloping sepals; pedicels ± 0. **FL:** sepals 4.5–5.5 mm, acute.
Uncommon. Sandy ridges; 2200 m. **DMtns (New York Mtns)**;
sw NV. Jun

  var. *subcongesta* (S. Watson) S. Watson (p. 259)   **LF** 10–30
mm, 0.5–1 mm wide, herbaceous, needle-like. **INFL:** cyme, ±
open; bracts scattered among fls; pedicels obvious. **FL:** sepals
4–5 mm, acute. Uncommon. Open, rocky slopes, flats, often on
volcanics; 1350–2750 m. CaRH, n SNH, **GB, DMtns**; to UT. ❀
DRN,SUN, IRR:1–3,6,7,14–16,18;DFCLT. Jun–Jul

A. kingii (S. Watson) M.E. Jones var. *glabrescens* (S. Watson)
Maguire (p. 259)   Per, tufted, green. **ST** 1–20 cm, ± dull,
glandular-hairy. **LF** 3–20 mm, 0.5–1.2 mm wide, needle-like,
herbaceous, gen sharp-pointed; vein 1. **INFL:** cyme, terminal;
fls few–many (fl 1–2 in alpine pls); pedicels 2–15 mm. **FL:** sepals
2.5–4 mm, in fr < 4.5 mm, acute to acuminate; petals 4–7 mm;
nectaries < 0.5 mm, rounded. **SEEDS** 2–5, 1.2–1.8 mm, elliptic-
oblong to ovate, compressed, reddish brown to dark purple;

tubercles low, rounded, often elongate. Rocky slopes, summits,
canyon floors; 2100–4050 m. CaRH, SNH, **SNE**; to OR, ID,
UT. High elevation pls 2–6 cm, called ssp. *compacta* (Cov.)
Maguire, with lvs 3–6 mm and sepals 2.5–3.5 mm, intergrade completely, do not deserve recognition. ❀ DRN,IRR,SUN:1–3,7,14–
16. Jun–Aug

A. macradenia S. Watson   DESERT SANDWORT   Per, tufted,
green. **ST** 20–40 cm, rounded, ± dull, sometimes glandular-hairy.
LF 20–60 mm, 0.5–2 mm wide, needle-like, herbaceous, blunt
to sharp-pointed; vein 1. **INFL:** cyme, terminal; fls several–many,
compact to open; pedicels 3–55 mm. **FL:** sepals 3–7.2 mm, in fr
< 8 mm, acute to acuminate; petals 6–11 mm; nectaries 2-lobed,
0.7–1.5 mm. **SEEDS** 4–9, 1.8–2.7 mm, ± spheric to ovate, compressed, reddish brown to blackish; tubercles low, rounded to
conic. Open woodlands, sagebrush flats, dry, rocky slopes; 1100–
2500 m. c&s SNH, SnJV, SnGb, SnBr, **SNE, DMoj**; to UT, AZ.
3 vars. and 1 ssp. in CA.

  ssp. *ferrisiae* Abrams (p. 259)   **LVS** ± ascending, 0.5–1 mm
wide. **INFL:** open; branches spreading, glabrous to moderately
glandular-hairy. **FL:** sepals 3–4.3 mm, in fr < 5.5 mm, glabrous
to sparsely glandular-hairy. Pine and oak woodlands, granitic
alluvium; 1450–2500 m. c&s SNH, **SNE**; to UT. May–Jul

  var. *macradenia* (p. 259)   **LVS** ± ascending, 0.8–1.2 mm
wide. **INFL** ± compact; branches erect to ascending, ± glabrous.
FL: sepals 4.5–7.2 mm, in fr < 8 mm, ± glabrous. Open woodlands, sagebrush flats, dry, rocky slopes, alluvial deposits, often
on carbonates; 1100–2200 m. SnJV, SnBr, SnGb, **SNE, DMoj**;
to UT, AZ. [Pls called var. *parishiorum* Robinson, with petals ±
= sepals and < 5 cauline lf pairs, intergrade completely and should
be considered a synonym] Apr–Jun

CERASTIUM   MOUSE-EAR CHICKWEED

Ann, per, erect to mat-forming; taproot or rhizomes present. **LF:** blade linear to ovate; vein 1. **INFL:** cyme, terminal
or axillary; fls few–many, open to tightly dense; pedicels 1–36+ mm. **FL:** sepals 5, 3.5–12 mm, free, lanceolate to

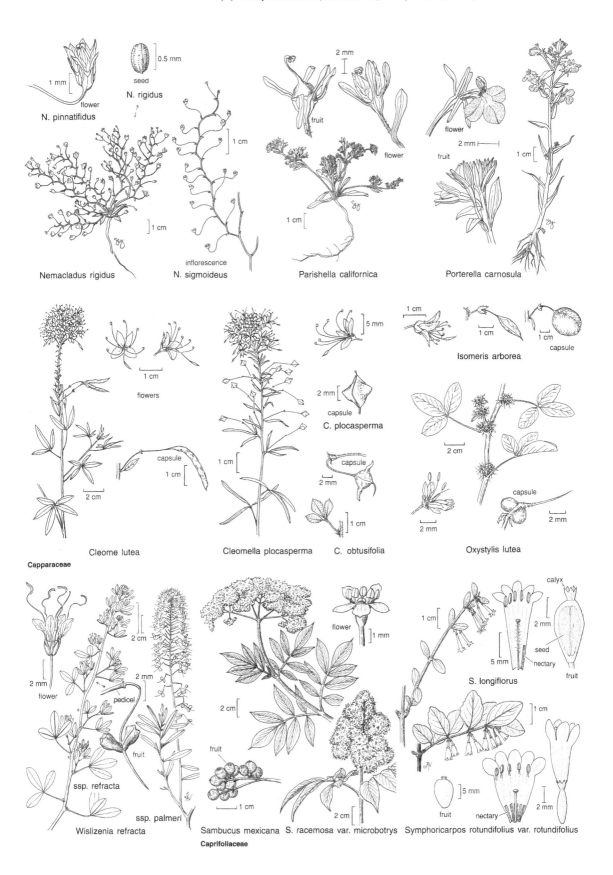

N. pinnatifidus

seed
N. rigidus

flower

Nemacladus rigidus

inflorescence
N. sigmoideus

fruit

flower

Parishella californica

flower

fruit

Porterella carnosula

flowers

capsule

Cleome lutea

Capparaceae

capsule
C. plocasperma

capsule

Cleomella plocasperma

capsule
C. obtusifolia

Isomeris arborea

capsule

capsule

Oxystylis lutea

flower

pedicel

fruit

ssp. refracta

ssp. palmeri

Wislizenia refracta

Caprifoliaceae

flower

fruit

Sambucus mexicana

S. racemosa var. microbotrys

calyx

seed
nectary

S. longiflorus

fruit

fruit

nectary

Symphoricarpos rotundifolius var. rotundifolius

ovate, hairy to glandular-hairy; petals 0 or 5, 2.5–15 mm, ± 2-lobed; stamens (5)10; styles 5, 0.5–3.3 mm. **FR**: capsule, cylindric, ± curved in upper 1/2; teeth 10, spreading to recurved. **SEEDS** several–many, pale brown to reddish brown. 60 spp.: worldwide. (Greek: horn, from fr shape)

C. beeringianum Cham. & Schldl. var. *capillare* Fern. & Wieg. (p. 259) Per, gen not fl first year, 1.5–10 cm, glandular-hairy. **STS** both vegetative (mat-forming) and fl (± erect). **LVS**: those on fl st 5–15 mm, lanceolate to elliptic; axillary lf clusters 0. **INFL**: bracts ± completely herbaceous; pedicels in fr 0.3–2.2 × sepals. **FL**: calyx 4.5–6 mm, ± glandular-hairy, with 0 long hairs > tip; scarious margin of outer sepals < 0.2 mm wide; petals 1.2–2.8 mm > sepals. **FR** 5.5–8.5 mm. **SEED** 0.7–0.8 mm. 2*n*=72. Moist, rocky areas, grassy meadows, open slopes; 2900–4300 m. c SNH, **W&I**; to WA, MT, Colorado. [ssp. *earlei* (Rydb.) Hultén] Jul–Aug

GYPSOPHILA BABY'S BREATH

Ann, bien, per, erect, taprooted or rhizomed. **LVS**: blade ± lanceolate; veins 1–3, often obscure. **INFL**: cyme, gen panicle-like, terminal; fls ± few–many; pedicels 1–30+ mm. **FL**: sepals 5, fused, glabrous or glandular-hairy, tube ± prominent, ± 1.3–4 mm, 0.8–2 mm diam, cup- to bell-shaped, round to angled in ×-section, white-scarious between teeth, veins ± 5, teeth 0.2–1 mm, < tube, lanceolate to triangular; petals 5, ± 2.2–9 mm, claw long, blade entire to notched; styles 2, 2–5 mm. **FR**: capsule, oblong to spheric; teeth 4, ascending to recurved. **SEEDS** 2–several, black. 125 spp.: temp Eurasia, n Afr. (Greek: gypsum lover, from habitat of 1 sp.) [Barkoudah 1962 Wentia 9:1–203]

1. Sepals, pedicels glabrous; largest lvs gen 2–9 mm wide . . . . *G. paniculata* var. *paniculata*
1' Sepals, pedicels glandular-hairy; largest lvs gen 10–35 mm wide . . . . *G. scorzonerifolia*

G. paniculata L. var. *paniculata* BABY'S BREATH Per 50–90 cm; rhizome stout. **LF**: blade 2–9 mm wide, ± lanceolate. **INFL** openly branched; fls many; pedicels glabrous. **FL**: calyx 1.5–2.1 mm, glabrous; petals 1.5–2 × calyx, white. 2*n*=26,28,34. Disturbed areas; 1200–2000 m. NCoRI, CaRH, n SNH, SnJV, SCoRO, SCo, MP, **DMoj**; native to e&c Eur, adjacent Asia. NOXIOUS WEED; infestations widely scattered.

G. scorzonerifolia Ser. Per 30–90 cm; rhizome stout. **LF**: blade 3–35 mm wide, linear-oblong to lanceolate. **INFL** ± open; fls ± few; pedicels glandular-hairy. **FL**: calyx 2–2.5 mm, glandular-hairy; petals 1.5–2 × calyx, white to light pink. 2*n*=34,68. Disturbed areas; ± 1200 m. **SNE (Inyo Co.)**; native to e Eur.

LOEFLINGIA

Ann, erect to prostrate, taprooted. **LF**: stipules 0.4–1.2 mm, bristle-like, scarious, entire, ± white; blade awl-like to oblong; vein 0–1. **INFL** axillary, sessile; fls 1–2. **FL**: sepals 5, 2.7–6 mm, ± free, ± lanceolate, glandular-hairy; petals 0 or rudimentary; stamens 3–5; styles 3, < 0.1 mm. **FR**: capsule, lanceolate to ovoid; valves 3, ± recurved at tip. **SEEDS** many, tan with reddish brown band on curved edge. 7 spp.: N.Am., Medit. (P. Loefling, Swedish botanist & explorer, 1729–1756) [Barneby & Twisselmann 1970 Madroño 20:398–408]

L. squarrosa Nutt. Pl 1–7 cm, much-branched at base, glandular-hairy, ± fleshy. **ST** stiff. **LF**: blade 2–7 mm, erect to ± recurved, bristly; base gen fused into a short scarious sheath; tip blunt to spine-tipped. **FL** cleistogamous; sepals spine-tipped, becoming hardened, margin often scarious. **FR** 0.5–0.8 × sepals, 3-angled. **SEED** 0.4–0.6 mm, minutely papillate on flat edge. Sandy, gravelly areas; < 1200 m. Teh, SnJV, SnFrB, SCo, PR, **GB, DMoj**; to OR, NE, TX. 2 vars. in CA.

var. *artemisiarum* (Barneby & Twisselm.) R. Dorn (p. 259) SAGEBRUSH-LIKE LOEFLINGIA **LF**: stipules 0.4–0.6 mm; blade 2–4 mm, oblong, erect to ± spreading; tip blunt or short-spined. **FL**: sepals 2.7–3 mm, in fr ± equal and tip ± straight, lateral spurs 0. **FR** gen 1.5–2.5 mm, 2–2.7 × longer than wide. UNCOMMON. Sand dunes, sandy flats; 700–1200 m. **GB, DMoj (se Kern, ne Los Angeles cos.)**; to OR, WY. [ssp. *a.* Barneby & Twisselm.]

MINUARTIA SANDWORT

Ann, per, erect to mat-forming, taprooted or rhizomed. **LF**: blade thread-like to awl-shaped or narrowly oblong; veins or ribs 1–3. **INFL**: cyme, terminal or axillary; fls 2–many, open to ± dense, or fl solitary; peduncles and pedicels 0.5–35+ mm. **FL**: hypanthium short, obscure; sepals 5, ± free, 1.9–7 mm, ± lanceolate to ovate, glabrous to glandular-hairy; petals 5 or 0, 0.7–10 mm, entire or notched; stamens arising from an obscure to prominent disk; styles 3, 0.3–2 mm. **FR**: capsule, narrowly ovoid to widely elliptic; teeth 3, ascending to recurved. **SEEDS** 1–many, reddish tan to reddish, purplish, or blackish brown. 120 spp.: arctic to Mex, n Afr, s Asia. (J. Minuart, Spanish botanist & pharmacist, 1693–1768) [McNeill 1980 Rhodora 82: 495–502]

1. Pl glabrous, ± densely cespitose, rhizomes and trailing systems 0; sepals 2–3.2 mm; petals 0 . . . . *M. stricta*
1' Pl densely glandular-hairy, mat forming, rhizomes elongate or st trailing; sepals 3.5–7 mm; petals present
. . . . *M. nuttallii*
2. Lvs prominently recurved; sepals 3-ribbed . . . . ssp. *fragilis*
2' Lvs straight or slightly recurved; sepals 1- or 3-ribbed, central rib prominent . . . . ssp. *gracilis*

M. nuttallii (Pax) Briq. Per, mat-forming, 2–20 cm, ± green, densely glandular-hairy; taproot > 3 mm diam; rhizomes and trailing st < 60+ cm. **STS** in fl ascending to erect. **LVS** 4–12(15) mm, ± 0.3–1.1 mm wide, > internodes, needle-like to awl-shaped, straight to recurved, ± rigid, ± evenly spaced; axillary lvs well developed. **FL**: sepals 3.5–7 mm, acute to acuminate, margin not incurved, ribs 1 or 3; petals 0.7–1.6 × sepals. **SEED** 1.5–2.2 mm; margin thick, reddish brown to dark brown. Sandy and rocky slopes and ridges, barren rock, chaparral, open pine woodland, often on serpentine; 650–3800 m. NW, CaRH, SNH, **GB**; OR, NV. [*Arenaria n.* Pax] 3 sspp. in CA.

ssp. *fragilis* (Maguire & A. Holmgren) McNeill (p. 259) **LF** prominently recurved. **FL**: sepals 3-ribbed; petals 0.8–1.1 × sepals. Uncommon. Basins, limestone talus; 1650–2400 m. **GB**; NV, OR. [*Arenaria n.* ssp. *f.* Maguire & A. Holmgren] May–Jul

ssp. *gracilis* (Robinson) McNeill (p. 259) **LF** straight or slightly recurved. **FL**: sepals 1-ribbed, sometimes obscurely 3-ribbed; petals 0.7–0.9 × sepals. Loose talus, sandy flats, gravelly areas, barren rock; 1500–3800 m. SNH, **SNE**; w NV. [*Arenaria n.* ssp. *g.* (Robinson) Maguire] ❀ TRY. Jul–Aug

M. stricta (Sw.) Hiern (p. 259) Per, cespitose, 0.8–2.5 cm, green, glabrous; taproot < 1.5 mm diam; rhizomes and stolons 0. **ST** decumbent to erect. **LVS** 2–9 mm, 0.3–0.6 mm wide, > internodes, needle-like, ± straight, flexible, mostly near base; axillary lvs well developed. **FL**: sepals 2–3.2 mm, acute to acuminate, margin not incurved, vein 1 or 3, often prominent; petals 0. **SEED** 0.5–0.6 mm; margin thick, reddish brown. $2n=26,30$. Uncommon. Granitic gravels, sandy wet spots, sedge meadows, alpine; 3500–3900 m. c&s SNH, **W&I**; circumboreal; to Colorado. [*Arenaria rossii* Richardson misapplied] ❀ TRY. Aug

SAGINA PEARLWORT

Ann, per, tufted to matted, taprooted. **LF**: blade linear to awl-shaped; vein 0–1. **INFL**: fl solitary, terminal or axillary; pedicels 2–30 mm. **FL**: sepals 4–5, free, 1.3–3.5 mm, lanceolate to ovate, glabrous to glandular-hairy; petals 0 or 4–5, 1–3 mm, entire or sometimes notched; stamens 4, 5, 8, or 10; styles 4–5, 0.1–0.6 mm. **FR**: capsule, ovoid; valves 4–5, spreading to recurved. **SEEDS** few–many, brown or reddish brown. 25 spp.: n temp, trop mtns. (Latin: fattening, once applied to *Spergula*, used as early forage) [Crow 1978 Rhodora 80:1–91]

S. saginoides (L.) Karsten (p. 259) Per (1)2–12 cm, glabrous; sterile basal rosettes often present. **ST** slender, ascending or sometimes decumbent. **LF** not fleshy; blade (3)5–15 mm, narrowly linear. **INFL**: pedicels 10–30 mm, thread-like, recurved in fr. **FL**: sepals 5, ± appressed in fr, 1.5–2.5 mm; petals 5, 3/4–1 × sepals; stamens gen 10. **FR** 1.5–2 × sepals. **SEED** 0.2–0.4 mm, obliquely triangular, ± compressed, smooth or slightly roughened, brown; back grooved. $2n=22$. Moist banks, streamsides, dry creeks; (100)1000–3800 m. KR, NCoRO, CaRH, SN, TR, PR, **GB**; to MT, WY, NM, Mex; circumboreal. [var. *hesperia* Fern.] ❀ TRY. May–Sep

SAPONARIA BOUNCING BET, SOAPWORT

Per, erect, rhizomed. **LF** petioled or not; blade oblanceolate to ovate; veins 3. **INFL**: cyme, terminal or axillary; fls 20–40+, dense; pedicels gen 0–3 mm. **FL**: sepals 5, fused, tube prominent, 15–20 mm, 4–8 mm diam, lanceolate to oblong, rounded, veins 20, obscure, teeth 5, 1.5–5 mm, < tube, triangular, tapered; petals 5, 25–40 mm, claw long, blade entire to ± obcordate, appendages 2; stamens fused with petals to ovary stalk; styles 2, 2–2.5 cm. **FR**: capsule, ± ovoid; stalk 2–3 mm; valves 4, ascending to recurved. **SEEDS** many, purplish to black. 30 spp.: Eurasia. (Latin: soap, because juice lathers with water)

S. officinalis L. (p. 259) Pl 3–9+ dm, ± glabrous. **ST** simple or branched above. **LF** 3–10 cm. **INFL**: peduncles ascending or erect, bracted. **FL**: calyx tube base depressed, indented, lobes unequal, thinly scarious, triangular-acuminate; petals 8–12 mm wide, obovate, pink, appendages 1–2 mm, linear-lanceolate. **SEED** 1.5–2 mm, ± spheric, notched, ± compressed, tubercled. $2n=28$. Roadsides, oak woodlands, streambeds, disturbed areas; < 1500 m. NW, CaRH, n SNF, SnFrB, SCoRO, SCo, PR, **GB**; native to s Eur. Jun–Sep

SCOPULOPHILA

Per, erect, dioecious, taprooted. **LF**: stipules 0.8–3.5 mm, triangular, scarious, jagged to ciliate, white; blade linear to lanceolate; vein ± 1. **INFL** axillary; fls 1–4, sessile. **FL** unisexual (appears bisexual); hypanthium in fr conic to urn-shaped, abruptly expanded above; sepals 5, free, 1.1–2.1 mm, elliptic to round, ± glabrous, margin wide, scarious, white; petals 0; stamens 5, sterile in pistillate fl, 1–1.5 mm, oblong, petal-like, arising from hypanthium rim; nectaries wide; ovary sterile in stamiante fl, style 3-branched in upper 1/3, ± 1.5 mm. **FR**: utricle, modified, ovoid; teeth 3, minute. **SEED** 1, tan. 2 spp.: sw US, Mex. Closely related to and possibly same as *Achyronychia*. (Greek: fond of high places, from habitat)

S. rixfordii (Brandegee) Munz & I.M. Johnston (p. 259) RIXFORD ROCKWORT Pl glabrous exc base; lf axils densely woolly. **STS** many, branched above, 10–30 cm. **LF**: blade 8–25 mm, ± fleshy. **INFLS** eventually many. **FL** 2.2–4.2 mm; hypanthium green, becoming brown, thickened, ± hard, ± angled; sepals erect to spreading, ± concave, often unequal, central portion linear to oblong, fleshy, green, margin much wider, entire to irregular, white, possibly deciduous. **SEED** 0.9–1.1 mm, ovoid, ± compressed, red dot near narrow end. Uncommon. Limestone outcrops; 1200–1550 m. SNE, **n DMoj**; w NV. Apr–Jul

SILENE CATCHFLY, CAMPION

Dieter H. Wilken

Ann, bien, per, ± erect, rarely dioecious, taprooted or rhizomed. **LVS** petioled or not; blade linear to oblanceolate; vein 1. **INFL**: cyme, gen terminal, sometimes axillary, open to dense; fls few–many, gen erect, gen with pedicels 5–40+ mm. **FL** gen bisexual; sepals 5, fused, tube prominent, 4–25 mm, 2–13 mm diam, cylindric to bell-shaped, rounded, hairs various or 0, veins gen 10+, lobes or teeth 1–13 mm, < tube, triangular to linear; petals 5, 6–48 mm, claw long, blade entire or 2–6-lobed, appendages 0–6 at junction of claw and blade; basal lobes present or 0; stamens gen fertile, fused with petals to stalk; ovary chamber 1 or ± incompletely 3–5, styles 3–5, 1–35 mm. **FR**: capsule, cylindric to ovoid; stalk 0–7 mm, gen glabrous; teeth 3, 6, or 10, ascending to recurved. **SEEDS** many, gray to red, brown, or black. 500 spp.: n hemisphere. (Greek: Probably from mythological Silenus, intoxicated foster-father of Bacchus, who was covered with foam, from sticky secretions of many spp.) [Hitchcock & Maguire 1947 Univ Wash Publ Biol 13:1–73; Showers 1987 Madroño 29–40]

1. Ann, taproot gen slender
  2. Calyx 4–9 mm, tube 10-veined, lobes 1–2 mm, ... *S. antirrhina*
  2' Calyx 18–26 mm, tube ± 30-veined, lobes 5–10 mm . . . . *S. conoidea*
1' Per, from few- to much-branched caudex
  3. Petal blade gen 4-lobed
    4. Calyx tube gen > 10 mm . . . . *S. bernardina*
    4' Calyx tube 4–5 mm . . . . ²*S. menziesii*
  3' Petal blade 2-lobed, sometimes short-toothed near base
    5. Basal and lower cauline lvs densely tufted, fleshy, ± linear, 1.5–3 cm, < 5 mm wide . . . . *S. sargentii*
    5' Basal lvs not densely tufted, not fleshy, lanceolate to elliptic, 2–9 cm, 2–20 mm wide
      6. Petals < 10 mm; calyx 5–7 mm . . . . ²*S. menziesii*
      6' Petals > 10 mm; calyx 10–15 mm . . . . *S. verecunda* ssp. *andersonii*

S. antirrhina L. (p. 259) Ann 12–80 cm. **ST** erect, glabrous or puberulent; upper internodes gen sticky. **LVS** gradually reduced upward, 1–3(6) cm, 3–5 mm wide; lower ± oblanceolate; upper linear to oblanceolate. **FL**: calyx 4–9 mm, gen glabrous, 10-veined, lobes 1–2 mm; petal claw glabrous, appendages 0, blade 2-lobed, white to pink; stamens incl; styles 3, ± incl. **FR** ± ovoid; stalk ± 1 mm. **SEED** < 1 mm, black. 2*n*=24. Open areas, burns; < 1800 m. CA-FP, **D (uncommon)**; to B.C., e US. Apr–Aug

S. bernardina S. Watson (p. 259) Per 15–55 cm; caudex gen few-branched. **ST** erect, puberulent to short-hairy, glandular above or throughout. **LVS** gradually reduced upward; lower 2–8 cm, 2–6 mm wide, linear to oblanceolate; upper 1–6 cm, 1–4 mm wide, ± linear. **INFL** axillary and terminal. **FL**: calyx 12–15 mm, glandular-puberulent, 10-veined, lobes 2–3.5 mm; petal claw ciliate at base, appendages 2, blade gen 4-lobed, white, pink, or purple; stamens slightly > petals; styles 3–4, ± = stamens. **FR** slightly elliptic; stalk 2–5 mm, puberulent. **SEED** 1.5–2 mm, brown. 2*n*=48. Rocky slopes, shrubland, coniferous forest, alpine; 1350–3600 m. KR, NCoR, CaR, SN, **GB, n DMtns**; OR, NV. [*S. montana* S. Watson ssp. *bernardina* (S. Watson) C. Hitchc. & Maguire, *maguirei* Bocq., *montana*, vars. *sierrae* C. Hitchc. & Maguire, *rigidula* (Robinson) Tiehm] ❀ TRY. Jun–Aug

S. conoidea L. Ann 50–80 cm. **ST** erect, puberulent below, glandular-puberulent above. **LVS** gradually reduced upward; lower 5–12 cm, 5–10 mm wide, lanceolate to oblanceolate; upper 1–8 cm, 3–8 mm wide, lanceolate. **FL**: calyx 18–26 mm, puberulent, sometimes glandular, ± 30-veined, lobes 5–10 mm; petal claw glabrous, appendages 2, blade entire, fine-toothed, or notched, white, pink, or purplish; stamens exserted; styles 3, slightly > petals. **FR** ovoid to conic; stalk 1–2.5 mm. **SEED** 1–1.5 mm, grayish brown. Uncommon. Disturbed, open areas; < 500 m. n SN, SCo, **DSon**; to WA, e US; native to Eur.

S. menziesii Hook. (p. 259) Per 5–20(50) cm, ± mat-like, gen dioecious (fls appear bisexual); caudex much-branched. **ST** decumbent to erect, puberulent to short-hairy, gen glandular above. **LVS** slightly reduced upward, 2–6 cm, 3–20 mm wide, lanceolate to elliptic. **INFL**: fls ascending to erect. **FL**: calyx 5–7 mm, puberulent, ± glandular, faintly 10-veined, lobes 1–2 mm; petal claw glabrous, appendages 2, blade 2–4-lobed, white; stamens deeply incl (pistillate fl) or ± = corolla (staminate fl); styles 3(4), << petals (staminate fl) or > petals (pistillate fl). **FR** ovoid; stalk 1–2 mm. **SEED** 0.5–1 mm, reddish brown. 2*n*=24,48. Coniferous forest, pinyon/juniper woodland; 900–2900 m. KR, NCoR, CaR, SN, SnBr, **GB**; to AK, Rocky Mtns. [sspp. *dorrii* (Kellogg) C. Hitchc. & Maguire, *menziesii*; var. *viscosa* (E. Greene) C. Hitchc. & Maguire] ❀ TRY. Jun–Jul

S. sargentii S. Watson (pl. 44) Per 10–15(20) cm; caudex much-branched. **ST** decumbent to erect, puberulent above, gen glandular above. **LVS** ± abruptly reduced upward; basal tufted, fleshy, 1.5–3 cm, 1–3 mm wide, oblanceolate; cauline 1–2.5 cm, 1–2 mm wide, ± linear. **FL**: calyx 9–15 mm, glandular-puberulent, 10-veined, lobes 2–3 mm; petal claw ciliate at base, appendages 2, blade 2-lobed, white to red-purple; stamens ± = petals; styles 3(4), < or = petals. **FR** ± ovoid; stalk 1.5–3 mm, woolly-puberulent. **SEED** 1–2 mm, brown. 2*n*=48. Subalpine forest, alpine; 2400–3800 m. SNH, **SNE (Sweetwater, White mtns)**; NV. Jul–Aug

S. verecunda S. Watson Per 10–55 cm; caudex branched. **ST** erect, densely puberulent, sometimes glandular above. **LVS** ± gradually reduced upward; lower 3–9 cm, 2–9 mm wide, gen lanceolate; upper 1–4.5 cm, 2–6 mm wide, linear to lanceolate. **FL**: calyx 10–15 mm, gen glandular-puberulent, 10-veined, lobes 2–5 mm; petal claw puberulent at base, appendages 2, blade 2-lobed, white to rose; stamens ± = petals; styles 3(4), ± = petals. **FR** oblong to ovoid; stalk 2–5 mm, puberulent. **SEED** 1–1.5 mm, dark brown. Open areas, chaparral, sagebrush scrub, oak woodland, pinyon/juniper woodland, coniferous forest; < 3400 m. c&s NCoR, SN, CW, TR, PR, **W&I, DMtns**; to UT, Baja CA. 3 sspp. in CA.

ssp. ***andersonii*** (Clokey) C. Hitchc. & Maguire **ST** ± scabrous. **LVS** ± thick, stiff; lower gen 4–8 cm; middle ascending to erect. **FL**: calyx short-hairy; petal claw ciliate throughout. 2*n*=48. Open areas, sagebrush scrub, pinyon/juniper woodland, coniferous forest; 1700–2700 m. SNH (e slope), **W&I, DMtns**; to UT. Jun–Jul

SPERGULARIA SAND-SPURREY

Ann, per, erect to sprawling; taprooted. **LF**: stipules 1–11 mm, lanceolate and acuminate to widely triangular, scarious, ± entire or splitting ± at tip, white to tan; blade thread-like to linear; vein 1. **INFL**: gen cyme, terminal, few–many-fld, open to dense; pedicels 0.5–28+ mm. **FL**: sepals 5, ± free, 1.5–10 mm, lanceolate to ovate, glabrous to glandular-hairy; petals 5, 0.6–9 mm, entire; stamens 2–10; styles 3, 0.3–1.9 mm. **FR**: capsule, ovoid; valves 3, spreading with tip recurved. **SEEDS** few–many, dark brown, reddish brown, or black. 40 spp.: worldwide. (Latin: derivative of *Spergula*) [Rossbach 1940 Rhodora 42:57–83,105–143,158–193,203–213]

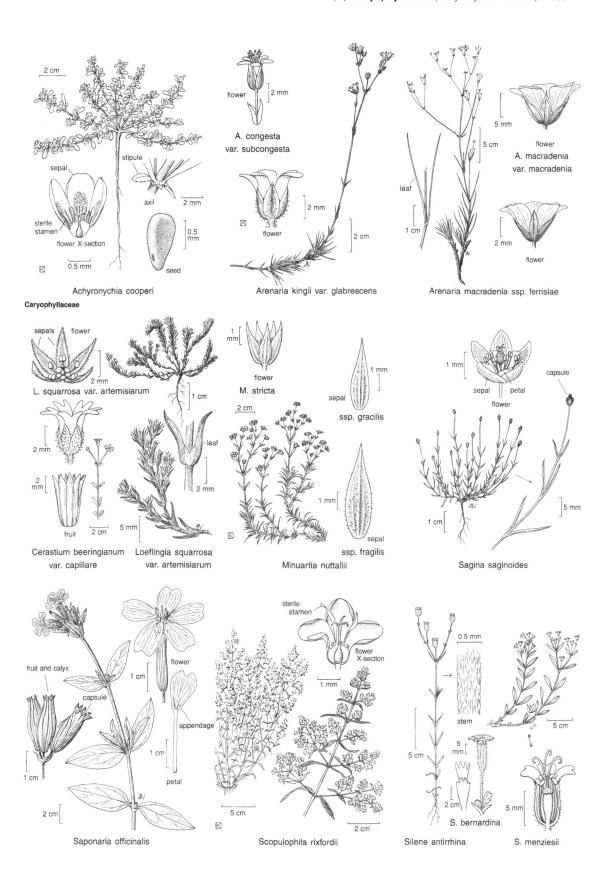

Achyronychia cooperi

A. congesta
var. subcongesta

Arenaria kingii var. glabrescens

A. macradenia
var. macradenia

Arenaria macradenia ssp. ferrisiae

Caryophyllaceae

L. squarrosa var. artemisiarum

Cerastium beeringianum
var. capillare

Loeflingia squarrosa
var. artemisiarum

M. stricta

ssp. gracilis

ssp. fragilis

Minuartia nuttallii

Sagina saginoides

Saponaria officinalis

Scopulophila rixfordii

S. bernardina

Silene antirrhina

S. menziesii

1. Pl strongly per, stout; stipules 4.5–11 mm; calyx lobes 4.5–5.5 mm in fl; stamens 9–10
.... *S. macrotheca* var. *leucantha*
1' Pl ann, delicate; stipules 1.2–3 mm; calyx lobes 1.8–4.5 mm in fl; stamens 2–5 .... *S. marina*

S. macrotheca (Hornem.) Heynh. Pl strongly per, stout. **ST**: lower main 0.8–3 mm diam. **LF** fleshy; axillary cluster 0 or 1–2+-leaved; stipules 4.5–11 mm, narrowly triangular, dull white to tan, ± conspicuous, tip long-acuminate. **INFL** simple or 1–3+ × compound or fl axillary, solitary; glandular-hairy. **FL**: sepals fused 0.5–1.8 mm, lobes 4.5–7 mm, < 8 mm in fr; petals white or pink to rosy; stamens 9–10; styles 0.5–3 mm. **FR** 4.6–10 mm, 0.8–1.4 × calyx. **SEED** 0.6–0.9 mm, ± reddish brown, gen winged; surface smooth, tubercled, or sculpture worm-like or of low rounded mounds, not papillate. 2*n*=36,72. Alkaline marshes, meadows, fields, salt flats, mud flats, coastal bluffs, outcrops; < 800 m. NCo, NCoRI, GV, CCo, SnFrB, SCoRO, SCo, ChI, **DMoj**; to B.C., Baja CA. 3 vars. in CA.

var. **leucantha** (E. Greene) Robinson (p. 265) Pl 10–40 cm. **FL**: calyx lobes 4.5–5.5 mm, < 6.5 mm in fr; petals white; styles 1.2–1.8 mm. **FR** 1.2–1.4 × calyx. **SEED** 0.7–0.8 mm. Alkaline soils, floodplains, vernal pools and meadows, marshy ground; < 800 m. GV, SnFrB, SCoRO, SCo, **DMoj**. Apr–Jun

S. marina (L.) Griseb. (p. 265) Ann, delicate. **ST**: lower main 0.6–2 mm. **LF** fleshy; axillary clusters 0; stipules 1.2–3 mm, widely triangular, dull white, inconspicuous, tip ± acute. **INFL** 1–3+ × compound or fl axillary, solitary; glandular-hairy. **FL**: sepals fused 0.5–1 mm, lobes 1.8–4.5 mm, < 4.8 mm in fr; petals white or pink to rosy; stamens 2–5; styles 0.4–0.6 mm. **FR** 2.8–6.4 mm, 1–1.5 × calyx. **SEED** 0.5–0.8 mm, light brown to reddish brown, gen wingless; surface smooth or slightly roughened, papillate or not. 2*n*=36. Mud flats, alkaline fields, sandy river bottoms, sandy coasts, salt marshes; < 700 m. NCo, NCoRO, c SNF, GV, CCo, SnFrB, SCo, ChI, PR, **D**; to WA, e US, S.Am; Eurasia. [var. *tenuis* (E. Greene) R. Rossbach] *S. salina* J.S. Presl & C. Presl may prove to be the correct name for this sp. Mar–Sep

STELLARIA CHICKWEED, STARWORT

Ann, per, erect to prostrate; taproot and rhizomes present. **LF**: petiole present or 0; blade linear to ovate; vein 1. **INFL**: cyme, terminal or axillary, few–many-fld, open to dense or umbel-like or fl axillary, solitary; peduncles, pedicels 0.8–50+ mm. **FL**: sepals gen 5, free, 1.5–5.5 mm, lanceolate to ovate, glabrous to glandular-hairy; petals 0 or 5, 0.8–7 mm, gen 2-lobed ± to base; stamens 10, sometimes fewer; styles 3(4–5 in *S. calycantha*), 0.2–2.8 mm. **FR**: capsule, ± ovoid to cylindric-oblong; teeth 6(8,10), ascending to recurved. **SEEDS** several–many, brown to yellowish, reddish, or purplish brown. 120 spp.: worldwide. (Latin: star, from fl shape) [Chinnappa & Morton 1991 Rhodora 93:129–135; Morton & Rabeler 1989 Canad J Bot 67:121–127]

1. Ann, from slender taproot; internode with line of hairs .... *S. media*
1' Per, from slender, white rhizomes; internode glabrous or with scattered hairs
  2. Sepals 0.7–1 × petals, 3–5.5 mm .... *S. longipes* var. *longipes*
  2' Sepals > 2 × petals, 1.5–3 mm .... *S. calycantha*

S. calycantha (Ledeb.) Bong. (p. 265) Per, prostrate to erect, 5–25 cm, often glabrous; rhizome white. **ST**: internodes glabrous or with wavy scattered hairs. **LVS** ± evenly spaced; blade 3–25 mm, elliptic to ovate; margin ± smooth, flat to wavy, shiny, sometimes ciliate. **INFL** terminal or axillary, 1–few-fld; bracts leafy; pedicels ascending to erect, in fr often recurved. **FL**: sepals 5, 1.5–3 mm, ovate to elliptic, ± acute, glabrous, margin scarious, veins gen obscure; petals (0)5, 0.3–0.5 × sepals. **SEED** 0.7–0.9 mm, reddish brown; surface ± smooth or minutely roughened. 2*n*=26. Mossy banks, bogs, dry creeks, wet meadows, shaded areas; 1700–3800 m. KR, NCoRO, CaRF, n&c SN, Teh, SnGb, MP, **W&I**; to AK, NM; ne Asia. [ssp. *interior* Hultén; var. *simcoei* (J. Howell) Fern.] Jun–Aug

S. longipes Goldie var. **longipes** (p. 265, pl. 45) Per, ascending to erect, 5–35 cm, gen glabrous; rhizomes white. **ST**: internodes glabrous or hairs scattered, wavy. **LVS** ± evenly spaced; blade 10–40 cm, linear to linear-lanceolate; margin smooth, flat, shiny, sometimes ciliate near base. **INFL** terminal or axillary, gen 1–7-fld, ± narrow; bracts leafy or ± scarious; pedicels ascending to erect, in fr ± straight. **FL**: sepals 5, 3–5.5 mm, lanceolate to ± ovate, ± acute, glabrous, margin widely scarious, in fr ± 3-ribbed; petals 5, 1–1.2 × sepals, 2-lobed > 1/2 to base. **SEED** 0.7–1 mm, reddish brown; surface minutely roughened. 2*n*=52,65,72, 78,84,91,104. Streambanks, moist to boggy meadows, seeps; 1250–3500 m. KR, NCoRI, CaRH, SN, SnBr, Wrn, **n SNE**, **W&I**; to MT, circumboreal. [var. *laeta* S. Watson] May–Aug

S. media (L.) Villars COMMON CHICKWEED Ann but often over-wintering, prostrate to erect, 7–50 cm; taproot slender. **ST**: internodes hairy in line. **LVS** ± evenly spaced; blade 8–45 mm, ± ovate; margin ± smooth, ± flat, shiny, often ciliate near base. **INFL** terminal or axillary, few-fld, ± dense; bracts leafy; pedicels spreading to erect, in fr curved to reflexed. **FL**: sepals 5, 3–4.5 mm, > 6 mm in fr, lanceolate to ovate, acute to obtuse, glabrous or ± hairy and glandular, margin ± widely scarious, ribs often 1 or 3 near base; petals 5, 0.7–0.9 × sepal. **SEED** 0.9–1.3 mm, reddish or purplish brown; surface papillate. 2*n*=40,42,44. Oak woodlands, meadows, disturbed areas; < 1300 m. NW, CaRH, c SNF, GV, CCo, SnFrB, SCo, ChI, **DSon**; native to sw Eur. Often a pernicious urban weed. Feb–Sep

VACCARIA COW-HERB, COCKLE

1 sp. (Latin: cow, from use as fodder or prevalence in pastures)

V. hispanica (Miller) Rauschert Ann (8)20–100 cm, glabrous, glaucous, taprooted. **LF** 2–12 cm; petiole present or 0; blade lanceolate to ovate, base rounded to cordate-clasping; veins 3–7. **INFL**: cyme, terminal, 10–70-fld +, ± flat-topped; bracts leafy; pedicels 5–40+ mm. **FL**: sepals 5, fused, glabrous, tube prominent, 7.5–17 mm, 1.5–9 mm diam, cylindric to urn-shaped, 5-angled or 5-keeled, veins 10, teeth 1.5–3 mm, < tube, ovate to triangular; petals 5, 15–25 mm, claw long, blade oblanceolate to obovate, entire or obcordate, pink to reddish; styles 2, 9–21 mm. **FR**: capsule, ovoid; stalk 0.5–1 mm; teeth 4, ascending to recurved. **SEEDS** many, 1.6–1.8 mm, ± spheric; tubercles fine, low, reddish brown to black. 2*n*=24,30,60. Fields, disturbed areas; < 2800 m. KR, NCoR, CaRH, c SNF, n SNH, Teh, ScV, CW, SCo, PR, **GB**; native to Eurasia, Medit. [*V.* *segetalis* (Necker) Asch.] May–Aug

CELASTRACEAE STAFF-TREE FAMILY

Barry A. Prigge

Shrub (sometimes climbing), tree, sometimes thorny, gen glabrous. **LVS** simple, opposite or alternate, ephemeral to persistent, subsessile or petioled; veins pinnate. **INFL**: cluster, cyme, raceme, panicle, or fl solitary, axillary or terminal, bracted. **FL** gen bisexual, radial, small; hypanthium ± cup-shaped; sepals 4–5; petals (0)4–5, free; stamens 4–5, alternate petals, attached below or to rim of disk; ovary superior or ± embedded in disk, 2–5-chambered, placentas axile or basal, style gen 1, short, stigma ± head-like, 2–5-lobed. **FR**: capsule, winged achene, berry, drupe, or nutlet, often 1-chambered. **SEED** gen 1 per chamber, arilled. 50 genera, 800 spp.: worldwide, esp se Asia; some orn (*Celastrus, Euonymus, Maytenus, Paxistima*). [Brizicky 1964 J Arnold Arbor 45:206–234]

MORTONIA

Shrub, erect, scabrous. **LVS** alternate, persistent, ascending, leathery, entire; margin gen thicker. **INFL**: panicle, terminal; fls many. **FL**: parts in 5's; hypanthium obconic; petals white; disk fused to hypanthium exc at top, fleshy, ± white, becoming red-purple; ovary superior, narrowly ovoid, stigma lobes 5, slender, spreading. **FR**: nutlets 5, oblong-cylindric, light brown. **SEED** 1 per nutlet, straw-colored, very difficult to separate from nutlet; aril 0. 5 spp.: sw US, Mex. (S.G. Morton, 19th century N.Am naturalist)

M. utahensis (Trel.) Nelson (p. 265) Pl 3–12 dm, coarsely scabrous. **ST**: twigs creamy white, turning gray. **LF**: petiole ±0–1 mm; blade 6–16 mm, ovate to round, from above transversely concave and longitudinally convex, base rounded to tapered, tip round to acute, sometimes with small point. **INFL** 8–65 mm, 6– 23 mm wide. **FL**: hypanthium 1.5–2 mm; sepals 1–2.3 mm, tips often acute, keeled; petals 2.2–3 mm, ovate. **FR** 5–7 mm, glabrous. Limestone slopes, canyon bottoms; 900–2100 m. **DMtns (Inyo, ne San Bernardino cos.)**; to sw UT. ❀ DRN,DRY:1–3,10;DFCLT. Mar–May

CERATOPHYLLACEAE HORNWORT FAMILY

Fosiée Tahbaz

Ann, per, submerged aquatic, monoecious. **ST** slender, well branched. **LVS** whorled, repeatedly forked; divisions thread-like or narrow, ± stiff, minutely serrate. **INFL**: fls solitary in axils, sessile. **FL** minute; perianth thin, many-parted; stamens 10–20, free; pistil gen 1, ovary sessile, chamber 1, ovule 1, style persistent. **FR**: achene, smooth to spiny. 1 genus.

CERATOPHYLLUM HORNWORT

The only genus; ± 3 spp.: worldwide. (Greek: horn leaf, from stiff lf divisions) Cult in pools, aquaria.

C. demersum L. (p. 265) **ST**: branches 5–25 dm. **LVS** 5–12 per whorl, 1–2.5 cm, ± prickly-serrate on 1 surface. **FR** ± 5 mm, widely elliptic, smooth to spiny-tubercled. 2*n*=24. Ponds, ditches, slow streams; gen < 2000 m. **CA**; worldwide. Jun–Aug

CHENOPODIACEAE GOOSEFOOT FAMILY

Dieter H. Wilken, except as specified

Ann to tree, sometimes monoecious or dioecious, glandular or with bead-like hairs that collapse with age, becoming scaly or powdery. **ST** often fleshy. **LVS** gen alternate, entire to lobed; veins gen pinnate. **INFL**: raceme, spike, catkin-like, or spheric cluster, or fl 1; bracts 0–few. **FL**: sepals 1–5, often 0 in pistillate fls, free or fused, gen persistent in fr; petals 0; stamens 0–5; ovary gen superior, chamber 1, ovule 1, styles 1–3. **FR**: gen utricle. **SEED** 1, vertical (fr compressed side-to-side) or horizontal (fr compressed top-to-bottom). 100 genera, 1300 spp.: worldwide, esp deserts, saline or alkaline soils; some cult for food (*Beta*, beets, chard; *Chenopodium*, quinoa). Recently treated in expanded Amaranthaceae [Downie et al 1997 Amer J Bot 84: 253–273] Key to genera adapted by Margriet Wetherwax.

1. Upper sts clearly jointed, ± fleshy; lvs scale-like
  2. Lvs alternate; shrub . . . . **ALLENROLFEA**
  2' Lvs opposite; per . . . . **SALICORNIA**
1' Upper sts not jointed, not fleshy; lvs gen with blades
  3. Lvs opposite, clasping; sepals strongly overlapping side-to-side; pls from rhizomes . . . . **NITROPHILA**
  3' Lvs mostly alternate, not clasping; sepals not overlapping; rhizomes 0
    4. Shrubs or subshrubs
      5. Herbage stellate-hairy, gen becoming rust-colored; lf margins inrolled . . . . **KRASCHENINNIKOVIA**

5' Herbage glabrous, puberulent, minutely scaly, or powdery, green or grayish; lf margins gen flat
  6. Pistillate fls gen ± enclosed by 2 tightly appressed or ± fused bractlets; pistillate calyx 0; lf blades gen flat
    7. Bractlets compressed, gen triangular, tip obtuse to acute; lvs minutely scaly . . . . [2]**ATRIPLEX**
    7' Bractlets thick, ± round, tip rounded; lvs glabrous or with soft, branched hairs . . . . **GRAYIA**
  6' Fls not ± enclosed by 2 bractlets (sometimes bisexual); lf blade gen ± cylindric (exc *Kochia, Suaeda*)
    8. St branches becoming spine-like; staminate fls in spikes with peltate bracts, calyx 0; pistillate
       fls 1–3, axillary, perianth cup-like, winged in fr . . . . **SARCOBATUS**
    8' St branches not spine-like; fls gen bisexual, axillary; calyx lobes rounded to winged
      9. Calyx lobes rounded in fr; ovary with neck-like extension; fl-st gen glabrous . . . . [2]**SUAEDA**
      9' Calyx lobes winged in fr; ovary top rounded to ± depressed, not neck-like; fl-st gen densely hairy
        . . . . [3]**KOCHIA**
4' Ann or per (sometimes woody below ground)
  10. Pistillate fls enclosed by 2 bracts; pistillate calyx 0; staminate calyx segments 3–5 . . . . [2]**ATRIPLEX**
  10' Fls sometimes bisexual, bracts 0–1 (if 2, not enclosing fl); calyx segments 1–5
    11. Lvs or infl bracts spine- or bristle-tipped
      12. Lvs and infl bracts cylindric, abruptly bristle-tipped; sepal tips winged in fr . . . . **HALOGETON**
      12' Lvs and infl bracts thread- to awl-like, acute; infl bract widely lanceolate, spine-tipped; sepal
        back winged in fr . . . . **SALSOLA**
    11' Lvs or infl bracts not spine- or bristle-tipped
      13. Calyx bilateral, 1 lobe > others . . . . [2]**SUAEDA**
      13' Calyx radial, lobes gen equal
        14. Calyx winged in fr
          15. Lf ± wavy-toothed; st finely tomentose; wing continuous around calyx, irregularly toothed
            . . . . **CYCLOLOMA**
         15' Lf entire; st glabrous to straight-hairy; wings 5, opposite calyx lobes . . . . [3]**KOCHIA**
        14' Calyx not winged in fr (fr winged or not)
          16. Calyx lobes tubercled or curved- to hooked-spiny
            17. Calyx densely and finely hairy, lobes curved- to hooked-spiny . . . . **BASSIA**
            17' Calyx glabrous to ciliate, lobe tips tubercled . . . . [3]**KOCHIA**
          16' Calyx lobes not appendaged
            18. Calyx tube > lobes . . . . [2]**CHENOPODIUM**
            18' Calyx tube 0 or << lobes
              19. Calyx lobes 4–5; stamens gen 5 . . . . [2]**CHENOPODIUM**
              19' Calyx lobes 1–3(5); stamens 1–3
                20. Fr ± elliptic, ± flat, winged; lvs subtending fls membranous, scarious-margined
                  . . . . **CORISPERMUM**
                20' Fr ovoid to ± spheric, not winged; lvs subtending fls herbaceous, often fleshy . . . . **MONOLEPIS**

ALLENROLFEA   IODINE BUSH

1 sp. (Robert Allen Rolfe, English botanist, 1855–1921)
A. occidentalis (S. Watson) Kuntze (p. 265) Shrub 50–200 cm, glabrous. **ST** much-branched, jointed; internodes 5–20 mm, green to ± glaucous, fleshy. **LVS** sessile, ± decurrent, scale-like, triangular. **INFL**: spike, 5–25 mm, cylindric, sessile; fls spirally arranged; bracts peltate. **FL** bisexual; calyx 1–1.5 mm, 4–5- lobed, enclosing and falling with fr; stamens 1–2, exserted; stigmas 2. **FR** ± 1 mm, ovoid. **SEED** red-brown. Saline soils, flats, bluffs; < 1300 m. SnJV, e SnFrB, SCoRI, n WTR, e PR, **s SNE**, **D**; to OR, ID, TX, n Mex. ❀ TRY. Jun–Aug

ATRIPLEX   SALTBUSH

Dean Taylor & Dieter H. Wilken

Ann (gen monoecious) to shrub (gen dioecious), often scaly. **LVS** gen alternate, gen entire; lower gen ± short-petioled; upper gen sessile, ± reduced. **STAMINATE INFL**: spike or spheric cluster; bracts 0. **PISTILLATE INFL**: clusters to spike- or panicle-like; bracts 2 per fr, free to fused, gen compressed, gen sessile. **STAMINATE FL**: calyx lobes 3–5; stamens 3–5. **PISTILLATE FL**: calyx ± 0; ovary ovoid to spheric, style branches 2. **SEED** gen erect. ± 250 spp.: temp to subtrop worldwide. (Latin: ancient name) Gen in alkaline or saline soils; some weedy; some accumulate selenium. Key to species adapted by Margriet Wetherwax.

1. Per, subshrub, or shrub
  2. Per; central st herbaceous to ± woody at base
    3. Sts gen erect; bract margin irregularly dentate to tubercled . . . . [2]***A. fruticulosa***
    3' Sts gen prostrate to decumbent; bract margin (sub)entire . . . . ***A. semibaccata***
  2' Subshrub or shrub, central st aboveground clearly woody

4. Fr bracts together 4-winged base to top . . . . *A. canescens*
  5. Lf blade 3–8 mm wide; fr bracts stalked, wings gen 3–6 mm wide . . . . ssp. *canescens*
  5' Lf blade 1–3 mm wide; fr bracts sessile, wings gen < 3.5 mm wide . . . . ssp. *linearis*
4' Fr bracts together not clearly 4-winged
  6. Lvs gen irregularly and sharply dentate . . . . *A. hymenelytra*
  6' Lvs gen entire
    7. Fr bracts 5–20 mm
      8. Fr bracts free, widely elliptic to ± round, few-toothed below middle, entire above middle
      . . . . *A. confertifolia*
      8' Fr bracts fused ± to middle, entire and spheric below middle, compressed and irregularly
      toothed above middle . . . . *A. spinifera*
    7' Fr bracts 2–6 mm
      9. Upper lvs gen sessile to clasping, base truncate to cordate . . . . *A. parryi*
      9' Upper lvs gen subsessile to short-petioled, base tapered to round
        10. Lf blade oblong to narrowly oblanceolate, gen 2–4 mm wide . . . . *A. polycarpa*
        10' Lf blade ovate to deltate, gen 10–50 mm wide . . . . *A. lentiformis*
        11. Twigs gen smooth, not striate . . . . ssp. *lentiformis*
        11' Twigs gen sharp-angled, striate . . . . ssp. *torreyi*
1' Ann
12. Lvs gen green on both surfaces, glabrous to sparsely powdery or fine-scaly
  13. Fr bracts gen ± free above middle; lf lanceolate to deltate, base tapered to hastate . . . . *A. phyllostegia*
  13' Fr bracts fused to near top; lf elliptic to lanceolate, bases tapered . . . . [2]*A. serenana* var. *serenana*
12' Lvs white to gray, densely and finely scaly (esp below)
  14. Fr bracts gen widest below middle, ovate, deltate, or diamond-shaped
    15. Lvs coarsely wavy-toothed; fr bracts hard . . . . *A. rosea*
    15' Lvs gen entire; fr bracts flexible
      16. Sts prostrate to decumbent, gen flexible; lvs gen opposite . . . . *A. parishii*
      16' Sts erect (branches widely spreading), generally rigid; lvs gen alternate . . . . *A. pusilla*
  14' Fr bracts (incl stalk) gen widest at or above middle, wedge-shaped to ± round
    17. Staminate fls in terminal spikes, pistillate fls in axillary clusters
      18. Sts prostrate to decumbent . . . . [2]*A. fruticulosa*
      18' Sts gen decumbent to ascending — mat-like . . . . [2]*A. serenana* var. *serenana*
    17' Staminate fls in axillary clusters (or staminate and pistillate fls mixed, gen in axillary clusters)
      19. Fr bracts round, ± dentate to cut . . . . *A. elegans*
        20. Fr bract margin 0.5–1 mm wide, dentate to cut . . . . var. *elegans*
        20' Fr bract margin 0.3–0.5 mm wide, minutely crenate to finely toothed . . . . var. *fasciculata*
      19' Fr bracts gen wedge-shaped to obovate or ± spheric, entire below middle, toothed above
        21. Fr bract tip gen truncate, 4–5-toothed, margin entire — gen SNE, DMoj (incl edges) . . . . . . *A. truncata*
        21' Fr bract tip acute to rounded in outline, margin toothed
          22. Lf margin coarsely serrate . . . . *A. suberecta*
          22' Lf margin entire, wavy . . . . *A. argentea*
          23. Upper lvs short-petioled; lowest lvs alternate; GB . . . . var. *argentea*
          23' Upper lvs sessile; lowest lvs opposite; D . . . . var. *mohavensis*

A. argentea Nutt. SILVERSCALE Ann 1.5–8 dm. **STS** decumbent to erect, densely branched, finely gray-scaly, peeling. **LF:** blade 7–40 mm, elliptic to deltate, gray-scaly, wavy-margined, base gen ± subhastate. **PISTILLATE INFL:** bracts in fr 4–8 mm, fused to near top, widely deltate to ± round, gen tubercled, margins green, toothed. **SEED** 1.5–2 mm, brown. Saline soils; < 1700 m. n SNH, GV, e SnFrB, e SCo, **GB, D**; to se OR, c U.S., TX, n Mex. 3 vars. in CA.

  var. *argentea* (p. 265) **STS** 3–8 dm, erect; branches ascending. **LVS:** blade 7–20 mm, elliptic to deltate; upper short-petioled. Habitats of sp.; < 1500 m. **GB**; to c US, n Mex. ❀ TRY. Jun–Sep

  var. *mohavensis* M.E. Jones (p. 265) **STS** 3–8 dm, erect; branches ascending. **LVS:** blade 7–40 mm, lanceolate to ovate, often curled toward st; lowest opposite. Habitats of sp.; < 1000 m. GV, e SnFrB, e SCo, **D**; to TX, n Mex. [ssp. *expansa* (S. Watson) H.M. Hall & Clements]. ❀ TRY. Jul–Nov

A. canescens (Pursh) Nutt. FOURWING SALTBUSH Shrub < 20 dm, erect; branches many, spreading to ascending. **LF:** blade 8–50 mm, linear to oblanceolate, densely white-scaly. **PISTILLATE INFL** terminal; bracts in fr 4–25 mm, gen fused to near top, ovoid to spheric, hard, wings 4, 3–6 mm wide, wavy to deeply sharp-dentate. **SEED** 1.5–2.5 mm. $2n=18,36$. Clay to gravelly flats, slopes, shrubland; < 2400 m. SNH (e slope), Teh, SCoRI, SCo, n TR, PR, **GB, D**; to w Can, SD, n Mex. [var. *laciniata* Parish a form of both CA sspp.] Sspp. intergrade.

  ssp. *canescens* (p. 265) **LF:** blade linear to oblanceolate. **PISTILLATE INFL:** bracts in fr 6–25 mm, stalked, wings gen 3–6 mm wide. Habitats and range of sp. [var. *macilenta* Jepson]. ❀ DRN,DRY,SUN:1,**2,3,7–13**,14–17,**18–23**,24;CVS. Jun–Aug

  ssp. *linearis* (S. Watson) H.M. Hall & Clements **LF:** blade linear. **PISTILLATE INFL:** bracts in fr sessile, 4–8 mm, wings gen < 3.5 mm wide. Sandy soils, dunes, flats; < 100 m. n WTR?, **DSon**; n Mex. ❀ TRY. May–Jul

A. confertifolia (Torrey & Frémont) S. Watson (p. 265) SHAD-SCALE Shrub < 10 dm, gen rounded, ± erect, much-branched; twigs gen spreading, stiff, becoming spine-like. **LF** short-petioled; blade 8–24 mm, elliptic to widely ovate, firm, densely gray-scaly. **PISTILLATE INFL** sometimes terminal; bracts in fr 5–12(20) mm, free, widely elliptic to ± round, smooth, entire to few-toothed. **SEED** 1.5–2 mm. 2*n*= 18,36,54. Alkaline flats, gravelly slopes in shrubland, pinyon/juniper woodland; < 2400 m. SNH (e slope), **GB, DMoj**; to OR, n-c US, n Mex. May hybridize with *A. canescens.* ❀ TRY. Apr–Jul

A. elegans (Moq.) D. Dietr. WHEELSCALE Ann ± 1–5 dm. **STS** decumbent to ascending, ± branched, finely scaly, becoming glabrous. **LF**: blade 5–25 mm, elliptic to oblanceolate, densely white-scaly below, entire to dentate, tapered to base. **PISTILLATE INFL**: bracts in fr 2–3.5 mm, fused to near top, round, smooth or with 1 low tubercle, toothed. **SEED** 1–1.5 mm, brown. Saline or alkaline soils, dry lakes; < 1000 m. **s SNE, D**; to TX, n Mex. Vars. intergrade.

var. ***elegans*** (p. 265) **LF**: blade entire to dentate. **PISTILLATE INFL**: bract in fr deeply dentate, margin 0.5–1 mm wide. ± saline creosote-bush scrub; < 800 m. **e D**; to TX, n Mex.

var. ***fasciculata*** (S. Watson) M.E. Jones (p. 265) **LF**: blade gen entire. **PISTILLATE INFL**: bract in fr minutely toothed, margin 0.3–0.5 mm wide. *n*=9. Habitats of sp. **s SNE, D**; AZ, n Mex. Mar–Jul

A. fruticulosa Jepson (p. 265) Per < 5 dm, monoecious. **ST** gen simple below; branches many, ± decumbent to erect, scaly, becoming glabrous. **LF**: blade 5–12(20) mm, narrowly lanceolate to elliptic, densely gray-scaly. **PISTILLATE INFL**: bracts in fr 3–5 mm, fused to middle or above, widely obovate to subspheric, ± hard, smooth to few-tubercled below middle, margin irregularly dentate to sharply tubercled. **SEED** ± 1.5 mm. Clay or alkaline soils, open sites, shrubland; < 700 m. Teh, s ScV, SnJV, SnFrB, SCoRI, **w DMoj**. ❀ TRY. Apr–Nov

A. hymenelytra (Torrey) S. Watson (p. 265, pl. 46) Shrub 4–10 dm, rounded, silver-scaly. **ST** simple below, erect; branches many, spreading to ascending. **LVS** petioled; blade 12–45 mm, widely ovate to round, thick, irregularly and sharply dentate. **PISTILLATE INFLS** sometimes terminal; bracts in fr short-stalked, 6–20 mm, free, round to ± reniform, entire to ± crenate. **SEED** ± 2 mm. Slopes, washes, shrubland; < 1500 m. se PR, **SNE, D**; to sw UT, AZ, n Mex. ❀ DFCLT. Jan–Apr

A. lentiformis (Torrey) S. Watson BIG SALTBUSH Shrub 8–30 dm, gen wider than tall. **ST** erect; branches many, spreading to ascending; twigs densely fine-scaly, becoming glabrous. **LVS**: blade 7–50 mm, ovate to deltate or hastate, densely gray-scaly. **PISTILLATE INFL** ± panicle-like, terminal; bracts in fr 2–6 mm, fused ± to middle, widely ovate to round, entire to minutely crenate. **SEED** ± 1.5 mm. Alkaline or saline washes, dry lakes, shrubland; < 1500 m. s SN, deltaic GV, SnJV, SCoRI, SCo, n WTR, **SNE, D**; to sw UT, n Mex.

ssp. ***lentiformis*** (p. 265) Monoecious or dioecious. **ST**: twigs smooth. **LF**: blade 15–50 mm. **PISTILLATE INFL**: bracts in fr 2.5–6 mm. *n*=9. Habitats and range of sp. Coastal and ChI pls with large lvs and frs have been called ssp. *breweri* (S. Watson) H.M. Hall & Clements. ❀ SUN,DRN:**2,3,7,14–24**&IRR:**8–13**;STBL; CVS. Jul–Oct

ssp. ***torreyi*** (S. Watson) H.M. Hall & Clements (p. 265) Gen dioecious. **ST**: twigs sharply angled, striate. **LF**: blade 7–30 mm, base sometimes hastate. **PISTILLATE INFL**: bracts in fr 2–4.5 mm. Alkaline clay soils, dry lakes, washes; 300–1500 m. **SNE, n DMoj**; to sw UT. [*A. torreyi* (S. Watson) S. Watson]. ❀ TRY. Jun–Oct

A. parishii S. Watson (p. 265) PARISH'S BRITTLESCALE Ann < 2 dm. **STS** prostrate to decumbent, gen flexible, white, scaly to densely woolly near tips. **LVS** gen opposite; blade 4–8 mm, ovate

to cordate, gen densely white-scaly, tip acute; lower gen sessile. **PISTILLATE INFL**: bracts in fr 2.5–3 mm, fused to near top, ovate or diamond-shaped, ± densely tubercled, entire to few-toothed. **SEED** ± 1–1.5 mm, reddish. **PRESUMED EXTINCT.** (Recently rediscovered in w Riverside Co.) Saline or clay soils; < 1900 m. SW (exc ChI), **w DMoj**; Baja CA. Jun–Oct

A. parryi S. Watson (p. 269) Shrub 2–5 dm, rounded, gen dioecious. **ST** erect; branches many, spreading to erect; twigs slender, stiff, becoming spine-like, densely scaly. **LF**: blade 5–20 mm, elliptic to widely ovate, densely white-scaly. **PISTILLATE INFL**: bracts in fr 2.5–4 mm, fused ± at middle or above, round to reniform, smooth, margin entire to low-tubercled above middle. **SEED** 1–1.5 mm. Alkaline soils, flats, dry lakes; < 1500 m. **s SNE, DMoj, n DSon**; NV. ❀ TRY. May–Aug

A. phyllostegia (Torrey) S. Watson (p. 269) ARROWSCALE Ann 1–4 dm. **STS** much-branched from base, faintly striate, green, glabrous to sparsely fine-scaly. **LF**: blades 10–40 mm, lanceolate to deltate, fleshy, brittle, base tapered to hastate. **PISTILLATE INFL**: bracts in fr 5–20 mm, fused at base, lanceolate to triangular, smooth or tubercled, base often ± hastate. **SEED** ± 1 mm, brown. *n*=9. Saline soils, meadows, flats; < 1500 m. SnJV, **SNE, DMoj**; to OR, UT. Apr–Aug

A. polycarpa (Torrey) S. Watson Shrub 5–20 dm, densely gray-scaly. **ST** erect; branches many, spreading to ascending; twigs slender, becoming ± spine-like. **LVS** gen subsessile; blade 3–25 mm, oblong to narrowly oblanceolate, ± thick. **PISTILLATE INFL** sometimes terminal; bracts in fr 2–3 mm, fused ± to middle, ± subspheric, smooth to tubercled, minutely crenate to toothed. **SEED** 1–1.5 mm. Alkaline flats, dry lakes; < 1500 m. SnJV and margins, n TR, e PR, **s SNE, D**; to UT, n Mex. ❀ STBL;may be INV. Jul–Oct

A. pusilla (Torrey) S. Watson (p. 269) Ann < 3 dm. **STS** 1–several from base, rigid, not brittle, gen scaly; tips reddish. **LVS**: blade 3–15 mm, elliptic to ovate, thick, ± fleshy, ± rounded, gen sparsely scaly; lower sessile. **STAMINATE INFL** terminal; fls 1–2. **PISTILLATE INFL**: fls 1–2; bracts in fr 1–2 mm, fused to near top, ovate, smooth, entire, tip acute. **SEED** 0.8 mm, brown. Alkaline soils, hot springs; 1300–2000 m. MP, **n SNE**; OR, NV. Jun–Sep

A. rosea L. TUMBLING ORACLE Ann 4–15 dm. **ST** erect; branches ascending, gen ± glabrous. **LVS** firm, persistent; blades 10–60 mm, 4–30 mm wide, densely fine-scaly below, greenish above, becoming red, coarsely wavy-toothed, base tapered. **PISTILLATE INFL**: bracts in fr 4–8 mm, fused ± to middle, diamond-shaped to deltate, hard, tubercled, dentate. **SEED** 2–2.5 mm, brown. *n*=9. Common. Open, disturbed places, fields; < 2400 m. CA-FP, **D (uncommon)**; to e N.Am; native to Eurasia. Jul–Oct

A. semibaccata R.Br. AUSTRALIAN SALTBUSH Per or sub-shrub < 3.5 dm, monoecious. **STS** several, 3–10 dm, ± spreading; branches spreading to ascending, ± white-scaly or becoming glabrous. **LVS**: blade 8–30 mm, oblong to narrowly elliptic, entire to wavy-toothed, ± scaly (esp below); upper short-petioled. **PISTILLATE INFL**: fls few; bracts in fr 3–6 mm, fused to middle or above, ovate to ± diamond-shaped, fleshy, reddish, net-veined, (sub)entire. **SEED** 1.5–2 mm. 2*n*=18. Waste places, shrubland, woodland; < 1000 m. ± CA-FP (exc CaR, n&c SN), **D**; to UT, TX, n Mex; native to Australia. Apr–Dec

A. serenana Nelson BRACTSCALE Ann 3–10 dm, mat-like. **STS** decumbent to ascending, sparsely scaly; tips flexible. **LVS** subsessile; blade 10–40 mm, elliptic to lanceolate, ± greenish, sparsely fine-scaly above, dentate. **STAMINATE INFL**: spikes 1–many in panicles (or clusters spheric, terminal). **PISTILLATE INFL**: bracts in fr 2–3.5 mm, fused to middle, ± round to wedge-shaped, smooth or tubercled, toothed above middle. Alkaline flats, coastal bluffs; < 2000 m. s SN, SnJV, SCoRO, SW, **SNE (naturalized), w DMoj, DSon**; Baja CA. 2 vars. in CA.

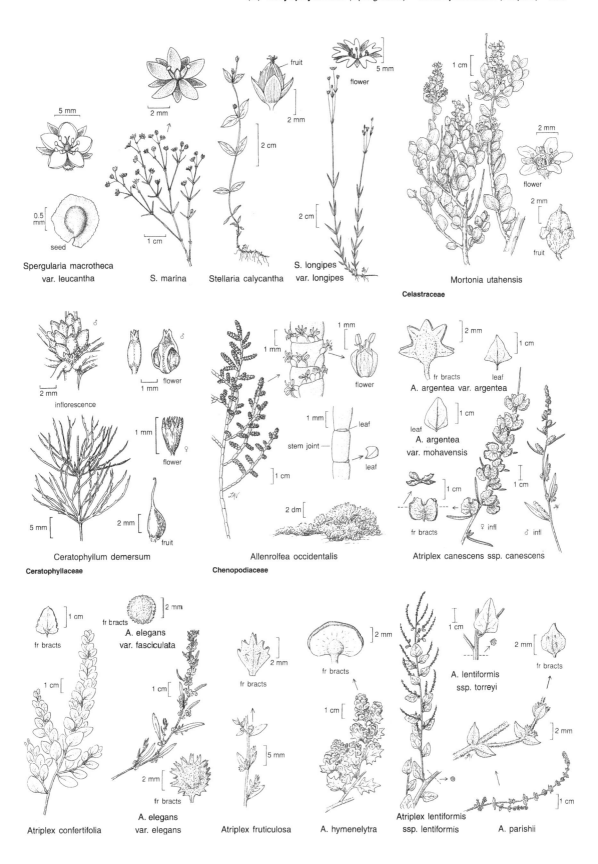

Spergularia macrotheca
var. leucantha

S. marina

Stellaria calycantha

S. longipes
var. longipes

Mortonia utahensis

Celastraceae

Ceratophyllum demersum

Ceratophyllaceae

Allenrolfea occidentalis

Chenopodiaceae

A. argentea var. argentea

A. argentea
var. mohavensis

Atriplex canescens ssp. canescens

Atriplex confertifolia

A. elegans
var. fasciculata

A. elegans
var. elegans

Atriplex fruticulosa

A. hymenelytra

Atriplex lentiformis
ssp. lentiformis

A. lentiformis
ssp. torreyi

A. parishii

var. *serenana* (p. 269) **LF**: blade 15–40 mm. **STAMINATE INFL**: panicle, elongate. **PISTILLATE INFL**: fr bract ± 1-veined. *n*=9. ± habitats and range of sp. May–Oct

A. spinifera J.F. Macbr. (p. 269) Shrub 3–20 dm, densely white-to gray-scaly. **ST** erect; branches many, ascending to erect; twigs spreading, ± stiff, becoming spine-like. **LVS**: blade 6–28 mm, elliptic to widely ovate, entire to 2-toothed below middle; upper short-petioled. **PISTILLATE INFL**: bracts in fr 5–20 mm, 3–11 mm wide, fused ± to middle, smooth, below middle spheric and densely scaly, above middle toothed and sparsely scaly. **SEED** 2–3 mm. Saline soils, flats, dry lakes; < 800 m. s SnJV, SCoRI, n WTR, **DMoj**. Apr–Jun

A. suberecta I. Verd. Ann 2–6 dm, sprawling. **STS** decumbent to ascending, densely scaly below. **LVS**: blades 12–30 mm,

ovate to ± diamond-shaped, ± fine-scaly below, ± glabrous above, coarsely serrate, base ± obtuse; upper short-petioled. **PISTILLATE INFL**: bracts in fr 2–5 mm, fused to middle, diamond-shaped, entire below middle, widest and 2–4-toothed above middle. *n*=9. Disturbed places, fields; < 300 m. SnJV, SCo, **DSon**; native to Australia.

A. truncata (Torrey) A. Gray WEDGESCALE Ann < 7 dm. **STS** 1–few, erect; branches ascending, gray-scaly. **LVS**: blade 10–40 mm, ovate to deltate, glabrous to sparsely gray-scaly, base truncate to ± lobed. **PISTILLATE INFL**: bracts in fr 2–4 mm, fused to near top, widely wedge-shaped, smooth to barely tubercled, entire, tip ± truncate to notched, 4–5-toothed. **SEED** 1–1.5 mm, brown. *n*=9. Alkaline soils, flats; 600–2500 m. SNH (e slope), TR (desert slopes), **SNE, DMoj**; to B.C., Colorado, NM. Jun–Sep

BASSIA

Ann, gen hairy. **ST**: axis gen erect; branches ascending to erect. **LVS** linear to lanceolate, reduced upward. **INFL**: spike; bracts lf-like; fls 1–few per axil. **FLS** gen bisexual; calyx lobes 5, incurved, hooked-spiny in fr; stamens gen 5; stigmas gen 2. **FR** ± depressed-spheric. **SEED** horizontal. ± 5 spp.: warm temp Eurasia. (Ferdinando Bassi, Italian botanist, 1710–1774) Perhaps best incl in *Kochia*.

B. hyssopifolia (Pallas) Kuntze (p. 269) Pl < 1 m. **LVS**: lower 5–60 mm, 1–3.5 mm wide, flat, often withered in fr. **INFL** 5–50 mm; bracts 2–5 mm, ± oblong. **FL**: calyx densely tan-woolly, base in fr leathery, spines ± 1 mm. **FR** 1–1.5 mm diam. **SEED**

dark brown. Disturbed sites, fields, roadsides; < 1200 m. **CA** (exc NW, SNH); widespread N.Am; native to Eurasia. Sometimes confused with *Kochia scoparia*. Jul–Oct

CHENOPODIUM PIGWEED, GOOSEFOOT

Ann or per, glabrous, glandular, or powdery. **ST**: branches 0 to gen ± spreading. **LVS** gen petioled, linear to deltate, entire to lobed, reduced upward; base gen tapered. **INFL**: spheric clusters, spikes, or panicle-like, gen dense; bracts gen 0; fls gen sessile. **FL**: calyx segments gen 5, fused or not, persistent, flat to keeled; stamens gen 5; ovary lenticular to spheric, stigmas 2–5. **SEED** vertical or horizontal, red-brown to black; wall very thin. ± 150 spp.: temp; some cult for food or grain. (Greek: goose foot, from lf shape of some) [Wahl 1954 Bartonia 27:1–46; Crawford 1975 Brittonia 27:279–288] Fr gen required for identification. Key to species adapted by Margriet Wetherwax.

1. Herbage ± glandular, strong-smelling . . . . ***C. botrys***
1' Herbage glabrous to minutely scaly or powdery, gen not strong-smelling
  2. Per from stout, fleshy caudex; calyx tube gen > lobes in fr . . . . ***C. californicum***
  2' Ann; calyx tube gen < lobes
    3. Most seeds vertical, or vertical and horizontal seeds ± equal in number; calyx lobes gen 3
      4. Upper infls axillary, ± spheric, < 10 mm diam
        5. Lower lf blades lanceolate to deltate, gen glabrous below; sepals thickened or ± fleshy in fr . . ***C. foliosum***
        5' Lower lf blades oblong to ovate, densely powdery below; sepals thin or membranous in fr —
          vertical and horizontal seeds gen ± equal in number . . . . ***²C. glaucum***
      4' Upper infls axillary or terminal, spike-like, 15+ mm
        6. Vertical seeds enclosed by calyx tube; style lobes ascending . . . . ***C. chenopodioides***
        6' Vertical seeds subtended by calyx tube; style lobes spreading
          7. Lf densely powdery below; vertical and horizontal seeds often ± equal in number . . . . ***²C. glaucum***
          7' Lf gen glabrous below; vertical seeds many more than horizontal . . . . ***C. rubrum***
    3' Most or all seeds horizontal; calyx lobes gen 5
      8. Lf bright green, deeply 3–5-lobed, base truncate to cordate, tip acuminate to long-tapered; fr
        1.5–3 mm diam . . . . ***C. simplex***
      8' Lf dull green to grayish, entire to toothed, base tapered to acute, tip rounded to acute; fr gen < 1.5 mm diam
        9. Lower lf blades 1-veined from base
          10. Lower lf blades > 4 mm wide; calyx lobes strongly keeled in fr; fr wall free from seed, easily
            detached with dissecting needle . . . . ***²C. desiccatum***
          10' Lower lf blades < 3.5 mm wide; calyx lobes weakly keeled in fr; fr wall adherent to seed, firm
            or fracturing when teased with dissecting needle . . . . ***C. leptophyllum***
        9' Lower lf blades 3-veined from base
          11. Fr wall free from seed, easily detached with dissecting needle
            12. Lf length < 1.5 × width
              13. Fr top visible between calyx lobes; fr 1–1.5 mm diam; lf blade thin . . . . ***C. fremontii***
              13' Fr top hidden by calyx lobes; fr ± 1 mm diam; lf blade thick, becoming ± leathery . . . . . . ***C. incanum***
            12' Lf length gen 1.5+ × width

14. Sts several from base, branches ± dense; fr ± 1 mm diam, top hidden by calyx lobes . . ²***C. desiccatum***
14' Sts 1–few from base, branches well spaced; fr 1–1.5 mm diam, top visible between calyx lobes
  15. Lower lvs entire, length 1.5–3 × width . . . . ***C. atrovirens***
  15' Lower lvs ± 1–2-lobed (-toothed), length 3–5 × width . . . . ***C. pratericola***
11' Fr wall adherent to seed, firm or fracturing when teased with dissecting needle
  16. Lower lf blade gen entire or 1–2-toothed at base
    17. Pl gen few-branched, slender; infl mostly spikes, ascending to erect; fr 1–1.5 mm diam . . . . . ***C. hians***
    17' Pl gen much-branched, ± wide; infl often rounded in fr; fr < 1 mm diam . . . . ***C. nevadense***
  16' Lower lf blade few–many-toothed above base
    18. Fr wall surface minutely honeycombed at 20×; calyx lobes widely keeled in fr . . . . ***C. berlandieri***
    18' Fr wall surface smooth or irregularly rough at 20×; calyx lobes smooth to narrowly keeled
      19. Lf gen shiny dark green above, glabrous to sparsely powdery below; seed sharply angled at
        equator . . . . ***C. murale***
      19' Lf dull green to grayish above, powdery below; seed obtusely angled or rounded at equator
        20. Fr top gen hidden by calyx lobes; terminal infl branches weak, gen straight; lower st branches
          0 or gen ascending . . . . ***C. album***
        20' Fr top gen visible between calyx lobes; terminal infl branches often stiffly ascending or erect;
          lower st branches often decumbent . . . . ***C. strictum*** var. ***glaucophyllum***

C. album L. (p. 269) PIGWEED, LAMB'S QUARTERS Ann 18–100+ cm. **LF:** blade 15–70 mm, lanceolate to ± deltate, entire to irregularly wavy-toothed, dull green above, powdery below. **FL:** sepals gen enclosing fr, gen keeled, powdery. **FR** 1–1.5 mm diam; wall ± rough at 20×, adherent to seed. **SEED** horizontal. 2*n*=54. Common. Disturbed places, fields, roadsides; < 1800 m. **CA**; ± temp worldwide; probably native to Eur. Often confused with *C. berlandieri*, and *C. strictum*. Pls with lf blade as wide as long and calyx lobes strongly keeled have been called *C. opulifolium* Schrader. Jun–Oct

C. atrovirens Rydb. (p. 269) Ann 7–60 cm. **LF:** blade 9–35 mm, oblong to widely ovate, entire or 1–2-toothed to -lobed at base, ± glabrous to sparsely powdery. **FL:** sepals not enclosing fr, barely keeled, sparsely powdery. **FR** 1–1.5 mm diam; wall free from seed. **SEED** horizontal. 2*n*=18. Open places, shrubland, woodland, coniferous forest; 300–3500 m. NW, CaR, SN, SnBr, PR, **GB, n DMoj**; to WY, c US, NM. Much like *C. pratericola*. Jul–Sep

C. berlandieri Moq. (p. 269) PITSEED GOOSEFOOT Ann 18–85+ cm. **LF:** blade 15–50 mm, lanceolate to ± deltate, gen irregularly wavy or toothed, powdery. **FL:** sepals enclosing fr, back strongly keeled, powdery. **FR** 1–1.5 mm diam; wall surface honeycombed at 20×, adherent to seed. **SEED** horizontal. 2*n*=36. Common. Open, often disturbed places; < 2700 m. **CA**; to e N.Am, n Mex. [vars. *sinuatum* (Murray) Murray, *zschackei* (Murray) Murray] Often confused with *C. album*.

C. botrys L. (p. 269) JERUSALEM OAK Ann 14–65 cm, strong-smelling. **LF:** blade 3–65 mm, short-stalked-glandular, ovate to elliptic and wavy to pinnately lobed below, oblong and gen entire above. **INFL:** branches arched or curved; fls short-pedicelled. **FL:** sepals weakly enclosing fr, ± flat, densely short-stalked-glandular. **FR** ± 0.5 mm diam; wall adherent to seed. **SEED** horizontal. 2*n*= 18. Disturbed places; < 2100 m. **CA**; to Can, e US; native to Eur. Jun–Oct

C. californicum (S. Watson) S. Watson (p. 269) Per 20–90+ cm; caudex stout, fleshy. **STS** several from base, decumbent to ascending. **LF:** blade 15–90+ mm, ± deltate, coarsely dentate to wavy-toothed, green to glaucous, base truncate to hastate, tip acute. **INFL:** axillary, ± spheric, < 10 mm diam; terminal spikes, 8–20 cm, interrupted. **FL:** calyx tube enclosing fr, lobes ± erect, flat, ± glabrous. **FR** 1.5–2 mm diam; wall adherent to seed. **SEED** vertical. Gen open sites, sandy to clay soils; < 2000 m. s NCo, NCoRO, c&s SNF, Teh, GV, CW, SW, **s SNE, w DMoj**; Baja CA. Mar–Jun

C. chenopodioides (L.) Aellen (p. 269) Ann 10–45 cm. **LF:** blade 8–32 mm, ovate to deltate, entire to irregularly few-toothed, ± glabrous. **FL:** calyx tube enclosing fr, > lobes, ± glabrous. **FR** 0.5–1 mm diam; wall free from seed. **SEEDS** mostly vertical; calyx of vertically seeded fls 3-lobed; calyx of horizontally seeded fls 4–5-lobed. Saline soils, drying ponds, mudflats; < 2500 m. NCoRO, SN, SnBr, PR, **GB**, expected elsewhere; to WA; native to S.Am. Jul–Oct

C. desiccatum Nelson (p. 269) Ann 12–35 cm. **STS** gen several from base. **LF:** blade 5–25 mm, oblong to elliptic, ± fleshy, margin entire, lower surface densely powdery, ± glabrous above. **FL:** sepals enclosing fr, keeled, powdery. **FR** ± 1 mm diam; wall free from seed. **SEED** horizontal. 2*n*=18. Uncommon. Open places, shrubland, coniferous forest; < 2900 m. SN, SnJV, TR, **GB, n DMoj**; to WY, Colorado. Jul–Sep

C. foliosum (Moench) Asch. (p. 269) Ann 5–60 cm. **LVS:** blades 7–38 mm, entire to irregularly toothed, ± glabrous; lower lanceolate to deltate, base 2-lobed to hastate; upper oblong to lanceolate, base tapered. **INFL** 3–5 mm diam, axillary, ± spheric, leafy-bracted. **FL:** sepals gen 3, ± enclosing fr, smooth, gen glabrous, becoming ± reddish; stamens 3–4. **FR** 1–1.5 mm diam; wall ± adherent to seed. **SEED** gen vertical. 2*n*=16,18. Open, gravelly or sandy soils; < 3700 m. NCoR, CaRF, c SNH, TR, **GB, w DMoj**; to w Can, c US; native to Eur. [*C. overi* Aellen] Pls with infl clusters 10+ mm diam, reddish, fleshy, leafy bracts 0, have been called *C. capitatum* (L.) Asch., reported from nw CA. Jun–Aug

C. fremontii S. Watson (p. 269) Ann 22–55 cm. **LF:** blade 8–40 mm, widely ovate to deltate, gen 1–3-lobed at base, sometimes entire, ± powdery below, glabrous above. **FL:** sepals not enclosing fr, keeled, ± powdery. **FR** 1–1.5 mm diam; wall free from seed. **SEED** horizontal. 2*n*=18. Gen shaded places, shrubland, coniferous forest; 700–3100 m. SN, TR, PR, **GB, DMtns**; to e N.Am, n Mex. Much like *C. incanum*. Jun–Oct

C. glaucum L. Ann 8–20 cm. **STS:** lower often prostrate; upper ascending. **LF:** blade 5–35 mm, oblong to ovate, margin entire to toothed, densely powdery below, glabrous to sparsely powdery above. **FL:** sepals gen 3, not enclosing fr, back flat, glabrous; stamens gen 3. **FR** ± 1 mm diam; wall free from seed. **SEEDS** vertical and horizontal. 2*n*=18. Open places, often saline soils, drying ponds, streambanks; < 2200 m. CaR, SNH (e slope), SnBr, **GB**, expected elsewhere; to Can, e US; native to Eurasia. [ssp. *salinum* (Standley) Aellen] Jul–Oct

C. hians Standley (p. 269) Ann 30–80 cm. **LF:** blade 8–30 mm, narrowly lanceolate to elliptic, entire, ± densely powdery below, glabrous to sparsely powdery above. **FL:** sepals not enclosing fr, smooth, powdery. **FR** ± 1 mm diam; wall adherent to seed. **SEED** horizontal. 2*n*=18. Open places, shrubland, woodland; 300–2700 m. SN, SCoRI, TR, PR, **GB**; to WY, NM. Much like *C. leptophyllum*.

C. incanum (S. Watson) A.A. Heller var. ***occidentale*** D.J. Crawford (p. 269) Ann 8–30 cm. **LF**: blade 6–22 mm, narrowly deltate to diamond-shaped, entire to weakly 2-lobed at base, powdery below, glabrous to sparsely powdery above. **FL**: sepals gen enclosing fr, strongly keeled, powdery. **FR** ± 1 mm diam; wall free from seed. **SEED** horizontal. Open places, sandy or gravelly soils; 700–2300 m. SNH (e slope), **GB**, **DMoj**; to ID, sw UT. [*C. fremontii* S. Watson var. *i.* S. Watson] Other vars. in Rocky Mtns, w Great Plains. Much like *C. fremontii*. Apr–Aug

C. leptophyllum Moq. Ann 20–60 cm. **LF**: blade 8–25 mm, linear to narrowly lanceolate, ± thin, entire, densely powdery below, less so above. **FL**: sepals enclosing fr or tips reflexed, ± keeled, gen densely powdery. **FR** ± 1 mm diam; wall adherent to seed. **SEED** horizontal. 2*n*=18. Open, gravelly soils, shrubland; 300–3200 m. s NCoRH, c&s SN, **W&I**, **n DMoj**; to e US, n Mex. Much like *C. hians*. Jul–Sep

C. murale L. Ann 15–70+ cm. **LF**: blade 15–60 mm, widely ovate to deltate, toothed, shiny dark green above, sparsely powdery below, base truncate to wedge-shaped. **FL**: sepals ± enclosing fr, keeled, powdery. **FR** 1–1.5 mm diam; equatorial margin sharply angled; wall adherent to seed. **SEED** horizontal. 2*n*=18. Common. Disturbed places, fields; < 2900 m. CA-FP, **D (uncommon)**; to Can, e US, n Mex; native to Eur. Most of year (esp spring)

C. nevadense Standley (p. 273) Ann 10–48 cm. **LVS**: blades 6–18 mm, entire, ± powdery (esp below), sometimes becoming ± glabrous; lower blades ovate to diamond-shaped; upper blades elliptic. **INFL** panicle-like, often rounded in fr. **FL**: sepals ± enclosing fr, back flat, densely powdery. **FR** < 1 mm diam; wall adherent to seed. **SEED** horizontal. Washes, shrubland; 1400–2000 m. c SNH (e slope), **SNE**; se OR, NV. Jun–Aug

C. pratericola Rydb. Ann 16–65 cm. **LVS**: blades 10–30 mm, thin, narrowly elliptic to lanceolate, densely powdery below, glabrous to sparsely powdery above; lower blades gen 1–2-lobed or -toothed. **FL**: sepals not enclosing fr, back keeled, moderately to densely powdery. **FR** 1–1.5 mm diam; wall free from seed. **SEED** horizontal. 2*n*=18. Open, dry places; < 2500 m. CaR, SN, GV, e SnFrB, SCoR, WTR, **SNE**, **D**; to c US. [*C. desiccatum* Nelson var. *leptophylloides* (Murray) Wahl] Much like *C. atrovirens*. Jun–Sep

C. rubrum L. Ann 20–58+ cm. **LF**: blade 15–90+ mm, lanceolate to widely ovate, toothed, glabrous to sparsely powdery below, ± glabrous above; base wedge-shaped. **INFL**: spike, axillary, 10–30 mm. **FL**: sepals gen 3, weakly enclosing fr, back flat, glabrous to sparsely powdery. **FR** 0.5–1 mm diam; wall ± adherent to seed. **SEEDS** vertical and horizontal. 2*n*=36. Open, saline places, drying mudflats; < 1100 m. NCoRO, CaR, Teh, GV, CCo, SnFrB, SCo, **GB**, **DMoj**; to Can, e US, Mex, Eurasia. [*C. humile* Hook.] Some forms may be introduced. Aug–Oct

C. simplex (Torrey) Raf. (p. 273) LARGE-SEEDED GOOSEFOOT Ann 35–150 cm, gen few-branched above middle. **LF**: blade 25–150+ mm, widely ovate to deltate, wavy-lobed (lobes 3–5), glabrous to sparsely powdery, base truncate to ± cordate. **FL**: sepals not enclosing fr, back flat, glabrous to sparsely powdery. **FR** 1.5–3 mm diam; wall free or adherent to seed, netted. **SEED** horizontal. 2*n*=36. UNCOMMON. Disturbed or open places, shrubland, coniferous forest; 1400–2400 m. n SN, MP, **W&I**; to e US. [*C. gigantospermum* Aellen] Jul–Oct

C. strictum Roth var. ***glaucophyllum*** (Aellen) Wahl Ann 45–120+ cm. **LVS** glabrous or sparsely powdery above; basal 25–40 mm, gen ovate, irregularly serrate; cauline 12–20 mm, lanceolate to narrowly ovate, entire. **FL**: sepals not enclosing fr, back flat or weakly keeled, sparsely powdery. **FR** ± 1 mm diam; wall gen adherent to seed. **SEED** horizontal. 2*n*=36. Open, disturbed places; < 1700 m. CaRF, SN, GV, SnFrB, SCoRO, SCo, SnBr, **DMtns (uncommon)**; to Can; native to e US. Perhaps best considered part of *C. album*. Var. *strictum* native to Eurasia. Aug–Oct

CORISPERMUM

Ann, gen erect. **ST**: branches 0–few, spreading to ascending. **LF** gen linear. **INFL**: spike; bracts lf-like. **FL** bisexual; sepals 0–5, scarious; stamens 1–5; ovary superior, chamber 1, stigmas 2. **FR** elliptic to obovate, ± flat. **SEED** vertical, gen black. 60 spp.: n temp. (Latin & Greek: leathery seed) [Maihle & Blackwell 1978 Sida 7:382–391]

C. hyssopifolium L. (p. 273) Pl 5–30 cm. **LF** 9–25 mm, 1–2.5 mm wide. **INFL** 1–4 cm, dense; bracts in fr 3–10 mm, margin scarious. **FR** 3–4 mm diam, round, dark green to black; wing white to translucent. Uncommon. Sandy soils, dunes; 100–1100 m. n SNE, **n DSon, n DMoj**; to AK, e N.Am, n Mex; native to Eurasia.

CYCLOLOMA   WINGED PIGWEED

1 sp. (Greek: circular wing, from calyx in fr)

C. atriplicifolium (Sprengel) J. Coulter (p. 273) Ann 12–75 cm, rounded. **ST** with many spreading branches, slender, striate, finely woolly, becoming glabrous. **LVS** gradually reduced upward; petiole 0–12 mm; blade 5–65 mm, lanceolate to ovate, ± wavy-toothed. **INFL** panicle-like, terminal, open in fr; bracts 0; fls sessile. **FL**: gen bisexual (pistillate); calyx enclosing fr, 2–3 mm diam in fr, winged, lobes 5, ± keeled; stamens 5; ovary densely and finely tomentose, style deeply 2–3-lobed. **FR** ± 2 mm diam. **SEED** 1.5–2 mm, horizontal, black. Fields, disturbed sites; < 800 m. GV, s SCo, w PR, **e DMoj**; native to c N.Am. May–Sep

GRAYIA   HOP-SAGE

1 sp. (Asa Gray, eminent Am botanist, Harvard University, 1810–1888)

G. spinosa (Hook.) Moq. (p. 273, pl. 47) Shrub gen < 1 m, gen dioecious; hairs simple or scaly. **ST**: branches many, stiff; bark straw-colored to gray; twigs becoming spine-like. **LF** 5–35 mm, gen elliptic to oblanceolate, flat, entire, tapered to short-petioled. **STAMINATE INFL** spike-like, terminal, 7–18 mm; bract ± lf-like; fls < 8 per axil. **PISTILLATE INFL** ± spike-like, axillary or terminal, 6–18 cm in fr; bracts 3–10 mm, ± lf-like; bractlets 2, 7–15 mm, fused, together sac-like, ± round, flat, winged, white to red-tinged. **STAMINATE FL**: calyx lobes 4; stamens 4. **PISTILLATE FL**: calyx 0; stigmas 2. **FR** gen 2–3 mm. 2*n*=36. Sandy to gravelly soils, shrubland, pinyon/juniper woodland; 500–2800 m. SNH (e slope), Teh, s SCoRI, WTR (n slope), **GB**, **DMoj**, **nw DSon**; to WA, MT, NM. Sometimes incl in *Atriplex*. ❀ TRY; also STBL. Mar–Jun

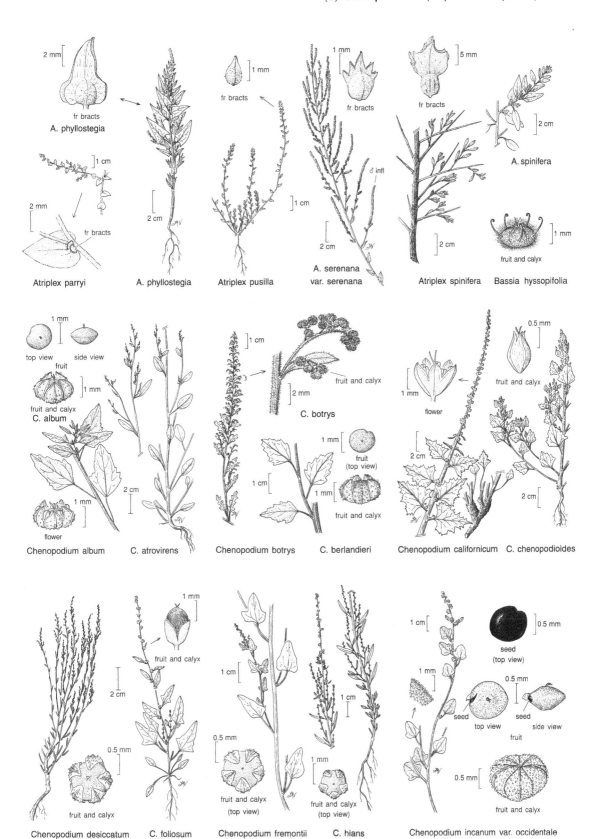

2 mm

fr bracts
A. phyllostegia

1 cm

2 mm

fr bracts

Atriplex parryi

2 cm

A. phyllostegia

1 mm

fr bracts

1 cm

Atriplex pusilla

1 mm

fr bracts
A. serenana

♂ infl

2 cm

A. serenana
var. serenana

5 mm

fr bracts

2 cm

A. spinifera

2 cm

1 mm

fruit and calyx

Atriplex spinifera

Bassia hyssopifolia

1 mm

top view   side view

fruit

1 mm

fruit and calyx
C. album

1 mm

flower

Chenopodium album

2 cm

C. atrovirens

1 cm

2 mm

C. botrys

fruit and calyx

1 mm

fruit
(top view)

1 mm

fruit and calyx

1 cm

Chenopodium botrys

C. berlandieri

0.5 mm

fruit and calyx

1 mm

flower

2 cm

2 cm

Chenopodium californicum   C. chenopodioides

2 cm

2 mm

1 mm

fruit and calyx

0.5 mm

fruit and calyx

Chenopodium desiccatum   C. foliosum

1 cm

0.5 mm

fruit and calyx
(top view)

Chenopodium fremontii

1 cm

1 mm

fruit and calyx
(top view)

C. hians

1 cm

0.5 mm

seed
(top view)

1 mm

0.5 mm

seed
top view

0.5 mm

seed
side view

fruit

0.5 mm

fruit and calyx

Chenopodium incanum var. occidentale

HALOGETON

Ann, glabrous to papillate. **ST** gen branched at base, prostrate to erect. **LVS** ± cylindric, fleshy, abruptly pointed to bristle- or spine-tipped. **INFL** axillary; fls densely clustered; bractlets 0–2. **FL** bisexual or pistillate; calyx lobes 5, gen enclosing fr, tip winged in fr; stamens 2–5; stigmas 2. **FR**: wall adherent to seed. **SEED** vertical or horizontal. 3 spp.: Eurasia. (Greek: salty neighbor, from habitat)

H. glomeratus (M. Bieb.) C. Meyer (p. 273) **STS** 6–25 cm, curved, leafy throughout. **LF** 4–22 mm, 1–1.5 mm wide, sessile, withered or deciduous in fr; bristle 1–2 mm, stiff. **INFL**: bracts 1.5–2 mm, ± glaucous; fls many, throughout st. **FL**: calyx lobes 1–2 mm, wings 2–3.5 mm, fan-like, membranous, veiny. **FR** 1–2 mm. Alkaline soils, open flats, shrubland; 800–1800 m. CaR, **GB**, **DMoj**; to ID, Colorado, NV; native to Eurasia. NOXIOUS WEED; TOXIC to livestock from concentrated oxalates. Summer

KOCHIA

Ann to subshrub, gen erect, gen short-hairy. **LVS** alternate or opposite, linear to lanceolate, flat to cylindric, fleshy or not. **INFL**: spike or clusters; bracts lf-like; fls 1–few per axil. **FL** gen bisexual; calyx lobes 5, incurved, keeled and tubercled to winged in fr; stamens 5; stigmas 2–3. **FR** ± compressed-spheric. **SEED** horizontal. ± 20 spp.: w N.Am, Eurasia. (Wilhelm D. Koch, German physician & botanist, 1771–1849)

1. Ann; lower cauline lvs short-petioled, gen 3–5-veined below middle . . . . *K. scoparia*
1' Per or subshrub; lvs sessile, vein 1 or obscure
  2. Sts many from base, gen simple, glabrous to finely white-tomentose; lvs gen overlapping . . . . . . *K. americana*
  2' Sts 1–few from base, gen branched throughout, gray- to brown-puberulent; lvs not overlapping . . . *K. californica*

K. americana S. Watson Subshrub 8–40 cm. **STS** many from base, gen simple, ascending to erect, gen finely white-tomentose, sometimes becoming glabrous. **LF** 5–20 mm, 1–2 mm wide, gen overlapping, ± cylindric, ± fleshy, glabrous to sparsely silky-hairy; vein obscure or 1. **INFL**: fls 1–3 per axil. **FL**: calyx lobes gen white-tomentose, wings in fr < 2 mm, < 4 mm wide. Alkaline soils, flats, dry lake margins; 600–2100 m. **GB**, **DMoj**; to OR, ID, NM. May–Aug

K. californica S. Watson (p. 273) Per or subshrub 20–60 cm. **STS** 1–few from base, erect, branched throughout, densely gray- to brown-puberulent. **LF** 3–12 mm, 1.5–3 mm wide, gen well spaced, gen flat, ± fleshy, silky-hairy; vein obscure or 1. **INFL**: fls 1–5 per axil. **FL**: calyx lobes densely short-hairy, wings in fr ± 1–2 mm, < 3 mm wide. Alkaline soils, flats; < 1000 m. s SnJV, **DMoj**; s NV. May–Sep

K. scoparia (L.) Schrader Ann 30–120 cm. **ST** simple to much-branched, glabrous to silky-hairy. **LF** 8–50 mm, 1–3.5 mm wide, flat, glabrous to short-soft-hairy, 3–5-veined below middle. **INFL**: fls gen 3–7 per axil. **FL**: calyx lobes short-hairy, with tubercles or wings < 1 mm, < 1 mm wide in fr. Disturbed places, fields, roadsides; < 1500 m. GV, n SnFrB, SCo, **GB**, **DMoj**, DSon, expected elsewhere; to e US, Eur; native to Asia. [vars. *culta* Farw., *subvillosa* Moq.] Aug–Oct

KRASCHENINNIKOVIA

Per or shrub, monoecious or dioecious, gen erect, stellate or long-hairy. **LF** linear to lanceolate, flat, entire. **INFL** spike-like, terminal; staminate fls above pistillate fls, bracts ± lf-like; pistillate fls few, clustered, subtended by 2 bractlets ± fused at base. **STAMINATE FL**: calyx lobes 4; stamens 4. **PISTILLATE FL**: calyx lobes 0; stigmas 2. **FR**: wall free or adherent to seed. **SEED** vertical. ± 8 spp.: n Medit, temp Asia, w N.Am. (Stephan P. Krascheninnikov, Russian botanist, 1713–1755)

K. lanata (Pursh) A.D.J. Meeuse & Smit (p. 273) WINTER FAT Shrub gen 5–10 dm, gen monoecious; hairs white, becoming ± rust-colored. **LF** 6–30 mm, 1.5–5 mm wide; margins inrolled. **INFL** 3–19 cm; staminate fls many; pistillate fls 1–4 in lower axils; bractlets densely hairy, 4–6 mm in fr. **STAMINATE FL**: calyx lobes 1–2 mm, densely hairy; stamens exserted. **PISTIL-** **LATE FL**: stigmas exserted. **FR** ± 2 mm, white-hairy. Rocky to clay soils, flats, gentle slopes; 100–2700 m. SNH (e slope), Teh, s SnJV, s SCoRI, WTR (n slope), **GB**, **DMoj**; to WA, n-c US, NM, n Mex. [*Ceratoides l.* (Pursh) J. Howell, *Eurotia l.* (Pursh) Moq., both invalid] ❀ DRN, SUN:1,**2,3,7–12**,13,**14,18–24**. Mar–Jun

MONOLEPIS POVERTY WEED

Ann, gen glabrous. **LVS** alternate, gen reduced upward. **INFL**: clusters gen axillary, 1–15+-fld; bracts lf-like. **FL** bisexual or pistillate; sepals 1–3; stamens 0–1; style branches 2. **FR**: wall pitted to tubercled, sometimes adherent to seed. **SEED** gen vertical. 3 spp.: w N.Am. (Greek: 1 scale, from sepal number in most spp.)

1. Sts gen much-branched; fls 1–3 per cluster, axillary and terminal; lf blade elliptic to ± oblong . . . . . . *M. pusilla*
1' Sts 2–many from base, not branched above; fls 4–15+ per cluster, axillary; lf blade (ob)lanceolate
  2. Lf blade 2-toothed to hastate; fr 1.5–2 mm, wall minutely pitted, adherent to seed . . . . *M. nuttalliana*
  2' Lf blades entire; fr ± 0.5 mm, wall minutely papillate, free from seed . . . . *M. spathulata*

M. nuttalliana (Schultes) E. Greene (p. 273) Pl 4–40 cm. **STS** 2–many from base, ascending to erect, fleshy. **LF** 10–45 mm, lanceolate, fleshy, 2-toothed to hastate. **INFL**: fls gen 5–15+ per cluster. **FL**: sepal oblanceolate to obovate. **FR** 1.5–2 mm; wall minutely pitted, adherent to seed. **SEED** dark brown. Open, disturbed, often wet places; < 3500 m. **CA** (exc NW); to c N.Am, n Mex. Often confused with *Chenopodium.*   Apr–Sep

M. pusilla S. Watson (p. 273)   Pl 4–14 cm. **STS** 1–5 from base, simple to ± intricately branched. **LF** 3–6 mm, elliptic to ± oblong. **INFL**: fls 1–3 per cluster. **FL**: sepals 1–3, oblanceolate. **FR** 0.5–1 mm; wall minutely tubercled, ± free from seed. **SEED** dark brown. Open places, shrubland; 1500–2400 m. n SNH (e slope), MP, **n SNE**; to ID, Colorado.   May–Aug

M. spathulata A. Gray (p. 273)   Pl 2–25 cm. **STS** 3–11 from base, decumbent to erect, fleshy. **LF** 3–120 mm, narrowly oblanceolate, fleshy, entire. **INFL**: fls 4–15+ per cluster. **FL**: sepal oblanceolate. **FR** ± 0.5 mm; wall minutely papillate, ± free from seed. **SEED** red-brown. Open places, shrubland, coniferous forest; 1400–2600 m. SNH, SnBr, **SNE**; to ID, NV, Baja CA. Jun–Sep

NITROPHILA

Per, glabrous, rhizomed. **ST** decumbent to erect; branches paired. **LVS** opposite, linear to ovate, fleshy, sessile to clasping. **INFL** axillary; bracts gen 2, unequal; fls 1–3 per cluster. **FL**: sepals 5(–7), enclosing fr, papery; sides overlapping; back ribbed; stamens 5, incl; stigmas 2, persistent in fr. **FR** ± 2 mm. **SEED** vertical, black. ± 8 spp.: temp Am. (Greek: soda loving)

1. Lf 3–4.5 mm, ± ovate; pl 3–10 cm . . . . *N. mohavensis*
1' Lf 5–16 mm, linear to oblong; pl 7–30 cm . . . . *N. occidentalis*

N. mohavensis Munz & Roos (p. 273)   AMARGOSA NITROPHILA Pl 3–10 cm. **ST** gen erect; internodes < lvs. **LF** 3–4.5 mm, ± ovate. **FL**: sepals 1.5–2 mm, pink, becoming white. RARE. Alkaline flats; 600–750 m. **n DMoj (Amargosa Desert)**; w NV. May–Jul

N. occidentalis (Nutt.) Moq. (p. 273)   Pl 7–30 cm. **ST** decumbent to erect; internodes = or > lvs. **LF** 5–16 mm, linear to oblong. **FL**: sepals 1–2 mm, white or pink, becoming white. Moist, alkaline soils; < 1500 m. CaRF, GV, SCo, w PR, **SNE**, DMoj; to e OR, UT, n Mex. May–Oct

SALICORNIA   PICKLEWEED

Ann to subshrub, glabrous. **ST** gen many-branched, jointed; internodes green to glaucous, fleshy. **LVS** opposite, sessile, ± decurrent; lf pairs gen fused at base, clasping or together ring-like. **INFL**: spike, terminal, cylindric, dense; bracts scale-like; fls gen 3 per axil, sessile to sunken in axis. **FL**: calyx bladder-like, slitted, ± deciduous in fr; stamens gen 2; stigmas 2–3. **FR**: wall free from seed. **SEED** vertical. ± 13 spp.: ± worldwide. (Greek: salt horn) [For alternate taxonomy see Scott 1977 J Linn Soc, Bot 75:357-374] Needs further study.

1. Spike 2–3 mm wide, distal internodes lacking fls . . . . *S. subterminalis*
1' Spike 3–5 mm wide, fld to tip . . . . *S. utahensis*

S. subterminalis Parish (p. 273)   Per 7–30 cm. **ST** spreading to erect; branches ascending; internodes 6–18 mm, 2–3 mm wide. **INFL** 10–40 mm, 2–3 mm wide; distal 5–14 nodes lacking fls; lower bracts gen obtuse to rounded; fls at same level. **SEED** ± 1 mm, glabrous. Salt marshes, alkaline flats; < 800 m. SnJV, CCo, SnFrB, SCo, ChI, **w DMoj, DSon**; n Mex.   ❀ STBL. Apr–Sep

S. utahensis Tidestrom   Per or subshrub 14–28 cm. **ST**: axis ascending to erect; branches ascending; internodes 10–40 mm, 3–6 mm wide. **INFL** 10–38 mm, 3–5 mm wide; lower bracts acute to obtuse; fls at same level. **SEED** puberulent. Alkaline soils; < 100 m. **ne DMoj (Death Valley)**; also UT. Differences between CA and UT pls need study.   ❀ STBL.

SALSOLA

Ann to subshrub. **ST** simple to much-branched. **LVS** gen reduced upward, thread-like to subcylindric, gen becoming thick, ridged, spine-tipped. **INFL** axillary; bracts 1–2; fls gen 1 per axil. **FL** bisexual; sepals 4–5, in fr thickened, persistent, gen tubercled to winged; stamens gen 5, exserted, style branches gen 2, exserted. **FR** spheric to obovoid; top ± depressed. **SEED** gen horizontal. ± 100 spp.: ± worldwide. (Latin: salty, from habitats)

1. Sepal wings 2.5–4.5 mm in fr; lvs yellow-green; branchlets often minutely papillate . . . . *S. paulsenii*
1' Sepal wings 0.5–2.5 mm in fr; lvs ± green; branchlets often short-stiff-hairy . . . . *S. tragus*

S. paulsenii Litv. (p. 273)   Ann gen < 50 cm, glabrous to minutely papillate. **ST** much-branched. **LF** 5–32 mm, thread-like, becoming yellow-green, thick; base becoming wide, leathery; tip spiny; margin at base white to translucent. **INFL**: bract subcylindric, spiny. **FL**: calyx 2.5–3.5 mm, wings in fr gen 2.5–4.5 mm, entire to irregularly, minutely toothed. Common. Disturbed places; 700–1800 m. n WTR, **SNE, DMoj**; to UT; native to se Eur, c Asia. May hybridize with *S. tragus*; needs further study.

S. tragus L. (p. 273)   RUSSIAN THISTLE, TUMBLEWEED   Ann < 1 m, gen rounded, glabrous to short-stiff-hairy. **ST** gen much-branched, in fr deciduous at base. **LF** 8–52 mm, thread-like, becoming rigid; base becoming wide, leathery, tip sharp-pointed to spiny, margin at base translucent. **INFL**: bract subcylindric, spiny. **FL**: calyx 2.5–3 mm, wings in fr 0.5–2.5 mm, gen minutely toothed to crenate. Common. Disturbed places; < 2700 m. **CA**; to e N.Am, Mex; native to Eurasia. [*S. australis* R. Br., *S. iberica* Sennen & Pau, *S. kali* L. var. *tenuifolia* Tausch., all misapplied] NOXIOUS WEED. Jul–Oct

SARCOBATUS   GREASEWOOD

1 sp. (Greek: fleshy bramble)   Recently treated in Sarcobataceae [Behnke 1997 Taxon 46:495–507; Angiosperm Phylogeny Group 1998 Ann Missouri Bot Gard 85:531–553]

S. vermiculatus (Hook.) Torrey (p. 277) Shrub 5–30 dm, often rounded, gen monoecious. **STS** yellowish to light gray; twigs short, spreading, rigid, often spine-tipped. **LVS** opposite below, alternate above, deciduous; blades 5–28 mm, ± linear, subcylindric, entire, fleshy, glabrous to hairy. **INFL**: spike, 5–19 mm, cylindric, dense, catkin-like; staminate fls gen many, spirally arranged, bracts peltate; pistillate fls 1–3, below staminate fls, bract lf-like. **STAMINATE FL**: perianth 0; stamens 2–3. **PISTILLATE FL**: perianth cup-like, in fr winged, 4–8 mm wide, persistent; ovary half-inferior. **FR** 4–5 mm, conic above wing, ± glabrous. Alkaline soils, dry lakes, washes, shrubland; 100–2100 m. **GB**, **DMoj**; to w Can, Great Plain, n Mex. [var. *baileyi* (Cov.) Jepson]. ❀ STBL. May–Aug

SUAEDA SEA-BLITE, SEEPWEED

Wayne R. Ferren, Jr.

Ann, per, shrubs, glabrous to hairy. **LVS** gen alternate; blade entire, sometimes cylindric or upper surface flat, fleshy, gen glaucous, tip acute or pointed. **INFL**: cyme; clusters sessile, gen arrayed in compound spikes; bracts lf-like or reduced; bractlets subtending fls 1–3, minute, membranous; fls 1–12. **FL** gen bisexual; calyx radial or bilateral, lobes 5, rounded, hooded, keeled, horned, or wing-margined; ovary ± lenticular, rounded, conic or with a neck-like extension, stigmas 2–3(5). **FR**: utricle, enclosed in calyx. **SEED** horizontal or vertical, lenticular or flat, of 2 kinds in some spp. 115 spp.: worldwide, saline and alkaline soils. (Arabic: suwayda, black; Bedouin name for *Suaeda asphaltica*) [Ferren & Whitmore 1983 Madroño 30:181–190] ❀ STBL.

1. Ann; sts 1–several, prostrate to erect; fresh lf ×-section ± uniformly green at 10×; calyx bilateral
   (1 lobe larger), lobes horned, ± keeled, ± wing-margined; stigmas 2, glabrous; seeds flat and dull
   . . . . *S. calceoliformis*
1' Subshrub or shrub; sts gen several from woody base; fresh lf ×-section ringed with dark green at 10×;
   calyx ± radial, lobes gen rounded, not horned or wing-margined; stigmas 3, hairy-papillate; seeds
   lenticular and shiny . . . . *S. moquinii*

S. calceoliformis (Hook.) Moquin (p. 277) HORNED SEA-BLITE Ann < 8 dm, glabrous, glaucous. **STS** prostrate to erect, 1–several; branches ascending or spreading, gen striped. **LVS** often tightly ascending, < 40 mm, linear, sessile; upper surface flat, green or reddish. **INFL** gen dense, branched; fls 3–5 per cluster; bracts subtending branches = lvs; bracts subtending fls < lvs, wider at base. **FL** bilateral, 1–4 mm incl horns; calyx lobes horned and ± keeled, ± wing-margined; ovary rounded to lenticular, stigmas gen 2, glabrous. **SEED** horizontal; lenticular form 0.8–1.7 mm, shiny, gen black; flat form 1–1.5 mm, dull, brown. Dry, saline or alkaline wetland soils; < 2200 m. **CA**; to AK, e Can, TX. [*S. depressa* (Pursh) S. Watson var. *erecta* S. Watson; *S. d.* var. *d.* misapplied] Yellow-green pls (gen GB) with lf-base ± tapered, infl gen open and slender, branches spreading, calyx lobes wing-margined, have been called *S. occidentalis* S. Watson. Jul–Oct

S. moquinii (Torrey) E. Greene (p. 277) BUSH SEEPWEED Subshrub, shrub 2–15 dm, glabrous or hairy, glaucous. **STS** spreading or erect, several, base gen woody; ann sts shiny, yellow-brown; branches spreading. **LVS** ascending to widely spreading, gen not overlapping; petiole 1 mm; blade 10–30 mm, subcylindric to flat, linear to narrowly lanceolate, base narrow, yellow-green or red. **INFL** gen open; clusters confined to upper sts; branches thin, 0.7–2 mm diam; fls 1–12 per cluster; bracts gen < lvs. **FL** bisexual or lateral pistillate, radial, 0.7–2 mm; calyx lobes rounded; ovary ± pear-shaped, stigmas 3, hairy-papillate. **SEED** horizontal or vertical, 0.5–1 mm, biconvex, shiny, black. Interior and desert (rarely coastal), alkaline and saline places; < 1600 m. GV, SnFrB, SW, **GB**, **D**; to w Can, TX, Mex. [*S. torreyana* S. Watson incl var. *ramosissima* (Standley) Munz; *S. fruticosa* (L.) Forsskal misapplied] May–Sep

CONVOLVULACEAE MORNING-GLORY FAMILY

Lauramay T. Dempster (except *Calystegia*)

Per (ann), gen twining or trailing. **LVS** alternate. **INFL**: cyme or fls solitary in axils; pedicels often with 2 bracts. **FL** bisexual, radial; sepals 5, ± free, overlapping, persistent, often unequal; corolla gen showy, gen bell-shaped, ± shallowly 5-lobed, gen pleated and twisted in bud; stamens 5, epipetalous; pistil 1, ovary superior, chambers gen 2, ovules gen 2 per chamber, styles 1–2. **FR**: gen capsule. **SEEDS** 1–4(6). 50 genera, 1,000 spp.: warm temp to trop; some cult as orn. Recently treated to include Cuscutaceae [Angiosperm Phylogeny Group 1998 Ann Missouri Bot Gard 85:531-553] Family description by L.T. Dempster, key to genera adapted by Margriet Wetherwax.

1. Lf < 1 cm, ± sessile, ± elliptic; st not twining . . . . **CRESSA**
1' Lf gen > 1 cm, gen petioled, blade gen reniform, hastate, or deeply lobed (sometimes oblanceolate);
   st often twining
   2. Stigma 1, head-like . . . . **IPOMOEA**
   2' Stigmas 2, ± linear
      3. Calyx 7 mm; corolla gen > 3 cm; stgma lobes cylindric or oblong, ± flattened . . . . **CALYSTEGIA**
      3' Calyx < 5 mm; corolla < 3 cm; stigma lobes cylindric or thread-like, not flattened . . . . . **CONVOLVULUS**

CALYSTEGIA MORNING-GLORY

Richard K. Brummitt

Per, subshrub from caudex or rhizome, glabrous to tomentose. **ST** very short to high-climbing, gen twisting and twining. **LF** gen > 1 cm, linear to reniform, often sagittate to hastate, rarely deeply divided. **INFL**: peduncle gen

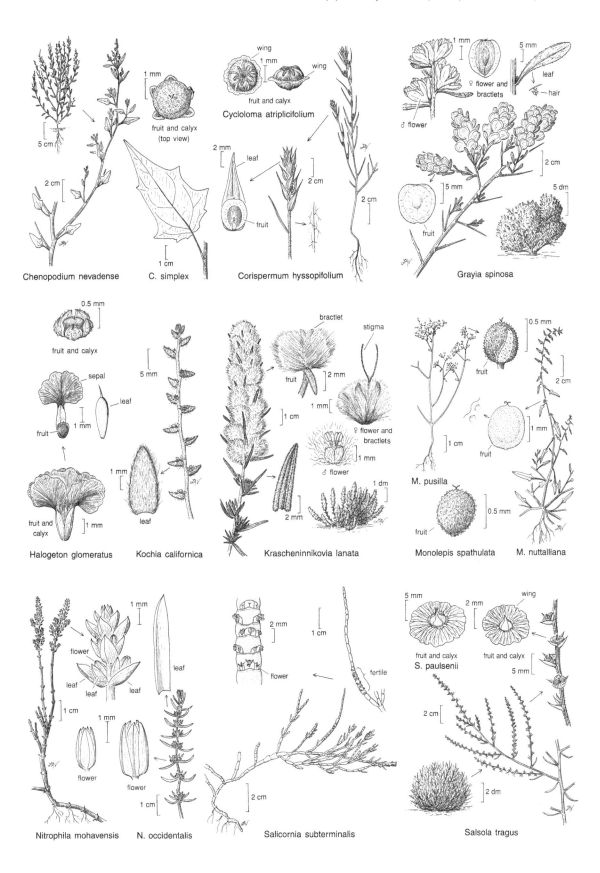

Chenopodium nevadense

C. simplex

Cycloloma atriplicifolium

Corispermum hyssopifolium

Grayia spinosa

Halogeton glomeratus

Kochia californica

Krascheninnikovia lanata

M. pusilla

Monolepis spathulata

M. nuttalliana

Nitrophila mohavensis

N. occidentalis

Salicornia subterminalis

Salsola tragus

S. paulsenii

1-fld; bractlets small and remote from calyx to large and concealing calyx, sometimes lobed. **FL** gen showy; corolla glabrous, white or yellow to pink or purple; ovary chamber 1 (septa gen incomplete), stigma lobes 2, gen swollen, cylindric or oblong, ± flattened. **FR** ± spheric, ± inflated. **SEEDS** gen ± 4. ± 25 spp.: temp, worldwide. (Greek: concealing calyx, from bractlets of some) [Brummitt 1980 Kew Bull 35(2):327–328] Intergradation common; intermediate forms often difficult to identify. Appears similar to *Convolvulus*, but anatomy suggests that the 2 genera are not very closely related.

1. St densely intertwined; subshrub; bractlets linear, attached 4–50 mm below calyx . . . . **C. longipes**
1' St twining; vine; bractlets ovate to oblong, attached 0–2 mm below calyx
  2. St < 40 cm, weakly climbing; bractlets 2.5–4 mm wide . . . . **C. peirsonii**
  2' St gen > 1 m, strongly climbing; bractlets > 6 mm wide . . . . **C. sepium** ssp. **limnophila**

C. longipes (S. Watson) Brummitt (p. 277) Subshrub from woody caudex, ± hemispheric, glabrous. **ST** stiffly erect or intertwining, 3–10 dm. **LF** < 6 cm, linear to narrowly triangular; lobes linear; sinus rounded; upper lvs less lobed. **INFL**: peduncle gen 1-fld, < 20 cm, >> subtending lf; bractlets 3–17 mm, 0.2–3 mm wide, linear, often with basal lobes, gen alternate, attached 4–50 mm below calyx. **FL**: sepals 8–11 mm; corolla 28–36 mm, white or cream to pale pink or lavender. Dry, rocky places, desert scrub; 600–1300 m. s CA-FP, **D**; NV, AZ. [*Convolvulus l.* S. Watson, *C. linearilobus* Eastw.] Intergrades with *C. peirsonii* in Los Angeles Co. ❀ DRY:7,8, **9**,10–12,**14,18–23**&SUN:**15–17,24**. May–Jul

C. peirsonii (Abrams) Brummitt (p. 277) PEIRSON'S MORNING-GLORY Per from rhizome, glabrous, ± glaucous. **ST** decumbent to weakly climbing, < 0.4 m. **LF**: blade < 2 cm at midrib, narrowly triangular; lobes sometimes = blade, broadest at tip, gen distinctly 2-tipped; sinus rounded to ± square. **INFL**: peduncle 2–8 cm, gen < subtending lf; bractlets attached 0–3 mm below calyx and overlapping it, 3–7 mm, 2.5–4 mm wide, elliptic to widely oblong, entire, flat. **FL**: calyx 9–13 mm; corolla 25–40 mm, white. UNCOMMON. Rocky slopes; 1000–1500 m. n SnGb, **adjacent DMoj (Antelope Valley)**. [*Convolvulus p.* Abrams] Intergrades with *C. longipes*. ❀ TRY. May–Jun

C. sepium (L.) R.Br. HEDGE BINDWEED Per from rhizome, glabrous to hairy. **ST** climbing, < 4 m. **LF**: blade gen 4–8 cm at midrib, lobed. **INFL**: peduncle < subtending lf; bractlets attached just below calyx and ± concealing it, entire, flat or keeled. **FL**: sepals gen 10–18 mm; corolla 30–70 mm, white or pink. 2n=20,22,24. Salt and freshwater marshes; < 500 m. Deltaic GV, SnFrB, SCoR, SCo, TR, **e DMoj**; temp regions worldwide. Highly variable, many geographic sspp. 2 sspp. in CA.

  ssp. **limnophila** (E. Greene) Brummitt (p. 277) Pl gen glabrous (densely hairy). **LF**: lobes ± abruptly spreading; sinus rounded to square; tip acute. **INFL**: bractlets 13–28 mm (= or > sepals), 8–18 mm wide, narrowly ovate, flat or keeled. **FL**: corolla 35–70 mm, white or pink-tinged. Marshes; < 500 m. Deltaic GV, SnFrB, SCoR, TR, **e DMoj (Amargosa River, 500 m)**; to e US, n Mex. [*Convolvulus l.* E. Greene; *C. s.* var. *repens* (L.) A. Gray] Intergrades with *C. peirsonii*. ❀ TRY.

CONVOLVULUS MORNING-GLORY, BINDWEED

Ann, per from caudex or rhizome, gen ± glabrous. **ST** gen trailing to high-climbing, gen twisting and twining. **LF** gen > 1 cm, gen petioled; blade gen cordate or hastate. **INFL**: bracts gen 2, below calyx. **FL** gen showy; corolla gen funnel-shaped, pleated, 5-angled or -lobed; stamens incl; ovary chambers 2, septa complete, stigma lobes 2, linear or thread-like, not flattened. **FR** spheric, ± inflated. **SEEDS** gen 4. (Latin, entwine) 250 spp.: gen temp. Not easily distinguished from *Calystegia*.

C. arvensis L. (p. 277) BINDWEED, ORCHARD MORNING-GLORY Per from deep persistent root, ± glabrous to minutely hairy. **ST** prostrate or twining. **LF** gen 2–3 cm; blade round, oblong or ovate, base hastate, tip gen rounded. **INFL**: peduncles gen 2.5–6 cm, 1–several-fld, in fr gen reflexed; bracts ± linear. **FL**: calyx ± 5 mm, lobes oblong; corolla 2–2.5 cm, open funnel-shaped, white, purplish outside, particularly at folds, margin entire. **FR** ± 8 mm. n=24. Abundant. Orchards, gardens, disturbed places; gen < 1500 m. **CA**; native to Eur. NOXIOUS WEED. May–Oct

CRESSA ALKALI WEED

Per, subshrub, canescent. **ST** prostrate to erect, not twining. **LF** entire. **INFL**: fls solitary in upper axils, appearing like a 1-sided, leafy raceme. **FL**: calyx ± erect, concealing corolla tube; corolla lobes ± = tube, exserted; ovary chambers 2, ovules 4, styles 2, stigmas head-like. 5 spp.: trop & subtrop. (Greek: a Cretan woman)

C. truxillensis Kunth (p. 277) ALKALI WEED **ST** 7–25 cm, much branched from base, densely silky-canescent. **LF** gen < 1 cm, ± sessile, ± elliptic. **INFL**: peduncle short, bracted. **FL** 5–8 mm; sepals elliptic; corolla white, persistent, lobes ovate, acute; stamens and styles exserted. **SEED** often 1, by abortion of others. 2n=28. Saline and alkaline soils; < 1200 m. CA-FP, **D**; to OR, TX, Mex. [var. *vallicola* (A.A. Heller) Munz; not *C. cretica* L.] ❀ DRY,SUN,ALKALINE,SALINE:**7–11**,13,**14**,15–17,**18–24**;INV. May–Oct

IPOMOEA MORNING-GLORY

Ann, per from rhizome or caudex. **ST** trailing to high-climbing. **LF** petioled; blade cordate, sometimes lobed. **INFL**: bracts 0. **FL**: corolla gen ± funnel-shaped, not or barely lobed; style 1, stigma head-like or of 2–3 spheric lobes. **FR**

Plate 1 Woodsia scopulina

Plate 2 Cheilanthes parryi

Plate 3 Juniperus osteosperma

Plate 4 Ephedra viridis

Plate 5 Pinus longaeva

Plate 6 Pinus monophylla

Plate 7 Justicia californica

Plate 8 Amaranthus fimbriatus

Plate 9 Rhus trilobata

Plate 10 Lomatium mohavense

Plate 11 Asclepias erosa

Plate 12 Sarcostemma cynanchoides ssp. hartwegii

Plate 13 Adenophyllum cooperi

Plate 14 Atrichoseris platyphylla

Plate 15 Baileya pleniradiata

Plate 16 Cirsium mohavense

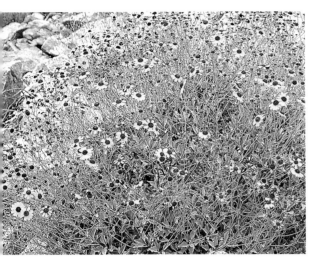

Plate 17 Encelia farinosa

Plate 18 Enceliopsis covillei

Plate 19 Glyptopleura marginata

Plate 20 Hymenoclea salsola var. salsola

Plate 21 Malacothrix glabrata

Plate 22 Monoptilon bellioides

Plate 23 Nicolletia occidentalis

Plate 24 Peucephyllum schottii

Plate 25 Townsendia scapigera

Plate 26 Trichoptilium incisum

Plate 27 Trixis californica var. californica

Plate 28 Xylorhiza tortifolia var. tortifolia

Plate 29 Chilopsis linearis ssp. arcuata

Plate 30 Cryptantha circumscissa

Plate 31 Tiquilia plicata

Plate 32 Caulanthus inflatus

Plate 33 Draba oligosperma var. oligosperma

Plate 34 Lesquerella kingii ssp. kingii

Plate 35 Stanleya elata

Plate 36 Echinocereus engelmannii

Plate 37 Ferocactus cylindraceus var. cylindraceus

Plate 38 Mammillaria dioica

Plate 39 Opuntia basilaris var. basilaris

Plate 40 Opuntia ramosissima

Plate 41 Nemacladus sigmoideus

Plate 42 Cleome lutea

Plate 43 Achyronychia cooperi

Plate 44 Silene sargentii

Plate 45 Stellaria longipes var. longipes

Plate 46 Atriplex hymenelytra

Plate 47 Grayia spinosa

Plate 48 Crossosoma bigelovii

Plate 49 Brandegea bigelovii

Plate 50 Chamaesyce albomarginata

Plate 51 Euphorbia incisa

Plate 52 Astragalus coccineus

Plate 53 Astragalus funereus

Plate 54 Lupinus arizonicus

Plate 55 Senna armata

Plate 56 Trifolium andersonii var. beatleyae

Plate 57 Fouquieria splendens
ssp. splendens

Plate 58 Gentiana newberryi var. tiogana

Plate 59 Ribes aureum var. aureum

Plate 60 Nama demissum var. demissum

Plate 61 Phacelia calthifolia

Plate 62 Krameria erecta

Plate 63 Hyptis emoryi

Plate 64 Salazaria mexicana

Plate 65 Salvia funerea

Plate 66 Pholisma arenarium

Plate 67 Mentzelia involucrata

Plate 68 Petalonyx thurberi ssp. thurberi

Plate 69 Eremalche rotundifolia

Plate 70 Sphaeralcea ambigua var. ambigua

Plate 71 Abronia villosa var. villosa
(with Oenothera deltoides ssp. deltoides)

Plate 72 Allionia incarnata

Plate 73 Forestiera pubescens

Plate 74 Camissonia brevipes ssp. brevipes

Plate 75 Orobanche cooperi

Plate 76 Arctomecon merriamii

Plate 77 Eschscholzia minutiflora

Plate 78 Ipomopsis congesta ssp. congesta

Plate 79 Langloisia setosissima ssp. punctata

Plate 80 Linanthus nuttallii ssp. pubescens

Plate 81 Loeseliastrum schottii

Plate 82 Centrostegia thurberi

Plate 83 Chorizanthe rigida

Plate 84 Eriogonum gracilipes

Plate 93 Coleogyne ramosissima

Plate 94 Petrophyton caespitosum ssp. caespitosum

Plate 95 Potentilla morefieldii

Plate 96 Prunus andersonii

Plate 97 Purshia mexicana var. stansburyana

Plate 98 Galium angustifolium ssp. borregoense

Plate 99 Thamnosma montana

Plate 100 Anemopsis californica

Plate 117 Fagonia laevis

Plate 118 Larrea tridentata

Plate 119 Washingtonia filifera

Plate 120 Carex douglasii

Plate 121 Agave utahensis

Plate 122 Allium atrorubens var. cristatum

Plate 123 Calochortus kennedyi var. kennedyi

Plate 124 Calochortus striatus

Plate 125 Hesperocallis undulata

Plate 126 Nolina parryi

Plate 127 Yucca schidigera

Plate 128 Achnatherum speciosum

spheric; valves 2–4. (Greek: worm-like) 500 spp.: trop and warm temp; some cult as orn or for food (*I. batatas*, sweet-potato).

I. triloba L. (p. 277) Ann. **LF** simple; blade 3–6 cm, cordate, ± acuminate, entire or 3-lobed, glabrous. **INFL**: peduncles 1–5-fld. **FL**: sepals ± 8 mm, ± narrowly ovate, acuminate; corolla 1.2–2 cm, narrowly bell-shaped, shallowly 5-lobed. $2n=30$. Fields, orchards, other disturbed places; –34–50 m. **DSon**; native to trop Am. NOXIOUS WEED.

CORNACEAE DOGWOOD FAMILY

James R. Shevock

Per, shrubs, trees, sometimes dioecious. **LVS** gen opposite, simple, gen entire, gen deciduous; stipules 0. **INFL**: cyme or racemes, gen umbel- or head-like, sometimes subtended by showy, petal-like bracts. **FL** gen small, gen bisexual; calyx gen 4-lobed; petals 0 or 4(5), free; stamens gen as many as and alternate petals; ovary inferior, chambers 1–4, each 1-ovuled, style simple, stigma 1–4-lobed. **FR**: gen drupe or berry. **SEEDS** gen 1–2. ± 12 genera, ± 100 spp.: esp n temp (also s trop, subtrop). Cult as orn (*Cornus, Aucuba*); some timber spp. Genera diverse; many have been treated as constituting families, but trend is to treat Cornaceae broadly. [Eyde 1987 Syst Bot 12:505–518]

CORNUS DOGWOOD

Per, shrubs, trees. **LVS** gen opposite or whorled, simple, gen deciduous; both ends gen tapered. **INFL** small, head- or umbel-like, and surrounded by showy bracts (or cyme, large, open, lacking showy bracts). **FL** gen minute; sepals 4, fused at base; petals 4; stamens 4, attached to receptacle; style 1, thread-like, stigma simple. **FR**: drupe; stone 1–2-chambered. ± 50 spp.: n temp (rare in s hemisphere); many cult as orn, some for autumn color. Some fr used for jam, syrup. (Latin: horn, from the hard wood) Divided by some into at least 6 genera.

C. sericea L. AMERICAN DOGWOOD Shrub gen 1.5–4 m. **ST**: branches reddish to purple, ± glabrous to minutely strigose; older sts grayish green, gen glabrous. **LVS** deciduous; blade gen 5–10 cm, lanceolate to ovate or elliptic, paler beneath, sparsely strigose, veins 4–7 pairs. **INFL**: cyme, strigose. **FL**: petals 2–4.5 mm; style 1–3 mm. **FR** 7–9 mm, white to cream; stone smooth to grooved on face, furrowed on sides. Many, gen moist habitats; < 2800 m. CA-FP, **W&I**; to AK, e N.Am, Mex. Highly variable complex with many local forms, treated broadly here. Sspp. intergrade widely. ❀ IRR: 1–3,**4–9**,10,**14–23**,24;rather INV. 2 sspp. in CA.

ssp. *sericea* (p. 277) Pl glabrous to ± strigose, not rough-hairy. **FL**: petals gen 2–3 mm; style 1–2 mm. **FR**: stone smooth on faces, furrowed on sides. Habitats of sp. CA-FP, **W&I** (uncommon in s CA). [*C. ×californica* C.A. Meyer, incl var. *nevadensis* Jepson; *C. stolonifera* Michaux] May–Jul

CRASSULACEAE STONECROP FAMILY

Ann, per, fleshy. **LVS** simple, basal and cauline, alternate or opposite, gen reduced upward. **INFL**: gen cyme, gen bracted. **FL**: sepals gen 3–5, gen ± free; petals gen 3–5, gen ± free; stamens = or 2 × sepals, free or epipetalous; pistils 3–5, simple (sometimes fused at base), ovary 1-chambered, placenta 1, parietal, ovules 1–many, style 1. **FR**: follicles 3–5. **SEEDS** 1–many, small. ± 30 genera, ± 1500 spp.: ± worldwide, esp dry temp; many cult for orn. Family description and generic key by Reid Moran.

1. Ann; lvs << 1 cm, opposite, pairs fused around stem; fls axillary, < 2 mm . . . . **CRASSULA**
1' Per; lvs > 1 cm, alternate, free; fls in axillary cymes, 2–16 (20) mm
  2. Lvs in basal rosette, entire; flowers bisexual, petals fused at base, > 1 cm . . . . **DUDLEYA**
  2' Lvs cauline, upper often dentate; flowers gen unisexual, petals free, < 1 cm . . . . **RHODIOLA**

CRASSULA

Reid Moran

Ann in CA, terrestrial, sometimes submersed and later stranded in dry ponds, glabrous in CA. **ST** decumbent to erect, branched or not. **LVS** basal and cauline, opposite; lf bases fused, ± sheathing. **INFL**: solitary in lf axils. **FL**: sepals 3–5, ± fused at base; petals free or ± fused at base; stamen number = sepal number; pistils 3–5. **FR**: follicles 3–5, gen erect. **SEEDS** 1–many, in CA gen < 0.5 mm, elliptic to elliptic-oblong, red-brown. ± 300 spp.: esp Afr (ann spp. ± worldwide). (Latin: diminutive of thick) [Moran 1992. Cact. Succ. J. 64:223-231]

C. connata (Ruiz & Pav.) A. Berger (p. 283) PYGMY-WEED **STS** erect, 2–6(10) cm, branched or not, in age red. **LF** 1–3(6) mm, ovate to oblong; tip obtuse, acute, or abruptly fine-pointed. **INFL**: fls (1)2 per lf pair, gen crowded; pedicel < 6 mm. **FL** 0.5–2 mm; sepals (3)4, 0.5–2 mm, lanceolate, tip acute to acuminate; petals gen < sepals, narrow-triangular. **FR**: follicles ascending, ovoid, tapered to styles. **SEEDS** 1–2, elliptic. $2n=16$. Open areas; < 1500 m. NW, SNF, GV, CW, SW, **DSon**; to OR, TX, n C.Am; also in w S.Am. [*Tillaea erecta* Hook. & Arn.] Locally abundant. Feb–May

DUDLEYA

Jim A. Bartel

Per, fleshy, glabrous. **ST** gen caudex or corm-like, gen vertical, branched or not, gen covered with dried lvs or lf bases. **LVS** in basal rosettes, evergreen or vernal and ephemeral. **INFL:** cyme, axillary, 1-sided; bracts lf-like, alternate. **FL:** sepals 5, fused below; petals 5, fused at base, erect to spreading above; stamens 10, epipetalous; pistils 5, ± fused below. **FR:** follicles 5, erect to spreading, many-seeded. **SEED** < 1 mm, narrowly ovoid, brown, striate. ±45 spp.: sw N.Am. (W.R. Dudley, western US botanist, 1849–1911) [Moran in Jacobsen 1960 Handb Succ Pl:344–359] Unopened follicles gen most reliable for posture (erect, spreading, etc).

1. Petals fused ± 1/2 length; pedicel in fr often sharply bent . . . . ***D. pulverulenta*** ssp. ***arizonica***
1' Petals fused < 1/3 length; pedicel in fr not sharply bent
  2. Lf 5–30 cm, 1–4 cm wide; peduncle 1.5–7.5 dm; petals fused 1-2 mm . . . . ***D. lanceolata***
  2' Lf 1–15 cm, 0.3–2.5 cm wide; peduncle 0.3–3.5(4) dm; petals fused 1–4 mm
    3. Petals pale yellow, darker yellow to red-tinged to red-lined on keel; pedicel 0.5–14 mm; w to sw edge DMoj
      4. Infl 1° branches 0 or gen 2–3, ascending, gen simple; pedicel gen 0.5–5 mm; petals red-lined on keel;
        sw edge DMoj . . . . ***D. abramsii*** ssp. ***affinis***
      4' Infl 1° branches 2–4, gen spreading, branched 0–3 ×; pedicel 2–14 mm; petals darker yellow to
        red-tinged on keel; s SNH, Teh, w edge DMoj . . . . ***D. calcicola***
    3' Petals greenish yellow to bright yellow, petals often red-tinged; pedicel 5–20 mm; DMtns . . . . ***D. saxosa***
      5. Caudex 1–3 cm wide; lf 4–15 cm; peduncle 1.0–3.5(4) dm; petals rarely red-tinged; DMtns . . . ssp. *aloides*
      5' Caudex 1–1.5 cm wide; lf 3–9 cm; peduncle 0.5–2.0 dm; petals gen red-tinged; n DMtns
        (Panamint Mtns) . . . . ssp. *saxosa*

D. abramsii Rose **ST:** caudex simple to gen branched, cespitose. **LF** 1.5–11 cm, 5–20 mm wide, oblong-lanceolate or tapered base to tip, rarely elliptic to oblanceolate, gen glaucous, upper surface gen ± flat; rosette 2–6 cm wide. **INFL:** 1° branches 0 or gen 2–3, gen simple, ascending; terminal branches 3–15 cm, 2–15-fld; lower bracts 4–40 mm; pedicel 0.5–5(11) mm. **FL:** sepals 2–5 mm, deltate; petals 8–13 mm, 1.5–3.5 mm wide, fused 1–4.5 mm, elliptic, acute, pale yellow, keel gen with fine, purple to red lines, margin often jagged. *n*=17. Rocky outcrops; 50–2600 m. s SCoRO, SW, **sw edge DMoj;** n Baja CA. 5 sspp. in CA.

ssp. *affinis* K. Nakai SAN BERNARDINO MOUNTAINS DUDLEYA **ST:** caudex 10–15 mm wide, gen simple. **LF** 2–4 cm, 7–15 mm wide, glaucous, oblanceolate to elliptic. **INFL:** peduncle 5–11 cm, 1–3 mm wide; lower bracts 5–6 mm. **FL:** petals fused 1.5–2.5 mm, keel gen red-lined. UNCOMMON. Outcrops, granitic or quartzite, rarely limestone; 1800–2600 m. SnBr, **sw edge DMoj.**

D. calcicola J. Bartel & J.R. Shevock LIMESTONE DUDLEYA **ST:** caudex 1–2 cm wide, cespitose; branches 0 to gen many. **LF** 1–10 cm, 3–13 mm wide, oblong-lanceolate or tapered base to tip, glaucous; tip acute to subacuminate; rosette 1–9 cm wide. **INFL:** 1° branches 2–4, gen spreading, branched 0–3 ×; peduncle 3–18 cm, 1.5–4 mm wide; terminal branches 1–6 cm, gen spreading, 2–8-fld; pedicel 2–14 mm. **FL:** sepals 2.5–6 mm, triangular-ovate to -lanceolate; petals 9–15 mm, 3.5–5 mm wide, fused 1–3 mm, lanceolate, narrowly acute, pale yellow, keel darker yellow to red-tinged. *n*=17. UNCOMMON. Open, rocky outcrops; 500–2600 m. s SNH, Teh, **w edge DMoj.** [*D. abramsii* spp. *c.* (J. Bartel & J.R. Shevock) K. Nakai] ❀ DRN,DRY:7,14–24;DFCLT.

D. lanceolata (Nutt.) Britton & Rose (p. 283) **ST:** caudex erect, 1–3 cm wide, branches 0 or few. **LF** 5–30 cm, 1–4 cm wide, 1.5–6 mm thick, oblong-lanceolate, glaucous or not; base 1–3 cm wide; tip acute to abruptly pointed. **INFL:** 1° branches 2–3, simple or forked 1 ×; peduncle 15–75 cm, 3–12 mm wide; terminal branches 2–25 cm, 2–20-fld; pedicel spreading, 2–12 mm, becoming erect. **FL:** sepals 3–6 mm, deltate-ovate; petals 10–16 mm, 3.5–5 mm wide, fused 1–2 mm, elliptic to oblanceolate, acute, yellow to gen red. *n*=34. Rocky slopes; 30–1250 m. SCoR, TR, PR, **DMtns;** n Baja CA. [*D. cymosa* ssp. *minor* (Rose) Moran] ❀ DRN,DRY:14, 15,**16,17,22–24**&IRR:**7,9,18–21**. Apr–Jul

D. pulverulenta (Nutt.) Britton & Rose Pl covered with dense, mealy powder to chalky wax. **ST:** caudex simple. **LF** oblong to oblong-obovate. **INFL:** 1° branches 3–10, simple or forked 1 ×; terminal branches twisted at base. **FL:** sepals acute to obtuse. *n*=17. Rocky places; < 1500 m. c&s CCo, SCoRO, SCo, TR, PR, **DMtns;** s NV, w AZ, nw Mex. 2 sspp. in CA.

ssp. *arizonica* (Rose) Moran **ST:** caudex 1–4 cm wide. **LVS** 15–25 per rosette, (3)5–15(17) cm, 1–5 cm wide, 2–4 mm thick, oblong to oblong-obovate; base 1–3.5 cm wide; tip gen long-acuminate. **INFL:** peduncle 1.5–6 dm, 2–6 mm wide; terminal branches ascending, 3–6, 4–27 cm, 3–6-fld; pedicel 5–15(20) mm, gen erect to ascending. **FL:** petals 9–15 mm, fused 4–8 mm, red or yellow, lobes 1.5–2 mm wide. Rocky slopes; 600–1500 m. **DMtns;** s NV, w AZ, nw Mex. [*D. a.* Rose] ❀ DRN:2,3,7, 10,11,14,18,**19–23**&DRY:15–17,**24**. May–Jul

D. saxosa (M.E. Jones) Britton & Rose **ST:** caudex short; branches 0–few. **LF** oblong-lanceolate, glaucous in youth; tip acute. **INFL** often reddish; 1° branches 2–3, simple or forked 1 ×; pedicel 5–20 mm. **FL:** sepals 4–8 mm, deltate, acute; petals 2.5–4 mm wide, oblong-lanceolate, acute. Rocky, gen n-facing slopes; 240–2200 m. PR, **DMtns;** c AZ, n Baja CA.

ssp. *aloides* (Rose) Moran (p. 283) **ST:** caudex 1–3 cm wide. **LF** 4–15 cm, 6–20 mm wide, 2–5 mm thick; base 10–25 mm wide. **INFL:** peduncle 1–3.5(4) dm, 4–9 mm wide; terminal branches wavy, 1–12 cm, 2–20-fld. **FL:** petals 8–15(20) mm, fused 1.5–3 mm, greenish yellow to yellow, rarely red-tinged. *n*=17. Rocky, shaded slopes; 240–1700 m. PR, **DMtns;** n Baja CA. [*D. alainae* Reiser, Banner dudleya] ❀ DRN,DRY:7,10–24;DFCLT. Apr–Jun

ssp. *saxosa* PANAMINT DUDLEYA **ST:** caudex 1–1.5 cm wide. **LF** 3–9 cm, 5–15 mm wide, 1.5–3 mm thick; base 5–15 mm wide. **INFL:** peduncle 0.5–2 dm, 2–4 mm wide; terminal branches not wavy, 1–4 cm, 2–9-fld. **FL:** petals 9–12 mm, fused 1–4 mm, bright yellow, gen red-tinged. *n*=68,85. UNCOMMON. N-facing, granitic or limestone slopes; 1100–2200 m. **n DMtns (w Panamint Mtns).** ❀ DRN,DRY:2,3,7,10,11,14–24;DFCLT. May–Jun

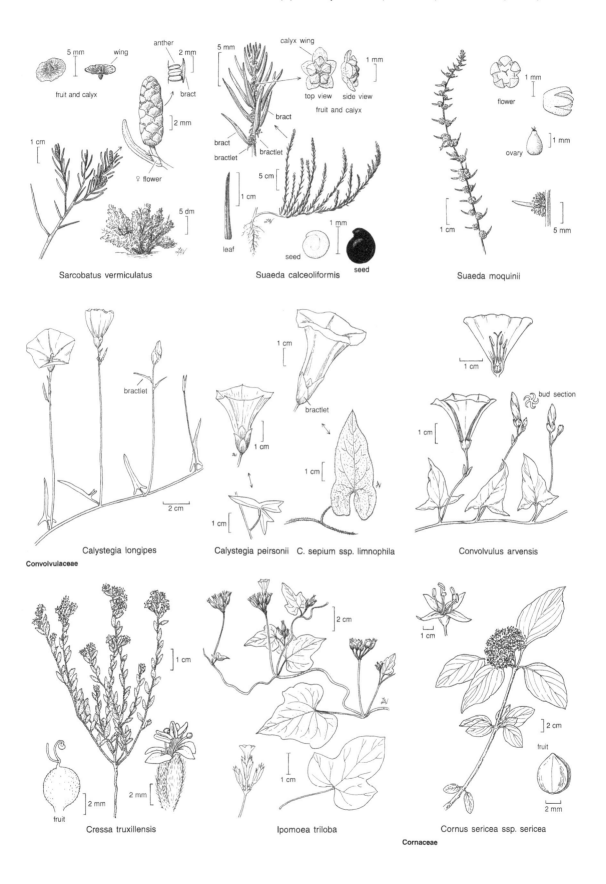

Sarcobatus vermiculatus

Suaeda calceoliformis

Suaeda moquinii

Calystegia longipes

Convolvulaceae

Calystegia peirsonii   C. sepium ssp. limnophila

Convolvulus arvensis

Cressa truxillensis

Ipomoea triloba

Cornus sericea ssp. sericea

Cornaceae

RHODIOLA

Reid Moran

Per from short, scaly caudex, glabrous, dioecious or not. **LVS** cauline, sessile, alternate, entire to toothed. **INFL** branched cyme. **FL**: sepals 4–5, fused below; petals 4–5, ± free; stamens 8 or 10, epipetalous; pistils 4–5, free or fused below, **FR** erect. **SEEDS** many, < 3 mm, brown, striate. 36 spp.: N temp. (Greek: rhodon, a rose, referring to the rose-scented roots)

R. integrifolia Raf. (p. 283) WESTERN ROSEROOT Pl 2–30 cm; caudex short, thick, fleshy, branching. **LF** 7–25 mm, oblanceolate to obovate, widest 3.5–5 mm below tip, flat, obtuse, sometimes dentate above middle. **INFL** ± 1–3 cm, 7–50-fld, dense. **FLS** 4(5)-parted, most or all unisexual; petals free, slightly spreading, 2–3.5 mm, oblong, fleshy, obtuse to acute, deep purple; anthers light brown to red-purple. **FR** 3–6 mm, erect. **SEED** 1.5–2 mm. *n*= 18. Cliffs, talus, alpine ridges; 1800–4000 m. KR, c SNF, SNH, Wrn, **W&I**; to AK, ne N.Am, Colorado, Eurasia. [*Sedum roseum* (L.) Scop. ssp. *integrifolium* (Raf.) Hultén] ❀ DRN,IRR:1,2, 4–7,15–17;DFCLT. May–July

CROSSOSOMATACEAE CROSSOSOMA FAMILY

James R. Shevock

Shrubs. **ST** gen glabrous; twigs or branchlets gen thorny. **LVS** gen deciduous, simple, small, gen alternate, entire; stipules minute or 0. **INFL**: fls solitary. **FL** gen bisexual, radial; hypanthium short, mostly forming thick nectary-disk; sepals gen 5(3–6), free; petals ephemeral, gen 5(3–6), free, gen white; stamens 4–50, attached to nectary-disk; pistils 1–9, simple, styles short, stigma head-like, ovules gen 2–many. **FR**: follicles 1–9. **SEEDS** brown to black, with an aril. 3 genera, 8 spp.: w US, Mex (*Apacheria* of AZ, NM has 1 sp.). [Thorne & Scogin 1978 Aliso 9:171–178]

1. Lf gen elliptic; stamens 15–50; fr 1–9, 8–20 mm, stalked . . . . **CROSSOSOMA**
1' Lf gen oblanceolate; stamens 4–10, fr 1–3,1–5 mm, sessile . . . . **GLOSSOPETALON**

CROSSOSOMA

Shrubs. **LVS** deciduous (or dry during dormant periods), gen ± narrowly elliptic; petiole 0–short. **INFL**: peduncle sometimes bracted. **FL**: petals 9–15 mm, rounded or oblong, gen white; stamens 15–50 in several series. **FR**: follicles 8–20 mm, cylindric. **SEEDS** gen > 2 per follicle, black, shiny, round to flat; aril conspicuous, fringed, yellowish. 2 spp.: s CA, e to AZ, s to Mex. (Greek: fringe body, from aril)

C. bigelovii S. Watson (p. 283, pl. 48) Shrub 1–2 m. **ST** much-branched; branchlets thorny. **LVS** clustered, 5–15 mm, gray-green. **FL**: sepals 4–5 mm, rounded; petals 9–12 mm, oblong, white to purplish, distinctly clawed. **FR**: follicles 1–3, 8–10 mm, ± straight. **SEEDS** gen 2–5 per follicle, ± 2 mm diam. Dry, rocky slopes, canyons; < 1250 m. e SnBr, **D**; NV, AZ, Baja CA. ❀ DRN:14,18–21&SHD:13; DFCLT. Feb–Apr

GLOSSOPETALON

Small shrubs, gen densely branched. **ST** greenish, glabrous to sparsely hairy, angled; branchlets thorny. **LVS** small, gen deciduous, gen oblanceolate to obovate, ± entire. **FL**: petals narrowly oblanceolate, white; stamens 4–10. **FR**: follicles 1–3, ovoid, sessile, gen striate, gen beaked. **SEEDS** gen 1–2 per follicle, gen brown; aril ± inconspicuous, gen whitish. ± 5 spp.: w US, n Mex, esp limestone in desert mtns. (Greek: tongue petal, from petal shape) [Holmgren 1988 Brittonia 40:269–274] Formerly called *Forsellesia* and assigned to Celastraceae.

1. Lf spine-tipped; stipules 0; fls terminal; fr < 1 mm . . . . ***G. pungens***
1' Lf tip rounded or short-pointed; stipules minute; fls axillary; fr 3–5.5 mm . . . . ***G. spinescens***

G. pungens Brandegee PUNGENT FORSELLESIA Shrub 5–20 cm, low, matted. **ST** not thorny. **LF** 6–10 mm, oblanceolate or narrowly elliptic, sharply spine-tipped; veins prominent below; stipules 0. **INFL**: fls terminal. **FL**: sepals acuminate, 2–3 spine-tipped; petals 6–8 mm; stamens 10 in 2 series, longer opposite sepals. **FR**: follicle 1, < 1 mm. **SEED** 1 per follicle, < 1 mm, light brown. RARE. Limestone cliffs; 1700–2000 m. **e DMtns (Clark Mtns)**; s NV (Sheep Mtns). [*Forsellesia p.* (Brandegee) A.A. Heller, incl var. *glabra* Ensign] May–Jun

G. spinescens A. Gray (p. 283) NEVADA GREASEWOOD Shrub < 2 m, ± erect. **ST**: tips ± thorny. **LF** 5–17 mm, oblong to obovate; veins ± inconspicuous below; stipules appearing as 2 minute bristles near petiole base. **INFL**: fls axillary. **FL**: sepals rounded, none spine-tipped; petals 3–7 mm; stamens 6–10. **FR**: follicles 1–2, 3–5.5 mm. **SEEDS** 1–2 per follicle, 2–3 mm, gen shiny brown. Limestone; 850–2200 m. s KR, s SNH (Piute Mtns, Kern Co.), SnBr (n base), **W&I, DMtns**; to WA, WY, TX, n Mex. Highly variable; if the 4 weak vars. are recognized, CA pls are var. *aridum* M.E. Jones. [*Forsellesia arida* (M.E. Jones) A.A. Heller; *F. nevadensis* (A. Gray) E. Greene; *F. stipulifera* (St. John) Ensign, stipule-bearing forsellesia] ❀TRY. Apr–May

CUCURBITACEAE GOURD FAMILY

Robert L. Schlising

Ann, per, gen monoecious; hairs often hardened by calcium deposits. **STS** trailing or climbing, 1–many; tendril gen 1 per node, often branched. **LVS** gen simple, alternate, gen palmately lobed, veined, petioled; stipule 0. **INFLS** at nodes; staminate fls in racemes, panicles, small clusters, rarely solitary; pistillate fls gen solitary. **FL** unisexual in CA, radial; hypanthium > ovary; calyx (apparently 0 or) gen 5-lobed; corolla rotate or cup-shaped, gen 5-lobed; stamens 3–5 (or appearing 1–3 from fusion), anthers often > filaments, twisted together; ovary ± inferior, chambers gen 5, placentas parietal, ± growing into chambers, styles 1–3, stigmas gen lobed, large. **FR**: berry (sometimes drying) or capsule (irregularly dehiscent), gen gourd- or melon-like. **SEEDS** 1–many. 100 genera, 700 spp.: esp trop; some cult (*Citrullus*; *Cucumis*; *Cucurbita*; *Sechium*, chayote). *Citrullus colocynthis* var. *lanatus* has been reported as a weed in DSon. Keys adapted by Margriet Wetherwax.

1. Corolla white, greenish, or cream, < 2 cm wide; staminate fls gen in racemes, panicles, or small clusters, pistillate fls solitary or in small clusters, gen at same nodes as staminate; fr gen prickly, sometimes irregularly dehiscent; seeds gen not flat
   2. Staminate fl < 3 mm wide; fr asymmetric, beak ± = body; seed 1; D . . . . **BRANDEGEA**
   2' Staminate fl > 3 mm wide; fr symmetric, beak < body or 0; seeds gen 3–12; DSon . . . . **MARAH**
1' Corolla yellow, gen 2–12 cm wide; staminate and pistillate fls 1–few, gen at different nodes; fr unarmed (weakly prickly in some *Cucumis),* gourd- or melon-like, indehiscent; seeds ± flat
   3. Corolla 3–12 cm wide, deeply cup-shaped, fused portion > 3 cm; per from large, tuber-like root
      . . . . **CUCURBITA**
   3' Corolla gen < 3 cm wide, shallowly cup-shaped or rotate, fused portion < 1 cm; ann . . . . **CUCUMIS**

BRANDEGEA

1 sp. (T.S. Brandegee, CA botanist & engineer, 1843–1925)

B. bigelovii (S. Watson) Cogn. (p. 283, pl. 49) Per from taproot. **ST** ± glabrous; tendril unbranched. **LF** round, cordate to ± square, gen deeply lobed; central lobe gen longest; upper surfaces dotted with white glands. **INFLS**: staminate fls in small axillary clusters; pistillate fls 1 per node. **FL**: corolla 1.5–3 mm wide, rotate or shallowly cup-shaped, cream. **FR** dry, indehiscent, asymmetric, ± prickly; body 5–6 mm, ± = beak. **SEED** 1. Canyons, washes; < 900 m. **D**; sw AZ, Mex. Mar–Apr

CUCUMIS CUCUMBER, MELON

Ann, per. **ST** gen scabrous; tendril unbranched. **LF** angled to ± palmately lobed; 1° lobes ± entire to irregularly lobed. **INFLS**: staminate fls 1–several per node; pistillate fls gen 1 at different nodes. **FL**: corolla 2–3 cm wide, rotate or shallowly cup-shaped, yellow, deeply 5-lobed, fused portion < 1 cm; anthers 3, free; styles 3–5, ± reniform. **FR** gourd- or melon-like, indehiscent, cylindric to round; rind firm, net-veined or hairy, prickles 0 or weak. **SEEDS** many, < 1 cm, ± flat; margin plain. ± 40 spp.: Afr, s Asia. (Greek: cucumber)

C. melo L. var. ***dudaim*** (L.) Naudin DUDAIM MELON s**FR** cylindric to ± round, orange, sometimes irregularly blotched or striped. Fields, roadsides; < 200 m. n SCo (Santa Barbara Co.), **se DSon (Imperial Co.)**; native to Afr. NOXIOUS WEED.

CUCURBITA GOURD, SQUASH

Ann, per (from large, fleshy, tuber-like root). **ST** smooth to scabrous; tendril gen branched. **LF** lanceolate to round, entire to deeply lobed. **INFL**: fls 1 per node, staminate and pistillate at different nodes. **FLS**: corolla > 3 cm wide (staminate gen < pistillate), deeply cup- or bell-shaped, yellow, fused portion 4–12 cm, lobes gen recurved; stigmas 3, each 2-lobed. **FR** gourd-like, indehiscent, ± round to ± flat; rind firm, smooth, rough, or grooved. **SEEDS** many, < ± 2 cm, ± ovate, ± flat; margin thick or raised. ± 30 spp.: warm Am. (Latin: gourd) [Rhodes et al. 1968 Brittonia 20:251–266]

1. Lf blade angled, finely toothed, or weakly lobed at base; tendril branched > 1 cm from base . . . ***C. foetidissima***
1' Lf blade deeply lobed; tendril branched ± from base
   2. Lf blade lobes ± linear-lanceolate, distinct nearly to petiole . . . . ***C. digitata***
   2' Lf blade lobes triangular or widely lanceolate, distinct ± 1/2 to petiole . . . . ***C. palmata***

C. digitata A. Gray (p. 283) FINGER-LEAVED GOURD Herbage ± scabrous, hairy; tendril branched ± from base. **LF** 3–9 cm; blade ± cordate in outline, green, main veins lighter, lobes gen 5, distinct nearly to petiole, ± linear-lanceolate, lateral 2 lobes gen coarsely toothed. **FL**: corolla 3–5 cm. **FR** 7–8 cm wide, round to oblong, dark green, ± mottled, with several, narrow, well defined, whitish stripes. **SEED** 10–11 mm, white. 2*n*=40. Uncommon. Sandy, open or shrubby places; < 1200 m. PR, **DSon**; to NM, Mex. ❀TRY. Aug–Oct

C. foetidissima Kunth (p. 283) CALABAZILLA Herbage coarsely scabrous, hairy; tendril branched gen > 1 cm above base. **LF** 15–30 cm; blade gen triangular-ovate, ± cordate or truncate at base, gray-green, ill-smelling, angled, finely toothed or weakly lobed at base. **FL:** corolla 9–12 cm. **FR** gen 7–8 cm wide, ± round, green, mottled, with coarse, white stripes. **SEED** 12–14 mm, white. $2n=40$. Sandy, gravelly places; < 1300 m. GV, CW, SW, **D**; to NE, TX, n Mex. Other localities possibly due to human transport. ❀ SUN, DRN:**7,14–16**,17,**18–24**&IRR:**8–13**;deciduous,GRNCVR. Jun–Aug

C. palmata S. Watson (p. 283) COYOTE MELON Herbage scabrous, hairy; tendril branched ± from base. **LF** 8–15 cm; blade ±

cordate, gray-green, main veins whitish, lobes gen 5, distinct ± half way to petiole, middle 3 widely triangular, gen without prominent teeth. **FL:** corolla 6–8 cm. **FR** 8–9 cm wide, round, dull green, mottled, with poorly defined, whitish stripes. **SEED** 10–14 mm, white. $2n=40$. Sandy places; < 1300 m. SnJV, CW, SW, **D**; AZ, Baja CA. ❀ DRN,SUN:**7**,14,**19–23**,24&IRR:**8,9**,10,11,**12,13**;deciduous, GRNCVR. Apr–Sep

MARAH MAN-ROOT, WILD CUCUMBER

Per, sometimes temporarily dioecious; tuber large. **ST** ± scabrous or hairy, becoming glabrous; tendril branched. **LF** ± round, cordate, ± 5–7-lobed. **INFLS:** staminate fls in racemes or panicles with nonglandular axes (or 1 fl per axil early in season); pistillate fl 1 per axil (gen same axil as staminate). **FL:** sepals 0; corolla 3–15 mm wide (wider in pistillate), cup-shaped to rotate, white or cream to yellowish green; stamens fused, anthers twisted together; stigma 1, ± hemispheric. **FR:** capsule, irregularly dehiscent, ± symmetric, 3–20 cm, round, ovate, or oblong, sometimes tapered to a beak, ± prickly. **SEED** gen > 1 cm. 7 spp.: w N.Am. (Latin: bitter, from taste of all parts) [Schlising 1969 Amer J Bot 56:552–561] Extremely variable in habit, lvs, sexual expression; presumed hybrids occur where spp. overlap. Sometimes incl in *Echinocystis*.

1. Corolla rotate, yellowish green, cream, or white; ovary and fr ± round, fr 4–5 cm, prickles sparse to dense, ± stiff . . . . **M. fabaceus**
1' Corolla ± cup-shaped, white; ovary and fr oblong, fr 5–12 cm, prickles ± dense, stiff
. . . . **M. macrocarpus** var. **macrocarpus**

M. fabaceus (Naudin) E. Greene (p. 283) CALIFORNIA MAN-ROOT Herbage gen not glaucous. **FL:** corolla rotate, yellowish green, cream or (esp inland) white. **FR** 4–5 cm, ± round; prickles sparse to dense, < 12 mm, ± stiff, unhooked. **SEEDS** 2–4, 18–24 mm, ovate to oblong, ± flat on sides or not. $2n=32$. Streamsides, washes, shrubby and open areas; < 1600 m. CA-FP (exc n NW, n CaR), **DMoj** ❀ DRN: **7**,8–10,**14–24**;INV. Feb–Apr

M. macrocarpus (E. Greene) E. Greene Herbage not glaucous.

FL: corolla shallowly cup-shaped, white. **FR** 5–12 cm, oblong, gen rounded at both ends (sometimes with sharp beak); prickles ± dense, stiff. **SEEDS** gen 4–12(24), 13–33 mm, ± round, oblong, or ovate, angled at tip or not. $2n=32,64$. Washes, shrubby or open areas; < 900 m. SW, **DSon**; Baja CA. 2 vars. in CA.

var. **macrocarpus** (p. 283) **FL:** staminate 8–13 mm wide. **SEED** 13–20 mm. Habitats of sp. SW mainland, **DSon**; Baja CA. Jan–Apr

CUSCUTACEAE DODDER FAMILY

Tania Beliz

Ann, parasitic vine. **ST** twining, ± thread-like, yellow-green to bright orange, gen glabrous. **LVS** 0 or scale-like, ± 2 mm, gen triangular to lanceolate. **INFL:** cyme or cluster (rarely fls solitary), gen head- or spike-like, axillary, sometimes bracted. **FL** bisexual, radial; calyx gen persistent, lobes gen 4–5, gen overlapped; corolla gen deciduous, < 6 mm, mostly white, tube gen appendaged opposite stamens, lobes 4–5; stamens 4–5, alternate corolla lobes; ovary superior, chambers 2(3), 2-ovuled, styles gen 2, stigma gen 1 per style, gen ± head-like. **FR:** capsule (circumscissile or irregularly dehiscent) or berry-like. 1 genus, ± 150 spp.: esp Am trop; some crop pests. Recently treated within Convolvulaceae [Angiosperm Phylogeny Group 1998 Ann Missouri Bot Gard 85:531-553]. Key to species adapted by Margriet Wetherwax.

CUSCUTA DODDER

(Arabic: ancient name)

1. Corolla appendages 0–0.1 mm . . . . **C. californica**
  2. Ovary and fr conic, top acute . . . . var. **apiculata**
  2' Ovary and fr obovoid, top depressed . . . . var. **californica**
1' Corolla appendages 0.7–2.5 mm
  3. Corolla funnel-shaped, tube longer than wide
    4. Anthers on filaments; fl 2–3 mm . . . . **C. salina** var. **salina**
    4' Anthers sessile; fl 4.5–5.5 mm . . . . **C. subinclusa**
  3' Corolla shallowly bell- to urn-shaped, tube ± shorter than wide
    5. Corolla appendage divisions gen 0–few, knob-like . . . . **C. denticulata**
    5' Corolla appendage divisions many, finger-like . . . . **C. indecora**
      6. Pedicel and calyx not papillate . . . . var. **indecora**
      6' Pedicel and calyx ± papillate . . . . var. **neuropetala**

C. californica Hook. & Arn. **INFL** spike-like. **FL** 2.6–4 mm; calyx persistent, lobes 5, spreading to recurved, lanceolate, acute

to acuminate, 0.5–1 × corolla tube; corolla persistent, shallowly bell-shaped, gland-dotted, lobes 5, ± 3–6 mm, reflexed to spreading, lanceolate, acute, appendages 0–0.1 mm; filaments 0.7–1.4 mm, anthers 0.2–1.1 mm; ovary 1–2 mm, gen obovoid, gland-dotted, top gen depressed < styles 0.7–3 mm. **FR** 1.5–2 mm, enveloped by perianth, gen obovoid; top gen depressed. On herbs and shrubs on roadsides, chaparral, grassland, yellow-pine forest; gen < 2500 m. CA-FP, **SNE, DSon**; to WA, Colorado, Mex. 4 vars. in CA.

var. *apiculata* Engelm. **FL**: corolla not papillate; ovary (and fr) conic, top acute. On herbs; probably < 500 m. **e DSon (near Colorado River)**.

var. *californica* (p. 283) **FL**: corolla not bulged between stamens, not papillate; ovary (and fr) obovoid, depressed. Habitats of sp. CA-FP, **SNE**; to WA, NV, Baja CA. May–Aug

C. denticulata Engelm. **INFL** loose, spike-like, few-fld; pedicels 0–3.3 mm. **FL** 2–4 mm; calyx persistent, lobes 5, 0.3–1.5 mm, finely toothed, acute to obtuse; corolla gen persistent, 2–3 mm, shallowly bell- to urn-shaped, tube gen shorter than wide, lobes 5, becoming reflexed, tube, widely ovate, obtuse, appendages 0.7–1.4 mm, divisions 0–few, knob-like; ovary ± 1–2 mm, conic, top acute. On herbs or esp shrubs in creosote-bush scrub, Joshua-tree woodland; gen < 1300 m. **W&I, D**; to UT, AZ, Baja CA. May–Oct

C. indecora Choisy (p. 283) **INFL** spike- or panicle-like; pedicels 0–4.4 mm. **FL** 3–5 mm; calyx 1–2 mm, lobes 5, not overlapped; ± 1/2 corolla tube, acute; corolla gen persistent, 3–4.5 mm, shallowly bell-shaped, tube gen shorter than wide, lobes 5, < tube, triangular, erect, tips incurved, appendages 1.5–2.5 mm, divisions many, finger-like; ovary 2–3 mm, ovoid-spheric, top with thickening, styles 3–4 mm. **FR** sometimes gland-dot-lined; top as ovary. Common. On herbs, often in moist fields,

roadsides; probably < 1500 m. NCo, NCoR, SN, GV, **D**; to c&se US, Mex; also Caribbean, S.Am.

var. *indecora* **INFL**: pedicel and calyx not papillate. n=15. Habitats and range of sp. [*C. jepsonii* Yuncker; *C. suaveolens* Ser. misapplied] Jul–Aug

var. *neuropetala* (Engelm.) Hitchc. **INFL**: pedicel and calyx ± papillate. Common. Habitats of sp. SNF, GV, **D**. Jul–Oct

C. salina Engelm. (p. 283) **INFL** spike-like; pedicels 0–2.5 mm. **FL** 2–4.5 mm; perianth parts ± acute, gland-dotted; calyx 1.3–2.7 mm, lobes 5, erect to spreading, triangular-lanceolate; corolla 3–5 mm, bell-shaped, tube longer than wide, lobes 5, erect to spreading, lanceolate, appendages 1.2–1.5 mm, divisions many, knob-like; filaments 0.1–0.7 mm, anthers 0.4–0.7 mm; ovary 1–3 mm, gland-dotted, thickening at top more conspicuous in fr, styles 0.4–1 mm. Common. On herbs in salty marshes, flats, ponds; gen < ± 100 m. NCo, KR, GV, CCo, SnFrB, SCo, **W&I**; to B.C., UT, AZ, Baja CA. 4 vars. in CA.

var. *salina* **FL** 2–3 mm; perianth not papillate. Inland salt flats; gen < ± 100 m. KR, GV, SnFrB, **W&I**; to B.C., UT, AZ, Baja CA. May–Sep

C. subinclusa Durand & Hilg. **INFL** spike- or head-like; pedicels 0–1 mm. **FL**: calyx persistent, 2–3.4 mm, lobes (4)5, ± 1.5–2.5 mm, lanceolate, acute to acuminate; corolla ± 4.5–5.5 mm, funnel-shaped, tube longer than wide, lobes spreading, 1–1.5 mm, < tube, triangular, tip often papillate outside, appendages 1.6–2.1 mm, spoon-shaped, divisions short, knob- to finger-like; anthers sessile, ± 1–2 mm; ovary 1–1.5 mm, obovoid to elliptic, top thickened, styles 1.1–1.5 mm. Common. Gen on shrubs, in forests near streams, rivers; < 1600 m. NCoR, SN, GV, SnFrB, SCoR, **SNE**. Jun–Oct

DATISCACEAE DATISCA FAMILY

William J. Stone

Per, tree, gen dioecious. **LVS** simple, alternate, gen pinnate. **INFL**: axillary spike, raceme, or cluster. **STAMINATE FL**: sepals 3–9; petals 6–8 or 0; stamens < 25. **PISTILLATE OR BISEXUAL FL**: sepals 3–8; petals 0; stamens ± functional; ovary inferior, chamber 1, placentas 3, parietal, styles 3. **FR**: capsule, opening at top between styles. **SEEDS** many, minute. 3 genera, 4 spp.: Asia, w N.Am.

DATISCA

Per, glabrous. **LVS** pinnately divided. **INFL**: small axillary clusters. **STAMINATE FL**: calyx very short, lobes 4–9, unequal; corolla 0; stamens 8–12, filaments short. **PISTILLATE FL**: calyx 3-toothed; stamens, if present, 2–4; ovary ovoid, 3-angled, styles 3, thread-like, 2-forked. 2 spp.: 1 Asia, 1 w N.Am. (Derivation unknown)

D. glomerata (C. Presl) Baillon (p. 291) DURANGO ROOT **STS** gen clustered, erect, branched above, 1–2 m. **LVS** alternate above but appearing opposite or somewhat whorled below, ± 15 cm, ovate to lanceolate, acuminate, unequally pinnate; petioles 2–3(4) cm. **STAMINATE FL**: calyx 2 mm; anthers 4 mm, ± sessile,

yellow. **PISTILLATE FL**: calyx 5–8 mm; styles ± 6 mm. **FR** ± 8 mm. **SEEDS** ± 1 mm, light brown, pitted in rows. Dry stream beds or washes in many communities; < 2000 m. CA-FP (exc GV), **D**; w NV, Baja CA. All parts toxic.

ELAEAGNACEAE OLEASTER FAMILY

Elizabeth McClintock

Shrub, tree, sometimes ± dioecious, often thorny, gen densely silvery-hairy throughout; hairs often scale-like. **LVS** simple, alternate or opposite, gen deciduous, entire; petiole gen short; stipules 0. **INFL** gen umbel-like, axillary; fls 1–few. **FL** radial; hypanthium rotate to salverform, lower part gen receptacle-like, persistent, with a disk, becoming fleshy; sepals (hypanthium lobes) gen 4, ± petal-like; petals 0; anthers 4 or 8, ± sessile; ovary superior (appearing inferior), chamber 1, style 1. **FR**: achene enclosed in fleshy hypanthium, the whole drupe- or berry-like. 3 genera, ± 45 spp.: N.Am, Eur, Asia, e Australia; esp temp, subtrop. [Graham 1964 J Arnold Arbor 45:274–278]

1. Lvs alternate; fls bisexual, stamens 4 . . . . **ELAEAGNUS**
1' Lvs opposite; fls unisexual (ssp. dioecious), stamens 8 . . . . **SHEPHERDIA**

ELAEAGNUS

Shrub, tree. **LVS** alternate. **FL** bisexual; hypanthium bell-shaped to salverform, lobes 4; stamens 4, barely exserted; disk flask-shaped, enclosing base of style; stigma ± elongate, on 1 side of style. ± 40 spp.: N.Am, s Eur, Asia. (Greek: olive, chaste-tree)

E. angustifolius L. OLEASTER, RUSSIAN OLIVE Pl < 7 m, sometimes thorny. **LF** 4–8 cm, lanceolate to oblong, more silvery on lower surface. **FL** 5–10 mm, ± as wide at top, fragrant; hypanthium ± dark yellow inside, tube ± = lobes. **FR** 10–20 mm, elliptic in outline, yellow. Uncommon. Disturbed, sometimes moist places; gen < 1500 m. SnJV, SnFrB, **SNE**, **DMoj**; native to temp Asia. Cult as orn. May–Jun

SHEPHERDIA

Shrub, dioecious. **LVS** opposite or some alternate. **STAMINATE FL**: hypanthium rotate; stamens 8; disk lobes alternate stamens. **PISTILLATE FL**: hypanthium urn-shaped, nearly closed at top; disk lobes 8, nearly meeting above ovary; stigma cap-like. 3 spp.: N.Am. (John Shepherd, 1764–1836, curator of Liverpool Botanic Garden)

S. argentea Nutt. (p. 291) BUFFALO BERRY Pl 2–6 m, much-branched, thorny. **LF** 2–6 cm, gen oblong, more silvery on lower surface. **FL**: hypanthium greenish yellow inside. **STAMINATE FL** < 2 mm, ± 4 mm wide at top; tube < lobes. **PISTILLATE FL** ± 2 mm, 1 mm wide; tube > lobes. **FR** 5 mm, elliptic in outline, red. Along streams, river bottoms, slopes; 1000–2000 m. e KR, c SNH, SCoR, WTR, SnBr, **n SNE**; to Can, c US. Fr sour, made into sauce eaten with buffalo meat along Overland Trail; sometimes cult as orn. ❀ SUN&DRN4–6,**10**&IRR:**1–3,7–10,14–22,23,24**;STBL. Apr–May

ELATINACEAE WATERWORT FAMILY

Gordon C. Tucker

Ann, per, in or near water; roots fibrous, from a taproot or not, gen from lower lf axils as well. **ST** gen soft. **LVS** simple, opposite, ± 4-ranked; stipules scarious. **INFL**: fls axillary or terminal, solitary or clustered. **FLS** small, inconspicuous, radial, bisexual; sepals and petals gen free, 3–5, equal in number; ovary spheric, styles 3–5, very short. **FR**: capsule, septicidal, ± spheric, ovoid, or depressed-ovoid, walls thin; chambers 2–5, each several–many-seeded. **SEED** very small; surface net-like or glossy. 2 genera, 50 spp.: ± worldwide. [Tucker 1986 J Arnold Arbor 67:471–483]

ELATINE WATERWORT

Ann, short-lived per, glabrous. **ST** erect underwater, ± prostrate on wet ground, branched or not; base not woody. **LVS** opposite, ± 4-ranked; petiole < 1/3 blade, flat, ± blade-like; blades narrowly elliptic to ± round, ± entire, bases wedge-shaped to ± rounded, tips rounded. **INFL**: fls 1(2) per node, 0–1 per upper lf axil. **FL**: sepals 3–4, widely elliptic, membranous, very pale green; petals gen as many as and ± = sepals, widely elliptic, membranous, pale greenish white; stamens 3, 6, 8, rarely 1, filaments ± 1/2 × to ± = petals, anthers widely ovoid; styles 3–4. **FR** ± spheric or depressed-ovoid; chambers 3–4, each 3–15-seeded; pedicel gen ± 0. **SEEDS** ± visible through fr wall, elliptic, straight or curved, brown to yellowish brown; surface net-like. ± 25 spp.: worldwide. (Greek: fir tree, from a Eur sp. that suggests such a pl in miniature) At least 20× magnification needed for pits on seeds.

1. Seed pits 10–15 per row; fls 1 per node; lf ovate to narrowly oblong . . . . *E. brachysperma*
1' Seed pits 16–35 per row; fls 1–2 per node; lf oblong-lanceolate . . . . *E. rubella*

E. brachysperma A. Gray (p. 291) **ST** decumbent to erect, 1–5(12) cm. **LF** ovate, narrowly oblong; petiole 1/4–1/3 blade or indistinct. **INFL**: fls 1 per node. **FL**: sepals 2 or, if 3, 1 < others; petals 3, equal, = and ± wider than sepals; stamens 3, opposite sepals. **FR**: chambers 3. **SEED** widely oblong, curved ± 15°; pits 10–15 per row, wider than long. Muddy shores, shallow pools; 50–500 m. **CA**; c&s US. [*E. obovata* (Fassett) H. Mason] Apr–Sep

E. rubella Rydb. **ST** prostrate to erect, 3–6(15) cm, often tinted reddish. **LF** oblong-lanceolate; petiole < 1/4 blade; tip blunt to notched. **INFL**: fls 1–2 per node. **FL**: sepals 2 or, if 3, 1 < others; petals 3, equal, widely elliptic; stamens 3, opposite sepals. **FR**: chambers 3. **SEED** narrowly oblong, straight or curved < 15°; pits 16–35 per row, ± as wide as long. Muddy shores, shallow vernal pools, ditches, rice fields; < 500 m. **CA**; widespread in w N.Am.

ERICACEAE HEATH FAMILY

Gary D. Wallace, except as specified

Per, shrub, tree. **ST**: bark often peeling distinctively. **LVS** simple, gen cauline, alternate, opposite, rarely whorled, evergreen or deciduous, often leathery, petioled or not; stipules 0. **INFL**: raceme, panicle, cyme, or fls solitary, gen bracted; pedicels often with 2 bractlets. **FL** gen bisexual, gen radial; sepals gen 4–5, gen free; petals gen 4–5, free or fused; stamens 8–10, free, filaments rarely appendaged, anthers awned or not, dehiscent by pores or slits; nectary gen at ovary base, disk-like; ovary superior or inferior, chambers gen 1–5, placentas axile or parietal, ovules

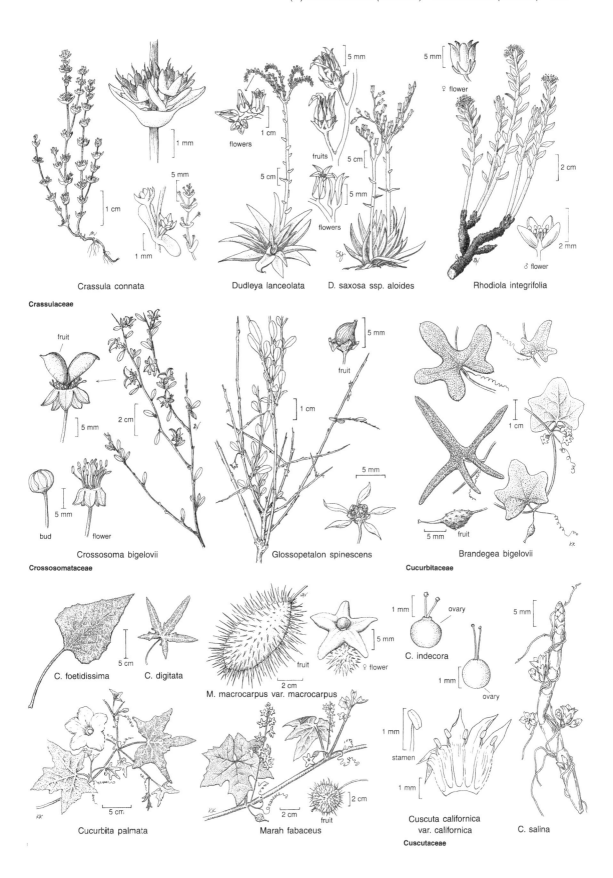

Crassula connata

Dudleya lanceolata

D. saxosa ssp. aloides

Rhodiola integrifolia

Crassulaceae

Crossosoma bigelovii

Glossopetalon spinescens

Brandegea bigelovii

Crossosomataceae

Cucurbitaceae

C. foetidissima

C. digitata

M. macrocarpus var. macrocarpus

C. indecora

Cucurbita palmata

Marah fabaceus

Cuscuta californica
var. californica

C. salina

Cuscutaceae

1–many per chamber, style 1, stigma head- to funnel-like or lobed. **FR**: capsule, drupe, berry. **SEEDS** gen many, sometimes winged. ± 100 genera, 3000 spp.: gen worldwide exc deserts; some cult, esp *Arbutus, Arctostaphylos, Rhododendron, Vaccinium*. [Wallace 1975 Wasmann J Biol 33:1–88; 1975 Bot Not 128:286–298] Subfamilies Monotropoideae, Pyroloideae, Vaccinioideae sometimes treated as families. Nongreen pls obtain nutrition from green pls through fungal intermediates.

1. Fruit a drupe, stones 2-10, free or fused; petals fused; anthers awned; dry sites . . . . **ARCTOSTAPHYLOS**
1' Fruit a septicidal capsule; petals free exc sometimes at base; anthers unawned; moist sites . . . . **LEDUM**

ARCTOSTAPHYLOS MANZANITA

Philip V. Wells

Shrubs, small trees. **ST** prostrate to erect; fire-resistant burl sometimes present at base; bark gen reddish, smooth or gray, rough, and shredded; hairs gen alike on twig, infl axis, bract. **LVS** alternate, spreading to ascending, evergreen; blade surfaces gen alike, sometimes convex, differing in color (stomata restricted to lower surface) or hairiness; margin flat to rolled. **INFL**: raceme or panicle-like, terminal; branches raceme-like; fls bracted; bracts lf-like, gen flat or scale-like, gen folded, keeled; immature infl present late summer through winter. **FL** radial; sepals gen 5, free, persistent; corolla gen 5-lobed, urn-shaped to ± spheric, white to pink; stamens gen 10, incl, filament base glabrous or hairy, anther 2-pored, awns 2, recurved; ovary superior, base surrounded by nectary disk, chambers 2–10, ovule 1 per chamber, style 1, stigma head-like. **FR**: drupe, berry-like, gen ± spheric; pulp gen thick, mealy; stones 2–10, free, separable, or strongly fused. ± 60 spp.: N.Am (esp CA) to C.Am, Eurasia. (Greek: bear berries) [Wells 1988 Madroño 35:330–341] Observation of hairs requires 10× magnification. Distribution of many spp. local; hybridization occurs in areas of overlap. ❀ Beautiful but mostly DFCLT due to fungus and often salinity and alkali. Avoid overhead watering in hot weather. CVS are the easier garden subjects. *A. parryana* Lemmon ssp. *deserticum* J.E. Keeley, L. Boykin & A. Massihi recently described, from western Borrego Palm Canyon [Madroño 44:265. 1997] Key to species adapted by Gary D. Wallace.

1. Infl a panicle, branches 4–8; pedicels finely glandular-puberulent; lvs 2–4 cm wide, white-glaucous,
   dull . . . . *A. glauca*
1' Infl a raceme, sometimes 1-branched; pedicels glabrous; lvs 1–1.8 cm wide, bright green, shiny
  2. Infl axis slender, not club-like near tip; fls, bracts evenly spaced; fr spheric; lvs obovate or oblanceolate
   . . . . *A. nevadensis*
  2' Infl axis stout, club-like near tip; fls, bracts crowded distally; fr depressed-spheric; lvs elliptic to
   lanceolate-elliptic . . . . *A. pungens*

A. glauca Lindley (p. 291) Shrub, tree-like, (< 1) 2–8+ m; burl 0. **ST**: twigs glabrous, ± glaucous, sometimes finely bristly. **LVS** erect; petiole 7–15 mm; blade 2.5–5 cm, 2–4 cm wide, oblong-ovate to ± round, base rounded, truncate, or slightly lobed, margin entire to toothed, surfaces alike, white-glaucous, dull, smooth. **INFL** open; branches 4–8; bracts 3–6 mm, scale-like, deltate to ± awl-like, weakly keeled; lowest bract 10–15 mm, lf-like, widely lanceolate; pedicel finely glandular-bristly, 8–10 mm in fr; immature axes 2–3 cm, spreading; bracts ± spreading, buds exposed. **FL**: ovary glandular. **FR** 12–15 mm wide, spheric or ovoid, sticky; pulp thick, leathery, not mealy; stones fused into 1 subspheric unit (± 10 mm wide), surface with vertical seams. *n*=13. Rocky slopes, chaparral, woodland; < 1400 m. ne SnFrB (Mount Diablo), SCoRI, TR, PR, sw DMtns (**Little San Bernardino Mtns**); Baja CA. [var. *puberula* J. Howell] ❀ DRN,SUN,DRY:7,11,14–24. Dec–Mar

A. nevadensis A. Gray (p. 291) Shrub < 0.6 m, mat- to mound-like; burl 0. **STS** spreading to decumbent, twigs finely tomentose. **LVS** ± erect; petiole 3–7 mm; blade 1–3 cm, 1–1.5 cm wide, obovate or oblanceolate, base ± wedge-shaped, tip ± obtuse, margin entire, fine ciliate, surfaces alike, bright green, shiny, ± puberulent, becoming glabrous, smooth. **INFL** spheric, dense; raceme, sometimes weakly 1-branched; axis, bract hairs like twig hairs or sometimes finely glandular-bristly; bracts 2–3 mm, scale-like, linear or linear-lanceolate, acuminate, sharp-pointed, lowest bract gen 5–10 mm, lf-like, linear-lanceolate; pedicel 3–5 mm, glabrous; immature axis 5–10 mm. **FL**: ovary glabrous. **FR** 6–8 mm wide, ± spheric, glabrous; stones ± separable. *n*=26. Rocky soils, coniferous forest; 900–3000 m. KR, NCoRH, CaRH, SNH, **n SNE (Sweetwater Mtns)**; to WA. ❀ DRN:1,2,4–6,15–17& IRR,SHD:3,7,14,18;DFCLT. May–Jul

A. pungens Kunth (p. 291) Shrub, erect, 1–3 m; burl 0. **ST**: twigs tomentose. **LVS** erect; petiole 4–8 mm; blade 1.5–4 cm, 1–1.8 cm wide, elliptic to lanceolate-elliptic, base obtuse to wedge-shaped, sometimes rounded, margin entire, surfaces alike, bright or dark green, shiny, ± finely tomentose, becoming glabrous, smooth. **INFL** spheric; raceme, sometimes 1-branched; bracts 2–4 mm, scale-like, ovate-deltate, acuminate, tip sharp-pointed; pedicel 5–10 mm, glabrous; immature axis 5–15 mm, stout, club-like, bracts recurved, crowded near tip. **FL**: ovary glabrous. **FR** 5–8 mm wide, glabrous. *n*=13. Rocky slopes, ridges, chaparral, coniferous forest; 900–2250 m. SCoRI (Monterey, San Benito cos.), SnBr, PR, **e DMtns**; to UT, TX, Mex. ❀ DRN, SUN,DRY: 2,3,**7**,9,10,14–17,**18**,19–23. Feb–Mar

LEDUM LABRADOR TEA

Shrub, gen hairy. **ST** decumbent to erect, rooting. **LVS** alternate, reflexed in age, evergreen, leathery; margin entire, rolled under or not. **INFL**: raceme, ± flat-topped, terminal, bracted; bractlets 1–2, deciduous; pedicels not jointed to fl. **FL**: sepals 5, fused; petals 5, free exc sometimes at base, when corolla ± rotate; stamens 8–10, anthers dehiscent by pores, unawned; ovary superior, chambers 5, placentas axile. **FR**: capsule, septicidal, dehiscent base to tip. **SEEDS** many per chamber, fusiform, unwinged. 2–3 spp.: n hemisphere. (Greek: for pl now known as *Cistus*) [Kron & Judd 1990 Syst Bot 15:57–68 (where incl in *Rhododendron*)]

L. glandulosum Nutt. (p. 291) WESTERN LABRADOR TEA **ST** gen erect, < 1.5 m; bark smooth; twigs puberulent to glandular. **LF** 1–3.5 cm, oblong to elliptic; margin not or ± rolled under; lower surface finely hairy, with sessile, flat glands. **FL** cream-yellow to whitish. Common. Boggy areas; < 3600 m. NCo, KR, s NCoRO, CaRH, SNH, CCo, SnFrB, n SCoRI, **SNE**; to w

Can, Colorado. [var. *californicum* (Kellogg) C. Hitchc.; ssp. *columbianum* (Piper) C. Hitchc. var. *australe* C. Hitchc.; ssp. *olivaceum* C. Hitchc.; *Rhododendron neoglandulosum* Harmaja] ❀ WET or IRR,DRN:1,2,**4–6,17**&SHD:7,14–16; acidic soil;CV. Jun–Aug

EUPHORBIACEAE   SPURGE FAMILY

Grady L. Webster, except as specified

Ann, per, shrub, tree, vine, monoecious or dioecious. **ST** gen branched, sometimes fleshy or spiny. **LVS** gen simple, alternate or opposite, gen stipuled, petioled; blade entire, toothed, or palmately lobed. **INFL**: cyme, panicle, raceme, spike; fls sometimes in clusters (dense, enclosed by involucre, fl-like in *Chamaesyce, Euphorbia*), terminal or axillary. **FL** unisexual, ± radial; sepals gen 3–5, free or fused; petals gen 0; stamens 1–many, free or filaments fused; ovary superior, chambers 1–4, styles free or fused, simple or lobed. **FR**: gen capsule. **SEEDS** 1–2 per chamber; seed scar appendage sometimes present, pad- to dome-like. 300 genera, 7500 spp.: ± worldwide esp trop; some cult (*Aleurites*, tung oil; *Euphorbia* ssp.; *Hevea*, rubber; *Ricinus*). [Webster 1967 J Arnold Arbor 48:303–430] Many spp. ± highly TOXIC. Key to genera adapted by Margriet Wetherwax.

1. Lvs opposite or 3-whorled
  2. Sts prostrate or erect, sap, milky; lf base oblique; infl dense, enclosed by involucre, ± fl-like, bisexual
  . . . . **CHAMAESYCE**
  2' Sts erect, sap clear; lf base not oblique; infl ± open, not enclosed by involucre, unisexual, not fl-like
  . . . . [2]**TETRACOCCUS**
1' Lvs alternate (sometimes opposite in *Euphorbia*)
  3. Infl dense, enclosed by involucre, ± fl-like, bisexual . . . . **EUPHORBIA**
  3' Infl ± open, not enclosed by involucre, not fl-like, unisexual
    4. St and lf hairs 2-branched or stellate, or 0
      5. Hairs 0 or 2-branched; petals present; filaments fused; seeds pitted . . . . **DITAXIS**
      5' Hairs stellate; petals 0; filaments free; seeds smooth
        6. Ovary 1-chambered; lvs prominently 3-veined, veins raised . . . . **EREMOCARPUS**
        6' Ovary 3-chambered; lvs pinnately veined, not raised
          7. Lf margin bluntly toothed; stipules persistent; staminate infl axillary . . . . **BERNARDIA**
          7' Lf margin entire; stipules inconspicuous; staminate infl terminal . . . . **CROTON**
    4' St and lf hairs simple
      8. Pistillate bract toothed, > fr . . . . **ACALYPHA**
      8' Pistillate bract entire, < fr
        9. Shrub; staminate infl cyme; ovules 2 per chamber . . . . [2]**TETRACOCCUS**
        9' Per; staminate infl spike or raceme; ovule 1 per chamber
          10. Lvs glabrous; stamens 2; base of pistillate fl bracts glandular; st sap milky . . . . **STILLINGIA**
          10' Lvs with stinging, nettle-like hairs; stamens 3–6; base of pistillate bracts not glandular; st sap
          clear . . . . **TRAGIA**

ACALYPHA

Ann, per, shrub, < 2 m, mostly monoecious; sap clear. **STS**: central erect, gen much-branched; lateral spreading to ascending. **LVS** simple, cauline, alternate, stipuled; hairs simple, sometimes glandular. **INFL**: spike, terminal or axillary; staminate bracts minute; pistillate bracts lf-like, toothed. **STAMINATE FL**: sepals 4; petals 0; stamens 4–8, filaments free or fused at base; nectary disk 0. **PISTILLATE FL**: sepals 3(–5); petals 0; nectary or 0; ovary 3-chambered, styles 3, deeply cut. **FR** ± spheric, smooth or ± lobed. **SEEDS** 1 per chamber, smooth to pitted; scar appendage minute. ± 400 spp.: trop, warm temp worldwide. (Greek: ancient name for a kind of nettle)

A. californica Benth. (p. 291)   Shrub < 1.5 m, hairy, ± glandular. **LF**: stipules 2–5 mm, linear; petiole < 1.5 cm; blade 1–2 cm, ovate to ± deltate, base truncate to ± lobed, margin crenate. **STAMINATE INFL** 1.5–4 cm, slender. **PISTILLATE INFL** < 2 cm; bracts together cup-like, hairy, margin glandular. **STAMI-**

NATE FL: sepals ± 0.5 mm, puberulent; stamens> sepals. **PISTILLATE FL**: sepals ± 1 mm, puberulent; ovary ± 1 mm diam, puberulent, styles reddish. **FR** 1–3 mm diam, puberulent. Rocky slopes, chaparral, oak woodland; 200–1300 m. PR, **w DSon**; Baja CA. ❀ DRN:15,16,18–21,**22–24**. Jan–Jun

BERNARDIA

Shrub, monoecious or dioecious; sap clear. **STS** erect, gen much-branched. **LVS** simple, cauline, alternate, stipuled; hairs simple or stellate. **STAMINATE INFL**: spike or raceme, axillary. **PISTILLATE INFL** terminal; fl sometimes solitary. **STAMINATE FL** sessile or short-pedicelled; calyx splitting into 3–4 segments; petals 0; stamens 3–25, filaments free; nectar disk minute or 0. **PISTILLATE FL** sessile; sepals 4–6; petals 0; nectar disk 0; ovary 3-chambered, styles 3, free, 2-lobed or -toothed. **FR** 3-lobed. **SEED** 1 per chamber; scar appendaged. 30–40 spp.: trop, subtrop Am. (Latin: Bernard de Jussieu, French taxonomist, 1699–1776)

B. myricifolia (Scheele) S. Watson (p. 291) Shrub < 2.5 m, dioecious, hairy. **LF**: stipules ± 1 mm, deciduous; petiole 1–5 mm; blade 0.5–3 cm, elliptic, tip obtuse or rounded, margin crenate. **STAMINATE INFL**: raceme; pedicel 3–4 mm. **PISTILLATE INFL**: fl 1, sessile. **STAMINATE FL**: stamens 12–15; nectar disk of small glands. **PISTILLATE FL**: sepals 5, ± 2 mm, unequal; ovary tomentose, styles jagged. **FR** 8–10 mm diam, tomentose. **SEEDS** 5 mm, smooth; back ribbed. Washes, rocky canyons; < 1200 m. **s DMoj, DSon**; to TX, Mex. [*B. incana* C. Morton] ❀ TRY. Apr–May, Oct–Nov

CHAMAESYCE PROSTRATE SPURGE

Daryl L. Koutnik

Ann, per, gen monoecious, glabrous to hairy; sap milky. **ST** prostrate to erect, < 5 dm; branches alternate. **LVS** cauline, opposite, short-petioled; stipules present; blade base gen asymmetric, veins dark green. **INFL** fl-like, gen 1 per node; involucre ± bell-shaped, bracts 5, fused; glands 4, distal appendages gen colorful, petal-like; fls central. **STAMINATE FLS** 3–many, gen in 5 clusters around pistillate fl, each fl a stamen. **PISTILLATE FL** 1, central, stalked; ovary chambers 3, ovule 1 per chamber, styles 3, separate or fused at base, divided to entire. **FR**: capsule, round to 3-angled or -lobed in ×-section. **SEED** gen 4-angled, smooth or sculptured. ± 250 spp.: dry temp, subtrop worldwide, esp Am. Often treated as subg. of *Euphorbia*. (Greek: ancient name for kind of prostrate plant) [Wheeler 1941 Rhodora 43:97–154, 168–286] ❀ STBL.

1. Infls in dense axillary cyme-like clusters; st erect — sparingly hairy . . . . *C. nutans*
1' Infls 1 per node, sometimes crowded on lateral branches; st prostrate to ascending (sometimes erect)
  2. Involucre, fr hairy; st, lvs gen hairy or becoming glabrous
    3. Per
      4. Involucre urn-shaped; staminate fls < 12 — DSon . . . . *C. arizonica*
      4' Involucre ± bell-shaped; staminate fls > 15
        5. Seed 3-angled, transversely 4–5-ridged, ridges rounded . . . . *C. pediculifera*
        5' Seed 4-angled, smooth to ± wrinkled
          6. St and lf hairs short, straight . . . . ²*C. polycarpa*
          6' St and lf hairs tomentose or long, appressed, dense
            7. Involucre glands red . . . . *C. melanadenia*
            7' Involucre glands green to yellow . . . . *C. vallis-mortae*
    3' Ann
      8. Gland appendage deeply 3–5-lobed; involucre urn-shaped . . . . *C. setiloba*
      8' Gland appendage scalloped or 0; involucre bell-shaped or obconic
        9. Gland appendage 0; seed smooth to slightly wrinkled . . . . ²*C. micromera*
        9' Gland appendage present, scalloped; seed transversely wrinkled . . . . *C. maculata*
  2' Involucre, fr, st, lvs glabrous (sometimes hairy)
    10. Stipules fused into wide, membranous scale . . . . *C. albomarginata*
    10' Stipules separate or fused below
      11. Per
        12. Glands round, appendage 0; lvs < 5 mm . . . . *C. parishii*
        12' Glands elliptic or oblong, appendage present; lvs > 3 mm
          13. Fr > 2 mm; stipules separate; seed > 2 mm . . . . *C. fendleri*
          13' Fr < 2 mm; lower stipules fused; seed < 2 mm . . . . ²*C. polycarpa*
      11' Ann
        14. Lvs linear, base symmetric; st prostrate to erect
          15. Glands 1–4, oval, appendage narrower than gland; fr and seed > 1.5 mm; seed smooth; st
            prostrate to ascending . . . . *C. parryi*
          15' Glands ± 4, round, appendage wider than gland or 0; fr and seed < 1.5 mm; seed transversely
            ridged; st erect . . . . *C. revoluta*
        14' Lvs lanceolate to round, base asymmetric; st prostrate to ascending
          16. Lf margin toothed, at least toward tip
            17. Pl hairy; seed transversely ridgid; appendage wider than gland . . . . ²*C. abramsiana*
            17' Pl glabrous; seed smooth to wrinkled; appendage narrower than gland . . *C. serpyllifolia* ssp. *serpyllifolia*
          16' Lf margin entire
            18. Seed flattened top to bottom, surface smooth; gland ovate . . . . *C. platysperma*
            18' Seed 3–4-angled, surface smooth, wrinkled, or transversely ridgid; glands disc-like or elliptic
              19. Gland appendage present; seed transversely ridgid . . . . ²*C. abramsiana*
              19' Gland appendage 0; seed smooth or wrinkled
                20. Fr < 2 mm; seed < 1 mm; staminate fls < 10 . . . . ²*C. micromera*
                20' Fr > 2 mm' seed > 1 mm; staminate fls > 30 . . . . *C. ocellata* ssp. *arenicola*

C. abramsiana (Wheeler) Koutnik Ann. **ST** prostate, hairy to subglabrous. **LF** 2–12 mm; stipules separate, 2–5-parted; blade ovate to elliptic-oblong, hairy to glabrous, tip obtuse, margin entire to finely toothed. **INFL** dense on short lateral branches; involucre < 1 mm, obconic, glabrous; gland < 0.5 mm, round to elliptic; appendage wider than gland, entire or shallowly 2-lobed, white. **STAMINATE FLS** 3–5. **PISTILLATE FL**: style divided 1/2 length. **FR** 1.5–2 mm, oblong, round, glabrous. **SEED** 1–1.5 mm, ovoid, transversely 4–6-ridged, white. Sandy flats; < 200 m. **DSon**; to AZ, Mex. [*Euphorbia a.* Wheeler] Sep–Nov

C. albomarginata (Torrey & A. Gray) Small (p. 291, pl. 50) RATTLESNAKE WEED Per. **ST** prostrate, glabrous. **LF** 3–8 mm; stipules fused, triangular, ciliate; blade round to oblong, glabrous, tip obtuse, margin entire. **INFL**: involucre < 2.5 mm, bell-shaped to obconic, glabrous; gland < 1 mm, oblong; appendage wider than gland, entire to slightly scalloped, white. **STAMINATE FLS** 15–30. **PISTILLATE FL**: style divided 1/2 length. **FR** 2–2.5 mm, ovoid, 3-angled, glabrous. **SEED** 1–2 mm, oblong, smooth, white. Common. Dry slopes; < 2300 m. s SnJv, SW, **D**; to UT, TX, Mex. [*Euphorbia a.* Torrey & A. Gray] Apr–Nov

C. arizonica (Engelm.) J.C. Arthur ARIZONA SPURGE Per. **ST** prostrate to erect, hairy. **LF** 2–10 mm; stipules gen separate, minute; blade ovate, hairy, tip acute, margin entire. **INFL**: involucre < 2 mm, urn-shaped, hairy; gland < 0.5 mm, ovate, appendage wider than gland, entire, white to pink. **STAMINATE FLS** 5–10. **PISTILLATE FL**: style divided 1/2 length. **FR** < 2 mm, spheric, lobed, hairy. **SEED** ± 1 mm, ovoid, transversely ridged, white to brown. RARE in CA. Sandy flats; < 300 m. **DSon**; to TX, Mex. [*Euphorbia a.* Engelm.] Mar–Apr

C. fendleri (Torrey & A. Gray) Small Per. **ST** decumbent, glabrous. **LF** 3–11 mm; stipules separate, linear, entire; blade ovate, glabrous, tip acute, margin entire. **INFL**: involucre < 2 mm, bell-shaped to obconic, glabrous; gland < 1 mm, elliptic; appendage narrower than gland, scalloped, white. **STAMINATE FLS** 25–35. **PISTILLATE FL**: style divided 1/2 length. **FR** 2–2.5 mm, ovoid, lobed, glabrous. **SEED** 2–2.5 mm, ovoid, smooth to slightly wrinkled, white. Uncommon. Dry slopes, woodland; 1500–2300 m. **W&I, DMtns**; to NE, TX, Mex. [*Euphorbia f.* Torrey & A. Gray] May–Oct

C. maculata (L.) Small (p. 291) SPOTTED SPURGE Ann. **ST** prostrate, hairy. **LF** 4–17 mm; stipules separate, fringed; blade ovate to oblong, hairy or becoming glabrous, tip acute to obtuse, margin finely toothed. **INFLS** dense on short, lateral branches; involucre < 1 mm, obconic, hairy; gland < 0.5 mm, elliptic; appendage width = gland width, scalloped,white to pink. **STAMINATE FLS** 4–5. **PISTILLATE FL**: style divided < 1/2 length. **FR** < 1.5 mm, ovoid, lobed, hairy. **SEED** < 1.5 mm, ovoid, transversely wrinkled, light brown. Waste places, gardens; < 200 m. CA-FP, **DSon**; native to e US. [*Euphorbia m.* L., *E. supina* Raf.] Apr–Oct

C. melanadenia (Torrey) Millsp. Per. **ST** decumbent to ascending, tomentose or becoming glabrous. **LF** 2–9 mm; stipules separate, linear; blade ovate, tomentose, tip acute, margin entire. **INFL**: involucre 1–1.5 mm, bell-shaped, tomentose; gland < 1 mm, oblong; appendage width = gland width, scalloped, white. **STAMINATE FLS** 15–20. **PISTILLATE FL**: style divided > 1/2 length. **FR** 1.5–2 mm, ovoid, lobed, tomentose. **SEED** 1–1.5 mm, ovoid, slightly wrinkled, white. Dry, stony slopes or flats; < 1300 m. SW, **DSon**; AZ, Baja CA. [*Euphorbia m.* Torrey] Dec–May

C. micromera (Engelm.) Wooton & Standley Ann. **ST** prostrate, glabrous to hairy. **LF** 2–7 mm; stipules fused below, separate above, triangular, ciliate; blade ovate to oblong, glabrous to hairy, tip acute to obtuse, margin entire. **INFL**: involucre < 1 mm, bell-shaped, glabrous to hairy; gland << 0.5 mm, round, red or pink; appendage 0. **STAMINATE FLS** 2–5. **PISTILLATE FL**: style divided 1/2 length. **FR** < 1.5 mm, spheric, angled, glabrous to hairy. **SEED** 1–1.5 mm, ovoid, smooth to slightly wrinkled, white to brown. Sandy places; < 1000 m. **D**; to UT, TX, n Mex. [*Euphorbia m.* Engelm.] Mostly Sep–Dec, also Apr–Jun

C. nutans (Lagasca) Small Ann. **ST** erect, sparingly hairy or becoming glabrous. **LF** 8–35 mm; stipules fused, triangular; blade oblong, glabrous to hairy, tip obtuse, margin finely toothed, glabrous to hairy. **INFLS** 1 per node or dense in axillary cyme-like clusters; involucre 1–2 mm, obconic, glabrous; gland < 0.5 mm, oblong; appendage wider than gland, entire, white to red. **STAMINATE FLS** 5–11. **PISTILLATE FL**: style divided 1/2 length. **FR** 2–2.5 mm, ovoid, lobed, glabrous. **SEED** 1–1.5 mm, ovoid, shallowly wrinkled, black to brown. Waste areas; < 300 m. GV, CW, SW, **DSon**; native to se US, S.Am. [*Euphorbia n.* Lagasca] Apr–Oct

C. ocellata (Durand & Hilg.) Millsp. Ann. **ST** prostrate, glabrous to hairy. **LF** 4–15 mm; stipules separate, thread-like; blade ovate to lanceolate, glabrous to hairy, tip acute to obtuse, margin entire, rolled down. **INFL**: involucre 1.5–2 mm, obconic to bell-shaped, glabrous to hairy; gland < 1 mm, round, appendage wider than gland or 0. **STAMINATE FLS** 40–60. **PISTILLATE FL**: style divided 1/2 length. **FR** 2–2.5 mm, spheric, lobed, glabrous to hairy. **SEED** 1.5–2 mm, ovoid, widely 3-angled, smooth to shallowly wrinkled, white. Sandy soils; < 800 m. CA-FP, **D**; to UT, AZ. [*Euphorbia o.* Durand & Hilg.] 3 sspp in CA.

 ssp. ***arenicola*** (Parish) Thorne **ST** glabrous. **LF** < 16 mm; blade ovate to lanceolate, tip acute, glabrous. **INFL**: involucre glabrous; gland appendage 0. **FR** glabrous. **SEED** smooth. Sandy places; < 800 m. **D**; to UT, AZ. [*Euphorbia o.* var. *a.* (Parish) Jepson] May–Sep

C. parishii (E. Greene) Millsp. Per. **ST** prostrate, glabrous. **LF** 2–4 mm; stipules separate, linear, ciliate; blade ovate, glabrous, tip abruptly pointed, margin entire, glabrous. **INFL**: involucre ± 1 mm, bell-shaped, glabrous; gland < 0.5 mm, round, yellow to red; appendage 0. **STAMINATE FLS** 40–50. **PISTILLATE FL**: style divided 1/2 length. **FR** < 2 mm, spheric, lobed, glabrous. **SEED** < 1.5 mm, ovoid, 4-angled, slightly wrinkled, white. Uncommon. Sandy washes; < 1000 m. **D**; NV. [*Euphorbia p.* E. Greene] Apr–Oct

C. parryi (Engelm.) Rydb. Ann. **ST** ascending to prostrate, glabrous. **LF** 5–28 mm; stipules separate, linear; blade linear, glabrous, tip acute to obtuse, margin entire. **INFL**: involucre 1.5–2 mm, bell-shaped, glabrous; glands 1–4, < 0.5 mm, oval; appendage narrower than gland, entire, white. **STAMINATE FLS** 40–55. **PISTILLATE FL**: style divided 1/2 length. **FR** 2 mm, spheric, lobed, glabrous. **SEED** ± 2 mm, ovoid, 3-angled, smooth, brown to white. Sand dunes; < 700 m. **DMoj**; to Rocky Mtns, TX, Mex. [*Euphorbia p.* Engelm.] May–Jun

C. pediculifera (Engelm.) Rose & Standley Per. **ST** prostrate to erect, hairy or becoming glabrous. **LF** 2–20 mm; stipules separate, thread-like; blade ovate to spoon-shaped, hairy to glabrous, tip acute, margin entire. **INFL**: involucre 1.5–2 mm, bell-shaped, hairy to glabrous; gland 0.5 mm, oblong, appendages unequal, entire to almost 0 on some glands, largest wider than gland. **STAMINATE FLS** 22–25. **PISTILLATE FL**: style divided to base. **FR** 2 mm, ovoid, lobed, hairy. **SEED** 1–1.5 mm, ovoid, white, 3-angled, transversely 4–5-ridged, ridges rounded. Uncommon. Dry slopes; < 500 m. **DSon**; AZ, Mex. [*Euphorbia p.* Engelm.] Jan–Apr

C. platysperma (S. Watson) Shinn. FLAT SEEDED SPURGE Ann. **ST** prostrate, glabrous. **LF** 5–10 mm; stipules separate, 2–3-lobed; blade oblong to obovate, glabrous, tip obtuse to rounded, margin entire. **INFL**: involucre 1.5–2 mm, bell-shaped, glabrous; gland 1 mm, ovate, glabrous; appendage 0. **STAMINATE FLS** ± 50. **PISTILLATE FL**: style divided to base. **FR** < 4.5 mm, widely ovoid, slightly lobed, glabrous. **SEED** 2.5–3 mm, ovoid, flattened top to bottom, smooth, white. Sandy soil; < 100 m. **DSon** (**Coachella Valley**); to sw AZ, n Sonora. [*Euphorbia p.* S. Watson] Not seen in CA since 1914.   May

C. polycarpa (Benth.) Millsp. (p. 291) Per. **ST** prostrate to ascending, glabrous to hairy. **LF** 1–10 mm; stipules separate, triangular; blade round to ovate, glabrous to hairy, tip acute to obtuse, margin entire. **INFL**: involucre 1–1.5 mm, bell-shaped, glabrous to hairy; gland < 1 mm, oblong, appendage wider to narrower than gland, entire to scalloped, white to red. **STAMINATE FLS** 15–32. **PISTILLATE FL**: style divided > 1/2 length. **FR** 1–1.5 mm, spheric, lobed, glabrous to hairy. **SEED** 1–1.5 mm, ovoid, smooth, white to light brown. Common. Dry, sandy slopes and flats; < 1000 m. SW, **D**; NV, Mex. [*Euphorbia p.* Benth.] DMoj pls with hairy sts, lvs, involucre have been called var. *hirtella* (Boiss.) Parish; relationship to glabrous pls needs further study. Most of year

C. revoluta (Engelm.) Small Ann. **ST** erect, glabrous. **LF** 3–26 mm; stipules separate, linear; blade linear, glabrous, tip acute to obtuse, margin entire, rolled down. **INFL**: involucre < 1.5 mm, obconic, glabrous; gland < 0.5 mm, round, appendage wider than gland or 0, entire, white. **STAMINATE FLS** 5–10. **PISTILLATE FL**: style divided 1/2 length. **FR** slightly < 1.5 mm, spheric, lobed, glabrous. **SEED** 1–1.5 mm, ovoid, 3-angled, transversely 2–3-ridged, white to gray. Uncommon. Rocky slopes; < 3100 m. PR, **D**; to Rocky Mtns, Mex. [*Euphorbia r.* Engelm.] Aug–Sep

C. serpyllifolia (Pers.) Small THYME-LEAVED SPURGE Ann. **ST** prostrate to ascending, glabrous to hairy. **LF** 3–14 mm; stipules separate, linear; blade ovate to oblong, glabrous to hairy, tip rounded, margin finely toothed. **INFL**: involucre ± 1 mm, bell-shaped, glabrous to hairy; gland < 0.5 mm, oblong; appendage narrower than gland, entire to scalloped, white. **STAMINATE FLS** 5–18. **PISTILLATE FL**: style divided 1/2 length. **FR** 1.5–2 mm, ovoid, lobed, glabrous to hairy. **SEED** 1–1.5 mm, ovoid,

smooth to wrinkled, white to brown. Common. Dry habitats; < 2500 m. **CA**; to B.C., e N.Am., Mex. [*Euphorbia s.* Pers.] 2 sspp in CA.

ssp. *serpyllifolia* **ST** glabrous. **LF** glabrous. **INFL**: involucre glabrous. **FR** glabrous. Common. Habitat and range of sp. Mostly Aug–Oct

C. setiloba (Torrey) Parish (p. 291) Ann. **ST** prostrate, hairy. **LF** 2–7 mm; stipules separate, thread-like; blade oblong to ovate, hairy, tip acute, margin entire. **INFL**: involucre < 1.5 mm, urn-shaped, hairy; gland < 0.5 mm, oblong, appendage wider than gland, 3–5-lobed, white. **STAMINATE FLS** 3–7. **PISTILLATE FL**: style divided to base. **FR** < 1.5 mm, spheric, lobed, hairy. **SEED** ± 1 mm, ovoid, wrinkled, white to brown. Uncommon. Sandy places; < 1500 m. **D**; to TX, Mex. [*Euphorbia s.* Torrey] Most of year

C. vallis-mortae Millsp. Per. **ST** prostrate to decumbent, tomentose or becoming glabrous. **LF** 4–6 mm; lower stipules fused; upper stipules separate, thread-like, tomentose; blade oblong to ovate, tomentose, tip obtuse, margin entire. **INFLS** gen clustered at branch tips; involucre < 2.5 mm, bell-shaped, tomentose; gland < 1 mm, oblong; appendage = to or wider than gland, entire to scalloped, white. **STAMINATE FLS** 17–22. **PISTILLATE FL**: style divided 1/2 length. **FR** 2 mm, ovoid, 3-lobed, tomentose. **SEED** < 2 mm, ovoid, smooth, white. Uncommon. Dry, sandy places; < 1300 m. **SNE**, **DMoj**. [*Euphorbia v-m.* (Millsp.) J. Howell] May–Oct

CROTON

Ann, per, shrub, tree, monoecious or dioecious; sap clear or colored. **STS** gen erect. **LVS** gen simple, cauline, alternate; hairs gen stellate. **INFL**: spike or raceme, gen terminal. **STAMINATE FL** gen pedicelled; sepals gen 5; petals 5 or 0; stamens 8–50(300), filaments free, bent inward in bud; nectar disk gen divided. **PISTILLATE FL**: pedicel short or 0, becoming longer in fr; sepals gen 5, entire to lobed; petals gen 0; nectar disk entire; ovary 3-chambered, styles 2-lobed or toothed. **FR** spheric or 3-lobed, smooth or tubercled. **SEEDS** 1 per chamber, smooth to ribbed or pitted; scar appendaged. 900–1000 spp.: trop, warm temp, worldwide. (Greek: from resemblance of seed to a tick) This genus now considered to include *Eremocarpus*. [Webster 1992 Novon 2:269-273]

1. Seeds 3.5–5.5 mm; pedicels in fr < 2 mm; staminate sepals ± 2–2.5 mm . . . . *C. californicus*
1' Seeds 6.5–7 mm; pedicels in fr 4–7 mm; staminate sepals 2.5–3 mm — sand dunes (se Imperial Co.) . . . . *C. wigginsii*

C. californicus Muell. Arg. (p. 291) Per or subshrub < 1 m, dioecious; hairs stellate, scale-like. **LF**: petiole 1–4 cm; blade 2–5.5 cm, elliptic to narrowly oblong, tip rounded to obtuse, margin entire. **INFL**: raceme. **STAMINATE FL**: pedicels 1–5.5(7) mm; petals 0; stamens 10–15, filaments hairy. **PISTILLATE FL**: pedicel < or = 1 mm, 1–1.5(3) mm in fr; sepals ± 2 mm, entire; styles 2-lobed, lobes 2-forked. **SEED** 3.5–5.5 mm, smooth. Sandy soils, dunes, washes; < 900 m. CCo, SCo, s ChI (Santa Catalina), **D**; AZ, Baja CA. [vars. *mohavensis* A. Ferg., *tenuis* (S. Watson) A. Ferg.] ❀ DRN:15–24; DFCLT. Mar–Oct

C. wigginsii Wheeler (p. 291) WIGGINS' CROTON Shrub or subshrub, < 1 m, dioecious; hairs stellate, scale-like. **LF**: petiole 1–4 cm; blade 2–8.5 cm, narrowly elliptic to linear-oblong, tip rounded to obtuse, margin entire. **INFL**: raceme. **STAMINATE FL**: pedicels 1–5.5(7) mm; petals 0; stamens 10–15, filaments hairy. **PISTILLATE FL**: pedicel < 2 mm, 4–7 mm in fr; sepals ± 2 mm, entire; styles 2-lobed, lobes 2-forked. **SEED** 6.5–7 mm, smooth. RARE in CA. Sand dunes; < 100 m. **se DSon (se Imperial Co.)**; AZ, nw Mex. Mar–May

DITAXIS

Ann, per, subshrub, gen monoecious; sap clear; hairs 0 or gen 2-forked, gen appressed. **STS** spreading to erect, 1–10 dm. **LVS** simple, alternate, stipuled. **INFL**: raceme, axillary; staminate fls gen above pistillate fls; axis gen densely appressed-hairy; bracts entire. **STAMINATE FL**: sepals 5, edges abutting in bud; petals 5; stamens 5–15, gen in 2 sets, some > others, filaments fused into a column, staminodes 0–3 at column tip. **PISTILLATE FL**: sepals 5, overlapping in bud; petals 5; nectar disk ± dissected; ovary 3-chambered, styles 3, 2-lobed. **FR** smooth. **SEEDS**: surface net-like to finely pitted; scar not appendaged. ± 50 spp.: trop, warm temp Am. (Greek: 2-ranked, from 2 sets of anthers)

1. Pl glabrous . . . . *D. californica*
1' Pl hairy, hairs 2-forked, appressed, sometimes simple or glandular
  2. Margin of stipules, bracts, and pistillate sepals with stalked glands; infl axis minutely spreading-hairy
    . . . . *D. claryana*

2' Margin of stipules, bracts, and pistillate sepals glabrous or faintly glandular, glands not stalked;
  infl axis appressed-hairy
  3. Subshrub, sts brittle; pistillate sepals ± = petals; fr appressed-hairy; style lobes expanded . . . . . ***D. lanceolata***
  3' Ann or per (sometimes woody at base), sts not brittle; pistillate sepals clearly > petals; fr ± spreading-
    hairy; style branches entire to barely toothed
    4. Seeds angled in ×-section, clearly pitted; lvs gen lanceolate, not densely hairy, entire to faintly toothed
      . . . . ***D. neomexicana***
    4' Seeds round in ×-section, ± striate, not clearly pitted; lvs gen widely elliptic, densely hairy,
      clearly toothed distally . . . . ***D. serrata***

D. californica (Brandegee) Pax & K. Hoffm. (p. 295) CALI-
FORNIA DITAXIS   Ann or per. **ST** 1.5–5 dm, glabrous. **LF** 1–5 cm;
stipules ± 1 mm, faintly gland-toothed; blade lanceolate to elliptic, glabrous, margin finely toothed. **STAMINATE FL**: sepals
1.5–2.5 mm; petals 2–2.5 mm; stamen column 1 mm. **PISTIL-
LATE FL**: sepals 2.5–4.5 mm, faintly gland-toothed; petals ±
1.5 mm, glabrous; ovary glabrous, styles gen free, lobe tips not
expanded. **FR** ± 3.5 mm. **SEED** ± 2 mm, ± angled, ± pitted.
RARE. Washes, canyons; 50–1000 m. **DMoj (Eagle Mtn), nw
DSon (Coachella Valley)**. Mar–May, Oct–Dec

D. claryana (Jepson) Webster (p. 295) GLANDULAR DITAXIS
Ann or per. **ST** 1–5 dm; some hairs simple and spreading, others
2-forked and appressed. **LF** 1–4 cm; stipules 1.5–3 mm, gland-
toothed; blade lanceolate, lower surface hairy, margin finely
gland-toothed. **STAMINATE FL**: sepals 3.5–5 mm, hairy; petals ± = sepals, glabrous or hairy; stamen column 1.5–2 mm.
PISTILLATE FL: sepals 3.5–5.5 mm, unequal, glabrous or
hairy; petals ± = sepals; ovary sparsely hairy, styles fused below,
lobe tips expanded. **FR** ± 4.5 mm. **SEED** ± 2 mm, angled, faintly
pitted. RARE in CA. Sandy soils, creosote-bush scrub; < 100
m. **DSon (Coachella Valley)**. [*D. adenophora* (A. Gray) Pax &
K. Hoffm. misapplied] Dec–Mar

D. lanceolata (Benth.) Pax & K. Hoffm. (p. 295) Subshrub.
STS gen erect, 1–5 dm, brittle, appressed-hairy. **LF** 2–6 cm;
stipules ± 1 mm, entire; blade lanceolate, densely hairy, margin
entire. **STAMINATE FL**: sepals 2.5–3 mm, hairy; petals 3–3.5

mm, back hairy; stamen column ± 1.5 mm. **PISTILLATE FL**:
sepals 3–4 mm, entire; petals ± = sepals, lanceolate to ovate,
back hairy; ovary densely appressed-hairy, styles gen free, lobe
tips expanded. **FR** 3–5 mm. **SEED** 2–2.5 mm, angled, pitted.
Rocky soils, slopes, canyons; < 600 m. **DMoj (Eagle Mtn)**,
DSon; AZ, Mex. Mar–May

D. neomexicana (Muell. Arg.) A.A. Heller (p. 295) Ann or
per. **ST** 1–3.5 dm, densely appressed-hairy. **LF** 1–3.5 cm; stipules
1–1.5 mm, entire; blade lanceolate, ± hairy, margin ± entire.
STAMINATE FL: sepals 2–2.5 mm; petals ± 2 mm, glabrous;
stamen column ± 1 mm. **PISTILLATE FL**: sepals 3–4 mm, back
hairy, margin ± entire, ± white; petals ± 2.5 mm, lanceolate, glabrous or appressed-hairy, hairs not exceeding petal tip; ovary stiff-
hairy; styles free, lobe tips not expanded. **FR** 3–4 mm. **SEED** ±
2 mm, angled, pitted. Slopes, creosote-bush scrub; < 300 m.
DMoj (s edge), DSon; to TX, Mex. Mar–Dec

D. serrata (Torrey) A.A. Heller (p. 295) Ann or per. **ST** 1–3.5
dm, densely appressed-hairy. **LF** 1–3 cm; stipules 1–1.5 mm,
entire; blade widely elliptic to ovate, densely hairy, margin clearly
toothed distally. **STAMINATE FL**: sepals 2–2.5 mm; petals ± 2
mm, glabrous; stamen column ± 1 mm. **PISTILLATE FL**: sepals 3–4 mm, back hairy; petals ± 2.5 mm, obovate, back clearly
appressed-hairy, hairs exceeding petal tip; ovary stiff-hairy, styles
free, lobe tips not expanded. **FR** ± 2 mm. **SEED** ± 2 mm, rounded,
± striate, not clearly pitted. Sandy or rocky soils, creosote-bush
scrub; < 200 m. **D**; AZ, Mex. Apr–Nov

EREMOCARPUS   TURKEY MULLEIN, DOVE WEED

1 sp. (Greek: solitary fr)

E. setigerus (Hook.) Benth. (p. 295) Ann < 2 dm, < 8 dm
wide, mound-like, monoecious; sap clear. **ST** much-branched
from base, spreading to ascending. **LVS** simple, cauline, alternate; stipules vestigial; petiole 1–5 cm; blade 1–6 cm, ovate, base
obtuse to wedge-shaped, margin entire, densely soft stellate-hairy,
3-veined, veins raised. **STAMINATE INFL**: cyme, terminal;
pedicel 2–3 mm. **PISTILLATE INFL** axillary, below staminate
infl; fls 1–3. **STAMINATE FL**: receptacle finely bristly; sepals
5–6; petals 0; stamens 6–10, free, exserted, filaments 1.5–2 mm;

nectary 0. **PISTILLATE FL**: sepals and petals 0; glands below
ovary 4–5; ovary 1-chambered, puberulent, style slender. **FR** ± 4
mm diam. **SEED** 1, 3–4 mm, smooth or ± ridged; scar not
appendaged. 2n=20. Dry, open, often disturbed areas; < 1000
m. CA-FP, **w D**; to WA. Seeds eaten by birds; herbage TOXIC
to livestock, esp in hay. ❀ SUN, DRN:**7–9,14**,15–17,**18–
24**;rather INV. May–Oct *This genus now treated as *Croton*.
[Webster 1992 Novon 2:269-273] ***Croton setigerus*** Hook.

EUPHORBIA   SPURGE

Daryl L. Koutnik

Ann, per, gen monoecious, glabrous or hairy. **ST** ascending to erect, < 1 m; branches forked, forks equal. **LVS** cauline,
gen alternate; stipules 0 or gland-like; petiole present or 0; lf base symmetrical. **INFL** fl-like or not, gen clustered;
clusters gen umbel-like or cyme-like; involucre ± bell-shaped; bracts 5, fused; glands gen 4, distal appendages gen 0;
fls central. **STAMINATE FLS** 5–many, gen in 5 clusters around pistillate fl. **PISTILLATE FL** 1, central, stalked;
ovary chambers 3, ovule 1 per chamber, styles 3, separate or fused at base, divided or entire. **FR**: capsule, round to 3-
angled or -lobed in ×-section. **SEED** round or angled in ×–section; surface smooth or sculptured, gen with a knob-like
structure at attachment scar. ± 1500 spp.: warm temp to trop, worldwide. See *Chamaesyce*. (Latin: Euphorbus,
Physician to the King of Mauritania, 1st century) [Wheeler 1936 Bull S Calif Acad Sci 35:127–147] ❀ STBL.

1. Involucre with petal-like appendages on glands; stipules present, gland-like, minute
  2. Lvs opposite, margin finely toothed; ann; seed with a knob at tip . . . . ***E. exstipulata***
  2' Lvs alternate, margin entire; woody; seed without a knob . . . . ***E. misera***
1' Involucre without petal-like appendages on glands; stipules 0

3. Infls few–many in cyme-like clusters at branch tips; involucre glands 1–3, cup-shaped; involucre,
fr hairy . . . . *E. eriantha*
3' Infl clusters umbel-like below, cyme-like above; involucre glands 4, ± flat; involucre, fr glabrous
  4. Per; seed sculpture low, net-like; fr rounded, lobes smooth; glands hornless, margin scalloped . . . . . . *E. incisa*
  4' Ann; seed dotted; fr 4-angled, 2-keeled on lobes; glands 2-horned, margin entire — waste places
    and gardens . . . . *E. peplus*

E. eriantha Benth. (p. 295) BEETLE SPURGE Ann. **ST** erect, 1.5–5 dm, glabrous or hairy and becoming glabrous. **LF** 2–7 cm, petioled; stipules minute, obscure; blade linear, hairy, becoming glabrous, tip acute to obtuse and abruptly pointed, margin entire. **INFL** 1 or few clustered at branch tips; involucre 1.5–2 mm, obconic, hairy; glands 1–3, ± 1.5 mm, round, cupped, lobes 5–7, curved over gland. **STAMINATE FLS** 23–36. **PISTILLATE FL:** style undivided. **FR** 4–5 mm, oblong, lobed, hairy. **SEED** 3.5–4 mm, flattened top to bottom, 4-angled, tubercled, white to gray. Canyons, rocky slopes; < 100 m. **DSon**; to TX, Mex. Mar–Apr

E. exstipulata Engelm. var. **exstipulata** CLARK MOUNTAIN SPURGE Ann. **ST** erect, < 2.5 dm, hairy. **LVS** opposite, 2–5 cm, petioled; stipules gland-like; blade linear to lanceolate, hairy, tip acute, margin finely toothed. **INFL** 1 per node; involucre 1–2 mm, bell-shaped, hairy; gland < 0.5 mm, oblong, appendage = to or wider than gland, 2–4-lobed. **STAMINATE FLS** 8–14. **PISTILLATE FL:** style divided > 1/2 length. **FR** 2.5–3.4 mm, spheric, lobed, hairy. **SEED** 2–3 mm, oblong, 4-angled, transversely 2–3-ridged, tubercled, gray to brown or white. RARE in CA. Rocky slopes; 1800–2000 m. **e DMtns (Clark Mtns)**; to TX, Mex. Other vars. in sw US, Mex.

E. incisa Engelm. (pl. 51) MOJAVE SPURGE Per. **ST** ascending to erect, 1–4 dm, glabrous to slightly hairy. **LF** 0.6–2 cm, sessile; blade obovate to elliptic, glabrous, tip acute to abruptly pointed, margin entire. **INFL:** involucre 2–3 mm, bell-shaped, glabrous;

gland 1–2 mm, crescent-shaped, margin scalloped. **STAMINATE FLS** < 20. **PISTILLATE FL:** style divided < 1/2 length. **FR** 4–5 mm, oblong, lobed, glabrous. **SEED** 2–3 mm, oblong, round, white to gray; surface low net-like to almost smooth. Rocky or sandy slopes; 1000–2300 m. **W&I, DMtns**; NV, AZ. Intergrades with *E. palmeri*, esp in gland shape. Mar–May

E. misera Benth. CLIFF SPURGE Shrub. **ST** erect, 5–10 dm, hairy, becoming glabrous. **LF** 0.4–1.5 cm, petioled; stipule thread-like; blade ovate to round, hairy, tip round, margin entire. **INFL** 1 at branch tip; involucre 2–3 mm, bell-shaped, hairy; glands 5, 1.5–2 mm, oblong, appendage = to gland width, scalloped, white. **STAMINATE FLS** 30–40. **PISTILLATE FL:** style divided 1/2 length. **FR** 4–5 mm, spheric, lobed, becoming glabrous. **SEED** 2.5–3 mm, ovoid, round, wrinkled, white to gray; knob 0. RARE in CA. Rocky slopes, coastal bluffs; < 500 m. SCo, s ChI, w DSon; Baja CA. In cult. Jan–Aug

E. peplus L. (p. 295) PETTY SPURGE Ann. **ST** ascending to erect, 1–4.5 dm, glabrous. **LF** 1–3.5 cm, petioled; blade obovate to ovate, glabrous, tip obtuse to notched, margin entire. **INFL:** involucre 1–1.5 mm, bell-shaped, glabrous; gland < 0.5 mm, crescent-shaped, 2-horned. **STAMINATE FLS** 10–15. **PISTILLATE FL:** style divided 1/2 length. **FR** ± 2 mm, spheric, glabrous, lobed; lobes 2-keeled. **SEED** 1–1.5 mm, oblong, 4-angled, dotted, white to gray. Common. Waste places, gardens; < 300 m. CA-FP, **DSon**; to Can,e US; native to Eur. Feb–Aug

STILLINGIA

Ann, per, < 2 m, monoecious; sap clear or milky. **STS** erect. **LVS** alternate, simple, entire or toothed; stipules minute, petioled; blade base gen with 2 glands. **INFL:** spike, axillary or terminal; bracts glandular. **STAMINATE FL:** calyx 2-lobed; petals 0, stamens 2; nectary disk 0. **PISTILLATE FL:** sepals 3 (overlapping in bud), reduced, or 0; petals 0; ovary 3-chambered, styles free, fused below, lobes 0. **FR** gen 3-lobed, separating into 3 1-seeded segments; central axis persistent. **SEEDS** pointed; scar not appendaged. 30 spp.: trop, warm temp. (Latin: Benjamin Stillingfleet, British botanist, 1702–1771) [Johnston & Warnock 1963 Southw Naturalist 8:100–106]

1. Lvs elliptic to ovate, sharply toothed, 3-veined; spikes axillary; seeds striate . . . . *S. spinulosa*
1' Lvs linear, entire or sparsely toothed below middle, 1-veined; spikes terminal; seeds smooth
  2. Spikes not projecting above lvs, open in lower 1/3, pistillate fls well separated; base of lf blade entire;
    seeds ± 2 mm . . . . *S. linearifolia*
  2' Spikes projecting above lvs, dense in lower 1/3, pistillate fls crowded; base of lf blade few-toothed;
    seeds 2.5–3 mm . . . . *S. paucidentata*

S. linearifolia S. Watson (p. 295) Per < 7 dm. **LF:** blade 1–4 cm, < 2 mm wide, linear, margin entire. **INFL** 2–7 cm; glands of pistillate bracts stalked, ± 1 mm. **PISTILLATE FLS** 3–6 per infl, well separated; styles ± 1 mm. **FR** ± 3.5 mm. **SEED** ± 2 mm, smooth. Dry slopes, washes; < 1500 m. SW, **D**; AZ, Mex. Mar–May

S. paucidentata S. Watson (p. 295) Per < 5 dm. **LF:** blade 3–8 cm, 1.5–5 mm wide, linear, margin few-toothed below middle, teeth ± spiny. **INFL** 2–7 cm; glands of pistillate bracts subsessile, ± 1–1.5 mm. **PISTILLATE FLS** 2–3 per infl, crowded; styles ±

2.5 mm. **FR** 4–4.5 mm. **SEED** 2.5–3 mm, smooth. Slopes, flats, creosote-bush scrub; < 1500 m. **DMoj, n DSon**; AZ. Apr–Jun

S. spinulosa Torrey (p. 295) Ann or per < 10 dm. **LF:** blade gen 2–4 cm, 5–12 mm wide, elliptic to ovate, margin sharply toothed. **INFL** 1–2 cm; glands of pistillate bracts stalked, ± 2 mm. **PISTILLATE FLS** 1–2 per infl, ± open; styles 3–3.5 mm. **FR** 4–5 mm. **SEED** 3–3.5 mm, striate, minutely roughened. Sandy soils, dunes, creosote-bush scrub; < 900 m. **D**; s NV, AZ. Mar–May

TETRACOCCUS

Shrub, gen 0.5–2 m, dioecious; sap clear. **ST:** axis erect; branches gen many, spreading to erect; twigs gen reddish, becoming gray, gen hairy, becoming glabrous; young lateral twigs short, sometimes becoming spine-like. **LVS** simple, cauline, alternate, opposite, or whorled in 3's, gen clustered at short, lateral branch tips; stipules 0; petiole < 2 mm; blade leathery, entire or toothed, base obtuse to acute. **STAMINATE INFL:** cyme, raceme, or panicle, axillary, sometimes clustered on short, lateral twigs, minutely bracted. **PISTILLATE INFL** axillary; fl 1. **STAMINATE FL:** sepals

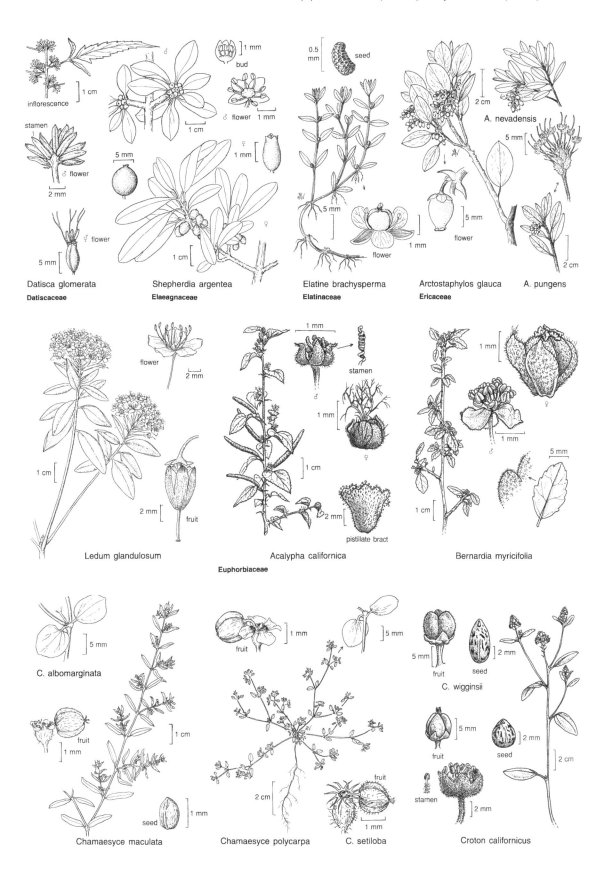

Datisca glomerata
Datiscaceae

Shepherdia argentea
Elaeagnaceae

Elatine brachysperma
Elatinaceae

Arctostaphylos glauca
Ericaceae

A. pungens

A. nevadensis

Ledum glandulosum

Acalypha californica

Bernardia myricifolia

Euphorbiaceae

C. albomarginata

Chamaesyce maculata

Chamaesyce polycarpa

C. setiloba

C. wigginsii

Croton californicus

4–10, 0.5–2 mm; petals 0; stamens 5–10, filaments glabrous or hairy; nectary disk ± minutely lobed. **PISTILLATE FL**: sepals 4–13, 2–5 mm; petals 0; nectary disk minutely lobed; ovary (2)3–5-chambered, styles = chambers, free, ± flattened, gen spreading. **FR** ± spheric, gen lobed, glabrous or short hairy, gen brown. **SEEDS** 1–2 per chamber, smooth, shiny; scar gen appendaged. 5 spp.: CA, AZ, Mex. (Latin: 4 seeds, from 4-lobed ovary in *T. dioicus*) [Dressler 1954 Rhodora 56:45–61] Key to species adapted by Margriet Wetherwax.

1. Llvs gen alternate, entire; pistillate pedicel gen < 3 mm; ovary gen 3-lobed, 3-chambered; styles gen 3;
    stamen filaments glabrous; se DMoj, DSon . . . . ***T. hallii***
1' Lvs gen opposite or 3-whorled, toothed; pistillate pedicel 8–15 mm; ovary 4-lobed, 4-chambered;
    styles gen 4; stamen filaments soft-hairy at base; DMtns . . . . ***T. ilicifolius***

T. hallii Brandegee (p. 295) **ST**: twigs sparsely short-strigose, becoming glabrous. **LVS** gen alternate, gen clustered on short, lateral twigs; blade 2–12 mm, oblanceolate to obovate, tip obtuse to rounded, margin entire. **STAMINATE INFL**: cyme; fls gen 1–5, gen clustered on short, lateral twigs; pedicel 3–5.5 mm. **STAMINATE FL**: sepals 4–6, ± round; stamens 4–8, filaments 1.5–2.5 mm, glabrous. **PISTILLATE FL**: pedicel 0.5–1(3) mm; sepals gen 5, 2–5 mm, ovate to deltate; ovary densely and finely gray-tomentose, chambers gen 3, sometimes 2 or 4, style 1.5–2 mm. **FR** 8–12 mm, 6–10 mm wide, finely tomentose. Rocky slopes, washes; < 1200 m. **se DMoj, DSon**; w AZ. ✤ TRY. Mar–May

T. ilicifolius Cov. & Gilman (p. 295) HOLLY-LEAVED TETRA-COCCUS **ST**: twigs sparsely and finely brown-tomentose, becoming glabrous. **LVS** gen opposite or 3-whorled; blade 15–30, ovate to widely elliptic, sometimes narrowly ovate, tip obtuse to acute, margin toothed, teeth 8–20 per lf. **STAMINATE INFL**: panicle, ± dense; fls ± sessile; pedicel << 0.5 mm. **STAMINATE FL**: sepals 7–9, ± linear to lanceolate; stamens 7–9, filaments 2–3 mm, base soft-hairy. **PISTILLATE FL**: pedicel 8–15 mm; sepals 5–8, 2–4 mm, widely lanceolate to ovate; ovary densely tomentose, chambers 4, style ± 3 mm. **FR** 8–9 mm, 6–8 mm wide, brown tomentose. RARE. Dry, rocky slopes; 600–1700 m. **n DMtns (Grapevine, Panamint mtns)**. In cult. May–Jun

TRAGIA NOSEBURN

Per < 0.5 m, monoecious; hairs stinging, nettle-like. **STS** spreading to erect, branched, sometimes twining. **LVS** gen simple, cauline, alternate; stipules persistent. **INFL**: raceme, terminal or opposite lf; staminate fls above pistillate fls. **STAMINATE FL**: sepals 3(–5); petals 0; stamens 3–6(50); nectary 0. **PISTILLATE FL**: sepals 4–8; petals 0; ovary 3-chambered, styles simple, ± fused at base. **FR** ± spheric. **SEEDS** 1 per chamber, smooth or ± rough; scar not appendaged. ± 100 spp.: trop, warm temp worldwide. (Latin: Tragus, name for Hieronymus Bock, German herbalist, 1498–1554) [Miller & Webster 1967 Rhodora 69:241–305]

T. ramosa Torrey (p. 295) Pl rough-hairy. **ST** 1–3 dm. **LF**: stipules 1–4.5 mm, lanceolate to ovate; petiole 2–20 mm; blade 1–2 cm, lanceolate to ovate, base truncate to ± lobed, margin coarsely, sharply toothed. **INFL** 0.5–1 cm, ± spreading; pedicels 1–2 mm; staminate fls 2–4; pistillate fl 1. **STAMINATE FL**: sepals 4–5, ± 1 mm, recurved; stamens 3–6, = sepals, filaments ± flattened. **PISTILLATE FL**: sepals 5, 1.5–2 mm; ovary < 2 mm

diam, puberulent to finely bristly, styles fused in lower 1/3. **FR** 3–4 mm, 6–8 mm wide, depressed-spheric, sparsely and finely bristly. **SEED** 2.5–3.5 mm, ± spheric. Dry, rocky slopes, shrubland, pinyon/juniper woodland; 900–1700 m. **DMtns (Clark, New York, Providence mtns)**; to c US, TX, Mex. [*T. stylaris* Muell. Arg.] Apr–May

FABACEAE [LEGUMINOSAE] LEGUME FAMILY

Ann to tree. **LVS** gen compound, alternate, stipuled; lflets gen entire. **INFL**: gen raceme, spike, umbel or head; fls sometime 1–2 in axils. **FLS** gen bisexual, gen bilateral; hypanthium gen flat or cup-like; sepals gen 5, fused; petals gen 5, free, or the 2 lower ± fused; stamens 1–many, often 10 with 9 filaments at least partly fused, 1 (uppermost) free; pistil 1, ovary superior, gen 1-chambered, ovules 1–many, style, stigma 1. **FR**: legume, sometimes incl a stalk-like base above receptacle, dehiscent, or indehiscent and breaking into 1-seeded segments, or indehiscent, 1-seeded, and achene-like. **SEEDS** 1–several, often ± reniform, gen hard, smooth. ± 650 genera, 18,000 spp.: worldwide; with grasses, requisite in agriculture and most natural ecosystems. Many cult, most importantly *Arachis*, peanut; *Glycine*, soybean; *Phaseolus*, beans; *Medicago*; *Trifolium*; and many orns. [Isely, Duane. 1998. Native and naturalized Leguminosae (Fabaceae) of the U.S. Provo, UT: Monte L. Bean Life Science Museum, Brigham Young University] Family description by Duane Isely, key to genera adapted by Margriet Wetherwax.

Key to Groups

1. Fl radial; calyx, corolla gen inconspicuous; petals gen fused, lobes not overlapping in bud; stamens
    10–many, often long-exserted; lf 2-pinnate (Mimosoideae) . . . . **Group 1**
1' Fl gen bilateral, sometimes ± radial; calyx, corolla gen conspicuous; petals overlapping in bud, free or
    2 lowermost ± fused; stamens 1–10, gen ± incl; lf 1- or 2-pinnate (sometimes simple)
    2. Fl slightly bilateral; sepals ± free; stamens exposed, visible without manipulation, free;
        lf 1- or 2-pinnate (Caesalpinioideae) . . . . **Group 2**
    2' Fl evidently bilateral, upper petal (banner) outside lateral ones (wings) in bud (petal position not evident
        in some *Dalea*); sepals fused basally; stamens gen hidden within lower petals (keel) (exposed in some
        Dalea), gen all or 9 filaments fused; lf 1-pinnate (lflets often 3), some palmately compound (esp
        Lupinus) or simple (Papilionoideae) . . . . **Group 3**

Group 1: Mimosoideae

1. Stamens 10; 1° lflets gen 1–2 pairs .... **PROSOPSIS**
1' Stamens > 10; 1° lflets gen several pairs
  2. Filaments free; infl a spike .... **ACACIA**
  2' Filaments fused below, free above; infl a head .... **CALLIANDRA**

Group 2: Caesalpinioideae

1. Lf 1-compound, pl unarmed or main lf axis gen > 2 cm, tip with 1 weak spine .... **SENNA**
1' Lf 2-compound, main axis gen < 2 cm, a strong spine, or subtended by a strong spine gen < 2 cm, main
    axis not a spine at tip
  2. Pl unarmed
    3. Shrub, > 2 m .... [2]**CAESALPINIA**
    3' Per, < 30 cm .... **HOFFMANNSEGGIA**
  2' Pl armed
    4. Pl with scattered prickles; sepals not all alike; fr indehiscent .... [2]**CAESALPINIA**
    4' Pl with thorns in lf axils (see lf scars); main lf axis, main 1° lflet axes falling with 2° lflets ... **CERCIDIUM**

Group 3: Papilionoideae

1. Shrub, tree
  2. Pl armed with prickles, spines, or thorns
    3. Lvs simple, sometimes small or falling early
      4. Pl gland-dotted; corolla blue to pink-purple; seed 1 .... [4]**PSOROTHAMNUS**
      4' Pl not gland-dotted; corolla reddish; seeds few-several .... **ALHAGI**
    3' Lvs all or mostly compound, gen persistent
      5. Pl gland-dotted; fr indehiscent; seed 1 .... [4]**PSOROTHAMNUS**
      5' Pl not gland-dotted; fr dehiscent (sometimes slowly so); seeds gen several
        6. Lf odd-pinnate; corolla white or pink; fr flat .... [2]**ROBINIA**
        6' Lf even-pinnate; corolla 2-colored; fr plump, gen narrowed between seeds .... **OLNEYA**
  2' Pl unarmed
    7. Filaments all fused; lvs palmately compound .... [2]**LUPINUS**
    7' Filaments all free or 9 fused, 1 (uppermost) free or 0; lvs gen pinnately compound
      8. Infl gen an umbel, fls sometimes 1–3; corolla gen yellow .... [4]**LOTUS**
      8' Infl a raceme; corolla not yellow
        9. Fl 15–25 mm; pl not gland-dotted; fr dehiscent; seeds several .... [2]**ROBINIA**
        9' Fl 6–10 mm; pl gland-dotted; fr indehiscent; seed 1(see also *Marina*) .... [4]**PSOROTHAMNUS**
1' Ann, per, subshrub
  10. Lf palmately compound; lflets gen 3–9
    11. Lflets gland-dotted; fr indehiscent or breaking transversely; seed 1
      12. Lvs mostly basal or clustered; lflets widely obovate to oblanceolate; fr incl in calyx exc for
          conspicuous beak .... **PEDIOMELUM**
      12' Lvs cauline; lflets obovate to linear; fr exserted from calyx .... **PSORALIDIUM**
    11' Lflets not gland-dotted; fr indehiscent or dehiscent through longitudal sutures; seeds gen several
      13. Filaments of all stamens fused at least in basal 1/2; lflets gen 5–9, entire .... [2]**LUPINUS**
      13' Filaments of 9 stamens fused, 1 (uppermost) free; lflets 3–5, entire, or margin toothed or wavy
        14. Lflets 5, lower 2 in stipular position, others palmate, stipules reduced to bumps or not apparent;
            infl an umbel; corolla yellow .... *Lotus corniculatus*
        14' Lflets gen 3, lower 2 not in stipular position, stipules gen papery or membranous, not reduced
            to bumps, rarely lflet-like, if so then infl not an umbel and corolla not yellow
          15. Lflet entire; fr not enclosed in corolla .... [3]**ASTRAGALUS**
          15' Lflet margin ± toothed or wavy; fr enclosed in persistent corolla .... **TRIFOLIUM**
  10' Lf pinnately to subpalmately compound (axis apparent beyond most lflets); lflets 3–many
    16. Lflets 3; fr (exc in *Phaseolus*) gen indehiscent
      17. Keel petals spirally coiled; lflet gen lobed — trailing ot twining vine .... **PHASEOLUS**
      17' Keel petals not spirally coiled; lflet margin entire, toothed, or waxy
        18. Lflet margin toothed or wavy — fr ovoid, reniform, or ± coiled, 1–several-seeded
          19. Fr spirally coiled, rarely only sickle-shaped; seeds > 2 .... [2]**MEDICAGO**
          19' Fr not spirally coiled; seeds 1–2
            20. Fr reniform, with curved ridges; corolla 2–3 mm .... [2]**MEDICAGO**
            20' Fr ovoid, with transverse ridges or a ± distinct network of lines; corolla 2.5–7 mm ..... **MELILOTUS**
        18' Lflet margins not toothed, not wavy
          21. Infl a raceme; fr exserted from calyx .... [3]**ASTRAGALUS**

21' Infl an umbel; fr exserted from calyx or incl exc for beak . . . . ⁴**LOTUS**
16' Lflets > 3 on all or most lvs; fr dehiscent
  22. Lf even-pinnate, main axis extending as a tendril
    23. Lflets 30–60; fr 15–20 cm, 2–3 mm wide, linear . . . . **SESBANIA**
    23' Lflets < 30; fr < 8 cm > 5 mm wide, ± oblong
      24. Style ± flat, puberulent on concave side ± 1/3–1/2 length; lflets ± rolled in bud . . . . **LATHYRUS**
      24' Style gen round in ×-section, puberulent all around tip; lflets folded in bud . . . . **VICIA**
  22' Lf odd-pinnate, main axis ending as a lflet
    25. Pl gland-dotted on sts, lflets, or both; fr indehiscent
      26. Fr several-seeded, long-exserted from calyx, gen prickly; lflets 6–10 mm wide . . . . . **GLYCYRRHIZA**
      26' Fr 1-seeded, incl in calyx, glandular; lflets gen < 6 mm wide
        27. Petals all arising from receptacle; stamens 10; infl sometimes head-like . . . . ⁴**PSOROTHAMNUS**
        27' Petals, exc banner, arising from side or top of column of fused filaments; stamens 5 or 9–10;
          infl not head-like
          28. Pl prostrate or decumbent . . . . ²**DALEA**
          28' Pl ascending or erect
            29. Infl a dense spike; stamens 5 . . . . ²**DALEA**
            29' Infl an open raceme; stamens 9–10 . . . . **MARINA**
    25' Pl not obviously gland-dotted; fr dehiscent or indehiscent
      30. Infl an umbel . . . . ⁴**LOTUS**
      30' Infl a spike or raceme
        31. Stigma or style tip finely hairy; stipules spiny . . . . **PETERIA**
        31' Stigma and style glabrous; stipules not spiny
          32. Keel tip rounded to acute, rarely short-beaked; fr 1-chambered or, if 2-chambered, septum
            arising from lower suture (rarely a narrow flange from upper suture also present) . . . ³**ASTRAGALUS**
          32' Keel tip distinctly beaked; fr gen 2-chambered, the partial or complete septum arising from
            upper suture . . . . **OXYTROPIS**

ACACIA

Elizabeth McClintock

Tree, shrub, armed or unarmed. **LVS** even-2-pinnate or, if simple, true blades 0, petioles and midribs blade-like (comprising phyllodia), gen alternate, gen evergreen; axes with prominent raised glands or not. **INFL**: heads, spheric, gen axillary, these solitary or in racemes or panicles, or fls in spikes. **FL** radial; sepals, petals inconspicuous; stamens many, conspicuous, exserted, free. **FR** gen dehiscent, sometimes tardily so, flat or ± cylindric. ± 1200 spp.: trop, subtrop, esp Australia. (Greek: sharp point) [Whibley 1980 Acacias of South Australia; Clarke et al. 1989 Systematic Botany 14:549–564] Australian spp. cult, sometimes naturalized and spreading in CA (seed arilled, stalk often elongated, encircing seed or not).

A. greggii A. Gray (p. 295) CATCLAW   Shrub, small tree 2.5–7 m, with curved prickles on st. **ST**: twig ± angled, hairy. **LVS** 2-pinnate, alternate, clustered on short-shoots or not, deciduous, gray-green; 1° lflets 2–3 pairs, separated, 1–1.5 cm; 2° lflets < 10 pairs, ± overlapped, < 6 mm, oblong. **INFL**: spikes 1 or more, clustered with lvs on short-shoots, gen > lf. **FL** light yellow. **FR** 5–15 cm, recurved or twisted, flat, narrowed between seeds, glaucous, brown. **SEED**: stalk 0. Common. Rocky slopes, flats, washes; 100–1400 m. **D**; to TX, Mex. ✿ SUN,DRN:**7–10,12,14–16**,17,18,**19–24**&IRR: **11,13**. Apr–Jun

ALHAGI

Duane Isely

Per, shrub; thorns axillary; rhizome spreading. **LVS** simple, small. **INFL**: raceme, axillary; main axis a thorn. **FL**: 9 filaments fused, 1 free. **FR** indehiscent, oblong, round in ×-section, narrowed between seeds. ± 3 spp.: Medit, w Asia. (Arabic: pilgrim)

A. pseudalhagi (M. Bieb.) Desv. (p. 295) CAMEL THORN   **ST** much branched, 3–10 dm, greenish. **LF** 7–20 mm, elliptic or obovate. **FL**: corolla 8–9 mm, red-purple. **FR** 1–3 cm, narrowed between seeds, becoming glabrous; stalk-like base short. 2*n*=16. Uncommon. Arid agricultural areas; esp < 500 m. GV, **s SNE**; **D**; sporadic to w TX; native to w Asia. [*A. camelorum* Fischer; *A. maurorum* Medikus] Desert forage, source of manna. NOXIOUS WEED. Most infestations have been eradicated.   Jun–Jul

ASTRAGALUS

Richard Spellenberg

Ann or per from crown, glabrous to hairy; hairs sometimes forked at base, branches parallel with lf surface, sometimes very unequal. **ST** 0 or prostrate to erect. **LVS** odd-1-pinnate; lflets gen jointed to midrib; stipules membranous, sometimes fused around st at st base. **INFL**: raceme, axillary, sometimes head- or umbel-like; fls 2–many. **FL** bilateral; calyx 5-lobed; banner outside wings in bud, keel blades with small protrusion at base locking into pit on adjacent

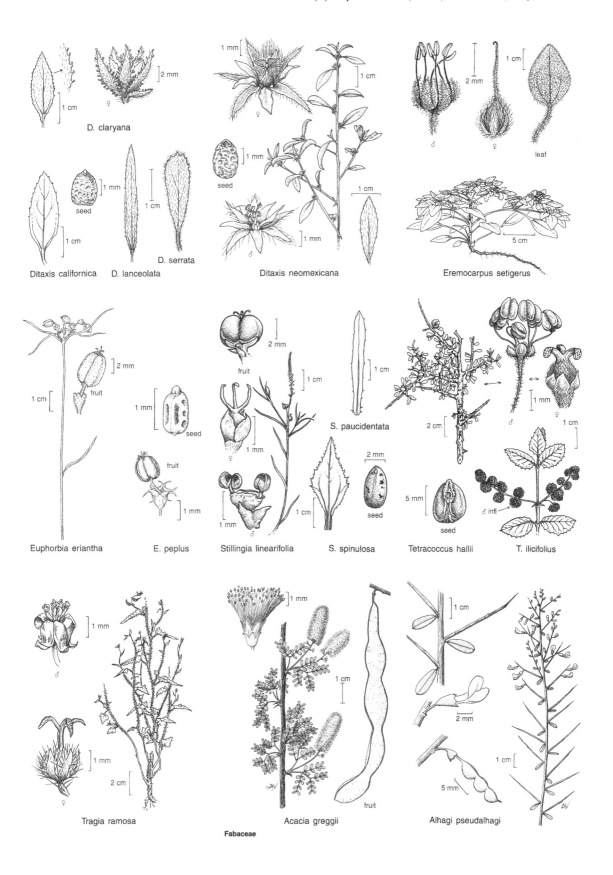

D. claryana

seed

Ditaxis californica   D. lanceolata   D. serrata

seed

Ditaxis neomexicana

leaf

Eremocarpus setigerus

fruit   seed   fruit

Euphorbia eriantha   E. peplus

fruit   seed   S. paucidentata

Stillingia linearifolia   S. spinulosa

Tetracoccus hallii   T. ilicifolius

Tragia ramosa   Acacia greggii   fruit   Alhagi pseudalhagi

Fabaceae

wing; 9 filaments fused, 1 free; ovary (and fr) gen sessile, style slender, stigma minute. **FR** gen 1- or ± 2-chambered, often mottled, gen becoming ± dry; placenta on upper suture. **SEEDS** 2–many, smooth, compressed, ± notched at attachment scar. > 2000 spp.: ± worldwide (380 in N.Am, 94 in CA incl many rare taxa). (Greek: ankle-bone or dice, perhaps from rattling of seeds within fr) [Barneby 1964 Mem NY Bot Gard 20:1–1188; Isely 1986 Iowa State J Res 61:157–289] Very difficult; both fl and fr needed for identification; many good spp. appear similar; some spp. complexes need study. Taxa near province boundaries may appear in > 1 key. Vars. keyed under spp. for simplicity; spp. with vars. so identified in key. Fr length incls beak and any stalk-like base unless fr body specified. Keys adapted by Martin F. Wojciechowski.

1. Corolla bright red, banner 35–41 mm; calyx 18–24 mm; fr 25–40, densely woolly . . . . *A. coccineus*
1' Corolla not red, banner < 35 mm; calyx < 18 mm; fr often < 25 mm, often not woolly
  2. Lflet ± 1 mm wide, spine tipped; raceme 1–3-fld, ± incl in axil; peduncle inconspicuous or 0; fl and
    fr 4–9 mm wide . . . . *A. kentrophyta* vars.
  2' Lflet gen > 1 mm wide, not spine-tipped; raceme often > 3-fld, well exserted from axil; peduncle
    conspicuous; fl 3–30 mm; fr 2–60 mm
   3. Plants of the Great Basin Floristic Province
    4. Terminal lflet not or only obscurely jointed to midrib, joint unlike that of lateral lflet; lateral lflets
      often reduced or 0, spaces between gen >> 1.5 × lflet width; lvs few, sparse . . . . . [3]*A. serenoi* var. *shockleyi*
    4' Terminal lflet clearly jointed to midrib, joint like that of lateral lflet; lateral lflets present, spaces
      between rarely > 1.5 x lflet width; lvs gen many, not sparse
     5. Ann, slender, rarely persisting into second season
      6. Fr not bladdery, not inflated, 2–3 mm wide; lflet upper surface ± appressed-hairy . . . . [2]*A. acutirostris*
      6' Fr inflated, 6–10 mm wide; lflet upper surface ± glabrous . . . . *A. geyeri* var. *geyeri*
     5' Per, often coarse
      7. Lf hairs forked at base
       8. Lflets 1–7, gen 3, crowded near lf tip; st gen 0; fls 1–8, ascending or spreading
         . . . . [2]*A. calycosus* var. *calycosus*
       8' Lflets 7–25; st 1.5–5.5 dm; fls 20–many, reflexed, overlapping . . . . *A. canadensis* var. *brevidens*
      7' Lf hairs simple
       9. Fr densely white-hairy, resembling a small ball of cotton; st 0–14 cm; pl hairs dense, of 2 kinds or not
        10. Sts forming a thickened crown covered by persistent lf bases; longer hairs of lf gen straight,
          some spreading; some fr hairs curly and short, some long and ± straight (see also *A. platytropis*)
          . . . . [2]*A. newberryi* var. *newberryi*
        10' Sts tufted or matted, lf bases not persistent; longer hairs of lf wavy, tangled; fr hairs all ±
          wavy or all straight . . . . [2]*A. purshii* vars.
       9' Fr sometimes silvery-hairy, but not resembling a small ball of cotton; st often > 10 cm; pl
        hairs sparse or dense, gen not of 2 kinds
        11. Calyx base strongly asymmetric, pedicel attached at lower side, upper ± pouched, petals white
         to yellowish; fr curved 1/4 to ± full circle, base stalk-like
         12. St hairs subappressed or incurved; fr 2.5–4.5 mm wide; lowest stipules not fused
          . . . . *A. curvicarpus* var. *curvicarpus*
         12' St hairs spreading or reflexed; fr 4–8 mm wide; lowest stipules fused around st into low sheath
          . . . . *A. gibbsii*
        11' Calyx base ± symmetric, pedicel attached ± at middle, upper side not conspicuously
         pouched; petals white, yellowish, or purplish pink; fr straight to curved ± full circle, base
         stalk-like or not
         13. Lower stipules fused around st into short sheath (often ruptured by st expansion or missing
          on specimens without base)
          14. Stalk-like fr base > 0.5 mm
           15. Banner < 6 mm; fr body narrow, 7–11 mm, triangular in ×-section . . . . *A. johannis-howellii*
          15' Banner > 8 mm; fr bladdery, > 15 mm, ± round in ×-section. . . . . *A. whitneyi* var. *whitneyi*
         14' Stalk-like fr base < 0.5 mm or 0 (see also *A. johannis-howellii*, 15.)
          16. Pls cespitose; st < 2 cm; fr bladdery . . . . *A. platytropis*
          16' Pls tufted or loosely matted; st > 5 cm; fr not bladdery
           17. Crown at surface; lowest few internodes above surface, densely white-tomentose; fr
            bluntly triangular in ×-section . . . . *A. andersonii*
           17' Crown below surface; lowest few internodes below surface, ± glabrous; fr oblong or
            bluntly triangular in ×-section
            18. Keel 6.7–8 mm; fr oblong in ×-section, not fully 2-chambered, ± narrowed at sutures
             . . . . *A. monoensis*
            18' Keel 10–12.2 mm; fr bluntly triangular in ×-section, fully 2-chambered (bilocular),
            not narrowed at sutures . . . . *A. sepultipes*
        13' Lower stipules not fused around st into sheath
         19. Fls 1–4 per infl, strongly ascending; corolla pink-purple; sts low, tufted or matted, softly
          gray-hairy . . . . *A. argophyllus* var. *argophyllus*

19' Fls gen > 4 per infl, strongly ascending to reflexed; corolla ± purple to white; sts low or not,
     tufted or not, rarely matted, ± glabrous to strigose or spreading-hairy
  20. Fr 1-chambered at mid-section
    21. Ovary, fr glabrous; fr bladdery, widely ovoid, stalk-like base 3–10 mm . . . . . . ²**A. oophorus** vars.
    21' Ovary, fr ± hairy; fr bladdery or only swollen, ovoid or incurved, stalk-like base 0 or 2–5 mm
      22. Fr base 2–5 mm, corolla pink-purple . . . . **A. inyoensis**
      22' Fr sessile; corolla white to purple
        23. St ascending or erect, wiry; lflets > 4 × longer than wide, spaces between >> width;
           keel > 10 mm . . . . ²**A. casei**
        23' St ± prostrate to decumbent, ± mat-forming; lflets < 2.5 × longer than wide, spaces
           between ± = or < width; keel < 10 mm
          24. Banner 10–15.5 mm; herbage, fr hairs stiff, straight, appressed,
             < 0.7 mm . . . . **A. iodanthus** var. **iodanthus**
          24' Banner 9–10 mm; herbage, fr hairs soft, wavy, spreading, ± 1 mm . . . . . ²**A. pseudiodanthus**
  20' Fr ± 2-chambered at mid-section
    25. Banner 4.8–6.1 mm; peduncles often in 2's or 3's in upper axils . . . . **A. lemmonii**
    25' Banner > 7 mm; peduncles gen solitary in upper axils
      26. Stipules 7–17 mm, whitish or straw-colored; fr, herbage hairs ± 1.5–2.5 mm, spreading;
         fr pendent . . . . **A. malacus**
      26' Stipules < 6 mm, often greenish; fr, herbage hairs 0 or gen < 1.5 mm, ± appressed;
         fr spreading or reflexed
        27. Crown below surface; fr curved at least 1/2 circle, wider than deep . . . . . . ²**A. pseudiodanthus**
        27' Crown at surface; fr at most slightly curved, gen ± as deep as wide . . . . . ⁵**A. lentiginosus** vars.
3' Plants of the Desert Floristic Province
  28. Terminal lflet not or only obscurely jointed to midrib, joint unlike that of lateral lflet; lateral
     lflets often reduced or 0, spaces between gen >> 1.5 × lflet width; lvs few, sparse
    29. Banner 10–14.2 mm; lflets 3–13; fr half-ovoid to ± spheric, bladdery . . . . . **A. magdalenae** var. **peirsonii**
    29' Banner 17–26 mm; lflets 5–11; fr plumply oblong, not bladdery . . . . ³**A. serenoi** var. **shockleyi**
  28' Terminal lflet clearly jointed to midrib, joint like that of lateral lflet; lateral lflets present,
     spaces between rarely > 1.5 × lflet width; lvs gen many, not sparse
    30. Hairs dense, forked at base; lflets 1–7, crowded near lf tip . . . . ²**A. calycosus** var. **calycosus**
    30' Hairs 0 to dense, not forked; lflets often > 7, gen not crowded near lf tip
      31. Ann (or fl 1st year and appearing so)
        32. Fl, fr strongly ascending in dense, head-like racemes; fr ± spheric, 2–4 mm . . . . **A. didymocarpus** vars.
        32' Fl, fr ± ascending to spreading in often open racemes; fr linear to ± ovate-inflated, > 4 mm
          33. Keel 10–21 mm
            34. Pl densely shaggy-hairy; st gen < 7 cm, prostrate, often nearly 0; fr 1-chambered . . . ²**A. tidestromii**
            34' Pl ± glabrous to strigose; st > 9 cm, prostrate to erect; fr 1- or 2-chambered
              35. Keel 10–13 mm; fr 2-chambered at middle . . . . ⁵**A. lentiginosus** vars.
              35' Keel 17–21 mm; fr 1-chambered . . . . ²**A. crotalariae**
          33' Keel 2–10.5 mm
            36. Fr bladdery, walls thinly papery, translucent
              37. Infl gen 10–35-fld
                38. Lflets 11–33, fr 1-chambered throughout . . . . **A. hornii** var. **hornii**
                38' Lflets 11–19; fr 2-chambered below beak . . . . ⁵**A. lentiginosus** vars.
              37' Infl gen 2–10-fld
                39. Corolla whitish; lflets 11–19; upper suture of fr convex; immature seeds 13–21
                   . . . . ²**A. allochrous** var. **playanus**
                39' Corolla purplish; lflets 7–19; upper suture of fr straight or convex; immature seeds 7–24
                  40. Keel 4.8–6 mm; lflet gen 11–19; upper suture of fr much less convex than lower; immature
                     seeds 7–14 . . . . **A. insularis** var. **hardwoodii**
                  40' Keel 5.9–6.6 mm; lflets gen 7–13; upper and lower sutures of fr ± equally convex;
                     immature seeds 19–24 . . . . **A. nutans**
            36' Fr linear or swollen, not bladdery, walls papery (but not esp translucent) or leathery
              41. Infl 20–40-fld . . . . ²**A. palmeri**
              41' Infl < 20-fld
                42. Fr 1-chambered, swollen, half-ovate or -elliptic in side view, ± ovate or round in ×-section
                  43. Pl, fr hairs appressed, ± straight, ± obscuring surface; immature seeds 3–7 . . . . **A. aridus**
                  43' Pl, fr hairs ascending or spreading, ± wavy, not obscuring surface; immature seeds 10–19
                     . . . . **A. sabulonum**
                42' Fr 2-, rarely 1-chambered, then often ± linear, curved, ± 2- or 3-sided in ×-section
                  44. Upper surface of lflets sparsely hairy; infl 1–6-fld; keel < 6 mm; fr often < 3 mm wide
                    45. Lflets of upper lvs gen blunt, notched at tip; fr quickly deciduous, maturing pale brown
                       . . . . ²**A. acutirostris**
                    45' Lflets of upper lvs gen acute at tip; fr persistent, maturing dark brown or blackish
                       . . . . **A. nuttallianus** vars.

44' Upper surface of lflets silvery-canescent; infls 3–16-fld; keel > 6 mm; fr often > 3 mm wide
  46. Fr 10–18 mm, 2.8–3.5 mm wide, incurved, ± 3-sided; immature seeds 8–11;
    Cushenbury Canyon (of ne SnBr, adjacent DMoj) . . . . ²*A. albens*
  46' Fr 13–32 mm, 3.5–8.5 mm wide, straight, ± 2-sided, or incurved, ± 3-sided; immature
    seeds 20–30; widespread in DMoj . . . . ²*A. mohavensis* vars.
31' Per
  47. Fr, gen ovary, glabrous (*A. nutans* also sometimes glabrous, under 47')
    48. Lflet hairs 0 or few, restricted to margins, midrib on 1 or both surfaces; fr base stalk-like
      49. Fls early ascending, then sometimes reflexed; keel 9.5–10.6 mm; fr ± ascending, ±
        2-chambered, base 5–12 mm . . . . *A. cimae* vars.
      49' Fls strongly ascending; keel 11–19 mm; fr erect, 1-chambered, base 2–7 mm
        . . . . *A. preussii* var. *preussii*
    48' Lflet hairs ± throughout 1 or both surfaces; fr base stalk-like or not
      50. Keel 6–9.4 mm; fr 2-chambered, at least at middle
        51. Pl often coarse, leafy, gen in open places; st not wiry; fr sessile, swollen, beak strongly
          compressed side-to-side, triangular in side view, 1-chambered . . . . ⁵*A. lentiginosus* vars.
        51' Pl slender, sparsely leafy, often in shelter of shrubs; st ± wiry; fr base stalk-like, 1–5 mm
          52. Fls 10–25 per infl, ascending; fr ascending, 3-sided . . . . *A. bernardinus*
          52' Fls 5–15 per infl, early ascending, then reflexed; fr pendent, ± 2-sided . . . . *A. jaegerianus*
      50' Keel 9.7–19 mm; fr 1- or 2-chambered
        53. Calyx glabrous; banner recurved ± 85°; fr bladdery, 1-chambered, base stalk-like
          . . . . ²*A. oophorus* var. *oophorus*
        53' Calyx strigose; banner recurved < 50°; fr ovoid or narrow, not bladdery, ± 2-chambered,
          base stalk-like or not
          54. Corolla pink-purple; lflets very narrow, upper surface often paler than lower (silvery vs
            greenish); fr erect, sessile; n Inyo Co. . . . . ³*A. serenoi* var. *shockleyi*
          54' Corolla white to cream; lflets narrow or wide, upper surface not paler than lower; fr
            ascending to pendent, stalk-like base 1–10 mm; w or sw edge of D
            55. Fls early spreading, then reflexed; fr pendent . . . . *A. trichopodus* var. *phoxus*
            55' Fls and fr spreading-ascending
              56. Banner 15–22 mm; fr 15–27 mm, swollen, ± compressed side-to-side; lflets
                sometimes > 17 . . . . *A. pachypus* vars.
              56' Banner 12.6–15.7 mm; fr 24–42 mm, ± linear, thin-walled, 3-sided; lflets 17–27
                . . . . *A. tricarinatus*
  47' Fr, ovary with at least some hairs (*A. nutans* sometimes glabrous)
    57. Pl conspicuously woolly- or wavy-hairy; sts < 9 cm, gen densely tufted
      58. Lflets > 21, at least on some lvs; infl 10–35-fld; se DMoj . . . . *A. tephrodes* var. *brachylobus*
      58' Lflets not > 21 on any lf (rarely to 23 in *A. layneae*); infl often < 10-fld
        (10–45-fld in *A. layneae*); widespread
        59. Keel 18–28 mm; fr 1-chambered, densely white-hairy
          60. Calyx with more black than white hairs; fr 25–50 mm; lf bases not notably persistent
            on crown; ne DMtns (e of Death Valley) . . . . *A. funereus*
          60' Calyx gen with more white than black hairs; fr 13–28 mm; lf bases persistent on crown;
            widespread, incl DMtns . . . . ²*A. newberryi* var. *newberryi*
        59' Keel gen < 20 mm; fr 2-chambered if densely white-hairy
          61. Infl 10–45-fld; corolla 2-colored, whitish with keel tip, wing tips, and sometimes banner
            tip purplish; fr hairs spreading — pls with deep rhizomes (± stless pls of *A. minthorniae*
            will key out here; see 65.) . . . . ²*A. layneae*
          61' Infl 3–16-fld; corolla pink-purplish or ± 2-colored (± as at 61.); fr hairs appressed or
            densely white-cottony
            62. Keel 11.5–20.8 mm; petals pink-purple; fr often ± 2-chambered, hairs spreading or
              curly, densely white-cottony . . . . ²*A. purshii* var. *tinctus*
            62' Keel 10–12 mm; petals whitish, tinged dull purple; fr 1-chambered, hairs gen appressed
              . . . . ²*A. tidestromii*
    57' Pl hairy to ± glabrous; sts gen > 9 cm (if < 9 cm, pl not woolly- or wavy-hairy), tufted or not
      63. St < 16 cm, from deep rhizome; corolla 2-colored, whitish with keel tip, wing tips, and
        sometimes banner tip purplish . . . . ²*A. layneae*
      63' St often > 15 cm, from crown at or slightly below surface; corolla 1- or 2-colored
        64. Lflet hairs 0 or few on margins, midrib on 1 or both sides; stalk-like fr base 0
          . . . . *A. preussii* var. *laxiflorus*
        64' Lflet hairs at least few, ± throughout 1 or both sides; stalk-like fr base 0 or < 3 mm
          65. St, lf hairs spreading, sometimes curly, 0.8–1.5 mm; fr shaggy-hairy, ± erect
            . . . . *A. minthorniae* var. *villosus*
          65' St, lf hairs appressed, if spreading then < 0.8 mm; fr often strigose or silky-canescent,
            erect to reflexed or pendent

66. St slender, wiry, incurved, < 15 cm; petioles persistent, wiry; infl 1–4-fld; — pls
of limestone, resembling an unkempt nest . . . . ***A. panamintensis***
66' St slender or coarse, if wiry then not incurved, often > 15 cm; petioles deciduous,
not wiry; infl often > 4-fld
67. Petals dingy white at least at base, dull lilac on margins or tips; wings ± twisted,
tips shortly fringed or notched; fr pendent, ± compressed side-to-side, 2-chambered
. . . . ***A. atratus*** var. ***mensanus***
67' Petals white to ± purple; wings plane or curved, tips not fringed, not notched; fr erect
to reflexed, ± 3-sided or round in ×-section, sometimes plump and widely grooved on
upper side, 1- or 2-chambered
68. Fr 2-chambered in middle, beak often 1-chambered; banner recurved 30–50°
69. Lflets gen > 11; fr ×-section ± round, upper suture gen ± sunken in a wide channel;
pl gen robust — fr beak widely triangular, 1-chambered . . . . ⁵***A. lentiginosus*** vars.
69' Lflets 3–11; fr ± 2- or 3-sided, upper suture raised; pl robust or delicate
70. Fr 10–18 mm, 2.8–3.5 mm wide, incurved, ± 3-sided; immature seeds 8–11;
Cushenbury Canyon (ne SnBr, adjacent DMoj) . . . . ²***A. albens***
70' Fr 13–32 mm, 3.5–8.5 mm wide, straight, ± 2-sided, or incurved, ± 3-sided;
immature seeds 20–30; widespread in DMoj . . . . ²***A. mohavensis*** vars.
68' Fr 1-chambered throughout; banner recurved 45–90°
71. Banner > 11 mm; calyx gen > 7 mm; fr stiffly papery or leathery
72. Keel < 14 mm; fr reflexed, 5–10 mm wide; SNE, DMtns (Inyo Co.) . . . . ²***A. casei***
72' Keel > 17 mm; fr ascending or spreading, 10–14 mm wide; DSon . . . . ²***A. crotalariae***
71' Banner < 11 mm; calyx 4–7 mm; fr bladdery, ± papery
73. Infl gen 10–40-fld; petals gen bright pink-purple; w edge DSon . . . . ²***A. palmeri***
73' Infl often < 10-fld; petals whitish or pink-purple; e DMoj, c&e DSon
74. Petals whitish; lflets 11–19; fr beak short, obscure . . . . ²***A. allochrous*** var. ***playanus***
74' Petals pink-purple; lflets 7–13; fr beak deltate, strongly compressed side-to-side
. . . . ***A. gilmanii***

A. acutirostris S. Watson (p. 303) Ann; hairs ± appressed, ±
curved. **ST** prostrate or ascending, 2–30 cm. **LF** 1–4 cm; lflets
7–15, 2–8 mm, ± oblong, tips gen notched. **INFL**: fls 1–6, ±
spreading. **FL**: petals ± white, banner 4.7–7 mm, recurved ± 45°,
keel 4.3–5.8 mm. **FR** 12–30 mm, 2–3 mm wide, gently curved,
± 3-sided, pale brown, thin-walled but not bladdery, quickly deciduous; chambers 2. Sandy or gravelly areas; 600–1500 m. **SNE,
DMoj**. Like and often growing with *A. nuttallianus* (which has
fr persistent, ultimately dark brown or blackish, curved most
strongly near base; at least upper lflets acute). Apr–May

A. albens E. Greene (p. 303) CUSHENBURY OR SILVERY-WHITE
MILK-VETCH Ann, sometimes ± per, delicate; hairs dense, appressed, flat, silvery. **ST** ± prostrate, 2–30 cm, loosely matted.
LF 1–5.5 cm; lflets 5–9, 2–10 mm, ovate to obovate, tips blunt,
± notched. **INFL**: fls 5–14, widely spreading or reflexed. **FL**:
petals pink-purple, banner 7.3–9.5 mm, recurved ± 40°, keel ± =
banner, > wings. **FR** 10–18 mm, 2.8–3.5 mm wide, crescent-
shaped, ± 3-sided (lower grooved), stiffly papery, densely strigose; chambers 2. RARE. Rocky areas; 1200–1800 m.
Cushenbury Canyon (ne SnBr, **adjacent DMoj**). Threatened by
limestone mining. Mar–May

A. allochrous A. Gray var. ***playanus*** (M.E. Jones) Isely (p.
303) Ann or per, leafy, thinly, minutely strigose. **ST** ± prostrate
to erect, 1–5 dm. **LF** 2–12 cm; lflets 11–19, 5–20 mm, ± oblanceolate or oblong, tips shallowly notched or obtuse. **INFL**: fls
4–10, early ascending, then spreading or reflexed. **FL**: petals whitish, banner 5–7 mm, recurved ± 45°, keel 4–6 mm. **FR** spreading, 15–30 mm, 12–20 mm wide, bladdery, papery, minutely strigose; beak short, obscure; chamber 1. 2*n*=22. Sandy flats; ±
800 m. **e DMoj**; to TX, Mex, where fl color, fr size, vary. [*A.
wootoni* E. Sheldon var. *w.*] In CA, var. *playanus* seems ± distinct from other members of *Inflatai*, which are geog isolated; in
AZ and NM it intergrades thoroughly with other vars. of *A.
allochrous*; entire complex needs study. Mar–Jul

A. andersonii A. Gray (p. 303) Per, loosely tufted; hairs spreading or ascending, wavy, grayish, on lower internodes dense, interwoven, whitish. **ST** decumbent to ascending, 7–20 cm. **LF** 2–
10 cm; lower stipules fused around st into scarious sheath; lflets
9–21, 3–14 mm, elliptic to ± oblanceolate. **INFL**: fls 12–26, early

ascending, then reflexed. **FL**: petals white, sometimes tinged dull
purple, banner 9.5–14.5 mm, recurved ± 45°, keel 6.6–9 mm.
FR spreading or pendent, 10–18 mm, 3–5 mm wide, widely oblong, ± 3-sided (lower narrow, shallowly grooved), stiffly papery;
hairs wavy, long; upper suture thick, raised; chambers 2. 2*n*=24.
Gen disturbed flats, slopes; 1300–2200 m. MP, **n&e-c SNE**; w
NV. Apr–Jul

A. argophyllus Torrey & A. Gray var. ***argophyllus*** (p. 303)
SILVERLEAF MILKVETCH Per, cespitose, from heavy crown; hairs
on lvs dense, appressed or ascending, silvery. **ST** ± prostrate, <
15 cm. **LF** 2–15 cm; lflets 9–21, 4–15 mm, ± elliptic or ovate,
tips acute or obtuse. **INFL**: fls 1–4, ascending. **FL**: petals bright
pink-purple, banner 22–24 mm, keel 17–20 mm. **FR** 15–25 mm,
7–12 mm wide, ± widely lanceolate, straight or curved, densely,
loosely strigose, early fleshy, then stiffly leathery; chamber 1.
RARE in CA. Heavy alkaline or saline soil; 1400–2350 m. MP
(e Lassen Co.), **SNE (nw Inyo Co.)**; to ID, WY, UT. Apr–Jun,
–Aug at higher elevations

A. aridus A. Gray (p. 303) Ann, silvery silky-canescent. **ST**
decumbent or ascending, 3–30 cm. **LF** 2–9 cm; lflets 7–17, 4–16
mm, ± oblong or oblanceolate, tips acute to notched. **INFL** fls
3–9, ascending. **FL**: petals whitish, pink-tinged, or ± tannish pink,
banner 3.3–6.5 mm, recurved ± 40°, keel 3.5–5 mm. **FR** ascending, 10–17 mm, 4–7 mm wide, ± half-elliptic in side view, round
in ×-section, grayish, thickly papery; hairs appressed; beak triangular, strongly compressed side-to-side; chamber 1. Sandy places;
< 350 m. **DSon**; AZ, nw Mex. Highly variable in stature: low,
slender (drier spring seasons) or robust, coarse (moist seasons).
Feb–Apr

A. atratus S. Watson var. ***mensanus*** M.E. Jones (p. 303) DAR-
WIN MESA MILKVETCH Per, wiry, loosely matted or scrambling
through shrubland, minutely strigose. **ST** 3–25 cm. **LF** 1.5–15
cm; lflets 7–15, well separated, 3–16 mm, linear to ± ovate, tips
acute to shallowly notched. **INFL**: fls 4–18, reflexed. **FL**: petals
dingy white at least at base, dull lilac on margins or tips, banner
9.8–13.4 mm, recurved ± 90°, blade pinched to appear fiddle-
shaped, wings irregularly toothed, ± twisted, keel 8.2–10 mm.
FR pendent, 16–22 mm, ± 4 mm wide, ± linear-oblong, ± compressed side-to-side, stiffly papery or leathery, minutely strigose;

upper suture more convex than lower; tip slightly reflexed; chambers 2. RARE. With pinyon pine; 1700–1850 m. **DMtns (n and w of Panamint Valley, Inyo Co.)**; other vars. in OR, ID, NV. May–Jul

A. bernardinus M.E. Jones (p. 303) Per, often twining among sagebrush, wiry, sparsely leafy; minutely strigose, esp above. **ST** 1–5 dm, slender. **LF** 3–14 cm; lflets 7–19, ± well separated, 4–20 mm, ± lanceolate, tips acute to notched. **INFL** ± among lvs; fls 10–25, well separated, ascending. **FL**: calyx tube 2.7–4.1 mm; petals pale to dark lilac, banner 7–10.2 mm, recurved 45–90°, keel 6.8–9.4 mm, < 0.5 mm < banner. **FR** ascending, 20–30 mm, 4–5 mm wide, narrowly oblanceolate in side view, straight or ± curved, 3-sided (lower shallowly channeled or ± flat), pale, papery, glabrous; chambers 2. Stony areas among desert shrubs, junipers; 900–2000 m. SnBr, **DMtns (New York, Ivanpah mtns)**. Apr–Jun

A. calycosus S. Watson var. *calycosus* (p. 303) TORREY'S MILKVETCH Per, ± stless, tufted, silvery-strigose; hairs forked at base. **LF** 1–7 cm; lflets 1–7, gen 3, crowded near lf tip, 5–19 mm, elliptic to obovate, tips gen obtuse or acute. **INFL**: fls 1–8, ascending or spreading. **FL**: petals ± white to bright purple, wing tips white, banner 10–13 mm, keel 7.4–9.4. **FR** ascending, 10–25 mm, 3–4 mm wide, oblong, ± 3-sided (lower narrow, grooved), strigose; chambers ± 2. 2*n*=22. Rocky areas, sagebrush shrublands to pine forests; 1500–3550 m. SNE, n DMtns; to ID, WY. Apr–Jul

A. canadensis L. var. *brevidens* (Gand.) Barneby (p. 303) Per from rhizome, leafy, ± strigose; hairs forked at base. **ST** 1.5–5.5 dm. **LF** 5–23 cm, stipules fused around st into sheath; lflets 7–25, 5–40 mm, widely lanceolate or elliptic, ± glabrous, tips obtuse or shallowly notched, small-pointed. **INFL** spike-like; fls 20–many, reflexed, overlapping. **FL**: petals cream, rarely tinged dull purple, banner 11.7–17.5 mm, recurved 40–90°, keel 8.9–13.6 mm. **FR** erect; body 10–15 mm, 3–4 mm wide, ± cylindric, stiffly papery, gen minutely strigose; beak recurved, ± 3 mm, stiff; chambers 2. Heavy soil where moist at least in spring; 1500–2450 m. MP, n SNE, adjacent CA-FP; to WA, WY, Colorado. Jun–Sep

A. casei A. Gray (p. 303) Per, slender, open and widely branched, wiry, sparsely leafy, minutely strigose. **ST** ascending or erect, 1–4 dm, often zig-zag. **LF** 3–10 cm, midrib rigid, tapered; lflets 5–15, well separated, 3–25 mm, linear to oblanceolate, tips gen obtuse or notched. **INFL**: fls 8–25, well separated, ultimately reflexed. **FL**: petals pink-purple, wing, keel tips white, rarely petals all white, banner 12–18 mm, recurved ± 45°, keel 10.6–13.3 mm. **FR** pendent; 20–55 mm, 5–10 mm wide, slightly incurved, ± half-lanceolate, wider than deep, minutely strigose, early pulpy, then tough, wrinkled; beak sharp, rigid, compressed side-to-side; sutures keel-like; chamber 1. Dry, gravelly soils or on dunes, with sagebrush or pinyon; 1200–2000 m. SNE, **DMtns (Inyo Co.)**; w NV. Apr–Jun

A. cimae M.E. Jones Per, ± coarse, ± fleshy; hairs ± 0. **ST** ± spreading, 2–25 cm. **LF** 4.5–11 cm; lflets 11–23, 5–20 mm, obovate to ± round, tips gen obtuse or notched. **INFL**: fls 10–25, early ascending, then sometimes reflexed. **FL**: petals reddish purple, white- or pale-tipped, banner 12–15 mm, keel 9.5–10.6 mm. **FR** ± ascending or spreading; body ± oblong in side view, bladdery or not, glabrous; stalk-like base 5–12 mm, tapered to narrow base; chambers ± 2. Calcareous soils, gen among sagebrush, sometimes pinyon pine; 1400–1850 m. **W&I(Inyo Mtns)**, **e DMtns**.

1. Fr 8–12 mm wide, sides drying stiffly leathery or woody . . . . var. *cimae*
1' Fr 13–21 mm wide, sides drying thinly papery . . . . var. *sufflatus*

var. *cimae* (p. 303) CIMA MILKVETCH **FR**: body 15–25 mm, gen incurved > 90° (so beak is perpendicular to infl axis), fleshy when green; stalk-like base 6–8 mm. RARE. Gen among sagebrush; 1400–1850 m. **e DMtns (e San Bernardino Co.)**. Apr–Jun

var. *sufflatus* Barneby **FR**: body 20–37 mm, straight to incurved 90° (so beak is ± parallel to infl axis), bladdery, becoming papery; stalk-like base 5–12 mm. Habitats of sp.; 1500–1850 m. **W&I (e slope Inyo Mtns)**. May

A. coccineus Brandegee (p. 303, pl. 52) SCARLET MILKVETCH Per, tufted, ± stless; hairs dense, whitish. **LF** 3–10 cm; lflets 7–15, 3–14 mm, ± oblanceolate, tips ± acute. **INFL**: fls 3–10, ascending. **FL**: calyx 18–24 mm; petals scarlet, banner 35–41 mm, recurved 20–30°, keel 35–40 mm. **FR** 25–40 mm, 10–12 mm wide, plump, ± narrowly ovoid in side view, often curved, ± strongly compressed top-to-bottom, early fleshy, then leathery, hairs dense, shaggy, white or tawny; chamber 1. 2*n*=22. Gravelly places, gen sagebrush scrub or pinyon woodland; 750–2450 m. SNE, DMtns; s NV, w AZ, n Baja CA. Longer fr, remnants of fl ± distinguish this in fr from *A. newberryi*. ✿ DFCLT;TRY. Mar–Jun

A. crotalariae (Benth.) A. Gray (p. 303) SALTON MILKVETCH Per (often fl 1st season and seemingly ann), bushy-clumped, coarse, ± ill-scented, ± strigose. **ST** ± erect, 1.5–6 dm, often hollow. **LF** 5–16.5 cm; lflets 9–19, 5–35 mm, ± obovate to round, flat, thick, tips notched. **INFL**: fls 10–25, ascending or spreading. **FL**: calyx 7.6–12.3 mm; petals bright reddish purple, sometimes white, banner 21–28 mm, recurved ± 40°, keel 17–21 mm. **FR**: body 20–30 mm, 10–14 mm wide, inflated, ovate in side view, sparsely to densely strigose, with fine, net-like pattern, drying thickly papery; stalk-like base 1–1.5 mm, stout; beak erect or incurved, tipped by persistent style-base; sutures ± raised; chamber 1. 2*n*=24. UNCOMMON. Sandy or gravelly areas; –60–250 m. DSon; w AZ, n Baja CA. Jan–Apr

A. curvicarpus (A.A. Heller) Macbr. var. *curvicarpus* (p. 303) Per, leafy; hairs ± appressed, spreading, or curly, often grayish. **ST** decumbent to ascending, 1.5–4 dm, often stout. **LF** 2.5–9 cm; lflets 7–21, 3–23 mm, obovate, tips obtuse or notched. **INFL**: fls 5–35, not crowded, reflexed. **FL**: petals early white, then pale yellow, banner 15–21 mm, recurved ± 45°, keel 11–15.2 mm. **FR** pendent; body 20–35 mm, 2.5–4.5 mm wide, narrowly oblong, curved 1/4 to ± full circle, ± strongly compressed side-to-side, sparsely wavy-hairy or rarely glabrous, drying stiffly papery; stalk-like base 9–20 mm, jointed to receptacle; chamber 1. Loose soil, often with sagebrush; 1300–2900 m. GB; to OR, ID, NV. Apr–Jul

A. didymocarpus Hook. & Arn. TWO-SEEDED MILKVETCH Ann, gen slender, ± minutely grayish strigose. **ST** prostrate to ± erect, 3–30 cm. **LF** 0.8–7.5 cm; lflets 9–17, 2–14 mm, linear to oblanceolate, tips notched. **INFL** head-like; fls 5–30, < 9 mm, erect or ascending. **FL**: calyx hairs ± mixed black, white; petals whitish, purple-tinged. **FR** ascending, ± incl in calyx, 2–4 mm, ± 2 mm wide, ± spheric, 2-lobed in ×-section, ± minutely strigose, rarely glabrous, coarsely wrinkled, drying stiffly papery; chambers 2. Open grassy, gravelly, or sandy areas; 0–1550 m. c&s SNF, Teh, GV, CW, SW, **D**; s NV, w AZ, nw Mex. Vars. intergrade. 4 vars. in CA.

1. Calyx hairs mostly black, lobes gen 0.8–1.5 mm, < (rarely =) tube; lflets ± glabrous on upper side; st ± erect . . . . var. *didymocarpus*
1' Calyx hairs mostly white, lobes gen 1.5–2.4 mm, > tube; lflets ± canescent on upper sides; st ± prostrate . . . . var. *dispermus*

var. *didymocarpus* (p. 303) **ST** ± erect, 25–45 cm; herbage ± green, sparsely ± strigose. **INFL**: fls 5–25. **FL**: calyx hairs mostly black, lobes gen < 1.5 mm; banner 2.8–6.1 mm, keel 2.4–4.5 mm, abruptly curved, tip bluntly pointed. 2*n*=24. Grassy areas; < 700 m. c&s SNF, Teh, GV, CW, SW (exc ChI), **DMoj**; s NV. [var. *daleoides* Barneby] Feb–May

var. *dispermus* (A. Gray) Jepson **ST** ± prostrate, 15–27 cm; herbage ± grayish hairy. **INFL**: fls 7–20. **FL**: calyx hairs mostly white, lobes gen > 1.5 mm; banner 3.4–5.4 mm, keel 3.4–4.5 mm, abruptly incurved, tip bluntly or sharply triangular. 2*n*=26. Sandy or gravelly areas; 30–1200 m. **D**. Feb–May

A. funereus M.E. Jones (p. 303, pl. 53) BLACK MILKVETCH Per; hairs grayish, dense, stiff, some short, ± wavy. **STS** prostrate, loosely tufted, 2–8 cm. **LF** 2.5–7 cm; lflets 7–17, ± crowded, 3–12 mm, obovate, tips blunt or notched. **INFL**: fls 3–10, ascending. **FL**: calyx hairs black or mixed black, white; petals pink-purple, banner 22–29 mm, recurved ± 40°, keel 21–28 mm. **FR** ascending, 25–50 mm, 10–15 mm wide, ± lanceolate in side view, straight, ± strongly compressed top-to-bottom at base, gently incurved near tip, leathery; hairs dense, 1.5–2.5 mm, wavy, white; chamber 1. RARE. Gravelly, clayey, or rocky areas; 1300–1500 m. **ne DMtns (Grapevine Mtns e of Death Valley)**. [*A. purshii* Hook. var. *f.* Jepson] Apr–May

A. geyeri A. Gray var. *geyeri* (p. 303) GEYER'S MILKVETCH Ann, rarely persistent into 2nd season, often very slender, minutely strigose. **ST** prostrate to ascending, 1–20 cm. **LF** 1.5–10 cm; lflets 3–13, ± well separated, 5–15 mm, ± linear to oblong, terminal often notably longest. **INFL** among lvs; fls 3–8, well separated, early ascending, then reflexed. **FL**: petals whitish, sometimes blushed with lilac, keel tip purple, banner 5.2–7.6 mm, recurved ± 45(–80)°, keel 3.8–4.8 mm. **FR** 15–25 mm, 6–10 mm wide, inflated, half-ovate in side view, thinly papery, minutely strigose; beak triangular; upper suture straight or ± concave; chamber 1. RARE in CA. Sandy areas; ± 1500 m. **e SNE**; to WA, WY, UT. Apr–Jul

A. gibbsii Kellogg (p. 303) Per, leafy; hairs gen ± spreading, minute, wavy, ± grayish. **STS** ± prostrate, many, 1.5–3.5 dm. **LF** 1.5–9.5 cm; lower stipules small, fused around st into sheath; lflets 7–19, 4–20 mm, obovate or oblong. **INFL**: fls 10–30, reflexed. **FL**: calyx base with pouch on upper side; petals yellowish, wings and banner ± equal, banner 14–18 mm, shallowly S-shaped, keel 12–15 mm. **FR** pendent; body 22–30 mm, 4–8 mm wide, oblong, incurved 1/4–1/2 circle, swollen but ± compressed side-to-side, densely wavy-hairy, early fleshy, then leathery; stalk-like base 7–22 mm; sutures prominent, raised, keel-like; chamber 1. Among sagebrush or pines; 1300–1650 m. n SNH, **GB**; w NV. May–Jul

A. gilmanii Tidestrom (p. 303) GILMAN'S MILKVETCH Per or ± ann; hairs ± spreading or curved, minute. **ST** ± ascending, 5–25 cm. **LF** 1.5–7.5 cm; lflets 7–13, 2–12 mm, ± oblanceolate, margins often purple, tips notched or not. **INFL** ± among lvs; fls 4–9, well separated, early ascending, then reflexed. **FL**: petals pink-purple, banner 6.1–8 mm, recurved ± 45°, keel 4.7–6.1 mm. **FR** 14–25 mm, 8–16 mm wide, ± ovoid, bladdery, papery; hairs minute, curved; chamber 1; immature seeds 8–12. UNCOMMON. Gravelly areas; 2000–3050 m. **n DMtns (Panamint Mtns)**; NV. May–Jun

A. hornii A. Gray var. *hornii* (p. 303) Ann, open and widely branched; hairs ± appressed or ascending. **ST** 3–12 dm, slender-solid or stout-hollow. **LF** 1.5–13 cm, often reflexed; lflets 11–33, 5–20 mm, ± elliptic. **INFL** dense, head-like; fls 10–35, spreading and ascending. **FL**: petals white to pale lilac, banner 7.8–10.2 mm, recurved ± 40°, keel 5.9–8.4 mm. **FRS** crowded, spreading, in cylindric or subspheric heads (2.5–3.5 cm wide), each fr 12–18 mm, 7–9 mm wide, ± ovoid, inflated, bladdery, papery; hairs spreading, coarse; beak prominent, pointed; chamber 1. Salty flats, lake shores; 60–150 (850 in w DMoj) m. s SnJV, WTR, **w edge DMoj**; w-c NV. May–Sep

A. insularis Kellogg var. *harwoodii* Munz (p. 303) HARWOOD'S RATTLEWEED or MILKVETCH Ann, ± grayish strigose. **ST** decumbent to ascending, 5–40 cm, slender. **LF** 2–12 cm; lflets 9–21, ± well separated, 4–20 mm, ± narrowly elliptic or oblong, tips gen notched. **INFL** among lvs; fls 4–9, well separated, early spreading, then reflexed. **FL**: petals pink-violet, banner 5.5–7.4 mm, recurved ± 50–60°, keel 4.8–6 mm. **FR** spreading or ± reflexed, 15–24 mm, 5–15 mm wide, half-ovoid, bladdery, papery, strigose; beak conspicuous; upper suture straight or ± convex; lower suture strongly convex; chamber 1; immature seeds 7–14. RARE in CA. Sandy or gravelly areas; 0–300 m. **DSon**; AZ, nw Mex. Jan–May

A. inyoensis Cov. (p. 303) INYO MILKVETCH Per, sparsely leafy, minutely strigose, grayish green. **STS** prostrate, loosely matted, 1–6 dm, slender, zig-zag. **LF** 1.5–4.5, widely spreading; lflets 9–21, crowded, 3–10 mm, narrowly obovate, tips blunt or notched. **INFL**: fls 6–15, well separated, spreading to reflexed. **FL**: petals pink-purple, banner 8.6–10.8 mm, recurved ± 45°, keel 8.2–9.6 mm. **FR** pendent; body 12–15 mm, 6–8 mm wide, lanceolate but strongly incurved 1/8–1/2 circle, wider than deep, ± 3-sided (lower side deeply grooved), stiffly leathery, minutely strigose; stalk-like base 2–5 mm, stout; chamber ± 1. 2*n*=22. Gravelly areas; 1500–2300 m. **W&I**. May–Jun

A. iodanthus S. Watson var. *iodanthus* (p. 303) HUMBOLDT RIVER MILKVETCH Per, sparsely strigose. **ST** ± prostrate, 5–40 cm. **LF** 2–7 cm; lflets 11–21, 3–18 mm, obovate or ± round, tips blunt or notched. **INFL**: fls 7–25, early crowded-ascending, then well separated and reflexed. **FL**: petals purple, whitish with purple keel tip, or cream, banner 10–15.5 mm, recurved ± 45°, ± = or > keel. **FR** pendent, 20–40 mm, 5–8.5 mm wide, incurved 1/4–full circle, wider than deep or ± 3-sided; chambers ± 2 in lower half, 1 in upper. Dry areas, sagebrush shrubland; 1200–2100 m. se MP, **SNE**; to OR, ID, UT. Apr–Jun

A. jaegerianus Munz (p. 307) LANE MOUNTAIN MILKVETCH Per, sparsely leafy; hairs minute, ± flat, scale-like. **ST** weak, often scrambling through shrubs, 3–7 dm. **LF** 2–5 cm, stiffly spreading or reflexed; lflets 7–15, ± well separated, 3–15 mm, narrow, hairier on upper surface. **INFL**: fls 5–15, well separated, early ascending, then reflexed, twisted to 1 side. **FL**: petals dull pale purplish with darker veins, fading cream, often dingy, banner 6.5–10 mm, recurved 50–75°, keel 6.4–8.5 mm. **FR** pendent; body 18–25 mm, 3–5 mm wide, plump, stiffly papery or leathery, glabrous; stalk-like base 3–5 mm; sutures thick, raised, ± wavy; chambers 2. RARE. Among desert shrubs, in sand or gravel; 900–1200 m. **c DMoj (near Barstow)**. Threatened by military activity, grazing, vehicles. Apr–Jun

A. johannis-howellii Barneby (p. 307) LONG VALLEY MILKVETCH Per, open and widely branched; hairs ± appressed to ascending. **ST** ± prostrate or decumbent, 3–20 cm, slender. **LF** 4–6 cm; lower stipules fused around st into sheath; lflets 13–23, 2–6 mm, narrowly obovate, upper surface glabrous. **INFL** ± among lvs; fls 6–12, very well separated, reflexed in age. **FL**: petals whitish, banner 5–5.5 mm, recurved 90°, keel 3.3–3.9 mm. **FR** pendent; body 7–11 mm, ± 3 mm wide, half-ellipsoid, 3-sided (lower side deeply, openly grooved), thinly papery, minutely strigose; stalk-like base 0.5–2.5 mm; chambers 2. 2*n*=22. RARE CA. Sandy areas, sagebrush shrubland; 2150 m. **SNE (sw Mono Co.)**. Threatened by grazing, mining. Jun–Aug

A. kentrophyta A. Gray Per, tufted or matted, gen spiny, strigose; hairs attached by base or on side. **ST** < 3 dm. **LF** 2–26 mm; stipules sometimes like lflets, gen minutely spine-tipped, at least lower fused around st into sheath; lflets 3–9, 1–17 mm, linear or narrowly lanceolate, ± spine-tipped. **INFL**: fls 1–3. **FL**: petals white to pink-purple, banner 3.9–9.2 mm, recurved ± 45°, keel 2.9–6.3 mm. **FR** 4–9 mm, 1.5–4 mm wide, compressed side-to-side, finely hairy; sutures prominent; chamber 1. Open areas, clay, gravel, talus, rock; 2280–3660 m. SNH, SNE; to s Can, WY, ND, NM.

1. Hairs forked at base (1 branch gen short, sometimes ± 0); immature seeds 2–4; pl not a cushion or mat
   .... var. *elatus*
1' Hairs simple; immature seeds 5–8; pl a cushion or mat
   2. Lflets 3 (sometimes 5 on lower lvs), rigid, spine-tipped; pl a dense, rounded cushion ...... var. *danaus*
   2' Lflets 5–9, ± soft, minutely spine-tipped; pl a ± flat mat .... var. *tegetarius*

var. *danaus* (Barneby) Barneby (p. 307) SWEETWATER MOUNTAINS MILKVETCH Pl a dense, rounded, very spiny cushion, strigose; hairs simple. **ST** gen < 5 cm. **LF** 4–20 mm; lflets 3 (sometimes 5 on lower lvs), 3–7 mm, stiff, spine-tipped. **FL**: petals pale purple or whitish with purple keel-tip, banner 4–5.6 mm, keel 3.3–4.1 mm. **FR:** immature seeds 5–8. UNCOMMON. Rocky places at and above timberline; 3000–3600 m. c&s SNH, **SNE.** Jul–Sep

var. *elatus* S. Watson (p. 307) SPINY-LEAVED MILKVETCH Pl not a cushion or mat, spiny, strigose; hairs forked at base (1 branch gen short, sometimes ± 0). **ST** ± erect, ± 1 dm, open and widely branched. **LF** 10–26 mm; lflets 3–7, stiff, spine-tipped. **FL:** petals gen white or faintly purple-veined, banner 4.8–6.2 mm, keel 3.7–4.1 mm. **FR:** immature seeds 2–4. RARE in CA. Open rocky areas; 3000 m. **W&I;** to WY, Colorado, NM. CA pls differ in aspects of habit, fr from e pls and may be taxonomically distinct. Jun–Sep

var. *tegetarius* (S. Watson) Dorn (p. 307) Pl a ± flat mat, ± spiny, ± strigose; hairs simple. **LF** 2–20 mm; lflets 3–9, ± soft, minutely spine-tipped. **FL:** petals gen purple, sometimes whitish, banner 3.9–6.2 mm, keel 2.9–4 mm. **FR:** immature seeds 5–8. Open, rocky areas; 2700–3350 m. **W&I;** to OR, MT, NM. [var. *implexus* (Canby) Barneby] ❀ DRN,DRY:1–3,14–16,18;DFCLT. Jun–Sep

A. layneae E. Greene (p. 307) Per from deep rhizome (gen overlooked); hairs coarse, gen grayish. **ST** erect, 2.5–16 cm. **LF** 4–16 cm; lflets 11–23, 5–23 mm, ovate to ± round. **INFL:** fls 10–45, early ascending, then spreading. **FL:** calyx hairs mostly black, some white; petals whitish with keel tip, wing tips purple, banner often lilac-blushed, banner 12.5–18 mm, recurved ± 50°, keel 10.4–16.5 mm. **FR** 20–65 mm, 3.5–8 mm wide, incurved 1/4–full circle, leathery; hairs spreading, wavy; chambers ± 2 in lower half. 2*n*=44. Sandy flats, washes; 450–1550 m. **W&I, DMoj;** s NV, nw AZ. Mar–Jun

A. lemmonii A. Gray (p. 307) Per, open and widely branched, sparsely strigose. **ST** ± prostrate, 1–5 dm, slender. **LF** 1–4.5 cm; lflets 7–15, 2–11 mm, narrowly elliptic, tips acute. **INFLS** in 2's or 3's in upper axils; fls 2–13, clustered at ends of peduncles, ascending. **FL:** petals whitish or dull lilac, banner 4.8–6.1 mm, recurved 45–85°, keel 3.4–4 mm. **FR** ± spreading, 4–7 mm, ± 2 mm wide, elliptic, 3-sided (lower narrow), papery; chambers ± 2. Meadows, lake shores; 1300–2200 m. **GB** (lower in MP), adjacent edge c SNH; s-c OR. May–Jul

A. lentiginosus Hook. FRECKLED MILKVETCH Per (sometimes fl 1st year or ann), moderately leafy, ± glabrous to silvery-strigose. **LF** 1–15 cm; lflets linear to widely ± ovate. **INFL:** fls 3–±50, ascending or spreading. **FL:** petals purplish, cream, whitish, or mixed purplish and whitish, keel 0.65–0.8 × banner, banner recurved 30–50°. **FR** ovoid or spheric, widely grooved above and below, gen ± bladdery, ± papery, deciduous; beak gen triangular, flat; chambers 2 below beak. 2*n*=22. Gen dry, open places; –30–3600 m. **CA;** to OR, WY, n Mex. Highly variable; vars. often very distinct, yet intermediates common; fl, fr both needed for identification. 19 vars in CA.

1. Lflets 3–5, linear-oblanceolate . . . . var. *piscinensis*
1' Lflets gen 7 or more, often wider than linear-oblanceolate
  2. Keel 8.4–15 mm; calyx often > 6.5 mm
    3. Herbage sparsely hairy, greenish, or if densely hairy then gen either fr sparsely hairy or calyx lobes < 1.4 mm
      4. St prostrate, > 60 cm; moist alkaline flats, extreme n DMoj . . . . var. *sesquimetralis*
      4' St ascending or spreading, < 50 cm; gen sand, widespread . . . . [3]var. *variabilis*
    3' Herbage hairs dense, silvery, grayish, or whitish; fr densely hairy; calyx lobes > 1.4 mm
      5. Fr slightly swollen, < 7 mm wide, upper suture concave in side view; e DMoj, s DSon . . var. *borreganus*
      5' Fr very swollen, often bladdery, > 8 mm wide, upper suture convex in side view; n DMoj, nw DSon
        6. Pl base not woody; DSon (Coachella Valley) . . . . var. *coachellae*
        6' Pl base ± woody; n DMoj (Eureka Valley) . . . . var. *micans*

2' Keel 5.5–9 mm (± 10 mm in var. *ineptus*); calyx rarely > 6.5 mm
  7. Lflet densely hairy, ashy-gray or silvery, at least on 1 side (see also 7')
    8. Fr beak declined or curved down, away from st; clayey, vernally moist alkaline flats, seeps — w margin DMoj . . . . var. *albifolius*
    8' Fr beak straight or curved upward, toward st; habitats various, gen exc clayey, alkaline flats, seeps
      9. Keel 5.6–8.5 mm; fr glabrous to sparsely hairy; SNE, e DMoj . . . . [2]var. *fremontii*
      9' Keel > (8)8.5 mm; fr sparsely to thinly to densely hairy; s SNE, s&w DMoj . . . . [3]var. *variabilis*
  7' Lflet sparsely hairy or ± glabrous, green (sometimes densely hairy, grayish in var. *ineptus*)
    10. Pls of n SNE (from c Mono Co.) . . . var. *floribundus*
    10' Pls of SNE, D
      11. Petals gen bright pink-purple; D
        12. Keel 6–8.5 mm; gen e DMoj . . . . . [2]var. *fremontii*
        12' Keel > 8 mm; gen w DMoj . . . . [3]var. *variabilis*
      11' Petals gen ± white, sometimes blushed with lilac, rarely purple; SNE
        13. Lf 1.5–5.5 cm, lflets 9–21, crowded; SNE . . . . var *ineptus*
        13' Lf 4–9 cm, lflets 13–27, ± well separated; W&I . . . . var. *semotus*

var. *albifolius* M.E. Jones Per. **ST** ± prostrate, 3–7 dm. **LF:** lflets 9–21, 3–18 mm, narrowly oblanceolate, densely strigose. **INFL** dense; fls 9–35; axis in fr 5–40 mm. **FL:** petals gen whitish with some purple, banner 8.2–11.5 mm, keel 6–8.5 mm. **FR** 10–17 mm, 8–14 mm wide, inflated, bladdery, thinly papery; beak curved down, 3–5 mm. Alkaline flats, seeps; 600–1500 m. **SNE, w DMoj (Los Angeles Co.).** Apr–Jul

var. *borreganus* M.E. Jones (p. 307) BORREGO MILKVETCH Ann (sometimes per), ± densely silvery-hairy. **ST** ascending, 1–3 dm. **LF** 6–16 cm; lflets 7–19, 4–21 mm, ± obovate. **INFL:** fls 13–50; axis in fr 4.5–26 cm. **FL:** petals pink-purple, banner 12–14.8 mm, keel 10–13 mm. **FR** 15–23 mm, 4.5–6 mm wide, swollen, not bladdery, papery, silky-hairy, in side view lanceolate or narrowly ovate, gently incurved; upper suture concave, tapered to short, ± triangular, tooth-like beak. UNCOMMON. Sand; 30–250 m. **e DMoj, s DSon;** nw Mex. [var. *coulteri* (Benth.) M.E. Jones] Mar–May

var. *coachellae* F. Shreve & Wiggins (p. 307) COACHELLA VALLEY MILKVETCH Ann or per, densely silvery-hairy. **STS** ascending, clumped, 1–3 dm. **LF** 5–11.5 cm; lflets 7–21, 5–17 mm, ± widely ovate. **INFL:** fls 11–25; axis in fr 3–10 cm. **FL:** petals pink-purple, banner 12.7–14.5 mm, keel 10.8–11.6 mm. **FR** 16–21 mm, 9–14 mm wide, greatly inflated, stiffly papery, grayish strigose; beak 3.5–6 mm. RARE. Sand; 0–350 m. **DSon (Coachella Valley).** Threatened by vehicles, development. Feb–May

var. *floribundus* A. Gray (p. 307) Per, gen short-lived. **ST** ± ascending, 2–5 cm, gen with several–many branches in lower half. **LF** 3–11 cm; lflets 11–19, 5–15 mm, obovate, ± glabrous. **INFL** dense; fls 10–40; axis in fr 1–4(10) cm. **FL:** petals white, ± lilac-tinged, banner 8.8–11.6 mm, keel 6.6–8 mm. **FR** 8–21 mm, 6–12 mm wide, bladdery-inflated, thinly papery, glabrous or sparsely strigose, straw-colored; beak 3–7 mm, incurved or erect. Often among sagebrush; 1150–1600 m. n SNH s MP, **SNE;** s-c OR, w-c NV. May–Jul

var. *fremontii* (A. Gray) S. Watson (p. 307) Ann or per, densely or sparsely hairy. **ST** decumbent to erect, 1–5 dm. **LF** 3–12 cm; lflets 9–15, 5–19 mm, obovate. **INFL** axis in fr 2.5–16 cm. **FL:** petals ± purple, banner 9.1–12.4 mm, keel 5.6–8.5 mm. **FR** 14–36 mm, 5–18 mm wide, gen bladdery, thinly papery, glabrous to thinly hairy; beak 2–10 mm, ± incurved. 2*n*=22. Open sand, gravel; 900–2900 m. **SNE, e DMoj;** s NV. At lower elevations in DMoj. Esp hairy in SNE. Apr–Jul, sometimes Sep–Oct

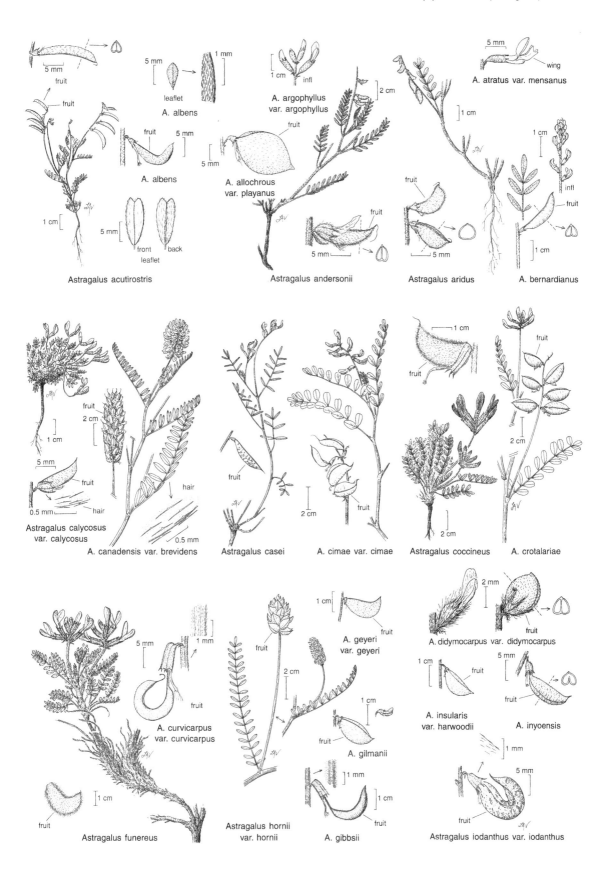

A. albens

leaflet

A. albens

fruit

A. acutirostris

front leaflet back

Astragalus acutirostris

A. argophyllus var. argophyllus

fruit

A. allochrous var. playanus

Astragalus andersonii

fruit

A. atratus var. mensanus

wing

fruit

Astragalus aridus

infl

fruit

A. bernardianus

Astragalus calycosus var. calycosus

fruit

hair

hair

A. canadensis var. brevidens

hair

fruit

Astragalus casei

fruit

A. cimae var. cimae

fruit

Astragalus coccineus

fruit

A. crotalariae

A. curvicarpus var. curvicarpus

fruit

fruit

Astragalus funereus

fruit

Astragalus hornii var. hornii

A. geyeri var. geyeri

fruit

fruit

A. gilmanii

fruit

A. gibbsii

A. didymocarpus var. didymocarpus

fruit

A. insularis var. harwoodii

fruit

fruit

A. inyoensis

fruit

Astragalus iodanthus var. iodanthus

var. *ineptus* (A. Gray) M.E. Jones (p. 307) Per; hairs ± ascending, grayish. ST decumbent, 1–3 dm. LF 1.5–5.5 cm; lflets 9–21, 2–10 mm, crowded, obovate. INFL: fls 10–21; axis in fr 10–25 mm. FL: calyx lobes > 1.2 mm; petals cream, banner 8.8–12.2 mm, keel 7.2–9.3 mm. FR 10–18 mm, 6–12 mm wide, bladdery, thinly papery, minutely strigose, rarely glabrous; beak erect or incurved. Open, gravelly places; 2000–3600 m. SNH (barely in n SNH), SNE. Intergrades with var. *semotus*. Jun–Aug

var. *micans* Barneby SHINING MILKVETCH Per; hairs dense, silvery or white, silky. STS clumped, ascending or erect, 2–4 dm. LF 4.5–9.5 cm; lflets 11–17, 5–14 mm, widely ± ovate. INFL: fls 12–35; axis in fr 4–10 cm. FL: petals white, tips blushed with pink-lavender, banner 12.2–14.3 mm, keel 9.6–10 mm. FR 15–20 mm, 8–10 mm wide, very swollen, stiffly papery, densely silky-hairy; beak 2.5–4 mm. RARE. Dunes; 900 m. **n DMoj (Eureka Valley)**. Threatened by vehicles. More strongly per, with longer hairs (1.1–2 mm) than similar pls of var. *variabilis*. Apr–Jun

var. *piscinensis* Barneby (p. 307) FISH SLOUGH MILKVETCH Per, sparsely leafy, ± canescent. ST prostrate, < 1 m. LF 3–4 cm; lflets 3–5, linear-oblanceolate, lateral 7–20 mm, terminal 14–32 mm. INFL: fls 5–12; axis in fr 1.5–4 cm. FL: petals lavender, banner 13 mm, keel 9 mm. FR 20–24 mm, 8–12 mm wide, strongly inflated, stiffly papery, densely strigose; beak 4.5–7 mm. RARE. Wet soil; 1300 m. **c SNE (Mono, Inyo cos., near Bishop)**.

var. *semotus* Jepson Per, ± strigose; hairs subappressed. STS loosely tufted, < 15 cm. LF 4–9 cm; lflets 13–27, 2–9 mm, ± well separated, ± narrowly elliptic, upper surface ± glabrous. INFL: fls 6–10; axis in fr 1–3 cm. FL: calyx lobes > 1.2 mm; petals whitish, banner 10.4–12 mm, keel 7.7–8.5 mm. FR 10–20 mm, 6–12 mm wide, bladdery, papery, sparsely strigose; beak 4–7 mm, incurved. 2*n*=22. Sandy or gravelly flats, hillsides; 2450–3350 m. **W&I**. Intergrades with var. *ineptus*. Jun–Aug

var. *sesquimetralis* (Rydb.) Barneby SODAVILLE MILKVETCH Per, sparsely strigose. ST prostrate, 6–8 dm. LF 2–5 cm; lflets 7–17, 6–18 mm, oblanceolate, upper surface ± glabrous. INFL: fls 5–12; axis in fr 1–2.5 cm. FL: petals purple, banner 12–14.5 mm, keel 9.3–9.5 mm. FR 12–26 mm, 5–12 mm wide, ± inflated, stiffly papery, sparsely strigose; beak 4–8 mm. **ENDANGERED** CA. Moist, alkaline flats; 950 m. **n DMoj (n Death Valley, e slope Last Chance Mtns)**; NV. May–Jun

var. *variabilis* Barneby Ann or per, gen robust, coarse; fl-st sometimes single, weak; hairs sparse, appressed, straight to ± dense, spreading, wavy. ST ascending, 1–4 dm. LF 2.5–13 cm; lflets 11–25, 4–17 mm, obovate. INFL: fls 10–30; axis in fr 4–17 cm. FL: petals purple, rarely white, banner 11.1–15 mm, keel 8.4–12.3 mm. FR 12–30 mm, 8–15 mm wide, bladdery, ± firmly papery, thinly to densely strigose or ± wavy-hairy; beak 3–9 mm, gently incurved. 2*n*=22. Sand; 140–1600 m. s SNF, Teh, s SnJV, **s-most SNE, w&s DMoj**. Intergrades with vars. *fremontii*, *micans* to n; *coachellae* to w. Mar–Jun

A. magdalenae E. Greene var. *peirsonii* (Munz & J. McBurney) Barneby (p. 307) PEIRSON'S MILKVETCH Ann or per, silvery-canescent. ST ascending or erect, 2–9 dm. LF 1–15 cm; midrib ± flattened; lflets 3–15, ± well separated, 2–8 mm, narrow, oblong, terminal not jointed to midrib. INFL: fls 5–20, ascending or spreading. FL: petals pink-purple, often white-tipped, banner 10–14.2 mm, recurved ± 40°, keel 8.5–10 mm. FR spreading, 20–35 mm, 10–20 mm wide, half-ovoid to ± spheric, bladdery, papery, finely strigose; chamber 1. **ENDANGERED** CA. Sand dunes; 50–250 m. **DSon**; reported from w AZ, n Baja CA. Seeds to 5 mm, largest of Am spp. of *Astragalus*. Dec–Apr

A. malacus A. Gray (p. 307) Per; hairs 1.5–2.5 mm, spreading. ST ascending or erect, 1–4 dm. LF 4–15 cm; lflets 7–21, 5–20 mm, elliptic to obovate, tips ± notched. INFL: fls 9–35, reflexed in age. FL: petals reddish violet, banner 15–21 mm, re-

curved ± 45°, keel 12.3–16 mm. FR pendent; body 18–38 mm, 5–6 mm wide, incurved, ± 3-sided, stiffly papery; hairs long; stalk-like base 1–3 mm, stout; chambers 2. Dry, rocky or heavy soils, with sagebrush, pinyon woodland; 1050–2350 m. e MP, SNE; to OR, ID, NV. Apr–Jun

A. minthorniae (Rydb.) Jepson var. *villosus* Barneby (p. 307) Per, robust or ± cespitose; hairs ± spreading, coarse. ST ascending or erect, 3–30 cm. LF 4–17 cm; lower stipules long-woolly; lflets 7–17, 8–25 mm, ± obovate. INFL: fls 7–35, ascending to reflexed. FL: petals cream (keel tip purple) or purplish (wing tips pale), banner 12–18 mm, recurved ± 45°, keel 9.5–13 mm. FR ± erect, 15–30 mm, 4–6 mm wide, ± compressed side-to-side but sides convex, leathery; hairs ± spreading; chambers 2. Rocky, calcareous hillsides, washes, gen with pinyon, juniper; 1350–2300 m. SnBr, **W&I, DMtns**; s NV. Apr–Jun

A. mohavensis S. Watson Ann or ± per, open and widely branched or tufted, grayish or silvery-strigose or canescent. ST weakly ascending, 5–35 cm. LF 2–12.5 cm; lflets 3–11, 3–18 mm, ovate to ± round, tips gen blunt, sometimes (rarely on upper) shallowly notched. INFL: fls 3–15, early ascending, then reflexed. FL: petals pink-purple, banner 7–12.5 mm, recurved ± 45°, keel 6.4–10.5 mm. FR reflexed, 13–32 mm, 3.5–8.5 mm wide, stiffly leathery, densely, minutely strigose; chambers ± 2 below, 1 near tip; immature seeds 20–30. Dry, rocky areas; 750–2300 m. **DMoj**; s NV.

1. Fr incurved 1/4–1/2 circle, ± 3-sided, upper suture raised, lower suture in wide, shallow groove; DMtns (immediately w of Death Valley) . . . . var. *hemigyrus*
1' Fr straight or incurved gen < 1/4 circle, ± compressed side-to-side or 3-sided, upper and lower sutures gen both raised; DMoj . . . . var. *mohavensis*

var. *hemigyrus* (Clokey) Barneby (p. 307) HALF-RING-POD, CURVED-POD, or MOJAVE MILKVETCH **PRESUMED EXTINCT** in CA; extant in s NV. Limestone; 1250–1600 m. **DMtns (Darwin Mesa, w of Death Valley, Inyo Co., in 1941)**; s NV.

var. *mohavensis* Gen limestone; 750–2300 m. **DMoj** Apr–Jun

A. monoensis Barneby (p. 307) MONO MILKVETCH Per from rhizome; crown below soil surface; hairs dense, silky, wavy. ST 7–20 cm, 2–6 cm buried, ± glabrous; above ground part decumbent, hairy. LF 7–30 mm; lower stipules, esp underground stipules, fused around st into sheath; lflets 9–15, crowded, 2–3 mm, ovate. INFL head-like; fls 6–12, spreading. FL: petals ± pale pink-tinged but drying ± yellow, banner 10–13 mm, recurved ± 50°, keel 6.7–8 mm. FR spreading or ascending, 15–20 mm, 6–9 mm wide, widely incurved-lanceolate, papery; hairs short, wavy; chambers ± 2; immature seeds 18–20. 2*n*=22. **RARE** CA. Open areas, pumice sand, or gravel; 2250–2400 m. **SNE (c Mono Co)**. Threatened by vehicles, road maintenance, grazing. Jun–Aug

A. newberryi A. Gray var. *newberryi* (p. 307) Per, ± cespitose; old lf bases persistent; hairs silky, longer ± straight, partly spreading. LF 1.5–15 cm; lflets 3–15, 5–20 mm, obovate, tips acute or blunt, sometimes notched. INFL: fls 3–8, ascending. FL: calyx with more white than black hairs; petals pink-purple or whitish, tipped with pink-purple, banner 21.5–30 mm, recurved ± 40°, keel 18.5–26 mm. FR 13–28 mm, 7–13 mm wide, ovoid, gen incurved; hairs dense, longer ones 2–4.5 mm, white, woolly, some curly, some ± straight; chamber 1. 2*n*=22. Rocky areas; 1300–2350 m. **SNE, ne DMoj**; to OR, UT, NM. Like *A. purshii* (which has hairs extremely fine, entangled). Apr–Jun

A. nutans M.E. Jones (p. 307) PROVIDENCE MOUNTAIN MILKVETCH Ann or ± per, minutely strigose. ST prostrate to erect, 6–15 cm. LF 2–8 cm; lflets 7–13, 5–15 mm, ± narrowly elliptic or obovate, tips acute or shallowly notched. INFL: fls 6–10, ascending to reflexed. FL: petals pink-purple, wing tips often paler, banner 7.8–10.4, recurved ± 90°, keel 5.9–6.6 mm. FR spreading, 15–25 mm, 11–15 mm wide, ovoid, bladdery, thinly papery, sparsely strigose, rarely ± glabrous; beak strongly compressed side-to-side, widely triangular; chamber 1; immature seeds 19–24; seeds suspended from flange 1–2.5 mm wide. $2n$=22. UNCOMMON. Sandy or gravelly places; 450–1950 m. **se DMtns, DSon.** Mar–Jun, sometimes Oct

A. nuttallianus A. DC. Ann, slender, minutely strigose. **ST** prostrate or weakly ascending, 4–45 cm. **LF** 1.5–6.5 cm; lflets 5–13, 2–10 mm, at least upper elliptic, acute at tips, lower sometimes blunt, notched at tip. **INFL** fls 1–4, 4–7 mm. **FL** corolla whitish, faintly lilac-tinged (rarely purplish). **FR** 10–20 mm, 2–3 mm wide, linear in side view, gently curved near base, ± straight toward tip, ± 3-sided (laterals ± convex, bottom grooved), maturing dark brown or blackish, glabrous or minutely strigose; chambers ± 2. Gravelly or sandy areas; 0–1950 m. **D;** to Colorado, n Mex. See note under *A. acutirostris*; vars. intergrade.

1. Calyx lobes gen 1.8–2.8 mm, hairs 0.7–1.2 mm,
    spreading, very stiff . . . . var. *austrinus*
1' Calyx lobes gen 1–1.7 mm, hairs 0.5–0.8 mm, ascending,
    not very stiff
    2. Lflets tips notched on lower lvs, acute on upper;
        calyx tube 1.4–1.7 mm . . . . var. *cedrosensis*
    2' Lflet tips acute; calyx tube 1.9–2.8 mm . . var. *imperfectus*

var. *austrinus* (Small) F. Shreve & Wiggins **LF** lflet tips acute. **FL** calyx tube ± 2.5 mm, lobes gen 1.8–2.8 mm, hairs spreading, 0.7–1.2 mm, very stiff, silvery. $2n$=22. Calcareous soils; gen 600–2150 m. Approaches **DSon** near AZ border, possibly a waif; to OK, TX, nw Mex. Mar–May

var. *cedrosensis* M.E. Jones **LF** lflet tips notched on lower lvs, acute on upper. **FL** calyx tube 1.4–1.7 mm, ± as wide, lobes 0.7–1.6 mm, hairs ascending, short, not very stiff. Rocky flats, washes; < 300 m. **DSon;** AZ, nw Mex. Dec–Apr

var. *imperfectus* (Rydb.) Barneby (p. 307) **LF** lflets tips acute. **FL** calyx tube 1.9–2.8 mm, ± 1/2 as wide, lobes 1–2 mm, hairs ascending, short, not very stiff. $2n$=22. Sandy or gravelly flats or washes; 300–1950 m. **e DMoj;** to UT, AZ. Mar–May

A. oophorus S. Watson Per, ± robust, ± glabrous. **ST** decumbent or ascending, 1–3 dm. **LF** 5–15 cm; lflets 7–21, 4–20 mm, ovate to ± round. **INFL** fls 4–10, not crowded, spreading. **FL** calyx glabrous; corolla red-purple with white wing tips, or cream, or white drying cream, banner 16–23 mm, recurved ± 85°, keel 10–16.5 mm. **FR** spreading or pendent; body 25–55 mm, 10–20 mm wide, widely ± ovoid, bladdery, glabrous; stalk-like base 3.5–10 mm; chamber 1. $2n$=24. Dry, open areas, often among sagebrush, pinyon; 1500–3100 m. **SNE, n DMtns;** w&c NV.

1. Petals cream, or white drying cream; lflets 7–11
    . . . . var. *lavini*
1' Petals red-purple, wing tips white; lflets > 11 on longer
    lvs . . . . var. *oophorus*

var. *lavinii* Barneby LAVIN'S MILKVETCH UNCOMMON. Habitats of sp.; ± 2450–3050 m. **SNE (Bodie Hills, Mono Co.);** w&c NV.

var. *oophorus* (p. 307) Habitats of sp.; 1500–3100 m. **SNE, n DMtns.** May–Jun

A. pachypus E. Greene Per, robust, rigid, bushy, sparsely leafy; hairs < 0.3 mm, appressed, ± scale-like, ± gray. **ST** ± erect, 2–8 dm, wiry. **LF** 2.5–16.5 cm; lflets 11–27, ± well separated, 3–34 mm, narrow. **INFL** fls 4–28, well separated, ascending. **FL** petals white or cream, banner 15–22 mm, recurved ± 45°, keel 10.7–15.3 mm. **FR** ascending or spreading; body 12–28 mm, 4–8 mm wide, straight or ± curved, compressed side-to-side and stiffly leathery when mature, gen glabrous; stalk-like base 4–8 mm, stout; beak short, sharp, rigid, persistent; chambers 2. Open slopes in grassland or shrubland; 500–1900 m. Teh, s SnJV, SCoRI, WTR, PR, **w edge D.**

1. Petals yellow when fresh, or dry, banner 15–17 mm;
    calyx tube 3.7–4.3 mm . . . . var. *jaegeri*

1' Petals white when fresh, drying yellow, banner
    15–22 mm; calyx tube 4–5.2 mm . . . . var. *pachypus*

var. *jaegeri* Munz JAEGER'S MILKVETCH **LF** lflets 15–25. RARE. Rocky or sandy areas; 500–750 m. n PR (incl SnJt), **nw edge DSon.** Dec–Jun

var. *pachypus* (p. 307) **LF** lflets 11–21. $2n$=22. Open areas or in shrubland, often on gravelly clay, shale, or sandstone; 500–1900 m. Teh, s SnJV, SCoRI, WTR, **w edge D.** San Diego Co. pls have smaller fls but are otherwise indistinguishable. Mar–Jul

A. palmeri A. Gray (p. 307) Ann or per, low, open and widely branched, sparsely to densely silvery-strigose. **ST** decumbent, 2–5 dm. **LF** 2–16 cm; lflets 9–21, 5–25 mm, widely ± elliptic. **INFL** fls gen 20–40, early dense, then ± well separated, spreading or widely ascending. **FL** petals gen pink-purple, sometimes cream with purple veins and petal tips, banner 7–10.3 mm, recurved 90°, keel 6.2–8.8 mm. **FR** 10–25 mm, 5–14 mm wide, moderately to very swollen (then ± bladdery), ± ovoid or half-ellipsoid, papery but not esp thin and translucent, thinly to densely strigose; beak ± erect, 1/5–1/3 × body, triangular; chamber 1. Sandy or rocky places; 150–1650 m. **w edge DSon,** adjacent foothills of SnBr, PR; Baja CA. [*A. vaseyi* S. Watson] Dec–Jun

A. panamintensis E. Sheldon (p. 311) Per, resembling an unkempt nest due to wiry, incurved branches and persistent petioles, silvery-canescent. **ST** 1–15 cm, slender. **LF** 1.5–12 cm; lflets 5–11, well separated, 2–14 mm, linear-elliptic, late deciduous, tips acute. **INFL** fls 1–4, ascending, well separated. **FL** petals pink-purple, banner 8–14 mm, recurved ± 45°, keel 7–9 mm. **FR** spreading or ascending, 8–18 mm, 3–5 mm wide, ± oblong-elliptic in side view, bluntly triangular in ×-section, densely strigose, becoming papery; chambers ± 2 in lower 1/2–2/3. In cracks, limestone ledges; 1200–2150 m. **n DMtns (Inyo Co.).** Apr–Jun

A. platytropis A. Gray (p. 311) BROAD-KEELED MILKVETCH Per, cespitose; hairs dense, silvery or gray. **ST** < 2 cm. **LF** 1–9 cm; lower stipules barely fused around st into sheath; lflets 5–15, 4–11 mm, elliptic to obovate, tips gen blunt. **INFL** head-like; fls 4–9. **FL** petals whitish or pale purple, banner 7.2–9.5 mm, recurved 30–45°, keel 7.8–8.6 mm. **FR** ascending, 15–33 mm, 10–18 mm wide, bladdery, ± strongly compressed top-to-bottom, strigose; chambers 2 by inflexion of lower suture, which meets seed-bearing flange in middle of fr, seeds therefore along center of partition. RARE in CA. Rocky areas above timberline; 2800–3550 m. **SNE, DMtns;** to ID, MT, NV. Jul–Aug

A. preussii A. Gray Per, robust, ill-scented, ± glabrous. **ST** ± erect, 1–3.5 dm. **LF** 3.5–18 cm; lflets 3–25, ± well separated, 2–27 mm, linear to ± round, hairs 0 or only on margins, midribs. **INFL** fls 4–22, ascending. **FL** petals pink-purple or ± pale, banner 14–24 mm, recurved ± 40°, keel 11–19 mm. **FR** erect or ascending; body 12–40 mm, 7–13 mm wide, inflated, oblong-ellipsoid, ± round in ×-section, stiffly papery, glabrous or minutely hairy; stalk-like base 0 or 2–7 mm; chamber 1. Alkaline clay flats, gravelly washes; 700–750 m. **DMoj;** to UT, AZ.

1. Fr base not stalk-like; infl open, axis in fr 4–23 cm;
    banner ± 14 mm . . . . var. *laxiflorus*
1' Stalk-like fr base 2–7 mm; infl ± dense, axis in fr
    1–9 cm; banner 17–24 mm . . . . var. *preussii*

var. *laxiflorus* A. Gray (p. 311) LANCASTER MILKVETCH UNCOMMON (possibly extinct in CA). Alkaline flats; 700 m. **sw DMoj (Antelope Valley);** s NV, sw UT, w AZ.

var. *preussii* DESERT MILKVETCH UNCOMMON. Clay flats; ± 750 m. **e DMoj (se Inyo, ne San Bernardino cos.);** to UT, AZ. Apr–Jun

A. pseudiodanthus Barneby (p. 311) TONOPAH MILKVETCH Per from underground crown; hairs soft, ± spreading, curved or curly. **STS** ± prostrate, loosely matted, 2–3 dm. **LF** 2.5–5 cm; lflets 7–19, ± crowded, 3–10 mm, ± obovate. **INFL** fls 7–25, early crowded, spreading, then ± well separated, reflexed. **FL:** petals reddish lilac, banner 9–10 mm, recurved ± 40°, keel 8.5–9.3 mm. **FR** reflexed, 12–24 mm, 5–8 mm wide, curved > 1/2 circle, wider than deep, early fleshy, then leathery; hairs spreading; chambers 1 or ± 2. RARE. Sand; 2050 m. **SNE (e Mono Co.);** w NV. [*A. iodanthus* S. Watson var. *p.* (Barneby) Isely] Possibly a sand ecotype of *A. iodanthus*. May–Jun

A. purshii Hook. PURSH'S MILKVETCH Per, sparsely to densely cespitose; herbage hairs gen 1–2.3 mm, extremely fine, cottony, entangled, silvery or gray. **ST** 0–14 cm. **LF** 1–15 cm; lflets 3–17, 2–20 mm, narrowly elliptic to ± round, tips blunt to notched. **INFL** ± among lvs; fls 1–11, ascending. **FL:** petals white, cream, pink-purple, or purple, banner 9–26 mm, recurved ± 40°, keel 8–21.2 mm. **FR** ascending, 7–27 mm, 4–13 mm wide, ovoid or widely lanceolate in side view; hairs gen very dense, gen 1.5–5 mm, white (fr resembling a cotton boll), all ± wavy or all straight; chambers 1 or 2. Dry flats, slopes; 450–3350 m. NW, CaR, SNH, Teh, SCoRI, SnBr, **GB, DMoj;** to Can, ND, Colorado. Locally and regionally variable. Like *A. newberryi* var. *n.* (which has longer, ± straight, partly spreading hairs). 4 vars. in CA.

1. Calyx < 10 mm; fr 7–17 mm; infl 1–5-fld . . . . var. *lectulus*
1' Calyx > 10 mm; fr 13–27 mm; infl 3–11-fld . . . var. *tinctus*

var. ***lectulus*** (S. Watson) M.E. Jones **ST** 0–10 cm. **LF** 1–5 cm; lflets 3–11, 2–10 mm. **INFL:** fls 1–5. **FL:** calyx 5.6–8.8 mm; petals pink or pale purple, banner 10.3–15 mm, keel 9.4–11.7 mm. **FR** 7.5–15 mm, 4–8 mm wide, incurved only near beak; chamber 1; immature seeds 24–32. Dry, open flats, slopes, often with juniper, pines, to rocky slopes above timberline; 1800–3350 m. SNH, SnBr, **w edge SNE.** May–Aug

var. ***tinctus*** M.E. Jones (p. 311) **ST** 0–10 cm. **LF** 2–11 cm; lflets 3–17, 2–14 mm. **INFL:** fls 3–11. **FL:** calyx 12–19 mm; petals pink-purple or purple, banner 14.6–25 mm, keel 11.5–20.8 mm. **FR** 13–27 mm, 5–13 mm wide; chambers often ± 2 (esp in **DMoj);** immature seeds 18–46. Gravelly, sandy flats, slopes, often with pines or sagebrush; 450–2900 m. KR, NCoRH, CaRH, n&s SNH, Teh, SCoRI, WTR, **GB, DMoj;** OR, NV. [var. *longilobus* M.E. Jones] ❀ DRN,DRY:1,2,16,18;DFCLT. Apr–Jun

A. sabulonum A. Gray (p. 311) Ann, coarse, leafy; hairs ± dense, ascending or spreading, ± wavy. **ST** erect or decumbent, 2–26 cm. **LF** 1.5–6.5 cm; lflets 5–15, 2–13 mm, oblanceolate, tips blunt ± notched. **INFL:** fls 2–7, well separated, spreading or ascending. **FL:** petals dingy cream, lilac-tinged, banner 5.2–7.2 mm, recurved 50–70°, keel 5–6.5 mm. **FR** 9–20 mm, 5–11 mm wide, ± ovoid, incurved (often abruptly so near tip), leathery; hairs ± dense, stiff, spreading, wavy; chamber 1. 2*n*=24. Sandy or gravelly areas; –30–200 m. **D;** to UT, NM. As high as 2000 m outside CA. Feb–Jul

A. sepultipes (Barneby) Barneby (p. 311) BISHOP MILKVETCH Per, open and widely branched, loosely clumped; hairs ± dense, ± appressed, silvery- or grayish silky. **ST** ascending, 1.5–3.5 dm; buried parts 1–5 cm, ± glabrous. **LF** 2–8 cm; lower stipules fused around st into sheath; lflets 7–17, 3–14 mm, obovate, tips notched or ± blunt. **INFL:** fls 10–30, spreading. **FL:** petals pale pink- or lilac-white, wing tips whitish, banner 12.7–17.5 mm, recurved ± 40°, keel 10–12.2 mm. **FR** spreading to reflexed, 15–20 mm, 3–6 mm wide, ± incurved, ± 3-sided, stiffly papery, hairs fine, shaggy; chambers 2; immature seeds 14–20. Dry, granitic sand, among sagebrush; 1800–1950 m. SNE, perhaps e edge c&s SNH. May–Jul

A. serenoi (Kuntze) E. Sheldon var. ***shockleyi*** (M.E. Jones) Barneby (p. 311) NAKED MILKVETCH Per, bushy-clumped, minutely, thinly strigose, often grayish. **ST** 1.5–4.5 dm. **LF** 5–15 cm; lflets 5–11, well separated, 5–30 mm, very narrow, upper surface more densely strigose, paler than lower, terminal lflet not or only obscurely jointed to midrib. **INFL:** fls 3–25, well separated, ascending. **FL:** petals ± purple with pale wing tips, strongly graduated, banner 17–26 mm, recurved 35–40°, keel

12.4–18.5 mm. **FR** erect, 17–31 mm, 7–12 mm wide, plumply oblong, leathery, glabrous; beak stout, sharp; chambers ± 2. 2*n*=22(24?). UNCOMMON. Open, dry, alkaline gravelly-clay soil, gen with sagebrush or pinyon; 1500–2250 m. **W&I, n DMtns (Inyo Co.);** NV. Pl with unilocular fr are *A. s.* var. *serenoi*, known only from NV. May–Jun

A. tephrodes A. Gray var. ***brachylobus*** (Gray) Barneby (p. 311) Per, tufted, coarse, greenish gray to silvery-silky; hairs appressed or ± ascending. **ST** prostrate, 0–8 cm. **LF** 4–16 cm; lflets 11–31, 4–17 mm, widely ± obovate. **INFL:** fls 10–35, ascending. **FL:** petals ± pink-purple, banner 14–24 mm, recurved ± 45°, keel 14.7–21 mm. **FR** ascending, 17–30 mm, 6–10 mm wide, plumply lanceolate, stiffly leathery; hairs appressed or ± spreading, rarely 0; chamber 1. Open, dry ground; 150 m. **se DMoj (near Needles, San Bernardino Co.);** to sw UT, AZ. Possibly a waif in CA. Apr–Jul

A. tidestromii (Rydb.) Clokey (p. 311) Per (or fl 1st year and appearing ann), tufted; hairs dense, ± stiff, often curly, entangled, shaggy. **ST** 0 or ± prostrate, < 7 cm. **LF** 3–15 cm; lflets 7–19, 4–17 mm, widely obovate. **INFL:** fls 5–16, well separated, spreading-ascending. **FL:** petals whitish, tinged with dull purple, wings, keel tipped dark purple, banner 12–17.7 mm, recurved ± 40°, keel 10–12 mm. **FR** ascending, 15–55 mm, 6–16 mm wide, ± lanceolate, curved 1/4–5/4 circle, ± 4-sided, compressed side-to-side at both ends, from top-to-bottom in middle, stiffly leathery, minutely strigose; beak long, narrowly triangular; chamber 1. Open, calcareous gravel; 600–1500 m. **DMoj (extreme e-c DMtns and mouth of Cushenbury Canyon near n edge SnBr);** s NV. Mar–May

A. tricarinatus A. Gray (p. 311) TRIPLE-RIBBED MILKVETCH Per, loosely tufted, finely strigose. **ST** ± erect, 5–25 cm, stiff. **LF** 7–20 cm; lflets 17–27, well separated, 3–12 mm, narrowly ± obovate, silvery-strigose on upper surface, greenish on lower. **INFL:** fls 5–15, well separated, spreading-ascending. **FL:** calyx 6.1–7.6 mm, tube 4.1–5 mm; petals cream, banner 12.6–15.7 mm, recurved ± 45°, keel 9.7–11 mm. **FR** ascending; body 24–42 mm, 3.5–5.5 mm wide, ± linear, 3-sided, thinly papery, glabrous; stalk-like base 1–2.5 mm, stout, jointed at top; upper suture a narrow ridge; chambers 2. RARE. Exposed rocky slopes, canyon walls; 450–550 (1250) m. e SnBr (Whitewater, Morongo Valley), adjacent edges D. Feb–May

A. trichopodus (Nutt.) A. Gray Per, robust, bushy-branched, ± minutely strigose, gen also spreading-hairy. **ST** ± erect, 2–10 dm. **LF** 2.5–20 cm; lower stipules rarely fused around st into sheath; lflets 15–39, 2–25 mm, ± lanceolate. **INFL:** fls 10–50, spreading or reflexed. **FL:** calyx 5–8.7 mm, tube 3.6–5.4 mm; petals ± cream, sometimes faintly lilac-blushed or -lined, banner 11.3–19 mm, recurved 40–45°, keel 8.6–13.7 mm. **FR** pendent; body 13–45 mm, 4.8–21 mm wide, linear-elliptic and compressed side-to-side to ovate and bladdery, persistent, thinly papery; hairs minute, sparse; dehiscence from tip, along top suture; stalk-like base 5–17 mm, slender, strigose or ± glabrous, joint between body base and stalk 0; chamber 1; immature seeds 10–30. Open, grassy areas, bluffs, rocky sites; 0–1220 m. SCoRO, s SCoRI, SCo, ChI, WTR, w PR, **w edge DMoj.** Like *A. asymmetricus, A. curtipes, A. oxyphysus,* but note calyx length; stalk length, hairs, position of joint; ovule number. 3 vars. in CA.

var. ***phoxus*** (M.E. Jones) Barneby **INFL:** fls 10–50. **FL:** banner 11.4–16.7 mm; keel 9.3–12.7 mm. **FR:** body 15–36 mm, 5–9 mm wide, compressed side-to-side, with ± flat or low-convex sides, glabrous or rarely minutely strigose. Gen inland, grassy or shrubby hillsides; 50–1220 m. s SCoR, n SCo (Santa Barbara, Ventura cos.), WTR, **w edge DMoj.** [*A. antiselli* A. Gray] Feb–Jun

A. whitneyi A. Gray Per, gen open and widely branched, silvery-hairy or not. **ST** ± ascending or erect, 4–40 cm. **LF** 1.5–10 cm; lower stipules fused around st into sheath; lflets 5–21, 2–21 mm, oblong to obovate. **INFL:** fls 3–16, ± well separated,

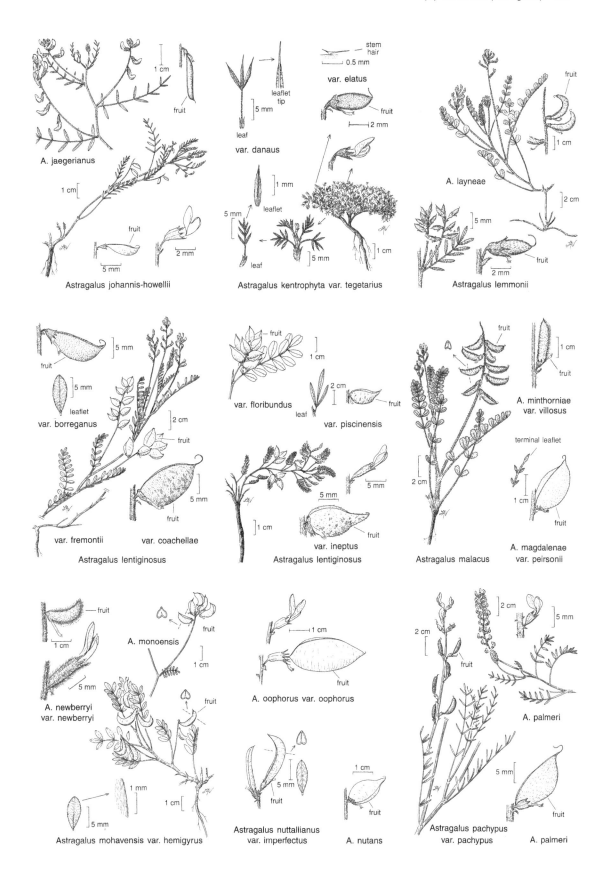

A. jaegerianus

Astragalus johannis-howellii

var. elatus

var. danaus

leaflet tip

leaf

stem hair
0.5 mm

fruit

leaflet

leaf

Astragalus kentrophyta var. tegetarius

A. layneae

fruit

Astragalus lemmonii

var. borreganus

leaflet

var. fremontii   var. coachellae

Astragalus lentiginosus

fruit

var. floribundus

leaf

var. piscinensis

fruit

var. ineptus

Astragalus lentiginosus

fruit

A. minthorniae var. villosus

terminal leaflet

fruit

A. magdalenae var. peirsonii

Astragalus malacus

fruit

A. monoensis

fruit

A. newberryi var. newberryi

fruit

Astragalus mohavensis var. hemigyrus

A. oophorus var. oophorus

fruit

Astragalus nuttallianus var. imperfectus

fruit

A. nutans

fruit

A. palmeri

fruit

Astragalus pachypus var. pachypus   A. palmeri

spreading or reflexed. **FL**: petals cream to pink-purple, banner 8.3–17.2 mm, recurved 50–80°, keel 7.3–13.8 mm. **FR** pendent; body 15–60 mm, 10–25 mm wide, bladdery, papery, glabrous to minutely strigose; stalk-like base 2–9 mm, slender; chamber 1. Open, sandy, gravelly, or rocky places; 800–3660 m. KR, NCoRH, SNH, w WTR, MP, **n SNE, W&I**; to WA, ID, NV. Intergrading complex. 4 vars. in CA.

var. ***whitneyi*** (p. 311) Pl greenish to grayish; hairs strigose. **LF** 3–11 cm, with spaces between lflets. **INFL**: fls 3–15. **FL**: petals pink or lilac, wing tips pale, banner 8.3–16.5 mm. **FR**: body 1.5–3 cm, glabrous; stalk-like base 3–15 mm. 2*n*=22. Open, rocky areas; 2050–3660 m. s SNF, SNH, n WTR, **n SNE, W&I**; w NV. May–Aug

CAESALPINIA

Elizabeth McClintock

Per, shrub, tree, armed or not, gland-dotted or not. **LVS** odd- or even-2-compound, alternate. **INFL**: gen raceme, axillary or terminal. **FL** ± bilateral; sepals ± free, overlapped above; stamens 10, ± exserted, free. **FR** dehiscent or not, inflated or flat. ± 200 spp.: trop, warm temp; some cult. (A. Caesalpini, Italy, 1519–1603) [Isely 1975 Mem New York Bot Garden 25(2):33–51]

C. virgata E.M. Fisher (p. 311) Shrub 0.5–2 m, hairy; branches slender, gen lfless, rush-like, green. **LVS** deciduous; 1° lflets 3, ternately arranged, lateral pair 0.5–1 cm, with 3–6 pairs of 2° lflets, terminal 1.5–4 cm, with 8–10 pairs of 2° lflets. **INFL** 5–15 cm, few-fld, puberulent. **FL**: sepals 5–6 mm; petals 6–8 mm, banner yellow with reddish marks, later entirely reddish; stamens ± 8 mm, slightly exserted, yellow. **FR** dehiscent, 1.5–2.5 cm, sickle-shaped, with sessile or short-stalked glands esp when young, margins ciliate. Uncommon. Gravelly or sandy desert gullies, washes, or canyon slopes; 100–500 m. **DSon**; to TX, Baja CA. [*Hoffmannseggia microphylla* Torrey] ✿ DRN,DRY, SUN:8–14,19–21;DFCLT. Mar–May

CALLIANDRA   FAIRYDUSTER, MOCK MESQUITE

Elizabeth McClintock

Per, shrub, tree, unarmed. **LVS** even-2-pinnate, alternate. **INFL**: gen head, axillary, few-fld. **FL** radial, purplish red or white; sepals, petals inconspicuous; stamens many, strongly exserted, filaments fused below, free above. **FR** dehiscent; 2 valves recurving; margins thickened. ± 200 spp.: trop Am, Madagascar, India. (Greek: beautiful stamens)

C. eriophylla Benth. (p. 311) FAIRYDUSTER Shrub < 30 cm; branches many, spreading from base. **LVS** deciduous; 1° lflets gen 2–4 pairs; 2° lflets gen 7–9 pairs, ± 2–3.5 mm. **FL** reddish purple; stamens 18–22 mm. **FR** 5 cm, flat. RARE in CA. Sandy washes, slopes, mesas; ± 1500 m. **DSon**; AZ, n Mex. In cult. Mar–Apr

CERCIDIUM   PALO VERDE

Elizabeth McClintock

Tree, shrub; branches with pointed tips; thorns in lf axils (see lf scars). **LVS** even-2-pinnate, alternate, falling early; 1° lflets gen 1 pair. **INFL**: raceme, axillary, < 7-fld. **FL** slightly bilateral; sepals ± free, all alike, reflexed; petals ± equal, clawed, yellow or cream-white; stamens 10, exserted, free. **FR** dehiscent or not, flat, narrowed between seeds or not. 4 spp.: deserts, se CA; AZ, nw Mex. (Greek: weaver's shuttle, from fr) [Carter 1974 Proc Calif Acad Sci 40(2):17–57]

1. Lf petioled; 2° lflets 1–3 pairs, 4–8 mm; fr not or very slightly narrowed between seeds
   . . . . ***C. floridum*** ssp. ***floridum***
1' Lf sessile; 2° lflets 4–8 pairs, 1–5 mm; fr narrowed between seeds . . . . ***C. microphyllum***

C. floridum A. Gray ssp. ***floridum*** (p. 311) BLUE PALO VERDE Tree gen < 8 m; branches spreading, ± zig-zagged, ± glabrous. **LF** blue-green; 1° lflets 1 pair, < 1 cm. **FL**: banner 9–15 mm, widely ovate, orange-dotted or not. **FR** 3–11 cm; tip beak-like. Uncommon. Washes, flood plains; ± 1100 m. **DSon**; to AZ, nw Mex. Fls gen 2 weeks before *C. microphyllum*. ✿ SUN,DRN:7,**8,9**,10,**14**, **19–23**,24&IRR:11,**12,13**; also STBL. Apr–May

C. microphyllum (Torrey) Rose & I.M. Johnston (p. 311) Shrub, small tree 3–4(9) m; branches gen ascending or spreading, broom-like, hairy. **LF** yellow-green; 1° lflets 1 pair, 3–6.5 cm. **FL**: banner < 10 mm, widely ovate, gen cream-white. **FR** < 11 cm; tip beak-like, gen ending in a spine. Uncommon. Rock slopes; ± 600 m. **DMoj**; to AZ, nw Mex. Branches used as livestock feed; seeds edible. Hybrids with *C. floridum* reported. ✿ SUN,DRN:**8,9**,10, **14**,15,16,**19–21**,22,23&IRR: 11,**12,13**; also STBL. Apr–May

DALEA

Duane Isely

Ann, per, unarmed, gland-dotted. **LVS** gen odd-1-pinnate; stipules inconspicuous, thread-like or glandular. **INFL**: spike in CA; bracts gen ± conspicuous. **FL**: calyx tube 10-ribbed; banner arising from receptacle, other petals from side or top of filament column; stamens 9–10 or 5, filaments fused; ovules 2. **FR** indehiscent, incl in or slightly exserted from calyx. **SEED** 1. ± 165 spp.: w US, Mex, s S.Am. (T. Dale, English botanist, 18th century) [Barneby 1977 Mem New York Bot Gard 27:135–582, 650–877] Incl spp. sometimes placed in *Petalostemon*; exc others found here in *Marina, Psorothamnus*. Key to species adapted by Margriet Wetherwax.

1. Pl (exc infl) glabrous, ascending; stamens 5 .... *D. searlsiae*
1' Pl hairy, gen prostrate, sometimes tiny; stamens 9–10
  2. Calyx 3–7 mm, lobes ± = tube; infl 8–15 mm wide; corolla > or ± = calyx; lflet margin entire, flat ... *D. mollis*
  2' Calyx 7–8 mm, lobes gen > tube; infl gen 14–16 mm wide; corolla < calyx; lflet margin gen shallowly
     lobed or wavy .... *D. mollissima*

D. mollis Benth. (p. 311) Ann, diminutive or mat-forming, hairy. **LF**: lflets 8–12, 3–7 mm, ± round to obovate-oblong, flat or folded, entire, flat. **INFL** 8–15 mm wide, ovoid or short-cylindric; bracts < ± 1 mm wide. **FL**: calyx 3–7 mm, lobes needle-like, ± = tube, shaggy-hairy; petals often > sepals, whitish or lavender; stamens 9–10. 2*n*=16. Common. Creosote-bush flats, washes, roadsides; < 800 m. D; to AZ, Mex. ❀ DRN,SUN:**10–13**. Mar–Jun

D. mollissima (Rydb.) Munz (p. 311) Ann or per, diminutive or mat-forming, hairy. **LF**: lflets 8–12, 3–10 mm, ± round to obovate-oblong, often folded, shallowly lobed, wavy. **INFL** gen 14–16 mm wide, ovoid or short-cylindric; bracts gen 1–1.5 mm wide. **FL**: calyx ± 7–8 mm, lobes needle-like, gen > tube, shaggy-hairy; petals < sepals, whitish or lavender; stamens 9–10. 2*n*=16. Common. Desert flats, washes; < 900 m. D; s NV, w AZ. ❀ DRN,SUN:**10–13**.

D. searlsiae (A. Gray) Barneby (p. 317) Per, glabrous. **STS** clustered, ascending, 3–5 dm. **LF**: lflets 5–7, obovate to oblong. **INFL** (minus corollas) ± 8–12 mm wide, oblong or narrowly oblong, initially compact, exposing axis in fr. **FL**: calyx 3.5–4.5 mm, tube recessed or slit on upper side, becoming leathery, puberulent; petals 5–7 cm, purple; stamens 5. Juniper/sagebrush scrub, slopes, bluffs; 1200–2000 m. **W&I (Inyo Mtns), DMtns**; to UT, AZ. [*Petalostemon s.* A. Gray] ❀ DRN,DRY:**1–3,10,11**. May–Jun

GLYCYRRHIZA   LICORICE

Duane Isely

Per, unarmed or ± prickly on axes, fr, sometimes glandular-hairy. **LVS** odd-1-pinnate; stipules deciduous; lflets gland-dotted. **INFL**: raceme, spike-like, axillary. **FL**: calyx lobes unequal, < or ± = tube; corolla white-yellow to blue; 9 filaments fused, 0 or 1 free. **FR** indehiscent, ellipsoid, prickly or rarely glabrous. ± 20 spp.: esp Eurasia. (Greek: sweet root) Several spp. cult.

G. lepidota Pursh (p. 317)   WILD LICORICE   Pl glabrous to very glandular-hairy. **LF**: lflets 9–19, lanceolate to narrowly ovate. **INFL** dense. **FL**: corolla 9–14 mm, yellowish or green-white. **FR** 12–15 mm. 2*n*=16. Moist, gen open, disturbed sites, such as creekbanks, roadsides; < 2400 m. **CA**; to c US, Can, Mex. ❀ DRN:**1–5,6,7,14–24**&IRR:**8–10**,11,12;INF;DFCLT;STBL. May–Jul

HOFFMANNSEGGIA

Elizabeth McClintock

Subshrub from spreading roots, unarmed. **LVS** 2-pinnate, ± basal. **INFL**: raceme, scapose. **FL** slightly bilateral; sepals ± free, equal; petals ± equal, yellow to orange-red; stamens 10, exserted, free, filaments often glandular. **FR** tardily dehiscent. **SEEDS** several. 28 spp.: Am, S. Afr. (J. Centurius, Count of Hoffmannsegg, Germany, 1766–1849) [B.B. Simpson 1998 Lundellia 2:14-54]

H. glauca (Ortega) Eifert (p. 317)   PIG-NUT, HOG POTATO   Pl erect, < 30 cm; stalked glands throughout; roots deep, tubered. **ST**: branches from base, slender. **LF** 5–12 cm; 1°lflets 5–11, odd-pinnate, 5–20 mm; 2° lflets 10–20, even-pinnate, 4–6 mm. **INFL** 5–15 cm, often glandular. **FL**: petals spreading, ± 1 cm, orange-red. **FR** 1.5–4 cm, ± curved; glands scattered, short-stalked, deciduous or not. Uncommon. Dry, alkaline flats in deserts, disturbed areas; < 900 m. SnJV, SCoRO, SCo, WTR, **D**; to TX, Mex, S.Am. [*H. densiflora* A. Gray] Aggressive weed, spreading by edible tubers. Apr–Jun

LATHYRUS   WILD PEA

Duane Isely

Ann or per, unarmed, glabrous or hairy, rarely glandular, gen rhizomed. **ST** sprawling, climbing, or erect; st angled, flanged, or winged. **LVS** even-1-pinnate; stipules persistent, upper lobe > lower; main axis ending as a tendril or short bristle; lflets 0–16, ± opposite or alternate, linear to widely ovate. **INFL**: raceme, gen axillary, 1–many-fld. **FL**: upper calyx lobes gen < and wider than lower; corolla 8–30 mm, pink-purple or pale, sometimes white or yellow; 9 filaments fused, 1 free; style flat, finely hairy on concave side. **FR** dehiscent, oblong, ± flat. ± 150 spp.: temp N.Am, Eurasia. (Ancient Greek name) [Broich 1987 Syst Bot 12:139–153] Some spp. variable, intergrading with others; some hybridization probable. Seeds of most alien spp. **TOXIC** to humans (esp young males) and livestock (esp horses). Key to species adapted by Margriet Wetherwax.

1. Infl gen 2-fld; lflets 1–1.4 cm; washes, desert scrub; ne DMoj (Grapevine Mtns, Inyo Co.) .. *L. hitchcockianus*
1' Infl gen 2–10-fld; lflets 2–8 cm; open slopes, pine forests; GB .... *L. lanszwertii* var. *lanszwertii*

L. hitchcockianus Barneby & Reveal Per, glabrous or puberulent. **ST** angled or flanged, not winged. **LF**: stipules small; lflets ± 4–6, 1–1.4 cm, linear to lanceolate; tendril coiled, sometimes branched. **INFL** gen 2-fld, open. **FL**: calyx tube > lobes; corolla 8–12 mm, lilac to purple. **FR** glabrous. Washes, desert scrub; 1500 m. **ne DMoj (Grapevine Mtns, Inyo Co.)**; w NV. Expected, but last collected in CA a century ago.

L. lanszwertii Kellogg Per, glabrous or puberulent. **ST** angled or ± flanged, not winged. **LF**: stipules small, gen narrow; lflets 4–10, opposite or subopposite, 2–8 cm, linear to lanceolate; tendril branched, coiled, reduced to bristle, or 0. **INFL** 2–10-fld. **FL**: calyx tube gen > upper lobes; corolla 7–16 mm. **FR** glabrous. 2*n*=14,28. Open, dry, gen coniferous woodlands, meadows; 200–2000 m. KR, NCoR, SNH, **GB**; to WA, WY, CO, NM. 3 vars. in CA.

var. ***lanszwertii*** (p. 317) Pl gen puberulent. **ST** climbing or ascending. **LF**: lflets elliptic-lanceolate to lanceolate; tendril gen coiled. **FL**: corolla 10–16 mm, purple to lavender. 2*n*=28. Open slopes, pine forests; 1200–2000 m. KR, NCoR, **GB**; to WA, ID, UT. ❀ TRY. May–Jul

LOTUS

Duane Isely

Ann, per, shrub, unarmed. **LVS** gen odd-1-pinnate (sometimes ± palmately compound, rarely some or most simple); stipules conspicuous or not; lflets 3–many, often irregularly arranged. **INFL**: umbel or 1–2-fld, axillary, gen peduncled, often bracted. **FL**: corolla gen yellow (sometimes white or pink), fading darker; 9 filaments fused, 1 free. **FR** dehiscent or not, exserted from calyx or not, ovoid to oblong, ± beaked. **SEEDS** 1–several. (Greek: derivation unclear) [Isely 1981 Mem New York Bot Garden 25:128–206] Spp. gen variable; intermediates may be hybrids. Key below separates natural groups. Key to species adapted by Margriet Wetherwax.

1. Lflets 5, 3 near axis tip, lower pair large, stipular in position; stipules gland-like, often not apparent
   . . . . ***L. corniculatus***
1' Lflets 3–15, lowest not stipular in position; stipules as above or not
   2. Stipules scarious, fragile; lflets 7–11, pinnately arranged, ± opposite . . . . . . ***L. oblongifolius*** var. ***oblongifolius***
   2' Stipules gland-like, bump-like, or conic, often not apparent; lflets 3–9, pinnately or ± palmately arranged, gen irregularly arrayed on lf axis
      3. Fr indehiscent, gen not flat, gen incl or moderately exserted from calyx (sometimes conspicuously exserted and curved most of length), gen with curved, 2–3 mm beak; gen per
         4. Lflets 3; infl 1–3-fld, ± sessile
            5. Lflet 2–5 mm; fr 6–9 mm; uncommon, sw DSon . . . . ***L. haydonii***
            5' Lflet 4–12 mm; fr 10–15 mm; widespread
               6. Calyx strigose . . . . ***L. procumbens*** var. ***procumbens***
               6' Calyx glabrous . . . . ²***L. scoparius*** var. ***brevialatus***
         4' Lflets > 3 on some or most lvs; infl 2–7-fld, sessile
            7. Lflets glabrous or finely strigose, gen green . . . . ²***L. scoparius*** var. ***brevialatus***
            7' Lflets conspicuously hairy, green, gray, or silvery
               8. St, esp near tip, with conspicuous, spreading or obliquely directed, often straight and stiff hairs < 0.5 mm; corolla 4–5 mm . . . . ***L. heermannii*** var. ***heermannii***
               8' St gen strigose or with ± spreading, often wavy hairs gen 0.2–0.4 mm; corolla 5–10 mm
                  . . . . ***L. nevadensis*** var. ***nevadensis***
      3' Fr dehiscent, gen flat, gen almost entirely exserted and straight, gen with straight or curved, 0.5–1.5 mm beak; ann or per
         9. Per or shrub-like; fr gen straight (sometimes curved throughout, rarely only at or near tip); wings gen > keel; stigma nearly glabrous or finely hairy
            10. Shrub-like, ascending, < 1.5 m, finely strigose; corolla 12–22 mm . . . . ***L. rigidus***
            10' Per, prostrate to low-ascending, < 0.3 m, silvery-silky or gray-puberulent; corolla 8–12 mm . . ***L. argyraeus***
               11. St prostrate, mat-forming; calyx lobes ± 2 mm . . . . var. ***argyraeus***
               11' St decumbent to low-ascending; calyx lobes 2–3 mm
                  12. Lflet oblanceolate to obovate, length ± 3–4 × width; New York Mtns . . . . var. ***multicaulis***
                  12' Lflet obovate, length ± 2 × width; Providence Mtns . . . . var. ***notitius***
         9' Ann; fr gen straight; wings < or ± = keel; stigma glabrous
            13. Fls gen 1 per lf axil; peduncle << 1 cm, bracts 0
               14. Calyx lobes 1–2 × tube; fr gen 3–4 mm wide; pl gen with soft, spreading hairs . . . . ***L. humistratus***
               14' Calyx lobes ± 0.8–1.2 × tube; fr gen 2.3–3 mm wide; pl inconspicuously strigose or with soft, spreading hairs . . . . ***L. wrangelianus***
            13' Fls 1–several per lf axil; peduncle gen > 1 cm, gen bracted
               15. Infl gen 2–4-fld (1st-formed sometimes 1–2-fld); lflets 3–7, obovate to ± round, terminal gen largest
                  . . . . ***L. salsuginosus***
                  16. Corolla 3.5–5 mm, keel > other petals; fr 1–1.5 cm, becoming narrowed between seeds
                     . . . . var. ***brevivexillus***
                  16' Corolla 6–10 mm, keel ± = other petals; fr 1.5–3 cm, not narrowed between seeds . . . var. ***salsuginosus***
               15' Infl gen 1-fld, lflets 1–5, lanceolate, elliptic, or obovate, ± equal
                  17. Calyx lobes >> tube; lflets gen 3 (often 1 on upper lvs), 10–20 mm . . . . ***L. purshianus***
                  17' Calyx lobes < tube; lflets 4–9 throughout, 3–10 mm . . . . ***L. strigosus***

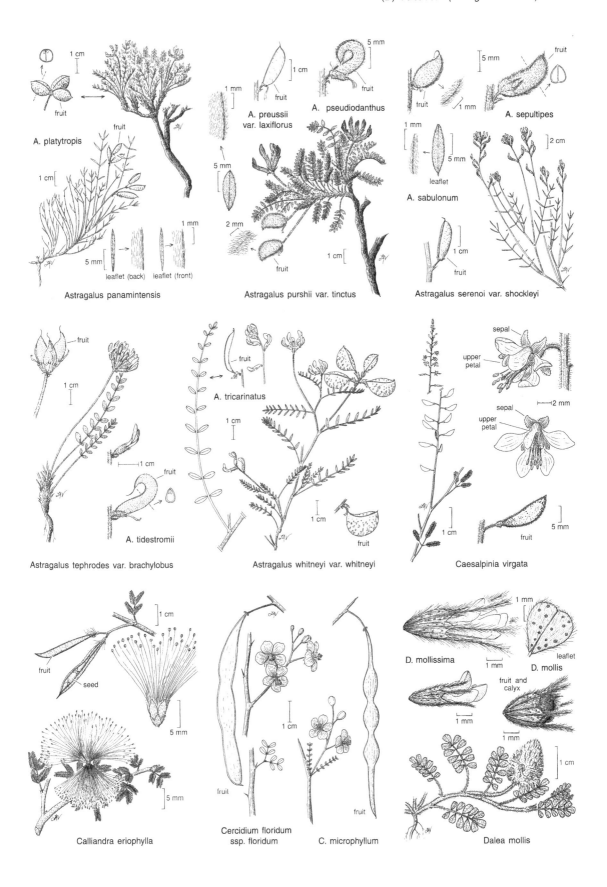

A. platytropis

A. preussii var. laxiflorus

A. pseudodanthus

A. sepultipes

A. sabulonum

Astragalus panamintensis

Astragalus purshii var. tinctus

Astragalus serenoi var. shockleyi

A. tidestromii

A. tricarinatus

Astragalus tephrodes var. brachylobus

Astragalus whitneyi var. whitneyi

Caesalpinia virgata

Calliandra eriophylla

Cercidium floridum ssp. floridum

C. microphyllum

D. mollissima

D. mollis

Dalea mollis

L. argyraeus (E. Greene) E. Greene Per, hairy, silvery or gray. **ST** prostrate to ascending. **LF** irregularly pinnate or ± palmate; stipules gland-like; lflets 3–5, 4–12 mm, oblanceolate or obovate. **INFL** 1–3-fld; peduncle bract small or 0. **FL:** calyx lobes ± = or < tube; corolla 7–12 mm, wings unequal, > keel; stigma finely puberulent. **FR** dehiscent, 1–2.5 cm, oblong, straight or slightly curved. **SEEDS** several. Mtn slopes, sometimes in pine/juniper woodland; 1200–2400 m. SnBr, SnJt, **DMtns**; to Mex. Vars. similar but geog distinct.

var. *argyraeus* **ST** gen prostrate, mat-forming. **LF** canescent; lflets obovate. **FL:** calyx lobes ± 2 mm; corolla 7–10 mm. **FR** 1.5–2 cm, gen straight. Open granitic slopes or with pine; 1500–2400 m. SnBr, SnJt, **DMtns**; to Mex. [ssp. *a.*] ❀ TRY. May–Aug

var. *multicaulis* (Ottley) Isely **ST** decumbent to ascending. **LF** green or canescent; lflets oblanceolate to obovate, length ± 3–4 × width. **FL:** calyx lobes 2.5–3 mm; corolla 8–12 mm. **FR** 2–2.5 cm, curved at tip. Uncommon. Pinyon/juniper woodland; 1200–1500 m. **DMtns (New York Mtns)**. [ssp. *m.* (Ottley) Munz] ❀ TRY

var. *notitius* Isely **ST** low-spreading or ascending. **LF** green or canescent; lflets obovate, length ± 2 × width. **FL:** calyx lobes 2–3 mm; corolla 8–12 mm. **FR** 1–2.5 cm, straight. Uncommon. Pinyon/juniper woodland; 1200–2000 m. **DMtns (Providence Mtns)**. ❀ TRY.

L. corniculatus L. (p. 317) BIRDFOOT TREFOIL Per, ± glabrous or strigose. **ST** decumbent or ascending. **LF:** stipules gland-like; lflets 5, 5–20 mm, linear to obovate, 3 ± palmately arrayed at axis tip, lower 2 stipular in position. **INFL** 3–8-fld; peduncle bracted. **FL:** calyx 2–3.5 mm, lobes ± = tube, not outcurved in bud; corolla 8–14 mm, bright yellow. **FR** dehiscent, 1.5–3.5 cm, narrowly oblong. **SEEDS** few–several. 2*n*=12,24. Probably naturalized in open, disturbed areas; < 1000 m. CA–FP, **GB**; to n US; native to Eurasia. In Eur, diploid *L. tenuis* Willd. is segregated; it seems indistinguishable in CA. Some pls TOXIC by production of cyanide-releasing compounds. Jun–Sep

L. haydonii (Orc.) E. Greene HAYDON'S LOTUS Per, finely strigose. **ST** ascending or sprawling, bushy, sparsely leafy, rush-like. **LF** subpalmate, tiny, deciduous; stipules gland-like or 0; lflets 3, 2–5 mm, elliptic. **INFL** 1–2-fld; peduncle < 3 mm or 0, bract 0. **FL:** calyx 2.5–3 mm, lobes < tube; corolla 4–5 mm, keel > other petals. **FR** indehiscent, exserted, ascending or reflexed, 6–9 mm, oblong, curved, beaked. **SEEDS** 1–2. RARE. Creosote-bush scrub to pinyon/juniper woodland; 600–1200 m. se PR, **sw DSon (esp San Diego Co.)**; Baja CA. Local, little known. Mar–Jun

L. heermannii (Durand & Hilg.) E. Greene Per (or fl 1st year and appearing ann). **ST** prostrate, often mat-forming; hairs spreading or obliquely directed, often straight and stiff, < 0.5 mm, esp near tip. **LF** subpalmate; stipules gland-like; lflets 4–6, irregularly arrayed, 4–16 mm, ovate or obovate; axis sometimes flat, ± blade-like. **INFL** 3–8-fld; peduncle gen < 5 mm, bracted. **FL:** calyx 2–4 mm, lobes < tube; corolla 4–7 mm, dark-tipped, wings > other petals. **FR** indehiscent, narrowly oblong, gen curved, tapered to long beak. **SEEDS** 1–2. Washes, riverbanks, coastal scrub, chaparral; < 2000 m. NCo, NCoRO, CCo, SCoRO, SCo, SnBr, PR, **DSon**. Vars. based esp on habitat, geography. 2 vars. in CA.

var. *heermannii* **FL:** corolla gen 4–5 mm; ovary gen strigose. Washes, riverbanks, chaparral; < 2000 m. SCo, SnBr, PR, **DSon**. Mar–Oct

L. humistratus E. Greene (p. 317) Ann, often ± fleshy, hairy. **ST** mat-forming to ascending. **LF** subpinnate or palmate; stipules gland-like; lflets gen 4, ± alternate, 4–12 mm, elliptic or obovate; axis flat, ± blade-like. **INFL** axillary, 1-fld, ± sessile. **FL:** calyx 3–6 mm, lobes 1–2 × tube; corolla 5–9 mm, yellow, wings ± = keel. **FR** dehiscent, ascending, 6–12 mm, gen 3–4 mm wide, oblong. **SEEDS** few. 2*n*=12. Abundant. Grassland, oak and pine woodland, desert flats and mtns, roadsides; < 1700 m. CA-FP, **D**; to sw UT, w NM, Mex. ❀ DRN,DRY:**7–10**,11,12,**14–24**;STBL. Mar–Jun

L. nevadensis (S. Watson) E. Greene Per, conspicuously hairy. **STS** mat-forming or ascending. **LF** subpinnate or palmate; stipules gland-like; lflets 3–5, irregularly arrayed, 4–12 mm, obovate to oblong, gen green. **INFL** 3–12-fld; peduncle gen bracted. **FL:** calyx 4–9 mm, lobes < tube; corolla 5–10 mm. **FR** indehiscent; body exserted, tapered-oblong, bent; beak slender, recurved. **SEEDS** 1–2. 2*n*=14. Oak, yellow-pine, lodgepole, and fir forests, open bracken meadows; 850–2750 m. KR, NCoR, CaR, SN, CCo, SCo, TR, PR, **DMtns**; to B.C., ID, NV. Forms intermediates with *L. argophyllus, L. scoparius*. Many named variants. 2 vars. in CA.

var. *nevadensis* (p. 317) **LF:** axis (incl petiole) gen 5–10 mm. **INFL** gen 5–12-fld. **FL** 5–10 mm, not blocky; banner upcurved 30–90°, gen drying orange-yellow. **FR** gen 1.8–2 mm wide. Pine and fir forests, bracken meadows, dry slopes; 850–2750 m. KR, NCoR, CaR, SN, CCo, SCo, w PR, **DMtns**; to B.C., ID, NV. [*L. douglasii* E. Greene] NW pls gen have larger fls (corolla 8–10 mm). ❀ **1–3**,7;STBL. May–Aug

L. oblongifolius (Benth.) E. Greene Per, glabrous or hairy. **ST** sprawling or ascending. **LF:** stipules scarious, fragile; lflets 3–11, ± opposite, 1–2.5 cm, gen elliptic to oblong. **INFL** 2–6-fld; peduncle bract 0 or just below umbel, simple or divided. **FL:** calyx 4–6 mm, lobes < tube; corolla 8–13 mm, gen whitish yellow, claws incl to slightly exserted from calyx tube. **FR** dehiscent, 2.5–5 cm, 1.5–2 mm wide, oblong. **SEEDS** few. Open, moist forests, river bottoms, marshy meadows, open pine woodland; 200–2600 m. CA-FP, **DMoj**; s OR, w&s NV, Mex. 2 vars. in CA.

var. *oblongifolius* **LF:** lflets 7–11, hairy or glabrous. **FL:** corolla 9–13 mm. Locally common. Open, moist forests, river bottoms, marshy meadows; 200–2400 m. Range of sp. [var. *nevadensis* (A. Gray) Munz] ❀ IRR;1–3,**7**,8–11,**14–24**;STBL. May–Sep

L. procumbens (E. Greene) E. Greene Per or stiff subshrub, puberulent or strigose, often gray. **ST** decumbent to erect, gen much-branched, < 1 m. **LF** subpalmate; stipules gland-like; lflets gen 3, 4–12 mm, oblanceolate to obovate. **INFL** 1–3-fld; peduncle 0 or < 3 mm, bract 0. **FL:** calyx 2–6 mm, strigose, lobes << or ± = tube; corolla 6–12 mm, yellow or with red, wings > keel. **FR** indehiscent, exserted, pendent, 1–1.5 cm, initially curved, then gen straight. **SEEDS** 2–3. Chaparral to pine forest, open slopes, ridges, flats, roadsides; < 2300 m. s SN, ScV, SCoR, TR, PR, **DMoj**. 2 vars. in CA.

var. *procumbens* (p. 317) **FL:** calyx 2–3 mm, lobes << tube; corolla 6–8 mm. Chaparral to Jeffrey-pine forest, sandy flats and slopes, roadsides; < 2300 m. ScV, SCoR, TR, PR, **DMoj**. Stiff subshrubs are found esp in Los Angeles Co. Apr–Jun

L. purshianus (Benth.) Ottley var. *purshianus* (p. 317) Ann, gen hairy. **ST** prostrate to erect, simple or openly branched, 0.5–6 dm. **LF** pinnate or ± simple; stipules gland-like; lflets gen 3, gen 1–2 cm, lanceolate to elliptic. **INFL** 1-fld; peduncle bracts simple. **FL:** calyx 3–6.5 mm, lobes> tube; corolla 5–9 mm, yellow to pink, wings ± = keel. **FR** dehiscent, widely spreading or pendent, 1.5–3 cm, oblong. **SEEDS** few. 2*n*=14. Coast, chaparral, mtn forests, water courses, and roadsides and other weedy areas; 0–2400 m. CA (exc DSon); to Can, c US, Mex. [var. *glaber* (Nutt.) Munz] Many races and ecological forms. ❀ **1–11**,14–24; STBL & forage. May–Oct

L. rigidus (Benth.) E. Greene (p. 317) Per, shrub-like, finely strigose. **STS** ascending, clustered, branched, 0.5–1.5 m. **LVS** irregularly pinnate to ± palmate, well spaced; stipules gland-like or 0; lflets 3–4, 0.5–1.5 cm, oblanceolate to obovate; axis (incl petiole) 1–8 mm. **INFL** 1–3-fld; peduncle 3–6 cm; bract near top or 0. **FL:** calyx 5–8 mm, lobes < tube; corolla 12–22 mm, wings > keel; stigma ± glabrous. **FR** slowly dehiscent, spreading or erect, 2–4 cm, oblong, ± straight, much exserted, gen glabrous. **SEEDS** several–many. 2*n*=14. Chaparral, desert flats, washes, foothills; < 1550 m. PR, **D**; to s NV, se UT, AZ, Baja CA. ❀ TRY. Mar–May

L. salsuginosus E. Greene   Ann, often fleshy, very small to robust, glabrous or strigose. **STS** clustered, prostrate or ascending. **LF** irregularly pinnate; stipules gen not apparent; lflets 3–7, ± alternate, 0.5–1.5 cm, obovate or ± round, terminal largest; axis flat, ± blade-like. **INFL** 2–4-fld (fr often only 1); peduncle bract lf- or lflet-like or perhaps 0. **FL**: calyx 2–4 mm, lobes < tube; corolla 3.5–10 mm, wings ± = or < keel. **FR** dehiscent, gen 1.5–3 cm, narrowly oblong, often curved; beak short, hooked. **SEEDS** few–many. 2*n*=14. Coastal scrub, foothill woodlands, washes, talus, deserts incl mtns; < 1850 m. CW, SCo, TR, PR, **D**; NV, AZ, Mex.

var. *brevixillus* Ottley   **FL**: calyx ± 2.5 mm; corolla 3.5–5 mm, keel > other petals. **FR**: 1–1.5 cm, becoming narrowed between seeds. Deserts, incl mtns; < 1850 m. **D**; NV, AZ, Mex. [ssp. *b.* (Ottley) Munz]

var. *salsuginosus*   **FL**: calyx 3–4 mm; corolla 6–10 mm, keel ± = other petals. **FR**: 1.5–3 cm, not narrowed between seeds. Coastal scrub, foothill woodlands, washes, talus; < 1200 m. CW, SCo, TR, PR, **DSon**; Mex. [ssp. *s.*] Mar–Jun

L. scoparius (Nutt.) Ottley   CALIFORNIA BROOM   Per, often shrubby, glabrous or finely strigose. **STS** clustered, gen ascending to erect (sometimes prostrate and mat-forming), bushy-branched, 0.5–2 m, greenish. **LVS** ± pinnate, well spaced, often deciduous; stipules gland-like or 0; lflets 3–6 (gen 3 on upper st), 6–15 mm, elliptic. **INFL** 2–7-fld; peduncle gen 0. **FL**: calyx 2.5–5 mm, glabrous, lobes < tube, not curved outward; corolla 7–11 mm. **FR** indehiscent, widely spreading or pendent, 1–1.5 cm, curved, long-beaked. **SEEDS** gen 2. 2*n*=14. Chaparral, roadsides, coastal sand, desert slopes, flats, washes; < 1500 m. NCo,

NCoR, n SNF, CCo, SnFrB, SCo, TR, PR, **DSon**; AZ, Mex. 2 vars. in CA.

var. *brevialatus* Ottley   **FL** gen 8–9 mm; keel > wings. Desert slopes, flats, washes; < 1400 m. SCo, TR, PR, **DSon**; AZ, Mex. [ssp. *b.* (Ottley) Munz].

L. strigosus (Nutt.) E. Greene (p. 317)   Ann, often fleshy, hairy or not. **ST** prostrate, gen branched from base. **LF**: stipules gland-like; lflets 4–9, gen alternate, 3–10 mm, oblanceolate to obovate; axis ± flat, ± blade-like. **INFL** 1–2-fld; peduncle gen bracted. **FL**: calyx 3–5.5 mm, lobes < tube; corolla opening or not, 5–10 mm, yellow, turning orange or reddish, wings gen > keel; stigma puberulent. **FR** dehiscent, 1–3.5 cm, gen ± curved only at or near tip. **SEEDS** several. 2*n*=14. Coastal scrub, chaparral, foothills, deserts, roadsides, other disturbed areas; < 2300 m. GV, CW, SW, **D**; AZ, Mex. [var. *hirtellus* (E. Green) Ottley; *L. tomentellus* E. Greene] Several variants often recognized (see Isely, pp. 193–198); pls in CA-FP gen ± strigose, with narrow lflets; pls in D fleshy, gen canescent, with wide lflets. Conspicuous in spring.   Mar–Jun

L. wrangelianus Fischer & C. Meyer   Ann, ± strigose or hairs soft, spreading. **ST** gen prostrate, branched at base. **LF**: stipules gland-like or ± 0; lflets gen 4, ± alternate, 4–15 mm, elliptic or obovate; axis flat, ± blade-like. **INFL** 1-fld; peduncle ± 0, slightly longer in fr. **FL** sometimes not opening; calyx 2.5–5 mm, lobes ± 0.8–1.2 × tube; corolla 5–9 mm, yellow, wings ± = keel. **FR** dehiscent, 10–18 mm, 2.2–3 mm wide, oblong. **SEEDS** few–several. 2*n*=12. Abundant. Coastal bluffs, chaparral, disturbed areas; 0–1500 m. CA-FP, probably naturalized in MP, **DSon** through agriculture. [*L. subpinnatus* Lagasca misapplied] Mar–Jun

LUPINUS LUPINE

Rhonda Riggins (ann) & Teresa Sholars (per to shrubs)

Ann to shrubs; cotyledons gen petioled, withering early. **ST** gen erect. **LVS** palmately compound in CA, gen cauline; stipules fused to petiole; lflets 3–17, gen oblanceolate, entire. **INFL**: raceme; fls spiraled or whorled; bracts gen deciduous. **FL**: calyx 2-lipped, lobes entire or toothed, gen appendaged between lobes; banner centrally grooved, sides reflexed, wing tips slightly fused, keel gen pointed; stamens 10, filaments fused, 5 long with short anthers, 5 short with long anthers; style brushy. **FR** dehiscent, gen oblong. **SEEDS** 2–12, gen smooth. ± 200 spp.: esp w N.Am, w S.Am to e US, also trop S.Am, Medit. (Latin: wolf, from mistaken idea that pls rob soil of nutrients) Some cult for fodder, green manure, edible seed, orn; some naturalized from CA in e N.Am, S.Am, Australia, s Afr; some (e.g., *L. arboreus, L. latifolius, L. leucophyllus*) have alkaloids (esp in seeds, frs, young herbage). TOXIC to livestock (esp sheep). [Barneby 1989 Intermountain Flora 3(B):237–267] Infl length does not incl peduncle. ❀ Many Lupine taxa need seed pre-treatment (scarification, stratification, inoculation) for successful germination.

1. Ann, rarely living > 1 growing season
  2. Cotyledons petioled, gen withering, gen deciduous; fr oblong; seeds 3–6, smooth
    3. Both margins of keel glabrous . . . . *L. concinnus*
    3' Lower (and often upper) margins of keel ciliate near claw
      4. Lflets 5–10 mm wide, upper surface glabrous; petals dark pink to magenta . . . . *L. arizonicus*
      4' Lflets 2–4 mm wide, upper surface hairy at least near margins; petals gen blue (pinkish) . . . . *L. sparsiflorus*
  2' Cotyledons sessile, persistent at pl base, disk- or cup-like or leaving a circular scar; fr gen ovoid;
    seeds gen 2, gen tubercled or wrinkled (fr oblong; seeds 2–6 in *L. odoratus*)
    5. Fls distinctly whorled; infl bracts reflexed; upper margins of keel ciliate near claw . . . . *L. microcarpus*
      6. Wings widely elliptic, persistent, becoming translucent, upper (and gen lower) margins ciliate near
        claw; lower margins of keel ciliate near claw; calyx appendages 1–2 mm . . . . var. *horizontalis*
      6' Wings linear to oblanceolate, early withering, not becoming translucent, upper (and rarely lower)
        margins gen ciliate near claw; lower margins of keel gen glabrous near claw; calyx appendages gen 0
        . . . . var. *microcarpus*
    5' Fls spiraled, not distinctly whorled; infl bracts straight; upper margins of keel glabrous
      7. Lvs basal; herbage sparsely hairy when young, gen glabrous at fl; pedicel 3–5 mm . . . . *L. odoratus*
      7' Lvs cauline, often crowded near base; herbage remaining densely to sparsely hairy or canescent;
        pedicel gen < 3 mm
        8. Herbage canescent; upper lip of calyx ± = lower; upper suture of fr wavy and densely stiff-ciliate;
          fr sides with short inflated hairs that become scale-like when dry . . . . *L. shockleyi*

8' Herbage sparsely to densely ± long-hairy; upper calyx lip < lower; upper suture of fr not obviously wavy and densely stiff-ciliate; fr sides with long, uninflated, wavy hairs
    9. Peduncle 0–1 cm; infl < lvs; fr oblong, narowed between seeds . . . . *L. pusillus* var. *intermontanus*
    9' Peduncle 2–5 cm; infl > lvs; fr ovoid, not narrowed between seeds
        10. Infl 1–2.5 cm; pedicel 0.5–1.5 mm; calyx appendages present; seed smooth . . . . *L. brevicaulis*
        10' Infl 3–8 cm; pedicel 2–3 mm; calyx appendages 0; seed wrinkled . . . . *L. flavoculatus*
1' Per to shrub
  11. Upper keel margin ciliate from claw to middle (glabrous near tip) — SNE . . . . *L. latifolius* var. *columbianus*
  11' Upper keel margin glabrous or ciliate middle or claw to tip
    12. Calyx spur 1–3 mm
      13. Wings with dense patch of hair on outside margin near tip; lf gen green, strigose . . . . *L. arbustus*
      13' Wings glabrous; lf gen silver-silky . . . . *L. argenteus* var. *heteranthus*
    12' Calyx spur 0–1 mm
      14. Banner back gen hairy (best seen in bud)
        15. Upper keel margin ± glabrous
          16. Lf green, glabrous to hairy; petals yellow . . . . *L. angustiflorus*
          16' Lf silver- to white-woolly; petals cream to pale yellow . . . . *L. padre-crowleyi*
        15' Upper keel margin ciliate
          17. Shrub or subshrub
            18. Pl < 2 dm, matted; fl 6–11 mm . . . . *L. breweri* var. *grandiflorus*
            18' Pl gen > 2 dm, not matted; fl 9–18 mm — grape-soda scented . . . . *L. excubitus*
              19. Shrub 10–15 dm, silver-appressed-hairy; petiole < 12 cm; SNE, DMoj and boundaries . . **var. *excubitus***
              19' Subshrub < 7 dm, silver-tomentose; petiole > 12 cm; DSon (se San Diego, sw Imperial cos.)
                . . . . [2]var. *medius*
          17' Per
            20. Pl < 1 dm — uncommon, SNE (Mono Co.) . . . . [3]*L. duranii*
            20' Pl > 1 dm
              21. Calyx not bulged or spurred
                22. Lf hairs dense, tomentose to woolly; lflets silvery; DSon . . . . [2]*L. excubitus* var. *medius*
                22' Lf hairs appressed, sometimes sparse; lflets green; SNE . . . . *L. pratensis* var. *eriostachyus*
              21' Calyx gen bulged or spurred < 1 mm . . . . *L. argenteus*
                23. Lvs basal and cauline; lower petioles gen 7–15 cm . . . . var. *montigenus*
                23' Lvs cauline; petioles 1–10 cm
                  24. St hairs appressed; lf gen appearing green; fl (8) 10–12 mm . . . . var. *argenteus*
                  24' St hairs spreading; lf gray to silver; fl 8–10 mm . . . . var. *palmeri*
      14' Banner back ± glabrous
        25. Lf upper surface glabrous, green . . . . *L. polyphyllus* var. *burkei*
        25' Lf upper surface hairy, greenish gray to silver
          26. Infl gen dense, bracts gen persistent; pedicels stout, short, gen < 3 mm
            27. Largest lflet > 40 mm . . . . *L. pratensis* var. *pratensis*
             27' Largest lflet < 40 mm . . . . *L. lepidus* (in part)
              28. Infl < lvs; pl matted — infl 2–3 whorls . . . . var. *utahensis*
              28' Infl gen > lvs; pl matted or not
                29. Cauline lvs 0 or clustered near the base (appearing basal)
                  30. Pl < 1 dm; infl ± head-like, 2–8 cm, gen < some lvs . . . . var. *lobbii*
                  30' Pl 1.2–3.5 dm; infl elongate, 4.5–11 cm, > lvs . . . . var. *sellulus*
                29' Cauline lvs gen spread along st
                  31. Lvs cauline; infl 5–30 cm, whorls > 7, ± crowded; meadows and vernally moist areas,
                    1500–3000 m . . . . var. *confertus*
                31' At least some lvs basal; infl 2–10 cm, whorls 3–7, well spaced; subalpine, 3000–4000 m
                  . . . . var. *ramosus*
      26' Infl gen ± open, bracts gen ± deciduous; pedicels slender, gen > 3 mm
        32. Upper keel margin gen glabrous
          33. Pl > 2 dm; fl 9–12 mm, keel gen upcurved; lf green, hairy . . . . *L. andersonii*
          33' Pl < 2 dm; fl 4–11 mm, keel ± straight; lf silvery-hairy
             34. Caudex above ground, pl forming a dense tuft; stipule 6–11 mm, pedicel 4–5 mm; lvs basal,
             petioles 2–8 cm; fl 8–11 mm, raceme 10–22-fld; rare, Mono Co. . . . . [3]*L. duranii*
             34' Caudex below ground, pl mat-forming; stipule 2–5 mm; pedicel 1–3(4) mm; lvs clustered
             near base, petioles 1–6 cm, fl 4–7 mm, raceme 4–14-fld; common . . . . *L. breweri* var. *bryoides*
        32' Upper keel margin ± ciliate
          35. Pls < 2 dm; fl 8–11 mm — rare, Mono Co. . . . . [3]*L. duranii*
          35' Pls > 2 dm; fl 5–18 mm
             36. Lf densely woolly; st hairs sharp; fl 10–18 mm, fragrant . . . . *L. magnificus*
             36' Lf hairy, but not woolly; st hairs not sharp; fl 5–15 mm, gen not fragrant
               37. Fl 5–7(10) mm; lvs ± appressed-silvery; SNE . . . . *L. argenteus* var. *meionanthus*

37' Fl 8–18 mm; lvs long-hairy; GB, DMtns
  38. Pl 4–7 dm; fl 13–15 mm; ne DMtns . . . . *L. holmgrenanus*
  38' Pl 1–4 dm; fl 10–12 mm; GB, DMtns (Mono Co.) . . . . *L. nevadensis*

L. andersonii S. Watson (p. 317) Per 2–9 dm, green, hairy. **ST** erect. **LVS** cauline; stipules 3–15 mm; petiole 2–6 cm; lflets 6–9, 20–60 mm. **INFL** 2–23 cm, open; peduncle 1–8.5 cm; pedicel 1.5–5 mm; fls ± whorled; bracts 2–10 mm, deciduous. **FL** 9–12 mm; calyx upper lip 5–7 mm, 2-toothed, lower lip 3–8 mm, 2–3-toothed; petals gen lavender to purple (yellowish), banner back glabrous, patch white turning purple, keel glabrous. **FR** 2–4.5 cm, silky. **SEEDS** 4–6, 4–6 mm, mottled tan, brown. Dry slopes; 1500–3000 m. NW, SNH, WTR, SnBr, **SNE**; s OR, w NV. ± indistinct from *L. albicaulis*. ❀ TRY. Jun–Sep

L. angustiflorus Eastw. (p. 317) Per 5–12 dm, green, glabrous to hairy. **ST** erect. **LVS** cauline; stipules 5–13 mm; petiole ± 1–5 cm; lflets 6–9, 20–60 mm. **INFL** 6–34 cm, open; peduncle ± 1–8 cm; pedicel 2–4 mm; bracts 3–7 mm, ± persistent. **FL** 8–10(12) mm; calyx upper lip 4–8 mm, 2-toothed, lower lip 4–9 mm, entire to 3-toothed; petals pale yellow to orange-yellow, banner back gen hairy, patch orange to yellow, keel glabrous, tip lavender. **FR** 2.5–4 cm, hairy. **SEED** 4.5–5.5 mm, speckled tan and brown. Gen volcanic soils; 1000–3500 m. CaRH, n&c SNH, **GB**. [*L. andersonii* var. *christinae* (A.A. Heller) Munz] ❀ TRY.

L. arbustus Lindley (p. 319) SPUR LUPINE Per 2–7 dm, green or gray-silky. **ST** erect. **LVS** cauline and sometimes basal; stipules 4–9 mm; petiole 2–16 cm; lflets 7–13, 20–70 mm. **INFL** 3–18 cm, open; peduncle 2–5 cm; pedicel 1–7 mm; fls whorled; bracts 3–6 mm, deciduous. **FL** 8–14 mm; calyx spur distinct, 1–3 mm, upper lip 2–4 mm, 2-toothed, lower lip 2.5–5 mm, 3-toothed; petals blue, purple, pink, white, or yellowish, banner back hairy, patch white, yellowish, or 0, wings with dense hair patch outside near tip, upper keel margins ciliate, lower keel margins glabrous. **FR** 2–3 cm, silky. **SEEDS** 3–6, 5–6 mm, tan. Open sagebrush shrubland or mixed-conifer forest; 1500–3000 m. SNH, SnGb, **GB**; to OR, ID, UT. [sspp. *calcaratus* (A.A. Heller) Dunn and *silvicola* (Kellogg) Dunn; var. *montanus* (Howell) Dunn] Like *L. argenteus* exc wing hairs. ❀ TRY. May–Jul

L. argenteus Pursh Per 1–15 dm, green-glabrous to silvery-hairy. **ST** erect. **LVS** basal to cauline; stipules 2–12 mm; petiole gen 1–15 cm; lflets 5–9, 10–60 mm, < 10 mm wide, glabrous to hairy above, hairy below. **INFL** 5–16(25) cm; peduncle 1–10 cm; pedicel 1–6 mm; fls whorled to not; bracts gen deciduous. **FL** 5–14 mm; calyx upper lip 4–8 mm, 2-toothed to entire, lower lip 4–8 mm, entire to 3-toothed, bulge or spur 0–3 mm (may be variable on 1 pl); petals blue, violet, or white, banner back gen hairy, patch yellowish to whitish to 0, upper keel margins ciliate, lower keel margins glabrous. **FR** 1–3 cm, hairy or silky. **SEEDS** 2–6, tan, brown, or red. Montane forest, sagebrush scrub; 1000–3500 m. CaR, SNH, **GB**, **DMtns**; to s Can, SD, NM. Highly variable; vars. intergrade. Var. *rubricaulis* (E. Greene) Welsh [*L. alpestris* Nelson] probably not in CA. ❀ TRY.

var. ***argenteus*** (p. 319) Pls 2–15 dm. **LVS** cauline, green, appressed-hairy. **FL** (8)10–12 mm; calyx bulge < 1 mm; petals blue or purple to white. Esp dry sagebrush scrub; 1000–2000 m. CaR, n SNH, **GB**, **DMtns**; to s Can, SD, NM. [var. *tenellus* (G. Don) D. Dunn; *L. sublanatus* Eastw.] Jun–Oct

var. ***heteranthus*** (S. Watson) Barneby (p. 319) Pls 2–8 dm. **LVS** basal or some cauline, densely silver-silky. **FL** 8–14 mm; calyx spur 1–2 mm; petals violet or blue to white, banner back silky, wings glabrous. Dry, open slopes; 1000–3000 m. **GB**; to OR, ID, UT. [*L. caudatus* Kellogg; *L. inyoensis* A.A. Heller] May–Sep

var. ***meionanthus*** (A. Gray) Barneby (p. 319) Per or subshrub 2–9 dm. **LVS** cauline, appressed-silvery to gray-greenish. **FL** 5–7(10) mm; petals dull blue to lilac, banner back glabrous, patch yellow. Dry banks; 1500–3500 m. n&c SNH, **SNE**; NV. [*L. m.* A. Gray] Jul–Aug

var. ***montigenus*** (A.A. Heller) Barneby Pls < 4 dm, densely silvery-hairy. **LVS** basal and cauline; lower petioles gen 7–15 cm. **FL** 9–12(14) mm; calyx bulge < 1 mm; petals blue to violet, banner patch yellow to cream. Dry, open montane forests, sagebrush scrub; 2500–3500 m. c SNH, **SNE**; NV. [*L. m.* A.A. Heller] Jul–Aug

var. ***palmeri*** (S. Watson) Barneby (p. 319) Pls 3–6 dm. **LVS** cauline, densely gray-spreading-hairy and silvery-silky. **FL** 8–10 mm; calyx bulge or spur < 1 mm; petals blue, banner back hairy. Dry, open montane forests; 2000–2500 m. SNH, **SNE**; to WA, UT, NM. Like var. *argenteus* exc hairs. [*L. p.* S. Watson] May–Jun

L. arizonicus (S. Watson) S. Watson (p. 319, pl. 54) ARIZONA LUPINE Ann 1–5 dm, short-appressed- and long-spreading-hairy. **LF**: petiole 2.5–7 cm; lflets 6–10, 10–40 mm, 5–10 mm wide, upper surface glabrous. **INFL** 4–24 cm; fls spiraled, sometimes appearing ± whorled; peduncle 2–6 cm; bracts 4–8 mm, gen persistent; pedicels 2–4 mm. **FL** 7–10 mm; calyx 3–6 mm, lips ± equal, upper lip deeply lobed; banner, wings dark pink to magenta, drying blue-purple or whitish, banner spot yellowish, becoming darker magenta, upper margins of keel glabrous, lower margins ciliate near claw. **FR** 1–2 cm, ± 5 mm wide, coarsely hairy, often on 1 side of infl axis. **SEEDS** 4–6. Sandy washes, open areas; < 1100 m. e DMoj, DSon; NV, AZ, Mex. [*L. concinnus* var. *brevior* (Jepson) D. Dunn] Robust pls have been called var. *barbatulus* (J. Thornber) I.M. Johnston. Locally common. ❀ TRY. Mar–May

L. brevicaulis S. Watson (p. 319) SAND LUPINE Ann < 1 dm, hairy; cotyledons disk-like, persistent. **LVS** cauline, crowded near base; petioles 2–5 cm; lflets 6–8, 10–15 mm, 3–6 mm wide, linear to oblanceolate, upper surface glabrous. **INFL** 1–2.5 cm, exceeding lvs; fls spiraled, dense; peduncle 2–5 cm; bracts 2–3 mm, straight, persistent; pedicels 0.5–1.5 mm. **FL** 6–8 mm; upper lip of calyx ± 3 mm, 2-toothed, lower lip ± 6 mm, appendages present; petals bright blue, banner spot white or yellow, keel glabrous. **FR** ± 1 cm, ± 5 mm wide, ovoid, hairy. **SEEDS** 1–2. Sandy washes, open areas; < 2300 m. MP (Lassen Co.), **W&I**, e DMoj; to OR, Colorado, Mex. ❀ TRY. May–Jun

L. breweri A. Gray Per or subshrub < 2 dm, matted or tufted, silvery-silky. **ST** prostrate; base ± woody. **LVS** cauline, clustered near base; stipules 2–5 mm; petiole 1–5(6) cm; lflets 5–10, 3–20 mm. **INFL** 1–10 cm, ± dense; peduncle 1–8 cm; pedicel 1–3(4) mm; bracts 3–5 mm, deciduous. **FL** 4–11 mm; calyx upper lip 4–7 mm, 2-toothed, lower lip 4–6 mm, entire to 3-toothed; petals blue to violet, banner back glabrous to densely hairy, patch white or yellow, keel straight, upper keel margins glabrous or ciliate, lower keel margins glabrous. **FR** 1–2 cm, silky. **SEEDS** 3–4, 3–4 mm, mottled tan, brown. Common. Gen open montane forest; 1000–4000 m. CA-FP, **SNE**; s OR, NV. 3 vars. in CA.

var. ***bryoides*** C.P. Smith (p. 319) Per. **LF**: lflets 3–5 mm. **INFL** < 2.5 cm; peduncle < 2 cm. **FL** 4–7 mm; keel and banner glabrous. Habitats of sp.; 2500–4000 m. s SNH, WTR, **SNE**; NV. ❀ TRY. Jul–Aug

var. ***grandiflorus*** C.P. Smith (p. 319) Subshrub, matted. **LF**: lflets 6–11 mm. **INFL** 2–10 cm; peduncle 2–8 cm. **FL** 6–11 mm; keel ciliate, banner back ± silky. Volcanic sand; 2000–3500 m. SN, SnBr, **SNE**. ❀ TRY. Jun–Aug

L. concinnus J. Agardh (p. 319) BAJADA LUPINE Ann 1–3 dm, hairy. **ST** erect or decumbent. **LF**: petiole 2–7 cm; lflets 5–9, 10–30 mm, 1.5–8 mm wide, sometimes linear. **INFL** 1.5–9 cm, often dense; fls spiraled, gen also in lower lf axils; peduncle 0–8 cm; bracts 2.5–4 mm, straight, persistent; pedicels 0.7–2 mm. **FL** 5–12 mm; calyx 3–5 mm, lips ± equal, upper lip deeply lobed;

petals pink to purple, rarely white, banner spot white or yellowish, keel gen glabrous. **FR** 1–1.5 cm, 3–5 mm wide, hairy. **SEEDS** 3–5. 2*n*=48. Common. Open or disturbed areas, burns; < 1700 m. c&s CW, SW, **D**; to UT, TX, Mex. [vars. *agardhianus* (A.A. Heller) C.P. Smith, *desertorum* (A.A. Heller) C.P. Smith, *optatus* C.P. Smith, *orcuttii* (S. Watson) C.P. Smith, and *pallidus* (Brandegee) C.P. Smith] Highly variable, gen self-pollinated, needs study; named vars. ± indistinct; pls in D with linear, coarsely hairy lflets and barely ciliate lower keel margins may be confused with *L. sparsiflorus*. ❀ DRN,DRY,SUN:**7–12**,13,**14–16**,17,**18–23**,24;STBL. Mar–May

L. duranii Eastw. (p. 319) MONO LAKE LUPINE Per 5–12 cm, robust, tufted, shaggy. **LVS** basal; stipules 6–11 mm; petiole 2–8 cm; lflets 5–8, 5–20 mm. **INFL** 2–6 cm; peduncle < 3.5 cm; pedicel 4–5 mm; whorls many, crowded; bracts 4–5 mm, ± deciduous. **FL** 8–11 mm; calyx upper lip 5–7 mm, deeply 2-toothed, lower lip 6–7 mm, ± entire; petals violet, banner back glabrous, patch cream or white, upper keel margins glabrous to sparsely ciliate, lower keel margins glabrous. **FR** 1–2 cm, 8–20 mm. **SEEDS** 3–5, white. RARE. Dry volcanic pumice, gravel; 2000–2500 m. **SNE (Mono Co.)**. [*L. tegeticulatus* var. *d.* (Eastw.) Barneby] Reports from Madera Co. questionable. May–Aug

L. excubitus M.E. Jones GRAPE SODA LUPINE Subshrub or shrub 2–20 dm, greenish to silver-hairy. **ST** prostrate to erect. **LVS** cauline, clustered at base or not, gen silver-hairy; stipules 5–20 mm; petiole 4–15 cm; lflets 7–10, 5–50 mm. **INFL** < 70 cm; peduncle < 30 cm; pedicel 2–7 mm; fls whorled or not; bracts 8–9 mm, deciduous. **FL** 9–18 mm, with distinctive sweet smell; calyx upper lip 6–8 mm, deeply notched, lower lip 6–8 mm, entire to 3-toothed; petals violet to lavender, banner back gen hairy, patch bright yellow (turning purple at fl), keel gen lobed near base, upper keel margins ciliate from middle to tip, lower keel margins glabrous. **FR** 3–5 cm, silky. **SEEDS** 5–8, mottled yellow-brown with lateral lines. Dry areas; < 3000 m. s SNH, Teh, SW, **SNE**, **D**. 5 vars. in CA.

var. *excubitus* (p. 319) Shrub 10–15 dm, silver-hairy. **INFL**: axis gen persistent in winter. **FL** 9–13 mm. Desert slopes, washes; < 2500 m. **DMoj**, adjacent s SNH, **SNE**. ❀ TRY. Apr–Jun

var. *medius* (Jepson) Munz (p. 319) MOUNTAIN SPRINGS BUSH LUPINE Subshrub < 7 dm, silver-tomentose. **FL** 10–13 mm. RARE. Desert washes; < 1000 m. **sw DSon (se San Diego, sw Imperial cos.)**. Mar–Apr

L. flavoculatus A.A. Heller (p. 319) Ann 0.5–2 dm, hairy; cotyledons disk-like, persistent. **LVS** cauline, crowded near base; petioles 3–6 cm; lflets 7–9, 10–20 mm, 5–8 mm wide, upper surface glabrous. **INFL** 3–8 cm, > lvs, gen < peduncle, dense; fls spiraled; peduncle 3–5 cm, 10 cm in fr; bracts 2–3 mm, straight, persistent; pedicels 2–3 mm. **FL** 7–10 mm; calyx upper lip 1–3 mm, deeply lobed, lower lip 4–5 mm, appendages 0; petals bright blue, banner spot yellow, keel blunt, glabrous. **FR** 0.5–1 cm, ± 5 mm wide, ovoid, hairy, often on 1 side of infl axis. **SEEDS** 1–2, wrinkled. Sand or gravel; < 2200 m. **W&I, e DMoj**; NV. ❀ TRY. Apr–Jun

L. holmgrenanus C.P. Smith (p. 319) HOLMGREN'S LUPINE Per 4–7 dm, long-hairy. **ST** erect. **LVS** basal and cauline; stipules 5–20 mm; lower petioles 2–17 cm, smaller upward; lflets 4–7, 15–50 mm. **INFL** 10–26 cm, open to ± dense; peduncle 3–10 cm; pedicel 6–10 mm; fls spiraled; bracts 8–10 mm, gen deciduous. **FL** 13–15 mm; calyx bulge or spur < 1 mm, upper lip 6–7 mm, 2-toothed, lower lip 7–9 mm, entire; petals violet, banner back glabrous, patch yellow, upper keel margin slightly ciliate, lower keel margins glabrous. **FR** 4–5 cm, hairy. **SEEDS** 5–7. RARE in CA. Dry desert slopes; 1500–2500 m. **ne DMtns (Last Chance, Grapevine mtns, Inyo Co.)**; w NV. May–Jun

L. latifolius J. Agardh Per 3–24 dm, green, glabrous to hairy. **ST** erect. **LVS** cauline; stipules 5–10 mm; petiole 4–20 cm; lflets 5–11, 40–100 mm, upper surface glabrous to hairy, lower surface ± hairy. **INFL** 16–60 cm, open; peduncle 8–20 cm; pedicel

2–12 mm; fls whorled or not; bracts 8–12 mm, deciduous. **FL** 8–18 mm; calyx upper lip 5–10 mm, entire to 2-toothed, lower lip 4–8 mm, entire or notched; petals blue or purple to white; banner back glabrous, patch gen white to yellowish turning purple, upper keel margins ciliate from claw to middle, lower keel margins gen ciliate. **FR** 2–4.5 cm, ± densely hairy. **SEEDS** 6–10, 3–4 mm, mottled dark brown. Common. Gen moist areas in woodlands, shady to open areas; < 3500 m. CA-FP (exc GV), **GB**; to B.C., UT, AZ, Baja CA. [*L. rivularis* Lindley in part] TOXIC: causes birth defects in livestock. 6 vars. in CA.

var. *columbianus* (A.A. Heller) C.P. Smith (p. 319) **ST** subglabrous to strigose. **FL** 10–14 mm; keel mostly covered by wings. Moist slopes, streamsides; 1000–3500 m. SN, **SNE**; to WA. ❀ TRY. May–Sep

L. lepidus Douglas DWARF LUPINE Per < 6 dm, matted, hairy. **ST** 0 or prostrate to ± erect. **LVS** gen basal; stipules 3–25 mm; petiole 2–10 cm; lflets 5–8, 5–40 mm. **INFL** < 30 cm, gen dense; peduncle < 14 cm; pedicel 1–3 mm; bracts 4–15 mm, persistent. **FL** 6–11 mm; calyx upper lip 3–7 mm, entire to 2-toothed, lower lip 4–7 mm, entire to 3-toothed; petals pink, violet, or blue, banner back glabrous, upper keel margins ciliate, lower keel margins glabrous. **FR** 1–2 cm, hairy. **SEEDS** 2–4, 2–4 mm, ± mottled tan or green to brown. Montane to alpine open places; 1500–4000 m. CaRH, SNH, SW, **GB**, **DMtns**; to B.C., MT, Colorado. Variable complex best characterized by habit, infl, bracts, habitats. Other vars. throughout w N.Am. 6 vars. in CA.

var. *confertus* (Kellogg) C.P. Smith (p. 319) Pl 25–60 cm, hairy. **ST** decumbent to erect. **LVS** cauline. **INFL** 5–30 cm, > lvs; whorls > 7, ± crowded; bracts 5–14 mm. **FL** 7–9 mm; banner patch yellowish fading brown to red. 2*n*=48. Common. Meadows, vernally moist areas; 1500–3000 m. CaRH, SNH, WTR, SnBr, **GB**, **DMtns**; NV. [*L. c.* Kellogg] ❀ TRY. Jun–Aug

var. *lobbii* (S. Watson) C. Hitchc (p. 319) Pl < 10 cm, hairy to shaggy. **ST** prostrate. **LVS** gen basal. **INFL** 2–8 cm, partly < lvs; bracts 5–6 mm. **FL** 6–10 mm; banner patch white. Dry rocks, meadows; 2000–3500 m. c SNH, **SNE**; to WA, NV. [*L. lobbii* S. Watson; *L. lyalli* A. Gray incl var. *danaus* (A. Gray) S. Watson] ❀ TRY. Jun–Aug

var. *ramosus* Jepson Pl 13–30 cm, shaggy. **ST** decumbent to erect. **LVS** basal to cauline; lflets gen 5–15 mm. **INFL** 2–10 cm, open; whorls 3–7, ± well spaced; bracts 4–9 mm. **FL** 7–9(12) mm, fragrant; banner patch white to yellow. 2*n*=48. Subalpine; 3000–4000 m. c&s SNH, **SNE**. [*L. hypoplasius* E. Greene] Jul–Aug

var. *sellulus* (Kellogg) Barneby (p. 319) Pl 12–35 cm. **ST** short, prostrate to ± erect. **LVS** subbasal; lflets 10–30 mm. **INFL** 4.5–11 cm, > lvs; bracts 4–8 mm. **FL** 8–9 mm; banner patch yellow to white turning red. 2*n*=48. Dry rocks, open woodlands; 500–3500 m. NW, CaRH, SN, **GB**; to OR, ID, NV. [*L. sellulus* Kellogg incl ssp. *ursinus* (Eastw.) Munz and var. *artulus* (Jepson) Eastw.] ❀ TRY;DFCLT. Jun–Aug

var. *utahensis* (S. Watson) C. Hitchc (p. 319) Pl 10–25 cm, short-lived, matted, densely hairy. **ST** very short. **LVS** appearing basal. **INFL** 3–6 cm, < lvs; whorls 2–3; bracts 8–15 mm. **FL** 7–10 mm; banner patch white, upper keel margins ciliate near tip. 2*n*= 48. Sand or rocks, with sagebrush, lodgepole pine; 1500–3500 m. **n W&I (White Mtns)**; to OR, MT, Colorado. [*L. caespitosus* Nutt.] ❀ TRY. Jun–Jul

L. magnificus M.E. Jones (p. 319) PANAMINT MOUNTAINS LUPINE Per 6–12 dm, white-woolly. **ST** erect, branched from base; long-sharp-stiff-hairy. **LVS** gen basal; stipules 10–24 mm; petiole 6–30 cm; lflets 5–9, 20–55 mm, 6–15 mm wide. **INFL** 20–50 cm; peduncle 10–50 cm; pedicel 2–8 mm; fls whorled or not; bracts 4–5 mm, deciduous. **FL** 10–18 mm, fragrant; calyx upper lip 5–9 mm, 2-toothed, lower lip 5–11 mm, entire; petals lavender to rose, banner back glabrous, patch yellow turning purple,

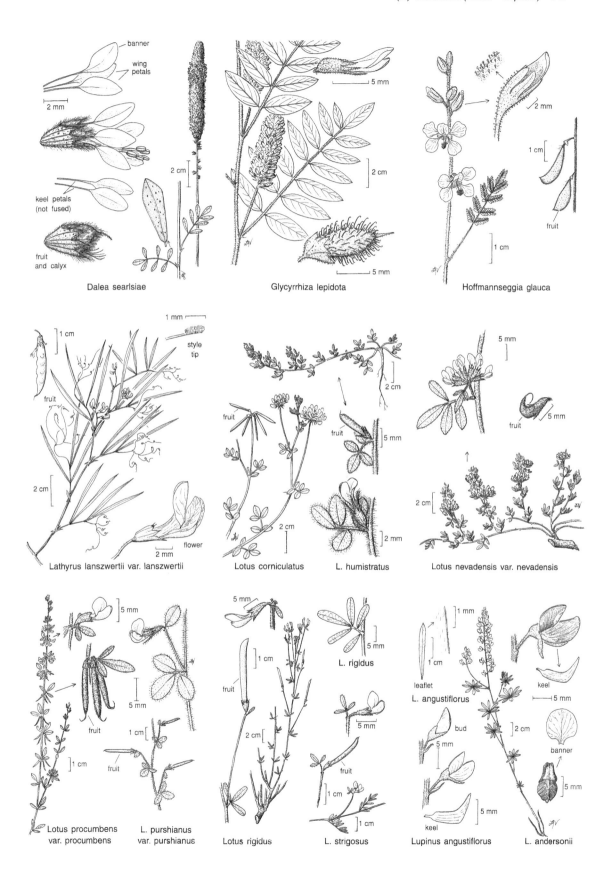

banner

wing petals

2 mm

keel petals (not fused)

fruit and calyx

Dalea searlsiae

5 mm

2 cm

2 cm

5 mm

Glycyrrhiza lepidota

2 mm

1 cm

fruit

1 cm

Hoffmannseggia glauca

1 cm

1 mm

style tip

fruit

2 cm

2 mm

flower

Lathyrus lanszwertii var. lanszwertii

2 cm

fruit

fruit

5 mm

2 cm

2 mm

Lotus corniculatus   L. humistratus

5 mm

fruit

5 mm

2 cm

Lotus nevadensis var. nevadensis

5 mm

fruit

5 mm

fruit

1 cm

fruit

Lotus procumbens var. procumbens

L. purshianus var. purshianus

5 mm

fruit

1 cm

2 cm

L. rigidus

5 mm

5 mm

fruit

1 cm

1 cm

Lotus rigidus   L. strigosus

1 mm

1 cm

leaflet

L. angustiflorus

bud

5 mm

keel

5 mm

keel

5 mm

2 cm

banner

5 mm

Lupinus angustiflorus   L. andersonii

upper keel margins ciliate from middle to tip, lower keel margins glabrous. **FR** 3–7 cm, densely hairy. **SEEDS** 5–8, 3–4 mm, tan. RARE. Desert slopes, washes; 1500–2500 m. **SNE, n DMoj**. Small-fld pls from Coso Mtns, Inyo Co., have been called var. *glarecola* M.E. Jones, Coso Mtn lupine; straight-keeled pls from SNE have been called var. *hesperius* (A.A. Heller) C.P. Smith, McGee Meadows lupine. May–Jun

L. microcarpus Sims CHICK LUPINE Ann 1–8 dm, sparsely to densely hairy; cotyledons disk-like, persistent, or leaving a circular scar. **LF**: petiole 3–15 cm; lflets 5–11, gen 9, 10–50 mm, 2–12 mm wide, sometimes linear, upper surface glabrous. **INFL** 2–30 cm; peduncle 2–30 cm; bracts 3.5–12 mm, reflexed, persistent; pedicels 0.5–5 mm. **FL** 8–18 mm; calyx upper lip 2–6 mm, lower lip 5–10 mm, appendages gen 0; petals white to dark yellow, pink to dark rose, or lavender to purple, wings gen ciliate on upper (less often lower) margins near claw, upper keel margins ciliate, lower keel margins less so or glabrous near claw. **FR** 1–1.5 cm, ± 10 mm wide, ovoid, hairy, erect to spreading, often on 1 side of infl axis or not. **SEEDS** 2, tan to brown, gen mottled, wrinkled or smooth. 2*n*=48. Abundant. Open or disturbed areas, sometimes seeded on roadbanks; < 1600 m. CA-FP, MP, **w DMoj**; to B.C., Baja CA, S.Am. Highly variable; vars. intergrade. 3 vars. in CA.

var. *horizontalis* (A.A. Heller) Jepson (p. 323) **INFL**: bracts short- to long-spreading-hairy. **FL**: calyx appressed- to spreading-hairy, appendages 1–2 mm; petals lavender to purple, becoming translucent, wings widely elliptic, persistent, ciliate on upper (and gen lower) margins near claw, lower keel margins ciliate near claw. **FR** gen spreading, gen not on 1 side of infl axis. Washes, sand or gravel; < 1500 m. s SnJV, **e DMoj**. [*L. h.* A.A. Heller incl var. *platypetalus* C.P. Smith; *L. arenicola* A.A. Heller; *L. densiflorus* Benth. var. *glareosus* (Elmer) C.P. Smith] ❀ TRY. Apr–May

var. *microcarpus* (p. 323) **INFL**: bracts (and calyx) long-shaggy-hairy. **FL**: calyx appendages gen 0; petals gen pink to purple (yellowish or white), wings linear to lanceolate, withering, upper (and rarely lower) margins gen ciliate near claw, lower keel margins gen glabrous near claw. **FR** ± erect, gen not on 1 side of infl axis. Habitats and range of sp. [*L. subvexus* C.P. Smith (incl vars.); *L. ruber* A.A. Heller; *L. densiflorus* Benth. vars. *austrocollium* C.P. Smith, *palustris* (Kellogg) C.P. Smith (as used by Munz), and *persecundus* C.P. Smith] ❀ SUN:3–6,**7–9,14–17**, 18,**19–24**. Apr–Jun

L. nevadensis A.A. Heller (p. 319) Per 1–4 dm, long-hairy. **ST** erect. **LVS** basal and cauline; stipules 8–10 mm; basal petioles < 14 cm, cauline < 4 cm; lflets 6–10, 20–50 mm. **INFL** 5–17 cm; peduncle 3–6 cm; pedicel 4–8 mm; fls spiraled; bracts 4–5 mm, deciduous. **FL** 10–12 mm; calyx upper lip 3–4 mm, 2-toothed, lower lip 4–5 mm, 3-toothed; petals blue, banner back glabrous, patch white to yellowish, keel strongly upcurved, upper keel margins ciliate, lower keel margins glabrous. **FR** 2.5–4 cm, densely hairy. **SEEDS** 3–4, light. Hillsides, valleys, with sagebrush; 1000–3000 m. **GB, DMtns**; OR, NV. ❀ TRY. Apr–Jun

L. odoratus A.A. Heller (p. 323) MOJAVE LUPINE Ann 1–3 dm, sparsely short-hairy when young, becoming glabrous; cotyledons disk-like, persistent. **LVS** basal; petioles 2–12 cm; lflets 7–9, 10–20 mm, 3–8 mm wide, sometimes obovate, bright green. **INFL** 4–13 cm; fls spiraled; peduncle 6–15 cm; bracts 2–4 mm, straight, persistent, tips sparsely ciliate; pedicels 3–5 mm. **FL** with violet odor, 7–10 mm; calyx tips gen glabrous, rarely ciliate, upper lip 3–3.5 mm, rounded or 2-toothed, lower lip 4–5 mm; petals deep blue-purple, banner spot white or yellow becoming magenta, keel glabrous. **FR** 1.5–2.5 cm, ± 8 mm wide, oblong, upper suture wavy, densely long-ciliate, sides glabrous or with a few short hairs that become scale-like when dry. **SEEDS** 2–6, wrinkled. Sandy flats, open areas; < 1600 m. **GB, DMoj**; NV, AZ. Pls having sparsely hairy peduncles, petioles, and lower lf surfaces have been called var. *pilosellus* C.P. Smith, may be confused with *L. flavoculatus*. ❀ TRY. Apr–May

L. padre-crowleyi C.P. Smith (p. 323) FATHER CROWLEY'S LUPINE Per 5–7.5 dm, silver- to white-woolly. **ST** erect, from a woolly mat. **LVS** basal and cauline; stipules 5–11 mm; petiole 2–3 cm; lflets 6–9, 25–75 mm. **INFL** 7–21 cm; peduncle 2–5.5 cm; pedicel 2–3.5 mm; fls ± whorled; bracts 4–9 mm, deciduous or not. **FL** 10–14 mm; calyx upper lip 5–7 mm, 2-toothed, lower lip 5.5–8 mm, 3-toothed; petals cream to pale yellow, banner back gen hairy, keel glabrous. **FR** 2–3 cm, silky. **SEEDS** 2–3, 4–5 mm, white, mottled black. RARE CA. Decomposed granite; 2500–4000 m. s SNH, **SNE (n Inyo, Tulare cos.)**. [*L. dedeckerae* Munz & Dunn] May–Jul

L. polyphyllus Lindley Per 2–15 dm, green, glabrous or sparsely hairy. **ST** erect, stout. **LVS** basal and cauline; stipules 5–30 mm; petioles 3–45 cm, upper shorter; lflets 5–17, 40–150 mm. **INFL** 6–40 cm, open; peduncle 3–13 cm; pedicel 3–15 mm; fls ± whorled; bracts 7–11 mm, deciduous. **FL** 9–15 mm; calyx lips entire, upper lip 4–7 mm, lower lip 4–7 mm; petals violet to lavender to pink to white, banner back glabrous, patch yellow to white sometimes turning red-purple, keel upcurved, gen glabrous. **FR** 2.5–4 cm, hairy. **SEEDS** 3–9. 2*n*=48. Moist areas to bogs; < 3000 m. CA-FP, **GB**; to B.C. 3 vars. in CA.

var. *burkei* (S. Watson) C. Hitchc.(p. 323) **LF**: lflets 5–11, glabrous above, sometimes sparsely hairy below. Wet places; 1500–3000 m. SNH, SnGb, SnBr, SnJt, **GB**; OR, ID, NV. [sspp. *bernardinus* (Abrams) Munz and *superbus* (A.A. Heller) Munz] ❀ SUN,IRR, DRN:1–3,6,**7**,14,**15–16,18**. May–Aug

L. pratensis A.A. Heller Per 3–7 dm, green, hairy. **ST** erect. **LVS** basal and cauline; stipules 5–20 mm; petiole 3–21(25) cm; lflets 5–10, 30–80(130) mm. **INFL** 5–28 cm, dense; peduncle 4–17 cm; pedicel 1–3 mm; bracts 5–10 mm, persistent. **FL** 10–12 mm; calyx upper lip 4–7 mm, 2-toothed, lower lip 5–6 mm, entire; petals violet to dark blue, banner back glabrous or hairy, patch orange to red, upper keel margins densely ciliate, lower keel margins glabrous. **FR** 1.5–2 cm, hairy to woolly. **SEEDS** 4–6, 3–4 mm, mottled tan, brown. 2*n*=48. Meadows, streambanks; 1000–3500 m. c&s SNH, **SNE**.

var. *eriostachyus* C.P. Smith **FL**: banner back hairy. Moist places; 1500–3500 m. **SNE (Big Pine Creek, Inyo Co.)** ❀ TRY

var. *pratensis* **FL**: banner back glabrous. Habitats and range of sp. ❀ TRY. May–Aug

L. pusillus Pursh var. *intermontanus* (A.A. Heller) C.P. Smith (p. 323) Ann < 1 dm, hairy; cotyledons disk-like, persistent. **LVS** cauline, crowded near base; petioles 3–6 cm; lflets 5–6, 10–20 mm, ± 5 mm wide, upper surface glabrous. **INFL** ± 3 cm, < lvs, dense; fls spiraled; peduncle 0–1 cm; bracts ± 3 mm, straight, persistent; pedicels < 1 mm. **FL** 6 mm; calyx upper lip ± 2.5 mm, lower lip ± 5 mm; petals pale blue, fading pinkish or whitish, banner spot white or yellow, keel glabrous. **FR** 1.5 cm, 6 mm wide, oblong, narrowed between seeds, hairy. **SEEDS** 2; surface wrinkled; margin ridged. 2*n*=48. Uncommon. Open, sandy areas; < 1600 m. **GB (Modoc, Inyo cos.)**; to WA, c US, AZ. May–Jun

L. shockleyi S. Watson (p. 323) DESERT LUPINE Ann 0.5–3 dm, canescent; cotyledons disk-like, persistent. **LVS** cauline, crowded near base; petioles 4–12 cm; lflets 8–10, 10–30 mm, 4–10 mm wide, upper surface glabrous. **INFL** 2–6 cm; fls spiraled; peduncle 1–10 cm; bracts 2–4 mm, straight, persistent; pedicels 1–3 mm. **FL** 4.5–6 mm; calyx 3–6 mm, lips ± equal; petals dark blue-purple, banner spot yellow, keel blunt, glabrous. **FR** 1.5–2 cm, 8–12 mm wide, ovoid, upper suture wavy, densely and stiffly long-ciliate, sides with short, inflated hairs that become scale-like when dry. **SEEDS** 2, wrinkled. Open, sandy areas; < 1200 m. **D**; NV, AZ. ❀ TRY. Apr–May

L. sparsiflorus Benth. (p. 323) COULTER'S LUPINE Ann 2–4 dm, short-appressed- and long-spreading-hairy. **LF**: petiole 3–4

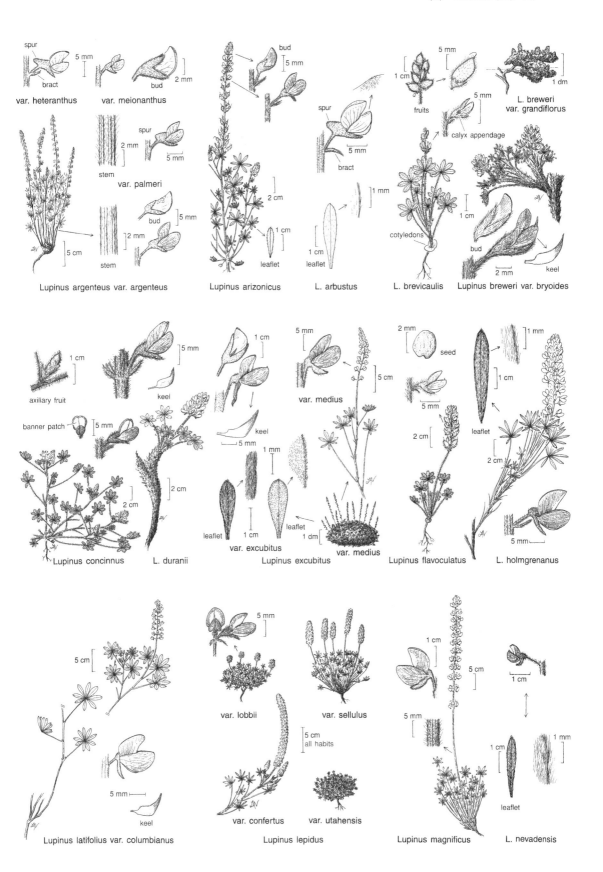

spur
bract
var. heteranthus
5 mm
2 mm
bud
var. meionanthus

stem
2 mm
spur
5 mm
var. palmeri

5 cm
stem
2 mm
bud
5 mm

Lupinus argenteus var. argenteus

bud
5 mm
spur
bract
5 mm

Lupinus arizonicus

2 cm
1 cm
leaflet

1 mm
1 cm
leaflet

L. arbustus

fruits
1 cm
5 mm
calyx appendage
5 mm

cotyledons
1 cm
bud
keel
2 mm

L. brevicaulis

5 mm
1 dm
L. breweri var. grandiflorus

Lupinus breweri var. bryoides

axiliary fruit
1 cm
keel
5 mm

banner patch
5 mm

2 cm
2 cm
Lupinus concinnus   **L. duranii**

1 cm
keel
5 mm
var. medius
5 mm

1 mm
leaflet
1 cm
leaflet
var. excubitus

5 cm
1 dm
var. medius

Lupinus excubitus

2 mm
seed
5 mm

2 cm

Lupinus flavoculatus

1 mm
1 cm
leaflet
2 cm

5 mm
L. holmgrenanus

5 cm

Lupinus latifolius var. columbianus
5 mm
keel

var. lobbii
5 mm

var. sellulus

5 cm
all habits

var. confertus   var. utahensis

Lupinus lepidus

1 cm
5 cm
5 mm
1 cm

1 cm
leaflet
1 mm

Lupinus magnificus   **L. nevadensis**

cm; lflets 7–11, 15–30 mm, 2–4 mm wide, linear to oblanceolate, upper surface hairy at least near margins. **INFL** 15–20 cm; fls spiraled, sometimes appearing ± whorled; peduncle 2–4 cm; bracts 3–5 mm, < buds, gen deciduous; pedicels 2–4 mm. **FL** gen 10–12 mm; calyx 3–6 mm, lips ± equal, upper lip deeply lobed; petals gen blue (pinkish), drying darker, banner spot whit-ish becoming magenta, lower (and often upper) margins of keel ciliate near claw. **FR** 1–2 cm, ± 5 mm wide, coarsely hairy. **SEEDS** 4–5. Washes, sandy areas; < 1300 m. s SCoR, SW, **DMoj**; to UT, AZ, n Mex. Locally common. Pls in DMoj (often with smaller fls) have been called ssp. *mohavensis* Dziekanowski & D. Dunn ❀ STBL.Mar–May

MARINA

Duane Isely

Per in CA, unarmed, hairy, gen gland-dotted. **LVS** odd-1-pinnate; stipules obscure; lflets opposite. **INFL**: raceme, terminal. **FL**: calyx lobes < tube; corolla gen 2-colored (blue-violet and white), petals (exc banner) arising from side of filament column; stamens gen 10, filaments fused; ovule 1. **FR** indehiscent, incl or slightly exserted. **SEED** 1. 38 spp.: esp Mex. (Marina, interpreter for Mexican conqueror Cortez, 16th century) [Barneby 1977 Mem New York Bot Gard 27:55–133; 608–649]

M. parryi (Torrey & A. Gray) Barneby (p. 323) Per 2–8 dm, strigose, gen gland-dotted. **LVS** well spaced; lflets 11–23, 0.5–6 mm, oblong obovate to orbicular to ± round, not folded, ± gland-dotted. **INFL** 2–10 cm. **FL**: corolla 5–7 mm. $2n$=20. Open desert washes, stony slopes, roadsides; < 800 m. **D**; AZ, Mex. [*Dalea p.* Torrey & A. Gray] Feb–Jun

MEDICAGO

Duane Isely

Ann or per, unarmed. **ST** prostrate to erect. **LVS** gen odd-1-pinnate, sometimes subpalmately compound; stipules somewhat fused with petiole, entire or deeply cut; lflets 3, gen toothed near tip. **INFL**: raceme, axillary or terminal, few–many-fld. **FL**: calyx lobes ± equal or not; corolla yellow or purple; 9 filaments fused, 1 free. **FR** indehiscent, gen coiled 1.5–5 turns, gen prickly. **SEEDS** 1–several. ± 55 spp.: Eurasia (esp Medit); several cult and naturalized world-wide. (Medea, source of alfalfa, which then bore Greek name Medice) [Small & Jomphe 1989 Can J Bot 67:3260–3294] Key to species adapted by Margriet Wetherwax.

1. Ann; fr 1-seeded, reniform; fl 2–3 mm, yellow . . . . *M. lupulina*
1' Per; fr several-seeded, spirally coiled, rarely sickle-shaped; fl 3–11 mm, purple or multicolored,
   rarely yellow . . . . *M. sativa*

M. lupulina L. BLACK MEDICK, YELLOW TREFOIL Ann, hairy, sometimes finely strigose. **ST** decumbent to ascending, gen 1–4 dm. **LF**: stipules entire or toothed; lflets gen 1–1.5 cm, obovate. **INFL** ovoid, 10–20-fld, dense. **FL**: calyx 1–1.5 mm; corolla 2–3 mm, yellow. **FR** 2–2.5 mm, reniform, black; veins strong, concentric; prickles 0. **SEED** 1. $2n$=16,32. Disturbed areas, sometimes in forests, mtns; < 2500 m. CA-FP, **GB**; most of US, widespread elsewhere; native to Eur. [var. *cupaniana* (Guss.) Boiss.] Apr–Jul

M. sativa L. (p. 323) ALFALFA, LUCERNE Per, ± glabrous or puberulent. **ST** decumbent to gen erect, 2–8 dm. **LF**: stipules entire to sharply toothed; lflets 1–1.5 cm, narrowly lanceolate to obovate. **INFL** spike-like, 8–25-fld, longer in fr. **FL**: calyx 4–4.5 mm; corolla 8–10 mm, purple or multicolored (i.e., violet, violet-green, greenish yellow, rarely yellow). **FR** gen coiled 2–3 turns (rarely only sickle-shaped), leathery; prickles 0. $2n$=16,32. Disturbed and agricultural areas; < 1500 m. CA-FP, **GB**; most of US exc se; native to Eurasia. [Isely 1983 Iowa State J Res 57:207–220] Cult; variable polyploid complex in US, incl genetic components from several spp.; often divided into several spp. or sspp. Apr–Oct

MELILOTUS SWEETCLOVER

Duane Isely

Ann or bien, unarmed. **ST** gen erect. **LVS** odd-1-pinnate; stipules gen narrow or bristle-like, bases fused to petiole; lflets 3, margin toothed or wavy. **INFL**: raceme, axillary or terminal, slender or short-cylindric, many-fld. **FL**: calyx lobes ± equal; corolla yellow or white; 9 filaments fused, 1 free. **FR** indehiscent, 2–4 mm, ovoid, compressed but thick, leathery, ridged or bumpy. **SEEDS** 1–2. 20 spp.: Eurasia, esp Medit; several spp. widely cult for soil improvement and naturalized. (Greek: honey-Lotus) [Isely 1954 Proc Iowa Acad Sci 61:119–131] TOXIC: inclusion in hay enhances production of mold toxins that may cause cattle death from hemorrhaging. Key to species by Dave Keil.

1. Corolla white; fr with network of lines . . . . *M. alba*
1' Corolla yellow; fr bumpy or with faint lines or irregularly cross-ridged
  2. Corolla 2.5–3 mm; ann, 1–6 dm; fr bumpy or with faint lines . . . . *M. indica*
  2' Corolla 4.5–7 mm; bien, 5–20 dm; fr irregularly cross-ridged . . . . *M. officinalis*

M. alba Medikus (p. 323) WHITE SWEETCLOVER Ann or bien, ± glabrous or strigose. **ST** erect, 0.5–2 m. **LF**: lflets 1–2.5 cm, elliptic-oblong to obovate, ± toothed. **INFL** slender; axis gen 3–8 cm when fls open. **FL**: calyx ± 2 mm; corolla 4–5 mm, white. **FR** 3–5 mm, ovoid, with network of lines. **SEED** 1. $2n$=16,24. Locally abundant. Open, disturbed sites; < 1500 m. **CA**; most of n US, adjacent Can; native to Eurasia. Indistinguishable from *M. officinalis* prior to fl. May–Sep

M. indica (L.) All. (p. 323) SOURCLOVER Ann, ± glabrous. **ST** spreading or erect, 1–6 dm. **LF**: lflets 1–2.5 cm, oblanceolate to wedge-shaped-obovate, gen sharply dentate. **INFL** slender, compact; axis gen 1–2 cm when fls open. **FL**: calyx 1–1.5 mm; corolla 2.5–3 mm, yellow. **FR** 2–3 mm, ovoid, bumpy or with obscure lines. **SEED** 1. 2*n*=16. Open, disturbed areas; < 1500 m. **CA**, most common s; to se US; native to Medit. Apr–Oct

M. officinalis (L.) Pall. (p. 323) YELLOW SWEETCLOVER Bien, ± glabrous to strigose. **ST** erect, 0.5–2 m. **LF**: lflets 1–2.5 cm, elliptic-oblong to obovate, ± toothed. **INFL** slender; axis gen 3–8 cm when fls open. **FL**: calyx 2–2.5 mm; corolla 4.5–7 mm, yellow. **FR** 3–5 mm, ovoid, irregularly cross-ridged. **SEED** 1. 2*n*=16. Open, disturbed sites; < 1500 m. **CA**; most of n US, adjacent Can; native to Eurasia. May–Aug

OLNEYA

Duane Isely

1 sp. (S.T. Olney, Am botanist, 1812–1878)

O. tesota A. Gray (p. 323) IRONWOOD Shrub or tree with stipular spines, canescent. **LVS** even-1-pinnate, alternate or clustered; stipular spines sometimes 0 or breaking off, leaving scar, sometimes persisting above lvs and appearing internodal; lflets 8–19, opposite or subopposite, obovate or elliptic, thick; axis extending beyond lflets, pointed. **INFL**: raceme, axillary, 2–4-fld. **FL**: corolla 10–12 mm, wings purple, other petals yellow-white to pink; 9 filaments fused, 1 free. **FR** slowly dehiscent, oblong or elliptic, plump, irregularly narrowed between seeds, persistent. **SEEDS** 1–3. 2*n*=18. Often abundant in washes; < 600 m. PR, **DSon**; AZ, Mex. Fls erratically. ❀ DRN,DRY:8,9,14,19,21,23, IRR:10,11,**12,13**;DFCLT. Apr–May

OXYTROPIS

Duane Isely

Per, unarmed, hairy. **LVS** odd-1-pinnate, basal, sometimes also cauline; stipules gen partly fused to petiole, initially forming a sheath, or free. **INFL**: raceme, gen scapose, spike- or head-like, or 1–2-fld; bracts gen persistent. **FL**: calyx lobes < tube; corolla pink-purple, white, or yellowish, keel tip beaked; 9 filaments fused, 1 free; style glabrous. **FR** ascending or reflexed, gen persistent, lanceolate or inflated, ± 2-chambered, septum arising from upper suture, ± incomplete. ± 300 spp.: Eurasia, N.Am. (Greek: sharp keel) [Barneby 1952 Proc Calif Acad Sci Series IV 27:177–309; Welsh 1991 Great Basin Nat 51:377-396] Seriously TOXIC: causes "staggers" in livestock, mostly outside CA.

1. Fl quickly reflexed, fr reflexed; lvs basal and gen 1–3 cauline; infl a loose raceme . . . . . . *O. deflexa* var. *sericea*
1' Fl and fr ascending or erect; lvs all basal; infl densely spike- or head-like
  2. Lflets 23–39; pl green, puberulent, sticky-glandular; corolla 12–18 mm . . . . *O. borealis* var. *australis*
  2' Lflets 7–17; pl silvery or gray, silky or tomentose, not glandular; corolla 7–10 mm
    3. Fls 2–12 per infl; fr ovoid-inflated, ± 7–9 mm wide, papery . . . . *O. oreophila* var. *oreophila*
    3' Fls 1–3 per infl; fr lanceolate, ± 5–6 mm wide, leathery . . . . *O. parryi*

O. borealis DC. var. *australis* Welsh (p. 323) Pl green, puberulent, glandular-sticky, sometimes tiny and cespitose. **LVS** basal; lflets 23–39, 3–10 mm, oblong-lanceolate to ovate, flat or folded. **INFL** spike-like, exserted, sometimes longer in fr; fls 4–many, ascending or erect. **FL**: corolla 12–18 mm, white to yellowish. **FR** ascending or erect, 8–14 mm, lanceolate to ovate in outline, somewhat inflated, papery, slightly 2-chambered; stalk-like base 0. Aspen meadows to alpine; 1200–3900 m. s SNH, **W&I**; circumboreal. Pls with fls purple are var. *viscida* (Nutt.) Welsh ❀ TRY Jul–Aug

O. deflexa (Pall.) DC. var. *sericea* Torrey & A. Gray (p. 323) BLUE PENDENT-POD OXYTROPE Pl green or gray. **LVS** basal, often also 1–3 cauline; stipules ± free; lflets 15–31, 3–20 mm, lanceolate to ovate, flat or folded. **INFL** spike-like; fls few–many, quickly reflexed; peduncle gen > 15 cm, often curved. **FL**: corolla 5–10 mm, dull white to pale lilac or blue, ± = or > calyx. **FR** reflexed, 3–4.5 cm, elliptic or oblong, membranous or papery, slightly 2-chambered; stalk-like base gen short. 2*n*=16. RARE in CA. Moist meadows, forest openings; ± 2800 m. **W&I (White Mtns, Mono Co.)**; to AK, NV, NM. Jun–Aug

O. oreophila A. Gray var. *oreophilla* (p. 323) Pl silvery or gray, silky, cespitose. **LVS** basal, congested; lflets 7–17, 2–10 mm, elliptic to oblong, folded. **INFL** head-like, incl or exserted; fls 2–12, ascending or erect. **FL**: corolla 7–10 mm, pink-purple, sometimes white. **FR** ascending or erect, 9–15 mm, 7–9 mm wide, ovoid-inflated, thinly papery, slightly 2-chambered; stalk-like base 0. 2*n*=16. Uncommon. Alpine; 3400–3800 m. SnBr, **W&I**; to UT, AZ. Variable; upper alpine pls only a few cm tall. Other vars not in CA. Jul

O. parryi A. Gray (p. 323) Pl gray, silky or tomentose, cespitose. **LVS** basal, clustered; lflets 11–15, 2–12 mm, ovate to oblong, folded. **INFL** head-like, gen exserted; fls 1–3, ascending or erect. **FL**: corolla 7–10 mm, purple. **FR** ascending or erect, 15–20 mm, 5–6 mm wide, lanceolate, leathery, nearly 2-chambered; stalk-like base 0. Near and above timberline; 3400–3800 m. SNE, **W&I**; to ID, Colorado, NM. ❀ TRY. Jun–Jul

PEDIOMELUM   BREADROOT

James W. Grimes

Per, unarmed, gland-dotted; hairs, stalked glands, or both; roots deep, woody, enlarged near ground surface. **ST**: main axis erect, nearly 0 to short; branches short, decumbent to ascending, sometimes underground. **LVS** ± palmately compound, ± basal (or cauline at branch tips); stipules at base of pl fused, those above free; lflets 5–7. **INFL**: basal, axillary, or terminal on branches, raceme with 1 sometimes tardily deciduous bract and 2–3 fls per node; pedicel sometimes very short. **FL**: calyx base swollen on top, tube enlarging in fr; corolla at least partly blue to purple; 9 filaments fused, 1 less so or free; ovary ± hairy, ovule 1, style tip curved to bent, stigma head-like. **FR** transversely dehiscent, beaked, hairy, rarely glandular. **SEED** elliptic, smooth or ridged. 22 spp.: N.Am. (Greek: plain apple) [Grimes 1990 Mem New York Bot Gard 61:1–114] *Bituminaria bituminosa* (L.) Stirton (*Psoralea b.* L.) possibly naturalized in SnBr; *Pediomelum mephiticum* (S. Watson) Rydb. incorrectly reported for CA.

P. castoreum (S. Watson) Rydb. (p. 323) **LF**: stipule 5–13.5 mm; petiole 6.8–15 cm; lflets 5–6, 25–42 mm, elliptic to oblanceolate. **INFL**: bract 3.5–8 mm. **FL** 9–13 mm; calyx 10–12 mm; banner 9–13 mm. **FR** ovate to elliptic in outline; body 6–8 mm; beak 8–11 mm, straight to curved, triangular. **SEED** 6 mm, reniform, ridged, gray. Open areas, roadcuts; < 1750 m. **DMoj (San Bernardino Co.)**; NV, AZ. [*Psoralea c.* S. Watson] ❀ TRY. Apr–May

PETERIA

Duane Isely

Per; spines stipular. **ST** decumbent to erect. **LVS** odd-1-pinnate; stipules spiny or 0; lflets 9–many. **INFL**: raceme, spike-like, terminal. **FL**: calyx tube sometimes bulged on 1 side near base, lobes < or > tube, upper pair fused 1/2 or more; corolla pink or white; 9 filaments fused, 1 free; style tip finely hairy around stigma. **FR** dehiscent, oblong, ± flat but plump, leathery; base stalk-like. 4 spp.: s US, Mex. (R. Peter, 19th century Kentucky botanist) [Porter 1956 Rhodora 58:344–354]

P. thompsoniae S. Watson (p. 323)   SPINE-NODED MILKVETCH Pl rhizomed; taproot often swollen. **ST** 2–6 dm. **LF**: lflets 13–21, elliptic or obovate. **INFL** often glandular-hairy. **FL**: calyx 11–15 mm, tube cylindric, gen darkly glandular-puberulent; corolla gen 15–20(25) mm; style hairs often hidden by pollen. **FR** 4–6 cm, glabrous. RARE in CA. Sandy alluvial fans; 800 m. **DMtns (California Valley, se Inyo Co.)**; to ID, UT, sw Colorado, AZ.

PHASEOLUS WILD BEAN

Duane Isely

Ann, per, vine, unarmed; hairs gen incl minute, hooked ones. **LVS** odd-1-pinnate; main axis extended beyond basal lflets, with stipule-like appendages, at least at tip; stipules persistent; lflets 3, entire or lobed. **INFL**: raceme; bracts persistent. **FL**: calyx lobes << tube; corolla incurved, sickle-shaped in bud, keel coiled 2–3 turns; 9 filaments fused, 1 free. **FR** dehiscent, linear to oblong, flat or round in ×-section. **SEEDS** few–several. ± 50 spp.: neotrop, warm regions; those of economic value on all continents. (Classical name, presumably for a bean) *P. vulgaris* L., *P. lunatus* L., and others may be found as waifs from cult.

P. filiformis Benth. (p. 333)   WRIGHT'S PHASEOLUS, SLENDER-STEMMED BEAN **LF**: lflets ovate-triangular in outline, gen lobed. **INFL**: peduncle 2–10 cm. **FL**: corolla pink-purple. **FR** 2.5–3.5 cm, 5–7 mm wide, oblong, gen curved, ± glabrous. UNCOMMON. Washes; ± 125 m. **DSon (Coachella Valley, Riverside Co.)**; to TX, n Mex. [*P. wrightii* A. Gray] ❀ TRY.

PROSOPIS MESQUITE

Elizabeth McClintock

Shrub, tree; stipular spines gen 2 per node; roots long, spreading. **LVS** even-2-pinnate, alternate, deciduous; 1° lflets gen 1–2 pairs, opposite; 2° lflets gen many, opposite. **INFL**: raceme, axillary, spike-like or spheric head, many-fld. **FL** radial, small, greenish white or yellow; calyx shallowly bell-shaped, lobes very short; petals gen inconspicuous; stamens 10, exserted, free; style exserted, gen appearing before stamens. **FR** indehiscent, ± flat, ± narrowed between seeds or tightly coiled, pulpy when young, then woody. **SEEDS** several. ± 44 spp.: esp Am (also sw Asia, Afr). (Greek: burdock, for obscure reasons) [Burkhart 1976 J Arnold Arbor 57:220–524; Holland 1987 Madroño 34:324–333] Used for timber, firewood, shade, orn, bee, human, and livestock food.

1. Fr not coiled; petals free . . . . *P. glandulosa* var. *torreyana*
1' Fr tightly coiled; petals fused
  2. Infl a spike-like raceme; lf hairy . . . . *P. pubescens*
  2' Infl a head; lf glabrous or short-hairy . . . . *P. strombulifera*

P. glandulosa Torrey var. **torreyana** (L. Benson) M. Johnston (p. 333)   MESQUITE, HONEY MESQUITE Shrub, tree < 7 m, ± glabrous; crown often wider than tall. **ST**: branches arched, crooked; spines 0.5–4 cm. **LF** glabrous; 1° lflets gen 1 pair, 6–17 cm; 2° lflets 7–17 pairs, 1–2.5 cm, oblong, length 7–9 × width. **INFL**: raceme, 6–10 cm, spike-like. **FL**: petals free, 2.5–3.5 mm. **FR** 5–20 cm, slightly narrowed between seeds, glabrous. **SEEDS** gen 5–18, 6–7 mm, oblong. Common. Grasslands, alkali flats, washes, bottomlands, sandy alluvial flats, mesas; 0–1700 m. SnJV, PR, **D**. ❀ SUN,DRN:7,9,10,**12,14–16**,17,18,**19–24**& IRR:**8,11,13**;also STBL. Apr–Aug

P. pubescens Benth.   SCREW BEAN, TORNILLO Shrub, tree < 10 m; crown gen ± narrow. **ST**: branches ascending; spines 4–12 mm. **LF** hairy; 1° lflets 1 or 2 pairs, 3–5 cm; 2° lflets 5–8 pairs, 2–10 mm, oblong, length 2–3 × width. **INFL**: raceme, 4–8 cm, spike-like. **FL**: petals fused, 2–3 mm. **FR** 3–5 cm, tightly coiled. **SEEDS** gen 3 mm, ovoid. 2*n*=28. Uncommon. Creek, river bottoms, sandy or gravelly washes or ravines; 100–1300 m. SnBr, **D**; sw US, n Mex. Fr used for food and as a coffee substitute.

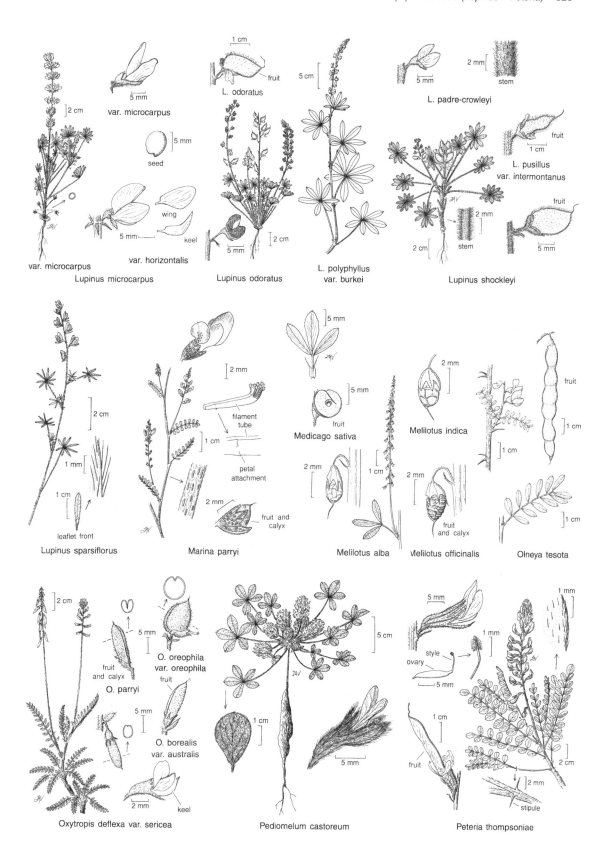

var. microcarpus

var. microcarpus

seed

wing

keel

var. horizontalis

Lupinus microcarpus

L. odoratus

fruit

1 cm

5 mm

Lupinus odoratus

2 cm

5 mm

L. polyphyllus
var. burkei

L. padre-crowleyi

2 mm

5 mm

stem

L. pusillus
var. intermontanus

fruit

1 cm

fruit

2 mm

stem

5 mm

2 cm

Lupinus shockleyi

Lupinus sparsiflorus

2 cm

1 mm

1 cm

leaflet front

Marina parryi

filament
tube

petal
attachment

2 mm

fruit and
calyx

2 mm

1 cm

Medicago sativa

5 mm

5 mm

fruit

2 mm

1 cm

Melilotus alba

Melilotus indica

2 mm

2 mm

fruit
and calyx

Melilotus officinalis

fruit

1 cm

1 cm

1 cm

Olneya tesota

2 cm

fruit
and calyx

O. parryi

5 mm

O. oreophila
var. oreophila

fruit

5 mm

O. borealis
var. australis

2 mm

keel

Oxytropis deflexa var. sericea

5 cm

1 cm

5 mm

Pediomelum castoreum

5 mm

style

ovary

5 mm

1 mm

1 mm

fruit

1 cm

2 cm

2 mm

stipule

Peteria thompsoniae

❀ SUN:7,**14**,15, 16,18,**19–21**,22,23&IRR:**8,9**,10,11,**12,13**;also STBL. Apr–Sep

P. strombulifera (Lam.) Benth.   Shrub < 3 m; crown not seen; roots long, spreading. **ST** ± zig-zag; spines 1–2 cm. **LF** glau-cous, glabrous or short-hairy; 1° lflets 1 pair, 1–3 cm; 2° lflets 2–8 pairs, 2–10 mm. **INFL**: head, ± 15 mm wide. **FL**: petals fused, 3–4 mm. **FR** 1.5–5 cm, tightly coiled. 2*n*=28. Uncommon. Disturbed places; ± 50 m. **se DSon (Bard, Imperial Co.)**; native to Argentina. Jul

PSORALIDIUM   SCURF-PEA

James W. Grimes

Per, unarmed, gland-dotted, glabrous to ± sparsely hairy; rhizomes or roots (or both) woody. **ST** erect, < 7.5 dm, green or yellow toward base. **LVS** palmately compound, cauline; stipules free; lflets 3–5. **INFL**: raceme, axillary, with 1 deciduous bract and 1–3 fls per node. **FL** pedicelled, 6–7 mm; calyx flaring back, tearing along 1 lateral sinus in fr; corolla white, yellow, or purple; 9 filaments fused, 1 less so or free; ovary glabrous to hairy, ovule 1, style tip bent, stigma head-like. **FR** indehiscent, ± spheric. **SEED** elliptic to round in outline. 4 spp.: N.Am. (Greek: diminutive of *Psoralea*) [Grimes 1990 Mem New York Bot Gard 61:1–114]

P. lanceolatum (Pursh) Rydb. (p. 333)   **LF**: petiole 9–21 mm; lflet 17–33 mm, linear to oblanceolate. **FL** 4–7 mm; calyx 2–2.5 mm; petals white to purple-blue. **FR** 4–6 mm, papillate-glandular to glandular, ± hairy. **SEED** 4–5 mm, smooth, shiny. Alluvial plains, sand; < 2500 m. **GB**; to c Can, c US. [*Psoralea l.* Pursh ssp. *scabra* (Nutt.) Piper] May–Jul

PSOROTHAMNUS

Duane Isely

Per, shrub, small tree, gen with thorns, gland-dotted. **ST** gen intricately branched. **LVS** odd-1-pinnate or simple. **INFL**: raceme, sometimes spike- or head-like, axillary or terminal; pedicels gen with bractlets. **FL**: calyx lobes gen unequal, upper pair often largest; petals all arising from receptacle, violet, blue, or 2-colored (purple and white); stamens 10, filaments partly fused; ovules gen 2. **FR** indehiscent, incl in or exserted from calyx, gen glandular. **SEED** 1. 9 spp.: sw US, Mex. (Greek: scabshrub) [Barneby 1977 Mem New York Bot Garden 27:21–54, 598–607]

1. Lf simple, linear or oblanceolate, rarely with 3 lflets
  2. Lvs persistent; infl axis not a thorn . . . . *P. schottii*
  2' Lvs deciduous by early summer; infl axis extended beyond fls as a thorn . . . . *P. spinosus*
1' Lvs all odd-1-pinnate or some simple
  3. Fr 2–4.5 mm, incl; infl a spike or raceme, dense, spheric or short-cylindric; pedicel lacking bractlets
    4. Terminal lflets of major lvs (or some of them) > lateral lflets; glands of twigs << 0.5 mm wide . . . . *P. emoryi*
    4' Terminal lflets of all lvs ± = lateral lflets; glands of twigs ± 0.5 mm wide . . . . *P. polydenius*
  3' Fr 7–10 mm, exserted; infl a ± open, sometimes spike-like raceme; pedicel with bractlets
    5. Fr glabrous, with many small glands forming longitudinal lines; D (e half of San Bernardino Co.) . *P. fremontii*
      6. Lflet linear, < ± 1 mm wide; DSon (Whipple Mtns, extreme se San Bernardino Co.) . . . . . . var. *attenuatus*
      6' Lflet narrowly elliptic to ovate or obovate, ± 1.5–3 mm wide; DMtns (e-c San Bernardino Co.)
      . . . . var. *fremontii*
    5' Fr glabrous or finely hairy, with few or several large, scattered glands; SNE, DMoj (exc e half of San Bernardino Co.) . . . . *P. arborescens*
      7. Lflets gen continuous with axis . . . . var. *simplicifolius*
      7' Lflets gen (or at least some) jointed to axis
        8. Calyx 7–9 mm, gen conspicuously hairy . . . . var. *arborescens*
        8' Calyx 5–7 mm, glabrous or sparsely hairy . . . . var. *minutifolius*

P. arborescens (A. Gray) Barneby   Shrub < 1 m, sometimes unarmed, glabrous to puberulent. **LF**: lflets gen 5–7, linear to ovate, 3–10 mm, terminal (sometimes all) often continuous with axis. **INFL**: raceme, open; pedicels with bractlets. **FL**: calyx 5–9 mm, lobes ± equal, gen < tube; corolla 6–10 mm, violet-purple. **FR** exserted, 7–10 mm, glabrous or finely hairy, with large, scattered glands. Desert mtns, slopes, canyons, flats, washes; 100–1900 m. SnBr, **SNE, DMoj**; s NV, Mex. Vars. similar morphologically, distinct geog. Close to *P. fremontii*, which might be considered to constitute two additional vars. ❀ TRY.

var. *arborescens* (p. 333) MOJAVE INDIGO BUSH   **LF**: lflets lanceolate to ovate, larger 2.5 mm or more wide, upper often continuous with axis. **FL**: calyx 7–9 mm, gen conspicuously hairy. UNCOMMON. Desert, incl mtns; 400–800 m. **sw DMoj (Kern, San Bernardino cos.)**; Mex. [*Dalea a.* (A. Gray) A.A. Heller; *D. fremontii* var. *saundersii* (Parish) Munz] Apr–May

var. *minutifolius* (Parish) Barneby (p. 333)   **LF**: lflets lanceolate to ovate, larger 2.5 mm or more wide, gen jointed to axis. **FL**: calyx 5–7 mm, glabrous or sparsely hairy. Desert mtn slopes, canyons, talus; 150–1900 m. **SNE (Mono Co.), n&c DMoj**; s NV. [*Dalea fremontii* Torrey var. *m.* (Parish) Benson]

var. *simplicifolius* (Parish) Barneby (p. 333)   **LF**: lflets oblanceolate, gen 1–2 mm wide, gen continuous with axis. **FL**: calyx 5–7 mm, puberulent. Lower desert mtn slopes, flats, washes; 100–1100 m. SnBr, **adjacent DMoj (s San Bernardino, adjacent Riverside cos.)**. [*Dalea californica* S. Watson] Apr–May

P. emoryi (A. Gray) Rydb. (p. 333)   Subshrub < 1 m, < 2 m wide, gen canescent, (sometimes becoming glabrous); glands of twigs << 0.5 mm wide. **LF**: lflets of middle lvs 5–9, 2–10 mm, narrowly oblong to obovate, terminal lflet gen > others. **INFL** spike-like, ovoid or spheric, dense; bractlets 0. **FL**: calyx 4–7 mm, lobes ± = or > tube; corolla 4–6 mm, 2-colored (purple and white), puberulent. **FR** incl, ± 2.5–3 mm. 2*n*=20. Desert flats, washes, dunes; < 700 m. **s DMoj, DSon;** sw AZ, Mex. *[Dalea e.* A. Gray] ❀ DRN:11,**12, 13.** Mar–May

P. fremontii (A. Gray) Barneby   Shrub < 1 m, gen silvery-strigose. **LF:** lflets 3–25 mm, linear to ovate, gen jointed to axis, exc uppermost. **INFL:** raceme, ± open; pedicels with bractlets. **FL:** calyx 5–9 mm, lobes unequal, < tube; corolla 7–9.5 mm, violet-purple. **FR** exserted, 7–10 mm, glabrous; small glands forming longitudinal lines. 2*n*=20. Granite and volcanic slopes, flats, canyons; 250–1350 m. **e DMtns, DSon;** to s UT, AZ. Like *P. arborescens* exc in fr, geography. ❀ TRY.

var. *attenuatus* Barneby   **LF:** lflets linear, < ± 1 mm wide. Uncommon. Habitats of sp.; 450–900 m. **e DSon (se San Bernardino Co.);** s NV, AZ.

var. *fremontii* (p. 333)   **LF:** lflets narrowly elliptic to ovate or obovate, ± 1.5–3 mm wide. Habitats of sp.; 250–1350 m. **e DMtns (ne San Bernardino Co.);** to s UT, AZ. *[Dalea f.* A. Gray] Apr–May

P. polydenius (S. Watson) Rydb.   Shrub < 1.5 m, widely spreading, finely strigose; hairs pointing down, or puberulent; glands of twigs ± 0.5 mm wide. **LF:** lflets 7–13, 1–4.5 mm, obovate to ±

round, terminal ± = lateral. **INFL:** raceme, spike-like, ovoid or short-cylindric, dense; pedicels < 1 mm, bractlets 0. **FL:** calyx 5–6 mm, lobes unequal, < tube; corolla 4.5–6 mm, pink-purple, persistent after fl. **FR** incl, < 4.5 mm. 2*n*=20. Locally abundant. Desert flats, hills; 900–2250 m. **SNE, DMoj;** to UT. *[Dalea p.* S. Watson] ❀ TRY. May–Sep

P. schottii (Torrey) Barneby   Shrub < 2 m, green or gray-strigose. **LF** simple, persistent, 1–3 cm, linear, gland-dotted. **INFL:** raceme, open, 5–15-fld; axis not extended beyond fls, not a thorn. **FL:** calyx 4–6 mm, lobes < tube; corolla 7–10 mm, bright blue. **FR** exserted; glands large, spaced. 2*n*=20. Slopes, benches, washes; 0–600 m. **DSon;** AZ, Mex. *[Dalea s.* Torrey, incl var. *puberula* (Parish) Munz] ❀ TRY. Mar–May

P. spinosus (A. Gray) Barneby (p. 333) SMOKE TREE   Shrub or tree 1.5–8 m, gray-canescent. **LF** simple, early deciduous, 0.5–2 cm, oblanceolate, thick, gland-dotted. **INFL:** raceme, gen dense, 5–15-fld; axis extended beyond fls, gen as a thorn; pedicels ± 1 mm. **FL:** calyx 4.5–5 mm, lobes < tube; corolla 6–8 mm, blue-purple. **FR** slightly exserted. 2*n*=10. Common. Desert washes; < 400 m. **D;** AZ, nw Mex. *[Dalea s.* A. Gray] ❀ DRN:**12,13.** Jun–Jul

ROBINIA   LOCUST

Duane Isely & Elizabeth McClintock

Shrub or tree, gen spreading from underground parts, gen with stipular spines not gland-dotted. **LVS** odd-1-pinnate, alternate, deciduous. **INFL:** raceme, axillary. **FL:** calyx bell-shaped, lobes 5; petals 5, white or pink, banner reflexed; 9 filaments fused, 1 free. **FR** flat or plump, dehiscent. 4 spp.: US. (J. & V. Robin, who introduced pls to Eur, 16–17th century) [Isely & Peabody 1984 Castanea 49:187–202]

1. Corolla pink; infl hairs coarse, glandular-hairy; probably native, DMoj . . . . ***R. neomexicana***
1' Corolla white; infl hairs fine, nonglandular; fr glabrous; naturalized, GB . . . . ***R. pseudoacacia***

R. neomexicana A. Gray (p. 333) DESERT LOCUST   Shrub or small tree. **INFL** pendent or not. **FL:** corolla 2–2.5 cm. **FR** thick, not winged. Uncommon. Canyons in pinyon/juniper woodland; < 1500 m. **e DMoj (Mid Hills, ne San Bernardino Co.);** to w TX, n Mex. ❀ 15–17,IRR:1,**2,3,7–10,**11,**14,18,19,**20–24. Apr–Aug

R. pseudoacacia L. (p. 333) BLACK LOCUST   Tree. **INFL** pendent. **FL:** corolla 1.5–2 cm. **FR** flat; upper margin narrowly winged. 2*n*=20. Locally abundant. Near abandoned houses, along roadsides, canyon slopes, streambanks; 50–1900 m. **CA-FP, GB;** native to e US. TOXIC: ingested seeds, lvs, bark may be fatal to humans and livestock. May–Jun

SENNA

Elizabeth McClintock

Per, shrub, tree, unarmed or spines weak. **LVS** even-1-pinnate, alternate; stipules sometimes small or ephemeral; lflets 2–10(18) pairs. **INFL:** raceme or panicle, axillary or terminal. **FL** gen slightly bilateral, gen showy; sepals ± free; petals free, gen yellow; stamens free, 7 fertile, 3 sterile, anthers gen > filaments, opening by terminal pores. **FR** dehiscent or not. **SEEDS** few–many. ± 260 spp.: esp Am trop, also warm temp, sometimes deserts. (Arabic: Sana) [Irwin & Barneby 1982 Mem New York Bot Gard 35:1–918] Some cult as orns. Dried lvs of some spp. cathartic.

1. St, lvs ± glabrous; lflets not close, not overlapped, 4–6 mm, exceeded by main lf axis . . . . ***S. armata***
1' St, lvs densely white-hairy; lflets close, overlapped, 1–2.5 cm, not exceeded by main lf axis . . . . ***S. covesii***

S. armata (S. Watson) H. Irwin & Barneby (p. 333, pl. 55) SPINY SENNA   Shrub, armed with weak spines, ± lfless most of year, ± glabrous. **ST** 0.5–1 m; branches from base, grooved, ending in a weak thorn or not, green. **LF:** stipules minute or 0; lflets 2–4 pairs, not overlapped, ± opposite, ephemeral, ± sessile, 4–6 mm, asymmetric, oblong; main axis elongating after lflets fall, weakly spine-tipped. **INFL** terminal, raceme-like (fls 1–2 per axil of upper lvs). **FL:** petals 8–12 mm, obovate, ± irregular, yellow to salmon-red. **FR** dehiscent, 2.5–4 cm, lanceolate, straight. **SEEDS** few. Uncommon. Sandy or gravelly washes; 200–1000 m. **D;** NV, AZ, Baja CA. [*Cassia a.* S. Watson] ❀ TRY. Mar–Jul

S. covesii (A. Gray) H. Irwin & Barneby COUES' CASSIA   Subshrub, unarmed, leafy, densely white-hairy. **ST** 3–6 dm. **LF:** stipules bristle-like, some persistent; lflets 2–3 pairs, overlapped, opposite, short-stalked, 1–2.5 cm, elliptic. **INFL:** raceme, axillary, 5–15 mm, few-fld. **FL:** petals ± 12 mm, oblong-obovate, prominently veined. **FR** dehiscent, 2–5 cm, oblong, ± straight. **SEEDS** several. RARE in CA. Dry, sandy desert washes, slopes; 500–600 m. **DSon;** NV, AZ, Baja CA. [*Cassia c.* A. Gray] Apr

SESBANIA

Duane Isely

Ann, unarmed. **LVS** even-1-pinnate; stipules gen deciduous; lflets gen many. **INFL**: raceme, axillary; bractlets sometimes appressed to calyx base. **FL**: calyx lobes subequal, < tube; corolla gen yellow, dark-spotted; 9 filaments fused, 1 free. **FR** gen dehiscent, linear or oblong, 4-sided. **SEEDS** few–many. ± 60 spp.: trop, warm regions. (Ancient Arabic name)

S. exaltata (Raf.) Cory (p. 333) COFFEE WEED, COLORADO RIVER HEMP **ST** erect, simple or branched, 1–3 m. **LF**: lflets 30–60, 1–2.5 cm, oblong. **INFL** 2–6-fld. **FL**: corolla 1–1.5 cm, yellowish or mottled. **FR** 15–20 cm, 2–3 mm wide, linear. **SEEDS** many. $2n= 12$. Along streams, other moist sites, often in cult or old fields; < 500 m. **e DSon**, probably elsewhere; to se US. 🌸 IRR:**8,9,11–14**; STBL. Apr–Oct

TRIFOLIUM CLOVER

Duane Isely

Ann or per, unarmed. **LVS** gen palmately compound; stipules conspicuous, partly fused to petiole; lflets gen 3, sometimes 5–9, ± serrate or dentate. **INFL**: raceme (often umbel-like), head, or spike, axillary or terminal, gen many-fld, often involucred, gen peduncled; fls bracted or not. **FL** spreading to erect, often becoming reflexed; corolla gen purple to pale lavender, sometimes yellow, persistent after fl; 9 filaments fused, 1 free. **FR** gen indehiscent, but often breaking, short, plump, gen incl in corolla; base often stalk-like. **SEEDS** 1–6. (Latin: 3 lvs) [Gillett 1980 Can J Bot 58: 1425–1558; Zohary & Heller 1984 Genus *Trifolium*] Key to species adapted by Margriet Wetherwax.

Group 1: Involucre obvious

1. Calyx or entire banner soon inflated in fr; involucre bracts gen ± free . . . . *T. fragiferum*
1' Calyx, corolla not inflated in fr; involucre bracts gen fused into a bowl- or wheel-shaped, deeply toothed or cut ring
  2. Involucre bowl-shaped, lobes few, ± entire, when pressed often hiding fls; seeds 1–2
    3. Calyx lobe bristle tips irregularly and secondarily forked; involucre margin wavy, finely toothed; pl gen glabrous . . . . *T. cyathiferum*
    3' Calyx lobe bristle tips unbranched; involucre lobes ± entire; hairs on pl gen fine, wavy . . . *T. microcephalum*
  2' Involucre wheel-shaped, lobes many, deeply cut, when pressed hiding only bases of fls; seeds1–4
    4. Ann; involucre 1–2.5 cm wide; seeds 1–2 . . . . *T. variegatum*
    4' Per; involucre 2–3 cm wide; seeds 2–4 . . . . *T. wormskioldii*

Group 2: Involucre inconspicuous

1. Lflets 3–7; infl 15–25 mm wide; pl silvery or gray, densely hairy or tomentose. . . . **T. andersonii** var. **beatleyae**
1' Lflets 3; infl 8–12 mm wide (sometimes wider in fr); pl not silvery or gray, glabrous to variously hairy, but not tomentose
  2. In fr, calyx inflated, infl becoming a red or brown fuzzy ball ± 2 cm wide, pl creeping and rooting or cespitose . . . . *T. fragiferum*
  2' In fr, calyx rarely inflated, infl then << 2 cm wide; pls various in habit, incl creeping and rooting or cespitose
    3. Fls gen 1–3 per infl, gen with tiny bracts; corolla white lavender-striate; pl gen very small
    . . . . *T. monanthum* var. *monanthum*
    3' Fls gen > 5 per infl unless pl starved; corolla various in color, incl white to lavender-striate; pl various in stature, incl very small
      4. Pedicels remaining < 1 mm, fls remaining erect-spreading
        5. Infl peduncled. . . . . *T. longipes* var. *nevadense*
        5' Infl ± sessile above a pair of reduced lvs . . . . *T. pratense*
      4' Pedicels often initially short but quickly 1–4 mm, some or all fls becoming reflexed
        6. Lvs all basal or 2–3 cauline, lflet lanceolate, thick, ± serrate; SNE — RARE
        . . . . *T. macilentum* var. *dedeckerae*
        6' Lvs cauline from ground level, lflet linear to obovate, thin, ± entire; GB, DMoj
          7. Pl creeping and rooting, turf-forming; all petioles from ground level ± equal; corolla white . . . . *T. repens*
          7' Pl not creeping or rooting, often cespitose; some or 0 petioles from ground level, lower > upper; corolla pink to purple
            8. Corolla 5–7 mm; calyx 4–6 mm, glabrous; lflet obovate to obcordate . . *T. gracilentum* var. *gracilentum*
            8' Corolla 10–18 mm; calyx 5–10 mm, ± puberulent; lflet linear to obovate . . . . *T. longipes* var. *shastense*

T. andersonii A. Gray Per, short-tufted or cushion-forming, soft-hairy or tomentose, silvery or gray. **ST** 0 (peduncle ascending or erect). **LVS** basal; stipules entire, persistent; lflets 3–7, 5–20 mm, oblanceolate to obovate, entire. **INFL** head-like, 1.5–2.5 cm wide; pedicel 0.5–1 mm. **FL:** calyx 8–10 mm, lobes slender, > tube, plumose, hairs ± 1 mm; corolla 10–15 mm, pink-purple or 2-colored. **SEEDS** 1–2. $2n=16$. Meadows, talus, washes, yellow-pine forest, rocky alpine slopes; 900–4000 m. n SNH, **GB**; se OR, NV. 2 vars. in CA.

var. *beatleyae* (J.M. Gillett) Isely (pl. 56) BEATLEY'S FIVE-LEAVED CLOVER **LF:** lflets gen 0.5–1.4 cm; longer petiole hairs 0.4–1.2 mm. **INFL** > lvs, gen exserted; peduncle 2–10 cm. RARE in CA. Washes, talus, pine forest to alpine slopes; 1300–4000 m. SNE **(Mono Co.)**; NV. [*T. monoense* E. Greene] May–Aug

T. cyathiferum Lindley (p. 333) Ann, miniature to robust, ± glabrous. **ST** ascending to erect. **LVS** cauline; stipules entire or deeply toothed, tip acuminate; lflets 1–3 cm, oblanceolate to obovate. **INFL** head-like, 6–20 mm wide, 3–many-fld; involucre bowl-shaped, margin wavy, finely toothed. **FL:** calyx 7–11 mm, lower and lateral lobes ± = tube, with bristle tips irregularly and secondarily forked; corolla ± = or < calyx, white or yellowish, pink-tipped. **FR:** stalk-like base short. **SEEDS** 1–2. $2n=16$. Gen spring-moist valleys, chaparral, roadcuts, ditches, forests; < 2500 m. NW, CaR, n&c SN, GV, **n SNE**; to B.C., ID, NV. May–Aug

T. fragiferum L. (p. 333) STRAWBERRY CLOVER Per, stoloned or cespitose, glabrous. **LVS** ± basal; stipules gen overlapping; petiole > blade; lflets 0.5–2 cm, elliptic to obovate. **INFL** head-like, 8–12 mm wide (in fr ± 2 cm wide, spheric, fuzzy, red or brown); involucre bracts tiny, ± free. **FL:** calyx in fr inflated, lobes linear, lower 2 bristle-like, > upper 3; corolla 5–6 mm, pink, quickly ± hidden by calyx. **SEEDS** 1–2. $2n=16$. Lawns, roadsides, etc., often in saline soil; < 1500 m. NCoR, ScV, SCoR, SW, MP, **W&I**, probably elsewhere; sporadic in much of US; native to Eur, Afr. Cult.

T. gracilentum Torrey & A. Gray Ann, glabrous or slightly hairy. **ST** prostrate to erect. **LVS** cauline; stipules ovate-tapered; lflets 5–15 mm. **INFL:** umbel, 1–1.5 cm wide, 3–many-fld, often turned to side; axis often ± exserted; involucre abortive; pedicels 1–2 mm; fls becoming recurved. **FL:** calyx 4–6 mm, glabrous, tube 10-veined, lobes ± 0.5 mm wide at base, > tube, shortly bristle-tipped; corolla 5–7 mm, pink to pink-purple. **FR** 4–6 mm, > petals. **SEEDS** 1–2. $2n=16$. Open, disturbed, moist or dry places, grassy areas near ocean, sometimes serpentine; < 1800 m. CA-FP, **DMoj**; to WA, AZ, Mex.

var. *gracilentum* (p. 333) **LF:** lflets obovate to obcordate, length 1.5–2.5 × width; lower stipules gen inconspicuous, < 1 cm. Open, disturbed, moist or dry places, sometimes serpentine; < 1800 m. CA-FP (exc ChI), **DMoj**; to WA, AZ. [var. *inconspicuum* Fern.] Apr–Jun

T. longipes Nutt. Per, rhizomed or not, cespitose or not, gen puberulent. **ST** decumbent to erect, sometimes ± 0. **LVS** basal and cauline; stipules oblong-lanceolate to ovate, < 2 cm; lflets 2–5 cm, linear to obovate. **INFL** head-like, incl or exserted, 1.5–3 cm wide; peduncle often bent or curved at tip. **FL:** calyx 5–10 mm, ± puberulent, lobes lanceolate to bristle-like, gen > tube; corolla 10–18 mm, dull white, purple, or 2-colored. **FR:** stalk-like base 0–1 mm. **SEED** gen 1. $2n=16,24,32,48$. Meadows, streambanks, forests, open slopes, sometimes on serpentine; 600–3000 m. KR, NCoR, CaR, SN, SnBr, SnJt, **GB**; to WA, MT, Colorado. Intergrading complex [Gillett 1969 Can J Bot 47:93–113]; differentiating characters often indiscernible on herbarium specimens. Var. *longipes* not present in CA. 4 vars. in CA.

var. *nevadense* Jepson (p. 333) **LF:** lflet length 2.5–6 × width in CA-FP, to 10 × width in GB. **INFL:** pedicel 0.3–1 mm; fls spreading to ascending. **FL:** calyx lobes 1–3 × tube; corolla white to purple. Dry or boggy meadows, open slopes, woodlands, subalpine; 1100–3000 m. KR, NCoR, CaR, SN, SnBr, SnJt, **GB**;

OR, NV. Pls very small at high elevation. ❀ DRN,IRR:**1–3,6**,7,15–17. Jun–Sep

var. *shastense* (House) Jepson Pl often cespitose. **LF:** lflet length 1.5–4 × width. **INFL:** pedicel 1–2 mm; some or most fls becoming reflexed. **FL:** calyx lobes 1–3 × tube; corolla lavender to purple, longer petals tapered to point. Coniferous forest to alpine slopes; 1400–2700 m. KR, CaR, **GB**; to WA. [var. *multipedunculatum* (Kenn.) Isely] ❀ TRY. Jun–Sep

T. macilentum E. Greene var. *dedeckerae* (J.M. Gillett) Barneby (p. 335) DEDECKER'S CLOVER Per, cespitose, glabrous. **ST** ± ascending. **LVS** basal, sometimes 2–3 cauline; basal stipules clasping st; lflets 0.5–4 cm, lanceolate, thick, ± serrate; petiole gen> blades. **INFL** head-like, 1.5–3 cm wide; pedicel 1–2 mm; fls soon reflexed. **FL:** calyx 5–8 mm, glabrous, lobes unequal, gen > tube, narrow or bristle-like; corolla 12–15 mm, pink or pale violet. **FR:** stalk-like base < 2 mm. **SEEDS** probably 2. RARE. Pinyon woodland to alpine crevices; 2100–3500 m. s SNH, SNE. [*T. dedeckerae* J.M. Gillett] Basal lvs sometimes ± deteriorated on herbarium specimens.

T. microcephalum Pursh (p. 335) Ann; hairs gen soft, wavy. **ST** decumbent to erect. **LVS** cauline; stipules often ± bristle-tipped; lflets 0.8–2 cm, gen obovate, tip notched. **INFL:** head, 7–many-fld, often bur-like in fr; involucre bowl-shaped, lobes ± entire. **FL:** calyx 4–6 mm, tube 10-veined, lobes ± = or > tube, entire, unbranched, bristle-tipped; corolla 4–7 mm, pink to lavender. **FR** rupturing corolla. **SEEDS** 1–2. $2n=16$. Streambanks, moist, disturbed areas, roadsides, serpentine, conifer forest; 200–2700 m. NW, SN, GV, SnFrB, SCo, ChI, SnBr, PR, **DMoj**; to B.C., ID, AZ. ❀ STBL. Apr–Aug

T. monanthum A. Gray Per, very small, often cespitose, glabrous or puberulent. **ST** slender or reduced. **LVS** gen basal; stipules lanceolate to ovate; lflets 2–12 mm, elliptic-oblanceolate to widely obovate. **INFL:** reduced head, incl or exserted from lvs, 1–6-fld; involucre vestigial or bracts 2–5, inconspicuous, ± free, 1–3 mm. **FL:** calyx 4–5 mm, lobes ± = tube, bristle-tipped; corolla 7–12 mm, white or lavender-striate. **FR** sometimes rupturing corolla. **SEEDS** 1–2. $2n=16$. Pinyon-pine belt upwards, mtn forests near streams, wet meadows with aspen or willows, coniferous woodlands; 1500–3800 m. CaR, SN, SCo, SnGb, SnBr, SnJt, **SNE**; NV. 2 vars. in CA.

var. *monanthum* (p. 335) **LF:** lflet length 2–4 × width, tip gen rounded or truncate. **INFL** 1–3-fld. Pinyon-pine belt upwards, wet meadows with aspen or willows, coniferous woodland; 1500–3800 m. CaR, SN, **SNE**; NV. [var. *eastwoodianum* J.S. Martin; var. *parvum* (Kellogg) L.F. McDermott] Jun–Aug

T. pratense L. (p. 335) RED CLOVER Per, gen hairy. **ST** ascending. **LVS** cauline; stipules bristle-tipped; lflets 1.5–3.5 cm, elliptic to obovate. **INFL** terminal, above pair of reduced lvs, head-like, 2–3 cm wide; peduncle 0. **FL:** calyx 4.5–7.5 mm, tube 10-veined, lobes ± = or > tube, bristle-like, sparsely plumose; corolla 11–15 mm, red-purple. **FR** circumscissile. **SEEDS** 1–2. $2n=14, 28$. Disturbed areas; esp < 1000 m. CA-FP, **GB**, esp n; US, esp n, Can; native to Eur. Important forage crop. Apr–Oct

T. repens L. (p. 335) WHITE CLOVER Per, gen ± glabrous. **ST** creeping, rooting. **LVS** cauline from ground level, clearly alternate or clustered (from lack of st elongation); stipules white-membranous; petioles> blades; lflets 0.5–2.5 cm, obovate. **INFL** umbel-like, 1–2.5 cm wide; pedicel 1–3 mm; fls becoming reflexed. **FL:** calyx 3–6 mm, lobes tapered, all or lower > tube; corolla 7–11 mm, white. **SEEDS** 3–4. $2n=16,28,48$. Agricultural, disturbed, urban areas; esp < 1500 m. CA-FP, **GB**, esp n; n US; native to Eurasia. Important forage crop. Apr–Dec

T. variegatum Nutt. Ann or possibly short-lived per, gen ± glabrous. **ST** prostrate to erect, wiry to fleshy. **LVS** cauline; lower stipules gen entire; upper stipules deeply cut; lflets gen obovate or wedge-shaped, sometimes narrower. **INFL** head-like, incl or exserted from lvs, 0.5–2.5 cm wide, 1–many-fld; involucre wheel-like, gen well developed. **FL**: calyx 3–10 mm, tube 10–many-veined, all or some lobes gen > tube, bristle-tipped; corolla 3.5–16 mm, lavender to purple, tips gen white. **FR**: stalk-like base short or 0. **SEEDS** 1–2. 2n=16. 50–2500 m. CA-FP, **SNE**; sporadic to B.C., MT, Colorado, AZ, Baja CA. Most variable of CA clovers; ± 30 names available; most conspicuous CA variants treated as phases below, with commonly used names indicated for each. Keel shape seems taxonomically insignificant (acute to beaked in phase 1); additional research needed. 5 phases in CA. Apr–Jul

1. Infl ± 10–15 mm wide, ± 8–10-fld; corolla ± 6–10 mm
   . . . . **phase 1**
1' Infl 15–25 mm wide, > 10-fld; corolla 10–16 mm
   . . . . **phase 2**

phase 1 (p. 335) Ann, very small to robust. **ST** decumbent to erect, sometimes mat- or tangle-forming, 0.5–5 dm. **INFL** ± 1–1.5 cm wide, ± 8–10-fld. **FL**: calyx 4–6 mm; corolla ± 6–10 mm. Open fields, wet forest meadows, roadsides; 50–2500 m. Range of sp. [*T. appendiculatum* Lojacono var. *rostratum* (E. Greene) Jepson] Gen considered "typical" *T. variegatum*; merging with phases 2 and 3. ❀ WET:1–4,**5–9,14–24**. Apr–Jun

phase 2 Ann or possibly short-lived per, gen robust. **ST** gen ascending, 1–5(10) dm. **INFL** 1.5–2.5 cm wide, many-fld. **FL**: calyx 6–10 mm; corolla 10–16 mm. Permanently wet or inundated sites, incl meadows, marshes, ditches; 200–2200 m. NCo, NCoR, SN, GV, CW, SW, **SNE**; sw OR. [var. *melananthum* (Hook. & Arn.) E. Greene; *T. appendiculatum* Lojacono var. *a.*] Commonly confused with *T. wormskioldii*. ❀ WET:1–3,**4–9,14–24**. Apr–Jul

T. wormskioldii Lehm. (p. 335) Per, cespitose or not, glabrous. **ST** decumbent or ascending. **LVS** gen basal; lower stipules bristle-tipped; upper stipules wide, toothed or sharply lobed; lflets 1–3 cm, narrowly elliptic to widely ovate. **INFL** head-like, incl or exserted from lvs, 2–3 cm wide; involucre wheel-like, segments or lobes many. **FL**: calyx 7–11 mm, lobes tapered, tips bristled; corolla 12–16 mm, pink-purple or magenta, tip white. **FR**: stalk-like base 0–1 mm. **SEEDS** 2–4. 2n=16,32. Beaches to mtn meadows, ridges, gen open moist or marshy places; < 3200 m. NW, CaR, SN, SnJV, CW, SCo, PR, **SNE**; to B.C., WY, NM, Mex. The only involucred per sp. in Pacific coast states; incl matted, rhizomed form (dry coastal sands); lush, long-stemmed form (lower to middle elevations); slender, often tiny form (middle to higher elevations). ❀ IRR orWET,SUN:1–3,**4–9,14–17**. May–Oct

VICIA VETCH

Duane Isely

Ann or per, unarmed. **ST** gen sprawling or climbing, ridged or angled. **LVS** even-1-pinnate; stipules with an upper (often toothed or lobed) and smaller lower segment; lflets 4–many, alternate to opposite (often on 1 pl), linear to ovate; main lf axis gen ending as a tendril. **INFL**: raceme or cluster, axillary; peduncle or pedicels present; bracts small or 0. **FL**: corolla gen lavender to purple, sometimes white or yellow; 9 filaments fused, 1 free; style gen round in ×-section, hairs tufted at tip. **FR** dehiscent, gen ± oblong, gen flat; base stalk-like or not. **SEEDS** 2 or more. ± 130 spp.: N.Am, Eurasia. (Latin: vetch) [Herman 1960 USDA Handb 168] Best separated from *Lathyrus* by style characters.

V. americana Willd. var. ***americana*** (p. 335) AMERICAN VETCH Per, hairy or glabrous. **ST** sprawling or short and erect. **LF**: stipules gen sharply lobed; lflets 8–16, 1–3.5 cm, widely elliptic, wedge-shaped, to narrowly oblong, tip acute, truncate, notched, or 1–5-toothed. **INFL** ± = subtending lf; fls 3–9, gen loosely spaced, on > 1 side of axis. **FL**: corolla 1.5–2.5 cm (length when pressed ± 2.5–3.5 × width), gen blue-purple to lavender. **FR** 2.5–3 cm, 5–7 mm wide, glabrous or hairy; stalk-like base 2–5 mm. 2n=14. Gen open, moist forest, along streams, disturbed areas; < 2400 m. CA-FP, **GB**; N. Am (exc se US). [ssp. *oregana* (Nutt.) Abrams; var. *linearis* S. Watson; var. *truncata* (Nutt.) Brewer; *V. californica* E. Greene] Attempts to use lflet form and hairs to define infraspecific taxa are untenable [see Gunn 1968 Iowa State Coll J Sci 42:171–214]. Often mistaken for a *Lathyrus*. ❀ STBL. Apr–Jun

FAGACEAE OAK FAMILY

John M. Tucker

Shrub or tree, monoecious, deciduous or evergreen. **LVS** simple, alternate, petioled; margin entire to lobed; stipules small, gen deciduous. **STAMINATE INFL**: catkin or stiff spike; fls many. **PISTILLATE INFL** 1–few-fld, gen above staminate infl; involucre in fr gen cup-like or lobed and bur-like, bracts many, gen overlapping, flat or cylindric. **STAMINATE FL**: sepals gen 5–6, minute; petals 0; stamens 4–12+. **PISTILLATE FL**: calyx gen 6-lobed, minute; petals 0; ovary inferior, style branches gen 3. **FR**: acorn (nut subtended by scaly, cup-like involucre) or 1–3 nuts subtended by spiny, bur-like involucre; nut maturing in 1–2 years. **SEED** gen 1. 7 genera, ± 900 spp.: gen n hemisphere. Wood of *Quercus* critical for pre-20th century ship-building, charcoal for metallurgy; some now supply wood (*Fagus, Quercus*), cork (*Q. suber*), food (*Castanea*, chestnut).

QUERCUS OAK

Evergreen or deciduous. **LF**: stipules small, gen early deciduous. **STAMINATE INFLS**: catkins, 1–several, slender, on proximal part of twig. **PISTILLATE INFL** axillary among upper lvs, short-stalked; fl gen 1. **STAMINATE FL**: calyx 4–6-lobed, minute; stamens 4–10. **PISTILLATE FL**: calyx minute, gen 6-lobed; ovary enclosed by involucre. **FR**: acorn, maturing in 1–2 years; nut enclosed by cup-like involucre with thin or tubercled scales. 2n=24 for all reports. ± 600 spp.: n hemisphere, to n S.Am, India. (Latin: ancient name for oak) Many more hybrids have been named but are not incl here. Reproduction of many spp. declining.

1. Acorn shell ± woolly inside; fr maturing in 2 years
  2. Twigs ± flexible, or if stiff, 3–4 mm in diam; shrub or small tree < 20 m; lf margin entire to spine-toothed; e DMtns . . . . ***Q. chrysolepis***

2' Twigs rigid, 1–3 mm in diam; shrub 2–6 m; lf margin clearly spine-toothed; s,w edge D . . . . *Q. palmeri*
1' Acorn shell glabrous inside; fr maturing in 1 year
  3. Lf 2-colored, upper surface gray- or yellow-green, lower surface whitish, densely finely tomentose,
    hairs obscuring lateral veins . . . . *Q. cornelius-mulleri*
  3' Lf 1- or 2-colored, upper surface gray-green. lower surface dull green, finely stellate-pubescent, hairs
    not obscuring lateral veins
    4. Acorn sessile, nut conic-ovoid, gradually tapered to tip, 20–30 mm, gen dark brown, lvs irregularly
      and weakly spine-toothed; sw edge DMoj . . . . *Q. john-tuckeri*
    4' Acorn stalked (stalk < 15 mm), nut cylindric-ovoid to elliptic in outline, abruptly tapered at tip,
      12–23 mm, gen yellow-brown; lvs regularly spine-toothed; e DMtns . . . . *Q. turbinella*

Q. chrysolepis Liebm. (p. 335) MAUL OAK, CANYON LIVE OAK
Tree < 20 m, sometimes shrub-like, evergreen; trunk bark becoming narrowly furrowed, scaly, pale gray; twigs golden-tomentose, becoming ± glabrous. **LF** (1.5)3–6 cm, leathery; petiole 3–10 mm; blade oblong to oblong-ovate, sometimes round-ovate, tip acute to abruptly pointed, margin entire or spine-toothed, upper surface dark green, lower surface golden-puberulent, becoming glabrous, dull, grayish. **FR** maturing in 2 years; cup 17–30 mm wide, 5–10 mm deep, saucer- to bowl-shaped, scales thick, flat to slightly tubercled, golden-tomentose; nut 25–30 mm, 14–20 mm wide, ± ovoid, tip rounded to pointed, shell woolly inside. Canyons, shaded slopes, chaparral, mixed-evergreen forest, woodland; 200–2600 m. CA-FP (exc GV), e DMtns; OR, AZ, Baja CA. Highly variable. Hybridizes with *Q. palmeri.*. Shrubs with lvs 2–4 cm have been called var. *nana* (Jepson) Jepson. ❀ DRN,SUN:**4–7,14–18,22,23**,24&IRR:1–3,8,9,10,11,**19–21**. Apr–May

Q. cornelius-mulleri K. Nixon & K. Steele (p. 335) MULLER'S OAK Shrub 1–2.5 m, evergreen, densely branched; twigs finely tomentose. **LF** 2.5–3.5 cm, leathery; petiole 2–5 mm; blade oblong, ovate, or narrowly obovate, tip acute to rounded, margin entire or 4–6-toothed, upper surface sparsely puberulent, dull, yellow- to gray-green, lower surface densely and finely tomentose, whitish, midrib yellow. **FR** maturing in 1 year; cup 12–20 mm wide, 5–8 mm mm deep, hemispheric to cup-shaped, scales ± flat, gray-canescent; nut 20–30 mm, elliptic in outline to widely conic, tip obtuse, puberulent, shell glabrous inside. Slopes, gen granitic soils, chaparral, pinyon woodland; 1000–1800 m. SnBr (n slope), PR (e slope), s DMtns (Little San Bernardino Mtns); Baja CA. ❀ DRN,SUN:**7,9,14**,15,16,**18–21**,22,23 &IRR:2,3,8, 10,11;STBL.

Q. john-tuckeri K. Nixon & C.H. Muller (p. 335) TUCKER'S OAK Shrub 2–5 m (sometimes tree-like, < 7 m), evergreen; young twigs finely tomentose. **LF** 1.3–2.8 cm; petiole 1–4 mm; blade oblong, elliptic, or obovate, base rounded to widely wedge-shaped, tip obtuse to rounded, margin irregularly spine-toothed, upper surface dull, gray-green, lower surface finely hairy, pale gray-green. **FR** maturing in 1 year; cup 10–15 mm wide, 5–7 mm deep, thin, obconic to hemispheric, scales flat to slightly tubercled; nut 20–30 mm, ovoid to conic, tapered to tip, shell glabrous inside. Slopes on desert borders, chaparral, pinyon/juniper woodland; 900–2000 m. Teh (e slope), SCoRI, WTR (n slope), SnGb (n slope), sw edge DMoj. [*Q. turbinella* E. Greene ssp. *californica* J. Tucker] ❀ DRN,SUN:**7,8,9,**11,**14–16,**17, **18–23**,24.

Q. palmeri Engelm. (p. 335) PALMER'S OAK Shrub 2–6 m, evergreen; twigs spreading, rigid. **LF** 1–3 cm, very stiff; petiole 2–5 mm; blade elliptic to round-ovate, tip gen spiny, margin wavy and spine-toothed, upper surface glabrous to sparsely puberulent, ± shiny, olive-green, lower surface densely glandular-puberulent when young, pale gray-green. **FR** maturing in 2 years; cup 10–25 mm wide, 6–12 mm deep, gen bowl-shaped, rim ± spreading, scales flat, densely hairy; nut 20–30 mm, ± ovoid, tip gen obtuse, shell densely woolly inside. Uncommon. Rocky slopes, flats; 700–1300 m. e NCoRI (Colusa Co.), nw SnJV (Alameda, Contra Costa cos.), SCoR (San Luis Obispo, Santa Barbara cos.), SnGb (n slope), e PR, DMtns (Little San Bernardino Mtns); AZ, Baja CA. [*Q. dunnii* Kellogg] Hybridizes with *Q. chrysolepis*. ❀ DRN,SUN:**7,8,9,**10,**14–16,**17,**18–23**,24&IRR:11;STBL. Apr–May

Q. turbinella E. Greene (p. 335) SHRUB LIVE OAK Shrub 2–5 m (sometimes tree-like, < 7 m) evergreen; twigs densely and finely tomentose. **LF** 1.5–3 cm; petiole 1–3 mm; blade oblong to elliptic, base rounded to subcordate, tip acute to obtuse, sometimes spiny, margin spine-toothed, upper surface dull, gray-green, lower surface with both appressed-stellate and glandular, yellowish hairs. **FR** maturing in 1 year; stalk < 15 mm; cup 9–12 mm wide, 4–6 mm deep, ± hemispheric, scales flat to slightly tubercled, thin; nut 12–23 mm, cylindric-ovoid to elliptic in outline, tapered abruptly to tip, yellow-brown, shell glabrous inside. Pinyon/juniper woodland; 1200–2000 m. e DMtns (New York Mtns); to Colorado, TX, Baja CA. ❀ DRN:**7,8,9,**10,11,**14,18–21**&SUN:5,15–17,22–24;STBL.

FOUQUIERIACEAE OCOTILLO FAMILY

William J. Stone

Shrubs, trees, spiny. **ST** branched near base, or trunk single, thick, fleshy. **LVS** simple, alternate, small, somewhat fleshy, glabrous, of 2 types: primary soon deciduous after rains, petiole long, it and midrib develop into persistent spine after blade drops; secondary lvs clustered in axil of developing spine. **INFL**: spike, raceme, or panicle, terminal; fls many. **FL** showy; sepals 5, unequal, overlapping, persistent; corolla tube cylindric, lobes 5, spreading, bright red or yellow; stamens 10–20, in 1–2 whorls, filaments free, epipetalous; pistil 1, ovary superior, incompletely 3-chambered, placenta axile at base, parietal above, ovules 3–6 per chamber. **FR**: capsule. **SEEDS** elliptic, angled. 1 genus (includes *Idria*), 11 spp.: sw US, Mex.

FOUQUIERIA OCOTILLO, CANDLEWOOD, BOOJUM

(P.E. Fouquier, French professor of medicine) [Henrickson 1972 Aliso 7:439–537]

F. splendens Engelm. ssp. ***splendens*** (p. 335, pl. 57) OCOTILLO **STS** branched near base, erect to outwardly arching or ascending, 6–100, 2–10 m, gen < 6 cm diam, cane-like, lfless most of year; bark gray with darker furrows; spines 1–4 cm. **LVS**: primary 1–5 cm, petioles 1–2.5 cm; secondary 2–6 per cluster, 1–2 cm, 4–9 mm wide, petioles 2–8 mm, blade spoon-shaped to obovate, tip rounded to notched. **INFL**: panicle, gen 10–20 cm, widely to narrowly conic. **FL**: corolla 1.8–2.5 cm, bright red. **FR** ± 2 cm. 2*n*=24. Dry, gen rocky soils; 0–700 m. **DSon**; to TX, c Mex, Baja CA. Sts used for fences, huts; bark for waxes, gums. ❀ DRN,DRY:**10–13**,19–21. Mar–Jul

FRANKENIACEAE FRANKENIA FAMILY

R. John Little

Subshrubs, shrubs, gen from rhizome; salt-secreting glands present. **ST** prostrate to erect, nodes swollen, often rooting; petioles or dead lvs persisting on older sts. **LVS** opposite, 4-ranked, ± clustered; blade entire, gen leathery or fleshy, glabrous to hairy, margins rolled under. **INFL**: cyme, axillary; fls 1–25. **FL** gen bisexual, radial; sepals 4–7, fused; petals 4–7, free, overlapping, clawed (together appearing salverform), white to blue-purple, petal blade with a scale-like appendage near base; stamens 3–12 in two whorls, outer shorter; ovary superior, chambers 1–4, style branches 1–4; ovules 1–many. **FR**: loculicidal capsule. **SEED** ivory to gold-brown. 1 genus, 90 spp.: temp saline and gypsum soils. (Whalen 1987 Syst Bot Monogr 17:1–93)

FRANKENIA FRANKENIA

(Possibly named for J. Franke, Swedish botanist born 1590 or for Johann Frankenius, colleague of Linnaeus)

F. salina (Molina) I.M. Johnston (p. 337) ALKALI HEATH Subshrub forming mats < 3 m diam. **ST** ± prostrate, 1–6 dm; twig glabrous to hairy. **LF** glabrous to densely hairy. **INFLS** in most upper axils. **FL**: calyx tube 4–9 mm; petals 5–14 mm, white to dark pink or blue-purple; stamens gen 6, 5–12 mm; style branches gen 3. **FR** 3–5 mm. **SEEDS** 1–20, 1–1.5 mm, ± ellipsoid. Salt marshes, alkali flats; < 750 m. GV, CCo, SCo, ChI, **SNE**, **DMoj**; to NV, Mex, S.Am. [*F. grandifolia* Cham. & Schldl.; *F. g.* var. *campestris* A. Gray] ❀ SUN,IRR:**8**,10,**11**,12,13,**14**, 15,**16–24**;GRCVR. Jun–Oct

GARRYACEAE SILK TASSEL FAMILY

Thomas F. Daniel

Shrub, small tree, dioecious. **LVS** simple, opposite, evergreen, petioled; blade ± leathery, flat to concave-convex, margin entire, flat, rolled under, or strongly wavy. **INFL** catkin-like, pendent; fls small, in axils of opposite, 4-ranked, basally fused bracts. **STAMINATE FLS** (1–)3(4) per bract, pedicelled; perianth parts 4, gen fused at tips; stamens 4, alternate perianth parts, filaments free, anthers 2-chambered. **PISTILLATE FLS** (1–)3 per bract; pedicel ± 0 or short; perianth 0 or vestigial with 2 small appendages; ovary inferior, chamber 1, styles 2(3). **FR**: berry, spheric to ovoid, green, fleshy, becoming dark blue, black, or whitish gray, dry, brittle, not or irregularly dehiscent. **SEEDS** gen 2. 1 genus, 14 spp.: w US, C.Am, Caribbean; some cult. [Dahling 1978 Contr Gray Herb 209:1–104]

GARRYA SILK TASSEL BUSH

(N. Garry, 1st secretary of Hudson Bay Co., friend of David Douglas, 1782?–1856) Intergradation among CA spp. suggests some may be unworthy of that status.

G. flavescens S. Watson (p. 337) Shrub < 3 m. **LF** 19–75 mm, 9–45 mm wide, 1.3–3.3 × longer than wide, flat to ± concave-convex, elliptic to obovate-elliptic; margin flat to wavy; lower surface hairs ± 0 to ± dense, ± coarse, straight to slightly curved, appressed toward lf tip. **FR**: hairs gen dense. Desert slopes, chaparral, pine/oak woodland; 650–2350 m. NW, SN, CW, SW, **DMtns**; to UT, AZ, n Baja CA. [var. *pallida* (Eastw.) Ewan] Variation in hairiness not related to geog in CA. ❀ DRN,DRY,SUN:**7**,8–11,**14**,15–17,**18–23**,24. Feb–Apr

GENTIANACEAE GENTIAN FAMILY

James S. Pringle (except as specified)

Ann, per. **ST** decumbent to erect, < 2 m. **LVS** simple, cauline (sometimes also basal), opposite or whorled, entire, sessile or basal ± petioled; stipules 0. **FL** bisexual, radial, parts in 4's or 5's, exc pistil 1; sepals fused, persistent; petals fused, persistent or deciduous; stamens epipetalous, alternate corolla lobes; ovary superior, chamber 1, placentas parietal, often intruding. **FR**: capsule, 2-valved. **SEEDS** many. ± 70 genera, ± 1200 spp.: worldwide; some cult (*Eustoma*, *Exacum*, *Gentiana*). [Wood & Weaver 1982 J Arnold Arbor 63:441–487]

1. Sinuses between corolla lobes with a variously shaped, sometimes fringed appendage, or ± truncate
   (base of sinus gen > 1 mm wide) . . . . **GENTIANA**
1' Sinuses between corolla lobes unappendaged and acutely tapered
   2. Corolla ± rotate, lobed ± to base, nectary pits prominent, margins fringed . . . . **SWERTIA**
   2' Corolla not rotate, with distinct tube, nectary pits 0 (but corollas sometimes bearing fringes or scales)
      3. Corolla widely bell-shaped, lobes >> tube . . . . **EUSTOMA**
      3' Corolla not widely but sometimes narrowly bell-shaped, lobes < or = tube

4. Corolla salverform or funnel-shaped, pink to white . . . . **CENTAURIUM**
4' Corolla ± funnel-shaped or narrowly bell-shaped, blue to violet, rarely white
 5. Corolla < 2 cm, lobes with fringes or scales on inner surface near base . . . . **GENTIANELLA**
 5' Corolla gen > 2 cm, lobes without fringes or scales on inner surface . . . . **GENTIANOPSIS**

CENTAURIUM CENTAURY

James C. Hickman

Ann, bien, erect, glabrous. **LVS** opposite, often ± sheathing. **INFL**: cyme or panicle-like. **FL**: calyx lobes gen 5, ± linear-keeled, appressed to corolla; corolla salverform or funnel-shaped, gen pink, lobes ± entire, scales 0, sinus appendages 0; stamens gen ± exserted, dehisced anthers spirally twisted; style thread-like, stigma oblong to fan-shaped. **FR** ± cylindric-fusiform. **SEED** gen < 0.5 mm, ± rounded, ± brown, netted. ± 50 spp.: worldwide, exc s Afr. (Latin: centaur, mythological discoverer of its medicinal properties) Variable and difficult; worldwide study needed. Key to species by James S. Pringle.

1. Pedicels mostly 10–65 mm, some or all distinctly > closed corollas; corolla lobes 3–7 mm, not very showy
 . . . . *C. exaltatum*
1' Pedicels 2–30 mm, all or most < closed corollas; corolla lobes 8–20 mm, ± showy
 2. Basal lvs gen rosetted; corolla lobes 8–12 mm, < 1.5 × as wide distally as at base . . . . *C. calycosum*
 2' Lvs all cauline; corolla lobes (2) 10–20 mm, > 2 × as wide distally as at base . . . . *C. venustum*

C. calycosum (Buckley) Fern. Gen bien?, gen 20–50 cm. **LVS**: basal often rosetted, gen 15–70 mm, gen oblanceolate. **INFL** open; pedicels 10–40 mm. **FL**: corolla lobes 8–12 mm, gen rose-purple; undehisced anthers 3–6 mm; stigma lobes 1–2 mm when fresh, separately stalked, fan-shaped, spreading. Uncommon. Damp places, esp riverbanks; 50–100 m. **e DSon**; to Colorado, TX, Mex. Intergrades with *C. exaltatum* & *C. venustum* in DMoj. ❀ TRY. Apr–May

C. exaltatum (Griseb.) Piper Ann 10–35 cm. **LF** 10–30 mm, ± oblong, acute. **INFL** open; pedicels 10–50 mm. **FL**: corolla lobes 3–7 mm, gen to inrolled and appearing ± linear, not overlapping, white to rose; undehisced anthers ± 1 mm; stigma lobes fan-shaped. Moist, gen alkaline scrub; gen < 1500 m. **SCoR, SnGb, PR, GB, DMoj, w edge DSon**; to e WA, UT. [*C. namophilum* Rev. et al. var. *nevadense* Broome, Nevada centaury] CA (not NV) pls that have been considered *C. n.* var. *namophilum*,

spring-loving centaury, **THREATENED** US, with denser infl, from e DMoj (Ash Meadows), are apparently all *C. exaltatum*. [Holmgren 1984 Intermtn Flora 4:5–6] Much variation apparently environmental. May–Aug

C. venustum (A. Gray) Robinson (p. 337) CANCHALAGUA Ann 3–50 cm. **LF** 5–25 mm, narrowly oblong to ovate. **INFL** dense or open; pedicels often short. **FL**: corolla throat ± white, lobes (2)10–20 mm, gen rose-purple (white); undehisced anthers 3–6 mm; stigma lobes 1–1.5 mm, separately stalked, fan-shaped, spreading. Common. Dry scrub, grassland, forest; < 1300 m. **e KR, NCoRI, s NCoRO, CaRF, SNF, e SnJV, SW** (exc n ChI, esp San Diego Co.), **DMoj**. [ssp. *abramsii* Munz] Highly variable; intergrades with *C. calycosum, C. exaltatum*. Much variation apparently environmental. ❀ SUN,IRR:2,3,**7–11,14–24**. May–Aug

EUSTOMA CATCHFLY GENTIAN

Ann, short-lived per, glabrous, ± glaucous. **ST** erect, branched. **LVS** basal and cauline, opposite, clasping. **INFL**: cyme or fls solitary. **FL**: parts in 5's; calyx lobed ± to base, lobes linear, narrowly keeled, acuminate; corolla > 2 cm, bell-shaped, lobes >> tube, scales 0, sinus appendages 0, nectary pits 0; nectaries at base of ovary; ovary sessile, style distinct, slender, stigma 2-lobed. 2 weakly differentiated spp.: N.Am, C.Am, West Indies. (Greek: beautiful mouth, from corolla tube)

E. exaltatum (L.) Don (p. 337) Pl 1.5–10 dm. **LVS**: basal 2–10 cm, spoon-shaped-obovate to elliptic-oblong, obtuse; cauline 1.5–9 cm, elliptic to lanceolate, ± obtuse, upper acute. **INFL**: pedicels 2–14 cm. **FL**: calyx 10–21 mm; corolla 2–4.5 cm, pale to deep violet-blue, rarely rose-violet or white, throat whitish

with dark purple blotches, lobes elliptic to obovate, rounded to subacute. $2n=\pm72$. Roadsides, alkaline marshes, other open, wet places; 100–600 m. **SCo, PR, DSon**; to se US, C.Am, West Indies. ❀ IRR&SUN:7–9, 13,14,16,17,19–22,**23,24**. Most of year

GENTIANA GENTIAN

Ann, per, gen glabrous. **ST** gen simple below infl. **LVS** cauline, opposite (sometimes also basal). **INFL**: compact cyme or fls solitary. **FL** parts in 4's or 5's; calyx tube obvious; corolla tube narrowly bell-shaped, lobes spreading, < tube, scales 0, sinuses between lobes ± truncate, unappendaged or gen with a variously shaped, sometimes fringed appendage, nectary pits 0 (nectaries on ovary stalk); ovary stalked, style short or indistinct, stigma 2-branched. ± 300 spp.: temp to subarctic and alpine Am, Eurasia. (Gentius, king of ancient Illyria, who might have found medicinal value in the pls)

1. Perennial; lf 6–50 mm; fl > 2 cm . . . . *G. newberryi* var. *tiogana*
1' Low annual; lf 2–6 mm; fl < 2 cm . . . . *G. prostrata*

G. newberryi A. Gray (p. 337) Per. **STS** arising laterally, below rosette, decumbent, 1–several, 0.5–10(23) cm. **LVS**: basal, lower cauline 8–50(75) mm, 2–25 mm wide, widely spoon-shaped to oblanceolate, often ± petioled, rounded or abruptly

pointed; upper cauline few, gen > internodes, 6–30(45) mm, 1–8(13) mm wide, oblanceolate to linear, acute. **INFL**: fls 1–5. **FL**: calyx (10)14–30 mm, lobes linear to narrowly ovate, acute; corolla 23–55 mm, lobes 7–17 mm, elliptic-obovate, short-acuminate, sinus appendages deeply divided into 2 triangular, ±serrate-jagged parts tapered to thread-like points. **SEED** winged. Wet mtn meadows; 1200–4000 m. KR, CaRH, SNH, **W&I**; OR, w NV. Size variation ± correlated with altitude, moisture; vars. intergrade in n SNH. 2 vars in CA.

var. *tiogana* (A.A. Heller) J. Pringle (pl. 58) St, lf, fl size often in lower to middle part of ranges for sp. **FL**: corolla white to pale blue exc exterior dark brownish purple on and below lobes. 2*n*=26. Wet mtn meadows; 1500–4000 m. SNH, **W&I**; w NV. [*G. tiogana* A.A. Heller]   ❀ WET&DRN:**1,2**,4–6,16,17& SHD:14,15. Jul–Sep

G. prostrata Haenke (p. 337) PIGMY GENTIAN Ann. **STS** ± prostrate to decumbent, 1–several, 0.5–8 cm. **LVS** basal and cauline similar, spreading, 2–6 mm, 1–4 mm wide, spoon-shaped-obovate to broadly oblanceolate, obscurely or not white-margined, abruptly pointed; cauline ± evenly spaced, gen 0.5–2 × internodes. **INFL**: fl 1. **FL**: calyx 3.5–10 mm, lobes triangular, abruptly pointed; corolla 7–18 mm, medium to deep blue, lobes 2.5–5 mm, ovate, acute to acuminate, sinus appendages triangular, ± entire or ± irregularly jagged toward tip, acute. **SEED** wingless. 2*n*=36. UNCOMMON. Wet mtn meadows; 3500–3800 m. **W&I (White Mtns)**; to AK, Colorado, Eurasia. Jul–Aug

GENTIANELLA

Ann, glabrous. **LVS** basal and cauline, opposite. **INFL**: cyme or fls solitary in axils or on branches. **FL**: parts in 4's or 5's; calyx tube < lobes; corolla < 2 cm, narrowly funnel-shaped, lobes spreading, < tube, subentire, with fringes or scales on inner surface near base, sinus appendages 0, nectary pits 0 (nectaries on lower part of corolla tube); ovary sessile, style indistinct or 0, stigma 2-branched, persistent. ± 250 spp.: ± worldwide, esp temp to alpine, exc Afr. (Latin: little *Gentiana*)   [Gillett 1957 Ann Missouri Bot Gard 44:195–269]

1. Fls in cymes and solitary in axils, parts gen in 5's; pedicels gen < internode below (subg. *Gentianella*)
. . . . ***G. amarella*** ssp. ***acuta***
1' Fls solitary, terminal on branches, parts gen in 4's; pedicels >> internode below (subg. *Comastoma*)
. . . . ***G. tenella*** ssp. ***tenella***

G. amarella (L.) Boerner ssp. *acuta* (Michaux) J.M. Gillett (p. 337) Pl 5–80 cm. **ST** erect, 1, often with long branches near base, simple or with shorter branches above. **LVS**: basal withering early; lower cauline < 45 mm, spoon-shaped to elliptic-oblong, crowded, upper cauline < 60 mm, oblong-lanceolate to ovate, gen << internodes. **INFL**: pedicels gen < 2.5(–5) cm. **FL**: calyx 5–10(18) mm, lobes 1–6 × tube, linear to lanceolate; corolla blue to rose-violet or white, lobes ovate-triangular, a fringe of hairs across each. 2*n*=18. Wet meadows, bogs; 1500–3500 m. KR, CaRH, SNH, SnBr, **W&I**; to AK, e N.Am, Baja CA, e Asia. [*Gentiana a*. L. var. *acuta* (Michaux) Herder]   ❀ TRY. Jun–Sep

G. tenella (Rottb.) Boerner ssp. *tenella* (p. 337) Pl < 15(–25) cm. **STS** decumbent to erect, 1–25, simple or branched near base. **LVS** mostly basal, < 20 mm, elliptic-oblong to spoon-shaped; cauline 1–4 pairs, smaller. **INFL**: pedicels 2–10 cm. **FL**: calyx 4–11 mm, lobes >> tube, ovate to lanceolate; corolla pale violet-blue to white, lobes ovate-oblong, with 2 fringed scales at base of each. 2*n*=10. Open, wet places; 3200–3900 m. c&s SNH, **W&I**; to AK, Colorado; circumpolar. [*Gentiana t*. Rottb.] ❀ TRY. Jul–Aug

GENTIANOPSIS   FRINGED GENTIAN

Ann, per, glabrous. **LVS** basal and cauline, opposite. **INFL**: fl 1 per st or branch. **FL**: parts in 4's; calyx tube distinct, lobes lanceolate, acuminate; corolla > 2 cm, funnel-shaped, blue, rarely white, lobes < or = tube, oblong to elliptic-obovate, obtuse or rounded, ± entire to conspicuously serrate, jagged, or fringed, sinus appendages 0, nectary pits 0 (nectaries on corolla tube near base); ovary stalked, style short or indistinct, persistent, stigma 2-branched. ± 15 spp.: temp N.Am, Eurasia. (Greek: resembling *Gentiana*)

G. holopetala (A. Gray) Iltis (p. 337) Ann or per from root-sprouts or rooting sts, 3–45 cm. **ST** ± decumbent to erect. **LVS**: basal, lower cauline < 70 mm, < 15 mm wide, spoon-shaped, rounded to acute; upper cauline few, lance-elliptic to linear, acute. **INFL**: peduncle (0.4)2.5–21 cm, gen > 2.5 × subtending internode, gen > whole st. **FL**: calyx (10)14–36 mm; corolla 20–55 mm. **SEED** papillate, obtuse. 2*n*=78. Wet meadows; 1800–4000 m. c&s SNH, **SNE**; w NV. [*Gentiana h*. (A. Gray) Holm] ❀ TRY. Jul–Sep

SWERTIA

Per (dying after fl in *S. albomarginata*, *S. fastigiata*, *S. parryi*, *S. puberulenta*, *S. radiata*; non-fl rosettes preceding fl-sts in these spp., accompanying fl-sts in others). **LVS**: basal ± petioled; cauline < basal, gen whorled or opposite. **INFL**: cyme or panicle of dense clusters. **FL**: parts in 4's (gen 5's in *S. perennis*, many spp. outside CA); calyx fused only near base, lobes lanceolate; corolla rotate, rarely bell-shaped, sometimes with fringed ridges or scales between stamen bases, lobes > tube, sinus appendages 0, nectary pits prominent, 1–2 per lobe, pit margin fringed; ovary sessile, style short, persistent, stigma 2-branched. ± 120 spp.: temp N.Am, Eurasia, Afr. (E. Sweert, Dutch herbalist, born 1552) [St. John 1941 Amer Midl Naturalist 21:1–29] *Frasera* sometimes segregated.

1. Cauline lvs whorled, uppermost sometimes opposite . . . . ***S. albomarginata***
1' Cauline lvs all opposite . . . . ***S. puberulenta***

S. albomarginata (S. Watson) Kuntze (p. 337) Pl 3–6 dm, glabrous or sts puberulent. **STS** 1–few. **LVS** white-margined; basal 2–9 cm, 5–10 mm wide, oblanceolate, tips acute; cauline whorled, upper sometimes opposite, linear-lanceolate, tips acuminate. **INFL** open; pedicels 5–50 mm. **FL**: calyx 5–12 mm; corolla 8–14 mm, with a low, fringed ridge between stamen bases, greenish white, often purple-dotted, lobes lance-oblong, abruptly acuminate; nectary pit 1 per corolla lobe, oblong, wider, 2-lobed at tip. Dry, open woodlands; 1500–2200 m. DMtns; to Colorado. [*Frasera a*. S. Watson]   ❀ DRN&SHD:2,6,15–17. May–Jul

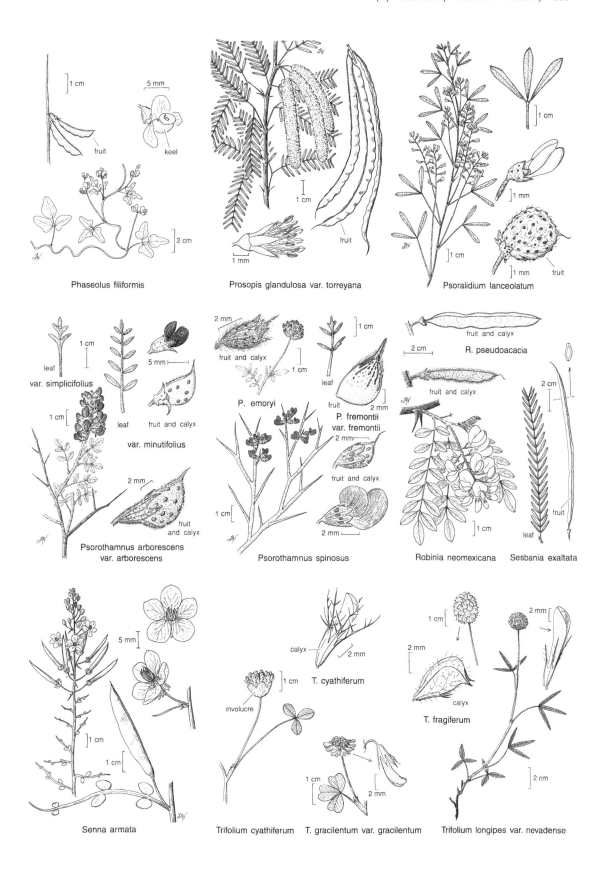

Phaseolus filiformis

Prosopis glandulosa var. torreyana

Psoralidium lanceolatum

var. simplicifolius

var. minutifolius

Psorothamnus arborescens var. arborescens

P. emoryi

P. fremontii var. fremontii

Psorothamnus spinosus

R. pseudoacacia

Robinia neomexicana   Sesbania exaltata

Senna armata

Trifolium cyathiferum   T. gracilentum var. gracilentum

T. cyathiferum

T. fragiferum

Trifolium longipes var. nevadense

S. puberulenta (Davidson) Jepson (p. 337) Pl 1–3 dm; sts, lower lf surfaces puberulent. **STS** 1–several. **LVS** narrowly white-margined; basal 2–12 cm, 6–10(17) mm wide, narrowly obovate to elliptic-oblong, tips obtuse or abruptly pointed; cauline opposite, upper oblong to lanceolate, tips acute. **INFL** open; pedicels 5–35 mm. **FL**: calyx 6–11 mm; corolla 7–13 mm, with a low, ± fringed ridge between stamen bases, greenish white, purple-dotted, lobes oblong-obovate, abruptly acuminate; nectary pit 1 per corolla lobe, oblong-obovate. Dry, open coniferous woodlands; 1700–3400 m. c&s SNH, **W&I**. [*Frasera p.* Davidson; *S. albomarginata* var. *purpusii* Jepson] Jun–Aug

GERANIACEAE GERANIUM FAMILY

Ann, per, or ± woody, gen hairy. **LVS** simple to compound, basal and cauline; cauline alternate or opposite, stipules present. **INFL**: cyme or umbel. **FL** bisexual, radial or ± bilateral; sepals 5, free, overlapping in bud; petals 5, free, with nectar glands at base; stamens gen 5 or 10; staminodes scale-like or 0; pistil 5-lobed, chambers 5, placentas axile, styles 5, fused to axis, columnar in fr, stigmas atop axis 5, free. **FR**: segments 5, dry, 1–2-seeded, separating from each other and then from column; fr body dehiscent on 1 side or not; part of style persistent atop ovary and separating with it, curved to tightly coiled when dry. 14 genera, ± 750 spp.: temp, ± trop. Some cult for orn, perfume oils. [Aeda 2000 Anales Jard. Bot. Madrid 58:39–82; Robertson 1972 J Arnold Arbor 53:182–201] Family description by M.S. Taylor. Keys adapted by Margriet Wetherwax.

1. Lf simple or pinnately compound; fertile stamens 5 . . . . **ERODIUM**
1' Lf palmately lobed; fertile stamens gen 10 . . . . *Geranium pusillum*

ERODIUM STORKSBILL, FILAREE

Mary Susan Taylor

Ann, per. **LVS** simple to pinnately compound; lower basal; upper opposite; blade lanceolate to reniform in outline, base cordate to truncate, short-hairy. **INFL**: umbel. **FL**: stamens 5, alternate 5 scale-like staminodes. **FR**: body indehiscent, fusiform, 1-seeded, base sharply pointed, top gen pitted, pits subtended by 1–2 furrows or not; style segment persistent to fr body stiffly hairy on side facing column. ±75 spp.: temp Am, Eurasia, n Afr, Australia. (Greek: heron, from bill-like fr) [Guittonneau 1972 Boissiera 20:1–154] Some cult for forage, dyes.

1. Lower lvs pinnately compound, lflets 9–13 . . . . *E. cicutarium*
1' Lower lvs simple, crenate to shallowly lobed . . . . *E. texanum*

E. cicutarium (L.) Aiton (p. 337) Ann. **ST** decumbent to ascending, 1–5 dm, ± glandular-hairy. **LVS** compound; lower 3–10 cm, blade > petiole, ovate to oblanceolate in outline, sparsely hairy; lflets 9–13, deeply dissected, ultimate segments 1–2 mm wide. **FL**: sepals 3–5 mm, tip bristly; petals ± = sepals, red-lavender, base gen purple. **FR**: body 4–7 mm, pit ± round, glabrous, subtended by 1 shallow furrow or not; style column 2–5 cm. $2n=40$. Open, disturbed sites, grassland, shrubland; < 2000 m. **CA**; widespread US; native to Eurasia. Feb–May

E. texanum A. Gray (p. 337) Ann, bien. **ST** prostrate to ascending, 1–5 dm, ± canescent. **LF** simple, 1.5–4 cm; blade gen < petiole, ovate, cordate, crenate to shallowly lobed, densely puberulent to strigose. **FL**: sepals 5–10 mm, tip strigose; petals unequal, 7–12 mm, lavender to purple. **FR**: body 5–8 mm, pit transversely elliptic, glabrous, furrow 0; style column 3–7 cm. $2n=20$. Dry, open sites, shrubland; < 1500 m. s SnJV, s SCo, **D**; to TX, n Mex. Mar–May

GERANIUM CRANESBILL, GERANIUM

Mary Susan Taylor

Ann, per. **LVS** palmately lobed or divided; upper alternate or opposite; blade gen round in outline, base gen cordate, ± hairy. **INFL**: cyme; fls (1)2. **FL**: sepals awned or not; stamens 10, outer 5 opposite petals, inner 5 alternate petals. **FR**: body dehiscent, gen ovoid, 1–2-seeded, base rounded; style column narrowed at top below free stigmas, forming a beak in fr; part of style persistent to fr body glabrous to puberulent on side facing column. 250–300 spp.: temp, trop mtns. (Greek: crane, from beak-like fr) [Jones & Jones 1943 Rhodora 45:5–26;32–53] Some orn, cult for oils. Native per vary regionally, are often difficult to separate, need further study. *G. californicum, G. viscosissimum* have been reported from Bodie, Mono Co.

G. pusillum L. (p. 337) Ann. **ST** prostrate to decumbent, 1–5 dm, often branched, puberulent. **LVS**: lower 10–25 cm; blades 2–5 cm wide, divided into 5–9 oblong to wedge-shaped segments, upper half of segments lobed or not. **FL**: pedicels 8–15 mm; sepals 3–4 mm, acute; petals ± = sepals, ± notched, pink to violet; fertile stamens 5–8. **FR**: body ± 2 mm, minutely strigose; style column 7–9 mm, beak < 1 mm. **SEED** smooth. Uncommon. Disturbed sites; < 500 m. NCo, s SNF, CCo, s MP, **s SNE**; to B.C., e N.AM; native to Eur. Jun–Sep

GROSSULARIACEAE GOOSEBERRY FAMILY

Michael R. Mesler & John O. Sawyer, Jr.

Shrub gen < 2 m. **ST** gen erect; nodal spines 0–9; internodal bristles gen 0; twigs gen hairy, gen glandular. **LVS** simple, alternate, gen clustered on short, lateral branchlets, petioled, gen deciduous; blade gen palmately 3–5-lobed,

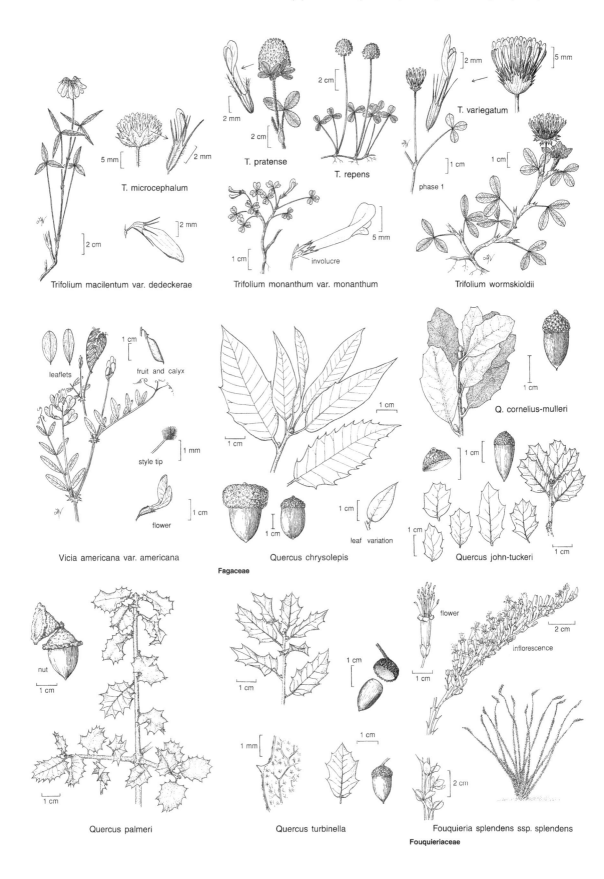

T. microcephalum

T. pratense

T. repens

T. variegatum

phase 1

Trifolium macilentum var. dedeckerae

involucre

Trifolium monanthum var. monanthum

Trifolium wormskioldii

leaflets

fruit and calyx

style tip

flower

Vicia americana var. americana

Quercus chrysolepis

leaf variation

Fagaceae

Q. cornelius-mulleri

Quercus john-tuckeri

nut

Quercus palmeri

Quercus turbinella

flower

inflorescence

Fouquieria splendens ssp. splendens

Fouquieriaceae

gen thin, gen dentate or serrate, base gen cordate. **INFL**: raceme, axillary, gen pendent, 1–25-fld; pedicel gen not jointed to ovary, gen hairy or glandular; bract gen green. **FL** bisexual, radial; hypanthium tube exceeding ovary; sepals gen 5, gen spreading; petals gen 5, gen < sepals, gen flat; stamens gen 5, alternate petals, gen inserted at level of petals (hypanthium top), anthers gen free, gen glabrous, tips gen rounded; ovary inferior, chamber 1, ovules many, styles gen 2, gen fused exc at tip, gen glabrous. **FR**: berry. 1 genus, 120 spp.: n hemisphere, temp S.Am. Some cult as food, orn. Hypanthium data refer to part above ovary; statements about ovary hairs actually refer to the hypanthium around the ovary. Formerly incl in Saxifragaceae.

RIBES CURRANT, GOOSEBERRY

The only genus. (Arabic: for pls of this genus)

1. Nodal spines 0 (subg. *Coreosma*)
  2. Sepals yellow . . . . *R. aureum* **var.** *aureum*
  2' Sepals white to pink . . . . *R. cereum*
    3. Bract tip wide, with several prominent teeth; styles gen hairy . . . . var. *cereum*
    3' Bract tip acute, with 1–3 shallow teeth on each side; styles glabrous . . . . var. *inebrians*
1' Nodal spines present (sometimes 0 on some shoots)
  4. Hypanthium disk- or saucer-shaped; fruits orange-red (subg. *Grossularioides*) . . . . *R. montigenum*
  4' Hypanthium cup- to tube-shaped; fruits purple to black (subg. *Grossularia*)
    5. Anthers exserted from petals . . . . *R. inerme*
    5' Anthers not exserted from petals
      6. Ovary hairs conspicuous, short and long, glandular and not . . . . *R. velutinum*
      6' Ovary hairs 0 or inconspicuous, gen short, gen nonglandular . . . . *R. quercetorum*

R. aureum Pursh GOLDEN CURRANT Shrub < 3 m. **ST**: nodal spines 0; internodes glabrous or puberulent. **LF**: blade firm, 15–50 mm, toothed or not, light green, gen glandular when young, gen glabrous when mature, base wedge-shaped to subcordate. **INFL** 5–15-fld. **FL**: hypanthium 6–10 mm, longer than wide; sepals 3–8 mm, yellow; petals 2–3 mm; styles fused base to tip. **FR** 6–8 mm, red, orange, or black, glabrous. 2*n*=16. Many habitats; < 3000 m. KR, NCoRI, CaR, SNH, SnJV, SnFrB, SCoR, SW, **GB**; to B.C., SD, NM. 2 vars in CA.

var. *aureum* (p. 337, pl. 59) **FL**: odor spicy; hypanthium 1.5–2 × sepals; sepals 5–8 mm; petals yellow turning orange. Many habitats; < 3000 m. KR, CaR, SNH, SnJV, **GB**; to B.C., SD, NM. ❀4,5,**6**,17; IRR:1–3,**7–10**,11,12,**14–16,18–24**. Apr–May

R. cereum Douglas WAX CURRANT **ST**: nodal spines 0. **LF**: odor spicy; blade 10–40 mm, round, shallowly lobed, finely toothed, upper surface glossy. **INFL** 3–7-fld. **FL**: hypanthium 6–8 mm, > 2 × longer than wide; sepals 1–2 mm, white to pink; petals < 1 mm, white to pink; stamens inserted below level of petals, anther tips with cup-like depression; styles fused ± to tip. **FR** 10–12 mm, red, glabrous to sparsely glandular. 2*n*=16. Dry montane to alpine slopes, among rocks, forest edges; 1500–4000 m. KR, CaRH, SNH, Teh, TR, SnJt, **GB**, **DMtns**; to B.C., c US, AZ.

var. *cereum* (p. 337) **LF**: hairs 0 to dense, glandular. **INFL**: bract tip wide, with several prominent teeth. **FL**: styles gen hairy. Habitats of sp. KR, CaRH, SNH, Teh, TR, SnJt, **GB**, **DMtns**; to B.C. ❀ DRN:4,5,6&IRR:1–3,7,14–18. Jun–Jul

var. *inebrians* (Lindley) C. Hitchc. **LF**: hairs dense, nonglandular. **INFL**: bract tip acute, with 1–3 shallow teeth on each side. **FL**: styles glabrous. Open, rocky areas; 2100–4000 m. s SNH, **W&I**; to ID, NE, NV, AZ. [*R. i.* Lindley] ❀ DRN&IRR:1,2. Jun–Jul

R. inerme Rydb. WHITE-STEMMED GOOSEBERRY **ST** scrambling; nodal spines 0–3; internodes gen glabrous. **LF**: blade 20–30 mm, coarsely toothed. **INFL** 1–5-fld. **FL**: hypanthium 2–3 mm, ± as long as wide; sepals reflexed, 3–4 mm, greenish white, purple at base or not; petals 1–2 mm, white; filaments exceeding petals by

< 2 mm. **FR** 7–11 mm, purple, glabrous. 2*n*=16. Forests, streamsides, meadow edges; 1200–3300 m. KR, NCoRO, CaRH, SNH, Wrn, **SNE**; to B.C., Rocky Mtns. [Sinnott 1985 Rhodora 87:189–286] 2 vars in CA.

var. *inerme* **LF**: hairs 0 to sparse, short, soft. **FL**: hypanthium and sepal hairs 0. Forests, streamsides, meadow edges; 1200–3300 m. KR, CaRH, SNH, Wrn, **SNE**; to B.C., Rocky Mtns. [*R. divaricatum* var. *i.* (Rydb.) McMinn] ❀ IRR:4,5,**6**,15–17&SHD:1–3,**7**, 14,18. May–Jun

R. montigenum McClatchie (p. 347) MOUNTAIN GOOSEBERRY **ST** spreading or decumbent; nodal spines 1–5; internodes ± bristly or not. **LF**: blade 1.5–2.5 cm, deep-lobed to -serrate, hairy, glandular. **INFL** gen > 5-fld; pedicel jointed to ovary. **FL**: hypanthium 1 mm, saucer-shaped; sepals 3–4 mm, green to greenish white; petals 1 mm, red. **FR** 4–5 mm, orange-red; bristles glandular. Many subalpine, alpine habitats; 2100–4800 m. KR, CaRH, TR, SnJt, Wrn, **n DMtns**; to B.C., ID, NV, AZ. ❀ DRN&IRR:1–3. Jun–Jul

R. quercetorum E. Greene (p. 347) OAK GOOSEBERRY **ST** arched; nodal spines 1(3); internodes puberulent. **LF**: blade 10–20 mm, dentate, gen hairy, glandular. **INFL** 2–3-fld. **FL**: hypanthium < 3 mm, longer than wide; sepals reflexed, 3 mm, yellow; petals 1 mm, cream; anthers not exserted from petals; styles fused ± to tip. **FR** 7–8 mm, black, ± glabrous. Oak woodlands, chaparral; < 1350 m. c&s SNF, Teh, SnFrB, SCoR, WTR, PR, **w edge D**; AZ, n Baja CA. ❀ DRN:7,8,9,11,**14,18–24**&SUN:5,**15–17**. Mar–May

R. velutinum E. Greene (p. 347) **ST** stout, arched; nodal spines 1(3). **LF**: blade 5–20 mm, crenate. **INFL** 1–4-fld. **FL**: hypanthium 2–3 mm, ± as long as wide; sepals 3 mm, white to yellow; petals 2 mm, white to yellow; anthers not exserted from petals; ovary hairs conspicuous, short and long, glandular and not. **FR** 6–7 mm, yellow becoming purple. Sagebrush steppe, juniper woodland, pine forest; 700–2500 m. KR, CaRH, SNH, Teh, TR, **GB**, **DMtns**; to UT, AZ. [var. *glanduliferum* (A.A. Heller) Jepson] ❀ DRN,DRY: 1–3,**7**,10. May–Jun

HALORAGACEAE WATER-MILFOIL FAMILY

Elizabeth McClintock

Ann, per, shrub, gen monoecious, gen aquatic. **LVS** cauline, opposite, alternate, or whorled; submersed blades with pinnate, thread-like divisions; aerial lvs simple, entire to divided. **INFL**: panicle, raceme, or spike; fls 1 or

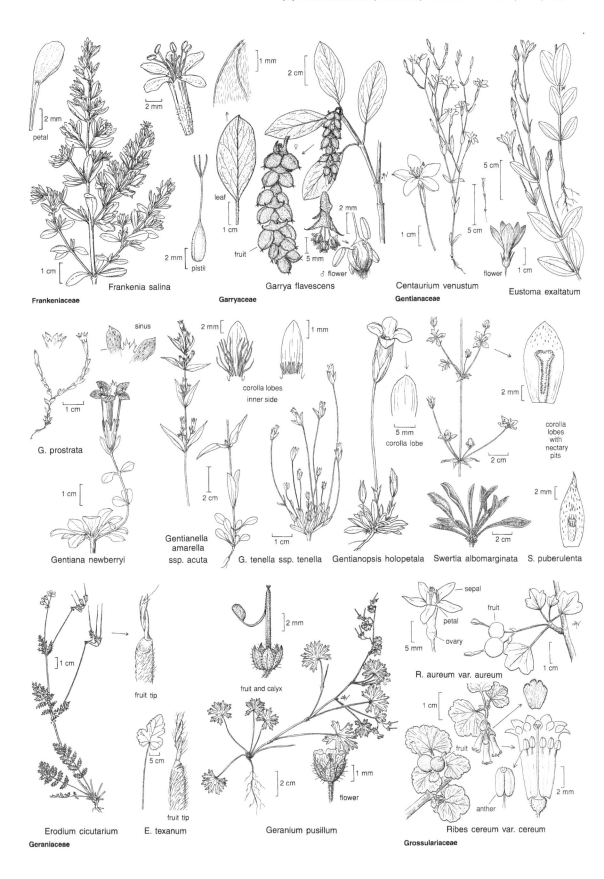

2 mm
petal

2 mm

1 cm

Frankenia salina

Frankeniaceae

2 mm
pistil

1 mm

2 cm

leaf
1 cm

fruit

2 mm

5 mm
♂ flower

♀

Garrya flavescens

Garryaceae

2 cm

5 cm

1 cm

Centaurium venustum

Gentianaceae

5 cm

flower
1 cm

Eustoma exaltatum

sinus

1 cm

G. prostrata

1 cm

Gentiana newberryi

2 mm

1 mm

corolla lobes
inner side

2 cm

**Gentianella
amarella
ssp. acuta**

1 cm

G. tenella ssp. tenella

5 mm
corolla lobe

Gentianopsis holopetala

2 cm

2 cm

Swertia albomarginata

2 mm

corolla
lobes
with
nectary
pits

2 mm

S. puberulenta

1 cm

fruit tip

5 cm

fruit tip

Erodium cicutarium

Geraniaceae

E. texanum

2 mm

fruit and calyx

2 cm

1 mm

flower

Geranium pusillum

sepal

petal

ovary

5 mm

fruit

1 cm

R. aureum var. aureum

1 cm

1 cm

fruit

anther

2 mm

Ribes cereum var. cereum

Grossulariaceae

clustered, short-pedicelled to ± sessile. **FL** gen unisexual (bisexual in *Haloragis*), small, biradial; calyx tube short, fused to ovary, lobes 2–4; petals gen 2–4; stamens 4 or 8, filaments gen short; ovary inferior, chambers 1–4, styles 2–4, separate, stigmas gen plumose. **FR** fleshy or nut-like, dehiscent or not. **SEEDS** gen 1 per chamber. 6–8 genera, ± 100 spp.: esp s hemisphere, some cult.

MYRIOPHYLLUM WATER-MILFOIL

Per from rhizomes, sometimes with overwintering bulblets, gen monoecious. **STS** simple or branched, gen open, gen green. **LVS**: submersed lvs whorled, 3–6 per node, pinnate divisions thread-like; emergent lvs opposite, lf- or bract-like, entire to pinnately divided. **INFL** spike-like, terminal, or fls clustered, axillary, gen emergent. **FL** unisexual; lower pistillate; middle sometimes bisexual; upper staminate; calyx lobes 4; petals gen 4, ephemeral on staminate fls, minute or 0 on pistillate fls; stamens gen 8; ovary 4-chambered, stigma plumose. **FR**: segments 4, nut-like. **SEED** 1 per chamber. ± 40 spp.: worldwide. (Greek: many lvs, from lf divisions) [Aiken & McNeill 1980 J Linn Soc Bot 80:213–222]

M. sibiricum V. Komarov (p. 347) Monoecious; bulblets sometimes present. **ST** > 1 m, whitish when dry. **LVS**: submersed lvs 1–3 cm, midrib and divisions linear, divisions < 15 mm, < 26 per lf. **INFL**: spike, 3–8 cm, emergent; lvs 1–3 mm, < fls, bract-like, oblanceolate to ovate, entire to coarsely toothed. Ponds, ditches, streams, lakes; < 2600 m. NCo, KR, CaR, n SN, SnJV, CCo, SnFrB, **GB**, **s DMoj (Mojave River)**; to B.C., e US, Eurasia. [*M. spicatum* L. ssp. *exalbescens* (Fern.) Hultén misapplied] ✹ TRY. Jun–Sep

HIPPOCASTANACEAE BUCKEYE FAMILY

William J. Stone

Shrub or tree. **LVS** opposite, gen 4-ranked, palmately compound. **INFL**: panicle or raceme, terminal; fls many. **FLS** showy, ± bilateral, some staminate; sepals 5, free or fused into tube, lobes unequal; petals 4–5, clawed, unequal; stamens 5–8, filaments long, slender; ovary chambers 3, ovules gen 2 per chamber. **FR**: capsule, spheric or slightly 3-lobed, leathery, roughly spiny to shiny. **SEEDS** large, shiny. 3 genera, 18 spp.: n hemisphere. [Hardin 1957 Brittonia 9:145–170, 173–194]. Recently treated in an expanded Sapindaceae [Gadek et al 1996 Amer J Bot 83:802–811; Angiosperm Phylogeny Group 1998 Ann Missouri Bot Gard 85: 531–553].

AESCULUS BUCKEYE, HORSE CHESTNUT

Shrub or tree, 4–30 m, < 15 m diam. **LVS** deciduous. **INFL**: pedicels jointed; staminate fls, if any, at top of infl; seed-producing fls gen at base. **FL** ill-smelling; sepals fused into tube; style of seed-producing fls long, thick, of sterile fls short. ± 15 spp.: n hemisphere; some cult. (Latin: name of some oak)

A. californica (Spach) Nutt. (p. 347) CALIFORNIA BUCKEYE Large shrub or tree, 4–12 m, broad, rounded. **LF**: lflets 5–7, 6–17 cm, oblong-lanceolate, finely serrate, acute to acuminate; petiole 1–12 cm. **INFL** panicle-like, erect, 1–2 dm, finely hairy; pedicel 3–10 mm. **FL**: calyx 5–8 mm, 2-lobed; petals 12–18 mm, white to pale rose; stamens 5–7, 18–30 mm, exserted, anthers orange. **FR** gen 1 (sometimes 2–9), 5–8 cm diam, borne at infl tip. **SEED** gen 1, 2–5 cm, glossy brown. *n*=20. Dry slopes, canyons, borders of streams; < 1700 m. c&s NW, n&c CW, s CaR, SNF, Teh, **sw DMoj**, scattered in GV near foothills. All parts TOXIC. Native Americans used ground seed as fish poison; nectar and pollen toxic to honeybees. ✹ 4–6,**7,14–17,19–24**;IRR:3,**8,9**,10,**18**;CVS. Gen deciduous Jun–Feb. May–Jun

HYDROPHYLLACEAE WATERLEAF FAMILY

Richard R. Halse, except as specified; Robert W. Patterson, Family Editor

Ann, per, shrub, gen hairy, gen taprooted. **ST** prostrate to erect. **LVS** simple to pinnately compound, basal or cauline, alternate or opposite; stipules 0. **INFL**: cyme (gen raceme-like and coiled) or fls solitary. **FL** bisexual, gen radial; calyx lobes gen 5, gen fused at base, gen persistent, enlarging in fr; corolla gen deciduous, rotate to cylindric, lobes gen 5, appendages in pairs on tube between filaments or 0; stamens gen 5, epipetalous, filament base sometimes appendaged, appendages scale-like; ovary gen superior, chamber 1, placentas 2, parietal, enlarged into chamber, sometimes meeting so ovary appears 2–5-chambered, styles 1–2, stigmas gen head-like. **FR**: capsule, gen loculicidal; valves gen 2. 20 genera, 300 spp.: esp w US; some cult (*Emmenanthe, Nemophila, Phacelia*). Recently treated to be included in an expanded Boraginaceae (also including Lennoaceae) [Angiosperm Phylogeny Group 1998 Ann Missouri Bot Gard 85: 531–553; Olmstead et al 2000 Mol Phylog Evol 16: 96-112].

1. Shrub; lvs evergreen . . . . **ERIODICTYON**
1' Ann, bien, or per; lvs deciduous
  2. Calyx lobes strongly unequal, outer 3 wide, cordate, enlarged and veiny in fr, inner 2 linear . . . **TRICARDIA**
  2' Calyx lobes equal to subequal, if unequal, not wide and cordate
    3. Styles 2
      4. Pl gen < 0.5 m; seeds surface variable, but not striate . . . . ²**NAMA**
      4' Pl 1–3 m; seeds striate . . . . **TURRICULA**

3' Style 1, gen 2-lobed
  5. Pl scapose; fls solitary on elongate peduncles . . . . **HESPEROCHIRON**
  5' Pl not scapose; st clearly present; fls 2–many per infl
    6. Per or bien, caudex gen well developed . . . . ²**PHACELIA**
    6' Ann, caudex 0
      7. Herbage sticky, scented; ovules borne on 2 sides of placenta . . . . **EUCRYPTA**
      7' Herbage gen not sticky, not strongly scented; ovules borne on 1 side of placenta
        8. Ovary chamber 1, placenta lining inner capsule wall; reflexed appendages gen present between
          calyx lobes (exc *Pholistoma membranaceum*)
          9. Ovary and fr glabrous to hairy, not bristly . . . . **NEMOPHILA**
          9' Ovary and fr bristly-hairy . . . . **PHOLISTOMA**
        8' Ovary chambers 2, placentas enlarged into chamber, meeting in middle; reflexed appendages
          between calyx lobes 0
          10. Stamens gen unequal, attached to corolla at different levels . . . . ²**NAMA**
          10' Stamens gen equal, attached ± at same level near corolla base
            11. Fls pendent; corolla yellow to cream, persistent, becoming paper-like . . . . **EMMENANTHE**
            11' Fls spreading to erect; corolla blue, purple, or white, gen deciduous (sometimes persistent
              and yellow) . . . . ²**PHACELIA**

EMMENANTHE

1 sp. (Greek: abiding fl, from persistent corolla)

E. penduliflora Benth. (p. 347) WHISPERING BELLS Ann, glandular, sticky, odorous. **ST** erect, simple to many-branched, 5–85 cm. **LVS** simple, basal, cauline, alternate; lower short-petioled; upper sessile, gen clasping, 1–12 cm, gen < 3 cm wide, toothed to deeply pinnately lobed. **INFL** terminal; pedicels 5–15 mm in fl, 12–25 mm in fr, thread-like, recurved. **FL**: calyx lobes 4–11 mm, 1–4 mm wide, lanceolate, glandular; corolla persistent, withering, enclosing fr, 6–15 mm, bell-shaped, white, yellow, or pink, hairy, glandular; stamens incl; ovary chambers 2, style incl, lobes 2, 1–4 mm. **FR** 7–10 mm, 2–4 mm wide, glandular. **SEEDS** 6–15, flat, oval, brown; surface honeycombed. *n*=18. Dry, open slopes, common after burns, disturbances; < 2200 m. NCoRH, NCoRI, c&s SN, SnJV, CW, SW, SNE, **D**; to NV, UT, AZ. 2 vars in CA.

var. *penduliflora* **FL**: corolla yellow to cream. Chaparral to creosote-bush scrub, rocky, sandy, decomposed granite, serpentine soils; < 2200 m. Range of sp. ❀ DRN,DRY:**7–10**,11–13,**14–16, 18–23**&SUN:6,17,**24**. Apr–Jul

ERIODICTYON YERBA SANTA

Shrub. **ST** erect; bark shredding. **LVS** simple, cauline, alternate, leathery; upper surface glabrous, shiny, sticky, or tomentose; lower surface tomentose. **INFL** gen open, terminal. **FL**: corolla funnel- to bell-shaped, white, lavender, or purple, hairy outside; stamens incl, filaments gen hairy; ovary chambers 2, styles 2, gen hairy. **FR** 1–3 mm wide; valves 4. **SEEDS** striate, dark brown or black. 9 spp.: sw US, Mex. (Greek: woolly net, from undersurface of some lvs) [Hannan 1988 Amer J Bot 75:579–588]

1. Lf linear, 2–11 mm wide, gen entire (or teeth few and well separated) . . . . *E. angustifolium*
1' Lf gen linear-lanceolate to narrowly oblong, 5–40 mm wide, gen toothed . . . . *E. trichocalyx*
  2. Twigs sparsely to densely hairy; lf densely white-tomentose below, lacking net-like pattern . . . . . var. *lanatum*
  2' Twigs gen glabrous; lf gen tomentose below, with net-like pattern . . . . var. *trichocalyx*

E. angustifolium Nutt. (p. 347) NARROW-LEAVED YERBA SANTA **ST** < 2 m; twigs glabrous, sticky to sparsely hairy. **LF** 2–10 cm, 2–11 mm wide, linear to lance-linear, sessile to short-petioled, entire to coarsely toothed, glabrous or sticky to sparsely hairy above, hairy between veins forming net-like pattern below; margin rolled under. **FL**: calyx lobes 1–4 mm, glabrous or sparsely hairy, ciliate; corolla 3–6 mm, white, densely hairy; styles 2–3 mm. **SEEDS** 1–8. *n*=14. UNCOMMON. Washes, slopes; 1500–1900 m. **e DMtns (New York, Grapevine Mtns)**; to NV, UT, AZ, Baja CA. Jun–Jul

E. trichocalyx A.A. Heller **ST** < 2 m; twigs glabrous to hairy. **LF** 3–14 cm, 0.5–4 cm wide, linear-lanceolate to narrowly oblong, short-petioled, entire to toothed, glabrous, sticky to sparsely hairy above, sparsely to densely tomentose below; margin rolled under. **FL**: calyx lobes 1–5 mm, glabrous to hairy, ciliate; corolla 4–13 mm, white to lavender, gen densely hairy; styles 1–6 mm. **SEEDS** 4–8. *n*=14. Slopes, mesas, ravines, grasslands, chaparral; 100–2800 m. SCo, SnGb, SnBr, PR, **D** (w edge); Baja CA.

var. *lanatum* (Brand) Jepson **ST**: twigs sparsely to densely hairy. **LF** gen densely white-tomentose below, lacking net-like pattern. Habitats of sp.; 300–1300 m. PR, **w edge DSon**; Baja CA. [*E. l.* (Brand) Abrams] ❀ TRY. Apr–Jun

var. *trichocalyx* (p. 347) **ST**: twig gen glabrous, sticky. **LF** gen tomentose below, with net-like pattern. Habitats and range of sp. ❀ DRN,SUN,DRY or IRR:7,8,9,11,**14**,15–17,**18–24**;INV,STBL. May–Aug

EUCRYPTA

Ann, glandular, sticky, scented. **ST** erect, much-branched. **LVS** simple, 1–3-pinnately toothed to dissected; lower cauline lvs opposite, petioled; upper lvs alternate, becoming smaller, sessile, clasping; petioles gen narrowly winged, ciliate. **INFL** terminal or axillary; pedicels thread-like, elongate in fr. **FL**: calyx < half-fused, bell-shaped, glandular, lobes oblong to spoon-shaped, ciliate; corolla bell-shaped, gen > calyx, with V-shaped transverse fold between each pair of filaments below throat; stamens incl, equal, equally attached; ovary chamber 1 (or appearing 5 from complex, enlarged placenta), ovules borne on both sides of placenta, style 1, stigmas 2. **FR** ovoid to spheric, bristly. **SEEDS** 5–15. 2 spp.: sw US. (Greek: well hidden, from seeds) [Constance 1938 Lloydia 1:143–152]

1. Calyx lobes spreading below fr; seeds of 2 kinds; lower lvs 2–3-pinnate
. . . . *E. chrysanthemifolia* var. *bipinnatifida*
1' Calyx lobes erect, enclosing fr; seeds alike; lower lvs 1-pinnate . . . . *E. micrantha*

E. chrysanthemifolia (Benth.) E. Greene **ST** erect to openly spreading, < 9 dm. **LVS**: lower 2–10 cm, 1–5 cm wide, petioles < 1/2 blade, widened, clasping, blade oblong to widely ovate, pinnate to deeply pinnately lobed, lobes 7–13, deeply 1–2-pinnately lobed, teeth obtuse; upper lvs smaller, narrower, less lobed, bases clasping. **INFL**: fls 4–15 per branch; pedicels gen recurved in fr. **FL**: calyx 2–4 mm; corolla 2–6 mm, lobes hairy on back; style < 3 mm. **FR** 2–4 mm wide, < spreading calyx. **SEEDS** 6–8, dark brown, of 2 kinds; some elliptic or round, disk-like, smooth; others oblong-ovoid, wrinkled. Canyons, chaparral, disturbed areas, slopes; 0–2300 m. s SNF, Teh, SnJV, CW, SW, **SNE**, **D**; NV, AZ, Baja CA. 2 vars in CA.

var. *bipinnatifida* (Torrey) Constance **ST** openly spreading. **LVS**: lower 2–7 cm, 1–4 cm wide, lobes 7–9. **INFL**: fls 4–8 per branch. **FL**: corolla 2–3 mm, 2–3 mm wide, = calyx, white or bluish; style < 1 mm. *n*=10,20. Cliffs, rocky slopes, washes, crevices; 30–2300 m. s SNF, Teh, SnBr, e PR, **SNE**, **D**; NV, AZ, Baja CA. Mar–May

E. micrantha (Torrey) A.A. Heller (p. 347) **ST** weak, < 3 dm, gen with stalked glands. **LVS**: lower 1–5 cm, < 2 cm wide, petiole short, widened to clasping base, blade oblong or ovate, deeply pinnately 7–9-lobed, lobes oblong or oblanceolate, straight or sickle-shaped, entire or few-toothed; upper lvs greatly reduced, lobed, toothed or entire. **INFL**: fls 4–12 per branch; pedicels gen erect in fr. **FL**: calyx 2–5 mm, gen black-glandular; corolla 2–4 mm, white or blue-purple, tube yellow; style 1–2 mm. **FR** 2–3 mm wide, < calyx. **SEEDS** 7–15, oblong, becoming incurved, worm-like, black or dark-brown, wrinkled. *n*=6,12. Canyons, hillsides, rocky crevices, washes, slopes; 60–2500 m. SnJt, **SNE**, **D**; to NV, UT, TX, Mex. Mar–Jun

HESPEROCHIRON

Per, scapose; root caudex-like. **LVS** simple, in basal rosette, spreading or ascending; blade tapered to petiole, gen entire, margins gen ciliate. **INFL**: fl solitary; peduncle erect or spreading, 1–10 cm, slender. **FL**: calyx lobes gen unequal, 2–9 mm, glabrous to hairy, ciliate; corolla tube gen densely hairy inside, throat gen yellow, lobes glabrous to hairy, white or bluish, gen tinged or marked with lavender or purple; stamens incl, gen unequal, filament base widened; ovary hairy, chamber 1, style 1, stigmas 2, 2–5 mm. **FR** 5–11 mm, ovoid, hairy. **SEEDS** many, ovoid, angular, reddish brown; surface honeycombed or pitted. 2 spp.: w US, n Mex. (Greek: evening or western centaur)

1. Corolla bell- or funnel-shaped; lvs gen hairy on both surfaces . . . . *H. californicus*
1' Corolla rotate; lvs gen glabrous at least on lower surface . . . . *H. pumilus*

H. californicus (Benth.) S. Watson (p. 347) **LVS** gen > 6, < 8 cm, < 3 cm wide, oblanceolate to elliptic or ovate; surfaces gen densely to sparsely spreading-hairy. **INFL**: fls gen > 5. **FL**: corolla < 3 cm, < 2 cm wide, bell- or funnel-shaped, lobes oblong, 3–10 mm. *n*=8. Wet meadows, flats, valleys; 1000–2900 m. KR, CaRH, SNH, Teh, WTR, SnBr, **GB**; to WA, MT, WY, UT, Baja CA. ❀ IRR or WET,SUN,DRN:1–3,7,18;DFCLT. May–Jul

H. pumilus (Griseb.) Porter (p. 347) **LVS** gen 2–10, 1–7 cm, < 2 cm wide, linear-oblong to oblanceolate or oblong; upper surface glabrous or hairy. **INFL**: fls gen 1–8. **FL**: corolla 5–15 mm, 7–30 mm wide, rotate, lobes 3–11 mm, rounded. *n*=8. Wet meadows, slopes, flats; 400–3000 m. KR, NCoRI, CaRH, SNH, Teh, WTR, **GB**, **n DMoj (Death Valley)**; to WA, MT, UT, AZ. ❀ DRN,DRYorIRR,SUN or part SHD: 1–3,7,15–17,**18**;DFCLT. Apr–Jul

NAMA PURPLE MAT

John D. Bacon

Gen ann, hairy. **LVS** cauline, gen alternate, simple; margin entire, wavy, crenate, or rolled under. **INFL**: clusters (gen terminal, leafy) or fls solitary or paired in axils, not coiled. **FL**: corolla salverform to bell-shaped; stamens gen attached to corolla at different levels, gen unequal, portion fused to corolla gen narrowly winged; scales at filament base 0. **FR** gen loculicidal, ovoid to elliptic. **SEEDS** gen many, small, reddish brown, brown, black or yellow. ± 55 spp.: sw US, trop Am, Hawaii. (Greek: a stream) [Hitchcock 1933 Amer J Bot 20:415–430, 518–534] Key to species by Robert W. Patterson.

1. Per; infls terminal heads . . . . *N. rothrockii*
1' Ann; infls not terminal heads
  2. Corolla < 3 mm, bowl- or bell-shaped with a distinct basal tube ± 0.5 mm; seeds < or = 4 . . . . *N. californicum*
  2' Corolla > 2 mm, if bowl- or bell-shaped then without a distinct basal tube; seeds > 4
    3. Styles fused > 1/2 length . . . . *N. densum*
     4. Styles > 3 mm . . . . *N. aretoies* var. *multiflorum*
     4' Styles < 2.5 mm . . . . *N. densum*
      5. Pl grayish, gen densely spreading-rough-hairy; style 0.3–1 mm . . . . var. *densum*
      5' Pl greenish, bristly-strigose; style gen 1–2.5 mm . . . . var. *parviflorum*
    3' Styles free
     6. Corolla limb < 4 mm wide
      7. Sepals glandular . . . . *N. dichotomum* var. *dichotomum*

7' Sepals not glandular
    8. Lf sessile; sepals gen pale greenish, not canescent; corolla slightly funnel-shaped or salverform;
      seeds gen oblong, surface gen cross-ridged . . . . *N. depressum*
    8' Lf petioled; sepals white- or gray-canescent at least in basal 1/3–1/2; corolla ± cylindric;
      seeds irregular, surface slightly net-like . . . . *N. pusillum*
  6' Corolla limb > 4 mm wide
    9. St ascending to erect; seeds fusiform, yellow to orange . . . . *N. hispidum* var. *spathulatum*
    9' St prostrate; seeds spheric to ovoid, brown to black . . . . *N. demissum*
      10. Lf petioled; pl gray-green . . . . var. *covillei*
      10' Lf sessile; pl gen green . . . . var. *demissum*

N. aretioides (Hook. & Arn.) Brand Hairs gen dense, coarse, appressed to spreading, gen swollen at base. **ST** prostrate, 3–12 cm, repeatedly forked. **LF** ± sessile, gen sickle-shaped. **INFL**: fls sessile. **FL**: sepals narrowly linear to lanceolate; corolla salverform; fused parts of filaments unwinged; style 1, 2-lobed. **FR** 2–4 mm. **SEED** gen compressed, irregularly elliptic-ovoid, brown to black, smooth to minutely cross-ridged, with prominent depressions. Dry, sandy or loamy areas; 1200–2300 m. KR, NCoRO, CaR, n SNH, **GB**; to ID, NV. 2 vars in CA.

  var. ***multiflorum*** (A.A. Heller) Jepson (p. 347) **LF** 7–30 mm, lanceolate to spoon-shaped. **FL**: corolla 9–18 mm, gen pink or purple; stamens 3–8 mm, attached 2–5 mm above corolla base; style 3–7 mm. **SEED** 0.6–0.8 mm, brown to black, minutely cross-ridged. 2*n*=14. Dry, sandy areas; 1200–2300 m. KR, NCoRO, CaR, n SNH, **GB**; NV. ☙ TRY;DFCLT. May–Jun

N. californicum (A. Gray) J. Bacon (p. 347) Gen puberulent to finely strigose; hair bases gen swollen. **ST** prostrate, forked, 3–10 cm. **LF** gen sessile, 5–14 mm, 1–4 mm wide, long tapered, oblanceolate, spoon-shaped, or elliptic. **INFL**: fls ± sessile. **FL**: sepals 2–5 mm, linear-lanceolate, silky-hairy; corolla 1–3 mm, bell-shaped, white to pale pink, tube ± 0.5 mm, limb 1–2 mm diam, lobes 0.4–0.8 mm, 0.5–0.8 mm wide; stamens 0.8–1.3 mm, attached < or = 0.5 mm above corolla base, free filament abruptly expanded above attachment; styles 0.5–1 mm. **FR** 2–2.5 mm. **SEEDS** < or = 4, 0.8–1 mm, ovoid, minutely cross-ridged, with prominent depressions. 2*n*=14. Dry, sandy areas; 900–2400 m. NCoRI, Teh, GV, e SnFrB, SCoRI, SW, **DMoj**; w NV. [*Lemmonia c.* A. Gray] Apr–Jun

N. demissum A. Gray Hairs gen dense, fine to coarse, gen mealy-glandular, bases swollen. **ST** prostrate, forked, 3–20 cm. **INFL**: pedicels 0–5 mm. **FL**: corolla funnel-shaped to salverform; stamens attached 2–4 mm above corolla base. **SEED** ± 0.5 mm, slightly cross-ridged, with depressions. Sandy or gravelly flats; < 1600 m. SNH, SW, SNE, **D**; to UT, AZ, Mex.

  var. ***covillei*** Brand Hairs 0.5–1.2 mm, soft, shaggy. **LF**: blade 5–13 mm, elliptic or spoon-shaped to diamond-shaped; petiole 1.5–5 mm, winged. **FL**: sepals 4–8 mm, linear to oblanceolate; corolla 8–12 mm, blue-violet to pink, limb 8–9 mm diam, lobes 2–4 mm, 3–4 mm wide; stamens 3–6 mm; styles 2–4 mm. **FR** 2–5 mm. **SEED** ovoid to spheric, black. Dry, sandy flats and slopes; < 500 m. **n DMoj (Death Valley region)**.

  var. ***demissum*** (p. 349, pl. 60) Hairs < 1 mm, finely strigose. **LF** 1–4 cm, linear to spoon-shaped, long-tapered; petiole 0. **FL**: sepals 3–8 mm, linear-lanceolate; corolla 7–14 mm, blue-purple to rose-pink, limb 5–9 mm diam, lobes 2–3 mm, 2–4 mm wide; stamens 4–7 mm; styles 3–6 mm. **FR** 3–4 mm. **SEED** elliptic-ovoid, brown to black. 2*n*=14. Sandy or gravelly flats, slopes; 500–1600 m. s SNH, SW, **SNE**, **D**; NV, AZ, Mex. [var. *deserti* Brand] Mar–May

N. densum Lemmon Hairs gen dense, stiff, gen appressed; bases gen swollen. **ST** prostrate, forked. **LF** gen long-tapered, lanceolate to oblanceolate; petiole 0. **INFL**: pedicels 0–1 mm. **FL**: sepals linear-lanceolate; corolla cylindric to funnel-shaped, white to pale purple; fused part of filament unwinged; style 2-lobed. **FR** 2–4 mm. **SEED** 0.6–0.9 mm, gen elliptic to ovoid, angled on underside, smooth to cross-ridged, brown to black. Sandy or gravelly flats, slopes; 1200–3400 m. NW, CaR, n&c SNH, **GB**; to WA, NV; also in WY, UT, Colorado.

  var. ***densum*** Pl grayish canescent; hairs gen dense, spreading, rough. **FL**: corolla 2–5 mm, gen narrowed below limb; stamens attached ± 1 mm above corolla base; style 0.3–1 mm. **SEED** gen smooth, gen shiny. 2*n*=14. Habitats & elevations of sp. NW, CaR, n&c SNH, **GB**; to WA, NV. May–Jul

  var. ***parviflorum*** (Greenman) C. Hitchc. (p. 349) Pl green; hairs gen sparse, bristly-strigose. **FL**: corolla 3–8 mm, gen not narrowed below limb; stamens attached 1–2 mm above corolla base; style (0.6)1–2.5 mm. **SEED** minutely cross-ridged, with prominent depressions. 2*n*=14,28. Habitats of sp.; 1200–1800 m. Range of sp. (exc SNH)

N. depressum A. Gray Pl puberulent; hairs pointed, appressed to ascending. **ST** prostrate, forked, 2–10 cm; lower 1/2 of st gen lfless. **LF** 2–16 mm, long-tapered, oblanceolate to spoon-shaped; petiole 0. **INFL**: fls pediceled. **FL**: sepals 3–5 mm, linear to slightly spoon-shaped; corolla 3–6 mm, funnel-shaped-salverform, pink or white, limb 2–4 mm diam, lobes 0.6–1.2 mm, 0.8–1.2 mm wide; stamens 2–3 mm, attached 1–2 mm above corolla base; styles 1–2 mm. **FR** 2–4 mm. **SEED** 0.4–0.7 mm, oblong to ovoid, brown, slightly cross-ridged, gen with prominent depressions. 2*n*=14. Dry, sandy or gravelly flats, slopes; 600–1600 m. s SNH, SNE, **DMoj**; NV. Apr–May

N. dichotomum (Ruiz & Pav.) Choisy var. ***dichotomum*** Pl short-glandular-strigose and short non-glandular-hairy; nonglandular hairs sometimes spreading. **ST** gen erect, simple or forked, 5–20 cm. **LF** 6–20 mm, long-tapered, linear to spoon-shaped; petiole 0. **INFL**: pedicels 1–2 mm, slender. **FL**: sepals 4–10 mm, linear to spoon-shaped; corolla 3–8 mm, cylindric to bell-shaped, white to bluish, limb 2–4 mm diam, lobes 1–2 mm, 1–1.5 mm wide; stamens 2–4 mm, attached 0.5–1.2 mm above corolla base, filament wider just above attachment; styles 1–3 mm. **FR** 2–6 mm. **SEED** 0.5–0.7 mm, irregularly oblong to ovoid, brown, cross-ridged, with prominent depressions. 2*n*=28. Uncommon. Granite or limestone slopes, ridgetops; 1900–2200 m. **e DMtns (New York Mtns)**; to TX, Mex.

N. hispidum A. Gray var. ***spathulatum*** (Torrey) C. Hitchc. Pl gen mealy-glandular; hairs gen dense, fine to bristly-strigose, base gen swollen. **ST** ascending to erect, simple or freely branched, 7–30 cm. **LF** 1–5 cm, linear to spoon-shaped, ± rolled under; base long-tapered; petiole 0. **INFL**: pedicels 0–4 mm. **FL**: sepals 4–10 mm, ± linear to oblanceolate; corolla 10–15 mm, salverform to narrowly bell-shaped, blue to purple-lavender, limb 5–14 mm diam, lobes 1–4 mm, 2–5 mm wide; stamens 2–6 mm, attached 1–4 mm from corolla base; styles 1–4 mm. **FR** 3–6 mm. **SEED** 0.3–0.7 mm, fusiform, yellow to orange; surface net-like. 2*n*=14. Dry, sandy or gravelly flats, slopes; < 600 m. SW, **D**; to TX, Mex. [var. *revolutum* Jepson] Mar–May

N. pusillum A. Gray Pl gen densely short-spreading-hairy; hairs slender, base gen swollen. **ST** prostrate, forked, 2–6 cm. **LF**: blade gen 2–11 mm, lanceolate to ovate, abruptly narrowed to winged petiole 1–6 mm. **INFL**: pedicels < or = 4 mm. **FL**: sepals 3–6 mm, linear to slightly spoon-shaped; corolla 3–5 mm, cylindric to slightly funnel-shaped, white to pale pink, limb < or = 2 mm diam, lobes < or = 0.5 mm, 0.5–1 mm wide; stamens 1–3 mm, attached 1–2 mm above corolla base; styles 1–2 mm. **FR** 2–4 mm. **SEED** ± 0.4 mm, irregular, brown-black; surface slightly net-like, with prominent depressions. 2*n*=14. Sandy to rocky flats; < 1700 m. SNE, **DMoj**; NV, AZ. Mar–May

N. rothrockii A. Gray (p. 349) Per, rhizomed, forming colonies, gen densely short-glandular-hairy and long-bristly-nonglandular-hairy; longer hairs gen swollen at base. **ST** erect, simple to few-branched, 20–30 cm, sticky. **LF** 2–6 cm, lanceolate to elliptic, margin crenate-dentate. **INFL**: heads, terminal; pedicels ± 1 mm. **FL**: sepals 8–15 mm, linear-lanceolate; corolla 13–18 mm, ± funnel-shaped, pink, purple, or pale blue, limb 8– 10 mm diam, lobes 4–6 mm, 3–6 mm wide; stamens 6–11 mm, attached 4–7 mm above corolla base; styles 7–11 mm. **FR** 3–6 mm. **SEED** 1–2 mm, ovoid, red-brown to brown, minutely pitted. $2n=34$. Sandy alluvial flats, gravelly granitic slopes, meadows; 1700–4000 m. c&s SNH, SnBr, **W&I**; NV; nw AZ. ❀ TRY;DFCLT. Jul–Aug

NEMOPHILA

Ann. **ST** simple to openly branched, prostrate to erect, fleshy, brittle, angled or winged, glabrous to gen bristly. **LVS** simple, cauline, opposite or alternate; petiole gen bristly-ciliate; blade pinnately toothed or lobed, gen bristly, upper gen reduced. **INFL**: fls solitary in axils or opposite lvs; pedicels longer in fr, recurved. **FL**: calyx bell-shaped to rotate, sinuses gen with spreading or reflexed appendages; corolla bell-shaped to rotate, white, blue, or purple, sometimes spotted or marked; stamens incl; ovary chamber 1, style 1, gen lobed 1/3–1/2. **FR** gen 2–7 mm wide, spheric to ovoid, hairy, gen enclosed by calyx. **SEEDS** ovoid, smooth, wrinkled or pitted, with a conic, colorless appendage at 1 end. 11 spp.: se US, w N.Am. (Greek: woodland-loving) [Constance 1941 Univ CA Publ Bot 19:341–398]

N. menziesii Hook. & Arn. BABY BLUE-EYES **LVS** opposite; lower 1–5 cm; petiole = blade, blade linear-oblong to ovate, 5–13-lobed, lobes obtuse, entire or 1–3-toothed; upper lvs nearly sessile, blade entire or less lobed (sometimes only toothed). **INFL**: pedicels 20–60 mm in fl, < 70 mm in fr. **FL**: calyx lobes 4–8 mm, appendages 1–4 mm; corolla 5–20 mm, 6–40 mm wide, bowl-shaped to rotate, bright blue with white center to white (and gen blue-veined and black-dotted); filaments = or > corolla tube, anthers 2–3 mm; style 2–7 mm. **FR** 5–15 mm wide. **SEEDS** 4–20, brown to black, wrinkled and tubercled. Meadows, fields, woodlands, roadsides, grasslands, canyons; 15–1900 m. CA-FP, **SNE, DMoj**; OR, Baja CA. Highly variable; vars. intergrade. 3 vars in CA.

1. Lower lvs 5–7-lobed; filaments > corolla tube;
    seeds 4–10 . . . . var. *integrifolia*
1' Lower lvs 6–13-lobed; filaments = corolla tube;
    seeds 10–20 . . . . var. *menziesii*

var. *integrifolia* Parish **LVS**: lower 5–7-lobed; upper entire to shallowly few-toothed, diamond-shaped, oblong or oblanceolate, sessile. **FL**: corolla 5–10 mm, 6–15 mm wide, blue, black-dotted at center or blue-veined; filaments > corolla tube. **SEEDS** 4–10. $n=9$. Grasslands, canyons, woodlands, burns, slopes; 100–1900 m. CCo, SCoR, SW, **SNE, DMoj**; Baja CA. ❀ TRY. Apr–May

var. *menziesii* (p. 349) **LVS**: lower 6–13-lobed; upper fewer, short-petioled, toothed or more narrowly lobed. **FL**: corolla 5–20 mm, 10–40 mm wide, bright blue with white center or blue-veined, center gen dotted; filaments = corolla tube. **SEEDS** 10–20. $n=9$. Meadows, grasslands, chaparral, woodlands, slopes, desert washes; 15–1600 m. CA-FP, **DMoj**. ❀ SUN:**4–6,15–17,24**&SHD:1–3,**7–9**,10–12,**14,18–23**;CVS. Feb–Jun

PHACELIA

Dieter H. Wilken, Richard R. Halse, & Robert W. Patterson

Ann, per, gen glandular-hairy, tap-rooted or from ± thick caudex. **LVS** gen alternate, simple to 2-pinnately compound, gen ± reduced upward. **INFL**: cyme, gen dense, coiled, gen 1-sided; pedicels gen short. **FL**: corolla rotate to bell-shaped, white to purple, tube base with scales free or fused to filaments; stamens gen attached at same level, equal; ovary chamber 1 (or 2 below middle), placentas parietal, enlarging and meeting in fr, style 2-lobed, gen hairy below lobes. **FR** oblong to spheric. **SEEDS** 1–many, oblong to spheric, gen brownish; back gen pitted or cross-furrowed. ± 175 spp.: Am; some cult for orn. (Greek: cluster, from the dense infl) [Halse 1981 Madroño 28:121–132; Heckard 1960 Univ Calif Publ Botany 32:1–126; Lee 1988 Syst Bot 13:16–20] Bristly hairs may cause severe dermatitis. CA pers often hybridize, difficult to separate. Bien and per spp. by Richard Halse. Key to species by Dieter Wilken.

Keys to Groups

1. Bien or per from ± woody taproot or branched caudex . . . . **Group 1**
1' Ann from slender taproot (some key in both the following groups)
  2. Most cauline lf blades deeply lobed to compound (some sinuses reaching midrib) . . . . **Group 2**
  2' Most cauline lf blades entire (lowest sometimes few-lobed at base, sinuses not reaching midrib) . . . . . **Group 3**

Group 1: bien or per

1. Ovules and seeds > 20 per fr . . . . *P. perityloides*
  2. Corolla 12–15 mm; pedicels in fr 10–30 mm, reflexed . . . . var. *jaegeri*
  2' Corolla 10–12 mm; pedicels in fr 5–15 mm, ascending to spreading . . . . var. *perityloides*
1' Ovules 4; seeds 1–4
  3. Lvs gen simple (or basal lobed to compound); lflets entire (if any)

4. Basal lvs dissected, segments 5–9 . . . . *P. heterophylla* ssp. *virgata*
4' Basal lvs simple to compound (lflets 3–15) . . . . *P. hastata*
  5. St decumbent to ascending, 5–20(30) cm; hairs mostly spreading; calyx lobes gen glandular . . . ssp. *compacta*
  5' St ascending to ± erect, 20–50 cm; hairs ± appressed; calyx lobes gen not glandular . . . . ssp. *hastata*
3' Lvs compound; lflets toothed or lobed . . . . *P. ramosissima*
  6. Lvs and sts below infl glandular
    7. St below infl with mostly long, coarse, stiff, bulb-based hairs (some hairs soft, spreading) . . . . . var. *latifolia*
    7' St below infl with long and short, soft, spreading hairs . . . . var. *ramosissima*
  6' Lvs and sts below infl not glandular
    8. St below infl glabrous or sparsely short-hairy . . . . var. *eremophila*
    8' St below infl densely puberulent . . . . var. *subglabra*

Group 2: Ann; most cauline lvs deeply lobed to compound

1. Corolla persistent, ± enclosing fr, gen white to yellow (or lobes purplish)
  2. Corolla tube puberulent inside; filaments puberulent . . . . *P. monoensis*
  2' Corolla tube gen glabrous inside; filaments glabrous
    3. Sepals and petals gen 5; corolla 2–3 mm . . . . *P. inyoensis*
    3' Sepals and petals gen 4; corolla < 2 mm . . . . *P. tetramera*
1' Corolla gen deciduous, variously colored
  4. Ovules 3–many per chamber; seeds gen 7+ per fr
    5. Corolla 2–5 mm, narrowly bell-shaped; stamens and styles 1–3 mm
      6. Upper st, infl axis, pedicels with dark, stalked glands . . . . *P. glandulifera*
      6' Upper st, infl axis, pedicels hairy, glandular or not, glands not esp dark
        7. Calyx lobes oblanceolate to spoon-shaped, 6–10 mm in fr, short-stiff-hairy, ± glandular . . . . *P. affinis*
        7' Calyx lobes gen linear to oblong, 3–6 mm in fr, minutely ciliate, not glandular . . . . *P. ivesiana*
    5' Corolla 6–20 mm, funnel- to widely bell-shaped; styles 2–7 mm; stamens 3–8 mm
      8. Corolla tube white to bluish; pedicels 3–8 mm; seeds pitted . . . . *P. douglasii*
      8' Corolla tube yellow; pedicels 1–4 mm; seeds cross-furrowed
        9. Lower lvs 1–2-compound; filaments puberulent . . . . *P. bicolor*
        9' Lower lvs deeply lobed to 1-compound; filaments glabrous . . . . *P. fremontii*
  4' Ovules 1–2 per chamber; seeds 1–4 per fr (2–8 in *P. austromontana*)
    10. Some lvs (esp upper cauline) entire or toothed; seeds 2–8 per fr . . . . *P. austromontana*
    10' Lvs lobed to compound, segments gen toothed or lobed; seeds 1–4 per fr
      11. Stamens and style branches incl; stamens 3–5 mm; style 3–5 mm . . . . *P. cryptantha*
      11' Some stamens (and gen style branches) exserted; stamens 4–15 mm; style 5–15 mm
        12. Seed cylindric or angled, inner surface not clearly ridged or grooved
          13. Calyx lobes 4–8 mm in fr, ± straight, ± enclosing fr; corolla ± blue
            14. Fr ± spheric, 2–3 mm; corolla deciduous; pedicels 1–3 mm in fr . . . . *P. distans*
            14' Fr ± ovoid, 3–4 mm; corolla ± persistent in fr; pedicels < 1 mm . . . . *P. tanacetifolia*
          13' Calyx lobes 7–12 mm in fr, ± curved, not enclosing fr; corolla lavender to violet
            15. Fr ± spheric, hairs few, stiff, gen > 1 mm, bulb-based . . . . *P. cicutaria* var. *hispida*
            15' Fr ± ovoid, ± puberulent (stiff hairs few, gen < 1 mm, not bulb-based) . . . . *P. vallis-mortae*
        12' Seed inner surface with central ridge separating 2 longitudinal grooves
          16. Sepals 1–1.5 mm > fr; pedicels ± thread-like, densely long-hairy, not glandular . . . . *P. pedicellata*
          16' Sepals ± = fr; pedicels stout, short-glandular-hairy
            17. Corolla limb white, scales ovate; calyx lobes ± 1 mm wide; fr ovoid, sparsely stiff-hairy . . . *P. amabilis*
            17' Corolla limb mostly blue to purple, scales linear; calyx lobes ± 1.5 mm wide; fr ovoid to spheric,
               ± puberulent, glandular . . . . *P. crenulata*
              18. Corolla tube white, limb lavender to blue; calyx lobes 3–4.5 mm in fr . . . . var. *minutiflora*
              18' Corolla blue to purple ± throughout; calyx lobes 3.5–5.5 mm in fr
                19. St glandular above middle; corolla 5–10 mm; stamens and style exserted 9+ mm . . . . . var. *ambigua*
                19' St glandular throughout; corolla 4.5–7 mm; stamens and style exserted 5.5–11 mm . . var. *crenulata*

Group 3: Ann; most cauline lvs entire to weakly lobed

1. Corolla ± persistent, ± enclosing fr
  2. Corolla tube puberulent inside; filaments puberulent . . . . *P. monoensis*
  2' Corolla tube gen glabrous inside; filaments gen glabrous
    3. Calyx lobes 5–7 mm in fr; seeds pitted; limestone slopes, woodlands . . . . [2]*P. saxicola*
    3' Calyx lobes 3.5–5 in fr; seeds cross-furrowed; alkaline flats, meadows
      4. Sepals and petals gen 5; corolla 2–3 mm . . . . *P. inyoensis*
      4' Sepals and petals gen 4; corolla < 2 mm . . . . *P. tetramera*
1' Corolla deciduous

5. Ovules 1–2(4) per chamber; seeds 1–4 per fr (2–8 in *P. austromontana*)
  6. Most cauline lvs crenate to lobed (lobes again gen ± toothed); inner seed surface with central ridge
    separating 2 longitudinal grooves
    7. Corolla 3–7 mm; stamens 2–5 mm; style 2–5 mm; stamens and style exserted 0–2 mm
      8. Corolla 5–7 mm, widely bell-shaped; stamens 3–5 mm; style 3–5 mm .... *P. anelsonii*
      8' Corolla 3–5 mm, bell-shaped; stamens 2–3 mm; style 2–3 mm .... *P. coerulea*
    7' Corolla 6–10 mm; stamens 10–15 mm; style 12–15 mm; stamens and style gen exserted ...... *P. crenulata*
      9. Corolla tube white, limb lavender to blue; calyx lobes 2–3 mm, 3–4.5 mm in fr .... var. *minutiflora*
      9' Corolla tube blue to ± purple throughout; calyx lobes 2.5–5.5 mm, 3.5–5.5 mm in fr
        10. St glandular above middle; corolla 5–10 mm; stamens exserted 9+ mm................ var. *ambigua*
        10' St glandular throughout; corolla 4.5–7 mm; stamens exserted 5.5–11 mm .... var. *crenulata*
  6. Most cauline lvs entire (or lobes entire, obtuse to acute); inner seed surface not ridged or grooved
    11. Upper st, infl axis, and pedicels not clearly glandular
      12. Calyx lobes densely white-hairy; filaments hairy .... *P. humilis* var. *humilis*
      12' Calyx lobes puberulent and stiffly ciliate, not white-hairy; filaments glabrous........ [2]*P. novenmillensis*
    11' Upper st, infl axis, and pedicels glandular
      13. Calyx lobes in fr 4–6 mm, short-stiff-hairy and ciliate, hairs < 1 mm .... *P. austromontana*
      13' Calyx lobes in fr 8–10 mm, puberulent and stiffly ciliate, some hairs > 1 mm .... [2]*P. novenmillensis*
5' Ovules 3–many per chamber; seeds gen 5+ per fr
  14. Lf blades gen longer than wide
    15. Ovules or seeds 20–80
      16. Corolla 3–4 mm, tube white, limb blue to violet; stamens 1–2 mm .... [2]*P. saxicola*
      16' Corolla 4–6 mm, tube yellow, limb white to lavender; stamens 2–4 mm
        17. Lvs cauline, gradually reduced upward; blade slightly lobed to toothed; fr glandular-puberulent
        .... *P. lemmonii*
        17' Lvs ± basal, ± abruptly reduced upwards; blade entire to barely toothed; fr short-hairy...... [2]*P. parishii*
    15' Ovules or seeds < 20
      18. Lower pedicels > upper (esp in fr), gen S-shaped or recurved in fr .... *P. curvipes*
      18' Pedicels gen equal, gen straight in fr
        19. Corolla 3–5 mm, scales lanceolate; stamens 2–3 mm, glabrous; seeds not cross-furrowed .. *P. barnebyana*
        19' Corolla 5–11 mm, scales 0; stamens 3–6 mm, short-hairy; seeds cross-furrowed .... *P. gymnoclada*
  14' Lf blades, esp lower, gen as wide as long
    20. Corolla 7–40 mm, scales 0 or fused to filament base and tooth-like, filament base slightly
      swollen; stamens (8)10–45 mm; style 8–45 mm
      21. Corolla limb violet to deep purple; filament teeth puberulent
        22. Corolla bell-shaped, 10–40 mm, uniformly purple; style 15–40 mm .... *P. minor*
        22' Corolla rotate to widely bell-shaped, 10–20 mm, throat with white spots below lobes;
        style 10–20 mm .... *P. parryi*
      21. Corolla limb white to bright blue; filament teeth glabrous
        23. Corolla 7–12 mm, white to blue; seeds 10–20 per fr .... *P. longipes*
        23' Corolla 10–40 mm, uniformly bright blue or throat white-spotted; seeds 40–80 per fr
          24. Calyx lobes 3–4 mm, 5–8 mm in fr; stamens 12–20 mm; corolla 10–18 mm, tube white,
          limb blue, throat white-spotted below sinuses .... *P. nashiana*
          24' Calyx lobes 6–8 mm, 9–11 mm in fr; stamens 20–45 mm; corolla 15–40 mm, ± bright blue
          throughout .... *P. campanularia*
            25. Corolla ± bell-shaped, 15–30 mm; petiole 1–10 cm; w DSon .... ssp. *campanularia*
            25' Corolla ± funnel-shaped, 25–40 mm; petiole 5–20 cm; DMoj, n DSon .... ssp. *vasiformis*
    20' Corolla 3–15 mm, scales linear to ovate, free from filaments; filament base not swollen,
      gen not toothed; stamens 2–10 mm; style 1–12 mm
      26. Seeds with 4–8 cross-furrows; calyx lobes 1.5–3 mm, 2–4 mm in fr; fr gen spheric
        27. Corolla 8–12 mm; stamens 5–6 mm; style 5–6 mm................................ *P. calthifolia*
        27' Corolla 4–7 mm; stamens 2–4 mm; style 2–3 mm
          28. Corolla cream; infl axis among lvs, finely glandular .... *P. neglecta*
          28' Corolla violet to purple; infl axis gen > lvs, ± dark-glandular .... *P. pachyphylla*
      26' Seeds not cross-furrowed; calyx lobes 2–5 mm, 4–9 mm in fr; fr ovoid to ± oblong
        29. Corolla 3–6 mm; style 1–2 mm
          30. Lf blade length gen > width, base tapered, margin entire to barely toothed; clay soils,
          dry lake margins .... [2]*P. parishii*
          30' Lf blade length < to = width, base truncate to slightly lobed, margin clearly toothed; gravelly
          to rocky soils, slopes and canyons
            31. Calyx lobes 6–8 mm in fr, some 1–2 mm wide; lf blade crenate to irregularly toothed ... *P. peirsoniana*
            31' Calyx lobes 4–6 mm in fr, gen < 1 mm wide; lf blade dentate to weakly lobed, lobes obtuse
            .... *P. rotundifolia*
        29' Corolla 6–15 mm; style 3–5 mm

32. Stem base ± woolly to spreading-hairy; corolla 10–15 mm . . . . *P. perityloides*
    33. Corolla 12–15 mm; pedicels in fr 10–30 mm, reflexed . . . . var. *jaegeri*
    33' Corolla 10–12 mm; pedicels in fr 5–15 mm, ascending to spreading . . . . var. *perityloides*
32' Stem base puberulent to finely hairy; corolla 6–12 mm
    34. Corolla narrowly bell-shaped, limb gen 3–5 mm wide; upper infl ± short-glandular-hairy
        (some hairs > 0.5 mm); gravelly or rocky slopes . . . . *P. mustelina*
    34' Corolla widely bell-shaped, limb gen > 5 mm wide; upper infl glandular-puberulent;
        clay flats . . . . *P. pulchella* var. *gooddingii*

P. affinis A. Gray (p. 349) Ann 6–30 cm. **ST** erect, simple to branched at base, short-hairy; hairs spreading to reflexed, glandular-puberulent above. **LF** 8–70 mm; blade > petiole, narrowly oblong, deeply lobed to compound, lobes entire, lflets ± lobed. **FL**: pedicel 1–2 mm, < 10 mm in fr; calyx lobes 4–5 mm, 6–10 mm in fr, oblanceolate, short-stiff-hairy, glandular; corolla 3–5 mm, narrowly bell-shaped, tube yellow, limb white to lavender, deciduous, scales ± 0; stamens 2–3 mm, glabrous; style 1–2 mm, gen glabrous. **FR** 4–6 mm, oblong to elliptic, puberulent below, sparsely short-glandular-hairy above. **SEEDS** 15–30, ± 1 mm; cross-furrows 5–8. *n*=11,12. Open, sandy or gravelly areas; < 3400 m. s SCoRI (Caliente Mtn), TR, PR, **W&I, D**; to UT, NM, Baja CA. Pls with spreading st hairs and glandular lower lvs have been called var. *patens* J. Howell. Mar–Jun

P. amabilis Constance (p. 349) SALINE VALLEY PHACELIA Ann 10–60 cm. **ST** decumbent to erect, few-branched, sparsely soft-hairy, glandular-puberulent. **LVS** 20–120 mm; blades > petioles, lower gen oblong, deeply lobed, lobes crenate to toothed, upper ± ovate, coarsely toothed. **FL**: pedicel 1–2 mm; calyx lobes 3–4 mm, 4–5 mm in fr, oblong, short-hairy, glandular; corolla 6–8 mm, bell-shaped, white, deciduous, scales ovate; stamens 9–15 mm, glabrous; style 12–15 mm, puberulent below. **FR** 3–4 mm, ovoid, sparsely stiff-hairy. **SEEDS** 2–4, 3–4 mm; back finely pitted; inner surface with central ridge separating 2 longitudinal grooves. **PRESUMED EXTINCT**. Gravelly soils, canyons; 500–700 m. s **W&I (Inyo Mtns)**. Apr–May

P. anelsonii J.F. Macbr. AVEN NELSON'S PHACELIA Ann 10–50 cm. **ST** erect, gen simple, short-glandular-hairy; gland-tipped hairs dark. **LVS** 15–80 mm; blades > petioles, lower oblong to oblanceolate, deeply lobed, lobes gen crenate, upper ± ovate, toothed. **FL** gen sessile; calyx lobes 3–4 mm, 3.5–5.5 mm in fr, oblanceolate, sparsely short-hairy; corolla 5–7 mm, widely bell-shaped, white, pale blue, or lavender, deciduous, scales ± linear; stamens 3–5 mm, glabrous; style 3–5 mm, puberulent. **FR** 2–3.5 mm, ovoid, glandular-puberulent. **SEEDS** 2–4, 2.5–3.5 mm; back pitted; inner surface with central ridge separating 2 longitudinal grooves. RARE in CA. Sandy or gravelly soils, creosote-bush scrub, woodland; 1200–1500 m. e **DMtns (New York Mtns)**; to sw UT. Apr–May

P. austromontana J. Howell (p. 349) Ann 5–27 cm. **ST** decumbent to ascending, many-branched, puberulent, sparsely stiff-hairy, ± glandular. **LF** 10–30 mm; blade = or > petiole, narrowly elliptic to oblanceolate, entire or few-lobed. **FL**: pedicel 1–2 mm; calyx lobes 3–5 mm, 4–6 mm in fr, unequal, linear to narrowly oblanceolate, short-hairy, glandular; corolla 3–6 mm, bell-shaped, pale blue to lavender, deciduous, scales lanceolate; stamens 2–4 mm, glabrous to papillate; style 3–4 mm, puberulent. **FR** 3–5 mm, ovoid, beaked, puberulent. **SEEDS** 2–8, 1–2 mm, pitted. *n*=9. Open, sandy to rocky areas; 1800–3000 m. c&s SN, TR, SnJt, **W&I, DMtns**; to sw UT. ❀ TRY. May–Jun

P. barnebyana J. Howell Ann 5–30 cm. **ST** erect, simple to branched at base, glandular-puberulent. **LVS** 10–20 mm, little reduced upward; blade < petiole, gen ovate, entire to obscurely toothed. **FL**: pedicel 3–5 mm; calyx lobes 2–4 mm, 4–6 mm in fr, narrowly oblanceolate, short-hairy, glandular; corolla 3–5 mm, narrowly bell-shaped, tube white to yellow, limb pale violet, deciduous, scales lanceolate; stamens 2–3 mm, glabrous; style 1–2 mm, puberulent. **FR** 3.5–5.5 mm, elliptic, puberulent. **SEEDS** 15–20, ± 1 mm, pitted. Limestone scree; 1600–2700 m. **W&I, ne DMtns (Clark Mtns)**; w NV. Closely related to *P. lemmonii*. May–Jul

P. bicolor S. Watson var. *bicolor* (p. 349) Ann 6–40 cm. **ST** decumbent to erect, gen branched at base, short-hairy, glandular-puberulent. **LVS** 20–60 mm, little reduced upward, 1–2-compound; lflets toothed to irregularly lobed. **FL**: pedicel 1–4 mm; calyx lobes 3–5 mm, 4–6 mm in fr, ± linear, short-hairy, glandular; corolla 8–18 mm, funnel- to bell-shaped, tube yellow, limb lavender to purple, deciduous, scales linear; stamens 5–8 mm, puberulent; style 3–7 mm, puberulent. **FR** 3–4 mm, ovoid; tip short-stiff-hairy. **SEEDS** 12–20, 1–1.5 mm; cross-furrows 7–11. *n*=13. Sandy or alkaline soils, shrubland; 700–3400 m. SNH (e slope), **GB, DMoj**; OR, NV. Other var. in e OR. ❀ TRY. May–Jun

P. calthifolia Brand (pl. 61) Ann 10–30 cm, fleshy, ± brittle. **ST** spreading to erect, few-branched, short-glandular-hairy; gland-tipped hairs black. **LVS** 10–30 mm, ± basal; blade ± = petiole, ± round, entire to crenate, base lobed. **FL**: pedicel 1–3 mm; calyx lobes 2–3 mm, 3–4 mm in fr, ± linear, short-hairy, glandular; corolla 8–12 mm, funnel- to bell-shaped, violet to purple, deciduous, scales linear; stamens 5–6 mm, puberulent; style 5–6 mm. **FR** 4–5 mm, ± spheric, sparsely puberulent; tip sparsely dark-glandular. **SEEDS** 30–50, 1–1.5 mm; cross-furrows shallow, gen 5–8. *n*=11,12. Sandy soils, gen in creosote-bush scrub; < 1000 m. **n DMoj**; w NV. ❀ TRY. Mar–May

P. campanularia A. Gray Ann 18–55 cm. **ST** erect, gen simple, short-hairy, ± glandular. **LF**: petiole 20–40 mm; blade < or > petiole, ovate to ± round, clearly toothed. **FL**: pedicel 7–20 mm; calyx lobes 6–8 mm, 9–11 mm in fr, oblong, hairy, glandular; corolla 15–40 mm, funnel- to bell-shaped, bright blue, deciduous, scales fused to filament base, short, truncate to toothed; stamens 20–45 mm, puberulent; style 20–45 mm, short-hairy. **FR** 8–15 mm, ovoid, beaked, short-glandular-hairy. **SEEDS** 40–80, 1–1.5 mm, pitted. Open, sandy or gravelly areas; < 1600 m. **DMoj, n&w DSon**. Vars. intergrade in s DMtns.

    ssp. *campanularia* **LF**: petiole 10–100 mm; blade 20–70 mm. **FL**: corolla 15–30 mm, ± bell-shaped; style parted 1/3–1/2 length. *n*=11. Habitats of sp. **w DSon**. ❀ DRN,SUN:**7,14–24**& IRR:**8,9**,10,11,**12**,13. Feb–May

    ssp. *vasiformis* G. Gillett (p. 349) **LF**: petiole 50–200 mm; blade 30–100 mm. **FL**: corolla 25–40 mm, ± funnel-shaped; style parted 1/4–1/2 length. *n*=11. Habitats of sp. **DMoj, n DSon**. ❀ TRY. Mar–May

P. cicutaria E. Greene (p. 349) Ann 18–60 cm. **ST** ascending to erect, simple to branched, stiff-hairy, glandular. **LF** 20–150 mm; blade gen > petiole, ovate to ± oblong, deeply lobed to compound, segments toothed. **FL**: pedicel 0.5–3 mm; calyx lobes 6–8 mm, 9–12 mm in fr, ± linear, long-hairy, glandular; corolla 8–12 mm, bell-shaped, lavender to yellowish, deciduous, scales elliptic; stamens 8–12 mm, glabrous; style 8–12 mm, short-hairy. **FR** 3–4 mm, ± spheric, sparsely short-hairy. **SEEDS** 2–4, 2–3 mm, pitted. Gravelly or rocky slopes, chaparral, oak/pine woodland, grassland; < 1400 m. SNF, Teh, SCoR, SW, **w D**; Baja CA. 3 vars in CA.

    var. *hispida* (A. Gray) J. Howell **INFL**: axis long-stiff-hairy; lower fls ± well separated; upper fls dense. **FL**: calyx lobes grayish, hairs stiff-hairy; corolla lavender. *n*=11. Rocky slopes, oak/pine woodland, grassland; < 1400 m. SCoR, SW, **w D**; Baja CA. ❀ TRY. Mar–Jun

P. coerulea E. Greene   Ann 12–40 cm. **ST** ascending to erect, simple to branched at base, puberulent to sparsely short-glandular-hairy. **LF** 15–70 mm; blade gen > petiole, oblong to oblanceolate, gen lobed toward base, crenate above middle. **FL**: pedicel 0.5–1 mm; calyx lobes 2–3 mm, 3–4 mm in fr, subequal, ± oblong, short-hairy; corolla 3–5 mm, bell-shaped, pale blue to pale purple, deciduous, scales narrow; stamens 2–3 mm, glabrous; style 2–3 mm, puberulent. **FR** 2.5–3.5 mm, spheric, sparsely puberulent. **SEEDS** gen 4, 2–3 mm; back pitted; inner surface with central ridge separating 2 longitudinal grooves. *n*=11. Open, sandy to rocky areas, gen in creosote-bush scrub; 1400–2000 m. **e DMoj**; to UT, TX, n Mex. Mar–May

P. crenulata Torrey   Ann 7–60 cm. **ST** gen erect, 0–few-branched at base, sparsely to densely short-hairy, glandular. **LVS** 20–80(120) mm, abruptly reduced upwards; blade > petiole, oblong to elliptic, crenate to deeply lobed. **FL**: pedicel 0.5–2 mm; calyx lobes 2–5 mm, 3–5.5 mm in fr, oblong, puberulent to short-hairy, glandular; corolla 4–10 mm, bell-shaped, blue to purple, throat ± white, deciduous, scales ± narrow; stamens 10–15 mm, glabrous; style 12–15 mm, glandular-puberulent. **FR** 2.5–4 mm, ovoid to spheric, puberulent. **SEEDS** gen 4, 2–3.5 mm; back finely pitted; inner surface with central ridge separating 2 longitudinal grooves. Sandy to gravelly washes, slopes; < 2200 m. se MP, **SNE, D**; to UT, AZ, nw Mex.

var. ***ambigua*** (M.E. Jones) J.F. Macbr.   **ST** gen densely stiff-hairy, glandular above. **FL**: calyx lobes 3–5 mm, 4–5 mm in fr; corolla 5–10 mm, blue to purple; stamens and style exserted 9+ mm. **FR** 3.3–3.5 mm, spheric. *n*=11. Habitats of sp.; < 1600 m. **D**; to sw UT, AZ, nw Mex. ❀ TRY. Mar–May

var. ***crenulata***   **ST** gen glandular- and stiff-hairy through-out. **FL**: calyx lobes 2.5–4 mm, 3.5–5.5 mm in fr; corolla 4.5–7 mm, blue to violet; stamens and style exserted 5.5–11 mm. **FR** 2.5–4 mm, ovoid. *n*=11. Habitats of sp.; < 2200 m. se MP, **SNE, e DMoj**; to UT. [var. *funerea* J. Voss] Mar–May

var. ***minutiflora*** (J. Voss) Jepson   **ST** gen densely stiff-hairy, glandular above. **FL**: calyx lobes 2–3 mm, 3–4.5 mm in fr; corolla ± 4 mm, tube white, limb lavender to blue; stamens and style exserted < 2 mm. **FR** ± 3 mm, ± spheric. *n*=11. Habitats of sp.: < 500 m. **DSon**; AZ, Baja CA. [*P. minutiflora* J. Voss] Mar–Apr

P. cryptantha E. Greene   Ann 16–50 cm. **ST** gen erect, 0–few-branched, puberulent to sparsely stiff-hairy, glandular. **LVS** 20–150 mm; blade > petiole, elliptic to ovate, lower gen compound, upper lobed to compound, segments toothed. **FL**: pedicel 2–4 mm; calyx lobes 4–7 mm, 8–10 mm in fr, ± linear, ciliate; corolla 4–7 mm, narrowly bell-shaped, bluish to lavender, deciduous, scales linear; stamens 3–5 mm, glabrous; style 3–5 mm, short-hairy. **FR** 4–5 mm, ± spheric, short-stiff-hairy. **SEEDS** gen 4, 1.5–3 mm, pitted. *n*=11. Gravelly or rocky slopes, canyons; < 1900 m. s SN, **SNE, n DMoj**; to UT. Mar–May

P. curvipes S. Watson   Ann 4–15 cm. **ST** spreading to ascending, many-branched, short-hairy, glandular above. **LVS** 10–40 mm; lower ± = upper; blade gen > petiole, elliptic to oblanceolate, entire. **FL**: pedicels 1–6 mm, lower < 25 mm in fr; calyx lobes 3–6 mm, 7–10 mm in fr, unequal, narrowly oblong, puberulent, stiffly ciliate; corolla 4–8 mm, rotate to bell-shaped, tube and throat white, lobes blue to violet, deciduous, scales linear; stamens 2–6 mm, short-hairy; style 2–4 mm, short-hairy. **FR** 4–5 mm, ovoid, puberulent and appressed-short-hairy. **SEEDS** 6–16, 1–1.5 mm, pitted. *n*=11. Sandy to rocky slopes, chaparral, oak/pine woodland, coniferous forest; 500–2700 m. s SN, Teh, TR, **SNE, DMoj**; to sw UT, nw AZ. Apr–Jun

P. distans Benth. (p. 349)   Ann 15–80 cm. **ST** decumbent to erect, simple to branched at base, puberulent, sparsely stiff-hairy,

finely glandular above. **LF** 20–100(150) mm; blade >> petiole, 1–2-compound, segments obtuse to toothed. **FL** gen subsessile; calyx lobes 3–4 mm, 4–5 mm in fr, gen unequal, oblong to oblanceolate, densely hairy, glandular; corolla 6–9 mm, funnel- to bell-shaped, whitish to blue, deciduous, scales ovate; stamens 8–12 mm, glabrous; style 7–12 mm, glabrous. **FR** 2–3 mm, ± spheric, puberulent. **SEEDS** 2–4, 2–3 mm, pitted. *n*=11. Common. Clay to rocky soils, slopes; < 2100 m. s NCoR, s ScV (Sutter Buttes), SnJV, CW, SW, **SNE, D**; s NV, n Mex. [*P. cinerea* Eastw.] ❀ SUN,DRN:**7,14–24**&IRR:3, **8,9**,10,**11–13**. Mar–May

P. douglasii (Benth.) Torrey   Ann 6–40 cm. **ST** spreading to erect, branched at base, short-hairy, glandular. **LF** 5–80 mm, ± basal; blade ± = petiole, oblanceolate to ovate, deeply lobed to compound, segments rounded to obtuse. **FL**: pedicel 3–8 mm; calyx lobes 2–5 mm, 4–7 mm in fr, oblanceolate, short-hairy; corolla 6–12 mm, widely bell-shaped, light blue to purplish, deciduous, scales lanceolate to ovate; stamens 3–7 mm, glabrous to puberulent; style 2–7 mm, puberulent, glandular or not. **FR** 5–7 mm, ovoid, short-hairy. **SEEDS** 8–20, 0.5–1 mm, pitted. *n*=11. Open, gen sandy areas; < 1700 m. s SNF, Teh, s ScV, SnJV, CW, WTR, **w DMoj**. ❀ TRY. Mar–May

P. fremontii Torrey (p. 349)   Ann 7–30 cm. **ST** decumbent to erect, branched at base, puberulent, gen glandular above. **LVS** 15–50 mm, ± basal; blade > petiole, oblong to oblanceolate, deeply lobed to compound, segments gen rounded. **FL**: pedicel 1–3 mm; calyx lobes 3–5 mm, 4–6 mm in fr, subequal, linear to oblanceolate, short-hairy; corolla 7–15(20) mm, funnel- to bell-shaped, tube and lower throat yellow, upper limb blue to violet, deciduous, scales linear; stamens 3–8 mm, glabrous; style 3–5 mm, short-hairy. **FR** 4–6 mm, ovoid, puberulent below, short-stiff-hairy above. **SEEDS** 10–18, 1–1.5 mm; cross-furrows 6–9. *n*=13. Sandy or gravelly soils, shrubland, grassland; < 2300 m. s SN, SnJV, e SnFrB, SCoRI, TR, **SNE, DMoj**; to sw UT, AZ. ❀ TRY. Mar–Jun

P. glandulifera Piper (p. 349)   Ann 5–25 cm. **ST** erect, 0–few-branched, short-hairy; gland-tipped hairs dark. **LVS** 10–50 mm; lower ± = upper; blade gen > petiole, oblong to oblanceolate, deeply lobed to compound, segments obtuse or toothed. **FL**: pedicel < 5 mm; calyx lobes 2–3 mm, 3–5 mm in fr, unequal, linear to oblanceolate, short-hairy; corolla 2–5 mm, narrowly bell-shaped, tube yellow, limb bluish, deciduous, scales 0; stamens 1–2.5 mm, glabrous; style 1–3 mm. **FR** 4–5.5 mm, elliptic, glabrous below, short-hairy above. **SEEDS** 7–14, 1–1.5 mm; cross-furrows 8–10. *n*=13. Gen sandy soils, shrubland, juniper woodland; 800–2500 m. CaR, **GB**; to e WA, WY. May–Jun

P. gymnoclada S. Watson   Ann 5–20 cm. **ST** spreading to ascending, branched at base, short-hairy, glandular-puberulent. **LVS** 10–40 mm, ± basal or reduced upward; blade < to > petiole, oblanceolate to ovate, wavy to obtusely lobed. **FL**: pedicel 1–7 mm; calyx lobes 3–5 mm, 5–8 mm in fr, ± linear, short-hairy; corolla 5–11 mm, funnel- to bell-shaped, tube yellow, limb lavender to purple, deciduous, scales 0; stamens 3–6 mm, short-hairy; style 2–5 mm, short-hairy. **FR** 3–4 mm, ± oblong, sparsely short-hairy. **SEEDS** 5–8, ovoid, 1–1.5 mm; cross-furrows 7–9. *n*=13. Clay to gravelly soils, gen in shrubland; 1700–2300 m. **n SNE (Mono Co.)**; OR, NV. ❀ TRY. May–Jun

P. hastata Lehm.   Per 5–50 cm. **ST** decumbent to ± erect, ± stiff-hairy, not glandular. **LVS** mostly basal; blade 15–120 mm, < to = petiole, lanceolate to widely elliptic, gen entire (or 2–4-lobed or compound with 3–5 lflets). **FL**: pedicels 0.5–3 mm; calyx lobes 3–7 mm, 5–9 mm in fr, linear to lanceolate; corolla 4–7 mm, urn- to bell-shaped, white to lavender, scales oblong; stamens 6–10 mm, exserted, glabrous to hairy; style 7–10 mm, exserted. **FR** 2–4 mm, ovoid, stiff-hairy. **SEEDS** 1–3, 1.5–2.5 mm, pitted in vertical rows. Flats, slopes, talus, shrubland, coniferous forest, alpine; 900–4000 m. KR, CaRH, SNH, SnBr, **GB**; to w Can, SD, Colorado. Ssp. intergrade.

ssp. ***compacta*** (Brand) Heckard (p. 349)   **ST** decumbent to

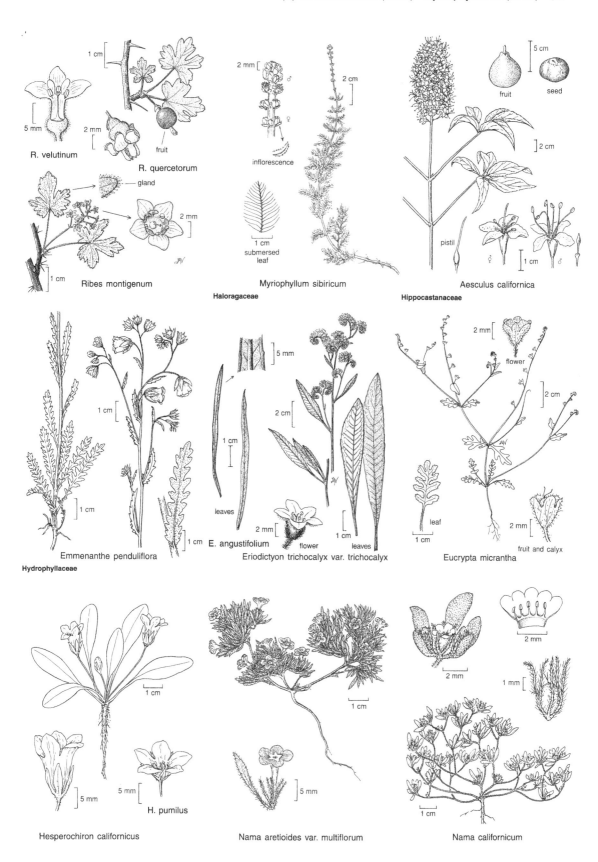

R. velitunum

R. quercetorum

fruit

gland

2 mm

1 cm

Ribes montigenum

inflorescence

submersed leaf

Myriophyllum sibiricum

Haloragaceae

fruit

seed

pistil

Aesculus californica

Hippocastanaceae

Emmenanthe penduliflora

Hydrophyllaceae

leaves

flower

E. angustifolium

Eriodictyon trichocalyx var. trichocalyx

flower

leaf

fruit and calyx

Eucrypta micrantha

H. pumilus

Hesperochiron californicus

Nama aretioides var. multiflorum

Nama californicum

ascending, 5–20(30) cm; hairs mostly stiff, spreading. **FL**: calyx lobes gen glandular. *n*=22,33. Sandy to rocky slopes, flats, talus, coniferous forest, alpine; 1800–4000 m. CaRH, SNH, Wrn, **SNE**; to WA, NV. [*P. frigida* E. Greene sspp. *frigida, dasyphylla* (J.F. Macbr.) Heckard] Jul–Sep

ssp. *hastata* (p. 349) **ST** ascending to ± erect, 20–50 cm; hairs ± appressed, some stiff, spreading. **FL**: calyx lobes gen not glandular. *n*=11,22. Sandy to rocky slopes, shrubland, coniferous forest; 900–2400 m. KR, CaRH, SNH, SnBr, **GB**; to w Can, SD, Colorado. [*P. oreopola* Heckard ssp. *simulans* Heckard] May–Jul

P. heterophylla Pursh ssp. *virgata* (E. Greene) Heckard (p. 349) Bien, weak per, 20–120 cm. **STS**: central erect, > lateral; lateral ascending, stiff-hairy, not glandular. **LVS** mostly basal; blade 50–150 mm, gen > petiole, lanceolate to ovate, basal dissected, segments 5–9, upper simple to dissected. **FL**: pedicels 0.5–1 mm; calyx lobes 3–6 mm, 6–10 mm in fr, oblong to lanceolate; corolla 4–7 mm, bell-shaped, white to lavender, scales oblong; stamens 8–10 mm, exserted, hairy; style 8–10 mm, exserted. **FR** 2.5–3 mm, ovoid, stiff-hairy. **SEEDS** 1–3, 1.5–2.5 mm, pitted in vertical rows. *n*=11,22. Slopes, flats, roadsides; 100–2900 m. NW, CaR, n&c SN, **GB**; to OR, ID, WY. ❀ SUN, DRN:1–3,6,**7,14–16**,17. May–Jul

P. humilis Torrey & A. Gray (p. 353) Ann 5–20 cm. **ST** gen erect, simple to branched at base, short-stiff-hairy, sparsely glandular. **LVS** 10–40 mm; lowest opposite; blade gen >> petiole, elliptic to ovate, entire. **FL** gen subsessile; calyx lobes 2–3 mm, 5–12 mm in fr, ± linear, densely white-hairy; corolla 4–7 mm, bell-shaped, lavender to violet, deciduous, scales ± deltate; stamens 4–8 mm, short-hairy; style 4–7 mm, glabrous to short-hairy. **FR** 2–5 mm, ovoid, short-hairy. **SEEDS** 1–4, 1.5–3 mm, finely pitted. Flats, meadows; 800–2300 m. SNH (e slope), Teh, **GB**; se OR, nw NV, also c WA. 2 vars in CA.

var. *humilis* **FL**: calyx lobes 5–8 mm in fr; stamens 4–6 mm, barely exserted. **SEEDS** 1.5–2.5 mm. *n*=11. Habitats of sp.; 1500–2300 m. SNH (e slope), **GB**; se OR, nw NV, also c WA. May–Jul

P. inyoensis (J.F. Macbr.) J. Howell (p. 353) INYO PHACELIA Ann 3–10 cm. **ST** decumbent to erect, branched at base, short-stiff-hairy, glandular-puberulent. **LVS** 5–20 mm, ± basal; blade > petiole, elliptic to obovate, entire to few-lobed. **FL**: pedicel 1–6 mm; calyx lobes 2–3 mm, 3.5–5 mm in fr, narrowly oblanceolate, short-hairy; corolla 2–3 mm, ± narrowly bell-shaped, pale yellow, persistent in fr, scales 0; stamens 1.5–3 mm, glabrous; style 1 mm, glabrous. **FR** 3–4 mm, oblong, puberulent. **SEEDS** 18–25, 0.5–1 mm; cross-furrows 5–8. *n*=12. UNCOMMON. Alkaline meadows; 1400–3200 m. **SNE**. May–Jul

P. ivesiana Torrey Ann 5–25 cm. **ST** spreading to ascending, many-branched at base, finely short-hairy to glandular-puberulent. **LVS** 10–60 mm; lower ± = upper; blade = to > petiole, deeply lobed to ± compound, segments oblong to ± deltate. **FL**: pedicel 1–5 mm; calyx lobes 2.5–4 mm, 3–6 mm in fr, ± linear, minutely ciliate; corolla 2–4 mm, narrowly bell-shaped, tube yellow, lobes white, deciduous, scales 0 or minute; stamens 1–2 mm, glabrous; style ± 1 mm, ± glabrous. **FR** 3–5 mm, ovoid, puberulent; tip short-stiff-hairy. **SEEDS** 10–15, 1–1.5 mm; cross-furrows 5–12. *n*=11. Open, sandy areas; < 1000 m. **D**; to WY, Colorado, AZ. Pls with ± deltate lf lobes, seeds having 5–7 cross-furrows, and *n*=23 have been called var. *pediculoides* J. Howell. Mar–Jun

P. lemmonii A. Gray Ann 7–20 cm. **ST** gen erect, simple to branched at base, glandular-puberulent. **LF** 10–40 mm; blade gen = or > petiole, ovate, > toothed to slightly obtusely lobed. **FL**: pedicel ± 1 mm; calyx lobes 2–4 mm, 5–7 mm in fr, subequal, oblong to oblanceolate, short-hairy, glandular; corolla 4–6 mm, narrowly bell-shaped, tube yellow, limb white to lavender, deciduous, scales linear or 0; stamens 2–3 mm, glabrous; style 2–3 mm, glandular-puberulent. **FR** 3–4 mm, ± oblong, glandular-puberulent. **SEEDS** 30–80, ± 0.5 mm. *n*=22,24. Sandy

washes, drying streambanks, slopes; 400–2300 m. ne PR, **SNE, DMoj, n DSon**; to UT. Mar–Jul

P. longipes A. Gray Ann 10–40 cm. **ST** decumbent to erect, simple to branched at base, sparsely stiff-hairy, short-glandular-hairy. **LVS** ± basal, 20–140 mm; blade gen < petiole, ovate to round, ± crenate to irregularly toothed. **FL**: pedicel 3–10(30) mm; calyx lobes 3–4 mm, 4–6 mm in fr, ± linear, short-hairy, glandular; corolla 7–12 mm, widely bell-shaped, white to blue, deciduous, scales fused to filament base; stamens 10–15 mm, puberulent; style 10–15 mm, puberulent. **FR** 5–8 mm, ovoid, short-glandular-hairy; tip stiff-hairy. **SEEDS** 10–20, 1–1.5 mm, pitted. *n*=11. Gravelly or rocky soils, chaparral, juniper woodland, coniferous forest; 400–1900 m. WTR, SnGb, **sw DMoj**. Apr–Jun

P. minor (Harvey) Thell (p. 353) Ann 20–60 cm. **ST** gen erect, 0–few-branched, short-glandular-hairy, sparsely stiff-hairy. **LF** 20–110 mm; blade < to = petiole, ovate to ± round, irregularly toothed. **FL**: pedicel 10–15(20) mm; calyx lobes 5–6 mm, 6–8 mm in fr, narrowly oblong, short-hairy, glandular; corolla 10–40 mm, bell-shaped, purple, deciduous, scales fused to filament base, elongate; stamens 15–35 mm, short-glandular-hairy; style 15–40 mm, short-hairy. **FR** 7–13 mm, ovoid, beaked, puberulent; tip short-stiff-hairy. **SEEDS** 30–80, ± 1 mm, pitted. *n*=11. Open areas, burns, slopes; < 1600 m. SCo, e WTR, SnGb, SnBr, PR, **w DSon**; Baja CA. ❀ TRY. Apr–Jun

P. monoensis R. Halse (p. 353) MONO COUNTY PHACELIA Ann 2–12 cm. **ST** spreading to ascending, branched throughout, short-hairy, ± glandular. **LF** 8–25 mm; blade gen = petiole, ± oblong to ovate, entire to lobed. **FL**: pedicel 1–2 mm; calyx lobes 2–4 mm, 4–6 mm in fr, narrowly oblanceolate, short-hairy; corolla 2–4 mm, gen narrowly bell-shaped, yellow, persistent in fr, scales 0; stamens 1.5–3 mm, puberulent; style < 1.5 mm, puberulent. **FR** 2.5–4 mm, ovoid, puberulent. **SEEDS** < 10, 1–1.7 mm; cross-furrows 8–11. *n*=12. RARE. Clay soils, alkaline meadows, shrubland; 1900–2900 m. **n SNE (Mono Co.)**; w NV. Jun

P. mustelina Cov. (p. 353) DEATH VALLEY ROUND-LEAVED PHACELIA Ann 6–30 cm. **ST** decumbent to ascending, many-branched, ± short-glandular. **LF** 10–40 mm; blade = or > petiole, ± round, irregularly toothed. **FL**: pedicel 1–3(5) mm; calyx lobes 2–4 mm, 5–6 mm in fr, narrowly oblanceolate, glandular-short-hairy; corolla 6–10 mm, narrowly bell-shaped, violet to purple, deciduous, scales lanceolate; stamens 2–4 mm, short-hairy; style 3–5 mm, sparsely short-hairy. **FR** 3–4 mm, ovoid, puberulent. **SEEDS** 20–60, ± 0.5 mm, finely pitted. *n*=12. RARE. Gravelly or rocky slopes, creosote-bush scrub, pinyon/juniper woodland; 1000–2100 m. **DMtns**; w NV. Mar–Jun

P. nashiana Jepson (p. 353) CHARLOTTE'S PHACELIA Ann 4–18 cm. **ST** ascending to erect, simple to branched at base, short-stiff-hairy; gland-tipped hairs black. **LVS** ± basal, 15–70 mm; blade gen < petiole, widely ovate to ± round, irregularly crenate to slightly lobed. **FL**: pedicel 5–10 mm; calyx lobes 3–4 mm, 5–8 mm in fr, oblong, short-hairy, glandular; corolla 10–18 mm, rotate to widely bell-shaped, tube white, throat blue with 5 ± white spots below sinuses, lobes bright blue, deciduous, scales fused to filament base, toothed; stamens 12–20 mm, short-hairy; style 10–20 mm. **FR** 7–14 mm, ovoid, beaked, short-glandular-hairy. **SEEDS** 40–80, ± 2 mm, pitted. *n*=11. RARE. Sandy to rocky, granitic slopes, gen in Joshua-tree or pinyon/juniper woodland; < 2200 m. s SNH, Teh (e slope), **w edge DMoj**. Apr–May

P. neglecta M.E. Jones (p. 353) Ann 3–20 cm. **ST** ascending to erect, 0–few-branched, ± short-glandular. **LVS** ± basal, 10–50 mm; blade = or > petiole, ± round, wavy to crenate. **FL**: pedicel 1–8 mm, < 12 mm in fr; calyx lobes 1.5–2 mm, 2–3.5 mm in fr, narrowly ovate, densely short-hairy, glandular; corolla 4–6 mm, cream-white, funnel- to bell-shaped, deciduous, scales linear; stamens 2–4 mm, puberulent; style 2–3 mm. **FR** 3–4 mm, spheric, puberulent. **SEEDS** > 60, ± 1 mm; cross-furrows 4–7. *n*=11. Clay or alkaline soils, flats, slopes; < 1000 m. **D**; s NV, AZ. Mar–May

P. novenmillensis Munz (p. 353) NINE-MILE CANYON PHACELIA

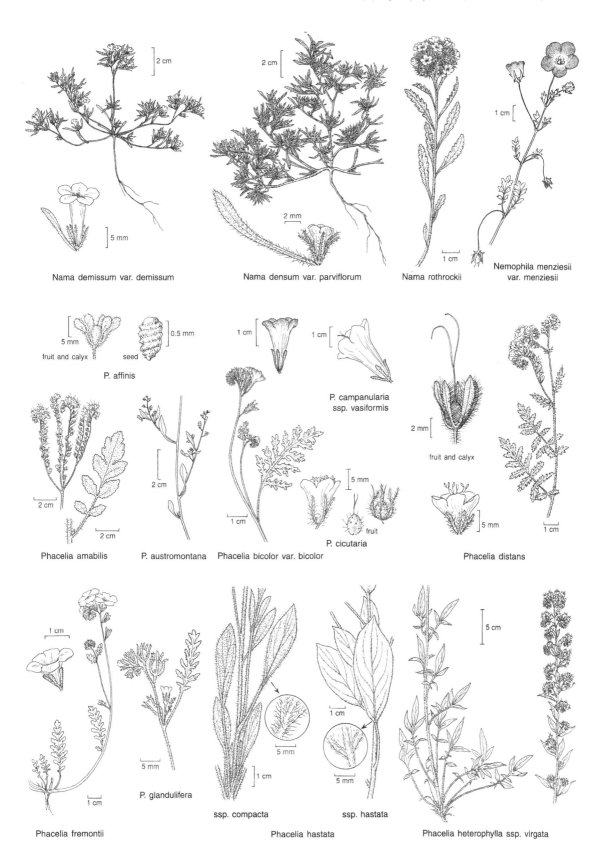

Nama demissum var. demissum

Nama densum var. parviflorum

Nama rothrockii

Nemophila menziesii var. menziesii

fruit and calyx   seed

P. affinis

P. campanularia ssp. vasiformis

fruit and calyx

Phacelia amabilis

P. austromontana

Phacelia bicolor var. bicolor

fruit

P. cicutaria

Phacelia distans

Phacelia fremontii

P. glandulifera

ssp. compacta

ssp. hastata

Phacelia hastata

Phacelia heterophylla ssp. virgata

Ann 5–10 cm. **ST** ascending to erect, 0–few-branched, short-soft-hairy, sparsely glandular-puberulent. **LF** 20–80(110) mm; blade gen < petiole, (ob)lanceolate to narrowly elliptic, entire; lower-most sometimes lobed to irregularly compound. **FL**: pedicel 1–3 mm; calyx lobes 2–4 mm, 8–10 mm in fr, subequal, linear, long-ciliate; corolla 3–4 mm, bell-shaped, lavender, deciduous, scales lanceolate; stamens 2–4.5 mm, glabrous; style 4–5 mm. **FR** 2–3 mm, ovoid, short-hairy. **SEEDS** 2–4, 1.5–2 mm, pitted. RARE. Open, sandy to gravelly soils, pinyon/juniper woodland, conifer-ous forest; 1900–2200 m. s SNH (e slope), **w edge DMoj**. May

P. pachyphylla A. Gray (p. 353) Ann 4–17 cm. **ST** erect, 0–few-branched, short-glandular. **LVS** ± basal, 20–50 mm; blade gen < petiole, ± round, entire to slightly crenate. **FL**: pedicel < 3 mm; calyx lobes 2–3 mm, 3–4 mm in fr, oblong, densely short-hairy, glandular; corolla 5–7 mm, funnel- to bell-shaped, violet to purple, deciduous, scales linear; stamens 2–4 mm, short-hairy; style 2–3 mm. **FR** 5–7 mm, ± spheric, short-hairy, some gland-stalked. **SEEDS** > 60, ± 1 mm; cross-furrows 6–8. *n*=11. Flats, ± alkaline soils, creosote-bush scrub; < 1000 m. **D**; Baja CA. Apr–May

P. parishii A. Gray (p. 353) PARISH'S PHACELIA Ann 5–15 cm. **ST** ascending to erect, branched at base, glandular-puberulent. **LVS** ± basal, 8–30 mm; blade > petiole, widely elliptic to ob-ovate, entire to barely toothed. **FL**: pedicel ± 1 mm; calyx lobes 3–5 mm, 6–8 mm in fr, unequal, ± linear to ovate, puberulent; corolla 4–6 mm, narrowly bell-shaped, tube yellow, limb lavender, deciduous, scales ± elliptic; stamens 2–4 mm, sparsely short-hairy; style 1–2 mm. **FR** 3–5 mm, ± oblong, short-hairy. **SEEDS** 20–40, 1–1.5 mm, finely pitted. RARE. Clay or alkaline soils, dry lake margins; 800–1200 m. **w DMoj (nw San Bernardino Co)**; NV. Apr–Jul

P. parryi Torrey Ann 10–70 cm. **ST** erect, 0–few-branched, glandular-puberulent, stiff-hairy. **LF** 10–120 mm; blade < or = petiole, oblong to ovate, irregularly toothed. **FL**: pedicel 10–20 mm; calyx lobes 4–6 mm, 6–8 mm in fr, ± linear, sparsely hairy, ± glandular; corolla 10–20 mm, rotate to widely bell-shaped, tube and lower throat white to light violet, limb violet to purple, de-ciduous, scales fused to filament base, square; stamens 10–20 mm, long-hairy; style 10–20 mm, short-hairy. **FR** 6–10 mm, ovoid, beaked, short-stiff-hairy, gland-dotted. **SEEDS** 40–60, ± 1 mm, pitted. *n*= 11. DSon plants are intermediate to *P. minor*. Open areas, burns, slopes, coastal-sage scrub, chaparral; < 2400 m. SCoRI, SW (exc ChI, w WTR), **w edge D**; Baja CA. ❀ TRY. Mar–Jun

P. pedicellata A. Gray (p. 353) Ann 12–50 cm. **ST** erect, 0–few-branched above base, short-stiff-hairy, glandular. **LF** 20–120 mm; blade gen = petiole, ovate to round, lower compound (lflets 3–7, rounded or toothed), upper lobed to compound (segments 3, gen rounded). **FL**: pedicel 1–2 mm; calyx lobes 3–4.5 mm, 4–5.5 mm in fr, narrowly oblanceolate, sparsely hairy, ciliate, glan-dular; corolla 5–7 mm, bell-shaped, pink to blue, deciduous, scales round; stamens 6–8 mm, glabrous; style 6–8 mm. **FR** 3–3.5 mm, ± spheric, puberulent. **SEEDS** gen 4, ± 3 mm; back pitted; inner surface with central ridge separating 2 longitudinal grooves. *n*=11. Sandy or gravelly washes, canyons; < 1400 m. **D**; s NV, AZ, Baja CA. Mar–May

P. peirsoniana J. Howell Ann 4–30 cm. **ST** erect, simple to branched above base, glandular-puberulent, sparsely short-stiff-hairy. **LF** 10–60 mm; blade gen < petiole, ± round, ± toothed. **FL**: pedicel 2–4 mm; calyx lobes 3–4 mm, 6–8 mm in fr, oblong, short-hairy; corolla 4–6 mm, narrowly bell-shaped, white to vio-let, deciduous, scales lanceolate; stamens 2–3 mm, sparsely short-hairy; style 1–2 mm, short-hairy. **FR** 4–6 mm, oblong, puberu-lent. **SEEDS** 20–50, ± 1 mm, pitted. *n*=12. Rocky slopes, can-yons, sagebrush scrub, pinyon/juniper woodland; 1500–2700 m. **n SNE**, w NV. May–Jun

P. perityloides Cov. (p. 353) Per 5–40 cm. **STS** spreading to pendent, glandular; base spreading-hairy to woolly. **LF**: petiole 3–50 mm; blade 5–25 mm, ± round, irregularly toothed to lobed. **FL**: pedicel 5–30 mm; calyx lobes 3–6 mm, 4–6 mm in fr, ob-long to oblanceolate; corolla 10–15 mm, narrowly funnel- to bell-shaped, tube yellowish or aging purple, limb white, scales lin-ear; stamens 3–6 mm, unequal, incl, glabrous; style 4–6 mm, incl, short-lobed. **FR** 2.5–4 mm, narrowly ovoid, hairy. **SEEDS** 50–200, ± 0.5 mm, angular, pitted. *n*=11. Crevices on cliffs, rocky, often calcareous slopes; 600–2300 m. **n W&I, n&e DMoj**; NV, AZ. Another var. in AZ.

var. **jaegeri** Munz **ST**: base spreading-hairy, not clearly woolly. **FL**: pedicels 10–30 mm in fr, reflexed; calyx lobes < 1 mm wide; corolla 12–15 mm. Habitats of sp.; 1900–2300 m. **e DMtns (Clark Mtn)**; NV. May–Jun

var. **perityloides** **ST**: base clearly woolly, some hairs spread-ing. **FL**: pedicels 5–15 mm in fr, spreading to ascending; calyx lobes 1–2 mm wide; corolla 10–12 mm. Habitats of sp.; 600–2200 m. **n W&I, n DMoj**. Mar–Jul

P. pulchella A. Gray var. **gooddingii** (Brand) J. Howell (p. 353) GOODDING'S PHACELIA Ann 5–20 cm. **ST** ascending to erect, branched throughout, glandular-puberulent. **LF** 5–40 mm; blade = or > petiole, ovate to ± round, ± toothed. **FL**: pedicel ± 2 mm; calyx lobes 4–5 mm, 6–9 mm in fr, oblanceolate, glandular-puberulent; corolla 6–12 mm, bell-shaped, tube yellow, limb lav-ender to violet, deciduous, scales ± lanceolate; stamens 3–5 mm, puberulent; style 4–5 mm, short-hairy. **FR** 3–5 mm, ± oblong, short-hairy. **SEEDS** (25)30–50, ± 0.5 mm, pitted. RARE in CA. Clay soils, flats; 800–1000 m. **n DMoj**; also nw AZ. Other vars. in NV, s UT, n AZ. Apr–Jun

P. ramosissima Lehm. Per 30–150 cm. **ST** prostrate to as-cending, many-branched, glabrous to densely hairy, glandular or not. **LF**: blade 40–200 mm, gen> petiole, oblong to widely ovate, compound; lflets ± sessile, elliptic to oblong, coarsely toothed or lobed, lobes often toothed. **FL** ± sessile; calyx lobes 4–6 mm, not gen longer in fr, oblanceolate to ± spoon-shaped; corolla 5–8 mm, funnel- to bell-shaped, white to lavender, scales ovate; sta-mens 7–10 mm, exserted, glabrous; style 7–10 mm, exserted. **FR** 3–4 mm, ovoid, sharply bristly. **SEEDS** 2–4, 2–3 mm, pit-ted. *n*=11. Many habitats; < 3100 m. **CA**; to WA, ID, AZ. Vars. difficult, need study. 6 vars in CA.

var. **eremophila** (E. Greene) J.F. Macbr. **ST** below infl gen glabrous, sometimes with few short hairs. Slopes, open places, coniferous forest; 600–2800 m. CaRH, SNH, Wrn, **SNE**; OR, NV. [var. *valida* M. Peck] Jul–Aug

var. **latifolia** (Torrey) Cronq. **ST** below infl glandular; hairs mostly long, coarse, stiff, bulb-based, some hairs soft, spread-ing. Slopes, canyons, washes, flats; 50–2500 m. Teh, SnFrB, SCoR, SCo, TR, PR, **n DMtns (Panamint Range)**; NV, AZ. [var. *suffrutescens* Parry] May–Aug

var. **ramosissima** (p. 353) **ST** below infl glandular; hairs short to long, soft, spreading. Slopes, ridges, washes, meadows; 100–2800 m. KR, NCoRH, NCoRI, SN, Teh, CW, WTR, SnBr, **DMtns (Panamint Range)**; to WA. May–Jul

var. **subglabra** M. Peck (p. 353) **ST** gen < 60 cm, not glan-dular below infl, densely puberulent; some hairs fine, long. Slopes, meadows, coniferous forest; 200–3100 m. SNH, Teh, SnGb, **SNE**; to OR, ID.

P. rotundifolia S. Wats. Ann 4–28 cm. **ST** decumbent to erect, few–many-branched, short-stiff-hairy, gen glandular. **LF** 10–40 mm; blade gen < petiole, ± round, toothed. **FL**: pedicel 1–4 mm; calyx lobes 2–4 mm, 4–6 mm in fr, narrowly oblanceolate, short-hairy; corolla 3–6 mm, narrowly bell-shaped, deciduous, tube pale yellow, limb white to violet, scales lanceolate; stamens 2–3 mm, glabrous; style 1–2 mm, sparsely short-hairy. **FR** 3.5–4.5 mm, oblong, puberulent. **SEEDS** 50–100, ± 0.5 mm, pitted. *n*=12. Rocky slopes, crevices, ledges, creosote-bush scrub & pinyon/juniper woodland; < 2000 m. **W&I, D**; to sw UT, AZ. Apr–Jun

P. saxicola A. Gray Ann 5–15 cm. **ST** ascending to erect, gen many-branched, short-stiff-hairy, glandular-puberulent. **LF** 3–10 mm; blade gen = petiole, narrowly oblanceolate to ovate, entire. **FL**: pedicel 1–2 mm; calyx lobes 3–4 mm, 5–7 mm in fr, subequal, linear to narrowly oblanceolate, short-hairy; corolla 3–4 mm, narrowly bell-shaped, tube white, limb blue to violet, ± persistent in fr, scales linear; stamens 1–2 mm, ± glabrous; style 1–2 mm. **FR** 2–3 mm, ovoid, short-stiff-hairy. **SEEDS** 20–50, ± 0.5 mm, pitted. Limestone slopes, woodland; 1000–2300 m. **SNE, n DMoj**; s NV, nw AZ. Apr–Sep

P. tanacetifolia Benth. (p. 353) Ann 15–100 cm. **ST** erect, 0–few-branched, ± short-stiff-hairy, glandular-puberulent. **LF** 20–200 mm; blade > petiole, ± oblong to ovate, gen compound, lflets toothed to lobed. **FL** ± sessile; calyx lobes 4–6 mm, 6–8 mm in fr, ± linear, densely long-hairy; corolla 6–9 mm, widely bell-shaped, blue, ± persistent in fr, scales narrow; stamens 9–15 mm, glabrous; style 11–15 mm, glabrous. **FR** 3–4 mm, ± ovoid, glabrous; tip puberulent to short-hairy. **SEEDS** 1–2, 2–3 mm, wrinkled, pitted. *n*= 11. Sandy to gravelly slopes, open areas; < 2000 m. s NCoRO, c&s SNF, Teh, s ScV (Sutter Buttes), SnJV, e SnFrB, SCoR, SW (exc ChI), **DMoj**; s NV, AZ. ❀ SUN,DRN: **7–12,14–24.** Mar–May

P. tetramera J. Howell Ann 2–15 cm. **ST** spreading to ascending, branched throughout, sparsely short-hairy, ± glandular. **LF** 5–30 mm; blade gen = petiole, oblong to ovate, entire to few-toothed. **FL**: pedicel 1–4 mm; calyx lobes 4, 1.5–3 mm, 3.5–4.5 mm in fr, narrowly oblanceolate, puberulent to short-hairy; corolla 1.3–2 mm, bell-shaped, whitish, persistent in fr, lobes 4, scales 0; stamens 1–2 mm, glabrous; ovules 12–24, style < 0.5 mm, glabrous. **FR** 2.5–4 mm, oblong to ± spheric, puberulent. **SEEDS** 5–12, ± 1 mm; cross-furrows 6–9. *n*=11. Alkaline flats, washes, meadows; 1500–2400 m. **SNE**; to e OR, UT. May–Jun

P. vallis-mortae J. Voss (p. 353) Ann 20–60 cm. **ST** ascending to erect, simple to branched, puberulent and sparsely stiff-reflexed-hairy. **LF** 15–80 mm; blade > petiole, ± oblong, compound, lflets toothed or slightly lobed. **FL**: pedicel ± 1 mm; calyx lobes 4–6 mm, 7–10 mm in fr, linear to narrowly elliptic, long-hairy; corolla 8–15 mm, funnel- to bell-shaped, lavender to violet, deciduous, scales 0; stamens 6–12 mm, glabrous; style 5–12 mm, short-glandular-hairy. **FR** 3–4 mm, ovoid, puberulent. **SEEDS** gen 4, 2.5–3 mm, pitted. *n*=11. Sandy to rocky soils, shrubland; 600–2400 m. w SnJV, **SNE, e DMoj, n DSon**; to sw UT, nw AZ. Pls from w SnJV with dark-veined corolla have been called var. *heliophila* (J.F. Macbr.) J. Voss; relationships to *P. cryptantha* or *P. cicutaria* need careful study. ❀ TRY. May–Jun

PHOLISTOMA

Ann, fleshy. **ST** many-branched, prostrate or reclined, brittle; angles ± glabrous, bristly, or gen with hooked prickles. **LVS** simple, cauline; lower opposite; upper alternate; petioles gen winged, clasping; blade pinnately lobed, uppermost reduced, short-petioled, gen deltate, 3-lobed, with small, sharp bristles on both surfaces. **INFL** terminal, axillary, opposite lvs, or fls solitary; pedicels present. **FL**: calyx lobes hairy, bristly-ciliate; corolla rotate, lobed to middle, lobes gen hairy; stamens incl, equal, equally attached; ovary chamber 1, bristly-hairy, style 1, 2-lobed in distal 1/2. **FR** spheric; bristles stout. **SEEDS** 1–8, spheric, brown, pitted or honeycombed. 3 spp.: CA. (Greek: scale mouth) [Constance 1939 Bull Torrey Bot Club 66:341–352]

1. Calyx sinus appendages present; calyx enclosing mature fr . . . . *P. auritum* var. *arizonicum*
1' Calyx sinus appendages 0; calyx rotate below mature fr . . . . *P. membranaceum*

P. auritum (Lindley) Lilja (p. 353) **ST** 1–15 cm. **LVS**: lower 4–16 cm, 1–8 cm wide, petiole widely winged, clasping, blade oblong to ovate-lanceolate, base cordate, tip acuminate, lobes 5–13, oblong or lanceolate, obtuse or acute, entire or 1–5-toothed. **INFL**: fls solitary or 2–6 in cymes; pedicels 1–3 cm. **FL**: calyx lobes 3–9 mm, ± lanceolate, sinus appendages 1–4 mm; corolla 3–15 mm, 5–30 mm wide, blue to purple with darker marks in throat; style 4–8 mm. **FR** 5–10 mm wide, enclosed in calyx. **SEEDS** 1–4. Ocean bluffs, talus slopes, woodlands, streambanks, canyons, desert scrub; 0–1900 m. NCoRI, s SN, c SNF, Teh, SnJV, CW, SW, **ne DSon**; to AZ. 2 vars in CA.

var. *arizonicum* (M.E. Jones) Constance ARIZONA PHOLISTOMA **LF**: lobes gen 5–11, very obtuse. **FL**: corolla 3–7 mm, = calyx, < 1 cm wide. RARE in CA. Desert scrub; 300–700 m. **ne DSon (Whipple Mtns)**; to AZ.

P. membranaceum (Benth.) Constance (p. 353) **ST** 5–90 cm, gen glaucous. **LVS**: lower 2–13 cm, 1–8 cm wide, petiole narrowly winged, not clasping, blade oblong to ovate, base cordate or truncate, tip obtuse, lobes 5–11, oblong, obtuse, entire or 1-toothed. **INFL**: cyme; fls gen 2–10; pedicel 5–20 mm. **FL**: calyx rotate in fr, lobes 1–3 mm, oblong, sinus appendages 0; corolla 3–6 mm, < 1 cm wide, white, gen purple spot on each lobe; style 1–2 mm. **FR** 2–4 mm wide. **SEEDS** 1–2. *n*=9. Beaches, bluffs, ravines, wooded slopes, desert washes; 40–1400 m. c&s SNF, Teh, SnJV, CW, SW, **D**; Baja CA. Mar–May

TRICARDIA

1 sp. (Greek: 3 hearts, from calyx)

T. watsonii S. Watson (p. 353) THREE HEARTS Per; herbage long-soft-hairy, becoming ± glabrous; taproot woody, gen topped by a branched caudex covered by persistent petiole bases of previous years. **STS** several from root crown, erect, 5–40 cm. **LVS** simple, gen in basal rosette, petioled, 2–9 cm, 5–25 mm wide; cauline alternate, smaller upward, lower short-petioled, upper sessile, 6–20 mm, 3–15 mm wide, entire. **INFL** loose, terminal; fls pedicelled. **FL**: calyx lobes 5, very unequal, outer 3 cordate, 5–9 mm in fl, in fr 9–25 mm, wide, scarious, veiny, green to purplish, inner 2 ± 4 mm in fl, linear, enlarged in fr; corolla 4–8 mm, bell-shaped to rotate, white to cream, gen marked lavender; stamuns incl, unequal, equally attached; ovary chamber 1, style 1, 3–4 mm, tip lobed. **FR** 7–9 mm, glabrous. **SEEDS** 4–8, oblong, brown, rough. *n*=8. Sandy or gravelly desert slopes, flats, mtns, gen in shelter of shrubs; 100–2300 m. SnBr, **SNE, D**; to NV, UT, AZ. ❀ TRY. Apr–Jun

TURRICULA

1 sp. (Latin: little tower)

T. parryi (A. Gray) J.F. Macbr. (p. 353) POODLE-DOG BUSH Subshrub, densely glandular, sticky, ill-scented. **ST** erect, stout, densely leafy, 1–3 m, branched from base. **LVS** simple, cauline, alternate, sessile; blade 4–30 cm, lanceolate, entire or toothed, margins of upper sometimes rolled down. **INFL** terminal, branched; fls short-pedicelled, densely clustered. **FL**: calyx lobes 3–6 mm, glandular, coarsely long-hairy; corolla 10–20 mm, shallowly lobed, funnel-shaped, glandular, hairy outside, blue, lavender, or purple; stamens incl, unequal; ovary chambers 2, styles 2, 4–7 mm. **FR**: valves 4, 3–4 mm, ovoid, glandular, hairy.

SEEDS many, oblong-ovoid, angular, shiny black, finely ridged, minutely net-sculptured. *n*=13. Gen disturbed areas, chaparral, dry granitic soils of slopes, ridges; 100–2300 m. s SN, s SnJV, s SCoRO, SW, **n DMtns (Panamint Mtns)**; Baja CA. ✿ DRN,DRY,SUN:1,2,7–9,14–24;DFCLT. Jun–Aug

HYPERICACEAE ST. JOHN'S WORT FAMILY

Jennifer Talbot

Ann, per, shrub, tree. **LVS** simple, cauline, opposite or whorled; stipules 0; blade often with black dots or embedded clear glands. **INFL**: cyme, panicle, or fl solitary, terminal or axillary. **FL** bisexual, radial; sepals persistent, gen 5, often fused at base, overlapping; petals gen 5, free; stamens gen many, free or ± fused into 3–5 clusters; pistil 1, ovary superior, chambers 1–3, placentas gen axile, style branches 3. **FR**: capsule, gen septicidal. **SEEDS** many, small. 10 genera, 400 spp.: worldwide.

HYPERICUM

Ann, per, shrub, glabrous. **LF** sessile, ± gland-dotted. **INFL**: gen cyme, gen terminal, bracted. **FL**: sepals 5; petals 5, deciduous or persistent, yellow; anthers sometimes black-dotted; ovary chambers 1 or 3, placentas 3, axile or parietal and projecting into chamber. 350 spp.: worldwide. (Greek name)

H. formosum Kunth var. ***scouleri*** (Hook.) J. Coulter (p. 357) Per from taproot or rhizome. **STS** erect, few from base, 2–7 dm, slender; sterile axillary branches gen < 2 cm. **LF** 1–3 cm, ovate to oblong; flat; base ± clasping; margin black-dotted, lower surface inconspicuously dotted. **INFL**: fls gen 3–25 per st. **FL**: sepals 3–4 mm, oblong to ovate, gen obtuse, black-dotted, margin glabrous; petals 7–12 mm, obovate, pale to bright yellow, black-dotted; stamens many, in 3 clusters, anthers black-dotted; styles 3–5 mm. **FR** 6–7 mm, 3-lobed. **SEED** < 1 mm, brown. Springs, meadows, moist places; 100–2500 m. NW, CaRH, SN, ScV, SnFrB, SCoRI, TR, PR, **GB**; to w Can, Colorado. Jun–Aug

KOEBERLINIACEAE JUNCO FAMILY

Staria S. Vanderpool

Shrub, ± lfless, thorny, unscented. **ST**: branches many, smooth, rigid, interlocking. **LVS** simple, alternate, scale-like, ephemeral. **INFL**: raceme, axillary, bracted. **FL** radial, bisexual, small; sepals 4, free; petals 4, free; stamens 8, filaments flat; ovary ± stalked, spheric, stigma minute, head-like. **FR**: berry; chambers 2, each 1–2-seeded.

KOEBERLINIA ALLTHORN

The only genus. 1 sp. (C.L. Koeberlin, German clergyman, botanist, born 1794)

K. spinosa Zucc. ssp. ***tenuispina*** (Kearney & Peebles) E. Murray (p. 357) CROWN-OF-THORNS Pl < 5 m, short-spreading-hairy. **ST**: branchlets 25–70 mm, pale green, tipped by 3–6 mm, black thorns. **LF** < 2 mm. **INFL** 3–15 mm; pedicels 3–6 mm. **FL**: sepals 1–2 mm, ovate, entire, greenish white; petals 3–4 mm, 0.5–1 mm wide, short-clawed with an obovate or oblanceolate limb, white; stamens 2.8–4 mm, anthers 0.8–1 mm; ovary 1–1.2 mm, stalk 0.3–0.5 mm, style 1–1.5 mm. **FR** 2.5–3.5 mm, black. RARE in CA. Creosote-bush scrub; 1600 m. **DSon (Chocolate Mtns)**; sw AZ, Mex (nw Sonora). Often incl in Capparaceae. Ssp. *spinosa* not in CA

KRAMERIACEAE RHATANY FAMILY

Beryl B. Simpson

Per, shrub, root parasite with chlorophyll. **ST** prostrate to erect, much branched. **LVS** gen simple, alternate, sessile; blade linear to ovate, hairy, sometimes glandular, tip abruptly pointed. **INFL**: fls gen solitary in axils; pedicel bracts 2. **FL** bisexual, bilateral; sepals 4–5, free, conspicuous; petals gen 5, 3 upper linear to clawed, held in ± upright "flag", 2 modified into glands flanking ovary; stamens gen 4, opening by pores; ovary superior, hairy, style slender, recurved. **FR** nut-like, bearing smooth or barbed spines. 1 genus, 17 spp.: Am, esp trop. [Simpson 1989 Fl Neotropica 49:1–109] Pollinating bees collect oils secreted by glandular petals.

KRAMERIA RHATANY

(Possibly named for J. Kramer, 1700's, Austrian army physician)

1. Sepals cupped; claws of flag petals fused; fr spines smooth or barbed along shaft . . . . ***K. erecta***
1' Sepals reflexed; flag petals free; fr spines barbed only at tip . . . . ***K. grayi***

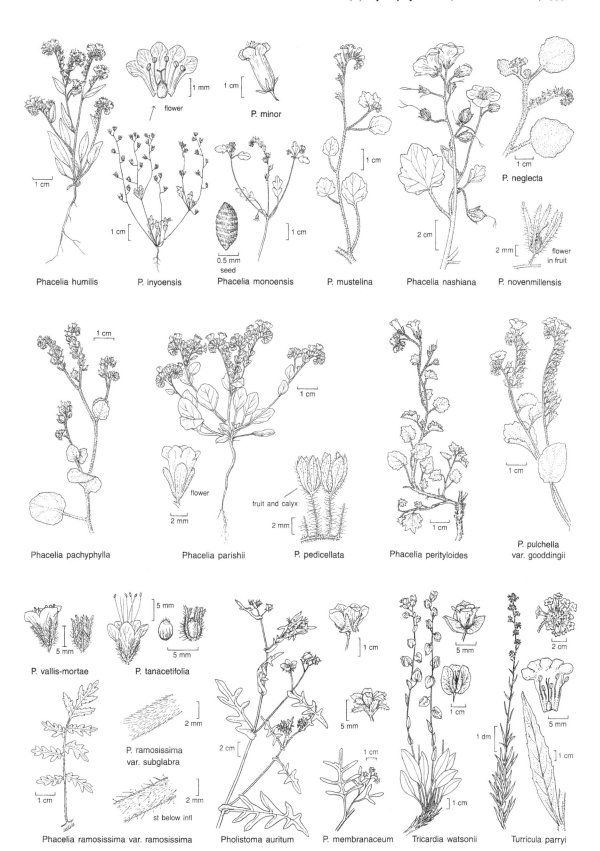

Phacelia humilis

P. inyoensis

flower

P. minor

Phacelia monoensis

seed

P. mustelina

Phacelia nashiana

P. neglecta

P. novenmillensis

flower in fruit

Phacelia pachyphylla

Phacelia parishii

flower

P. pedicellata

fruit and calyx

Phacelia perityloides

P. pulchella var. gooddingii

P. vallis-mortae

P. tanacetifolia

P. ramosissima var. subglabra

st below infl

Phacelia ramosissima var. ramosissima

Pholistoma auritum

P. membranaceum

Tricardia watsonii

Turricula parryi

K. erecta Schultes (p. 357, pl. 62) PIMA RHATANY, PURPLE HEATHER Shrub, ± strigose to canescent or ± silky-hairy. **ST** < 1 m; branches often ascending, tips blunt. **LF** ± linear. **FL**: buds ovate, barely curved; sepals cupped, pink; flag petal claws ± fused, blades triangular, green and pink; glandular petals pink, outer face glandular-blistered near margin. **FR** cordate, somewhat flat; spines smooth or barbs scattered. Dry, rocky ridges, slopes; < 1200 m. e PR (Santa Rosa Mtns), **D**; to NV, TX, n Mex. [*K. glandulosa* Rose & Painter; *K. parvifolia* Benth.; *K. p.* var. *imparata* J.F. Macbr.] ❀ TRY. Mar–May

K. grayi Rose & Painter (p. 357) WHITE RHATANY Shrub, densely canescent or silky-hairy. **ST** < 1 m; branches ± spreading, tips spiny. **LF** narrowly lanceolate. **FL**: buds curved upward; sepals reflexed, deep purple-red; flag petals free, blade oblanceolate, base green, tip pink or purple; glandular petals purple, outer face covered with blister-like glands. **FR** ± spheric; spines barbed only at tip. Dry, rocky or sandy places, esp on lime soils; < 1400 m. **D**; to NV, TX, n Mex. [*K. canescens* A. Gray] ❀ TRY. Apr–May

LAMIACEAE MINT FAMILY

Dieter H. Wilken, except as specifed

Ann, per, shrub, glabrous to hairy, gen aromatic. **STS** gen erect, gen 4-angled. **LVS** gen simple to deeply lobed, opposite, gen gland-dotted. **INFL**: cyme, gen clustered around st, head-like, separated by evident internodes (terminal in *Monardella*) or collectively crowded, spike-like to panicle-like (sometimes raceme or fls 2–12); subtended by lvs or bracts; fls sessile or pedicelled. **FL** gen bisexual; calyx gen 5-lobed, radial to bilateral; corolla gen bilateral, 1–2-lipped, upper lip entire or 2-lobed, ± flat to hood-like, sometimes 0, lower lip gen 3-lobed; stamens gen 4, gen exserted, paired, pairs unequal, sometimes 2, staminodes 2 or 0; ovary superior, gen 4-lobed to base, chambers 2, ovules 2 per chamber, style 1, arising from center at junction of lobes, stigmas gen 2. **FR**: nutlets 4, gen ovoid to oblong, smooth. ± 200 genera, 5500 spp.: worldwide. Many cult for herbs, oils (*Lavandula*, lavender; *Mentha*, mint; *Ocimum*, basil; *Rosmarinus*, rosemary; *Thymus*, thyme), some cult as orn (in CA *Cedronella*, *Leonotis*, *Phlomis*). [Cantino & Sanders 1986 Syst Bot 11:163–185]

1. Fertile stamens 2; staminodes sometimes present
  2. Corolla 4-lobed, not 2-lipped; lobes sub-equal; fr compressed, edge corky . . . . **LYCOPUS**
  2' Corolla evidently 2-lipped, lobes of upper lip very different from lobes of lower lip; fr not compressed
    3. Fertile anther sacs 1 or 2, on thread-like appendage hinged or joined to filament tip; sterile anther
      sac reduced, modified, or 0 . . . . **SALVIA**
    3' Fertile anther sacs 2, attached to filament tip side by side
      4. Shrub, rounded to mound-like, gen 5–10 dm . . . . **POLIOMINTHA**
      4' Ann, per, or subshrub, gen < 5 dm
        5. Calyx ± radial, lobes subequal . . . . **MONARDA**
        5' Calyx ± 2-lipped, lobes of upper and lower lips unequal . . . . **HEDEOMA**
1' Fertile stames 4; staminodes 0
  6. Corolla ± radial to slightly 2-lipped, lobes or lips equal in length
    7. Infls 2+, axillary, gen subtended by 2 lvs or bracts, each head-like or collectively spike-like . . . . . **MENTHA**
    7' Infl gen 1, head-like, terminal, subtended by involucre-like whorl of bracts . . . . **MONARDELLA**
  6' Corolla clearly bilateral, gen 1–2-lipped, lobes and lips unequal
    8. Calyx 2-lobed, lobes lip-like, entire
      9. Shrub; calyx becoming bladder-like, 1–2 cm in fr . . . . **SALAZARIA**
      9' Per from rhizomes; calyx not becoming bladder-like, gen < 1 cm in fr — calyx back transversely
        ridged or with dome-like bump . . . . **SCUTELLARIA**
    8' Calyx 5- or 10-lobed, radial to 2-lipped (lips 2–3-lobed)
      10. Calyx 10-lobed, lobe tips recurved or hooked; stamens incl in corolla tube . . . . **MARRUBIUM**
      10' Calyx 5-lobed, lobe tips straight; stamens exserted, hidden under upper lip to > corolla lobes
        11. Stamens reclining on lower corolla lip; lip pouched; twigs densely stellate-hairy — DSon . . . . . **HYPTIS**
        11' Stamens ascending under or ± parallel to and exceeding upper corolla lip; lip ± flat, not
          pouched; twigs glabrous to hairy, not stellate
          12. Nutlets fused laterally below middle; corolla 1-lipped, lip entire, ± reflexed, > other 4 lobes,
            or lip 5-lobed, ± straight
            13. Lvs crenate to deeply lobed; corolla lip 5-lobed, straight; fr ± smooth, puberulent at top
              . . . . **TEUCHRIUM**
            13' Lvs entire; corolla lip entire, ± reflexed; fr irregularly ridged, puberulent to hairy throughout —
              stamens >> corolla lobes . . . . **TRICHOSTEMA**
         12' Nutlets separate to base; corolla 2-lipped, upper lip entire, notched, or lobed; lower lip 1–3-lobed
          14. Infls axillary, gen separated by evident internodes subtended by bracts or lvs . . . . ²**STACHYS**
          14' Infls terminal, collectively ± dense, spike-like, sometimes interrupted by 1–3 internodes below
            15. 2 or 4 stamens clearly exceeding corolla . . . . **AGASTACHE**
            15' Stamens ascending under upper lip, not clearly > lower corolla lobes
              16. Calyx 2-lipped, upper lip 3-toothed, lower lip 2-lobed, acuminate . . . . **PRUNELLA**
              16' Calyx ± radial, 5-lobed, lobes ± equal . . . . ²**STACHYS**

AGASTACHE HORSEMINT

Deborah Engle Averett

Per, erect, gen < 1 m, aromatic. **LVS** petioled; blade ± lanceolate to triangular, margin crenate to coarsely serrate. **INFL**: spike of sessile clusters, dense; bracts 1–several at base, lf-like, lanceolate. **FL**: calyx 5-lobed, 2-lipped, turning pink before fr, lobes acuminate; corolla 5-lobed, rose to rose-purple, 2-lipped, lower longer, broader, upper 2-lobed; stamens 4, in 2 pairs (1 pair longer), exserted, anther sacs spreading; style 2-lobed, exserted. **FR** ± 2 mm, oblong, brown, smooth; with small hairs at tip. $2n=18$. 22 spp.: N.Am, Mex, Asia. (Greek: many spikes)

A. urticifolia (Benth.) Kuntze (p. 357) **LF** 3–8 cm, 1.5–7 cm wide. **FL**: corolla rose to rose-purple. Common. Gen woodlands, but many habitats; 400–3000 m. NCoR, CaR, n&c SNF, SNH, SCoRO, SnBr, Wrn, **n** SNE; to B.C., Rocky Mtns. [*A. glaucifolia* A.A. Heller] ❀ DRN, IRR:**1,2,4–6**,17&SHD:**3,7**,14–16. Jun–Aug

HEDEOMA MOCK PENNYROYAL

Ann, per, subshrub, aromatic; hairs short, spreading to recurved. **STS** decumbent to erect, branched at base. **LF** short-petioled to sessile; blade ovate to linear, entire or toothed. **INFLS** axillary at upper st nodes, each head-like, subtended by lvs; bracts minute. **FL**: calyx 2-lipped, upper lip 3-lobed, lower lip 2-lobed, lobes acuminate, sharp-pointed, tube swollen or pouched below middle; corolla 2-lipped, upper lip > lower, entire to 2-lobed, lower lip 3-lobed; stamens 2, under upper lip or exserted, staminodes minute or 0; style unequally 2-lobed. **FR**: nutlets pitted, glaucous, gelatinous when wet. 38 spp.: N.Am., S.Am. (Greek: ancient name for strongly aromatic mint) [Irving 1980 Sida 8: 218–295]

1. Lf blades linear to ± oblong; calyx lobes in fr converging, throat ± closed; sts becoming glabrous below
. . . . *H. drummondii*
1' Lf blades ovate to ± round; calyx lobes in fr spreading to reflexed, throat ± open; sts puberulent below
. . . . *H. nanum* ssp. *californicum*

H. drummondii Benth. (p. 357) Per 15–45 cm. **ST** puberulent, becoming glabrous below. **LF**: blade 5–11 mm, 1–4 mm wide, linear to ± oblong, tip gen obtuse. **INFL**: fls 3–7; pedicels 2–3.5 mm. **FL**: calyx 5–6 mm, lobes converging, throat ± closed in fr; corolla 7–9 mm, lower lip length = width. $2n=34,36$. Rocky, gravelly soils; 1400–1700 m. **e DMtns (New York Mtns)**; to MT, NE, Mex.

H. nanum (Torrey) Briq. ssp. *californicum* W.S. Stewart (p. 357) Per 10–25 cm. **ST** puberulent. **LF**: blade 3–8.5 mm, 2–4.5 mm wide, ovate to ± round, tip acute. **INFL**: fls 3–5; pedicel 3–4.5 mm. **FL**: calyx 4.5–5.5 mm, lobes spreading or ± reflexed, throat ± open in fr; corolla 8–9 mm, lower lip length < width. $2n=36$. Rocky, often limestone outcrops; 900–2100 m. **DMtns**; NV, nw AZ. May–Jun

HYPTIS

Ann, per, shrubs, glabrous to densely, gen stellate-hairy. **STS** erect to spreading, branched. **LVS** gen petiolate. **INFLS** axillary; each cluster ± dense, subtended by lvs or bracts. **FL**: calyx gen 5-lobed, lobes equal, acute to long-acuminate; corolla 2-lipped, upper lip 2-lobed, flat, lower lip 3-lobed, central lobe reflexed, ± pouched; stamens 4, fertile, curved, reclining on lower lip, exserted; style lobes ± equal. **FR**: nutlets angled, ridged or smooth. ± 350 spp.: warm temp, trop Am. (Greek: turned back, from lower lip position) Some spp. used for food (seeds), oil, wood, fiber.

H. emoryi Torrey (p. 357, pl. 63) DESERT-LAVENDER Shrub 1–3 m. **ST**: branches spreading to erect; twigs densely stellate, becoming glabrous. **LF**: petiole 3–7 mm; blade gen ovate to ± round, crenate. **INFLS**: lower subtended by lvs, upper by bracts; bracts << lvs, linear to elliptic; pedicel 1–3 mm. **FL**: calyx 4–5 mm, densely stellate, lobes ± acuminate; corolla 5–6 mm, violet. Gravelly, sandy washes, canyons, desert shrubland; < 1000 m. **s DMoj, DSon**; AZ, nw Mex. ❀ DRN,DRYorIRR:8–14&SUN:15–17,19–24;DFCLT. Jan–May

LYCOPUS BUGLEWEED

Per from rhizomes, glabrous or hairy. **STS** erect, branched or not. **LVS** short-petioled to sessile, gen ovate to lanceolate; margin toothed to deeply lobed or cut below middle. **INFLS** axillary, each head-like, subtended by lvs. **FL**: calyx gen 5-lobed, lobes ± equal, obtuse to short-awned; corolla slightly bilateral, not 2-lipped, gen 4-lobed, lobes ± unequal, odd lobe notched or entire; stamens 2, exserted, staminodes 2, minute, club-shaped; style exserted. **FR**: nutlets ± compressed, edge corky-thickened, truncate or rounded. 14 spp.: temp N.Am, Eurasia, 1 spp. in Australia. (Greek: wolf foot, from French common name) [Henderson 1962 Amer Midl Naturalist 68:95–135]

1. Lvs gen short-petioled, deeply lobed to cut in lower half; fr 1–1.5 mm, top rounded . . . . *L. americanus*
1' Lvs gen subsessile to sessile, serrate; fr 1.5–2 mm, top ± truncate . . . . *L. asper*

L. americanus W.C. Barton (p. 357) Rhizomes ± slender, not thickened at tip. **STS** erect, 2–8 dm, gen glabrous; nodes short-hairy. **LF** 2–8(10) cm, short-petioled, oblong to lanceolate, lobed to cut esp in lower half, glabrous to puberulent on veins. **FL**: calyx lobes awl-like, short-awned; corolla 2–3 mm, ± = calyx, white. **FR**: nutlet 1–1.5 mm, top rounded, smooth. $2n=22$. Moist areas, marshes, streambanks; < 1000 m. CA-FP, **SNE**; to B.C., e N.Am. Aug–Sep

L. asper E. Greene (p. 357) Rhizomes thicker and tuber-like near tip. **STS** erect, 3–8(10) dm, puberulent to short-hairy. **LF** 2.5–7(9) cm, ± sessile, lanceolate to narrowly elliptic, serrate, glabrous to puberulent. **FL:** calyx lobes awl-like, acuminate to short-awned; corolla 3–5 mm, slightly > calyx, white. **FR:** nutlet 1.5–2 mm; top truncate, sometimes minutely toothed. 2*n*=22. Uncommon. Moist areas, marshes, streambanks; < 1300 m. Deltaic GV, SnFrB, **GB**; to w Can, Great Plains. [*L. lucidus* Benth. misapplied]. Jun–Oct

MARRUBIUM HOREHOUND

Per. **STS** gen erect, gen branched, tomentose. **LVS** petioled to subsessile, gen ovate to round, crenate or toothed, **INFLS** gen axillary, each head-like, gen subtended by lvs. **FL:** calyx 10-lobed in CA, lobes spreading or recurved, sharp-pointed; corolla 2-lipped, upper lip entire to 2-lobed, lower lip 3-lobed; stamens 4, fertile, lower pair gen > upper pair, incl in tube; style incl, lobes ± equal. **FR:** nutlet top truncate. 30 spp.: Eur. (Latin: based on ancient Hebrew word for bitter juice) Some spp. cult for folk medicine, flavorings, some toxic.

M. vulgare L. **STS** ascending to erect, 1–6 dm. **LF:** petiole < blade; blade 1.5–5.5 cm, widely ovate to ± round, base rounded to ± lobed, margin crenate. **FL:** calyx 4–6 mm; teeth 10, short-soft-hairy; corolla > calyx, lips ± equal. 2*n*=34. Disturbed sites, gen overgrazed pastures; < 600 m. CA-FP, **DMtns (uncommon)**; widespread worldwide; native to Eur. Formerly cult for tea, flavoring. Spring & summer

MENTHA MINT

Per from rhizomes, glabrous to hairy. **STS** gen ascending to erect, gen branched. **LF** petioled to sessile, elliptic, ovate, or lanceolate, toothed to lobed. **INFLS** axillary, each head-like and subtended by lvs, or collectively spike- or panicle-like and by bracts. **FL:** calyx ± radial, gen 10-veined, lobes equal or unequal; corolla ± 2-lipped, lips gen equal, upper lip notched, lower lip 3-lobed; stamens 4, ± equal, gen exserted; style lobes unequal. 25 spp.: temp. N.Am, Eurasia. (Latin: ancient name for mint) [Tucker, Harley, & Fairbrothers 1980 Taxon 29:233–255] Cult for oils, flavoring, herbs. Many cult and naturalized populations derived from hybridization, gen complexly polyploid, some sterile, reproducing vegetatively.

1. Internodes between upper infls evident, gen > 1 cm; infls subtended by ovate or elliptic lvs . . . . ***M. arvensis***
1' Internodes between upper infls gen inconspicuous, gen < 6 mm; infls gen subtended by linear, lanceolate, or awl-like bracts
   2. Cauline lf base rounded to obtuse, tip acute to acuminate, lower surface gen glabrous or white-soft-hairy
    . . . . ***M. spicata***
   2' Cauline lf base slightly lobed, tip gen rounded, lower surface stellate-tomentose . . . . ***M. suaveolens***

M. arvensis L. (p. 357) **ST** 1–5(8) dm, puberulent to short-hairy. **LF:** 1.5–5(8) cm; lower short-petioled; cauline gen subsessile; blade ovate to elliptic, base tapered, tip gen acute, crenate to serrate, lower surface (esp veins) short-hairy. **INFLS** axillary, each head-like, subtended by spreading lvs; bracts minute or 0. **FL:** calyx 1.5–3 mm, short-hairy; corolla 4–7 mm, white, pink, or violet; stamens > corolla lobes. 2*n*= 24,54,72,90. Moist areas, streambanks, lake shores; < 2400 m. CA-FP, **GB**; circumboreal. [var. *villosa* (Benth.) S.R. Stewart]. Some pls sterile; some pls naturalized from Eur. Jul–Oct

M. spicata L. var. *spicata* SPEARMINT **ST** 3–10(12) dm, glabrous. **LF** gen 1–6 cm, ± sessile; blade ovate to lanceolate, base rounded to obtuse, tip acute to acuminate, gen serrate, lower surface gen glabrous. **INFLS** densely clustered, subtended by linear-lanceolate bracts, collectively spike-like. **FL:** calyx 1.5–2.5 mm, gen glabrous; corolla 3–4 mm, white, pink, or lavender; stamens exserted but not > corolla lobes. 2*n*=36,48. Moist areas, marshes, lake shores; < 1650 m. CA-FP, **W&I, DMtns (uncommon)**, cult elsewhere; to e US; native to Eur. Pls with lower blade surface white-soft-hairy are var. *longifolia* L. [*M. longifolia* (L.) Hudson], native to Eur. Jul–Oct

M. suaveolens Ehrh. **ST** 5–10 dm, soft-hairy. **LF** 1–4 cm, subsessile; blade ovate, oblong, or broadly elliptic, base slightly lobed, tip gen rounded, crenate to serrate, lower surface stellate-tomentose. **INFLS** densely clustered at upper nodes, subtended by linear or awl-like bracts, collectively spike-like. **FL:** calyx 1–1.5 mm, short-hairy; corolla 2–3 mm, white or pinkish; stamens > corolla lobes. Moist areas, ditchbanks; < 1200 m. NW, SN, CW, SCo, **SNE**, expected elsewhere; native to s Eur. [*M. rotundifolia* (L.) Hudson misapplied] Jul–Oct

MONARDA BEE BALM

Ann, per, gen short-hairy. **STS** erect, gen branched. **LF** petioled to sessile. **INFLS** axillary, each head-like; lower subtended by lvs; upper by bracts. **FL:** calyx 5-lobed; corolla 2-lipped, upper lip entire or ± 2-lobed, hood-like, arched, lower lip gen 2–3-lobed, central lobe gen > lateral lobes; stamens 2, ascending under upper lip, = or > upper lip; style unequally lobed. 16 spp.: N.Am. (Nicolas Monardes, Spanish physician & botanist, 1493–1588) [Scora 1967 Univ Calif Publ Bot 41:1–71] Some cult for fls, tea.

M. pectinata Nutt. (p. 357) Ann. **ST** 1.5–3.5 dm; hairs short, ± curled down. **LF** short-petioled to subsessile; blade 1.5–4 cm, gen oblong to lanceolate, entire to serrate, ± glabrous to finely strigose esp on veins. **INFL:** subtending lvs gradually reduced upward, uppermost ± 4–7 mm, ovate. **FL:** calyx tube 6–8 mm, throat densely puberulent within, lobes 2–4 mm, long-acuminate; corolla 12–25 mm, white to pink, lower lip sometimes purple-spotted. 2*n*=18,36. Washes, rocky slopes, pinyon/juniper woodland; 1150–1500 m. **e DMtns (New York Mtns)**; to w Great Plains, n Mex. ❀ TRY. Jul–Sep

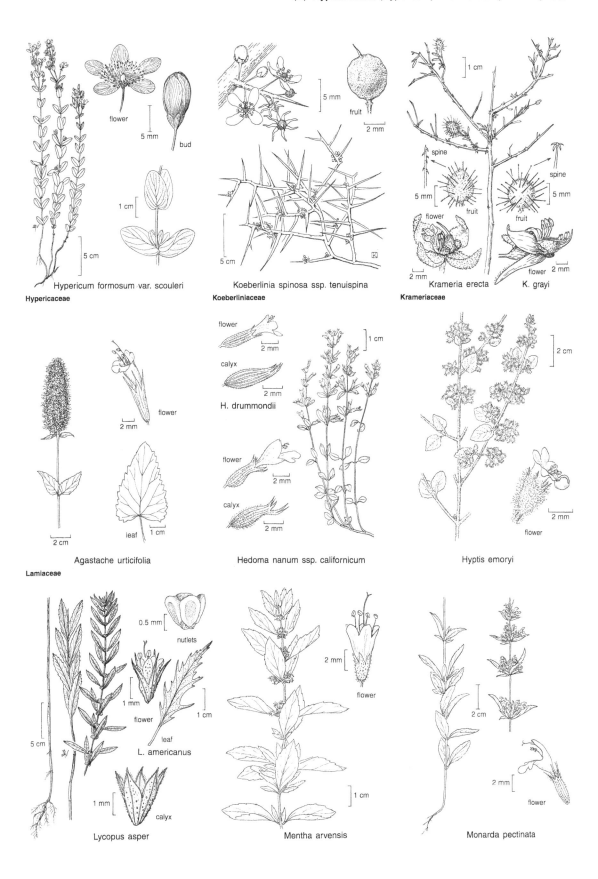

Hypericum formosum var. scouleri
Hypericaceae

Koeberlinia spinosa ssp. tenuispina
Koeberliniaceae

spine

spine

flower   fruit

fruit

flower

flower

Krameria erecta   K. grayi
Krameriaceae

flower

Agastache urticifolia
Lamiaceae

leaf

flower

calyx

H. drummondii

flower

calyx

Hedoma nanum ssp. californicum

Hyptis emoryi

flower

nutlets

flower

leaf

L. americanus

calyx

Lycopus asper

flower

Mentha arvensis

flower

Monarda pectinata

MONARDELLA

James D. Jokerst

Ann, per, ± gland-dotted. **LVS** entire to serrate. **INFL**: heads 1 or more per main st, sometimes arrayed in spikes or panicles; bracts in 2–3 series (outer series (0)1–2 pairs, ± like lvs, 0–several mm below heads, erect to reflexed; middle series 2–4 pairs, like lvs to papery or leathery, sometimes straw-colored to purple, erect in cup-like involucre to reflexed; inner series 0–few pairs, membranous, linear-lanceolate). **FL**: calyx 5-lobed, gen < 12 mm; corolla white to purple, upper lip erect, 2-lobed, lower lip recurved, 3-lobed; stamens 4; style unequally 2-lobed. ± 20 spp.; w N.Am. (Latin: small *Monarda*) [Epling 1925 Ann Missouri Bot Gard 12:1–106] Complex; hybrids common, often outnumbering non-hybrids; head width and bract orientation given for unpressed specimens. ❀ often DFCLT. Many spp., sspp., or populations have exacting soil requirements. Key to species adapted by Bruce Baldwin.

1. Ann; heads several–many per main st, arrayed in panicles — DMoj . . . . ***M. exilis***
1' Per; heads gen 1 per main st (if > 1, arrayed in spikes or racemes in *M. robisonii*)
  2. Heads gen several per main st; middle bracts scarious, lanceolate to ovate, reflexed to ascending,
    gen not in cup-like involucre — DMtns . . . . ***M. robisonii***
  2' Heads gen 1 per main st; middle bracts scarious or papery, elliptic to ovate or round, erect, in cup-like
    involucre (not always apparent on pressed material)
    3. Outer bract series present, reflexed, 25–40 mm; middle bracts scarious or membranous, greenish,
      sometimes ± purple-tinged; calyx lobes woolly; W&I . . . . ***M. odoratissima*** ssp. *pallida*
    3' Outer bract series 0; middle bracts papery, whitish to straw or purple, sometimnes rose- or
      purple-tinged, less often grayish; calyx lobes stiff-hairy; GB, DMoj
      4. St green, dark gray, or appearing glaucous; lf elliptic, narrowly ovate, or ovate; GB . . . . ***M. glauca***
      4' St silvery to ash-gray with densely matted minute hairs; lf linear to lanceolate; SNE, DMoj
      . . . . ***M. linoides*** ssp. *linoides*

M. exilis E. Greene   Ann; 1° branches gen basal or below middle of main axis. **LF** lanceolate to narrowly ovate. **INFL**: bracts gen narrowly ovate, green or whitish, purple-tinged or not, lateral veins 0 or few, cross-veins 0, margins, tips white-scarious, abruptly acuminate. **FL**: calyx lobe tips erect, acute, white; corolla 10 mm, white; stamens slightly exserted, tissue between pollen chambers notched. Desert scrub, washes; 600–1100 m. s SnJV, **DMoj**. May–Jun

M. glauca E. Greene (p. 363)   Per. **ST** green, dark gray, or appearing glaucous, with gland-tipped bristles or not. **LF** elliptic to ovate, entire. **INFL**: outer bracts 0; middle bracts elliptic to ovate, papery, finely short-hairy, ciliate, rose or purple (or grayish), outer of middle series in cup-like involucre, sometimes lf-like at tips. **FL**: calyx lobes sparsely to densely stiff-hairy; corolla 10–20 mm, purple or red-purple. Rocky openings, sagebrush scrub to alpine forest; 1000–3500 m. KR, NCoRH, CaRH, SNH, **GB**. [*M. odoratissima* ssp. *g.* (E. Greene) Epling] Highly variable: small-headed, matted pls of SNH, W&I, AZ, NV, Rocky Mtns have been called *M. o.* ssp. *parvifolia* (E. Greene) Epling; pls intermediate to *M. linoides* with erect, lanceolate lvs, internodes 1.5–2 × lvs, bracts scarious below, deep rose above, of SNE, e SNH, e MP, NV, AZ, Rocky Mtns have been called *M. rubella* E. Greene. Hybridizes with *M. o.* ssp. *pallida*, *M. linoides*. ❀ TRY. Jun–Aug

M. linoides A. Gray (p. 363)   Per, erect, open. **ST** 10–50 cm, woody below, silvery to ash-gray with densely matted hairs; internodes gen > lvs. **LF** 10–40 mm, linear to narrowly ovate, entire, greenish, silvery, or ash-gray. **INFL**: head 20–30 mm wide; outer bracts 0; middle bracts ovate, acuminate, papery, hairy, ciliate, whitish, straw-colored, rose, or purple, outer of middle series erect, in cup-like involucre, sometimes lf-like at tips. **FL**: calyx lobes stiff-hairy; corolla 12–15 mm, whitish, lavender, or pale purple; stamens exserted. *n*=21. Desert scrub, pinyon/juni-
per woodland, open conifer forest, subalpine; 900–3100 m. s SNH, Teh, SnGb, SnBr, PR (and **adjacent w D**), SNE, **DMoj**; NV, Baja CA. 4 sspp in CA.

  ssp. *linoides*   **ST** silvery. **LF** 10–40 mm, linear to lanceolate, silvery. **INFL**: bracts = or > calyces, white to rose. Habitats, elevations of sp. s SNH, e SnBr, SnJt, **SNE**, **DMoj**; NV. Hybridizes with *M. glauca*. Jun–Aug

M. odoratissima Benth.   Per, tufted or not. **ST** erect, greenish, hairy, with gland-tipped bristles or not. **LF** 5–45 mm, lanceolate to ovate, entire, sparsely to densely hairy, green to ash-gray, often purple-tinged. **INFL**: head 10–25 mm wide; outer bracts reflexed or spreading, like lvs; middle bracts erect, in cup-like involucre, lanceolate to ovate, acute to obtuse, hairy, ciliate, scarious, greenish, sometimes lavender- to rose- or purple-tinged. **FL**: calyx hairy, lobes woolly; corolla 10–20 mm, white, lavender, or purple. *n*=21. Sagebrush scrub, montane forest; 600–3100 m. KR, NCoRH, CaRH, SNH, nw MP, **W&I**; to WA, NV. 2 sspp in CA.

  ssp. *pallida* (A.A. Heller) Epling (p. 363)   Pl greenish or ash-gray. **LF** 20–45 mm, both surfaces green-glabrous to ash-gray-hairy. **INFL**: outer bracts reflexed, 25–40 mm, linear-lanceolate. **FL**: calyx woolly; corolla gen white (sometimes lavender- or purple-tinged). Montane forest, rocky slopes; 1000–3100 m. KR, NCoRH, CaRH, SNH, **W&I**; NV. ❀ DRN:15–17&SHD:1–3,7&IRR:8,9,**14**,18–23;dfclt. Jul–Sep

M. robisonii Epling (p. 363)   ROBISON'S MONARDELLA   Per, erect, open. **ST** grayish, hairs spreading, wavy. **LF** 10–40 mm, lanceolate to narrowly ovate, ash-gray, hairy. **INFL**: heads 10–20 mm wide, gen several arrayed in ± terminal spikes or racemes, sometimes solitary; bracts narrowly acute, scarious, pink-tinged or not, aging brown. **FL**: calyx lobes narrowly acute; corolla pale rose. RARE. Desert scrub, pinyon/juniper woodland; 1100–1500 m. **DMtns**; Baja CA? Hybridizes with *M. linoides*. Jun

POLIOMINTHA

Shrub. **STS** spreading to erect, branched throughout, densely strigose, gen grayish. **LVS** short-petioled to subsessile; blades gen narrow, entire. **INFLS** axillary, gen subtended by lvs. **FL**: calyx ± radial, 15-veined, lobes subequal; corolla 2-lipped, lips ± equal, upper lip ± flat, lower lip 3-lobed, central lobe notched; stamens 2, ± exserted, staminodes short; style unequally lobed. 4 spp.: sw US, n Mex. (Greek: hoary white mint)

P. incana (Torrey) A. Gray (p. 363) Pl 5–10 dm, rounded or mound-like. **LF** subsessile, gen reduced upward; blade 5–18 mm, 2–4 mm wide, oblong-elliptic to narrowly linear. **INFL**: fls 2–6; pedicel 1–2 mm. **FL**: calyx densely short-hairy, tube 3–5 mm, puberulent within, lobes 1–2 mm; corolla 8–10 mm, light blue to lavender, upper lip 2–3.5 mm, lower 3–4 mm, minutely purple-dotted. Uncommon. Sandy soils, rocky slopes, 1600–1700 m. **s DMoj (Cushenbury Springs)**; to Colorado, TX, n Mex. ❀TRY; DFCLT. Jun–Jul

PRUNELLA SELF-HEAL

Per, glabrous to hairy, gen with bisexual fls only, sometimes with only pistillate fls. **STS** prostrate to erect, sometimes rooting at lower nodes. **LVS** basal and cauline, gen petioled; blade gen entire. **INFLS** densely clustered, collectively ± spike-like, terminal; bract gen wide, abruptly acuminate. **FL**: calyx 2-lipped, upper = lower, upper lip 3-toothed, lower 2-lobed; corolla finely hairy inside, 2-lipped, lower lip 3-lobed, upper lip ± entire, hood-like, ± enclosing stamens; stamens 4, lower pair > upper, filament minutely toothed below anther. **FR**: nutlets obovoid. 4 spp.: temp, esp Eurasia. (Latin: from early German name for pl used to treat chest pains)

P. vulgaris L. **ST** 1–5 dm, glabrous to short-hairy. **LVS**: lower petioled, petiole 5–30 mm; upper subsessile; blade 2–7 cm, gen 1–4 cm wide, ovate, elliptic, or lanceolate, base gen wedge-shaped. **INFL** 2–6.5 cm; bract margin ciliate, reddish. **FL**: calyx 7–11 mm, dark green to purplish; corolla 12–15 mm in bisexual fls, 8–11 mm in pistillate fls, bluish violet, sometimes pink or white. 2*n*= 28,32. Moist areas; < 2400 m. CA-FP, SNE; circumboreal. 2 vars in CA.

var. *lanceolata* (Barton) Fern. (p. 363) **STS** gen decumbent to erect. **LVS**: cauline blade length 3 × width. Moist areas, gen coniferous forest, woodland; < 2400 m. CA-FP, **SNE**; N.Am, e Asia. [var. *atropurpurea* Fern.] ❀ IRR or WET:1,2,**4–7,15–17, 22–24**&SHD:3,**8,9**,10,**14,18–21**;can be INV. May–Sep

SALAZARIA BLADDER SAGE

1 sp. (Don Jose Salazar y Larrequi, Mexican astronomer, US-Mexican Boundary Survey)

S. mexicana Torrey (p. 363, pl. 64) Shrub, 5–10(15) dm, ± rounded, branched. **ST**: lateral branches spreading, rigid, tips becoming spine-like; twigs ± canescent. **LF** short-petioled to subsessile; blade 3–15(20) mm, 2–8 mm wide, gen ovate to elliptic, base rounded, margin entire, glabrous to puberulent. **INFLS** axillary at distal 3–10 nodes; fls 2; axis finely glandular-puberulent; bracts 0. **FL** 2-lipped; calyx lobes ± equal, entire, purplish, becoming 1–2 cm, bladder-like in fr; corolla 15–25 mm, upper lip ± entire, white to light violet, lower lip ± 3 lobed, violet to purple; stamens 4, gen enclosed by upper lip, lower stamen pair < upper pair, anthers ciliate. **FR**: nutlets, widely ovoid, short-stalked, tubercled. Sandy to gravelly slopes, washes, shrubland, woodland; < 1800 m. **s SNE, D**; to UT, TX, n Mex. ❀ DRN,IRR:7,**8–12**,13,**14**,18,**19–21**&SUN: 15,16,22,23. Mar–Jun

SALVIA

Deborah Engle Averett & Kurt R. Neisess

Ann, per, shrub. **LF** entire, lobed, or toothed, gen not spine-tipped. **INFL**: clusters gen many-fld, gen head-like, gen spheric, gen involucred, gen surrounding nodes in gen ± spike-like, gen interrupted panicles, or fls 1–several per lf axil. **FL**: calyx gen 2-lipped, upper lip entire or of 3 gen shallow, sometimes spine-tipped lobes, lower lip gen of 2 gen spine-tipped lobes; corolla 2-lipped, upper lip 2-lobed to entire, lower lip with 3 spreading lobes (middle often expanded); fertile stamens 2, attached in throat, anther sacs 1–2 per stamen (if 2, then separate on thread-like structure, 1 fertile, > other); style forked at tip. ± 900 spp.: ± worldwide, esp trop, subtrop Am. (Latin: to save, from medicinal use) ❀ All spp. are excellent bee fodder and have edible seeds (a traditional food of native Californians).

1. Ann
  2. Corolla 15–25 mm; bract 2–5 cm; pl white-woolly . . . . ***S. carduacea***
  2' Corolla 6–8 mm; bract < ± 1 cm; pl short-hairy . . . . ***S. columbariae***
1' Per to shrub
  3. Lf spine-tipped
    4. Lf white-woolly, short-petioled, gen deciduous; spines in 1–2 pairs or 0 on lf margins; calyx
      lobes triangular, white-woolly . . . . ***S. funerea***
    4' Lf green-tomentose, sessile or short-petioled, persistent; spines in 2–7 pairs on lf margins; calyx
      lobes lanceolate, not white-woolly . . . . ***S. greatae***
  3' Lf not spine-tipped
    5. Lower corolla lip > 2 × upper; corolla white to lavender
      6. Lf grayish from ± bristly hairs or gen greenish; infl gen < 5 dm . . . . ***S. eremostachya***
      6' Lf grayish velvety from minute appressed hairs; infl gen > 10 dm
        7. Lvs lanceolate, base tapered; calyx lobes barely or not spine-tiped; corolla white with lavender . . ***S. apiana***
        7' Lvs oblong-ovate, base ± truncate to tapered; calyx lobes spine-tipped; corolla white . . . . ***S. vaseyi***
    5' Lower corolla lip < 2 × upper; corolla blue or blue-violet to rose, rarely white
      8. Width of lower middle corolla lobe < 1/2 length; lf puckered, teeth rounded, hairs minute, moderately
        dense to sparse, spreading . . . . ***S. mohavensis***
      8' Width of lower middle corolla lobe > or = length; lf not puckered, ± entire, scaly, or hairs very dense,
        appressed

9. Corolla blue-violet to rose, gen 13–23 mm; bract 10–20 mm; lf 20–50 mm . . . . *S. pachyphylla*
9' Corolla blue, rarely purple, rose, or white, gen 6–13 mm; bract 5–12(14) mm; lf 4–30 mm . . . . . . *S. dorrii*
10. Bract, calyx lower surface with soft, shaggy hairs or scaly, margin hairs long . . . . var. *pilosa*
10' Bract, calyx glabrous to scaly, margin hairs gen short . . . . var. *dorrii*

S. apiana Jepson (p. 363) WHITE SAGE Per, subshrub. **ST** < 1 m. **LVS** gen basal, 4–8 cm, widely lanceolate; base tapered; teeth minute, rounded; hairs dense, minute, simple, appressed. **INFL**: clusters few-fld, in ± spike-like clusters, these in ± raceme-like, interrupted panicles; bracts < to > calyx, linear-lanceolate, re-curved. **FL**: calyx 8–10 mm, lobes barely or not spine-tipped, upper lip entire; corolla tube 12–22 mm, white with lavender, upper lip < 2 mm, lower lip 4–5 mm, upcurved, blocking throat; stamens and style exserted. **FR**: nutlet 2.5–3 mm, light brown, shiny. *n*=15. Common. Dry slopes, coastal-sage scrub, chaparral, yellow-pine forest; gen < 1500 m. SCo, TR, PR, **w edge D**; Baja CA. ❀ DRN,DRY,SUN:**7**,8,9,11,**14–16**,17,**18–24**;also STBL. May–Aug

S. carduacea Benth. (p. 363) THISTLE SAGE Ann, 1–10 dm, white-woolly. **LVS** basal, subsessile, 3–10(30) cm, oblanceolate, 1-pinnately dissected; margin wavy, short-spiny. **INFL** scapose; clusters 1.5–3 cm wide, 1–4 per fl st; bracts 2–5 cm, lanceolate, spiny. **FL**: calyx 10–17 mm, lobes spine-tipped, upper lip 3-lobed; corolla tube 15–25 mm, lavender, rarely blue or white, upper lip 2-lobed, lower lip > 2 × upper; stamens exserted. **FR**: nutlet 2.5 mm, tan to gray, flecked. *n*=16. Common. Sandy or gravelly soils; < 1400 m. Teh, SnJV, e SnFrB, SCoRI, SW, **w D**; n Baja CA. ❀ DRN,SUN,DRY:**7–9,11**,14–16,**18–21**,22–24. Mar–May

S. columbariae Benth. (p. 363) CHIA Ann, 1–5 dm; hairs gen sparse, short. **LVS** basal, 2–10 cm, oblong-ovate, 1–2-pinnately dissected; lobes irregularly rounded, minutely bristly. **INFL** ± scapose; clusters gen 1–2 per fl st; bracts < ± 1 cm, ± round, awn-tipped. **FL**: calyx 8–10 mm, purple-tipped, upper lip unlobed but 2(3)-awned; corolla tube 6–8 mm, pale to deep blue, upper lip entire to shallowly 2-lobed, 2–3 mm, lower lip ± 2 × upper; stamens and style exserted. **FR**: nutlet, 1.5–2 mm, tan to gray. *n*=13. Common. Dry, disturbed sites, chaparral, coastal-sage scrub; gen < 1200 m. **CA** (exc KR, CaR, n SN); to UT, AZ. [var. *ziegleri* Munz] ❀ DRN,SUN,DRY:**7–10**,11,12,**14–24**. Mar–Jun

S. dorrii (Kellogg) Abrams Shrub, spreading to mat-forming, 10–70 cm, densely white-scaly throughout. **LF** 4–30 mm, linear to spoon-shaped, ± entire. **INFL**: clusters gen 12–30 mm wide; bracts 5–12(14) mm, ± round. **FL**: calyx gen 6–11 mm, blue, purple, or rose, upper lip gen entire, rounded, lobes of lower lip acute, not spine-tipped; corolla tube gen 6–13 mm, blue, rarely purple, rose, or white, upper lip 2-lobed, 2–3 mm, < lower; stamens and style exserted. **FR**: nutlet, 1.8–3.5 mm, gray to reddish brown. *2n*=30. Common. Dry, mostly rocky places; 1000–4000 m. nw CaRH, s SNH (e slope), Teh, **GB**, **n DMoj**; to WA, ID, UT, AZ. Highly variable; vars. intergrade. 3 vars in CA.

var. **dorrii** (Kellogg) Abrams (p. 363) **LF** 6–20 mm, widest 2–13 mm from base, abruptly narrowed to petiole. **INFL**: bract and calyx glabrous to scaly, margin hairs gen short. Dry flats, slopes; 800–3100 m. nw CaRH, s SNH (e slope), Teh, **GB, n DMoj**; to OR, ID, UT, AZ. [ssp. *argentea* (Rydb.) Munz; ssp. *gilmanii* (Epling) Abrams] ❀ TRY. May–Jul

var. **pilosa** (A. Gray) J.L. Strachan & Rev. **LF** 4–32 mm, widest 1–15 mm from base, abruptly narrowed to petiole. **INFL**: bract and calyx lower surfaces with soft, shaggy hairs or scaly, upper surface glabrous to minutely hairy, margin hairs long. Desert slopes, washes; 900–1900 m. s SNH (e slope), Teh, **GB, DMoj**; NV, AZ. ❀ TRY;DFCLT.

S. eremostachya Jepson (p. 363) DESERT SAGE Shrub, erect, 60–80 cm, finely branched. **LF** 1.5–3.3 cm, linear, puckered; base truncate; teeth minute, rounded. **INFL**: clusters 1–3 per fl st; bracts 5–12 mm, lanceolate to ovate, rose to purple, papery. **FL**: calyx 6–9 mm, upper lip minutely 3-lobed; corolla tube white to pale lilac, 10–17 mm, upper lip 2-lobed, 1–4 mm, lower lip 4–8 mm; stamens and style exserted. **FR**: nutlet 3 mm, yellow-brown. *2n*=30. UNCOMMON. Dry, rocky, gravelly places, lower pinyon/juniper; 700–1400 m. **w edge DSon**; n Baja CA. Locally common. Mar–May

S. funerea M.E. Jones (p. 363, pl. 65) DEATH VALLEY SAGE Shrub, 5–12 dm, densely branched, densely white-woolly. **LF** 9–20 mm, ± ovate, short-petioled, gen deciduous; spines 1 at tip, in 1–2 pairs or 0 on margins. **INFL**: fls gen 3 per lf axil. **FL**: calyx 4.5–6 mm, lobes 5, subequal, triangular, spine-tipped; corolla tube 12–16 mm, violet, rarely blue, upper lip 2-lobed, 2–2.5 mm, lower lip almost 2 × upper; stamens and style incl. **FR**: nutlet ± 3 mm, brown, smooth. *2n*=64. UNCOMMON. Dry washes, canyons; 0–360 m. **ne DMoj (Death Valley, Amargosa and Panamint mtns)**. Mar–Jun

S. greatae Brandegee (p. 363) OROCOPIA SAGE Shrub, < 1 m; hairs glandular, tangled. **LF** 9–20 mm, ± ovate, sessile or short-petioled, green-tomentose, persistent; spines 1 at tip, in 2–7 pairs on margins. **INFL**: clusters many-fld. **FL**: calyx 9–11 mm, upper lip of 3 shallow, spine-tipped lobes; corolla tube 9–11 mm, lavender to rose, upper lip 2-lobed, 2–2.5 mm, lower lip 4–5 mm; mature stamens and style exserted. **FR**: nutlet, 2–3 mm, flat, keeled, gray to brown. *2n*=±30. RARE. Alluvial slopes; 30–240 m. **se DSon (Orocopia, Chocolate Mtns)**. Mar–Apr

S. mohavensis E. Greene (p. 363) Shrub, gen < 1 m; hairs minute, simple. **LF** 1.5–2 cm, lance-oblong to ovate, puckered; base tapered to ± truncate; teeth small, rounded; some hairs glandular. **INFL**: clusters subspheric, gen 1 per fl st; bracts gen 1–1.5 cm, >> calyx, ovate to ± round, pale, papery. **FL**: calyx 7–12 mm, minutely glandular-hairy, sometimes tinged blue, upper lip entire to minutely 3-lobed; corolla tube 18–20 mm, tinged sky-blue, upper lip entire, slightly < lower, 4–6 mm; stamens and style exserted. **FR**: nutlet, 2–3 mm, tan to brown. *n*=15. Dry, rocky slopes, blackbush scrub, pinyon/juniper woodland; 300–1500 m. **DMtns**; AZ. Locally common. ❀ TRY;DFCLT. Jul–Oct

S. pachyphylla Munz (p. 363) Shrub, prostrate, 20–80 cm, rooting at nodes, scaly. **LF** 20–50 mm, obovate to spoon-shaped; margins wavy, ± entire. **INFL**: clusters 20–50 mm wide, 2–5 per fl st; fl st ± hidden between clusters; bracts 1–2 cm, papery, ovate to oblong-elliptic, greenish to purple or rose. **FL** 8–13 mm; calyx = bracts in color, upper lip entire, obtuse to abruptly soft-pointed, 3–6 mm, lower lip 1–3 mm, lobes acuminate, not spine-tipped; corolla tube gen 13–23 mm, blue-violet to rose, upper lip entire, 4–6 mm, < lower lip; stamens and style exserted. **FR**: nutlet, 2.5–3 mm, tan to brown. *2n*=30. Dry slopes, pinyon/juniper to yellow-pine forest; 1400–2500 m. s SNH, Teh, SnBr, PR, **DMtns**; NV, AZ, n Baja CA. ❀ TRY;DFCLT. Jul–Oct

S. vaseyi (Porter) Parish (p. 363) Subshrub, < 1.5 m. **LVS**: basal > cauline, 2–6 cm, oblong-ovate; base ± truncate to tapered; teeth minute, rounded; hairs dense, minute, appressed. **INFL**: clusters 1.5–3 cm wide; bracts < 2 cm, lanceolate. **FL**: calyx 8–14 mm, lobes spine-tipped; corolla tube 13–20 mm, white, upper lip shallowly 2-lobed, 2–4 mm, lower lip 6–9 mm; stamens and style exserted. **FR**: nutlet, 2.5–3 mm, light brown. *n*=15. Dry, rocky desert slopes, canyons, creosote-bush scrub; < 800 m. **w edge DSon**; n Baja CA. Locally common. Apr–Jun

SCUTELLARIA SKULLCAP

Richard G. Olmstead

Per, gen hairy, sometimes glandular, from rhizomes or tubers. **STS** erect, branched or not. **LVS** basal and cauline; lower gen petioled; cauline becoming ± sessile upward. **INFL**: fl gen 1 per lf axil (or appearing as a bracted raceme). **FL**: calyx 2-lipped, lips ± equal, enclosing nutlets, back of upper lip dome-like or transversely ridged, gen with concave depression behind ridge; corolla 2-lipped, white to violet-blue, upper lip < lower lip, ± entire, hood-like, lower lip 3-lobed; stamens 4, pairs ± equal, enclosed by upper corolla lip, anthers ciliate, lower two 1-chambered; disk below ovary gen green-yellow. **FR** gen ovoid, gen minutely papillate, brown or black. ± 300 spp.: gen temp world-wide. (Latin: tray, from calyx dome or ridge) [Olmstead 1990 Contr Univ Michigan Herb 17:223–265] Key to species adapted by Margriet Wetherwax.

1. Corolla 12–14 mm, white to light yellow, lower lip sometimes blue- to violet-mottled; drier gravelly soils
.... *S. bolanderi* ssp. *austromontana*
1' Corolla 6–8 mm, blue to violet-blue, lower lip gen white-patched; marshes, wet meadows ...... *S. lateriflora*

S. bolanderi A. Gray Pl 30–100 cm; rhizomes slender, tips ± swollen. **ST**: hairs 1–2 mm, spreading, often gland-tipped. **LVS**: basal petioles 2–10 mm; upper cauline blades ovate to cordate, crenate (rarely entire), base truncate to ± lobed, tip rounded. **FL**: pedicel 2–3 mm; calyx 3–5 mm, ridged; corolla 13–19 mm, white, lower lip blue-mottled, inner surface long-soft-hairy. **FR** brown to black. Gravelly soils, streambanks, oak or pine woodland; 300–2000 m. SN, SCoRI, SW, s DMoj. 2 sspp in CA.

ssp. ***austromontana*** Epling (p. 363) SOUTHERN SKULLCAP **LF** length > 2 × width. **FL**: corolla 12–14 mm. RARE. Habitat like sp.; 600–2000 m. SnBr, PR, **s DMoj**.

S. lateriflora L. (p. 363) BLUE SKULLCAP Pl 20–60 cm; rhizomes slender. **ST** glabrous or hairs sparse, hairs << 0.5 mm, ascending to upcurled, gen not glandular. **LVS**: basal gen 0; lower cauline petioles 10–20 mm; upper cauline blades ovate to lanceolate, ± dentate, base rounded to truncate, tip acute. **INFL**: raceme or spike, bracted; bracts < 8 mm. **FL**: pedicel 1–3 mm; calyx 1.5–3 mm, upper lip back dome-like; corolla 6–8 mm, blue, lower lip blue, inner surface glabrous or sparsely soft-hairy. **FR** ± spheric, brown. RARE in CA. Marshes, wet meadows; < 500 m. n SnJV, **SNE (Saline Valley)**; to B.C., e US. ❀ WET:4–9,10,11,**14–16**,17, 18,**19–23**,24. May–Jul

STACHYS HEDGE NETTLE

Barrett H. Anderson & Barry D. Tanowitz

Ann, per, hairy, often glandular; rhizome slender or 0. **ST** decumbent to erect, 1–25 dm. **LVS** 1.5–18 cm; lower gen petioled; upper ± sessile; blades oblong to ovate, serrate to crenate. **INFL**: spike of ± sessile clusters, gen terminal, interrupted or continuous, bracted. **FL**: calyx bell-shaped, radial to ± 2-lipped, veins 5–10, lobes 5, erect or spreading, triangular, tips sharp; corolla white, pink, red, magenta, or purple, tube narrow, with internal ring of hairs gen above base, perpendicular to oblique to tube axis, sometimes narrowed on lower surface, upper lip erect or gen parallel to tube axis, concave, entire, rarely notched, gen hairy, lower lip perpendicular to tube axis or reflexed, 3-, rarely 2-lobed, glabrous to hairy. **FR** oblong to ovoid, brown to black. ± 300 spp.: temp (exc Australia); some cult for orn. (Greek: ear of corn, from infl) [Epling 1934 Fedde Rep Sp Nov Regni Veg 80:1–75]

1. Lvs and infl with dense, cobwebby hairs; corolla gen white to pink .... *S. albens*
1' Lvs and infl with soft or stiff hairs, sometimes dense, but not cobwebby; corolla pale pink to purple
.... *S. ajugoides* var. *rigida*

S. ajugoides Benth. **ST** branched or not. **LF**: petiole 1.5–6 cm. **INFL** > 5 cm, interrupted; bracts sometimes long, leafy. **FL**: calyx tube 3–6 mm, hairs 0 or soft to stiff, sometimes glandular; corolla tube 6–11 mm, sometimes spurred or sac-like, ring of hairs > 2 mm from base, upper lip 2.5–5 mm, lower lip 4.5–8.5 mm. Gen moist places, sometimes dry hillsides, many communities; < 2500 m. **CA**; to B.C., Baja CA. 2 vars in CA.

var. *rigida* Jepson & Hoover (p. 367) **ST** gen erect to ± decumbent, 6–10 dm, ± glabrous to soft- or stiff-hairy, sometimes glandular. **LF**: blade 5–9 cm, ovate to lanceolate, glabrous or soft-hairy to densely felt-like, sometimes glandular, base rounded to ± cordate, tip acute to obtuse. **INFL**: clusters 6–16-fld. **FL**: corolla pink to magenta or purplish, tube 6–10 mm, ring of hairs strongly oblique. Moist to ± dry places; < 2500 m. **CA (very**

uncommon in D); to WA, Baja CA. [*S. r.* Benth. incl ssp. *lanata* Epling, ssp. *quercetorum* (A.A. Heller) Epling, ssp. *rivularis* (A.A. Heller) Epling; *S. emersonii* Piper; *S. mexicana* Benth] ❀ STBL. Jul–Aug

S. albens A. Gray (p. 367) **ST** erect, 5–25 dm, often branched. **LF**: petiole < 5 cm; blade 3–15 cm, widely ovate, crenate to serrate, gen with dense, cobwebby hairs, base ± cordate, tip acute to obtuse. **INFL** 10–30 cm, ± interrupted, clusters 6–12-fld; hairs dense, cobwebby. **FL**: calyx tube 4–5.5 mm, hairs dense, cobwebby; corolla white to pinkish, tube 6–9 mm, ring of hairs > 2 mm from base, oblique, upper lip 3.5–5.5 mm, lower lip 6–8 mm. Wet, swampy to seepy places; < 3000 m. NCoR, c&s SN, Teh, GV, CCo, SW, **W&I, rarely D**. ❀ WET or IRR:1,2,4,**5–9**,**14–17**,**19–23**,24&SHD:3,10, **18**;INV. May–Oct

TEUCRIUM

Ann, per, glabrous to short-hairy. **STS** ascending to erect, branched or not. **LVS** petioled, crenate to deeply lobed, lobes oblong. **INFLS** in CA axillary at upper nodes; fls gen 2, subtended by lvs or bracts. **FL**: calyx ± radial, 10-veined, 5-lobed, lobes subequal; corolla 1-lipped, tube split above, lip 5-lobed, ± flat, distal lobe > lateral lobes, tip rounded, lateral lobe tips acute to obtuse; stamens 4, lower pair gen > upper; style lobes gen equal. ± 100 spp.: worldwide, esp Medit. (Greek: ancient name) [McClintock & Epling Brittonia 5:491–510]

1. Ann, 1–3 dm, simple to branched at base; pedicels < 5 mm; corolla lip 4–8 mm, white to bluish,
   purple-spotted . . . . ***T. cubense*** ssp. ***depressum***
1' Per, 5–10 dm, gen branched throughout; pedicels 8–25 mm; corolla lip 10–17 mm, white, gen
   violet-streaked . . . . ***T. glandulosum***

T. cubense Jacq. ssp. ***depressum*** McClint. & Epling (p. 367)
ST simple to branched at base. **LF** gen withering in fr; lower
2–4 cm, blade ovate to obovate, crenate to lobed; upper 0.5–
1.5 cm, gen deeply 3-lobed. **INFL**: pedicel 1–5 mm. **FL**: calyx
tube 1–3 mm, lobes 3–6 mm, bristle-tipped; corolla 7–15 mm,
slightly puberulent inside, white to bluish, purple-spotted; filament glabrous. Sandy soils, washes, fields; < 400 m. **DSon**; to
TX, Baja CA. Other 4 sspp. in s US, Caribbean, Mex, S.Am.
Mar–May

T. glandulosum Kellogg (p. 367) **ST** gen branched throughout. **LF** 1–4 cm, ± persistent, gen deeply 3-lobed. **INFL**: pedicel
8–25 mm. **FL**: calyx tube 2–4 mm, lobes 4–8 mm, gen acute;
corolla 15–21 mm, densely puberulent inside, white, violet-
streaked; filament short-hairy below middle. Uncommon. Rocky
slopes, canyons; 400–500 m. **ne DSon (Whipple Mtns)**; AZ,
Baja CA. ✤ TRY. Apr–May

TRICHOSTEMA BLUECURLS

Harlan Lewis

Ann, shrub, strong-scented. **ST** hairy, often glandular. **LF** simple; blade linear to ovate, entire. **INFL**: cymes (racemes
in *T. lanceolatum*), axillary. **FL**: calyx lobes 5, equal or uppermost 1 narrower; corolla blue or lavender, tube straight
or curved upward, sometimes abruptly near throat, incl to much exserted from calyx, lobes 5, lowest a gen reflexed lip;
stamens 4, attached near throat, gen much exserted, ascending between upper corolla lobes, gen arched. **FR**: nutlets 4,
joined in basal ± 1/3, puberulent to hairy, irregularly ridged. ± 17 spp.: N.Am. (Greek: hair, stamen) [Lewis 1945
Brittonia 5:276–303]

T. austromontanum Harlan Lewis (p. 367) Ann < 5 dm. **ST**:
short hairs appressed, long hairs spreading, some hairs glandular. **LF**: petiole indistinct or < 5 mm; blade 2–5 cm, elliptic, length
> 4 × width. **FL**: calyx lobes > 2 × tube, widest at base, acute, ±
equal; corolla tube 1.5–3 mm, curved gradually upward, ± = ca-

lyx, lower lip 1.8–3 mm; stamens 3–5.5 mm, exserted, arched.
Uncommon. Drying margins of lakes, meadows, streams; 1000–
2500 m. Teh, TR, PR, **SNE**; Baja CA. 2 sspp in CA.

ssp. ***austromontanum*** **ST** < 5 dm. **LF** < 50 mm, ± = or <
internode above. *n*=14. Habitat and range of sp. Jul–Sep

LENNOACEAE LENNOA FAMILY

George Yatskievych

Ann, per, root parasite lacking chlorophyll. **ST** (actually a peduncle) fleshy, underground, gen unbranched, white or
± brown. **LVS** scale-like, alternate. **INFL**: panicle, spike, or head. **FL** bisexual, ± radial; calyx lobes 4–10; corolla
lobes 4–10; stamens as many as corolla lobes, epipetalous, incl; ovary superior, chambers 10–32, placentas axile,
style 1, stigma lobes 5–9. **FR**: capsule, circumscissile, hidden by persistent perianth. **SEEDS** in a ring, 1 per chamber, ± reniform, flat, brown. 2 genera, 4 spp.: sw US to n S.Am, nowhere common; some historically harvested for
food. [Yatskievych & Mason 1986 Syst Bot 11:531–548] Recently treated in an expanded Boraginaceae (also
including Hydrophyllaceae) [Angiosperm Phylogeny Group 1998 Ann Missouri Bot Gard 85: 531–553; Olmstead
et al 2000 Mol Phylog Evol 16: 96-112].

PHOLISMA

Per. **ST** < 1.5 m. **LF** 5–25 mm, linear to triangular, glandular. **FL** 7–10 mm. **FR** ± circumscissile below middle. 3 spp.:
s CA, w AZ, nw Mex. (Greek: scale, from scaly stem)

1. Infl a dense panicle or spike; calyx lobes glandular-puberulent (hairs < 0.5 mm) . . . . ***P. arenarium***
1' Infl a concave head; calyx lobes appearing feathery (glandular hairs 1–1.5 mm) . . . . ***P. sonorae***

P. arenarium Hook. (p. 367, pl. 66) **ST** 3–8 dm, 1–2 cm diam.
FL: calyx lobes linear to spoon-shaped; corolla lavender to bluish purple, margin white, exterior minutely puberulent; ovary
chambers 10–20. 2*n*=36. Uncommon. Sandy soil, coastal dunes,
chaparral, desert; < 1900 m. CCo, SCo, PR, **D**; w AZ, nw Mex.
Parasitic on *Croton, Eriodictyon*, various shrubby Asteraceae.
Apr–Jul, Oct

P. sonorae (A. Gray) G. Yatskievych SAND FOOD **ST** 5–15
dm, 0.5–2 cm diam. **FL**: calyx lobes linear; corolla pink to purple,
margin white, exterior glabrous; ovary chambers 12–32. 2*n*=36.
RARE. Dunes, sandy areas; < 0–200 m. **DSon (se Imperial
Co.)**; w AZ, nw Mex. Parasitic on *Eriogonum, Tiquilia, Ambrosia, Pluchea.* [*Ammobroma s.* A. Gray] Threatened by off-road
vehicles. Apr–May

LENTIBULARIACEAE BLADDERWORT FAMILY

Lawrence R. Heckard

Ann, per, carnivorous, of moist or aquatic habitats. **ST** a caudex (*Pinguicula*) or filamentous with alternate or
whorled branch systems in place of lvs (*Utricularia*). **LVS** 0 or in basal rosette, simple. **INFL**: raceme, spike, 1-fld

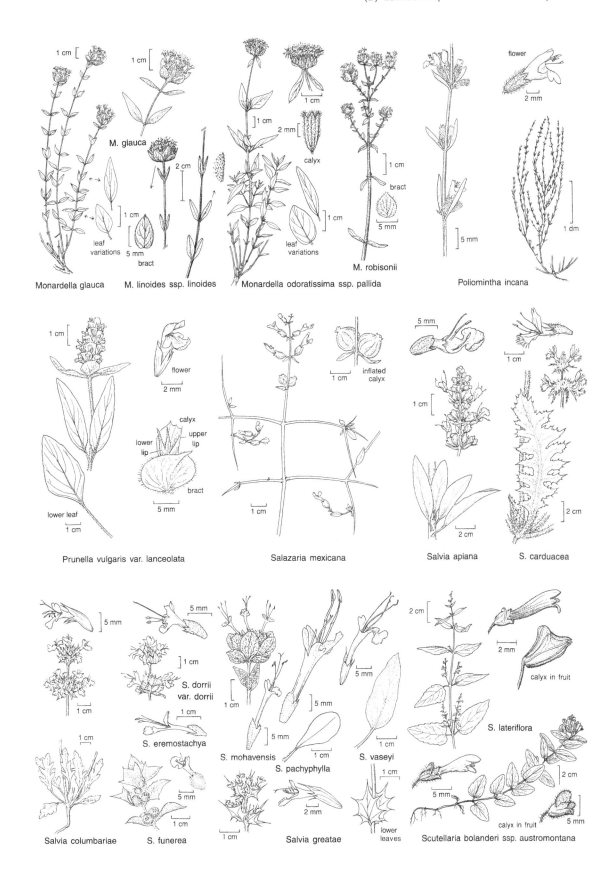

Monardella glauca

M. linoides ssp. linoides

Monardella odoratissima ssp. pallida

M. robisonii

Poliomintha incana

Prunella vulgaris var. lanceolata

Salazaria mexicana

Salvia apiana

S. carduacea

S. dorrii var. dorrii

S. eremostachya

Salvia columbariae

S. funerea

S. mohavensis

Salvia greatae

S. pachyphylla

S. vaseyi

S. lateriflora

Scutellaria bolanderi ssp. austromontana

or scapose. **FL** bisexual; calyx 2-lipped, gen 4–5-lobed; corolla 2-lipped, 5-lobed, spurred at base, lower lip arched and gen pouched upward, blocking throat; fertile stamens 2, epipetalous; ovary superior, chamber 1, placenta gen free-central; stigma unequally 2-lobed, ± sessile. **FR**: capsule, round, gen 2-valved (sometimes irregularly dehiscent). **SEEDS** gen many, small.   4 genera, 200 spp.: worldwide, esp trop.

UTRICULARIA BLADDERWORT

Ann, per, gen aquatic. **ST** submersed or creeping, uncoiling at tip, sometimes with claw-like appendages, with gen alternate green branch systems ("lvs") of ± linear or thread-like segments; bladders borne on st or "lvs" trap small organisms when entrance hairs triggered; terminal, overwintering buds dispersed in some spp. **LVS**: true lvs 0. **INFL**: raceme (rarely 1-fld), bracted, emergent. **FL**: calyx lips entire; corolla gen yellow, lower lip entire or 3-lobed, gen blocking throat, hairy, red-spotted, upper lip ± entire. **FR** gen circumscissile (or opening irregularly). ± 180 spp.: worldwide, esp. trop. (Latin: little bag, from bladders) [Ceska & Bell 1973 Madroño 22: 74–84] ❀ Pure water ponds; DFCLT.

U. vulgaris L. (p. 367) COMMON BLADDERWORT **STS** shallowly floating; winter buds 1–2 cm, bristly; "lvs" dense, 2–9 cm, 1–2-parted at base, each part several times unequally pinnately dissected; ultimate segments dense, 20–150, thread-like, not flat, margin bristly; bladders near "lf" base > those near tip. **INFL** 5– 20-fld; peduncle 1–4 dm, stout, < 2 mm wide; pedicel recurved in fr. **FL**: corolla 1–2 cm; lower lip ± = upper, slightly > curved, cylindric, pointed spur. **SEED** unwinged. $2n=±40$. Quiet water; < 2700 m. NW, CaR, SN, SnFrB, SnBr, **GB**, **w DMoj**; to AK, e N.Am, Mex; circumboreal. Jul–Sep

LINACEAE FLAX FAMILY
Niall F. McCarten

Ann, per, shrub. **ST** gen erect, branched, glabrous, hairy, or glandular. **LVS** gen cauline, alternate, opposite, or whorled, simple, sessile, gen linear to ovate; stipules glandular or 0. **INFL**: raceme or cyme, axillary, open to dense. **FL** gen bisexual, radial, nodding in bud; sepals 4–5, free, glabrous, hairy, or margins gland-toothed; petals 4–5, free, blue, white, yellow, or pink, ephemeral; stamens 4–5, alternate petals, gen appendaged; staminodes alternate stamens or 0; ovary superior, chambers 2–5 but becoming 4–10 by growth of false septa, styles 2–5. **FR**: gen capsule. 13 genera, 300 spp.: worldwide, esp temp; some cult (*Linum usitatissimum*, flax, linseed; *L. bienne*, *L. grandiflorum*, orn). [Robertson 1971 J Arnold Arbor 52: 649–665]

LINUM FLAX

Ann, per. **ST** 5–90 cm. **LVS** alternate, opposite, or whorled, erect, glabrous or hairy; margins entire or gland-toothed. **INFL**: raceme or cyme. **FL**: sepals 5, margins gen translucent; petals 5, 8–25 mm; stamens 5; staminodes 5 or 0; ovary chambers 10, styles 5, free or not, stigmas > styles in width (spheric or ± elongate) or = styles in width (± linear). **FR** 5–10 mm. **SEEDS** 10, gen gelatinous when wet. ± 200 spp.: temp & subtrop, esp Medit. *Linum usitatissimum* has been reported from D as a garden escape. (Latin: flax)

1. Petals yellow to orange; margins of lvs, bracts, and sepals gland-toothed; styles fused to near tip . . . *L. puberulum*
1' Petals blue to white; margins of lvs, bracts, and sepals glabrous; styles free . . . . *L. lewisii*
2. Petals 6–7 mm; styles < 6 mm . . . . var. *alpicola*
2' Petals 10–15 mm; styles gen > 6 mm . . . . var. *lewisii*

L. lewisii Pursh Per. **ST** glabrous. **LF** 10–20 mm, linear to lanceolate, glabrous. **INFL**: pedicels 10–30 mm. **FL**: sepals 4–6 mm, ovate; petals 6–15 mm, blue, sometimes white; styles free, stigmas ± spheric. **FR** 5–8 mm. **SEED** 3.5–4.5 mm. Dry, open ridges, slopes; 400–3400 m. CA-FP, **n SNE**, **W&I**, **D**; to AK, Rocky Mtns, TX, Mex.

var. *alpicola* Jepson **FL**: petals 6–7 mm; styles < 6 mm. Alpine ridges; 2000–3400 m. SNH, **W&I**; to UT.

var. *lewisii* (p. 367) **FL**: petals 10–15 mm; styles > 6 mm. Open slopes; 400–3000 m. CA-FP, **n SNE**, **D**; OR. ❀ DRN, SUN:1,**4–6,15–17**& IRR:2,3,7,8,9,**14,18–24**. May–Sep

L. puberulum (Engelm.) A.A. Heller (p. 367) Per. **ST** puberulent. **LF** 5–10 mm, linear; margin gland-toothed. **INFL**: bract margins gland-toothed. **FL**: sepals 4–7 mm, < or = fr, lanceolate, margins gland-toothed; petals 10–15 mm, yellow to orange; styles fused to near tip. **FR** 3.5–4 mm. **SEED** ± 2.5 mm. $n=15$. Dry ridges; 1000–2500 m. **e DMtns**; to Colorado, TX. ❀ TRY. May–Jul

LOASACEAE LOASA FAMILY
Barry Prigge

Ann to shrub; hairs needle-like, stinging, or rough. **LVS** alternate in CA, gen ± pinnately lobed; stipules 0. **INFL** various. **FL** bisexual, radial; sepals gen 5, gen persistent in fr; petals gen 5, free or fused to each other or to filament tube; stamens 5–many, filaments thread-like to flat, sometimes fused at base or in clusters; petal-like staminodes sometimes present; pistil 1, ovary inferior, chamber gen 1, placentas gen 3, parietal, style 1. **FR**: gen capsule (utricle). **SEEDS** 1–many. 15 genera, ± 200 spp.: esp Am (Afr, Pacific). [Ernst & Thompson 1963 J Arnold Arbor 44:138–142]

1. Subshrub; stamens 5; stigma 1; fr utricle, seed 1 . . . . **PETALONYX**
1' Herbs or subshrubs; stamens gen many; stigma lobes 3 or 5; fr capsule, seeds gen many
  2' Some hairs needle-like, stinging; petals fused to each other or filament tube; stigma lobes and placentas 5
  . . . . **EUCNIDE**
  2' Hairs not needle-like (often barbed-rough); petals free; stigma lobes and placentas 3 . . . . **MENTZELIA**

EUCNIDE ROCK NETTLE

Ann to subshrub; hairs gen needle-like and stinging (or barbed). **LF** widely ovate to ± round, toothed to ± lobed; base widely tapered to cordate. **INFL**: cyme, bracted. **FL**: petals fused below middle or to filament tube; stamens many, epipetalous or fused at base into short tube; ovary club-shaped to spheric, placentas 5, stigma lobes 5, gen appressed. **FR** obovoid to spheric, nodding or reflexed, clearly pedicelled, dehiscent from top by 5 valves. **SEEDS** many, < 1 mm, ± oblong, grooved or ribbed. 8 spp.: sw US, n Mex. (Greek: strongly nettle-like)

1. Ann; petals fused, tube 8–15 mm, lobes 2–5 mm, greenish; stamens epipetalous, sessile or filaments
   < 1.5 mm. . . . . *E. rupestris*
1' Subshrub; petals mostly free, 30–50 mm, fused at base to short filament tube, whitish to pale yellow;
   stamens fused at base, filaments 20–30 mm . . . . *E. urens*

E. rupestris (Baillon) H.J. Thompson & W.R. Ernst ROCK NETTLE Ann < 30 cm. **LF**: petiole < 6 cm; blade 1–8 cm, gen round, toothed to weakly lobed, green and shiny above, base widely tapered to cordate. **FL**: pedicel in fr < 2 cm, reflexed; sepals 3.5–7 mm, 2–3 mm wide, tip obtuse; corolla tube 8–15 mm, narrowly cylindric, with ring of hairs at base, gen greenish, lobes 2–5 mm, erect, green; stamens epipetalous, sessile or filaments < 1.5 mm; stigma lobes below or = anthers. **FR** 7–15 mm, cylindric. RARE in CA. Crevices, cliffs; 500–600 m. **s DSon (Painted Gorge, Imperial Co.)**; AZ, n Mex (incl islands).

[*Sympetaleia r.* (Baillon) A. Gray] Dec–Apr
E. urens (A. Gray) C. Parry (p. 367) Subshrub 30–100 cm. **LF**: petiole < 5 cm; blade < 10 cm, gen ovate, ± irregularly toothed, gray-green and dull above, base truncate to cordate. **FL**: pedicel in fr < 1.5 cm, ascending to erect; sepals 15–22 mm, 5–7 mm wide, tip acuminate; petals ± free, 30–50 mm, spreading, whitish to pale yellow; stamens fused at base, tube 2–5 mm, filaments 10–20 mm; stigma lobes slightly above anthers. **FR** 10–20 mm, club-shaped to obconic. n=21. Cliffs, rocky slopes, washes; < 1400 m. **D**; to sw UT, AZ, n Mex. Apr–Jun

MENTZELIA BLAZING STAR

Ann to shrub; hairs gen barbed-rough (smooth). **LF** linear to ovate, gen ± lobed; basal in rosettes, gen petioled; cauline gen sessile, ± reduced upward. **INFL**: gen cyme (or fl 1); bract gen 1 per fl, gen green. **FL**: sepals lanceolate to deltate, gen persistent; petals 5, free, gen yellow; stamens gen many, ± free, gen unequal, inner filaments gen thread-like; staminodes 0 or 5–many, outer often petal-like; ovary gen cylindric, placentas gen 3, style thread-like, stigma 3-furrowed or -lobed. **FR** gen tapered to base, sometimes curved. **SEEDS** gen many, grain- to prism-like (triangular in ×-section), angled, or lenticular and winged (important to identification). ± 50 spp.: w US, ± trop Am. [Darlington 1934 Ann Missouri Bot Garden 21:103–226] Key to species adapted by Margriet Wetherwax.

1. Bien or per (sometimes fls 1st year); seeds lenticular and winged or fusiform and ribbed lengthwise
  2. Seeds ± fusiform, 3-ribbed lengthwise . . . . *M. torreyi*
  2' Seeds lenticular, winged
    3. Petal-like staminodes 5
      4. Petal tips obtuse to rounded; sandy soils . . . . *M. multiflora* ssp. *longiloba*
      4' Petal tips acute; gypsum-clay soils. . . . . *M. pterosperma*
    3' Petal-like staminodes 0
      5. Petals > 30 mm; fr gen > 25 mm . . . . *M. laevicaulis*
      5' Petals < 20 mm; fr gen < 25 mm
        6. Lvs entire, narrow. . . . . *M. polita*
        6' Lvs toothed or lobed
          7. Fr > 12 mm . . . . *M. inyoensis*
          7' Fr < 10 mm . . . . *M. oreophila*
1' Ann; seeds grain- or prism-like, or narrowed and folded near middle
  8. Fr > 4 mm diam; seeds narrowed and folded near middle, end beak-like, ashy white; filament tip
   gen wide and minutely 2-toothed
    9. Petals 8, < 12 mm; filament tips acute, teeth 0. . . . . *M. reflexa*
    9' Petals 5, gen > 15 mm; filament tips 2-toothed, anther gen on thread-like stalk between teeth
      10. Fl bract white-scarious, margin green . . . . *M. involucrata*
      10' Fl bract green ± throughout (exc sometimes base)
        11. Upper lvs and bracts ± cordate, clasping; fr sessile, 14–25 mm, erect. . . . . *M. hirsutissima*
        11' Upper lvs and bracts not clasping; fr pedicelled, gen 9–15 mm, often reflexed
          12. Anther stalk gen < teeth; seeds narrowed at middle; e DSon, DMtns . . . . *M. tricuspis*
          12' Anther stalk gen > teeth; seeds narrowed above and below middle; c DMoj . . . . *M. tridentata*
  8' Fr gen < 4 mm diam; seeds grain- or prism-like; filament tip ± thread-like, not toothed
   13. Seeds 1-rowed above mid-ovary, prism-like, angles grooved

14. Basal lvs lobed; sepals 2–7 mm; petals 3–10 mm; style 3–7 mm . . . . ***M. affinis***
14' Basal lvs entire or toothed; ; sepals 1–4 mm; petals 2–6 mm; style 1–3.5 mm . . . . ***M. dispersa***
13' Seeds 2–3-rowed above mid-ovary, grain-like, angles rounded or sharp, not grooved
  15. Fl bracts toothed or lobed; axillary frs straight, erect
    16. Fl bracts wide, ± concealing fls and frs, mostly white-scarious; infl dense . . . . ***M. congesta***
    16' Fl bracts narrow, not concealing fls and frs, whitish below middle; infl open to ± dense
      17. Styles 1.5–3 mm; cauline lvs gen entire (or 2-lobed at base) . . . . ***M. montana***
      17' Styles 3–6 mm; cauline lvs gen lobed . . . . ***M. veatchiana***
  15' Fl bracts gen entire; axillary frs often curved 90°+
    18. Petals gen > 8 mm
      19. Seed not folded at 1 end, angles acute, surface pointed-papillate . . . . ***M. jonesii***
      19' Seed with flap-like fold at 1 end, angles rounded, surface rounded-papillate
        20. Sepals > 8 mm; petals 12–24 mm; style 7– 15 mm . . . . ***M. eremophila***
        20' Sepals < 8 mm; petals 7–15 mm; style 4– 8 mm . . . . ***M. nitens***
    18' Petals gen < 8 mm
      21. Seed surface smooth . . . . ***M. desertorum***
      21' Seed surface papillate
        22. Seed angles acute, surface pointed-papillate . . . . ***M. albicaulis***
        22' Seed angles rounded, surface rounded-papillate . . . . ***M. obscura***

M. affinis E. Greene (p. 367) Ann, 5–47 cm, erect. **LVS** 1–17 cm; basal toothed or lobed; cauline entire to lobed. **INFL**: bract lanceolate to obovate, entire or toothed. **FL**: sepals 2–7 mm; petals 3–10 mm, base with orange spot, tip rounded, notched, or toothed; stamens < 6.5 mm; style 3–6.5 mm. **FR** 12–28 mm, 1–3 mm wide, straight or arched < 90°. **SEED** 1–2 mm, ± 1 mm wide, prism-like, smooth; angles grooved; sides ± flat. Sandy grassland, woodland, creosote-bush scrub; s SN, SnJV, se SnFrB (Mount Hamilton), SCoRI, TR, SnJt, **D**; AZ, Baja CA. ❀ DRN,DRY, SUN:2,7,9,**10**,11,12,14–16,18–24;DFCLT. Apr–May

M. albicaulis Hook (p. 367) Ann 5–42 cm. **ST** decumbent to erect, white. **LVS** 1–11 cm; lower lobed (comb-like); upper entire to lobed (comb-like). **INFL**: bract lanceolate to ovate, entire or with 2–4 small teeth or angles, base rarely faintly whitish. **FL**: sepals 1–5 mm; petals 2–7 mm, base sometimes with yellow spot; stamens 3–5 mm; style 2–5 mm. **FR** 8–28 mm, 1.5–3.5 mm wide, straight, sometimes curved < 180°. **SEED** 1–1.5 mm, grain-like, pointed-papillate; angles acute. *n*=27,36. Gravel fans, washes, shrubland to pinyon/juniper woodland; 500–2300 m. Teh, SnGb, SnBr, **GB**, **D**; to B.C., Colorado, Baja CA. Pls with *n*=27 have been called *M. mojavensis* H.J. Thompson & Joyce Roberts. Mar–Jul

M. congesta Torrey & A. Gray (p. 367) Ann 7–40 cm. **ST** branched from base, erect, tan. **LF** 1–9 cm, entire to lobed. **INFL** dense; bract ± fused and appressed to ovary, ± entire, widely wedge-shaped, tip truncate to 3-toothed, lower 3/4 white-scarious. **FL**: sepals 1–4 mm; petals 3–9 mm, base with orange spot; stamens ± = style; style 1.5–5 mm. **FR** 5–12 mm, 2–3 mm wide, gen straight. **SEED** ± 1 mm, grain-like, ± sharp-angled, papillate, checkered or not; sides concave. Disturbed slopes, pine forest, sagebrush scrub, pinyon/juniper woodland; 1500–2700 m. SNH (e slope), Teh, WTR, SnGb, PR, **SNE**, **DMtns**; to ID. [var. *davidsoniana* (Abrams) J.F. Macbr.] May–Jul

M. desertorum (Davidson) H. J. Thompson & Joyce Roberts Ann, 10–41 cm, erect. **LVS** 1–12 cm; basal toothed or lobed; cauline entire to lobed. **INFL**: bract lanceolate to ovate, gen entire. **FL**: sepals 2–4 mm; petals 2.5–6 mm, base sometimes with orange spot; stamens = style; style 2–4 mm. **FR** 12–27 mm, 1–2.5 mm wide, straight to curved < 180°. **SEED** ± 1 mm, grain-like, rounded to slightly angled, smooth. *n*=9. Sandy flats in creosote-bush scrub; < 700 m. **D**; Baja CA.

M. dispersa S. Watson (p. 367) Ann, 7–48 cm, erect. **LVS** < 10 cm; basal entire to lobed; cauline entire to toothed. **INFL**: bract ovate or tapered from middle to tip, margin entire to lobed. **FL**: sepals 1–3.5 mm; petals 2–6 mm; stamens < 4.5 mm; style 1–3.5 mm. **FR** 7–25 mm, 1–2.5 mm wide, straight or curved < 30°. **SEED** 0.5–1.5 mm, 0.5–1 mm wide, prism-like; angles gen

sharp, grooved; sides smooth, shiny. *n*=9,18. Sandy or rocky soils; < 2500 m. CA-FP, **GB**; to WA, n-c US, NM. May–Aug

M. eremophila (Jepson) H.J. Thompson & Joyce Roberts (p. 367) Ann 8–43 cm. **ST** erect, tan to whitish, spreading in age. **LVS** 1–10 cm; basal lobed; cauline entire or lobed. **FL**: bract ovate to triangular, entire or 2-toothed at base. **FL**: sepals 8–16 mm; petals 12–24 mm; stamens 3–10 mm; style 7–15 mm. **FR** 19–40 mm, 2–3.5 mm wide, gen curved < 270°. **SEED** ± 1 mm, grain-like, minutely round-papillate, ± checkered; angles rounded; tip ± folded, flap-like. *n*=9. Washes, roadsides, flats, creosote-bush scrub; 700–1100 m. **s DMoj**. Sometimes confused with *M. nitens*. ❀ TRY.

M. hirsutissima S. Watson Ann 6–31 cm. **ST** branched at base, spreading. **LF** 1–11 cm, toothed to lobed. **INFL**: bract ovate to triangular, 2–5-toothed or -lobed, ± clasping, whitish at base. **FL**: sepals 9–18 mm; petals 12–28 mm, pale yellow; stamens 4–12 mm; filaments gen minutely 2-toothed or -shouldered at tip; style 6–13 mm. **FR** 14–25 mm, 5–8 mm wide, straight. **SEED** ± 2.5 mm, ±1.5 mm wide, beaked, narrowed above and below middle, minutely papillate, ashy white. Washes, fans, slopes; creosote-bush scrub; < 600 m. **DSon**; Baja CA. [var. *stenophylla* (Urb. & Gilg) I.M. Johnston, hairy stickleaf]. ❀ TRY. Apr–May

M. involucrata S. Watson (p. 367, pl. 67) Ann 7–35 cm. **LVS** 2–18 cm, gen irregularly lobed. **INFL**: bracts 4–5 per fl, (ob)ovate, white-scarious, margin green, 3–10-toothed or -lobed. **FL**: sepals 7–23 mm; petals 13–62 mm, cream-yellow with orange veins; stamens 4–26 mm, filament 2-toothed at tip; style 8–30 mm. **FR** 14–22 mm, 5–10 mm wide, straight. **SEED** 2–3 mm, ± 2–2.5 mm wide, ± compressed, beaked, narrowed above and below middle, rough, ashy white. Washes, fans, steep slopes, creosote-bush scrub; < 900 m. **D**; n Mex. Pls from n DSon, with petals > 30 mm, have been called var. *megalantha* I.M. Johnston. ❀ TRY. Jan–May

M. inyoensis B. Prigge Per 15–40 cm. **LVS** 4–11 cm, lobed. **INFL**: bract gen linear, entire, sometimes 2-lobed at base. **FL**: sepals 4–12 mm; petals 11–18 mm, clawed; stamens 5–15 mm; outer filaments < 1.5 mm wide; style 10–13 mm. **FR** gen 12–16(25) mm, 6–8 mm wide, cylindric. **SEED** 2–3 mm, ± 2 mm wide, lenticular, faintly fine-pebbled, winged. *n*=11. Rocky slopes, canyons; 1900–2000 m. **n W&I (White Mtns.)**; NV.

M. jonesii (Urb. & Gilg) H.J. Thompson & Joyce Roberts Ann, 10–40 cm, gen sprawling. **LF** 10–14 cm, lobed or toothed. **INFL**: bract triangular-ovate, entire. **FL**: sepals 2–10 mm; petals 6–22 mm, base sometimes with orange spot; stamens 3–10 mm, outer > inner; style 4–10 mm. **FR** 15–38 mm, 2–4 mm wide at flared tip, curved < 180° or S-shaped. **SEED** ± 1 mm, grain-like, pointed-papillate; angles acute. *n*=18,27. Washes, fans, flats, roadsides; 200–1500 m. **W&I**, **D**; NV, AZ. Closely related to *M. nitens*. ❀ TRY.

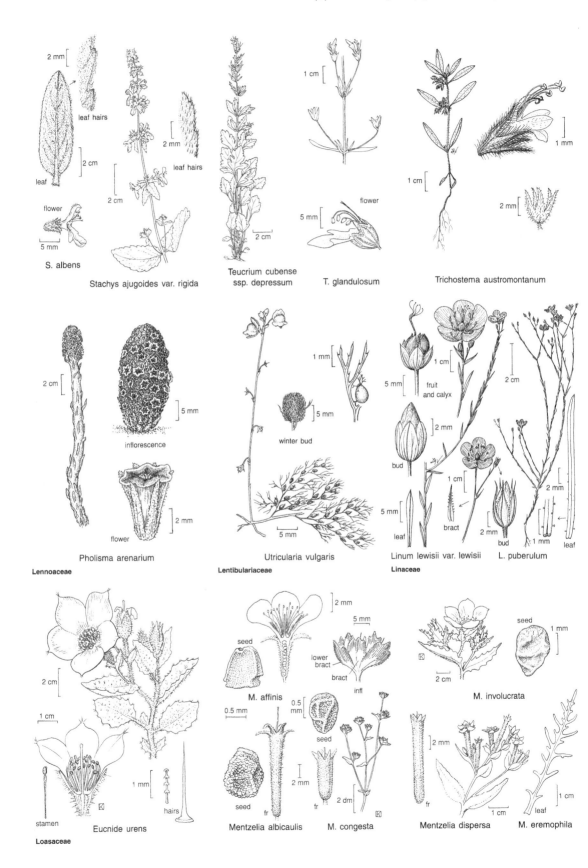

S. albens

Stachys ajugoides var. rigida

Teucrium cubense
ssp. depressum

T. glandulosum

Trichostema austromontanum

Pholisma arenarium

Lennoaceae

Utricularia vulgaris

Lentibulariaceae

Linum lewisii var. lewisii

L. puberulum

Linaceae

Eucnide urens

Loasaceae

M. affinis

Mentzelia albicaulis

M. congesta

M. involucrata

Mentzelia dispersa

M. eremophila

M. laevicaulis (Hook.) Torrey & A. Gray (p. 373) Per 22–100 cm. **ST** erect, branching above middle. **LVS** gen lobed; basal 19–24 cm; cauline 2–10 cm. **INFL**: bract linear to lanceolate, entire, toothed at base to deeply 4–5-lobed. **FL**: sepals 15–46 mm; petals 30–80 mm; stamens 15–55 mm, outer filaments < 2 mm wide; style 2.5–7 mm. **FR** 15–44 mm, 8–13 mm wide, straight. **SEED** 3–4 mm, 2–2.5 mm wide, lenticular, winged, fine-pebbled. *n*=11. Sandy to rocky slopes, washes, roadcuts; < 2700 m. **CA** (exc GV, DSon); to B.C., MT, WY, NV. ❀ DRN,DRY,SUN:1,**2**,**3**,**7**,9,12,14–24; DFCLT. Jun–Oct

M. montana (Davidson) Davidson Ann, 4–48 cm, erect. **LF** < 13 cm, entire or base 2-lobed. **INFL**: bract (ob)ovate, 2–6-toothed below middle, base whitish. **FL**: sepals 1–4 mm; petals 2–7 mm, base sometimes with orange spot; stamens 2–7 mm; style 1.5–3 mm. **FR** 6–20 mm, 2–3 mm wide, sometimes cylindric. **SEED** 1–1.5 mm, ± 1 mm wide, grain-like, rough; angles sharp. *n*=18. Open slopes, flats, sagebrush scrub, coniferous forest; 1200–2600 m. CaR, SN, WTR, SnJt, **n** SNE, **W&I**, **n DMtns**; to OR, Colorado, TX, n Mex. Intermediate to *M. congesta, M. veatchiana.*

M. multiflora (Nutt.) A. Gray ssp. *longiloba* (J. Darl.) Felger Bien or per, 15–100 cm, branched at base, erect. **LVS** 1–15 cm; basal crenate or shallowly lobed; cauline entire to toothed. **INFL**: bract 0 or linear to lanceolate, gen entire. **FL**: sepals 6–12 mm; petals 10–20 mm, golden yellow, tip obtuse to rounded; staminodes 5, 9–20 mm, petal-like, tip acute to serrate; stamens 3–13 mm, outer > and wider than inner; style 5–13 mm. **FR** 8–17 mm, 7–10 mm wide, cup-shaped. **SEED** 3–4 mm, 2.5–3.5 mm wide, lenticular, winged, whitish. Sandy creosote-bush scrub; < 700 m. **s DMoj, DSon**; AZ, n Mex. Ssp. *multiflora* e of CA. ❀ TRY. Apr–Jun

M. nitens E. Greene (p. 373) Ann 12–34 cm. **ST** glabrous to sparsely hairy, whitish. **LVS** 6–15 cm; basal toothed or lobed; cauline entire to lobed. **INFL**: bract 0 or lanceolate and entire. **FL**: sepals 3–8 mm; petals 7–15 mm, base sometimes with orange spot; stamens < or = style; style 4–8 mm. **FR** 13–26 mm, 2–3.5 mm wide, curved > 180°. **SEED** ± 1 mm, grain-like, round-papillate, checkered; end folded, flap-like; angles rounded. *n*=9. Sandy washes, slopes; 500–1800 m. **SNE, n DMoj**; s NV. [*M. californica* H.J. Thompson & Joyce Roberts] Closely related to *M. eremophila.* ❀ TRY. Apr–Jun

M. obscura H.J. Thompson & Joyce Roberts (p. 373) Ann 8–45 cm. **LVS** toothed to lobed; basal 7–22 cm; cauline 1–7 cm. **INFL**: bract ovate, entire. **FL**: sepals 2–6 mm; petals 3–12 mm, base sometimes with faint orange spot; stamens 2–7 mm, outer > inner; style 2–6 mm. **FR** 11–31 mm, 1.5–3 mm wide, slightly flared at tip, curved 90°–250°. **SEEDS** ± 1 mm, grain-like, round-papillate; angles rounded. *n*=18. Washes, slopes, roadsides, shrubland, Joshua-tree woodland; 200–1500 m. **D**; to UT, AZ, n Mex. Much like *M. albicaulis* (exc seeds).

M. oreophila J. Darl. (p. 373) Per 15–40 cm. **STS** few, 0–many branched, white, peeling. **LVS** < 9 cm, ± wavy-dentate; upper clasping. **INFL**: bract linear to lanceolate, entire. **FL**: sepals 4–11 mm, becoming inrolled; petals 7–17 mm; staminodes 5, petal-like, gen 5–8(14) mm; stamens 3–9 mm, outer > inner; style 5–9 mm. **FR** 5–10 mm, 6–11 mm wide, cup-shaped. **SEED** 2–3.5 mm, lenticular, winged, finely pebbled. *n*=11. Washes, limestone soils, talus; 900–1800 m. **W&I, DMtns**; to UT, w AZ. [*M. leucophylla* Brandegee misapplied] Pls with non-clasping lvs have been called *M. puberula* J. Darl. May–Jun

M. polita Nelson (p. 373) Per, 12–31 cm, rounded; caudex thick. **ST** white, peeling. **LF** < 7 cm, linear to lanceolate, entire. **INFL**: bract narrowly lanceolate, green, entire. **FL**: sepals 5–7

mm; petals 8–14 mm, white to pale yellow; stamens 3–7 mm, outer filaments 1–1.5 mm wide; style 5–7 mm. **FR** 5–8 mm, 5–8 mm wide, cup- or urn-shaped. **SEED** ± 3 mm, ovate, lenticular, winged, fine-pebbled. *n*= 11. Limestone, gypsum soils; 1200–1500 m. **e DMtns (Clark Mtns)**; s NV.

M. pterosperma Eastw. Per (may fl 1st year) 6–42 cm. **LVS** 1–13 cm; basal entire to short-toothed; cauline toothed (or lobed near base). **INFL**: bract narrowly lanceolate, entire. **FL**: sepals 6–10 mm; petals 9–24 mm, tips acute; staminodes 5, petal-like, slightly < petals; stamens 4–15 mm, outer filaments 1–3 mm wide; style 6–17 mm. **FR** 8–14 mm, 6–10 mm wide, cup- to barrel-shaped. **SEED** ± 3–4 mm, elliptic in outline, lenticular, winged, fine-pebbled. *n*=11. Gypsum clay soils; 1100 m. **e DMtns (Clark Mtns)**; to Colorado, nw AZ.

M. reflexa Cov. (p. 373) Ann, 2–20 cm, rounded, branched at base. **LF** 1–10 cm, toothed. **INFL**: bract lanceolate to ovate, lobed. **FL**: sepals 5–7 mm; petals 8, 6–12 mm, pale yellow; stamens 3–8 mm, filaments slightly expanded and acute at tip; style 5–6.5 mm. **FR** 9–13 mm, 5–7 mm wide, cylindric to barrel-shaped. **SEEDS** 2–2.5 mm, ± pear-shaped, narrowed and folded near middle, minutely papillate, ashy white; tip narrow, flap-like. *n*=10. Washes, rocky slopes, roadsides; < 1200 m. **W&I, DMoj**. Mar–May

M. torreyi A. Gray (p. 373) Per, 10–16 cm, much-branched. **LF** 2–4 cm, entire or with 2–4 linear lobes, rolled under. **INFL**: bract linear, base sometimes 2-lobed, inrolled. **FL**: sepals 3–6 mm; petals 9–15 mm, pale yellow, hairy above; stamens 7–10 mm; style 8–12 mm. **FR** 4–8 mm, 1–6 mm wide, urn-shaped. **SEED** 2–2.5 mm, ± fusiform, ± spirally 3-ribbed; 1 end acute, other truncate. Sandy or alkaline soils, slopes, shrubland, pin-yon woodland; 1200–2100 m. **SNE**; to OR, ID. Jun–Aug

M. tricuspis A. Gray (p. 373) Ann, 14–27 cm, spreading to erect. **LF** 2–12 cm, wavy or toothed. **INFL**: bract lanceolate to widely ovate, sessile, 2–8-toothed, base sometimes whitish. **FL**: sepals 7–17 mm; petals 11–50 mm, cream-white; stamens 7–17 mm, filaments 2-toothed at tip, teeth linear, anther stalk between teeth 1–1.5 mm (gen < teeth), thread-like; style 10–12 mm. **FR** 9–15 mm, 5–8 mm wide, cylindric to barrel-shaped, gen pedicelled, reflexed. **SEED** 2–2.5 mm, narrowed at middle, beaked, minutely papillate, whitish. *n*=10. Sandy or gravelly slopes or washes in creosote-bush scrub; 200–400 m. **e DSon, DMtns**; to UT, AZ. ❀ TRY. Mar–May

M. tridentata (Davidson) H.J. Thompson & Joyce Roberts Ann, 10–25 cm, spreading to erect. **LF** 1–9 cm, wavy or toothed. **INFL**: bract lanceolate to widely ovate, sessile, 2–10-toothed, base sometimes whitish. **FL**: sepals 5–13 mm; petals 10–40 mm, cream-white; stamens 6–15 mm, filaments 2-toothed at tip, teeth linear, anther stalk between teeth 1–1.5 mm (gen > teeth), thread-like; style 9–12 mm. **FR** 9–18 mm, 5–8 mm wide, cylindric to barrel-shaped, gen pedicelled, erect to reflexed. **SEED** 1.5–2.5 mm, narrowed above and below middle, beaked, minutely papillate, whitish. *n*=10. Creosote-bush scrub; 700–1000 m. **c DMoj**. [*M. tricuspis* var. *brevicornuta* I.M. Johnston] ❀ TRY. Apr–May

M. veatchiana Kellogg (p. 373) Ann, 3–45 cm, erect. **LVS** 1–18 cm; basal deeply lobed, sinuses or lobes toothed; cauline toothed to lobed. **INFL**: bract (ob)ovate, gen toothed, base whitish. **FL**: sepals 1–5 mm; petals 4–8 mm, (yellow-)orange; stamens 3–7 mm, outer > inner; style 3–6 mm. **FR** 8–28 mm, gen straight (or curved < 70°). **SEED** 1–2 mm, grain-like, pointed-papillate; angles acute. *n*=27. Sandy grassland, shrubland, oak/pine woodland; 400–1900 m. **n SNH, Teh, SnJV, SCoRI, TR, PR, GB, D**; OR, NV, AZ, Baja CA. Apr–Jun

PETALONYX

Subshrub; hairs gen rough with whorls of fine barbs. **LF** linear to ± round, entire to toothed; base tapered to cordate. **INFL**: raceme, gen terminal; bracts 3 per fl, outer 1 > inner 2. **FL**: sepals ± deciduous; petals free or claws adherent, fused below blades; stamens gen 5, free; ovary ± ovoid, placenta 1, stigma 1. **FR**: utricle, ± ovoid, gen 5-veined or

-ribbed, erect. **SEED** 1, 1.5–2.5 mm, ± fusiform, gen smooth. n=23 for all spp. 5 spp.: sw US, nw Mex. (Greek: petal claw) [Davis & Thompson 1967 Madroño 19:1–32]

1. Petals free; lvs linear to narrowly (ob)lanceolate . . . . *P. linearis*
1' Petal claws adherent below, fused above; lvs lanceolate to widely ovate
  2. Lvs petioled, cauline ± equal . . . . *P. nitidus*
  2' Lvs sessile, cauline reduced upwards . . . . *P. thurberi*
    3. Hairs soft, spreading; stamens 4–7.5 mm . . . . ssp. *gilmanii*
    3' Hairs rough, appressed downward; stamens 6–10 mm . . . . ssp. *thurberi*

P. linearis E. Greene Pl 15–100 cm. **LF** gen sessile, 10–25 mm, linear to narrowly (ob)lanceolate, obtuse to acute, entire to irregularly toothed. **INFL** 4–10 cm; outer bract 5–8 mm, ovate to ± round; inner bracts 3–4 mm, ovate, ± cordate, acute to notched, lobed; pedicels 1–2 mm. **FL**: petals 2–5.5 mm, free, white; stamens 3–7 mm, barely exserted; style ± 3–6 mm. Sandy or rocky canyons, gen in creosote-bush scrub; < 1000 m. **se DMoj, DSon**; sw AZ, nw Mex. ❀ TRY. Mar–May

P. nitidus S. Watson (p. 373) Pl 15–45 cm. **LF** widely tapered to base, 15–40 mm, gen ovate, acute to acuminate, serrate to coarsely few-toothed. **INFL** 3–4.5 cm; outer bract 5–13 mm, narrowly ovate; inner bracts 1–5 mm, elliptic to ovate, truncate, crenate; pedicels 1–2 mm. **FL**: petals 5–11 mm, cream, claws adherent, fused in upper 1/4; stamens 7–14 mm, well exserted; style 8–15 mm. Sandy or rocky canyons in creosote-bush scrub, Joshua-tree woodland, pinyon/juniper woodland; 1000–2100 m. **W&I, DMtns**; to sw UT, nw AZ. ❀ TRY. May–Jul

P. thurberi A. Gray (p. 373) Pl < 100 cm. **LF** clasping, 4–45 mm, lanceolate to deltate-ovate, acute to acuminate, entire to few-toothed. **INFL** 1–4 cm; outer bract 4–7.5 mm, deltate-ovate; inner bracts 2–3 mm, lanceolate to ovate, tapered to cordate, acute to ± acuminate, crenate or lobed; pedicels < 1 mm. **FL**: petals 2.5–6.5 mm, cream, claws adherent, fused in upper 1/5; stamens 4–10 mm, well exserted; style 3–11 mm. Sandy or gravelly dunes, washes, canyons in creosote-bush scrub; < 1200 m. **D**; NV, AZ, nw Mex.

  ssp. *gilmanii* (Munz) W.S. Davis & H.J. Thompson DEATH VALLEY SANDPAPER PLANT Hairs soft, spreading. **FL**: stamens 4–7.5 mm. RARE. Sandy washes, dunes; < 1200 m. **n DMoj**. [*P. g.* Munz] May–Jun, Sep–Nov

  ssp. *thurberi* (pl. 68) Hairs rough, appressed downward. **FL**: stamens 6–10 mm. Habitats and range of sp. ❀ TRY. May–Jul

LYTHRACEAE LOOSESTRIFE FAMILY

Elizabeth McClintock

Ann, per, shrubs, trees. **ST** angled or cylindric. **LVS** simple, entire, gen opposite, sometimes alternate or whorled. **INFL**: raceme, spike, or panicle, terminal, or axillary clusters with 1–several fls. **FL** bisexual, gen radial; hypanthium cylindric to bell-shaped, gen membranous, persistent in fr; sepals 4–6, gen persistent, appendages 3–5 or 0, alternate sepals; petals, stamens inserted on inner hypanthium; petals 4–6 or 0, alternate sepals, deciduous; stamens gen = or 2 × petals, incl or exserted; ovary superior, chambers 2–6, style gen slender, stigma head-like. **FR**: capsule, opening by valves from top, splitting sometimes irregular or 0. **SEEDS** 3–many. ± 25 genera, 450 spp.: temp, trop, gen in wet habitats. Some orn or cult for medicine, dyes. [Graham 1964 J Arnold Arbor 45:235–250]

1. Hypanthium cylindric; fls 1–2 per lf axil . . . . **LYTHRUM**
1' Hypanthium bell- or urn-shaped, becoming globose in fruit; fls 3–5 per lf axil . . . . **AMMANNIA**

AMMANNIA

Ann. **ST** prostrate to erect, glabrous. **LVS** opposite, 4-ranked, sessile, linear to oblanceolate, gen with basal ear-like lobes. **INFL**: cluster, axillary; fls 3–10; bractlets 2, inconspicuous. **FL** radial; hypanthium bell- to urn-shaped, ± spheric in fr; sepals 4, appendages < or ± = sepals, thick, horn-like; petals (0)4; stamens 4(8), incl or exserted. **FR** ± spheric, irregularly dehiscent. **SEEDS** many, ± 1 mm. ± 25 spp.: temp, trop. (Paul Ammann, Germany, 1634–1691) [Howell 1985 Wasmann J Biol 43:72–74]

1. Lower peduncles > 3 mm; petals rose-purple; fr 3–5 mm diam . . . . *A. coccinea*
1' All peduncles < 1 mm or 0; petals pale lavender; fr 4–6 mm diam . . . . *A. robusta*

A. coccinea Rottb. (p. 373) **ST** decumbent to erect, 1–10 dm, solitary or branched at base. **LF** 2–8 cm, 2–15 mm wide, linear to narrowly lanceolate. **INFL** compact; fls 3–5; lower peduncles 3–5(9) mm. **FL**: hypanthium urn-shaped; petals 3–4(5) mm, obovate, deep rose-purple; stamens 4(7), exserted in fl, anthers deep yellow. **FR** 3–5 mm diam. n=33. Wet places, drying ponds, lake & creek margins; < 300 m. CaRF, c&s SNF, GV, SnFrB, SW, **DSon**; to e US, C.Am. [*A. auriculata* Willd. misapplied] Jun–Aug

A. robusta Heer & Regel (p. 373) **ST** decumbent to erect, 1–10 dm, solitary or branched. **LF** 1.5–8 cm, 5–15 mm wide, gen linear-lanceolate. **INFL** compact; fls 1–3(5), sessile. **FL**: hypanthium urn-shaped, often 4-ridged; appendages ± = sepals in fr; petals 3–4(8) mm, obovate, pale lavender; stamens 4(5–12), exserted in fl, anthers pale yellow. **FR** 4–6 mm diam. n=17. Wet places, drying pond and ditch margins; < 500 m. NCoR, s SNF, GV, CW, SCo, s ChI (Santa Catalina Island), **DSon**; to c US, Mex. [*A. coccinea* Rottb. misapplied] Jun–Aug

LYTHRUM

Ann, per. **ST** prostrate to erect, sometimes 4-angled. **LVS** opposite, alternate, or whorled, linear to ovate, sessile or short-petioled. **INFL**: raceme- or spike-like; fls gen 1–2 per axil; bracts 2 per node. **FL** radial to slightly bilateral, sometimes of 2–3 forms (heterostyly); hypanthium cylindric to bell-shaped; sepals 4–6, appendages < to > sepals; petals 4–6 or 0; stamens 4–6 or 12, incl or exserted; styles < to > stamens. **FR** gen cylindric, gen dehiscent by 2 valves. **SEEDS** many, < 1 mm. ± 35 spp.: temp. (Greek: blood, from fl color) [Stuckey 1980 Bartonia 47:3–20]

L. californicum Torrey & A. Gray (p. 373) CALIFORNIA LOOSE-STRIFE Per, heterostylous. **ST** erect, 2–6 dm, gen branching above; branches spreading to ascending, glabrous. **LVS**: lower opposite; upper alternate, 1–7 cm, linear to linear-lanceolate, ± glaucous. **INFL** terminal, ± spike-like; bracts gen linear; pedicels 1–2 mm. **FLS** of 2 style forms; hypanthium cylindric, 4–7 mm; sepals narrowly deltate, < 1 mm, ± = appendages; petals 4–8 mm, purple; stamens gen 6, incl; style incl or exserted. **FR** ovoid, ± = hypanthium. 2*n*=20. Marshes, pond and stream margins; < 2200 m. s NCoRI, SNF, s SNH, GV, CW, SW, **W&I, D**; to c US, n Mex. ✤ INV;STBL. Apr–Sep

MALVACEAE MALLOW FAMILY

Steven R. Hill, except as specified

Ann, per, shrubs, trees, gen stellate-hairy; juice sticky; inner bark tough, fibrous. **LVS** alternate, simple, petioled; blade gen palmately veined or lobed, stipules present. **INFL** often leafy; whorl or involucre of bractlets often subtending calyx. **FL** gen bisexual, radial; calyx lobes 5, margins abutting in bud; petals 5, free (fused at base to filament tube, so falling together); stamens many, filaments fused into a tube surrounding style, tube fused in turn to petal bases; pistil 1, ovary superior, chambers gen 5 or more, style branches, stigmas gen 1 or 2 × as many as chambers. **FR** of 5–many disk- or wedge-shaped segments, loculicidal capsule, or berry. 100 genera, 2000 spp.: worldwide, esp warm regions; some cult (e.g., *Abelmoschus*, okra; *Alcea*; *Gossypium*, cotton; *Hibiscus, Malvaviscus*). Recently treated to include Sterculiaceae [Angiosperm Phylogeny Group 1998 Ann Missouri Bot Gard 85:531–553; Alverson et al 1999 Amer J Bot 86:1474–1486; Bayer et al 1999 Bot J Linn Soc 129:267–303]. Mature fr important for identification.

1. Fr a capsule, chambers 5; seeds 2–many per chamber; fls showy, solitary in axils near st tips . . . . . . **HIBISCUS**
1' Fr of 5–40 segments that gen separate from axis and each other; seeds gen 1–4 per chamber; infl various
  2. Stigmas linear, on inner side of style branches; fr segments indehiscent; seed 1 per segment
    3. Lvs gen entire to shallowly lobed, not much reduced toward st tip; bractlets subtending calyx 3, gen forming an involucre; fls 2–6 per axil; fr segments ± 11–15, length gen ± = height, beak 0 . . . . **MALVA**
    3' Lvs gen deeply lobed, gen much reduced toward st tip; bractlets subtending calyx 0(–3); infl raceme or spike-like; fr segments 5–10, length < height, small beak gen present . . . . **SIDALCEA**
  2' Stigmas head-like; fr segments dehiscent or not; seeds 1 or more per segment
    4. Fr segment with 2 scarious wings at top; bractlets 0; subshrub . . . . **HORSFORDIA**
    4' Fr segment not 2-winged; bractlets 0–3; herb or shrub
      5. Seeds gen 1–9 per fr segment
        6. Corolla orange, rarely orange-pink to red; bractlets 0; fr segment beaked, dehiscent from top to bottom; seeds 2–9 per fr segment . . . . **ABUTILON**
        6' Corolla white to rose, lavender, or red-orange; bractlets gen 1–3, gen deciduous; fr segment not beaked, smooth and dehiscent at top, net-veined and indehiscent at base; seeds 1–2 per fr segment . . ²**SPHAERALCEA**
      5' Seed gen 1–2 per fr segment
        7. Fr segment dehiscent throughout into 2 valves; infl of dense, axillary clusters; shrub or subshrub . . . . **MALACOTHAMNUS**
        7' Fr segment at least partly indehiscent; infl gen raceme-like, or fls solitary in axils; ann or per
          8. Fr segment smooth and dehiscent at top, strongly net-veined and indehiscent at base . . . ²**SPHAERALCEA**
          8' Fr segment ± uniform, indehiscent, gen not strongly net-veined
            9. Ann; pl with scattered stellate hairs; lvs 3–5-lobed or crenate; petals white to rose-purple . . . . **EREMALCHE**
            9' Per; pl densely stellate-canescent or scaly; lvs wide-reniform to triangular; petals yellowish . . . . **MALVELLA**

ABUTILON INDIAN MALLOW

Ann, per, shrub, stellate-canescent, tomentose, or bristly-hairy. **ST** decumbent to erect. **LF**: blade crenate or toothed, cordate, lobes gen 0. **INFL**: fls solitary in axils or in leafy panicles; bractlets subtending calyx 0. **FL**: petals yellow to reddish; anthers borne at top of filament tube; stigmas head-like. **FR** ± cylindric to ± spheric; segments smooth-sided, beaked, walls firm, sometimes woody, gen not separating, dehiscent ± to base. **SEEDS** 2–9 per fr segment. 200 spp.: warm regions. (Arabic name) [Borssum Waalkes 1966 Blumea 14:159–177]

1. Petals 10–20 mm, orange; fr segments gen 7–10, = or < calyx . . . . *A. palmeri*
1' Petals 3–6 mm, orange-pink to red; fr segments gen 5, >> calyx . . . . *A. parvulum*

A. palmeri A. Gray (p. 373) Subshrub. **ST** 15–20 dm; hairs stellate (and either long and soft or bristly). **LF**: blade 2–5 cm wide, ± round, cordate, velvety; lobes 3, obscure. **FL**: calyx 9–15 mm, = or > mature fr; petals 10–20 mm, orange. **FR**: segments gen 7–10, ± 10 mm tall, bristly or densely soft-hairy, beaks 1.5–2 mm, ± erect. Uncommon. Dry, gen e-facing mtn slopes, creosote-bush scrub; 600–800 m. **DSon**, adjacent PR; AZ, n Mex. ✤ DRN,DRY,SUN: 8,9,11,12,**13**. Apr–May

A. parvulum A. Gray (p. 373) Per. **ST** 1–4 dm, slender, much-branched, ± decumbent from woody root, stellate-canescent. **LF**: blade 1–4.5 cm wide, ovate, cordate; hairs scattered, stellate. **FL**:

calyx 3–5 mm, << mature fr; petals 3–6 mm, orange-pink to red. **FR**: segments gen 5, 7–9 mm, stellate-puberulent, beaks 1–2 mm, ± erect. Uncommon. Arid, rocky slopes, shadscale scrub; 900– 1300 m. **DMtns (Providence Mtns)**; to s Colorado, w TX, n Mex. Apr–May

EREMALCHE

David M. Bates

Ann, some pls with only pistillate fls, gen stellate-hairy. **ST** prostrate to erect. **LF**: blade toothed to lobed or parted. **INFL**: fls solitary in axils or in terminal clusters; pedicel longer in fr; bractlets subtending calyx 3, linear to thread-like. **FL**: calyx lobes > tube, acuminate; petals gen > calyx, white to purplish (drying darker); filament column incl; stigmas head-like. **FR**: segments 9–36, separating, indehiscent, unarmed, glabrous, lateral walls fragile, margins and outer wall ridged or net-veined. **SEED** 1 per fr segment. 3 spp.: sw US, nw Mex. (Greek: lonely mallow, from desert habitats)

1. Lvs palmately lobed; petals 4–4.5 mm, uniformly white or pale pinkish purple; fr segments
   1.4–1.8 mm. . . . . *E. exilis*
1' Lvs ± crenate; petals 15–30 mm, bases bright purple; fr segment 2.8–3.5 mm . . . . *E. rotundifolia*

E. exilis (A. Gray) E. Greene (p. 373) **ST** prostrate to decum-bent, < 40 cm, finely stellate-hairy. **LF** gen 1–2.5 cm wide, 3–5-lobed; lobe tips entire or 3-toothed. **INFL**: fls sometimes near st base; bractlets 3–7 mm. **FL**: calyx 4–7 mm, lobes 3–5 mm, 1.5–2.5 mm wide; petals 4–5.5 mm, white or pale pinkish purple. **FR**: segments 9–13, 1.4–1.8 mm, ± wedge-shaped in ×-section, margins rounded, cushion-like, outer wall cross-ridged. 2*n*=20,40. Desert scrub; < 1500 m. **W&I, D**; AZ, n Baja CA. [*Malvastrum e.* A. Gray] Mar–May

E. rotundifolia (A. Gray) E. Greene (p. 373, pl. 69) DESERT FIVE-SPOT **ST** erect, 8–60 cm, sometimes branched from base; hairs gen simple, bristly. **LF** 1.5–6 cm wide, round-reniform, crenate. **INFL**: fls gen > lvs; bractlets 6–10 mm. **FL**: calyx 9.5–14 mm, lobes 5.5–11 mm, 3.5–7 mm wide; petals 15–30 mm, pinkish purple, each with a bright purple basal blotch. **FR**: seg-ments gen 25–35, 2.8–3.5 mm, wafer-like, margins sharp, outer wall net-veined. 2*n*=20. Dry desert scrub; –50–1200 m. **D**; NV, AZ. [*Malvastrum r.* A. Gray] ❀ TRY. Mar–May

HIBISCUS ROSE-MALLOW, HIBISCUS

Ann, per, shrubs, trees. **INFL**: fls gen solitary in axils; bractlets subtending calyx 3–many, slender, persistent, forming an involucre. **FL** gen showy; filament tube 5-toothed, anthers scattered on upper 1/2 below tip. **FR**: capsule; chambers 5. **SEEDS** 2–many per chamber. 200 spp.: warm regions. (Greek: name used by Dioscorides for marshmallow) [Fryxell 1980 Techn Bull USDA 1624:1–53]

H. denudatus Benth. (p. 377) PALE FACE Subshrub, tomen-tose. **ST** 3–6 dm, slender. **LF**: blade 1–3 cm, ovate, finely toothed. **FL**: peduncle < 1 cm; sepals fused only at base; petals 1–2.5 cm, white to lavender, base gen purple. **FR** < calyx. **SEED** reniform, densely silky. Desert scrub of mesas, canyons; < 800 m. **DSon**; to w TX, n Mex. Feb–May

HORSFORDIA

Subshrub, densely stellate-tomentose or -scabrous. **ST** erect. **LF**: petiole stout; blade lanceolate to round, cordate, ± entire or irregularly fine-toothed. **INFL**: panicle or fls solitary in axils; bractlets subtending calyx 0. **FL**: petals yellow-orange to red-lavender; stigmas head-like. **FR**: segments 8–12, upper portion dehiscent, with 2 spreading, scarious wings; lower portion seed-bearing, indehiscent, firm, net-veined. **SEEDS** 1–3, white-puberulent. 4 spp.: sw US, Mex. (F.H. Horsford, Vermont, botanical collector) [Fryxell 1985 Syst Bot 10:268–272]

1. Petals pink to red-lavender, 10–21 mm; fr segments 10–12, wings >> (± 3 ×) lower part; harshly
   canescent . . . . *H. alata*
1' Petals yellow or orange, 8–10 mm; fr segments 8–9, wings gen = lower part; velvety rusty- or
   yellow-hairy . . . . *H. newberryi*

H. alata (S. Watson) A. Gray (p. 377) **ST** 10–40 dm, few-branched, harshly stellate-canescent. **LF**: blade 2–9 cm, slightly sticky, acute to acuminate, ± cordate, finely toothed. **FL**: pedicels 2–4 mm in fl, 6–10 mm in fr; calyx lobes 3–5 mm, wide-ovate, acuminate; petals 10–21 mm, pink to red-lavender, drying blu-ish. **FR**: segments 10–12, 7–8 mm; wings ± lanceolate, 3 × length of lower part. **SEED** 1 per segment. Uncommon. Rocky can-yons, creosote-bush scrub, washes; 100–500 m. **DSon**; s AZ, n Mex. ❀ TRY. Mar–Apr, Nov–Dec

H. newberryi (S. Watson) A. Gray (p. 377) **ST** erect, 10–30 dm, branched above base; hairs velvety, yellow to rusty, stellate. **LF**: blade 3–15 cm, shallowly cordate, entire to crenate. **FL**: pedicels 5–10 mm in fl, < 20 mm in fr; calyx lobes 2–3 mm, ± triangular, acute; petals 8–10 mm, yellow or pale orange. **FR**: segments 8–9, ± 6 mm; wings ovate, ± = lower part. **SEEDS** 2–3 per segment. Uncommon. Creosote-bush scrub; 100–800 m. PR, **w DSon**; to sw AZ, n Mex. Mar–Apr, Nov–Dec

MALACOTHAMNUS BUSH MALLOW

David M. Bates

Subshrubs, shrubs; hairs sparse to dense, stellate (stalked or sessile), simple, and glandular. **ST** erect; branches sometimes spreading. **LF**: blade toothed, lobes 0 or 3–7. **INFL** head-like to panicle-like, composed of axillary clusters (each a cyme) variously arrayed; clusters few–many-fld, loose to dense, congested or well separated; bractlets subtending calyx 3. **FL**: petals > calyx, gen pale pinkish purple or white (often purplish when dry); filament column incl; stigmas head-like. **FR** disk-like; segments 7–14, 2–5 mm, separating, each dehiscing into 2 valves, unarmed, smooth, top hairy. **SEED** 1 per fr segment. 11 spp.: CA, nw Mex. (Greek: soft shrub) Spp. represent major morphological variants; they are all interfertile and sometimes intergrade in areas of proximity. Variation between populations (esp in hairs, infl, and fls) is high and of ± complex pattern within most spp.

1. Herbage sparsely to densely pubescent but not woolly; calyx lobes joined in bud, in fl ± = tube, triangular to ovate . . . . ***M. fasciculatus***
1' Herbage ± white-woolly; calyx lobes free in bud, in fl > to >> tube, gen narrowly triangular-acuminate
. . . . ***M. fremontii***

M. fasciculatus (Torrey & A. Gray) E. Greene (p. 377) CHAP-ARRAL MALLOW Pl 10–50 dm; hairs sparse to dense, white to tawny. **ST**: branches gen slender. **LF**: blade 2–6(11) cm, ovate to ± round, gen thin; upper surface gen sparsely hairy; lobes 0–7, angular to rounded, not overlapping (sinuses open). **INFL** spike- to openly panicle-like; clusters gen many-fld, dense and sessile to widely spreading; bractlets 1–8 mm, < 1 mm wide. **FL**: calyx 4–11 mm, lobes 1.8–8, ± = tube, triangular to ovate, acute to short-acuminate, densely hairy, 2–3 often joined when fl open. 2*n*=34. Coastal-sage scrub, chaparral; gen < 600 m (< 2450 in PR, Santa Rosa Mtns). NCoRI (Mendocino Co.), interior SnFrB, SCoRO, SW, **sw edge DMoj**; n Baja CA. [var. *catalinensis* (Eastw.) Kearney; var. *laxiflorus* (A. Gray) Kearney; var. *nesioticus* (Robinson) Kearney (Santa Cruz Island bush mallow, **ENDANGERED** CA); var. *nuttallii* (Abrams) Kearney; *M. arcuatus* (E. Greene) E. Greene (arcuate bush mallow); *M. hallii* (Eastw.) Kearney (Hall's bush mallow); *M. mendocinensis* (Eastw.) Kearney (Mendocino bush mallow, **PRESUMED EXTINCT**); *M. parishii* (Eastw.) Kearney (Parish's bush mallow)]

Highly variable, with many indistinct and intergrading local forms. Apr–Jul

M. fremontii A. Gray (p. 377) Pl 5–20 dm; hairs ± woolly, white. **ST**: branches gen coarse, stiff. **LF**: blade 4–6(11) cm, ovate to wider than long, thin or thick, upper surface hairy, lobes obscurely 5. **INFL** spike- to panicle-like; clusters many-fld, sessile to peduncled, gen well separated; bractlets 3–15 mm, gen < 0.5 mm wide. **FL**: calyx 5.5–13 mm, not angled in bud, lobes free in bud, 3–10 mm, gen > or > tube, gen narrowly triangular-acuminate, soft-hairy; petals light purple. 2*n*=34. Chaparral to pine woodland; 60–1300 m (n CA), 450–2300 m (s CA), 1700–2800 m (SNE). NCoRH, NCoRI, SNF, Teh, e SnFrB, TR, SnJt, **SNE, DMtns (Panamint Mtns)**. [ssp. *cercophorus* (Robinson) Munz; *M. helleri* (Eastw.) Kearney (Heller's bush mallow); *M. howellii* (Eastw.) Kearney; *M. orbiculatus* (E. Greene) E. Greene] s CA pls are gen less woolly ("*M. orbiculatus*" form), may intergrade with *M. fasciculatus*. ✿ DRN:6,**7,14,18**,19–23&SUN:15,**16, 17**,24. Apr–Sep

MALVA MALLOW

Ann, bien, per, gen ± glabrous. **ST** prostrate to erect, gen < 1 m. **LF** petioled; blade round to reniform, gen crenate, lobes 0 or shallow. **INFL**: fls 1–several in axils; bractlets subtending calyx gen 3, free. **FL**: calyx lobes ± = tube; petals white, pink, or purple; anthers borne along filament tube, not clustered; stigmas linear, on inner side of style branches. **FR**: segments gen 6–15, indehiscent; beak 0. **SEED** 1 per segment. 100 spp.: Eur, Asia, Afr. (Greek: mallow) Some spp. reportedly TOXIC to livestock from selenium or nitrate concentration.

1. Calyx not much enlarged in fr; fr segment back smooth, margins rounded; petals 8–13 mm; st gen decumbent . . . . ***M. neglecta***
1' Calyx much enlarged in fr; fr segment back wrinkled, margins thin-winged; petals 4–5 mm; st erect
. . . . ***M. parviflora***

M. neglecta Wallr. (p. 377) COMMON MALLOW, CHEESES Ann, bien. **ST** decumbent, 2–6 dm, gen densely stellate-hairy. **LF**: blade 2–6 cm, lobes 0 (or 5–7, obscure), crenate. **INFL**: fls 3–6 per axil; pedicel in fr < 25 mm; bractlets 3–5 mm, ± broadly linear. **FL**: calyx gen 4–6 mm, not much enlarged in fr, lobes acuminate; petals 8–13 mm, pale lilac or white, claws hairy; stamentube hairy. **FR**: segments ± 15, puberulent, back round, smooth or weakly ridged, not net-veined, margins rounded. Disturbed places; < 3000 m. n&c **CA**; native to Eurasia. [sometimes identified as *M. rotundifolia* L. or *M. verticillata* L. var. *crispa* L.] The latter would key here, but has glabrous stamen-tube, 8–11 fr

segments; cult for salad, likely waif. May–Oct

M. parviflora L. (p. 377) CHEESEWEED, LITTLE MALLOW Ann. **ST** erect, 2–8 dm, widely branching, ± stellate near tips, glabrous below. **LF**: blade 2–8 cm wide, 5–7-angled to -lobed, teeth ± rounded. **INFL**: fls 2–4 per axil, crowded; pedicels in fr gen < 10 mm; bractlets 1–2 mm, linear. **FL**: calyx ± 3 mm in fl, enlarging and spreading in fr, net-veined, lobes widely ovate; petals 4–5 mm, white to pink, glabrous. **FR**: segments ± 11, back wrinkled, net-veined, glabrous to hairy, margins thin-winged, finely toothed. Common. Disturbed places; < 1500 m. CA-FP, **D**; native to Eurasia; widespread weed. Fls nearly all year.

MALVELLA ALKALI-MALLOW

Per. **ST** prostrate to decumbent. **LF**: blade gen asymmetric, gen silvery-stellate. **INFL**: pedicel ± jointed at tip, gen recurved in fr; bractlets subtending calyx 1–3, linear, sometimes deciduous. **FL**: calyx lobes ± = tube; petals stellate-hairy in bud, cream-white to yellow; stigmas head-like. **FR**: segments gen 6–10, indehiscent, slightly inflated, beak 0. **SEED** 1 per segment. 4 spp.: Am, Medit. (Greek & Latin: small mallow) [Fryxell 1974 Southw. Naturalist 19:97–103]

M. leprosa (Ortega) Krapov. (p. 377) ALKALI-MALLOW, WHITE-WEED Per. **ST** decumbent, 1–4 dm, densely white-stellate; some hairs bristly, some scale-like. **LF**: blade 1–3 cm, reniform, round,

or triangular, asymmetric, toothed, margin wavy. **INFL**: fls gen 1–3 per axil; bractlets 3, 3 mm. **FL**: calyx 6–10 mm; petals 10–15 mm, cream-white to yellow. **FR**: segments 6–10, 3 mm, gen

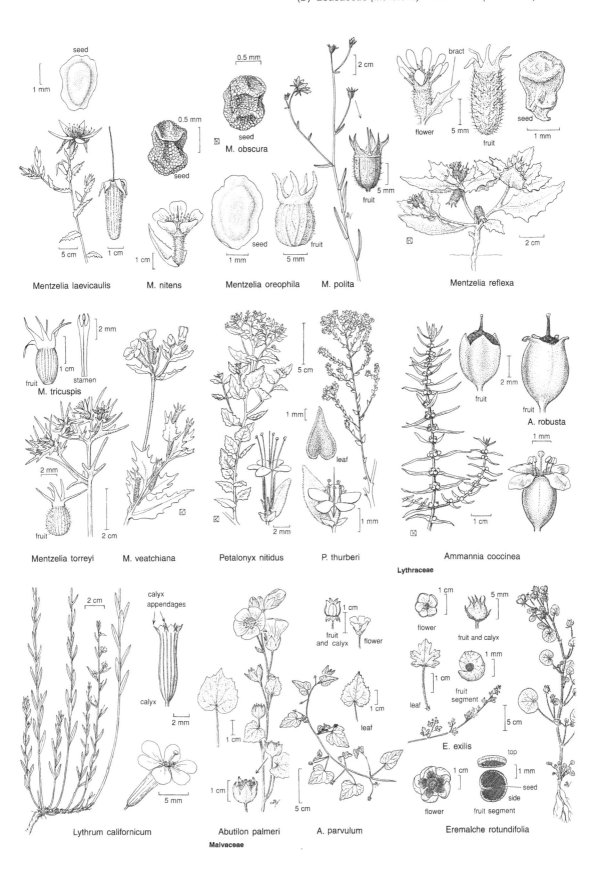

Mentzelia laevicaulis

M. nitens

M. obscura

Mentzelia oreophila

M. polita

Mentzelia reflexa

M. tricuspis

Mentzelia torreyi

M. veatchiana

Petalonyx nitidus

P. thurberi

A. robusta

Ammannia coccinea

Lythraceae

Lythrum californicum

Abutilon palmeri

A. parvulum

E. exilis

Eremalche rotundifolia

Malvaceae

net-veined on sides. Valleys, orchards, gen in saline soil; < 1000 m. **CA** (esp GV); to WA, ID, TX, Mex, S.Am. [*Sida l.* (Ortega)

Schumann var. *hederacea* (Hook.) Schumann] TOXIC to sheep, perhaps other livestock.

SIDALCEA CHECKER MALLOW, CHECKERBLOOM

Ann, per, sometimes from long, creeping rhizomes. **ST** gen erect or base ± decumbent. **LVS** gen mostly from near st base; lowest blades gen crenate to shallowly lobed, upper blades gen deeply lobed (gen ± compound). **INFL** gen spike- or panicle-like, gen more open in fr; bracts at pedicel base 2, gen stipule-like; bractlets subtending calyx gen 0(–3). **FL**: calyx lobes = or > tube; petals purple or rose-pink to white; stamen-tube with gen 2 series of ± fused filaments near tip; stigmas linear, on inner side of style branches. **FR**: segments gen 5–10, indehiscent, gen ± beaked, walls thin. **SEED** 1 per fr segment. ± 25 spp.: w N.Am. (Greek: combination of 2 names for mallow) [Hitchcock 1957 Univ Wash Publ Biol 18:1–96] Highly variable and difficult, with many local forms; some pls will not key with certainty. Additional work warranted.

1. Roots fleshy; rhizome and caudex 0; infl open
  2. Calyx minutely stellate; fr segment back slightly net-veined; Inyo Co. . . . . *S. covillei*
  2' Calyx with minute stellate and longer bristly hairs; fr segment back smooth; San Bernardino, San
    Diego cos. . . . . *S. neomexicana*
1' Roots not fleshy, caudex woody; infl open or densely fld
  3. Lowest lvs deeply lobed, lobes gen subdivided; infl loosely 3–9-fld; st and calyx stellate but not
    bristly-hairy . . . . *S. multifida*
  3' Lowest lvs gen crenate to simply lobed; infl dense, spike-like, > 10-fld; st and calyx gen both stellate-
    and bristly-hairy . . . . *S. oregana* ssp. *spicata*

S. covillei E. Greene (p. 377) OWENS VALLEY CHECKERBLOOM Per; roots ± fleshy, simple to clustered; rhizome and caudex 0. **ST** 2–6 dm; lower st finely stellate to coarsely bristly, hairs finer upward. **LF**: blade fleshy, glaucous, densely stellate. **INFL** open, finely stellate. **FL**: calyx 5–8 mm, uniformly fine-stellate; petals 10–15 mm, pink-lavender. **FR**: segment ± 2.5 mm, sparsely glandular-puberulent, back net-veined, sides strongly so. 2n=20. **ENDANGERED** CA. Alkaline flats; 1100–1300 m. **SNE (Owens Valley, Inyo Co.).** Threatened by lowering of water table, grazing. May–Jun

S. multifida E. Greene (p. 377) Per from woody caudex; very gray-glaucous; rhizomes 0. **ST** 1–6 dm, appressed-stellate. **LF**: blade fleshy, deeply 7-lobed, lobes of all lvs gen deeply ternate, segments linear to oblong. **INFL** open, glabrous to finely stellate; fls gen 3–9. **FL**: calyx 7–10 mm in fl, uniformly finely stellate-hairy; petals 10–25 mm, rose-pink. **FR**: segment 3.5–4 mm, net-veined and pitted, glandular-puberulent on back. 2n=20. Dry places in sagebrush scrub or pine forest; 2000–2500 m. c&s SNH, SNE; NV. May–Jul

S. neomexicana A. Gray (p. 377) Per from clustered, fleshy roots. **ST** 2–9 dm, glabrous to sparsely hairy. **LF**: blade fleshy. **INFL** slender, loosely many-fld, glabrous to stellate. **FL**: calyx gen 5–8 mm, with few small stellate and longer hairs on swollen pads, lobes acuminate, veins gen prominent; petals 6–18 mm,

rose. **FR**: segment ± 2 mm, smooth to very lightly net-veined on sides, nearly glabrous. 2n=20. Uncommon. Alkaline springs and marshes; gen < 1500 m. SCo, PR, SnGb, SnBr, sw DMoj; to NM, n Mex. [ssp. *thurberi* (A. Gray) C. Hitchc.] Apr–Jun

S. oregana (Torrey & A. Gray) A. Gray Per from woody taproot. **ST** 3–15 dm, rarely rooting at base; lower st coarsely stellate to long-bristly. **LVS** ± basal; lower blades crenate to deeply lobed; upper blades ± compound, segments entire to deeply lobed. **INFL** ± spike-like, dense to open; pedicels gen 1–3 mm. **FL**: calyx gen ± 5 mm in fl, lobes lanceolate, glabrous to densely and uniformly stellate or bristly; petals 5–18 mm, rose-pink. **FR**: segment 2–3 mm, smooth to lightly net-veined and pitted. Meadows, marshes, other wet places; < 3000 m. NW, CaR, SN, **GB**; to WA, WY, UT. None of the vars. (within sspp.) of C. Hitchc. are recognized. 5 sspp. in CA.

  ssp. ***spicata*** (Regel) C. Hitchc. (p. 377) **ST** 3–8 dm; lower st gen long-bristly. **INFL** spike-like, dense, often branched. **FL**: calyx ± 6 mm in fr, gen densely bristly and stellate (or stellate only). **FR**: segment 2.5–3 mm, gen smooth to lightly net-veined, glandular-puberulent. 2n=20,40. Meadows, streamsides; 1100–3000 m. KR, n NCoRH, CaRH, n&c SNH, **GB**; OR, w NV. [*S. setosa* C. Hitchc., in part, incl ssp. *s.*, Edgewood checkerbloom] Variable; most readily recognized ssp. ✿WET or IRR,DRN:1–5,**6,7,14–17**,18. Jun–Aug

SPHAERALCEA GLOBEMALLOW

John C. La Duke

Ann, per, stellate-hairy. **LF** petioled; blade linear-lanceolate to triangular, entire to deeply dissected. **INFL**: panicle or raceme-like (unbranched exc for clusters in axils). **FL**: petals obovate, red-orange, white, or lavender; filament tube glabrous or stellate-hairy, anthers yellow or purple; stigmas head-like. **FR** breaking into 9–17 segments; upper part of segment dehiscent, smooth; lower part indehiscent, strongly net-veined, 1–2-seeded. **SEEDS** gray, black, or brown. ± 50 spp.: warm Am, s Afr. (Greek: globe mallow, from fr shape) [Kearney 1935 Univ Calif Publ Bot 19:1–128] Polyploidy and intermediates common.

1. Ann (obviously so) . . . . *S. coulteri*
1' Per (perhaps in fl during first year)
  2. Lvs linear-lanceolate . . . . *S. angustifolia*
  2' Lvs rounded, triangular, or nearly compound
    3. Lvs deeply divided (nearly compound) . . . . *S. rusbyi* var. *eremicola*
    3' Lvs triangular or rounded (often lobed, but not approaching compound)
      4. Infl an open panicle (sometimes dense in bud) . . . . *S. ambigua* (in part)

5. Petals red-orange . . . . var. *ambigua*
5' Petals lavender to pink . . . . var. *rosacea*
4' Infl raceme-like or a dense panicle
  6. Lower lf surface markedly ridged . . . . *S. ambigua* var. *rugosa*
  6' Lower lf surface ± smooth
    7. Hairs coarse; lvs gray-green; fr segment 4.5–5 mm, dehiscent part ± 60% of segment
    . . . . *S. emoryi* var. *emoryi*
    7' Hairs soft; lvs yellow-green; fr segment 2.5–3 mm, dehiscent part < 20% of segment . . . . *S. orcuttii*

S. ambigua A. Gray APRICOT MALLOW Pl canescent. **ST** erect, 5–10 dm. **LF**: blade 15–50 mm, ± triangular, weakly 3-lobed, green or yellow-green, 3-veined, base wedge-shaped, truncate, cordate, margin crenate, wavy. **INFL**: tip not leafy. **FR**: segments 9–13, < 6 mm, < 3.5 mm wide, truncate-cylindric, dehiscent part < 3.5 mm, rounded, 60–75% of segment. **SEEDS** 2 per segment, brown, glabrous, ± hairy. Desert scrub; 150–2500 m. **s SNE, D**; to UT, AZ, Mex.

  var. *ambigua* (p. 377, pl. 70) **LF** not markedly wrinkled. **INFL** open, sometimes raceme-like; fls in clusters or solitary. **FL**: petals red-orange to apricot; filament tube 3–9 mm, hairy, anthers yellow-purple. **FR**: dehiscent portion ± 60% of segment. Habitat and range of sp. [var. *monticola* (Kearney) Kearney; *S. parvifolia* Nelson misapplied to CA pls] ❀ DRN, SUN,DRY:2,3,**7–12**,13,**14**,15–17, **18–23**,24. Feb–Jul

  var. *rosacea* (Munz & I.M. Johnston) Kearney PARISH MALLOW **LF** not markedly wrinkled. **INFL** open. **FL**: petals lavender to pink; filament tube ± 11 mm, ± glabrous, anthers purple-gray. **FR**: dehiscent portion ± 60% of segment. Desert scrub; 150–800 m. **D (esp DSon)**; AZ, Mex. ❀ DRN,SUN,DRY:7,**8–14**,15–17, **18–23**,24. Mar–Jul, Oct

  var. *rugosa* (Kearney) Kearney (p. 377) **LF** markedly wrinkled. **INFL** dense. **FL**: petals red-orange; filament tube ± 5 mm, ± glabrous, anthers yellow. **FR**: dehiscent portion 75% of segment. Habitat and range of sp. ❀ DRN,SUN,DRY:2,3,**7–12**,13, **14**,15–17,**18–23**,24. Mar–Sep, Dec

S. angustifolia (Cav.) G. Don (p. 377) Pl canescent. **ST** erect, < 30 dm. **LF**: blade 15–48 mm, ± linear-lanceolate, sometimes hastately lobed, 3–5-veined, light green-gray, base tapered, margin entire to wavy-crenate. **INFL** raceme-like; tip gen leafy. **FL**: pedicel ± = calyx; petals 7–9 mm, red-orange; filament tube 5.5–7 mm, hairy, anthers yellow. **FR**: segments 9–13, 4–7 mm, 1.5–2 mm wide, truncate-conic; dehiscent part erect, 3–4 mm, ± 75% of segment. **SEEDS** 2, brown-black, hairy. Desert scrub; –6 –500 m. **DMoj, n DSon**; to Kansas, TX, n Mex. [var. *cuspidata* A. Gray; var. *oblongifolia* (A. Gray) Shinn.; *S. emoryi* var. *nevadensis* Kearney] Relationship to *S. emoryi* merits further study. ❀ TRY. Mar–Oct

S. coulteri (S. Watson) A. Gray (p. 381) Ann; hairs few, long, soft. **ST** sprawling to erect, 1.5–15 dm, slender. **LF**: blade 15–45 mm, sometimes wider than long, ± triangular or cordate, thin, soft, gray-green, lobes 3 or 5, rounded, coarsely toothed. **INFL** gen raceme-like (fls clustered in axils); tip gen leafy. **FL**: pedicel > calyx; petals < 11 mm, salmon-orange; filament tube ± 5 mm, hairy, anthers yellow. **FR**: segments ± 15, 1.5–2 mm, 2–2.5 mm wide, ± hemispheric; dehiscent part ± 1 mm, flat, ± 30% of segment, projecting toward fr axis. **SEED** 1 per segment, brown, glabrous or ± hairy. Uncommon. Dry, sandy places; < 300 m. **s DSon**; AZ, Mex. ❀ TRY. Mar–May

S. emoryi Torrey var. *emoryi* (p. 381) Pl coarsely canescent. **ST** erect, < 21 dm. **LF**: blade 25–55 mm, ovate-triangular, 3-lobed, 3–5-veined, gray-green, base cordate, tip ± truncate to acute, margin crenate. **INFL** raceme-like below, compact panicle of clusters above. **FL**: calyx 6–8 mm; petals 10–12 mm, red-orange to lavender; filament column ± 6 mm, anthers yellow. **FR**: segments 10–16, 4.5–5 mm, 2.5 mm wide, truncate-conic; dehiscent part acute, < 3 mm, ± 60% of segment. **SEEDS** 1–2 per segment, brown or black. Fields, roadsides; < 600 m. **s DMoj, DSon**; NV, AZ, Mex. [var. *arida* (Rose) Kearney; var. *variabilis* (Cockerell) Kearney] Intergrades with *S. angustifolia*. ❀ TRY. Feb–Jul, Oct

S. orcuttii Rose (p. 381) CARRIZO MALLOW Taproot large. **ST** erect, 5–12 dm, ± yellow-canescent. **LF**: blade 30–50 mm, rounded to triangular, ± 3-lobed, thick, prominently 3-veined, yellow-green to pale green, base tapered to truncate, margin entire to ± wavy. **INFL** raceme-like or with branches near lfless tip; fls many per axil. **FL**: pedicel < calyx; petals 10–12 mm, red-orange; filament tube 5–6 mm, hairy, anthers yellow. **FR**: segments 12–17, 2.5–3 mm, 2–3 mm wide, ± hemispheric; dehiscent part < 1 mm, round, < 20% of segment. **SEED** 1 per segment, brown, glabrous or ± hairy. Dry, sandy, ± alkaline desert scrub; –20–900 m. **s DSon**; AZ, Mex. ❀ TRY. Feb–Sep

S. rusbyi A. Gray var. *eremicola* (Jepson) Kearney (p. 381) RUSBY'S DESERT MALLOW, PANAMINT MALLOW **ST** erect, ± 3 dm. **LF**: blade ± palmately compound, 15–20 mm, broadly ovate, 5-veined, light green, base truncate to cordate, margin entire. **INFL**: panicle; tip ± leafy. **FL**: pedicels < calyx; sepals 13–15 mm; petals < 20 mm, red-orange; filament tube ± 9 mm, hairy, anthers yellow. **FR**: segments ± 13, ± 5 mm, 2 mm wide, ± truncate-spheric; dehiscent part ± 3 mm, ± 60% of segment. **SEEDS** 1–2 per segment, black-gray, glabrous to ± hairy. RARE. Desert scrub; 1300–1500 m. **n DMtns (Death Valley region, e Inyo Co.; Clark Mtns, ne San Bernardino Co.).** May

MARTYNIACEAE UNICORN-PLANT FAMILY

Lawrence R. Heckard

Ann, per, glandular-hairy, gen strongly scented. **LVS** simple, opposite or alternate; stipules 0; petiole long. **INFL**: raceme, terminal, bracted; bractlets 2, just below fl. **FL** bisexual; sepals 5, ± unequal; corolla 2-lipped, gen 5-lobed; stamens epipetalous, gen 2 long, 2 short, 1 vestigial; ovary superior, 1-chambered, placentas 2, parietal, each 2-lobed, style >> ovary, curved, stigma 2-lobed, flat, gen closing when touched. **FR**: capsule, drupe-like; outer layer fleshy, deciduous; inner layer ultimately exposed, woody; beak incurved, splitting to form 2 horns (claws). 3 genera, 15 spp.: gen ± trop Am; some cult. Placed by some authors in Pedaliaceae (Sesame Family). [Bretting & Nilsson 1988 Syst Bot 13:51–59]

PROBOSCIDEA UNICORN-PLANT, DEVIL'S CLAW

Ann, per; taproot branched or tuberous. **ST** prostrate to spreading, gen < 1 m. **LF**: blade broadly ovate to round or triangular, palmately veined (gen palmately lobed), base cordate. **INFL**: bractlets < calyx. **FL**: calyx 1–2 cm, gen

5-lobed and split to base on lower side (or sepals free); corolla 2–5 cm, bell- to funnel-shaped, showy, tube cylindric, gen < 1 cm, bent downward, throat 10–30 mm, limb with 5 flaring lobes, throat and lower limb with colored lines ("nectar guides"). **FR**: body 5–10 cm, fusiform; surface sculptured or spiny throughout, crested with branched projections gen only along upper suture; beak (claws) 1.5–3 × body. **SEED** 8–13 mm, angled, gen black, corky. 8 spp.: Am. (Greek: beak) Dispersed by attachment of fr claw to animals. Key to species adapted by Margriet Wetherwax.

1. Largest lvs 3–7 cm wide; corolla mostly yellow with maroon-brown flecks, fragrant; fr body ± 1 cm
   thick, lanceolate in outline; per; taproot fusiform, tuber-like, wider than stem at base . . . . *P. althaeifolia*
1' Largest lvs 5–15 cm; corolla white to pink, not fragrant; fr body 2–3 cm thick, narrowly ovate in
   outline; ann; taproot diam ± = stem diam . . . . *P. parviflora*

P. althaeifolia (Benth.) Decne. (p. 381) DESERT UNICORN-PLANT Per; root fusiform, tuber-like, yellow. **ST** decumbent. **LF**: blade 3–7 cm wide, broadly ovate to round or triangular, gen palmately 3–5-lobed, crenate. **INFL** 5–50-fld, gen overtopping lvs. **FL** fragrant; corolla bright yellow to orangish, maroon-brown-streaked on lower lobe and spotted in 2 rows along upper side of throat; nectar guides orange-yellow. **FR**: body ± 1 cm thick, upper crest with distal teeth sometimes extended into 2 slender, accessory beaks; sometimes also crested along lower suture. **SEED** 6–7 mm. UNCOMMON. Sandy places; < 1000 m. **DSon**; to TX, n Mex; Peru. [*P. altheaefolia* (Benth.) Decne.] ❀ SUN,DRN,DRY: 7,14–17&IRR:**8,9**,10,11, **12,13**,18–24. Summer

P. parviflora (Wooton) Wooton & Standley ssp. *parviflora* (p. 381) Ann; barely ill-smelling. **LF**: blade gen 5–15 cm wide, ± broadly ovate-triangular, entire to shallowly 3–7-lobed or -toothed. **INFL** 4–10-fld, < or barely overtopping lvs. **FL** not fragrant; corolla gen white to pink, purplish mottled, throat with 2 lines of purplish spots or not; upper corolla lip purplish splotched; nectar guides yellow; anther 2–3 mm; stigma not closing with touch. **FR**: body 2–3 cm thick, narrowly ovoid. 2*n*=30. Uncommon. Disturbed, dry places; < 1000 m. SW, **D**; to TX, Mex. Pls from D with white seed, fr crests > 5 cm long and > 5 mm high, horns > 18 cm have been called var. *hohokamiana* Bretting: cult by sw native Americans for black basket fibers from claws. ❀ TRY. Summer

MOLLUGINACEAE CARPET-WEED FAMILY

Wayne R. Ferren, Jr.

Ann, per, shrub, glabrous or hairy. **ST** prostrate to erect. **LVS** simple, gen basal and cauline, alternate, opposite, or whorled, rarely fleshy; stipules 0, conspicuous, or small and deciduous. **INFL**: cyme, cluster, or fl solitary, axillary. **FL** gen bisexual, small, radial; calyx persistent, sepals 4–5, gen free; corolla 0 or small; stamens 5–10, sometimes petal-like, attached to hypanthium, filaments free or fused at base; nectary a ring; ovary superior, chambers 1–10, placentas gen axile, styles 1 or 3–5, gen free. **FR**: gen capsule, gen loculicidal. **SEEDS** 1 or more per chamber, sometimes with arils. 14 genera, 95 spp.: gen trop, subtrop, esp Afr.

MOLLUGO CARPET-WEED, INDIAN CHICKWEED

Ann, per, glabrous to puberulent. **ST** prostrate to ascending, slender; branches many. **LVS** gen whorled, linear to oblanceolate; stipules 0. **INFL**: umbel, cluster, or fl solitary; pedicel slender. **FL**: sepals 5, ± petal-like, persistent, midrib gen green, margins white-scarious; petals 0; stamens 3, alternate ovary chambers or 5, alternate sepals (rarely 10); ovary ± ovate, chambers 3–5, styles 3–5, linear. **FR** thin, incl in calyx. **SEEDS** many, reniform, smooth, ridged, or tubercled, red-brown; aril short. 20 spp.: trop, subtrop. (Greek: soft pl)

1. St ± erect; lvs linear, gen 1 mm wide, glaucous; infl a peduncled umbel, fls 1–several; stamens 5 . . *M. cerviana*
1' St prostrate; cauline lvs narrowly oblanceolate, 1–8 mm wide, not glaucous; infl a sessile cluster
   of stalked fls; stamens gen 3 . . . . *M. verticillata*

M. cerviana (L.) Ser. (p. 381) Ann, glabrous, glaucous. **ST** ± erect, < 20 cm. **LVS** in whorls of 4–10, 3–15 mm, gen 1 mm wide, linear. **INFL**: umbel; fls 1–several; peduncle = or > slender pedicels; bracts minute, margins translucent. **FL**: sepals 5, 1–1.5 mm; stamens 5, alternate sepals, 1 mm; stigmas 3, sessile. **FR** ± round. **SEED** 0.3–0.4 mm, faintly keeled and net-sculptured, brown. 2*n*=18. Uncommon. Seasonal pools, sandy washes, flats, slopes; < 1700 m. SnJt, **D**; to TX, Mex, trop; native to Old World. Sep–Mar

M. verticillata L. (p. 381) Ann, mat-forming, < 50 cm diam, glabrous. **ST** prostrate, forked unequally. **LVS** in whorls of 3–6, < 40 mm, 1–8 mm wide, unequal; blade ± oblanceolate. **INFL**: sessile cluster (peduncle 0); fls 1–5; pedicel 5–15 mm, slender. **FL**: sepals 5, 1.5–2.5 mm, oblong; stamens 3–5, 3 mm, alternate ovary chambers; stigmas 3, ± sessile. **FR** ovoid, ± > sepals. **SEED** 0.6 mm, ± smooth, gen with several ridges, dark reddish brown, shiny. 2*n*=64. Common. Moist, exposed, disturbed wetland margins, roadsides, fields; < 1000 m. CA-FP, **SNE**; N.Am; native to trop Am. May–Nov

NYCTAGINACEAE FOUR O'CLOCK FAMILY

Richard Spellenberg

Per, shrub, tree, glabrous or hairy. **ST** often forked. **LVS** opposite, sessile or petioled, pairs gen unequal; blade gen entire. **INFL** gen forked; of spikes, clusters, or umbels, each unit sometimes with a calyx-like involucre. **FL** bisexual, radial; perianth of 1 whorl, petal-like, bell- to trumpet-shaped, base hardened, tightly surrounding ovary in fr, lobes 4–5, gen notched to ± bilateral; stamens 1–many; ovary superior (appearing inferior because of hardened perianth base), style 1. **FR**: achene or nut, smooth, wrinkled, or ribbed. 30 genera, 300 spp.: warm regions, esp Am; some orn (*Bougainvillea*; *Mirabilis*, four o'clock). Keys adapted by Margriet Wetherwax.

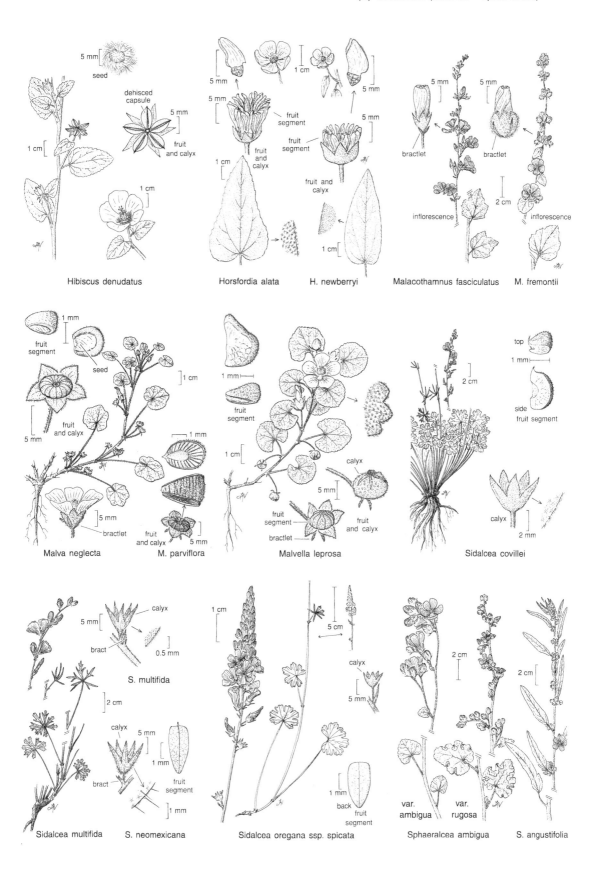

Hibiscus denudatus

Horsfordia alata

H. newberryi

Malacothamnus fasciculatus

M. fremontii

Malva neglecta

M. parviflora

Malvella leprosa

Sidalcea covillei

S. multifida

Sidalcea multifida

S. neomexicana

Sidalcea oregana ssp. spicata

Sphaeralcea ambigua

S. angustifolia

1. Stigma linear, incl in perianth; infl a head or umbel; fr gen wing
  2. Fr wing thick or 0, not continuous above fr body; receptacle conic, ± smooth . . . . **ABRONIA**
  2' Fr wing membranous, continuous above fr body; receptacle flat or ± conic, studded with peg-like
     pedicels . . . . **TRIPTEROCALYX**
1' Stigma ± spheric, gen exserted; infl various; fr gen unwinged
  3. Bracts 3–5, fused or free, forming a calyx-like involucre; fls 1–many per involucre, but infl often
    appearing as 1 fl
    4. Fls 3 per involucre, blooming simultaneously, each subtended by an involucral bract; fr bilateral . . **ALLIONIA**
    4' Fls 1–16 per involucre, blooming sequentially; involucre gen 5-lobed; fr radial . . . . **MIRABILIS**
  3' Bracts 1–3, free, not forming an involucre; infl not fl-like
    5. Fr wing thin, membranous; perianth 30–40 mm, trumpet-shaped . . . . **SELINOCARPUS**
    5' Fr wing 0; perianth < 30 mm or > 70 mm, trumpet-shaped or not
      6. St erect, > 7 mm diam; fr glabrous, ridges 10, fine, inconspicuous . . . . **ANULOCAULIS**
      6' St gen trailing, gen << 5 mm diam; fr sometimes hairy, ridges or ribs 5, prominent
        7. Perianth > 70 mm, trumpet-shaped; fr glabrous, oblong, 6–10 mm . . . . **ACLEISANTHES**
        7' Perianth < 30 mm; fr glabrous or hairy, club-shaped, < 3.5 mm . . . . **BOERHAVIA**

ABRONIA   SAND VERBENA

Ann, per, gen glandular. **ST** prostrate to ascending, gen ± red. **LF** gen fleshy, petioled. **INFL**: head or umbel; fls opening together or outer first; receptacle conic, ± smooth. **FL**: perianth salverform to trumpet-shaped, gen fragrant, lobes 4–5; stamens 4–5, incl; stigma linear, incl. **FR**: body fusiform; lobe-like wings (0)2–5 (if present, prominent, opaque, thick, not continuous above fr body). 25 spp.: w N.Am. (Greek: graceful) [Galloway 1975 Brittonia 27:328–347] Closely related to *Tripterocalyx*.

1. Sts densely tufted; montane per . . . . *A. nana* ssp. *covillei*
1' Sts elongate, aboveground; gen desert ann or per
  2. Fr wings 0; thick, prominent angles present or not; ovary chamber extends into angle
    3. Perianth white to pale magenta; fr angles gen 5, inflated and flat on top . . . . *A. turbinata*
    3' Perianth gen bright pink; fr smooth or with few low ridges. . . . . ²*A. villosa* var. *villosa*
  2' Fr with 2–5 thin, conspicuous wings; ovary chamber does not extend into wing
    4. Fr wings gen 2; fr outline broadly cordate . . . . *A. pogonantha*
    4' Fr wings gen 3–5; fr cordate, obconic, or ovoid
      5. Lvs thin, widely scalloped; fr beak inconspicuous, not hardened . . . . *A. gracilis*
      5' Lvs thin or thick, ± entire; fr beak evident, hardened . . . . *A. villosa*
        6. Fr wings pronounced, exceeding top of fr body; fl gen > 20 mm . . . . var. *aurita*
        6' Fr wings 0 or not exceeding top of fr body; fl gen < 20 mm . . . . ²var. *villosa*

A. gracilis Benth. Ann, glandular-hairy. **ST** prostrate, < 1 m. **LF**: petiole 10–35 mm; blade 20–45 cm, thin, oblong or elliptic, edges scalloped or wavy. **INFL**: peduncle 2–10 cm; bracts 6–10 mm, ovate; fls 13–24. **FL**: perianth tube 6–17 mm, ± white or pink, limb 7–16 mm wide, light magenta to purplish red. **FR** 6–11 mm, glandular-hairy; wings 5, thin, broadly rounded, extending above top of fr body, base of fl tube not forming a hard beak at top of fr body. Sandy soil; < 500 m. SCo (San Diego Co.), s DSon; nw Mex. Intergrades with *A. villosa*. ✿ sand,DRN, DRY,SUN:**24**.

A. nana S. Watson ssp. **covillei** (Heimerl) Munz   Per, densely tufted. **ST** < 6 cm. **LF** glaucous; petiole 1–4 cm; blade 5–20 mm, ovate to ± round. **INFL**: peduncle < 10 cm, scapose; bracts 6–8 mm, lanceolate; fls > 6. **FL**: perianth white, tube 11–15 mm, limb 6–8 mm wide. **FR** 7–8 mm, with 5 thin wings rounded at top. Dry sandy places; 1600–2800 m. **DMtns**; sw NV. Pls from e DMoj have wider bracts and longer fls, approach ssp. *nana*. ✿ TRY. Jun–Aug

A. pogonantha Heimerl (p. 381) Ann, glandular-hairy. **ST** decumbent to ascending, 10–55 cm. **LF**: petiole 0.8–4 cm; blade 1.5–5.5 cm, 1–3 cm wide, ovate to oblong-ovate. **INFL**: peduncle 2–7 cm; bracts 4–9 mm, lanceolate to broadly ovate; fls 12–24. **FL**: perianth tube 1–2 cm, ± red or green, limb 6–8 mm wide, white or pink. **FR** 4–6 mm, 2–3-winged, outline cordate; wing interior spongy. Sand, desert communities; < 1550 m. s SnJV,

DMoj **(and adjacent margins of SN)**; w NV. ✿ TRY. Apr–Jul

A. turbinata S. Watson (p. 381) Ann, rarely per, glabrous or sparsely glandular-hairy. **ST** ascending to erect, < 50 cm. **LF**: petiole 1–5 cm, slender; blade 1–5 cm, broadly ovate to round. **INFL**: peduncle 3–9 cm; bracts 3–10 mm, lanceolate to ovate; fls 15–35. **FL**: perianth tube 6–18 mm, ± green or pink, limb 5–8 mm wide, white or pale magenta. **FR** 3–7 mm, as wide as long, with gen 5 wings inflated and truncate at top, glandular-hairy at top, glabrous below; wings hollow inside (some fr with 2 wings folded together). Dry, sandy soil, desert scrub; 900–2500 m. **SNE**, **DMoj**; se OR, w NV. ✿ TRY. May–Jul

A. villosa S. Watson (p. 381) Ann, glandular-hairy. **ST** prostrate to ascending, < 80 cm. **LF**: petiole 0.5–5 cm; blade 1–5 cm, 1–4.5 cm wide, triangular-ovate to ± round. **INFL**: peduncle 2–10 cm; bracts 3–11 mm, lanceolate to narrowly ovate; fls 15–35. **FL**: perianth tube 1.3–3.5 cm, ± pink, limb 6–18 mm wide, pale to bright magenta. **FR** 5–10 mm; base of fl tube hardened as a beak on top of fr body; wings 3–5, thin, rounded or angled, or 0. Sandy places in creosote-bush or coastal-sage scrub; < 1600 m. SCo, **D**; s NV, sw AZ, nw Mex. Hybridizes with *A. gracilis*.

  var. **aurita** (Abrams) Jepson **FL**: perianth tube 2–3.5 cm, limb > 1.5 cm wide. **FR**: body nearly smooth; wings wide, extending well above body. Coastal-sage scrub, chaparral, etc.; < 1600 m. c&s SCo, **w DSon**. ✿ sand,DRN,DRY,SUN:**13,18–24**. Mar–Aug var. *villosa* (pl. 71) **FL**: perianth tube 1.3–2 cm, limb gen < 1.5 cm wide. **FR**: body prominently wrinkled, appearing pitted; wings barely extending above fr body. Creosote-bush scrub; < 1000 m. **D**; s NV, sw AZ, nw Mex. ✿ sand,DRN,DRY,SUN:**13**. Feb–Jul

ACLEISANTHES TRUMPETS

Per from thick taproot, minutely hairy. **ST** prostrate to ascending. **LF** sessile or petioled; blade < 8 cm, ± firm. **INFL**: bracts 1–3, free, not forming an involucre; fl gen 1. **FL** nocturnal or some cleistogamous; perianth trumpet-shaped, white, tube slender; stamens 5 (2 in cleistogamous fls), slightly exserted; stigma ± spheric, exserted slightly beyond anthers. **FR** oblong, with rounded ridges, glabrous; wings 0. 7 spp.: esp Chihuahuan Desert, ne Mex. (Greek: without closure, from absence of involucre) Pollinated by hawkmoths.

A. longiflora A. Gray (p. 381) ANGEL TRUMPETS   **ST** < 1 m. **LF** petioled; blade < 4 cm, lanceolate to triangular, acute to acuminate, margins ± wavy. **FL**: perianth 7–17 cm, tube ± green or pale purple, limb 1–3 cm wide, white. **FR** 6–10 mm, oblong, ridges 5, 2 parallel grooves between adjacent ridges. Dry places, gen on limestone; 10–2500 m. **e DSon (Maria Mtns, e Riverside Co.)**; to TX, n Mex. ✿ TRY. May

ALLIONIA WINDMILLS, TRAILING FOUR O'CLOCK

Ann, short-lived per, glabrous to densely glandular-hairy. **ST** trailing, < 1 m. **LF** petioled; blade < 4 cm, oval to oblong, paler below. **INFL**: involucres clustered, each involucre resembling 1 fl; bracts 3, ± 1/2 fused, hairy; fls 1 per bract, blooming together. **FL**: stamens gen 4(–7), exserted; stigma ± spheric, exserted. **FR** bilateral. 2 spp.: Am. (C. Allioni, Italian botanist, 1725–1804)

A. incarnata L. (p. 381, pl. 72) **FL**: perianth 3–15 mm, 3-lobed, oblique, red-purple, tube funnel-shaped, limb longest above subtending bract. **FR** 3–4.5 mm, compressed, bilateral; margin strongly incurved, entire or with 3–5 irregular teeth; outer surface convex; inner surface concave, with 2 rows of sticky glands. Creosote-bush scrub; 0–1500 m. **D**; to Colorado, TX, S.Am. ✿ DRN,DRY,SUN: 10,11,**17,18**. Apr–Sep

ANULOCAULIS RINGSTEM

Per from thick caudex. **ST** little-branched, erect, > 7 mm diam; internodes with sticky brown ring. **LVS** few, ± on lower st half, petioled; blade oblong to round, thick. **INFL** openly branched; fls in heads, racemes, or umbel-like clusters; bracts 1–3, small, not forming an involucre. **FL**: perianth funnel-shaped; stamens 3 or 5, gen long-exserted; stigma ± spheric, exserted. **FR** finely 10-ribbed, glabrous. 5 spp.: esp Chihuahuan Desert, ne Mex. (Latin: ring stem, from sticky internodal rings)

A. annulatus (Cov.) Standley (p. 381) **ST** < 1.5 m. **LF**: blade 3–10 cm, oblong to ovate-triangular, stiff-hairy, hairs with enlarged, dark, glandular base. **INFL**: head-like umbels on long peduncles. **FL**: perianth ± 8 mm, tube ± green, hairy, limb pale pink. **FR** 4–5 mm, thick-fusiform, gray-brown. Rocky slopes, canyons; < 1200 m. **ne DMoj (Death Valley region)**. [*Boerhavia a.* Cov.] Apr–May

BOERHAVIA SPIDERLING

Ann, per. **ST** prostrate to erect; internode often with sticky area. **LF** petioled; blade 1–6 cm, paler beneath, often brown-dotted. **INFL** openly branched; unit a raceme, umbel, or head; bracts 1–3, free, not forming an involucre. **FL**: perianth < 30 mm, bell-shaped, closing by afternoon; stamens 1–5; stigma ± spheric, gen exserted. **FR** < 3.5 mm, club-shaped; ridges 4–5; wings 0. ± 30 spp.: warm regions. *B. spicata* Choisy (fr ribs narrow, acute) is widespread in sw US, expected as agricultural weed near Mex border. (H. Boerhaave, Dutch botanist, l668–1738)

1. Perianth red-violet; per; fr hairy . . . . *B. coccinea*
1' Perianth pale pink or white; ann; fr glabrous
  2. Peduncle long, infl umbel-like or fl 1
    3. Ultimate 2 cm of branches with 0–1 branches, ending in umbel-like infl; fr length 3 × width . . . *B. intermedia*
    3' Ultimate 2 cm of branches with 2–4 appressed side branches, each with 1–2 fls; fr length 2 × width
    . . . . *B. triquetra*
  2' Peduncle short, infl a raceme-like spike
    4. Bracts deciduous, << fr; fr ridges gen 5; st puberulent . . . . *B. coulteri*
    4' Bracts ± persistent, ± = fr; fr ridges gen 4; st glandular-hairy . . . . *B. wrightii*

B. coccinea Miller (p. 381) SCARLET SPIDERLING   Per. **ST** prostrate or sprawling, < 15 dm; glandular-hairy. **LF**: blade broadly ovate, blunt, slightly wavy. **INFL**: head. **FL** 2 mm; perianth red-violet. **FR** 2.5–3.5 mm, hairy; tip rounded; wide and smooth between ribs. Dry, disturbed places; < 1000 m. s SnJV, SCo, **DSon**; to se US, nw S.Am. ✿ DRN,DRY,SUN:**8,9**,11,12,**13**. Apr–Jul

B. coulteri (Hook.) S. Watson (p. 387) Ann. **ST** decumbent to erect, < 8 dm, puberulent. **LF**: blade lanceolate to ovate-triangular, ± acute. **FL** < 1.5 mm; perianth white to pale pink. **FR** 2.5 mm, glabrous; narrow (nearly closed) and wrinkled between wide ribs; tip rounded. Gravelly hillsides, washes; < 1150 m. **DMtns (Clark Mtns), DSon**; to AZ, Mex. Sep–Nov

B. intermedia M.E. Jones (p. 387) Ann. **ST** ascending or erect, 2–5 dm; hairs fine, sparse. **LF**: blade broadly lanceolate or ovate, acute to obtuse. **INFL** openly branched; fls in small umbel-like clusters at branch tips. **FL** 1.5–2 mm; perianth white to pale pink. **FR** 2–2.7 mm, glabrous; wrinkled between sharp ribs; tip truncate. Gravelly washes, flats; < 1300 m. **e DMtns (Clark,**

Kingston mtns), e DSon; to TX, Mex. [*B. erecta* L. var. *i.*(M.E. Jones) Kearney & Peebles] *B. erecta* L. (fr 3–4.5 mm) is widespread in trop Am, expected as agricultural weed near Mex border. Aug–Oct

B. triquetra S. Watson (p. 387) Ann. **ST** branched from base, ascending to erect, < 6 dm, slender, puberulent. **LF**: blade narrowly lanceolate to oblong, obtuse to acute. **INFL** openly branched; peduncle 2–6 cm, slender; fls 1–2. **FR** 1–1.3 mm; perianth white to pale pink. **FR** 2–2.5 mm, glabrous; coarsely wrinkled between 3–5 wide, smooth, sharp ribs; tip truncate.

Sandy or rocky areas; < 1600 m. e PR (n base Santa Rosa Mtns), **DMtns (Little San Bernardino Mtns), DSon**; w AZ, nw Mex. Sep–Dec

B. wrightii A. Gray (p. 387) Ann. **ST** branched from base, erect, < 7 dm, glandular-hairy. **LF**: blade lanceolate to oblong-ovate, acute. **INFL** loose, spike-like. **FL** 1–2 mm, subtended by wide, reddish, persistent bracts, 2–3 mm in fr. **FR** 2–2.5 mm, nearly as wide, wrinkled between the 4 wide ridges. Dry, sandy places; < 1400 m. **D**; to NV, TX, Mex. Aug–Dec

MIRABILIS FOUR O'CLOCK

Per, subshrub. **ST** repeatedly forked, decumbent to erect. **LF** gen petioled. **INFL** forked; calyx-like involucres densely clustered or solitary in axils, bell- to saucer-shaped; fls 1–16 per involucre, blooming sequentially. **FL**: perianth funnel- to bell-shaped, lobes 5; stamens 3–5, gen exserted; stigma ± spheric, gen exserted. **FR** ± round to club-shaped, smooth to 5-ribbed; wing 0. ± 60 spp.: Am, Himalayas. (Latin: wonderful) Fls open in evening, close in morning. Spp. intergrade; *Hermidium, Oxybaphus* sometimes segregated, but intergrade with other spp.; careful study needed. [Pilz l978 Madroño 25:113–132; for revised taxonomy see Spellenberg and Rodríguez Tijerina 2001 Sida (in press)]

1. Fr strongly 5-ribbed, < 3 mm wide, gen with prominent warts or coarse wrinkles between ribs; involucre in fr enlarged, brown, papery
    2. Lf ± sessile, ± linear . . . . ***M. coccinea***
    2' Lf obviously petioled, blade lanceolate or wider
        3. St branched gen near top, ascending to erect, gen glabrous (or finely strigose) in lower half; lf length > width, tip gen acute . . . . ***M. oblongifolia***
        3' St much-branched from near base, decumbent to ascending, hairy or glandular throughout; lf length gen = width, tip obtuse to acute . . . . ***M. pumila***
1' Fr smooth to moderately 5- or 10-ribbed or -angled, often > 3 mm wide, sometimes with low wrinkles or warts; involucre in fr little changed
    4. Fls 3–16 per involucre, bracts either > 22 mm or free
        5. Perianth ± 15 mm; involucral bracts 15–30 mm, free to fused ± 1/2 length . . . . ***M. alipes***
        5' Perianth 40–60 mm; involucral bracts > 22 mm, fused > 1/2 length . . . . ***M. multiflora***
            6. Involucral bracts obtuse; fr warty, gelatinous when wet . . . . var. ***glandulosa***
            6' Involucral bracts acute; fr smooth or lightly warty with 10 slender, tan ribs, not gelatinous when wet
            . . . . var. ***pubescens***
    4' Fl 1 per involucre; bracts < 15 mm, fused
        7. Involucre > 10 mm, lanceolate lobes > tube; perianth white; lvs ascending . . . . ***M. tenuiloba***
        7' Involucre < 10 mm, ± ovate lobes < tube; perianth white to magenta; lvs widely spreading
            8. Perianth pink to purple-red; lf puberulent or glandular-hairy (youngest lvs with conic-based hairs); fr gen very lightly dotted or wrinkled, sometimes smooth . . . . ***M. californica***
            8' Perianth white to pale pink; lf finely glandular-hairy; fr gen lightly dotted or wrinkled . . . . ***M. bigelovii***
                9. Sts and lvs with long-spreading, glandular hairs . . . . var. ***bigelovii***
                9' Sts and lvs with short, reflexed hairs, gen also ± glandular . . . . var. ***retrorsa***

M. alipes (S. Watson) Pilz (p. 387) **ST** decumbent to erect, 2–4 dm; glaucous, glabrous, or sparsely hairy upward. **LF**: blade 2–7 cm, broadly ovate to round, fleshy, glabrous or sparsely short-hairy. **INFL**: involucre 1 per upper axil, peduncled, ± cup-shaped, glabrous; bracts gen 5–7, free to fused ± half length, 15–30 mm, broadly ovate; fl 1 per bract; pedicel fused to bract. **FL**: perianth ± 15 mm, funnel-shaped, magenta (rarely creamy white). **FR** 5.5–7 mm, elliptic, glabrous; ribs 10, slender, tan. Dry slopes, flats; 1200–2000 m. **W&I, DMtns (Panamint Mtns)**; to w Colorado [*Hermidium a.* S. Watson] TRY. May–Jun

M. bigelovii A. Gray **ST** ascending to erect, < 8 dm, gen glandular-hairy. **LF**: blade 1–4 cm, ovate to ± reniform, glandular-hairy. **INFL**: involucres clustered near ends of branches, bell-shaped; bracts 5, 5–6 mm, > 1/2 fused, lobes ovate; fl 1 per involucre. **FL**: perianth 8–12 mm, widely funnel-shaped, white to pale pink (esp in w part of range). **FR** ± 3 mm, ± spheric to ovoid, lightly dotted or wrinkled, glabrous. Rocky places; < 2300 m. **W&I, D**; to UT, AZ, nw Mex. Vars. intergrade, esp in e D. Poorly distinguished morphologically and geographically from *M. californica*; these may prove to be infraspecific entities in a more broadly conceived species.

var. ***bigelovii*** (p. 387) **ST**: hairs glandular, spreading. **FL**: perianth white, sometimes pink. **FR** gen ± ovoid, lightly dotted, often mottled. Habitat of sp.; < 200 m. **W&I, D, esp e D**. [var. *aspera* (E. Greene) Munz] DRN,DRY,SUN:7–10,**11–13**,14–16. Mar–Jun, Oct–Nov

var. ***retrorsa*** (A.A. Heller) Munz **ST** scabrous above with short, reflexed hairs, often also ± glandular. **FL**: perianth white to pale pink. **FR** ± spheric, with ± prominent pale lines. Rocky places; < 2300 m. **W&I, D**; to UT, AZ, nw Mex. DRN,DRY, SUN:7–10,**11–13**, 14–16. Mar–Jun

M. californica A. Gray (p. 387) WISHBONE BUSH **ST** trailing to ascending, < 8 dm, somewhat woody, grayish when old, scabrous or ± glandular-hairy. **LF**: blade 1–3.5 cm, ovate, puberulent or glandular-hairy. **INFL**: involucres clustered near ends of branches, bell-shaped; bracts 5, 5–8 mm, > 1/2 fused, lobes ovate; fl 1 per involucre. **FL**: perianth 5–14 mm, broadly funnel-shaped, pink to purple-red (white). **FR** ± 5 mm, ovoid, gen lightly dotted or wrinkled, glabrous. Common. Grassy areas, chaparral, dunes,

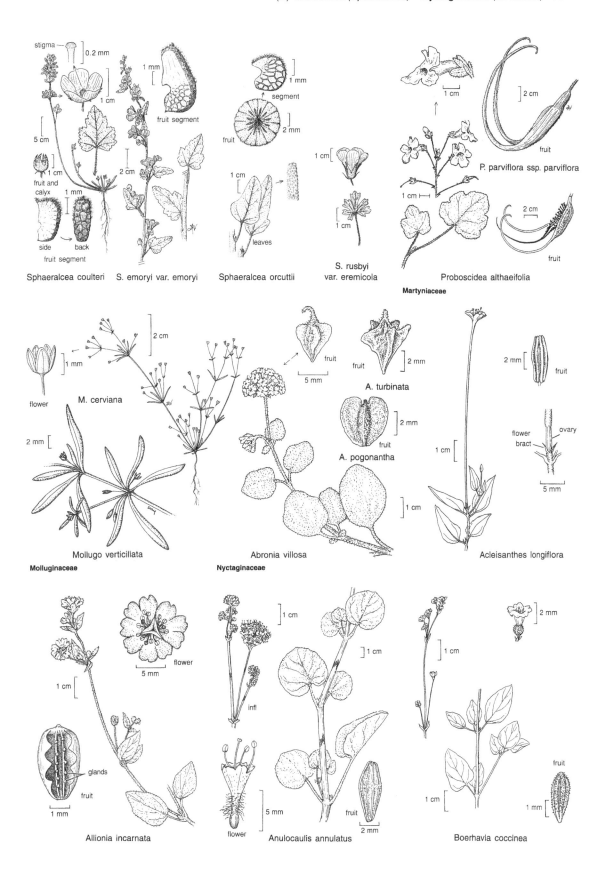

Sphaeralcea coulteri   S. emoryi var. emoryi   Sphaeralcea orcuttii   S. rusbyi var. eremicola   Proboscidea althaeifolia

P. parviflora ssp. parviflora

Martyniaceae

M. cerviana

Mollugo verticillata

Molluginaceae

Abronia villosa

Nyctaginaceae

A. turbinata

A. pogonantha

Acleisanthes longiflora

Allionia incarnata

Anulocaulis annulatus

Boerhavia coccinea

dry, rocky areas and washes; < 1000 m. CCo, SCoR, SW, **w edge D**; Baja CA. [*M. laevis* (Benth.) Curran var. *cedrosensis* (Standley) Munz, var. *laevis* misapplied] Poorly distinguished morphologically and geographically from *M. bigelovii*; these may prove to be infraspecific entities in a more broadly conceived species. ❀ DRN,DRY,SUN:**8,9,14–24**. Mostly Dec–Jun

M. coccinea (Torrey) Benth. & Hook. **ST** ascending to erect, < 6 dm, glabrous, glaucous. **LF** ± sessile; blade 2–12 cm, linear, fleshy, glabrous. **INFL** loosely forked; involucre bell-shaped, short-hairy, enlarged and papery in fr; bracts 5, 5–8 mm in fr, ± 1/2 fused; fls 1–3 per involucre. **FL**: perianth 15–20 mm, ± salverform (tube narrowly funnel-shaped), bright red. **FR** ± 5 mm, club-shaped; ribs 5, coarsely wrinkled between ribs; hairs fine. Dry, rocky slopes, washes; 1300–1800 m. **DMtns**; to w NM, nw Mex. [*Oxybaphus c.* Torrey] ❀ TRY. May–Jul

M. multiflora (Torrey) A. Gray (p. 387) **ST** ascending to erect, 3–8 dm. **LF**: blade 3–12 cm, round to ovate, fleshy, glandular-hairy or becoming glabrous. **INFL**: involucre 1 per upper axil, bell-shaped, ± glabrous to minutely glandular-hairy; bracts 5, 22–35 mm, 1/2–3/4 fused; fls 6 per involucre. **FL**: perianth 40–60 mm, narrowly funnel-shaped, magenta. **FR** 6–11 mm, elliptic. Dry, rocky or sandy places; < 2500 m. s SnJV, **W&I, D**; to w Colorado, AZ, Baja CA.

var. ***glandulosa*** (Standley) J.F. Macbr. **INFL**: involucral bracts obtuse. **FR** faintly warty, gelatinous when wet; ribs inconspicuous. Habitat of sp.; 900–2500 m. **W&I, n DMoj (Inyo Co.)**; to w Colorado. ❀ DRN,DRY,SUN:2,**10**,11.

var. *pubescens* S. Watson **INFL** involucral bracts gen acute. **FR** ± smooth, 10-ribbed, not gelatinous when wet. Habitat of sp.; 50–2100 m. s SnJV, **W&I, D**; to sw UT, nw Mex. [*M. froebellii* (Behr) E. Greene; *M. f.* var. *glabrata* (Standley) Jepson] ❀ DRN,DRY,SUN: 7,**8–10**,11–16,18,**19–23**. Apr–Aug

M. oblongifolia (A. Gray) Heimerl **ST** branched near top,

ascending to erect, 1–5 dm; glabrous or strigose. **LF**: blade < 4 cm, narrowly 3-angled, ± finely hairy. **INFL** broadly forked; involucre cup-shaped, glandular-hairy, enlarged and papery in fr; bracts 5, ± 8 mm in fr, 1/2–2/3 fused; fls 5 per involucre. **FL**: perianth 10–12 mm, broadly funnel-shaped, magenta. **FR** 3–5 mm, club-shaped, finely warty between the 5, wide, wrinkled or warty ribs. Dry, rocky areas; 1500–2500 m. **DMtns (Ivanpah, New York mtns)**; to Colorado, TX, n Mex. [*Oxybaphus comatus* (Small) Weath.] Variable complex, with many named forms; in need of critical study. CA pls have ± narrow lvs and branch ± at base (like *M. linearis* var. *decipiens* (Standley) Welsh of s Rocky Mtns). Exc for distribution of hairs, CA pls also resemble *M. pumila* from DMtns. The concept for both these taxa in California is tentative and is likely to change. ❀ TRY. May–Jun

M. pumila (Standley) Standley (p. 387) **ST** trailing to ascending, < 5 dm, short-hairy, often glandular. **LF**: blade 2–6 cm, triangular to broadly ovate, fleshy, hairy like sts. **INFL**: involucres in axils or narrow clusters, cup-shaped, densely glandular-hairy, enlarged and papery in fr; bracts 5, 7–8 mm in fr, ± 2/3 fused; fls 3 per involucre. **FL**: perianth 8–10 mm, broadly funnel-shaped, pale pink, hairy. **FR** ± 5 mm, club-shaped, shallowly wrinkled between 5 wide ribs. Dry, rocky places; 1400–2500 m. SnBr, SnJt, **W&I, DMtns**; to NV, NM, nw Mex. [*Oxybaphus p.* (Standley) Standley] The concept for both this species in California is tentative and is likely to change. ❀ TRY. Jun–Aug

M. tenuiloba S. Watson LONG-LOBED FOUR O'CLOCK **ST** trailing to erect, < 5 dm, glandular-hairy. **LF** ascending; blade 2.5–5 cm, narrowly to broadly triangular, glandular-hairy. **INFL**: involucres ± densely clustered in upper axils, narrowly bell-shaped, spreading glandular-hairy; bracts 5, 11–13 mm, < 1/2 fused, lobes narrowly lanceolate; fl 1 per involucre. **FL**: perianth 12–15 mm, funnel-shaped, whitish, lightly hairy. **FR** ± 5 mm, ovoid, smooth, blackish brown, glabrous. UNCOMMON. Rocky slopes in desert scrub; < 500 m. **w DSon (Imperial, Riverside, San Diego cos.)**; Baja CA. ❀ TRY. Mar–May

SELINOCARPUS MOONPOD

Per, shrub. **ST** forked. **LF** sessile or petioled; blade ± fleshy. **INFL**: bracts 1–3, free, not forming an involucre; fls gen solitary in axils. **FL** nocturnal or cleistogamous; open perianth trumpet-shaped; stamens 4–8; stigma ± spheric, gen exserted. **FR**: wing thin, membranous. 8 spp.: N.Am., Afr. (Greek: parsley fruit) (Fowler & Turner 1977 Phytologia 37(3): 177–208]

S. nevadensis (Standley) Fowler & B. Turner (p. 387) DESERT WING-FRUIT Ann; hairs appressed and divergent, white, also glandular-puberulent. **ST** prostrate to erect, < 3 dm. **LF** petioled, < 26 mm; blade ovate to round. **FL**: perianth 30–40 mm, ± 10 mm wide, tube ± green, limb white. **FR** 5–7 mm; wings 5, 2 mm. RARE in CA. Dry, rocky areas; 1250 m. **DMtns (ne Kingston Range, se Inyo Co.)**; to s NV, sw UT, nw AZ. [*S. diffusus* A. Gray var. *n.* Standley] Jun–Sep

TRIPTEROCALYX

Per from large taproot. **ST** much-branched. **LF** petioled; blade < 8 cm, fleshy, margin often wavy. **INFL**: head; bracts 5–10, green, fls blooming sequentially across head; receptacle flat to ± conic, studded with peg-like pedicels. **FL**: perianth trumpet-shaped, nocturnal, lobes 4–5; stamens 3–5, incl; stigma linear, incl. **FR**: wings 3–5, wide, continuous above and below fr body, thin, transparent, conspicuously net-veined. 3 spp.: arid N.Am. (Greek & Latin: 3-winged cup, from fr) [Galloway 1977 Brittonia 27:328–347] Closely related to *Abronia*.

T. micranthus (Torrey) Hook. (p. 387) **ST** < 6 dm, glandular-sticky or scabrous. **LF**: blade 1–6 cm, narrowly ovate to elliptic, glabrous to glandular-hairy, hairs denser on lower surface. **FL**: perianth 6–18 mm. **FR**: wings glabrous exc on edges. Dunes; 800–2450 m. **se DMoj (Kelso, San Bernardino Co.)**; to MT, SD, NM [*Abronia m.* Torrey] ❀ TRY. Apr–May

OLEACEAE OLIVE FAMILY

Dieter H. Wilken

Shrub, tree, or vine, some dioecious. **LVS** alternate or opposite, deciduous or evergreen, simple to pinnately compound. **INFL** various; fl sometimes solitary. **FL** sometimes unisexual, gen radial; calyx gen minute, tube cup-shaped, lobes 4–15; petals (0)4–6, gen fused; stamens gen 2, epipetalous; pistil 1, ovary superior, chambers 2, placentas axile, ovules 2–4 per chamber, style 1, stigma gen 2-lobed. **FR**: drupe, capsule, or winged achene. **SEED**

1 per chamber. ± 25 genera, 900 spp.: ± worldwide; some cult for orn (*Forsythia*; *Jasminum*, jasmine; *Ligustrum*, privet; *Syringa*, lilac) or food (*Olea*). [Wilson & Wood 1959 J Arnold Arbor 40:369–384]

1. Lf pinnately compound, lflets (1)3–7; fr winged . . . . **FRAXINUS**
1' Lf simple; fr capsule or drupe
  2. Lvs gen alternate; fr a deeply 2-lobed capsule . . . . **MENODORA**
  2' Lvs opposite or clustered; fr a drupe . . . . **FORESTIERA**

FORESTIERA

Shrub, gen dioecious. **LVS** simple, opposite or clustered, gen deciduous, short-petioled. **INFL**: clusters, axillary; staminate fls subsessile; pistillate fls pedicelled. **FL**: calyx minute, minutely ± 4-lobed, deciduous; corolla 0. **STAMINATE FL**: stamens 1–4; pistil vestigial. **PISTILLATE FL**: stamens 0; ovules 2 per chamber, stigma 1–2-lobed. **FR**: drupe. ± 20 spp.: Am. (Charles Le Forestier, French physician & naturalist, early 19th century)

F. pubescens Nutt. (p. 387, pl. 73) DESERT OLIVE Shrub 5–25 dm. **ST**: bark smooth, grayish; twigs short, spine-like, puberulent, becoming glabrous. **LF**: blade 15–40 mm, lanceolate to elliptic, leathery, entire to minutely toothed, glabrous. **INFL** gen appearing before lvs. **STAMINATE FL**: stamens 3–6 mm. **FR** 5–8 mm, elliptic in outline, purple-black, ± glaucous. 2*n*=46. Streambanks, canyons, washes; 100–1800 m. s SNF, c&s SNH (e slope), Teh, e SnFrB, SCoRO (e slope), SCoRI, TR, PR, **s SNE, DMoj**; to Colorado, TX, n Mex. [*F. neomexicana* A. Gray] ❀ 1,**7,14–17,22–24**&IRR:2,3, **8–12**,13,**18–21**. Mar–Apr

FRAXINUS ASH

Shrub or tree, gen dioecious. **ST**: bark smooth to furrowed, gen gray; twigs gen puberulent, becoming glabrous. **LVS** opposite, deciduous, odd-pinnate, petioled, gen thin, gen glabrous; lflet dark green above, pale below, gen glabrous, base and tip rounded to acute. **INFL**: clusters or panicles, axillary, often peduncled; fls pedicelled. **FL**: calyx 1–2 mm, shallowly ± 4-lobed; petals 0, 2, or 4, free or fused. **STAMINATE FL**: stamens gen 2; pistil vestigial. **PISTILLATE FL**: stamens 0; ovules 2 per chamber. **FR**: achene, winged. **SEED** gen 1. ± 65 spp.: temp N.Am, Eurasia, trop Asia. (Latin: ancient name) [Little 1952 J WA Acad Sci 42:369–380]

1. Lvs gen appearing simple (lflet 1, sometimes 3–5 on twigs); twigs gen 4-angled; fr body flat; n&e DMtns . . . . *F. anomala*
1' Lflets 3–7; twigs gen cylindric; fr body ± cylindric; SNE, DMoj . . . . *F. velutina*

F. anomala S. Watson (p. 387) SINGLE-LEAF ASH Shrub or tree < 6 m. **ST**: twigs gen 4-angled. **LF** 2–10 cm; petiole glandular-puberulent; lflets gen 1(–5), 1–6 cm, stalked, narrowly ovate to ± round, thick, entire to irregularly crenate. **FL** gen bisexual; petals 0. **FR** 13–25 mm, 5–10 mm wide; body flat, winged from near base. 2*n*=46. Washes, rocky slopes, shrubland, pinyon/juniper woodland; 1100–2400 m. **n&e DMtns**; to Colorado, NM. ❀ DRN,IRR:1,**2,3,7**,8–10,14–16,**18**,19–23. Apr–May

F. velutina Torrey (p. 387) VELVET ASH Tree < 10 m, dioecious. **ST**: twigs cylindric. **LF**: 10–25 cm; petiole puberulent, often becoming glabrous; lflets (3)5–7, 3–10 cm, short-stalked, lanceolate to ovate, entire to ± serrate, sparsely puberulent to glabrous below. **FL**: petals 0. **FR** 15–30 mm, 4–8 mm wide; body subcylindric, winged from ± middle. 2*n*=46,92. Canyons, streambanks, woodland; 200–1600 m. s SN, SCo, TR, PR, **s SNE, DMoj**; to sw UT, TX, n Mex. [var. *coriacea* (S. Watson) Rehder] Apparently hybridizes with and difficult to separate from *F. latifolia* in s SN, w DMoj. ❀ 4–6,17&IRR:1,**2,3,7–11**,12,**14–16,18–23**,24;CVS. Mar–Apr

MENODORA

Per to shrub, gen glabrous. **LVS** opposite to alternate, simple, gen entire, sessile to short-petioled. **INFL** appearing after lvs; axillary clusters or terminal panicle. **FL** bisexual; calyx lobes 5–15, ± linear, persistent; corolla ± rotate to funnel-shaped, lobes 4–6; ovules 2–4 per chamber, style slender, stigmatic lobes 2, ± spheric. **FR**: capsule, dehiscent by valves, circumscissile, or ± indehiscent, 2-lobed to near base. **SEEDS** 4–8. ± 25 spp.: Am, s Afr. (Greek: perhaps half-moon spear, from appearance of fr on stiff pedicel) [Steyermark 1932 Ann Missouri Bot Gard 19:87–176]

1. Shrub; branches many, spreading to ascending, branchlets short, stout, becoming spiny; corolla white, often purple- or brown-tinged; fr indehiscent . . . . *M. spinescens*
1' Per or subshrub; branches 0–few, ascending to erect, slender, not becoming spiny; corolla yellow; fr circumscissile
  2. Herbage rough-puberulent to scabrous; calyx lobes 8–11; upper lf length < 4 × width . . . . *M. scabra*
  2' Herbage ± glabrous; calyx lobes 5–8; upper lf length > 5 × width . . . . *M. scoparia*

M. scabra A. Gray (p. 387) Subshrub, scabrous to rough-puberulent. **STS** 3–many from base, 11–28 cm, ascending to erect. **LVS** gen alternate, 7–20 mm, oblong to ovate, reduced and wider upward. **INFL** terminal. **FL**: calyx lobes 8–11, 3–6 mm, rough-hairy; corolla tube 4–6 mm; anthers and stigma exserted. **FR** circumscissile; lobes 5.5–8 mm. 2*n*=44. Rocky soils, canyons; 1200–1800 m. **e DMtns (Clark, Eagle, New York mtns)**; to Colorado, TX, n Mex. May

M. scoparia A. Gray Per to subshrub, ± glabrous. **STS** gen many from base, 30–60 cm, gen erect. **LVS** gen alternate, 5–25 mm, linear to obovate, reduced and narrower upward. **INFL** terminal. **FL**: calyx lobes 5–8, 3–6 mm, glabrous; corolla tube 3–5 mm; anthers and stigma exserted. **FR** circumscissile; lobes 5–7 mm. 2*n*=22. Rocky slopes, canyons; 600–2000 m. **e&s DMoj, w DSon**, e PR; to TX, n Mex. May–Jul

M. spinescens A. Gray   Shrub < 90 cm, sparsely puberulent. **ST** intricately branched; branchlets short, stout, becoming spiny. **LVS** alternate or clustered, 3–11 mm, oblong to obovate, fleshy. **INFL** axillary. **FL**: calyx lobes 5–7, 3–5 mm, sparsely rough-hairy; corolla tube 4–9 mm; anthers and stigma barely exserted. **FR** indehiscent or breaking apart irregularly; lobes 5–8 mm. Rocky slopes, canyons; 900–2300 m. SnBr (n slope), **s SNE**, **DMtns**; s NV, nw AZ. ✿ TRY. Apr–May

ONAGRACEAE   EVENING PRIMROSE FAMILY

Warren L. Wagner, except as specified

Peter H. Raven, Family Coordinator

Ann to tree. **LVS** basal or cauline, alternate, opposite, or whorled, gen simple and toothed (to pinnately compound); stipules 0 or gen deciduous. **INFL**: spike, raceme, panicle, or fls solitary in axils; bracted. **FL** gen bisexual, gen radial, opening at dawn or dusk; hypanthium sometimes prolonged beyond ovary (measured from ovary tip to sepal base); sepals gen 4(2–7); petals gen 4 (or as many as sepals, rarely 0), often "fading" darker; stamens gen 4 or 8 (2), anthers 2-chambered, opening lengthwise, pollen gen interconnected by threads; ovary inferior, chambers gen 4 (sometimes becoming 1), placentas axile or parietal, ovules 1–many per chamber, style 1, stigma 4-lobed (or lobes as many as sepals), club-shaped, or hemispheric. **FR**: capsule, loculicidal (sometimes berry or indehiscent and nut-like). **SEEDS** sometimes winged or hair-tufted. 15 genera, ± 650 spp.: worldwide, esp w N.Am; many cult (*Clarkia, Epilobium, Fuchsia, Gaura, Oenothera*). [Munz 1965 N.Am Fl II 5:1–278]

1. Sepals persistent; hypanthium not prolonged beyond ovary; petals gen easily deciduous; moist habitats
   . . . . **LUDWIGIA**
1' Sepals deciduous; hypanthium gen conspicuously (exc *Gayophytum*) prolonged beyond ovary;
   petals not gen easily deciduous, various habitats
   2. Fr indehiscent, nut-like, short . . . . **GAURA**
   2' Fr dehiscent, not nut-like, long
      3. Ovary chambers 2; hypanthium barely prolonged beyond ovary; st branches hair-like . . . **GAYOPHYTUM**
      3' Ovary chambers 4; hypanthium well developed or not; st branches gen not hair-like
         4. Seeds hair-tufted at 1 end or sepals erect at fl; pollen gen shed in groups of 4 . . . . **EPILOBIUM**
         4' Seeds not hair-tufted; sepals reflexed singly, in pairs, or 3 adherent and reflexed to 1 side;
            pollen gen shed singly
            5. Stigma ± head-like or hemispheric . . . . **CAMISSONIA**
            5' Stigma 4-lobed . . . . **OENOTHERA**

CAMISSONIA   SUN CUP

Ann, per, from taproot or lateral roots. **LVS** basal, cauline, or both, alternate, simple to 2-pinnate. **INFL** bracted; spike, raceme, or fls solitary in axils. **FL** radial, gen opening at dawn (rarely at dusk); sepals 4, reflexed (sometimes 2–3 remaining adherent); petals 4, yellow, white, lavender, often with darker basal spots, gen fading purplish or reddish; stamens (4)8, longer ones opposite sepals, anthers gen attached at middle (or base), pollen grains 3-angled exc in polyploid taxa (visible with hand-lens); ovary chambers 4, stigma ± head-like or hemispheric, gen > anthers and cross-pollinated (or ± = anthers and self-pollinated). **FR** straight to coiled, gen sessile. **SEEDS** in 1–2 rows per chamber. 62 spp.: w N.Am (esp CA-FP), 1 S.Am. (L.A. von Chamisso, French-born German botanist, 1781–1838) [Raven 1969 Contr US Natl Herb 37:161–396] Polyploidy and self-pollination have predominated in evolution of genus. Previously incl in *Oenothera* ("*O.*" in synonyms).

1. Ovary tip (below hypanthium, above seeds) slender, sterile; st ± 0 (sect. *Tetrapteron*)
   2. Ann; fr winged; lf narrowly oblanceolate, minutely serrate; petals 2–3.5 mm . . . . *C. palmeri*
   2' Per; fr not winged; lf lanceolate to narrowly elliptic; subentire to irregularly pinnately lobed; petals
      5–23 mm
      3. Pl hairs dense, spreading or appressed; hypanthium 4–8.5 mm; stigma > anthers
         . . . . *C. tanacetifolia* ssp. *tanacetifolia*
      3' Pl ± glabrous; hypanthium 1.5–3 mm; stigma ± = or slightly > anthers . . . . *C. subacaulis*
1' Ovary without a slender, sterile tip; st present (or pl immature)
   4. Seed wing thick, with club-shaped hairs; petals white, bases yellow (sect. *Chlismiella*) . . . . . . *C. pterosperma*
   4' Seed wing 0; petals white, yellow, lavender, rarely cream or red
      5. Fr pedicelled, not coiled or twisted or wavy; seeds in 2 rows per chamber
         6. Hypanthium 4.5–40 mm; pl with cauline, simple lvs (rosette 0 or poorly developed) (sect. *Lignothera*)
            7. Hypanthium 18–40 mm; infl open; sepals 8–15 mm . . . . *C. arenaria*
            7' Hypanthium 4.5–14 mm; infl compact; sepals 3–9 mm . . . . *C. cardiophylla*
               8. Hairs gen spreading (sometimes some glandular); hypanthium 4.5–12 mm . . . . ssp. *cardiophylla*
               8' Hairs gen glandular (sometimes spreading, longer, nonglandular); hypanthium 9–14 mm . . . ssp. *robusta*
      6' Hypanthium 0.4–8 mm; pl gen with well developed basal rosette, sometimes also with cauline lvs;
         lvs gen 1-pinnate (sect. *Chylismia*)
         9. Petals lavender, bases gen yellow, lavender-dotted; rosette poorly developed . . . . *C. heterochroma*

9' Petals yellow or white (sometimes fading purple or red or bases purple- or red-dotted); rosette
  well developed
  10. Fr gen < 2 mm wide, ± same width throughout
    11. Sepals 1.5–4 mm; hypanthium 1–1.5 mm; stigmas ± = anthers; infl erect, buds drooping;
       lflets < 30 mm. . . . . *C. walkeri* ssp. *tortilis*
    11' Sepals 5–9 mm; hypanthium 3–8 mm; stigmas > anthers; infl nodding; lflets gen < 10 mm or 0
       . . . . *C. brevipes*
      12. Fl buds reflexed; petals gen fading red. . . . . ssp. *arizonica*
      12' Fl buds not reflexed; petals not fading red
        13. Pl hairs spreading; sepals in bud with free subterminal tips; petals not red-dotted . . . . . ssp. *brevipes*
        13' Plants strigose (rarely also with spreading hairs below); sepals in bud without free tips;
           petals gen red-dotted. . . . . ssp. *pallidula*
  10' Fr > 2 mm wide, wider toward tip
    14. Pedicel and fr becoming reflexed; petals yellow . . . . *C. munzii*
    14' Pedicel and fr spreading or ascending; petals white or yellow . . . . *C. claviformis*
      15. Pl spreading-hairy below; free tips of sepals in bud conspicuous; petals yellow . . . . ssp. *peirsonii*
      15' Pl strigose or glabrous below; free tips of sepals in bud 0 or inconspicuous (exc ssp. *funerea*);
         petals white or yellow
        16. Petals yellow
          17. Pl strigose above, also glandular-hairy or not; free tips of sepal in bud 0 or short . . . . . . ssp. *yumae*
          17' Pl glabrous above or glandular-hairy; free tips of sepal in bud 0. . . . . ssp. *lancifolia*
        16' Petals white (rarely pale yellow in ssp. *claviformis*)
          18. Lateral lflets well developed
            19. Pl strigose above (rarely with some glandular hairs) . . . . ssp. *aurantiaca*
            19' Pl ± glabrous above (or with some glandular hairs). . . . . ssp. *claviformis*
          18' Lateral lflts gen 0 or poorly developed
            20. Free tips of sepals in bud conspicuous; pl gen strigose above . . . . ssp. *funerea*
            20' Free tips of sepals in bud 0 or minute; pl ± glabrous above or gen with some glandular
               hairs . . . . ssp. *integrior*
5' Fr sessile (exc. some *C. kernensis),* often coiled or twisted and wavy; seeds in 1 row per chamber
  21. Petals white; fl opening at dusk (sect. *Eremothera)*
    22. Fr of ± same width throughout; fls at lower nodes 0
      23. Sepals 1.5–2.5 mm; petals 1.8–3 mm; stigma ± surrounded by anthers . . . . *C. chamaenerioides*
      23' Sepals 4–6 mm; petals 3.5–7 mm; stigma exceeding anthers . . . . *C. refracta*
    22' Fr enlarged at base and tapering to the tip; fls at lower nodes present or 0
      24. Rosette present at time of first fl (spring); bracts not lf-like, ± inconspiduous; pl ± glabrous
         (or hairs not spreading). . . . . *C. boothii*
        25. Fr 2–3.8 mm wide at base, woody, curved outward, but not downward . . . . ssp. *condensata*
        25' Fr 1–1.6 mm wide at base, not woody, curved downward. ssp. *desertorum*
      24' Rosette gen withered by time of first fl (late spring, summer); bracts lf-like, conspicuous;
         pl hairs often spreading
        26. Pl strigose, also with rarely spreading or glandular hairs; seeds all minutely pitted . . . . . ssp. *alyssoides*
        26' Pl hairs spreading, some glandular; seeds of 2 kinds, minutely pitted and coarsely papillate
          27. Pl gen 15–40 cm; cauline lf narrowly ovate or lanceolate, serrate . . . . ssp. *boothii*
          27' Pl 5–20 cm; cauline lf narrowly lanceolate to lanceolate, ± entire to minutely serrate. . . ssp. *intermedia*
  21' Petals yellow; fl opening near dawn
    28. Pl straight, slender, erect; lf sharply pinnately lobed; petal bases ± red-flecked; fr reflexed
       (sect. *Eulobus*) . . . . *C. californica*
    28' Pl not simultaneously straight, slender, and erect; lf entire, serrate or dentate; petals bases with
       0–2 red dots; fr not reflexed
      29. Fr 4-angled (at least when dry), not swollen by seeds; fls gen from lower (through upper)
         nodes; lf gen narrowly elliptic-lanceolate; seeds dull, 1–1.5 mm (sect. *Holostigma*) . . . . *C. pallida*
        30. Petals 6.5–13 mm; hypanthium 3.8–4.2 mm; style 6.5–10.5 mm . . . . ssp. *hallii*
        30' Petals 2–6.5 mm; hypanthium 1–3 mm; style 2.1–6.5 mm . . . . ssp. *pallida*
      29' Fr cylindric, ± swollen by seeds; fls 0 at lower nodes; lf gen linear to narrowly elliptic or
         narrowly oblanceolate; seeds shiny, gen < 1 mm (sect. *Camissonia*)
        31. Sepals all separating and reflexed when fl opens
          32. Petals 8–18 mm; sepals 5–11 mm; stigma exceeding anthers . . . . *C. kernensis*
            33. Pl somewhat open, sparsely spreading-hairy, sparsely glandular-hairy; lvs clustered at
               base; pedicel 3–15 mm. . . . . ssp. *gilmanii*
            33' Pl compact, densely spreading-hairy, sparsely glandular-hairy; lvs clustered at base; pedicel
               0–5 mm. . . . . ssp. *kernensis*
          32' Petals 1.8–4 mm; sepals < 3.8 mm; stigma ± surrounded by anthers
            34. Rosette 0; lf subentire; pl ± glabrous or minutely strigose, gen sparsely glandular-hairy,
               rarely with a few spreading nonglandular hairs . . . . *C. parvula*

34' Rosette gen ± 0; lf serrate; pl hairs gen spreading, some glandular
　35. Hypanthium 1.3–3 mm; sepals 2.2–3.8 mm; fr (18)26–50 mm . . . . *C. pubens*
　35' Hypanthium 0.8–1.6 mm; sepals 1.2–2 mm; fr 18–32 mm . . . . *C. pusilla*
31' Sepals gen remaining adherent in pairs and reflexed when fl opens
　36. Stigma exceeding anthers; sepals 3–8(12) mm; petals (3.5)5–15 mm　..　*C. campestris* ssp. *campestris*
　36' Stigma surrounded by anthers; sepals 1.6–4 mm; petals 2.1–4.5 mm . . . . *C. strigulosa*

C. arenaria (Nelson) Raven (p. 391)　Ann or bushy per, erect; hairs spreading, in infl a few glandular. **ST** < 180 cm. **LF**: petiole < 60 mm; blade < 60 mm, cordate-deltate, teeth coarse or larger and smaller. **INFL** open, nodding. **FL** opening at dusk; hypanthium 18–40 mm; sepals 8–15 mm; petals 8–20 mm, yellow. **FR** 30–44 mm, ascending, cylindric, ± straight; pedicel 2–5 mm. **SEEDS** in 2 rows per chamber, 0.5–0.7 mm, brown. 2*n*=14. Sandy washes, rocky slopes, desert scrub; < 0–430 m. **DSon**; sw AZ, n Mex (Sonora). [*O. a.* (Nelson) Raven] Gen cross-pollinated. ❀ TRY.

C. boothii (Douglas) Raven　Ann, gen reddish; rosette gen ± 0 (to well developed); hairs minutely strigose and spreading (some glandular, esp in infl). **ST** erect, 3–65 cm, peeling. **LVS** 20–100(130) mm, lanceolate to narrowly elliptic or narrowly ovate, sparsely minutely dentate or serrate; lower oblanceolate or not. **INFL** nodding; fls gen 0 at lower nodes. **FL** opening at dusk; hypanthium 4–8 mm; sepals (2.7)4–8 mm; petals 3–7.5 mm, gen white (red) fading reddish. **FR** 8–35 mm, 1–3.8 mm wide, cylindric exc base wider than tip, ± curved outward to very wavy and twisted, persistent, tardily dehiscent. **SEEDS** in 1 row per chamber, 1.4–2.1 mm, gen of 2 kinds (minutely pitted in rows and pale brown; coarsely papillate and dark brown). 2*n*=14. Shrubby or open, dry areas, gen desert; < 2400 m.　s SN, CW, WTR, **GB**, **D**; to WA, ID, UT, nw Mex. [*O. b.* Douglas] Cross-pollinated. 6 sspp. in CA.

　ssp. ***alyssoides*** (Hook. & Arn.) Raven　PINE CREEK EVENING PRIMROSE　Pl: hairs gen densely, minutely strigose (esp in infl, hairs rarely spreading or glandular). **ST** 3–35 cm. **LVS** 10–40 mm, lanceolate to narrowly ovate; lower oblanceolate or not. **INFL**: bracts lf-like; fls sometimes present at lower nodes. **FR** 1–1.4 mm wide, gen very wavy and twisted. **SEEDS** all alike. RARE in CA. Sandy slopes, flats, gen sagebrush scrub; 600–1700 m. MP, **W&I**; to NV, w UT, s ID. [*O. a.* Hook. & Arn.] Intergrades widely with sspp. *boothii, intermedia* in NV; much like ssp. *desertorum.* May–Aug

　ssp. ***boothii*** (p. 391)　BOOTH'S EVENING PRIMROSE　Pl: hairs spreading and glandular. **ST** gen 15–40 cm. **LF** 30–80 mm, lanceolate to narrowly ovate, serrate. **INFL**: bracts lf-like. **FR** 1.4–2 mm wide, gen very wavy and twisted. UNCOMMON. Sandy flats, steep loose slopes, Joshua-tree and pinyon-juniper woodlands; 900–2400 m. **SNE**; to WA, nw AZ. Intergrades widely with sspp. *alyssoides, intermedia* in NV. Jun–Aug

　ssp. ***condensata*** (p. 391) (Munz) Raven　Pl stout, subglabrous (exc infl minutely strigose or glandular-hairy); rosette well developed. **ST** 5–20(30) cm. **LF** 25–100(130) mm, gen lanceolate to oblanceolate, subentire to minutely dentate. **INFL**: bracts inconspicuous. **FR** 2–3.8 mm wide, curved outward. Sandy slopes, washes, desert scrub; –70–1200 m. **D**; to s NV, s UT, w AZ, nw Mex. [*O. b.* ssp. *c.* (Munz) Munz] Intergrades widely with ssp. *desertorum.*

　ssp. ***desertorum*** (Munz) Raven (p. 391)　Pl: rosette well developed; hairs sparse, minutely strigose (or also glandular, esp in infl). **ST** 10–35 cm. **LF** gen 10–40 mm, lanceolate to narrowly ovate (or lower oblanceolate), entire to minutely dentate. **INFL**: bracts inconspicuous. **FR** 1–1.6 mm wide, gen curved downward. Sandy or gravelly slopes, washes, gen creosote-bush scrub; 450–2000 m. s SNH, **s SNE, DMoj.** [*O. b.* sspp. *d.* Munz, *inyoensis* Munz] Intermediate between sspp. *condensata, decorticans.*

　ssp. ***intermedia*** (Munz) Raven (p. 391)　Pl: hairs dense, spreading (and glandular, esp in infl). **ST** 5–20 cm. **LF** gen < 25 mm, ± lanceolate (or lower oblanceolate), minutely serrate. **INFL**: bracts lf-like; fls sometimes present at lower nodes. **FR** 1–1.4

mm wide, gen curved outward or ± wavy and twisted. Sandy soils, sagebrush scrub; 1500–2150 m. **SNE, DMtns**; NV. [*O. b.* ssp. *i.* Munz] Intermediate between sspp. *alyssoides, boothii*; ± uniform.

C. brevipes (A. Gray) Raven　Ann, strigose or hairs spreading. **ST** 3–75 cm. **LVS** gen basal, simple to 1-pinnate; terminal lflet < 65 mm, lateral lflets gen < 10 mm or 0. **INFL** nodding. **FL**: hypanthium 3–8 mm; sepals 5–9 mm, tips in bud gen free and subterminal; petals 3–18 mm, yellow, bases ± red-dotted; stamens ± equal. **FR** ascending to spreading, 18–92 mm, cylindric, straight or curved; valves with strong midrib; pedicel 2–20 mm. **SEEDS** in 2 rows per chamber, 1–1.5 mm. 2*n*=14. Sandy or rocky slopes, washes, creosote-bush scrub, Joshua-tree woodland; –70–1800 m. **D**; to w&s NV, sw UT. [*O. b.* A. Gray] Cross-pollinated.

　ssp. ***arizonica*** (Raven) Raven (p. 391)　Pl: hairs spreading. **FL**: bud reflexed; hypanthium 3–5 mm; sepal tips in bud free or not; petals 3–8 mm, gen red-dotted, gen fading red. **FR** 18–60 mm; pedicel 2.5–5 mm. Rocky slopes and flats; 70–300 m. **se DSon** (Imperial Co.); sw AZ. [*O. b.* ssp. *a.* Raven] Hybridizes with *C. claviformis* ssp. *yumae.*

　ssp. ***brevipes*** (p. 391, pl. 74)　Pl: hairs spreading. **FL**: bud not reflexed; hypanthium 4–8 mm; petals 6–18 mm, gen not red-dotted, not fading red. **FR** 20–92 mm; pedicel 5–20 mm. Sandy slopes, washes, alluvial fans (moister than other sspp.); –70–1800 m. **D**; to w&s NV, sw UT. Intergrades with ssp. *pallidula*; hybridizes with *C. claviformis, C. munzii.* ❀ TRY. Mar–May

　ssp. ***pallidula*** (Munz) Raven (p. 391)　Pl strigose (hairs rarely spreading below). **FL**: bud not reflexed; hypanthium 4–5 mm; sepal tips in bud not free; petals 7–12 mm, gen red-dotted, not fading red. **FR** 20–42 mm; pedicel 2–10 mm. Dry flats, desert pavement, gen with *Larrea, Ambrosia*; 70–1100 m. **D** (**se Inyo, ne Imperial cos.**); to sw UT, nw AZ. [*O. b.* ssp. *pallidula* (Munz) Raven] ❀ TRY.

C. californica (Torrey & A. Gray) Raven (p. 391)　Ann, subglabrous (or lvs minutely strigose, infl sparsely glandular-hairy); rosette well developed. **ST** straight, slender, erect, 2–180 cm, ± glaucous or green. **LVS** much reduced upward, < 300 mm, narrowly elliptic, irregularly and sharply pinnately lobed. **INFL**: fls widely spaced. **FL**: hypanthium 0.6–1.5 mm, closed by fleshy disk; sepals 3.9–8 mm; petals 6–14 mm, yellow, base ± red-flecked. **FR** reflexed, 45–110 mm, cylindric (drying 4-angled), ± straight. **SEEDS** in 1 row per chamber, 1.3–1.6 mm, olive with purple spots. 2*n*=14, 28. Open places in coastal-sage scrub, chaparral, desert scrub; < 1300 m. s NCoRO (Sonoma Co.), SnJV (Fresno Co.), SCoR, SW, **D**; sw AZ, nw Mex. [*Eulobus c.* Torrey & A. Gray; *O. leptocarpa* E. Greene] Self-pollinated. ❀ SUN,DRY,DRN:**7–12,14–17**,18,**19–24**. Apr–Jun

C. campestris (E. Greene) Raven　MOJAVE SUN CUP　Ann, slender; rosette gen ± 0; hairs 0, coarse, or glandular. **ST** decumbent or erect, 5–25(50) cm, peeling. **LF** 5–30 mm, linear to narrowly elliptic or narrowly oblanceolate, minutely to coarsely serrate. **INFL** nodding. **FL**: hypanthium 1.5–5 mm; sepals 3–8(12) mm, remaining adherent in pairs; petals (3.5)5–15 mm, yellow fading reddish, base with (1)2 red dots; stigma exceeding anthers. **FR** 20–43 mm, 0.7–2 mm wide, cylindric, alternately narrow and swollen by seeds, straight or wavy, subsessile. **SEEDS** in 1 row per chamber, 0.8–1.6 mm, shiny, minutely pitted. 2*n*=14. Open sandy flats, desert scrub, noncoastal grasslands; 0–2000 m. SNF, GV, CW, e SW, **DMoj**. [*O. c.* E. Greene] Cross-pollinated. Sspp. intergrade extensively. 2 sspp. in CA.

　ssp. ***campestris***　**ST** gen erect. **LF** linear to narrowly elliptic or narrowly oblanceolate, minutely serrate. Habitats, range of sp. [*O. c.* ssp. *parishii* (Abrams) Munz; *O. dentata* Cav. var.

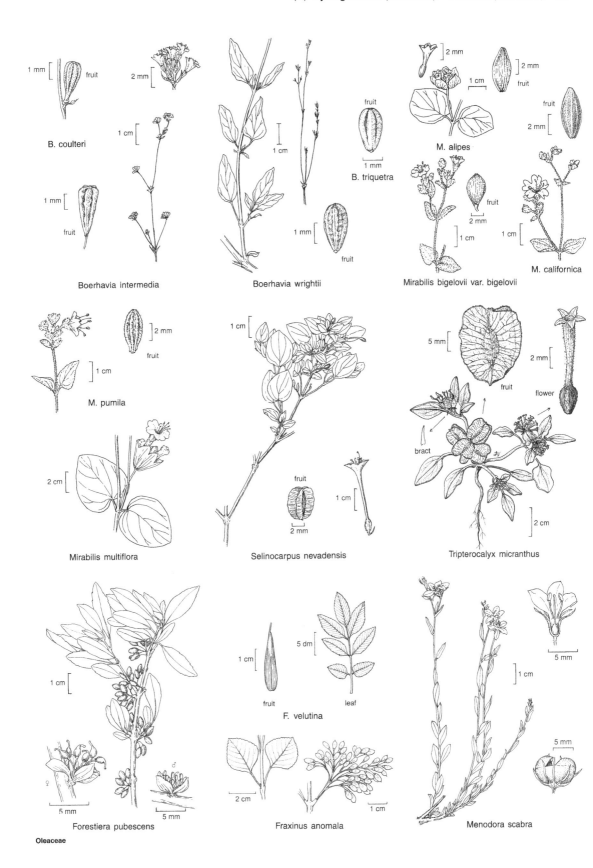

B. coulteri

fruit

Boerhavia intermedia

Boerhavia wrightii

B. triquetra

fruit

M. alipes

Mirabilis bigelovii var. bigelovii

M. californica

M. pumila

fruit

Mirabilis multiflora

Selinocarpus nevadensis

fruit

flower

bract

Tripterocalyx micranthus

F. velutina

fruit

leaf

Forestiera pubescens

Fraxinus anomala

Menodora scabra

Oleaceae

johnstonii Munz; *O. cruciata* (S. Watson) Munz misapplied]. ❀ TRY. Mar–May

C. cardiophylla (Torrey) Raven Ann, per; rosette 0; hairs spreading, glandular or not. **ST** < 100 cm. **LF**: blade < 55 mm, ovate to rounded-cordate, irregularly dentate; petiole < 75 mm. **INFL** dense, nodding. **FL** opening at dusk; hypanthium 4.5–14 mm; sepals 3–9 mm; petals 3–12 mm, yellow or cream. **FR** 20–55 mm, ascending, cylindric, ± straight; pedicel 1–18 mm. **SEEDS** in 2 rows per chamber, 0.5–0.7 mm, brown. 2*n*=14. Sandy washes, slopes, rocky walls, creosote-bush scrub; 0–1400 m. **D**; sw AZ, nw Mex. [*O. c.* Torrey] Gen cross-pollinated. 3 sspp., 2 in CA.

ssp. **cardiophylla** (p. 391) Pl: sometimes some hairs glandular. **LF** cordate. **FL**: hypanthium 4.5–12 mm; petals 3–12 mm. Habitats of sp.; 0–600 m. **c&s DMoj (San Bernardino Co.)**, **DSon**; s AZ, nw Mex. ❀ TRY. Mar–May

ssp. **robusta** (Raven) Raven (p. 391) Pl: hairs both glandular and not. **LF** cordate-round. **FL**: hypanthium 9–14 mm; petals 7–11 mm. Habitats of sp.; 600–1400 m. **n DMoj (Inyo Co.)**. [*O. c.* ssp. *r.* Raven] ❀ TRY.

C. chamaenerioides (A. Gray) Raven Ann, gen reddish, glandular-hairy (infl also minutely strigose); rosette gen ± 0. **ST** erect, 8–50 cm, peeling. **LF** < 80 mm, narrowly elliptic to narrowly lanceolate, sparsely minutely dentate. **INFL** nodding. **FL** opening at dusk; hypanthium 1.6–2.3 mm; sepals 1.5–2.5 mm; petals 1.8–3 mm, white fading reddish. **FR** spreading, 35–55 mm, 0.8–0.9 mm wide, cylindric, ± straight. **SEEDS** in 1 row per chamber, 0.9–1 mm, minutely pitted in rows. 2*n*=14. Sandy slopes, flats, desert scrub; ± –50–1300 m. **W&I, D**; to UT, TX, nw Mex. [*O. c.* A. Gray] Self-pollinated. Related to *C. refracta*. Mar–Jun

C. claviformis (Torrey & Frémont) Raven Ann. **ST** 3–70 cm. **LVS** gen basal, gen 1-pinnate; terminal lflet 8–90 mm, lanceolate to cordate; lateral lflets < 25 mm or 0. **INFL** nodding. **FL** gen opening at dusk; hypanthium 1–6.5 mm; sepals 2–8 mm, tips in bud free and subterminal or not; petals 1.5–8 mm, yellow or white; stamens ± equal. **FR** ascending or spreading, 8–38 mm, wider to tip, straight or curved; pedicel 4–40 mm. **SEEDS** in 2 rows per chamber, 0.6–1.5 mm. 2*n*=14. Sandy or rocky slopes or washes; –70–2000 m. **PR, GB, D**; to OR, ID, UT, AZ, nw Mex. [*O. c.* Torrey & Frémont] Cross-pollinated; most complex, widespread sp. in genus; 11 sspp., 8 in CA.

ssp. **aurantiaca** (Munz) Raven (p. 391) Pl strigose (rarely glandular-hairy above). **LF**: terminal lflet < 30 mm, narrowly ovate; lateral lflets well developed, like terminal. **FL**: hypanthium 3–5 mm; sepal tips not free in bud (rarely inconspicuously so); petals 2.5–8 mm, white gen fading purple, bases rarely purple-dotted. Sandy flats, washes, creosote-bush scrub; –70–900 m. **D**; s NV, w AZ, n Baja CA. [*O. c.* ssp. *a.* (Munz) Raven] Intergrades extensively with sspp. *peirsonii, yumae*, and those with white petals; sometimes hybridizes with *C. brevipes, C. munzii*. ❀ TRY.

ssp. **claviformis** Pl glabrous or strigose below, ± glabrous or glandular-hairy above. **LF**: terminal lflet < 60 mm, narrowly ovate, purple-dotted or not; lateral lflets gen large. **FL**: hypanthium 3–5.5 mm; sepal tips free in bud, conspicuous; petals 3.5–8 mm, white (pale yellow) gen fading purple, bases purple-dotted or not. Alluvial slopes, flats, creosote-bush scrub; 850–1700 m. **DMoj and edges**. Intergrades widely and gradually with sspp. *aurantiaca, funerea*; hybridizes with *C. brevipes* ssp. *b.* ❀ TRY. Mar–May

ssp. **funerea** (Raven) Raven (p. 391) Pl gen strigose (densely so at least below). **LF**: terminal lflet < 80 mm, ovate, gen cordate; lateral lflets gen 0. **FL**: hypanthium 3–5.5 mm; sepal tips free in bud, conspicuous, subterminal; petals 3.5–7.5 mm, white gen fading purple. Dry slopes, flats, creosote-bush scrub; –70–900 m. **n DMoj (Eureka, Saline, Death valleys)**. [*O. c.* ssp. *f.* Raven] Intergrades with sspp. *aurantiaca, claviformis*; sometimes hybridizes with *C. munzii, C. brevipes* ssp. *b.*

ssp. **integrior** (Raven) Raven Pl strigose below, gen glandular-hairy or subglabrous above. **LF**: terminal lflet < 70 mm, ± ovate, base gen subcordate; lateral lflets 0–few, small. **FL**: hypanthium 3–6 mm; sepal tips not free in bud (or inconspicuously so); petals 4.5–8 mm, white fading purple, bases purple-dotted or not. Dry flats, desert scrub; 1200–2000 m. **SNE**; c OR, NV. [*O. c.* ssp. *i.* Raven] Intergrades with sspp. *aurantiaca, cruciformis*; hybridizes with *C. brevipes*.

ssp. **lancifolia** (A.A. Heller) Raven (p. 391) Pl strigose below, glabrous and glaucous above. **LF**: terminal lflet < 50 mm, lanceolate; lateral lflets 0–few, small. **FL**: hypanthium 3.5–6 mm; sepal tips not free in bud; petals 3.5–7 mm, yellow, bases gen red-dotted. Sandy soils, sagebrush scrub; 1200–1700 m. **SNE, DMtns**. [*O. c.* ssp. *l.* (A.A. Heller) Raven]

ssp. **peirsonii** (Munz) Raven (p. 391) Pl spreading-hairy, rarely strigose or glandular. **LF**: terminal lflet < 90 mm, narrowly ovate; lateral lflets gen large. **FL**: hypanthium 2.5–4.5 mm; sepal tips in bud free, conspicuous; petals 4.5–7 mm, gen yellow (white). Sandy flats, creosote-bush scrub; –70–300 m. **PR, DSon (Imperial Co.)**; n Baja CA. [*O. c.* ssp. *p.* (Munz) Raven] Intergrades with ssp. *aurantiaca*.

ssp. **yumae** (Raven) Raven Pl strigose (gen densely so), sometimes also glandular-hairy above. **LF**: terminal lflet < 65 mm, lanceolate; lateral lflets reduced or not. **FL**: hypanthium 2.5–4 mm; sepal tips not free in bud (or inconspicuously so); petals 4–5.5 mm, yellow, sometimes fading reddish. Dunes or sandy flats, creosote-bush scrub; 0–300 m. **se DSon (se Imperial Co.)**; sw AZ, nw Mex. [*O. c.* ssp. *y.* Raven] Probably derived from ssp. *aurantiaca* × ssp. *peeblesii* (Munz) Raven in AZ; intergrades with ssp. *aurantiaca*.

C. heterochroma (S. Watson) Raven (p. 391) SHOCKLEY'S EVENING PRIMROSE Ann, glandular-hairy (or subglabrous and glaucous above). **ST** 10–100 cm. **LVS** gen basal, < 70 mm, ovate; base gen cordate. **INFL** erect, longer in fr. **FL**: hypanthium 2–5 mm; sepals 1.5–3.5 mm, tips not free in bud; petals 2–6 mm, lavender, bases gen yellow, lavender-dotted. **FR** 7–13 mm, erect, club-shaped, straight; pedicel 2–5 mm. **SEEDS** in 2 rows per chamber, 1–1.2 mm, brown. 2*n*=14. Alluvial slopes, rock slides, creosote-bush scrub to pinyon/juniper woodland; 600–2100 m. **SNE, n DMtns (Grapevine Mtns)**; NV. [*O. h.* S. Watson, incl ssp. *monoensis* (Munz) Raven] Cross-pollinated. ❀ TRY. May–Jun

C. kernensis (Munz) Raven Ann, robust; rosette gen ± 0; hairs dense, spreading (some glandular, or ± 0, esp in infl). **ST** erect, 5–30 cm. **LF** 10–38(55) mm, gen narrowly elliptic, sparsely serrate. **INFL** nodding. **FL**: hypanthium 2.2–3.8(5.5) mm; sepals 5–11 mm, free; petals 8–18 mm, yellow fading reddish, bases with 2 large red dots. **FR** 22–37 mm, 1.5–1.7 mm wide, cylindric, ± swollen by seeds, straight or wavy; pedicel 0–15 mm. **SEEDS** in 1 row per chamber, 1.1–1.2 mm, shiny, minutely pitted. 2*n*=14. Sandy slopes, flats, washes, sagebrush scrub, Joshua-tree and pinyon/juniper woodland; 700–1800 m. se SNH, **s SNE (esp Inyo Co.), n&w DMoj**; s NV. [*O. k.* Munz] Cross-pollinated. Related to *C. parvula, C. pubens, C. pusilla*. Sspp. intergrade extensively.

ssp. **gilmanii** (Munz) Raven (p. 391) Pl open; hairs ± 0–few, glandular and spreading. **ST** < 30 cm. **LVS** not esp clustered at base. **FR**: pedicel 0–5 mm. Washes, slopes; 760–1800 m. Range of sp. [*O. k.* sspp. *g.* (Munz) Munz, *mojavensis* Munz] ❀ TRY.

ssp. **kernensis** (p. 391) KERN COUNTY EVENING PRIMROSE Pl compact; hairs dense, spreading, few glandular. **ST** 5–15(22) cm. **LVS** clustered at base. **FR**: pedicel 3–15 mm. UNCOMMON. Sandy slopes, flats, gen in sagebrush scrub or Joshua-tree woodland; 850–1800 m. se SNH, **w DMoj (Piute Mtns, El Paso Mtns, Grapevine Canyon, Kern Co.)**. Often locally abundant. ❀ TRY. May

C. munzii (Raven) Raven Ann, strigose. **ST** 8–50 cm. **LVS** gen basal, 1-pinnate; terminal lflet < 60 mm, ovate; lateral lflets well developed, like terminal. **INFL** nodding. **FL:** hypanthium 2–3 mm; sepals 4–7 mm, tips not free in bud; petals 3–10 mm, yellow, bases red-dotted; stamens ± equal. **FR** 8–24 mm, wider toward tip, ± straight; pedicel (and fr) reflexed in age, 8–28 mm. **SEEDS** in 2 rows per chamber, 0.8–1.6 mm, pale brown. $2n=14$. Slopes, washes, mtns; 600–1600 m. **ne DMoj**; s NV. [*O. m.* Raven] Cross-pollinated. Sometimes hybridizes with *C. brevipes* ssp. *b.* and *C. claviformis* ssp. *a.*

C. pallida (Abrams) Raven Ann, rosetted, grayish, densely strigose, in infl also glandular. **ST** decumbent and branched or erect and simple, < < 60 cm. **LVS** 10–30 mm; cauline narrowly elliptic-lanceolate, subentire to minutely dentate; petiole < 2 mm. **INFL** nodding. **FL:** hypanthium 1–4 mm; sepals 1.5–8 mm; petals 2–13 mm, yellow fading reddish, bases with 1–3 red dots. **FR** 10–25 mm, 1.1–1.2 mm wide, ± 4-angled, straight to 3-coiled. **SEEDS** in 1 row per chamber, 1–1.5 mm, minutely pitted in rows, dull brownish black. $2n=14$. Desert slopes, flats, washes, creosote-bush scrub to pinyon/juniper woodland; 30–1800 m. s SnJV (Kern Co.), n slope SnBr, **D**; NV, AZ. [*Sphaerostigma p.* Abrams] Gen self-pollinated. Sspp. intergrade.

ssp. **hallii** (Davidson) Raven **FL:** hypanthium 3.8–4.2 mm; sepals 4.8–8 mm; petals 6.5–13 mm; style 6.5–10.5 mm. Habitats of sp. n slope SnBr, **s DMoj**, **n DSon**. [*O. h.* (Davidson) Munz] Sometimes cross-pollinated. ❀ TRY.

ssp. **pallida** **FL:** hypanthium 1–3 mm; sepals 1.5–5.5 mm; petals 2–6.5 mm; style 2.1–6.5 mm. Habitats and range of sp. [*O. abramsii* J.F. MacBr.]

C. palmeri (S. Watson) Raven (p. 391) Ann, strigose (few hairs spreading). **ST** ± 0, peeling. **LF** 15–55 mm, narrowly oblanceolate, minutely serrate. **INFL** nodding. **FL:** hypanthium 0.8–1.3 mm, closed by fleshy disk; sepals 1.6–2.3 mm; petals 2–3.5 mm, yellow; anthers attached at base; sterile tip of ovary 5–12 mm. **FR** 5–7 mm, 4-angled, 4-winged near tip, ± straight, leathery, tardily dehiscent. **SEEDS** in 2 rows per chamber, 1.2–2 mm, papillate, tan with brown spots. $2n=14$. Desert flats, sagebrush scrub; 600–1400 m. s GV, SCoRI, WTR (Tejon Pass), PR (Jacumba), **W&I**, **DMoj**; OR, NV. [*O. p.* S. Watson] Self-pollinated. Apr–May

C. parvula (Torrey & A. Gray) Raven (p. 391) Ann, slender, ± glabrous or minutely strigose (some hairs gen glandular, rarely few spreading); rosette 0. **ST** erect, < 30 cm, wiry. **LF** 10–30 mm, linear, subentire. **INFL** nodding. **FL:** hypanthium < 2 mm; sepals < 2.2 mm, free; petals < 3.6 mm, yellow fading reddish. **FR** ± 20–30 mm, 0.6–0.9 mm wide, cylindric, ± swollen by seeds, straight or wavy, subsessile. **SEEDS** in 1 row per chamber, 0.7–0.8 mm, shiny, minutely pitted. $2n=28$. Sandy soils, gen sagebrush scrub; 1000–2000 m. **GB**; to WA, WY, Colorado. [*O. p.* Torrey & A. Gray; *O. contorta* Douglas var. *flexuosa* (Nelson) Munz] Self-pollinated. Related to *C. kernensis, C. pubens, C. pusilla.*

C. pterosperma (S. Watson) Raven (p. 391) Ann, slender; hairs bristly, in infl also glandular. **ST** erect, 2–14 cm, peeling. **LF** 3–30 mm, narrowly lanceolate to oblanceolate, entire. **INFL** nodding. **FL:** hypanthium 1–2 mm; sepals 1.5–2.5 mm; petals 1.5–2.5 mm, white fading purplish, bases yellow. **FR** 12–28 mm, 0.6–0.8 mm wide, ascending or spreading, cylindric, ± straight, slightly swollen by seeds; pedicel 4–8 mm. **SEEDS** in 2 rows per chamber, 1–1.5 mm; wing thick; hairs club-shaped. $2n=14$. Uncommon. Well drained, gen volcanic slopes, pinyon/juniper woodland or sagebrush scrub; 1400–2400 m. **s W&I (Inyo Mtns)**, **n DMtns (Panamint Mtns)**; to OR, UT. [*O. p.* S. Watson] Self-pollinated. May–Jun

C. pubens (S. Watson) Raven Ann; rosette gen ± 0; hairs gen glandular, some spreading. **ST** erect, < 38 cm. **LF** 15–45 mm, narrowly lanceolate, wavy-serrate. **INFL** nodding. **FL:** hypanthium 1.3–3 mm; sepals 2.2–3.8 mm, free; petals 2.2–4 mm, yellow fading reddish, bases with 1–few red dots. **FR** (18)26–50 mm, 0.8–1.2 mm wide, cylindric, ± swollen by seeds, straight or wavy; pedicel 0–2 mm. **SEEDS** in 1 row per chamber, 0.7–0.8 mm, shiny, minutely pitted. $2n=28$. Sandy soils, gen sagebrush

scrub or pinyon/juniper woodland; 1000–3000 m. **GB**, **n DMoj (scattered)**; w NV. [*O. p.* (S. Watson) Munz] Self-pollinated. Related to *C. kernensis, C. parvula, C. pusilla.*

C. pusilla Raven (p. 397) Ann, slender; rosette gen ± 0; hairs glandular, gen also spreading. **ST** erect, 2–22 cm. **LF** 10–30 mm, linear, minutely serrate. **INFL** nodding. **FL:** hypanthium 0.8–1.6 mm; sepals 1.2–2 mm, free; petals 1.8–3.1 mm, yellow fading reddish, bases with 2 red dots. **FR** 18–32 mm, 0.6–0.9 mm wide, cylindric, ± swollen by seeds, straight or wavy; pedicel 0–2 mm. **SEEDS** in 1 row per chamber, 0.7–0.8 mm, shiny, minutely pitted. $2n=14$. Sandy soils, gen sagebrush scrub; 760–3000 m. n slope SnBr (Cactus Flats), **GB, DMoj (scattered)**; to WA, ID, UT. [*O. contorta* Douglas var. *flexuosa* (Nelson) Munz misapplied] Self-pollinated. Related to *C. kernensis, C. parvula, C. pubens.*

C. refracta (S. Watson) Raven (p. 397) Ann, gen reddish, sparsely minutely strigose, esp in infl also glandular; rosette ± 0. **ST** erect, 6–45 cm, peeling. **LVS** < 60 mm, narrowly elliptic to narrowly lanceolate (lowest gen oblanceolate), sparsely minutely dentate. **INFL** nodding. **FL** opening at dusk; hypanthium 4–6 mm; sepals 4–6 mm; petals 3.5–7 mm, white fading reddish; 70–100% of pollen grains 4–5-angled. **FR** 20–50 mm, 0.7–1 mm wide, cylindric, straight or wavy. **SEEDS** in 1 row per chamber, 0.9–1.5 mm, minutely pitted in rows. $2n=14$. Sandy slopes, flats, desert scrub; –30–1300 m. **D**; to s NV, sw UT, w AZ. [*O. r.* S. Watson; *O. deserti* M.E. Jones] Cross-pollinated. ❀ TRY. Mar–May

C. strigulosa (Fischer & C. Meyer) Raven (p. 397) Ann, slender, minutely strigose (hairs glandular or not, toward base also coarse, spreading); rosette ± 0. **ST** decumbent or erect, < 50 cm, wiry, peeling. **LF** 8–35 mm, linear to very narrowly elliptic, minutely serrate. **INFL** nodding. **FL:** hypanthium 1.6–2.7 mm; sepals 1.6–4 mm, remaining adherent in pairs; petals 2.1–4.5 mm, yellow fading reddish, bases with 0–2 red dots. **FR** 15–45 mm, 0.8–1.3 mm wide, cylindric, ± swollen by seeds, straight or wavy, subsessile. **SEEDS** in 1 row per chamber, 0.6–0.8 mm, shiny, minutely pitted. $2n=28$. Open, sandy soils of dunes, grassland, desert scrub; 0–2100 m. s edge s SNH, Teh, CW, SW, n ChI (Santa Rosa Island), **w DMoj**; n Baja CA. [*O. s.* (Fischer & C. Meyer) Torrey & A. Gray; *O. dentata* Cav. misapplied] Self-pollinated. Related to S.Am *C. dentata*; hybridizes with *C. kernensis* ssp. *k.* Mar–May

C. subacaulis (Pursh) Raven (p. 397) Per, ± fleshy, ± glabrous; taproot thick. **ST** ± 0. **LF:** blade 20–220 mm, lanceolate to narrowly elliptic, subentire to irregularly pinnately lobed, rarely sparsely and minutely strigose; petiole 10–120 mm. **FL** erect; hypanthium 1.5–3 mm, closed by fleshy disk; sepals 4.1–13 mm; petals 5–16 mm, yellow; anthers attached at base; ovary with sterile tip 15–80 mm. **FR** 11–28 mm, linear-ovoid, 4-angled, barely swollen by seeds, ± straight or slightly curved, leathery; pedicel 0–10 mm. **SEEDS** in 2 rows per chamber, 1.3–1.9 mm, pitted, pale brown. $2n=14$. Moist meadows, gen clay soils; 450–2600 m. SN, **GB**; to WA, MT, Colorado. [*O. s.* (Pursh) A.O. Garrett incl var. *taraxacifolia* (S. Watson) Jepson] Gen cross-pollinated. ❀ TRY.

C. tanacetifolia (Torrey & A. Gray) Raven Per; taproot woody, new shoots from lateral roots; hairs gen ± dense (sparse), short, spreading or appressed. **ST** 0. **LF** 65–320 mm, narrowly elliptic, deeply and irregularly pinnately lobed; petiole 10–80 mm. **INFL** erect. **FL:** hypanthium 4–6.5(8.5) mm, closed by fleshy disk; sepals 5.5–13 mm; petals 8–23 mm, yellow; anthers attached at base; sterile tip of ovary 14–55 mm. **FR** 7–25 mm, long-tapered, swollen by seeds, leathery, ± straight or slightly curved, disintegrating irregularly. **SEEDS** in 2 rows per chamber, 1.5–2 mm, pitted in rows, pale brown. Open fields, moist slopes, clay soils; 700–2500 m. CaR, SN, **GB**; to WA, ID, NV. [*O. t.* Torrey & A. Gray] Crosspollinated. 2 sspp. in CA.

ssp. **tanacetifolia** (p. 397) Pl: hairs gen dense (sparse), spreading or appressed. **FL:** < 5% of pollen grains 4-angled. $2n=14,28$. Open fields, moist slopes, clay soils; 700–2500 m. CaR,

SN, **GB**; to WA, ID, NV [*C. breviflora* (Torrey & A. Gray) Raven misapplied: not in CA] ✸ SUN:**15–17**&IRR:1–3,7,14,18–24. May–Jul

C. walkeri (Nelson) Raven ssp. *tortilis* (Jepson) Raven Ann, short-lived per; hairs spreading, in infl also glandular or bristly. **ST** 10–60 cm. **LVS** gen basal, simple to 2-pinnate, 30–220 mm, gen purple-dotted; lateral lflets < 30 mm. **INFL** erect; buds drooping. **FL**: hypanthium 1–1.5 mm; sepals 1.5–4 mm, gen purple-dotted, in bud tips free, subterminal; petals 3–6 mm, yellow. **FR** ascending or spreading, 10–45 mm, 1.2–1.8 mm wide, cylindric, straight or curved; valves gen twisted; pedicel 5–30 mm. **SEEDS** in 2 rows per chamber, 0.6–1.2 mm. $2n=14$. Rocky places near cliffs, along ephemeral streams, creosote-bush scrub to pinyon/juniper woodland; 600–1800 m. **W&I, n DMoj (Inyo, ne San Bernardino cos.)**; to s NV, sw UT. [*O. scapoidea* Torrey & A. Gray var. *t.* Jepson; *O. w.* (Nelson) Raven ssp. *t.* (Jepson) Raven] Self-pollinated. Ssp. *w.* outside CA. Mar–May

EPILOBIUM FIREWEED, WILLOW HERB

Peter C. Hoch

Ann to subshrub. **LVS** gen opposite below (or clustered in axils), gen ± fine-toothed; veins gen obscure. **INFL**: gen raceme, bracted. **FL** radial or ± bilateral; sepals 4, erect; petals 4, gen notched; stamens 8, anthers attached at middle, pollen grains gen shed in 4's, gen cream-yellow; ovary chambers 4, stigma gen club-like. **FR** straight, cylindric to club-like. **SEEDS** gen in 1 row per chamber, gen with white, deciduous hair-tuft. 171 spp.: worldwide exc trop. (Greek: upon pod, from inferior ovary) [Raven 1976 Ann Missouri Bot Gard 63:326–340] Incl *Boisduvalia*, *Zauschneria*. Most taxa polyploid; many with anthers ± = stigma self-pollinated; many hybrids. **Epilobium angustifolium* now treated in *Chamerion*. See Hoch 1999 Flora of Japan IIc:241; Baum et al 1994 Syst Bot 19:363–388.

1. Lvs alternate; hypanthium 0; petals entire; stamens subequal, in 1 whorl; pollen bluish gray, grains shed singly . . . . **E. angustifolium** ssp. **circumvagum**
1' Lvs opposite at least at base; hypanthium 0.3–34 mm; petals notched; stamens in 2 unequal whorls; pollen cream, grains gen shed in 4's
  2. Ann; lvs opposite only near base; lower st peeling; seed hair-tuft 0 . . . . **E. torreyi**
  2' Ann to subshrub; lvs gen opposite up into infl; lower st gen not peeling; seeds hair-tufted
    3. Hypanthium 17–34 mm; corolla red-orange, ± bilateral . . . . **E. canum** ssp. **latifolium**
    3' Hypanthium gen 0.5–2.6(16) mm; corolla white to purple
      4. Ann; st peeling below; lvs narrowly lanceolate, upper often clustered . . . . **E. brachycarpum**
      4' Per from caudex; st not peeling; lvs narrowly lanceolate to ovate, not clustered
        5. St ± erect, with offset rosettes or fleshy underground shoots
          6. Lf veins conspicuous; st with rosette or fleshy shoot; seed ridged . . . . **E. ciliatum** ssp. **ciliatum**
          6' Lf veins ± obscure; st with fleshy shoot; seed papillate . . . . **E. saximontanum**
        5' Sts ± ascending, clumped or cespitose; leafy or thread-like stolons sometimes tipped by fleshy bulblets
          7. Pl subglabrous (exc scattered infl hairs)
            8. Pl 2– 8.5 dm, clumped; stolons short, scaly; pedicels 5–25 mm; infl crowded (bracts ± = internodes) . . . . **E. glaberrimum** ssp. **glaberrimum**
            8' Pl < 4 dm, matted; stolons thread-like; pedicels 20–65 mm; infl open (bracts << internodes) . . . . **E. oregonense**
          7' Pl variously hairy, often in decurrent lines on stem
            9. Stolons thread-like, tipped by fleshy bulblets; lf ± narrowly lanceolate; seed 1.4–2.2 mm, hairs persistent . . . . **E. leptophyllum**
            9' Stolons short, leafy; lf gen lanceolate or wider; seed 0.7–2.1 mm, hair-tuft gen deciduous
              10. Pl cespitose, < 2 dm; lf 8–28 mm; fr 17–40 mm . . . . **E. anagallidifolium**
              10' Pl loosely clumped, 1–5 dm; lf 15–55 mm; fr 40–100 mm
                11. Pedicel in fr 5–15 mm; fr 40–65 mm; seed 0.9–1.2 mm, papillate; petals gen pink to rose-purple . . . . **E. hornemannii**
                11' Pedicel in fr 20–45 mm; fr 50–100 mm; seed 1.1–1.6 mm, netted; petals gen white . . **E. lactiflorum**

E. anagallidifolium Lam. (p. 397) Per < 2 dm, cespitose, ascending to erect, ± strigose in decurrent lines; stolons short. **LVS** 8–25 mm; basal spoon-shaped to oblong, glabrous; upper elliptic to lanceolate, sparsely strigose, obtuse; petiole 1–6 mm. **INFL** ± nodding, sometimes glandular. **FL**: hypanthium 0.6–1.4 mm; sepals 1.5–5 mm; petals 2.5–9 mm, pink to rose-purple. **FR** 17–36 mm, subglabrous; pedicel 10–55 mm. **SEED** 0.7–1.4 mm, netted; hairs persistent. $2n=36$. Moist alpine slopes, meadows, streambanks; 2100–3700 m. KR, CaRH, SNH, SnBr, **n W&I**; circumboreal. ✸ TRY. Jul–Sep

E. angustifolium L. ssp. *circumvagum* Mosq. (p. 397) FIREWEED Per < 30 dm, gen strongly colonial, ± glabrous to densely strigose above. **LVS** alternate, 15–200 mm, lanceolate; midrib strigose below; veins ± conspicuous; petiole 2–7 mm. **INFL** dense, gen canescent; bracts small, linear. **FL** nodding in bud; hypanthium 0 (exc as greenish disk); sepals 7–16 mm; petals 10–25 mm, gen deep pink to magenta, entire; stamens subequal, < pistil, maturing before stigma, pollen bluish gray, shed singly; stigma 4-lobed. **FR** 40–100 mm, gray-hairy; pedicel 7–20 mm. **SEED** 1–1.3 mm, fusiform, irregularly netted; hairs persistent. $2n=72$. Common. Open places, gravel bars, roadsides, esp after fires; < 3300 m. NCo, KR, NCoRO, CaRH, SNH, SnBr, **W&I, ne DMtns**; circumboreal. Ssp. *angustifolium* ($2n=36$), farther n & higher, might be expected in CA. ✸ SUN: **4–6**&IRR:**1–3,7,14–18**,19–21;STBL;INV [CVS non-CA]. **Chamerion angustifolium* (L.) Holub ssp. *circumvagum* (Mosq.) P. Hoch Jul–Sep

E. brachycarpum C. Presl (p. 397) Ann 2–20 dm, glabrous and peeling below, strigose and gen glandular-hairy above. **LVS** gen early deciduous, 10–50 mm, linear to narrowly elliptic, acuminate, gen folded along midrib, ± glabrous; petiole 0–4 mm. **FL**: hypanthium 1.5–8(16) mm; sepals 2–8 mm; petals 2–15 mm,

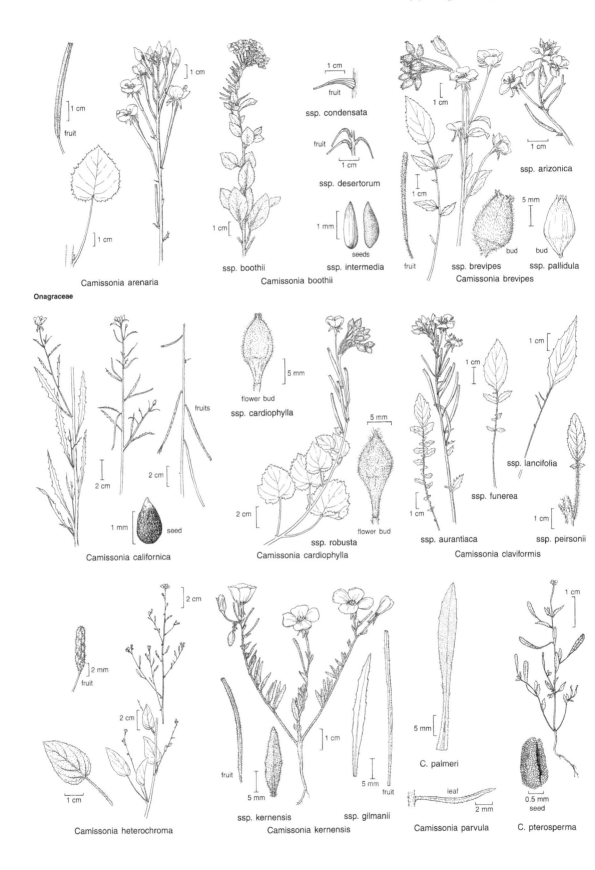

Onagraceae

Camissonia arenaria

ssp. condensata

ssp. desertorum

ssp. boothii   ssp. intermedia
Camissonia boothii

ssp. arizonica

ssp. brevipes   ssp. pallidula
Camissonia brevipes

ssp. cardiophylla

ssp. robusta
Camissonia cardiophylla

ssp. lancifolia

ssp. funerea

ssp. aurantiaca   ssp. peirsonii
Camissonia claviformis

Camissonia californica

Camissonia heterochroma

ssp. kernensis   ssp. gilmanii
Camissonia kernensis

C. palmeri

Camissonia parvula   C. pterosperma

white to rose-purple; stamens < or = pistil; stigma sometimes 4-lobed. **FR** 15–35 mm, glabrous or glandular; pedicel 3–20 mm. **SEED** 1.4–2.7 mm, papillate. $2n$=24. Common. Dry, open woodland, grassland, roadsides; < 3300 m. CA-FP (exc ChI), MP, **W&I**; to B.C., SD, NM, also e Can; introduced in s S.Am. [*E. paniculatum* Torrey & A. Gray incl vars. *laevicaule* (Rydb.) Munz, *tracyi* (Rydb.) Munz] Highly variable, esp fl size. KR pls with large fls, pollen shed singly, have been called *E. p.* var. *jucundum* (Rydb.) Trel. ❀ SUN,DRN:**4–6**& IRR:1–3,7,8,9,**14–24**. Jun–Sep

E. canum (E. Greene) Raven CALIFORNIA FUCHSIA, ZAUSCHNERIA Per (clumped with basal scaly shoots) to subshrub 1–9 dm, ± densely spreading-hairy and gen glandular. **LF** subsessile, 5–50 mm, linear to ovate, green to grayish, sometimes strongly toothed. **FL** red-orange; hypanthium 20–34 mm; sepals 8–15 mm; petals 8–17 mm; stamens << pistil; stigma 4-lobed. **FR** 20–35 mm, ± beaked, hairy; pedicels 0–2 mm. **SEED** 1.5–2.3 mm, low-papillate. Dry slopes, ridges; < 3000 m. CA-FP (exc NCo, NCoRH), **DMtns**; to OR, WY, NM, n Mex. Hummingbird-pollinated. Sspp. intergrade, esp in s CA; ssp. *garrettii* (Nelson) Raven has been reported from DMtns. 2 sspp. in CA.

ssp. *latifolium* (Hook.) Raven (p. 397) Per 1–5 dm, gen glandular. **LVS** opposite, widely lanceolate to ovate, gen green. $2n$=60. Habitats of sp.; 500–3000 m. KR, CaR, c&s SNF, SNH, Teh, SnJV, TR, **DMtns**; to sw OR, w NM, nw Mex. [*Zauschneria californica* ssp. *l.* (Hook.) Keck]. ❀ SUN,DRN:**4–6,17**&IRR:1–3,7,8–10,**14–16,18–24**;rather INV;CVS;some forms GRCVR (deciduous);also STBL. Aug–Sep

E. ciliatum Raf. Per < 19 dm, loosely clumped, with basal rosettes or fleshy bulblets, gen strigose in lines or spreading-hairy. **LF** 1–15 cm, narrowly lanceolate to ovate; veins conspicuous; petiole 0–8 mm. **INFL** densely strigose, ± spreading-hairy, gen glandular. **FL**: hypanthium 0.5–2.6 mm; sepals 2–7.5 mm; petals 2–14 mm, white to rose-purple; stamens < or = pistil; stigma club- or head-like. **FR** 15–100 mm, hairy; pedicels 0–30 mm. **SEED** 0.8–1.9 mm, ridged. $2n$=36. Abundant. Disturbed places, moist meadows, streambanks, roadsides; < 4100 m. **±CA**; most of N.Am, e Asia, s S.Am, introduced in Australasia, Eur, w Asia. 3 sspp. in CA.

ssp. *ciliatum* (p. 397) Rosettes well developed. **LF** lanceolate. **INFL** openly branched, not leafy. **FL**: petals 2–6(9) mm, white to pink. Common. Habitats and range ± of sp. [*E. adenocaulon* Hausskn. incl vars. *holosericeum* (Trel.) Munz, *occidentale* Trel., *parishii* (Trel.) Munz] Jul–Oct

E. glaberrimum Barbey Per < 8.5 dm, clumped, ± glabrous, glaucous; stolons short, scaly. **LF** subsessile, 10–70 mm, narrowly lanceolate to narrowly ovate, clasping. **FL**: hypanthium 0.7–2.3 mm; sepals 1.6–7.5 mm; petals 2.5–12 mm, gen pink to rose-purple (white). **FR** 20–75 mm, sometimes sparsely hairy; pedicel 5–25 mm. **SEED** 0.7–1.2 mm, papillate. $2n$=36. Gravel bars, scree, roadsides, moist rocky areas; 600–3800 m. KR, NCoRH, CaRH, SNH, TR, PR, Wrn, **n W&I**; to w Can, MT, WY, n Mex. 2 sspp. in CA.

ssp. *glaberrimum* (p. 397) **ST** 2–8.5 dm, gen branched above. **LF** 20–70 mm, (narrowly) lanceolate. **FL**: sepals 3–7.5 mm; petals 5–12 mm. **FR** 45–75 mm; pedicel 5–25 mm. **SEED** 0.7–1 mm. Well drained, gravelly soils, streambanks, roadsides;

600–3000 m. KR, NCoRH, CaRH, SNH, TR, PR, Wrn, **n W&I**; to w Can, MT, WY, n Mex. Jul–Aug

E. hornemannii Reichb. ssp. *hornemannii* (p. 397) Per 1–4.5 dm, loosely clumped, ± strigose in lines (esp above); stolons short, leafy. **LF** 15–55 mm, lanceolate to ovate (narrower above), gen ± glabrous, obtuse to acute; petiole 0–8 mm. **INFL** glandular. **FL**: hypanthium 1–2.2 mm; sepals 2–4.5 mm; petals 3–9 mm, (white) pink to rose-purple. **FR** 40–65 mm, hairy; pedicel 5–15 mm. **SEED** 0.9–1.2 mm, papillate. $2n$=36. Moist meadows, streambanks; 1500–3400 m. KR, NCoRH, CaRH, SNH, SnBr, SnJt, Wrn, **n W&I**; ± circumboreal. Jul–Aug

E. lactiflorum Hausskn. (p. 397) Per 1.5–5 dm, gen clumped, minutely strigose in lines (esp above); stolons short, leafy. **LF** 20–55 mm, narrowly ovate to narrowly lanceolate (narrower above), ± ciliate, obtuse to acute; petioles ± winged, 0–12 mm. **INFL** nodding to erect, glandular. **FL**: hypanthium 1–2.2 mm; sepals 2–3.6 mm; petals 3–6.8 mm, white (pink). **FR** 50–100 mm, hairy; pedicel 20–45 mm. **SEED** 1.1–1.6 mm, netted. $2n$=36. Moist meadows, streambanks, talus; 1400–3300 m. KR, CaRH, SNH, **n W&I**; to AK, MT, Colorado, also e N.Am, n Eurasia. Jul–Aug

E. leptophyllum Raf. Per 1–10 dm, ± densely strigose; stolons thread-like, tipped with fleshy, winter bulblets. **LF** subsessile, 20–75 mm, linear to narrowly elliptic. **INFL** ± glandular. **FL**: hypanthium 0.8–1.5 mm; sepals 2.5–4.5 mm; petals 3.5–7 mm, white to pink. **FR** 35–80 mm, gray-hairy; pedicel 1–4 mm. **SEED** 1.5–2.2 mm, papillate; hairs persistent. $2n$=36. Uncommon. Boggy meadows, damp places; 2000 m. **s SNE**; to B.C., MT, Colorado; native to e N.Am.

E. oregonense Hausskn. (p. 397) Per < 4 dm, often matted, delicate, (sub)glabrous, ± purplish above; stolons sprawling, thread-like, with minute rounded lvs. **LF** sessile, clasping, 5–25 mm, round (below) to linear (above); tip obtuse to round. **INFL** very open (bracts << internodes), sparsely glandular. **FL**: hypanthium 0.8–1.5 mm; sepals 1–4 mm; petals 2–8 mm, white to rose-purple; stigma sometimes cylindric. **FR** 20–50 mm; pedicel 20–65 mm, gen > fr. **SEED** 1–1.4 mm, papillate. $2n$=36. ± boggy areas; 1700–3600 m. KR, CaRH, SNH, SnBr, SnJt, Wrn, **SNE**; to WA, w MT, WY. Jul–Aug

E. saximontanum Hausskn. (p.397) Per < 6 dm, from fleshy, scaly, underground shoots, ± strigose in decurrent lines. **LF** subsessile, gen clasping, 10–65 mm, narrowly elliptic to (ob)ovate. **INFL** glandular. **FL**: hypanthium 0.8–1.4 mm; sepals 1.2–3.5 mm; petals 2.2–7 mm, white (pink). **FR** 20–70 mm, hairy; pedicels 0–5 mm. **SEED** 1–1.8 mm, low-papillate. $2n$=36. Moist montane meadows, streambanks, ± disturbed roadsides; 1400–3500 m. c&s SNH, **SNE**; to e OR sw Can (also e Can), MT, NM.

E. torreyi (S. Watson) P. Hoch & Raven (p. 397) Ann 1–6.2 dm, grayish spreading-hairy, peeling below. **LVS** opposite only near base (where glabrous), subsessile, 5–45 mm, ± linear-lanceolate, hairy. **INFL** glandular. **FLS** gen cleistogamous; hypanthium 0.4–1 mm; sepals 0.7–2 mm; petals 1–3 mm, pink or white; stigma rarely 4-lobed. **FR** sessile, 8–13 mm, cylindric, beaked, flexible; axis disintegrating. **SEED** 1–1.5 mm, netted; hairs 0. $2n$=18. Streambanks, moist slopes; < 2600 m. NW (exc NCo), CaR, SN, GV, SnFrB, SCoRO, MP, **sw DMoj**; to B.C., ne NV. [*Boisduvalia stricta* (A. Gray) E. Greene] May–Jul

GAURA

Ann, bien, per, from woody caudex, rhizome, or taproot. **LVS** basal and cauline, alternate, sessile; margin gen wavy-dentate. **INFL**: spike, terminal, bracted. **FL** gen bilateral, opening at dusk or dawn; sepals gen 4, gen widely opening; petals gen 4, white or yellow, often fading reddish or purplish; stamens 8, filaments gen with paired teeth at base, anthers attached at middle; ovary chambers gen 4 (in fr 1), stigma deeply lobed, gen elevated above anthers (pl then cross-pollinated). **FR** indehiscent, ± erect, nut-like, gen 4-angled or -winged; walls woody; base stalk-like or not. **SEEDS** gen 3–4, gen 2–3 mm, ovoid, gen flat-sided, yellowish to pale brown. 21 spp.: temp N.Am (esp TX), C.Am. (Greek: proud, from showy fls of some) [Raven & Gregory 1972 Mem Torrey Bot Club 23:1–96]

1. Fr not linear; stalk-like base 2–4 mm, thick; per, not mat-forming (sect. *Campogaura*) .... *G. coccinea*
1' Fr ± linear; stalk-like base 2–8 mm, slender; rhizomatous per, forming large mats (sect. *Stipogaura*) .. *G. sinuata*

G. coccinea Pursh (p. 397) WILD HONEYSUCKLE, LINDA TARDE
Per, gen minutely strigose and with long spreading hairs, or ±
glabrous; caudex woody, branched below ground. **ST** 10–120
cm. **LF** 10–70 mm, linear to narrow-elliptic, entire to coarsely
wavy-serrate. **INFL:** bracts 2–5 mm. **FL:** hypanthium 4–13 mm;
sepals 5–10 mm; petals 3–8 mm. **FR** erect or spreading, 4–9 mm,
4-angled; stalk-like base short, thick, > 1/2 diam of widest part.
SEED 1.5–3 mm. 2*n*=14,42,56. Dry slopes, gen limestone,
Joshua-tree or pinyon/juniper woodland; 900–1600 m. **DMtns**
(naturalized in Teh, SW); to w Can, c US, Mex. [var. *glabra*
(Lehm.) Torrey & A. Gray] Though native, may become a NOX-
IOUS WEED. ❀ TRY;INV. Apr–Jun

G. sinuata Ser. (p. 397) WAVY-LEAVED GAURA Per, forming
large mats, rhizomed; hairs gen sparse or 0. **ST** 20–60 cm,
branched, sparsely minutely strigose and with long, spreading
hairs. **LF** 10–110 mm, linear to narrow-oblanceolate, slightly
wavy-dentate. **INFL:** bracts 1–5 mm. **FL:** hypanthium 2.5–5 mm;
sepals 7–14 mm; petals 7–14.5 mm. **FR** erect, 4–16 mm, ± linear, narrowly 4-winged; stalk-like base 2–8 mm, slender, tapered.
SEEDS 1–4, 2–3 mm. 2*n*=28. Light sandy loam of cult fields;
< 1000 m. GV, CW, SW, **DMoj**; native to OK, TX; widely natu-
ralized, esp se US. [*G. villosa* Torrey incl var. *mckelveyae*
Munz, misapplied] NOXIOUS WEED, limited by self-sterility.
Jun–Sep

GAYOPHYTUM

Harlan Lewis

Ann. **ST** gen erect, < 1 m, slender; hairs 0 to dense, rarely glandular. **LVS** cauline, alternate (or lowest subopposite),
entire, petioled or not, narrow-lanceolate. **INFL:** fls axillary, pedicelled or not, opening at dawn. **FL:** hypanthium
inconspicuous; sepals 4, staying fused in 2's or all coming free; petals 4, 0.5–8 mm, white, with 1–2 yellow or greenish
spots at base, fading pink or red; stamens 8, those opposite sepals larger, pollen ± yellow; ovary chambers 2, stigma
gen not beyond anthers, gen touching them, gen ± spheric. **FR:** capsule, ± cylindric or flat; valves 4, all gen coming
free, gen equal. **SEEDS** few–many, gen all maturing, gen appressed to septum, alternate or subopposite between
chambers, in each chamber gen in 1 row and gen not overlapped, 0.5–2.3 mm, ovoid, glabrous or hairy, brown or gray
mottled with brown; appendages 0. ± 9 spp.: w N.Am, 2 S.Am. (C. Gay, French author of Flora of Chile, 1800–1873)
[Lewis & Szweykowski 1964 Brittonia 16:343–391] Self-compatible; taxa with petals < 3 mm self-pollinated.

1. Seeds in each chamber overlapped, gen in 2 rows
  2. Petals 0.7–1.5 mm; pedicel gen > fr . . . . *G. ramosissimum*
  2' Petals 1.5–3 mm; pedicel gen < fr . . . . *²G. diffusum* ssp. *parviflorum*
1' Seeds in each chamber not overlapped, in 1 row
  3. Branches ± throughout, gen 2–8 nodes between . . . . *G. decipiens*
  3' Branches at base or not, many above, gen 0–1 node between . . . . *²G. diffusum* ssp. *parviflorum*

G. decipiens Harlan Lewis & J. Szweykowski **ST** < 50 cm;
branches ± throughout, gen 2–8 nodes between. **LVS** 1–3 cm,
gen ± reduced above. **INFL:** 1st fl gen 1–5 nodes above base.
FL: petals 1.1–1.8 mm; larger stamens 0.8–1.5 mm; ovary hairy.
FR 6–8 mm, > pedicel, ± not flat, slightly knobby. **SEEDS** 10–25,
subopposite, glabrous to dense-puberulent. 2*n*=14. Pinyon/juni-
per woodland, pine or fir forest; 1800–4200 m. SNH, SnGb, SnBr,
Wrn, **W&I, DMtns (Panamint Mtns)**; ± w US. May–Sep

G. diffusum Torrey & A. Gray **ST** < 60 cm; branches at base
or not, gen forked above. **LVS** 1–6 cm, gen reduced above. **INFL:**
1st fl 1–20 nodes above base. **FL:** petals 1.2–7 mm; larger stamens 0.9–6 mm; ovary hairy, stigma beyond anthers or not, hemispheric or not. **FR** 3–15 mm, sessile or gen > pedicel, cylindric,
slightly to very knobby. **SEEDS** 3–18, alternate or subopposite,
in each chamber sometimes in 2 rows and overlapped, glabrous
to densely puberulent. 2*n*=28. Common. Open montane forest,
sagebrush scrub; 800–3700 m. NW, CaR, SN, TR, PR, **GB**; w

US to B.C., Baja CA. Complex from several 2*n*=14 spp.; sspp.
may intergrade locally. 2 sspp. in CA.

  ssp. *parviflorum* Harlan Lewis & J. Szweykowski (p. 397)
FL: petals 1.2–3 mm; larger stamens 0.9–2 mm; stigma not beyond anthers, ± spheric. **FR:** seeds in each chamber sometimes
in 2 rows and overlapped. Common. Habitats, elevations of sp.
NW, CaR, SN, TR, PR, **GB**. Variable; most small-fld pls assigned by Munz, others to *G. nuttallii* belong here; may occur
with any member of genus. May–Sep

G. ramosissimum Torrey & A. Gray Pl herbage ± glabrous.
ST < 50 cm; branches throughout, forked exc at base. **LVS** 1–4
cm, much reduced above. **INFL:** 1st fl 5–15 nodes above base.
FL: petals 0.7–1.5 mm; larger stamens ± 0.5 mm; ovary glabrous or puberulent. **FR** 3–9 mm, gen < pedicel, cylindric, slightly
knobby. **SEEDS** 10–30, in each chamber in 2 rows, overlapped,
glabrous. 2*n*=14. Sagebrush scrub; 500–3000 m. CaRH, n SN,
GB; ± w US. May–Sep

LUDWIGIA FALSE LOOSESTRIFE, WATER PRIMROSE

Peter C. Hoch

Ann to subshrub, sometimes floating or rooting at nodes. **LVS** alternate to opposite, simple; stipules gen deciduous.
INFL: spike; fls 1 per bract. **FL** radial; hypanthium 0; sepals 4–5(7), persistent; petals (0)4–5(7), white to yellow;
stamens 4 or 10(12), pollen gen shed singly in CA; stigma club-shaped to spheric. **FR** dehiscing irregularly; wall thick
or thin. **SEEDS** free or embedded in fr wall. 82 spp.: ± worldwide. (C.G. Ludwig, German botanist & physician,
1709–1773) [Raven 1963 Reinwardtia 6:327–427] Many polyploids. Key to species adapted by Margriet Wetherwax.

1. Lvs opposite; sepals 4; petals 4, 1–3 mm; stamens 4; fr 4–10 mm, erect; seeds free from fr wall . . . . . *L. repens*
1' Lvs alternate; sepals 5(6); petals 5(6), 7–24 mm; stamens 10(12) in 2 unequal sets; fr 10–40 mm,
  reflexed; seeds embedded in fr all . . . . *L. peploides* ssp. *peploides*

L. peploides (Kunth) Raven Per, matted, floating, or creeping. **ST** 1–30 dm, prostrate to erect, simple or branched. **LVS** <

10 cm, alternate, ± clustered; blade oblong to round, subentire,
glabrous to spreading-hairy above. **FL:** sepals 5(6), 3–12 mm;

petals 5(6), 7–24 mm; stamens 10(12) in 2 unequal sets, anthers 0.5–2.2 mm. **FR** reflexed; pedicel 6–90 mm; body cylindric, ± 5-angled, hard, subglabrous to spreading-hairy. **SEED** 1–1.5 mm, embedded in inner fr wall. 2*n*=16. Ditches, streambanks, lakeshores; < 900 m. NCo, NCoRO, SNF, GV, CCo, SnFrB, SCo, WTR, **sw DMoj**; to OR, se US, S.Am, Eurasia, Australia. [*Jussiaea repens* L. misapplied] May be serious wetland or agricultural weed. 2 sspp. in CA.

ssp. *peploides* (p. 397) Pl (sub)glabrous. **LF**: tip gen not glandular; lower petioles 3–8 mm. **FR**: body 10–25 mm. 2*n*=16. Ditches, shores, streambanks; < 900 m. NCo, NCoRO, SNF, GV, CCo, SnFrB, SCo, WTR, **sw DMoj**; to OR, TX, S.Am, in-troduced into Australia. [*J. r.* var. *p.* (Kunth) Griseb. misapplied] May–Oct

L. repens Forster Per, matted. **ST** 1–3 dm, decumbent, rooting at nodes, ± branched, subglabrous. **LVS** opposite, < 5 cm; blade narrowly elliptic to ± round, entire, subglabrous to densely and minutely strigose. **FL**: sepals 4, 1.8–5 mm; petals 4, 1–3 mm, yellow; stamens 4, anthers 0.4–0.9 mm, pollen shed ± in 4's. **FR** erect; pedicel 0–3 mm; body 4–10 mm, oblong to narrowly obconic, sometimes ± hairy. **SEED** 0.6–0.8 mm, free of fr wall. 2*n*=32. Muddy or sandy streambanks, ponds, ditches; < 900 m. SnBr, PR, **sw DMoj**; to se US, Caribbean, n Mex, introduced into s Asia, Japan. [*L. natans* Elliott var. *stipitata* Fern. & Griscom]. ❀ INV;STBL. Jul–Sep

OENOTHERA   EVENING PRIMROSE

Ann, bien, per, gen from taproot. **LVS** basal or cauline, alternate, gen pinnately toothed to lobed. **INFL**: spike, raceme-like, or fls in axils of upper, reduced lvs. **FL** radial, gen opening at dusk; sepals 4, reflexed in fl (sometimes 2–3 remaining adherent); petals 4, yellow, white, rose, or ± purple, gen fading orangish to purplish, tip notched or toothed; stamens 8, anthers attached at middle; ovary chambers 4, stigma deeply lobed, gen > anthers and cross-pollinated (or ± = anthers and self-pollinated). **FR** cylindric to 4-winged, straight to curved, gen sessile (base sometimes seedless, stalk-like). **SEEDS** in gen 2?(1–3) rows per chamber, or clustered. 119 spp.: Am, some widely naturalized. (Greek: wine-scented) [Dietrich & Wagner 1988 Syst Bot Monogr 24:1–91] Many spp. self-pollinated; some of these have chromosome peculiarities (ring of 14 in meiosis) and ± 50% pollen fertility; they yield genetically ± identical offspring; they are identified as permanent translocation heterozygotes.

1. Petals white; seeds in 1 row per chamber (fr not tubercled) or in 2 rows per chamber (fr tubercled)
  2. St not peeling; buds erect; fr 4–9 mm wide, tubercled; seeds in 2 rows per chamber, with an obvious
    cavity on 1 side (sect. *Pachylophus*) . . . . ***O. caespitosa***
    3. Fr 10–34 mm, lanceolate to elliptic-ovate, gen S-shaped, stalk-like base 0–1 mm; petals fading
      rose to deep purple; margin of seed-cavity lobed . . . . ssp. ***crinita***
    3' Fr 25–68 mm, cylindric, ± straight, stalk-like base 0–55 mm; petals fading lavender to pink;
      margin of seed-cavity entire . . . . ssp. ***marginata***
  2' St peeling; buds nodding; fr 1.5–5 mm wide, not tubercled; seeds in 1 row per chamber, cavity 0,
    embryo filling seed coat (sect. *Anogra*)
    4. Ann; new shoots from lateral roots 0; free sepal tips in bud 0–1 mm . . . . ***O. deltoides***
      5. Spreading hairs ± 2 mm above; fr base 3–5 mm wide . . . . ssp. ***cognata***
      5' Spreading hairs 1–1.5 mm above; fr base 2–9 mm wide . . . . ssp. ***deltoides***
    4' Per; new shoots gen from lateral roots; free sepal tips 0 or < 1 mm . . . . ***O. californica***
      6. Pl hairs dense, short, appressed, and long, wavy; roots fleshy; new rosettes at st tips . . . . . . ssp. ***eurekensis***
      6' Pl hairs 0 to dense, short, appressed and some long, spreading; roots not fleshy; new rosettes not
        forming at st tips
        7. Cauline lf gen ± pinnately lobed; pl ± grayish green . . . . ssp. ***avita***
        7' Cauline lf subentire to deeply wavy-dentate; pl green to slightly grayish . . . . ssp. ***californica***
1' Petals yellow; seeds in 2 rows per chamber (fr not tubercled, but hairs sometimes with red blister-like bases)
  8. St erect; bien; fls in gen dense spikes; seeds angled, irregularly pitted (sect. *O.* subsect. *Oenothera*)
    9. Hypanthium 60–135 mm . . . . ***O. longissima***
    9' Hypanthium 20–55 mm . . . . ***O. elata*** ssp. ***hirsutissima***
  8' St ± 0, decumbent or erect; ann or short-lived per; fls few, in upper axils; seeds not both angled
    and irregularly pitted
    10. Fr lanceolate to ovate, 4–11 mm wide; seeds 2.4–3.5 mm, ± wrinkled; pollen ± 100% fertile
      (sect. *Eremia*) . . . . ***O. primiveris***
      11. Lf grayish green; petals (22)29–40 mm; stigmas above anthers . . . . ssp. ***bufonis***
      11' Lf gen green; petals 6–25(28) mm; stigmas ± not above anthers . . . . ssp. ***primiveris***
    10' Fr cylindric, 2–4 mm wide; seeds 1–1.8 mm, pitted; pollen ± 50% sterile
      12. Bud nodding; free tips of sepal in bud 0.1–1 mm (sect. *O.* subsect. *Nutantigemma*) . . . . ***O. pubescens***
      12' Bud curved upward; free tips of sepal in bud 0.3–3 mm (sect. *O.* subsect. *Raimannia*) . . . . ***O. laciniata***

O. caespitosa Nutt. FRAGRANT EVENING PRIMROSE Per, rosetted; caudex woody, new shoots gen from lateral roots; hairs glandular and sometimes also coarse and non-glandular. **ST** sprawling, < 2 dm, or ± 0. **LF** 1.7–36 cm, oblanceolate to narrowly elliptic, gen irregularly dentate to lobed. **INFL**: fls in axils. **FL**: hypanthium 30–165 mm; sepals 16–50 mm, tips in bud not free; petals 16–56 mm, white. **FR** 10–68 mm, 4–9 mm wide, cylindric to elliptic-ovate, tubercled. **SEED** obovate to ± triangular, papillate or netted, 1 side with a cavity sealed by a depressed, gen splitting membrane. 2*n*=14,28. Open desert scrub, pinyon/juniper woodland, coniferous and bristlecone-pine for-ests; 1100–3400 m. MP (likely), **SNE**, **D**; w US. Cross-polli-nated. 5 intergrading sspp., 2 in CA.

ssp. *crinita* (Rydb.) Munz CAESPITOSE EVENING PRIMROSE Pl loosely to densely cespitose. **FL**: hypanthium 30–85 mm; petals fading rose to purple. **FR** 10–34 mm, lanceolate to elliptic-ovate, gen S-shaped; stalk-like base 0–1 mm. **SEED** 2.9–3.5 mm; cavity margin lobed. UNCOMMON. Calcium soils in bristlecone-pine forest, pinyon/juniper woodland, desert scrub; 1150–3370 m. MP (likely), **SNE**, **D**; w US. [var. *c.* (Rydb.) Munz] 2 intergrading forms differ in elevation, habit, lf size, petal color; more study needed. ❀ TRY. Jun–Sep ssp. *marginata* (Hook. & Arn.) Munz (p. 397) Pl loosely cespitose. **FL:** hypanthium 65–165 mm; petals fading lavender to pink. **FR** 25–68 mm, cylindric, ± straight; stalk-like base 0–55 mm. **SEED** 2.2–3.4 mm; cavity margin entire. Rocky or sandy sites in granite, limestone, or sandstone soils, pinyon/juniper woodland to pine forest; < 2400 m. MP (likely), **SNE**, **D**; w US. [var. *m.* (Hook. & Arn.) Munz] Variable. ❀ DRN,DRYorIRR, SUN:1–3,**7–14,18–21**. Apr–Aug

O. californica (S. Watson) S. Watson Per, rosetted when young, glabrous to densely minutely strigose, also sometimes with longer, spreading hairs; roots gen not fleshy, gen new shoots from laterals. **ST** decumbent or ascending, 1–8 dm, peeling. **LVS:** cauline 1–6 cm, lanceolate or deltate-ovate, entire to pinnately lobed. **INFL:** fls in upper axils; buds nodding. **FL:** hypanthium 20–40 mm; sepals 15–30 mm, free tips in bud 0–1 mm; petals 15–35 mm, white fading pink. **FR** 20–80 mm, 2–3 mm wide, cylindric, straight or curved. **SEEDS** in 1 row per chamber, 1.4–3 mm, obovate, smooth. 2*n*=14,28. Sandy or gravelly areas, dunes, desert scrub to pinyon/juniper or ponderosa-pine woodlands; < 2500 m. CW, SW, SNE, **D**; to sw UT, nw AZ, nw Baja CA. Cross-pollinated.

ssp. *avita* Klein (p. 397) Pl ± grayish green; hairs dense, short, appressed and also longer, spreading. **ST:** new rosettes not formed at tips. **LVS:** cauline oblong to lanceolate, gen ± pinnately lobed. **FR** 20–80 mm. 2*n*=14. Sandy or gravelly areas, desert scrub to pinyon/juniper or ponderosa-pine woodlands; 800–2500 m. SNE, **D**; ± sw US. [*O. a.* (Klein) Klein] ❀ TRY.

ssp. *californica* (p. 397) Pl green to slightly grayish; hairs 0, or dense, short, appressed, also sometimes long, spreading. **ST:** new rosettes not formed at tips. **LVS:** cauline oblong to lanceolate, subentire to deeply wavy-dentate. **FR** 30–55 mm. 2*n*=28. Sandy or gravelly areas, open coastal-sage scrub, chaparral, oak woodland; < 1900 m. CW (San Luis Obispo Co. s), SW, **s DMtns (Little San Bernardino Mtns)**; nw Baja CA. [*O. c.* var. *glabrata* Munz] Intergrades with ssp. *avita.* ❀ TRY. Apr–Jun

ssp. *eurekensis* (Munz & Roos) Klein (p. 397) EUREKA DUNES EVENING PRIMROSE Pl: roots fleshy; hairs dense, short, appressed, also long, spreading, wavy. **ST:** new rosettes formed at tips. **LVS:** cauline deltate-ovate, entire to dentate. **FR** 30–70 mm. 2*n*=14. **RARE** CA, **ENDANGERED** US. Dunes, gen with *Psorothamnus polydenius*; 900–1200 m. **n-most DMoj (Eureka Valley, ne Inyo Co.)**. [*O. avita* (Klein) Klein ssp. *e.* (Munz & Roos) Klein] Populations few, large.

O. deltoides Torrey & Frémont DEVIL'S LANTERN, LION-IN-A-CAGE, BASKET EVENING PRIMROSE Ann, per, loosely rosetted; hairs 0, curly, or straight, also sometimes glandular. **ST** decumbent or erect, 2–10 dm, stout, spongy, peeling. **LVS:** cauline 2–15 cm, ± diamond-shaped-obovate to oblanceolate, subentire to pinnately lobed. **INFL:** fls in upper axils; buds nodding. **FL:** hypanthium 20–40 mm; sepals 8–30 mm, free tips in bud 0–9 mm; petals 10–40 mm, white fading pink. **FR** 20–60(80) mm, 1.5–4(5) mm wide, cylindric, gen curved, ± twisted. **SEEDS** in 1 row per chamber, 1.5–2 mm, obovate, smooth. 2*n*=14. Sandy, often dunes; < 1800 m. SnJV, ne SnFrB, MP, **D**; to OR, UT, AZ, nw Mex. Gen cross-pollinated. 5 sspp., 4 in CA.

ssp. *cognata* (Jepson) Klein (p. 397) Ann, short-lived per; hairs spreading, ± 2 mm above. **ST** gen 2–4(6) dm, branched from base. **LVS:** upper subentire to wavy-dentate, rarely pinnately lobed. **FL:** bud tip obtuse; sepal tips not free; petals 25–40 mm. **FR:** base 3–5 mm wide. Sandy soils, grassland; < 700 m. SnJV, **w DMoj**. Sometimes self-pollinated. ❀ TRY.

ssp. *deltoides* (p. 397, pl. 71) Ann; hairs spreading, 1–1.5 mm above, also minutely strigose or not. **ST** gen 2–10 dm, ± branched from base. **LVS:** upper dentate. **FL:** bud tip obtuse to acute; sepal free tips 0–1 mm; petals 15–38 mm. **FR:** base 2–3 mm wide. Sandy soils, incl dunes; gen < 1100 m. **D**; s NV, w AZ, nw Mex. ❀ TRY. Mar–May

O. elata Kunth Bien, short-lived per, densely minutely strigose and (esp in infl) glandular; hairs also long, appressed to spreading, sometimes with red, blister-like bases. **ST** erect, 4–25

dm. **LVS:** cauline 4–25 cm, oblanceolate to lanceolate or elliptic, gen dentate to subentire. **INFL:** spike. **FL:** hypanthium 20–48(55) mm; sepals 27–48 mm, free tips in bud 1–7 mm; petals 25–52 mm, yellow fading reddish orange. **FR** 20–65 mm, 4–7 mm wide, narrowly lanceolate, ± straight. **SEED** 1–2 mm, angled, irregularly pitted. 2*n*=14. Moist places; 0–2800 m. **CA**; w N.Am to C.Am. Gen cross-pollinated. 3 sspp., 2 in CA.

ssp. *hirsutissima* (S. Watson) W. Dietr. (p. 397) **ST** 10–25 dm. **FL:** sepals green or red-flushed, hairs glandular or not, also spreading and sometimes with red, blister-like bases, free tips in bud 3–6 mm; anthers 8–15 mm. Moist places, gen inland; < 2800 m. **CA**; w US, nw Mex. [*O. hookeri* Torrey & A. Gray sspp. *angustifolia* (Gates) Munz, *grisea* (Bartlett) Munz, *venusta* (Bartlett) Munz] Several intergrading forms. ❀ SUN:4–6,**14–17**,&IRR:1–3,**7–12**,13,**18–24**;INV,STBL.

O. laciniata Hill Ann, short-lived per, rosetted, minutely strigose; hairs above gen also long, spreading, and also glandular. **ST** decumbent to erect, 0.5–5 dm. **LVS:** cauline 2–10 cm, narrowly oblanceolate to ± elliptic or oblong, subentire to pinnately lobed. **INFL:** fls in upper axils; buds curved upward. **FL:** hypanthium 12–35 mm; sepals 5–15 mm, free tips in bud 0.3–3 mm; petals 5–22 mm, yellow fading orange. **FR** 20–50 mm, 2–4 mm wide, cylindric. **SEED** 1–1.8 mm, ± spheric or widely elliptic, pitted. 2*n*=14. Open, gen sandy, gen urban, disturbed places; gen < 500 m. CW, SW, **DMoj**; native to e US, widely naturalized, but occurrences scattered. Self-pollinated. Permanent translocation heterozygote. May–Jul

O. longissima Rydb. Bien, rosetted, minutely strigose; hairs also gen long, spreading, with red, bristle-like bases, sometimes some glandular. **ST** erect, 6–30 dm. **LVS:** cauline 5–22 cm, narrowly oblanceolate to ± elliptic, subentire to dentate. **INFL:** spike. **FL:** hypanthium 60–135 mm; sepals 23–47 mm, free tips in bud 2–6 mm; petals 28–65 mm, yellow fading reddish orange. **FR** 25–55 mm, 4–9 mm wide, narrowly lanceolate, ± straight. **SEEDS** 1–2 mm, angled, irregularly pitted. 2*n*=14. Seasonally moist places in creosote-bush scrub, pinyon/juniper woodland; 1000–1700 m. **e DMtns (New York Mtns)**; to Colorado, AZ. [ssp. *clutei* (Nelson) Munz] Gen cross-pollinated. ❀ TRY. Jul–Sep

O. primiveris A. Gray Ann, rosetted, minutely strigose, in infl gen glandular; hairs also coarse, with red, blister-like bases or not. **ST** gen 0 (sometimes erect or ascending, < 3.5 dm). **LF** 4–28 cm, oblanceolate, wavy-dentate to 1–2-pinnately lobed. **INFL:** fls in axils. **FL:** hypanthium 20–72 mm; sepals 7–30 mm, free tips in bud 0; petals 6–40 mm, yellow fading reddish orange to purple. **FR** 10–60 mm, 4–8 mm wide, lanceolate to ovate, straight, curved, or S-shaped. **SEED** 3–3.5 mm, irregularly obovate to oblanceolate, papillate, 1 side coarsely wrinkled in distal 1/2, other side with thick, U-shaped area forming groove and small cavity at tip. 2*n*=14. Sandy flats, low hills, dune margins, arroyos; 30–1400 m. **D**; ± sw US, n Mex. Self- or cross-pollinated. 2 intergrading sspp.

ssp. *bufonis* (M.E. Jones) Munz (p. 401) **LF** grayish green. **FL:** petals (22)29–40 mm. Habitats of sp. **D**; to UT, w AZ, nw Mex. Gen cross-pollinated. ❀ TRY.

ssp. *primiveris* **LF** gen green. **FL:** petals 6–25(28) mm. Uncommon. Habitats of sp. **D**; to TX, nw Mex. [ssp. *caulescens* (Munz) Munz] Self-pollinated. ❀ TRY. Mar–May

O. pubescens Sprengel (p. 397) Ann, bien, rosetted, minutely strigose; hairs also long, spreading, and gen some glandular. **ST** decumbent to erect, 0.5–8 dm. **LVS:** cauline 2–8 cm, narrow-oblanceolate to lanceolate or elliptic, pinnately lobed to subentire. **INFL:** fls in upper axils; buds nodding. **FL:** hypanthium 15–50 mm; sepals 5–25 mm, free tips in bud 0.1–1 mm; petals 5–35 mm, yellow fading orange. **FR** 20–45 mm, 2–4 mm wide, cylindric. **SEED** 1–1.5 mm, subspheric, pitted. 2*n*=14. Open places; ± 600 m. **c DMoj (Newberry Springs, San Bernardino Co.)**; to NM, S.Am. [*O. laciniata* Hill ssp. *p.* (Sprengel) Munz] Permanent translocation heterozygote. CA pls possibly introduced from AZ. May–Jul

OROBANCHACEAE BROOM-RAPE FAMILY

Lawrence R. Heckard

Ann, per, non-green root parasites; roots modified into absorptive structures; pl an erect, fleshy, mostly underground st (peduncle) with terminal infl. **LF**: true lvs 0. **INFL**: spike, raceme, or panicle; bracts alternate, scale-like. **FL** bisexual; calyx cylindric or cup-shaped, lobes 0–5, persistent; corolla ± 2-lipped, lobes gen 5; stamens 4, epipetalous in 2 pairs (sometimes a 5th vestigial); ovary superior, chamber 1, placentas gen 2–4, parietal, simple or lobed, stigma gen 2–4-lobed, gen bowl- to funnel-shaped. **FR**: capsule, loculicidal; valves 2–4. **SEEDS** many, small, angled; surface netted. 14 genera, 200 spp.: esp n temp. [Thieret 1971 J Arnold Arbor 52:404–432] Recently treated to include hemiparasitic genera of Scrophulariaceae (e.g., *Castilleja, Cordylanthus, Orthocarpus, Pedicularis, Triphysaria*) [Olmstead et al 2001 Mol Phylogen Evol 16:96–112].

OROBANCHE BROOM-RAPE

Ann, per, gen glandular-puberulent above; root attachment sometimes tuber-like. **ST** simple or branched. **INFL** gen ± spike-like (lower fls often short-pedicelled or on short branches), gen dense; fls gen > 20; bracts gen lanceolate to deltate (wider on peduncle); bractlets 0 or 2. **FL**: calyx lobes gen 4–5; corolla glandular-puberulent (hairs short and tack-shaped or long-stalked), gen lacking ring of hairs at stamen bases, upper lip erect to reflexed, gen 2-lobed, lower lip 3-lobed, spreading, yellow-lined; anthers glabrous to hairy; stigma lobes 2, spreading or peltate. **FR** 2-valved; placentas gen 2 or 4, often lobed. **SEED** < 0.7 mm. 140 spp.: worldwide, esp Medit. (Greek: vetch strangler, from parasitic habit) [Heckard 1973 Madroño 22:41–70]. Key to species adapted by Margriet Wetherwax.

1. Fls 5–20; pedicels 3–15 cm, scapose; bractlets on pedicel 0 (sect. *Gymnocaulis*) . . . . ***O. fasciculata***
1' Fls > 20; pedicels 0 or < 5 cm, not scapose; bractlets on pedicel 2
  2. Infl and fls ± dark purple; calyx 8–12 mm . . . . ***O. cooperi***
  2' Infl and fls gen buff to pinkish, corolla lips white to pink lavender with darker veins; calyx gen 10–20 mm
    3. Corolla 20–50 mm, lips 10–14 mm, widely flaring (see also *O. corymbosa* for pls of GB)
    . . . . ***O. californica*** ssp. ***feudgei***
    3' Corolla 15–30 mm, lips 4–10 mm, upper erect, lower spreading
      4. Infl 3–4 cm, branched, forming a convex flat-topped cluster; anthers densely hairy throughout
      . . . . ***O. corymbosa***
      4' Infl > 4 cm, of long, gen unbranched, ± raceme-like units; anthers glabrous to hairy along dehisced
      margin . . . . ***O. parishii*** ssp. ***parishii***

O. californica Cham. & Schldl. Pl 4–35 cm, pale to dark purple aboveground, glandular-puberulent. **STS** 1 or clustered, slender to stout, branched below or throughout. **INFL** long or branched and ± flat- to convex-topped; pedicels 0–4 cm, shorter upward. **FL**: calyx 12–20 mm, gen pale to mauve, lobes linear-triangular,> tube; corolla 20–50 mm, purplish or white to rose with darker veins, moderately glandular-puberulent, lips 10–14 mm, gen widely flaring; upper lip obtuse to rounded; anthers woolly; placentas 4, stigma lobes 2, triangular, margins recurved. Gen dry, ± rocky soils, on herbs, gen Asteraceae; < 2500 m. CA-FP, **GB**; to B.C., Baja CA. [*O. grayana* G. Beck misapplied] 5 sspp. in CA.

  ssp. ***feudgei*** (Munz) Heckard (p. 401) Pl 10–30 cm. **ST** gen branched above, stout. **INFL** ± flat- or round-topped; branches gen < 9 cm. **FL**: corolla 25–35 mm, tube ± stout, abruptly expanded above sinus, > 4 mm wide at constriction, forming ± hump-back throat 8–10 mm wide, lips moderately recurved, whitish to yellowish, purple-tinged and veined. 2*n*=48. Dry washes, mtn slopes, flats, on *Artemisia tridentata*; 700–2500 m. s SNH, Teh, TR, PR, **s SNE**; Baja CA. [*O. grayana* G. Beck var. *f.* Munz] May–Jul

O. cooperi (A. Gray) A.A. Heller (p. 401, pl. 75) Pl 10–40 cm, gen dark purplish aboveground, glandular-puberulent; root attachment sometimes a coral-like thickening. **STS** simple or branched, often forming large clumps, stout, little enlarged at base. **INFL** 4–5 cm wide; lower pedicels < 5 cm, upper 0. **FL**: calyx 8–12 mm, lobes > tube, triangular, acuminate; corolla 18–32 mm, purplish, hairs long-stalked, gen glandular, tube lacking ring of hairs, lips 5–10 mm, upper lobes 6–10 mm, > lower, obtuse; anthers gen hairy; stigma lobes 2, thin, recurved. 2*n*=24,48,72. Sandy flats, washes, on Asteraceae (gen *Ambrosia, Hymenoclea, Encelia*) (weed on tomatoes, DSon, in 1960's); < 500 m. **D**; to UT, AZ, Baja CA. [*O. ludoviciana* vars. *c.* (A. Gray) G. Beck, *latiloba* Munz] An undescribed form (probably best a ssp.), 2*n*=96, with smaller, shorter-lobed corolla and peltate, bowl-shaped stigma occurs on same hosts, over range

of sp. Jan–May

O. corymbosa (Rydb.) Ferris (p. 401) Pl 3–17 cm, pale to purple-tinged aboveground, glandular-puberulent. **ST**: branches clustered, stout, gen thickened at base. **INFL** 3–4 cm, branched, gen round-topped, few-fld; lower pedicels 5–30 mm below, upper 0. **FL**: calyx 12–18 mm, lobes > tube, linear-triangular; corolla 20–30 mm, sparsely short-glandular, lips 5–8 mm, purplish to pink, veins darker, upper lobes rounded; anthers woolly; stigma 2-lobed, peltate. 2*n*=48,96. Openings in sagebrush scrub, gen on *Artemisia tridentata*; 1200–2800 m. n CaR, SNH, **GB**, **n DMtns (Panamint Mtns)**; to B.C., MT, UT. [*O. californica* var. *co.* (Rydb.) Munz] Closely related to *O. californica*, esp ssp. *feudgei.* Jun–Aug

O. fasciculata Nutt. (p. 401) CLUSTERED BROOM-RAPE **STS** 1 or clustered, 5–20 cm, branched or not. **INFL**: raceme, ± flat-topped, gen 5–20-fld; bracts > 6, glandular-puberulent; pedicels 3–15 cm, shorter upwards; bractlets 0. **FL**: calyx lobes 3–7 mm, gen < tube, deltate, gen ± acuminate; corolla 15–30 mm, curved, becoming erect, yellow to purple-tinged, lobes rounded to narrowly acute; anthers gen hairy; stigma 2-lobed, recurved. 2*n*=48. Dry, gen ± bare places, gen on shrubs (esp *Artemisia, Eriodictyon, Eriogonum*); < 3300 m. CA-FP, **GB**, **DMtns**; to Yukon, c N.Am, n Mex. [vars. *franciscana* D.B. Achey, *lutea* (C. Parry) D.B. Achey] Apr–Jul

O. parishii (Jepson) Heckard Pl 5–26 cm, ± yellowish white. **ST** gen simple, stout, glandular-puberulent. **INFL**: bracts narrowly ovate, with > 5 conspicuous, parallel veins. **FL**: calyx 10–20 mm, ± narrowly triangular, pale; corolla 15–30 mm, buff to pinkish, lips 4–8 mm, lobes rounded, veins reddish; anthers glabrous to hairy. Bare, sandy to rocky soils, gen on shrubs; < 2800 m. SNH, Teh, CCo, SW, **W&I**, **DMtns**; Baja CA. [*O. californica* Cham. & Schldl. var. *p.* Jepson] 2 sspp. in CA.

  ssp. ***parishii*** Pl 15–26 cm. **INFL** 5–14 cm. **FL**: calyx lobes 10–16 mm; corolla 20–25 mm, lips 6–8 mm, spreading; anthers glabrous or hairy; stigma lobes wide, spreading. 2*n*=48.

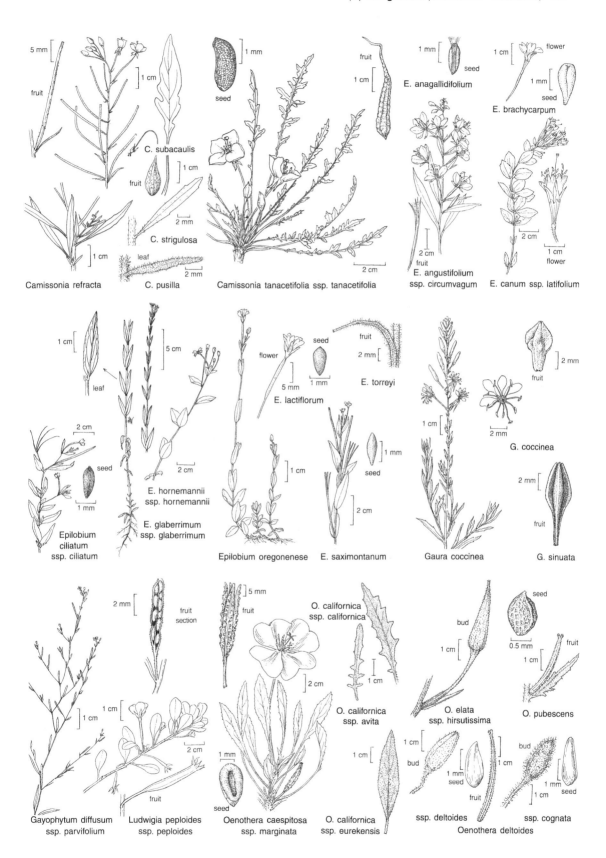

5 mm

fruit

1 cm

C. subacaulis

fruit

1 cm

C. strigulosa

2 mm

leaf

2 mm

C. pusilla

Camissonia refracta

seed

1 mm

Camissonia tanacetifolia ssp. tanacetifolia

2 cm

fruit

1 cm

E. anagallidifolium

seed

1 mm

1 cm

flower

E. brachycarpum

seed

1 mm

2 cm

fruit

E. angustifolium
ssp. circumvagum

2 cm

1 cm

flower

E. canum ssp. latifolium

1 cm

leaf

2 cm

seed

1 mm

Epilobium
ciliatum
ssp. ciliatum

5 cm

2 cm

E. hornemannii
ssp. hornemannii

E. glaberrimum
ssp. glaberrimum

Epilobium oregonenese

flower

5 mm

E. lactiflorum

seed

1 mm

fruit

2 mm

E. torreyi

seed

1 mm

1 cm

2 cm

E. saximontanum

1 cm

Gaura coccinea

fruit

2 mm

2 mm

G. coccinea

2 mm

fruit

G. sinuata

2 mm

fruit
section

1 cm

Gayophytum diffusum
ssp. parvifolium

1 cm

2 cm

fruit

Ludwigia peploides
ssp. peploides

5 mm

fruit

2 cm

1 mm

seed

Oenothera caespitosa
ssp. marginata

O. californica
ssp. californica

1 cm

O. californica
ssp. avita

1 cm

O. california
ssp. eurekensis

bud

1 cm

O. elata
ssp. hirsutissima

1 cm

bud

fruit

seed

1 mm

ssp. deltoides

Oenothera deltoides

seed

0.5 mm

1 cm

fruit

O. pubescens

bud

1 cm

seed

1 mm

ssp. cognata

Uncommon. Openings in chaparral, scrub, gen on shrubs; < 2800 m. s SNH, Teh, SW, **W&I**, **DMtns**; Baja CA. [*O. californica* var. *p.* Jepson] Separation from *O. ludoviciana* Nutt. var. *arenosa* (Suksd.) Cronq. blurred in GB.   May–Jul

OXALIDACEAE OXALIS FAMILY
Robert Ornduff

Ann to tree. **LVS** compound (palmate, pinnate, or lflet 1), alternate, often ± basal in rosettes or in clusters at st or rhizome tips, gen petioled; stipules gen 0; lflets gen sessile. **INFL**: cyme, sometimes umbel- or raceme-like, or fls solitary, gen in axils; peduncle bracted. **FL** gen bisexual, radial; sepals 5, free or fused at base; petals 5, free or fused above base; stamens 10 or 15, fused below, of 2 lengths; pistil 1, ovary superior, chambers 3–5, placentas axile, styles 1–5, gen ± free. **FR**: gen capsule, loculicidal. **SEEDS** gen with aril. 8 genera, 575 spp.: esp temp. Often heterostylous.

OXALIS

Ann, per, shrub; roots fibrous or woody; bulbs, tubers, or rhizomes often present. **ST** sometimes 0 or very short. **LF** petioled: stipules 0 or small; lflets 3, gen ± obcordate in CA, gen entire, gen green. **FL**: petals clawed; stamens 10; ovary chambers 5, styles 5, free, erect or curved. **FR** cylindric to spheric, explosively dehiscent. **SEEDS** flat, often ridged; aril translucent. ± 480 spp.: esp temp. (Greek: sour) [Eiten 1963 Amer Midl Nat 69:257–309; Lourteig 1975 Phytologia 42:57–197] Gen heterostylous; many (esp aliens in CA exc *O. laxa*) fine orn; some noxious weeds; contained oxalates may be toxic to livestock.

O. corniculata L. (p. 401) Per; taproot ± fleshy; bulbs 0. **ST** creeping, rooting at nodes, < 30 cm, ± hairy. **LVS** cauline; petiole < 7 cm; lflets < 2 cm, often maroon. **INFL**: cyme, ± umbel-like or not, 2–5-fld; pedicel < 1 cm. **FL**: sepals < 4.5 mm; petals gen < 8 mm, yellow. **FR** 6–25 mm, cylindric, ± angled. Abundant. Lawns, gardens; gen < 2500 m. **CA**; probably native to Old World. Fls ± all year. Pernicious urban weed. Possibly toxic in quantity to sheep. Most of year

PAPAVERACEAE POPPY FAMILY
Curtis Clark

Ann to small tree; sap often colored, often milky. **LVS** basal, cauline, or both, gen toothed, lobed, or dissected; cauline gen alternate; stipules 0. **INFL**: cyme, raceme, or panicle (terminal), or fl solitary. **FL** bisexual, gen radial; sepals 2–4, sometimes shed ± at fl; petals gen 4 or 6 (or more), sometimes in 2 unlike pairs; stamens 4–many; ovary gen 1, superior, chamber gen 1, stigma lobes 0–many, ovules 1–many. **FR**: gen capsule, dehiscent by valves or pores, gen septicidal. 40 genera, 400 sp.: n temp, n trop, s Afr; some cult (*Papaver, Dicentra, Eschscholzia*). Petal length incls any spur or pouch. *Hunnemannia fumariifolia* Sweet (*Eschscholzia*-like garden per with free sepals) an uncommon waif in CA. *Corydalis, Dicentra, Fumaria* formerly treated in Fumariaceae. Key to genera adapted by Margriet Wetherwax.

1. Fl bilateral; petals 4, not alike, upper spurred at base; lvs dissected (Fumarioideae) .... **CORYDALIS**
1' Fl radial; petals 4 or 6, alike, not spurred at base; lvs entire or dissected (Papaveroideae)
  2. Lvs entire to minutely toothed, cauline opposite; petals gen 6 .... **PLATYSTEMON**
  2' Lvs toothed to dissected (exc entire in *Canbya*), cauline alternate; petals 4 or 6
    3. Pl spiny-armed; lvs gen cauline .... **ARGEMONE**
    3' Pl unarmed; lvs basal, cauline, or both
      4. Hairs gen ± 5–15 mm; lvs gen basal — ne DMoj .... **ARCTOMECON**
      4' Hairs 0–3 mm; lvs basal, cauline, or both
        5. Lvs entire; petals 6; stamens 6–9 — small annual; D .... **CANBYA**
        5' Lvs dissected; petals gen 4; stamens 12–many .... **ESCHSCHOLZIA**

ARCTOMECON

Per; hairs gen ± 5–15 mm; taproot stout; sap colorless. **LVS** gen basal, obovate or wedge-shaped; teeth rounded. **INFL**: fls solitary, terminal. **FL**: bud nodding; sepals 2–3, long-hairy; petals 4 or 6, free, obovate, white or yellow, gen persistent after pollination; stamens many, free; placentas 3–6, style 1, stigma lobes 3–6. **FR** ovate to oblong, dehiscent from tip. **SEEDS** few, oblong, wrinkled, black. 3 spp.: sw US. (Greek: bear poppy, from long hairs)

A. merriamii Cov. (p. 401, pl. 76) WHITE BEAR POPPY Per 2–5 dm. **ST** glaucous. **LF** 25–75 cm. **FL**: sepals 15–20 mm; petals 25–40 mm, white. **FR** 25–35 mm. **SEEDS** 1.5–2 mm. RARE in CA. Rocky slopes; 900–1400 m. **ne DMoj**; s NV. *A. californica* Torrey & Frémont (petals yellow) in sw NV, undocumented in CA. Apr–May

ARGEMONE PRICKLY POPPY

Ann, per, armed; sap yellow or orange, milky. **LVS** gen cauline, oblanceolate to ovate, toothed or deeply pinnately lobed. **INFL**: fls solitary, terminal. **FL**: sepals 2(3), prickly, with pointed appendage below tip, shed at fl; petals 4 (or 6), free, obovate, crinkled, white, shed after pollination; stamens 12–many, free; placentas 4–6, style 0 or 1, stigma lobes 4–6. **FR** ovate to lanceolate, prickly, dehiscent at tip by slits. **SEEDS** many, 1–2.5 mm, round to ovate, net-ridged, brown or black. ± 30 spp.: N.Am, S.Am, Hawaii. (Greek: ocular cataract, supposedly remedied by sap of other pls with this name)

1. Sap orange; lvs gen less prickly on upper surface; fr 25–30 mm; stamens 100–120; seed 1–1.5 mm
.... *A. corymbosa*
1' Sap yellow; lvs ± equally prickly on both surfaces; fr 35–55 mm; stamens 150–250; seed 2–2.5 mm ... *A. munita*

A. corymbosa E. Greene Per 4–8 dm; sap orange. **LF** 8–15 cm, prickly on margins and veins, gen less so on upper surface. **FL**: petals 20–35 mm; stamens 100–120. **FR** 25–30 mm. **SEED** 1–1.5 mm. Dry slopes, flats; 400–1100 m. **SNE, DMoj**. ❀ TRY. Apr–May

A. munita Durand & Hilg. (p. 401) CHICALOTE Ann or per 6–15 dm; sap gen yellow (rarely red in NCoRH). **LF** 5–15 cm, ±

equally prickly on margins and veins of both surfaces. **FL**: petals 25–40 mm; stamens 150–250. **FR** 35–55 mm. **SEED** 2–2.5 mm. Open areas; 70–3000 m. NW (exc NCo), CW, SW, **GB, D**; n Baja CA. [sspp. *argentea* Ownbey, *robusta* Ownbey, prickly poppy, *rotundata* (Rydb.) Ownbey,] TOXIC but not gen eaten. ❀DRN,SUN,DRY:1–3,**7**,8,**9**,10,11,**14**,15–17,**18–21**,22–24. Aug

CANBYA

Ann; sap colorless. **LVS** ± basal, linear-oblong, entire. **INFL**: fls solitary, terminal. **FL**: sepals 3; petals 6, free, elliptic, white, persistent after pollination; stamens 6–9, free; placentas 3, style 0, stigma lobes 3, linear, radiating from below ovary top. **FR** ovate, dehiscent from tip. **SEEDS** many, shiny, brown. 2 spp.: CA, OR, NV. (W.M. Canby, Delaware botanist, 1831–1904)

C. candida C. Parry (p. 401) Pl 10–30 mm, tufted, ± glabrous. **LF** 5–9 mm, fleshy. **INFL**: peduncle 10–20 mm. **FL**: petals 3–4 mm, closing over fr. **FR** 2–2.5 mm. **SEED** 0.6 mm. Sandy places; 600–1200 m. **w DMoj**, adjacent SN. ❀ TRY. Apr–May

CORYDALIS

Ann to per, glabrous, glaucous; sap colorless. **LVS** deeply pinnately dissected. **INFL**: raceme or panicle. **FL** bilateral; sepals 2, shed at fl; petals 4, yellow or white to pink, persistent after pollination, outer 2 petals free, not alike, keeled (upper spurred at base), inner 2 petals adherent at tips, oblanceolate, crested on back; stamens 6, ± fused in 2 sets, opposite outer petals; ovary lanceolate, placentas 2, style 1, stigma lobes 4–8. **FR** gen cylindric to oblong, dehiscent from tip. **SEEDS** several–many, 2–2.5 mm, round-reniform, smooth or rough, black. ± 100 spp.: n hemisphere, s Afr (some orn). (Greek: crested lark)

C. aurea Willd. (p. 401) Ann or bien, 10–40 cm, glaucous. **LVS** several–many, 3–18 cm. **INFL**: raceme, few-fld. **FL**: petals 9–11 mm, yellow, spur 4–5 mm. **FR** 18–25 mm, cylindric. Open areas; 1500–2300 m. **GB**; to AK, e US, n Mex. ❀ TRY. May–Aug

ESCHSCHOLZIA

Ann, per; taproot sap colorless or clear orange. **LVS** basal and gen some cauline, ± linear-dissected. **INFL**: cyme, 1–many-fld. **FL**: receptacle funnel-shaped, tip cupped around ovary base, sometimes spreading-rimmed below petals; sepals 2, fused, shed as a unit at fl; petals gen 4 (exc doubled fls), free, obovate or wedge-shaped, gen yellow to orange (white or pink), shed after pollination leaving crown-like membrane (formerly called "inner rim"); stamens 12–many, free; placentas 2, style 0, stigma lobes 4–8, spreading, linear. **FR** cylindric, dehiscent from base. **SEEDS** many, 1–2 mm, round to ovate, net-ridged, bur-like, or pitted, tan, brown, or black. 12 spp.: w N.Am. (J.F. Eschscholtz, Russian naturalist, 1793–1831)

1. Per from heavy taproot or ann; receptacle rim 0.5–5 mm; cotyledons gen 2-lobed .... [2]*E. californica*
1' Ann; receptacle rim gen 0–0.3(2) mm; cotyledons entire
   2. Receptacle rim 0.5–2 mm; cotyledons sometimes 2-lobed .... [2]*E. californica*
   2' Receptacle rim 0–0.3 mm; cotyledons entire
      3. Lvs gen basal; lf segments acute .... *E. glyptosperma*
      3' Lvs basal and cauline; lf segments acute or obtuse
         4. Lf segments widened to tip, gen obtuse, < 2 mm; lvs grayish or bluish green; petals 2–26 mm; seeds gen oblong to elliptic .... *E. minutiflora*
         4' Lf segments not widened to tip, acute, > 1.5 mm; lvs bright green or yellow-green; petals 8–30 mm; seeds gen round .... *E. parishii*

E. californica Cham. (p. 401) CALIFORNIA POPPY Ann (or per from heavy taproot), erect or spreading, 5–60 cm, glabrous, sometimes glaucous. **LF** segments obtuse or acute. **FL**: bud erect, acute to long-pointed, glabrous, sometimes glaucous; receptacle obconic; petals 20–60 mm, yellow, bases gen orange-spotted. **FR** 3–9 cm. **SEED** 1.5–1.8 mm wide, round to elliptic, net-ridged, brown to black. 2*n*=12. Grassy, open areas; 0–2000 m. CA-FP, **w SNE, DMoj**; to s WA, NV, NM, nw Baja CA. Highly variable (> 90 taxa have been described). Ann pls with entire cotyledons

from DMtns have been called ssp. *mexicana* (E. Greene) C. Clark. CA state fl. TOXIC but rarely eaten. ❀ SUN:1–5,**6–12**,**14**–24&IRR: **13**;CVS.Feb–Sep

E. glyptosperma E. Greene (p. 401) Ann, erect, 5–25 cm, glabrous, sometimes glaucous. **LVS** basal; segments acute. **INFL**: fl 1. **FL**: bud gen nodding, acute, glabrous, sometimes glaucous; receptacle obconic; petals (10)12–25 mm, yellow. **FR** 4–7 cm. **SEED** 1.2–1.8 mm wide, round, minutely pitted, tan to brown.

$2n=14$. Desert washes, flats, slopes; 50–1500 m. **D**; to s NV, sw UT, w AZ. ✿ TRY. Mar–May

E. minutiflora S. Watson (p. 401, pl. 77) Ann, erect or spreading, 5–35 cm, glabrous, gray- or blue-glaucous. **LF**: segments obtuse. **FL**: bud nodding, short-pointed, glabrous, sometimes glaucous; receptacle obconic; petals 3–26 mm, yellow, base sometimes orange-spotted. **FR** 3–6 cm. **SEED** 1–1.4 mm wide, gen oblong to elliptic, net-ridged, brown to black. $2n=12,24,36$. Desert washes, flats, slopes; 0–2000 m. **SNE, D**; to s NV, sw UT, w AZ, nw Mex. Variable. Pls from n&c DMoj with petals 6–18 mm & $2n=24$ have been called ssp. *covillei* (E. Greene) C. Clark;

pls from w DMoj (ne Kern Co.) with 10–26 mm & $2n=12$ have been called ssp. *twisselmannii* C. Clark & Faull. [*E. parishii* has been misapplied to the pls called ssp. *t.*] ✿ TRY. Mar–May

E. parishii E. Greene Ann, erect, 5–30 cm, glabrous, bright green to yellow-green, sometimes glaucous. **LF**: segments > 1.5 mm, not widened to tip, acute. **FL**: bud nodding, long-pointed, glabrous, sometimes glaucous; receptacle obconic; petals 8–30 mm, yellow. **FR** 5–7 cm. **SEED** 1–1.4 mm wide, gen round, net-ridged, tan to brown. $2n=12$. Desert slopes, hillsides; 0–1200 m. **s DMoj**, **DSon**; nw Mex. ✿TRY. Mar–Apr

PLATYSTEMON

1 sp. (Greek: wide stamen)

P. californicus Benth. (p. 405) CREAM CUPS Ann 10–30 cm, shaggy-hairy; sap colorless. **LVS** basal and cauline, opposite, 2–8 cm, linear to lanceolate or narrowly oblong, entire. **INFL**: fls solitary, terminal; peduncle 10–20 cm. **FL**: sepals 3, hairy; petals 6, free, elliptic, white to yellowish, 8–16 mm, persistent after pollination; stamens > 12, free, filaments flattened; longitudinal ovary segments (carpels) gen 9–18, fused, separating in fr. **FR**: segments 10–16 mm, linear, narrowed between seeds, breaking into 1-seeded units. **SEED** 1 mm, elliptic to reniform, smooth, brown. Open grasslands, sandy soils, burns; < 1000 m. CA-FP, **w D**; to OR, UT, AZ, Baja CA. Highly variable. [vars. *crinitus* E. Greene, *horridulus* E. Greene, *nutans* Brandegee, *ornithopus* (E. Greene) Munz]. ✿ DRN,DRY,SUN:6, **7–9**,10–12,**14–24** Mar–May

PHILADELPHACEAE MOCK ORANGE FAMILY

Charles F. Quibell

Shrub, subshrub. **ST** < 3 m, gen erect; bark gen peeling as thin sheets or narrow strips. **LVS** simple, opposite, deciduous or not, ± hairy; stipules 0; blade ± round to narrowly elliptic, entire or toothed. **INFL**: cyme, raceme, panicle, or fl 1, terminal or axillary, gen bracted. **FL** bisexual, radial; sepals 4–7, free, spreading or erect; petals 4–7, free, ± round to narrowly elliptic, gen white; stamens 10–12 in 2 whorls or many and clustered, filament base linear or wide and flat; pistil 1, ovary superior to 2/3 inferior, chambers 2–8, ovules 1–2 or many per chamber, placentas axile or parietal, styles 3–8, free or fused at base. **FR**: capsule, loculicidal or septicidal; styles persistent or not. **SEEDS** gen many, small to minute. 7 genera, 130 spp.: temp, subtrop n hemisphere; some cult for orn (*Carpenteria, Deutzia, Philadelphus*) Recently treated within Hydrangeaceae [Soltis et al 1995 Amer J Bot 82:504–514]. Key to genera adapted by Margriet Wetherwax.

1. Fls < 8 mm wide, odorless; sepals ± 1.5 mm, ± deciduous; petals 3–4 mm . . . . **FENDLERELLA**
1' Fls gen > 15 mm wide, fragrant; sepals 3–5 mm, persistent; petals 5–8 mm
  2. Stamens 10, alternating long and short; filament base wide, flat; stigma terminal, < 0.5 mm . . . . . . **JAMESIA**
  2' Stamens > 20, ± equal, filament base ± linear; stigma linear, elongated along style, > 1 mm
  . . . . **PHILADELPHUS**

FENDLERELLA

Shrub < 8 dm. **ST**: bark whitish, peeling as thin sheets or strips; twigs strigose. **LVS** deciduous, leathery, ± sessile. **INFL**: cymes clustered, terminal, dense to open; fls (1)3–11. **FL**: odorless; sepals 5; petals 5, white; stamens 10, alternating long and short, filament base wide, flat; ovary half-inferior, chambers 3, placentas axile, ovule 1 per chamber, styles 3, persistent, spreading in fr, stigma terminal. **FR** ± cylindric, septicidal. **SEED** fusiform, red-brown. 3 spp.; sw US, n Mex. (August Fendler, plant collector, 1813–1883)

F. utahensis (S. Watson) A.A. Heller (p. 405) YERBA DESIERTO **LF**: blade 8–16 mm, 3–6 mm wide, ovate to elliptic, strigose, 3-veined from base, margin entire, ± rolled under. **INFL** 12–18 mm, short-peduncled. **FL**: sepals ± 1.5 mm, linear-lanceolate, sparsely strigose; petals 3–4 mm, oblong-obovate. **FR** ± 4 mm. **SEED** ± 2 mm. UNCOMMON. Limestone soils, crevices, slopes; 1300–2800 m. **DMtns**; to Colorado, n Mex. ✿ TRY;DFCLT. Jun–Aug

JAMESIA CLIFFBUSH

1 sp. (Edwin P. James, naturalist, 1797–1861) [Holmgren & Holmgren 1989 Brittonia 41:335–350]

J. americana Torrey & A. Gray var. ***rosea*** C. Schneider (p. 405) Shrub < 1 m; herbage gen densely hairy. **ST**: bark gen gray, peeling as narrow strips. **LVS** deciduous; petiole 2–6 mm; blade 1.5–4 cm, 1–2 cm wide, widely ovate to ± round, pinnately veined, margin toothed, upper surface green, finely strigose, lower surface pale gray-white, densely hairy. **INFL**: cyme, terminal; fls (1)3–11. **FL** 1.2–1.5 cm wide, slightly fragrant; sepals 5, 3–4 mm, gray-strigose; petals 5, 5–8 mm, elliptic to obovate, gen pink; stamens 10, alternating long and short, filament base wide, flat; ovary half-inferior, chambers 3–5, 1 in fr, placentas parietal, ovules many, styles 3–5, > sepals, persistent, spreading in fr, stigma terminal. **FR** 1–1.3 cm, conic, septicidal. **SEEDS** many,

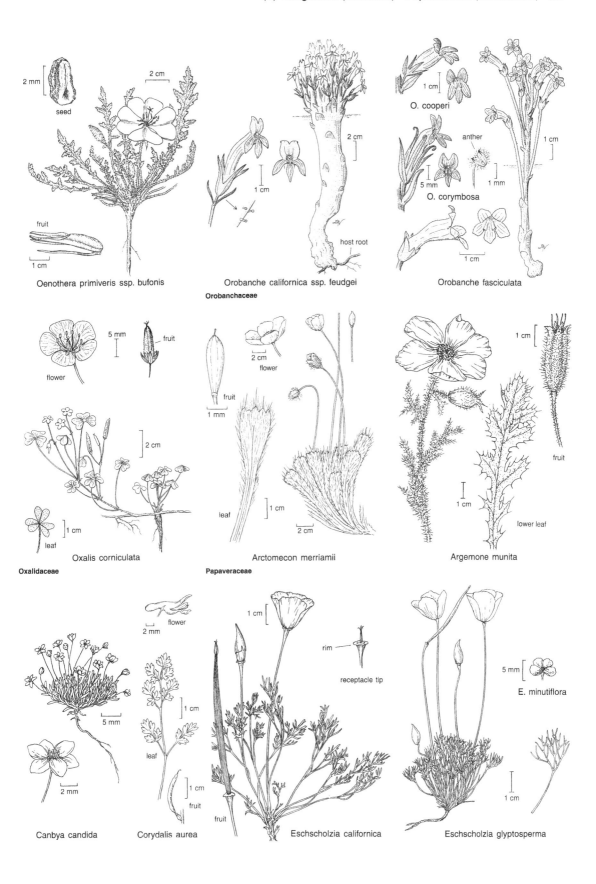

2 mm

seed

2 cm

fruit

1 cm

Oenothera primiveris ssp. bufonis

1 cm

2 cm

host root

Orobanche californica ssp. feudgei

Orobanchaceae

1 cm

O. cooperi

anther

5 mm

1 mm

O. corymbosa

1 cm

1 cm

Orobanche fasciculata

flower

5 mm

fruit

leaf

1 cm

Oxalis corniculata

Oxalidaceae

fruit

2 cm

flower

1 mm

leaf

1 cm

2 cm

Arctomecon merriamii

Papaveraceae

1 cm

fruit

1 cm

lower leaf

Argemone munita

flower

2 mm

Canbya candida

2 mm

flower

2 mm

1 cm

leaf

1 cm

fruit

Corydalis aurea

1 cm

rim

receptacle tip

fruit

Eschscholzia californica

5 mm

E. minutiflora

1 cm

Eschscholzia glyptosperma

fusiform, brown. Rocky slopes, cliffs; 2250–3700 m. c&s SNH, **W&I, n DMtns (Grapevine Mtns)**; w NV. [var. *californica* (E.

Small) Jepson] Other vars. in GB, Rocky Mtns.   ❀ TRY;DFCLT. Jul–Aug

PHILADELPHUS   MOCK ORANGE

Shrub < 3 m. **ST**: bark red-brown, aging gray, peeling as narrow strips or narrow rectangles; twigs glabrous to hairy. **LVS** deciduous, petioled; blade 3-veined from base, ± glabrous to hairy, margin entire to toothed. **INFL**: raceme, panicle, or fl one, terminal, ± open. **FL** fragrant; sepals 4–5, glabrous to hairy; petals 4–5, white; stamens gen many, clustered, filaments linear, fused at base, ovary inferior to half-inferior, chambers 4–5, placentas axile, ovules many, style 1, branches gen 4 above middle, stigma linear along style branch. **FR** becoming woody, gen loculicidal. **SEEDS** many, gen fusiform, gen brown. ± 65 spp.: temp Amer, Eurasia. (Greek: for Ptolemy Philadelphus, Greek king of Egypt, 309–247 B.C.) [Hu 1956 J Arnold Arbor 37:15–90]

P. microphyllus A. Gray (p. 405)   LITTLELEAF MOCK ORANGE **LF**: petiole 5–18 mm; blade 8–25 mm, 3–8 mm wide, narrowly ovate to elliptic, margin entire, ± rolled under. **INFL**: fls 1–3. **FL** 1–1.5 cm wide; sepals 3–5 mm, petals 6–8 mm, widely elliptic. Rocky slopes, cliffs; 1200–2750 m. e PR (SnJt, Santa Rosa Mtns),

W&I, e DMtns; to TX, n Mex. [ssp. *stramineus* (Rydb.) C. Hitchc.] Pls with lf blades 8–12 mm, rough-hairy above have been called ssp. *pumilus* (Rydb.) C. Hitchc. [Hitchcock 1943 Madroño 7:35–56] ❀ DRN,IRR:1–3,**7**,10,14–16,**18**. May–Jul

PLANTAGINACEAE   PLANTAIN FAMILY

Lauramay T. Dempster

Ann, per, gen scapose. **LVS** gen basal, simple, with longitudinal ribs; stipules 0. **INFL**: spike, gen terminal, gen dense; fls few–many, each subtended by 1 bract. **FL** unisexual or bisexual, gen radial; calyx deeply 4-lobed, lobes gen overlapping, persistent, margin mostly scarious; corolla salverform or cylindric, lobes 4, spreading to erect, scarious, persistent, colorless; stamens 2–4, alternate corolla lobes, epipetalous; ovary superior, 2–4-chambered, ovules several per chamber, style 1, stigma long, hairy. 3 genera, ± 270 spp.: worldwide, esp temp; some weedy, some (esp *P. afra*, psyllium) cult for laxative. Recently treated to include Callitrichaceae, and most non-parasitic California genera of Scrophulariaceae (but not *Mimulus, Scrophularia, Verbascum*); = Veronicaceae sensu Olmstead et al [Angiosperm Phylogeny Group 1998 Ann Missouri Bot Gard 85:531-553; Olmstead et al 2001 Mol Phylog Evol 16: 96-112]

PLANTAGO   PLANTAIN

Ann, per. **ST** decumbent to erect. **FL** gen bisexual; corolla radial, sometimes bilateral. **FR**: capsule, circumscissile ± at or below middle. **SEEDS** 2–many, gelatinous when wet. ± 250 spp.; worldwide. (Latin: sole of foot) Key to species adapted by Margriet Wetherwax.

1. Per; glabrous exc peduncle
   2. Pl taprooted; lf oblanceolate to narrowly elliptic-ovate, tapered gradually to wide petiole . . . . **P. eriopoda**
   2' Pl with clustered fibrous roots; lf widely elliptic to ± cordate, narrowed abruptly to petiole . . . . **P. major**
1' Ann; ± densely hairy throughout
   3. Bracts ± = sepals, ovate to round, not exserted . . . . **P. ovata**
   3' Bracts 1–2× sepals, linear, lower exserted . . . . **P. patagonica**

P. eriopoda Torrey (p. 405)   Per, from stout caudex, taprooted below, gen glabrous exc peduncle. **LF** 5–25 cm, oblanceolate to narrowly elliptic-ovate, tapered gradually to wide petiole, margin gen ± wavy or minutely dentate. **INFLS** gen 1–6, 10–55 cm incl peduncle; spike 8–18 cm, narrowly cylindric, becoming arched, dense, loose near base; bract ± = sepals, appressed to sepals, widely ovate, not exserted. **FL**: corolla lobes ± 1 mm, spreading, ovate-lanceolate; stamens 4. **SEEDS** 2–4, ± 2 mm. 2*n*=24. Moist and alkaline places; < 1000 m. KR, n CaR, **SNE (Mono Co.)**; to Can, c US.   Jul–Aug

P. major L. (p. 405)   COMMON PLANTAIN   Per (rarely ann), gen glabrous exc peduncle; caudex short, with fibrous roots. **LF** blade 5–18 cm, widely elliptic to somewhat cordate, narrowed ± abruptly to petiole; margin entire or ± finely dentate, rough to touch. **INFLS** gen 3–7, 5–60 cm incl peduncle; spike gen 3–20 cm, linear-cylindric, becoming ± loose; bract ± = sepals, ovate, not exserted. **FL**: corolla lobes 0.5 mm, spreading, ovate-lanceolate; stamens 4. **SEEDS** 5–16, < 1 mm. 2*n*=6,12,24. Disturbed areas; < 2200 m. CA-FP (exc SN), **GB, DMtns (uncommon)**; to e US; native to Eur. Highly variable; named vars. (incl *pilgeri* Domin, *scopulorum* Fries & S.P. Broberg) indistinct.   Apr–Sep

P. ovata Forsskal (p. 405) Ann, ± densely silky-hairy. **LF** 2–17 cm, linear or rarely oblong, entire or teeth few, minute. **INFLS** few–many, gen 2–27 cm incl peduncle; spike 0.5–3.5 cm, gen short-cylindric, dense, woolly; bract ± = sepals, ovate to round, not exserted. **FL**: corolla lobes 1.3–2.8 mm, spreading, round-ovate, tips obtuse; stamens 4. **SEEDS** 2, 2–2.5 mm. 2*n*=8. Sandy or gravelly soils, creosote-bush scrub, Joshua-tree woodland, sagebrush scrub, coastal strand; < 1400 m. CW, SW, **SNE, D**; to UT, TX, Baja CA; also Medit. [*P. insularis* Eastw. incl var. *fastigiata* (E. Morris) Jepson] May be alien, naturalized very early from Medit. Feb–Apr

P. patagonica Jacq. (p. 405)   Ann, ± densely hairy. **LF** 2–10 cm, linear or very narrowly oblanceolate, entire. **INFLS** often many, 2–18 cm incl peduncle; spike 1–6 cm, nearly cylindric, a little wider at base, dense, ± densely woolly; bract 1–2 × sepals, linear, lower exserted. **FL**: corolla lobes 1.5–2 mm, spreading (or 1 erect), ovate, tips ± acute; stamens 4. **SEEDS** 2, ± 3 mm. 2*n*=20. Sandy, rocky, or grassy slopes, pinyon/juniper or Joshua-tree woodland, chaparral; 500–2200 m. SnBr, SnJt, **DMoj**; to B.C., n-c US, TX, Mex; also Argentina. [*P. purshii* Roemer & Shultes var. *oblonga* (E. Morris) Shinners] ❀ TRY. Apr–Jun

PLATANACEAE PLANE TREE, SYCAMORE FAMILY

Elizabeth McClintock

Tree, gen monoecious, wind-pollinated. **ST**: branches irregular below, spreading to erect above; bark irregularly colored, scale-like, peeling; twigs dense-hairy. **LVS** simple, alternate, deciduous, gen palmately 3- or 5-lobed, -veined; stipules gen lf-like, free or fused around st, shed before lvs; petiole at base dilated, hollow, ± covering bud; blade dense-hairy, glabrous in age, hairs stellate. **INFL**: heads 1–6, ± evenly spaced on axis, spheric, many-fld, sessile or on pendent peduncles, gen unisexual; staminate breaking apart in age; pistillate persistent; bracts subtending heads, fls. **FLS** unisexual; calyx cup-shaped, scale-like, entire or 3–6(8)-lobed or sepals ± free. **STAMINATE FL**: petals 3–6, fleshy or scale-like, minute or vestigial; stamens 3–6(8), alternate petals, anthers subsessile, axis above anther expanded, disc-like; pistils vestigial. **PISTILLATE FL**: petals 3–6, minute, or gen 0; staminodes often 3–4; pistils (3)5–9, ovaries superior, 1-chambered, gen 1-ovuled, style 1. **FR**: spheric head of small, hairy, basally bristly achenes; style persistent. 1 genus, ± 8 spp.: n temp; some orn, esp *P.* ×*acerifolia* (Aiton) Willd., London plane tree; some cult for wood, veneer. [Ernst 1963 J Arnold Arbor 44: 206–210]

PLATANUS

(Greek: probably broad, from lvs)

P. racemosa Nutt. (p. 405) WESTERN SYCAMORE **ST** 10–35 m; base < 1 m wide; bark smooth, pale. **LF**: stipules 2–3 cm; petiole 3–8 cm; blade ± 10–25 cm, ± round, lobes 3 or 5, acute to acuminate, entire. **INFL**: heads 3–5, ± 1 cm. **FR**: heads 2–3 cm, sessile or not. 2*n*=42. Common. Streamsides, canyons; < 2000 m. c&s SNF, Teh, GV, CW, SW, **w DSon**; Baja CA. Pls in PR, with some pistillate infls with peduncles 5–15 mm have been called var. *wrightii* (S. Watson) L. Benson. ❀ IRR,SUN:1–3,7–9,**11**,14–17,**18–21**,22–24; susceptible to sycamore anthracnose. Feb–Apr

POLEMONIACEAE PHLOX FAMILY

Robert W. Patterson, Family Editor

Ann, per, shrub, vine. **LVS** simple or compound, cauline (or most in basal rosette), alternate or opposite; stipules 0. **INFL**: cymes, heads, or fls solitary. **FL**: calyx gen 5-ribbed, ribs often connected by translucent membranes that are gen torn by growing fr; corolla gen 5-lobed, radial or bilateral, salverform to bell-shaped, throat often well defined; stamens gen 5, epipetalous, attached at same or different levels, filaments of same or different lengths, pollen white, yellow, blue, or red; ovary superior, chambers gen 3, style 1, stigmas gen 3. **FR**: capsule. **SEEDS** 1–many, gelatinous or not when wet. 19 genera, 320 spp.: Am, n Eur, n Asia; some cult (*Cantua, Cobaea* (cup-and-saucer vine), *Collomia, Gilia, Ipomopsis, Linanthus, Phlox*). *See also revised taxonomy of Porter and Johnson 2000 Aliso 19 (1): 55–91; Porter 1998 Aliso 17:83-85.

1. Lvs pinnately compound, lflets lanceolate to ovate . . . . **POLEMONIUM**
1' Lvs simple, entire to deeply lobed (deep lobes gen linear)
  2. Lvs 0 (exc cotyledons); involucral bracts fused near base; filaments 0 . . . . **GYMNOSTERIS**
  2' Lvs present; involucral bracts (if present) not fused; filaments present
    3. Calyx membrane 0; calyx sinus pleated to expanded in fr . . . . **COLLOMIA**
    3' Calyx membrane present between ribs; calyx sinus not pleated or expanded in fr
      4. Lvs opposite, gen cauline
        5. Stamens attached at different levels . . . . **PHLOX**
        5' Stamens attached at same level . . . . **LINANTHUS**
      4' Lvs alternate, sometimes in basal rosette
        6. Lvs deeply palmately lobed . . . . **LEPTODACTYLON**
        6' Lvs not deeply palmately lobed
          7. Lvs and calyx lobes bristle-tipped
            8. Corolla bilateral . . . . **LOESELIASTRUM**
            8' Corolla radial
              9. Basal teeth of upper lvs reduced to 2–3 bristles . . . . **LANGLOISIA**
              9' Lf bristles solitary — lvs holly-like . . . . ⁴GILIA
          7' Lvs and calyx lobes not bristle-tipped
            10. Infl a dense spiny-bracted head; calyx lobes gen unequal
              11. Infl woolly . . . . **ERIASTRUM**
              11' Infl gen hairy, not woolly . . . . **NAVARRETIA**
           10' Infl not a dense, spiny-bracted head; calyx lobes equal
            12. St scapose; lvs abruptly reduced above well developed basal rosette . . . . ⁴GILIA
            12' St not clearly scapose; lvs ± gradually reduced above basal lvs
              13. Fls densely clustered, collectively head-like, ± subsessile
                14. Lower lvs 1–2-pinnately lobed, axis and lobes ± linear, not clearly flat; corolla blue or
                  throat yellow, purple-spotted . . . . ⁴GILIA

14' Lower lvs entire or 1-pinnately to -palmately toothed to lobed, axis and lobes narrowly
elliptic to oblanceolate, ± flat; corolla white, tube light yellow . . . . [2]**IPOMOPSIS**
13' Fls not densely clustered, 1 per node or few-fld cyme, collectively raceme- or panicle-like;
fls gen pedicelled
15. Upper lvs ± palmately lobed, distal lobe gen > lateral lobes; lf and sepal tips gen rounded
at 20× (if pointed, not translucent); seeds black . . . . **ALLOPHYLLUM**
15' Upper lvs entire or pinnately lobed, distal lobe gen = or > lateral lobes; lf and sepal tips ±
sharp-pointed at 20×; seeds light brown to grayish
16. Corolla gen bell- or funnel-shaped; throat color clearly different from tube color . . . . [4]**GILIA**
16' Corolla gen salverform; tube color gen uniform (red, pink, lavender, or white) — corolla
lobes gen mottled . . . . [2]**IPOMOPSIS**

ALLOPHYLLUM

Alva G. Day

Ann, hairy, ± glandular; glands minute. **ST** erect, leafy, gen branched. **LVS** alternate, dark green, gen deeply pinnately
lobed, ± palmately lobed upward; lobes linear to lanceolate, blunt-tipped, central lobe widest. **INFL**: fls gen in clusters. **FL**: calyx lobes blunt-tipped, tube narrowly membranous between ribs, ribs translucent below in fr, membrane
splitting; corolla funnel-shaped, radial or bilateral, throat narrow, tapered, lobes narrowly obovate; stamens attached
in tube; stigmas 3. **FR** spheric, < calyx; valves gen falling. **SEEDS** 1–3 per chamber, concave, black or brown,
gelatinous when wet; ends rounded. 4 spp.: w N.Am. (Greek: other leaf) [Grant & Grant 1955 El Aliso 3:93–110]

A. gilioides (Benth.) A.D. Grant & V. Grant **ST** < 40 cm, puberulent; hairs gland-tipped in infl. **LF**: lobes 0–11, linear to narrowly lanceolate. **INFL** open or dense; fls 2–8 in clusters. **FL**:
corolla < 10 mm, dark blue-purple, lobes 1–3 mm; stamens nearly
equal, incl to slightly exserted; style incl. **SEEDS** 1 per chamber,
black. Open, sandy, gen damp or grassy areas; 200–2900 m.
NCoR, CaR, SN, SnFrB, SCoR, TR, PR, **n SNE**, **DMtns**; s OR,
w NV, AZ, Baja CA.

ssp. *violaceum* (A.A. Heller) A.G. Day (p. 405) **ST** gen ±
15 cm, slender. **LVS** ± cauline, entire or 3–7-lobed; lobes 0.5–3
mm wide. **INFL** open; fls solitary or 2–3 in clusters; pedicels
gen elongating in fr. **FL**: corolla 5–8 mm. Habitat and range of
sp. (exc not in CaR, OR); 1200–2900 m. ***A. violaceum* (A.A.
Heller) A.D. Grant & V. Grant May–Jul

COLLOMIA

Dieter H. Wilken

Ann, per. **ST** hairy, glandular. **LVS** alternate, simple, entire to gen pinnately lobed, linear to ovate; basal short-
petioled; cauline sessile. **INFL**: heads or clusters, terminal (or fls 1–3 in axils). **FL**: calyx green, becoming straw-
colored, membranous in age, gen bell-shaped, sinuses in fr pleated to expanded; corolla salverform to funnel-shaped.
FR ovoid to elliptic. **SEEDS** 3 per chamber, oblong, gen gelatinous, brown. $2n$=16. 15 spp.: N.Am; also in s S.Am.
(Greek: glue, from wet seed surface) Self-compatible; ann spp. self-pollinating; per spp. gen cross-pollinating. [Wilken
et al 1982 Biochem Syst Ecol 10:239–243]

1. St branched, branches spreading; fls in axillary and terminal clusters . . . . *C. tinctoria*
1' St gen simple, erect; infl head-like, terminal
   2. Corolla yellow to orange, gen > 15 mm (< 5 mm in cleistagamous fls); pollen gen blue . . . . *C. grandiflora*
   2' Corolla white to pink, < 15 mm; pollen gen white . . . . *C. linearis*

C. grandiflora Lindley Ann. **ST** erect, simple or branched
above in robust pls. **LVS**: basal lanceolate, toothed; cauline lanceolate to linear, entire, gen glabrous above, glaucous, slightly
glandular below. **INFL**: head, terminal; fls sessile. **FL**: calyx 7–
10 mm, lobes lanceolate; corolla 15–30 mm (< 5 mm in cleistogamous fls), yellow to orange; pollen gen blue. Open areas; 600–
2500 m. CA-FP, **n SNE**; to B.C., ID, Colorado, AZ; naturalized
in Eur. Fully cleistogamous pls have single sts < 10 cm with 3–
7 fls at tip. ❀ SUN, DRN:1–5,**6,7**,8,**9,10,14–24**. Apr–Jun

C. linearis Nutt. (p. 405) Ann. **ST** erect, gen simple, branched
above in robust pls. **LVS**: basal lanceolate, toothed; cauline lanceolate to linear, entire, gen glabrous above, glaucous, slightly

glandular below. **INFL**: bracted head, terminal; fls 7–20, sessile.
FL: calyx 4–7 mm; corolla 8–15 mm, white or pink; pollen gen
white. Open areas; 600–3300 m. KR, CaR, SN, SnBr, **GB**; to
AK, e N.Am, AZ; naturalized elsewhere. May–Aug

C. tinctoria Kellogg (p. 405) Ann. **ST** 2–8 cm, branched diffusely. **LF** linear-lanceolate, glandular, entire. **INFL** axillary; fls
gen 2–3. **FL**: calyx 5–7 mm, lobes long tapered, awned; corolla
8–11 mm, slender, tube maroon to violet, lobes pink; stamens
attached at same level, 1–2 anthers exserted, stigma exserted.
Gravelly to rocky, open areas; 600–2800 m. KR, NCoRH, CaRH,
n&c SNH, WTR, **GB**; to WA, ID, NV. May–Aug

ERIASTRUM

Robert W. Patterson

Ann, per, glabrous to woolly. **ST** gen erect. **LVS** cauline, alternate, pinnately lobed or simple; lobes gen linear or
lanceolate. **INFL**: heads, bracted, gen densely woolly; bracts lf-like; fls sessile. **FL**: calyx lobes unequal; corolla
radial or bilateral, funnel-shaped to salverform; stamens equal or not, attached at same level, anthers gen sagittate,

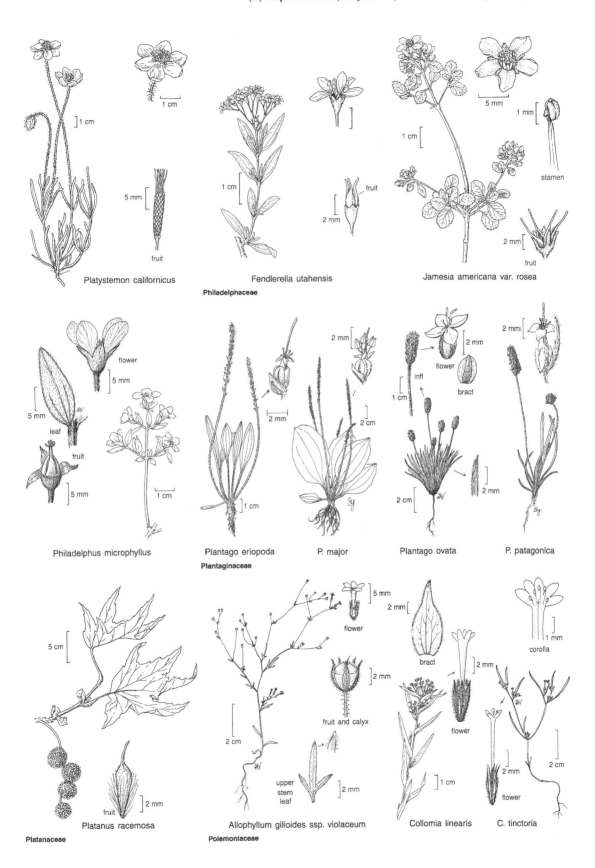

Platystemon californicus

fruit

5 mm

1 cm

1 cm

Fendlerella utahensis

Philadelphaceae

1 cm

fruit

2 mm

Jamesia americana var. rosea

5 mm

1 mm

stamen

2 mm

fruit

Philadelphus microphyllus

flower

5 mm

5 mm

leaf

fruit

5 mm

1 cm

Plantago eriopoda

Plantaginaceae

2 mm

2 mm

P. major

2 cm

Plantago ovata

infl

1 cm

flower

bract

2 mm

2 mm

2 cm

2 mm

P. patagonica

2 mm

Platanus racemosa

Platanaceae

fruit

2 mm

5 cm

Allophyllum gilioides ssp. violaceum

Polemoniaceae

flower

5 mm

fruit and calyx

2 mm

upper
stem
leaf

2 mm

2 cm

Collomia linearis

bract

2 mm

flower

2 mm

1 cm

C. tinctoria

corolla

1 mm

2 mm

flower

2 mm

2 cm

pollen white or blue; style incl or exserted. **SEEDS** 1–few per chamber. 13 spp.: w N.Am. (Greek, woolly star) [Harrison 1972 Brigham Young Univ Sci Bull, Biol Ser 16:1–26]

1. Per . . . . *E. densifolium*
  2. Lvs not recurved, blades gen equally wide at base, tip, lobes not spine-tipped . . . . ssp. *elongatum*
  2' Lvs strongly recurved, blade wider at base that at tip, lobes spine-tipped . . . . ssp. *mohavense*
1' Ann
  3. Corolla 15–23 mm
    4. Corolla bilateral; stamens unequal, bent toward lower lip . . . . *E. eremicum*
    4' Corolla radial; stamens equal . . . . *E. pluriflorum*
  3' Corolla < 15 mm
    5. Stamens extended to tip of corolla lobes or beyond; corolla lobes bright blue . . . . *E. sapphirinum*
    5' Stamens not extended as far as tip of corolla lobes; corolla lobes pale blue, white, pink, or yellow
      6. Stamens equal . . . . *E. sparsiflorum*
      6' Stamens unequal
        7. Stamens attached at top of throat; calyx 5–6 mm . . . . *E. diffusum*
        7' Stamens attached at base of throat or lower; calyx 6–10 mm . . . . *E. wilcoxii*

E. densifolium (Benth.) H. Mason Per. **ST** erect or spreading, nearly glabrous to woolly. **LF** 10–50 mm; lobes gen 2–16, glabrous to woolly. **FL**: corolla funnel-shaped, blue or white, 15–35 mm, lobes 5–15 mm; stamens equal, attached in throat or upper tube, exserted. Dunes, dry river beds, open slopes; 0–2900 m. s SN, s CW, SW, **SNE, w DMoj**; Baja CA. 5 sspp. in CA.

  ssp. ***elongatum*** (Benth.) H. Mason (p. 413) **LF**: lobes gen 2–8, 5 mm, slightly to moderately woolly. **INFL** terminal, axillary; bracts gen 1–5-lobed; fls gen < 15. **FL**: corolla tube 14–18 mm. 2*n*=14. Dry places; < 1800 m. s SCoRO, SW, **w edge DMoj**; Baja CA. ❀ DRN, SUN:7–12,14,**19–24**. Jun–Sep

  ssp. ***mohavense*** (Craig) H. Mason **LF**: axis 1–4 mm wide, strongly recurved; blade base wider than tip; lobes 2–8, spine-tipped, densely woolly. **FL**: corolla tube 15–17 mm. 2*n*=14. Sandy slopes, flats; 800–2800 m. **SNE, w DMoj**. In cult. Jun–Oct

E. diffusum (A. Gray) H. Mason Ann. **ST** erect or spreading, 3–20 cm. **LF** 10–25 mm; lobes gen 2–4 near base, nearly glabrous to woolly. **FL**: corolla gen funnel-shaped, tube 3–5 mm, blue or yellow, lobes 3–5 mm, light blue or cream; stamens attached in upper throat, unequal, exserted. 2*n*=14. Open areas, sandy soil; < 2000 m. **W&I, D**; to Colorado, NM, TX. Despite name, not always openly or diffusely branched. Mar–May

E. eremicum (Jepson) H. Mason ssp. *eremicum* (p. 413) Ann. **ST** 5–30 cm. **LF** 5–55 mm; lobes 2–6, nearly glabrous to woolly. **FL**: corolla gen bilateral, tube 4–9 mm, blue or yellow, throat gen yellow, lobes 4–8 mm, light to dark blue; stamens attached in lower throat or upper tube, unequal, bent toward lower corolla lip, exserted. 2*n*=14. Open areas in sandy soils; < 1800 m. **W&I, D**; to NV, UT, AZ. Ssp. *yageri* (M.E. Jones) H. Mason occurs in AZ. Apr–Jun

E. pluriflorum (A.A. Heller) H. Mason (p. 413) MANY-FLOWERED ERIASTRUM Ann. **ST** 2–32 cm. **LF** 20–50 mm, thread-like, woolly, entire or 2–10-lobed. **FL**: corolla salverform, tube 12–16 mm, white or yellow, throat yellow, lobes 4–9 mm, bright blue; stamens attached in sinuses, gen equal, exserted. 2*n*=14. UNCOMMON. Chaparral, woodlands, pine forests; < 2000 m. c&s SNF, SnJV, SnFrB, e SCoR, **w DMoj**. [ssp. *sherman-hoytae* (Craig) H. Mason] ❀ TRY. May–Jul

E. sapphirinum (Eastw.) H. Mason (p. 413) Ann. **ST** 5–40 cm; hairs minute, glandular. **LF** 5–55 mm, thread-like, nearly glabrous to woolly, entire or 2-lobed near base. **FL**: corolla funnel-shaped, bilateral from unequal sinuses, tube 5–6 mm, blue or yellow, throat yellow, lobes 5–10 mm, bright blue; stamens attached in lower throat or upper tube, equal, exserted, gen extended beyond corolla lobes. 2*n*=14. Many communities; 700–2700 m. s CA-FP, **w D**. [sspp. *ambiguum* (M.E. Jones) H. Mason, *dasyanthum* (Brand) H. Mason, *gymnocephalum* (Brand) H. Mason] ❀ TRY. Jun–Aug

E. sparsiflorum (Eastw.) H. Mason Ann. **ST** 5–30 cm. **LF** 5–35 mm, thread-like, woolly, entire or 2-lobed near base. **FL**: corolla salverform or narrow funnel-shaped, tube 4–6 mm, yellow, throat yellow, lobes 2–3 mm, bright blue, pink, yellow, or cream; stamens attached in lower throat or upper tube, equal, slightly exserted or incl. 2*n*=14. Desert slopes; < 2400 m. s SNH, Teh, **SNE, DMtns**; OR, NV, Baja CA. [ssp. *harwoodii* (Craig) H.K. Harrison] Jun–Jul

E. wilcoxii (Nelson) H. Mason Ann. **ST** 3–30 cm. **LF** 15–30 mm, thread-like, densely woolly, entire or 2–4-lobed near base. **FL**: corolla gen salverform, tube 5–8 mm, yellow, throat yellow, lobes 4–5 mm, blue; stamens attached in lower throat or upper tube, unequal, filaments 1–5 mm, exserted or opening. 2*n*=14. Desert flats, slopes; < 2700 m. **SNE, n DMtns**; to ID, WY. Jun–Aug

GILIA

Alva G. Day

Ann, per, gen erect. **ST** glabrous, hairy, glandular, or cobwebby. **LVS** simple, gen alternate; basal gen in rosette, toothed, pinnately lobed, or entire; cauline gen reduced; lf tips, calyx lobes acute, acuminate, or needle-like. **INFL**: fls solitary or clustered, 1–many in axils of bracts. **FL**: calyx membranous between ribs, membrane splitting or expanding; corolla > calyx, lobes gen ovate. **FR** gen ovoid; chambers 3, valves separating from top. **SEEDS** 3–many, brown, gen gelatinous when wet. ± 70 spp.: w N.Am, S.Am. (Felipe Gil, 18th century Spanish botanist) This genus is polyphyletic and currently being revised: Porter and Johnson 2000 Aliso 19 (1): 55–91. *Aliciella humillima* (A. Brand) J.M. Porter recently recorded from Diaz Lake, Inyo Co.

1. Cobwebby hairs present; pl ± scapose; basal rosette of lvs gen present (sect. *Arachnion*)
  2. Calyx, pedicel densely glandular (glands minute, stalked), rarely cobwebby; fls, frs gen in clusters
    . . . . ***G. brecciarum***
    3. Stamens slightly exserted; corolla 7–11 mm, upper throat yellow with purple veins . . . . ssp. ***brecciarum***

3' Stamens long-exserted; corolla 10–20 mm, upper throat white with five yellow spots and white veins
. . . . ssp. *neglecta*
2' Calyx, pedicel slightly glandular (glands stalked or not) to glabrous or cobwebby; fls, frs loosely
  dispersed, sometimes clustered
  4. Cauline lvs clasping or expanded at base (except *G. diegensis* ± so)
    5. Corolla 6–12 mm, 1–2 × calyx
      6. St green (not glaucous), slightly cobwebby or glandular below middle . . . . **G. modocensis**
      6' St glaucous, glabrous below middle
        7. Lobes of basal lvs gen lanceolate, entire; pollen blue . . . . **G. diegensis**
        7' Lobes of basal lvs gen roundish, toothed; pollen white . . . . **G. sinuata**
    5' Corolla 10–35 mm, 2–7 × calyx . . . . **G. latiflora**
      8. St glaucous and glabrous below middle
        9. Corolla purple to mid-throat; fls in loose clusters . . . . [2]ssp. *daveyi*
        9' Corolla purple to base of throat; fls not in clusters . . . . ssp. *latiflora*
      8' St cobwebby below middle
        10. Corolla tube stout, throat > 7 mm wide, corolla length gen 3 × lobe length . . . . [2]ssp. *daveyi*
        10' Corolla tube slender, throat < 7 mm wide, corolla length gen 4–6 × lobe length . . . . ssp. *elongata*
  4' Cauline lvs not clasping or expanded at base
    11. Calyx glabrous or cobwebby, not gland-dotted
      12. Stamens long-exserted, gen = or > corolla lobes, calyx membrane purple-tinged or -spotted
        13. Basal lvs densely cobwebby . . . . **G. salticola**
        13' Basal lvs glabrous or slightly cobwebby . . . . **G. aliquanta**
          14. Corolla tube incl, lobe length = tube + throat . . . . ssp. *aliquanta*
          14' Corolla tube exserted, lobe length gen = 1/2 tube + throat . . . . ssp. *breviloba*
      12' Stamens slightly exserted, < corolla lobes; calyx membrane not purple-tinged or -spotted
        15. Lobes of basal lvs spreading; corolla throat white and yellow
          16. Corolla 4–6.5 mm, throat < or = lobes, white with five yellow spots at base; calyx lobes short-
            pointed . . . . **G. clokeyi**
          16' Corolla 7–12 mm; throat = or > lobes, bright yellow to near top of throat; calyx lobes acuminate
            . . . . **G. ophthalmoides**
        15' Lobes of basal lvs gen ascending; corolla throat yellow in lower 1/2, blue above . . . . **G. ochroleuca**
          17. St gen cobwebby or glandular below infl (or if glabrous, not glaucous) . . . . ssp. *exilis*
          17' St glabrous and glaucous below infl
            18. Corolla 8–14 mm; style > stamens . . . . ssp. *bizonata*
            18' Corolla 4–6 mm; style < or = stamens . . . . ssp. *ochroleuca*
    11' Calyx gland-dotted (sometimes cobwebby in young pls)
      19. Infl gen showy; corolla 8–29 mm, throat without purple spots
        20. Corolla throat yellow or white . . . . **G. leptantha** ssp. *transversa*
        20' Corolla throat yellow in lower 1/2, blue above . . . . **G. cana**
          21. Adjacent pedicels subequal, spreading
            22. Corolla throat narrowly conic, tube 2–4 × calyx . . . . ssp. *speciformis*
            22' Corolla throat cup-shaped, tube gen < 2 × calyx . . . . ssp. *triceps*
          21' Adjacent pedicels unequal, ± ascending
            23. Fr narrowly ovoid, gen < 5 mm; basal lvs cobwebby-matted . . . . ssp. *cana*
            23' Fr widely ovoid, gen > 5 mm; basal lvs cobwebby, not matted
              24. Corolla tube length = throat; stamens unequal, longest well exserted . . . . ssp. *bernardina*
              24' Corolla tube length 3–6 × throat; stamens equal, slightly exserted . . . . ssp. *speciosa*
      19' Infl not showy; corolla 4–11 mm, throat often with purple spots below each lobe (*G. inconspicua*
        complex)
        25. Longest stamens well exserted (exceeding mid-corolla lobe) . . . . **G. inconspicua**
        25' Longest stamens slightly exserted (not reaching mid-corolla lobe)
          26. Lf lobes < 1 mm wide, ascending; fr exserted from calyx, valves not detaching; seeds < 1 mm
            . . . . **G. minor**
          26' Lf lobes 1–2 mm wide, spreading; fr incl in calyx, valves detaching; seeds ± 1.5 mm
            27. Branches decumbent; corolla spots 0, lobe tips obtuse . . . . **G. malior**
            27' Branches erect to spreading; corolla gen with purple spots in throat, lobe tips acute . **G. transmontana**
1' Cobwebby hairs 0; pl scapose or not; basal rosette of lvs present or 0
  28. Corolla tube very short or merged with throat; stamens attached near corolla base or mid-throat;
    pollen yellow, rarely white . . . . (sect. *Giliastrum*)
    29. Principal lvs broad, holly-like; pl > 10 cm; corolla pink, veins in each lobe ± 20, with few inter-
      connections; calyx gland-dotted
      30. Ann; lf teeth tips spine-like, < 2 mm . . . . **G. latifolia**
      30' Per; lf teeth tips needle-like, 3–6 mm . . . . **G. ripleyi**
    29' Principal lvs narrower, not holly-like; pl gen < 9 cm; corolla yellow or white, veins in each lobe
      3–10, with 0 interconnections; calyx not gland-dotted

31. Fls in dense clusters; pl < 3 cm; lf 2–5 mm; st finely translucent-hairy; calyx membrance ciliate; corolla lobe with red spot at base . . . . ***G. maculata***

31' Fls dispersed, not in dense clusters; pl > 3 cm; lf 6–20 mm; st glandular-puberulent; calyx membrance glabrous; corolla lobe without red spot

    32. Lf narrowly linear; corolla yellow . . . . ***G. filiformis***

    32' Lf oblong or narrowly oblanceolate; corolla yellow and white (bluish)

        33. Corolla 7–9 mm, with 2 purple spots below each lobe, tube well exserted from calyx . . . ***G. campanulata***

        33' Corolla 2–7 mm, not purple-spotted, tube incl in calyx

            34. Lower lvs white-hairy; corolla 4–7 mm; seeds 10–15 per chamber . . . . ***G. inyoensis***

            34' Lower lvs gen glabrous; corolla 2–3 mm; seeds 1 per chamber . . . . ***G. tenerrima***

28' Corolla tube well defined below well expanded throat; stamens attached in upper corolla throat or gen near sinuses of lobes; pollen blue or white

    35. Pl hairy on lf, st; upper lvs lobed or toothed; pollen blue (sect. *Saltugilia*)

        36. Calyx not gland-dotted; fr narrowly ovoid, 1–2 × calyx; seeds 7–13 per chamber . . . . ***G. australis***

        36' Calyx gland-dotted; fr widely ovoid, incl in calyx; seeds 2–6 per chamber

            37. Hairs on basal lvs opaque-white, sharply bent, non-glandular; corolla throat with a purple spot below each lobe . . . . ***G. stellata***

            37' Hairs on basal lvs translucent, straight or curved, glandular; corolla throat without spots . . ***G. scopulorum***

    35' Pl glandular-puberulent; upper lvs gen entire; pollen white

        38. Lower lvs gen 2-pinnately lobed; corolla tube glandular-puberulent, lobes wavy-margined

        . . . . ***G. hutchinsifolia***

        38' Lower lvs gen toothed or 1-pinnately lobed; corolla tube glabrous, lobes not wavy-margined

            39. Tips of calyx lobes thickened, sinus membrane narrowly U-shaped, attached below lobe tip; corolla tube thread-like, well exserted from calyx

                40. Corolla 5-pointed, each lobe truncate with acute tip, throat clearly expanded . . . . ***G. leptomeria***

                40' Corolla 15-pointed, each lobe 3-toothed, throat indistinct . . . . ***G. triodon***

            39' Tips of calyx lobes not thickened, sinus membrane widely V-shaped, attached near lobe tip; corolla tube stout, incl or exserted from calyx

                41. Corolla 2–4 mm, throat gen incl in calyx; pedicels in fr recurved or spreading; fr spheric, gen < calyx

                . . . . ***G. micromeria***

                41' Corolla 4–7 mm, throat exserted from calyx; pedicels in fr nearly erect; fr ovoid, gen > calyx

                    42. Basal lvs ± prostrate-spreading, lanceolate, dentate or lobed, upper surface glabrous, lower glandular

                    . . . . ***G. lottiae***

                    42' Basal lvs semi-erect, strap-shaped, pinnately lobed, glandular on both surfaces . . . . ***G. subacaulis***

G. aliquanta A.D. Grant & V. Grant **ST**: 8–16 cm, branches spreading from base, glabrous or slightly cobwebby below, glandular in infl. **LVS**: basal in semi-erect cluster, 1–3 cm, pinnate; axis 1–3 mm wide; lobes 2–7 mm, gen > axis width. **INFL** loose; pedicels 1–11 mm in unequal pairs. **FL**: calyx 3–5 mm, glabrous or cobwebby, bright purple or purple-spotted, membrane puckered between ribs; corolla 6–12 mm, tube purple, lobes lavender, obovate, tips rounded; stamens unequal, exserted; style exserted. **FR** = or > calyx, widely ovoid, detaching at base. Rocky slopes, washes; 700–1600 m. s SN, n SnGb, n SnBr, **DMoj**; s NV.

ssp. ***aliquanta*** (p. 413) **FL**: corolla tube incl in calyx, throat spreading, tube + throat = lobes, lobes widely obovate; longest stamens gen > corolla lobes; stigmas touching or > anthers. 2*n*=18. Habitat and range of sp.; 700–1300 m. Apr–May

ssp. ***breviloba*** A.D. Grant & V. Grant (p. 413) **FL**: corolla tube exserted from calyx, throat narrow, tapered, tube + throat > 2 × lobes, lobes narrowly obovate; stamens < lobes; stigmas touching anthers. Uncommon. Habitat of sp.; 800–1600 m. **DMoj**; s NV. Mar–May

G. australis (H. Mason & A.D. Grant) V. Grant & A.D. Grant **ST**: 20–45 cm, branches ± erect, hairy below. **LVS**: basal in loose rosette, 2–7 cm, 2-pinnate, translucent-hairy; axis linear; lobes short-acuminate; upper linear, entire or lobed at base. **INFL** loose; pedicels 5–20 mm in unequal pairs, glands flat-topped. **FL**: calyx gen glabrous, green with purple marks or not, rib width < membrane, lobes acuminate; corolla 5–10 mm, 2–3 × calyx, lavender to white, tube incl, throat yellow-spotted, lobes 1–1.5 × throat; stamens, style exserted. **FR** 4–6 mm, 1–2 × calyx, narrowly ovoid. **SEEDS** 7–13 per chamber. 2*n*=18. Open, sandy areas; 600–1200 m. TR, PR, **DMtns**. ****Saltugila australis*** (H. Mason & A.D. Grant) L.A. Johnson Mar–Jun

G. brecciarum M.E. Jones Pl sometimes with skunk-like odor. **ST** 8–35 cm, densely cobwebby near base. **LVS**: basal in semi-erect cluster, 2–5 cm, 2-pinnate; lobes > axis width; upper cauline lobes gen finger-like, middle lobe widest. **INFL**: fls (and gen frs) clustered; pedicel glands dense, minute, black, stalked. **FL**: calyx lobes acute, spreading, ribs thick, densely glandular, width > membrane, membrane gen purple-tinged. **FR** 4–7 mm, widely ovoid. Open, sandy places; 500–2300 m. c&s SNH, Teh, SCoRI, TR (n edges), **DMoj**; se OR, NV, UT. Sspp. intergrade somewhat in s CA. 3 sspp. in CA.

ssp. ***brecciarum*** (p. 413) **ST**: branches spreading. **FL**: corolla 7–11 mm, tube incl, purple, throat exserted, narrowly V-shaped, purple below, yellow and purple-veined above, lobes pinkish lavender with white base; stamens, style slightly exserted. 2*n*=18. Sandy flats in open woodlands or shrubland; 1200–2300 m. s SNH, Teh, n edges SnGb, **GB, DMtns (Cottonwood Mtns)**; to se OR, NV, UT. Apr–Jun

ssp. ***neglecta*** A.D. Grant & V. Grant (p. 413) **ST** spreading to erect. **FL**: corolla showy, 10–20 mm, tube purple, gen exserted, throat widely V-shaped, lower half purple, white above, yellow spotted, lobes white, tips lavender; stamens, style long-exserted. 2*n*=18. Sandy desert flats; 650–2100 m. c&s SNH, Teh, **DMoj**. [ssp. *argusana* A.D. Grant & V. Grant] ❀ TRY. Mar–May

G. campanulata A. Gray **ST** 3–12 cm; branches spreading, gland-dotted. **LVS**: basal few, rosette 0; lower cauline narrowly oblanceolate, entire or 3–7-toothed, with white, jointed hairs; upper cauline spreading or recurved, entire, gland-dotted. **INFL**: fls 2 per st; pedicels 3–11 mm, thread-like, glandular. **FL**: calyx lobes acuminate, free nearly to base, membrane bordering lobes to near tip; corolla 7–9 mm, 2–3 × calyx, bell-shaped, yellow with white lobes, 2 purple lines below lobes, tube = 1 mm, throat

> lobes; stamens attached near base of tube, incl; style incl. **FR** 2–3 mm, < calyx. **SEEDS** 7–8 per chamber, gelatinous when wet. 2*n*=18. Open, sandy flats; 900–2100 m. **SNE**; NV **Linanthus campanulatus* (A. Gray) J.M. Porter & L.A. Johnson. ❀ TRY. May

G. cana (M.E. Jones) A.A. Heller Pl sometimes with skunk-like odor. **ST**: 9–32 cm, branches 1–several from base, cobwebby near base, gen glandular above. **LVS**: basal in rosette, 1–2-pinnate, cobwebby; axis < 3 mm wide; lobes gen ascending, toothed on both sides, teeth short-pointed or acuminate. **INFL** spreading, showy; fls many. **FL**: calyx gland-dotted, lobes acute to acuminate; corolla tube purple, throat bluish in upper half, yellow below, lobes pink; stamens exserted, to below middle of corolla lobes; style > stamens. **FR** < calyx, ovoid to spheric; valves detaching. Open, gravelly or sandy flats or washes; 800–3100 m. SnBr, s slope s SNH, **W&I**, **DMoj**; sw NV; nw AZ. ❀ TRY.

ssp. **bernardina** A.D. Grant & V. Grant **LVS**: basal densely cobwebby. **INFL**: fls in 2's or 3's; pedicels ± ascending, very unequal. **FL**: calyx cobwebby to slightly glandular; corolla 17–23 mm, tube exserted, throat widely conic, lobes 4–6 mm wide; stamens unequal, longest exserted. **FR** 5–9 mm, widely ovoid. **SEEDS** 4–6 per chamber. 2*n*=18. Habitat of sp.; 800–1500 m. n SnBr, **sw edge DMoj**. Some populations intergrade with *G. leptantha* ssp. *transversa*. Apr–May

ssp. **cana** (p. 413) **LVS**: basal gen cobwebby-matted. **INFL**: pedicels ascending, narrowly spreading, very unequal. **FL**: calyx cobwebby or glandular; corolla 14–26 mm, tube well exserted, throat narrowly conic, lobes 2–3 mm wide; stamens gen unequal, longest exserted to above middle of corolla lobes. **FR** 3–6 mm. **SEEDS** 2–4 per chamber. 2*n*=18. Rocky, sandy, or granitic slopes; 1800–3100 m. e slope s SNH, **DMtns (Coso Mtns)**. [*G. latiflora* ssp. *cosana* A.D. Grant & V. Grant] Jun–Aug

ssp. **speciformis** A.D. Grant & V. Grant (p. 413) **LVS**: basal 2–5 cm, densely cobwebby. **INFL**: fls 4–8 per bract; pedicels slender, in spreading, subequal pairs. **FL**: corolla 15–29 mm, tube 2–4 × calyx, throat narrowly conic, lobes 3–9 mm wide; stamens equal, exserted to below middle of corolla lobes. **FR** 5–8 mm, widely ovoid to spheric. **SEEDS** 4–6 per chamber. Gen basalt gravel or sand; 800–1200 m. **e DMtns**; sw NV, AZ. Apr–May

ssp. **speciosa** (Jepson) A.D. Grant & V. Grant (p. 413) **LVS**: basal 2–4 cm, densely cobwebby. **INFL**: fls 4–8 per bract; pedicels slender, spreading, unequal, nearly glabrous. **FL**: calyx glabrous or glandular; corolla 20–32 mm, tube 3–4 × throat, throat short, narrow, lobes 4–9 mm wide; stamens equal, exserted to base of lobes. **FR** 5–9 mm. **SEEDS** 10–20 per chamber. 2*n*=18. Habitat of sp.; 200–1200 m. e slope s SNH, **w edge DMoj**. Locally abundant. Intergrades with ssp. *cana*. Apr–May

ssp. **triceps** (Brand) A.D. Grant & V. Grant (p. 413) **LVS**: basal 1–5 cm, slightly cobwebby. **INFL**: fls many; pedicels slender, spreading, nearly equal. **FL**: corolla 8–23 mm, tube < 2 × calyx, throat cup-shaped, lobes ± 3 mm wide; stamens exserted to below middle of corolla lobes. **FR** 4–7 mm, widely ovoid to spheric. **SEEDS** 4–6 per chamber. 2*n*=18. Sandy flats, gen on limestone; 800–1600 m. **W&I**, **DMoj**; sw NV, sw AZ. Apr–May

G. clokeyi H. Mason (p. 413) **ST** 8–17 cm, cobwebby below middle, glandular above. **LVS** 1–3 cm, gray-green, cobwebby; basal in rosette, 1–2-pinnate, lobes short-pointed, spreading; upper cauline palmate, middle lobe widest. **INFL** loose; pedicels paired, unequal, spreading, thread-like. **FL**: calyx 2–4 mm, glabrous (or cobwebby in early fls), lobe short-pointed, tip thick; corolla 4–6.5 mm, tube gen incl, throat exserted, white with 5 yellow spots at base, lobes 1–3 mm, acute, white, gen blue-streaked; stamens, style slightly exserted. **FR** 3–6 mm, spheric, detaching at base. 2*n*=18. Open, rocky slopes or sandy washes, gen on limestone; 400–2000 m. **e DMoj**, **DMtns**; to Colorado, NM.

G. diegensis (Munz) A.D. Grant & V. Grant **ST** 10–40 cm, glabrous, glaucous below middle, glandular above. **LVS**: basal in prostrate rosette, 1–7 cm, strap-shaped, toothed or pinnate, lobes spreading, cobwebby; cauline shorter, ±clasping, lanceolate,

serrate. **INFL** cluster in fl, loose in fr; pedicels unequal. **FL**: calyx 3–5 mm; corolla 8–12 mm, tube incl or slightly exserted, purple, lower throat purple, upper yellow, lobes 2–5 mm, lavender or white; stamens unequal, exserted to mid-lobe, pollen blue; style gen = or > stamens. **FR** 4–7 mm, < calyx. **SEEDS** many. 2*n*=18. Sandy areas, open forest or shrubland; 500–2200 m. SnGb, SnBr, PR, **s edge DMoj**, **w edge DSon**; Baja CA. Apr–Jun

G. filiformis A. Gray Pl glabrous or sparsely gland-dotted. **ST**: 3–15 cm, branches few–many, thread-like, glaucous. **LVS** 1–3 cm, linear, entire; basal rosette 0 or ephemeral; cauline spreading. **INFL**: fls gen 2 per st; pedicels 2–16 mm. **FL**: calyx 2–4 mm, tube < 1 mm, lobes acuminate, spreading in fr, margins membranous to near tip; corolla 4–7 mm, yellow, lobes 3–5 mm, oblong; stamens attached near base; stamens, style well exserted, < corolla lobes. **FR** 2–4 mm. **SEEDS** many, gelatinous when wet. 2*n*=18. Open, rocky canyon slopes or washes; 300–1800 m. **W&I**, **DMoj**; to UT, AZ. **Linanthus filiformis* (A. Gray) J.M. Porter & L.A. Johnson Mar–May

G. hutchinsifolia Rydb. Pl densely glandular-puberulent (studded with sand grains), glandular-hairy below, with skunk-like odor. **ST** 5–40 cm, branched from base or above. **LVS**: lower 2–8 cm, in semi-erect cluster, 1–2-pinnate, axis linear, lobes > 2 × axis width; upper linear, entire. **INFL**: fls gen 2 per st; pedicels unequal. **FL**: calyx 3–4 mm, lobes linear, membrane attached well below lobes, U-shaped between lobes in fr; corolla 7–14 mm, tube well exserted, glandular-puberulent, throat green-spotted, yellow above, lobes 2–4 mm, gen wavy-margined, lavender; stamens, style slightly exserted. **FR** 3–6 mm, = or > calyx. **SEEDS** many, not gelatinous when wet. 2*n*=18. Sandy or gravelly flats, slopes, dunes; 400–1800 m. **GB**, **DMoj**; to UT, AZ. **Aliciella hutchinsifolia* (Rydb.) J.M. Porter ❀ TRY. Apr–May

G. inconspicua (Smith) Sweet (p. 413) **ST**: 8–32 cm, branches ascending or spreading, cobwebby below infl, black-glandular above. **LVS**: basal in rosette, 1-pinnate, cobwebby, lobes 2–10 mm, = or > axis width, linear (or rounded and short-pointed), entire or toothed, ascending. **INFL**: cluster in fl, loose in fr; fls 2–4 per st; pedicels unequal. **FL**: calyx glabrous or gland-dotted (or cobwebby in early fls); corolla 6–11 mm, 3–4 × lobes, tube and throat gen yellow, lobes lavender with purple spot at base; stamens and style exserted; longest stamens > middle of corolla lobe. **FR** 5–8 mm, < calyx, oblong-ovoid; valves detaching. **SEEDS** 4–6 per chamber. 2*n*=36. Rocky or sandy sagebrush slopes, washes; 1200–2000 m. **GB**; to WA, ID, NV. Apr–Jun

G. inyoensis I.M. Johnston (p. 413) **ST**: 3–10 cm, branches spreading, finely glandular, white-hairy below. **LVS** white-jointed-hairy; basal few, 4–8 mm, oblanceolate to obovate, entire or toothed; cauline 3–6 mm, upper entire, spreading or recurved. **INFL**: fls 2 per st; pedicels 4–8 mm, thread-like, spreading. **FL**: calyx 2–4 mm, tube < 1 mm, membrane bordering lobes to near tip; corolla 4–7 mm, tube and throat yellow, incl in calyx, lobes white, obovate, short-pointed; stamens, style slightly exserted. **FR** 2–4 mm. **SEEDS** 10–15 per chamber, gen not gelatinous. Common. Open, sandy flats in pine forest or sagebrush shrubland; 1900–2600 m. s SNH, **SNE**; NV. [var. *breviuscula* Jepson] **Linanthus inyoensis* (I.M. Johnston) J.M. Porter & L.A. Johnson Apr–Jul

G. latiflora (A. Gray) A. Gray BROAD-FLOWERED GILIA Pl ± scapose. **LVS**: basal in prostrate rosette, 2–7 cm, cobwebby, strap-shaped, toothed or lobed, lobes spreading; cauline shorter, clasping, entire or lobed at base, tapered. **INFL** gen loose; pedicels unequal. **FL** showy, fragrant; calyx slightly glandular, or early fls cobwebby; corolla 10–35 mm, tube purple, upper throat, base of lobes white, tips lavender; stamens exserted, < lobes; style > stamens. **FR** gen > calyx. Gen deep sandy soils; 700–1500 m. SCoRI, n TR, **s&w DMoj**. Forms showy displays. Sspp. variable within populations, intergrading where ranges overlap. 4 sspp. in CA.

ssp. **davyi** (Milliken) A.D. Grant & V. Grant (p. 413) **ST** 10–30 cm, glabrous, glaucous below middle, or cobwebby. **INFL**: cluster in fl, loose in fr. **FL**: calyx 4–7 mm; corolla 18–24 mm,

3× lobe length, tube exserted, throat gen long-tapered, purple to middle or higher. **FR** 6–9 mm. Common. Open, sandy flats; 700–1200 m. SCoRI, **s&w DMoj**. [*G. latiflora* ssp. *excellens* (Brand) A.D. Grant & V. Grant] ✿ DRN,SUN:**7–12,14–16**,17,**18–23**,24 Mar–May

ssp. *elongata* A.D. Grant & V. Grant (p. 413) **ST** 13–32 cm, cobwebby below middle. **FL:** calyx 4–6 mm; corolla 21–35 mm, 4–6 × lobe length, tube > 3 × calyx, white-veined, throat narrowly tapered, yellow, white. **FR** 6–9 mm. Open, sandy slopes; 700–1200 m. **w DMoj (Rand, El Paso ranges, Black Rock Hills)**. ✿ TRY. Apr–May

ssp. *latiflora* (p. 413) **ST** 10–33 cm, glabrous, glaucous below middle. **INFL** loose. **FL:** calyx 4–5 mm; corolla 15–22 mm, tube slightly exserted, white-veined, throat widely expanded, yellow at base. **FR** 6–9 mm. 2*n*=18. Common. Open, sandy flats; 700–1100 m n base SnGb and SnBr, **sw DMoj**. ✿ TRY. Apr–May

G. latifolia S. Watson BROAD-LEAVED GILIA Ann. Pl strong-scented. **ST** 10–30 cm, branches spreading, glandular-hairy below. **LVS:** basal in rosette, petioled, ovate, holly-like, appressed-hairy; upper lvs reduced, needle-like. **INFL:** many-fld loose cluster, gland-dotted; pedicels spreading, longer in fr. **FL:** calyx fused in lower 1/2, lobes fine-pointed; corolla 7–11 mm, tube white, lobes bright pink above, whitish or dull pink below; stamens attached in lower throat, unequal, longest slightly exserted, pollen white. **FR** equal or > calyx, ovoid. **SEEDS** many, not gelatinous. 2*n*=36. Common. Rocky slopes, washes; < 1800 m. **W&I, D**; to AZ, NV, UT, Baja CA. Like *G. ripleyi*. **Aliciella latifolia* (S. Watson) J.M. Porter ssp. *latifolia* (ssp. *imperialis* (S.L. Welsh) J.M. Porter restricted to UT) Apr–May

G. leptantha Parish **ST** spreading or erect, cobwebby below middle, glandular above. **LVS:** basal in rosette, 1-pinnate, cobwebby, axis 1–2 mm wide, linear, lobes toothed on both sides, teeth short-pointed or acuminate. **INFL** clustered or loose with 2–6 fls above a bract; pedicels unequal, barely spreading. **FL:** calyx slightly glandular (or cobwebby in early fls); corolla tube yellow or purple with yellow veins, throat gen yellow, lobes pink to lavender; stamens unequal, longest gen > corolla lobes. Open, rocky or sandy areas; 700–2800 m. s SNH, SCoR, n slope SnGb, SnBr, **sw edge DMoj**. 4 sspp. in CA.

ssp. *transversa* A.D. Grant & V. Grant **ST** 16–40 cm, densely glandular above rosette. **LVS** basal, lower cauline 2–8 cm. **FL:** glandular or cobwebby; corolla 13–17 mm, tube 1–1.5 × calyx, throat widely expanding, lobes 3–6 mm wide, obovate; stamens unequal, longest well exserted, < corolla lobes; style > stamens. **FR** 3–8 mm, < or = calyx, ovoid. 2*n*=18. Rocky or sandy soil, gen near streams; 900–1800 m. n slope SnGb, SnBr, **sw edge DMoj**. ✿ TRY Apr–Jul

G. leptomeria A. Gray (p. 413) GREAT BASIN GILIA **ST:** 7–23 cm, branches spreading, 1–several, thread-like, glandular-puberulent. **LVS:** basal in rosette, 1–6 cm, lanceolate or strap-shaped, toothed or lobed, lobes rounded and short-pointed, gen < axis width; cauline linear, entire. **INFL:** fls 1–3 per st. **FL:** calyx 2–3 mm, lobe tips thickened, membrane U-shaped; corolla 4–7 mm, tube 1.5–3 × calyx, thread-like, purple, throat yellow, lobes truncate, short-pointed, white, outside purple; stamens, style slightly exserted, pollen white. **FR** 3–5 mm, = calyx, narrowly ovoid. **SEEDS** many, not gelatinous. 2*n*=34,36. Common. Open, sandy or rocky areas; 800–2100 m. **GB, DMtns (uncommon)**; to OR, ID, Colorado, NM. **Aliciella leptomeria* (A. Gray) J.M. Porter Apr–Jun

G. lottiae A.G. Day (p. 413) **ST** 5–43 cm, branches gen spreading from base, glandular-puberulent. **LVS:** basal in flat rosette, 2–11 cm, widely strap-shaped, fleshy, serrate or pinnate, lobes spreading; cauline linear, entire. **INFL** clustered, nodding, becoming erect. **FL:** calyx membrane V-shaped, attached near lobe tip; corolla 6–7 mm, tube slightly exserted, stout, throat > tube, white, yellow-spotted, lobes lanceolate to ovate, pink, lavender, or white; stamens, styles slightly exserted. **FR** 3–5 mm, < 1.5 ×

calyx, ovoid, tip pointed. **SEEDS** many, not gelatinous. 2*n*=50. Sandy soils, sagebrush scrub; 400–2100 m. **GB, e DMtns (Clark Mtns)**; to WA, ID, UT. **Aliciella lottiae* (A.G. Day) J.M. Porter

G. maculata Parish (p. 413) LITTLE SAN BERNARDINO MOUNTAINS GILIA Pl hairy. **ST** 1–3 cm. **LF** 2–5 mm, entire, gen oblong or oblanceolate. **INFL:** cluster, dense; fls ± sessile. **FL:** calyx tube < 1 mm, lobe margins widely membranous, membrane ciliate; corolla 3–5 mm, white, lobes 1–2 mm, wavy, recurved, red-spotted at base; stamens, style slightly exserted; pollen yellow. **FR** < calyx, widely ovoid. **SEEDS** 6–16, not gelatinous. 2*n*=18. RARE. Sandy flats; 900–1100 m. **s DMtns (Little San Bernardino Mtns)**. [*Linanthus m.* (Parish) Milliken] Threatened by development. Apr–May

G. malior A. G. Day & V. Grant **ST** 5–20 cm; lower branches decumbent, spreading, cobwebby below infl, gland-dotted above. **LVS:** lower in basal rosette, 1-pinnate, axis, lobes 1–2 mm wide, linear, cobwebby, lobes spreading, sometimes toothed; upper palmate to entire above. **INFL:** cluster in fl, loose in fr; pedicels unequal. **FL:** calyx gen slightly glandular; corolla 6–11 mm, tube and throat purple or partly yellow above, throat tapered, lobes lavender, obovate, tip obtuse; stamens, style slightly exserted. **FR** incl in calyx; tip rounded; valves detaching. 2*n*=36. Open, sandy, rocky flats; 700–2000 m. Teh, s SnJV, SCoRI, SNE, **DMoj**; w NV, s OR. Intermediate between *G. minor*, *G. aliquanta*.

G. micromeria A. Gray (p. 413) **ST** 4–14 cm; branches many, spreading, thread-like, slightly puberulent. **LVS:** basal in rosette, 1–6 cm, pinnate, lobes spreading, length 0.5–1 × axis width; upper linear, entire. **INFL** loose, nodding in bud; pedicels spreading or recurved in fr. **FL:** calyx ± 2 mm, membrane attached near lobe tips, V-shaped; corolla ± 3 mm, tube gen incl, throat pale yellow, lobes ovate, entire, white or lavender; stamens, style slightly exserted; pollen white. **FR** ± 3 mm, ± = calyx, spheric. **SEEDS** 2–8 per chamber, not gelatinous. 2*n*=18. Uncommon. Rocky or sandy soil; 1200–1670 m. **GB** (Modoc, **Inyo cos.**); to OR, Colorado. **Aliciella micromeria* (A. Gray) J.M. Porter Apr–Jun

G. minor A.D. Grant & V. Grant (p. 413) **ST:** lower branches decumbent, 6–20 cm, cobwebby below infl. **LVS:** basal in rosette, 1-pinnate, axis and lobes < 1 mm wide, linear, cobwebby, lobes ascending; upper glandular. **INFL** clustered in fl, loose in fr; pedicels very unequal. **FL:** calyx glandular (or cobwebby in early fls); corolla 4–8 mm, ± 2 × calyx, tube incl, purple, throat purple or yellow with purple-veins, tapered, lobes ovate, acute, lavender; stamens, style slightly exserted. **FR** 2 × calyx, narrowly ovoid; valves partly separating, not detaching. 2*n*=18. Firm sand, gen at base of shrubs; 300–1100 m. SCoR, **DMoj**; AZ. Mar–Apr

G. modocensis Eastw. **ST** 10–34 cm; branches gen several, spreading from below, slightly cobwebby or glandular near base. **LVS:** basal in rosette, strap-shaped, pinnate, lobes short, spreading, toothed; cauline clasping, gen deeply lobed. **INFL** clustered in fl, loose in fr; pedicels glandular; glands black. **FL:** calyx gland-dotted, rib width > membrane, lobes erect or spreading in fr; corolla 7–10 mm, tube incl, purple, throat yellow, or midveins purple, lobes 2–3 mm, lavender; stamens, style, slightly exserted. **FR** < or = calyx. 2*n*=36. Open, rocky areas, pinyon/juniper woodland; 400–2300 m. Teh, WTR, SnBr, **GB, DMtns**; OR, ID, NV. Apr–Jun

G. ochroleuca M.E. Jones **ST:** branches gen spreading from below, 5–30 cm, glabrous or cobwebby, gen glandular in infl. **LVS:** lower in rosette, 1–2-pinnate or entire, axis, lobes gen linear, ascending; cauline palmate; infl lvs gen entire. **INFL:** pedicels thread-like, widely spreading, gen in subequal pairs, glandular. **FL:** calyx 2–4 mm, ribs and lobes linear, membrane colorless; corolla tube purple, throat well expanded, blue in upper half, yellow below, lobes pink to white; stamens slightly exserted. **FR** 2–5 mm; valves detaching. **SEEDS** 3–15. Open, gen sandy places; 200–2500 m. SCoRI, TR, PR, **W&I, w D**; Baja CA. 4 sspp. in CA.

ssp. *bizonata* A.D. Grant & V. Grant   Pl gray-green. **ST** glabrous below calyces, gen glabrous, glaucous below infl. **LF** cobwebby or glabrous, 1–2-pinnate; lobes 1–15 mm. **FL**: calyx glabrous, rarely cobwebby, lobes acute; corolla 8–14 mm, tube gen incl, < throat; stamens < style. **FR** spheric. 2*n*=18. Common. Flats; 300–2200 m. SCoRI, TR (n slope), **w DMoj**.

ssp. *exilis* (A. Gray) A.D. Grant & V. Grant (p. 413) Pl yellow -green. **ST**: branches sub-erect, gen cobwebby below. **LF** cobwebby, entire or 1-pinnate. **INFL** glabrous or glandular. **FL**: calyx slightly cobwebby, lobes acute to acuminate; corolla 8–11 mm, tube slightly exserted, > throat; style slightly exserted; stigmas near or just above anthers. **FR** spheric to ovoid. 2*n*=18. Common. Flats; 200–1800 m. s SnBr, PR, **w D**; Baja CA.

ssp. *ochroleuca* (p. 413)   Pl gray-green. **ST**: branches spreading, < 30 cm, glabrous, glaucous below infl, glabrous beneath fls. **LF** slightly cobwebby, 1–2-pinnate, lobes 3–10 mm, ascending; infl lvs palmate. **FL**: calyx glabrous; corolla 4–6 mm, tube incl, throat = tube; style slightly exserted, stigmas near anthers. **FR** spheric. 2*n*=18. Desert sand; 700–1500 m. **W&I, w DMoj**. Apr–Jun

G. ophthalmoides Brand (p. 417)   **ST**: branches sub-erect, 15–30 cm, densely cobwebby below middle, glandular above. **LVS**: lower in rosette, densely cobwebby, 1–2-pinnate, axis linear, lobes spreading, gen toothed (or lobed) on both sides. **INFL**: pedicels thread-like, unequal. **FL**: calyx glabrous, lobes acuminate, reddish; corolla 7–12 mm, tube 1.5–2 × calyx, slender, purple, throat narrow, long-tapered, bright yellow, lobes 1–3 mm, pink; stamens, style slightly exserted. **FR** 4–6 mm, gen < calyx; valves detaching. **SEEDS** 3–4 per chamber. 2*n*=36. Open, rocky soil, gen pinyon/juniper woodland; 1100–2600 m. **n SNE, W&I, DMtns**; to Colorado, AZ. ❀ TRY.   May–Jun

G. ripleyi Barneby (p. 417)   RIPLEY'S GILIA   Per. **ST** 10–30 cm, densely glandular-hairy. **LVS**: basal, lower in cluster, 3–6 cm, glandular-hairy, obovate, veins raised beneath, holly-like, tips of teeth needle-like; upper lvs < 1 cm, gland-dotted, needle-like. **INFL** loose; fls many, gland-dotted. **FL** opening in evening; calyx 4–6 mm, tube 2 mm, lobes acuminate, margins membranous to near tip; corolla 7–10 mm, tube 1/2 corolla, white, throat white, lobes 3–5 mm, ovate, both surfaces pink, veins many; stamens attached near base, incl or longest slightly exserted; style slightly exserted. **FR** 3–4 mm. **SEEDS** many, not gelatinous. 2*n*=18. RARE in CA. Limestone cliffs; 900–1400 m. **n DMtns (Inyo Co.)**; NV (where rare). *Aliciella ripleyi* (Barneby) J.M. Porter   May–Jun

G. salticola Eastw. (p. 417)   **ST**: branches spreading, 4–20 cm, densely cobwebby. **LVS**: basal in rosette, densely cobwebby, 1–2-pinnate, axis and lobes 2–3 mm wide, entire or toothed. **INFL** loose-clustered, black-gland-dotted. **FL**: calyx glabrous or cobwebby, lobes acuminate, outcurved in fr, membrane purple-spotted or colorless; corolla 6–15 mm, tube incl, yellow, throat yellow, lobes 3–6 mm, gen > throat, bright pinkish lavender; stamens, style exserted, > lobes. **FR** 4–6 mm, < calyx. **SEEDS** 10–15. 2*n*=18. Open, volcanic or granitic areas; 1500–2700 m. n&c SNH, w MP, **nw SNE (Mono Co.)**; w NV. [*G. leptantha* Parish ssp. *s.* (Eastw.) A.D. Grant & V. Grant] May–Jun

G. scopulorum M.E. Jones (p. 417)   **ST**: branches ± erect, 10–30 cm, glandular-hairy. **LVS** glandular-hairy; hairs translucent; lower lvs in semi-erect rosette, 1–2-pinnate, axis linear below, winged above, merging with lobes, lobes toothed, teeth short-pointed; upper minute, 2-toothed. **INFL** loose; pedicels thread-like, spreading in unequal pairs, gland-dotted, glands black, flat-topped, short-stalked. **FL**: calyx slightly glandular, ribs green, width > membrane; corolla 9–17 mm, 2–4 × calyx, tube purple, throat < tube, yellow, lobes ovate, acute, lavender to pink; stamens, style slightly exserted. **FR** 5–6 mm, < or = calyx, widely ovoid. **SEEDS** many. 2*n*=18,36. Semi-shaded, rocky ravines; < 1200 m. **s SNE, e DMoj, DSon**; to UT, AZ. Apr–May

G. sinuata Benth. (p. 417)   **STS** 1 or several spreading from base, 13–34 cm, glabrous, glaucous below middle. **LVS**: basal in prostrate rosette, cobwebby, pinnate, lobes spreading, gen toothed; cauline reduced, clasping with expanded base, toothed or entire, tapering to tip. **INFL** glandular; fls in clusters; pedicels longer in fr, unequal, 1–9 mm. **FL**: calyx 3–5 mm; corolla 7–12 mm, tube exserted, purple, white-veined, lobes 2–3 mm, lavender, pink, or white; stamens, style slightly exserted; pollen white. **FR** 4–7 mm. 2*n*=36. Open, sandy flats; 150–1800 m. **GB, DMoj**; to WA, ID, Colorado, AZ. Apr–Jun

G. stellata A.A. Heller (p. 417)   STAR GILIA   Pl hairy below; hairs white, sharply bent. **ST**: branches gen several from base, 10–40 cm, gland-dotted above. **LVS**: basal in rosette, grayish, 1–2-pinnate, axis linear, lobes toothed on both sides, teeth short-pointed; upper cauline minute, entire or 2-toothed. **INFL** loose; fls 4–8 per st; pedicels spreading, in unequal pairs, glabrous or gland-dotted. **FL**: calyx hairy or glandular, glands black, stalked; corolla funnel-shaped, tube incl or slightly exserted, throat yellow with purple spots, lobes pink or white; stamens, style slightly exserted. **FR** 5–7 mm, widely ovoid. **SEEDS** 3–6 per chamber. 2*n*=18. Common. Sandy desert flats, washes; < 1700 m. **W&I, D**; to UT, AZ, Baja CA. Mar–May

G. subacaulis Rydb. (p. 417)   **ST** 5–30 cm, glandular-puberulent; branches gen spreading from base. **LVS**: basal 1–7 cm, in ± erect cluster, glandular-puberulent, pinnate, axis narrow, lobes widely obtuse to pointed; cauline entire, linear. **INFL**: cluster, nodding, becoming erect. **FL**: calyx lobe short; corolla 4–7 mm, tube incl, stout, throat > tube, well expanded, yellow-spotted, lobes ovate, white, lavender-streaked below; stamens exserted; style gen exserted. **FR** 3–5 mm, 1.5–2 × calyx, ovoid; tip pointed. 2*n*=16. Sandy soils, sagebrush scrub, pinyon/juniper woodland; 500–2500 m. **SNE, DMoj**; to OR, WY, AZ *Aliciella subacaulis* (Rydb.) J.M. Porter & L.A. Johnson

G. tenerrima A. Gray (p. 417)   **ST** spreading, 5–22 cm, thread-like, glandular-hairy below, leafy. **LVS** 1–2 cm, glabrous, linear to narrowly lanceolate, entire or 1–2-lobed; basal few. **INFL** loose; fls 2–3 per st; pedicels gland-dotted, spreading or recurved in fr. **FL**: calyx tubular below, lobe margins membranous to tip; corolla 2–3 mm, 2 × calyx, tube, throat white, incl, lobes bluish; stamens, style slightly exserted. **FR** 1–3 mm, spheric; valves spreading, detaching. **SEED** 1 per chamber, gelatinous when wet. 2*n*=36. Creeks, rocky canyon slopes; 500–2800 m. SNH, **SNE**; to OR, WY, UT. . *Lathrocasis tenerrima* (A. Gray) L.A. Johnson

G. transmontana (H. Mason & A.D. Grant) A.D. Grant & V. Grant   **ST**: branches ascending or spreading, 10–32 cm, slightly cobwebby below middle. **LVS** ± cobwebby; basal in rosette, 1-pinnate, axis and lobes linear, lobes 3–11 mm, spreading, entire or few-toothed. **INFL** loose; fls 2–3 per st; pedicels unequal, glandular. **FL**: calyx glabrous or slightly glandular; corolla 5–8 mm, tube incl, purple, lower throat yellow, upper throat white with purple spots below lobes, lobes > throat, ovate, acute, lavender; stamens, style slightly exserted. **FR**: tip pointed or round; valves detaching. 2*n*= 36. Rocky or sandy desert slopes or washes; 600–2000 m. **DMoj**; to UT. Mar–May

G. triodon Eastw. (p. 417)   Pl glandular-puberulent. **ST** gen 1, branched above; branches spreading, 5–13 cm, thread-like. **LVS**: basal in rosette, 5–20 mm, oblanceolate to obovate, toothed or lobed, lobes spreading, short-pointed, length < axis width; upper lvs linear, entire. **INFL** loose; fls 1–3 per st. **FL**: calyx 2–3 mm, lobes short-pointed, tips thickened, membrane U-shaped; corolla 5–7 mm, tube 1.5–2 × calyx, thread-like, purple, throat narrower than tube, lobes 3-toothed, base yellow, tips white; stamens, style slightly exserted; pollen white. **FR** 3–4 mm, narrowly ovoid. **SEEDS** many, not gelatinous. 2*n*=18. Open, sandy or rocky areas, sagebrush scrub, juniper woodland; 1200–1700 m. **e DMtns (Clark Mtns)**; to Colorado, NM. Like *G. leptomeria* exc corolla throat, lobes, lf shape. *Aliciella triodon* (Eastw.) Brand

GYMNOSTERIS

Dieter H. Wilken

Ann. **ST** erect, gen solitary, glabrous. **LVS** 0. **INFL** head-like; bracts lf-like, in involucre; fls 1–5. **FL**: calyx urn-shaped, scarious, abruptly awned, puberulent; corolla salverform, white to lavender, throat gen yellow; stamens attached in throat, filaments 0. **FR** ovoid. **SEEDS** 1–5 per chamber, angled. 2 spp.: w US. (Greek, naked st) [Wherry 1944 Am Mid Nat 31: 216–231] Self-pollinated.

G. parvula A.A. Heller (p. 417) **ST** < 7 cm. **INFL**: bracts 2–13 mm, lanceolate to ovate, glabrous. **FL**: calyx < 4 mm; corolla < 7 mm, tube = calyx, lobes gen oblong, acuminate, white or tinged pink. Gravelly, sandy areas, gen in meadows or wet depressions in shrublands; 2400–3700 m. n&c SNH, **SNE**; to OR, WY, Colorado. *G. nudicaulis* (Hook. & Arn.) E. Greene, with corolla 8–12 mm, occurs in w NV, may be expected in SNE. May–Jun

IPOMOPSIS SCARLET GILIA

Dieter H. Wilken

Ann, per. **ST** gen branched at base. **LVS** alternate, simple, gradually smaller upward, entire to pinnately or palmately lobed; lobes gen small-pointed at tip. **INFL**: clusters, lateral and 1-sided or terminal and open to head-like, rarely solitary and pedicelled in lower axils. **FL**: calyx gen bell-shaped, tube and sinuses membranous, lobes gen small-pointed at tip; corolla bell-shaped or salverform, radial or bilateral, white to red or lavender. **SEEDS** slender, angled, slightly winged, white to light brown. 30 spp.: w N.Am, se U.S., s S.Am. (Greek, striking appearance) [Grant & Wilken 1988 Bot Gaz 149:443–449] Per spp. cross-pollinated, ann spp. gen self-pollinated. Distinguished from *Gilia* by infl, lf morphology, chromosome number, flavonoid chemistry.

1. Infl of lateral clusters
  2. Corolla white to pink or lavender, tube 25–45 mm; calyx lobes long-tapered, glandular, hairy; n SNE
    . . . . *I. tenuituba*
  2' Corolla red, tube 10–30 mm; calyx lobes narrowly acuminate, glabrous or glandular; SNE, DMtns
    3. Corolla tube 20–30 mm; stamens exserted; widespread . . . . *I. aggregata* ssp. *formosissima*
    3' Corolla tube 10–20 mm; stamens included; DMtns . . . . *I. arizonica*
1' Infl terminal, open, and few-fld, or head-like and bracted, or fls solitary in lower axils
  4. Corolla 15–28 mm, red . . . . *I. tenuifolia*
  4' Corolla gen < 10 mm, white to deep pink
    5. Ann
      6. Lf entire to slightly toothed, linear to elleptic; lower fls 1–3 in axil . . . . *I. depressa*
      6' Lf gen pinnately lobed, gen oblanceolate; fls terminal . . . . *I. polycladon*
    5' Per . . . . *I. congesta*
      7. Lf pinnately lobed . . . . ssp. *congesta*
      7' Lf palmately lobed
        8. Lf lobes gen 3; st < 1 dm; alpine . . . . ssp. *montana*
        8' Lf lobes gen 5; st 1–3 dm; montane . . . . ssp. *palmifrons*

I. aggregata (Pursh) V. Grant   Per, dying after flowering once. **ST** erect, glabrous or glandular, slightly hairy. **LVS**: basal 3–5 cm, pinnately 9–11-lobed, withered at fl; cauline 5–7-lobed, glabrous to puberulent. **INFL** 1-sided; clusters lateral, compact; fls 1–7. **FL**: calyx lobes deltate to acuminate; corolla tube 20–30 mm, salverform, gen red with yellow mottling in throat and bases of lobes, lobes acute to acuminate; stamens attached at different levels, exserted; style exserted. Openings in shrublands, woodlands; 1100–3300 m. KR, CaRH, n&c SNH, **GB**; to B.C., Colorado, Mex. 8 sspp., 2 in CA, ssp. *aggregata* in Rocky Mtns.

  ssp. *formosissima* (E. Greene) Wherry (p. 417) **LF**: lobes acute. **INFL**: fls 3–7. **FL**: calyx lobes 3–4 mm, narrowly acuminate; stamens exserted, pollen white, light yellow, or slightly bluish. Shrublands, montane; 1100–2500 m. KR, CaRH, n&c SNH, **GB**; to WA, Colorado, n Mex. ❀ DRN:1–4,6,15–17 & IRR:7,14,18;DFCLT. Jun–Sep

I. arizonica (E. Greene) Wherry (p. 417) Per, dying after flowering once. **ST** erect, glabrous or glandular, slightly hairy. **LVS**: basal 3–5 cm, pinnately 7–11-lobed; cauline hairy. **INFL** 1-sided; fls 5–13 on upper third of axis. **FL**: calyx lobes 1–3 mm, acuminate, glabrous to glandular-puberulent; corolla 11–20 mm, salverform, red, lobes 5–10 mm, flared to reflexed; stamens attached at same level, incl; style incl. Open, sandy to rocky areas in canyons; 1500–3100 m. **DMtns**; s NV, n AZ. [*I. aggregata* ssp. *a.* (E. Greene) V. Grant & A.D. Grant] ❀ TRY.

I. congesta (Hook.) V. Grant   Per. **ST** decumbent to erect, 1–3 dm, glabrous to densely puberulent. **LF** 1–4 cm, gen hairy, entire or pinnately to palmately 3–5-lobed. **FL**: calyx 3–5 mm; corolla gen > calyx, tube gen yellow, lobes gen oblong, white; stamens attached at same level, exserted; style incl. Dry, open shrublands, woodlands, alpine; 1200–3700 m. KR, n CaR, n&c SNH, MP, **W&I**; to OR, SD, Colorado. 7 sspp., esp GB.

  ssp. *congesta* (pl. 78) **ST** 1–2 dm, branched upward. **LVS**: lower pinnately 3–5-lobed, hairy. Valleys, basins; 1200–2150 m. KR, n CaR, n&c SNH, MP, **n SNE**; to OR, w Great Plains. ❀ TRY. Jun–Jul

  ssp. *montana* (Nelson & Kenn.) V. Grant Pl cespitose. **ST** < 1 dm. **LF** gen palmately 3-lobed, hairy. Alpine; 2100–3700 m. n&c SNH, **GB**; to OR, NV. Jul–Sep

  ssp. *palmifrons* (Brand) A.G. Day (p. 417) **ST** 1–3 dm. **LF** palmately 5-lobed, sparsely hairy. Gen woodlands, shrublands of foothills, montane; 1500–2750 m. **GB**; to OR, ID. Intergrades with ssp. *montana* in n CaR, MP.

I. depressa (A. Gray) V. Grant   Ann. **ST** decumbent, < 1 dm, glandular-puberulent. **LVS**: cauline < 2 cm, linear to elliptic,

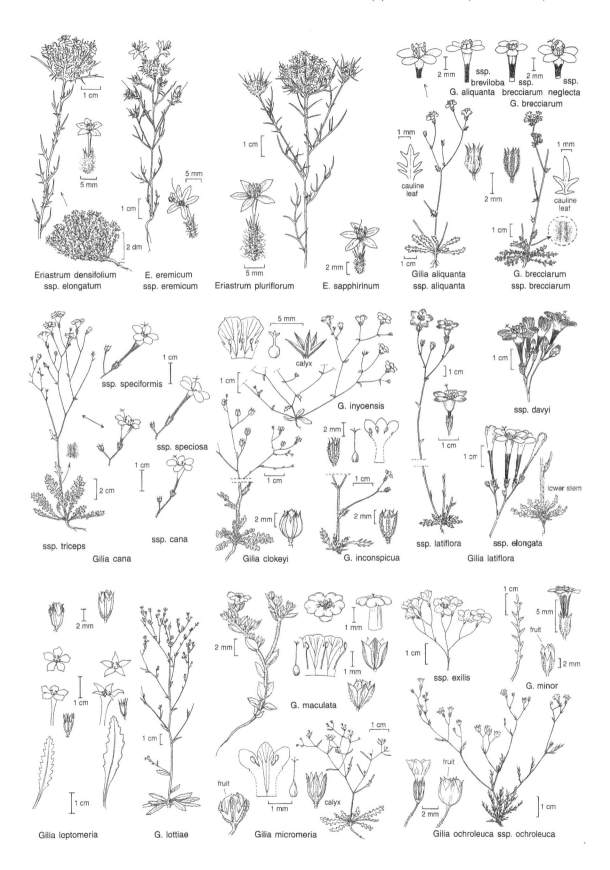

Eriastrum densifolium
ssp. elongatum

E. eremicum
ssp. eremicum

Eriastrum pluriflorum

E. sapphirinum

ssp.
breviloba
G. aliquanta

ssp.
brecciarum

ssp.
neglecta

G. brecciarum

cauline
leaf

Gilia aliquanta
ssp. aliquanta

cauline
leaf

G. brecciarum
ssp. brecciarum

ssp. speciformis

ssp. speciosa

ssp. triceps

ssp. cana

Gilia cana

calyx

G. inyoensis

Gilia clokeyi

G. inconspicua

ssp. davyi

ssp. latiflora

lower stem

ssp. elongata

Gilia latiflora

Gilia leptomeria

G. lottiae

G. maculata

fruit

calyx

Gilia micromeria

ssp. exilis

fruit

G. minor

fruit

Gilia ochroleuca ssp. ochroleuca

acute, entire to toothed; upper crowded below infl. **INFL** compact, terminal; bracts lf-like, hairy to canescent; fl 1–3 per axil. **FL:** calyx lobes acuminate, canescent; corolla 5–7 mm, lobes < 2 mm, white; stamens attached at same level, unequal, slightly exserted; style incl. Sandy soils of gentle slopes, flats; 1000–1550 m. **SNE, n DMoj;** to UT. ***Loeseliastrum depressum** (A. Gray) J.M. Porter & L.A. Johnson Apr–Jun

I. polycladon (Torrey) V. Grant (p. 417) Ann. **ST** decumbent to prostrate, < 1 dm, glandular-puberulent. **LVS** < 2 cm, entire to toothed or pinnately 5–7-lobed; basal, cauline crowded below infl, terminal lobe = axis width, glabrous above, puberulent below. **INFL** head-like, terminal; bracts oblanceolate, lobed, pointed at tip. **FL:** corolla 3–6 mm, salverform, lobes < 2 mm; stamens attached at same level, filaments = anthers; style incl. **FR** < 5 mm, ellipsoid. Sandy, gravelly soils; 900–2150 m. **SNE, DMoj;** to TX, n Mex. Apr–Jun

I. tenuifolia (A. Gray) V. Grant (p. 417) SLENDER-LEAVED IPOMOPSIS Per. **ST** 1–4 dm, sparsely puberulent. **LVS:** basal 5–35 mm, pinnately 3–5-lobed, lobes remote, linear; cauline en-

tire. **INFL** open, terminal or in upper axils; fls 1–7. **FL:** calyx lobes < 3 mm; corolla 15–28 mm, bell-shaped, slightly bilateral, tube 5–10 mm, throat 6–11 mm, lobes 5–7 mm, oblong, notched, red with white mottling on lobes, throat; stamens, style exserted. RARE in CA. Gravelly to rocky slopes, canyons; 100–1200 m. se PR (s San Diego Co), **DSon;** Baja CA. [*Loeselia t.* A. Gray] Mar–May

I. tenuituba (Rydb.) V. Grant (p. 417) Per, dying after flowering once. **ST** erect, glabrous or glandular, slightly hairy. **LVS:** basal 3–6 cm, pinnately 9–17-lobed, withered at fl; cauline gen puberulent. **INFL** 1-sided; fls 3–7, lower well spaced on axis. **FL:** calyx lobes tapered, glandular, hairy; corolla tube 25–45 mm, salverform, lobes white to pink or lavender, slightly speckled at base; stamens attached at different levels, incl, pollen white to yellow, rarely blue; style incl to slightly exserted. Gravelly to rocky slopes; 2400–3050 m. n-c SNH, e MP, **n SNE;** to Colorado. [*I. aggregata* ssp. *attenuata* (A. Gray) V. Grant & A.D. Grant misapplied to CA pls] Hybridizes with *I. aggregata* in SNH. ❀ TRY. Mar–May

LANGLOISIA

Steven L. Timbrook

1 sp. (Rev. A.B. Langlois, Louisiana botanist) [Timbrook 1986 Madroño 33:157–174]

L. setosissima (Torrey & A. Gray) E. Greene Ann, bristly, gen hairy; hairs branched, nonglandular. **ST** erect, gen naked below, leafy above. **LVS** alternate, simple, linear or oblanceolate, 3–5-toothed at tip, each tooth with 1 bristle; basal teeth of upper lvs reduced to cluster of 2–3 bristles. **INFL:** clusters, terminal, head-like; bracts lf-like; pedicels 0–short. **FL:** calyx lobes equal, bristle-tipped; corolla funnel-shaped; stamens attached at or below sinuses, equal, exserted, pollen white to blue; style exserted. **FR** oblong-lanceolate, triangular in ×-section; outer wall of valve flat. **SEEDS** gelatinous when wet. Dry, gen sandy places; gen < 1800 m. **SNE, D;** to NV, AZ, n Mex; also in e OR, w ID. Self-compatible; gen cross-pollinated. Sspp. intergrade.

1. Corolla lobes 1/2 to nearly = tube, purple dotted; stamens > 3 mm . . . . ssp. *punctata*
1' Corolla lobes 1/3–1/2 tube, unmarked or streaked with purple, seldom dotted; stamens < 3 mm . . ssp. *setosissima*

ssp. *punctata* (Cov.) S. Timbrook (p. 417, pl. 79) LILAC SUNBONNET **FL:** corolla white to light blue with purple dots, lobes 1/2 to nearly = tube, gen 2 yellow spots in middle of each;

stamens > 3 mm. Common. Desert washes, flats, slopes, gravelly to sandy soils; < 1800 m. **SNE, DMoj;** to NV. Smaller-fld populations in w ID, e OR. [*L. p.* (Cov.) Goodd.] ❀ TRY. Mar–Jun

ssp. *setosissima* (p. 417) BRISTLY LANGLOISIA **FL:** corolla lavender to blue, unmarked or with purple lines, lobes 1/3–1/2 tube; stamens < 3 mm. Common. Desert washes, flats, slopes, gravelly to sandy soils; gen < 1800 m. **SNE, D;** to NV, AZ, n Mex. Feb–Jun

LEPTODACTYLON

Robert W. Patterson & Paul A. Meyers

Per, open or cespitose. **ST** decumbent to erect. **LVS** cauline with clustered axillary lvs, alternate or opposite, simple, deeply lobed; lobes linear, gen spine-tipped, palmate or pinnate. **INFL** gen terminal; fls gen sessile. **FL:** calyx membrane wider than ribs, lobes 4–6, linear; corolla funnel-shaped or salverform, lobes 4–6; stamens attached at same level, anthers at throat, pollen yellow; style incl. 7 spp.: w N.Am. (Greek: narrow finger) [Gordon-Reedy 1989 Madroño 37:28–42]

L. pungens (Torrey) Rydb. (p. 417) Pl gen hairy, glandular or not. **ST** 1–3 dm. **LVS** alternate, 3–7-lobed; middle lobe gen longest, spine-tipped. **FL** gen opening in evening; calyx lobes gen unequal, spine-tipped, membrane extended along lobes; corolla funnel-shaped, tube + throat 7–15 mm, lobes 7–10 mm, obovate, white or pink with purplish shading on outer surface; stamens attached at throat. $2n=18$. Open, rocky areas in montane, subal-

pine forests, alpine fell-fields; 1700–4000 m. **CA;** to B.C., Rocky Mtns. [ssp. *hallii* (Parish) H. Mason; ssp. *hookeri* (Douglas) Wherry; ssp. *pulchriflorum* (Brand) H. Mason] Some proposed sspp. sort well elsewhere, but not in CA. Further study warranted. ❀ DRN,DRY, SUN:1–3,6,7,14–18. ***Linanthus pungens** (Torrey) J.M. Porter & L.A. Johnson May–Aug

LINANTHUS

Robert W. Patterson

Ann, per. **ST** gen erect, gen branched from base. **LVS** cauline, opposite, entire or palmately 3–9-lobed; lobes linear to narrowly lanceolate or spoon-shaped. **INFL** head-like, open, or fl solitary; bracts lf-like; fls sessile or pedicelled. **FL:** calyx tubular, or lobes nearly free, bordered by translucent membrane; corolla funnel-shaped, salverform, or bell-shaped; stamens attached at same level, pollen yellow. 41 spp.: w N.Am, Chile. (Greek: flax flower) [Patterson 1977 Madroño 24:36–48; Bell et al 1999 Syst Bot 24:632–644]

1. Per
  2. Corolla tube gen < calyx; seeds < 2 mm . . . . ***L. nuttallii*** ssp. ***pubescens***
  2' Corolla tube gen > calyx; seeds > 2 mm . . . . ***L. pachyphyllus***
1' Ann
  3. Corolla salverform, tube >> calyx, < 1 mm wide; infl a dense, bracted head . . . . ***L. breviculus***
  3' Corolla not salverform, if so, infl not a dense, bracted head
    4. Corolla gen open only in evening, without red marks near throat; calyx membrane wider than ribs
      5. Calyx glandular-hairy . . . . ***L. jonesii***
      5' Calyx not glandular-hairy
        6. Calyx hairy on inner surface; on gypsum soils . . . . ***L. arenicola***
        6' Calyx not hairy on inner surface; gen not on gypsum soils
          7. Lf not lobed; corolla tube without hairy pads where filaments . . . . ***L. bigelovii***
          7' Lf gen 3–7-lobed; corolla tube with hairy pads where filaments attach . . . . ***L. dichotomus***
    4' Corolla open in daytime; calyx membrane narrower than ribs, if wider then corolla with red marks near throat
      8. Fls on thread-like pedicel
        9. Corolla tube, filaments glabrous; corolla gen < calyx . . . . ***L. harknessii***
        9' Corolla tube, filaments, or both hairy; corolla gen > calyx
          10. Filaments glabrous, attached above ring of hairs . . . . ***L. aureus***
            11. Corolla lobes bright yellow . . . . ssp. ***aureus***
            11' Corolla lobes white . . . . ssp. ***decorus***
          10' Filaments hairy at base, or if glabrous then attached at ring of hairs inside corolla
            12. Corolla 4–15 mm; tube << lobes . . . . ***L. liniflorus***
            12' Corolla 2–4 mm; tube = lobes . . . . ***L. septentrionalis***
      8' Fls sessile
        13. Calyx lobes fused by membrane; corolla lobes yellow . . . . ***L. lemmonii***
        13' Calyx lobes free to base; corolla lobes not yellow
          14. Corolla bell-shaped, 5–6 mm, white with 2 red marks at base of lobes . . . . ***L. demissus***
          14' Corolla funnel-shaped, 6–12 mm, blue-purple or cream with purple kidney-shaped crest at base of lobes . . . . ***L. parryae***

L. arenicola (M.E. Jones) Jepson & V. Bailey SAND LINANTHUS Ann. **ST** 1–8 cm, glabrous to slightly hairy. **LF**: lobes 3–12 mm, upper surface hairy, lower surface glabrous. **INFL**: cyme. **FL** opening in evening; calyx 4–5 mm, lobes unequal, membrane 2/3 calyx length; corolla 5–7 mm, funnel-shaped, light yellow with purple throat; stamens attached in lower throat. RARE in CA. Saline flats in gypsum soils; 800–1400 m. **DMoj**; NV. Mar–Apr

L. aureus (Nutt.) E. Greene (p. 417) Ann. **ST** thread-like, glabrous or hairy, sometimes glandular. **LF**: lobes 3–6 mm, linear. **INFL**: fl solitary; peduncle 5–15 mm, thread-like. **FL**: calyx 4–6 mm, membrane wider than ribs; corolla funnel-shaped, tube 3–5 mm, ring of hairs inside and outside, lobes 5–7 mm, oblanceolate; stamens attached in throat; stigmas 3–4 mm, exserted. Desert flats; < 2000 m. **D**; to NM, Baja CA. Both sspp. occur in gen same range but rarely occur together. ****Leptosiphon aureus*** (Nutt.) J.M. Porter & L.A. Johnson

  ssp. ***aureus*** **FL**: corolla tube, lobes bright yellow, throat gen brighter yellow or maroon. $2n=18$. Habitat and range of sp. ****Leptosiphon aureus*** (Nutt.) J.M. Porter & L.A. Johnson ssp. ***aureus*** ❀ TRY. Mar–Jun

  ssp. ***decorus*** (A. Gray) H. Mason **FL**: corolla tube, lobes white or cream, throat maroon. $2n=18$. Habitat and range of sp. ****Leptosiphon aureus*** (Nutt.) J.M. Porter & L.A. Johnson ssp. ***decorus*** (A. Gray) J.M. Porter & L.A. Johnson ❀ TRY.

L. bigelovii (A. Gray) E. Greene Ann, glabrous. **ST** erect, 5–20 cm. **LF** simple, 10–30 mm, linear. **INFL**: cyme. **FL** opening in evening; calyx 8–12 mm, membrane much wider than ribs; corolla funnel-shaped, tube 4–5 mm, lobes 7–8 mm, cream or white, purplish shading on back of lobes; stamens attached in upper tube. $2n=18$. Deserts, dry areas; gen < 1700 m. SCoRI, WTR, **D**. Mar–May

L. breviculus (A. Gray) E. Greene Ann, hairy. **ST** 10–25 cm. **LF**: lobes 3–10 mm, linear. **INFL** head-like; fls sessile. **FL**: calyx 7–8 mm, membranes as wide as ribs; corolla salverform, tube 15–25 mm, maroon, glabrous, throat purple, lobes 4–6 mm, white, pink, or blue; anthers gen incl. $2n=18$. Deserts, dry montane

areas; < 2400 m. SnGb, SnBr, **DMoj**. ❀ TRY. May–Aug

L. demissus (A. Gray) E. Greene (p. 417) Ann. **ST** decumbent, 2–10 cm, hairy or glandular. **LF**: lobes 6–10 mm, hairy or glabrous. **INFL**: clusters, terminal, bracted; fls gen sessile. **FL**: calyx 3–4 mm, calyx lobes obscure, lobes membrane margined; corolla bell-shaped, tube 1–2 mm, white or yellow green, throat, 4–6 mm, white or light yellow, lobes 2–3 mm, white with 2 purple lines at base; stamens incl. $2n=18$. Limestone soils, desert pavement, sandy areas; < 1700 m. **DMoj**; to UT, AZ. Mar–May

L. dichotomus Benth. (p. 427) EVENING SNOW Ann, glabrous. **ST** 5–20 cm, glaucous. **LF**: lobes 3–7, 10–22 mm, linear. **INFL**: cyme. **FL** gen opening in evening; calyx 8–14 mm, membrane much wider than ribs; corolla funnel-shaped, tube 7–10 mm, purple, throat white or cream, lobes 10–16, white with light purple shading on back; stamens attached in lower tube, incl, filaments dilated into hairy pad at base. $2n=18$. Common. Drying open areas, esp serpentine; < 1700 m. CA-FP, **W&I**, **D**; AZ, NV. On pls from SnFrB s, fls open in evening; on pls from SnFrB n [ssp. *meridianus* (Eastw.) H. Mason], fls open during daylight. ❀ DRN,DRY,SUN: **7**,8–12,14–24;fragrant;DFCLT. Apr–Jun

L. harknessii (Curran) E. Greene (p. 427) Ann. **ST** 5–15 cm, thread-like, gen glabrous. **LF**: lobes 5–15 mm, thread-like, glabrous or hairy. **INFL**: fls solitary, bracted; peduncles 5–20 mm, thread-like. **FL**: calyx 1–3 mm, membranes 1/2 calyx length, glabrous; corolla funnel-shaped, white or pale blue, gen incl in calyx; stamens incl. $2n=18$. Open flats; 1000–3200 m. NCoR, CaR, SNH, **GB**; to WA. [ssp. *condensatus* H. Mason, Plaskett Meadows linanthus] ****Leptosiphon harknessii*** (Curran) J.M. Porter & L.A. Johnson Jun–Aug

L. jonesii (A. Gray) E. Greene Ann, glandular-hairy. **ST** 3–15 cm. **LF** 10–20 mm, linear, curved outward, entire. **INFL**: cyme. **FL** opening in evening; calyx 7–8 mm, glandular, membrane wider than ribs; corolla funnel-shaped, tube 5–6 mm, white, throat yellow, lobes 3–5 mm, white or cream-yellow, light purple shading on back; stamens incl. Sandy flats, washes; < 900 m. **D**; AZ, Baja CA. Mar–May

L. lemmonii (A. Gray) E. Greene Ann, hairy. **ST** 5–15 cm, gen glandular. **LF**: lobes 2–5 mm, linear. **INFL** terminal, bracted; fls sessile. **FL**: calyx 4–5 mm, membrane 1/2 calyx length; corolla funnel-shaped, tube 1–3 mm, light yellow, ring of hairs on upper inner surface, throat yellow or maroon, lobes 2–3 mm, yellow or cream; stamens incl. $2n$=18. Dry, open areas gen away from coast, chaparral, woodlands, deserts; < 1900 m. SW, **w DSon**; Baja CA. *****Leptosiphon lemmonii*** (A. Gray) J.M. Porter & L.A. Johnson .Apr–Jun

L. liniflorus (Benth.) E. Greene Ann. **ST** 10–50 cm, glabrous or hairy. **LF**: lobes 1–3 cm, linear. **INFL**: cyme or panicle; peduncles 1–3 cm. **FL**: calyx 3–5 mm, hairy, membrane wider than ribs, extended along lobes; corolla funnel-shaped, white, tube 1–2 mm, throat wide, 2–4 × tube, lobes 8–10 mm, veins gen purple; stamens exserted, filaments hairy at base. $2n$=18. Woodlands, open areas, common on serpentine soil, deserts; < 1700 m. n&c CA-FP, **w DMoj**; to WA. [ssp. *pharnaceoides* (Benth.) H. Mason] *****Leptosiphon liniflorus*** (Benth.) J.M. Porter & L.A. Johnson Apr–Jun

L. nuttallii (A. Gray) Milliken (p. 427) Per, hairy. **STS** many, 10–20 cm. **LF**: lobes gen 5, linear. **INFL**: cluster, terminal, bracted; fls gen sessile. **FL**: calyx 8–9 mm, membrane gen obscure, 1/2 calyx length; corolla funnel-shaped, tube incl in calyx, white or light yellow, throat yellow, lobes 4–5 mm, white; stamens incl or slightly exserted; stigmas slightly exserted. Dry flats, openings in forest, sometimes on serpentine; 500–3500 m. KR, NCoRO, NCoRH, CaR, **GB**; to B.C., Rocky Mtns, n Mex. Gen self-pollinated. 3 sspp. in CA. *****Leptosiphon nuttallii*** (A. Gray) J.M. Porter & L.A. Johnson

ssp. *pubescens* R. Patt. (pl. 80) **LF**: lobes 5–10 mm, densely hairy. $2n$=18. Dry flats, openings in forests; 2800–3500 m. **SNE**; to NV.. *****Leptosiphon nuttallii*** (A. Gray) J.M. Porter & L.A. Johnson ssp. *pubescens* (R. Patt.) J.M. Porter & L.A. Johnson

L. pachyphyllus R. Patt. Per, hairy. **ST** 10–20 cm. **LF**: lobes gen 5, 10–16 mm, linear. **INFL**: cluster, terminal, bracted; fls gen sessile. **FL**: calyx 9–10 mm, densely hairy, membrane gen obscure, 1/2 calyx length; corolla funnel-shaped, tube 12–15 mm, white or pale yellow, throat yellow, lobes 6–9 mm, white; stamens gen exserted; stigmas slightly exserted. $2n$=36. Open, wooded areas; 1700–2500 m. SNH, **SNE**. Gen self-pollinated. Distinguished from *L. nuttallii* by gen larger features, chromosome number. *****Leptosiphon pachyphyllus*** (R. Patt.) J.M. Porter & L.A. Johnson

L. parryae (A. Gray) E. Greene Ann. **ST** decumbent or very short-erect concealed by lvs, 2–10 cm, glandular-hairy. **LF**: lobes 5–15 mm, linear, hairy. **INFL** crowded leafy; fls sessile. **FL**: calyx 6–8 mm, tube obscure, membrane extended along lobes; corolla funnel-shaped, white or blue-purple, tube 1 mm, throat 1–2 mm, lobes 8–12 mm, gen jagged at tip, dark purple kidney-shaped arch at base; stamens incl. $2n$=18. Sandy, open, flat areas; < 2000 m. s SNF, s SnJV, SCoRI, WTR, **DMoj**. Some populations have pls with both corolla colors. ❀ TRY. Mar–May

L. septentrionalis H. Mason Ann. **ST** 5–30 cm, thread-like, glabrous or hairy. **LF**: lobes 5–20 mm, thread-like. **INFL**: fls solitary, bracted; peduncles 10–25 mm, thread-like. **FL**: calyx 1–3 mm, membrane 2/3 calyx length, extended along lobes; corolla funnel-shaped, tube 2–3 mm, white, ring of hairs at or above stamen attachment, throat yellow, lobes 1–2 mm, white or pale blue; stamens slightly exserted. Common. Sagebrush shrublands, pinyon/juniper woodlands; 2000–3000 m. **GB**; to w Can, Colorado. *****Leptosiphon septentrionalis*** (H. Mason) J.M. Porter & L.A. Johnson May–Jul

LOESELIASTRUM

Steven L. Timbrook

Ann, bristly and gen soft; hairs unbranched. **ST** erect, gen naked below, leafy above. **LVS** alternate, simple, 1–4 cm, linear to oblanceolate, toothed; teeth with 1 bristle. **INFL**: bracts lf-like; pedicels glandular. **FL**: calyx lobes equal, bristle-tipped, glandular-hairy; corolla 2-lipped, white to deep pink, upper lip gen 3-lobed with maroon arches above throat, lower lip 2- lobed, less marked; stamens attached at or below sinuses, unequal, curved, exserted, pollen yellow; style exserted. **FR** 2–5 mm, ovoid, 3-lobed in × -section; outer wall of valve indented between walls separating chambers. **SEEDS** gelatinous when wet. 2 spp.: sw US, n Mex. (Latin: like *Loeselia*) [Timbrook 1986 Madroño 33:157–174] Self-compatible; gen cross-pollinated.

1. Corolla 11–21 mm, upper lip 3/4 to > tube; longer stamens > upper lip; calyx (exc bristles) 1/2–3/4 corolla tube . . . . ***L. matthewsii***
1' Corolla 8–15 mm, upper lip gen 1/2–3/4 tube; longer stamens < upper lip; calyx (exc bristles) > 3/4 corolla tube . . . . ***L. schottii***

L. matthewsii (A. Gray) S. Timbrook (p. 427) DESERT CALICO **FL**: corolla strongly bilateral, white to deep rose-purple, upper lip 5–11 mm, 3/4 to > tube, lobe bases with bright maroon arches and white blotch, lower lip 4–7 mm, lobe tips gen truncate, 3-toothed or notched, gen with inward directed projections in sinuses on either side of middle lobe of upper lip; longer stamens > upper lip. Common. Desert washes, flats, slopes, sandy to gravelly soils; gen < 1800 m. **SNE, DMoj, w DSon** (e San Diego Co.); NV. [*Langloisia m.* (A. Gray) E. Greene] ❀ TRY. Mar–Jun

L. schottii (Torrey) S. Timbrook (p. 427, pl. 81) **FL**: corolla weakly to moderately bilateral, gen white to pink, upper lip 3–7 mm, gen 1/2–3/4 tube, lobe bases maroon-streaked, lower lip 2–5 mm, lobe tips gen acute; longer stamens < upper lip. Common. Desert washes, flats, slopes, sandy to gravelly soils; gen < 1800 m. sw SnJV, **SNE, D**; to sw UT, w AZ, n Mex. Weakly bilateral forms are self-pollinated. [*Langloisia s.* (Torrey) E. Greene] Mar–Jun

NAVARRETIA

Alva G. Day

Ann, gen erect; branches spreading or ascending, hairy, glandular or puberulent. **LVS** simple, alternate, gen deeply pinnately lobed or entire. **INFL**: head; bracts pinnately to palmately toothed or lobed, spine-tipped; fls sessile or subsessile. **FL**: calyx membranous between ribs, lobes 4–5, entire or toothed, unequal, spine-tipped; corolla lobes 4–5; stigmas 2 or 3. **FR** gen ovoid, chambers 1–3. **SEEDS** 1–many per chamber, free or stuck together, brown, gelatinous when wet. ± 30 spp.: w N.Am, also in Argentina, Chile. (F. Navarrete, Spanish physician, 1700's)

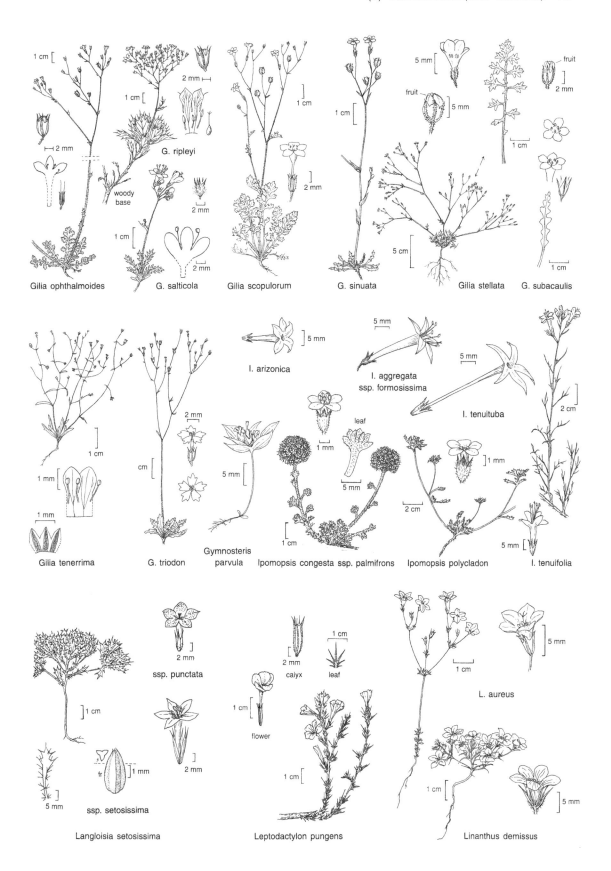

1 cm

2 mm

1 cm

G. ripleyi

2 mm

2 mm

woody base

1 cm

2 mm

5 mm

fruit

5 mm

fruit

2 mm

1 cm

Gilia ophthalmoides

G. salticola

Gilia scopulorum

G. sinuata

Gilia stellata

G. subacaulis

1 cm

2 mm

1 cm

1 mm

1 mm

I. arizonica

5 mm

I. aggregata ssp. formosissima

5 mm

5 mm

I. tenuituba

2 cm

2 mm

cm

5 mm

leaf

1 mm

5 mm

1 mm

2 cm

5 mm

Gilia tenerrima

G. triodon

Gymnosteris parvula

Ipomopsis congesta ssp. palmifrons

Ipomopsis polycladon

I. tenuifolia

1 cm

ssp. punctata

2 mm

2 mm

calyx

1 cm

leaf

1 cm

5 mm

1 cm

1 cm

5 mm

L. aureus

1 cm

fr

1 mm

flower

2 mm

1 cm

1 cm

5 mm

5 mm

ssp. setosissima

Langloisia setosissima

Leptodactylon pungens

Linanthus demissus

N. breweri (A. Gray) E. Greene (p. 427) Pl gen densely branched, as wide as high. **ST** 3–8 cm, white-puberulent, brown or red. **LF** gray-green, pinnate; axis < 0.5 mm wide; lobes needle-like, entire or forked. **INFL**: outer bracts like lvs but axis wider, shorter. **FL**: calyx 6–9 mm, sinus U-shaped, lobes needle-like, hairy inside; corolla 6–7 mm, yellow, lobes 1 mm; stamens, style exserted; stigmas 3, minute, appearing as 1. **FR** obovoid, dehiscing from tip. **SEEDS** 1–2 per chamber. Open, wet areas, meadows, streamsides; 1200–3300 m. SNH, **GB**; to WA, ID, Colorado, AZ. ❀ TRY. Jun–Aug

PHLOX

Robert W. Patterson & Dieter H. Wilken

Ann, per. **ST** prostrate to erect, or tufted to cushion-like. **LVS** cauline, opposite, simple, sessile, linear-lanceolate to elliptic, entire. **FL**: corolla salverform; stamens attached at different levels, some stamens unequal. ± 60 spp.: Am, Siberia. (Greek: flame, ancient name for *Lychnis* of Caryophyllaceae) [Cronquist 1984 in Intermountain Flora, V.4]

1. Ann . . . . *P. gracilis*
1' Per
  2. Pl erect, branches open, not matted; fls gen 3–15 per infl . . . . *P. stansburyi*
  2' Pl ± matted to tightly cushion-like; fls 1–3 per infl
    3. Pl tightly cushion-like (sometimes difficult to determine on dried specimens) . . . . *P. condensata*
    3' Pl ± matted, not tightly cusion-like
      4. Glandular hairs present
        5. Glandular throughout herbage; GB . . . . *P. cespitosa*
        5' Glandular only on calyces; SNE . . . . *P. pulvinata*
      4' Glandular hairs 0
        6. Lvs hairy only at base . . . . *P. diffusa*
        6' Lvs hairy throughout . . . . *P. hoodii* var. *canescens*

P. cespitosa Nutt. Per, ± matted, glandular-hairy. **LF** 4–8 mm, linear-lanceolate, sharp-tipped. **INFL** terminal; fls 1–3, sessile. **FL**: calyx 7–9 mm; corolla pink or pale lavender to white, tube ± 10 mm; style 4–6 mm. Dry areas, sagebrush scrub, juniper woodland; 1500–2000 m. **GB**; to s WA. [*P. douglasii* Hook. ssp. *rigida* (Benth.) Wherry] ❀ TRY;DFCLT. Apr–May

P. condensata (A. Gray) E. Nelson (p. 427) Per, tightly cushion-like. **LF** 3–5 mm, lanceolate, coarsely ciliate; upper surface gen concave; lower surface with 2 elongate grooves. **INFL** terminal; fls solitary, sessile. **FL**: calyx 5–6 mm, gen glandular-puberulent; corolla white or pale pink, tube 8–10 mm. 2*n*=28. Dry, open, rocky areas, esp limestone, travertine; 2000–4000 m. SNH, SnBr, **SNE**; NV. [*P. covillei* E. Nelson; *P. cespitosa* Nutt. var. *condensata* A. Gray] ❀ TRY;DFCLT. Jun–Aug

P. diffusa Benth. (p. 427) Per, ± matted, ± glabrous to hairy, not glandular. **ST** decumbent. **LF** 10–15 mm, linear-lanceolate or ± awl-like not sharp-tipped, nearly glabrous exc base densely white-woolly. **INFL** terminal; fls solitary; peduncle short. **FL**: calyx 8–10 mm, hairy, membrane not keeled; corolla white to pink or blue, tube 9–13 mm. 2*n*=28. Dry, open areas; 1100–3600 m. **CA** (exc SW, D); w N.Am. [*P. azurea* G. Smith] ❀ DRN,SUN:1,2–6, 15–17&IRR:3,7,14,18–21;DFCLT. May–Aug

P. gracilis (Hook.) E. Greene (p. 427) Ann; upper parts gen glandular-hairy. **ST** decumbent and highly branched to simple and erect, < 20 cm. **LVS** opposite (uppermost alternate), 10–30 mm, oblanceolate to lanceolate. **FL**: calyx 5–10 mm; corolla tube 8–12 mm, yellowish, lobes 1–2 mm, gen truncate or notched, bright pink to white. 2*n*=14. Dry to moist areas; < 3300 m. **CA**; to B.C., MT, Colorado, Mex; also in S.Am. [*Microsteris g.* (Hook.) E. Greene, incl ssp. *humilis* (E. Greene) V. Grant] **Microsteris gracilis* (Hook.) E. Greene Mar–Aug

P. hoodii Richardson ssp. *canescens* (Torrey & A. Gray) Wherry Per, ± matted, ± glabrous to woolly, not glandular. **ST** glabrous. **LF** ± awl-like, gen hairy. **INFL** terminal; fls solitary, sessile. **FL**: calyx 7–8 mm, woolly near base of lobes; corolla white to lilac, tube 10–12 mm. Open, rocky areas, sagebrush scrub, pinyon/juniper woodland; 1500–2700 m. n SNH, **GB**; MT, UT, n AZ. [ssp. *lanata* (Piper) Munz] ❀ TRY;DFCLT. May–Jul

P. pulvinata (Wherry) Cronq. Per, cespitose. **LF** 4–10 mm, lanceolate, coarsely ciliate; upper surface gen flat. **INFL** terminal; fls solitary, sessile. **FL**: calyx 7–8 mm, gen glandular-puberulent; corolla white or pale pink, tube 8–10 mm. Dry, rocky areas, subalpine forest, alpine fell-fields; 3300–4300 m. SNH, **SNE**; to ID, Colorado. [*P. cespitosa* Nutt. var. *p.* (Wherry) Cronq.] Jul–Aug

P. stansburyi (Torrey) A.A. Heller (p. 427) Per. **ST** openly branched from base, gen growing up through shrubs. **LF** 1–3 cm, linear-lanceolate, > internodes, glandular-hairy to long-soft-hairy. **INFL**: fls few; peduncles 5–25 mm. **FL** 8–12 mm; calyx glandular-hairy, membrane gen keeled; corolla pink to white, tube 12–33 mm; style nearly = corolla tube. Dry areas in sagebrush scrub, pinyon/juniper woodland; 1700–3000 m. **SNE, DMtns**; to UT, NM. [*P. longifolia* Nutt. var. *puberula* E. Nelson; *P. viridis* E. Nelson ssp. *compacta* (Brand) Wherry] ❀ TRY;DFCLT. Apr–Jun

POLEMONIUM

Dieter H. Wilken

Ann, per. **STS** decumbent to erect, 1–10, 1–10 dm, glandular, hairy. **LVS** pinnately compound, alternate; basal petioled; cauline sessile upward; lflets entire to divided, glabrous to glandular. **INFL** open to head-like. **FL**: calyx bell-shaped, membranous in age but not separated into membrane and ribs, glandular; corolla rotate to funnel-shaped, white to blue or purple; stamens attached at same level, filaments hairy at base. **FR** ovoid to spheric. **SEEDS** 3–10 per chamber, gen 1–3 mm, elliptic to ovate in outline, slightly gelatinous when wet, brown to black. ± 20 spp.: Am, Eurasia. (Greek: perhaps from Polemon, Athenian philosopher, or polemos, strife or war) [Grant 1989 Bot Gaz 150:158–169] Pers gen cross-pollinated, anns self-pollinated.

1. Lflets deeply 3–5-lobed (appearing ± whorled); infl head-like . . . . *P. chartaceum*
1' Lflets entire; infl gen open
  2. St erect, 5– 10 dm; lvs cauline; infl open . . . . *P. occidentale* ssp. *occidentale*
  2' St cespitose, gen 1–2 dm; lvs ± basal; infl crowded . . . . *P. pulcherrimum* var. *pulcherrimum*

P. chartaceum H. Mason (p. 427) MASON'S SKY PILOT Per, hairy. **STS** erect, 3–8, < 3 dm. **LVS** basal, narrowly oblong in outline; petioles sheathing, membranous at base, brownish; lflets > 13, < 5 mm, deeply 3–5-lobed; lobes gen elliptic, glandular, appearing as whorled lflets. **INFL** head-like; pedicels < 5 mm. **FL**: calyx 6–8 mm, lobes = tube, rounded; corolla 10–14 mm, funnel-shaped, tube > lobes, white, throat blue or yellow, lobes blue; stamens and style exserted. UNCOMMON. Rocky slopes, talus, 2600–4200 m. KR, **SNE (Sweetwater Mtns), W&I**. Probably related to both *P. eximium* (c&s SNH) and *P. elegans* E. Greene (WA, B.C.); warrants further study, esp because of interrupted range. ❀ DFCLT. Jul–Aug

P. occidentale E. Greene ssp. *occidentale* (p. 427) Per; rhizomes short. **ST** erect, 5–10 dm. **LVS** cauline, 1–4 dm; lower with 13–17 lflets; upper with 9–13 lflets; lflets elliptic to lanceolate, green above, glaucous below, glabrous. **INFL** open; pedicels < 5 mm. **FL**: calyx 4–10 mm, lobes = tube, acute; corolla 10–15 mm, bell-shaped, gen blue, throat gen yellow, lobes >> tube; stamens incl; style exserted. **FR** < 4 mm. Moist areas, meadows, woodlands; 900–3300 m. KR, SN, SnBr, **SNE**; to B.C., Colorado. (other sspp occur elsewhere in the US) [*P. caeruleum* L. ssp. *amygdalinum* (Wherry) Munz] ❀ DRN:4–6&IRR,SHD:1–3,7,14–18. Jun–Aug

P. pulcherrimum Hook. Per, cespitose, gen from slender rhizome. **STS** decumbent to erect, 4–10, < 3 dm; hairs soft. **LVS** ± basal, abruptly smaller upward; lflets 9–21, 4–8 mm, ovate to round, glandular, hairs small. **INFL** crowded or head-like; pedicels < 10 mm. **FL**: calyx 4–8 mm, lobes = tube; corolla 5–8 mm wide, rotate to bell-shaped, tube white, throat often yellow, lobes = tube, blue to white; stamens and style ± incl. Subalpine to alpine talus; 2400–3700 m. KR, NCoRH, n&c SNH, MP, **n SNE**; to AK, WY. 2 vars. in CA.

  var. ***pulcherrimum*** Pl gen 1–2 dm; herbage sparsely hairy and glandular. **FL**: corolla blue to purple. Talus; 2400–3700 m. KR, n&c SNH, **n SNE**; to AK, WY. ❀ TRY;DFCLT. Jun–Aug

POLYGALACEAE MILKWORT FAMILY

Thomas L. Wendt

Ann, per, shrub, tree, vine, some non-green, dependent on fungi for nutrition; hairs unbranched. **LVS** simple, gen alternate (rarely opposite or whorled); veins pinnate; margin gen ± entire; stipules gen 0. **INFL**: raceme, spike, or panicle. **FL** bisexual, gen bilateral (appearing ± pea-fl-like) or ± radial; sepals 5, free or fused, lateral (inner) pair often larger and petal-like (wings); petals 5 or 3, individually fused to stamen tube, ± similar or different with 1 lower keel petal, 2 strap-like upper petals, and 0 or 2 small lateral petals; stamens 3–10, ± fused, tube open on upper side; ovary chambers 1–8, ovule 1 per chamber, style 1 or 0. **FR**: capsule, drupe, or nut, sometimes winged. **SEED** often with aril. 18 genera, 800 spp.: esp trop, subtrop, very few cult. [Blake 1924 N Amer Fl 25:305–379]

POLYGALA MILKWORT

Ann, per, shrub, tree, vine; roots gen with wintergreen odor. **INFL**: raceme or spike, sometimes grouped and panicle-like. **FL** bilateral; lateral 2 sepals enlarged as wings; petals 3 or 5, keel petal often with cylindric beak or fringed crest at tip; stamens 6–8, anthers dehiscent at tip, appearing 1-chambered; nectary disc or gland present; ovary chambers 2, stigma 2- lobed. **FR**: capsule. **SEED** fusiform or ovoid, black, gen hairy, gen with prominent white aril on one end. ± 500 spp.: trop, temp. (Greek: much milk, some Eur spp. said to increase milk flow in cows) [Wendt 1979 J Arnold Arbor 60:504–514] Key to species adapted by Margriet Wetherwax.

1. Fls 7.5–13.5 mm; wings pink; twigs glabrous to puberulent . . . . *P. heterorhyncha*
1' Fls 2.5–5.3 mm; wings cream or greenish; twigs densely hairy
  2. Pedicels, outer sepals, and lvs with spreading hairs . . . . *P. acanthoclada*
  2' Pedicels glabrous; outer sepals glabrous, ciliate, or sparsely hairy near tip; lf hairs incurved or appressed
  . . . . *P. intermontana*

P. acanthoclada A. Gray (p. 427) Subshrub, shrub; hairs spreading. **ST** weak and sprawling to erect, gen not highly branched, < 10 dm; twig hairs dense, white. **LF** 5–25 mm, oblanceolate to narrowly elliptic or obovate, hairy. **INFL** thorn-tipped; fls 1–15; pedicels 1.5–5.8 mm, hairy. **FL** 3–5.3 mm; outer sepals hairy, wings cream; beak of keel-petal 0 or minute. **FR** 4–6 mm. *n*=9. Desert scrub, Joshua-tree or pinyon/juniper woodlands, often loose, sandy or gravelly soils; 900–1700 m. **DMoj (Eagle, New York mtns, Lucerne Valley)**; to s UT, AZ. May–Aug

P. heterorhyncha (Barneby) T. Wendt (p. 427) NOTCH-BEAKED MILKWORT Per, subshrub, forming thorny mats 1–2 dm high, < 7 dm diam. **ST** glabrous to puberulent, ± glaucous. **LF** 4–20 mm, ovate, elliptic, or obovate, often ± scabrous; base tapered to rounded. **INFL** strongly thorn-tipped. **FL** 7.5–13.5 mm; sepal wings pink; beak of keel-petal 1.4–4 mm, prominently notched on lower side, yellow. **FR** 4.2–7.8 mm incl stalk; veins promi-nent. **SEED** 3–4.4 mm, incl hairs; seed body widely elliptic to round in ×-section, hairier near aril end; aril glabrous. *n*=18. RARE. Desert scrub; 900–1600 m. **DMtns (Funeral Mtns, Inyo Co)**; s NV. [*P. subspinosa* var. *h.* Barneby] Apr–May

P. intermontana T. Wendt (p. 427) Subshrub, shrub, stiffly branched, ± open, < 10 dm high (or forming thorny mats 1.5–5 dm high, gen < 10 dm diam). **ST**: twig hairs dense, white, appressed to irregularly ascending. **LF** 3–25 mm, linear to obovate; hairs incurved or appressed. **INFL** thorn-tipped; fls 1–7; pedicels 2.5–9 mm, glabrous. **FL** 2.5–5.2 mm; outer sepals glabrous or ciliate, sometimes with incurved hairs near tip, wings cream or greenish; beak of keel-petal 0 or minute. **FR** 3.5–5.8 mm. *n*=9. Pinyon/juniper woodland; 2600–2700 m. **SNE (n Mono Co)**; to UT, n AZ. [*P. acanthoclada* var. *intricata* Eastw.] May–Aug

POLYGONACEAE BUCKWHEAT FAMILY

James C. Hickman

Ann to trees, some dioecious. **ST**: nodes often swollen. **LVS** simple, basal or cauline, alternate, opposite, or whorled, gen entire; stipules 0 or obvious and fused into a gen scarious sheath around st. **INFL**: small cluster, axillary or arrayed in cymes or panicles; involucres sometimes subtending 1–many fls. **FL** gen bisexual, small, ± radial; perianth gen 5–6-lobed, base ± tapered, often jointed to pedicel; stamens 2–9, often in 2 whorls; ovary superior, styles gen 3, gen fused at base. **FR**: achene, gen enclosed by persistent perianth, gen 3-angled, ovoid, and glabrous. 50 genera, 1100 spp.: worldwide, esp n temp; some cult for food (*Fagopyrum*; *Rheum*, rhubarb; *Rumex*, sorrel) or orn (*Antigonon*, coral-vine; *Muehlenbeckia*; *Polygonum*). [Ronse Decraene & Akeroyd 1988 Bot J Linn Soc 98:321–371; Reveal et al 1989 Phytologia 66:83–414] Family key, keys and treatments of the 15 eriogonoid genera are by James L. Reveal, who is gratefully acknowledged.

1. Stipules obvious, sheathing st; lvs cauline (or some basal); node ± swollen; fls often in clusters but not involucred (Polygonoideae)
  2. Lf blades reniform; perianth lobes 4; styles 2; fr flat . . . . **OXYRIA**
  2' Lf blades not reniform; perianth lobes (4)5 or 6; styles (2) 3; fr gen 3-angled
    3. Perianth lobes gen 5, tubercles 0 in fr . . . . **POLYGONUM**
    3' Perianth lobes 6; sometimes tubercled in fr . . . . **RUMEX**
1' Stipules 0; lvs often basal; node gen not swollen; fls gen in involucres of ± fused bracts (Eriogonoideae)
  4. Lvs cauline, opposite, entire to deeply notched, appearing 2-lobed; involucral bracts in fr large, scarious, 2-winged, net-veined — pls sprawling ann . . . . **PTEROSTEGIA**
  4' Lvs gen basal or cauline alternate, not deeply notched; involucre ± tubular, gen spiny , rarely 0
    5. Involucral bracts, if any, essentially free
      6. Shrub . . . . **DEDECKERA**
      6' Ann
        7. Fls obscured in woolly hairs; lf blade linear to oblanceolate, petiole indistinct; perianth greenish white to dark red; stamens 3 . . . . **NEMACAULIS**
        7' Fls not obscured; lf blade obovate to round, petiole obvious; perianth yellow; stamens 9
          8. Perianth thinly hairy, involucral bracts 0; lf-like bracts 3 per node . . . . **GILMANIA**
          8' Perianth hairy at base; involucral bracts 5; lf-like bracts 2 per node . . . . **GOODMANIA**
    5' Involucral bracts obviously fused (even if only at base), forming a funnel-shaped to cylindric involucre
      9. Involucral lobes not awned-tipped . . . . **ERIOGONUM**
      9' Involucral lobes awned-tipped
        10. Involucre ± funnel-shaped, awns gen 4; fls 2–10 per involucre . . . . **OXYTHECA**
        10' Involucre cylindric; awns 3–6; fls 1–2 per involucre
          11. Bracts 2–3 per infl node . . . . **CHORIZANTHE**
          11' Bracts 1 per infl node
            12. Bracts not encircling st; involucre awned at base and tip; fls 2 per involucre . . . . . . **CENTROSTEGIA**
            12' bracts encircling st; involucre awned at tip only; fls 1 per involucre . . . . **MUCRONEA**

CENTROSTEGIA

1 sp. (Greek: spurred cover, from involucre) [Reveal 1989 Phytologia 66:199–220]

C. thurberi Benth. (p. 427, pl. 82) THURBER'S SPINE-FLOWER Ann, spreading, 3–30 cm, glandular. **LVS** basal; stipule 0; ± petioled; blade 5–40 mm, oblanceolate, glabrous. **INFL** 6–50 cm diam; bract 1 per node, 1–10 mm, 3-lobed, awns 1–2 mm; involucre 1 per node, sessile, 2–8 mm, cylindric above, swollen at base, upper awns 5, < 1 mm, basal awns 3, < 2 mm, curved; fls 2 per involucre. **FL**: perianth 2–3.5 mm, white to pink, hairy, lobes 6; stamens 9. **FR** 2–2.5 mm, brown. *n*=19. Common. Desert scrub, other dry sandy places; 300–2400 m. s SnJV, e SCoRI, TR, **SNE**, **D**; to NV, UT, AZ, nw Mex. [*Chorizanthe t.* (Benth.) S. Watson] Apr–Jun

CHORIZANTHE SPINEFLOWER

Ann, per, glabrous or hairy, sometimes glandular. **ST** gen scapose (made up of infl axes). **LVS** basal (rarely some cauline); stipule 0; blade gen ± oblanceolate. **INFL** open or of few heads, sometimes 1-sided; bracts gen opposite, lf-like to scale-like; involucres 1–several per axil, sessile, tube cylindric to bell-shaped, gen ± cross-ridged or net-veined, bracts (and ribs) 3–6, awns straight or hooked; fls 1–2 per involucre. **FL**: perianth white to red or yellow, lobes 6, entire to fringed or toothed; stamens 3–9. **FR** 1.5–4.5 mm, gen ± brown, glabrous. 50 spp.: temp w N.Am, sw S.Am. (Greek: divided fl, from perianth) [Reveal & Hardham 1989 Phytologia 66(2):98–198]

1. Involucral bracts (or ribs) and awns 3 or 5
  2. Involucral bracts and awns 3
    3. Involucral tube cylindric, prominently cross-ridged, glabrous; stamens 6; D . . . . *C. corrugata*
    3' Involucral tube 3-angled, funnel-shaped, not cross-ridged, hairy; stamens 9; SNE, D . . . . *C. rigida*
  2' Involucral bracts and awns 5

4. Awns straight; perianth 2.5–3.5 mm long; stamens attached at base of perianth tube . . . . *C. spinosa*
4' Awns hooked; perianth 1.5–2.5 mm long; stamens attached near top of perianth tube . . . . *C. watsonii*
1' Involucral bracts (or ribs) and awns 6
  5. Perianth lobes unequal, inner gen < and narrower than outer; stamens 9
    6. Pl grayish, canescent; involucral tube 1.5–2 mm; nw edge DSon . . . . *C. parryi* var. *parryi*
    6' Pl reddish, thinly hairy; involucral tube 3.5–5 mm; SNE . . . . *C. xanti* var. *xanti*
  5' Perianth lobes all equal in length and width; stamens 3 . . . . *C. brevicornu*
    7. Lvs ± linear, petiole indistinct . . . . var. *brevicornu*
    7' Lvs oblanceolate, petiole distinct . . . . var. *spathulata*

C. brevicornu Torrey BRITTLE SPINEFLOWER **ST** ± erect, 5–50 cm, thinly hairy. **LF**: blade 10–40 mm, ± linear to oblanceolate. **INFL** breaking at nodes; involucral tube 3–5 mm, cylindric, green, thinly strigose, bracts 6, awns hooked. **FL**: perianth 2–4 mm, gen white, glabrous, lobes equal, ± linear, entire; stamens 3, attached near top of perianth tube. Common. Desert scrub, sagebrush scrub, juniper woodland; < 3000 m. **SNE, D**; to se OR, s ID, w UT, AZ, nw Mex.

  var. *brevicornu* (p. 427) **ST** 5–50 cm. **LF**: blade 15–40 mm, ± linear; petiole indistinct. **INFL**: involucral tube prominently ribbed. *n*=19–21,23. Desert scrub; < 2300 m. **s SNE, D**; s NV, UT, AZ, nw Mex. Mar–Jun

  var. *spathulata* (Rydb.) C. Hitchc. **ST** 5–30 cm. **LF**: blade 10–20 mm, oblanceolate; petiole distinct. **INFL**: involucral tube obscurely ribbed. *n*=19. Sagebrush scrub, pinyon/juniper woodland; 700–3000 m. **SNE, DMtns**; to se OR, s ID. [ssp. *s.* (Rydb.) Munz] Jun–Jul

C. corrugata (Torrey) Torrey & A. Gray (p. 427) **ST** erect, 3–50 cm, thinly tomentose. **LF**: blade 5–20 mm, ovate to rounded. **INFL**: involucral tube 3–4 mm, cylindric, green to tan, prominently cross-ridged, ± glabrous, bracts 3, 2–4.5 mm, gen lanceolate, awns hooked. **FL**: perianth 2–2.5 mm, white, thinly hairy; stamens 6, attached near top of perianth tube. *n*=19. Common. Dry soils in desert scrub; –70–920 m. **D**; s NV, w AZ, nw Mex. Feb–May

C. parryi S. Watson **ST** prostrate to ascending, 2–30 cm, strigose. **LF**: blade 5–40 mm, oblanceolate to oblong. **INFL**: involucral tube 1.5–2 mm, urn-shaped, canescent, bracts 6, awns straight or hooked. **FL**: perianth 2.5–3 mm, white, sparsely hairy, lobes equal or not, gen nearly entire; stamens 9. Uncommon. Sandy places, gen in coastal or desert scrub; 200–1200 m. SCo, e TR, **nw edge DSon**. 2 vars. in CA.

  var. *parryi* **INFL**: involucral awns hooked. **FL**: perianth lobes unequal. Habitat of sp.; 300–1200 m. c&e SCo, e TR, **nw edge DSon**. Apr–Jun

C. rigida (Torrey) Torrey & A. Gray (p. 427, pl. 83) SPINY-HERB **ST** erect, densely branched, 2–15 cm, greenish, hairy.

LF long-petioled; blade gen 5–10 mm, oblanceolate to elliptic, hairy; rosette withering early. **INFL**: main bracts like lvs but larger (blade gen 10–35 mm) and wider, later bracts many, spine-like, < 30 mm; involucral tube 2–3 mm, funnel-shaped, bracts 3, 2–10 mm (1 much the longest), lanceolate, awns straight, **FL**: perianth < 2 mm, incl, yellow, densely hairy; stamens 9, attached near top of perianth tube. *n*=19,20. Common. Desert scrub; < 2000 m. **SNE, D**; to sw UT, w AZ, nw Mex. Dead spiny skeletons persist. Mar–May

C. spinosa S. Watson (p. 429) MOJAVE SPINEFLOWER **ST** prostrate to ascending, 3–40 cm, grayish, thinly hairy. **LF**: blade 3–20 mm, oblong. **INFL**: bracts gen 3 per node, ± lanceolate, awns straight; involucral tube 2–2.5 mm, urn-shaped, bracts gen 5, one >> others (to 10 mm, ± lanceolate), awns straight. **FL**: perianth 2.5–3.5 mm, gen white, glabrous; stamens 9, attached at base of perianth tube. **FR** black. *n*=22. UNCOMMON. Desert scrub; 6–1300 m. **w DMoj**. Apr–Jul

C. watsonii Torrey & A. Gray WATSON'S SPINEFLOWER **ST** ± erect, 2–15 cm, canescent. **LF**: blade 3–20 mm, gen thinly tomentose. **INFL**: involucral tube 3–4.5 mm, cylindric, green, hairy, bracts 5, largest 2–6 mm, ± lf-like, other four 1–2 mm, awns hooked. **FL**: perianth 1.5–2.5 mm, yellow, thinly hairy; stamens 9, attached near top of perianth tube. Desert and sagebrush scrub and woodland; 300–2200 m. sw SnJV, n edge TR, **GB, DMoj**; to NV, sw UT, nw AZ. Apr–Jul

C. xanti S. Watson **ST** ± erect, 3–30 cm, reddish, thinly hairy. **LF**: blade 3–15 mm, oblong to ovate, thinly hairy above, densely tomentose below. **INFL**: lower bracts ± oblanceolate, ± lf-like, persistent, awns straight; involucres ± loosely clustered, tube 3.5–5 mm, cylindric, smooth, hairs slender, curly, bracts 6, awns hooked. **FL**: perianth 4.5–6 mm, rose to red, hairy, lobes unequal, gen entire; stamens 9. Foothill and pinyon/juniper woodland, desert scrub; 100–1600 m. s SN, Teh, s SCoRI, n&e TR, SnJt, **SNE (s Mono Co.)**. 2 vars. in CA.

  var. *xanti* (p. 429) **INFL**: involucre thinly hairy. *n*=19,21. Common. Habitat and range of sp., exc desert scrub in e SnBr, SnJt. ❀ TRY. Apr–Jun

DEDECKERA

1 sp. (Mary C. DeDecker, e CA botanist, 1909–2000) [Reveal & Howell 1976 Brittonia 28:245–251]

D. eurekensis Rev. & J. Howell (p. 429) JULY GOLD Shrub, 2–10 dm, 5–20 dm diam, minutely hairy. **LVS** cauline; stipule 0; petiole 2–5 mm; blade 7–15 mm, oblanceolate. **INFL** 1–8 cm; bracts 2–5 per node, free; most nodes with 1 ± sessile fl and 1–several branches; branches and bracts reduced and fls more per node upward; true involucre 0. **FL**: perianth gen 2–4 mm, yellowish, lobes 6; stamens 9. **FR** 2–3.5 mm, light reddish brown, puberulent at tip. *n*=14. RARE CA. Limestone outcrops; 1220–2200 m. **W&I, DMtns (Last Chance, Panamint Mtns)**.

ERIOGONUM WILD BUCKWHEAT

Ann to shrub. **LVS** gen ± basal (clustered on low sts or cauline), petioled, gen ± tomentose below (often shedding above); stipule 0. **INFL** openly cyme-like, umbel-like, or head-like, gen ± scapose; bracts (any whorled, lf-like structures on infl) 3–many per node, lf-like to scale-like; involucres gen 1 per node, gen ± obconic, lobes (or short teeth) gen 3–10, gen erect; fls gen many per involucre, pedicelled. **FL**: perianth white, yellow, or red, lobes 6, gen ± oblong to obovate; stamens 9. **FR** brown to black, glabrous to hairy. ± 250 spp.: N.Am. (Greek: woolly knees, from hairy nodes of some) [Reveal 1989 Phytologia 66:295–414] Largest dicot genus in CA; apparently currently differentiating; many taxa ± indistinct, descriptions for several new taxa from DMoj are currently in press. Better habitat data needed. Many are excellent bee fodder. ❀ Most are attractive and easy to grow with good drainage.

Key to Groups

1. Ann
  2. Involucre ± ribbed or angled, sessile, gen erect . . . . **Group 1**
  2' Involucre unribbed, unangled, gen stalked, gen reflexed . . . . **Group 2**
1' Per to shrub
  3. Perianth base ± stalk-like, ± as wide as pedicel and jointed to it; bracts gen lf-like, ± reduced
    above, 2–10 per node . . . . **Group 3**
  3' Perianth base not stalk-like; bracts gen scale-like throughout, 3 per node . . . . **Group 4**

Group 1: Ann; involucre ± ribbed, sessile, gen erect

1. Most or all involucres appearing terminal; infl ± repeatedly forked or ± evenly branched
  2. Perianth cream; infl narrow, erect, branches stout; SNE . . . . *E. ampullaceum*
  2' Perianth yellow; infl spreading, open, branches threadlike; DMoj . . . . *E. mohavense*
1' Most involucres appearing lateral; infl cyme-like, ± unevenly branched
  3. Involucres 3–4 mm — SNE, DMoj . . . . *E. davidsonii*
  3' Involucres < 1.5 mm
    4. Perianth minutely hairy — rare, DMtns (Cottonwood Mtns) . . . . *E. puberulum*
    4' Perianth (sub)glabrous (sometimes minutely glandular)
      5. Infl axes glabrous
        6. Perianth white to pink, 1–2 mm, gen glandular . . . . *E. baileyi* var. *baileyi*
        6' Perianth yellow, ± 1 mm, glabrous . . . . *E. brachyanthum*
      5' Infl axes ± tomentose
        7. Perianth glandular; uncommon . . . . *E. baileyi* var. *praebens*
        7' Perianth glabrous; common
          8. Perianth yellow to red; infl branches gen curved ± inward . . . . *E. nidularium*
          8' Perianth white to pink or pale yellow; infl branches gen curved ± outward . . . . *E. palmerianum*

Group 2: Ann; involucre unribbed, stalked, gen reflexed

1. Lvs basal and cauline (or lower bracts lf-like)
  2. Lf linear, < 2 mm wide, rolled under; infl axes glandular; W&I . . . . *E. spergulinum* var. *reddingianum*
  2' Lf lanceolate to ovate, > 2 mm wide, flat; infl axes not glandular; GB, D
    3. Perianth soft-wavy-hairy — w DSon . . . . [2]*E. ordii*
    3' Perianth glabrous or glandular
      4. Stamens long-exserted . . . . *E. angulosaum*
      4' Stamens incl
        5. Outer perianth lobes flat, margins wavy; DMoj . . . . *E. gracillimum*
        5' Outer perianth lobes with a pouch-like bulge or cup, margins not wavy; GB, D
          6. Outer perianth lobes cupped at base; GB, D . . . . *E. maculatum*
          6' Outer perianth lobes cupped at tip; w DMoj . . . . *E. viridescens*
1' Lvs basal, at most a few ± cauline on lowermost st
  7. Lvs glabrous to coarsely hairy or hairy in patches (not tomentose); perianth often yellow (sometimes
    white to red)
    8. Perianth glabrous — SNE . . . . *E. esmeraldense*
    8' Perianth subglabrous to hairy
      9. Perianth subglabrous to sparsely soft-hairy
        10. Infl axes sparsely hairy — w DSon . . . . [2]*E. ordii*
        10' Infl axes subglabrous or sparsely glandular
          11. Perianth white, ± 2 mm; sw DMtns . . . . *E. apiculatum*
          11' Perianth pink to red, ± 0.5 mm; n W&I . . . . *E. parishii*
      9' Perianth densely and often coarsely hairy
        12. Perianth whitish; infl glandular throughout — n&ne DMtns . . . . *E. glandulosum*
        12' Perianth yellow; infl glabrous or glandular only at base or nodes
          13. Involucre teeth 4 . . . . *E. trichopes*
          13' Involucre teeth 5
            14. Infl axes slender throughout, nodes glandular — e DMoj . . . . *E. contiguum*
            14' Infl axes wider below nodes, glabrous (pls rarely 1st-year-fl per; see also Group 4) . . . . . . *E. inflatum*
7' Lvs ± tomentose on one or both surfaces; perianth gen white to reddish (sometimes yellow)
  15. Outer perianth lobes ± cordate at base (magnification needed)
    16. Involucres erect
      17. Infl wide, flat-topped, gen < 30 cm long; ne DMoj . . . . *E. bifurcatum*
      17' Infl sometimes narrow, not flat-topped, 20–120 cm long; s DMoj, DSon . . . . *E. deflexum* var. *rectum*
    16' Involucres reflexed

18. Infl axes glandular . . . . *E. brachypodum*
18' Infl axes glabrous, exc sometimes at very base
  19. Perianth yellow
    20. Involucres sessile; infl ± umbrella-like; SNE. . . . . *E. hookeri*
    20' Involucres on spreading or recurved stalks; infl ± open, not umbrella-like; D . . . . [2]*E. thomasii*
  19' Perianth white to reddish
    21. Infl in several tiers; involucres ± 1 mm; perianth base swollen — ne DMoj. . . . . *E. rixfordii*
    21' Infl not tiered; involucre 1–3 mm; perianth base not swollen . . . . *E. deflexum* (in part)
      22. Involucre 2–3 mm, stalk 3–15 mm; main infl axes often stout or gen wider below nodes . . var. *baratum*
      22' Involucre 1–2 mm, stalk 0–5 mm; infl axes slender throughout . . . . var. *deflexum*
15' Outer perianth lobes obtuse to truncate at base
  23. Perianth minutely hairy
    24. Perianth white to red, ± hair-tufted inside . . . . *E. thurberi*
    24' Perianth yellow or if whitish, not hair-tufted inside
      25. Outer perianth lobe bases swollen . . . . [2]*E. thomasii*
      25' Outer perianth lobe bases not swollen
        26. Involucre minutely glandular . . . . *E. pusillum*
        26' Involucre glabrous . . . . *E. reniforme*
  23' Perianth glabrous
    27. Involucres ± sessile
      28. Sand; SNE . . . . [2]*E. deflexum* var. *nevadense*
      28' Rocky slopes; n DMtns (Death Valley region, Inyo Co.). . . . . *E. hoffmannii*
        29. Lf margin flat; pl spreading, < 50 cm; perianth lobes narrowly spoon-shaped . . . . var. *hoffmannii*
        29' Lf margin strongly wavy; pl erect, < 100 cm; perianth lobes narrowly ovate . . . . var. *robustius*
    27' Involucres on ± recurved or reflexed stalks
      30. Infl glandular ± throughout — s W&I, n DMtns . . . . *E. eremicola*
      30' Infl glabrous
        31. Outer perianth lobes wavy-margined — W&I . . . . *E. cernuum* var. *cernuum*
        31' Outer perianth lobes flat
          32. Involucre obconic, ± 1.5 mm; SNE . . . . [2]*E. deflexum* var. *nevadense*
          32' Involucre bell-shaped, ± 2 mm; n SNE . . . . *E. nutans*

Group 3: Per to shrub; perianth base ± stalk-like, jointed to pedicel; bracts gen ± lf-like below, ± reduced above

1. Perianth base ± indistinctly stalk-like, often ± wider than pedicel and funnel-like
  2. Lvs subglabrous and green — W&I, Panamint Mtns. . . . . *E. latens*
  2' Lvs woolly below, **white- to brown-** tomentose above
    3. Infl densely umbel-like; involucres terminal; bracts lf-like; stalk-like perianth base short, not angled . . *E. lobbii*
    3' Infl open, cyme-like; most involucres lateral; bracts scale-like; stalk-like perianth base long, sharply
      3-angled . . . . *E. saxatile*
1' Perianth base distinctly stalk-like, ± as wide as pedicel
  4. Perianth hairy . . . . *E. caespitosum*
  4' Perianth glabrous . . . . *E. umbellatum*
    5. Infl a simple umbel (only subtending rays bracted, rays unbranched)
      6. Perianth bright yellow, gen weakly striped
        7. Low dense mat; infl gen spreading; uncommon, subalpine rocks . . . . var. *covillei*
        7' Subshrub; infl ± erect; common, Jeffrey-pine forest, sagebrush shrublands . . . . var. *nevadense*
      6' Perianth cream to pale yellow, or pink to red, gen strongly striped
        8. Perianth gen cream to pale yellow; W&I . . . . var. *dichrocephalum*
        8' Perianth gen pink to reddish brown; s W&I, n DMtns. . . . . [2]var. *versicolor*
    5' Infl a compound umbel (some rays branched, or bracts whorled near mid-axis or mid-ray)
      9. Lf blade ± glabrous on both surfaces . . . . var. *chlorothamnus*
      9' Lf blade hairy at least below
        10. Perianth pink to reddish brown, conspicuously dark-striped . . . . [2]var. *versicolor*
        10' Perianth whitish to yellow, not conspicuously striped
          11. Perianth whitish to pale yellow; e DMtns (San Bernardino Co.). . . . . var. *juniporinum*
          11' Perianth bright yellow; SNE, DMtns . . . . var. *subaridum*

Group 4: Per to shrub; perianth base not stalk-like; bracts gen scale-like throughout

1. St jointed, breaking into short-cylindric sections; fls 150–200 per involucre — n DMtns. . . . . . . *E. intrafractum*
1' St not jointed, not breaking into sections; fls 7–100 per involucre
  2. Pl cushion-like or matted
    3. Inner and outer perianth lobes distinctly different (outer round, wider than inner, often notched)

4. Perianth balloon-like from bulging outer perianth lobes; uncommon — n DMtns . . . . *E. gilmanii*
4' Perianth not balloon-like; common . . . . *E. ovalifolium*
  5. Perianth bright yellow . . . . var. *ovalifolium*
  5' Perianth white to cream or purple
    6. Lf blade 2–8 mm; infl axis < 5 cm; perianth gen 2–3 mm; W&I . . . . var. *nivale*
    6' Lf blade 10–60 mm; infl axis 4–20 cm; perianth gen 4–6 mm; GB . . . . var. *purpureum*
3' Inner and outer perianth lobes similar
  7. Perianth yellow to reddish
    8. Involucre and infl axis ± tomentose; shale or gravel; 1300–2100 m — e Mono Co.
    . . . . *E. ochrocephalum* var. *alexanderae*
    8' Involucre and infl axis glandular; dry, volcanic outcrops; 1500–4000 m
      9. Lf blade 10–25 mm; involucre teeth 5; 1500–2500 m . . . . *E. beatleyae*
      9' Lf blade 4–12 mm; involucre teeth 5–8; 2300–4000m . . . . *E. rosense*
  7' Perianth white (yellowish) to rose
    10. Infl axis glabrous . . . . *E. kennedy* var. *purpusii*
    10' Infl axis hairy or glandular
      11. Perianth and fr glabrous; n W&I . . . . *E. gracilipes*
      11' Perianth and fr long-hairy; SNE, n DMtns . . . . *E. shockleyi* var. *shockleyi*
2' Pl gen ± erect, gen not cushion-like or matted (*E. wrightii* var. *subscaposum* and *E. microthecum*
  vars. loosely matted)
12. Subshrub or shrub
  13. Gen most involucres sessile, lateral; infl branches long
    14. Branches fragile; involucres lateral esp near infl tips . . . . [2]*E. nummulare*
    14' Branches stout; involucres lateral throughout infl (see 32' for vars) . . . . [2]*E. wrightii*
  13' Most involucres terminal; infl branches ± short
    15. Involucres clustered at nodes . . . . *E. fasciculatum*
      16. Lf greenish, linear, tightly rolled under, subglabrous above . . . . var. *flavoviride*
      16' Lf grayish, oblanceolate, slightly rolled under, ± tomentose above . . . . var. *polifolium*
    15' Involucres solitary (1 per node)
      17. Infl with definable main axis (± unevenly branched, side branches shorter than main axis)
        18. Infl ± tiered; outer perianth lobes narrowed to base . . . . *E. plumatella*
        18' Infl gen not tiered; outer perianth lobe bases ± cordate . . . . [2]*E. heermannii*
        19. Infl axes deeply grooved — limestone cliffs . . . . var. *sulcatum*
        19' Infl axes round, not grooved
          20. Infl axes thinly tomentose . . . . var. *floccosum*
          20' Infl axes ± glabrous
            21. Perianth 1.5–2.5 mm; limestone cliffs . . . . var. *argense*
            21' Perianth 2.5–3 mm; gravel . . . . var. *humilius*
      17' Infl without definite main axis, branches ± even on length
        22. Perianth and fr hairy — s DSon . . . . *E. deserticola*
        22' Perianth and fr glabrous
          23. Large, ± erect shrub, < 200 cm; lf blade elliptic to oblong, 10–30 mm
            24. Infl branches stout, ± spreading (see 18' for vars.) . . . . [2]*E. heermannii*
            24' Infl open, branches fragile — SNE . . . . [2]*E. nummulare*
          23' Low, spreading shrub or subshrub, < 60 cm; lf blade linear to narrowly elliptic, 3–20 mm
            25. Lf 4–6 mm, linear, strongly rolled under — e DMtns . . . . *E. ericifolium*
            25' Lf 3–25 mm, gen ± oblanceolate, nearly flat (upper, smaller lvs sometimes rolled under)
            . . . . *E. microthecum*
              26. Largest lvs gen 15–25 mm, nearly flat
                27. Perianth yellow . . . . var. *ambiguum*
                27' Perianth white to pink
                  28. Lvs flat; sts, infl thinly tomentose to glabrous; Inyo Co. northward . . . . var. *laxiflorum*
                  28' Lvs rolled under; sts, infl densely woolly to tomentose; s Inyo Co. . . . . var. *simpsonii*
              26' Largest lvs gen 3–15 mm, gen distinctly rolled under
                29. Pl shrubby, rounded, gen 30–60 cm . . . . var. *panamintense*
                29' Pl subshrubs, decumbent, 4–15 cm
                  30. Infl compact, main axes gen < 1 cm; lf gen thinly hairy above; n SNE . . . . var. *alpinum*
                  30' Infl ± open, main axes gen 2–4 cm; lf gen ± densely tomentose above; s W&I . . . var. *lapidicola*
12' Per (at most somewhat woody above caudex)
  31. Most involucres terminal, gen stalked (exc *E. nudum*); infl branches either ± short or ± equal
    32. Lf blade gen 40–150 mm, lanceolate — long-hairy and green on both surfaces . . . . *E. elatum* var. *elatum*
    32' Lf blade gen < 40 mm; oblong to oblanceolate or ovate
      33. Lvs coarsely short-hairy; involucres 1–1.5 mm . . . . *E. inflatum*
        34. Infl slender; uncommon; s DMoj, DSon . . . . var. *deflatum*
        34' Infl main axes wider below major nodes; common; GB, D . . . . var. *inflatum*

33' Lvs tomentose (at least below); involucres 2–5 mm . . . . *E. nudum*
   35. Perianth glabrous, white; w SNE . . . . var. *deductum*
   35' Perianth hairy, yellow; s SNE, w DMoj . . . . var. *westonii*
31' Most involucres lateral, sessile; infl branches long, ± wand-like, ± unequal
   36. Lf blade elliptic to broadly ovate or round, 1.5–4 cm
      37. Lf blade elliptic to oblong — n W&I, ne DMtns (Cottonwood Mtns). . . . . *E. rupinum*
      37' Lf blade broadly ovate to round
         38. Lf blade round; involucre bell shaped, 2–4 mm; s W&I, n DMtns (Panamint Mtns) . . . . . *E. mensicola*
         38' Lf blade broadly elliptic to ovate or obovate; involucre funnel shaped, 3–5 mm; W&I, n DMtns
           . . . . *E. panamintense*
   36' Lf blade linear to elliptic, < 1.5 cm . . . . [2]*E. wrightii*
      39. Pl loosely matted, 10–30 cm; W&I . . . . var. *subscaposum*
      39' Pl not matted, 15–100 cm; D
         40. Pl 30–100 cm; lf woolly; infl axils stout, rigid, densely gray-woolly; s DMoj, w DSon. . . var. *nodosum*
         40' Pl 15–50 cm; lf tomentose (esp below); infl gen slender, glabrous to white- or green-tomentose;
           DMoj . . . . var. *wrightii*

E. ampullaceum J. Howell (p. 429) [Group 1] Ann 5–30 cm. **LVS** basal; blade 5–15 mm, ± round, tomentose (esp below). **INFL** cyme-like, erect, slender, ± repeatedly forked; bracts scale-like; involucres sessile (few lateral), ± erect, 1.5–2 mm, ± ribbed, teeth 5. **FL**: perianth 1–1.5 mm, cream. **FR** 1–1.3 mm, glabrous. Uncommon. Sand; 1700–2200 m. **n SNE**; w NV. Aug–Sep

E. angulosum Benth. (p. 429) [Group 2] Ann 10–90 cm. **LVS** basal and cauline on lower st; blade 5–40 mm, ± lanceolate, tomentose (esp below). **INFL** cyme-like; axes gen grooved or angled; bracts ± lf-like below (reduced upward), and scale-like; involucres on slender stalks, 1.5–3 mm, bell-shaped, unribbed, sparsely hairy, teeth 5, rounded. **FL**: perianth 1.5–1.8 mm, white to rose, minutely glandular, lobes ± cupped, inner narrower. **FR** 1–1.3 mm, glabrous. Common. Sand or clay; 150–1900 m. c&s SNF, Teh, SnJV, CW, TR, s PR, **w DMoj**. ❀ TRY. May–Nov

E. apiculatum S. Watson (p. 429) [Group 2] Ann 20–90 cm, ± glandular. **LVS** basal; blade 5–40 mm, ± oblanceolate, coarsely hairy and glandular. **INFL** cyme-like, slender, unevenly branched; involucres on thread-like stalks, ± 1.5 mm, unribbed, ± glabrous, teeth 4. **FL**: perianth 1.5–2.5 mm, ± white, short-white-hairy, outer lobes abruptly soft-pointed. **FR** ± 1.5 mm, glabrous. Uncommon. Granite sand; 1100–2700 m. c&s PR (Santa Rosa, Palomar, Cuyamaca mtns), **sw DMtns (Little San Bernardino Mtns)**. Jul–Aug

E. baileyi S. Watson [Group 1] Ann 10–50 cm. **LVS** basal; blade 5–20 mm, ± round, tomentose. **INFL** cyme-like, ± unevenly branched; bracts scale-like; involucres sessile (many lateral), 1–1.5 mm, ± ribbed, teeth 5. **FL**: perianth 1–2 mm, white to pink, subglabrous or minutely glandular. **FR** 1–1.5 mm, glabrous. Common. Sand or gravel; 650–2900 m. SNF, Teh, s SCoR, TR, **GB, w DMoj**; to WA, ID, UT.

  var. **baileyi** (p. 429) **INFL** glabrous. Common. Habitats and range of sp. May–Sep

  var. **praebens** (Gand.) Rev. **INFL** tomentose. Uncommon. Sand; 1300–2900 m. **GB**; w NV. [var. *divaricatum* (Gand.) Rev.]

E. beatleyae Rev. (p. 429) BEATLEY'S BUCKWHEAT [Group 4] Per 8–15 cm; mats 10–50 cm diam. **LVS** ± basal; blade 10–25 mm, ± oblanceolate, densely tomentose. **INFL** head-like; axis erect, glandular; bracts scale-like, ± hidden; involucres 3–6 per head, 3–4 mm, bell-shaped, rigid, glandular and thinly hairy, teeth 5. **FL**: perianth 2.5–4 mm, pale to reddish yellow, subglabrous, stalk-like base 0. **FR** 2.5–3.5 mm, glabrous. UNCOMMON. Dry volcanic outcrops; 1500–2500 m. **n SNE**; NV. Much like *E. ochrocephalum*.

E. bifurcatum Rev. (p. 429) FORKED BUCKWHEAT [Group 2] Ann 5–40 cm. **LVS** basal; blade 5–30 mm, round, tomentose (esp below). **INFL** cyme-like, 30–150 cm diam, spreading, flat-topped; bracts scale-like; involucres slender-stalked (esp lower), 2–2.5 mm, unribbed, glabrous, teeth 5. **FL**: perianth 1.5–2 mm, white to reddish, outer lobe bases cordate, inner lobes narrower. **FR** 2–2.5 mm, glabrous. *n*=20. RARE. Sand; 700–800 m. **ne DMoj (s Inyo, ne San Bernardino cos.)**; s NV.

E. brachyanthum Cov. [Group 1] Ann 5–40 cm. **LVS** basal; blade 5–20 mm, ovate to round, tomentose (esp below). **INFL** ± cyme-like, densely short-branched, gen rounded, glabrous; bracts scale-like; involucres sessile (most lateral), ± 1 mm, angled, glabrous, teeth 5. **FL**: perianth ± 1 mm, yellow, glabrous, inner lobes narrower. **FR** ± 1 mm, glabrous. Sand; 600–2000 m. **SNE, DMoj**; NV. Locally common. ❀ STBL. May–Sep

E. brachypodum Torrey & A. Gray (p. 429) [Group 2] Ann 5–50 cm, 20–100 cm diam, minutely glandular. **LVS** basal; blade 10–50 mm, round-cordate, tomentose (esp below). **INFL** cyme-like, stout, gen ± flat-topped; bracts scale-like; involucres on slender, reflexed stalks, 1–2.5 mm, minutely glandular, unribbed, teeth 5. **FL**: perianth 1–2.5 mm, white to reddish, outer lobes gen bulged and cordate at base, inner narrower. **FR** 1–2 mm, glabrous. *n*=20. Sand or gravel; 150–2400 m. **SNE, DMoj**; to sw UT, nw AZ. Locally common. ❀ STBL.

E. cernuum Nutt. NODDING BUCKWHEAT [Group 2] Ann 5–60 cm, ± slender. **LVS** basal; blade 5–25 mm, ± round, white-tomentose (esp below). **INFL** cyme-like, glabrous, glaucous; bracts scale-like; involucres gen on reflexed stalks, 1–2 mm, unribbed, glabrous, teeth 5. **FL**: perianth 1–2 mm, white to pinkish, glabrous, outer lobes ± wavy-margined. **FR** 1.5–2 mm, glabrous. Sand or gravel; 1300–3300 m. c SNH (e slope), **GB**; to WA, w&c Can, c US, NM.

  var. **cernuum** (p. 429) **INFL**: involucre stalk 2–25 mm, ascending, spreading, or reflexed. Uncommon. Sand or gravel; 2100–3300 m. c SNH (e slope), **W&I**; range of sp. outside CA. Jun–Sep

E. caespitosum Nutt. (p. 429) [Group 3] Per from much-branched woody caudex, < 10 cm; mats 20–70 cm diam; some fls staminate. **LVS** ± basal on caudex branches; blade 2–5(15) mm, ± elliptic, rolled under, tomentose. **INFL**: bracts 0; involucre 1 per axis, 2–3.5 mm, lobes 6–9, long, reflexed, tomentose. **FL**: perianth 2.5–5 mm (staminate fl) or 6–10 mm (bisexual fl), yellow to reddish, hairy, stalk-like base short, distinct. **FR** 3.5–5 mm; tip sometimes slightly hairy. Sand or gravel; 1200–3100 m. **GB**; to OR, MT, Colorado. Locally common. ❀ TRY; DFCLT. May–Jul

E. contiguum (Rev.) Rev. (p. 429) ANNUAL DESERT TRUMPET [Group 2] Ann 3–30 cm, spreading. **LVS** basal; blade 4–20 mm, round-cordate, sparsely and coarsely hairy. **INFL** cyme-like, dense, slender; nodes glandular; bracts scale-like; involucres on ± erect, thread-like stalks, 1–2 mm, unribbed, glabrous, teeth gen 5. **FL**: perianth 1–2.5 mm, yellow, coarsely yellow-hairy. **FR** 1.5–2 mm, glabrous. *n*=16. RARE in CA. Sandy flats; 50–1000 m. **e DMoj (se Inyo, ne San Bernardino cos.)**; sw NV.

E. davidsonii E. Greene [Group 1] Ann 5–40 cm. **LVS** basal; blade 10–20 mm, round to reniform, wavy-margined, tomentose (esp below). **INFL** cyme-like, unequally branched; bracts scale-like; involucres sessile (most lateral), 3–4 mm, cylindric, ribbed, glabrous, teeth 5. **FL:** perianth 1.5–2 mm, white to pink or red, glabrous, inner lobes narrower, shorter. **FR** 2–2.5 mm, glabrous. Volcanic or granitic soils; 200–2800 m. SW, **SNE, DMoj**; to sw UT, n AZ, n Baja CA. [*E. molestum* S. Watson var. *d.* (E. Greene) Jepson] ❀ DRN,DRY,SUN:1–3,7,8–10,**14–24**. Jun–Sep

E. deflexum Torrey FLAT-TOPPED BUCKWHEAT [Group 2] Ann 5–200 cm. **LVS** ± basal; blade 10–40 mm, round to reniform, white-woolly below, tomentose above. **INFL** cyme-like, ± unevenly branched, gen ± widely spreading, glabrous; bracts scale-like; involucres on gen reflexed stalks, 1–3 mm, ± unribbed, teeth 5. **FL:** perianth 1–3 mm, white to pinkish, glabrous, outer lobes gen cordate at base, inner ± narrow. **FR** 1.5–3 mm, glabrous. *n*=20 (all vars.). Sandy desert scrub; < 2900 m. Teh, s SnJV, TR, **SNE, D**; to w UT, w NM, nw Mex.

var. ***baratum*** (Elmer) Rev. (p. 429) **INFL** sometimes narrow; main axes often stout or wider below main nodes; involucre stalk 3–15 mm, reflexed; involucre 2–3 mm. **FL:** perianth 1.5–2 mm, outer lobe bases truncate. Dry slopes; 900–2900 m. Teh, s SnJV, TR, **W&I, n&w DMoj**; s NV. [ssp. *b.* (Elmer) Munz] Locally common. ❀ TRY. Jul–Oct

var. ***deflexum*** (p. 429) **INFL:** axes ± slender; involucre stalk 0–5 mm, reflexed; involucre 1–2 mm. **FL:** perianth 1–2 mm, outer lobe bases cordate. Sand; < 2000 m. **W&I, D**; to s UT, AZ, nw Mex. Locally common. ❀ TRY. May–Oct

var. ***nevadense*** Rev. **INFL:** axes ± slender; involucre stalk 0–5 mm, reflexed; involucre ± 1.5 mm. **FL:** perianth 1–2 mm, outer lobe bases truncate. Sand; 1100–2200 m. **SNE**; to w UT. Locally common. Intergrades with var. *deflexum* in n DMoj. ❀ TRY.

var. ***rectum*** Rev. **INFL** sometimes narrow; axes slender to stout; involucre 1.5–2 mm, erect, stalk 0–5 mm. **FL:** perianth 1–2 mm. Uncommon. Sand; < 1000 m. **s DMoj, DSon**. [*E. insigne* S. Watson misapplied to CA pls]

E. deserticola S. Watson (p. 429) [Group 4] Shrub 60–150 cm, 100–300 cm diam, openly branched. **LVS** cauline; blade 5–15 mm, ± ovate, densely tomentose. **INFL** cyme-like, open, fragile; bracts scale-like; involucres gen sessile (many lateral), 1.5–2 mm, gen ± hairy, teeth 4, very short. **FL:** perianth 2.5–3.5 mm, yellow, hairy, stalk-like base 0, lobe tips truncate. **FR** 3–4 mm, gen hairy. Uncommon. Sand; ~65–100 m. **s DSon (Imperial Co.)**; sw AZ, nw Sonora. Sep–Dec

E. elatum Benth. [Group 4] Per 40–150 cm, 5–15 cm diam. **LVS** basal; blade gen 40–250 mm, ± lanceolate, gen tomentose (esp below). **INFL** cyme-like; bracts scale-like or barely lf-like; involucres 1–5 per cluster, gen short-stalked, 2.5–4 mm, teeth 5. **FL:** perianth 2.5–3 mm, white to red, base hairy, not stalk-like. **FR** 3.5–4 mm, glabrous. Sand or gravel; 600–3100 m. **n CA**; to WA, ID, c NV. 2 vars. in CA.

var. ***elatum*** (p. 429) **INFL:** axes and involucres glabrous. *n*=20. Common. Habitats and range of sp. ❀ DRN,SUN:1–3,6,**7**, 14,**15–17**,18–21. Jun–Sep

E. eremicola J. Howell & Rev. (p. 429) WILD ROSE CANYON BUCKWHEAT [Group 2] Ann 8–25 cm, ± spreading. **LVS** basal; blade 3–25 mm, ± round, white-woolly below, tomentose above. **INFL** cyme-like, open, finely glandular; bracts scale-like; involucres on reflexed stalks (at least below), 1–2 mm, unribbed, teeth 5. **FL:** perianth 2–2.5 mm, white, yellow, or red, glabrous. **FR** 2–2.3 mm, glabrous. RARE. Sand or gravel; 2200–3100 m. **s W&I (Inyo Mtns), n DMtns (Panamint Mtns)**.

E. ericifolium Torrey & A. Gray var. ***thornei*** Rev. & Henrickson THORNE'S BUCKWHEAT [Group 4] Subshrub 4–15 cm, 10–45 cm diam. **LVS** cauline; blade 4–6 mm, linear, rolled under, tomentose (esp below). **INFL** ± cyme-like, ± dense; bracts scale-like; involucres sessile or short-stalked, 1.5–2 mm, glabrous,

teeth 5. **FL:** perianth 1.5–2 mm, white to pink, glabrous, stalk-like base 0. **FR** 2 mm, glabrous. **ENDANGERED** CA. Copper-rich gravel; ± 1800 m. **e DMtns (New York Mtns)**. Other vars. in AZ. Very difficult to distinguish from *E. microthecum* complex.

E. esmeraldense S. Watson var. ***esmeraldense*** [Group 2] Ann 5–50 cm. **LVS** basal; blade 5–40 mm, obovate to round, sparsely and coarsely hairy. **INFL** cyme-like, gen spreading, glabrous, glaucous; bracts scale-like; involucres on slender, spreading to reflexed stalks, 0.8–1.8 mm, unribbed, glabrous, teeth 5. **FL:** perianth 1–3 mm, white to red, glabrous. **FR** 1.4–2.5 mm, glabrous. Sand; 1500–3200 m. c SNH (e slope), **SNE**; NV. Locally common. Another var. in c NV. Jul–Sep

E. fasciculatum (Benth.) Torrey & A. Gray CALIFORNIA BUCKWHEAT [Group 4] Shrub 10–200 cm, 50–300 cm diam. **LVS** cauline, clustered at nodes; blade 6–18 mm, linear to oblanceolate, leathery, ± rolled under, ± white-tomentose below. **INFL:** head or umbel of heads, gen ± canescent; lowest bracts lf-like, upper scale-like; involucres gen several per head, 2.5–4 mm, angled, teeth 5. **FL:** perianth 2.5–3 mm, white to pinkish, stalk-like base 0. **FR** 1.8–2.5 mm, glabrous. Common. Dry slopes, washes, canyons in scrub; < 2300 m. c&s CA-FP, **SNE, D**; to UT, AZ, nw Mex. 4 vars. in CA.

var. ***flavoviride*** Munz & I.M. Johnston Pl ± rounded. **LF** linear, tightly rolled under, light (or yellow-) green, glabrous above. **INFL:** involucre 2–3 mm, gen glabrous. **FL:** perianth gen hairy inside. *n*=20. Dry slopes, washes; 200–1300 m. **D**; Baja CA. ❀ TRY. Mar–May

var. ***polifolium*** (Benth.) Torrey & A. Gray (p. 429) Pl ± rounded. **LF** oblanceolate, slightly rolled under, ± tomentose above. **INFL** ± tomentose; involucre 2.5–3.5 mm, hairy. **FL:** perianth hairy. *n*=20. Dry slopes, washes; 60–2300 m. s SN, SCoRI, e SCo, TR, e PR, **SNE, D**; to sw UT, w AZ, nw Mex. ❀ DRN,DRY,SUN: **7**,8–11,**14–16**,17,**18–24**;STBL. Apr–Nov

E. gilmanii S. Stokes (p. 429) GILMAN'S BUCKWHEAT [Group 4] Per from a shredding woody caudex, 1–5 cm; mats < 20 cm diam. **LVS** ± rosetted on low sts; blade 2–4 mm, elliptic, tomentose. **INFL** dense, gen ± head-like; bracts scale-like; involucres gen several per head, ± sessile, gen 3–4 mm, rigid, teeth 5. **FL:** perianth ± balloon-like, 3.5–5 mm, reddish, glabrous, lobes 3–5 mm, outer shorter, ± round, cupped, inner narrow, stalk-like base 0. **FR** 2.5–3 mm, glabrous. UNCOMMON. Gravel slopes; 1800–2000 m. **n DMtns (Panamint, Last Chance mtns)**. Aug–Sep

E. glandulosum (Nutt.) Benth. (p. 429) [Group 2] Ann 5–40 cm, ± widely spreading. **LVS** basal; blade 5–15 mm, widely elliptic to round, sparsely and coarsely hairy. **INFL** cyme-like, flat-topped, glandular ± throughout; bracts scale-like; involucres on slender, ± reflexed stalks, ± 1 mm, unribbed, glabrous, teeth 5. **FL:** perianth 1–2 mm, white to pink, stiffly white-hairy, lobes ± widely lanceolate. **FR** 1–1.3 mm, glabrous. Sand or gravel; 800–1600 m. **n&ne DMtns**; sw NV. [*E. carneum* (J. Howell) Rev.] Locally common.

E. gracilipes S. Watson (p. 429, pl. 84) [Group 4] Per 5–10 cm; mats 5–20 cm diam. **LVS** clustered on low sts; blade 5–20 mm, ± elliptic, white-tomentose. **INFL** head-like; axis glandular; bracts scale-like; involucres 5–7 per head, sessile, 2–3 mm, flexible, ± hairy, teeth 5. **FL:** perianth 2–3 mm, white to rose, glabrous, stalk-like base 0. **FR** 2–2.5 mm, glabrous. Uncommon. Dry granite; 2900–4000 m. c&s SNH (e slope), **n W&I (White Mtns)**; w NV. ❀ TRY;DFCLT. Jun–Aug

E. gracillimum S. Watson (p. 433) [Group 2] Ann 5–50 cm. **LVS** basal and seemingly cauline; blade 5–40 mm, ± elliptic, white-tomentose (esp below), margins wavy, ± rolled under. **INFL** cyme-like, slender; bracts lf- and scale-like below, reduced upward; axes gen ± grooved; involucres on slender, spreading stalks, ± 2 mm, bell-shaped, nearly unangled, finely glandular, teeth 5. **FL:** perianth ± 2 mm, white to rose, subglabrous, margins gen ± wavy. **FR** ± 1 mm, glabrous. *n*=20. Common. Clay to gravel; 500–1300 m. s SN, SCoR, TR, **DMoj**. ❀ STBL. Apr–Sep

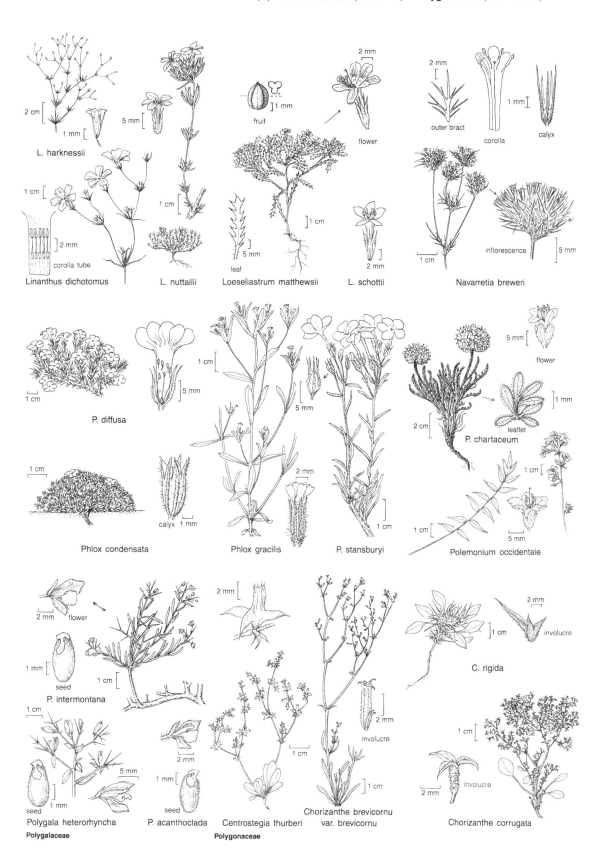

L. harknessii

Linanthus dichotomus

L. nuttallii

fruit

flower

leaf

Loeseliastrum matthewsii

L. schottii

outer bract

corolla

calyx

inflorescence

Navarretia breweri

P. diffusa

Phlox condensata

Phlox gracilis

P. stansburyi

flower

leaflet

P. chartaceum

Polemonium occidentale

flower

seed

P. intermontana

seed

Polygala heterorhyncha

Polygalaceae

seed

P. acanthoclada

Centrostegia thurberi

Polygonaceae

involucre

Chorizanthe brevicornu
var. brevicornu

involucre

C. rigida

involucre

Chorizanthe corrugata

E. heermannii Durand & Hilg. [Group 4] Shrub 10–200 cm, 10–250 cm diam, gen rounded. **LVS** cauline; blade 5–40 mm, ± linear to spoon-shaped, gen ± tomentose. **INFL** cyme-like, gen dense; bracts scale-like; axes many, rigid, ± widely spreading; involucres sessile, 1–2 mm, glabrous, teeth 5. **FL**: perianth 1.5–4 mm, white or yellowish to rose, glabrous, outer lobes ± obovate-cordate, inner ± oblanceolate, stalk-like base 0. **FR** 2–2.5 mm, glabrous. Dry rocks, washes; 600–2800 m. c&s SNH (e slope), Teh, SCoR, TR, **SNE, esp DMtns**; to UT, n AZ. Highly variable complex. 6 vars. in CA.

var. *argense* (M.E. Jones) Munz Pl 10–45 cm, 20–80 cm diam, ± low, spreading. **LF**: blade 5–10 mm, oblanceolate to spoon-shaped. **INFL**: axes round, glabrous. **FL**: perianth 1.5–2.5 mm, white. Uncommon. Limestone outcrops, cliffs; 1200–2800 m. **W&I, DMtns**; s NV.

var. *floccosum* Munz CLARK MOUNTAIN BUCKWHEAT Pl 30–60 cm, 40–80 cm diam. **LF**: blade 5–15 mm, ± linear, rolled under. **INFL**: axes round, thinly tomentose. **FL**: perianth 2–3 mm, yellowish white. UNCOMMON. Dry slopes, washes; 900–2400 m. **e DMtns**; s NV. ❀ TRY. Aug–Oct

var. *humilius* (S. Stokes) Rev. Pl 30–70 cm, 50–120 cm diam, low, spreading. **LF**: blade 8–15 mm, oblanceolate to spoon-shaped. **INFL** ± open to dense; axes round, glabrous. **FL**: perianth 2.5–3 mm, white. Gravel; 1000–2000 m. c&s SNH (e slope), **SNE, n DMtns**; NV.

var. *sulcatum* (S. Watson) Munz & Rev (p. 433). Pl 10–40 cm, 15–60 cm diam. **LF**: blade 4–12 mm, narrowly (ob)lanceolate to spoon-shaped. **INFL**: axes deeply grooved, minutely scabrous. **FL**: perianth 1.5–2 mm, yellowish white. Uncommon. Limestone outcrops; 700–2700 m. **DMtns**; to sw UT, nw AZ. Sep–Oct

E. hoffmannii S. Stokes [Group 2] Ann 10–100 cm. **LVS** ± basal; blade 10–50 mm, ± round, tomentose (esp below). **INFL** cyme-like, open; bracts scale-like; main axes gen dominant; involucres ± sessile (many lateral), 1–2 mm, unribbed, teeth 5. **FL**: perianth 1.5 mm, white to reddish, glabrous; stamen incl, glabrous. **FR** 2 mm, glabrous. *n*=20. Dry slopes; 300–1700 m. **n DMtns (Death Valley region, Inyo Co.)**.

var. *hoffmannii* HOFFMANN'S BUCKWHEAT Pl 10–50 cm, spreading. **LF**: blade flat. **FL**: perianth lobes narrowly spoon-shaped. UNCOMMON. Dry slopes; 1000–1700 m. **n DMtns (w slope Panamint Mtns)**. Aug–Sep

var. *robustius* S. Stokes ROBUST HOFFMANN'S BUCKWHEAT Pl 30–100 cm, erect. **LF**: blade strongly wavy-margined. **FL**: perianth lobes narrowly ovate. UNCOMMON. Dry slopes; 300–750 m. **ne DMtns (Black, Funeral mtns)**. Aug–Sep

E. hookeri S. Watson (p. 433) [Group 2] Ann 10–60 cm. **LVS** basal; blade 10–50 mm, ± round, white-tomentose (esp below). **INFL** cyme-like, gen umbrella-like; bracts scale-like; involucres sessile, reflexed, 1–2 mm, widely bell-shaped, unribbed, teeth 5. **FL**: perianth ± 2–4 mm, yellow or reddish, outer lobes round, bases ± cordate, inner lobes narrower. **FR** 2–2.5 mm, glabrous. *n*=20. Uncommon. Sand or gravel; 1000–2500 m. **SNE, DMtns (Panamint Mtns)**; to ID, WY, Colorado, NM.

E. inflatum Torrey & Frémont DESERT TRUMPET [Group 4] Ann or per 10–150 cm. **LVS** basal; blade 5–50 mm, oblong to reniform, ± coarsely short-hairy. **INFL** cyme-like, glabrous, glaucous; main axes gen wider below nodes; bracts scale-like; involucres on erect or spreading, thread-like stalks, 1–1.5 mm, teeth 5. **FL**: perianth 1–3 mm, yellow, coarsely white-hairy, lobes narrowly ovate. **FR** 2–2.5 mm, glabrous. *n*=16. Dry sand or gravel; –15–2000 m. **GB, D**; to Colorado, NM, nw Mex.

var. *deflatum* I.M. Johnston Per, long-lived, not fl 1st year, 50–150 cm. **INFL**: branches many, slender. Habitats of sp.; < 1100 m. **D**; s NV, w AZ, nw Mex. Young pls much like var. *inflatum*. ❀ TRY.

var. *inflatum* (p. 433) Ann or per, often fl 1st year, 2–80 cm. **INFL**: branches few–many; main axes wider below nodes.

Common. Habitats and range of sp ❀ DRN,DRY,SUN:1–3,7–9,**10–13**,14,19–24. Mar–Jul, Sep–Oct

E. intrafractum Cov. & C. Morton (p. 433) JOINTED BUCKWHEAT [Group 4] Per 60–150 cm. **LVS** basal; blade 25–70 mm, oblong-ovate, ± woolly. **INFL** ± cyme-like; axes few, stout, brittle, finally breaking into napkin-ring-like segments; bracts ± lf-like at first node, scale-like above; branches 2–3, long, wand-like; involucres gen 3 per node, sessile, 2.5–3 mm, bell-shaped, splitting as fls mature, teeth 5. **FLS** many per involucre; perianth 1.5–3 mm, yellow to red, bristly, lobes oblanceolate, stalk-like base 0. **FR** 2–2.5 mm; tip bristly. *n*=20. UNCOMMON. Dry limestone; 1100–1600 m. **n DMtns (Grapevine, Panamint mtns)**. May–Aug

E. kennedyi S. Watson [Group 4] Per 5–15 cm; cushions 10–50 cm diam. **LVS** densely clustered on low sts; blade 2–12 mm, elliptic to oblong, often rolled under, tomentose. **INFL** head-like, gen ± tomentose; bracts scale-like, subtending involucres; involucres 3–8, sessile, 1.5–4 mm, rigid, ± hairy, teeth 5–7. **FL**: perianth 1.5–3.5 mm, white to rose, glabrous, lobes ± widely elliptic, stalk-like base 0. **FR** 2–4 mm, glabrous. Dry gravel; 1200–3600 m. c&s SN, TR, **SNE, nw DMtns**. 5 vars. in CA.

var. *purpusii* (Brandegee) Rev. **LF**: blade 2.5–6 mm, oblong. **INFL**: axis 4–10 cm; involucre 1.5–2 mm, glabrous. **FL**: perianth 2–2.5 mm. **FR** 2.5–3 mm. Uncommon. Dry granite sand; 1200–2500 m. c&s SN (e slope), SNE, **nw DMtns (Argus, Coso mtns)**.

E. latens Jepson [Group 3] Per from thick woody caudex, 10–50 cm; mats 10–20 cm diam. **LVS** basal; blade 10–35 mm, gen ± round, ciliate, subglabrous, green. **INFL** ball-like (branches and bracts ± hidden); axis stout, ± glabrous; involucres gen many per head, 6–8 mm, lobes 5–8, ± spreading. **FL**: perianth 3–6 mm, cream to pale yellow, stalk-like base indistinct, angled, hairy. **FR** 3–5 mm, glabrous. Uncommon. Granite sand; 2500–3400 m. c&s SNH (e slope), **W&I, n DMtns (Panamint Mtns)**; w NV. ❀ DRN, SUN,IRR:1–3,7,14,**15–17**,18. Jul–Aug

E. lobbii Torrey & A. Gray (p. 433) [Group 3] Per from thick, sparsely branched, woody caudex, gen 3–15 cm, 5–40 cm diam. **LVS** basal; blade 10–50 mm, ± round, woolly below, ± tomentose above. **INFL** densely umbel-like (sometimes compound); axis often ± prostrate, slender, tomentose; bracts lf-like, subtending branches; involucre 1 per ultimate ray, 4–12 mm, lobes 6–10, long, ± spreading. **FL**: perianth 5–7 mm, white to rose, dark-striped, glabrous, stalk-like base very short, ± indistinct. **FR** 5–8 mm, glabrous. Common. Open, rocky slopes and ridges; 1400–3500 m. NW, CaR, SN, **GB**; OR, w NV. ❀ DRN,SUN:1–3,6,7, 14–18;DFCLT. Jun–Aug

E. maculatum A.A. Heller (p. 433) [Group 2] Ann 5–40 cm. **LVS** basal (some seemingly cauline); blade 5–40 mm, lanceolate to obovate, densely white-tomentose (esp below); margin sometimes wavy or rolled under. **INFL** cyme-like, open; axes ± angled, ± tomentose; lower bracts lf- and scale-like, reduced upward; involucres on thread-like, spreading stalks, 1–2 mm, cup-shaped, unribbed, glandular, teeth 5. **FL**: perianth 1–2.5 mm, white to red or yellow, striped rose to purple, glandular, outer lobes wider, base ± cupped, inner lobes longer. **FR** 1–1.5 mm, glabrous. *n*=20. Common. Gravel to clay soils; 100–2500 m. s SN, TR, **GB, D**; to WA, UT, Baja CA. ❀ TRY. Apr–Nov

E. mensicola S. Stokes [Group 4] Per 15–30 cm, 10–25 cm diam. **LVS** basal; blade 10–15 mm, round ± densely tomentose. **INFL** cyme-like, slender, ± unevenly branched; bracts scale-like; involucres sessile (many lateral), 2–4 mm, bell-shaped, tomentose, teeth 5. **FL**: perianth 3–4 mm, white or brownish, glabrous, stalk-like base). **FR** 2.5–3 mm, glabrous. Open gravel; 1800–2500 m. **s W&I, n DMtns (Panamint Mtns)**; NV. Jul–Aug

E. microthecum Nutt. [Group 4] Subshrub or shrub 5–150 cm, 60–160 cm diam. **LVS** cauline; lower ± clustered; blade 3–25 mm, gen ± oblanceolate, gen ± rolled under, tomentose (esp below). **INFL** cyme-like, gen ± flat-topped; bracts scale-like; involucres terminal, 1.5–4 mm. **FL**: perianth 1.5–4 mm, white to

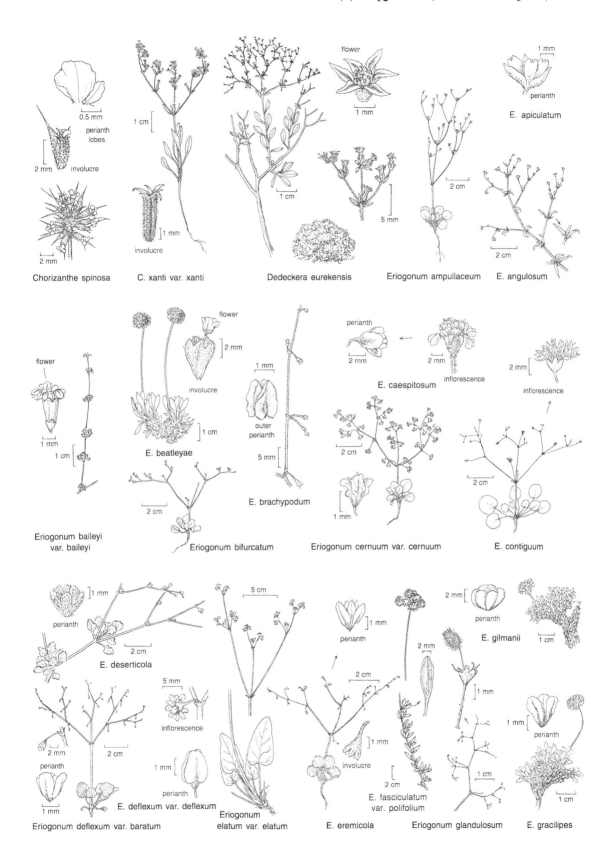

Chorizanthe spinosa

C. xanti var. xanti

Dedeckera eurekensis

Eriogonum ampullaceum

E. angulosum

E. apiculatum

Eriogonum baileyi var. baileyi

E. beatleyae

Eriogonum bifurcatum

E. brachypodum

E. caespitosum

Eriogonum cernuum var. cernuum

E. contiguum

E. deserticola

Eriogonum deflexum var. baratum

E. deflexum var. deflexum

Eriogonum elatum var. elatum

E. eremicola

E. fasciculatum var. polifolium

Eriogonum glandulosum

E. gilmanii

E. gracilipes

rose or yellow, glabrous, stalk-like base 0. **FR** 1.5–3 mm, glabrous. Dry places; 1100–3300 m. SN, SnGb, SnBr, **GB, DMoj**; to WA, MT, WY, Colorado, NM. [Reveal 1971 Brigham Young Univ Sci Bull Biol Ser 13(1):1–45] Highly variable; vars. intergrading, often indistinct, differentiated esp by geog. 8 vars. in CA, typical var. in OR, ID.

var. *alpinum* Rev. Pl 4–10 cm. **LF**: blade 3–9 mm, oblanceolate, gen rolled under, thinly hairy above. **INFL**: main branches gen < 1 cm. **FL**: perianth white to rose. Uncommon. Rocks; 2500–3300 m. c & n SNH, **n SNE**.

var. *ambiguum* (M.E. Jones) Rev. (p. 433) Pl 5–50 cm. **LF**: blade 8–25 mm, largest nearly flat. **FL**: perianth yellowish. Common. Rocks; 1100–3000 m. SN (e slope), **GB**; NV. ❀ TRY.

var. *lapidicola* Rev. (p. 433) INYO MOUNTAINS BUCKWHEAT Pl 5–15 cm. **LF**: blade 3–7 mm, gen ± densely tomentose above. **INFL**: main branches gen 2–4 cm. **FL**: perianth whitish to rose or orange. RARE. Rocks; 2600–3100 m. **s W&I (Inyo Mtns)**; NV.

var. *laxiflorum* Hook. Pl 10–50 cm. **LF**: blade 5–25 mm, gen nearly flat. **FL**: perianth white to pink. Rocks; 1200–3100 m. SN (e slope), **GB, DMtns**; range of sp. outside CA.

var. *panamintense* S. Stokes (p. 433) PANAMINT MOUNTAINS BUCKWHEAT Pl 30–60 cm. **LF**: blade gen 5–8(15) mm, strongly rolled under. **FL**: perianth brownish. RARE. Rocks; 1900–2800 m. **s W&I (Inyo Mtns), n DMtns (Panamint Mtns)**. Jul–Oct

var. *simpsonii* (Benth.) Rev. Pl 10–150 cm. **LF**: blade 5–25 mm, upper smaller, strongly rolled under. **INFL** ± open, not very flat-topped. **FL**: perianth white to pink. Sand or gravel; 1300–2200 m. **s SNE, DMoj**; to Colorado, NM. ❀ TRY.

E. mohavense S. Watson (p. 433) [Group 1] Ann 5–30 cm. **LVS** basal; blade 4–20 mm, oblong to round, wavy-margined, tomentose (esp below). **INFL** cyme-like, evenly branched, slender, subglabrous; involucres sessile (few lateral), 1.5–2 mm, glabrous, ribbed, teeth 5. **FL**: perianth < 1 mm, yellow, gen glabrous. **FR** 1–1.3 mm, glabrous. Uncommon. Sand; 700–1300 m. **DMoj**. May–Aug

E. nidularium Cov. (p. 433, pl. 85) [Group 1] Ann 5–30 cm. **LVS** basal; blade 5–20 mm, ± round. **INFL** cyme-like, densely and ± unevenly branched, often curved ± inward; axis thinly tomentose; bracts scale-like; involucres sessile (most lateral), ± 1 mm, ribbed, teeth 5. **FL**: perianth gen 2–3 mm, yellow to red, glabrous, lobes fan-shaped. **FR** ± 1 mm, glabrous. Common. Sand or gravel flats, washes; 300–2300 m. **GB, D**; to se OR, s ID, UT, AZ. ❀ STBL. Apr–Oct

E. nudum Benth. [Group 4] Per, 10–200 cm. **LVS** basal or ± spaced on lower st; blade 10–70 mm, oblanceolate to ovate, gen tomentose (esp below). **INFL** cyme- to head-like; axes glabrous, sometime wider below major nodes; bracts scale-like (lowest sometimes lf-like); involucres 1–10 per cluster, 2–5 mm, teeth 5–8. **FL**: perianth 1.5–4 mm, white, yellow, or red, subglabrous, stalk-like base 0. **FR** 1.5–3.5 mm, glabrous. Abundant. Dry open places; < 3800 m. **CA (exc ChI)**; to WA, NV, nw Mex. Vars. difficult, intergrading. 13 vars. in CA.

var. *deductum* (E. Greene) Jepson (p. 433) **LVS** basal, gen 10–20 mm; margins gen flat. **INFL** slender, glabrous; involucres gen 1–3 per node. **FL**: perianth 2.5–3.5 mm, white, glabrous. Open places; 1000–3200 m. SN, **w SNE**; w NV. Jul–Aug

var. *westonii* (S. Stokes) J. Howell **LVS** basal, 10–30 mm, often tomentose above; margins ± flat. **INFL** slender or stout, glabrous; involucres 1–4 per node, 3–5 mm, gen glabrous. **FL**: perianth 2.5–3 mm, yellow, obviously hairy. *n*=20. Common. Dry places; 300–2800 m. s SN, Teh, s CW, TR, **s SNE, w DMoj**. Intergrades with var. *pubiflorum*. ❀ TRY. Jun–Sep

E. nummulare M.E. Jones (p. 433) [Group 4] Shrub 15–100 cm, 30–150 cm diam. **LVS** cauline; blade 10–30 mm, gen ± oblanceolate, tomentose (esp below). **INFL** cyme-like, open; axes thick,

fragile; bracts scale-like; involucres sessile (some lateral, esp near branch tips), 2–3 mm, teeth 5. **FL**: perianth 1.5–3 mm, white, glabrous, stalk-like base 0. **FR** 2–3 mm, glabrous. Sand; 1100–2600 m. s MP, **SNE, DMtns (Last Chance Mtns)**; to w UT. [*E. kearneyi* Tidestrom, Kearney's buckwheat, incl var. *monoense* (S. Stokes) Rev.]

E. nutans Torrey & A. Gray (p. 433) NODDING BUCKWHEAT [Group 2] Ann 5–30 cm, reddish. **LVS** basal; blade 5–20 mm, round, densely tomentose (esp below). **INFL** cyme-like, open; bracts scale-like; involucres on slender, ± reflexed stalks, ± 2 mm, bell-shaped, subglabrous, teeth 5. **FL**: perianth 2–3 mm, white to red, glabrous, outer lobes ± widely elliptic, inner narrower. **FR** ± 2 mm, glabrous. *n*=40. UNCOMMON. Sand or gravel; ± 3000 m. e MP (ne Lassen Co.), **n SNE (e Mono Co.)**; to se OR, w UT. Scattered but expected more widely. Grayish pls from NV [var. *glabrum* Rev., glabrous nodding buckwheat] may be roadside waifs in SNH. May–Jul

E. ochrocephalum S. Watson [Group 4] Per 10–40 cm; mats 5–15 cm diam. **LVS** clustered on low sts; blade 10–35 mm, lanceolate to obovate, tomentose. **INFL** head-like; involucres 5–8 per head, 3.5–5 mm, ± bell-shaped, rigid, thinly tomentose, ribbed, teeth 6–8. **FL**: perianth 2–3 mm, yellow, glabrous, lobes oblong, stalk-like base 0. **FR** 1.5–3 mm, glabrous. Uncommon. Dry clay or rocky slopes; 1200–2400 m. **GB**; to sw OR, s ID, w NV. Much like *E. beatleyae*. 2 vars. in CA.

var. *alexanderae* Rev. (p. 433) ALEXANDER'S BUCKWHEAT **INFL**: axis tomentose; involucre gen tomentose. RARE in CA. Shale or gravel; 1300–2100 m. **n SNE (e Mono Co.)**; w NV.

E. ordii S. Watson [Group 2] Ann 5–70 cm, erect, gen glabrous. **LVS** basal; blade 20–80 mm, narrowly oblanceolate to obovate, subglabrous to hairy in patches. **INFL** cyme-like, slender; lowest bracts gen lf-and scale-like, reduced upward; involucres on thread-like stalks, 1–1.5 mm, glabrous, teeth 4. **FL**: perianth 1–2.5 mm, white to red, soft-hairy. **FR** ± 2 mm, glabrous. Uncommon. Barren clay; 15–1500 m. s SNF, Teh, SCoRI, n WTR, **w DSon**; nw AZ? Mar–Jun

E. ovalifolium Nutt. [Group 4] Per 15–35 cm; mats 3–40 cm diam. **LVS** ± basal on dense caudex; petiole long, slender; blade 2–60 mm, elliptic to ± round, tomentose to woolly. **INFL** gen head-like, < 30 cm; bracts scale-like, ± hidden; involucres 1–15 per head, 2–6.5 mm, rigid, hairy, teeth 5. **FLS** sometimes unisexual; perianth 2.5–7 mm, white to purple or yellow, glabrous, outer lobes obovate to round, inner narrower, gen notched, stalk-like base 0. **FR** 2–3 mm, glabrous. *n*=20. Common. Dry sand or gravel; 1200–4100 m. KR, CaRH, SNH, ne SnBr, **GB, DMtns**; to s Can, MT, Colorado, NM. Vars. ± intergrade. 5 vars. in CA.

var. *nivale* (Canby) M.E. Jones (p. 433) **LF**: blade 2–8 mm, gen round, ± unmargined. **INFL**: axis < 5 cm; involucres 3–5, 2–4 mm. **FL**: perianth gen 2–3 mm, white to cream. Common. Alpine sand or gravel; 1500–4100 m. CaRH, SNH, **n SNE, W&I**; to B.C., w UT. ❀ DRN,SUN:1–3,**7,15–17**. Jul–Aug

var. *ovalifolium* (p. 433) **LF**: blade 10–60 mm, gen ± obovate, ± unmargined. **INFL**: axis 4–20 cm; involucres 3–15, 4–6.5 mm. **FL**: perianth gen 4–6 mm, yellow. Common. Sand or gravel; 1200–2900 m. SNH (e slope), **GB, DMtns**; to OR, MT, Colorado. ❀ DRN,SUN:1–3,**7,15–17**,22–24. May–Jul

var. *purpureum* (Nelson) Durand **LF**: blade 5–20 mm, gen ± obovate, ± unmargined. **INFL**: axis 4–20 cm; involucres 3–15 per cluster, 4–6.5 mm. **FL**: perianth gen 4–6 mm, white to cream or purplish. Common. Sand or gravel; 1200–2800 m. KR, CaRH, SNH (e slope), **GB**; range of sp. [var. *ovalifolium* misapplied] ❀ TRY.

E. palmerianum Rev. (p. 433) [Group 1] Ann 5–30 cm. **LVS** basal; blade 5–15 cm, gen round, tomentose (esp below). **INFL** cyme-like, ± open; branches few, tips often outcurved; involucres sessile (many lateral), ± 1 mm, thinly hairy, ± angled, teeth 5. **FL**: perianth 1.5–2 mm, white to pink or pale yellow, glabrous. **FR** ± 1.7 mm, glabrous. Sand or gravel; 600–3400 m. **SNE, D**; to Colorado, NM. Locally common. ❀ TRY.

E. panamintense C. Morton (p. 433) [Group 4] Per 15–30 cm, 10–25 cm diam. **LVS** basal; blade 15–40 mm, elliptic to ovate or obovate, ± densely tomentose. **INFL** cyme-like, slender, ± unevenly branched; bracts lf-like below, reduced upward; involucres sessile (few lateral), 3–5 mm, funnel-shaped, tomentose, teeth 5. **FL**: perianth 3.5–5 mm, white or brownish, glabrous, stalk-like base 0. **FR** 2.5–3 mm, glabrous. Open gravel; 1500–2900 m. **W&I, DMtns**; NV. May–Oct

E. parishii S. Watson (p. 433) [Group 2] Ann 10–50 cm, densely and finely branched. **LVS** basal; blade 20–60 mm, ± oblanceolate, coarsely hairy. **INFL** cyme-like, open, slender, glandular above nodes; bracts scale-like; involucres on thread-like, spreading stalks (many seemingly lateral), ± 0.5 mm, glabrous, unribbed, teeth 4. **FL**: perianth ± 0.5 mm, pink to red, subglabrous. **FR** 1–1.3 mm, glabrous. *n*=20. Common. Granite sand; 1200–3200 m. s SN, TR, PR, **n W&I (White Mtns)**; AZ, n Baja CA. ❀ TRY. Jul–Sep

E. plumatella Durand & Hilg. (p. 433) [Group 4] Shrub 30–60 cm, 30–60 cm diam. **LVS** cauline; blade 6–15 mm, gen obovate, ± densely white-tomentose. **INFL** cyme-like, open, appearing ± layered; main axes obvious; branches spreading-recurved; bracts ± scale-like; involucres sessile, ± 2 mm, glabrous, teeth 5. **FL**: perianth 2–2.5 mm, white to pale yellow, glabrous, outer lobes spreading, obovate, inner erect, narrower, stalk-like base 0. **FR** 2.5–3 mm, glabrous. Common. Dry slopes, washes; 300–1800 m. **D**; to sw UT, nw AZ. [var. *jaegeri* (Munz & I.M. Johnston) Munz] ❀ TRY. Aug–Oct

E. puberulum S. Watson (p. 433) DOWNY BUCKWHEAT [Group 1] Ann 2–30 cm. **LVS** basal; blade 5–15 mm, obovate to round, sparsely hairy. **INFL** cyme-like, minutely silky; lowest bracts ± strap-like, upper scale-like; involucres sessile (most lateral), ± 1 mm, barely angled, minutely silky, lobes 4, deep. **FL**: perianth 1–1.5 mm, white to red, minutely hairy. **FR** 1–1.5 mm, glabrous. RARE in CA. Gravelly pinyon/juniper woodland; 1300–2900 m. **n DMtns (Cottonwood Mtns, Death Valley region, Inyo Co.)**; to sw UT.

E. pusillum Torrey & A. Gray (p. 433) [Group 2] Ann 5–30 cm. **LVS** basal; blade gen 5–15 mm, gen round, white-woolly (esp below). **INFL** cyme-like, slender, ± glabrous; bracts scale-like; involucres on slender straight stalks, ± 2 mm, cup-shaped, unribbed, minutely glandular, teeth 5. **FL**: perianth 1–2.5 mm, yellow becoming reddish, glandular, lobes ± obtuse at base. **FR** ± 0.7 mm, glabrous. *n*=16. Common. Sand or gravel; 150–2600 m. SCoRI, WTR, **GB, D**; to OR, ID, NV. ❀ STBL. Mar–Jul

E. reniforme Torrey & Frémont (p. 433) [Group 2] Ann 5–40 cm. **LVS** basal; blade 5–20 mm, round-reniform, ± densely white-tomentose. **INFL** cyme-like, spreading, slender, ± glabrous, glaucous; bracts scale-like; involucres on slender, ± curved stalks, ± 2 mm, unribbed, glabrous, glaucous, teeth 5. **FL**: perianth ± 1 mm, yellow, becoming reddish, minutely glandular, outer lobes wider, ± obtuse. **FR** ± 1 mm, glabrous. *n*=16. Common. Sand or gravel; < 1500 m. **D**; s NV, w AZ, nw Mex. ❀ STBL. Mar–Jun

E. rixfordii S. Stokes (p. 433) RIXFORD'S BUCKWHEAT [Group 2] Ann 15–40 cm. **LVS** basal; blade 5–30 mm, round-cordate, densely white-tomentose (esp below). **INFL** cyme-like, of pagoda-like tiers of branches, stout, glabrous, glaucous; bracts scale-like; involucres ± sessile, reflexed, ± 1 mm, unribbed, glabrous, teeth 5. **FL**: perianth ± 1.5 mm, white to reddish, outer lobes narrowly ovate, bases ± cordate. **FR** 1.5 mm, glabrous. *n*=20. UNCOMMON. Sand or gravel; 150–1600 m. **ne DMoj (Death Valley)**; sw NV.

E. rosense Nelson & Kenn. (p. 433) [Group 4] Per gen < 10 cm; mats 5–20 cm diam, ± glandular throughout. **LVS** clustered

on low sts; blade 4–15 mm, ± oblanceolate, coarsely tomentose (esp below). **INFL** head-like; bracts scale-like; involucres 3–5, 3–3.5 mm, rigid, sparsely hairy, teeth 5–8. **FL**: perianth 2–3 mm, bright yellow to reddish, stalk-like base 0. **FR** 1.5–2 mm, glabrous. *n*=20. Uncommon. Dry granite or volcanic outcrops; 2300–4000 m. s CaRH (Lassen Peak), n&c SNH, **n SNE**; w NV. [*E. anemophilum* E. Greene misapplied]

E. rupinum Rev. [Group 4] Per 30–50 cm, 20–40 cm diam. **LVS** basal; blade 20–40 mm, elliptic to oblong, thinly tomentose. **INFL** cyme-like, slender, ± unevenly branched; bracts lf-like below, reduced upward; involucres sessile (many lateral), 3–4 mm, funnel-shaped, thinly tomentose, teeth 5. **FL**: perianth 2.5–3.5 mm, white or cream, glabrous, stalk-like base 0. **FR** 2–3 mm, glabrous. Open gravel; 1800–3000 m. **n W&I, DMtns (Cottonwood Mtns)**; NV.

E. saxatile S. Watson (p. 435) [Group 3] Per 10–20(40) cm; mats 5–20 cm diam. **LVS** clustered on low sts; blade 3–25 mm, elliptic to round, densely tomentose. **INFL** cyme-like, ± stout, ± tomentose; involucres sessile (most lateral), 2–4 mm, ± tomentose, teeth 5–6. **FL**: perianth 3–7 mm, white to rose or yellowish, glabrous, outer lobes narrower, stalk-like base long, sharply 3-angled, ± indistinct. **FR** 3.5–4 mm, glabrous, winged. *n*=20. Common. Decomposed granite or volcanic rocks; 900–3400 m. s SN, CW, SW, **SNE, DMtns**; s NV. ❀ DRN,DRY,SUN:1–3,7,14, **15–18**,19–24. May–Jul

E. shockleyi S. Watson var. ***shockleyi*** (p. 435) SHOCKLEY'S BUCKWHEAT [Group 4] Per gen < 6 cm; mats 5–100 cm diam. **LVS** clustered on low sts; blade 2–12 mm, oblanceolate to elliptic, tomentose. **INFL** head-like, ± erect; bracts scale-like; involucres 2–6 per head, 2–6 mm, bell-shaped, ± tomentose, teeth 5–10. **FL**: perianth 2.5–4 mm, white or yellowish, densely soft-hairy toward base, stalk-like base 0. **FR** ± 3 mm, hairy. UNCOMMON. Dry limestone gravel in pinyon/juniper woodland; 1700–2700 m. **SNE, n DMtns (Last Chance Mtns)**; to s ID, Colorado, NM.

E. spergulinum A. Gray [Group 2] Ann 5–40 cm. **LVS** basal; blade 3–30 mm, linear (< 2 mm wide), ciliate. **INFL** cyme-like, prostrate to erect, slender; involucres on ± erect, thread-like stalks, 0.5–1 mm, unribbed, glabrous, teeth 4. **FL**: perianth 1–3 mm, white, striped darker. **FR** ± 2 mm, glabrous. Dry sand (esp granitic); 1200–3500 m. NCoRH, CaRH, SN, n SCoRO, TR, **W&I**; to OR, ID, NV. 3 vars. in CA.

var. ***reddingianum*** (M.E. Jones) J. Howell **INFL** erect, 8–40 cm; axes glandular. **FL**: perianth ± 2 mm. Common. Sand or gravel; 1400–3500 m. NCoRH, CaRH, SN, n SCoRO, TR, **GB**; to OR, ID, NV. Jun–Aug

E. thomasii Torrey (p. 435) [Group 2] Ann 5–30 cm. **LVS** basal; blade 5–20 mm, round-cordate, densely tomentose (esp below). **INFL** cyme-like, open, ± glabrous, glaucous; bracts scale-like; involucres on slender, spreading stalks, ± 1 mm, cup-shaped, unribbed, glabrous, teeth 5. **FL**: perianth 0.5–1 mm (2 mm in fr), yellow becoming white or rose, minutely hairy, outer lobes sometimes swollen-cordate at base. **FR** ± 1 mm, glabrous. *n*=20. Sand or gravel; – 61–1300 m. **D**; s NV, w AZ, nw Mex. Locally common. ❀ STBL. Mar–Jun

E. thurberi Torrey (p. 435) [Group 2] Ann 5–40 cm. **LVS** basal; blade 8–45 mm, elliptic to round, tomentose (esp below). **INFL** cyme-like, ascending; axes glandular; involucres on slender stalks, ± 1.5 mm, unribbed, minutely glandular, teeth 5. **FL**: perianth ± 1.5 mm, white becoming reddish, minutely glandular, lobes ± hair-tufted inside, outer lobes ± fan-shaped, bases ± obtuse. **FR** ± 0.7 mm, glabrous. Sand or gravel; 100–1300 m. e PR, **s DMoj, DSon**; AZ, nw Mex. ❀ STBL. Apr–Jul

E. trichopes Torrey (p. 435) [Group 2] Ann 10–180 cm. **LVS** basal; blade 5–40 mm, oblong to reniform, gen ± wavy-margined, glabrous to thinly hairy. **INFL** cyme-like; base ± hairy; axes often wider below main nodes; bracts scale-like; involucres on thread-like stalks, 0.5–2.5 mm, glabrous, teeth gen 4. **FL**: perianth 1–2.5 mm, yellow becoming reddish, coarsely and minutely white-hairy, lobes ovate. **FR** ± 2 mm, glabrous. *n*=16. Open clay, sand, gravel, or serpentine; –60–1300 m. SCoRI, n WTR, **D**; to UT, NM, nw Mex. Apr–Aug

E. umbellatum Torrey SULFUR FLOWER [Group 3] Per to shrub, 10–200 cm, 10–200 cm diam. **LVS** clustered on low sts; blade 3–40 mm, gen ± elliptic, gen densely tomentose (esp below). **INFL** umbel- to head-like, erect, slender; bracts lf-like, subtending rays (if ray 1, appearing whorled near mid-axis), rarely alternate on main axis); involucre gen 1 per ray, 1–6 mm, ± tomentose, lobes 6–12, long, reflexed. **FL**: perianth 2.5–12 mm, gen ± yellow becoming reddish (cream to purple), glabrous, lobes ± obovate, stalk-like base long, distinct. **FR** 2–5 mm, glabrous. Abundant. Dry, open, often rocky places; 200–3700 m. **CA (exc coast)**; to w Can, Colorado, NM. Extremely variable and difficult. Many vars. intergrade, best dispositions unclear; more study needed. 17 vars. in CA.

var. *chlorothamnus* Rev. Gen shrub, glabrous. **LF**: blade gen ± oblanceolate. **INFL** compound; some rays again branched. **FL**: perianth 3–6 mm, bright yellow. Uncommon. Sagebrush scrub; 1600–2900 m. s SNH, **SNE, nw DMoj**. Intergrades with var. *subaridum*.

var. *covillei* (Small) Munz & Rev. Low mat. **LF**: blade 2–6 mm wide, ± narrowly elliptic, sometimes remaining densely tomentose above. **INFL** simple, gen spreading. **FL**: perianth 2–5 mm, bright yellow. Uncommon. Subalpine rocks; 2600–3600 m. c&s SNH (esp Tulare, Inyo cos.), **W&I**. Intergrades with var. *nevadense*. Jul–Sep

var. *dichrocephalum* Gand. **LF**: blade ± elliptic, gen subglabrous below. **INFL** simple, subglabrous. **FL**: perianth 4–8 mm, whitish to pale yellow, gen conspicuously striped. Sand or gravel; 1500–3700 m. Wrn, **W&I**; to OR, MT, WY. ❀ TRY.

var. *juniporinum* Rev. JUNIPER BUCKWHEAT Subshrub. **INFL** compound, ± tomentose; rays gen branched again. **FL**: perianth 4–6 mm, whitish to pale yellow. RARE in CA. Pinyon/juniper woodlands, sagebrush shrublands; 1300–2500 m. **e DMtns (e San Bernardino Co.)**; NV. Like var. *subaridum*.

var. *nevadense* Gand. (p. 435) Subshrub. **LF**: blade ± thinly tomentose below. **INFL** simple, gen thinly tomentose. **FL**: perianth 4–7 mm, bright yellow. Common. Jeffrey-pine forest, sagebrush shrublands; 1500–3200 m. SN (e slope), **GB, nw DMoj**; s OR, c NV. [var. *umbellatum* (Rocky Mtns) misapplied] ❀ TRY.

var. *subaridum* S. Stokes (p. 435) Subshrub. **LF**: blade narrowly elliptic or oblanceolate. **INFL** compound; rays gen again branched (or with small bracts whorled near middle). **FL**: perianth 3–7 mm, bright yellow. Common. Sagebrush shrublands, rocks; 1300–2800 m. se SN, TR, **SNE, DMtns**; to Colorado, AZ. Closely related to var. *chlorothamnus*.. ❀ TRY. Jul–Aug

var. *versicolor* S. Stokes Mat, gen 10–40 cm diam. **INFL** simple or rays sometimes branched again. **FL**: perianth 3–6 mm, pink to reddish brown, conspicuously dark-striped. Uncommon. Sagebrush shrublands; 1800–3200 m. **s W&I (Inyo Mtns)**, **DMtns**; s NV. Jul–Sep

E. viridescens A.A. Heller (p. 435) [Group 2] Ann 5–30 cm. **LVS** basal and seemingly cauline on lower st, 5–30 mm, gen ± elliptic, white-tomentose (esp below). **INFL** cyme-like, open, ± tomentose; axes gen round; bracts lf- and scale-like below, reduced upward; involucres on spreading, slender stalks, 2–3 mm, cup-shaped, unribbed, minutely glandular, teeth 5. **FL**: perianth 1–2 mm, white to rose, minutely glandular, outer lobes obovate, ± cupped at tip. **FR** ± 1 mm, glabrous. *n*=20. Sand, gravel, or clay; < 1700 m. SnJV, SCoR, TR, **w DMoj**. Locally common. ❀ TRY. May–Oct

E. wrightii Benth. [Group 4] Per to shrub, 15–100 cm, 10–150 cm diam. **LVS** cauline or clustered on low sts; blade 1–30 mm, linear to widely elliptic, gen densely tomentose (esp below). **INFL** cyme- to head-like; branches long, wand-like, glabrous to densely woolly; bracts scale-like; involucres sessile (most lateral), 1–4 mm, glabrous to densely woolly, teeth 5. **FL**: perianth 1.5–4 mm, white to pink or rose, glabrous, lobes obovate, stalk-like base 0. **FR** 1–3 mm, glabrous. Dry gravel or rocks; 50–3500 m. NW, s CaR, SN, CW, SW, **W&I, D**; to s NV, TX, Mex. Variable; vars. intergrade. 6 vars. in CA.

var. *nodosum* (Small) Rev. (p. 435) Shrub 30–100 cm, 30–150 cm diam, densely gray-woolly. **LF**: blade 8–12 mm, ± flat. **INFL** open; branches stout; involucre 1.5–2.5 mm. **FL**: perianth 3–4 mm. **FR** 2.5–3 mm. Uncommon. Dry gravel or rocks; 150–1600 m. e PR, **s-most DMoj**, w DSon; n Baja CA.

var. *subscaposum* S. Watson Subshrub 5–30 cm; mats 10–50 cm diam, loose, glabrous to tomentose. **LF**: blade 5–12 mm, ± flat. **INFL** dense to open, gen slender; involucre gen 2–4 mm. **FL**: perianth 2–3 mm. **FR** 2–2.5 mm. *2n*=34. Common. Dry gravel or rocks; 200–3400 m. SN, CW, TR, e PR, **W&I, DMtns (Panamint Mtns)**. ❀ DRN,SUN:1–3,7,14,**15,16,**17, **18–21,**22–24. Jul–Oct

var. *wrightii* Subshrub 15–50 cm, 10–50 cm diam, gen tomentose. **LF**: blade 5–15 mm, ± flat. **INFL** open, slender; involucre 2–2.5 mm. **FL**: perianth ± 3 mm. **FR** 2.5–3 mm. Uncommon. Dry gravel or rocks; 30–2300 m. **DMoj**; to TX, n Mex. Aug–Oct

GILMANIA

1 sp. (M. French Gilman, CA botanist, 1871–1944) [Reveal 1989 Phytologia 66:236–245]

G. luteola (Cov.) Cov. (p. 435, pl. 86) GOLDEN CARPET Ann, ± prostrate, 3–20 cm diam, thinly hairy. **LVS** basal (bracts appear as cauline lvs); stipule 0; petiole < 3 cm; blade 1–1.5 cm, ± obovate, ± glabrous. **INFL** 5–30 cm diam; bracts clustered at nodes, lf-like; involucre 0; fls several–many per bract axil; pedicels 2–7 mm. **FL**: perianth 1–2 mm, yellow, thinly hairy, lobes 6; stamens 9. **FR** 1.5–2 mm, brownish. RARE. Barren alkaline scrub; < 500 m. **DMoj (Death Valley, Inyo Co.)**. Mar–Apr

GOODMANIA

1 sp. (George J. Goodman, Oklahoma botanist, 1904–) [Reveal & Ertter 1976 Brittonia 28:427–429]

G. luteola (C. Parry) Rev. & B. Ertter (p. 435) Ann, spreading, 1–8 cm, thinly hairy. **LVS** basal; stipule 0; petiole 3–20 mm; blade 2–6 mm, ± round, lower surface tomentose, upper surface thinly hairy. **INFL** 2–10 cm diam; bracts lf-like to linear and awned; involucral bracts 5, free, 3–8 mm, unequal, linear, awned; fls several–many per involucre. **FL**: perianth 0.8–1 mm, yellow, base hairy, lobes 6; stamens 9. **FR** 1–1.2 mm, brownish. *n*=20. Uncommon. Grassland, alkaline and desert scrub; < 2200 m. s SnJV, **SNE, w DMoj**; NV [*Oxytheca l.* C. Parry] May–Aug

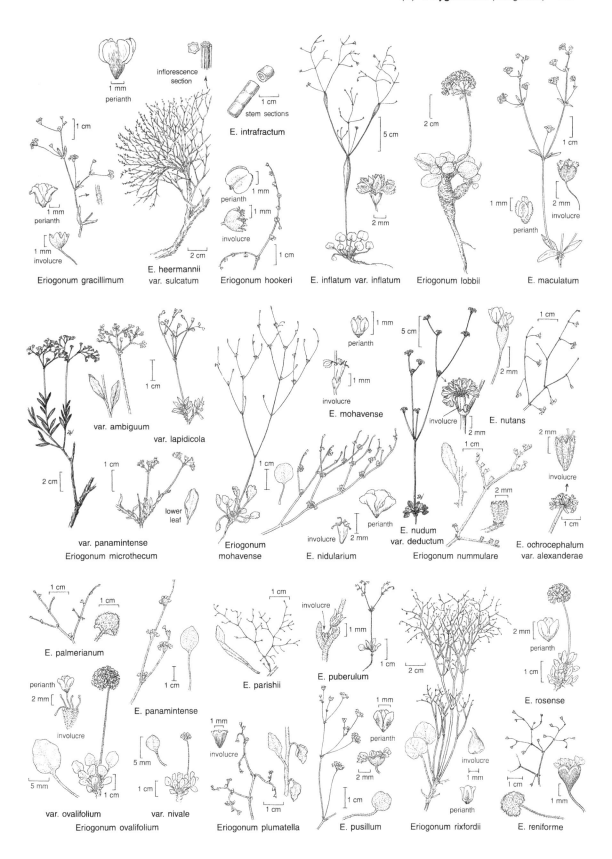

inflorescence section

1 mm perianth

E. intrafractum

1 cm stem sections

5 cm

2 cm

2 mm

1 cm

1 cm

1 mm perianth

1 mm perianth

1 mm involucre

perianth 1 mm

involucre 1 mm

2 mm

1 cm

1 mm perianth

2 mm involucre

Eriogonum gracillimum

E. heermannii var. sulcatum

Eriogonum hookeri

E. inflatum var. inflatum

Eriogonum lobbii

E. maculatum

var. ambiguum

var. lapidicola

1 cm

1 mm perianth

1 mm involucre

E. mohavense

5 cm

1 mm perianth

2 mm involucre

1 cm

2 mm

2 mm involucre

2 cm

var. panamintense
Eriogonum microthecum

1 cm

lower leaf

1 cm

involucre 2 mm

perianth

Eriogonum mohavense

E. nidularium

involucre 2 mm

E. nudum var. deductum

Eriogonum nummulare

E. nutans

1 cm

2 mm

E. ochrocephalum var. alexanderae

1 cm

1 cm

E. palmerianum

1 cm

involucre 1 mm

1 cm

1 cm

E. panamintense

perianth 2 mm

involucre

5 mm

1 cm

var. ovalifolium

Eriogonum ovalifolium

5 mm

1 cm

var. nivale

1 cm

E. parishii

involucre 1 mm

1 cm

E. puberulum

2 cm

1 mm involucre

1 cm

Eriogonum plumatella

1 mm

perianth

2 mm

1 cm

E. pusillum

Eriogonum rixfordii

2 mm perianth

1 cm

E. rosense

involucre 1 mm

perianth

1 cm

1 mm

E. reniforme

MUCRONEA

Ann, glandular. **LVS** basal; stipule 0; petiole indistinct; blade oblanceolate. **INFL** open; bract 1 per node, 5–25 mm, lobes 3, awned; involucre gen 1 per node, sessile, cylindric, lobes 2–4, awned, awns straight, often unequal; fl gen 1 per involucre. **FL**: perianth white to pink, hairy, lobes 6; stamens 6–9. **FR** 2–3 mm, lanceolate in outline, brown to black. 2 spp.: CA. (Latin: sharp-pointed, from bracts) [Reveal 1989 Phytologia 66:199–220]

M. perfoliata (A. Gray) A.A. Heller (p. 435) PERFOLIATE SPINEFLOWER Pl 2–30 mm; non-glandular hairs nearly 0. **LF** gen 30–60 mm. **INFL**: bracts encircling st, gen 10–25 mm, gen lobed < halfway to base; involucre 1 per node, 3–8 mm, asymmetric or ± bilateral, lobes 4, awns < 1.5 mm; fl 1 per involucre. **FL**: perianth lobes gen jagged. *n*=19. Chaparral, Joshua-tree woodland, etc., in sandy soils; 150–1900 m. s SNF, Teh, SnJV, SCoRI, n WTR, **w DMoj**. [*Chorizanthe p.* A. Gray] ❀ TRY. Apr–Jun

NEMACAULIS

1 sp. (Latin: thread st) [Reveal & Ertter 1980 Madroño 27:101–109]

N. denudata Nutt. WOOLLY-HEADS Ann, prostrate to erect, 4–40 cm, woolly. **LVS** basal; stipule 0; petiole indistinct; blade 5–80 mm, linear to oblanceolate. **INFL** open, 20–80 cm diam; branches wiry; bracts 3 per node, 1–5 mm, gen linear; involucral bracts many, 1–4 mm, oblanceolate, upper surface very woolly; fls 5–30 per involucre. **FL**: perianth 0.8–1.5 mm, greenish white to dark red, glabrous or glandular, lobes 6; stamens 3. **FR** ± 1 mm, brown to black. Coastal strand, desert scrub, etc., in sandy soils; < 400 m. SW, **DMoj (Devil's Playground), DSon**; AZ, nw Mex. 2 vars. in CA.

var. ***gracilis*** (p. 435) Goodman & L. Benson **FLS** gen obscured by surrounding wool; outer perianth lobes linear. Coastal strand, desert scrub, etc., in sandy soils; < 400 m. SW, **DMoj (Devil's Playground), DSon**; AZ, nw Mex. ❀ TRY. Mar–May

OXYRIA MOUNTAIN SORREL

1 sp. (Greek: sour, from acidic taste)

O. digyna (L.) Hill (p. 435) Per, glabrous, becoming red-tinged; caudex thick, scaly. **ST** erect, < 5 dm. **LVS** ± basal, alternate, petioled; stipules fused, loosely sheathing st above nodes, breaking; blade << petiole, reniform, fleshy, sour-tasting. **INFL**: panicle, erect, ± open. **FL** nodding; perianth lobes 4, < 2 mm; stamens 6; styles 2. **FR** flat, conspicuously winged. 2*n*=14. Alpine rock crevices, talus; 1800–4000 m. KR, CaRH (Mt. Lassen), SN, SnBr, SnJt, Wrn, **SNE**; circumboreal. ❀ DRN,IRR:1; DFCLT. Jul–Sep

OXYTHECA

Ann, glandular, glaucous. **LVS** basal; stipule 0; petiole indistinct; blade linear to obovate, hairy. **INFL** open; bracts gen 3 per node, gen linear-lanceolate, gen fused at base, awned; involucre 1 per node, gen stalked, narrowly funnel-shaped to bell-shaped, teeth 3–36, awned; fls 2–20 per involucre. **FL**: perianth white to rose or pale yellow, hairy, lobes 6; stamens 9. **FR** golden brown to dark brown. w N.Am., s S.Am. (Latin: sharp box, from involucre) [Ertter 1980 Brittonia 32:70–102]

1. Involucral bracts 5 (6), awns 3, 5, or 6 — w DMtns (Little San Bernardino Mtns) . . . . ***O. trilobata***
1' Involucral bracts 4, awns 4
  2. Bracts fully fused and completely surrounding st; involucres sessile . . . . ***O. perfoliata***
  2' Bracts fused at base only; involucres ± stalked
    3. Stalks of lower involucres 0.5–15 mm; bract awns < 0.5 mm; GB . . . . ***O. dendroidea*** ssp. ***dendroidea***
    3' Stalks of lower involucres 0.5–2 mm; bract awns > 1 mm; se W&I (Santa Rosa Hills, Inyo Co.) . . ***O. watsonii***

O. dendroidea Nutt. ssp. ***dendroidea*** (p. 435) Pl 5–40 cm. **LF** 1–4.5 cm, gen narrowly oblanceolate, hairy, sparsely glandular. **INFL** sparsely glandular; bract awns 0–0.5 mm; branches sometimes recurved; involucre narrowly funnel-shaped, glabrous, involucral bracts 4, 1–2 mm, < half fused, awns 0.5–3 mm; fls 2–6 per involucre. **FL**: perianth 1–2 mm, white to pink, strigose, lobes entire. **FR** 1.5–2 mm. *n*=20. Dry sandy places in scrub or pine forest: 1200–2200 m. **GB**; to WA, WY, NV. Jun–Sep

O. perfoliata Torrey & A. Gray (p. 435) Pl 6–20 cm, nearly glabrous or sparsely glandular. **LF** 1–6 cm, oblanceolate to oblong, ciliate. **INFL** glandular on lower half of internodes; bracts fused around node, fused unit 1–2.5 cm diam; involucre sessile, narrowly funnel-shaped, involucral bracts 4, 2–5 mm, < half fused, awns 2–3 mm; fls 5–10. **FL**: perianth 1.5–2.5 mm, white to greenish yellow, hairy, lobes entire. **FR** 1.5–2 mm. *n*=20. Common. Creosote-bush or pinyon scrub, sandy to rocky soils; 200–2000 m. s SnJV, **GB, DMoj**; to NV, sw UT, nw AZ. Apr–Jul

O. trilobata A. Gray (p. 435) Pl 7–50 cm. **LF** 1–9 cm, gen linear to oblanceolate, sparsely strigose and glandular. **INFL** sparsely glandular near base of internodes; involucre widely funnel-shaped, sparsely glandular, involucral bracts 5–6, << half fused, awns sometimes 3, 0.3–2 mm; fls 3–10 per involucre. **FL**: perianth 2.5–4 mm, white to pink, hairy and glandular, lobes deeply divided into 3 ± regular, overlapping lobes. **FR** 1.2–2 mm. *n*=20. Common. Sandy places; 700–2800 m. SnGb, SnBr, PR, **w DMtns (Little San Bernardino Mtns)**; Baja CA. Jul–Sep

O. watsonii Torrey & A. Gray (p. 435) WATSON'S OXYTHECA Pl 5–25 cm. **LF** 0.7–5 cm, oblanceolate to obovate, gen sparsely strigose and glandular. **INFL** glandular at or near nodes; bract awns 1–3 mm; involucre narrowly funnel-shaped, glabrous, involucral bracts 4, 1.5–2 mm, ± half fused (exc awns), awns 2.5–3 mm; fls 2–7 per involucre. **FL**: perianth 1–1.5 mm, greenish white to pink, hairy; lobes entire. **FR** 1–1.5 mm. *n*=20. RARE in CA. Sandy places; 1200–2000 m. **se W&I (Santa Rosa Hills, Inyo Co.)**; NV. Name misapplied to *O. parishii* var. *goodmaniana*. May–Jul

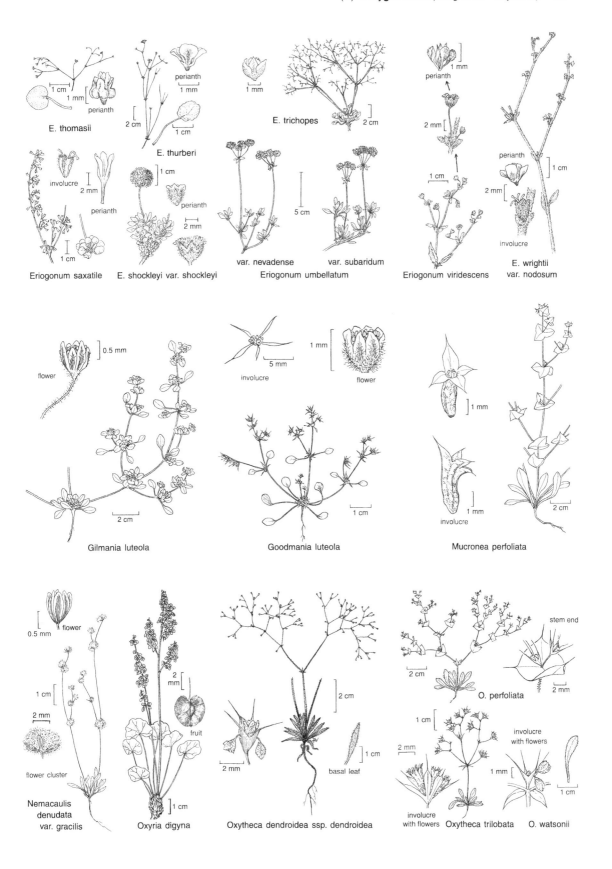

E. thomasii

E. thurberi

E. trichopes

perianth

Eriogonum saxatile

involucre

E. shockleyi var. shockleyi

perianth

var. nevadense   var. subaridum
Eriogonum umbellatum

perianth

Eriogonum viridescens

perianth

involucre

E. wrightii
var. nodosum

flower

Gilmania luteola

involucre   flower

Goodmania luteola

involucre

Mucronea perfoliata

flower

flower cluster

Nemacaulis
denudata
var. gracilis

fruit

Oxyria digyna

basal leaf

Oxytheca dendroidea ssp. dendroidea

stem end

O. perfoliata

involucre
with flowers

involucre
with flowers   Oxytheca trilobata   O. watsonii

POLYGONUM KNOTWEED, SMARTWEED

Ann, per, shrub, vine. **ST** prostrate to erect, or climbing, or floating, < 3 m. **LVS** gen cauline, alternate, sessile or petioled; stipules fused, sheathing st above nodes, gen scarious or membranous; blade sometimes obviously jointed to stipule sheath. **INFL**: unit a 1–8-fld cluster, these arrayed singly or in head-like to open panicles. **FL**: perianth lobes gen 5; stamens 3–8, filaments gen wider at base. **FR** gen ovoid, 3-angled, sometimes round, flat, indented; shiny to dull, brown to black. ± 300 spp.: worldwide, esp n temp. (Greek: many knees, from swollen nodes of some spp.) [Ronse Decraene & Ackeroyd 1988 Bot J Linn Soc 98:321–371] Segregate genera (e.g., *Bistorta, Fallopia, Persicaria*) are sometimes recognized. Key to species adapted by Margriet Wetherwax.

1. Lf 0.5–8 cm, ± sessile, gen linear to (ob)lanceolate; filaments not thread-like, much wider at base
   2. St ± angled (esp just below nodes), ribs between angles 0 or obscure . . . . *P. douglasii* ssp. *douglasii*
   2' St ± round, obviously and regularly 8–16-ribbed
      3. Fl clusters scattered, some or all in axils of full-sized lvs; st prostrate to erect, < 1.5 m . . . . . . *P. arenastrum*
      3' Fl clusters ± restricted to branch tips, in axils of bracts; st erect, < 6 dm . . . . *P. argyrocoleon*
1' Lf gen 5–35 cm, petioled, blade gen lanceolate or wider; filaments thread-like
   4. St unbranched, erect; gen subalpine; lvs ± basal, blade gen oblong, strap-shaped; perianth lobes nearly
      free . . . . *P. bistortoides*
   4' St branched, gen ± sprawling; lower elevations; lvs cauline, blade rarely oblong; perianth lobes fused
      to nearly free
      5. Perianth gland-dotted; infl open, slender; fr 2–3-angled . . . . *P. punctatum*
      5' Perianth not gland-dotted; infl open or dense; fr 2-angled or ± flat
         6. Perianth greenish to light pink; infl spikes ± 5 mm wide
            7. Perianth gen light pink, opening, veins at fruiting inconspicuous; fr 3-angled; per from extensive
            rhizome . . . . *P. hydropiperoides*
           7' Perianth greenish remaining closed. veins at fruiting prominent, forked, recurved; fr flat. gen
            indented on both sides; ann . . . . *P. lapathifolium*
         6' Perianth deep pink; infl spikes 7–20 mm wide
            8. Ann; ± terrestrial; st ± erect; lf narrowly lanceolate, 0.5–2 cm . . . . *P. persicaria*
            8' Per from rhizome; st decumbent or floating; lf widely lanceolate, 2–4 cm . . . . *P. amphibium*
               9. St terrestrial or emergent in fl; lf blade lanceolate, tip ± acuminate; infl > 4 cm . . . . var. *emersum*
               9' St floating in fl; lf blade oblong, tip round to acute; infl < 3.5 cm . . . . var. *stipulaceum*

P. amphibium L. WATER SMARTWEED Per from rhizome, terrestrial, emergent, or floating. **ST** prostrate to erect, rooting at nodes, stout. **LF** < 35 cm, petioled; stipule veiny, brown, fracturing, bristles 0 exc in terrestrial sts with flared stipules; blade widely lanceolate, oblong, or ovate, base tapered to cordate, tip acuminate to round. **INFL** 1–15 cm, dense. **FL**: perianth 4–6 mm, deep pink; filaments thread-like. **FR** 2.5–3 mm, round, flat, brown, shiny. 2*n*=96,98. Common. Shallow lakes, streams, shores; < 3000 m. CA-FP, **w DMoj**; to e N.Am, Eurasia. Variable, in part due to environment. Vars. intergrade.

var. **emersum** Michaux (p. 439) Rhizome coarse. **ST** gen erect, terrestrial or emersed in fl. **LF**: blade lanceolate, ± hairy, tip ± acuminate. **INFL** 4–10 cm, narrowly cylindric or conic. Habitat of sp.; < 1500 m. Range of sp. [*P. coccineum* Muhlenb.] Prostrate, non-fl, terrestrial sts hairier. Jun–Oct

var. **stipulaceum** Coleman (p. 439) Rhizome slender. **ST** prostrate, gen floating in fl. **LF**: blade < 10 cm, gen oblong, ± glabrous, tip round to acute. **INFL** < 3.5 cm, ± ovoid. Habitat of sp.; < 3000 m. CA-FP, **SNE**. Shore growth form has spreading hairs, flared stipules. ❀ Water or WET,SUN:**1–9,14**,15–24. Jul–Sep

P. arenastrum Boreau (p. 439) COMMON KNOTWEED. DOORWEED Ann or weak per. **ST** prostrate to erect, < 1.5 m, round, ribbed. **LVS** gen 0.5–2 cm, gen smaller upward and on side branches, jointed to stipule sheath, sessile; blade linear to narrowly elliptic. **INFL**: fls 2–8 in nearly all axils. **FL**: perianth 2–3.2 mm, fused 30–50% its length, green, margin white or pink, base in fr often prominently veiny. **FR** 2–3 mm, at maturity gen 2 sides convex, 1 ± concave, dark brown, ± dull. 2*n*=40,60. Abundant. Disturbed places; < 2500 m. **CA**; to e N.Am, native to Eur. [*P. montereyense* Brenckle] Highly variable, tolerant of trampling. *P. aviculare* L. (perianth 3.1–4.9 mm, fr 2.8–3.8 mm: Styles 1962 Watsonia 5:177–214) is very closely related, apparently undocumented in CA. Jul–Aug

P. argyrocoleon Kunze (p. 439) Ann. **ST** erect, < 6 dm, round, ribbed. **LF** gen 2–4 cm, sessile; blades linear to lanceolate, reduced gradually upward, then abruptly at infl. **INFL**: branches raceme-like, restricted to st tips, ± 15 cm; fls 4–8 per axil; bracts < fls or falling early. **FL**: perianth ± 2 mm, exserted from stipule, margin red. **FR** < 2 mm, widely ovoid, 3 sides ± equal, brown, shiny. 2*n*= 40. Fields, disturbed places; < 1000 m. **CA (esp s)**; native to sw Asia. Jun–Oct

P. bistortoides Pursh (p. 439) WESTERN BISTORT Per from short, bulb-like rhizome. **ST** erect, unbranched, gen 20–70 cm. **LVS** ± basal or reduced rapidly upward, gen 15–40 cm, petioled; blade oblong or strap-shaped to narrowly lanceolate, leathery, ± glaucous, esp below. **INFL** long peduncled, dense, ± oblong. **FL**: perianth 4–5 mm, white or pink, lobes nearly free; stamens exserted, filaments thread-like. **FR** 3–4.5 mm, ± obovoid, brown, shiny. Common. Wet mtn meadows; 1500–3000 m. CA-FP, **n SNE** (uncommon in coastal freshwater marshes; 0–20 m. NCo, SnFrB); to AK, e N.AM. Variable. ❀ WET,SUN:**1–6**,7,15–18;INV Jun–Aug

P. douglasii E. Greene Ann. **ST** gen erect, 3–80 cm, ± angled; ribs 0 or obscure. **LF** < 8 cm, sessile; blade linear, elliptic, or oblanceolate, smaller upward, vein 1. **INFL** 5–20 cm, open or dense. **FL**: perianth 2–5 mm, opening or not, pink or white, margins and midribs gen red or green; stamens gen 8 (3–8, may vary within pl). **FR** 2–5 mm, black, shiny. 2*n*=40. Common. Open, slopes, dry meadows; < 3500 m. **CA** (esp mtns, coast); to Can, e N.Am. Variable intergrading complex; sspp. maintained in mixed populations merge over geog range. Gen self-pollinating. Homopteran insect parasites shorten st, broaden lvs, enlarge and sterilize fls in infected pls. Further study warranted. 5 sspp. in CA.

ssp. **douglasii** (p. 439) **ST** decumbent at high elevation, 3–80 cm. **LVS** often falling before fls appear, < 8 cm, linear to oblanceolate, abruptly reduced to scale-like bracts. **INFL** spike-like, open. **FL**: perianth 3–4 mm, gen closed (< 2.5 mm if open).

FR 3–4 mm, reflexed; enfolding perianth with a minute, curved stalk at base. Common. Drying places; 1000–3000 m. **CA**; to Can, e N.Am. [var. *latifolium* (Engelm.) E. Greene] Jun–Sep

P. hydropiperoides Michaux (p. 439) WATERPEPPER Per from extensive, slender rhizomes. **ST** ascending to erect, < 1 m. **LF** gen ± 10 cm, petioled; stipule strigose, margin bristly; blade linear to lanceolate, strigose (esp margin, midrib). **INFL**: branches spike-like, ± erect, gen 6–10 cm, 2–5 mm wide; bract margin bristly. **FL**: perianth ± 3 mm, light pink; filaments thread-like. **FR** ± 2 mm, 3-angled, dark brown, shiny. 2*n*=40. Wet banks; < 1500 m. CA-FP, **DSon (Colorado River)**; to Mex, e N.AM. [var. *asperifolium* E. Stanford] Often confused with *P. persicaria*. Jun–Oct

P. lapathifolium L. (p. 439) WILLOW WEED Ann. **ST** ascending to erect, < 1.5 m. **LF** < 20 cm, petioled; stipule veiny, esp at base, ± glabrous, clear brown; blade ± lanceolate, often hairy below. **INFL**: branches spike-like, 3–8 cm, ± 5 mm wide, gen drooping, dense. **FL**: perianth ± 2 mm, greenish to pink, lobes gen 4, veins in fr prominent, those of outer lobes forked, hooked near tips; filaments thread-like; styles 2. **FR** ± 2 mm, round, flat, indented on both sides, brown, shiny. 2*n*=22. Common. Moist places; < 1500 m. **CA**; to e N.Am. Highly variable, some forms are induced by environment. Merits study. Jun–Oct

P. persicaria L. (p. 439) LADY'S THUMB Ann. **ST** ± erect, < 1 m, sometimes ± swollen above nodes. **LF** < 20 cm, petioled; stipule strigose, margin bristly; blade narrowly lanceolate, gen with dark central spot, gen gland-dotted below. **INFL** spike-like, dense, < 3 cm, gen interrupted below; peduncles glabrous; bracts ± strigose-bristly. **FL**: perianth 2 mm, deep pink (rarely white). **FR** ± 2 mm, ± flat to equally 3-angled, brown to black, shiny. 2*n*=22,40,44. Common. Moist urban places; < 1500 m. **CA**; native to Eur. [*P. fusiforme* E. Greene] Variable. Hybrids with *P. lapathifolium* have narrower infl and ± flat, brown frs. Often confused with *P. hydropiperoides*. Jun–Nov

P. punctatum Elliott (p. 439) Ann or per from slender rhizome. **ST** decumbent to erect, < 1 m. **LF** < 15 cm, ± sessile; stipule margin entire or bristly; sessile or petiole short; blade narrowly lanceolate. **INFL**: branches gen 5–8 cm, slender, interrupted below. **FL**: perianth gen green, dotted with green glands, margin gen white; filaments thread-like. **FR** 2–3-angled, black, shiny. Common. Shallow water, shores; < 1500 m. **CA**; to WA, e N.Am, S.Am. Highly variable. Planted as waterfowl food. Jul–Oct

PTEROSTEGIA

1 sp. (Greek: winged cover, from involucre) [Reveal 1989 Phytologia 66:228–235]

P. drymarioides Fischer & C. Meyer (p. 439) Ann, prostrate or sprawling, (mats < 12 dm diam), thinly hairy, monoecious. **LVS** cauline, opposite, petioled; stipule 0; blade 3–20 mm, gen fan-shaped, entire to deeply notched and appearing 2-lobed. **INFL**: fls 1–2 per node, 1 or both gen staminate; only pistillate fls involucred; involucre growing with fr to 1–3 mm, net-veined, cream to pink or rose, wings 2, outer surface rounded, inner hollowed. **FL**: perianth 0.9–1.2 mm, pale yellow to pink or rose, sparsely hairy, lobes 5–6; stamens 6. **FR** ± 1.5 mm, brownish. *n*=14. Common. Shady, often moist places; < 1600 m. **CA**; to s OR, s NV, sw UT, w AZ, nw Mex. Mar–Jul

RUMEX DOCK

Ann, bien, per, gen from stout taproot, some dioecious, glabrous. **ST** gen erect, gen unbranched below infl, < 2 m, ± ridged, gen red-brown in fr; nodes ± swollen. **LVS** gen ± basal, alternate, petioled; stipules fused, sheathing st above nodes, fracturing; blade < 50 cm. **INFL**: bracted clusters gen arrayed in erect panicles. **FL** gen bisexual, < 3 mm, gen green; perianth lobes 6, persistent, outer 3 in fr ± inconspicuous, inner 3 in fr enlarged, hardened, ± veiny, covering fr, midrib often expanded into a tubercle; stamens 6; stigmas 3, fringed. **FR** brown, shiny. (Latin: sorrel) [Mitchell 1978 Brittonia 30:293–296] Hybrids common. Mature inner perianth lobes gen required for identification. Some cult for vegetable greens. TOXIC in quantity to livestock; seldom eaten. Key to species adapted by Margriet Wetherwax.

1. Dioecious (rarely some fls bisexual); lvs all basal; st erect, slender, gen < 4 dm; perianth ± red in fl, tubercles 0 in fr . . . . ***R. paucifolius***
1' Fls gen bisexual; lvs basal and cauline; st decumbent to erect, gen stout, gen > 7 dm; perianth gen green or pink in fl, tubercles various or 0 in fr
  2. Lvs cauline; main st prostrate to ± erect, gen with smaller branches at most nodes, branches often producing fls after main st . . . . ***R. salicifolius***
    3. Tubercles 0 or inconspicuous; inner perianth lobes triangular, with short, triangular teeth near base
    . . . . var. ***denticulatus***
    3' Tubercles 1 or 3 per fl, ± obvious; inner perianth lobes lanceolate to round, teeth gen 0
      4. 1 tubercle obvious, width > 1/3 width of inner perianth lobe . . . . var. ***salicifolius***
      4' 3 tubercles obvious, ± equal, width << 1/3 width of inner perianth lobe
        5. Some lvs oblong to elliptic; GB — shores, ± salty lake beds . . . . var. ***lacustris***
        5' Lvs linear to lanceolate; CA . . . . var. ***triangulivalvis***
  2' Lvs basal and cauline (basal rosette often withered when fr mature); st ± erect, gen unbranched below infl
    6. Tubercles 0 or inconspicuous; perianth lobes gen pink — margin of inner perianth lobe entire
      7. Lf blade fleshy, base tapered; inner perianth lobe 8–15 mm . . . . ***R. hymenosepalus***
      7' Lf blade ± leathery, base truncate to cordate; inner perianth lobe 4–10 mm . . . . ***R. occidentalis***
    6' Tubercles 1 or 3 per fl, gen obvious; perianth lobes gen green
      8. Margin of inner perianth lobe entire
        9. Inner perianth lobe ± 5 mm, tubercle < 1/3 width of lobe; infl gen narrow, interrupted to continuous, ± leafy . . . . ***R. crispus***
        9' Inner perianth lobe ± 3 mm, tubercle 1/3 width of lobe or wider; infl open, interrupted below, nearly lfless . . . . [2]***R. violascens***

8' Margin of inner perianth lobe with 0.5–4 mm teeth
  10. Ann, bien
    11. Widest inner perianth lobe narrowly ovate, teeth 2–4 mm, fine; infl dense, leafy; widespread
         . . . . **R. maritimus**
    11' Widest inner perianth lobes widely ovate or ± equilaterally triangular, teeth < 1 mm or obscure;
         infl ± open, nearly lfless; esp D . . . . ²**R. violascens**
  10' Per
    12. Largest lf 20–75 cm, blade lance-ovate, ± cordate; inner perianth lobe narrowly ovate-triangular
         . . . . **R. obtusifolius**
    12' Largest lf < 12 cm, fiddle-shaped; inner perianth lobe ovate . . . . **R. pulcher**

R. crispus L. (p. 439) CURLY DOCK Per. **ST** < 15 dm, stout. **LVS** basal and cauline, < 50 cm; blade lanceolate, margin strongly curled, esp near base. **INFL** narrow, interrupted to continuous, ± leafy. **FL**: inner perianth lobe ± 5 mm, ovate to round, base cordate, teeth 0, tubercles 3, ovate, unequal, < 1/3 width of lobe. **FR** 2 mm. $2n=60$. Abundant. Disturbed places; < 2500 m. **CA**; N.Am, native to Eurasia. Variable. Most of year

R. hymenosepalus Torrey (p. 439) CANAIGRE, WILD-RHUBARB Per from cluster of tuber-like roots. **ST** 6–12 dm, stout, ± fleshy, smooth. **LVS** basal and cauline, < 6 dm, fleshy; blade lanceolate to oblong, base tapered, tip acuminate, margin curled. **INFL** dense. **FL**: inner perianth lobe 8–15 mm, round-cordate, ± pink, teeth 0, tubercles 0. **FR** 4–7 mm. $2n=40,100$. Dry sandy places; 0–2000 m. SW, **DMoj**; to WY, TX, Baja CA. ❀DRN,SUN:7–14**16**,17–24. Jan–May

R. maritimus L. (p. 439) GOLDEN DOCK Ann, bien. **ST** sometimes branched, < 10 dm, slender, or stout and hollow. **LVS** gen cauline, < 20 cm; blade linear to ovate, base cordate to tapered, margin ± curled. **INFL** dense, interrupted below, leafy. **FL**: inner perianth lobe 2–3.5 mm, narrowly ovate, teeth variable, gen 2–4 mm, fine, tubercles 3, ± equal, linear and inconspicuous to ovate-round and as wide as lobes. **FR** 1–2 mm. $2n=40$. Wet, ± salty places; 0–2000 m. **CA**; to AK, e N.Am, S.Am, Eurasia. [*R. fueginus* Philippi; *R. persicarioides* L.] Highly variable in size, color, lvs, inner perianth lobes (all traits of which vary within fls); apparently inseparable into geog races. ❀ TRY STBL. May–Sep

R. obtusifolius L. (p. 439) BITTER DOCK Per. **ST** < 15 dm, stout. **LVS** basal and cauline, < 75 cm; blade lance-ovate, cordate, flat, entire. **INFL** open, interrupted below, ± lfless. **FL**: inner perianth lobe 3–5 mm, narrowly ovate-triangular, teeth gen ± 0.5 mm (in some fls 1–2 mm or 0), tubercle 1 (or 3, unequal). **FR** 2–3 mm. $2n=40$. Moist places; < 1500 m. **CA**; native to w Eur. [ssp. *agrestis* (Fries) Danser] Hybridizes with *R. crispus*, other Eur spp. Jun–Dec

R. occidentalis S. Watson (p. 439) WESTERN DOCK Per. **ST** 5–20 dm, stout, ± red. **LVS** basal and cauline, < 60 cm, ± leathery; blade lanceolate, base truncate to cordate, margin ± curled. **INFL** dense above, ± interrupted below. **FL**: inner perianth lobe 4–10 mm, round to ovate, cordate, often pink, teeth 0, tubercles 0. **FR** 3–4 mm. $2n=140,200$. Uncommon. Wet, ± salty places; 0–1500 m. **CA**; to AK, e N.Am. [*R. fenestratus* E. Greene] Variable. Aug–Sep

R. paucifolius S. Watson Per, dioecious (rarely some fls bisexual). **STS** clustered on taproot, sometimes branched, < 4(7) dm, slender. **LVS** gen ± basal, 2–9 cm; blade linear to lanceolate. **INFL** ± open. **FL**: perianth ± red, inner lobe 3–4 mm, > fr faces, teeth 0, tubercles 0. **FR** ± 1.3 mm. $2n=14,28$. Moist places; 1500–4000 m. CaR, SN, MP, **W&I**; to B.C., Alberta, Colorado. [var. *gracilescens* Rech.f.] ❀ TRY STBL. Aug–Sep

R. pulcher L. (p. 439) FIDDLE DOCK Per. **ST** < 1 m, slender. **LVS** basal and cauline, < 12 cm; blade lanceolate, basal cordate, often narrowed below middle, forming a "fiddle" shape. **INFL** interrupted; branches spreading. **FL**: inner perianth lobe 3–5 mm, ovate, teeth slender to base, some 1.5 mm, tubercles 3, ovate, unequal or inconspicuous. **FR** 2–3 mm. $2n=20,40$. Common. Disturbed places; gen < 1500 m. **CA**; native to Medit. Variable. May–Sep

R. salicifolius J.A. Weinm. WILLOW DOCK Per. **ST** prostrate to ± erect, < 1 m; branches from most nodes often producing fls later than main axis. **LVS** cauline, gen crowded below, 3–20 cm; blade linear to ovate, margin entire, flat or ± curled. **INFL** dense to open. **FL**: inner perianth lobe 2–5 mm, narrowly ovate to ± round, teeth gen 0, tubercles variable. **FR** 2–3 mm. Common. Moist places; 0–3500 m. **CA**; to AK, Rocky Mtns, Mex. [Hickman 1984 *Madroño* 31:249–252] Highly variable, even within pls; intergrading complex, warrants detailed study. 6 vars. in CA.

  var. **denticulatus** Torrey (p. 439) **ST** ascending to ± erect. **LF**: blade linear to lanceolate. **INFL** ± open, interrupted, 15–30 cm. **FLS** gen both bi- and unisexual on 1 pl; inner perianth lobe 3–4 mm, ± triangular, teeth near base, short, irregular, tubercles inconspicuous or 0. $2n=20$. Moist places; 0–3500 m. **CA** (esp mtns, coast); to Yukon, MT, Colorado, n Mex. [*R. californicus* Rech.f.; *R. utahensis* Rech.f.] Except for tubercles, very much like var. *salicifolius*. ❀ TRY STBL. May–Sep

  var. **lacustris** (E. Greene) J. Hickman (p. 439) **ST** ± decumbent if terrestrial, erect and emersed if aquatic. **LF**: blade gen widely lanceolate, some elliptic or oblong, ± thickened. **INFL** dense, continuous, ± 5 cm. **FL**: inner perianth lobe 2–5 mm, tubercles 3, gen << 1/3 width of lobe, narrowly lanceolate, ± equal. Beds, shores of ± salty lakes; 1000–2500 m. **GB** (esp MP); s OR, NV. [*R. l.* E. Greene] Growth form variable with water level: emergent pls gen unbranched, lfless below initial water level. ❀ TRY STBL. Jul–Sep

  var. **salicifolius** (p. 439) **ST** decumbent to ± erect. **LF**: blade linear to lanceolate. **INFL** open, ± interrupted, 15–30 cm. **FL**: inner perianth lobe 2–3 mm, tubercle 1, > 1/3 width of lobe. $2n=20$. Moist places, esp coastal, montane; 0–3000 m. CA-FP, **W&I**; NV, Baja CA. ❀ TRY STBL. May–Sep

  var. **triangulivalvis** (Danser) J.C. Hickman (p. 439) **ST** ± erect. **LF**: blade linear to lanceolate. **INFL** dense, ± continuous, gen 15–30 cm. **FL**: inner perianth lobe 3–5 mm, tubercles 3, < 1/3 width of lobe, ± equal, lanceolate. $2n=20$. Many habitats; 300–2500 m. **CA** (esp SN); to B.C., e N.Am, Mex, naturalized in Eur. [*R. t.* (Danser) Rech.f.] ❀ TRY STBL. Jul–Sep

R. violascens Rech.f. (p. 439) MEXICAN DOCK Ann, bien. **ST** < 1 m, gen stout (appearing per). **LVS** basal and cauline, gen 6–10 cm; blade lance-linear to elliptic, base cordate to tapered, margin ± curled. **INFL** ± open, interrupted below, 10–20 cm, nearly lfless. **FL**: inner perianth lobe ± 3 mm, widely ovate to ± equilaterally triangular, teeth < 1 mm, sometimes obscure, tubercles 3, ovoid, unequal. **FR** ± 2 mm. Uncommon. Wet, ± salty places; < 500 m. s ScV, SnJV, PR, **D**; to TX, Mex. Variable. Mar–Aug

PORTULACACEAE PURSLANE FAMILY

Ann or per, gen fleshy. **STS** gen glabrous. **LVS** simple, alternate or opposite, sometimes stipuled. **INFL** various. **FL** bisexual, radial; sepals gen 2(–8), free or fused at base; petals 3–18, free or ± fused; stamens 1–many, free or

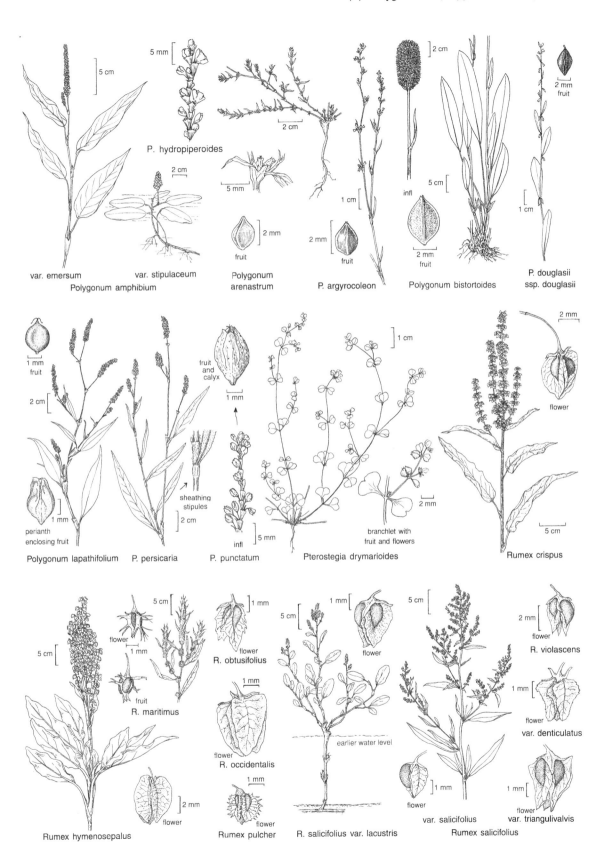

5 cm

5 mm

P. hydropiperoides

2 cm

var. emersum

var. stipulaceum

Polygonum amphibium

5 mm

2 cm

5 mm

2 mm
fruit

Polygonum
arenastrum

2 cm

1 cm

2 mm
fruit

P. argyrocoleon

2 cm

infl

5 cm

2 mm
fruit

Polygonum bistortoides

2 mm
fruit

P. douglasii
ssp. douglasii

1 cm

1 mm
fruit

2 cm

1 mm
perianth
enclosing fruit

Polygonum lapathifolium

fruit
and
calyx

1 mm

sheathing
stipules

2 cm

P. persicaria

infl

5 mm

P. punctatum

1 cm

2 mm

branchlet with
fruit and flowers

Pterostegia drymarioides

2 mm

flower

5 cm

Rumex crispus

5 cm

flower

1 mm

fruit

1 mm

R. maritimus

1 mm

flower

R. obtusifolius

1 mm

flower

R. occidentalis

1 mm

flower

Rumex pulcher

1 mm

flower

1 mm

earlier water level

R. salicifolius var. lacustris

5 cm

5 cm

flower

2 mm

flower

Rumex hymenosepalus

1 mm

flower

var. salicifolius

Rumex salicifolius

2 mm

flower

R. violascens

1 mm

flower

var. denticulatus

1 mm

flower

var. triangulivalvis

inserted on corolla; ovary superior or partly inferior, chamber 1, placenta free-central or basal; styles 2–8, gen fused at base. **FR**: capsule, circumscissile or 2–3-valved. **SEEDS** 1–many, gen black, gen shiny. ± 20 genera, ± 400 spp.: gen temp Am, Australia, s Afr; some cult (*Lewisia, Portulaca, Calandrinia*). [Bogle 1969 J Arnold Arbor 50:566–598] Family description and key to genera by Dieter H. Wilken & Walter A. Kelley.

1. Fr circumscissile
  2. Sepals free, 2–8; ovary superior; fr circumscissile near base . . . . **LEWISIA**
  2' Sepals fused at base, 2-lobed; ovary partly inferior; fr circumscissile near middle . . . . **PORTULACA**
1' Fr 2–3-valved, dehiscing from tip
  3. Infl umbel or panicle of head-like clusters, dense; sepals in fr gen round to reniform, ± scarious
     . . . . ²**CALYPTRIDIUM**
  3' Infl cyme, raceme, panicle, or fls solitary, gen open; sepals in fr gen not round or reniform, gen not scarious
    4. Style 2-branched (or stigmas 2, sessile); petals fused at base, tube expanding in fr, cap-like; fr 2-valved
     . . . . ²**CALYPTRIDIUM**
    4' Styles 3, sometimes fused at base; petals ± free, tube not expanding in fr, not cap-like; fr gen 3-valved
    5. Cauline lvs alternate . . . . **CALANDRINIA**
    5' Cauline lvs opposite
      6. Lvs mostly basal; cauline lvs gen 2, free or fused (2-lobed to disk- or cup-like) . . . . **CLAYTONIA**
      6' Lvs ± gradually reduced upward; cauline lvs gen 4+, not fused . . . . **MONTIA**

CALANDRINIA

Walter A. Kelley

Ann, per, ± fleshy, glabrous or glaucous. **STS** several–many, prostrate to erect, 3–45 cm. **LVS** simple, alternate; blade linear to spoon-shaped, cylindric or flat. **INFL**: raceme or panicle; bracts scarious or lf-like. **FL**: sepals 2, overlapping, persistent in fr; petals 3–7, gen 5, red or white; stamens 3–15; style 3-branched. **FR**: capsule, 3-valved from tip. **SEEDS** many, ovate to ± elliptic, gen black, smooth, finely tuberculate, or with a fine, net-like pattern, short-hairy or not. 150 spp.: w Am, Australia. (J.L. Calandrini, Switzerland, born 1703) [Kelley 1973 MS Thesis, CA State Univ, Northridge]

1. Petals white; infl a panicle of umbel-like clusters . . . . *C. ambigua*
1' Petals red-purple; infl a raceme . . . . *C. ciliata*

C. ambigua (S. Watson) Howell (p. 445) Ann, glabrous. **ST** spreading to erect. **LF** 1.5–6 cm, linear to ± spoon-shaped, ± cylindric. **INFL**: panicle of umbel-like clusters, compact; bracts scarious; pedicels 1–3 mm, straight in fr. **FL**: sepals 2–5 mm, glabrous, margins white-scarious; petals 3–5, 2–5 mm, white; stamens 5–10. **FR** < calyx. **SEEDS** 6–15, 1–2 mm wide, ± elliptic to ovate, at 10× shiny black, smooth. Common. Sandy to silty soil; < 1000 m. **D**; AZ, nw Mex. Jan–Apr

C. ciliata (Ruiz & Pav.) DC. (p. 445) RED MAIDS Ann, ± glabrous. **ST** spreading. **LF** 1–10 cm, linear to oblanceolate, flat.

INFL: raceme, elongate; bracts lf-like; pedicels 4–25 mm, gen straight in fr. **FL**: sepals 2.5–8 mm, glabrous to ciliate; petals gen 5, 4–15 mm, red; stamens 3–15. **FR** not > calyx by 3 mm or more. **SEEDS** 10–20, 1–2.5 mm wide, elliptic, at 10× shiny black, at 30× black, glabrous, with a very fine, net-like pattern. 2n=24. Common. Sandy to loamy soil, grassy areas, cult fields; < 2200 m. CA-FP, w MP, **s SNE, n DMtns (Coso Range)**; to NM, C.Am; also in nw S.Am. [var. *menziesii* (Hook.) J.F. Macbr.; *C. micrantha* Schldl.] Variable vegetatively; uniform in fl, fr, and seed. ✿ SUN:4–6,**7–9,14–24**&IRR:1–3,**10–12**,13. Mostly Feb–May

CALYPTRIDIUM PUSSYPAWS

Dieter H. Wilken & Walter A. Kelley

Ann, per, ± fleshy, from taproot or fibrous roots, gen glabrous. **STS** gen several, gen spreading to ascending. **LVS** in basal rosette or basal and cauline, simple, oblanceolate to spoon-shaped. **INFL**: raceme, panicle, or umbel, scapose, leafy, or bracted; fls gen on 1 side of axis, deciduous or persistent in fr; bracts gen < sepals, scarious. **FL**: sepals 2, ovate to reniform, gen scarious or scarious-margined, persistent in fr; petals 2–4, minute, < sepals, tips adherent and cap-like in fr, falling as 1 unit; stamens 1–3; style 0 or 1, thread-like, stigmas gen 2. **FR**: capsule, gen translucent, 2-valved, gen compressed, oblong to ± round. **SEEDS** 1–many, black, gen shiny. 8 spp.: w Am. (Greek: cap, from petals in fr) [Hinton 1975 Brittonia 27:197–208; Thomas 1956 Leafl W Bot 8:9–11] Observation of fl, seeds requires 20× magnification.

1. Style thread-like; fr widely elliptic to ± round
  2. Infl axillary, 2+ per rosette; bracts subtending infl ovate to deltate, < sepals . . . . *C. monospermum*
  2' Infl terminal, 1 per rosette; bracts subtending infl ± round, ± = sepals . . . . *C. umbellatum*
1' Style 0, stigmas sessile; fr ovate to narrowly oblong
  3. Fls pedicelled, pedicels 1–3 mm; petals 2; stamen 1 . . . . *C. roseum*
  3' Fls subsessile; petals gen 3; stamens gen 3
    4. Fr > 2 × sepal length . . . . *C. monandrum*
    4' Fr < 2 × sepal length . . . . *C. parryi* var. *nevadense*

C. monandrum Nutt. (p. 445) Ann, 1.5–18 cm; taproot slender. **STS** spreading to decumbent, leafy. **LVS** basal and cauline; basal 1–5 cm, withering in fr; cauline 1–2(4) cm. **INFL**: raceme or panicle, 1–4 cm, gen open, axillary; bracts narrowly ovate; fls subsessile, gen persistent in fr. **FL**: sepals 1–2 mm, ovate to deltate, fleshy, becoming membranous, narrowly white-margined; petals 3, 1–3 mm, pink to reddish; stamen gen 1; stigmas sessile. **FR** 3–6 mm, oblong. **SEEDS** 4–10. Sandy soils, open sites, burned areas, shrub-land; < 2150 m. s SN, Teh, s SnJV, SCoRI, SW, **SNE, D**; NV, AZ. ✿ TRY. Mar–Jun

C. monospermum E. Greene (p. 445) Per, < 50 cm; caudex short, thick; taproot slender to thick. **STS** gen spreading to ascending, scapose to leafy. **LVS**: basal rosette 1; basal lvs 1.5–6 cm; cauline lvs 0.8–3 cm. **INFL**: umbel, gen compound, ± open, simple and dense in small pls, axillary, 2+ per rosette, 1–10 cm diam; bracts ovate to deltate; fls subsessile to short-pedicelled, persistent in fr. **FL**: sepals 3–8 mm, ± round, clearly scarious; petals 4, 3–7 mm, rose to white; stamens 3, pink to rose; style ± 1 mm, thread-like. **FR** 2–3.5 mm, widely elliptic to ± round. **SEEDS** 1–3. Open, sandy or gravelly soils, coniferous forest; 600–3200 m. KR, NCoRH, CaR, SN, TR, **SNE**; s OR, NV, Baja CA.

C. parryi A. Gray Ann, 2–11 cm; taproot slender. **STS** spreading to ascending, leafy. **LVS** basal and cauline, 1–3 cm; basal withering in fr. **INFL**: raceme or panicle, open to dense, 1–3.5 cm, axillary; bracts ovate to elliptic; fls subsessile, persistent to deciduous in fr. **FL**: sepals 2–5 mm, equal to unequal (outer > and wider than inner), ovate, round, or reniform, scarious to membranous; petals 3, 1.5–3 mm, gen white; stamens gen 3; stigmas

sessile. **FR** 3–7 mm, ovate to oblong. **SEEDS** 10–15. Open areas, chaparral, oak or pinyon/juniper woodland, coniferous forest; 700–3350 m. s SNH, s SnFrB, n SCoRI, TR, **SNE, n DMtns**; w NV, AZ. (1 other var. in s AZ, n Mex.) 3 vars. in CA.

var. ***nevadense*** J. Howell **INFL**: fl deciduous in fr. **FL**: outer sepal ± reniform, margin white-scarious. **SEED**: margin fine-tubercled. Pinyon/juniper woodland; 1550–2500 m. **SNE, n DMtns**; w NV. Jun–Jul

C. roseum S. Watson (p. 445) Ann, 1.5–10 cm; taproot slender or fibrous. **STS** spreading to ascending, leafy. **LVS** basal and cauline, 0.5–4 cm, gen persistent in fr. **INFL**: raceme or panicle, open, 1–4 cm, axillary; bracts ovate to elliptic; fl persistent to ± deciduous in fr; pedicels 1–3 mm. **FL**: sepals 2–3 mm, ± round to reniform, white-margined; petals 2, ± 1 mm, white; stamen 1; stigmas sessile. **FR** 2–3 mm, ovate to oblong. **SEEDS** 5–11. Gravelly soils, coniferous forest, sagebrush shrublands; 1500–3750 m. e slope SNH, **n SNE, W&I**; to ID, NV. Jun–Aug

C. umbellatum (Torrey) E. Greene (p. 445) Per, < 6 dm; caudex short, thick; taproot slender to thick. **STS** gen spreading to ascending, scapose. **LVS**: basal rosettes gen 2+; lf 1.5–7 cm. **INFL**: umbel, gen compound, simple in small pls, 1–7 cm diam, terminal, gen 1 per rosette, dense; bracts subtending infl < or = sepals, ± round; fls subsessile to short-pedicelled, persistent in fr. **FL**: sepals 3–8 mm, ± round, clearly scarious; petals 4, 3–8 mm, white; stamens 3, yellow to red; style ± 1 mm, thread-like. **FR** 2–3 mm, widely elliptic to ± round. **SEEDS** 1–8. Open, sandy to rocky soils, coniferous forest, alpine; 1500–4300 m (100–200 m in SnFrB). KR, NCoRH, CaR, SN, sw SnFrB (Santa Cruz Mtns), **GB**; to MT, w WY.. ✿ TRY;DFCLT. May–Aug

CLAYTONIA

Kenton L. Chambers

Ann or per, from stolon, rhizome, tuber, or taproot, glabrous, ± fleshy. **LVS** entire; basal 0–many, rosetted; cauline gen 2, gen opposite, free to fully fused into ± 2-toothed disk or cup surrounding st. **INFL**: raceme, terminal, 1-sided; pedicels reflexed, becoming erect in fr. **FL**: petals 5, pink or white; stamens 5, epipetalous; ovary chamber 1, placentas basal, style 1, stigmas 3. **FR**: capsule; valves 3, margins rolling inward and forcibly expelling seeds. **SEEDS** 3–6, gen black, gen clearly appendaged. 28 spp.: N.Am, e Asia. (John Clayton, colonial Am botanist, born 1686) [Miller 1978 Syst Bot 3:322–341] Some spp. formerly placed in *Montia.*

1. Per; fl-sts from spheric tuber (tuber, if present, sometimes bearing a slender taproot)
2. Cauline lvs sessile (if short-petioled, blade linear-lanceolate); tuber fibrous-rooted only; in this region
   in DMtns (Panamint Mtns) . . . . ***C. lanceolata***
2' Cauline lvs with petiole ± = ovate blade; tuber often from slender taproot; in this region in SNE . . . . ***C. umbellata***
1' Ann or per from rhizomes; taproot 0 or slender
   3. Per; rhizomes long, gen branched; petals >> sepals — in this region in SNE . . . . ***C. nevadensis***
   3' Ann; petals ± = sepals
      4. Basal adult lvs linear or blade > 3× longer than wide, spoon-shaped, tapered gradually to petiole;
         petals 1–4 mm . . . . ***C. parviflora***
         5. Cauline lf-pair ± disk-like; widespread, esp DMtns . . . . ssp. ***utahensis***
         5' Cauline lvs linear, free or ± fused on 1 side; in this region in SNE, DMtns . . . . ssp. ***viridis***
      4' Basal adult lf blades < 3 × longer than wide, ovate to deltate, base abruptly tapered to cordate; petals
         gen < 5 mm
         6. Basal lvs gen many, prostrate to spreading, smaller inward, all of the blades widely diamond-shaped
            to deltate; cauline lf-pair often unequally fused; petals minute; in this region in SNE . . . ***C. rubra*** ssp. ***rubra***
         6' Basal lvs few–many, spreading to erect, ± equal; blades of seedling lvs ± linear, adult lvs ovate to
            diamond-shaped; cauline lvs gen equally fused; widespread, esp DMtns . . . ***C. perfoliata*** ssp. ***intermontana***

C. lanceolata Pursh (p. 445) WESTERN SPRING BEAUTY Per; caudex 0; tuber 1–3 cm wide, brownish, spheric; rhizomes and stolons 0. **ST** 5–15 cm, erect. **LVS**: basal 0–2, 5–8 cm, blade 1–3 cm, elliptic, base wedge-shaped, tip acute, petiole thread-like; cauline 1–7 cm, ± linear to ovate, gen ± sessile. **INFL** gen short-stalked or sessile, 1-bracted at base; fls 3–15. **FL**: sepals 3–7 mm; petals 5–12 mm, white or pinkish (base sometimes yellow). **FR** 3.5–4.5 mm. **SEED** 2–2.5 mm, round, shiny. 2*n*=16,24,32, many other numbers. Gravelly woodlands, meadows; 1500–2600 m. KR, NCoRH, CaRH, n&c SNH, SnGb, MP, **DMtns**

(**Panamint Mtns**); to w Can, MT, NM. [vars. *peirsonii* Munz & I.M. Johnston, Peirson's spring beauty; *sessilifolia* (Torrey) Nelson] Much variability apparently environmental; needs study. ✿ TRY;DFCLT. May–Jul

C. nevadensis S. Watson (p. 445) Per; caudex long, 1–3 mm diam, horizontal, white or yellowish, continuous with fleshy, much-branched rhizomes; bulb-like offshoots and stolons 0. **ST** 2–10 cm, spreading or erect. **LVS**: basal 2–15 cm, blade 1–5 cm, elliptic to widely ovate, base wedge-shaped, tip gen obtuse, petiole linear, often buried; cauline < 2 cm, free, ovate, sessile, obtuse. **INFL** short-stalked or sessile, gen dense; bracts 0–1; fls 2–8. **FL**: sepals 3–8 mm; petals 6–10 mm, pinkish. **FR** 3–4 mm. **SEED** 1.5–2 mm, round, shiny, smooth. $2n=14$. Subalpine streams, springs, melting snowbeds, in gravel or sand; 2200–3500 m. KR, CaRH, SNH, **SNE**; OR. Jul–Sep

C. parviflora Hook. Ann. **ST** 1–30 cm, spreading to erect. **LVS**: basal 1–18 cm, linear to narrowly oblanceolate, blade 1–7 cm, > 3 × longer than wide, gradually tapered to petiole, obtuse to acute; cauline free (< 6 cm, linear) or disk-like (< 5 cm diam, round or squarish). **INFL** stalked or sessile, dense or open, 1-bracted at base; fls 3–40. **FL**: sepals 1.5–4 mm; petals 2–6 mm, white or pinkish. **FR** 1.5–4 mm. **SEEDS** 1.2–2.3 mm, ovate to round, shiny, smooth. Vernally moist, often disturbed places in sun or shade; < 2300 m. CA-FP, **GB**, **DMtns**; to B.C., MT, n Mex. 3 sspp. in CA.

ssp. **utahensis** (Rydb.) John M. Miller & Chambers (p. 445) **LVS**: cauline pair fused, disk-like. **INFL** stalked or sessile, open or dense. **FL**: sepals 1.5–4 mm; petals 2–6 mm; stamens maturing ± with stigmas. **SEED** 1.2–2.3 mm. $2n=24,36,48$. Habitats and range of sp. [*Montia perfoliata* (Donn) Howell forma *parviflora* (Hook.) J. Howell, var. *utahensis* (Rydb.) Munz; *C. parviflora* ssp. *parviflora*, in part, Jepson Manual, 1993, as to plants of CA deserts] Variable; intergrades with other members of *C. perfoliata* complex. Self-pollinating.

ssp. **viridis** (Davidson) John M. Miller & Chambers (p.445) Herbage green. **LVS**: cauline free, linear, often curved, spreading or erect. **FL**: sepals 1.5–2 mm; petals 2–3.5 mm; stamens maturing ± with stigmas. **SEED** 1.2–1.5 mm. $2n=24,36$. Shrub- or woodland, sometimes dry; < 1800 m. s SN, SCoR, TR, PR, **SNE**, **DMtns**; to n Mex. [*Montia spathulata* var. *v.* Davidson] Intergrades with ssp. *u.*, *C. rubra*. Self-pollinating. Apr–Jun

C. perfoliata Willd. MINER'S LETTUCE Ann. **ST** 1–40 cm, spreading to erect. **LVS**: basal 1–25 cm, blade < 4 cm, < 3 × longer than wide, elliptic to reniform, tip rounded to acute, petiole linear; cauline pair fused, disk-like, < 10 cm diam, round or squarish (or free on 1 side). **INFL** stalked or sessile, open or dense, 1-bracted at base; fls 5–40. **FL**: sepals 1.5–5 mm; petals 2–6 mm, white or pinkish. **FR** 1.5–4 mm. **SEED** 1.2–2.7 mm, ovate to round, shiny, smooth. Common. Vernally moist, often

shady or disturbed sites; < 2000 m. CA-FP, **GB**, **DMtns**; to B.C., MT, C.Am. Highly variable; sspp. difficult because of environmental plasticity, genetic mixing among polyploids, and geog overlap of distinct, self-pollinating forms. 2 sspp. in CA.

ssp. **intermontana** John M. Miller & Chambers (p. 445) **LVS**: basal elliptic to round-deltate, tip obtuse to acute; cauline pair gen round or ± obtuse-angled. $2n=24,36,48$. Habitats and range (exc s AZ to C.Am) of sp. [*Montia perfoliata* (Donn) Howell; *C. p.* ssp. *perfoliata*, in part, Jepson Manual, 1993, as to plants of CA deserts] Polyploids are derived from hybridization with *C. parviflora*, *C. rubra*. ❀DRN:4&SHD:**5–7,9,10, 14–24**&IRR:1,2,8; rather INV.

C. rubra (Howell) Tidestrom Ann. **ST** 1–15 cm. **LVS**: basal 1–8 cm, blade < 2 cm, elliptic to widely deltate, base truncate to wedge-shaped, tip gen obtuse, petiole linear; cauline pair fused or partly free on 1 side, < 4 cm wide, gen round or with 2 squarish corners. **INFL** sessile, ± dense, 1-bracted at base; fls 3–30. **FL**: sepals 1.5–3 mm; petals 2–3.5 mm, white or pinkish. **FR** 2–3 mm. **SEED** 1–2 mm, elliptic, shiny, smooth. Vernally moist dunes, coniferous forest, shrubland, in sun or shade; < 2500 m. NW, CaRH, SNH, Teh, w CW, **GB**; to B.C., SD, Colorado. Intergrades with *C. parviflora*, *C. perfoliata*. 2 sspp. in CA.

ssp. **rubra** (p. 445) **LVS**: basal diamond-shaped or deltate, widest below middle, base often truncate; cauline ± free (at least on 1 side). $2n=12$. Habitats and range (exc NCo, CCo) of sp. [*Montia perfoliata* (Donn) Howell var. *depressa* (A. Gray) Jepson misapplied] Apr–Jun

C. umbellata S. Watson Per from oblong to spheric, brownish tuber 1–5 cm diam; caudex, rhizomes, and stolons gen 0; taproot gen 2–5 mm diam. **ST** 5–25 cm, erect, mostly underground. **LVS**: basal 0–3, 5–25 cm, blade 1–3 cm, widely elliptic to ovate, base acute or truncate, tip obtuse, petiole mostly underground; cauline 1.5–5 cm, blade 1–3 cm, widely elliptic, obtuse, abruptly tapered to linear petiole. **INFL** ± sessile, dense, 1–2-bracted at base; fls 2–12. **FL**: sepals 4–7 mm; petals 6–12 mm, pinkish. **FR** 5–6 mm. **SEED** 2.5–3 mm, round, shiny. $2n=16$. Uncommon. Talus slopes, stony flats, crevices; 1900–3500 m. KR, **GB**; OR, NV. Jul–Aug

LEWISIA

Lauramay T. Dempster

Per, gen from short, thick, ± branched taproot, topped by short, sometimes very thick caudex at or below ground level, sometimes from spheric corm. **ST**: aerial parts restricted to infl. **LVS** gen in basal rosette, simple, entire or not; base wide; margin gen ± translucent. **INFL** ± scapose; sts 1–many, gen lfless but bracted, sometimes disjointing in age, 1–many-fld. **FL**: sepals 2–8, free, persistent; petals 4–18, variously colored, overlapping in bud; stamens 5–many; styles 2–8, fused at base, stigmas 2–8, thread-like. **FR**: capsule, translucent, spheric or ovoid, circumscissile near base. **SEEDS** 2–many, dark, gen shiny, smooth or finely tuberculate. ± 20 spp.: w N.Am. (Captain Meriwether Lewis, 1774–1809, of Lewis & Clark Expedition) [Elliott 1966 Bull Alpine Gard Soc 34] ❀ DRN&IRR:pots and rock gardens only;DRY when dormant; DFCLT. Key to species adapted by Margriet Wetherwax.

1. Sepals 6–8, entire, becoming scarious in age, petal-like .... **L. rediviva**
1' Sepals 2, margin jagged or toothed, persistent, not petal-like
  2. Sepal margins gland-toothed; bracts in 1–2 irregular pairs, above middle of st, ± ovate, gland-toothed; stamens 6 .... **L. glandulosa**
  2' Sepal margins ± jagged; bracts 2, at or below middle of st, ± widely lanceolate, entire or dentate; stamens ± 8 .... **L. pygmaea**

L. glandulosa (Rydb.) Dempster (p. 445) Root + caudex elongate, gen widest at top, tapered or branched below. **LVS** many, in dense rosette, gen 2–10 cm, thread-like to narrowly lanceolate, entire, gen persistent after withering, tapered to fleshy, expanded base; tip obtuse. **INFL**: sts many, 1–3.5 cm, 1- or often 2-fld; fls not or barely exserted from lvs; bracts in 1–2 irregular pairs, above middle of st, ± ovate, gland-toothed. **FL**: sepals 2, ± 1/2 × corolla, roundish or truncate, margin reddish gland-toothed; petals 6–8, ± 8 mm, obovate, white, pink, or reddish, tip irregular, gen acuminate; stamens 6; stigmas 4. Granite sand, rock cracks, wet meadows; 3000–4000 m. c&s SNH, **SNE**. [*L. pygmaea* ssp. *g.*

(Rydb.) Ferris] Jul–Sep

L. pygmaea (A. Gray) Robinson Root + caudex gen largest at top, tapered below or short-fusiform or corm-like. **LVS** gen several, in rosette, 2–8 cm, thread-like to linear-lanceolate, fleshy, entire, tapered to expanded base; tip blunt. **INFL**: sts several–many, 1–5 cm, each gen 1-fld (less often several-fld); fls gen incl in lvs; bracts 2, at or below middle of st, ± widely lanceolate, entire or dentate. **FL**: sepals 2, ± 1/2 × corolla, ± ovate, margin ± jagged or toothed, rarely glandular, tip pointed, rounded, or truncate; petals 6–9, 5–10 mm, obovate, white, pink, or red, often striped; tip ± jagged; stamens ± 8; stigmas ± 4.   *n*=±33.   Rocky slopes, wet granite sand or gravel, moist meadows, along streams; 1700–4500 m.   KR, NCoRH, CaRH, SNH, WTR (Mt. Pinos), SnBr, Wrn, **W&I**; to B.C., Rocky Mtns.   [*L. sierrae* Ferris]   Ill-defined, probably of hybrid origin; intergrades and apparently hybridizes with *L. glandulosa.*   Jul–Aug

L. rediviva Pursh (p. 445)   BITTER ROOT   Root + caudex short, thick, ± expanded above; roots radiating, fleshy. **LVS** many, in rosette, 0.5–5 cm, linear, thick, entire, barely tapered at base; tip blunt. **INFL**: sts several–many, 2–6 cm, each 1-fld, disjointing near middle, leaving ring of 5–many scarious, awl-like bracts;

fls exserted from lvs. **FL**: sepals 6–8, ± 3/4 × corolla, petal-like, scarious in age, widely obovate, entire; petals 10–19, 12–25 mm, obovate-oblong, white or pink, tip obtuse-notched; stamens 40–47; stigmas 6–8.   *n*=14.   Rocky, sandy ground, talus, serpentine, clay, granite, shale, open woodlands and sagebrush shrublands with pine, oak, or juniper; 60–3000 m.   KR, NCoRH, NCoRI, SN, SnFrB, SCoRI, TR, SnJt, MP, **SNE, DMtns (Panamint Mtns)**; to Rocky Mtns.   Roots once food for native Americans, bitter if taken after fl.   Smaller pls of TR, PR, SNE have been called var. *minor* (Rydb.) Munz.   Mar–Jun

MONTIA

Kenton L. Chambers

Ann or per, glabrous, ± fleshy, sometimes ± aquatic, sometimes matted. **LVS**: cauline > 2, alternate or opposite, entire. **INFL**: raceme, 1-sided; lowest fl gen bracted; pedicels recurved, erect in fr. **FL**: petals (3–)5, equal or 2 larger; stamens (3–)5, filaments fused to corolla-base; ovary chamber 1, placentas basal, style 1, stigmas 3. **FR**: capsule; valves 3, margins rolling inward and forcibly expelling seeds. **SEEDS** 1–3, gen black, smooth to tubercled, fleshy-appendaged or not.   12 spp.: Am, Siberia, Australia.   (Giuseppe Monti, Italian botanist, 1682–1760) [McNeill 1975 Can J Bot 53:789–809]   Sometimes divided into 9 genera.

1. Per with stolon-like sts bearing pink bulblets; petals 5–9 mm, >> sepals . . . . *M. chamissoi*
1' Ann; sts often matted;   petals 1–2 mm, ± = sepals . . . . *M. fontana*

M. chamissoi (Sprengel) E. Greene (p. 445)   TOAD LILY   Per 2–30 cm, from leafy stolons, erect to prostrate or floating, often tufted or matted; rhizomes scaly, with pink, overwintering bulblets. **LVS** opposite, 5–50 mm, oblanceolate. **INFLS** gen few, some axillary; fls 2–8. **FL**: sepals 1.5–2 mm, spheric; petals 5–9 mm, equal, pink or white; stamens 5. **FR** 1–1.5 mm. **SEED** 1–1.3 mm; tubercles low, rounded; appendage flattened.   2*n*=22.   Wet, sandy or loamy soil, seeps, wet meadows; 1100–3700 m.   KR, NCoRH, CaR, SNH, TR, PR, **GB**; to AK, WY, NM, also c Can, n-c US.   Jun–Aug

M. fontana L. (p. 445)   WATER CHICKWEED, BLINKS   Ann 1–30 cm, prostrate to erect, sometimes floating, often tufted or mat-

ted, rooting from lower nodes. **LVS** opposite, ± sessile, 3–20 mm, linear to widely oblanceolate; tip acute to obtuse; base tapered. **INFLS** many, some axillary; fls 1–8, lowest 1–2-bracted. **FLS** often cleistogamous; sepals 1–2 mm, round, truncate; petals 1–2 mm, ± unequal, white; stamens 3. **FR** 1–2 mm. **SEED** 0.5–1.2 mm, gen ± rough with acute tubercles; appendage round or flat.   2*n*=18,20.   Common.   Ponds, streams, vernal pools, seeps, ditches; < 3200 m.   CA-FP, **GB**; to AK, e N.Am, ± worldwide.   Highly variable.   [*M. hallii* (A. Gray) E. Greene; *M. verna* Necker]   Gen self-pollinated.   Pls from s SNH, TR, SNE, with seed smooth have been called ssp. *variabilis* Walters [*M. funstonii* Rydb., Funston's montia].   Jul–Aug

PORTULACA

Walter A. Kelley

Ann, ± fleshy. **STS** several–many, spreading to erect. **LVS** simple, alternate or opposite, upper 2–5 forming involucre; blade linear, ovate, or spoon-shaped, flat or cylindric. **INFL**: fls solitary or in clusters at st tips. **FL**: sepals 2, fused at base, lower portion fused to ovary and persistent in fr; petals 4–6, inserted on calyx, yellow; stamens 4–20; ovary ± inferior, 1-chambered; style 3–6-branched. **FR**: capsule, circumsissle. **SEEDS** many, reniform, ± tuberculate.   100 spp.: warm regions worldwide.   (Probably Latin: small gate or door, from capsule lid) [Matthews & Levins 1985 Sida 11(1):45–61]

1. Lf axils with many long (to 8 mm) hairs; involucre hairy . . . . *P. halimoides*
1' Lf axils glabrous, scarious, or with few hairs; involucre glabrous . . . . *P. oleracea*

P. halimoides L. (p. 449)   **ST** spreading to ascending, 1–6 cm. **LF** 3–15 mm, ± linear, ± cylindric. **INFL**: fls in clusters of 2–10 at st tips. **FL**: sepals 1–2.5 mm, gen reddish, not keeled; petals 1–2.5 mm, yellow, drying reddish; stamens 4–15; style 3–5-branched. **FR** 1–3 mm wide. **SEED** 0.5–0.8 mm wide, black or metallic silver.   Uncommon.   Sandy washes, flats; 1000–1200 m. **DMtns (Little San Bernardino Mtns, New York Mtns)**; to TX, n Mex.   [*P. parvula* A. Gray]   CA pls reported as *P. mundula* I.M. Johnston (desert portulaca, UNCOMMON) appear to be-

long here; additional study needed.

P. oleracea L. (p. 449)   COMMON PURSLANE   **ST** spreading. **LF** 5–30 mm, ovate to spoon-shaped, gen flat. **INFL**: fls solitary or in clusters of 2–5 at st tips. **FL**: sepals 3–5 mm, green or reddish, gen keeled; petals 3–5 mm, yellow; stamens 5–20; style 4–6-branched. **FR** 3–8 mm wide. **SEED** 0.6–1 mm wide, dark brown to black.   2*n*=36,54.   Disturbed soil; < 1400 m.   CA; native to Eur.   [*P. retusa* Engelm.]   May–Sep

PRIMULACEAE   PRIMROSE FAMILY

Anita F. Cholewa & Douglass M. Henderson

Ann, per, subshrub, glabrous to glandular-hairy. **LVS** simple, basal or cauline, alternate, opposite, or whorled, sessile or petioled; stipules 0. **INFL** sometimes scapose. **FL** bisexual, radial; parts gen in 4's or 5's; calyx deeply lobed, often persistent; corolla lobes spreading to reflexed; stamens epipetalous, opposite corolla lobes; ovary gen superior, 1-chambered, placenta basal or free-central, style 1, stigma head-like. **FR**: capsule, circumscissile or 2–6-

valved. **SEEDS** small, few–many. ± 25 genera, 600 spp.: esp n hemisphere; several orn (*Cyclamen, Dodecatheon, Primula*). [Channell & Wood 1959 J Arnold Arbor 40:268–288]

1. Fls in lf axils . . . . **GLAUX**
1' Infl scapose, ending in an umbel
  2. Corolla lobes reflexed . . . . **DODECATHEON**
  2' Corolla lobes spreading to erect
    3. Corolla lobes 1–3 mm . . . . **ANDROSACE**
    3' Corolla lobes 5–10 mm . . . . **PRIMULA**

ANDROSACE   ROCK-JASMINE, FAIRY-CANDELABRA

Ann, per, gen < 12 cm. **LVS** in basal rosette. **INFL**: umbel 1 per scapose peduncle, terminal, subtended by involucral bracts. **FL**: parts in 5's; calyx tube scarious; corolla salverform, tube narrowed at top, lobes acute to obtuse at tip; stamens incl, filaments ± 0 or short, anthers oblong; ovary superior, spheric, style short. **FR** 5-valved, spheric. ± 100 spp.: n temp, arctic, esp Asia. (Greek: uncertain sea-plant) [Robbins 1944 Amer Midl Nat 32:137–163]

A. septentrionalis L. ssp. ***subumbellata*** G. Robb. (p. 449) Ann or weak per, 1–6 cm, hairy; peduncles 1 to gen several. **LF** 5–20 mm, linear-lanceolate, tapered to petiole, entire to finely dentate. **INFL**: involucre bracts 1.7–3 mm, gen ± < 0.5 mm wide, linear-lanceolate to lanceolate; pedicels 0.5–5 cm. **FL**: calyx (2.5)3–4 mm, glabrous or puberulent at base, tube > lobes, scarious between ridges, lobes widely lanceolate to triangular, gen reddish, tips acute to obtuse; corolla = or > calyx, white. Dry, rocky sites; 2700–3600 m. c&s SNH, SnBr, SNE; to B.C., Rocky Mtns. [ssp. *puberulenta* (Rydb.) G. Robb.] Jul–Aug

DODECATHEON   SHOOTING STAR

Per, glabrous or glandular-hairy; roots fleshy-fibrous. **LVS** basal. **INFL**: umbel, 1 per scapose peduncle, terminal, few–many-fld, subtended by bracts. **FL** nodding; parts in 4's or 5's; sepals reflexed, later erect, persistent; corolla tube short, lobes reflexed; stamens exserted, filaments very short, wide, often fused, anthers erect, ± lanceolate, surrounding style; ovary superior, style slender, ± exserted from anthers. **FR** ± 5-valved or circumscissile, ovate to spheric. ± 14 spp.: gen N.Am. (Greek: 12 gods, presumably the Olympians) [Thompson 1953 Contr Dudley Herb 4:73–154] Polyploid group; spp. sometimes intergrade.

1. Stamens 4 . . . . *D. alpinum*
1' Stamens 5
  2. Corolla tube not covering anther bases, filaments exposed . . . . *D. pulchellum*
  2' Corolla tube gen covering anther bases so filaments not visible . . . . *D. redolens*

D. alpinum (A. Gray) E. Greene (p. 449) ALPINE SHOOTING STAR Pl gen glabrous. **LF** 2–20 cm; blade linear to linear-oblanceolate, narrowed gradually to petiole. **INFL** 1–10-fld. **FL**: parts gen in 4's; corolla lobes 8–16 mm, magenta to lavender; filaments free, 0.5 mm, black, anthers 4.7–8.5 mm, tissue at base wrinkled, black; stigma enlarged. **FR** with valves. *n*=22. Boggy meadows, streambanks; 2400–3400 m. KR, NCoRH, SNH, n WTR (Mt. Pinos), SnBr, SnJt, Wrn, **SNE**; to OR, UT, AZ. [ssp. *majus* H.J. Thompson] ❀ bog and pure water:1–4,6;DFCLT. Jul–Aug

D. pulchellum (Raf.) Merr. (p.449) Pl glabrous. **LF** 4–25 cm; blade oblanceolate to ovate, gen narrowed ± gradually to petiole, gen entire. **INFL** 2–15-fld. **FL**: parts in 5's; corolla lobes gen 9–14 mm, magenta to lavender; filament tube 1.5–3.5 mm, dark maroon to black, anthers 3–5.5 mm, tissue near base smooth to longitudinally wrinkled (when dry), maroon to black; stigma not much enlarged. **FR** with valves. 2*n*=44,88. Wet meadows; 1200–2200 m. **GB, DMtns**; to AK, e US, Mex. [ssp. *monanthum* (E. Greene) H.J. Thompson] The name *D. pauciflorum* E. Greene may have priority. ❀ bog and pure water:1–4,6;DFCLT. Apr–May

D. redolens (H.M. Hall) H.J. Thompson (p. 449, pl. 87) Pl densely glandular-hairy. **LF** 20–40 cm; blade oblanceolate, narrowed gradually to petiole, entire. **INFL** 5–10-fld. **FL**: parts in 5's; corolla tube covering anther bases, lobes 15–25 mm, magenta to lavender; filaments free, gen < 1 mm, anthers 7–11 mm, tissue near base transversely wrinkled, dark maroon to black; stigma enlarged. **FR** with valves. Moist sites; 2400–3600 m. c&s SNH, SnGb, SnBr, SnJt, **SNE**, **DMtns**; to NV, w UT. ❀ WET:1–4,6;DFCLT. Jul–Aug

GLAUX   SEA-MILKWORT

1 sp.: n temp, arctic. (Greek: bluish green)

G. maritima L. (p.449) Per, fleshy, erect, tufted. **ST** 5–40 cm. **LVS** cauline, opposite, sessile, 4–20 mm, linear to oblong, entire. **INFL**: fls solitary in axils, ± sessile. **FL** 3–4 mm; parts in 5's; calyx corolla-like, lobes ± 2 × tube, ovate or oblong, white to reddish or lavender; corolla 0; stamens 5, alternate calyx lobes, filaments awl-like, anthers cordate; ovary superior. **FR** 5-valved, < calyx, ovoid. **SEEDS** few, elliptic in outline, ± flat. 2*n*=30. Coastal salt marsh, saline meadows. NCo, KR, deltaic GV, CCo, SCo, e-most MP, **SNE**; to e N.Am, Eurasia. ❀ TRY:STBL. May–Jul

PRIMULA   PRIMROSE

Per, rhizomed, sometimes stoloned. **LVS** basal or crowded on branches near ground, sessile. **INFL**: umbel 1 per scapose peduncle, terminal, subtended by bracts. **FL**: parts in 5's; calyx tube angled; corolla funnel-shaped or salverform, lobes spreading or erect, entire or notched at tip; stamens incl, filaments short, anthers oblong, obtuse; ovary superior. **FR** 5-valved, elliptic to ovate. **SEEDS** many, peltate, dotted. 200 spp.: gen n temp. (Latin: diminutive of first, from early fl) Spp. often heterostylous.

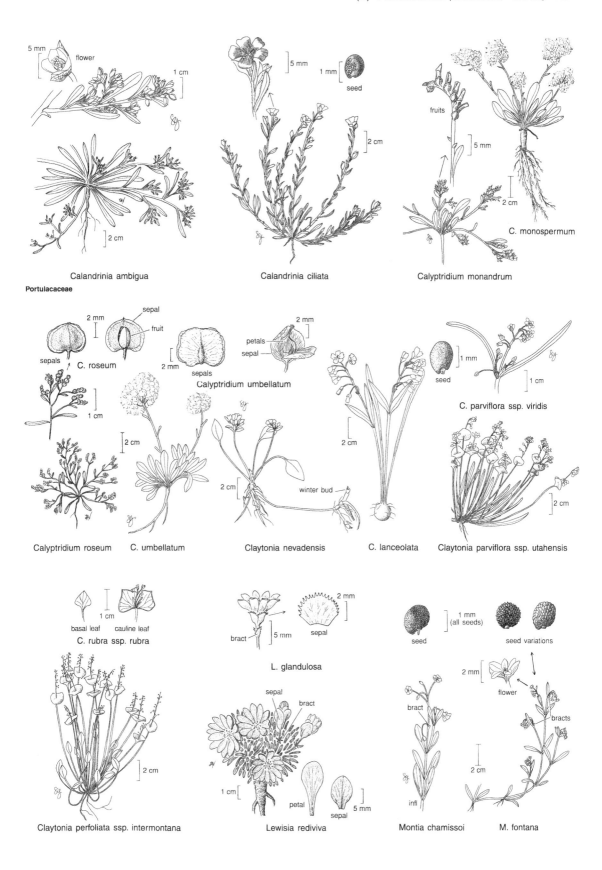

Calandrinia ambigua

Calandrinia ciliata

C. monospermum

Calyptridium monandrum

Portulacaceae

C. roseum

Calyptridium umbellatum

C. parviflora ssp. viridis

Calyptridium roseum

C. umbellatum

Claytonia nevadensis

C. lanceolata

Claytonia parviflora ssp. utahensis

C. rubra ssp. rubra

L. glandulosa

Claytonia perfoliata ssp. intermontana

Lewisia rediviva

Montia chamissoi   M. fontana

P. suffrutescens A. Gray (p. 449, pl. 88) SIERRA PRIMROSE Subshrub. **ST** creeping, branched; base woody. **LF** 15–35 mm, ± spoon-shaped, glabrous; base tapered; tip gen rounded, dentate. **INFL** glandular-hairy; peduncle 3.5–12 cm; pedicels 2–several, < 1.5 cm. **FL**: calyx 4–8 mm, lobes lanceolate; corolla magenta with yellow throat, tube 7–10 mm, = lobes. **FR** ± = calyx, ovoid. $2n=44$. Gen in rock crevices; gen 2000–4200 m. KR, SNH, SNE. ❀ DRN,IRR: 1–4,6;DFCLT. Jul–Aug

RAFFLESIACEAE RAFFLESIA FAMILY

George Yatskievych

Per, st or root parasites lacking chlorophyll, gen monoecious or dioecious. **ST** reduced to thread-like tissue underground or inside host st. **LF** scale-like. **INFL**: short spike or fl solitary. **FL** gen unisexual, radial, fleshy; sepals 4–16, often fused at base; petals 0; stamens 5–many, fused to style axis forming a column that is gen expanded at tip into a disk with stigmatic areas or stamens along or under margin; ovary ± inferior, chambers 1–several, placentas parietal. **FR** various, gen fleshy. **SEEDS** many, minute. ± 8 genera ± 50 spp.: worldwide, esp trop. Poorly known taxonomically. *Rafflesia* spp. have world's largest fls (1 m diam); *Pilostyles* fls are < 2 mm diam.

PILOSTYLES

Per, st parasites. **ST** appearing 0. **LVS** reduced to bracts subtending fl. **INFL**: fl solitary. **FL** minute; sepals 4–5. **STAMINATE FL**: anthers many, sessile on column under margin of disk. **PISTILLATE FL**: ovary inferior, chamber 1, stigmas in ring along disk margin. **FR**: fleshy capsule. 20 spp.: trop Am, Afr, Australia, sw Asia (esp Iran). (Latin: hair pillar, from the central column) Only fls and bracts are visible on surface of host st.

P. thurberi A. Gray (p. 449, pl. 89) THURBER'S PILOSTYLES **FL** < 2 mm, brown or maroon; bracts 4–7, 1–1.5 mm, overlapping, round or ovate; sepals like bracts; disk < 1 mm diam; stamens in ring of ± 3 rows. UNCOMMON. Open desert scrub; 0–300 m. s DSon **(Riverside, San Diego, Imperial cos.)**; to NV, AZ, TX, Mex. Parasitic on *Psorothamnus*, esp *P. emoryi*. Jan

RANUNCULACEAE BUTTERCUP FAMILY

Dieter H. Wilken, except as specified

Ann, per, sometimes aquatic. **LVS** gen basal and cauline, gen alternate, simple or compound; petioles at base gen flat, sometimes sheathing or stipule-like. **INFL**: cyme, raceme, panicle, or fls solitary. **FL** gen bisexual, radial; sepals gen 5, free, early deciduous or withering in fr, gen green; petals 0–many, free; stamens gen 10–many; pistils 1–many, ovary superior, chamber 1, style 1, gen ± persistent in fr as beak, ovules 1–many. **FR**: achene, follicle, berry, or utricle-like, 1–many-seeded. ± 60 genera, 1700 spp.: worldwide, esp n temp, trop mtns; many orn (*Adonis, Aquilegia, Clematis, Consolida, Delphinium, Erianthis, Helleborus*), some highly TOXIC (*Aconitum, Actaea, Delphinium, Ranunculus*). [Duncan & Keener 1991 Phytologia 70:24–27]

1. Fl bilateral; sepals not alike, uppermost spurred or hooded, gen > others
  2. Upper sepal hooded, gen enclosing upper 2 petals & stamens, not spurred . . . . **ACONITUM**
  2' Upper sepal flat to curved but not hooded, gen not enclosing petals & stamens . . . . **DELPHINIUM**
1' Fl radial; sepals gen alike
  3. Lvs simple, thread-like to narrowly oblanceolate; sepal spurs 1–2.5 mm; stamens gen 5 or10 . . **MYOSURUS**
  3' Lvs gen dissected or compound, sometimes simple in Ranunculus; sepals not spurred (petals spurred
    in *Aquilegia*, spurs gen 10–23 mm long); stamens gen many
    4. Petals spurred; pistils, frs gen 5; fr a follicle . . . . **AQUILEGIA**
    4' Petals 0 (sepals then petal-like) or not spurred
      5. Cauline lvs gen opposite or in 1–2 whorls
        6. Per from tuber; cauline lvs few, dissected; infl terminal; fr cluster oblong to elliptic . . . . **ANEMONE**
        6' Woody vine; cauline lvs many, 1–2-pinnate; infl gen axillary; fr cluster gen spheric . . . . **CLEMATIS**
      5' Cauline lvs alternate or 0
        7. Perianth parts in 2 whorls; petals gen 5, white to yellow . . . . **RANUNCULUS**
        7' Perianth parts in 1 whorl; sepals greenish white to purplish; petals 0 . . . . **THALICTRUM**

ACONITUM MONKSHOOD

Per from rhizome or tuber; roots fibrous or fleshy. **STS** 1–few, gen erect, gen simple. **LF** gen palmately lobed; deep lobes 3–7, toothed to lobed; cauline gradually reduced upward. **INFL**: gen raceme, terminal, bracted; pedicels spreading to upcurved. **FL** bilateral; sepals 5, petal-like, upper 1 > others, hooded, gen enclosing upper 2 petals and stamens, tip gen rounded to beaked; petals 2–5, upper 2 clawed, blades gen inflated, tip spurred, lower 3 << sepals, scale-like, or 0; pistils gen 3. **FR**: follicles. **SEED** angled or winged, dark brown to black. ± 100 spp.: temp N.Am, Eurasia. (Greek: ancient name) [Brink 1980 Amer J Bot 67:263–273] Most spp. highly TOXIC, causing death in livestock, humans.

A. columbianum Nutt. (p. 449) Pl 3–15(20) dm. **ST** erect, less often reclining or twining above; upper axils (incl infl) with deciduous bulblets or not. **LF** 3–17 cm, 5–14 cm wide; deep lobes 3–5, wedge- to diamond-shaped, toothed to irregularly cut or lobed above middle. **INFL** 5–55 cm, open. **FL**: sepals deep bluish purple or white to yellow-green, upper 10–15(20) mm, beak

3–8 mm, lateral 8–18 mm, round to reniform, lower 7–12 mm, lanceolate to ovate; upper petals blue to whitish, spur < blade, lower petals 0. **FR** glabrous to puberulent, glandular or not. n=8. Streambanks, moist areas, meadows, coniferous forest; 600–2900 m. NW (exc NCo), CaR, SNH, MP, **n SNE (Sweetwater Mtns)**; to B.C., SD, NM. [*A. geranioides* E. Greene, *A. leibergii* E. Greene] Pls with bulblets (KR, n SNH) have been called var. *howellii* (Nelson & J.F. Macbr.) C. Hitchc. [*A. hanseni* E. Greene, *A. viviparum* E. Greene] ❀ WET,DRN:1–3,**4–6**,7, 14–18;DFCLT. Jul–Aug

ANEMONE ANEMONE

Per from stout, simple to branched caudex, rhizome, or tuber. **STS** 1–several, erect, gen simple. **LVS** simple to 1-ternate, blade or lflet toothed to dissected; basal lvs rosetted, petioled, in fl or fr withered or persistent; cauline lvs gen 2–3, in 1–2 whorls, petiole 0 to short. **INFL** terminal; peduncles 1–5, erect, 1-fld, in fr elongated. **FL** radial; receptacle in fr elongated; sepals 5–8(10), petal-like; petals 0; pistils many, styles in fr gen persistent, gen glabrous to puberulent. **FR**: achenes, densely clustered. ± 100 spp.: temp worldwide. (Greek: fl shaken by wind) Some spp. cult for orn. Pls with long, plumose styles sometimes separated as *Pulsatilla*.

A. tuberosa Rydb. (p. 449) Pl 12–35 cm; tuber atop slender caudex. **STS** 1–few, glabrous. **LVS** dissected; segments oblong to ovate, 4–8 mm wide; basal few, petioles 5–8 cm, glabrous; cauline petioles ± 0 to short. **INFL**: fls 1–5. **FL**: sepals 5–8, 6– 14 mm, reddish, lower surface soft-hairy. **FR**: cluster oblong to elliptic, (15) 20–30 mm, densely woolly; styles 1–2 mm. Rocky slopes, ledges; 900–1900 m. **e DMtns**; to UT, NM. ❀ TRY; DFCLT. Apr–May

AQUILEGIA COLUMBINE

Per from thick, simple to branched caudex. **STS** 1–few, ascending to erect, branched to not, scapose to not, glabrous to glandular-hairy. **LVS**: basal 1–3-ternate, petiole gen long; cauline 0–few, gen much reduced, deeply 3-lobed to 1–2-ternate, petiole short to ± 0; segments gen wedge-shaped to obovate, upper surface green to pale green, lower surface pale green to glaucous. **INFL**: few-fld raceme or fl solitary, terminal; axis and pedicels glabrous to glandular; fls often pendent. **FL** radial; sepals 5, petal-like, spreading to slightly reflexed; petals 5, gen with spur projecting between sepals; pistils gen 5. **FR**: follicles, glabrous to glandular. **SEEDS** smooth, shiny, brown to black. ± 70 spp.: temp N.Am, Eurasia. (Derivation uncertain, perhaps Latin: eagle, from spurs, or water-drawer, from habitats) [Munz 1946 Gentes Herb 7:1–150] Many spp. and hybrids cult as orn; natural hybrids common.

A. formosa Fischer (p. 449) Pl 20–80(150) cm. **LVS**: basal and lower cauline 2–3-ternate, petioles 5–30(40) cm, segments 7–45 (130) mm; upper cauline gen simple to deeply 3-lobed. **INFL**: fl nodding, in fr erect. **FL**: sepals 12–20(25) mm, red; petal blade 1–8 mm, yellow, spur (4)10–23 mm, ± straight to ± incurved, tube red, tip 1.5–4 mm wide, mouth 4–8 mm wide, ± round, ± 90° to fl axis, red to yellowish; stamens 10–18 mm. $2n$=14. Streambanks, seeps, moist places, chaparral, oak woodland, mixed-evergreen or coniferous forest; < 3300 m. CA-FP (exc GV, SCo, ChI), **GB**, **DMtns**; to AK, MT, Baja CA. [vars. *hypolasia* (E. Greene) Munz, *pauciflora* (E. Greene) H.S. Boothman, *truncata* (Fischer & C. Meyer) Baker; forma *anomala* J. Howell] Lf variation needs study; pls with pale green to glaucous, gen 3-ternate lvs (DMtns), have been called *A. shockleyi* Eastw. ❀ IRR,DRN:**4–6,15–17,24**& SHD:1–3,7,8,9,**14, 18–23**. Jul

CLEMATIS VIRGIN'S BOWER

Frederick B. Essig

Gen woody vine, sometimes dioecious. **LVS** gen 1–2-pinnate, cauline, gen opposite; petiole gen twining; lflets ovate to lanceolate, often irregularly 2–3-lobed, coarsely toothed. **INFL**: panicle to 1-fld, axillary or terminal. **FL** radial; sepals gen 4, free, petal-like, gen lanceolate, white to cream (or brightly colored elsewhere); petals 0; stamens many, free; pistils gen many, simple. **FR**: many achenes, each gen with an elongate, feathery style. 250 spp.: worldwide. (Greek: twig) Worldwide revision badly needed but CA spp. distinct.

1. Infl several–many-fld, appearing June–September; sepal hairy on both surfaces . . . . *C. ligusticifolia*
1' Infl 1–3-fld, appearing January–June; sepal hairy only on lower surface . . . . *C. pauciflora*

C. ligusticifolia Nutt. (p. 453) VIRGIN'S BOWER, YERBA DE CHIVA **LF**: lflets 5–15, irregularly lobed or toothed, largest on pl 2–8 cm. **INFL** several–many-fld, axillary. **FL**: sepals 6–10 mm, both surfaces hairy; stamens 25–40, 5–9 mm, ± = sepals; pistils 35–65. **FR**: body hairy. Along streams, wet places; < 2400 m. **CA**; to B.C., SD, NM, nw Mex. Fls June–September. ❀ IRR:**1–7, 14–17, 22–24**&SHD:**8,9,18–21**.

C. pauciflora Nutt. (p. 453) ROPEVINE **LF**: lflets 3–9, ± 3-lobed, toothed, largest on pl gen 1–3 cm. **INFL** 1–3-fld, axillary. **FL**: sepals 6–12 mm, only lower surface hairy; stamens 30 –50, 6–12 mm, ± = sepals; pistils 25–50. **FR**: body glabrous. Dry chaparral; < 1300 m. SW, **DMtns (Little San Bernardino Mtns)**; Baja CA. Fls January–June. ❀ DRN,DRY:**7,14–17,22–24**&SHD: **8,9,18–21**.

DELPHINIUM LARKSPUR

Michael J. Warnock

Per; root gen < 10 cm, ± fibrous or fleshy; buds gen obscure. **ST** gen 1, erect, gen unbranched; base gen ± as wide as root, gen firmly attached to root, gen ± reddish or purplish. **LVS** simple, basal and cauline, petioled; blades gen palmately lobed, deep lobes gen 3–5, gen < 6 mm wide, gen also lobed; lower lvs gen dry, often 0 in fl; cauline merging into bracts upward. **INFL**: raceme or somewhat branched, terminal; fls gen 10–25; pedicels gen ± spreading.

FL bilateral; sepals 5, petal-like, gen spreading, gen ± dark blue, uppermost spurred; petals 4, << sepals, upper 2 with nectar-secreting spurs enclosed in uppermost sepal, lower 2 clawed, with blades gen 4–8 mm, notched, gen ± perpendicular to claws, gen colored like sepals, gen obviously hairy; pistils 3(–5). **FR** aggregate of 3(–5) erect follicles, gen 2.5–4 × longer than wide. **SEED** dark brown to black, often appearing white, gen winged when immature, gen without inflated collar; coat cell margins gen straight. (Latin: dolphin, from bud shape) [Lewis & Epling 1954 Brittonia 8:1–22] Hybrids common, esp in disturbed places. Root length here includes coarse but not thread-like parts. Most spp. highly TOXIC, attractive and causing many deaths to cattle, less often to horses, sheep. ❀ Exc as noted, successful in cult only within natural range and habitat. Lowland ssp.: DRY. Upland spp.: winter chilling required.

1. Sepals red or pink
  2. Infl hairs glandular; pedicels ascending; st ± glabrous . . . . *D. purpusii*
  2' Infl hairs nonglandular; pedicels spreading; st gen curled-puberulent . . . . *D. cardinale*
1' Sepals blue, purple, or white.
  3. Lower sts and lower petioles with spreading hairs > 1 mm . . . . *D. hansenii* ssp. *kernense*
  3' Lower sts and lower petioles glabrous or with spreading hairs < 1 mm
    4. Cauline lf lobes (> 60% to base) 5; lobes > or = 6 mm wide.
      5. Green lvs mostly on lower 1/3 of st in fl . . . . *D. polycladon*
      5' Green lvs mostly on upper 2/3 of st in fl . . . . *D. glaucum*
    4' Cauline lf lobes (> 60% to base) > = 7; lobes < 6 mm wide.
      6. Fr > 3× longer than wide; seed coat cell margins straight (at 10×) . . . . *D. andersonii*
      6' Fr < 3× longer than wide; seed coat cell margins wavy (at 10×)
        7. Sepals ± sky blue, reflexed . . . . *D. parishii* ssp. *parishii*
        7' Sepals darker, duller blue, not reflexed . . . . *D. parishii* ssp. *subglobosum*

D. andersonii A. Gray (p. 453) Root gen > 10 cm, distally branched. **ST** 20–90 (gen 30–60) cm; base often narrower than root, but firmly attached to root, ± glabrous. **LVS** mostly on lower 1/2 of st, ± glabrous; lobes 7–30, < 4 mm wide. **INFL** cylindric; pedicels 8–68 mm, 5–25 cm apart, ± S-shaped, glabrous to puberulent. **FL**: sepals dark blue, lateral 9–16 mm, spur 11–18 mm. **FR** 17–32 mm, gen > 4 × longer than wide. **SEED**: coat smooth, shiny, ± translucent, inflated. 2*n*=16. Talus, dry soils in sagebrush scrub; 1300–2700 m. ne SNH, **GB**; to se OR, c ID, s UT. [ssp. *cognatum* (E. Greene) Ewan; *D. decorum* var. *nevadense* S. Watson; not *D. menziesii* DC.] ❀ DRN,DRY;DFCLT. May–Jun

D. cardinale Hook. (p. 453) CARDINAL or SCARLET LARKSPUR Root gen > 15 cm, distally branched. **ST** 30–270 cm, gen curled-puberulent, base sometimes narrower than root, but firmly attached to root. **LVS**: basal present in fl or not, ± glabrous, lobes 5–27. **INFL**: pedicels 15–55 mm, 6–80 (gen > 15) mm apart, puberulent. **FL**: sepals red, gen ± forward-pointing, lateral 11–15 mm, spur 15–24 mm; lower petals flattened, blades 4–5 mm, with hairs few, short, yellow, obscure to naked eye. **FR** 12–18 mm, ± straight. **SEED** bumpy. 2*n*=16. Slopes (often talus) in chaparral; 300–1500 m. SCoR, SW, **w edge DSon**; Baja CA. Gen hummingbird pollinated. ❀DRN,DRY:15–17&SHD: 7,14,18–23 Feb–Jul

D. glaucum S. Watson (p. 453) MOUNTAIN LARKSPUR Root gen > 15 cm, distally branched; buds gen prominent (exc on herbarium specimens). **ST** gen 2 or more per root system, 80–300 (gen > 150) cm,l ower glabrous, glaucous. **LVS** gen glabrous; lobes sharply cut at tips, often > 6 mm at widest. **INFL** gen branched; fls gen > 50; lower bracts lf-like; pedicels 10–48 mm, 3–25 mm apart, glabrous to puberulent. **FL**: sepals ± forward-pointing to spreading, purplish blue, lateral 8–14(21) mm, spur 10–19 mm. **FR** 9–20 mm. **SEED** bumpy. 2*n* =16. Wet thickets, streamsides; 1600–3200 m. KR, SNH, SnGb, SnBr, **n SNE**; to AK, NV. May hybridize with *D. polycladon*. ❀WET, DRN:1,4–6&SHD: 2,3. Jul–Sep

D. hansenii (E. Greene) E.Greene HANSEN'S LARKSPUR **ST** 25–180 (gen 40–80) cm; base puberulent to hairy. **LF** hairy, esp on lower surface; lobes 3–18; petioles hairy. **INFL**: fls often > 25; pedicels 3–57 mm, gen < 8 mm apart, ± ascending, puberulent. **FL**: sepals spreading to forward-pointing, lateral 7–13 mm, spur 6–16 mm; lower petal blades gen hairier on inner lobes. **FR** 8–20 mm, often < 3 × longer than wide. **SEED** finely prickly. Oak woodland or open chaparral; 60–3000 m. NCoRI, CaRF, SNF,

c&s SNH, Teh, GV, **w edge DMoj**. 3 sspp. in CA.

ssp. *kernense* (Davidson) Ewan **ST** 34–110 cm; base puberulent. **LVS** gen basal but dry in fl. **INFL**: pedicels 7–25 (gen > 10) mm apart. **FL**: sepals white to dark blue-purple, lateral 7–13 mm, spur 8–16 mm. 2*n* =16,32. Open oak woodland, chaparral; 800–1900 m. s SN, Teh, **w edge DMoj**. ❀ DRN, DRY; DFCLT. Apr–May

D. parishii A. Gray Root often > 15 cm, branched. **ST** 17–100 (gen < 60) cm; base often narrower than root, but firmly attached to root, glabrous to puberulent. **LVS** mostly basal or mostly cauline in fl, glabrous to puberulent; lobes 3–20, gen < 6 mm at widest. **INFL**: fls 6–75; pedicels 3–48 mm, 3–25 mm apart, glabrous to puberulent. **FL**: sepals dark blue to white or pink, lateral 7–13 mm, spur 8–14 mm; lower petal blades 3–6 mm. **FR** 8–21 mm, gen < 3 × longer than wide. **SEED** ± winged; coat inflated; cell margins wavy. Desert scrub, with scattered trees or not; 300–2500 m. Teh, TR, **SNE, D**; to sw UT, AZ, n Baja CA. 3 sspp. in CA.

ssp. *parishii* (pl. 90) PARISH'S or DESERT LARKSPUR **ST** 17–100 cm. **LVS**: basal 3–5-lobed but gen 0 in fl; cauline 3–15-lobed. **INFL**: pedicels 8–25 mm apart. **FL**: sepals ± sky blue, lateral reflexed, 8–12 mm, 3–6 mm wide, spur 8–15 mm; lower petal blades 3–6 mm. **FR** 9–21 mm. 2*n*=16. Desert scrub, juniper woodland; 300–2500 m. Teh, e TR, **SNE, D**; to sw UT, AZ, n Baja CA. ❀ DRN,DRY; DFCLT. Mar–May

ssp. *subglobosum* (Wiggins) Harlan Lewis & Epling **ST** 19–78 cm. **LVS**: basal gen present in fl (cauline much reduced); lobes 7–12. **INFL**: pedicels 8–17 mm apart. **FL**: sepals spreading, dark blue, lateral 9–13 mm, 5–7 mm wide, spur 12–14 mm; lower petal blades 4–6 mm. **FR** 8–11 mm. UNCOMMON. Chaparral or desert scrub; 600–1300 m. **w DSon**; n Baja CA. ❀ DRN,DRY;DFCLT. Mar–Apr

D. polycladon Eastw. HIGH MOUNTAIN LARKSPUR Root gen > 15 cm, distally branched; buds prominent (exc on herbarium specimens). **ST** 15–160 (gen 80–120) cm, glabrous, often 2 or more per root system. **LVS** mostly on lower third of st, glabrous; lobes often > 6 mm at widest. **INFL** often ± 1-sided; fls 3–35; pedicels 10–150 mm, ± S-shaped, 10–80 mm apart, glabrous to puberulent. **FL**: lateral sepals 12–18 mm, spur 11–22 mm. **FR** 13–20 mm. **SEED** ± striate. 2*n*=16. Streamsides, wet talus; 2200–3600 m. SNH, **W&I**. Hybridizes with *D. glaucum*. ❀ DRN,DRY; DFCLT. Jul–Sep

D. purpusii Brandegee (p.453) PURPUS', ROSE-FLOWERED, or

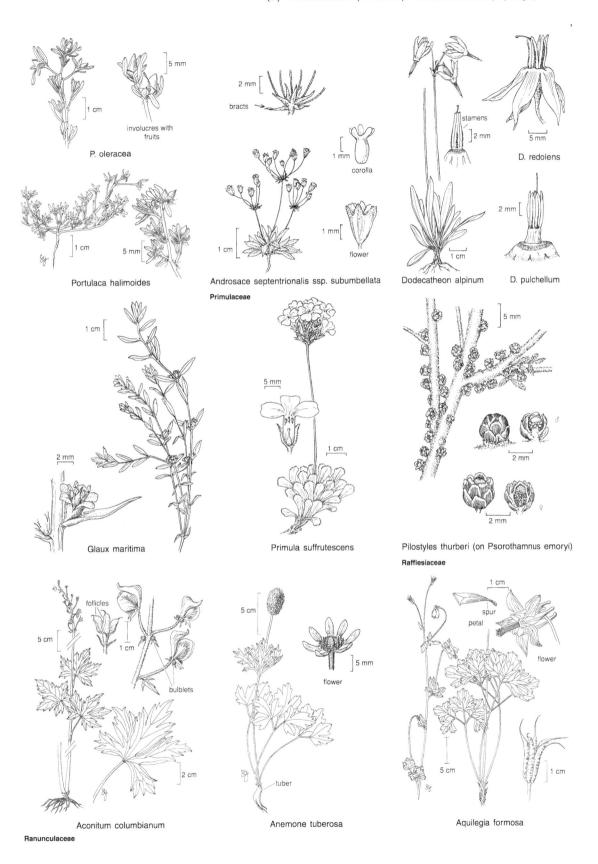

P. oleracea

involucres with fruits

Portulaca halimoides

Androsace septentrionalis ssp. subumbellata

Primulaceae

corolla

flower

bracts

D. redolens

stamens

Dodecatheon alpinum

D. pulchellum

Glaux maritima

Primula suffrutescens

Pilostyles thurberi (on Psorothamnus emoryi)

Rafflesiaceae

Aconitum columbianum

follicles

bulblets

Anemone tuberosa

flower

tuber

Aquilegia formosa

spur

petal

flower

Ranunculaceae

KERN COUNTY LARKSPUR Root often > 15 cm, distally branched. **ST** 30–120 cm; base gen narrower than root, not firmly attached to root, ± glabrous. **LVS** mostly on lower half of st, ± puberulent; lobes > 6 mm at widest or not. **INFL** narrow; pedicels 5–48 mm, 8–50 mm apart, ± ascending, hairs glandular, yellow. **FL**: sepals gen reflexed, pink to deep pink, lateral 10–16 mm, spur 10–19 mm; lower petal blades 3–4 mm, ± glabrous, angled > 130° to claw. **FR** 11–29 mm. **SEED** shiny; coat inflated, ± clear, winged. 2*n*=16. UNCOMMON. Talus, cliffs: 300–1300 m. s SN, w edge DMoj. ❀ DRN, DRY, SHD; DFCLT. Mar–May

MYOSURUS MOUSE-TAIL

Ann from fibrous roots, gen glabrous. **STS** 1–many, ascending to erect, slender, gen tufted. **LVS** simple; basal, ± sessile, thread-like to narrowly oblanceolate, entire. **INFL** scapose; peduncle 1-fld, in fr gen > 1 cm. **FL** bisexual, radial; receptacle in fr much elongated, cylindric; sepals 5–7, spurred, white to green, fading brownish; petals 0 or 3–5, white to greenish or yellowish, gen early deciduous; stamens 5–many; pistils many. **FR**: achenes, glabrous to puberulent; keel on outer surface, in depression or not, beak (continuation of keel) ± ascending to erect-appressed. 10–15 spp.: temp Am, Eurasia, New Zealand. (Greek: mouse tail, from receptacle in fr) [Campbell 1952 El Aliso 2:389–403; Stone 1959 Evolution 13:151–174] Fr needed for identification. *M. apetalus* Gay var. *borealis* Whittemore has been collected se of Masonic in n SNE.

1. Fr beak ± ascending, 0.5–1 mm, gen > body; stamens gen 5 . . . . ***M. minimus***
1' Fr beak gen erect to erect-appressed, gen < 0.5 mm, gen << or = body; stamens gen 10 . . . . ***M. cupulatus***

M. cupulatus S. Watson (p. 453) Pl 2–14 cm. **LF** 1–7 cm, linear to narrow-oblanceolate. **INFL** in fr > lvs. **FL**: receptacle in fr 5–40 mm; sepals 1.5–3 mm, spur 1–2.5 mm; petals 5, 1.5–2.5 mm; stamens gen 5. **FR**: body not compressed, ± as long as wide; keel in a depression; beak ± ascending, 0.5–1 mm. Wet places, shrubland, pinyon/juniper woodland; 1000–1700 m. e DMtns; to NM. Apr–May

M. minimus L. (p. 453) Pl 2–12 cm. **LF** 0.5–6 cm, thread-like to narrowly oblanceolate. **INFL** in fr < to > lvs. **FL**: receptacle in fr 10–40 mm; sepals 1–3 mm, spur ± = sepal; petals 5, 1.5–3 mm; stamens gen 10. **FR**: body ± compressed laterally or not; keel in a slight depression or not; beak gen erect to erect-appressed, gen < 0.5 mm. Wet places, vernal pools, marshes; < 1500 m. NCoR, CaRF, SNF, GV, SnFrB, s SCoRO, SCo, SnJt, MP, **W&I**; to B.C., e US, Eurasia. [ssp. *major* (E. Greene) G.R. Campbell, var. *filiformis* E. Greene]

RANUNCULUS BUTTERCUP

Ann, per, sometimes from stolons or caudices, terrestrial or aquatic; roots gen fibrous. **ST** prostrate to erect. **LVS** basal and gen cauline, gen reduced upwards, gen glabrous; petiole base flat, stipule-like or not; basal and lower cauline petioles gen long; blades simple to dissected or compound, entire to toothed. **INFL**: cyme, axillary or terminal, 1–few-fld. **FL** radial; sepals gen 5, gen early deciduous, gen glabrous, gen green to yellowish; petals gen 5, gen > sepals, gen white to yellow, shiny; nectar gland near petal base, pocket-like or with flap-like scale; anthers yellow; pistils gen many. **FR**: achene, gen compressed, beaked, gen glabrous; walls thick. ± 250 spp.: temp world-wide, trop mtns; some orn. (Latin: (Pliny) little frog, from gen wet habitats)

1. Pl gen aquatic, st gen submersed or floating (sometimes stranded)
  2. Most or all lvs submersed, gen 2–6-dissected . . . . ***R. aquatilis*** var. ***capillaceus***
  2' Most lvs floating or emergent, simple, entire . . . . ***R. hydrocharoides***
1' Pl gen terrestrial (if aquatic st gen emergent)
  3. Ann; basal lf segments linear; fr body < beak, with 2 lateral bulges . . . . ***R. testiculatus***
  3' Per; basal lf entire to lobed or dissected; fr body gen > beak, lenticular to ovoid, not bulged
    4. Basal lvs deeply lobed to dissected
      5. Petals white, red-tinged or not; fr utricle, body 8–15 mm, fr cluster spheric . . . . ***R. andersonii***
      5' Petals yellow; fr achene, body 1.5–2 mm, fr cluster ovoid to subcylindric
        6. Per; sepals 3.5–7 mm; petals 5–12 mm; fls few, terminal; st solid . . . . ***R. eschscholtzii*** var. ***oxynotus***
        6' Ann; sepals 2–3 mm; petals 2–5 mm; fls many, axillary, terminal; st hollow below . . . . ***R. sceleratus***
    4' Basal lvs entire to 3-lobed
      7. Basal lvs ovate to reniform, margins crenate; pls gen from stolons; fr cluster cylindric
       . . . . ***R. cymbalaria*** var. ***saximontanus***
      7' Basal lvs ± oblong to round, margins entire; stolons 0; fr cluster spheric
        8. Upper cauline lvs linear to narrowly lanceolate, entire; sts gen 3–5 from base . . . . ***R. alismifolius*** var. ***alismellus***
        8' Upper cauline lvs 0 or oblong to ovate, entire to 2–3-lobed; sts gen 1, sometimes few-branched above
         . . . . ***R. glaberrimus*** var. ***ellipticus***

R. alismifolius Benth. Per 4–50 cm; roots ± thick, fleshy. **STS** decumbent to erect, gen 3–5 from base, gen branched from base, glabrous. **LVS**: basal and lower cauline petioles 2–7 cm, blades 2–12 cm, ± oblong to ovate, entire, base gen tapered; upper cauline lvs linear to narrowly lanceolate, entire. **FL**: receptacle glabrous; sepals 3–5 mm, ± reflexed, glabrous to puberulent; petals 5–7 in CA, 5–15 mm, 2–5.5 mm wide. **FRS** 12–many; cluster spheric; body 1.5–2.5 mm, sides 1–2 mm wide, smooth to sparsely puberulent, back rounded; beak 0.5–1 mm, ± straight. Wet places, streambanks, meadows, coniferous forest; 1300–3600 m. KR, NCoRH, CaRH, SNH, SnBr, SnJt, MP, **n W&I**; to B.C., ID, WY, NV. Vars. intergrade, difficult. 2 vars. in CA.

var. ***alismellus*** A. Gray (p. 453) Pl 4–20(30) cm. **LF**: basal blade 2–4 cm, gen ovate. **FL**: petals 5–7 mm. Habitats of sp.; 1400–3600 m. KR, NCoRH, CaRH, SNH, SnBr, SnJt, **n W&I**; to WA, NV. ❀ WET or IRR,DRN:1,2,7&SUN:4–6,15–17. Jun–Jul

R. andersonii A. Gray (p. 453) Per 8–18 cm, scapose, glabrous. **LVS** basal; petioles 3–6 cm, blades 2–3 cm, cordate to reniform, 1-ternate, lflets gen 2-dissected, segments lanceolate to oblong. **FL:** receptacle puberulent; sepals 6.5–8 mm, persistent, white, red-tinged or not; petals 10–21 mm, 8–15 mm wide, white, red-tinged or not. **FRS** 14–many; cluster spheric; body 8–15 mm, very compressed, sides 6–10 mm wide, smooth, margin wing-like; beak ± 0.5 mm, ± straight. Rocky slopes, gravelly soils, shrubland, pinyon/juniper woodland; 900–2300 m. SNH (e slope), MP, **W&I, n DMtns (Grapevine, Panamint Mtns)**; to OR, ID, NV. ✿ TRY; DFCLT. Jun

R. aquatilis L. Per (5)20–80 cm, aquatic, often mat-forming. **ST** submersed or floating, branched throughout, rooting at lower nodes, glabrous. **LVS** cauline, most or all submersed; submersed blades 10–40 mm, 3–6-dissected, cordate to reniform, segments thread-like, petioles < 1.5 cm; floating or emergent gen 0 or few, like submersed or not. **FL** floating or emergent; receptacle hairy; sepals 2–5 mm; petals 4–10(14) mm, 1–2.5 mm wide, white, base yellow or not. **FRS** 10–50; cluster spheric; body 1–2 mm, sides 1–1.5 mm, with transverse, broken, wavy ridges, glabrous to sparsely puberulent, back ± rounded; beak < 0.5 mm, ± straight. Ponds, lake margins, marshes, rivers; < 2900 m. CA-FP (exc ChI), **GB**; to AK, e N.Am, Mex. Vars. intergrade, difficult to separate. ✿ WET,SUN:1–3,**4–7**,8,9,14,**15–18**,19–24. 3 vars. in CA.

var. *capillaceus* (Thuill.) DC. (p. 453) **LVS** gen = or < internodes; submersed lvs 4–6-dissected, segments thread-like, petioles below gen flat, wide, above gen thread-like; floating or emergent gen 0 or like submersed. **INFL:** pedicel ± straight in fr. **FRS** 10–30. Habitats, elevations of sp. CA-FP (exc ChI), **GB**; to AK, e N.Am, Mex. [var. *harrisii* L. Benson] Apr–Aug

R. cymbalaria Pursh var. *saximontanus* Fern. (p. 453) Per 3–20(30) cm, often scapose, gen from stolons. **ST** gen erect, simple; hairs 0–sparse. **LVS:** petioles 1.5–8 cm, blades 0.5–2.5 cm, ovate to reniform, gen crenate. **FL:** receptacle glabrous to puberulent; sepals 2–5 mm; petals 5–8, 4–8 mm, 2–3 mm wide. **FRS** many; cluster cylindric; body 1–2 mm, sides ± 1 mm wide, striate to ridged, glabrous, back rounded; beak 0.5–1 mm, ± straight. 2*n*=16. Meadows, streambanks, marshes, pond margins; 200–3200 m. NW (exc NCo), CaR, SN, e SCo, TR, PR, **GB, DMoj**; to w Can, SD, Mex, also S.Am. Other vars. in Rocky Mtns, Asia. ✿ WET:1,**2**,3,7,8,10,11,14,18–23&SUN: 4,5,**6,15,16**,17,24. Jun–Aug

R. eschscholtzii Schldl. Per 5–25 cm, scapose or not. **ST** gen erect, glabrous; branches 0–few, at base. **LVS:** basal and lower cauline (if present) petioles 1–8 cm, blades 1–3 cm, round to reniform, with 3–5(7) lobes > 1/2 way to base, lobes entire to few-toothed or -lobed; upper cauline lvs 0 or 1–2, sessile, deeply 3-lobed to 1-ternate, lobes or lflets oblong to elliptic, entire. **FL:** receptacle glabrous; sepals 3.5–7 mm, hairs 0–sparse; petals 5–12 mm, 5–9 mm wide. **FRS** 17–many; cluster narrowly ovoid to subcylindric; body 1.5–2 mm, ± plump, sides ± 1 mm wide, glabrous, back rounded; beak 0.5–1 mm, ± straight. 2*n*=56. Meadows, rocky slopes, ledges; 1800–4000 m. KR, CaR, SNH, SnBr, SnJt, **GB**; to AK, MT, Colorado, AZ. Some plants difficult to separate from typical var. of AK and w Canada. 2 vars. in CA.

var. *oxynotus* (A. Gray) Jepson (p. 453) **LVS:** basal and lower cauline (if present) with 3–5(7) lobes > 1/2 way to blade

base, middle lobe with 1 tooth or lobe or gen entire, tips obtuse to rounded. Habitats and elevations of sp. CaR, SNH, SnBr, SnJt, **GB**; NV. ✿ TRY;DFCLT. Jul–Aug

R. glaberrimus Hook. Per 5–26 cm, scapose or not; roots fleshy. **STS** gen 1 from base, decumbent to erect, glabrous; branches gen 0–few, above. **LVS** gen basal; basal and lower cauline (if present) petioles 2–8 cm, blades 2–5 cm, elliptic to round, gen tapered at base, entire to 3-lobed; upper cauline lvs 0 or oblong to ovate, gen sessile, entire or gen deeply 2–3-lobed. **FL:** receptacle glabrous; sepals 4–7 mm, reflexed, hairs 0–sparse; petals gen 5, 6–15 mm, 4–8 mm wide. **FRS** many; cluster spheric; body ± 2 mm, sides ± 1.5 mm wide, glabrous to sparsely puberulent, back faintly keeled; beak ± 0.5 mm, ± straight. Open areas, meadows, rocky soils in coniferous forest, juniper woodland, sagebrush scrub, shrubland; 900–3500 m. CaR, n SNH, MP, **n SNE**; to B.C., SD, NM. 2 vars. in CA.

var. *ellipticus* (E. Greene) E. Greene (p. 453) **LVS:** basal and lower cauline (if present) elliptic to obovate, gen entire; upper cauline lvs entire or gen deeply 2–3-lobed, terminal lobe gen > lateral. Coniferous forest, shrubland; 900–3500 m. n SNH (e slope), MP, **n SNE**; to B.C., NE, NM. ✿ DRN,IRR:1–3,7,14,18&SUN:6,15,16;DFCLT. Apr–Jun

R. hydrocharoides A. Gray (p. 453) FROG'S-BIT BUTTERCUP Per 5–25 cm, aquatic or terrestrial, scapose or not. **ST** submersed, floating, or prostrate to decumbent, glabrous; branches 0–few. **LVS** most floating or emergent; basal and lower cauline (if present) petioles 2–10(15) cm, blades simple, 0.6–3 cm, ovate to ± cordate, entire; upper cauline lvs 0 or much smaller. **FL:** receptacle glabrous; sepals 2–4 mm; petals 5–8, 2.5–6 mm, 1–3 mm wide. **FRS** 9–many; cluster spheric; body 1.5–4 mm, plump, ± 1 mm wide, sides smooth, back rounded; beak 1–1.5 mm, ± straight. RARE in CA. Marshes, small streams; 1100–2700 m. c SNH, **s SNE (Mono Co.)**; to NM, Mex. Jul

R. sceleratus L. Ann 15–50 cm; roots ± fleshy. **ST** erect, simple to branched, hollow below; hairs 0–sparse. **LVS:** basal and lower cauline petioles 3–9(12) cm, blades 1.5–3 cm, cordate to reniform, deeply 3-lobed to 1-ternate, lobes or lflets toothed to deeply lobed; upper cauline lvs ± sessile. **FL:** receptacle gen puberulent; sepals 2–3 mm; petals 2–5 mm, 1–2 mm wide. **FRS** many; cluster ovoid to subcylindric; body 1–2 mm, sides < 1 mm wide, smooth or transversely ridged, back slightly keeled; beak < 1 mm. 2*n*=32. Shallow water, lake or pond margins, streambanks; < 2000 m. CaR, GV, MP, **e SNE (Deep Springs Lake, Inyo Co.)**; to AK, MN, AZ, to e N.Am, Eur. [var. *multifidus* Nutt.] Jun–Sep

R. testiculatus Crantz (p. 453) Ann 1–6(10) cm, scapose. **ST** decumbent to erect; branches 0–few, at base. **LVS** basal, tomentose; petioles < 3 cm, blades 0.5–3 cm, ovate, 2–3-dissected, segments linear, ± entire. **FL:** receptacle tomentose; sepals 2–8 mm, persistent in fr, tomentose; petals 2–5, 3–8 mm, 1–4.5 mm wide, early deciduous. **FRS** many; cluster ovoid to spheric; body 1.5–2.5 mm, plump, with 2 lateral bulges, finely tomentose or glabrous; beak 3–4 mm, spine-like, ± straight. Waste areas, overgrazed pastures, shrubland; 1400–2500 m. CaR, s SNH, **GB**, expected elsewhere; to WA, Colorado; native to Eurasia. [*Ceratocephala t.* (Crantz) Roth] TOXIC to livestock. Apr–Jun

THALICTRUM MEADOW-RUE

Per from caudex or rhizomes, dioecious or fls bisexual, gen glabrous. **STS** 1–few, gen erect; branches 0 or few. **LVS** gen 1–4-ternate, basal or basal and cauline, gen reduced upwards, petioled; segments wedge-shaped, fan-shaped, or ± round; upper surface gen green; lower surface pale green. **INFL:** raceme or panicle, axillary or terminal, gen erect; pedicels gen erect in fr; bracts simple to 1-ternate. **FL** radial; sepals 4–5, gen green, petal-like or not, often early deciduous; petals 0; stamens 8–many, gen > sepals, anthers gen narrowly oblong, tip gen abruptly pointed, filaments gen thread-like; pistils 2–20. **FR:** achenes, compressed laterally to not, ribbed or veined, beaked. ± 80 spp.: temp N.Am, Eurasia, Afr; some orn, medicinal. (Greek: name given by Dioscorides, Greek physician-botanist) [Boivin 1944 Rhodora 46:337–377, 391–445, 453–487]

1. Pl 5–15(20) cm; basal lvs 1.5–6 cm; infl raceme; pedicel recurved in fr . . . . *T. alpinum*
1' Pl 60–200 cm; basal lvs 4–46 cm; infl panicle; pedicel erect in fr
  2. Fl unisexual; sepals gen 4; filaments thread-like . . . . *T. fendleri* var. *fendleri*
  2' Fl bisexual; sepals gen 5; filaments flat . . . . *T. sparsiflorum*

T. alpinum L. Pl 5–15(20) cm; fls bisexual. **LVS** gen basal, 1.5–6 cm; segments 4–10 mm, glabrous; upper surface dull green; lower surface glaucous, tip rounded. **INFL**: raceme, scapose; pedicels recurved in fr. **FL**: sepals 1.5–2.5 mm, purplish; stamens 7–12(15). **FRS** 1–6, ± pendent (due to recurved pedicel); body 2–3.5 mm, slightly compressed laterally, side obliquely ± ovate to ± obovate, with 2–3 curved ribs. 2*n*=14. Meadows, often moist, gravelly soils, gen alpine; 2900–3700 m. n&c SNH, **n W&I**; to AK, NM, arctic Eurasia. ❀ IRR,DRN,SUN:**1**,2–7,15, 16,18; DFCLT. Jun–Aug

T. fendleri A. Gray Pl 60–200 cm, gen dioecious. **LVS** basal and cauline, 7–46 cm; segments 8–20 mm, glabrous to finely glandular-puberulent, tip acute to rounded. **INFL**: panicle, leafy to bracted above. **FL**: sepals gen 4, 2–5 mm, greenish white to purplish; stamens 15–28. **FRS** 7–20, spreading to ascending; body 4–8 mm, side with 1–3 ± curved ribs, 0 or several wavy veins. Moist, open to shaded places, woodland, forest; < 3200 m. CA-FP (exc GV, ChI), **GB**; to WA, WY, TX, n Mex. 2 vars. in CA difficult, need study.

var. *fendleri* (p. 453) **LF**: lower surface (esp upper lvs) gen finely glandular-puberulent (at 20 ×). **FR**: body ± compressed laterally throughout, side obliquely ± ovate to ± obovate, with 2–3 ribs, 0 veins. 2*n*=28,56,70. Habitats of sp.; 900–3200 m. KR, CaR (very uncommon), SN, SnFrB, SCoRO, TR, PR, **GB**; to OR, WY, TX, n Mex. ❀ DRN:**4–6**&IRR:**17**&SHD:1–3,7,9,**14–16,18**,19–21,**22–24**. May–Aug

T. sparsiflorum Fischer & C. Meyer Pl 60–180 cm; fls bisexual. **LVS** basal (few) and cauline, 4–30 cm; segments 12–20 mm, finely glandular-puberulent, tip obtuse to rounded. **INFL**: panicle, gen leafy. **FL**: sepals gen 5, 2.5–4 mm, greenish white; stamens 10–20, anthers ovate to oblong, tip obtuse, filaments flat. **FRS** 6–22, ± spreading; body 4–6 mm, strongly compressed laterally, side gen semi-circular to crescent-shaped, veined or weakly ribbed; beak 1–1.5 mm, straight. 2*n*=42. Uncommon. Moist places, streambanks, coniferous forest; 1400–3300 m. CaR, c&n SN, SnBr, SnJt, **n SNE (Sweetwater, White mtns)**; to AK, Colorado, Asia. ❀ TRY. Jul–Aug

RESEDACEAE MIGNONETTE FAMILY

Thomas F. Daniel

Ann, bien, per, shrub. **LVS** simple, alternate; stipules small, tooth- or gland-like; blade entire to deeply lobed. **INFL**: raceme, terminal, spike-like. **FL** gen bisexual, small, asymmetric, 1 per bract; sepals 2–8; petals 0–8; disk sometimes present; stamens 3–50+, gen on disk, anthers 2-chambered; pistils gen ± compound, 2–7-parted, gen open at top, ovary superior, sessile or short-stalked, gen 1-chambered, stigmas beak-like. **FR**: capsule, gaping at top, or berry. **SEEDS** few–many, reniform. 6 genera, 70 spp.: n&e hemispheres, esp Medit. [Abdallah & de Wit 1978 Meded Landbouwhogeschool 78(14):99–416]

OLIGOMERIS

Ann, per. **LF** sessile, entire. **FL**: sepals 2–6, margins white; petals 2 (rarely more), entire to shallowly lobed; disk 0; stamens 3–12; stigmas 3–5. **FR**: capsule. 3 spp.: w N.Am, e hemisphere. (Greek: few parts)

O. linifolia (M. Vahl) J.F. Macbr. (p. 453) Ann, ± fleshy, glabrous. **ST** erect, < 45 cm. **LF** 8–45 mm, 0.5–2 mm wide, linear to ± oblanceolate. **INFL**: bracts triangular to awl-like. **FL** 1–2 mm; sepals 4; petals 2, whitish; stamens 3; pistil barely compound, 8-lobed. **FR** 2–3 mm, 4-parted, depressed-spheric. **SEEDS** black, shiny. *n*=24. Rocky slopes, open dunes, ocean bluffs, roadsides, alkaline places; < 850 m. SW, **D**; to TX, Mex; also in Eurasia. Feb–Jul

RHAMNACEAE BUCKTHORN FAMILY

John O. Sawyer, Jr. (except *Ceanothus*)

Shrub, vine, tree, gen erect, often thorny. **LVS** simple, gen alternate, often clustered on short-shoots, gen petioled, gen stipuled; blade often 1–3-ribbed from base. **INFL**: cyme, panicle, or fls solitary in axils. **FL** gen bisexual, radial; hypanthium subtending, surrounding, or partly fused to ovary; sepals 4 or 5; petals 0, 4, or 5, clawed; stamens 4 or 5, alternate sepals, attached to hypanthium top, each gen fitting into a petal concavity; ovary superior or partly inferior, chambers 2–5, each 1–2-ovuled, style lobes or parts 1–3. **FR**: capsule, drupe. 55 genera, 900 spp.: esp trop, subtrop; some cult (*Ceanothus*; *Colletia*, anchor-plant; *Gouania*; *Phylica*; *Rhamnus*; *Ventilago*; *Ziziphus*). [Brizicky 1965 J Arnold Arbor 45:439–463]

1. Fls showy; petals, sepals, pedicels blue, pink, or white . . . . **CEANOTHUS**
1' Fls inconspicuous; petals, if any, green or white; sepals and pedicels green or gray
  2. Fr a capsule . . . . **COLUBRINA**
  2' Fr a drupe
    3. Fr of 2–4 separate stones; widespread . . . . **RHAMNUS**
    3' Fr of 1 stone; DSon, e PR
      4. Petals 0; hypanthium below and not surrounding ovary . . . . **CONDALIA**
      4' Petals 5; hypanthium surrounding base of ovary . . . . **ZIZIPHUS**

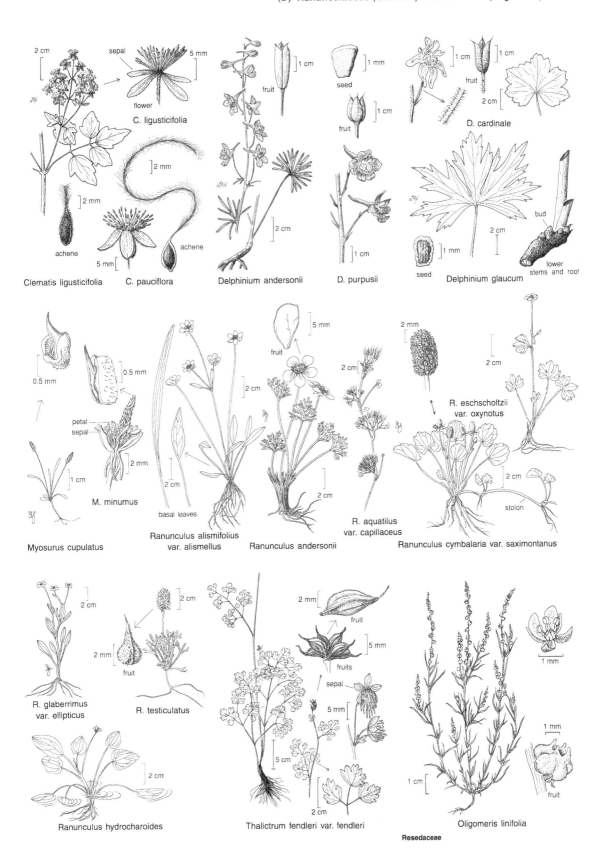

C. ligusticifolia

flower

sepal

fruit

seed

fruit

D. cardinale

fruit

achene

achene

Clematis ligusticifolia

C. pauciflora

Delphinium andersonii

D. purpusii

seed

bud

lower stems and root

Delphinium glaucum

petal

sepal

M. minumus

Myosurus cupulatus

basal leaves

Ranunculus alismifolius var. alismellus

fruit

Ranunculus andersonii

R. eschscholtzii var. oxynotus

R. aquatilus var. capillaceus

stolon

Ranunculus cymbalaria var. saximontanus

fruit

R. glaberrimus var. ellipticus

R. testiculatus

fruit

fruits

sepal

Thalictrum fendleri var. fendleri

Ranunculus hydrocharoides

fruit

Oligomeris linifolia

Resedaceae

CEANOTHUS CALIFORNIA LILAC

Clifford L. Schmidt

Shrub, small tree, prostrate to erect, thorny or not. **ST**: branches gen arranged as lvs. **LVS** alternate or opposite, deciduous or evergreen, petioled; blade 1–3-ribbed from base, margin entire or not. **INFL**: gen panicle-like aggregations of umbel-like, 3-fld clusters. **FL** gen < 5 mm; hyphanthium surrounding fleshy disk below ovary base, in fr thick, not splitting; sepals gen 5, lanceolate-deltate, incurved, colored like petals, persistent; petals gen 5, hooded, white to deep blue; stamens gen 5, opposite petals; ovary superior, 3-lobed, chambers 3, each 1-ovuled, style parts 3. **FR**: capsule, ± spheric, 3-valved. **SEEDS** 3, ± 3 mm, 1 surface convex. 45 spp.: N.Am, esp w. (Greek: thorny pl) [Rensselaer & McMinn 1942 *Ceanothus* Santa Barbara Bot Gard 1–308] Hybridization common (named hybrids not recognized here); hybrid forms may not key adequately.

1. Lvs opposite, 1 principal vein from blade base . . . . ***C. greggii*** var. ***vestitus***
1' Lvs alternate, 3 principal veins from blade base
  2. Twigs rigid, gen thorny; lvs < 3 cm; branches gen gray . . . . ***C. cordulatus***
  2' Twigs flexible, gen not thorny; lvs < 8 cm; branches gen not gray . . . . ***C. velutinus*** var. ***velutinus***

C. cordulatus Kellogg MOUNTAIN WHITETHORN Pl spreading to ascending, < 1.5 m, thorny. **ST**: twigs round, yellow-green, puberulent, becoming light gray. **LVS** alternate, evergreen, < 3 cm; stipules deciduous; petiole < 6 mm; blade ovate to elliptic, 3-ribbed from base, tip acute to obtuse, margin gen entire, upper surface light to gray-green, glabrous to puberulent, lower paler, glabrous to puberulent. **INFL** panicle-like, < 4 cm. **FL** white. **FR** < 5 mm, rough, 3-lobed; valves ± crested. 2*n*=24. Rocky ridges, open pine forests; 900–2900 m. KR, NCoRO, SN, SnGb, SnBr, SnJt, **DMtns (Panamint Mtns)**; OR, NV, n Baja CA. ☸ DRN,DRY,SUN:1–3,6,7,14–16,18;STBL. May–Jul

C. greggii A. Gray Pl erect, < 2 m. **ST**: twigs round, gray, to-mentose, not changing color. **LVS** opposite, evergreen; stipules persistent; petiole < 3 mm; blade esp variable in shape on individual pls, 1-ribbed from base, firm, tip obtuse to truncate, margin entire to dentate, upper surface (in our vars.) gen ± cupped, both surfaces gray-canescent. **INFL** raceme-like, < 2 cm. **FL** white. **FR** 3–5 mm; horns near middle or 0; ridges 0. Dry slopes; 300–2300 m. Teh, TR, PR, **W&I, DMtns**; to UT, TX, n Mex. 2 vars. in CA.

var. ***vestitus*** (E. Greene) McMinn (p. 459) MOJAVE CEANOTHUS **LF** < 1.5 cm; blade elliptic-ovate to oblanceolate, entire to toothed, upper surface gray-green. Desert slopes, Joshua-tree and pinyon/juniper woodlands, sagebrush scrub; < 2300 m. Teh, TR, **W&I, DMtns**; to UT, AZ. ☸ DRN,SUN:2,3,7,10,15–17&IRR:11, 14,**19–21**;STBL. May–Jun

C. velutinus Dougl. **ST**: twigs round, brown, ± puberulent, becoming dark brown. **LVS** alternate, evergreen, aromatic, < 8 cm; stipules deciduous; petiole < 18 mm; blade widely elliptic to ovate-elliptic, 3-ribbed from base, tip obtuse, margin gland-toothed, upper surface dark green, ± shiny, glabrous, lower paler. **INFL** panicle-like, < 12 cm. **FL** white. **FR** 3–4 mm, 3-lobed. Open, wooded slopes; < 3000 m. KR, NCoRO, CaRH, SNH, n SnFrB, Wrn, **SNE (exc W&I)**; to B.C., SD.

var. ***velutinus*** (p. 459) TOBACCO BRUSH Pl ascending-erect, < 2 m; crown rounded. **LF**: lower surface canescent. **FR** ± smooth; valve crests minute or 0. Inland; 1000–3000 m. . KR, CaRH, SNH, Wrn, **SNE (exc W&I)**. ☸ DRN:1,2,6,7,15–17;DFCLT. Apr–Jul

COLUBRINA

Tree, shrub. **ST**: branches opposite and alternate, rigid; twigs spreading, thorn-tipped or not, gen hairy. **LVS** in part clustered on short-shoots, evergreen or deciduous; stipules deciduous; blade 3–5-ribbed from base, entire but gen with round, marginal glands. **INFL** umbel-like, few-fld; axis sometimes thorn-tipped. **FL**: hyphanthium filled with nectar in fl, lower part adhering to developing fr, upper deciduous; sepals 5; petals 5, oblanceolate, = sepals; stamens 5; ovary chambers 3, each 1-ovuled, styles 3. **FR**: capsule, shallowly 3-valved. 31 spp.: warm places, worldwide. (Latin: from French for serpent tree) [Johnston 1971 Brittonia 23:2–53]

C. californica I.M. Johnston (p. 459) LAS ANIMAS COLUBRINA Shrub < 3 m. **LVS** deciduous; blade 12–30 mm, oblong to obovate, dull gray-green, hairs silky, denser on lower surface, base rounded or wedge-shaped, tip rounded to ± notched, sometimes with a small point. **INFL** 5–10 mm, 3–12-fld, very dense; pedicel 1–2 mm, 2–4 mm in fr. **FLS** appearing after rain; hyphanthium 3 mm wide. **FR** 8–10 mm, persistent 3–6 months. UNCOMMON. Creosote-bush scrub; < 1000 m. **DSon**; AZ, Mex. Local. ☸ DRN,DRY, SUN:8,9,**10–12**,13,14,19–23. Apr–May

CONDALIA

Shrub. **ST**: branches alternate, rigid, 3-ranked; twigs spreading, thorn-tipped. **LVS** clustered on short-shoots, deciduous; stipules deciduous; petioles ± 0 or short; blade obovate, 1-ribbed from base, entire. **INFL**: fls solitary or in clusters on short-shoots. **FL**: hyphanthium below, not surrounding ovary, becoming flat after fl, persistent below fr; sepals 5, deciduous; petals gen 0; stamens 5; ovary spheric, strongly narrowed at base, chambers 2, each 1-ovuled, style 1. **FR**: drupe, stone 1. **SEED** tightly held in stone. 18 spp.: arid Am. (A. Condal, Spanish physician, 1745–1804) [Johnston 1962 Brittonia 14:332–368]

C. globosa I.M. Johnston var. ***pubescens*** I.M. Johnston (p. 459) **ST** < 4 m; bark smooth, gray; twigs 3–13 cm, thorn-tipped, pale olive or purple, short-hairy. **LVS** in clusters of 2–7; blade 3–12 mm, slightly thickened; stipules brown. **INFL** 1–8-fld. **FL**: hyphanthium 1–1.5 mm, olive or purple, short-hairy; sepals 1 mm, olive; stamens < sepals; pistil purple. **FR** 3–5 mm, black, juicy, bitter or sweet. Uncommon. Creosote-bush scrub; < 1000 m. **DSon**; AZ, Mex. ☸ DRN,DRY,SUN:**10,12**,14–16,**22–24**&IRR:8,9,11,13,**19–21**. Mar–Apr

RHAMNUS BUCKTHORN

Shrub, small tree. **ST**: branches alternate, flexible; twigs sometimes thorn-tipped. **LVS** sometimes clustered on short-shoots, deciduous or evergreen, petioled; stipules deciduous; blade 1-ribbed from base, entire or not. **INFL**: umbel or fls solitary, axillary. **FL** bisexual or unisexual, gen < 3 mm; hypanthium at base fused to, developing around ovary in fr, above base deciduous; sepals 4 or 5; petals 0, 4, or 5; stamens 4 or 5; ovary appearing partly inferior, chambers 2–4, each 1–2-ovuled, style lobes 2–4. **FR**: drupe, 2–4-stoned. 125 spp.: temp, few trop. (Greek: name for pls of this genus) [Wolf 1938 Rancho Santa Ana Bot Gard Monogr 1] Some species treated by some authors in the segregate genus *Frangula*. [Kartesz & Gandhi 1996 Phytologia 76: 441-457] Some of value in medicine or as dyes.

1. Terminal bud covered by scales; lf margin spiny; petals 0; fr red . . . . *R. ilicifolia*
1' Terminal bud not covered by scales; lf margin not spiny; fr black
  2. Lf blade ± glabrous on lower side; fr -stoned . . . . *R. californica* ssp. *californica*
  2' Lf blade tomentose on lower side,; fr 3-stoned . . . . *R. tomentella*
    3. Lf margin dentate to dentate-serrate, tip abruptly pointed . . . . ssp. *cuspidata*
    3' Lf margin serrate to ± entire. tip acute to rounded . . . . ssp. *ursina*

R. californica Eschsch. CALIFORNIA COFFEEBERRY Shrub < 5 m. **ST**: bark bright gray or brown; twigs glabrous to finely hairy; terminal bud not covered with scales. **LVS** evergreen; petiole 3–10 mm; blade 20–80 mm, ovate to elliptic, thick, base acute to rounded, tip acute, rounded, or truncate, margin serrate to entire, sometimes rolled under, both surfaces ± glabrous. **INFL** 5–60-fld; pedicels < 20 mm. **FL** bisexual; hypanthium 1–2 mm; sepals 5; petals 5; style incl. **FR** 10–15 mm, black. Coastal-sage scrub, chaparral, woodlands, forests; < 2300 m. NW, CW, SW, **DMtns (Providence Mtns)**; sw OR. 2 sspp. in CA.

  ssp. ***californica*** **ST**: twigs red. **LF**: blade narrowly to widely elliptic, base and tip acute, upper surface gen dark green, lower bright green or yellow, veins prominent. **FR** 2-stoned. 2*n*=24. Non-serpentine; < 2000 m. NW, CW, SW (incl Santa Cruz Island), **DMtns (Providence Mtns)**. ✣ DRN, DRY:**14–17,22–24**&SHD,IRR:**7,8,9,18–21**;CVS,incl.GRCVR. May–Jul

R. ilicifolia Kellogg (p. 459) HOLLY-LEAF REDBERRY Shrub < 4 m. **ST**: bark gray; branches ascending, stiff; twigs glabrous to densely hairy; terminal bud scales 3 mm. **LVS** evergreen; petiole 2–10 mm; blade 20–40 mm, ovate to round, thick, base and tip rounded, margin spiny, surfaces glabrous or hairy, lower concave, veins not prominent. **INFL** 1–6-fld; pedicels 2–4 mm. **FL** unisexual; hypanthium 2 mm; sepals 4; petals 0; style exserted. **FR** 2-stoned, 8 mm, red. 2*n*=24. Chaparral, montane forests; < 2000 m. CA-FP, **DMtns**; AZ, Baja CA. [*R. crocea* ssp. *i.* (Kellogg) C. Wolf] ✣ DRN,DRY:**7,14–17**,20–23&SHD:**18,19**. Mar–Jun

R. tomentella Benth. HOARY COFFEEBERRY Shrub < 6 m. **ST**: bark gray or red; twigs velvety; terminal buds not covered with scales. **LVS** evergreen; petiole 3–10 mm; blade 20–70 mm, ovate to narrowly elliptic, thick, base acute to rounded, tip acute, rounded, or with a small point, margin entire to toothed, sometimes rolled under, lower surface tomentose, veins prominent. **INFL** 5–60-fld; pedicels < 20 mm. **FL** bisexual; hypanthium 1–2 mm; sepals 5; petals 5. **FR** 3-stoned, 10–15 mm, black. Chaparral, woodlands; < 2300 m. s KR, NCoR, CaRF, SN, ScV, SnFrB, SCoR, SW, **SNE, DMoj**; to NM, Baja CA. Several sspp. distinguishable, but intermediates are common. 4 sspp. in CA.

  ssp. ***cuspidata*** (E. Greene) J.O. Sawyer (p. 459) **ST**: twigs red, hairs of 2 lengths. **LF**: blade 20–60 mm, elliptic, upper surface green, tip abruptly narrowed to a point, margin dentate to dentate-serrate. Chaparral, montane woodlands; 600–2300 m. c&s SN, Teh, TR, nw PR, SnJt, **SNE, DMoj**. [*R. californica* ssp. *cuspidata* (E. Greene) C. Wolf] Pls at high elevations sometimes deciduous. ✣ DRN,DRY, SUN:**2,3,7,14**,15–21. Apr–Jul

  ssp. ***ursina*** (E. Greene) J.O. Sawyer **ST**: twigs gray, tomentose. **LF**: blade 30–70 mm, elliptic or ovate, upper surface green, tip acute to rounded, margin serrate to ± entire. Woodlands; 1200–2100 m. **DMtns (Clark, New York, Providence Mtns)**; to NV, NM. [*R. californica* ssp. *ursina* (E. Greene) C. Wolf] ✣ DRN, DRY,SUN:**2,3,7,14**,15–21. May–Jul

ZIZIPHUS

Tree, shrub, vine. **ST**: branches alternate, flexible, sometimes 2–3-ranked; twigs thorn-tipped. **LVS** in part clustered on short-shoots, deciduous or evergreen, petioled; stipules sometimes unequal spines; blade ovate to oblong, 1–3-ribbed from base, ± entire to serrate. **INFL**: cyme or small panicle. **FL**: hypanthium surrounding base of ovary; sepals 5; petals 5, < or = sepals; stamens 5; ovary broadly attached at base, chambers 2, each 1-ovuled, styles 2. **FR**: drupe, stone 1. 100 spp.: gen trop. (Latin: reason for application obscure) [Johnston 1963 Amer J Bot 50:1020–1027]

1. Fr 7–10 mm, blue-black, not beaked . . . . *Z. obtusifolia* var. *canescens*
1' Fr 10–25 mm, brown, beaked . . . . *Z. parryi* var. *parryi*

Z. obtusifolia (Torrey & A. Gray) A. Gray var. ***canescens*** (A. Gray) M. Johnston (p. 459) GRAYTHORN **ST** < 3 m; bark gray; twigs 1–8 cm, thorn-tipped, hairs short, dense, white. **LVS** deciduous; blade 2–20 mm, ovate or oblong, firm, gray, margin entire or teeth 2–10, glandular; stipules brown. **INFL** 10–30-fld. **FL**: hypanthium 1.5–2 mm, olive, glabrous to tomentose; sepals = petals, yellowish; stamens < petals; pistil olive. **FR** juicy. Uncommon. Creosote-bush scrub; < 1000 m. **DSon**; AZ, Mex. [*Condalia lycioides* (A. Gray) Weberb. var. *c.* (A. Gray) Trel.] ✣ DRN,DRY,SUN:**8,9,10–12**,13,14,16,19–23. Apr–Jun

Z. parryi Torrey var. ***parryi*** (p. 459, pl. 91) LOTEBUSH **ST** < 4 m; bark gray to brown; twigs 1.3–3 cm, thorn-tipped, with 1 node, 1 short-shoot, glabrous. **LVS** deciduous; blade 10–25 mm, elliptic to obovate, membranous, olive, margin ± entire; stipules brownish, membranous. **INFL** 2–5-fld. **FL**: hypanthium 2–2.2 mm, purplish green, glabrous; sepals < petals, yellow; stamens < petals; pistil brownish. **FR** dry. Uncommon. Chaparral; 500–1500 m. **w edge DSon**; Mex. [*Condalia p.* (Torrey) Weberb.] ✣ DRN,DRY,SUN:**8,9,10–12**,13,14,19–23. Feb–Apr

ROSACEAE ROSE FAMILY

Ann to tree. **LVS** simple to pinnately to palmately compound, gen alternate; stipules free to fused, persistent to deciduous. **INFL**: cyme, raceme, panicle, or fls solitary. **FL** gen bisexual, radial; hypanthium free or fused to ovary, saucer- to funnel-shaped, often with bractlets alternate with sepals; sepals gen 5; petals gen 5, free; stamens (0)5–many, pistils (0)1–many, simple or compound; ovary superior to inferior, styles 1–5. **FR**: achene, follicle, drupe, pome, or blackberry- to raspberry-like. **SEEDS** gen 1–5. 110 genera, ± 3000 spp.: worldwide, esp temp. Many cult for orn and fr, esp *Cotoneaster, Fragaria, Malus, Prunus, Pyracantha, Rosa,* and *Rubus.* [Robertson 1974 J Arnold Arbor 55:303–332,344–401,611–662] Family description, key to genera by Barbara Ertter and Dieter H. Wilken.

1. Ann, bien, or per herb
  2. Style tapered to fr, hooked or ± hairy — lvs 1-pinnate . . . . **GEUM**
  2' Style jointed to fr, neither hooked nor hairy
    3. Lf palmate to ternate
      4. Stamens 10–20; petals gen >> 1 mm (exc 1—2.5 in *P. biennis*); lflets 3–7, gen > 3-toothed . . ²**POTENTILLA**
      4' Stamens 5; petals ± 1 mm; lflets 3, 3-toothed at tip — low, matted per, > 1900 m . . . . **SIBBALDIA**
    3' Lf pinnate
      5. Hypanthium cup-like, ± flat-bottomed; filaments gen ± flat, often forming a tube; petals ± white or pinkish
        6. Stamens 10; lflets 4–14 per side, toothed to deeply lobed . . . . **HORKELIA**
        6' Stamens 20; lflets 25–35 per side, deeply 5–10-lobed . . . . **HORKELIELLA**
      5' Hypanthium either shallow or not flat-bottomed; filaments slender, spreading; petals white to yellow
        7. Pls hanging clumps on vertical rocks — lflets 2–6 per side, gen ± separate . . . . ²**IVESIA**
        7' Pls matted to erect, not hanging
          8. Lf ± cylindrical; lflets 5–50 per side, divided nearly to base, overlapping; stamens 5–20; petals 2–5 mm, narrowly oblanceolate to obovate . . . . ²**IVESIA**
          8' Lf gen flat; lflets 2–13 per side, toothed to deeply divided, separate or overlapping; stamens 20–25; petals 4–20 mm, obcordate to elliptic . . . . ²**POTENTILLA**
1' Shrub or subshrub
  9. Lvs pinnately compound
    10. Lvs gen 2-pinnate, glandular, strong-smelling; pistils 4–5; fr a follicle — pl not prickly; petals white . . . . **CHAMAEBATIARIA**
    10' Lvs 1-pinnate, neither glandular nor strong-smelling; pistils many; fr an achene
      11. Pl not prickly; lflets entire; petals yellow; fr not enclosed in fleshy hypanthium . . . . . . *Potentilla fruticosa*
      11' Pl gen ± prickly; lflets toothed; petals pink; fr enclosed in fleshy hypanthium . . . . **ROSA**
  9' Lvs ± simple, sometimes deeply lobed
    12. Ovary inferior, chambers gen 2–5; fr a pome
      13. Lvs ± evenly distributed on branches and twigs, blade elliptic to round, petiole well defined . . . . **AMELANCHIER**
      13' Lvs ± clustered on short lateral branches, blade ± oblanceolate, tapered to base, ± subsessile . . . . **PERAPHYLLUM**
    12' Ovary superior (sometimes hidden in funnel- or urn-shaped hypanthium), chamber gen 1; fr not a pome
      14. Lf margin ± rolled under; style persistent, gen ± hairy to plumose in fr
        15. Lf entire, linear to narrowly (ob)lanceolate; petals 0 . . . . **CERCOCARPUS**
        15' Lf deeply lobed to toothed, often ± wedge-shaped; petals 5
          16. Hypanthium bractlets present; petals 10–25 mm, white . . . . **FALLUGIA**
          16' Hypanthium bractlets 0; petals 6–8 mm, white to yellow . . . . **PURSHIA**
      14' Lf margin not rolled under; style gen not persistent, not plumose (exc. *Coleogyne ramossisima*)
        17. Lf entire
          18. Mat-forming subshrub; lvs in rosettes — infl ± spikelike; gen on limestone . . . **PETROPHYTON**
          18' Erect shrub; lvs often clustered but not in rosettes
            19. Lvs opposite; sepals 4; petals 0; fr an achene . . . . **COLEOGYNE**
            19' Lvs alternate; sepals 5; petals 5; fr a drupe . . . . *Prunus emarginata*
        17' Lf ± toothed
          20. Lf finely serrate, petals 4–9 mm; fr a drupe . . . . **PRUNUS**
          20' Lf ± coarsely toothed and lobed; petals < 4 mm; fr a follicle or achene
            21. Lf pinnately veined; infl a panicle; petals 1.5–2 mm; fr an achene . . . . **HOLODISCUS**
            21' Lf palmately 3-veined from base; infl umbel-like; petals 3–4 mm; fr a follicle . . . . . **PHYSOCARPUS**

AMELANCHIER SERVICE-BERRY

Dieter H. Wilken

Shrub or small tree. **ST**: bark gray- to red-brown; twigs gen short. **LVS** alternate or clustered, simple, deciduous; stipules deciduous. **INFL**: racemes or clusters; fls 3–16+. **FL**: hypanthium bell- to urn-shaped; sepals persistent; petals ascending to erect, white; stamens ± 10–20; ovary inferior, 2–5-chambered, styles 2–5. **FR**: pome, berry-like, gen spheric, blue black. ± 10 spp.: temp N.Am, Eurasia, n Afr. Fr of some spp. used by native Americans for food. (Latin: from old French common name) [Jones 1946 Illinois Biol Mongr 20(2):1–126] Variation in w N.Am needs further study.

A. utahensis Koehne (p. 459) UTAH SERVICE-BERRY Shrub 1–4 m. **ST**: twigs glabrous to white-tomentose. **LF**: blade 13–45 mm, 10–45 mm wide, entire to serrate from below middle, dull green below, darker above, glabrous to minutely tomentose; lateral veins 12–24. **INFL** 1–4 cm; fls 3–8. **FL**: petals 5–11 mm, 1–4 mm wide, elliptic to wedge-shaped; ovary top glabrous to tomentose, styles 2–4(5). **FR** 4–10 mm diam. Open, rocky slopes, shrubland, pinyon/juniper woodland, coniferous forest; 200–3400 m. NW, CaR, SN, CW, SW, **SNE, DMtns**; to OR, MT, TX, Baja CA. Variable; pls from n DMtns gen < 2 m, with twigs and lvs glabrous, have been called var. *covillei* (Standley) N.H. Holmgren; pls with petals > 9 mm, lf blades with < 18 lateral veins, have been called *A. pallida* E. Greene. ❀ DRN:**2**,4,5,**6,15–17**&IRR:**1**,3,**7**,8–10,**14,18,**19–21. Apr–May

CERCOCARPUS MOUNTAIN-MAHOGANY

Richard Lis

Shrub or small tree, evergreen. **ST**: trunk < 80 cm diam; bark gen gray to reddish brown; twigs short. **LVS** gen clustered, simple; stipules often deciduous; blade ± thin to leathery, entire to toothed, upper surface gen ± glabrous. **INFL**: clusters; fls 1–12. **FL**: hypanthium funnel-like, tube persistent in fr, rim cup-like, deciduous; petals 0; stamens 10–45, in 2–3 rows on hypanthium rim, anthers glabrous or hairy; pistil 1, free from hypanthium tube, ovary superior, 1-ovuled, style terminal, persistent in fr, straight or becoming twisted, plumose. **FR**: achene, cylindric, hairy, incl in hypanthium. 13 spp.: w N.Am, Mex. (Greek: tailed fr) [Lis 1992 Int J Pl Sci 153:258–272] Key to species adapted by Margriet Wetherwax.

1. Lf blade gen 3–10 mm, margins inrolled to midrib . . . . ***C. intricatus***
1' Lf blade 10–40 mm, margin plane to inrolled (but not to midrib, midrib and veins visible)
. . . . ***C. ledifolius*** var. ***intermontanus***

C. intricatus S. Watson (p. 459) Shrub 1–3 m, intricately branched. **LF**: petiole 0–1 mm; blade gen 0.3–1 cm, linear, thick-leathery, entire, inrolled to midrib, glabrous to gray-white-hairy below. **INFL**: fls 2–3, sessile. **FL**: hypanthium 3–5 mm in fl and fr, rim 1–3 mm diam; stamens 10–15, anthers glabrous; style slightly exserted, stigma straight. **FR** 5–7 mm; style 1–2 cm. Dry, rocky outcrops, slopes, pinyon/juniper woodland; 1400–3000 m. s SNH, **SNE, DMtns**; to Colorado, AZ. ❀ DRN,DRY,SUN:1–3,7,9,**10**,11, **14**,15,16,**18**,19–24;DFCLT. May

C. ledifolius Nutt. CURL-LEAF MOUNTAIN-MAHOGANY Shrub or tree 1–10 m, much-branched. **LF**: petiole 0–6 mm; blade 1–4 cm, narrowly (ob)lanceolate, resinous, entire, inrolled or plane, ± glabrous to densely woolly below. **INFL**: fls 1–3, axillary, ± sessile. **FL**: stamens 15–25, anthers glabrous; style exserted, stigma straight. **FR** 4–13 mm. Deep soils, rocky slopes; 1200–2800 m. KR, NCoRH, CaRH, SNH, Teh, TR, n PR, **GB, DMtns**; to WA, MT, AZ, Baja CA. Hybridizes with *C. intricatus.* 2 vars. in CA.

var. ***intermontanus*** N. Holmgren (p. 459) Pl < 8 m; trunk < 8 dm diam. **LF**: petiole < 6 mm; blade 1–4 cm, ± elliptic-lanceolate, glabrous to sparsely woolly below, midrib and veins visible. **FL**: hypanthium 5.5–9 mm, 6–10 mm in fr, rim 4–5 mm diam. **FR** 8–13 mm; style 4–7 cm. 2n=18. Pinyon/juniper woodland, sagebrush scrub; 1500–2800 m. KR, NCoRH, CaRH, SNH, Teh, TR, n PR, **GB, DMtns**; to WA, WY, AZ, Baja CA. [Hybrids with *C. intricatus* are var. *intercedens* C. Schneider] ❀ DRN,SUN:**1**,4–6,15–17& IRR:**2,3,7**,8–10,**14,18–21**,22,23.

CHAMAEBATIARIA FERN BUSH, DESERT SWEET

Thomas J. Rosatti

1 sp. (Greek: *Chamaebatia*-like)

C. millefolium (Torrey) Maxim. (p. 459, pl. 92) Shrub < 20 dm, densely branched, strong-smelling, gen stellate-hairy, glandular. **LF** odd-(1)2-pinnate, 2–8 cm, oblong; stipules entire; 1° lflets 31–43; 2° lflets (0)11–35, < 2 mm, sessile, sometimes with a sessile gland-tip, lobes or teeth 0–3(5). **INFL**: panicle or raceme, 3–10 cm. **FL**: bractlets on hypanthium 0; sepals 3–5 mm, lanceolate; petals ± 5 mm, ± round, white; pistils 4–5, ovaries superior, ± fused below, styles free. **FR**: follicles 3–5 mm, leathery, dehiscent on inner suture and upper half of outer suture. **SEEDS** few, 2.5–3.5 mm, narrowly fusiform, yellowish. Dry, rocky sagebrush scrub, pinyon/ juniper woodland, pine forest; 900–3400 m. KR, CaR, SN (e slope), **GB, ne DMtns**; to OR, WY, AZ. ❀ DRN,SUN:1,**2,3,7,14–16**,17,**18**,19–21. Jun–Aug

COLEOGYNE BLACKBUSH

Thomas J. Rosatti

1 sp. (Greek: sheath female, from hypanthium enclosing pistil)

C. ramosissima Torrey (p. 459, pl. 93) Shrub 3–20 dm, much-branched, ± strigose, thorny. **LVS** in opposite clusters, 5–15 mm, linear-oblanceolate, ± thick, entire; stipules persistent. **INFL**: fls solitary at st tips; bract (closely subtending fl) ± linear, base 2-lobed. **FL** gen yellow (sepals often reddish outside); hypanthium bell-shaped, leathery, sheath at top enclosing pistil, 4–5 mm; sepals 4, 7–8 mm, erect, elliptic, persistent, inner 2 widely scarious-margined; petals 0; stamens 30–40; pistil 1, ovary superior, style lateral, long-hairy esp below, persistent. **FR**: achene, ± 3–4 mm, ± crescent-shaped, brown, glabrous. Dry, open slopes, creosote-bush scrub, pinyon/juniper woodland; 600–1600 m. **SNE, D (esp DMtns)**; to Colorado, NM. Apr–Jun

FALLUGIA   APACHE PLUME

Margriet Wetherwax

1 sp. (V. Fallugi, Italian abbot)

F. paradoxa (D. Don) Endl. (p. 465)   Shrub < 2 m, ± erect. **ST** much-branched; bark grayish white-tomentose, peeling. **LVS** alternate to clustered, 7–15 mm, ovate to wedge-shaped, lobed; lobes 3–7, deeply pinnate, linear, obtuse, rolled under, densely hairy above, rusty-scaly below; stipules lanceolate, deciduous. **INFL** terminal; fls 1–3; bracts 5, linear, alternate sepals. **FLS**: hypanthium hemispheric, silky-hairy; sepals 5–8 mm, ovate, acute to long-acuminate, tomentose; petals 10–25 mm, ± round, white; stamens many; pistils many, ovary superior, chamber 1, style 1, persistent, 30–50 mm, plumose, purplish. **FR**: achene, 3–5 mm, silky-hairy. 2n=18. Dry, ± rocky slopes in pinyon/juniper woodland; 1000–2200 m. **e DMtns**; to Colorado, w TX, n Mex. ❀ DRN,SUN:1,2,7–9,**14**, **18–23**,24&DRY:5,**15,16**,17. May–Jun

GEUM

Barbara Ertter

Per; glands inconspicuous. **ST** ascending to erect. **LVS** gen basal, odd-1-pinnate; lflets lobed, unevenly toothed, often alternately large and small. **INFL**: gen cyme, open. **FL**: hypanthium shallow; bractlets gen 5; sepals 5; petals 5; stamens > 20; pistils many, ovaries superior, continuous to style at top. **FR**: achene ± flat; style long, persistent. 40–50 spp.: gen n temp, arctic. (Latin: ancient name)

1. Style in fr 2.5–5 mm, not plumose, hooked; fls gen 3–10, erect, rotate; petals yellow, deciduous;
   bractlets < 3 mm . . . . *G. macrophyllum*
1'. Style in fr 20–35 mm, plumose, not hooked; fls 1–3, nodding, ± cupshaped; petals cream or tinged
   pink, persistent; bractlets 5–14 mm . . . . *G. triflorum*

G. macrophyllum Willd. (p. 465)   BIGLEAF AVENS   Pl tufted, drying greenish. **ST** gen 20–100 cm. **LF** gen 10–45 cm; main lflets gen 2–4 per side, gen << terminal lflet; terminal lflet gen 8–10 cm, ± bluntly cordate-reniform, 3-lobed < 3/4 to base, irregularly toothed. **INFL** gen 3–10-fld; pedicels straight. **FL** rotate; bractlets gen ± 2 mm, linear, often some missing; sepals reflexed, 3–5 mm; petals 3–7 mm, obovate-round, shallowly notched, yellow, deciduous. **FR**: achene body 2.5–3.5 mm; style 2.5–5 mm, hooked below gen deciduous tip, glabrous to glandular. n=21. Meadows, streambanks; < 3300 m. NW, CaRH, SNH, SnBr, **GB**; to n&e N.Am. Vars. indistinct in CA. Some pls from s SNH approach *G. aleppicum.* ❀ **4–6**&IRR:1,2,**7,15–17**&SHD:3,**14,18–24**;rather INV. May–Aug

G. triflorum Pursh (p. 465)   OLD MAN'S WHISKERS   Pl in patches, rhizomed, ± gray-green. **ST** gen 20–50 cm. **LF** gen 5–20 cm; lflets wedge-shaped, gen 2–3-lobed ±1/2 to base, lobes deeply few-toothed, main lflets 3–6 per side, largest 1–3 cm, ± = terminal lflet. **INFL** 1–3-fld; pedicels curved (straight in fr). **FL** ± cup-shaped; bractlets 5–14 mm, linear-oblanceolate; sepals erect, 6–12 mm; petals 7–13 mm, ± elliptic, outcurved, cream or pink-tinged, persistent. **FR**: achene body 2.5–5 mm; style < 35 mm incl gen persistent tip, not hooked, plumose. n=21. Dry meadow edges, sagebrush scrub, open yellow-pine forest; 1300–3200 m. c KR (Marble Mtns), CaRH, n&c SNH, **GB**; to n&e N.Am. [*G. canescens* (E. Greene) Munz; *G. ciliatum* Pursh] Vars. indistinct in CA. ❀ 1,**2**,6&IRR:3, **7,14–16**,17,**18–21**,22–24;GRNCVR. May–Jul

HOLODISCUS

Richard A. Lis

Shrub, ± hairy. **ST** 3–60 dm; bark reddish, in age gray, shredding. **LVS** simple, alternate, 0.3–12 cm, thin to leathery, toothed; base truncate to gen ± wedge-shaped; lower surface strongly veined; stipules 0; petiole distinct or not. **INFL**: panicle, ± terminal, dense, 2.5–25 cm, ± conic, many-fld, persistent; pedicel slender, bractlets 1–3. **FL**: hypanthium 3–5 mm wide, saucer-shaped, prominent nectary-disk below inner rim; sepals 5, 1–2 mm; petals 5, 1.5–2 mm, ± ovate, gen white; stamens 15–20, wider at base; pistils 5, ovaries superior, 2-ovuled, hairs dense, bristle-like, persistent in fr, style 1 mm, persistent, stigma ± 2-lobed. **FR**: achenes 5, 1–1.5 mm, often with sessile glands. 5 spp.: w N.Am, C.Am, n S.Am. (Greek: whole disk) Spp. highly variable; lvs of peg-like sts best for identification; complexity in c SNH evidently from local climatic variation, hybridization.

H. microphyllus Rydb.   Pl 0.3–1 m, ± dense. **ST**: peg-like sts very predominant. **LVS** of peg-like sts gen 0.3–2 cm, obovate, others 1–3 cm, ovate to ± round; teeth of 1 size, those of lvs on peg-like sts near tip, rarely to middle, those of others gen to middle; petiole gen indistinct, winged or not. **INFL** gen mixed with lvs, 2–8 cm, 1.5–5 cm wide. **FL**: sepal lower surface glabrous to densely short- to long-hairy, upper surface glabrous. Rocky places, outcrops; 600–4000 m. NCoRH, SNH (2700–4000 m), SnGb (1800–3300 m), SnBr, SnJt, **GB**, **DMtns**; to OR, ID, WY, Colorado, AZ, Baja CA. Lvs of peg-like sts from pls of wet places more often ovate to round than obovate. 2 vars. in CA.

var. *microphyllus* (p. 465)   ROCK SPIRAEA   **ST** hairy to long-hairy; glands 0 or obscured by hairs. **LF** ± long-hairy on 1 or both surfaces; hairs not greatly longer on margins and veins; glands 0 or often obscured by hairs. **FL**: sepals hairy to long-hairy; glands 0. Habitats of sp.; 1200–4000 m. Range of sp. [var. *sericeus* Ley] ❀ TRY. Jun–Aug

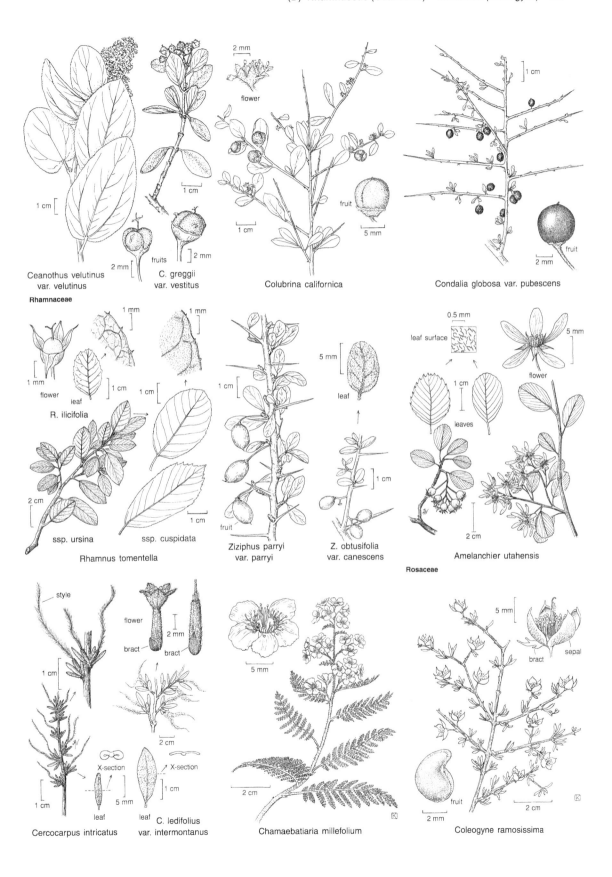

Ceanothus velutinus
var. velutinus

C. greggii
var. vestitus

fruits

Rhamnaceae

Colubrina californica

flower

fruit

Condalia globosa var. pubescens

fruit

R. ilicifolia

flower

leaf

ssp. ursina

ssp. cuspidata

Rhamnus tomentella

Ziziphus parryi
var. parryi

fruit

Z. obtusifolia
var. canescens

leaf

leaf surface

leaves

flower

Amelanchier utahensis

Rosaceae

style

flower

bract

bract

X-section

X-section

leaf

leaf C. ledifolius
var. intermontanus

Cercocarpus intricatus

Chamaebatiaria millefolium

flower

bract

sepal

fruit

Coleogyne ramosissima

HORKELIA

Barbara Ertter

Per, gen ± glandular, gen resinous-smelling; caudex gen branched. **ST** gen ascending to erect. **LVS** gen basal, odd-1-pinnate, gen ± flat; cauline alternate, reduced upward; uppermost lateral lflets gen ± fused with terminal. **INFL**: cyme, open or of dense clusters; pedicels gen straight. **FL**: hypanthium a ± flat-bottomed cup, width ± 2 × length; bractlets 5, gen 2/3 sepals; sepals 5, often reflexed; petals 5, gen ± = sepals, blunt, white; stamens 10, filaments flat, often forming a tube; pistils 2–many, ovaries superior, styles jointed below fr tip, ± thicker at base. **FR**: achene. 19 spp.: w N.Am. (J. Horkel, German pl physiologist, 1769–1846) Data apply to basal lvs, pressed hypanthia.

1. Lvs flat, lflets gen 5–15 mm, separated, 4–8 per side, toothed; fls in head-like clusters . . *H. fusca* ssp. *parviflora*
1' Lvs gen ± cylindrical; lflets gen 2.5–4 mm, crowded, 10–14 per side, divided > 3/4 to base; fls in ±
   flat-topped clusters . . . . *H. hispidula*

H. fusca Lindley Pl gen tufted (± matted), green to grayish. **ST** gen 10–60 cm. **LF** gen 4–15 cm; lflets 3–15 per side, separated to ± crowded, gen 4–20 mm, narrowly wedge-shaped to ± round, ± 5–10-toothed to lobed, hairs sparse to dense. **INFL**: clusters 1–several, gen ± head-like, gen 5–30-fld; pedicels gen 1–3 mm. **FL**: hypanthium width gen 2–3.5 mm, ± 1–2 × length; bractlets < 0.5 mm wide, linear; sepals gen 2–3 mm; petals 2–6 mm, ± wedge-shaped; filaments 0.1–1.5 mm, bases 0.2–1 mm wide, anthers gen ± 0.5 mm; pistils gen 10–20, styles 0.8–1.5 mm. **FR** 1–1.8 mm. Dry meadow edges, open forest, volcanic or granitic soils; 1000–3300 m. e KR, CaRH, SNH, **GB**; to WA, WY, NV. Sspp. need study. 3 sspp. in CA.

ssp. *parviflora* (Nutt.) Keck (p. 465) **LF** gen 4–15 cm; lflets gen 4–8 per side, separated, gen 5–15 mm, wedge-shaped to ± round, ± 5-toothed 1/4–1/2 to base, hairs sparse to dense. **INFL**: clusters gen 5–20-fld. **FL**: petals 2–4 mm; filaments 0.2–1 mm, gen longer than wide, anthers ± 0.4 mm; styles ± 1 mm. **FR** ± 1.2 mm. *n*=14. Habitats of sp.; 1000–3300 m. e KR, n CaRH, SNH, MP (exc Wrn), **n SNE**; to OR, ID, NV. [ssp. *pseudocapitata* (Rydb.) Keck] ❀ DRN,IRR:1–7,15–18;DFCLT. Jul–Aug

H. hispidula Rydb. WHITE MOUNTAINS HORKELIA Pl matted, gen grayish. **ST** < 25 cm. **LF** 3–10 cm, gen ± cylindric; lflets gen 10–14 per side, crowded, gen 2.5–4 mm, divided > 3/4 to base into 3–7 oblanceolate lobes, hairs many. **INFL** ± dense or open, gen 3–15-fld; pedicels gen 1–6 mm. **FL**: hypanthium width 3–4 mm, ± 2 × length, inner wall ± hairy; bractlets < 0.5 mm wide, linear to lanceolate; sepals 2.5–4 mm; petals 3–5 mm, oblong to oblanceolate; filaments 0.5–2 mm, bases ± 0.5 mm wide, anthers ± 0.5 mm; pistils 10–20, styles ± 2 mm. **FR** ± 1.5 mm. RARE. Dry flats; 3000–3400 m. **n W&I (White Mtns)**. In cult. Jan–Aug

HORKELIELLA

Barbara Ertter

Per, tufted, ± glandular; odor ± resinous. **ST** ascending to erect, 15–50 cm. **LVS** gen basal, odd-pinnately compound; lflets many, ± overlapped, toothed to palmately divided, segments ± oblanceolate. **INFL**: ± cyme; pedicels straight. **FL**: hypanthium cup-like, flat-bottomed; bractlets 5; sepals 5, often reflexed; petals 5, white or pinkish, midvein often reddish; stamens 20, filaments ± flat, often an erect tube; pistils many, ovaries superior, style jointed below fr tip, ± rough, thick. **FR**: achene, ± 1.5 mm. 2 spp.: CA. (Latin: small *Horkelia*) Fl ± like that of *Horkelia*.

H. congdonis (Rydb.) Rydb. (p. 465) Pl green, gen glandular-sticky. **LVS**: basal gen 8–25 cm, lflets 25–35 per side, gen 2–6 mm, 5–10-lobed; cauline 2–7. **INFL** ± dense; pedicels gen ± 2 mm. **FL** ± 15 mm wide; sepals 4–8 mm; petals 3–6 mm, oblong-oblanceolate; anther ± 1 mm, filaments 1.5–3 mm, central longest; style 2–5 mm. Meadows in sagebrush flats; 1500–2900 m. c SNH (e slope), **w SNE**. [*Ivesia purpurascens* (S. Watson) Keck ssp. *c.* (Rydb.) Keck] ❀ DRN,IRR,SUN:1–3,**7,15–16**,17. Jul–Aug

IVESIA

Barbara Ertter

Per, glandular; odor resinous. **LVS** gen basal, odd-1-pinnate, gen ± cylindric; cauline reduced; lflets gen overlapped, gen divided ± to base. **INFL**: cyme. **FL**: hypanthium shallow or deep; bractlets (0)5, gen < sepals; sepals gen 5; petals gen 5, acute to rounded; stamens gen < or = 20; pistils 1–many, ovaries superior, style jointed below fr tip, base ± rough-thickened. **FR**: achene. 30 spp.: w N.Am. (E. Ives, Yale Univ. pharmacologist, 1779–1861) [Ertter 1989 Syst Bot 14:231–244] Lf and lflet data are for basal lvs.

1. Pls hanging clumps on vertical rocks; lflets 2–6 per side, separate to ± overlapped; infl open; petals yellow
   2. Lflets gen lobed nearly to base, ± overlapped, 4–6 per side; stamens 20; bractlets 5 . . . . *I. jaegeri*
   2' Lflets toothed or lobed gen < 3/4 to base, gen ± separate, 2–4 per side; stamens 5–40; bractlets 5 or 0
      3. Stamens 15–40; bractlets 5 . . . . *I. saxosa*
      3' Stamens 5–10; bractlets gen 0
         4. Hypanthium width < to = length, receptacle stalked in pistil-bearing portion; n DMtns (Inyo Co.)
            . . . . *I. arizonica*
         4' Hypanthium width > 2 × length; receptacle not stalked; e DMtns (Kingston Mtns.) . . . . *I. patellifera*
1' Pls matted to erect, not hanging; lflets 5–50 per side, overlapped; infl open or head-like; petals white
   or yellow

5. Petals white; stamens 20; lflets 30–50 per side; moist alkaline sites < 2000 m . . . . *I. kingii*
5' Petals yellow; stamens 5; lflets < 35 per side; ± dry and rocky sites, gen > 2000 m
   6. Infl gen ± open between separate fls; pl matted from much-branched caudex; lflets gen 5–10 per side
   . . . . *I. shockleyi*
   6' Infl of 1–few gen head-like clusters; pl rosetted or tufted from simple to few-branched caudex; lflets
      10–35 per side
      7. Hypanthium length = or > width; petals narrowly oblanceolate; pistils gen 2–4; fr ± 2 mm, mottled brown
      . . . . *I. gordonii*
      7' Hypanthium length ± < width; petals ± obovate; pistils gen 5–15; fr 1–1.5 mm, not mottled . . *I. lycopodioides*
         8. Lflet lobes ± 1 mm, ± round, ± glabrous, not bristle-tipped; petals 2–3 mm . . . . ssp. *lycopodioides*
         8' Lflet lobes 1–3 mm, ± obovate, moderately to densely hairy, gen bristle-tipped; petals 3–5 mm
         . . . . ssp. *scandularis*

I. arizonica (J. Howell) B. Ertter var. *arizonica* (p. 465) YEL-LOW PURPUSIA Pls hanging clumps, green. **ST** 5–10 cm. **LF** gen 5–10 cm, flat; sheathing bases gen glabrous; lflets 2–4 per side, separated, 5–15 mm, ± round, ± evenly toothed or lobed < 3/4 to base; cauline lvs 1–3. **INFL** open, gen < 10-fld; pedicels 5–30 mm, often ± S-shaped in fr. **FL** 5–10 mm wide; hypanthium length 1–2 × width; bractlets gen 0; petals 2–3 mm, oblanceolate to elliptic, yellow, ± = sepals; stamens 5; pistils 2–10; receptacle stalked in pistil-bearing portion (unique in *Ivesia*). **FR** 1.5–2 mm, ± ridged, pale. UNCOMMON. Limestone crevices; 1200–3100 m. **n DMtns (Inyo Co.)**; s NV, nw AZ. Var. *saxosa* (Brandegee) B. Ertter [*Purpusia s.* Brandegee], rock purpusia, (NV) has white petals. May–Aug

I. gordonii (Hook.) Torrey & A. Gray (p. 465) Pl tufted, green; caudex 0–few-branched. **ST** ascending to erect, 5–20 cm. **LF** gen 3–8 cm; sheathing bases ± glandular; lflets gen 10–16 per side, overlapped but distinct, lobes 4–8, gen 2–5 mm, oblanceolate to obovate; cauline lf gen 1. **INFL**: cluster gen 1, 10–30 mm wide, head-like, 10–20-fld; pedicels gen < 3 mm, straight. **FL** ± 5 mm wide; hypanthium length = or > width; petals 2–3 mm, narrowly oblanceolate, yellow; stamens 5; pistils gen 2–4. **FR** ± 2 mm, ± smooth, mottled brown. Open, dry, rocky ridges, slopes; 1800–3500 m. KR, NCoRH, n&c SNH, Wrn, **n SNE (Sweetwater Mtns)**; to WA, WY, Colorado. ❀ TRY;DFCLT. Jul–Aug

I. jaegeri Munz & I.M. Johnston JAEGER'S IVESIA Pl hanging to ± matted, green; caudex branched. **ST** 3–15 cm. **LF** 2–10 cm; sheathing bases sparsely hairy; lflets 4–6 per side, lobes 3–6, 2–5 mm, oblanceolate to obovate; cauline lvs ± 2. **INFL** open, gen < 10- fld; pedicels 6–30 mm, ± S-shaped in fr. **FL** 5–10 mm wide; hypanthium length < l/2 width; petals gen 2 mm, narrowly oblanceolate, yellow, < or = sepals; stamens 20; pistils 3–8. **FR** 1–2 mm, ± ridged, pale. RARE. Limestone crevices; 2100–3600 m. **e DMtns (Clark Mtn)**; sw NV. Jun–Jul

I. kingii S. Watson var. *kingii* ALKALI IVESIA Pl rosetted, glabrous or short-appressed-hairy, glaucous or not; caudex gen simple. **ST** decumbent to ascending, 15–40 cm. **LF** 7–15 cm; sheathing bases gen strigose; lflets 30–50 per side, overlapped but distinct, lobes < 4, 2–6 mm, oblanceolate to obovate; cauline lvs 4–13. **INFL** open; clusters gen < 10, 10–20 mm wide, loosely head-like, gen < 5-fld; pedicels 3–25 mm, straight. **FL** ± 10 mm wide; hypanthium length ± l/2 width; petals 3–5 mm, ± obovate, white, > sepals; stamens 20; pistils 2–6. **FR** 2–2.5 mm, smooth, light brown. RARE in CA. Moist alkaline clay; 1200–2000 m. **n SNE (Mono Co.)**; to UT. Var. *eremica* (Cov.) B. Ertter, Ash Meadows mousetails, w NV, has branched caudex, denser hairs, more tightly overlapped lflets. Jun–Jul

I. lycopodioides A. Gray CLUB-MOSS IVESIA Pl rosetted, green; caudex gen simple. **ST** decumbent to erect, 3–30 cm. **LF** 1–15 cm; sheathing bases ciliate; lflets 10–35 per side, overlapped but distinct, lobes 4–10, 1–8 mm, oblanceolate to round; cauline lf gen 1. **INFL**: cluster gen 1, 10–20 mm wide, ± dense or head-like, gen < 15-fld; pedicels gen < 5 mm, straight. **FL** 8–15 mm wide; hypanthium length ± < width; petals 2–5 mm, ± obovate, yellow, > sepals; stamens 5; pistils gen 5–15. **FR** 1–1.5 mm, smooth, pale. Wet meadows to alpine rocks; 2300–4000 m. SNH, SNE. Sspp. intergrade. 3 sspp. in CA.

  ssp. *lycopodioides* **ST** 3–15 cm. **LF** 1–7 cm; lobes ± 1 mm, ± round, ± glabrous, bristle-tip gen 0. **FL**: hypanthium length ± 1/2 width; petals 2–3 mm, gen < 2 mm wide, obovate; filaments ± 1 mm; styles 1–2 mm. 2*n*=28. Rocky areas; 3000–4000 m. n&c SNH, **n SNE (Sweetwater Mtns)**. ❀ DRN,IRR, SUN:1,16,17; DFCLT. Jul–Aug

  ssp. *scandularis* (Rydb.) Keck **ST** gen 5–15 cm. **LF** gen 3–8 cm; lobes 1–3 mm, ± obovate, moderately to densely hairy, bristle-tip gen 0.5–1 mm. **FL**: hypanthium length ± < 1/2 width; petals 3–5 mm, gen > 2 mm wide, widely obovate; filaments ± 1.5 mm; styles 2–3 mm. Vernally moist, open, rocky areas; 3000–4000 m. c&s SNH, **W&I**. Jul–Aug

I. patellifera (J. Howell) B. Ertter (p. 465) KINGSTON MOUNTAINS IVESIA Pls hanging clumps, green. **ST** 10–20 cm. **LF** gen 5–12 cm, flat; sheathing bases glabrous; lflets 2–3 per side, separated, 5–20 mm, ± round, evenly toothed or lobed < 1/2 to base, cauline lvs 0–2. **INFL** open, few–many-fld; pedicels 5–30 mm, gen ± S-shaped in fr. **FL** 5–10 mm wide; hypanthium length < 1/2 width; bractlets gen 0; petals 2–3 mm, narrowly oblanceolate, yellow, < sepals; stamens 5–10; pistils 4–10. **FR** 1.5–2 mm, ± ridged, pale. RARE. Granite crevices; 1400–2100 m. **e DMtns (Kingston Mtns)**. [*Potentilla p.* J. Howell, Kingston Mtns cinquefoil] Jun & Oct

I. saxosa (E. Greene) B. Ertter (p. 465) Pls hanging clumps, green. **ST** 5–30 cm. **LF** gen 5–15 cm, flat; sheathing bases ± glabrous; lflets 2–4 per side, separated, gen 5–15 mm, ± round, shallowly and evenly toothed to unevenly lobed ± to base; cauline lvs 2–4. **INFL** open, few–many-fld; pedicels 7–30 mm, gen ± S-shaped in fr. **FL** 7–10 mm wide; hypanthium length < 1/2 width; petals 2–4 mm, oblanceolate to obovate, yellow, < sepals; stamens 15–40; pistils 3–20. **FR** 1–1.5 mm, ± ridged, pale. Granitic or volcanic crevices; 900–3000 m. s SNH, SnBr, PR, **SNE**, **DMtns**; n Baja CA. [*Potentilla s.* E. Greene incl ssp. *sierrae* Munz] Stamen number, lf lobing, hairiness variations apparently not useful taxonomically. Apr–Aug

I. shockleyi S. Watson var. *shockleyi* Pl matted, green; caudex much-branched. **ST** spreading, < 15 cm. **LF** gen 2–8 cm; sheathing bases strigose; lflets gen 5–10 per side, lobes 3–5, 1–5 mm, oblanceolate to obovate; cauline lf 1, gen ± bract-like. **INFL** open, gen < 10-fld; pedicels 4–10 mm, S-shaped in fr. **FL** 5–10 mm wide; hypanthium length 1/2 width; petals ± 2 mm, oblanceolate, yellow, < or = sepals; stamens 5; pistils 2–5(6). **FR** ± 2 mm, smooth, light brown. Rocky areas; 2700–4000 m. n&c SNH, **W&I**; se OR, c&n NV. *I. cryptocaulis* (Clokey) Keck (sw NV) has open, buried caudex. ❀ DRN,IRR,SUN:1–3,16,17;DFCLT. May–Jul

PERAPHYLLUM WILD CRAB APPLE

Thomas J. Rosatti

1 sp. (Greek: very leafy)

P. ramosissimum Nutt. (p. 465) Shrub 1–3 m, much-branched. **LVS** ± clustered on short lateral branches, deciduous, 2–4 cm; stipules soon deciduous; petiole 0–short; blade ± oblanceolate, pointed, strigose, entire to minutely gland-toothed. **INFL**: fls 1–3, pedicelled, on short lateral branches. **FL**: hypanthium ± funnel-shaped, bractlets 0; sepals spreading to reflexed, 3–5 mm, hairy at least inside, persistent; petals spreading, 6–8 mm, ± obovate, white to rose; stamens ± 15–20, free; ovary 1, inferior, 4(6)-chambered, styles 2(3). **FR**: pome, 8–10 mm, spheric, yellowish to reddish. Dry washes, sagebrush scrub, pinyon/juniper woodland, pine forest; 1200–2500 m. e CaRH, n&c SNH (e slope), **GB, DMtns (Panamint Mtns)**; to OR, Colorado, NM. ❀ DRN,SUN:1,**2**,15–17&IRR:**3,7**,14,**18**,19–21. Apr–May

PETROPHYTON ROCK SPIRAEA

Thomas J. Rosatti

Shrub, matted, scapose. **LVS** crowded, evergreen, gen ± oblanceolate, entire. **INFL** ± spike-like. **FL**: bractlets on hypanthium 0; sepals persistent; petals white; stamens 20–40; pistils gen 5, simple, ovary superior, hairy, styles thread-like. **FR**: follicles, dehiscing along both sutures. **SEEDS** 1–several, linear. ± 4 spp.: w N.Am. (Greek: rock pl)

P. caespitosum (Nutt.) Rydb. (p. 465) Pl 3–8 dm wide; rosettes many. **STS** very stout. **LF** 1–3-veined below. **INFL** 4–14 cm; peduncle 3–10 cm, bracted. **FL**: sepals ± 1.5 mm, narrowly ovate, acute; petals ± 1.5 mm, gen obtuse; style ± 3 mm. **FR** ± 2 mm. **SEEDS** 1–2, ± 1.5 mm, linear to obovoid, brown, smooth. Limestone soils, pinyon/juniper woodland, coniferous forest; 1200–3000 m. s SNH, **W&I, DMtns**; to Rocky Mtns. 2 sspp. in CA.

  ssp. ***caespitosum*** (pl. 94) **LF** 5–12 mm, densely hairy. Limestone ledges, rocks, often in pinyon/juniper woodland; 1500–3000 m. **W&I, DMtns**; to Rocky Mtns. ❀ DRN,SUN:1,2,7,14–16;DFCLT. May–Sep

PHYSOCARPUS NINEBARK

Thomas J. Rosatti

Shrub, gen ± stellate-hairy. **LVS** stipuled, petioled, deciduous; blade ovate to ± round in CA, gen palmately 3–7-lobed, lobes crenate to serrate. **INFL** ± umbel-like, bracted. **FL**: hypanthium bell-shaped, bractlets 0; sepals persistent; petals rounded, white; stamens 20–40; pistils 1–5, free or ± fused, ovary superior, style thread-like, stigma head-like. **FR**: follicles, inflated, often opening along both sutures. **SEEDS** 2–4, ovoid; coat hard, shiny. ± 10 spp.: N.Am, Asia. (Greek: bladdery fr)

P. alternans (M.E. Jones) J. Howell (p. 467) Pl 5–15 dm. **LF**: petiole 5–10 mm; blade 5–20 mm, gen densely hairy, lobes 3–7, gen crenate. **FL**: sepals ± 3 mm; petals 3–4 mm; stamens alternately short and long; pistil gen 1. **FR** hairy. Dry, rocky pinyon/juniper woodland; 1800–3100 m. **W&I, n DMtns**; to UT. [sspp. *annulatus* J. Howell and *panamintensis* J. Howell] ❀ TRY; DFCLT. Jun–Jul

POTENTILLA CINQUEFOIL

Barbara Ertter

Ann to shrub; odor resinous or 0. **LVS** gen basal, odd-1-pinnate to 1-palmate or 1-ternate; lflets ± toothed or lobed, terminal gen ± = lateral. **INFL**: cyme, gen ± open; pedicels gen ± straight. **FL**: hypanthium ± shallow; bractlets 5; sepals gen 5, ± triangular; petals gen 5, gen = or > sepals, gen ± widely obcordate, gen yellow; stamens gen 20; pistils gen many, styles gen jointed near tip. **FR**: achene. 200–500 spp.: n temp. (Latin: diminutive of powerful, for reputed medicinal value) [Clausen, Keck, & Hiesey 1940 Carn Inst Wash Pub 520:26–195]

1. Shrub; fr densely hairy . . . . *P. fruticosa*
1' Herb; fr glabrous (often nested among receptacle hairs)
  2. Fls solitary on peduncles at nodes of stolons; lf pinnate, main lateral lflets alternating with reduced lflets, densely hairy . . . . *P. anserina*
  2' Stolons 0; fls in cymes; lf palmate, ternate, or pinnate, alternating reduced lflets 0, variously hairy to glabrous
    3. Styles attached below middle of fr, widest near middle; lf pinnate, terminal lflet > others . . . . . *P. glandulosa*
      4. St nonglandular, gen 20–60 cm; petals cream; lflet teeth >10 . . . . ssp. *nevadensis*
      4' St gen glandular, gen 5–20 cm; petals pale yellow; lflet teeth < 10 . . . . ssp. *pseudorupestris*
    3' Styles attached just below tip of fr, slender throughout or widest near base; lf palmate, ternate, or pinnate (if pinnate, terminal lflet not >> others)
      5. Ann, bien or short-lived per; lflets 3, basal lvs often withered or fallen in fl
        6. St-base hairs gen ± dense, < 1 mm, many glandular; petals 1–2.5 mm, stamens 10; fr whitish, ± 0.6 mm . . . . *P. biennis*
        6' St-base hairs sparse, to 2 mm, glandless; petals 3–4 mm; stamens 15–20; fr light brown, ± 1 mm . . . . *P. norvegica*
      5' Per; lflets > 3, basal lvs gen present in fl
        7. Lvs palmate; basal central lflet 2–6 cm; st ± ascending, gen 20–50 cm; infl few–many-fld, flat-topped; pedicels straight . . . . *P. gracilis*
        8. Lflet lobed > 3/4 to midvein, lobes often narrowest at base, gen tomentose below . . . . var. *elmeri*

8' Lflet toothed < 1/2 to midvein, teeth widest at base, equally hairy above and below . . . . var. *fastigiata*
7' Lvs pinnate to palmate (if palmate then basal central lflet 5–15 mm); st prostrate to ascending,
    gen < 25 cm; infl gen < 10-fld, gen not flat-topped, pedicels often ± recurved in fr
    9. Pl of moist meadows < 2000 m; lvs pinnate, lflets 5–13 per side . . . . *P. millefolia*
    9' Pl of rocky alpine barrens > 2700; lvs pinnate or palmate, lflets 5 or 2–5 per side
        10. Lflets densely white-tomentose below, green and strigose above; styles > 1 mm, gen slender
            11. Lf palmate; lflets gen 5, distally toothed ± 1/2 to midvein . . . . *P. concinna*
            11' Lf ± pinnate; lflets 2–4 per side, often assymetrically split as well as lobed ± 3/4 to midvein
                . . . . *P. morefieldii*
        10' Lflet surfaces similarly hairy (but often denser below); styles ± 1 mm, tapered from rough-
        thickened base
            12. Lf ± pinnate; lflets 5–11, lobed ± 2/3 to midvein . . . . *P. pensylvanica*
            12' Lf gen ± palmate; lflets ± 5, lobed > 3/4 to midvein . . . . *P. pseudosericea*

P. anserina L. (p. 467) Pl tufted from stolons, nonglandular. **LF** pinnate, 3–50(75) cm; main lflets 5–10 per side (reduced lflets alternating), 10–50 mm, ± elliptic to oblanceolate, toothed 1/2–1/3 to midvein, densely hairy (at least below). **INFL:** fls solitary from stolon nodes. **FL:** hypanthium 4–7 mm wide; petals 7–20 mm; filaments 1–3.5 mm, anthers ± 1 mm; styles ± 2 mm, slender. **FR** ± 2 mm, rough, dark red-brown. $2n$=28,42. Wet, often brackish areas; 0–2500 m. NCo, s SNH, CCo, SCo, SnBr, **GB**; circumboreal. [Rousi 1965 Ann Bot Fenn 2:47–112] 2 sspp. in CA.

ssp. ***anserina*** **LF** 3–15 cm, gen densely hairy above; main lflets 10–25 mm. **INFL:** pedicel gen 2–7 cm. **FL:** petals 7–10 mm; pistils 10–50. $2n$=28,42. Habitats of sp.; 1200–2500 m. s SNH, SnBr, **GB**; circumboreal. [var. *sericea* Hayne] ❀ TRY. May–Oct

P. biennis E. Greene (p. 467) Ann or bien, taprooted, glandular. **ST** ascending to erect, 10–70 cm, ± spreading-hairy. **LF** ternate; basal often withered in fl; cauline gen 2–10 cm, lflets 3, 5–30 mm, ± obovate, toothed ± 1/4 to midvein, ± hairy. **INFL** few–many-fld. **FL:** hypanthium 2–4 mm wide; petals 1–2.5 mm, < sepals, oblanceolate-elliptic; stamens 10, filaments 0.5–1 mm, anthers ± 0.2 mm; style ± 0.7 mm, tapered from rough-thickened base. **FR** ± 0.6 mm, smooth, whitish. Moist shores; 1500–3100 m. SNH, TR, **GB, DMtns**; w N.Am. May–Aug

P. concinna Richardson (p. 467) ALPINE CINQUEFOIL Pl tufted from branched caudex; glands 0 or hidden. **ST** prostrate to decumbent, 3–15 cm, ± ascending-hairy. **LVS** palmate; basal gen 1.5–5 cm, lflets gen 5, central lflet 5–20 mm, narrowly obovate, distal 1/2 few-toothed ± 1/2 to midvein, densely white-tomentose below, ± green and strigose above. **INFL** gen 1–5-fld; pedicels often ± recurved in fr. **FL:** hypanthium 2.5–5 mm wide; petals 2.5–5.5 mm; filaments 1–2.5 mm, anthers 0.6–0.9 mm; styles 1.5–2.5 mm, slender. **FR** 1.5–2 mm, smooth, light brown. RARE in CA. Rocky meadows, ridges; ± 3170 m. **n W&I (White Mtns)**; w N.Am. [var. *divisa* Rydb.; *P. beanii* Clokey] Pls intermediate to *P. gracilis* var. *pulcherrima* (Lehm.) Fern. (otherwise unknown from CA) also in n W&I. May–Jul

P. fruticosa L. (p. 467) SHRUBBY CINQUEFOIL Shrub < 1 m; glands 0 or hidden. **LF** ± pinnate, 1–3 cm; lflets 2–3 per side, 5–20 mm, linear to narrowly elliptic, entire, ± hairy. **INFL:** fls 1–2 at end of twigs. **FL:** hypanthium 5–8 mm wide; petals 5–10 mm; stamens 20–25, filaments 2–3 mm, anthers ± 1 mm; styles ± 2 mm, attached near middle of fr, narrowly club-shaped. **FR** ± 1.5 mm, densely hairy (unique in CA). n=7,14. Meadows, rocks; 2000–3600 m. KR, CaRH, SNH, Wrn, **W&I**; circumboreal. ❀ DRN,SUN:**4–6**& IRR:1,7,15–17&WET:2,3,14,18–24;DFCLT; CV. Jun–Aug

P. glandulosa Lindley Pl gen ± tufted from loosely branched caudex; glandular hairs often many. **ST** ± erect, 5–90 cm, spreading-hairy. **LVS** pinnate; basal 3–30 cm, terminal lflet largest (0.5–12 cm), lateral gen 3–5 per side, ± obovate, toothed 1/4–1/2 to midvein, ± hairy. **INFL** gen 2–30-fld. **FL:** hypanthium 3–6 mm wide; petals 3–10 mm, gen widely ovate, yellow to white; stamens ± 25, filaments 1–3.5 mm (longest opposite sepals), anthers 0.6–1.2 mm; styles ± 1–2.5 mm, attached below middle of fr, ± fusiform and rough. **FR** ± 1 mm, smooth or ± ridged, golden to reddish brown. n=7. Common. Many habitats; < 3800 m. CA-FP (exc GV), **GB**; w N.Am. Often confused with *Horkelia*; despite much work, sspp. remain poorly defined. 8 sspp. in CA.

ssp. ***nevadensis*** (S. Watson) Keck (p. 467) **ST** gen 20–60 cm, nonglandular, short-hairy. **LVS:** basal gen 5–20 cm, sheathing base gen glabrous, terminal lflet gen 20–40 mm, lateral lflet teeth gen 10–20, gen single. **INFL:** branch angle gen 10–20°. **FL:** bractlets gen 0.5–1 mm wide; petals 4–8 mm, > sepals, cream; styles ± 1 mm. Gen ± moist, often rocky places; 1800–3700 m. SNH, TR, SnJt, **SNE**; NV. ❀ IRR,DRN:4–6&SHD:1,**2**,3,7, 14,**15–17**. Jun–Aug

ssp. ***pseudorupestris*** (Rydb.) Keck **ST** gen 5–20 cm, gen glandular-hairy. **LVS:** basal gen 3–8 cm, sheathing base gen appressed-hairy, terminal lflet gen 5–20 mm; lateral lflet teeth < 10, gen single. **INFL:** branch angle gen ± 40°. **FL:** bractlets ± 1 mm wide; petals 4–7 mm, > sepals, cream to pale yellow; styles ± 1–1.5 mm. Rocky areas; 2400–3800 m. NW, CaRH, SNH, Wrn, **W&I**; to MT. [*P. p.* Rydb.] CA pls possibly separable; further study needed. Jul–Aug

P. gracilis Hook. Pl tufted from short, thick rhizome; glands gen 0 or hidden. **ST** ± ascending, gen 20–100 cm, strigose to spreading-hairy. **LVS** palmate; basal gen 6–30 cm, lflets gen 5–7, central lflet ± oblanceolate, toothed or lobed ± throughout, hairy esp below. **INFL** few–many-fld. **FL:** hypanthium gen 3–5 mm wide; petals gen 4–10 mm; filaments 1.5–2.5 mm, anthers 0.6–1.6 mm; styles 1.5–2 mm, slender. **FR** 1–1.5 mm, smooth, light brown. $2n$=52–109. Common. Meadows, open forests; 120–3500 m. NW (exc sw), CaR, SNH, Teh, TR, PR, **GB**; w N.Am. Variation complex; many but not all pls are assignable to the following extremes. 4 vars. in CA.

var. *elmeri* (Rydb.) Jepson (p. 467) Hairs gen appressed. **ST** gen 20–50 cm. **LVS:** basal central lflet 20–60 mm, gen tomentose below, lobed > 3/4 to midvein, lobes often narrowest at base, entire. **FL:** petals 5–8 mm. n=21. Dry meadows; 1280–3050 m. SNH, Teh, TR, SnJt, **SNE**; w N.Am. [*P. flabelliformis* Lehm. var. *inyoensis* Jepson; *P. pectinisecta* Rydb.] May–Jul

var. *fastigiata* (Nutt.) S. Watson (p. 467) Hairs spreading to appressed. **ST** gen 20–50 cm. **LVS:** basal central lflet 20–60 mm, ± equally hairy above and below, toothed < 1/2 to midvein, teeth widest at base. **FL:** petals gen 4–7 mm. $2n$=52–109. Common. Gen open forests, dry meadows; 800–3500 m. Range of sp. [ssp. *nuttallii* (Lehm.) Keck; vars. *glabrata* (Lehm.) C. Hitchc. and *permollis* (Rydb.) C. Hitchc.] Jun–Aug

P. millefolia Rydb. (p. 467) Pl rosetted from thick taproot, sometimes glandular. **ST** prostrate to decumbent, 5–20 cm, spreading- to appressed-hairy. **LVS** pinnate; basal 2–15 cm, lflets 5–13 per side, overlapped, 5–20 mm, narrowly 3–10-lobed > 1/2 to base, ± glabrous to hairy. **INFL** gen < 10-fld; pedicels gen ± recurved in fr. **FL:** hypanthium 3–6 mm wide; petals 4–8 mm; filaments gen 2–3.5 mm, anthers ± 1 mm; pistils 10–30, styles 2–3 mm, slender. **FR** 1.5–2 mm, smooth, ± tan. Vernally wet meadows; 900–2000 m. CaRH, ne-most SNH, **GB**; OR, NV. Variable; further study needed. Spreading-hairy pls have been called var. *klamathensis* (Rydb.) Jepson. ❀ DRN,IRR,SUN:1–3,6,7,15–18;DFCLT. Jun–Jul

P. morefieldii B. Ertter (p. 467, pl. 95) MOREFIELD'S CINQUEFOIL Pl tufted from few-branched caudex; glands 0 or hidden. **ST** prostrate to decumbent, 5–15 cm, ± ascending-hairy. **LVS** pinnate; basal 2–6 cm, lflets 2–4 per side, ± overlapped, 5–20 mm, ± oblanceolate, few-lobed ± 3/4 to midvein, often ± asymmetrically split ± to base, white-tomentose below, ± green and strigose above. **INFL** gen 5–15-fld; pedicels often ± recurved in fr. **FL**: hypanthium 2.5–5 mm wide; petals 4–6 mm; filaments ± 1–2 mm, anthers 0.5–1 mm; styles ± 1.5–2 mm, ± slender. **FR** ± 1.8 mm, smooth or ± ridged, pale brown. RARE. Low areas in alpine calcareous (or granite?) rocks; 3300–4000 m. **n W&I**.

P. norvegica L. (p. 467) Ann to short-lived per from taproot, nonglandular. **ST** ascending to erect, 10–70 cm; hairs spreading, sparse and long below, denser and shorter above. **LVS** gen ternate; basal often withered or fallen in fl; cauline gen 3–12 cm, lflets gen 3, 15–50 mm, oblanceolate, toothed ± 1/3 to midvein, ± hairy. **INFL** several–many-fld. **FL**: hypanthium 4–10 mm wide; petals 3–4 mm, < sepals; stamens 15–20, filaments 0.5–2 mm, anthers ± 0.3 mm; styles ± 0.8 mm, tapered from rough-thickened base. **FR** ± 1 mm, veined, light brown. *n*=28,35; 2*n*=56,63,70. Moist, disturbed areas; < 2300 m. c SNH, ScV, **SNE**; N.Am, native to Eurasia. [ssp. *monspeliensis* (L.) Asch. & Graebn.] Jun–Sep

P. pensylvanica L. (p. 467) Pl tufted from simple or branched caudex, inconspicuously glandular. **ST** ± ascending, 8–25 cm, short-spreading- and long-ascending-hairy. **LVS** ± pinnate; basal gen 3–10 cm, lflets 2–5 per side, 8–25 mm, ± oblanceolate, ± evenly 10-lobed ± 2/3 to midvein, densely hairy below, gen greener and less hairy above. **INFL** ± 5–10-fld. **FL**: hypanthium 3.5–5.5 mm wide; petals ± 4 mm; filaments 0.5–1.5 mm, anthers ± 0.6 mm; styles ± 1 mm, tapered from rough-thickened base. **FR** ± 1 mm, ± smooth, pale brown. 2*n*=28. Alpine rocky barrens; 2700–3800 m. c SNH, **W&I**; N.Am, Eurasia. Highly variable outside CA; w N.Am pls have been called var. *strigosa* Pursh. Jun–Aug

P. pseudosericea Rydb. (p. 467) Pl tufted or matted from ± branched caudex; glands 0 or hidden. **ST** decumbent to ascending, gen 5–10 cm, ± strigose. **LVS** gen ± palmate; basal gen 2–4 cm, lflets ± 5, central lflet gen 5–15 mm, ± obovate, 3–10-lobed > 3/4 to midvein, tomentose below, densely white-strigose above. **INFL** gen ± 5-fld. **FL**: hypanthium 3–4 mm wide; petals ± 3 mm; filaments 0.5–1 mm, anthers ± 0.4 mm; styles ± 1 mm, tapered from rough-thickened bases. **FR** ± 1 mm, ± smooth, pale brown. Rocky flats, slopes; 3200–4300 m. c&s SNH, **n SNE (Sweetwater Mtns), W&I**. Jul–Aug

PRUNUS

Dieter H. Wilken

Shrub or tree. **ST**: bark gray to red-brown. **LVS** gen alternate, simple, gen glabrous; stipules deciduous. **INFL**: raceme or umbel-like cluster, often on short branchlets. **FL**: hypanthium cup- to urn-shaped; sepals spreading to reflexed; stamens 15+, gen in 2+ whorls; pistil 1, ovary superior, chamber 1, ovules 2, style 1, stigma subspheric. **FR**: drupe, gen ovoid to spheric. ± 400 spp.: temp N.Am, Eurasia, n Afr; many cult for wood, orn, edible fr; some persisting near human habitation (*P. armeniaca*, apricot; *P. avium*, sweet cherry; *P. cerasus*, sour cherry; *P. domestica*, plum; *P. laurocerasus*, laurel cherry; *P. lusitanica*, portugal laurel; *P. mahaleb*; *P. persica*, peach). Seeds of many spp. ± TOXIC from production of hydrocyanic acid.

1. Infl a raceme, fls 10–many; frs on naked infl axis, glabrous
  2. Raceme short, ± flat-topped; fls 3–12 . . . . *P. emarginata*
  2' Raceme longer than wide; fls gen many. . . . . *P. virginiana* var. *demissa*
1' Infl a cluster, often umbel-like, fls 1–5; frs from short, leafy branchlet or on leaf axil, gen puberulent
  3. Lf blade ovate to round, 7–22 mm wide, base obtuse to ± cordate . . . . *P. fremontii*
  3' Lf blade narrowly elliptic to oblanceolate, 1-**7** mm wide, base tapered
    4. Petals reddish to salmon; lf blade 4–7 mm wide, finely serrate . . . . *P. andersonii*
    4' Petals white to yellowish; lf blade 1–3 mm wide, gen entire . . . . *P. fasciculata* var. *fasciculata*

P. andersonii A. Gray (p. 467, pl. 96) DESERT PEACH Shrub < 2 m. **ST** much-branched; twigs rigid, becoming spine-like. **LVS** gen clustered, deciduous; petiole 0–7 mm; blade 9–30 mm, elliptic to oblanceolate, finely serrate, base tapered, tip gen acute. **INFL**: fls 1–2; pedicel 4–7 mm. **FL**: sepals puberulent inside; petals 5–9 mm, reddish. **FR** 10–14 mm, subspheric, densely puberulent, red-orange; pulp dry. Rocky slopes, flats, shrubland, coniferous forest; 900–2600 m. SNH (e slope), **GB, n DMtns (Last Chance Range)**; w NV. ❀ DRN,SUN:1,2,**3**,**7**,8–10,**18**–**21**,22–23&DRY:**14**,15–17. Mar–Apr

P. emarginata (Hook.) Walp. (p. 467) BITTER CHERRY Shrub or tree 1–10 m, often forming dense thickets. **LVS** gen clustered, deciduous; petiole 3–12 mm; blade 20–65 mm, gen elliptic to obovate, crenate-serrate, base obtuse to tapered. **INFL**: raceme, ± flat-topped; fls 3–12; pedicel 3–12 mm. **FL**: hypanthium and sepals glabrous to puberulent; petals 4–8 mm, white. **FR** 7–14 mm, ovoid to spheric, glabrous, red to purple; pulp ± fleshy. Rocky slopes, canyons, chaparral, mixed-evergreen or coniferous forest; < 2800 m. CA-FP (exc GV, ChI), **GB**; to B.C., MT, WY. ❀ DRN:**4–6**& IRR:1,**2**,3,**7**,8–10,14,**15–18**,19–23

P. fasciculata (Torrey) A. Gray (p. 467) DESERT ALMOND Shrub 1–2 m. **ST** much-branched; twigs rigid, becoming ± spine-like. **LVS** gen clustered, deciduous, subsessile; blade 7–15 mm, narrowly oblanceolate, gen entire, base tapered, tip acute to obtuse.

INFL: fls 1–3, subsessile. **FL**: sepals glabrous to sparsely puberulent; petals 2–4 mm, white to yellowish. **FR** 8–15 mm, ovoid to spheric, densely puberulent, gray to red-brown; pulp dry. Slopes, canyons, washes, shrubland, woodland; < 2200 m. s SNH, Teh, s CCo, s SCoRI, n TR, e PR, **D**; to UT, Baja CA. 2 vars. in CA.

  var. *fasciculata* **LF**: surface puberulent. Creosote-bush scrub, Joshua-tree or pinyon/juniper woodland; 700–2200 m. s SNH, Teh, s SCoRI, n TR, e PR, **D**; to UT, Baja CA. ❀ DRN,SUN,DRY:1,2,**3**,**7**,8–11,**14**,**18**,19–23. Mar–May

P. fremontii S. Watson (p. 467) DESERT APRICOT Shrub 1–4 m. **ST** much-branched; twigs rigid, becoming spine-like. **LVS** gen clustered, deciduous; petiole 2–7 mm; blade 12–30 mm, ovate to round, serrate, base obtuse to ± cordate, tip obtuse to rounded. **INFL**: fls 1–5; pedicel 1–3 mm. **FL**: petals 4–8 mm, white. **FR** 8–15 mm, ovoid to spheric, puberulent, yellowish; pulp dry. Rocky slopes, canyons, shrubland, pinyon/juniper woodland; 200–1200 m. e PR, **w DSon**; Baja CA. ❀ DRN:3,**7**,8,9,**14**,**18**,**19** &SUN:**15**,**16**,**20**–**24**. Feb–Mar

P. virginiana L. var. *demissa* (Nutt.) Torrey (p. 467) WESTERN CHOKE-CHERRY Shrub or small tree < 6 m. **LVS** deciduous; petiole 10–25 mm; blade 50–100 mm, elliptic to obovate, finely serrate, base rounded to ± cordate, tip gen acute. **INFL**: raceme; fls many; pedicel 5–8 mm. **FL**: petals 4–7 mm, white. **FR** 6–14 mm, ovoid to spheric, glabrous, dark red to black; pulp ± fleshy.

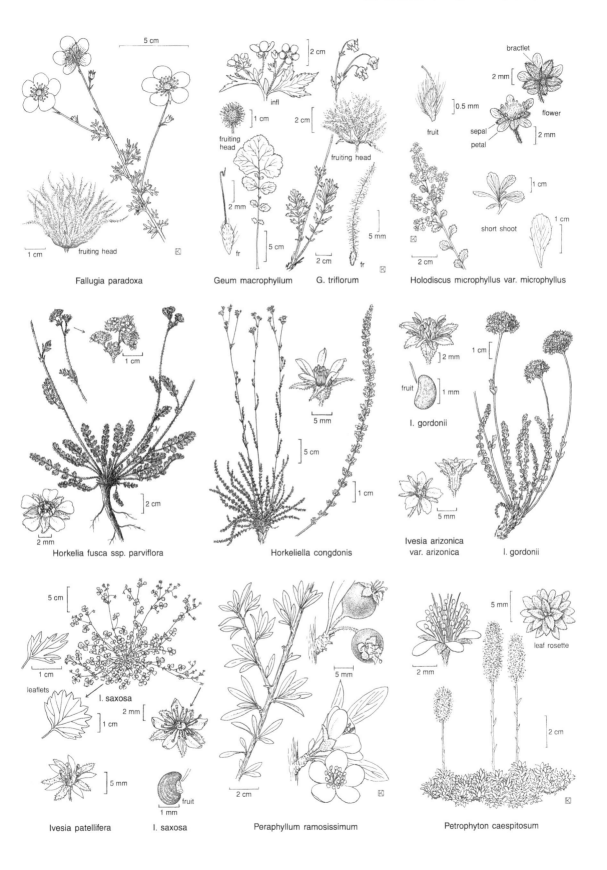

Fallugia paradoxa

Geum macrophyllum

G. triflorum

Holodiscus microphyllus var. microphyllus

Horkelia fusca ssp. parviflora

Horkeliella congdonis

I. gordonii

Ivesia arizonica
var. arizonica

I. gordonii

Ivesia patellifera

I. saxosa

Peraphyllum ramosissimum

Petrophyton caespitosum

$2n=16$. Rocky slopes, canyons, shrubland, oak/pine woodland, coniferous forest; 100–2900 m. CA-FP (exc coast, GV), **GB**; to B.C., c US, TX, n Mex. [var. *melanocarpa* (Nelson) Sarg.] Var. *virginiana* of e N.Am, gen a tall tree, sometimes cult. ❀ DRN,SUN or part SHD: **4–6,14–17**&IRR:1,**2**,3,7,8–10,**18–20**,21–24;STBL; rather INV. May–Jun

PURSHIA ANTELOPE BUSH

Thomas J. Rosatti

Shrub or small tree. **LVS** ± clustered on short lateral branches, mostly deciduous, gen deeply lobed, ± strongly rolled under, gen with ± sunken glands above; bases persistent, overlapping, sheathing st. **INFL**: fls solitary on side-branch tips. **FL**: hypanthium ± funnel-shaped, sometimes partly glandular, bractlets 0; stamens ± 25; pistils 1–5(12), simple, styles persistent, ± hairy. **FR**: achene, ± fusiform to oblong. ± 5 spp.: w N.Am. (Frederick T. Pursh, N.Am flora author, 1774–1820) [Koehler & Smith 1981 Madroño 28:13–25; Henrickson 1986 Phytologia 60:468]

1. Pistils 4–5(12); styles in fr 2–6 cm, plumose; lf lobes (3)5–9, from just below middle
.... ***P. mexicana*** var. ***stansburyana***
1' Pistils 1(–3); styles in fr < 1 cm, canescent at least below tip; lf lobes 3(–5), gen from above middle
.... ***P. tridentata***
 2. Lf upper surface sparsely nonglandular-hairy, sessile or sunken glands few–many; twig hairs mostly glandular.. var. ***glandulosa***
 2' Lf upper surface densely nonglandular-hairy, sessile or sunken glands 0–few; twig haris mostly nonglandular.. var. ***tridentata***

P. mexicana (D. Don) Welsh var. ***stansburyana*** (Torrey) Welsh (p. 471, pl. 97) Shrub 3–30 dm. **LF**: lobes (3)5–9, from just below middle, entire to again lobed. **FL**: hypanthium ± 5 mm; sepals 4–6 mm, ± ovate; petals 6–8 mm, widely ovate, white to cream; pistils 4–5(12). **FR** glabrous to becoming so; styles 2–6 cm, plumose. *n*= 9. Dry Joshua-tree or pinyon/juniper woodland; 1100–2500 m. **W&I**, **DMtns**; to Colorado, NM, n Mex. [*Cowania m.* D. Don var. *stansburiana* (Torrey) Jepson] Hybridizes with *P. tridentata* var. *glandulosa*. ❀ DRN,SUN: 1,2,**3**,**7**,9,**10**,11,**14**,**18–21**&DRY:15,16,22,23. May–Jul

P. tridentata (Pursh) DC. Shrub 10–50 dm. **LF**: lobes 3(–5), gen from above middle, gen entire. **FL**: hypanthium ± 3–4 mm; sepals ± 3 mm, ± oblong; petals 6–8 mm, ± obovate, cream to yellow; pistils 1(–3). **FR** canescent; style < 1 cm, canescent at least below tip. Dry sagebrush scrub, chaparral, Joshua-tree or pinyon/juniper woodland, coniferous forest; 700–3400 m. KR, NCoRH, CaR, SNH (e slope), Teh, n TR and e PR (**D edge**), **GB**, **DMtns**; to B.C., MT, NM. Vars. intergrade.

var. ***glandulosa*** (Curran) M.E. Jones (p. 471) Pl greenish; twig hairs mostly glandular. **LF** 5–10 mm, sparsely tomentose and greenish, sessile or sunken glands few–many above. $2n=18$. Chaparral, Joshua-tree or pinyon/juniper woodland; 700–3000 m. c&s SNH (e slope), Teh, n TR and e PR (D edge), **SNE**, **DMtns**;NV,AZ. [*P. glandulosa* Curran] ❀ DRN,SUN: 1,2,**3**,**7**,9–11&DRY:**14**,15,16,**18**,19–21. Apr–Jun

var. ***tridentata*** (p. 471) Pl grayish; twig hairs mostly nonglandular. **LF** 5–30 mm, ± densely white-tomentose, sessile or sunken glands 0–few above. $2n=18$. Sagebrush scrub, coniferous forest, juniper woodland; 900–3400 m. KR, NCoRH, CaR, SNH (e slope), **GB**; to B.C., MT, NM. ❀ DRN,SUN: 1,**2**,**3**,**7**&DRY:**14**,15,16,18–21. May–Jul

ROSA

Barbara Ertter

Shrub to vine, often thicket-forming, gen prickly. **LVS** gen odd-pinnately compound; stipules gen attached to petiole, gen gland-margined. **INFL**: gen ± cyme or fls solitary. **FL**: hypanthium urn-shaped; bractlets 0; sepals 5, often with long expanded tip; petals gen 5 (exc cultivars), gen pink in CA (white to red or yellow); stamens gen> 20; pistils gen many, ovaries superior, jointed to gen hairy styles. **FR**: bony achenes enclosed in fleshy, gen reddish hypanthium (hip). 100+ spp.: gen n temp. (Latin: ancient name) Spp. hybridize freely; other non-natives established locally.

R. woodsii Lindley var. ***ultramontana*** (S. Watson) Jepson (p. 471) INTERIOR ROSE Shrub, loose to thicket-forming, gen 5–30 dm. **ST** gray- or red-brown; prickles few–many, gen ± slender, gen ± straight. **LF**: lflets (sub)glabrous; terminal lflet ± 10–40 mm, ± elliptic, tip ± obtuse, margins single-toothed, glandless. **INFL** gen 1–5-fld; pedicels gen ± 10–20 mm, gen ± glabrous, glandless. **FL**: hypanthium gen 3–5 mm wide at fl, glabrous, neck 2–4 mm wide; sepals glandless, gen entire (or with simple, linear lobes), tip gen ± = body, entire; petals gen 15–20 mm; pistils gen > 10. **FR** 5–12 mm wide. Gen ± moist areas; 800–3400 m. CaRH, SNH, SnGb, SnBr, **GB**, **DMtns**; to B.C., MT, NV. [vars. *glabrata* (Parish) Cole, *gratissima* (E. Greene) Cole] Some DMtns pls have thick-based, curved prickles. Var. *woodsii* in c US. ❀ DRN,SUN:4,5,**6** &IRR: **15–17**&part SHD:1,**2**,3,**7**,**14**,18–24; rather INV; also STBL. May–Jul

SIBBALDIA

Barbara Ertter

Per, low, ± matted; caudex branched. **LVS** gen basal, gen 1-ternate in CA. **INFL**: cyme. **FL**: hypanthium shallow; bractlets gen 5; sepals gen 5; petals gen 5; stamens 4–10; pistils few–many, ovaries superior, jointed to slender style on side. **FR**: achene. ± 20 spp.: n hemisphere, esp Himalayan. (R. Sibbald, Scottish naturalist, physician, 1641–1722) [Dixit & Panigrahi 1981 Proc Indian Acad Sci 90:253–272]

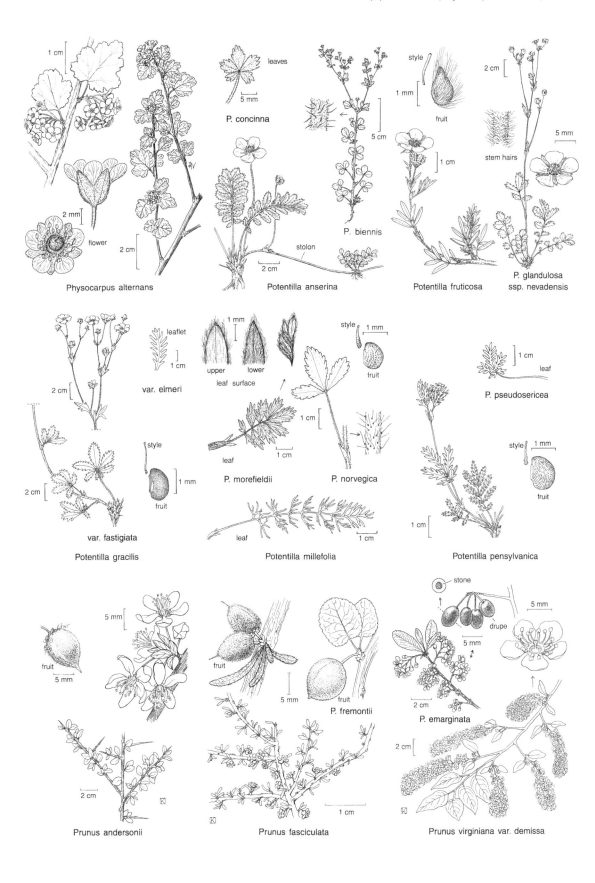

1 cm

leaves

P. concinna

5 mm

style

1 mm

fruit

style

2 cm

5 mm

stem hairs

2 mm

flower

2 cm

Physocarpus alternans

5 cm

stolon

2 cm

P. biennis

Potentilla anserina

1 cm

Potentilla fruticosa

P. glandulosa
ssp. nevadensis

leaflet

1 cm

var. elmeri

2 cm

1 mm

upper   lower
leaf surface

style

1 mm

fruit

1 cm

P. pseudosericea

leaf

2 cm

style

1 mm

fruit

var. fastigiata

Potentilla gracilis

leaf

1 cm

P. morefieldii

1 cm

P. norvegica

style

1 mm

fruit

Potentilla pensylvanica

1 cm

leaf

1 cm

Potentilla millefolia

fruit

5 mm

5 mm

Prunus andersonii

2 cm

fruit

5 mm

P. fremontii

fruit

5 mm

Prunus fasciculata

1 cm

stone

5 mm

drupe

5 mm

5 mm

P. emarginata

2 cm

2 cm

Prunus virginiana var. demissa

S. procumbens L. (p. 471) Pl hairs appressed, gen ± sparse. **ST** 2–15 cm, ± spreading. **LF**: petiole 1–7 cm; lflets 5–25 mm, ± wedge-shaped, teeth gen 3, at tip. **INFL**: pedicels gen 3–10 mm, straight. **FL**: bractlets linear, < sepals; sepals 2–4 mm, triangular; petals ± 1 mm, widely oblanceolate, yellow; stamens 5. **FR** ± 1 mm, smooth, brown, often retained in disintegrating fl. *n*=7. Moist rocky areas; 1900–3700 m (lowest in n). KR, CaRH, SNH, SnBr, Wrn, **SNE**; to ne N.Am, Eurasia. ❀ TRY;DFCLT. Jun–Aug

RUBIACEAE MADDER FAMILY

Lauramay T. Dempster

Ann, per, shrub, vine, tree. **LVS** gen opposite, entire; stipules gen on st, sometimes lf-like (then lvs apparently whorled and stipules considered lvs), adjacent pairs sometimes fused. **INFL**: cyme, panicle, cluster, or fl solitary, gen terminal and ± axillary. **FL** gen bisexual; calyx gen ± 4-lobed, sometimes 0; corolla gen radial, 4-lobed; stamens epipetalous, alternate corolla lobes, gen incl; ovary gen inferior, chambers gen 2 or 4, style 1, ± fused if 2. **FR**: 2 or 4 nutlets or a berry, drupe, or capsule. ± 500 genera, 6000 spp.: world-wide, esp trop; many cult (incl *Coffea*, coffee; *Cinchona*, quinine; many orn). [Dempster 1979 Fl CA 4(2):1–47] Key to genera adapted by Margriet Wetherwax.

1. Lvs opposite; fls sessile; corolla funnel-shaped . . . . **DIODIA**
1' Lvs whorled; fls pedicelled; corolla gen rotate . . . . **GALIUM**

DIODIA BUTTONWEED

Ann. **LVS** opposite; stipules fused to lf bases, small, fringed. **INFL**: fls 1–several in axils, sessile. **FL**: corolla funnel-shaped, lobes gen 4. **FR**: 2 nutlets; calyx teeth persistent. 50 spp.: warm Am, Afr. (Greek: thoroughfare, from habitats)

D. teres Walter ROUGH BUTTONWEED, POOR JOE (p. 471) **ST** erect, simple or branched, 7–25 cm, 4-angled. **LF** 2.5 cm, lanceolate, tip sharp; petiole and blade base transparent, fringed, forming a cup ± enclosing infl. **FL**: calyx lobes 4, stiff, unequal; corolla 3 mm, white or pink. **FR**: hairs stiff. Sandy places; 50 m. **DSon (sw Imperial Co)**; to e US, Mex.

GALIUM BEDSTRAW, CLEAVERS

Ann, per, sometimes ± shrubby, often ± dioecious, glabrous or hairy, often scabrous. **ST** when young 4-angled. **LVS** in whorls of 4 or more, incl lf-like stipules. **INFL**: panicle, or axillary clusters of 1–many fls. **FL** bisexual or unisexual (with sterile stamens or pistils); calyx 0; corolla gen rotate, sometimes ± bell-shaped, gen greenish, fading yellow or white, sometimes reddish, lobes gen 4; ovary 2-lobed, styles 2, ± fused basally. **FR**: 2 nutlets or 1 berry. ± 400 spp.: worldwide, esp temp. (Greek: milk, from use of some spp. in its curdling) Hairiness of ovary and fr gen ± equal on a single pl; staminate pls often identified only by association with pistillate. Key to species adapted by Bruce Baldwin.

1. Ann
  2. Corolla lobes gen 3
    3. Lvs in whorls of 2–4; fls gen solitary in axils . . . . *G. bifolium*
    3' Lvs in whorls of 4–6; fls several in axils . . . . ²*G. trifidum* var. *pacificum*
  2' Corolla lobes gen 4
    4. Lvs in whorls of 6–8 . . . . *G. aparine*
    4' Lvs in whorls of 2–4 . . . . *G. proliferum*
1' Per or shrub
  5. Corolla gen 3-lobed; nutlets glabrous . . . . ²*G. trifidum* var. *pacificum*
  5' Corolla gen 4-lobed; nutlets long- or short- straight-hairy
    6. Pl with staminate, pistillate, and sometimes bisexual fls; woody at base only
      7. Lf ovate to orbicular, not sharp at tip; infl narrow, open between dense clusters . . . . *G. parishii*
      7' Lf linear to oblanceolate, acute or sharp at tip; infl open . . . . *G. wrightii*
    6' Pl with staminate or pistillate fls only; habit various
      8. Lf ± linear, gen widest above middle . . . . *G. angustifolium*
        9. Corolla hairy; DSon (Borrego Desert) . . . . ssp. *borregoense*
        9' Corolla gen glabrous or not more hairy than st or lf; DMtns . . . . ssp. *gracillimum*
      8' Lf gen ovate, gen widest below middle
        10. Pl woody at base and above; lf sharp at tip . . . . *G. stellatum* var. *eremicum*
        10' Pl woody only at base; lf sharp at tip or not
          11. Corolla gen bell-shaped or cupped at base
            12. Herbage glabrous
              13. Lvs at each node ± equal, lanceolate to ovate, tapered to a pointed tip; DMtns . . . . ²*G. argense*
              13' Lvs at each node ± unequal, the 2 larger ovate to orbicular, rounded abruptly to a pointed tip; GB
              . . . . ²*G. multiflorum*
           12' Herbage hairy, sometimes scabrous
             14. Lf rounded abruptly to a pointed tip, gen ± arched . . . . ²*G. multiflorum*
             14' Lf tapered to a pointed tip, gen not arched . . . . *G. hilendiae*
              15. Pl gen tall, stiff; internode 2–6 × lf; herbage ± hairy, gen scabrous; lf tip ± sharp to touch
              . . . . ssp. *carneum*

15' Pl gen ± low, flexible; internode 1–4.5 × lf; herbage with long hairs; lf tip not sharp
  16. Corolla throat wide; lf ovate to ± round . . . . ssp. *hilendiae*
  16' Corolla throat slender; lf lanceolate, ovate, or elliptic . . . . ssp. *kingstonense*
11' Corolla gen rotate, sometimes shallowly bell-shaped
  17. Pl glabrous . . . . [2]*G. argense*
  17' Pl hairy, at least in part
    18. Corolla glabrous or velvety with magnification . . . . *G. hypotrichium*
    19. Corolla gen rotate, greenish, yellow or pink-tinged; herbage velvety with magnification
    . . . . ssp. *hypotrichium*
    19' Corolla shallowly bell-shaped, reddish; herbage gray with ± dense fine hairs . . . . . ssp. *tomentellum*
    18' Corolla hairy
      20. Herbage hairy . . . . *G. munzii*
      20' Herbage glabrous below hairy infl
        21. Pl flexible; lf 8–23 mm, gen laanceolate to ovate, tip acute but not sharp to touch. . *G magnifolium*
        21' Pl stiff; lf 2–10 mm, ovate, tip sharp to touch . . . . *G. matthewsii*

G. angustifolium Nutt. NARROW-LEAVED BEDSTRAW Per, low, tufted or with elongate sts, woody at least at base, dioecious, glabrous to hairy. **ST** 6–100 cm. **LVS** in whorls of 4, ± strap-shaped, 3-veined. **INFL**: panicle, individual clusters with few–many fls. **FL**: corolla rotate, gen yellowish. **FR**: nutlets, gen > pedicel; hairs dense, long, straight, spreading. Mtn areas where roots sheltered; 15–2650 m. s SN, SCoRO, SW, **DMtns, DSon**; Baja CA. 8 sspp. in CA.

ssp. *borregoense* Dempster & Stebb. (pl. 98) BORREGO BED-STRAW Pl glabrous. **ST** 35–50 cm, slender; ridges ± covering surfaces between. **LF** < 8 mm. **INFL** many-fld, ± dense. **FL**: corolla hairs longer than those of herbage. RARE. Among boulders; 350–1250 m. **DSon (Palm Canyon, Hellhole Canyon, Pinyon Mtn Valley; San Diego Co.)**.

ssp. *gracillimum* Dempster & Stebb. Pl very slender throughout, gen glabrous exc lvs scabrous. **ST** ± 40 cm, woody; ridges ± as wide as surfaces between. **LVS** gen 4–15 mm, ephemeral. **INFL** few-fld, ± open. **FL**: corolla hairs gen 0. **FR**: nutlets, < 2 mm, > hairs. $2n=22$. Shaded places among granite boulders in canyons, on outcrops; 130–1550 m. **DMtns (Providence, Little San Bernardino mtns)**.

G. aparine L. (p.471) GOOSE GRASS Ann, climbing or prostrate, or sometimes low, erect; herbage adhesive by small, hooked prickles. **ST** 3–9 dm, weak, brittle. **LVS** in whorls of 6–8, 13–31 mm; lowest petioled, ± round; upper sessile, ± narrowly oblanceolate. **INFL**: fls few on branchlets in most axils. **FL** bisexual; corolla rotate, whitish. **FR**: nutlets; hairs many, short, hooked. $2n=20,22,42,44,63,64,66,\pm86,88$. Grassy, half-shady places, weedy in gardens; 30–1500 m. CA-FP, **SNE, DMoj**; to AK, e Coast, perhaps native to Eur. Mar–Jul

G. argense Dempster & Ehrend. Per, erect, 20–50 cm, stiff, open, dioecious, glabrous. **LVS** in whorls of 4, 8–17 mm, lanceolate to ovate, tapered to sharp tip. **INFL**: clusters terminal on branchlets in upper, ± lfless part of pl. **FL**: corolla ± bell-shaped to rotate, cream. **FR**: nutlets; hairs long, straight, spreading. $2n=44$. Loose, stony slopes; 1250–2400 m. **DMtns (Argus, Nelson ranges; Inyo Co.)**.

G. bifolium S. Watson (p. 471) LOW MOUNTAIN BEDSTRAW Ann, erect, gen 5–15 cm, glabrous; branches few or 0. **LVS** in whorls of 4, in 2 unequal pairs, or uppermost often opposite, 10–21 mm, lanceolate or narrowly elliptic. **INFL**: fls gen solitary on slender branchlets; pedicels nodding. **FL** bisexual; corolla white, lobes gen 3, ovate, ascending. **FR**: nutlets; hairs short, hooked. Open coniferous forest, gravelly slopes, meadows; 1500–3700 m. KR, NCoRI, SNH, WTR, MP, **n SNE (Bodie Hills), DMtns (Cottonwood Mtns)**; to B.C., MT, Colorado. Jun–Sep

G. hilendiae Dempster & Ehrend. HILEND'S BEDSTRAW Per, dioecious, hairy, sometimes scabrous; base woody. **ST** 13–38 cm. **LVS** in whorls of 4, lance-ovate to orbicular. **INFL**: panicle, ± terminal, leafy. **FL**: corolla ± bell-shaped, gen ± pink. **FR**: nutlets; hairs long, straight. Rocky places; 1200–3400 m. **DMoj**; c NV.

ssp. *carneum* (Hilend & J. Howell) Dempster & Ehrend.

PANAMINT MOUNTAINS BEDSTRAW **ST** ± erect, gen 13–38 cm, stiff; internodes 2–6 × > lvs. **LF** lanceolate to ovate, tip tapered to acute or sharp to touch. **FL**: corolla throat open. $2n=44$. UNCOMMON. Rocky slopes, open flats; 1650–3400 m. **DMtns (Panamint Mtns)**. [*G. munzii* var. *c*. Hilend. & Howell] May–Jul

ssp. *hilendiae* Pl hairs many, long. **ST** 13–33 cm, not stiff; internodes 1.5–4.5 × > lvs. **LF** ovate to ± round; tip tapered to acute but not sharp to touch. **FL**: corolla throat open. Dry canyons, rocky places; 1300–2700 m. **DMoj**.

ssp. *kingstonense* (Dempster) Dempster & Ehrend. KINGSTON MOUNTAINS BEDSTRAW Pl often matted; hairs many, long. **ST** < 35 cm, slender, weak; internodes 2–4 × > lvs. **LF** lance-elliptic to ovate; tip obtuse or acute but not sharp to touch. **FL**: corolla pink, bell-shaped, throat slender, ± = lobes. $2n=44$. RARE. Rocky places; 1200–2100 m. **DMtns (Kingston Mtns, ne San Bernardino Co.)**; c NV. [*G. munzii* var. *k*. Dempster]

G. hypotrichium A. Gray ALPINE BEDSTRAW Per, low or ± dwarfed, dioecious; base ± woody. **ST** 2.5–23 cm. **LVS** in whorls of 4, ± crowded, ovate to round, gen ± fleshy; tip obtuse or acute but not sharp to touch. **INFL** leafy; fls few on branchlets. **FL**: corolla rotate or ± bell-shaped. **FR**: nutlets; hairs long, straight, yellowish. Talus, rocky places; 1950–4200 m. c&s SNH, **SNE, DMtns**. 4 sspp. in CA.

ssp. *hypotrichium* (p. 471) Pl densely branched, velvety with magnification. **ST** 2.5–12 cm. **LF** 3–8 mm, ovate to nearly round. **FL**: corolla ± rotate, pale yellow or pink-tinged. $2n=22,44$. Ridges, talus; 3000–4200 m. c&s SNH, **SNE (White, Sweetwater mtns)**. May–Aug

ssp. *tomentellum* Ehrend. TELESCOPE PEAK BEDSTRAW Pl ± densely branched, gray; hairs dense, fine. **ST** < 9 cm. **LF** 5–9 mm, ovate. **FL**: corolla shallowly bell-shaped, reddish. $2n=22$. RARE. Talus; 3300–3550 m. **DMtns (Telescope Peak, Panamint Mtns)**. Jun–Aug

G. magnifolium (Dempster) Dempster Per, erect, 10–45 cm, dioecious, glabrous exc upper bracts; base woody. **LVS** in whorls of 4, 8–23 mm, gen lanceolate to ovate; tip tapered to acute. **INFL**: panicle, open; fls solitary or in clusters in axils. **FL**: corolla rotate, whitish, with long hairs outside. **FR**: nutlets; hairs few, short, straight. Rocky slopes in juniper belt; 800–2000 m. **DMtns (Clark Mtn)**; NV, UT.

G. matthewsii A. Gray (p. 471) MATTHEWS' BEDSTRAW Per, erect, 13–30 cm, open, dioecious, glabrous exc upper bracts; base woody. **STS** gen many, slender, tangled or not; internodes 2–7 × > lvs. **LVS** in whorls of 4, 2–10 mm, ovate, leathery; tip sharp. **INFL**: panicle; fls in terminal clusters on branches. **FL**: corolla rotate, yellowish, with long hairs outside. **FR**: nutlets; hairs long, straight. $2n=22$. Dry slopes, washes; 1100–3000 m. SNH, **W&I, DMoj**. May–Aug

G. multiflorum Kellogg (p. 471) KELLOGG'S BEDSTRAW Per, erect, 15–35 cm, dioecious, glabrous or hairy; base woody. **LVS**

in whorls of 4, in 2 unequal pairs; the larger 8–15 mm, broadly ovate, 3-veined, base round, tip ± obtuse but often with a small point. **INFL**: panicle of few-fld clusters, terminal. **FL**: corolla ± bell-shaped, whitish or pinkish. **FR**: nutlets; hairs long, straight. $2n=22$. Rocky places among sagebrush; 1300–2900 m. n SNH, **GB, DMtns**; NV. [f. *hirsutum* (A. Gray) Ehrend.; *G. watsonii* (A. Gray) A.A. Heller] May–Aug

G. munzii Hilend & J. Howell (p. 471) MUNZ'S BEDSTRAW Per, erect, 10–30 cm, dioecious; hairs coarse; base woody. **ST** flexible. **LVS** in whorls of 4, in 2 unequal pairs; the larger 6–19 mm, gen ovate, ± 3-veined, tip tapered, ± sharp to touch. **INFL**: panicle, open; clusters axillary, few-fld. **FL**: corolla rotate, whitish or pink, hairy outside. **FR**: nutlets; hairs long, straight. $2n=44$. UNCOMMON. Cool n- or e-facing slopes, shady canyon bottoms; 1100–2250 m. **DMtns**; to s UT. May–Jul

G. parishii Hilend & J. Howell (p. 471) PARISH'S BEDSTRAW Per, erect or matted, gen velvety; base woody. **ST** 5–40 cm. **LVS** in whorls of 4, in 2 unequal pairs, 1–9 mm, ovate to orbicular, gen 3-veined. **INFL** open between dense clusters; pedicel ± 0. **FL** bisexual or unisexual (sometimes on individual pls); corolla rotate, gen red or pink, sometimes yellow. **FR**: nutlets; hairs long, straight. $2n=22$. Steep slopes, talus, at base of boulders; 1675–3400 m. SnGb, SnBr, SnJt, **DMtns**; s NV. Jun–Aug

G. proliferum A. Gray DESERT BEDSTRAW Ann, erect, 5–30 cm, ± hairy, often scabrous. **LVS** in whorls of 2–4, 3–10 mm, gen lanceolate, sessile (lowermost broad, petioled); tip obtuse or abruptly soft-pointed. **INFL**: fls gen solitary in axils, subsessile. **FL** bisexual; corolla rotate, yellowish. **FR**: nutlets; hairs short, hooked. Very uncommon. Rocky banks, limestone ledges; 1100–

1400 m. **DMtns**; to AZ, Mex. Apr–May

G. stellatum Kellogg var. *eremicum* Hilend & J. Howell (p. 471) Shrub, ± erect, 30–75 cm, dioecious, scabrous. **ST** stout, brittle; branches many, spreading. **LVS** in whorls of 4, gen 4–8 mm, lanceolate to needle-like, gray-green; tip sharp to touch. **INFL**: panicle, axillary, leafy; fls few–many. **FL**: corolla rotate, whitish. **FR**: nutlets; hairs dense, long, straight, white. $2n=22,44$. Rocky slopes; 130–1600 m. **D**; to NV, nw Mex. Mar–Apr

G. trifidum L. (p. 477) Per, rarely ann, minutely scabrous. **ST** 10–50 cm, slender, weak, tangled. **LVS** in whorls of 4–6, 4–19 mm, linear to elliptic or narrowly ovate-obovate, petioled; tip rounded. **INFL**: few-fld clusters; pedicel slender. **FL** bisexual; corolla rotate, white or pinkish, gen 3-lobed. **FR**: nutlets, spheric, hard, smooth, black when dry; hairs 0. Wet places; < 2800 m. CA-FP, **GB**; to AK, Mex. 2 vars. in CA.

var. *pacificum* Wieg. Pl straggling. **ST** 10–50 cm. $2n=24$. Wet places; < 2400 m. CA-FP, **GB**; to AK, Mex. [*G. cymosum* Wieg.; *G. trifidum* L. var. *subbiflorum* Wieg.] ❀ WET:**4–6,17,24**&SHD:2,3,**7,8,9,14–16,18–23**. Jul–Aug

G. wrightii A. Gray WRIGHT'S BEDSTRAW Per, erect or straggling, ± dioecious, glabrous exc just above base; base woody. **ST** 15–50 cm. **LVS** in whorls of 4, < 8 mm, gen linear to oblanceolate, attachment broad; tip acute or ± sharp to touch. **INFL**: panicle, terminal, open. **FL** bisexual or unisexual on individual pls; corolla rotate, gen red or pink. **FR**: nutlets; hairs few, long, straight. $2n=22$. RARE in CA. Shady canyons among rocks; 1600–2000 m. **DMtns (Clark Mtn)**; to NM, w TX, Mex (Coahuila, Sonora, Baja CA). May–Jun, Aug–Sep

RUTACEAE RUE FAMILY

James R. Shevock

Per, shrubs, trees, very aromatic, sometimes thorny. **LVS** gen alternate, simple to pinnately compound (sometimes reduced to spines), prominently oil-gland-dotted; stipules 0. **INFL**: cyme, raceme, or fls solitary, gen bracted. **FL** gen bisexual, gen strongly aromatic; sepals gen 5, free or fused at base, gen persistent; petals gen 5, free or fused at base, gen whitish or greenish; stamens gen 2–4 × petal number; ovary gen superior, gen lobed, chambers gen 4–5, ovules gen many. **FR**: berry, drupe, winged achene, or capsule, gen aromatic. **SEEDS** gen oily. ± 150 genera, ± 1500 spp.: esp trop, warm temp, esp s Afr, Australia; used or cult for food (*Citrus*, 50 spp.), perfume, medicine, timber, orn (*Choisya, Skimmia*, etc). Some TOXIC: oils may promote localized sunburn or produce dermatitis.

THAMNOSMA TURPENTINE-BROOM

Subshrub or shrub. **LF** very small, ephemeral. **INFL**: panicle (raceme-like or fls scattered along sts). **FL** bisexual; calyx persistent, 4-lobed; petals 4, erect in fl; stamens 8, in 2 series; ovary stalked, deeply 2-lobed, style thread-like. **FR**: capsule, deeply 2-lobed, leathery, opening at tip. 6 spp.: sw N.Am, s Afr. (Greek: bush odor)

T. montana Torrey & Frémont (p. 477, pl. 99) **ST** 3–6 dm, broom-like, yellowish green, thickly covered with blister-like glands, gen lfless. **FL**: sepals ± 2 mm, ± round, greenish; petals 8–12 mm, ± elliptic, ± leathery, purplish, tip rolled out; style ± exserted, ovules 8–9 per chamber. **FR**: lobes ± 5 mm thick, ± spheric. **SEEDS** 1–3, ± 4 mm, reniform, whitish. Dry slopes; < 700 m. **D**; to NM, Mex. ❀DRN,DRY:**8–13,18–21**&SUN:7,**14–16,22–24**;DFCLT. Mar–May

SALICACEAE WILLOW FAMILY

Shrub, tree, gen dioecious (rarely monoecious). **ST**: trunk < 40 m; wood soft; bark smooth, bitter; buds scaly. **LVS** simple, alternate, deciduous; stipules gen deciduous, often large. **INFL**: catkin, gen appearing before lvs; each fl subtended by disk or 1–2 nectary glands and 1 bract. **FL**: perianth 0. **STAMINATE FL**: stamens 1–many. **PISTILLATE FL**: pistil 1, ovary superior, chamber 1, stigma lobes 2–4. **SEEDS** many; hairs fine, white, cottony. **FR**: capsule; valves 2–4. 2 genera, 340 spp.: gen temp (exc Australia, Malay Archipelago) moist places; many cult. Hybridization common; identification often difficult. Family description, key to genera by John O. Sawyer, Jr.

1. Lf length not >> (often ± =) width; catkin pendent, bract cut into narrow segments; staminate fl with
    8–many stamens; pistillate fl with cup-like disk; winter bud scales > 3 . . . . **POPULUS**
1' Lf length gen >> width; catkin erect, bract entire; staminate fl with 1–8 stamens; pistillate fl with 1–2
    glands; winter bud scale 1 . . . . **SALIX**

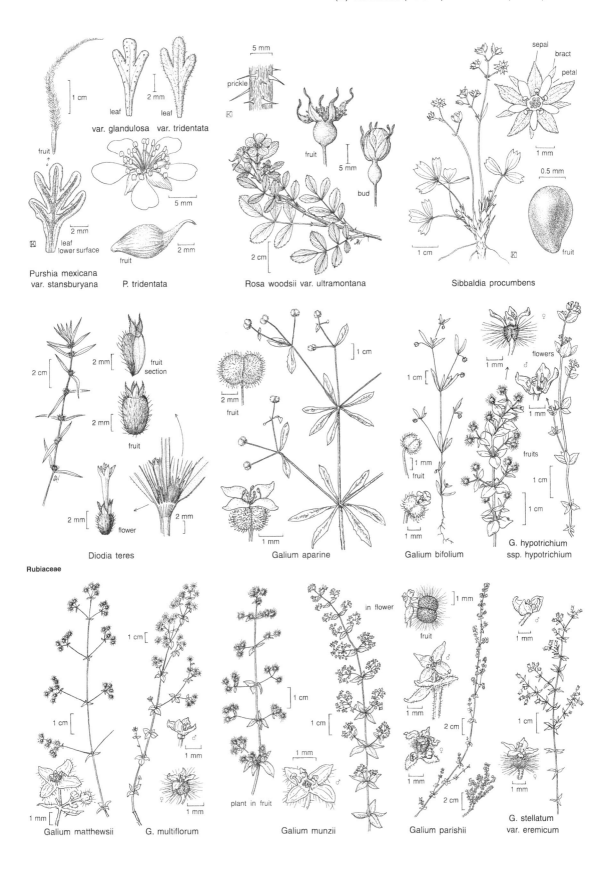

Purshia mexicana
var. stansburyana

P. tridentata

Rosa woodsii var. ultramontana

Sibbaldia procumbens

Diodia teres

Galium aparine

Galium bifolium

G. hypotrichium
ssp. hypotrichium

Rubiaceae

Galium matthewsii

G. multiflorum

Galium munzii

Galium parishii

G. stellatum
var. eremicum

POPULUS POPLAR, COTTONWOOD, ASPEN

John O. Sawyer, Jr.

Tree. **ST**: trunk < 40 m; young bark smooth, pale yellow-green to gray; older bark furrowed, brown to gray; twigs with swellings below lf scars; winter bud gen resinous, scales > 3. **LVS** gen glabrous (juvenile lvs may differ from adult lvs); blade 3–11 cm, elliptic to triangular, veins pinnate or subpalmate, tip gen elongate. **INFL**: catkin pendent, 3–8 cm; bract cut into narrow segments; fls sessile, on a cup- or saucer-like disk. **STAMINATE FL**: stamens 8–60. **PISTILLATE FL**: style short, stigmas 2–3(4), large, scalloped to 2-lobed. **FR** spheric to conic; valves 2–3(4), 3–12 mm. 40 spp.: n hemisphere. (Latin: name for pls of this genus)

1. Lf blade white-tomentose below .... *P. alba*
1' Lf blade glabrous or hairy below
  2. Lf blade lanceolate .... *P. angustifolia*
  2' Lf blade ovate, nearly round, or triangular
    3. Lf blade triangular, margin coarsely scalloped............................... *P. fremontii* ssp. *fremontii*
    3' Lf blade ovate to nearly round, margin finely scalloped
      4. Lf blade narrowly to widely ovate, 3–7 cm; petiole 1/3–1/2 blade length, round in ×-section
      .... *P. balsamifera* ssp. *trichocarpa*
      4' Lf blade widely ovate to nearly round, gen 2–4 cm, petiole 2/3 to = blade length, laterally compressed
      in ×-section .... *P. tremuloides*

P. alba L. WHITE POPLAR **ST**: crown wide; trunk < 20 m; twigs and winter buds white-tomentose. **LF**: petiole 1/3–1/2 blade length; blade 3–9 cm, 3–5-lobed, base rounded to slightly cordate, tip acute, margin entire to toothed, upper surface blue-green, glossy, lower white-tomentose. Disturbed places near settlements; 600–1800 m. KR, CaR, **GB**, expected elsewhere; native to c Eur, c Asia. Persisting primarily by clonal root-sprouting.

P. angustifolia James (p. 477) NARROW-LEAVED COTTONWOOD **ST**: crown slender; trunk < 15 m; twigs glabrous; winter buds very gummy. **LF**: petiole 1/3 blade length, lower side round, upper widely channeled; blade 4–11 cm, lanceolate, base wedge-shaped, tip acute or tapered, margin finely scalloped, surfaces yellow-green, glabrous. 2*n*=38. RARE in CA. Streamsides; 1200–1800 m. **SNE**; to OR, Rocky Mtns, n Mex. [*P.* ×*acuminata* Rydb. misapplied] In cult.

P. balsamifera L. ssp. *trichocarpa* (Torrey & A. Gray) Brayshaw (p. 999) BLACK COTTONWOOD **ST**: crown wide; trunk < 30 m; twigs brown, becoming gray; winter buds finely ciliate, very resinous, fragrant when opening. **LF**: petiole 1/3–1/2 blade length, lower side round, upper channeled, with pair of glands at junction with blade; blade 3–7 cm, narrowly to widely ovate, base round to cordate, tip acute to tapered, margin finely scalloped, upper surface green, lower glaucous, often stained with brown resin. 2*n*=38. Scattered. Alluvial bottomlands, streamsides; < 2800 m. CA-FP, **GB**; to AK, n Rocky Mtns, UT, n Baja CA. [*P. t.* Torrey & A. Gray] [Brayshaw 1965 Canad Field-

Naturalist 79:91–95] ❀ IRR or WET,SUN:1,**2–7**,8,9,14,**15–18**, 19–23,**24**; susceptible to galls. Feb–Apr

P. fremontii S. Watson ssp. *fremontii* (p. 477) ALAMO or FRE-MONT COTTONWOOD **ST**: crown wide; trunk < 20 m; twigs yellow, becoming gray, glabrous to hairy; winter buds resinous. **LF**: petiole 1/2 to = blade length, laterally compressed near blade; blade 3–7 cm, deltate, base slightly cordate to flat, tip ± tapered, margin coarsely scalloped, surfaces yellow-green, glabrous to hairy, often stained with milky resin. Scattered. Alluvial bottomlands, streamsides; < 2000 m. **CA (exc MP)**; to c Rocky Mtns, n Mex. [var. *arizonica* (Sarg.) Jepson, var. *macdougalii* (Rose) Jepson] [Eckenwalder J Arnold Arbor 58:193–208] ❀ IRR or WET, SUN:1,**2,3**, 4–6,**7–24**; susceptible to mistletoe; INV;STBL. Mar–Apr

P. tremuloides Michaux (p. 477) QUAKING ASPEN **ST**: crown slender; trunk < 15 m, highly clonal; twigs greenish white, glabrous; winter buds shiny. **LF**: petiole 2/3 to = blade length, laterally compressed; blade 2–4(7) cm, widely ovate or wider, base rounded to cordate, tip tapered, margin finely scalloped, surfaces glabrous, upper green, lower glaucous. 2*n*=38. Streamsides, moist openings and slopes in montane and subalpine forests, woodlands, sagebrush steppe; 1800–3000 m. KR, NCoRH, CaR, SNH, SnBr, **GB**; to AK, e N.Am, Mex. ❀IRR,SUN,DRN: **1,2**,3,**4–7**,14–17,**18**,19–21;INV; also STBL. Apr–Jun

SALIX WILLOW

George W. Argus

Shrub, tree, dioecious; bud scale 1, not sticky, margins gen fused (or free, overlapping). **ST**: twigs gen flexible and not glaucous. **LF**: blade linear to widely obovate, entire to toothed, gen ± hairy. **INFL**: dense catkin emerging before, with, or after lvs, sessile or on a short leafy shoot; bract subtending each fl. **FL**: perianth 0. **STAMINATE FL**: stamens 1–8. **PISTILLATE FL**: ovary stalked or sessile, style 1 or 0, stigmas 2, each sometimes 2-lobed; nectaries 1–several, gen rod-like, gen between infl axis and fl. **FR**: valves 2. ± 400 spp.: ± worldwide, esp n temp, arctic. (Latin: ancient name) [Argus 1986 Syst Bot Monog 9:1–170; Dorn 1976 Canad J Bot 54:2769–2789] Difficult, highly variable. Not all specimens will key easily; sprout shoots and other extreme forms are not incl in keys, may require field comparison for identification. Studies of variation, hybridization needed.

1. Pl vegetative (without reproductive structures) .... **Key 1**
1' Pl with reproductive structures
  2. Pl pistillate .... **Key 2**
  2' Pl staminate .... **Key 3**

Key 1 (vegetative plants)

1. Tree; bud scale margin free and overlapping
  2. Lf blade not glaucous below; twigs yellowish or yellow-gray . . . . **S. goodddingii**
  2' Lf blade glaucous below; twigs yellow-brown or red-brown . . . . **S. laevigata**
1' Shrub or tree; bud scale margin fused
  3. Petiole with glands at base of blade
    4. Mature lf blade shiny to highly glossy above
      5. Lf blade apex long-acuminate . . . . **S. lucida**
      5' Lf blade apex acute
        6. Twigs gray- to red-brown; lf blades soft-shaggy-hairy above, becoming glabrous . . . . [2]**S. boothii**
        6' Twigs yellowish, yellow-brown, or violet; lf blades tomentose or long silky-hairy above, becoming
          glabrous . . . . [3]**S. eastwoodiae**
    4' Mature lf blade dull above
      7. Shrub < 4 m; lvs not glaucous below; twigs erect, flexible or sometimes brittle at base; petiole
        glabrous or long-soft-wavy-hairy . . . . [3]**S. eastwoodiae**
      7' Tree 7–25 m; lvs glaucous below; twigs pendent or erect, brittle at base; petiole tomentose or silky-hairy
        8. Stems erect, spreading, or pendent; twigs brownish to golden-yellow; twigs and young lvs densely
          silky-hairy; petiole silky-hairy; stipules acute at tip . . . . **S. alba**
        8' Stems long-pendent; twigs gray- to yellow-brown; twigs and young lf blade sparsely long-soft,
          -wavy-hairy, becoming glabrous; petiole tomentose; stipules acuminate at tip . . . . [2]**S. babylonica**
  3' Petiole lacking glands at base of blade
    9. Lf blade not glaucous below
      10. Mature lf blades soft-shaggy-hairy, becoming glabrous, shiny to highly glossy above; young lvs
        shaggy-hairy . . . . [2]**S. boothii**
      10' Mature lf blades tomentose, silky-hairy, becoming glabrous, dull above; young lvs densely tomentose
        or silky-hairy . . . . [3]**S. eastwoodiae**
    9' Lf blade glaucous below
      11. Lvs dull above
        12. Trees 7–25 mm, twigs long-pendent, brittle at base; stipule tip ± long-acuminate . . . . [2]**S. babylonica**
        12' Trees or shrubs 0.2–9 m, twigs erect or decumbent, flexible at base; stipule acute to rounded at tip
          13. Lvs glabrous to sparsely hairy below . . . . [2]**S. lutea**
          13' Lvs silky-hairy below . . . . **S. orestera**
      11' Lvs shiny or highly glossy below
        14. Lf blades linear, length 10–38× width, margins entire to sharply serrate, teeth well-separated . . **S. exigua**
        14' Lf glades wider, length 3–11× width; margins entire, gland-dotted, remotely serrate, finely serrate,
          or crenate
          15. Pl sometimes prostrate, < 1 m; mature lf blade 20–43 mm; 2500–4000 m . . . **S. planifolia** ssp. *planifolia*
          15' Pl erect, 1.2–10 m; mature lf blade 35–125 mm; gen 900–3600 m
            16. Stipules 0
              17. Twigs often glaucous; petiole 2–9 mm long; hairs on lower lf surface straight . . . . [2]**S. geyeriana**
              17' Twigs not glaucous; petiole 3–16 mm long; hairs on lower leaf surface wavy . . . . [2]**S. lasiolepis**
            16' Stipules lf-like or vestigial
              18. Stipule apex obtuse to rounded; lf white-hairy; petiole glabrous or sparsely hairy . . . . [2]**S. lutea**
              18' Stipule apex acute; lf white or white-and-rusty-hairy; petiole densely hairy
                19. Twigs glaucous
                  20. Lf blade 32–74 mm, persistently long-hairy; petiole 3–9 mm; stipules 0 or vestigial . . . [2]**S. geyeriana**
                  20' Lf blade 44–110 mm, becoming glabrous or sparsely short-hairy; petiole 5–16 mm;
                    stipules vestigial on young lvs, lf-like on mature lvs . . . . [2]**S. lemmonii**
                19' Twigs not glaucous
                  21. Twigs long-soft-wavy-hairy, becoming glabrous; young lvs entire to finely serrate . . . [2]**S. lasiolepis**
                  21' Twigs gen glabrous; young lvs entire . . . . [2]**S. lemmonii**

Key 2 (pistillate plants)

1. Ovary hairy
  2. Fl bract deciduous, tawny, sparsely to moderately hairy
    3. Slender shrub, clonal by root-sprouting; bud scale margins fused; lf blade margins entire to finely
      serrate (teeth well separated); petiole glands 0; ovary stalk 0.2–0.9 mm . . . . [2]**S. exigua**
    3' Tree, not clonal by root-sprouting; bud scale margins free and overlapping; lf blade margins
      entire to finely serrate; petiole glands present; ovary on stalk 1.2–3.2 mm . . . . [2]**S. gooddingii**
  2' Fl bract persistent, brown or tawny, gen moderately hairy
    4. Twigs gen conspicuously glaucous
      5. Catkins subspheric, 6–20 mm; styles 0.1–0.4 mm long; young lvs rusty- or white- and rusty hairy;
        twigs gen conspicuously glaucous; stipules gen 0 . . . . **S. geyeriana**

5' Catkins cylindric, 13–50 mm; styles 0.3–1.0 mm; young lvs white-hairy; twigs gen not glaucous or
    only thinly so; stipules lf-like or vestigial . . . . *S. lemmonii*
4' Twigs gen not glaucous, rarely thinly so
  6. Catkins appearing before lvs; petiole glabrous; stipules 0 or vestigial . . . . *S. planifolia* ssp. *planifolia*
  6' Catkins appearing after the lvs; petiole hairy; stipules lf-like
    7. Lf blade hairy becoming glabrous, not glaucous below, base rounded, rarely acute . . . . *S. eastwoodiae*
    7' Lf blade silky-hairy, glaucous below (surface obscured by hairs), base acute. . . . . *S. orestera*
1' Ovary glabrous
  8. Fl bract deciduous after fl, tawny, sparsely hairy, tip often ragged
    9. Margins of bud scale free and overlapping; tree
      10. Lf blade not glaucous below; twigs yellowish; lf margins on leafy catkin shoots finely serrate;
        young lvs white-hairy; ovary beak abruptly tapering to style . . . . ²*S. gooddingii*
      10' Lf blade glaucous below; twigs red- to yellow-brown; lf margins on leafy catkin shoots entire;
        young lvs white- or white-and-rusty-hairy; ovary beak gradually tapering to style . . . . *S. laevigata*
    9' Margins of bud scale fused; tree or shrub
      11. Petiole glands 0 . . . . ²*S. exigua*
      11' Petiole with glands near base of blade
        12. Tree 10–25 m; stipules not conspicuously glandular; fr 3.5–5 mm long; twigs yellowish to
          gray- or red-brown . . . . *S. alba*
        12' Tree or shrub < 10 m; stipules conspicuously glandular; fr 6–11 mm; twigs brownish . . . . *S. lucida*
  8' Fl bract persistent after fl, tawny to black, gen hairy, tip entire
    13. Tree; twigs strongly pendent; ovary stalk < nectary. . . . . *S. babylonica*
    13' Shrub; twigs erect; ovary stalk > nectary
      14. Catkins appearing before the lvs, sessile
        15. Margins of young lvs on vegetative shoots or at base of catkins glandular-dotted or finely serrate
          . . . . ²*S. boothii*
        15' Margins of young lvs on vegetative shoots or at base of catkins entire, not glandular . . . . *S. lasiolepis*
      14' Catkins appearing with the lvs, on distinct leafy shoots
        16. Young lvs hairy; styles 0.3–1.0 mm long; lvs not glaucous below . . . . ²*S. boothii*
        16' Young lvs glabrous; styles 0.2–0.6 mm long; lvs glaucous below. . . . . *S. lutea*

Key 3 (staminate plants)

1. Stamens > 2 per fl
  2. Margins of bud scales fused; anthers 0.7–1 mm; petiole gen glandular-lobed at base of blade . . . . *S. lucida*
  2' Margins of bud scales free and overlapping; anthers 0.4–0.6 mm; petiole only glandular-dotted at
    base of blade
    3. Lf blade not glaucous below; twigs yellowish to yellow-gray; margins of lvs on catkin shoots
      finely serrate; young lvs white-hairy . . . . *S. gooddingii*
    3' Lf blade glaucous below; twigs yellow- or red-brown; margins of lvs on catkin shoots entire; young
      lvs glabrous or white- or white-and-rusty-hairy. . . . . *S. laevigata*
1' Stamens 2 per fl
  4. Tree; twigs brittle at base
    5. Sts erect or weakly pendent, twigs gen yellow, sometimes gray- or red-brown; twigs, young lvs
      densely silky-hairy . . . . *S. alba*
    5' Sts pendent; twigs yellow-brown, long-soft-wavy-hairy becoming glabrous; young lvs glabrous or
      sparsely silky-hairy. . . . . *S. babylonica*
  4' Shrubs; twigs flexible at base
    6. Catkins appearing before lvs, sessile or subtended by greenish bracts
      7. Fl bracts tawny to light brown
        8. Stipules 0 or vestigial; catkin 6–12 mm; filaments free; anther 0.4–0.5 mm. . . . . ⁴*S. geyeriana*
        8' Stipules lf-like (vestigial on young lvs); catkin 13–35 mm; filaments free or fused at base; anther
          0.5–0.9 mm. . . . . ³*S. lemmonii*
      7' Fl bracts dark brown to blackish
        9. Lf blade not glaucous below. . . . . *S. boothii*
        9' Lf blade glaucous below
          10. Pl < 1 m, low, often prostrate; petiole glabrous; mature lf blade 20–45 mm . . *S. planifolia* ssp. *planifolia*
          10' Pl 1.2–10 m, erect; petiole tomentose or velvety; mature blade 35–125 mm
            11. Fl bract widest at its broadly rounded tip, densely hairy with short, wavy or straight, ± equally
              long hairs . . . . *S. lasiolepis*
            11' Fl bract widest at middle, sparsely to moderately hairy with straight hairs of unequal length
              . . . . ³*S. lemmonii*
    6' Catkins appearing with lvs, on distinct, leafy shoots
      12. Young lvs rusty- or white-and-rusty hairy
        13. Catkins subspheric, 6–20 mm; fl bract tawny to light brown; anthers 0.4–0.5 mm long; stipules 0
          . . . . ⁴*S. geyeriana*

13' Catkins cylindric 13–50 mm; fl bract dark brown; anthers 0.5–0.9 mm long; stipules lf-like or
     vestigial . . . . **³S. lemmonii**
12' Young lvs white-hairy only
  14. Filaments glabrous
    15. Twigs gen glaucous; stipules 0, stipule scars none or minute . . . . **⁴S. geyeriana**
    15' Twigs not glaucous; stipules lf-like, or at least stipule scars distinct
      16. Catkins plump, 16–45 mm; margins of young lvs gland-dotted; lf blade not glaucous below; twigs
          violet to brownish . . . . **²S. eastwoodiae**
      16' Catkins slender, 15–75 mm; margins of young lvs finely serrate; lf blade glaucous below; twigs
          yellowish to yellow-gray . . . . **S. lutea**
  14' Filaments hairy
    17. Catkins subspheric, 6–20 mm; stipules 0 . . . . **⁴S. geyeriana**
    17' Catkins cylindric, 12–70 mm; stipules gen lf-like (0 or vestigial)
      18. Fl bract tawny; nectaries 2; lvs linear, 2–8 mm wide, base wedge-shaped; stipules 0 or vestigial
          . . . . **S. exigua**
      18' Fl bract brown; nectary 1; lvs narrowly oblong, narrowly elliptic or elliptic to oblanceolate;
          7.5–37 mm wide, base acute to rounded; stipules lf-like
        19. Lf blade not glaucous below; lvs on catkin shoots short, slender, finely serrate (rarely only
            gland-dotted) . . . . **²S. eastwoodiae**
        19' Lf blade glaucous below (surface obscured by hair); lvs on catkin shoots entire or gland-dotted
            . . . . **S. orestera**

S. alba L. WHITE WILLOW Tree < 25 m. **ST** erect, spreading, or pendent; twigs yellowish to gray- or red-brown, sometimes ± brittle at base, silky, becoming glabrous. **LVS:** stipules lf-like; petiole with glands; young lvs silky; mature blade 63–115 mm, lanceolate to elliptic, acute (base acute or wedge-shaped), finely serrate, silky, becoming glabrous, lower surface glaucous, upper surface dull. **INFL** appearing with lvs, 30–60 mm, on leafy shoot 3–25 mm; fl bract tawny, glabrous or sparsely hairy, tip rounded; pistillate bract deciduous after fl. **STAMINATE FL:** stamens 2; axillary and outer nectaries present. **PISTILLATE FL:** ovary glabrous, style ± 0.2 mm, stalk 0.2–0.8 mm. 2*n*=76. Disturbed places, gen near settlements; gen < 1000 m. **CA;** to e N.Am; native to Eur. Mostly cult as orn, but many CVS and hybrids ± naturalized. Var. *vitellina* (L.) Stokes (twigs bright yellow or golden, spreading) most common.

S. babylonica L. (p. 477) WEEPING WILLOW Tree < 20 m. **ST** pendent ± to ground; twigs gray- to yellow-brown, long-soft-wavy-hairy, becoming glabrous, brittle at base. **LVS:** stipules lf-like; petiole gen with glands; young lvs glabrous or sparsely silky; mature blade 70–140 mm, linear-lanceolate, acuminate (base wedge-shaped), gen finely sharp-serrate, silky, becoming glabrous, lower surface glaucous. **INFL** appearing with lvs, 18–35 mm, on leafy shoots 4–10 mm (staminate ± sessile); fl bract tawny. **STAMINATE FL:** stamens 2. **PISTILLATE FL:** ovary glabrous, stigma shrub-like, style ± 0.2 mm, stalk 0–0.3 mm. 2*n*=76. Disturbed places, around settlements; gen < 1000 m. **CA;** to US; native to Asia. Mostly cult as orn, but many CVS and hybrids ± naturalized, incl hybrids with *S. alba*

S. boothii Dorn (p. 477) BOOTH'S WILLOW Shrub < 6 m. **ST:** twigs gray- to red-brown, glabrous or shaggy-hairy. **LVS:** petiole gen with glands; young lvs shaggy-hairy; mature blade 26–102 mm, lanceolate to widely elliptic, acute (base rounded to acute), entire to finely serrate, soft-shaggy-hairy, becoming glabrous, upper surface shiny, lower surface not glaucous. **INFL** appearing before or with lvs, 10–70 mm, sessile or on leafy shoots < 15 mm (margins of shoot-lvs gland-dotted or finely serrate); fl bract gen dark brown, hairs wavy. **STAMINATE FL:** stamens 2. **PISTILLATE FL:** ovary glabrous, style 0.3–1 mm, stalk 0.5–2.5 mm. 2*n*=76. Uncommon. Wet subalpine meadows, shores; 1700–3200 m. KR, CaRH, n SNH, Wrn, **W&I;** to w Can, Colorado. [*S. pseudocordata* Andersson; *S. myrtillifolia* Andersson misapplied] Jun–Jul

S. eastwoodiae A.A. Heller (p. 477) SIERRA WILLOW Shrub < 4 m. **ST:** twigs yellowish or red-brown to violet, glaucous or not, shaggy-hairy, becoming glabrous. **LVS:** stipules lf-like, early-deciduous; petiole gen with glands, glabrous or long-hairy; young lvs hairy; mature blade 27–99 mm, lanceolate to widely elliptic or oblanceolate, acute (base rounded to acute), entire to finely short-slender-serrate, hairy, becoming glabrous, not glaucous below. **INFL** appearing with lvs, 16–45 mm, on leafy shoots 2–20 mm; margin of shoot lvs finely short-slender-serrate or gland-dotted; fl bract brown or tawny. **STAMINATE FL:** stamens 2. **PISTILLATE FL:** ovary silky, style 0.6–1.5 mm, stalk 0.2–1.6 mm. 2*n*=76. Alpine and subalpine meadows, streams, talus; 1600–3800 m. KR, CaRH, SNH, **W&I;** to OR, MT, WY. [*S. commutata* Bebb incl var. *denudata* Bebb, misapplied] ❀ WET:1–6,18. Jun–Jul

S. exigua Nutt. (p. 477) NARROW-LEAVED WILLOW Shrub < 7 m, clonal by root-sprouting; **ST:** twigs brownish, silky or long-soft-wavy-hairy, becoming glabrous. **LVS:** stipules 0 or vestigial; petiole short; young lvs silky; mature blade 50–124 mm, linear, acuminate (base wedge-shaped), entire to sharply serrate (teeth well separated), silky, glaucous below or not. **INFL** appearing with or after lvs, sometimes branched, 22–70 mm, on leafy shoots 5–110 mm; fl bract tawny; pistillate bract deciduous after fl. **STAMINATE FL:** stamens 2. **PISTILLATE FL:** ovary glabrous, silky, or sparsely soft-shaggy-hairy, stigma deciduous after fl, style 0–0.2 mm, stalk 0.2–1 mm. 2*n*=38. Common. Streamsides, marshes, wet ditches; < 2700 m. **CA;** to AK, e N.Am, AZ, Mex. [var. *stenophylla* (Rydb.) C. Schneider; *S. hindsiana* Benth. incl vars. *leucodendroides* (Rowlee) C. Ball and *parishiana* (Rowlee) C. Ball] Pls with spreading hairs on lvs and twigs, slender stigmas 0.6–1 mm, and ± entire lvs from throughout CA have been called *S. hindsiana*; these features vary independently; the type of *S. h.* does not share them all.. ❀ WET,SUN:1–5,6–24;STBL. Mar–May

S. geyeriana Andersson (p. 477) GEYER'S WILLOW Shrub < 5 m. **ST:** twigs yellowish to brownish, gen glaucous, tomentose or velvety, becoming glabrous, sometimes brittle at base. **LVS:** stipules 0 or vestigial; petiole short; young lvs silky; mature blade 32–74 mm, lanceolate, elliptic, or oblanceolate, acuminate (base wedge-shaped), entire, flat, ± persistently white- or white-and-rusty-hairy, upper shiny, lower glaucous. **INFL** appearing with or just before lvs, 6–20 mm, subspheric, sessile or on leafy shoots < 8 mm; fl bract tawny to brown. **STAMINATE FL:** stamens 2. **PISTILLATE FL:** ovary white- or white-and-rusty-hairy, style 0.1–0.4 mm, stalk 1–2.8 mm. 2*n*=38. Subalpine streams, meadows; 1450–3600 m. s CaRH, n SNH, s SNH (esp Kern Plateau), SnBr, **GB;** to B.C., MT, NM. [var. *argentea* (Bebb) C. Schneider] ❀ WET:1–3,4,5,6,7;STBL. May–Jun

S. gooddingii C. Ball (p. 477) GOODDING'S BLACK WILLOW Tree < 30 m. **ST:** twigs yellowish, velvety or soft-shaggy-hairy, becoming glabrous, sometimes brittle at base; bud scale margins free, overlapping. **LVS:** petiole with glands; young lvs hairy;

mature blade 67–130 mm, narrowly lanceolate or narrowly ovate, acuminate (base wedge-shaped), finely serrate, glabrous or becoming so, not glaucous below. **INFL** appearing with lvs, 22–65 mm, on leafy shoot 4–25 mm (margins of shoot-lvs finely serrate); fl bract tawny; pistillate bract deciduous after fl. **STAMINATE FL**: stamens 4–8. **PISTILLATE FL**: ovary glabrous or hairy, style 0.1–0.3 mm, stalk 1.2–3.2 mm. $2n=38$. Common. Streamsides, marshes, seepage areas, washes, meadows; gen < 500 m (–23–1600 m). NCoRI, CaRF, SNF, GV, SCo, PR, **D (esp GV, D)**; to TX, Mex. [var. *variabilis* C. Ball] ❀ WET:5,6,**7–9**,10,11,**12–14**,15–17,**18–24**;STBL. Mar–Apr

S. laevigata Bebb (p. 477) RED WILLOW Tree < 15 m. **ST**: twigs red- to yellow-brown, hairy, becoming glabrous, brittle at base; bud scale margins free, overlapping. **LVS**: stipules vestigial on early lvs, lf-like later; petiole gen with glands; young lvs glabrous or white- or white-and-rusty-hairy; mature blade 67–150 mm, lanceolate to widely elliptic, acuminate (base rounded to acute), indistinctly finely crenate, glabrous, shiny to highly glossy above, glaucous below. **INFL** appearing with or after lvs, 35–110 mm, on leafy shoots 3–35 mm (shoot lvs entire); fl bract tawny; pistillate bract deciduous after fl. **STAMINATE FL**: stamens 5. **PISTILLATE FL**: ovary glabrous, style 0.2–0.6 mm, stalk 1–3.4 mm. $2n=38$. Common. Riverbanks, seepage areas, lake shores (subalkaline or brackish), canyons, ditches; 0–1700 m. **CA** (exc MP, DSon); s OR, n NV, AZ, Mex, n C.Am. [var. *araquipa* (Jepson) C. Ball] Mar–May

S. lasiolepis Benth. (p. 477) ARROYO WILLOW Shrub, small tree, < 10 m. **ST**: twigs yellowish to brownish, tomentose, long-soft-wavy-hairy or velvety, becoming glabrous, gen brittle at base. **LVS**: petiole 3–16 mm, tomentose to velvety; young lvs hairy; mature blade 35–125 mm, ± lanceolate-elliptic to oblanceolate, acute (base wedge-shaped), entire to irregularly serrate, slightly rolled under, gen white- or white-and-rusty-tomentose, becoming glabrous, shiny above, glaucous below. **INFL** appearing before lvs, 15–70 mm, sessile or on leafy shoots < 10 mm (shoot lvs entire); fl bract dark brown, ± densely straight- or wavy-hairy, tip broadly rounded. **STAMINATE FL**: stamens 2. **PISTILLATE FL**: ovary glabrous, style 0.1–0.6 mm, stalk 0.5–2.4 mm. $2n=76$. Abundant. Shores, marshes, meadows, springs, bluffs; < 2800 m. **CA**; to WA, ID, TX, Mex. [vars. *bracelinae* C. Ball, *sandbergii* (Rydb.) C. Ball; *S. lutea* Nutt. var. *nivaria* Jepson; *S. tracyi* C. Ball, Tracy's willow] Highly variable. ❀ WET,SUN:1–5,**6–9**,10–13,**14–24**;STBL;INV. Feb–Apr

S. lemmonii Bebb (p. 481) LEMMON'S WILLOW Shrub < 4 m. **ST**: twigs yellowish to brownish, gen thinly glaucous, gen glabrous, sometimes brittle at base. **LVS**: stipules lf-like (vestigial on young lvs); young lvs silky; mature blade 44–110 mm, oblong-lanceolate to elliptic, acuminate (base acute), entire or finely serrate, white- or white-and-rusty-hairy, becoming glabrous, glaucous below. **INFL** appearing with or just before lvs, 13–50 mm, sessile or on leafy shoots < 6 mm (shoot lvs entire); fl bract tawny to dark brown. **STAMINATE FL**: stamens 2. **PISTILLATE FL**: ovary silky, style 0.3–0.8 mm, stalk 1.2–2.1 mm. $2n=76$. Streams, wet meadows, burns in subalpine pine forests; 1400–3500 m. e KR, CaRH, SNH, SnBr, MP, **W&I**; to B.C., ID, NV. See *S. geyeriana*. ❀ WET or IRR,SUN:**1–3**,4,5,**6,7**;STBL. May–Jun

S. lucida Muhlenb. SHINING WILLOW Shrub, tree, < 10 m. **ST**: twigs brownish, glabrous or soft-shaggy-hairy, sometimes brittle at base. **LVS**: stipules glandular-lobed; petiole with glands; young

lvs glabrous or white- or white-and-rusty-hairy; mature blade 53–170 mm, lanceolate, long-acuminate (base acute to rounded), finely serrate, glabrous or becoming so, upper surface shiny, lower glaucous or not. **INFL** appearing with lvs, 20–90 mm, on leafy shoots 8–45 mm; fl bract tawny; pistillate bract deciduous after fl. **STAMINATE FL**: stamens 3–5. **PISTILLATE FL**: ovary glabrous, style 0.2–0.6 mm, stalk 0.8–2 mm. $2n=76$. Wet places; < 3200 m. **CA**; to AK, w Can, NV.

1. Lf blade not glaucous below, with stomata above
.... ssp. *caudata*

1' Lf blade glaucous below, at least largest lacking stomata above .... ssp. *lasiandra*

  ssp. *caudata* (Nutt.) E. Murray Wet meadows, lakeshores; 1500–3200 m. SNH, SnBr, GB; to B.C., NV. [*S. c.* (Nutt.) A.A. Heller var. *bryantiana* C. Ball and N.F. Bracelin] ❀ TRY. May

  ssp. *lasiandra* (Benth.) E. Murray (p. 481) Common. Wet meadows, shores, seepage areas; 0–2700 m. **CA** (less common in s CA, D). [*S. lasiandra* Benth. incl vars. *abramsii* C. Ball, *lancifolia* (Andersson) Bebb] ❀ WET:1–5,**6–10**,11–13,**14–24**;STBL. Mar–May

S. lutea Nutt. (p. 481) YELLOW WILLOW Shrub, small tree, < 7 m. **ST**: twigs yellow-gray or yellow-brown, glabrous or hairy. **LVS**: young lvs glabrous or silky; mature blade 42–116 mm, lanceolate to oblanceolate, acuminate (base acute to rounded), entire, gland-dotted to finely crenate, glabrous to sparsely hairy, dull above, glaucous below. **INFL** appearing just before or with lvs, 15–75 mm, sessile or on leafy shoots < 8 mm; fl bract brown, glabrous or sparsely curly-hairy. **STAMINATE FL**: stamens 2. **PISTILLATE FL**: ovary glabrous, style 0.2–0.6 mm, stalk 1–3.4 mm. $2n=38$. Creek margins, wet meadows; 900–3100 m. c&s SNH (esp e slope), SnBr, SnJt, **GB**, **w DMoj**; to w&c Can, NE. [var. *watsonii* (Bebb) Jepson] ❀ WET,SUN:**1–3**,4–6,**7–9**,10–12,**14**,15–17,**18–21**,22–24;STBL. May–Jun

S. orestera C. Schneider (p. 481) GRAY-LEAVED SIERRA WILLOW Shrub < 2 m. **ST**: twigs yellowish to brownish, glaucous or not, silky or short-shaggy-hairy, becoming glabrous. **LVS**: stipules lf-like; young lvs silky; mature blade 35–95 mm, lanceolate to oblanceolate, acute (both ends), entire, silky, dull above, glaucous below or this obscured by hairs. **INFL** appearing with lvs, 12–55 mm, on leafy shoots 3–15 mm (shoot lvs entire or gland-dotted); fl bract dark brown. **STAMINATE FL**: stamens 2. **PISTILLATE FL**: ovary silky, style 0.6–1 mm, stalk 0.8–2 mm. Alpine and subalpine meadows, wet places; 2100–4000 m. SNH, **W&I**; w NV. [*S. commutata* Bebb var. *rubicunda* Jepson misapplied] ❀ TRY;STBL. Jun–Jul

S. planifolia Cham. ssp. *planifolia* (p. 481) TEA-LEAVED WILLOW Shrub < 1 m. **ST**: twigs brownish, glabrous or silky. **LVS**: stipules 0 or vestigial; young lvs glabrous or white- or white-and-rusty-silky; mature blade 20–43 mm, ± elliptic, acute (both ends), entire to finely serrate, glabrous or silky, glossy above, glaucous below. **INFL** appearing before lvs, 10–45 mm, sessile, rarely on leafy shoots < 3 mm; fl bract brown to black. **STAMINATE FL**: stamens 2. **PISTILLATE FL**: ovary white- or white-and-rusty-silky, style 0.5–1.4 mm, stalk 0.3–0.8 mm. $2n=76, 57$. Subalpine meadows, riverbanks; 2500–4000 m. c SNH, **SNE**; to n&e N.Am, NM. [*S. phylicifolia* L. var. *monica* (Bebb) Jepson] ❀ TRY. Jun–Aug

SANTALACEAE SANDALWOOD FAMILY

William J. Stone

Herbs, shrubs, trees, green root-parasites. **LVS** entire; stipules 0. **INFL**: gen small cymes in upper axils. **FL**: calyx lobes 4–5 parted, fused side-to-side in bud, persistent; petals 0; stamens opposite sepals, inserted on fleshy disk; ovary superior to fully inferior, chamber 1, ovules 2–4, suspended from top of free-central placenta, style 1, thread-like, stigma head-like. **FR**: drupe or nut. **SEED** 1, spheric or ovoid. ± 26 genera, 250 spp.: gen trop. Recently treated to include Viscaceae [Angiosperm Phylogeny groupd 1998 Ann Missouri Bot Gard 95:531–303]

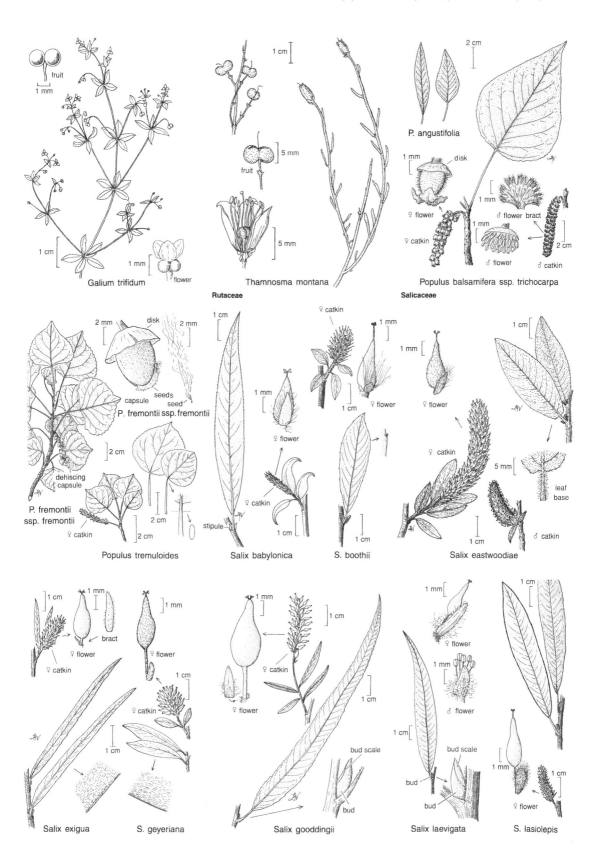

fruit
1 mm

1 cm

1 mm
flower

Galium trifidum

1 cm

fruit
5 mm

5 mm

Thamnosma montana

Rutaceae

P. angustifolia

2 cm

1 mm
disk

♀ flower
♀ catkin

1 mm
♂ flower bract

1 mm
♂ flower

2 cm
♂ catkin

Populus balsamifera ssp. trichocarpa

Salicaceae

2 mm
disk

2 mm

seeds
capsule
seed
P. fremontii ssp. fremontii

2 cm

P. fremontii
ssp. fremontii
dehiscing
capsule
♀ catkin

2 cm

2 cm

Populus tremuloides

1 cm

1 mm
♀ flower

♀ catkin

stipule

Salix babylonica

♀ catkin
1 mm

1 cm
♀ flower

♀ flower

S. boothii

1 mm
♀ flower

♀ catkin

5 mm
leaf
base

♂ catkin

1 cm

Salix eastwoodiae

1 cm

1 cm
1 mm
bract
♀ flower
♀ catkin

1 mm
♀ flower

1 cm
♀ catkin

1 cm

Salix exigua   S. geyeriana

1 mm

1 cm
♀ catkin

♀ flower

1 cm

bud scale
bud

Salix gooddingii

1 mm
♀ flower

1 mm
♂ flower

1 cm

bud scale
bud
bud

Salix laevigata

1 cm

1 mm
♀ flower

1 cm

S. lasiolepis

COMANDRA BASTARD TOAD-FLAX

Per, subshrub, smooth; rhizome extensive, horizontal. **ST** 7–40 cm, green, blue-green, or grayish, striate. **LVS** alternate, ± sessile. **FL** subtended by bractlet; calyx tube bell- or urn-shaped, lobes (4–)5, ovate to oblong; anther base with a tuft of hair. **FR** drupe-like, crowned by persistent calyx. **SEED** spheric. 4 spp.: 3 Am, 1 Eur. (Greek: hair, man, for hairy attachment of stamens) [Piehl 1965 Mem Torrey Bot Club 22(1):1–97]

C. umbellata (L.) Nutt. ssp. *californica* (Rydb.) Piehl (p. 481) Sub-shrub. **STS** many from rhizome, 10–40 cm, leafy, glaucous, branched. **LF** 15–55 mm, lanceolate, acute to sharp-pointed, sometimes paler below. **FL** 2–3.5 mm; sepals lanceolate to narrowly ovate, whitish; anthers ± 0.5 mm. **FR** 5–7.5 mm, oblong-ovoid; calyx tube sometimes forms "neck" above fr. Gen dry, ± rocky areas; 300–3000 m. KR, CaR, SN, **SNE**, **n DMoj?**; to B.C., sw NV, nw AZ. [*C. californica* Rydb.] Intergrades with ssp. *pallida* (A. DC.) Piehl in OR, WA, NV, AZ. ❀ TRY. Apr–Aug

SAURURACEAE LIZARD'S-TAIL FAMILY

Elizabeth McClintock

Per, from rhizomes or stolons. **LVS** simple, alternate; stipules joined to petiole. **INFL**: spike, dense, many-fld, terminal, sometimes subtended by petal-like involucre bracts and so resembling a single fl. **FLS** small, bisexual; perianth 0; stamens 6, 8 (or 3); ovary superior but sometimes embedded in infl axis, compound, 1-chambered or carpels fused only at base, styles 3–4, distinct. **FR**: capsule, ± fleshy, dehiscent at tip or ± berry-like. **SEEDS** many or 1, spheric or ovate. 5 genera, 7 spp.: e Asia, N.Am. [Wood 1971 J Arnold Arb 52:479–485]

ANEMOPSIS

1 sp. (Greek: anemone-like, from infl) [Howell 1971 Wasmann J Biology 29:97–100]

A. californica (Nutt.) Hook. & Arn. (p. 481, pl. 100) YERBA MANSA Rhizome creeping, thick, woody. **ST** 15–50 cm, gen naked, hollow, glabrous or hairy. **LVS**: basal several, blade 5–15 cm, elliptic to oblong or base sometimes cordate, petiole 10–20 cm; cauline few, ovate, gen subsessile to clasping, sometimes subtending 1–3 short-petioled lvs. **INFL** 1.5–4 cm, conic; involucre bracts 5–8, 1–3 cm, petal-like, white, often tinged reddish. **FL** (exc lowermost) subtended by a 5–6 mm white bract; stamens appearing to arise from infl axis. Common. Saline or alkaline soil, wet or moist areas, seeps, springs; 75–1700 m. sw ScV, SnJV, SnFrB, SCoR, SCo, n&s ChI, PR, **W&I**, **DMoj**; to Utah, w TX, nw Mex. [*A. c.* var. *subglabra* Kelso] Pls aromatic, once used to treat diseases of skin, blood. ❀ IRR or WET:3,6,**7–10,14–24**&SHD:**11,12**;deciduous GRCVR. Mar–Sep

SAXIFRAGACEAE SAXIFRAGE FAMILY

Patrick E. Elvander

Per or subshrub from caudex or rhizome, gen ± hairy. **ST** often ± leafy on lower half, rarely trailing and leafy throughout. **LVS** gen simple, basal or sometimes cauline, gen alternate, gen petioled; veins ± palmate. **INFL**: panicle, gen ± scapose. **FL** gen bisexual, gen radial; hypanthium free to ± fused to ovary; calyx lobes gen 5; petals gen 5, free, gen clawed, gen white; stamens gen 5 or 10; pistils 2 and simple or 1 and compound (chambers 1–2, placentas 2–4, axile or parietal), ovary superior to inferior, sometimes more superior in fr, styles gen 2. **FR**: 2 follicles or 2–4-valved capsule. **SEEDS** gen many, small. 40 genera, 600 spp.: esp n temp, arctic, alpine; some cult (*Bergenia*, *Darmera*, *Heuchera*, *Saxifraga*, *Tellima*, *Tolmiea*). [Soltis 1988 Syst Bot 13:64–72]

1. Fl solitary, showy; staminodes 5, lobed; placentas 4 (Parnassioideae) . . . . **PARNASSIA**
1' Fls several, not showy; staminodes 0; placentas 2–3 (Saxifragoideae)
　2. Stamens 5; styles and placentas 2; petals not lobed . . . . **HEUCHERA**
　2' Stamens 10; styles and placentas 3; petals lobed . . . . **LITHOPHRAGMA**

HEUCHERA ALUMROOT

Rhizome scaly; bulblets 0. **LVS** basal, sometimes a few cauline; blade ovate, base cordate to reniform, lobes and teeth shallow, irregular. **INFL** gen raceme-like; bracts gen scale-like. **FL** radial to ± bilateral; hypanthium partly fused to ovary; calyx lobes equal or not; petals 0 or 5, gen equal; stamens 5, gen equal; pistil 1, ovary > half inferior, chamber 1, placentas 2, parietal. **FR**: capsule. 50 spp.: N.Am. (J.H. von Heucher, German professor of medicine, 1677–1747) [Rosendahl, Butters, & Lakela 1936 Minn Stud Pl Sci 2:1–180] A very difficult genus, highly variable at many levels and needing much additional research.

1. Fl radial; styles and most or all stamens incl in full fl; calyx lobes equal, yellow with pink tips . . . . . *H. duranii*
1' Fl ± bilateral; styles and all stamens exserted in full fl; calyx lobes unequal, usually whitish to pink-red with green tips . . . . *H. rubescens* var. *alpicola*

H. duranii Bacigal. (p. 481) DURAN'S ALUMROOT **LF**: petiole 2.5–5 cm; blade 1–2 cm, round-reniform, shallowly 5–9-lobed. **INFL** 14–20 cm, dense toward tips, ± short-glandular. **FL** radial; part of hypanthium fused to ovary 0.8 mm, < free part, together with calyx lobes 2–2.5 mm; calyx lobes yellow with pink tips; petals 2.5–3 mm, ± = calyx lobes, lanceolate to oblanceolate;

stamens < calyx lobes, incl; mature styles < 1 mm, incl. RARE in CA. Rocky areas; 2500–3500 m. **SNE (Sweetwater Mtns, White Mtns)**; w NV. Probably a depauperate high elevation ecotype of *H. parvifolia* Nutt. In cult. Jul–Aug

H. rubescens Torrey **LF**: petiole 1–15 cm; blade 8–60 mm, broadly ovate to ± round, ± deeply 5–9-lobed. **INFL** 7–55 cm, often 1-sided, open or dense, cylindric to conic, glandular. **FL** ± bilateral; part of hypanthium fused to ovary 0.9–2.5 mm, ± = long side of free part, together with calyx lobes 3–6 mm; calyx lobes gen unequal, whitish to pink-red with green tips, becoming redder; petals 3–6 mm, > calyx lobes, narrowly oblanceolate to thread-like; stamens > calyx lobes, exserted; mature styles gen

>> 1.5 mm, exserted. Dry, rocky areas; 1500–4000 m. SN, PR, **GB (exc n MP), DMtns**; to OR, ID, Colorado, TX, n Mex. Highly variable. Closely related to *H. parishii* and *H. mexicana* of Mex; hybridizes with other spp. Vars. intergrade. Needs monographic study. ❀ DRN:**4,5**& IRR:**1,2,**15–17,24&SHD:**3,6,**7,14,18–23. 4 vars. in CA.

var. ***alpicola*** Jepson **INFL**: ± open, ± narrow, ± 1-sided. **FL**: hypanthium gen ± as long or longer than wide; part of hypanthium fused to ovary together with calyx lobes gen 4–6 mm; petals oblanceolate; styles strongly papillate throughout. Habitat of sp. s SN, **SNE, DMtns**. [var. *pachypoda* (E. Greene) C. Rosend., F.K. Butters, & Lakela] May–Jul

LITHOPHRAGMA WOODLAND STAR

Rhizome slender, scaleless, bearing bulblets. **LVS** basal and cauline, reduced, sometimes opposite, more deeply lobed upward; blade round, base cordate to reniform, ± lobed, gen toothed. **INFL**: raceme; bracts scale-like or 0. **FL**: hypanthium gen partly fused to ovary; petals gen lobed or toothed; stamens 10; pistil 1, ovary superior to ± inferior, chamber 1, placentas 3, parietal, styles 3. **FR**: capsule, valves 3. 12 spp.: w N.Am. (Greek: rock hedge, from habitats) [Taylor 1965 U Calif Publs Bot 37:1–122] Generic names ending in "phragma" are considered of neuter, not feminine, gender.

L. glabrum Nutt. (p. 481) **LF**: basal blade deeply 3-lobed or ± palmately compound (lflets 3), lobes and lflets lobed, teeth ± sharp-tipped. **INFL** 8–25 cm; fls 1–7; pedicels 3–6 mm; lower bracts often with axillary bulblet. **FL**: hypanthium spheric to bell-shaped, part fused to ovary gen < free part; petals 3–7 mm, ovate, deeply 3–5-lobed; ovary < half inferior. **SEED** spiny. 2*n*=14,28. Dry, gravelly places; < 3500 m. KR, CaRH, SNH, MP, **n SNE**; to B.C., SD, Colorado. [*L. bulbiferum* Rydb.] Apr–Jul

PARNASSIA GRASS-OF-PARNASSUS

Pl glabrous; caudex without scales or bulblets. **LVS** basal; blade ovate, entire, base tapered to cordate or reniform. **INFL**: fl solitary; bract gen 1, sessile, lf-like. **FL**: hypanthium minute, free of ovary; stamens 5, staminodes 5, toothed to fringed or divided; pistil 1, ovary superior, chamber 1, placentas 4, parietal, styles inconspicuous, stigmas 4. **FR**: capsule, valves 4. **SEED** ± winged. 25 spp.: n temp, arctic. (Mount Parnassus, Greece)

P. parviflora DC (p. 481) **LF** 2–7 cm; blade 1–3.5 cm, ± ovate, base gen ± truncate. **INFL** 10–35 cm; bract below middle of peduncle, ± elliptic. **FL**: calyx lobes 4–7 mm, elliptic, entire, gen erect to spreading; petals 5–13 mm, ovate-elliptic, entire; staminodes 3.5–6 mm, lobes < 12, tips spheric. **FR** 4–13 mm. Uncommon. Rocky seeps; < 3000 m. **SNE**; to e Can. [*P. palustris* L. var. *p.* (DC) J. Boivin]. Jul–Oct

SCROPHULARIACEAE FIGWORT FAMILY

Lawrence R. Heckard, Family Coordinator

Ann to shrubs, gen glandular, some green root-parasites. **ST** gen round. **LVS** gen alternate, simple, gen ± entire; stipules gen 0. **INFL**: spike to panicle, gen bracted, or fls 1–2 in axils. **FL** bisexual; calyx lobes gen 5; corolla gen strongly bilateral, gen 2-lipped (upper lip gen 2-lobed, lower lip gen 3-lobed); stamens gen 4 in 2 pairs, gen incl, a 5th (gen uppermost) sometimes present as a staminode; pistil 1, ovary superior, chambers gen 2, placentas axile, style 1, stigma lobes gen 2. **FR**: capsule, gen ± ovoid, loculicidal or septicidal. **SEED**: coat sculpture often characteristic. ± 200 genera, 3000 spp.: ± worldwide; some cult as orn (e.g., *Antirrhinum, Mimulus, Penstemon*) or medicinal (*Digitalis*). Recently treated to include only *Buddleja, Scrophularia,* and *Verbascum* in CA; other CA genera moved to Orobanchaceae (*Castilleja, Cordylanthus, Orthocarpus, Pedicularis, Triphysaria*), Phrymaceae (*Mimulus*), and Plantaginaceae (=Veronicaceae sensu Olmstead et al) [Angiosperm Phylogeny Group 1998 Ann Missouri Bot Gard 85:531-553; Olmstead et al 2001 Mol Phylog Evol 16: 96-112] Key to genera by Margriet Wetherwax.

1. Lvs alternate, not all basal
  2. Stamens 5; corolla ± radial, 5-lobed; fr septicidal . . . . **VERBASCUM**
  2' Stamens 2 or 4; corolla ± radial to 2-lipped; fr loculicidal or dehiscing by terminal slits or pores
    3' Corolla base sac-like; fr dehiscent by terminal slits or pores
      4. Lvs palmately veined, long-petioled, entire to dentate . . . . **MAURANDYA**
      4' Lvs not palmately veined, ± sessile, gen entire
        5. Sts twining or decumbent . . . . **²ANTIRRHINUM**
        5' Sts erect
          6. Stamens 4, anther sacs 2 per stamen; fr asymmetric, dehiscing by 1–2 pores near tip . **²ANTIRRHINUM**
          6' Stamens 2, anther sacs 1 per stamen; fr symmetric, dehiscent by irregular slit at tip . . . . **MOHAVEA**
    3. Corolla base not sac-like; fr loculicidal
      7. Corolla upper lobe not forming a beak or hood . . . . **²VERONICA**
      7' Corolla upper lip forming a beak or hood
        8. Corolla hood rounded; anther sacs 2 per stamen, equal; lvs gen toothed (or divisions gen > 7)
        . . . . **PEDICULARIS**

8' Corolla beak ± straight; anther sacs 1 per stamen or if 2, then unequal; lvs entire or divisions 3–7
  9. Beak tip open; stigma expanded (head-like or 2-lobed); seed attached at base . . . . **CASTILLEJA**
  9' Beak tip closed (opening directed downward); stigma unexpanded (dot-like); seed attached at side
    10. Calyx sheath-like (sometimes ± notched); bracts grading from lf-like to calyx-like; lower corolla
    lip 3-lobed; infl various; gen ± low elevation . . . . **CORDYLANTHUS**
    10' Calyx unequally 4-lobed; bract 1 per fl; lower corolla lip 3-toothed; infl spike-like; montane
    . . . . **ORTHOCARPUS**
1' Lvs gen opposite, rarely whorled or all basal
  11. Stamens 2, staminodes 0 or 2
    12. Corolla ± radial, gen 4-lobed, tube < lobes; calyx gen 4-lobed; staminodes 0 . . . . ²**VERONICA**
    12' Corolla 2-lipped, 5-lobed, tube > lobes; calyx gen 5-lobed; staminodes 2, forming corolla throat
    ridges . . . . **LINDERNIA**
  11' Stamens 4, staminode 0 or 1
    13. Calyx tube well developed
      14. Pl aquatic; lvs all basal; stolons present . . . . **LIMOSELLA**
      14' Pl gen of drier habitats; lvs gen opposite; stolons gen 0
        15. Stigma head-like; calyx tube not prominently ribbed or pleated . . . . **COLLINSIA**
        15' Stigma lobes 2, flat; calyx tube gen prominently ribbed or pleated . . . . **MIMULUS**
    13' Calyx segments ± free
      16. Staminode 0, filaments 4
        17. Pl gen aquatic; lvs spoon-shaped to round, entire; corolla white to pink . . . . **BACOPA**
        17' Pl gen of dry habitats; lvs lanceolate, toothed; corolla violet to purple . . . . **STEMODIA**
      16' Staminode 1, filaments 5
        18. St square, staminode attached near base of upper lip . . . . **SCROPHULARIA**
        18' St round; staminode attached near base of corolla throat
          19. Fertile filament bases densely hairy, attached to corolla at 1 level . . . . **KECKIELLA**
          19' Fertile filament bases glabrous, attached to corolla at different levels . . . . **PENSTEMON**

ANTIRRHINUM SNAPDRAGON

David M. Thompson

Ann, per, glabrous to hairy. **ST** vine-like, ascending or erect, often clinging by twining pedicels or branchlets. **LVS** gen opposite below, alternate above, gen reduced upward; veins pinnate. **INFL**: raceme or fls solitary in axils. **FLS** often cleistogamous; uppermost calyx lobe gen largest; corolla tube of opening fls truncate or with rounded sac-like extension at base, lower lip base gen swollen and closing mouth; staminode 0. **FR** ovoid to spheric; chambers gen dehiscent by 1–2 pores near tip, lower chamber gen larger, upper sometimes indehiscent. **SEEDS** many, gen with tubercles or netted ridges. 36 spp.: w N.Am, w Medit. (Greek: nose-like, from corolla shape) [Thompson 1988 Syst Bot Monogr 22:1–142] *For alternate taxonomy see Ghebrehiwet et al 2000 Pl Syst Evol 220:223–239; Sutton 1988 A revision of the tribe Antirrhineae. British Museum (Natural History), London, & Oxford University Press, Oxford.

1. Pedicels 3–10 cm, twining . . . . *A. filipes*
1' Pedicels < 1.3 cm, not twining
  2. Lvs all glandular-hairy . . . . *A. cyathiferum*
  2' Lower lvs glabrous
    3. Calyx lobes unequal; corolla 5–7 mm or cleistogamous . . . . *A. kingii*
    3' Calyx lobes equal; corolla lobes 9–12 mm, opening . . . . *A. coulterianum*

A. coulterianum Benth. (p. 481) Ann, glabrous below infl, hairy in infl. **ST** erect but weak, often clinging to other pls or debris. **LVS**: basal rosette often present (unique in genus). **INFL**: raceme, terminal; pedicels 1–5 mm, lowest gen subtended by twining branchlets. **FL**: calyx lobes equal; corolla 9–12 mm, white to lavender; lower stamens gen exserted. **FR**: upper chamber indehiscent. *n*=15. Among shrubs in deserts, gen on burns elsewhere; 0–1700 m. s SCoRO, SW (exc ChI), **nw edge DSon**; n Baja CA. ❀ DRN,DRY:**7,14**,15–17,**18–24**&part SHD:**8,9**,10–12;also STBL. **Sairocarpus coulterianus* (A. DC.) D.A. Sutton Apr–Jul

A. cyathiferum Benth. (p. 481) DEEP CANYON SNAPDRAGON Ann, hairy. **ST** erect, self-supporting. **LVS** not reduced upward. **INFL**: fls solitary; pedicels 1–6 mm, in fr bending down, subtending branchlets 0. **FL**: calyx lobes equal; corolla 8–9 mm, cream and purple, veins purple. **FR**: chambers equal, opening by irregular bursting near tip. **SEED**: wing large, cup-shaped. *n*=13. RARE in CA. Washes, rocky slopes; 0–800 m. **w DSon (Deep Canyon, Riverside Co.)**; s AZ, nw Mex. **Pseudorontium cyathiferum* (Benth.) Rothm. Jan–Apr

A. filipes A. Gray (p. 485, pl. 101) Ann, glabrous. **ST** vine-like, climbing. **INFL**: fls solitary; pedicels 3–10 cm, twining, subtending branchlets gen 0. **FL**: calyx lobes equal; corolla of opening fls 10–13 mm, yellow and gold with maroon flecks on lower lip. **FR**: chambers equal, fragile, opening by irregular bursting on sides. **SEED**: ridges 4–6, thick, parallel. *n*=15. On shrubs, debris, gen in washes; 0–1400 m. **D**; to sw UT, w AZ, nw Mex. **Neogaerrhinum filipes* (A. Gray) Rothm. Early Mar–mid-May

A. kingii S. Watson (p. 485) Ann, gen glabrous below infl (infl sometimes sparsely hairy). **ST** erect but weak, often clinging to other pls or debris. **INFL**: raceme or fls solitary; upper pedicels gen 1–4 mm, lowest gen subtended by twining branchlets. **FL**: calyx lobes unequal; corolla of opening fls 5–7 mm, white, veins violet. *n*=15. Uncommon. Washes, scree; 500–2300 m. SNE, **DMtns (esp ne San Bernardino Co.)**; to se OR, w UT, nw AZ. **Sairocarpus kingii* (S. Wats.) D.A. Sutton Late Apr–mid Jul

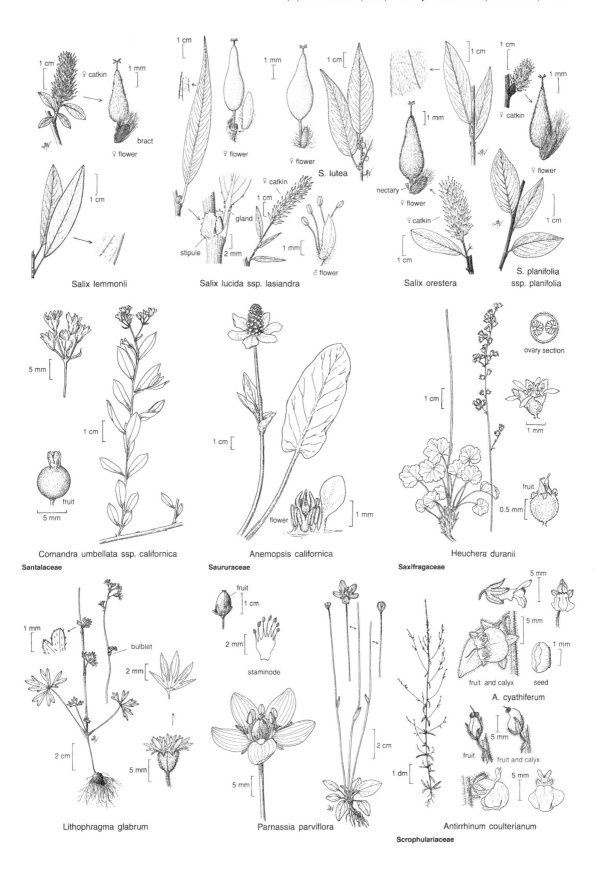

Salix lemmonii

Salix lucida ssp. lasiandra

♀ catkin
♀ flower
bract
♀ flower
gland
stipule
♀ catkin
S. lutea
♂ flower

Salix orestera
nectary
♀ flower
♀ catkin

S. planifolia
ssp. planifolia
♀ catkin
♀ flower

Comandra umbellata ssp. californica
Santalaceae
fruit

Anemopsis californica
Saururaceae
flower

Heuchera duranii
Saxifragaceae
ovary section
fruit

Lithophragma glabrum
bulblet

Parnassia parviflora
fruit
staminode

Antirrhinum coulterianum
Scrophulariaceae
fruit and calyx
seed
A. cyathiferum
fruit
fruit and calyx

BACOPA WATER-HYSSOP

John L. Strother

Ann, ± per. **ST** prostrate to erect, gen < 6 dm. **LVS** cauline, opposite, gen < 4 cm, gen narrowly ± obovate to ± round. **INFL**: fls 1–3 per lf axil, sessile or pedicelled. **FL**: sepals 4–5; corolla (3–)5-lobed, gen white to pink, throat yellow; stamens (2–)4(–5); ovary subtended by nectary, stigma weakly 2-lobed. **FR**: capsule, spheric, loculicidal or septicidal. **SEEDS** > 30, 0.1–0.3 mm. ± 100 spp.: trop, warm temp. (Presumed to be aboriginal name) [Barrett & Strother 1978 Syst Bot 3:408–419]

1. Lf narrowly ± obovate, vein 1; pedicel bractlets 2, near tip . . . . ***B. monnieri***
1' Lf obovate to ± round, veins > 6, palmate; pedicel bractlets 0 . . . . ***B. eisenii***

B. eisenii (Kellogg) Pennell (p. 485) **ST** prostrate to ascending. **LF** 12–34 mm, obovate to ± round; veins > 6, palmate. **INFL**: pedicel 15–51 mm, gen > lf, stout; bractlets 0. **FL**: sepals 5, outer 4.3–6.6 mm, widely ovate to ± round; corolla 10–14 mm. 2*n*=56. Rice fields, muddy places, wet soil, or floating; < 100 m (< 1200 in SNE). GV, s CW, **SNE**; NV. ❀ TRY. May–Oct

B. monnieri (L.) Wettst. (p. 485) **ST** prostrate to ascending. **LF** gen 5–25 mm, narrowly ± obovate; vein 1. **INFL**: pedicel 8–40 mm, gen > lf; bractlets 2, near tip. **FL**: sepals 5, outer 5–7 mm, lanceolate to ovate, corolla 8–10 mm. 2*n*=64. Wet soil or in shallow water; < 100 m. **DSon (e Riverside Co.)**; to s US; native to trop. Recent alien, probably naturalized. All year

CASTILLEJA PAINTBRUSH, OWL'S-CLOVER

T.I. Chuang & Lawrence R. Heckard

Ann to subshrub, green root-parasites. **LVS** sessile, entire to dissected. **INFL** spike-like; bracts becoming shorter, wider, more lobed than lvs, tips gen colored. **FL**: calyx gen unequally 4-lobed, gen colored like bract tips; corolla upper lip beak-like, tip open, lower lip gen reduced, 3-toothed to -pouched; anther sacs 2, unequal; stigma entire to 2-lobed, gen exserted. **FR** loculicidal, ± ovoid, ± asymmetric. **SEED** gen ± brown, attached at base; coat netted, net-like walls sometimes aligned ladder-like. ± 200 spp.: esp w N.Am. (Domingo Castillejo, Spanish botanist) [Chuang & Heckard 1991 Syst Bot 16:644–666] Highly variable within & between populations. Hybridization & polyploidy common; polyploid forms may have separate ranges or be ± identifiable within populations by minor characters. Biologically consistent taxa very difficult to define. ❀ TRY with host; usually DFCLT. Key to species adapted by Margriet Wetherwax.

1. Infl not very reddish (gen green, white, yellow, or purplish); fl ± short (gen bee-pollinated); lower lip
  3-pouched, teeth gen ± erect, white or green to lavender, upper lip << 3 × lower (subg. *Colacus*)
   2. Per
    3. Bract lobes acuminate, green-margined; corolla pale yellow blotched purplish; seed coat loose-fitting
      . . . . ***C. nana***
    3' Bract lobes gen rounded (acute), white-margined; corolla pale yellow-green throughout; seed coat
      tight-fitting . . . . ***C. pilosa***
   2' Ann (sect. *Oncorhynchus*)
    4. Bracts green throughout; corolla beak straight, puberulent . . . . ***C. tenuis***
    4' Bracts purplish or whitish; corolla beak hooked, densely shaggy-hairy . . . . ***C. exserta***
     5. Corolla white to yellow or rose to purple-red; widespread . . . . ssp. ***exserta***
     5' Corolla bright rose-red with orange-tipped lower lip; w DMoj . . . . ssp. ***venusta***
1' Infl ± red or yellow-orange; fl long (gen hummingbird-pollinated); lower lip of small, incurved, gen
  dark green teeth, upper lip > 3 × lower (subg. *Castilleja*)
   6. Ann; st simple; wet places — lvs and bracts entire, ± narrowly lanceolate . . . . ***C. minor*** ssp. ***minor***
   6' Per; sts gen clustered; gen drier places (exc *C. miniata*)
    7. Pl hairs branched
     8. Whole pl densely white-woolly; calyx ± undivided on sides . . . . ***C. foliolosa***
     8' Herbage green, calyces white-woolly; calyx sinuses deeper on sides than in back and front . . ***C. plagiotoma***
    7' Pl hairs not branched or 0
     9. Calyx divided 2/3 in front, 1/3 in back; corolla (incl lower lip) gen curved out through front calyx
       sinus; lvs linear, gen rolled upward . . . . ***C. linariifolia***
     9' Calyx divided ± 1/3–1/2 in front and back; corolla (incl lower lip) gen not curved out through front
       calyx sinus; lvs ± lanceolate, entire to lobed
      10. Pl not glandular below infl; lf margin not wavy
       11. Lvs gen 3–5-lobed; calyx divided 1/4–1/3 in back and front; gen dry sagebrush or desert scrub
         . . . . ***C. angustifolia***
       11' Lvs gen entire; calyx divided ± 1/2 in back and front; moist places . . . . ***C. miniata*** ssp. ***miniata***
      10' Whole pl very glandular-sticky; lf margin ± wavy . . . . ***C. applegatei***
       12. St 10–25 cm; lf lobes 0–3; calyx 13–15 mm, divided to ± 1/4 on sides; subalpine . . . . ssp. ***pallida***
       12' St. 30–80 cm; most lvs entire; calyx 15–25 mm, gen divided < 1/5 on sides; gen below subalpine
        13. Calyx lobes obtuse to rounded . . . . ssp. ***martinii***
        13' Calyx lobes acute . . . . ssp. ***pinetorum***

C. angustifolia (Nutt.) G. Don (p. 485) DESERT
PAINTBRUSH Per 15–45 cm, few-branched, gray-green, ± bristly, nonglandular. **LF** 20–70 mm, linear-lanceolate; lobes 0–5,
widely spreading. **INFL** 4–15 cm; bracts 20–30 mm, lobes 3–5,
bright red to yellowish orange (violet). **FL**: calyx 15–25 mm,
divided 1/4–1/3 in back and front, ± 1/7 on sides, long-
nonglandular- and short-glandular-hairy, lobes obtuse to rounded;
corolla 20–35 mm, beak ± = tube, yellowish green, back puberulent, margins reddish, lower lip 2–3 mm, dark green, incl; stigma
2-lobed. **FR** 10–15 mm. **SEED** 1.5–2 mm; coat deeply netted,
most walls ladder-like. 2*n*= 24,48. Dry sagebrush scrub, pin-
yon/juniper woodland; 1000–3000 m. ne SnBr, **GB**, **DMoj**; to
OR, MT, WY, Colorado, NM. [*C. martinii* Abrams ssp. *ewanii*
(Eastw.) Munz (2*n*=24)] Type from ID; earliest epithet in the
widespread *C. chromosa* Nelson complex. More study needed.
May–Sep

C. applegatei Fern. Per 10–80 cm, few-branched, green to
dusty, gen very short-glandular-sticky-hairy and long-
nonglandular-hairy. **STS** gen ± few. **LF** 20–70 mm, ± lanceolate,
gen wavy-margined; lobes 0–3. **INFL** 5–20 cm; bracts 15–25
mm, lobes 0–7, bright red to yellowish. **FL**: calyx 13–25 mm,
divided 1/3–1/2 in back and front, ± 1/8 on sides, lobes acute to
rounded; corolla 20–40 mm, beak ± = tube, back puberulent,
margins reddish, lower lip 1–3 mm, dark green, incl or exserted;
stigma slightly 2-lobed, well exserted. **FR** 8–15 mm. **SEED** 1–
1.5 mm; coat shallowly netted, loose-fitting, side walls ladder-
like. Dry, open forest or scrub; 300–3600 m. KR, NCoRH,
NCoRI, CaR, SN, e SnFrB, SCoRI, SW, **GB**, **DMtns**; to OR, ID,
NV, n Baja CA. Highly variable complex (other sspp. outside
CA), unique in combination of glandular-sticky herbage and wavy
lf margins. 4 sspp. in CA.

  ssp. *martinii* (Abrams) Chuang & Heckard (p. 485) **ST** 30–
80 cm. **LF**: lobes 0–3. **FL**: calyx 15–25 mm, divided ± 1/3 in
back and front, 1/4–1/6 on sides, lobes ovate, obtuse or rounded;
corolla 25–40 mm, beak 12–18 mm. 2*n*=24,48,72. Dry chaparral, open yellow-pine forest, sagebrush scrub; 300–2800 m.
NCoRI, e SnFrB, SCoRI, SW, **SNE**, **DMtns**; n Baja CA, s NV.
[*C. m.* Abrams incl var. *clokeyi* (Pennell) N. Holmgren; *C. roseana*
Eastw.] May–Sep

  ssp. *pallida* (Eastw.) Chuang & Heckard (p. 485) **STS** many,
10–25 cm. **LF**: lobes gen 3. **FL**: calyx 13–15 mm, divided > 1/3
in back and front, ± 1/4 on sides, lobes acute; corolla 16–22 mm,
beak 7–10 mm. 2*n*=48. Dry, rocky slopes and flats, red-fir forest to alpine barrens; 2400–3600 m. SNH, **SNE**. [*C. breweri*
Fern. incl var. *p.* Eastw.] Jun–Aug

  ssp. *pinetorum* (Fern.) Chuang & Heckard (p. 485) **ST** 30–
60 cm. **LF**: lobes 0–3. **FL**: calyx 16–25 mm, divided 1/3–
1/2 in back and front, 1/8–1/5 on sides, lobes lanceolate, acute;
corolla 25–35 mm, beak 11–16 mm. 2*n*=24,48. Open coniferous forests, dry sagebrush scrub; 700–2700 m. KR, NCoRH,
CaR, SN, **GB**, **n DMtns**; to OR, ID

C. exserta (A.A. Heller) Chuang & Heckard PURPLE OWL'S
CLOVER Ann 10–45 cm, glandular-puberulent, stiff-hairy. **LF** 10–
50 mm; lobes 5–9, ± thread-like. **INFL** 2–20 cm, 2–4 cm wide;
bracts 10–25 mm, white to purplish red, lobes 5–9 (lowest pair
often again 2–4-lobed), ± linear (tips wider). **FL**: calyx 10–22
mm, divided 1/2 in front, 1/3 on sides, 2/3 in back; corolla 12–30
mm, beak 6–7 mm, shaggy-hairy, tip hooked, lower lip 4–6 mm,
pouches 3–8 mm wide, 3–4 mm deep; filaments puberulent;
stigma ± incl. **FR** 10–15 mm. **SEED** 1–2 mm; coat deeply netted, loose-fitting. Open fields, grassland; < 1600 m. NW, SNF,
GV, CW, SW, **w DMoj**; AZ, nw Mex. [*Orthocarpus purpurascens*
Benth.] Highly variable. Very showy in spring. 3 sspp. in CA.

  ssp. *exserta* (p. 485) **INFL**: upper bracts gen < 5 mm wide,
lobes > 2 mm, tipped white, pale yellow, or purplish red. **FL**:
corolla colored like bracts. 2*n*=24. Habitats and range of sp.
[*O. p.* incl var. *pallidus* Keck] Mar–May

  ssp. *venusta* (A.A. Heller) Chuang & Heckard **INFL**: upper bracts gen < 5 mm wide, lobes > 2 mm, tipped deep pink or

purple. **FL**: corolla pouch orange-tipped. Dry sand, washes; 600–
900 m. **w DMoj**. [*O. p.* var. *ornatus* Jepson]

C. foliolosa Hook. & Arn. (p. 485) WOOLLY PAINTBRUSH Per or
subshrub 30–60 cm, felt-like with white to gray, much-branched
hairs. **ST** much-branched, with short, axillary shoots. **LF** 10–50
mm, ± linear; lobes 0–3, tips obtuse. **INFL** 3–20 cm; bracts 15–
25 mm, lobes 0–5, orange-red (yellow-green). **FL**: calyx 15–18
mm, divided 1/3–2/5 in back and front, entire or barely notched
on sides, swollen in fr; corolla 18–25 mm, beak ± = tube, exserted, back puberulent, margins pale, lower lip 2 mm, dark green,
incl; stigma club-shaped, slightly 2-lobed. **FR** 10–15 mm. **SEED**
1.5–2 mm; coat deeply netted, most walls ladder-like. 2*n*=24.
Dry, open, rocky slopes, edges of chaparral; < 1800 m. NCoR,
SNF, CW, s ChI (Santa Catalina Island), **sw edge DMoj**; n Baja
CA. ❀ DRN,DRY,SUN, with host:7,14–17, 22–24;DFCLT.
Mar–Jun

C. linariifolia Benth. (p. 485, pl. 102) Per 30–100 cm, few-
branched, yellowish to gray-green, gen becoming purplish, glabrous to slightly puberulent. **LF** 20–80 mm, gen linear, rolled
upward; lobes 0–3. **INFL** 5–20 cm, open below; bracts 15–30
mm, lobes gen 3, narrow, < calyx, bright red to yellow. **FL**: calyx
20–35 mm, divided 1/3 in back, 2/3 in front, ± 1/8 on sides,
puberulent, lobes acute; corolla 25–45 mm, beak ± = tube,
yellow-green, back puberulent, margins red, lower lip 2–3 mm,
dark green; stigma slightly 2-lobed. **FR** 10–15 mm. **SEED** 1.5–
2 mm; coat shallowly netted, loose-fitting, side walls ladder-like.
2*n*=24,48. Dry plains, rocky slopes, sagebrush shrub or pinyon/
juniper woodland; 1000–3000 m. CaR, TR, **GB**, **DMtns**; to OR,
MT, NM. Jun–Sep

C. miniata Hook. Per 40–80 cm, few-branched, green or becoming purplish, glabrous, to long-soft-hairy above. **LF** 30–60
mm, ± lanceolate, entire; tip acute. **INFL** 3–15 cm; bracts 15–35
mm, lobes 0–5, bright red to yellowish. **FL**: calyx 10–30 mm,
divided 1/2–2/3 in back, more deeply in front, ± 1/4 on sides,
lobes acute; corolla 15–40 mm, beak ± = tube, yellow-green,
back puberulent, margins red, lower lip 1–2 mm, dark green,
slightly exserted; stigma slightly 2-lobed. **FR** 6–12 mm. **SEED**
1.5–2 mm; coat shallowly netted, loose-fitting, inner walls membranous, persistent. Moist places, esp meadows, bogs; < 3500
m. NW, CaR, SNH, c-w CW, SW, **GB**; to AK, Rocky Mtns. 2
sspp. in CA.

  ssp. *miniata* (p. 485) **ST** stout. **INFL** gen bright red (orange-red). **FL**: calyx 15–30 mm; corolla 20–40 mm. **FR** 10–12
mm. 2*n*=24,48,72,96,120. Common. Wet montane meadows,
streambanks; gen 1500–3500 m. Range of sp. ❀ DRN,IRR,
SUN,with host:1,2,4–7,14–17;DFCLT May–Sep

C. minor (A. Gray) A. Gray Ann, ± simple, ± slender, 30–150
cm, green to grayish, variously hairy. **LF** 40–100 mm, linear-
lanceolate, entire. **INFL** 10–40 cm, narrow, open below; bracts
20–50 mm, lf-like, entire, tips red, ± long-tapered; pedicels < 10
mm below (0 above). **FL**: calyx 14–28 mm, divided 2/3–3/4 in
back and front, ± 1/8 on sides, lobes narrow, acute;
corolla 15–35 mm, beak < tube, back ± puberulent, margins pale,
lower lip 2–3 mm, yellowish, exserted from calyx, spreading;
stigma slightly 2-lobed. **FR** 10–15 mm. **SEED** 1–1.5 mm; coat ±
deeply netted, most walls ladder-like. 2*n*=24. Wet places; <
2300 m. NCoR, c&s SNF, SnJV, CCo, n SnFrB, SCo, **GB**; to
WA, NM. 2 sspp. in CA.

  ssp. *minor* (p. 485) Hairs long, soft, mostly glandular. **FL**:
corolla 15–20(30) mm. Alkaline marshes; 1200–2200 m. **GB**;
to WA, NM. [*C. exilis* Nelson] Jul–Sep

C. nana Eastw. (p. 485, pl. 103) Per 5–25 cm, spreading-hairy,
mostly nonglandular. **LF** 10–35 mm, linear-lanceolate; lobes 0–
5. **INFL** 3–13 cm; bracts 15–30 mm, lobes 3–5, acuminate, yellow-green or purplish, green-margined. **FL**: calyx 12–20 mm,
subequally divided ± 1/2, lobes linear-lanceolate;
corolla 15–20 mm, beak 4–6 mm, pale yellow blotched purplish,
lower lip 3–5 mm, pouches shallow; stigma ± exserted, dark. **FR**
8–12 mm. **SEED** 1–1.5 mm; coat shallowly netted, loose-fitting,

with ladder-like thickenings. Dry, ± alpine barrens; 2400–3700 m. SNH, **W&I**. Jul–Aug

C. pilosa (S. Watson) Rydb. (p. 485) Per 8–35 cm, often decumbent, spreading-hairy, nonglandular. **LF** 10–50 mm, linear-lanceolate; lobes 0–3. **INFL** 3–20 cm; bracts 10–30 mm, lobes 3–5, pale green or purplish, white-margined, central lobe truncate or rounded. **FL**: calyx 10–20, subequally divided ± 1/2, lobes linear-lanceolate; corolla 13–22 mm, beak 4–6 mm, pale yellow-green, lower lip 3–5 mm, pouches shallow; stigma ± exserted, dark. **FR** 8–12 mm. **SEED** ± 1 mm; coat deeply netted, tight-fitting, walls ladder-like. 2*n*=24. Dry sagebrush scrub to alpine barrens; 1200–3400 m. n&c SNH, **GB**; to e OR, c ID, w WY. [ssp. *jussellii* (Eastw.) Munz; *C. psittacina* (Eastw.) Pennell] Highly variable; forms intergrade; needs further study. Jun–Aug

C. plagiotoma A. Gray (p. 485) MOJAVE PAINTBRUSH Per 30–60 cm, gray-green, becoming ± maroon, puberulent (esp lvs); hairs branched. **LF** 20–50 mm, ± linear; lobes 3–5. **INFL** 3–20 cm; bracts 13–20 mm, white-woolly lobes 3–5, central lobe wide, truncate, green. **FL**: calyx 12–18 mm, pale yellow, white-woolly,

divided 1/8 in back, 1/4 in front, ± 1/2 on sides, front lobes ± 2 mm > back lobes; corolla 12–20 mm, beak ± = tube, incl, yellowish, back puberulent, margins pale, lower lip 1 mm, pale green; stigma barely exserted, head-like. **FR** ± 10 mm. **SEED** 1–1.5 mm; coat deeply netted, tight-fitting, most walls ladder-like. UNCOMMON. Dry sagebrush scrub, pinyon woodland; 300–2500 m. s SN, s SnJV, SCoRI, TR, **DMoj**. Apr–Jun

C. tenuis (A.A. Heller) Chuang & Heckard (p. 485) Ann 10–45 cm, spreading-hairy, glandular above. **LF** 10–40 mm, linear-lanceolate; lobes 0–3(5). **INFL** 5–25 cm, 1–3 cm wide; bracts 10–30 mm, ovate, green, lobes 3–7, ± lanceolate. **FL**: calyx 8–12 mm, divided 1/3 in front and on sides, 1/2 in back; corolla 12–20 mm, white or yellow, beak 4–5 mm, straight, puberulent, lower lip 3–4 mm, pouches 2–4 mm wide, ± 2 mm deep; stigma well incl. **FR** 6–9 mm. **SEED** ± 1 mm; coat ± deeply netted, loose-fitting. 2*n*=24,48. Moist flats, meadows; 1000–2800 m. NCoR, CaR, SNH, MP, **n SNE**, rare in SnBr and PR; to AK. [*Orthocarpus hispidus* Benth.] May–Aug

COLLINSIA

Elizabeth Chase Neese

Ann, often glandular, sometimes brown-staining. **LVS** opposite; lower petioled. **INFL** bracted, often interrupted; fls 1–many in lf axils. **FL**: calyx 5-lobed; corolla ± pea-like, gen glabrous outside, tube short, throat ± angled to tube, ± pouched on upper side, lips gen ± = throat, upper lobes 2, ± reflexed, gen paler, lower lobes 3, lateral spreading, central lobe keeled, enclosing stamens and style; stamens 4, attached unequally near throat base; staminode gland-like. **FR** septicidal and loculicidal (valves 2-lobed). **SEEDS** gen few, ± oblong, gen plump; inner surface ± hollow. ± 18 spp.: N.Am, esp CA. (Zaccheus Collins, 1764–1831, Philadelphia botanist) Late-season fls gen atypically small. Key to species adapted by Margriet Wetherwax.

1. Fls crowded in whorls; lower pedicels < calyx; corolla 9–14 mm, lavender to purple
 .... *C. bartsiifolia* var. *davidsonii*
1' Fls 1–5 per node, not very crowded; lower pedicels > calyx; corolla 4–9 mm, lavender-blue or blue and white
 2. St, lvs ± thick, fleshy; infl conspicuously glandular; corolla 7–9 mm, lavender-blue .... *C. callosa*
 2' St slender; lvs thin; infl glabrous to sparsely glandular; corolla 4–8 mm, gen white with lower lobes blue
 .... *C. parviflora*

C. bartsiifolia Benth. Pl 7–35 cm. **LF** gen 1–4 cm, ± oblong, thickish, obtuse, crenate, rolled under, gen finely hairy. **INFL** interrupted, ± finely glandular or shaggy; whorls dense; pedicels < calyx. **FL**: calyx lobes ± blunt; corolla glabrous outside, gen whitish to pinkish lavender (purplish), veiny when pressed, throat longer than wide, hairy inside, lips ± equal; upper lobes ± oblong, toothed, often back-to-back, lateral lobes obovate, notched; upper filaments hairy, spur 0–0.5 mm. **SEEDS** many, ± plump. *n*=7. Open sandy places; < 1300 m. NW, CaRF, SNF, CW, **w DMoj**. 2 vars. in CA.

 var. *davidsonii* (Parish) V. Newsom **FL**: corolla 9–14 mm, gen pale lavender to purple. **SEEDS** ± 1.5 mm. Habitats of sp.; 500–1300 m. s SNF, SCoRI, **w DMoj**. ❀ TRY. Apr–Jun

C. callosa Parish (p. 485) Pl 4–25 cm, stout, fleshy. **LVS** gen < 3 cm, oblong to ovate, obtuse, gen entire, ± rolled under; middle and upper lvs clasping. **INFL** open, glandular; bracts > 2 mm; fls

1–3 per node, pedicels > calyx, not reflexed. **FL**: calyx in fr gen > 5 mm wide, base ± truncate, lobes ± = fr; corolla gen 7–9 mm, lavender-blue. **SEEDS** 6–8, ± 2 mm. Disturbed rocky slopes, open chaparral, sagebrush scrub, pinyon/juniper or pine woodland; 1000–2300 m. s SNH, Teh, TR, **nw DMtns**. Apr–Jun

C. parviflora Lindley (p. 485) BLUE-EYED MARY Pl 3–40 cm. **LF** ± linear-lanceolate, obtuse, gen rolled under. **INFL** open, leafy, glabrous to sparsely and finely glandular; pedicels > calyx, gen 1 per node below, 3–5 per node above, often reflexed. **FL**: calyx > 2/3 corolla, ± = fr, lobes sharply acute to acuminate; corolla 4–8 mm, throat barely angled to tube, tube and throat white, narrowed to lips, pouch angular, ± hidden by calyx, upper lip whitish or blue-tipped, main lobes ± 1 mm wide, oblong, gen blue (purplish). **SEEDS** gen 4. *n*=7. Common. Moist, ± shady places in mtns; 800–3500 m. CA-FP, **GB**; to B.C., e Can, Colorado. ❀ TRY. Apr–Jun

CORDYLANTHUS BIRD'S-BEAK

T.I. Chuang & Lawrence R. Heckard

Ann, green root-parasites, gen much-branched. **LVS** sessile, 0–11-lobed. **INFL**: spike (subtended by outer bracts) or fls solitary (each subtended by outer bracts) but often clustered; outer bracts ± lf-like; inner bract calyx-like (formerly confused with calyx). **FL**: calyx gen divided to base in front, sheath-like, tip gen entire or shallowly notched; corolla ± club-shaped, upper lip beak-like, enclosing anthers and style, tip closed, lower lip ± = upper, pouched, middle lobe gen tightly rolled under; stamens gen 4, anther sacs 1–2 per stamen, unequal; style bent near tip, stigma unexpanded, ± exserted downward from closed beak tip. **FR** loculicidal. **SEEDS** gen 10–20, attached at side; coat netted or ridged, tight-fitting. 18 spp.: w N.Am. (Greek: club-shaped fl) [Chuang & Heckard, 1986 Syst Bot Monogr 10:1–105] Close to *Orthocarpus*, distinguished by infl and calyx. Gen fls late summer. Key to species adapted by Margriet Wetherwax.

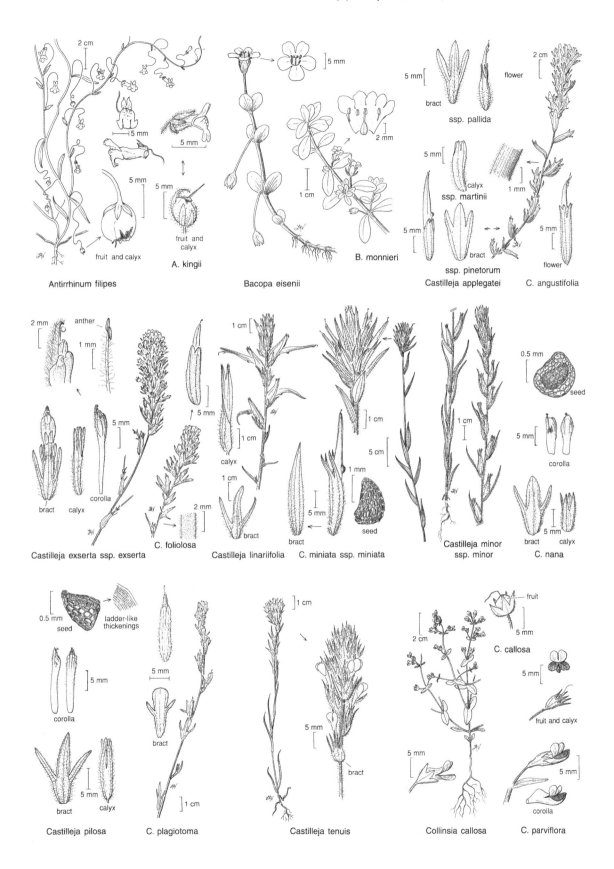

2 cm

5 mm

5 mm

fruit and calyx

Antirrhinum filipes

fruit and calyx

A. kingii

5 mm

5 mm

fruit and calyx

1 cm

2 mm

5 mm

Bacopa eisenii

B. monnieri

bract

flower

5 mm

ssp. pallida

5 mm

calyx

1 mm

ssp. martinii

5 mm

bract

ssp. pinetorum

Castilleja applegatei

2 cm

5 mm

flower

C. angustifolia

2 mm   anther

1 mm

5 mm

bract   calyx

corolla

Castilleja exserta ssp. exserta

5 mm

2 mm

C. foliolosa

1 cm

calyx

1 cm

bract

Castilleja linariifolia

1 cm

1 cm

1 mm

5 cm

5 mm

bract

seed

C. miniata ssp. miniata

1 cm

Castilleja minor
ssp. minor

0.5 mm

seed

5 mm

corolla

5 mm

bract   calyx

C. nana

0.5 mm
seed

ladder-like
thickenings

5 mm

corolla

5 mm

bract   calyx

Castilleja pilosa

5 mm

bract

1 cm

C. plagiotoma

1 cm

5 mm

5 mm

bract

Castilleja tenuis

2 cm

fruit

5 mm

C. callosa

5 mm

fruit and calyx

5 mm

5 mm

corolla

Collinsia callosa

C. parviflora

1. Lf linear to lanceolate, entire; infl a spike, 20–150 mm; middle lobe of lower corolla lip erect, not rolled under; saline or alkaline places (subg. *Hemistegia*)
  2. Stamens 4; inner bract entire or notched . . . . *C. maritimus* ssp. *canescens*
  2' Stamens 2; inner bract 3–7-lobed . . . . *C. tecopensis*
1' Lf thread-like to linear, entire to subpalmately 5-lobed; infl a spike < 30 mm or fls solitary, in loose clusters; middle lobe of corolla lip tightly rolled under; nonsaline places (subg. *Cordylanthus*)
  3. Calyx divided to base; corolla ± dusty yellow . . . . *C. ramosus*
  3' Calyx tube 1–4 mm; corolla gen pink to dull purple-yellow
    4. Pl canescent, not glandular-sticky; outer bracts 3–7-lobed . . . . *C. eremicus* ssp. *eremicus*
    4' Pl densely glandular-sticky; outer bracts gen 3- or 5-lobed
      5. Infl a spike, 20–30 mm; corolla lips ± equal; inner bract pinnately lobed; SNE . . . . . . *C. kingii* ssp. *helleri*
      5' Fls solitary, scattered, 1–4 per loose cluster; lower corolla lip << upper; inner bract gen entire; e DMtns
      . . . . *C. parviflorus*

C. eremicus (Cov. & C. Morton) Munz Pl 10–80 cm, (yellow-) green, often tinged red, canescent. LF 10–40 mm, thread-like or linear, entire to 5-lobed. INFL: spike, 15–25 mm, 3–14-fld; outer bracts 3–7, 5–20 mm, lobes 3–7; inner bract 10–18 mm. FL: calyx 10–18 mm, tube 1–3 mm; corolla 10–20 mm, purplish, pinkish, or yellow-green, often yellow-tipped, throat base blotched maroon, pouch 4–6 mm wide, soft-white-hairy; stamens 4, anther sacs 2. SEED 1.5–2 mm, ± ovoid, deeply netted, pale brown. 2*n*=26. Dry rocks; gen 1800–3000 m. s SNH, n SnBr, **n DMtns**. 2 sspp. in CA.

  ssp. **eremicus** (p. 491) INFL: outer bracts ± sparsely scabrous, tips wider, ± maroon-thickened. FL: calyx tube 2–3 mm; corolla lavender to pinkish, blotched purple. SEED papillate between nets. Sagebrush scrub, pinyon/juniper woodland; gen 1800–2800 m. n SnBr (near Cushenbury, 1000 m), **n DMtns**. [*C. ramosus* Benth. ssp. *e.* (Cov. & C. Morton) Munz; *C. bernardinus* Munz] Aug–Oct

C. kingii S. Watson ssp. **helleri** (Ferris) Chuang & Heckard (p. 491, pl. 104) Pl 10–60 cm, gray-green, tinged red, densely glandular-sticky. LF gen 10–25 mm, linear. INFL: spike, 20–30 mm, 1–4- fld; outer bracts 3–6, 10–15 mm, gen 3-lobed; inner bract 10–15 mm, 3–5-lobed, glandular. FL: calyx 15–20 mm, tube 2.5 mm; corolla 15–25 mm, rosy-lavender to dull purple-red or -yellow, hairy; stamens 4, anther sacs 2. SEED 2–2.5 mm, ± ovoid, deeply netted, pale brown, papillate between nets. 2*n*=26. Open pinyon/juniper woodland or sagebrush scrub; 1300–3100 m. s MP, **SNE**; NV. [*C. h.* (Ferris) J.F. Macbr.] Ssp. *k.* in NV, UT. Jul–Sep

C. maritimus Benth. Pl 10–40 cm, gray-green, glaucous, often tinged purple and salt-encrusted, gen ± short-hairy. LF 5–25 mm, ± linear-lanceolate, entire. INFL: spike, 20–90 mm, many-fld; outer bract lf-like; inner bract 15–30 mm. FL: calyx 15–25 mm; corolla 15–25 mm, white to cream, puberulent, lips pale to brownish or purplish red, middle lobe of lower lip erect; stamens 4, anther sacs 2 (lower pair) or 1 (upper pair). SEEDS 10–40, 1–3 mm, ± reniform, deeply netted, dark brown. 2*n*=30. Coastal salt marshes (< 10 m), inland alkaline flats (1200–1900 m). n NCo, n CCo, SCo, **SNE**; to OR, UT, n Baja CA. 3 sspp. in CA.

  ssp. **canescens** (A. Gray) Chuang & Heckard (p. 491) ST: branches gen many, ± erect, upper gen > central spike. INFL dense; inner bract gen entire. SEEDS 25–40, 1–1.5 mm. Inland alkaline flats; 1200–1900 m. **SNE**; to s OR, UT. [*C. c.* A. Gray] ✿STBL. Jun–Sep

C. parviflorus (Ferris) Wiggins (p. 491) PURPLE BIRD'S-BEAK Pl 20–60 cm, gray-green, tinged red, glandular-sticky and long-hairy. LF 5–30 mm, linear. INFL fls solitary, 1–4 per loose cluster; outer bract gen 1, 5–15 mm, 3–5-lobed, tips obtuse, ciliate, densely glandular; inner bract 10–12 mm. FL: calyx 10–15 mm, tube 1–1.5 mm; corolla 15–20 mm, pink to lavender, pouch ± 7 mm wide, often dark-veined; stamens 4, anther sacs 2. SEED 1.5–2 mm, ± ovoid, shallowly netted, dark brown, densely papillate between nets. 2*n*=26. RARE in CA. Dry sagebrush scrub or pinyon/juniper/Joshua-tree woodland; 700–2200 m. **e DMtns (New York, Providence mtns)**; to se NV, sw UT, nw AZ, also s-c ID. Aug–Oct

C. ramosus Benth. (p. 491) Pl 10–90 cm, gray-green or tinged red, ± canescent. LF 10–40 mm, ± thread-like, entire to 5-lobed. INFL: fls solitary; clusters spike-like, 15–25 mm, 3–7-fld; outer bract 1 per fl, 10–20 mm, entire to 7-lobed, sometimes bristly, segments ± thread-like; inner bract 10–20 mm, entire. FL: calyx 10–15 mm; corolla 10–20 mm, dusty yellow, often marked maroon, ± puberulent; stamens 4, anther sacs 2. SEED ± 2 mm, ± ovoid, deeply netted, light brown. 2*n*=24. Rocky to alkaline soils in sagebrush scrub; 1200–2800 m. n CaRH (Shasta Valley), **GB**; to OR, WY, Colorado. [ssp. *setosus* Pennell] Jul–Aug

C. tecopensis Munz & Roos (p. 491) TECOPA BIRD'S-BEAK Pl 10–60 cm, grayish or tinged purple, sparsely puberulent, glaucous. LF 5–15 mm, linear-lanceolate, entire. INFL: spike, 20–150 mm, loose; outer bract lf-like; inner bract 10–15 mm, lobed near middle. FL: calyx 10–13 mm; corolla 10–15 mm, pale lavender, densely puberulent, middle lobe of lower lip erect; stamens 2 (upper pair vestigial); style puberulent. SEEDS 8–10, 2–3 mm, ± reniform, deeply netted, light brown. 2*n*=28. RARE. Alkaline meadows and flats; 100–900 m. **s SNE, n DMoj**; w NV. Aug–Oct

KECKIELLA

Noel H. Holmgren

Subshrub or shrub. ST wand-like to much-branched. LVS drought-deciduous, ± opposite or in 3's, gen short-petioled. INFL: panicle or spike-like; bracts small. FL: calyx lobes 5, ± equal; corolla not purplish (whitish to red), short-glandular outside, upper lip hooded, lobes short, rounded, external in bud, lower lip rounded, lobes often reflexed; filaments densely nonglandular-hairy at base, anthers small, glabrous, anther sacs gen spreading flat at dehiscence; staminode well developed, glabrous to densely bearded; nectary a disk; stigma head-like. FR ovoid, septicidal and sometimes also loculicidal at tip. SEEDS many, irregularly angled or ± winged. 7 spp.: esp CA. (David D. Keck, CA botanist, 1903–1995) [Straw 1966 Brittonia 18:80–95] Segregated from *Penstemon*.

1. Corolla 15–23 mm, tube + throat < upper lip . . . . *K. antirrhinoides* var. *microphylla*
1' Corolla 10–12 mm, tube + throat > upper lip . . . . *K. rothrockii* var. *rothrockii*

K. antirrhinoides (Benth.) Straw **ST** spreading to erect, 6–25 dm; young sts canescent (rarely glabrous). **LVS** ± opposite, in axillary clusters on older sts; blade 5–20 mm, (ob)lanceolate to narrowly (ob)ovate, base tapered, margin gen entire. **INFL** finely short-hairy and sparsely glandular. **FL**: corolla 15–23 mm, yellow (drying blackish), tube + widely expanded throat 6–10 mm, upper lip 8–15 mm; anther sacs 1.1–1.8 mm; staminode densely yellow-hairy, exserted. Scrub, woodland; 100–1600 m. SnBr, PR, **D**; AZ, Mex. 2 vars. in CA.

var. *microphylla* (A. Gray) N. Holmgren (p. 491, pl. 105) Pl canescent. **FL**: calyx 5.5–9 mm, lobes lanceolate, tips acute to acuminate; anther sacs 1.1–1.5 mm. Juniper/pinyon woodland, Joshua-tree scrub; 400–1800 m. **D**; AZ. [ssp. *m*. (A. Gray) Straw] ❀ TRY.

K. rothrockii (A. Gray) Straw Pl wide, low, 3–6 dm. **ST** densely short-hairy; young sts green. **LVS** subopposite or in 3's, subsessile; main lvs 5–16 mm, (ob)lanceolate to widely obovate, entire or finely serrate towards tip. **INFL** short-hairy. **FL**: calyx 4–7 mm, sometimes glandular, lobes lanceolate; corolla 11–16 mm, brownish yellow, pale yellow, or cream, purple- or reddish brown-lined, tube + expanded throat 7–11 mm, upper lip 4–5 mm; anther sacs 0.8–1.1 mm, barely spreading at dehiscence; staminode glabrous distally, ± incl. Sagebrush steppe, juniper/pinyon woodland; 1900–3200 m. SnJt, **SNE**, **DMtns**; NV. 2 vars. in CA.

var. *rothrockii* (p. 491) **LF** canescent. **FL**: corolla 10–12 mm, becoming glabrous. Habitats and elevations of sp. **SNE**, **DMtns**; NV. ❀ TRY.

LIMOSELLA MUDWORT

Margriet Wetherwax

Ann, sometimes partly submersed, with stolons. **LVS** basal, erect; stipules gen present. **INFL** scapose; fl 1. **FL**: calyx bell-shaped, lobes 5, ovate, acute; corolla radial, bell-shaped, lobes 5, upper surfaces sparsely papillate; stamens 4; style gen subterminal, stigmas gen fused, head-like. **FR** elliptic to spheric, chambers 2 below, 1 above. **SEEDS** many, minute. ± 15 ssp.: worldwide. (Latin: mud seat, from habitat)

L. acaulis Sess & Mociño (p. 491) Cespitose, gen mat-forming. **LF**: 1–6 cm, 0.5–2 mm wide, flat, linear to ± spoon-shaped; petiole gen not differentiated; stipules gen ear-shaped, transparent. **INFL**: pedicels < 2/3 lf length, spreading in fr. **FL**: calyx lobes ± = tube; corolla 2–3 mm, white to lavender, lobes rounded; stamens attached at 1 level; style 0.5–1 mm, erect to slightly curved. **FR** 3–5 mm, spheric. Wet, muddy places; < 3300 m. SNH, GV, SnFrB, SW, **GB**; to NM, Mex. ❀ TRY. May–Oct

LINDERNIA FALSE PIMPERNEL

Margriet Wetherwax

Ann, bien. **ST** openly branched. **LVS** opposite, entire to finely dentate. **FL**: calyx radial, sepals 5, ± free; corolla 2-lipped, upper lip erect, 2-lobed, lower lip > upper, spreading, 3-lobed, throat with 2 yellow, hairy ridges; fertile stamens 2 (lower stamen pair antherless, forming corolla throat ridges and with free, forked, filament-like tips); stigmas 2, flat. ± 50 spp.: esp trop, subtrop. (Franz B. von Lindern, German physician, botanist, 1682–1755)

L. dubia (L.) Pennell Ann, glabrous. **ST** < 38 cm, spreading to erect. **LF** 1–37 mm; blade lanceolate to ovate, entire to finely dentate, tapered to round-clasping at base. **INFL**: pedicels 0.5–28 mm. **FL**: sepals < 7 mm, linear; corolla 7–10 mm, white to bluish or lavender. **FR** < or = sepals. **SEED** length 1.5–3 × width, yellow. Wet places; < 1600 m. KR, SN, GV, SnFrB, **SNE**; N.Am, S.Am. 2 vars. in CA.

var. *dubia* (p. 491) **LF** 1–37 mm, slightly reduced above; blade lanceolate, lanceolate to ovate; lower lvs (or all) tapered at base. **INFL:** pedicels 0.5–27 mm, < or slightly > subtending lvs. **FL**: corolla 9–10 mm, pale lavender, darker distally. **FR** < or = calyx. **SEED** length 2–3 × width, pale yellow. Streamsides; 100–700 m. KR, SN, GV, SnFrB, **SNE**; to WA, e US, Mex. Jul–Sep

MAURANDYA

David M. Thompson

Per, glabrous to hairy. **ST** prostrate, erect, pendent, or climbing by twining sts or petioles. **LVS** gen alternate (lowest opposite on seedlings), entire to irregularly bristly-dentate; veins palmate. **INFL**: fls solitary in lf axils. **FL**: calyx lobes 5, ± equal; corolla tube with sac-like extension at base, tube-throat floor gen with 2 longitudinal folds, lower lip base gen not swollen, gen not closing mouth. **FR** ovoid to ± spheric, often oblique; chambers 1–2, equal or not, each gen dehiscent by 2–3 pores or irregularly near tip. **SEEDS** many, ± tubercled, pitted, or winged. 20 spp.: sw US, Mex. (C.P. Maurandy, botany professor, Cartagena, Spain, ± 18th century) *For alternate taxonomy see see Ghebrehiwet et al 2000 Pl Syst Evol 220:223–239; Elisens 1985 Syst Bot Monogr 5:1–97.

1. Pl glabrous; lf hastate to sagittate, entire; sts or petioles twining; corolla ± reddish to violet
. . . . *M. antirrhiniflora* ssp. *antirrhiniflora*
1' Pl hairy; lf ± round to reniform, irregularly bristly-dentate; st erect to pendent, neither sts nor petioles twining; corolla yellowish to white . . . . *M. petrophila*

M. antirrhiniflora Willd. ssp. *antirrhiniflora* (p. 491) VIOLET TWINING SNAPDRAGON Pl glabrous; sts or petioles twining. **LF** hastate to sagittate, entire; **INFL**: pedicel 12–47 mm. **FL**: calyx lobes entire; corolla ± reddish to violet, tube-throat 13–17 mm, floor without longitudinal folds, lower lip swollen at base, ± closing throat. **FR**: chambers unequal. $2n=24$. RARE in CA.

Desert flats, washes; 0–2600 m. e DMtns (**Providence Mtns, e San Bernardino Co.**); to TX, Mex. In cult. *Maurandella antirrhiniflora* (Willd.) Rothm. Apr–May

M. petrophila Cov. & C. Morton ROCK LADY (p.491) Pl hairy; neither sts nor petioles twining. **ST** erect to pendent. **LF** ± round to reniform, irregularly bristly-dentate. **INFL**: pedicel 1–4 mm. **FL**: calyx lobes irregularly bristly-dentate; corolla tube-throat 20–24 mm. **FR**: chamber 1 (septum incomplete). **RARE** CA. Limestone crevices of canyons; 1200–1400 m. **n DMoj (Titus, Fall canyons, Death Valley region, Inyo Co.).** **Holmgrenanthe petrophila* (Cov. & C. Morton) Elisens Apr–Jun

MIMULUS MONKEYFLOWER

David M. Thompson

Ann to shrub, glabrous to hairy. **ST** gen erect. **LVS** opposite, gen ± sessile, gen toothed or gen entire, reddish or gen green. **INFL**: raceme, bracted, or fls gen 2 per axil. **FL** sometimes cleistogamous; calyx gen green, lobes 5, gen << tube, equal or not, gen uppermost largest; corolla gen deciduous, white to red, maroon, purple, gold or yellow, limb width measured at widest point looking into fl, lower lip base sometimes swollen, ± closing mouth, tube-throat floor gen with 2 longitudinal folds; pollen chambers spreading; placentas 2, axile or parietal; stigma lobes gen lf-like, gen incl. **FR** gen ovoid to fusiform, gen upcurved if elongate, gen ± fragile, loculicidal near tip (sometimes hard, indehiscent); chambers 1–2. **SEEDS** many, gen < 1 mm, ovoid, yellowish to dark brown. ± 100 spp.: w N.Am, Chile, e Asia, s Afr, New Zealand, Australia. (Latin: little mime or comic actor, from face-like corolla limb of some) [Grant 1924 Ann Missouri Bot Gard 11:99–388]

1. Pedicel gen < calyx; corolla persistent (subg. *Schizoplacus*)
  2. Subshrub; calyx 2–4 cm (sect. *Diplacus*) . . . . *M. aurantiacus*
  2' Ann; calyx < 2 cm
    3. Corolla radial, salverform, maroon and ± white, throat floor without longitudinal folds (sect.
      Mimulastrum) . . . . *M. mohavensis*
    3' Corolla bilateral, gen not maroon and white, throat floor gen with 2 longitudinal folds
      4. Fr hard, indehiscent while st alive; corolla lobes each with 1 round magenta-purple spot at base
        (sect. *Oenoe*) . . . . *M. rupicola*
      4' Fr fragile, gen promptly dehiscent; corolla lobes without round basal spots (sect. *Eunanus*)
        5. All fl nodes with 1 fl or fr developing . . . . *M. fremontii*
        5' At least some fl nodes (gen upper) with 2 fls or frs developing
          6. Stigma ± exserted
            7. Lvs and calyx hairy; corolla ± magenta or yellow . . . . *M. mephiticus*
            7' Lvs and calyx ± puberulent; corolla ± magenta . . . . *M. nanus*
          6' Stigma incl
            8. Calyx puberulent, often purplish throughout, upper lobe rounded, ± with small point at tip, fr incl
            . . . . *M. parryi*
            8' Calyx hairy, sometimes purplish on ribs, upper lobe long-tapered; fr exserted . . . . *M. bigelovii*
              9. Lower internodes gen > upper; lf elliptic, sometimes acuminate but not abruptly so . . . . . var. *bigelovii*
              9' Lower internodes ± = upper; lf gen wide-ovate to ± round, abruptly sharp-pointed . . . . . var. *cuspidatus*
1' Pedicel gen > calyx; corolla deciduous (subg. *Mimulus*)
  10. Calyx in fr strongly and asymmetrically swollen, uppermost lobe longest, lowest 2 gen upcurved;
    corolla yellow, lower lip base gen swollen, ± closing mouth (sect. *Simiolus*)
    11. Fls gen > 5 in a definite, bracted raceme; ann or per; corolla tube-throat 2–40 mm; fls cleistogamous
      or opening . . . . *M. guttatus*
    11' Fls 1–5 per st, solitary in axils of upper lvs, not in bracted raceme; per; corolla tube-throat 17–45 mm;
      fls opening . . . . *M. tilingii*
  10' Calyx in fr not (or symmetrically) swollen, lobes equal, not upcurved; corolla yellow or not, lower
    lip base not swollen, not closing mouth
    12. Per; corolla tube-throat 4–5 cm (sect. *Erythranthe*) . . . . *M. cardinalis*
    12' Ann or per; corolla tube-throat < 2.5 cm
      13. Calyx lobes ± = tube; pl hairs dense, long, soft, wavy (sect. *Mimuloides*) . . . . *M. pilosus*
      13' Calyx lobes << tube; pl glabrous, puberulent, or hairs ± straight (sect. *Paradanthus*)
        14. Per; stolons or rhizomes present
          15. Calyx glabrous; lf palmately 3-veined . . . . *M. primuloides* ssp. *primuloides*
          15' Calyx hairy; lf pinnately 5- or more veined . . . . *M. moschatus*
        14' Ann; stolons or rhizomes 0
          16. Calyx tube ± hairy
            17. Lf palmately veined; calyx 8–10(13 in fr) mm; corolla pinkish to white . . . . *M. parishii*
            17' Lf ± pinnately veined; calyx 3–8 mm; corolla yellow . . . . *M. floribundus*
          16' Calyx tube glabrous to puberulent
            18. Corolla lobes lavender-magenta or yellow
              19. Calyx lobes ciliate . . . . *M. rubellus*
              19' Calyx lobes glabrous . . . . *M. suksdorfii*
            18' Corolla lobes, or at least some, purple or maroon
              20. Lowest corolla lobe comprising almost the entire lower lip, yellow . . . . *M. shevockii*
              20' Lobes (3) of lower lip ± equal, purple
                21. Corolla tube-throat 5–8 mm . . . . *M. androsaceus*
                21' Corolla tube-throat 10–22 mm . . . . *M. palmeri*

M. androsaceus E. Greene (p. 491) Ann, 0.5–9 cm, minutely puberulent. **LF** 3–13 mm, ± lanceolate to oblong or ovate, narrowly sheathing st. **FL:** pedicel 7–27 mm, ± spreading (in fr ascending at tips); calyx 3.5–6.5 mm, tube gen minutely puberulent, lobes equal, 0.2–1 mm, ± glabrous; corolla red-purple, tube-throat 5–8 mm, limb 4–6 mm wide; placentas axile. **FR** 4–5 mm. Uncommon. Moist runoff areas on gentle slopes; < 2100 m. NCoRI, Teh, CW, WTR, SnBr, e PR (Santa Rosa Mtns, Riverside Co.), **w edge DMoj**. Mar–Jun

M. aurantiacus Curtis (p. 491) Subshrub, shrub, glabrous to hairy. **ST** 10–150 cm; main lf axils often with clusters of smaller lvs. **LF** 20–80 mm, linear to obovate; edges gen rolled under; upper surface glabrous, often sticky. **FL:** pedicel 3–30 mm; calyx 20–37 mm, not swollen at base, glabrous to hairy, lobes unequal, 3–10 mm, acute to acuminate; corolla persistent, white to buff, yellow, orange, or red, tube-throat 25–60 mm; placentas parietal. **FR** 12–20 mm, splitting only along upper suture. *n*=10. Common. Rocky hillsides, cliffs, canyon slopes, disturbed areas, borders of chaparral, open forest; < 1600 m. CA-FP, **nw edge DSon**. [sspp. *australis* (McMinn) Munz, *lompocensis* (McMinn) Munz] Highly complex, with many intergrading, hybridizing, local forms. ❀ DRN,IRR fall through spring & DRY when dormant:**7**,8,9,**14**,18,**19–23**&SUN:4,5,**15–17,24**;CVS. Mar–Jun

M. bigelovii (A. Gray) A. Gray Ann, 2–25 cm, densely hairy. **LF** 7–35 mm, elliptic to round, often abruptly acuminate. **FL:** pedicel 1–4 mm; calyx 6–13 mm, sometimes purplish but only on ribs, hairy, lobes spreading, unequal, 1.5–6 mm, long-tapered; corolla persistent, rose to dark magenta, tube-throat 12–22 mm, throat floor gold, mouth gen with 2 lateral, dark maroon patches; placentas parietal. **FR** 7–13 mm, exserted. *n*=8. Rocky desert slopes, margins of washes; 120–2300 m. **s SNE, D**; to sw UT, w AZ. Variable; vars. intergrade.

  var. *bigelovii* (p. 491) **ST:** lower internodes gen > upper. **LVS** elliptic, sometimes acuminate but not abruptly so, upper narrower, longer-tapered than lower. Habitats of sp.; 120–1700 m. **D**; s NV, w AZ. ❀ DRN,SUN:6,**7,14**,15–17&IRR:**8–12**,13,**18–24**. Late Feb–Jun

  var. *cuspidatus* A.L. Grant **ST:** lower internodes ± = upper (all reduced in severe drought). **LVS** gen widely ovate to ± round, abruptly sharp-pointed, upper lvs as wide or wider, often more abruptly sharp-pointed than lower lvs. Habitats and elevations of sp. **s SNE, n D**; to sw UT, nw AZ. [var. *panamintensis* Munz; *M. spissus* A.L. Grant] ❀ TRY. Late Feb–Jun

M. cardinalis Benth. (p. 491) Per, rhizomed, hairy. **ST** 25–80 cm, often decumbent or ascending. **LVS** 20–80 mm, oblong to obovate, palmately 3–5-veined; upper clasping st. **FL:** pedicel 50–80 mm; calyx 20–30 mm, hairy, lobes equal, 4–5 mm, acute to acuminate; corolla orange to red, tube-throat 40–50 mm, upper lip arched forward, lower lip reflexed; anthers, stigma exserted but arched in upper lip; placentas axile. **FR** 16–18 mm. *n*=8. Moist to wet places along streams, seepage areas; < 2400 m. CA-FP, **W&I, DMtns (Panamint Mtns)**; to se OR, sw UT, w-c NV, w NM, n Baja CA. ❀ WET or IRR:1–3,**7–9**,10,11,**14,18–23**&SUN:4,5,**6,15–17,24**;CV. Apr–Oct

M. floribundus Lindley (p. 491) Ann, hairy, often ± slimy. **ST** 3–50 cm, often decumbent or ± climbing. **LF:** petiole 0–20 mm; blade 5–45 mm, lanceolate to ovate, base gen rounded to cordate, veins ± pinnate. **FL:** pedicel 5–30 mm, not reflexed in fr; calyx 3–8 mm, ± hairy, lobes equal, 1–2 mm, ± acute; corolla yellow, tube-throat 6–15 mm; placentas axile. **FR** 4–7 mm. *n*=16. Crevices, seeps around granite outcrops, near streams; < 2500 m. CA-FP (esp c&s SNF), **W&I, DMtns (Panamint Mtns)**; to B.C., SD, n Mex. [ssp. *subulatus* A.L. Grant; *M. arenarius, M. dudleyi* A.L. Grant] Many minor, ± indistinct forms (if corolla tube-throat > 15 mm, see *M. moschatus*). ❀ DRN,SUN: 4–6&IRR:1–3,7,8–10,**14–24**. Apr–Jul

M. fremontii (Benth.) A. Gray (p. 491) Ann, 1–20 cm. **LVS** 2–30 mm, gen narrowly elliptic (to obovate); lowest glabrous, upper hairy. **INFL:** fl 1 per node. **FL:** pedicel 1–4 mm; calyx 5–14 mm, wide-ribbed, swollen, often reddish, often ± whitish hairy,

lobes ± equal, 0.5–3.5 mm, acute to acuminate; corolla persistent, magenta to red-purple (rarely yellow), ± darker at mouth, tube-throat 10–23 mm, throat floor ± glabrous, folds gold; placentas parietal. **FR** 7–13 mm. *n*= 8. Sandy, disturbed areas among shrubs, often on banks, benches along streams; 75–2100 m. SCoR, SW, **s DMoj**; n Baja CA. [*M. subsecundus* A. Gray, one-sided monkeyflower] ❀ TRY. Late Mar–Jun

M. guttatus DC. (p. 491) Ann or rhizomed per, 2–150 cm, glabrous to hairy. **LVS** abruptly reduced to sessile bracts; petioles 0–95 mm; blades 4–125 mm, ovate to round, often crenate, base often irregularly small-lobed or dissected. **INFL:** raceme, gen > 5-fld; bracts ovate to cordate, fused at base or not, not glaucous. **FL** cleistogamous or opening; pedicel 10–80 mm; calyx 6–30 mm, asymmetrically swollen in fr, glabrous to hairy, lobes unequal, lowest 2 upcurved in fr; corolla yellow, tube-throat 2–40 mm; placentas axile. **FR** 5–12 mm. *n*=14,15,16,24,28. Common. Wet places, gen terrestrial, sometimes emergent or floating in mats; < 2500 m. **CA**; to AK, w Can, Rocky Mtns, n Mex. [sspp. *arenicola* Pennell, *arvensis* (E. Greene) Munz, *litoralis* Pennell, *micranthus* (A.A. Heller) Munz; *M. glabratus* Kunth ssp. *utahensis* Pennell, Utah monkeyflower; *M. microphyllus* Benth., small-lvd monkeyflower; *M. nasutus* E. Greene; *M. whipplei* A.L. Grant, Whipple's monkeyflower] Exceedingly complex: local populations may be unique but their forms intergrade over geog or elevation; variants not distinguished here. ❀ WET or IRR:1–3,**7–9**,10–12,**14,18–23**&SUN:**4–6,15–17,24**;occas.INV. Mar–Aug

M. mephiticus E. Greene (p. 491, pl. 106) Ann, 1–15 cm, hairy. **LF** 6–30 mm, ± linear to narrowly lanceolate; tip obtuse to rounded. **FL:** pedicel 2–7 mm; calyx 5–11 mm, hairy, lobes ± unequal, 1–3 mm, acute to long-tapered; corolla persistent, ± magenta or yellow (often mixed in a population), tube-throat 11–19 mm, lower lip base with 3 maroon lines and maroon spots at mouth; stigma and sometimes anthers ± exserted; placentas parietal. **FR** 6–11 mm. *n*=8. Bare, sandy or gravelly areas, often around granite outcrops; gen 1500–3500 m. SN, s MP, **SNE**; NV. [*M. coccineus* Congdon; *M. densus* A.L. Grant] Corolla more often magenta (gradually darker), pl more tufted, anthers more often exserted from low to high elevations. Early May–early Sep

M. mohavensis Lemmon (p. 495) MOJAVE MONKEYFLOWER Ann, 2–10 cm, ± puberulent. **LF** 7–27 mm, narrowly elliptic, red-purple. **FL:** pedicel 2–5 mm; calyx 7–15 mm, enlarged in fr, red-purple, ± puberulent along veins, lobes spreading, ± unequal, 2–4 mm, acuminate, ciliate; corolla radial, salverform, persistent, tube-throat 9–15 mm, tube-throat, limb at base ± solid maroon, veins maroon, fading into white border, appearing << 0.5 mm wide; placentas parietal. **FR** 8–13 mm. *n*=7. RARE. Gravelly banks of desert washes; 600–1000 m. **DMoj (near Barstow, San Bernardino Co.)**. Early Apr–mid-May

M. moschatus Lindley (p. 495) MUSK MONKEYFLOWER Per, rhizomed, ± glabrous to densely slimy-hairy, often musk-scented. **ST** 5–30 cm, prostrate to ascending. **LF:** petiole 0–15 mm; blade 10–60 mm, oblong to ovate, pinnately 5+-veined. **FL:** pedicel 10–50 mm; calyx 8–12 mm, ± glabrous to hairy, lobes equal, 2–5 mm; corolla yellow, tube ± cylindric, 2–4 mm wide (exc mouth), tube-throat 15–26 mm, throat floor deeply grooved; placentas axile. **FR** 4–9 mm. *n*=16. Common. Seeps, streambanks, often in partial shade; < 2900 m. CA-FP, **n SNE (Sweetwater Mtns)**; to B.C., Rocky Mtns; naturalized in ne US, Chile, Eur. [var. *moniliformis* (E. Greene) Munz] Variable. ❀ WET or IRR:**1**,2,3,**7**,8–10,**14–18**,19–24&SUN:**4–6**;occas. INV. Jun–Aug

M. nanus Hook. & Arn. (p. 495) Ann, 1–10 cm, ± puberulent. **LF** 7–25 mm, oblanceolate to obovate; lower surface ± purple; tip obtuse to rounded. **FL:** pedicel 1–4 mm; calyx 6–10 mm, puberulent, lobes ± unequal, 1–4 mm, acute to ± long-tapered, mouth in fr slightly oblique; corolla persistent, magenta to purplish, tube-throat 11–19 mm, throat floor gen with 2 gold stripes surrounded by and often dotted deeper magenta; stigma ± exserted; placentas parietal. **FR** 6–12 mm. *n*=8. Sandy runoff areas above streamlets or in bare openings among shrubs; gen

1000–2300 m. NW, CaRH, **GB**; to WA, MT, WY. Intermediates with *M. mephiticus* found in e GB. ❀ WET,DRN,SHD:1–3,4–7,14,**15–17**,18,24. Mid May–late Jul

M. palmeri A. Gray Ann, 1–28 cm, ± puberulent. **LF** 3–28 mm, ± linear to ovate (< 12 mm wide). **FL:** pedicel 10–65 mm; calyx 4–10 mm, glabrous to puberulent, lobes equal, 0.5–1.5 mm, rounded to truncate, with small point at tip, glabrous to densely ciliate; corolla purple with variable marks, not lipped, tube-throat 10–22 mm; placentas axile. **FR** 3.5–9 mm. Sandy washes, disturbed areas; < 2100 m. s SNF, Teh (and **adjacent w DMoj**), n SCoRO, TR, PR; n Baja CA. [*M. diffusus* A.L. Grant, Palomar monkeyflower]. ❀ TRY. Mar–Jun

M. parishii E. Greene (p. 495) Ann, 3–85 cm, hairy. **LF** 8–75 mm, oblanceolate to ovate, veins 3, palmate. **FL:** pedicel 15–18 mm; calyx 8–10 (13 in fr) mm, hairy, lobes equal, 1–2 mm; corolla pinkish to white, tube-throat 9–14 mm; placentas axile. **FR** 6–10 mm. Uncommon. Wet, sandy streamsides; < 2100 m. s SN, SW (and adjacent w D), **n&e DMtns (Granite, New York, Panamint mtns)**; n Baja CA. May–Aug

M. parryi A. Gray (p. 495) Ann, 1–17 cm, densely puberulent. **LF** 3–26 mm, ± linear to oblanceolate; tip rounded to acute. **FL:** pedicel 1–5 mm; calyx 6–12 mm, membranous, often purplish throughout, puberulent, lobes spreading, upper lobe 1.5–3 mm, rounded, tip ± small-pointed, other lobes gen 0.5–2 mm, acuminate; corolla persistent, magenta, tube-throat 11–18 mm; placentas parietal. **FR** 5.5–10 mm, incl. *n*=8. Very uncommon. Steep hillsides, along washes; 1200–2600 m. **W&I (esp Inyo Mtns)**; to sw UT, nw AZ. ± common outside CA, where corollas yellow. Mid May–Jul

M. pilosus (Benth.) S. Watson (p. 495) Ann, 2–35 cm, densely long-soft-wavy-hairy. **LF** 10–30 mm, lanceolate to oblong. **FL:** pedicel gen 10–15 mm; calyx 6–7 mm, densely hairy, lobes unequal, 3–4 mm, ± = tube, obtuse; corolla yellow, tube-throat 7–8 mm; placentas axile. **FR** 4–7 mm. Moist, sandy areas, esp running or dry streamlets, disturbed areas; < 2600 m. CA-FP, **SNE, DMtns**; to WA, UT, AZ, Baja CA. ❀ IRR,DRN,SUN:1–6,7,8–10,**14–18**,19–24. Apr–Aug

M. primuloides Benth. Per; rhizomes or stolons forming mats of ± distinct rosettes or tufted pls; forming bulblets in autumn. **ST** 0.5–12 cm, glabrous. **LF** 7–50 mm, linear to obovate; upper surface glabrous to densely long-hairy; veins 3, palmate. **FL:** pedicel 10–120 mm, stiffly erect; calyx 5–12 mm, glabrous, lobes equal, 0.5–1.5 mm; corolla yellow, tube-throat 8–20 mm, base of each lower lip lobe gen with reddish spot; placentas axile. **FR** 6–7 mm. Wet meadows, seeps, streamsides; 600–3400 m. NW, CaR, SN, SnBr, SnJt, **GB**; to WA, NV. 2 sspp. in CA.

ssp. ***primuloides*** (p. 495) **ST** 0.5–4 cm. **LVS** gen ± spreading in ± distinct rosettes, 7–35 mm, oblong to obovate. **FL:** calyx 5–9 mm. *n*=17. Habitats & range of sp. Smaller, hairier pls sometimes found with others (e.g., Echo Summit, El Dorado Co.) have been called var. *pilosellus* (E. Greene) Smiley; such variation ± continuous. ❀ WET or IRR,DRN:**1**,2,3,7,14,18&SUN:4–

6,15–17; DFCLT. Jun–Aug

M. rubellus A. Gray (p. 495) Ann, 2–32 cm, minutely puberulent. **LVS** 3–31 mm, lanceolate to ovate; lower ± petioled; upper sessile, narrowly sheathing st. **FL:** pedicel 5–18 mm, ascending in fr; calyx 4.5–9 mm, gen glabrous, ribs brownish, lobes equal, 0.5–1.5 mm, rounded, often with small point at tip, ciliate; corolla yellow or lavender-magenta, tube-throat 6–10 mm, limb 3–4 mm wide; placentas axile. **FR** 3.5–7 mm. Gen in and along washes; 800–2600 m. **W&I, DMtns**; to WY, NM, n Baja CA. Apr–Jun

M. rupicola Cov. & A.L. Grant (p. 495) DEATH VALLEY MONKEYFLOWER Per, 1–17 cm, puberulent. **LF** 18–60 mm, oblanceolate, not ciliate. **FL:** pedicel 1–3 mm; calyx 8–17.5 mm, puberulent, lobes unequal, 1.5–7 mm, long-tapered; corolla persistent, pinkish to white with 1 large, magenta-purple spot at center of each lobe base, tube-throat 17–35 mm; placentas parietal. **FR** 3–8 mm, ovoid-oblong, slightly curved, hard, indehiscent. **SEEDS** few, 1–2 mm. *n*=8. UNCOMMON. Limestone crevices; 300–1800 m. **n DMtns (Cottonwood, Funeral, Grapevine, Last Chance, n Panamint ranges)**. Late Feb–early Jun

M. shevockii Heckard & Bacigal. (p. 495) KELSO CREEK MONKEYFLOWER Ann, 2–12 cm, minutely puberulent. **LF** 3–10 mm, lanceolate to ovate, clasping st; pairs at nodes fused or not. **FL:** pedicel 10–22 mm; calyx 4–7 mm, with reddish spots or solid red, ± puberulent, lobes equal, 0.5–1 mm, gen rounded; corolla tube-throat 8–12 mm, upper lip appearing 4-lobed (lateral pair small), maroon-purple, lower lip appearing as 1 lobe, notched, yellow with maroon dots at base; anthers and stigma exserted; placentas axile. **FR** 5–6 mm. *n*=16. RARE. Alluvial fans, dry streamlets, gen granitic soils; 900–1300 m. s SNF (Cortez, Cyrus canyons, Kern Co.), **w edge DMoj (Kelso Creek)**. Apr–May

M. suksdorfii A. Gray (p. 495) Ann, 0.5–10 cm, minutely puberulent. **LVS** 4–23 mm, ± linear to ovate; upper sessile, narrowly sheathing st. **FL:** pedicel 2–8 mm, spreading and ± S-curved in fr; calyx 3–5.5 mm, glabrous, lobes equal, 0.3–1 mm, rounded or with small point at tip; corolla yellow, tube-throat 4–6 mm; placentas axile. **FR** 2.5–5 mm. Moist, gen clay soils in ± full sun; 1100–4000 m. CaRH, SNH, WTR, SnBr, SnJt, **n DMtns (Grapevine Mtns, Inyo Co.)**; to WA, MT, WY, Colorado, AZ. May–Aug

M. tilingii Regel (p. 495) Per, 2–35 cm, rhizomed, glabrous to ± hairy. **LF:** petiole 0–25 mm; blade 5–30 mm, elliptic to ± round; pairs at nodes not fused. **INFL:** fls 1–5 per st, solitary in axils of upper lvs, not in bracted raceme. **FLS** opening; pedicel 10–90 mm; calyx 7–25 mm, asymmetrically swollen in fr, glabrous to puberulent, lobes unequal, lowest 2 upcurved in fr; corolla yellow, tube-throat 17–45 mm; placentas axile. **FR** 5–10 mm. *n*=14,15,24,25, 28. Seeps, streamsides, wet meadows; 1400–3400 m. NW, CaRH, SNH, SnBr, SnJt, **W&I**; to AK, MT, Colorado. Intergrades with *M. guttatus* in some areas. ❀ WET or IRR:**1–3,7**,14,**18**,19–23&SUN: 4,5,**6**,15–17. Jul–Sep

MOHAVEA

David M. Thompson

Ann, hairy. **ST** erect; branches 0 or few. **LVS** alternate, widely lanceolate, entire; veins pinnate. **INFL** crowded; fls solitary in axils. **FL:** calyx lobes 5, ± equal; corolla tube with sac-like extension at base, lips flaring, ± fan-shaped, lower lip base swollen, closing mouth; stamens 2, incl, staminodes 2. **FR** obliquely ovoid, fragile; chambers dehiscent by 1–2 large pores near tip. **SEEDS** ovate, flat, smooth; wing incurved, ± cup-shaped. 2 spp.: sw US, n Mex. (Mojave River, where first collected by John Frémont)

1. Corolla dark yellow; fl 15–20 mm . . . . ***M. breviflora***
1' Corolla yellowish to white; fl 25–35 mm . . . . ***M. confertiflora***

M. breviflora Cov. (p. 495, pl. 107) Pl 5–20 cm. **INFL:** pedicel 2–5 mm. **FL** 15–20 mm; corolla dark yellow, lower lip lobed to within 2–3 mm of swollen base, maroon-spotted only on swollen base. **FR** 8–10 mm. *n*=15. Gravelly desert slopes, washes; 100–

1400 m. **n&e DMoj**; s NV, nw AZ. ❀ TRY;DFCLT. Mar–Apr

M. confertiflora (Benth.) A.A. Heller (p. 495, pl. 108) GHOST FLOWER Pl 10–40 cm. **INFL:** pedicel 5–10 mm. **FL** 25–35 mm;

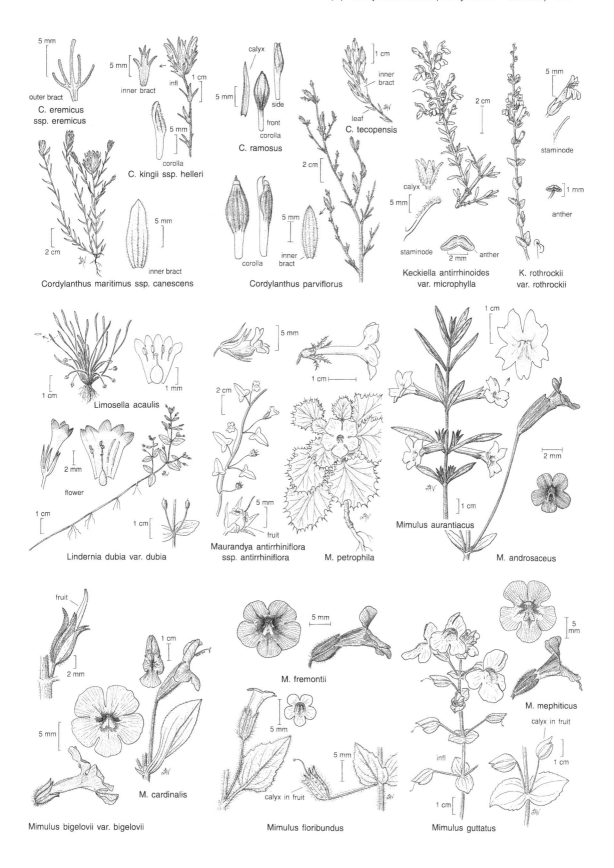

5 mm

outer bract

C. eremicus
ssp. eremicus

5 mm

inner bract

infl

1 cm

5 mm

corolla

C. kingii ssp. helleri

calyx

5 mm

side

front

corolla

C. ramosus

1 cm

inner
bract

leaf

C. tecopensis

2 cm

calyx

5 mm

staminode

anther

2 mm

Keckiella antirrhinoides
var. microphylla

5 mm

staminode

anther

1 mm

K. rothrockii
var. rothrockii

2 cm

inner bract

Cordylanthus maritimus ssp. canescens

5 mm

2 cm

corolla

5 mm

inner
bract

Cordylanthus parviflorus

Limosella acaulis

1 cm

1 mm

2 mm

flower

1 cm

1 cm

Lindernia dubia var. dubia

5 mm

2 cm

1 cm

5 mm

fruit

Maurandya antirrhiniflora
ssp. antirrhiniflora

M. petrophila

1 cm

2 mm

1 cm

Mimulus aurantiacus

2 mm

M. androsaceus

fruit

1 cm

2 mm

5 mm

Mimulus bigelovii var. bigelovii

M. cardinalis

5 mm

M. fremontii

5 mm

5 mm

calyx in fruit

Mimulus floribundus

5 mm

M. mephiticus

infl

1 cm

calyx in fruit

1 cm

Mimulus guttatus

corolla yellowish to white, lower lip lobed to within 6–8 mm of swollen base, maroon-spotted on swollen base and limb. **FR** 10–12 mm. *n*= 15. Gravelly or sandy desert slopes, washes; 0–1100 m. **D (exc Inyo Co.)**; s NV, w AZ, nw Mex. ❀ TRY;DFCLT. Mar–Apr

ORTHOCARPUS

T.I. Chuang & Lawrence R. Heckard

Ann, green root-parasites. **LVS** sessile, entire to 3-lobed. **INFL**: spike; bracts entire to 5-lobed, tips gen colored. **FL**: calyx 4-lobed, deepest sinus in back; corolla club-shaped, upper lip beak-like, tip closed, enclosing anthers and style, lower lip shorter, ± 3-pouched, gen 3-toothed; stamens 4, anther sacs 2, unequal; style and stigma slender. **FR** loculicidal, ± ovoid, gen ± notched. **SEEDS** gen 8–15, often ± curved, ± keeled, attached at side; coat netted or ridged, tight-fitting. 9 spp.: w N.Am. (Greek: straight fr) [Chuang & Heckard 1992 Syst Bot 17:560–582] Close to *Cordylanthus*; other spp. formerly placed here are in *Castilleja* (Owl's-clovers). ❀ TRY with host; DFCLT. Key to species adapted by Margriet Wetherwax.

1. Bracts grading into upper lvs, uniformly ± green (or uppermost purple-tinged), all 3 lobes triangular-lanceolate; infl densely glandular-puberulent; corolla golden-yellow . . . . *O. luteus*
1' Bracts differing abruptly from upper lvs, tips purplish or white; lateral 2–4 lobes much narrower than ovate or oblong central lobe; infl glabrous or sparsely glandular-puberulent; corolla purplish-pink
   . . . . *O. cuspidatus* ssp. *cryptanthus*

O. cuspidatus E. Greene Pl 10–40 cm, puberulent to scabrous, sparsely glandular, becoming ± purple-tinged. **ST** simple to much-branched, gen slender. **LVS** 10–50 mm, ± lanceolate; lower entire; upper with 3 deep, linear lobes. **INFL** 2–10 cm, dense; bracts differing abruptly from lvs, 10–20 mm, ± ovate, with 2 narrow, ± basal lobes, central lobe 7–15 mm wide, tip abrupty pointed, uppermost purplish pink on distal 1/3. **FL**: calyx 7–10 mm, divided 2/3 in back, 1/2 in front, 1/4 on sides; corolla 10–25 mm, exserted or not, lips purplish pink, densely puberulent, beak nearly straight, 4–10 mm, 0–4 mm > lower lip, lower lip ± pouched, teeth 1–2 mm, triangular, densely puberulent; stigma well incl. **FR** 6–8 mm. **SEED** dark brown. 2*n*=28. Open slopes or sagebrush; 700–3200 m. NW, CaR, SN, **GB**; to OR, n-c US. 3 sspp. in CA.

ssp. ***cryptanthus*** (Piper) Chuang & Heckard (p. 495) **FL**: corolla 9–14 mm, ± hidden by bract, beak 3–5 mm, 0–1 mm > lower lip, pouches 1–2 mm deep; upper medial anther sac 0.7–1 mm. Drying meadows, open sagebrush scrub; 1500–3200 m. SN, **GB**; to OR, n-c US. [*O. copelandii* Eastw. var. *cryptanthus* (Piper) Keck]

O. luteus Nutt. (p. 495, pl. 109) Pl 10–40 cm, glandular- and longer-nonglandular-hairy, yellow-green, often becoming ± purple-tinged. **ST** gen simple, slender. **LF** 15–50 mm, ± linear, entire or upper deeply 3-lobed. **INFL** 5–20 cm, densely glandular-puberulent; bracts grading into upper lvs, 10–20 mm, ± green, 2 lateral lobes below middle, narrowly triangular, central lobe ± lanceolate, 2–5 mm wide. **FL**: calyx 5–8 mm, divided 3/4 in back, 1/3 in front, 1/4 on sides; corolla 10–15 mm, golden-yellow, exserted, puberulent (esp beak), lips ± equal, beak 2–4 mm, tip minute, downward-projecting, cylindric, lower lip moderately pouched, teeth 0.5 mm, blunt, incurved; stigma incl. **FR** 5–7 mm. **SEEDS** 20–35, yellowish to dark brown. 2*n*=28. Moist fields, sagebrush scrub, mtn meadows; 1500–3000 m. **GB**; to B.C., n-c US, NM. Jul–Aug

PEDICULARIS LOUSEWORT

Linda Ann Vorobik

Per, green root-parasites. **STS** decumbent to erect, gen 1–several from gen short caudex. **LVS** alternate, gen ± basal, gen < infl, crenate to divided, gen reduced upward; petiole gen < blade. **INFL**: raceme, spike-like; bracts (at least lower) gen ± like upper lvs; pedicels 1–6 mm. **FL**: calyx lobes (2,4) gen 5, uppermost gen shortest (all gen < tube), lateral fused in pairs; corolla white or yellow to red or purple, upper lip hood-, beak-, or trunk-like, lower lip 3-lobed, narrow to fan-shaped, central lobe gen smallest; fertile stamens 4, gen glabrous, anthers gen incl; stigma head-like, gen exserted. **FR** loculicidal, gen ± ovoid, asymmetric, opening mostly on upper side. **SEED** smooth or netted. ± 500 spp.: cool wet n temp, circumboreal, S.Am. (Latin: lice, from belief that ingestion by stock promoted lice infestation) [Macior 1977 Bull Torrey Bot Club 104: 148–154]

1. Lf deeply lobed to compound; calyx lobes 5; fr ovoid . . . . *P. attollens*
1' Lf simple, ± toothed; calyx lobes gen 2 (4 in some pls); fr ± lanceolate in outline
   2. Upper corolla lip hooded, not beaked; lf crenate; sts gen simple . . . . *P. crenulata*
   2' Upper corolla lip extended in a long, curved-down beak; lf serrate to dentate; sts gen branched above
      . . . . *P. racemosa*

P. attollens A. Gray (p. 495) LITTLE ELEPHANT'S HEAD **ST** 6–60 cm, tomentose above. **LVS**: basal 3–20 cm, ± linear; segments 17–41, linear, toothed. **INFL** 2–30 cm; bracts tomentose, < fls. **FL**: calyx 4–6 mm, tomentose; corolla 4.5–7 mm, light pink to purple, marked darker, glabrous, upper lip 3.5–6 mm, trunk-like, curved upward, beak 3–5 mm, lower lip 3–5 mm, ± fan-like; anthers ± 1 mm, bases obtuse. **FR** 6–10 mm. **SEED** 2.5–4 mm, finely netted. Wet meadows, streamsides, bogs; 1500–3900 m. KR, CaR, SNH, MP, **W&I**; OR. ❀ very DFCLT. Jun–Sep

P. crenulata Benth. (p. 495, pl. 110) SCALLOP-LEAVED LOUSEWORT **ST** 12–40 cm, tomentose above. **LVS** basal and cauline, 3–11 cm, ± linear; margins ± doubly crenate, thick, rolled under, wavy, white. **INFL** 2–12 cm, gen simple; bracts gen < fls. **FL**: calyx 8–12 mm, tomentose, lobes 2(4); corolla 20–26 mm, ± club-like, white in CA, glabrous, upper lip 11–15 mm, hooded, lower lip 7–12 mm, ± fan-like, margins ± wavy, lobes ± equal; anthers 2–3 mm, bases acute. **FR** 10–20 mm, lanceolate in outline. **SEED** 1.5–2 mm, smooth. Wet meadows, streambanks; 2100–2300 m. **SNE (Convict Creek, Mono Co)**; to NV, WY, Colorado. [forma *candida* J.F. Macbr.] Jun–Jul

P. racemosa Hook. (p. 495) LEAFY LOUSEWORT **STS** few–many, ± decumbent, branched above, 12–80 cm, subglabrous.

LVS cauline, 2–10 cm, ± narrowly lanceolate, singly to doubly toothed. **INFL** 1–5 cm; lower bracts = or > fls. **FL**: calyx 4.5–8 mm, glabrous, lobes 2(4); corolla 10–16 mm, whitish, yellowish, or purplish, glabrous, upper lip 5–7.5 mm, beak strongly incurved, lower lip 5–9 mm, fan-like; anthers 1.5–2.5 mm, bases acute. **FR** 10–16 mm, lanceolate in outline. **SEED** 1.5–3 mm, smooth. Open coniferous forests; 900–2300 m. KR, CaR, n SNH, **SNE**; to w Can, WY, NM. Pls e of CA-FP with linear lvs and whitish to yellowish corollas have been called ssp. *alba* Pennell. ❀ very DFCLT. Jun–Aug

PENSTEMON BEARDTONGUE

Noel H. Holmgren

Per to shrub. **LVS** gen opposite, entire to toothed; upper sessile. **INFL**: panicle or raceme; bracts gen small. **FL**: calyx lobes 5, ± equal; corolla tube ± cylindric or lower side expanded, ± 2-lipped, gen pink or blue to purple (some red, yellow, or white), upper lip 2-lobed, external in bud; anther sacs 2, gen spreading ± flat at dehiscence; staminode attached near base of corolla tube, well developed, gen hairy on upper side; nectaries 2, at bases of upper stamens; stigma head-like. **FR**: capsule, septicidal and sometimes also loculicidal at tip. **SEEDS** gen many, irregularly angled. 250 spp.: N.Am., esp w US. (Latin & Greek: almost thread, from stamen-like staminode) [Holmgren 1984 In Cronquist et al. Intermtn Fl 4:370–457] Largest genus of fl pls endemic to N.Am. See also *Keckiella*.

1. Anther sacs not dehiscing the full length, valves opening only partially . . . . **Group 1**
1' Anther sacs dehiscing the full length and across the connective, valves gen spreading widely . . . . **Group 2**

Group 1

1. Anther sacs dehiscing at proximal end, leaving distal portion indehiscent.
  2. Corolla red to orange, lips long, lower lip strongly reflexed . . . . *P. rostriflorus*
  2' Corolla lavender, bluish, or purplish, lips short, lower lip projecting or spreading, not reflexed.
    3. Lvs well distributed on sts, narrowly oblong to oblanceolate, 7–25 mm wide; calyx 7–12 mm . . . *P. papillatus*
    3' Lvs gen basal, the few cauline lvs ± linear, 1–3 mm wide; calyx 3–5 mm . . . . *P. scapoides*
1' Anther sacs dehiscing at distal end only, proximal portion indehiscent.
  4. Corolla red, nearly radial, lobes ± equal, tube gradually expanding into throat, 4–7(9) mm wide at widest
    point (when pressed) . . . . *P. eatonii*
    5. Pl glabrous throughout . . . . var. *eatonii*
    5' Pl short-hairy . . . . var. *undosus*
  4' Corolla blue or blue-violet, distinctly 2-lipped, tube ± abruptly expanded into throat, 6–13 mm wide
    at widest point (when pressed)
    6. Calyx 3–4(5) mm; corolla throat 6–8 mm wide when pressed, floor hairy; anther sacs 1.6–2.2 mm
    . . . . *P. pahutensis*
    6' Calyx (4)6–13 mm; corolla throat 7.5–13 mm wide when pressed, floor glabrous; anther sacs
    (1.8)2–3 mm . . . . *P. speciosus*

Group 2

1. Anther sacs densely woolly; sts distinctly woody . . . . *P. davidsonii*
1' Anther sacs glabrous; sts herbaceous or woody only at very base
  2. Herbage (sts and lvs below inflorescence) hairy
    3. Staminode exserted, coiled at tip; corolla abruptly and widely expanded into throat . . . . *P. barnebyi*
    3' Staminode ± included, not coiled at tip; corolla gradually expanded into throat
      4. Calyx at fl 6.5–11 mm
      5. Anther sacs 0.6–0.7 mm, valves spreading widely; calyx 6.5–8.5 mm . . . . *P. calcareus*
      5' Anther sacs 1.1–1.4 mm, valves not spreading widely; calyx at fl 8–11 mm . . . . *P. monoensis*
      4' Calyx at fl 2.5–6(7) mm
        6. Herbage glandular-hairy; corolla glandular-hairy on floor; pls not mat-forming
        . . . . *P. deustus* var. *pedicellatus*
        6' Herbage nonglandular-hairy; corolla not glandular-hairy on floor; pls mat-forming
          7. Basal lvs (15)20–75 mm, petioled, cauline lvs sessile; sts erect or ascending; anther sacs
          0.5–0.8 mm . . . . *P. humilis*
          7' Lvs similar throughout, 5–20 mm, 1.8–5(6) mm wide, all petioled; plants low, fl-sts prostrate to
          ascending; anther sacs 0.8–1.3 mm . . . . *P. thompsoniae*
  2' Herbage glabrous
    8. Infl glabrous; lf margins entire to weakly toothed
      9. Corolla 8–17 mm
        10. Openly branched shrub; lvs narrowly linear, gen rolled upward, 0.6–1.4(2) mm wide . . . . *P. thurberi*
        10' Per herb, fl-sts gen little-branched; lvs lanceolate to ovate, flat to folded, 3.3–22 mm wide
          11. Lvs thick; anther sacs 1–1.2 mm; corolla floor glabrous . . . . ²*P. patens*
          11' Lvs ± thin; anther sacs 0.5–1 mm; corolla floor hairy
            12. St base gen buried in sand and lower cauline lvs scale-like; staminode glabrous . . . . *P. albomarginatus*

12' St base above ground and lower lvs well developed; staminode densely hairy
.... *P. rydbergii* var. *oreocharis*
9' Corolla 17–34 mm.
  13. Anther sacs 0.6–1.2 mm; lvs (ob)lanceolate, 5–20 mm wide; staminode included, glabrous or
     with short, papillate hairs at apex
    14. Corolla gen < 17mm, moderately expanded, lavender, magenta, or violet .... [2]*P. patens*
    14' Corolla gen > 17 mm, cylindric or narrowly funnel-shaped, red
      15. Corolla lobes projecting, not spreading, not glabrous; lvs cordate-clasping .... *P. centranthifolius*
      15' Corolla lobes spreading, glandular; lvs not cordate-clasping....................... *P. utahensis*
  13' Anther sacs 1.6–2.1 mm; lvs linear to narrowly lanceolate, 3.3–7(9) mm wide; staminode exserted,
    densely hairy... *P. fruticiformis*
    16. Corolla 22–24 mm, throat 8–10 mm wide when pressed, glandular hairy outside; calyx lobes
     ovate, gen (4.5)5–7.5 mm.. var. *amargosae*
    16' Corolla 24–28 mm, throat 10–14 mm wide when pressed, glabrous outside; calyx lobes broadly
     ovate to ± round, gen 4.5–6 mm... var. *fruticiformis*
8' Infl glandular-hairy; lf margins toothed (exc. *P. heterodoxus*)
  17. Upper lf pairs fused around st
    18. Anther-sacs 0.8–1.3 mm, valves spreading widely
      19. Corolla (17)20–25 mm, throat (5)6–9 mm wide when pressed, lower lip prominently lined; anther
       sacs 1.1–1.3 mm .... *P. pseudospectabilis*
      19' Corolla 15–20 mm, throat 4–6 mm wide when pressed, lower lip lacking prominent lines; anther
       sacs 0.8–1.1 mm .... *P. stephensii*
    18' Anther-sacs 1.3–2.2 mm, valves not spreading widely (exc. *P. bicolor*)
      20. Corolla 18–24(27) mm, throat 6–11 mm wide when pressed; anther sacs 1.3–2 mm; staminode
       rarely exserted .... *P. bicolor*
      20' Corolla 25–32 mm, throat 12–20 mm wide when pressed; anther sacs 1.8–2.2 mm; staminode
       exserted.. *P. palmeri*
  17' Lvs free, not fused around st
    21. Corolla 10–16 mm; lvs not thick; sts short, 0.5–2 dm tall .... *P. heterodoxus* var. *heterodoxus*
    21' Corolla 18–32 mm; lvs thick; sts 2–12 dm tall.
      22. Lvs linear to narrowly lanceolate, 1.5–7 mm wide; anther sacs 1.8–2.4 mm .... *P. incertus*
      22' Lvs lanceolate, ovate, or widely ovate, 10–60 mm wide; anther sacs 0.8–1.9 mm.
        23. Corolla gradually expanded into throat, throat 4.5–8 mm wide when pressed .... *P. clevelandii*
        24. Lvs entire to moderately serrate; staminode glabrous or sparsely hairy .... var. *clevelandii*
        24' Lvs sharply serrate; staminode densely hairy.................................. var. *mohavensis*
        23' Corolla ± abruptly expanded into throat, throat 6–16 mm wide when pressed .... *P. floridus*
        25. Corolla 21–27 mm, less abruptly expanded into throat, 6–11 mm wide when pressed, mouth
         perpendicular to tube, tube 6–8 mm .... var. *austinii*
        25' Corolla 24–30 mm, ± abruptly expanded into throat, 10–16 mm wide when pressed, mouth
         oblique to tube, tube 7–12 mm .... var. *floridus*

P. albomarginatus M.E. Jones (p.497) WHITE-MARGINED BEARDTONGUE Per 15–35 cm, glabrous; crown buried in sand. **LVS**: lower-most scale-like; upper 15–50 mm, oblanceolate, white-margined, entire or weakly dentate. **INFL** glabrous. **FL**: calyx 3–5 mm, lobes ovate-elliptic, white-margined, finely serrate; corolla 13–17 mm, pink to purple, glabrous (exc floor hairy); anther sacs ± 0.9–1 mm, spreading flat; staminode glabrous. 2*n*=16. RARE. Loose desert sand, gen on stabilized dunes; 700–900 m. **DMoj**; NV, AZ. In cult. Mar–May

P. barnebyi N. Holmgren (p. 497) BARNEBY'S BEARDTONGUE Per 6–30 cm; hairs short, backward-pointing. **LVS** 20–75 mm; basal lvs well developed; upper cauline lvs lanceolate, (sub)entire. **INFL** glandular. **FL**: calyx 4.5–12 mm, lobes ± lanceolate; corolla 10–14 mm, abruptly expanded to throat on lower side, 4–5 mm wide when pressed, violet, blue distally, glandular outside, throat white, dark-lined, floor ± yellow-hairy; anther sacs 0.7–0.9 mm, spreading flat; staminode much exserted, densely orange-yellow-hairy. RARE in CA. Limestone gravel or silt in sagebrush scrub or juniper/pinyon woodland; 1500–2500 m. **W&I**; NV.

P. bicolor (Brandegee) Clokey & Keck Per < 150 cm, glabrous. **LVS** thick; upper cauline 4–11 cm, ovate, bases fused around st, sharply serrate. **INFL** glandular. **FL**: calyx 4–6 mm, lobes ± ovate; corolla 18–24(27) mm, throat 6–11 mm wide when pressed, cream to magenta, strongly lined, glandular outside, floor long-whitish-hairy; anther sacs 1.3–2 mm, spreading flat; staminode

incl, densely yellow-hairy. Gravelly or rocky soils, creosote-bush or blackbush scrub, Joshua-tree woodland; 700–1500 m. **ne DMtns (Castle Mtns)**; NV. ❀ TRY;DFCLT.

P. calcareus Brandegee (p. 497) LIMESTONE BEARDTONGUE Per 7–25 cm; hairs fine, backward-pointing, ashy (short, densely glandular on upper st and infl). **LVS**: upper cauline 20–60 mm, widely lanceolate, entire to shallowly dentate. **FL**: calyx 6.5–8.5 mm, lobes ± lanceolate; corolla 13–17 mm, cylindric to funnel-shaped, bright pink to rose-purple, glandular outside, throat 4–6 mm wide when pressed, floor nearly glabrous; anther sacs 0.6–0.7 mm, spreading flat; staminode densely yellow-hairy, ± incl. RARE in CA. Limestone crevices, rocky slopes in juniper/pinyon woodland, Joshua-tree scrub; 1200–1600 m. **n DMtns**. Apr–May

P. centranthifolius (Benth.) Benth. (p. 497) SCARLET BUGLER Per 30–120 cm, glabrous, glaucous. **LVS** thick; middle cauline gen largest, 40–100 mm, lanceolate to ovate, cordate-clasping, entire. **FL**: calyx 3.5–6.5 mm, lobes ovate to ± round; corolla 20–30 mm, cylindric, lobes projecting, not spreading, bright red, glabrous (incl floor); anther sacs 0.8–1.2 mm, spreading flat; staminode glabrous. 2*n*=16. Dry, open or wooded places, gen in chaparral or oak woodland; < 1800 m. NCoR, n SNF, GV (margins), CW, SW, **sw edge DMoj**; Mex. Common hybrids with *P. spectabilis* have been called *P.* ×*parishii*. ❀ SUN,DRY,DRN; **7,9,14–21,**22,23;may be DFCLT. Apr–Jun

P. clevelandii A. Gray Per 30–70 cm, much-branched, glabrous.

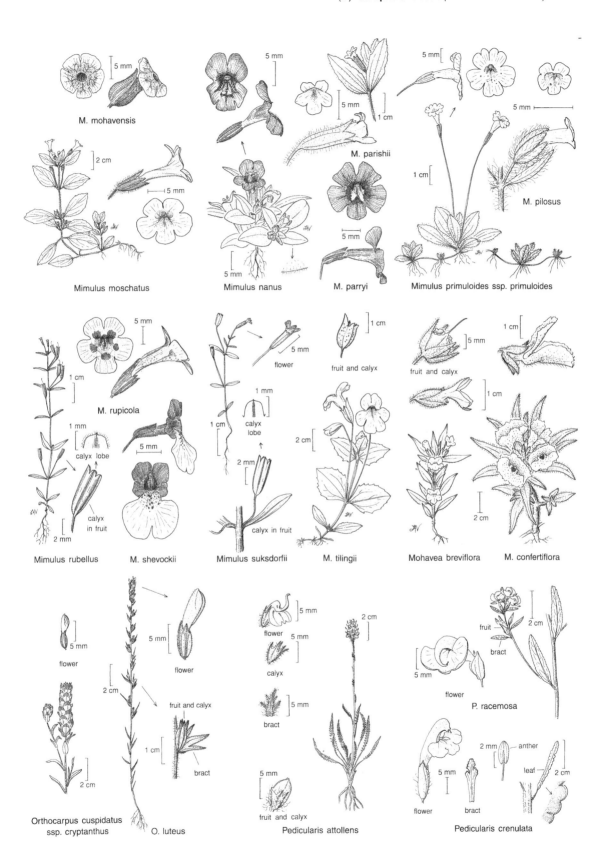

M. mohavensis

M. parishii

M. pilosus

Mimulus moschatus

Mimulus nanus

M. parryi

Mimulus primuloides ssp. primuloides

M. rupicola

flower

fruit and calyx

fruit and calyx

calyx lobe

calyx in fruit

calyx lobe

calyx in fruit

Mimulus rubellus

M. shevockii

Mimulus suksdorfii

M. tilingii

Mohavea breviflora

M. confertiflora

flower

flower

fruit and calyx

bract

Orthocarpus cuspidatus ssp. cryptanthus

O. luteus

flower

calyx

bract

fruit and calyx

Pedicularis attollens

fruit

bract

flower

P. racemosa

anther

leaf

flower

bract

Pedicularis crenulata

LF thick, 20–60 mm, ± ovate, entire to serrate. **INFL** glabrous or glandular. **FL:** calyx 3.5–6.5 mm, lobes ovate to ± round; corolla 18–24 mm, gradually expanded into throat, 4.5–8 mm wide when pressed, pink, magenta, or reddish purple, unlined, glandular outside and inside (floor lacks nonglandular hairs); anther sacs dehiscing full length; staminode incl, glabrous to densely hairy. Rocky hillsides, rock crevices in creosote-bush scrub, juniper/pinyon woodland, chaparral; 400–1500 m. PR, **s DMtns, w DSon**; Mex. 3 vars. in CA.

var. *clevelandii* (p. 497) **LVS** entire to moderately serrate; upper cauline lf bases free, cordate-clasping to rounded. **FL:** anther sacs 0.8–1.3 mm, spreading flat; staminode glabrous or sparsely hairy. 2*n*=16. Habitats and elevations of sp. PR, **w DSon**; Mex. Variable. ❀ DRN,DRY:7,14–16,**18–21**,22,23. Mar–May

var. *mohavensis* (Keck) McMinn (p. 497) **LVS** sharply serrate; upper cauline lf bases free, widely wedge-shaped. **FL:** anther sacs 0.9–1.1 mm, valves spreading but not flat; staminode densely hairy. Habitats of sp. **s DMtns (Little San Bernardino, Granite mtns)**. [ssp. *m.* Keck] Mar–May

P. davidsonii E. Greene var. *davidsonii* (p. 497) Subshrub < 10 cm, mat-forming, short-hairy. **LVS** ± basal (much reduced upward); main blades 5–30 mm, elliptic to (ob)ovate, (sub)entire, glabrous, green. **INFL** glandular; pedicels short. **FL:** calyx 7–13 mm, lobes linear to lanceolate; corolla 20–36 mm, blue-violet to blue-purple, floor ± white-shaggy-hairy; anther sacs 0.9–1.3 mm, valves spreading flat, white-woolly; staminode incl, densely pale yellow-hairy. 2*n*=16. Montane to alpine outcrops, talus; 2000–3600 m. KR, CaRH, SNH, Wrn, **n SNE (Sweetwater Mtns)**; OR. ❀ DRN, IRR:15–17&SHD:1–6,7,14–21;DFCLT. Jul–Aug

P. deustus Lindley Per gen < 40 cm, subglabrous to glandular. **LVS** mostly basal, dentate; cauline lvs 10–50 mm. **INFL** ± glandular. **FL:** corolla 8–15 mm, cream-white, dark-lined, glandular outside and on floor; anther sacs 0.5–0.7 mm, valves spreading flat; staminode glabrous or sparsely hairy distally. 2*n*=16. Open, rocky, gen volcanic places; 600–3000 m. KR, NCoRH, CaR, SNH, MP, **n SNE (Sweetwater Mtns)**; to WA, MT, WY, UT. 2 vars. in CA.

var. *pedicellatus* M.E. Jones (p. 497) **LF** lanceolate. **FL:** calyx 2.5–5 mm, lobes lanceolate, acute; corolla 10–15 mm. Sagebrush scrub, juniper/pinyon woodland, yellow-pine and montane forests; 900–3000 m. CaR, SNH, MP, **n SNE (Sweetwater Mtns)**; OR, NV, UT. [ssp. *heterander* (Torrey & A. Gray) Pennell & Keck] ❀ DRN,DRY,SUN:1, **2,3,7**,14,**15–17**,18–21. May–Jul

P. eatonii A. Gray (p. 497) Per 40–100 cm. **LVS:** cauline 30–90 mm, widely lanceolate to ovate, entire. **FL:** calyx 3.5–6 mm, lobes ovate; corolla 24–33 mm, cylindric, obscurely 2-lipped, lobes subequal, barely spreading, scarlet, glabrous; anther sacs 1.4–2.8 mm, dehiscing in distal 1/2–3/4, short-hairy on sides; staminode glabrous to sparsely hairy at tip. 2*n*=16. Dry sagebrush scrub, juniper/pinyon woodland, yellow-pine forest; 1500–2800 m. SnBr, **DMtns**; to Colorado, NM. ❀ SUN,DRY:1–3,**7**,10,15–21.

var. *eatonii* Pl glabrous. **DMtns**; to Colorado, NM. Mar–Jul

var. *undosus* M.E. Jones Pl short-hairy. Habitats and range of sp. [ssp. *u.* (M.E. Jones) Keck]

P. floridus Brandegee (p. 497) Per 50–120 cm, glabrous, glaucous. **LVS** thick; upper cauline lvs lanceolate or ovate, cordate-clasping, gen dentate (upper-most sometimes subentire). **INFL** glandular. **FL:** calyx 4.2–6.2 mm, lobes ovate; corolla 21–30 mm, throat narrowed towards mouth, rose-pink, strongly lined, glandular outside and inside (floor lacks nonglandular hairs); anther sacs 1.3–1.9 mm, spreading flat; staminode incl, glabrous. 2*n*=16. Gravelly washes, canyon floors, in sagebrush scrub and juniper/pinyon woodland; 1000–2400 m. **SNE, n DMtns**; NV. Vars. intergrade in s White Mtns.

var. *austinii* (Eastw.) N. Holmgren (p. 497) **FL:** corolla 21–27 mm, gradually expanded into throat, 6–11 mm wide when pressed, tube 6–8 mm, mouth perpendicular to tube. Habitats and elevations of sp. **W&I, n DMtns**; NV. [ssp. *a.* (Eastw.) Keck] ❀ TRY; DFCLT. May–Jun

var. *floridus* (p. 497) **FL:** corolla 24–30 mm, ± abruptly expanded into throat, 10–16 mm wide when pressed, tube 7–12 mm, mouth oblique to tube. Habitats of sp.; 1600–2400 m. **SNE**; NV. ❀ TRY;DFCLT. May–Jul

P. fruticiformis Cov. Shrub 30–60 cm, much-branched below, gen wider than tall. **ST:** young sts glabrous, gen glaucous. **LVS** thick, 25–65 mm; upper lvs ± narrowly lanceolate, (sub)entire, gen folded lengthwise or rolled inward. **INFL** glabrous. **FL:** corolla pale pink to whitish, limb sometimes ± lavender, strongly lined, floor shaggy-hairy; anther sacs, 1.6–2.1 mm, dehiscing full length, valves barely spreading; staminode exserted, densely hairy. Gravelly washes, canyon floors in creosote-bush scrub, juniper/pinyon woodland; 1000–1800 m. **s SNE, n DMtns**; w NV.

var. *amargosae* (Keck) N. Holmgren (p. 497) DEATH VALLEY BEARDTONGUE **FL:** calyx (4.5)5–7.5 mm, lobes ovate; corolla 22–24 mm, throat 8–10 mm wide when pressed, glandular-hairy outside. UNCOMMON. Creosote-bush scrub; 1000–1200 m. **ne DMtns (Kingston Mtns)**; w NV. [ssp. *a.* Keck]

var. *fruticiformis* (p. 497) **FL:** calyx 4.5–6.5 mm, lobes widely ovate to ± round; corolla 24–28 mm, throat 10–14 mm wide when pressed, glabrous exc on floor. 2*n*=16. Habitats and range of sp. (exc NV). ❀ TRY;DFCLT. May–Jun

P. heterodoxus A. Gray Per 5–65 cm, mat-forming, ± glabrous. **LVS** entire, sometimes folded lengthwise; basal many; cauline narrowly lanceolate to ovate, sometimes clasping. **INFL** glandular. **FL:** calyx 2.5–6 mm, lobes narrowly oblong to obovate; corolla 10–16 mm, cylindric to moderately expanded, deep blue-purple, glandular outside, floor yellow-brown-hairy; anther sacs 0.5–1 mm, dehiscing full length but barely spreading; staminode moderately yellow-hairy. Montane to alpine slopes, meadows, scree; 1100–3900 m. KR, CaRH, SN, **n SNE, W&I**; NV. 3 vars. in CA.

var. *heterodoxus* (p. 497) Pl 5–20 cm. **LF** 5–50 mm, 2–10 mm wide. **INFL** of 1(2) clusters, glandular. 2*n*=16. Montane and sulapine habitats of sp.; 2000–3900 m. KR, SN, **n SNE, W&I**; NV. ❀ DRN,IRR:1–3,7,14&SUN:15,16;usually DFCLT. Jul–Aug

P. humilis A. Gray var. *humilis* (p. 497) Per 5–35 cm, gen mat-forming, ± short-(sometimes ashy-)hairy. **LVS** entire; basal lvs many, (15)20–75 mm long, (2)4–32 mm wide, (ob)ovate, petioled; cauline lvs lanceolate to obovate, sessile, clasping. **INFL** glandular. **FL:** calyx 3–6 mm, lobes lanceolate to ovate; corolla 11–15 mm, cylindric narrowly funnel-shaped, blue with lighter floor, dark-lined, glandular outside, floor ± yellow- or white-hairy; anther sacs 0.5–0.8 mm, dehiscing full length, valves barely spreading; staminode orange- to yellow-hairy. 2*n*=16. Open montane to subalpine forests, sagebrush scrub, juniper/pinyon woodland; 1500–3000 m. CaRH, n&c SNH, **GB**; to OR, WY, Colorado. Small-fld, small-lvd pls from n CA, s OR, nw NV have been called *P. cinereus* Piper, gray beardtongue, which intergrades fully with *P. h.* May–Jul

P. incertus Brandegee (p. 497) Shrub 20–100 cm, rounded; young sts glabrous, glaucous. **LVS** thick; largest lvs at mid st, 40–70 mm, linear to narrowly lanceolate, gen rolled inward, entire. **INFL** glandular. **FL:** calyx 4.5–7.3 mm, lobes ovate, glabrous to glandular; corolla 23–32 mm, throat 8–12 mm wide when pressed, violet to purple (limb bluish), unlined, glandular outside, glabrous inside exc small hair patch on floor; anther sacs 1.8–2.4 mm, dehiscing full length, valves barely spreading; staminode incl, densely hairy. Gen sandy soil along washes, canyon slopes, in sagebrush scrub, Joshua-tree and juniper/pinyon woodlands; 1000–1700 m. **s SNH, Teh, SnBr, nw DMtns**. ❀ TRY;DFCLT. May–Jun

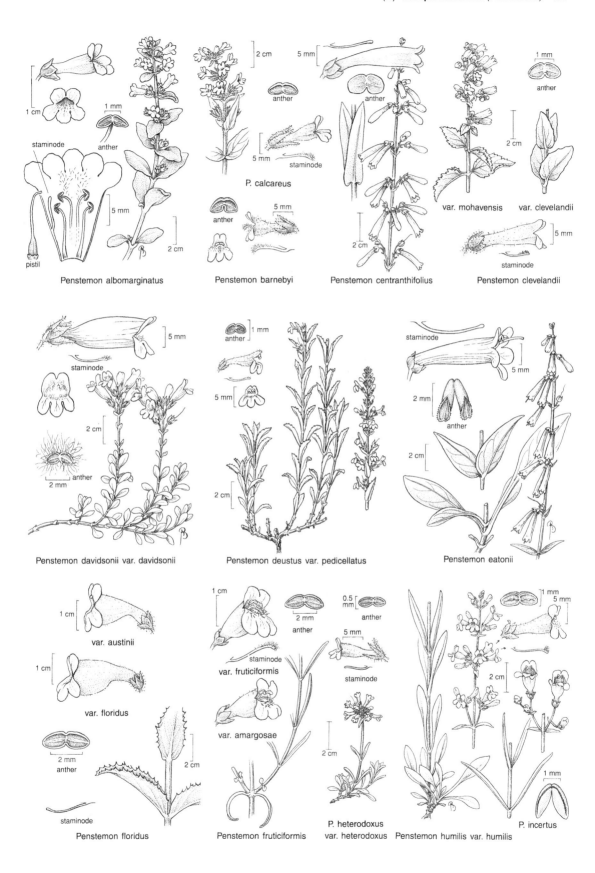

P. calcareus

Penstemon albomarginatus

Penstemon barnebyi

var. mohavensis var. clevelandii

Penstemon centranthifolius

Penstemon clevelandii

Penstemon davidsonii var. davidsonii

Penstemon deustus var. pedicellatus

Penstemon eatonii

var. austinii

var. floridus

Penstemon floridus

var. fruticiformis

var. amargosae

Penstemon fruticiformis

P. heterodoxus
var. heterodoxus

Penstemon humilis var. humilis

P. incertus

P. monoensis A.A. Heller (p. 501) Per 7–30 cm; hairs dense, ashy, backward-pointing. **LVS** gen cauline, 50–120 mm, ± (widely) lanceolate, entire to toothed. **INFL** glandular. **FL**: calyx 8–11 mm, lobes gen lanceolate (ovate); corolla 15–20 mm, cylindric to narrowly funnel-shaped, glandular outside, pink to red-violet, throat 4–6 mm wide when pressed, floor white to pale pink to white, sparsely hairy; anther sacs 1.1–1.4 mm, dehiscing full length, valves barely spreading; staminode ± incl, yellow-hairy. Sandy and gravely washes and hills, sagebrush scrub, Joshua-tree and juniper/pinyon woodland; 1300–1800 m. **W&I**. ❀ TRY;DFCLT. Apr–May

P. pahutensis N. Holmgren (p. 501) PAHUTE BEARDTONGUE Per 15–35 cm, glabrous, sometimes glaucous. **LVS**: cauline 30–100 mm, linear to narrowly lanceolate, entire. **INFL** glabrous. **FL**: calyx 3–4(5) mm, lobes widely ovate; corolla 17–30 mm, bluish lavender, throat 6–8 mm wide when pressed, floor yellow- or white-hairy; anther sacs 1.6–2.2 mm, ± S-shaped, dehiscing in distal 2/3, glabrous; staminode densely golden-yellow-hairy. $2n=16$. RARE in CA. Sagebrush scrub, juniper/pinyon woodland; 1900–2300 m. **ne DMtns (Grapevine Mtns)**; NV.

P. palmeri A. Gray var. *palmeri* (p. 501, pl. 111) Per 50–200 cm, glabrous, glaucous. **LVS** thick; upper cauline lvs 40–120 mm, triangular-clasping, ± dentate. **INFL** glandular. **FL** fragrant; calyx 4–7 mm, lobes ± ovate; corolla 25–32 mm, abruptly expanded into throat, 12–20 mm wide when pressed, pale (lavender-)pink, strongly lined, glandular outside and inside, floor sparsely long-whitish hairy; anther sacs 1.8–2.2 mm, dehiscing full length, valves barely spreading; staminode exserted, densely long-spreading-yellow-hairy. $2n=16,32$. Washes, roadsides, canyon floors, in creosote-bush scrub to juniper/pinyon woodland; 1100–2300 m. **e DMoj**; to UT, AZ. [*P. bryantae* Keck] ❀ DFCLT. May–Jun

P. papillatus J. Howell (p. 501) INYO BEARDTONGUE Per 20–40 cm, ashy-hairy, woody-branched below. **LVS** gen cauline, 25–40 mm long, 7–25 mm wide, narrowly elliptic to oblanceolate, gen widest at base, cordate-clasping, entire. **INFL** glandular. **FL**: calyx 7–12 mm, lobes narrowly lanceolate; corolla 24–30 mm, blue-violet, glandular outside, floor glabrous; anthers sacs 1.7–1.9 mm, dehiscing only across common tip, inner margins glabrous; staminode pale-yellow-hairy. UNCOMMON. Rocky openings of juniper/pinyon woodland and montane forest communities; 2000–2700 m. c&s SNH, **SNE**. Jun–Jul

P. patens (M.E. Jones) N. Holmgren (p. 501) Per 15–40 cm, glabrous, glaucous. **LVS** thick; basal lvs well developed; cauline lvs 25–90 mm long, 5–20 mm wide, lanceolate, entire. **INFL** glabrous. **FL**: calyx 3–7 mm, lobes widely ovate; corolla 13–20 mm, lavender to magenta, glabrous; anther sacs 1–1.2 mm, dehiscing full length, valves barely spreading; staminode minutely orange- to yellow-hairy. $2n=16$. Sagebrush scrub, juniper/pinyon woodland, yellow-pine forest; 1900–3000 m. c&s SNH, **s SNE**; s NV. [*P. confusus* M.E. Jones ssp. *patens* (M.E. Jones) Keck] ❀ TRY. May–Jun

P. pseudospectabilis M.E. Jones (p. 501) Shrub 30–100 cm. **ST**: young sts glabrous, glaucous. **LVS** ± thin; upper cauline lvs 30–90 mm, widely triangular-ovate, ± serrate, bases fused around st; upper lf pairs ± disk-like. **INFL** gen glandular. **FL**: calyx 3.5–7.5 mm, lobes ovate; corolla (17)20–25 mm, gradually expanded to throat, (5)6–9 mm wide when pressed, reddish pink, strongly lined, glandular outside and inside (floor lacks nonglandular hairs); anther sacs 1.1–1.3 mm, spreading flat; staminode incl, glabrous. $2n=16$. Gravelly or rocky desert washes, canyon floors, in creosote-bush scrub, juniper/pinyon woodland; 100–1400 m. **s DMtns, ne DSon**; AZ. ❀ DRN,DRY,SUN:10,12–21;DFCLT. Mar–May

P. rostriflorus Kellogg (p. 501, pl. 112) Per 30–100 cm, glabrous or sts finely hairy below; clumps large, woody-branched. **LVS** gen cauline, 20–70 mm, linear to lanceolate, entire. **INFL** glandular. **FL**: calyx 4–8 mm, lobes lanceolate to narrowly ovate; corolla 22–33 mm, upper lip forming hood over anthers, lower

lip strongly reflexed, (orange-)red sparsely glandular outside, floor glabrous; anther sacs 1.8–2.5 mm, dehiscing across common tip 1/4–1/3 their length, glabrous on sides; staminode glabrous. $2n=16$. Dry sagebrush or Joshua-tree scrub, juniper/pinyon woodland, montane forest; 1600–2800 m. c&s SNH, SW, **SNE**, **DMtns**; to Colorado, NM. [*P. bridgesii* A. Gray] Jun–Aug

P. rydbergii Nelson var. *oreocharis* (E. Greene) N. Holmgren (p. 501) Per 20–60 cm, gen glabrous (or finely hairy on sts below). **LVS** entire; basal and lower cauline lvs well developed, oblanceolate; upper cauline lvs 25–70 mm, lanceolate. **FL**: calyx 3–5 mm, lobes narrowly oblong to (ob)ovate; corolla 9–14 mm, blue to purple, glabrous (exc white- to yellow-hairy floor); anther sacs 0.5–0.8 mm, dehiscing full length; staminode densely golden-yellow-hairy. $2n=16$. Moist meadows, streambanks, gen in montane to subalpine forests; 1200–3100 m. CaR, SNH, **GB**; OR, NV. [*P. o.* E. Greene] ❀ DRN,IRR,SUN:1,2,4–6,**7**,14–16;DFCLT. May–Aug

P. scapoides Keck (p. 501) Per 15–60 cm, subscapose, woody-branched below. **ST** glabrous, glaucous. **LVS** gen dense, in basal mat, entire, densely hairy; basal lvs 1.5–6 mm, ovate to ± round; cauline lvs few, narrowly linear to narrowly oblanceolate. **INFL** glandular. **FL**: calyx 3–5 mm, lobes oblong to widely ovate; corolla 25–34 mm, pale lavender to purple or blue, glandular outside, floor yellow-hairy; anther sacs 1.4–1.7 mm, dehiscing across common tip 1/2 their length, inner margins glabrous; staminode pale yellow-hairy. $2n=16$. Sagebrush, juniper-pinyon, and bristlecone-pine woodland; 2000–3200 m. **W&I, n DMtns (Last Chance Mtns)**. ❀ TRY;DFCLT. Jun–Jul

P. speciosus Lindley (p. 501) Per 5–60 cm, short-hairy on sts. **LVS**: upper cauline 35–90 mm, lanceolate, clasping, sometimes folded lengthwise, entire, gen glabrous. **INFL** glabrous to short-hairy, rarely glandular. **FL**: calyx (4)6–13 mm, lobes lanceolate to ovate; corolla 25–37 mm, ± abruptly expanded into throat, 7–13 mm wide when pressed, sky-blue (gen white inside), glabrous; anther sacs 1.8–3 mm, ± S-shaped, dehiscing in distal 2/3, glabrous; staminode glabrous or tip hairy. $2n=16$. Open sagebrush scrub to subalpine forest; 1200–3300 m. KR, CaR, SNH, WTR, SnGb, **GB**, **n DMtns**; to WA, ID, UT. [ssp. *kennedyi* (Nelson) Keck] ❀ DRN, IRR,SUN:1–3,7,14–16;DFCLT. May–Jul

P. stephensii Brandegee (p. 501) STEPHENS' BEARDTONGUE Shrub 30–150 cm, glabrous. **LVS** ± thin; upper cauline lvs 25–50 mm long, 1.8–5(6) mm wide, triangular-ovate, pairs fused at base, finely and sharply serrate. **INFL** glabrous to glandular. **FL**: calyx 3–5.2 mm, lobes ovate to ± round; corolla 15–20 mm, rose to magenta, unlined, throat 4–6 mm wide when pressed, glandular outside and inside (floor lacks nonglandular hairs); anther sacs 0.8–1.1 mm, spreading flat; staminode incl, glabrous. RARE. Rocky slopes, washes, rock crevices in creosote-bush scrub, juniper/pinyon woodland; 1200–1500 m. **DMtns**. Apr–Jun

P. thompsoniae (A. Gray) Rydb. (p. 501) Per 5–15 cm, ± prostrate, gen matted; hairs appressed, backward-pointing, scale-like, ashy. **LF** 5–20 mm, ± obovate, entire, petioled. **FL**: calyx 4–7.5 mm, lobes lanceolate, some hairs gen glandular; corolla 10–18 mm, violet to blue, glandular outside, floor pale yellow-hairy; anther sacs 0.8–1.3 mm, dehiscing full length but barely spreading; staminode densely orangish- to golden-yellow-hairy. White calcareous soils in juniper/pinyon woodland; 1500–2700 m. **e DMtns (New York, Clark mtns)**; to UT, AZ. ❀ TRY;DFCLT. May–Jun

P. thurberi Torrey (p. 501) THURBER'S BEARDTONGUE Shrub 20–80 cm, ± round, glabrous. **LVS** cauline, narrowly linear; main blades 10–45 mm, gen rolled upward, nearly round in ×-section, entire. **INFL** glabrous. **FL**: calyx 2.4–4.5 mm, lobes ovate; corolla 8–15 mm, ± funnel-shaped (lips obliquely spreading), lavender, rose, or blue-purple, glabrous outside, floor hairy; anther sacs 0.7–0.9 mm, spreading flat; staminode glabrous. $2n=16$. UNCOMMON. Sandy and gravelly slopes and mesas, in chaparral, creosote-bush or Joshua-tree scrub, juniper/pinyon woodland; 500–1200 m. PR, **DMoj**; to NV, NM, Mex. May–Jun

P. utahensis Eastw. (p. 501) Per 15–50 cm, glabrous, glaucous. **LVS** thick, gen folded lengthwise; basal lvs well developed; upper cauline lvs 15–55 mm long, 5–20 mm wide, ± lanceolate, entire. **INFL** glabrous. **FL**: calyx 2.5–5 mm, lobes widely ovate; corolla 17–25 mm, cylindric to funnel-shaped (limb slightly oblique, lobes spreading), pink to red (or purplish), glandular outside and inside; anther sacs 0.6–1.1 mm, spreading flat; staminode glabrous or tip minutely hairy. Sagebrush scrub, juniper/pinyon woodland; 1200–2500 m. **e DMtns (Kingston, New York mtns)**; to UT, AZ. ❀ DRN,SUN, DRY:2,3,7,10,14–16,18–21;DFCLT. Apr–May

SCROPHULARIA FIGWORT

Margriet Wetherwax

Per or shrub, long-soft-shaggy- and short-stout-glandular-hairy. **ST** square. **LVS** opposite, 4-ranked, lanceolate to triangular-ovate, petioled. **INFL**: panicle; pedicels > bracts; bracts lanceolate, long-tapered. **FL**: calyx lobes 5, obtuse to rounded; corolla 5-lobed, 2-lipped, greenish yellow to blackish, tube urn-shaped to spheric, lowest lobe gen recurved; fertile stamens 4, incl, anther sacs 1, staminode attached near base of upper lip, appressed to upper side of corolla; style slender, gen reflexed at fl, stigma head-like; nectary at base of ovary, fleshy. **FR** septicidal. **SEEDS** many, oblong-ovoid, ridged. ± 150 spp.: n temp, esp Eurasia. (Latin: scrofula, a disease supposedly cured by some spp.) [Shaw 1962 Aliso 5: 147–178]

S. desertorum (Munz) R.J. Shaw (p. 501) Per, minutely glandular. **STS** clustered, 7–12 dm. **LF**: blade 10–13 cm, lanceolate, base wedge-shaped, narrowly acute, serrate; petiole 7–10 cm. **INFL** long. **FL**: calyx lobes 2–3 mm, triangular-ovate, obtuse to rounded; corolla 7–9 mm, ± spheric, upper half maroon, lower half cream, lowest lobe recurved; staminode club-shaped, longer than wide, light maroon, with 2 nectar drops gen conspicuous at base. **FR** 5–9 mm, ovoid, long-tapered. Dry slopes; 1000–3000 m. SN, **SNE, n DMtns**; NV. ❀ DFCLT. Apr–Aug

STEMODIA

Margriet Wetherwax

Per. **STS** 1–many, ascending to erect. **LVS** gen opposite, clasping, serrate. **INFL** raceme; bracts 2 below calyx. **FL**: sepals 5, free; corolla 5-lobed, 2-lipped, violet to purple, tube 4-angled, upper lip 2-lobed, arched, lower lip reflexed; stamens 4, in 2 pairs, anther sacs well separated by thick tissue; stigma lobes 2, flat. **FR** ± cylindric, loculicidal. **SEEDS** many, netted. ± 20 spp.: trop Am. (Latin: from stemodiacra: stamens with 2 tips)

S. durantifolia (L.) Sw. (p. 505) Pl glandular-hairy. **ST** 10–50 cm. **LVS** 2–3 per node, 2–4 cm (smaller upward), lanceolate, widest at base, ± dentate, ± sessile. **INFL** open; pedicels 2–12 mm; bracts ± = sepals. **FL**: sepals 5–7 mm, lanceolate, long-tapered; corolla 7–10 mm. **FR** ± 5 mm, ovoid-cylindric. Wet sand or rocks, drying river beds; < 300 m. SnJt, **DSon**; AZ, Mex. ❀ TRY. Most of year

VERBASCUM MULLEIN

Margriet Wetherwax

Bien in CA (lf rosette 1st year, stout fl-st 2nd year). **ST** erect, simple or branched just below infl. **LVS** basal and cauline, alternate, reduced upward. **INFL**: raceme or panicle, bracted. **FL**: calyx ± radial, deeply 5-lobed; corolla ± radial, rotate, 5-lobed; stamens 5, lowest 2 filaments > upper 3, all or upper hairy; stigmas fused, head-like. **FR**: capsule, septicidal. **SEEDS** small, wingless, many. ± 360 spp.: Eurasia. (Latin: from root for bearded) Lvs used medicinally in cigarets for asthmatics.

V. thapsus L. (p. 505) WOOLLY MULLEIN Bien, densely stellate-hairy. **ST** 30–200 cm, gen simple. **LVS**: basal 8–50 cm, oblanceolate, gen entire, short-petioled; cauline 5–30 cm, lanceolate, gen entire, long-decurrent. **INFL**: raceme, dense; bracts 12–18 mm; pedicels < 2 mm, gen fused to st. **FL**: calyx 7–9 mm, lobes lanceolate; corolla 15–25(30) mm wide, yellow; upper 3 filaments white- or yellow-hairy, lower 2 glabrous to sparsely hairy. **FR** 7–10 mm, ovoid. 2*n*=32,34,36. Disturbed areas; < 2200 m. CA-FP, **n SNE**; native to Eurasia. Jun–Sep

VERONICA SPEEDWELL, BROOKLIME

Margriet Wetherwax

Ann, per. **ST** erect or prostrate. **LVS** opposite. **INFL**: raceme, terminal or axillary, or fls solitary in axils; bracts small, alternate. **FL**: sepals gen 4(5), ± free, gen unequal; corolla ± rotate, 4-lobed, upper lobe wide (formed by fusion of upper pair), blue or violet to white; stamens 2, exserted; stigma head-like. **FR**: capsule, flattened perpendicular to septum, gen obcordate, loculicidal and septicidal. ± 250 spp.: n temp, esp Eurasia. (Possibly named for Saint Veronica)

1. Most racemes terminal; herbage gen hairy; ann or per
  2. Tap-rooted ann; glandular-hairy; lvs linear to spoon-shaped; style 0.1–0.4 mm . . . ***V. peregrina*** ssp. ***xalapensis***
  2' Rhizomed per, ± hairy; lvs elliptic to widely ovate; style 2–3 mm . . . . ***V. serpyllifolia*** ssp. ***humifusa***
1' Racemes axillary, opposite; herbage glabrous; per

3. Lvs sessile (lower rarely petioled); fr rounded, notched < 0.1 mm . . . . *V. anagallis-aquatica*
3' Lvs petioled; fr obcordate
    4. Lvs lanceolate to ± ovate, obtuse to acute; pedicel 5–10 mm; widespread . . . . *V. americana*
    4' Lvs obovate, rounded; pedicel 3–6 mm; SNE (near Bridgeport, Mono Co.) . . . . *V. beccabunga*

V. americana (Raf.) Benth. (p. 505) AMERICAN BROOKLIME Per, rhizomed, glabrous. **ST** gen decumbent, rooting at lower nodes, branched, 5–60(100) cm. **LF** 5–50 mm, lanceolate to ovate, ± serrate, acute to obtuse, petioled. **INFL** axillary; fls 10–25; bracts linear; pedicels 5–10(13) mm. **FL:** sepals ± 3 mm, oblanceolate, acute; corolla 7–10 mm, violet-blue, dark-lined; style 1.5–3(4)mm. **FR** 3–4 mm, ± round, entire to barely notched. **SEED** 0.5 mm, flat. 2*n*=36. Common. Moist to wet soil, springs, slow streams, meadows, lakeshores; < 3300 m. CA-FP, **SNE**, **DMtns (uncommon)**. May–Aug

V. anagallis-aquatica L. WATER SPEEDWELL Per, rhizomed, glabrous. **ST** gen decumbent, rooting at lower nodes, simple to many-branched from base, 10–60(100) cm. **LF** 20–80 mm, elliptic to ovate, clasping to cordate, entire to serrate, light green, sessile (exc lowest ± short-petioled). **INFL** axillary, glabrous to ± glandular-puberulent; fls gen > 30; bracts linear-lanceolate; pedicels 4–8 mm, upcurved. **FL:** sepals 3–5.5 mm, lanceolate to elliptic; corolla 5–10 mm, pale lavender-blue, violet-lined; style 1.5–3 mm. **FR** 2.5–4 mm, at least as wide, rounded, barely notched. **SEED** 0.5 mm, flat. 2*n*=18,36. Wet meadows, streambanks, slow streams; < 2500 m. CA-FP, **W&I**, **DMtns (uncommon)**; widely naturalized in N.Am, S.Am; native to Eur. May–Sep

V. beccabunga L. Per, rhizomed, glabrous. **ST** gen prostrate to ascending, rooting at nodes, 15–55 cm. **LF** 10–50 mm, obovate, widely obtuse, ± crenate, petioled. **INFL** axillary; fls 6–15; bracts small; pedicels 3–5 mm. **FL:** sepals 2.5–3.5 mm, unequal, oblanceolate; corolla 5–7 mm, blue to purple; style 1.5–2 mm. **FR** ± 3 mm, ± spheric; notch 0.2–0.3 mm. **SEED** 0.5 mm, flat. 2*n*=16,18, 36. Wet meadows, slow streams; < 2000 m. **SNE (near Bridgeport)**; to se US; native to Eurasia. May–Aug

V. peregrina L. ssp. ***xalapensis*** (Kunth) Pennell (p. 505) PURSLANE SPEEDWELL Ann, taprooted, gen glandular-hairy. **ST** erect, gen branched, 5–30 cm. **LF** 5–25 mm, linear to spoon-shaped, entire to ± serrate; lower ± petioled. **INFL** terminal, open; bracts lanceolate, > pedicels; pedicels 0.5–2 mm. **FL:** sepals 3–6 mm, ± equal, lanceolate; corolla 2–3 mm, whitish; style 0.1–0.4 mm. **FR** 3–4 mm, obovate; notch 0.2–0.5 mm. **SEED** 0.5 mm, flat. 2*n*=52. Moist places; < 3100 m. CA-FP, **W&I**; to w Can, Mex, S.Am. Glabrous ssp. *p.* is widespread in e N.Am, Eur. Apr–Aug

V. serpyllifolia L. ssp. ***humifusa*** (Dickson) Syme (p. 505) Per, rhizomed, ± hairy. **ST** erect at tips only, 5–30 cm. **LF** 10–25 mm, elliptic to widely ovate, obtuse, entire to crenate. **INFL** terminal, glandular-hairy; pedicels 2.5–7 mm. **FL:** sepals 2.5–4 mm, ± equal, oblong to ovate; corolla 6–7 mm, bright blue; style 2–3 mm. **FR** 2.8–3.7 mm, wider than long, ± glandular-hairy; notch 0.3–0.8 mm. **SEED** 0.5 mm, flat. 2*n*=14,28. Moist streambanks, lakeshores, meadows; 1700–3000 m. CaR, SN, SnBr, SNE; to AK, ne US, NM; also in S.Am, Eurasia. [var. *h.* Dickson]

SIMAROUBACEAE QUASSIA OR SIMAROUBA FAMILY

Elizabeth McClintock

Shrub, tree, gen dioecious. **ST** often thorny; bark often bitter. **LVS** gen alternate, simple, entire, or pinnately compound with subentire lflets. **INFL:** panicle, raceme, or fls solitary. **FL** inconspicuous; sepals gen 5, gen fused at base, gen erect; petals gen 5, free, gen spreading; stamens gen 10–15, gen inserted on a disk, filaments often with a basal scale; pistils gen 2–5, ovaries superior, 1-ovuled (if pistil 1, chambers gen 2–5, 1-ovuled), styles free or partly fused. **FR:** winged achene, drupe, berry, or capsule. ± 25 genera, ± 150 spp.: trop, warm temp; some cult. [Brizicky 1962 J Arnold Arbor 43:173–186] Bark, lvs used in medicine.

CASTELA

Shrub or small tree, appearing ± lfless, dioecious; thorns large, branched. **LF** simple, sometimes scale-like. **FL:** calyx lobes 4–8; petals 4–8; stamens 8–24; ovaries 4–8, adherent near middle, style bases fused, tips spreading. **FR:** drupe, dry, 4–8 per fl, spreading. ± 15 spp.: sw & s-c US, to S.Am. (René R.L. Castel, French botanist, poet, editor, opera librettist, 1759–1832) [Moran & Felger 1968 Trans San Diego Soc Nat Hist 15:31–40]

C. emoryi (A. Gray) Moran & Felger (p. 505) CRUCIFIXION THORN Pl gen < 1 m, intricately branched; young parts densely puberulent. **LF** scale-like, entire, ephemeral and rarely seen. **INFL:** panicle, much branched, 2.5–5 cm, stiff. **FL** 6–8 mm diam. **FR** ± 6 mm, flat-topped; base ± rounded, sometimes persisting several years. **SEED** 1. RARE in CA. Dry, gravelly washes, slopes, plains; ± 650 m. **D**; AZ, nw Mex. [*Holacantha e.* A. Gray] Two other desert pls have same common name. Jun–Jul

SIMMONDSIACEAE JOJOBA FAMILY

William J. Stone

Shrub, evergreen, dioecious, much branched, unusual secondary growth. **ST:** bark smooth. **LVS** opposite, simple, small, leathery; base jointed; stipules 0. **INFL:** staminate fls in axillary clusters; pistillate fls gen solitary. **FL:** small, radial; sepals gen 5, overlapping, becoming larger in female, disk 0; corolla 0; stamens 8–12, free, anthers elongate with longitudinal slits; ovary superior, chambers 3, styles 3, stigmas long, feathery. **FR:** capsule, loculicidal. **SEED** 1. 1 genus: sw U.S., Mex. Sometimes placed in Buxaceae.

SIMMONDSIA JOJOBA, GOAT-NUT, PIG-NUT

1 sp. (F.W. Simmonds, English botanist, died exploring Trinidad in 1804)

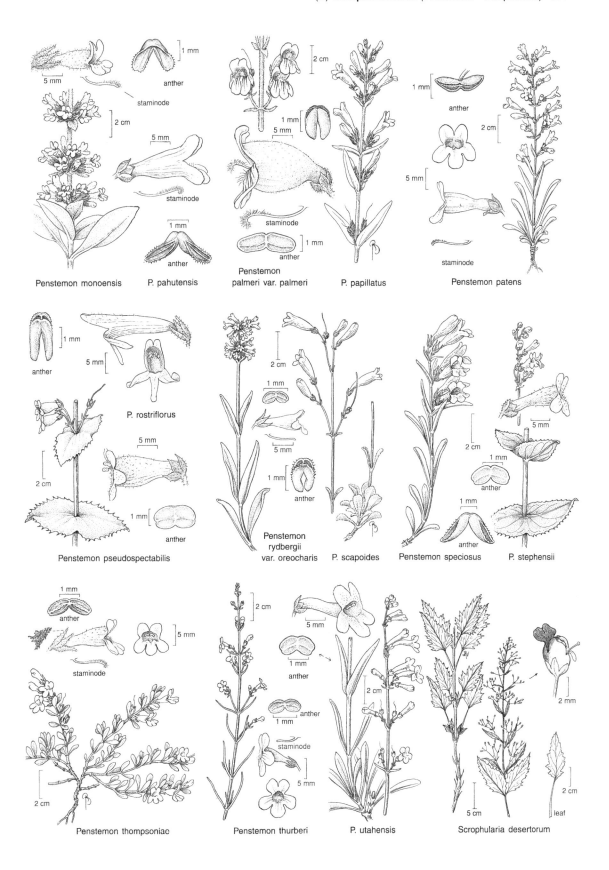

Penstemon monoensis   P. pahutensis

Penstemon palmeri var. palmeri   P. papillatus   Penstemon patens

P. rostriflorus

Penstemon pseudospectabilis   Penstemon rydbergii var. oreocharis   P. scapoides   Penstemon speciosus   P. stephensii

Penstemon thompsoniac   Penstemon thurberi   P. utahensis   Scrophularia desertorum

S. chinensis (Link) C. Schneider (p. 505, pl. 113) **ST** 1–2 m; young growth ± hairy; branches stiff. **LF** 2–4 cm, oblong-ovate, dull green, ± canescent-puberulent, subsessile. **INFL**: peduncles 3–10 mm. **FL**: sepals in staminate fls 3–4 mm, greenish, in pistillate fl becoming 10–20 mm. **FR** < 2.5 mm, nut-like, ovoid, tough, leathery, obtusely 3-angled. **SEED** large, contains liquid wax. 2*n*=26. Locally common. Arid areas; < 1500 m. w PR, SnJt, **DSon**; n Mex. [*S. californica* Nutt.] Important as forage pl, seed wax is source of substitute for sperm oil, fr edible. ❀ DRN:**8,9**,10, 11,**12**,13,**19–21**&SUN:**14**,18,**22–24**&DRY:7,15–17;CVS. Mar–May

SOLANACEAE NIGHTSHADE FAMILY

Michael Nee

Ann to shrub. **LVS** gen simple, gen alternate, gen petioled; stipules 0; blade entire to deeply lobed. **INFL** various. **FL** bisexual; calyx lobes gen 5; corolla ± radial, cylindric to rotate, lobes gen 5; stamens 5, alternate corolla lobes; ovary superior, gen 2-chambered, style 1. **FR**: berry or capsule, 2–5-chambered. 75 genera, 3000 spp.: worldwide, esp ± trop; many alien weeds in CA; many cult for food, drugs, or orn (potato, tomato, peppers, tobacco, petunia); many TOXIC. Keys adapted by Thomas J. Rosatti.

1. Fr a capsule
  2. Fr wall prickly . . . . **DATURA**
  2' Fr wall not prickly
    3. Corolla narrowly urn-shaped; seeds flat, narrowly winged . . . . **ORYCTES**
    3' Corolla funnel-shaped; seeds angled but not flat, narrowly winged
      4. Fls in racemes or panicles (only lower bracts lf-like) . . . . **NICOTIANA**
      4' Fls solitary in axils of lvs or lf-like bracts . . . . **PETUNIA**
1' Fr a berry
  5. Corolla narrowly funnel-shaped . . . . **LYCIUM**
  5' Corolla ± rotate to widely bell-shaped
    6. Anthers opening by pores or short slits near tip . . . . ²**SOLANUM**
    6' Anthers opening by slits from tip to near base
      7. Anthers gen > filaments . . . . ²**SOLANUM**
      7' Anthers gen < filaments
        8. Calyx in fr not bladder-like; corolla tube tomentose between filament bases . . . . . . **CHAMAESARACHA**
        8' Calyx in fr bladder-like; corolla tube hairy but not tomentose between filament bases . . . . . . **PHYSALIS**

CHAMAESARACHA

Per; hairs ± scale-like. **ST** decumbent, branched. **LF** entire to ± deeply pinnately lobed. **INFL**: cluster, axillary, 1–5-fld. **FL**: calyx in fr ± enlarged, open at top; corolla rotate, tomentose near base between stamens; anthers free, gen < filaments, opening by slits; style 1. **FR**: berry, spheric, partly enclosed by calyx. **SEEDS** ± flat, reniform. ± 9 spp.: esp sw US, Mex. (Greek: low *Saracha*, a S.Am genus in family) [Averett 1973 Rhodora 75:325–365]

1. Lf ± linear to lanceolate, ± entire to ± deeply lobed; corolla 10–15 mm wide; e DMoj . . . . *C. coronopus*
1' Lf ovate, entire; corolla 15–25 mm wide; GB . . . . *C. nana*

C. coronopus (Dunal) A. Gray (p. 505) **STS** many from base, 10–50 cm. **LF** 20–65 mm, 1–10 mm wide. **INFL**: pedicels ± 1 cm, in fr < 2 cm, reflexed. **FL**: calyx 3–5 mm, in fr 5–10 mm, lobes in fr 2 mm; corolla dirty or greenish white; filaments 3 mm. **FR** 5–8 mm wide. **SEEDS** 3 mm. *n*=12,24,36. Dry, clay soil; ± 1500 m. e DMtns (**New York Mtns**); to UT, TX, n Mex. May–Jul

C. nana (A. Gray) A. Gray (p. 505) **STS** 1–several from base, 5–25 cm. **LF** 15–50 mm. **INFL**: pedicels 8–18 mm, in fr < 3 cm, recurved. **FL**: calyx ± 5 mm, in fr < 10 mm, lobes in fr 2–3 mm; corolla whitish; filaments 5 mm. **FR** 1–1.2 cm wide. **SEEDS** 1.5–2 mm. *n*=12. Sandy soils, slopes, coniferous forest; 1500–2800 m. CaR, SNH, **GB**; OR, NV. [*Leucophysalis nana* (A. Gray) Averett] ❀ DRN,DRY:1–3,**7**,14,18&SUN:15–17;GRNCVR May–Jul

DATURA JIMSON WEED, THORN-APPLE

Ann to subshrub, ± glabrous or hairs simple, ill-smelling. **LF** entire to deeply lobed. **INFL**: fls solitary in branch forks. **FL**: calyx circumscissile near base, leaving a ± rotate collar in fr; corolla funnel-shaped, white or purplish, lobes 5(10); filaments attached below middle of corolla tube; ovary 2- or 4-chambered. **FR**: capsule, leathery or ± woody, gen prickled; valves 2–4, or indefinite. **SEEDS** ± flat, black, brown, grayish brown, or tan. ± 13 spp.: warm regions, esp Mex. (Hindu: ancient name) All spp. HIGHLY TOXIC; several orn, some source of drugs.

1. Corolla 10–16 cm, glabrous; calyx 5–9 cm, 5-winged toward base; seeds black with white outgrowth
   near attachment scar . . . . *D. discolor*
1' Corolla 15–20 cm, pubescent; calyx 8–12 cm, ribbed towards base; seeds tan . . . . *D. wrightii*

D. discolor Bernh. Ann < 5 dm. **ST** grayish hairy. **LF** 6–12 cm, 4–10 cm wide, widely ovate, coarsely toothed. **FL** erect; calyx 5–9 cm, 5-winged toward base, lobes 1–1.5 cm; corolla 10–16 cm, glabrous, white with purple markings in tube, lobes shallow; filaments 4.5 cm, anthers 5–6 mm; style 10–14 cm. **FR** 4-valved, nodding, 35 mm wide, puberulent; prickles < 2 cm.

SEEDS 3–3.5 cm, coarsely wrinkled, black with a white outgrowth near attachment scar. *n*=12. Sandy, gravelly soils, washes; < 500 m. **DSon**; Mex. Apr–Oct

D. wrightii Regel (p. 505) Ann or per 5–15 dm. **ST** whitish puberulent. **LF** 7–20 cm, ovate, entire or coarsely lobed. **FL** erect to nodding; calyx 8–12 cm, ribbed toward base, lobes ± 2 cm; corolla 15–20 cm, puberulent, white, lobes 1–2 cm, tips long, narrow; filaments 13–15 cm, anthers 12–15 mm; style 15–18 cm.

FR irregularly valved, nodding, 25–30 mm wide, puberulent; prickles 5–12 mm. **SEEDS** 5 mm, flat, tan; margin grooved. *n*=12. Sandy or gravelly open areas; < 2200 m. NCoRI, c&s SNF, Teh, GV, CW, SW, **D**; to UT, TX, Mex. Sometimes cult for showy fls; may have been introduced by early Spanish; may be the same as *D. inoxia* J.S. Miller [*D. meteloides* A.DC.], native to Mex. ❀ DRN,SUN:**7–10,14–24**&IRR:11–13;occas INV. Apr–Oct

LYCIUM BOX THORN

Shrub 1–4 m, gen with leafy thorns, glabrous, hairy, or glandular. **LVS** alternate or clustered, entire, small, fleshy. **INFL**: clusters; fls 1–several. **FL**: calyx cylindric to bell-shaped, lobes 2–5; corolla funnel-shaped or rotate, whitish, greenish, or purplish, lobes 4–5; stamens attached at different levels in corolla. **FR**: berry, 2-chambered, dry to fleshy. **SEEDS** 2–many. ± 100 spp.: warm, dry areas worldwide. (Latin: Lycia, ancient country of Asia Minor) [Hitchcock 1932 Ann Missouri Bot Gard 19:179–374]

1. Calyx lobes 1/2 to > tube (or gen > 2 mm)
  2. St glandular-puberulent; fr yellow to orange, with 2 cross-grooves above middle . . . . ***L. cooperi***
  2' St glabrous; fr greenish purple, not cross-grooved . . . . ***L. pallidum*** var. ***oligospermum***
1' Calyx lobes < 2/3 tube (or gen < 2 mm
  3. Corolla lobes 1/3 to = tube . . . . ***L. brevipes*** var. ***brevipes***
  3' Corolla lobes < 1/3 tube
    4. Pl glandular
      5. Calyx 4–8 mm, lobes triangular, 1–2 mm . . . . ***L. fremontii***
      5' Calyx 2.5–5 mm, lobes oblong-ovate, 2–4 mm . . . . ***L. parishii***
    4' Pl ± glabrous
      6. Corolla lobe margin glabrous to finely straight-ciliate; lf ± linear-oblanceolate . . . . ***L. andersonii***
      6' Corolla lobe margin woolly-ciliate; lf narrowly oblanceolate to obovate . . . . ***L. torreyi***

L. andersonii A. Gray (p. 505, pl. 114) Pl ± glabrous; branches stiffly spreading to erect. **LF** 3–15 mm, ± linear, ± elliptic in ×-section. **FL**: calyx 1.5–3 mm, cup-shaped, lobes (2)4–5, ± 0.8 mm; corolla narrowly funnel-shaped, whitish, often violet-tinged, tube 5–10 mm, lobes 1–1.5 mm; stamens ± incl to slightly exserted, unequal, attached 1/3 from base. **FR** 3–8 mm, red or orange, juicy. **SEEDS** many. Gravelly or rocky slopes, washes; < 1900 m. s Teh, s SnJV, SCo, n WTR, PR, **SNE**, **D**; to UT, NM, nw Mex. [var. *deserticola* (C. Hitchc.) Jepson] ❀ STBL. Mar–May

L. brevipes Benth. Pl gen glandular-puberulent; branches spreading. **LF** 5–15 mm, ± obovate. **FL**: calyx 2–6 mm, bell-shaped, lobes 2–4(6); corolla funnel-shaped, tube 6–10 mm, lavender to whitish, lobes 3–5 mm, ovate; stamens exserted, attached ± at middle of tube. **FR** ± 10 mm, red. **SEEDS** many. Coastal bluffs, slopes; < 600 m. s ChI, **w DSon**; nw Mex. ❀ SUN,DRN:7,**8**,**9**,**14**,15,16,18,**19–24**&IRR:**12,13**;also STBL. 2 vars. in CA.

  var. ***brevipes*** **FL**: calyx lobes 1/3 to = tube, triangular to linear. Habitats and elevations of sp. s ChI (San Clemente Island), **w DSon**; nw Mex. Intergrades with *L. torreyi* in w DSon. Mar–Apr

L. cooperi A. Gray (p. 505) Pl glandular-puberulent; branches rigidly ascending to erect, leafy. **LF** 1–3 cm, oblanceolate to obovate. **FL**: calyx 8–15 mm, narrowly bell-shaped, lobes 4–5, 1.5–3 mm, 1/2 to = tube; corolla narrowly funnel-shaped, greenish white, lavender-veined, outside glabrous or puberulent, tube 9–12 mm, lobes ovate-triangular; stamens barely exserted, attached ± at middle of tube. **FR** 5–9 mm, yellow to orange, with 2 cross-grooves above middle. **SEEDS** several. Sandy to rocky flats, washes; < 2000 m. SNH (e slope), s SnJV, **SNE**, **DMoj** (incl w margins); to UT, AZ. ❀ STBL. Mar–May

L. fremontii A. Gray (p. 509) Pl glandular-hairy; branches spreading to ascending. **LF** 10–25 mm, narrowly obovate. **FL**:

calyx 4–8 mm, cylindric, lobes 5, 1–2 mm, < tube, triangular; corolla 10–15 mm, narrowly funnel-shaped, violet or whitish, purple-veined, lobes < 1/3 tube; stamens mostly incl, unequal, attached in lower 1/3 of tube. **FR** 6–8 mm, red, juicy. **SEEDS** many. Alkaline soils, flats; < 500 m. PR (e slope), **s DSon**; s AZ, nw Mex. ❀ DRN,SUN:**14**,15,16,18,**19–24**&IRR:**8,9**,10,11,**12,13**; also STBL. Mar–Apr

L. pallidum Miers var. ***oligospermum*** C. Hitchc. (p. 509) Pl very thorny, glabrous; branches many, spreading to ascending. **LF** 10–50 mm, oblong to narrowly obovate, glaucous. **FL**: calyx 5–8 mm, bell-shaped, lobes 5, ± = tube; corolla narrowly funnel-shaped, greenish white, veins purple, tube 8–12 mm; stamens exserted, attached at middle of tube. **FR** 8–10 mm, fleshy, firm, greenish purple. **SEEDS** 5–7. Flats, washes, slopes; < 1200 m. **DMoj**, **n DSon**; s NV. Var. *p.* (corolla tube 12–18 mm, seeds 15+) expected in e DMoj. ❀ TRY. Mar–May

L. parishii A. Gray PARISH'S DESERT-THORN Pl glandular-hairy; intricately spreading-branched. **LF** 5–30 mm, oblanceolate. **FL**: calyx 2.5–5 mm, bell-shaped, lobes 5, 2–4 mm, ± = tube, oblong-ovate; corolla narrowly funnel-shaped, purple, tube 2.5–6 mm, lobes ± 1 mm, ovate; stamens slightly exserted, attached ± at middle of tube. **FR** 4–6 mm, red. **SEEDS** 7–12. RARE in CA. Sandy to rocky slopes, canyons; < 1000 m. e SCo, **w DSon**; AZ, nw Mex. Mar–Apr

L. torreyi A. Gray Pl nearly glabrous (exc lf clusters hair-tufted); twigs slender, ± climbing. **LF** 1–5 cm, ± oblanceolate. **FL**: calyx 2.5–4.5 mm, lobes 5, 1–1.5 mm; corolla 8–15 mm, narrowly funnel-shaped, greenish lavender or whitish, lobes 3–4 mm, lanceolate to ovate; stamens slightly exserted, attached at middle of tube. **FR** 6–10 mm, red or orange, juicy. **SEEDS** 8–30. Washes, streambanks; < 700 m. **s DMoj, DSon**; to UT, TX, n Mex. ❀ TRY. Mar–May

NICOTIANA TOBACCO

Ann to small tree. **LVS** entire. **INFL**: raceme or panicle, terminal. **FL**: calyx 5-lobed, ± enlarging, not fully enclosing fr; corolla gen radial, gen funnel-shaped to salverform; stamens 5, equal or 1 smaller. **FR**: capsule. **SEEDS** many, minute, angled. ±60 spp.: gen Am; *N. tabacum* L. widely cult. (J. Nicot, 1530–1600, said to have introduced tobacco to Eur) [Goodspeed 1954 Chron Bot 16:1–536] Seriously TOXIC to livestock.

1. Shrub or small tree; herbage glabrous, glaucous . . . . *N. glauca*
1' Ann or per; herbage gen glandular-hairy
  2. Cauline lvs clasping; per, base often slightly woody; corolla open during day . . . . *N. obtusifolia*
  2' Cauline lvs not clasping; ann, corolla closed during day
    3. Cauline lvs petioled . . . . *N. attenuata*
    3' Cauline lvs gen ± sessile (exc lowest)
      4. Corolla tube 15–20 mm, limb 8–10 mm wide; filaments attached ± at 1 level below middle of corolla tube; calyx lobes very unequal . . . . *N. clevelandii*
      4' Corolla tube 25–50 mm, limb 20–50 mm wide; filaments attached at various levels above middle of corolla tube; calyx lobes ± unequal . . . . *N. quadrivalvis*

N. attenuata Torrey (p. 509) Ann 5–15 dm, ± glandular. **LVS** 3–10 cm, petioled; basal elliptic to ovate; cauline gradually reduced. **INFL**: bracts < 30 mm, linear. **FL**: calyx 6–10 mm, pock-marked, lobes < tube, ± equal, triangular, acute; corolla ± salverform, greenish, pink-tinged, tube 20–27 mm, limb 4–8 mm wide, white; stamens unequal, filaments attached below middle of tube. **FR** 8–12 mm. Open, well drained slopes; 200–2800 m. **CA** (exc coast); to B.C., MT, NM, nw Mex. ❦ TRY. May–Oct

N. clevelandii A. Gray (p. 509) Ann 2–6 dm, slender, ± glandular. **LVS** 3–18 cm; lowermost lvs petioled, elliptic to ovate; upper lvs sessile, lanceolate. **INFL**: bracts < 30 mm, linear to lanceolate. **FL**: calyx 8–10 mm, lobes 4, very unequal (1 > tube), linear; corolla ± salverform, greenish white, tube 15–20 mm, limb 8–10 mm wide; stamens unequal, filaments attached below middle of corolla tube. **FR** 4–6 mm. Gen sandy washes, shrubland; < 500 m. SCo, ChI (Santa Catalina, Santa Cruz islands), **DSon**; AZ, nw Mex. Mar–Jun

N. glauca Graham TREE TOBACCO Gen tree < 6 m, gen glabrous, gen glaucous; wood soft. **LF** 5–21 cm, petioled, gen ± ovate. **INFL**: bracts < 5 mm, linear. **FL**: calyx ± 10 mm, lobes < tube, ± unequal, triangular; corolla 30–35 mm, ± cylindric, yellow; stamens ± equal, filaments attached below middle of tube. **FR** 7–15 mm. Open, disturbed flats or slopes; < 1100 m. NCoRI,

c&s SNF, GV, CW, SW, **D**; to s US, Mex, Afr, Medit; native to S.Am. Spring, summer

N. obtusifolia Martens & Galeotti (p. 509) Per from woody base, 2–8 dm, glandular. **LVS** 2–10 cm; lower lvs short-petioled, (ob)ovate; upper lvs ± narrowly ovate, clasping. **INFL**: bracts < 20 mm, linear to lanceolate. **FL**: calyx 10–15 mm, lobes ± = tube, ± equal, narrowly triangular; corolla ± funnel-shaped, greenish or dull white, tube and throat 15–26 mm, limb 8–10 mm wide; stamens unequal, filaments attached near base of tube. **FR** 8–10 mm. Gravelly or rocky washes, slopes; < 1600 m. **s SNE, D**; to UT, TX, Mex. [*N. trigonophylla* Dunal] Mar–Jun

N. quadrivalvis Pursh Ann 3–20 dm, glandular. **LVS** 4–15 cm; lower lvs short-petioled, elliptic to ovate; upper lvs sessile, reduced. **INFL**: bracts < 35 mm, ± linear. **FL**: calyx 15–20 mm, 10-ridged, lobes narrowly lanceolate, 1 sometimes > tube; corolla ± salverform, white with green or violet tinge, tube 25–50 mm, limb 20–50 mm wide; stamens unequal, filaments attached at different levels above middle of tube. **FR** 15–20 mm. Open, well drained washes, slopes; < 1500 m. CA-FP, **uncommon in SNE, DMoj**; to WA, c US. [*N. bigelovii* (Torrey) S. Watson] Widely cult by w Am native people. ❦ DRN,SUN:**7–10,14–24**. May–Oct

ORYCTES

1 sp. (Greek: a digger)

O. nevadensis S. Watson (p. 509) NEVADA ORYCTES Ann 5–20 cm, branched from slender taproot, leafy, sticky; some hairs scale-like. **LF** 1–3 cm, linear to ovate, entire to shallowly lobed, drying ± wavy; petiole 5–10 mm, narrowly winged. **INFL**: umbels in axils, few-fld. **FL**: calyx 2–3 mm, < 10 mm in fr; corolla 5–8 mm, narrowly urn-shaped, purplish, lobes small; stamens

attached at base of tube, unequal, some slightly exserted; style = stamens. **FR**: capsule, 2-valved, 6–7 mm wide, spheric. **SEEDS** 10–15, round, flat; body 2 mm wide; wing 0.5 mm wide. RARE in CA. Sandy soils, dunes; 1200–1500 m. **s SNE (Inyo Co.)**; w NV. Seriously threatened by grazing. May

PETUNIA

Ann or per, sticky-glandular. **ST**: main branches from base, with ± long internodes. **LVS** subopposite near fls, entire. **INFL**: fls solitary in axils of lvs or lf-like bracts. **FL**: calyx divided nearly to base, lobes ± lf-like, esp in fr; corolla funnel-shaped, 5-lobed; stamens ± equal or 1 short, 2 medium, 2 long. **FR**: capsule. **SEEDS** many, minute, spheric to angled. ± 30 spp.: gen S.Am; some cult for orn, sometimes waifs. (Native Am: petun, name for tobacco) *Our species currently treated in *Calibrachoa*. Wijsman 1990 Acta Bot Neerl 39:101-102.

P. parviflora A.L. Juss. (p. 509) **ST** prostrate to decumbent, < 4 dm, rooting at nodes; axillary branches short, leafy. **LF** 5–14 mm, ± oblanceolate, fleshy, ± subsessile. **FL**: calyx lobes 3–6 mm, < 11 mm in fr, elliptic to narrowly obovate; corolla 4–6 mm, purplish, tube whitish. **FR** 2–3 mm. **SEEDS** pale brown.

Open washes, dry streambeds; < 1300 m. CW, SCo, n ChI (Santa Rosa Island), PR, **DSon**; to sw UT, se US, Mex; also S.Am; waif elsewhere in US. ***Calibrachoa parviflora*** (A.L. Juss.) D'Arcy Apr–Aug

PHYSALIS GROUND-CHERRY

Ann or rhizomed per; hairs sometimes branched. **LVS** sometimes ± opposite, entire to pinnately lobed. **INFL**: fls 1–few per axil, pedicelled. **FL**: calyx 5-lobed, enlarged and persistent in fr; corolla rotate to widely bell-shaped, yellowish, often dark-spotted inside; stamens 5, filaments inserted on hairy band in corolla tube, anthers free, gen <

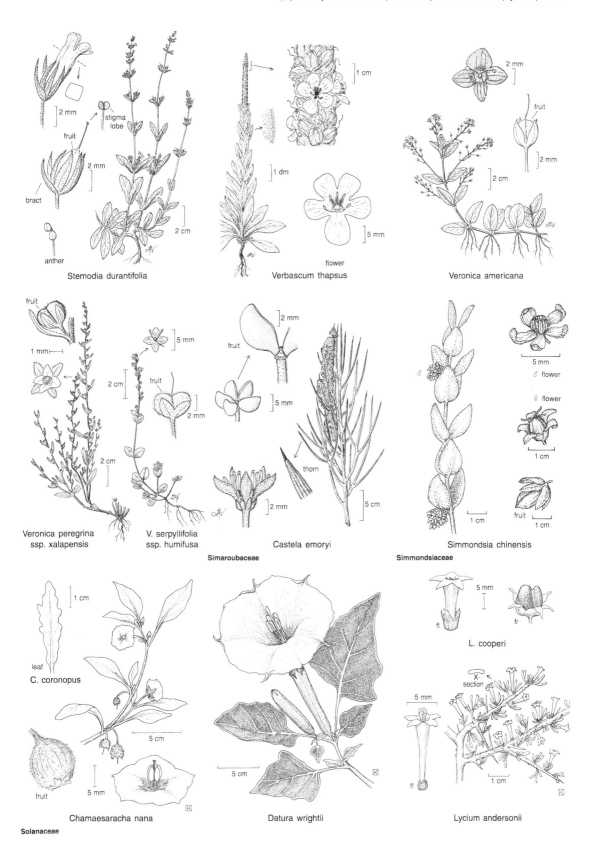

Stemodia durantifolia

Verbascum thapsus

flower

Veronica americana

Veronica peregrina
ssp. xalapensis

V. serpyllifolia
ssp. humifusa

Simaroubaceae

Castela emoryi

thorn

Simmondsiaceae

♂ flower

♀ flower

fruit

Simmondsia chinensis

C. coronopus

leaf

Chamaesaracha nana

Solanaceae

Datura wrightii

L. cooperi

X-
section

Lycium andersonii

filaments, opening by slits; style gen straight. **FR**: berry. **SEEDS** many, 2–2.5 mm, ± spheric to reniform. ± 85 spp.: Am, Eurasia, Australia. (Greek: bladder, from calyx in fr) [Sullivan 1985 Syst Bot 10:426–444] Some spp. cult for edible or orn fr. Unripe fr often TOXIC. Needs further study in w US.

1. Fl erect; corolla purple; style curved . . . . *P. lobata*
1' Fl gen nodding; corolla yellow; style straight
  2. Lf with branched hairs; per . . . . *P. hederifolia* var. *fendleri*
  2' Lf glabrous or with unbranched hairs; ann or per
    3. Per or subshrub from rhizomes
      4. Pedicel > fl or calyx in fr . . . . *P. crassifolia*
      4' Pedicel < fl or calyx in fr . . . . *P. hederifolia* var. *palmeri*
    3' Ann from taproot
      5. Pl ± densely hairy . . . . *P. pubescens*
      5' Pl glabrous to sparsely hairy
        6. Corolla pale yellow with dark yellow center, 15–23 mm wide; lf deeply toothed; anthers 3 mm
        . . . . *P. acutifolia*
        6' Corolla evenly yellow throughout, 7–8 mm wide; lf entire to ± weakly toothed; anthers 1–2 mm
        . . . . *P. lancifolia*

P. acutifolia (Miers) Sandw. (p. 509) Ann 2–10 dm, branched; hairs simple, short, appressed. **LF** 4–12 cm, lanceolate to ± ovate, tapered to base, teeth < 7 mm, prominent, slender. **INFL**: pedicels 15–25 mm, in fr < 40 mm. **FL**: calyx 3–4.5 mm, in fr 20–25 mm, spheric, with 10 equal veins; corolla 15–23 mm wide, rotate, pale yellow with darker yellow center; anthers 3 mm, yellow and blue-green. Waste places, roadsides; < 200 m. s SnJV, SCo, **DSon**; to TX, n Mex. [*P. wrightii* A. Gray] Jul–Oct

P. crassifolia Benth. (p. 509, pl. 115) Per or subshrub < 8 dm; hairs simple, dense, short, gen glandular. **ST** often zigzag, ridged. **LF** 1–3 cm, gen ovate, fleshy, entire or ± wavy; petiole ± = blade. **INFL**: pedicels 15–30 mm, in fr > fls. **FL**: calyx 4–7 mm, in fr 20–25 mm, weakly angled; corolla 15–20 mm diam, widely bell-shaped, yellow; anthers 2–3 mm, yellow. Gravelly to rocky flats, washes, slopes; < 1300 m. PR, s SNE, **D**; NV, AZ, n Mex. [var. *versicolor* (Rydb.) Waterf.; *P. greenei* Vasey & Rose, Greene's ground-cherry] ❀ TRY;DFCLT. Mar–May

P. hederifolia A. Gray Per or subshrub 1–8 dm, from fleshy rhizome; hairs sometimes branched or glandular. **LF** 2–4 cm, ovate, entire to coarsely toothed, gen gray-green; base tapered or subcordate. **INFL**: pedicels 3–5(10) mm, in fr < 15 mm. **FL**: calyx 6–7 mm, in fr 20–30 mm, ± spheric, with 10 green veins; corolla 10–15 mm, widely bell-shaped, yellow, gen with 5 purple-brown spots at base inside. Gravelly to rocky slopes; 700–1800 m. s PR, **s SNE**, **DMoj**; to TX, n Mex.

  var. ***fendleri*** (A. Gray) Cronq. Hairs gen branched, nonglandular. *n*=12. Habitats of sp; 900–1800 m. **s SNE (incl Inyo Mtns)**, **DMoj**; to TX, n Mex. [var. *cordifolia* (A. Gray) Waterf.; *P. f.* A. Gray] May–Jul

  var. ***palmeri*** (A. Gray) C. Hitchc. Hairs unbranched, many glandular. *n*=12. Habitats of sp.; 700–1600 m. s PR, **DMtns**; to UT, AZ, Baja CA.

P. lancifolia Nees Ann < 8 dm; hairs simple, few, minute, appressed. **LF** 3–13 cm, lanceolate, tapered to base, entire or teeth < 3 mm. **INFL**: pedicels 10–25 mm, in fr < 40 mm. **FL**: calyx 2–3 mm, in fr 20–25 mm, 10-veined; corolla 7–8 mm wide, widely bell-shaped, yellow; anthers 1–2 mm, blue-tinged. Wet places, fields, waste places; < 200 m. GV, SnFrB, SCoRI, **DSon**; to TX, c Mex; native to S.Am. [*P. angulata* L. var. *l.* (Nees) Waterf.] Jun–Sep

P. lobata Torrey (p. 509) LOBED GROUND-CHERRY Per < 5 dm, decumbent to spreading, few-branched, glabrous to minutely papillate. **LF** 1–7 cm, lanceolate to ovate, entire to lobed, tapered to base. **INFL**: pedicels 3–4.5 mm, not gen longer in fr. **FL**: calyx 3–4.5 mm, in fr 15–20 mm; corolla 15–20 mm wide, rotate, purple, tube white inside; anthers 1–2.5 mm, yellowish. *n*=11,22. RARE in CA. Granitic soils, dry lake margins; 500–800 m. **se DMoj**, **ne DSon**; to KS, TX, n Mex.

P. pubescens L. Ann < 8 dm; hairs simple, ± dense, spreading, most with small glands. **LF** 3–9 cm, widely ovate to ± cordate, entire to coarsely toothed. **INFL**: pedicels 3–12 mm, in fr ± longer. **FL**: calyx 5 mm, in fr 20–40 mm, sharply 5-angled with ribs between; corolla ± 10 mm wide, widely bell-shaped, yellow, tube with 5 dark spots inside; anthers 1.5–2 mm, blue. Waste places, cult fields; < 1500 m. s NCoRO, SnJV, **s SNE**, **D**, expected elsewhere; native to e US. 2 vars. in CA.

  var. *grisea* Waterf. **LF** 5–9 cm, ± thick, often drying reddish brown; teeth many. *n*=12. Habitats and elevations of sp. **s SNE**, **D**; native to c&e US. [*P. pruinosa* L.; *P. neomexicana* Rydb. misapplied to CA pls] Often cult. Aug–Oct

SOLANUM NIGHTSHADE

Ann to shrub or vine, often glandular, sometimes prickly. **LVS** alternate to subopposite, often unequal, entire to deeply pinnately lobed. **INFL**: panicle or umbel-like, often 1-sided. **FL**: calyx ± bell-shaped; corolla ± rotate, white to purple; anthers free, > filaments, oblong or tapered, opening by 2 pores or short slits near tip; ovary 2-chambered, style 1, stigma head-like. **FR**: berry, gen spheric (or dry, capsule-like). **SEEDS** many, compressed, gen reniform. ± 1500 spp.: worldwide, esp trop Am. (Latin: quieting, from narcotic properties) [Symon 1981 J Adelaide Bot Gard 4:1–367] Many cult for food (incl potato, *S. tuberosum*), orn; many TOXIC. *S. rostratum* has been reported from the Panamint Mtns.

1. Hairs, or at least some, stellate; anthers tapered, opening only by terminal pores; pl often prickly
  . . . . *S. elaeagnifolium*
1' Hairs simple or branched, not stellate; anthers oblong, opening by terminal pores that become short slits;
  pl gen not prickly
  2. Lf very deeply lobed . . . . *S. triflorum*
  2' Lf entire to shallowly lobed (or deeply 2-lobed below middle)
    3. Corolla lobes shallow (< tube + throat), gen spreading . . . . *S. xanti*

3' Corolla lobes deep (= or > tube + throat), often reflexed
  4. Calyx enlarged and enclosing fr base; fr yellowish or green . . . . *S. sarrachoides*
  4' Calyx not enlarged, not enclosing fr base; fr black
    5. Anthers 1.4–2.2 mm . . . . *S. americanum*
    5' Anthers gen 2.5–4 mm . . . . *S. douglasii*

S. americanum Miller (p. 509) Ann to subshrub 3–8 dm, ± glabrous or hairs short, curved or ± appressed. **LF** 2–15 cm, ovate, entire to coarsely wavy-toothed. **INFL** umbel- or ± raceme-like. **FL**: calyx 1–2 mm, lobes in fr recurved; corolla 3–6 mm wide, deeply lobed, white; anthers 1.4–2.2 mm; style 2.5–4 mm. **FR** 5–8 mm diam, greenish or black. **SEED** 1–1.5 mm. *n*=12. Open, often disturbed places; < 1000 m. CA-FP, **DMoj (uncommon)**; to Can, e US, Mex. [*S. nodiflorum* Jacq.] Apr–Nov

S. douglasii Dunal Per or subshrub < 20 dm, much-branched; hairs gen simple, < 1 mm, ± curved, white. **LF** 1–9 cm, ovate, entire to coarsely and irregularly toothed. **INFL** gen umbel-like. **FL**: calyx ± 2 mm; corolla ± 10 mm wide, deeply lobed, white with greenish spots at base; anthers 2.5–4 mm; style 4–5 mm, puberulent below. **FR** 6–9 mm diam. **SEED** 1.2–1.5 mm. *n*=12. Dry shrubland, woodland; < 1000 m. s NCo, Teh, CW, SW, **DMoj**; n Mex. Most of year

S. elaeagnifolium Cav. (p. 509) WHITE HORSE-NETTLE Per < 10 dm, from rhizome, forming colonies, gen prickly, densely scaly-stellate (hair rays fused at center). **LF** 2–15 cm, oblong, entire to barely lobed, ± yellowish. **INFL** raceme-like. **FL**: calyx 5–8 mm, lobes ± = tube; corolla 20–30 mm wide, purple or blue; anthers 8–10 mm, ± equal. **FR** 8–15 mm diam, orange, often persistent. **SEED** 2.5–3 mm. *n*=12. Common. Dry, disturbed places, fields; < 1200 m. CA (exc NCo, KR, GB); to S.Am; native to c US, n Mex. NOXIOUS WEED. May–Sep

S. sarrachoides Sendtner Ann 1–9 dm, decumbent, sticky; hairs spreading, some glandular. **LF** 2–6 cm, ovate, entire to irregularly toothed or shallowly lobed. **INFL** gen raceme-like, few-fld. **FL**: calyx 2–2.5 mm, in fr 4–6 mm, enclosing fr base; corolla 3–5 mm wide, white, lobes > tube; anthers 1.5–2 mm. **FR** 6–7 mm diam, yellowish. **SEED** ± 2 mm, yellow. *n*=12. Disturbed areas; gen < 1000 m. NCo, KR, CaRF, GV, w CW, SCo, n ChI (Santa Cruz Island), **SNE**; to Can, e US, Mex; native to S.Am. May–Oct

S. triflorum Nutt. Ann 1–5 dm, decumbent; hairs ± curved or spreading, sometimes glandular. **LF** 2–5 cm, oblong to ovate, very deeply lobed. **INFL** umbel-like; fls 2–3. **FL**: calyx 2.5–3 mm, enlarged in fr, enclosing fr base; corolla 7–9 mm wide, white, lobes ± = to > tube; anthers ± 3 mm. **FR** 8–12 mm diam, green. **SEEDS** many, 2.5–3 mm, yellow. *n*=12. Dry shrubland, juniper woodland; 100–2300 m. s SNH (e slope), SCo, **GB**, **n DMoj**; to c US; native to S.Am. Jun–Sep

S. xanti A. Gray (p. 509) Per to subshrub 4–9 dm, much-branched, ± hairy, sometimes glandular. **LF** 2–7 cm, lanceolate to ovate, ± entire to 1–2-lobed at base; base obtuse to subcordate. **INFL** umbel-like, sometimes branched. **FL**: calyx 4–5 mm; corolla 15–30 mm wide, dark blue or lavender, lobes < tube; anthers 4–5.5 mm. **FR** 10–15 mm diam, greenish. **SEED** 1.5–2 mm. Shrubland, oak/pine woodland, coniferous forest; < 2700 m. CA-FP (exc CaR, GV), **n DMtns**; Baja CA. [vars. *hoffmannii* Munz (Hoffmann's nightshade), *intermedium* Parish, *montanum* Munz; *S. tenuilobatum* Parish (narrow-lvd nightshade)]. 🌺 DRN:2,3,7,9,14,18–21&SUN:15,16,**17,22–24**;CV. Feb–Jun

STERCULIACEAE CACAO FAMILY
R. David Whetstone & T.A. Atkinson

Per to tree; hairs stellate (or scale-like, peltate). **LVS** cauline, alternate, simple or palmately compound, evergreen, petioled; stipules gen deciduous. **INFL**: gen complex clusters, cymes, or fls solitary (in axils, opposite lvs, or on a spur branch); whorl of bracts often subtends calyx (esp if petals 0). **FL** bisexual, radial; sepals 5, gen fused at base; petals 0 or 5, clawed, sometimes fused to filament tube; stamens 5 (sometimes alternate 5 staminodes), filaments fused below into tube; ovary superior, sometimes on a stalk that may be fused to filament tube, chambers gen 5, style 1. **FR**: capsule. 60 genera, 700 spp.: gen trop, subtrop; some cult for orn (*Fremontodendron*) or for drugs and food (*Cola*, *Theobroma*, chocolate). [Brizicky 1966 J Arnold Arbor 47:60–74] Recently treated in an expanded Malvaceae [Alverson et al 1999 Amer J Bot 86:1474–1486; Bayer et al 1999 Bot J Linn Soc 129: 267–303].

AYENIA

Shrub; taproot stout. **ST** erect, much-branched at base, 1–4 dm; twig hairs stellate or 0. **LF** ovate-obovate, unlobed, serrate. **INFL**: fls 1(–2) in axils. **FL** gen < 3 mm wide; sepals ± spreading, narrowly ovate; petal claw thread-like, coiled, limb ± obcordate, incurved, parachute-like, sinus with an anther below and a stalked, gland-like appendage above; filament tube ± cup- or urn-shaped at top, surrounding ovary, with 5 short, stalked anthers bent outward and downward (each stalk inserted in sinus, thereby attached to petal), staminodes 5, < stamens; ovary (and fr) stalked above receptacle. **FR**: chambers 1-seeded. ± 50 spp.: warm Am. (Louis de Noailles, Duc d'Ayen, 1739–1777) [Cristobal 1960 Opera Lilloana 4:1–230]

A. compacta Rose (p. 509) AYENIA **FL**: sepals ± 1.5 mm; petals 2–3 mm, claws ± 2 mm. **FR** ± 5 mm, spheric, straw-colored, with ± cylindric, purplish protuberances. **SEED** black. RARE in CA. Dry, rocky canyons; < 500 m. **e DMtns (Providence Mtns)**, **w&c DSon (incl Eagle Mtns)**; Baja CA. [*A. californica* Jepson] Mar–Apr

TAMARICACEAE TAMARISK FAMILY
Dieter H. Wilken

Shrubs, trees, much-branched, often in saline habitats. **ST**: trunk bark rough. **LVS** alternate, sessile, entire. **INFL**: racemes or spikes; bracts scale-like. **FL**: sepals 4–6, gen free, overlapping; petals 4–5, overlapping, gen attached below nectary; stamens 4–10, attached to disk-like nectary; ovary 1-chambered, placentas parietal or basal, ovules 2–many, styles 2–5. **FR**: capsule, loculicidal. **SEEDS** many, hairy. ± 5 genera, 100 spp.: Eurasia, Afr, esp Medit.

TAMARIX TAMARISK

STS green, glabrous; twigs jointed, slender, often drooping. **LVS** on twigs, gen overlapping, awl- to scale-like, gen excreting salt. **INFLS** gen in panicle-like clusters on current or previous year's twigs. **FL**: sepals gen 5, persistent; petals gen 5, deciduous to ± persistent, white to reddish; stamens gen 5, filaments alternate or confluent with nectar disk lobes; nectar disk 4–5-lobed; placentas basal, styles 3. **SEEDS** hairy-tufted. (Latin: Tamaris River, Spain) [Baum 1967 Baileya 15:19–25] Invasive weeds with deep roots that lower water table, esp along streams, irrigation canals. Most CA spp. cult for orn, windbreaks; some may hybridize.

1. Twigs-lvs not overlapping, strongly clasping, abruptly pointed; sepals ± round . . . . ***T. aphylla***
1' Twigs-lvs overlapping, not clasping, acute to long-acuminate; sepals elliptic to ovate
  2. Sepals, petals, and stamens gen 4 . . . . ***T. parviflora***
  2' Sepals, petals, and stamens gen 5
    3. Nectar disk lobes longer than wide, confluent with filaments . . . . ***T. gallica***
    3' Nectar disk lobes wider than long, alternate filaments
      4. Lvs oblong to narrowly lanceolate, acute; sepals entire . . . . ***T. chinensis***
      4' Lvs ovate, acute to acuminate; sepals minutely toothed . . . . ***T. ramosissima***

T. aphylla (L.) Karsten (p. 515) ATHEL Tree < 12 m. **LVS** on twigs not overlapping, ± 2 mm, strongly clasping, abruptly pointed. **INFL**: spike 2–6 cm; bract clasping, triangular, tip acuminate. **FL**: sepals 1–1.5 mm, ± round, tip obtuse, entire, outer < inner; petals ± 2 mm, oblong to elliptic; nectar disk lobes wider than long; stamens alternate disk lobes. Uncommon. Washes, roadsides; < 200 m. SnJV, e SCo, **D**; to TX; native to India, Afr. May–Jul

T. chinensis Lour. Tree < 10 m. **LF** 1.5–3 mm, oblong to narrowly lanceolate, acute. **INFL**: raceme 1.5–7 cm, dense; bract linear to lanceolate, acute to acuminate; pedicel ± 1 mm. **FL**: sepals 0.5–1.5 mm, ovate, acute; petals 1.5–2 mm, oblong to elliptic; nectar disk lobes wider than long; stamens alternate disk lobes. Uncommon. Canyons, riverbanks, roadsides; < 200 m. SCo, **D**; to s Can, Mex; native to Asia. Possibly *T. ramosissima.*

T. gallica L. Shrub or tree < 8 m. **LF** 1.5–2 mm, linear to narrowly lanceolate, acute. **INFL**: spike 2–5 cm; bract oblong to lanceolate, acute to acuminate. **FL**: sepals 0.5–1 mm, ovate, acute, outer slightly < inner; petals 1.5–2 mm, elliptic to ovate; nectar disk lobes longer than wide, confluent with filaments. Uncommon. Washes, flats, roadsides; < 300 m. s NCoR, SnJV, SnFrB,

SCo, **n DMoj (Death Valley)**; to TX; native to s Eur. Pls with outer sepals narrower than inner, petals sometimes > 2 mm, have been called *T. africana* Poiret; native to Afr. Jun–Aug

T. parviflora DC. (p. 515) Shrub or tree 1.5–5 m. **LF** 2–2.5 mm, ± linear, long-acuminate. **INFL**: spike 1–4 cm; bract > pedicel, triangular, tip obtuse to blunt. **FL**: sepals 4, 1–1.5 mm, elliptic to ovate, entire to finely toothed, outer slightly > and wider than inner, acute to abruptly pointed, inner obtuse; petals 4, ± 2 mm, ± oblong; stamens gen 4; nectar disk 4-lobed, lobes longer than wide, confluent with filaments. Common. Washes, slopes, sand dunes, roadsides; < 800 m. s NCoR, s SNF, Teh, GV, CCo, SnFrB, SCoRI, SCo, **SNE**, **D**; to ne N.Am; native to se Eur. [*T. tetrandra* Pallas misapplied] Mar–Apr

T. ramosissima Ledeb. (p. 515) Shrub or tree < 8 m. **LF** 1.5–3.5 mm, ovate; tip acute to acuminate. **INFL**: spike 1.5–7 cm; bract > pedicel, triangular, acuminate. **FL**: sepals 0.5–1 mm, ± ovate, obtuse to acute; petals 1–2 mm, elliptic to oblanceolate; nectar gland lobes wider than long; stamens alternate disk lobes. Common. Washes, streambanks, ditches; < 800 m. Teh, SnJV, SCo, WTR, **SNE**, **D**; to c US, AZ; native to e Asia. [*T. pentandra* Pallas] Possibly *T. chinensis.* Apr–Aug

ULMACEAE ELM FAMILY

Dieter H. Wilken

Shrub or tree, often monoecious. **LVS** simple, gen alternate, gen 2-ranked, short-petioled; stipules deciduous; blade base often oblique, veins pinnate. **INFL**: cyme, clustered, axillary; fls 1–few; bracts 0. **FL** radial; sepals 4–6, free to fused; corolla 0; stamens 4–6, opposite sepals; ovary superior, chamber 1, ovule 1, style branches gen 2. **FR**: drupe or winged nutlet. ± 15 genera, ± 150 spp.: temp to trop; some cult for orn (*Celtis, Ulmus, Zelkova*), some used for wood, fibers (esp *Ulmus*). [Elias 1970 J Arnold Arbor 51: 18–40] *Celtis* recently treated in Celtidaceae [Angiosperm Phylogeny Group 1998 Ann Missouri Bot Gard 95:531–303; Wiegrefe & Sytsma 1998 Pl Syst Evol 210:249–270]

1. Fr a spheric drupe; lf blade entire to serrate, lower surface yellowish green, clearly net-veined . . . . **CELTIS**
1' Fr a compressed, winged nutlet; lf blade srrate (sometimes doubly so), lower surface pale green, not
  clearly net-veined . . . . **ULMUS**

CELTIS HACKBERRY

Shrub or tree; fls often bisexual and unisexual on same pl. **ST**: trunk bark deeply furrowed. **LF**: blade entire to serrate, base oblique. **INFL** gen stalked. **FL**: sepals gen 5–6, gen free to near base; stamens 4–6, exserted; style branches elongate, spreading to reflexed. **FR**: drupe, ± spheric, persistent. **SEED** 1, spheric. ± 60 spp.: esp trop and dry temp.

C. reticulata Torrey (p. 515) NET-LEAF HACKBERRY Pl 1–8 m. **ST**: trunk bark corky, checkered between furrows; twigs puberulent. **LF**: blade 2–8 cm, lanceolate to ovate, leathery, entire to serrate, lower surface yellowish green, clearly net-veined, base obtuse to ± cordate, tip obtuse to acuminate, scabrous. **FL**: pedicel in fr 4–15 mm. **FR** 5–12 mm diam, brownish to purple; pulp thin. Uncommon. Canyons, seeps, washes; 500–1700 m. s SNF, Teh, SnBr, PR, **s SNE**, **e DMtns**; to WA, KS, TX, n Mex. Sometimes also cult, persisting (ScV, SNE, D). ❀ SUN,DRN:4–6,**14–17,22–24**&IRR:1,**2,3,7–13,18–21**.

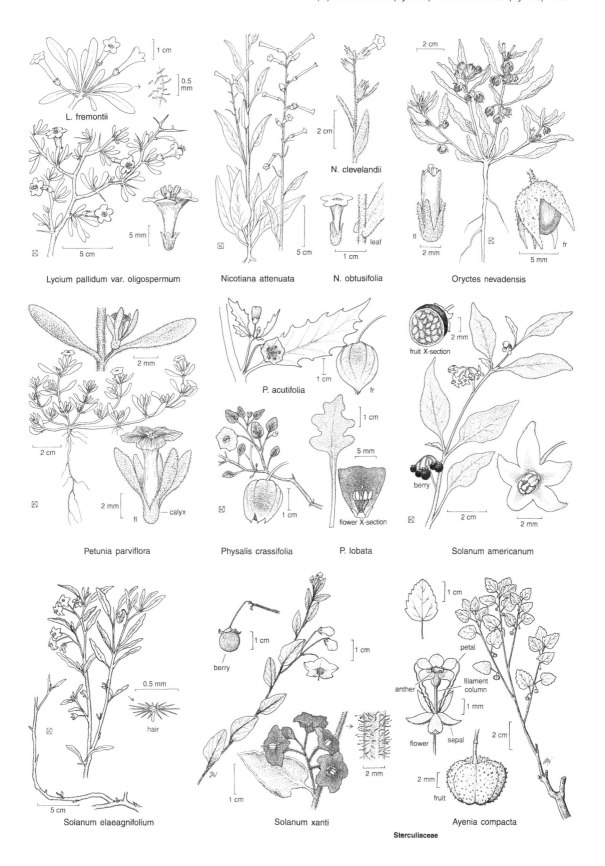

L. fremontii

Lycium pallidum var. oligospermum

Nicotiana attenuata

N. clevelandii

N. obtusifolia

leaf

Oryctes nevadensis

fl

fr

Petunia parviflora

fl — calyx

P. acutifolia

fr

Physalis crassifolia

P. lobata

flower X-section

fruit X-section

berry

Solanum americanum

Solanum elaeagnifolium

hair

Solanum xanti

berry

hair

Ayenia compacta

petal

anther

filament column

flower

sepal

fruit

Sterculiaceae

ULMUS ELM

Shrub or tree. **ST**: trunk bark scaly to furrowed, gray to brown. **LF** serrate (or doubly so), base gen oblique, obtuse to cordate. **INFL** sessile to short-stalked. **FL** bisexual; calyx gen bell-shaped, lobes 5–9; stamens 5–9, exserted; style branches spreading. **FR**: nutlet, compressed, clearly winged. ± 20 spp.: n temp; some widely cult (e.g. *U. americana*, american elm, *U. ×hollandica*, dutch elm).

U. pumila L. SIBERIAN ELM Tree gen < 10 m. **ST**: twigs glabrous to sparsely puberulent. **LF**: blade 3–9 cm, narrowly elliptic to lanceolate, margin gen singly serrate, glabrous (lower surface vein axils puberulent). **INFL**: fls appearing before lvs in spring. **FL**: calyx lobes short, unequal. **FR** 10–15 mm, 10–12 mm wide; body 2–4 mm wide; wing translucent, whitish. Waste places, roadsides, washes; < 1800 m. SNF, GV, **SNE, n DMoj**; to c US; native to c Asia. Cult; reproducing by seeds and root-sprouts. Spring

URTICACEAE NETTLE FAMILY

Dennis W. Woodland

Ann to (soft-wooded) trees, glabrous or stinging-hairy, monoecious or dioecious. **LVS** alternate or opposite, gen stipuled, petioled, often with embedded crystals. **INFL** various, axillary. **FLS** gen unisexual, small, greenish; sepals gen 4–5, free to fused; petals 0. **STAMINATE FL**: stamens gen 4–5, opposite sepals, in bud incurved, then springing out. **PISTILLATE FL**: ovary 1, superior, chamber 1, style 0–1, stigma 1, gen hair-tufted. **FR**: gen achene (drupe). 50 genera, 700 spp.: worldwide; some cult (*Boehmeria*; *Pilea*, clearweed) [Miller 1971 J Arnold Arb 52:40–68] Wind-pollinated. Keys adapted by Margriet Wetherwax.

1. Stinging hairs 0; ann; lvs alternate . . . . **PARIETARIA**
1' Stinging hairs present; per; lvs opposite . . . . **URTICA**

PARIETARIA PELLITORY

Ann or per, not stinging-hairy. **LVS** alternate, 1–8 cm, lanceolate to round, entire; stipules 0; crystals round. **INFL** head-, spike-, or panicle-like, gen few-fld; fls subtended by involucre of 1–3 bracts. **FL**: sepals 4, fused below. **STAMINATE FL**: stamens 4. **FR** ovoid, shiny. 20–30 spp.: worldwide. (Latin: wall, from habitat of some) [Hinton 1969 Sida 3:293–297]

1. Lf blade ovate, base gen truncate to cordate, lowest veins often from basal margin; calyx lobes 2–3 mm
. . . . *P. hespera* var. *hespera*
1' Lf blade linear to ovate, base ± tapered, lowest veins gen from midrib; calyx lobes 1.5–2 mm . . *P. pensylvanica*

P. hespera B.D. Hinton Ann 2–55 cm, decumbent to erect, sometimes matted. **LF**: blade 5–20 mm, ovate to round, base gen truncate to cordate, tip notched to short-acuminate, lowest veins often from basal margin. **FL**: calyx lobes gen 2–3 mm, acute or acuminate. **FR** hidden between calyx lobes, 0.9–1.3 mm, ovate, tan to ± brown, tip obtuse. Chaparral, deserts, dry woodlands, roadsides, often moist shade, sandy or rocky soils; < 1400 m. SnFrB, SW, **D**; to UT, NM, nw Mex. [*P. floridana* Nutt. misapplied] 2 vars. in CA.

var. *hespera* (p. 515) **LF**: blade ovate, wider than long. **FL**:
calyx lobes erect, acute, dark reddish brown. ± habitats and range of sp. Feb–Jun

P. pensylvanica Willd. (p. 515) Ann < 6 dm, decumbent to erect. **LF**: blade 7–90 mm, linear-lanceolate to ovate, base tapered to obtuse, tip gen acuminate to long-tapered, lowest veins gen from midrib. **FL**: calyx lobes ± erect, 1.5–2 mm, acute, dark reddish brown. **FR** hidden between calyx lobes or not, 0.9–1.5 mm, 0.5–1 mm wide, ovate, ± (reddish) brown, tip obtuse. $2n=16$. Dry ledges, slopes, shady waste areas; < 2500 m. CA-FP, **DSon**; US. Mar–Jul

URTICA STINGING NETTLE

Ann to shrub, weak, stinging-hairy or not, monoecious or dioecious. **LVS** opposite, lanceolate to cordate, toothed, prominently 3–5-veined from base; crystals round to elongate. **INFL** head-, raceme-, or panicle-like. **STAMINATE FL**: sepals 4, ± free, green, sharp-bristly; stamens 4. **PISTILLATE FL**: sepals 4, ± free, outer 2 < inner 2. **FR** lenticular to deltate, enclosed by 2 inner sepals. ± 50 spp.: esp temp. (Latin: to burn, from stinging hairs) [Woodland 1982 Syst Bot 7:282–290]

U. dioica L. Per 10–30 dm, from rhizome, ± erect. **LF**: blade gen > 40 mm, narrowly lanceolate to widely ovate, base tapered to cordate. **INFL** spike-, raceme-, or panicle-like, 1–7 cm, gen > petiole, with only staminate or pistillate fls. **FR** ovate. Streambanks, margins of deciduous woodlands, moist waste places; < 3000 m. CA-FP, **W&I, DMtns**; US, Can, n Mex, Eurasia. Ssp. *d.* dioecious, native to Eurasia; naturalized in N.Am incl CA. 3 sspp. in CA.

ssp. *holosericea* (Nutt.) Thorne (p. 515) HOARY NETTLE Gen monoecious. **ST** 10–30 dm, gray-green; nonstinging hairs (st and lower lf surface) moderate to dense. $2n=26$. Habitats of sp.; < 3000 m. CA-FP (± exc NW), **W&I, DMtns**; w US, n Mex. [*U. h.* Nutt.; *U. serra* Blume misapplied]. Jun–Sep

VALERIANACEAE VALERIAN FAMILY

Lauramay T. Dempster

Ann, per, sometimes strongly scented; odor gen disagreeable. **LVS** simple, pinnately lobed, or compound; petioles sometimes sheathing; basal ± whorled; cauline opposite, petioled to sessile. **INFL**: cyme, panicle, or head-like, gen ± dense. **FLS** gen bisexual; calyx fused to ovary top, limb 0 or highly modified (if present, lobes gen 5–15, coiled inward, becoming plumose, pappus-like, spreading in fr); corolla radial to 2-lipped, lobes gen 5, throat gen > lobes, >> tube, base gen spurred or swollen, tube slender, long or short; stamens gen 1–3, epipetalous; ovary inferior, chamber gen 1 (sometimes 3 but 2 empty or vestigial). **FR**: achene, smooth, ribbed, or winged. ± 17 genera, 300 spp.: gen temp, worldwide exc Australia. Some spp. cult (*Centranthus*), some medicinal (*Valeriana*). [Ferguson 1965 J Arnold Arbor 46:218–225] Recently treated in an expanded Caprifoliaceae [Donoghue et al 1992 Ann Missouri Bot Gard 79:333–345].

VALERIANA VALERIAN

Ann, per (in CA) from rhizome or short underground caudex, glabrous to soft-hairy. **ST** gen erect, 1–several. **LVS**: basal simple or pinnately lobed, tapered to petiole; cauline subsessile to ± clasping, pinnately lobed, distal lobe gen > others. **INFL**: cyme, clustered, ± dense to open, terminal or axillary. **FL**: calyx lobes 5–15, gen ± coiled inward, becoming plumose, spreading and persistent in fr; corolla ± funnel-shaped, white or pink, lobes ± equal, throat >> tube, sometimes swollen near base, tube slender, sometimes obscured by swollen throat; stamens 3; ovary ± 1-chambered. **FR** gen compressed, gen 6-veined vertically. ± 200 spp.: temp worldwide exc Australia. (Latin: strength, from use in folk medicine, or after Valerian, a Roman emperor) [Meyer 1951 Ann Missouri Bot Gard 38:377–503]

V. pubicarpa Rydb. (p. 515) Pls gen with bisexual fls, gen glabrous. **ST** 1.1–7 dm. **LVS** 4–30 cm; basal simple, blade elliptic to spoon-shaped; cauline few, reduced, simple or few-lobed, margin ± entire. **FL**: corolla ± 6 mm, white, sometimes pinkish, throat ± 2 × lobe length. **FR** 3.5–5 mm, lanceolate in outline. Moist, rocky slopes, coniferous forest; 2100–3200 m. **W&I**; to OR, MT, UT.

VERBENACEAE VERVAIN FAMILY

Dieter H. Wilken

Ann to tree, gen hairy. **LVS** cauline, opposite, gen toothed; stipules 0. **INFL**: raceme, spike, or head, gen elongated in fr; bract gen 1 per fl. **FL** bisexual; calyx gen 4–5-toothed; corolla 4–5-lobed, radial to bilateral, salverform to 2-lipped; stamens 4–5, epipetalous (if 4, gen in unequal pairs); ovary superior, 2- or 4-lobed, gen 2-chambered, style 1, often with 2 unequal lobes, only 1 stigmatic, lateral. **FR**: 2 or 4 nutlets, drupe-like, or capsule. ± 90 genera, ± 1900 spp.: esp Am trop. Some cult (*Clerodendron, Lantana, Verbena, Vitex*); some weedy worldwide (*Lantana*); some used for wood (*Tectona*, teak).

1. Shrub . . . . **ALOYSIA**
1' Ann to per
  2. Calyx 2–4-toothed, ± compressed; corolla 2-lipped; nutlets 2 . . . . **PHYLA**
  2' Calyx 5-toothed, cylindric; corolla gen radial; nutlets 4 . . . . **VERBENA**

ALOYSIA

Shrub, strong-smelling. **LF**: blade lanceolate to ovate. **INFL**: raceme or spike. **FL**: calyx 4-toothed; corolla 5-lobed, ± 2- lipped; stamens 4, in unequal pairs; ovary 2-chambered, ovules 2, style unlobed, stigma ± spheric, terminal. **FR**: nutlets 2. ± 35 spp.: Am. Some used for food flavoring, tea. (Maria Luisa Teresa, Queen of Spain or Luis Antonio Bourbon, Prince of Asturias, Spain, both 18th century).

A. wrightii Abrams (p. 515) OREGANILLO Shrub < 2 m, ± rounded. **ST**: branches many; twigs brown, angles white. **LF**: petiole < 4 mm; blade 4–17 mm, ovate to ± round, crenate, lower surface densely hairy. **INFL**: spike 1.5–6 cm; axis densely and finely tomentose; bract slightly < calyx, lanceolate. **FL**: calyx 1.5–3 mm, puberulent; corolla 2.5–3.5 mm, white, lobes rounded, upper 2 larger. **FR** < 2 mm. Uncommon. Rocky, often limestone, slopes, Joshua-tree or pinyon/juniper woodland; 900–1600 m. **s&e DMtns**; to TX, n Mex. [*Lippia w.* Torrey invalid] ❀ TRY;DFCLT. Aug–Oct

PHYLA

Per, gen mat-like. **ST**: central gen stolon-like; branches decumbent to erect, glabrous or ± strigose. **LVS** opposite or clustered, strigose to appressed-hairy; hairs forked. **INFL**: spike, ± spheric, becoming cylindric in fr, dense; bracts ovate to wedge-shaped. **FL**: calyx ± compressed, 2–4-toothed; corolla ± 2-lipped, tube gen > calyx; stamens 4, in unequal pairs; ovary 2-chambered, ovules 2, style lobes 2, stigma lateral. **FR**: nutlets 2. ± 15 spp.: warm temp, subtrop Am. (Greek: clan or tribe, from clustered fls) ❀ IRR or WET:7,**8,9**,10–12,**13,14**,18,**19–23**&SUN:4–6,**15–17,24**;turf-like GRNCVR; fls attract bees.

1. Lf widest below middle, 15–25 mm wide, teeth 11–21 . . . . *P. lanceolata*
1' Lf gen widest at or above middle, 5–10 mm wide, teeth 5–11 . . . . *P. nodiflora*
  2. Lf gen 4–5 × longer than wide, narrowly wedge-shaped . . . . var. *incisa*
  2' Lf gen 2–4 × longer than wide, oblanceolate to elliptic . . . . var. *nodiflora*

P. lanceolata (Michaux) E. Greene (p. 515) **ST**: internodes gen 3–10 cm; branches 15–50 cm. **LF**: blade 25–60 mm, lanceolate to ovate; margin serrate from below middle, teeth 11–21. **INFL** 7–18 mm; peduncle 4–9 cm. **FL**: corolla white or pale blue to purplish. $2n=32$. Wet places, marshes, ditches; < 400 m. GV, CCo, SnFrB, SCo, **D**; to e N.Am, n Mex. [*Lippia l.* Michaux] May–Nov

P. nodiflora (L.) E. Greene **ST**: internodes gen < 4 cm; branches gen < 15 cm. **LF**: blade 5–30 mm, ± ovate to wedge-shaped; margin gen serrate from above middle; teeth 5–11. **INFL** 6–10 mm; peduncle 1.5–9 cm. **FL**: corolla white to reddish. $2n=36$. Gen wet places; < 400 m. NW (exc KR, NCoRH), GV, CCo, SnFrB, SCo, ChI (Santa Cruz, Catalina islands), PR, **DSon**; to e N.Am, n Mex.

var. *incisa* (Small) Mold. (p. 515) **LF** gen 4–5 × longer than wide; blade narrowly wedge-shaped. Wet places, pond margins, ditches; < 400 m. SnJV, SCo, **DSon**; to c US, n Mex. [*Lippia i.* (Small) Tidestrom] Jul–Nov

var. *nodiflora* (p. 515) **LF** gen 2–4 × longer than wide; blade elliptic to oblanceolate. Wet places, ditches, fields; < 300 m. NW (exc KR, NCoRH), GV, CCo, SnFrB, SCo, ChI (Santa Cruz, Santa Catalina islands), PR, **DSon**; warm temp, trop ± worldwide. [*Lippia n.* vars. *canescens* (Kunth) Kuntze, *reptans* (Kunth) Kuntze] Densely matted pls with lvs < 1 cm, widely naturalized from S.Am, have been called *Phyla cespitosa* (Rusby) Moldenke] May–Nov

VERBENA

Ann to shrub. **STS** often 4-angled; hairs gen short, stiff. **LVS** reduced upward; blade entire to pinnately lobed. **INFL**: spikes, often in panicle-like clusters, gen terminal, gen very long in fr. **FL**: calyx 5-ribbed, 5-toothed, hairs gen strigose or appressed; corolla 4–5-lobed, ± radial to bilateral and 2-lipped; stamens 4, paired; ovary 4-chambered, ovules 4, style 1, lobes 2, 1 tooth-like, 1 with ± spheric stigma. **FR**: nutlets 4, gen oblong. ± 250 spp.: temp, trop Am, Medit Eur. (Latin: ancient name) [Umber 1979 Syst Bot 4:72–102; Barber 1982 Syst Bot 7:433–456]

1. Calyx 6–9.5 mm; corolla 8–14 mm, limb 6–10 mm wide . . . . *V. gooddingii*
1' Calyx 2–4 mm; corolla 2–5 mm, limb < 5 mm wide
  2. Fl bract 4–8 mm, 2–3 mm > calyx; most sts prostrate to decumbent . . . . *V. bracteata*
  2' Fl bract 2–3 mm, gen < calyx; most sts ascending to erect . . . . *V. menthifolia*

V. bracteata Lagasca & J.D. Rodriguez (p. 515) Ann or bien 8–30 cm. **STS** few–many from base, prostrate to decumbent; hairs sparse, spreading. **LF** 1–3(6) cm, ± oblanceolate, coarsely serrate to lobed, rough-hairy; base tapered to ± flat petiole. **INFL**: spikes 1–3 per st branch, 2–10 cm in fr, 6–10 mm diam, gen dense; fl bract 4–8 mm. **FL**: calyx 2–4 mm; corolla 4–5 mm, white to lavender or blue. **FR** 1–2 mm. $2n=14$. Open, disturbed places, pond or lake margins; < 2200 m. CaR, GV, SCoR, SW, **GB, D**; to B.C., e N.Am, n Mex. May–Oct

V. gooddingii Briq. (p. 515) Per (sometimes fl 1st year) 10–45 cm. **STS** 3–10+ from base, decumbent to erect; hairs soft, spreading. **LF** 1–4 cm, lanceolate to ovate, obtusely toothed, short-soft-hairy; base 3–5-lobed, tapered to ± flat petiole. **INFL**: spike gen 1 per st branch, 2–6 cm in fr, 10–15 cm diam, dense; fl bract 4.5–6 mm. **FL**: calyx 6–9.5 mm, hairs spreading; corolla 8–14 mm, purplish blue. **FR** 2–3 mm. $2n=30$. Sandy soils, washes, rocky slopes; 1200–2000 m. **e DMoj, ne DSon**; to UT, N Mex. [var. *nepetifolia* Tidestrom; *Glandularia g.* (Briq.) Solbrig] ❀DRN,SUN:**10**& IRR:**7–9,11–16**,17,**18–23**,24. Apr–Jun

V. menthifolia Benth. Bien or per 30–75 cm. **STS** 1–3 from base, ascending to erect, gen sparsely strigose. **LF** 2–4(6) cm, ovate, gen deeply 1–2-lobed near base, coarsely serrate, sparsely strigose; base tapered to ± flat petiole. **INFL**: spikes 1–3 per st, 6–30 cm in fr, < 0.5 mm diam, open (frs not overlapping); fl bract 2–3 mm. **FL**: calyx 2.5–3 mm; corolla 2–3 mm, purple. **FR** 1–1.5 mm. Open, gen dry places, shrubland; < 300 m. SCo, w PR, **DSon**; to TX, n Mex. Apr–Jun

VIOLACEAE VIOLET FAMILY

R. John Little

Ann to shrub or vine (gen per in CA). **LVS** basal, cauline, or both, gen alternate, entire to compound; stipules gen small. **INFL**: head, raceme, panicle, or fls solitary; peduncle bractlets 2. **FL** gen bisexual, gen bilateral; sepals 5, free to slightly fused, gen persistent; petals 5, free, lowest gen spurred or pouched at base; stamens gen 5, alternate petals, filaments short, wide, anthers surrounding ovary, adherent or fused, often with nectaries at base, often with membranous appendage at tip; ovary superior, chamber 1, placentas 3, parietal, ovules gen many, style 1. **FR**: gen capsule, 3-valved, gen explosively dehiscent. **SEEDS** gen appendaged. 15 genera, 600 spp.: gen temp, worldwide; some cult as orn; some Eur spp. medicinally useful as emetics, diuretics, purgatives. [Brizicky 1961 J Arnold Arbor 42:321–333]

VIOLA VIOLET

Ann or per < 35 cm, glabrous to hairy. **LF** entire to compound. **INFL**: fl gen solitary, axillary. **FL** bilateral; sepals subequal, appendaged at base; petals unequal, lowest spurred or pouched at base, lateral 2 equal, gen spreading, often hairy near base, upper 2 equal, erect; lower 2 stamens with nectaries projecting into spur. **FR**: capsule, ovoid to oblong. (Latin: ancient name) [Clausen 1964 Madroño 17:173–197] Cleistogamous fls gen present. Seeds often dispersed by ants that feed on seedappendages. Key to species adapted by Margriet Wetherwax.

1. Petals blue-violet, without yellow . . . . ***V. sororia*** ssp. ***affinis***
1' Petals yellow to ± orange, or white with yellow base and spur
  2. Lf canescent above; pl gen 6–13 cm; peduncle 30–100 mm; uncommon — SNE, DMoj . . . . ***V. aurea***
  2' Lf glabrous to hairy (not caulescent) above; pl 1–35 cm; peduncle 15–170 mm; widespread . . . . . ***V. purpurea***
    3. Pl 4–35 cm; st gen not buried, clearly branched and elongated above by late summer; lower petal
      (incl spur) gen > 11 mm . . . . ssp. ***mohavensis***
    3' Pl 1–12 cm; st gen buried, little branched or elongated by late summer; lower petal (incl spur) gen
      < 11 mm . . . . ssp. ***venosa***

V. aurea Kellogg (p. 515) Pl 6–13 cm. **ST** decumbent or erect from woody taproot, gray-tomentose. **LVS** simple, canescent; blade base tapered to truncate; basal 1–6, petiole 40–70 mm, blade 12–50 mm, oblong to round, crenate to shallowly and irregularly serrate, tip obtuse; cauline petiole 14–55 mm, blade 15–37 mm, ± lanceolate to ovate, gen dentate-serrate. **INFL:** peduncle 30–100 mm, gen canescent. **FL:** petals yellow, lowest (incl spur) 8–13 mm, lower 3 veined dark brown, lateral 2 with thick hairs, at least upper 2 purple or brown outside. **FR** ± 6 mm, puberulent. *n*=6. Uncommon. Dry, sandy slopes; 1000–1800 m. **SNE, DMoj;** w NV. Apr–Jun

V. purpurea Kellogg Pl 1–35 cm. **ST** appearing ± early, sometimes elongating, ascending to erect from woody taproot, gen hairy. **LVS** simple, entire to toothed, often purplish (esp below), ± hairy; basal 1–5, petiole 20–145 mm, blade 10–50 mm, ovate to round, tapered to cordate at base, often fleshy; cauline blade < basal, lanceolate to ovate, base tapered to cordate. **INFL:** peduncle 15–170 mm. **FL:** petals deep lemon-yellow, lowest (incl spur) 6–17 mm, lower 3 veined purple-brown, lateral 2 bearded, upper 2 purplish outside. **FR** 5–12 mm, puberulent. n=6,12. Chaparral, dry forest, timberline communities, sagebrush or desert scrub; 400–3100 m. NW, CaR, SN, CW, SW, **GB, DMtns (Panamint Mtns);** to WA, WY, Colorado, AZ, n Baja CA. Variable; sspp. intergrade; needs study. ❀ TRY;DFCLT. 5 sspp. in CA.

ssp. ***mohavensis*** (M. Baker & J. Clausen) J. Clausen (p. 515) Pl 8–22 cm. **LVS:** basal gen erect, blade regularly dentate, base tapered to truncate; cauline blade deeply toothed, base tapered, tip acute. **INFL:** peduncle < 140 mm. *n*=6. Desert or sagebrush scrub, dry yellow-pine forest; 1000–2400 m. TR, PR, **SNE;** sw NV. [*V. aurea* ssp. *m.* M. Baker & J. Clausen] Variable. Apr–Jun

ssp. ***venosa*** (S. Watson) M. Baker & J. Clausen (p. 515) Pl 4–11 cm. **LVS:** basal erect to spreading, blade often fleshy, base tapered to subcordate; cauline blade gen crenate to deeply toothed, base gen tapered. **INFL:** peduncle < 70 mm. *n*=6. Dense shade of forests or shrublands; 1300–3100 m. **GB, DMtns (Panamint Mtns);** to WA, WY, Colorado. [sspp. *atriplicifolia* (E. Greene) M. Baker & J. Clausen, *geophyta* M. Baker & J. Clausen]

V. sororia Willd. ssp. ***affinis*** (Le Conte) R.J. Little (p. 515) LECONTE VIOLET Pl 3–25 cm from short thick rhizome with long fibrous roots, ± glabrous. **ST** 0. **LVS** basal, simple; petiole 50–230 mm; blade 10–45 mm, gen widely ovate to reniform, toothed, tip acute to obtuse. **INFL:** peduncle 50–250 mm. **FL:** petals deep blue-violet, lowest (incl spur) 10–19 mm (spur ± 3 mm, as wide as long), lower 3 dark-veined, white-bearded. **FR** 5–10 mm, glabrous. *n*=27. Shady, moist ground; 300–2300 m. KR, SnBr, Wrn, **SNE.** [*V. nephrophylla* E. Greene] ❀ IRR,DRN: **4–6,15–17**&SHD:1–3,**7**,8, 9,**14**,18–24;occas. INV. May–Jun

VISCACEAE MISTLETOE FAMILY
Frank G. Hawksworth and Delbert Wiens

Per, shrub, gen ± green, parasitic on aboveground parts of woody pls, dioecious or monoecious. **ST** brittle; 2° branches gen many. **LVS** simple, entire, opposite, 4-ranked, with blade or lvs scale-like (then each pair gen fused). **INFL:** spikes or open cymes, gen axillary, sometimes terminal; bracts opposite, 4-ranked, scale-like, each pair fused. **FL** unisexual, radial, 2–4 mm; perianth parts in gen ± 1 series. **STAMINATE FL:** perianth parts 3–4(7); anthers gen ± sessile, opposite and gen on perianth parts. **PISTILLATE FL:** perianth parts gen 2–4; ovary inferior, 1-chambered, style unbranched, stigma ± obscure. **FR:** berry, shiny, gelatinous. **SEEDS** 1(–2), without thickened coat. 7 genera, ±450 spp.: trop, gen n temp. [Kuijt 1982 J Arnold Arbor 63:401–410] Sometimes incl in Loranthaceae; parasitic on pls in many other families. Frs gen dispersed by birds or seeds explosively ejected. All parts of most members may be TOXIC. Recently treated in Santalaceae [Angiosperm Phylogeny groupd 1998 Ann Missouri Bot Gard 95:531–303] Keys adapted by Margriet Wetherwax.

1. St gen < 12 cm; angled, at least when young, olive-green to brown; lf < 1 mm, scale-like; berry
  2-colored, seeds dispersed by explosion . . . . **ARCEUTHOBIUM**
1' St gen > 20 cm; rounded, green, less often reddish; lf 10–47 mm, or scale-like; berry 1-colored,
  seeds dispersed by consumption (by birds) . . . . **PHORADENDRON**

ARCEUTHOBIUM DWARF MISTLETOE

Per, shrub, glabrous, dioecious. **ST** gen < 20 cm, ± angled, at least when young, yellow, straw, yellow-green, olive-green, green, brown, reddish, purple; 2° branches gen not whorled, gen in ± 1 plane. **LF** scale-like. **INFL:** spikes, many-fld, open or ± interrupted, short-peduncled; fls gen opposite, 4-ranked, less often whorled. **STAMINATE FL:** perianth parts 3–4(7); anthers ± 1-chambered. **PISTILLATE FL:** perianth parts 2, persistent. **FR** gen 2–5 mm, ± compressed-spheric, 2-colored (1 color below, 1 above), dispersed by explosion, seeds projected < 15 m; pedicel short, recurved. ± 45 spp.: temp and trop n hemisphere. (Greek: juniper, life) Most important timber pathogens, causing annual loss of many millions of dollars; most spp. cause abnormal branching (witches' brooms) in hosts. [Hawksworth & Wiens 1972 USDA Handbook No. 401]

A. divaricatum Engelm. (p. 525) PINYON DWARF MISTLETOE **ST** 7–12 cm, 2–4 mm wide at base, olive-green to brown. **FL** August–September. **SEED** mature September–October. *n*=14. Woodlands, on *Pinus edulis, P. monophylla* (on *P. quadrifolia* in n Baja CA, expectedly so in s CA); 1400–2300 m. c&s SNH, TR, PR, **SNE, DMtns;** to Colorado, TX, Baja CA. [*A. campylopodum* f. *d.* (Engelm.) Gill]

PHORADENDRON MISTLETOE

Shrub, woody at least at base, glabrous or hairy, dioecious in CA. **ST** gen > 20 cm, rounded, green, less often reddish. **LF** with blade or lf scale-like (then each pair fused). **INFL**: spikes, many-fld, open or ± interrupted, short-peduncled; fls sunken into axis. **FL**: perianth parts gen 3. **STAMINATE FL**: anthers 2-chambered. **PISTILLATE FL**: perianth parts persistent. **FR** ± 3–6 mm, ± spheric, 1-colored, white, pink, or reddish, maturing (in temp) in 2 seasons, dispersed by consumption (by birds); pedicel 0. ± 200 spp.: temp, trop Am. (Greek: tree thief) *P. tomentosum* (DC.) Engl. collected in Texas for sale nationally in Christmas trade; other spp. similarly important locally. [Wiens 1974 Brittonia 16:11–54]

1. Lf scale-like, << 1 mm
  2. St 4–10 dm, canescent, esp at tip, reddish to green; spike with 2–3(7) fertile internodes; on *Acacia*,
     Cercidium, Larrea (rarely), *Olneya, Prosopsis*; fl Jan–Mar; < 1200 m . . . . **P. californicum**
  2' St glabrous, ± green or yellow-green; spike with 1–2 fertile internodes; on *Calocedrus, Juniperus*;
     fl June–July; 1300–2600 m . . . . **P. juniperinum**
1' Lf with blade, gen 10–47 mm
  3. Lf length gen > 3 × width, 10–15 mm; st glabrous; spike with gen 1 fertile internode, pistillate 2-fld;
     on conifers . . . . **P. densum**
  3' Lf length gen > 1.5 × width, 15–47 mm; st gen short-hairy, at least at tip; spike with 2–5(7) fertile
     internodes, pistillate 6–15(20)-fld; on woody dicots
    4. Lf 30–42 mm, 15–23 mm wide, gen glabrous, if hairy not very densely so; fr glabrous; staminate
      fertile internodes 2–5(7), gen 30–35-fld, pistillate 2–4(6), 6–10(20)-fld; on woody dicots other than
      Quercus; fl Dec–Mar; < 1200 m . . . . **P. macrophyllum**
    4' Lf 15–47 mm, 10–25 mm wide, very densely short-hairy; fr short-hairy near tip; staminate fertile
      internodes 2–4, gen 25–30-fld, pistillate 2(–3), gen 10–15-fld; gen on *Quercus*; fl July–Sept; 60–2100 m
      . . . . **P. villosum**

P. californicum Nutt. (p. 525, pl. 116) DESERT MISTLETOE **ST** 4–10 dm, gen pendent in age, reddish to green, canescent, esp at tip, ± glabrous in age; internodes 13–28 mm. **LF** scale-like. **STAMINATE INFL**: fertile internodes 2–3(6), 4–10-fld. **PISTILLATE INFL**: fertile internodes 2–3(7), 2–3-fld. **FL** January–March. **FR** ± 3 mm, white to reddish pink, ± glabrous. *n*=14. Deserts, on *Acacia, Cercidium, Larrea* (rarely), *Olneya, Prosopis*; < 1200 m. **D**; NV, AZ, Baja CA. [vars. *distans* Trel., *leucocarpum* (Trel.) Jepson]

P. densum Trel. DENSE MISTLETOE **ST** 3–5 dm, gen ± erect, green, glabrous; internodes 6–17 mm. **LF** gen 10–15 mm, 2–5 mm wide, gen oblanceolate-oblong, ± sessile. **STAMINATE INFL**: fertile internodes 1(–2), 6–13-fld. **PISTILLATE INFL**: fertile internodes 1(–2), 2-fld. **FL** June–August. **FR** ± 4 mm, white to straw or pinkish, glabrous. *n*=14,27. Juniper/pinyon woodlands, on *Cupressus, Juniperus* (reported on *Pinus monophylla*, in Mount Pinos area, Ventura Co.); 200–2300 m. KR, NCoR, CaR, CW, TR, PR, **GB, DMtns**; to s OR, c AZ, nw Mex. [*P. bolleanum* (Seemann) Eichler ssp. *d.* (Trel.) Fosb.; *P. b.* var. *d.* (Trel.) Wiens]

P. juniperinum A. Gray JUNIPER MISTLETOE **ST** 2–4 dm, gen erect, gen woody only at base, yellow-green, glabrous; internodes 5–10(12) mm. **LF** scale-like. **STAMINATE INFL**: fertile internodes 1(2), 5–9-fld. **PISTILLATE INFL**: fertile internode 1, 2-fld. **FL** June–July. **FR** ± 4 mm, pinkish white, glabrous. *n*=14. Juniper/pinyon woodlands, on *Juniperus*; 1700–2600 m. CaRH,

SNH, SnBr, MP, **W&I, DMtns**; to OR, Colorado, TX, Mex. [var. *ligatum* (Trel.) Fosb.]

P. macrophyllum (Engelm.) Cockerell BIG LEAF MISTLETOE **ST** > 1 m, ± erect, green, short-hairy, esp near tip, ± glabrous in age; internodes 22–59 mm. **LF** 30–42 mm, 15–23 mm wide, obovate to elliptic-round, ± petioled to ± not, gen glabrous, if hairy not very densely so. **STAMINATE INFL**: fertile internodes 2–5(7), gen 30–35-fld. **PISTILLATE INFL**: fertile internodes 2–4(5), 6–10(20)-fld. **FL** December–March. **FR** ± 4–5 mm, white, pink-tinged or not, glabrous. *n*=14. On woody dicots other than *Quercus* (esp *Alnus, Fraxinus, Juglans, Platanus, Populus, Prosopis, Robinia, Salix*); < 1200 m. NCoRO, NCoRI, SNF, GV, CW, SCo, TR, PR, **D**; to Colorado, w TX, Baja CA. [*P. flavescens* (Pursh) Nutt. var. *m.* Engelm.; *P. tomentosum* ssp. *m.* (Engelm.) Wiens]

P. villosum (Nutt.) Nutt. (p. 525) OAK MISTLETOE **ST** ± 1 m, ± erect, gray-green, gen densely short-hairy when young, ± glabrous in age; internodes 15–38 mm. **LF** 15–47 mm, 10–25 mm wide, obovate-elliptic, ± petioled or not, very densely short-hairy. **STAMINATE INFL**: fertile internodes 2–4, gen 25–30-fld. **PISTILLATE INFL**: fertile internodes 2(–3), gen 10–15-fld. **FL** July–September. **FR** ± 3–4 mm, pinkish white, short-hairy near tip. *n*=14. Oak woodlands, on gen *Quercus*, less often other associated woody dicots (e.g., *Adenostoma, Arctostaphylos, Rhus, Umbellularia*); 60–2100 m. KR, NCoR, SNF, ScV, SCoR, SCo, TR, PR, **DMtns (Little San Bernardino Mtns)**; to n OR, TX, Mex. [*P. flavescens* var. *v.* (Nutt.) Engelm.]

VITACEAE GRAPE FAMILY

Michael O. Moore

Woody vines; tendrils opposite lvs; fls sometimes unisexual. **LVS** gen many, cauline, simple or compound, alternate, petioled, deciduous; stipules gen deciduous. **INFL** cyme or panicle, gen opposite lf, peduncled. **FL** radial; sepals gen reduced, gen fused, lobes 5 or 0; petals gen 5, free, reflexed and falling individually (or adherent at tips, ± erect, and falling as unit), reddish or yellowish; stamens gen 5, opposite petals; nectaries 0 or between stamens as ± free glands; ovary 1, superior, chambers gen 2(–4), style 1 or 0, stigma inconspicuous or head-like. **FR**: berry. **SEEDS** 1–6, large. 15 genera, ± 800 spp.: esp warm regions; some cult (*Cissus*, grape ivy; *Parthenocissus*; *Vitis*). [Moore 1991 Sida 14:339–367]

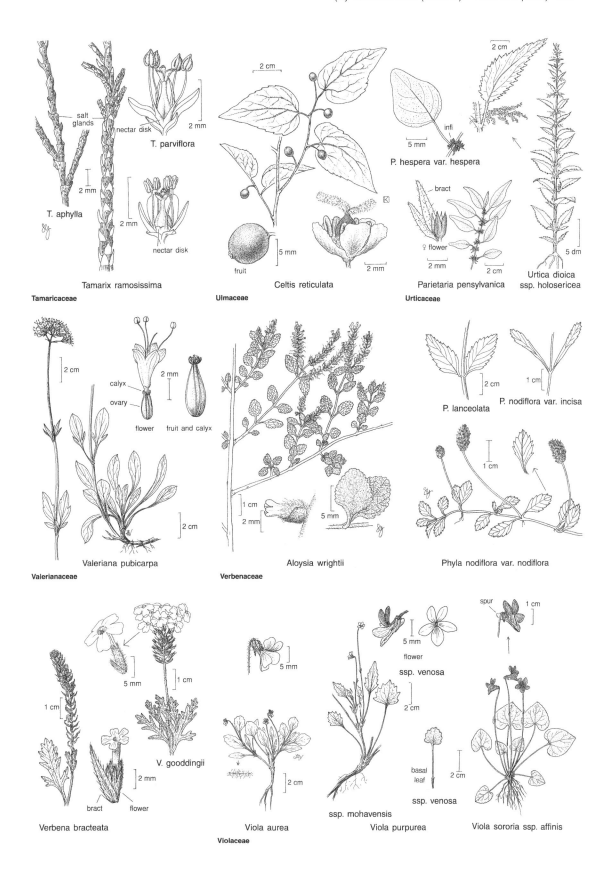

salt glands

nectar disk

T. parviflora

2 mm

2 mm

T. aphylla

2 mm

nectar disk

Tamarix ramosissima

Tamaricaceae

2 cm

5 mm

fruit

2 mm

Celtis reticulata

Ulmaceae

5 mm

infl

P. hespera var. hespera

bract

♀ flower

2 mm

2 cm

Parietaria pensylvanica

Urticaceae

2 cm

5 dm

Urtica dioica
ssp. holosericea

2 cm

calyx

ovary

2 mm

flower   fruit and calyx

Valeriana pubicarpa

Valerianaceae

1 cm

2 mm

5 mm

Aloysia wrightii

Verbenaceae

2 cm

P. lanceolata

1 cm

P. nodiflora var. incisa

1 cm

1 cm

Phyla nodiflora var. nodiflora

5 mm

1 cm

1 cm

bract   flower

2 mm

V. gooddingii

Verbena bracteata

5 mm

2 cm

Viola aurea

Violaceae

5 mm

flower

ssp. venosa

2 cm

basal
leaf

ssp. venosa

ssp. mohavensis

Viola purpurea

spur

1 cm

2 cm

Viola sororia ssp. affinis

VITIS GRAPE

ST: bark peeling; st center brown, clearly partitioned at nodes; tendrils not tipped with adhering disks. **LF** simple; blade crenate to serrate. **INFL**: panicle. **FL** unisexual or bisexual; calyx greenish, lobes 0 or shallow; petals adherent at tips, yellowish; stamens 3–9, gen erect; nectaries ± free glands. **PISTILLATE FL**: stamens reflexed or 0. **FR** 4–20 mm wide, spheric to ovoid, glaucous or not. **SEEDS** 1–4, obovoid, with a round structure opposite attachment scar that is either raised or sunken. 65 spp.; temp, subtrop. (Latin: vine) [Olmo & Koyama 1980 Proc 3rd Intl Symp Grape Breeding 33–41] Key to species adapted by Margriet Wetherwax.

1. Fr purple, very glaucous, when 3–4-seeded gen > 8 mm wide; seed structure opposite attachment scar
   gen raised; stipules gen < 3.5 mm; st tomentose when young, becoming less so . . . . *V. californica*
1' Fr black, slightly to not glaucous, when 3–4-seeded gen < 8 mm wide; seed structure opposite attachment
   scar gen sunken; stipules gen > 3.5 mm; st densely tomentose when young, ± remaining so . . . . *V. girdiana*

V. californica Benth. (p. 525) CALIFORNIA WILD GRAPE **ST** to-mentose when young, becoming less so; nodal partitions gen 3–4 mm thick. **LF**: stipules gen < 3.5 mm; blade lobes 0–3, shallow, margin crenate to slightly serrate, lower surface ± tomentose. **FL** unisexual. **FR** gen > 8 mm wide, spheric, purple, glaucous; skin separating from pulp. **SEED**: round structure opposite attachment scar gen raised. 2*n*=38. Streamsides, springs, canyons; < 1000 m. NW, CaRF, SNF, GV, CW, **SNE**; OR. ❀ 4,5,6;IRR:**7–9**,10–12,**14–24**;CVS. May–Jun

V. girdiana Munson (p. 525) DESERT WILD GRAPE **ST** ± densely tomentose; nodal partitions gen 2–3 mm thick. **LF**: stipules gen > 3.5 mm; blade lobes 0 or 3–5 and shallow, margin gen serrate, lower surface tomentose to densely so. **FL** unisexual. **FR** gen < 8 mm wide, spheric, black, not or slightly glaucous; skin separating from pulp. **SEED**: round structure opposite attachment scar gen sunken. 2*n*=38. Streamsides, canyon bottoms; < 1250 m. SW, **s SNE**, **D**; Baja Ca. Intergrades with (perhaps best a var. of) *V. californica*. ❀ 4–6;IRR:**7–16**,17,**18–24**. May–Jun

ZYGOPHYLLACEAE CALTROP FAMILY

Duncan M. Porter

Herb, shrub, often armed; caudex present or not. **ST** branched; nodes often angled, swollen. **LVS** 1-compound, opposite; stipules persistent or deciduous; lflets entire. **INFL**: fls 1–2 in axils. **FL** bisexual; sepals 5, free, persistent or deciduous; petals 5, free, gen spreading, sometimes twisted and appearing propeller-like; stamens 10, sometimes appendaged on inside base; ovary superior, chambers 5–10, ovules 1–several per chamber, placentas axile. **FR**: capsule or splitting into 5–10 nutlets. 26 genera, ± 250 spp.: widespread esp in warm, dry regions; some cult (*Guaiacum*, lignum vitae; *Peganum*, harmal (NOXIOUS and illegal); *Tribulus*, caltrop (pernicious)). *Peganum harmala* L. has been reported as a pernicious weed near Daggett, San Bernardino Co. [Porter 1972 J Arnold Arbor 53:531–552]

1. Lflets 2
   2. Lflets fused at base . . . . **LARREA**
   2' Lflets free at base . . . . **ZYGOPHYLLUM**
1' Lflets 3 or more
   3. Lflets 3, palmate, spine-tipped; stipules spine-tipped . . . . **FAGONIA**
   3' Lflets 6–18, pinnate, not spine-tipped; stipules not spine-tipped
      4. Fr tuberculed, nutlets 10 . . . . **KALLSTROEMIA**
      4' Fr spiny, nutlets 5 . . . . **TRIBULUS**

FAGONIA

Per, shrub. **ST** < 1 m, spreading, angled or ridged. **LF** palmately compound; stipules stiff, spine-tipped; lflets 3, spine-tipped, terminal largest. **INFL**: fls solitary in axils. **FL**: sepals deciduous; petals clawed, twisted, propeller-like, purple to pink, deciduous. **FR**: capsule, deeply 5-lobed, obovoid, ± septicidal; style persistent; peduncle reflexed. **SEED** 1 per chamber. ± 18 spp.; sw N.Am, Chile, Medit, sw Afr. (G.C. Fagon, French physician to Louis XIV, 1638–1718)

1. St ascending to erect, scabrous; glands only on youngest herbage, << 0.1 mm wide; stipules curved; lflets
   lanceolate . . . . *F. laevis*
1' St prostrate, not scabrous; glands also on older herbage, ± 0.15 mm wide; stipules straight; lflets elliptic
   to ovate . . . . *F. pachyacantha*

F. laevis Standley (p. 525, pl. 117) Shrub < 1 m, intricately branched. **LF**: lflets 3–9 mm, gen < petiole, 1–4 mm wide. **FL** ± 1 cm wide. **FR** 4–5 mm wide, minutely strigose or hairy, gen with some glands, rarely glabrous; style 1 mm, wider at base. Rocky hillsides to sandy washes; 0–700 m. **D**; to sw UT, nw Mex. [*F. californica* Benth. ssp. *l.* (Standley) Wiggins] Pls with minute, glandular hairs on frs have been called *F. longipes* Standley. ❀ TRY. Mar–May, Nov–Jan

F. pachyacantha Rydb. (p. 525) Per; caudex woody. **LF**: lflets < 25 mm, ± = or > petiole, < 9 mm wide. **FL** ± 1.5 cm wide. **FR** 5 mm wide, hairy, gen with some glands; style 2–3 mm, not or barely wider at base. Flat, sandy or rocky habitats; 0–500 m. **DSon**; to sw AZ, Mex. [*F. californica* Benth. var. *glutinosa* Vail] ❀ TRY. Mar–May, Nov–Jan

KALLSTROEMIA

Ann, unarmed. **ST** spreading radially, prostrate to ascending, < 1 m. **LF** even-1-pinnate; stipules narrow, green. **INFL**: fls solitary in axils. **FL**: sepals deciduous or persistent; petals yellow to orange, withering but not deciduous. **FR** 10-lobed, splitting into 10 tubercled nutlets; style persistent; peduncle reflexed. **SEED** 1 per chamber. 17 spp.: warm and trop Am. (A. Kallstroem, obscure contemporary of Scopoli, author of genus) [Porter 1969 Contr Gray Herb 198:41–153]

1. Peduncle < subtending lf; petal 3– 5 mm; sepals deciduous; style < fr body . . . . *K. californica*
1' Peduncle > subtending lf; petal 6–30 mm; sepals persistent; style > fr body
  2. Sepals >> fr body, spreading; petal 15–30 mm, yellow to orange, darker at base . . . . *K. grandiflora*
  2' Sepals ± = to slightly > fr body, appressed; petals 6–12 mm, orange . . . . *K. parviflora*

K. californica (S. Watson) Vail (p. 525) **ST** prostrate to decumbent, < 0.7 m, glabrous to strigose. **LF**: stipules 1.5–5 mm; lflets 6–12. **FL**: peduncle < subtending lf; sepals deciduous; petals 3–5 mm, yellow. **FR** ovoid; body > style, 3–5 mm wide. RARE in CA. Flat, sandy or disturbed areas; 0–600 m. D; to TX, Mex. Last collected in CA in 1948. Aug–Oct

K. grandiflora A. Gray **ST** decumbent to ascending, < 1 m; some hairs densely silky, others sharply bristly. **LF**: stipules 4–10 mm; lflets 8–16. **FL**: peduncle > subtending lf; sepals persistent, margins strongly inrolled, >> fr body, spreading; petals 15–30 mm, yellow to orange, darker at base. **FR** 2–3 mm wide, ovoid; style 3 × fr body. Uncommon. Sandy roadsides; 300–900 m.

DSon (near Desert Center, Riverside Co.; Jacumba, San Diego Co; expected elsewhere); native AZ to TX, w Mex. Waif in CA. Sep–Oct

K. parviflora Norton (p. 525) **ST** prostrate to decumbent, < 1 m, strigose to glabrous. **LF**: stipules 5–7 mm; lflets 6–10. **FL**: peduncle > subtending lf; sepals persistent, ± = to slightly > fr body, appressed; petals 6–12 mm, orange. **FR** 4–6 mm wide, ovoid; style to 3 × fr body. Uncommon. Sandy roadsides, slopes; 1500–2000 m. c PR (near Warner Hot Springs, San Diego Co.), **e DMtns (Clark Mtns, San Bernardino Co.)**, expected elsewhere; native AZ to c US, c Mex. Waif in CA. Aug–Oct

LARREA CREOSOTE BUSH

Shrub, unarmed. **ST** branched, erect to prostrate, < 4 m, reddish becoming gray; nodes swollen, darker; hairs 0 or appressed. **LF**: stipules persistent; lflets 2, fused at base. **INFL**: fls solitary in axils. **FL**: sepals deciduous, unequal, overlapping; petals clawed, twisted, propeller-like, yellow, deciduous; stamen appendages bract-like, coarsely toothed. **FR** 5-lobed, spheric, short-stalked, hairy, splitting into 5 hairy, 1-seeded nutlets. 5 spp.: warm, dry Am. (J.A. Hernandez de Larrea, Spanish clergyman)

L. tridentata (DC.) Cov. (p. 525, pl. 118) **LF**: lflets < 18 mm, < 8.5 mm wide, obliquely lanceolate to curved; deciduous awn between lflets < 2 mm. **FL** < 2.5 cm wide; sepals ovoid, appressed-hairy; petal claw brownish; stamens > appendages; ovary hairs dense, straight, stiff, silvery (reddish brown in fr); style 4–6 mm, persistent on young fr. **FR** 4.5 mm wide. Common. Desert scrub; < 1000 m. **SNE, D**, (uncommon in Teh, SCo, SnJt);

to sw UT, TX, c Mex. [var. *arenaria* L. Benson, Algodones creosote bush; *L. divaricata* ssp. *t.* (DC.) Lowe & Felger] Closely related to s S.Am *L. d.* Clones may live 10,000 years, longer than any other living pls known; resinous odor characteristic; dominant shrub over vast areas of desert. ❀ DRN,SUN:7,**8–10,14**, 18,**19–21**,22,23&IRR:**11–13**; also STBL. Apr–May

TRIBULUS PUNCTURE VINE, CALTROP

Ann. **ST** prostrate, spreading radially, < 1 m. **LF** even-1-pinnate; stipules ± lf-like. **INFL**: fls solitary in axils. **FL**: sepals deciduous; petals yellow, deciduous. **FR** 5-lobed, splitting into 5 nutlets, each with 2–4 stout spines; style deciduous; peduncle reflexed. **SEEDS** 3–5 per chamber. ± 12 spp.: esp dry Afr. (Greek: caltrop, weapon used to impede cavalry, from armed fr)

T. terrestris L. (p. 525) **ST** ± silky or appressed-hairy, sharply bristly to glabrous. **LF**: stipules 1–5 mm; lflets 6–12. **FL** < 5 mm wide; peduncle < subtending lf. **FR** 5 mm, < 1 cm wide, ± flat, hairy, gray or yellowish; spines 4–7 mm, spreading, hairy to glabrous. Roadsides, railways, vacant lots, other dry, disturbed ar-

eas; gen < 1000 m. **CA**; to WY, e US, c Mex; native to Medit. First collected in CA in 1902; long a pernicious weed, now partly controlled by introduced weevils. TOXIC to livestock in vegetative condition, frs cause mechanical injury. Apr–Oct

ZYGOPHYLLUM BEAN-CAPER

Per, unarmed; caudex woody. **ST** branched, fleshy. **LF**: stipules ± lf-like or membranaceous; lflets 2 or more, free, fleshy. **INFL**: fls 1–2 in axils. **FL**: sepals often unequal, deciduous or not; petals ± erect, white, yellow, or orange, base sometimes orange or red. **FR**: capsule, cylindric to spheric, 5-angled, septicidal. **SEEDS** 1–few per chamber. ± 70 spp.: Eurasia, s Afr, Australia. (Greek: yoke lf, from sometimes oblique lflets)

Z. fabago L. (p. 525) **ST** erect, glabrous. **LF**: lflets < 4.5 cm, 3 cm wide, obliquely ovate; awn between lflets 1 mm. **FL** 6–7 mm; sepals ± = petals; stamens exserted, appendages < stamens, ± linear, divided at tip. **FR** 25–35 mm, oblong-cylindric; style

persistent, thread-like, < 7 mm; peduncle reflexed. Uncommon. Disturbed areas; < 1000 m. NCoRO, SnJV, **D**; native to Medit, c Asia. [var. *brachycarpum* Boiss.] Forms large colonies. Summer

ALISMATACEAE WATER-PLANTAIN FAMILY

Charles E. Turner

Ann, per from corms, stolons, rhizomes, or tubers, aquatic (± emergent or on mud), gen bisexual; roots fibrous. **ST**: caudex short. **LVS**: basal, simple, palmately veined, sometimes floating; submersed blades gen linear to ovate; emergent blades linear to sagittate. **INFL** gen scapose, umbel- to panicle-like; fls whorled, in interrupted clusters. **FLS** bisexual or unisexual, radial; sepals 3, gen green, gen persistent; petals 3, gen > sepals, white or pink; stamens 6–many; pistils 6–many, gen simple. **FR**: achene, gen compressed, beaked. ± 12 genera, 75–100 spp.: esp n hemisphere. [Rogers 1983 J Arnold Arbor 64: 383–420]

SAGITTARIA ARROWHEAD, TULE POTATO, WAPATO

Ann, per, gen monoecious; roots partitioned. **LVS**: petiole unangled; submerged blades tapered to base; floating or emergent blades gen sagittate (or linear to ovate). **INFL**: lowest node gen with 3 pistillate fls; staminate fls above. **FLS** gen unisexual; sepals 3–10 mm, reflexed to appressed in fr; petals gen entire. **PISTILLATE FL**: receptacle convex; pistils many, in spheric cluster. **STAMINATE FL**: stamens many. **FR**: body gen 2–3.5 mm, strongly compressed, back winged or ridged; beak gen lateral, spreading or erect. ± 20 spp.: worldwide, esp Am. (Latin: arrow, from lf shape) [Bogin 1955 Mem NY Bot Gard 9:179–233] Some spp. weedy; tubers of some used for human and wildlife food. Key to species adapted by Margriet Wetherwax.

1. Emergent lf blades 5–15 cm, lower lobes < terminal lobe; fr beak ascending to ± erect, < 0.5 mm ... ***S. cuneata***
1' Emergent lf blades 6–30 cm, lower lobes ± = terminal lobe; fr beak spreading, 1–2 mm .... ***S. latifolia***

S. cuneata E. Sheldon (p. 525) Per; tuber oblong, ± white or bluish. **LVS**: emergent blades 5–15 cm, sagittate, lower lobes < terminal lobe. **PISTILLATE FL**: pedicel ascending in fr; sepals reflexed in fr. **STAMINATE FL**: filaments glabrous. **FR**: beak < 0.5 mm, ascending to ± erect. 2*n*=22. Ponds, slow streams, ditches; < 2500 m. NW, CaR, SN, SnBr, **GB**; to s Can, n US, TX. ❀ TRY. Jun–Aug

S. latifolia Willd. (p. 525) Per, sometimes dioecious; tubers oblong, ± white or bluish. **LVS**: emergent blades 6–30 cm, sagittate, lower lobes ± = terminal lobe. **PISTILLATE FL**: pedicel ascending in fr; sepals reflexed in fr. **STAMINATE FL**: filaments glabrous. **FR**: beak 1–2 mm, spreading. 2*n*=22. Ponds, slow streams, ditches; < 1500 m. CA-FP, **GB**; to s Can, e US, n S.Am. ❀ SUN,WET (fresh water):1–5,**6–9**,10–12,**14–24**. Jul–Aug

ARACEAE ARUM FAMILY

Elizabeth McClintock

In CA per, terrestrial or aquatic, from short, gen erect caudex, often monoecious; elsewhere shrub, vine, or growing on other pl. **STS** sometimes above ground in addition to caudex. **LVS** simple or compound, basal (or cauline and 2-ranked). **INFL**: gen spike, fleshy, gen ill-smelling; fls gen many, bisexual or pistillate below, staminate above; bract subtending spike 1, gen showy (petal-like), gen > spike, sometimes sheathing. **FL**: perianth parts 4 or 6, free or fused; stamens gen 0, 4, or 6, free or fused; ovary superior to half-inferior and sunken in infl axis, chambers 1–3, stigma ± sessile. **FR**: berry. **SEEDS** 1–many. ± 110 genera, 1800 spp.: gen trop, subtrop. Some cult for food (*Colocasia*, taro) or orn (*Philodendron, Anthurium*). [Wilson 1960 J Arnold Arbor 41:47–63] Recently treated to include Lemnaceae [French et al 1995 Monocotyledons: Systematics and Evolution, eds. Rudall et al, Royal Bot Gard, Kew]. Needle-like crystals in most tissues cause intense irritation when chewed; those of *Dieffenbachia*, dumb-cane, may induce temporary speechlessness.

PISTIA WATER-LETTUCE

1 sp. (Greek: liquid, from aquatic habitat)

P. stratiotes L. Per, aquatic, floating, monoecious, spreading from stolons; roots many. **ST** short, corm-like. **LVS** basal, whorled, sessile; blade < 20 cm, < 7 cm wide, widely wedge-shaped, tip truncate or notched, velvety-hairy. **INFL**: spike < bract, fused to bract (exc tip); bract ± 1.5 cm, hairy outside, upper half open, ± flared, reflexed, lower half closed. **FLS**: perianth 0. **STAMINATE FLS** several, whorled, sessile; stamens 2–8. **PISTILLATE FL** 1; ovary ± superior, chamber 1, style stout, curved, persistent. **FR** green. **SEEDS** many. Ditches; < 50 m. **e DSon (Colorado River)**, expected elsewhere; to se US, trop worldwide.

ARECACEAE [PALMAE] PALM FAMILY

Elizabeth McClintock

Shrub, tree, evergreen, monoecious, dioecious, or fls bisexual. **ST**: trunk gen ± erect, unbranched. **LVS** splitting to be palmately or pinnately dissected or compound, alternate, forming a terminal crown, large; base sheathing; petiole often long. **INFL**: gen large panicle, axillary; peduncle sheathed by 1 or more large bracts; fls many, gen ± sessile. **FL** gen small, ± radial; sepals and petals gen 3, sometimes similar, fused at base or free; stamens gen 6; pistils 1 or

3, ovaries superior, gen 3, (if 1, chambers gen 3), styles free or fused. **FR**: often a drupe. **SEED** 1. ± 200 genera, 3,000 spp.: trop, subtrop; many cult, esp for orn. [Uhl & Dransfield 1987 Genera Palmarum] Used for food (fats, oils, frs, seeds) and building materials.

1. Lf blade pinnately compound, ± elongate; fl unisexual . . . . **PHOENIX**
1' Lf blade palmately dissected, ± round; fl bisexual . . . . **WASHINGTONIA**

PHOENIX DATE PALM

Tree, dioecious. **LVS** pinnately compound; bases persistent on trunk; lflets folded longitudinally with margins upward, lower sometimes smaller, spine-like. **INFL** within crown, < lvs. **FL**: perianth yellowish; calyx 3-lobed; petals gen free; ovaries 3, free, simple. ± 12 spp.: Afr, Asia. (Greek: name for date palm, of uncertain meaning)

P. dactylifera L. DATE, DATE PALM **LF** gen < 7 m. **FR** 2.5–5 cm, oblong-ovate, brown, pulp thick. Uncommon. Near habitations, adjacent moist areas; < 200 m. SCo, **D**; native to n Afr. Abundantly cult; fr (commercial date) pulp sweet, edible.

WASHINGTONIA FAN PALM

Tree. **LVS** persistent as a "skirt", palmately divided; segments folded longitudinally with margins upward; petiole margins and lower blade gen spiny. **INFL** within crown, > lvs. **FL** bisexual; perianth white; calyx lobes 3, ± erect; corolla lobes 3, reflexed; ovary 3-lobed, chambers 3. 2 spp.: deserts of s CA, AZ, n Mex. (George Washington, 1st president, US, 1732–1799) [Bailey 1936 Gentes Herb 4: 52–82] Widely cult as orn, CA to Florida.

W. filifera (L. Linden) H.A. Wendl. (p. 525, pl. 119) CALIFORNIA FAN PALM **ST**: trunk thick, robust, < 20 m; base gen not swollen. **LF** gray-green; petiole 1–2 m; blade 1–2 m, divided nearly to middle, segments 40–60, margins with thread-like fibers, tips ± reflexed. **FR** oblong or ovate, black. UNCOMMON (but locally abundant). Groves, moist places, seeps, springs, stream sides; < 1200 m. DSon; se AZ, n Baja CA. Reportedly naturalized along Kern River (Kern Co.) and near springs in DMoj (Death Valley National Monument); expected elsewhere. ❀ SUN:**14–16**,17,**22,23**,24& IRR:**8,9**,10,**11–13,18–21**. Jun

CYMODOCEACEAE MANATEE-GRASS FAMILY

Robert F. Thorne

Per from long, slender, jointed rhizomes, dioecious, marine aquatic, glabrous. **ST** erect, sometimes short. **LVS** simple, cauline, alternate or opposite, 2-ranked; sheath open; ligule present; blade 0.5–15 mm wide, ± linear, ± flat or cylindric, margin entire. **INFL**: fls 1–2 (or bracted cyme), axillary. **FLS** minute; perianth 0. **STAMINATE FL**: stamens 2, filaments 0 or short and ± fused; pollen filament-like, sticking together in string-like bodies < 1 mm wide. **PISTILLATE FL**: pistils 1–2, simple, ovary superior, chamber 1, ovule 1, style 1, lobes 0–3. **FR**: utricle. **SEED** 1. 5 genera, ± 13 spp.: trop, subtrop worldwide.

HALODULE

STS erect, several, short. **LVS** alternate to ± opposite, narrowly linear; tip gen 3-toothed. **FLS** 1 per node, ± enclosed by lf sheaths. **STAMINATE FL**: anther sacs attached to filament at different places. **PISTILLATE FL** short-stalked; pistils 2, simple, attached to receptacle at slightly different places, ovary asymmetric, style 1, simple, thread-like, from side of ovary. 1–2 spp.: trop, subtrop marine. (Greek: perhaps from ancient name meaning under salty water) Poorly known exc to students of algae. Sometimes treated in Zannichelliaceae.

H. wrightii (Asch.) Asch. SHOAL-GRASS **LF** gen 10–20 mm, 0.5–3 mm wide. Aquatic; –70 m. DSon (Salton Sea, where presumed extirpated), expected in warm, coastal, marine habitats; native to coastal Mex and Caribbean Sea.

CYPERACEAE SEDGE FAMILY

Raymond Cranfill, except as specified

Ann or per, often rhizomed, often of wet open places, gen monoecious; roots fibrous, hairy. **ST** gen 3-sided. **LVS** often 3-ranked; sheath gen closed; ligule gen 0; blade (0) various, parallel-veined. **INFL**: spikelets variously clustered; fls gen sessile in axil of fl bract. **FL** small, gen wind-pollinated; perianth 0 or bristle-like; stamens gen 3, anthers attached at base, 4-chambered; ovary superior, 1-chambered, 1-ovuled, style 2–3-branched. **FR**: achene, gen 3-sided. ± 110 genera, 3600 spp.: worldwide, esp temp. [Tucker 1987 J Arnold Arbor 68:361–445] Difficult: taxa differ in technical characters of infl and fr. Key to genera adapted by Margriet Wetherwax.

1. Fl and fr enveloped in ± closed bract (perigynium) . . . . **CAREX**
1' Fl and fr not enveloped in a perigynium
  2. Gen only uppermost fls bisexual and fruiting (lower staminate)
    3. Perianth bristles 0; fr not whitish and bony . . . . **CLADIUM**

3' Perianth bristles present; fr white, bony . . . . **SCHOENUS**
2' Fls gen bisexual only, fruiting (exc sometimes lowest 1–3)
  4. Fl bracts 2-ranked
    5. Spikelet 1–many-fld, gen not shed a a unit . . . . **CYPERUS**
    5' Spikelet 1–2-fld, shed as a unit . . . . **KYLLINGA**
  4' Fl bracts spiraled
    6. Perianth bristles 0 . . . . **FIMBRISTYLIS**
    6' Perianth bristles gen present
      7. Fr top with tubercle of different color and texture; st lvs 0 . . . . **ELEOCHARIS**
      7' Fr top gen beaked, tubercle 0; st lvs gen 1–several . . . . **SCIRPUS**

CAREX SEDGE

Joy Mastrogiuseppe

Per, cespitose or from rhizomes, gen monoecious. **ST** gen sharply 3-angled, gen solid. **LVS** 3-ranked, gen glabrous exc gen scabrous on midrib, margin; sheath closed, back (blade side of st) green, ribbed, front gen thin, translucent, forming gen U-shaped mouth at top. **INFL**: spikelets gen several–many, arrayed in raceme, panicle, or head-like cluster, each 1–many-fld, gen subtended by a spikelet bract. **FLS** unisexual, each subtended by 1 fl bract; perianth 0. **STAMINATE FL**: stamens gen 3. **PISTILLATE FL** enclosed by perigynium (sac-like bract); perigynium body 2–3-sided or round, wall gen delicate; perigynium beak tip open, often notched; style 1, gen deciduous, stigmas 2–4, exserted. **FR** 2–4-sided. (Latin: cutter, from sharp lf and st edges) [Standley 1985 Syst Bot Monogr 7:1–106] Fully mature perigynia needed for identification, so are described under "**FR**" (long-persistent perigynia are often atypical); perigynium "front" faces spikelet axis; "fr" refers to achene body (excluding beak). "Shredding" lower lf sheath fronts become a network or fringe of veins; some others shred longitudinally only. Difficult because of many spp. and minute key characters; longer key statements and descriptions are designed to enhance both ease and probability of correct identification. Group descriptions are assumed in specific descriptions. ❀ Many spp. especially those with rhizomes are INVASIVE. This is one of the most effective genera for knitting moist or wet soil.

Key to Groups

1. Spikelet 1 per infl; staminate fls above pistillate if both in a spikelet; bristle-like axis sometimes in
  perigynium . . . . **Group 1**
1' Spikelets > 1 per infl (sometimes in a dense, head–like cluster); staminate fls often below pistillate if
  both in a spikelet; bristle-like axis in perigynium 0
  2. Dioecious . . . . [2]**Group 4**
  2' Monoecious
    3. Perigynium faces hairy, at least near tip; stigmas 3; fr 3-sided
      4. Perigynia 3–15 per spikelet; mouth of lf sheath finely toothed; lateral pistillate spikelets 1–2, < 1.5 cm
      . . . . *C. rossii*
      4' Perigynia 25–45 per spikelet; mouth of lf sheath entire; lateral, pistillate spikelets gen > 2, > 1.5 cm . . *C. pellita*
    3' Perigynium faces glabrous; stigmas 2–3; fr 2–3-sided
      5. Stigmas 3; fr 3-sided
        6. Style tough, persistent, not jointed to fr; perigynium veins many, each gen rib-like; staminate
        spikelets 1–5 . . . . *C. utriculata*
        6' Style delicate, deciduous, breaking easily at joint to fr; perigynium veins 0–many, gen 2 rib-like;
        staminate spikelet 0–1 . . . . **Group 2**
      5' Stigmas 2; fr 2-sided
        7. Spikelets gen > 1.5 cm, sessile or not, ± cylindric, lower pistillate (or tip staminate), upper gen
        staminate . . . . **Group 3**
        7' Spikelets gen < 1.5 cm, sessile, ovoid to elliptic, each gen both staminate and pistillate
          8. Perigynia 1–3 per spikelet, 2.2–3 mm; infl 3–4 mm wide; terminal spikelet often with a conspicuous,
          narrow, staminate tip . . . . *C. disperma*
          8' Perigynia > 3 per spikelet, 2–8 mm; infl > 5 mm wide; terminal spikelet tip, if staminate, not
          conspicuous and narrow
            9. Spikelets staminate at tip, pistillate at base; perigynium ribs 0 or 2, sometimes with wing-like
            margins . . . . [2]**Group 4**
            9' Spikelets staminate at base, pistillate at tip; perigynium ribs 2, with distinct, gen serrate wings
            at least on upper body and beak . . . . **Group 5**

Group 1

Some dioecious. **LF**: blade flat or folded, less often rolled. **INFL**: spikelets 1(2) per infl, each staminate at tip, pistillate below (or staminate on separate spikelet); spikelet bract gen 0; pistillate fl bract gen not lf-like, < 1 cm (exc awn, if present). **PISTILLATE FL**: stigmas gen 3. **FR**: perigynium 2–7 mm, gen glabrous, beak tip narrowly white-margined; bristle-like axis gen present within perigynium; fr gen 3-sided.

1. Stigmas 2; fr 2-sided; bristle-like axis in perigynium present; wet places . . . . **C. capitata**
1' Stigmas 3; fr 3-sided; bristle-like axis in perigynium present or 0; moist to dry places
   2. Perigynium 3-sided to round, thick, ± = fr, faces gen hairy at least just below beak; style exserted,
     base black, persistent . . . . **C. filifolia** var. *erostrata*
   2' Perigynium flat, thin, > fr, faces glabrous; style not exserted, base not black, not persistent
     3. Perigynium 2.5–4.1 mm, 0.9–2 mm wide, fr width > distance from fr to perigynium side; spikelet
       3–6 mm wide . . . . **C. subnigricans**
     3' Perigynium 4–7 mm, 2.1–4.8 mm wide, fr width < distance from fr to perigynium side; spikelet
       6–10 mm wide . . . . **C. breweri** var. *breweri*

Group 2

Monoecious. **LF** gen < st; blades flat or folded, basal gen not minute; lower sheath fronts sometimes shredding to network or fringe of veins. **INFL**: terminal spikelet gen staminate, gen > 2 mm wide, gen oblong; lower spikelets pistillate or with staminate tips, gen erect, gen 1–4 cm, at least lowest ones gen stalked; lowest spikelet bract ± lf like, gen < or = infl, gen sheathless, gen ascending. **PISTILLATE FL**: stigmas 3. **FR**: perigynia gen 5–50 per spikelet, gen appressed to ascending, flat or 3-sided, faces gen glabrous, 0–many but gen 2-ribbed, beak tip gen notched, teeth gen glabrous; fr 3-sided.

1. Upper body of perigynium conspicuously papillate
   2. Pistillate fl bract ± = perigynium; perigynium 2.5–3.4 mm, body at least 1/4 empty above fr; lvs
     gray-green; dry rocky soils, often on summits . . . . **C. albonigra**
   2' Pistillate fl bract < perigynium; perigynium 1.6–3 mm, body ± filled by fr; lvs green; meadows
     3. Terminal spikelet 6–14 mm, staminate at base, pistillate at tip; pistillate fl bract << perigynium;
       perigynium 1–1.3 mm wide, without stiff hairs on or just below beak . . . . **C. norvegica**
     3' Terminal spikelet 10–30 mm, pistillate throughout or staminate at base or tip; pistillate fl bract
       < perigynium; perigynium 1.1–1.6 mm wide, gen with few stiff hairs on or just below beak
       . . . . [2]**C. parryana** var. *hallii*
1' Upper body of perigynium not or inconspicuously papillate
   4. Pistillate fl bract gen < 1 mm wide, > but width << perigynium; st gen < 30(50) cm; 2400–4100 m . . . **C. helleri**
   4' Pistillate fl bract gen > 1.2 mm wide, length < or >, width < or >, color = perigynium or not; st
     10–100 cm; 1800–3800 m
     5. Perigynium not flattened around fr, 1.6–3 mm, gen with few stiff hairs on or just below beak
       . . . . [2]**C. parryana** var. *hallii*
     5' Perigynium strongly flattened around fr, 2.5–4.5 mm, without stiff hairs on or just below beak . . **C. heteroneura**
      6. Perigynium > 3.5 mm, flat margin (0.3)0.5–0.7 mm wide, fr base to perigynium base 0.5–1.1 mm
       . . . . var. *epapillosa*
      6' Perigynium gen < 3.5 mm, flat margin < 0.4 mm wide; fr base to perigynium base 0–0.4 mm
       . . . . var. *heteroneura*

Group 3

Monoecious, gen rhizomed. **LF**: blades gen flat, basal gen not minute; bladeless sheaths gen < 7 mm wide at midlength, gen not keeled. **INFL** > 2.5 cm; spikelets gen 1 per node, terminal 1–3 gen staminate, gen < 5 cm, lower 3 or more gen pistillate, gen erect, at least lowest 1 gen stalked; lowest spikelet bract gen lf-like, gen not sheathing; pistillate fl bract gen > 0.5 mm wide, rarely lf-like, gen acute, gen purple to black, gen glabrous. **PISTILLATE FL**: stigmas 2. **FR**: perigynium gen ascending, gen abruptly narrowed above, gen papillate, gen glabrous, 2-ribbed, gen green, beak gen ± purple or brown, tip gen notched < 0.1 mm, teeth gen glabrous; fr 2-sided, sometimes deeply indented on 1 or both sides.

1. Lowest spikelet bract sheath > 4.5 mm; lower spikelet stalks often from near pl base; perigynium green
   or white to orange, ± fleshy, papillate
   2. Lower pistillate fl bract not appressed against, often falling before fully expanded perigynium;
     staminate portion of terminal spike 0.9–2 mm wide; perigynium body gen strongly ribbed, tip gen
     not conspicuously papillate at 10×, fully mature body orange when fresh, ± squashed below . . . . **C. aurea**
   2' Lower pistillate fl bract closely appressed against, falling after fully expanded perigynium; staminate
     portion of terminal spike (1.8)2–3.5 mm wide; perigynium body veined but gen not ribbed, tip
     conspicuously papillate at 10×, fully mature body gold or white when fresh, gen not squashed . . . . . **C. hassei**
1' Lowest spikelet bract sheath < 4 mm; lower spikelet stalks not from near pl base; perigynium brown,
   purple, or green with red or purple dots or blotches, not fleshy, papillate or not
   3. Perigynium with strong veins on faces, wall often ± tough, resisting puncture, beak 0.2–0.6 mm, tip
     notched > 0.1 mm, teeth minutely hairy; fresh lf blade gen blue-glaucous . . . . **C. nebrascensis**
   3' Perigynium without veins on faces, wall thin, soft, easy to puncture, beak gen < 0.4 mm, tip
     notched < 0.1 m, glabrous; fresh lf blade gen not blue-glaucous

4. Lowest spikelet bract gen ± = infl; infl gen open, lowest internode 20–180 mm; perigynium ±
   red-dotted or -blotched, weakly papillate; fr shiny . . . . *C. aquatilis* var. *aquatilis*
4' Lowest spikelet bract gen < 1/2 infl; infl ± dense, lowest internode 5–45 mm; perigynium gen
   purple above, strongly papillate; fr dull . . . . *C. scopulorum* var. *bracteosa*

Group 4

Sometimes dioecious. **LF**: basal blades minute. **INFL** gen dense, > 5 mm wide; spikelets each staminate at tip, pistillate below, gen < 1.5 cm, ovate to oblong, sessile; lowest spikelet bract gen bristle-like, gen < infl. **PISTILLATE FL**:
style gen < 1.5 mm, stigmas 2, gen < 4 mm. **FR**: perigynia > 3 per spikelet, planoconvex, gen widest above base, gen
abruptly narrowed above, glabrous, gen stalked, margin of upper body, beak gen serrate, beak > 0.2 mm, tip gen
notched; fr 2-sided.

1. Pl staminate
  2. Pl cespitose; spikelets > 1 per lower node or branch; lf sheath front gen red-dotted . . . . ^{2}*C. alma*
  2' Pl rhizomed; spikelets gen 1 per node throughout infl; lf sheath front gen not red-dotted
    3. Anther with narrow, < 0.1 mm tip; lf blade < 1.5 mm wide; rhizome 1–2 mm thick; infl 0.5–20 mm
    . . . . ^{2}*C. eleocharis*
    3' Anther with 0.1–1 mm awn; lf blade 1–5 mm wide; rhizome 1—5 mm thick; infl < 50 mm
      4. Rhizome 1–2 mm thick, brown; anther awn 0.2–1 mm, glabrous; lf sheath mouth with thick rim;
        infl (7)10–25 mm wide . . . . ^{2}*C. douglasii*
      4' Rhizome 1.5–5mm thick, dark brown; anther awn 0.1–0.4 mm, glabrous or hairy; lf sheath mouth
        fragile or with thick rim; infl < 15 mm wide
        5. Anther awn gen hairy at 20×, slender, tip tapered or obtuse; lf sheath mouth often with thick rim,
          without brown stripe . . . . ^{3}*C. praegracilis*
        5' Anther awn glabrous at 20×, stout, tip obtuse; lf sheath mouth fragile, often with brown stripe across
        . . . . ^{2}*C. simulata*
1' Pl bisexual or pistillate
  6. Perigynium wall filled with pithy tissue below
    7. Pl cespitose; spikelets > 1 per lower node or branch; infl (0.8)1–2 cm wide, spikelets > (often >>) 10
      8. Perigynium 3.5–4.3 mm, body ovate, widest ± at base; lower lf sheath front gen not cross-wrinkled,
        back not white-spotted . . . . ^{2}*C. alma*
      8' Perigynium 2–3.7(4) mm, body diamond-shaped, widest ± at middle; lower lf sheath front gen
        cross-wrinkled, back often white-spotted . . . . ^{2}*C. dudleyi*
    7' Pl rhizomed; spikelet 1 per node throughout infl; infl gen < 1 cm wide; spikelets gen < 10
      9. Perigynium > 2.6 mm, ± dull, beak > 1/4 body . . . . ^{3}*C. praegracilis*
      9' Perigynium < 2.6 mm, shiny, beak gen < 1/4 body . . . . ^{2}*C. simulata*
  6' Perigynium wall not filled with pithy tissue
    10. Spikelets > 1 per lower node or branch; spikelets >, often >> 10; lower lf sheath fronts often cross-
      wrinkled; pl cespitose . . . . ^{2}*C. dudleyi*
    10' Spikelet 1 per node throughout infl (if > 1 per lower node or branch, pl rhizomed); spikelets gen <
      10; lower lf sheath fronts not cross-wrinkled; pl cespitose or rhizomed
      11. Perigynium beak, upper body entire, body ± shiny
        12. Pl < 6 cm; rhizome long, gen wavy; lf > infl, blade 0.7–1.5 mm wide; perigynium 2.8–3.5 mm, >
          pistillate fl bract; dry places . . . . *C. incurviformis* var. *danaensis*
        12' Pl 5–30 cm; rhizome short, straight; lf < infl, blade 2–4 mm wide; perigynium 3.3–4.8 mm, <
          pistillate fl bract; places wet from snow-melt . . . . *C. vernacula*
      11' Perigynium beak, upper body serrate, body shiny or dull
        13. Perigynium conspicuously spreading, > pistillate fl bract (exc *C. occidentalis*)
          14. Perigynium back bulged so that marginal ribs are on front, not sides, very shiny . . . . *C. vallicola*
          14' Perigynium back not bulged (marginal ribs are on sides), dull or ± shiny
            15. Perigynium 2.5–3.5 mm, ± = pistillate fl bract; infl ± open . . . . ^{2}*C. occidentalis*
            15' Perigynium 3.4–5 mm, > pistillate fl bract; infl dense (sometimes lowest spikelet separate) . . *C. hoodii*
        13' Perigynium appressed to ascending, < or = pistillate fl bract
          16. Pl cespitose or short-rhizomed, bisexual; perigynium base narrowed for > 0.4 mm below fr
          . . . . ^{2}*C. occidentalis*
          16' Pl long-rhizomed, often unisexual; perigynium base narrowed for < or > 0.4 mm below fr
            17. Rhizome > 2 mm thick; lf blade flat or V-shaped . . . . ^{3}*C. praegracilis*
            17' Rhizome < 2.1 mm thick; lf blade margin gen inrolled (sometimes ± folded)
              18. Stigmas 4–6 mm, persistent, style 1.8–3.5 mm, exserted; perigynium 3.5–4.6 mm; infl on
                pistillate pl 1.3–2.7 cm wide . . . . ^{2}*C. douglasii*
              18' Stigmas < 4 mm, deciduous, style < 1.5 mm, incl; perigynium 2.5–3.5 mm; infl on pistillate or
                bisexual pl < 1 cm wide . . . . ^{2}*C. eleocharis*

Group 5

Monoecious, cespitose. **ST** hollow. **LF**: blades gen flat, basal minute; ligule gen < 2.5 mm. **INFL** dense or ± open; spikelets each staminate at base, pistillate above, gen erect, gen < 15 mm, ovate to fusiform, sessile; lowest spikelet bract gen ± like pistillate fl bract; pistillate fl bract gen < perigynium, in width < or ± = perigynium, gen brown, gen acute. **PISTILLATE FL**: stigmas 2. **FR**: perigynia > 3 per spikelet, gen ascending, gen ovate, planoconvex or flat, glabrous, back veins gen < 9, wings from base to near beak tip, gen serrate, beak tip notched; fr 2-sided, gen at base of perigynium body.

1. Lower 2–3 spikelet bracts elongate, ± leaf-like, at least lowest >> infl (sometimes broken off or damaged late in season); infl dense . . . . *C. athrostachya*
1' Lower 2–3 spikelet bracts gen like pistillate fl bracts or bristle-like, < infl, lowest sometimes lf-like and > infl; infl dense or open
  2. Perigynium shallowly scoop-shaped, wings incurved to front
    3. Perigynium 0.8–1.2 mm wide; beak tips inconspicuous in infl; lf blade 0.5–2 mm wide, gen inrolled
    . . . . *C. leporinella*
    3' Perigynium 1–2.5 mm wide; beak tips conspicuous in infl; lf blade 1.5–4.5 mm wide, flat . . . . . . ²*C. abrupta*
  2' Perigynium not scoop-shaped, wings not incurved to front
    4. Perigynium body much wider than fr, flat exc over fr or planoconvex due to air within, beak tip to fr top ± > 1/2 perigynium length
      5. Perigynium beak tip cylindric and entire for < 0.3 mm, red-brown, very tip white, perigynium wing minutely crinked at least above . . . . *C. straminiformis*
      5' Perigynium beak tip cylindric and entire for > 0.3 mm, green to black, perigynium wing gen not crinkled
        6. Perigynium 4.2–6.5 mm, beak tip to fr top gen > 2.5 mm; fr body (1.2)1.4–2.3 mm; pl often < 20 cm, st often decumbent; 2400–4200 m . . . . *C. haydeniana*
        6' Perigynium 2.9–5 mm, beak tip to fr top gen < 2.5 mm, fr body 1–1.6 mm; pl gen > 20 cm, st erect; 1400–3400 m . . . . *C. microptera*
    4' Perigynium body as wide as fr, planoconvex, not flat, beak tip to fr top < or > 1/2 perigynium length
      7. Lowest 2 infl internodes together < 1/3 infl length . . . . ²*C. abrupta*
      7' Lowest 2 infl internodes together > 1/3 infl length
        8. Fr 1–2.1 mm, 0.7–1.2 mm wide, 0.2–0.5 mm thick
          9. Lf blade folded or margin downcurved, 0.5–2.5 mm wide; lower pistillate fl bracts ± > 3.5 mm; gen rocky slopes, subalpine to alpine; 2700–3900 m . . . . *C. phaeocephala*
          9' Lf blade flat, 1.2–3.7 mm wide; lower pistillate fl bracts gen < 3.5 mm; seasonally wet meadows 100–3500 m . . . . *C. subfusca*
        8' Fr 1.8–3 mm, 1.1–1.8 mm wide, 0.5–0.8 mm thick
          10. Lf blade flat; fr 2.2–3 mm; dry to wet meadows, grasslands, 600–3200 m . . . . *C. petasata*
          10' Lf blade folded or margin downcurved; fr 1.7–2.1(2.6) mm; dry rocky slopes, summits, 900–3700 m . . . . *C. tahoensis*

C. abrupta Mack. (p. 527) ABRUPT-BEAKED SEDGE (Group 5) **ST** erect, gen > 20 cm. **LF**: blade 1.5–4.5 mm wide. **INFL** gen dense, 10–22 mm, brown; lowest 2 internodes together < 1/3, gen ± 1/5 total infl length; pistillate fl bract gen < perigynium, obtuse, brown or coppery. **FR**: perigynium 3.6–5.4 mm, 1–2.5 mm wide, sometimes shallowly scoop-shaped with wings incurved to front, flat margin incl wing < 0.3 mm wide, body ± planoconvex, veined both sides, brown, upper margin green to gold, beak gen < 1 mm, tip cylindric, entire and unmargined for gen > 0.6 mm, gen dark brown, 1.6–2.2 mm to fr top; fr 1.2–1.8 mm, 0.7–1.2 mm wide. Gen moist meadows, open forest; 1400–3300 m. KR, NCoR, CaRH, SN, TR, PR, MP, **n SNE (Sweetwater Mtns), DMtns**; to OR, WY, NV. ❀ TRY.

C. albonigra Mack. (p. 527) (Group 2) Cespitose. **ST** 10–30 cm. **LF**: blade 2.5–5 mm wide, gray-green. **INFL** gen head-like; terminal spikelet pistillate above, staminate at base; pistillate fl bract ± = perigynium, obtuse to with a small point, dark purple, margin, tip ± white. **FR**: perigynium not conspicuous in infl, 2.5–3.4 mm, 1.3–2 mm wide, ± flat, gen papillate above, dark purple, empty space above fr at least 1/4 body, beak 0.1–0.5 mm; fr 1.3–2 mm, 0.7–1.3 mm wide, < perigynium body. 2*n*=52. ± Dry, rocky soil, often summits; 3000–3900 m. c&s SNH, **W&I**; to w Can, Colorado. ❀ TRY.

C. alma L. Bailey (p. 527) (Group 4) Cespitose, rarely dioecious. **LF**: blade 3–6 mm wide; sheath front red-dotted. **INFL** open or dense, 2.5–15 cm, 1–2 cm wide; spikelets >> 10, > 1 per lower node or branch; spikelet bracts conspicuous; pistillate fl bract > or ± = perigynium, ± white or widely white-margined,

awned. **STAMINATE FL**: anther > 1.9 mm, awn < 0.1 mm. **FR**: perigynium 3.5–4.3 mm, 1.5–2.3 mm wide, 0.7–1.1 mm thick, body ovate to deltate, tapered or abruptly narrowed above, gold to dark brown, veins on back few, on front 0–few, wall rounded over fr, filled with pithy tissue below, beak 0.6–1.4 mm, serrate or ciliate; fr 1.5–2.5 mm, 0.9–1.8 mm wide, ovate. Springs, streambanks; 120–2400 m. c&s SNH, SnJV, CCo, SCo, TR, PR, **DMtns**; NV, AZ, Baja CA. ❀ TRY.

C. aquatilis Wahlenb. (Group 3) **INFL** gen open; lowest spikelet bract lf-like, > 1/2 infl; pistillate fl bract tip often white. **PISTILLATE FL**: style often exserted. **FR**: perigynium faces unveined, weakly papillate, beak tip notched < 0.1 mm, glabrous; fr 1.1–1.8 mm, 0.7–1.6 mm wide, shiny. 2*n*=72,74,76,>80. Wet places; < 3200 m. NW, CaRH (Butte Co.), SNH, CCo, SnBr, Wrn, **W&I**; to AK, e Can; Eur. 2 vars. in CA.

  var. ***aquatilis*** **LF**: blade 3–8 mm wide; sheath front not purple-dotted, mouth white or pale brown. **INFL**: lowest internode 2–18 cm; lateral spikelets 1–10 cm, 3–7 mm wide, stalks 0 or gen erect, < 4.2 cm, tip gen pistillate; lowest spikelet bract gen ± = infl. **FR**: perigynium 2–3.6 mm, 1.2–2.3 mm wide, ± red-dotted or -blotched, beak 0.1–0.2 mm, thickened. 2*n*=72,76, 78,80. Wet places; < 3200 m. KR, NCoR, SNH, SnBr, Wrn, **W&I**; to AK, e Can; n Eur. ❀ TRY:STBL.

C. athrostachya Olney (p. 527) (Group 5) **LF**: blade 1.5–4 mm wide. **INFL** dense, 10–20 mm, green to light brown; axis ± erect; spikelets ± indistinct; lowest 2–3 spikelet bracts ascending, >> infl, gen < 1.8 mm wide, lf-like, base expanded ± around st; pistillate fl bract < perigynium, often short-awned. **FR**: perigynium appressed to ascending, 2.8–4.8 mm, 0.8–1.8 mm wide, lanceolate or narrowly ovate, ± stalked, green to light brown, flat margin incl wing 0.1–0.2 mm wide, veins weak, < 8 on back, 0–4 on front, beak tip gen cylindric, unwinged and entire for > 0.4 mm, gen green above, very tip white, 2.1–2.5 mm to fr top; fr 1.1–1.7 mm, 0.7–1.1 mm wide. Common. Seasonally moist meadows, marshes; 100–3200 m. KR, NCoR, CaR, SN, SnBr, MP, **W&I**; w N.Am. ❀ TRY.

C. aurea Nutt. (p. 527) (Group 3) **LF**: blade 2–4 mm wide. **INFL**: terminal spikelet sometimes pistillate above, 0.9–2 mm wide in staminate portion, at least lowest non-basal spikelet separate, lateral ones erect to nodding, 4–20 mm, 3–5 mm wide; lowest spikelet bract sheath > 4.5 mm, mouth U-shaped; lower staminate fl bracts 2–4 mm; pistillate fl bract <, not appressed against, often falling before fully expanded perigynium, acute to narrow-awned, white to red-brown. **FR**: perigynia gen 4–10 per spikelet, ± ascending to spreading, 1.8–3 mm, 1–2 mm wide, spheric in spikelet, gen round or wide-tapered at base, gen sessile, gen strongly ribbed, green at full size but turning orange just before falling (rare in genus), translucent, fleshy, when dry orange or gold to purplish, upper part often white, lower ± squashed, body tip wide, blunt, gen not conspicuously papillate at 10×, beak ± 0, tip unnotched, often red-brown; fr 1.3–2 mm, 1–1.6 mm wide, beak < 0.1 mm. $2n=52$. Wet places; 1100–3300 m. NCoRH, CaRH, SNH, TR, MP, n SNE(Bodie), **W&I, DMtns**; to B.C., ne N.Am, UT, NM. Dried, immature perigynia ± indistinguishable from *C. hassei*. ❀ TRY:STBL.

C. breweri Boott var. *breweri* (p. 527) (Group 1) Rhizomed. **ST** 10–25 cm, > lvs. **LF**: blade 1 mm wide, rolled, quill-like. **INFL**: spikelet 6–10 mm wide, wide-conic; pistillate fl bract golden-brown to black, 3-veined, very wide, margin wide, whitish. **FR**: perigynia 10–40 per spikelet, ascending to spreading, 4–7 mm, 2.1–4.8 mm wide, very flat, thin, golden brown, a bristle-like axis within; fr 1.7–2.3 mm, 0.8–1 mm wide (width < distance from fr to perigynium side), << perigynium body, 3-sided. Dry gravelly, to sandy, open areas; 2300–3700 m. CaRH, SNH, **W&I**; to WA. ❀ TRY: STBL.

C. capitata L. (p. 527) (Group 1) Loosely cespitose. **ST** 10–35 cm, > lvs. **LF**: blade < 1 mm wide, rolled, quill-like. **INFL**: spikelet bisexual; spikelet bract 0; pistillate fl bracts brown, with wide, white margin. **PISTILLATE FL**: stigmas < 3. **FR**: perigynia 6–25 per spikelet, ascending to spreading, 2–3.5 mm, 1.3–2 mm wide, flat around fr, finely veined on back, pale green, a bristle-like axis within, edge sharp to base, entire, faces glabrous, beak tip dark; fr 1–1.8 mm, 0.5–1.2 mm wide, < perigynium body, 2-sided. $2n=50$. Gen wet places; 1900–3900 m. CaRH, SNH, **W&I**; to AK, ne N.Am, S.Am, Eurasia. ❀ TRY:STBL.

C. disperma Dewey (p. 527) Rhizomed. **LF**: blade 0.7–2 mm wide. **INFL** open, 1.5–2.5 cm, 3–4 mm wide; spikelets 2–4, lower << internodes between; staminate tip of terminal spikelet often very narrow, conspicuous; pistillate fl bract white, acuminate. **FR**: perigynia 1–3 per spikelet, 2.2–3 mm, 1.1–1.6 mm wide, light- to yellow-green, veined both sides, wall filled with pithy tissue to near beak, beak 0.2–0.3 mm, margin entire, tip white-margined; fr 1–2 mm, 0.9–1.3 mm wide. $2n=70$. Wet streamsides, lake margins; 1100–3400 m. SNH, **W&I**; to n Can, e N.Am, Eurasia. ❀ TRY: STBL.

C. douglasii Boott (p. 527, pl. 120) (Group 4) Gen dioecious; rhizome 1–2 mm thick, brown. **LF**: blade 1–2.5 mm wide, folded or inrolled at margin, thick; sheath mouth with thick rim. **STAMINATE INFL** dense, < 3 cm, 7–25 mm wide; lowest spikelet bract like pistillate fl bract. **PISTILLATE INFL** dense, 1.5–3.5 cm, 1.3–2.7 cm wide; spikelets gen < 10; pistillate fl bract > perigynium, gold, minutely pointed. **STAMINATE FL**: filament exserted, anther conspicuous, ± 3.5 mm, awn 0.2–1 mm, glabrous. **PISTILLATE FL**: style 1.8–3.5 mm, exserted; stigmas 4–6 mm, conspicuous, persistent. **FR**: perigynium appressed, 3.5–4.6 mm, 1.3–1.8 mm wide, many-veined, often obscurely, both sides, tapered to tip, gold to medium brown, beak 0.9–1.8 mm, ± = body, tip with narrow white margin; fr 1.4–1.9 mm, 1–1.5 mm wide. Dryish sandy, gravelly, or alkaline areas; 300–3800 m. KR, SNH, SnJV, SCoR, TR, PR, **GB, DMtns (Cottonwood Mtns)**; to w&c Can, c US, NM. Distinctive. ❀ TRY:STBL.

C. dudleyi Mack. (p. 527) (Group 4) Cespitose. **LF**: blade 3–6 mm wide; sheath back often white-spotted, front gen cross-wrinkled, red-dotted or not; ligule 0–9 mm. **INFL** 1.5–6.5 cm, 0.8–1.5 cm wide; spikelets >> 10, > 1 per lower node or branch; spikelet bracts conspicuous; pistillate fl bract gen red-brown, sometimes white, often awned. **FR**: perigynium ascending to spreading, 2–4 mm, 1.3–2.1 mm wide, pale gold, translucent, back veined, front veined or not, rounded over, with narrow flat margin around fr, body ± diamond-shaped, lower wall rarely filled with pithy tissue, beak 0.8–1.5 mm, conic to ± linear, gen < 0.6 mm wide 0.2 mm above fr, serrate or ciliate, tip reddish; fr 1.5–1.9 mm, 1.1–1.6 mm wide, 0.7–1.1 mm thick. At least seasonally wet places; 30–600 m. NCo, NCoR, GV, SN, CW, WTR, MP, **DMtns**; OR. ❀ TRY.

C. eleocharis L. Bailey (p. 527) SPIKERUSH SEDGE (Group 4) Sometimes dioecious; rhizome 1–2 mm thick, brown. **LF**: blade 0.5–1.5 mm wide, margin inrolled. **INFL** dense, 0.5–2 cm, 5–10 mm wide; spikelets < 10; lowest spikelet bract like pistillate fl bract to bristle-like; pistillate fl bract > perigynium, gold to brown, minutely pointed. **STAMINATE FL**: anther with narrow, < 0.1 mm tip. **PISTILLATE FL**: style < 1.5 mm, incl; stigmas < 4 mm, deciduous. **FR**: perigynium ascending, 2.5–3.5 mm, 1.5–1.8 mm wide, thick-walled, white to black, back veined, front unveined, beak 0.4–1.2 mm, < body, tip unnotched, white-margined; fr 1.5–2 mm, 1.25–1.7 mm wide. $2n=60$. RARE in CA. Dry areas in sagebrush shrubland, coniferous forest; 3500–4100 m. **W&I**; scattered w N.Am; also Eurasia. Sometimes considered *C. stenophylla* Wahlenb.

C. filifolia Nutt. (Group 1) Cespitose. **LF**: blade 0.2–0.7 mm wide, rolled, quill-like. **INFL**: lower pistillate fl bracts pale redbrown, white- or yellowish brown-margined, obtuse to awned, clasping perigynium base. **PISTILLATE FL**: style exserted, base black, persistent. **FR**: perigynium appressed to ascending, 1.9–3.7 mm, 1.3–2.1 mm wide, 3-sided to round, a bristle-like axis within, faces hairy at least just below beak, unveined, whitish to gold; fr 1.6–3 mm, 1.2–1.8 mm wide, ± = perigynium body. Meadows, dryish areas with subsurface moisture; 1500–3700 m. SNH, SnBr, **W&I**; w N.Am. 2 vars. in CA.

var. *erostrata* Kük. (p. 527) **ST** 5–25 cm, < or > lvs. **INFL**: longest pistillate fl bracts (exc awns) gen < 2.5 mm, gen < perigynia. **FR**: perigynium 1.9–3 mm, tapered to body tip, sometimes ruptured by fr, beak 0 or < 0.3 mm; fr 1.6–2.2 mm. Meadows; 1500–3700 m. SNH, SnBr, **W&I**; s OR, w NV. [*C. exserta* Mack.] ❀ TRY.

C. hassei L. Bailey (p. 527) (Group 3) **LF**: blade 2–4 mm wide. **INFL**: terminal spikelet sometimes pistillate at tip, staminate part 1.8–3.5 mm wide, lowest non-basal spikelet separate, lateral ones erect to nodding, 3–5, 7–25 mm, 3.5–4.5 mm wide, long-stalked, lower gen from near pl base; lowest spikelet bract >> infl, sheath > 4.5 mm, mouth U-shaped; lower staminate fl bracts 3–6 mm; pistillate fl bract <, appressed against, falling after fully expanded perigynium, obtuse to awned, red-brown, margin, tip white. **FR**: perigynia gen 10–20 per spikelet, ascending to spreading, 2.1–3.1 mm, 1.2–1.6 mm wide, ± obovate, stalked, veined but gen not ribbed, when fresh white to gold, ± translucent, fleshy, when dry greenish white or pale gold, sometimes gold at base, gen not squashed, body wide-obovate or -elliptic, base tapered, tip wide, blunt, conspicuously papillate at 10×, beak < 0.2 mm, tip often red-brown, unnotched; fr 1.5–2 mm, 1–1.5 mm wide, beak < 0.1 mm. Wet places; < 2700 m. NCo, KR, SNH, CCo, SnFrB, SnGb, SnBr, **SNE, DMtns**; to B.C.; also in NV. [*C. aurea* var. *androgyna* Olney] See note under *C. aurea*. ❀ TRY:STBL.

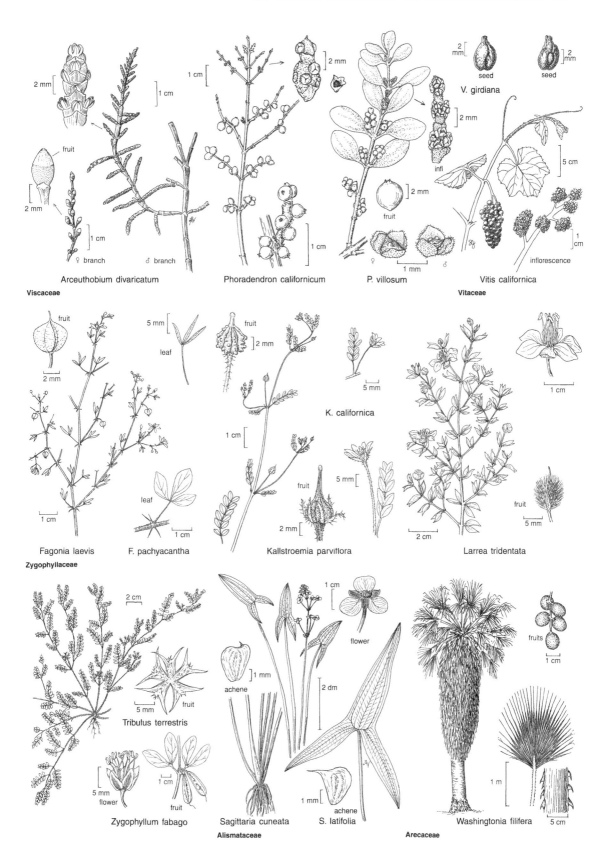

2 mm

1 cm

fruit

2 mm

1 cm

♀ branch   ♂ branch

Arceuthobium divaricatum

Viscaceae

1 cm

2 mm

1 cm

Phoradendron californicum

2 mm

infl

2 mm

fruit

♀   ♂

1 mm

P. villosum

2 mm

seed

2 mm

seed

V. girdiana

5 cm

1 cm

inflorescence

Vitis californica

Vitaceae

fruit

2 mm

5 mm

leaf

Fagonia laevis

leaf

1 cm

F. pachyacantha

Zygophyllaceae

fruit

2 mm

K. californica

5 mm

1 cm

fruit

2 mm

5 mm

Kallstroemia parviflora

1 cm

2 cm

fruit

5 mm

Larrea tridentata

2 cm

fruit

5 mm

Tribulus terrestris

5 mm

flower

1 cm

fruit

Zygophyllum fabago

Alismataceae

1 cm

flower

achene

1 mm

2 dm

achene

1 mm

Sagittaria cuneata

S. latifolia

fruits

1 cm

1 m

5 cm

Washingtonia filifera

Arecaceae

C. haydeniana Olney (p. 527) (Group 5) **ST** often decumbent, often < 20 cm. **LF**: blade 1.5–4 mm wide, flat. **INFL** dense, 9–18 mm, ± spheric, gen dark brown, less often greenish; spikelets gen indistinct; pistillate fl bract < perigynium, dark brown to blackish. **FR**: perigynium 4.2–6.5 mm, 1.5–2.6 mm wide, dark brown, margin sometimes green, flat margin incl wing 0.3–0.7 mm wide, back ± veined, front gen unveined, beak tip cylindric for > 0.3 mm, ± minutely serrate, dark brown to black, 2.3–3.8 mm to fr top; fr 1.2–2.3 mm, 0.8–1.1 mm wide. 2*n*=82. Uncommon. Rocky slopes, flats, moist soil; 2400–4200 m. c&s SNH, **W&I**, Wrn; OR; also in B.C., Rocky Mtns. [*C. nubicola* Mack.]

C. helleri Mack. (p. 527) (Group 2) Cespitose. **ST** 15–50 cm. **LF**: blade 2–3.5 mm wide. **INFL** gen head-like; lowest spikelet sometimes separate; terminal spikelet pistillate above; lower spikelets sessile; pistillate fl bract gen < 1 mm wide, > but width << perigynium, red-brown to purple, tip long-tapered. **FR**: perigynium 2.5–3.8 mm, 1.5–2.8 mm wide, gen red-brown to purple at least above, occasionally weakly papillate above, beak 0.2–0.5 mm; fr 1.4–1.8 mm, 0.7–1.1 mm wide. Dry, rocky or gravelly slopes; 2400–4100 m. CaRH, SNH, **SNE (Sweetwater Mtns)**, **W&I**; NV (Elko Co.). ❀ TRY.

C. heteroneura W. Boott (Group 2) Cespitose. **ST** 25–100 cm. **INFL** ± dense, not head-like; terminal spikelet gen pistillate above; lowest spikelets gen erect, gen conspicuously 2-colored; pistillate fl bract gen > 1.2 mm wide, dark purple, awned or not. **FR**: perigynium ascending, green or purplish above, 3-sided but flat around fr, body elliptic to obovate, ± rounded at tip, beak 0.2–0.5 mm. Wet meadows, forest openings, rocky slopes; 1800–3800 m. KR, CaRH, SNH, SnGb, SnBr, SnJt, Wrn, **SNE (Sweetwater Mtns)**, **W&I**; to WA, Rocky Mtns.

var. *epapillosa* (Mack.) F. Herm. **LF**: blade 3.5–7 mm wide. **INFL**: pistillate fl bract midstripe inconspicuous. **FR**: perigynium 3.5–4.5 mm, 1.7–3.2 mm wide, flat margin (0.3)0.5–0.7 mm wide; fr 1.1–1.9 mm, 0.7–1.0 mm wide, base to perigynium base 0.5–1.1 mm. Habitats of sp. KR (Medicine Lake, Siskiyou Co.), CaRH, n&c SNH, Wrn, **SNE (Sweetwater Mtns)**; to WA, Rocky Mtns. [*C. epapillosa* Mack.] ❀ TRY.

var. *heteroneura* (p. 527) **LF**: blade 2–4 mm wide. **INFL**: pistillate fl bract midstripe pale. **FR**: perigynium 2.5–3.9 mm, 1.4–2.5 mm wide, flat margin < 0.4 mm; fr 1–2 mm, 0.7–1.4 mm wide, base to perigynium base < 0.4 mm. Meadows, forest openings, rocky slopes; 1800–3500 m. KR, CaRH, SNH, SnGb, SnBr, SnJt, Wrn, **SNE (Sweetwater Mtns)**, **W&I**; to WY. ❀ TRY.

C. hoodii Boott (p. 527) (Group 4) Cespitose. **LF**: blade 1.5–3.5 mm wide. **INFL** dense, 1–2 cm, 8–15 mm wide; spikelets 4–8, indistinct exc sometimes lowest; lower spikelet bracts gen < spikelets; pistillate fl bract < perigynium, acute to long-acuminate, red-brown, midrib green, margin white. **FR**: perigynium spreading, 3.4–5 mm, 1.4–2 mm wide, sessile, coppery brown, margin green, veins gen 0 both sides, beak 0.8–1.8 mm, tip notched; fr 1.7–2.1 mm, 1.3–1.7 mm wide. 2*n*=60. Rocky or gravelly slopes, meadow edges; 1200–3400 m. KR, NCoR, CaRH, SNH, SnBr, Wrn, **W&I**; to w Can, SD, Colorado. ❀ TRY.

C. incurviformis Mack. var. *danaensis* (Stacey) F. Herm. (p. 527) (Group 4) Rhizome long, gen wavy. **ST** < 6 cm. **LF** > infl; blade 0.75–1.5 mm wide. **INFL** dense, 6–9 mm, 5–8 mm wide; spikelets < 10, indistinct; lowest spikelet bract like pistillate fl bract; pistillate fl bract acuminate, brown, margin white. **FR**: perigynium spreading, 2.8–3.5 mm, 1–1.6 mm wide, > pistillate fl bract, brown, darker near beak, shiny, sessile, veins many both sides, margin entire above, beak 0.7–1.5 mm, entire, tip unnotched, narrowly white-margined; fr 1.4–1.7 mm, 0.8–1.3 mm wide, round to square. Open, dry gravelly or rocky slopes; 3700–4000 m. c&s SNH, **W&I**; also in Colorado. [*C. danaensis* Stacey] ❀ TRY:STBL.

C. leporinella Mack. (p. 527) (Group 5) **LF**: blade 0.5–2 mm wide, gen ± rolled. **INFL** gen open, 15–35 mm; spikelets distinct, gen fusiform; lowest spikelet bract sometimes bristle-like; pistillate fl bract gen covering perigynium, red-brown, white-

margined. **FR**: perigynium appressed, 3.2–4.2 mm, 0.8–1.2 mm wide, planoconvex, gold, veined both sides, shallowly scoop-shaped, wings incurved to front, flat margin incl wing ± 0.1 mm wide, beak tips inconspicuous in infl, cylindric and entire for > 0.4 mm, gold to brown, 1.5–2.2 mm to fr top; fr 1.4–1.9 mm, 0.7–1 mm wide. Moist meadows; 2100–4000 m. KR, CaRH, SNH, Wrn, **SNE (Sweetwater Mtns)**, **W&I**; to WA, UT. ❀ WETorIRR,SUN:1,**6,15,16**,17.

C. microptera Mack. (p. 527) (Group 5) **ST** gen > 20 cm, erect. **LF**: blade 2–6 mm wide. **INFL** dense, 12–25 mm, green to less often dark brown; spikelets ± indistinct; lowest spikelet bract like pistillate fl bract, << infl, or sometimes lf-like and > infl; pistillate fl bract, < perigynium, brown. **FR**: perigynium ascending-spreading, 2.9–5 mm, 1–2.2 mm wide, flat but sometimes planoconvex due to air within, veined both sides, front sometimes only at base, green to gold, center often brown, flat margin incl wing 0.2–0.5 mm wide, beak 1–1.5 mm, tip cylindric for gen 0.3–0.5 mm, ± serrate, green or brown, gen 1.6–2.5 mm to fr top; fr 1–1.6 mm, 0.7–1.4 mm wide. 2*n*=80,82,90. Common. Meadows; 1500–3400 m. KR, NCoRH, CaRH, SNH, TR, PR, Wrn, **SNE (Sweetwater Mtns)**, **W&I**, **DMtns**; w N.Am. [*C. festivella* Mack.] Highly variable.

C. nebrascensis Dewey (p. 531) (Group 3) **LVS** gen tufted at st base; blade 3–12 mm wide, thick, gen strongly blue-glaucous. **INFL**: lateral spikelets 3–6 cm, 5–9 mm wide, gen 2-colored, pressing ± flat; lowest spikelet bract gen > infl, base often dark-margined; pistillate fl bract obtuse to awned, tip or awn glabrous, margin often white. **FR**: perigynia 30–150 per spikelet, ascending, 2.6–4 mm, 1.6–2.5 mm wide, sessile, not papillate, veined, gen ± tough, often red-dotted, beak 0.2–0.6 mm, tip brown or purple, notched > 0.1 mm, teeth minutely hairy; fr 1.3–2 mm, 0.9–1.8 mm wide. 2*n* =66,68. Meadows, swamps; < 2500 m. KR, CaRH, SNH, GV, WTR, SnJt, **GB**, **DMtns**; w US. ❀ WET,SUN:1–3,**4–7**, 8,9,**14–18**&STBL.

C. norvegica Retz. (p. 531) SCANDINAVIAN SEDGE (Group 2) Short-rhizomed. **ST** 10–70 cm. **LF**: blade 1.5–3.5 mm wide. **INFL** gen ± open; terminal spikelet 6–14 mm, pistillate above, lateral spikelets sessile or short-stalked; pistillate fl bract << perigynium, obtuse to with a small point, purple-black or brownish black, margin, tip white. **FR**: perigynium 2–3 mm, 1–1.3 mm wide, 3-sided, papillate, green or gold to dark purple, conspicuous in infl, beak 0.3–0.4 mm; fr 1.2–1.7 mm, 0.7–1 mm wide. RARE in CA. Wet meadows; 2900 m. **W&I (White Mtns)**; to AK, ID, UT, NM, n-c US; circumboreal.

C. occidentalis L. Bailey (p. 531) (Group 4) Cespitose or short-rhizomed. **LF**: blade 1.5–2.5 mm wide. **INFL** ± open, 1.5–3 cm, 6–8 mm wide; spikelets gen < 10, ± distinct, lower ± separate; lowest spikelet bract < infl, like pistillate fl bract; pistillate fl bract brown, short-awned. **FR**: perigynium ascending to spreading, 2.5–3.5 mm, 1.5–1.9 mm wide, ± = pistillate fl bract, green to brown, veins ± 0, beak 0.6–1.2 mm, tip notched; fr 1.3–2 mm, 1.1–1.7 mm wide. Dry woodlands, meadows; ± 1900 m. SnBr, **W&I**; to WY, NM. ❀ TRY.

C. parryana Dewey var. *hallii* (Olney) D.F. Murray (p. 531) PARRY'S SEDGE (Group 2) Short-rhizomed. **ST** 10–60 cm. **LF**: blade 2–4 mm wide. **INFL** ± open; spikelets pistillate or terminal one staminate at base or tip, terminal spikelet 10–30 mm, lateral sessile or short-stalked; pistillate fl bract < perigynium, gen dark purple or brown, margin white. **FR**: perigynium 1.6–3 mm, 1.1–1.6 mm wide, gen brown, upper body gen papillate, hairs on, just below beak few, stiff, beak 0.1–0.5 mm; fr 1.3–2 mm, 1–1.5 mm wide, ± = perigynium body. RARE in CA. Meadows; ± 3200 m. **W&I (White Mtns)**; to n Can, ID, UT. [*C. hallii* Olney]

C. pellita Willdenow (p. 531) WOOLLY SEDGE Rhizomed. **ST** 30–100 cm. **LF**: blade 1.5–5 mm wide, flat, glabrous, slender tip short, basal blades minute; ligule of upper lvs > 2 mm, thin, membranous; sheath mouth densely hairy or not but not ciliate, lower fronts often not shredding. **INFL**: lowest internode 0.5–0.9 mm

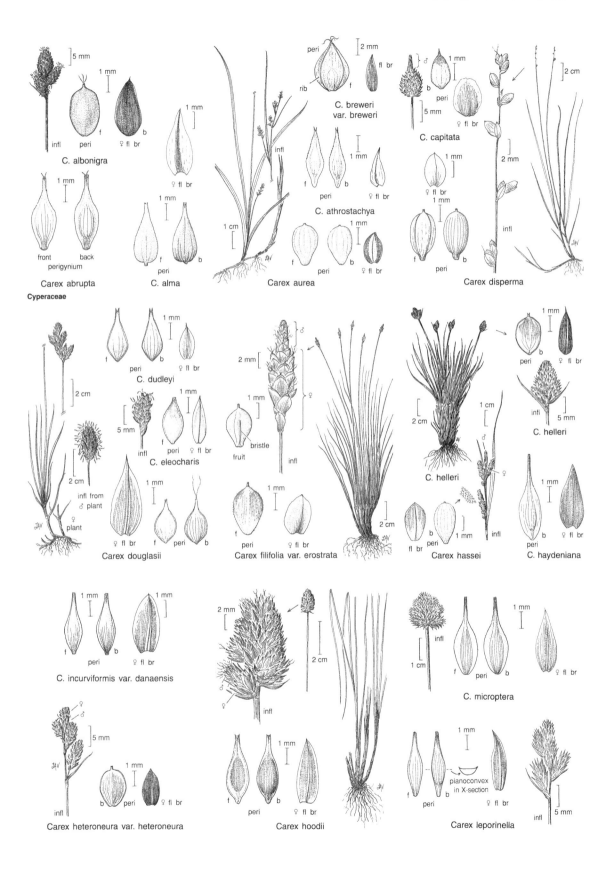

C. albonigra

front back
perigynium
Carex abrupta

C. alma

Carex aurea

C. breweri
var. breweri

C. athrostachya

C. capitata

Carex disperma

Cyperaceae

C. dudleyi

C. eleocharis

Carex douglasii

Carex filifolia var. erostrata

C. helleri

C. helleri

Carex hassei

C. haydeniana

C. incurviformis var. danaensis

C. microptera

Carex heteroneura var. heteroneura

Carex hoodii

Carex leporinella

wide; pistillate spikelets gen sessile; pistillate fl bract > perigynium, ± ciliate, purplish or brown, with very narrow white margin, awned. **FR**: perigynia 25–75 per spikelet, 2.8–5 mm, 1.5–2 mm wide, thick-walled, weakly many-ribbed, gen purplish, beak 0.5–1.2 mm, teeth ± outcurved, 0.2–0.9 mm; fr 1.5–2.1 mm, 1–1.5 mm wide, persistent style base gen straight. Gen marshy places; 60–3300 m. NCo, KR, CaRH, SNH, CCo, SnFrB, SnGb, SnBr, **GB**, **DMoj**; to B.C., e N.Am. [*C. lanuginosa* Michaux misapplied; *C. lasiocarpa* var. *lanuginosa* Kük. misapplied] ❀ TRY:STBL.

C. petasata Dewey (p. 531) LIDDON'S SEDGE (Group 5) **LF**: blade 2–5 mm wide. **INFL** open, 2–6 cm, often whitish; 2nd lowest internode gen 4–11 mm; spikelets 3–8, distinct, base long-tapered; lowest spikelet bract sometimes bristle-like; pistillate fl bract ± covering perigynium, gen white or widely white-margined. **FR**: perigynium appressed, 5.7–8.1 mm, 1.5–2.5 mm wide, body planoconvex, tapered above, cream-white to brown, flat margin incl wing 0.2–0.5 mm wide, veins 5–8 both sides, beak tip cylindric, ± entire, red-brown, very tip white, 2.8–4.6 mm to fr top; fr 2.2–3 mm, 1.1–1.8 mm wide. Uncommon. Dry to wet meadows, grasslands; 600–3200 m. e KR, CaRH (Lassen Co.), n SNH (Alpine Co.), MP, **W&I**; to B.C., Rocky Mtns. [*C. liddonii* Boott]

C. phaeocephala Piper (p. 531) (Group 5) **LF**: blade 0.5–2.5 mm wide, folded or margin downcurved. **INFL** dense or open, 10–35 mm, > 10 mm wide; 2nd lowest internode 2–4.5 mm; spikelets 2–7, distinct, ascending, club-shaped or fusiform; pistillate fl bract ± = perigynium in length, width, dark brown to orangish, white margin < 0.3 mm. **FR**: perigynium ascending, 3.5–5.8 mm, 1.2–2.5 mm wide, 2.2–4 × longer than wide, gen ± planoconvex, gen gold to brown, sometimes cream-white, green-margined, sometimes translucent, flat margin incl wing 0.2–0.6 mm wide, back veined, front veined or not, beak 0.5–1.2 mm, tip cylindric and ± entire for 0.4–1 mm, gen red-brown, white at very tip, with white flap along middle of back, tip to fr top 1.5–2.7 mm; fr 1.1–1.9 mm, 0.7–1.1 mm wide, 0.2–0.5 mm thick. 2*n*=84. Often rocky soils; 2700–3900 m. CaRH, SN, Wrn, **SNE (Sweetwater Mtns)**, **W&I**; to w Can, Colorado. ❀ TRY:STBL.

C. praegracilis W. Boott (p. 531) (Group 4) Sometimes dioecious; rhizome 2–5 mm thick. **LF**: blade 1.5–3 mm wide, flat or V-shaped; sheath mouth with ± thick rim. **INFL** gen dense, 1–5 cm, 6–10 mm wide; spikelets gen < 10, sometimes > 1 per lower node or branch; pistillate fl bract ± dull, gold to brown, white-margined or not, minutely pointed to awned. **STAMINATE FL**: filament exserted; longest anther awns 0.1–0.2 mm, slender, gen minutely hairy at 20×. **PISTILLATE FL**: style gen incl. **FR**: perigynium ascending, 2.8–4 mm, 1.3–1.6 mm wide, widest above base, tapered above, dark brown, gen dull, veins several on back, 0 on front, lower wall sometimes filled with pithy tissue, base narrowed for < 0.3 mm below fr, beak 0.6–1.5 mm, serrate, tip unnotched, very tip white; fr 1.2–1.9 mm, 1–1.4 mm wide. Common. Often alkaline, ± moist places; < 2700 m. CA-FP, **W&I**, **DMtns**; w&c N.Am, S.Am. ❀ TRY:STBL.

C. rossii Boott (p. 531) Cespitose. **ST** 3–40 cm, < or > lvs; glabrous or scabrous on angles only just below infl, sometimes scabrous nearly to base. **LF**: blade 1–4 mm wide, stiff or not, ± flat, bright or gray-green. **INFL**: staminate spikelet of taller infls 5–10 mm, 1–2.5 mm wide; pistillate spikelets 1–3; pistillate fl bract green or reddish tinged, awned or not. **FR**: perigynia 3–15 per spikelet, 2.4–5 mm, 1–1.7 mm wide, beak tip notched in front gen > 0.1 mm; fr 1.3–2.4 mm, 1.1–1.7 mm wide. Dry forests, meadows; < 3800 m. NW, CaRH, SNH, SnGb, SnBr, MP, **W&I**, **DMtns**; to n Can, Colorado, AZ, scattered to n-c US. [*C. novae-angliae* Schwein. var. *r.* (Boott) L. Bailey; *C. brevipes* W. Boott] ❀ TRY.

C. scopulorum Holm var. **bracteosa** (L. Bailey) F. Herm. (p. 531) (Group 3) Loosely cespitose. **LF**: blade 3–6 mm wide; sheath front purple-dotted. **INFL** ± dense, gen dark purple; lowest internode 5–45 mm; lateral spikelet 8–30 mm, 4–6 mm wide, gen erect, stalk 0–15 mm, tip gen pistillate; lowest spikelet bract

< 1/2 infl, ± like pistillate fl bract; pistillate fl bract obtuse to acute. **FR**: perigynia 8–50 per spikelet, ascending to spreading, 2–4 mm, 1.2–2.3 mm wide, unveined, papillate, gen purple above, beak 0.1–0.3 mm, glabrous, tip notched < 0.1 mm; fr 1.2–1.8 mm, 0.9–1.5 mm wide, dull. 2*n*=72,76,78,80. Wet places; 1200–3400 m. KR, NCoRO, NCoRH, CaRH, SNH, Wrn, **SNE (Sweetwater Mtns)**; to B.C., Rocky Mtns [*C. gymnoclada* Holm] ❀ TRY:STBL.

C. simulata Mack. (p. 531) (Group 4) Often dioecious; rhizome long, 1.5–3 mm thick, dark brown. **LF**: blade 2–5 mm wide; sheath mouth fragile, often with brown stripe across. **INFL** dense, 1.2–3.5 cm, 6–10 mm wide; spikelets < 10; lowest spikelet bract < infl, like pistillate fl bract; pistillate fl bract brown, white-margined, lower ± awned. **STAMINATE FL**: filaments exserted, longest anther awns 0.1–0.2 mm, stout, glabrous at 20×, tip obtuse. **PISTILLATE FL**: style exserted. **FR**: perigynium 1.7–2.6 mm, 1.3–1.6 mm wide, brown, shiny, veins few both sides, lower wall filled with pithy tissue, beak 0.2–0.6 mm, gen < 1/4 body, serrate, tip unnotched, white-margined; fr 1.1–2.3 mm, 0.7–1 mm wide. Moist soil; < 3300 m. KR, CaRH, SNH, CCo, SnFrB, **GB**; to WA, MT, UT, NM. ❀ TRY:STBL.

C. straminiformis L. Bailey (p. 531) MOUNT SHASTA SEDGE (Group 5) **LF**: blade 2–4 mm wide. **INFL** dense, 15–25 mm; spikelets distinct, ± spheric; pistillate fl bract < perigynium, red-gold to brown. **FR**: perigynium ascending to spreading, 4–5.8 mm, 1.8–3.4 mm wide, body green or gold, flat exc over fr, flat margin incl wing 0.4–1 mm wide, edge minutely crinkled at least above, veins many at least on back, weak, beak tip cylindric and entire for < 0.3 mm, red-brown, very tip white, 2.2–3 mm to fr top; fr 1.3–2.3 mm, 1–1.6 mm wide. Common. Rocky or gravelly soils; 2000–3800 m. KR, NCoRI, CaRH, SNH, MP, **SNE (Sweetwater Mtns)**, **W&I**; to WA; also in ID, UT. ❀ TRY.

C. subfusca W. Boott (p. 531) (Group 5) **LF**: blade 1.2–3.7 mm wide. **INFL** open or dense, 11–30 mm; spikelets distinct; lowest spikelet bract bristle-like or not, > infl or not; lower pistillate fl bracts, < or ± = perigynium, gen ± < 3.5 mm, pale brown to reddish. **FR**: perigynium appressed to spreading, 2.5–4.3 mm, 0.9–1.9 mm wide, planoconvex, whitish green, pale gold, or light brown, upper margin green or gold, flat margin incl wing 0.1–0.3 mm wide, at least back veined, beak tip flat or cylindric, ± serrate, light to medium brown, very tip white, 1.1–2 mm to fr top; fr 1–1.6 mm, 0.7–1.2 mm wide. 2*n*=84. Common. Seasonally wet meadows; 100–3500 m. KR, NCoR, CaRH, SN, SnFrB, SCoR, TR, PR, Wrn, **DMtns**; w N.Am. [*C. teneriformis* Mack.] ❀ TRY.

C. subnigricans Stacey (p. 531) (Group 1) Rhizomed. **ST** 5–20 cm, > lvs. **LF**: blade 1 mm wide, rolled, quill-like. **INFL**: spikelet bisexual, 3–6 mm wide; spikelet bract 0; pistillate fl bract wide, gen 1-veined, light brown. **FR**: perigynia 10–40 per spikelet, ascending, 2.5–4.1 mm, 0.9–2 mm wide, gold to golden brown, a bristle-like axis within, faces glabrous; fr 1.3–1.8 mm, 0.7–1 mm wide (width > distance from fr to perigynium side), < perigynium body. Meadows, gen dry, rocky slopes; 2600–3800 m. SNH, **SNE (Sweetwater Mtns)**, **W&I**; to ID, UT. ❀ TRY:STBL.

C. tahoensis F.J. Smiley (p. 531) (Group 5) **LF**: blade 0.7–2 mm wide, folded or margin downcurved. **INFL** open or dense, 15–35 mm, 5–20 mm wide; spikelets 3–7, distinct, appressed to ascending, < 5 mm wide; pistillate fl bract ± covering perigynium, white margin 0.1–0.6 mm wide. **FR**: perigynium appressed, 3.7–7(7.5) mm, 1.3–2.7 mm wide, planoconvex, gold to brown, flat margin incl wing 0.1–0.5 mm wide, back veined, front gen with 5–8 veins to fr top, beak tip cylindric and ± entire for 0– 0.7 mm, reddish, white at very tip, with white flap along middle of back, tip to fr top 0.6–4.4 mm; fr 1.7–2.1(2.6) mm, 1.1–1.6 mm wide, 0.5–0.7 mm thick. Uncommon. Open, rocky slopes; 2900–3700 m. SNH, **SNE (Sweetwater Mtns)**; scattered in montane w NA. [*C. eastwoodiana* Stacey; *C. phaeocephala* f. *eastwoodiana* (Stacey) F. Herm.]

C. utriculata Boott (p. 531)   Long-rhizomed. **ST** 30–120 cm, ± = lvs, angles blunt. **LF**: ligule length < or ± = width. **INFL**: lateral spikelets erect, sessile or short-stalked; pistillate fl bract green with reddish margin, awn < body or 0. **FR**: perigynia 50–200 per spikelet, in many dense rows, ascending-spreading to reflexed, 3.9–7 mm, 1.3–3 mm wide, abruptly narrowed to 1–2 mm beak, shiny, gold to brown, teeth erect, 0.1–0.8 mm; fr 1.1–2 mm, 0.9–1.3 mm wide. 2*n*=72,76,82. Common. Wet places, shallow water; < 3400 m. KR, NCoR, CaRH, SNH, CCo, SnFrB, SnBr, **SNE**; N.Am, Eurasia. [*C. rostrata* Stokes] ❀ TRY:STBL.

C. vallicola Dewey (p. 533)   (Group 4) Cespitose, sometimes long-rhizomed. **ST** 15–60 cm. **LF**: blade 0.5–2 mm wide. **INFL** 1–3 cm, 4–8 mm wide; spikelets < 10; lowest spikelet bract like pistillate fl bract or bristle-like; pistillate fl bract < perigynium, white, minutely pointed. **FR**: perigynium spreading, 2.5–4 mm, 1.5–2.2 mm wide, sessile, green to brown, very shiny, back bulged so that marginal ribs on front rather than sides, ± veined,

front unveined, beak 0.6–1 mm, teeth reddish; fr 1.6–2.5 mm, 1.4–2.1 mm wide. Moist to ± dry slopes, montane; < 2700 m. **SNE (Sweetwater Mtns)**; to MT, SD, UT. ❀ TRY:STBL.

C. vernacula L. Bailey (p. 533)   (Group 4) Cespitose; rhizomes short, straight. **ST** 5–30 cm. **LF** < infl; blade 2–4 mm wide. **INFL** dense, 8–16 mm, 8–16 mm wide; spikelets < 10, indistinct; lowest spikelet bract like pistillate fl bract; pistillate fl bract dark brown with narrow white tip, shiny, long-tapered or awned. **FR**: perigynium ascending to spreading, 3.3–4.8 mm, < pistillate fl bract, 1.2–1.8 mm wide, brown, often green-margined, ± shiny, back veined, front veined or not, margin entire above, beak 0.8–1.7 mm, entire; fr 1.2–1.6 mm, 0.7–1.1 mm wide, ovate. Open, often rocky areas, wet from snow melt; 1800–4000 m. CaRH, SNH, Wrn, **SNE (Sweetwater Mtns), W&I**; scattered in montane w US. ❀ TRY.

CLADIUM   TWIG RUSH

Per. **ST** rounded, leafy, hollow. **INFL**: spikelets in terminal and axillary, panicle-like clusters; fl bracts spiraled, basal sterile, reduced; uppermost fls bisexual, others staminate. **FL**: perianth bristles 0; stamens 2; style 2–3-branched. **FR** weakly 2-sided, truncate at base; style base persistent; tubercle 0.   ± 40 spp.: worldwide, esp Australia.   (Greek: branch, from infl)

C. californicum (S. Watson) O'Neill (p. 533)   SAW-GRASS   Pl 10–20 dm, stout, coarse, from long rhizomes, forming dense bunches. **LF** 7–10 mm wide, flat; margin sharply serrate. **FR** purple-brown; tip 2–2.5 mm, buff. 2*n*=36. Uncommon. Gen alkaline marshes, swamps; < 2000 m. s CCo, SCoRO, SCo, WTR, **D**; NV, AZ. [*C. mariscus* R.Br. var. *c*. S. Watson] ❀ TRY;STBL. Jun–Sep

CYPERUS   NUTSEDGE, GALINGALE

Gordon C. Tucker

Ann or per, glabrous. **STS** gen > 1, erect, 2–100 cm, 3-angled or round. **LVS** basal; blades 0 or linear. **INFL**: bracts 1–9, lf-like, spreading or erect; rays < 10 cm; spikelets flat to ± cylindric; fl bracts 2-ranked, 1 per fl, 2–36 per spikelet. **FL** bisexual; perianth 0; stigmas gen 3. **FR** (ob)ovoid, gen 3-angled, brown, gen not beaked.   (Greek: name for 1 sp.) [Tucker 1994 Syst Bot Monogr 43:1–213]   Mature fr gen needed for identification.

1. Fr lenticular; stigmas 2
  2. Fr face next to spikelet axis . . . . *C. laevigatus*
  2' Fr edge edge next to spikelet axis. . . . . *C. niger*
1' Fr 3-angled; stigmas 3
  3. St 1; stolons with tubers; fr seldom maturing
    4. Fl bract yellow to brown . . . . *C. esculentus*
    4' Fl bract ± dark red to purple . . . . *C. rotundus*
  3' St gen > 1, clumped or not; stolons, tubers 0; fr maturing
    5. Tip of fl bract outcurved, bristle-like; stamen 1 . . . . *C. squarrosus*
    5' Tip of fl bract not outcurved, acute, obtuse, or rounded, sometimes with a small point but not bristle-like; stamens 3
      6. Spikelet flat; fl bract reddish . . . . *C. parishii*
      6' Spikelet slightly flat or subcylindric; fl bract brown, light reddish brown, or yellowish
        7. Fl bract 1.3–1.5 mm; fr 0.7–1.1 mm . . . . *C. erythrorhizos*
        7' Fl bract 2–4 mm; fr 1.5–2.4 mm
          8. Ann; st base not corm-like; fl bract persistent, light brown splotched reddish, midvein green, elliptic to ovate; spikelet axis joined at base of each bract, spikelet falling apart as it matures . . *C. odoratus*
          8' Per (often fl first season); st base corm-like; fl bract persistent or not, yellowish, lanceolate; spikelet axis continuous, spikelet gen falling as a unit as it matures . . . . *C. strigosus*

C. erythrorhizos Muhlenb. (p. 533)   Ann 5–100 cm; roots reddish. **INFL**: bracts 3–11, 5–70 cm; spikelets 20–150, 3–11 mm, linear, subcylindric, light (reddish) brown, in open, ovoid-cylindric spikes, these 1–6 per ray; fl bracts 6–30 per spikelet, 1.3–1.5 mm, overlapped, tips obtuse, lateral veins 0. **FR** 0.7–1 mm, light gray to brown, glossy, sides unequal. Ditches, riverbanks, shores; 0–500 m. **CA**; US, n Baja CA. Jul–Oct

C. esculentus L. (p. 533)   Per 20–80 cm; stolons with tubers < 15 mm wide. **ST** 1. **INFL**: bracts 3–7; rays 5–10, 20–200 mm;

spikelets 5–30 mm, linear, brown, in ± open, widely elliptic spikes; fl bracts 2–3 mm, overlapped, ovate to elliptic, (yellowish) brown. **FR** 1–1.6 mm, elliptic, seldom maturing. Croplands, disturbed places; 0–1000 m. **CA**; worldwide weed. Jun–Oct

C. laevigatus L. (p. 533)   Per 1–60 cm; rhizomes 1–2 mm thick, horizontal. **STS** clumped or not, 1–30 mm apart. **LF**: blades gen 0(–7 cm). **INFL**: bracts 1–3, 1–10 cm; spikelets 1–14, 4–12 mm, elliptic to oblong-lanceolate, subcylindric, greenish white to reddish or dark brown, in open, ± ovoid heads; fl bracts 8–24 per spikelet, 1.5–2 mm, obovate to round, overlapped, lateral veins 0. **FL**: stigmas 2. **FR** 1.2–1.8 mm, oblong-elliptic to ovate, lenticular, dark brown or black. Alkaline or brackish, wet soils, hot springs, permanent pools in arroyos; 30–1000 m. SW, **GB**, **D**; to TX, Mex, scattered in trop, warm temp worldwide. ❀ TRY;STBL. Jul–Dec

C. niger Ruiz & Pav. (p. 533) Per 1–50 cm; rhizomes short, 1 mm thick. **INFL**: bracts 2–3, 1–15 cm; rays 0–2, 0–5(40) mm; spikelets 3–25, 3–9 mm, linear-oblong, flat, black to (rarely light) brown, in gen dense, ovoid, head-like spikes 5–15 mm wide; fl bracts 3–9 per spikelet, 1.5–2.2 mm, ovate, light brown to black, lateral veins 0. **FL**: stigmas 2. **FR** 1.2–1.4 mm, elliptic to ovoid, lenticular, brown. Marshes, swamps, moist roadsides; 0–1500 m. NCoR, CaRF, SNF, n SNH, GV, CW, SCo, PR, **SNE**, S.Am. ❀ TRY;STBL. Jul–Nov

C. odoratus L. (p. 533) Ann 10–50 cm. **INFL**: bracts 5–9, 5–24 cm; rays 6–12, 10–100 mm; spikelets linear, cylindric or slightly flat, in ± ovoid spikes; fl bracts 6–24 per spikelet, 2.1–3.5 mm, elliptic to ovate, light brown splotched reddish, midvein conspicuous. **FR** 1.5–1.9 mm, slightly flat front-to-back. Wet disturbed soils; 0–500 m. GV, SCo, **D**; trop and warm temp. [*C. ferax* Rich.] Jul–Oct

C. parishii Britton (p. 533) Ann 5–25 cm, cespitose. **INFL**: bracts 2–5, 3–20 cm; rays 1–6, 20–70 mm; spikelets 5–30, 6–22 mm, linear, flat, red, reddish purple, or reddish brown, in open, spheric spikes; fl bracts 8–10 per spikelet, gen ± 3 mm, elliptic, reddish, deciduous, 7–9-veined, midvein green, gen prominent.

FR 1–1.3 mm, widely elliptic, (dark purplish) brown. Streambanks, roadsides; 0–800 m. SW, **D**; to AZ.

C. rotundus L. PURPLE NUTSEDGE Per 10–40 cm; stolons with tubers < 1 cm thick. **INFL**: bracts 2–5, 5–25 cm; rays 5–10, 10–150 mm; spikelets 3–12, 4–40 mm, linear to ± lanceolate, flat, reddish purple, in open, elliptic spikes 1–5 cm wide; fl bracts 6–36 per spikelet, 2–3.4 mm, ovate, 7–9-veined. **FR** 1.4–1.9 mm, elliptic, black, seldom maturing. Disturbed soils, croplands; 0–250 m. GV, SCo, **DSon**; trop and warm temp; native to Eurasia. Often considered world's worst weed. Jul–Nov

C. squarrosus L. (p. 533) Ann 1–10(20) cm. **INFL**: bracts 1–4, 1–15 cm; rays 0–6, 4–40 mm; spikelets 1–30, 2–20 mm, widely lanceolate to oblong, flat, greenish, yellowish, or reddish brown, in open, ovoid spikes 5–15 mm wide; fl bracts 2–8 per spikelet, 1.2–2 mm, oblong-lanceolate, 5–11-veined, bristle-like tip 0.5–1.3 mm. **FL**: stamen 1. **FR** 0.7–1.1 mm, ± obovoid, light brown to black. Moist, sunny, disturbed places, esp pond margins, riverbanks; 0–1500 m. **CA**; temp and trop ± worldwide. [*C. aristatus* Rottb.] Jun–Nov

C. strigosus L. (p. 533) FALSE NUTSEDGE Per 5–70 cm. **ST** corm-like at base. **INFL**: bracts 3–6, 5–30 cm; rays 0–7, 20–150 mm; spikelets 15–50, 6–30 mm, linear, subcylindric, in spikes 1–5 cm wide; fl bracts 3–11 per spikelet, 2.8–4 mm, lanceolate, yellowish, 7–11-veined. **FR** 1.8–2.4 mm, narrowly elliptic. Moist soils, pond margins, ditches, roadsides; 0–1000 m. **CA**; US, s Can. Jul–Oct

ELEOCHARIS SPIKERUSH

Ann or per. **ST** gen round, ridged and grooved, gen solid. **LVS** basal, 1(–4); base sheathing; blade gen 0. **INFL**: spikelet solitary, terminal, erect; spikelet bract 0; fl bracts gen spiraled. **FLS** bisexual; perianth bristles 0–6, persistent, barbs gen recurved; stamens (1–)3; style 2–3-branched, base bulb-like, persistent. **FR** 2–3-sided or round; top tubercled. ± 250 spp.: worldwide. (Greek: marsh grace) St shape best seen in fresh material (or just below spikelet); drying exaggerates ridges and grooves. *E. montevidensis* has been reported from n DMtns. Key to species by S. Galen Smith.

1. Fr (purplish) black, shiny — tubercle hat-like, gen < 1/5 fr body . . . . *E. geniculata*
1' Fr yellowish brown, shiny or dull
  2. Style 2-branched; fr 2-sided . . . . *E. macrostachya*
  2' Style 3-branched; fr 2–3-sided . . . .
    3. Spikelet 2–9-fld
      4. St gen 2–7 cm, not rhizomed; fr 1–1.3 mm, gen = perianth bristles; spikelet 2–3.5 mm . . . . *E. parvula*
      4' St gen 10–40 cm, rhizomed; fr 2–3 mm, gen < perianth bristles; spikelet 4–7 mm . . . . *E. quinqueflora*
    3' Spikelet gen 10–many-fld
      5. Fr tubercle base slightly narrowed; fr body appearing 2-sided; st gen ascending, tip not rooting . . *E. parishii*
      5' Fr tubercle base not narrowed; fr body weakly 2–3-sided; st tip often rooting to form new pl . . . *E. rostellata*

E. geniculata (L.) Roemer & Schultes (p. 533) Ann < 4 dm. **LF**: base brown, becoming straw-colored above; tip oblique, 1-toothed. **INFL**: spikelet 2–8 mm, much wider than st, ovate to ± round, 10–many-fld; fl bracts ovate-obtuse, greenish with dark brown sides, lower deciduous. **FL**: style 2-branched. **FR**: perianth bristles gen > fr; body ± 0.7–1 mm, obovate, strongly 2-sided, (purplish) black, shiny; tubercle < 1/5 fr body, wide, hat-like, base weakly narrowed. Wet soil, esp marshes, streambanks; < 1000 m. SW, **D**. [Wilson 1990 Cyper Newsl 7:6–7] Mar–Dec

E. macrostachya Britton (p. 533) Per 5–10 dm, rhizomed. **LF** loosely sheathing, base purplish, becoming straw-colored above; tip truncate, often 1-toothed. **INFL**: spikelet 5–25 mm, not much wider than st, gen 10–many-fld, elongate, tip acute; fl bract lanceolate-acute, brown to purplish. **FL**: style 2-branched. **FR**: perianth bristles unequal, < to > fr body; body ± 1.5–2.5 mm, obovate, strongly 2-sided, yellowish brown; tubercle ± conic, base narrowed. *n*=5,8,16,18,19,21,23. Marshes, pond margins, vernal pools, ditches; < 2500 m. **CA**; temp, montane w hemisphere. [*E. palustris* (L.) Roemer & Schultes, in part] Complex, polyploid, needs study. ❀ TRY;also STBL.

E. parishii Britton (p. 533) Per 1–3 dm, not glaucous; rhizome long, reddish. **LF** purplish brown, becoming straw-colored above; tip truncate, 1-toothed. **INFL**: spikelet 10–15 mm, much wider than st, ± linear-lanceolate, gen 10–many-fld, tip acute to acuminate; fl bract ± dark brown, tip short, translucent. **FL**: style 3-branched. **FR**: perianth bristles 6–7, unequal, = or > fr; body ± 1 mm, obovate, appearing 2-sided, yellowish to light brown, shiny; tubercle short, conic, base barely narrowed. *n*=5. Moist, often sandy openings; gen < 2000 m. **CA** (exc MP); to OR, NV, NM, n Mex. [*E. montevidensis* vars. *disciformis* (Parish) V. Grant and *p*. (Britton) V. Grant] ❀ TRY;GRNCVR;also STBL. May–Sep

E. parvula (Roem. & Schult.) Bluff, Nees & Schauer (p. 533) SMALL SPIKERUSH Per < 1 dm; rhizome 0. **ST** often spongy when fresh, flattened in pressing. **LF** membranous. **INFL**: spikelet 2–3.5 mm, wider than st, ovate, 2–9-fld; fl bract ovate, greenish to brownish, tip obtuse to acute. **FL**: style 3-branched. **FR**: perianth bristles gen = fr; body 1–1.3 mm, obovate, strongly 3-sided, straw-colored, shiny; tubercle base not narrowed. 2*n*=8,10. UNCOMMON. Wet, gen saline flats, marshes; < 2500 m. NCo, **GB**, **D**; circumboreal. [var. *coloradoensis* (Britton) Beetle] Pls from interior N.Am sometimes separated. Jul–Aug

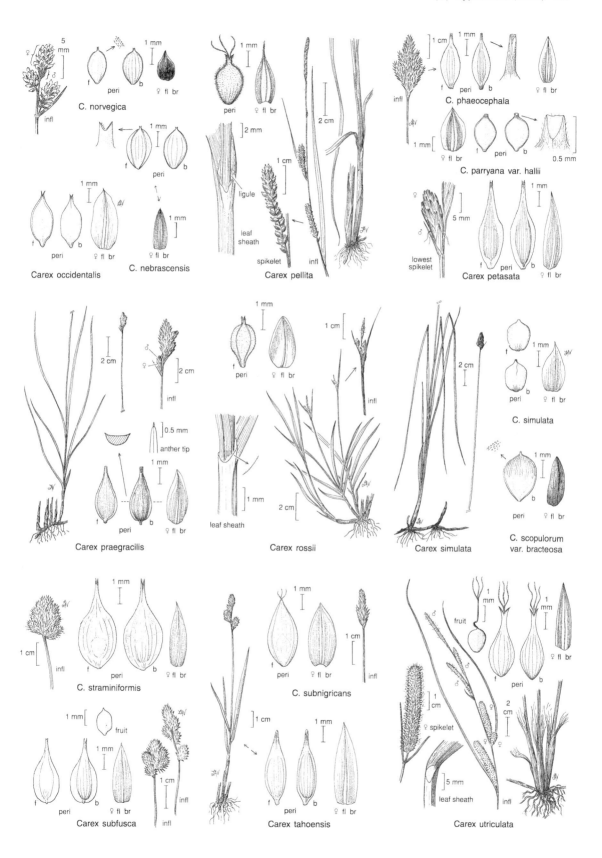

C. norvegica

Carex occidentalis

C. nebrascensis

Carex pellita

ligule

leaf sheath

spikelet

infl

C. phaeocephala

C. parryana var. hallii

lowest spikelet

Carex petasata

Carex praegracilis

anther tip

Carex rossii

leaf sheath

C. simulata

Carex simulata

C. scopulorum var. bracteosa

C. straminiformis

fruit

Carex subfusca

infl

C. subnigricans

Carex tahoensis

fruit

spikelet

leaf sheath

infl

Carex utriculata

E. quinqueflora (Hartmann) O. Schwarz Per 1–4 dm, rhizomed. **LF** straw-colored to brownish; tip truncate. **INFL**: spikelet 4–7 mm, gen wider than st, ovate, 2–7-fld; fl bract ovate, dark brown to blackish, margin scarious, tip obtuse to acute. **FL**: style 3-branched. **FR**: perianth bristles gen > fr, barbs irregular; body 2–3 mm, obovate to fusiform, weakly 2–3-sided, yellow-brown, weakly netted; tubercle base not narrowed, tip dark. Uncommon. Meadows; 1500–3700 m. NW, CaR, SN, SnBr, SnJt, MP, **nSNE**; circumboreal; also s Chile. [*E. pauciflora* S. Wats.]

E. rostellata (Torrey) Torrey (p. 533) Per 2–15 dm, rhizomed.

ST spongy, flattened in pressing; tip often arched, rooting to form new pl. **LF** rigid, straw-colored to greenish; tip truncate to oblique. **INFL**: spikelet 8–20 mm, wider than st, ovate, gen 10–many-fld; fl bracts light brown, often purple-spotted, lowest rounded, upper acute. **FL**: style 3-branched. **FR**: perianth bristles ± = fr; body 2–3 mm, obovate, weakly 2–3-sided, olive, shiny, weakly netted; tubercle deltate, not flat, base not narrowed. Alkaline marshes, sinks, springs; < 2000 m. SnFrB, SW, **SNE, D**; to Caribbean, Mex, also w S.Am. Indicator of saline, calcareous soils. ❀ TRY;GRNCVR;also STBL. May–Aug

FIMBRISTYLIS

Ann or per, cespitose. **ST** round, slender, solid, ridged and grooved. **LVS** several, basal; bases sheathing, margins wide, scarious; blades 0 to linear, flat, folded, or upcurled. **INFL**: spikelets in terminal, panicle-like clusters; fl bracts spiraled, deciduous; **FLS** bisexual; perianth bristles 0; stamens gen 3; style 2–3-branched, gen fringed below branches, widest at base, deciduous. **FR** 2–3-sided or round, surface cells ± square or horizontally elongate. ± 200 spp.: worldwide. (Latin: fringed style) [Kral 1971 Sida 4:57–227] Key to species adapted by Margriet Wetherwax.

1. Per; spikelets gen long-peduncled; fl bracts densely hairy; W&I, DMoj . . . . ***F. thermalis***
1′ Ann; spikelet subsessile; fl bracts glabrous; e DSon . . . . ***F. vahlii***

F. thermalis S. Watson (p. 533) HOT-SPRINGS FIMBRISTYLIS Per < 15 dm, rhizomed. **ST** > lf, round below, flat above. **LVS** spiraled; blades linear, flat or ± upcurled. **INFL**: spikelet gen long-peduncled, elongate, tip acute; fl bracts densely hairy, fruiting bracts narrowly ovate, pale brown, midrib greenish, exserted, tip acute. **FL**: style 2-branched. **FR** ± 1.5 mm, obovate, 2-sided, dark brown, shiny. 2*n*=20. RARE in CA. Wet, alkaline soils, gen near hot springs; > 500 m. **W&I, DMoj**; s NV, AZ. [*F. spadicea* (L.) M. Vahl misapplied] Like *F. caroliniana* (Lambert) Fern. of se US. Aug–Sep

F. vahlii (Lam.) Link (p. 533) Ann < 1 dm. **ST** > lf, round. **LVS** spiraled; blade ± thread-like, upcurled. **INFL**: spikelets in tight clusters, subsessile, elongate, tips acute; fl bracts glabrous, fruiting bracts lanceolate, straw-colored to pale green, midrib dark green, exserted, tip acute. **FL**: style 2-branched. **FR** ± 0.6 mm, obovate to round, pale brown. 2*n*=20. Wet soil, often mud flats, silty levees; < 500 m. SnJV, **e DSon**; to se US, C.Am. Perhaps alien. Jul–Nov

KYLLINGA

Gordon C. Tucker

Per, glabrous. **ST** erect, 2–100 cm. **LVS** basal, linear. **INFL**: bracts 2–4; spikelets 50–100+, 1–2-fld, slender-elliptic, breaking apart just above 2 basal fl bracts at maturity, in 1–3, sessile, dense, ovoid, light green spikes; fl bracts 2-ranked, 4–5 per spikelet, 2 at base << others, 3rd subtending a bisexual fl, 4th (and rarely 5th) ± = 3rd but sterile or subtending staminate fl. **FL**: perianth 0; stamens 1–3; stigmas 2. **FR** lenticular, ± smooth, brown. (P. Kylling, Danish botanist, died 1696) [Tucker 1984 Rhodora 86:507–538]

K. brevifolia Rottb. (p. 533) Pl rhizomed. **STS** 1–6, 2–55 cm. **INFL**: longest bract erect; spikelets 2.2–3.2 mm, oblong-lanceolate, in 1–3 spikes 4–7 mm. **FR** 1–1.3 mm, ± elliptic, ± stalked.

Lawns, ditches, croplands; 0–300 m. GV, SW, **D**; trop and warm temp; native to trop Am. Jul–Sep

SCHOENUS

Per. **LVS** basal, erect, stiff, sheathing. **INFL** subtended by 2 unequal, lf-like bracts; spikelet bracts 1–2, sheathing; spikelets in terminal head-like clusters, flat; fl bracts 2-ranked, only uppermost fruiting. **FLS** bisexual and staminate; perianth bristles 3–6, gen < 1 mm, deciduous; stamens 3; style 3-branched. **FR** obovoid; tubercle gen 0. ± 80 spp.: worldwide, esp Australia. (Greek: a rush)

S. nigricans L. (p. 539) BLACK SEDGE Pl 2–7 dm, densely cespitose; rhizomes 0. **LF** > 1/2 st, ± 1 mm wide, entire to minutely serrate, uprolled; ligule wide, very dark brown. **FR** 3-sided, with a ± white, bony covering. 2*n*=44,54. Uncommon. Marshes,

swamps, springs, gen alkaline soils; < 2000 m. SnBr, **DMoj**; Caribbean, Eurasia, n Afr. Used as roof thatch in Ireland. ❀ TRY. Aug–Sep

SCIRPUS

S. Galen Smith, A.E. Schuyler, & William J. Crins

Ann or per, rhizomed or not; roots fibrous. **ST** gen erect, 3-angled or cylindric, solid. **LVS** basal or cauline, alternate, 3-ranked; sheaths closed; ligule 0 or present; blades linear, sometimes vestigial and scale-like. **INFL** panicle- or head-like; bracts 1–several, lf- or ± st-like; spikelets 1–many, gen many-fld; fl bracts spiraled, gen ovate, scale-like. **FL** bisexual; perianth bristles gen < and hidden by fl bracts, gen ± straight, gen slender, gen stiff, gen persistent on fr, gen finely spined or fringed, sometimes 0 or vestigial; stamens gen 3; style 1, ribbon- or thread-like, stigmas 2–3, gen exserted. **FR**: achene, gen obovoid, lenticular or ± 2–3-angled, gen beaked, not tubercled. ± 200 spp.: gen wet sites,

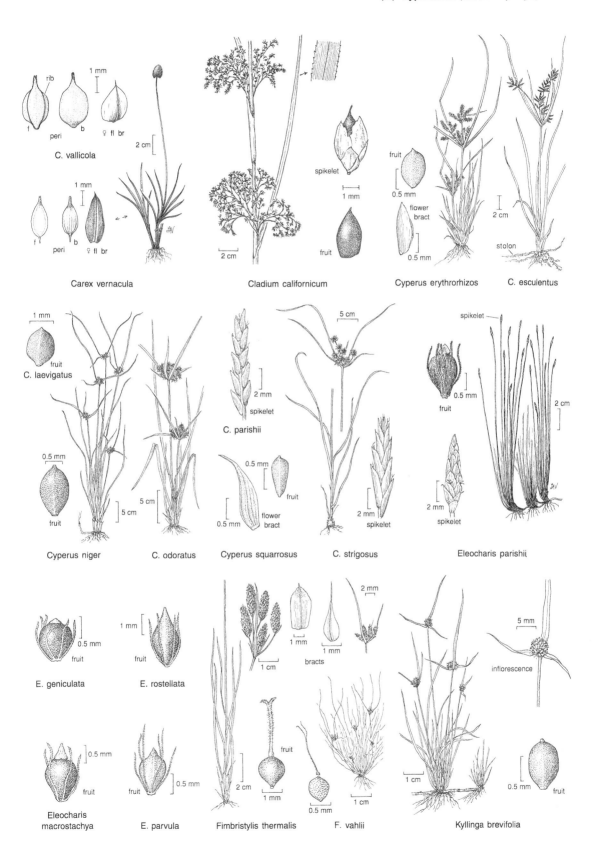

C. vallicola

Carex vernacula

Cladium californicum

Cyperus erythrorhizos

C. esculentus

C. laevigatus

Cyperus niger

C. odoratus

C. parishii

Cyperus squarrosus

C. strigosus

Eleocharis parishii

E. geniculata

E. rostellata

Eleocharis macrostachya

E. parvula

Fimbristylis thermalis

F. vahlii

Kyllinga brevifolia

worldwide. Some spp. mistaken for *Eleocharis*. *CA plants treated as *Amphiscirpus, Bolboschoenus, Schoenoplectus* [Browning et al 1995 Brittonia 47:433–445; Smith 1995 Novon 5:97–102].

1. Well developed infl bracts gen 2+, thin, lf-like, gen spreading; lvs ± cauline, blades gen > sheaths, thin
  2. Spikelet 5–10 mm wide; fr 2.5–3.5 mm; fl bract scabrous, tip notched, awned; tubers present . . . ***S. maritimus***
  2' Spikelet 1–3.5 mm wide; fr ± 0.5–1.5 mm; fl bract glabrous, tip entire, ± abruptly pointed; tubers 0
  . . . . ***S. microcarpus***
1' Well developed infl bract gen 1, thick, ± st-like at least above, erect, spreading or bent to side; lvs ±
  basal; blades 0 or gen < sheaths (exc. *S. nevadensis, S. pungens*), thin to st-like
  3. St 3-angled, at least above
    4. Infl panicle-like, dense to open, spikelets 20–many; perianth bristles on fr flat, thick, fringed with
    soft hairs; st often cylindric below, < 4 m . . . . ***S. californicus***
    4' Infl head-like, or rarely with 1–2 short branches, spikelets 1–12(20); perianth bristles on fr slender,
    minutely spiny; st 3-angled throughout, gen < 2.2 m
      5. Fl bract tip with awn < 0.5 mm; uppermost lf blade gen < sheath; st sides deep-concave . . . . ***S. americanus***
      5' Fl bract tip with awn 0.5–1.5 mm; uppermost lf blade > sheath; st sides ± flat to shallow-concave
      . . . . ***S. pungens***
  3' St cylindric throughout
    6. Infl head-like; spikelets 1–10; fl bract tip acute to obtuse, notch 0 . . . . ***S. nevadensis***
    6' Infl panicle-like; spikelets 3–100+; fl bract tip short-awed, notched (tip sometimes worn off in age)
      7. Fl bract gen prominently spotted, awn 0.5–1 mm, gen contorted; spikelets gen 1–7 per cluster
      . . . . ***S. acutus*** var. ***occidentalis***
      7' Fl bract faintly or not spotted, awn gen < 0.5 mm, ± straight; spikelets gen solitary or 2– 4 per cluster
      . . . . ***S. tabernaemontani***

S. acutus Bigelow var. ***occidentalis*** (S. Watson) Beetle (p. 539) TULE Per 150–400 cm; rhizome long. **STS** erect, 4–12 mm wide in middle, cylindric. **LVS** ± basal; sheaths prominent; blades 1–2, << sheaths, flat, glabrous. **INFL** panicle-like; branches stiff to flexible, erect to arched; spikelets 3–many, in clusters of 1–7 at branch tips, 7–10(24) mm, 3–4.5 mm wide; main bract 1–4(11) cm, gen erect, stiff, st-like at least above; fl bract ± 4 mm, finely prickly to papillate above middle, ± colorless to straw-colored, or orange to dark red-brown above, gen prominently spotted, margin woolly, tip notched, awn 0.5–1 mm, gen contorted. **FL**: perianth bristles gen 6, < or = fr; stigmas 2 or 2+3. **FR** 2–2.5 mm, obtusely 2–3-angled, smooth, gray-brown; beak 0.1–0.3 mm. Marshes, lakes, streambanks; < 2500 m. **CA**; temp N.Am. Hybridizes with *S. californicus, S. tabernaemontani*; some pls in mtns with < 40 spikelets per infl, stigmas 2, fr 2-angled, perianth bristles ± = fr may be var. *acutus* (gen e N.Am). *****Schoenoplectus acutus*** (Bigelow) Á. & D. Löve var. ***occidentalis*** (S. Wats.) S.G. Smith 🌣 STBL. May–Aug

S. americanus Pers. (p. 539) Per 50–220 cm; rhizome long. **STS** erect to arched, 3–7 mm wide in middle, 3-angled; sides deeply concave. **LVS** ± basal, < 1/2 st; blades 1–2, gen < sheaths, flat, folded below, subcylindric above, glabrous. **INFL** head-like; spikelets 2–12(20), 5–11 mm, 4–5 mm wide; bract 1, 1–4 cm, gen erect, stiff, less often bent to side, ± st-like; fl bract 3–3.5 mm, glabrous, orange- to red-brown, margin short-woolly, tip notch < 0.5 mm, awn < 0.5 mm, ± straight. **FL**: perianth bristles 2–7, 1/2 to > fr; stigmas 2 or 2+3. **FR** ± 2–3 mm, obtusely 2-angled, or back low-ridged, smooth, dark brown. *n*=39. Marshes, ponds; < 2000 m. **CA** (exc SNH); to e N.Am, S.Am. [*S. olneyi* A. Gray] Hybridizes with *S. pungens*. 🌣 STBL. *****Schoenoplectus americanus*** (Pers.) Schinz & R. Keller May–Aug

S. californicus (C. Meyer) Steudel(p. 539) Per 200–400 cm; rhizome long. **STS** erect, < 10 mm wide in middle, 3-angled throughout or cylindric below. **LVS** ± basal; sheaths < 1/6 st; blades 0 or 1–2, < sheaths, flat, glabrous. **INFL** panicle-like; spikelets 20–many, clustered at branch tips, 5–11 mm, ± 3 mm wide; branches flexible, arched; main bract 3–8 cm, erect, stiff, ± st-like (at least above); fl bract 2.5–3 mm, glabrous, orange-brown, midrib thick, margin woolly, tip notch short, awn < 0.5 mm, gen bent. **FL**: perianth bristles 2–4, ± = fr, flat, thick, fringed; stigmas 2. **FR** ± 2 mm, 2-angled, smooth, dark brown; beak ± 0.2 mm. Marshes; < 200 m. NCo, GV, CCo, SnFrB, SCo, **e D (Colorado River)**; to s US, S.Am, oceanic islands.

*****Schoenoplectus californicus*** (C.A. Meyer) J. Sojak 🌣 STBL. Jun–Sep

S. maritimus L. (p. 539) Per 80–150 cm; rhizome long with tubers < 2 cm wide. **STS** erect, ± 3–8 mm wide, sharply 3-angled; sides smooth, angles minute-scabrous. **LVS** ± cauline; sheath top opposite blade with V-shaped veinless area; upper blades >> sheaths, gen 3–12 mm wide when dry, flat to V-shaped, midrib and margin fine-scabrous. **INFL**: spikelets 4–many, 10–30 mm, 5–10 mm wide, often in 1 sessile, dense cluster or < 1/2 in clusters of 1–15 at branch tips; 2–3 bracts >> infl, spreading, lf-like, longest gen 2–5 mm wide when dry; fl bract not strongly appressed, 5–7 mm, pale grayish to orange-brown, ± translucent, scabrous, tip notch ± 0.5 mm, awn 1.5–3 mm, ± contorted. **FL**: perianth bristles gen 6, gen < 1/2 fr, gen deciduous; stigmas 2. **FR** 2.5–3.5 mm, ± compressed, ± weakly 2–3-angled, smooth, shiny, dark brown; sides convex; tip truncate, beak 0.1–0.2 mm. 2*n*=90. Marshes; < 2500 m. NW, GV, CW, SW, MP, **D**; almost worldwide. [var. *paludosus* (Nelson) Kük. *****Bolboschoenus maritimus*** (L.) Palla 🌣 STBL.

S. microcarpus C. Presl(p. 539) Per 70–160 cm; rhizome long. **STS** erect, clumped or not, 5–7 mm wide, ± 3-angled. **LVS** basal and cauline; sheaths prominent; blades < to = sheaths, flat, glabrous, margin scabrous. **INFL** ± panicle-like; spikelets 50+, gen in head-like clusters of 4–12 at branch tips, 2–8 mm, 1–3.5 mm wide; longest bract 1.2–4 cm, ascending to erect, lf-like; fl bract 1–3 mm, glabrous, dark brown, tip ± abruptly pointed. **FL**: perianth bristles gen 4, < to > fr, fine-toothed throughout; stigmas gen 2. **FR** ± 0.5–1.5 mm, gen 2-angled, smooth. Marshes, wet meadows, streambanks, pond margins; < 2900 m. **CA**; to AK, e US, Asia. 🌣 IRRorWET:1,**2–10,14–24**;also STBL;INV. May–Aug

S. nevadensis S. Watson (p. 539) Per 10–45 cm; rhizomes long, some vertical. **STS** erect to arched, well spaced to clumped, < or = 1 mm wide, cylindric, stiff, wiry. **LVS** basal and cauline; lower sheaths coarsely fibrous; upper blades >> sheaths, ± strongly curved, C-shaped to ± cylindric, wiry. **INFL** head-like; spikelets 1–10, 10–20 mm, ± 5 mm wide; bract gen 1, 1–10 cm, erect to spreading, stiff, ± st-like; fl bract ± 4 mm, ± glabrous, brown, tip acute to obtuse. **FL**: perianth bristles 1–4, << to 1/2 fr; stigmas 2. **FR** ± 2.5 mm, 2-angled, shiny, brown, pitted; tip ± rounded, beak 0. Wet places, saline soils; 1000–2500 m. **GB**; to B.C., Great Plains, S.Am. *****Amphiscirpus nevadensis*** (S. Watson) Oteng-Yeboah 🌣 STBL. Jun–Aug

S. pungens Vahl COMMON THREESQUARE Per 20–200 cm; rhizome long. **STS** erect to strongly arched, 2–6 mm wide in middle, sharply 3-angled; sides flat to ± concave. **LVS** ± basal, ± 1/2 to = st; blades 2–6, gen > sheaths, ± flat to subcylindric below, subcylindric above, glabrous or minutely scabrous near tip. **INFL** head-like; spikelets 1–5(10), 7–13 mm, 4–5 mm wide; main bract 3–11(20) cm, gen erect, ± st-like, next bract 5–12 mm; fl bract 3.5–5 mm, glabrous, orange- to dark red-brown, margin short-woolly, tip notch ± 0.5–1 mm, awn 0.5–1.5 mm, ± straight or tip contorted. **FL**: perianth bristles gen 2–7, gen 1/2 to = fr; stigmas 3. **FR** ± 2–3.5 mm, obtusely 3-angled, ± compressed, smooth, dark brown. Marshes, lake shores; < 2000 m. NCo, SN, SnJV, CW, SCo, **GB**, **DMoj**; N.Am, S.Am, Eur, s Pacific. [*S. americanus* Pers. var. *longispicatus* Britton]*Schoenoplectus pungens* (Vahl) Palla ❀ STBL. May–Aug

S. tabernaemontani C. Gmelin (p. 539) Per 100–200(300) cm; rhizome long. **STS** erect, densely clustered, 4–8 mm wide in middle, cylindric. **LVS** ± basal; sheaths prominent; blades 1–2, << sheaths, flat, glabrous. **INFL** panicle-like; branches flexible, arched; spikelets 12–125, solitary or in clusters of 2–4 at branch tips, 5–10(13) mm, 3–4 mm wide; main bract 2–5 cm, gen erect, stiff, ± st-like (at least above); fl bract 2.5–3 mm, orange-brown, faintly or not spotted, midrib minutely spiny-papillate, margin woolly, tip ± rounded, notched, awn gen < 0.5 mm, ± straight. **FL**: perianth bristles 6, = fr; stigmas gen 2. **FR** ± 2–2.5 mm, ± obtusely 2-angled, smooth, brown; beak < 0.2 mm. *n*=21. Marshes, lakes, streambanks; < 200 m. NW, SnJV, SnFrB, SCo, **DSon**; N.Am, S.Am, Eurasia, South Pacific. [*S. validus* Vahl] *Schoenoplectus tabernaemontani* (Gmelin) Palla ❀ TRY. Aug–Sep

HYDROCHARITACEAE WATERWEED FAMILY

Robert F. Thorne

Ann, per, aquatic, freshwater or marine, bisexual, monoecious, or dioecious. **LVS** basal or cauline, alternate, opposite, or whorled, gen ± sheathing at base, glabrous. **INFL**: cyme; fls 1–few, subtended by ± sheathing, entire or lobed bract; staminate fls sometimes deciduous, free-floating. **FL** radial; perianth tube 0 or much elongated, peduncle-like in fl; sepals (0)3, green; petals (0)3, colored or white; stamens (1)3–many, gen in 1+ series; ovary inferior, chamber 1, placentas parietal or basal, ovules 1–many; style lobes gen 3, linear, lobed or notched. **FR**: achene or berry-like and dehiscing irregularly, linear to spheric, submersed. ± 17 genera, ± 130 spp.: worldwide; some cult for aquaria, others noxious weeds. Keys adapted by Margriet Wetherwax.

1. Fl sessile, 1–few per axil; perianth 0; fr achene with net-like surface . . . . **NAJAS**
1' Fl with perianth tube peduncle-like; sepals green, petals colored or white; fr berry-like and dehiscing irregularly, surface not net-like
  2. Lvs 2–4 cm, 2–5 mm wide, gen 3–6 per node; staminate fls 2–3, petals 8–10 mm . . . . **EGERIA**
  2' Lvs < 1.5(2) cm, < 3 mm wide, opposite or 2–3 per node; staminate fl 1, petals gen < 5 mm
    3. Lf margin minutely and finely serrate, midrib on lower surface ± smooth . . . . **ELODEA**
    3' Lf margin clearly and sharply serrate, midrib on lower surface ± rough . . . . **HYDRILLA**

EGERIA BRAZILIAN WATERWEED

Per, rooted in mud, submersed or floating, dioecious (CA pls staminate). **ST** slender, sometimes branched or not. **LVS** opposite below, crowded and whorled above, 3–6 per whorl, sessile. **STAMINATE INFL**: fls 2–3, clustered, peduncled; bract slender. **PISTILLATE INFL**: fl 1, axillary; bract slender, ± tubular. **STAMINATE FL** floating; perianth tube elongated, slender, peduncle-like, persistent on pl; stamens 9. **PISTILLATE FL** floating; perianth tube elongated, slender, peduncle-like, persistent on pl; stigmas 3, 3–4-lobed. 2 spp.: S.Am. (Latin: a mythical water nymph)

E. densa Planchon **ST** 2–3 mm thick. **LF** 2–4 cm, 2–5 mm wide, narrowly oblong; margin finely serrate. **STAMINATE FL**: perianth tube < 6 cm; petals 8–10 mm, white, ± round. **PISTILLATE FL**: perianth tube < 6 cm; petals 6–7 mm, white, ± round. 2*n*=48. Streams, ponds, sloughs; < 2200 m. n&s SNF, SnJV, SnFrB, SnJt, **SNE** (expected elsewhere); to e US, Eur; native to S.Am. Staminate pls widely cult for aquaria. [*Elodea d.* (Planchon) Caspary] Jul–Aug

ELODEA

Per, rooting at nodes, gen submersed, dioecious or some fls bisexual. **ST** slender, gen branched. **LVS** opposite or whorled, gen 3 per whorl, sessile; blade margin minutely, finely toothed. **INFL**: fls solitary, axillary, sessile or peduncled; bract gen notched. **STAMINATE FL** floating, deciduous or not; perianth tube elongated, slender, peduncle-like; stamens 3–9. **PISTILLATE FL** floating; perianth tube elongated, slender, peduncle-like; style slender, stigmas 3, simple or 2-lobed. **FR** cylindric to ovoid. **SEEDS** several. ± 12 spp.: temp and trop Am. (Greek: of marshes) [St. John, 1965 Rhodora 67:1–35,155–180]

1. Lf 1–4 mm wide, tip obtuse or abruptly pointed; staminate fl not deciduous, not free-floating; staminate sepals 3.5–5 mm, petals ± 5 mm; pistillate sepals ± 2–3 mm . . . . *E. canadensis*
1' Lf 0.3–2 mm wide, tapered to acute tip; staminate fl deciduous, free-floating; staminate sepals ± 2 mm, petals < 1 mm or 0; pistillate sepals ± 1 mm . . . . *E. nuttallii*

E. canadensis Rich.(p. 539) COMMON WATERWEED **LVS** gen crowded at st tips; middle and upper 3 per whorl; blade gen 9–15 mm, 1.5–3 mm wide, linear, tip obtuse or abruptly pointed.

STAMINATE INFL: bract ovoid to elliptic, swollen; fl not deciduous, not free-floating. **STAMINATE FL**: sepals 3.5–5 mm, ± 4 mm wide; petals ± 5 mm. **PISTILLATE FL**: sepals ± 2–3 mm. **SEED** glabrous. 2*n*=24,48. Shallow water, ditches, sloughs, ponds, lakes; 300–2600 m. NCoRO, CaRH, SNH, GV, SnFrB, SnGb, SnBr, **GB**; to B.C., e US; naturalized in Eur. Jul–Aug

E. nuttallii (Planchon) H. St. John (p. 539) NUTTALL'S WATERWEED **LVS** gen not crowded at st tips; middle and upper 3(4) per whorl; blade gen 6–13 mm, 0.3–1.5(2) mm wide, linear

to narrowly lanceolate, tapered to acute tip. **STAMINATE INFL**: bract ovoid, abruptly pointed; fl deciduous, free-floating. **STAMINATE FL**: sepals ± 2 mm, ± 1.5 mm wide; petals gen 0 or ± 0.5 mm. **PISTILLATE FL**: sepals ± 1 mm. **SEED** short-soft-hairy. 2*n*=48. Shallow water, streams, lakes, ponds, ditches; 500–2800 m. KR, SNH, ScV, SnBr, **GB**; to WA, c US. Jul–Aug

HYDRILLA

1 sp. (Greek: growing in water)

H. verticillata (L.f.) Caspary (p. 539) Per, rooting at nodes, submersed, dioecious. **STS**: horizontal, tuber-like; erect sts elongate, much-branched. **LVS** whorled, gen 4–8 per whorl, sessile, 1–2 cm, 1.5–2 mm wide, oblong, margin clearly, sharply toothed, mid-rib ± keeled below, rough, with tooth-like, conic bumps, tip gen sharp-pointed. **STAMINATE INFL**: fl 1, bract at base of elongated perianth tube, spheric. **PISTILLATE INFL**: fl 1, bract at base of elongated perianth tube, ± cylindric. **STAMINATE FL** deciduous, free-floating; perianth parts 6 in 2 whorls, 3–5 mm; stamens 3. **PISTILLATE FL** persistent, floating; perianth parts 6 in 2 whorls, 3–5 mm; stigmas 3. **FR** 5–6 mm, ± fusiform. Ditches, canals, ponds, reservoirs, lakes; < 200 m. NCoRI, n&c SNF, ScV, SCo, **D**; to s US, C.Am.; native to Eurasia. NOXIOUS WEED

NAJAS

Ann, aquatic, submersed, sometimes mat-like, monoecious or dioecious. **STS** several, much-branched, slender. **LVS** simple, cauline, opposite or appearing ± whorled; sheath gen wider than blade, expanded abruptly at junction with blade; blade gen linear, margin entire to spiny. **INFL** axillary; fls 1–few, clustered, inconspicuous. **STAMINATE FL**: perianth 0; stamen 1, anther opening irregularly, subsessile, subtended by 2 minute involucres, inner membranous, flask-shaped, outer cup-like, tubular, or with free scales. **PISTILLATE FL**: perianth part 0–1; ovary 1, chamber 1, ovule 1, style short, stigmas 2–4, linear. **FR**: achene, fusiform; outer wall thin, ± translucent. ± 50 spp.: ± worldwide. (Greek: water nymph) [Shaffer-Fehre 1991 Bot J Linn Soc 107:189–209]

1. Lf entire to minutely toothed (at 20× magnification), flexible, 0.5–1 mm wide, light to olive-green
or reddish; internodes not spiny; fr gen < 3 mm, < 2 mm wide; monoecious . . . . *N. guadalupensis*
1' Lf coarsely spine-toothed, stiff, 1–3 mm wide, bright green; internodes often minutely spiny; fr gen <<
5 mm, 2–3 mm wide; dioecious . . . . *N. marina*

N. guadalupensis (Sprengel) Magnus (p. 539) COMMON WATER-NYMPH Monoecious. **ST** < 6 dm, gen much-branched. **LVS** 1–2.5 cm, 0.5–1 mm wide, gen ± evenly spaced, flexible; sheath rounded to obtuse at junction with blade, few-toothed; blade margin minutely few-toothed, tip acute. **STAMINATE FL** 2–3 mm; anther 4-chambered. **PISTILLATE FL** 2–3 mm; stigmas 2–3. **FR**: surface dull, clearly net-like, pitted. 2*n*=12,36,42,48,54,60. Ponds, lakes, irrigation ditches, reservoirs; < 1200 m. NCoRO, GV, CW, SCo, **D**; to OR, e US, S.Am. Jul

N. marina L. (p. 539) HOLLY-LEAVED WATER-NYMPH Dioecious. **ST** < 4 dm, branched from base; internodes often minutely spiny. **LVS** 1–4 cm, 1–3 mm wide, ±evenly spaced, stiff; sheath rounded at junction with blade, entire or minutely few-toothed; blade margin and lower surface coarsely spine-toothed. **STAMINATE FL** 3–4 mm; anther 4-chambered. **PISTILLATE FL** 3–4 mm; stigmas 3. **FR**: surface smooth (finely net-like at 20× magnification). Ponds, lakes, marshes, rivers; < 1000 m. NCoR, s SNF, CCo, SCoR, SCo, SnBr, PR, **D**; to n-c US, Baja CA, Eurasia, Pacific Islands. Jul–Aug

IRIDACEAE IRIS FAMILY

Elizabeth McClintock, except as specified

Per, bulbed, cormed, or rhizomed. **ST** gen erect. **LVS** gen basal (a few cauline), 2-ranked, ± linear, gen grass-like, gen sharply folded along midrib; bases overlapping, sheathing. **INFL**: spike, raceme, panicle, ± terminal, or fls solitary; bracts ± like lf bases, sheathing. **FL** gen bisexual, gen radial; hypanthium fused to ovary; perianth parts gen fused into tube above ovary, gen petal-like, in 2 series of 3, outer (sepals) gen ± like inner (petals); stamens 3, gen attached to sepals, filaments fused below into a tube or not; ovary inferior, 3-chambered, placentas gen axile, style 1, each of 3 branches entire or 2-branched or -lobed, petal-like or not, with stigma on under surface instead of at tip. **FR**: capsule, loculicidal. **SEEDS** few–many. 80 genera, ± 1500 spp.: worldwide, esp Afr; many cult (e.g., *Iris, Gladiolus, Crocus, Freesia*). Key to genera adapted by Margriet Wetherwax.

1. Style branches petal-like, conspicuous, covering stamens; sepals unlike petals in shape, size, and position . . . **IRIS**
1' Style branches not petal-like, not conspicuous, not covering stamens; sepals ± like petals . . . . **SISYRINCHIUM**

IRIS IRIS

Anita F. Cholewa & Douglass M. Henderson

Per; rhizome creeping or ± tuber-like. **INFL**: fls 1–many. **FL**: perianth parts clawed, sepals (wider, spreading or reflexed) unlike petals (gen narrower, erect); style branches ± petal-like, arching over stamens, each with flat, scale-like stigma on surface facing stamen, just below gen 2-lobed tip (crest). **SEEDS** compressed, pitted. Perhaps 150 spp.: gen n temp. (Greek: rainbow, from fl colors) [Lenz 1958 Aliso 4:1–72; Clarkson 1959 Madroño 15:115–122] ❀ Pacific Iris hybrids; CVS.

I. missouriensis Nutt. (p. 539) WESTERN BLUE FLAG Rhizome 20–30 mm diam. **ST** 2–5 dm, sometimes branched. **LF** 3–9 mm wide; base sometimes purplish. **INFL**: fls 1–2; lowest 2 bracts opposite, gen scarious, outermost 4–8 cm, 5–7 mm wide. **FL**: perianth pale lilac to whitish with lilac-purple veins, tube < 12 mm, sepals < 6 cm, 2 cm wide; style branches < 25 mm, crests averaging 8 mm, stigmas 2-lobed. 2*n*=38. Moist, grassy places; 900–3400 m. NCoR, SN, SCoRI, TR, PR, **GB**; much of w N.Am, n Mex. ❀ IRR:1,**2–7,9**,10,11,14,**15**,16,**18**,19–21. May–Jun

SISYRINCHIUM

Anita F. Cholewa & Douglass M. Henderson

Per; rhizomes compact. **STS** single or tufted, gen ± flat and winged or rounded, sometimes with lf-bearing nodes well above basal lvs, each with 1 or more fl-branches. **LF** narrow, grass-like. **INFL**: fls in umbel-like clusters; bracts 2, equal in length or not, margins translucent. **FL** ephemeral; perianth reddish purple, bluish, violet, yellow, rarely white, parts ± alike, but outer gen wider; filaments ± completely free to ± completely fused. **SEEDS** ovoid, smooth or pitted. ± 70 spp.: w hemisphere. (Name used by Theophrastus for Iris-like pl) [Henderson 1976 Brittonia 28:149–176]

1. St branched
   2. Translucent margin of inner bract not extending beyond green tip; perianth gen blue-violet, rarely
      pale blue or white . . . . *S. bellum*
   2' Translucent margin of inner bract extending above green tip as 2 rounded teeth; perianth gen pale blue
      . . . . *S. funereum*
1' St simple, unbranched
   3. Inner bracts with translucent margins wider at tip or extending beyond as rounded teeth; outer
      perianth parts widely wedge-shaped . . . . *S. halophilum*
   3' Inner bract with uniformly narrow translucent margins; outer perianth parts elliptic to oblanceolate
      . . . . *S. idahoense* var. *occidentale*

S. bellum S. Watson (p. 539) BLUE-EYED-GRASS **STS** gen tufted, < 64 cm, sometimes 5.3 mm wide, almost always with lf-bearing nodes. **INFL**: translucent margins of inner bract wider just below tip, not extending above tip. **FL**: perianth 10.5–17 mm, deep bluish purple to blue-violet, or pale blue, rarely white, tips truncate to notched, with a small point. Common. *n*=16. Open, gen moist, grassy areas, woodlands; gen < 2400 m. **CA**; OR. [*S. eastwoodiae* E. Bickn.; *S. greenei* E. Bickn.; *S. hesperium* E. Bickn.] Gen self-incompatible. ❀ DRN:4,5,**6,7**,8,**9,14–18,20–24**&IRR:3,**19**&SHD:10,11;CVS. Mar–May

S. funereum E. Bickn. **STS** tufted, < 70 cm, almost always with lf-bearing nodes, pale green, glaucous. **INFL**: translucent margins of inner bract widest at tip, extending above tip as 2 rounded teeth. **FL**: perianth 10–15 mm, gen pale blue, tips truncate to sometimes notched, with a small point. *n*=16. Gen strongly alkaline margins of wet areas; < 800 m. **ne DMoj (Death Valley region)**; adjacent NV. Self-incompatible. ❀ IRR, ALKALINE:**8–14,19–23**. Feb–Apr

S. halophilum E. Greene (p. 539) **STS** tufted, < 40 cm, gen without lf-bearing nodes, glaucous. **INFL**: translucent margins of inner bract widest at tip, extending above tip as 2 rounded teeth. **FL**: perianth gen 9–12 mm, gen medium blue to blue-violet, outer parts gen widely tapered at base, tips truncate to rarely notched, with a small point. *n*=16. Gen moist, alkaline meadows; < 2550 m. **SNE, adjacent DMoj**; to UT. [*S. leptocaulon* E. Bickn.] Self-incompatible. ❀ IRR,ALKALINE:**8–14**,15–17,**18–23**. May–Jun

S. idahoense E. Bickn. (p. 539) **STS** tufted, < 45 cm, sometimes with 1 or more lf-bearing nodes, green to glaucous. **INFL**: translucent margins of inner bract ± uniformly narrow. **FL**: perianth 8–17 mm, gen blue to blue-violet, outer parts gen narrowly elliptic or oblanceolate, tips rounded to deeply notched, with a small point. Open, moist, grassy places; 900–3150 m. KR, CaR, SN, **GB**; to B.C., MT, Colorado. Highly variable, self- or cross-pollinating. Vars. intergrade. 2 vars. in CA.

var. *occidentale* (E. Bickn.) D. Henderson **ST**: upper margin gen entire. **INFL**: margins of outer bract fused gen in basal 4.5 mm. **FL**: perianth gen < 13 mm. *n*=32. Habitat of sp. SN, **SNE**; to e WA, MT, Colorado. [*S. occidentale* E. Bickn.] ❀ IRR,DRN:**6,7**,14–16 &SUN:1–3.**4,5**,17.

JUNCACEAE RUSH FAMILY

Janice Coffey Swab

Ann, per, gen from rhizomes. **ST** round or flat. **LVS** gen mostly basal; sheath margins fused, or overlapping and gen with 2 ear-like extensions at blade junction; blade round, flat, or vestigial, glabrous or margin hairy. **INFL**: head-like clusters or single fls, variously arranged; bracts subtending infl 2, gen lf-like; bracts subtending infl branches 1–2, reduced; bractlets subtending fls gen 1–2, gen translucent. **FL** gen bisexual, radial; sepals and petals similar, persistent, green to brown or purplish black; stamens gen 3 or 6, anthers linear, persistent; pistil 1, ovary superior, chambers gen 1 or 3, placentas 1 and basal or 3 and axile or parietal, stigmas gen > style. **FR**: capsule, loculicidal. **SEEDS** 3–many, often with white appendages on 1 or both ends. 9 genera, 325 spp.: temp, arctic, trop mtns. Fls late spring to early fall.

1. Lvs stiff, glabrous, sheath open; fr 3-chambered (rarely 1 in minute pls); seeds many . . . . **JUNCUS**
1' Lvs flexible, margins gen hairy, sheath closed; fr 1-chambered; seeds 3 . . . . **LUZULA**

JUNCUS RUSH

Ann, per; rhizome (if any) gen with scale-like lvs. **ST** gen cylindric or flat. **LF**: blade well developed and cylindric or flat, or reduced to small point; crosswalls often present (pull fresh blade apart lengthwise to see or slide lf between fingers to feel); appendages often present at blade-sheath junction. **INFL** gen terminal (appearing lateral when pushed aside by lowest infl bract); bractlets 0–2. **FLS**: stamens gen 3 or 6 (2 in some very small ann taxa); pistil 1, ovary chambers 1 or 3, placentas axile or parietal, stigmas gen 3(2). **SEEDS** many. 225 spp.: worldwide, esp n hemisphere. (Latin: to join or bind, from use of sts) [Ertter 1986 Mem NY Bot Gard 39:1–90] Key to species by Barbara Ertter.

1. Ann, gen < 10 cm; lvs narrow, gen < 1 mm wide, flat or channeled, crosswalls 0
  2. Fls 1 per node in a branched infl; stamens gen 6; lvs mostly cauline . . . . *J. bufonius*
    3. Perianth 4–7 mm long . . . . var. *bufonius*
    3' Perianth < 4 mm long . . . . var. *occidentalis*
  2' Fls solitary or 2–7 in clusters in an unbranched terminal infl; stamens 2– 4; lvs all basal
    4. Seeds with distinct longitudinal ridges; fls gen > 1 per st . . . . *J. tiehmii*
    4' Seeds with longitudinal and horizontal ridges equally faint; fls 1 per st
      5. Perianth > fr, curving inward, gen < 2.3 mm; fls gen 2-merous; st hair-like, 0.1—0.2 mm thick; fr
        aging bright reddish . . . . *J. bryoides*
      5' Perianth < or = fr, erect to recurved, gen > 2.3 mm long; fls gen 3-merous; st stouter, gen > 0.2 mm
        thick; fr aging orange-red . . . . *J. hemiendytus* var. *hemiendytus*
1' Per, gen > 10 cm; lvs cylindrical or flat, crosswalls 0, present, or incomplete
  6. Lvs rigidly sharp-pointed, cylindric; infl of small clusters subtended by one bract; bractlet 0
    7. Perianth segments ± rounded, broadly margined, 2–4 mm long; sepals thin . . . . *J. acutus* ssp. *leopoldii*
    7' Perianth segments acute to acuminate, narrowly margined, 5–6 mm long; sepals firm . . . . *J. cooperi*
  6' Lvs ± flexible, not rigidly sharp-pointed, flat or cylindric; infl ± open; fls solitary, each subtended by
    2 bractlets
    8. Fls 1 per node; infl appearing to be lateral, infl bract st-like; lvs gen reduced to basal sheaths
      (subg. *Genuini*)
      9. Fls gen 1–3 per infl; seed with conspicuous appendages; ± dry exposed rocky areas > 2000 m . . . . . *J. parryi*
      9' Fls gen 2–many per infl; seed appendage 1and minute, or 0; wet or dry areas, gen < 2200 m
        (exc. *J. mexicanus*)
        10. Plants forming dense tufts; stamens 3, filaments > anthers . . . . *J. effusus* var. *pacificus*
        10' Plants rhizomatous, forming open or dense stands; stamens 6, filaments << anthers
          11. Lf blades gen absent; st gen cylindric, not twisted, 3.5–11 dm . . . . *J. balticus*
          11' Lf blades present on at least some lvs; st flattened, twisted, 1–6 dm . . . . *J. mexicanus*
    8' Fls in headlike clusters; infl bract lf-like; lf blades gen present
      12. Lf blade flat
        13. Lf blade oriented with edge toward st, bases overlapping (*Iris*-like), crosswalls gen incomplete
          14. Anthers large, conspicuous, >> filaments . . . . *J. macrandrus*
          14' Anthers inconspicuous, gen < or = filaments
            15. Lf sheath margins membranous, prolonged into small ear-like appendages . . . . *J. saximontanus*
            15' Lf sheath appendages 0 or inconspicuous
              16. Infl gen of few, many-fld clusters; perianth dark brown to black; stamens gen 3; fr abruptly
                tapered to short beak; . . . . *J. ensifolius*
              16' Infl gen of many, few-fld clusters; perianth green to brown; stamens 6; fr gradually tapered
                to beak . . . . *J. xiphioides*
        13' Lf blade oriented with flat side toward st, crosswalls 0
          17. Lf sheath appendages 0 or obscure . . . . *J. orthophyllus*
          17' Lf sheath appendages well developed
            18. Infl clusters 1–9, 3–12-fld; sepals = to > petals . . . . *J. longistylis*
            18' Infl clusters 8–25, 3–5-fld; sepals < petals . . . . *J. macrophyllus*
      12' Lf blade cylindric — bases not overlapping; crosswalls gen complete
        19. Infl cluster gen 1, gen > 12-fld . . . . *J. mertensianus*
        19' Infl clusters 1–many, < 10 fld
          20. Anthers >> filament length . . . . *J. nevadensis*
          20' Anthers < to > filaments
            21. Lf, st surfaces conspicuously wrinkled . . . . *J. rugulosus*
            21' Lf , st surfaces not wrinkled
              22. Infl clusters many (<150) . . . . *J. dubius*
              22' Infl clusters gen < 25
                23. Infl gen ± spreading, clusters < 1 cm broad, 10–25-fld; perianth segments 2.5–4 mm, not
                  rigidly pointed; sepals . . . . *J. nodosus*
                23' Infl ± dense, clusters 1–1.5 cm broad, gen > 25-fld; perianth segments 3.5–5.6 mm,
                  rigidly pointed; sepals > petals . . . . *J. torreyi*

J. acutus L. ssp. ***leopoldii*** (Parl.) Snog. (p. 543) Per, cespitose, 50–140 cm; rhizome much-branched. **ST** rigid, hardened, cylindric. **LVS** basal, 40–120 cm, rigid; tip hard, sharp; sheath appendages firm. **INFL** appearing lateral, gen ± open; lowest bract cylindric, resembling st, < to = infl; branches uneven; clusters 2–4-fld, each subtended by 1 obvious, clasping bract; bractlets 0. **FL**: perianth 2–4 mm, margins membranous, sepals thin, obtuse, petals ± rounded; anthers 6, > filaments, reddish brown. **FR** gen ± 2 × perianth, nearly spheric, shiny brown. **SEED** irregular; appendages small. 2*n*=48. Moist saline places (salt marshes, alkaline seeps); gen < 300 m. CCo, SCo, s ChI, **DSon**; to AZ, Baja CA, S.Am, s Afr. [var. *sphaerocarpus* Engelm., spiny rush] ✤ IRRorWET:5,7,**8,9, 12–17,19–24**;also STBL;good barrier plant. May–Jun

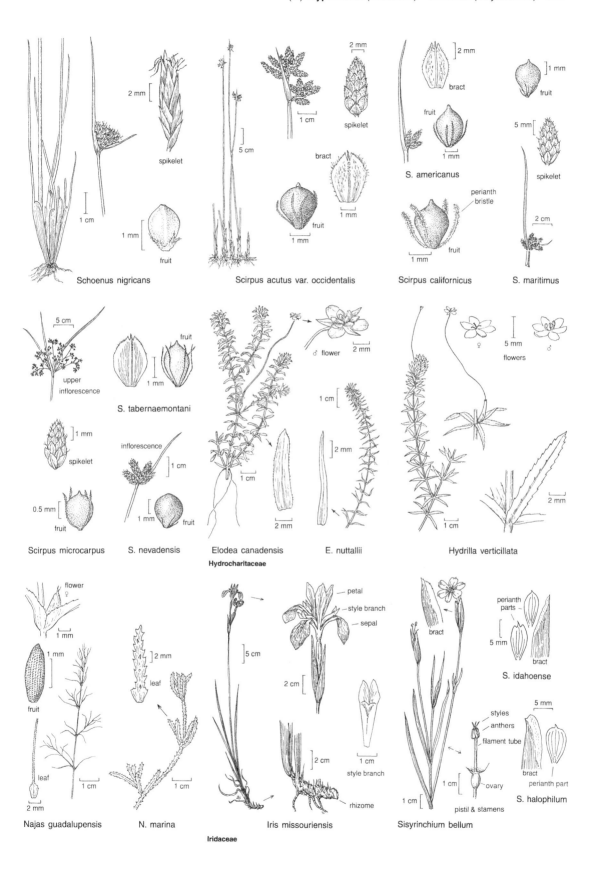

Schoenus nigricans

Scirpus acutus var. occidentalis

S. americanus

Scirpus californicus

S. maritimus

S. tabernaemontani

Scirpus microcarpus

S. nevadensis

Elodea canadensis

E. nuttallii

Hydrilla verticillata

Hydrocharitaceae

Najas guadalupensis

N. marina

Iris missouriensis

Sisyrinchium bellum

S. idahoense

S. halophilum

Iridaceae

J. balticus Willd. (p. 543) Per 35–110 cm; rhizome scaly, creeping, gen unbranched, slender to stout. **ST** 1–6 mm wide, gen cylindric. **LVS** basal; blades 0; sheaths variable, 2–15 cm. **INFL** appearing lateral, ± open; lowest bract cylindric, resembling st, gen >> infl; fls 5–50 or more; bractlets 2, membranous. **FL**: perianth segments 3–6 mm, sepals ± = to > petals, scarious margins of petals wider than those of sepals; stamens 6, filaments << anthers. **FR** ± = perianth; beak small but obvious. **SEED** 0.4–0.8 mm; appendages 0. 2n=40,80. Moist to rather dry places; gen < 2200 m. **CA**; to AK, e N.Am, S.Am, Eurasia. [var. *montanus* Engelm.] Highly variable, intergrading complex needing study. ✿ IRR,DRN:1–3,**4–7**,8–13,**14–24**;alsoSTBL;INV. May–Aug

J. bryoides F.J. Herm. (p. 543) Ann, densely cespitose, 0.1–2.2 cm, reddish brown in fr. **ST** hair-like (0.1–0.2 mm wide). **LVS** basal, 1/4 to = st; sheath appendages 0. **INFL**: fl 1 per st; bracts 1–2, < 1 mm. **FL**: perianth segments 4–8, 1–2.8 mm (sepals < to = petals), incurved over fr; stamens 2–4, filaments > anthers. **FR** < perianth, ovoid to elliptic; chambers 2–3. **SEED** 0.3–0.5 mm, not striate; appendages small. Common. Wet places, washes, meadows, granitic seeps; 600–3600 m. SNH, SCoRO, TR, PR, **W&I**; to OR, ID, Colorado, Baja CA. Gen self-pollinating. May–Aug

J. bufonius L. TOAD RUSH Ann, gen branched from base, 2–30 cm. **ST** gen ± 1 mm wide. **LVS** ± cauline, 1–3 per st, 0.5–1.5 mm wide. **INFL**: fls 1–few in small clusters, ± throughout pl; lowest bracts lf-like; bractlets 1–2.5 mm. **FL**: perianth segments 2–7 mm (sepals gen > petals), acuminate (or petals obtuse); stamens 6, filaments < to > anthers. **FR** < perianth, oblong to obovoid; tip acute. **SEED** 0.3–0.6 mm, ovoid to elliptic; appendages 0 or small. Moist (sometimes saline) open or disturbed places; < 3200 m. **CA**; ± worldwide. Detailed study of variation in N.Am pls needed. 3 vars. in CA.

var. *bufonius* (p. 543) Pl relatively large in all features. **INFL**: fls solitary at nodes. **FL**: perianth 4–7 mm, petals acuminate. **FR** < petals. 2n=108. Moist (sometimes saline) open or disturbed places; < 3200 m. **CA**; ± worldwide. ✿ STBL. Apr–Sep

var. *occidentalis* F.J. Herm. Pl relatively small in all features. **INFL**: fls solitary at nodes. **FL**: perianth 3–4 mm, petals acuminate. **FR** < petals. Dry pools, streamsides; < 3200 m. **CA**; to OR, ID, TX. [gen misidentified as *J. sphaerocarpus* Nees of Eur]

J. cooperi Engelm. (p. 543) Per, ± cespitose, 40–80 cm; rhizome short, thick, much-branched; roots large, spongy. **LVS** basal; blades short, stiff, cylindric, tips sharp. **INFL** appearing lateral; lowest bract cylindric, resembling st, > infl, tip sharp; branches unequal; clusters 2–10-fld; bracts within infl obvious, > cluster; bractlets white. **FL**: perianth segments 5–6 mm (sepals > petals), pale greenish straw, tips acute, firm; stamens 6, large, filaments < anthers. **FR** slightly > perianth, narrowly oblong, 3-angled. **SEED** with a conspicuous white ridge; appendages unequal. Uncommon. Alkaline places; < 600 m. **D**; NV. May

J. dubius Engelm. (p. 543) Per, gen densely matted, 15–70 cm; rhizome stout. **LVS**: basal blades 0; cauline lvs few, sheath appendages 4–6 mm, prominent, clear, blade cylindric, blade crosswalls complete but obscure. **INFL**: lowest bract gen 1–2 cm, inconspicuous; branches spreading; clusters gen many, 4–10-fld; bracts and bractlets clear. **FL**: perianth segments 2.5–3 mm, ± equal, bristle-tipped; stamens 6, filaments << anthers. **FR** barely > perianth, 3-sided. **SEED** obovate; appendages minute. Wet places; < 2000 m. CaRF, c&s SNF, SnFrB, SCoRO, PR, **DMoj**; OR. ✿ TRY;GRCVR. Apr–Aug

J. effusus L. Per, cespitose, 60–130 cm; rhizome stout, branched. **ST** 1–3.5 mm wide at base. **LVS** basal; sheaths 5–15 cm; blades 0. **INFL** appearing lateral, open; lowest bract cylindric, resembling st, >> infl; fls gen many, single; bractlets 2 per fl. **FL**: perianth segments 1.8–4.2 mm, ± equal, gen ascending, very pale to dark; stamens 3, filaments > anthers. **FR** ± = peri-

anth, obovoid, ± truncate. **SEED** 0.5 mm; appendage 1, minute. 2n=40. Wet places; < 2500 m. **CA (exc DSon)**; to B.C., e US, Mex, Eurasia. 4 vars. in CA.

var. *pacificus* Fern. & Wieg. (p. 543) **LF**: sheath uniformly dark, edges overlapping at top. **FL**: perianth 2.5–4.2 mm, firm when fresh, rigid when dry, green to pale brown. Many moist habitats; < 2500 m. **CA (exc DSon)**; to B.C. A large, s CA variant with dark lf sheaths may warrant status as var. ✿ IRRor WET:1–3,**4–7**,8,9, **14–24**. Jun–Aug

J. ensifolius Wikström (p. 543) Per 20–60 cm; rhizome slender, creeping. **LVS** mostly basal; bases overlapping; sheath appendages obscure; blade flat, with edge toward st, 2–5 mm wide, gen curved, crosswalls incomplete, tip long. **INFL**: lowest bract ± 1/2 infl length, gen curved; clusters 2–7, many-fld, hemispheric. **FL**: perianth segments ± 3 mm, ± equal, dark brown to black; stamens gen 3, ± 1/2 perianth length, filaments > anthers. **FR** barely > perianth, oblong, abruptly short-beaked. **SEED** ± 0.5 mm, widely fusiform; appendages 0. 2n=40. Common. Wet places; < 2800 m. NW, CaR, SN, CCo, SnBr, **W&I**; to AK, w Can, UT, Mex. ✿ STBL. Jul–Aug

J. hemiendytus F.J. Herm. Ann, densely cespitose, 0.1–3.2 cm. **ST** 0.1–0.5 mm wide. **LVS** basal, < 1.8 cm; sheath appendages 0. **INFL**: fl gen 1 per st; bracts 0–2, < 2 mm, ovate to rounded, not sheathing st. **FL**: perianth segments 4–6, 1.9–3.5 mm (sepals = or > petals), dull; stamens 2–3, filaments > anthers, (anthers < 0.7 mm). **FR** < to > perianth, obovoid to oblong; chambers 2–3. **SEED** 0.3–0.55 mm, not striate; appendages small. Damp open areas, esp vernally wet places; 400–3400 m. NW, SN, SnBr, **GB**; to WA, ID, UT. Gen self-pollinating. 2 vars. in CA.

var. *hemiendytus* (p. 543) **ST** 0.1–0.3 mm wide, not wider below fl. **INFL**: bracts 1–2. **FR** > perianth. 2n=32. Damp open areas, esp vernally wet places; 400–3200 m. NW, SN, SnBr, **GB**; to WA, ID, UT. Jun–Jul

J. longistylis Torrey (p. 543) Per, loosely cespitose, 20–60 cm. **ST** slender. **LVS** mostly basal; sheath appendages obvious, 0.5–2 mm, often truncate; blade with flat side toward st, < 1/2 st length, 1–3 mm wide, crosswalls 0; cauline lvs 1–3. **INFL**: lowest bract 1–2 cm, membranous; clusters 1–9, 3–12-fld. **FL**: perianth segments 5–6 mm, ± equal (or sepals > petals), middle dark green, scarious margin wide; stamens 6, filaments < anthers. **FR** < perianth, 3-angled, brown; beak short. **SEED** 0.5 mm, narrow; appendages short. 2n=40. Moist places in conifer forests; 1800–2900 m. SNH, TR, PR, Wrn, **W&I**; to B.C., e N.Am, NM. Jul–Aug

J. macrandrus Cov. (p. 543) Per, cespitose, 30–50 cm; rhizome creeping. **LVS** mostly basal; bases overlapping; sheath appendages 0; blade flat, with edge toward st, 2–3 mm wide, crosswalls gen incomplete. **INFL**: lowest bract << infl; clusters 3–many, 3–5-fld. **FL**: perianth 3–4 mm, dark purple-brown; stamens 6, filaments << conspicuous anthers. **FR** < perianth, widely oblong; beak abruptly pointed. **SEED** squarish; appendages small. Wet places, montane conifer forests; 1200–2800 m. SNH, TR, **W&I**. Jul–Aug

J. macrophyllus Cov. (p. 543) Per 20–100 cm; rhizome spreading. **LVS** mostly basal; sheath appendages 1.5–3 mm, membranous; blade with flat side toward st, often = st, 1.5–3 mm wide, somewhat channeled, midrib prominent; cauline lvs 1–2, thick. **INFL**: lowest bract lf-like, < infl; branches ascending; clusters 8–25, 3–5-fld. **FL**: perianth segments 5–6 mm (sepals < petals), margins membranous; stamens 6, filaments << anthers. **FR** gen < to ± = perianth, obovoid, short-beaked, shiny brown. **SEED** plump; appendages 0. Uncommon. Wet slopes; < 2600 m. SCoRO, TR, **DMtns**; to AZ, Baja CA. May–Aug

J. mertensianus Bong. (p. 543) Per 15–45 cm; rhizome vertical, stout. **ST** flat. **LF**: cauline lvs 2–3; sheath appendages prominent, rounded to acute, opaque; blade ± flat with edge toward st, crosswalls complete but gen obscure. **INFL**: lowest bract short-sheathing, wide, narrow tip > infl; cluster gen 1, gen > 12-fld; bractlets short-awned, dark brown, opaque. **FL**: perianth segments 3–4 mm, soft, shiny brownish black, sepals narrowly acuminate, petals acute; stamens 6, filaments < to > anthers. **FR** ± = perianth, oblong; tip notched. **SEED** ± 0.5 mm, ovate-lanceolate; appendages minute. $2n$=40. Common. Alpine and subalpine meadows, streambanks, lake margins; 1200–3500 m. KR, NCoRI, CaRH, SNH, Wrn, **W&I**; to AK, SD, Colorado. Important forage for sheep. ❀ WET:**1–3**,18. Jul–Aug

J. mexicanus Willd.(p. 543) Per 10–60 cm; rhizome heavy. **ST** erect or gen spirally twisted, flattened, slender. **LVS** basal; sheaths loose, appendages short, firm; upper sheaths gen bearing 5–20 cm blades that resemble st (blades less common farther n). **INFL** appearing lateral, ± compact; lowest bract 1–25 cm, cylindric, resembling st; fls 2–many. **FL**: perianth segments 3–5.5 mm (sepals > petals) gen acuminate, color variable, midstripe varying in width and intensity, margins gen clear; stamens 6, filaments << anthers. **FR** < to > perianth, ovoid, 3-angled; beak 0.3 mm. **SEED** 0.5–0.7 mm; appendages 0. Common. Coast to montane meadows; < 3800 m. **CA** (exc GV); to WA, Colorado, TX, S. Am. ❀ STBL. May–Aug

J. nevadensis S. Watson (p. 543) Per gen 10–50 cm; rhizome creeping. **ST** slender. **LF**: sheath appendages 2–3 mm, membranous; blade gen < 2 mm wide, cylindric, crosswalls complete, obvious or obscure. **INFL** variable; lowest bract inconspicuous; clusters gen 1–4 (many), < 10-fld. **FL**: perianth segments 3–3.5 mm, ± equal, dark brown; stamens 6, filaments << anthers. **FR** < to ± = perianth, oblong, shiny brown; tip abruptly beaked. **SEED** 0.6 mm, ± spheric, brown; appendages 0. Common. Mtn meadows, streambanks; 1200–3300 m. KR, NCoRH, NCoRI, CaRH, SNH, SnGb, SnBr, **GB**; to B.C., WY. [*J. mertensianus* ssp. *gracilis* (Engelm.) F.J. Herm.; *J. phaeocephalus* Engelm. var. *gracilis* Engelm.] Important forage for cattle and horses. Jul–Aug

J. nodosus L. (p. 543) KNOTTED RUSH Per 15–60 cm; rhizome creeping, slender, tuber-bearing. **ST** slender. **LF**: sheath appendages 0.5–1 mm, rounded, firm; blade cylindric, upper > infl, crosswalls complete, prominent. **INFL**: lowest bract = to > infl; clusters 2–20, 10–25-fld, spreading, spheric. **FL**: perianth segments 2.5–4.1 mm, acuminate, ± equal; stamens 6, filaments > anthers. **FR** > perianth, slender, sharply 3-angled, long-tapered. **SEED** 0.5 mm; appendages small. $2n$=40. UNCOMMON. Streambanks, lake shores, wet meadows; < 1700 m. se SNH, **W&I, n DMtns**; scattered across US, s Can. Jul–Sep

J. orthophyllus Cov. (p. 543) Per 20–50 cm; rhizome scaly, hard, dark, creeping. **LVS** mostly basal; sheath appendages 0 or minute; blade with flat side toward st, << to ± = st, 2–6 mm wide, crosswalls 0; cauline lvs 1–3, above middle. **INFL** open; lowest bract inconspicuous; branches papillate; clusters 1–many, 6–10-fld, hemispheric; bractlets prominent, ± 1/2 perianth. **FL**: perianth segments 5–6 mm, minutely papillate, bristle-tipped, middle green, margins brown, sepals < petals; stamens 6, filaments < anthers. **FR** slightly < perianth; tip flat; beak small. **SEED** 0.5 mm, brown; appendages minute. Inland wet places, esp meadows, streambanks in forest; 1200–3500 m. NCoRH, CaRH, SNH, TR, **DMtns**; to WA, ID, NV. Jul–Aug

J. parryi Engelm. (p. 543) Per, densely cespitose, 6–30 cm. **LVS** basal; sheaths many, appendages 0 (or obscure, membranous); blades on inner lvs only, 3–8 cm, thread-like. **INFL** appearing lateral; lowest bract gen> infl, cylindric; fls 1–3, sometimes sessile; bractlets 2 per fl, dissimilar. **FL**: perianth segments 5–7 mm (sepals > petals), pale brown, tips acuminate, veins ob-

vious in lower part; stamens 6, filaments << anthers. **FR** gen > perianth, 3-angled; tip acute. **SEED** 2 mm, ovate; appendages > seed body. Dry granitic slopes; 2000–3800 m. KR, SN, SnBr,**W&I**; to sw Can, Colorado. Jul–Aug

J. rugulosus Engelm. (p. 543) WRINKLED RUSH Per, often densely matted, 15–70 cm; rhizome stout, horizontal; surfaces prominently wrinkled. **LF**: basal blades 0; sheath appendages 4–6 mm, prominent, clear; cauline blades cylindric, crosswalls complete but obscure. **INFL**: lowest bract gen inconspicuous; branches spreading; clusters gen many (<150), 4–8-fld; bracts and bractlets clear. **FL**: perianth segments brown to brownish red, clear margins wide, sepals 2.5 mm, ± bristle-tipped, petals 2.5–3 mm, petal veins prominent at base; stamens 6, filaments ± = anthers. **FR** > perianth, 3-angled, long-beaked, bright brownish red. **SEED** 0.4 mm, plump, brown; appendages minute. $2n$=40. Common. Wet places; < 2100 m. s SNF, CCo, TR, PR, **DMtns**. ❀ TRY;STBL.

J. saximontanus Nelson (p. 543) Per 30–60 cm; rhizome stout, creeping. **LF**: bases overlapping; sheath margins membranous, appendages small but distinct; blade flat, with edge toward st, 2–5 mm wide, crosswalls gen incomplete. **INFL**: lowest bract short; clusters few–many, 4–25-fld. **FL**: perianth segments 2.5–3 mm (sepals > petals), lanceolate, pale to dark brown; stamens 6 (3), ± 2/3 perianth length, filaments > inconspicuous anthers. **FR** < to ± = perianth, oblong. **SEED** 0.4–1 mm; appendages minute to 0.3 mm. Wet places, montane conifer forest; 1500–2900 m. KR, CaRH, SNH, SnGb, SnBr, **W&I**; to AK, TX, Mex. [forma *brunnescens* (Rydb.) F. J. Herm] Jul–Aug

J. tiehmii B. Ertter (p. 549) Ann, densely cespitose, 0.5–6 cm. **ST** 0.1–0.2 mm wide. **LVS** basal, < 2.5 cm; sheath appendages 0. **INFL**: fls 1–7 per st; bracts 0.6–1.5 mm, acute or acuminate. **FL**: perianth segments 4–6, 1–2.9 mm, ± equal, pale green or pink; stamens 2–3, filaments > anthers, anthers < 0.7 mm. **FR** gen > perianth, obovoid to oblong, color of perianth. **SEED** 0.3–0.5 mm, obviously striate; appendages small. $2n$= 34. Bare, moist granitic sand of seeps, streamsides, meadows; 300–3100 m. SN, SCoR, TR, PR, **n SNE (Bodie Hills)**; to OR, ID, NV, Baja CA.

J. torreyi Cov. (p. 549) Per 30–100 cm; rhizome thin, creeping, bearing narrow tubers. **ST** stout. **LF**: sheath appendages prominent, thin; blade cylindric, crosswalls complete. **INFL**: lowest bract > infl; clusters 1–20, 25–80-fld, crowded. **FL**: perianth segments slender, tapered to rigid points, sepals 4–5.6 mm, petals 3.5–5 mm; stamens 6, filaments > anthers. **FR** ± > perianth, thin, 3-angled, tapered to point, with seeds only below middle. **SEED** 0.4–0.5 mm, oblong; appendages 0. $2n$=40. Meadows, moist woodlands; 0–1800 m. SW, **GB, D**; to s Can, much of US. Jul–Sep

J. xiphioides E. Meyer (p. 549) Per 40–80 cm; rhizome stout, creeping. **LF**: sheath appendages obscure; blade 3–12 mm wide, flat, with edge toward st, curved, crosswalls gen incomplete, tip long. **INFL**: lowest bract < 1/2 infl length; clusters many, 3–10-fld. **FL**: perianth segments 3–3.5 mm, very narrow (revealing fr between), green to brown; stamens gen 6, filaments = to > inconspicuous anthers. **FR** = to > perianth, oblong, gradually tapered to beak. **SEED** ± 0.5 mm; appendages minute. $2n$=40. Wet places; 0–2100 m. NW, CW, SW, GV, Teh, SNF, s SNH, **W&I, DMtns**; OR, AZ, Baja CA. ❀ STBL;INV. May–Oct

LUZULA HAIRY WOOD RUSH

Per; rhizome often short, vertical. **ST** round. **LVS** mostly basal, reduced upward; sheath closed; margin and sheath opening gen with long, soft hairs; blades flat or channeled, veins indistinct, tips often thick. **INFL**: head-like clusters or panicles of separate fls; bractlets 1–3, margins often hairy. **FL**: stamens 6; pistil 1, chamber 1, placenta basal. **SEEDS** 3, plump, elliptic, often with a distinct ridge, sometimes attached to placenta by tuft of hairs. 80 spp.: worldwide, esp n hemisphere. (Latin: light; Italian: glowworm) When present, fleshy seed appendage (outer seed coat) adapts large seeds to ant dispersal.

L. spicata (L.) DC. (p. 549) Pl densely cespitose, 3–33 cm. **ST**: bases thick, extending several cm into soil, reddish. **LF**: sheath opening densely hairy; blade erect, 2–15 cm, 1–4 mm wide, linear, channeled, tip not thickened. **INFL**: panicle of dense, nodding, spike-like clusters, each 1–25 mm, often interrupted by 10–70 mm; lowest bract gen > infl; bracts and bractlets = to > fls, clear, margin hairy, tip narrow and extended. **FL**: perianth segments 2–2.5 mm (sepals > petals), bristle-pointed, brown with clear margins or very pale throughout. **FR** gen < perianth, round with ± acute tip, pale to dark brown. **SEED** 1 mm, brown; appendage 0.2 mm. 2*n*=24. Alpine slopes, subalpine forests; 2900–3700 m. SNH, Wrn, **SNE (Sweetwater Mtns)**; circumpolar, to AK, high mtns of N.Am, Eurasia. [ssp. *saximontana* Á. Löve & D. Löve; var. *nova* Smiley] Jul–Aug

JUNCAGINACEAE ARROW-GRASS FAMILY

Robert F. Thorne

Ann or per from rhizomes, submersed or emergent, sometimes dioecious or pls with some unisexual fls. **ST** short, erect, ± scapose. **LVS** basal or cauline, alternate, gen narrowly cylindric; sheath open, gen liguled. **INFL**: spike or raceme (terminal) or fls solitary in axils; bracts 0. **FL**: perianth parts gen 6 in 2 whorls (exc 0–1 in *Lilaea*), free, greenish; stamens gen 1, 3, or 6, filament short, ± fused to inner perianth parts, anthers elongate, dehiscing outward; pistil 1 (simple) or seemingly so (ovaries 3, 4, or 6, fused to central axis, each with 1 chamber and 1 style), ovule 1 per chamber, style short and plumose or long and thread-like. **FR**: follicle or nutlet. 5 genera, ± 20 spp.: temp and circumboreal. Keys adapted by Margriet Wetherwax.

1. Perianth parts 0 or 1; stamen 1; ovary chamber (and seed) 1; fls bisexual and staminate (many on emergent spike) and pistillate (few, axillary, submersed) . . . . **LILAEA**
1' Perianth parts 6; stamens gen 6; ovary chambers (and seeds) 6; fls bisexual (on spike or raceme) . . . . **TRIGLOCHIN**

LILAEA

1 sp. (A. Raffeneau-Delile, French botanist, 1778–1850)

L. scilloides (Poiret) Hauman (p. 549) FLOWERING-QUILLWORT Ann from fibrous roots, aquatic, gen emergent. **ST** erect, short, obscure. **LVS** basal, tufted, 5–20(45) cm, 1–5 mm wide; sheath 3–10 cm, translucent; blade tip acute to pointed. **INFL**: bisexual and staminate fls in erect, gen emergent spike, 6–20 cm; pistillate fls solitary in submersed axils, enclosed by lf sheaths. **BISEXUAL FL**: perianth part 1, 2–3 mm; stamen 1; ovary 1, chamber 1, ovule 1, style short. **STAMINATE FL**: perianth part 1, 2–3 mm; stamen 1. **PISTILLATE FL**: perianth 0; ovary 1, chamber 1, ovule 1, style 6–20 cm, thread-like, stigma floating. **FR**: nutlet, 2–10 mm, ribs 25–30; tip beaked. 2*n*=12. Vernal pools, ditches, streams, ponds, lake margins: < 1700 m. NCo, NCoRI, SN, GV, CW, SCo, PR, **GB**; to w Can, MT, Mex, Chile; naturalized in Australia. Sometimes treated in Lilaeaceae. ❁ WET,fresh water margins & mud;SUN:**1–3**,4,5,**6–9**,10–12,**14–24**. Mar–Oct

TRIGLOCHIN ARROW-GRASS

Per from rhizomes in CA, terrestrial or aquatic. **ST** erect, short, obscure. **LVS** basal, ± tufted; sheath membranous; ligule entire to 2-lobed. **INFL**: raceme, scapose, narrowly cylindric, glabrous; pedicels short. **FL** bisexual; perianth parts 3–6, gen green, inner surface concave; stamens (1)3–6, subsessile, anthers wide; ovaries 3 or 6 (if 6, 3 sometimes sterile), 1-chambered, ± fused to central axis, style short, stigma papillate. **FR**: follicles 3–6, separating from axis. **SEED** 1, linear, ± flat or angled. ± 12 spp.: temp and circumboreal. (Greek: 3-pointed, from frs of some) TOXIC when fresh from cyanogenic compounds.

1. Lf < 20 cm, blade ± cylindric, ± 1 mm wide; ligule deeply 2-lobed, 0.5–1.5 mm . . . . . . *T. concinna* var. *debilis*
1' Lf < 70 cm, blade flat, gen 1.5–5 mm wide; ligule entire to slightly notched, (1)1.5–5 mm . . . . *T. maritima*

T. concinna Burtt Davy ARROW-GRASS Per from spreading to ascending rhizomes, < 3(4.5–6) dm. **LF** < 20 cm, ± 1 mm wide, subcylindric, ± fleshy; ligule 0.5–1.5 mm, 2-lobed. **INFLS** 2 or more per pl, > lvs; pedicels < 5 mm, ascending. **FL**: perianth parts gen 6, ± 1.5 mm; stamens gen 6; fertile ovaries 6. **FRS** 6 per fl, < 5 mm, oblong-ovoid, falling from axis. 2*n*=48,96. Salt marshes, alkaline meadows, seeps, mudflats, stream and lake margins; < 2500 m. NCo, s SNH, CCo, SCo, **GB**, **DMoj**; to Can, Baja CA. 2 vars in CA.

var. *debilis* (M.E. Jones) J. Howell (p. 549) Pl 3–6 dm. Alkaline meadows, seeps, mudflats, stream and lake margins; < 2500 m. s SNH, **GB**, **DMoj**; to Can, Colorado. ❁ TRY. May–Oct

T. maritima L.(p. 549) SEASIDE ARROW-GRASS Per from short, thick rhizomes, gen > 3 dm. **LF** < 70 cm, 1.5–5 mm wide, flat; ligule 1–5 mm, entire to slightly notched. **INFLS** 2 or more per pl, gen > lvs; pedicels 1–3(5) mm, ascending. **FL**: perianth parts 6, ± 2 mm; stamens gen 6; fertile ovaries 6. **FRS** 6 per fl, 3–5(7) mm, oblong to ovoid, falling from axis. 2*n*=12,24,30,36,48,60,96,120, 144. Uncommon. Saline, brackish, or alkaline marshes; < 2800 m. NCo, SN, CCo, SCo, SnBr, **GB**; circumboreal. ❁ WET or IRR, alkaline to fresh water margins & mud; SUN:**1–9**,**14–24**;STBL. Apr–Aug

LEMNACEAE DUCKWEED FAMILY

Wayne P. Armstrong

Per, floating aquatics, small, clonal, in dense populations; new pls produced in budding pouch at base or along

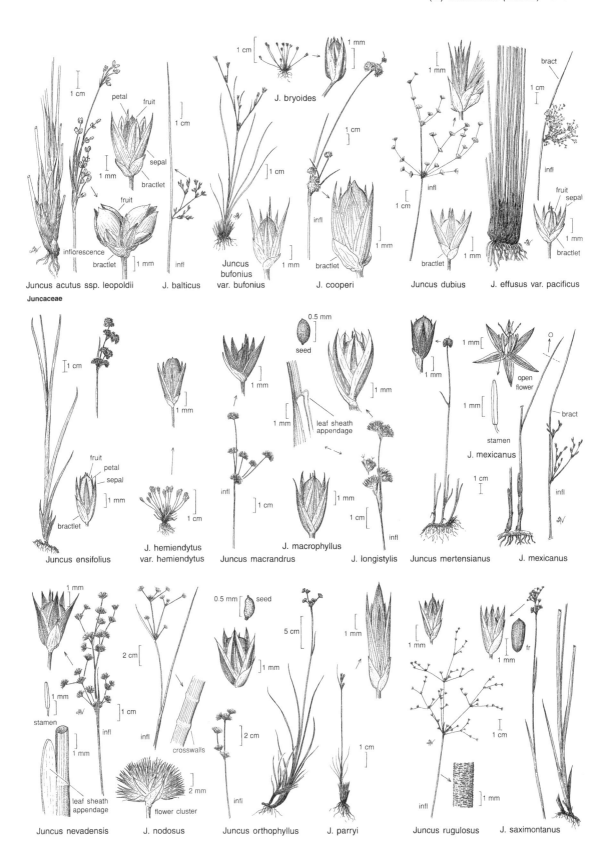

Juncus acutus ssp. leopoldii

J. balticus

J. bryoides

Juncus bufonius var. bufonius

J. cooperi

Juncus dubius

J. effusus var. pacificus

Juncaceae

Juncus ensifolius

J. hemiendytus var. hemiendytus

Juncus macrandrus

J. macrophyllus

J. longistylis

Juncus mertensianus

J. mexicanus

Juncus nevadensis

J. nodosus

Juncus orthophyllus

J. parryi

Juncus rugulosus

J. saximontanus

margins; may overwinter on bottom as dense, rootless, starch-filled daughter pl (winter bud); roots 0–many. **PL BODY** 0.4–10 mm, flat and tongue-shaped to spheric. **FL** 1, rarely seen, minute, appearing like 2–3 unisexual fls, often sheathed by minute membrane; perianth 0; stamens 1–2; pistil 1, simple, maturing before stamens. **FR** achene-like, sometimes winged. **SEEDS** 1–3, smooth or ribbed. Spp. best separated by chemistry and fr; clones vary; magnification, backlighting gen needed to identify vegetative pls. 4 genera, 34 spp.: worldwide; orn in pools, aquaria. [Landolt 1986 Veröff Geobot Inst ETH Stiftung Rübel Zürich 71] Recently treated in Araceae [French et al 1995 Monocotyledons: Systematics and Evolution, eds. Rudall et al, Royal Bot Gard, Kew]. ❀ May be used in still ponds, 0–2500 m, depending on individual tolerances; very invasive.

1. Roots 0 . . . . *Lemna trisulca*
1' Roots present
  2. Root 1;winter buds gen 0 . . . . **LEMNA**
  2' Roots gen 5–16; winter buds often produced in autumn . . . . **SPIRODELA**

LEMNA DUCKWEED

Pls gen in clusters of 2–many; root gen 1, gen 2–6 mm, sheath near base gen not winged. **PL BODY** gen 2–5 mm, flat, gen widely elliptic to oblong, pale to dark green, often ± red; veins 1–5 (visible in backlight); winter buds gen 0. **FLS** in 2 lateral pouches, sheathed by minute membrane. **FR** gen unwinged. **SEED** ribbed, gen smooth between ribs. 13 spp.: worldwide. (Greek: lake or swamp)

1. Pl body 6–10 mm, on tapered stalk, often connected in branched chains of 10–30 . . . . *L. trisulca*
1' Pl body 1–5 mm, appearing sessile, single or in clusters of 2–8
  2. Vein 1 per pl body (sometimes obscure); pl body gen 1–2.5 mm . . . . *L. minuta*
  2' Veins 3–5 per pl body; pl body 2–5 mm
    3. Root sheath with 2 obvious, wing-like appendages at base . . . . *L. aequinoctialis*
    3' Root sheath not winged
      4. Pl body widely elliptic to round, tip assymetric, lower surface gen much swollen or rounded . . . . . . *L. gibba*
      4' Pl body widely elliptic, tip symmetric, lower surface flat
        5. Upper surface gen smooth; lower surface green; winter buds 0 . . . . *L. minor*
        5' Upper surface gen with midline row of minute bumps; lower surface becoming ± red; winter buds
          dark green or brown in autumn . . . . *L. turionifera*

L. aequinoctialis Welw. Root-sheath winged. **PL BODIES** in pairs, 2–3.5 mm, obovate-elliptic, light green; base asymmetric; tip symmetric; upper surface tip and root node with minute bumps; lower surface light green, smooth. **SEED** cross-lined between ribs. Freshwater in hot regions; < 200 m. GV, CCo, SCo, **DSon**; worldwide. [*L. paucicostata* Hegelm.; not *L. perpusilla* Torrey] Fall, Dec

L. gibba L. (p. 549) **PL BODIES** gen in pairs or 3's, 3–6 mm, widely elliptic to round, glossy green or yellow-green mottled red; base symmetric; tip asymmetric; upper surface barely ridged, midline bumps gen 0; lower surface gen swollen, with enlarged air spaces often bordered in red. **FR** strongly winged. Common. Fresh or brackish water; gen < 1500 m. **CA**; worldwide. Variable: vegetative, non-swollen forms appear much like *L. minor*. Gen replaced > 1500 m by *L. turionifera*. Summer

L. minor L. **PL BODIES** in pairs or 3's, 2–5 mm, obovate-elliptic, glossy-green; base and tip gen symmetric; upper surface smooth, midline bumps gen 0; lower surface air spaces ± obscure. Common. Freshwater; < 2000 m. **CA**; worldwide. [*L. m.* var. *minima* Chevall.] Aug

L. minuta Kunth **PL BODIES** gen in pairs (esp when crowded, in full sunlight), 1–2.5 mm, widely elliptic to oblong, pale green; base and tip gen symmetric; margin clearly thinner than center; vein 1, obscure, < 2/3 distance from root node to tip of pl body (gen < region of visible air spaces between cells). **SEED** cross-lined between ribs. Common. Freshwater; < 2200 m. **CA** (exc MP); w US, S.Am, Eur, n Asia [*L. minima* Hegelm., *L. minuscula* Herter] Needs study. Aug

L. trisulca L. Root often 0. **PL BODY** 6–10 mm, lanceolate to oblong, transparent green, on a long, tapered stalk, connected in branched chains of 8–30; base and tip symmetric; surfaces smooth. Meadows, mtn streams; < 3000 m. CaR, SN, SCo?, SnBr, **GB**, **DMoj?**; nearly worldwide. Gen forming dense, tangled masses below water surface. Aug

L. turionifera Landolt **PL BODIES** gen in pairs or 3's, 2–5 mm, widely elliptic to obovate, glossy green; base and tip gen symmetric; upper surface shiny green with row of minute bumps at midline; lower surface often ± red; winter buds 1–2 mm, dark green or brown. Freshwater; < 3000 m. **CA**; N.Am, Asia. Like *L. minor*, exc for winter buds. Aug

SPIRODELA DUCKMEAT

Pls gen in clusters of 2–5; roots 1–16, some passing through minute scale on lower surface. **PL BODY** 3–10 mm, oblong to round, flat; upper surface dark shiny green, lower gen red-purple; veins 3–12 (visible in backlight); young pls with minute scale-like lf on each side at base; winter buds often produced. **FLS** in 2 lateral budding pouches, sheathed by minute membrane. **SEED** ribbed. 3 spp.: worldwide. (Greek: visible thread, from roots)

S. polyrrhiza (L.) Schleiden (p. 549) Roots 5–16, 1–2 passing through minute scale. **PL BODY** 5–10 mm, round-ovate, symmetric; veins 7–12; upper surface smooth; winter buds produced in autumn. Freshwater; 0–2500 m. **CA**; nearly worldwide. Aug

LILIACEAE LILY FAMILY

Dale W. McNeal, except as specified

Per to trees, from membranous bulb, fibrous corm, scaly rhizome, or erect caudex. **ST** gen underground. **LVS** gen basal, often withering early, alternate, gen ± linear. **INFL** various, gen bracted. **FL** gen bisexual, gen radial; perianth often showy, segments gen 6 in two petal-like whorls (outer sometimes sepal-like), free or fused at base; stamens 6 (or 3 + gen 3 ± petal-like staminodes), filaments sometimes attached to perianth or fused into a tube or crown; ovary superior or inferior, chambers 3, placentas gen axile, style gen 1, stigmas gen 3. **FR**: gen capsule, loculicidal or septicidal (berry or nut). ± 300 genera, 4600 spp.: esp ± dry temp and subtrop; many cult for orn or food; some TOXIC. Recently treated in restricted sense (here including *Fritillaria*), with recognition of Agavaceae (*Agave*, *Yucca*), Alliaceae (*Allium*), Asparagaceae (*Asparagus*), Calochortaceae (*Calochortus*), Convallariaceae (*Nolina*, *Smilacina*), Hesperocallidaceae (*Hesperocallis*), Melanthiaceae (*Veratrum*, *Zigadenus*), Themidaceae (*Androstephium*, *Dichelostemma*, *Muilla*) [Dahlgren et al 1985 Families of the Monocotyedons, Springer-Verlag; Chase et al 1995 Monocotyledons: Systematics and Evolution, eds. Rudall et al, Royal Bot Gard, Kew; Angiosperm Phylogeny Group 1998 Ann Missouri Bot Gard 85:531–553]. *Asparagus officinalis* L. has been reported from SNE.

1. Shrub- or tree-like, coarse — lvs large, ± sword-like, persistent
 2. Ovary inferior; perianth (greenish) yellow . . . . **AGAVE**
 2' Ovary superior; perianth white to purplish
  3. Fls bisexual and unisexual, < 1cm; lvs not spine-tipped . . . . **NOLINA**
  3' Fls bisexual, > 3 cm; lvs spine-tipped . . . . **YUCCA**
1' Per (± fleshy if coarse) herbs
 4. Pl from a caudex or rhizome; lvs alternate, > 1 cm broad, gen not withering at fl
  5. Pl < 1 m; lvs finely veined; infl < 1.5 dm; fr a berry . . . . **SMILACINA**
  5' Pl > 1 m; lvs coarsely veined; infl gen 2–7 dm; fr a capsule . . . . **VERATRUM**
 4' Pl from a scaly bulb or corm; lvs basal or whorled, if alternate < 1 cm broad, gen withering at fl
  6. Infl a scapose, bracted umbel
   7. Perianth parts ± free at base, not forming an obvious tube; filaments ± free
    8. Pls with onion odor; pedicels not each subtended by scarious bractlets . . . . **ALLIUM**
    8' Onion odor 0; pedicels subtended by scarious bractlets above main bracts . . . . **MUILLA**
   7' Perianth parts fused at base, forming an obvious tube; filaments forming a crown-like tube
    9. Anthers in 2 ± dissimilar series; lvs 2–3; capsules 4–6 mm . . . . **DICHELOSTEMMA**
    9' Anthers in a single series, all alike; lvs < 5; capsules 10–15 mm . . . . **ANDROSTEPHIUM**
  6' Infl various (not a scapose, bracted umbel)
   10. Outer perianth parts narrower and ± sepal-like . . . . **CALOCHORTUS**
   10' Outer and inner perianth parts similar
    11. Lvs mostly cauline, 2–3 per node; fls nodding, purplish brown, mottled yellow or white . **FRITILLARIA**
    11' Lvs mostly basal (cauline obviously reduced); fls spreading or erect, white or cream
     12. Infl dense, raceme or panicle; fls gen erect, perianth parts distinct or ± fused to ovary base,
      1–2 cm, saucer-shaped; styles 3 . . . . **ZIGADENUS**
     12' Infl open, raceme; fls spreading, perianth parts fused into a tube, 4–6 cm, funnel-shaped; style 1
      . . . . **HESPEROCALLIS**

AGAVE

Katy K. McKinney

Shrub-like, often producing rosettes without seeds, blooms once and dies. **LVS** in basal rosette, long-lived, < 50 cm, sessile, linear to ovate, fleshy, glabrous; lateral teeth and tip spine-like. **INFL**: panicle or raceme-like, scapose, gen bracted, gen 2–4 m. **FL**: perianth segments 6 in 2 petal-like whorls, 3–10 cm; stamens 6; ovary inferior, chambers 3. **FR**: capsule, ± ovoid, loculicidal. **SEEDS** many, flat, black. ± 300 spp.: warm and trop Am. (Greek: noble, from imposing stature) [Gentry 1982 Agaves of N.Am, Univ AZ Press] Key to species by Dale McNeal.

1. Bracts on peduncle 0; infl ± raceme-like; fr 15–35 mm . . . . *A. utahensis*
1' Bracts on peduncle obvious; infl a panicle; fr 35–50 mm . . . . *A. desertii*

A. deserti Engelm. (p. 549) Pl gray-glaucous; caudex 30–50 cm, unbranched. **LF** 25–40 cm, gen lanceolate; marginal teeth 2–8 mm, slender, widely spaced. **INFL**: peduncle bracts triangular, scarious; fls in pedicelled in small clusters. **FL**: perianth 40–60 mm, light yellow, tube 4–6 mm, lobes equal; nectary disk thick; filaments attached near base of tube, 25–35 mm; ovary 22–40 mm, neck slightly narrowed. **FR** 35–50 mm, short-stalked, beaked. Rocky slopes, washes in desert scrub; < 1500 m. **s DMtns**, **DSon**; AZ, Baja CA. ❀DRN:3,7,**8–10**,11,**12**,13,**14**,**19–21**&SUN,DRY:15,16,18,22,23. May–Jul

A. utahensis Engelm. (pl. 121) UTAH AGAVE Pl blue-glaucous; caudex 18–30 cm, branched. **LF** 15–30 cm, gen narrowly lanceolate; lateral teeth 2–4 mm, blunt, detachable. **INFL** raceme-like; peduncle bracts 0; fls 2–8 per cluster. **FL**: perianth 25–31 mm, yellow, tube 2.5–4 mm, lobes equal; filaments attached near base of tube, 18–20 mm; ovary 10–25 mm. **FR** 15–35 mm, short-beaked. UNCOMMON. Shadscale scrub, Joshua-tree woodland; 900–1500 m. **n&e DMtns**; to UT, AZ. [var. *eborispina* (Hester) Breitung, ivory-spined agave; var. *nevadensis* Engelm., Clark Mtn agave] ❀DRN,DRY,SUN:3,7–9,**10–12**,13,14,**18**,19–21. May–Jul

ALLIUM ONION, GARLIC

Per with onion odor, taste; bulb solitary, reforming each year, divides at base into daughter bulbs, outer bulb coats gen brown or gray, cell sculpture gen important to identification, inner bulb coats gen white. **ST** scapose, cylindric or flat. **LVS** basal, 1–5, linear, cylindric, channeled, or flat, gen withering from tip before fl. **INFL**: umbel; bracts gen 2–4, conspicuous, ± fused, scarious. **FL**: perianth segments 6, in 2 petal-like whorls; stamens 6, epipetalous, filaments wide at base, fused into a ring; ovary superior, sometimes with 3 or 6 crests, chambers 3, ovules gen 2 per chamber, style 1, stigma entire or 3-lobed. **FR**: capsule, loculicidal. **SEED** obovoid, black, sculpture net-like, smooth, or granular. 500 spp.: worldwide, esp CA. (Latin: garlic) Bulb coat characters available at several couplets in the key, but collecting of bulbs for identification purposes is strongly discouraged.

1. Lvs 2+ per st; ovary crests 3 or 6, prominent or not
    2. Ovary crests 6, prominent, margins minutely dentate; fls rose-purple — bulb sculpture ± prominent, rectangular, walls very wavy . . . . *A. bisceptrum*
    2' Ovary crests 3, minute, papillate or not; fls white, midveins green or reddish
        3. St 3–12 cm, lvs ± broad, flat, ± sickle shaped; crests and upper ovary smooth . . . . *A. parvum*
        3' St 10–35 cm; lvs narrow, flat or channeled; crests and upper ovary densely papillate — bulb sculpture obvious, meshes prominent, walls thick, ± wavy, polygonal . . . . *A. lacunosum*
            4. St 10–25 cm; infl dense, pedicel 0.7–1.5 × fl . . . . var. *kernensis*
            4' St 15–25 cm; infl open, pedicel 1.5–3.5 × fl . . . . var. *davisiae*
1' Lf 1 per st, cylindric; ovary crests 6, prominent
    5. Stigma ± entire
        6. Fls gen white or fading to pink with a darker pink midvein, perianth parts lance-linear to lanceolate, widely spreading from base, recurved at tip; outer bulb coat sculpture clearly net-like with ± long, twisted meshes . . . . *A. nevadense*
        6' Fls gen deep red-purple or pink, perianth parts broadly lanceolate to ovate, erect; outer bulb coat sculpture 0, only 2–3 vertical cells above root pad, or indistinct and polygonal
            7. Perianth parts 12–18 mm; pedicels stout, gen < fls . . . . *A. parishii*
            7' Perianth parts 8–12 mm; pedicels slender, > fls . . . . *A. atrorubens*
                8. Perianth parts gen deep red-purple (white), lanceolate, appearing long-acuminate . . . . var. *atrorubens*
                8' Perianth parts pale pink, dark-veined, ovate, acute to acuminate . . . . var. *cristatum*
    5' Stigma lobes 3, often slender and recurved
        9. Ovary crest margins entire, to finely and irregularly dentate; inner perianth parts dentate near tip
        . . . . *A. denticulatum*
        9' Ovary crest margins gen deeply cut; perianth parts entire . . . . *A. fimbriatum*
            10. Perianth parts deep purple-red, tips gen recurved-spreading . . . . var. *fimbriatum*
            10' Perianth parts white, pink, or light lavender, tips erect . . . . var. *mohavense*

A. atrorubens S. Watson Bulb 10–16 mm, ovoid to ± spheric; outer coat red-brown, sculpture 0 or 2–3 rows of vertical cells above root pad; inner coats pink to white. **ST** 5–17 cm. **LF** < 2 × st, cylindric, tip tightly coiled when fresh. **INFL**: fls 5–50; pedicels 6–20 mm. **FL** 8–12 mm; perianth parts lanceolate to ovate, entire, red-purple or pink (rarely white), inner narrower, = or > outer; ovary crests 6, prominent, triangular, tip entire or notched. Rocky or sandy soil; 1200–2100 m. **GB, DMtns**; to UT, AZ.

var. ***atrorubens*** **FL**: perianth parts ± narrowly lanceolate, ± acute, red-purple (rarely white), tip margins rolled inward, so appearing long-acuminate. *n*=7. Common. Rocky or sandy soil; 1200–2100 m. **GB, DMtns**; to UT, AZ.. May–Jun

var. ***cristatum*** (S. Watson) D. McNeal (pl. 122) **FL**: perianth parts ± ovate, acute, pale pink with deep pink midveins, tip margins flat. *n*=7. Uncommon. Sandy soils; 1200–2100 m. **SNE, DMtns**; w NV. [var. *inyonis* (M.E. Jones) F. Ownbey & Aase; *A. nevadense* S. Watson var. *c*. (S. Watson) F. Ownbey] May–Jun

A. bisceptrum S. Watson var. ***bisceptrum*** (p. 549) Bulb 10–15 mm, ovoid to ± spheric; daughter bulbs gen clustered at bulb base; outer coat brown to gray, sculpture ± square, walls very wavy. **STS** 1–3, 10–35 cm. **LF** ± = st, widely channeled. **INFL**: fls 15–40; pedicels 10–25 mm. **FL** 6–10 mm; perianth parts lanceolate, acuminate, entire, rose-purple, papery in fr; ovary crests 6, low, ± triangular, minutely dentate. *n*=7. Meadows, aspen groves; 2000–2900 m. **GB**; to ID, UT. May–Jul

A. denticulatum (Traub) D. McNeal Bulb 10–14 mm, ovoid to ± spheric; outer coats red-brown, sculpture 0 or 2–3 rows of vertical cells above root pad; inner coats pale brown to white. **ST**

5–18 cm. **LF** 1.5–2 × st, cylindric. **INFL**: fls 5–30; pedicels 5–20 mm. **FL** 9–17 mm; perianth parts minutely dentate near tip, rose-purple; ovary crests 6, prominent, entire to finely and irregularly dentate. *n*=7. Dry slopes; 900–1600 m. s SN, Teh, w **DMoj**. [*A. fimbriatum* var. *d*. Traub] Apr–May

A. fimbriatum S. Watson Bulb 10–17 mm, ovoid to ± spheric; outer coats red-brown, sculpture 0 or 2–3 rows of vertical cells above root pad; inner coats pale brown to white. **LF** 1.5–2 × st, cylindric. **INFL**: fls 6–75; pedicels 6–20 mm. **FL** 6–12 mm; perianth parts lanceolate to ovate, entire, dark red-purple to white; ovary crests 6, prominent, finely dentate to deeply cut. Dry slopes and flats; 300–2700 m. s NCoR, s SNF, Teh, CW, SW, **SNE, D**; n Baja CA. 3 vars. in CA.

var. ***fimbriatum*** (p. 549) **ST** 10–20 cm. **INFL**: fls 6–35. **FL**: perianth parts dark red-purple; ovary crests finely dentate to deeply cut. *n*=7. Common. Dry slopes and flats; 300–2700 m. s NCoR, s SNF, Teh, CW, SW, **D** [*A. anserinum* Jepson] ❀ DRY,DRN,SUN:3,7–9,11,14–24;DFCLT. Mar–Jul

var. ***mohavense*** Jepson **ST** 10–25 cm. **INFL**: fls 12–60. **FL**: perianth parts white, pink, or light lavender; ovary crests deeply cut, sometimes with additional outgrowths on ovary. *n*=7. Common. Dry slopes and flats; 700–1400 m. **SNE, w DMoj**. Pls from n base SnBr intermediate to var. *f*. are placed here provisionally. Apr–May

A. lacunosum S. Watson (p. 549) Bulb 1–2 cm, ovoid; outer coats often many, thickly surrounding bulb, yellow-brown, sculpture ± square, walls obscurely wavy. **ST** 10–35 cm. **LVS** 2, 0.7–2 × st, ± cylindric or flat. **INFL**: fls 5–45; pedicels 5–25 mm. **FL**

4–9 mm; perianth parts oblanceolate to narrowly ovate, entire, white or pale pink, midveins darker; ovary crests 3, minute, 2-lobed, central, crests and upper ovary densely papillate. Common. Dry, open hillsides; 50–2100 m. s SNF, Teh, SnFrB, SCoR, SCo, n ChI, WTR, SnBr, **SNE, DMoj.** 4 vars. in CA.

var. *davisiae* (M.E. Jones) D. McNeal & F. Ownbey **INFL** open; bracts 2; fls 10–35; pedicels 10–25 mm. **FL** 6–8 mm. *n*=7. Uncommon. Open, sandy slopes, ridges; 600–2100 m. WTR, SnBr, **SNE, DMoj.** [*A. d.* M.E. Jones] ✿ SUN,DRN,DRY:3,7–9,11,14–16,18–21;DFCLT. Apr–May

var. *kernensis* D. McNeal & F. Ownbey **INFL** dense; bracts 2; fls 10–45; pedicels 10–15 mm. **FL** 5–7 mm. *n*=7. Uncommon. Open sandy slopes; 700–1300 m. s SNF, Teh, **w DMoj.**

A. nevadense S. Watson (p. 549) Bulb 9–15 mm, ovoid; daughter bulbs gen 1–2, stalked, at bulb base; outer coat brown to gray, sculpture ± elongate, intricately twisted; inner coats white or pink. **ST** 5–15 cm. **LF** 1.5–2 × st, cylindric, tip tightly coiled before withering. **INFL**: fls gen 5–25; pedicels 6–17 mm. **FL** 7–12 mm; perianth parts spreading, lanceolate to ovate, ± recurved at tip,

entire, white or fading to pink, midveins dark pink; ovary crests 6, prominent, triangular, entire or tip notched. *n*=7,14. Uncommon. Sandy or gravelly slopes; 1300–1700 m. **DMtns**; to OR, ID, AZ. Apr–Jun

A. parishii S. Watson PARISH'S ONION Bulb 10–15 mm, ovoid; outer coats red-brown, sculpture 0, very obscure, or square; inner coats pink. **ST** 5–25 cm. **LF** 1, < 2 × st, cylindric. **INFL**: fls 6–25; pedicels 5–15 mm. **FL** 12–18 mm; perianth parts spreading, lanceolate, entire, pale pink, midveins darker; ovary crests 6, entire or finely and irregularly dentate. *n*=7. UNCOMMON. Open, rocky slopes; 900–1400 m. **DMoj**; w AZ. Populations scattered. Apr–May

A. parvum Kellogg Bulb 10–25 mm, oblique to ovoid; outer coat brown to gray, sculpture 0, obscure, or ± square. **ST** 3–12 cm. **LVS** 2, 1.5–3 × st, flat, sickle-shaped. **INFL**: fls 5–30; pedicels 3–12 mm. **FL** 6–12 mm; perianth parts erect, oblong to elliptic, entire, white or pink, midveins wide, dark; ovary crests 3, minute, round, central. *n*=7. Common. Stony clay slopes, talus; 1200–2800 m. KR, CaRH, n&c SNH, MP, **n SNE**; to e OR, ID, UT. Apr–Jul

ANDROSTEPHIUM

Glenn Keator

Per from spheric, fibrous-coated corm. **LVS** basal, linear, channeled. **INFL** umbel-like, scapose, straight; bracts papery; pedicels unjointed, erect. **FL**: perianth segments 6 in 2 petal-like whorls, tube funnel-shaped, lobes narrowly oblong; stamens 6, filaments fused into a crown-like tube with toothed lobes between anthers, anthers attached near middle; ovary superior, sessile, chambers 3, style persistent. **FR**: capsule, subspheric, obtusely 3-angled, loculicidal. **SEEDS** several per chamber, flat, black. 3 spp.: sw US. (Greek: stamen crown, from fused filaments)

A. breviflorum S. Watson (p. 549) **LF** 10–30 cm, 2 mm wide, scabrous. **INFL** 10–30 cm, scabrous near base; bracts lanceolate; pedicels 15–30 mm; fls 3–12. **FL**: perianth white to light violet drying yellow-brown, tube 5–7 mm, lobes 10–14 mm; filaments 8–10 mm, tube ± funnel-shaped, appendages ± 2 mm, anthers ± 3 mm. **FR** 10–15 mm, deeply 3-lobed. Open desert scrub; 700–1600 m. **e D**; to w Colorado.

CALOCHORTUS

Peggy Fiedler & Bryan Ness

Bulb coat gen membranous. **LVS** gen linear to lanceolate; basal lf 1; cauline lvs 0–several, smaller upwards. **INFL** often ± umbel-like; fls 2–many. **FL** spheric and closed to nearly rotate; sepals gen < petals, gen ± lanceolate (ovate), gen nearly glabrous; petals gen widely wedge-shaped, gen hairy inside, nectary near base; filaments ± flat, anthers gen attached at base; style 1, stigmas 3. **FR**: capsule, septicidal, gen ± oblong, gen 3-angled or -winged; chambers 3. **SEEDS** many in 2 rows per chamber, gen flat, gen netted, gen ± yellow. ± 65 spp.: w N.Am, C.Am; many cult. Bulbs of some eaten by native Americans. Nectary shape and hairs important to identification. (Greek: beautiful grass) [Ness 1989 Syst Bot 14:495–505] Sect. *Calochortus* by Bryan Ness.

1. Nectary not round, not depressed, not surrounded by a fringed membrane (but sometimes 1 of these)
 2. Nectary ± arched, ± triangular or sagittate; sepals gen > petals . . . . *C. macrocarpus*
 2' Nectary oblong to crescent shaped; sepals gen < petals
  3. Petals spotted, not conspicuously veined; sts twining or straggling; common . . . . *C. flexuosus*
  3' Petals conspicuously purple-veined, never spotted; sts ± erect; rare . . . . *C. striatus*
1' Nectary ± round, depressed, surrounded by conspicuous fringed membrane
 4. Fls orange, red, or yellow; st sometimes twisted; median green stripe on outer surface of petal absent
  . . . . *C. kennedyi*
  5. Petals gen orange (e DMoj) to red (W&I, w DMoj) . . . . var. *kennedyi*
  5' Petals ± yellow; n & e DMtns . . . . var. *munzii*
 4' Fls white, white gen tinged with lilac, or lavender; st not twisted; median green stripe on outer surface
  of petal present
  6. Sepals not spotted; bracts gen 3–8 cm; anthers red-brown — SNE (Alabama Hills, Inyo Co.) . . *C. excavatus*
  6' Sepals dark-spotted near base; bracts gen 2–5 cm; anthers yellow, maroon, blue, purple, or red
   7. Petals with red or purple arch above nectary, nectary surrounded by yellow spot; SNE . . . . . . *C. bruneaunis*
   7' Petals without red or purple arch above nectary, nectary surrounded by red or purple spot;
   DMtns (Panamint Mtns) . . . . *C. panamintensis*

C. bruneaunis Nelson & J.F. Macbr. (p. 553) **ST** 10–40 cm, gen simple; base bulblet-bearing. **LVS**: basal 10–20 cm, withering; upper cauline inrolled. **INFL**: fls 1–4, erect; bracts 2–4 cm. **FL**: perianth bell-shaped; sepals 10–40 mm, dark-spotted near base; petals 20–40 mm, narrowly obovate, white, tinged lilac, striped green outside, with dark red or purple arch above nectary, ± glabrous, nectary surrounded by yellow spot, round, depressed, densely short-hairy, encircled by fringed membrane. **FR** 3–7 cm, erect, linear-lanceolate, angled. *n*=7. Dry shrubs or grass in pinyon/juniper woodland; 1700–3000 m. **s SNE**; to OR, MT, NV, UT. [*C. nuttallii* Torrey var. *b.* (Nelson & J.F. Macbr.) F. Ownbey] ❀ DFCLT. May–Aug

C. excavatus E. Greene (p. 553) INYO COUNTY STAR-TULIP **ST** 10–30 cm, slender, simple; base gen bulblet-bearing. **LVS**: basal 10–20 cm, gen persistent. **INFL** ± umbel-like; fls 1–6, erect; bracts 3–8 cm, paired. **FL**: perianth widely bell-shaped; sepals 20–30 mm, not spotted; petals 30–40 mm, white, base marked dark purple, green-striped outside, sparsely short-hairy near nectary, nectary round, depressed, encircled by fringed membrane, densely short-branched-hairy; anthers red-brown. **FR** erect, narrowly lanceolate, angled. RARE. Grassy meadows in shadscale scrub; 1300–2000 m. **w SNE (Mono, Inyo cos.)**. Threatened by groundwater development. Apr–May

C. flexuosus S. Watson (p. 553) **ST** 10–20 cm, branched, gen wavy, ± sprawling; bulblets 0. **LVS**: basal 1–2, withering. **INFL**: fls 1–6, erect; bracts 1–3 cm. **FL**: perianth bell-shaped; segments purple-spotted, yellow-banded; sepals 20–30 mm; petals 30–40 mm, white, lilac-tinged, sparsely short-hairy near nectary, nectary 1 crescent, not depressed, densely short-hairy. **FR** erect, 3–4 cm, stout, angled. *n*=7. Dry, rocky sites; 600–1700 m. Creosote-bush or sagebrush scrub. **e D**; to Colorado. ❀ DFCLT. Apr–May

C. kennedyi Porter (p. 553) **ST** 10–20(50) cm, gen simple, sometimes twisted; bulblets gen 0. **LVS**: basal 10–20 cm, glaucous, channeled, withering. **INFL** ± umbel-like; fls 1–6, erect; bracts 2–4 cm, base widest. **FL**: perianth bell-shaped; segments often dark-spotted near base; sepals 20–30 mm; petals 30–50 mm, yellow to red, sparsely club-like-hairy near nectary, nectary round, depressed, encircled by fringed membrane, densely simple- or forked-hairy. **FR** erect, 4–6 cm, ± lanceolate, angled, striped. *n*=8. Heavy or rocky soil in creosote-bush scrub, pinyon/juniper woodland; 600–2200 m. **n TR, W&I, DMoj**; NV, AZ.

var. ***kennedyi*** (pl. 123) **FL**: petal orange (esp e DMoj) or red (W&I, w DMoj). Heavy or rocky soil in creosote-bush scrub, pinyon/juniper woodland; 600–2200 m. **n TR, W&I, DMoj**; NV, AZ.. ❀ DFCLT. Apr–Jun

var. ***munzii*** Jepson **FL**: petal yellow; 600–2200 m. **n&e DMtns (Panamint, Clark, Providence mtns)**. ❀ DFCLT.

C. macrocarpus Douglas (p. 553) **ST** 20–50 cm, stout, gen simple; base gen bulblet-bearing. **LVS**: basal 5–10 cm, withering; cauline inrolled, tips curled. **INFL** ± umbel-like; fls 1–6, erect; bracts 3–5 cm. **FL**: perianth bell-shaped; sepals 40–50 mm, gen > petals; petals 35–60 mm, narrowly obovate, purple, green-striped outside, gen purple-banded and bearded above nectary, nectary slightly depressed, ± triangular-sagittate, densely slender-hairy. **FR** erect, 4–5 cm, ± lanceolate, angled. *n*=7. Common. Sagebrush scrub, yellow-pine forest, gen in volcanic soil; 1300–2000 m. **n CaR, GB**; to B.C. ❀ DFCLT. Jun–Aug

C. panamintensis (F. Ownbey) Rev. PANAMINT MARIPOSA LILY **ST** 40–60 cm, gen simple; base bulblet-bearing. **LVS**: basal 10–20 cm, withering; upper cauline inrolled. **INFL**: fls 1–4, erect; bracts 2–4 cm. **FL**: perianth bell-shaped; sepals 10–40 mm, dark-spotted near base; petals 20–40 mm, narrowly obovate, white tinged lilac, green-striped outside, not spotted, ± glabrous, nectary in red or purple spot, round, depressed, encircled by fringed membrane, densely short-hairy. **FR** erect, ± 7 cm, ± linear, angled. *n*=7. RARE. Dry pinyon/juniper woodland; 2500–3200 m. **n DMtns (Panamint Mtns)**. [*C. nuttallii* Torrey var. *p.* F. Ownbey] Jun–Jul

C. striatus Parish (p. 553, pl. 124) ALKALI MARIPOSA LILY **ST** 1–5 cm; bulblets 0. **LVS**: basal 10–20 cm, gen withering. **INFL** ± umbel-like; fls 1–5, erect; bracts 1–3 cm. **FL**: perianth bell-shaped, base narrowed; sepals 10–20 mm; petals 20–30 mm, irregularly toothed above, white to lavender, purple-veined, sparsely hairy near nectary, nectary not depressed, oblong, densely simple-hairy. **FR** erect, 4–5 cm, linear, angled. RARE. Alkaline meadows, moist creosote-bush scrub; 800–1400 m. **w DMoj**; w NV. Threatened by grazing, urbanization. In cult. Apr–Jun

DICHELOSTEMMA

Glenn Keator

Per from spheric, fibrous-coated corm; cormlets gen sessile. **LVS** basal, 2–5, narrowly lanceolate, gen keeled, entire, glabrous, sometimes withered by fl. **INFL** umbel- or raceme-like, gen dense (pedicels < fl); axis gen curved to twining, cylindric; bracts 2–4, ± papery. **FL**: perianth tube cylindric to bell-shaped, lobes 6 in 2 petal-like whorls; staminodes gen 0 (stamen-like in 1 sp.); stamens 3 (6 in 1 sp.), filaments fused to perianth and into a crown-like tube, free filaments gen ± 0, anthers attached at base; style 1, stigma 3-lobed. **FR**: capsule, gen not stalked, gen ovoid, 3-angled, loculicidal. **SEEDS** sharply angled, black. *n*=9. 5 spp.: w US, esp n CA (Greek: toothed crown, from stamen appendages) [Keator 1992 Four Seasons 9:24–39]

D. capitatum (Benth.) A.W. Wood BLUE DICKS Cormlets sessile and on stolons. **LVS** 2–3, 10–40 cm, barely keeled. **INFL** head- or umbel-like, dense; axis < 65 cm; bracts widely lanceolate, whitish to dark purple; fls 2–15. **FL**: perianth blue, blue-purple, pink-purple, or white, tube 3–12 mm, narrowly cylindric to short-bell-shaped, lobes gen ascending, 7–12 mm; stamens 6, crown segments 4–6 mm, deeply notched, lanceolate, white, angled inward, slightly reflexed at tip, outer filaments wider at base, outer anthers 2–3 mm, inner free filaments ± 0, inner anthers 3–4 mm; ovary sessile, style 4–6 mm. Open woodlands, scrub, desert, grassland; 0–2300 m. **CA**; to OR, UT, NM, n Mex. [*D. pulchellum* (Salisb.) A.A. Heller; *Brodiaea p.* (Salisb.) E. Greene] Variable; sspp. intergrade. Key to subspecies by Dale McNeal.

1. Bracts gen dark purple; pedicels gen < bracts; fls gen > 6
.... ssp. *capitatum*

1' Bracts whitish or purple striped; pedicels gen > bracts;
fls gen < 6 .... ssp. *pauciflorum*

ssp. ***capitatum*** (p. 553) **INFL**: bracts gen dark purple (or paler and striped dark purple); pedicels 1–15 mm, gen < bracts; fls 6–15. **FL**: perianth lobes gen ascending. *n*=9,18,27,36,45. Open woodlands, scrub, desert, grassland; 0–2300 m. **CA**; to OR, UT, NM, n Mex. [*D. lacuna-vernalis* L. Lenz, vernal pool brodiaea] ❀ DRN,DRY,SUN:1–3,5,6,**7–10**,11,12,**14–24**. Mar–May

ssp. ***pauciflorum*** (p. 553) (Torrey) Keator **INFL**: bracts whitish; pedicels 6–35 mm, > bracts; fls 2–5. **FL**: perianth lobes widely spreading. *n*=9,27. Deserts, open scrub; 300–2100 m. **SNE, D**; to UT, NM, n Mex. [*Brodiaea pulchella* (Salisb.) E. Greene var. *pauciflora* (Torrey) C. Morton] ❀ DRN,DRY,SUN:2,3,**7–10**,11,**12,14–24**.

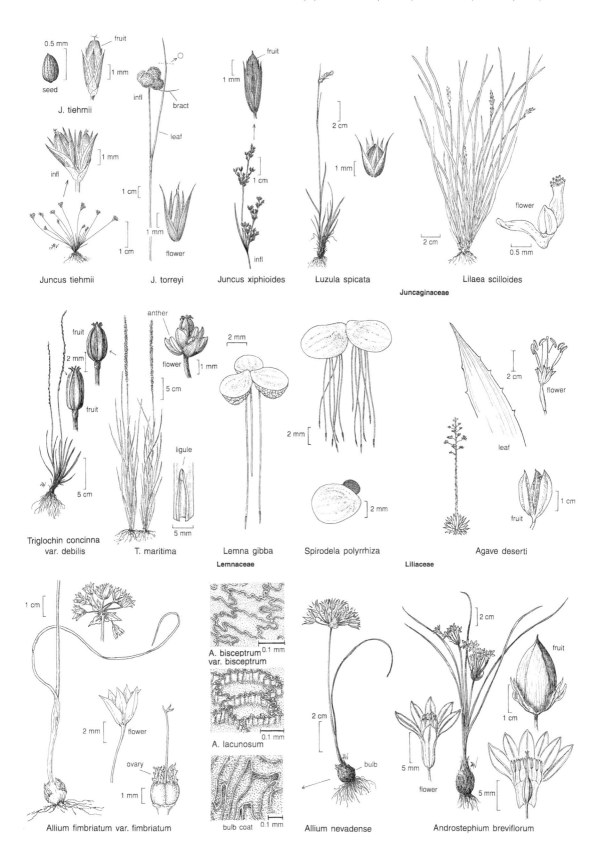

J. tiehmii

Juncus tiehmii

J. torreyi

Juncus xiphioides

Luzula spicata

Lilaea scilloides

Juncaginaceae

Triglochin concinna var. debilis

T. maritima

Lemna gibba

Lemnaceae

Spirodela polyrrhiza

Agave deserti

Liliaceae

Allium fimbriatum var. fimbriatum

A. bisceptrum var. bisceptrum

A. lacunosum

bulb coat

Allium nevadense

Androstephium breviflorum

FRITILLARIA FRITILLARY

Bryan D. Ness

Per; bulb with 1–several large, fleshy scales, 0–many small scales. **ST** erect, simple (0 in non-fl pls). **LVS** alternate (or whorled below), sessile, linear to ± ovate (1 "bulb-lf" in non-fl pls). **INFL**: raceme; bracts lf-like. **FL** gen nodding, bell- or cup-shaped; perianth segments 6, of 2 similar whorls; nectaries 6, on perianth parts; stamens 6, incl, inserted at perianth base, anthers attached ± near middle; ovary ± sessile, style 1, entire or 3-branched. **FR**: capsule, loculicidal, thin walled, ± rounded, 6-angled, or winged, chambers 3. **SEEDS** many, 2 rows per chamber, flat, brownish. ± 100 spp.: n temp. (Latin: dicebox, from fr shape) [Turrill & Sealy 1980 Hooker's Icones Plantarum 34:1–275] Bulbs of some eaten by Indians of N.Am. ❀ DRN: for pots or rock gardens; DRY when dormant. Most are very DFCLT.

F. atropurpurea Nutt. (p. 553) Bulb; large scales 2–5; small scales 45–50. **ST** 1–6 dm. **LVS** 2–3 per node, 4–12 cm, linear to lanceolate. **FL** nodding; perianth widely open, parts 1–2.5 cm, oblong to ± diamond-shaped, purplish brown mottled yellow or white; nectary indistinct, covering most of perianth part, elliptic, yellow with dark reddish dots; style divided > 1/2. **FR** acutely angled. Common. Lf mold under trees; 1000–3200 m. KR, CaR, SN, **GB**, **DMtns**; to OR, MT, NM. Apr–Jul

HESPEROCALLIS DESERT LILY

1 sp. (Greek: western beauty)

H. undulata A. Gray (p. 553, pl. 125) Per; bulb 4–6 cm, ovoid, deep. **ST** 30–180 cm, gen simple, ± scapose. **LVS** mostly basal (cauline reduced), 20–50 cm, 8–15 mm wide, blue-green; margins wavy, white. **INFL**: raceme; bracts conspicuous, ± ovate, papery; pedicels ± 1 cm. **FL** jointed to pedicel, 4.5–6 cm, funnel-shaped; perianth parts 6, fused below, petal-like, white with silver-green midstripe, lobes 3–4 cm, ± oblanceolate, spreading; stamens 6, attached to perianth; ovary superior, chambers 3, style slender, stigma slightly 3-lobed. **FR**: capsule, 12–16 mm, 3-lobed, loculicidal. **SEED** ± 5 mm, flat, black. *n*=24. Sandy flats; < 800 m. **D**; w AZ. ❀ DRN, sand,SUN:12,13;DFCLT. Mar–May

MUILLA

Glenn Keator

Per from small, fibrous-coated corm. **LVS** few–several, basal, narrow, gen ± cylindric. **INFL** umbel-like; axis stiff, straight, cylindric; bracts gen 3, papery, acuminate; pedicels slender, unjointed, erect. **FL**: perianth segments 6 in 2 petal-like whorls, barely fused at base, lobes gen equal, lanceolate or oblong, widely spreading, midribs 2-veined; stamens 6, filaments thread-like to winged, anthers attached at middle; style short, club-shaped, persistent, stigma 3-lobed. **FR**: capsule, subspheric, 3-angled, loculicidal. **SEEDS** irregularly angled, black. ± 6 spp.: sw US, n Mex. (Anagram of *Allium*, from superficial resemblance) [Shevock 1984 Aliso 10:621–627] Key to species by Dale McNeal.

1. Filaments thread-like or slightly wider at base; anthers blue, green, or purple . . . . *M. maritima*
1' Filaments much wider at base, forming a crown; anthers yellow
   2. Infl 3–15 cm; perianth lobes 3–4 mm . . . . *M. coronata*
   2' Infl 15–50 cm; perianth lobes 6–8 mm . . . . *M. transmontana*

M. coronata E. Greene (p. 553) CROWNED MUILLA **LF** < 18 cm. **INFL** 3–15 cm; pedicels 10–30 mm; fls 3–10. **FL**: perianth lobes 3–4 mm, greenish outside, white or faded blue inside; stamens 2–4 mm, filaments petal-like, ± translucent, much wider at base and fused into cylindric crown, anthers yellow. **FR** 3–7 mm. *n*=7. UNCOMMON. Open desert scrub, woodland, in heavy soils; 1000–1600 m. s SNH (e slope), **SNE**, **n&w DMoj**; w NV. Mar–Apr

M. maritima (Torrey) S. Watson (p. 553) COMMON MUILLA **LF** 10–60 cm. **INFL** 10–50 cm; pedicels 10–50 mm; fls 4–20. **FL**: perianth lobes 3–6 mm (inner slightly wider), greenish white, striped brown; stamens 2–5 mm, filaments thread-like or slightly wider at base, anthers blue, green, or purple. **FR** 5–8 mm. *n*=7,8, 10. Grassland, open scrub, woodland, in alkaline, granitic, or serpentine soils; 0–2300 m. c&s NW, CW, SW, uncommon in c SNF, GV, **w D**; Baja CA. ❀ DRN,DRY,SUN:3,7,14–14;may be DFCLT. Mar–Jun

M. transmontana E. Greene **LF** 20–60 cm. **INFL** 15–50 cm; pedicels 2–3 cm; fls 12–30. **FL**: perianth lobes 6–8 mm, white aging lilac; stamens 4–6 mm, filaments wider and fused at base into cup, anthers yellow. **FR** 8–10 mm. *n*=10. High desert scrub, open forest; 1200–2300 m. **GB**; w NV. ❀ DFCLT. Jun

NOLINA BEARGRASS

James C. Dice

Tree-like subshrubs, dioecious or some fls bisexual. **STS** thick and woody or ± underground. **LVS** densely rosetted, 6–20 dm, stiff, sword-like; bases much expanded, white, fleshy. **INFL**: panicle or raceme-like, scapose, bracted, < 4 m. **FL**: perianth parts 6 in 2 petal-like whorls, ± white, < 6 mm; stamens 6, filaments slender; ovary superior, 3-chambered, style and 3 stigmas short, ovules 2 per chamber. **FR**: capsule, papery. **SEEDS** 1–3 per fr, ovoid. ± 25 spp.: s US, Mex. (P.C. Nolin, 18th century French agriculturist) [Munz & Roos 1950 Aliso 2:217–238; Hess & Dice 1995 Novon 5:162–164] Key to species by Dale McNeal.

1. Old lf margins fibrous-shedding; bracts deciduous; seeds gray . . . . *N. bigelovii*
1' Lf margins minutely and persistently serrate; bracts persistent, large, papery; seeds reddish brown . . . *N. parryi*

N. bigelovii (Torrey) S. Watson (p. 553) **ST** 10–25 dm, simple to several-branched aboveground. **LVS** 50–150 per rosette, 15–45 mm wide just above expanded base, ± glaucous; base 7–15 cm, 5–11 cm wide; margin minutely serrate when young, shredding-fibrous when mature. **INFL** 15–35 dm; axis 20–40 mm diam at base; bracts ± lanceolate, gen early deciduous. **SEEDS** 2.5–3.5 mm, grayish white. *n*=19. Rocky slopes and ridges; 300–1500 m. **DMtns, DSon**; s NV, w AZ, nw Mex. Scattered and local. ❀ DRN,SUN:2,3,**7,10,14**,15,16,**18–24**&IRR:**8,9**,11–13. May–Jun

N. parryi S. Watson (p. 553, pl. 126) **ST** 3–21 dm, simple to several-branched aboveground (sometimes belowground). **LVS** 65–200 per rosette, 20–40 mm wide just above expanded base, gen green; base 9–23 cm, 5–17 cm wide; margins minutely serrate. **INFL** 14–38 dm; axis 26–90 mm diam at base; bracts large, ± deltoid, papery, persistent. **SEEDS** 3–4 mm, reddish brown. *n*=19,20. Dry slopes and ridges; 900–2100 m. s SNH (Kern Plateau), e SnBr, PR, **DMtns, DSon**. [ssp. *wolfii* Munz] Scattered, local. ❀ SUN,DRN:1–3,**7,12,14–16**,17,**18–24**&IRR:**8,9,11,13**. May–Jun

SMILACINA FALSE SOLOMON'S SEAL

Per; rhizome creeping. **ST** erect, scaly below, leafy above. **LVS** alternate, > 5. **INFL**: panicle or raceme, terminal. **FL** small; perianth parts 6, petal-like, white; stamens 6, attached to perianth; ovary superior, chambers 3, style 1, stigma ± 3-lobed. **FR**: berry. **SEEDS** 1–3. ± 25 spp.: n temp. (Greek: little smilax) CA spp. now treated in *Maianthemum* [LaFrankie 1986 Taxon 35:584–589].

S. stellata (L.) Desf. (p. 553) Rhizome slender. **ST** 30–70 cm, straight or ± zigzag above, glabrous to puberulent. **LF** 5–17 cm, (ob)lanceolate to elliptic, acuminate, puberulent below, sessile, clasping. **INFL**: raceme (rarely branched at lowest node), gen 2–8 cm; fls 5–15. **FL** 4–6 mm; perianth parts oblong to lanceolate, spreading; stamens < perianth. **FR** 7–10 mm, spheric, reddish purple to black. **SEED** brown. *n*=18. Moist woodlands, streambanks, open slopes; < 2400 m. CA-FP, **W&I**; to B.C., e N.Am. [var. *sessilifolia* (Baker) L. Henderson] ❀DRN:4&IRR: 1,2,**5,6**&SHD:3,**7**,8–10,14,**15–17**,18–24 Apr–Jun

VERATRUM CORN LILY, FALSE HELLEBORE

Per, coarse, leafy; rhizome thick. **ST** erect, 1–2 m, simple, hollow. **LVS** many, alternate, lanceolate to widely ovate, gen acute, clasping, coarsely veined, reduced upward. **INFL**: panicle; fls many. **FL** bisexual or staminate; perianth parts 6, petal-like, free, widely spreading, white or greenish to red-brown, nectary glands 1–2 near base; stamens 6, attached to perianth; ovary slightly inferior, chambers 3, styles 3, short, stigmas long. **FR**: capsule, septicidal. ± 25 spp.: n temp. (Latin: dark roots) Alkaloids used medicinally and TOXIC to both livestock and humans.

V. californicum Durand var. *californicum* (p. 553) **LVS** ovate; lower 20–40 cm, tomentose-ciliate, lower surface curly-hairy, upper surface glabrous or veins sparsely short-hairy. **INFL** erect, gen 30–60 cm, tomentose; branches ascending or spreading; pedicels 1–6 mm. **FL** 10–15 mm; perianth parts elliptic to obovate, white or greenish, glabrous to sparsely woolly below, entire to ± dentate, glands 1–2, Y-shaped, green; stamens 1/2–2/3 perianth length; ovary glabrous. **FR** 2–3 cm, narrowly ovoid. **SEED** 10–12 mm, ± winged. *n*=16. Streambanks, moist meadows, forest edges; ± 1000–3500 m. CA-FP, **GB**; to WA, MT, Colorado, Mex. ❀WET:1,**2**,6,7. Jul–Aug

YUCCA SPANISH BAYONET

Katy K. McKinney & James C. Hickman

Subshrub or tree-like, sometimes dying after fr. **LVS** rosetted (basal or elevated on branches), 2–15 dm, linear, stiff, sword-like, stoutly spine-tipped; bases ± expanded; edges gen curved up. **INFL**: panicle, dense; fls pendent. **FL** 2–13 cm; perianth parts 6 in 2 petal-like whorls, gen ± fused, ± white, fleshy, waxy; stamens 6, filaments ± thick, fleshy; ovary superior, style short, stigma 3-lobed, concave or dome-like. **FR**: gen capsule. **SEEDS** ± many in 2 rows per chamber, black, often flat. ± 40 spp.: esp dry sw N.Am. (Haitian: yuca, or manihot, because young infls sometimes roasted for food) Pollinated at night by small moths that simultaneously lay eggs in ovary.

1. Lvs 30–150 cm, margins conspicuously fibrous-shredding
  2. Aboveground st ± 0; perianth 5–13 cm . . . . *Y. baccata*
  2' Aboveground st conspicuous, gen few-branched; perianth 3–5 cm . . . . *Y. schidigera*
1' Lvs 20–35 cm, margins minutely serrate, not fibrous-shredding
  3. Pl tree-like, (rosettes aerial); lvs 20–35 cm, ± dark green . . . . *Y. brevifolia*
  3' Aboveground st ± 0 (rosettes basal); lvs gen 40–100 cm, ± gray-green . . . . *Y. whipplei*

Y. baccata Torrey (p. 563) **STS** ± 0 aboveground; rosettes ± open, solitary or in small clumps. **LF** 50–75 cm, ± dark green; expanded base ± 10 cm, 5 cm wide, reddish; margins strongly fibrous-shredding. **INFL** 6–8 dm, heavy, purple-tinged; peduncle < 2 dm. **FL**: perianth 5–13 cm, bell-shaped, reddish brown outside, ± white inside, segments lanceolate (outer narrower), fused below; pistil 5–8 cm. **FR** 15–17 cm, fleshy when young, eventually pendent. Uncommon. Dry Joshua-tree woodland; 800–1300 m. **e DMtns**; to UT, TX. [var. *vespertina* McKelvey]

Reported to hybridize with *Y. schidigera*. ❀ DRN,SUN,DRY: 2,3,7,8,9,**10**,11,14–16,**18–21**,22,23. May–Jun

Y. brevifolia Engelm. (p. 563) JOSHUA TREE Tree-like, 1–15 m, gen openly branched (sometimes clumped or only low-branching); rosettes at st tips, dense. **LF** 20–35 cm, dark green; expanded base 2–4 cm, 4–5 cm wide, ± white; margins minutely serrate. **INFL** 3–5 dm, heavy; peduncle gen ± 1 dm. **FL**: perianth 4–7 cm, ± bell-shaped, cream to greenish, segments ± widely lanceolate, fused ± to middle; filaments thick throughout; pistil ± 3 cm, stigma cavity surrounded by lobes. **FR** spreading to erect. Desert flats, slopes; 500–2000 m. s SNH (e slope), Teh, **e SNE**, **DMoj**; to sw UT, w AZ. [vars. *herbertii* (J.M. Webber) Munz, *jaegeriana* McKelvey] Growth form variable. ❀ DRN,SUN, DRY:2,**3**,7,9,**10**,11,12,14–16,**18–21**,22,23. Apr–May

Y. schidigera K.E. Ortgies (p. 563, pl. 127) MOHAVE YUCCA Shrub- or tree-like 1–5 m, 1–few-branched; rosettes at st tips, not very dense. **LF** 30–150 cm, yellowish or bluish green; expanded base 2–8 cm, 4–11 cm wide, ± white; margins coarsely fibrous-shredding. **INFL** 6–12 dm, heavy; peduncle 1–5 dm. **FL**: perianth 3–5 cm, narrowed at base, gen cream; pistil 2–3 cm, stigma cavity surounded by lobes. **FR**: capsule, ± pendent. Chaparral, creosote-bush scrub; < 2500 m. s SW (San Diego Co.), **s DMoj, nw DSon**; NV, AZ, n Baja CA. ❀ DRN,SUN,DRY:

2,3,7–9,**10**,11,12,**14**,15,16,**18–23**,24. Apr–May

Y. whipplei Torrey (p. 563) OUR LORD'S CANDLE Pl dies after fr. **STS** ± 0 aboveground; rosettes 1–many, very dense. **LF** gen 40–100 cm, flat or ± 3-angled, ± gray-green; expanded base ± 4–7 cm, 4–7 cm wide, ± white to greenish; margins minutely serrate. **INFL** 2–40 dm, not appearing heavy; peduncle 15–35 dm; branches and fls very many. **FL**: perianth ± 3 cm, ± spheric, white, gen purple-tipped; filaments linear below, tip angled, club-like; pistil 1–2 cm, stigmas domed, clear-papillate. **FR** spreading to erect. Chaparral, coastal or desert scrub; < 2500 m. s SN (esp e slope), s SCoR, SW, **w edge DMoj**; n Baja CA. Growth form highly variable; branched pls from desert edge have been called ssp. *cespitosa* (M.E. Jones) A.L. Haines. ❀ DRN,SUN,DRY:1–3,7,8–12,**14–16**,17,**18–23**,24. Warrants treatment in segregate genus *Hesperoyucca* Trel., as *H. whipplei* (Torrey) Trel. Apr–May

ZIGADENUS DEATH CAMAS

Per from bulb or rhizome. **ST** ± scapose. **LVS** many, ± basal (reduced upward), linear, gen folded, ± curved. **INFL**: raceme or panicle. **FL** bisexual, staminate, or sterile; perianth parts 6, petal-like, free or ± fused to ovary base, white to yellowish in CA, glands 1–2 near base; stamens 6, free to ± attached to perianth; ovary chambers 3, styles 3. **FR**: capsule, septicidal. **SEEDS** many. ± 15 spp.: temp N.Am, Asia. (Greek: yoke-gland, from gland shape of some) All taxa should be considered highly TOXIC to livestock (gen unpalatable) and humans from alkaloids (esp in bulbs); caused serious illness to some members of Lewis & Clark Expedition.

1. Stamens < perianth; perianth segments ± equal . . . . ***Z. brevibracteatus***
1' Stamens = or > perianth; outer perianth segments < inner
  2. Infl a well developed panicle; fls of main axis bisexual, fls of branches often staminate; stamens 1–2 mm > perianth . . . . ***Z. paniculatus***
  2' Infl raceme-like (rarely branched below); fls all bisexual; stamens ± = perianth . . ***Z. venenosus*** var. ***venenosus***

Z. brevibracteatus (M.E. Jones) H.M. Hall (p. 563) Bulb 25–40 cm diam, ± spheric; outer coats dark brown to black. **ST** 30–50 cm, glabrous. **LF** 15–30 cm, 5–10 mm wide, scabrous-ciliate. **INFL**: gen panicle, 10–35 cm, open; fls of branches mostly bisexual or staminate; pedicels spreading, 10–35 mm. **FL** 5–8 mm; perianth parts ovate to elliptic, ± obtuse, short-clawed, glands greenish, indistinct exc ridge at lower edge; stamens ± 2/3 perianth; styles spreading or recurved. **FR** 12–20 mm, oblong. *n*=11. Sandy desert; 600–1800 m. Teh, WTR, **DMoj**. ❀ DRN,DRY, SUN:2,3,7,9, **11**,12,**14**,15,16. Apr–May

Z. paniculatus (Nutt.) S. Watson (p. 563) Bulb 30–50 mm, 8–30 mm wide, ovoid; outer coats dark brown to black. **ST** 20–70 cm, glabrous. **LF** 20–50 cm, 6–16 mm wide, scabrous-ciliate. **INFL**: panicle, 10–30 cm, densely fld at tip; fls bisexual (or sterile or staminate on branches); pedicels ascending, 5–25 mm. **FL** 3–6 mm; perianth parts unequal, outer 3–5 mm, ovate, acute to subacuminate, short-clawed, inner 3.5–6 mm, triangular-ovate,

acute, longer-clawed, glands yellowish green, indistinct; stamens 1–2 mm > perianth; styles erect or ± spreading. **FR** 8–20 mm, cylindric. *n*=11. Dry sagebrush scrub to coniferous forest; 1200–2300 m. NW, CaR, n SNH, **GB**; to WA, MT, Colorado, NM. ❀ DRN,SUN,DRY:1,**2**,3,**7**,10,14–16,18–21. May–Jun

Z. venenosus S. Watson var. ***venenosus*** (p. 563) Bulb 12–25 mm diam, widely ovate; outer coats ± brown. **ST** 15–70 cm, glabrous. **LF** 10–40 cm, 4–10 mm wide, scabrous-ciliate. **INFL** raceme-like (sometimes with basal branches), 5–25 cm; pedicels erect or ascending, 10–20 mm. **FLS** bisexual, 4–6 mm; perianth parts unequal, ± ovate, obtuse, outer 4–5.5 mm, gen subsessile, inner ± 4.5–6 mm, claw ± 1 mm, glands yellowish green; upper margin a heavy, irregular ridge; stamens = or > perianth; styles erect. **FR** 8–14 mm, cylindric. *n*=11. Moist meadows to dry rocky hillsides; < 2600 m. NW, CaR, SN, CW, WTR, PR, **GB**, **DMtns**; to B.C., c US, AZ, n Baja CA. ❀ SUN:**4–6**&IRR:1,**2**,3, **7**,**14–21**,22–24. Mostly May–Jul

ORCHIDACEAE ORCHID FAMILY

Dieter H. Wilken & William F. Jennings

Per, terrestrial in CA, some nongreen, gen from rhizomes. **LVS** linear to ± round or scale-like, gen sessile. **INFL**: gen raceme or spike, bracted. **FL** bisexual, bilateral, sometimes spurred; sepals gen 3, gen petal-like, gen free, uppermost gen erect; petals 3, lowest different ("lip"); stamen gen 1, fused with style and stigma into column, pollen gen sticky, gen removed as sessile anther sacs; ovary inferior, gen twisted 180° (so lip appears to be lowest perianth segment), 1-chambered, placentas 3, parietal; stigmas 3, gen under column tip. **FR**: capsule. **SEEDS** very many, minute. ± 800 genera, ± 18,000 spp.: esp trop (worldwide exc deserts). Many cult for orn, esp *Cattelya, Cymbidium, Epidendrum, Oncidium, Paphiopedalum; Vanilla planifolia* frs used as source of food flavoring. [Luer 1975 Orchids US and Can, NY Bot Garden] Nongreen pls derive nutrition through fungal intermediates.

1. Pl nongreen, sts red to yellow-brown; lvs bract-like . . . . **CORALLORHIZA**
1' Pl green; lvs 1–many, well developed
  2. Sepals 12–20 mm, uppermost ± flat, greenish to yellowish, veins purple; infl leafy . . . . **EPIPACTIS**
  2' Sepals 3–6 mm, uppermost concave, hood-like, white to green, veins not colored; infl naked or bracted . . . . **PLATANTHERA**

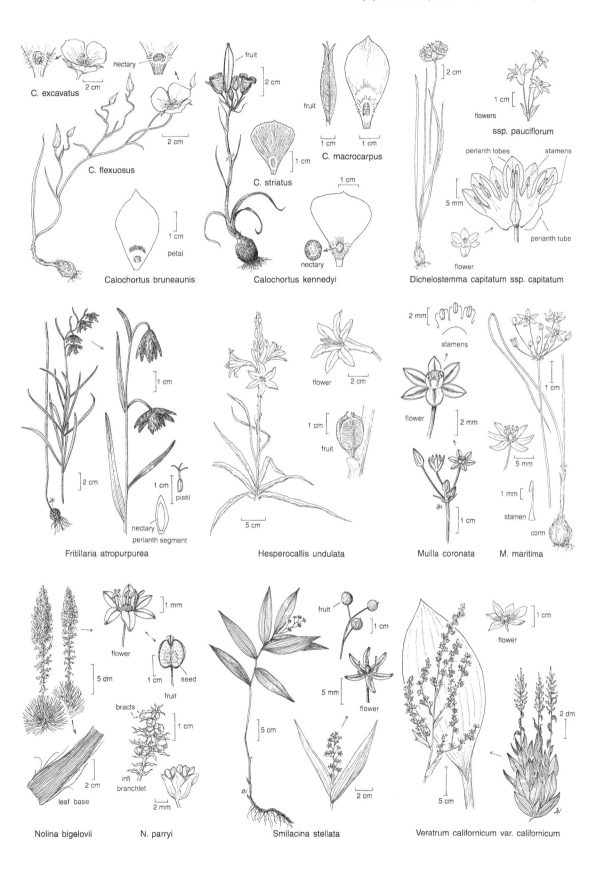

C. excavatus

nectary

2 cm

C. flexuosus

2 cm

Calochortus bruneaunis

petal

1 cm

fruit

2 cm

C. striatus

1 cm

nectary

Calochortus kennedyi

fruit

C. macrocarpus

1 cm   1 cm

1 cm

2 cm

flowers

ssp. pauciflorum

1 cm

perianth lobes   stamens

5 mm

flower

perianth tube

Dichelostemma capitatum ssp. capitatum

Fritillaria atropurpurea

1 cm

2 cm

1 cm

nectary

perianth segment

pistil

Hesperocallis undulata

flower

2 cm

1 cm

fruit

5 cm

2 mm

stamens

flower

2 mm

1 cm

Muilla coronata

1 cm

5 mm

1 mm

stamen

corm

M. maritima

Nolina bigelovii

5 dm

leaf base

2 cm

flower

fruit

seed

1 cm

1 cm

bracts

1 cm

infl
branchlet

2 mm

N. parryi

Smilacina stellata

fruit

1 cm

flower

5 mm

5 cm

2 cm

flower

1 cm

2 dm

5 cm

Veratrum californicum var. californicum

CORALLORHIZA CORALROOT

Pls yellowish green to purplish; rhizome branches many, short, scaly, together coral-like. **ST** ± scapose. **LVS** bract-like, ± sheathing. **INFL**: raceme; fl bract << fl, often scale-like. **FL**: sepals ± alike, oblong to (ob)lanceolate, gen curved over column and lip, gen 3-veined, lower gen fused at base; lateral petals = or > lip, spreading or curved toward lip; lip simple to 3-lobed, spreading to reflexed, sometimes short-spurred; column gen convex above, concave below, curved over lip. **FR** pendent. ± 10 spp.: N.Am, C.Am. (Greek: coral root) Albino pls require careful comparison of sepals and lips for identification.

C. maculata Raf.(p. 563) SPOTTED CORALROOT Pl 17–55 cm. **ST** red to yellow-brown. **FL**: sepals 5.5–10 mm, lower spreading, gen colored like sts; lateral petals gen deep pink to red, sometimes dark-spotted; lip 5–7 mm, white, clearly red- to purple-spotted, base with 2 lateral lobes, tip finely crenate or toothed; spur < 2.5 mm; column 3–5 mm, yellowish, purple-spotted. **FR** 15–20 mm. $2n=42$. Shaded coniferous forest, in decomposing lf litter; < 2800 m. NW, CaR, SN (exc Teh), sw SnFrB, SnGb, SnBr, PR, **W&I**; to B.C., e US, NM. Jun–Aug

EPIPACTIS

LVS gradually reduced upward, lanceolate to widely ovate, often ± ribbed. **INFL** ± 1-sided, open; fls 4+; fl bract ± lf-like. **FL**: sepals ± alike, lanceolate to ovate, lower spreading to descending; lateral petals ascending or curved forward, shape and color ± like sepals; lip abruptly narrowed at ± middle, concave to pouch-like below middle, ± flat or grooved above middle; column curved over lip, convex above. **FR** spreading to pendent. ± 25 spp.: N.Am, Eurasia, n Afr. (Greek: ancient name)

E. gigantea Hook. (p. 563) STREAM ORCHID Pl 30–70(100) cm. **LVS** 5–15 cm, lanceolate to widely elliptic. **INFL**: fl bract lanceolate to oblong. **FL**: sepals 12–20 mm, green, purple-veined; lateral petals 13–15 mm; lip 14–20 mm, lower half deeply concave, greenish to yellowish, veined red-purple, upper half grooved, yellow, reddish tinged or veined below; column 5–9 mm. **FR** 20–28 mm. $n=20$. Seeps, wet meadows, streambanks; < 2600 m. CA-FP (exc GV, s ChI), **SNE, D**; to B.C., SD, TX, Mex. ⚘ IRRorWET,DRN:**4–6,15–17,24**&SHD:1–3,7,8–10,**14**,18–21,**22,23**;CV. Mar–Oct

PLATANTHERA BOG-ORCHID

Caudex cylindric to fusiform. **LVS** basal and cauline; basal 1–2 gen < lower cauline; cauline linear to elliptic, gradually reduced upward. **INFL** gen spike; fl bracts lf-like. **FL**: perianth white to greenish; sepals ± equal, upper gen hood-like, lower free, gen spreading; lateral petals gen erect; lip pendent to upcurved, spurred from back; column ± erect, tip (often the stigma) separating pollen sacs. **FR** ascending to erect. ± 85 spp.: temp N.Am, Eurasia. (Greek: wide anther)

1. Perianth white to cream; lip abruptly wider below middle, ± oblong above middle . . . . ***P. leucostachys***
1' Perianth green to yellow-green; lip ± gradually tapered to tip
  2. Column gen 1.5–2 mm; lip gen 4–7 mm; spur tip blunt; infl gen 5–15 cm, dense . . . . ***P. hyperborea***
  2' Column gen 2.5–4 mm; lip gen 6–10 mm; spur tip acute; infl gen ± 15–40 cm, open (esp below) . . ***P. sparsiflora***

P. hyperborea (L.) Lindley (p. 563) GREEN-FLOWERED BOG-ORCHID Pl 15–100 cm. **LVS**: cauline 4–12 cm, 6–30 mm wide. **INFL** gen 5–15 cm, ± dense; lower bracts 18–30 mm. **FL**: perianth green to yellowish green; sepals 3–6 mm; lip 4–7 mm, narrowly lanceolate, descending to slightly upcurved; spur slightly < lip, ± cylindric, slightly curved, tip blunt; column 1.5–2 mm. $2n=42,84$. Wet coniferous or subalpine forest; 1800–3200 m. KR, SNH (e slope), **n W&I**; to AK, ne N.Am, Colorado, Iceland, Japan. [*Habenaria h.* (L.) R. Br.] Jun–Aug

P. leucostachys Lindley (p. 563) WHITE-FLOWERED BOG-ORCHID Pl 15–100 cm. **LVS**: cauline 5–25 cm, 9–30 mm wide. **INFL** 5–35 cm, gen dense; lower bracts 9–25 mm. **FL**: perianth white to cream; sepals 4–8 mm; lip 5–10 mm, abruptly wider below middle, pendent; spur 5–15 mm, 1–2 × lip in CA, gen cylindric, ± curved; column gen < 1/2 upper sepal. $2n=42$. Wet, gen open places, meadows; < 3400 m. CA-FP (exc GV), **GB, n DMtns (Panamint Mtns)**; to AK, MT, UT. [*Habenaria dilatata* (Pursh) Hook. var. *l.* (Lindley) Ames] Pls with spur ± = lip and barely curved have been incorrectly called *P. dilatata* (Pursh) Hook., of n N.Am. ⚘ IRR,DRN:1,4–7,14–17;DFCLT. May–Sep

P. sparsiflora (S. Watson) Schltr. (p. 563) SPARSE-FLOWERED BOG-ORCHID Pl 25–55 cm. **LVS**: cauline 4–15 cm, 5–30 mm wide. **INFL** gen 15–40 cm, ± open; lowest fls gen not overlapping. **FL**: perianth green; sepals 5–9 mm; lip 6–10 mm, ± linear; spur ± = lip, ± cylindric, slightly curved, tip acute; column 2.5–4 mm. Wet meadows, streambanks, coniferous forest; 300–3400 m. NW (exc NCoRI), SNH, TR, PR, **SNE**; to WA, Colorado, NM. [*Habenaria s.* S. Watson] ⚘ DFCLT. May–Sep

POACEAE [GRAMINEAE] GRASS FAMILY

James P. Smith, Jr., except as specified

Ann to bamboo-like; roots gen fibrous. **ST** gen round, hollow; nodes swollen, solid. **LVS** alternate, 2-ranked, gen linear; sheath gen open (margins adjacent or overlapping but not fused together) or sometimes closed (margins fused together for part of their length); ligule membranous or hairy, at blade base. **INFL** various (of gen many spikelets). **SPIKELET**: glumes gen 2; florets (lemma, palea, fl) 1–many; lemma gen membranous, sometimes glume-like; palea gen ± transparent, ± enclosed by lemma. **FL** gen bisexual, minute; stamens gen 3; stigmas gen 2, gen plumose. **FR**: achene-like grain. 650–900 genera; ± 10,000 spp.: worldwide; greatest economic importance of any family (wheat, rice, maize, millet, sorghum, sugar cane, forage crops, orn, weeds; thatching, weaving, building materials). [Hitchcock 1951 Manual grasses US, USDA Misc Publ 200; Clayton & Renvoize 1986 Kew Bull Add

Series 13] See Glossary for illustrations of general family characteristics. Gen wind-pollinated. Key to genera by J. Travis Columbus.

1. Pl > 2 m and st > 5 mm diam
  2. Infl with long hairs (> 2 mm)
    3. Spikelet base long-hairy . . . . ²**SACCHARUM**
    3' Spikelet base not long-hairy
      4. Lemma long-hairy; spikelet axis glabrous . . . . **ARUNDO**
      4' Lemma glabrous; spikelet axis long-hairy . . . . **PHRAGMITES**
  2' Infl without long hairs
    5. Spikelet > 10 mm; glumes narrow, awl-like, not enclosing florets . . . . *Leymus condensatus*
    5' Spikelet (exc awn, if present) gen << 10 mm; glumes gen elliptic to ovate, gen enclosing florets
    . . . . ²**SORGHUM**
1' Pl < 2 m or st < 5 mm diam
  6. Spikelets enclosed in spiny, bur-like involucre . . . . **CENCHRUS**
  6' Spikelets not concealed in bur-like involucre
    7. Spikelet or spikelet cluster subtended by 1+ bristles
      8. Spikelet or spikelet cluster subtended by 30–50 bristles; most bristles plumose, > 2 cm . . **PENNISETUM**
      8' Spikelet or spikelet cluster subtended by 1–12 bristles; bristles glabrous or scabrous, < 2 cm . . . . **SETARIA**
    7' Spikelet or spikelet cluster not subtended by bristles (1–2 bristle-like structures subtending each
      sessile spikelet in *Andropogon* are actually stalks bearing a vestigial spikelet)
      9. Lf blades gen < 1.5 cm, conspicuously 2-ranked; pl creeping, per; spikelets at shoot tips,
        inconspicuous; salt marshes . . . . **MONANTHOCHLOË**
      9' Lf blades gen > 1.5 cm, gen not conspicuously 2-ranked; pl gen noncreeping, ann or per; spikelets
        at shoot tips or not, gen conspicuous; various habitats
      10. Florets gen replaced by leafy bulblets; st base ± bulb-like . . . . *Poa bulbosa*
      10' Florets not replaced by leafy bulblets; st base not bulb-like
        11. Spikelets all attached (sessile or stalked) to main axis of infl; branches not present in infl
          12. Infls of 2 kinds (in overall morphology), gen unisexual, gen on separate pls (dioecious) or
            the same pl (monoecious); stolons present, rhizomes 0; staminate spikelet florets 5–10(20);
            staminate spikelet lemma veins 3(5); staminate spikelet lemma awn 0 or 1, gen < 1 mm;
            pistillate spikelet lemma awns 3, 6–12 cm . . . . ²**SCLEROPOGON**
          12' Infls of 1 kind (in overall morphology), gen bisexual (unisexual, pls gen dioecious in *Distichlis*);
            stolons, rhizomes present or 0; spikelet florets 1–20+; lemma veins 1–11 or indiscernible;
            lemma awns 0 or 1–3, gen < 5 cm
            13. Awns 5–10 cm, twisted, conspicuously bent, puberulent, red-brown . . . . **HETEROPOGON**
            13' Awns 0 or not as above
              14. Spikelets 3 per node, of 2 kinds (in overall morphology), falling as a cluster, central
                bisexual, laterals gen staminate or sterile
                15. Glumes awn-like; lateral spikelets gen short-stalked; spikelet cluster not hairy-tufted at
                  base; infl axis breaking at nodes in fr; ann to per; rhizomes 0 . . . . **HORDEUM**
                15' Glumes not awn-like; lateral spikelets sessile; spikelet cluster hairy-tufted at base; infl axis
                  not breaking at nodes in fr; per, rhizomed . . . . **PLEURAPHIS**
             14' Spikelets gen < 3 per node (2–17 in *Leymus cinereus*), gen of 1 kind (in overall morphology),
               gen not falling as a cluster, gen bisexual
              16. Floret 1 per spikelet
                17. Lemma awns 3 . . . . *Aristida californica*
                17' Lemma awn 0 or 1
                  18. Lower glume awns 2–3 . . . . ²**LYCURUS**
                  18' Lower glume awn 0 or 1
                    19. Ann; ligule hairy; glume awn 0; spikelet breaking below glumes . . . . ²**CRYPSIS**
                    19' Per; ligule membranous; glume gen awned; spikelet breaking above glumes . . . . . . ²**PHLEUM**
              16' Florets gen > 1 per spikelet
                20. Ligule hairy; lemma veins 3 or awns 3
                21. Lemma awns 3 . . . . ²**BOUTELOUA**
                21' Lemma awn 0 or 1
                  22. Ann; lemma central vein glabrous or puberulent below middle . . . . **MUNROA**
                  22' Per; lemma central vein densely hairy below middle
                    23. Lemma awn 1 . . . . ²**ERIONEURON**
                    23' Lemma awn 0 . . . . ²**TRIDENS**
             20' Ligule membranous (sometimes fringed); lemma veins indiscernible or 5–11, awn 0 or 1
              (sometimes 3 in *Elymus*)
              24. Lower glume 0 exc spikelet at infl tip . . . . **LOLIUM**
              24' Lower glume present in all spikelets (sometimes 0 in *Leymus salinus* ssp. *mojavensis*,
                sometimes minute in *Vulpia myuros*)

25. Spikelets short-stalked (stalks sometimes < 0.5 mm)

 26. Lemma awn 0

  27. Per, rhizomed; spikelets unisexual; lemma veins 9–11 — gen dioecious .... [2]**DISTICHLIS**

  27' Ann; rhizomes 0; spikelets bisexual; lemma veins 5 .... *Puccinellia simplex*

 26' Lemma awn 1

  28. Lemma awned from below tip; lf sheath closed .... *Bromus hordeaceus*

  28' Lemma awned from tip; lf sheath open

   29. Per .... *Festuca brachyphylla*

   29' Ann .... [2]**VULPIA**

25' Spikelets sessile

 30. Ann; lemma stiff-ciliate on keel and exposed margin .... **SECALE**

 30' Per; lemma gen not ciliate

  31. Middle of infl axis internodes < 3 mm .... **AGROPYRON**

  31' Gen all infl axis internodes > 3 mm

   32. Glumes narrow, awl-like (< 2 cm)................................ **LEYMUS**

   32' Glumes awn-like (> 2 cm) to broad, not awl-like

    33. Glume tips truncate or obtuse, awn 0; lemma awn gen 0 .... **ELYTRIGIA**

    33' Glume tips acute to gen awned, or glumes awn-like; lemma gen awned

     34. Rhizomes 0... **ELYMUS**

     34' Rhizomes present

      35. Glumes not curved to the side (symmetric), tip acute (excluding awn, if present)

       .... *Elytrigia repens*

      35' Glumes curved slightly to the side (asymmetric), tip acuminate ..... **PASCOPYRUM**

11' Some spikelets attached to branches of infl; branches of infl sometimes ± inconspicuous (short, appressed, or axis concealed by spikelets), sometimes present only near base of infl

 36. Spikelets within infl clearly of 2 kinds, gen falling as a cluster of 1 gen larger, bisexual spikelet and 1–6 sterile (sometimes vestigial) or staminate spikelets

  37. Spikelets gen in clusters of 2–7, all or most stalked

   38. Spikelet clusters spreading or drooping; sterile spikelet > fertile spikelet (exc awn), linear; fertile spikelet glumes not winged, lemma awn 1............................ **LAMARCKIA**

   38' Spikelet clusters ascending; sterile spikelet gen < fertile spikelet, variously shaped; fertile spikelet glumes winged from keel, lemma awn 0....................... *Phalaris paradoxa*

  37' Spikelets regularly arranged in pairs (trios at branch tips), both attached to 1 node, 1 sessile and 1 stalked (stalked spikelet vestigial in *Andropogon*)

   39. Infl gen not long-hairy.. [2]**SORGHUM**

   39' Infl with hairs ± 5 mm

    40. Infls many per st, each closely subtended by a lf .... **ANDROPOGON**

    40' Infl 1 per st, distantly subtended by a lf .... **BOTHRIOCHLOA**

 36' Spikelets within infl all ± alike (although sometimes smaller or vestigial at base or tip of infl or branches), gen not falling as a cluster

  41. Infl densely long-hairy (hairs gen > 4 mm)

   42. Awns hairy on basal segment .... *Achnatherum speciosum*

   42' Awns 0 or glabrous

    43. Awns 0; infl 1–3 dm, < 3 cm wide .... **IMPERATA**

    43' Awns present; infl 2.5–6 dm, > 3 cm wide .... [2]**SACCHARUM**

  41' Infl not densely long-hairy

   44. Lemma awns 9, plumose .... **ENNEAPOGON**

   44' Lemma awns < or = 3, gen not plumose

    45. Florets 2 per spikelet; lemma of lower floret unawned, lemma of upper floret awned below tip; awn twisted to recurved, ± 1 mm .... **HOLCUS**

    45' Florets 1–20+ per spikelet; lemmas gen all awned or not; awn (if present) from back or tip, gen straight or bent, length variable

     46. Spikelets with 1 well developed floret; additional, conspicuously different (gen much reduced, often vestigial, or with thinner lemma) floret(s) sometimes present above or below, then gen 1–2

      47. Florets 2 per spikelet, upper well developed, gen hard and shiny at maturity, gen compressed front-to-back; lower floret lemma thin, glume-like; spikelet breaking just below glumes (lower glume sometimes 0 in *Eriochloa*, *Paspalum*, spikelet thereby appearing to have 1 floret)

       48. Ligule 0 .... **ECHINOCHLOA**

       48' Ligule present

        49. Spikelet base with a disc- or cup-like ring .... **ERIOCHLOA**

        49' Spikelet base without a disc- or cup-like ring

         50. Infl with 2° branches .... **PANICUM**

         50' Infl with 1° branches only

          51. Ann.. **DIGITARIA**

51' Per... **PASPALUM**
47' Floret gen 1 per spikelet, gen neither hard nor shiny at maturity, gen compressed
    side-to-side or not compressed; spikelet gen breaking above glumes
  52. Infl cylindric or ovoid, dense; spikelets compressed side-to-side; glumes completely
     enclosing floret, keel gen ciliate
    53. Glume awn 0; lemma awned from back; spikelet breaking below glumes ... **ALOPECURUS**
    53' Glume gen awned; lemma awn gen 0; spikelet breaking above glumes .... **²PHLEUM**
  52' Infl variously shaped, open or dense; spikelets compressed side-to-side or not compressed;
    glumes gen not completely enclosing floret(s), keel gen not ciliate
    54. Some or all awns > 1 mm, persistent (sometimes deciduous in *Achnatherum*,
      Piptatherum), or infl branches falling with spikelets
     55. Florets gen 2 or 3 per spikelet
      56. Infl not umbel-like .... **²BOUTELOUA**
      56' Infl umbel-like .... **CHLORIS**
     55' Floret 1 per spikelet
      57. Lemma awns 3 .... **ARISTIDA**
      57' Lemma awn 1
       58. Lower glume awns 2–3 .... **²LYCURUS**
       58' Lower glume awn 0 or 1
        59. Lemma awned below tip (sometimes just below)
         60. Palea 0 or < 1/3 lemma .... **²AGROSTIS**
         60' Palea ± = lemma
          61. Ann; lemma awned near tip...................................... **APERA**
          61' Per; lemma awned near base .... **CALAMAGROSTIS**
        59' Lemma awned from tip
         62. Glume awn 1; spikelet breaking below glumes .... **POLYPOGON**
         62' Glume awn gen 0; spikelet breaking above glumes
          63. Lemma (exc awn) < glumes, gen hard at maturity, margins gen overlapping;
           awn at maturity with 1–2 bends and persistent or ± straight and deciduous
           64. Lemma awn < 11 mm, at maturity ± straight, deciduous
            65. Lemma gen hairy, hairs > 2.5 mm...................... **²ACHNATHERUM**
            65' Lemma gen glabrous .... **PIPTATHERUM**
           64' Lemma awn > 13 mm, at maturity with 1–2 bends, persistent
            66. Lemma awn gen < 65 mm; callus < 2 mm; palea < lemma, gen flexible
              .... **²ACHNATHERUM**
            66' Lemma awn > 65 mm; callus 2–5 mm; palea = lemma, hard.... **HESPEROSTIPA**
          63' Lemma (exc awn) gen > glumes, gen not hard at maturity, margins not
           overlapping; awn at maturity ± straight, persistent........... **²MUHLENBERGIA**
54' Awns 0 or < 1 mm, persistent; infl branches persistent, not falling with spikelets
  67. Glumes 0 .... **LEERSIA**
  67' Glumes present
    68. Spikelet ± circular in outline, breaking below glumes; glumes wrinkled near keel
      .... **BECKMANNIA**
    68' Spikelet not circular in outline, gen breaking above glumes; glumes not wrinkled near keel
     69. Spikelets all attached to 1° branches, gen sessile, overlapping
      70. Infl umbel-like .... **CYNODON**
      70' Infl not umbel-like, branches attached at different points along axis .... **SPARTINA**
     69' Spikelets not all attached to 1° branches (some attached to infl axis or to 2° branches),
      stalked, sometimes overlapping
      71. Lf sheath closed; vestigial florets present above fertile floret .... *Melica imperfecta*
      71' Lf sheath open; vestigial florets above fertile floret 0, sometimes present below
       fertile floret
       72. Vestigial florets 1–2, below fertile floret, awl- or chaff-like .... **²PHALARIS**
       72' Vestigial florets 0
        73. Lemma < glumes; lemma veins gen 5, often obscure................ **²AGROSTIS**
        73' Lemma gen > lower glume; lemma veins 1 or 3, gen obvious
         74. Lemma veins 3 .... **²MUHLENBERGIA**
         74' Lemma veins 1
          75. Ann; infl < 7 cm; spikelet breaking below glumes.................. **²CRYPSIS**
          75' Per; infl > 8 cm; spikelet breaking above glumes.............. **SPOROBOLUS**
46' Spikelets with 2+ well developed florets; additional, conspicuously different floret(s)
  not present, although upper often slightly or gradually reduced or sometimes vestigial
  76. Lemma awn attached to lemma back, sometimes near tip
    77. Lemma awned at or below middle
     78. Lemma awn > 20 mm .... **AVENA**

78' Lemma awn < 10 mm . . . . **DESCHAMPSIA**
77' Lemma awned above middle, near tip
  79. Florets gen > 3 per spikelet; lf sheath closed; infl gen open, variously shaped in
    outline . . . . **²BROMUS**
  79' Florets 2 or 3 per spikelet; lf sheath open; infl dense, cylindric to narrowly elliptic in
    outline . . . . **TRISETUM**
76' Lemma awn 0 or attached to lemma tip, sometimes between lemma teeth or lobes,
  sometimes to lemma back a minute distance below tip in *Festuca*, *Lolium*
  80. Stolons present; rhizomes 0; florets 5–10(20) per spikelet, staminate; lemma veins 3(5)
    . . . . **²SCLEROPOGON**
  80' Stolons gen 0; rhizomes sometimes present; florets 2–20+ per spikelet, gen bisexual;
    lemma veins 3, 5–11, or indiscernible
    81. Lemma veins 3
      82. Spikelets all attached to 1° branches
        83. Infl umbel-like . . . . **DACTYLOCTENIUM**
        83' Infl not umbel-like, branches attached at different points along axis . . . . . **LEPTOCHLOA**
      82' Spikelets not all attached to 1° branches (some attached to infl axis or to 2° branches)
        84. Rhizomes present . . . . *Muhlenbergia asperifolia*
        84' Rhizomes 0
          85. Lemma awned
            86. Lemma awns 3, central ciliate at least at base . . . . **BLEPHARIDACHNE**
            86' Lemma awn 1, glabrous . . . . **²ERIONEURON**
          85' Lemma awn 0
            87. Lemma glabrous . . . . **ERAGROSTIS**
            87' Lemma back densely hairy on lower half . . . . **²TRIDENS**
    81' Lemma veins 3, 5–11, or indiscernible
      88. Lower glume 0 on most spikelets . . . . *Lolium perenne*
      88' Lower glume present (sometimes minute in *Vulpia myuros*)
        89. Spikelet +/- as long as wide, < 5 mm; lemma papery to translucent, rounded at tip,
          awn 0; infl open; ann . . . . **BRIZA**
        89' Spikelet gen longer than wide, gen > 5 mm; lemma gen membranous, gen acute or
          acuminate, awn 0 or present; infl open or dense; ann or per
          90. St > 10 dm; glumes awl-like; spikelets sessile or short-stalked . . . . . *Leymus condensatus*
          90' St gen < 10 dm; glumes not awl-like; spikelets gen stalked
            91. Lemma veins 9–11; rhizomes present; spikelets unisexual . . . . **²DISTICHLIS**
            91' Lemma veins 5–9; rhizomes sometimes present; spikelets sometimes unisexual
              92. Rhizomes present, thick; ligule hairy; lf stiff, sharply pointed; lemma densely
                long-hairy near lower margin . . . . **SWALLENIA**
              92' Rhizomes gen 0; ligule gen membranous; lf not stiff, not sharply pointed; lemma
                gen glabrous near lower margin
                93. Glumes not alike, lower ± linear, upper widely obovate, gen 3–4 × wider than
                  lower . . . . **SPHENOPHOLIS**
                93' Glumes ± alike in shape, upper < 3 × wider than lower
                  94. Ligule hairy; lemma 9-veined, toothed or notched; ann . . . . **SCHISMUS**
                  94' Ligule membranous (sometimes fringed); lemma 5–7-veined, gen entire; ann or per
                    95. Spikelets arranged in dense, gen 1-sided clusters at branch tips; lemma gen
                      stiff-ciliate on keel; lf sheath closed . . . . **DACTYLIS**
                    95' Spikelets not arranged in dense, 1-sided clusters at branch tips; lemma not
                      stiff-ciliate on keel; lf sheath closed or open
                      96. Infl gen cylindric, dense, ± shiny, axis puberulent; florets 2–3(4) per spikelet
                      . . . . **KOELERIA**
                    96' Infl variously shaped, gen not dense or shiny, axis gen glabrous; florets
                      2–10+ per spikelet
                        97. Upper vestigial florets gen > 1, densely clustered and appearing as 1 floret;
                        lower glume broadly translucent on upper portion and along margin; lf sheath
                        closed; per; rhizomes 0 . . . . **MELICA**
                      97' Upper vestigial floret 0 or 1; lower glume gen membranous or narrowly
                      translucent near tip and along margin; lf sheath closed or open; ann or per;
                      rhizomes 0 or present
                        98. Lemma awn gen 1; lowest lemma of spikelet gen > 5 mm; lf blade gen
                        not prow-tipped
                        99. Lf sheath closed; lemma gen 2-toothed . . . . **²BROMUS**
                        99' Lf sheath open (closed in *Festuca rubra*); lemma gen entire
                          100. Per . . . . **FESTUCA**
                          100' Ann . . . . **²VULPIA**

98' Lemma awn 0; lowest lemma of spikelet gen < 5 mm; lf blade gen prow-tipped
   101. Lemma acute; lemma veins ± converging at tip; pls of various habitats,
     sometimes in saline meadows . . . . **POA**
   101' Lemma truncate, obtuse, or acute; lemma veins ± parallel, not converging
     at tip; pls of saline meadows and flats, mineral springs . . . . . **PUCCINELLIA**

ACHNATHERUM NEEDLEGRASS

Mary E. Barkworth

Per, tufted. **ST** gen erect. **LF**: ligule membranous, sometimes long-ciliate, blade gen flat. **INFL** panicle-like, gen narrow; branches gen ascending. **SPIKELET**: glumes > floret (exc awn), tapered below midpoint; axis breaking above glumes; floret 1, gen cylindric; callus blunt or sharp; hairs stiff; lemma stiffly membranous to hard, evenly hairy or glabrous above, awned from tip; awn > 10 mm, persistent, with 1–2 bends, or < 10 mm, readily deciduous, ± straight; palea < lemma, hairy, veined. ± 75 spp.: temp worldwide. (Greek: awned scale, from lemma) Segregated mostly from *Stipa*; see also *Hesperostipa*. Key to species adapted by Margriet Wetherwax.

1. Basal segment of awn hairy, some hairs > 1 mm; callus sharp
  2. Lowest hairs on awn > upper lemma hairs
    3. Awn with 1 bend; uppermost ligule 0.3–1 mm, long ciliate . . . . **A. speciosum**
    3' Awn with 2 bends; uppermost ligule 2–7 mm, glabrous . . . . **A. thurberianum**
  2' Lowest hairs on awn < upper lemma hairs
    4. Palea 2/3–3/4 lemma length . . . . **A. nevadense**
    4' Palea 1/3–1/2 lemma length . . . . **A. occidentale**
     5. Lower awn segments hairy, upper glabrous . . . . ssp. **californicum**
     5' All awn segments hairy . . . . ssp. **occidentale**
1' Basal segment of awn scabrous; callus sharp or blunt
  6. Hairs of lemma tip < 2 mm
    7. Distal segment of awn wavy . . . . **A. aridum**
    7' Distal segment of awn straight . . . . **A. lemmonii** ssp. **lemmonii**
  6' Hairs of lemma tip 2–7 mm
    8. Awn 3–11 mm, straight, readily deciduous
     9. Infl open, branches and pedicels widely spreading . . . . **A. hymenoides**
     9' Infl dense, branches and pedicels ascending . . . . **A. webberi**
    8' Awn 13–35 mm, with 1 or 2 bends, persistent
     10. Hairs of lemma tip 1–7 mm, > hairs at lemma midpoint; awn with 1 bend . . . . **A. parishii**
     10' Hairs of lemma tip 1.5–3.5mm, = hairs at lemma midpoint; awn with 2 bends . . . . **A. pinetorum**

A. aridum (M.E. Jones) Barkworth MORMON NEEDLEGRASS **ST** 3.5–8.5 dm. **LF**: blade 0.9–3 mm wide. **INFL** 5–17 cm, often partly enclosed by uppermost lf sheath. **SPIKELET**: glumes 8–15 mm, lower > upper; floret 4–6.5 mm; callus sharp; lemma 1.4–2.4 × palea length, hairs short, awn 40–80 mm, distal segment wavy; palea 2–3.2 mm. RARE in CA. Outcrops, shrubsteppe, pinyon/juniper; 1200–1550 m. **e DMoj**; to Colorado, AZ. [*Stipa a.* M.E. Jones] May–Jun

A. hymenoides (Roemer & Schultes) Barkworth (p. 563) INDIAN RICEGRASS **ST** 2.5–7 dm. **LF**: blade rolled, < 1 mm diam. **INFL** ± open; branches, spikelet stalks widely spreading. **SPIKELET**: glumes 5–9 mm, ± equal, base swollen; floret 3–4.5 mm; callus sharp; lemma ± spheric, hairs 2.5–6 mm, awn 3–6 mm, deciduous. 2*n*=46,48. Dry, well drained, often sandy soil, desertshrub, sagebrush shrubland, pinyon/juniper; < 3400 m. CaR, SN, SW, **GB**, **D**; to B.C., Great Plains, Mex. [*Oryzopsis h.* (Roemer & Schultes) Ricker] Hybrids with other spp., incl *O.* ×*bloomeri* (Bolander) Ricker, have narrower florets and awns 12–16 mm. Used by Native Americans for food; highly palatable to livestock. ❀ DRN,DRY,SUN:1,**2,3,7**,8–11,**14**,15,16,**18–21**,22–24;STBL. Apr–Jul

A. lemmonii (Vasey) Barkworth ssp. **lemmonii** (p. 563) **ST** 1.5–9 dm. **LF**: basal ligules 0.5–1.2 mm; blade 0.5–1.5 mm wide. **INFL** 7–21 cm. **SPIKELET** somewhat laterally compressed; glumes 7–11.5 mm, ± equal; floret 5.5–7 mm; callus blunt; lemma 1.1–1.3 × palea length, hairs short, tip lobe ± 0.1 mm, thick, awn 16–30 mm. 2*n*=34,36. Sagebrush shrubland, coniferous forest; < 2300 m. KR, NCoR, CaR, SN, Teh, **GB**; to B.C., ID, UT. ssp. *pubescens* (Crampton) Markworth occurs in Tehama and Lake

cos. Barkworth 1993 Phytologia 74:1-25. [*Stipa l.* (Vasey) Scribner, incl var. *pubescens* Crampton, pubescent needlegrass; *S. columbiana* Macoun] ❀ DRN,SUN:1,**2**,3–6,**7**,9,**14–18**,19–24;STBL. May–Jul

A. nevadense (B. Johnson) Barkworth **ST** 2–9 dm, hairy below nodes. **LF**: blade 1–3 mm wide, gen inrolled. **INFL** 6–20 cm. **SPIKELET**: glumes 8–14 mm; floret 5.5–8.5 mm; callus sharp; lemma ± 1.3 × palea length, tip hairs < 1.5 mm, > lower awn hairs; awn 20–35 mm, bent twice, lower 2 segments hairy. 2*n*=68. Sagebrush shrubland, open woodlands; < 3100 m. **SNE**; NV. [*Stipa n.* B. Johnson] ❀ DFCLT.

A. occidentale (Thurber) Barkworth **ST** 2–12 dm. **LF** glabrous to hairy; sheath gen ciliate at top; blade 0.5–1 mm wide, gen rolled. **INFL** 8–30 cm. **SPIKELET**: glumes 9–15 mm, ± equal; floret 5.5–7.5 mm; callus sharp; lemma 1.6–3 × palea length, hairs short; awn 15–40 mm, bent twice, lower 2 segments hairy. 2*n*=36. Open, dry sites, sagebrush shrubland, coniferous forest, alpine; 150–3400 m. CA-FP (exc GV, CW), **SNE**; to WA, ID, UT. 3 sspp.in CA.

  ssp. **californicum** (Merr. & Burtt Davy) Barkworth (p. 563) **SPIKELET**: lemma tip hairs < awn hairs; upper awn segment smooth to rough, not hairy. Sagebrush shrubland, coniferous forest; 150–3100 m. KR, CaR, SN, SnBr, PR, **GB**; to WA, ID, NV. Intergrades with ssp. *pubescens*. [*Stipa c.* Merr. & Burtt Davy] ❀ DRN:**1**,15,16,24SHD:**2**,3,7,**14,18**,19–23;STBL. May–Aug

  ssp. **occidentale** (p. 563) **SPIKELET**: lemma tip hairs gen = awn hairs; all awn segments hairy. Coniferous forest, alpine; gen 3000–3400 m. SN, PR, **SNE**. *[Stipa o.* Thurber] ❀ DRN,SHD:**1–3**; STBL. Jun–Aug

A. parishii (Vasey) Barkworth (p. 563) **ST** 3–8 dm. **LF**: basal sheath ciliate at top, hairs 2–3 mm; blade 1.5–3 mm wide. **INFL** 11–15 cm. **SPIKELET**: glumes unequal, lower 11–21 mm, upper 2–3 mm; floret 4–8 mm; callus blunt; lemma 1.4–2 × palea length, hairs < 5 mm, tip hairs longest, awn 15–35 mm, bent once. Dry, rocky slopes, shrubland, pinyon/juniper woodland; 900–2700 m. s SN, TR, PR, **W&I, DMtns** ; to UT, AZ, Baja CA. *[Stipa coronata* var. *depauperata* (M.E. Jones) A. Hitchc.] ❀DRN,SUN:1,**2,3,7**,8–10,**14–18**,19–24;STBL. May–Aug

A. pinetorum (M.E. Jones) Barkworth **ST** 1–5 dm. **LF**: blade 0.5–1 mm wide. **INFL** 4.5–20 cm; branches appressed. **SPIKELET**: glumes 7–11 mm, ± equal; floret 3.5–5.5 mm; callus blunt to sharp; lemma 1–1.5 × palea length, hairs at midlength 1.5–3.5 mm, at tip < 5 mm, awn 13–25 mm, bent twice. 2*n*=32. Rocky soil, pinyon/juniper woodland, coniferous forest; 1900–3810 m. SN, **W&I, DMtns**; to ID, NM. *[Stipa p.* M.E. Jones] ❀ TRY. Jun–Aug

A. speciosum (Trin. & Rupr.) Barkworth (p. 563, pl. 128) DESERT NEEDLEGRASS **ST** 3–6 dm. **LF**: basal sheath hairy; blade gen rolled, < 1 mm diam. **INFL** 10–15 cm, gen partly enclosed by uppermost lf sheath. **SPIKELET**: glumes 14–20 mm,

± equal; floret 8–9 mm; callus sharp; lemma 1.2–2.1 × palea length, densely short-hairy near base, glabrous near tip, awn 35–40 mm, bent once, hairs of lower segment < 8 mm. 2*n*=66,68,±74. Rocky slopes, canyons, washes; < 2200 m. s SN, SCoRO, SW, **s SNE, D**; to Colorado, Mex, S.Am. *[Stipa s.* Trin. & Rupr.] ❀ DRN,SUN:1,**2,3,7**,8,9,**10**,11,**14–18**,19–21,**22–24**;also STBL. Apr–Jun

A. thurberianum (Piper) Barkworth **ST** 3–7.5 dm, hairy. **LF**: ligule 1.5–8 mm; blade 0.5–2 mm wide. **INFL** 7–15 cm. **SPIKELET**: glumes 10–15 mm, 2.5–4 mm wide, lower > upper; floret 6–9 mm; callus sharp; lemma 1.1–1.4 × palea length, tip hairs 2 × lower hairs; awn 32–56 mm, bent twice, lower segments long-hairy. 2*n*=34. Canyons, foothills, sagebrush shrubland, juniper woodland; 1500–2600 m. SN, WTR, **GB**; to WY, Colorado. *[Stipa t.* Piper] ❀ DRN:1,**2,3,7,10**,11,12,**14–24**;STBL. Jun–Jul

A. webberi (Thurber) Barkworth **ST** 1–3.5 dm. **LF**: blade gen rolled, < 0.5 mm diam, stiff. **INFL** 2.5–7 cm, dense, often partly enclosed by upper lf sheath. **SPIKELETS**: glumes 6–10 mm, ± equal; floret 4.5–6 mm, callus blunt; lemma evenly, densely long-hairy, hairs 2.5–3.5 mm, awn 5–11 mm, straight, readily deciduous; palea ± = lemma. 2*n*=32. Dry, open flats, rocky slopes, gen with sagebrush; 1500–3000 m. SN, MP, **W&I**; to OR, Colorado. *[Oryzopsis w.* (Thurber) Benth.] ❀ DRN,SUN: 1,**2,3,7**,8,9,**10**,11,**14–24**; STBL. Jun–Jul

AGROPYRON CRESTED WHEATGRASS

Mary E. Barkworth

Per. **ST** erect, gen tufted. **LF**: sheath sometimes appendaged; ligule membranous; blade flat or rolled. **INFL** spike-like, dense; axis not breaking apart in fr; spikelets 2-ranked, strongly overlapping, 1 per node, gen spreading. **SPIKELET**: glumes gen ± equal, = or < lower floret, lanceolate, keeled, acute to short-awned; axis breaking above glumes and between florets; lemma firm, acute to awn-tipped; anthers 2–5 mm. 12 spp.: Medit, e Eur, c Asia. (Greek: ancient name for wild wheat) [Barkworth & Dewey 1985 Amer J Bot 72:767–776] See also *Elymus, Elytrigia, Pascopyrum, Pseudoroegneria* for spp. sometimes treated here.

A. desertorum (Fischer) Schultes (p. 563) DESERT CRESTED WHEATGRASS **ST** 2.5–10 dm. **LF**: sheath, blade gen glabrous, sometimes soft-hairy or scabrous. **INFL** 3–10 cm; internodes 1–3 mm; spikelets spreading to ascending. **SPIKELET** 7–11 mm; glumes < lower floret, translucent margin ± 0.4 mm wide, gen 3-veined; florets 3–8; lemma 5–9 mm, gen 5-veined, awn < 3.5

mm. 2*n*=28. Disturbed areas, roadsides; 600–1500 m. CaRF, n SNH, **GB**; to Great Plains; native to e Eur. Pls with infl internodes < 1 mm, spikelets spreading are called *A. cristatum* (L.) Gaertner. Pls with spikelets appressed, lemma awn < 1 mm, are called *A. fragile* Roth *[A. sibiricum* Willd.] All forms (and hybrids) are planted for erosion control.

AGROSTIS BENT

M. J. Harvey

Ann or per, gen tufted, sometimes from rhizomes or stolons. STS gen erect. **LF**: sheath gen smooth, glabrous; ligule membranous; blade flat to rolled. INFL panicle-like, densely cylindric to openly ovate. SPIKELET: glumes gen subequal, back gen glabrous, vein gen finely scabrous, 1-veined, gen acute; floret 1, < glumes, gen breaking above glumes; callus glabrous to densely hairy; lemma gen 5-veined, veins not converging, sometimes extended as short teeth, awned from back or not; palea 0 to ± = lemma, translucent; anthers gen 3. ± 200 spp.: esp temp Am, Eurasia. (Greek: pasture) [Carlbom 1967 PhD OR State Univ] Some cult in pastures, lawns. *A. gigantea* has not been reported from the desert areas but may be expected. This heat-resistant European weed should be looked for near habitations and irrigated areas.

1. Mature infl branches widely spreading; spikelets not crowded, gen well spread on same branch
  2. Infl ± 2 × longer than wide, < 1/3 st; st lvs 3 or more . . . . **A. idahoensis**
  2' Infl ± as long as wide, ± 1/2 st; st lvs 1 or 2 . . . . **A. scabra**
1' Mature infl narrow, branches gen ascending; spikelets crowded and often overlapping on same branch
  3. Spikelet breaking below glumes, falling as 1 unit; older sts decumbent, rooting . . . . **A. semiverticillata**
  3' Floret breaking above persistent glumes; sts stoloniferous or tufted
    4. Pl from stolons, palea slightly < lemma . . . . **A. stolonifera**
    4' Pl tufted or loosely matted; palea 0 or < 1/3 lemma
      5. Pl tufted, rhizomes 0; glumes acute to narrowly acuminate; lemmas awnless or awn < 3.5 mm . . **A. exarata**
      5' Pl from rhizomes < 10 cm, loosely matted; glumes acute; lemmas awnless or awn < 0.5 mm . . . . . **A. pallens**

A. exarata Trin. (p. 563) Per 8–100 cm. **LVS**: ligule 2.5–4 mm; lower blades 4–15 cm, 2–7 mm wide, flat. **INFL** 5–30 cm, oblong to ± ovate in outline, ± open to dense, sometimes interrupted near base; 1° branches 1–2 cm, ascending to ± appressed. **SPIKELET**: glumes 1.5–3.5 mm, acute to narrowly acuminate; callus hairs < 0.5 mm; lemma 1–2 mm, sometimes awned above middle, awn < 3.5 mm, straight to bent; palea < 1/3 × lemma; anthers 0.3–0.6 mm. 2*n*=28,42,56. Common. Moist or disturbed areas, open woodland, coniferous forest; < 2000 m. CA-FP, **W&I, DMtns (Panamint Mtns)**; to AK, SD, Mex. [*A. ampla* A. Hitchc.; *A. longiligula* Vasey vars. *l.* and *australis* J. Howell] Pls with dense, narrowly cylindric infl and awned lemmas have been called var. *monolepis* (Torrey) A. Hitchc. ❀ STBL. Jun–Aug

A. idahoensis Nash (p. 563) Per 8–30 cm. **LVS**: ligule 1–3 mm; lower blades 1–5 cm, 0.5–2 mm wide, flat, often inrolled with age. **INFL** 3–13 cm, lanceolate to ovate in outline, ± open; 1° branches gen ascending, lower 1–4 cm, axes thread-like. **SPIKELET**: glumes 1.5–2.5 mm; callus glabrous or hairs < 0.3 mm; lemma 1–2 mm, awn 0; palea minute, << lemma; anthers 0.3–0.5 mm. Common. Open, wet meadows, coniferous forest; < 3500 m. NW, CaR, SN, n SnFrB, SnBr, SnJt, **W&I**; to AK, MT, NM. [*A. tenuis* Vasey] Jul–Aug

A. pallens Trin. Per 10–70 cm, sometimes from rhizomes < 10 cm. **LVS**: ligule 1.5–3 mm; lower blades 1.5–5 cm, 1–6 mm wide, flat to inrolled. **INFL** 5–20 cm, lanceolate to narrowly ovate in outline, ± open; 1° branches gen ascending, lower 2–5 cm. **SPIKELET**: glumes 2–3 mm; callus hairs minute; lemma 1.5–2.5 mm, sometimes awned near tip, awn < 0.5 mm, straight; palea 0 or minute, << lemma; anthers 0.7–1.8 mm. 2*n*=42,56. Common. Open meadows, woodland, forest, subalpine; 200–3500 m. CA-FP, **GB**; to B.C., MT. [*A. lepida* A. Hitchc.; *A. diegoensis* Vasey] Geog and ecological variation need study. ❀ DRN,part SHD:**1,2,3,4–7,**8,9,**14–24**;decid.GRNCVR. Jun–Aug

A. scabra Willd. (p. 563) Per 20–75 cm. **STS** ascending to erect. **LVS** mostly basal; ligule 2–5 mm; lower blades 4–14 cm, 1–3 mm wide, flat, finely scabrous. **INFL** 8–25 cm, ovate in outline, open; 1° branches spreading below, ascending above, lower 4–11 cm, axes thread-like, branched 1–2 × above middle, often breaking at base in fr. **SPIKELET**: glumes 1.5–3 mm; callus hairs minute, sparse; lemma 1.5–2 mm, sometimes awned from below middle, awn < 2 mm, ± straight; palea 0 or minute, << lemma; anthers 0.4–0.7 mm. 2*n*=42. Open roadsides, meadows, coniferous forest; 1000–3100 m. KR, NCoR, SN, TR, SnJt, **SNE**; to AK, ne US. Stunted alpine pls have been called var. *geminata* (Trin.) Swallen. ❀ SUN:4,5,**6**&IRR:1–3,7,14,**15–17**,18–24;may be INV. Jul–Sep

A. semiverticillata (Forsskal) C. Chr.(p. 563) Gen per 10–75 cm, decumbent to long-trailing, rooting at nodes. **LVS**: ligule 1–6 mm; lower blades 3–18 cm, 2–10 mm wide, flat. **INFL** 2–15 cm, narrowly lanceolate to elliptic in outline, dense, often interrupted; 1° branches erect to ± appressed, lower 1–4 cm. **SPIKELET** breaking below glumes, falling as 1 unit; glumes 1.5–2.5 mm, back finely scabrous; callus glabrous; lemma 1–1.5 mm, awn 0; palea slightly < lemma; anthers 0.5–0.7 mm. 2*n*=28. Common. Disturbed areas, wet areas, ponds, ditches, streambanks; < 2000 m. CA; to s US; native to Eur. [*A. viridis* Gouan] May–Jun

A. stolonifera L. (p. 563) CREEPING BENT Per 8–60 cm, decumbent to erect, often mat-like; stolons 5–100 cm. **LVS**: ligule 2–5 mm; lower blades 2–10 cm, 2–5 mm wide, flat. **INFL** 3–15 cm, elliptic to lanceolate in outline, ± dense; 1° branches ascending to ± erect, lower gen 2–6 cm. **SPIKELET**: glumes 1.5–3 mm; callus hairs minute, sparse; lemma 1.5–2 mm, awn 0; palea slightly < lemma; anthers 1–1.5 mm. 2*n*=28. Ditches, lake margins, marshes; < 1000 m. NW, CaR, n SN, CW, SW (exc ChI), **W&I, DMtns**; to s Can, s US; native to Eur. [*A. alba* L. vars. *major* (Gaudin) Farw. and *palustris* (Hudson) Pers.; *A. alba* var. *a.* misapplied in part]. Jun–Sep

ALOPECURUS FOXTAIL

William J. Crins

Ann, per, cespitose or from stolons. **ST** decumbent to erect, 1–8 dm; nodes visible, brown. **LVS**: ligule 1–6 mm, membranous, truncate to acute, gen scabrous; blade flat, glabrous or scabrous. **INFL** panicle-like, gen cylindric, dense; branches short. **SPIKELET** ± compressed, breaking below glumes, falling as 1 unit; glumes ± equal, gen = spikelet, membranous, gen keeled, keel and lateral veins gen stiff- or appressed-hairy, margins free or fused near base, tip obtuse, acute, or short-awned, 3-veined; floret 1; lemma membranous, margins keeled, sometimes fused near base, truncate to acute, 3–5-veined, awned on back below middle, awn straight or abruptly bent gen at lemma tip; palea gen 0; anthers 0.5–4 mm. **FR** glabrous. ± 35 spp.: temp N.Am, Eurasia. (Greek: fox tail) [Rubtzoff 1961 Leafl West Bot 9:165–180]

A. aequalis Sobol. (p. 569) SHORT-AWN FOXTAIL Per. **ST** 0.9–4.7 dm. **LF**: ligule 2–5.5 mm; blade 2.5–10 cm, 1–5(8) mm wide. **INFL** 1–7.5 cm, 3–6(9) mm wide. **SPIKELET**: glumes 2–3(3.5) mm; lemma awn straight, exceeding lemma body by 0–2 mm; anthers 0.5–1 mm. 2*n*=14. Common. Wet meadows, shores; 100–3500 m. CA-FP, **GB**; to subarctic, e N.Am, Eurasia. May–Jul

ANDROPOGON BLUESTEM

Kelly W. Allred

Per, cespitose. **STS** erect, branched; nodes gen hairy. **LVS** cauline; ligule membranous, minutely ciliate; blade flat or folded. **INFL** panicle-like with 2 or more spike-like branches, solitary or compactly clustered, partly enclosed in lf sheaths; axes breaking apart with age; spikelet sessile, subtended by hairy, naked stalk and axis segment, falling with stalk and axis segment as 1 unit. **SPIKELET**: glumes ± = florets, lanceolate, ± translucent; florets 2, lower vestigial, obscure, upper fertile, lemma translucent, awned, palea << lemma or 0; stamens 1–3. **FR** oblong, brownish or purplish. ± 100 spp.: warm temp, trop. Some spp. cult for forage, revegetation. (Greek: man beard, from hairy staminate spikelets) [Campbell 1983 J Arn Arbor 64:171–254]

A. glomeratus (Walter) Britton, Sterns, & Pogg. var. ***scabriglumis*** C.S. Campbell (p. 569) SOUTHWESTERN BUSHY BLUESTEM **ST** 0.8–1.5 m. **LF**: sheath scabrous; ligule 1–2.2 mm; lower blades 3–6 dm, 3.5–6 mm wide. **INFLS** many, compactly clustered, plume-like; branches gen 2–4. **SPIKELET** 4–4.5 mm; lower glume keel gen scabrous at base; callus hairs 1–2 mm; awn 0.5–2 mm. 2*n*=20,40. Moist, open, disturbed areas, seeps; < 600 m. TR, **DMoj**, naturalized in NCo, NCoRO, n SNF, SNH, ScV, SCo; to NM, Baja CA. ❀ IRRorWET,SUN:5,**7–11**,12,**14–16**,17,18,**19–24**. Sep–Mar

APERA

Dieter H. Wilken

Ann. **STS** erect, solitary to tufted, gen glabrous. **LVS** basal and cauline; ligule membranous, acute to toothed at tip; blade flat to rolled, gen ridged. **INFL** panicle-like; branches spike-like, minutely scabrous. **SPIKELET**: glumes subequal, > floret, membranous, lower 1-veined, upper 1–3-veined; floret 1, bisexual, breaking above glumes; lemma rounded on back, membranous, short bristly at base, minutely scabrous and awned near tip, obscurely 3–5-veined, awn straight, flexible; palea ± = lemma. ± 3 spp.: temp Eur, Asia. Cult for revegetation. (Greek: not maimed, alluding to occasional presence of vestigial florets)

A. interrupta (L.) P. Beauv. **STS** gen < 3, 1–6 dm, sometimes tufted. **LF**: ligule 1–4 mm; blade 1–3 mm wide, minutely scabrous above. **INFL** 5–15 cm; branches ascending; glumes 1.5–3 mm; lemma 1.5–2 mm, minutely scabrous at tip, awn 5–8 mm. $2n$=14, 28. Disturbed sites; 1220–1700 m. MP (w of Alturas, Modoc, Co.), **DMtns (Furnace Creek, e Inyo Co.)**; reported elsewhere in w US; native to Eur, Asia.

ARISTIDA THREE-AWN

Kelly W. Allred

Ann, per, cespitose. **ST** ascending to erect. **LVS** basal and cauline; basal often tufted; ligule hairy; blades flat or inrolled. **INFL** raceme-like or panicle-like; branches spike-like. **SPIKELET**: glumes narrowly lanceolate, thin, 1-veined, awn gen 0; floret 1, breaking above glumes; lemma ± fusiform, hard when mature, 3-veined, tip beak-like or not, awned at tip, awns 3, equal or unequal; palea < lemma, enclosed by lemma, transparent. **FR** narrowly fusiform. ± 300 spp.: worldwide, arid warm temp. Some spp. noxious. (Latin: awn) [Allred 1992 Great Basin Nat 52:41–52]

1. Lower sts densely hairy . . . . *A. californica*
1' Lower sts glabrous
  2. Ann . . . . *A. adscensionis*
  2' Per
    3. Base of 1° infl branches abruptly spreading or ascending from main axis
      4. Lower infl branches ascending, upper appressed . . . . *A. purpurea* var. *parishii*
      4' All infl branches spreading . . . . *A. ternipes* var. *hamulosa*
    3' Base of 1° infl branches appressed to main axis, tips sometimes spreading or ascending . . . . *A. purpurea*
      5. Most awns 4–10 cm
        6. Upper glume gen > 16 mm; lemma ± linear . . . . var. *longiseta*
        6' Upper glume gen < 16 mm; lemma gradually tapered to tip . . . . [2]var. *purpurea*
      5' Most awns < 3.5 cm
        7. Mature infl branches and spikelet stalks slender, curving to drooping, appearing S- or U-shaped
        . . . . [2]var. *purpurea*
        7' Mature infl branches and spikelet stalks gen stiff, straight, erect to spreading
          8. Tip of lemma gen < 0.2 mm wide; awns delicate, gen < 0.2 mm wide at base . . . . var. *nealleyi*
          8' Tip of lemma gen > 0.2 mm wide; awns stiff, gen > 0.2 mm wide at base
            9. Infl gen 3–14 cm; lf blade gen < 10 cm . . . . var. *fendleriana*
            9' Infl gen 15–30 cm; lf blade gen > 10 cm
              10. Infl dense, reddish, fading to straw-colored; lower branches with 8–18 spikelets . . . . var. *parishii*
              10' Infl somewaht loose, tan to brown, fading to straw-colored; lower branches with 2–10 spikelets
              . . . . var. *wrightii*

A. adscensionis L. (p.569) SIX-WEEKS THREE-AWN Ann. **ST** branched below, 0.5–8 dm. **LF**: blade < 10 cm, inrolled. **INFL** 2–22 cm, narrow. **SPIKELET**: lower glume 5–7 mm, upper 8–10 mm; lemma 6–13 mm, awns ±equal, 7–23 mm. $2n$=22. Dry, open places, rocky sites, shrubland ; < 1400 m. CCo, SCo, s ChI, WTR, PR, **D**; to MO, TX, S.Am. Some pls (Riverside Co.) have very short lateral awns. Feb–Jun

A. californica Thurber var. *californica* CALIFORNIA THREE-AWN Per, ± bushy. **ST** much-branched, gen 1–4 dm, densely hairy. **LF**: sheath << internodes; blade < 5 cm, gen inrolled. **INFL** 2–6 cm. **SPIKELET**: lower glume 4–8 mm, upper 9–12 mm; lemma 5–7 mm, narrow beak at tip 8–26 mm, awns 2–4.5 cm, beak and awns breaking from lemma. $2n$=22. Dry sandy sites, shrubland; < 150 m. **D**; AZ, nw Mex. ❀ TRY. Apr–May

A. purpurea Nutt. Per. **ST** gen erect, unbranched, 10–80 cm. **LF**: blade 5–25 mm, 1–2 mm wide, gen inrolled. **SPIKELET**: glumes thin, lower 4–12 mm, upper 7–25 mm; awns 0.5–6 cm, equal or central slightly longer. $2n$=22,44,66,88. Sandy to rocky soils, slopes, plains; < 2000 m. SCo, SnBr, PR, **SNE, D**; to sw

Can, Great Plains, n Mex.

  var. *fendleriana* (Steudel) Vasey FENDLER THREE-AWN **LVS** gen basal; blade gen < 10 cm. **INFL** 3–14 cm, narrow. **SPIKE-LET**: lower glume 5–8 mm, upper 10–15 mm; lemma 8–14 mm, awn gen 1.8–4 cm, 0.2–0.3 mm wide at base. $2n$=22,44. Dry, rocky slopes, shrubland; 1000–2000 m. SnBr, PR, SNE, **DMoj**; to MT, Great Plains, n Mex. [*A. f.* Steudel] ❀ STBL. May–Jul

  var. *longiseta* (Steudel) Vasey (p. 569) RED THREE-AWN **LVS** basal or cauline; blade 4–16 cm. **INFL** 5–15 cm; branches stiff and erect to delicate and drooping. **SPIKELET**: lower glumes 8–12 mm, upper gen 16–25 mm; lemma 12–16 mm, 0.4–0.8 mm wide just below awns, awn 4–10 cm, 0.2–0.5 mm wide at base, stiff. $2n$=22,44,66,88. Dry slopes, plains, shrubland; 300–1600 m. SnBr, **D**; to sw Can, n Mex. [*A. l.* Steudel] ❀ TRY. Mar–May

  var. *nealleyi* (Vasey) K.W. Allred (p. 569) NEALLEY THREE-AWN **LVS** gen basal; blade 5–25 cm. **INFL** 8–18 cm, narrow, light brown; branches gen erect. **SPIKELET**: lower glume 4–7

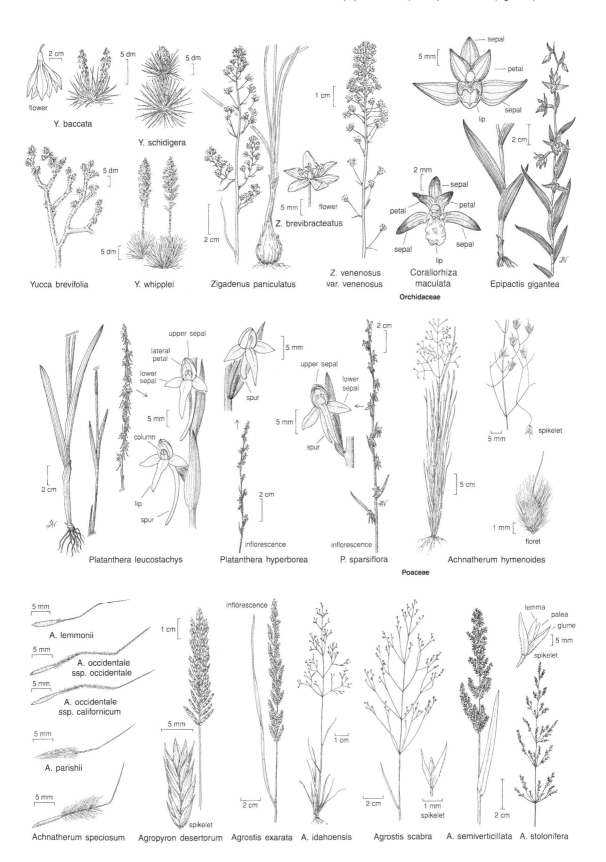

Y. baccata

Y. schidigera

Yucca brevifolia

Y. whipplei

Zigadenus paniculatus

Z. brevibracteatus

flower

Z. venenosus var. venenosus

Corallorhiza maculata

Epipactis gigantea

Orchidaceae

Platanthera leucostachys

Platanthera hyperborea

P. sparsiflora

Achnatherum hymenoides

Poaceae

A. lemmonii

A. occidentale ssp. occidentale

A. occidentale ssp. californicum

A. parishii

Achnatherum speciosum

Agropyron desertorum

Agrostis exarata

A. idahoensis

Agrostis scabra

A. semiverticillata

A. stolonifera

mm, upper 8–14 mm; lemma 7–13 mm, 0.1–0.2 mm wide just below awns; awns 1.5–2.5 cm, gen 0.1 mm wide at base, delicate. 2*n*=22,44. Dry slopes, plains, shrubland; 200–2000 m. SCo, SnBr, PR, **D**; to s UT, OK, n Mex. [*A. glauca* (Nees) Walp.] Mar–May, Sep

var. ***parishii*** (A. Hitchc.) K.W. Allred (p. 569) PARISH THREE-AWN **LVS** cauline; blade > 10 cm, gen flat. **INFL** 15–24 cm, narrow; lower branches sometimes stiffly spreading, reddish when young. **SPIKELET**: lower glume 7–11 mm, upper 10–15 mm; lemma 10–13 mm, 0.2–0.3 mm wide just below awns, awns 2–3 cm, 0.2–0.3 wide at base. Dry slopes, plains, chaparral, shrubland; 300–1300 m. SCo, SnBr, **D**; s NV, Baja CA. [*A. parishii* A. Hitchc.] Apr–Jun

var. ***purpurea*** PURPLE THREE-AWN **LVS** gen cauline; blade 5–25 cm. **INFL** 5–20 cm, gen nodding, purplish; branches gen delicate, wavy or drooping. **SPIKELET**: lower glume 4–9 mm, upper 7–16 mm; lemma 6–12 mm, 0.1–0.3 mm wide just below awns, awns gen 2–6 cm, 0.1–0.3 mm wide at base. 2*n*=22,44,66,88. Dry slopes, shrubland; 250–800 m. SCo, SnBr,

PR, **DMtns**; to Arkansas, n Mex. ❀ TRY. May–Jul

var. ***wrightii*** (Nash) K.W. Allred (p. 569) WRIGHT THREE-AWN **LVS** cauline, 10–25 cm, flat to inrolled. **INFL** gen 14–30 cm; branches gen erect. **SPIKELET**: lower glume 5–10 mm, upper 10–16 mm; lemma 8–14 mm, 0.2–0.3 mm wide just below awns, awns gen 2–3.5 cm, 0.2–0.3 wide at base. 2*n*=22,44,66. Sandy to rocky slopes, plains, shrubland; 500–1800 m. PR, **D**; to s UT, OK, n Mex. [*A. w.* Nash] ❀ DRN,SUN,DRY: 3,7,8,**9**,**10**, 11,12,**14–24**. Mar–Jun, Sep

A. ternipes Cav. var. ***hamulosa*** (Henrard) J.S. Trent (p. 569) HOOK THREE-AWN Per, sometimes bushy. **STS** few, prostrate to erect, 25–80 cm. **LF**: blade 5–40 cm, flat to inrolled, base sparsely long-hairy. **INFL** 15–40 cm, open; primary branches spreading. **SPIKELET**: glumes ± equal, 9–15 mm; lemma 10–15 mm, gen straight at tip, awns equal to unequal, central awn 10–25 mm, lateral 6–23 mm. 2*n*=44. Dry hills and slopes; 100–1350 m. GV, SCo, TR, PR, **DMoj**; to TX, C.Am. [*A. h.* Henrard] ❀ DRN,SUN:2,3,**7**,8,9,**10**,11,12,**14–16**,17,**18–24**;also STBL. May–Nov

ARUNDO

Kelly W. Allred

Per; rhizomes thick. **ST** erect, cane-like. **LVS** cauline; sheaths > internodes, glabrous; ligule thinly membranous, fringed; blade flat or folded, glabrous, margin scabrous. **INFL** panicle-like, dense, silvery to purplish. **SPIKELET** laterally compressed; glumes > florets, 3–5-veined; axis glabrous; florets 4–5, breaking above glumes and between florets; lemmas 3–7-veined, hairy, awn ± 0; palea < lemma; 3 spp.: warm temp, trop. (Latin: a reed grass)

A. donax L. (p. 569) GIANT REED **ST** < 8 m; nodes glabrous; internodes < 4 cm thick. **LF**: blade < 1 m, 2–6 cm wide. **INFL** 3–6 dm, plume-like; branches ascending. **SPIKELETS** 10–14 mm; glumes 10–13 mm, thin, brownish or purplish; lemmas 8–12 mm,

tip 2-toothed, hairs < 8 mm, silky; palea 3–5 mm, hairy at base; anthers 2.5–3 mm. 2*n*=110. Moist places, seeps, ditchbanks; < 500 m. c SNF, CCo, SCo, SnGb, **D**; native to Eur. Mar–Sep

AVENA OATS

Dieter H. Wilken

Ann. **STS** erect, 1–6, ± glabrous. **LVS** basal and cauline; ligules 2–5 mm, membranous, rounded at tip; blade flat. **INFL** panicle-like, open. **SPIKELETS** gen stalked, ± pendent; glumes subequal, gen > florets, membranous, 5–7-veined, gen glabrous; axis prolonged behind upper floret, vestigial floret at tip; florets 2–3, upper floret > lower, bisexual, breaking above glumes and between florets or not breaking; lemmas hard, glabrous to hairy below awn, awned at or slightly below middle, tip 2-forked, 5–7-veined, awn stiff, bent to straight, slightly to strongly coiled below bend; palea ± < lemma. ± 10–15 spp.: temp Eur, Asia. Cult for grain, hay. (Latin: oats)

1. Forks at lemma tip bristly, bristles > 2 mm; glumes 5–7-veined . . . . ***A. barbata***
1' Forks at lemma tip < 1 mm; glumes 7–9-veined . . . . ***A. fatua***

A. barbata Link (p. 569) SLENDER WILD OAT **ST** 3–6 dm. **LF**: blade 2–6 mm wide, glabrous to minutely scabrous (sometimes ciliate). **SPIKELET**: glumes 18–30 mm, 5–7-veined; lemmas 12–18 mm, gen soft-hairy below awn, forks at tip 2–6 mm, bristly, awn 20–40 mm, bent, twisted below bend. 2*n*=28. Disturbed sites; 40–1200 m. CA-FP, MP, **DMoj**; reported elsewhere in US; native to s Eur. Mar–Jun

A. fatua L. (p. 569) WILD OAT **ST** 3–12 dm. **LF**: blade 4–12 mm wide, minutely scabrous. **SPIKELET**: glumes 18–25 mm, 7–9-veined; lemmas 14–20 mm, gen glabrous on back to soft-hairy in lower third, forks at tip < 1 mm, awn 25–40 mm, bent, twisted below bend. 2*n*=42. Disturbed sites; 25–1220 m. CA-FP, MP, **DMoj**; to AK, e Can, most of US; native to Eur. *A. occidentalis* Durand, with 4 florets, has been reported from CA. Apr–Jun

BECKMANNIA

Ann, per. **STS** gen erect. **LVS** cauline; sheath glabrous; ligule membranous, acute, entire or irregularly cut; blades flat. **INFL** panicle-like; branches short, spike-like, ascending to appressed; spikelets 2-ranked, overlapping. **SPIKELET** ± round in outline; glumes ± equal, strongly compressed, wrinkled, 3-veined, winged; axis breaking below glumes, falling as 1 unit; florets 1–2, bisexual; lemma 5-veined, acuminate; palea narrow, < lemma. 2 spp.: N.Am, Eur. (J. Beckmann, German botanist, 1739–1811) [Reeder 1953 Bull Torrey Bot Club 80:187–196]

B. syzigachne (Steudel) Fern. (p. 569) **ST** 2.5–10 dm. **LF**: ligule 4–8 mm; blades 6–21 cm, 4–10 mm wide. **INFL** 9–28 cm, narrow, ± dense; branches 0.5–4 cm, ascending to appressed. **SPIKELET** 1.5–3 mm; glumes glabrous to finely scabrous, tip

short-pointed; floret 1; lemma 2–3 mm, lanceolate, acuminate; palea < lemma, 2-toothed. 2*n*=14. Pond margins, streams, marshes; < 1500 m. NW, CaR, n SN, n CCo, SnFrB, MP, **n SNE**; to AK, e US. May–Jul

BLEPHARIDACHNE EYELASH GRASS

Ann, per. **LVS** short, tufted. **INFL** panicle-like, dense, short, barely elevated above lvs. **SPIKELET** bisexual; glumes > florets, thin, 1-veined; axis breaking apart above glumes, falling as 1 unit; florets 4, lower 2 sterile or staminate, 3rd bisexual, uppermost sterile, ± vestigial; lemma ciliate, back rounded, 3-veined, deeply 2-lobed to middle, lobes short-awned, tip awned between lobes, awn = lobes; palea ciliate or glabrous. 4 spp.: sw N.Am, Argentina. (Greek: eyelash + chaff, from ciliate lemma)

B. kingii (S. Watson) Hackel (p. 569) KING'S EYELASH GRASS
Per, tufted. **ST** gen < 10 cm. **LF:** blade 1–3 cm, < ± 1 mm wide; sheath margins translucent; blade curved, ± stiff, sharp-pointed. **INFL** < 2 cm, head-like, terminal, exceeded by upper blades, straw-colored or purplish. **SPIKELET** compressed; glumes 7–8 mm, ± equal, acuminate; callus soft-hairy; lemmas ± 6 mm, margins ciliate, base soft-hairy; sterile palea narrower and < lemma; fertile palea = lemma. 2*n*=14. RARE in CA. Pinyon/juniper woodland; 1350–1600 m. **SNE (Inyo, Mono cos.)**; to s UT. [Hunziker & Anton 1979 Brittonia 31:446–453] May

BOTHRIOCHLOA

Kelly W. Allred

Per, cespitose. **ST** gen erect; nodes gen short-hairy. **LVS** basal and cauline; ligule membranous; blade flat or folded. **INFL** panicle-like, branching 1–several ×; branches spike-like, long-soft-hairy; axes grooved, breaking apart with age. **SPIKELETS** paired; lower sessile, bisexual; upper stalked, staminate or sterile; pair with subtending axis segment falling as 1 unit. **SESSILE SPIKELET:** glumes > florets, membranous; florets 2, lower vestigial, obscure, upper fertile; lemma translucent, tip awned. **FR** oblong to fusiform. ± 35 spp.: worldwide, warm temp, trop. (Greek: pit, from pitted glumes of some spp.) [Allred & Gould 1983 Syst Bot 8:168–184] Cult for forage, revegetation.

B. barbinodis (Lagasca) Herter (p. 573) CANE BLUESTEM **STS** 6–12 dm, clumped. **LVS** basal and cauline; blade 20–30 cm. **INFL** 7–14 cm, gen branching 2 ×, primary branches 4–9 cm. **SPIKE-LET:** upper spikelet stalk 3–4 mm. **SESSILE SPIKELET** > stalked spikelet, 4.5–7.5 mm, gen tan; lower glume sometimes pitted on back; awn 2–3 cm. 2*n*=180. Dry slopes; < 1200 m. SCo, s ChI, WTR, SnGb, PR, **DMtns**; w to OK, TX, Mex. [*Andropogon b.* Lagasca] ❀ TRY. Feb–Sep

BOUTELOUA GRAMA

J. Travis Columbus

Ann, per, gen cespitose. **ST** solid, gen glabrous. **LVS** gen basal; ligule gen < 1 mm, gen hairy; blade flat to inrolled, upper surface gen ± short-hairy, often ciliate near ligule, hairs long, bulbous-based. **INFL** gen panicle-like; branches spike-like, 1 per node, persistent or deciduous in fr; spikelets 2-rowed on 1 side of axis, overlapping. **SPIKELET** sessile or short-stalked, ± cylindric to laterally compressed; glumes gen unequal, gen lanceolate, 1-veined, upper glume firmer than lower; axis (if infl branch persistent) breaking between glumes and lower floret; florets gen 2–3, lower floret bisexual, > upper, upper florets gen vestigial, sterile; lemmas 3-veined, gen 3-awned, awns straight, scabrous; palea ± = lemma. ± 40 spp.; Am. (Claudio (born 1774) and Esteban (born 1776) Boutelou, Spanish botanists, horticulturists) [Gould 1979 Ann Missouri Bot Gard 66:348–416] Many spp. important for forage.

1. Ann
　2. Spikelets 1–4 per infl branch; branch axis prolonged beyond terminal spikelet node 5+ mm; infl branches deciduous in fr, spikelets falling with branch . . . . ***B. aristidoides***
　2' Spikelets gen > 7 per infl branch; branch axis terminated by spikelet; infl branches persistent, spikelet axis breaking between glumes and lower floret
　　3. Tip of lower floret lemma 2-lobed, central awn from sinus; base of awned upper floret hairy-tufted
　　 . . . . ***B. barbata***
　　3' Tip of lower floret lemma tapered to central awn; base of upper floret glabrous . . . . ²***B. trifida***
1' Per, rhizomes or stolons sometimes present
　4. Infl branches 13–60, deciduous in fr, spikelets falling with branch . . . . ***B. curtipendula***
　4' Infl branches < 9, persistent in fr, spikelet axis breaking between glumes and lower floret
　　5. Lower st internodes hairy; infl branch axis slightly prolonged beyond terminal spikelet node . . . . ***B. eriopoda***
　　5' Lower st internodes glabrous or minutely scabrous; infl branch axis terminated by fertile or vestigial spikelet
　　　6. Base of awned upper floret hairy-tufted; tip of lower floret lemma 2-lobed, central awn from sinus . . ***B. gracilis***
　　　6' Base of upper floret glabrous; tip of lower floret lemma tapered to central awn . . . . ²***B. trifida***

B. aristidoides (Kunth) Griseb. var. ***aristidoides*** (p. 573)
NEEDLE GRAMA Ann. **ST** decumbent to erect, 0.5–3.5 dm. **LF:** blade < 7 cm, < 2 mm wide. **INFL:** branches 4–16, 8–25 mm, appressed to reflexed, deciduous in fr; branch axis exceeding terminal spikelet node > 5 mm, base densely short-hairy; spikelets 1–4 per branch, gen appressed, falling with branch. **SPIKELET:** upper glume 5–7 mm, glabrous or hairy, acute; florets (1)2; lower floret lemma ± = upper glume, glabrous or hairy, lobes 0, awns 0 or 2, < 1 mm; upper floret base hairy-tufted, lobes gen 0 between awn bases, awns 2–7 mm, gen unequal. 2*n*=40. Dry, open, sandy to rocky slopes, flats, washes, disturbed sites, scrub, woodland; < 1800 m. e PR, **e&s DMoj, DSon**; to UT, TX, s Mex, S.Am. Another var. in AZ, NM, n Mex. ❀ STBL. Apr–Sep

B. barbata Lagasca var. *barbata* (p. 573) SIX WEEKS GRAMA Ann. **ST** prostrate to erect, 0.5–3 dm. **LF**: blade < 6 cm, < 2 mm wide. **INFL**: branches 2–8, 6–25 mm, spreading to appressed, persistent in fr; branch axis terminated by spikelet, base glabrous or puberulent; spikelets 7–40 per branch, spreading to ascending, breaking apart between glumes and lower floret. **SPIKE-LET**: upper glume 1–3.5 mm, glabrous or hairy, tip gen notched, gen awned from sinus < 1 mm; florets 2–3, lower floret lemma ± = upper glume, hairy below middle, tip 2-lobed, awns < 3.5 mm, ± equal, central awn from sinus; base of middle or, if only 2 florets, upper floret hairy-tufted, lobed between awn bases, awns 1–3.5 mm, ± equal; uppermost floret (if present) < 1 mm, awn 0. $2n=20,40$. Gen open, sandy to rocky slopes, flats, washes, roadsides, disturbed sites, scrub, woodland, pine forest; < 1700 m. SnJV, e PR, **D**; to Colorado, TX, s Mex. Other vars. in AZ, NM, n Mex. ❀ STBL. Jul–Dec

B. curtipendula (Michaux) Torrey (p. 573) SIDE-OATS GRAMA Per, sometimes rhizomed. **ST** gen erect, 2–9 dm. **LF**: blade < 25 cm, < 4 mm wide. **INFL**: branches 13–60, 5–20 mm, gen pendent, deciduous in fr; branch axis slightly exceeding terminal spikelet node, base puberulent; spikelets 1–13 per branch, ascending to appressed, falling with branch. **SPIKELET**: upper glume 3–10 mm, glabrous or scabrous, acute or awned < 0.5 mm; florets 1–2; lower floret lemma < to ± = upper glume, glabrous or sparsely hairy, lobes 0, awns 2, < 2 mm; base of upper floret (if present) glabrous, tip 2-lobed, central awn from sinus, < 6 mm, lateral awns < 4 mm. $2n=20,28,35,40–103$. Dry, rocky slopes, crevices, sandy to rocky drainages, scrub, woodland; < 1900 m. s ScV (Yolo Co. as roadside waif), PR (Santa Rosa, Cuyamaca mtns), **e&s DMtns**; to s Can, e US, S.Am. Pls with erect sts and rhizomes 0 have been called var. *caespitosa* Gould & Kapadia. ❀ DRN,SUN:2,3,**7,14–16**,17,**18, 24**&IRR:**8–12,19–23**;alsoSTBL. May–Aug

B. eriopoda (Torrey) Torrey (p. 573) BLACK GRAMA Per, sometimes stoloned. **ST** decumbent to erect, 1.5–6 dm; inter-nodes, esp lower, hairy. **LF**: blade < 10(15) cm, < 2 mm wide. **INFL**: branches 2–7, 10–40 mm, spreading to appressed, persistent in fr; branch axis slightly exceeding terminal spikelet node, base densely hairy; spikelets 6–18 per branch, ascending to appressed, breaking apart between glumes and lower floret. **SPIKELET**: upper glume 4–8 mm, gen glabrous, sharply acute; florets 2; lower floret lemma ± = to > upper glume, base hairy-tufted, glabrous or sparsely hairy above, lobes 0, central awn 1.5–4 mm, lateral awns < 2 mm; upper floret base hairy-tufted, lobes gen 0 between awn bases, awns 4–8 mm, ± equal. $2n=20,21,28$. Dry, open, sandy to rocky slopes, flats, washes, scrub, woodland; 900–1900 m. e DMtns; to WY, OK, n Mex. Jun–Sep

B. gracilis (Kunth) Griffiths (p. 573) BLUE GRAMA Per, gen short-rhizomed. **ST** decumbent to erect, 1–6 dm. **LF**: blade < 15 cm, < 2 mm wide. **INFL**: branches 1–4(6), 10–50 mm, spreading to appressed, persistent in fr; branch axis terminated by vestigial spikelet, base hairy; spikelets 20–80 per branch, spreading to ascending, breaking apart between glumes and lower floret. **SPIKELET**: upper glume 3.5–6 mm, vein gen with long, bulbous-based hairs, acute or awned < 0.5 mm; florets 2–3; lower floret lemma ± = upper glume, base hairy-tufted, back hairy, tip 2-lobed, awns 1–3 mm, unequal, central awn from sinus; base of middle or, if only 2 florets, upper floret hairy-tufted, lobed between awn bases, awns 2.5–6 mm, ± equal; uppermost floret (if present) < 2 mm, awn 0. $2n=20, 21,28,35,40,42,60,61,77,84$. Sandy to rocky slopes, flats, drainages, scrub, woodland, pine forest; < 2300 m. SnBr, **e DMtns (Ivanpah, New York, Clark mtns)**, waif elsewhere; to s Can, e US, s Mex, S.Am. ❀ DRN,SUN:**7,15,16**,17&IRR:2,3,**8,9,14,18–24**; GRCVR;also STBL. May–Aug

B. trifida S. Watson (p. 573) RED GRAMA Per, sometimes fl 1st year, sometimes short-rhizomed. **ST** ascending to erect, 1–3 dm. **LF**: blade < 5 cm, < 1.5 mm wide. **INFL**: branches 1–7, 10–35 mm, ascending to appressed, persistent in fr; branch axis terminated by spikelet, base puberulent to hairy; spikelets 8–32 per branch, ascending, breaking between glumes and lower floret. **SPIKELET**: upper glume 2–5 mm, glabrous, acute or notched, awned from sinus < 1 mm; florets 2(3), lower floret lemma < upper glume, glabrous or hairy, tip tapered to central awn, awns 2–8 mm, ± equal; upper floret base glabrous, lobes 0 between awn bases, awns 2–8 mm, ± equal. $2n=20,28$. RARE in CA. Dry, rocky, often calcareous slopes, crevices, scrub; 700–2000 m. **e DMtns**; to UT, TX, c Mex. May–Sep

BRIZA QUAKING GRASS

Dieter H. Wilken

Ann, per. **STS** ascending to erect, 1(–6). **LVS** basal to cauline; ligules membranous to translucent; blades flat. **INFL** panicle-like, open. **SPIKELETS** erect to pendent, ± compressed, subconic to ovoid; glumes subequal, papery, rounded at tip, 3–9-veined; florets 6–20, breaking above glumes and between florets; lemma width > length, papery to translucent, rounded at tip, 7–9-veined; palea ± = lemma. 10–12 spp.: Eur, n Afr. (Greek: a kind of grain)

B. minor L. (p. 573) Ann. **STS** 1–8, 10–50 cm. **LF**: ligule 3–6 mm, blade 3–10 mm wide. **INFL** 3–20 cm. **SPIKELETS** gen > 15 per infl, 2–5 mm, erect, subconic, truncate at base; glumes 1.5–4 mm, 3–5-veined; florets 4–6; lemmas 1–2 mm. $2n=10$. Shaded or moist, open sites; 20–600 m. NCo, NCoRO, CCo, SnFrB, n SCoRO, n&c SNF, s ScV, n SnJV, **DSon (Rancho Mirage)**; to OR, e US; native to s&w Eur. Apr–Jul

BROMUS BROME

Dieter H. Wilken & Elizabeth L. Painter

Ann to per. **LVS** basal and cauline; sheath closed, gen hairy; ligule gen < 5 mm, membranous, entire to fringed; blade flat to inrolled. **INFL** gen panicle-like, open to dense; spikelet stalk gen stiff, rigid. **SPIKELET** strongly compressed to cylindric; axis breaking above glumes and between florets; glumes unequal, gen < lower floret, lower gen 1–3-veined, upper 3–7-veined, back rounded to keeled, tip acute; lemmas faintly 5–9-veined, tip gen 2-toothed, short-pointed to straight-awned from between teeth; palea gen < lemma. ± 150 spp.: temp worldwide. (Greek: ancient name) [Stebbins 1947 Contr Gray Herb 165:42–55; Wagnon 1952 Brittonia 7:415–480] Native spp. need careful study. Key to species by Elizabeth L. Painter.

1. Spikelet strongly compressed, glumes and lemmas clearly keeled
  2. Lemma awn 0–2.5 mm, lower glume gen 5–7-veined . . . . ***B. catharticus***
  2' Lemma awn 3–15 mm, lower glume 3–5-veined

3. Ann; lemma body ± = upper glume, back glabrous to scabrous, short-hairy between outer 1–2 veins
 and margin . . . . **B. arizonicus**
3' Per; lemma body gen > upper glume, back evenly glabrous to densely short-hairy
 . . . . **B. carinatus** var. **carinatus**
1' Spikelet subcylindric to compressed, lemmas (sometimes glumes) slightly compressed or with back rounded
 4. Lemma awns gen 15–55 mm, teeth gen 3–7 mm; lower glume gen 1-veined
  5. Lemma awn bent, gen twisted 1time below middle . . . . **B. trinii**
  5' Lemma awn gen straight, not bent, not twisted
   6. Infl dense, spikelets all ascending to erect, > lower branches or talks . . . . [2]**B. madritensis**
    7. St, lf sheath gen glabrous, infl narrowly ovoid to oblong, most spikelet stalks visible . . . . . ssp. **madritensis**
    7' St, lf sheath gen puberulent, infl ovoid; most spikelet stalks obscure . . . . ssp. **rubens**
   6' Infl ± open, lower spikelets (or branches) spreading to drooping, spikelets gen < lower branches or stalks
    8. Lemma body 18–30 mm, teeth 3–7 mm; awn 30–55 mm . . . . **B. diandrus**
    8' Lemma body 9–13 mm, teeth 1–3 mm; awn 8–18 mm . . . . [2]**B. tectorum**
 4' Lemma awns gen 0–15 mm, teeth gen 0–3 mm; lower glume 1–3(5)-veined
  9. Lemma awn gen < 3 mm
   10. Infl gen nodding, lower branches drooping; spikelet ± compressed; glumes, lemmas puberulent
    . . . . [2]**B. anomalus**
   10' Infl erect, branches ascending to erect; spikelet ± cylindric; glumes, lemmas gen glabrous (to
    short hairy) . . . . **B. inermis** ssp. **inermis**
  9' Lemma awns gen 3–25 mm
   11. Ann; lemma teeth gen 0.5–4 mm, acute
    12. Lower glume 1-veined, upper glume 3-veined
     13. Infl dense; branches ascending to erect, stalks gen < 10 mm . . . . [2]**B. madritensis** (see 6. for sspp.)
     13' Infl gen open (to ± dense), branches spreading to nodding, stalks gen > 10 mm . . . . [2]**B. tectorum**
    12' Lower glume 3–5-veined, upper glume 5–7-veined
     14. Infl ± dense; spikelets all ascending to erect, stalks gen 0–5 mm . . . . **B. hordeaceus**
     14' Infl open; lower spikelets spreading to nodding, some stalks or branches > 10 mm
      15. Spikelet gen soft-hairy; lower glume 7–10 mm, upper glume 8–12 mm . . . . **B. arenarius**
      15' Spikelet gen glabrous; lower glume 4.5–7 mm, upper glume 5–8 mm . . . . **B. japonicus**
   11' Per; lemma tip gen rounded to minutely lobed
    16. Lemma glabrous or evenly hairy throughout . . . . [2]**B. anomalus**
    16' Lemma unevenly hairy, back gen glabrous above middle, puberulent below middle or between
     outer veins and margin . . . . **B. ciliatus**

B. anomalus Fourn. (p. 573) Per 30–90 cm, often tufted. **LF** glabrous to soft-hairy; blade 1.5–5 mm wide. **INFL** 6–20 cm, open, gen nodding. **SPIKELET** ± compressed; glumes rounded, gen puberulent, lower 5–8 mm, 1–3-veined, upper 6–11 mm, 3-veined; florets 6–11; lemma body 7–13 mm, back rounded, 5–7-veined, puberulent to densely short-hairy, tip rounded to minutely lobed, awn 1–3.5 mm. $2n=14$. Uncommon. Dry, open places, slopes, coniferous forest; 3000–3500 m. **W&I** (reported from n SNH, SnBr); to w Can, TX, Mex. [*B. porteri* (J. Coulter) Nash] CA pls need study. Jul–Aug

B. arenarius Labill. (p. 573) Ann 15–60 cm. **LF** soft-hairy; blade 2–5 mm wide. **INFL** 4–19 cm, open; branches nodding to spreading, threadlike; spikelet stalks wavy to S-shaped. **SPIKE-LET** ± compressed, gen short-soft-hairy; glumes with flat to rounded back, lower 7–10 mm, gen 3-veined, upper 8–12 mm, 5–7-veined; florets (3)6–9; lemma body 7–11 mm, back rounded, 5–7-veined, teeth 0.5–2 mm, awn (6)9–14 mm. Open, disturbed places; < 2000 m. CA-FP, **D**; to OR, AZ; native to Australia. Apr–Jul

B. arizonicus (Shear) Stebb. (p. 573) Ann 40–90 cm. **LF**: glabrous to sparsely short-hairy; blade 3–9 mm wide, margin clearly ciliate below middle. **INFL** 5–30 cm, ± open; branches ascending to erect, lower ± spreading in fr. **SPIKELET** ± strongly compressed; glumes keel-like, glabrous, lower 8–13 mm, 3–5-veined, upper 9–15 mm, 5–9-veined; florets 5–9; lemma body 9–13 mm, keel-like, 7–9-veined, back glabrous to scabrous, short-hairy between margin and outer 1–2-veins, awn 7–15 mm. $2n=84$. Open places, grassland, shrubland; < 1000 m. SnJV, s SCoR, SCo, ChI, **D**; AZ, Baja CA. Mar–May

B. carinatus Hook. & Arn. CALIFORNIA BROME Per 45–150 cm, sometimes fl 1st yr. **LF**: sheath glabrous to soft-hairy; blade 3–12 mm wide, glabrous, scabrous, or soft-hairy. **INFL** 5–20 cm, ± dense, gen becoming open in fr; lowest branches gen spread-

ing to ascending; upper branches ascending to erect. **SPIKE-LET** strongly compressed; glumes keel-like, glabrous to short-soft-hairy, lower 6.5–12 mm, gen 3-veined, upper 9–15 mm, 5–7(9)-veined; florets 7–11; lemma body 12–17 mm, 7–9-veined, keel-like, glabrous to densely short-hairy, awn 3–15 mm. $2n=56$. Open shrubland, woodland, coniferous forest; < 3500 m. **CA** (exc GV, DSon); to AK, TX, n Mex. Pls gen self-pollinating, florets often cleistogamous; forms formerly recognized as spp. widespread, often occurring together.

var. **carinatus** (p. 573) **INFL**: spikelets not crowded or over-lapping; branches and stalks gen > spikelets. Habitat and range of sp. [*B. breviaristatus* Buckley; *B. marginatus* Steudel; *B. polyanthus* Scribner, Great Basin brome] ✿ STBL;CV. Apr–Aug

B. catharticus Vahl RESCUE GRASS Ann or short-lived per, 30–100 cm. **LF**: sheath gen short-soft-hairy; blade 3–10 mm wide, glabrous to scabrous or short-soft-hairy. **INFL** 10–30 cm, ± open, gen erect; lower branches spreading to nodding; upper branches ascending. **SPIKELET** strongly compressed, glabrous to mi-nutely scabrous; glumes keel-like, lower 7–12 mm, 5–7(9)-veined, upper 8–13 mm, 7–9-veined; florets 5–13; lemma body 9–12 mm, keel-like, 7–11-veined, keeled, tip acute, awn 0–2.5 mm. $2n=42$. Open, gen disturbed places, fields; < 1500 m. **CA**; to s Can, e US, Eur; native to S.Am. [*B. haenkeanus* (C. Presl) Kunth, *B. unioloides* Kunth, *B. willdenovii* Kunth; Pinto-Escobar 1981 Bot Jahrb Syst 102:445–457] Apr–Nov

B. ciliatus L. (p. 573) FRINGED BROME Per 35–90 cm, gen tufted. **LF** glabrous to soft-hairy; blade 4–11 mm wide. **INFL** 8–21 cm, open; lower branches gen nodding; upper branches spread-ing to ascending. **SPIKELET** compressed; glumes rounded, gla-brous, lower 5–8 mm, 1(3)-veined, upper 6–11 mm, gen 3-veined; florets 5–10; lemma body 7.5–14 mm, back rounded, gen 5-veined, glabrous, puberulent between outer veins and margin, tip minutely lobed, awn 2–6 mm; anthers 1–2.5 mm. $2n=14,28$.

Meadows, coniferous forest; 1100–3200 m. SNH, SnBr, **n W&I**; to AK, ne N.Am, n Mex. Pls from s SNH, SnBr, with anthers 2–3 mm, 2*n*=28, have been called *B. richardsonii* Link. Jul–Aug

B. diandrus Roth RIPGUT GRASS Ann 15–80 cm. **LF** gen soft-hairy; blade 2–7 mm wide, margin scabrous. **INFL** 6–25 cm, ± open; lower branches gen nodding; upper spreading to ascending. **SPIKELET** ± compressed, glabrous to scabrous; lower glume 12–25 mm, gen 1-veined; upper glume 18–30 mm, gen 3-veined; florets 5–8; lemma body 18–30 mm, 5–7-veined, back rounded, tip with teeth 3–7 mm, awn 30–55 mm. 2*n*=28,42,56. Open, gen disturbed places, fields; < 2000 m. **CA**; to B.C., S.Am; native to Eur. [*B. rigidus* Roth incl var. *gussonei* (Parl.) Cosson & Durieu misapplied] Apr–Jun

B. hordeaceus L. (p. 573) SOFT CHESS Ann 11–65 cm. **LF** gen soft-hairy; blade 1.5–5 mm wide. **INFL** 2.5–13 cm, gen ± dense; branches ascending to erect; spikelet > stalk. **SPIKELET** ± compressed, glabrous to short-soft-hairy; lower glume 5–8 mm, gen 3-veined, upper glume 6–9 mm, 5–7-veined; florets 5–10; lemma 6.5–10 mm, back rounded, 5–7-veined, teeth 0.5–1.5 mm, awn 4–10 mm; palea gen < lemma; anthers gen < 1 mm. 2*n*=14,28. Open, often disturbed places; < 1000 (2100) m. **CA (uncommon D)**; Am; native to Eurasia. [*B. mollis* L., *B. racemosus* L., and *B. scoparius* L. misapplied] Pls with awn outcurved in fr, ± flat near base, have been called ssp. *molliformis* (Godron) Maire [*B. molliformis* Godron]; pls with anthers 3–5 mm have been called *B. arvensis* L. Apr–Jul

B. inermis Leysser ssp. *inermis* SMOOTH BROME Per 45–100+ cm, rhizomed. **LF** glabrous to sparsely scabrous; blade 5–12 mm wide, margin ± scabrous. **INFL** 9–20 cm, ± dense; branches gen ascending to erect. **SPIKELET** ± cylindric, glabrous to minutely scabrous; lower glume 5–9 mm, gen 1-veined; upper glume 6–10 mm, gen 3-veined; florets 5–10; lemma body 8–12 mm, 5–7-veined, back rounded, tip obtuse to minutely lobed, awn 0–2.5 mm. 2*n*=28,42,± 56. Meadows, ditches, fields; < 2700 m. SNH (e slope), SCoRO, SCo, PR, **GB**; to e US; native to Eur. Cult widely for forage, revegetation after fire. Ssp. *pumpellianus* (Scribner) Wagnon with puberulent lemmas is native to nw N.Am, Rocky Mtns. May–Aug

B. japonicus Murr. Ann 17–85 cm. **LF**: sheath soft-hairy; blade 1.5–6 mm wide, glabrous to sparsely soft-hairy. **INFL** 3–26 cm, ± open; branches spreading to ascending, lower sometimes nodding; spikelet stalks thread-like, flexible. **SPIKELET** subcylindric to slightly compressed, gen glabrous; glumes with

rounded back, lower 4.5–7 mm, 3–5-veined, upper 5–8 mm, 5–7-veined; florets 3–9; lemma 7–10 mm, gen glabrous, back rounded, faintly 7–9-veined, teeth fused or free, < 1 mm, awn 5–11 mm, straight to slightly curved. Open, disturbed places; < 2300 m. **CA**; Am; native to Eurasia. [*B. commutatus* Schrad.] Pls with infl branches nodding and awns slightly curved are uncommon. May–Jul

B. madritensis L. FOXTAIL CHESS Ann 10–50 cm. **LF** glabrous to short-soft-hairy; blade 1–4 mm wide. **INFL** 3–11 cm, ± dense; branches ascending to appressed. **SPIKELET** cylindric to slightly compressed, glabrous to puberulent; lower glume 5–11 mm, 1-veined, upper 9–15 mm, 3-veined; florets 3–9; lemma body 10–18 mm, back gen rounded, 3–5-veined, teeth 2–4 mm, awn 10–25 mm. 2*n*=28. Open, gen disturbed places; < 2200 m. **CA**; to B.C., e US, n Mex; native to Eur.

ssp. *madritensis* (p. 573) **ST** gen glabrous. **LF**: sheath gen glabrous. **INFL** 4–11 cm, narrowly ovoid; most branches visible. Habitats of sp. CA-FP, **D (uncommon)**; s OR, Baja CA; native to Eur. Apr–Jan

ssp. *rubens* (L.) Husnot (p. 573) **ST** gen puberulent. **LF**: sheath gen puberulent. **INFL** 3–8 cm, ovoid, dense; branches (exc lowest) obscure. Habitats and range of sp. [*B. rubens* L.] Mar–Jun

B. tectorum L. (p. 573) CHEAT GRASS, DOWNY BROME Ann 5–40 cm. **LF**: sheath gen densely soft-hairy; blade 1–5 mm wide, ± glabrous to densely soft-hairy, gen long-ciliate near base. **INFL** 6–22 cm, open to ± dense; branches spreading to nodding. **SPIKELET** subcylindric to slightly compressed; glumes glabrous to short-hairy, lower 5–8 mm, 1-veined, upper 7–12 mm, 3-veined; florets 3–7; lemma body 9–13 mm, 5–7-veined, glabrous to short hairy, tip with 2 teeth 1–3 mm, awn 8–18 mm. Open, disturbed places; < 2200 m. **CA**; Am; native to Eurasia. [var. *glabratus* Shear] May–Jun

B. trinii Desv. (p. 573) CHILEAN CHESS Ann 20–90 cm, often tufted. **LF**: sparsely to densely long-soft-hairy; blade 2–9 mm wide. **INFL** 8–30 cm, ± open; branches gen ascending. **SPIKELET** ± compressed; glumes rounded, gen glabrous, lower 8–16 mm, 1-veined, upper 10–18 mm, 3-veined; florets 3–9; lemma body 6–15 mm, back rounded, densely silky-hairy, teeth 3–5 mm, thread-like awn 14–22 mm, bent, twisted below middle. 2*n*=42. Open, sandy or gravelly soils; < 1700 m. NCoRI, CaRF, SN, SnJV, CW, SW, **SNE, D**; to Colorado, n Mex; probably native to w S.Am. [var. *excelsus* Shear] Mar–May

CALAMAGROSTIS REED GRASS

Craig W. Greene

Per, gen from rhizomes. **STS** 1–15 dm, gen not branched; ± smooth; nodes gen 2–4. **LVS** gen basal and cauline; sheath smooth or scabrous; ligule membranous; blade flat to inrolled. **INFL** panicle-like, open to dense; branches ± drooping to appressed; spikelets ascending to appressed. **SPIKELET**: glumes subequal, gen lanceolate, acute to acuminate, lower gen 1-veined, upper 3-veined; floret 1, breaking above glumes; axis prolonged beyond floret, hairy; callus hairy; lemma < glumes, awned from below middle to near base, tip gen 4-toothed, veins 3–5, awn straight to twisted, bent; palea ± = lemma, thin. ± 100 spp.: cool temp (esp moist montane); some forage value. (Greek: reed grass) [Greene 1980 Ph.D. Thesis Harvard University] Hybridization, polyploidy, and asexual seed set contribute to taxonomic difficulty. Key to species adapted by Margriet Wetherwax.

1. Awn exserted 2–4 mm beyond glume tips, twisted, bent; lemma 4.5–8 mm, callus hairs < 1/4 lemma length . . . . ***C. purpurascens***
1' Awn ± = glume tip, straight; lemma 2–4 mm, callus hairs 1/2–3/4 lemma length . . . . ***C. stricta*** ssp. ***stricta***

C. purpurascens R.Br. (p. 573) Cespitose; rhizomes short. **ST** 1–8 dm; nodes 2–3. **LF**: collar glabrous to short-hairy; ligule 2–6 mm; blade 2–5 mm wide, flat, lower surface smooth, upper gen soft-hairy. **INFL** 4–15 cm, dense, narrow; branches < 3.5 cm. **SPIKELET**: glumes 4.5–8 mm, scabrous; axis 1–2 mm, hairs 1–2 mm; callus hairs < 1/4 lemma length; lemma ± = glumes, awned near base; awn > lemma, exserted 2–4 mm beyond glume tips, twisted, bent. 2*n*=28,40–58,84. Subalpine, alpine rocky slopes, sandy soils; 1300–4000 m. CaRH, SN, **n SNE, W&I**; to

AK, Siberia, Greenland. Some pls set seed asexually. ✿ DRN,IRR:**1**,2,3,7,14,18–24&SUN:4–6,**15,16**,17. Jul–Sep

C. stricta (Timm) Koeler Loosely cespitose. **ST** 2–24 dm. **LF**: sheath smooth; ligule 1–5.5 mm; blade 2–5 mm wide, gen inrolled, lower surface gen smooth, upper surface smooth to scabrous. **INFL** 5–20 cm, dense, narrow; branches < 1.5–5+ cm, ascending to appressed. **SPIKELET**: glumes 2–6 mm, smooth to scabrous; axis ± 1 mm, hairs = callus hairs; callus hairs 1/2 to

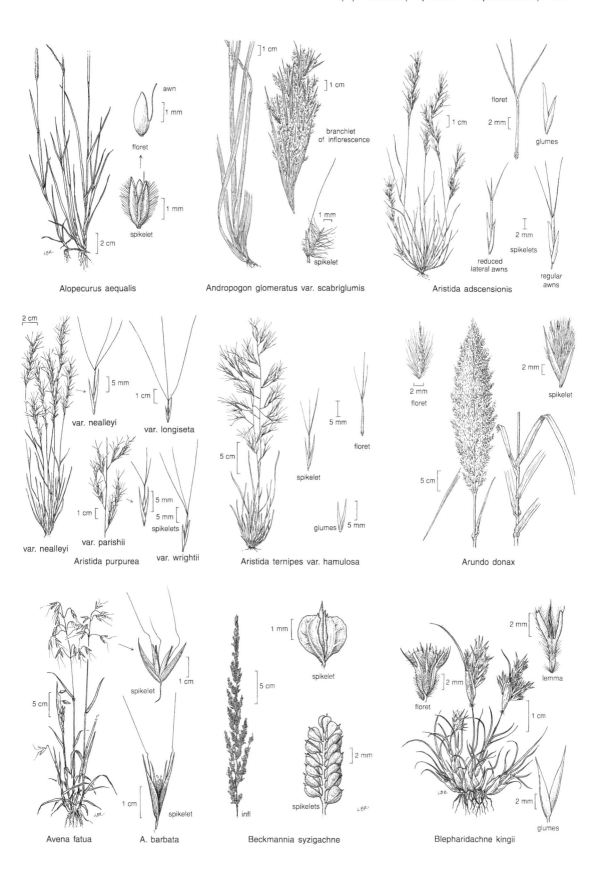

Alopecurus aequalis

Andropogon glomeratus var. scabriglumis

Aristida adscensionis

var. nealleyi
Aristida purpurea

Aristida ternipes var. hamulosa

Arundo donax

Avena fatua

A. barbata

Beckmannia syzigachne

Blepharidachne kingii

= lemma length; lemma 2–5 mm, fine-scabrous, awned at or below middle; awn ± = glume tip, gen straight. Mtn slopes, meadows, coastal marshes; < 3400 m. NW, CaR, SNH, CCo, **W&I**; to AK, ne N.Am, Eurasia. 2 sspp in CA, intergrading.

ssp. *stricta* **ST** 2–9 dm. **LF**: ligule 1–3.5 mm; blade gen inrolled, upper surface smooth to scabrous. **INFL** 5–12 cm; long-

est branches < 4 cm. **SPIKELET**: glumes 2–4.5 mm, gen thin, margin sometimes translucent; callus hairs 1/2–3/4 lemma; lemma 2–4 mm; awn straight, slender; anthers gen fertile. 2n=28,42,56,±70. Coniferous forest, meadows, slopes; 1500–3350 m. SNH, **W&I**; to AK, ne N.Am, Eurasia.

CENCHRUS SANDBUR

Robert Webster

Ann, per. **STS**: internode solid to spongy inside. **LVS** basal and cauline; sheath gen smooth; ligule short-hairy or membranous, ciliate; blade flat or folded. **INFL**: main axis straight or wavy; spikelets in groups of 1(–8), gen enclosed by bur-like involucre, bracts bristle- or spine-like, fused; involucre and enclosed spikelets falling as 1 unit; spikelet sessile to ± embedded in short axis. **SPIKELET** ± compressed; glumes strongly unequal, lower 1-veined, upper ± = florets; florets 2, lower floret sterile or staminate, lemma gen 5-veined, palea gen present, upper floret fertile, lemma thick, ± hard, palea ± = lemma. ± 20 spp.: warm temp Am, Afr, s Asia. (Greek: ancient name) [Delisle 1963 Iowa State Coll J Sci 37:259–351] Key to species adapted by Margriet Wetherwax.

1. Lower involucre bracts whorled, ± flexible, upper bracts spiny . . . . *C. echinatus*
1' Lower involucre bracts not clearly whorled, all bracts spiny
  2. Involucre bracts 8–40; longest bract < 5 mm . . . . *C. incertus*
  2' Involucre bracts > 50; longest bract > 5 mm . . . . *C. longispinus*

C. echinatus L. SOUTHERN SANDBUR Ann. **ST** 1–5 dm. **LF**: sheath 3–7 cm; ligule ± 1–1.5 mm; blade 6–20 cm, 3.5–11 mm wide, upper surface glabrous or hairy. **INFL** 3.5–8 cm; main axis wavy; involucre bracts 40–60, fused. **SPIKELET** 5–6.5 mm, ± 1.5–2 mm wide, lanceolate to ovate, green; lower glume ± 1–3 mm, upper ± 4–6 mm; lower floret sterile, lemma acute. 2n=68,70. Disturbed places, fields; < 150 m. s ScV (Solano Co.), SCo (San Diego), **DSon**; native to s US, Mex, C. and S.Am. NOXIOUS WEED. Oct

C. incertus M. Curtis COAST SANDBUR Ann. **ST** decumbent, 1–5 dm. **LF**: sheath 2.5–7 cm; ligule 0.5–1.5 mm; blade 4–12 cm, 2.5–5 mm wide, upper surface glabrous. **INFL** 2–5 cm; main axis wavy; involucre bracts 8–40, fused. **SPIKELET** 3.5–5.5 mm, ± 1–2.5 mm wide, ovate, green; lower glume 1–3.5 mm,

upper glume 3–5 mm; lower floret staminate, lemma 5–7-veined, acute. 2n=34. Disturbed places, fields; < 800 m. GV, SCo, **DMoj (near Daggett)**; OR; native to s US, Mex, C. and S. Am. NOXIOUS WEED.

C. longispinus (Hackel) Fern. (p. 573) MAT SANDBUR Ann. **ST** 1–6 dm. **LF**: sheath 2.5–8 cm; ligule ± 1 mm; blade 4–12 cm, 2.5 mm wide, upper surface glabrous. **INFL** 3–5 cm; main axis wavy; involucre bracts 50–75, fused. **SPIKELET** ± 6–7.5 mm, 2–3 mm wide, lanceolate to ovate, green; lower glume 1.5–4 mm, upper 4.5–6 mm; lower floret staminate, lemma acute. 2n=34. Disturbed places, fields; < 900 m. s ScV, SnJV, SCo, MP (Lassen Co.), **DMoj**; native to c&e US. [*C. pauciflorus* Bentham misapplied]. NOXIOUS WEED. Jul–Sep

CHLORIS

Dennis Anderson

Ann, per, cespitose or from rhizomes. **ST** decumbent to erect, 2–30 dm. **LF**: ligule membranous or hairy-tufted; blade gen 10–40 cm, 0.5–1.5 cm wide, flat. **INFL** gen umbel-like; branches 2–30, sometimes in distinct whorls, each raceme- or spike-like branch with 2 rows of overlapping spikelets on 1 side of axis. **SPIKELET**: glumes unequal, < florets, 1–3-veined; axis breaking above glumes; lower 1–2 florets fertile, upper 1–3 sterile, < 1/2 lower floret length; fertile floret lemma ovate to lanceolate, back glabrous, midvein hairy, 3-veined, awn 1; palea < lemma, translucent, obscure. **FR** ± fusiform, 3-angled. ± 50 spp.: warm temp, trop worldwide. (Greek: mother of Nestor, goddess of flowers)

C. virgata Sw. Ann. **ST** gen 1–5 dm. **LF**: sheath glabrous to hairy near collar; ligule glabrous to hairy; blade < 30 cm, 1.5 cm wide. **INFL** umbel-like; branches ± erect, 5–10 cm. **SPIKELET** 2.5–4.5 mm; lower glume 1.5–2.5 mm, 0.2–0.5 mm wide, acute; upper glume 2.5–4.5 mm, 0.3–0.5 mm wide, acute; fertile floret 1, 2.5–4 mm, 0.5–1.5 mm wide, ovate, obovate, or elliptic, keel

hairy near tip, margin gen hairy-tufted, awn 2.5–15 mm; sterile floret 1, 1.5–3 mm, 0.5–0.8 mm wide, awn 3–9.5 mm. **FR** 1.5–2 mm. 2n=20,26, 30,40. Disturbed areas; < 200 m. GV, SCo, **D**; to s Great Plains, n Mex; native to warm temp regions worldwide. Apr–Sep

CRYPSIS PRICKLE GRASS

Ann. **STS** prostrate, ascending, or erect. **LF**: ligule hairy; blade gen short, linear to narrowly lanceolate. **INFL** panicle-like, dense, cylindric and exserted or head-like and enclosed by enlarged sheaths. **SPIKELET** bisexual, strongly compressed, falling as 1 unit; glumes, lemma acute or short-pointed; glumes < or = floret, gen lanceolate, keeled, strongly 1-veined; floret 1; lemma membranous, 1-veined; palea gen 2-veined, gen splitting with age; stamens gen 3. 8 spp.: Medit Eur, Asia, c Afr. (Greek: concealment, from partly hidden infl) [Hammel & Reeder 1979 Syst Bot 4:267–280]

C. alopecuroides (Piller & Mitterp.) Schrader (p. 573) Pl gen purple to black. **STS** ascending, 5–75 cm; branches few. **LF**:

blade 5–12 cm, green or ± glaucous; sheath glabrous. **INFL** 1.5–6.5 cm, 4–6 mm wide, cylindric. **SPIKELET** 2–3 mm; glume keel hairy, lower glume < upper; lemma gen > glumes; palea 2-veined. 2*n*=16. Bottom-lands, reservoir and river margins; <

800 m. KR, NCoRI, CaRH, CCo (Marin Co.), **SNE**; to WA, e US; native to Eur.

CYNODON

Per, mat-like, from rhizomes or stolons. **STS** ± branched. **LF**: blade short, flat, narrow, fleshy. **INFL** umbel-like; branches 2–20, spike-like, with 2 rows of overlapping spikelets along 1 side of axis; spikelets sessile. **SPIKELET** bisexual, strongly compressed; glumes ± equal, 1-veined, awn 0; floret 1, rarely 2, upper floret vestigial, breaking above glumes; lemma keeled, 3-veined, awn 0; palea = lemma, 2-veined. 8–10 spp.; trop, warm temp Eurasia, Afr. (Greek: dog tooth, from hard scales on rhizomes) [Harlan et al 1970 Okla State Univ Agric Exp Sta Bull B–673] Key to species adapted by Margriet Wetherwax.

1. Infl branches 3–20, gen 3–8 cm . . . . *C. dactylon*
1' Infl branches 2, gen 1–2 cm . . . . *C. transvalensis*

C. dactylon (L.) Pers. (p. 573) BERMUDA GRASS Per from rhizomes or stolons. **ST** gen erect, 1–4 dm, flat. **LF**: ligule white-hairy; blade < 6 cm, glabrous or upper surface hairy. **INFL**: branches 2.5–5 cm, gen 4–7. **SPIKELET** ± 2 mm; glumes ± 1.5 mm, gen purplish; lemma ± 2 mm, boat-shaped, acute, keel and margins hairy; palea keels glabrous. 2*n*=36. Disturbed sites; < 900 m. CA-FP, **W&I, D**; warm temp, trop; native to Afr. Cult for lawns, forage. TOXIC: important pollen source in hay fever; may produce contact dermatitis. Jun–Aug

C. transvalensis Burtt Davy AFRICAN BERMUDA GRASS Per from rhizomes and stolons. **ST** prostrate to ascending, gen < 10 cm. **LF**: ligule membranous, short; blade gen 10–30 cm, ± 1.5 mm wide, gen yellow-green, hairy. **INFL**: branches 1–2 cm, gen 2, sometimes 1 or 3. **SPIKELET** 2–4 mm; glumes unequal, < 3/4 spikelet length; lemma 2–4 mm, elliptic, keel ± hairy; palea keels glabrous. 2*n*=18. Uncommon. Roadsides; < 100 m. **DSon (Imperial Co.)**; native to Afr.

DACTYLIS ORCHARD GRASS

Dieter H. Wilken

1 sp. (Greek: finger, from crowded spikelets at infl tips) [Stebbins & Zohary 1959 Univ Calif Publ Bot 31:1–40]

D. glomerata L. (p. 577) Per from short rhizomes. **STS** (1)2–5, 3–11 dm. **LVS** basal and cauline; sheath closed, keeled; appendages 0; ligules 4–9 mm, ± translucent, fringed to toothed at obtuse tip, glabrous to minutely soft-hairy; blades 3–6 mm wide, slightly scabrous. **INFL** 1–1.5 dm, panicle-like; branches spike-like, stiffly erect to spreading. **SPIKELETS** crowded, gen on 1 side of branch tips, short-stalked to subsessile, laterally com-pressed; florets 2–4, gen bisexual (upper sometimes staminate), breaking above glumes and between florets; glumes and lemmas short-awned at tip, minutely ciliate to short-hairy on back; glumes 3–6 mm, subequal; lemmas 4–7 mm, margin translucent, 5-veined; palea slightly < lemma, 2-forked at tip. 2*n*=14,21,27–31,42. Disturbed, often moist sites; < 2500 m. CA-FP, **GB**; to Can, US; native to Eurasia. Cult for forage, hay. May–Aug

DACTYLOCTENIUM

Ann, per. **STS** decumbent to erect. **LF**: blade flat. **INFL** umbel-like; branches 2–11, spike-like, with 2 rows of over-lapping, sessile spikelets gen along 1 side of axis; axis tip ± naked. **SPIKELET** bisexual, compressed; glumes un-equal, wide, 1-veined, lower persistent, acute, upper deciduous, short-awned, awn gen curved; axis breaking apart above glumes and between florets; florets 3–5; lemma membranous, keeled, 3-veined, lateral veins gen faint, tip acuminate or short-awned, awn gen curved; palea ± = lemma. 13 spp.: Afr. (Greek: finger + a small comb, from spikelet arrangement) [Fisher & Schweickerdt 1941 Ann Natal Mus 10:47–77]

D. aegyptium (L.) Willd. (p. 577) CROWFOOT GRASS Ann, gen mat-like. **ST** 2–4(10) dm, rooting at lower nodes. **LF**: blade < 15 cm, 4–5 mm wide, margin ciliate, hairs swollen at base. **INFL**: branches gen 4–5, 1–5 cm. **SPIKELET** ± 4 mm; upper glume keeled, 1.5–2 mm; florets 3; lemma widely ovate. 2*n*=20,36,40,45,48. Disturbed places; < 300 m. SnJV (Kern Co.), SCo, PR, **DSon**; e&s US, trop Am, Asia; native to Afr.

DESCHAMPSIA HAIRGRASS

Dieter H. Wilken and Elizabeth Painter

Ann, per. **STS** erect, solitary to densely clumped. **LVS** basal to cauline; ligule narrow, decurrent to sheath, glabrous to minutely hairy; blades flat to inrolled. **INFL** panicle- to spike-like, open to narrow. **SPIKELET**: glumes and lemmas shiny; glumes equal to ± subequal, > lower floret; axis prolonged beyond upper floret, bristly (sometimes with vesti-gial floret at tip); florets 1–3, bisexual, breaking above glumes and between florets; callus soft-hairy; lemmas rounded, 2–4-toothed at truncate tip, faintly 3–7-veined, awned at or below middle, awn straight to bent; palea ± = lemma. 30–40 spp.: temp Am, Eurasia, New Zealand, Antarctica. (J. L-Deslongchamps, France, born 1774)

1. Ann . . . . ***D. danthonioides***
1' Per
  2. Infl open, > 1 cm wide; basal lf blades 1–4 mm wide . . . . ***D. cespitosa*** ssp. ***cespitosa***
  2' Infl narrow, < 1 cm wide; basal lf blades ± 1 mm wide . . . . ***D. elongata***

D. cespitosa (L.) P. Beauv. TUFTED HAIRGRASS Per. **STS** densely clumped, 2–10 dm. **LVS** gen basal, tufted, glabrous to scabrous; ligule 3–8 mm, acute to obtuse, entire to toothed at tip; blades gen 8–20 cm, 1–4 mm wide, flat to inrolled. **INFL** narrow to open; lower branches erect to drooping. **SPIKELET**: glumes and tips of lemmas purplish; glumes 3–7 mm, ± subequal, lanceolate, acute, lower 1-veined, upper 3-veined; florets gen 2; callus hairs gen < 1/3 lemma length; lemmas 2–4 mm, gen 4-toothed at tip, faintly 5-veined, awned below middle, awn 2–6 mm, straight to slightly bent. 2*n*=26–28. Wet sites, meadows, streambanks, coastal marshes, forests, alpine; < 3900 m. NW, CaR, SN, CCo, SnFrB, TR, Wrn, **n SNE, W&I**; to AK, e N.Am, S.Am, Eurasia, New Zealand. Sspp. intergrade, 2 in CA.

ssp. **cespitosa** (p. 577) **INFL** open; lower branches spreading to drooping. **SPIKELETS** solitary to clustered on exposed branchlets. Meadows, streambanks, coastal marsh, forests, alpine; < 3820 m. NW, CaR, SN, CCo, SnFrB, TR, Wrn, **n SNE, W&I**; to AK, Can, e US, Eurasia. Variable in pl and spikelet size, awn length.. ❀ SUN or part SHD:**4–6,17**&IRR:1–3,7,8–10, **14–16,18–24**; also STBL. Jul–Aug

D. danthonioides (Trin.) Munro (p. 577) ANNUAL HAIRGRASS Ann. **STS** gen solitary to loosely clumped, 1.2–6 dm. **LVS** gen basal, glabrous; ligule 2–4 mm, acute to acuminate, entire; blades 1–9 cm, 1–2 mm wide, gen inrolled. **INFL** narrow to open; lower branches gen ascending. **SPIKELET**: glume and lemma tips sometimes purplish; glumes 4–9 mm, equal, lanceolate, acute to acuminate, 3-veined; florets gen 2; callus hairs < 1/2 lemma length; lemmas 2–3 mm, gen 2–3-toothed at tip, faintly 1–3-veined, awned below middle, awn 4–9 mm, slightly bent. 2*n*=26. Moist to drying, open sites, meadows, streambanks, temporary ponds; < 2700 m. CA-FP, MP, **DMtns (uncommon)**; to AK, AZ, Baja CA, S.Am. ❀ SUN,IRR:1–3,**4–10**,11,12,**14–24**. Mar–Aug

D. elongata (Hook.) Munro(p. 577) SLENDER HAIRGRASS Per. **STS** densely clumped, 1–7 dm. **LVS** gen basal, tufted, glabrous; ligule 2–8 mm, acute to acuminate, entire; blades 4–8 cm, ± 1 mm wide, flat to inrolled. **INFL** narrow; branches spike-like, ascending. **SPIKELET**: glumes and lemmas green to tan, tips sometimes purplish; glumes 3–5 mm, ± subequal, narrow-lanceolate, acute to acuminate, 3-veined; florets gen 2; callus hairs ± 1/2 lemma length; lemmas 2–3 mm, 2–4-toothed at tip, faintly 5-veined, awned near middle, awn 1–5 mm, gen straight. 2*n*=26. Wet sites, meadows, lakeshores, shaded slopes; 100–3100 m. CA-FP, **GB**; to AK, WY, Baja CA, S.Am. ❀ IRR or WET:1–3,**4–6,17**&partSHD:7,8,9,**14–16,18–24**. May–Aug

DIGITARIA

Robert Webster

Ann, per. **STS** decumbent to erect. **LVS** basal and cauline; ligule membranous, ciliate or not; blade gen flat. **INFL** umbel- to panicle-like; 1° branches ± spike-like, spreading to ascending; spikelets gen many per branch, 2–3 per node, short-stalked to subsessile, on one side of axis. **SPIKELET** compressed, falling as 1 unit; glumes unequal, upper glume < or = spikelet, appressed-hairy, clearly 3–5-veined, veins minutely ridge-like; florets 2, lower floret sterile, lemma texture like upper glume, upper floret fertile, lemma ± thin, flexible, back facing away from infl axis, margin flat, tip gen obtuse; palea ± = lemma. ± 200 spp.: warm temp, trop, worldwide. (Latin: finger, from infl branch arrangement) [Webster 1987 Sida 12:209–222]

D. sanguinalis (L.) Scop. (p. 577) Ann. **ST** 2–7 dm; nodes 2–5. **LF**: sheath 2.5–15 cm, hairy; ligule 1–3 mm; blade 3–17 cm, 2–14 mm wide, upper surface soft-hairy. **INFL**: branches 3–9 cm; spikelets many, gen 2 per node, stalk ± 0.5 mm. **SPIKELET** ± 2.5–3 mm, ± 1 mm wide, lanceolate to ovate, purple in fr; lower glume < 0.5 mm, upper glume ± 1/2–3/4 spikelet length; lemma of lower floret 5–7-veined, acuminate to acute. 2*n*=36. Disturbed places, fields, roadsides; < 500 m. NCo, s CaRF, SNF, CW, SCo, **SNE, DMoj (Panamint Valley)**, expected elsewhere; OR, e US; native to Eur. Jun–Sep

DISTICHLIS SALTGRASS

Per from scaly rhizomes, gen dioecious. **STS** ascending to erect, stiff, glabrous, solid in ×-section. **LVS** 2-ranked; ligule membranous, fringed; blade flat or ± rolled, pointed, glabrous, gen deciduous at collar. **INFL** panicle- or raceme-like; spikelets stalked. **SPIKELET** unisexual; pistillate gen = staminate, compressed; glumes unequal, firm, awns 0, lower 3(5)-veined, upper 5–7(9)-veined; axis breaking above glumes and between florets; florets 3–20; lemma wide, 9–11-veined, awn 0; palea < or > lemma; keels minutely hairy. 3–6 spp.: Am, 1 sp. in Australia. (Greek: 2-ranked, from lf arrangement) [Beetle 1943 Bull Torr Bot Club 70:638–650]

D. spicata (L.) E. Greene (p. 577) SALTGRASS Per; rhizomes stout, yellowish; stolons sometimes present. **STS** erect, 1–5 dm, **LF**: blade gen 2–10 cm, 1–4 mm wide, stiff. **INFL** panicle-like, 2–8 cm, narrowly ± cylindric to elliptic in outline, ± dense. **SPIKELET** 6–20(25) mm, straw-colored to purplish; glumes 1.5–4 mm; florets 5–15; lemmas 3–6 mm; palea ± = lemma, keel ciliate in pistillate florets. 2*n*=40. Salt marshes, moist, alkaline areas; < 1000 m. **CA**; s Can, US. [vars. *divaricata* Beetle, *nana* Beetle, *stolonifera* Beetle, *stricta* (Torrey) Beetle] ❀ SUN:**4–7,14–17,21–24**&IRR:3,**8,9**,10,11,**12,13,18–20**;GRNCVR;INV; STBL;tolerates salt & alkali. Apr–Jul

ECHINOCHLOA

Robert Webster

Ann, sometimes per. **STS** decumbent to erect; internode spongy inside. **LVS** basal and cauline; sheath gen glabrous; ligule gen 0; blade gen flat, upper surface gen glabrous. **INFL** panicle-like, ± dense; branches gen ascending to appressed, axis gen glabrous; spikelets gen many, 1–2 per node, short-stalked to subsessile, ± on 1 side of axis. **SPIKELET** falling as one unit, ovoid to compressed; glumes unequal, lower < upper, short-bristly to hairy, gen green to purplish, upper glume gen awned; florets 2, lower floret sterile or staminate, lemma gen like glumes, upper floret fertile, lemma leathery or hard, shiny to dull, margin inrolled, tip abruptly pointed, palea free from lemma. ± 35 spp.: warm temp, subtrop, worldwide. (Greek: hedgehog grass, from bristly spikelet) [Gould, Ali, & Fairbrothers 1972 Amer Midl Nat 87:36–59] Key to species adapted by Margriet Wetherwax.

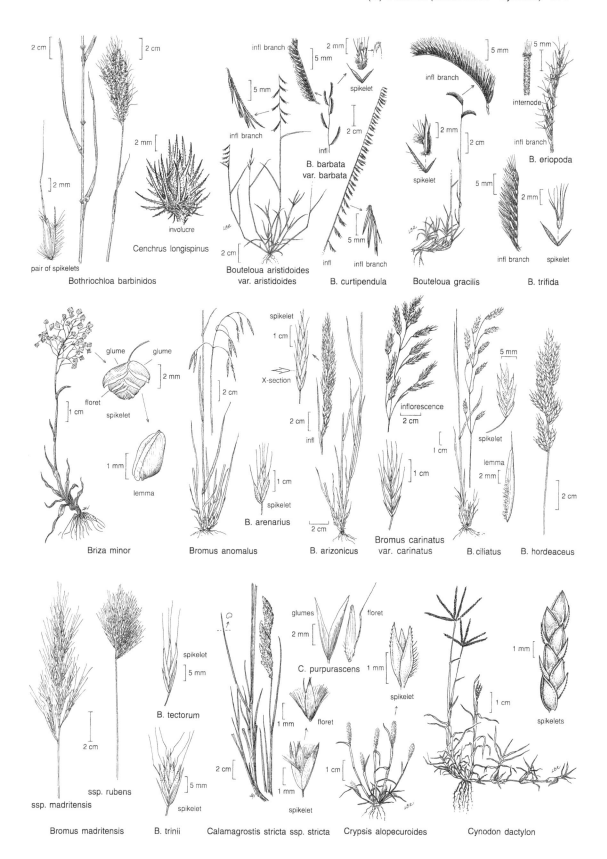

2 cm | 2 cm

2 mm

2 mm

involucre

Cenchrus longispinus

pair of spikelets

Bothriochloa barbinodis

infl branch

2 mm | 5 mm

spikelet

5 mm

infl branch

2 cm

infl

**B. barbata
var. barbata**

2 cm

Bouteloua aristidoides
var. aristidoides

5 mm

infl   infl branch

B. curtipendula

infl branch

2 mm

2 cm

spikelet

Bouteloua gracilis

5 mm

internode

infl branch

B. eriopoda

5 mm

2 mm

infl branch   spikelet

B. trifida

glume   glume

2 mm

floret

spikelet

1 cm

1 mm

lemma

Briza minor

2 cm

B. arenarius

Bromus anomalus

spikelet

1 cm

X-section

2 cm

infl

1 cm

spikelet

B. arizonicus

inflorescence

2 cm

1 cm

**Bromus carinatus
var. carinatus**

5 mm

spikelet

1 cm

lemma

2 mm

B. ciliatus

2 cm

B. hordeaceus

glumes   floret

2 mm

C. purpurascens

1 mm

spikelet

spikelet

5 mm

B. tectorum

1 mm

floret

1 mm

spikelet

1 mm

spikelets

1 cm

5 mm

spikelet

2 cm

ssp. rubens

ssp. madritensis

Bromus madritensis

B. trinii

2 cm

1 mm

spikelet

Calamagrostis stricta ssp. stricta

1 cm

Crypsis alopecuroides

Cynodon dactylon

1. 1° infl branches gen 1–3 cm; spikelet 2–3 mm, ± 1–1.5 mm wide; lemma of lower floret acute . . . . . . *E. colona*
1' 1° infl branches gen 3–7 cm; spikelet 3–4 mm, ± 1.5–2 mm wide; lemma of lower floret acuminate
  2. Lemma of upper floret abruptly pointed . . . . *E. crus-galli*
  2' Lemma of upper floret tapered to a point . . . . *E. muricata*

E. colona (L.) Link Ann. **ST** decumbent to erect, 1–9 dm. **LF**: sheath 4–9 cm; blade 3–22 cm, 3–7.5 mm wide. **INFL** 3–13 cm; 1° branches 1–3 cm, axis glabrous; spikelets 2 per node, stalk < 1 mm. **SPIKELET** 2–3 mm, 1–1.5 mm wide, ovate to elliptic; lower glume 1–1.5 mm, 3–5-veined, upper glume ± = spikelet; florets ± equal; lower floret staminate or sterile, lemma 5-veined, acute. Wet places, fields; < 100 m. s SnFrB, **DSon**, expected elsewhere; to e US; native to Eurasia. Jul–Sep

E. crus-galli (L.) P. Beauv. (p. 577) Ann. **ST** erect, 2.5–15 dm. **LF**: sheath 3–7 cm; blade 0.5–30 cm, 6–20 mm wide. **INFL** 6–10 cm; 1° branches 3–7 cm, axis glabrous or hairy; spikelet 1 per node, stalk 0.5–1 mm. **SPIKELET** 3–4 mm, ± 1.5–2 mm wide, lanceolate to elliptic; lower glume ± 1–1.5 mm, 3–5-veined, upper glume ± = spikelet; florets subequal, lower sterile, lemma 5–7-veined, acuminate. Waste places, often wet sites, fields, roadsides; < 1500 m. **CA** (esp CA-FP); worldwide; native to Eurasia, Afr. [*var. zelayensis* (Kunth) A. Hitchc.] Jul–Oct

E. muricata (P. Beauv.) Fern. Ann, sometimes rooting at lower nodes. **ST** erect, 5–10 dm. **LF**: sheath 6–20 cm; blade 7–20 cm, 4–11 mm wide. **INFL** 6–25 cm; 1° branches 3–6 cm; spikelets 1–2 per node, stalk < 1 mm. **SPIKELET** ± 3–4 mm, ± 1.5–2 mm wide, elliptic; lower glume ± 1 mm, 3–5-veined, upper glume ± = spikelet; lower floret slightly < upper, sterile, lemma 5-veined, acuminate. Waste places, often wet sites, fields; < 1000 m. **CA** (esp CA-FP); to e US; native to Eurasia.

ELYMUS

Mary E. Barkworth

Per, sometimes from rhizomes. **ST** gen bent at base or erect, gen tufted. **LF**: sheath appendaged, appendages sometimes small, fragile; ligule membranous, truncate to obtuse; blades flat, folded, or rolled. **INFL** spike-like, open to dense; axis gen not breaking apart in fr; spikelets ± 2-ranked or not, 1–4 per node, gen ascending. **SPIKELET**: glumes lanceolate to awl-like, sometimes 0, awned from tip or not; florets 1–7; lemma gen > glumes, gen rounded, tip acute to awned, awn straight or curved outward; anthers 1–6 mm. 150 spp.: temp worldwide. (Greek: ancient name for millet). [Barkworth 1998 Phytologia 83:302–311; Barkworth & Dewey 1985 Amer J Bot 72:767–776; Wilson 1963 Brittonia 15:303–323] See *Agropyron, Elytrigia, Leymus* for spp. sometimes treated here. Some spp. hybridize; hybrids with *Hordeum, Leymus* also occur. Key to species adapted by Margriet Wetherwax.

1. Glumes gen awn- to awl-like, base gen < 1.5 mm wide, awn 16–110(200) mm
  2. Lower spikelets with 2 glumes; glumes entire . . . . *E. elymoides* ssp. *brevifolius*
  2' Lower spikelets with 2 glumes and 1+ glume-like, reduced florets; glumes entire to divided into 2+ awns
    3. Glumes divided into 3+ outward-curving awns; sheath appendages evident, 0.5–1.5 mm . . . . *E. mutisetus*
    3' Glumes entire or divided into 2 outward-curving awns; sheath appendages obscure, < 0.5(1) mm . . *E. elymoides*
      4. All glumes entire, awn gen < lemma awn . . . . ssp. *californicus*
      4' Some glumes divided, awn(s) gen > lemma awn . . . . ssp. *elymoides*
1' Glumes flat, base gen > 1.5 mm wide, awn < 25 mm
  5. Infl with 2 spikelets at all or most nodes . . . . *E. glaucus* ssp. *glaucus*
  5' Infl with 1 spikelet at all or most nodes
    6. Lemma awns curving out at maturity . . . . *E. scribneri*
    6' Lemma awns straight . . . . *E. trachycaulus* ssp. *trachycaulus*

E. elymoides (Raf.) Swezey (p. 577) SQUIRRELTAIL **ST** 1–6.5 dm. **LF**: sheath glabrous to long-hairy, appendages < 1 mm; blade 1–6 mm wide, flat, folded, or rolled. **INFL** 2.5–15 cm (exc awns), breaking apart with age; internodes 3–10 mm; spikelets gen 2 per node. **SPIKELET** 12–20 mm; glumes 35–85 mm, awn-like, base narrow, thick, gen spreading, sometimes with 1–2 short awns at base; lemma awn 30–90 mm, spreading; anthers ± 2 mm. $2n=28$. Dry, open areas; 600–4200 m. KR, CaR, SN, TR, PR, **GB**, D; to Great Plains, TX, n Mex. [*Sitanion hystrix* (Nutt.) J.G. Smith] Hybrids with *E. trachycaulus* have been called *E. macounii* Vasey [*Agropyron saundersii* (Vasey) A. Hitchc.] Hybrids with *E. glaucus* have been called *E.* ×*hansenii* Scribner [*Sitanion h.* (Scribner) J.G. Smith]. (see also *E. multisetus*). 4 sspp.in CA.

  ssp. *brevifolius* (J.G. Smith) Barkworth **INFL**: spikelets gen 2 per node. **SPIKELET**: glumes entire, awn 50–110(125) mm; lowest floret fertile, not glume-like; fertile florets 1+; lemma awn 50–105 mm. $2n=28$. Habitat of sp.; 600–3000 m. SnBr, PR, MP, **W&I**; to OR, Great Plains, n Mex. [*Sitanion longifolium* J.G. Smith] ❀ SUN,DRN:1–3,**7,8,9,10**,11,**14–24**. Jul–Aug

  ssp. *californicus* (J.G. Smith) Barkworth **INFL**: spikelets gen 2 per node. **SPIKELET**: glumes entire, awn 16–60 mm; lowest floret gen sterile, glume-like, fertile florets 1+; lemma awn 25–70 mm. $2n=28$. Habitat of sp.; 800–4200 m. KR, CaR, SN, SnGb, SnBr, SnJt, **SNE**; to WA, MT, UT. [*Sitanion hystrix* (Nutt.) J.G. Smith var. *c*. (J. G. Smith) F.D. Wilson] ❀ SUN,DRN:**1–3**,7,10,14,**15–17**,18–24. Jul–Aug

  ssp. *elymoides* **INFL**: spikelets gen 2 per node. **SPIKELET**: some glumes divided, awns 2, 35–85 mm; lowest floret gen sterile, glume-like, fertile florets 1+; lemma awn 25–75 mm. $2n=28$. Habitat of sp.; 800–4000 m. SnFrB, TR, SnJt, **GB**, **D**; to WA, WY, Colorado. [*Sitanion hystrix* (Nutt.) J. G. Smith var. *h*.] ❀ SUN,DRN:1,**2,3**,7–11,14–24. Jul–Aug

E. glaucus Buckley (p. 577) BLUE WILDRYE Sometimes with short stolons. **ST** 6–14 dm. **LF**: sheath glabrous or hairy, appendages ± 2 mm; blade 4–12 mm wide, gen flat. **INFL** 6–16 cm (exc awns), not breaking apart with age; internodes 4–8 mm; spikelets gen 2 per node. **SPIKELET** 8–16 mm; glumes 6.5–19 mm, short-awned; lower florets concealed; lemma 8.5–14 mm, awn < 30 mm, gen straight; anthers 1.5–3 mm. $2n=28$. Open areas, chaparral, woodland, forest; < 2500 m. **CA**; to AK, Great Plains, n Mex. Hybridizes with *E. elymoides, E. stebbinsii, E. trachycaulus*. ❀ DRN,SUN:**4–6,15–17**&IRR:1,**2,3**,7,8,9, **10**,11,**14,18–24**. 3 sspp.in CA.

  ssp. *glaucus* **LF**: sheath, blade ± glabrous or scabrous. **SPIKELET**: lemma awn 10–30 mm. $2n=28$. Habitats and range of sp. Jun–Aug

E. multisetus (J.G. Smith) Burtt Davy   BIG SQUIRRELTAIL   **ST** 1.5–6 dm. **LF**: sheath glabrous, appendages gen 0.5–1.5 mm; blade 1.5–5 mm wide, flat or rolled. **INFL** 3–17 cm (exc awns), breaking apart with age; internodes 4–8 mm; spikelets gen 2 per node. **SPIKELET** 8–11 mm; glumes divided near base into 3–5 awns, 25–200 mm, distal half curving outward with age; lowest florets like glumes, vestigial; lemma 8–10 mm, tip gen 2-lobed, lobes awn-like, < 20 mm, awn between lobes 25–100 mm; anthers 1–2 mm. 2*n*=28. Open, sandy to rocky areas; < 3200 m. CA; to WA, Rocky Mtns. [*Sitanion jubatum* J.G. Smith] Hybrids with *E. trachycaulus* have been called *Agropyron saundersii* (Vasey) A. Hitchc. or *E. aristatus* Merr. Hybrids with *E. glaucus* have been called *Sitanion hansenii* (Scribner) J.G. Smith. (see also *E. elymoides*).   ❀ SUN,DRN:**1,2**,3–6,**7,14–24**; also STBL;may be INV. May–Jul

E. scribneri (Vasey) M.E. Jones (p. 577)   SCRIBNER'S WHEATGRASS   **ST** 1.5–5.5 dm, becoming prostrate. **LF**: sheath short-hairy, appendages < 1 mm; blade 1.5–3.5 mm wide, rolled or flat. **INFL** 3.5–8 cm, breaking apart with age; internodes 3.5–

8 mm; spikelet 1 per node. **SPIKELET** 9–15 mm; glumes 4–7 mm, awn 10–25 mm, curving strongly outward; lemma 7–10 mm, awn curving outward; anthers 1–1.6 mm. 2*n*=28. RARE in CA. Rocky areas, alpine; 2900–4200 m. **n SNE, W&I**; to B.C., Rocky Mtns. [*Agropyron scribneri* Vasey] Hybridizes with *E. trachycaulus, E. elymoides.*   Jul–Aug

E. trachycaulus (Link) Shinn. (p. 577)   SLENDER WHEATGRASS   **ST** 3–15 dm. **LF**: sheath glabrous to hairy, appendages < 1 mm; blade 2–8 mm wide, gen flat. **INFL** 4–25 cm, not breaking apart with age; internodes 4–15 mm; spikelet 1 per node. **SPIKELET** 10–20 mm; glumes 6–12 mm, acute to short-awned; lemma 6–13 mm, glabrous to hairy, awn 1–30 mm; anthers 1–2.5 mm. 2*n*=28. Dry to moist, open areas, forest, woodland; < 3400 m. **CA (exc GV)**; to AK, e US. Hybrids with *Hordeum jubatum* have been called *E. macounii* Vasey. 2sspp.in CA.

ssp. ***trachycaulus*   SPIKELET**: lemma awns < 7 mm. 2*n*= 28. Habitat of sp.: < 3300 m. **CA (exc GV)**; to AK, e US. [*Agropyron t.* (Link) Malte]   ❀ TRY. Jun–Aug

ELYTRIGIA

James K. Jarvie & Mary E. Barkworth

Per, gen from rhizomes. **STS** erect, sometimes tufted. **LF**: sheath appendaged; ligule membranous; blade flat or rolled. **INFL** spike-like; axis gen not breaking apart in fr; spikelets 2-ranked, strongly overlapping, ± appressed to axis, 1 per node. **SPIKELET**: glumes thick, midvein gen prominent and scabrous at least above middle, tip truncate, obtuse, acute or short-awned; axis breaking above glumes and between florets; lemma gen awnless. 25 spp.: Medit Eur, Asia. (Greek: from combination of *Elymus* and *Triticum*) [Jarvie 1990 PhD dissertation UT State Univ] Some spp. cult for forage, erosion control; some serious weeds. See *Agropyron, Elymus*. For revised taxonomy see Barkworth, M.E. 2000 Changing perceptions of the Triticeae in S.W.L. Jacobs and J. Everett (Editors), Proceedings of the Third International Symposium on Grass Systematics and Evolution, p. 110–120. CSIRO,Canberra, Australia. Key to species adapted by Margriet Wetherwax.

1. Pls without rhizomes; glume tip truncate; glumes weakly keeled . . . . ***E. pontica*** ssp. ***pontica***
1' Pls strongly rhizomed; glume tip gen acute to obtuse; glumes keeled
  2. Glumes truncate to obtuse, not awn-tipped; blades stiff, distance between veins < or = vein width
     . . . . ***E. intermedia*** ssp. ***intermedia***
  2' Glumes acute to awn-tipped; blades drooping, distance between veins >> vein widths . . . . ***E. repens***

E. intermedia (Host) Nevski ssp. *intermedia*   INTERMEDIATE WHEATGRASS   **ST** 5–11.5 dm. **LF**: sheath glabrous or ciliate; blade 2–8 mm wide; veins many, weakly ribbed. **INFL** 8–21 cm. **SPIKELET** 11–18 mm; glume tips truncate to obtuse; florets 3–10; lemmas 7.5–10 mm, gen glabrous, sometimes rough-hairy, awn gen 0; anthers 5–7 mm. 2*n*=42. Open areas, slopes; < 2100 m. KR, NCoRI (Yolo Co.), CaR, SN, WTR, SnBr, **DMtns**; to B.C., Great Plains; native to Eurasia. [*Agropyron i.* (Host) P. Beauv.] Pls with hairy lf blades, rough-hairy lemmas have been called ssp. *barbulata* (Schur) Á. Löve. [*Agropyron trichophorum* (Link) Richter]

E. pontica (Podp.) Holub ssp. *pontica* (p. 577)   TALL WHEATGRASS   Rhizomes 0. **STS** 5–22 dm, tufted. **LF**: sheath ciliate below middle; blade 2–6.5 mm wide, gen rolled; veins < or =

8, strongly ribbed. **INFL** 10–42 cm. **SPIKELETS** 13–30 mm; florets 6–12; glume tips truncate; lemmas 9–12 mm, awn 0; anthers 2.5–6 mm. 2*n*=70. Disturbed, often alkaline areas; < 1600 m. **CA (exc NW)**; to B.C., e US; native to se Eur, w Asia. [*Agropyron elongatum* (Host) P. Beauv. in part]

E. repens (L.) Nevski (p. 577)   QUACKGRASS   **ST** 5–10 dm. **LF**: sheath gen glabrous or lowermost soft-hairy; blade 2–14 mm wide, veins of 2 kinds, some faint, others strongly ribbed, widely spaced. **INFL** 8–20 cm. **SPIKELETS** 9–16 mm; florets 3–8; glumes gen with awn 0.5–4 mm, otherwise acute; lemmas 6–12 mm, tapering to point or awn 0.5–10 mm; anthers 4–5.5 mm. 2*n*=42. Weed in disturbed areas, cult fields; < 1800 m. CA-FP, **GB**; to e US; native to Eurasia. [*Agropyron r.* (L.) P. Beauv.] NOXIOUS WEED.

ENNEAPOGON   PAPPUS GRASS

Dieter H. Wilken

Per, gen cespitose. **STS** ascending to erect. **LVS** basal and cauline; ligule hairy; blade flat to inrolled. **INFL** panicle- to spike-like, narrow, gen compact. **SPIKELET**: glumes ± subequal, 3–9-veined; florets 3–6, breaking above glumes and weakly between florets, lower 1–3 florets fertile, bisexual, upper gen sterile, gradually reduced; lemmas < glumes, elliptic to ovate, firmly membranous, rounded on back, gen 9-veined, awned at truncate tip, awns 9, plumose; fertile palea slightly > lemma. ± 30 spp.: warm temp N.Am, Afr, Asia, Australia. (Greek: nine beards, from 9 plumose awns) [Renvoise 1968 Kew Bull 22: 393–401]

E. desvauxii P. Beauv. (p. 577)   NINE-AWNED PAPPUS GRASS   **ST** 1–4 dm; nodes dense, short-hairy; internodes 2–5 cm. **LF** soft-hairy; sheath ciliate; ligule hairs < 1 mm; blade > 2 × internode, < 2 mm wide, ± inrolled. **INFL** spike-like, 3–6 cm, grayish. **SPIKELET**: glumes minutely soft-hairy on back, strongly veined, lower 3–5 mm, upper 4–6 mm; florets 3, lower 1 fertile,

upper 2 sterile; lemmas 1–3 mm, awns 2–5 mm, exserted. 2*n*=20. RARE in CA. Rocky slopes, crevices, calcareous soils, desert woodland; 1275–1825 m. **e DMoj (DMtns)**; to Colorado, w TX, n Mex, S.Am. Lower sheaths sometimes enclose cleistogamous spikelets that disperse with st parts. Aug–Sep

ERAGROSTIS LOVEGRASS

John R. Reeder

Ann, per, often glandular; glands often wart-like, circular, pitted. **LF**: sheath margin hairy on sides just below collar; ligules ciliate. **INFL** gen panicle-like, open or dense, sometimes spike-like, often glandular. **SPIKELET** laterally compressed; glumes ± unequal, acute or acuminate, 1(3)-veined; florets 3–many, axis breaking above glumes and between florets (or persistent with glumes and lemmas deciduous, paleas remaining attached or not); lemma keeled or rounded, acute or obtuse, 3-veined, veins gen obvious; palea ± = lemma. **FR** lens-shaped or elliptic, sometimes grooved, gen red-brown. ± 300 spp.: trop, warm temp. (Greek: eros, love, agrostis, a kind of grass) [Koch 1974 Ill Biol Monogr 48:1–74]

1. Per; anthers 0.5–1+ mm; spikelet axis breaking tardily above glumes and between florets
  2. Sts simple, erect, not weak or trailing; spikelets 1.5–2 mm wide; fr > 1 mm . . . . **E. curvula**
  2' Sts abruptly bent, branching, some becoming decumbent and rooting at nodes; spikelets ± 1 mm wide;
    fr gen < 0.8 mm . . . . **E. lehmanniana**
1' Ann; anthers gen 0.2–0.4 mm; spikelet axis not breaking apart, persistent, glumes and lemmas
  deciduous, paleas deciduous or remaining attached
  3. Fr with one or both ends truncate; surface checkered, with an evident groove on side opposite embryo;
    spikelets linear to linear-lanceolate, ± 1 mm wide . . . . **E. mexicana** ssp. *virescens*
  3' Fr elliptic, ovate, or sometimes ± spheric, tip rounded; surface smooth, not grooved; spikelet shape various
    4. Pls without conspicuous glands or glandular areas — spikelet stalks appressed to branches
    . . . . **E. pectinacea** var. *pectinacea*
    4' Pls with conspicuous glands or glandular areas on lf sheaths, blade margins, infl axis and branches,
      spikelet stalks, or lemma keel
      5. Spikelets 2.5–3 mm wide; spikelet stalks gen not glandular . . . . **E. cilianensis**
      5' Spikelets < 2 mm wide; spikelet stalks gen with 1 or 2 glands . . . . **E. minor**

E. cilianensis (All.) Janchen (p. 581) STINKGRASS Ann. **STS** spreading or decumbent, sometimes abruptly bent, often branching, << 6 dm; glands gen present below st nodes. **LF**: sheath glabrous, long-hairy below collar, keel glandular; blade 10–20 cm, 2–8 mm wide, flat or inrolled, margin with wart-like glands. **INFL** < 20 cm, 5–6+ cm wide, ± compact, gen gray-green. **SPIKELET** 2.5–3 mm wide, linear to ovate; glumes 1.5–2 mm, ± equal, midvein often glandular; axis not breaking apart, paleas persistent; florets (5)10–45; lemma 2–2.5 mm, lateral veins prominent, midvein gen glandular; anthers ± 0.2 mm. **FR** ± 0.6 mm, ± spheric. $2n=20$. Disturbed soils; < 1500 m. **CA**; to e US; native to Eur. [*E. megastachya* (Koeler) Link] Jun–Oct

E. curvula (Schrader) Nees var. *curvula* (p. 581) WEEPING LOVEGRASS Per. **ST** erect, densely tufted, unbranched, 4–12 dm, glabrous. **LF**: sheaths < internodes, glabrous or hairy, gen with long hairs on collar and inside upper sheath margin; ligule ± 1 mm; blade (1)2–3(5) dm, inrolled, long-tapered, distally thread-like, scabrous. **INFL** < 35 cm, 15 cm wide, open, gen nodding; branches flexible; lower axils long-hairy; spikelets short-stalked. **SPIKELET** 5–8 mm, gen gray-green; glumes acute, 1-veined, lower 1.5–2 mm, upper 2.5–3 mm; axis breaking apart tardily; lemmas obtuse to ± acute, 3-veined, lemma of lowest floret 2–3 mm; anthers 1–1.5 mm, purple. **FR** ± 1.5 mm, ± ovoid or oblong, light brown; embryo ± 1/2 fr length, dark brown to ± black. $2n=20,40,50,60$. Roadsides, near gardens; < 500 m. CaRF, GV, SCoRO, SnBr, PR, **DMoj**; to s US; native to s Afr. Orn, cult for erosion control. Aug–Oct

E. lehmanniana Nees (p. 581) LEHMANN LOVEGRASS Per, glabrous. **STS** decumbent to erect, often abruptly bent at lower nodes, often stolon-like, branched, 3–6 dm; nodes glabrous. **LF**: sheaths gen < internodes, glabrous or sparsely hairy near collar; ligule ± 0.5 mm; blade (5)8–15 cm, 1–3 mm wide, flat to inrolled, tapered to a rigid point. **INFL** 6–20 cm, < 10 cm wide, open; lower branches loosely spreading. **SPIKELET** 4–8 mm, ± 1 mm wide; glumes unequal, lower ± 1 mm, acute, upper 1–2 mm, obtuse; axis breaking apart tardily; florets 6–10, linear, gray-green; lemmas membranous, obtuse, lowest floret lemma ± 1.5 mm; anthers ± 0.8 mm. **FR** 0.6–0.8 mm, oblong, pale; embryo ± 1/2 fr length, dark brown or black. $2n=40,60$. Roadsides; < 1000 m. SCo, **DMoj**; to TX, n Mex; native to s Afr. Cult for erosion control.

E. mexicana (Hornem.) Link Ann. **ST** widely spreading to erect, gen 1.5–10 dm, gen with ring of glandular depressions below nodes. **LF**: sheath glabrous or papillate-soft-hairy on upper margins, often with glandular depressions on veins; blade 5–25 cm, 4–7 mm wide, flat, sometimes hairy below, midvein rarely glandular. **INFL** (5)10–35 cm, open; axis below nodes, branches, spikelet stalks gen sparsely glandular. **SPIKELET** ovoid to linear, gray-green to reddish; glumes ± 2 mm, upper slightly > lower, lanceolate; axis not breaking apart; florets 5–15; lemmas 1.5–2.5 mm, ovate, acute, gen glabrous; palea slightly < lemma, persistent. **FR** pear-shaped, elliptic, or ± rectangular; surface checkered, with shallow to deep groove on side opposite embryo. Disturbed, gen open sites; < 2000 m. **CA**; to OK, TX, S.Am. 2 sspp.in CA.

ssp. *virescens* (J. Presl) Koch & E. Sánchez (p. 581) **INFL** not glandular. **SPIKELET** ± 1 mm wide, linear to linear-lanceolate. **FR** elliptic or pear-shaped. Disturbed soils in fields, sandy river banks, etc.; < 2000 m. **CA**; NV, S.Am. [*E. orcuttiana* Vasey] May–Oct

E. minor Host (p. 581) Ann. **STS** erect, becoming prostrate, often branching at base, < 6 dm. **LF** gen glabrous; sheath long-soft-hairy near collar; blades like *E. cilianensis*. **INFL** gen 3.5–15 cm, 2.5–6 cm wide, gen gray-green; spikelet stalk with 1 or 2 glands near middle. **SPIKELET** ± 2 mm wide, linear to ovate; glumes not glandular; axis not breaking apart, paleas persistent; florets 8–12; lemma not glandular. **FR** ± 0.5 mm, subspheric to elliptic. $2n=40$. Disturbed soils; < 1400 m. NCoRO, n&c SNF, GV, SCo, MP, **DMoj**; e US, native to Eur. [*E. poaeoides* Roemer & Schultes] Jun–Sep

E. pectinacea (Michaux) Nees Ann. **ST** erect, sometimes abruptly bent at base, 1.5–6(7.5) dm. **LF**: sheaths gen < internodes, glabrous exc white-hairy-tufted at collar margins; blade 2–15(20) cm, 1.5–3(5) mm wide, flat, glabrous. **INFL** (5)10–25 cm, 3–12(15) cm wide, open; primary branches spreading or ascending, straight, alternate or opposite; spikelet stalks spreading or appressed. **SPIKELET** 5–8(10) mm, 1.2–2 mm wide, gen linear; glumes thin, lower 0.5–1 mm, upper 1–1.5 mm; axis not breaking apart, paleas persistent; florets 5–15(20); lemma 1.5–2 mm, membranous, gray-green or red-tinged near tip, veins prominent. **FR** ± 1 mm, oblong to slightly pear-shaped, not grooved.

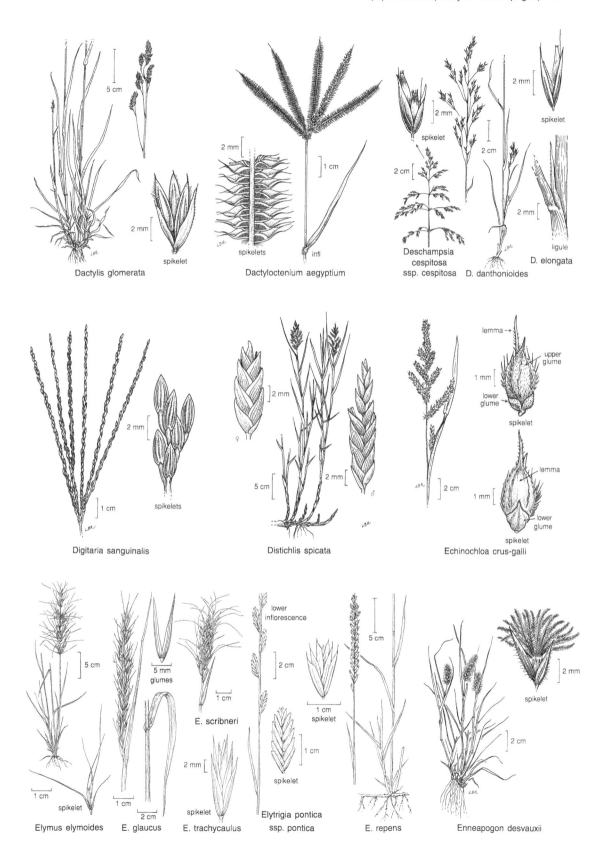

Dactylis glomerata

Dactyloctenium aegyptium

Deschampsia cespitosa ssp. cespitosa   D. danthonioides   D. elongata

Digitaria sanguinalis

Distichlis spicata

Echinochloa crus-galli

Elymus elymoides

E. glaucus

E. scribneri

E. trachycaulus

Elytrigia pontica ssp. pontica

E. repens

Enneapogon desvauxii

Open disturbed sites, fields; < 1400 m. **CA**; widespread US, to Caribbean, C.Am. 2 vars. in CA.

var. *pectinacea* (p. 581) **INFL**: spikelet stalks appressed to branches, diverging < 20°. 2*n*=60. Habitat and range of sp. [*E. diffusa* Buckley] Jul–Oct

ERIOCHLOA

Robert Webster

Ann, per. **STS** decumbent to erect. **LVS** basal and cauline; sheath glabrous or hairy; ligule gen < 1 mm, membrane hairy-fringed; blade gen flat. **INFL** panicle-like, ± dense; 1° branches spreading to appressed; 2° branches appressed; spikelets many, 1–2 per node, short-stalked to subsessile, on one side of axis. **SPIKELET** lanceolate, ± compressed, gen green, falling as 1 unit; glumes strongly unequal, lower glume gen 0, fused to spikelet base to form a disc- or cup-like ring between stalk and upper glume, upper glume ± = spikelet; florets 2, lower floret sterile, gen acuminate, palea 0, upper floret fertile, lemma firm or hard, gen wrinkled, margin inrolled, tip short-pointed to awned. ± 30 spp.; warm temp, trop, worldwide. (Latin: woolly grass) [Shaw & Webster 1987 Sida 12:165–207]

1. Upper floret awnless . . . . *E. acuminata*
1' Upper floret tapered to short awn, awn ± 1–2.5 mm
  2. Spikelets 2 per node at middle of branch axis; upper lf blade surface glabrous; glume awn ± 1–3 mm; upper floret lemma abruptly pointed, point < 0.5 mm . . . . *E. aristata*
  2' Spikelets gen 1 per node; upper lf blade surface short-hairy; glume awn ± 1 mm; upper floret lemma awn slightly < 1 mm . . . . *E. contracta*

E. acuminata (C. Presl) A. Hitchc. var. ***acuminata*** (p. 581) Ann. **ST** 3–12 dm; nodes 2–5. **LF**: sheath 4–8 cm, glabrous or short-hairy; blade gen 5–12 cm, 5–12 mm wide, upper surface glabrous or short-hairy. **INFL**: main axis 7–16 cm; 1° branches 1–5 cm; spikelets gen 2 per node, 1 per node near the axis tip, stalk = or < 1 mm. **SPIKELET** 4–6 mm, ± 1–1.5 mm; lower floret lemma gen 5-veined; upper floret ± 0.8 × lower floret length, lemma acuminate. Seasonal streams, irrigated fields, orchards; < 200 m. SW, **DSon**; to s US, Baja CA. [*E. gracilis* (Fourn.) A. Hitchc.] Other var. native to s AZ, NM, TX, n Mex. Aug–Sep

E. aristata Vasey var. ***aristata*** AWNED CUP GRASS **ST** 4–10 dm; nodes 3–10. **LF**: sheath 4–13 cm, glabrous; blade gen 6–20 cm, 6–20 mm wide, upper surface gen glabrous. **INFL**: main axis 5–20 cm; 1° branches 2–3.5 cm; spikelets 1–2 per node; stalk 0.5–2 mm. **SPIKELET** 4–7 mm (exc awn), ± 1–1.5 mm wide; glume awn ± 1–3 mm; lower floret lemma 3–7-veined; upper floret ± 0.6 × lower floret length. RARE in CA. Seasonal streams, riverbanks; < 100 m. **DSon (Imperial Co.)**; to s US, n Mex. Other var. in C.Am. Jun–Nov

E. contracta A. Hitchc. (p. 581) Ann. **ST** 2–10 dm; nodes 2–5. **LF**: sheath 4–8 cm, glabrous to short-hairy; blade gen 8–12 cm, 2–8 mm wide, upper surface gen short-hairy. **INFL**: main axis 6–20 cm; 1° branches 1.5–4.5(6) cm; spikelets gen 1 per node; stalk = or < 1 mm. **SPIKELET** 3.5–4.5 mm, ± 1–2 mm wide, becoming purple; lower glume ± = spikelet; lower floret lemma 3–7-veined; upper floret ± 0.7 × lower floret length. Seasonal streams, ditches, irrigated fields; < 100 m. Deltaic GV (Solano Co.), SW, **DSon**; native to c&s US.

ERIONEURON

Per, tufted or mat-like, sometimes from short stolons. **STS** spreading to erect. **LVS** gen basal; ligule short-hairy; blade ± stiff, pointed. **INFL** raceme- or panicle-like, head-like, ± dense. **SPIKELET** compressed; glumes ± equal, membranous, gen acuminate, glabrous, 1-veined; axis breaking apart above glumes and between florets; florets 4–20, lower bisexual, upper staminate or sterile; lemma wide, back rounded, gen densely soft-hairy below middle, 3-veined, tip 2-lobed or not, short-awned from tip or not; palea densely hairy below middle; stamens 1 or 3. 5 spp.: sw US, Mex, S.Am. (Greek: woolly nerve, from lemma, palea hairs) [Tateoka 1961 Amer J Bot 48:565–573]

1. Infl stalked, > terminal lvs; lemma minutely 2-toothed, awn < 2 mm . . . . *E. pilosum*
1' Infl ± subsessile, < some terminal lvs; lemma deeply 2-lobed, awn 2.5–4 mm . . . . *E. pulchellum*

E. pilosum (Buckley) Nash (p. 581) HAIRY ERIONEURON Per, cespitose. **STS** erect, 1–3(4) dm. **LF**: sheath margin long-soft-hairy at collar; blade 3–6 cm, 1–1.5 mm wide, flat or folded, margin white. **INFL** raceme- or panicle-like, 1.5–4 cm, 1–2 cm wide, stalked and elevated above terminal lf cluster. **SPIKELET** 1–1.5 cm; glumes 3–7 mm, tan or purplish, awn 0; lemma 4–7 mm, tip minutely 2-toothed, awn < 2 mm. 2*n*=16. RARE in CA. Rocky slopes, ridges, pinyon/juniper woodland; 1500–2000 m. **SNE, e DMtns**; to KS, TX, Mex. [*Tridens p.* (Buckley) A. Hitchc.] May–Jun

E. pulchellum (Kunth) Tateoka (p. 581) FLUFF GRASS Per, mat-like, with clusters of lvs from short stolon-like sts. **STS** spreading to erect, < 1 dm. **LF**: sheath margin short-soft-hairy at collar; blade 2–6 cm, inrolled. **INFL** ± panicle-like, 1–2.5 cm, 1–1.5 cm wide, short-stalked to subsessile and gen < terminal lvs. **SPIKELET** 6–9 mm; glumes 6–9 mm, tan or purplish, short-awned; lemma 2.5–5 mm, 2-lobed from near middle, awn 2.5–4 mm. 2*n*=16. Sandy to rocky slopes, flats, desert shrubland, woodland; 300–1700 m. **D**; to Colorado, TX, c Mex. [*Tridens p.* (Kunth) A. Hitchc.] Feb–May

FESTUCA FESCUE

Susan G. Aiken

Per, gen cespitose, gen ± glabrous; bisexual, dioecious in *F. kingii*. **ST** erect. **LVS** ± basal; sheath gen persisting; collar gen glabrous; ligule gen < 1 mm, membranous, truncate, minutely fringed; blade flat or rolled, basal lobes gen 0. **INFL** panicle-like; branches dense and appressed to open and spreading. **SPIKELET**: glumes < lowest floret, unequal, lower 1–3-veined, upper 3–5-veined; axis breaking above glumes and between florets, florets 2–10, gen bisexual; lemma base gen glabrous, 5-veined (rarely 3- or 7-veined), not converging at tip; awn gen terminal, straight, glabrous; palea ± = lemma; stamens 3. **FR** free from palea, gen beakless. (Latin: ancient name) [Frederiksen 1982 Nord J Bot 2:525–536]

1. Florets unisexual, pl dioecious — spikelets staminate or pistillate; reduced, sterile pistils sometimes present (subg. *Leucopoa*); rhizomes short; lemma awnless . . . . ***F. kingii***
1' Florets gen bisexual (subg. *Festuca*)
  2. Basal lobes of lf blade prominent, ± clasping st
    3. Basal lobes hairy; lemma awn gen 0.5–1.5 mm . . . . ***F. arundinacea***
    3' Basal lobes glabrous; lemma awn << 0.5 mm or 0 . . . . ***F. pratensis***
  2' Basal lobes of lf blade, if any, not clasping st
    4. Lf sheath closed, ± reddish, gen with downward-pointing hairs; rhizomes often present . . . . ***F. rubra***
    4' Lf sheath open at least half its length, gen green, glabrous; rhizomes 0
      5. Infl 0.8–2.5 cm, gen unbranched; lf blade 1–6 cm; spikelet 3.5—5.5 mm; anthers 1–1.5 mm
      . . . . ***F. brachyphylla*** ssp. ***coloradensis***
      5' Infl 1–5 cm, branched only at lowest node; anthers; lf blade 3–10 cm; spikelet 5–7.5 mm; anthers
      1.5–2 mm . . . . ***F. saximontana*** var. ***purpusiana***

F. arundinacea Schreber (p. 581) TALL FESCUE **ST** 8–20 dm, robust; nodes visible. **LF**: sheath shredding with age; ligule 0.5–1 mm; blade 25–70 cm, 4–10 mm wide, flat or loosely rolled, hairy, ± rigid, prominently ribbed above, basal lobes ± clasping st, hairy. **INFL** 15–35 cm; branches many, spreading. **SPIKELET** 8–16 mm; lower glume 3–6 mm, upper 4–9 mm; florets 3–8; lemma 6–12 mm, scabrous near tip, often tinged purple, awn 0.5–2 mm; anthers 3–4 mm; ovary tip glabrous. 2*n*=28,42,70. Disturbed places; < 2700 (gen < 1000) m. CA-FP, **W&I**; to e N.Am; native to Eur. May–Jun

F. brachyphylla Schultes & Schultes f. ssp. ***coloradensis*** S. Frederiksen (p. 581) **STS** 0.4–2 dm, densely clumped; nodes ± concealed. **LF**: ligule < 0.5 mm; blade 1–6 cm, < 0.5 mm wide, folded. **INFL** 0.8–2.5 cm, gen unbranched, narrow. **SPIKELET** 3.5–5.5 mm; lower glume 2–3 mm, upper 2.5–4.5 mm; florets 2–4; lemma 3–4.5 mm, slightly scabrous near tip, awn ± 1–1.5 mm; anthers ± 1–1.5 mm; ovary tip glabrous. Rocky places, subalpine or alpine; 2800–4300 m. c&s SNH, **n SNE**, **W&I**. Other sspp. to Arctic, e N.Am. [*F. brachyphylla* Schultes & Schultes f. ssp. *breviculmis* S. Frederiksen] ✿ DRN,IRR, SUN:1,2,16,17;DFCLT. Jul–Sep

F. kingii (S. Watson) Cassidy (p. 581) Dioecious; rhizomes short. **STS** 3–8 dm, tufted; nodes visible. **LF**: sheath glabrous to densely hairy, conspicuously persisting, gen reddish brown with age; ligule 0.5–3.5 mm, back puberulent; blade 3–30 cm, 1.5–7 mm wide, ± flat, stiffly erect, glaucous. **INFL**: branches appressed, some with spikelets near base. **STAMINATE SPIKELET** 5–10(12) mm; lower glume 3–5.5 mm, upper 4–6.5 mm; florets 2–5; lemma 5–8 mm, finely scabrous, awn 0; anthers 3.5–5 mm; pistil reduced or 0. **PISTILLATE SPIKELET** 5–8 mm; lower glume 3–5 mm, upper 3.5–6 mm; florets 2–5; lemma 5–8 mm, finely scabrous, awn 0; stamens sterile or 0; ovary tip hairy. **FR** plump, minutely beaked. 2*n*=56. Dry, sandy places, sagebrush plains to subalpine forest; > 2000 m. s SNH, SnBr, **GB**; to

OR, Great Plains. [*Hesperochloa kingii* (S. Watson) Rydb., *Leucopoa k.* (S. Watson) W.A. Weber] ✿ TRY. Jun–Aug

F. pratensis Hudson (p. 581) MEADOW FESCUE **STS** 3–13 dm, loosely clumped; nodes visible. **LF**: sheath shredding with age; ligule < 0.5 mm; blade 10–30 cm, 2–7 mm wide, flat or loosely rolled, basal lobes ± clasping st, glabrous. **INFL** 10–25 cm, narrow, branched only at lowest node. **SPIKELET** 12–15.5 mm; lower glume 2.5–4 mm, upper 3.5–5 mm; florets 4–10; lemma 6–8 mm, awn gen 0; anthers 2–4.5 mm; ovary tip glabrous. Disturbed places; gen < 2000 m. CA-FP, **GB**, less common in SW; to e N.Am; native to Eur. Grown for forage. [*F. elatior* L.] May–Jul

F. rubra L. (p. 581) RED FESCUE Rhizomes gen present, sometimes very short. **STS** 3–8 dm, ± clumped, decumbent at base; nodes visible. **LF**: sheath closed, ± reddish, shredding with age, hairs ± downward-pointing; ligule < 0.5 mm; blade 5–30 cm, < 3 mm wide, ± folded. **INFL** 5–20 cm, ± open; branches ± ascending. **SPIKELET** 9–12 mm; lower glume 2.5–3.5 mm, upper 3.5–5.5 mm; florets 3–10; lemma 5–7 mm, sometimes scabrous near tip, awn < 4 mm, scabrous; anthers 2.5–4 mm; ovary tip glabrous. 2*n*=14,28,42,56,70,128. Sand dunes, grassland, subalpine forest; gen < 2500 m. NW, CaR, n&c SN, CW, TR, **W&I**; worldwide. Some commercial cultivars much naturalized in CA. ✿ IRR:1,**4–6,15–17,24**&partSHD:2,3,7,8,9,**14,18–23**;CVS:GRNCVR. May–Jul

F. saximontana Rydb. var. ***purpusiana*** (St.-Yves) S. Frederiksen & L.E. Pavlick **STS** 0.5–2.5 dm, clumped; nodes ± concealed. **LF**: ligule < 0.5 mm; blade 3–10 cm, < 0.5 mm wide, folded, scabrous near tip. **INFL** 1–5 cm, very narrow, branched only at lowest node. **SPIKELET** ± 5–7.5 mm; lower glume 2–3.5 mm, upper 3.5–5 mm; florets 2–5; lemma ± 3.5–5.5 mm, scabrous near awn, awn ± 1–2 mm; anthers 1.5–2 mm; ovary tip glabrous. 2*n*=42. Alpine, subalpine summits, dry granitic gravel, talus fields, sagebrush scrub; gen > 3000 m. c&s SNH, SnBr, **W&I**.

HESPEROSTIPA

Mary E. Barkworth

Per, cespitose. **ST** erect, unbranched. **LF**: blade upper surface conspicuously ridged, gen inrolled. **INFL** panicle-like, narrow. **SPIKELET**: glumes tapered from near base to acute tip, awn 0; axis breaking above glumes; floret 1, 7–25 mm, narrowly cylindric; callus 2–5 mm, sharp, densely stiff-hairy; lemma hard, margins overlapping at maturity, upper portion fused, awn 6.5–18 cm, bent twice, lower segments twisted, last segment not twisted; palea = lemma, hard, 2-veined, veins terminating at tip. 4 spp.: N.Am. (Greek: western stipa) Segregated from *Stipa*; most closely related to *Piptochaetium, Nassella*.

H. comata (Trin. & Rupr.) Barkworth NEEDLE-AND-THREAD **ST** 1–11 dm. **INFL** 10–28 cm. **SPIKELET**: lower glume 18–35 mm, upper 1–3 mm shorter; floret 7–13 mm; lemma evenly hairy, hairs ± 1 mm, white; awn 65–195 mm. Well-drained soils; 200–3500 m. SN, TR, PR, **GB**, **DMtns**; to Yukon, Great Plains, Mex. [*Stipa c.* Trin. & Rupr.]. 2 sspp.in CA.

ssp. ***comata*** (p. 581) **SPIKELET**: awn 75–195 mm, distal segment wavy or curly. 2*n*=38,44,46. Grassland, sagebrush shrubland; 200–3500 m. SN, TR, PR, **GB**, **DMtns**, introduced elsewhere; to Yukon, c US, Mex. ✿ DRN,SUN:1,**2,15–17**&IRR:**3,7**,14,18–24; also STBL. Jun–Jul

HETEROPOGON

Kelly W. Allred

Ann, per, cespitose. **ST** erect, densely clumped. **LVS** cauline; sheaths flattened, gen < internodes; ligule truncate, thinly membranous; blade flat or folded. **INFL** ± spike-like. **SPIKELETS** on 1 side of axis, paired; spikelets of lower 1–4 pairs ± equal, sterile or staminate; spikelets of upper pairs strongly unequal, lower spikelet sessile, bisexual, awned, upper stalked, staminate or sterile, awn 0, pair breaking below sessile spikelet. **SESSILE SPIKELET**: glumes hard, short-hairy, tightly enclosing florets; florets 2, lower vestigial, obscure, upper fertile; lemma transparent, fragile. **STALKED SPIKELET**: glumes, lemmas thinly membranous. 6 spp.: warm temp, trop. (Greek: different beard, from awned and awnless spikelets)

H. contortus (L.) Roemer & Schultes (p. 581) Per. **ST** 2–8 dm. **LF**: blade 6–24 cm, 4–8 mm wide. **INFL** 4–8 cm. **SESSILE SPIKELET**: glume 5–8 mm, dark brown, sharp-pointed, callus hairs red-brown, awn 5–10 cm, puberulent, red-brown. **STALKED SPIKELET** staminate. 2*n*=60. Uncommon. Rocky slopes, washes, open areas; < 800 m. **DSon**, naturalized in SCo (San Diego Co.); Baja CA, to s US, worldwide. NOXIOUS.

HOLCUS VELVET GRASS

Dieter H. Wilken

Per, cespitose or from rhizomes, glabrous to velvety soft-hairy. **LVS** gen basal; ligule membranous, truncate, puberulent; blade flat. **INFL** panicle-like, ± congested. **SPIKELETS** laterally compressed, breaking below glumes, falling as 1 unit; glumes ± subequal, lower 1-veined, upper 3-veined; florets gen 2, lower floret bisexual, upper staminate or sterile; callus hairy; lemmas ± 2 mm, shiny, membranous, faintly 3–5-veined, lemma of lower floret awnless, lemma of upper floret awned near 2-lobed tip; palea ± = lemma. ± 8 spp.: temp Eurasia, Afr. (Latin: a grass)

H. lanatus L. (p. 581) COMMON VELVET GRASS **STS** clumped, ascending to erect, 6–20 dm; nodes and internodes soft-hairy. **LF**: ligule 1–3 mm; blade 5–18 cm, 4–9 mm wide. **INFL** 7–15 cm. **SPIKELET**: glumes 3–6 mm, purplish, short-hairy on back, keel long-hairy; lemmas 3–4 mm, awn twisted to recurved, ± 1 mm. 2*n*=14. Moist sites, roadbanks, cult fields, meadows; 100–2300 m. CA-FP, **GB, DMoj**; to AK, Can, widespread US, S.Am; native to Eur. Cult for forage, hay.

HORDEUM BARLEY

Mary E. Barkworth

Ann, per, sometimes from short rhizomes. **ST** decumbent to erect, gen abruptly bent at base. **LF**: sheath glabrous or hairy, short-appendaged or not; blade flat or ± rolled. **INFL** spike-like, dense; axis breaking apart at nodes in fr; spikelets 2-ranked, strongly overlapping, 3 per node, spikelets of 2 kinds. **CENTRAL SPIKELET** bisexual, gen sessile; with 1 stalked or sessile floret; glumes awn-like, gen > floret; lemma awned. **LATERAL SPIKELETS** 2, sterile or staminate, gen short-stalked; with 1 sessile floret; glumes awn-like, > floret, lemma gen awned. 32 spp.: temp worldwide exc Australia. (Latin: ancient name for Barley) [Baum & Bailey 1990 Canad J Bot 68:2433–2442] Key to species adapted by Margriet Wetherwax

1. Per
  2. Central spikelet glumes 7–17 mm; gen straight with age . . . . *H. brachyantherum* ssp. *brachyantherum*
  2' Central spikelet glumes 22–80 mm, spreading with age
    3. Glumes 22–26 mm, flat near base; lemma awn of central spikelet 17–20 mm . . . . ²*H. arizonicum*
    3' Glumes 35–80 mm, not flat; lemma awn of central spikelet 25–90 mm . . . . ²*H. jubatum*
1' Ann
  4. Lf sheath appendages well developed, 1–4 mm; lateral florets exc awns 8–15 mm . . . . *H. murinum*
    5. Central spikelet sessile to subsessile, stalk < 0.5 mm; central floret (exc awn) gen = lateral florets;
      palea of lateral floret ± glabrous . . . . ssp. *murinum*
    5' Central spikelet stalk 1–2 mm; central floret (exc awn) < lateral florets; palea of lateral floret hairy
      6. Central floret slightly < lateral florets; anthers of lateral florets > 2 × anthers of central floret . . . ssp. *glaucum*
      6' Central floret << lateral florets; anthers of central and lateral florets ± equal . . . . ssp. *leporinum*
  4' Lf sheath appendages < 2 mm or 0; lateral florets exc awns < 5(7.5) mm
    7. Lemma awn of central spikelet 16–90 mm
      8. Glumes ± straight with age . . . . ²*H. marinum* ssp. *gussoneanum*
      8' Glumes spreading with age
        9. Central spikelet glumes flat near base . . . . ²*H. arizonicum*
        9' Central spikelet glumes not flat . . . . ²*H. jubatum*
    7' Lemma awn of central spikelet 3–18 mm
      10. Lemma awn of lateral spikelet 3–8 mm . . . . ²*H. marinum* ssp. *gussoneanum*
      10' Lemma awn of lateral spikelet < 1 mm . . . . *H. depressum*

H. arizonicum Covas Ann, bien, sometimes per. **STS** 2.1–7.5 dm, gen erect, tufted; nodes glabrous. **LF**: lower sheaths hairy, upper sheaths glabrous, appendages < 2 mm or 0; blade < 4 mm wide, scabrous. **INFL** 5–12 cm, 6–10 mm wide, pale green. **CENTRAL SPIKELET**: glumes 22–26 mm, flat near base, spreading with age; lemma awn 17–20 mm. **LATERAL SPIKELET**: stalks curved; glumes 20–26 mm; floret 6–7.5 mm, narrow; lemma tip tapered. 2*n*=42. Uncommon. Wet places, irrigated fields; < 300 m. **DSon**; AZ.

H. brachyantherum Nevski (p. 585) Per. **STS** < 9 dm, loosely

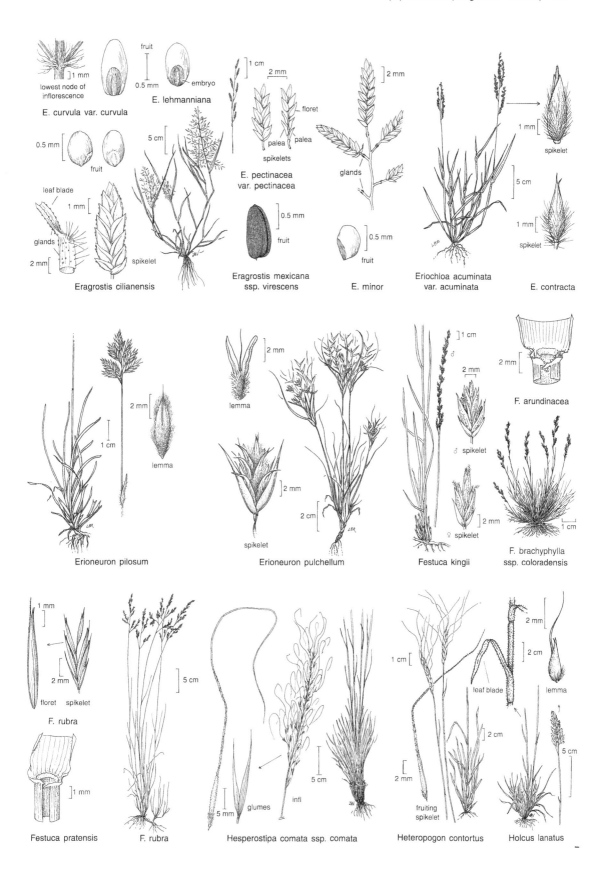

lowest node of inflorescence
E. curvula var. curvula

fruit
0.5 mm
embryo
E. lehmanniana

0.5 mm
fruit

leaf blade
1 mm
glands
2 mm
spikelet
Eragrostis cilianensis

5 cm

1 cm
2 mm
floret
palea palea
spikelets
E. pectinacea
var. pectinacea

0.5 mm
fruit
Eragrostis mexicana
ssp. virescens

2 mm
glands

0.5 mm
fruit
E. minor

2 mm

Eriochloa acuminata
var. acuminata

1 mm
spikelet

5 cm

1 mm
spikelet
E. contracta

lemma

2 mm

1 cm
lemma
Erioneuron pilosum

lemma
2 mm

2 mm
spikelet

2 cm
Erioneuron pulchellum

1 cm
2 mm
♂ spikelet
2 mm
♀ spikelet
Festuca kingii

2 mm
F. arundinacea

1 cm
F. brachyphylla
ssp. coloradensis

1 mm
2 mm
floret spikelet
F. rubra

1 mm
Festuca pratensis

5 cm
F. rubra

5 mm
glands
5 cm
infl
Hesperostipa comata ssp. comata

1 cm
leaf blade
2 cm
2 mm
fruiting spikelet
Heteropogon contortus

2 mm
2 cm
lemma
5 cm
Holcus lanatus

to densely tufted; nodes glabrous. **LF**: sheath glabrous to sparsely hairy; appendages 0; blade 1.5–9 mm wide, glabrous or hairy. **INFL** 8–10 cm, green to ± purplish. **CENTRAL SPIKELET**: glumes 7–19 mm, sometimes flat at base, gen straight with age; floret 5.5–10 mm; lemma awn 3.5–22 mm. **LATERAL SPIKE-LETS**: glumes 6.5–19 mm, straight or spreading with age, lower glume sometimes flat at base; floret < 7 mm or vestigial; lemma awn < 7.5 mm; anthers 0.9–5.5 mm. 2*n*=14,28,42. Meadows, pastures, streambanks; < 3400 m. CA-FP, **GB**; to AK, Rocky Mtns, Mex, Eurasia. 2 sspp.in CA.

ssp. **brachyantherum** **ST** gen robust. **LF**: blade < 19 cm, < 9 mm wide, glabrous or sparsely short-hairy. **CENTRAL SPIKELET**: glumes 7–17 mm, gen straight with age; lemma awn < 4.5 mm; anthers 1–3.5 mm. 2*n*=28,42. Habitat and range of sp. May–Aug

H. depressum (Scribner & J.G. Smith) Rydb. (p. 585) LOW BARLEY Ann. **STS** 1–5.5 dm, erect, loosely tufted; nodes glabrous. **LF**: basal sheaths hairy, appendages 0; blade < 4.5 mm wide, sparsely to densely hairy. **INFL** 2.2–7 cm, 4.8 mm wide, pale green or reddish. **CENTRAL SPIKELET**: glumes 5.5–20 mm, sometimes flattened to 0.5 mm wide at base; floret 5–9 mm; lemma awn 3–12 mm. **LATERAL SPIKELETS** staminate or sterile, sometimes bisexual; glumes 5–20 mm, lower glume ± flat at base; floret vestigial; lemma awn < 1 mm; anthers 0.5–1.5 mm. 2*n*=28. Moist sites, vernal pools, gen alkaline soils; < 1800 m. **CA (exc mtns)**; to WA, ID. Apr–May

H. jubatum L. (p. 585) FOXTAIL BARLEY Per or ann. **STS** 2–6 dm, bent at base or erect, densely tufted. **LF**: sheath glabrous to hairy; blade < 5 mm wide, scabrous to short-hairy. **INFL** 3–10 cm, breaking apart in fr, whitish green to light purple. **CENTRAL SPIKELET**: glumes (10)35–80 mm, not flat at base, strongly spreading with age; floret 5.5–8 mm; lemma awn 25–90 mm. **LATERAL SPIKELETS** staminate or sterile; glumes 35–80 mm, not flat at base; floret 4–5 mm; lemma awn 2–7 mm. 2*n*=28. **CA**; to AK, e US, Mex. [var. *caespitosum* (Scribner) A. Hitchc.] Scabrous spikelet clusters can cause mechanical injury to animals. May–Jul

H. marinum Hudson ssp. **gussoneanum** (Parl.) Thell. MEDITERRANEAN BARLEY Ann. **ST** 1–5 dm, bent at base or erect; nodes glabrous. **LF**: basal sheaths ± hairy, appendages < 2 mm or 0; blade 1–6 mm wide. **INFL** 1.5–7 cm, 5–20 mm wide, green to purple, breaking apart at nodes in fr; central and lateral spikelets falling together. **CENTRAL SPIKELET**: glumes 14–26 mm, not flat at base, ± straight with age; floret 5–8 mm; lemma awn 6–18 mm. **LATERAL SPIKELETS** sterile; glumes 10–24 mm, rounded to slightly flat; floret < 5 mm; lemma awn 3–8 mm. Dry to moist, disturbed sites; < 1500 m. **CA** (exc mtns); to B.C., ID, AZ; native to Eur. [*H. hystrix* Roth, *H. geniculatum* All.] Apr–Jun

H. murinum L. Ann. **ST** 1–11 dm, erect, sometimes ± prostrate; nodes glabrous. **LF**: basal sheaths glabrous to ± hairy, appendages 1–4 mm; blade 2–5 mm wide, glabrous, scabrous or sparsely long-hairy. **INFL** 3–8 cm, 7–16 mm wide, green to glaucous, sometimes reddish or brown in fr. **CENTRAL SPIKE-LET**: glumes 11–25 mm, flattened at base, gen ciliate; floret 8–14 mm; lemma awn 20–40 mm. **LATERAL SPIKELETS**: glumes 11–35 cm, base flattened; floret 8–15 mm; lemma awn 20–50 mm. Moist, gen disturbed sites; gen < 1000 m. **CA**; to B.C., e US, n Mex; native to Eur.

ssp. **glaucum** (Steudel) Tzvelev (p. 585) Summer ann. **ST** 1.5–4 dm. **INFL** green to glaucous, gen brown in fr. **CENTRAL SPIKELET** stalked; floret ± < lateral florets; lemma awn < awn of lateral floret. 2*n*=14. Habitat and range of sp. [*H. g.* Steudel, *H. stebbinsii* Covas] Apr–May

ssp. **leporinum** (Link) Arcang. Winter ann. **ST** 3–11 dm. **INFL** gen green. **CENTRAL SPIKELET** stalked; floret << lateral florets; lemma awn slightly < awn of lateral floret. 2*n*=28,42. Habitat and range of sp. [*H. leporinum* Link] Apr–Jun

ssp. **murinum** Winter ann. **ST** 3–6 dm. **INFL** green. **CENTRAL SPIKELET** sessile or subsessile; stalk < 0.5 mm; floret gen = lateral florets; lemma awn slightly < awn of lateral floret. 2*n*=28. Habitat and range of sp. Apr–Jun

IMPERATA

Kelly W. Allred

Per with rhizomes. **ST** erect, solid in ×-section. **LVS** cauline; ligule membranous, truncate; blade flat. **INFL** panicle-like, ± cylindric; branches appressed, many, short, spike-like, densely silky. **SPIKELETS** in pairs, stalked, ± round in ×-section, breaking below glumes, falling as 1 unit; glumes unequal, lower < upper, thinly membranous; florets 2, lower vestigial, obscure, upper bisexual, < glumes; lemma reduced, transparent or 0, awn 0; palea << lemma, ± vestigial; stamens 1–2; style exserted, stigmas plumose. 8 spp.: warm temp, trop. (F. Imperato, Italian naturalist, 1500's)

I. brevifolia Vasey (p. 585) SATINTAIL Rhizomes hard, scaly. **ST** 0.7–1.5 m. **LF**: ligule densely ciliate; blade 15–50 cm, 4–15 mm wide, narrow at collar. **INFL** 1–3 dm, plume-like, densely white-silky-hairy, appearing speckled from adherent brown anthers and stigmas; hairs 8–15 mm. **SPIKELET**: glumes 2.5–5 mm, faintly 5-veined. Wet springs, meadows, streamsides, flood plains; < 500 m. SnJV, SCo, SnGb, SnBr, **DMoj**, cult elsewhere; to TX, n MEX. Sep–May

KOELERIA

Dieter H. Wilken

Ann, per. **STS** erect. **LVS** basal to cauline; ligule membranous, glabrous to minutely ciliate, toothed at obtuse to truncate tip; blade narrow, flat to inrolled. **INFL** panicle-like, gen compact, narrow. **SPIKELET** laterally compressed; glumes unequal, upper > and wider than lower, keeled, acute, lower 1-veined, upper faintly 3–5-veined; axis prolonged beyond fertile floret, bristly (sometimes with vestigial floret at tip); florets 2–5, bisexual, breaking above glumes and between florets; lower lemmas gen > glumes, awned or not, 5-veined; palea ± < lemma, tip minutely 2-forked. ± 30 spp.: temp N.Am, Eurasia. (G.L. Koeler, Germany, born 1765)

K. macrantha (Ledeb.) J. A. Schultes (p. 585) JUNEGRASS Per, cespitose. **STS** 2–7 dm, glabrous to puberulent. **LVS** gen basal, tufted, glabrous to puberulent; ligule 1–2 mm; blade 3–20cm, 1–2(3) mm wide, gen ridged. **INFL** 2–15 cm, 1–2 cm wide,cylindric to narrowly conic, axis and branches puberulent. **SPIKELET** 4–6 mm, ± shiny, tan (sometimes purplish); florets2–3(4); glumes and lower lemmas minutely scabrous on back; lower glume ± 3 mm, upper ± 5 mm; lemmas 3–5 mm, acute to small-pointed at tip. 2n=14. Dry, open sites, clay to rocky soils, shrubland, woodland, coniferous forest, alpine; < 3500 m. NW, CaR, SN, CW, TR, PR, **GB, DMtns**; to AK, e Can, c&e US, n Mex. [*K. cristata* (L.) Pers., an illegitimate name; *K. pyramidata* (Lam.) P. Beauv. misapplied] ❀ DRN:4–6,14–17&IRR,part SHD:1–3,7, 8–10,18–24; alsoSTBL. May–Jul

LAMARCKIA GOLDENTOP

Lynn G. Clark

1 sp.: native to Medit, naturalized in similar climates worldwide (J.B. Lamarck, French botanist, 1744–1829)

L. aurea (L.) Moench (p. 585) Ann, cespitose, glabrous. **ST** gen erect, 7–40 cm. **LVS** cauline, ± evenly distributed; ligule 3–7 mm, membranous, glabrous, tip ± irregularly cut; blade 2.5–9 cm, 2.5–7 mm wide, flat. **INFL** panicle-like, terminal, 2–8 cm, dense, golden yellow to purplish; axis short-white-hairy in branch axils; spikelets short-stalked, with 1 fertile and 1–3 sterile spikelets in spreading to drooping clusters, each cluster gen falling as 1 unit. **FERTILE SPIKELET**: glumes 2.5–4 mm, ± equal, gen = spikelet; florets 2; lower floret fertile, 2.5–3 mm, lemma awned from near tip, awn 6–7 mm, straight; upper floret sterile, ± 0.5 mm, awn 4–5 mm. **STERILE SPIKELET** 6–9 mm,> glumes, linear; glumes > lower floret; florets 5–8; lemmas ± overlapping, 1.5–2 mm, obtuse, tip ± fringed, awn 0 (stalk base sometimes with a reduced, sterile spikelet that is like the fertile in size and shape). *n*=7. Open ground, moist seeps, rocky hillsides, sandy soil; < 660 m. **CA**; AZ; native to Medit. Somewhat weedy. Feb–May

LEERSIA CUTGRASS

Dieter H. Wilken

Per from long rhizomes. **STS** gen solitary, decumbent to erect. **LVS** cauline; ligule membranous; blade flat to folded. **INFL** panicle-like, open; lateral branchlets, spikelet stalks arched to wavy. **SPIKELET** laterally compressed; glumes 0; floret 1, bisexual, falling as 1 unit; lemma, palea firmly membranous; lemma strigose on back, awnless, 5-veined; palea ± = lemma. 17 spp.: trop, warm temp Am, Eurasia. (J.D. Leers, Germany, born 1727) [Pyrah 1969 Iowa State Coll J Sci 44:215–270]

L. oryzoides (L.) Sw. (p. 585) RICE CUTGRASS **STS** 1–1.5 m; nodes short, soft-hairy. **LF**: sheath glabrous to minutely scabrous; ligule ± 1 mm, truncate; blade 10–28 cm, 8–14 mm wide, margin strongly scabrous, with downward-pointing teeth. **INFL** 12–20 cm; lower branches ± spreading. **SPIKELET** 4–5 mm, oblong to narrowly elliptic; lemma 4–5 mm, width 3–5 × palea width, strigose on back. *2n*=48. Marshes, streams, ponds; < 700 m. NW, ne SCo, **SNE (Owens Valley)**; to B.C., e N.Am, Eurasia. Lateral infls enclosed by sheath, gen cleistogamous. ❀ STBL. Aug–Oct

LEPTOCHLOA SPRANGLETOP

Ann, per. **STS** spreading to erect. **LVS** gen cauline; ligule membranous, ± entire to jagged, sometimes ciliate; blade flat. **INFL** panicle-like; branches spike-like; spikelets short-stalked or sessile. **SPIKELETS** compressed or ± cylindrical; glumes equal or unequal, 1(3)-veined, short-awned or not; axis breaking apart above glumes and between florets; florets gen 2–12; lemma back rounded or keeled, glabrous or hairy, 3-veined, tip obtuse or minutely 2-lobed, awn gen 0; stamens 2 or 3. 40 spp.: warm temp, trop. (Greek: slender grass, from slender infl) [McNeill 1979 Brittonia 31:399–404] For revised taxonomy see Snow 1998 Novon 8:77–80. Key to species adapted by Margriet Wetherwax.

1. Spikelets 1–3 mm, compressed; florets 2–4; lemma strongly keeled . . . . ***L. mucronata***
1' Spikelets 4–12 mm, ± cylindric; florets 6–12; lemma rounded
  2. Lemma gen acute, awned, awn 0.5–3(5) mm . . . . ***L. fascicularis***
  2' Lemma obtuse to truncate, abruptly pointed . . . . ***L. uninervia***

L. fascicularis (Lam.) A. Gray (p. 585) BEARDED SPRANGLETOP Ann. **STS** spreading to erect, 3–10 dm. **LF**: sheath gen glabrous; ligule 2–5 mm, membranous, jagged; blade 10–50 cm, 1–5 mm wide. **INFL** 1–3 dm; branches spreading to ascending, lower 8–15 cm. **SPIKELET** 6–12 mm, ± cylindric in ×-section; glumes 2–4 mm, upper > and wider than lower; florets 6–12; lemma 3.5–5 mm, back rounded, densely hairy below middle, awn at tip 1–3(5) mm. *2n*=20. Marshes, wetlands, sometimes wet disturbed areas; < 1200 m. GV, **GB**, **D**; to e US, Mex, S.Am. Fls gen self-pollinating. Jun–Oct

L. mucronata (Michx.)Kunth. (p. 585) RED SPRANGLETOP Ann. **STS** decumbent to erect, 1–10 dm, sometimes reddish or purple. **LF**: sheath papillate; ligule 1–2 mm, entire to jagged and hairy; blade 5–30 cm, 3–10 mm wide. **INFL** 3–20 cm; branches gen ascending, lower 5–10 cm. **SPIKELET** 1–2 mm, compressed; glumes 1–2 mm, ± equal, purplish; florets (2)3–4; lemma 1–2 mm, keeled, veins hairy, tip minutely 2-lobed, awn 0. *2n*=20. Wet sites, drying ponds; < 100 m. **DSon (Imperial Co.)**; to s US, Mex, S.Am. Sep–Dec

L. uninervia (C. Presl) A. Hitchc. & Chase MEXICAN SPRANGLETOP Ann. **STS** erect, sometimes few-branched, 3–10 dm. **LF**: sheath glabrous or scabrous; ligule 2–6 mm, entire to jagged; blade 10–45 cm, 1–4 mm wide. **INFL** 1–3 dm; branches gen ascending, lower 3–6 cm. **SPIKELET** 5–7 mm, ± cylindric in ×-section; glumes 1–2 mm, upper > and wider than lower; florets 6–9; lemma 2–3 mm, back rounded, tip obtuse to truncate, abruptly short-pointed, awn 0. *2n*=20. Ditches, drying ponds, disturbed wet places; gen < 1000 m. s SNF, SnJV, SW, **GB**, **D**; to s US, Mex, S.Am. Mar–Dec

LEYMUS

Mary E. Barkworth

Per, gen from rhizomes. **LF**: ligule membranous; blade flat or rolled, strongly ribbed above, glabrous or hairy. **INFL** gen spike-like (panicle-like in *L. condensatus*), dense, gen > upper cauline lvs; some nodes with 2+, gen sessile spikelets. **SPIKELET**: glumes < spikelet, lanceolate, membranous, flexible or narrowly lanceolate to awl-like, stiff; axis breaking above glumes and between florets; florets 2–7; lemma acute to short-awned; palea < lemma; anthers 2.5–7 mm. 31 spp.: N.Am., Eurasia. (Anagram of *Elymus*) [Barkworth & Atkins 1984 Amer J Bot 71:609–625]

Some spp. important in revegetation, often on saline soils. Sometimes treated in *Elymus*. Key to species adapted by Margriet Wetherwax.

1. Pls gen > 7 dm; blades often > 10 mm wide
  2. Infl unbranched; spikelets sessile on central axis . . . . *L. cinereus*
  2' Infl often branched at lowest nodes; some spikelets stalked, stalks 1–5 mm . . . . *L. condensatus*
1' Pls gen < 10 dm; blades 1–5 mm wide
  3. Pl cespitose, rhizomes 0; DMtns . . . . *L. salinus* ssp. *mojavensis*
  3' Pl from rhizomes; GB . . . . *L. triticoides*

L. cinereus (Scribner & Merr.) Á. Löve (p. 585) Pl cespitose; rhizomes 0 or short. **ST** 7–21 dm, gen glabrous; lowest node often hairy. **LF:** ligule 2.5–6.5 mm; blade 3–12 mm wide, upper surface scabrous. **INFL** 9–19 cm, unbranched; spikelets 2–17 per node, sessile. **SPIKELET:** glumes 8–18 mm, awl-like; lemma 6.5–12 mm, acute to awn-tipped, glabrous to short-hairy. 2*n*=28,56. Streamsides, canyons, roadsides, sagebrush shrubland, open woodlands; < 3000 m. CaR, SN, Teh, ScV, TR, **GB, DMtns**; to Can, Colorado. [*Elymus c.* Scribner & Merr.] Hybridizes with *L. triticoides*. ❀ TRY. Jun–Aug

L. condensatus (C. Presl) Á. Löve Pl cespitose; rhizomes 0 or short. **ST** 11–30 dm, glabrous. **LF:** blade 10–28 mm wide. **INFL** 17–44 cm, panicle-like; lower nodes short-branched; spikelets sessile or stalked. **SPIKELETS:** glumes 6–16 mm, awl-like; lemmas 7–14 mm, glabrous to hairy, acute to awn-tipped, awn < 4 mm. 2*n*=28,56. Dry slopes, open woodland; < 1500 m. CW, SW, **DMoj**; Mex. [*Elymus c.* C. Presl] Hybridizes with *L. triticoides*. ❀ DRN:**7,14–24**&IRR:**8–12**;CV. Jun–Aug

L. salinus (M.E. Jones) Á. Löve ssp. ***mojavensis*** Barkworth & R.J. Atkins Pl cespitose. **ST** 4–14 dm, gen glabrous. **LF:** ligule 0.5–1 mm; blade 1–5 mm wide, upper surface evenly hairy. **INFL** 4–14 cm; spikelets gen 1 at lower, upper nodes, 2 at central nodes. **SPIKELET:** glumes 0 or < 13 mm, awl-like; lemma 7–13 mm, gen glabrous, acute or awn-tipped, awn < 2.5 mm. Hillsides; 1350–2000 m. **DMtns**; to ID, w Colorado, n AZ. [*Elymus s.* M.E. Jones] Other ssp. Great Basin. ❀ TRY. May–Jun

L. triticoides (Buckley) Pilger (p. 585) Pl from rhizomes. **ST** 4.5–13 dm, glabrous to hairy. **LF:** ligule 0.2–1.3 mm; blade 2.5–4 mm wide, upper surface finely scabrous. **INFL** 5–20 cm, narrow; spikelets 1–3 per node. **SPIKELET:** glumes 5–16 mm, awl-like; lemmas 5–12 mm, glabrous to puberulent, gen awn-tipped, awn gen ± 3 mm. 2*n*=28. Moist, often saline, meadows; < 2300 m. **CA** (exc D); to WA, Rocky Mtns, TX. [*Elymus t.* Buckley] Hybridizes with *L. condensatus, L. cinereus.* ❀ 4,**5,6,15–17**&IRR:1,**2,3,7–10**,11,**14,18–24**;GRNCVR;also INV;CVS. Jun–Jul

LOLIUM RYEGRASS

Dieter H. Wilken

Ann, per. **STS** solitary to loosely clumped, ascending to erect. **LVS** basal and cauline; appendages acute; ligules membranous, obtuse to truncate; blade flat to folded. **INFL** spike-like (sometimes panicle-like). **SPIKELETS** gen 2-ranked, ± laterally compressed, narrow edge facing infl axis, sessile; glume 1, on outside edge of spikelet, > lower floret, 3–9-veined; florets 5–20, bisexual, breaking above glume and between florets; lemma ± membranous, rounded on back, awnless to awned at acute to obtuse tip, 5-veined; palea ± = lemma. 8 spp.: Eurasia. (Latin: ancient common name for ryegrass) [Terrell 1966 Bot Rev 32:138–164]

1. Glume < rest of spikelet, lower lemmas firmly membranous, flat to rounded at base . . . . *L. perenne*
1' Glume gen = or > rest of spikelet exc awns, lower lemmas hard, ± swollen at base . . . . *L. temulentum*

L. perenne L. (p. 585) PERENNIAL RYEGRASS Per. **STS** ascending to erect, (3)5–8 dm, gen glabrous. **LF:** ligule 1–3 mm; blade 4–25 cm, 2–5 mm wide, folded in bud. **INFL** (8)12–30 cm, spike-like (sometimes panicle-like, branches spike-like, spreading). **SPIKELET** 10–15 mm; glume < rest of spikelet, 5–10 mm, linear-lanceolate; lemmas 5–7 mm, ± lanceolate, awnless. 2*n*=14,28. Disturbed sites, abandoned fields, lawns; < 1000 m. CA-FP, **DMtns (Panamint Mtns)**; to AK, e N.Am; native to Eur. May–Sep

L. temulentum L. (p. 585) DARNEL Ann. **STS** ascending to erect, 4–9 dm, glabrous to scabrous. **LF:** ligule < 2 mm; blade 5–20 cm, 2–7 mm wide, folded in bud. **INFL** 7–28 cm, spike-like. **SPIKELET** (15)20–30 mm; glume = or > rest of spikelet exc awns, 10–17(20) mm, lanceolate to oblong; lemmas 5–8 mm, awnless to awned at tip, awn 8–18 mm. 2*n*=14. Open, disturbed sites; 150–1750 m. CA-FP, MP, **D**; to B.C., e US; native to Medit. TOXIC: grain can contaminate flour or birdseed; severe poisoning rare. Apr–Jun

LYCURUS

Per. **STS** erect, ± solid in ×-section. **LVS** gen basal, gray-green; sheath compressed, keeled; ligule membranous, long tapered, entire; blade flat to folded. **INFL** panicle-like; branches short, ± appressed, dense. **SPIKELETS** stalked, paired; pair with subtending axis segment falling as 1 unit; lower spikelet staminate or sterile, upper bisexual; glumes unequal, < lemma, 2–3-veined, awned; floret 1; lemma firm, long-tapered, awned; palea ± = lemma. 3 spp.: sw US, Mex, S.Am. (Greek: wolf tail, from infl shape) [Reeder 1985 Phytologia 57: 283–291]

L. setosus (Nutt.) C. Reeder (p. 585) WOLFTAIL Per, tufted. **ST** decumbent to erect, 2–6 dm, compressed, puberulent. **LF:** ligule 3–7 mm, decurrent to sheath; blade 3–10 cm, 1–2 mm wide, margin and midvein ± white. **INFL** 3–6 cm, ± 5 mm wide, cylindric. **SPIKELET** compressed; glumes 1–2 mm, ± equal, < flo- ret, awns 3–7 mm, lower glume 2-veined, awns 2–3, upper glume 1-awned; floret 1; lemma 3–4 mm, 3-veined, awn 2–7 mm. 2*n*=40. RARE in CA. Joshua-tree woodland, pinyon/juniper woodland; ± 500 m. **e DMtns (New York Mtns)**; to OK, Mex. Previously misidentified as *L. phleoides* Kunth.

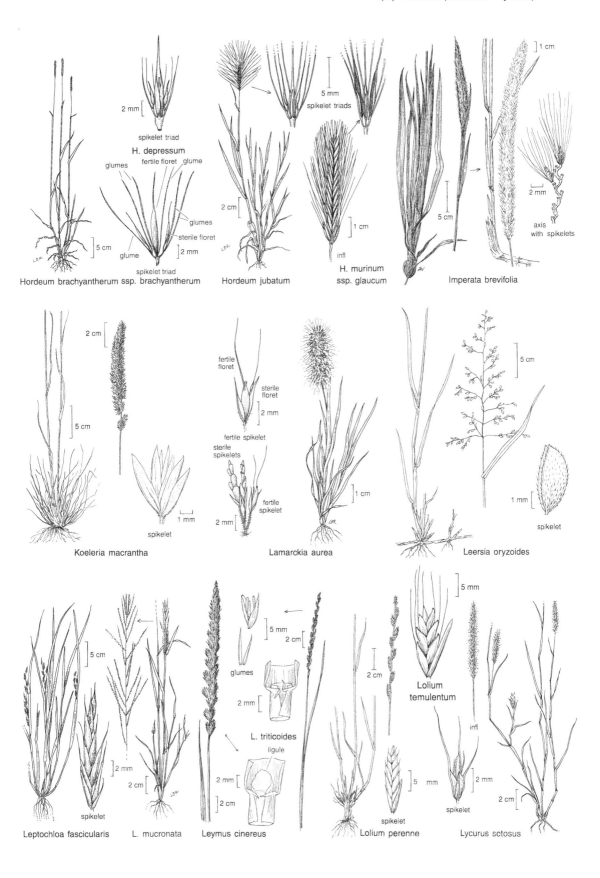

spikelet triad

2 mm

H. depressum

glumes   fertile floret   glume

glumes

sterile floret

glume

spikelet triad

2 mm

5 cm

Hordeum brachyantherum ssp. brachyantherum

5 mm

spikelet triads

2 cm

infl

Hordeum jubatum

5 mm

spikelet triads

1 cm

infl

**H. murinum
ssp. glaucum**

5 cm

1 cm

2 mm

axis
with spikelets

Imperata brevifolia

2 cm

5 cm

1 mm

spikelet

Koeleria macrantha

fertile
floret

sterile
floret

2 mm

fertile spikelet

sterile
spikelets

fertile
spikelet

2 mm

1 cm

Lamarckia aurea

5 cm

1 mm

spikelet

Leersia oryzoides

5 cm

2 mm

spikelet

Leptochloa fascicularis

2 mm

2 cm

L. mucronata

5 mm

2 cm

glumes

2 mm

L. triticoides

ligule

2 mm

2 cm

Leymus cinereus

5 mm

2 cm

2 mm

**Lolium
temulentum**

2 cm

5 mm

spikelet

Lolium perenne

infl

2 mm

spikelet

2 cm

Lycurus setosus

MELICA MELIC, ONIONGRASS

Mary E. Barkworth

Per; rhizomes, corms gen 0. **STS** gen erect, gen densely clumped. **LVS** ± basal; sheath closed to near top, glabrous to short-hairy; ligule thin, membranous, tip obtuse to truncate, gen jagged; blade gen 2–5 mm wide, flat, veins inconspicuous. **INFL** raceme- or panicle-like, gen narrow. **SPIKELET**: glumes papery, back rounded, tip rounded, translucent, lower glume 3–5-veined, upper 1–3-veined; axis gen breaking above glumes; lower florets fertile, 1–7, uppermost florets sterile, ± densely clustered at axis tip; lemma ± like glumes, prominently 5–7-veined, veins not converging, base ± red; palea < lemma. ± 80 spp.: gen temp, exc Australia. (Latin: honey, or old Italian name for pl with sweet sap) [Boyle 1945 Madroño 8:1–26] Key to species adapted by Margriet Wetherwax.

1. Glumes deciduous; spikelets falling as 1 unit, 1–2 per infl branch (sect. *Melica*)..................**M. stricta**
1' Glumes persistent; spikelet axis breaking above glumes, spikelets > 2 per lower infl branch (sect. *Bromelica*)
  2. Fertile florets 3–6 in some or all spikelets...**M. frutescens**
  2' Fertile florets 1–2 in all spikelets...**M. imperfecta**

M. frutescens Scribner Rhizomes short to long; corms 0. **ST** 4–20 dm, gen branching at basal nodes. **LF**: blade 5–9 cm, 2–4 mm wide. **INFL** 12–40 cm, narrow. **SPIKELET** 12–18 mm; lower glume 7–12 mm, upper 9–15 mm, translucent margin 1–2 mm wide; fertile florets 3–6, sterile cluster tapered, concealed by uppermost fertile floret; lemma 8–11 mm, acute; palea 1/2–3/4 lemma length. *n*=9. Dry slopes, chaparral, woodland; 300–1500 m. s SCoRI, TR, PR, **D**; AZ, Baja CA. ❀ TRY. Mar–May

M. imperfecta Trin. (p. 591) **ST** 5–11 dm. **LF**: ligule 3–6 mm; blade 1–6 mm wide. **INFL** 5–36 cm, narrow to wide. **SPIKELET** 3.5–7 mm; glumes 2–6 mm, ± equal; fertile florets 1–2, sterile cluster 0.5–4 mm, acute to obtuse; lemma 3–7 mm, acute or obtuse, gen glabrous or minutely scabrous. *n*=9. Dry, rocky hillsides, chaparral, woodland; < 1500 m. c&s SN, SnFrB, SCoR, SW, **DMoj**; Baja CA. ❀**15–17**;partSHD:**7**,11&IRR:**14, 18–24**;also STBL. Apr–May

M. stricta Bolander (p. 591) **ST** 1–9 dm. **INFL** 3–30 cm, very narrow; spikelets 1–2 per branch. **SPIKELET** open, appearing V-shaped; glumes 6–18 mm, ± equal, spreading, upper half translucent; axis falling as 1 unit; fertile florets 2–5, sterile cluster 2–7 mm; lemma 8–16 mm, tip obtuse to acute, awn 0; palea 1/2–3/4 lemma length; anthers 1–3 mm. *n*=9. Open sites, coniferous forest, rocky areas in alpine; 1200–3350 m. KR, NCoRO, SN, Teh, TR, **n SNE, W&I, DMtns**; to OR, UT. ❀ DRN,SUN:1–3,6,**7**,10,14–24;DFCLT. Jun–Aug

MONANTHOCHLOË

Dieter H. Wilken

Per, mat-like, dioecious. **STS**: central prostrate to decumbent, widely creeping, rooting at nodes; lateral ascending to erect. **LVS** cauline, tufted on prostrate sts, densely clustered on erect branches; sheath open, persistent, fibrous with age; ligule ring-like, ciliate; blade gen oblong, thick, obtuse to pointed. **SPIKELET** gen 1, sessile to short-stalked, concealed by upper lf blades, unisexual; glumes 0; florets 3–5, lower 1–2 fertile, upper sterile; florets of staminate spikelet breaking apart, lemmas firmly membranous; florets of pistillate spikelet weakly breaking apart or not, lower lemmas hard, strongly enfolding fl and fr; palea ± = lemma, membranous. 2 spp.: warm temp N.Am, Cuba, Mex; also Argentina. (Greek: one-flower grass)

M. littoralis Engelm. (p. 591) **STS**: prostrate 3–8 dm; lateral 5–23 cm. **LVS**: upper ± 2-ranked; sheath 5–12 mm, scarious; blade 4–12 mm. **SPIKELET** 5–11 mm; staminate floret sessile to short-pedicelled; pistillate floret sessile, gen concealed by lvs. Salt marshes; ± 0 m. SCo, Chl, **DSon (Salton Sea)**; to s US, Cuba, Mex. ❀ STBL. May–Jun

MUHLENBERGIA MUHLY

Paul M. Peterson

Ann, per, sometimes mat-like, often rhizomed. **ST** decumbent to erect, ± clumped. **LVS** basal and cauline; sheath open; ligule membranous, entire to irregularly toothed, sometimes with 1 large tooth on each side; blade flat to rolled. **INFL** panicle-like, narrow to open; branches spreading to appressed. **SPIKELET**: glumes subequal, gen 1-veined, short-pointed to awned, upper glume sometimes 3-veined; florets 1, sometimes 2, breaking above glumes; lemma short-pointed to awned, glabrous to hairy, 3-veined; palea < to = lemma. **FR** ± fusiform, reddish brown, gen falling with lemma and palea. ± 160 spp.: temp Am, s Asia. (H.L.E. Muhlenberg, Pennsylvania botanist, 1753–1815) [Reeder 1981 in Gould and Moran 1981 San Diego Soc Nat Hist Memoir 12:67–78] *M. glauca* (Nees) Mez. collected once in Jamacha, San Diego Co. 1894 ❀ STBL.

1. Ann
  2. Lemma awned, awn 1–3 cm, lemma 2.5–6 mm
    3. Lemma gen 4.5–6 mm; infl branches closely appressed, < 1.5 cm wide; glumes gen 1–2 mm ... **M. appressa**
    3' Lemma gen 2.5–4.5 mm; infl branches spreading to ascending, 1–5 cm wide; glumes gen < 1 mm
    ....[2]**M. microsperma**
  2' Lemma abruptly pointed, point < 1 mm, or awn 0; lemma < 2.5 mm
    4. Infl narrow, < 1 cm wide; branches closely appressed; sts often rooting at lower nodes ....[2]**M. filiformis**

4' Infl open, 1.5–8 cm wide; branches reflexed to ascending; sts not rooting at lower nodes
   5. Infl branches stiffly reflexed ± 90° from axis; glumes glabrous; ligule teeth 2, 1 on each side . . . *M. fragilis*
   5' Infl branches ascending < 80° from axis; glume tip short-hairy; ligule irregularly short-toothed
    . . . . *M. minutissima*
1' Per
  6. Lemma awn 0 or short-pointed, point < 1 mm
    7. Infl 5–14 cm wide, open, ovoid, branches spreading . . . . *M. asperifolia*
    7' Infl < 1.3 cm wide, narrow, cylindric, branches ascending to appressed
      8. Sts 5–15 dm, erect, densely clumped; infl 15–60 cm; blade 10–50 cm long . . . . *M. rigens*
      8' Sts 0.2–4 dm, erect to decumbent, loosely clumped; infl 1–12 cm; blade 1–5 cm long
        9. Rhizomes 0, stolons often present; lower nodes not swollen or knot-like; glumes obtuse, ± toothed at
          10× . . . . [2]*M. filiformis*
        9' Rhizomes creeping, ± scaly; lower nodes often swollen or knot-like; glumes acute to ± short-awned
          . . . . *M. richardsonis*
  6' Lemma awn 1–30 mm
    10. Infl open, 6–15 cm wide, branches spreading . . . . *M. porteri*
    10' Infl gen narrow, < 5 cm wide, branches ascending to appressed
      11. Rhizome gen scaly, creeping
        12. Blade gen flat, 2–4 mm wide; lemma base with hairs = lemma; anther yellow, gen 0.5–1.5 mm . . . *M. andina*
        12' Blade gen rolled, < 2 mm wide; lemma base with hairs << lemma; anther purple, 1.5–3 mm
          13. Lemma, palea short-soft-hairy on lower half; sts loosely clumped, decumbent . . . . [2]*M. arsenei*
          13' Lemma base sparsely short-hairy, palea glabrous to ± scabrous; sts erect, ± rooting at lower nodes
            . . . . [2]*M. pauciflora*
      11' Rhizomes 0
        14. Glumes gen < 1.5 mm, obtuse; cleistogamous spikelets gen present in lowermost st axils . . [2]*M. microsperma*
        14' Glumes gen 1.5–3.5 mm, acute, acuminate, or awned; cleistogamous spikelets 0
          15. Lemma, palea short-soft-hairy on lower half; sts loosely clumped, decumbent . . . . [2]*M. arsenei*
          15' Lemma base sparsely short-hairy, palea glabrous to ± scabrous; sts erect, ± rooting at lower nodes
            . . . . [2]*M. pauciflora*

M. andina (Nutt.) A. Hitchc. FOXTAIL MUHLY Per; rhizome scaly, creeping. **ST** 2.5–8.5 dm. **LF**: ligule 0.5–1.5 mm, truncate, ciliate; blade 4–16 cm, 2–4 mm wide, flat. **INFL** 4–15 cm, 5–15 mm wide, narrow; branches appressed, loosely fld. **SPIKELET**: glumes 2–4 mm, acuminate or short-awned; lemma 2–3.5 mm, hairs at base = lemma, awn 1–7 mm; anthers 0.5–1.5 mm, yellow. 2*n*=20. Canyons, streambanks, wet meadows; < 3100 m. KR, NCoRI, SN, SCoRI, SnBr, **SNE, DMtns**; to Can, Colorado, w TX. Jul–Sep

M. appressa C.O. Goodd. Ann. **ST** 1–4 dm. **LF**: ligule 1.5–3 mm, truncate to obtuse, decurrent to sheath, toothed; blade 1–5 mm, 1–2 mm wide, flat or folded. **INFL** 4–14 cm, 0.5–1.5 cm wide, narrow, loosely fld; branches 0.5–1.5 cm, appressed. **SPIKELETS** in lower branch axils cleistogamous, enclosed by tightly rolled sheath; glumes 1–2 mm, obtuse to acute; lemma 4.5–6 mm, hairs at base short-appressed between veins, awn 1–3 cm; anther 0.5–1 mm, purple. Uncommon. Open canyon bottoms and rocky slopes; 20–1600 m. s ChI (San Clemente Island), **DMtns (Providence Mtns)**; s AZ, Baja CA.

M. arsenei A. Hitchc. (p. 591) TOUGH MUHLY Per; rhizomes ± short. **ST** 1.5–4 dm, decumbent at base. **LF**: ligule 1–2 mm, acuminate, toothed, ± decurrent to sheath, with 1 large tooth on each side; blade 1–5 cm, < 2 mm wide, rolled. **INFL** 4–12 cm, < 3 cm wide, narrow; branches ascending to appressed, loosely fld. **SPIKELET**: glumes 2–3 mm, acute, ± short-awned, awn < 1 mm; lemma 3.5–5 mm, short-soft-hairy on lower half, awn 4–12 mm; palea short-soft-hairy on lower half; anther 1.6–3 mm, purple. RARE in CA. Limestone rock outcrops, slopes; 1400–1860 m. **DMtns (Clark Mtns)**; to se UT, n NM, n Baja CA. Aug–Sep

M. asperifolia (Nees & Meyen) L. Parodi (p. 591) SCRATCHGRASS Per; rhizomes shiny, scaly; ± stoloned. **ST** decumbent to erect, 1–6 dm. **LF**: ligule 0.2–1 mm, truncate, minutely ciliate; blade 2–6 cm, 1–2.8 mm wide, flat or folded. **INFL** 6–17 cm, ovoid, open; branches 5–14 cm, spreading. **SPIKE-LET**: glumes 0.5–1.5 mm, acute; florets 1–2; lemma 1–2 mm, glabrous ± short-awned; anther 1–1.2 mm, purple. 2*n*=20,22,28. Moist, often alkaline meadows, seeps, hot springs; 120–2150 m. **CA**; w N.Am, also in s S.Am. Jul–Oct

M. filiformis (Thurber) Rydb. PULL-UP MUHLY Ann, often short

stoloned. **ST** decumbent, loosely clumped, rooting at lower nodes, 0.2–3 dm. **LF**: ligule 1–2.5 mm, obtuse to acute, margin serrate; blade 1–4 cm, 1–2 mm wide, flat or rolled. **INFL** 1–6 cm, < 1 cm wide, cylindric, narrow; branches closely appressed. **SPIKELET**: glumes 0.5–1.2 mm, obtuse, ± toothed at 10×; lemma 1.5–2 mm, short-awned, awn < 1 mm; anther 0.5–1 mm, purple. 2*n*=18. Moist meadows, seeps, streambanks; 150–3350 m. NW, SN, SnBr, SnJt, **SNE**; to B.C., Great Plains, NM, Mex. ❀WET:**1–3**&SUN:**6,7,14–16,18**. Jun–Aug

M. fragilis Swallen (p. 591) DELICATE MUHLY Ann. **ST** erect or spreading, 1–3.5 dm. **LF**: ligule 1–3 mm, decurrent to sheath, with 1 large tooth at each side; blade 1–6 cm, 1–2 mm wide, flat, margin, midvein strongly white-thickened. **INFL** 10–30 cm, 3.5–8 cm wide, ovoid, open; branches thread-like, stiffly spreading to reflexed ± 90° from central axis. **SPIKELET**: glumes 0.5–1 mm, obtuse to acute, glabrous; lemma ± 1 mm, gen glabrous, margin, midvein sometimes short-hairy; anther < 0.5 mm, purple. 2*n*=20. RARE in CA. Open, ± disturbed, limestone gravelly wash; ± 1600 m. e **DMtns (Clark, New York mtns)**; to w TX, Mex. In cult.

M. microsperma (DC.) Trin. (p. 591) LITTLESEED MUHLY Ann, short-lived per. **ST** 1–6 dm. **LF**: ligule 1–2 mm, decurrent to sheath, truncate to obtuse, toothed; blade 2–6 cm, 1–2.5 mm wide, flat or loosely rolled. **INFL** 5–20 cm, 1–5 cm wide; branches spreading to ascending, loosely to densely fld. **SPIKELETS** in lower branch axils cleistogamous, enclosed by tightly rolled sheath; glumes 0.5–1 mm, obtuse; lemma 2.5–4.5 mm, short-soft-hairy at base, awn 1–3 cm; anther 0.5–1 mm, purple. 2*n*=20,40,60. Open ± disturbed sites; < 1650 m. CCo, SCoRO, SW, **D**; to sw UT, AZ, Mex, also in S.Am. Mar–May

M. minutissima (Steudel) Swallen (p. 591) Ann. **ST** ascending to erect, 0.2–3 dm. **LF**: ligule 1–2 mm, truncate to obtuse, short-toothed; blade 0.5–4 cm, 1–2 mm wide, flat. **INFL** 1–20 cm, narrowly ovoid, open; branches 1.5–5 cm, ascending < 80° from central axis. **SPIKELET**: glumes 0.5–1 mm, obtuse, tip short-hairy; lemma 1–1.5 mm, margins, midvein short-hairy; anther 0.5–1 mm, purple. 2*n*=60,80. Open, ± disturbed, sandy slopes, seeps; 400–2300 m. KR, n&c SNH, SnBr, SnJt, **SNE**; to WA, MT, w TX, Mex. Jul–Oct

M. pauciflora Buckley (p. 591) FEW-FLOWERED MUHLY Per; rhizomes ± short, knot-like. **ST** erect, 3–5 dm, wiry, rooting at lower nodes; lower nodes knot-like. **LF**: ligule 1–2.5 mm, decurrent to sheath, with 1 large tooth on each side; blade 5–8 cm, 0.5–1.5 mm wide, flat to ± folded. **INFL** 5–12 cm, < 3 cm wide, narrow; branches ascending to appressed, loosely fld. **SPIKELET**: glumes 1.4–3.2 mm, acuminate to ± short-awned, awn < 1 mm; lemma 4–5 mm, hairs at base short-appressed, awn 5–20 mm; palea glabrous to ± scabrous; anther 1.8–2 mm, purple. RARE in CA. Rocky slopes, ledges, canyons; 1755 m. **e DMtns (New York Mtns)**; to s Colorado, w TX, Mex. In cult.

M. porteri Beal (p. 591) Per. **ST** 2.5–8 dm, wiry; lower nodes knot-like. **LF**: ligule 1–2.5 mm, truncate, decurrent to sheath, toothed; blade 2–8 cm, 1–2 mm wide, flat to ± folded. **INFL** 4–15 cm, 6–15 cm wide, ovoid, open; branches thread-like, spreading. **SPIKELET**: glumes 2–3 mm, acuminate; lemma 3–4.2 mm, hairy below middle, awn 2–10 mm; anther 1.5–2.3 mm, purple to yellow. 2*n*=20,23,24,40. Among boulders or shrubs, rocky slopes, cliffs; 610–1680 m. SnBr, SnJt, **SNE, DMoj**; to Colorado, w TX, Mex. Jun–Oct

M. richardsonis (Trin.) Rydb. (p. 591) MAT MUHLY Per; rhizome scaly, matted. **ST** decumbent to erect, 0.5–4 dm; lower nodes often swollen or knot-like. **LF**: ligule 1–2.5 mm, acute to truncate, decurrent to sheath; blade 1–5 cm, 1–2 mm wide, flat to ± rolled. **INFL** 1–12 cm, 1–4 mm wide, cylindric, narrow; axis gen obscured by appressed branches. **SPIKELET**: glumes 0.8–1.8 mm, acute to ± short-awned; lemma 2–3 mm, glabrous, ± scabrous at tip, ± short-awned, awn < 0.5 mm; anther 1.2–1.5, yellow to purple. 2*n*=40. Open sites, ± moist meadows, talus slopes, along streams; 1220–3670 m. KR, CaRH, SNH, SCoRO, TR, SnJt, **GB, DMtns**; to Can, ne US, Mex. ❀ IRR,SUN:**1–3,6,11,14–16,18,21**;GRCVR. Jun–Aug

M. rigens (Benth.) A. Hitchc. (p. 591) DEERGRASS Per. **ST** densely clumped, 5–15 dm. **LF**: ligule 0.5–2 mm, truncate, ± ciliate; blade 10–50 cm, 1.5–6 mm wide, flat. **INFL** 15–60 cm, 5–12 mm wide, cylindric, narrow; branches appressed, densely fld. **SPIKELET**: glumes 1.8–3 mm, acute or obtuse, ± scabrous; lemma 2.5–3.5 mm, base sparsely short-hairy, ± abruptly pointed; anther 1.3–1.7, yellow to purple. 2*n*=40. Sandy to gravelly places, canyons, stream bottoms; < 2150 m. CaRH, SN, GV, SCoRO, SCo, TR, SnJt, **SNE, DMoj**; to TX, Mex. ❀**6,14–17,22–24**;IRR:1–3,**7–11,18–21**. Jun–Sep

MUNROA

Dieter H. Wilken

Ann, multi-branched. **STS** spreading to erect. **LVS** tightly clustered; ligule short-hairy; blade stiff, flat or folded, margin white. **INFL** ± panicle-like; spikelets subsessile, clustered, ± concealed by terminal lf clusters. **SPIKELET**: glumes gen narrow, acute, 1-veined; axis gen breaking above glumes and between florets; florets bisexual; lemmas gen lanceolate, thick, pointed or short-awned, 3-veined; palea ± = lemma; lower spikelets with glumes subequal, slightly < lower floret, florets 3–4; upper spikelets with glumes strongly unequal, << florets, lower glume sometimes 0, florets 2–3. 3 spp.: temp N.Am, s S.Am. (W. Munro, English agrostologist, 1818–1880) [Parodi 1934 Revista Mus La Plata Secc. Bot 34:171–193]

M. squarrosa (Nutt.) Torrey (p. 591) FALSE BUFFALOGRASS Pl mat-like, gen < 20 cm wide. **ST** gen 3–10 cm; internodes ± scabrous, sometimes puberulent; lower internodes evident, upper crowded, concealed by lvs. **LF**: sheath stiff-hairy near collar, margins ± ciliate; ligule hairs < 1 mm; blades 1–2 cm, 1–3 mm wide. **SPIKELETS** 6–8 mm; lower gen sessile; upper short-stalked; glumes 2–4 mm; lemmas 3–5 mm, outer veins short-stiff-hairy below middle, awn < 2 mm; anthers 1–1.5 mm. 2*n*=16. RARE in CA. Open, gravelly or rocky places; 1500–1800 m. **DMtns (Clark Mtn.)**; to Great Plains, TX, n Mex.

PANICUM MILLET, PANICGRASS

Robert Webster

Ann, per. **STS** gen erect; internode solid to hollow inside. **LVS** basal and cauline; sheath glabrous or hairy; ligule short-hairy or membranous, ciliate, hairs gen > membrane. **INFL** panicle-like, gen open; 1° branches spreading to ascending; 2° branches spreading to appressed; spikelets many, 1–2 per node, gen stalked, on one side of axis or not, stalk tip expanded, one side concave. **SPIKELET** falling as 1 unit, ± compressed, gen green to purplish; glumes gen unequal, lower gen < upper, free, clasping, upper glume ± = spikelet, membranous, ± thin; florets 2, lower sterile or staminate, lemma texture like glumes, upper floret fertile, lemma leathery to hard, firm, gen shiny, smooth to rough, margin inrolled, tip blunt, palea enclosed by lemma margin. ± 450 spp.: trop to warm temp, worldwide. (Latin: ancient name for millet) [Spellenberg 1975 Brittonia 27:87–95] Some spp. cult for food. Key to species adapted by Margriet Wetherwax.

1. Lvs of 2 forms, basal < and gen wider than cauline, rosette well developed, gen persistent
. . . . *P. acuminatum* var. *acuminatum*
1' Lvs gradually reduced upward, basal rosette not well developed
  2. Per
    3. Spikelet 2.5–3 mm, glabrous; upper floret lemma with flat margin; ligule membranous, ciliate . . *P. antidotale*
    3' Spikelet 5–7.5 mm, short-hairy; upper floret lemma with inrolled margin; ligule short-hairy . . *P. urvilleanum*
  2' Ann
    4. Spikelet 4.5–5.5 mm; lf base minutely lobed . . . . *P. miliaceum*
    4' Spikelet 2–3.5 mm; lf base not lobed
      5. Spikelet axis not elongated, not clearly visible; lower glume 1–1.5 mm . . . . *P. capillare*
      5' Spikelet axis visible, ± elongated between glumes and between florets; lower glume 1.5–2.5 mm . . . *P. hirticaule*

P. acuminatum Sw. Per. **ST** 2–8 dm. **LF**: sheath 2–7 cm, gen short-hairy; ligule 2–5 mm, hairy; blade 3–15 cm long, 2–13 mm wide, upper surface glabrous or short-soft-hairy. **INFL** 4–9 cm; 1° branches 2–4 cm, axis glabrous or short-hairy; spikelet 1 per node, stalk 0.5–3 mm. **SPIKELET** 1–2.5 mm, slightly < 1 mm wide, elliptic to slightly obovate; lower glume ± 0.5–1 mm, 1-

veined, tip acute; lower floret sterile, lemma 7-veined, tip rounded, palea vestigial; upper floret slightly < lower floret. 2*n*=18. Moist places, marshes, streambanks; < 2600 m. NW, CaR, SN, GV, CCo, SnFrB, SCoRO, SW, **W&I**; to Can, e US, C. & n S.Am. 2 vars in CA.

var. ***acuminatum*** (p. 591) **ST** 1–6 dm. **LF**: sheath hairy; ligule 2–4 mm; blade 5–10 cm, 5–12 mm wide, upper surface short-soft-hairy. **INFL** 5–8 cm. **SPIKELET** 1.5–2.5 mm, obovate to elliptic; lower glume 0.5–1 mm. Habitat and range of sp. [*P. huachucae* Ashe, *P. occidentale* Scribner, *P. pacificum* A. Hitchc. & Chase, *P. shastense* Scribner & Merr., *P. thermale* Bolander, Geyser's panicum, **ENDANGERED** CA] Jun–Aug

P. antidotale Retz. BLUE PANICGRASS Per from rhizomes. **ST** erect, 1–25 dm; nodes 5–8. **LF**: sheath 4–8 cm, glabrous; ligule 1.5–2.5 mm, densely hairy; blade 15–30 cm, 4–12 mm wide, upper surface glabrous. **INFL** 13–28 cm; 1° branches 9–18 cm, glabrous; spikelets 1–2 per node, stalk < 2.5 mm. **SPIKELET** 2.5–3 mm, ± 1 mm wide, ovate to elliptic, brown, green, or purple; lower glume 1.5–2.5, 5-veined; lower floret staminate, lemma 7-veined, tip acute, palea ± = lemma; upper floret ± = lower floret. 2*n*=18. Open, gen disturbed areas, fields; < 100 m. **s DMoj**, **DSon**; to TX, n Mex; native to India. NOXIOUS WEED.

P. capillare L. (p. 591) WITCHGRASS Ann. **ST** 2–9 dm. **LF**: sheath 4–8 cm, short-hairy; ligule 0.5–1.5 mm, tip hairy; blade 6–22 cm, 6–20 mm wide, upper surface short-soft-hairy to roughly hairy. **INFL** 15–40 cm; 1° branches 12–30 cm, axis sparsely short-hairy; spikelet 1 per node; stalk 2–6 mm. **SPIKELET** 2–3.5 mm, ± 1 mm wide, lanceolate to elliptic; lower glume ± 1–1.5 mm, 3-veined; lower floret sterile, lemma 7-veined, tapered to a tail-like tip, palea gen 0; upper floret ± 0.8 × lower floret length. 2*n*=18. Open places, fields; < 1500 m. **CA**; to Can, e US. Jul–Sep

P. hirticaule C. Presl Per. **ST** 1–8 dm. **LF**: sheath 2–6 cm, axis glabrous to short-hairy; ligule ± 0.5–1.5 mm; blade 7–15 cm, 3–12 mm wide, upper surface gen sparsely short-hairy. **INFL** 5–12 cm; 1° branches 3–8 cm, glabrous; spikelets 1–2 per node, stalk 0.5–1 mm. **SPIKELET** ± 2.5–3 mm, ± 1 mm wide, lanceolate to ovate, green; axis between glumes and florets visible; lower glume ± 1.5–2.5 mm, gen 5-veined; lower floret sterile, lemma 7-veined, acuminate to acute, palea vestigial or 0; upper floret 0.7–0.8 × lower floret length. Sandy soils, open sites, creosote-bush scrub; < 1400 m. **DSon**; to TX, n Mex. ❀ STBL. Jul–Oct

P. miliaceum L. (p. 591) BROOM CORN MILLET Ann. **ST** 2–10 dm. **LF**: sheath 3.5–8 cm, short-hairy; ligule 1–3 mm, membranous; blade 10–20 cm, 6–25 mm wide, upper surface velvety to roughly hairy. **INFL** 10–40 cm; 1° branches 4–18 cm, axis glabrous; spikelet 1 per node, stalk 2–10 mm. **SPIKELET** 4.5–5.5 mm, ± 2–2.5 mm wide, elliptic, green to brown; lower glume 2.5–4.5 mm, 3–5-veined; lower floret sterile, lemma 11-veined, acuminate to acute, palea vestigial; upper floret 0.8 × lower floret length. 2*n*=36. Disturbed places, fields, roadsides; < 1000 m. GV, SnFrB, SW, **DSon**; to e US; native to Eurasia. Cult for seed, food. Aug–Oct

P. urvilleanum Kunth Per from stolons. **ST** 4–10 dm. **LF**: sheath 12–35 cm, hairy; ligule 1–1.5 mm; blade 20–45 cm, 4–7 mm wide, upper surface short-hairy. **INFL** 20–35 cm; 1° branches 8–13 cm, glabrous; spikelet 1 per node, stalk 1.5–10 mm. **SPIKELET** 5–7.5 mm, 2–2.5 mm wide, elliptic, green; lower glume 4.5–6.5 mm, 5–7-veined; lower floret staminate, lemma 9-veined, acute, palea ± = lemma; upper floret 0.8–0.9 × lower floret length, lemma margin inrolled. 2*n*=36. Sandy soils, dunes; < 900 m. e SCo, **D**; AZ, N Mex, S.Am. ❀ STBL. Mar–May

PASCOPYRUM

Mary Barkworth

1 sp. (Latin & Greek: pasture wheat) [Barkworth & Dewey 1985 Amer J Bot 72:767–776]

P. smithii (Rydb.) Á. Löve (p. 591) Per, from rhizomes. **ST** erect, (2)4–9 dm, glaucous. **LVS** basal and cauline, glaucous; ligule membranous; sheath appendages 1–2 mm, clasping; blade flat, rolled when dry. **INFL** 4–15 cm, spike-like; spikelets 2-ranked, overlapping, 1(2) per node. **SPIKELET**: glumes subequal, < or = lower floret, glabrous to rough, tapered from middle to acute tip, 3-veined in lower half, midvein curving slightly to side; axis above glumes and between florets; florets 4–11; lemma 8–14 mm, glabrous to hairy, awn < 5(15) mm; anthers 2–4.5 mm. 2*n*=56. Uncommon. Dry, alkaline soils, flats; 1500–2000 m. **GB, DMtns (New York Mtns)**; to s Can, Great Plains. [*Agropyron s.* Rydb.] Polyploid derived from hybrid between *Elymus* and *Leymus*. Jun–Aug

PASPALUM

Robert Webster

Per in CA, gen from rhizomes or stolons. **STS** decumbent to erect; internode solid to hollow inside. **LVS** basal and cauline; sheath glabrous or hairy; ligule gen membranous. **INFL** panicle-like; axis gen glabrous; 1° branches raceme-to spike-like, spreading to appressed; spikelets many, 1–2 per node, gen short-stalked, on one side of axis. **SPIKELET** falling as 1 unit, compressed, gen green; glumes 1–2, lower glume minute or 0, upper ± = spikelet; florets 2, lower floret gen sterile, palea vestigial or 0, upper floret fertile, lemma firm, thick, sometimes hard, back facing infl axis, smooth or striate, margin inrolled, tip blunt. ± 300 spp.: trop, warm temp worldwide. (Greek: ancient name) [Gould 1975 The Grasses of Texas pp. 500–527, Texas A&M Press] Key to species adapted by Margriet Wetherwax.

1. Margins of upper glume and lower floret lemma glabrous . . . . ***P. distichum***
1' Margins of upper glume and lower floret lemma ciliate . . . . ***P. dilatatum***

P. dilatatum Poiret (p. 591) DALLIS GRASS Per from short rhizomes. **ST** decumbent to erect, 2.5–14 dm; nodes 2–6. **LF**: sheath 6–30 cm, glabrous to hairy; ligule 2–8 mm; blade 9–35 cm, 4–10 mm wide, upper surface glabrous, sometimes hairy at base. **INFL**: main axis 3–20 cm; 1° branches 3–6, 4–12.5 cm; spikelets many, 2 per node, stalk 1–1.5 mm. **SPIKELET** 3–4 mm, 2–2.5 mm wide, elliptic, green to purple; lower glume 0; lower floret lemma 5–9-veined, tip acute to rounded; upper floret 0.7–0.9 × lower floret length. 2*n*=40,50. Moist places, ditches, roadsides; < 400 m. CA-FP, **DMoj**; to WA, e US, Eur; native to S.Am. May–Nov

P. distichum L. (p. 591) Per from stolons and rhizomes. **ST** decumbent to erect, 0.8–6 dm; nodes 5–15. **LF**: sheath 3–20 cm, glabrous; ligule < 1.5 mm; blade 2–22 cm, 2–7 mm wide, upper surface glabrous. **INFL**: main axis < 1.5 cm; 1° branches 2–3, 1.5–5.5 cm; spikelets many, 1–2 per node, stalk < 0.5 mm. **SPIKELET** 2.5–3.5 mm, ± 1.5 mm wide, obovate to elliptic;

lower glume < 2.5 mm, 1-veined; lower floret lemma 3–5-veined, tip acute; upper floret slightly < lower floret. 2*n*=40,60. Moist places, marshes, ditches; < 1650 m. CA-FP (exc mtns), **GB, n DMoj**; to WA, e US, S.Am. ❀ STBL. Jun–Oct

PENNISETUM

Robert Webster

Ann, per. **STS** prostrate to erect and tufted; internode spongy inside, sometimes hollow. **LVS** basal and cauline; sheath gen glabrous; ligule short-hairy or membranous, ciliate. **INFL** gen panicle-like, dense, ± cylindric (raceme-like, spikelets few in *P. clandestinum*), spikelets many, short-stalked to sessile, clustered, gen 1–4 per cluster, subtended by 5–50 flexible bristles; spikelet cluster and bristles falling as 1 unit. **SPIKELET** compressed; glumes 1–2, lower glume < upper or 0, upper ± = spikelet; florets 2, lower floret gen sterile, lemma like glumes, upper floret fertile, lemma firm, ± thick or hard, smooth or scabrous, gen dull, margin flat to inrolled, tip blunt. ± 80 spp.: warm temp, trop Eurasia, Afr. (Latin: feather bristle) Some spp. cult for orn, food.

P. villosum R.Br.(p. 597) Per, cespitose. **ST** erect, 1.6–7.5 dm. **LF**: sheath 4–10 cm, glabrous; ligule ± 1–1.5 mm; blade 5–40 cm, 2–4.5 mm wide, upper surface glabrous to hairy. **INFL** 4–8 cm; 1° branches (incl bristles) 3–5 cm, glabrous; bristles subtending cluster 30–50. **SPIKELET** 9–11 mm, ± 1.5 mm wide, lanceolate, green to white; lower glume < 1 mm, veins not visible; upper glume ± 0.5 × spikelet length; lower floret sterile or staminate, lemma 7–9-veined, tip acuminate, palea ± = lemma or 0; upper floret ± = to slightly > lower floret. Waste places, urban roadsides; < 100 m. SW, **DSon**, expected elsewhere; to TX; native to Afr. Jun–Aug

PHALARIS

Dennis Anderson

Ann, per, cespitose or from rhizomes. **ST** gen erect, 2–20 dm. **LF**: sheath open; ligule membranous, truncate; blade gen 2–5 dm, 1–2 cm wide, flat. **INFL** panicle-like, gen cylindric to ovoid, dense; branches ascending to appressed, obscure. **SPIKELET** gen fertile, sometimes also sterile in *P. paradoxa*, compressed; glumes equal, > florets, sometimes with wing-like keel, 3–5-veined; axis gen breaking above glumes, gen falling as 1 unit; florets 2–3, lower 1–2 vestigial or 0, upper 1 fertile; upper floret lemma gen ovoid, glabrous or appressed-hairy, shiny, faintly 5-veined, awn 0; palea < lemma, translucent. **FR** ± fusiform. ± 15 spp.: temp N.Am, Eurasia. (Greek: ancient name for grass with shiny spikelets) Key to species adapted by Margriet Wetherwax.

1. Fertile spikelet surrounded by 5–6 sterile spikelets, falling together, sterile spikelets = fertile spikelet
   or club-shaped; fertile floret lemma ± glabrous; lower florets < 0.2 mm . . . . *P. paradoxa*
1' Spikelets all fertile, gen breaking apart separately; fertile floret lemma gen hairy; lower florets gen >
   0.2 mm, gen 1/3 upper floret length
   2. Lower floret gen 1 . . . . *P. minor*
   2' Lower florets 2
      3. Infl 7–40 cm, 2–11 cm wide; lower florets awl-like . . . . *P. arundinacea*
      3' Infl 1.5–4 cm, 1.5– 2 cm wide; lower florets wide, chaff-like — gen < 1/2 upper floret . . . . . . *P. canariensis*

P. arundinacea L. (p. 597) REED CANARY GRASS Per from distinct rhizomes. **ST** 5–15 dm. **INFL** 7–40 cm, 2–11 cm wide, cylindric, interrupted near base; branches spreading in fl, appressed in fr. **SPIKELET**: glumes 3.5–7.5 mm, wing 0, tip acute; lower florets 2, 1–2 mm, awl-like, hairy; upper lemma 3–4.5 mm, ± 1.5 mm wide, narrowly lanceolate, glabrous to sparsely hairy. **FR** 1.5–2 mm, < 1 mm wide. 2*n*=14,27–31,35,42. Wet streambanks, moist areas, grassland, woodland; < 1600 m. CA-FP, **W&I**; temp N.Am, Eurasia. Cult for forage. May–Aug

P. canariensis L. (p. 597) CANARY GRASS Ann. **ST** 3–10 dm. **INFL** 1.5–4 cm, 1.5–2 cm wide, ovoid to oblong. **SPIKELET**: glumes 7–10 mm, 1.5–2.5 mm wide, wing-like above middle, tip widely acute; lower florets 2, 2.5–4.5 mm, 0.5–1 mm wide, chaff-like; upper lemma 5–7 mm, 1.5–2.5 mm wide, ovoid, densely hairy. **FR** ± 4 mm, ± 1.5 mm wide. 2*n*=12. Disturbed areas; < 300 m. NCo, CCo, SnFrB, SCo, **DMtns**; to Great Plains; native to Medit Eur. Apr–Jun

P. minor Retz. (p. 597) Ann. **ST** 2–10 dm. **INFL** 1–6 cm, 1–2 cm wide, oblong-ovoid. **SPIKELET**: glumes 4–6.5 mm, 1–1.5 mm wide, glabrous, wing entire or toothed; lower floret 1, 1–2 mm, gen awl-like; upper lemma 3–4 mm, 1–2 mm wide, widely ovate, hairy. **FR** ± 2–2.5 mm, ± 1.5 mm wide. 2*n*=28,29. Disturbed areas; < 300 m. CA-FP, **DSon**; native to Medit. Apr–Jul

P. paradoxa L. (p. 597) Ann. **ST** 2–10 dm. **INFL** 3–9 cm, 1–2 cm wide, oblong; base tapered; tip ± truncate to acuminate; spikelets in clusters, cluster falling as 1 unit; fertile spikelet gen surrounded by 5–6 sterile, short-stalked spikelets; sterile spikelet vestigial or = fertile. **FERTILE SPIKELET**: glumes 5.5–8 mm, ± 1 mm wide, gen glabrous, wing lobed to toothed, tip acute to acuminate; lower florets vestigial, knob-like; upper lemma 2.5–3.5 mm, 1–1.5 mm wide, narrowly ovoid, gen glabrous or sparsely hairy near tip. **FR** ± 2.5 mm, ± 1 mm wide. 2*n*=14. Disturbed areas; < 250 m. GV, SW, **DSon**; native to Medit Eur. Small pls gen have many, vestigial sterile spikelets. May–Aug

PHLEUM TIMOTHY

Dieter H. Wilken

Ann, per. **STS** ascending to erect. **LVS** basal and cauline; appendages 0 or small, acute to obtuse; ligule membranous to translucent, obtuse to truncate; blade gen flat, margin minutely scabrous. **INFL** panicle-like, cylindric to ovoid, dense; branches spike-like, short. **SPIKELET** ± sessile, strongly laterally compressed; glumes subequal, membranous, keel gen stiff-ciliate (comb-like), pointed to awned at obtuse to truncate tip, 3-veined; floret 1, breaking above glumes, bisexual; lemma gen awnless at wide, truncate tip, 3–7-veined; palea ± = lemma. 15 spp.: temp Am, Eurasia. (Greek: a marsh reed)

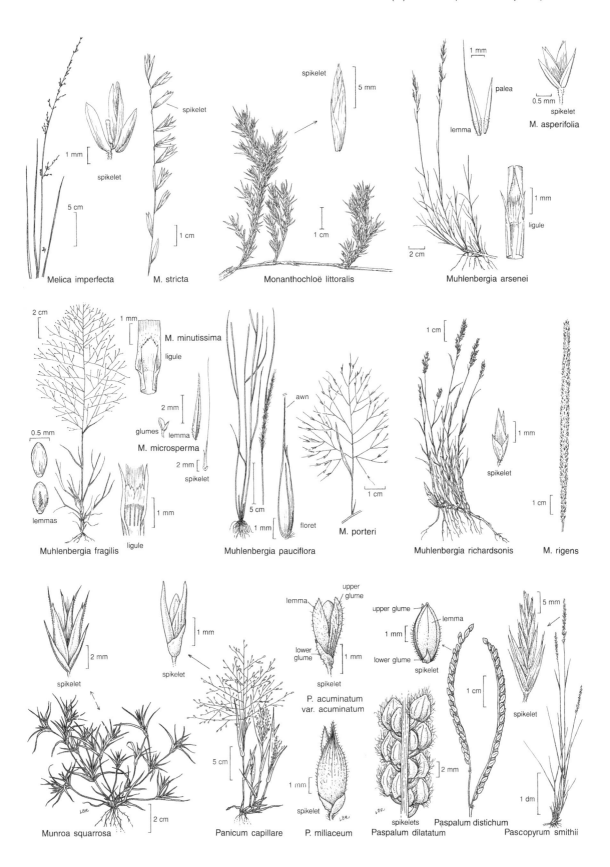

Melica imperfecta

M. stricta

Monanthochloë littoralis

M. asperifolia

Muhlenbergia arsenei

Muhlenbergia fragilis

M. minutissima

M. microsperma

Muhlenbergia pauciflora

M. porteri

Muhlenbergia richardsonis

M. rigens

Munroa squarrosa

Panicum capillare

P. miliaceum

P. acuminatum var. acuminatum

Paspalum dilatatum

Paspalum distichum

Pascopyrum smithii

1. Infl 1–6 cm, ovoid to cylindric, awn 2–3 mm . . . . *P. alpinum*
1' Infl 4–18 cm, cylindric, awn < 2 mm . . . . *P. pratense*

P. alpinum L. (p. 597) MOUNTAIN TIMOTHY Per. **STS** gen clumped, 2–6 dm. **LF**: basal loosely tufted; cauline blade 2–12 cm, 3–8 mm wide. **INFL** 1–6 cm, 7–12 mm wide, ovoid (often cylindric at low elevations). **SPIKELET**: glumes 2–5 mm, scabrous on back, awned at wide acuminate tip, awn 2–3 mm; lemma 2–3 mm, puberulent on back. $2n=14,28$. Wet meadows, streambanks, coniferous forest, alpine; 1500–3350 m (NCo, NCoR: < 1500 m). NCo, KR, n&c NCoR, CaRH, SNH, SnBr, SnJt, **SNE**; to AK, e Can, ne US, Mex; also in S.Am, Eur. ❀ STBL. Jul–Aug

P. pratense L. CULTIVATED TIMOTHY Per. **STS** solitary to loosely clumped, 5–10 dm; base gen swollen. **LF**: basal few, gen spreading; cauline blade 4–20 cm, 3–6 mm wide. **INFL** 4–18 cm, 5–8 mm wide, cylindric. **SPIKELET**: glumes 2–3 mm, lower 1/2 scabrous on back, awned at acuminate tip, awn gen < 2 mm; lemma 1–2.5 mm, veins puberulent. $2n=14,21,28$. Disturbed sites, roadsides, cult fields; < 2100 m. CA-FP, **GB**; N.Am, Mex; native to Eurasia. Widely cult for forage, hay. May–Jun

PHRAGMITES

Kelly W. Allred

Per with thick rhizomes or stolons, forming dense stands. **ST** tall, erect. **LVS** cauline; sheaths open; ligule short, membranous, truncate or hairy; blade flat or folded, gen deciduous. **INFL** panicle-like. **SPIKELET**: glumes unequal, lower> upper, 1–3-veined; axis long-soft-hairy; florets 1–10, breaking above glumes and between florets; lower florets sterile or staminate, upper bisexual; lemmas lanceolate, glabrous, gen 3–5-veined; palea << lemma; stamens gen 2–3. 3 spp.: temp, trop. (Greek: growing in hedges) [Clayton 1968 Taxon 17:168–169]

P. australis (Cav.) Steudel (p. 597) COMMON REED **ST** 2–4 m. **LF**: blade gen 20–45 cm, 1–5 cm wide, margins scabrous, gen breaking at collar. **INFL** 15–50 cm, plume-like, oblong to obovoid, purplish to whitish. **SPIKELETS** 10–16 mm; lower glume 3–7 mm, upper glume 5–10 mm; florets 2–10. $2n=36,44,46,48,49–52, 54,72,84,96$. Pond and lake margins, sloughs, marshes; < 1600 m. **CA**; worldwide. [*P. communis* Trin., incl. var. *berlandieri* (Fourn.) Fern.] Perhaps most widely distributed of all vascular seed pls. ❀ IRR or WET,SUN: 1,**2,3**,4,5,**6–11**,12,**14**,15–17,**18–23**,24;STBL;INV. Jul–Nov

PIPTATHERUM

Mary E. Barkworth

Per, cespitose. **ST** erect, unbranched. **LF**: ligule membranous, gen truncate; blade flat to inrolled, often tapered. **SPIKELET**: glumes ± equal, obtuse to acute, veins evident; axis breaking above glumes; floret 1, gen ± compressed; callus blunt; lemma margins widely separated with age, awn gen readily deciduous; palea ± = lemma, texture like lemma. 30 spp.: arid temp, subtrop Eurasia, Afr. (Greek: falling awn) Segregated from *Oryzopsis*.

P. micranthum (Trin. & Rupr.) Barkworth (p. 597) SMALL-FLOWERED RICEGRASS **ST** 3–6.5 dm. **LF**: basal ligule 0.4–1.5 mm. **INFL** 8–13 cm; lowest branches becoming reflexed. **SPIKELET**: glumes 2.5–3.5 mm; floret 1.5–2.5 mm; lemma gen glabrous, sometimes sparsely but evenly short-hairy, awn 4–8 mm, weakly bent. $2n=22$. RARE in CA. Gravel benches, rocky slopes, creek banks; 700–2950 m. **W&I, DMtns**; to B.C., Rocky Mtns. [*Oryzopsis m.* (Trin. & Rupr.) Thurb.] Jun–Sep

PLEURAPHIS

J. Travis Columbus

Per, cespitose, rhizomed. **ST** ascending to erect, solid; nodes gen hairy. **LF**: ligule membranous, ciliate-fringed; blade firm, flat to inrolled, sharply acute. **INFL** spike-like, gen cylindric; spikelets in clusters ± equal, 3 per node; clusters wedge-shaped, overlapping, appressed to ascending, hairy-tufted at base, falling as 1 unit; axis wavy; glumes of cluster together involucre-like. **CENTRAL SPIKELET** subsessile, appressed to or nearest infl axis; glumes equal, < floret, oblanceolate, keeled, ciliate, tip deeply 2-lobed, lobes lanceolate, awns 3–9, 1 from ± mid-keel, others terminal; florets gen 1–2, lower floret bisexual, upper floret (if present) bisexual or staminate; lemma lanceolate, 3-veined, gen ciliate, tip gen 2-lobed, gen 1-awned ± from sinus; palea ± = lemma. **LATERAL SPIKELETS** sessile; glumes < to ± = florets, ciliate, lower glume asymmetric with 1 awn from ± middle near margin, gen 2-lobed, lobes unequal; florets 1–4, gen staminate; lemma 3-veined, tip gen ciliate; palea ± = lemma. 3 spp.: w US, n Mex. (Greek: side needle, from awn position on lower glume of lateral spikelets) [Reeder & Reeder 1988 Madroño 35:6–9]

1. St internodes glabrous to puberulent . . . . *P. jamesii*
1' St internodes sparsely to densely hairy . . . . *P. rigida*

P. jamesii Torrey (p. 597) GALLETA Pl 1.5–4(6.5) dm, unbranched above base. **ST** ± 1 mm diam; internodes glabrous or puberulent; node hairs ± straight. **LVS** gen basal; glabrous or scabrous, long-ciliate near ligule; ligule membrane 1–3 mm, gen appendaged; blade < 13(21) cm, 2–3 mm wide, upper surface sometimes with short, ± straight, hairs. **INFL** 3–7 cm; spikelet clusters 6–9 mm. **CENTRAL SPIKELET**: glume margin hairs < 0.5 mm; lower lemma awn < 2.5 mm. **LATERAL SPIKE-LETS**: lower glume 1-awned; upper glume tip unlobed or 2-lobed, awn 0–0.4 mm, margin hairs < 0.5 mm; lemma tip unlobed or 2-lobed, awn 0–0.4 mm, margin hairs < 0.2 mm. $2n=36,38,72$. Dry, sandy to rocky slopes, flats, scrub, woodland; 1000–2500 m. **SNE, n&e DMtns**; to WY, TX. [*Hilaria j.* (Torrey) Benth.] ❀ SUN,DRN:1,2,3,7,**8–10, 14–16**,17,**18–23**,24&IRR: **11–13**;alsoSTBL. May–Jun

P. rigida Thurber (p. 597) BIG GALLETA Pl 3.5–10 dm, branched above base, gen bush-like. **ST** 1.5–3.5 mm diam at base; internodes sparsely to densely curly-hairy; node hairs curly. **LVS** gen cauline, gen sparsely to densely curly-hairy, esp near and sometimes overlapping ligule; ligule membrane < 1 mm, appendages 0; blade < 10 cm, 2–4 mm wide. **INFL** 4–10 cm; spikelet clusters 7–11 mm. **CENTRAL SPIKELET**: glume margin hairs 0.5–3 mm; lower lemma tip sometimes 4-lobed, awns 3 ± from sinuses, central awn 2–5.5 mm. **LATERAL SPIKELETS**: lower glume with 1+ subsidiary lobes, or larger lobe tip fringed, awns 2–4; upper glume tip with 2+ lobes or fringed, awns 1–3, 0.4–2.5 mm, margin hairs 0.5–2 mm; lemma tip 2-lobed, awn 1 ± from sinus, 0.4–2 mm, margin hairs 0.2–1 mm. 2*n*=18,36,±108. Common. Dry, open, sandy to rocky slopes, flats, and washes, sand dunes, scrub, woodland; < 1600 m. PR, **e&s DMoj**, **DSon**; to UT, n Mex. [*Hilaria r.* (Thurber) Scribner] Important forage; some pls from e DMoj and Ord Mtn. (San Bernardino Co.), with ± straight internode hairs, intermediate to *P. jamesii*. ✿ SUN,DRN:2,3,7,**8–10,14–16**,17,18,**19–21**,22–24&IRR:**11–13**;also STBL. Feb–Jun

POA BLUEGRASS

Robert J. Soreng

Ann, per, some ± dioecious. **ST** 0.3–12 dm. **LF**: sheath open to closed (best observed on upper st lf); ligule thin, flexible; blade grooved above on both sides of midvein, flat, folded, or inrolled, gen smooth or scabrous on veins, gen prow-tipped. **INFL** panicle-like; branches appressed to drooping. **SPIKELET** gen compressed, breaking between florets; glumes 2, similar, gen < lowest lemma, awnless; florets gen 2–6; callus indistinct, often with obvious tuft of long cobwebby hairs; lemma gen keeled to base, of same texture as glumes, awnless, veins gen 5, ± converging near tip; palea well developed, keels gen scabrous; fertile anthers 0.2–4.5 mm; ovary glabrous. ± 500 spp.: temp and cool regions. (Greek: ancient name) [Soreng 1991 Syst Bot 16:507–528] CA is center of diversity in N.Am. Spikelet features best observed on lowest florets of spikelet.

1. Florets mostly replaced by leafy bulbets; st base swollen . . . . *P. bulbosa*
1' Florets normal, rarely replaced by bulbets; st base not swollen
  2. Anthers < 1,1 mm; pl cespitose
    3. Lemmas without a web at the base
      4. Ann; widespread . . . . *P. annua*
      4' Per; alpine
        5. Anthers < 0.7 mm . . . . *P. lettermanii*
        5' Anthers > 0.7 mm . . . . ²*P. keckii*
    3' Lemmas with a web at the base
      6. Ann . . . . *P. bigelovii*
      6' Per
        7. Pl densely tufted . . . . ²*P. pattersonii*
        7' Pl loosely tufted . . . . *P. leptocoma*
  2' Anthers often > 1.2 mm; pl cespitose or rhizomatous
    8. Pl rhizomatous (hybrids between *P. pratensis* and *P. secunda* will key here)
      9. St and nodes distinctly compressed in ×-section; sheath closed < 1/4 length . . . . *P. compressa*
      9' St and nodes round in ×-section or only slightly compressed; sheath closed > 1/4 length
        10. Lemmas without a web at base, keel and marginal veins glabrous to sparsely hairy; pl unisexual
         . . . . *P. wheeleri*
        10' Lemmas with a web at base, keel and marginal veins hairy; pl bisexual . . . . *P. pratensis*
    8' Pl cespitose (sometimes stoloniferous in *P. palustris*)
      11. Lemmas glabrous
        12. Pl unisexual; basal blades densely short-hairy on upper surface; sheath closed > 1/4 length . . . *P. cusickii*
        13. Infl branches slender, longest gen > 17 mm . . . . ssp. *cusickii*
        13' Infl branches stout, longest gen < 15 mm . . . . ssp. *pallida*
        12' Pl bisexual; basal blades glabrous or indistinctly scabrous on upper surface; sheaths closed < 1/4 length
        14. Lemma distinctly keeled; pls alpine . . . . ²*P. keckii*
        14' Lemma weakly keeled; pls widespread . . . . *P. secunda* ssp. *juncifolia*
      11' Lemmas hairy
        15. Pl unisexual; basal blades densely short-hairy on upper surface; uppermost st-lf blade gen
          highly reduced; sheath closed ± 2/3 length . . . . *P. fendleriana* ssp. *longiligula*
        15' Pl bisexual; basal blades glabrous or indistinctly scabrous on the upper surface; upper st-lf blades
          not highly reduced in length; sheaths open 4/5 length to near base
        16. Lemma with a web at the base
          17. Pl < 15 cm; alpine . . . . ²*P. pattersonii*
          17' Pl > 20 cm; along streams and in adjacent meadows . . . . *P. palustris*
        16' Lemma without a web at the base
          18. Lemma distinctly keeled; upper spikelet internodes < 0.9 mm . . . . *P. glauca* ssp. *rupicola*
          18' Lemma weekly keeled; upper spikelet internodes gen > 1.2 mm . . . . *P. secunda* ssp. *secunda*

P. annua L.(p. 597) ANNUAL BLUEGRASS Ann, bien, cespitose or with stolons, 0.2–2 dm. **LF**: sheath open ± 2/3 length; ligule 1–5 mm, rounded to obtuse; blade gen 1–3 mm wide, soft, gen flat, bright or yellowish green. **INFL** 1–10 cm, triangular, 1.2–1.6 × longer than wide, open in fr; branches spreading, smooth, with spikelets only in top 1/2. **SPIKELET**: axis ± hidden; callus glabrous; lemma 2.5–4 mm, smooth, veins soft-hairy or glabrous; palea keels hairy. **FL**: bisexual or upper 1–2 pistillate; anther 0.6–1 mm. 2*n*=28. Abundant. Disturbed moist ground, lawns, etc.; gen < 2000 m. **CA** (esp near coast); ± worldwide, native to Eur. Fl winter and spring, continuously when moisture permits.

P. bigelovii Vasey & Scribner Ann, cespitose, gen 1.5–4 dm. **LF**: sheath open 1/2–3/4 length; ligule 1.5–6 mm, truncate to obtuse, minutely scabrous; blade 1.5–5 mm wide, soft, gen flat, abruptly prow-tipped. **INFL** 5–15 cm, ± linear; branches appressed, with spikelets from near base. **SPIKELET**: callus cobwebby; lemma 2.5–4 mm, veins (sometimes between) hairy; palea keels hairy. **FL**: anthers 0.2–1 mm. 2*n*=28. Uncommon. Shady places in desert scrub, yellow-pine forests; < 1500 m. SW, **D**; to TX, nw Mex. Fls early spring.

P. bulbosa L. (p. 597) BULBOUS BLUEGRASS Per, densely cespitose, 1.5–6 dm. **ST**: bases ± bulbous. **LF**: sheath open to near base; ligule 2–4 mm, obtuse; blade 1–2 mm wide, soft, flat or folded, soon withering. **INFL** 3–10 cm, ovate to lanceolate; branches gen ascending, smooth. **SPIKELET**: florets gen formed into leafy bulblets. **FL**: anthers 1.2–1.5 mm in fertile lemmas. 2*n*= 21–42. Disturbed places; gen < 2000 m. **CA**; ± worldwide temp; native to Eur. Sporadic in CA. Spring

P. compressa L. CANADIAN BLUEGRASS Per from long, stout rhizomes, 1.5–6 dm. **ST** (incl nodes) ± flattened, keeled, wiry; nodes obviously exposed. **LF**: sheath open 3/4 length to near base; ligule 1–3 mm, rounded; blade 1.5–4 mm wide, soft to ± firm, flat or folded. **INFL** 2–9 cm, lanceolate to ovate, dense (or sparse and interrupted); branches ± ascending, short, densely scabrous on angles. **SPIKELET**: callus glabrous or ± cobwebby; lemma 2.3–3.5 mm, keel and marginal veins hairy. **FL**: anthers 1.3–1.8 mm. 2*n*=42 most often. Moist, often disturbed low ground; < 1800 m. **CA**; ± worldwide temp; native to Eur. Summer

P. cusickii Vasey Per, ± densely cespitose, 1–6 dm, ± dioecious. **LF**: sheath open 1/5–3/4 length; ligule gen 1–6 mm (on sterile sts < 2 mm, truncate, scabrous); blade longest at mid-st, on sterile sts gen 0.5–1 mm wide, ± firm, inrolled (sometime also folded), upper surface finely hairy. **INFL** 2–12 cm, lanceolate to ovate, gen dense; branches ascending to appressed, slender, smooth or scabrous. **SPIKELET**: callus gen glabrous; lemma keeled, gen glabrous (rarely keel sparsely hairy), smooth or scabrous; palea keels scabrous. **FL**: fertile anthers 2–3.5 mm. Moist to dry meadows, sagebrush scrub, montane forest; 1500–3600 m. KR, SNH, **GB**; to s Can, ND, Colorado. 4 sspp.in CA.

ssp. **cusickii** **LVS**: basal tuft dense; sheath open 1/2–3/4 length; 0–1 nodes barely exposed. **INFL** 3–12 cm; branches slender, obviously scabrous, longest gen > 17 mm. **SPIKELET**: lemma 3.5–7.5 mm, glabrous or scabrous. **FL** gen staminate or pistillate. 2*n*=28. Moist meadows, dry slopes in sagebrush scrub or montane forest; 1500–2500 m. e KR, n SNH, **GB**; to WA, ID, n NV. [*P. hansenii* Scribner]. ✿ IRR:**1–3**. May–Jul

ssp. **pallida** R. Soreng **LVS**: basal tuft dense; sheath open 1/2–4/5 length; 0–1 nodes barely exposed. **INFL** 3–5 cm, ± dense; branches appressed, obviously scabrous, stout, longest gen < 15 mm. **SPIKELET**: lemma 5–6.5 mm, glabrous or scabrous. **FL** pistillate in CA. 2*n*=56,59. Uncommon in CA. High montane to lower alpine dry meadows, ridges; 2000–3500 m. c&s SNH, **W&I**; w Can, ND, Colorado. [*P. subaristata* Beal, not Philippi] ✿ DRY:**1–3**. Early summer

P. fendleriana (Steudel) Vasey ssp. **longiligula** (Scribner & Williams) R. Soreng (p. 597) LONGTONGUE MUTTON GRASS Per, densely cespitose, gen with short rhizomes, 1.5–7 dm, dioecious. **LF**: sheath open ± 1/3 length; ligule 1–18 mm, truncate to acuminate; blade 1.5–4 mm wide, firm, folded, inrolled, uppermost st-lf blade gen ± vestigial, sterile st blade upper surface gen finely hairy. **INFL** gen 2–12 cm, lanceolate to ovate, dense; branches scabrous. **SPIKELET**: callus glabrous; lemma 3.5–6 mm, keel and marginal veins (sometimes between) hairy. **FL** gen pistillate; fertile anthers 2–4 mm. 2*n*=56. Mtn slopes, sagebrush scrub to subalpine; 2000–3200 m. s SNH, **GB**, **DMtns**; to B.C., SD, n

Mex. [*P. longiligula* Scribner & Williams] ✿ DRY:**1–3**; STBL. Spring–early summer

P. glauca M. Vahl. ssp. **rupicola** (Nash) W.A. Weber TIMBERLINE BLUEGRASS Per, densely cespitose, all current shoots flowering, gen 0.5–1.5 dm. **LF**: sheath open > 4/5 length; ligule 0.5–3 mm, truncate and finely scabrous at margin to acute and smooth; blade 1–2 mm wide, soft, flat or folded, abruptly ascending or spreading. **INFL** 1–5 cm, lanceolate to ovate; branches ascending to appressed, gen < 1.5 cm, gen scabrous on angles. **SPIKELET**: upper internodes < 1 mm, not elongated; glumes 3-veined, upper 2.5–3.5 mm (< 3/4 length of lower); callus glabrous; lemma 2.5–3.5 mm, veins and base hairy. **FL**: anthers 1.2–1.8 mm. 2*n*=42–70. Dry alpine slopes, ridges; 3300–4000 m. c&s SNH, **W&I**; to sw Can, NM. [*P. rupicola* Nash]. Summer

P. keckii R. Soreng Per, densely cespitose, gen 0.3–1 dm. **LF**: sheath open 4/5–9/10 length; ligule of uppermost sterile st lf < 2.5 mm, smooth or sparsely scabrous; blade 1–2 mm wide, firm, folded, inrolled. **INFL** 1.5–6 cm, lanceolate to narrowly ovate. **SPIKELET**: callus glabrous; lemma 3–5 mm, glabrous or sparsely short-hairy. **FL**: anthers 0.7–2 mm. High alpine, often in open ground; > 3000 m. SNH, **SNE (White, Sweetwater Mtns)**. Formerly mistaken for *P. suksdorfii* (Beal) Piper of WA. Summer

P. leptocoma Trin. ssp. **leptocoma** BOG BLUEGRASS Per, loosely cespitose, 1–7 dm. **LF**: sheath open 1/2–3/4 length; ligule 1.5–4 mm, truncate to obtuse, smooth; blade 1–4 mm wide, soft, gen flat. **INFL** 4–15 cm, open; branches ascending in fl, ± drooping in fr, with spikelets in top 1/3. **SPIKELET**: lower glume 1-veined; callus cobwebby; lemma 3–4 mm, keel and marginal veins sparsely hairy. **FL**: anthers 0.2–1 mm. 2*n*=42. Moist subalpine, lower alpine meadows; 1800–3200 m. KR, CaRH, n&c SNH, **GB**; to AK, MT, NM. Summer

P. lettermanii Vasey Per, cespitose, delicate, gen 0.2–0.9 dm. **LF**: sheath open 3/4–9/10 length; ligule 1–4 mm, truncate to acute, smooth; blade 0.5–1.5 mm wide, soft, flat or folded. **INFL** gen 1–3 cm, narrowly lanceolate. **SPIKELET**: glumes (at least upper) > lowest lemma; callus glabrous; lemma 2.5–3 mm, glabrous or very sparsely short-hairy. **FL**: anthers 0.2–0.7 mm. 2*n*=14. High alpine, in sandy soil around boulders; > 3500 m. s SNH, **W&I**; to sw Can, Colorado. ✿ IRR:**1**. Summer

P. palustris L. FOWL BLUEGRASS Per, cespitose or with stolons, gen 2.5–12 dm. **LF**: sheath open 3/4 length to near base; ligule 1–3 mm, acute to rounded; blade 1.5–6 mm wide, soft, gen flat, often > sheath, slenderly prow-tipped, base closely ascending. **INFL** gen 10–30 cm, eventually open, lanceolate to narrowly triangular, many-fld; branches ascending to spreading in fr, scabrous on angles. **SPIKELET**: lower glume gen 3-nerved; callus cobwebby; lemma gen 2–3 mm, keel and marginal veins hairy. **FL**: anthers 0.8–1.4 mm. 2*n*=28. Disturbed ground in moist forests or sagebrush scrub, meadows, along streams; 1500–2000 m. NW, CaRH, SnGb, SnBr, **SNE**; cool temp; native to Eur.. Spring–early summer

P. pattersonii Vasey PATTERSON'S BLUEGRASS Per, densely cespitose, gen 0.5–1.5 dm; sterile shoots many. **LF**: sheath open 4/5 length to near base, basal sheaths persisting; ligule 1–3.5 mm, acute, smooth; blade 1.5–2.5 mm wide, folded, ± inrolled, closely ascending. **INFL** 2–6 cm, lanceolate, dense; branches appressed, smooth or sparsely scabrous. **SPIKELET**: glumes ± equal, upper 3.5–4.2 mm, gen = first lemma; callus gen cobwebby; lemma 3.5–4 mm, keel and marginal veins gen hairy; palea keels sparsely scabrous above, some sparsely short-hairy below. **FL**: anthers gen 0.7–1.2 mm. 2*n*=42,84. RARE in CA. High alpine open ground; gen > 3300 m. SNH, **W&I**; to Colorado. Mid–summer

P. pratensis L. ssp. **pratensis** (p. 597) KENTUCKY BLUEGRASS Per from long, stout rhizomes, tufted or loose, gen 2–7 dm. **LF**: sheath open 1/2–3/4 length; ligule 1–4 mm, truncated to rounded, smooth to minutely scabrous at margin; blade gen 2–4 mm wide, soft to ± firm, flat or folded. **INFL** gen 6–15 cm, ovate to triangular; branches gen spreading, smooth or scabrous. **SPIKELET**: callus densely long-cobwebby; lemma 3–4 mm, keel and marginal veins hairy. **FL**: anthers 1.2–2 mm. $2n=21$–117. Common. Many disturbed or stable habitats, incl saline or alkaline soils; 0–3500 m. **CA**; n temp; native to Eur. Widely planted as lawn or pasture grass. Fls spring to early summer. Ssp. *agassizensis* (Boivin & D. Löve) Taylor & McBride is widespread and possibly native in CA, has dense infl < 6 cm with smooth branches, lf firm and folded, sterile shoot blade upper surface often sparsely soft-hairy; ssp. *angustifolia* (L.) Arcang. is probably introduced in CA, has long, folded sterile shoots gen < 0.5 mm wide, sometimes hairy as in ssp. *agassizensis*, narrowly triangular open infl with with smooth ascending branches. Spring–early summer

P. secunda J.S. Presl (p. 597) Per, densely cespitose, 1.5–10 dm. **LF**: sheath open 3/4 length to near base; ligule 0.5–10 mm, truncate to acuminate, sometimes scabrous; blade gen 0.5–3 mm wide, soft to firm, flat to folded or inrolled. **INFL** 2–25 cm, often ± 1-sided, gen linear to lanceolate, gen dense; branches gen appressed to ascending (gen spreading only in fl), ± scabrous. **SPIKELET** ± cylindric or little compressed; upper internodes gen > 1.2 mm; callus glabrous or with a ring of short hairs; lemma 3.5–5 mm, weakly keeled to rounded across lower back, glabrous to ± evenly short-hairy across body (rarely soft-hairy only on veins), smooth to scabrous. **FL**: anther 1.5–3 mm. $2n=42$–106 (mostly high polyploids). Common. Many habitats; 0–3800 m. **CA**; to AK, Rocky Mtns, nw Mex, also in S.Am. Many ecological forms; sspp. tend to intergrade.

ssp. *juncifolia* (Scribner) R. Soreng Pl 3–12 dm. **LF**: ligule 0.5–6 mm, truncate to acuminate, scabrous (those on lateral shoots 0.5–2 mm); blade gen < 1.5 mm wide, ± firm, tightly folded and inrolled, retaining shape, often glaucous. **INFL** gen 6–25 cm. **SPIKELET**: lemma 3.5–6 mm, glabrous (rarely sparsely short-hairy on keel and marginal veins near base); palea keels scabrous. $2n=\pm63$ most often. Sagebrush shrubland to lower montane forest, often in alkaline depressions; 900–3000 m. CaRH,

SNH (esp e slope), **GB**; to s Can, ND, NM. Several important ecotypes: 1) "*P. ampla* Merr." (big bluegrass; pl 6–12 dm; lf blade flat, gen glaucous; non-alkaline); 2) "*P. juncifolia* Scribner" (alkali bluegrass: lf blade folded and inrolled, ligule truncate; often alkaline); 3) "*P. nevadensis* Scribner" (Nevada bluegrass: lf blade folded and inrolled, ligule long, acute; often alkaline). "*P. pratensis* × *P. secunda*" [*P. fibrata* Scribner] is a set of hybrids between *P. s.* ssp. *j.* and *P. pratensis*; $2n=63$–64; low, often saline meadows; 800–2000 m; range of ssp. *j.* ✿ DRY:**1–9,14–24**;salt tolerant. Early Summer

ssp. *secunda* ONE-SIDED BLUEGRASS Pl 1.5–10 dm. **LF**: ligule 2–10 mm, acute to acuminate, smooth or sparsely scabrous; blade 0.5–3 mm wide, soft, gen flat, soon withering, sometimes glaucous, basal often thread-like. **INFL** gen 2–15 cm. **SPIKELET**: lemma 4–5 mm, base ± evenly short-hairy (sometimes nearly glabrous); palea keels and between them gen hairy in lower 1/2. $2n=\pm84$ most often. Common. Habitat and range of sp. [*P. canbyi* (Scribner) Howell; *P. gracillima* Vasey, *P. incurva* Scribner & Williams, *P. sandbergii* Vasey, *P. scabrella* (Thurb.) Vasey] Many ecological forms have been named, all intergrade completely, probably do not warrant taxonomic recognition. ✿ DRY:**1–9,14–24**;some forms tolerate IRR. Spring–early summer (mid-summer in subalpine)

P. wheeleri Vasey (p. 597) Per from short rhizomes, ± tufted, gen 3.5–8 dm, dioecious. **LF**: sheaths open 1/3–2/3 length, lower often finely reflexed-scabrous or short-hairy; lower ligules 0.5–2 mm, truncate to rounded, scabrous; blade 1.5–3.5 mm wide, soft to ± firm, flat or folded; sterile st blades folded and ± inrolled, upper surface gen ± finely hairy. **INFL** gen 5–12 cm, ovate, sparse; branches gen 1–4, ± ascending to spreading, ± scabrous. **SPIKELET**: callus glabrous; lemma 3–6 mm, glabrous to sparsely hairy on keel and marginal veins, gen sparsely scabrous. **FL** pistillate (anthers ± vestigial). $2n=56,63,70$–91. Common (esp SN). Mtns, open forest in rich soil; 1300–3800 m. KR, NCoRH, CaRH, SNH, **GB**; to s Can, Colorado. Formerly mistaken for *P. nervosa* (Hook.) Vasey. ✿ DRY:**1–3**;REVEG. Summer

POLYPOGON BEARD GRASS

Steven A. Conley

Ann, per. **STS** decumbent to erect, simple. **LF**: sheath open, loosely enclosing st, glabrous; ligule thinly membranous, obtuse to truncate, minutely ciliate to toothed; blades ± cauline, flat, scabrous, veins minutely prickly at 10×. **INFL** panicle-like, oblong to narrowly ovoid in outline, interrupted to compact, dense. **SPIKELETS** breaking below glumes and part of stalk, falling as 1 unit; glumes ± equal, 1-veined, entire or 2-lobed, awned at tip or between lobes, awn straight; floret 1; lemma ± 0.5 × glumes, translucent, 5-veined, tip toothed, awn < glume awn; palea slightly < lemma, transparent; anthers tightly enclosed by lemma and palea. **FR** oblong, smooth, enclosed by lemma, palea. 18 spp.: warm temp Eurasia, Afr, S.Am. Some spp. orn. (Greek: much bearded)

1. Glume awn gen < 3.5 mm: per; infl interrupted to compact .... *P. interruptus*
1' Glume awn gen > 3.5 mm; ann; infl compact to dense .... *P. monspeliensis*

P. interruptus Kunth DITCH BEARD GRASS Per. **ST** 5–9 dm, clumped. **LF**: ligule 2–8(13) mm, obtuse to truncate, minutely hairy; blade 0.5–19.5 cm, 3–6 mm wide. **INFL** 1.5–18 cm. **SPIKELET**: stalk < 1 mm; glumes 1.5–3 mm, scabrous; awns 1.5–4.5 mm, purplish; lemma 1–2 mm, awn 0.5–3 mm. $2n=42$. Common. Streambanks, ditches; < 1300 m. **CA**; to Great Plains, sc US; native to S.Am. May–Aug

P. monspeliensis (L.) Desf. (p. 597) ANNUAL BEARD GRASS Ann. **ST** 2–10 dm. **LF**: ligule 5–6 mm, irregularly toothed, minutely hairy; blade 1–20.5 cm, 4–6 mm wide. **INFL** 1–17 cm, plume-like, densely fld. **SPIKELET**: glumes 1–2.5 mm, minutely bristly, awn 2–10 mm; lemma 0.5–1.5 mm, awn 0.5–4.5 mm. $2n=28,35$. Common. Moist places, along streams, ditches; < 2100 m. **CA**; N.Am; native to s&w Eur. Apr–Aug

PUCCINELLIA ALKALI GRASS

Jerrold I Davis

Ann, per; stolons and rhizomes gen 0. **STS** decumbent to erect. **LVS** basal and cauline; sheath open ± to base; ligule thinly membranous, acute to truncate, sometimes toothed. **INFL** panicle-like; lower branches reflexed to erect; spikelets stalked. **SPIKELET** bisexual; glumes < lowest floret, lower glume gen 1-veined, upper 3-veined; florets 2–9; lemma gen firm, back gen rounded, sometimes weakly keeled near tip, margin entire to scabrous-serrate near tip at 10×, glabrous to weakly puberulent at base, gen faintly 5-veined; awn 0; palea ± = lemma. 50 spp.: temp to arctic, N.Am, Eurasia. (B. Puccinelli, Italy, 1808–1850) [Davis 1983 Syst Bot 8:341–353] Gen on wet saline or alkaline soils; some spp. difficult to separate without hand lens. Key to species adapted by Margriet Wetherwax.

1. Ann; remains of previous years growth not present
  2. Lowest lemma ± 2 mm; lemma tip obtuse to truncate . . . . *P. parishii*
  2' Lowest lemma 3.5–4 mm; lemma tip acute . . . . *P. simplex*
1' Per; previous years lvs persistent
  3. Lemma tip obtuse to truncate
    4. Lowest floret lemma 1.5–2 mm; lower infl branches spreading to reflexed in fr . . . . *P. distans*
    4' Lowest floret lemma 2–4 mm; lower infl branches erect to reflexed in fr . . . . ²*P. nuttalliana*
  3' Lemma tip acute
    5. Cauline lf blade inrolled, 1–2 mm wide when flat; lemma margin near tip entire to uniformly
      scabrous-serrate at 10×; infl gen < 10 cm . . . . *P. lemmonii*
    5' Cauline lf blade flat to inrolled, 1–4 mm wide when flat; lemma margin near tip scabrous-serrate at
      10×; infl gen > 10 cm . . . . ²*P. nuttalliana*

P. distans (Jacq.) Parl. EUROPEAN ALKALI GRASS Per. **LF**: cauline blade flat to inrolled, 1–7 mm wide when flat. **INFL** 2.5–22 cm; lower branches spreading to reflexed in fr; spikelet stalk scabrous. **SPIKELET**: lemma tip widely obtuse to truncate, margin near tip scabrous-serrate, lowest lemma 1.5–2 mm; anthers of lowest floret 0.5–1 mm. 2*n*=14,42,56,42. Saline meadows and flats; < 2700 m. CA-FP, **GB**; to AK, ne N.Am; native to Eurasia. Jun–Jul

P. lemmonii (Vasey) Scribner (p. 603) LEMMON'S ALKALI GRASS Per. **LF**: cauline blade inrolled, 1–2 mm wide when flat. **INFL** 2–10(18) cm; lower branches ascending to reflexed in fr; spikelet stalk scabrous. **SPIKELET**: lemma tip acute, margin near tip entire to scabrous-serrate, lowest lemma 2.5–3.5 mm; anthers of lowest floret 1–2 mm. 2*n*=14. Saline meadows and flats; 700–2000 m. KR, CaR, **GB**; to OR, ID, WY. ❀ TRY; alsoSTBL. May–Aug

P. nuttalliana (Schultes) A. Hitchc. NUTTALL'S ALKALI GRASS Per. **LF**: cauline blade flat to inrolled, 1–4 mm wide when flat. **INFL** (3.5)10–22 cm; lower branches erect to reflexed in fr; spikelet stalk scabrous. **SPIKELET**: lemma tip acute to obtuse, margin near tip scabrous-serrate, lowest lemma (2)2.5–3.5 mm; an-

thers of lowest floret 0.5–1.5 mm. 2*n*=28,42,56. Saline meadows and flats; < 2300 m. CA-FP, **GB**; to AK, e N.Am. [*P. airoides* (Nutt.) S. Watson & J. Coulter] ❀ STBL; valuable forage grass. Jun–Sep

P. parishii A. Hitchc. PARISH'S ALKALI GRASS Ann. **LF**: cauline blade gen inrolled, < 1 mm wide when flat. **INFL** 1–8 cm; lower branches erect to reflexed in fr; spikelet stalk scabrous (sometimes sparsely scabrous). **SPIKELET**: lemma veins hairy in lower half, tip obtuse to truncate, margin near tip scabrous-serrate, lowest lemma ± 2 mm; anthers of lowest floret ± 0.5 mm. 2*n*=14. RARE. Mineral springs; ± 1000 m. **w DMoj (sw San Bernardino Co.)**; to NM. Apr–May

P. simplex Scribner (p. 603) Ann. **LF**: cauline blade gen inrolled, 1–2 mm wide when flat. **INFL** 1–18 cm; lower branches erect in fr; spikelet stalk ± glabrous to scabrous, often faintly short-soft-hairy. **SPIKELET**: lemma back short-hairy, rounded to weakly keeled at tip, acute, margin near tip entire to sparsely scabrous, lowest lemma 3.5–4 mm; anthers of lowest floret < 0.5 mm. 2*n*=56. Saline flats and mineral springs; < 800 m. Teh, GV, SnFrB, **w DMoj**; to UT. Mar–May

SACCHARUM

Kelly W. Allred

Per with rhizomes. **STS** tall, erect. **LVS** cauline; sheaths > internodes; ligule short, membranous or hairy; blades flat or folded. **INFL** panicle-like, silky-hairy. **SPIKELET** bisexual, in pairs; lower spikelet sessile; upper spikelet stalked; pair falling with subtending axis segment as 1 unit or stalked spikelet deciduous; glumes, lemmas long-tapered; glumes > florets, 3–5-veined; florets 2, lower vestigial, obscure, upper fertile; lemma awned or awn 0; palea < lemma or 0. 35–40 spp.: trop, subtrop, se Asia. Some orn, sugarcane (*S. officinarum*) widely cult for sugar. (Latin: sugar)

S. ravennae (L.) Murray RAVENNA GRASS **STS** densely tufted, gen 2–4 m. **LF**: ligule < 1 mm, thin; blade < 12 mm wide, gen densely hairy near ligule, strongly serrate. **INFL** plume-like, 2.5–6 dm. **SPIKELET** 3.5–7 mm; stalked spikelet deciduous; glumes

lanceolate, base densely silky-hairy; lemma awned, awn 3–5 mm. 2*n*=20. Ditchbanks, marshes; < 300 m. **s DSon (Imperial Co.)**; native to Eurasia. [*Erianthus r.* (L.) P. Beauv.]

SCHISMUS MEDITERRANEAN GRASS

Kelly W. Allred

Ann, bien, cespitose. **ST** erect to prostrate. **LVS** gen basal, tufted, glabrous; ligule short-hairy; blades flat or inrolled. **INFL** panicle-like, dense. **SPIKELET** ± laterally compressed; glumes lanceolate, membranous, 5–7-veined; florets 3–8, bisexual, breaking above glumes and between florets; lemma 9-veined, ciliate proximally, toothed to notched, awn ± 0; palea < to > lemma. 5 spp.: warm temp, subtrop, Eurasia, Afr. (Greek: split, from notched lemma) [Conert and Türpe 1974 Abh. Senckenberg Naturf. Ges. 532:1–81]

1 Lower lemmas gen > 2,5 mm, teeth narrowly triangular; palea of lowest floret acute, < lemma . . . . . *S. arabicus*
1' Lower lemmas gen < 2,5 mm, teeth obtuse to widely triangular, palea of lowest floret obtuse, > lemma
. . . . *S. barbatus*

S. arabicus Nees (p. 603) Ann. **ST** gen 5–20 cm. **LF**: blade < 1 mm wide, thread-like. **SPIKELET**: glumes 4.5–6.5 mm; lemma 2.5–4 mm, teeth ± 0.3 × lemma, narrowly triangular; palea 2–3 mm, gen << lemma. 2*n*=12. Dry, open, gen disturbed areas;

< 1300 m. SnJV, CW, s ChI, **W&I**, **D**; to TX, AZ; native to Eurasia. Mar–May

S. barbatus (L.) Thell. (p. 603) Ann. **ST** gen 2–16 cm. **LF**:

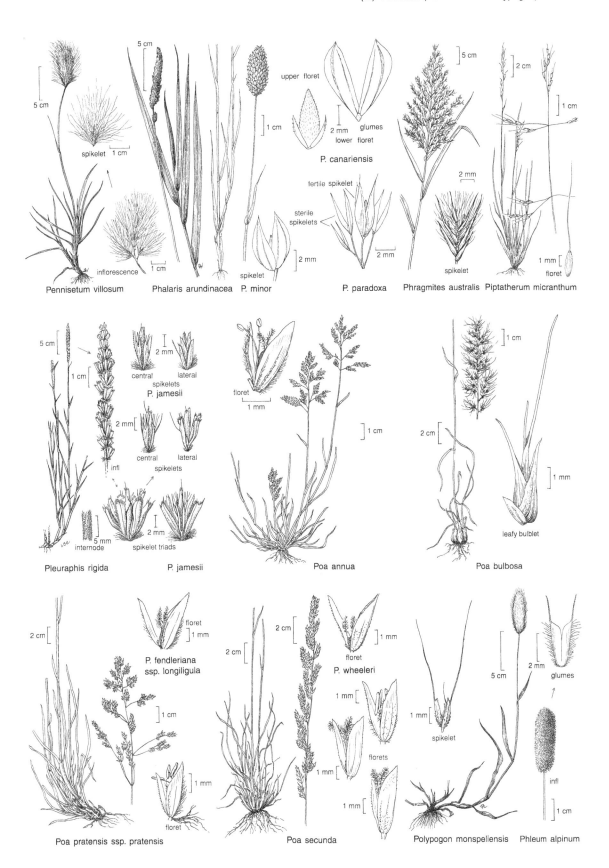

5 cm

spikelet 1 cm

inflorescence 1 cm

Pennisetum villosum

5 cm

Phalaris arundinacea

1 cm

spikelet

2 mm

P. minor

upper floret

2 mm glumes
lower floret

P. canariensis

fertile spikelet

sterile
spikelets

spikelet 2 mm

P. paradoxa

5 cm

2 mm

spikelet

Phragmites australis

2 cm

1 cm

1 mm floret

Piptatherum micranthum

5 cm

1 cm

2 mm

central lateral
spikelets

P. jamesii

2 mm

central lateral
spikelets

infl

2 mm

spikelet triads

internode 5 mm

Pleuraphis rigida

P. jamesii

floret

1 mm

1 cm

Poa annua

1 cm

2 cm

1 mm

leafy bulblet

Poa bulbosa

2 cm

floret

1 mm

P. fendleriana
ssp. longiligula

1 cm

floret

1 mm

Poa pratensis ssp. pratensis

2 cm

2 cm

floret

1 mm

P. wheeleri

1 mm

florets

1 mm

1 mm

Poa secunda

1 mm

spikelet

Polypogon monspeliensis

5 cm

2 mm glumes

infl

1 cm

Phleum alpinum

blade < 2 mm wide, thread-like. **SPIKELET**: glume 4–5 mm; lemma 2–2.5 mm, teeth ± < 0.2 × lemma, obtuse to widely triangular; palea 1.5–2.5 mm, gen = > lemma. 2*n*=12. Dry, open, gen disturbed areas; < 1200 m. Teh, SnJV, SW, **D**; to TX, n Mex; native to s Eur, Afr. Mar–Apr

SCLEROPOGON

1 sp. (Greek: hard beard, from firm awns) [Reeder & Toolin 1987 Phytologia 62:267–275]

S. brevifolius Philippi (p. 603) BURRO GRASS Per, from short stolons, mat-like, gen dioecious or monoecious, rarely bisexual. **ST** erect, 1–2 dm. **LVS** gen basal, densely tufted; sheath smooth; ligule short-hairy; blade 2–8 cm, 1–3 mm wide, flat, firm, sharp-pointed. **STAMINATE INFL** 3–7 cm, raceme- or panicle-like. **PISTILLATE INFL** 10–20 cm, spike-like. **STAMINATE SPIKELET** 1–2.5 cm, compressed; glumes 3–8 mm, ± equal, lanceolate, 1(3)-veined, awn 0; axis breaking above glumes and falling as 1 unit; florets 5–10(20); lemma ± = glumes, 3(5)-veined, short-awned. **PISTILLATE SPIKELET** 8–15 cm incl awns; body exc awns ± cylindric, gen subtended by 1 bract; bract ± = lower glume; glumes 3-veined, acute, awn 0, lower glume 1–2 cm, upper 1.5–3 cm; axis breaking above glumes and ± between florets; florets 3–5, lower with sharp-pointed callus, upper reduced to awns; lemma 7–11 mm, narrow, awns 3, 6–12 cm, ± spreading to ascending. 2*n*=40. RARE in CA. Open creosote-bush shrubland; ± 1600 m. **e DMtns (New York Mtns)**; to Colorado, TX, Mex, w S.Am.

SECALE RYE

Mary E. Barkworth

Ann, per. **ST** gen erect. **LF**: sheath appendaged; ligule membranous; blade gen flat. **INFL** spike-like, dense, ± flat; axis sometimes breaking at nodes in fr; spikelets 2-ranked, 1 per node, sessile, not sunken. **SPIKELET**: glumes narrow, rigid, keeled, vein gen 1; florets 2, fertile, sessile and side-to-side, sometimes with vestigial floret between; lemma with keel near margin, keel and margins ciliate, veins 5, tip tapered, awn straight, scabrous. 5 spp.: Eurasia. (Latin: ancient name for rye)

S. cereale L. (p. 603) Ann, sometimes bien. **ST** 6–12.5 dm, glabrous exc below infl. **LF**: sheath glabrous, appendages ± 1 mm; blade 3–10 mm wide. **INFL** 8–17 cm, gen not breaking apart. **SPIKELET**: glumes 6–17 mm, keeled; lemma 10–16 mm, awn 2–7 cm; anthers 7.5–8.5 mm. 2*n*=14,16,27–29. Disturbed slopes, roadsides; < 1800 m. n SNH, sw SnFrB, s MP, **DMoj**, expected elsewhere; native to sw Asia. May–Aug

SETARIA

Robert Webster

Ann, per. **STS** gen erect; internode solid to hollow inside. **LVS** basal and cauline; ligule short-hairy or membranous, ciliate. **INFL** panicle-like, dense, gen cylindric; 1° branches spreading to appressed; spikelets many, gen clustered on one side of short 2° branch, short-stalked to subsessile, subtended by 1–15 bristles; bristles gen scabrous. **SPIKELET** falling as 1 unit, gen elliptic; glumes unequal; florets gen 2, ± equal, lower floret sterile or staminate, palea gen < lemma, upper floret fertile, firm, gen hard, rough, margin inrolled, tip blunt. ± 100 spp.; warm temp, trop Eurasia, Afr. (Latin: bristly) [Rominger 1962 Illinois Biol Monogr 29:1–132] Some spp. cult for food. Key to species adapted by Margriet Wetherwax.

1. Spikelet or spikelet cluster subtended by 1–3 bristles; lower floret palea gen < 2/3 × lemma
  2. Short stiff hairs on bristles pointed downward; infl 1° branches 11–20 mm . . . . *S. verticillata*
  2' Short stiff hairs on bristles pointed toward tip; infl 1° branches 3–10 mm . . . . *S. viridis*
1' Spikelet or spikelet cluster subtended by 4–15 bristles; lower floret palea gen = lemma
  3. Spikelet gen < 2.8 mm, ± 1–1.5 mm wide; per . . . . *S. gracilis*
  3' Spikelet ± 2.8–3.5 mm, ± 1.5–2 mm wide; ann . . . . *S. pumila*

S. gracilis Kunth (p. 603) Per, cespitose, from short rhizomes. **ST** erect, 7–12 dm; base with hard, knot-like swellings. **LF**: sheath 4–9 cm, glabrous; ligule < 1 mm; blade < 25 cm, 2–8 mm wide, upper surface glabrous. **INFL** 3–8 cm; 1° branches 3–8 mm; axis glabrous; bristles 4–12; spikelet stalk << 0.5 mm. **SPIKELET** 2–3 mm, ± 1–1.5 mm wide; lower glume 1–1.5 mm, 3-veined, upper glume 0.5–0.8 × spikelet length; lower floret sterile or staminate, lemma 5–7-veined, tip acute, palea ± = lemma. Open areas, grassland, chaparral; < 400 m. GV, CW, SCo, **SNE (very uncommon), DMoj**; to e US, C. & S.Am. [*S. geniculata* (Lam.) P. Beauv. misapplied] ❀ STBL. May–Sep

S. pumila (Poiret) Roemer & Schultes Ann. **ST** 2–13 dm. **LF**: sheath 4–9 cm, glabrous; ligule ± 0.5–1 mm; blade 5–30 cm, 3–10 mm wide, upper surface glabrous. **INFL** 2–6 cm; 1° branches 5–10 mm, axis short-hairy; bristles 4–12; spikelet stalk << 0.5 mm. **SPIKELET** 3–3.5 mm, ± 1.5–2 mm wide; lower glume 1–1.5 mm, 3–5-veined, upper glume 0.5–0.7 × spikelet length; lower floret staminate, lemma 5-veined, tip acute, palea = lemma. Gen moist sites, fields; < 1200 m. e KR, CaRF, SNF, GV, CW, SCo, **SNE, DMoj**; to s Can, e US, Mex; native to Eur. [*S. lutescens* (Weigel) Hubb., *S. glauca* (L.) P. Beauv.] Jun–Oct

S. verticillata (L.) P. Beauv. Ann. **ST** decumbent to erect, 1.6–10 dm. **LF**: sheath 2–10 cm, glabrous; ligule 1–2 mm; blade 5–25 cm, 3–12 mm wide, upper surface glabrous to hairy. **INFL** 3–13 cm; 1° branches 1.1–2 cm, axis glabrous; bristles 1–2; spikelet stalk << 0.5 mm. **SPIKELET** ± 1.7–2.5 mm, 1–1.5 mm wide; lower glume 0.5–1.5 mm, 1–3-veined, upper glume ± = spikelet length; lower floret sterile, lemma 5–7-veined, tip acute to rounded, palea vestigial. Waste places, fields, roadsides; < 200 m. SnJV, SW, **DSon**; native to Eur. [*S. carnei* A. Hitchc. misapplied] May–Jul

S. viridis (L.) P. Beauv. (p. 603) Ann. **ST** decumbent to erect, 2–10 dm. **LF**: sheath 5–15 cm, scabrous; ligule 1–2 mm; blade 8–20 cm, 3–12 mm wide, upper surface glabrous. **INFL** 2–15 cm; 1° branches 3–10 mm, glabrous or hairy; bristles 1–3; spikelet stalk < 0.5 mm. **SPIKELET** ± 2 mm, ± 1 mm wide; lower glume ± 1 mm, 3-veined, upper glume = spikelet length; lower floret sterile, lemma 5-veined, tip acute, palea vestigial. $2n=18$.

Waste places, fields, roadsides; < 300 m. CA-FP, **DSon**; widespread N.Am; native to Eurasia. Jun–Aug

SORGHUM

Kelly W. Allred

Ann, per, cespitose or with rhizomes. **ST** erect; internodes gen solid. **LVS** cauline; sheaths gen < internodes; ligule membranous; blades flat or folded. **INFL** panicle-like, open to compact. **SPIKELETS** in pairs (trios at branch tips); lower sessile, bisexual; upper 1(2) stalked, staminate; pair with subtending axis segment falling as 1 unit. **SESSILE SPIKELET** ovoid; lower glume leathery, shiny, glabrous to puberulent, enclosing upper glume, florets; florets 2, lower sterile, upper fertile; lemma membranous, fertile lemma awned, awn bent, twisted; palea < lemma. ± 20 spp.: trop, subtrop, Afr. Cult for food, forage, sugar. (Italian: Sorgho) [de Wet 1978 Amer J Bot 65:477–484]

1. Ann . . . . ***S. bicolor***
1' Per with rhizomes . . . . ***S. halepense***

S. bicolor (L.) Moench (p. 603) SORGHUM, MILO, SUDAN GRASS Ann. **ST** erect, 1–2 m. **LF**: blade 3–10 cm wide. **INFL** 1–4 dm, open to compact; branches ± spreading to stiffly erect. **SESSILE SPIKELET** 4–9 mm; lemma 4–5 mm, awn 5–10 mm or 0. $2n=20$. Disturbed areas, roadsides, fallow fields; < 600 m. NCo, NCoR, GV, CCo, SCo, **D**; widely cult, native to Afr. [*S. lanceolatum* Stapf, *S. sudanense* (Piper) Staph, *S. virgatum* (Hackel) Stapf]

S. halepense (L.) Pers. (p. 603) JOHNSONGRASS Per with rhizomes. **ST** erect, 0.5–2 m. **LF**: blade 0.5–2 cm wide. **INFL** 1–5 dm, 5–25 cm wide, gen open; lower branches spreading to ascending. **SESSILE SPIKELETS** 4–6.5 mm; lemma 4–5 mm, awn 9–10 mm or 0. $2n=40$. Disturbed areas, ditchbanks, roadsides; < 800 m. NW, CaRF, SNF, GV, CW, SW, **DSon**; native to Medit. May–Aug

SPARTINA CORD GRASS

John R. Baird & John W. Thieret

Per, gen with rhizomes. **ST** erect, unbranched. **LVS** basal and cauline; sheath open, > internodes, glabrous, margin of sheath opening sometimes with long, shaggy hairs; ligule a fringe of hairs, 0.5–2 mm; blade flat to inrolled, long tapered, upper surface ridged. **INFL** panicle-like; each spike-like branch with 2 rows of overlapping spikelets on lower side of axis. **SPIKELET** laterally compressed, sessile, breaking below glumes, falling as 1 unit; glumes firmly membranous, obtuse to acuminate or with a small, sharp point, unequal, upper 1–3(5)-veined, gen > floret, lower 1-veined, < floret; floret 1; lemma 1(3)-veined, acute, awned or not, firmly membranous; palea ± = lemma, 2-veined. 15 spp.: Am, Eur, Afr. (Greek: a cord) [Spicher & Josselyn 1985 Madroño 32:158–167]

S. gracilis Trin. (p. 603) ALKALI CORD GRASS Rhizome 3–5 mm wide. **ST** gen solitary, 1.8–10 dm, 2–5 mm wide at base, slender, internodes firm. **LF**: blade 15–27 cm, 2.5–6 mm wide at base, gen inrolled when fresh, ridges on upper surface ± 5 per mm. **INFL** 4–25 cm, 5–12 mm wide, compact; branches 2–12, overlapping (often for only 1/2 their length, or lowest spike rarely separated), appressed, 1.5–8 cm, 2–6 mm wide. **SPIKELET** 6–

11 mm; glume and lemma keels ciliate (hairs gen 0.3–1 mm) at least near tip; lower glume 3–7 mm; upper glume 5–11 mm; lemma 6.5–10 mm. $2n=40,42$. UNCOMMON. Alkaline lake shores, stream banks, meadows, marshes; 1000–2100 m. **SNE, n DMoj**; to n&e Can, KS, NM. ✿ WET,SUN:1,**2,3**;STBL, salt tolerant. Jun–Aug

SPHENOPHOLIS WEDGEGRASS

Dieter H. Wilken

Ann, per. **STS** solitary to clumped, erect. **LVS** cauline, glabrous to hairy; ligule ± membranous, obtuse to truncate; blade flat to inrolled. **INFL** panicle-like, open to compact; lower branches spreading to erect. **SPIKELET** subsessile to stalked, ± laterally compressed; glumes unequal, lower < upper, lower linear to lanceolate, 1-veined, upper obovate, 3–5-veined; florets 2–3, bisexual; lemmas membranous, rounded on back, compressed at acute to obtuse tip, weakly 5-veined, awned or not; palea < lemma. 4 spp.: N.Am, Caribbean. (Greek: wedge scale, from upper glume shape) [Erdman 1965 Iowa State J Sci 39:289–336]

S. obtusata (Michaux) Scribner (p. 603) PRAIRIE WEDGEGRASS Per (sometimes fl first year). **ST** 2–8 dm. **LF**: sheath glabrous to scabrous; ligule 1–4 mm, tip jagged; blade 5–8 cm, flat. **INFL** 4–12 cm, gen erect, compact; lower branches ± ascending. **SPIKE-LET** 2–5 mm; lower glume ± 2 mm, ± linear, keel minutely sca-

brous, acute; upper glume 2–3 mm, widely obovate, obtuse; lemmas 2–3 mm, lower > glumes, acute to obtuse. $2n=14$. UNCOMMON. Wet meadows, streambanks, ponds; 300–2000 m. n SNH (Amador Co.), s SNH (Fresno Co.), ne SCo (Santa Ana River), sc PR (Cuyamaca Mtns),**W&I** ; N.Am, Caribbean. Apr–Jun

SPOROBOLUS DROPSEED, SACATON

Michael Curto

Ann, per. **STS** gen ascending to erect, 2–20 dm, gen tufted, ± solid in ×-section. **LVS** gen basal; cauline few, ascending or curving away; distal sheath margin and collar glabrous or hairy; ligule < 1 mm, hairy or membranous, fringed; blade flat to inrolled, gen glabrous or scabrous, sometimes short-soft-hairy. **INFL** terminal, also sometimes axillary, panicle-

or spike-like, gen partly enclosed by sheath; branches spreading or appressed. **SPIKELET** < 6 mm, gen pale to gray-green or purplish; glumes gen unequal, upper < or > lemma, membranous to translucent, 1-veined; floret bisexual, gen breaking above glumes; lemma texture gen like glumes, 1(3)-veined; palea < or > lemma. **FR** 1–3 mm, gen falling from floret, gen gelatinous when wet. ± 150 spp.: Am, Eurasia, Afr. (Greek: to throw seed, from deciduous seeds)

1. Anthers 1–2.5 mm; glume and lemma backs ± rounded, midvein glabrous at 20×; lowest infl branches
   gen > 6 cm .... *S. airoides*
1' Anthers < or = 1 mm; glume and lemma backs ± keeled, midvein scabrous at 20×; lowest infl branches gen < 6 cm
   2. Infl ± spike-like, cylindric, dense; 1° branches gen < 3 cm, appressed, obscure .... *S. contractus*
   2' Infl oblong to pyramid-like, ± open, sometimes enclosed by sheath; 1° branches gen > 3 cm, gen
      some or all spreading
      3. 1° infl branches ± spike-like, 2° and 3°, branches ± appressed, ± obscure .... *S. cryptandrus*
      3' 1° infl branches ± raceme-like, 2° and 3°, branches spreading, evident .... *S. flexuosus*

S. airoides (Torrey) Torrey (p. 603) ALKALI SACATON Per. **STS** tufted, 3–20 dm. **LF**: distal sheath margin glabrous to short-hairy; collar margin glabrous or sparsely long-hairy; ligule < 0.5 mm, fringed; blade 12–40 cm, 2–4(6) mm wide. **INFL** gen terminal, 1–6 dm; base 4–25 cm wide, panicle-like, pyramid-shaped, open to dense; branches ascending to spreading, lowest gen > 6 cm, 2° gen evident, spreading. **SPIKELET** 1–3 mm; glume, lemma backs rounded; base purplish; glumes unequal, narrowly lanceolate, tip acute to obtuse, lower 0.5–2 mm, upper > 0.5× lemma; lemma 2–3 mm, ovate to narrowly lanceolate; palea ± = lemma; anthers 1–1.8 mm. **FR** ± 1 mm. 2*n*= 80,90,108,126. Seasonally moist, alkaline areas; < 2100 m. SNF, Teh, s ScV, SnJV, s SCoRO, SW, **GB, D**; to e WA, c&s US, Mex. Pls from s SCo, s PR, s DSon with sts 1–2 m, infl base 12–25 cm wide, 1° branches ascending, 2° and 3° branches appressed, spikelets dense have been called var. *wrightii* (Scribner) Gould. ✿ SUN,IRR:1–3,7–24;alsoSTBL. Apr–Oct

S. contractus A. Hitchc. (p. 603) SPIKE DROPSEED Per. **STS** tufted, 4–12 dm. **LF**: distal sheath margin glabrous to short-hairy; collar margin glabrous to densely long-hairy; ligule < 1 mm, hairy; blade 4–30 cm, 2–8 mm wide. **INFL** gen terminal, 1.5–5 dm; base < 1 cm wide, spike-like, cylindric, dense; branches gen < 3 cm. **SPIKELET** 2–3 mm; glumes, lemma keeled; glumes unequal, lower ± 1 mm, ± awl-like, upper > 0.5 × lemma, narrowly lanceolate, acute; lemma 2–3 mm, ovate to narrowly lanceolate, acute; palea ± = lemma; anthers 0.2–1 mm. **FR** ± 1 mm. 2*n*=36. Washes, rocky slopes, shrubland, pinyon/juniper woodland; < 1900 m. **W&I, e D**; to Colorado, TX, Mex. ✿ SUN,DRN: 1,**2**,3,**7**,14–16,18–23. Aug–Oct

S. cryptandrus (Torrey) A. Gray (p. 603) SAND DROPSEED Per. **STS** tufted, 3–10 dm. **LF**: distal sheath margin glabrous to short-hairy; collar margin glabrous to densely long-hairy; ligule < 0.5 mm, hairy; blade 3–15 cm, 1–5 mm wide. **INFL** gen terminal, 0.8–3(4) dm; base 2–6(10) cm wide, panicle-like, pyramid-shaped or oblong in outline, ± open, sometimes completely enclosed by sheath; lower 1° branches gen > 3 cm, ± spike-like; 2° and 3° branches appressed. **SPIKELET** 1.5–2.5 mm; glumes, lemma keeled; glumes unequal, lower 0.5–1 mm, upper > 0.5 × lemma, narrowly lanceolate, acute; lemma 1.5–2.5 mm, ovate to narrowly lanceolate, acute; palea ± = lemma; anthers 0.2–1 mm. **FR** ± 1 mm. 2*n*=18,36,38,72. Rocky to sandy washes, slopes, shrubland, woodland; 350–2800 m. e-c SNH, SnBr, PR, **SNE, D** (used for revegetation elsewhere); widespread s Can, US, n Mex. ✿ SUN,DRN:1–3,**7**,9,**10,11**,12–24;also STBL. May–Oct

S. flexuosus (Vasey) Rydb. MESA DROPSEED Per. **STS** tufted, 3–10 dm. **LF**: distal sheath margin short-hairy; collar margin densely long-hairy; ligule < 0.5 mm, fringed; blade 5–20 cm, 1–6 mm wide. **INFL** terminal, 1–3 dm; base 4–14 cm wide, panicle-like, pyramid-shaped or oblong in outline, ± open; lower 1° branches gen > 3 cm; 2° and 3° branches evident, spreading. **SPIKELET** 1.5–2.5 mm; glumes, lemma keeled; glumes unequal, lower 0.5–1 mm, ± awl-like, upper > 0.5 × lemma, narrowly lanceolate, acute; lemma 1.5–2.5 mm, ovate to narrowly lanceolate, acute; palea ± = lemma; anthers 0.2–1 mm. **FR** ± 1 mm. 2*n*=36, 38. Rocky to sandy washes, slopes, shrubland; < 1200 m. **SNE, D**; to s UT, w TX, N Mex. ✿ SUN,DRN: 2,3,**7**,9,**10–12**,13–24. May–Oct

SWALLENIA

1 sp. (Jason Swallen, American agrostologist, 1903–) [Henry 1979 Fremontia 7(2):3–6]

S. alexandrae (Swallen) Söderstrom & Decker (p. 603) EUREKA VALLEY DUNE GRASS Per, tufted, from thick, scaly rhizomes; nodes ± woolly. **STS** ascending to erect, branching, 1.5–3.5 dm, stiff, ridged, gen glabrous. **LVS** gen cauline; sheath margins with soft-shaggy hairs near collar; ligule ± 1 mm, densely hairy; blade 5–14 cm, 3–6 mm wide, stiff, sharp-pointed, strongly veined. **INFL** panicle-like, 4–10 cm; branches spike-like, short, appressed; axis hairy. **SPIKELET** < 1.5 cm, persistent; glumes 9–14 mm, ± equal, > florets, wide, glabrous, 7–11-veined, awn 0; florets 3–7; lemma 7–9 mm, back rounded, lower margin with soft-shaggy hairs, 5–7-veined, awn 0; palea ± = lemma, margin hairs like lemma. **FR** 4 mm, 2 mm wide, falling from floret. 2*n*=20. **ENDANGERED** US; **RARE** CA. Sand dunes; 900–1200 m. **W&I (Eureka Valley, ne Inyo Co.)**. [*Ectosperma a.* Swallen] Apr–Jun

TRIDENS

Per, ± tufted. **STS** gen erect. **LF**: ligule fringed or short-hairy; blade flat or inrolled. **INFL** panicle-like, open to dense, or spike-like. **SPIKELET** ± compressed; glumes ± equal, membranous, lower 1-veined, upper 1–3(5)-veined; axis breaking above glumes and between florets; lemma wide, thin, back rounded, veins hairy below middle, 3-veined, tip notched or 2-toothed, gen short-pointed; palea gen < lemma, glabrous or minutely hairy. 18 spp.: N.Am. (Latin: 3-toothed, from lemma tip) [Tateoka 1961 Amer J Bot 48:565–573]

T. muticus (Torrey) Nash (p. 603) SLIM TRIDENS Pl densely tufted. **ST** 2–5 dm. **LF**: sheath hairy, esp near collar; ligule short-hairy; blade 3–25 cm, 1–3 mm wide, gen inrolled, ± fine scabrous, sometimes sparsely hairy. **INFL** 4–20 cm, 3–8 mm wide, narrow; branches short, appressed; spikelets subsessile to short-stalked. **SPIKELET** 8–11 mm, ± cylindric; glumes 5–6 mm, 1-veined; florets 6–8, strongly overlapping, pale to light purple; callus densely hairy; lemma ± 5 mm, veins densely hairy below middle, tip entire to minutely notched; palea keels densely hairy. 2*n*=40. Dry, rocky, gen limestone soils, creosote-bush shrubland, pinyon/juniper woodland; 900–2000 m. **SNE, DMtns**; to Colorado, TX, Mex. ✿ DRN,SUN:1,**2,3,7**,8,9,**10**,11,**14–23**,24. Apr–May, Oct–Nov

TRISETUM

Dieter H. Wilken

Ann, per. **STS** ascending to erect. **LVS** basal and cauline; ligule membranous, obtuse to truncate, toothed, tip ciliate or not; blade flat to inrolled. **INFL** panicle- to spike-like, open to compact, cylindric to narrowly conic. **SPIKELET**: glumes ± unequal, gen = or < lower floret, keeled, acute, lower 1-veined, upper 3-veined; axis stiff- to soft-hairy, gen prolonged behind upper floret, bristly or with vestigial floret; florets 2–3, bisexual, breaking above glumes and between florets (sometimes below glumes); callus short-hairy; lemmas ± keeled, tip 2-bristled or not, awned on back near tip or not, awn straight to bent; palea = or < lemma. 50–70 spp.: temp, trop mtns. (Latin: three bristle) Some spp. intergrade; needs critical study in w N.Am.

T. spicatum (L.) Richter (p. 605) Per, cespitose. **STS** 0.5–4 dm, densely clumped. **LVS** mostly basal, tufted; ligule 1–3 mm; cauline blade gen 1–4 mm wide. **INFL** spike-like, 2–8(10) cm, dense, cylindric to narrowly elliptic in outline; lower branches erect, ± appressed; axis hairy, hidden by spikelets. **SPIKELETS** on branches from base to tip; glumes lanceolate, acute, lower 5– 6 mm, upper 6–7 mm; lemmas 4–5 mm, awn 4–8 mm. $2n=14,28,42$. Open, dry to moist sites, meadows, streambanks; 1400–4000 m (KR: 150–2000 m). KR, CaRH, SN, **n SNE**, **W&I**; temp Am, Eurasia. ❀ DRN,SUN:4,5,6&IRR:**1**,2,**7**,**14–18**,19–24. Jul–Aug

VULPIA

Susan G. Aiken & Robert I. Lonard

Ann. **STS** < 8 dm, solitary or loosely clumped, ascending to erect, unbranched. **LVS** gen cauline; sheath < internode; ligule < 1 mm, membranous, minutely fringed; blade < 15 cm, 0.5–2.5 mm wide, flat or rolled when dry; basal lobes 0. **INFL** panicle-like, narrow, dense or open. **SPIKELET**: glumes unequal, lanceolate, lower sometimes minute, upper 3-veined; axis breaking above glumes and between florets; florets 2–10, gen cleistogamous; lemma back round, 3–5-veined, awn < 22 mm, straight; palea ± = lemma, tip forked; stamen gen 1, < 1 mm; ovary tip glabrous. **FR** 3–6.5 mm, ± linear, ± sticking to palea. (J.S. Vulpius, pharmacist-botanist of Baden, Germany) [Lonard & Gould 1974 Madroño 22:217–230] ❀STBL.

1. Lower glume < 1/2 upper glume length, sometimes minute . . . . ***V. myuros*** var. ***hirsuta***
1' Lower glume > 1/2 upper glume length
  2. Florets 5–10, closely overlapping; spikelet axis hidden; lemma awn < 5 mm . . . . ***V. octoflora***
    3. Lemma back prominently long-scabrous near tip, margins long-scabrous . . . . var. ***hirtella***
    3' Lemma back glabrous or barely scabrous, margins often short-scabrous . . . . var. ***octoflora***
  2' Florets gen 2–4, loosely overlapping; spikelet axis visible, each internode > 1mm; lemma awn gen
    2.5–12 mm
    4. Lowest infl branches appressed to erect; branch axil not thickened . . . . ***V. bromoides***
    4' Lowest infl branches spreading or reflexed; branch axis thickened. ***V. microstachys***
      5. Spikelet glabrous or only scabrous . . . . var. ***pauciflora***
      5' Spikelet obviously hairy
        6. Glumes and lemma hairy . . . . var. ***ciliata***
        6' Glumes glabrous; lemma hairy . . . . var. ***microstachys***

V. bromoides (L.) S.F. Gray (p. 605) **ST** < 5 dm, glabrous or hairy. **INFL** 1.5–15 cm, dense, clearly > uppermost lf; branches 1–3 per node; spikelet stalk flat or winged. **SPIKELET** 5–10 mm; glumes 3.5–8 mm, ± equal; florets 4–7; lemma 5.5–8 mm, awn 2.5–12 mm, scabrous. **FR** ± 3.5–5 mm. $2n=14$. Uncommon. Dry, disturbed places, coastal-sage scrub, chaparral; gen < 1500 m. CA-FP, **DMoj**; to e US; native to Eur. [*Festuca b.* L., *F. dertonensis* (All.) Asch. & Graebner] Apr–Jun

V. microstachys (Nutt.) Munro **ST** 1.5–7.5 dm, glabrous. **INFL** 2–24 cm, ± open, at least lower branches spreading or reflexed; branches l per node, < 7 cm; spikelet stalk angular. **SPIKELET** 5.5–10 mm; lower glume 2–5.5 mm, upper 3.5–7.5 mm; florets 2–4; lemma 4–9.5 mm, awn 3.5–12 mm. **FR** ± 4–6 mm. $2n=42$. Disturbed, open, gen sandy soils; gen < 1500 m. **CA**; to WA, ID, Baja CA. 4 vars. in CA.

  var. *ciliata* (Beal) Lonard & Gould **SPIKELET**: glumes hairy; lemma hairy. Locally abundant. Open, gen disturbed places, sandy soils, hillsides, forest; gen < 1000 m. **CA** (esp CA-FP). [*Festuca eastwoodae* Piper, *F. grayi* (Abrams) Piper] Apr–Jun

  var. ***microstachys*** (p. 605) **SPIKELET**: glumes glabrous; lemma hairy. Dry hillsides, coarse, sandy soils, crumbling serpentine or shale, open woodlands; gen < 1000 m. **CA** (esp CA-

FP); to WA, ID. [*Festuca m.* Nutt., *F. arida* Elmer] Apr–Jun

  var. *pauciflora* (Beal) Lonard & Gould **SPIKELET**: glumes glabrous or scabrous; lemma glabrous or scabrous. Common. Dry, open, wooded hillsides, sandy, often disturbed places; gen < 1500 m. **CA-FP**, **D**; to WA, Baja CA. [*F. m.* Nutt. var. *simulans* (Hoover) Hoover, *F. pacifica* Piper, *F. reflexa* Buckley] Apr–Jun

V. myuros (L.) C. Gmelin (p. 605) **ST** < 7.5 dm, glabrous or scabrous only near infl. **INFL** 4–25 cm, < 2 cm wide, ± dense; base often enclosed in sheath at maturity; branches 1–3 per node; spikelet stalk < 1 mm, slender. **SPIKELET** 5–11.5 mm; lower glume gen < 1.5 mm, upper ± 2.5–5.5 mm; florets 3–6; lemma 4.5–6.5, awn 5–15 mm. **FR** 3.5–4.5 mm. $2n=42$. Common. Gen open places, sandy soils; < 2000 m. CA-FP, **D**; worldwide; probably native to Eur. 2 vars. in CA.

  var. *hirsuta* Hack. **SPIKELET**: lemma margin ciliate near tip, awn of lowermost floret 9.5–22 mm. Open places, hillsides, washes; gen < 1500 m. CA-FP, **uncommon in D**; worldwide; native to Eur. [*Festuca megalura* Nutt.] Apr–Jun

V. octoflora (Walter) Rydb. (p. 605) **ST** < 6 dm, glabrous or hairy. **INFL** 0.4–16 cm, 0.5–2 cm wide, dense; branches 1 per node. **SPIKELET** 4.5–10 mm; lower glume ± 2–4.5 mm, upper

± 2.5–7 mm; florets 5–10; lemma ± 3–5 mm; awns 0.5–5 mm. **FR** 2–3.5 mm. $2n=14$. Sandy to rocky soils, open sites; < 2000 m. **CA**; widespread Am, Eur. [*Festuca o.* Walter]

var. ***hirtella*** (Piper) Henrard **SPIKELET**: lemma back prominently long-scabrous near tip. Sandy to rocky soils, open, gen disturbed areas, desert scrub to pinyon woodland, esp burned areas; gen < 1500 m. **CA** (esp s CA-FP); to B.C., Colorado, TX, Baja CA. [*Festuca o.* ssp. *hirtella* Piper]

var. ***octoflora*** **SPIKELET**: lemma back glabrous or barely scabrous, margins often scabrous. Sandy soils, washes, hills, chaparral; gen < 2000 m. s CA-FP, **D**; widespread Am, Eur. Apr–Jun

POTAMOGETONACEAE PONDWEED FAMILY

Robert F. Thorne

Ann, per, aquatic (gen fresh to alkaline water), glabrous, from rhizomes or small, bulb-like, winter buds. **STS** erect, simple to branched, cylindric or flattened. **LVS** simple, cauline, alternate or in subopposite pairs; submersed thread-like to round, sessile or petioled; floating, if present, elliptic to ovate, petioled, leathery; sheath open, continuous with petiole or ± free from lf base, gen with stipules, stipules sometimes fused, ligule-like. **INFL**: spike or head-like, axillary or terminal, gen emergent, peduncled; bracts 0. **FL** bisexual; perianth parts 0 or 4, clawed; stamens 2 or 4, if 4, each fused to base of perianth part, sessile or filament short, wide, anthers open to outside; pistils 4, ovary 1-chambered, ovule 1, style short or stigmas sessile. **FR**: drupe. **SEED** 1. 3 genera, ± 95 spp.: worldwide. Recently treated to include Zannichelliaceae [Angiosperm Phylogeny Group 1998 Ann Missouri Bot Gard 85:531–553].

POTAMOGETON PONDWEED

Ann, per, from rhizomes or small, bulb-like, winter buds. **ST** simple or branched, cylindric or flattened, rooting at lower nodes. **LVS** simple, cauline, gen alternate, gen flat, gen green, margin gen entire; submersed lvs sessile or petioled, linear to round, tip rounded to acuminate, veins 1–35; floating lvs, if any, elliptic to ovate, gen petioled, leathery; stipules free or fused, sheath-like below lf junction, free or fused (ligule-like) above lf junction. **INFL**: cylindric spike or head-like, axillary or terminal, floating to emergent. **FL**: inconspicuous; perianth parts 4, clawed, greenish; stamens 4, attached to base of perianth, anthers gen sessile; ovule attached at chamber base, style short or stigma sessile. **FR** gen obovate, sessile, floating. ± 90 spp.: mostly temp n hemisphere. (Greek: river neighbor, from aquatic habitat) [Haynes 1974 Rhodora 76:564–649; 1985 Sida 11:173–188; Wieglet 1988 Feddes Repert 99:249–266] Key to species adapted by Margriet Wetherwax.

1. Pls with both submersed and floating lvs; floating lvs gen elliptic to ovate, gen leathery; submersed lvs linear to ovate, membranous
  2. Submersed lvs linear, < 1 cm wide . . . . ²*P. nodosus*
  2' Submersed lvs lanceolate to ovate, gen 1–7 cm wide, sometimes linear
    3. Submersed lvs with 8+ veins, long-petioled . . . . ²*P. nodosus*
    3' Submersed lvs gen with 7 veins, all or lowermost sessile
      4. Margin of submersed lvs entire . . . . **P. gramineus**
      4' Margin of submersed lvs finely serrate near tip . . . . ²*P. illinoensis*
1' Pls with all lvs submersed
  5. Lvs lanceolate, elliptic, oblong, or ovate, sessile, base tapered to clasping (short-petioled in *P. illinoensis*)
    6. Fr with long slender beak 2–3 mm; st branched above; infl < 2 cm . . . . **P. crispus**
    6' Fr short-beaked; st much-branched; infl < 6 cm . . . . ²*P. illinoensis*
  5' Lvs linear
    7. Stipules fused to lf base, sheath-like below, sheath 1+ cm
      8. Lvs 2–5 mm wide; stipules fused above junction with blade, ligule-like . . . . **P. latifolius**
      8' Lvs linear to thread-like, < 1 mm wide; stipules free above junction with blade . . . . **P. pectinatus**
    7' Stipules free from lf or fused to lf base for 1–2 mm
      9. Keel of fr back wing-like, wavy; glands at lf base 0 or faint . . . . **P. foliosus**
        10. Stipules coarsely veined and ciliate, rigid, breaking apart into fibers; infl gen interrupted; fr pale green, 1.4–1.7 mm, 1.1–1.2 mm wide, keel on back gen < 0.2 mm high, beak < or = 0.2 mm . . . . var. ***fibrillosus***
        10' Stipules finely veined, not coarsely ciliate, decaying early; infl gen not interrupted; fr olive to greenish brown, 1.5–2.7 mm, 1.2–2.2 mm wide, keel on back gen 0.2–0.4 mm high, beak 0.2–0.6 mm . . . . var. ***foliosus***
      9' Fr back rounded, not keeled; glands at base of some or all lvs 2, prominent . . . . **P. pusillus**
        11. Fr sides concave, tip beaked at top of central axis; infl clearly interrupted, with 2–4 whorls; stipules fused below middle . . . . var. ***pusillus***
        11' Fr sides rounded, tip oblique, beak to 1 side of central axis; infl not clearly interrupted, with 1–2 whorls; stipules free, flat or margins inrolled . . . . var. ***tenuissimus***

P. crispus L. (p. 605) CRISPATE-LEAVED PONDWEED Per from slender rhizomes. **ST** branched above, < 90 cm, somewhat flattened. **LVS** all submersed, sessile, 4–8 cm, 0.5–0.8 cm wide, oblong, thick, crisped or wavy; tip rounded; margin finely serrate; stipules gen < 1 cm, ± free, becoming fibrous. **INFL**: spike, < 2 cm. **FR** 4–6 mm; back clearly keeled; beak 2–3 mm, slender, erect or curved. Uncommon. $2n=36,42,50,52,72,78$. Shallow water, ponds, reservoirs, streams; < 2100 m. NCoR, GV, CCo,

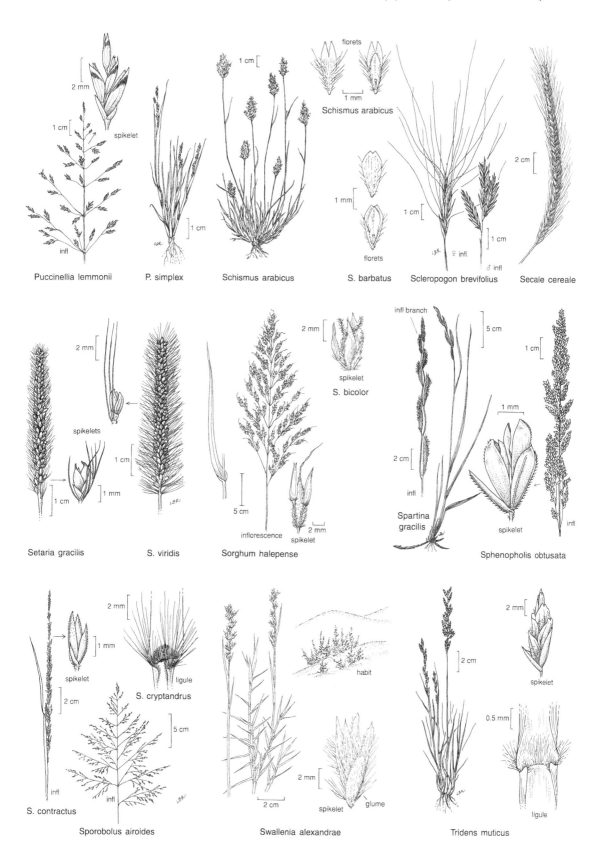

Puccinellia lemmonii

P. simplex

Schismus arabicus

Schismus arabicus

florets

S. barbatus

florets

Scleropogon brevifolius

♀ infl

♂ infl

Secale cereale

Setaria gracilis

spikelets

S. viridis

S. bicolor

spikelet

Sorghum halepense

inflorescence

spikelet

Spartina gracilis

infl branch

infl

spikelet

infl

Sphenopholis obtusata

S. contractus

spikelet

infl

S. cryptandrus

ligule

Sporobolus airoides

infl

Swallenia alexandrae

habit

spikelet

glume

Tridens muticus

spikelet

ligule

SnFrB, SCo, ChI, SnGb, SnBr, **DMoj**; ± worldwide; native to Eurasia. Jul–Sep

P. foliosus Raf. LEAFY PONDWEED Per, from densely matted, slender rhizomes. **ST** many-branched, somewhat flattened. **LVS** all submersed, 1–10 cm, < 0.3 cm wide, linear; veins < 7; stipules fused, sheath-like. **INFL** subspheric to short spike, < 1 cm; peduncle club-shaped. **FR**: back keeled; sides rounded to slightly concave. Shallow water, ponds, lakes, streams, irrigation ditches; < 2300 m. **CA**; to AK, e Can, C.Am.

var. ***fibrillosus*** (Fern.) R. Haynes & Rev. FIBROUS PONDWEED **ST** < 60 cm. **LF** 2–4 cm, 1–2 mm wide; base gen with 2 glands; tip acute; stipules < 12 cm, persistent, clearly veined, becoming fibrous. **INFL** head-like or short spike, interrupted; peduncle 4–12 mm. **FR** 1.4–1.7 mm, 1.1–1.2 mm wide, pale green; back with keel < 0.2 mm high; beak ± 0.2 mm. RARE in CA. Shallow water, small streams; < 1300 m. n NCo (Crescent City), w MP, **s SNE**; to WA, ID, WY.

var. ***foliosus*** (p. 605) **ST** < 100 cm, flattened. **LF** 1–10 cm, 0.3–2.5 mm wide; base without glands; tip acute to abruptly pointed; stipules < 2 cm, early deciduous, sometimes persistent as delicate fibers. **INFL** subspheric to head-like; peduncle 3–10 mm. **FR** 1.5–2.7 mm, 1.2–2.2 mm wide, olive to greenish brown; back with wavy, wing-like keel 0.2–0.4 mm high; beak 0.2–0.6 mm. 2n=28. Common. Habitats of sp.; < 2300 m. NW, CaRH, SNF, n SNH, GV, CW, SW (exc ChI), **GB**, **D**; to AK, e Can, C.Am. ❀ TRY. Jul–Oct

P. gramineus L. (p. 605) GRASS-LEAVED PONDWEED Per from matted rhizomes. **ST** with many short branches, < 100 cm, cylindric. **LVS**: submersed lvs gen sessile, 2–11 cm, 1–12 mm wide, linear to oblanceolate, tip acute or long-tapered; floating lvs gen on short, axillary branches, with petiole > blade, blade 1.5–7 cm, 1–3 cm wide, elliptic to ovate, tip ± obtuse; stipules < 3 cm, free, persistent. **INFL**: spike, 1–4 cm. **FR** 1.5–3 mm; back faintly keeled; beak recurved. n=26. Uncommon. Shallow water, ponds, lakes, bogs; 900–2750 m. KR, NCoRI, CaRH, SNH, SnFrB, SnBr, **GB**; to AK, Greenland, AZ, Eurasia. Hybridizes with *P. alpinus, P. illinoensis, P. natans, P. nodosus, P. richardsonii*. Jul–Aug

P. illinoensis Morong (p. 605) SHINING PONDWEED Per from rhizomes. **ST** gen much-branched, < 150 cm, slender, cylindric. **LVS**: submersed lvs sessile or petioles < 2 cm, blade 6–20 cm, 15–50 mm wide, elliptic to oblanceolate, margin finely serrate near tip, tip acute to short-pointed; floating lvs, if any, with blades 4–12 cm, 2–6 cm wide, widely elliptic to oblong-elliptic, gen > petiole; stipules 2.5–7 cm, free, persistent. **INFL**: spike, gen < 6 cm. **FR** ± 4 mm; back keeled; beak slightly below tip. n=52. Lakes, ponds, streams; 400–2350 m. NCoR, CaR, SNH, SnJV, CCo, SnFrB, SCoRO, SnGb, SnBr, PR, **GB**; to B.C., e Can, TX, Baja CA; also C.Am. and Caribbean. Hybridizes with *P. gramineus, P. nodosus, P. richardsonii*. ❀ TRY. Jun–Aug

P. latifolius (Robb.) Morong (p. 605) NEVADA PONDWEED Per from rhizome. **ST** much-branched, < 60 cm, subcylindric, whit-ish. **LVS** all submersed, sessile, 2–10 cm, 0.2–0.5 cm wide, linear; tip rounded or acute; veins 3–7; stipules < 3 cm, fused to blade, sheath-like below, ligule-like above, becoming deeply cut. **INFL**: spike, interrupted below, whorls 2–5, < 3 cm. **FR** < 3 mm, somewhat box-shaped, back faintly keeled. Very uncommon. Shallow, alkaline water, ponds, lakes; < 1450 m. c SnJV, **GB**, **DMoj**; to UT, TX, AZ. May be locally abundant. Jul–Aug

P. nodosus Poiret (p. 605) LONG-LEAVED PONDWEED Per from rhizomes. **ST** simple to branched above, gen < 300 cm, subcylindric. **LVS**: submersed lvs long-petioled, 2–15 cm, 10–40 mm wide, lowermost sessile, blade linear to elliptic-lanceolate, tapered at both ends; floating lvs long-petioled, 5–10 cm, < 5 cm wide, elliptic to ovate; base tapered to rounded; tip rounded; stipules 3–9 cm, free, breaking apart early. **INFL**: spike, < 5 cm. **FR** 3–5 mm, clearly beaked; back rounded; sides flat. n=26. Shallow water, lakes, ponds, ditches, streams; 100–2750 m. NCoR, s SNF, n SNH, GV, CCo, SnFrB, SCo, SnBr, PR, **GB**, **DMoj**; worldwide (exc Australia). Hybridizes with *P. gramineus, P. illinoensis*. May–Aug

P. pectinatus L. (p. 605) FENNEL-LEAF PONDWEED Per from matted rhizomes; tubers also present. **ST** many-branched, < 80 cm, ± cylindric. **LVS** all submersed, < 15(35) cm, < 1 mm wide, gen thread-like; tip acute to acuminate; stipules 2–5 cm, fused to blade, sheath-like below, sheath open, tips free. **INFL**: spike, interrupted in fr, whorls 2–6; peduncle 3–25 cm. **FR** 2.5–5 mm, plump; back rounded; beak very short. Common. Ponds, lakes, marshes, streams; < 2400 m. **CA**; worldwide (exc S.Am). Often weedy in reservoirs, irrigation canals; important food for waterfowl. May–Jul

P. pusillus L. (p. 605) SMALL PONDWEED **ST** < 100 cm, cylindric. **LVS**: all submersed, thread-like to linear; base gen with 2 prominent glands, veins < 7; stipule tips gen becoming finely fibrous. **INFL** ± interrupted, < 1.5 cm. **FR** < 2.5 mm; back rounded; beak short. Shallow water, ponds, lakes, reservoirs, ditches, vernal pools, slow streams; < 2700 m. NCo, KR, n SNF, SNH, GV, SnFrB, SCoR, SW (exc ChI), **GB**, **DMoj**; circumboreal.

var. ***pusillus*** **ST** gen many-branched, < 100 cm. **LF** < 7 cm, gen (0.5)2–3 mm wide; tip acute; veins 3–5; stipules 6–15 mm, fused below middle, breaking apart early. **INFL**: spike, 6–15 mm, gen interrupted; whorls 2–4; peduncle 1.5–8 cm. **FR**: sides concave; tip beaked at top of central axis. Common. Habitat and distribution of sp. May–Jun

var. ***tenuissimus*** Mert. & Koch **ST** simple to many-branched, < 80 cm. **LF** < 6 cm, gen < 1.5(2.5) mm wide; tip acute to obtuse; stipules 3–9 mm, free, flat or margins inrolled. **INFL** head-like to subcylndric spike; whorls 1–2; peduncle 0.5–4 cm. **FR**: sides round; tip beak to 1 side of central axis. n=13. Uncommon. Shallow water, mostly cold, acidic lakes, ponds; < 2100 m. NCo, KR, n SNF, SNH, SnJV, SCoRO, SnBr, **GB**; circumboreal. [*P. berchtoldii* Fieber] Jul–Aug

RUPPIA DITCH-GRASS

Per from slender rhizomes. **STS** gen many-branched, thread-like, rooting at lower nodes. **LVS** cauline, alternate, < 10 cm, < 1 mm wide; tip acute; stipules < 15 mm, ± completely fused to lf base, sheath-like below lf junction, ± open. **INFL** terminal; fls 2, sessile; peduncle elongated, straight or coiled in fr. **FL** minute; perianth 0; stamens 2, anthers sessile; pistils gen 4(2–8), simple, ovule attached at top of chamber, stigma sessile, peltate. **FR** ovoid, oblique, long-stalked. 2+ spp.; temp worldwide. (Heinrich Bernhard Ruppius, German botanist, 1688–1719) [Setchell 1946 Proc Calif Acad Sci 25:469–478]

R. cirrhosa (Petagna) Grande (p. 605) **LF** < 1 mm wide; tip acute. **INFL**: peduncle in fr 3–30 cm, coiled or flexible. 2n=40. Marshes, ponds, sloughs; < 2045 m. NCo, NCoRI, Teh, SnJV, CW, SCo, SnBr, **GB**, **D**; ± worldwide. [*R. spiralis* Dumort.] Possibly indistinct from *R. maritima*; needs further study. Mar–Aug

TYPHACEAE CATTAIL FAMILY

Per from long rhizomes, colonial, glabrous, gen aquatic (submersed to emergent), monoecious. **ST** erect and stiff or submersed and floating above, cylindric, solid. **LVS** basal and cauline, alternate, ± 2-ranked, spongy or stiff; sheath

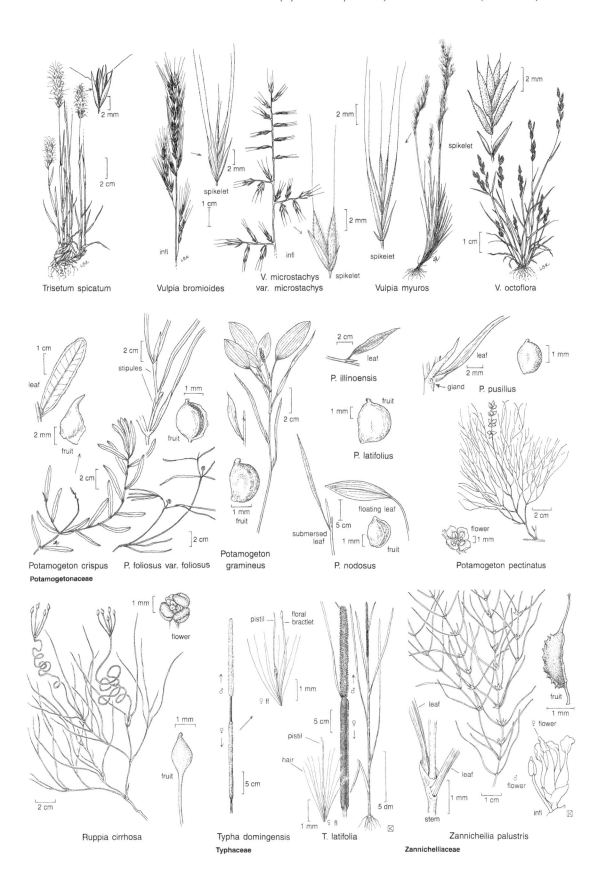

Trisetum spicatum

Vulpia bromioides

V. microstachys
var. microstachys

Vulpia myuros

V. octoflora

Potamogeton crispus

Potamogetonaceae

P. foliosus var. foliosus

Potamogeton
gramineus

P. illinoensis

P. latifolius

P. nodosus

P. pusillus

Potamogeton pectinatus

Ruppia cirrhosa

Typha domingensis

Typhaceae

T. latifolia

Zannichellia palustris

Zannichelliaceae

open; blade linear, flat, keeled, or triangular in ×-section, spongy. **INFL** spike-like (cylindric, dense) or head-like (spheric), terminal or axillary; staminate above pistillate, gen on same axis; fls subtended by 1, minute bract. **STAMINATE FL**: perianth parts 0 or 1–6 and scale-like; stamens 1–8. **PISTILLATE FL**: perianth parts 0 or 1–6 and flattened; ovary 1, chambers 1–2(3), ovules 1–2(3). **FR**: achene; wall thin, splitting in water. 2 genera, ± 25 spp.: worldwide. Family description by R.F. Thorne.

TYPHA CATTAIL

S. Galen Smith

Per from tough rhizomes, emergent or terrestrial, colonial, glabrous, monoecious. **STS** erect, simple, hard. **LF**: sheath open; blade linear, C-shaped in ×-section below, flat above. **INFL** spike-like, terminal, cylindric; staminate fls above, pistillate fls below; fls 1000+, staminate mixed with many papery scales; pistillate pedicels clustered on short, peg-like stalk. **STAMINATE FL**: perianth 0; stamens 2–7 on slender stalk; filaments slender, gen deciduous in fr. **PISTILLATE FLS** fertile and sterile; perianth 0; pedicel slender, long-hairy; ovary 1-chambered, ovule 1, style long, thread-like, stigma 1; sterile ovary truncate to rounded. **FR** minute, fusiform, falling with pedicel and hairs; wall thin, splitting in water. ± 8–13 spp.: worldwide. Rhizomes, pollen of some spp. used for food; lvs used for caning. (Greek: ancient name) [Smith 1987 Arch Hydrobiol Beih 27:129–138] All N.Am spp. hybridize.

1. Naked axis between staminate and pistillate fls gen 1–8 cm; pistillate spike bright yellow to orange-brown, 15–25 mm wide in fr; pistillate stalk ± 0.5 mm, peg-like; pistillate bractlets ± = or > pedicel hairs; stigmas linear, gen deciduous in fr . . . . ***T. domingensis***
1' Naked axis between staminate and pistillate fls gen 0; pistillate spike green when fresh, becoming brown, ± 28–36 mm wide in fr; pistillate stalk 1.5–3 mm, hair-like; pistillate bractlets 0; stigmas widely lanceolate, gen persisitent in fr . . . . ***T. latifolia***

T. domingensis Pers. SOUTHERN CATTAIL (p. 605) Pl 15–40 dm. **LF**: sheath tapered to blade, lobes 0 to membranous, ear-like; blade 6–18 mm wide when fresh, 5–15 mm wide when dry, gland-dotted on inside near base. **INFL** ± = lvs; naked axis between staminate and pistillate fls (0)1–8 cm; staminate bractlets irregularly branched, straw-colored to cinnamon-brown; pistillate stalk ± 0.5 mm, peg-like, spike < 35 cm, 15–25 mm wide in fr, bright yellow- to orange-brown, bractlet > pedicel hairs, tip acute to acuminate. **STAMINATE FL**: pollen grains single. **PISTILLATE FL**: stigma linear; sterile ovary rounded to ± truncate, gen < pedicel hairs, pale brownish, hair tips straw-colored to orange-brown. *n*=15. Marshes; < 1500 m. NCo, NCoRO, GV, CW, SW, **GB**, **D**; warm temp, trop worldwide. Jun–Jul

T. latifolia L. (p. 605) BROAD-LEAVED CATTAIL Pls 15–30 dm. **LF**: sheath tapered to blade or shouldered; blade 10–29 mm wide when fresh, (5)10–20 mm wide when dry, glands 0. **INFL** ± = lvs; naked axis between staminate and pistillate fls 0 (sometimes 4–8 cm); staminate bractlets simple, hair-like, whitish; pistillate stalk 1.5–3 mm, hair-like, spike 5–25 cm, 28–36 mm wide in fr, green when fresh, brown in fr, bractlets 0. **STAMINATE FL**: pollen grains in groups of 4. **PISTILLATE FL**: stigma widely lanceolate; sterile ovary rounded, << pedicel hairs, hair tips whitish. *n*=15. Common. Marshes, ponds, lakes; < 2000 m. **CA**; temp N.Am, C.Am, Eurasia, Afr. Jun–Jul

ZANNICHELLIACEAE HORNED-PONDWEED FAMILY

Robert F. Thorne

Per from slender, creeping rhizome, glabrous, aquatic, submersed, monoecious or dioecious. **ST** thread-like, weak. **LVS** alternate below to ± whorled and linear above; sheath fused to blade or free, stipule-like. **STAMINATE INFL** gen below pistillate infl, axillary; fls 1–3. **PISTILLATE INFL** terminal, short-stalked; fls 1–9, sessile, subtended by cup-like bract. **STAMINATE FL**: perianth parts 0 or 3 and scale-like; stamens 1–3, filaments free or fused; anther 1–2-chambered. **PISTILLATE FL**: perianth parts 0; ovary superior, chamber 1, ovule 1, style 1, simple or 3-lobed. **FR**: achene or nutlet. 4 genera; 9–10 spp.: ± worldwide. Recently treated in Potamogetonaceae [Angiosperm Phylogeny Group 1998 Ann Missouri Bot Gard 85:531–553].

ZANNICHELLIA HORNED-PONDWEED

Per, monoecious. **ST** gen few-branched. **LVS** ± opposite, clustered and ± whorled at upper nodes, stipuled. **STAMINATE FL**: stamen 1, filament slender, anther 2-chambered. **PISTILLATE FL**: style short, stigma peltate. **FR**: nutlet, stalked, compressed, tip beaked; back curved, ridged, toothed; front ± straight. 1–2 spp.: ± worldwide. (Gian G. Zannichelli, Venetian botanist, 1662–1729)

Z. palustris L. (p. 605) **ST** 3–10 dm. **LF** 2–10 cm; blade < 1 mm wide, tip acute. **FR** 2–4 mm; beak 1–1.5 mm, flask-shaped. $2n$=12,24,28,32,36. Streams, ponds, ditches, lakes; < 2200 m. **CA**; ± worldwide. Mar–Nov

INDEX

The following index includes all common names used in *The Jepson Desert Manual*, scientific names of families and genera, and synonyms. It does not include scientific names of accepted species or infraspecific taxa, which are arranged alphabetically in the text. Italics are not used for scientific names in the index.

Page numbers for all common names, synonyms, and miscellaneous entries are given in regular type. Accepted scientific names of families are followed by the page number of the beginning of the family description. Scientific names of described genera are often followed by two or more page numbers. The first page number indicates the page on which the genus description begins. This is followed by a page number in *italics* that refers to the first illustration plate for the genus.

Photographers

Anonymous: 119

Bruce G. Baldwin: 2, 7, 8, 14, 20, 24, 26, 27, 28, 47, 49, 61, 63, 67, 69, 70, 71, 74, 79, 81, 90, 93, 99, 101, 107, 114, 115, 128

George Becker: 43

John Game: 1, 3, 4, 5, 6, 16, 22, 25, 29, 30, 31, 33, 34, 37, 38, 41, 44, 45, 50, 51, 52, 54, 56, 57, 58, 59, 62, 64, 68, 72, 77, 78, 80, 82, 83, 84, 87, 88, 89, 92, 94, 95, 96, 102, 103, 104, 105, 106, 108, 109, 110, 112, 113, 116, 117, 118, 120, 122, 124, 125, 127

Lawrence R. Heckard: 39, 66, 75

Steve Junak: 48

Scott N. Martens: 36, 40, 60

Donald Myrick: 19, 23

G. Norvell: 18

Robert Ornduff: 12, 55

G. T. Robbins: 32, 121

James T. Vale: 9, 11, 13, 15, 17, 21, 35, 42, 73, 85, 86, 91, 97, 98, 111, 123

Charles S. Webber: 10, 46, 76, 126

Dieter Wilken: 100

Martin F. Wojciechowski: 53, 65

Hierarchical Outline of Geographic Subdivisions

CA-FP
California Floristic
Province

NW Northwestern California
    NCo North Coast
    KR Klamath Ranges
    NCoR North Coast Ranges
        NCoRO Outer North Coast Ranges
        NCoRH High North Coast Ranges
        NCoRI Inner North Coast Ranges

CaR   Cascade Ranges
    CaRF Cascade Range Foothills
    CaRH High Cascade Range

SN   Sierra Nevada
    SNF Sierra Nevada Foothills
        n SNF northern Sierra Nevada Foothills
        c SNF central Sierra Nevada Foothills
        s SNF southern Sierra Nevada Foothills
    SNH High Sierra Nevada
        n SNH northern High Sierra Nevada
        c SNH central High Sierra Nevada
        s SNH southern High Sierra Nevada
    Teh Tehachapi Mountains

GV   Great Central Valley
    ScV Sacramento Valley
    SnJV San Joaquin Valley

CW   Central Western California
    CCo Central Coast
    SnFrB San Francisco Bay Area
    SCoR South Coast Ranges
        SCoRO Outer South Coast Ranges
        SCoRI Inner South Coast Ranges

SW   Southwestern California
    SCo South Coast
    ChI Channel Islands
        n ChI northern Channel Islands
        s ChI southern Channel Islands
    TR Transverse Ranges
        WTR Western Transverse Ranges
        SnGb San Gabriel Mountains
        SnBr San Bernardino Mountains
    PR Peninsular Ranges
        SnJt San Jacinto Mountains

GB
Great Basin Province

MP   Modoc Plateau
    Wrn Warner Mountains
    MP exc Wrn Modoc Plateau except Warner Mountains

SNE   East of Sierra Nevada
    W&I White and Inyo Mountains
    SNE exc W&I East of Sierra Nevada except White and Inyo Mountains

D
Desert Province

DMoj Mojave Desert
    DMtns Desert Mountains
    DMoj exc DMtns Mojave Desert except Desert Mountains

DSon Sonoran Desert (also known as Colorado Desert)